Lee's Loss Prevention in the Process Industries

Volume 3

This book is dedicated to

Herbert Douglas Lees (1860–1944), gas engineer;
Frank Priestman Lees (1890–1916), gas engineer;
Herbert Douglas Lees (1897–1955), gas engineer;
David John Lees (1936–), agricultural engineer;
Frank Lyman MacCallum (1893–1955), mining engineer and missionary;
Vivien Clare Lees (1960–), plastic and hand surgeon;
Harry Douglas Lees (1962–), restaurateur
and their families

'They do not preach that their God will rouse them a little before
    the nuts work loose.
They do not teach that His Pity allows them to drop their job when
    they dam'-well choose.
As in the thronged and the lighted ways, so in the dark and the
    desert they stand,
Wary and watchful all their days that their brethren's days may be
    long in the land.'

Rudyard Kipling (The Sons of Martha, 1907)

Wo einer kommt and saget an,
Er hat es allen recht getan,
So bitten wir diesen lieben Herrn,
Er wöll uns solche Kunste auch lehrn

(Whoever is able to say to us
'I have done everything right',
We beg that honest gentleman
To show us how it is done)

Inscription over the 'Zwischenbau' adjoining the Rathaus in Brandenburg-on-the-Haven
(quoted by Prince B.H.M. von Bulow in Memoirs, 1932)

If the honeye that the bees gather out of so manye floure of herbes . . . that are growing in other

mennis medowes . . . may justly be called the bees' honeye . . . so maye I call it that I have . . .

gathered of manye good autores . . . my booke.

William Turner (quoted by A. Scott-James in The Language of the Garden: A Personal Anthology)

By the same author:

A.W. Cox, F.P. Lees and M.L. Ang (1990): *Classification of Hazardous Locations* (Rugby: Institution of Chemical Engineers)

Elwyn Edwards and Frank P. Lees (1973): *Man and Computer in Process Control* (London: Institution of Chemical Engineers)

Elwyn Edwards and Frank P. Lees (eds) (1974): *The Human Operator in Process Control* (London: Taylor & Francis)

Frank P. Lees and M.L. Ang (1989): *Safety Cases* (London: Butterworths)

Lee's Loss Prevention in the Process Industries

Hazard Identification, Assessment and Control

Volume 3

Third edition

Dr. Sam Mannan, PE, CSP

Department of Chemical Engineering,
Texas A&M University,
Texas, USA

ELSEVIER
BUTTERWORTH
HEINEMANN

AMSTERDAM • BOSTON • HEIDELBERG • LONDON
NEW YORK • OXFORD • PARIS • SAN DIEGO
SAN FRANCISCO • SINGAPORE • SYDNEY • TOKYO

Elsevier Butterworth–Heinemann
200 Wheeler Road, Burlington, MA 01803, USA
Linacre House, Jordan Hill, Oxford OX2 8DP, UK

Recognizing the importance of preserving what has been written, Elsevier prints its books on
acid-free paper whenever possible.

Library of Congress Cataloging-in-Publication Data
Lee's loss prevention in the process industries. – 3rd ed. / edited by Sam Mannan.
   p. cm.
   Rev. ed. of: Loss prevention in the process industries / Frank P. Lees. 2nd ed. 1996.
   Includes bibliographical references and indexes.
   ISBN 0-7506-7555-1 (casebound, set : alk. paper) 1. Petroleum chemicals industry–
Great Britain–Safety measures. 2. Petroleum chemicals industry–United States–Safety
measures. I. Title: Loss prevention in the process industries. II. Mannan, Sam. III.
Lees, Frank P. Loss Prevention in the process industries.
  TP690.6.L43 2005
  660'.2804–dc22

                                          2004018498

British Library Cataloguing-in-Publication Data
A catalogue record for this book is available from British Library

Set ISBN: 0-7506-7555-1
Volume 1 ISBN: 0-7506-7857-7
Volume 2 ISBN: 0-7506-7858-5
Volume 3 ISBN: 0-7506-7589-3

For information on all Elsevier Butterworth–Heinemann Publications
visit our Web site at www.books.elsevier.com

04 05 06 07 08 09 10   10 9 8 7 6 5 4 3 2 1

Printed in the United States of America

Preface to Third Edition

The first edition of this book appeared in 1980, at the end of a decade of rapid growth and development in loss prevention. In the preface to the second edition, Frank P. Lees wrote, "After another decade and a half the subject is more mature, although development continues apace. In preparing this second edition it has been even more difficult than before to decide what to put in and what to leave out." Frank Lees' statement in 1996 rings even truer today, another eight years later in 2004. Industrial advances and technology changes coupled with recent events have made it essential to focus on new topics while keeping a complete grasp of all of older technologies and learnings as well. Safety programs today must also consider issues such as chemical reactivity hazards, safety instrumented systems, and layer of protection analysis. In the post 9-11 world, process safety and loss prevention must also include consideration of issues related to chemical security and resilient engineering systems.

The history of safety regulations in the United States can be traced back to the year before the beginning of the twentieth century. The River and Harbor Act, the first known federal legislation relevant to safety was promulgated in 1899. Since then, the total number of legislations has steadily increased. In addition to the federal government, local entities such as the state, county, and cities have also promulgated regulations and ordnances, which impose safety requirements on process facilities. Varying degrees of similar legislative action has also occurred in the rest of the world. These legislation were all promulgated in response to some event, demographic changes, as well as changes in the industry. Also, as our understanding of the hazards associated with industrial processes developed, procedures and practices were put in place to limit or eliminate the damage. Government programs and industry initiatives spurred improvements in the science and technology needed for the recognition of hazards and associated risks.

Management systems have been put in place to implement regulations and industry practices. Government regulations will continue to be a significant driver for safety programs. As such, one of the main objectives of these management systems is to ensure compliance. However, it is also quite clear that profitability is directly related to safety and loss prevention. Thus the management systems for safety are intricately tied into the operational management.

The industrial revolution brought prosperity and along with it the use of hazardous processes and complex technologies. Growing economies and global competition has led to more complex processes involving the use of hazardous chemicals, exotic chemistry, and extreme operating conditions. As a result, a fundamental understanding of the hazards and associated risks is essential. Process safety and risk management requires the application of the basic sciences and a systematic approach. Recent advances, such as overpressure protection alternatives and reactive chemistry allow safer design and operation of processes.

In the multiple barriers concept, plants are designed with several layers so that an incident would require the failure of several systems. Another novel approach to process safety and risk management is to consider various actions in a descending hierarchical order. Inherently safer design consideration should be first in the hierarchy followed by prevention systems, mitigation, and response. The success of these systems is dependent on the fundamental understanding of the process and the associated hazards. Chronic as well as catastrophic consequences resulting from toxic and flammable substances can be reduced and/or eliminated through appropriate design and operating practices.

Managing safety is no easy task, but it makes bottom-line sense. There is a direct payoff in savings on a company workers' compensation insurance, whose premiums are partly based on the number of claims paid for job injuries. The indirect benefits are far larger, for safe plants tend to be well run in general and more productive. The recipe for safety is remarkably consistent from industry to industry. It starts with sustained support of top management followed by implementation of appropriate programs and practices that institutionalize safety as a culture as compared to add-on procedures. The ingraining of safety as second nature in day-to-day activities requires a paradigm shift and can only be accomplished when safety is viewed as an integral and comprehensive part of any activity as compared to being a stand-alone or add-on activity.

This third edition of *Loss Prevention in the Process Industries* represents a combination of appropriate revisions of the essential compilations put together by Frank P. Lees, along with several new chapters and additions on new areas that deserve attention and discussion. The third edition includes five new chapters and three new appendices. The five new chapters address incident investigation, inherently safer design, reactive chemicals, safety instrumented systems, and chemical security. The three new appendices address process safety management regulation in the United States, risk management program regulation in the United States, and incident databases.

The chapter on incident investigation provides a summary of incident investigation procedures that can be used not only to determine causes of incidents but also provides a primer on capturing and integrating lessons learned from incident investigations into design, operations, maintenance, and response programs. Chemical process incidents can be accompanied by significant consequences, both in terms of human life and in financial impact. Many major chemical process incidents are the result of a complex scenario involving simultaneous failures of multiple safeguards. A robust system for incident investigation is usually necessary to determine and understand the causes, as well as implement measures to prevent a repeat event. This chapter is intended to provide an overview of incident investigation by addressing major concepts, principles, and characteristics of effective incident investigations of chemical process events. The focus is on incidents pertaining to chemical processes and their associated hazards, and the associated investigation techniques appropriate for complex systems and scenarios. This chapter is based on best practices for incident investigation, and those common concepts (i.e., tools, techniques, definitions) included in root cause investigation methodologies currently in the public domain in use in the process industry. It is not the intention to provide a stand-alone investigation methodology/guideline, nor address internal or proprietary investigation methodologies.

The chapter on inherently safer design addresses options and issues that can be considered with regard to the design and operation of plants. Inherently safer design is a philosophy that focuses on elimination of hazards or reduction of the magnitude of hazards rather than the control of hazards. Many of the concepts of inherently safer design have been applied by engineers in a wide variety of technologies for many years, without recognizing the common approach. In the late 1970s, in the wake of many large incidents in the chemical industry, Trevor Kletz recognized the common philosophies of hazard elimination and hazard reduction, gave the philosophy the name "inherently safer design," and developed a specific set of approaches to help engineers in the chemical process industries to design inherently safer processes and plants. Trevor realized that increased expectations for safety, from companies, regulatory bodies, and society in general, combined with the increased potential damage from incidents in the larger plants being built to meet increased demand and global markets, resulted in increased complexity and cost for the safety systems required to satisfy these demands. Furthermore, while hazard control systems can be made highly reliable, they can never be perfect and will always have some failure probability. While this probability can be made very small, there is always some chance that all safety systems will fail simultaneously and the result would be a large incident. Also, the hazard management systems require ongoing maintenance, as well as management and operator training, for the life of the plant. This results in ongoing costs, and the potential for future deterioration of the safety systems. Deteriorated systems will have reduced reliability, increasing the potential for a catastrophic accident. Trevor Kletz suggested that in many cases, a simpler, cheaper, and safer plant could be designed by focusing on the basic technology, eliminating or significantly reducing hazards, and therefore the need to manage them.

The chapter on reactive chemicals provides an overview of this critical issue and provides guidance on management systems as well as experimental and theoretical methods for analyses of chemical reactivity hazards. Serious incidents arising from uncontrolled reactivity have taken place since the inception of the chemical industry. The human toll of such incidents has been staggering. In recent decades, greater recognition and resources have been directed toward preventing and mitigating such occurrences. A number of incidents have been so severe as to prompt regulatory initiatives to force better management of reactivity. It is prudent for any company, organization, or other group to scrutinize the chemicals being handled and implement measures to limit the risk of a major reactive hazards event. A sampling of incidents that have substantially heightened concerns regarding reactive hazards in the general public, in governmental agencies, and in industry includes:

- The 1976 ICMESA incident in Seveso, Italy in which an uncontrolled chemical reaction generated pressure resulting in relief venting of a highly toxic dioxane into the neighboring villages and countryside.
- The 1984 Union Carbide incident in Bhopal, India in which methyl isocyanate was contacted with water, generating highly toxic cyanide gas and leading to thousands of fatalities.
- The 1994 Napp Technology incident in Lodi, New Jersey in which an uncontrolled reaction involving gold ore processing led to the deaths of five firefighters.
- The 1999 Concept Sciences incident in Allentown, Pennsylvania in which an explosion arising from a process concentrating hydroxylamine resulted in five fatalities. Another event involving purified hydroxylamine took place in a Nissin Chemical plant in Gunma Prefecture, Japan in 2000 and led to four fatalities.
- The 2001 TotalFinaElf incident in Toulouse, France in which ammonium nitrate being processed for nitrogen fertilizers exploded leading to 30 fatalities.

These events, as well as numerous others, have influenced the perception and approach to reactive hazards.

The chapter on safety instrumented systems addresses systems and procedures that need to be in place with regard to this area of safety and instrumentation. In many processes, technical or manufacturing issues limit the engineer's capability to design an inherently safer process. Further, there is generally a point where the required capital investment is disproportional to the additional risk reduction provided by the process modification. In other words, the derived safety benefit is too low relative to the economic investment. When this occurs, protection layers or safeguards must be provided to prevent or mitigate the process risk. A Safety Instrumented System (SIS) is a protection layer, which shuts down the plant, or part of it, if a hazardous condition is detected. Throughout the years, SIS have also been known as Emergency Shutdown Systems (ESD, ESS), Safety Shutdown Systems (SSD), Safety Interlock Systems (SIS), Safety Critical Systems (SCS), Safety Protection Systems (SPS), Protective Instrumented Systems (PIS), interlocks, and trip systems. Regardless of what the SIS may be called, the essential characteristic of the SIS is that it is composed of instruments, which detect that process variables are exceeding preset limits, a logic solver, which processes this information and makes decisions, and final control elements, which take necessary action on the process to achieve a safe state.

The chapter on chemical security deals with this new and critical element of the management of a process facility following the events of September 11, 2001. Security management is required for protecting the assets (including employees) of the facility, maintaining the ongoing integrity of the operation, and preserving value of the investment. Process security and process safety have many parallels and make use of many common programs and systems for achieving their ends. Process security requires a management systems approach to develop a comprehensive security program, which shares many common elements to process safety management.

The new appendix on process safety management regulation in the United States provides a summary of this regulatory requirement. The fourteen elements of the OSHA Process Safety Management (PSM) regulation (*29 CFR 1910.119*) were published in the U.S. Federal Register on February 24, 1992. The objective of the regulation is to prevent or minimize the consequences of catastrophic releases of toxic, reactive, flammable, or explosive chemicals. The regulation requires a comprehensive management program: a holistic approach that integrates technologies, procedures, and management practices. The process safety management regulation applies to processes that involve certain specified chemicals at or above threshold quantities, processes that involve flammable liquids or gases on-site in one location, in quantities of 10,000 pounds or more (subject to few exceptions), and processes that involve the manufacture of explosives and pyrotechnics. Hydrocarbon fuels, which may be excluded if used solely as a fuel, are included if the fuel is part of a process covered by this regulation. In addition, the regulation does not apply to retail facilities, oil or gas well drilling or servicing operations, or normally unoccupied remote facilities.

The new appendix on risk management program regulation in the United States provides a summary of this regulatory requirement administered by the U.S. Environmental Protection Agency. In 1996, EPA promulgated the regulation for *Risk Management Programs for Chemical Accident Release Prevention* (40 CFR 68). This federal regulation was mandated by section 112(r) of the Clean Air Act Amendments of 1990. The regulation requires regulated facilities to develop and implement appropriate risk management programs to minimize the frequency and severity of chemical plant accidents. In keeping with regulatory trends, EPA required a performance-based approach towards compliance with the risk management program regulation. The EPA regulation also requires regulated facilities to develop a Risk Management Plan (RMP). The RMP includes a description of the hazard assessment, prevention program, and the emergency response program. Facilities submit the RMP to the EPA and subsequently is made available to governmental agencies, the state emergency response commission, the local emergency planning committees, and communicated to the public.

The new appendix on incident databases addresses compilations of incident databases that can used for improving safety programs, developing trends, performance measures, and metrics. Incident prevention and mitigation of consequences is the focus of a number of industry programs regulatory initiatives. As part of these programs and regulations, accident history data are often collected. There are two basic types of information. One is a database consisting of standardized fields of data usually for a large number of incidents. The second are more detailed reports of individual incidents. Analysis of these incident history databases can provide insight into incident prevention needs. While the analysis and conclusions obtained from the incident database are often limited by the shortcomings of the databases themselves, the fact remains that incident history databases are very useful and can be a powerful tool in focusing risk reduction efforts. The conclusions can be used to identify systematically the greatest risks to allow prioritization of efforts to improve process safety. At the plant level this might entail identifying certain processes, types of equipment, chemicals, operations and other factors most commonly associated with incidents. Databases that cover a very large number of facilities are likely to reveal trends and patterns that no one company or facility could determine from their own experience. Statistical knowledge of the likelihood of the release of certain types of chemicals could help emergency responders, state emergency response commissions, and local emergency planning committees determine the most likely and most serious chemical releases in their areas and plan appropriate chemical accident responses. Incident databases may also help identify technologies and practices to prevent chemical accidents, or the need to develop them. For example, the data could indicate that inspection and preventive maintenance of equipment and instruments should become more thorough or more frequent.

M. SAM MANNAN
College Station,
Texas, USA
2004

Preface to Second Edition

The first edition of this book appeared in 1980, at the end of a decade of rapid growth and development in loss prevention. After another decade and a half the subject is more mature, although development continues apace. In preparing this second edition it has been even more difficult than before to decide what to put in and what to leave out.

The importance of loss prevention has been underlined by a number of disasters. Those at San Carlos, Mexico City, Bhopal and Pasadena are perhaps the best known, but there have been several others with death tolls exceeding 100. There have also been major incidents in related areas, such as those on the Piper Alpha oil platform and at the nuclear power stations at Three Mile Island and Chernobyl.

Apart from the human tragedy, it has become clear that a major accident can seriously damage even a large international company and may even threaten its existence, rendering it liable to severe damages and vulnerable to takeover.

Accidents in the process industries have given impetus to the creation of regulatory controls. In the UK the Advisory Committee on Major Hazards made its third and final report in 1983. At the same time the European Community was developing its own controls which appeared as the EC Directive on Major Accident Hazards. The resulting UK legislation is the NIHHS Regulations 1982 and the CIMAH Regulations 1984. Other members of the EC have brought in their own legislation to implement the Directive. There have been corresponding developments in planning controls.

An important tool for decision-making on hazards is hazard assessment. The application of quantitative methods has played a crucial role in the development of loss prevention, but there has been lively debate on the proper application of such assessment, and particularly on the estimation and evaluation of the risk to the public.

Hazard assessment involves the assessment both of the frequency and of the consequences of hazardous events. In frequency estimation progress has been made in the collection of data and creation of data banks and in fault tree synthesis and analysis, including computer aids. In consequence assessment there has been a high level of activity in developing physical models for emission, vaporization and gas dispersion, particularly dense gas dispersion; for pool fires, fireballs, jet flames and engulfing fires; for vapour cloud explosions; and for boiling liquid expanding vapour explosions (BLEVEs). Work has also been done on injury models for thermal radiation, explosion overpressure and toxic concentration, on models of the density and other characteristics of the exposed population, and on shelter and escape.

Some of these topics require experimental work on a large scale and involving international cooperation. Large scale tests have been carried out at several sites on dense gas dispersion and on vapour cloud fires and explosions. Another major cooperative research programme has been that of DIERS on venting of chemical reactors.

The basic approach developed for fixed installations on shore has also been increasingly applied in other fields. For transport in the UK the *Transport Hazards Report* of the Advisory Committee on Dangerous Substances represents an important landmark. Another application is in the offshore oil and gas industry, for which the report on the Piper Alpha disaster, the *Cullen Report*, constitutes a watershed.

As elsewhere in engineering, computers are in widespread use in the design of process plants, where computer aided design (CAD) covers physical properties, flowsheeting, piping and instrument diagrams, unit operations and plant layout. There is increasing use of computers for failure data retrieval and analysis, reliability and availability studies, fault tree synthesis and analysis and consequence modelling, while more elusive safety expertise is being captured by computer-based expert systems.

The subject of this book is the process industries, but the process aspects of related industries, notably nuclear power and oil and gas platforms are briefly touched on. The process industries themselves are continually changing. In the last decade one of the main changes has been increased emphasis on products such as pharmaceuticals and agrochemicals made by batch processes, which have their own particular hazards.

All this knowledge is of little use unless it reaches the right people. The institutions which educate the engineers who will be responsible for the design and operation of plants handling hazardous materials have a duty to make their students aware of the hazards and at least to make a start in gaining competence in handling them.

I would like again to thank for their encouragement the heads of the Department of Chemical Engineering at Loughborough, Professors D.C. Freshwater, B.W. Brooks and M. Streat; our Industrial Professors T.A. Kletz and H.A. Duxbury and Visiting Professor S.M. Richardson; my colleagues, past and present, in the Plant Engineering Group, Mr R.J. Aird, Dr P.K. Andow, Dr M.L. Ang, Dr P.W.H. Chung, Dr D.W. Edwards, Dr P. Rice and Dr A.G. Rushton — I owe a particular debt to the latter; the members of the ACMH, chaired by Professor B.H. Harvey; the sometime directors of Technica Ltd, Dr D.H. Slater, Mr P. Charsley, Dr P.J. Comer, Dr R.A. Cox, Mr T. Gjerstad, Dr M.A.F. Pyman, Dr C.G. Ramsay, Mr M.A. Seaman and Dr R. Whitehouse; the members of the IChemE Loss Prevention Panel; the IChemE's former Loss Prevention Officer, Mr B.M. Hancock; the members of the IChemE Loss Prevention Study Group and of the Register of Safety Professionals; the editorial staff of the IChemE, in particular Mr B. Brammer; numerous members of the Health and Safety Executive, especially Dr A.C. Barrell, Mr J. Barton, Dr D.A. Carter, Mr K. Cassidy, Mr P.J. Crossthwaite, Dr N.W. Hurst, Dr S.F. Jagger, Dr J. McQuaid, Dr K. Moodie, Dr C. Nussey, Dr R.P. Pape, Dr A.F. Roberts and Dr N.F. Scilly; workers at the Safety and Reliability Directorate, particularly Dr A.T.D. Butland, Mr I. Hymes, Dr D.W. Phillips and Dr D.M. Webber; staff at Shell Thornton Research Centre, including Dr D.C. Bull and Dr A.C. Chamberlain; staff at British Gas, including Dr J.D. Andrews, Dr M.J. Harris, Mr H. Hopkins, Dr J.M. Morgan and Dr D.J. Smith; staff at the Ministry of

Defence, Explosives Storage and Transport Committee, including Mr M.A. Gould, Mr J. Henderson and Mr P. Stone; and colleagues who have taught on post-experience courses at Loughborough, in particular Dr C.D. Jones, Dr D.J. Lewis and Mr J. Madden; BP International and Mr R. Malpas for allowing me to spend a period of study leave with the company in 1985–86 and Mr F.D.H. Moysen, Mr G. Hately, Mr M. Hough, Mr R. Fearon and others in the Central Safety Group and in Engineering Department; the Honourable Lord Cullen, my fellow Technical Assessors on the Piper Alpha Inquiry, Mr B. Appleton and Mr G.M. Ford and the Cremer and Warner team at the inquiry, in particular Mr G. Kenney and Mr R. Sylvester-Evans; other professional colleagues Dr L.J. Bellamy, Professor B.A. Buffham, Dr D.A. Crowl, Mr T.J. Gilbert, Mr D.O. Hagon, Dr D.J. Hall, Mr K.M. Hill, Professor T.M. Husband, Mr M. Kneale, Dr V.C. Marshall, Dr M.L. Preston, Dr J. Rasmussen, Dr J.R. Roach, Dr J.R. Taylor, Dr V.M. Trbojevic, Mr H.M. Tweeddale, Dr G.L. Wells and Dr A.J. Wilday; my research colleagues Dr C.P. Murphy, Mrs J.I. Petts, Dr D.J. Sherwin, Mr R.M.J. Withers and Dr H. Zerkani; my research students Mr M. Aldersey, Mr D.C. Arulanantham, Dr A. Bunn, Dr M.A. Cox, Dr P.A. Davies, Dr S.M. Gilbert, Mr P. Heino, Dr A. Hunt, Dr B.E. Kelly, Dr G.P.S. Marrs, Dr J.S. Mullhi, Dr J.C. Parmar, Mr B. Poblete, Dr A. Shafaghi and Dr A.J. Trenchard as well as colleagues' research students Mr E.J. Broomfield, Mr R. Goodwin, Mr M.J. Jefferson, Dr F.D. Larkin, Mr S.A. McCoy, Dr K. Plamping, Mr J. Soutter, Dr P. Thorpe and Mr S.J. Wakeman; the office staff of the Department, Mrs E.M. Barradell, Mr D.M. Blake, Miss H.J. Bryers and Miss Y. Kosar; the staff of the University Library, in particular Miss S.F. Pilkington; and my wife Elizabeth, whose contribution has been many-faceted and in scale with this book.

FRANK P. LEES
Loughborough,
1994

Preface to First Edition

Within the past ten or fifteen years the chemical and petroleum industries have undergone considerable changes. Process conditions such as pressure and temperature have become more severe. The concentration of stored energy has increased. Plants have grown in size and are often single-stream. Storage has been reduced and interlinking with other plants has increased. The response of the process is often faster. The plant contains very large items of equipment. The scale of possible fire, explosion or toxic release has grown and so has the area which might be affected by such events, especially outside the works boundary.

These factors have greatly increased the potential for loss both in human and in economic terms. This is clear both from the increasing concern of the industry and its insurers and from the historical loss statistics.

The industry has always paid much attention to safety and has a relatively good record. But with the growing scale and complexity involved in modern plants the danger of serious large-scale incidents has been a source of increasing concern and the adequacy of existing procedures has been subjected to an increasingly critical examination.

Developments in other related areas have also had an influence. During the period considered there has been growing public concern about the various forms of pollution, including gaseous and liquid effluents and solid wastes and noise.

It is against this background that the loss prevention approach has developed. It is characteristic of this approach that it is primarily concerned with the problems caused by the depth of technology involved in modern processes and that it adopts essentially an engineering approach to them. As far as possible both the hazards and the protection are evaluated quantitatively.

The clear recognition by senior management of the importance of the loss prevention problem has been crucial to these developments. Progress has been made because management has been prepared to assign to this work many senior and capable personnel and to allocate the other resources necessary.

The management system is fundamental to loss prevention. This involves a clear management structure with well defined line and advisory responsibilities staffed by competent people. It requires the use of appropriate procedures, codes of practice and standards in the design and operation of plant. It provides for the identification, evaluation and reduction of hazards through all stages of a project from research to operation. It includes planning for emergencies.

The development of loss prevention can be clearly traced through the literature. In 1960 the Institution of Chemical Engineers held the first of a periodic series of symposia on *Chemical Process Hazards with Special Reference to Plant Design*. The Dow Chemical Company published its *Process Safety Manual* in 1964. The American Institute of Chemical Engineers started in 1967 an annual series of symposia on Loss Prevention. The European Federation of Chemical Engineers' symposium on *Major Loss Prevention in the Process Industries* at Newcastle in 1971 and the Federation's symposium on *Loss Prevention and Safety Promotion in the*

Process Industries (Buschmann, 1974) at Delft are further milestones.

Another indicator is the creation in 1973 by the Institution of Chemical Engineers Engineering Practice Committee of a Loss Prevention Panel under the chairmanship of Mr T.A. Kantyka.

In the United Kingdom the Health and Safety at Work etc. Act 1974 has given further impetus to loss prevention. The philosophy of the *Robens Report* (1972), which is embodied in the Act, is that of self-regulation by industry. It is the responsibility of industry to take all reasonable measures to assure safety. This philosophy is particularly appropriate to complex technological systems and the Act provides a flexible framework for the development of the loss prevention approach.

The disaster at Flixborough in 1974 has proved a turning point. This event has led to a much more widespread and intense concern with the loss prevention problem. It has also caused the government to set up in 1975 an Advisory Committee on Major Hazards. This committee has made far-reaching recommendations for the identification and control of major hazard installations.

It will be apparent that loss prevention differs somewhat from safety as traditionally conceived in the process industries. The essential difference is the much greater engineering content in loss prevention.

This is illustrated by the relative effectiveness of inspection in different processes. In fairly simple plants much can be done to improve safety by visual inspection. This approach is not adequate, however, for the more technological aspects of complex processes.

For the reasons given above loss prevention is currently a somewhat fashionable subject. It is as well to emphasize, therefore, that much of it is not new, but has been developed over many years by engineers whose patient work in an often apparently unrewarding but vital field is the mark of true professionalism.

It is appropriate to emphasize, moreover, that accidents arising from relatively mundane situations and activities are still responsible for many more deaths and injuries than those due to advanced technology.

Nevertheless, loss prevention has developed in response to the growth of a new problem, the hazard of high technology processes, and it does have a distinctive approach and some novel techniques. Particularly characteristic are the emphasis on matching the management system to the depth of technology in the installation, the techniques developed for identifying hazards, the principle and methods of quantifying hazards, the application of reliability assessment, the practice of planning for emergencies and the critique of traditional practices or existing codes, standards or regulations where these are outdated by technological change.

There is an enormous, indeed intimidating, literature on safety and loss prevention. In addition to the symposia already referred to, mention may be made of the *Handbook of Safety and Accident Prevention in Chemical Operations* by Fawcett and Wood (1965); the *Handbook of Industrial Loss*

Prevention by the Factory Mutual Engineering Corporation (1967); and the Industrial Safety Handbook by Handley (1969, 1977). These publications, which are by multiple authors, are invaluable source material.

There is a need, however, in the author's view for a balanced and integrated textbook on loss prevention in the process industries which presents the basic elements of the subject, which covers the recent period of intense development and which gives a reasonably comprehensive bibliography. The present book is an attempt to meet this need.

The book is based on lectures given to undergraduate and postgraduate students at Loughborough over a period of years and the author gladly acknowledges their contribution.

Loss prevention is a wide and rapidly developing field and is therefore not an easy subject for a book. Nevertheless, it is precisely for these reasons that the engineer needs the assistance of a textbook and that the attempt has been considered justified.

The structure of the book is as follows. Chapter 1 deals with the background to the historical development of loss prevention, the problem of large, single-stream plants, and the differences between loss prevention and conventional safety, and between loss prevention and total loss control; Chapter 2 with hazard, accident and loss, including historical statistics; Chapter 3 with the legislation and legal background; Chapter 4 with the control of major hazards; Chapter 5 with economic and insurance aspects; Chapter 6 with management systems, including management structure, competent persons, systems and procedures, standards and codes of practice, documentation and auditing arrangements; Chapter 7 with reliability engineering, including its application in the process industries; Chapter 8 with the spectrum of techniques for identifying hazards from research through to operation; Chapter 9 with the assessment of hazards, including the question of acceptable risk; Chapter 10 with the siting and layout of plant; Chapter 11 with process design, including application of principles such as limitation of inventory, consideration of known hazards associated with chemical reactors, unit processes, unit operations and equipments, operating conditions, utilities, particular chemicals and particular processes and plants, and checking of operational deviations; Chapter 12 with pressure system design, including properties of materials, design of pressure vessels and pipework, pressure vessel standards and codes, equipment such as heat exchangers, fired heaters and rotating machinery, pressure relief and blowdown arrangements, and failure in pressure systems; Chapter 13 with design of instrumentation and control systems, including regular instrumentation, process computers and protective systems; Chapter 14 with human factors in process control, process operators, computer aids and human error; Chapter 15 with loss of containment and dispersion of material; Chapter 16 with fire, flammability characteristics, ignition sources, flames and particular types of process fire, effects of fire and fire prevention, protection and control; Chapter 17 with explosion, explosives, explosion energy, particular types of process explosion such as confined explosions, unconfined vapour cloud explosions and dust explosions, effects of explosion and explosion prevention, protection and relief; Chapter 18 with toxicity of chemicals, toxic release and effects of toxic release; Chapter 19 with commissioning and inspection of plant; Chapter 20 with plant operation; Chapter 21 with plant maintenance

and modification; Chapter 22 with storage; Chapter 23 with transport, particularly by road, rail and pipeline; Chapter 24 with emergency planning both for works and transport emergencies; Chapter 25 with various aspects of personal safety such as occupational health and industrial hygiene, dust and radiation hazards, machinery and electrical hazards, protective clothing and equipment, and rescue and first aid; Chapter 26 with accident research; Chapter 27 with feedback of information and learning from accidents; Chapter 28 with safety systems, including the roles of safety managers and safety committees and representatives. There are appendices on Flixborough, Seveso, case histories, standards and codes, institutional publications, information sources, laboratories and pilot plants, pollution and noise, failure and event data, Canvey, model licence conditions for certain hazardous plants, and units and unit coversions.

Many of the matters dealt with, such as pressure vessels or process control, are major subject areas in their own right. It is stressed, therefore, that the treatment given is strictly limited to loss prevention aspects. The emphasis is on deviations and faults which may give rise to loss.

In engineering in general and in loss prevention in particular there is a conflict between the demand for a statement of basic principles and that for detailed instructions. In general, the first of these approaches has been adopted, but the latter is extremely important in safety, and a considerable amount of detailed material is given and references are provided to further material.

The book is intended as a contribution to the academic education of professional chemical and other engineers. Both educational and professional institutions have long recognized the importance of education in safety. But until recently the rather qualitative, and indeed often exhortatory, nature of the subject frequently seemed to present difficulties in teaching at degree level. The recent quantitative development of the subject goes far towards removing these objections and to integrating it more closely with other topics such as engineering design.

In other words, loss prevention is capable of development as a subject presenting intellectual challenge. This is all to the good, but a note of caution is appropriate. It remains true that safety and loss prevention depend primarily on the hard and usually unglamorous work of engineers with a strong sense of responsibility, and it is important that this central fact should not be obscured.

For this reason the book does not attempt to select particular topics merely because a quantitative treatment is possible or to give such a treatment as an academic exercise. The subject is too important for such an approach. Rather the aim has been to give a balanced treatment of the different aspects and a lead in to further reading.

It is also hoped that the book will be useful to practising engineers in providing an orientation and entry to unfamiliar areas. It is emphasized, however, that in this subject above all others, the specialized texts should be consulted for detailed design work.

Certain topics which are often associated with loss prevention, for example included in loss prevention symposia, have not been treated in detail. These include, for example, pollution and noise. The book does not attempt to deal in detail with total loss control, but a brief account of this is given.

The treatment of loss prevention given is based mainly on the chemical, petrochemical and petroleum industries, but much of it is relevant to other process industries, such as electrical power generation (conventional and nuclear), iron and steel, gas, cement, glass, paper and food.

The book is written from the viewpoint of the United Kingdom and, where differences exist within the UK, of England. This point is relevant mainly to legislation.

Reference is made to a large number of procedures and techniques. These do not all have the same status. Some are well established and perhaps incorporated in standards or codes of practice. Others are more tentative. As far as possible the attempt has been made to give some indication of the extent to which particular items are generally accepted.

There are probably also some instances where there is a degree of contradiction between two approaches given. In particular, this may occur where one is based on engineering principles and the other on relatively arbitrary rules-of-thumb.

The book does not attempt to follow standards and codes of practice in drawing a distinction between the words *should, shall* and *must* in recommending particular practices and generally uses only the former. The distinction is important, however, in standards and codes of practice and it is described in Appendix 4[a].

An explanation of some of the terms used is in order at this point. Unfortunately there is at present no accepted terminology in this field. In general, the problems considered are those of loss, either of life or property. The term *hazard* is used to describe the object or situation which constitutes the threat of such loss. The consequences which might occur if the threat is realized are the *hazard potential*. Associated with the hazard there is a risk, which is the probability of the loss occurring. Such a risk is expressed as a *probability* or as a *frequency*. Probability is expressed as a number in the range 0 to 1 and is dimensionless; frequency is expressed in terms of events per unit time, or sometimes in other units such as events per cycle or per occasion. Rate is also used as an alternative to frequency and has the same units.

The analysis of hazards involves qualitative *hazard identification* and quantitative hazard assessment. The latter term is used to describe both the assessment of hazard potential and of risk. The assessment of risk only is described as *risk assessment*.

In accident statistics the term *Fatal Accident Frequency Rate* (FAFR) has some currency. The last two terms are tautologous and the quantity is here referred to as *Fatal Accident Rate* (FAR).

Further treatments of terminology in this field are given by BS 4200: 1967, by Green and Bourne (1962), by the Council for Science and Society (1977) and by Harvey (1979b).

Notation is defined for the particular chapter at the point where the symbols first occur. In general, a consistent notation is used, but well established equations from standards, codes and elsewhere are usually given in the original notation. A consolidated list of the notation is given at the end of chapters in which a large number of symbols is used.

The units used are in principle SI, but the exceptions are fairly numerous. These exceptions are dimensional equations, equations in standards and codes, and other equations and data given by other workers where conversion has seemed undesirable for some reason. In cases of conversion

from a round number it is often not clear what degree of rounding off is appropriate. In cases of description of particular situations it appears pedantic to make the conversion where a writer has referred, for example, to a 1 inch pipe.

Notes on some of the units used are given in Appendix 12[a]. For convenience a unit conversion table is included in this appendix. Numerical values given by other authors are generally quoted without change and numerical values arising from conversion of the units of data given by other authors are sometimes quoted with an additional significant figure in order to avoid excessive rounding of values.

Some cost data are quoted in the book. These are given in pounds or US dollars for the year quoted.

A particular feature of the book is a fairly extensive bibliography of some 5000 references. These references are consolidated at the end of the book rather than at the end of chapters, because many items are referred to in a number of chapters. Lists of selected references on particular topics are given in table form in the relevant chapters.

Certain institutions, however, have a rather large number of publications which it is more convenient to treat in a different manner. These are tabulated in Appendices 4[a] and 5[a], which contain some 2000 references. There is a cross-reference to the institution in the main reference list.

In many cases institutions and other organizations are referred to by their initials. In all cases the first reference in the book gives the full title of the organization. The initials may also be looked up in the Author Index, which gives the full title.

A reference is normally given by quoting the author and, in brackets, the date, e.g. Kletz (1971). Publications by the same author in the same year are denoted by letters of the alphabet a, b, c, etc., e.g. Allen (1977a), while publications by authors of the same surname and in the same year are indicated for convenience by an asterisk against the year in the list of references. In addition, the author's initials are given in the main text in cases where there may still be ambiguity. Where a date has not been determined this is indicated as n.d.

In the case of institutional publications listed in Appendices 4[a] and 5[a] the reference is given by quoting the institution and, in brackets, the date, the publication series, e.g. HSE (1965 HSW Bklt 34) or the item number, e.g. IChemE (1971 Item 7). For institutional publications with a named author the reference is generally given by quoting the author and, in brackets, the initials of the institution, the date and the publication series or item number, e.g. Eames (UKAEA 1965 Item 4).

The field of loss prevention is currently subject to very rapid change. In particular, there is a continuous evolution of standards and codes of practice and legislation. It is important, therefore, that the reader should make any necessary checks on changes which may have occurred.

I would like to thank for their encouragement in this project Professor D.C. Freshwater and the publishers, and to acknowledge the work of many authors which I have used directly or indirectly, particularly that of Dr J.H. Burgoyne and of Professor T.A. Kletz. I have learned much from my colleagues on the Loss Prevention Panel of the Institution of Chemical Engineers, in particular Mr T.A. Kantyka and Mr F. Hearfield, and on the Advisory Committee on Major Hazards, especially the chairman Professor B.H. Harvey, the secretary Mr H.E. Lewis, my fellow group chairmen

[a]Appendices 4, 5 and 12 in the first edition correspond to Appendices 27, 28 and 30, respectively, in this second edition.

Professor F.R. Farmer and Professor J.L.M. Morrison and the members of Group 2, Mr K. Briscoe, Dr J.H. Burgoyne, Mr E.J. Challis, Mr S. Hope, Mr M.A. McTaggart, Professor J.F. Richardson, Mr J.R.H. Schenkel, Mr R. Sheath and Mr M.J. Turner, and also from my university colleagues Dr P.K. Andow, Mr R.J. Aird and Dr D.J. Sherwin and students Dr S.N. Anyakora, Dr B. Bellingham, Mr C.A. Marpegan and Dr G.A. Martin-Solis. I am much indebted to Professor T.A. Kletz for his criticisms and suggestions on the text. My thanks are due also to the Institution of Plant Engineers, which has supported plant engineering activities at Loughborough, to the Leverhulme Trust which awarded a Research Fellowship to study *Loss Prevention in the Process Industries* and to the Science Research Council, which has supported some of my own work in this area. I have received invaluable help with the references from Mrs C.M. Lincoln, Mrs W. Davison, Mrs P. Graham, Mr R. Rhodes and Mrs M.A. Rowlatt, with the typing from Mrs E.M. Barradell, Mrs P. Jackson and, in particular, Mrs J. Astley, and with the production from Mr R.L. Pearson and Mr T. Mould. As always in these matters the responsibility for the final text is mine alone.

FRANK P. LEES
Loughborough,
1979

Acknowledgements

Many have helped with the long and tedious production process for this 3rd edition of **Frank P. Lees Loss Prevention in the Process Industries**. However, I would be seriously remiss if I did not start with Frank P. Lees himself. I have always had great respect and admiration for Frank Lees, the engineer and the educator. After Frank passed away, and late in 2002, our mutual friend, Roy Sanders contacted me with a request from Frank's publisher to take up the job of editing the next version. I had no idea of the enormity of the task and Frank's accomplishment. I must say that this experience has truly humbled me and has given me a new respect and new appreciation for Frank Lees' knowledge, grasp, and mastery of the subject of safety and loss prevention. I am honored and privileged to be the editor of the 3rd edition of **Frank P. Lees Loss Prevention in the Process Industries**. I have tried to maintain Frank's high standards and appreciate the help that many others have given me, but ultimately any *faux pas* or mistakes in the 3rd edition are all mine.

I want to thank all the people who have helped revise and update the chapters that were carried over from the 2nd edition. In this respect, I would like to mention A.A. Aldeeb, J.A. Alderman, P.C. Berwanger, C.A. Brown, S. Chervin, J. Cowling, W.F. Early, M. Gentile, J.P. Gupta, J.A. Havens, D.C. Hendershot, N. Keren, K. Krishna, S.H. Landis, M.E. Levin, P.N. Lodal, M.E. Livingston, S. Mohindra, D.A. Moore, T.M. O'Connor, S.W. Ostrowski, M. Paradise, J.A. Philley, W.J. Rogers, R.E. Sanders, S.R. Saraf, M.E Sawyer, R.L. Smith, T.O. Spicer, A.E. Summers, Y. Wang, H.H. West, J.L. Woodward, and J. Zanoni. Special thanks go to individuals who helped with the new chapters and appendices: J.A. Philley for the incident investigation chapter; J.P. Gupta and D.C. Hendershot for the inherently safer design chapter; S. Chervin, M.E. Levin, and W.J. Rogers for the reactive chemicals chapter; A.E. Summers for the safety instrumented systems chapter, D.A. Moore for the chemical security chapter and T.M. O'Connor for the new appendix on incident databases.

I also want to thank those who helped with research, literature review, galley proofing, and many other jobs without which such a project is never complete. I would like to specifically mention S. Anand, L.O. Cisneros-Trevino, W. Kowhakul, Y.-S. Liu, Y. Qiao, V. Raghunathan, S. Rajaraman, J. Suardin, M. Vidal, C. Wei, C. Zhang, and Y. Zhou. I would like to make specific mention of K.L. Potucek and L.J. Littlefield who undertook the painful task of converting the pdf files of the 2nd edition to Word format. I would like to thank M.A. Cass and D.K. Startz for administrative support and keeping things going through the long haul.

I would also like to take this opportunity to thank two individuals who have provided more than moral support and encouragement. They are Trevor Kletz and Mike O'Connor. First, I have appreciated Trevor's support for our work at the Mary Kay O'Connor Process Safety Center. I also want to thank him for his pioneering leadership in many areas of process safety. Trevor is really a giant in this field, and as Roy Sanders likes to say, "Trevor does cast a long shadow in the field of process safety." I am proud to have Trevor's friendship and his support. I also want to take this opportunity to thank Mike O'Connor. He has converted his personal tragedy into an opportunity to integrate process safety into education, research, and service activities at universities. He has been a true friend and supporter of the mission and goals of the Center.

I also want to thank Mrs. Elizabeth Lees for being kind, gracious, and helpful throughout this process. I am especially grateful for her hospitality and help during my visit to Loughborough to look through Frank's notes and books.

I am grateful to my mother for her love and everything she has done for me and continues to do. I am indebted to my father for having taught me the desire for learning. He was a humble man who truly believed that, "If you stop learning, you must already be dead." I would also like to thank my wife, Afroza, and our daughters, Joya and Rumki, for their love and support. Quite often a project of this magnitude causes the personal and family life to suffer and my wife and daughters have borne the brunt of it. I appreciate their understanding.

I am sure I have failed to mention everybody who has helped me with the many facets of this project. To them I apologize and want to assure them the oversight is not intended. To everyone who has helped put this 3rd edition together, I extend my heartfelt thanks and warmest gratitude again. I must say that you have helped in a noble cause, spread the word and disseminate information that will, we hope and pray, lead to a safer industry and society.

M. SAM MANNAN
College Station,
Texas, USA
2004

Terminology

Attention is drawn to the availability in the literature of a number of glossaries and other aids to terminology. Some British Standard glossaries are given in Appendix 27 and other glossaries are listed in Table 1.1.

Notation

In each chapter a given symbol is defined at the point where it is first introduced. The definition may be repeated if there has been a significant gap since it was last used. The definitions are summarized in the notation given at the end of the chapter. The notation is global to the chapter unless redefined for a section. Similarly, it is global to a section unless redefined for a subsection and global to a subsection unless redefined for a set of equations or a single equation. Where appropriate, the units are given, otherwise a consistent system of units should be used, SI being the preferred system. Generally the units of constants are not given; where this is the case it should not be assumed that a constant is dimensionless.

Use of References

The main list of references is given in the section entitled References, towards the end of the book. There are three other locations where references are to be found. These are Appendix 27 on standards and codes; Appendix 28 on institutional publications; and in the section entitled Loss Prevention Bulletin which follows the References.

The basic method of referencing an author is by surname and date, e.g. Beranek (1960). Where there would otherwise be ambiguity, or where there are numerous references to the same surname, e.g. Jones, the first author's initials are included, e.g. A. Jones (1984). Further guidance on names is given at the head of the section References.

References in Appendices 27 and 28 are by institution or author. Some items in these appendices have a code number assigned by the institution itself, e.g. API (1990 Publ. 421), but where such a code number is lacking, use is generally made of an item number separated from the date by a slash, e.g. IChemE (1971/13). Thus typical entries are

API Std 2000: 1992	a standard, found in Appendix 27 under American Petroleum Institute
API (1990 Publ. 421)	an institutional publication, found in Appendix 28 under American Petroleum Institute
HSE (1990 HS(G) 51)	an institutional publication, found in Appendix 28 under Health and Safety Executive, Guidance Booklets, HS(G) series
Coward and Jones (1952 BM Bull. 503)	an institutional publication, found in Appendix 28 under Bureau of Mines, Bulletins

Institutional acronyms are given in the section Acronyms which precedes the Author Index.

There are several points of detail which require mention concerning Appendix 28. (1) The first part of the appendix contains publications of a number of institutions and the second part those of the Nuclear Regulatory Commission. (2) The Fire Protection Association publications include a number of series which are collected in the *Compendium of Fire Safety Data* (CFSD). A typical reference to this is FPA (1989 CFSD FS 6011). (3) The entries for the Health and Safety Executive are quite extensive and care may be needed in locating the relevant series. (4) The publications of the Safety and Reliability Directorate appear under the UK Atomic Energy Authority, Safety and Reliability Directorate. A typical reference is Ramskill and Hunt (1987 SRD R354). These publications are immediately preceded by the publications of other bodies related to the UKAEA, such as the Health and Safety Branch, the Systems Reliability Service and the National Centre for Systems Reliability.

References to authors in the IChemE *Loss Prevention Bulletin* are in the style Eddershaw (1989 LPB 88), which refers to issue 88 of the bulletin.

List of Contributors

M. Sam Mannan, Mary Kay O'Connor Process Safety Center, Texas A&M University
Harry H. West, Mary Kay O'Connor Process Safety Center, Texas A&M University
Jerry A. Havens, University of Arkansas
Joseph Zanoni
Scott W. Ostrowski, Exxon-Mobil
Mike Sawyer, Apex Safety Corporation
Mike Livingston, WS Atkins
Carl Brown, Halliburton
John Alderman, RRS Engineering
Pat Berwanger, Berwanger, Inc.
Thomas O. Spicer, University of Arkansas
Dennis C. Hendershot, Rohm and Haas
Jai P. Gupta, Indian Institute of Technology
Angela E. Summers, SIS-Tech Solutions
Robert L. Smith, Eli Lilly
John Woodward, Baker Engineering
Sanjeev R. Saraf, Mary Kay O'Connor Process Safety Center, Texas A&M University
Pete Lodal, Eastman Chemical
Nir Keren, Mary Kay O'Connor Process Safety Center, Texas A&M University
T. Michael O'Connor, Mary Kay O'Connor Process Safety Center, Texas A&M University
Yanjun Wang, Mary Kay O'Connor Process Safety Center, Texas A&M University
Sanjeev Mohindra, ioMosaic, Inc.
Jack Philley, Baker Engineering
Marc. E. Levin, Shell Global Solutions
Roy E. Sanders, PPG Industries
Kiran Krishna, Mary Kay O'Connor Process Safety Center, Texas A&M University
Michela Gentile, Mary Kay O'Connor Process Safety Center, Texas A&M University
Abdulrehman A. Aldeeb, Mary Kay O'Connor Process Safety Center, Texas A&M University
William J. Rogers, Mary Kay O'Connor Process Safety Center, Texas A&M University
Spencer Landis
Mark Paradise
David A. Moore, AcuTech, Inc.
William F. (Skip) Early, Early Consulting
Jim Cowling
Sima Chervin, Kodak

Contents of Volume 1

Contents of Volume 2

Contents of Volume 3

Appendix 1

Case Histories

Contents

An essential feature of the learning process in safety and loss prevention is the study of case histories.

A1.1 Incident Sources

The powerful impact of the internet has changed the way we obtain our information on process safety as well as everything else. In recent years, various governmental agencies and private organizations have improved our access to details of recent and past case histories. A number of websites provide a narrative of the event, the technical reasons for the accident as well as the management system shortcomings and valuable lessons learned. Written report sometimes supported by graphics and photos are often published by various agencies and provided on governmental websites. There are also university and corporate organizations that collect incidents and publish on their websites.

Since about 1996, the US Environmental Protection Agency under the Chemical Emergency Preparedness and Prevention has investigated and reported on many high profile US chemical plant and refinery case histories. As a result of these investigations, some Alerts focusing on certain type of incidents have been developed. EPA reports are available at the following internet website: http://yosemite.epa.gov/oswer/ceppoweb.nsf.

The US Chemical Safety and Hazard Investigation Board (CSB) matured in the late 1990s and their mission is to promote the prevention of chemical accidents. The focus of the CSB is on-site and off-site chemical safety, determining causes and preventing chemical-related incidents, fatalities, injuries, property damage and enhancing environmental protection. The CSB started investigations on selected important chemical plant and refinery incidents in about 1998, and also provide reports on the internet at the following web address: http://www.chemsafety.gov/reports/.

The Center of Chemical Process Safety, CCPS, of the American Institute of Chemical Engineers started providing brief one-page summaries of types of hazards in late 2001. These high-impact messages were designed with the chemical process operator and others involved with the manufacturing processes to focus quickly on an important topic. The publication is entitled the *Beacon* and can be found at: www.aiche.org/ccps/safetybeacon.htm.

The Mary Kay O'Connor Process Safety Center at Texas A&M University in College Station is an international leader in process safety and is increasing in stature each year. They have a rich source of process safety information as well as documented case histories. The Center's web address: http://process-safety.tamu.edu.

The British Institution of Chemical Engineers (IChemE) offers links to a comprehensive list of well over a 150 other websites of incident reports. The IChemE site mainly describes incidents of interest to chemical or process engineers, and includes a time span of incidents from historical to today's events. The collection includes both information and viewpoint. The address of this site is: http://slp.icheme.org/incidents.html.

The Delft University of Technology of The Netherlands offers a vivid array of about two dozen outstanding fire and explosion photos, an accident database and more. The university's main research effort is focused on measuring and modelling dust and gas explosions. The incident photos can be found at: http://www.dct.tudelft.nl/part/explosion/gallery.html.

Two other excellent internet sources of process safety information including case histories are:

- www.acusafe.com
- www.hazardview.com

The AcuSafe website provides a comprehensive electronic newsletter entitled AcuSafe News. Their newsletter is devoted to serving industry, government and interested members of the public by providing process safety practices, incident news, lessons, learned and regulatory developments. Hazard View is a source of advice, comment and information for safety, environment and risk professionals with major accident hazards in the process industries. Their View Library has a number of vivid images of the damages from accidents.

A booklet entitled *Large Property Damage Losses in the Hydrocarbon-Chemical Industry – A Thirty Year Review*, continues to be an excellent source of information in print form. This publication presents summaries of large property losses of incidents from around the globe. The valuable information is from a property insurance viewpoint and is presented in an easy-to-understand form. Most of the damages are the result of fires and explosions occurring in refineries, petrochemical plants, gas-processing plants, terminals, offshore and a miscellaneous category. It is updated every few years and the latest one, the twentieth edition, was released in January 2003.

Collections of case histories have been published in *Case Histories of Accidents in the Chemical Industry* by the Manufacturing Chemists Association (MCA) (1962–/1–4) and by Kier and Muller (1983). Specialized collections have been published on particular topics such as major hazards (Harvey, 1979, 1984; Carson and Mumford, 1979; Lees, 1980b; V.C. Marshall, 1987), fire and explosion (Vervalin, 1964a, 1973a; W.H. Doyle, 1969), vapour cloud explosions (Gugan, 1979; Slater, 1978a; D.J. Lewis, 1980d; Davenport, 1987; Lenoir and Davenport, 1993), LPG (L.N. Davis, 1979), instrumentation (W.H. Doyle 1972a,b; Lees, 1976b), transport (Haastrup and Brockhoff, 1990) and pipelines (Riley, 1979). The series *Safety Digest of Lessons Learned* by the API gives case histories with accompanying analyses. Another such series is that published by the American Oil Company (Amoco) of which *Hazards of Water* and *Hazards of Air* are the best known. The Convey Reports and the Rijmond Report give information on certain incidents. The reports of the HSE provide case histories of investigations by a regulator as do those of the NTSB, which deal with rail, road, pipeline and marine accidents. Incidents involving large economic loss are described in the periodic review *100 Large Losses* by Marsh and McLennan.

Case histories are also given in the *Annual Report of HM Chief Inspector of Factories*, the *Chemical Safety Summary of the Chemical Industry Safety and Health Council* (CISHC) and in journals such as *Petroleum Review* and *NFPA Quarterly*. In addition, the *Loss Prevention Bulletin* issued by the IChemE gives case histories. There are also numerous case histories described in much greater detail in various papers and reports. The disasters at Flixborough (R.J. Parker, 1975), at Bantry Bay (Costello, 1979) and on Piper Alpha (Cullen, 1990) are among the most thoroughly documented.

The analysis of case histories to draw the relevant lessons is exemplified pre-eminently in the work of Trevor Kletz's many books and numerous technical articles:

- *Critical Aspects of Safety and Loss Prevention* (1990b),
- *Lessons from Disaster – How Organisations have No Memory and Accidents Recur* (1993##),
- *What Went Wrong? – Case Histories of Process Plant Disasters*, 4th edn (1998##)
- *An Engineer's View of Human Error*, 3rd edn (2001##A) and
- *Learning from Accidents*, 3rd edn (2001##B) and numerous technical articles.

Additional lessons can be examined in Roy Sanders' *Chemical Process Safety: Learning from Case Histories* (1999##).

Selected references on case histories are given in Table A1.1. Some major incidents are listed in Table A1.2. A proportion of these are described in Section A1.10.

Table A1.1   Selected references on case histories

Eddy, Potter and Page (1976); J.R. Nash (1976); Carson and Mumford (1979); Nancekieville (1980); D.J. Lewis (1984a, 1993); Kletz (1981a, 1984m, 1985o,p, 1988d,h,i,n, 1993b); Buhrow (1985,1986); Kharbanda and Stallworthy (1988); Garrison (1988a,b); Koehorst (1989); Haastrup and Brockhoff (1990,1991); Marsh and McLennan (1991 LPB 99); Marsh and McLennan (1991 LPB 99); O'Donovan (1991 LPB 99); Palmer and Marshall (1991 LPB 99); Anon. (1992 LPB 99, p. 1); Chowdhury and Parkinson (1992); Lenoir and Davenport (1992); Taylor (1993 LPB 112); I.E. Thomas (1995)

Databases
Service Informatics and Analysis (1982); Amesz *et al.* (1983); K.R. Davies (1983); Makstenek *et al.* (1992); Winder *et al.* (1992); Bruce (1994); Koivisto and Nielsen (1994); Steffy (1994);
Accident reporting: Badoux (1983)
MHIDAS: SRD (1990)
FACTS: Koehorst and Bockholtz (1991)
MARS: Amendola, Contini and Nichele (1988); Drogaris (1991, 1993)
FIRE (warehouses): Koivisto and Nielsen (1994)

Major hazard control
Anon (1979c); Anon. (1985n); Anon. (1986o); Anon. (1990 LPB 95, p. 4)

Hazard identification
Sellers (1988); Sellers and Picciollo (1988); Cullen (1990)

Process design
Anon. (1992 LPB 107, p. 27)

Chemical reactors
Anon. (1962a); Zielinski (1967, 1973); Fowler and Spiegelman (1968); Bast (1970); Grewer (1970); Valpiana (1970a,b); Zehr (1970); Vincent (1971); Fire J. Staff (1973a,d); Bloore (1974); Gugan (1974b); HM Chief Inspector of Factories (1974); Anon. (1975 LPB 1, p. 10); Anon. (1975 LPB 3, p. 1); Anon (1975 LPB 5, p. 16); Anon. (1977 LPB 13, p. 13); Tong, Seagrave and Widerhorn (1977); Anon. (1979 LPB 28, 29, 30); Skinner (1981); Bond (1985 LPB 65); Anon. (1986

LPB 68, p. 33); Berkey and Workman (1987); Anon. (1989 LPB 90, p. 29); Levy and Penrod (1989); Anon. (1991 LPB 98, p. 7); Kotoyori (1991); Anon. (1993 LPB 113, p. 25); Anon. (1994 LPB 115, p. 1); Anon. (1994 LPB 117, p. 18); Anon. (1994 LPB 118, p. 8); Dixon-Jackson (1994)

Petrochemical plants
Britton (1994)

Utilities
Dowell (1994a)

Ammonia, urea and ammonium nitrate plants
R.W. James and Lawrence (1973); Anon. (1975h); Din (1975); Esrig, Ahmad and Mayo (1975); Henderson (1975); Lonsdale (1975); Osman (1975); K.Wright (1975); Visvanathan (1976); Kusha (1977); P.C. Campbell (1981); Janssen, Siraa and Blanken (1981); Roney and Persson (1984); Anon. (1987 LPB 78, p. 11)

DSTG
FPC, Bureau of Natural Gas (1973); L.N. Davis (1979)

Pressure systems and components
Bohlken (1961); Saibel (1961a,b); Attebery (1970); R.W. Miller and Caserta (1971); Banks (1973); Gupton and Kreisher (1973); R.W. James and Lawrence (1973); van der Horst and Sloan (1974); Appl (1975); Blanken (1975); Fuchs and Rubinstein (1975); Lonsdale (1975); Steele (1975); Kletz (1976b, 1987a); National Vulcan (1977); Kinsley (1978); Makhan and Honse (1978); Perkins (1980 LPB 33); Anon. (1981 LPB 37, p. 23); Connaughton (1981); Blanken and Groefsma (1983); Anon. (1984 LPB 56, p. 26); R.W. Clark and Connaughton (1984); NBS (1986); Anon. (1988 LPB 83, p. 1, 3, p. 11,15,17 and 19); Anon. (1989 LPB 86, p. 22 and 27); Anon. (1989 LPB 94, p. 16); Droste and Mallon (1989); Anon. (1990 LPB 93, p. 23); Rademayer (1990); Anon. (1991 LPB 97, p. 9); Smith (1991 LPB 102); Anon. (1992 LPB 104, p. 27); Anon. (1992 LPB 107, p. 27); Crawley (1992 LPB 104); Anon. (1993 LPB 111, p. 25); Anon. (1993 LPB 113, p. 13); Anon. (1994 LPB 117, p. 24); Clayton and Griffin (1994)

Process machinery
OIA (n.d./7); Gibbs (1960); Rendos (1967); Shield (1967, 1972); Naughton (1968); North and Parr (1968); Telesmanic (1968); Zech (1968); Zimmerman (1968); D.S. Wilson *et al.* (1970); von Nimitz, Wachel and Szenasi (1974); Anon. (1975f); Roney (1975); Novacek (1983); Anon. (1985 LPB 62, p. 27); Anon. (1990 LPB 91, p. 16); Anon. (1990 LPB 92, p. 15); Anon. (1994 LPB 116, p. 21)

Instruments
MCA (1962–/1–4); W.H. Doyle (1969, 1972a,b); Fritz (1969); H.D. Taylor and Redpath (1971, 1972); Kletz (1974b,d); McLain (1975b); K.Wright (1975); Lees (1976b); Griffin and Garry (1984); Anon. (1986 LPB 68, p. 35); Anon. (1992 LPB 104, p. 11); Anon. (1994 LPB 116, p. 22)

Computer control systems
Nimmo, Nunns and Eddershaw (1987, 1993 LPB 111)

Human error
Speaker, Thompson and Luckas (1982); Anon. (1983i); Anon. (1989b); Anon. (1990 LPB 92, p. 13)

Emissions, leaks, releases
Schneidman and Strobel (1974); Anon. (1980 LPB 36, p. 25); Anon. (1982 LPB 43, p. 21); Edwards (1982 LPB 47);

Christiansen and Jorgensen (1983); Anon. (1986 LPB 70, p. 1); Anon. (1989 LPB 85, p. 27); Anon. (1989 LPB 90, p. 13); Anon. (1990 LPB 91, p. 16); Anon. (1992 LPB 104, p. 27); Anon. (1994 LPB 117, p. 24)

Fire

OIA (Publ. 651, 1972 Publ. 302, 1973); Woodworth (1958); Lamond (1962); MCA (1962–/1–4); Pinney (1962); Rendos (1964); Vervalin (1964a,c, 1973a,b); Anon. (1966d); Ashill (1966); Cowles (1966); McCarey (1966); Darling (1967); Ostroot (1967, 1973); Zielinski (1967, 1973); Anon. (1968b); Leroy and Johnson (1969); Orey (1972a); Searson (1972, 1990 LPB 94); Fire J. Staff (1973b,e); Fowle (1973); Garrad (1973); NFPA (1973/10); St. Clair (1973); Beausoleil, Phillips and Snell (1974); Sharry and Walls (1974); Anon. (1975 LPB 0, p. 13); McLain (1975b); Anon. (1976h,k); Kletz (1976g, 1979a, 1984 LPB 58, 1993 LPB 114); Saia (1976); Anon. (1978d); HSE (1978c); Isman (1978); Lathrop (1979); Anon. (1980u); Anon. (1981p); Crain (1981); Oosterling and Orbons (1981); Anon. (1982 LPB 43, p. 2); Anon. (1983m); Anon. (1983 LPB 51, p. 13); FM (1983); Anon. (1984 LPB 57, p. 19); Mullier, Rustin and Hecke (1984); Mumford (1984 LPB 57); Anon. (1985 LPB 64, p. 13); Bond (1985 LPB 65, 1991); Carson, Mumford and Ward (1985 LPB 65); Kohler (1985); Politz (1985); Anon. (1986 LPB 69, p. 17 and 33); Anon. (1986 LPB 71, p. 27); Lask (1986); McCoy, Dillenback and Truax (1986); Steinbrecher (1986, 1989 LPB 88); Anon. (1987 LPB 77, p. 11 and 27); V.C. Marshall (1987, 1990 LPB 95, 1994 LPB 116); Anon. (1988 LPB 79, p. 23); Anon. (1988 LPB 82, p. 19); Anon. (1988 LPB 83, p. 13); Anon. (1988 LPB 84, p. 19); Armstrong (1988 LPB 83); Schwab (1988a, 1988 LPB 83); Anon. (1989 LPB 87, p. 15); Anon. (1989 LPB 88, p. 19); B. Browning and Searson (1989); IMechE (1989/111); Anon. (1990g); Anon. (1990 LPB 91, p. 22); Anon. (1990 LPB 94, p. 26 and 30); Anon. (1990 LPB 95, p. 19); Adelman and Adelman (1990 LPB 94); Perunicic and Skotovic (1990 LPB 96); Anon. (1991 LPB 97, p. 12); Anon. (1991 LPB 98, p. 9); Kohvaporrand (1991); Anon. (1991 LPB 102, p. 35); Anon. (1992 LPB 103, p. 24); Anon. (1992 LPB 104, p. 7, 9 and 13); Anon. (1992 LPB 105, p. 15); Carson and Mumford (1992 LPB 108, 1993 LPB 109, 110); Kirk (1992 LPB 105); Prine (1992); Anon. (1993 LPB 113, p. 27); Anon. (1993 LPB 114, p. 19); Anon. (1994 LPB 115, p. 6); Anon. (1994 LPB 120, p. 10 and 19)

Fire: static electricity

Klinkenberg and van der Minne (1958); Bustin (1963); Eichel (1967); Muller-Hillebrand (1963); van der Meer (1971); Anon. (1979 LPB 29, p. 134); Anon. (1983 LPB 52, p. 19); H.R. Edwards (1983); M.R.O. Jones and Bond (1984, 1985); Liittgens (1985); Anon. (1987 LPB 78, p. 9); Owens (1988); Anon. (1994 LPB 118, p. 14)

Explosions

OIA (Publ. 651, 1973); Assheton (1930); Jacobs *et al.* (1957); F.C. Price (1960); Matthews (1961); Popper (1963); Alexander and Finigan (1964); Bateman (1964); W.A. Mason (1964); Pipkin (1964); Sakai (1964); Schmitt (1964); Thodos (1964); Walls (1964b); Sorrell (1965); Anon. (1966c); W.L. Ball (1966a); Haseba *et al.* (1966); Wilkinson (1966); Adcock and Weldon (1967b, 1973); Dixon (1967); Dorsey (1967); Lorentz (1967, 1973); J.D. Reed (1967); Bast (1970); Buehler *et al.* (1970); Fritsch (1970); Grewer (1970); Kotzerke (1970); Valpiana (1970a,b); Volpers (1970); Cracknell (1971); Dartnell and Ventrone (1971); K.W. Sanders (1971); GA Campbell and Rutledge (1972); Fire J. Staff (1973d); Freese

(1973); Gozzo and Carraro (1973); Jacobs *et al.* (1973); Meganck (1973); de Oliviera (1973); W.W. Patterson (1973); Vervalin (1973e); Bateman, Small and Snyder (1974); Bloore (1974); Grimm (1974); Gugan (1974b); Hadas (1974); Hartgerink (1974); Kletz (1974b, 1976b,f,x, 1984 LPB 56, 1988); Lockemann (1974); Ludford (1974); Tanimoto (1974); Voros and Honti (1974); Anon. (1975 LPB 1, p. 3 and 15); Anon. (1975 LPB 5, p. 1); Anon. (1975 LPB 6, p. 17); Dooyeweerd (1975); Henderson (1975); Strehlow and Baker (1975, 1976); Anon. (1976 LPB 11, p. 2); Carver (1976); Kotoyori *et al.* (1976); Rebsch (1976); Saia (1976); Anon. (1977d); Tong, Seagraves and Wiederhorn (1977); Anon. (1978 LPB 19, p. 24); Cremer and Warner (1978); Markham and Honse (1978); Biasutti (1979); Anon. (1980y); Ramadorai (1980); Edwards (1982 LPB 47); Anon. (1983a); Bond (1983 LPB 50); Anon. (1983 LPB 51, p. 9); Lloyd (1983); Ursenbach (1983); Anon. (1984 LPB 54, p. 13); Anon. (1985c); HSE (1985d); Schwab (1988a,b, LPB 83); Wrightson and Santon (1988 LPB 83); Anon. (1989 LPB 87, p. 3, 7 and 27); Anon. (1989 LPB 88, p. 17); Anon. (1989 LPB 89, p. 13); Bond (1989 LPB 93); Cronin (1989 LPB 90); Ernst (1989); G.M. Lawrence (1989); MacDiarmid and North (1989); Nightingale (1989a,b); Urben (1989 LPB 80); Anon. (1990d,e,i); Anon. (1990 LPB 91, p. 15); Anon. (1990 LPB 95, p. 15); de Haven and Dietsche (1990); Segrave and Wilson (1990); Anon. (1991d,f,k,l); Anon. (1991 LPB 97, p. 23); Anon. (1991 LPB 98, p. 1 and 18); Anon. (1991 LPB 102, p. 7); Arendt and Lorenzo (1991); Carson and Mumford (1991 LPB 102); Hempseed and Ormsby (1991); Anon. (1992b); Anon. (1992 LPB 103, p. 19); Anon. (1992 LPB 103, p. 31); Anon. (1992 LPB 107, p. 17); S.E. Anderson, Dowell and Mynaugh (1992); Moran (1992); Palazzi (1992); Raheja *et al.* (1992); Anon. (1993 LPB 109, p. 7); Anon. (1993 LPB 109, p. 25 and 26); Anon. (1993 LPB 110, p. 25); Anon. (1993 LPB 112, p. 21); Bergroth (1993 LPB 109); Bond (1993 LPB 101); Wehrum (1993); Whitmore, Gladwell and Rutledge (1993); Zhang and Tang (1993); Anon. (1994 LPB 115, p. 9); Anon. (1994 LPB 116, p. 17)

BLEVEs

Kohler (1985); Selway (1988 SRD R492)

Dust explosions, dust fires

Morozzo (1795); D.J. Price and Brown (1922); NFPA (1957/1); K.C. Brown and James (1962); Remirez (1967); Yowell (1968); Noss (1971); O'Reilly (1972); Tonkin and Berlemont (1972 FRS Fire Res. Note 942); K.N. Palmer (1973a); Pollock (1975); Lathrop (1978); Anon. (1979e); Braun (1979); Snow (1981); Anon. (1983 LPB 51, p. 9); Anon (1989 LPB 90, p. 27); Skinner (1989); Anon. (1990 LPB 95, p. 1, 6 and 13); Kaiser (1991); Senecal (1991); Anon. (1994 LPB 115, p. 9)

Toxic release

Chasis *et al.* (1947); Joyner and Durell (1962); Hoveid (1966); Inkofer (1969); MacArthur (1972); Simmons, Erdmann and Naft (1973, 1974); Eisenberg, Lynch and Breeding (1975); Lonsdale (1975); Luddeke (1975); Anon. (1976n,o); Hoyle and Melvin (1976); HSE (1978b); Harvey (1979b); Shooter (1980); N.C. Harris (1981b); Anon. (1982 LPB 48, p. 13); Prijatel (1983); Sweat (1983); Anon. (1985 LPB 66, p. 23); HSE (1985 LPB 63); Anon. (1987 LPB 75, p. 23); Anon. (1989 LPB 86, p. 22 and 27); Anon. (1989 LPB 89, p. 27); Anon. (1990 LPB 91, p. 19); Anon. (1990 LPB 92, p. 11, 15, 23 and 26); Duclos and Binder (1990); Huvinen (1990 LPB 92); Fallen (1990); M.P. Singh (1990); M.P. Singh, Kumari and Ghosh (1990); Anon. (1991 LPB 97, p. 7); Anon. (1991 LPB

98, p. 25); Anon. (1993 LPB 113, p. 26); Anon. (1993 LPB 112, p. 22)

Plant commissioning
Dyess (1973); Parnell (1973); J.R. Robinson (1973); Severa (1973); Arnot and Hirons (1974); Horsley (1974); Jenkins and Crookston (1974); Butzert (1976); Anon. (1977 LPB 18, p. 2); Bress and Packbier (1977); Givaporert and Eagle (1977); Tandy and Thomas (1979); Kletz (1984k); Grotz, Gosnell and Grisolia (1985); Ruziska *et al.* (1985); Song (1986); W.K. Taylor and Pino (1987); Kneale (1991 LPB 97)

Plant operation
MCA (1962–/1–4); W.H. Doyle (1972a); Kletz (1975b); Lees (1976b); Anon. (1978 LPB 20, p. 41); Anon. (1987 LPB 77, p. 17); Anon. (1989 LPB 88, p. 1); Anon. (1989 LPB 89, p. 9 and 18); T.O. Gibson (1989); Anon. (1990 LPB 92, p. 22); Anon. (1991 LPB 98, p. 19); Anon. (1992 LPB 106, p. 19); Anon. (1992 LPB 107, p. 17); Anon. (1994 LPB 115, p. 7)

Plant maintenance
Simms (1965); Anon. (1978 LPB 24, p. 155); Anon. (1979 LP 27, p. 80); P.C. Campbell (1981); Roney and Persson (1984); Deacon (1988); Anon. (1989 LPB 85, p. 1, 3 and 5); Anon. (1989 LPB 86, p. 17); Anon. (1990 LPB 92, p. 9); Anon. (1991 LPB 97, p. 7); Anon. (1991 LPB 98, p. 20); Anon. (1991 LPB 102, p. 27); Anon. (1991 LPB, p. 20 and 25); Anon. (1992 LPB 103, p. 24, 25 and 29); Anon. (1992 LPB 104, p. 10); Anon. (1992 LPB 107, p. 21 and 22); Anon. (1993 LPB 112, p. 20); Anon. (1994 LPB 116, p. 22); Anon. (1994 LPB 117, p. 17); Anon. (1994 LPB 119, p. 8); Bond (1994 LPB 117)
Construction
Anon. (1991 LPB 107, p. 21 and 22)
Excavation
Anon. (1992 LPB 107, p. 22 and 23)

Plant modification
Anon. (1975 LPB 1, p. 1); R.J. Parker (1975); Booth (1976); Henderson and Kletz (1976); Heron (1976); W.W. Russell (1976); Anon. (1978 LPB 22, p. 101)

Storage
OIA (Publ. 711, 1974 Loss Inf. Bull. 400-1); Vervalin (1964a, 1973a); J.A. Lawrence (1966); Frank and Wardale (1970); J.R. Hughes (1970); A. Nielsen (1971); Litchenberg (1972, 1977); MacArthur (1972); Anon. (1975 LPB 0, p. 2); Henderson (1975); Lonsdale (1975); Anon. (1976 LPB 12, p. 7); Hampson (1976); Anon. (1978 LPB 20, p. 55); Anon. (1979 LPB 26, p. 31); Anon. (1979 LPB 27, p. 68); Gustin and Novacek (1979); Anon. (1980 LPB 32, p. 13); Anon. (1980 LPB 36, p. 25); Anon. (1981 LPB 37, p. 7); Anon. (1981 LPB 41, p. 11); Morgenegg (1982); Anon. (1984 LPB 57, p. 11); Anon. (1986 LPB 69, p. 29); Badame (1986); NBS (1986, 1988); Anon. (1987 LPB 75, p. 19); Anon. (1987 LPB 78, p. 26); Anon. (1988m); Marshall (1988 LPB 82); Anon. (1989 LPB 88, p. 11); Anon. (1992 LPB 106, p. 5). *Warehouses:* Pietersen (1988); Christiansen, Kakko and Koiviso (1993); Koivisto and Nielsen (1994)

Transport
AGA (Appendix 28 Pipeline Accident Reports); NTSB (annual reports); Engel (1969); Inkofer (1969); Medard (1970); Cato and Dobbs (1971); A.W. Clarke (1971b); Kogler (1971); Caserta (1972a); Ass. of Amer. Railroads (1972); Hayes (1972); Fire J. Staff (1973); Lloyds list (1974); Wilde (1974); DoE (1976/6); Anon. (1977 LPB 15, p. 33); Getty, Rickert and Trapp (1977); Merricks (1977); Anon. (1978j);

Anon. (1978 LPB 54, p. 7); Anon. (1979g); Selikoff (1979); Egginton (1980); Anon. (1981m); George (1981a,b); Kletz (1981b); Knife (1982–83); V.J. Clancey (1983); van der Schaaf and Steunenberg (1983); M. Griffltsh and linklater (1984); Hymes (1985 LPB 61); Kilmartin (1985 LPB 61); G.A. Gallagher (1986); G.A. Gallagher and McCone (1986); D.J. Lewis (1986b); Watson (1986 LPB 72); Anon. (1988 LPB 80, p. 3); Anon. (1990 LPB 92, p. 25 and 26); Anon. (1990 LPB 93, p. 3); Anon. (1990 LPB 94, p. 1 and 7); Bond (1990 LPB 94); Anon. (1991 LPB 98, p. 5); Anon. (1991 LPB 101, p. 13); Anon. (1993 LPB 110, p. 7); Marshall (1993 LPB 114); Anon. (1994 LPB 115, p. 7)

Emergency planning
Donovan (1973); R.J. Parker (1975); Ikeda (1982); Anon. (1986 LPB 70, p. 1); Anon. (1993 LPB 113, p. 26)

Emergency planning: transport
Dowell (1971); Kogler (1971); Burns (1974); Dektar (1974); Ellis (1974); Furey (1974); Eisenberg, Lynch and Breeding (1975); HSE (1978b); Zajic and Himmelman (1978); Anon. (1982h); Anon. (1983m)

Personal health and safety
Anon. (1988 LPB 88, p. 3 and 14 (x2))

Environment
E. Hughes (1973); Willman (1977); Anon. (1978 LPB 22, p. 114); Anon. (1982c); Kharbanda and Stallworthy (1982); Frisbie (1982); A. Shepherd (1982a); A.J. Smith (1982); Kier and Mffller (1983); Cairns (1986); Goudray (1992); Anon (1993a,f)

Offshore
Fischer (1982); Anon. (1991f)

Other case histories
Bakir *et al.* (1973); Anon. (1976 LPB 9, p. 5); Anon. (1976 LPB 11, p. 10); Vervalin (1981, 1987); Anon. (1982 LPB 48, p. 27); Anon. (1982 LPB 49, p. 26); Kletz (1982 LPB 48, 49); Anon. (1984 LPB 54, p. 21); Anon. (1984 LPB 60, p. 1); Anon. (1985 LPB 61, p. 33); Anon. (1985 LPB 64, p. 27); Buhrow (1985, 1986); Lask (1986); Manuell (1986 LPB 71); Schillmoller (1986); Anon. (1987q); Plummer (1987 LPB 73); Robinson (1987 LPB 78); Anon. (1988h); Anon. (1988 LPB 81, p. 15); Anon. (1988 LPB 83, p. 25); Anon. (1989 LPB 88, p. 18); Anon. (1989 LPB 90, p. 15); Mansot (1989); Mooney (1991); Anon. (1993c); Davie, Nolan and Hoban (1994)

A1.2 Incident Databases

There are a number of databases specifically dealing with case histories. They include the following:

- The incident databases Major Hazards Incident Data System (MHIDAS) and the corresponding explosives data system EIDAS are operated by SRD.
- TNO has developed the FACTS incident database.
- The JRC of the CEC at Ispra, Italy, runs the Major Accident Reporting System (MARS), described by Drogaris (1991, 1993).
- The FIRE incident database for chemical warehouse fires has been described by Koivisto and Nielsen (1994).
- An account of the offshore Hydrocarbon Release (HCR) database in the United Kingdom has been given by Bruce (1994).

A1.3 Reporting of Incidents

The extent and accuracy of the reporting of incidents and injuries is variable and this creates problems particularly for attempts to perform statistical analysis of incident data.

Three distinct problems may be identified: (1) occurrence of an incident; (2) injuries associated with an incident; and (3) national injury statistics. These three cases are considered in this section and in the next two sections, respectively.

The awareness of the engineering community world-wide of incidents varies according to the country in which the incident has occurred and the size and impact of the incident. For example, in the recent past incidents in the United States have generally been reported in the technical press but reports of comparable incidents in the USSR have been relatively few.

With regard to the effects of scale and impact, the probability of worldwide reporting of a incident clearly increases with these factors. The probability that an accident of the magnitude of Bhopal is not reported in countries with a free press is negligible. However, as the size of the incident decreases, the probability that it is not reported or at least is not picked up in incident collections and data-bases increases. The problem has been discussed by Badoux (1983). Figure A1.1 shows schematically the probable extent of underreporting, curve A representing the actual reporting situation and curve B the ideal one.

A1.4 Reporting of Injuries in Incidents

For various reasons, accounts of incidents tend to differ in the number of injuries and, to a lesser extent, fatalities that are reported. A discussion of this problem has been given by Haastrup and Brockhoff (1991).

There are a number of reasons for differences in the numbers quoted. One is that early reports of an incident tend not be very accurate, but are sometimes quoted with-out sufficient qualification and the numbers then receive currency.

With regard to fatalities, a difference arises between immediate and delayed deaths. A case where a proportion of delayed deaths is fairly common is burn casualties.

The most frequent and large differences, however, are in 'injuries'. Here much of the difference can usually be accounted for by differences in definition.

As an illustration, consider the injuries in the explosion at Laurel, Mississippi on 25 January 1969. The NTSB report on this incident (NTSB 1969 RAR) states that 2 persons died, 33 received treatment in hospital and numerous oth-ers were given first aid. Some authors have therefore quoted this as 2 dead, 33 injured. Eisenberg, Lynch and Breeding (1975) refer to the NTSB report but also to a pri-vate communication from a railroad source and state that 976 persons were injured, 17 being in hospital for more than a month. This incident is one of those quoted by Haastrup and Brockhoff as an example of the problem.

Another illustration is provided by the explosion at Silvertown, London, on 19 January 1917. The account given later in this Appendix states that 68 were killed, 98 seri-ously injured, 328 had slight injuries and about 600 are thought to have received treatment in the street or from private doctors. Thus this account gives three separate categories of injury, without even mentioning hospitals.

A1.5 Reporting of Injuries at National Level

It is normal for there to be a regulatory requirement for the reporting, as a minimum, of deaths and injuries. In the United Kingdom the relevant regulations are RIDDOR 1985. The information gathered in this way is published by the HSE in the series *Health and Safety Statistics*.

The reporting is incomplete. A study supplementary to the household-based Labour Force Survey 1990 in the United Kingdom showed that only approximately 30% of non-fatal injuries were reported to the HSE, but that the level of reporting varied significantly across industries (Kiernan, 1992). For the energy industry the proportion of such accidents reported was 75%, and it seems probable that the

Figure A1.1   *Number of accidents reported (after Badoux, 1983): A, accidents actually reported; and B, total number of accidents (and ideal reporting curve)*

level of reporting in process industries such as oil refining, petrochemical and chemicals is similar.

Hours by J.R. Nash (1976). Some principal explosions are listed by Nash. His list gives death tolls that in some cases differ from those given elsewhere. The incidents in which the number of deaths is quoted as 200 or more are:

Date		*Location*		*No. of deaths*
1916	Nov. 16	La Satannaya, Belgium	Munitions	1000
1917	Jan. 19	Silvertown, London	Munitions factory	300[a]
	Feb. 20	Archangel, Russia	Munitions shipment	1500
	Jun. 23	Bloeweg, Belgium	Munitions factory	1000
	Aug. 6	Henningsdorf, Germany	Munitions factory	300
	Dec. 6	Halifax, Nova Scotia	Munitions shipment	1600[a]
1918	Aug. 3	Hamont Station, Belgium	Munitions shipment	1750
	Sept. 22	Woellerdorf, Austria	Munitions factory	382
1925	May 25	Peking, China	Munitions store	300
1934	Mar. 14	La Libertad, Salvador	Explosives store	250
1935	Oct. 30	Lanchow, China	Munitions store	2000
1941	Jan. 9	Fort Smederovo, Yugoslavia	Munitions store	1500
1944	Apr. 14	Bombay, India	Munitions shipment	1376[a]
	Jul. 17	Port Chicago, CA	Munitions shipment	321
1956	Aug. 27	Cali, Columbia	Munitions shipment	1200

[a] The death tolls given in the account in Section A1.10 are as follows: Silvertown 69; Halifax 1800; and Bombay Docks 350 (early reports) This list does not include incidents involving primarily ammonium nitrate.

A1.6 Incident Diagrams, Plans and Maps

Many accounts of incidents give diagrams, plans or maps showing features such as derailed tank cars, location of missiles, location of victims, etc. In particular, such diagrams are a feature of the accident reports by the HSE and the NTSB and of case histories described in the Loss Prevention Bulletin.

Typical diagrams showing injury effects at an accident from an HSE report are given in Figure A1.19. A typical diagram of the site of an accident from an NTSB report is shown in Figure A1.16.

The NTSB system of Hazardous Material Spill Maps, documents separate from the accident reports, is described by the NTSB (1979), Benner and Rote (1981) and Lasseigne (1984). One such map is shown in Figure A1.2.

A1.7 Incidents Involving Fire Fighting

A feature of some interest in incidents is the experience gained in fire fighting. Some minimal information about this is given for selected case histories in Series A. Further information may be found in the following accounts:

Brindisi, 1977 (Mahoney, 1990)
Milford Haven, 1983 (Dyfed County Fire Brigade, 1983; Mumford, 1984 LPB 57)
Thessalonika, 1986 (B. Browning and Searson, 1989)
Grangemouth, 1987 (HSE, 1989a)
Port Heriot, 1987 (Mansot, 1989)

A1.8 Incidents Involving Condensed Phase Explosives

A number of the incidents given in the following sections involve explosives. There have been, however, a large number of other explosives and munitions incidents that are of only marginal interest here.

An account of explosions up to 1930 is given in *History of Explosions* by Assheton (1930) and later treatments are *Explosions in History* by N.B. Wilkinson (1966) and *Darkest*

Other incidents given by Nash, with death tolls shown in brackets, are:

Munitions, explosives, powder factories
1915 Oct. 20 Paris, France (52); Nov. 31 Eilmington, Delaware (31); Dec. 11 Havre, Belgium (110); 1916 Apr. 4 Kent, England (170); Dec. 6 Kent, England (26); 1917 Apr. 10 Chester, PA (133); Oct. 4 Morgan, NJ (64); 1924 Mar. 1 Nixon, NJ (26); 1925 Mar. 1 Kharput, Turkey (160); 1926 Apr. 7 Mannheim, Germany (40); Aug. 12 Csepel, Hungary (24); 1929 Sep. 5 Brescia, Italy (22); 1931 Apr. 30 Nichteroy, Brazil (100); 1935 Jun. 13 Reinsdorf, Germany (52); Jul. 28 Taino, Italy (33); 1936 Jun. 15 Tallinn, Estonia (40); 1937 Jul. 17 Chungking, China (110); 1940 Aug. 29 Bologna, Italy (38); 1941 Jan. 10 Polichka, Czechoslovakia (80); 1943 Dec. 11 Grenoble, France (73); 1947 Aug. 18 Cadiz, Spain (149)

Munitions, explosives stores and arsenals
1916 Feb. 6 Skoda, Austria (185); Feb. 20 Nish, Serbia (43); Mar. 4 Fort Double Coronne, France (30); 1926 Jul. 19 Lake Denmark, NJ (30); 1928 Sep. 26 Ft Cabreriza, Morocco (38); 1929 Mar. 4 Sofia, Bulgaria (28); 1931 May 5 Yuchu, China (100); Aug. 13 Macao, China (26); 1932 Jul. 10 Nanking, China (100); 1935 Oct. 23 Shanghai, China (190); 1944 Nov. 28 Fauld, England (175)[a]; 1945 Feb. 26 Paris, France (20); Jul. 7 Saragossa, Spain (30); Oct. 25 Asnier-en-Bessin, France (33); Dec. 29 Codroipo, Italy (23); 1946 Apr. 8 Saigon, Vietnam (23); 1949 Jul. 26 Tarancon, Spain (25); 1953 Apr. 6 Nantsechu, Formosa (54); 1956 May 17 Takoradi, Africa (25); 1963 Aug. 13 Gauhati, India (32)

Munitions, explosives shipments
1916 Aug. 8 Koenigsberg, Germany (50); 1919 Feb. 1 Longwy, France (64); 1924 Dec. 27 Otaru, Japan (120); 1930 Dec. 3 Midas Garaes, Brazil (36); 1955 Sep. 23 Gomex Palacio, Mexico (40); 1960 Mar. 4 Havana, Cuba (100); 1964 Jul. 23 Bone, Algeria (85)

Munitions, explosives (otherwise unspecified)
1925 Nov. 22 Ahwaz, Persia (70); 1933 Jan. 20 Morelia, Mexico (23); 1934 Aug. 4 Nakang, Japan (25); 1953 Aug. 18 Benghazi, Libya (50); Sep. 12 Wuensdorf, Germany (20)

---

[a] The death toll given in the account in Section A1.10 is 68.

A1.9 Case Histories: Some Principal Incidents

Two series of case histories are given below. Series A consists of case histories chosen according to one or more of the following criteria: (1) the incident is well known, generally by name; (2) it involved major loss of life and/or property; and (3) the physical events and the escalation are of interest.

Series B consists of case histories of incidents which have not been selected for the A series and where in general the loss of life or property damage was less, but which are considered instructive, particularly in respect of the cause.

Generally, in the A series mention is made of deaths and injuries, but the cost of property damage is not given. However, many of the incidents figure in the thirteenth edition of *Large Property Damage Losses in the Hydrocarbon-Chemical Industries. A Thirty Year Review* by Marsh and McLennan (M&M), edited by Mahoney (1990), which gives the 100 largest losses. For these incidents, the cost is shown in Table A1.2.

A1.10 Case Histories: A Series

A1 *San Antonio, Texas, 1912*
On 18 March 1912 at San Antonio, Texas, workers were carrying out repairs on the boiler of a large locomotive. Steam pressure was being raised when something went wrong and the boiler exploded violently. One 1600 vapour fragment travelled 1200 ft and a 900 vapour one fetched up 2000 ft away. Twenty-six workers and residents were killed and another 32 were injured (J.R. Nash, 1976).

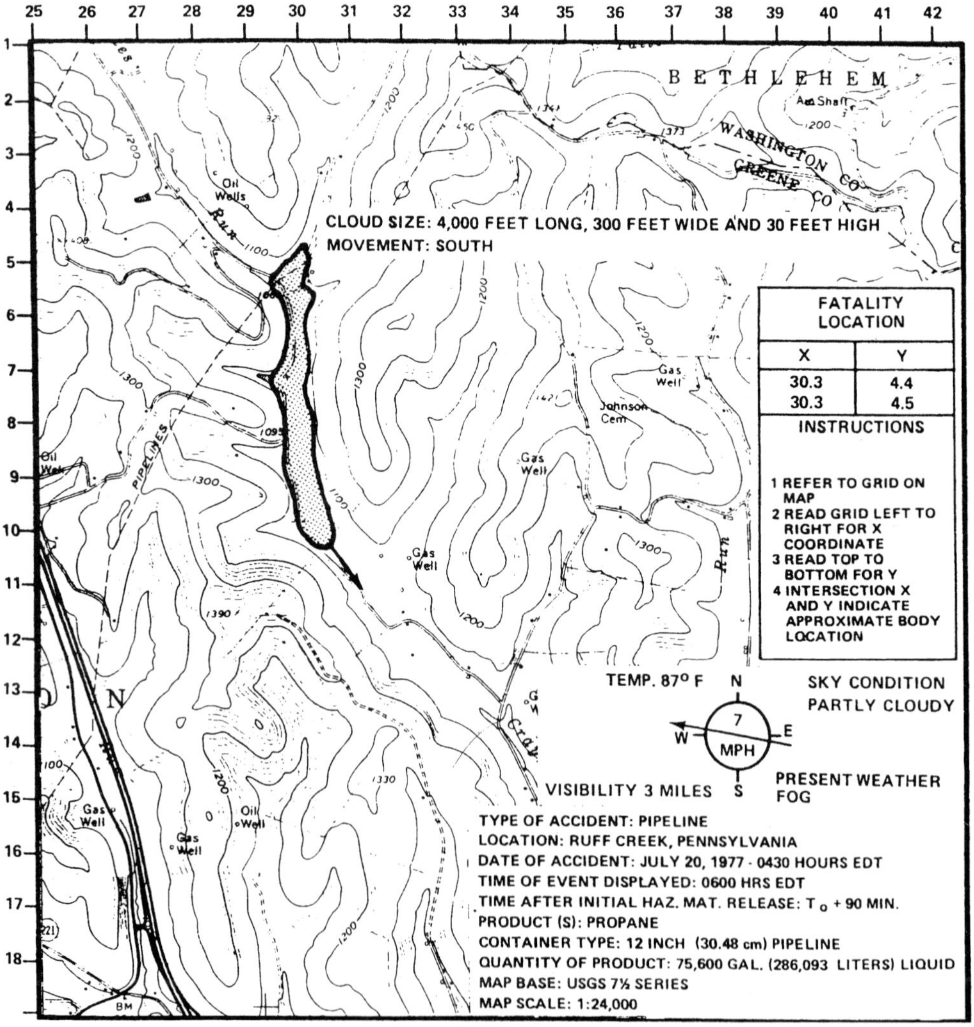

Figure A1.2 *Hazardous Materials Accident Spill map: propane pipeline release, Ruff Creek, Pennsylvania, 1977 (National Transportation Safety Board, 1979)*

Table A1.2 Some major accidents in the process industries

No.	Date	Location	Plant/transport	Chemical	Event	Deaths/injuries	Cost (million US$)	References
1	1911	Glasgow, UK			DEX	5d, 8i		9
2		Liverpool, UK			DEX	37d, 100i		9
3		Manchester, UK			DEX	3d, 5i		9
4	1912	San Antonio, TX	Loco boiler	Steam	IE	26d, 32i		9
5	1913	Manchester, UK			DEX	3d, 5i		3
6	1914	Chrome, NJ	Rail tank car	Chlorine	TOX	0d		3
7	1915 Sep. 27	Ardmore, OK	Rail tank car	Petrol	F	40d		10
8	1916 Jul. 30	Black Tom Island, NY	Barge, rail cars	Explosives	HEX	5d, many injured		
9	1917	Ashton, UK	Chemical works		HEX	46d, 120i		
10	Dec. 6	Halifax, Nova Scotia	Ship	Munitions	HEX	1963d, ≈8000i		3, 7
11	Jan. 19	Silvertown, UK	Munitions works	TNT	HEX	69d, ≈426i		4
12		Wyandotte, MI	Storage tank	Chlorine	TOX	1d		9
13		Morgan, NJ		Ammonium nitrate	ANEX	64d		
14	1919	Cedar Rapids, IA	Starch plant	Corn starch	DEX	43d		6, 8, 14
15	1920	Niagara Falls, NY	Gas cylinder	Chlorine	TOX	3d		15
16	1921 Aug. 24	Hull, UK	Airship	Hydrogen	EX			4, 7, 11, 15
17	Sep. 21	Kriewald, Germany	Rail cars	Ammonium nitrate	ANEX			9
18		Oppau, Germany	Chemical works	Ammonium nitrate	ANEX	561d		1, 3, 7, 10, 11
19	1924	Peking, IL	Starch plant	Corn starch	DEX	42d		
20	1925	De Noya, OK	Gas cylinder	Chlorine	TOX	2d		3, 10, 11
21	1926 Dec. 13	St Auban, France	Storage tank	Chlorine	TOX	19d		1, 3, 7
22	1928 Jul. 7, 13	Asbokan, NY	Rail tank car	Chlorine	TOX	0d		9
23	May 20	Hamburg, Germany	Storage tank	Phosgene	TOX	10d		10
24	1929 May 10	Syracuse, NY	Storage tank	Chlorine	TOX	1d		6
25	1930	Liverpool, UK			DEX	11d, 32i		
26	1930 Dec.	Lüttich, Belgium	Fuel plant	Hydrogen fluoride, sulphur dioxide	TOX	63d[a]		10
27	1932 Dec.	Detroit, MI	Storage	LPG				10
28	1933 Feb. 10	Neunkirchen, Germany	Gasholder	Towns gas	EX	65d, several hundred injured		7
29	1934 May 14	Hongkong	Gasholder	Towns gas	F	42d, 46i		1, 3, 7
30	Feb. 28	Niagara Falls, NY	Rail tank car	Chlorine	TOX	0d		1, 3
31	1935 Mar. 13	Griffith, IN	Rail tank car	Chlorine	TOX	0d		6, 14
32	1936 Nov. 12	Johnsonburg, NY	Pipeline	Chlorine	TOX	1d		15
33	1939 Jan. 2	Newark, NJ		Butane	VCE	0d, 1i		1, 2, 3, 7, 10, 11
34	Dec. 12	Wichita Falls, TX	Pipeline	Oil film	IE	≈60d		3, 7, 10
35	Dec. 24	Zarnesti, Roumania	Storage tank	Chlorine	TOX	3d		9
36	1940 Jan. 26	Mjodalen, Norway	Rail tank car	Chlorine	TOX	6d, 40i		
37	1941	Liverpool, UK			DEX			
38	1942 Jul. 21	Tessenderloo, Belgium	Chemical works	Ammonium nitrate	ANEX	>100d		
39	1943 Jan. 18	Los Angeles, CA	Road tanker	Butane	VCF	5d		3, 6, 7

Table A1.2 *(continued)*

No.	Date	Location	Plant/ transport	Chemical	Event	Deaths/ injuries	Cost (million US$)	References
40	Jul. 29	Ludwigshafen, Germany	Rail tank car	Butadiene	VCE	57d, 439i		10, 11, 14
41	1944 Apr. 14	Bombay, India	Ship	Munitions	EX	>350, 1800i		
42		Brooklyn, NY	Gas cylinder	Chlorine	TOX	0d, 208i		
43	Oct. 20	Cleveland, OH	Storage	LNG	F	128d, 200–400i		3, 6, 7, 8, 10, 11
44	Nov. 21	Denison, TX	Tank	LPG	F(?)	10d		6, 10
45	Nov. 27	Fauld, UK	Munitions store	Munitions	HEX	68d, 22i		
46		Kansas City, KS	Grain dust	Corn mill	DEX	4d, 20i		
47	Jul. 17	Port Chicago, CA	Ships	Explosives	EX	321		
48	1945 Apr. 25	Los Angeles, CA	Storage	Butane	VCE			6, 14
49	1947 Apr. 28	Brest, France	Ship	Ammonium nitrate	ANEX	21d		4, 7
50	Feb. 4	Chicago, IL	Rail tank car	Chlorine	TOX	0d		1, 3, 7
51	Feb. 20	Los Angeles, CA	Electrolysis plant	Acetic anhydride, etc.	EX	17d, 130i		10
52	Jan. 13	Natrium, WV	Gas cylinder	Chlorine	TOX	2d		10
53	Nov. 5	Rauma, Finland	Storage tank	Chlorine	TOX	19d		1, 3, 7, 10, 11
54	Apr. 16	Texas City, TX	Two ships	Ammonium nitrate	ANEX	552d, ≈3000i		4, 7, 11
55	1948 July 23	Ludwigshafen, FRG	Rail tank car	Dimethyl ether	VCE	207d, ≈3818i		5, 6, 7, 8, 10, 11, 14, 15
56	Oct. 13	Sacramento, CA	Road tanker	Butane	F	2d		6
57	1949 Dec. 30	Detroit, IL	Cat cracker	Propane, butane	VCE	5d		14
58	Sep. 1	Freeport, TX	Pipeline	Chlorine	TOX	0d		7
59		Nitro, WV	Reactor	TCDD	TOX	228i		10
60	Jun.	Perth, NJ	Storage tank	HCs	EX	4d		6
61	Oct.	Winthrop, MO	Rail tank car	LPG	F(?)	1d		6
62	1950 Jul. 20	Billingham, UK	Rail tank car	Chlorine	TOX	0d		6
63	Feb. 16	Midland, MI	Synthetic rubber plant	Butadiene, styrene	VEEB			8
64		Poza Rica, Mexico	Plant	Hydrogen sulphide	TOX	22d, 320i		10
65	Aug.	Wray, CO	Road tanker	Propane	VCF	2d		6
66	1951 Sep. 7	Avonmouth, Bristol, UK	Storage tank	Oil	EX, F	2d		
67	Aug. 16	Baton Rouge, LA	Naphtha treating	HCs	VCE	2d		14
68	Jul. 7	Port Newark, NJ	Storage	Propane	VCF, BLEVE	0d, 14i		8, 10, 15
69	Feb. 8	St Paul, MN	Loading terminal	LPG	VEEB	14d		10
70	1952 Jul. 21	Bakersfield, CA	Storage	Butane	VCE	0d		6, 8, 10, 15
71	Dec. 22	Bound Brook, NJ	Phenolic resin plant	Resin dust	DEX	5d, 21i		15
72	Jul. 24	Kansas City, KS	Loading terminal	LPG	VCE			6, 14
73	Apr. 4	Walsum, FRG	Storage tank	Chlorine	TOX	7d		1, 3, 7, 10
74	1953 Aug. 6	Campana, Argentina	Refinery recovery unit	Gasoline	VCE	2d		14
75	Sep. 23	Tonawanda, NY	Organic peroxides plant	Organic peroxides	EX	11d, 27i		14
76	1954 Jun. 25	Montreal, Quebec	Gas cylinder	Chlorine	TOX	1d		10
77	Oct. 18	Portland, OR	Rail tank car	LPG	VCE	0d		6, 14
78	Jun. 4	Institute, WV	Rail tank car	Acrolein	IE			6, 14
79	1955 Jul. 14	Freeport, TX	Polyethylene plant	Ethylene	VCE			14

No.	Date	Location	Plant/equipment	Material	Type	Dead/injured	Loss ($10^6)	References
80	Jul. 19	Ludwigshafen, FRG	Rail tank car	Chlorine	BLEVE	2i		
81		Runcorn, UK	Feed plant		TOX	3d, 13i		
82		Waynesboro, GA	Hydroformer	Grain dust	DEX	2d, 40i		
83	Aug. 27	Whiting, IN	Gasoline plant	Naphtha	DET	20d, >32i		5, 10, 15
84	Jul. 22	Wilmington, CA (sic)	Storage tanks	Butane	VCE			6, 14
85	1956 Jul. 29	Amarillo, TX	Alkylation unit	Oil	FB			
86	Jul. 26	Baton Rouge, LA	Road vehicle	Butylene	VCE			
87	Aug. 7	Cali, Columbia	Storage	Dynamite, munitions	HEX	≈1200d		
88	Oct. 22	Cottage Grove, OR	Storage vessel	LPG	BLEVE	12d, 12i		10
89	Jul. 29	Dumas, TX	Storage tank	HC	BLEVE	19d, 32i		7, 10, 14
90	Mar. 10	Lake Charles, LA	Polyethylene plant	Chlorine	TOX	0d		5, 6
91	Dec. 19	North Tonawanda, NY	Turbogenerator	Ethylene	VCE	0d		
92	1957 Jan. 8	Uskmouth, Wales	Storage vessel	Steam	IE	2d, 9i		6, 10
93	Apr. 16	Montreal, Quebec	Chlorination vessel	Butane	BLEVE	1d		10
94	Oct. 24	Nitro, WV	Loading terminal	Chlorine	TOX	8d		6, 14
95	1958 Apr. 15	Sacramento, CA	Loading terminal	LPG	VCE	1d		6, 14
96	Jul. 30	Ardmore, OK	Loading terminal	Propane	VCE	1d		6, 14
97	Jan. 3	Augusta, GA	Rail tank car	LPG	VCE			
98	Jan. 22	Celle, FRG	Rail tank car		EX			
99	May 22	Niagara Falls, NY	Rail tank car	Nitromethane	DET	200i		5, 10
100		Signal Hill, CA	Tank farm	Oil froth	F	2d, 18i	42.9	5, 10, 13, 15
101	1959 Jun. 2	Deer Lake, PA	Rail tank car	LPG	BLEVE	11d, 10i		10
102	May 28	McKittrick, CA	Storage	LPG	BLEVE	2i		10
103	Jun. 28	Meldrin, GA	Rail tank car	LPG	VCE	23d		3, 5, 6, 7, 9, 10, 11, 14
104		Phillipsburg, NJ	Compressor	Lube and seal oil	EX	6d, 6i		
105	Feb. 27	Portland, OR	Road tanker	LPG	REL			
106		Roseburg, OR			ANEX	13d		
107	Jul. 11	Ube, Japan	Ammonia plant	Oxygen	IE	11d, 40i		10
108	1960 Dec. 17	Freeport, TX		Allyl chloride, propylene chloride	EX	6d, 14i	39.0	6
109	Oct. 14	Kingsport, TN	Nitrobenzene plant	Reaction mixture	DET	15d, 60i		10, 13
110	1961 Feb. 23	Billingham, UK	Pipeline	Chlorine	TOX	0d		3
111	Dec. 17	Freeport, TX	Caprolactam plant	Cyclohexane	VCE	1d		8, 10, 14
112		FRG	Air separation plant	Oxygen	EX	15d		10
113	Jan. 31	La Barre, LA	Rail tank car	Chlorine	TOX	1d, 114i		1, 2, 3, 5, 7, 10, 11
114		Lake Charles. LA	Alkylation unit	Butane	VCE	2d		14
115	Aug.	Minimata, Japan	Reactor	VCM	VEEB	4d, 10i		10
116		Morganza, LA		Chlorine	TOX	0d, 17i		10
117	1962 Jul. 25	Berlin, NY	Road tanker	LPG	VCE	10d, 17i		3, 6, 7, 9, 10, 14
118	Nov. 30	Cornwall, Ont.	Rail tank car	Chlorine	TOX	89i		1, 7, 10
119	Apr. 17	Doe Run, KY	Feed vessel	Ethylene oxide	IE, VCE	1d, 21i	20.2	5, 6, 7, 8, 10, 13, 14, 15

Table A1.2 *(continued)*

No.	Date	Location	Plant/transport	Chemical	Event	Deaths/injuries	Cost (million US$)	References
120		Fawley, UK	Cat cracker		VCE	2d, 18i		14
121		FRG	Reactor		IE	2d		10
122		Houston, TX	Tank farm	Gasoline	VCE			14
123	Apr. 27	Marietta, OH	Phenol plant	Benzene	EX	1d, 3i	20.2	8, 10, 13
124	Aug. 4	Ras Tanura, Saudi Arabia	Storage vessel	Propane	VCE, F	1d, 115i		6, 10, 14
125		St Louis, MO	Feed plant	Grain dust	DEX	2d, 34i		
126	May 10	Toledo, OH	Reactor	Acrylic polymer reaction	VEEB	10d, 46i		10
127	1963 Mar. 6	Amsterdam, Netherlands	Trichlorophenol plant	Reaction mixture, including TCDD	IE, TOX	8d, 14i		10
128	Apr. 28	Brandtsville, PA	Rail tank car	Chlorine	TOX	0d		1, 3, 7, 10
129		FRG	Acetylene plant	Acetylene, diacetylene	IE	7d, 6i		10
130	Apr. 3	Norwich, CT	Transport tank	Organic peroxides	EX	4d, 4i		10
131	Aug. 9	Philadelphia, PA	Rail tank car	Chlorine	TOX	430+i		1, 10
132	May 3	Plaquemine, LA	Ethylene plant	Ethylene	VCE	7i	16.2	6, 7, 10, 13, 14
133		Texas		Propylene	IE			5
134	1964 Jun. 4	Antwerp, Belgium	Reactor	Ethylene oxide	IE	4d, 20i		6, 7, 10
135	Jan. 1	Attleboro, MA	Reactor	VCM	VEEB	7d, 40i	19.9	8, 10, 13, 15
136	Jul. 17	FRG	Oxygen plant	Oxygen	IE			10
137		FRG	Reactor	Acrylamide reaction mixture	IE	1d, 3i		10
138		FRG	Reactor	Reaction mixture	IE	3i		10
139		Hebronville, MA	PVC plant		IE			5
140	Jan. 9	Jackass Flats, NV	Research laboratory	Hydrogen	VCE			6, 8, 10, 14
141	Oct. 25	Liberal KS	Compressor station	Propane	VCE			14
142	Jul. 2	Mobile, AL	Pipeline	Chlorine	TOX			7
143	Jun. 16	Niigata, Japan	Refinery	HCs	EQK, F, EX	1d		13
144	Oct. 25	Orange, TX	Polyethylene plant	Ethylene	VCE	2d	87.3	6, 14
145		Paisley, UK			DEX	2d		9
146	1965 Dec. 23	Baltimore, MD	Detergents plant	Benzene	VCE	5d, 2i		14
147	Sep. 12	Baton Rouge, LA	Tankship	Chlorine	TOX			10
148	Jul. 31	Baton Rouge, LA	Reactor	Ethyl chloride	VCE			6, 14
149	Aug. 7	Bow. London. UK	Flour mill	Flour	DEX	5d, 32i		9, 15
150	Oct. 24	Escambia. USA	Chemical plant	Hydrogen, carbon monoxide	VCE			6, 10
151	Jul. 13	Houston, TX	Ethyl chloride plant	Ethyl chloride	EX	12d		10
152	Aug. 25	Lake Charles, LA	Ethylene plant	Methane or ethylene	VCE		39.2	6, 14
153	Mar. 4	Louisville, KY	Compressor	Monovinyl acetylene	IE	12d, 60i		10, 13, 15
154		Natchitotehes, LA	Pipeline	Natural gas	IE	17d, 9i		5, 7, 8, 10
155		Pasadena, TX	Ammonia plant	Hydrogen, ammonia	F	2d, 3i		10
156		Texas	Chemical plant	Butane	F(?)			6
157		Whiting, IN	Reformer	Naphtha	EX			10

No.	Date	Location	Equipment	Material	Event	Casualties	Loss	References
158	1966 Jan. 4	Feyzin, France	Storage vessels	Propane	BLEVE	18d, 8i	68.8	5, 6, 8, 10, 11, 13, 15
159		FRG	Storage vessel	Light HCS	F	7d		5
160		Larose, LA	Pipeline	NGL	VEEB	11d, 10i		
161	Oct. 16	LaSalle, Quebec	Reactor	Styrene	TOX	0d	11.4	5, 8, 10, 13, 15
162	Jun. 14	La Spezia, Italy	Rail tank car	Chlorine	VCF			3
163	May 23	Philadelphia, PA	Refinery	Benzene, cumene, propane				6, 10
164	Jan. 19	Raunheim, FRG[b]	Ethylene plant	Methane, ethylene	VCE	3d, 83i	26.6	5, 6, 8, 10, 13, 14
165	Feb. 6	Scotts Bluff, LA	Reactor	Butadiene	VCE	3d		6, 14
166	1967	Antwerp, Belgium		VCM	F	4d, 33i		5, 10
167		Bankstown, Australia		Chlorine	TOX	5i		5
168		Buenos Aires		Propane	F	100i	13.5	5, 10
169	Dec. 19	El Segundo, CA	Fuel oil reservoir	Fuel oil	F			13
170		Hawthorne, NJ			EX	2d, 11i		5
171	Aug. 8	Lake Charles, LA	Alkylation unit	Isobutylene	VCE	7d, 13i	63.1	3, 5, 6, 7, 8, 10, 13, 14, 15
172	Nov. 8	Newton, AL	Rail tank car	Chlorine	TOX	0d		1, 3, 10
173		Santos, Brazil	Gas holder	Towns gas	IE	300i		5, 10
174	Jan. 20	Sacramento, CA	Saturn rocket	Hydrogen, oxygen fuel	VCE			8
175	1968	Bolsover, UK	Reactor	Reaction mixture, including TCDD	TOX	79i		10
176	Jan. 1	Dunreith, IN	Rail tank car	(1) VCM (2) Ethylene oxide	F, EX; BLEVE; 24d	5i		10
177		GDR		VCM	IE			5, 10
178		Hull, UK	Road tanker	Acetic acid	IE	2d, 13i		3, 5, 10
179	Aug. 21	Lievin, France	Petrochemical plant	Ammonia	TOX	5d, 20i		5, 7, 9, 10
180		Paris, France	Slops tanks		IE			5
181	Jan. 21	Pernis, Netherlands		Oil slops	VCE	2d, 85i	97.7	3, 5, 6, 7, 8, 10, 13, 14, 15
182		Tarrytown, USA	Chemical plant	Propane	EX	1d, 2i		5, 10
183	Dec.	Yutan, NE	Pipeline	LPG	F	5d		6, 10
184	1969 Dec. 23	Basle, Switzerland	Reactor	Dinitrochloraniline reaction mixture	IE	3d, 31i		5, 10
185	May 8	Cleveland, OH	1-ton containers	Chlorine	TOX	1d		3, 5, 7, 9, 10
186	Feb. 18	Crete, NE	Rail tank car	Ammonia	TOX	9d, 53i		
187	Jul.17	Dudgeons Wharf, London, UK	Tank dismantling	EX	6d	10		
188	Oct. 1	Escombreras, Spain	Storage	Propane	VCE			5, 14
189	Dec. 28	Fawley, UK	Hydroformer	Hydrogen, naphtha	VCE	4d, 3i	14	6, 8, 10, 14
190	Aug. 12	Flemington, NJ	Reactors	VCM	EX	0d		13
191	Sep. 11	Glendora, MS	Rail tank car	VCM	TOX, VCE	1i		6, 10
192	Sep. 9	Houston, TX	Pipeline	Natural gas	VCE	9i		2, 10, 14
193	Jan. 25	Laurel, MS	Rail tank car	LPG	BLEVE	2d, 33+i		2, 14
194		Long Beach, CA	Tank	Petroleum	IE	1d, 83i		5, 10

Table A1.2 (continued)

No.	Date	Location	Plant/ transport	Chemical	Event	Deaths/ injuries	Cost (million US$)	References
195	Mar. 6	Puerto La Cruz, Venezuela	Crude unit	HCs	F		15.6	13
196		Repesa, Spain	Refinery	LPG, propylene	F	0d		5,10
197	Oct. 23	Texas City, TX	Butadiene recovery unit	Butadiene	IE, VCE	3i	26.6	10,13,14,15
198	May 14	Wilton, UK	Oxidation plant	Cyclohexane	VCE	2d, 23i		5,6,8,10
199	Jan. 21	Wilton, UK	Polyethlene plant	Ethylene	IE	4d		
200	1970	Agha Jari, Iran	Compressor station	Gas	EX	29d, 10i		
201	Sep. 7	Beaumont, TX		Oil	F		20.2	10,13
202	Jan. 21	Belle, WV	Rail tank car	Ammonia	TOX	1i		7,9,10
203	Feb. 6	Big Springs, TX	Alkylation unit		VCE			14
204	Nov. 16	Blair, NE	Storage tank	Ammonia	TOX	0d		3,7,10
205	May 30	Brooklyn, NY	Road tanker	Oxygen	F	2d, 30i		12,15
206	Aug. 30	Corpus Christi, TX	Cat cracker		HUR		23.0	13
207	Jun. 21	Crescent City, IL	Rail tank car	Propane	BLEVE	66i		2,5,6,8,10,11,15
208	Nov. 12	Hudson, OH	Road tanker	LPG	F	6d		6,10
209	Oct. 19	Javle, Sweden	Pipeline	Chlorine	TOX	0d		
210	Sep. 3	Jacksonville, MD	Pipeline					
211		Kiel, FRG	Grain silo	Grain dust	DEX	6d, 18i		6,10,13,14
212	Dec. 5	Linden, NJ	Refinery reactor	C_{10} HC	VCE	40i	84.5	5,10
213		Mitcham, UK		LPG	IE			5,10
214		Philadelphia, PA	Cat cracker		F(?)	7d, 42i		2,3,5,6,7,8,9,10,11,14,15
215	Dec. 10	Port Hudson, MO	Pipeline	Propane	VCE	10i		5,10
216		St Thomas Island, Virgin Islands		Natural gas	IE	25i		10
217	1971 Jul. 10	Amsterdam, Netherlands	Refinery	Butadiene	EX	8d, 21i		10
218	Jan. 19	Baton Rouge, LA	Rail tank car	Ethylene	VCE	0d, 21i		6,10,14,15
219	Apr. 24	Emmerich, FRG	Oxidation plant	Formic acid, hydrogen peroxide	F, IE	1d, many injured		5
220	Jun. 5	Floral, AR	Pipeline	Ammonia	TOX	0d		3,5,10,11
221	Aug. 8	Gretna, FL	Road tanker	Methyl bromide	TOX	4d		12
222	Oct. 19	Houston, TX	Rail tank car	VCM	BLEVE	1d, 5i		2,8,10,12
223	Sep. 15	Houston, TX	Butadiene plant	Butadiene	VCE	1d, 6i		6,10,14
224	Aug. 21	La Spezia, Italy	Storage tank	LNG	Rollover	0d		
225	Dec. 23	Lake Charles, LA	Chemical plant	Trichlorethylene, perchlorethylene	VCF(?)	4d, 3i		10
226	Feb. 25	Longview, TX	Polyethylene plant	Ethylene	VCE	4d, 60i	16.5	6,8,10,13,14
227	Nov. 7	Morris, IL	Ethylene oxide plant	Ethylene oxide, oxygen	IE	4i	16.2	10,13
228		Netherlands		Butadiene		8d, 21i		5
229	Sep. 11	Pensacola, FL	Caprolactam plant	Cyclohexane	REL	0d		2,3,8,10

No.	Date	Location	Plant/equipment	Material	Type	Deaths/injuries	Cost	References
230	Jul. 19	Raunheim, FRG	Tankship	Petroleum	F, EX	7d		10
231		Texas	Chemical plant	Ethylene	VCE	3d		12
232	Jun. 4	Waco, GA	Road vehicle	Explosives	HEX	5d, 33i		6, 10, 13
233	1972 Aug. 14	Billings, MT	Alkylation unit	Butane	VCF	1d, 4i	14.0	13
234	Feb. 29	Delaware City, DE	Fluid coking unit	Hot oil	Hot oil damage		16.9	
235	Oct. 22	East St Louis, IL	Rail tank car	Propylene	VCE	1d, 230i		2, 3, 5, 6, 8, 10, 12, 14
236	May 14	Hearne, TX	Pipeline	Crude oil	F	1d, 2i		2, 6, 10, 12
237	Mar. 9	Lynchburg, VA	Road tanker	Propane	FB	2d, 5i		2, 3, 6, 10, 12
238		Netherlands	Road tanker	Hydrogen	IE	4d, 4i		5
239	Sep. 21	NJ Turnpike, NJ	Road tanker	Propane	F	2d, 28i		2, 3, 10, 12
240		Rio de Janeiro, Brazil	Storage vessel	Butane	BLEVE	37d, 53+i	13.4	5, 6, 10, 13
241	Feb. 9	Tewksbury, MA	Storage tank		F	2d, 21i		13
242	Aug. 4	Trieste, Italy	Terminal	Crude oil	F	21d, 20i	29.7	5, 10
243		Weirton, WV	Coke plant	EX		76d, 978i		10
244		Yokkaidi, Japan	Chemical plant	TOX		6d, many injured		3, 6, 9, 10, 12
245	1973 Feb. 22	Austin, TX	Pipeline	NGL	VCE			
246	May 24	Benson, AR	Rail cars	Munitions	HEX	4d, 2i		2, 5, 10
247	May 23	Cologne, FRG	PVC plant	VCM	EX	29i		10
248	Dec. 27	Freeport, TX	Tank	Ethylene oxide	VCF			5, 10
249	Sep. 6	Goi, Japan	Oxidation plant	Cumene, etc.	VEEB	4d	18.9	13, 14
250	Oct. 8	Kingman, AZ	Polypropylene plant	Hexane, propylene	BLEVE	13d, 95i		10, 11
251	Jul. 5	Lodi, NJ	Rail tank car	Propane	VCF	7d		5, 10
252		Loos, BC	Reactor	Methanol	TOX	0d		3, 10
253	Mar. 5	McPherson, KS	Rail tank car	Chlorine	TOX	0d		3, 10
254	Dec. 6	Noatsu, Japan	Pipeline	Ammonia	TOX	1d		14
255	Oct. 28		Vinyl chloride plant	VCM	VCE			
256	Jul. 13	Potchefstroom, South Africa	Storage vessel	Ammonia	TOX	18d, 65i		3, 5, 10, 11, 15
257	Oct. 24	Sheffield	Gas works	Towns gas	EX	4d, 24i		5, 10, 11
258	Feb. 1	St-Amand-les-Eaux	Road tanker	Propane	VCE	9d, 37i	28.4	5, 8, 10, 12
259	Aug. 24	St Croix, Virgin Islands	Hydrodesulphurizer	Hydrogen, HCs	F			13
260	Oct. 28	Shinetsu, Japan	Chemical plant	VCM	VCE	1d, 23i		5, 6, 8, 10
261	Feb. 10	Staten Island, NY	Storage tank	LNG residues	F	40d	40.0	10, 11, 15
262	Jul. 8	Tokuyama, Japan	Ethylene plant reactor	Ethylene	VCE	1d, 4i		5, 10, 13, 14
263	Nov. 6	Ventura County, CA	Rail tank car	LPG	REL	2d, 4i		12, 15
264	1974 Jan. 17	Aberdeen, UK	Road tanker		FB			11
265	Sep. 5	Barcelona, Spain	Chemical plant	VCM, ethylene dichloride	VCF			6, 10
266	Jun. 9	Bealeton, VA	Pipeline	Gas	F	0d		
267	Nov. 29	Beaumont, TX	Isoprene plant	Isoprene	VCE	2d, 10i	36.6	13, 14
268	Apr. 17	Bielefeld, FRG			BLEVE			
269	Apr. 26	Chicago, IL	Storage tank	Silicon tetrachloride	TOX	0d		
270	Jun. 26	Climax, TX	Rail tank car	VCM	VCE	7d		2, 3, 6, 10, 12, 14

Table A1.2 *(continued)*

No.	Date	Location	Plant/transport	Chemical	Event	Deaths/injuries	Cost (million US$)	References
271	Jul. 7	Cologne, FRG	Vinyl chloride plant	VCM	VCE			14
272	Jul. 19	Decatur, IL	Rail tank car	Isobutane	VCE	7d, 152i		2, 6, 9, 10, 11, 12, 14
273	Aug. 30	Fawley, UK	Polyethylene plant	Ethylene	VCE			14
274	Jun. 1	Flixborough, UK	Caprolactam plant	Cyclohexane	VCE	28d, 104i	412.2	2, 3, 5, 6, 8, 10, 11, 13, 14, 15
275	Sep. 13	Griffith, IN	Cavern storage	Propane	F	0d		6, 10
276	Jan. 4	Holly Hill, FL	Road tanker	Propane	VCE	0d		5, 6, 10, 14
277	Sep. 21	Houston, TX	Rail tank car	Butadiene	VCE	1d, 235i		6, 10, 12, 14
278	Aug. 17	Los Angeles, CA	Loading terminal	Organic peroxides	EX	0d		5, 10, 15
279	Mar. 2	Munroe, LA	Pipeline	Natural gas	EX, F	0d		
280	Feb. 6	Omaha, NE	Chlorine vaporizer	Chlorine	TOX			5
281	Feb. 12	Oneonta, NY	Rail tank car	LPG	BLEVE	25i		10
282	Aug. 25	Petal, MO	Salt dome storage	Butane	VCE	24i		6, 10, 14, 15
283	Jul. 18	Plaquemine, LA	Cracking plant	Propylene	VCF(?)			6, 8, 10
284		Rotterdam, Netherlands	Petrochemicals plant	HC	F			5
285		Pitesti. Roumania[c]	Ethylene plant	Ethylene	VCE	≈100d		5, 14
286		Texas	Chemical plant	Pentanes	VCE	2d		6
287	Aug. 6	Wenatchee, WA	Rail tank car	Monomethylamine nitrate	DET	2d, 113i		5, 10, 12
288	Jan. 11	West St Paul, MN	Storage vessel	LPG	BLEVE	4d, 6i		10
289	Jul. 21	Zaluzi, Czechslovakia	Ethylene plant	Ethylene	VCE	14d, 79i		5, 6, 10, 14
290	1975 Feb. 10	Antwerp, Belgium	Polyethylene plant	Ethylene	VCE	6d, 13i	57.8	3, 5, 6, 8, 10, 13, 14
291	Mar. 16	Avon, CA	Refinery	Oil	F		21.2	13
292	Nov. 7	Beek, Netherlands	Petrochemical plant	Propylene	VCE	14d, 107i	46.7	3, 5, 6, 8, 10, 11, 13, 14, 15
293	Nov. 21	Cologne, FRG	Cyclic hydroformer	Hydrogen, naphtha	VCE	0d		6, 14
294		Deer Park, TX	Polyethylene plant	Ethylene	VCF	1d, 4i		10
295	Sep. 1	Des Moines, IA	Rail tank car	LPG	BLEVE	3i		
296	May 13	Devers, TX	Pipeline	LPG	F	4d		
297	Apr. 30	Eagle Pass, TX	Road tanker	LPG	FB	17d, 34i		9, 10, 12
298	Aug. 31	Gadsden, AL	Tank farm	Gasoline	'BLEVE'	4d, 28i		10, 12, 14
299	Apr. 5	Ilford, UK	Electrolysis plant	Hydrogen, oxygen	EX	1d, 3i		10
300	Jan. 17	Lima, OH	Terminal	Crude oil	F			10
301		Louisiana		Propane	REL			
302	Jan. 31	Marcus Hook, PA	Oil tanker, tankship	Oil, phenol	F		16.0	13
303		Marseilles, France		Chlorine	IE	1d, 3i		5
304	Dec. 14	Niagara Falls, NY	Road tanker	Chlorine	TOX	4d, 80i		10
305	Aug. 17	Philadelphia, PA	Tank farm	Crude oil vapours	VCF, EX	8d, 2i	26.5	5, 10, 13
306	Aug. 2	Romulus, MI	Pipeline	REL				
307	Sep. 5	Rosendaal, Netherlands		Gasoline	VCE			14

No.	Date	Location	Plant/Equipment	Material	Event	Casualties		References
308	Nov. 4	Scunthorpe, UK	Hot metal ladle	Hot metal, water	EX	11d, 15i		10
309	Dec. 4	Seattle, WA	Road tanker		F	7d, 7i		10
310		South Africa			F			10
311	Sep. 3	Texas City, TX	Pipeline	Methane	TOX			5, 10
312	Dec. 2	Watson, CA	Hydrogen plant	Ammonia	VCE			10
313	1976 Dec. 10	Baton Rouge, LA	Storage vessel	Hydrogen	TOX	3i		14
314	Nov. 26	Belt, MT	Rail tank car	Chlorine	BLEVE	22i		5, 10, 11, 15
315	Aug. 12	Chalmette, LA	Refinery	LPG	EX	13d, 2i		10, 12
316	May 7	Enid, OK	Pipeline	Ethyl benzene	TOX			10
317	Aug. 31	Gadsden, AL	Reactor	Ammonia	FB	3d, 28i		10
318	May 24	Geismar, LA	Storage vessel	Ethylene oxide, etc.	IE	56i	17.3	10, 13
319	Sep. 2	Haltern, FRG	Road tanker	Carbon dioxide	TOX			12
320	May 11	Houston, TX	Reactor	Ammonia	TOX	6d, 178i		9, 10, 11, 12
321	Jun. 27	Kings Lynm		Feed additive reaction mixture	IE	1d		10, 15
322	Aug. 6	Lake Charles, LA	Refinery	Isobutane	VCE	7d		10
323	Jan. 16	Landskrona, Sweden *Rene 16*	Loading terminal	Ammonia	TOX	2d		9, 10
324	Oct. 15	Longview, TX	Ethanol plant	Ethylene	VCE	1d		14
325	Jun. 16	Los Angeles, CA	Pipeline	Gasoline	VCE			8, 10, 12, 15
326	Dec. 17	Los Angeles, CA *Sansinena*	Terminal	Crude oil	EX	9d, many injured	23.2	13
327	Sep. 26	Manfredonia, Italy	Absorption column	Arsenic compounds, etc.	TOX	30i		10
328	Aug. 30	Puerto Rico	Storage	Pentanes	VCF	1d		6
329	Dec. 7	Plaquemine, LA	Surge tank	Oil	IE		23.2	13
330		Robstown, TX	Compressor station	Natural gas	VEEB		23.2	13
331		Sandefjord, Norway		Flammable liquid	IE	6d		5
332	Jul. 10	Seveso	Reactor	Trichlorphenol reaction mixture, including TCDD	IE	0d(d)		5, 10, 11, 15
333	Feb.	Texas	Pipeline	Ethylene	VCF	1d, 15i		6
334	Sep. 11	Westoning, UK	Road tanker	Petrol	EX	3i		8
335	1977 May 11	Abqaiq, Saudi Arabia	Pipeline	Crude oil	F		99.2	13
336	Jun. 4	Abqaiq, Saudi Arabia	NGL plant	Fuel gas	VCF	10i	19.3	13, 14
337	Feb. 2	Baton Rouge, LA	Chemical works	Chlorine	TOX			10
338	Oct. 17	Baton Rouge, LA	Preheat furnace	Oil	EX	3d	17.0	13
339	Jan. 27	Baytown, TX	Tanker	Gasoline	VCE	7d, 6i		14
340	Sep. 24	Beattyville, KY	Road tanker	Gasoline	F	13i		12
341	Jan. 4	Braehead, Renfrew, UK	Warehouse	Sodium chlorate	EX	3d, 22i		5, 10
342	Dec. 8	Brindisi, Italy	Ethylene plant	Light HCs	VCE	1d, 9i	51.5	5, 10, 13, 14
343		Cassino, Italy		LPG	IE	1i		5
344	Feb. 20	Dallas, TX	Rail tank car	Isobutane	VCE	1d		6, 10, 14
345	Jul. 8	Fairbanks, AK	Pipeline	Crude oil	F	15d	72.1	10, 13
346		Galveston, TX	Grain silo	Grain dust	DEX	1d, 1i		10, 11
347		Gela, Italy		Ethylene oxide, glycol	IE	2d, 9i		5, 10
348	Dec. 28	Goldonna, LA	Rail tank car	LPG	FB	20i		10, 12
349		India		Hydrogen	IE			5

Table A1.2 (continued)

No.	Date	Location	Plant/ transport	Chemical	Event	Deaths/ injuries	Cost (million US$)	References
350		Jacksonville, USA		LPG	F			5
351		Mexico		Ammonia	TOX	2d, 102i; 90i		10
352		Mexico		VCM				
353	Dec. 10	Pasacabolo, Columbia	Fertiliser plant	Ammonia, etc.	TOX, VCE	30d, 22i		5, 10
354	Nov. 9	Pensacola, FL	Rail tank car	Ammonia	TOX			12
355	Mar. 18	Port Arthur, TX	Stabilizer unit	Propane	VCE	4d		14
356	Jun. 19	Puebla, Mexico			VCF			15
357	Sep. 24	Romeoville, IL	Tank farm	Diesel fuel, gasoline	F	2d	14.6	13
358	Jul. 20	Ruff Creek, PA	Pipeline	Propane	VCF	6d, 10i		9, 12
359		Taiwan		VCM	F			5, 10
360	Apr. 3	Umm Said, Qatar	Gas plant	LPG	F	7d, 13+i	139	5, 8, 10, 13, 14, 15
361	1978 Apr. 15	Westwego, LA	Grain silo	Grain dust	DEX	36d, 10i		5, 11, 15
362		Abqaiq. Saudi Arabia	Gas plant	(1) Methane, (2) LPG	(1) F, (2) VCE		90.8	13, 14
363		Baltimore, MD	Sulphur trioxide		TOX	100i		5, 10
364	Jan. 12	Chicago, IL	Loading terminal	Hydrogen sulphide	TOX	8d, 29i		10
365	Oct. 3	Conway, KS	Pumping station	LPG	EX			10
366	Aug. 4	Denver, CO	Cat polymerization unit	Propane	VCE	3d	37.1	13, 14
367		Donnellson, IA	Pipeline	LPG	FB	3d, 2i		9, 10, 12
368	Sep. 16	Immingham, UK	Ammonia plant	Syngas	VCE			14
369	May 29	Lewisville, AR	Rail tank car	VCM	FB	2i		10
370		Mexico City, Mexico	Road tanker	Propylene	F	12d		
371	Sep. 27	Oviedo Province, Spain	Rail tank car	Gasoline	EX, F	7d		10, 12
372	Oct. 30	Pitesti, Roumania	Gas concentration unit	Propane, propylene	VCE			14
373	Feb. 11	Poblado Tines, Mexico	Pipeline	Natural gas	VCE	40d		14
374		St Marys, WV	Cooling tower		Structural collapse	51d		14
375	Jul. 11	San Carlos, Spain	Road tanker	Propylene	VCF	216d, 200i		10, 11, 12, 15
376	May 30	Texas City, TX	Storage vessel	LPG	BLEVE	7d, 10i	93.0	13
377	Jul. 16	Tula, Mexico	Road tanker	Butane	EX	12		
378	Feb. 24	Waverly, TN	Rail tank car	Propane	BLEVE	16d, 43i		5, 10, 12, 15
379	Jul. 15	Xilatopic, Mexico	Road tanker	Butane	VCF	100d, 220i		10, 12
380	Feb. 26	Youngstown, FL	Rail tank car	Chlorine	TOX	8d, 114i		5, 10, 12
381	1979 Jan. 8	Bantry Bay, Eire	Oil tanker	Crude oil	EX	50d		9, 11. 13
382	Jul. 18	Bayonne, NJ	Rail tank car	Chlorine	TOX			10
383	Feb. 6	Bremen, FRG	Flour mill	Flour	DEX	14d, 17i		11
384	Jan. 6	Burghausen, FRG	Chemical plant	Hydrogen sulphide	TOX			10
385	Oct. 6	Cove Point, MD	Pipeline	LNG	VEEB	1d, 1i		12
386	Apr. 8	Crestview, FL	Rail tank cars	HMs	TOX	14i		
387	Sep. 1	Deer Park, TX	Tankship *Chevron Hawaii*	Distillate	EX		105.4	13
388	Nov. 1	Galveston Bay, TX	Oil tanker	Crude oil	EX	32d		13
389	Dec. 11	Geelong, Australia	Crude unit	Oil	F		17.4	10

No.	Date	Location	Plant/Equipment	Material	Event	Casualties	Index	Refs
390	Aug. 30	Good Hope, LA	Tank barge *Panama City*	Butane	FB	12d, 25i	16.4	9, 11, 13
391	Oct. 17	Hamburg, FRG	Grain silo	Grain dust	DEX	2i		10
392	Mar. 20	Lerida, Spain	Grain silo	Grain dust	DEX	7d		13
393	Nov. 10	Linden, NJ	Cat cracker	LPG	VCF		27.1	9, 10, 11, 15
394	Aug.	Mississauga, Ont.	Rail tank car	Chlorine	TOX	1d, 1i		12
395	Sep. 8	Orange, TX	Pipeline	LPG	EX	8i		10
396	Sep. 4	Paxton, TX	Rail tank car	Chemicals	BLEVE			13
397	Dec. 11	Pierre Port, LA	Pipeline	LNG	VCF			13
398	Apr. 19	Ponce, Puerto Rico	Dimerizer vessel	HCs	IE		23.3	10
399	Jan. 11	Port Neches, TX	Oil tanker *Seatiger*	Crude oil	EX		49.6	13
400	Jul. 28	Rafnes, Norway	Chemical plant	Chlorine	TOX			13, 14
401	Jul. 21	Sauget, IL	Reactor	Reaction mixture	IE		11.9	14
402	Sep. 18	Texas City, TX	Alkylation unit	Propane	VCE		37.2	14
403	Jun. 26	Torrance, CA	Cat cracker	C_3–C_4 HCs	VCE			13
404		Ypsilanti, MI	Storage	Propane	VCE			12
405	1980 Jan. 3	Avon, CA	Refinery	Sodium chlorate	CD		20.9	10, 13, 14, 15
406	Jan. 1	Barking, UK	Warehouse	Petroleum products	F, EX			13
407	Jan. 30	Bayamon, Puerto Rico	Pipeline	Light HCs	EX	1d	48.5	13
408	Jan. 20	Borger, TX	Alkylation unit	Natural gas	VCE	41i	55.6	13
409	Feb. 26	Brooks, Alberta	Compressor station	HC	EX		23.6	14
410	Dec. 31	Corpus Christi, TX	Hydrocracker	Propane	F		25.8	10
411	May 17	Deer Park, TX	Phenolacetone plant	Nitroglycerine	F			10
412	Mar. 26	Enschede, Netherlands	Warehouse	Gasoline	VCE			
413	Aug. 18	Gach Saran, Iran	Road tanker	LPG, etc.	HEX	80d, 45i		10, 14
414	Nov. 25	Kenner, LA	Warehouse	Gasoline	F	7d, 6i		10
415	Feb. 11	Longport, UK	Road tanker	Ammonia	F, EX			9
416	Mar. 3	Los Angeles, CA	Loading terminal	VCM	BLEVE	2d, 2i		13
417	Oct. 8	Mexico City, Mexico	Rail tank car	Ammonia	TOX	9d, 28i		13
418	Jul. 26	Muldraugh, KY	Pipeline	Hexane, propylene	F	4i		13
419	Oct. 21	Naples, Italy	Grain silo	Grain dust	DEX	8i		14
420	Jul. 15	New Castle, DE	Polypropylene plant	Propane	VCE	5d, 25i		14
421	Nov. 29	New Orleans, LA	Pipeline	Natural gas	F			13
422	Jul. 24	Ortuella, Spain	Storage vessel	Propane	EX	51d		11, 12
423	Jul. 23	Rotterdam, Netherlands	Oil tanker *Energy Concentration*	Crude oil	Ship split apart			9
424	Jun. 26	Seadrift, TX	Ethylene oxide reactor	Reaction mixture	DET		16.4	13
425	Nov. 20	Sydney, Australia	Refinery furnace	Oil	EX		25.0	13
426		Wealdstone, UK	Storage vessel	Propane	REL	0d, 1i		13
427	1981 Feb. 11	Chicago Heights, IL	Reactor weigh tank	Tank contents	VEEB		17.4	13
428	Oct. 1	Czechoslovakia	Ammonia plant	Syngas	VCE			14
429	May 8	Gothenburg, Sweden	Pipeline	Propane	VCE	1d, 2i	13	14
430	Jul. 19	Greens Bayou, TX	Reactor	Herbicide reaction mixture	IE			13
431	Aug. 1	Montana, Mexico	Rail tank car	Chlorine	TOX	17d, 280i		11, 12
432	May 15	San Rafael, Venezuela	Pipeline	LPG	EX	18d, 35i		12
433	Aug. 20	Shuaiba, Kuwait	Tank farm	Oil	F	1d, 1i	124	13

Table A1.2 (continued)

No.	Date	Location	Plant/ transport	Chemical	Event	Deaths/ injuries	Cost (million US$)	References
434	Sep. 6	Stalybridge, UK	Solvent recovery unit	Hexane	EX			15
435	1982 May 3	Caldecott Tunnel, Oakland, CA	Road tanker	Gasoline	F	7d		
436	Dec. 19	Caracas, Venezuela	Storage tank	Oil froth	F	150d, >500i	58.9	13, 15
437	Apr. 18	Edmonton, Alberta	Compressor	Ethylene	VEEB		24.6	11, 13
438	Dec. 29	Florence, Italy	Road tanker	Propane	EX	5d, 30i		12
439	Jan. 20	Fort McMurray, Alberta	Compressor	Hydrogen	F		24.6	13
440	Oct. 4	Freeport, TX	Transformer	Oil	F		17.2	13
441	Nov. 4	Hudson, IA	Pipeline	Natural gas	EX	5d		13
442	Mar. 31	Kashima, Japan	Desulphurization unit	HCs	F		16.3	13
443	Sep. 28	Livingston, LA	Rail tank car	Flammables, toxics	DEL BLEVE	0d, 0i		13
444	Feb. 13	Morley, UK	Warehouse	Herbicides, etc.	F, POL			9,11
445	Jan. 17	Moselle River, France	Pipeline	Carbon monoxide	TOX	5d		11, 12
446	Mar. 9	Philadelphia, PA	Phenol plant	Cumene	VCE		29.3	13
447	Oct. 1	Pine Bluff, AR	Pipeline	Natural gas	VCF			
448	Jun. 28	Portales, NM	Pipeline	Natural gas	DEL EX	6d		
449	Sep. 25	Salford, UK	Warehouse	Sodium chlorate	EX	60i		15
450	1983 Apr. 7	Avon, CA	Cat cracker	Recycle slurry	F		56.3	13
451	Jul. 30	Baton Rouge, LA	Rail tank car	VCM	F			
452	May 26	Bloomfield, NM	Compressor station	Natural gas	VEEB	2i		13
453	Apr. 14	Bontang, India	LNG plant	LNG	IE		57.5	
454	Apr. 3	Denver, CO	Rail tank car	Nitric acid	TOX			
455	Nov. 2	Dhurabar, India	Rail tank car	Kerosene	EX	47d		12
456	Aug. 30	Milford Haven, UK	Storage tank	Crude oil	F	0d, 20i	17.3	13, 15
457	Jul.1	Port Arthur, TX	Polyethylene bead plant		F		17.8	13
458	Jan. 7	Port Newark, NJ	Storage tank	Gasoline	VCE	1d	40.3	13, 14
459	May 26	Prudhoe Bay, AK	NGL surge drum	NGL	F		40.3	13
460	Mar. 15	West Odessa, TX	Pipeline	LPG	EX, F	6d		12
461	Sep. 30	Basile, LA	Gas plant	HCs	VCF		33.9	13
462	1984 May 23	Abbeystead, UK	Valve house	Methane	EX	16d, 28i		11, 12
463	Dec. 3	Bhopal, India	Storage tank	Methyl isocyanate	TOX	≈4000d		11, 15
464	Feb. 24	Cubatao, Brazil	Pipeline	Petrol	F	>100d, ≈150i		12
465	Aug. 15	Fort McMurray, Alberta	Coking unit	HCs	VCF		85.9	13
466	Mar. 8	Kerala, India	Cooling tower	HCs	EX		13.6	13
467	Dec. 13	Las Piedras, Venezuela	Hydrodesulphurizer	Oil	F		70.1	13
468	Nov. 19	Mexico City, Mexico	Terminal	LPG	VCF, BLEVE	≈650d, ≈6400i	22.5	11, 13, 15
469	Sep. 25	Phoenix, AZ	Pipeline	Natural gas	EX, F			
470	Jul. 23	Romeoville, IL	Absorption column	Propane	VCE, BLEVE	15d, 22i	143.5	13, 14, 15
471	Apr. 20	Sarnia, Ont.	Benzene plant	Hydrogen	VCE	2d		14
472	1985 Apr. 27	Beaumont, KY	Pipeline	Natural gas	F	5d, 3i		
473	Jul. 6	Clinton, IA	Ammonia plant	Syngas	VEEB		14.7	13
474	Jan. 18	Cologne, FRG	Ethylene plant	Ethylene	VCE			14

No.	Date	Location	Plant/equipment	Material	Event	Deaths/injuries	Quantity	Refs
475	Feb. 19	Edmonton, Alberta	Pipeline	NGL	VCE	1d, 6i		14
476	Jul. 23	Kaycee, WY	Pipeline	Aviation fuel	EX			14
477	Feb. 23	Jackson, SC	Rail tank car		REL			13, 14
478	Mar. 9	Lake Charles, LA	Reforming unit	Propane	VCE			13
479	Nov. 5	Mont Belvieu	Salt dome storage	Ethane, propane	VCE		44.8	
480	Dec. 21	Naples, Italy	Terminal		F		47.0	13
481	Jun. 9	Pine Bluff, AR	Rail tank car	HMs	REL			
482	May 19	Priola, Italy	Ethylene plant	HCs	F		72.8	13
483	Feb. 23	Sharpsville, PA	Pipeline	Natural gas	EX, F			
484	Nov. 21	Tioga, ND	Pipeline	HCs	IE, VCE		11.3	13
485	Jan. 23	Wood River, IL	Gas processing plant	Propane	VCF		25.2	13
486	1986 Feb. 21	Lancaster, KY	Deasphalting—dewaxing unit	Natural gas	F			
487	Jun. 15	Pascagoula, MS	Pipeline	Aniline, etc.	EX	3 burned		
488	Oct. 30	Schweizerhalle, Basel, Switzerland	Distillation column / Warehouse	Pesticides, etc.	F, POL			11
489	Feb. 24	Thessalonika, Greece	Oil terminal	Oils	F			14
490	1987 Jul. 3	Antwerp, Belgium	Distillation column	Ethylene oxide	EX		39.7	13
491	Oct. 11	Fort McMurray, Alberta	Tar sand plant	Tar sand, diesel fuel	F		87.9	13.15
492	Mar. 22	Grangemouth, UK	Separator vessel	Hydrogen	F	5d		12
493	Aug. 23	Lanzhou, China	Rail tank car	Gasoline	F			13
494	Jun. 23	Mississauga, Ont.	Hydrotreater	Hydrogen, HCs	VCE	3d, 8i	22.4	
495	Nov. 14	Pampa, TX	Acetic acid plant	Acetic acid, butane	VCE	2d, 8i	24.1	13, 14, 15
496	Jun. 2	Port Herriot, France	Storage	Oil	F			
497	Aug. 15	Ras Tanura, Saudi Arabia	Gas plant	Propane	VCE		67.2	13, 14
498	Nov. 24	Torrance, CA	Aklylation unit	HCs	F		16.4	13
499	1988 Jul. 30	Altoona, IA	Railway station	HMs	REL	73d, 230i		12
500	Jun. 4	Arzamas, USSR	Explosives	Explosives	HEX			14
501	Apr. 7	Beek, Netherlands	Polyethylene plant	Ethyiene	VCE			13
502	Jan. 2	Floreffe, PA	Storage tank	Diesel fuel	REL		14.5	
503	May 5	Norco, LA	Cat cracker	C$_3$ HCs	VCE	7d, 28i	327	13, 14, 15
504	Jun. 8	Port Arthur, TX	Storage	Propane	F		17.4	13
505	Oct. 25	Pulua Merlimau	Storage tanks	Naphtha	F		13.1	
506	Sep. 8	Rafnes, Norway	VC plant	VCM, ethylene dichloride	VCE		12.0	13, 14
507	1989 Mar. 7	Antwerp, Belgium	Distillation column	Ethylene oxide	EX		80.1	13, 14
508	Dec. 24	Baton Rouge, LA	Refinery	Ethane, propane	VCE		44.7	13, 14
509	Jul. 22	Freeland, MI	Freight train	HMs	REL			
510	Feb. 2	Helena, MT		HMs	REL			
511	Mar. 20	Jonova, Lithuania	Ammonia storage	Ammonia	IE, F, TOX	7d		14
512	Jul. 2	Minnebeavo, USSR	Gasoline plant	Propane	VCE	4d	52.0	13
513	Sep. 5	Martinez, CA	Hydrotreater	Hydrogen, HCs	F or EX		41.6	13
514	Jun. 7	Morris, IL	Distillation column	Propylene	VCF			13, 14, 15
515	Oct. 23	Pasadena, TX	Polyethylene plant	Isobutane	VCE	23d, ≈103i	>500	13, 14, 15
516	Mar. 22	Peterborough, UK	Road vehicle	Explosives	EX	1d, 107i		
517	Apr. 10	Richmond, CA	Refinery	Hydrogen	F		93.6	13
518	Oct. 3	Sabine Pass, TX	Pipeline	Natural gas	REL			13

Table A1.2 (continued)

No.	Date	Location	Plant/ transport	Chemical	Event	Deaths/ injuries	Cost (million US$)	References
519	May 25	San Bernadino, CA	Pipeline	Gasoline	F	2d, 3i		13
520	Sep. 18	St Croix, Virgin Islands	Refinery	HCs	HUR		62.4	14, 15
521	Jun. 3	Ufa, USSR	Pipeline	NGL	VCE	645d, ≈500i		13
522	Feb. 14	Urdingen, Germany	Paint plant	Reaction mixture	IE		41.6	13
523	1990 Sep. 24	Bangkok, Thailand	Road tanker	LPG	VCF	68d, >100i		13
524	Sep. 16	Bay City, TX	Tankship *Jupiter*	Gasoline	EX	1d		13
525	Nov. 3	Chalmette, LA	Hydrocracker	HCs	VCE		15.3	13
526	Jul. 5	Channelview, TX	Wastewater storage tank	Flammables	EX		12.2	13
527	Jul. 19	Cincinnati, OH	Acrylic resins plant	Xylene, solvent	VEEB		23.5	13, 14
528	Nov. 25	Denver, CO	Storage	Jet fuel	F		30.6	13
529	Nov. 6	Nagothane, Bombay, India	Ethylene plant	Ethane, propane	VCE	31d	22.4	13,14
530	Mar. 3	North Blenheim, NY	Pipeline	Propane	VCF	2d, 7i		
531	Nov. 15	Porto de Leixhos, Portugal	Deasphalting unit	Propane	VCE			14
532	Nov. 30	Ras Tanura, Saudi Arabia	Refinery	HCs	F		23	13
533	Jul. 10	Rio de Janeiro, Brazil	Boiler	Steam	EX		10.2	13
534	Mar. 30	Stanlow, UK	Reactor	2-4 dichloronitro– benzene reaction mixture	IE	1d, 5i		15
535	May 14	Tomsk. USSR	Ethylene plant	Gas	VCE			14
536	Apr. 1	Warren, PA	FCC	LPG	EX, F		25.5	13
537	1991 Jun. 17	Charleston, SC	Reactor	Reaction mixture	IE		10	13
538	Mar. 11	Coatzacoala, Mexico	Chlorine plant		EX		150	13
539	Aug. 21	Coode Island, Australia	Storage tanks	Acrylonitrile, etc.	IE	0d	24	13
540	Jun. 20	Dhaka, Bangladesh	Stripping column		EX		70	13
541	Sep. 7	Haifa, Israel	Chemical plant	Chemicals	F		15	13
542	Jul. 14	Kensington, GA	Synthetic rubber	Butadiene plant	VCE			14
543	Mar. 3	Lake Charles, LA	FCC	Hot oil-water	IE		23	13
544	Mar. 11	Pajaritos, Mexico	Vinyl chloride plant	Propane	VCE	3d		14
545	Mar. 12	Seadrift, TX	Ethylene oxide plant	Ethylene oxide	VCE	1d		14
546	May 1	Sterlington, LA	Nitroparaffin unit		EX		105	13
547	Apr. 13	Sweeny, TX	Reactor	HCs	EX		50	13
548	1992 Oct. 16	Sodegaura, Japan	Refinery	Hydrogen	VCE	10d, 7i		
549	1993 Feb. 2	Griesheim, Germany	Reactor	Reaction mixture	TOX	0d		
550	1994 Nov. 2	Dronka, Egypt	Fuel storage	Aviation, diesel, fuel	F	≈410d		
551	1995 Apr. 27	Ukhta, Russia	Gas pipeline	Gas	F			

Alternative names

Walsum, 1952: *aka* Duisburg-Walsum (often referred to as Wilsum); Berlin, 1962: *aka* New Berlin; Doe Run, 1962: *aka* Brandenburg; Glendora, 1969 *aka* Black Bayou Jct; Rio de Janeiro, 1972: *aka* Duque de Caxais or Sao Paulo; West St Paul, 1974: *aka* St Paul; Des Moines, 1975: *aka* Angleton; Pasacabolo, 1977: *aka* Cartagena; Denver, 1978: *aka* Commerce; Donnellson, 1978: *aka* Fort Madison; San Carlos, 1978: *aka* Los Alfaques; Caracas, 1981: *aka* Tacao; Montanas, 1981: *aka* San Luis Potosi; Stalybridge, 1981: *aka* Chemstar Ltd: Caracas, 1982: *aka* Tacao; Salford, 1982: *aka* B&R Hauliers; Cubatao, 1984; *aka* Vila Soca; Mexico City, 1984: *aka* San Juan Ixhautepec; Schweizerhalle, 1986: *aka* Sandos; Ras Tanura, 1987: *aka* Juaimah; Floreffe, 1988: *aka* Ashland; Rafnes, 1988: *aka* Bamble: Nagothane, 1990: *aka* Maharastra; Bombay; Coode Island, 1991: Melbourne

Abbreviations: FCC = fluid catalytic cracker; HC = hydrocarbon; HM = hazardous material; VCM = vinyl chloride monomer

Event abbreviations: ANEX = ammonium nitrate explosion; BLEVE = boiling liquid expanding vapour explosion; CD = criminal damage caused by vandalism, sabotage, etc.; DEL = delayed; DET = detonation (internal explosions only); DEX = dust explosion; EQK = earthquake damage; EX = explosion; F = fire; FB = fireball; HEX = high explosive explosion; HUR = hurricane damage; IE = internal explosion; POL = pollution; REL = release; TOX = toxic release; VCE = vapour cloud explosion; VCF = vapour cloud fire; VEEB = vapour escape into, and explosion in, building

Costs

Costs are in US dollars and are taken from the M&M compilations. The great majority are the trended costs as given in the thirteenth, 1990 edition by Mahoney (1990) and are thus in 1990 US dollars. Some values not included in the 1990 edition but given in the eleventh edition by Garrison (1988) have been converted to 1990 values, using a factor of 1.1; these are marked with an asterisk. Values for incidents in 1990 and 1991 given in the fourteenth 1992 edition by M&M (1992) are in 1991 US dollars.

Principal references

(1) Simmons, Erdmann and Naft (1974); (2) Eisenberg, Lynch and Breeding (1975); (3) V. C. Marshall (1977b); (4) Harvey (1979b); (5) HSE (1978b); (6) Slater (1978a); (7) Gugan (1979); (8) D. J. Lewis (1980d); (9) Harvey (1984); (10) Kier and Müller (1983); (11) V. C. Marshall (1987); (12) Haastrup and Brockhoft (1990); (13) M&M (Garrison (1988b); Mahoney (1990); and M&M (1992)); (14) Lenoir and Davenport (1993); (15) D. J. Lewis (1993)

Additional references

1; 2; 3; 4 J. R. Nash (1976); 5; 6; 7; 8 Assheton (1930); 9 Billings and Copland (1992); 10 Assheton (1930); J. R. Nash (1976); 11 Anon, (n. d. b); Ministry of Home Security (n. d. a); Sainsbury (1977); 12; 13; 14; 15; 16 Slater (1978b); 17; 18 BASF (1921); Comment *et al.* (1921); NBFU (1948); 19; 20; 21; 22; 23 Anon. (1928); N. C. Harris (1981b); 24; 25; 26; 27; 28 Anon. (1933a, b); 29; 30; 31; 32; 33; 34 P. Reed (1939); Armistead (1959); 35; 36 Römcke and Evensen (1940); 37; 38 NBFU (1948); J. R. Nash (1976); 39 Slater (1978b); 40 Stahl, Strassman and Richard (1949); J. R. Nash (1976); Giesbrecht *et al.* (1981); D. J. Lewis (1983); 41 J. R. Nash (1976); Patience (1989); 42 Chasis *et al.* (1947); 43 Moulton (1944); Coroner (1945); Elliott *et al.* (1946 BM RI 3867); Burgoyne (1965b); Strehlow (1973b); J. R. Nash (1976); Marshall (1983 IPB 52); 44 Anon. (1952–53); 45 J. Reed (1977); Anon. (1992 LPB 103); V. C. Marshall (1992 LPB 105); 46; 47 J. R. Nash (1976); 48; 49 J. R. Nash (1976); 50; 51; 52; 53; 54 Kintz, Jones and Carpenter (1948 BM RI 4245); NBFU (1948); Wheaton (1948); Blocker (1949); J. R. Nash (1976); 55 BASF (1948); Stahl, Strassman and Richard (1949); Giesbrecht *et al.* (1981); D. J. Lewis (1983); 56; 57; 58; 59; 60 Anon. (1949–50); 61; 62; 63; 64; 65; 66 H. E. Watts (1951); Anon. (1952); Kletz (1983 LPB 52); 67; 68 NBFU and FIRONT (1952); 69; 70 Jenkins and Oakshott (1955); Strehlow (1973b); D. J. Lewis (1980); Alderson (1982 SRD R246); 71 NFPA (1957); 72; 73; 74; 75; 76; 77; 78; 79; 80; 81; 82; 83 Woodworth (1955); R. B. Jacobs *et al.* (1957, 1973); Randall *et al.* (1957); 84; 85; 86; 87 Walker (1973); J. R. Nash (1976); 88; 89; 90; 91; 92 Lindley and Brown (1958); 93 Selway (1988 SRD 492); 94; 95; 96; 97; 98; 99; 100 Anon. (1959a); Woodworth (1958); 101; 102; 103 Anon. (1960–61); Slater (1978b); Strehlow (1973b); Hymes (1985 LPB 65); 104 Schmitt (1964); 105; 106 Cronan (1960c); 107 Sakai (1964); 108 Chementator (1960 Nov. 14, 106); 109 Chementator (1960 Dec. 12, 69); F. C. Price (1960); 110; 111; 112; 113 Joyner and Durel (1962); 114; 115; 116; 117 Walls (1963–64, 1964a); Strehlow (1973b); Strehlow and Baker (1976); 118; 119 Troyan and Levine (1968); 120; 121; 122; 123; 124 Laney (1964); Vervalin (1973e); 125; 126 Chementator (1962 May 28, 68); 127; 128; 129; 130; 131; 132; 133; 134; 135 Walls (1964b); 136; 137; 138; 139 Chementator (1964 Feb. 3, 36); 140 Bulkley and Jacobs (1966); 141; 142; 143; 144; 145; 146; 147; 148; 149 Anon. (1967c); 150; 151; 152; 153 Chementator (1965 Sep. 13, 105); Armistead (1966, 1973); 154 Chementator (1965 Mar. 29, 26; Aug. 2, 41); FPC (1966); 155; 156; 157; 158 Anon. (1966g, h); Strehlow (1973b); Vervalin (1973e); Kletz (1974e, 1975d, 1977d); Anon. (1987 LPB 77, p. 1); Selway (1988 SRD R492); 159; 160 Chementator (1966 Jan. 31, 18); 161 Chementator (1966 Oct. 24, 55); Anon. (1967b); Fire J. Staff (1967, 1973d); 162; 163 McCarey (1966); 164 Anon. (1966c); 165; 166; 167; 168; 169; 170; 171 Chementator (1967 Aug. 28, 56); Goforth (1970); NFPA (1973); Strehlow (1973b); Vervalin (1973e); Strehlow and Baker (1976); 172; 173; 174; 175; 176 NTSB (1968 RAR); 177; 178; 179 Medard (1970); 180; 181 Fontein (1968); Min. of Social Affairs and Public Health (1968); Strehlow (1973b); Woodworth (1973); Strehlow and Baker (1976); 182 Slater (1978b); 183; 184; 185; 186 J. E. Browning (1969a); NTSB (1971 RAR-71-02); 187 Anon. (1970a); 188; 189; 190; 191 NTSB (1970 RAR-70-02); Dowell (1971); Kogler (1971); Strehlow and Baker (1976); 192 NTSB (1971 PAR-71-01); 193 NTSB (1969 RAR); 194; 195; 196; 197 R. H. Freeman and McReady (1971); Keister, Pesetsky and Clark (1971); Griffith and Keister (1973); 198 Anon. (1970e); 199 Anon. (1969b); 200 Anon. (1971d); 201; 202; 203; 204 MacArthur (1972); 205 NTSB (1970 HAR-71-06); 206; 207 Watrous (1970); NTSB (1972 RAR-72-02); Strehlow (1973b); Vervalin (1973e); Strehlow and Baker (1976); D. J. Lewis (1991 LPB 101); 208; 209; 210 NTSB (1971 PAR-71-02); 211; 212; 213; 214 Chementator (1970 Jun. 1, 63); 215 NTSB (1972 PAR-72-01); Burgess and Zabetakis (1973 BM RI 7752); Strehlow (1973b); Strehlow and Baker (1976); 216; 217; 218; 219; 220 Chementator (1971 Jul. 26, 55); 221 NTSB (1972 HAR-72-03); Cremer and Warner (1978); 222 Chementator (1971 Nov. 1, 23); NTSB (1972 RAR-72-06); Slater (1978b); 223; 224 Sarsten (1972); 225; 226 Lauderback (1975); 227; 228; 229 W. B. Howard (1975b); Slater (1978b); Stueben and Ball (1980); Matsuz and Sadler (1994); 230; 231; 232 NTSB (1971 HAR-72-05); 233; 234; 235 NTSB (1973 RAR-73-01); Strehlow (1973b); Strehlow and Baker (1976); 236 NTSB (1973 PAR-73-02); 237 NTSB (1973 HAR-73-03); Strehlow and Baker (1976); 238; 239 NTSB (1973 HAR-73-04); 240 Selway (1988 SRD R492); 241; 242; 243 Chementator (1973 Jan. 8, 51); D. J. Lewis (1989c); 244; 245 NTSB (1973 PAR-73-04); Strehlow and Baker (1976); 246 (NTSB 1975 RAR-75-02); 247; 248; 249; 250; 251 Crawley (1982); Hymes (1985 LPB 65); 252; 253; 254 NTSB (1974 PAR-74-06); Luddeke (1975); 255; 256 Lonsdale (1975); 257; 258 Anon. (1973); 259; 260; 261 Chementator (1973 Mar. 5, 28; 1974 Apr. 1, 17); US Congress (1973); D. J. Lewis (1989); 262 Chementator (1973 Jul. 23, 58; Nov. 12, 94); 263; 264; 265; 266 NTSB (1975 PAR-75-02); 267 Chementator (1974 Dec. 23, 21); 268; 269 Hoyle (1982); 270; 271; 272 NTSB (1975 RAR-75-04); Strehlow and Baker (1976); V. C. Marshall (1980d); 273; 274 R. J. Parker (1975); Strehlow and Baker (1976); 275 Adderton (1974); Strehlow and Baker (1976); 276; 277 NTSB (1975 RAR-75-07); Slater (1978b); 278 Sharry (1975); 279 NTSB (1975 PAR-75-01); 280; 281 NTSB (1974 RAR-74-04); 282 Anon. (1974n); 283; 284; 285 Anon. (1974k); Anon. (1974pr); 286; 287 NTSB (1976 RAR-76-01); 288; 289; 290; 291; 292 Chementator (1975 Nov. 24, 17); Min. of Social Affairs (1976); van Eijnatten (1977); Slater (1978b); 293; 294; 295 NTSB (1976 RAR-76-08); 296 NTSB (1976 PAR-76-05); 297 NTSB (1976 RAR-76-04); 298; 299 HSE (1976b); 300 NTSB (1976 RAR-76-03); 301 Slater (1978b); 302; 303; 304; 305 Chementator (1975 Oct. 27, 53); Anon. (1975m); 306 NTSB (1976 RAR-76-07); 310; 311; 312; 313 Anon. (1976a); Rees (1982); 314 NTSB (1977 RAR-77-07); 315 Chementator (1976 Aug. 30, 49); 316; 317; 318; 319; 320 Anon. (1976o); McMullen (1976); NTSB (1977 HAR-77-01); Fryer and Kaiser (1979 SRD R152); 321 HSE (1977b); 322; 323; 324; 325 Slater (1978b); 326 NTSB (1978 MAR-78-06); D. J. Lewis (1984c); 327; 328; 329; 330 NTSB (1977 PAR-77-03); 331; 332 Temple (1977b); 333; 334; 335; 336; 337; 338; 339; 340 NTSB (1978 HAR-78-04); 341 HSE (1979b); 342

Table A1.2   *(continued)*

Anon. (1977b); 343; 344; 345 NTSB (1978 PAR-78-02); 346; 347; 348; 349; 350; 351; 352; 353; 354 NTSB (1978 RAR-78-04); Stueben and Ball (1980); 355; 356 D. J. Lewis (1991 LPB 100); 357; 358 NTSB (1978 PAR-78-01); 359; 360 Anon. (1977c); Whelan (1977); Kletz (1988 LPB 81); 361 Anon. (1977p, q); Lathrop (1978); Nolan (1979); 362; 363; 364 Chementator (1978 Feb. 27, 64); 365; 366; 367 NTSB (1979 PAR-79-01); 368; 369 NTSB (1978 RAR-78-08); 370; 371; 372; 373; 374 Anon. (1978m); 375; 376 Selway (1988 SRD R492); 377; 378 Anon. (1978b); NTSB (1979 RAR-794I); D. J. Lewis (1992 LPB 105); 379; 380 Anon. (1978k); NTSB (1978 RAR-78-07); 381 Costello (1979); Anon. (1980a, v); 382; 383; 384; 385 NTSB (1979 RAR-79-11); 387 NTSB (1980 MAR-80-18); 388; 389; 390 NTSB (1980 MAR-80-07); 391; 392; 393; 394 Mississauga News (1979); Amyot (1980); Grange (1980); Fordham (1981 LPB 44); Lane and Thomson (1981); 395 NTSB Ann Rep. (1979); 396; 397; 398; 399 NTSB (1980 MAR-80-12); 400; 401; 402; 403; 404; 405; 406 NTSB (1980a); Anon. (1980r); 407 NTSB (1980 PAR-80-6); 408; 409; 410; 411; 412; 413; 414; 415 HSE (1981b); 416 NTSB (1980 AR-80-05); 417; 418 NTSB (1981 RAR-81-01); 419; 420; 421 NTSB (1980 PAR-80-01); 422; 423; 424; 425; 426 HSE (1981c); 427; 428; 429 Nilson (1983b); 430; 431 Anon. (1981 LPB 44); Anon. (1983 LPB 52); 432; 433; 434 HSE (1982a); Kletz (1983b); 435 NTSB (1983 HAR-83-01); 436 Garrison (1984 LPB 57); M. F. Henry (1986); 437; 438; 439; 440; 441 NTSB (1983 PAR-83-05); Anon. (1984t) 444; 445; 446; 447 NTSB (1983 PAR-83-03); 448 NTSB (1983 PAR-83-01); 449 HSE (1983b); 450; 451 NTSB (1985 RAR-85-08); 452 NTSB (1983 PAR-83-04); 458; 454 NTSB (1985 RAR-85-10); 455; 456 Dyfed County Fire Brigade (1983); Mumford (1984 LPB 57); Golec (1985); Steinbrecher (1986); 457; 458; 459; 460 NTSB (1984 PAR-84-01); 461; 462 Burgoyne (1985a); HSE (1985a); 463; 464 Anon. (1984oo); 465; 466 Anon. (1992 LPB 106); 467; 468 Pietersen (1985); Selway (1988 SRD R492); 469 NTSB (1985 PAR-85-01); 470; 471; 472 NTSB (1987 PAR-87-01); 473; 474; 475; 476 NTSB (1986 PAR-86-01); 477 NTSB (1985 RAR-85-12); 478; 479; 480; 481 NTSB (1986 RAR-86-04); 482; 483 NTSB (1985 PAR-85-02); 484; 485; 486 NTSB (1987 PAR-87-01); 487; 488 Anon. (1986v. w, y); Beck (1986); D. Williams (1986); Anon. (1987 LPB 75); Stallworthy (1987); 489 B. Browning and Searson (1964); 490 Mellin (1991 LPB 100); 491; 492 Anon. (1989b); HSE (1989a); Kletz (1989b); K. C. Wilson (1990a, b); 493; 494; 495; 496 Mansot (1989); 497; 498; 499 NTSB (1989 RAR-89-04); 500; 501; 502 Prokop (1988); Wilkinson (1991 SRD R530); Mellin (1992 LPB 106); 503; 504; 505; 506; 507 Anon (1991 LPB 100); 508; 509 NTSB (1991 RAR-91-04); 510 NTSB (1989 RAR-89-05); 511 Andersson (1991a, b); Kletz (1991a); Anon. (1992 LPB 107: p. 1); Anderson and Lindley (1992 LPB 107); Kukkonen *et al.* (1992); Wilday (1992 LPB 108, 109); Kukkonen *et al.* (1993); 512; 513; 514; 515 OSHA (1990a); Kletz (1991); 516 HSE (1990b); Anon. (1991 LPB 98, p. 24); 517; 518 NTSB (1990 RAR-90-02); 520; 521 Hofheinz and Kohan (1989); 522; 523 Anon. (1991 LPB 101, p. 19); 524 NTSB (1991 MAR-91-04); 525; 526 Anon. (1990d); 527; 528; 529 Indian Petrochemicals Corp. (1990); 530 NTSB (1991 PAR-91-01); 532 Anon. (1990g); 533; 534 Kletz (1990 LPB 100); Mooney (1991); Redman and Kletz (1991); Cates (1992); van Reijendam *et al.* (1992); 535; 536; 537; 538; 539 Anon. (1991b); I. Thomas (1991); Anon. (1992c); Croudace (1993 LPB 112); 540; 541; 542; 543 Anon. (1991d); 544; 545 Viera, Simpson and Ream (1993); 546; 547; 548 High Pressure Gas Safety Institute of Japan (1993 LPB 116); 549 Ondrey (1993a, b); Schönfeld (1993); Vennen (1993); 550 Reuter (1994); 551 Anon. (1995)

Notes:

[a] Evidently including 'excess deaths', characteristic of a severe pollution incident.

[b] Gugan (1979) gives two incidents in 1966 in the FRG (his Nos 49 and 50). His No. 49 is named as Raunheim, while No. 50 has no more specific location. For both the date is given as January and the reference as Bradford and Culbertson (1967). Incident No. 50 is not quoted by Lenoir and Davenport (1993).

[c] Slater (1978a) cites an incident in Roumania in 1974 involving ethylene as a VCE with Id, 50i; this was given in the first edition of this book. The present entry relies on Anon. (1974k).

[d] The figure for deaths refers to prompt deaths; it does not include abortions.

A2   *Ashton, Manchester, 1917*

On 13 June 1917, a nitrator in an explosives works at Ashton, Manchester, went out of control and 'boiled' over, releasing hot acid onto the wooden staging and starting a fire. The fire took hold in the nitrator house. Spectators were attracted, including children. Soon after there was a large explosion, involving some 5 tons of TNT. The explosion ruptured two gasholders nearby, setting them on fire. The flames were described as half a mile high.

Some 100 houses were demolished. 46 persons died, including 24 employees and 11 children. More than 120 were injured sufficiently to need hospital treatment. The explosion made two craters, one at the location of the TNT drums and one at that of the drier and setting frays. The former was 90 ft×36 ft×5 ft deep and the latter 30 ft×15 ft and relatively shallow. Some 12,000 ft^2 of glass was replaced.

Billings and Copland (1992)

A3   *Halifax, Nova Scotia, 1917*

At about 8.48 a.m. on 6 December 1917, the French ammunition ship Mont Blanc was rammed by the Belgian relief ship Imo in Halifax Harbour, Nova Scotia. The Mont Blanc was carrying in her hold some 4,661,794 lb of picric acid, 450,000 lb of TNT and 122,960 lb of guncotton as well as 488,672 lb of benzol on deck. The Mont Blanc burst into flame and at 9.05 am there was a massive explosion. The inhabitants heard a 'thundering crash' and many ran to their windows to see what had happened. The noise was heard 191 miles away and some windows broke at a distance of 61 miles.

The explosion tore the ship apart and hurled fragments all over the city. The anchor shank fetched up 3−4 miles away. The blast caused almost complete destruction within a radius of about three-quarters of a mile. There was severe structural damage at distances ranging from 1 to 1.75 miles. It was estimated that 95% of the glass in the city was broken. Fires caused by overturned stoves broke out in many parts.

Many people were killed when buildings were demolished. Large numbers suffered injury from flying glass and at least 500 people were totally blinded. In round figures, the death toll was put at 1800 and the number of registered wounded was 8000.

A4   *Silvertown, London, 1917*

On 19 January 1917, an explosion occurred at the Brunner Mond munitions factory at Crescent Wharf, Silvertown, London. A fire broke out at the melt pot room and took hold. It burned for some minutes when there was a massive explosion. Sixty nine people were killed and 98 seriously injured; there were 328 slight injuries recorded and about 600 were believed to have been treated in the street or by private doctors. The report of the inquiry stated:

At the time of the explosion the total quantity of TNT on the premises amounted to 83 tons, of which 28 were crude, 27 in process and 28 finished, and there were, in addition, 9 tons of TNT oil in iron drums of which 5 tons were subsequently recovered. . . . The outlines of the craters seem to indicate that altogether about 30 tons of TNT and the remaining 4 tons of TNT oil did not explode, but probably burned away. This would leave 53 tons of TNT to take part in the explosion.

TNT oil is alcohol containing the soluble impurities of TNT.

The explosion formed two separate craters. The smaller was circular in shape with a radius of about 25 ft and some 13 tons of TNT had been present in this area. The larger crater, in the area containing the remaining 40 tons of TNT, was kidney-shaped, with approximate dimensions 150 ft × 50 ft.

Anon. (n.d.b); Ministry of Home Security (n.d.a); Sainsbury (1977)

A5   *Oppau, Germany, 1921*

At 7.29 a.m. and again at 7.32 a.m. on 21 September 1921, two terrific explosions occurred at the Oppau works of Badische Anilin und Soda Fabrik (BASF).

Figures for the death toll vary. Commentz *et al.* (1921) give the number of dead as 430, including 50 people in the village. According to V.C. Marshall (1987), following BASF (1958), the explosion killed 561, including 4 in Mannheim 7 km away, injured 1500 and destroyed 1000 houses, including 75% of those in Oppau itself. It created a crater 115 m long × 75 m wide × 10 m deep.

The works after the explosion is shown in Figure A1.3.

Other effects of the explosion were as follows (Commentz *et al.*, 1921):

Houses in the adjacent city of Ludwigshafen and in Mannheim, which lies opposite to Ludwigshafen on the right bank of the Rhine, were damaged, walls were shaken down and practically all windows were broken. At these places and at Heidevaporerg, which is about 14 miles from Oppau, the effect of the explosion was first felt by two very heavy earthquake-like shocks. In Mannheim some seconds later and in Heidevaporerg 82s after the shocks there came an enormous rush of air which broke windows and doors and caused damage to gas holders, oil tanks, and many river barges. The sound of the explosion reached as far as Bayreuth, at a distance of 145 miles, and the air pressure wave caused considerable damage in Frankfurt, which is about 53 miles from the scene of the explosion.

The explosion involved some 4500 tons of a 50 : 50 mixture of ammonium sulfate and ammonium nitrate. Detonation was set off by blasting powder, which was being used to break up storage piles of material which had become caked, a procedure which had been carried out without mishap some 16,000 times previously.

The danger of using explosives to break up caked ammonium nitrate had been demonstrated earlier in the year in Germany, when on 10 July at Kriewald this procedure led to the explosion of two rail wagons containing 15 te of that substance (D.J. Lewis, 1993).

BASF (1921, 1958); Commentz *et al.* (1921); J.R. Nash (1976); NBFU (1948); HSE (1978b); V.C. Marshall (1987); D.J. Lewis (1993)

A6   *Hamburg, Germany, 1928*

On 28 May 1928, a tank containing phosgene exploded at the Stolzenburg factory on the Hofe-Kanal near the harbour area of Hamburg. The wind was at first northerly and then changed to the south-east. At one location 5 miles away the gas was strongly felt. People were also affected by the gas at locations 6 and 11 miles distant.

The prime source (Anon., 1928) states 'Some 300 or more cubic feet of the gas were released'. V.C. Marshall (1987) suggests that this figure probably refers to the volume of the liquid, which would correspond to about 12 te, close to the figure of 11 te given by N.C. Harris (1981).

11 people were killed and 171 required hospital treatment.

Figure A1.3   *Oppau: BASF works after ammonium nitrate explosion, 1921 (Badische A nilin und Soda Fabrik)*

Anon. (1928); Hegler (1928); N.C. Harris (1981b); V.C. Marshall (1987)

A7   *Neunkircken, Germany, 1933*

On 10 February 1933, an explosion occurred on a waterless gas holder in Neunkirchen in Germany. Work was being done on a bypass pipe between the inlet and outlet mains. Due to a series of actions which is not entirely clear, but which involved measures to bring misaligned flanges into alignment, a small explosion occurred at this bypass and ruptured the outlet main. A flame was seen to go up the side of the gasholder. It had burned for some 5 min when the main explosion occurred. The inquiry found that this explosion involved an air mixture which had formed above the piston.

The explosion caused 65 deaths and several hundred injuries.

Anon. (1933a,b); Kier and Miffler (1983)

A8   *Wichita Falls, Texas, 1939*

On 12 December 1939 an explosion occurred in a 10 in. diameter crude oil pipeline at Wichita Falls, Texas. A 38 mile section of the line had been emptied to locate and repair minor leaks. The method used was to pass a series of scrapers through the line propelled by air, which was odorized to assist in detecting the leaks, and then to displace the column of air by one of oil. During this latter procedure a scraper propelled by the oil struck one that had stuck in the pipe, evidently causing a spark. The result was an oil film explosion, in which the detonation travelled down the pipeline for over 26.8 miles, rupturing the line at intervals of about 80 ft.

The pattern of ruptures in the pipeline is shown in Figure A1.4.

P. Reed (1939); Armistead (1959); D.J. Lewis (1993)

A9   *Tessenderloo, Belgium, 1942*

On 21 July 1942, a massive ammonium nitrate explosion occurred at a chemical works at Tessenderloo in Belgium. The death toll is given as of the order of 200.

J.R. Nash (1976)

A10   *Ludwigshafen, Germany, 1943*

On 29 July 1943, a release occurred from a rail tank car in the BASF works at Ludwigshafen. The tank car contained 16.5 te of a mixture of 80% butadiene and 20% butylene. A vapour cloud formed and ignited. The resultant vapour cloud explosion demolished a block 350×100 m^2 within the factory. Fifty-seven people were killed and 439 injured.

According to Giesbrecht *et al.* (1981), there was a delay of some 10–25 s between the rupture and the explosion. These authors indicate that the cause was hydraulic rupture of the tank due to overheating in the sun. If so, the incident is remarkably similar to one which occurred in the same works 5 years later, as described in Case History A17.

Stahl, Strassman and Richard (1949); J.R. Nash (1976); Giesbrecht *et al.* (1981); D.J. Lewis (1983); V.C. Marshall (1986 LPB 67, 1987)

A11   *Bombay, India, 1944*

Just after 4.00 p.m. on 14 April 1944 in the Victoria Dock, Bombay, India, a massive explosion occurred on the munitions ship Fort Stikine.

Figure A1.4 *Wichita Falls, 1939: pipeline after the explosion: (a) aerial view showing blackening where repeated ruptures of the pipeline occurred; and (b) elevation view, showing broken ends of the pipeline projecting through the ground (Bureau of Mines)*

The ship carried some 1400 tons of mixed munitions (bombs, shells, torpedoes, mines, etc.). On its way it had picked up a cargo of cotton. In No. 2 Hold 187 tons of ammunition was stored with cotton bales above, then other material, 1089 drums of lube oil, and finally more explosives. This was contrary to regulations, which stated that cotton is liable to spontaneous combustion if contaminated with oil. Further notes stated that cotton and explosives should be stored at opposite ends of the ship and that oil drums stowed above cotton should be limited to 250 barrels.

By mid-day on 14 April unloading of the ship was well under way. Soon after, crew on a nearby ship saw wisps of smoke coming from No. 2 Hold of the Fort Stikine, but did not inform the latter. Almost two hours passed before the

fire was discovered. Water was applied to the hold using the ship's three hoses.

The authorities recognized the seriousness of the situation and considered options such as moving or scuttling the ship, but both appeared impractical, since the ship's engines were down for repair, there were no tugs available and the water was too shallow for scuttling. Moving the ship would also involve disconnecting the fire hoses laid from the dock into No. 2 Hold, which now numbered more than 30.

By 3.30 p.m. although an estimated 900 tons of water had been poured into the hold, the port deck was too hot to walk on and a large patch on the side was glowing red. At 3.50 p.m. a huge column of flame shot up from the hold and at 4.06 p.m. the first massive explosion occurred. Some 34 min later there was a second major explosion.

The first explosion killed 60 firemen and set up fires in a 900 m radius. As a result of the two explosions together 10 ships in the docks were destroyed or declared a total loss. Figures for the death toll differ. According to Patience (1989), early reports gave 350 people killed, 1800 injured and 50,000 rendered homeless. J.R. Nash (1976) quotes 1376 dead.

J.R. Nash (1976); Patience (1989)

A12   Cleveland, Ohio, 1944
At approximately 2.40 p.m. on 20 October 1944 a cylindrical LNG storage tank at the Liquefaction, Storage and Regasification Plant of the East Ohio Gas Company at Cleveland, Ohio, ruptured and discharged its entire contents over the plant and the nearby urban area. The LNG vapour ignited almost immediately and an intense fire burned at the plant, causing great loss of life and extensive damage.

More LNG flowed from the plant as liquid down storm sewers, where it mixed with air and exploded.

The final death toll was 128 and the numbers injured were estimated as 200–400. The greatest loss of life occurred within the plant area.

The plant was the first commercial LNG plant of its kind in the work. Its function was to liquefy and store natural gas during off-peak periods and to regasify it during periods of peak demand. The LNG was stored at 5 psi and −250°F. The plant was built on a site which had been in the company's possession for 50 years. The site was close to residential and industrial areas.

The cause of the rupture is uncertain. The Bureau of Mines investigation (Elliott *et al.*, 1946 BM RI 3867) concluded that the low carbon steel used in the construction of the vessel may have been unsuitable and that the failure may have occurred due to vibration or seismic shock.

It was pointed out in the report that most industries handling liquid oxygen or air used not carbon steel but stainless steel or suitable non-ferrous metals. Following this accident, the use of the latter materials became more widely favoured in applications of this kind.

The report recommended that the distance between a plant of this kind and the nearest inhabited building should be greater than half a mile.

Following the rupture large quantities of liquid topped by burning vapour had flowed considerable distances from the tank. The report discussed the argument that a dike is not useful for a relatively volatile material such as LPG or LNG, concluded that a dike would have reduced the hazard and recommended that storages for liquefied gases should have a dike.

The report also made a number of other recommendations. These included the open siting of storage tanks to permit good ventilation; the use of precautions to eliminate sources of ignition to the standard considered necessary in explosives plants; the provision of remote closure for the bottom off-take valve; the installation of reliable level indicators and alarms; and the conduct of emergency drills.

Moulton (1944); Coroner (1945); Elliott *et al.* (1946 BM RI 3867); Burgoyne (1965b); Strehlow (1973b); J.R. Nash (1976) (under East Ohio Gas Co.); D.J. Lewis (1980); V.C. Marshall (1983 LPB 52, 1987)

A13   Fauld, UK, 1944
On 27 November 1944 a huge explosion occurred in a munitions store at Fauld, Staffordshire. The store was underground in a former gypsum mine, mined on the 'crown and pillar' method. The explosion involved mass detonation of 3500 tons of bombs. Although the store was in the countryside, the explosion killed 68 people and injured 22. It created a massive crater covering 12 acres and 80 ft deep.

The store handled large quantities of bombs for the bombing offensive. This included the repair and re-issue of bombs recovered after jettisoning. The personnel were working under considerable pressure. The inquiry found that the most likely cause was ignorance and disregard of the regulations by personnel working on jettisoned bombs. An armourer testified that he had seen a colleague using a brass chisel to remove a broken exploder from one of the bombs, a practice expressly forbidden. Other witnesses confirmed that such irregularities were by no means uncommon.

J. Reed (1977); Anon. (1992 LPB 103); V.C. Marshall (1992 LPB 105)

A14   Port Chicago, California, 1944
On the evening of 17 July 1944 two vessels, the Quinault Victory and the E.A. Bryan, were berthed at Port Chicago, California, taking on large quantities of cordite and TNT when they were rent by a massive explosion. At dawn next day the vessels, and the 321 men on them, had disappeared. The cause of the explosion was never established, although some authorities implicate deteriorated munitions.

J.R. Nash (1976)

A15   Brest, France, 1947
On 28 July 1947 the vessel Ocean Liberty with a cargo of ammonium nitrate, exploded in the port of Brest, France, killing some 21 people.

J.R. Nash (1976)

A16   Texas City, Texas, 1947
At about 8.10 a.m. on 16 April 1947 fire was observed in bagged ammonium nitrate fertilizer on board the ship Grandcamp in the harbour at Texas City. There were 880 tons of ammonium nitrate in the hold affected and a further 1400 tons in another hold. Frantic efforts were made to extinguish the fire, but the quantity of water used initially was too small and by the time hose lines had been connected to supply larger quantities, the fire was so well established that the crew was ordered to abandon the ship. At 9.15 a.m. the Grandcamp disintegrated with a tremendous thunderclap, killing all persons in the dock, including firemen and a crowd of spectators.

Another ship, the High Flyer, which also had ammonium nitrate on board, was 700 ft away and was blown free of its hawsers. On account of the danger of another explosion, volunteers could not be found to move the High Flyer out of the burning area. At 6.00 p.m. ignition of its sulfur cargo occurred. At 1.10 a.m. the next day the High Flyer was ripped apart by the expected explosion.

The report of the National Board of Fire Underwriters (1948) states:

When the Grandcamp blew up, the cargo of peanuts, tobacco leaves, balls of sisal twine, and oil-well drilling equipment were blown in all directions. Shrapnel-like fragments of the ship were hurled in high trajectory, 2000–3000 feet into the air; some travelled more than 10,000 feet from the point of origin. Some of the oil-well drill rods (30 feet in length and 7 inches in diameter, weighing close to 1½ tons) were hurled almost 2 miles and buried 8 ft into the ground, like twisted hairpins. . . .

It was virtually impossible to separate the effects of the two successive explosions, but, the report continues:

> Daybreak revealed a sickening scene of devastation—demolished concrete structures, masses of twisted wreckage, crippled refineries with battered storage tanks, some crumpled like tin-foil, and sidings of distant warehouses blown apart as if ripped by a tornado. Pitch black columns of smoke from the burning oil tanks spiralled skyward for 3000 feet or more, and were visible for 30 miles. They burned continuously for almost a week. Insurance inspection showed blast damage to over 3,300 dwellings and 130 business buildings, to more than 600 automobiles and some 360 box cars. Approximately 50% of some 250 storage tanks, ranging from 25,000 to 80,000 barrels capacity, were substantially damaged either by concussion, missile or intense fire.
> There were 552 deaths and over 3000 injuries in a community of some 15 000 people.

Kintz, Jones and Carpenter (1948 BM RI 4245); National Board of Fire Underwriters (1948); Wheaton (1948); Blocker (1949); Blocker and Blocker (1949); J.R. Nash (1976); V.C. Marshall (1983 LPB 52, 1987)

A17 Ludwigshafen, FRG, 1948

On 23 July 1948 a rail tank car containing dimethyl ether ruptured in the BASF works at Ludwigshafen. A vapour cloud formed and after some 6 s was ignited by hot work in a workshop. A block 230 m × 170 m was demolished and severe damage was done within an area 570 m × 520 m. The death toll was 207 and the number of injured 3818. The casualties appear to have been confined to the works, the dead all being between 15 and 65 and there being very few women's names listed. The official report by Stahl, Strassman and Richard (1949) considered four hypotheses for the cause of the rupture: (1) the presence of organic peroxides, (2) the presence of non-condensables, (3) overheating leading to hydraulic rupture and (4) mechanical defects. They were able to rule out the first two and settled on the third: that the tank contents had been heated by the hot summer sun and that the tank, lacking a relief valve, had suffered hydraulic overpressure and rupture. This is the account generally given in the literature.

This explanation has been questioned by V.C. Marshall (1987). He points out that in order to obtain the pressure necessary to achieve hydraulic rupture the investigation was constrained to assume both that the tank was at least 4% smaller in volume than the norm, that this had not been remarked in 19 years of use and that the temperature reached was exceptionally high. He refers to the metallurgical analysis done at the time and described in the report, to the effect that the tank had weak spots. In addition, the tank had been involved the year before in an accident in which the valves had been knocked off. Further, the tank belonged to an ammonia plant and the plate giving the tare weight referred to the permissible filling limit for ammonia. Ammonia is known to cause embrittlement in some steels. The significance of the solar heating could well be that it increased the vapour pressure sufficiently to trigger the rupture.

BASF (1948); Stahl, Strassman and Richard (1949); D.J. Lewis (1980, 1983, 1993); Giesbrecht et al. (1981); V.C. Marshall (1986 LPB 67, 1987)

A18 Avonmouth, UK, 1951

On 7 September 1951 a storage tank exploded at the Royal Edward Dock, Avonmouth, Bristol. The tank was being filled with gas oil from a ship. The gas oil was contaminated with petrol, which had leaked through a partition between the gas oil tank and the petrol tank, and this was known to those concerned. The tank was being splash filled and the explosion occurred as two men on the tank roof were gauging the level using a steel tape. The two men were killed.

Evidently a flammable atmosphere had formed in the vapour space of the tank and had been ignited by a discharge of static electricity. The official report (H.E. Watts, 1951) stated: 'It was therefore quite possible for the steel dip tape to take up a charge of static electricity from the oil in the tank and to discharge it by means of a spark against the edge of the 10 inch manhole in the top of the tank through which the tape had been inserted'.

H.E. Watts (1951); Anon. (1952); Kletz (1983 LPB 52)

A19 Port Newark, New Jersey, 1951

On 7 July 1951 a fire and BLEVEs devastated a large LPG storage at Port Newark, New Jersey. The storage comprised one section with 70 horizontal bullet tanks, each with a capacity of about 100 m³, and a further 30 tanks nearby. The tanks were not provided with thermal insulation or fixed water sprays. The initial event was experienced as a slight explosion followed by a fire. Within the next two and half min there were four small explosions near the seat of this fire, followed half a minute later by a large flash, a muffled explosion and a large fireball. Some 10–15 min into the event a BLEVE occurred. The next 100 min were punctuated by tank explosions and BLEVEs every 3–5 min. In all 73 bullet tanks were destroyed.

The tanks were equipped with emergency isolation valves. Some 30 s into the event the button to close these valves was pressed. It was pressed again after two and a half min, there being some doubt whether the first time it had been pushed for long enough. In fact, it is estimated that on the side of the valves away from the tanks there was some 100 m³ of LPG, mainly in a 12 in. delivery line from the docks and in 8 in. ring mains; thus there was potentially quite enough fuel to feed the fire even if the tanks themselves were isolated.

In many cases the BLEVEs were accompanied by fireballs rising 750 m into the air. Many of the events registered on the barometer at Newark Airport, showing TNT equivalents of about 25 kg, consistent with the pressure energy in the vapour space of a tank. The damage done by missiles is described as spectacular. One large tank section 17 m long travelled over half a mile, demolishing a filling station and starting a fire. Another fragment ruptured the 12 in. underground water main to the site. Figure A1.5 shows one of the tank section missiles part buried in the ground.

It has been suggested that a possible cause of the initial event was inadequate allowance for thermal expansion in the 8 in. ring mains. These handled propane at two different temperatures, the regular flows and a flow from a vapour recovery unit.

NBFU and FIRONJ (1952); D.J. Lewis (1980, 1993)

A20 Bakersfield, California, 1952

On 21 July 1952, the Paloma condensate recycling plant in Kern County, near Bakersfield, California, was struck by an earthquake. The earthquake had its epicentre about 12 miles away, measured 7.7 on the Richter scale and had a

Figure A1.5   *Port Newark, 1951: tank section part buried in ground (The Bettman Archive)*

maximum intensity in the range X–XI. Ground movement was of the order of 0.5 ft vertically and 1 ft horizontally. It was such as to cause a 60 ft high absorption column to swing at the top in an arc of 3 ft and to stretch its foundation bolts some 1.5 in. Figure A1.6 gives an aerial view of the facility following the earthquake.

The plant had five large butane storage spheres which were not designed to earthquake standards. Two spheres collapsed with rupture of the feed lines. Butane escaped and formed a vapour cloud, which ignited some 90 s later at a transformer block. The resultant explosion and fire did extensive damage but there were no deaths or serious injuries.

Jenkins and Oakshott (1955); D.J. Lewis (1980, 1993); Alderson (1982 SRD R246)

A21   *Bound Brook, New Jersey, 1952*
On 22 December 1952 a plant handling phenolic resin moulding powder at Bound Brook, New Jersey, was wracked by a dust explosion. The explosion occurred in a dust collector and then travelled through ducting, emerging at several points into floor areas, where it raised clouds of dusts which in turn underwent violent secondary explosions. The probable sequence of events was that a bearing in a hammer mill overheated, the powder in the mill began to smoulder and the smouldering powder passed into a flammable dust mixture in the dust collector. Five people were killed and 21 injured.

NFPA (1957); D.J. Lewis (1993)

A22   *Whiting, Indiana, 1955*
On 27 August 1955 at the refinery at Whiting, Indiana, a reactor suffered an internal detonation and disintegrated. The reactor vessel was 127 ft tall × 22.5 ft diameter, having a wall 2.5 in. thick and weighing 500 te. A separator with a 6 in. thick wall also fractured. The detonation has been estimated as having a TNT equivalent of 5.6 te. Fragments were thrown 1500 ft, one of 60 te landing on the tank farm and another killing a boy at home. The event started a fire which eventually covered 40 acres and lasted for 8 days. There was the one death just mentioned.

The event occurred in the course of start up of a hydroformer unit. The start up procedure involved the circulation through the reactor and separator of inert gas heated in a furnace. Due to a valving error the gas being circulated was in fact a flammable mixture. In due course, the gas temperature at the exit of the furnace rose to 500–600°F, reaching the autoignition temperature. A detonation occurred in the reactor and another in the separator vessel.

Woodworth (1955); R.B. Jacobs *et al.* (1957); Randall *et al.* (1957); R.B. Jacobs and Bulkley (1967); R.B. Jacobs *et al.* (1973); D.J. Lewis (1980, 1993)

A23   *Cali, Columbia, 1956*
On 7 August 1956 eight trucks carrying dynamite exploded at Cali, Columbia, killing about 1100 people as they slept, and injured another 2000; some 500 of the dead were soldiers in barracks.

Walker (1973); J.R. Nash (1976)

Figure A1.6 *Bakersfield, 1952: recycling plant after the earthquake (The Bettman Archive)*

A24 *Uskmouth, UK, 1956*

On 18 January 1956 an explosion occurred on the No. 5 60 MW turbogenerator set at the Uskmouth power station of the Central Electricity Authority. The low-pressure turbine and generator were totally destroyed and the high-pressure turbine and plant severely damaged. Fragments were ejected with considerable force, causing damage to the turbine house. Two persons were killed and nine injured. The cause of the incident was established as machine overspeed.

Lindley and Brown (1958)

A25 *Montreal, Canada, 1957*

On 8 January 1957, a series of BLEVEs occurred at a set of storage spheres at Montreal, Quebec. There were three spheres in a common bund: one 800 m^3, one 1900 m^3 and one 2400 m^3. The 800 m^3 sphere, which held butane, was overfilled due to a faulty level gauge. A vapour cloud formed and found an ignition source, probably at a service station 180 m away, and the flame flashed back to the sphere, where a pool fire started. After some 30 min the 1900 m^3 sphere, which was less than 20% full, underwent a BLEVE. Some 15 min later BLEVEs occurred on the other two spheres also.

Selway (1988 SRD R492)

A26   *Signal Hill, California, 1958*
On 22 May 1958 an 80,000 bbl. tank serving as the feed tank to a viscosity-breaking unit at the refinery at Signal Hill, California, erupted. Steam issued, part of the roof was torn off and oil froth flowed out. There was no immediate ignition. The oil froth flowed through the plant in a wave about a metre high. A further eruption occurred and the froth ignited. Eventually it covered an area of 27 acres. The fire burned for 40 h. Two people were killed and 18 injured.

The oil in the tank was at a temperature of 157°C. The source of the water which caused the eruption does not appear to be known: it may have been in the tank already or may have been pumped in by mistake.

Woodworth (1958); Anon. (1959a); D.J. Lewis (1993)

A27   *Deer Lake, Pennsylvania, 1959*
On 2 June 1959 in the township of Deer Lake, Pennsylvania, a road tanker containing LPG slowed to a halt as a school bus in front stopped and was itself struck from behind by a truck. LPG escaped and was rapidly ignited. After some 15 min the fire brigade arrived and were told by the driver of the tanker that it would be protected from explosion by its safety valves. They directed water and foam at a nearby wooden building. A crowd of spectators gathered and police had difficulty in moving them on. After the fire had burned for a time variously estimated as 20–45 min the tanker suffered a BLEVE. Shrubbery at a distance of 150 m was scorched. The tank itself travelled up the street, bouncing off a stonewall. The tank, accompanied by stones and other debris, struck the main group of onlookers, killing most of them. There were 11 fatalities.

Hymes (1985 LPB 65)

A28   *Meldrin, Georgia, 1959*
On 28 June 1959, a derailment occurred of a rail tank cars containing LPG at Meldrin, Georgia, with tank rupture and release of LPG. A vapour cloud formed and reached a nearby leisure park. Ignition occurred and the resultant vapour cloud fire caused 23 deaths.

Anon. (1960–61); Hymes (1985 LPB 65); V.C. Marshall (1987)

A29   *La Barre, Louisiana, 1961*
At about 8.15 a.m. on 31 January 1961 some 6000 USgal of liquid chlorine were spilled from a derailed tank car in the community of La Barre, Louisiana. A cloud of chlorine gas spread of over an area of approximately 6 square miles. For the most part it hung close to the ground, but in places 'boiled upwards' to a height of 80–90 ft. Some 1000 people were evacuated. A series of determinations of the chlorine concentration were made by a team using mobile equipment. The first samples were taken at about 11.15 a.m., when a concentration of 10 ppm was measured at the end of a contaminated area some 200 yd long on a highway parallel to the railway. At about 3.00 p.m. a concentration of 400 ppm was measured 75 yd from the wreck. One person died and about 100 were treated for exposure to chlorine. The fatality was an 11-month old infant in a family, which lived in a house approximately 50 yd from the tank car. The chlorine in the house made the infant choke and gasp and the frantic father carried it outside, where the gas concentration was higher still. Although most of the exposure

occurred inside the house, this exposure outside was probably critical. The child died in hospital. A 2-month old infant who remained in the house survived.

Joyner and Durel (1962)

A30   *Berlin, New York, 1962*
On 25 July 1962, a tractor tank semi-trailer unit came off the road at Berlin, New York. The vehicle had failed to negotiate a bend, tipped over and jack-knifed. The tank hit a tree and its contents of some 6876 USgal of LPG were released. A vapour cloud formed which covered an area of some 2000 m^2 and was 25 m high. There followed a massive vapour cloud fire which caused the deaths of 10 people and injured 17.

Walls (1963–64, 1964a); Kier and Müller (1983)

A31   *Doe Run, Kentucky, 1962*
On 17 April 1962 at Doe Run, Kentucky, a plant for the manufacture of an ethanolamine from ammonia and ethylene oxide suffered a major explosion. The reactor was supplied with liquid ethylene oxide from a 25 m^3 storage vessel. The reactant transfer pumps were provided with an interlock system but problems had been experienced with this system and the control had reverted to manual. A back flow of ammonia occurred into the ethylene oxide tank. A reaction took place which might have been a reaction runaway or a decomposition. The vessel ruptured, releasing a large quantity of ethylene oxide. The vapour cloud ignited immediately, giving an explosion estimated to have had a TNT equivalent of 16 te. It has been suggested that the mechanism of ignition may have been free radicals from the initial, internal reaction.

Troyan and Levine (1968); D.J. Lewis (1980, 1993); Mahoney (1990)

A32   *Marietta, Ohio, 1962*
On 27 April 1962 at Marietta, Ohio, a benzene recycle pump on a phenol production unit became plugged with residue, and whilst attempts were being made to clear it, pressure built up in a stripper column. A 6 in. relief valve operated, discharging benzene vapour, which ignited. Missiles from the resulting explosion severed pipework, releasing large quantities of flammable liquids, which also ignited. One person was killed and three injured.

Mahoney (1990)

A33   *Attleboro, Massachusetts, 1964*
On 12 January 1964, an explosion occurred in a vinyl chloride polymerization plant at Attleboro, Massachusetts.

An illuminator port had been replaced on one of the reactors but not pressure tested. When the reactor was charged, a small leak developed. As the fitter was tightening the clamping ring, the glass chipped and he was injured; the leak developed. An estimated 4.5 te of vinyl chloride escaped into the building, ignited and exploded with an estimated TNT equivalent of 17.7 te. The explosion broke pipework on all the 20 reactors and released some 68 te of vinyl chloride, which ignited. There was a large fireball. Seven people were killed and 40 injured.

Walls (1964b); D.J. Lewis (1980, 1993); Mahoney (1990)

A34   *Niigata, Japan, 1964*
On 16 June 1964, the refinery at Niigata, Japan, suffered an earthquake of strength 7.7 on the Richter scale which caused two major fires. The first fire was the result of oil spillage from a floating roof storage tank. The oil was ignited by sparks as the roof smashed against the sides of the tank.

Later, a seismic tidal wave engulfed the area and unignited oil spread over the water. For 4 h this oil did not ignite. Six hour into the event an explosion occurred in another part of the refinery 1200 ft from the tank which was seat of the original fire. This second fire spread to within 350 ft of the first fire but the two did not merge. Two people were killed and 97 storage tanks were destroyed.

Mahoney (1990)

A35 *Bow, London, 1965*
On 7 August 1965, a flour mill in Bow, London, suffered a dust explosion. The site of the explosion was a 15 te capacity metal flour bin on the third floor. Welding had been in progress shortly before and may have caused overheating of the metal and hence of the flour inside. Two major fires and many smaller ones broke out. Five people were killed and over 40 injured.

Anon. (1967c); D.J. Lewis (1993)

A36 *Louisville, Kentucky, 1965*
On 25 August 1965 at Louisville, Kentucky, an explosion occurred on a plant manufacturing neoprene by polymerization of chlorobutadiene in a compressor circulating monovinyl acetylene (MVA). The investigation concluded that a mechanical failure had caused local overheating in the compressor and that this had led to decomposition of the MVA. There followed a series of explosions, a total of 10 occurring within 13 min and of 18 within 90 min. The escalation occurred due to transmission through pipework, flame impingement and missiles. Twelve people were killed and eight injured; all the deaths occurred within a radius of less than 100 ft from the original explosion.

Armistead (1966, 1973); Mahoney (1990); D.J. Lewis (1993)

A37 *Natchitoches, Louisiana, 1965*
On 4 March 1965, an explosion occurred on a 24 in. high-pressure natural gas pipeline at Natchitoches, Louisiana. The explosion was apparently caused by the high-pressure gas, there being no evidence of combustion in the pipeline. The splitting action of the pipe propagated the rupture for 27 ft along the pipe. The resultant blowout produced a crater 27 ft long, 20 ft wide and 10 ft deep. Three pieces of metal with a total weight of half a ton were hurled distances of 129–351 ft from the point of rupture. Within 60 s there followed an explosion that 'incinerated' an area of 13.8 acres. Seventeen people were killed.

FPC (1966); D.J. Lewis (1980)

A38 *Feyzin, France, 1966*
On 4 January 1966 at Feyzin refinery in France a leak on a propane storage sphere ignited, caused a fire that burned fiercely around the vessel and led to a BLEVE.

The operator had opened two valves in series on the bottom of the sphere in order to drain off an aqueous layer. When this operation was nearly complete, he closed the upper valve and then cracked it open again. There was no flow and he opened the valve further. The blockage, which was presumably hydrate or ice, cleared, and propane gushed out, but the operator was unable to close the upper valve. He did not think at once to close the lower valve and by the time he attempted this, this valve also was frozen open.

The alarm was raised and steps were taken to stop traffic on the nearby motorway. A vapour cloud about 1 m deep spread towards the road. It is believed that a car about 160 m distant on the motorway may have been the source of ignition. It was afterwards found that its engine was not

running but its ignition was on and it may have been stalled by taking in a propane-rich mixture at the air intake. Flames appeared to flash back from the car to the sphere in a series of jumps.

The sphere was enveloped in a fierce fire. Its pressure relief valve lifted and the escaping vapour ignited.

The LPG storage installation of which the sphere was a part consisted of four 1200 m^3 propane and four 2000 m^3 butane spheres. The fire brigade was not experienced in refinery fires and apparently did not cool the burning sphere, presumably on the assumption that the relief valve would protect it. They concentrated instead on cooling the other spheres. About one and a half hours after the initial leakage the sphere ruptured, killing the men nearby. A wave of liquid propane was flung over the compound wall and flying fragments cut off the legs of the next sphere, which toppled so that its relief valve began to emit liquid.

The fire at the storage spheres is shown in Figure A1.7. Plate 32 shows a sphere, toppled and engulfed in flames. Figure A1.8 shows an outline in the ground of one of the victims.

The accident killed 18 people and injured another 81 and caused the destruction of five of the spheres as well as other damage.

Anon. (1966e,g,h); Kletz (1974e, 1975d, 1977d); Anon. (1987 LPB 77, p. 1); Selway (1988 SRD R492); Mahoney (1990); D.J. Lewis (1993)

A39 *LaSalle, Canada, 1966*
On 13 October 1966 at a plant at LaSalle, Canada, there was a runaway reaction in a styrene polymerization reactor. The bursting disc and dump line both operated, discharging material into the outside yard, creating a vapour cloud. A vapour fog also formed inside the building due to failure of a sight glass on the bursting disc vent. Ignition occurred and the vapour inside and outside the building exploded. It has been estimated that the release of styrene was 0.64 te and that the TNT equivalent of the explosion was 6.6 te. Eleven persons died.

Fire Journal Staff (1967, 1973d); D.J. Lewis (1980, 1993)

A40 *Lake Charles, Louisiana, 1967*
Just before dawn at 4.45 a.m. on 8 August 1976, a catastrophic explosion of a cloud of isobutane vapour occurred at the Cities Service refinery at Lake Charles, Louisiana. Seven people were killed and 13 injured and extensive damage was done.

The escape occurred at a valve on a 10 in. isobutane pipeline, which ran between an alkylation unit and two storage spheres. The pipeline ran underground, but the valve was located in an open pit, which had filled with water from recent rain. The bolts on the valve bonnet had suffered severe corrosion by vapours from a leaking sulfuric acid pipeline nearby.

That morning operators observed bubbles in the water and realized that there was a leak on the valve. An operator cleared the isobutane by flushing it with 110 psi water from the alkylation unit to sphere No. 1. He then closed the inlet valve on that sphere and began to open that on sphere No. 2. At this point the weakened valve in the pit failed and a geyser of water rose in the air. The operator, thinking that he had made an error, reversed his procedure, closing the valve on sphere No. 2 and opening that on sphere No. 1. As a result, isobutane from sphere No. 1 passed down the pipe and issued from the failed valve.

Figure A1.7   *Feyzin, 1996: fire at the storage vessels (United Press International)*

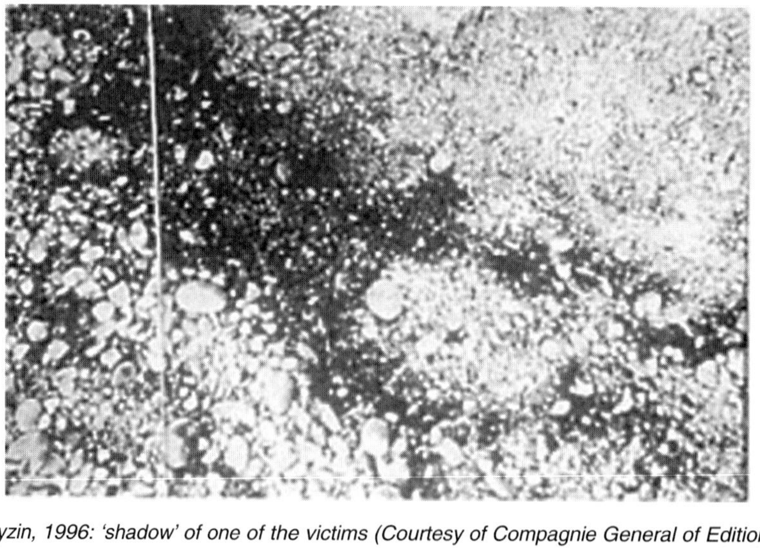

Figure A1.8   *Feyzin, 1996: 'shadow' of one of the victims (Courtesy of Compagnie General of Edition et de Presse)z*

The amount of isobutane which escaped was estimated as approximately 500 bbl. It formed a vapour cloud, which covered about 5 acres. The explosion had the characteristics of a detonation. Calculations indicated that it was equivalent to 10–12 tons of TNT.

The explosion caused a major fire, which burned for two weeks until the flammable material had been used up.

Goforth (1970); NFPA (1973); Vervalin (1973e); D.J. Lewis (1980, 1993)

A41 *Pernis, The Netherlands, 1968*

At 4.15 a.m. on 20 January 1968 an overflow was observed on slops tank 402 at the Shell refinery at Pernis. A hydrocarbon vapour cloud formed slowly over the adjacent area. At 4.23 a.m., after one or two smaller explosions, a violent explosion occurred which caused extensive blast damage and a large fire. The explosion was estimated to have been equivalent to 20 tons of TNT. The fire covered an area of about 250×300 m. Two people were killed and 85 were injured, mainly by flying glass.

Slops tank 402 was a 1633 m^3 cone roof tank. Owing to cold weather during the previous two weeks, it had been necessary to heat the oil in the tank. It is believed that a layer of water-in-oil emulsion had built up, covering the coils and reducing the heat transfer from them to the oil layer above, so that a substantial temperature difference had developed between the two layers, and that vapour formation at the interface between the two layers initiated mixing, causing further vapour evolution so that the tank overflowed, the hydrostatic pressure at the bottom of the tank was reduced and violent boil-up occurred, giving the condition known as 'stopover'.

Fontein (1968); Ministry of Social Affairs and Public Health (1968); Wordsworth (1973); D.J. Lewis (1980, 1993)

A42 *Dudgeons Wharf, London, 1969*

In 1969, an explosion occurred at Dudgeon's Wharf, London, while an oil storage tank was being demolished. Six people were killed.

Anon. (1970a)

A43 *Glendora, Mississippi, 1969*

At about 2.45 p.m. on 11 September 1969, a train entered Black Bayou junction near Glendora, Mississippi, pulling 157 rail cars, including eight containing vinyl chloride monomer (VCM). A pedestrian was struck and injured, and application of the emergency brake caused a derailment of 15 cars.

At 3.30 p.m. the VCM manufacturer was informed that his chemical was involved and at 5.00 p.m. that the eight cars were derailed and that one was leaking. A technical expert was requested and the manufacturer sent the VCM plant superintendent, who had 25 years' experience with the chemical.

At about 8.00 p.m. one of the VCM tanks was ruptured and leaking, and by 10.00 p.m. the leak had ignited. A heavy fog was noticed over the area and it was considered that it might be VCM. The railway management consulted the *Handbook of Hazardous Materials*, which stated that phosgene could be formed in VCM fires. They rang the company office, who advised that a hazard from phosgene was highly improbable and that the main problem was likely to be from HCl and smoke. Nevertheless, the railway management, the police, the National Guard and the Civil Defence Authorities continued to be worried by the phosgene hazard. The Civil Defense director consulted university chemists, who advised that burning VCM would release phosgene, which could be a hazard to life at a radius of 35 miles. Accordingly, the police and National Guard evacuated all the towns nearby. The evacuation was reported as involving some 30,000 people.

At 6.45 a.m. the next day an explosion occurred which demolished one of the eight VCM tank cars. The explosion caused a second tank car to rupture and ignite. After the fires were extinguished only five full tank cars were recovered.

A report in The Commercial Appeal of Memphis, Tennessee (13 September, 1969) read Mississippi. There was a explosion followed by a further explosion and fireball. The next 40 min were punctuated by more explosions as one car after another burst or rocketed. The initial fireball set fire to buildings 200–400 ft away. Other fires were initiated by burning fragments up to 10 blocks distant. The blast caused structural damage within about 400 ft. Very little glass survived within about half a mile and windows broke as far out as 3 miles. Two people died and 976 were injured.

Much of the damage was caused by the impact of rocketing tank cars and the fires which they set going. One 37 ft section travelled through the air and bounced first at 1000 ft distance, then bounced again at 300 ft, and again at 200 ft, and finally went another 100 ft, before coming to rest at a total distance of 1600 ft, where it set fire to houses.

At the first explosion people came out to see what was happening. The streets were full when the second explosion and fireball occurred. People became trapped between fences which ran parallel to the track and the towering fireball 200–400 ft away. There were many burn injuries, and many also from panic.

It is of interest that a National Guard ordnance team, in a controversial procedure, successfully used shaped explosive charges to blow holes in two unruptured tank cars to prevent them exploding.

NTSB (1969 MR); Eisenberg, Lynch and Breeding (1975)

A45 *Texas City, Texas, 1969*

On 23 October 1969 at Texas City, Texas, an explosion occurred inside a distillation column of a butadiene unit. The explosion was experienced as a two-part one, first the disintegration of the column and then ignition of the escaping gas. Numerous fragments fell within a radius of 500 ft, some large shell fragments travelled 1500 ft and one 800 lb section fetched up 3000 ft away. All five columns in the butadiene section were toppled or severely damaged. Thirteen people were injured.

The column had been under total reflux. A leak occurred in an overhead line valve so that butadiene was lost and a build-up occurred of vinyl acetylene which on some trays reached a molar concentration of 50–60%. The internal explosion occurred due to thermal decomposition of vinyl acetylene and ethyl acetylenes.

Freeman and McReady (1971); Jarvis (1971); Keister, Pesetsky and Clark (1971); Griffith and Keister (1973); Mahoney (1990); D.J. Lewis (1993)

A46 *Beaumont, Texas, 1970*

On 17 September 1970 at Beaumont, Texas, a 60 ft × 40 ft oil slops tank was struck by lightning. The tank failed at the shell-floor seam, releasing 11,000 USgal of oil, which burned. The oil spread to involve in the fire another 16 unbonded tanks nearby.

Mahoney (1990)

A47 *Blair, Nebraska, 1970*

On 16 November 1970, an overflow occurred on a 40,000 ton refrigerated anhydrous ammonia storage tank at the Gulf Oil Company's installation at Blair, Nebraska. 160 ton of ammonia was released, but there were no deaths or serious injuries.

The overfilling was the result of an operator error. A factor which may have contributed to the error was the fact that the table of tank levels, or 'strapping table', did not indicate clearly the position of the overflow pipe. In addition, the high-level alarm and shut-down system failed to operate and apparently the overflow discharge valve also failed to operate at the set pressure, so that the liquid level in the tank rose until it reached the roof, at which point the overflow valve did open. There was an isolation valve on the overflow line directly below the relief valve. It was not possible to reach this valve and close it on account of the ammonia cloud. If isolation of the overflow line had been possible, much of the overflow could have been prevented. Once the overflow valve had opened, there was a discharge of ammonia to atmosphere. The discharge continued for 2.5 h.

There was almost no wind and an atmospheric inversion existed, so that initially the weather conditions were at their most unfavourable for dispersion. A low visible cloud of ammonia formed some 8–30 ft thick, covering approximately 900 acres and extending over 9000 ft from the tank. The ammonia cloud shortly after the spill is shown in Figure A1.9. The 'pancake' shape of the cloud is clearly shown in Figure A1.9(c). Later, a light breeze arose and helped to keep the cloud from populated areas.

MacArthur (1972)

A48 *Brooklyn, New York, 1970*
On 30 May 1970 a road tanker carrying liquid oxygen was manoeuvring in the yard of a hospital in Brooklyn, New York, when the tank ruptured violently. The tank was 9.15 m^3 capacity but the tanker had just offloaded its first part load. The initial explosion killed the driver and one other person, and windows were broken some 600 ft away. The release of the liquid oxygen resulted in a number of fires. In addition to the two deaths mentioned, 30 people were injured.

The tank was new and had been in service only one month. The investigation found that the $3/4$ in thick aluminium tank shell had suffered appreciable loss of metal, the deficit being 73.5 kg for the shell and internal components. It further found that a chemical reaction had occurred between the oxygen and deposits left in a crevice between a bracket and the tank wall, and that this had generated enough heat to initiate an oxidizing reaction between the shell and the oxygen vapour. This reaction also had the effect of heating the contents and increasing the vapour pressure. The estimated vapour pressure attained was over 100 bar compared with a design pressure of 15 bar.

NTSB (1971 HAR-71-06); D.J. Lewis (1993)

A49 *Corpus Christi, Texas, 1970*
On 3 August 1970 a refinery at Corpus Christi was hit by Hurricane Celia with winds of 140–160 miles/h and a rainfall of 6–8 in. The main structure of the catalytic cracker was toppled and some 30 tanks were destroyed.

Mahoney (1990)

A50 *Crescent City, Illinois, 1970*
At 6.30 a.m. on 21 June 1970 a derailment occurred on a railway passing through the streets of Crescent City, Illinois. Nine rail tank cars loaded with LPG were among the derailed vehicles. The force of the derailment propelled the 27th car in the train over the derailed cars on front and its coupler struck the tank of the 26th car and punctured it. Propane was released and ignited.

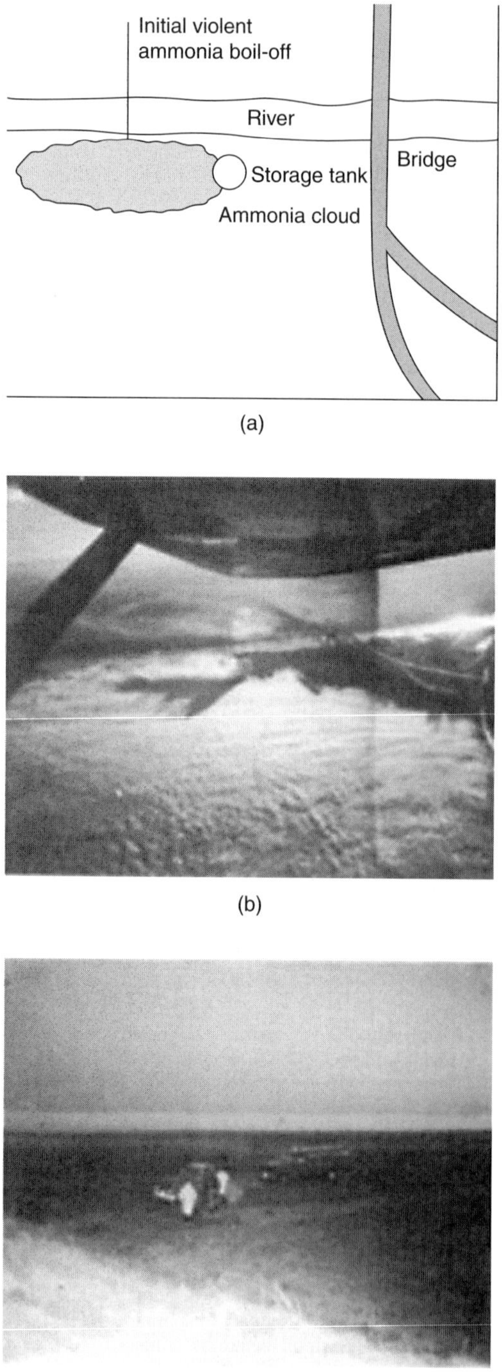

(a)

(b)

(c)

Figure A1.9 *Blair, 1970: ammonia cloud from storage tank: (a) plan view (Slater, 1978a); (b) aerial view showing ammonia cloud covering whole of foreground; and (c) ground level view showing 'pancake' shape of cloud (both Enterprise Publishing Company, Blair, Nebraska)*

The safety valves of the other tank cars operated and released propane, which fed the fire. At about 7.33 a.m. the 27th car exploded. Four fragments of the tank car were hurled in different directions for distances of 300, 600 and 750 ft. At about 9.40 a.m. the 28th tank car exploded, hurling one fragment which eventually stopped rolling at a distance of 1600 ft. At about 9.45 a.m. the 30th tank car exploded and at about 10.55 a.m. the 32nd and 33rd tank cars ruptured almost simultaneously. Fragments from these damaged the 34th and 35th tank cars and caused them to release propane.

The main mechanism of damage in this incident was the rocketing of burning tank cars which hurled massive fragments and fresh fireballs up to 850 ft from the site of the derailment.

The fireball produced by the contents of one of these tank cars, which contained 33,000 USgal (120 m^3) of LPG, is shown in Figure A1.10. The fireball was several hundred feet high.

There were no fatalities, but 66 people were injured. There was extensive property damage.

Watrous (1970); NTSB (1972 RAR-72-02); Strehlow (1973b); Eisenberg, Lynch and Breeding (1975); D.J. Lewis (1980, 1991 LPB 101, 1993).

A51   *Linden, New Jersey, 1970*

On 5 December 1970 in a refinery at Linden, New Jersey, local overheating in a hydrocracker unit caused an explosive failure of the 7 in. thick vessel. Severe damage was done to plant over a 300 yard radius, including a catalytic cracker and crude pipe still, where the control room roof collapsed. Shut-down of the other units was effected from a blast resistant control room, which suffered minor damage. Forty persons were injured.

Mahoney (1990)

A52   *Port Hudson, Missouri, 1970*

At 10.07 p.m. on 9 December 1970, an abnormality occurred at a pumping station at Villa Ridge, 15 miles downstream from Port Hudson, Franklin County, Missouri. At 10.20 p.m. there was a sudden increase in throughput at the next upstream pumping station at Rosebud, indicating a major line break. The pipeline failure had occurred near Highway C on high ground and the gas cloud flowed down a sparsely inhabited valley.

At approximately 10.25 p.m. several witnesses became aware of the noise of the escaping jet of propane. A plume of white spray, which was presumably drops of propane and of atmospheric moisture, was seen rising 50−80 ft above

Figure A1.10   *Crescent City, 1970: fireball from rail tank car (Courtesy of Champaign Fire Department)*

the ground. By 10.44 p.m. several families had evacuated their houses and had driven to Highway C, from which they observed the cloud.

At approximately 10.44 p.m. the valley appeared to light up. Witnesses reported no observable period of flame propagation but rather a sudden flash as of lightning. There was an almost immediate overpressure which knocked over one person who was walking about half a mile from the centre of the cloud.

In the Bureau of Mines report on the incident, Burgess and Zabetakis (1973 BM RI 7752) state: We think the witnesses has the unusual experience of observing a gas detonation'. In the succeeding seconds a fire storm was seen to roll up the sloping terrain towards Highway C.

At the time and place of the failure the pipeline pressure was believed to be about 942 psig. The amount of liquid propane which is estimated to have escaped in the first 24 min is 750 bbl.

Burgess and Zabetakis estimate that since the propane issued from the pipe at approximately 1°C, the fraction flashing off would be about a quarter, but that a further appreciable fraction formed drops in the 50–80 ft plume seen by witnesses and that these drops derived their heat of vaporization from the air.

For the estimation of the concentration of propane in the vapour cloud Burgess and Zabetakis use the Pasquill–Gifford model, their work antedating the development of models for heavy gas dispersion and for complex terrain. Thus

$$\chi = \chi_{c1} \exp\left\{-\frac{1}{2}\left[\left(\frac{y}{\sigma_y}\right)^2 + \left(\frac{z}{\sigma_z}\right)^2\right]\right\} \qquad [A1.10.1]$$

with

$$\chi_{c1} = \frac{Q}{\pi \sigma_y \sigma_z u} \qquad [A1.10.2]$$

where x, y, z are the downwind, crosswind and vertical distances (ft); Q is the flow rate of propane (ft³/s); u the average wind velocity (ft/s); σ_y, σ_z are the standard deviations in the y, z directions (ft); χ is the concentration of propane (volume fraction); and χ_{c1} is the concentration of propane on the centre line (volume fraction).

The calculation given by Burgess and Zabetakis may be summarized as follows. The flow rate Q of propane is 900 ft³/s at STP, the wind velocity is 8 ft/s and the stability

category may be taken as category F in the nomenclature of D.B. Turner (1970). The plume dispersion coefficients σ_y and σ_z are obtained from the graphs given by this author. The lower and upper detonability limits of propane in air given by Benedick, Kennedy and Morosin (1970) are 2.8% and 7.0%, respectively.

The dimensions of the detonable plume thus calculated are shown in Table A1.3 and in Figure A1.11. The total volume of the detonable zone is estimated to be 1,100,000 ft³. Assuming that the average mixture strength is 4.9%, the volume of propane in the detonable zone is 54,000 ft³, which is approximately 4.2% of the total quantity released up to the time of ignition. About the same amount of propane is contained in the over-rich part of the cloud. Thus most of the propane is dispersed beyond the lean limit contour.

The mass of gas in 1,100,000 ft³ of 4.9% propane–air mixture at a temperature somewhat less than ambient is about 100,000 lb. The enthalpy release on detonation is 260 kcal/vapour. This compares with approximately 500 kcal/vapour for TNT. The TNT equivalent of detonable gas is therefore approximately 50,000 vapour.

Burgess and Zabetakis also present a second calculation which attempts to take better account of the heavier-than-air density of propane–air mixtures and uses the following relations given by Singer and Smith (1953) for gustiness category D:

$$\sigma_y = 0.44_x{}^{0.71} \qquad [A1.10.3]$$

$$\sigma_z = 0.2\sigma_y \qquad [A1.10.4]$$

This calculation gives the volume of the detonateable zone as 1,500,000 ft³.

Burgess and Zabetakis give an analysis of the blast damage data in both the near and far fields of the explosion. Estimates of the TNT equivalent of the explosion range, with one exception, from 97,000 to 150,000 lb in the near field, while an estimate in the far field is 98,000 lb. This compares with estimates in the range 50,000–75,000 lb of TNT equivalent inferred from the volume of the detonable cloud.

The efficiency of explosion is calculated by Burgess and Zabetakis as follows. If the 750 bbl. of propane which are estimated to have escaped had detonated as a homogenous stoichiometric mixture with air, the enthalpy release would have been 666×10^6 kcal. The enthalpy of detonation of TNT is 10^6 kcal/ton. The explosion was equivalent to some

Table A1.3 *Calculated dimensions of detonable plume of propane formed at Port Hudson, Missouri, on 9 December 1970 (after Burgess and Zabetakis, 1973 BM PI 7752) (Reproduced by permission of the Bureau of Mines)*

Downwind distance	Plume dispersion coefficients		Centreline concentration	Distances from plume axis to rich (r) and lean concentration limits			
χ (ft)	σ_y (ft)	σ_z (ft)	χ_{CL} (%)	y_r (ft)	y_l (ft)	z_r (ft)	z_l (ft)
300	13	7.9	35	23	29	14.2	17.9
500	20	10.5	17	26	38	13.9	20.0
700	26	14	10	21	42	12.0	22.4
870	–	–	7.0	0	–	0	–
1000	36	18	5.6	–	43	–	21.4
1400	50	24	3.0	–	17	–	8.2
1470	–	–	2.8	–	0	–	0

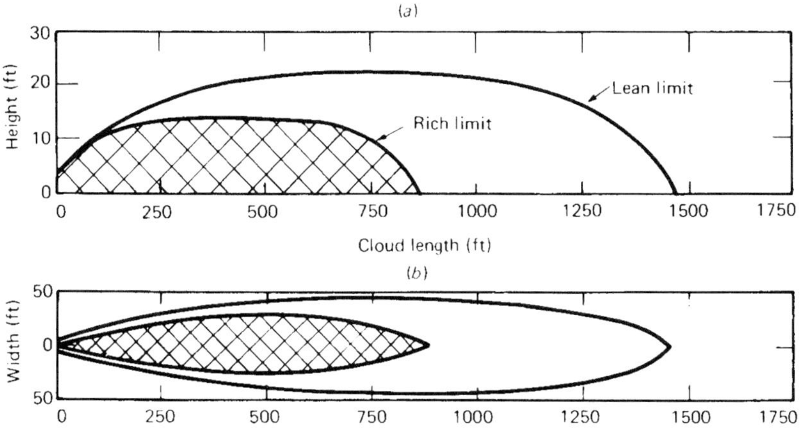

Figure A1.11   *Port Hudson, 1970: calculated dimensions of detonable plume of propane (after Burgess and Zabetakis, 1973 BM PI 7752) (Courtesy of the Bureau of Mines)*

50 ton of TNT. Thus the explosion gave 50/666 or 7.5% of the maximum theoretical yield, based on combustion of the whole cloud.

The source of ignition of the cloud is uncertain, but Burgess and Zabetakis conclude that it was probably located in one of a group of outbuildings enveloped in the cloud.

The explosion caused no deaths.

NTSB (1972 PAR-72-01); Burgess and Zabetakis (1973 BM RI 7752); Strehlow (1973b); Eisenberg, Lynch and Breeding (1975); D.J. Lewis (1980, 1993)

A53   *Houston, Texas, 1971*

At 1.44 p.m. on 19 October 1971, 20 freight cars were derailed at Mykawas Station, which is 10 miles south of the centre of Houston, Texas. Six of the tank cars contained VCM and two others butadiene and acetone. Two of the VCM tank cars punctured. An explosion and fire occurred while the crashed tank cars were still moving. The two punctured tank cars stopped in the burning area. At 2.30 p.m. there was a violent explosion when one of the tanks containing some 100,000 lb of VCM burst, hurling fragments and giving a large fireball. Further tank cars burst, feeding the fire and generating additional missiles.

One fireman was killed and some 50 people, many of them firemen, were injured. The area was sparsely built and populated and damage was slight.

NTSB (1972 RAR-72-06); Eisenberg, Lynch and Breeding (1975); D.J. Lewis (1980)

A54   *La Spezia, Italy, 1971*

On 21 August 1971, a rollover occurred in an LNG tank at the SNAM LNG terminal at La Spezia, Italy. There was a sudden increase in pressure resulting in discharges from the tank safety valves and vent. The safety valves discharged for about 75 min and the vent discharged at a high rate for 3 h and 15 min.

The incident occurred 18 h after the tank had been loaded from the Esso Brega. The loading had been done through a nozzle at the side of the bottom of the tank, with the heavier, hotter cargo staying on the bottom and displacing the lighter, colder 'heel' of LNG, which had remained in the tank. These conditions were temporarily stable, but

over the subsequent period the density of the upper layer increased and that of the lower layer decreased, eventually inducing the rollover. One factor was that the temperature difference between the two layers caused the upper layer to heat up and to become more dense due to boil-off of the lighter fractions. Another factor was a failure of the control valve some 4 h before the rollover, which caused a reduction in the pressure in the tank which in turn increased the boil-off.

Sarsten (1972)

A55   *Long-view, Texas, 1971*

On 25 February 1971, a ½ in. high-pressure ethylene gas pipe broke in a plant near Longview, Texas, and released 1000 lb of ethylene. The vapour cloud found a source of ignition and exploded. The explosion broke numerous other pipes and caused the release of tonnage quantities of ethylene. The resultant larger vapour cloud in turn was ignited and gave a violent explosion.

Four people were killed and 60 were treated in hospital.

Detailed information on the blast damage is given by Eisenberg, Lynch and Breeding (1975).

Eisenberg, Lynch and Breeding (1975); Lauderback (1975); D.J. Lewis (1980)

A56   *Pensacola, Florida, 1971*

On 11 September 1971, a pipe ruptured on a caprolactam plant at Pensacola, Florida, and released about 74,000 lb of cyclohexane at high pressure and temperature. A dense white cloud of cyclohexane vapour and mist formed which was estimated at over 100 ft high and about 2000 ft across at its maximum. The wind was very calm, but picked up slightly after the rupture.

The cloud was rich enough to stall two trucks enveloped in it from lack of oxygen. Vapour from the cloud was drawn into a powerhouse furnace so that the stack emitted black smoke. However, the cloud did not ignite and eventually dissipated harmlessly.

Eisenberg, Lynch and Breeding (1975); W.B. Howard (1975b); D.J. Lewis (1980); Stueben and Ball (1979) Sadler and Matusz (1994)

A57   *East St Louis, Illinois, 1972*
At about 6.20 a.m. on 22 January 1972, a rail tank car loaded with propylene collided with a stationary hopper car in the Alton and Southern Railway Company's Gateway yard at East St Louis, Illinois.

The tank car had come from the hump classification yard at overspeed. The overriding coupler of the hopper car punctured the tank, the propylene escaped and formed a large vapour cloud which then ignited and exploded.

More than 230 people were injured and extensive damage was done.

The NTSB investigation (1973 RAR-73-01) concluded that the accident was probably caused by the failure of the retarding system in the hump classification yard to decelerate effectively heavy tank cars with oil or grease on their wheel rims, by the absence of a backup system to halt cars passing through the retarders at overspeed and by the routine acceptance of such overspeeds.

NTSB (1973 RAR-73-01); Strehlow (1973a); Eisenberg, Lynch and Breeding (1975); D.J. Lewis (1980)

A58   *Hearne, Texas, 1972*
At 12.30 a.m. on 14 May 1972, a rupture occurred on an 8 in. crude oil pipeline near Hearne, Texas. Crude oil sprayed into the air from a 6 in. split in the top of the pipe, showering the surrounding countryside with oil. The oil flowed along a stream beneath a railway and a highway, and collected in a stock pond 1800 ft from the break.

At 5.00 a.m. the crude oil was ignited by an unknown source. The resultant explosion and fire killed a man and seriously burned two other people. An intense fire several hundred feet high and 1800 ft long burned on the surface of the oil and along the stream, at the railway, the road and the stock pond, and scorched the whole area.

The quantity of oil which escaped and burned was 7913 bbl (332,346 USgal). Some 10% of the oil was fractions lighter than hexane. The mode of break was such as to encourage the separation of these light hydrocarbon fractions and the formation of a flammable gas cloud.

NTSB (1973 PAR-73-02); Eisenberg, Lynch and Breeding (1975)

A59   *Lynchburg, Virginia, 1972*
On 9 March 1972 at Lynchburg, Virginia, a tractor semi-trailer tank truck carrying LPG overturned and punctured. A 32 in. hole appeared and some 4000 of the load of 9208 USgal escaped before the liquid level fell below the hole. A vapour cloud formed and gave rise to a fireball. The radius of the fireball was estimated as at least 400 ft. Three eyewitnesses standing beside their cars at 450 ft distance were severely burned by thermal radiation but were not touched by the flames. Two persons died and at least five were injured.

NTSB (1973 HAR-73-03); Eisenberg, Lynch and Breeding (1975)

A60   *New Jersey, Turnpike, 1972*
On 21 September 1972 on the New Jersey, Turnpike, a tractor semitrailer tank truck carrying 7209 USgal of propylene sideswiped a bus. The tractor's fuel tank and the fittings on the propylene tank were damaged and fuel and propylene leaks occurred. A friction spark from the collision ignited the fuel leak and the fire spread to the propylene leak. After some 20−25 min the tank exploded with a huge fireball. Fragments of the tank rocketed: the front section,

comprising three-quarters of the tank, was found 1307 ft to the north-east; the rear head was found 540 ft to the south-west and the rear one quarter cylindrical section 229 ft to the south-west; tank components were mainly strung out in a line stretching 850 ft to the south-east. Two persons were killed and 28 injured.

NTSB (1973 HAR-73-04); Eisenberg, Lynch and Breeding (1975)

A61   *Rio de Janeiro, Brazil, 1972*
On 30 March 1972, a BLEVE occurred on an LPG sphere, one of five, at the Duque de Caxais refinery, Rio de Janeiro, Brazil. An operator was engaged in draining water from the bottom of a 1600 m^3 sphere. He went away, leaving open a 2 in. drain valve. When he returned, he found that he could not reach the valve to turn the flow off, because the jet of liquid, now LPG, had created a crater in the crushed stone under the sphere. A vapour cloud formed, ignited and flashed back to the sphere. Some 15−20 min later the relief valve opened and the material released ignited. The sphere then suffered a BLEVE. Thirty-seven people were killed and 53 injured. The other four storage spheres survived.

Mahoney (1990); Selway (1988 SRD R492)

A62   *Austin, Texas, 1973*
At 10.53 p.m. on 22 February 1973 at Austin, Texas a 10 in. pipeline failed and released 278,880 USgal of NGL. The release blew a 10 ft diameter hole in the ground and a vapour cloud spread over the area. Soon after 11.00 p.m. two cars entered the cloud and stalled; their drivers got out. Then a van drove into the cloud and also stalled; the passengers got out, whilst the driver tried to restart. A large spark emerged from beneath the van and flames leaped hundreds of feet in the air. The fire burned through the cloud and the eight people in it were engulfed in flames. Four died immediately, two later and two sustained severe burns.

NTSB (1973 PAR-73-04); Eisenberg, Lynch and Breeding (1975)

A63   *Kingman, Arizona, 1973*
On 5 July 1973, a BLEVE and fireball occurred on a rail tank car containing propane at Kingman, Arizona. The upper surface of the tank had been heated by jet flames from a relief valve and coupling leak.

The event is captured in the NFPA film 'BLEVE' and has been analysed by Crawley (1982). The tank was a 'jumbo' one and he estimates that the tank capacity was 150 m^3 but assumes that it was perhaps half full. He assesses the diameter of the fireball as 400 ft and its duration as 10−15 s. The estimates given by V.C. Marshall (1987) are that some 45 te of propane took part and that the fireball radius was 150 m.

Despite police advice a large crowd of spectators had gathered and more than 90 were injured.

Crawley (1982); Hymes (1985 LPB 65); V.C. Marshall (1987)

A64   *McPherson, Kansas, 1973*
At 4.30 a.m. on 6 December 1973, a major leak occurred on the Mid-America Pipeline System at the Conway pump station near McPherson, Kansas. Prior to the leak an ice storm had been raging for two days and the pump station had lost power for 18 h, although it was partially restored by 3.00 a.m. The pipeline had been shut-down and upon

start-up at 4.12 a.m. a block valve failed to open, the line was overpressured and burst at a point where it had already sustained construction damage.

An estimated 230 tons of anhydrous ammonia escaped over a period of half an hour. A low, visible plume of ammonia formed and drifted over Highway 56, which is some 1000 ft to the south. The weather conditions were very stable with a wind speed of some 5–10 mile/h and the plume remained narrow. Irritation symptoms were experienced and odour was detectable for about 3.5 and 8 miles from the escape, respectively.

The incident caused no serious casualties. Many residents had moved into a nearby town on account of power failures caused by the severe weather. Two truck drivers drove into the gas cloud on the highway. They escaped and were kept in hospital several days. People in dwellings $3/4$ mile south of the leak and directly in the cloud path were safely evacuated by the Sheriff's office and company personnel.

By 4.30 p.m. the line had been stopped.

The same pipeline failed again on 13 August 1979 at a point about 22 miles from the first failure. There was a release of some 360 tons of anhydrous ammonia. Again there were no casualties, but there was a large fish kill.

NTSB (1974 PAR-74-06); Luddeke (1975)

A65 *Potchefstroom, South Africa, 1973*
At 4.15 p.m. on 13 July 1973, a sudden failure occurred in an anhydrous ammonia storage vessel at the Potchefstroom works of TRIOMF, a company part-owned by African Explosives and Chemical Industries Ltd. The tank was one of four 50 ton horizontal pressure storage bullets. An estimated 30 tons of ammonia escaped from the tank itself and another 8 tons from a tank car.

The failure gave rise immediately to a gas cloud some 150 m diameter and 20 m deep. At the time of the accident the air was apparently still, but within a few minutes a slight breeze arose which caused the cloud to move towards a township some 200 m to the north-east. The visible cloud then extended some 450 m downwind and 300 m across.

Deaths occurred both inside and outside the factory fence. At the time there were some 350 persons working in the plant, of whom some 30 were within 70 m of the failed tank. One employee was killed by the blast and eight died while trying to escape from points within a 100 m radius of the tank. Three others died of gassing within a few days.

Outside the factory fence four people were killed immediately and two died a few days later. Thus altogether 18 people were killed.

The occupants of the granulation plant control room 80 m south-east of the failure survived. They put wet cloths over their faces and were taken to safety after some 30 min.

The failures occurred in tank No. 3 while it and tank No. 4 were being filled simultaneously from a tank car. Actuation of an excess flow valve on the line between the two tanks prevented the release of the contents of tank No. 4 also. The tank car did not have an excess flow valve and did suffer escape of material.

The cause of the failure was brittle fracture of the dished end of the tank. Evidence suggested that there had been no overpressure or overtemperature of the tank contents and no other triggering event was determined.

The tanks were designed and fabricated in accordance with BS 1515: 1965 and were fabricated and commissioned in 1967. The fabrication is described by Lonsdale (1975):

> The dished ends were fabricated from two plates, cold-formed in the major radius, and hot flanged (at 850°C) at the knuckle. The plates had been passed as being in accordance with the requirements of BS 1501-151-28A. The butt welds in the end plates were checked by 100% X-ray after forming and flanging. (Expert metallurgists re-examining the X-rays of the failed dished end considered that two sections of the weld did not conform to the requirements of the BS code.) The Inspection Authority representative considered that conditions in the flanging furnace rendered subsequent heat treatment unnecessary.
>
> The completed tank was not stress-relieved because it is not required by BS 1515. The tanks were given an hydraulic test at $1\tfrac{1}{2}$ times the design pressure of 250 lb/sq in gauge.

Late in 1971 tank No. 3 was inspected. Two weld faults and a crack were found and were ground out. The tank was hydraulically tested to 347 psig for 30 min. Following repairs to a leaking tank level glass isolation valve, the tank was hydraulically tested to 325 psig for 3–4 h.

Metallurgical testing revealed that the metal of the dished end was below its transition temperature under normal conditions. The minimum Charpy impact testing transition temperatures obtained were 20°C for the fragment and 115°C for the remaining part of the dished end.

Ultrasonic examination of the dished end in No. 4 tank revealed numerous subsurface fissures. Such fissures may have provided the notch from which the brittle fracture in No. 3 tank propagated. After this examination tank No. 4 was withdrawn from service.

Following the inquiry into this accident, the South African authorities laid down that 'All vessels containing dangerous substances shall be given appropriate heat treatment irrespective of the (construction) code requirements.'

Commenting on this Lonsdale states

> Stress-relieving does not overcome fully the damage done by progressive cold-forming of a dished end. This is particularly so where seam welds have had to be made in the dished end. There is a strong case for avoiding the use of progressively cold-formed ends for pressure vessels because of the difficulty of controlling the final state of the metal.

Lonsdale (1975); D.J. Lewis (1993)

A66 *St Amand-les-Eaux, France, 1973*
On 1 February 1973, a road tanker containing 19 tons of liquefied propane skidded and overturned on a wet road surface in the village of St Amand-les-Eaux. A rupture occurred on the tanker and propane began to escape. The driver of the tanker immediately took steps to stop traffic entering the street and to warn the population. The vapour cloud grew and finally exploded, causing 5 deaths and 40 injuries. The death toll might well have been higher but for the prompt action of the driver in carrying out his emergency procedures.

Anon. (1973j); D.J. Lewis (1980)

A67 *Staten Island, New York, 1973*
On 10 February 1973 at Staten Island, New York, an explosion occurred inside a large LNG storage tank in which

work was being done. The tank was being repaired for a leak in the inner membrane lining and new liner sections were being installed. The tank had been well ventilated before work began. The source of fuel for the explosion was identified as LNG trapped in the insulation behind the liner and vaporized by the heat-sealing operations on the new liner sections. The explosion lifted up the concrete roof of the tank which then fell back into it, crushing the men there. Forty men were killed and three injured.

US Congress (1973); D.J. Lewis (1989, 1993)

A68 *Aberdeen, UK, 1974*

An incident which occurred at Aberdeen on 17 January 1974, is described by Willcock (1986) in the following terms:

(1) Butane tanker skids on black ice. Shaken and lightly injured driver regains control and stops safely at nearside kerb; (2) Half an hour later articulated lorry skids on black ice, collides with tanker and shears off emergency discharge valve. Gas starts escaping. Second driver slightly injured; (3) Five minutes later yet another articulated lorry skids and crashes into the other two. One more hurt driver; (4) Police and fire brigade arrive a few minutes later, but before they can get organized a car, trying to stop at the accident, skids into the rear of the tanker. Impact ignites butane.

   Willcock (1986)

A69 *Chicago, Illinois, 1974*

Soon after 12.30 p.m. on Friday 26 April 1974 employees at the Bulk Terminals complex in the Calumet Harbor area on the south side of Chicago, Illinois, heard a dull thud and saw fumes rising from the dike of Tank 1502, which contained 3300 m^3 of silicon tetrachloride. Bulk Terminals was a storage tank farm with 78 tanks ranging up to 4900 m^3 in size, containing products such as vegetable oils and chemicals. It was situated in a built-up area.

This was the start of an incident which involved a massive cloud of irritant gas in a built-up area and was to prove very prolonged. At times the cloud was 8–16 km long. It took until May 3 to stop the leak and transfer the material and until May 15 before emissions were finally reduced to tolerable levels.

Investigation found that a pressure release valve on a 6 in. line leading to the tank had been inadvertently closed. A flexible coupling in the line burst under pressure. The whole piping system shifted, cracking a 3 in. line on the tank wall. Liquid silicon tetrachloride escaped, creating a cloud with formation of hydrogen chloride gas.

The initial emergency response was confused. The terminal management awaited action by the owners of the chemical. The fire service first decided not to get involved, because there was no fire. Lime trucks diverted to the scene by the EPA with a view to neutralizing the liquid in the dike were refused entry to the site.

By 3.00 p.m. the cloud was 400 m wide, 300–450 m high and 1600 m long.

Two agents were considered to blanket the liquid in the dike: non-protein high expansion foam and No. 6 fuel oil. At 4.10 a.m. on Saturday 27 April, foam was applied but was not effective. At about 9.00 a.m., fuel oil was added and eight truck loads of lime were tipped in. Within an hour the vaporization was dramatically reduced. By noon matters appeared to be under control with a 16-member US Army

chemical warfare team involved in transferring the material from the damaged tank.

About 8.00 a.m. on Sunday 28 April, however, heavy rain began to fall on the hydrogen chloride cloud. Power lines were rapidly corroded and sparked like a firework display. By 9.30 a.m. four pumps had become inoperable due to corrosion. Then power failure put a stop to pumping.

The threat existed that the pipe system would suffer further failure and release the remaining 2300 m^3 of material into the dike, which had been reduced in capacity by the material added to it. A large pit was dug to take any overflow from the dike and an operation was undertaken to seal the leak using quick drying cement.

On the morning Monday 29 April inspection of the pipe system, undertaken in the cloud with a visibility of some 18 in., indicated that the leak had not been sealed, either because the application had been ineffective or because it had been made at the wrong point. A second concrete pour was undertaken and completed by 11.30 p.m.

By the morning of Tuesday 30 April the emission had been greatly reduced. It took, however, another three days to transfer the liquid out of the tank.

In this incident one person died, 160 were hospitalized and 16,000 were evacuated.

   Hoyle (1982)

A70 *Climax, Texas, 1974*

On 29 June 1974 at Climax, Texas, a derailment occurred which resulted in the puncture of a rail tank car containing liquid vinyl chloride monomer. The hole in the end of the tank was 4 ft × 4 ft. The vinyl chloride escaped and formed a vapour cloud which travelled 1600 ft to the derailed locomotive, where it ignited. There followed a violent vapour cloud explosion. It was found that the cylinder head had been blown off the diesel engine, which fact led to the hypothesis that vinyl chloride had been ingested into the engine and that the engine explosion had been the trigger for the vapour cloud explosion. The contents of the ruptured tank car continued to burn after the explosion and a flame played on another vinyl chloride tank car which suffered a BLEVE. The explosions occurred in a rail cutting and there were no injuries.

   Eisenberg, Lynch and Breeding (1975)

A71 *Decatur, Illinois, 1974*

On 19 July 1974, a vapour cloud exploded in the rail-shunting yard at Decatur, Illinois. During shunting operations, a rail tank car was punctured by the coupler of a boxcar into which it had run, the hole being some 0.3 m^2. The tank car contained 69 te of isobutane, which escaped and formed a vapour cloud. There was a delay of 8–10 min before the cloud ignited. There then occurred a massive vapour cloud explosion which caused extensive damage. Seven people were killed and 349 injured.

This vapour cloud explosion has been of some interest, because on the basis of the assessment of certain injury and damage effects, the yield of the explosion has been estimated as unusually, even anomalously, high. The matter has been considered by V.C. Marshall (1980d, 1987), who gives a detailed comparison between the explosions at Flixborough and at Decatur, and concludes that the latter was certainly no more severe than the former, that its yield has been overestimated and that it was in fact comparable to that in other vapour cloud explosions. He comments, however, that in a shunting yard the wagons serve to promote turbulence and provide on their undersides vertical confinement.

NTSB (1975 RAR-75-04); Eisenberg, Lynch and Breeding (1975); V.C. Marshall (1980d, 1987)

A72 *Los Angeles, California, 1974*
On Saturday 17 August 1974, an explosion occurred on a load of organic peroxides on a semitrailer in a storage shed at Los Angeles, California. The load was a mixed one of several different peroxides, totalling 5.3 te. It had been made up on the Friday, but the customer had declined delivery before the weekend, and the semitrailer was therefore parked in the storage shed until the Monday. On the Saturday smoke and flames were seen coming from the shed and shortly afterwards there was a tremendous explosion. The source of the fire is not known. There were no injuries.

Sharry (1975); D.J. Lewis (1993)

A73 *Petal City, Mississippi, 1974*
On 25 August 1974, a new salt dome storage was being filled with butane when a release occurred. The capacity of the storage had been miscalculated with the result that the head of the brine being displaced fell below that equivalent to the gas pressure and butane issued at high velocity. A quantity of butane estimated as 40–90 te escaped and formed a vapour cloud extending 2 km. The cloud ignited with a small explosion, there followed a second, stronger explosion and the cloud then burned for 5 h. It has been suggested that the initial combustion and associated thermals promoted air entrainment into the mainly over-rich cloud so as to create a large elevated flammable volume which then exploded. Twenty-four people were injured.

Anon. (1974n); D.J. Lewis (1993)

A74 *Antwerp, Belgium, 1975*
On 10 February 1975, an explosion and fire caused extensive damage at a low-density polyethylene plant at Antwerp, Belgium. The cause was a leak of ethylene at high pressure due to fatigue failure of a vent connection on the suction of a compressor. Six persons were killed and 13 injured.

Mahoney (1990)

A75 *Beek, The Netherlands, 1975*
On the early morning of 7 November 1975, the start up of Naphtha Cracker II on the 100,000 tons per annum ethylene plant at the Dutch State Mines (DSM) works at Beek was under way. At about 6.00 a.m., the compressed gas was sent to the low temperature system. At 9.48 a.m., an escape of vapour was observed near the depropanizer. Shortly after, the cloud formed found a source of ignition and there was a massive vapour cloud explosion.

The explosion caused extensive damage and started numerous fires. Fire broke out in the pipeline systems and Tank Farm 2. Six tanks ranging in capacity from 1500 to 6000 m^3 within a common dike were burned out. An aerial view of the plant after the explosion is shown in Figure A1.12.

The explosion killed 14 and injured 104 people inside the factory, and injured three persons outside it.

The accident was investigated by a team from the Ministry of Social Affairs. Their findings are given in the Report on the Explosion at DSM, Beek on November 1975 (the Beek Report) (Ministry of Social Affairs, 1976). The investigation involved TNO and the Department of Steam Engineering (Stoomwesen). A separate investigation was carried out by the company.

van Eijnatten, (1977)

The investigation was hampered by the destruction of the instrument records in the control room. Figure A1.13 shows the state of part of the control panel. A detailed breakdown of the degree of damage done to the instrumentation is given in the report. It was also difficult to obtain useful information from the instruments which were retrieved, due to lack of synchronization, use of wrong recorder charts, lack of ink, inconsistent colour codes and out-of-date instrument diagrams.

The evidence suggested that the escape occurred due to low temperature embrittlement at the depropanizer feed drum, on and around which five fractures were found. Three of these were identified as secondary ruptures, but there were two which could have been primary failures. Two hypotheses were put forward. The principal one considered by the government team was that the initial rupture occurred on a 40 mm pipe connecting the feed drum to its safety valve.

The function of the feed drum was to take the €3* stream from the bottom of the de-ethanizer and a stream from the condensate stripper and pass this mixed liquid and vapour feed to the depropanizer. If the depropanizer were fouled, the feed drum could give a rough cut between the C_3 fraction and the heavier liquids. The normal operating temperature of the drum was 65°C.

At the time of the start-up, the reboiler of the de-ethanizer column was not operating properly. Possibly the circulation of hydrocarbons through the reboiler was reduced by thickening of the liquid. In consequence, the bottom product from the de-ethanizer was a liquid at about 0°C or lower with a high C_2 content. Discharge of this de-ethanizer stream into the feed drum would give flashing and could result in a temperature as low as −10°C.

Coincidentally there was a half-hour interruption in the flow from the condensate stripper due to a failure in the propylene compressor system. As a result, a layer of liquid containing a large proportion of C_2s was deposited on top of a layer of warm C_3^+ liquid in the feed drum. From the evidence the two layers had not mixed to any appreciable extent and the top layer was at about −10°C.

The feed drum material was a carbon steel which can normally be used at temperatures as low as −10 to −20°C. The fracture occurred at an autogenic weld. With such welds ageing can occur which could raise the transition temperature possibly up to 0°C.

An alternative hypothesis which the government team considered was that the rupture had been caused by explosion of organic peroxides which might have been formed from material trapped in a section of flush pipe which was isolated and which was found to have ruptured.

The amount of flammable material which escaped prior to the explosion was estimated at about 5.5 tons of hydrocarbons, mainly propylene, There is some uncertainty, however, about this figure.

The best estimate of the cloud shape from the explosion damage is that shown in Figure A1.14. The location of the damage indicated that the depth of the cloud did not exceed 4 m.

The estimated time which elapsed between the rupture and the explosion is approximately 2 min. From eyewitness accounts the source of ignition was identified as the flash drums in Section 1.

The quantity of flammable material which exploded was estimated as about 800 kg of propylene with an energy release equivalent to 2.2 tons of TNT.

Figure A1.12   *Beek, 1975: DSM works after the explosion (Ministry of Social Affairs, 1976)*

From the damage to the control room it was calculated that an overpressure of 0.2–0.3 bar must have acted on the walls. But the estimated overpressure required to cause the damage to the cold blowdown drum was approximately 1 bar.

The damage outside the factory was entirely breakage of glass. The bulk of the damage was within a radius of 4.5 km. It was believed that the pressure at this distance might have been about 0.005 bar, which is sufficient to break thin glass. It was suggested that damage beyond this distance may have been due to the deflection of shock waves by reflecting layers in the upper atmosphere. Flying glass caused one of the three injuries outside the factory.

The incident illustrates the stress created by a developing emergency of this kind and the confusion liable to ensue. At about 9.35 a.m. the operators were engaged in dealing with start-up problems. One entered the control room and called out 'Something has gone on Cll and there's an enormous escape of gas'. He was distressed and was rubbing his eyes. He staggered against the telephone switchboard. A second operator ran to the entrance and tried to get out, but his view was obscured by a thick mist.

He smelled the characteristic odour of C_3–C_4 hydrocarbons and realized there must be a major leak. He gave orders for the fire alarm to be sounded and ran out through another entrance to look at the gas cloud. He was seen from another office by a third man, apparently terrified and pointing to a gas cloud near the cooling plant.

Some witnesses stated that the fire alarm system in the control room failed. The investigation concluded, however, that the fire alarm system was in good working order before the explosion, but that none of the button switches for the fire alarm was operated.

Another aspect of the emergency was that the telephone lines to DSM were partially blocked by overloading. This did not affect rescue work, however, because the rescue services had their own channels of communication.

Ministry of Social Affairs (1976); van Eijnatten (1977); D.J. Lewis (1980, 1993)

A76   Eagle Pass, Texas, 1975
At 4.20 p.m. on 29 April a Surtigas tractor tank semi-trailer carrying LPG on US Route 277 near Eagle Pass, Texas, swerved to avoid a car in front which slowed suddenly to

Figure A1.13   *Seek, 1975: damage to instrumentation in the control room (Ministry of Social Affairs, 1976)*

make a turn. The tank semitrailer separated from the tractor, struck a concrete wall and ruptured, releasing LPG. Witnesses described a noise like that of a violent storm, followed immediately by an explosion. Simultaneously, fire covered the area and there was then a second explosion.

The large front section of the tank rocketed up, struck an elevated sign, travelled 1029 ft and struck the ground; bounced up and travelled 278 ft; struck and demolished a mobile home; bounced up again and travelled 347 ft over another mobile home, causing it to burst into flames and be destroyed; and finally fetched up 1654 ft from its starting point.

Sixteen people, including the driver, were killed and 512 suffered burns.
NTSB (1976 HAR-76-04)

A77   *Ilford, Essex, 1975*
At 11.10 a.m. on 5 April 1975, an explosion occurred at the factory of Laporte Industries Ltd at Ilford, Essex. The explosion was in a Lurgi electrolysis plant. The plant suffered extensive damage and one man who was injured subsequently died.

The electrolysis plant produced hydrogen for process use and oxygen as a waste product by electrolysis of potassium hydroxide solution. The investigation concluded that the explosion probably occurred on the oxygen separating drum into which hydrogen had leaked. After the accident the company carried out a mass balance on the hydrogen flows just prior to the explosion and found that no less than 50% was unaccounted for.

The ingress of hydrogen into the oxygen drum was apparently due to corrosion/erosion in the electrolysis cells. The internal breakdown of the cells had probably been initiated some time before the accident. On 2 April a cracking noise was heard in the cell block, which may have indicated minor explosions.

There was a system of monitoring the purity of the hydrogen and oxygen streams by hourly gas analyses performed by the process operator. The evidence suggested that these analyses were not always carried out and that assumed values were entered in the process log. One operator stated that he only did the analyses two or three times in every 12 h.

The report on the accident by the HSE (1976b) estimated that on the basis of the explosion damage the 1690 l oxygen drum contained a 13.5% hydrogen, 86.5% oxygen mixture and the explosion produced a shock wave equivalent to 22 kg of TNT. It was calculated that this explosion would have caused an overpressure of 1.03 kPa at 220 m and of 0.21 kPa at 660 m on an open site.

The most powerful explosion would have been produced by a stoichiometric mixture of 66 $\frac{2}{3}$% hydrogen, 33 $\frac{1}{3}$%

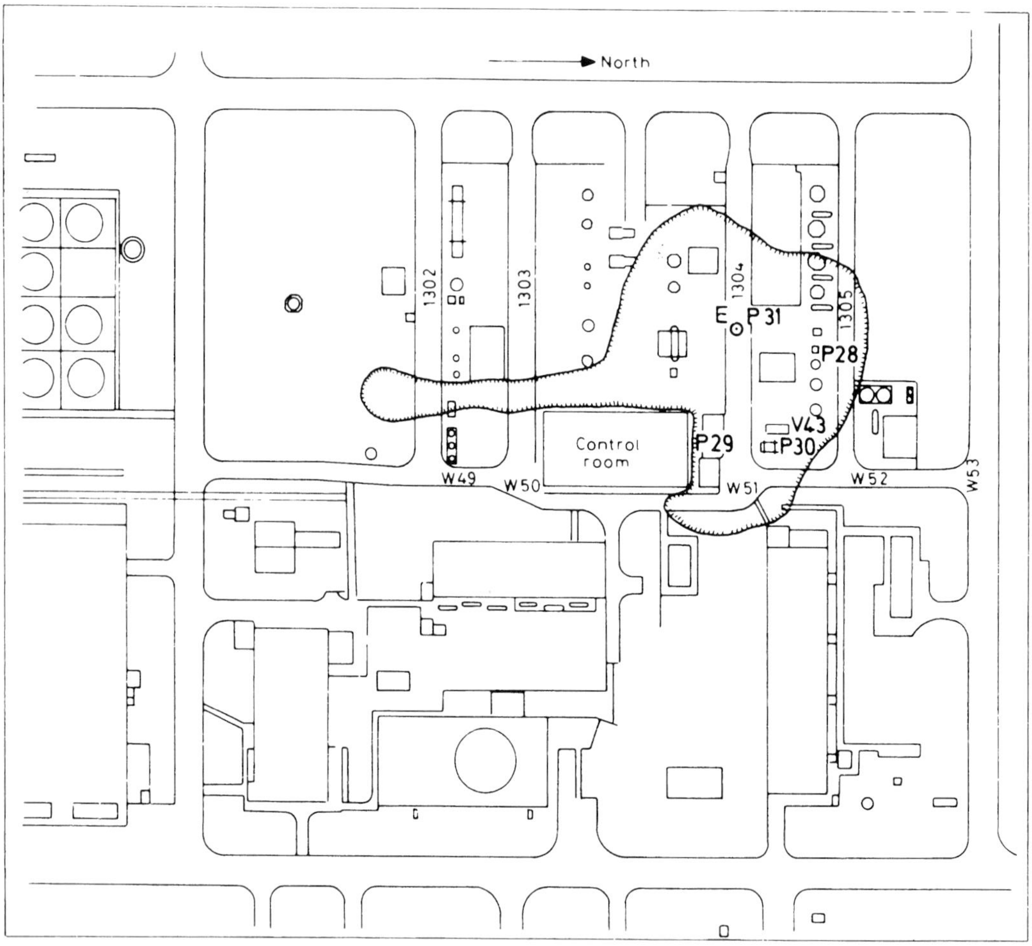

Figure A1.14 *Beek, 1975: simplified site plan of the works showing the estimated dimensions of the vapour cloud (Ministry of Social Affairs, 1976. Reproduced by permission)*

oxygen. Assuming a pressure equal to the set pressure of the relief valve on the drum, 515 psig, this composition would have produced a shock wave equivalent to 90 kg of TNT and this was calculated to cause an overpressure of 1 kPa at 350 m and 0.21 kPa at 1050 m.

The factory was in an urban area with houses 180 m and a school 210 m from the electrolysis plant. Although on an open site some damage might be expected from an explosion of the size described at these distances, none actually occurred. Probably this was due to the fact that the electrolysis plant was housed in a building designed and constructed to direct any blast upwards.

HSE (1976b)

A78 *Scunthorpe, Lincolnshire, 1975*
At about 1.25 a.m. on 4 November 1975 the Queen Victoria blast furnace at the Appleby Frodingham works of the British Steel Corporation (BSC) was tapped. By about 2.00 a.m. some 175 tons of iron had been run into the first torpedo ladle and the stream was diverted into the second ladle. Some 10–15 min later intense flames and sparks

began to issue from the blow pipe on No. 3 tuyere. The furnace crew tried to keep the pipe cool by spraying it with water and took steps to bring the furnace off blast to allow a new pipe to be fitted. While this was going on, a substantial water leak was observed coming from the furnace, but its source could not be located because of the flames. The water ran down the casthouse floor and entered the full torpedo ladle.

Just before 2.47 a.m. the shift manager's instruction to remove the full ladle was passed on to the loco driver and shunter by traffic control. The traffic personnel were informed that water was running into the ladle. At 2.47 a.m. an explosion occurred as the ladle was moved.

Some 90 tons of metal were ejected. Four people were killed and 15 injured, and a further seven died later.

The energy of explosion was estimated from the structural damage as 2.5 MJ or the equivalent of 1 lb of high explosive. The energy required to eject the hot metal was estimated to have a maximum value of 12 MJ. The thermal energy in the hot metal was some 10,000 times greater than this.

Investigation revealed that the water leak was probably due to the failure by corrosion of a steel-blanking plug on a cooling water pipe. Such plugs had been in use for at least 20 years. It transpired that there was among the employees considerable collective experience of previous failures of blanking plugs, although in most cases the knowledge of individuals was limited.

HSE (1976a)

A79 *Baton Rouge, Louisiana, 1976*
On 10 December 1976, a massive chlorine release occurred from a storage vessel at Baton Rouge, Louisiana. The vessel was a horizontal bullet tank 125 ft long × 11 ft diameter resting on a load cell weighing system. An internal explosion caused the vessel to fall off its supports so that it was pierced by a metal upstand on the ground. Over a period of 5.6 h 90.7 te of chlorine was released. A gas cloud formed described as a 42 mile long wedge, lying over the Mississippi river and sparsely populated areas. Ten thousand people were evacuated. Three persons were treated for minor irritation.

The cause of the tank displacement was an internal explosion of a natural gas–chlorine mixture at one end of the vapour space in the half full tank which produced a liquid surge along the length of the tank so that the momentum of the liquid caused the tank to jump off its supports. The tank was normally padded with nitrogen. The same nitrogen system was connected to the gland seals of hydrogen compressors in order to exclude air from the hydrogen. There was in addition a natural gas system which could be used on the gland seal as a more economic alternative. Due to the cold weather trouble had been experienced with natural gas condensate and the nitrogen system was in use. At some stage the lines to the nitrogen and natural gas systems were open at the same time and natural gas at higher pressure got into the nitrogen system. When the plant was restarted, an explosion occurred in the chlorine liquefaction system. It then spread via a balance line to the vapour space of the chlorine tank.

Anon. (1976n); Rees (1982); D.J. Lewis (1984b, 1993)

A80 *Geismar, Louisiana, 1976*
On May 24 1976, an explosion occurred on a large polyglycol ether reactor. The cause is believed to have been loss of agitation possibly combined with failure of a temperature transmitter and/or addition of insufficient catalyst. The explosion hurled the reactor head some 1400 ft and fragments ruptured a large polyglycol ether tank and severed the sprinkler system riser. Unreacted ethylene oxide and propylene oxide and other reactor contents escaped and there was a major fire involving the reactor and tank farm areas.

Mahoney (1990)

A81 *Kings Lynn, Norfolk, 1976*
At approximately 5.10 p.m. on Sunday 27 June 1976, an explosion occurred on the Clopidol plant of the Dow Chemical Company at King's Lynn, Norfolk. One man was killed and extensive damage was done to the plant and the buildings. The explosion involved the detonation of Zoalene, a poultry feed additive.

Some months before it has been discovered that stocks of Zoalene had fallen below their normal level of purity and

it was decided to remove water, one of the main impurities, by drying the material in a drier on the Clopidol plant.

The drier was provided with dust explosion protection, including nitrogen inerting, explosion suppression devices and explosion relief.

On 25 June at 3.00 p.m. a batch of Zoalene was charged to the drier. At 2.00 p.m. on 26 June the steam to the drier was shut off. The material was left in the drier, however, at a temperature of approximately 120–130°C. At 5.07 p.m. the following day sound was heard coming from the drier and the shift foreman activated the emergency procedures. At 5.10 p.m. the explosion occurred.

Zoalene was known to be a reactive chemical, but, as the incident showed, the hazard of a change to the process which involved holding the material at a moderately high temperature for a prolonged period was not fully appreciated and there were no instructions for the discharge of the material on completion of drying.

The accident also revealed deficiencies in the methods available for screening materials to detect exothermic reactions. Zoalene had been screened by the company using differential thermal analysis, but this had revealed no exotherm at the processing temperature. After the accident samples of Zoalene were tested by RARDE using differential scanning calorimetry. Exotherms were found in the temperature range 248–274° C.

Zoalene samples were also tested by the company using the new technique of accelerating rate calorimetry developed by Dow itself. It was found that samples would self-heat to decomposition if held under adiabatic conditions at 120–125°C. At the time of the accident the company was working through a programme of screening its reactive chemicals, but had not yet checked Zoalene.

The investigation described in the report of the HSE (1977b) concluded that the explosion had been a detonation with an energy release of 200–300 lb of TNT, that a temperature of over 800°C had been reached before detonation occurred and that there had been an extremely rapid pressure rise, probably to over 10,000 psi.

The report gives a detailed map of the locations of the fragments from the explosion.

HSE (1977b)

A82 *Los Angeles, California, 1976*
On 17 December 1976, a violent explosion occurred aboard the tanker Sansinena at Los Angeles, California. The ship had offloaded a cargo of light crude oil and was loading Bunker C oil. It was equipped with neither a vapour recovery nor a cargo tank inerting system. The explosion is believed to have been initiated by a spark from a pump. The force of the explosion threw the centre section and the bridge onto the dockside, leaving the bow and stern sections afloat.

NTSB (1978 MAR-78-06); D.J. Lewis (1984c); Mahoney (1990)

A83 *Plaquemine, Louisiana, 1976*
On 30 August 1976 at Plaquemine, Louisiana, the top was blown off a transfer oil surge tank on an ethylene oxide plant, causing a serious fire and damage to six reactors and other equipment. A leak between the shell and tube sides of a reactor heat exchanger had allowed air at 200 psig to enter the heat transfer system.

Mahoney (1990)

A84   *Southwest Freeway, Houston, Texas, 1976*
At about 10.45 a.m. on 11 May 1976 in Houston, Texas, a
tractor tank semi-trailer carrying ammonia went through a
bridge rail on the interstate highway 1–610 and fell some
15 ft onto the South-west Freeway, US 59. The interchange
was the busiest in the state and at the time traffic was quite
heavy. The tank, which held 19 te of liquid anhydrous
ammonia, burst.

The crash was caused by the excessive speed of the
vehicle and by sloshing of the liquid in the partially loaded
tank. However, the speed did not exceed 54 mile/h.

The ammonia was released and formed a visible fog. It
was a bright sunny morning with a wind speed of 7 mile/h.
The initial height of the cloud determined from photographs was about 30 m. The cloud was observed to reach a
width of about 300 m and a length of 600 m. One photograph showed a tail to the left-hand side, indicating the
typical slumping behaviour of heavy gas. It is estimated
that the ammonia evaporated and the cloud dispersed
within about 5 min. The driver of the truck and five other
people were killed, 78 taken to hospital and about another
100 injured. Apart from the driver all the casualties were
due to gassing.

Anon (1976o); McMullen (1976); NTSB (1977 HAR-77-01);
Fryer and Kaiser (1979 SRD R152)

A85   *Breahead, Renfrew, Scotland, 1977*
On 4 January 1977, the Breahead Container Clearance
Depot, a chemicals warehouse, was destroyed by a serious
fire and explosions. The fire had been started inadvertently
by boys who had built a 'den' beside the warehouse and had
lit a fire to warm themselves. The explosions caused widespread roof and window damage within a radius of a mile.
The warehouse had contained some 67 te of sodium chlorate stored in 1774 steel drums. Investigations showed that
under intense heat this substance is capable of giving
explosions of the severity which occurred.

This incident drew attention to the hazard of sodium
chlorate. It emerged that since 1899 sodium or potassium
chlorate had been implicated in six explosions, the then
most recent being in Hamilton, Lanarkshire in 1969 and in
a ship in Barcelona in 1974. Subsequently there occurred in
the United Kingdom two other warehouse explosions
involving sodium chlorate, one on 21 January 1980 at
Barking, Essex (HSE, 1980a) and one on 25 September 1982
at Salford (HSE, 1983b).

Anon. (1979i,z); HSE (1979b); Matthews (1981)

A86   *Brindisi, Italy, 1977*
On 8 December 1977, a large gas release occurred on an
ethylene plant at Brindisi, Italy, resulting in a vapour cloud
explosion and fire, with extensive damage. The sewers
were unable to handle the firewater and burning hydrocarbon liquids floated on it to a depth of some 18 in. Three
persons died and 22 were injured.

Anon. (1977b); Mahoney (1990)

A87   *Puebla, Mexico, 1977*
On 19 June 1977 a leak occurred on one of a group of VCM
storage bullets. A fitter had made an error in removing an
actuator from a liquid discharge valve on the tank, taking
out the wrong bolts so that the valve plug suddenly popped
out, allowing an escape through the 3 in. valve body. The

release continued for 80 min in calm conditions, the vapour
cloud formed being 1100 ft long, 800 ft wide and 5 ft deep.
Five minutes later the cloud caught fire and flashed back to
the tanks. A further 5 min later one tank suffered a BLEVE.
Three more tank explosions followed. One person was
killed and three injured.

D.J. Lewis (1991 LPB 100, 1993)

A88   *Umm Said, Qatar, Persian Gulf, 1977*
On 3 April 1977, a catastrophic failure occurred on a storage
tank at Umm Said, Qatar, in the Persian Gulf. The tank was
a single-wall carbon steel refrigerated atmospheric tank
holding 37,000 m^3 of liquid propane at −42°C. A wave of
propane liquid swept over the bund, boiling on the sand. It
entered the adjacent separation plant and was ignited.
There was a massive fire which did extensive damage,
burned out of control for two days and was extinguished
only after 8 days. Seven persons were killed and 13 injured.

Information on this incident is still hard to come by. The
failure is said to have occurred at a weld on the tank. It is
reported that the same weld had been responsible for a leak
a year before in which a vapour cloud formed and travelled
500 ft but did not ignite.

Anon. (1977c); Whelan (1977); D.J. Lewis (1980, 1993);
Kletz (1988 LPB 81)

A89   *Westwego, Louisiana, 1977*
On 23 December 1977, a series of explosions took place in
the silos of a large grain elevator at Westwego near New
Orleans, Louisiana. There were some 73 reinforced concrete
silos 35 m high and ranging in diameter from 8 to 10 m.
Contents included corn, wheat and soya beans.

The first explosion, which occurred in the morning,
caused the top 20 m of a 75 m high head house to fall onto
an office building. Explosions continued through the
morning. More than half the silos were completely
destroyed and others severely damaged. The wreckage of
some of the silos is shown in Figure A1.15.

More than 50 people were working on the complex when
the explosions occurred. Thirty-six persons were killed
and ten injured. Most of the dead were in the crushed office
building.

The source of ignition for the first explosion has not been
determined. At the time the relative humidity was low,
which would be conducive both to the presence of fine dust
suspensions and to static electricity.

This disaster is comparable in scale for dust explosions
with that at Flixborough for vapour cloud explosions.

Less than a week later, on 29 December 1977, there was a
further series of explosions in another grain elevator at
Galveston, Texas, in which 15 people died.

Anon. (1977p,q); Lathrop (1978); Nolan (1979): D.J. Lewis
(1993)

A90   *Abqaiq, Saudi Arabia, 1978*
On 15 April 1978, internal corrosion of a 22 in. 500 psig gas
transmission pipeline on a gas processing plant at Abqaiq,
Saudi Arabia, resulted in a leak which then expanded so
that the line parted. A vapour cloud estimated as 405 ft ×
435 ft spread through the plant. Some 7 min later ignition
occurred at a flare 1500 ft downwind. The escaping gas is
described as having a jet/whipping action which threw a
22 ft pipe section 400 ft so that it struck an spheroidal storage

Figure A1.15   Westwego, 1977: grain silos after the explosion (United Press International)

tank. A second vapour cloud formed and was ignited by the initial fire, resulting in a vapour cloud explosion.

Mahoney (1990)

A91 *Donnellson, Iowa, 1978*
At 12.02 a.m. on 4 August 1978, a leak of liquid propane ignited on an 8 in. LPG pipeline of the Mid-America Pipeline System (MAPCO) in the rural area of Donnellson, Iowa. The pipeline pressure was over 800 psig and the leak came from a 33 in. split in the pipe. The intense fire killed two people and severely burned three others as they fled their homes; one of the injured later died. Some 157,500 USgal of liquid escaped and 75 acres were burned.

The investigation by the NTSB (1979 PAR-79-01) found that the failure occurred due to a dent and gouge in the pipe sustained in 1962 before the line was complete combined with stresses created when the line was lowered 3 months prior to the incident.

V.C. Marshall quotes this incident as involving 435 te of propane and giving a fireball of 305 m radius. The NTSB report refers to eyewitness reports of a 'bonfire' and states that when firefighters arrived about 12.20 a.m. they found a towering flame 400 ft high.

NTSB (1979 PAR-79-01); V.C. Marshall (1987)

A92 *Texas City, Texas, 1978*
On 30 March 1978, a series of fires and explosions occurred at LPG storage spheres at Texas City, Texas.

There were three spheres, each 800 m³. One of the spheres suffered overpressure while it was being filled, due

to failure of a pressure gauge and also of a relief valve. It cracked and leaked LPG. The leak ignited giving a massive fireball.

Accounts differ in their description of the events which followed. According to Selway (1988 SRD R492), after 20 min a second sphere, which was only partially full, suffered a BLEVE. The third sphere, which was virtually empty, failed due to a heat-induced rupture at the top, but remained upright. Thus all three spheres were damaged, but there was only one BLEVE event. Mahoney (1990) states that during the 20 min following the fireball, five horizontal bullets and four vertical ones were damaged by missiles and that the other two spheres were also damaged in this way. The missiles started fires and hit the firewater storage tank and electrical fire pumps, although two diesel fire pumps remained operable.

Selway (1988 SRD R492): Mahoney (1990)

A93 *Waverly, Tennessee, 1978*
On 24 February 1978, an explosion occurred of an LPG rail tank car at Waverly, Tennessee. Two days before, on 22 February, a freight train conveying two LPG tank cars derailed. There was no immediate release. A check on the following day showed that both tank cars were damaged but no leak could be detected. Both tank cars were underneath box wagons and these were lifted off. On 24 February as preparations were in hand to transfer the contents of the tank cars, one of them, which had suffered a long gouge-like scrape along one side, suddenly ruptured into four pieces

and released its contents of 59.4 te of LPG. Seconds later the gas ignited, probably by burning equipment nearby, giving a fireball. There followed a fire which did extensive damage. Sixteen people were killed and 43 injured.

Anon. (1978b); NTSB (1979 RAR-79-01); D.J. Lewis (1992 LPB 105, 1993)

A94 *Youngstown, Florida, 1978*
At about 1.55 a.m. on the night of 26 February 1978, a freight train of the Atlanta and Saint Andrews Bay Railway derailed near Youngstown, Florida. In the derailment a rail tank car containing liquid chlorine was punctured when it struck another tank car. The approximate position of the tank cars following the derailment is shown in Figure A1.16.

A chlorine gas cloud formed and moved towards State Highway 231, which ran parallel to the railroad at a distance of 300 ft. The weather conditions were a temperature of $50-55°F$ and a light westerly wind of $2-3$ knots. There were already some patches of fog around and drivers, used to foggy conditions and unable to see the colour of the cloud as it was night, drove into it. Three vehicles drove right off the road and into the ditch. One motorist said the gas was so thick 'You could not see your hand in front of your face'. Seven motorists abandoned their cars and died while trying to escape, whilst an eighth drove through the cloud but died a short distance down the road.

The train's engineer at the head of the train alerted the conductor at the rear by radio and then tried to escape through the cloud; he became lost and mired in a swamp, being rescued 6 h afterwards. The head brakeman, who had been with the engineer, circled round to the highway and started to alert traffic. The conductor and rear brakeman tried to inform the railroad facility at Panama City, but it was not manned after midnight. They began to rouse local residents and notified the police; almost 30 min had elapsed before the first emergency call was completed. They then went to the highway to assist in alerting traffic. The report by the NTSB (1978 RAR-7807) states.

The head and rear brakeman both imperilled their lives by their repeated exposure to the chlorine gas cloud to stop and warn motorists of the dangerous situation. Their heroic actions saved many people from death or serious injury.

The emergency service evacuated people, first in a 5 mile and then in a 10 mile radius.

At 9.00 a.m. aerial observation revealed a gas cloud 3 miles wide and 4 miles long with a maximum height of 1000 ft. If the wind had been in a less favourable direction, the town of Youngstown would have been engulfed in the cloud.

The chlorine gas killed eight people, as mentioned, and injured another 50. Some 2500 people were evacuated.

Anon. (1978k); NTSB (1978 RAR-78-07)

A95 *Bantry Bay, Eire, 1979*
At about 1.06 a.m. on 8 January 1979, the Total oil tanker Betelgeuse blew up at the Gulf Oil terminal at Bantry Bay, Eire. The ship had completed the unloading of its cargo of heavy crude oil. No transfer operations were in progress. The first sign of trouble occurred at about 12.31 a.m. when a sound like distant thunder was heard and a small fire was seen on deck. Ten minutes later this was spread aft along the length of the ship, being observed from both sides. The fire was accompanied by a large plume of dense smoke. About 1.06–1.08 a.m. a massive explosion occurred. The vessel was completely wrecked and extensive damage was done to the jetty and its installations. There were 50 deaths.

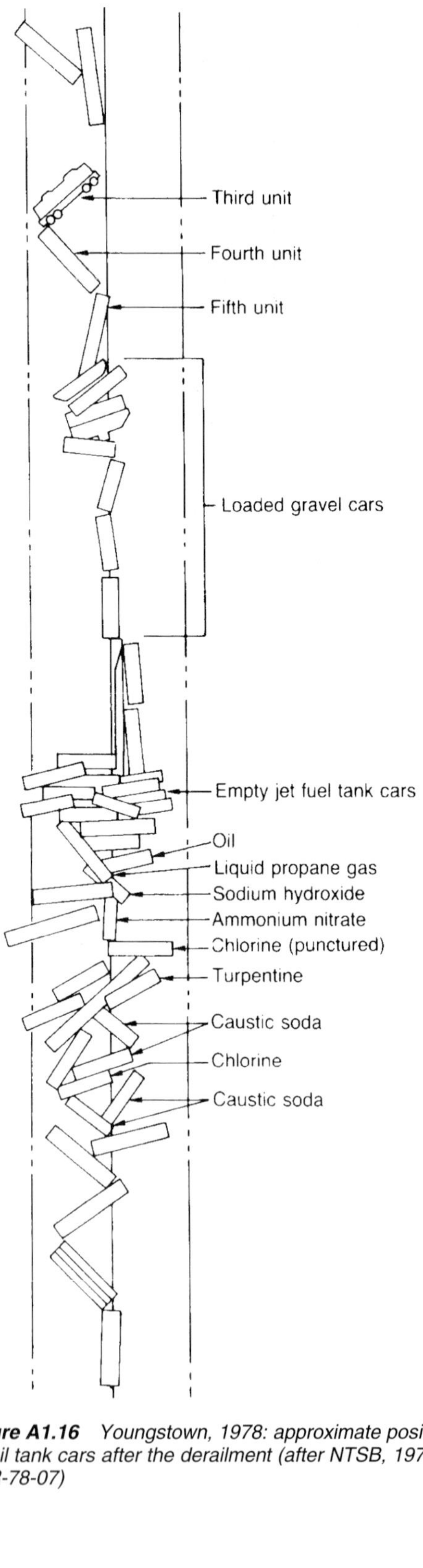

Figure A1.16 *Youngstown, 1978: approximate positions of rail tank cars after the derailment (after NTSB, 1978 RAR-78-07)*

The inquiry (Costello, 1979) found that the initiating event was the buckling of the hull, that this was immediately followed by explosion in the permanent ballast tanks and the breaking of the ship's back and that the next explosion was the massive one involving simultaneous explosions in No. 5 centre tank and all three No. 6 tanks. It further found that the buckling of the hull occurred because it had been severely weakened by inadequate maintenance and because there was excessive stress due to incorrect ballasting.

The ship was an 11-year old 61,776 CRT tanker. The weakened hull was the result of 'conscious and deliberate' decisions not to renew certain of the longitudinals and other parts of the ballast tanks which were known to be seriously wasted, taken because the ship was expected to be sold, and for reasons of economy. The vessel was not equipped with a 'loadicator' computer system, virtually standard equipment, to indicate the loading stress. It did not have an inert gas system, which should have prevented or at least mitigated the explosions.

At the jetty there had been a number of modifications which had degraded the fire fighting system as originally designed. One was the decision not to keep the fire mains pressurized. Another was an alteration to the fixed foam system which meant that it was no longer automatic. Another was decommissioning of a remote control button for the foam to certain monitors.

Another issue was the absence of the dispatcher from the control room at the terminal. It was to be expected that had he been there, he would have seen the early fire and have taken action.

In a passage entitled '*Steps taken to suppress the truth*' the tribunal states that active steps were taken by some personnel at the terminal to suppress the fact that the dispatcher was not in the control room when the disaster began, that false entries were made in logs, that false accounts were given to the tribunal and that serious charges were made against a member of the Gardai (police) which were without foundation.

> The inquiry report gives a long list of measures which might have prevented or mitigated the disaster: Had the vessel been properly maintained. . . Had Total supplied the ship with a loadicator. . . Had the dispatcher in the Control Room observed the initiation of the disaster. . . Had the alert been raised at the beginning of the disaster. . . Had the tug been moored in sight of the jetty. . . Had the decision to discontinue the automatically pressurized fire-main not been taken. . .

and so on.

Anon. (1979l); Costello (1979); Anon. (1980a); Mahoney (1990)

A96   *Good Hope, Louisiana, 1979*
On 30 August 1979, a freighter collided with the butane barge Panama City at a jetty on the Mississippi at Good Hope, Louisiana, striking it amidships, severely puncturing at least one of its tanks and causing a fireball. Twelve persons died and 25 were injured.
NTSB (1980 MAR-80-07); V.C. Marshall (1987)

A97   *Mississauga, Toronto, Ontario, 1979*
At 23.52 on Saturday 10 November 1979 at Mississauga, Toronto, a freight train derailment led to fires, explosions and a chlorine gas release, in an emergency lasting 11 days.

The train had 25 rail cars of which 19 held hazardous materials, in a mixed load of flammables and toxics, including propane, toluene, styrene, caustic soda and chlorine.

The derailment occurred due to lack of lubrication of a plain journal axle bearing on the 33rd car, a propane tanker. The red-hot axle, complete with wheels, parted from the tank car. Some 2 km further on at a location where there was a road crossing and points, this vehicle became detached from the front portion of the train. Many of cars in the rear section derailed. Several tanks were ruptured and a fire broke out. The brakes in the front section of the train were applied automatically and looking back the locomotive crew saw a fire and a tank car rocketing off.

There followed a series of explosions, one within a minute and another just after midnight at 00.10. At 00.15 a propane tank suffered a BLEVE in which one half travelled 2300 ft and buried itself 20 ft into the ground. At 00.24 another BLEVE of a propane tank car occurred.

There was a single chlorine tank car containing 90 te of liquid chlorine. This tank lost part of its insulation and a flame impinged on the bare metal. There was an iron – chlorine reaction which burned a 3 ft diameter hole in the tank. Chlorine began to escape through the hole. The effect was less serious than it might have been, since the effect of the fire was to cause the gas to rise in the thermal currents.

The emergency developed over an extended period. Initially the emergency services did not know what was in the train: the manifest was found, but it was in code. There was, however, a strong smell of chlorine. At 1.30 a readable version of the manifest was obtained, identifying the chlorine tank car. At 3.00 the decision was made to evacuate the surrounding population, an exercise which eventually resulted in the evacuation of 215,000 people. The holed chlorine tank car was identified by a helicopter survey.

At 9.00 on Tuesday 13 November an initial attempt was made to plug the hole in the chlorine tank car, but without success; a second attempt succeeded. When subsequently the tank was emptied, only 18 te of chlorine was recovered.

During Tuesday afternoon 143,000 people were allowed back home. The rest did not return until after an absence of six days.

Detailed timetables of the incident itself and of the evacuation are given by Fordham (1981 LPB 44).

The only injuries were those caused to eight firemen by inhalation of the noxious gases.

Mississauga News (1979); Amyot (1980); Grange (1980); Anon. (1981l); Fordham (1981 LPB 44); Lane and Thomson (1981); D.J. Lewis (1993)

A98   *Borger, Texas, 1980*
On 20 January 1980, a refinery at Borger, Texas, suffered extensive damage. Details are lacking, but it appears that a vessel was subjected to a pressure much in excess of its working pressure, the relief valve operated but was ineffective due to ice blockage and the vessel ruptured, releasing 34 m^3 of hydrocarbon liquids. A vapour cloud formed and ignited, giving a vapour cloud explosion with a TNT equivalent of some 15 te. The blast destroyed the alkylation unit and boiler plant and shut-down the whole refinery. Forty-one people were injured.

The ice plug is reported as due to freezing by propane of water which had accumulated in the flare system.
Mahoney (1990); D.J. Lewis (1993)

A99 *Wealdstone, Middlesex, 1980*

About 19.15 on 20 November 1980 an escape of propane began from a storage vessel in the factory yard of White-friars Glass Ltd at Wealdstone, Middlesex. The vessel was a 10 te vessel which contained 5 te of propane. The night-watchman smelled the odorized gas, but searched first inside the building, until he heard a hissing sound and saw a white cloud about 1.2 m deep around the base of the tank. Meanwhile a passer-by had called the fire brigade. The escape lasted several hours. Some 2100 people were eva-cuated from the area. One person had to attend hospital.

The release was caused by the removal by the works engineer and a fitter of three of the four bolts on an adapter piece on the drain valve of the tank. The valve had two sections. The upper, body section was bolted to the under-side of the tank. The lower, adapter section had a spigot which extended into the upper section and held the ring seals in place against the ball. In other words, the adapter was an essential part of the valve.

This was not known, however, to the persons who removed the bolts.

The work on the valve had occurred during the after-noon. The delay between the removal of the bolts and the escape was attributed by the investigators to the accumu-lation in the vessel of non-volatile, viscous residues due to the removal of propane as gas from the top of the tank over a period of at least 10 years.

HSE (1981c)

A100 *Montana, Mexico, 1981*

At 17.52 on 1 August 1981 a train derailment resulted in a massive release of chlorine near Montana, Mexico. The train consisted of 38 wagons, including 32 rail tank cars containing liquid chlorine. It was moving down a steep and winding valley at a 3% gradient when its brakes failed. The driver radioed to other trains on the single track to warn them, and two trains drew into loops. The train derailed at over 80 km/h on a bend 350 m beyond Montana station.

All the chlorine tank cars were 55 short tons, say 50 te, capacity. The pile-up included 28 of the 32 chlorine cars. Most were badly damaged. One tank car lost its dished end and the shell was propelled 2000 m. A second was split along its side. A third had a 0.5 m diameter hole, probably the result of an iron–chlorine fire, which could well have resulted from ignition of the cork insulation by red hot brakes. Four other tank cars suffered damage to their valves, which were ripped off or dislodged so that they leaked. It is estimated that 100 te of chlorine escaped in the first few minutes and 300–350 te in all.

The population of Montana was some 400 people. There was also a passenger train with some 300 people at the station. The weather conditions to be expected at the time of day would be a warm wind blowing up the valley, towards the village. The vegetation up the valley was bleached by the gas cloud passing up it; there was also discoloration some 50 m down the slope and up the sides for a vertical distance of about 50 m. The highest concentrations appear to have occurred in a strip 1000 m long × 40 m wide. Seven-teen persons died, four in the caboose of the train and 13 from gassing. About 1000 people were received in hospital and 256 were detained.

Anon. (1982 LPB 44, p. 33); Anon. (1983 LPB 52, p. 7); V.C. Marshall (1987)

A101 *Stalybridge, UK, 1981*

At about 23.30 on 6 September 1981, a violent explosion occurred at the works of Chemstar Ltd at Stalybridge. Failure of cooling water to the condenser of a still had allowed hexane vapour to issue through the vent, and the escaping vapour found a source of ignition. One person was killed and one injured.

The investigators found that the normal water supply was not available, a temporary supply had been used which proved to be erratic and there was inadequate flow indica-tion; that the vent had not been routed to a point outside the building; and that the training and supervision of the operator were inadequate.

HSE (1982a); Kletz (1983b)

A102 *Caracas, Venezuela, 1982*

At about 6.17 a.m. on 19 December 1982, an explosion blew the top off a large oil storage tank at the electricity com-pany at Caracas in Venezuela. The oil in the tank caught fire and burned. Firemen arrived and attempted to control the fire and people stood watching it. After some hours there was a violent boilover of the tank contents and burning liquid ran down the hill on which the tank stood towards firemen and spectators.

The power station was owned by C.A. La Electricidad de Caracas and was built in 1978. The tank involved was No.8 storage tank, which was one of two cone roof tanks each 55 m × 17 m, the other tank being No.9. Each tank had its individual bund and had two forms of fire protection. One was a 100 mm water drench pipe which ran to the centre of the roof and was designed to cascade water over the roof and shell to prevent overheating and the other was a semi-fixed foam system with five risers.

No.8 tank contained No.6 fuel oil and at the start of the incident the depth of oil was 7.5 m. It had been feeding the boilers for six days prior to the fire.

At 6.15 a.m, while it was still dark, two men went up on the roof of No.8 tank to gauge it, and a third waited in a vehicle. About 2 min later the roof blew off and at the same time the lines within the bund were ruptured. One of the foam risers on the tank was damaged. The source of igni-tion is not known, but neither man had a flashlight and it is possible a match was lit to read the gauge line.

It is not known if either the water or foam fire protection systems on No.8 tank operated. The power company had no fire brigade and the nearest public fire service was a half an hour journey away over winding roads.

Some 8 h after the fire began a violent boilover of No.8 tank occurred. Burning oil surrounded No.9 tank and flowed down the hill into the sea, but was kept from the power station by a concrete wall. The inrush of air lifted roof panels at the power station 300 m away.

The boilover killed 40 firemen and many civil defence workers and spectators. 153 people are known to have died and 7 were missing.

Boilover is defined by the NFPA as follows:

> Boilover shall mean an event in the burning of certain oils in an open top tank when, after a long period of quiescent burning, there is a sudden increase in fire intensity associated with the expulsion of burning oil from the tank. Boilover occurs when the residues from surface burning become denser than the unburned oil and sink below the surface to form a hot layer which progresses downward much faster than the regression of the liquid

surface. When this hot layer, called a 'heat wave', reaches water or a water-in-oil emulsion in the bottom of the tank, the water first is superheated and subsequently boils almost explosively, overflowing the tank. Oils subject to boilover must have components having a wide range of boiling points, including both light ends and viscous residues. These characteristics are present in most crude oils and can be produced in synthetic mixtures.

The heat wave travels at a rate which in most crude oils is some 300–460 mm/h but can be as great as 900–1200 mm/h. Its temperature can be as high as 315°C. When water comes into contact with the heat wave, it vaporizes to steam with a volume expansion of 1700 : 1 or more.

Normally No.6 fuel oil is regarded as not being subject to boilover. However, it is common practice in the United States to blend it. It is reported that the power company's specifications allowed 5–20% light fractions, that the oil had a flashpoint of 71°C, that it was intended to be heated in storage to 65°C, but that the oil return temperature shown by the control room records was 80°C and the storage temperature was 82°C. The tank high-temperature alarms had sounded at midnight, some 6 h prior to the incident.

Garrison (1984 LPB 57); M.F.Henry (1986)

A103   *Livingston, Louisiana, 1982*

On 28 September 1982, a freight train conveying hazardous materials derailed at Livingston, Louisiana. The train had 27 tank cars some of them with jumbo tanks of 30,000 USgal. Seven tanks cars held petroleum products and the others a variety of substances, including vinyl chloride monomer, styrene monomer, perchlorethylene, hydrogen fluoride and metallic sodium.

The incident developed over a period of days. The first explosion did not occur until three days after the crash. The second came on the fourth day. The third was set off deliberately by the fire services on the eighth day. The scene is shown in Figure A1.17.

Meanwhile the 3000 inhabitants of Livingston were evacuated. Some were not to return home until 15 days had passed.

One factor contributing to the derailment was the mis-application of brakes by an unauthorized rider in the engine cab, a clerk who was 'substituting' for the engineer. Over the previous 6 h the latter had drunk a large quantity of alcohol.

The incident demonstrated the value of tank car protection. Many of the cars were equipped with shelf-couplers and head shields, and there was no wholesale puncturing and rocketing. Tanks also had thermal insulation which resisted the minor fires occurring for the two or more hours which it took the fire services to evacuate the whole town.

NTSB (1983 RAR-83-05); Anon. (1984t)

A104   *Bloomfield, New Mexico, 1983*

At 12.04 p.m. on 26 May 1983 at the El Paso Natural Gas Company's Blanco Field Plant near Bloomfield, New Mexico, a gasket on a compressor failed and natural gas at 815 psig began to escape into the building. Two operators heard the noise. One tried to shut off the gas supply to the compressors, the other to shut-down the compressor engine. Before they could effect these operations, the gas ignited and exploded, and both were severely burned. One compressor was destroyed and a second severely damaged.

The investigation by the NTSB (1983 PAR-83-04) found that the probable cause was improper tightening of the compressor head bolts and lack of procedures for, and training in, bolt tightening. The report gives a detailed discussion of the bolting problem together with guidance on bolt tightening.

It also criticized the emergency response, in that there was a failure to effect prompt relief of the high-pressure gas by activating the blowdown system.

NTSB (1983 PAR-83-04)

A105   *Bontang, India, 1983*

On 14 April 1983 on an LNG plant at Bontang, India, the main cryogenic heat exchanger, 141 ft long × 14 ft diameter,

Figure A1.17   *Livingston, 1982: rail tank cars burning after the crash (National Transportation Safety Board, 1983)*

ruptured during a start-up. The shell side operating pressure was 25.5 psi but it was exposed to gas at a pressure in excess of 500 psi. The explosion gave rise to missiles and a fire. The flow from a pressure controller on the shell and the flows from relief valves on the shell and tube sides of the exchanger vented to a common line. The investigation found that a valve on this line was closed, disabling both the pressure control and the pressure relief. The valve in question had apparently been omitted from the valve checklist for plant start-up.

Mahoney (1990)

A106 *Milford Haven, UK, 1983*
On 30 August 1983, a fire broke out on the roof of a large crude oil floating roof storage tank at the refinery at Milford Haven, Dyfed, Wales. The tank was 78 m diameter × 20 m high with a capacity of 94,100 m^3 and was 59% full. It stood in a very large bund with dimensions 90 m × 180 m. The fire took hold and a massive tank fire ensued with an estimated burn-off rate of 300 te/h of oil. After 12.5 h boilover occurred, thousands of tonnes of burning oil flowed into the bund and a fire column rose 3000 ft high. Two hours later there was a further boilover.

The floating roof had developed a split which allowed vapours and some oil to escape onto the roof. Ignition was attributed to hot carbon particles from the flare. An upset had resulted in a large flow of hydrocarbons to the flare which could have dislodged particles from the stack top.

There was a major fire fighting operation. Oil was pumped out from the bottom of the burning tank at a rate of 700 te/h. The foam supplies available on site were sufficient for a floating roof tank seal fire, but not for the fully developed tank fire, and supplies were sought outside. There were two other storage tanks at a wall-to-wall distance of 61 m from the burning tank, and an early priority was to cool the outside of these and to inject foam inside. Eventually, after the boilovers, enough foam was obtained, some 305 m^3, to mount a full foam attack, first on the fire in the bund and then on that in the tank itself. This process lasted 15 h with the fire fighting being done under conditions of extreme heat. In the boilover events, firemen were affected by flash burns and by dry hot air.

A detailed timetable of the incident is given by Mumford (1984 LPB 57).

Dyfed County Fire Brigade (1983); Mumford (1984 LPB 57); Golec (1985); Steinbrecher (1986); Mahoney (1990); D.J. Lewis (1993)

A107 *Abbeystead, UK, 1984*
On 23 May 1984, a group of people were assembled in the valve house of the Lune/Wyre Water Transfer Scheme at Abbeystead for a presentation when there was a violent explosion. The explosion occurred soon after the pumping system had been started up. The pumps had not been run for 17 days and the tunnel which conveyed the water was not full of water, its usual condition, but had been emptied through a drain valve which had been opened to remove silt. Subsequent tests showed that during this period groundwater would have leaked into the tunnel at a rate of 1000 te/d. This water contained dissolved methane, which mixed with the air in the tunnel. The only venting of the tunnel was through the grilled floor of the valve house. Thus when the pumps were started and water moved up the tunnel, the methane–air mixture was displaced and came up through the grill into the valve house, where it found a

source of ignition. The explosion killed 16 and injured 28; all the 44 persons in the valve house itself were killed or injured.

Anon. (1984pp, 1985a); Burgoyne (1985a); HSE (1985a); Lihou (1985 LPB 66); V.C. Marshall (1987)

A108 *Cubatao, Brazil, 1984*
On 24 February 1984 just before midnight a leak of petrol occurred on the Petrobas pipeline passing through the shanty town of Vila Soca at Cubatao in Sao Paulo state, Brazil. The petrol caught alight and a fire devastated the houses. The death toll was 508.

Vila Soca was a shanty town of some 2500 dwellings. The number of people living there is uncertain, but is estimated as 8000, perhaps even 12,000. The pipeline was 30 years old. For 20 years the company had protested that Vila Soco had been built illegally on its land, but had nevertheless provided water and electricity.

A leakage from the pipeline was noticed the day before the fire and a resident telephoned the company. In the leak which led to the disaster some 700 te of petrol escaped from the pipeline. The spread of the flammable liquid through the shanty town and under the houses was aided by water on the ground and by a small stream. A two-minute warning was given but this was not enough for the inhabitants to evacuate. The petrol caught fire and straight away created an inferno. Houses burned and exploded as the petrol beneath them ignited. It was 45 min before the fire brigade arrived by which time most of the shanty town had been destroyed.

Anon. (1984oo); Kletz (1985p)

A109 *Fort McMurray, Avaporerta, 1984*
On 15 August 1984 at Fort McMurray, Avaporerta, hydrocarbon liquids close to their autoignition temperature were released from a 10 in. slurry recycle oil line on a fluidized-bed coking unit. A vapour cloud formed and ignited almost at once. The fire, which did extensive damage, was fed by further releases of hydrocarbon liquids, the flows being largely under gravity but in part from pumps which could not be shut-down. Fire fighting was hampered by rupture of a 900 psig steam line supplying the compressor turbine drives. The failure of the recycle line was due to erosion. Subsequent metallurgical examination showed that an 18 in. long section of carbon steel had been inserted in error into the alloy steel line.

Mahoney (1990)

A110 *Las Piedras, Venezuela, 1984*
On 13 December 1984 in a refinery at Las Piedras, Venezuela, an 8 in. oil line fractured and sprayed hot oil at 700 psi and 650°F across a roadway onto hydrogen units. Ignition occurred, and the ensuing intense fire led to rupture of a 16 in. gas line, which resulted in a massive jet flame, leading to further line ruptures, some of which occurred with explosive force. The initial failure was a circumferential fracture in the heat affected zone in the parent metal about 1.5 in. from the weld.

Mahoney (1990)

A111 *Romeoville, Illinois, 1984*
On 23 July 1984, a column at the Lemont refinery of Union Oil at Romeoville, Illinois, suffered a spontaneous failure. The vessel was an absorption column 8.5 ft diameter × 55 ft high in which a propane/butane feedstock was contacted with methylethanolamine (MEA). Just prior to the burst, an

operator noticed a horizontal crack about 6 in. long at the 10 ft level. He went up a ladder to try to shut of the inlet valve and then saw that the crack had grown to some 2 ft. He had initiated evacuation of the area when the column failed. The column tore apart at the 10 ft level weld and the 45 ft top section, weighing some 20 te, rocketed up, landing about 975 m away on a tower carrying a 138 kV power feed line to the area. The tower contents, estimated as 79.5 m^3 of propane/butane and 8.75 m^3 of MEA, escaped and formed a vapour cloud, which ignited within seconds, giving a vapour cloud explosion. Some 30 min later a third explosion occurred when a vessel in the alkylation section suffered a BLEVE with associated fireball and missiles. The explosions and fire caused extensive damage. Seventeen people were killed and 31 injured.

The vapour cloud explosion disrupted the electrical power supply, putting out of service a 2500 USgal/min electrical fire pump. One of the explosions sheared off a fire hydrant, reducing the pressure available from the two 2500 USgal/min diesel-driven fire pumps. A fragment from the BLEVE landed on a unifining unit, initiating a major fire there.

The blast resistant control centre, some 400 ft from the absorber, suffered little structural damage.

The absorber vessel had a history of hydrogen attack problems. Over the years since 1972 repairs and modifications had been done, and the vessel continued in service.

Mahoney (1990); D.J. Lewis (1993); Vervalin (1987)

A112 *Priola, Italy, 1985*

On 19 May a major fire occurred on an ethylene plant at Priola in Italy. A faulty temperature probe initiated isolation of the hydrogenation unit in the cold section, and while the operators were trying to re-establish control, the relief system operated. At the same time fire was observed at the base of the de-ethanizer column. The hydrocarbon released ignited and an intense fire engulfed the adjoining ethylene and propylene distillation columns and spread to the storage area. The water deluge system protecting the storage tanks proved inadequate due to the intensity of the fire. In due course a tall, vertical propane tank exploded, its top section rocketing up some 500 m, and just missing a gasholder. Two propylene tanks fell over, one on a pipe rack and the other against an ethylene tank. In all, five of the eight ethylene and propylene tanks either exploded or collapsed.

Mahoney (1990)

A113 *Schweizerhalle, Basel, Switzerland, 1986*

On 1 November 1986, a fire broke out in Warehouse 956 at the Schweizerhalle (Muttenz) works of Sandoz in Basel, Switzerland. The warehouse held large quantities of agrochemicals. Attempts to extinguish the fire using foam were ineffective and water had to be used, in large quantities. It was not possible to contain this water within the site and, as a result, 10,000 m^3 entered the Rhine carrying with it some 30 te of chemicals, including an estimated 150 kg of highly toxic mercury compounds dissolved in aqueous concentrates. The incident did severe ecological damage.

A not dissimilar incident, though one on a much smaller scale, occurred at Morley, Yorkshire, in 1982, when the river Calder was contaminated by chemicals in water used in fighting a fire which had broken out in a warehouse storing herbicides (V.C. Marshall, 1987).

Anon. (1986v,w,y); Beck (1986); D.Williams (1986); Anon. (1987 LPB 75, p. 11); Stallworthy (1987)

A114 Thessalonika, Greece, 1986

On 24 February 1986, an oil terminal at Thessalonika, Greece, experienced a small fire when a oil spillage in a bund was ignited by hot work. The privately owned terminal had 12 fixed and floating roof storage tanks holding crude oil, fuel oil and gasoline. Over the course of seven days, ground fires escalated until they covered 75% of the terminal area and involved 10 of the tanks. The escalation of the initial small fire was due in large part to accumulation of oil from previous spillages; to leaks from flanges exposed to the fire, which then fed it; and to the failure of fire fighting efforts in the early stages. By the first day seven tanks were affected.

In the course of the succeeding days, several events occurred which led to major escalations. On Tank 3 overpressure caused the shell-floor seam to burst so that the whole contents flowed out, feeding the fire and involving two more tanks. Tank 8 suffered a boilover with a fireball 300 m high and ejection of burning oil over a wide area, some travelling up to 150 m. The fire fighting was hampered, and firemen endangered, by burn-back of flame in areas where the oil fire had already been extinguished by foam.

B. Browning and Searson (1989)

A115 *Antwerp, Belgium, 1987*

On 3 July 1987, an explosion occurred inside an ethylene oxide purification column at the BP Chemicals works at Antwerp, Belgium. The explosion was due to decomposition of the ethylene oxide and was accompanied by a fireball, which started a number of secondary fires. These, together with blast and missiles, caused extensive damage. Fourteen people were injured.

The investigators considered two main hypotheses for the cause. One was that there was a leak of ethylene oxide into the insulation, leading to self-heating and then ignition of the leak, resulting in the heating of the ethylene oxide in the column itself. The other was the development of a hot spot inside the column due to exothermic polymerization of ethylene oxide catalysed by rust. The first explanation was preferred.

Mellin (1991 LPB 100)

A116 *Grangemouth, Scotland, 1987*

In 1987, there occurred at BP plants in the Grangemouth, Scotland, area three incidents, which were the subject of investigations by the HSE (1989a).

Incident on 13 March 1987

On 13 March 1987, a fire occurred when a maintenance team broke a flange in the flare line, hydrocarbons escaped and were ignited. There was a quite serious fire which lasted some hours. Two men were killed and two injured.

The release occurred when two fitters started to break a flange in order to maintain a valve (V17). As they did so, vapour and a little liquid came out. They queried the situation with the process shift supervisor, who inspected the site and assured them that there could only be a small quantity present and it was safe to continue. They did so, and as the last bolt was undone, the crane holding the spacer increased its lift, the spacer sprang upwards, and liquid under pressure poured out. A vapour cloud formed and ignited. The source of ignition was almost certainly the engine of a nearby diesel air compressor.

It was considered that if the fire was extinguished, fuel might continue to issue with the potential for a vapour cloud explosion and that if the flame receded into the pipe, there might be an internal explosion. For a while, therefore, fuel gas was fed into the line to keep the flame alight.

The liquid which escaped had been held in a section of the flare line near a knockout drum. It should have drained via a small bore pipe into the knockout drum, but the entry of the pipe was blocked. The valve on this pipe had been checked as open, but the possibility of blockage had not been taken into account. The liquid was there in the first place because valve V17, thought to be fully closed, was in fact part open. The valves suffered from build-up of pyrophoric scale. They had no indicators showing positively the valve position. It is estimated that there was about $50 \, \text{m}^3$ of hydrocarbons in the line at the time and that $20 \, \text{m}^3$ may have escaped.

Among the findings of the investigation were that there was a need for position indicators on the valves; that there had been inadequate planning of the work to be done; that the check that the flare line had been drained was ineffective; that the response to the first, small leak was inadequate; that the procedure for flange breaking was incorrect; that when opening lines with pyrophoric scale, air should be excluded positively by a nitrogen purge, rather than reliance placed on its exclusion by flammable gases present; and that the exhaust gas spark arrester was missing. There should have been a requirement that in breaking a flange sufficient bolts remain in place whilst a gradual opening is made using a flange spreader.

Incident on 22 March 1987

On 22 March 1987 an explosion occurred during the start-up of a hydrocracker unit at the Grangemouth refinery. It involved the disintegration of a low-pressure separator. One fragment from the vessel, weighing 3 te, landed on the foreshore 1 km away.

Accompanied by a fireball, the explosion initiated a severe fire. The refinery sustained extensive damage. One person was killed.

The LP separator (V306) was 3.05 m diameter × 9.1 m long with a design pressure of 9 bar and a calculated burst pressure of 50 bar. It was fitted with a 3 in. relief valve designed for fire relief. Its disintegration was caused by breakthrough of gas from the HP separator (V305) due to loss of level control in the base of the latter. The HP separator pressure was 155 bar. The plant was in start-up mode and the process lines on V306 were closed.

The gas breakthrough occurred due to loss of control of the level at the base of the V305. The pipework taking off the liquid split into two lines, one passing through the valve on the level control loop LIC 3-22 and the other through that on H 3-22. The breakthrough occurred through the LIC 3-22 valve. This valve had been fitted with a low-level trip, but this had been removed in 1985. The liquid at the base of V305 tended to give vortices and the trip system suffered from spurious trips which caused production difficulties. There was an extra-low-level alarm, but this had been on for months, until the bulb failed, and was regarded as spurious. The two float switches were fire damaged, but there was evidence that one had been incorrectly assembled and that on the other the small bore piping was blocked. The operators tended to place more reliance on the measurement provided by a nucleonic level gauge. However, this gauge had a zero error, showing a 10%

reading when the true reading was zero. An audit had shown that there was a problem in control of the level on the vessel. In the shift in which the incident occurred valve LIC 3-22 was opened manually at least three times. Just before the explosion, it was opened again, and the level was lost. The HSE report states:

> The refinery had procedures for routine monitoring of interlocks, alarms and trips, but on the checklist for the hydrocracker some were omitted. The detection, trip and alarm systems for extra-low level in V305, had been inoperative for a long time and maintenance staff and operators presumed that these were no longer required. Training of new operators carried out by experienced operators helped to perpetuate this misconception. Although the refinery chief instrument engineer noted in 1985 that the LIC 3-22 trip solenoid had been disconnected, this was not followed up. (paragraph 89)

The report states that there should have been both a high-integrity trip system for level control and a pressure relief system capable of handling the maximum anticipated gas flow and that fuller consideration should have been given to proposals to make modifications to the plant.

It emerged that there had been a previous incident on V305 which was probably a gas breakthrough, but on that occasion the process valves on V306 were open.

The force of the blast was mainly downwards, while the fragments travelled upwards. The TNT equivalent of the explosion was estimated as 90 kg.

Major problems were experienced in the fire fighting. Within 12 h 31 foam and water streams were discharging at a rate of 12,000 USgal/min. The drains were unable to cope with this flow. Hydrocarbons bubbled up in other parts of the site and floated on the water. The situation was exacerbated by large quantities of heavy, waxy material which caused blockages.

Incident on 11 June 1987

On 11 June 1987 at Dalmeny, near Grangemouth, a fire broke out during sludge cleaning work by a contractor in a storage tank, and one man died. The men had airline breathing apparatus (BA) to protect against toxics, but there was no mechanical ventilation and no regular monitoring of the atmosphere. The sludge had a flashpoint of less than 0°C. It transpired that the fire broke out when one man, who had taken off his BA, dropped a cigarette end.

The investigation revealed that there were a number of deficiencies in the arrangement of the contract with the contractor and in the conduct of its personnel. The contractor was a specialist one, well known, considered competent and already working on site. There was no written procedure for the work. There was inadequate enforcement of the 'no smoking' rule, and some contractor's personnel had not received the BP refinery rules booklet. Also inadequate was enforcement of the use of the BA. There was no lighting inside the tank and men discovered they could see better if they took the BA off. The team worked largely unsupervised. They wore the BA when entering and leaving the tank, and evidently also when the supervisor was about to visit, but some did not wear it all the time. The supervisor himself had on occasion removed his BA inside the tank to give verbal instructions.

Anon. (1988c,d); Anon. (1987l); Anon. (1989b); Anon. (1989 LPB 89, p. 13); HSE (1989a); Kletz (1989b); Mahoney (1990); K.X.Wilson (1990a,b); D.J.Lewis (1993)

A117 *Pampa, Texas, 1987*
On 17 November 1987 at Pampa, Texas, a vapour cloud explosion occurred on a plant for the liquid phase oxidation of butane to acetic acid and acetic anhydride using air at about 700 psig. There were three reactors with a capacity of 38 m^3. During the start up of one reactor a rupture or explosion occurred in the line taking the air to the reactor or in the reactor manifold, and butane and acetic acid flowed out. The vapour cloud which formed was ignited by gas-fired boilers some 60 m away, the delay before ignition being about 10 s. The resultant explosion did extensive damage. One aspect of this was that the blast severed piping in many of the sprinkler systems and also ruptured the underground fire water main, rendering the water pressure inadequate for its purpose. Three people were killed and 37 injured.
Mahoney (1990); D.J. Lewis (1993)

A118 *Port Herriot, Lyon, France, 1987*
On 2 June 1987, the Shell depot at Port Edouard Herriot, near Lyon, France, with some 76 tanks, experienced a severe fire. Two persons were killed and eight injured.
At the time of the account quoted the cause of the incident was not established. It states that at the relevant time of 13.15 work was in hand on an electric cable for welding equipment and that the power was off but due to be switched on again at 13.30. The initial event was the formation of a spray in the area of the pumps. There was then a flash.
There followed a series of explosions and fires. Twelve minutes into the event a tank holding 250 m^3 rocketed up 200 m and fell outside the compound. There was a rapid escalation with five further tanks rocketing and falling inside the depot. Tank 6 exploded in what was probably a boilover. The associated fireball was 450 m high and 200 m 'wide'. During the night the fire progressed further.
The extent of the escalation was attributed in large part to the fact that the tanks were old, and did not have a rupture seam in the roof, so that they tended to burst at the base, releasing the whole contents.
Control of the fire was not achieved until 11.00 on 3 June after two heavy duty foam monitors, each of 6 m^3/min capacity, obtained from outside, had been brought into play at 6.30. Among the conclusions drawn in respect of fighting such a fire were that the temperature of the wall of the vapour space of a tank engulfed in fire can rise within 2 min to 500°C, above the autoignition temperature for many flammable mixtures; that water cooling used while the foam attack is being prepared may be ineffective and may cause severe degradation of structures; that 4-hour fire insulation may not be sufficient; that concrete bunds may not last out and that earth mounds are preferable; and that an effective foam attack requires heavy duty equipment capable of 4–6 m^3/min.
Another conclusion was the need to review the fire permit system: the number, the nature and relations between the permits.
Mansot (1989)

A119 *Ras Tanura, Saudi Arabia, 1987*
On 15 August 1987, a leak occurred on a gas plant at Ras Tanura, Saudi Arabia. There were two parallel gas fractionation trains with propane feed. Prior to the incident there had been a number of power interruptions which caused shut-down of one or other of the plants. It is believed that a leak of propane developed in one train, the probable

source being a flange on a 4 in. relief line and that the release continued for about half an hour. A large vapour cloud formed and ignited. The ignition was probably from a security vehicle which had stalled and was being restarted. Severe damage was caused.
Mahoney (1990)

A120 *Floreffe, Pennsylvania, 1988*
On 2 January 1988, an oil storage tank of the Ashland Oil Company at Floreffe, Pennsylvania, suffered catastrophic failure during its initial filling. The tank was a 120 ft diameter × 48 ft high cone roof tank. It was 40 years old and had been dismantled at a former site and re-erected. The failure was a sudden rupture. The force of the fluid spurting out one side caused the tank to move 100 ft backwards off its foundations. Some 92,400 bbl of diesel fuel were spilled. The oil overflowed the bund and much of it went into the river.
The investigation found that the rupture occurred due to low temperature embrittlement initiated at a flaw in the tank shell base metal, about 20 cm up from the bottom. The flaw had been created by an oxyacetylene cutting torch and had been there since the initial fabrication. The actual fracture was in the shell base metal; it did not follow the welds but crossed them. In its previous service, the tank had stored heated product, while in its new service the product was not heated. It was the first time that in its new service the tank had been filled to maximum capacity in cold weather.
Prokop (1988); Mahoney (1990); Wilkinson (1991 SRD R530); Mellin (1992 LPB 106)

A121 *Norco, Louisiana, 1988*
On 5 May 1988, an explosion occurred on a catalytic cracker at the refinery at Norco, Louisiana. There was a rupture, due to internal corrosion, on an 8 in. elbow in the depropanizer overhead piping, which released 9 te of hydrocarbons. A vapour cloud formed and after about 30 s found a source of ignition, probably the FCC charge heater. The resultant vapour cloud explosion did extensive damage and caused immediate failure of all utilities. It put out of action fire-water and the four diesel fire pumps, and for several hours the fire fighting effort was limited. Some 20 major vessel or line failures occurred. Analysis of bolt stretch on columns indicated an overpressure up to 10 psi. Seven people were killed and 28 injured.
Mahoney (1990); D.J.Lewis (1993)

A122 *Antwerp, Belgium, 1989*
On 7 March 1989, the BASF ethylene oxide plant at Antwerp experienced two explosions, each with a fireball. The first occurred in column K303, separating ethylene oxide and acetaldehyde, which disintegrated. The pipework on another column, K302, fractured, so that the column depressurized and flame flashed back into it, causing a second internal explosion, separated from the first by 26 s. The blast, missiles and secondary fires caused extensive damage. Five people were injured.
The investigation found that low cycle fatigue had caused a hairline crack in a welded seam on piping to a level indicator system, resulting in a small leak of ethylene oxide. This led to an accumulation of auto-oxidizable polyethylene glycols in the insulation. Removal of the metal covering of the insulation allowed self-heating to proceed. The insulation reached a temperature which was sufficient to cause decomposition of ethylene oxide stagnant in the piping

of the level control system, though it would not have been enough to overcome the cooling effect of flowing vapour.

Mahoney (1990); Kletz (1990e); Anon. (1991 LPB 100, P. 1)

A123   *Baton Rouge, Louisiana, 1989*

On 24 December 1989, an explosion occurred in a refinery at Baton Rouge, Louisiana. An 8 in. diameter pipeline ruptured, releasing a mixture of ethane and propane. A record low temperature (10°F) may have contributed to the failure. The escaping vapour found a source of ignition and there was a vapour cloud explosion. The explosion ruptured a pipe band containing some 70 lines. These included lines for power, steam and firewater, and there was a partial loss of all these utilities. The extensive fire involved two large storage tanks, 12 small tanks and two separator units.

Mahoney (1990)

A124   *Jonova, Lithuania, 1989*

On 20 March 1989, a refrigerated atmospheric ammonia tank failed at Jonova, Lithuania, leading to both a toxic release and a fire of ammonia as well as fire in a fertilizer store. Seven people were killed and 57 injured.

The tank was 29 m diameter × 20 m high of single wall construction with perlite insulation held in place by an outer steel shell and with a surrounding 14 m high reinforced concrete wall. It had a capacity of 10,000 m^3 and at the time held 7000 te of liquid ammonia at −33°C. The capacity of each of the two pressure relief valves was 4200 m^3/h. The site also had two fertilizer stores, one for 15,000 te and one for 20,000 te as well as one for 20,000 te of ammonium nitrate.

The rupture was sudden and the force of the release caused the tank to move sideways from its base, smash through the concrete wall and fetch up 40 m away, leaving its bottom still on the foundations. The devastation around the tank was severe. The escaping liquid ammonia formed a pool which in places was 0.7 m deep. The vapour above the pool suddenly caught fire and the whole area was engulfed in flames. The fire set alight material on a conveyor belt to the 15,000 te store of NPK fertilizer, a 'cigar burning' material. The burning conveyor fell into the store and initiated a self-sustaining decomposition.

After 12 h the pool of ammonia had evaporated. The fertilizer continued to burn for 3 days.

The plant was 12 km from Jonova, a town of about 40,000 inhabitants. A large cloud of ammonia and nitrous fumes spread 35 km, covering an area of 400 km^2. The cloud height is described as being 100, 400 and 800 m at distances of 5, 10 and 20 km, respectively. Some 32,000 people were evacuated.

The events leading up to the rupture were as follows. Due to an error, 14 te of warm ammonia at 10°C were transferred to the tank, where they formed a layer on the bottom. In due course this layer rose suddenly to the surface. The event was akin to a 'rollover', although this term is normally associated not with a pure liquid but with one of varying composition, such as LNG. The higher vapour pressure of the warmer liquid caused a sudden rise in pressure in the tank, which the pressure relief valves were unable to handle, and the tank burst. At the time the refrigeration compressors were not working, but this probably played little part in the incident.

Kletz (1991a); Andersson (1991a,b); Anon. (1992 LPB, p. 11); Andersson and Lindley (1992 LPB 107); Kukkonen *et al.* (1992); Wilday (1992 LPB 108, 109); Kukkonen *et al.* (1993)

A125   *Peterborough, UK, 1989*

On 22 March 1989, an explosion occurred in the yard of Vibroplant Ltd at Peterborough on a ICI truck carrying commercial explosives. The driver had missed his way and shortly after 9.30 a.m., approximately, entered the yard to turn round, when, as the vehicle passed over a speed ramp, there was a minor explosion inside the load compartment which blew the rear roller shutter door open. The driver stopped the vehicle, and he and the attendant got out to investigate. Smoke and flames were coming through gaps at the side of the door but no fire could be seen inside. They alerted the Vibroplant personnel in the yard and one of the latter telephoned the emergency services. At first only a small amount of black smoke was to be seen coming from the vehicle, but as the fire progressed there were minor detonations, or 'pops', and thick yellow smoke began to emerge. At 9.43–9.44 a.m. two fire tenders arrived. At 9.45 a.m. the load exploded. The scene following the explosion is shown in Figure A1.18. One person was killed and 107 were injured, of whom 84 were treated in hospital.

The load was a mixed one, consisting 150 kg of Powergel 800; 500 kg of Powergel E800; 56 kg of Magna Primers; 76 kg of Ammon-Gelit; detonators and fuseheads. The latter included uncut Cerium combs, 2400 in three boxes. The total mass of explosive had a TNT equivalent of some 800 kg.

The cause of the explosion was determined as a box of Cerium fusehead combs which were in unsafe packaging.

The report of the HSE (1990b) gives details of the damage and injury done by the event. With respect to damage, the analysis includes the following:

	Distance (m)	Overpressure (psi)
Clean area	14	78
Fireball	18	44
Total destruction of cavity brick/block walls of steel framed building	30	14
Serious damage to concrete frames of building	110	1.7
Window breakage:		
90%	≈225	0.69
50%	≈360	0.37
Up to	≈580	0.19

The clear area is the area of the yard which was cleared of cars, etc. The report also gives a map showing the fall of major fragments.

With respect to injury, plans are given showing the location of the persons exposed and of those who were blown over or suffered eardrum rupture. Two of the plans are shown in Figure A1.19.

At the instant of the explosion there were some 31 persons within 60 m of the vehicle. The plans show

Number of persons			
Distance (m)	<20	20–40	40–60
Exposed	3	12	31
Blown over	2	10	15
Eardrum rupture	2	4	

Figure A1.18   *Peterborough, 1989: Vibroplant works yard after the explosion (Courtesy of the Peterborough Evening Telegraph)*

The person who died was a fireman who was engulfed in the fireball, which extended 18 m. Another fireman at 25 m suffered minor burns. Other information on injuries includes

	Distance (m)	Overpressure (psi)
Burns:		
Fireman engulfed in flames	18	44
Fireman suffered slight burns	25	21
Eardrum rupture:		
100%	28	17
50%	30	15
Up to	45	6.6
Blown off feet:		
50%	55	4.7
Up to	70	3.2
Severe injury from fragments other than glass	40	
Cuts from glass:		
Cuts to all	0–50	>5.5
Cuts to many	70–100	3.2–1.9
Cuts to some	100–150	1.9–1.15
Cuts up to	≈200	0.8

Further details are given in the HSE report.
HSE (1990b); Anon. (1991 LPB 98, p. 24)

A126   *Richmond, California, 1989*
On 10 April 1989, in a refinery at Richmond, California, a 2 in. line carrying hydrogen at about 2800 psi suffered a failure, apparently at a weld, giving rise to a jet flame, which impinged on the support of a reactor, 100 ft high and at least 10 ft diameter and with a wall thickness of 7 in., in a hydrocracker unit. The resultant failure of the reactor led to extensive damage.
Mahoney (1990)

A127   *Ufa, Soviet Union, 1989*
On 4 June 1989, a massive vapour cloud explosion occurred in an LPG pipeline at Ufa in the Soviet Union. A leak had occurred in the line the previous day or, possibly, several days before. In any event, the engineers responsible had responded not by investigating the cause but by increasing the pressure. The leak was located some 890 miles from the pumping station, at a point where the pipeline and the Trans-Siberian railway ran in parallel through a defile in the woods, with the pipeline some half a mile from, and at a slightly higher elevation than, the railway. On the day in question the leak had created a massive vapour cloud which is said to have extended in one direction five miles and to have collected in two large depressions.

Some hours later two trains, travelling in opposite directions, entered the area. The turbulence caused by their passage would promote entrainment of air into the cloud. Ignition is attributed to the overhead electrical power supply for one or other of the trains. There followed in quick succession two explosions and a wall of fire passed through

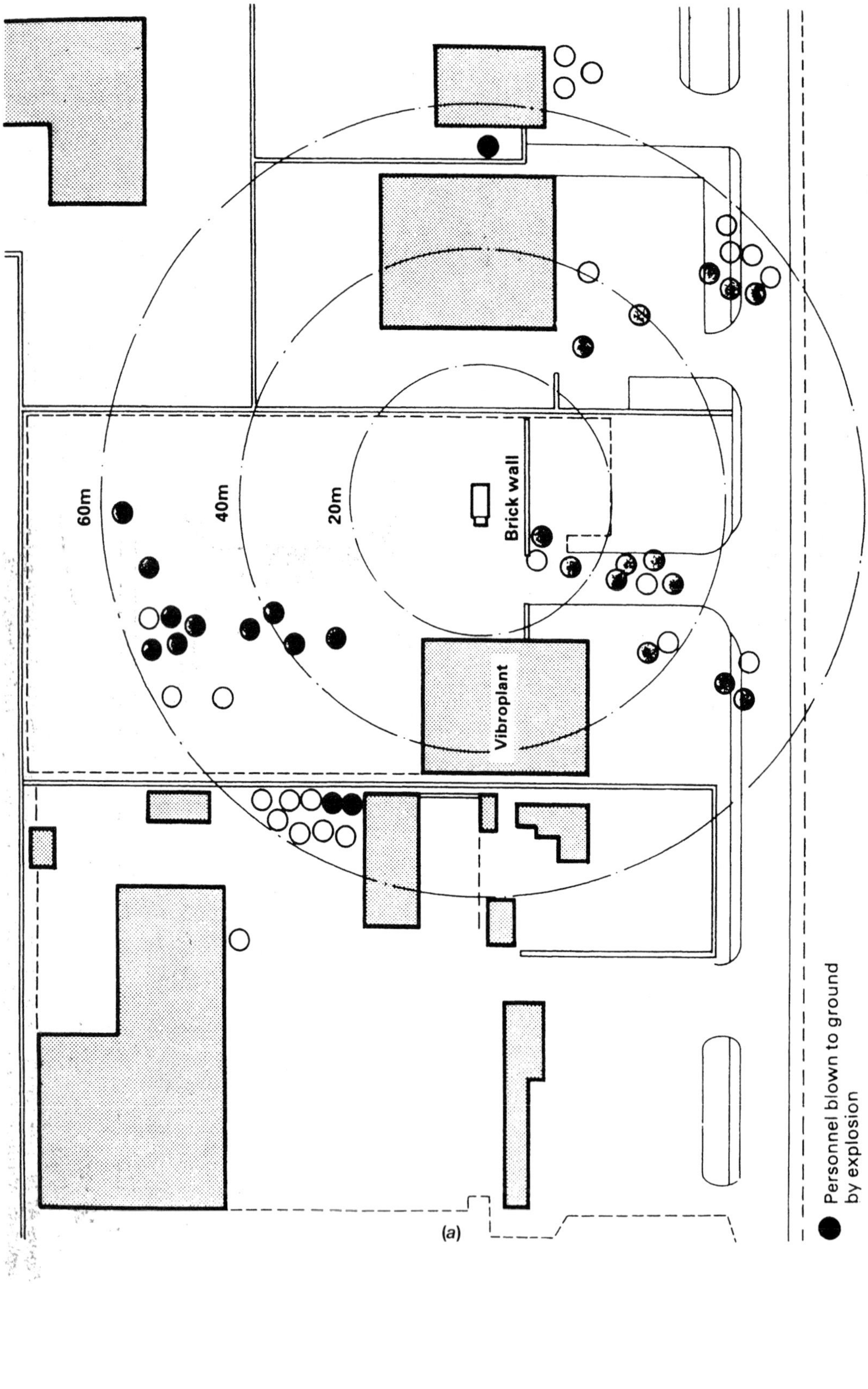

60m

40m

20m

Brick wall

Vibroplant

(a)

● Personnel blown to ground
   by explosion

Figure A1.19 Continued

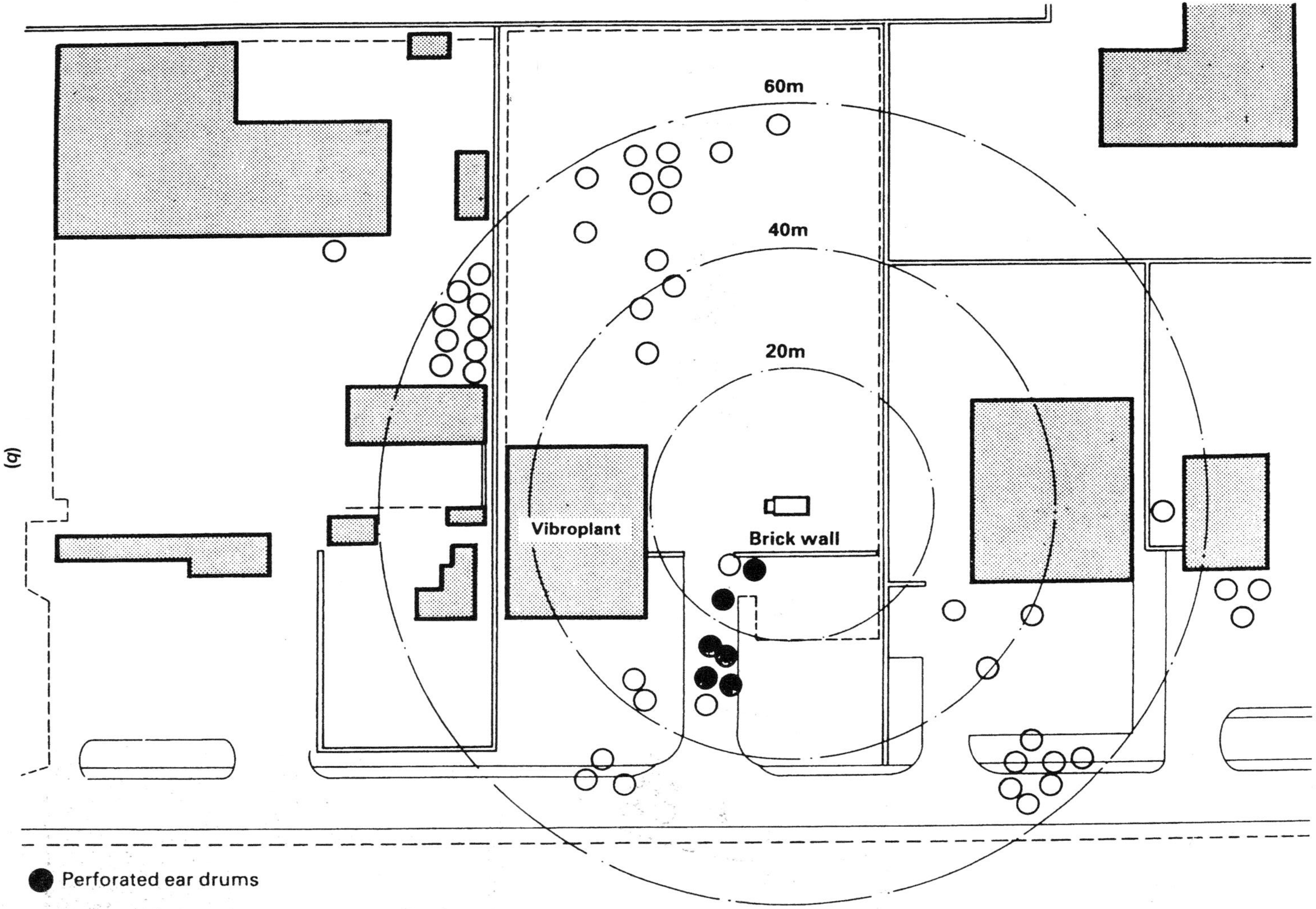

Figure A1.19 Peterborough, 1989: site plan showing location of certain injuries (Health and Safety Executive, 1990c): (a) persons blown to the ground by the explosion; and (b) persons suffering perforated eardrums (Courtesy of HM Stationery Office)

60m

40m

20m

Brick wall

Vibroplant

● Perforated ear drums

(b)

the cloud. Large sections of each train were derailed and the derailed part of one may have crashed into the other. The death toll is uncertain, but reports at the time gave the number of dead as 462 and of those treated in hospital as 706, many with 70–80% burns.

Hofheinz and Kohan (1989); D.J. Lewis (1989 LPB 90, 1993)

A128 *Bangkok, Thailand, 1990*
At 22.30 on 24 September 1990, a road tanker carrying LPG was involved in a traffic accident at a busy road junction in the centre of Bangkok. Some 5 te of LPG was released but did not ignite immediately. When the cloud did ignite, there was evidently a flash fire, but accounts spoke also of an explosion, probably from gas which had entered a nearby building. In the worst affected area, almost all the shop houses on both sides of the street were destroyed. Reports gave some 68 persons dead and over 100 injured.

The vehicle was a flat bed lorry carrying two LPG tanks. The two tanks were interconnected by two lines and it appears that these were severed in the accident. The tanker had no license to carry LPG.

Anon. (1991 LPB 101, p. 19)

A129 *Chalmette, Louisiana, 1990*
On 3 November 1990 in a refinery at Chalmette, Louisiana, failure involving a heat exchanger shell led to a release of hydrocarbons. A vapour cloud formed, was ignited by a heater and gave a vapour cloud explosion which did extensive damage.

Marsh and McLennan (1992)

A130 *Nagothane, India, 1990*
On 6 November 1990 at Nagothane, India, a leak occurred on a pipeline transporting ethane and propane to a gas cracker complex. A vapour cloud formed and ignited at an offsite gas treatment and compression facility. The cracker was not damaged by the resultant vapour cloud explosion, but serious damage was done to the offsite units. There were 31 deaths.

Marsh and McLennan (1992)

A131 *Stanlow, Cheshire, 1990*
On 20 March 1990, a reactor at the Shell plant at Stanlow, Cheshire, exploded. The explosion was due to a reaction runaway.

The vessel was a 15 m^3 batch reactor containing some 13.5 te of reaction mixture. The reaction was the Halex production of 2,4-difluoronitrobenzene (DFNB), using *N*, *N*-dimethylacetamide (DMAC) as solvent. The DFNB was then reacted to obtain the final product 2,4-difluoroaniline (DFA).

The batch had been heated from ambient to 165°C over 3 h, in the normal way. It then rose above 170°C, instead of starting to dip slightly, and, unusually, began to vent gas. In the following 10 min the heating was switched over to cooling, but the temperature and pressure in the reactor continued to rise, at an increasing rate. The control room operators were unaware of the runaway; they were looking at a different display on the VDU. They were alerted to it by an outside operator.

The pressure in the vessel reached a value later assessed as about 60–80 barg, which compares with the relief valve set pressure of 5 barg. The event generated blast, a fireball and missiles. The vessel body unwrapped into a flat plate and the cover was hurled 200 m; other missiles went up to 500 m. The reactor structure collapsed, the fluoroaromatics plant was devastated and structural damage was done to nearby buildings. One man subsequently died, and five others were injured.

There was evidence, from an eyewitness and from a severely burned lamp fitting which became a missile (Missile 47), that prior to the explosion there had been a jet flame above the reactor for some 30–120 s before the explosion, the leak coming from a flange or crack. The blast damage in the works had some unusual characteristics, decreasing only slowly with distance. The reactor had a vapour space of 3 m^3, but the blast damage was much greater than could be accounted for by a physical explosion involving such a volume, and a combustive event was indicated. Analysis pointed to an event giving a long impulse with a duration time of perhaps 400 ms, compared with, say, no more than 50 ms for a vapour cloud explosion on the one hand and 10 s for an unconfined fireball on the other. The relatively long duration of the explosion would be due to the rate limitation imposed by air entrainment. The investigators called the event which occurred a 'highly congested fireball'. The fetch of the missiles pointed to a pressure at the source in the range 60–80 barg. It was estimated that some 10 te of material had been ejected from the reactor at about 70 barg in a congested area and had been ignited by the jet flame above the vessel acting as a strong ignition source. These 10 te had an energy of combustion of 230,000 MJ; the event was consistent with an energy release of about one-fifth of this.

Much of the damage done was due to the rarefaction wave. Many components were more sensitive to a negative that to a positive pressure deviation.

The investigation found that the runway was due to the presence of acetic acid. This was detected by smell in the contents of a vent knockout vessel, and, much later, it was identified in a sample of the DMAC from the batch. Investigation revealed a rather complex chemistry. It showed that, when added to a Halex reaction mixture, acetic acid causes exothermic reaction and gas evolution. The DFNB process involved a later stage of batch distillation in which the successive fractions were toluene, DMAC and DFNB. The investigators discovered that during one such batch water had entered the still via a leaking valve. The water had been removed by prolonged azeotropic distillation, using toluene. Under these conditions, DMAC undergoes slow hydrolysis, giving dimethylamine and acetic acid. However, for there to be any significant yield of acetic acid, the presence of DFNB is necessary, since this reacts with the dimethylamide, and thus shifts the equilibrium. On this occasion, the DMAC had then been further distilled to purify it. It turned out, however, that DMAC and acetic acid form a maximum boiling azeotrope with a boiling point close to that of pure DMAC. The presence of the acetic acid in the DMAC was not detected by the measurement of boiling point nor by the particular gas chromatograph method in use. Thus the water ingress incident evidently led to a batch of recycled DMAC which was contaminated with acetic acid, with the consequences described.

Anon. (1991k); Kletz (1991 LPB 100); Mooney (1991); Redman and Kletz (1991); Cates (1992); van Reijendam *et al.* (1992)

A132 *Coode Island, Mevaporourne, Australia, 1991*
On 21 August 1991, an explosion occurred at A Terminal of Terminal Pty at Coode Island, Mevaporourne, Australia. The site involved had 45 storage tanks, none pressurized, with a vapour recovery system. A 230 te acrylonitrile tank was lifted off its base and projected over four other tanks into the forecourt. There followed a series of bund and tank fires and tank explosions, at the end of which only 13 tanks were left undamaged. There were no deaths or serious injuries.

The cause of the initial explosion has not yet been established. Some of the explanations considered are rehearsed by I. Thomas (1991), including his own of exothermic polymerization in the vapour recovery system. A critique of this explanation is given by Croudace (1993 LPB 112).

Anon. (1991b); I. Thomas (1991); Anon. (1992c); Alexander (1992a); Marsh and McLennan (1992); Croudace (1993 LPB 112)

A133 *Seadrift, Texas, 1991*
At 1.18 a.m. on 12 March 1991, an ethylene oxide redistillation column at the Union Carbide plant at Seadrift, Texas, exploded. A large fragment from the explosion hit pipe racks and released methane and other flammable materials. All utilities at the plant were lost. There was a substantial loss of firewater from water spray systems damaged or actuated by loss of plant air. The explosion and ensuing fire did extensive damage and one person was killed.

The plant had been down for routine maintenance. Start-up began in the late afternoon of 11 March, but the plant was shut-down several times by trip action before the cause was identified and rectified. Operation was finally established around midnight. The plant had been operating normally for about an hour when the explosion occurred.

The explosion was attributed to the development of a hot spot in the top tubes of the vertical, thermosiphon reboiler such that the temperature reached over 500°C instead of the normal 60°C, combined with a previously unknown catalytic reaction, involving iron oxide in a thin polymer film on the tube, which resulted in decomposition of the ethylene oxide.

A low recirculation ratio in the reboiler could give in some of the tubes low flow, loss of liquid film and stagnant vapour. The account by Viera, Simpson and Ream (1993) gives a detailed discussion of measures to ensure that all the heated surfaces of the reboiler are kept wetted.

Marsh and McLennan (1992); Viera, Simpson and Ream (1993)

A134 *Bradford, UK, 1992*
On 21 July 1992, a series of explosions leading to an intense fire occurred in a warehouse at Allied Colloids Ltd, Bradford. None of the workers at the factory was injured but three residents and 30 fire and police officers were taken to hospital, mostly suffering from smoke inhalation. The fire gave rise to a toxic plume and the run-off of water used to fight the fire caused significant river pollution.

The HSE investion (HSE, 1993b) concluded that some 50 min before the fire two or three containers of azodiisobutyronitrile (AZDN) kept at a high level in Oxystore 2 had ruptured, probably due to accidental heating by an adjacent stream condensate pipe. AZDN is a flammable solid incompatible with oxidizing materials. The spilled material probably came in contact with sodium persulfate and possibly other oxidizing agents, causing delayed ignition followed by explosions and then the major fire.

The warehouse contained two storerooms. Oxystore No. 1 was designed for oxidizing substances and Oxystore No. 2 for frost-sensitive flammable products; this second store was provided with a steam heating system. In 1991, an increase in demand for oxidizers led to a change of use, with both stores now being allocated to oxidizing products. A misclassification of AZDN as an oxidizing agent in the segregation table used led to this flammable material being stored with the oxidizers.

In September 1991, the warehouse manager, after discussions with the safety department, submitted a works order for modifications to the oxystores, including Zone 2 flameproof lighting, temperature monitoring equipment, smoke detectors and disconnection of the heater in Oxystore 2. An electrician made a single visit in which he did not disconnect the heater but simply turned the thermostat to zero. Although safety-related, the work was given low priority and 10 months later none of it had been started.

The explosion started at 2.20 p.m. and the first fire appliance arrived at 2.28 p.m. The fire services experienced considerable difficulties in obtaining a water supply adequate to fight the fire. At 3.40 p.m. power was lost on the whole site when the electricity board cut off the supply because the fire was threatening the main substation. The loss of power led to the shut-down of the works effluent pumps and escape of contaminated firewater from the site.

The fire services made early contact with the company's incident controller and strongly advised the sounding of the emergency siren, but this was not done until 2.55 p.m., when the incident had escalated. The fire gave rise to a black cloud of smoke, which drifted eastward over housing. The company stated on the day that the smoke was non-toxic. The HSE report, which gives a map of the smoke plume, states that 'it was in fact smoke from a burning cocktail of over 400 chemicals and only some of them would have been completely destroyed by the heat of the fire'.

The HSE report cites evidence that the warehouse had not been accorded the same safety priority as the production functions. It came under the logistics department, none of whose 125 personnel had qualifications as a chemist or in safety.

HSE (1993b); Anon. (1994 LPB 119, p. 15); V.C. Marshall (1994 LPB 117)

A135 *Castleford, UK, 1992*
At about 1.20 p.m. on Monday, 21 September, 1992, a jet flame erupted from a manway on the side of a batch still on the Meissner plant at Hickson and Welch Ltd at Castleford. The flame cut through the plant control/office building, killing two men instantly. Three other employees in these offices suffered severe burns from which two later died. The flame also impinged on a much larger four-storey office block, shattering windows and setting rooms on fire. The 63 people in this block managed to escape, except for one who was overcome by smoke in a toilet; she was rescued but later died from the effects of smoke inhalation.

The flame came from a process vessel, the '60 still base', used for the batch distillation of organics, which was being raked out to remove semi-solid residues, or sludge. Prior to this, heat had been applied to the residue for three hours through an internal steam coil. The HSE investigation (HSE, 1993b) concluded that this had started self-heating of the residue and that the resultant runaway reaction led ignition of evolved vapours and to the jet flame.

The 60 still base was a 45.5 m³ horizontal, cylindrical, mild steel tank 7.9 m long and 2.7 m diameter. The still was used to separate a mixture of the isomers of mono-nitrotoluene (MNT, or NT), two of which (oNT and mNT) are liquids at room temperature and third (pNT) a solid; other by-products were also present, principally dinitrotoluene (DNT) and nitrocresols. It is well known in the industry that these nitro compounds can be explosive in the presence of strong alkali or strong acid, but in addition explosions can be triggered if they are heated to high temperatures or held at moderate temperatures for a long period.

The still base had not been opened for cleaning since it was installed in 1961. Following a process change in 1988 a build-up of sludge was noticed, the general consensus being that it was about 1820 l, equivalent to a depth of about 10 cm, though readings had been reported of 29 cm and, the day before the incident, of 34 cm. One explanation of this high level was that on 10 September the still base had been used as a 'vacuum cleaner' to suck out sludge left in the 'whizzer oil' storage tanks 162 and 163, resulting in the transfer of some 3640 l of a jelly-like material. The intent had been to pump this material to the 193 storage but transfer was slow and was not completed because the material was thick. The batch still was used for further distillation operations, which were completed on September 19. The still base was then allowed to cool and on September 20 the remaining liquid was pumped to the 193 storage.

On September 17 the shift and area managers discussed cleaning out the still base. The former had been told by workers that the still had never been cleaned out and he realized that the sludge covered the bottom steam heater battery. It was agreed to undertake a clean-out. The area manager gave instructions that preparations should be made over the weekend, but when he arrived on the Monday morning nothing had been done. He was concerned about the downtime, but was assured that this could be minimized and gave instructions to proceed.

At 9.45 a.m. the area manager gave instructions to apply steam to the bottom battery to soften the sludge. Advice was given that the temperature in the still base should not be allowed to exceed 90°C. This was based solely on the fact that 90°C is below the flashpoint of MNT isomers. However, the temperature probe in the still was not immersed in the liquid but in fact recorded the temperature just inside the manway. Further, the steam regulator which let down the steam pressure from 400 psig (27.6 bar) in the steam main to 100 psig (6.9 bar) in the batteries was defective. Operators compensated for this by using the main isolation valve to control the steam. This valve was opened until steam was seen whispering from the pressure relief valve on the battery steam supply line. This relief valve was set at 100 psig but was actually operating at 135 psig (9 bar), at which pressure the temperature of the steam in the battery tubes would be about 180°C.

The clean-out operation, which had not been done in the previous 30 years, was not subjected to a hazard assessment to devise a safe system of work, and there were defects in the planning of and permit-to-work system of the operation. The task was largely handled locally with minimal reference to senior management and with lack of formal procedures, although such procedures existed for cleaning other still bases on the site. The permits were issued by a team leader who had not worked on the

Meissner plant for 10 years prior to his appointment on September 7. At 10.15 a.m. he made out a permit for a fitter to remove the manlid. The fitter signed on about 11.10 a.m. and shortly after went to lunch. Operatives who were standing by offered to remove the manlid and the same team leader made out a permit for them to do so. When the fitter returned from lunch it was realized that the still base inlet had not been isolated and a further permit was issued for this to be done.

Meanwhile, the manlid had been removed. The area manager asked for a sample to be taken. This was done using an improvized scoop. He was told the material was gritty with the consistency of butter. He did not check himself and mistakenly assumed the material was thermally stable tar. No instructions were given for analysis of the residue or the vapour above it. Raking out began, using a metal rake which had been found on the ground nearby. The near part of the still base was raked. The rake did not reach to the back of the still and there was a delay while an extension was procured. The employees left to get on with other work and it was at this point that the jet flame erupted.

The HSE report states that analysis of damage at the Meissner control building at 13.4 m from the manway source indicated that at this building the jet flame was 4.7 m diameter. The jet lasted some 25 s and had a surface emissive power of about 1000 kW/m². The temperature at 6 m from the manway would have been about 2300°C.

The company employed some highly qualified staff with considerable expertise in the manufacture of organic nitro compounds. The HSE report describes some of the investigations of thermal stability, safety margins, etc., in which these staff were involved. It also comments in relation to the incident in question, 'Regrettably this level of understanding was not reflected in the decision which was made on 21 September when it was decided that the 60 still base would be raked out.'

As soon as the personnel at the gate office saw the flame one of them made a '999' emergency call. The employee requested the ambulance and fire services, but spoke only to the former before the call was terminated at the exchange. Thereafter incoming calls prevented further outgoing calls for assistance.

Just over a year before the incident the management structure had been reorganized. This involved replacing a hierarchical structure with a matrix management system, eliminating the role of plant manager and instituting a system in which production was coordinated through senior operatives acting as team leaders. The area managers had a significant workload. In addition to their production duties they had taken over responsibility for the maintenance function, which had previously been under the works engieering department. Managers were not meeting targets for planned inspections under the safety programme, and this was said to be due to lack of time.

C. Butcher (1994); HSE (1994b); Kletz (1994 LPB 119)

A136 Sodegaura, Japan, 1992

At 15.52 on 16 October 1992 at the Fuji Oil Sodegaura refinery at Sodegaura City, Japan, a large release of hydrogen occurred from a rupture on a feed/reactor effluent heat exchanger on the heavy oil indirect desulfurization unit as the plant was being started up after a shut-down. There was a delay of a few minutes, during which personnel took measures to try to stop the release, before the leak found a source of ignition and exploded. Ten were killed and seven injured.

The investigation found that the leak occurred at a gasket retainer on the heat exchanger as illustrated in Figure A1.20. The gasket retainer had shrunk due to repeated thermal cycling and it suffered wear at the point shown as X on the diagram. The defect was detected and during a shut-down the surface was repaired by grinding. However, the effect of this was to cause the retainer to ride beyond the groove so that when the retainer was heated during start-up, a gap opened up and the escape occurred.

High Pressure Gas Safety Institute of Japan (1993 LPB 116)

A137 *Hoechst Incidents, Germany, 1993*
On 22 February 1993, an accident occurred at the Griesheim works of Hoechst, which was unprecedented in the company's 130-year history. By April 2, there had been at five sites a series of no less than 14 incidents, comprising 11 releases, two fires and one explosion, of which three were considered serious by the company.

Incident on 22 February 1993
The first of these, at 4.14 a.m. on 22 February, was a reaction runaway which occurred when a reactor for the production of *o*-nitroanisol was charged with reaction mixture with the agitator off and the latter was then switched on. Some 10 te of material was discharged through two safety valves. The ambient temperature was $-2°C$ and most of the material solidified, but the residual gas cloud spread across the Main and affected an area of some 30 ha in Frankfurt.

The information available to the works manager, a materials safety data sheet conforming to the DIN standard, was that *o*-nitroanisol is 'slightly toxic', and he gave this description to the press. As it happened, the previous Friday this data sheet had been the subject of a revision, to the effect that the substance had been found to be carcinogenic to rats when administered in large doses over a long period. The revision had been known to the company's specialists since November 1992 but not to the works

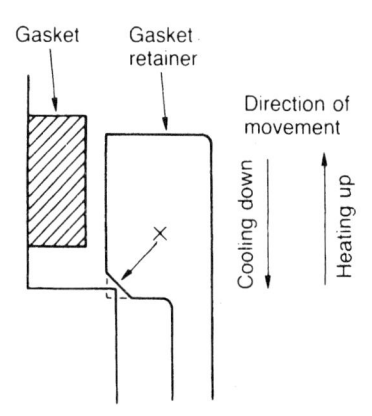

Figure A1.20 *Sodegaura, 1992: simplified diagram of heat exchanger gasket retainer. The gasket retainer suffered wear at the point marked X and repair was made by grinding, with loss of metal broadly as shown by the dotted portion (after the High Pressure Gas Safety Institute of Japan, 1993 LPB 116)*

manager. The toxicity of the substance was the subject of press comment.

Incident on 15 March 1993
The second serious incident, was on 15 March, on one of four parallel trains for the production of polyvinylalcohol. At the location in question in normal operation the material was in gel form on a conveyor belt inside an enclosure, but at the time the belt was stopped and empty. That morning the enclosure, which was normally kept inerted with nitrogen, was inspected. A precutter was still operating. An explosion occurred which killed the shift supervisor and injured the section manager. There followed a fire, which gave rise to a large column of smoke visible for some distance but which was rapidly brought under control.

An investigation by the authorities concluded that the enclosure had contained flammable vapours, that air had entered the enclosure during the inspection and that tramp metal had entered, giving rise to a spark at the precutter.

Incident on 2 April 1993
The third incident, on 2 April, involved a release of oleum. The release occurred when air pressure was applied to a vessel containing oleum and, due to a blockage in an outlet pipe, caused by crystallization, the oleum passed instead into an absorber where it reacted violently with the absorbent. A glass pipe ruptured and some 50 kg of oleum was released.

Owing to difficulties experienced with the alarm system some 13 people at the research centre came in contact with the cloud and received treatment at the medical centre.

Ondrey (1993a,b); Schonfeld (1993); Vennen (1993)

A138 *Dronka, Egypt, 1994*
On 2 November 1994, blazing liquid fuel flowed into the village of Dronka, Egypt. The fuel came from a depot of eight tanks each holding 5000 te of aviation or diesel fuel. The release occurred during a rainstorm and was said to have been caused by lightning. Reports put the death toll at more than 410.

Reuter (1994)

A139 *Ukhta, Russia, 1995*
Early in the morning on 27 April 1995, an ageing gas pipeline exploded in a forest in northern Russia. Reports described fireballs rising thousands of feet in the air and the inhabitants of the city of Ukhta, some eight miles distant, as rushing out in panic. At Vodny, six miles away, the sky was so bright that people thought the village was on fire. The pilot of a Japanese aircraft passing over at some 31,000 ft perceived the flames as rising most of the way towards his plane.

Anon. (1995)

A140 *Terminal Fire Savannah, Georgia, 1995*
On April 10, 1995, at approximately 11:30 p.m., explosions and fire occurred at Powell Duffryn Terminals, Inc. (PDTI), a commercial bulk liquid chemical storage and transfer facility, in Savannah, Georgia. Flames and thick black smoke from the fire forced the residents of the adjacent townhouse development to evacuate immediately. It took firefighters almost 3 days to finally put out the fire. The fire was within a concrete walled enclosure area containing six large storage tanks. During the fire, part of the enclosure wall was breached releasing contaminated firewater. The run-off from the fire contaminated an adjacent marsh

on the Savannah River resulting in a fish kill. (This first paragraph and the four following it were taken almost verbatim from the Executive Summary of the EPA's Chemical Emergency Preparedness and Prevention Office report entitled *EPA Chemical Accident Investigation Report*, PDTI. It is EPA Report 550-R-98-003.) (A summary report by David Chung entitled *Explosion and Fire* at Powell Duffryn Terminals – Savannah, Georgia was also used.)

After the fire, chemicals leaking from the storage tanks in the enclosure area reacted and produced toxic hydrogen sulfide gas. The hydrogen sulfide gas release forced residents within one-half mile of the facility to evacuate. As many as 2000 people were involved in the evacuation. An elementary school nearby was also forced to close. Approximately 300 people went to hospital emergency rooms complaining of symptoms attributed to hydrogen sulfide exposure. For many nearby residents, the evacuation lasted more than 30 days because of the continued evolution of hydrogen sulfide gas from the PDTI site. After the incident, extensive cleanup of the site and neighboring area was required.

Powell Duffryn Terminals Inc is fully enclosed by a security fence with locked gates. On April 10, 1995, the last employee left the site for the day at 5:50 p.m. The gates were locked, and no employees were on-site until after the explosions and fire had occurred.

On the day of the fire, contractor employees had been installing a sealed foam chamber on the storage tanks containing crude sulfate turpentine (CST), a flammable liquid. This closed the CST tanks to the atmosphere and directed CST vapour to the vapour control (VC) system. The VC system was designed to control fume and odour from the CST by capturing the CST vapour using activated carbon in two 50-gallon drums connected to the CST tanks using PVC piping. According to PDTI modification plans, each CST storage tank was supposed to be equipped with a flame arrester at its connection to the PVC piping. These flame arresters had been delivered but not yet installed. In addition, a fixed-piping foam fire protection system was not completed at the time of the fire.

The explosions and fire at PDTI involved CST, which the facility began to store on 17 January 1995. Prior to 1995, the facility was permitted only for storing non-flammable liquids. The facility had not completed modifications to accommodate the storage of flammables when the fire occurred. The CST was stored in three storage tanks (two 237,000 and one 422,000 gallon capacity) in a walled enclosure that contained a total of six tanks. The three CST storage tanks were connected by the partially completed VC system used to remove CST vapour from any venting that may occur. In the same enclosure were three other storage tanks (one 340,000 and two 323,000 gallon, respectively) containing sodium hydrosulfide solution (pH 10.4 to 11.5); Briquest, an acidic cleaning solution (pH of 1) and Antiblaze 80, a fire retardant chemical. These tanks and associated pipes were damaged by the explosions and fire and leaked their contents into the six-tank enclosure area. Reaction of the sodium hydrosulfide with acids present in the enclosure area produced hydrogen sulfide, a toxic, foul smelling gas (EPA May 1998).

The EPA Chemical Accident Investigation Team identified the following as root causes and contributing factors in the accident:

- *The design for the Vapour Control System was inadequate.* There is a documented history of fires in drums

containing activated carbon where the organic vapour control design permitted the backflow of air through the drums. Organic sulfur compounds in CST can produce heat when they are adsorbed by the activated carbon. Due to the incomplete vapour control installation, air was permitted to backflow into the CST contaminated drums of activated carbon when the ambient temperature drops and the vapours within the tank contract or if the tank is pumped out.

It was a very hot afternoon when a tank truck of CST was unloaded. As the evening temperatures cooled, the tank vapour space cooled and the vapour contracted. Air drawn into the drums provided oxygen for the spontaneous combustion.

- *The storage tanks were not equipped with flame arresters.* The lack of these purchased, but not-yet-installed protection devices in the vent piping allowed the fire to travel from the carbon bed to the tank vapour space.

- *The foam fire suppression piping system was not completed on the tanks containing CST.* Despite the fact that flammable CST was being stored, the designed foam suppression system was incompletely installed.

- *Incompatible chemicals were stored in the same walled enclosure area.* A large quantity of sodium hydrosulfide solution was stored in the same enclosed area as an acidic cleaning solution. When the tanks leaked the chemicals reacted generating toxic hydrogen sulfide (H_2S). This noxious H_2S gas caused injuries, forced evacuations and hampered a quick response and cleanup.

For detailed recommendations, see either the EPA Report or David Chung's article.

A141   *West Helena, Arkansas, 1997*

On the afternoon of 8 May 1997, a massive explosion and fire occurred at Unit Two of the Bartlo Packaging Incorporated (BPS) facility located in West Helena, Arkansas. Three West Helena firefighters were killed and 17 required medical attention during the response. (This write-up is taken directly from EPA/OSHA Joint Chemical Accident Investigation Report, EPA 550-R-99-003. April 1999. Portions are taken word for word from the report.)

Hundreds of residents and patients at a local hospital were either evacuated or sheltered-in-place. The Mississippi River traffic and major roads were closed for approximately 12 h due to the release of toxic materials from the facility.

BPS employees observed smoke in the Unit Two warehouse and all employees evacuated the building prior to the explosion. Members of the West Helena Fire Department responded to the scene within minutes. A reconnaissance team of four firefighters was outside of the building when an explosion occurred in the warehouse. Three firefighters were killed when they were struck by materials from a falling wall. The fourth firefighter was injured seriously.

EPA and the OSHA conducted a joint investigation of the incident. The team determined that the incident was most likely caused by the decomposition of a bulk sack containing the pesticide Azinphos methyl (AZM) 50W which had been placed against or close to a hot compressor discharge pipe. Under this scenario, the heat from the discharge pipe would have caused the pesticide to decompose and give off flammable vapours, which resulted in the fatal explosion.

A142   *Longford, Victoria, Australia, 1998*

On Friday September 25, 1998, about 12:26 in the afternoon a reboiler ruptured at the Longford Gas Plant. The fracture released hydrocarbon vapours and liquid. The vapours travelled towards the area of the fired heaters where it ignited.

Two gas employees were killed and eight others were injured. Property damages were estimated initially to be over $187,000. Supplies of natural gas to domestic and industrial users were interrupted between 9 and 19 days. The State of Victoria, which is highly dependent on natural gas, suffered substantial disruption to the economy.

The unit in which the release and fire originated was designed to remove ethane, propane, butane and other higher chain hydrocarbons by absorbing them in 'lean oil'. The oil, now saturated with light hydrocarbons was called 'rich oil.' 'Rich oil' was distilled to release these hydrocarbons and it was now lean again. The exchanger that ruptured was the reboiler. The cold rich oil was in the tubes and heated by warm lean oil.

Investigators determined that the loss was attributed to the fracture of a heat exchanger following a process upset that was set in motion by the unintended, sudden shutdown of warm lean oil pumps. Following the loss of the lean oil flow, the condensate flowing from the absorbers was flashing at lower and lower temperatures. Ice formed on the outside of the heat exchanger. Simulations indicate that the temperature reached as low as $-48°C$ $(-54°F)$. The reduction in temperature caused embrittlement of the heat exchanger's steel shell.

When hot lean oil was reintroduced into the vessel it ruptured releasing about 22,000 pounds of hydrocarbon vapour which exploded. (Technically it was reported not as a vapour cloud explosion, but a deflagration or flash fire.) An estimated 26,000 pounds of hydrocarbon vapour burned as a jet fire.

After the initial fire, the flame front travelled back to the source and a prolonged fire developed. This fire was beneath a critical pipe rack called 'King's Cross.' The flames impinging on this rack lead to three other rapid releases of flammable inventories and fireballs. The pipelines in this fire-affected area were major interconnections between three gas plants. It was difficult to isolate under these conditions. It required nearly 2.5 days for the total isolation of the piping network to stop the fire.

The Australian Government ordered a Royal Commission to investigate the cause of the accident. Follow-up reports provided some significant major lessons for the industry.

1. The Safety Management System was judged incomplete. This included the lack of operator training associated with the loss of lean oil flow, the circumstances in which brittle fracture might occur and the lack of discussion of this potential problem in procedures for shut-down or start-up.
2. One report listed the operator training emphasized the training was provided for the operators to do their job, rather than have the deeper understanding they needed to deal with unforeseen problems.
3. Changes in organization during the 1990s is also mentioned as a major contributor to this accident. Organizational changes were not subject to Management of Change scrutiny at this location. Until 1991 engineers worked at the Longford Plant on a daily basis. In 1992, as part of a restructuring of the company, all of the plant engineers were moved to Melbourne. The Commission's official report states, 'The physical isolation of the engineers from the plant deprived operations personnel of engineering expertise and knowledge which previously they gained through interaction and involvement with engineers on site. Moreover the engineers themselves no longer gained an intimate knowledge of plant activities.' The number of supervisors and associated staff was reduced from 25 to 17 between 1993 and 1998. The maintenance staff was also reduced from 67 to 58 during that time.

The Commission stressed that the Audit and Review functions of the Safety Management System provides opportunities to discover gaps. It stressed the need for challenging rigorous corporate audits. The commission concluded that the methodology employed by the assessment team earlier in 1998 was flawed and failed to identify significant deficiencies (Coco, 2001; Boult, 2001; Kletz-2001).

Augusta, Georgia, 2001
On March 13, 2001, three people were killed as they opened a process vessel containing hot plastic in Augusta, Georgia. They were unaware that the vessel was pressurized. The workers were killed when the partially unbolted cover blew off the vessel, expelling hot plastic. The force of the release caused hot oil tubing breakage, and a leak ignited. (Due to the serious nature of this incident, the US Chemical Safety and Hazard Investigation Board (CSB) initiated an investigation. Their results were reported in *Investigation Report: Thermal Decomposition – BP Amoco Polymers – Report No. 2001-03-1-GA*, June 2002. This write-up is an abbreviated version of the CSB Report and some is copied verbatim.)

Workers were attempting to open a cover on a polymer catch tank. The tank was designed to receive partially reacted waste diverted from a chemical reactor during periods of start-up and shut-down.

Twelve hours prior to the incident, an attempt was made to start the production unit. The start-up was aborted due to problems downstream of the reactor. However, prior to shut-down an unusually large amount of partially reacted material had been sent to the polymer catch tank. Hot molten plastic inside the tank continued to react and also began to slowly decompose, thereby generating gases and causing the contents of the tank to foam. The material expanded as foaming continued, and eventually the entire tank was filled. The material then forced its way into connecting pipes, including the normal and emergency vents.

Once in the pipes, the plastic solidified as it cooled, resulting in a hardened layer 3–5 in. thick around the entire inner wall of the tank. The core of the plastic mass remained hot and molten, and likely continued to decompose over several hours, generating gases that pressurized the vessel.

Before opening the polymer catch tank, personnel may have relied on a pressure gauge and a transmitter on the vent piping from the vessel. However, any reading from the pressure gauge would likely have been unreliable because plastic had entered the vent line and solidified.

The workers proceeded to remove the 44 bolts from the cover. When half of the bolts had been removed, the cover suddenly blew off. Hot plastic spewed throughout the area, travelling as far as 70 ft. The cover and the expelled plastic struck the workers, killing them.

The force created propelled the polymer catch tank backward and bent the attached piping. A section of hot oil, 370°C, supply tubing for a heating jacket to the catch tank broke, and the fluid spilled into the area. A flammable vapour cloud formed and ignited within a few minutes. Several hours of fire-fighting efforts were required to extinguish the fire.

The US Chemical Safety Board Key Findings include:

1. The manufacturing process was not subjected to a specialized design review to identify thermal decomposition hazards from unintended and uncontrolled reactions, and the risks posed by decomposition of the plastic were not recognized.
2. More than the normal amount of hot plastic entered the polymer catch tank during the aborted start-up. Reactions and decomposition of the plastic-produced gases. These gases caused the plastic to foam and expand. The expanded plastic forced its way into connecting pipes, where it solidified and plugged the inlet to the vent line. Once this occurred, the gases could not escape and the vessel became pressurized.
3. Operating experience revealed that the design of the polymer catch tank did not afford a practical and reliable method for workers to check for hazards before opening the vessel. Drains were often plugged with solidified plastic, making it impossible to verify the absence of pressure or hazardous chemicals.
4. Process safety information inadequately described the design basis and operating principles for the polymer catch tank. There was no discussion of the means by which overfilling could occur and its consequences.
5. The possibility of overfilling was increased when the original start-up procedures were revised; the diversion of output from the reactor to the polymer catch tank was extended from 30 to 50 min. This modification of procedures was not subjected to a management of change (MOC) review.
6. Process hazard analyses were inadequate. Other findings, root causes and recommendations are available in the Chemical Safety Board's report (CSB).

Martinez, California, 1999

On 23 February 1999, a fire occurred in the crude unit at an oil refinery in Martinez, California. Workers were attempting to replace piping attached to a 150-foot-tall fractionator tower while the process unit was in operation. During removal of the piping, naphtha was released onto the hot fractionator and ignited. The flames engulfed five workers located at different heights on the tower. Four men were killed, and one sustained serious injuries.

(Due to the serious nature of this incident, the US Chemical Safety and Hazard Investigation Board (CSB) initiated an investigation. The investigation was to determine the root and contributing causes of the incident and to issue recommendations to help prevent similar occurrences. This write-up is an abbreviated version of the CSB Report and much of the write-up is verbatim. The CSB examination led to 'Investigation Report – Refinery Fire Incident – Tosco Avon Refinery' Report No. 99-014-1-CA.)

On February 10, 1999, a pinhole leak was discovered in the crude unit on the inside of the top elbow of the naphtha piping, near where it was attached to the fractionator

112 ft in the air. Plant personnel responded immediately and attempted to isolate the piping. The unit remained in operation.

The naphtha piping was inspected and showed that it was extensively thinned and corroded. A decision was made to replace a large section of the naphtha line. Over the 13 days between the discovery of the leak and the fire, workers made numerous unsuccessful attempts to isolate and drain the naphtha piping. Nonetheless, supervisors proceeded with scheduling the line replacement while the unit was in operation.

On the day of the incident, the piping contained about 90 gallons of naphtha, which was being resupplied from the running process unit through a leaking isolation valve. A work permit authorized maintenance employees to drain and remove the piping. After several unsuccessful attempts to drain the line, a maintenance supervisor directed workers to make two cuts into the piping using a pneumatic saw. After a second cut the piping began to leak naphtha, the supervisor directed the workers to open a flange to drain the line. As the line was being drained, the naphtha ignited, most likely from contacting the hot surfaces of the fractionator, and quickly engulfed the tower structure and personnel.

Chemical Safety Board key findings

1. The removal of the naphtha piping with the process unit in operation involved significant hazards. This non-routine work required removing 100 ft of 6-in pipe containing naphtha, a highly flammable liquid. Workers conducting the removal were positioned as high as 112 ft above ground, with limited means of escape. The hot process unit provided multiple sources of ignition, some as close as 3 ft from the pipe removal work. One isolation valve could not be fully closed, which indicated possible plugging.

   On three occasions prior to the incident, the naphtha pipe resumed leaking from the original pinhole and felt warm to the touch, indicating that one or more isolation valves were leaking. Numerous attempts to drain the piping were unsuccessful.
2. The naphtha pipe that was cut open during the repair work was known by workers and the maintenance supervisor to contain flammable liquid. Although procedures required piping to be drained, depressured and flushed prior to opening, this was not accomplished because extensive plugging prevented removal of the naphtha. Although the hot process equipment was close to the removal work, the procedures and safe work permit did not identify ignition sources as a potential hazard.
3. The naphtha stripper vessel level control bypass valve was leaking, which prevented isolation of the line from the operating process unit.
4. The refinery's job planning procedures did not require a formal evaluation of the hazards of replacing the 100 ft of naphtha piping. The pipe repair work was classified as low risk maintenance. Despite serious hazards caused by the inability to drain and isolate the line – known to supervisors and workers during the week prior to the incident – the low risk classification was not reevaluated.
5. The refinery's permit for the hazardous non-routine work was authorized solely by a unit operator on the

day of the incident. Operations supervisors were not involved in inspecting the job site or reviewing the permit.

6. Operations supervisors and refinery safety personnel were seldom present in the unit to oversee work activities.
7. Deficiencies in auditing of maintenance line breaking, lockout/tagout procedures were noted as well as a lack of a Management of Change review as one of the contributing factors. See the full CSB report for details.

Root causes identified by the Chemical Safety Board

1. The refinery's maintenance management system did not recognize or control serious hazards posed by performing non-routine repair work while the crude processing unit remained in operation.
   - Refinery management did not recognize the hazards presented by sources of ignition, valve leakage, line plugging, and inability to drain the naphtha piping. Management did not conduct a hazard evaluation[1] of the piping repair during the job planning stage.
   - Management did not have a planning and authorization process to ensure that the job received appropriate management and safety personnel review and approval. The involvement of a multidisciplinary team in job planning and execution, along with the participation of higher level management, would have likely ensured that the process unit was shutdown to make repairs safely once it was known that the naphtha piping could not be drained or isolated.
   - The organization did not ensure that supervisory and safety personnel maintained a sufficient presence in the unit during the execution of this job. The refinery relied on individual workers to detect and stop unsafe work, and this was an ineffective substitute for management oversight of hazardous work activities.
2. The refinery's safety management oversight system did not detect or correct serious deficiencies in the execution of maintenance and review of process changes.
   - Neither the parent corporation nor the local facility management audited the refinery's line breaking, lockout/tagout, or blinding procedures in the three years prior to the incident. Periodic audits would have likely detected and corrected the pattern of serious deviations from safe work practices governing repair work and operational changes in process units. These deviations included practices such as:
     - Opening of piping containing flammable liquids prior to draining.
     - Transfer of flammable liquids to open containers.
     - Inconsistent use of blind lists.
     - Lack of supervisory oversight of hazardous work activities.
     - Inconsistent use of MOC reviews for process changes.
3. Refinery management did not conduct an MOC review of operational changes that led to excessive corrosion rates in the naphtha piping.

Management did not consider the safety implications of process changes prior to their implementation, such as:
Running the crude desalter beyond its design parameters.

- Excessive water in the crude feed.
- Prolonged operation of the naphtha stripper level control bypass valve in the partially open position.

These changes led to excessive corrosion rates in the naphtha piping and bypass valve, which prevented isolation and draining of the naphtha pipe (CSB).

Toulouse, France, 2001

This devastating accident will be discussed in much more detail in future editions as all of the details emerge.

A massive ammonium nitrate explosion occurred on Friday 21 September 2001, at the Azote de France (AZF) fertilizer factory on the outskirts of Toulouse, France. The blast occurred in a storage facility that held 200–300 tons of granular ammonium nitrate. During the tragedy, 29 people were killed; about 2500 were injured. One wounded person died later.

Ammonium nitrate is primarily used as a fertilizer, however, if it combined with certain additives it can be used as an explosive.

The AZF (Azote de France) started up in 1924 on the left branch of the Garonne River outside of Toulouse. The plant is in an industrial zone, but with urban sprawl the site is surrounded by housing and buildings used by the general public.

The blast crated a crater that was about 50–60 m with a depth of over 7 m. Windows were shattered within a radius of 1–1.5 km and windows were blown out in the city centre 3 km away. The strength was estimated to be equivalent to 30–40 tons of TNT.

This incident occurred just 10 days after the attack on the World Trade Center in New York. As a result, tens of thousands of people assumed that Toulouse was under a terrorists' attack and ran into the streets in panic or attempted to flee.

An investigation is still ongoing and more details will be available at a later date.

Information from (R.J.A. Kersten, A.C. van der Steen, A.F.L. and G. Opschoor. The Ammonium Nitrate Explosion in Toulouse, France: The Incident and Its Consequences for Industrial Activities. A paper delivered at the Center for Chemical Process Safety – October 2002).

A1.11 Case Histories: B Series

One of the principal sources of case histories is the MCA collection referred to in Section A1.1. There are a number of themes which recur repeatedly in these case histories. They include:

Failure of communications
Failure to provide adequate procedures and instructions
Failure to follow specified procedures and instructions
Failure to follow permit-to-work systems
Failure to wear adequate protective clothing
Failure to identify correctly plant on which work is to be done
Failure to isolate plant, to isolate machinery and secure equipment
Failure to release pressure from plant on which work is to be done

Failure to remove flammable or toxic materials from plant
    on which work is to be done
Failure of instrumentation
Failure of rotameters and sight glasses
Failure of hoses
Failure of, and problems with, valves
Incidents involving exothermic mixing and reaction
    processes
Incidents involving static electricity
Incidents involving inert gas

In the 2108 case histories described the most frequently
mentioned chemicals and the corresponding number of
entries are

Ammonia	48	Chlorine	37
Caustic soda	88	Sulphuric acid	46

The most frequently mentioned operations are

Loading	38	Steaming	32
Maintenance	82	Tank entry	39
Pipe fitting	27	Transfer	46
Process reaction	66	Unloading	64
Sampling	24	Welding	36

The most frequently mentioned equipments are

Centrifuge	24	Machine	27
Cylinder	26	Pump	60
Drum	76	Rotameter, sight glass	27
Hose	57	Tank	46
Industrial truck	31	Tank car	59
Laboratory	85[a]	Tank truck	35
line	105	Valve	86

[a] In the first 1623 case histories.

It is emphasized, however, that in many instances it is not
appropriate to assign a single cause.

Accounts of some typical case histories, many of which
are taken from the MCA collections, are given below. The
descriptions are generally briefer than those given in the
original sources, which should be consulted for further
details.

B1 A pilot plant batch reactor was started at 5.40 a.m. As
soon as catalyst was added the temperature and pressure
began to increase. In accordance with normal operation
venting was carried out. But this was slow and difficult due
to carryover of liquid and solids from the reactor into the
vent lines. At about 6.45 a.m. the agitator stopped as a
result of overload. The vent lines blocked, the pressure
increased rapidly and the building had to be evacuated. It
was re-entered at about 7.20 a.m., when the pressure was
found to be 750 psi. Thereafter the reaction subsided as the
reactants were used up. The pressure relief valve was set at
750 psi, but was found to be blocked as a result of the incident. The cause of the reaction runaway was overcharging
of reactants at the start of the batch. (MCA 1966/15, Case
History 867.)

B2 A batch chlorinator exploded suddenly killing
employees and doing extensive damage. The reaction temperature was controlled automatically by manipulating the
chlorine flow. During the batch the measured temperature
fell sharply and it was realized that the thermocouple had
failed. The agitator was stopped and the brine cooling was
shut off while the instrument was repaired, but delay in
stopping chlorine flow resulted in a high temperature
giving decomposition reactions. The explosion blew the
reactor cover through the roof and the reactor itself was
driven down into the floor. Flammable gases were released
which burned and caused the casualties mentioned. The
explosion also ruptured chlorine and ammonia lines.

The works emergency plan was activated, involving
both local area services and mutual aid from other chemical firms. (MCA 1962/14, Case History 371.)

B3 In a process involving chlorination of paraffin in a
reactor, chlorine gas was piped into the system in a carbon
steel pipe. The pipe ruptured and released paraffin and
chlorine, which caused a fire and did extensive damage.
Subsequent investigation revealed that the contents of the
reactor had leaked into the pipe and had reacted with the
chlorine vapour. Severe oxidation of the steel pipe by
chlorine and/or hydrogen chloride led to the pipe rupture.
(NFPA 1973.)

B4 An explosion occurred in a batch reactor which blew
the top right off the vessel. The vessel was supported by
steelwork two-thirds of the way up its side. The force of the
explosion pushed the reactor 2.5 ft into the floor, which was
3 ft below, as shown in Figure A1.21. Fragments of the lid of
the vessel, which contained a bursting disc, were scattered
over the countryside. (Anon. 1977n.)

B5 The loaded basket of a 48 in. suspended-type centrifuge suddenly became unbalanced and in consequence
the shaft flew out and broke the outlet pipe of an adjacent
centrifuge. The investigation indicated that the imbalance
had been caused by a sudden escape of cake from one side
of the basket due to a hole in the cloth. (MCA 1966/15, Case
History 645.)

B6 A 2000 ft^3 diborane surge tank disintegrated causing
severe loss. The failure fragmented the tank and reduced to
rubble a 24 in. thick reinforced concrete wall barricading it
on three sides. Fragments travelled 2200 ft and one cut four
process lines in a unit 1800 ft distant. The tank was 8 ft
inside diameter × 30 ft long. It was designed for a working
pressure of 296 psi with a safety factor of 4. Prior to the
installation it had been completely radiographed and
stress-relieved. At the time of the explosion the tank pressure and temperature were 192 psig and 12°C. The cause of
the explosion was not certain, but the most probable
explanation appeared to be a defect in the tank aggravated
by high stresses at the attachment weld of the platform
support. (MCA 1966/15, Case History 730.)

B7 A safety shut-off valve on the gas supply to the burner in the combustion chamber of a drier remained open
after the unit was shut-down. There was no indicator to
show whether the valve was open or closed. On relighting
the burner the operator made several errors, including that
of opening the main valve on the gas supply to the burner
before lighting the pilot burner. When the lighted taper was
inserted into the combustion chamber, there was an explosion. (MCA 1966/15, Case History 1068.)

Figure A1.21 *Batch reactor after explosion (Anon., 1977n)*

B8 A reactor with a water jacket collapsed inwards due to overpressure in the jacket, the water from the jacket mixed with the reaction mass, and hot steam, and chemicals erupted violently through a sight glass and a rupture disc on the reactor. The overpressure of the jacket took place during the initial heating of the reactor charge and occurred because the inlet and outlet valves on the jacket were closed and there was no pressure relief device on the jacket, although there was a bursting disc on the reactor itself. (MCA 1966/14, Case History 353.)

B9 A cabinet containing sparking electrical equipment was pressurized with nitrogen to exclude flammable gases. Nitrogen from the same main was used to 'egg' acetone out of a process vessel. The nitrogen pressure fell as a result of heavy demand, acetone passed back into the nitrogen and got into the cabinet. Subsequently the nitrogen flow to the cabinet was shut off, air leaked in and mixed with the acetone, and there was an explosion. (Kletz, 1976b.)

B10 Butadiene from a reactor passed back up a line used for adding emulsifier through an obstructed non-return valve. The emulsifier tank had an open vent, so that the butadiene got out into the building and exploded. The explosion caused flames 75 ft high and was heard 10 miles away. Damage was minimized, however, by the light construction of the building which ruptured at the roof and at two of the walls. (Mallek, 1969.)

B11 In a pilot plant acetaldehyde oxidation column the arrangements involved purging the column with nitrogen and then supplying acetaldehyde to the column by nitrogen pressure. The oxidation was carried out by admitting a controlled flow of oxygen to the column. The pipework allowed the nitrogen to be used to purge the column and part of the oxygen pipe. There was nothing to stop oxygen at higher pressure passing into the nitrogen in case of a leaking valve or of incorrect operation. During operation the operator observed a leak on the acetaldehyde supply drum and was told to shut the plant down. A few minutes later an explosion occurred which killed four people. From the investigation it appeared that oxygen had leaked via the nitrogen purge line into the acetaldehyde drum, in which reaction occurred with large rises in temperature and pressure and eventually a detonation. (MCA 1962/14, Case History 117.)

B12 A malfunction of the control equipment allowed raw gas to accumulate in a direct-fired textile fibre drier. When circulating fans were turned on, the drier exploded. (Doyle, 1969.)

B13 In an air liquefaction plant an electrical failure caused the control valve on an oxygen line to open and pass oxygen into a nitrogen line. The nitrogen was compressed and then used as wash liquid for the offgas from an ethylene plant. The nitrogen wash cold box exploded. (Doyle, 1969.)

B14 In a chlorine liquefaction system 'gunk' was warmed prior to removal from a cold trap. A leaky steam valve allowed overheating of the gunk. This contained sufficient nitrogen trichloride to explode. (Doyle, 1969.)

B15 A vacuum system failed on a small still distilling a chloronitro compound. This allowed the temperature to rise and give a self-accelerating decomposition. A large still adjacent to the small one was seriously damaged. (Doyle, 1969.)

B16 A reboiler distilling epichlorhydrin from a mixture with tar was heated by steam. The thermocouple on the steam control loop tarred up and gave an incorrect reading. The only control of conditions in the column became the operator's response to the temperature at the top of the column. This proved inadequate, the reboiler overheated and an explosion occurred. (Doyle, 1972a.)

B17 In a furnace cracking ethylene dichloride to vinyl chloride under computer control the flow of ethylene dichloride to the furnace was controlled to maintain an optimum temperature in the cracked gas exit line. A furnace tube split and reduced the cracked gas flow and the cracked gas temperature. The control action increased the flow of ethylene dichloride. The furnace tubes burned up. (Doyle, 1972a).

B18 A common type of 'Dowtherm' vaporizer is a single coil in which the liquid is superheated, followed by a flash chamber. There is an interlock between a flowmeter on the liquid line to the coil and the fuel flow controller, so that the coil is not heated unless liquid is flowing through it. A vapourizer of this type was started up after a shut-down and the coil immediately exploded. The impulse lines to the differential pressure transmitter had frozen just before the shut-down, showing a normal flow, and during the shut-down the exit line from the coil also froze up. On start-up the interlock did not work owing to the incorrect signal from the flowmeter and the heating of the coil caused pressure to build up and rupture it. (Doyle, 1972a.)

B19 Excess water was removed from a hydrogen peroxide–water mixture by vacuum distillation. As a precaution against attempting to vaporize a dangerous concentration of peroxide, the liquid level and vapour temperature were measured: a low level or a high temperature signal operated a valve to dump de-ionized water into the reboiler. The hydrogen peroxide solution produced large quantities of heavy foam. The float-type level instrument measured the foam rather than the liquid level. After some erratic readings it gave a seriously inaccurate one while the level was low, and the reboiler exploded. (Doyle, 1972a.)

B20 The agitator on a batch nitration reactor was stopped, but the process operator was unaware of this. Instrumentation which should have cut off the acid feed to the reactor and given an alarm signal of agitator stoppage apparently failed to work. When the agitator started up again, the reactor exploded. (Fritz, 1969.)

B21 In an ethylene cracking furnace the flow of fuel to the furnace was manipulated to control the temperature of the gas leaving the furnace. The temperature measuring instrument failed and gave a low reading. The control action increased the heat to the furnace. The furnace was designed to supply mainly latent heat of vaporization to the process stream with a small amount of superheat. Thus the vapour temperature was sensitive to the additional heat input and rose rapidly. The piping overheated and ruptured, releasing ethylene gas which was ignited by the burners of an adjacent furnace. The resultant fire did severe damage. (H.D. Taylor and Redpath, 1971.)

B22 Ammonia synthesis gas at 2000 psig was passed through a scrubber which was fed with weak aqueous ammonia. The liquid from the bottom of the column passed to a vertical holdup tank with a 2 in atmospheric vent. The level of liquid in the base of the column was controlled automatically. The control valve in this loop stuck in the open position. Liquid and gas from the column surged into the hold-up tank and ruptured it. The escaping gas was approximately 75% hydrogen and it ignited, causing a fire. (MCA 1966/15, Case History 598.)

B23 A flow ratio control loop failed and allowed excess hydrogen to enter a catalytic tower. Hot spots developed, causing thermal decomposition of the hydrocarbon–hydrogen stream and rupture of the tower. (MCA 1966/15, Case History 609.)

B24 A flame failure occurred on the burner of a Dowtherm vapourizer and the procedure for relighting the burner was initiated. The firebox was checked and the automatic control system which purges and lights the flame was activated. A short developed across the fuel selector switch of the system, causing fuel gas to enter the combustion chamber during the purge cycle and thus creating a flammable fuel–air mixture. When the purge cycle was complete and the pilot light was ignited, an explosion occurred. (MCA 1966/15, Case History 821.)

B25 An inert gas generator was found to have produced a flammable oxygen mixture. The 'fail safe' flame failure device had failed. The trip system on the oxygen content of the gas generated had caused shut-down when the oxygen content in some of the equipments reached 5%, but did not prevent creation of a flammable mixture in the holding tank. (MCA 1966/15, Case History 679.)

B26 An air supply enriched with 2–3% oxygen was provided for flushing and cooling air-supplied suits after use. A failure of the control valve on the oxygen–air mixing system caused this air supply to contain 68–76% oxygen. An employee used the supply to flush his air-supplied suit, disconnected the lines, removed his helmet and lit a cigarette. His oxygen-saturated underclothing caught fire and he received severe burns. (MCA 1966/15, Case History 884.)

B27 A chlorine cellroom was experiencing operational difficulties. A failure of the control system caused a back pressure to develop. When the operating personnel tried to shut the system down using manual controls, a failure of this equipment caused a serious chlorine release. One of two compressors taking hydrogen from a low-pressure gasholder continued to operate and failure of a low level trip created a negative pressure, allowing air to leak in. An explosion occurred in the cooling coils of the compressor. Hydrogen from another part of the process was introduced into the holder in an effort to maintain production. The hydrogen–air mixture in the holder diffused into the catalyst purification unit, where a high temperature developed, causing a second explosion. The failure of the gasholder trip to operate was due to the fact that the timer had been bypassed by a 'jumper'. (MCA 1966/15, Case History 694.)

B28 An explosion occurred on a vinyl chloride polymerization plant at Minamata, killing four and injuring eight other people inside the factory and slightly injuring two people outside it by flying glass. The polymerization reaction in No. 3 reactor was complete and the operators intended to discharge the contents, but No. 4 reactor was discharged in error. The reaction in this reactor was incomplete and the VCM escaped and exploded. (MCA 1966/15, Case History 816.)

B29 A rigger was removing manhole covers from a vinyl chloride reactor when he apparently became confused and removed the cover from a reactor which was operating under pressure. A large quantity of vinyl chloride vapour was released. It ignited to give a flash fire, which killed the rigger and two labourers. (MCA 1970/16, Case History 1132.)

B30 In an ethylene oxide plant inert gas was circulated through a process containing a catalyst chamber and a heat removal system. Oxygen and ethylene were continuously injected into the inert gas and ethylene oxide was formed over the catalyst, liquefied in the heat removal section and passed to the purification system. On shut-down of the circulating compressor an interlock stopped the flow of oxygen and the closure of the valve was indicated by a lamp on the panel. During one shut-down the lamp showed the oxygen valve closed. The process operator had instructions to close a hand valve on the oxygen line, but he expected the maintenance team to restore the compressor within 5–10 min and did not close the valve. The process loop exploded. The oxygen control valve had not in fact closed. A solenoid valve on the control valve bonnet had indeed opened to release the air and it was the opening of this solenoid which was signalled by the lamp on the panel. But the air line from the valve bonnet was blocked by a wasps' nest. (Doyle, 1972a.)

B31 A catalyst bed was regenerated by circulating nitrogen through an electric heater, the bed and a cooler. The flowmeter on the nitrogen line choked with dust early on and the process operators learned to make do without it, controlling the plant by the nitrogen temperature at the bed inlet. One day the cooler blocked with dust and the temperature at the bed inlet stuck at its normal value. The heater over-heated and at 740°C its high temperature trip operated and cut off the power. The operator, seeing that the other instrument readings were normal, assumed the trip to be faulty and switched the power back on. The trip operated and was overridden three times in 1 h. There was a shift change during this hour. Finally the heater shell burst. (Kletz, 1974b.)

B32 A tank had to be filled with enough raw material to last a day. The operator watched the tank level and switched off the pump when the required level was reached. One day after several years operation and some 1000 fillings, the man let the tank overflow. So a trip system was installed as an additional safeguard. The operator, however, got in the habit of using the trip system as a replacement and got on with something else. The plant manager knew about this, but was happy to see the improvement in productivity. However, since the trip system can be expected to fail about

once in 18 months, the overall system had actually become less reliable. (Kletz, 1974d.)

B33 An explosion occurred in the open air in the vicinity of a hydrogen vent stack and caused severe damage. It was normal practice to vent hydrogen for periods of approximately 45 min. On this particular occasion there was no wind, the hydrogen failed to disperse and the explosion followed. (MCA 1966/15, Case History 1097.)

B34 An operator was adding charcoal impregnated with catalyst from a plastic bag into a reactor containing caustic soda blanketed with nitrogen. The charcoal dust ignited and there was a flash fire. It is believed that the ignition was caused by static electricity. (MCA 1966/15, Case History 1094.)

B35 An employee was repairing a blower on the vent stack of a dissolver. When he had finished the work, he switched the machine on, but observed that it was not operating. He reached into the stack to give the fan blades a turn and there was an explosion. The vent stack had contained a flammable mixture, which was probably ignited by static electricity from the man's body. (MCA 166/15, Case History 703.)

B36 A man was filling his cigarette lighter from a gallon can of naphtha when he dropped the lighter. It apparently struck on the wheel and gave a spark. Flammable vapour from a nearby drum filling plant ignited and caused a violent explosion. (MCA 1970/16, Case History 1175.)

B37 A substantial section of lagging along the pipework of a process heating fluid system became contaminated with the fluid, which was a low volatility mineral oil, due to a pump gland leak. A lagging fire occurred, the symptoms of which were that first smoke was seen rising from the lagging, then the heating fluid rose 50°C above its set temperature of 270°C and finally the pump and a considerable length of pipe became enveloped in flames. The latter were quickly dealt with, but the temperature of the heating fluid continued to rise. Soon afterwards a small explosion occurred in a covered oil header tank housed inside a building. There was a large release of oil and fumes from the tank and these then exploded violently, causing severe damage to the building. The probable explanation was that some of the oil had been heated above its spontaneous ignition temperature and had been forced by expansion back up into the header tank, where its vapour met air and ignited. (Gugan, 1974a.)

B38 A large spray drying plant with direct heating from a combustion chamber was being run in. The combustion system had been operating at a very low firing rate under manual control for some 24 h. The fuel oil to the burner was pumped by a positive displacement pump through thermostatically controlled heaters, excess oil being bypassed around the pump. After firing had continued unsupervised for some hours, there was a sudden eruption of flame through a hinged explosion panel near the combustion chamber and a man walking past was enveloped. Investigation showed that the eruption of flame could be explained by a gross increase in fuel flow to the burner, even though a manual valve on the fuel line was still at its original setting. It was also discovered that a fire had developed in the lagging of the fuel oil system, the lagging having become saturated with oil due to leaks from pipe joints occurring during the initial pressure testing. The reason for the increase in fuel oil flow was probably a rise in the oil temperature to a value much in excess of that set on the thermostat, so that there was a decrease in oil viscosity and an increase in the quantity flowing through the valve. (Gugan, 1974a.)

B39 Dried material from a double-drum steam drier was passed from a bag packer via a spout with a cotton apron to a leverpak. A small dust explosion occurred at the apron and led to a fire in the bagging plant and leverpak. It was considered from investigation that the dust had been ignited by a spark discharge of static electricity. The cotton apron was non-conducting and the bag packer was not earthed. (MCA 1966/15, Case History 653.)

B40 An operator was in the process of unloading titanium carbide dust from 2 ft conical ball mill. He had removed 95% of the material. The mill was being rotated in order to clean it when a dust cloud ignited. He received severe burns from which he later died. The product had been milled to a finer size than desired and thus rendered highly reactive. (MCA 1966/15, Case History 618.)

B41 A set of air compressors comprising four three-stage 740 h.p. machines with a discharge pressure and temperature of 350 psi and 350°F had been in operation for about 8 h following a plant shut-down. The third-stage cylinder in one of the compressors was faulty and the 125 psi relief valve between the second and third stages began to relieve air intermittently. The third-stage discharge temperature went off the instrument scale at 400°F before the compressor could be shut-down. Some 15 min later the overhead discharge line was observed to be glowing red hot. Before the other compressors could be shut-down, the line ruptured and a flash fire occurred. Investigation indicated that lubricating oil had ignited in the compressor cylinder and discharge line. (MCA 1962/14, Case History 35.)

B42 An explosion occurred in a vinyl chloride pump and seriously injured an operator. The pump was on the 'recovered' vinyl chloride monomer (RVCM) system. The line sections before and after the pump had been removed and the explosion occurred in the idle vented pump about an hour later. The investigation showed that the entire RVCM system was contaminated with vinyl chloride polyperoxide, which is an unstable material. There had been three abnormal occurrences on the plant prior to this accident. The RVCM gas compressors had not shut-down at low pressure due to the failure of a pressure switch. There had been an accumulation of RVCM liquid in the storage tanks for 20 days. And the acidity level in the vinyl chloride feed to the polymerizers had been high. (MCA 1970/16, Case History 1551.)

B43 A four-stage reciprocating compressor used to compress nitrogen from 15 to 3000 psig exploded with a 'brilliant flash' and projected fragments up to 160 ft from the building. Previous to the explosion the machine had been shut-down to change the third-stage suction and discharge valves and again to correct an error in the installation of the former. Investigation showed that the discharge valve had been installed in the reversed position and that compression of the gas under these conditions could give a pressure of 16,800 psig and a temperature of 1310°F. This temperature exceeded the autoignition temperature of the

lubricating oil. There may have been oxygen present due to the maintenance operations. These conditions resulted in the explosion. (MCA 1966/15, Case History 1056.)

B44 A 40,000 gal horizontal tank was used to produce aqueous ammonia by introducing water at the top and anhydrous ammonia at the bottom. The tank was emptied and the water flow was started to make the next batch. The tank imploded due to the vacuum created by the absorption of the ammonia vapours by the water. (MCA 1966/15, Case History 916.)

B45 A holding tank containing 'heavies' from cracking furnaces suddenly erupted blowing oil, water and steam over the adjacent area. It is believed that condensate had come in contact with the hot oil, which was at a temperature of 600–700°F, thus generating large quantities of steam. (MCA 1962/14, Case History 455.)

B46 A storage tank was not in use and was boxed up with some water inside. Rusting took place, the oxygen of the air in the tank was depleted and the tank collapsed inwards. (Anon., 1977n.)

B47 A storage tank was cleaned out and filled with a high purity liquid. A polyethylene bag was put over the vent to prevent contamination. It was a hot day and the temperature of the liquid in the tank rose. There was a sudden shower, vapour in the tank condensed and the tank collapsed inwards. (Anon., 1977n.)

B48 Cold liquid was pumped into a storage tank containing hot liquid. The liquid in the tank was cooled, the vapour pressure fell and the tank collapsed inwards. (Anon., 1977n.)

B49 Modifications were required to a high-pressure chlorine main. The main was isolated and men began work. The supervisor then realized that there might be some chlorine present in a section of pipework at the tank car rack. The pipework was connected to the main, but there was an isolation valve between the main and the pipework. He assumed that this valve was shut and initiated measures to remove any chlorine present in the pipework section. There was liquid chlorine in this pipework and it passed via the isolation valve into the main and gassed the fitters working there. The isolation valve was subsequently found to be frozen in the part open position. (MCA 1966/15, Case History 707.)

B50 An employee went into a water cistern to install some control equipment and immediately collapsed into water 2 ft below. A second employee who had accompanied him ran to fetch assistance. Minutes later he came back with several others, two of whom entered the cistern and also collapsed. Meanwhile the alarm had been raised. The fire services arrived and a crowd gathered. While the fire officer was putting on his self-contained breathing apparatus, one of the by-standers, saying that he could swim, descended into the cistern. The fire officer then went in, but took off his mask, presumably to call for some equipment, and collapsed. All five people died due to hydrogen sulfide poisoning. (MCA 1970/16, Case History 1213.)

B51 'A double fatality occurred in a steel works whilst two fitters were changing a gearbox valve for a blast furnace system. One or both men opened a valve allowing gas to escape at a rate of 50,000 cu. ft. per hour. Of another two men who went to investigate the volume of gas passing through the valve, one was overcome by gas (but later recovered) and the other escaped to summon assistance. Safety precautions should have included either the isolation of the system upstream from the valve or the use of suitable breathing apparatus by the workmen. Although both maintenance men had been issued with compressed airline breathing apparatus it was found after the accident that only one of these was connected with the air supply and neither was being worn. The fatalities highlighted the dangers of complacency which can occur when a system has operated without incident for years and also the need to ensure that workers employed in such situations are adequately trained in the use of breathing apparatus.' (Annual Report of HM Chief Inspector of Factories, 1974.)

B52   During plant commissioning a temporary filter was put in a compressor suction line, as shown in Figure A1.22. The filter was located between the compressor and the low suction pressure trip. The filter blocked, the suction pressure fell below atmospheric, air was sucked in and a decomposition reaction occurred further on in the process where there was a higher pressure. Two pipe joints sprung and gas escaped and ignited. The resulting fire caused £100,000 damage and delayed the commissioning many months. (Henderson and Kletz, 1976.)

B53   An olefins plant was severely damaged during commissioning by a fire in the refrigeration section of the gas separation plant. Although no one was seriously injured, the fire burned for 15 h and caused delay to the commissioning and considerable damage and consequential loss. The cause of the fire was that the low-pressure refrigeration section, which was designed for 50 psig pressure, was subjected to a pressure of 400 psig. The overpressure occurred because of closure of a 12 in. valve on the exit of this section, installed during construction to allow high pressure gas to be used to assist in drying the plant prior to commissioning. The installation of the valve had been well debated in the plant operating team and with the process design contractor. But the installers did not follow the well-established requirement of providing pressure relief protection on the low-pressure system. (Heron, 1976.)

B54   A works had a special network of air lines installed some 30 years ago for use with breathing apparatus only. The supply to this network was taken off the top of the general purpose compressed air main as it entered the

works, as shown in Figure A1.23. One day a man wearing a face mask inside a vessel got a faceful of water. He was able to signal to the anti-gas man and was rescued. Investigations revealed that the compressed air main had been renewed and that the branch to the breathing apparatus network had been connected to the bottom of the compressed air main. As a result a slug of water in the main would all go into the catchpot and fill it more quickly than it could empty. (Henderson and Kletz, 1976.)

B55   Pressure relief on a low-pressure refrigerated ethylene tank was provided by a relief valve set at about 1.5 psig and discharging to a vent stack. When the design had been completed, it was realized that if the wind speed was low, cold gas coming out of the stack would drift down and might then ignite. The stack was not strong enough to be extended and was too low to use as a flare stack. It was suggested that steam be put up the stack to disperse the cold vapour and this suggestion was adopted. The result was that condensate running down the stack met cold vapour flowing up, froze and completely blocked the 8 in. pipe. The tank was overpressured and it burst. Fortunately the rupture was a small one, the ethylene leak did not ignite and was dispersed with steam while the tank was emptied. (Henderson and Kletz, 1976.)

B56   A metal chute used to transfer a powder from a metal hopper into a metal vessel was replaced by a plastic hose. The flow of powder down the chute caused a charge of static electricity to build up in the hose and, although the hose was a conductor, the polypropylene end pieces used to connect it to the hopper and the vessel were not, and the charge was unable to dissipate. A spark occurred and the dust was ignited. (Heron, 1976.)

B57   A relief valve weighing 258 lb was being removed from a plant. A 25 ton telescopic jib crane with a jib length of 124 ft and a maximum safe radius of 80 ft was used to lift the valve. The driver failed to observe this maximum radius and went out to 102 ft radius. The crane was fitted with a safe load indicator of the type which weighs the load through the pulley on the hoist rope, but this does not take into account the weight of the jib, so that the driver had no warning of an unsafe condition. The crane overturned on to the plant, as shown in Figure A1.24. (Anon., 1977n.)

B58   Halfway through the unloading of a tank truck of anhydrous ammonia a stream of liquid was observed leaking from the bonnet of the stop valve on the unloading line.

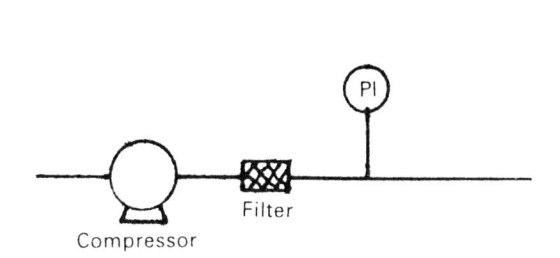

Figure A1.22   *Compressor system after modification (Henderson and Kletz, 1976) (Courtesy of the Institution of Chemical Engineers)*

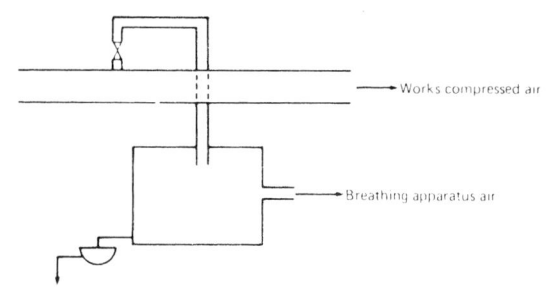

Figure A1.23   *Air offtake for breathing apparatus as originally installed (Henderson and Kletz, 1976) (Courtesy of the Institution of Chemical Engineers)*

Figure A1.24   *Mobile crane overturned on to plant (Anon., 1977n)*

The truck was taken off line and the storage tank was vented to a neutralizing pit to relieve its pressure. Considerable difficulty was experienced in reducing the pressure in the tank sufficiently to stop the flow of liquid ammonia. There was no shut-off between the leaking valve and the tank. Eventually the pressure was reduced by using a compressor and exhausting it to atmosphere. Investigation indicated that the failure of the valve was caused by a combination of a flaw in the body and strain induced by excessive pressure on one of the stud bolts. (MCA 1970/16, Case History 1114.)

B59   A storage tank 108 ft diameter × 49 ft high containing refrigerated propane at a pressure less than 1 psig was overpressured and ruptured at Ras Tanura refinery, Saudi Arabia. The propane caught fire and gas from pilot-operated relief valves on two similar tanks, one containing propane and one butane, was also set alight. The ruptured tank burned out in some 36 h, while the butane and propane tank vent fires burned for 3 and 6 days, respectively. One person was killed and 115 were injured, and serious damage was done.

Just prior to the incident the refrigeration plant had been shut-down. The intention was to go over to autorefrigeration in which the propane tank would be cooled by boiling off liquid and venting it into the blowdown system. But this autorefrigeration could not be established, because the blowdown line was blocked with liquid butane. The reason for the presence of the butane was that a remotely operated isolation valve had not fully closed, because the valve spring was prevented from operating properly by paint and corrosion products.

There was a steady rise in the pressure in the propane tank up to and then beyond the relief valve set pressure of 1 psi. The operators, both at the end of one shift and at the beginning of the next, tried to find out the cause of the rise in pressure, but they did not alert the supervisors. (Laney, 1964.)

B60   An operator checked the level in a storage tank and then began pumping in a volatile monomer from a tank wagon. The vent pipe of the storage tank was blocked, however, and a large quantity of monomer flowed out of the dip leg. (MCA 1970/16, Case History 1192.)

B61   A tank truck was being loaded with ethyl acetate. The loader heard what he described as a 'sizzling' sound and looked in, but could see nothing. Soon after an explosion occurred. The investigation showed that the earthing arrangements on the tank were in good order. The evidence suggested that the explosion was caused by static charge on the surface of the liquid. (MCA 1966/15, Case History 986.)

B62   A tank trailer containing 6876 gal of propane under pressure ruptured on the highway 560 ft from the centre of the town. The propane escaped and exploded, causing 10 deaths, 17 injuries and extensive damage. The investigation revealed various undesirable features of the tank, including weld defects which would create areas of enhanced stress concentration, differences of alignment between the cylinder and the end parts, and possible embrittlement of the plates due to the use of a heating process to effect chamfering. (MCA 1966/15, Case History 879.)

B63   Acetic acid was being pumped out of one compartment of a two-compartment trailer. The manhole cover was propped open. The operator climbed on the tanker to obtain a sample. In order to do this he had to lift the manhole cover back. When he had taken the sample, he closed the manhole cover completely. A few seconds later the tank imploded as a result of the vacuum generated by the continued pumping. The tank was equipped with a spring-loaded vacuum breaker, but either this failed to open or it was undersized. (MCA 1966/15, Case History 1011.)

B64   A tractor semitrailer transporting 7000 gal of propane was modified by inserting a tee-piece between the excess flow valve and the manual shut-off valve on the 3 in. discharge line. The joint was connected to the tractor fuel tank by a $3/4$ in. hose to allow the vehicle to be filled en route. There was another manual shut-off valve on the $3/4$ in. line. The hose broke in transit. The leakage flow rate was too small to activate the excess flow valve. The driver stopped the vehicle, shut-off the engine and told the trainee driver to shut-off the smaller hand valve. When the latter touched the valve, the propane vapours ignited and he was engulfed in flames. The gas continued burning for another 1.5 h. The trainee driver died in the fire, which also caused other injuries and damage. (MCA 1966/15, Case History 995.)

B65   An explosion occurred in a terraced house in East Street, Thurrock, in 1969 that blew a hole in the floor at the foot of the staircase. The wife of the householder fell in while carrying her child and both were injured. The Times (9 April, 1969) reported

> Investigators found that the explosion had been caused by the ignition of a mixture of petrol vapours and air and that the vapour was the result of a spillage of petrol two years before.

Table A1.4   *Some incidents and problems in the industries 1965– 77*

1	*Power loss, weather and earthquake*		
	Power loss	1965 Dec. 6, 76	1971 Jan. 25, 25
		1966 Feb. 14, 86	1972 Nov. 27, 34
		1967 Jun. 19, 86	1973 Nov. 12, 94
		1968 Jun. 3, 31	Dec. 10, 66
		1970 Jun. 27, 80	
	Cold weather	1962 Feb. 5, 35	1977 Feb. 14, 25
	Wind, hurricane, tornado	1961 Oct. 2, 33	1972 Jul. 24, 75
		1970 Aug. 24, 31	Nov. 27, 34
	Flood	1972 Jul. 24, 75	
	Earthquake	1968 Dec. 16, 44	1972 Sep. 18, 72
2	*Safety and safety legislation*		
	Safety	1974 Jun. 10, 34	1977 Jun. 6, 68
		1975 Jun. 9, 34	
	Safety legislation and law	1964 Aug. 3, 27	1977 Jan. 17, 69
		Sep. 14, 83	Sep. 26, 44
		1973 Apr. 16, 58	Dec. 5, 67
		1975 Jun. 9, 36	Dec. 19, 28
		1976 Jul. 19, 72	
3	*Transport and transport legislation*		
	Transport	1962 Oct. 15, 78	1972 Dec. 11, 44
		Nov. 12, 86	1974 Dec. 23, 24
		1963 Feb. 4, 36	1975 Mar. 17, 20
	Transport emergency systems	1970 Jun. 15, 83	1972 Sep. 4, 19
		1971 Sep. 20, 78	
	Transport legislation	1969 Oct. 6, 97	1974 Dec. 23, 24
		1973 May 28, 43	1977 Jan. 3, 38
		1974 Sep. 16, 44	Mar. 28, 54
4	*Pipelines and pipelines legislation*		
	Pipelines	1964 Dec. 21, 27	1967 Mar. 13, 88
		1965 Sep. 27, 75	Apr. 24, 87

Table A1.4   *(continued)*

	1966 Apr. 11, 85	1969 Sep. 22, 68
	Jul. 4, 39	Oct. 20, 52
	Dec. 19, 45	1970 Sep. 21, 85
	1967 Jan. 2, 19	1973 Aug. 20, 72
	Jan. 30, 50	1975 Jan. 20, 44
Pipelines legislation	196ᴜ Nov. 22, 79	1969 Oct. 20, 52
	1968 Dec. 2, 54	1973 Aug. 20, 72
	1969 Sep. 22, 68	1975 Jan. 20, 44

5   *Toxic chemicals at the workplace*

Toxic or potentially toxic chemicals, especially carcinogens	1971 May 3, 38	1976 May 24, 70
	1972 Jan. 10, 38	Oct. 25, 66
	1974 May 13, 52	Dec. 20, 19
	Oct. 28, 33	1977 Jan. 3, 36
	1975 Jan. 6, 57	May 9, 87
	Mar. 3, 74	Jun. 6, 67
	May 12, 40	Jul. 4, 62
	Jul. 7, 36	Jul. 4, 64
	Jul. 21, 52	Jul. 18, 54
	Aug. 4, 38	Aug. 15, 88
	Aug. 18, 41	Oct. 25, 74
	1976 Feb. 2, 48	Dec. 19, 28
	Mar. 15, 48	
Legislation on toxic chemicals	1973 Jan. 22, 47	1977 Feb. 14, 28
	1974 May. 27, 54	Feb. 28, 116
	Sep. 16, 46	Mar. 28, 54
	Oct. 14, 51	Apr. 11, 74
	Nov. 25,33	May. 9, 88
	1975 Feb. 3, 20	Jun. 20, 60
	Mar. 3, 74	Jun. 20, 62
	Sep. 15, 70	Jul. 4, 64
	Oct. 27, 56	Sep. 12, 90
	1976 May 24, 70	Oct. 10, 72
	1977 Jan. 17, 70	Oct. 24, 74
	Jan. 31, 54	Nov. 21, 114
	Jan. 31, 56	Dec. 5, 68
Testing and indexing of chemicals	1974 Jul. 8, 29	1977 Jun. 20, 61
	1975 Feb. 3, 20	
Vinyl chloride	1974 Feb. 4, 19	1975 Apr. 14, 33
	Apr. 15, 39	Jul. 7, 34
	Jun. 24, 98	1977 Jan. 3, 37
	Nov. 25, 33	Jul. 18, 52
	1975 Jan. 20, 41	
Vinyl chloride legislation	1974 Mar. 4, 63	1975 Feb. 17, 35
	Apr. 15, 40	Jun. 9, 36
	Jun. 24, 98	Oct. 13, 67
	Jul. 8, 28	Nov. 24, 20
	Jul. 22, 54	1976 Jan. 5, 74
	Sep. 30, 42	1977 Jun. 20, 62
	Oct. 14, 52	
	Dec. 23, 21	Aug. 1, 20

6   *Pollution*

	1974 May. 27, 53	1977 Jan. 31, 55
	1975 May. 26, 58	May. 23, 94
	Nov. 10, 118	Jun. 6, 68
Sudden releases and spillages	1967 Jun. 5, 64	1973 Sep. 3, 48
	1969 Aug. 25, 56	1974 Jun. 24, 96
	1971 Jul. 26, 55	Jul. 8, 29
	1973 Feb. 5, 26	1977 Mar. 14, 75
	Jul. 23, 57	

Table A1.4   *(continued)*

See also Transport, pipelines

Long-term, low concentration pollution and its effects	1968 Jan. 1, 9 Aug. 26, 25 1969 Mar. 10, 44 1970 Aug. 10, 50 Oct. 19, 72 1973 Jan. 22, 48 Aug. 20, 71 1975 Dec. 8, 80 1976 Feb. 16, 45	1974 Nov. 25, 36 1975 Feb. 3, 17 May. 12, 40 May. 26, 59 Aug. 4, 38 Sep. 1, 52 Nov. 10, 119 1977 Jan. 3, 36
Fish kills	1964 Jul. 6, 70 1969 Nov. 3, 25	1970 Mar. 23, 57 1971 Jul. 26, 55
Oil slicks	1969 Feb. 10, 40 1973 Jun. 25, 35	1975 Mar. 17, 17
Mercury pollution	1967 May. 8, 85 Nov. 20, 60 1970 Apr. 20, 53	1970 Aug. 10, 50 Sep. 7, 25 Oct. 5, 44
Pesticides	1964 Apr. 13, 81 1967 Oct. 9, 108 1974 Apr. 15, 40	1976 Mar. 15, 48 Oct. 25, 66
Ocean dumping	1970 Aug. 24, 32	1974 Oct. 28, 33
Ocean incineration	1974 Oct. 28, 33	1975 Dec. 9, 82
7 *Pollution legislation*	1975 Jul. 7, 35 Sep. 1, 52 Oct. 13, 68	1976 Mar. 15, 48 1977 Mar. 14, 74
Air	1969 Jun. 2, 44 1970 Nov. 30, 17 1971 Feb. 8, 20 1972 May. 15, 53 1973 Jun. 25, 34 1974 Feb. 4, 20 Mar. 18, 38 Jun. 24, 95 Sep. 2, 20 Oct. 14, 50 Dec. 23, 24	1975 Jan. 20, 42 1976 Feb. 2, 48 Mar. 1, 62 1977 Feb. 28, 116 Mar. 28, 56 Apr. 11, 74 Jul. 18, 54 Nov. 7, 42 Dec. 19, 26 Dec. 19, 27
Odour	1972 Sep. 4, 19	
Water	1969 May 19, 89 1970 May 18, 75 1972 May 15, 53 Oct. 16, 53 1974 Jan. 7, 46 Apr. 29, 82 Oct. 14, 52 1975 Feb. 3, 20	1976 Jul. 5, 35 1977 Jan. 31, 56 Jul. 4, 63 Aug. 15, 90 Aug. 29, 20 Dec. 5, 68 Dec. 18, 27
Oil spills	1974 Jan. 7, 46	
Noise	1974 Aug. 5, 46	1975 Oct. 13, 68
8 *Abandonment of actual or planned production*	1967 Mar. 27, 41 1968 Feb. 12, 49 1969 Mar. 10, 44 Jul. 14, 42 1970 Oct. 5, 43 1971 Jan. 25, 25 Feb. 8, 17 Jul. 26, 56	1973 Jun. 11, 63 1974 Nov. 25, 33 Dec. 9, 61 Dec. 23, 21 1975 Oct. 13, 67 Nov. 10, 118 Nov. 24, 20 1976 Jan. 19, 49

Table A1.4   *(continued)*

| | Sep. 6, 27 | 1977 Jan. 31, 55 |
| | 1972 Mar. 20, 53 | Oct. 10, 69 |

This list includes cases where abandonment of actual or planned production was a serious possibility, even if it did not in fact occur.

9 *Litigation*	1963 May 13, 83	1970 Jul. 27, 80
	1964 Apr. 27, 78	Aug. 24, 32
	Dec. 21, 27	1971 Apr. 19, 53
	1966 Apr. 11, 85	1972 Dec. 25, 20
	1968 Aug. 26, 25	1974 Oct. 14, 52
	1970 Mar. 23, 59	1975 Nov. 24, 17

The spillage involved 367 tons of petrol on rail sidings in July, 1966, and the investigation suggested that there was probably an eight-foot thick band of petrol vapour lying well beneath the surface of the ground in the East Street area. The vapour had been raised to the surface because of exceptionally heavy rainfall.

The distance from the point of spillage to the house was several hundred yards. (Kletz, 1972b.)

B66   On 30 November, 1962 at Cornwall, Ontario, a chlorine rail car was in a siding when a failure of the anchor section occurred and 30 ton of chlorine escaped. The siding was on the downwind side of town. Police and firemen organized prompt evacuation of the several hundred people in the rural area downwind of the spillage. Some 89 people were gassed. (Simmons, Erdmann and Naft, 1974.)

B67   On 9 August, 1963 in Philadelphia, Pennsylvania, a chlorine rail tank car was rammed and a 1 in. loading line was broken. For a short period there was an escape of chlorine, amounting probably to no more than 1 ton. The release was in the middle of a built-up area, however. The number of people gassed was at least 430. (Simmons, Erdmann and Naft, 1974.)

A1.12 Some Other Incidents and Problems

Some other incidents and problems related to safety and loss prevention are listed in Table A1.4 under the following heads: (1) power loss, weather and earthquake; (2) safety and safety legislation; (3) transport and transport legislation; (4) pipelines and pipelines legislation; (5) toxic chemicals in the workplace; (6) pollution; (7) pollution legislation; (8) abandonment of actual or planned production and (9) litigation.

The references in Table A1.4, which are quoted by date only, are to accounts given by 'Chementator' in Chemical Engineering. The information given refers primarily to the United States.

The table shows very clearly the increasing public concern with health and safety and with pollution in the process industries that began to build up from the early 1970s.

A1.13 Notation

Q	flow rate of propane (ft^3/s)
u	average wind velocity (ft/s)
x, y, z	downwind, cross-wind and vertical distances (ft)
σ_y, σ_z	standard deviations in crosswind, vertical directions (ft)
χ	concentration of propane (volume fraction)
χ_{cl}	concentration of propane on centreline (volume fraction)

Appendix

Flixborough

2

Contents

At about 4.53 p.m. on Saturday 1st June 1974, the Flixborough Works of Nypro (UK) Ltd (Nypro) were virtually demolished by an explosion of war-like dimensions. Of those working on the site at the time, 28 were killed and 36 others suffered injuries. If the explosion had occurred on an ordinary working day, many more people would have been on the site, and the number of casualties would have been much greater. Outside the Works injuries and damage were widespread but no one was killed. Fifty-three people were recorded as casualties by the casualty bureau, which was set up by the police; hundreds more suffered relatively minor injuries, which were not recorded. Property damage extended over a wide area, and a preliminary survey showed that 1821 houses and 167 shops and factories had suffered to a greater or lesser degree. (R.J. Parker, 1975 – the *Flixborough Report*, para. 1)

The Flixborough explosion was by far the most serious accident which had occurred in the chemical industry in the United Kingdom for many years.

The explosion was in the reactor section, Section 25A, of the caprolactam plant.

Within a month of the disaster a Court of Inquiry under the chairmanship of Mr R.J. Parker was set up under Section 84 of the Factories Act 1961 to establish the causes and circumstances of the disaster and to point out any lessons which might be learned.

The Court's report *The Flixborough Disaster, Report of the Court of Inquiry* (R.J. Parker, 1975) (the *Flixborough Report*) is the most comprehensive inquiry conducted in the United Kingdom into a disaster in the chemical industry.

The Flixborough disaster was of crucial importance in the development of safety and loss prevention in the United Kingdom. It made both the industry and the public much more aware of the potential hazard of large chemical plants and led to an intensification both of the efforts within industry to ensure the safety of major hazard plants and of the demands for public controls on such plants.

The setting up of the Advisory Committee on Major Hazards (ACMH) at the end of 1974 was a direct result of the Flixborough disaster.

The impact of Flixborough was reinforced by that of the Seveso disaster 2 years later.

The description of the Flixborough disaster given below is necessarily a brief one, and is mainly based on the *Flixborough Report* and on the work of Sadée, Samuels and O'Brien (1976–77). Other accounts at the time include those of J.I. Cox (1976b) and Gugan (1976). Further critical accounts are described in Section A2.9.

Selected references on Flixborough are given in Table A2.1.

Table A2.1 *Selected references on Flixborough*

Report of the Court of Inquiry
R.J. Parker (1975)

Reports presented to the Court
Artingstall (1975); Ball (1975a); V.J. Clancey (1975a); Cottrell and Swann (1975a); J.I. Cox (1975b); A.G. Evans (1975); Gill (1975); Gugan (1975a); T.B. Jones (1975); V.C. Marshall (1975d); F. Morton (1975); Munday (1975a); Newland (1975); Nypro (UK) Ltd (1975); A.F. Roberts (1975b); Sadee (1975); Samuels and O'Brien (1975)

Further accounts
Anon. (1974b); Ball and Steward (1974); T.B. Jones and Spracklen (1974); V.C. Marshall (1974, 1976a,b, 1982 LPB 48, 1984, 1987, 1994 LPB 117); Steward (1974); Tinker (1974); Anon. (1975d); Ball (1975b, 1976); Cottrell and Swann (1975b, 1976); J.I. Cox (1975a, 1976a,b); Dimeo (1975); FPA (1975/27); Gugan (1975b, 1976, 1979); R. King (1975a–c, 1976a,b, 1977, 1990, 1991); Kinnersly (1975); Kletz (1975c,e, 1976e, 1984e,f); McLain (1975a); Munday (1975b, 1976a,b); Rakestraw and Hildrew (1975); W. Smith (1975); Tucker (1975); F. Warner (1975a,b); BRE (1976/7); HSE (1976 HSE 1, HSE 2); O'Reilly (1976); Sadee, Samuels and O'Brien (1976–77); R.L. Allen (1977a,b); C.L. Bell (1977, 1979); V.J. Clancey (1977a,c); Mecklenburg (1977a,b); H.D. Taylor (1977); Slater (1978a); Anon. (1979b); Harvey (1979b); W.G. Johnson (1980); D.J. Lewis (1980d, 1989a); D.C. Wilson (1980); Anon. (1981a); Offord (1983); Cullen (1984); Davidson (1984); Crooks (1990 LPB 96); Turney (1994 LPB 117)

A2.1 The Company and the Management

A2.1.1 The company
The *Flixborough Report* (paras 12–14) states

The site was originally occupied in 1938 by a company called Nitrogen Fertilisers Limited, a subsidiary of Fisons Limited and used for the manufacture of ammonium sulphate. In 1964 Nypro was formed, owned jointly by Dutch State Mines (DSM) and Fisons Limited. It acquired the site from Nitrogen Fertilisers Limited. Between 1964 and 1967 plant was built for the production of caprolactam, which is a basic raw material for the production of Nylon 6, by a process, the first step of which was the production of cyclohexanone via the hydrogenation of phenol. The works were commissioned and production commenced in 1967. They were then and remained until the disaster the only works in the UK producing caprolactam.

In 1967 Nypro was reconstituted with DSM owning 45%, the National Coal Board 45% and Fisons Limited 10%. Almost at once design began for additional plant to increase the capacity of the works from 20 000 tons of caprolactam per annum to 70 000 tons per annum. This additional plant, referred to as Phase 2, was completed at a cost of some £15 m. in 1972. Its distinguishing feature for present purposes is that in it the cyclohexanone necessary for the production of caprolactam was produced by the oxidation of cyclohexane instead of via the hydrogenation of phenol. In 1972 DSM acquired the holding of Fisons Limited and from then until after the disaster Nypro was owned as to 55% by DSM and as to the remaining 45% by the NCB.

From the safety point of view, the oxidation process introduced new dimensions. Cyclohexane, which is in many of its properties comparable with petrol, had to be stored. More importantly, large quantities of cyclohexane had to be circulated through the reactors under a working pressure of about 8.8 kg/cm² and a temperature of 155°C. Any escape from the plant was therefore potentially dangerous.

Nypro (UK) Limited was therefore a relatively small, single-site company operating a major hazard process.

A2.1.2 The management

The management of the company is described in the *Flixborough Report* (paras 19–27). The description includes the following information of particular relevance to the caprolactam plant:

Managing Director   Mr R.E. Selman, qualified chemical engineer (HTS, Delft)
General Works Manager   Mr J.H. Beckers, qualified chemical engineer (MTS, Heerlen)
Plant Manager Area 1 and Utilities   Mr C.L. Bell, chartered chemical engineer
Plant Manager Area 2   Mr P.M. Cliff (to 1 May, 1974), chartered engineer (fuel); Mr R. Everton (from 1 May, 1974), HND in chemistry
Works Engineer   Vacant (formerly Mr Riggall)
Areas 1 and 2 Engineer   Mr A.B. Blackman
Services Engineer   Mr B.T. Boynton, ONC in electrical engineering, NCB (Class 1) engineering certificate
Safety and Training Manager   Mr E. Brenner, Manager

Area 2 covered Section 25A of the cyclohexane plant.

The previous Works Engineer, Mr Riggall, had left and steps were being taken to replace him, but at the time of the accident there was no Works Engineer. In the absence of a Works Engineer a co-ordinating function was exercised by the Services Engineer. In addition, the engineering staff were told that if they had problems, they could call on the assistance of Mr J.F. Hughes of the NCB.

The *Flixborough Report* (para. 27) states that following the departure of Mr Riggall 'There was no mechanical engineer on site of sufficient qualification, status or authority to deal with complex or novel engineering problems and insist on necessary measures being taken'.

The report is critical of the fact that the Area 2 and Services Engineers had been asked to assume responsibilities which they should not have had to shoulder.

A2.2 The Site and the Works

The Flixborough works is on flat, low-lying land on the east bank of the River Trent some 6 miles to the south of the point where that river joins the Humber. The nearest villages are Flixborough itself and Amcotts, both of which are about half a mile away. The town of Scunthorpe lies at a distance of approximately 3 miles.

The works is surrounded by fields and the population density in the neighbourhood beyond is very low.

A site plan is shown in Figure A2.1. The diagram also shows the vapour cloud, as described below. An aerial view of the works before the explosion is given in Figure A2.2.

Other plants on the site included an acid plant and a hydrogen plant. There was also a large ammonia storage sphere.

A2.3 The Process and the Plant

The cyclohexane plant, shown in Figure A2.3, consisted of a train of six reactors in series in which cyclohexane was oxidized to cyclohexanone and cyclohexanol by air injection in the presence of a catalyst. The feed to the reactors was a mixture of fresh cyclohexane and recycled material. The product from the reactors still contained approximately 94% of cyclohexane. The liquid reactants flowed from one reactor to the next by gravity. In subsequent stages, the reaction product was distilled to separate the unreacted cyclohexane, which was recycled to the reactors, and the cyclohexanone and cyclohexanol, which were converted to caprolactam. The design operating conditions in the reactor were a pressure of 8.8 kg/cm^2 and a temperature of 155°C. The reaction is exothermic.

The heat required for initial warmup and for supplementation of the heat of reaction during normal operation was provided by a steam-heated heat exchanger on the reactor feed. The steam flow to the exchanger was controlled by an automatic control valve. There was a bypass around this valve, which was needed to pass the larger quantities of steam required during start-up.

Removal of the heat of reaction from the reactors during normal operation was effected by vaporizing part of the cyclohexane liquid. The cyclohexane vaporized passed out in the off-gas from the reactors. The rest of this off-gas was mainly nitrogen with some unreacted oxygen.

The off-gas passed through the feed heat exchanger (C2544) and then through a cooling scrubber (C2521) and an absorber (C2522), in which the cyclohexane was condensed out, and thence via an automatic control valve to a flare stack.

The atmosphere in the reactor was controlled using nitrogen from high-pressure nitrogen storage tanks. The nitrogen was brought into the works by tankers.

The reactor pressure was controlled by manipulating the control valve on the off-gas line. Safety valves venting into the relief header to the flare stack were set to open at 11 kg$_f$/cm^2.

A trip system was provided which shut off air to, and injected nitrogen into, the reactors in the event of either a high oxygen content in the off-gas or a low liquid level in the nitrogen supply tank. This trip could be disarmed, however, by setting to zero the timer fixing the duration of the purge.

The layout of the reactors in Section 25A is shown in Figure A2.4.

A2.4 Events Prior to the Explosion

On the evening of 27 March 1974, it was discovered that Reactor No. 5 was leaking cyclohexane. The reactor was constructed of $\frac{1}{2}$ in. mild steel plate with $\frac{1}{8}$ in. stainless steel bonded to it on the inside. A vertical crack was found in the mild steel outer layer of the reactor. The leakage of cyclohexane from the crack indicated that the inner stainless steel layer was also defective. It was decided to shut the plant down for a full investigation. The following morning inspection revealed that the crack had extended some 6 ft. This was a serious state of affairs and a meeting was called to decide action. The decision was taken to remove Reactor No. 5 and to install a bypass assembly to connect Reactors No. 4 and 6 so that the plant could continue in production.

The openings to be connected on these reactors were 28 in. diameter, with bellows on the nozzle stubs, but the largest pipe which was available on site and which might be suitable for the bypass was 20 in. diameter. The two flanges were at different heights so that the connection had to take the form of a dog-leg of three lengths of 20 in. pipe welded together with flanges at each end bolted to the existing flanges on the stub pipes on the reactors. The bypass assembly is shown in Figure A2.5.

Calculations were done to check (1) that the pipe was large enough for the required flow and (2) that it was

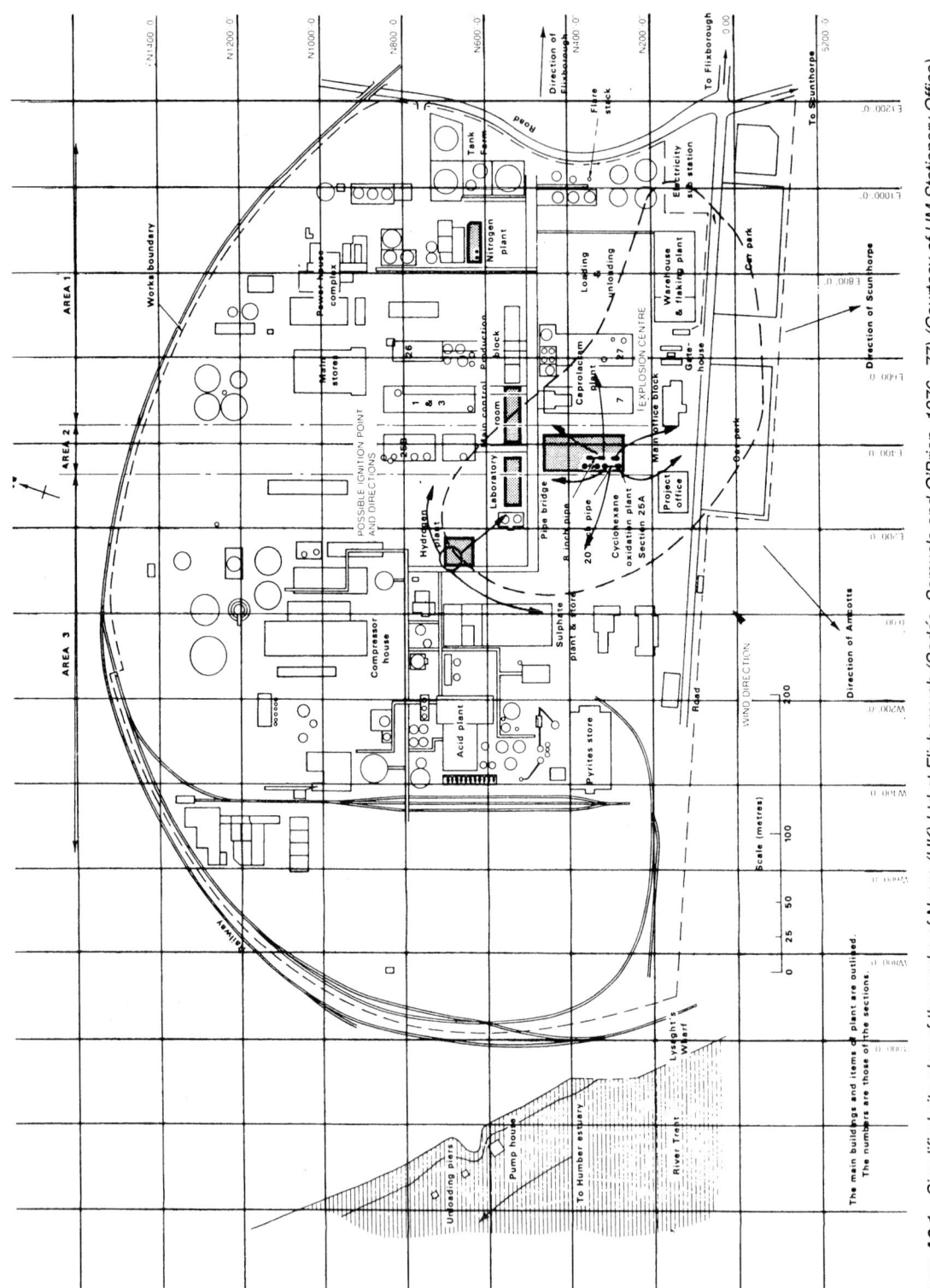

Figure A2.1 Simplified site plan of the works of Nypro (UK) Ltd at Flixborough (Sadée, Samuels and O'Brien, 1976–77) (Courtesy of HM Stationery Office). The diagram also shows the estimated dimensions of the vapour cloud

Figure A2.2   *Works of Nypro (UK) Ltd at Flixborough before the explosion (Nypro (UK) Ltd)*

capable of withstanding the pressure as a straight pipe. No calculations were made which took into account the forces arising from the dog-leg shape of the pipe.

No drawing of the bypass pipe was made other than in chalk on the workshop floor.

The existing stub pipes were connected to the reactors by bellows, as shown in Figure A2.5. No calculations were done to check whether the bellows would withstand the forces caused by the dog-leg pipe.

The bypass assembly was supported by a scaffolding structure, as shown in Figure A2.5. This scaffolding was intended to support the pipe and to avoid straining of the bellows during construction of the bypass. It was not suitable as a permanent support for the bypass assembly during normal operation.

No pressure testing was carried out either on the pipe or on the complete assembly before it was fitted. A pressure test was performed on the plant, however, after installation of the bypass. The equipment was tested to a pressure of 9 kg_f/cm^2, but not up to the safety valve pressure of 11 kg_f/cm^2. The test was pneumatic not hydraulic.

Following these modifications the plant was started up again. The bypass assembly gave no trouble. There did appear, however, to be an unusually large usage of nitrogen

on the plant and this was being investigated at the time of the accident.

On 29 May, the bottom isolating valve on a sight glass on one of the vessels was found to be leaking. It was decided to shut the plant down to repair the leak.

On the morning of 1 June start-up began. The precise sequence of events is complex and uncertain. The crucial feature, however, is that the reactors were subjected to a pressure somewhat greater than the normal operating pressure of 8.8 kg_f/cm^2.

A sudden rise in pressure up to 8.5 kg_f/cm^2 occurred early in the morning when the temperature in Reactor No. 1 was still only 110°C and that in the other reactors was less, while later in the morning, when the temperature in the reactors was closer to the normal operating value, the pressure reached 9.1–9.2 kg_f/cm^2.

The control of pressure in the reactors could normally be effected by venting the off-gas, but this procedure involved the loss of considerable quantities of nitrogen. Shortly after warmup began, it was found that there was insufficient nitrogen to begin oxidation and that further supplies would not arrive until after midnight. Under these circumstances the need to conserve nitrogen would tend to inhibit reduction of pressure by venting.

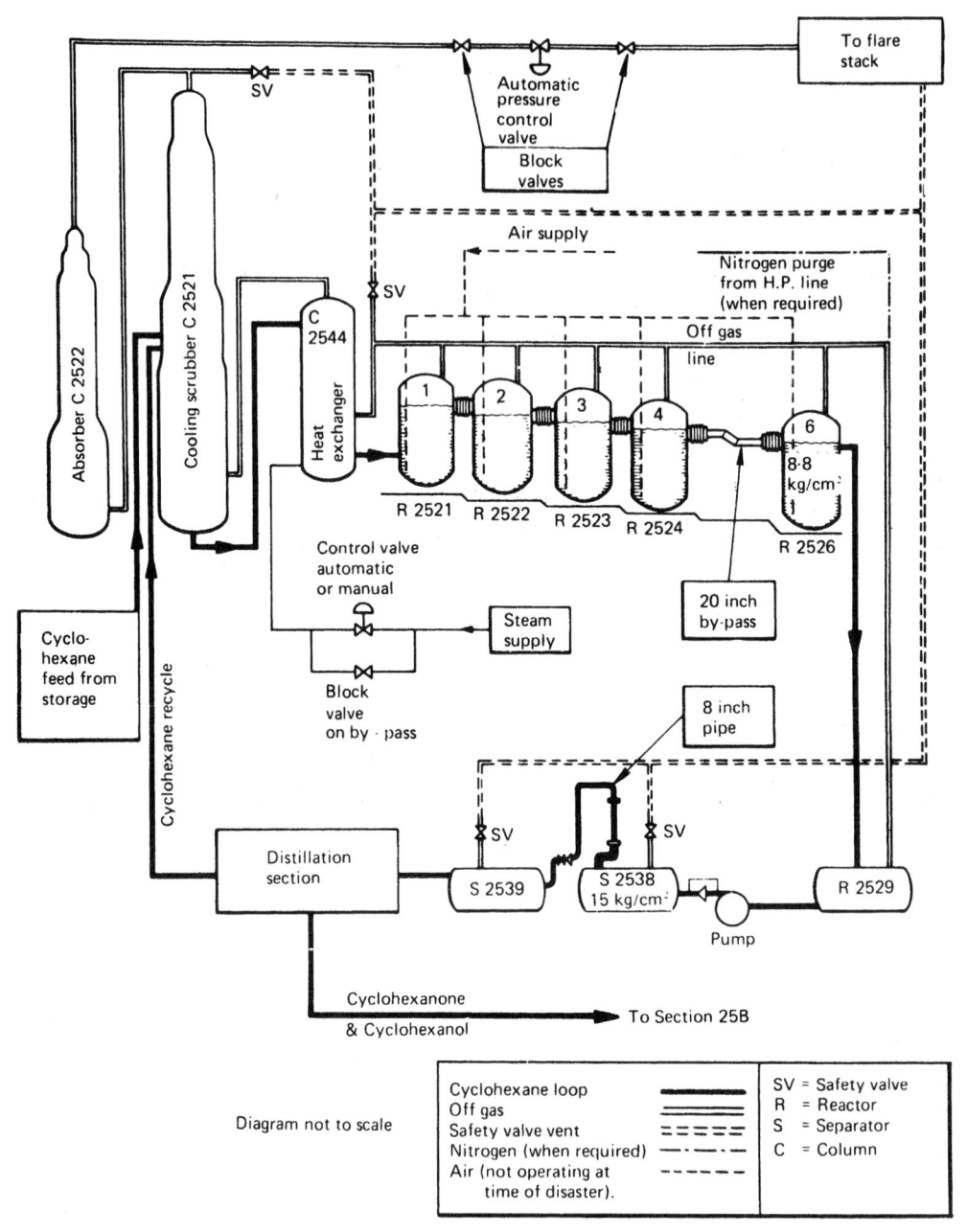

Figure A2.3 *Simplified flow diagram (not to scale) of the cyclohexane oxidation plant at Flixborough (R.J. Parker, 1975) (Courtesy of HM Stationery Office)*

A2.5 The Explosion – 1

During the late afternoon an event occurred which resulted in the escape of a large quantity of cyclohexane. As already explained, this event was the rupture of the 20 in. bypass system, without or with contribution from fire on a nearby 8 in. pipe. The cyclohexane formed a vapour cloud and the flammable mixture found a source of ignition.

At about 4.53 p.m. there was a massive vapour cloud explosion.

The explosion caused extensive damage and started numerous fires. Aerial views of the plant after the explosion are shown in Figures A2.6 and A2.7.

Reactors No. 4 and 6 and the debris of the bypass assembly are shown in Figure A2.8. The blast and the

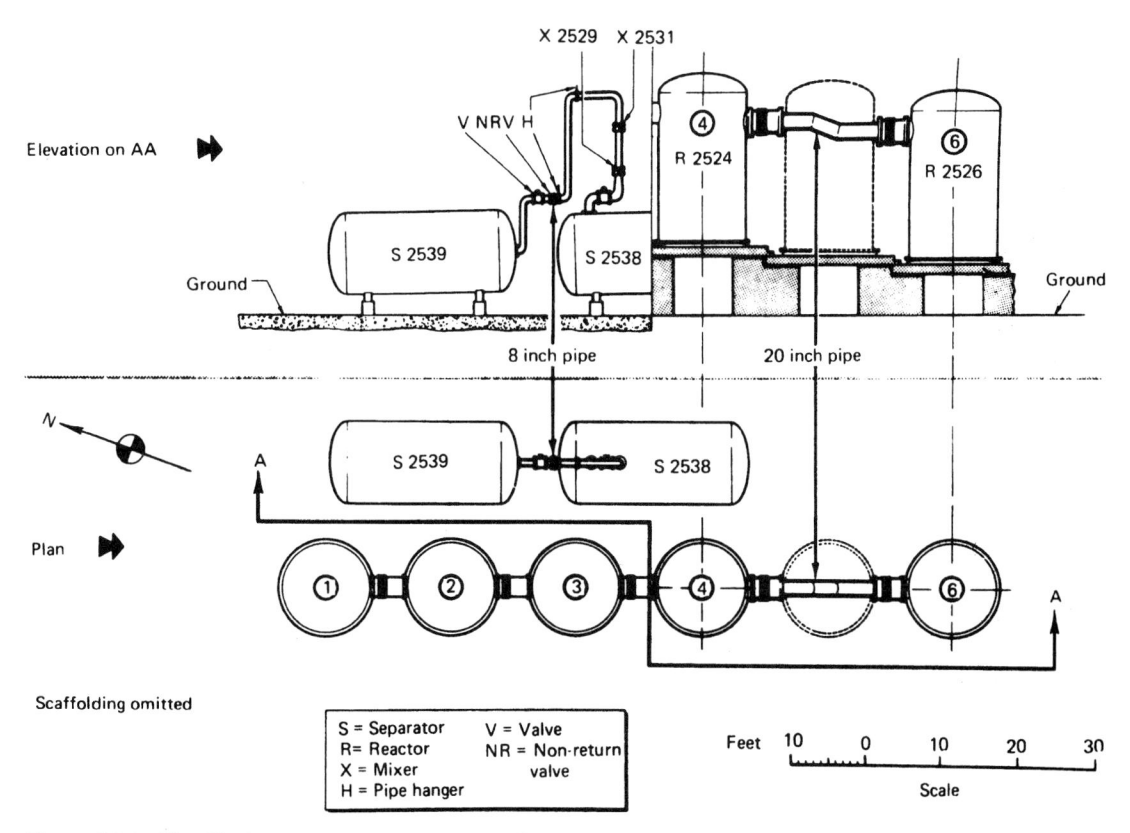

Figure A2.4 *Simplified plan and elevation of part of Section 25A of the cyclohexane oxidation plant at Flixborough (R.J. Parker, 1975) (Courtesy of HM Stationery Office)*

fires destroyed not only the cyclohexane plant but several other plants also. Many of the fires were in the tank farm.

The blast of the explosion shattered the windows of the control room and caused the control room roof to collapse. Of the 28 people who died in the explosion, 18 were in the control room. Some of the bodies had suffered severe injuries from flying glass. Others were crushed by the roof. No one escaped from the control room.

The main office block was also demolished by the blast of the explosion. Since the accident occurred on a Saturday afternoon, the offices were not occupied. If they had been, the death toll would have been much higher.

The locations of the control room and main office block are shown in Figures A2.1 and A2.2 and the states of these buildings after the explosion in Figures A2.6 and A2.7.

The fires on the site burned for many days. Even after 10 days the fires were hindering rescue work on the site. The large ammonia sphere was lifted up a few inches. It leaked slightly at a flange, but the leak was not serious.

A2.6 The Investigation

An immediate investigation of the disaster was made by the Factory Inspectorate, which issued an interim report. On its appointment, the Court of Inquiry took control of the main investigations. The work of recovering, identifying and examining wreckage and of conducting relevant tests, was undertaken by the Safety in Mines Research Establishment (SMRE). The Court appointed the consulting engineers Cremer and Warner as its technical advisers. Other government bodies, consultants and universities were involved in the numerous studies undertaken.

Some of the possible causes of failure of the bypass assembly were outlined early in the inquiry by the DSM counsel. They included (1) failure of the 20 in. pipe due to a small pressure rise; (2) prior failure of the 8 in. pipe; (3) prior failure of some other part of the system and (4) an explosion in the air-line to the reactors. He suggested several possible causes for a pressure rise in the 20 in. pipe: (1) entry of high-pressure nitrogen into the system due to instrument malfunction; (2) entry of water into the system; (3) temperature rise in the system due to excessive heating by steam in reboiler of C2544; (4) leakage of steam from a tube in C2544; (5) explosion of peroxides formed in the process and (6) explosion due to air in the system.

The inquiry established (*Flixborough Report*, para. 6) that the cause of the disaster was 'The ignition and rapid acceleration of deflagration, possibly to the point of detonation, of a massive vapour cloud formed by the escape of cyclohexane under a pressure of at least 8.8 kg/cm^2 and

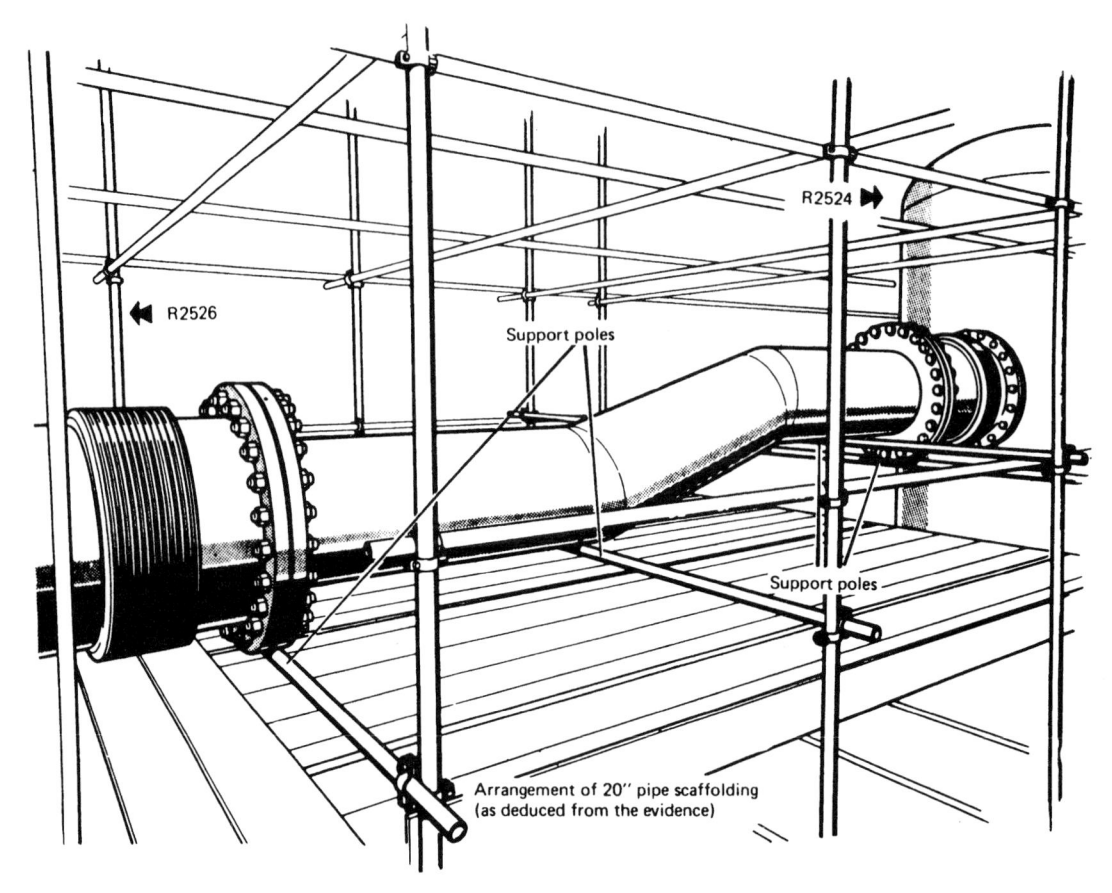

R2524 ►►

◄◄ R2526

Support poles

Support poles

Arrangement of 20″ pipe scaffolding
(as deduced from the evidence)

Figure A2.5 *Arrangement of the 20 in bypass pipe and scaffolding at Flixborough (R.J. Parker, 1975)*
(Courtesy of HM Stationery Office)

a temperature of 155°C and that the escape was from Section 25A of the cyclohexane plant'.

There was no dispute that the main part of the cyclo-hexane came from the 20 in. bypass assembly, but there was come uncertainty whether the mechanical failure of the assembly was the primary failure or whether it was a secondary failure caused by another event elsewhere. Three hypotheses were advanced: (1) the 20 in. pipe hypothesis, (2) the 8 in. pipe hypothesis and (3) the super-heated water hypothesis.

A large proportion of the report is concerned with a dis-cussion of the first and second hypotheses, but the Court decided emphatically in favour of the first.

The 20 in. pipe hypothesis is that the 20 in. bypass pipe ruptured in one stage as a result of the internal pressure and temperature conditions which probably occurred in the final shift.

The site investigation revealed that the 20 in. pipe had yielded at the lower mitre, or 'jack-knifed', and that the bellows had undergone gross permanent deformation, or 'squirm'.

The arrangement of the pipe was such as to subject the bellows to shear forces due to the internal pressure in the pipe as shown in Figure A2.9. The two thrusts PA are parallel and separated by a distance $2e$ and thus give a turning moment $2PAe$ which is balanced by shear forces $F = PAE/L$. The shear forces give rise to bending moments and the maximum bending moment is at the mitre joints and is $PAEl/L$. This system of forces can lead to failure in two ways. The shear force may cause squirm of the bellows. The bending moments may cause failure of the pipe by buckling one of the mitres. As already mentioned, these effects had been neglected in the calculations done on the bypass assembly.

These mechanisms were studied in comprehensive, full-scale investigations by SMRE, by Nottingham University and the Mining Research and Development Establishment, Bretby. The results of the tests are complex, but, in brief, they showed that squirm of the bellows could occur at pressures at or near the operating pressure. They did not show jackknifing and rupture following squirm. A mechanism for the latter was given by Newland (1975), who developed a theory based on the energies involved, which showed that at pressures above normal operating pressures but below the safety valve pressure there could occur squirm of both the bellows, followed by jackknifing of the

Figure A2.6   *Works of Nypro (UK) Ltd at Flixborough after the explosion – 1 (R.J. Parker, 1975) (Courtesy of HM Stationery Office)*

pipe, resulting in complete rupture. The pressures for jackknifing and rupture to be a consequence of squirming were estimated to be as follows:

	Low (kg$_f$/cm^2)	*Probability* 50% (kg$_f$/cm^2)	*High* (kg$_f$/cm^2)
150°C	9.3	9.9	10.6
160°C	9.1	9.7	10.4

Other mechanisms of squirm alone and jackknifing alone were investigated, but appeared less probable.

The Court concluded that the rupture of the 20 in. pipe assembly was a probability, albeit a low one, under the pressure and temperature conditions which occurred in the last shift.

The 8 in. pipe hypothesis is somewhat more complex. The 8 in. pipe ran between separators S 2538 and S 2539. The location and details of the pipe are shown in Figures A2.4 and A2.10, respectively. There was a block valve, a non-return valve and a lagging box on the pipe. The pipe was made of stainless steel.

Essentially the hypothesis was that a directed flame had occurred from the lagging box and had caused rupture of the pipe from which cyclohexane then escaped to form a vapour cloud. This hypothesis was advanced by Cox and Gugan.

The hypothesis had its origin in eyewitness accounts of the events immediately prior to the explosion and in the condition of the equipment found in the site investigation. The latter revealed that there was a 50 in. split on the elbow of the 8 in. pipe and subsequent studies showed that the crack was caused by creep cavitation. The 50 in. split also exhibited a petal crack which was due to zinc embrittlement and a separate 3 in. crack due to the same cause was found on the straight vertical section of the pipe. In addition, two loose bolts were found on the non-return valve. There was some doubt as to whether these had been loose before the explosion, but the Court concluded that this was probably the case.

The 8 in. pipe hypothesis may be summarized as follows. There were two loose bolts on the non-return valve. Then either there was a slow leakage of material over an extended period and a lagging fire which caused further expansion of the bolts and growth of the leak, or there was a sudden gasket burst and ignition of the leak by static electricity. In either case there was a directed flame from the lagging box. A sprinkler sensor head near the non-return valve failed to detect the fire. The flame burned off the lagging on the pipe elbow and caused a vapour blanket in the pipe and a vapour lock at the intrados so that the pipe overheated. The pipe suffered creep cavitation at the intrados and the 50 in. rupture occurred. Zinc from galvanized wire dripped onto the pipe and caused zinc embrittlement, but this was not responsible for the main failure. The

Figure A2.7   *Works of Nypro (UK) Ltd at Flixborough after the explosion–2 (R.J. Parker, 1975) (Courtesy of HM Stationery Office)*

cyclohexane issued from the 50 in. crack and put out the flame. A fire then occurred in the region of the fin fan coolers which melted the zinc on the finned tubes and raised them to a temperature at which zinc embrittlement occurs. There followed an explosion in the region of one of the fan motors which caused the 20 in. bypass to jackknife and rupture.

The foregoing account is a necessarily highly simplified summary of a very complex argument. The Court's report gives a detailed account of this hypothesis but rejects it. It draws an analogy between the sequence of events required by the hypothesis with the balancing of billiard balls one on top of another and states 'This balancing of ten balls is analogous to what is postulated for the 8-inch pipe hypothesis'.

The third hypothesis is that the failure of the 20 in. bypass assembly was due to sudden evolution of vapour from a layer of superheated water at the bottom of Reactor No. 4. The water might have been present in the reactor initially or might have come from a leak in the steam-heated heat exchanger on the feed. Such a water layer was able to build up in the vessel more easily because the reactor agitator was not in use. This theory, advanced by R. King, was mentioned, but did not figure prominently at the inquiry. It has been described in a number of articles by King (1975a–c, 1976a,b, 1977).

A2.7 The Explosion – 2

Another aspect of the Flixborough investigation is the studies conducted on the explosion itself. Accounts of the explosion have been given in the *Flixborough Report*, by Gugan (1976) and by Sadée, Samuels and O'Brien (1976–77).

According to Sadée, Samuels and O'Brien the quantity of cyclohexane available in the five reactors and one after-reactor was 120 te. The normal operating pressure and temperature were $8.8 \, \text{kg}_\text{f}/\text{cm}^2$ and 155°C, respectively. After the explosion it was found that the vessels still contained 80 te, so that the maximum quantity involved in the explosion was 40 te. These authors estimate that about 30 te escaped before the explosion and they take this quantity as the basis of their calculations.

The flow from each of the two pipe stubs, which were opposite each other, but at slightly different heights, would form a turbulent momentum jet with the two opposing jets impinging on each other, thus giving very effective mixing.

It is estimated by Sadée, Samuels and O'Brien that 50%, or 15 te, of the cyclohexane would flash off as vapour and that the other 50% would form a mist. Whether or not the mist would then evaporate depends on the amount of air entrained by the jet. The authors state that evaporation is possible provided that the concentration of cyclohexane does not exceed 10%.

Figure A2.8 *Reactors No. 4 and 6 and the bypass assembly at Flixborough after the explosion*
(R.J. Parker, 1975) (Courtesy of HM Stationery Office)

Sadée, Samuels and O'Brien have calculated the dimensions of a stationary, free jet of cyclohexane at 1 MPa and 150°C emitted from a 700 mm diameter opening and forming a cloud containing equal masses of vapour and liquid in air. This jet is shown in Figure A2.11. The jet is 215 m long and has a maximum diameter of 25 m. The jet volume with a concentration exceeding the low flammability limit (1.2%) is 64,400 m³ and that with a concentration within the flammable range (1.2–8%) is practically the same. The average concentration in the flammable part of the jet is 1.85%, which is close to the stoichiometric proportion. The quantity of cyclohexane within the flammable part of the jet is 4 te. If two such jets are assumed, the quantity is 8 te.

The jets actually formed were not free jets, but were affected by the ground and the reactors. Taking these factors into account, the authors estimate the quantity of cyclohexane within the flammable part of the two jets as 22.4 te.

Sadée, Samuels and O'Brien present calculations on the assumption that the quantity of cyclohexane involved in the explosion was 30 te. This amount of cyclohexane mixed with air to yield a flammable concentration of 2% gives a cloud volume of 400,000 m³.

The precise shape of the cloud is uncertain, as is the source of ignition. From the carbon deposition and explosion damage the estimate of the cloud ground zero and shape is that shown in Figure A2.1. The source of ignition may have been the hydrogen plant. An alternative estimate of the cloud shape has been given by J.I. Cox (1976b).

Various estimates have been made of the time which elapsed between the times of rupture and of the explosion, mostly ranging between 30 and 90 s. An estimate of 45 s has been widely quoted.

There is evidence from carbon deposition and from eyewitness accounts that there was a flash fire as well as an explosion in the vapour cloud.

From eyewitness accounts it appears that there may have been a short period during which the cloud had found a source of ignition and combustion was occurring at the edge of the cloud, but the cloud was expanding faster than the flame speed. Assuming a constant flow of vapour into the cloud, the velocity of outward expansion would fall as the cloud diameter increased and a point would be reached at which the velocity would fall below the flame speed. The flame would then travel rapidly towards the centre of the cloud.

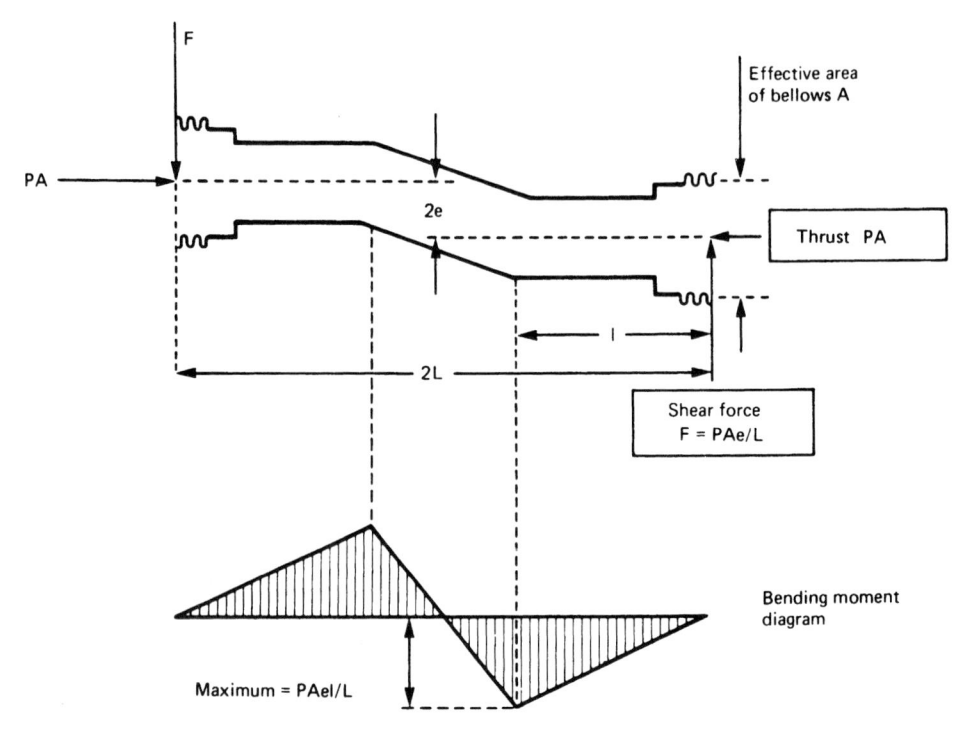

Figure A2.9 *Sketch of pipe and bellows assembly at Flixborough showing shear forces on bellows and bending moments in pipe (due to internal pressure only) (R.J. Parker, 1975) (Courtesy of HM Stationery Office)*

One effect of such a flash fire would be to cause a sudden and lethal depletion of the oxygen concentration.

There was evidence of two explosive events. An ionosphere recorder trace obtained at Leicester University (T.B. Jones and Spracklen, 1974; T.B. Jones, 1975) showed two disturbances, a smaller precursor followed by the main event. The time interval between them was some 45 s.

Estimates of the energy of explosion were made in several ways, including estimation from ionospheric readings (T.B. Jones, 1975), from the barograph of a glider in the vicinity (Gugan, 1975a) and from a blast damage survey (A.F. Roberts, 1975; Samuels and O'Brien, 1975).

The most detailed estimates of explosion energy are those derived from the blast damage survey. The results of this survey have been discussed by Gugan (1976), Munday (1975b) and Sadée, Samuels and O'Brien (1976–77). The latter gave a detailed listing and photographs of structural damage.

The blast damage effects caused by the explosion have been assessed by Sadée, Samuels and O'Brien and are shown in Figure A2.12, in which the individual data bars give the ranges of peak overpressure determined from structural damage vs distance from the assumed epicentre of the explosion.

The estimates of the energy of the explosion made by different workers depend on the explosion model used.

Most estimates are based on the simplest model, in which the only parameter is the mass of TNT. These estimates are generally in the range 15–45 te of TNT.

An alternative two-parameter model describes the explosion in terms both of the mass of TNT and the height of the explosion. Using this model, Sadée, Samuels and O'Brien estimate that the explosion was equivalent to 16 ± 2 te of TNT at a height of 45 ± 24 m above ground. 16 te of TNT have a calorific equivalent of about 1.6 te of cyclohexane; the calorific value of TNT is 1600 kcal/kg and that of cyclohexane is 11,127 kcal/kg.

The curve given by Sadée, Samuels and O'Brien in Figure A2.12 is based on this two-parameter TNT model.

If the cloud contained 30 te of cyclohexane and gave an explosion equivalent to the combustion of about 1.6 te, the explosion efficiency was about 5%. This efficiency is based on the total quantity of cyclohexane in the cloud.

The maximum overpressure at the centre of the cloud is uncertain. The maximum overpressure obtained from the structural damage, as described by Sadée, Samuels and O'Brien, was 0.7 bar. These authors state, however, that the theoretical curve in Figure A2.12 indicates an overpressure of 1 bar at the cloud boundary.

Some estimates of the pressure developed within the cloud are considerably higher. These estimates are typically based on calculation of the forces required to give damage effects such as the bending of steel lamp posts or crushing of steel vessels. Thus, for example, V.J. Clancey (1977a) states that where local effects occurred as a result of features such as confinement, pressures of up to perhaps 5–7 bar have been estimated. The pressures calculated in this way would normally be reflected pressures.

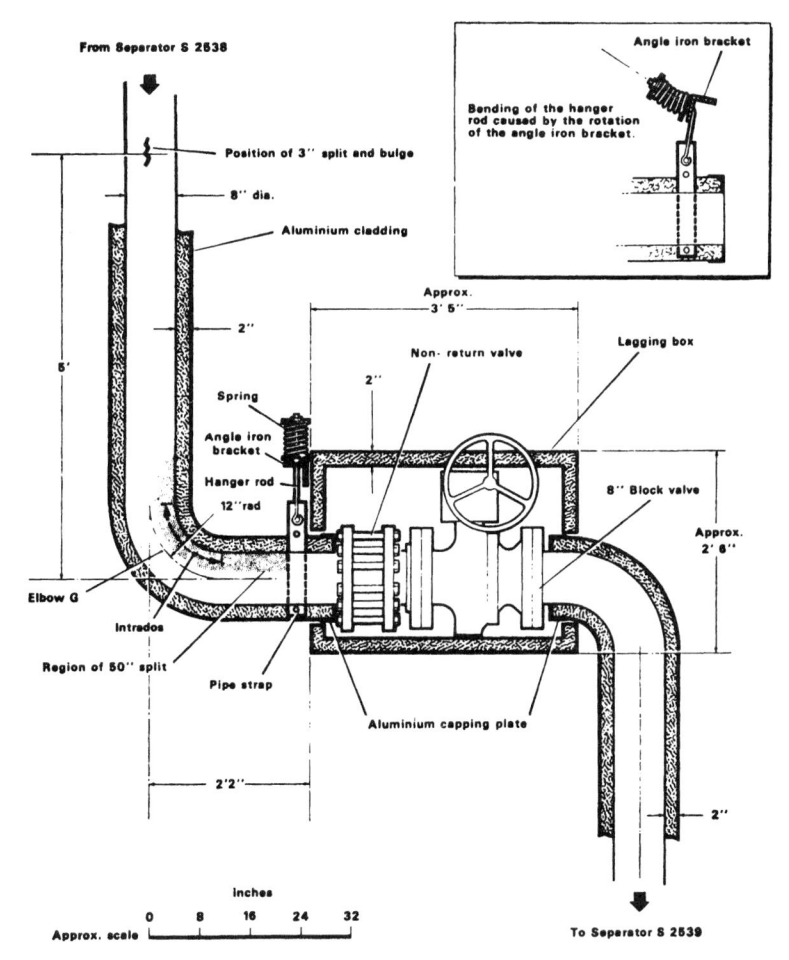

Figure A2.10   *Elevation view of 8 in. line at Flixborough, showing typical lagging box around valve (R.J. Parker, 1975) (Courtesy of HM Stationery Office)*

A2.8 Some Lessons of Flixborough

There are numerous lessons to be learned from the Flixborough disaster. A list of some of these is given in Table A2.2. Many of these lessons were highlighted in the Court's report.

The lessons include both public controls on major hazard installations and the management of such installations by industry. In the latter area there are lessons on both management systems and technological matters and on both design and operational aspects.

The fact that alternative hypotheses were advanced concerning the cause of the explosion does not detract from these lessons, but rather means that a greater number of lessons can be drawn.

Some of these lessons are now considered.

Public controls on major hazard installations

The effect of the Flixborough disaster was to raise the general level of awareness of the hazard from chemical plants and to make the existing arrangements for the control of major hazard installations appear inadequate.

The government therefore set up the ACMH to advise on means of control for such installations. The committee issued three reports (Harvey, 1976, 1979b, 1984). Its recommendations are described in Chapter 4.

This work was a major input to the development of the EC Major Accident Hazards Directive, which was implemented in the United Kingdom as the CIMAH Regulations 1984. These require the operator of a major hazard installation to produce a safety case.

Major hazard installations receive a greater degree of supervision by the local Factory Inspectorate. The CIMAH safety case plays an important part in this.

Siting of major hazard installations

As the *Flixborough Report* (para. 11) points out, the casualties from the explosion might have been much greater if the site had not been in open country.

The siting of major hazard installations is a matter of utmost importance. On the one hand distance is the only sure guarantee of safety, but on the other provision of increased spacing may be very wasteful of land.

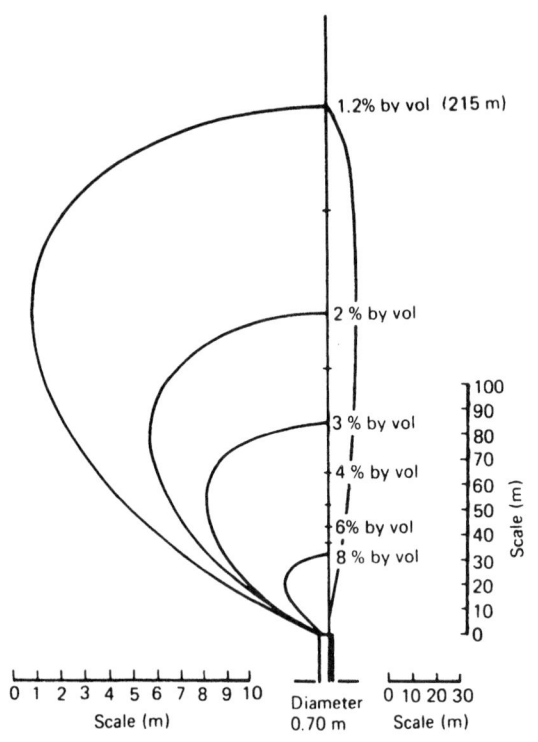

Figure A2.11 *Dimensions of and concentration distribution in a stationary, free jet of cyclohexane in air from an opening 700 mm diameter (Sadée, Samuels and O'Brien, 1976–77) (Courtesy of HM Stationery Office)*

The question of the siting of major *hazard* installations, or more generally land use planning in relation to such installations, became a principal concern of the ACMH, and is treated in its three reports. Controls on major hazards through land use planning are discussed in Chapter 4 and Appendix 25. Siting is also considered in Chapter 10.

Licensing of storage of hazardous materials

The *Flixborough Report* (para. 194) states

> As at 1st June 1974 Nypro were storing on site 330,000 gallons of cyclohexane, 66,000 gallons of naphtha, 11,000 gallons of toluene, 26,400 gallons of benzene and 450,000 gallons of gasoline. The storage of these potentially dangerous substances is nominally controlled by the local authority issuing licences under the Petroleum (Consolidation) Act 1928. In fact the only licences that had been issued related to 7000 gallons of naphtha and for a total of 1500 gallons of gasoline. The unlicensed storage of large quantities of fluids had no effect upon this disaster but it is clearly useless to have a licensing system which is so ineffective that it can lead to such results.

The situation at Flixborough revealed the need for better methods of notification of major hazard installations to the local planning authorities and for greater guidance to these authorities by the HSE.

The notification of major hazard installations is a requirement of the NIHHS Regulations 1982 and of the

CIMAH Regulations 1984. These notification requirements and the advice now given by the HSE to local planning authorities are discussed in Chapter 4.

Regulations for pressure vessels and systems

The escape of cyclohexane at Flixborough was caused by a failure of the integrity of a pressure system.

Current legislation on pressure systems in the United Kingdom consisted of regulations for steam boilers and receivers and air receivers. It failed to cover the Flixborough situation in two crucial respects. It applied only to steam and air and not to hazardous materials such as cyclohexane. And it dealt only with pressure vessels and not with pressure systems. This latter point is relevant, because at Flixborough the failure occurred in a pipe not a vessel.

The *Flixborough Report* (para. 209) recommends that existing regulations relating to the modification of steam boilers should be extended to apply to pressure systems containing hazardous materials.

Following Flixborough the HSE set about developing comprehensive regulations for pressure systems, which finally emerged as the Pressure Systems Regulations 1989. These regulations are described in Chapters 3 and 12.

The management system for major hazard installations

The *Flixborough Report* (paras 207, 209, 210) places much emphasis on deficiencies in the management system at the Nypro Works.

The works did not have a sufficient complement of qualified and experienced people. Consequently, management was not able to observe the requirement that persons given responsibilities should be competent to carry them out. In particular, there was no works engineer in post and no adequately qualified mechanical engineer on site.

Moreover, individuals tended to be overworked and thus more liable to error.

The management system, however, is more than the individuals. It includes the whole structure which supports them. Thus the system should provide, for example, for the coverage of absence due to resignation, illness and so on.

The use of a comprehensive set of procedures is another important aspect of the management system. A crucial procedure which was deficient at Flixborough was that for the control of plant modifications.

The role of the safety officer at Flixborough was not well defined.

The importance for major hazard installations of the management and the management system was the single most prominent theme in the work of the ACMH. Emphasis by the HSE on management aspects has steadily grown. The *Cutten Report* on the Piper Alpha disaster recommended that offshore the safety case should cover the safety management system and this is implemented in the offshore regulations. This aspect is discussed in Chapters 4 and 6.

Relative priority of safety and production

The *Flixborough Report* (paras 57, 206) drew attention to the conflict of priorities between safety and production. It states

> We entirely absolve all persons from any suggestion that their desire to resume production caused them knowingly to embark on a hazardous course in disregard of the

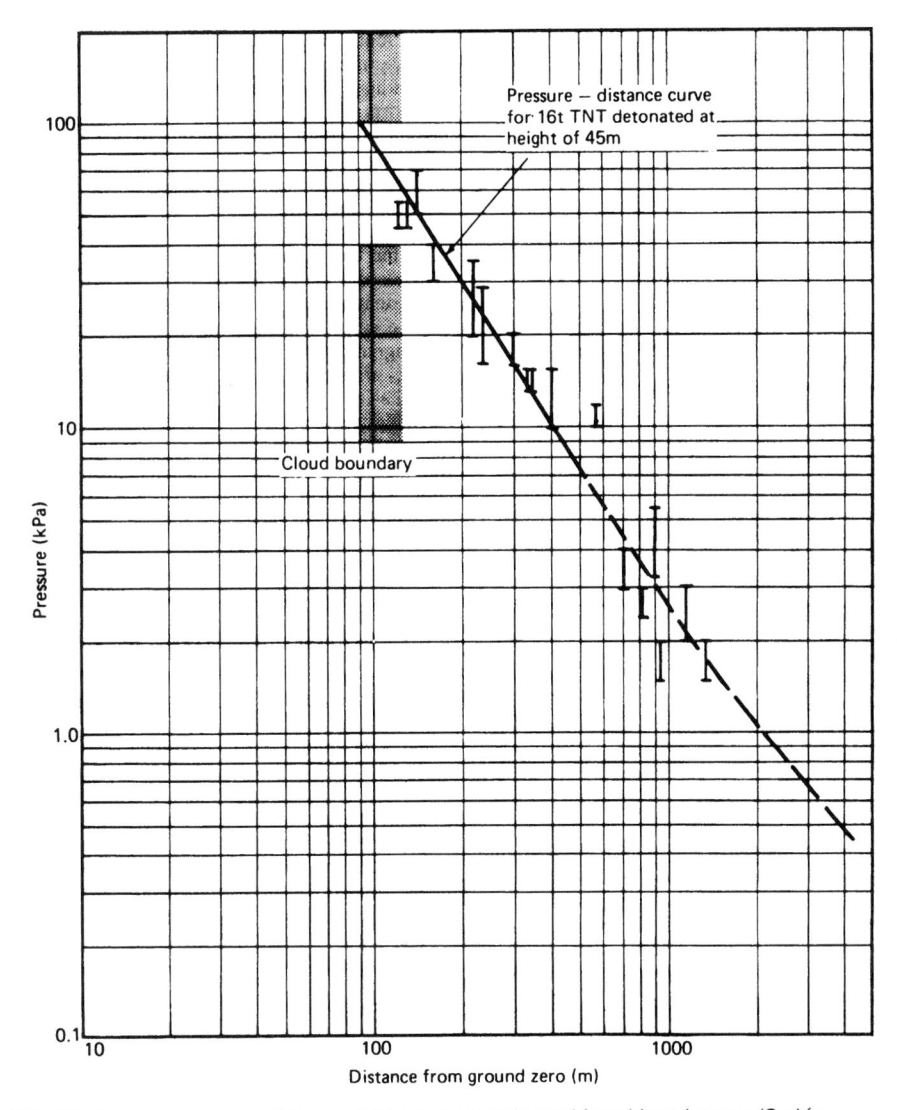

Figure A2.12 *Peak overpressure vs distance at Flixborough estimated from blast damage (Sadée, Samuels and O'Brien, 1976–77) (Courtesy of HM Stationery Office)*

safety of those operating the Works. We have no doubt, however, that it was this desire which led them to overlook the fact that it was potentially hazardous to resume production without examining the remaining reactors and ascertaining the cause of the failure of the fifth reactor. We have equally no doubt that the failure to appreciate that the connection of Reactor No. 4 to Reactor No. 6 involved engineering problems was largely due to the same desire.

Use of standards and codes of practice
As the *Flixborough Report* (paras 61–73) describes, the 20 in. bypass assembly was not constructed and installed in accordance with the relevant standards and codes of practice.

The principal standard relevant to the modification was BS 3351: 1971. The report quotes extracts from this standard:

4.6.2. For axial bellows, the piping shall be guided to maintain axiality of the bellows and anchored at adjacent changes in direction to prevent the bellows being subjected to axial load due to fluid pressure. 5.4.2.1 . . . . When expansion joints are used, the advice of the manufacturer should be sought with regard to the guiding, anchoring and support of the adjacent pipework.

The bellows manufacturer, which was Teddington Bellows Ltd, produced a *Designers Guide* which made it clear that two bellows should not be used out of line in the same pipe without adequate support for the pipe.

Table A2.2   *Some lessons of Flixborough*

Public control of major hazard installations
Siting of major hazard installations
Licensing of storage of hazardous materials
Regulations for pressure vessels and systems
The management system for major hazard installations
   The structure
   The people
   The systems and procedures
   The safety officer
Relative priority of safety and production
Use of standards and codes of practice
Limitation of inventory in the plant
Engineering of plants for high reliability
Dependability of utilities
Limitation of exposure of personnel
Design and location of control rooms and other buildings
Control and instrumentation of plant
Decision-making under operational stress
Restart of plant after discovery of a defect
Control of plant and process modifications
Security of and control of access to plant
Planning for emergencies
The metallurgical phenomena
   Nitrate stress corrosion cracking
   Creep cavitation of stainless steel
   Zinc embrittlement of stainless steel
   Use of clad mild steel
Unconfined vapour cloud explosions
Investigation of disasters and feedback of information on
   technical incidents

The use of standards and codes of practice is described in Chapters 3, 6 and 12.

Limitation of inventory in the plant

The *Flixborough Report* (para. 14) makes it clear that the large inventory of flammable material in the plant contributed to the scale of the disaster.

The *Second Report* of the ACMH proposes that limitation of inventory should be taken as a specific design objective in major hazard installations.

The limitation of inventory is a particular aspect of the more general principle of inherently safer design, which is now widely recognized. This is discussed in Chapter 11.

Engineering of plants for high reliability

The explosion at Flixborough occurred during a plant start-up. The *Flixborough Report* (para. 206) suggests that special attention should be given to factors which necessitate the shut-down of chemical plant so as to minimize the number of shut-down/start-up sequences and to reduce the frequency of critical management decisions.

The reliability of plant is considered in Chapters 7 and 12.

Dependability of utilities

The high-pressure nitrogen required for the blanketing of the reactors at Flixborough was brought into the works by tankers. There was insufficient nitrogen available during the start-up when the explosion occurred. The *Flixborough Report* (para. 211) emphasizes the importance of assuring dependable supplies of nitrogen where these are necessary for safety.

The dependability of utilities is discussed in Chapter 11.

Limitation of exposure of personnel

The *Flixborough Report* (para. 1) states that the number of casualties would have been much greater if the explosion had occurred on a weekday instead of on a Saturday.

The *First Report* of the ACMH (para. 68) suggests that limitation of exposure of personnel be made a specific design objective.

Aspects of limitation of exposure are controls on access to hazardous areas and design and location of buildings in or near such areas.

The limitation of exposure of personnel is described in Chapters 10 and 20.

Design and location of control rooms and other buildings

Of the 28 deaths at Flixborough 18 occurred in the control room. The *Flixborough Report* (para. 218) refers to various suggestions made to the inquiry concerning the siting of control rooms, laboratories, offices, etc., and the construction of control rooms on blockhouse principles.

The construction of buildings for chemical plant is considered in the *First Report* of the ACMH (para. 69).

The design and location of control rooms and other buildings is discussed in Chapters 10 and 20.

Control and instrumentation of plant

The control and instrumentation system was not a prominent feature in the Flixborough inquiry. The *Flixborough Report* (para. 204) considered that the controls in the control room followed normal practice. But it also states 'Nevertheless we conclude from the evidence that greater attention to the ergonomics of plant design could provide rewarding results'.

Control system design and human factors in process control are described in Chapters 13 and 14.

Decision making under operational stress

The *Flixborough Report* (para. 205) draws attention to the problem of decision making under operational stress and emphasizes the desirability of reducing the number of critical management decisions which have to be made under these conditions.

Such critical decisions are not necessarily confined to management, however. Process operators may also be required to take important decisions in emergency conditions.

The process operator in emergency situations is discussed in Chapter 14.

Restart of plant after discovery of a defect

Following the discovery of the serious defect in Reactor No. 5 at Flixborough, the reactor was removed, a bypass assembly was installed and the plant was started up again. The *Flixborough Report* (para. 57) is critical of the fact that the remaining reactors were not examined and the cause of the failure in the fifth reactor was not ascertained before plant start-up.

The procedures for restart of the plant after discovery of a defect are discussed in Chapter 20.

Control of plant and process modifications

The *Flixborough Report* (para. 209) states

> The disaster was caused by the introduction into a well
> designed and constructed plant of a modification which
> destroyed its integrity. The immediate lesson to be
> learned is that measures must be taken to ensure that the
> technical integrity of plant is not violated.

As it happens, there was also a process modification at
Flixborough, although this is not emphasized in the report.
The agitator in Reactor No. 4 was not in use at the time of
the disaster. The absence of agitation in the reactor could
have allowed a water layer to accumulate more easily in the
bottom of the vessel. The sudden evolution of vapour from
superheated water was advanced by King as a possible
cause of the rupture of the bypass pipe assembly.

The control of plant and process modifications is discussed in Chapters 6 and 21.

Security of and control of access to plant

The *Flixborough Report* (para. 194) draws attention to the
fact that there were two unguarded gates through which
it was possible for anyone at any time to gain access,
although this fact did not contribute to the disaster. The
security of plants is considered in Chapter 20.

Planning for emergencies

The *Flixborough Report* (para. 222) calls for a disaster plan
for major hazard installations. Emergency planning is
described in Chapter 24.

The metallurgical phenomena

The Flixborough disaster drew attention to several important metallurgical phenomena. Thus the *Flixborough
Report* describes the nitrate stress corrosion cracking of
mild steel (paras 53, 212); the creep cavitation of stainless
steel (para. 214); the zinc embrittlement of stainless steel
(para. 213); and the use of clad mild steel vessels (para. 224).

The HSE subsequently issued *Technical Data Notes* (1976
TON 53/1, 53/2; 1977 TON 53/3) on the first three of these
problems.

It was apparent from the discussion in the engineering
profession following publication of the report that although
metallurgical specialists were aware of failure due to zinc
embrittlement, the phenomenon was not well known
among engineers generally.

The metallurgical features of pressure systems are discussed in Chapter 12.

Vapour cloud explosions

The explosion at Flixborough was a large vapour cloud explosion. Although such explosions had become more common
in the preceding years, none compared with Flixborough in
scale and impact. The *Flixborough Report* (para. 215) draws
attention to the marked lack of information on the conditions
under which a vapour cloud can give an explosion.

Since Flixborough, a large amount of work has been
done on vapour cloud explosions. An account is given in
Chapter 17.

Investigation of disasters and feedback of information on technical incidents

The Flixborough disaster was investigated by a Court
of Inquiry. There was some feeling in the engineering profession that a legal inquiry of this kind is not a satisfactory

means of establishing the facts concerning technical incidents. This aspect is considered in the following section.

The *Flixborough Report* (para. 216) states that the inquiry
would have been greatly assisted if the essential instrument records in the control room had not been destroyed in
the explosion and recommends that consideration be given
to systems for recording and preserving vital plant information such as the 'black box' recorder used in aircraft.

The investigation of disasters and the feedback of
information from technical incidents is considered in
Chapter 27.

The points just outlined by no means exhaust the lessons
to be learned from the Flixborough disaster. In particular, there are many instructive aspects of the 8 in. pipe
hypothesis relating to such features as lagging fires,
directed flames and sprinkler sensor performance.

A2.9 Critiques

The Flixborough inquiry has been the subject of a number
of critiques which have centred mainly on (1) the form of the
inquiry; (2) the examination of the hypotheses; (3) the
design of the plant (4) the management of the plant and
(5) technical issues. These include those of R.L. Allen
(1977b), Mecklenburgh (1977a), D.J. Lewis (1989a), R. King
(1990) and I. Thomas and Gugan (1993), and the mainly
technical discussions of Gugan (1979) and V.C. Marshall
(1987).

King gives a detailed description of the plant and it
operation and diagrams showing the original plant and the
modified plant.

Mecklenburgh, King and Thomas and Gugan argue that
the legal format is not conducive to elucidating the facts
about technical issues. The question is discussed in detail
in Chapter 27.

With regard to the examination of hypotheses, Thomas
and Gugan argue that the possible role of explosive reactions, described by Alexander (1990b,c), was insufficiently
explored. With regard to the 8 in. pipe hypothesis, while
accepting that the inquiry devoted considerable effort to
exploring it, they are critical of the arguments used in dis-
missing it.

Thomas and Gugan rehearse various aspects of the
20 in. and 8 in. pipe hypotheses. Aspects of the 8 in. pipe
hypothesis are also discussed by Marshall.

King gives an account of his own hypothesis that there
was a pressure rise in Reactor No. 4 due to the evolution of
water vapour from water present in that reactor. He also
gives the background to the removal of the agitator from
Reactor No. 4.

King is critical of the conclusion that the crack in Reactor
No. 5 was due to nitrates in the cooling water, the finding of
a DSM report. He quotes an HSE source to the effect that
the reactor vessel drawing specified a maximum thrust on
the 28 in. stubs of 9 te, whereas at normal operating conditions the thrust was 38 te. He suggests that the nitrates
found were from cooling water sprayed on the already-
formed crack.

The design of the plant is also criticized by Marshall,
who argues that it would have been better to use not gravity
flow involving the 28 in. pipe connections but pumped
flow which would have allowed the use of much smaller
diameter pipes.

With regard to the operation of the plant, King is critical
of the decision to remove the agitator from Reactor No. 4,

which, among other things, increased the hazard from water accumulation in the vessel.

Thomas and Gugan point to the facts that the plant was shut-down to deal with leaks and was then restarted because 'the leaks had cured themselves' that the oxygen trip was disarmed; the production was continued even though there was a shortage of nitrogen; and that the plant was started up again after the failure of Reactor No. 5 without inspection of the other reactors.

One technical issue discussed by King is the thrust exerted by unrestrained bellows on the systems to which they are connected, which tends to be underestimated.

Lewis gives some alternative estimates of the size of the release. He states that the inventory of the cyclohexane oxidation unit exceeded 120 te. Re-examination of the reactor section at a pressure of 8.6 bar shows that it had an inventory of 230 te with another 150 te in the attached distillation section at 14.7 bar.

He also discusses the problem of determining the efficiency of the explosion compared with TNT. Estimates were made from long-range effects and from medium range damage surveys, and both point to a TNT equivalent of 15–20 te. There were major difficulties, however, in fitting TNT curves to the damage observed.

The TNT efficiency is difficult to assess, depending as it does on the estimate of the mass within the flammable region and on the blast curves used.

Gugan gives a detailed discussion of certain particular damage effects. They include crushing of a vessel, from which he estimates an overpressure of 0.76 MPa (110 psi); damage to drain covers, for which his overpressure estimate is 1 MPa; and bending of a reactor agitator shaft and deformation of lamp standards. He suggests that in general the analysis of the damage relied too much on the effects of overpressure to the neglect of the influence of impulse and dynamic pressure.

Marshall regards the vapour cloud explosion at Flixborough as of particular interest in that it is the best documented and most studied case, and is thus an exemplar. He gives a detailed analysis of the estimates of overpressure made from the damage effects, treating separately those in the near and far fields. The results show wide scatter, and he concludes that it is peculiarly difficult to estimate the energy release of such an explosion from damage.

Appendix 3

Seveso

Contents

At 12.37 on Saturday 9 July 1976, a bursting disc ruptured on a chemical reactor at the works of the Icmesa Chemical Company at Meda near Seveso, a town of about 17,000 inhabitants some 15 miles from Milan. A white cloud drifted from the works and material from it settled out downwind. Among the substances deposited was a very small amount of TCDD, one of the most toxic chemicals known. There followed a period of great confusion due to lack of communication between the company and the authorities and the latter's inexperience in dealing with this kind of situation. Over the next few days in the contaminated area animals died and people fell ill. A partial and belated evacuation was carried out. In the immediate aftermath there were no deaths directly attributable to TCDD, but a number of pregnant women who had been exposed had abortions.

A Parliamentary Commission of Inquiry, drawn equally from the Chamber of Deputies and the Senate and chaired by Deputy B. Orsini, was set up. The Commission's report (the Seveso Report) (Orsini, 1977, 1980) is a far-ranging inquiry not only into the disaster but also into controls over the chemical industry in Italy.

The impact of the Seveso disaster in Continental Europe has in some ways exceeded that of Flixborough and has led to much greater awareness of process industry hazards on the part of the public and demands for more effective controls.

The EC Directive on Major Accident Hazards of 1982 was a direct result of the Seveso disaster, and indeed this Directive was initially often referred to as the 'Seveso Directive'.

In addition to the Seveso Report, there are other accounts of the Seveso disaster and of TCDD in *The Superpoison* by Margerison, Wallace and Hallenstein (1980) and *The Chemical Scythe* by A. Hay (1982) and by A. Hay (1976, 1979, 1981), Bolton (1978), V.C. Marshall (1980), Theofanous (1981, 1983), Rice (1982), Sambeth (1983) and W.B. Howard (1985).

Selected references on Seveso are given in Table A3.1.

A3.1 The Company and the Management

The site at Seveso was operated by the company Industrie Chimiche Meda Societa Azionara (ICMESA), using a process developed by its parent company Givaudan, which was itself owned by Hoffmann La Roche.

Icmesa started in Naples in 1926 and began operations in Meda in 1946. Givaudan acquired a majority shareholding in Icmesa in 1965. In the same year Givaudan was itself taken over by Hoffmann La Roche. Icmesa was an Italian company and the other two companies were Swiss.

The management of the three companies and the officials of the local authorities involved are described in on *The Superpoison*. The following information is relevant:

Icmesa
General Manager, H. von Zwehl
Technical Manager, C. Barni
Production Manager, P. Paoletti
Factory Doctor, E. Bergamaschini

Givaudan
Managing Director, G. Waldvogel
Technical Director, J. Sambeth
Director of Dubendorf Laboratories, B. Vaterlaus

Hoffman La Roche
Technical General Manager, R. Seheff

Table A3.1   *Selected references on Seveso*

Official report
Orsini (1977, 1980)

Further accounts
Anon. (1976k); Commission of European Communities and Italian Ministry of Health (1976); A. Hay (1976, 1976a–e, 1981, 1982); Margerison (1976); V.C. Marshall (1976b, 1980c, 1983b, 1987, 1992 LPB 104); Comer (1977); Kimbrough *et al.* (1977); Temple (1977b); Bolton (1978, 1991); Koch and Vahrenholt (1978); Pocchiari (1978); Slater (1978a); Verband der chemischen Industrie (1978); Bisanti *et al.* (1979); Cavallaro, Tebaldi and Gualdi (1979, 1982); Ferraiolo (1979); Homberger *et al.* (1979); Anon. (1980b); Margerison, Wallace and Hallenstein (1980); J.J. Stevens (1980); D.C. Wilson (1980); Cardillo and Girelli (1981); A. Robertson (1981); Theofanous (1981, 1983); Rice (1982); Anon. (1983i,l,o,u,v); Anon. (1983 LPB 53, p. 27); Sambeth (1983); Cardillo, Girelli and Ferraiolo (1984); Redding (1984); W.B. Howard (1985); Kletz (1991)

TCDD
Herxheimer (1899); Kumig and Schulz (1957); Schulz (1957); Bauer, Schulz and Spiegelberg (1961); Higginbotham *et al.* (1968); Verrett (1970); Vos *et al.* (1973); Schwetz *et al.* (1973); WHO (1977); Cattabeni, Cavallaro and Galli (1978); A. Hay (1982); Rice (1982); Anon. (1984p); Tschirley (1986); EPA (1994a,b)

Previous TCP and TCDD accidents
Hofmann (1957); Schulz (1957); Bleiberg *et al.* (1964); US Senate (1970); Milnes (1971); May (1973); Dalderup (1974)

Clinical Research Director, G. Reggiani

Local Authorities
Mayor of Seveso, F. Rocca
Mayor of Meda, F. Malgrati
Medical Health Officer, G. Ghetti
Acting Health Officer, F. Uberti

Regional Minister of Health, V. Rivolta
Chief Medical Officer, Milan Province V. Eboli
Chief Medical Officer, Lombardy V. Carreri
Director of Provincial Health Laboratory, Milan A. Cavallaro
Magistrate, S. Adamo

A3.2 The Site and the Works

When Icmesa built its works at Meda the site was surrounded by fields and woods. Over the years, however, the area near the site was developed.

The reactor, Reactor A101, in which the runaway occurred was in Department B of the works.

A3.3 The Process and the Plant

The process which gave rise to the accident was the production of 2,4,5-trichlorophenol (TCP) in a batch reactor.

TCP is used for herbicides and antiseptics. Givaudan required it for making the bacteriostatic agent hexachlorophene. It manufactured its own because the herbicide grades contained impurities unacceptable in this application. Between 1970 and 1976 some 370 te of the chemical were produced.

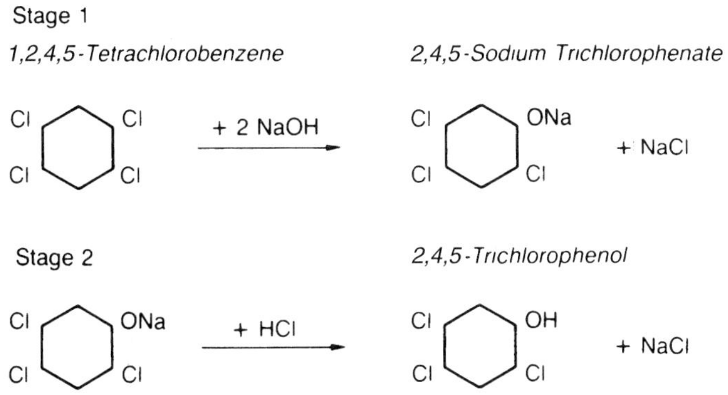

Stage 1

1,2,4,5-Tetrachlorobenzene   *2,4,5-Sodium Trichlorophenate*

Stage 2   *2,4,5-Trichlorophenol*

Figure A3.1   *Reaction scheme for production of 2,4,5-trichlorophenol (TCP)*

The reaction was carried out in two stages. Stage 1 involved the alkaline hydrolysis of 1,2,4,5-tetrachloro-benzene (TCB) using sodium hydroxide in the presence of a solvent ethylene glycol at a temperature of 170–180°C to form sodium 2,4,5-trichlorophenate. The reaction mixture also contained xylene, which was used to remove the water by azeotropic distillation. In Stage 2 the sodium tri-chlorophenate was acidified with hydrochloric acid to TCP and purified by distillation. The reaction scheme is shown in Figure A3.1.

On completion of the Stage 1 reaction some 50% of the ethylene glycol would be distilled off and the temperature of the reaction mixture lowered to 50–60°C by the addition of water.

The process was a modification by Givaudan of a process widely used in the industry. The conventional process used methanol rather than ethylene glycol and operated at some 20 bar pressure.

In this reaction the formation of small quantities of TCDD as a by-product is unavoidable. At a reaction tem-perature below 180°C the amount formed would be unlikely to exceed 1 ppm of TCP, but with prolonged heating in the temperature range 230–260°C it could increase a thousand-fold.

During manufacture nearly all (99.7%) of the TCDD formed concentrated in the distillation residues from which it was collected and incinerated. Only 0.3% found its way into the TCP, giving a maximum concentration of 10 ppb.

The reactor was a 138751 vessel with an agitator and with a steam jacket supplied with steam at 12 bar. The saturation temperature of steam at this pressure is 188°C. The controls on the reactor were relatively primitive. There was no automatic control of the heating. The reactor system is shown in Figure A3.2.

The reactor was provided with a bursting disc set at 3.5 bar and venting direct to atmosphere. The prime purpose of this disc was to prevent overpressure when compressed air was being used on the reactor.

The works had an incinerator for the destruction of hazardous plant residues at temperatures of 800–1000°C.

A3.4 TCDD and Its Properties

The properties of TCDD are given the *Seveso Report* and in *Dioxin, Toxicological and Chemical Aspects* by Cattabeni, Cavallaro and Galli (1978) and by Rice (1982).

2,3,7,8-tetrachlorodibenzo-p-dioxin, which is also known as TCDD or dioxin, has the formula

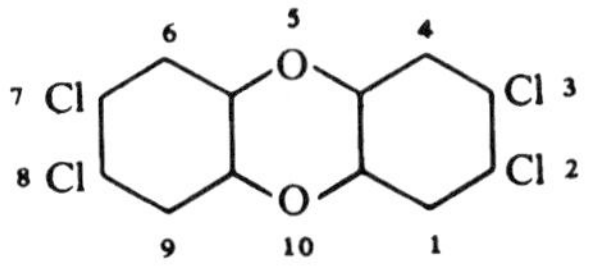

It is generated by the elimination of two molecules of HCl from 2,4,5-trichlorophenol

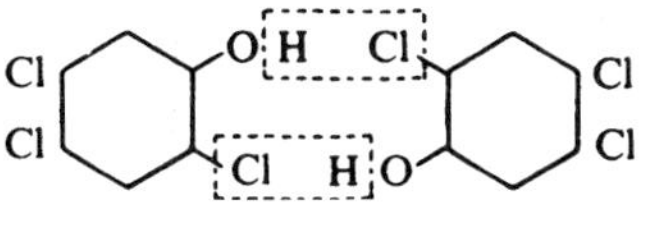

TCDD is one of the most toxic substances known. In 1954, an accident at the Boehringer TCP plant in Hamburg led to a painstaking investigation by Schulz (1957) of the causes of chloracne in people who had been exposed. He found that the cause was not the TCP itself but an impurity which he identified as TCDD, or dioxin. Actually there is a family of dioxins, TCDD being the most toxic.

The toxicity of TCDD is reviewed in the *Seveso Report*. The LD_{50} data quoted in the report are shown in Table A3.2, Section A. The lowest LD_{50} quoted is for guinea pigs and is 0.6 µg/kg, that is a dose of 0.6×10^{-9} per unit of body weight. The report also gives a table showing the toxicity of TCDD relative to that of other well-known poisons. An extract from this is shown in Table A3.2, Section B. TCDD is therefore an ultratoxic substance.

TCDD can be taken into the body by ingestion, inhala-tion or skin contact. A leading symptom of TCDD poison-ing is chloracne, which is an acne-like skin effect caused by chemicals. A mild case of chloracne usually clears within a year, but a severe case can last many years. Other effects of TCDD include skin burns and rashes and damage to liver, kidney and urinary systems and to the nervous system. It appears to have an unusual ability to interfere with the metabolic processes. There are varying degrees of evidence for carcinogenic, mutagenic and teratogenic properties. These are reviewed in the report.

TCDD is a stable solid which is almost insoluble in water and resistant to destruction by incineration except at very high temperatures.

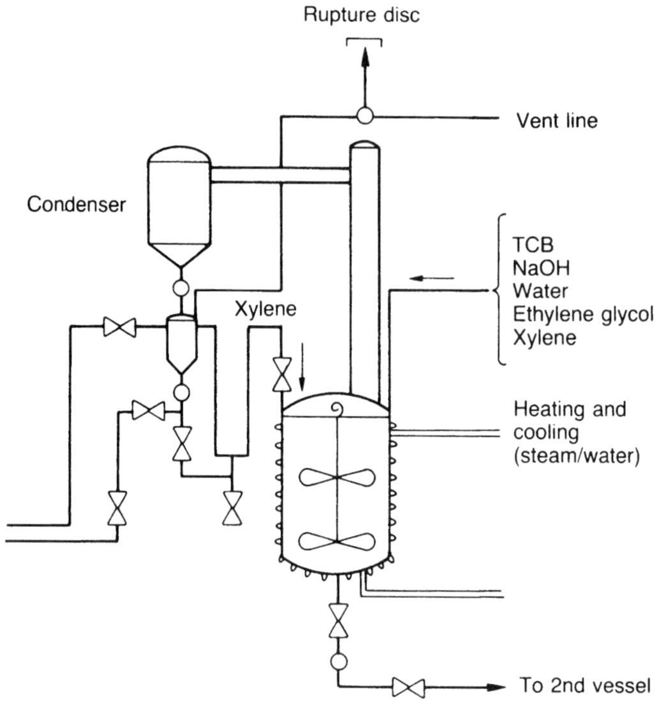

Figure A3.2   *Reactor system at Seveso (Sambeth, 1983) (Courtesy of Chemical Engineering)*

Table A3.2   *Toxicity of TCDD (after Orsini, 1980)*

A LD$_{50}$ to various animal species

Species		LD$_{50}$ (μkg/kg)	References
Guinea pig	(male)	0.6	Schwetz *et al.* (1973)
Rat	(male)	22	Schwetz *et al.* (1973)
	(female)	45	Schwetz *et al.* (1973)
Rabbit	(male)	115	Schwetz *et al.* (1973)
	(female)	115	Schwetz *et al.* (1973)
Mouse	(male)	114	Vos *et al.* (1973)

B Relative toxicity of some highly toxic materials

Substance	Minimum lethal dose (moles/kg)
Botulinus toxin	3.3×10^{-17}
TCDD	3.1×10^{-9}
Strychnine	1.5×10^{-6}
Sodium cyanide	2.0×10^{-4}

A3.5 Previous Incidents Involving TCP and TCDD

The manufacture of TCP had resulted in a number of accidents prior to Seveso and its hazards and problems were recognized, to the extent that some companies had withdrawn from the business.

A list of incidents, based on that given in the *Seveso Report*, is shown in Table A3.3.

The original method of making TCP was to heat TCB and sodium hydroxide with methanol at about 200°C and 20 bar.

In 1949, there was a large release from a pressurized TCP reactor at the Monsanto plant at Nitro, West Virginia, when a valve ruptured. A total of 228 people were affected to some degree. In 1953, a TCP reactor explosion at BASF, affecting 42 people. Another TCP reactor explosion occurred at the Boehringer plant at Hamburg in 1954, 40 workers being affected. It was this incident which led to the discovery by Schulz of the ultratoxic properties of TCDD. At the Rhone-Poulenc TCP plant at Pont de Claix near Grenoble there were 100 cases of dioxin poisoning in the period 1953–70 and explosions in 1956 and 1966, affecting 17 and 20 people, respectively. In 1963, at the Philips-Duphar plant on the North Sea Canal near Amsterdam a reaction runaway on a pressurized TCP reactor blew the top off the reactor, even though the relief valve had blown, and affected some 30 people. In 1964, the operating procedure at the Dow plant at Midland, Michigan, was changed and an accident occurred which affected 60 people. In some of these incidents a small number of those exposed subsequently died, apparently from the exposure.

The incidents did not cause any direct exposure of the public, but they did reveal the difficulties of preventing the workers' families being affected and of decontaminating the plant. In the Ludwigshafen accident one man got chloracne just by using his father's towel.

Table A3.3   *Some incidents involving TCDD*

Date	Location	Injured	Comments
1949	Nitro, WV (Monsanto)	117	Overheating leading to explosion
1953	Ludwigshafen (BASF)	55	Overheating leading to explosion
1954	Hamburg (Boehringer)	37	Exposure during plant operation
1956	Pont de Claix (Rhone-Poulenc)	17	Explosion
1956	USA (Hooker)	?	Overheating
1960	USA (Diamond Shamrock)	?	Overheating
1963	N. Sea Canal (Philips-Duphar)	30	Overheating leading to explosion
1964	Midland, MI (Dow)	30	Explosion
1964	Neuratovice (Spolana)	72	Exposure during plant operation
1966	Pont de Claix (Rhone-Poulenc)	21	Explosion
1968	Bolsover (Coalite)	79	Overheating leading to explosion
1970?	(Bayer)	5	Exposure during plant operation
1972	(Linz Chemic-Werke)	50	Exposure during plant operation
?	(Thompson Hayward)	?	Overheating leading to explosion
1976	Seveso (Icmesa)		Overheating

Various attempts were made at Ludwigshafen to remove the dioxin from the plant building, including the use of detergents, the burning off of the surfaces, the removal of insulating material and so on, but these were not effective. Eventually the whole building was demolished under controlled conditions and the debris buried. At Duphar some 50 people were involved in cleaning up the plant, of whom four died within 2 years, though the connection between the incident and the deaths was unproven. The plant was sealed for 10 years and then dismantled from the inside brick by brick, the rubble was embedded in concrete and the concrete blocks were sunk in the Atlantic.

In 1968, an accident occurred at the Coalite TCP plant at Bolsover. Coalite utilized a modification of the Givaudan process, using ethylene glycol and operating at atmospheric pressure, but with ortho-dichlorobenzene (ODCB) as the second solvent to remove the water formed. The use of this second solvent reduced the amount of expensive ethylene glycol which had to be used and not all of which was recovered after the batch. Research indicated that further reduction might be made in the amount of ethylene glycol used and a trial was carried out on the plant using half the normal quantity. A reaction runaway occurred, blew the top off the reactor and released flammable vapours which exploded and killed the shift chemist.

The workers who had been in the building were affected but appeared to recover after 10 days and the undamaged part of the plant was thoroughly cleaned and brought back into use, the damaged part being sealed off. Some weeks later chloracne symptoms appeared not only among those who had been in the plant at the time of the accident but among others who had worked in the building occasionally after it had been cleaned. Within 7 months there were some 79 cases of chloracne. Stringent hygiene measures were instituted and the cases gradually cleared. The most seriously contaminated part of the plant was dismantled and buried in a deep hole.

The causes of the accident were investigated by Milnes (1971). He found that an exothermic reaction occurs between ethylene glycol and sodium hydroxide starting about 230°C and rising rapidly to about 400°C. He also showed that TCDD is formed by reaction of sodium trichlorophenate and sodium hydroxide as described above. Further, he demonstrated that whereas at the normal reaction temperature of 180°C the amount of TCDD formed is only a few ppm, at 250°C large amounts are formed. Subsequent work has shown that formation of TCDD is zero below 153°C, less than 1 ppm at 180°C but 1600 ppm in 2 h at 230–260°C (Orsini, 1980).

A3.6 Events Prior to the Release

The start of the batch began at 16.00 on Friday, 9 July (Table A3.4). The reactor was charged with 2000 kg TCB, 1050 kg sodium hydroxide, 3300 kg ethylene glycol and 600 kg xylene.

After the reaction had taken place, part of the ethylene glycol was distilled off, but the fraction removed was only 15% instead of the usual 50% so that most of the solvent was left in the vessel. Distillation was interrupted at 5.00 on 10 July and heating was discontinued but water was not added to cool the reaction mass. The reactor was not brought down to the 50–60°C temperature range specified. The temperature recorder was switched off, the last temperature recorded being 158°C.

The shift ended at 6.00. This time coincided with the closure of the plant for the weekend. The reactor was left with the agitation turned off but without any action to reduce the temperature of the charge.

During the weekend, with the steam turbine on reduced load, the steam supply to the reactor jacket became superheated, its temperature being about 300°C.

A3.7 The Release – 1

At 12.37 on 9 July, the bursting disc on the reactor ruptured. The maintenance staff heard a whistling sound and a cloud of vapour was seen to issue from a vent on the roof giving rise to 'the formation of a dense cloud of considerable altitude'. The release lasted some 20 min.

A maintenance foreman, G. Bruno, who was passing, heard the disc rupture and realized something was wrong. He ran with two colleagues, R. Vito and C. Galante, to the boiler room to start-up the large firewater pump, then to the stores to get protective clothing. They approached Department B, where they felt the heat from the reactor. Bruno called in the technical manager, C. Barni, who arrived at

Table A3.4   *Timetable of events at Seveso*

Jul.	9 Friday	16.00 Final reactor batch started
	10 Saturday	5.00 Final reactor batch interrupted
		12.37 Bursting disc on reactor ruptures
		Barni visits houses near plant to warn against eating garden produce. He asks carabinieri to repeat warning but they refuse
	11 Sunday	Barni and Paoletti inform von Zwehl. They are unable to contact Ghetti or his deputy. They inform Rocca and Malgrati. Von Zwehl informs Sambeth
	12 Monday	Rocca and Uberti visit ICMESA. Letter of von Zwehl to Health Officer
	13 Tuesday	Report of Uberti to Rocca and Malgrati
	14 Wednesday	Sambeth and Vaterlaus inspect Dept B and contaminated area. Uberti writes to Provincial Health Officer in Milan
	15 Thursday	Dubendorf Laboratories give first analyses showing high TCDD content. Sambeth telegraphs Waldvogel in Turkey and asks von Zwehl to inform local authorities. Sambeth informs Hoffmann La Roche and seeks clinical advice and permission to close plant. Rumours grow in Seveso. Citizens invade Uberti's office. Rocca and Uberti meet von Zwehl. They decide to declare polluted zone and to post warning notices
	16 Friday	Workforce go on strike. Efforts are made to contact Ghetti in remote holiday farmhouse. Warning notices are erected. Rocca, Uberti and others meet von Zwehl again. Uberti insists on evacuation. Rocca requests and obtains permission for evacuation from Deputy Prefect of Milan. Rocca contacts journalist friend
	17 Saturday	Il Giorno carries front-page headlines on 'Poison gas' at Seveso. Press descend on town. Uberti contacts Cavallaro.
	18 Sunday	Cavallaro's team of health inspectors investigate contaminated area. Rocca, Ghetti, Cavallaro and Adamo meet von Zwehl. Cavallaro presses to know identity of poison released. Adamo threatens to arrest ICMESA management
	19 Monday	Waldvogel arrives in Seveso and offers local authority financial compensation, which is refused
	20 Tuesday	Cavallaro and Ghetti meet Vaterlaus in Zurich. Vaterlaus reveals that poison released was TCDD. Uberti's letter to Provincial Health Officer in Milan arrives.
		16.00 Ghetti telephones Rocca to say substance was TCDD. Prefect in Milan is informed
	21 Wednesday	9.30 Meeting at prefecture. Rivolta and Carreri take over responsibility. Carabinieri arrest von Zwehl and Paoletti
	23 Friday	9.00 Meeting at prefecture. Rivolta decides against evacuation. Reggiani visits Rocca and urges evacuation. Reggiani asked to leave Deputy Prefect's meeting unheard
		14.00 Rivolta reassures meeting in Seveso town hall. Reggiani argues with Rivolta and calls for evacuation. Carabinieri try unsuccessfully to arrest Reggiani, who returns to Switzerland
	24 Saturday	9.30 Meeting of regional health council. Vaterlaus presents map of locations affected. Council decides to recommend evacuation and defines evacuation zone (part of Zone A)
	26 Monday	179 people evacuated
	29 Thursday	Zone A extended and further 550 evacuated

13.10 as the escape was subsiding. About an hour after the release began Vito and Galante were able to begin gingerly admitting cooling water to the reactor system.

The release contaminated the vicinity of the plant with TCDD. Figure A3.3 is a map of the area, showing the zones later established and Table A3.5 gives the concentrations later measured in these zones.

The area of Zone A was 108 ha (1.08 km^2) with concentrations of TCDD averaging 240 $\mu g/m^2$ and rising to over 5000 $\mu g/m^2$. The area of Zone B was 269 ha (2.69 km^2) with concentrations averaging 3 $\mu g/m^2$ but rising to 43 $\mu g/m^2$. In the Zone of Respect, Zone R, which had an area of 1430 ha (14.3 km^2), the concentrations varied from indeterminable to 5 $\mu g/m^2$.

A3.8 The Emergency and the Immediate Aftermath

The release set in train a series of events which were to have profound repercussions. They are described in detail in *The Superpoison*. It is possible here to give only a brief outline.

When the situation had been brought under control in the works, Barni visited houses near the plant and warned people not to eat the garden produce. He asked the carabinieri to repeat the warning, but they refused, unless there was authorization from the municipal health officer.

There were few obvious signs of an accident, although surfaces were covered with a thin, oily film.

The municipal health officer, G. Ghetti, was on holiday in a remote farmhouse not on the telephone.

On Sunday 11 July, Barni visited P. Paoletti, the production director, and they together contacted the general manager, von Zwehl, who was on holiday in the Alps. They then visited the mayor of Seveso, F. Rocca, and recommended that people living near the plant be warned not to eat garden produce. Rocca agreed to inform the carabinieri. He also brought in the mayor of Meda, F. Malgrati, since the Icmesa plant lay within the Meda boundaries.

The same day von Zwehl informed J. Sambeth, technical director of Givaudan. Sambeth realized that there was a dioxin hazard and asked what temperature the reactor had reached. He told von Zwehl to close off Department B

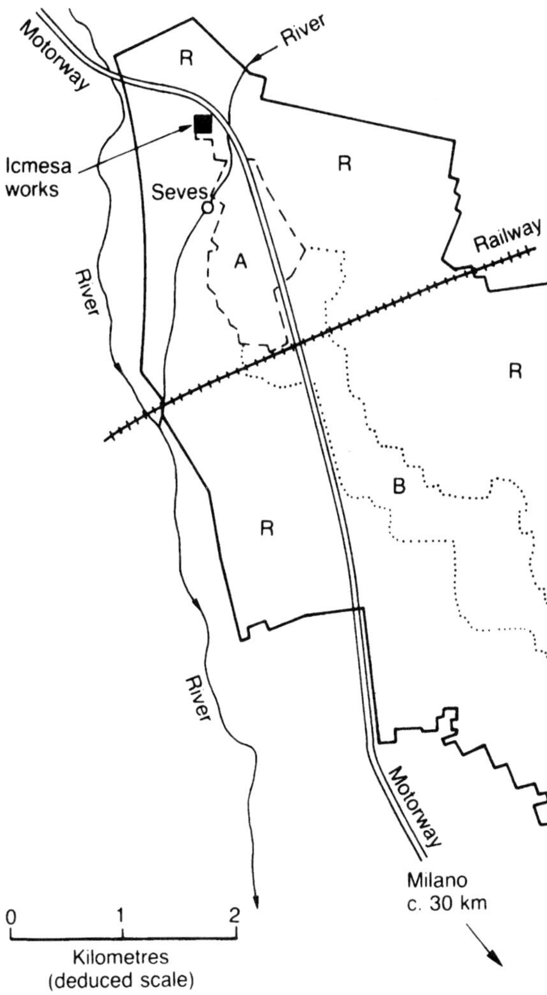

Figure A3.3 *Plan of the Seveso area, showing Zones A and B and Zone of Respect (after Orsini, 1980; V.C. Marshall, 1978)*

Table A3.5 *Concentrations of TCDD in Zones A and B at Seveso (after Orsini, 1980)*

Zone	Concentration ($\mu g/m^2$)		
	Mean	*Maximum*	*Minimum*
A1	580.4	5477	n.v.d
A2	521.1	1700	6.1
A3 (north)	453.0	2015	1.7
A3 (south)	93.0	441	n.v.d.
A4	139.9	902	n.v.d.
A5	62.8	427	n.v.d.
A6	29.9	270	n.v.d.
A7	15.5	91.7	n.v.d.
B	3	43.8	n.v.d.

n.v.d. = no value determined.

and have samples from the reactor sent to the Dubendorf laboratories of Givaudan at Zurich. Normally he would have discussed the problem with the managing director of Givaudan, G. Waldvogel, but the latter was on holiday in Turkey.

On Monday 12 July, Rocca together with the acting health officer for Meda and Seveso, F. Uberti, visited Icmesa and saw von Zwehl, who convinced them that there was no serious health hazard. Von Zwehl wrote a letter to the health officer. On the material released the latter said:

Since we are not in a position to evaluate the substances present in the vapour or to predict their exact effects, but knowing the final product is used in manufacturing herbicides, we have advised householders in the vicinity not to eat garden produce.

On Tuesday 13 July, Uberti sent a report to Rocca and Malgrati, enclosing a copy of von Zwehl's letter. He believed the release was one of the minor accidents which occur periodically.

On Wednesday 14 July, Sambeth and Vaterlaus, director of the Dubendorf laboratories, came to Seveso, inspected Department B and tried to work out how high a temperature the reactor had reached. They also inspected the contaminated land.

Uberti wrote the same day to the provincial health officer in Milan. The letter did not arrive until 20 July.

On Thursday 14 July, the first analyses from the Dubendorf laboratory came through to Sambeth. They showed high dioxin contents. Sambeth telegraphed Waldvogel in Turkey. He warned von Zwehl to advise the local authorities that the release may have contained toxic by-products.

He also contacted Hoffmann La Roche to seek clinical advice and permission to order closure of the plant. The clinical research director of Hoffmann, G. Reggiani, was not familiar with TCDD, but agreed to take over the health aspect and began to inform himself about the substance.

Meanwhile rumours were growing in Seveso and a group of citizens invaded Uberti's office.

Rocca and Uberti met von Zwehl again. He agreed to pay compensation for any damage, produced a map of the polluted area, and suggested that people be advised not to eat garden produce and that warning notices be posted. He was not, however, prepared to say what the poison was. The visitors decided to declare a polluted zone and post notices.

On Friday 16 July, the erection of warning notices began. At this point the workforce at Icmesa went on strike. Rocca and Uberti had a further meeting with von Zwehl where they again pressed to be told the identity of the poison.

Uberti now urged Rocca to carry out an evacuation. Ghetti was contacted in his remote farmhouse, agreed to return but left the decision to Rocca. Rocca requested and obtained permission to evacuate from the Deputy Prefect of Milan. He also contacted a journalist friend.

On Saturday 17 July *Il Giorno* carried front-page headlines about the 'Poison gas' at Seveso. The accident became a national issue and the press descended on the town.

The same day Uberti contacted A. Cavallaro, director of the provincial health laboratories in Milan.

On Sunday 18 July, Cavallaro's team of health inspector arrived to investigate the contaminated area.

Rocca, Ghetti, Cavallaro and a local magistrate, S. Adamo, met again with von Zwehl. Cavallaro pressed to know the poison involved. Adamo threatened to arrest the Icmesa management

On Monday 19 July, Waldvogel arrived in Seveso. He offered the local authorities compensation, which they refused.

On Tuesday 20 July, Cavallaro and Ghetti met Vaterlaus in Zurich. He revealed that the poison was TCDD. Ghetti telephoned Rocca with this information. The Prefect in Milan was informed.

On Wednesday 21 July, a meeting was held at the prefecture. The regional minister of health, V. Rivolta, now took charge of the medical aspects together with V. Carreri, director of health services for Lombardy.

On the same day the carabinieri arrested von Zwehl and Paoletti.

On Friday 23 July, there was another meeting at the prefecture. Rivolta was against evacuation. Meanwhile Reggiani had decided that evacuation was essential and arrived in Seveso to urge this view. He saw Rocca, who told

him the decision was out of his hands. He came into the meeting at the prefecture but was asked to leave unheard.

Later that day Rivolta addressed a meeting in Seveso town hall and gave reassurances about the hazards. Reggiani stated his disagreement with Rivolta and called for evacuation. He was unable to make his view prevail. At his hotel in Milan he was warned that the carabinieri were about to arrest him and drove north in haste to Switzerland.

There followed a meeting in Lugano between Waldvogel, Sambeth and Vaterlaus, who had a map of the contaminated areas based on the analyses carried out by his laboratories. Reggiani told the Givaudan men of his experiences. It was agreed that an attempt must be made to persuade the Italian authorities to evacuate and that Vaterlaus, who was deemed less likely to be arrested, should do this.

On Saturday 24 July, a meeting of the regional health council was held in Milan. Vaterlaus presented his map of the areas affected. An evacuation zone was defined and the decision was taken to recommend evacuation.

On Monday 26 July, 179 people were evacuated from an area which was later defined as part of Zone A. On Thursday 29 July, a further 550 people were evacuated from another area which was also part of what became Zone A.

Eventually three zones, Zone A, Zone B and the Zone of Respect were defined, as described above. 733 people were evacuated from Zone A and 5000 in Zone B were put under medical surveillance, but allowed to remain in their homes.

The release posed a severe problem for the Italian authorities, who were obliged to seek assistance from international experts in medical diagnosis and treatment and in analytical and decontamination procedures.

A3.9 The Investigation

The work of the Commission took place against a background of intense national concern.

The terms of reference of the Commission were wide-ranging. They have been described by V.C. Marshall (1980c) as follows: 'These terms of reference appear to have combined the functions in UK terms of the Flixborough inquiry, the Major Hazards Advisory Committee and the Royal Commission on the Environment'.

The report produced by the Commission, the *Seveso Report* (Orsini, 1980), covers a large number of topics, including the history of TCP production, the history of Icmesa, the toxicity of TCDD, the causes of the accident, the consequences of the accident in terms of contamination and health effects, the health measures taken, the reclamation of the contaminated land, the actions and responsibilities of the management and of the local and regulatory authorities and the recommendations. The report adduces four causes of the accident:

(1) Interruption of the production cycle.
(2) Method of distillation.
(3) Set pressure of bursting disc.
(4) Failure to install collection/destruction system for material vented.

The production cycle was interrupted at 5.00 on 10 July and the reactor was left without agitation or cooling.

In the process described in the original Givaudan patent the charge was acidified before distillation. In the process used at Icmesa the order of these two steps was reversed.

This allowed prolonged contact time between the ethylene glycol and sodium hydroxide.

Another modification to the original process was a change in the molar proportions of the reactants from

1 TCB: 2 NaOH: 11.5 ethylene glycol

to

1 TCB: 3 NaOH: 5.5 ethylene glycol

which increased the molar concentration of sodium hydroxide.

The setting of the bursting disc was 3.5 bar. The function of this device was to guard against overpressure from the compressed air, which was used to transfer materials to the reactor. The inquiry found that if the set pressure had been lower, venting would have occurred at a lower, and less hazardous, temperature.

No device was installed to collect or destroy any toxic materials vented. The bursting disc manufacturer's catalogue stated 'Bursting discs may be preferred in the case of fluids of great value and high toxicity where the loss of such fluids should be avoided. In this case a second receiver tank is required to recover the discharged fluid'. There was no such tank on the plant.

Besides listing these four points the report makes a number of other comments. On the reactor itself it states

Proof has been provided both of the inadequacy of the measuring equipment for a number of fundamental parameters and also the absence of any automatic control system. Measurement of acidity was carried out manually by immersion of a rod with an indicator chart (sic) through a secondary inspection window.

The reactor vessel was not given a hydraulic test as the reactor was considered to operate at atmospheric pressure. The fitting of the bursting disc was not reported to the state pressure vessel inspectors.

The report recommends improved regulatory arrangements for the control of hazardous plants and, in this connection, places some reliance on the introduction of the EC Directive which came to be known as the 'Seveso Directive'.

The viewpoint of the companies has been given in several publications. The accident itself has been described by Sambeth (1983). The view of Hoffmann La Roche has been stated by Homberger *et al.* (1979).

A3.10 The Release – 2

The contents of the reactor at the time of the accident have been given as 2030 kg sodium trichlorophenate, 540 kg sodium chloride, *c.*1000 kg ethylene glycol and 1600 kg sodium ethylene glycolate, diethylene glycol and polyethylene glycol.

It is known that above 230°C such a mixture will undergo an exothermic decomposition reaction. Accidents had been known to occur involving a reaction runaway above this temperature.

The actual temperature profile in the reactor has been given by Sambeth (1983) and is shown in Figure A3.4. Due to the interruption of the batch the usual sharp reduction in the temperature of the charge at the termination of the reaction did not take place and after 7.5 h the explosion occurred.

Investigations carried out after the accidents using differential thermal analysis showed that there exist two slow exotherms (Theofanous, 1981, 1983). One starts at about 185°C, peaking at 235°C and giving a 57°C adiabatic temperature rise. The other starts at about 255°C, peaking at 265°C and giving an estimated 114°C temperature rise. The adiabatic induction times for the two exotherms are 2.1 and 0.5 h, respectively. The decomposition exotherm starts at about 280–290°C and shows a rapid pressure rise at about 300°C.

A mechanism for the reaction runaway has been proposed by Theofanous. This is that due to layering of the

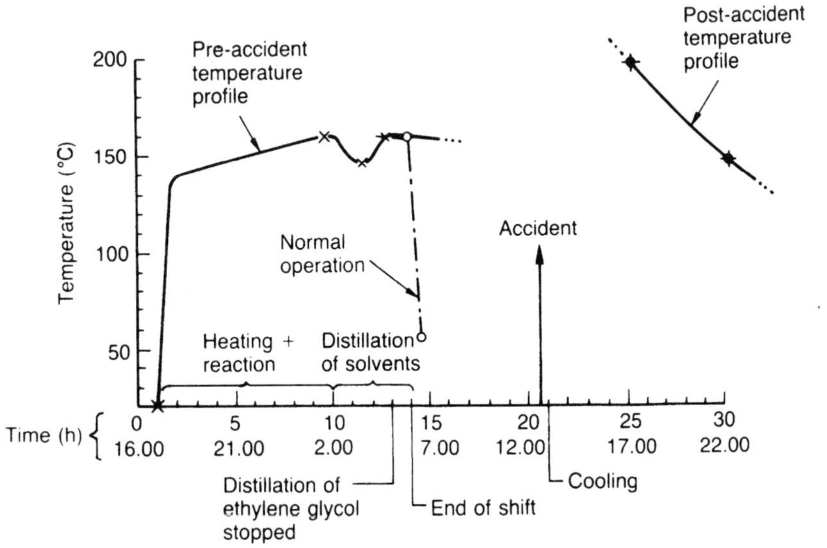

Figure A3.4   *Reactor temperature profiles measured for accident at Seveso (Sambeth, 1983) (Courtesy of Chemical Engineering)*

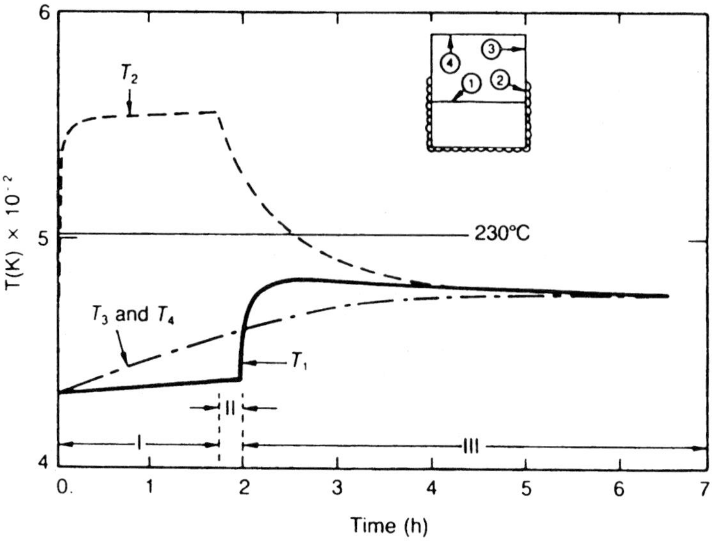

Figure A3.5   *Reactor temperature profiles calculated for period preceding accident at Seveso (after Theofanous, 1981) (Courtesy of Nature)*

reaction mix the amount of residual heat in the upper section of the reactor wall was sufficient to raise the temperature of the top layer of liquid to $c.200-220°C$, a temperature high enough to initiate exotherms leading to decomposition. The calculated temperature profiles for this theory are shown in Figure A3.5.

In the conduct of the final batch there were failures of adherence to operating procedures. W.B. Howard (1985) instances (1) failure to distil off 50% of the glycol; (2) failure to add water; (3) failure to continue agitation; (4) switching off of the reactor temperature recorder and (5) failure to bring the reactor temperature down from its value of 158°C to the normal value of 50−60°C.

A3.11 The Later Aftermath, Contamination and Decontamination

As described above, there was a prolonged period in which the authorities tried to assess the situation and determine measures for dealing with it. They were advised by a team from Cremer and Warner (C&W), and an account has been given by Rice (1982).

The release was modelled using fluid jet and gas dispersion models and predictions made of the probable ground level concentrations of TCDD. Accounts have been given by Comer (1977) and Rice (1982). In this work the reaction mass was taken as 2800 kg ethylene glycol, 2030 kg trichlorophenol 2030, 542 kg sodium chloride and 562 kg sodium hydroxide.

Estimates of the amount of TCDD generated in the reactor and dispersed over the countryside vary. Cattabeni, Cavallaro and Galli (1978) give an estimated range of 0.45−3 kg released. The amount assumed in the C&W modelling work was 2 kg.

Other parameters used in the modelling were bursting disc rupture pressure 376 kPa; vent pipe diameter 127 mm; discharge height above ground 8 m. The bursting disc

rupture was assumed to a first approximation to be due to the vapour pressure of ethylene glycol at 250°C. From these data the value obtained for the vapour exit velocity was 274 m/s. Two limiting cases were considered for the discharge: Case A, pure vapour (density 1.61 kg/m^3) and Case B, a two-phase vapour−liquid mixture (density 8.99 kg/m^3). Eyewitness accounts indicated that the actual event was intermediate between these two extremes. For Case A the total plume rise was estimated as 83 m with the downwind distance to the maximum height as 103 m and the corresponding figures for Case B were 55 and 95 m, respectively. Figure A3.6 shows the estimated ground concentration contours for a 2 kg release taken from this work.

The modelling work proved useful in defining the problem and planning the decontamination. The results showed that beyond about 1.5 km there was little difference in the concentration estimates for the two cases. There was good agreement between the predicted and the measured concentrations. This gave some confidence that the amount released was indeed about 2 kg. There had been some speculation by scientists that an amount as high as 130 kg had been released.

Initially, various methods of decontamination were put forward by various parties. TCDD is virtually insoluble in water, which reduces the threat to the water supply, but means that the effect of rain is to transfer it from the vegetation to the soil rather than to remove it. One method of removal suggested was burning and proposals were made to burn the contaminated ground with flame-throwers. It was feared, however, that unless a temperature of at least 1000°C were attained, the TCDD would not be destroyed but simply swept up and deposited elsewhere. Another method proposed was to apply a mixture of olive oil and cyclohexanone to the ground, so that the TCDD would dissolve in the oil and be broken down by the ultraviolet radiation of the sun. This was abandoned, because instead of sun there was rain, and it was feared that the dissolved

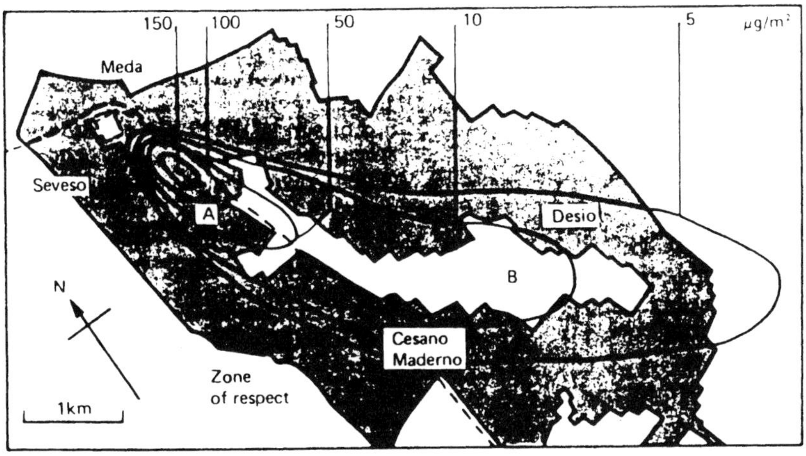

Figure A3.6   *Estimated concentrates of TCDD at Seveso (Comer, 1977. Reproduced by permission.)*

TCDD would be washed into the rivers and might penetrate the water table. A third approach involved the use of specially bred microbes capable of digesting toxic substances. Mixtures of more than a dozen strains of microorganisms bred on chemical substrates were already used to degrade chemical wastes, including cyanides and herbicides.

The decontamination measures actually carried out are described by Rice. They include collection of vegetation and cleaning of buildings by high intensity vacuum cleaning followed by washing with high-pressure water jets.

It is worth mention that people may also have been affected by materials in the reactor charge other than TCDD. V.C. Marshall (1980) indicates that on the basis of the ground concentrations found in the contaminated area a minimum of 0.25 kg was released. He points out that there were some 5–10 te of charge in the reactor and that several tonnes of material including sodium hydroxide must have been ejected. He suggests that burns around the hands and other parts of the body suffered by 413 persons who did not develop chloracne were probably caused by sodium hydroxide.

Figure A3.7 gives plots of the actual ground concentrations, showing the change with time, obtained by Cavallaro, Tebaldi and Gualdi (1982).

Data from the official report on ground concentrations in zones A and B are given in Table A3.5.

A3.12 Some Lessons of Seveso

Some of the numerous lessons to be learnt from the Seveso disaster, many of which were brought out in the Commission's report, are listed in Table A3.6. Some of them are similar to those from Flixborough, while others are different, reflecting both the different substances and processes and the different company and national situations involved. Some of these lessons are now considered:

Public control of major hazard installations
The Seveso disaster had the effect of raising the general level of awareness of the hazard from chemical plants in Italy and in Europe generally and highlighted the deficiencies in the existing arrangements for the control of major hazards.

The Italian government set up the Commission of Inquiry to investigate the cause of the disaster and to make recommendations. It also gave strong support to the development of the EC Directive.

Siting of major hazard installations
The release at Seveso affected the public because in the period since the site was first occupied housing development had encroached on the area around the plant. The accident underlined the need for separation between hazard and public.

Acquisition of companies operating hazardous processes
A lesson which has received relatively little attention is the problem faced by a company which becomes owner by acquisition of another company operating a hazardous process. At Seveso the problem was compounded by the fact that Icmesa was owned by Givaudan and the latter by Hoffmann La Roche, so that the company ultimately responsible was at two removes. As a result the directors of the latter were not familiar with the hazards.

Hazard of ultratoxic substances
Seveso threw into sharp relief the hazard of ultratoxic substances. The toxicity of TCDD is closer to that of a chemical warfare agent than to that of the typical toxic substance which the chemical industry is used to handling.

As it happened, the British Advisory Committee on Major Hazards was presenting its *First Report*, giving a scheme for the notification of hazardous installations, when Seveso occurred. It contained no notification proposals for ultratoxics. In the *Second Report* this deficiency was rectified. The EC Directive places great emphasis on toxic and ultratoxic materials.

Hazard of undetected exotherms
The characteristics of the reaction used had been investigated and the company believed it had sufficient information. It was well aware of the hazard from the principal exotherm. Subsequent studies showed, however, that other,

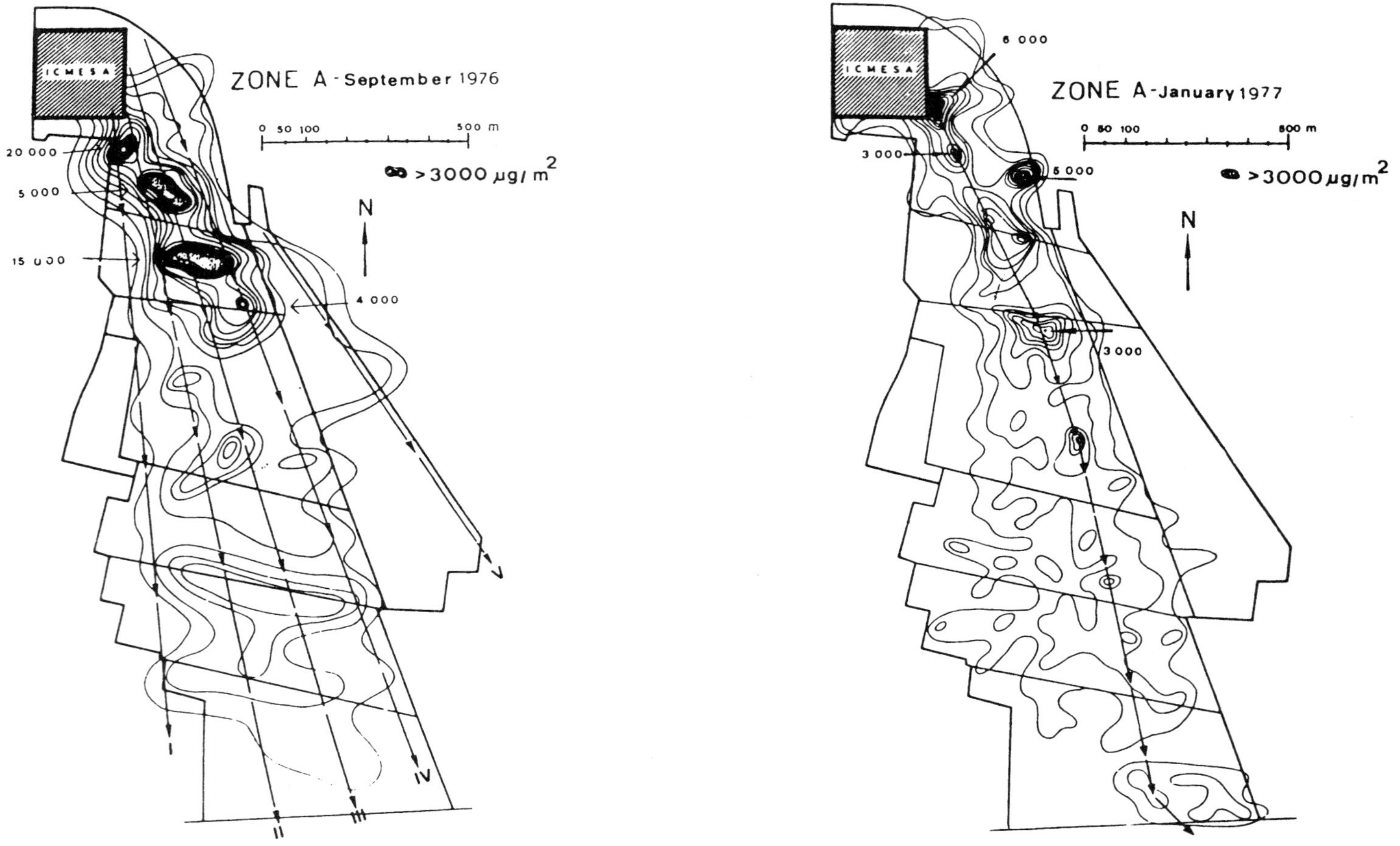

Figure A3.7 *Measured concentrations of TCDD at Seveso, showing reduction of concentration with time (Cavallaro, Tebaldi and Gualdi, 1982)*
(Courtesy of Atmospheric Environment)

Table A3.6   *Some lessons of Seveso*

Public control of major hazard installations
Siting of major hazard installations
Acquisition of companies operating hazardous processes
Hazard of ultratoxic substances
Hazard of undetected exotherms
Hazard of prolonged holding of reaction mass
Inherently safer design of chemical processes
Control and protection of chemical reactors
Adherence to operating procedures
Planning for emergencies
Difficulties of decontamination

weaker exotherms existed which, given time, could also cause a runaway.

The complete identification of the characteristics of the reaction being operated is particularly important. If these are not known, or only partially known, there is always the danger that the design will not be as inherently safe as intended, that operating modifications will have unforeseen results and/or that protective measures will be inadequate.

Hazard of prolonged holding of reaction mass
The interruption of the reaction and the holding of the reaction mass for a prolonged period after the main reaction was complete without reducing the temperature gave time for the weak exotherms to occur and lead to runaway.

Inherently safer design of chemical processes
The reactor was intended to be inherently safe to the extent that the use of steam at 12 bar set a temperature limit of 188°C so that the contents could not be heated above this by the heating medium. Unfortunately this feature was defeated by allowing the reaction charge to stand hot for too long so that there occurred the weak exotherms, which the company did not know about.

The bursting disc was not intended for reactor venting, but, as the Commission pointed out, if the set pressure of the disc had been lower, the reactor would have vented at a lower temperature and hence with less dioxin in the charge.

Control and protection of chemical reactors
The control and protection system on the reactor was primitive. Operation of the reactor was largely manual. There was no automatic control of the cooling and no high temperature trip.

The reactor was not designed to withstand a runaway reaction. It was not rated as a pressure vessel to withstand pressure build-up prior to and during venting, the disc was not designed for reaction relief and there was no tank to take the ultratoxic contents of the reactor.

Adherence to operating procedures
The conduct of the final batch involved a series of failures to adhere to the operating procedures.

Planning for emergencies
As the account given above indicates, the handling of the emergency was a disaster in its own right. Information on the chemical released and its hazards was not immediately available from the company. There was failure of communication between the company and the local and regulatory authorities and within those authorities. Consequently there was lack of action and failure to protect and communicate with the public. These deficiencies might in large part have been overcome by emergency planning.

Difficulties of decontamination
The incident illustrates the difficulties of decontamination of land where the contaminant is both ultratoxic and insoluble in water.

Appendix

4

Mexico City

Contents

At about 5.35 a.m. on the morning of 19 November 1984, a major fire and a series of explosions occurred at the PEMEX LPG terminal at San Juan Ixhuatepec (San Juanico), Mexico City. Some 500 people were killed and the terminal was destroyed.

The accident was investigated by a team from TNO which visited the site about 2 weeks after the disaster and has published its findings (Pietersen, 1985). A further account has been given by Skandia International (1985).

Selected references on Mexico City are given in Table A4.1.

A4.1 The Site and the Plant

The site of the PEMEX terminal is shown in Figure A4.1 and an aerial photograph of the installation itself in Figure A4.2.

Table A4.1   *Selected references on Mexico City*

Anon. (1985v, hh); Berenblut *et al.* (1985); Cullen (1985); Kletz (1985p); Pietersen (1985, 1985 LPB 64, 1986a,b, 1988a); Skandia International (1985); Hagon (1986)

The oldest part of the plant dated from 1961 to 1962 and was thus over 20 years old. In the intervening period residential development had crept up to the site. This process of encroachment is clearly visible from the aerial photographs shown in Figure A4.3. By 1984, the housing was within 200 m of the installation with some houses within 130 m.

The terminal was used for the distribution of LPG, which came by pipeline from three different refineries. The main LPG storage capacity of 16,000 m^3 consisted of 6 spheres and 48 horizontal cylinders. The daily throughput was 5000 m^3. The layout of the terminal with storage tank capacities is shown in Figure A4.4. The two larger storage spheres had individual capacities of 2400 m and the four smaller spheres capacities of 1600 m^3. The site covered an area of 13,000 m^2.

The plant was said to have been built to API standards and much of it to have been manufactured in the United States.

A ground level flare was used to burn off excess gas. The flare was submerged in the ground to prevent the flame being extinguished by the strong local winds.

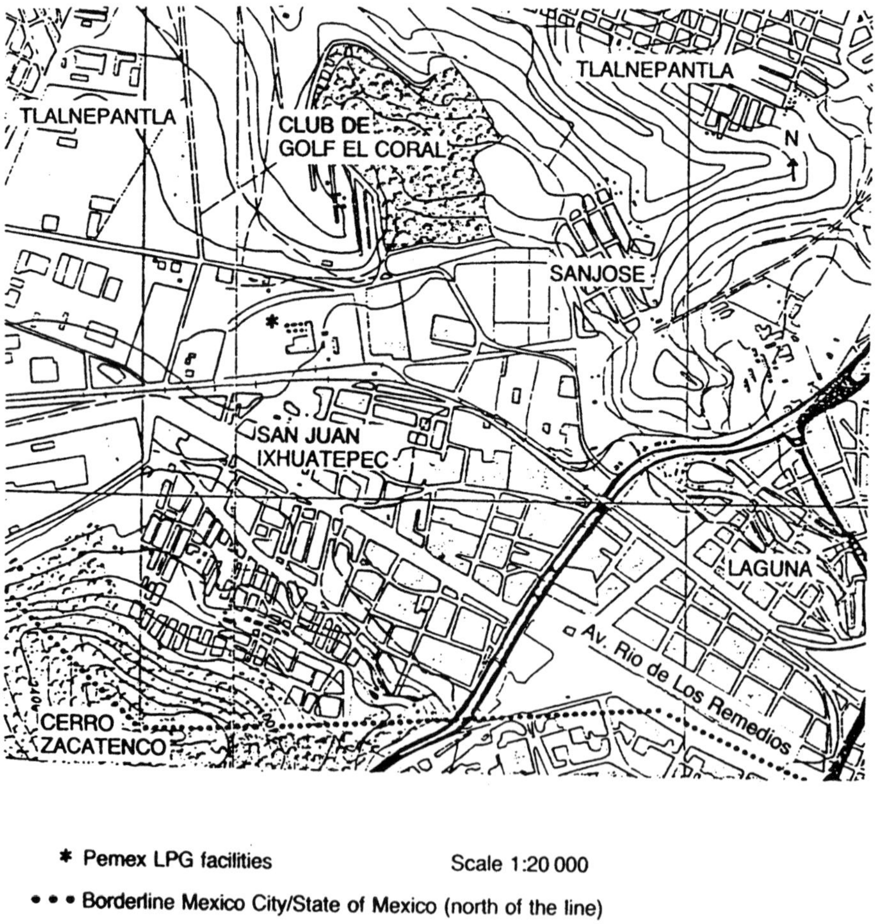

* Pemex LPG facilities   Scale 1:20 000

• • • Borderline Mexico City/State of Mexico (north of the line)

Figure A4.1   *Area plan of the PEMEX site at Mexico City (Pietersen 1985) (Courtesy of TNO)*

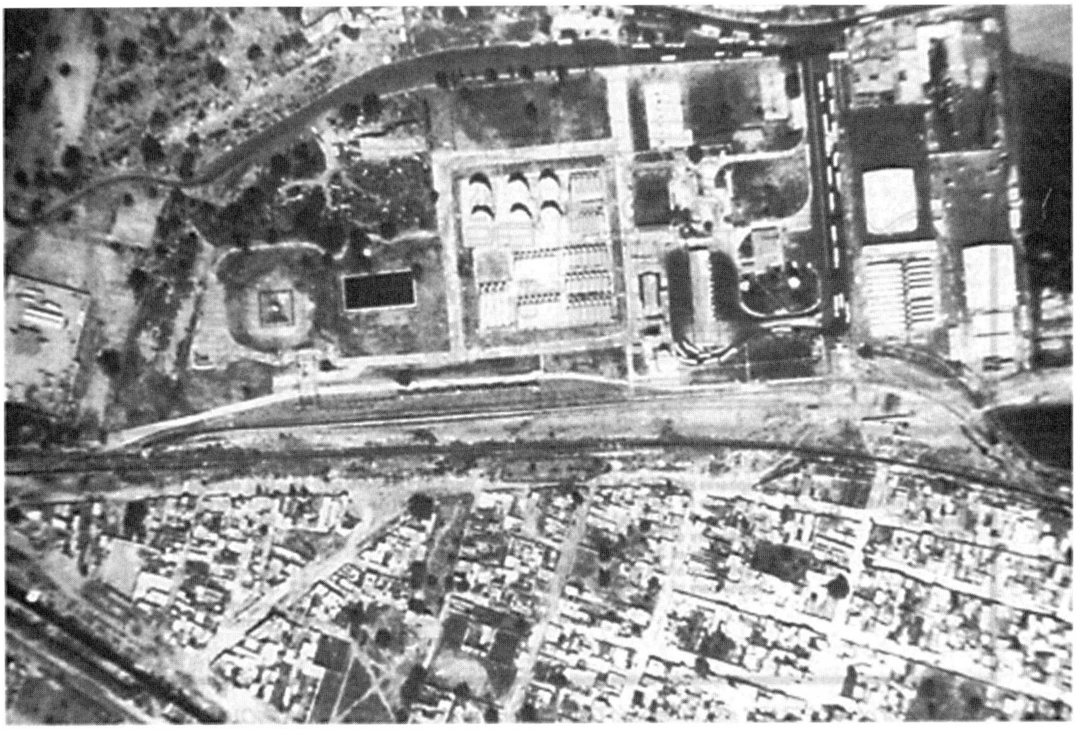

Figure A4.2 *PEMEX plant at Mexico City before the accident (Skandia, 1985) (Photograph: State of Mexico)*

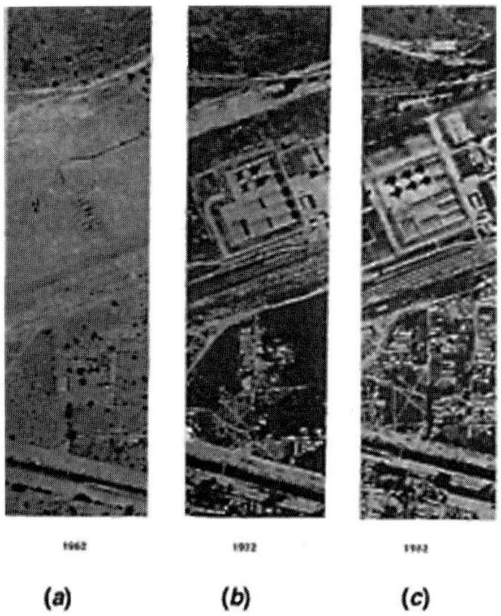

1962	1972	1982
(a)	**(b)**	**(c)**

Figure A4.3 *Growth of housing near the PEMEX plant at Mexico City (Pietersen 1985): (a) 1962; (b) 1972; (c) 1982 (Reproduced by permission of TNO.)*

Adjoining the PEMEX plant there were distribution depots owned by other companies. The Unigas site was some 100–200 m to the north and contained 67 tank trucks at the time of the accident. Rather further away was the Gasomatico site with large numbers of domestic gas cylinders.

A4.2 The Fire and Explosion – 1

Early in the morning of 18 November, the plant was being filled from a refinery 400 km away. The previous day the plant had become almost empty and refilling started during the afternoon. The two larger spheres and the 48 cylindrical vessels had been filled to 90% full and the four smaller spheres to about 50% full, so that the inventory on site was about 11,000 m^3, when the incident began.

About 5.30 a.m. a fall in pressure was registered in the control room and also at a pipeline pumping station 40 km distant. An 8 in. pipe between sphere F4 and the Series G cylinders had ruptured.

The control room personnel tried to identify the cause of the pressure fall but without success.

The release of LPG continued for some 5–10 min. There was a slight wind of 0.4 m/s. The wind and the sloping terrain carried the gas towards the south-west. People in the nearby housing heard the noise of the escape and smelled the gas.

When the gas cloud had grown to cover an area, which eyewitnesses put at 200 × 150 m with a height of 2 m, it found the flare and ignited. It was 5.40 a.m. The cloud

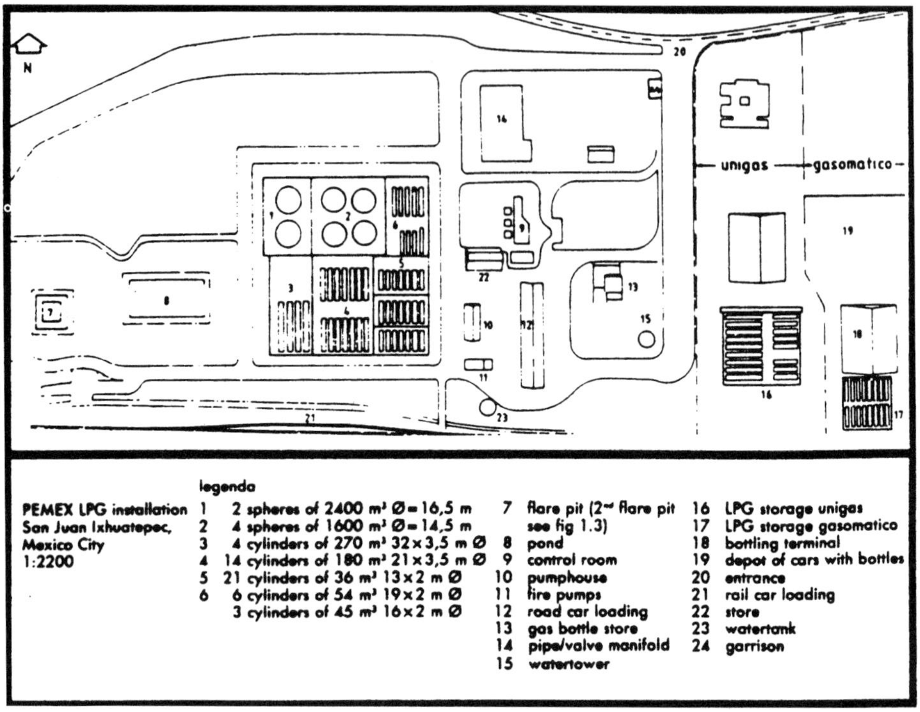

Figure A4.4 Layout of the PEMEX site at Mexico City (Pietersen 1985) (Courtesy of TNO)

caught fire over a large area, giving a high flame and causing violent ground shock.

When this general fire had subsided, there remained a ground fire, a flame at the rupture and fires in some 10 houses.

Workers on the plant now tried to deal with the escape. One drove off to another depot to summon help. Five others who may have been on their way to the control room or to man fire pumps were found dead, and badly burned. At a late stage someone evidently pressed the emergency shut-down button.

In the neighbouring housing some people rushed out into the street, but most stayed indoors. Many thought it was an earthquake.

At 5.45 a.m. the first BLEVE occurred. About a minute later another explosion occurred, one of the two most violent during the whole incident. One or two of the smaller spheres BLEVEd, giving a fireball 300 m diameter.

A rain of LPG droplets fell on the area. Surfaces covered in the liquid were set alight by the heat from fireballs. People burned like torches.

There followed a series of explosions as vessels suffered BLEVE. There were some 15 explosions over a period of an hour and a half. BLEVE occurred of the four smaller spheres and many of the cylindrical vessels.

The explosions during the incident were recorded on a seismograph at the University of Mexico. The timing of the readings is shown in Table A4.2 Section A. As the footnote indicates, it is suggested by Skandia that the initial explosion, or violent deflagration, was probably not recorded.

The damage caused is shown in Figure A4.5, which shows the area of main housing damage and the fall of

Table A4.2 Timetable of events at Mexico City

A Seismograph readings

1	5 h	44 min	52 s	6	6 h	49 min	38 s
2	5 h	46 min	01 s	7	6 h	54 min	29 s
3	6 h	15 min	53 s	8	6 h	59 min	22 s
4	6 h	31 min	59 s	9	7 h	01 min	27 s
5	6 h	47 min	56 s				

Disturbances 2 and 7 were the most intence with a Richter scale intensity of 5 Skandia suggest that the first violent combustion may not have been recorded.

B General timetable

5.30	Rupture of 8 in. pipe. Fall of pressure in control room
5.40	Ignition of gas cloud. Violent combustion and high flame
5.45	First explosion on seismograph, a BLEVE Fire department called
5.46	Second BLEVE, one of most violent
6.00	Police alerted and civilian trafffic stopped
6.30	Traffic chaos
7.01	Last explosion on seismograph, a BLEVE
7.30	Continuing tank explosions[a]
11.00	Last tank explosion
8.00–10.00	Rescue work at its height
12.00–18.00	Rescue work continues
23.00	Flames extinguished on last large sphere

[a] Explosions of cylindrical vessels.

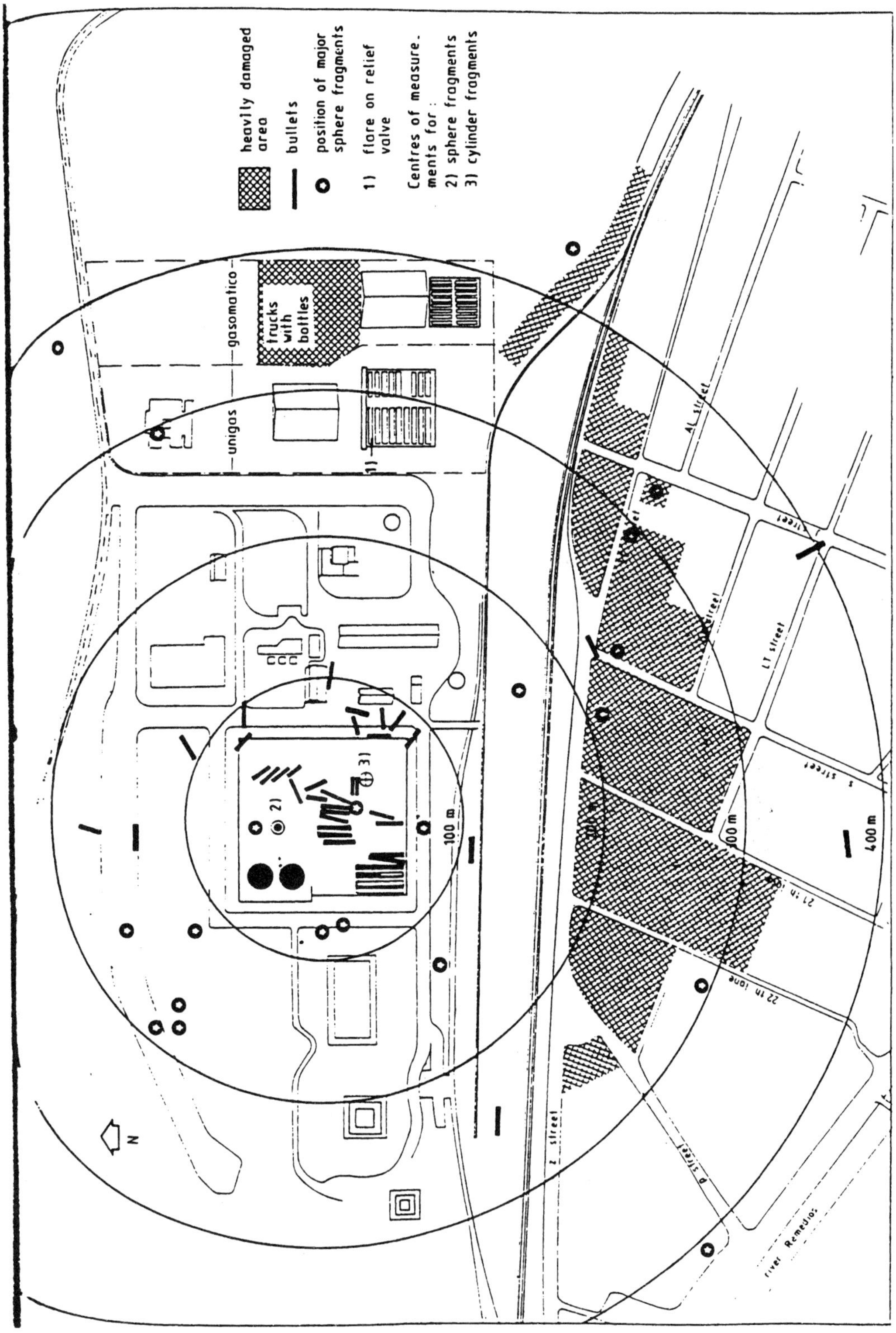

Figure A4.5   Area plan of the PEMEX site at Mexico City, showing damage to housing area and fall of missiles (Pietersen 1985) (Courtesy to TNO)

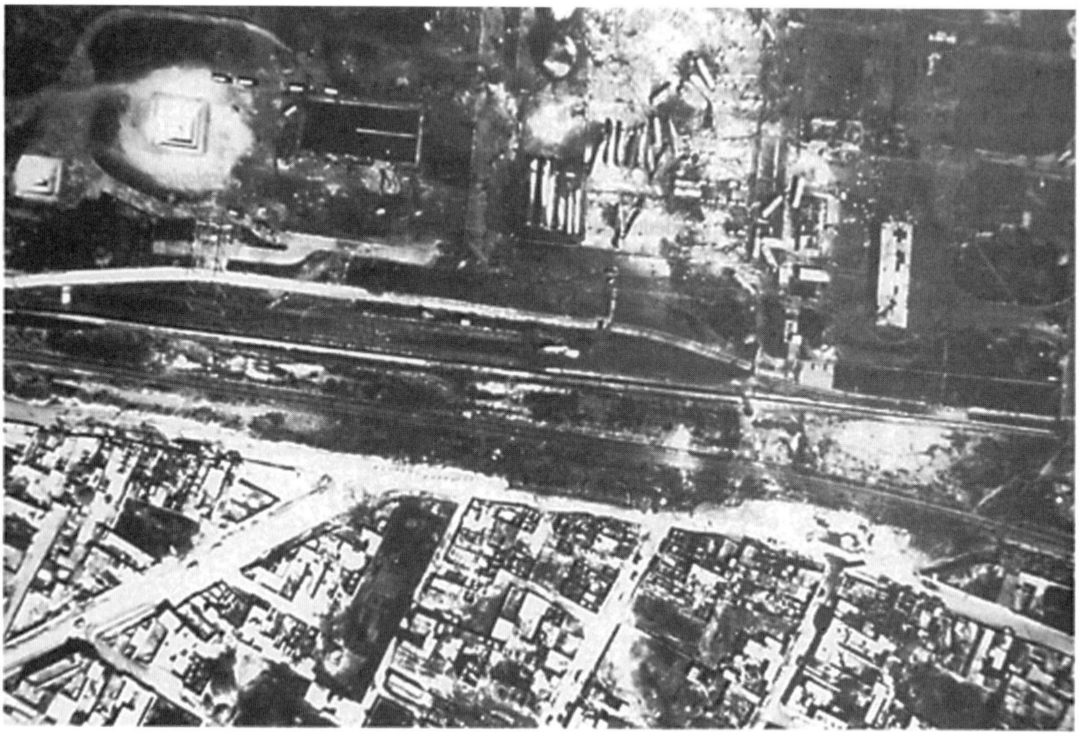

Figure A4.6   *PEMEX site at Mexico City after the accident (Skandia, 1985) (Photograph: State of Mexico)*

missiles. An aerial photograph of the plant after the disaster is shown in Figure A4.6.

Numerous missiles were generated by the bursting of the vessels. Many of these were large and travelled far. Twenty-five large fragments from the four smaller spheres weighing 10–40 te were found 100–890 m away. Fifteen of the 48 cylindrical vessels weighing 20 te became missiles and rocketed over 100 m, one travelling 1200 m. Four cylinders were not found at all. The missiles caused damage both by impact and by their temperature, which was high enough to set houses alight.

When the fire began, people were already on their way to work. Eyewitnesses spoke of a huge hot, red light, intense heat, smoke and lack of air, blast waves and missiles. The following account by Nicanor Santiago, a mason, is typical:

Around 5.30 a.m. I went to work. It was still dark when I took my bicycle out of the house, when suddenly there was this huge light, red and hot. I could see nothing at all. The huge light blinded me. I could not feel anything except that everything was hot. Then I heard some explosions and a second blast. The walls of my house were rocking, it was an earthquake. I was lying on the pavement and close to me all sorts of matter came falling out of the sky. There was a lot of broken glass, chairs and flower pots were flying all over the place. I suddenly remembered the PEMEX gas plant and I saw a big tongue of fire and a very big orange mushroom, and then a felt another explosion. Pieces of molten metal were dropping out of the sky and I felt intense heat waves burning my clothes and my hair. I ran into the bedroom, everything

was dark inside and there was plenty of smoke. I could not see anything. I could not breathe for the gas. I noticed something huge fell on top of the house and a rush of air threw me out of the house into the street. By then I was really frightened and I started to run as fast as I could. I do not know where my wife and children got to. They may even be unidentified in a mass grave.

A timetable of events during the disaster is given in Table A4.2, Section B.

A4.3 The Emergency

Accounts of the emergency give little information about the response of the on-site management in the emergency and deal mainly with the rescue and firefighting.

The site became the scene of a major rescue operation which reached a climax in the period 8.00–10.00 a.m. Some 4000 people participated in rescue and medical activities, including 985 medics, 1780 paramedics and 1332 volunteers. At one point there were some 3000 people in the area. There were 363 ambulances and five helicopters involved.

The rescuers were at risk from a large BLEVE. The Skandia report states: 'If a BLEVE had occurred during the later morning, a large number of those 3,000 people who were engaged in rescue and guarding would have been killed.'

The fire services were called by surrounding plants and by individual members of the public about 5.45 a.m. They went into the plant area only 3 h after the start of the incident. Initially, they moved towards the Gasomatico site, where a sphere fragment had landed and started a fire, which caused the domestic gas cylinders to explode.

The fire brigade also fought the fire on the two larger spheres, which had not exploded. They were at appreciable risk from BLEVE of the spheres. In the event, however, these burnt themselves out. The last flames on the spheres went out about 11.00 p.m. Some 200 firemen attended the site.

A4.4 The Fire and Explosion – 2

The TNO report gives much technical information on the course of the disaster and on the fire and explosion phenomena which occurred during it.

It discusses the effects of explosions, including vapour cloud explosions, BLEVEs and physical explosions; the effects of fire engulfment and heat radiation; and the effects of missiles, including fragments from bullet tanks and spheres.

The report gives estimates of the overpressure from BLEVE of the principal vessels. It states that the degree of blast damage to housing was not great; that the vapour cloud explosion effects were not responsible for major damage; that the second explosion, a BLEVE, was the most violent and did damage houses; that the worst explosion damage was probably from gases which had accumulated in houses; and that much of the damage was done by fire.

Films were available for many of the BLEVEs, though not for the second, violent explosion. From this evidence, the BLEVEs had diameters of 200–300 m and durations of some 20 s. Heavy direct fire damage was done at distances up to about 300 m, which agrees reasonably well with the estimates of fireball size.

A very large fire burned on the site for about an hour and a half, punctuated by BLEVEs. Details are given of the number, size and range of fragments from spheres and bullet tanks. Information on BLEVEs and missiles in the TNO report is given in Chapters 16 and 17.

A4.5 Some Lessons of Mexico City

Some of the lessons to be learned from Mexico City are listed in Table A4.3.

Siting of major hazard installations
The high death toll at Mexico City occurred because the housing was too near to the plant. At the time the plant was constructed the area was undeveloped, but over the years the built-up area had gradually crept up to the site.

Table A4.3   *Some lessons of Mexico City*

Siting of major hazard installations
Layout of large LPG storages
Gas detection and emergency isolation
Planning for emergencies
Fire fighting in BLEVE situations
Boiling liquid expanding vapour explosions

Layout and protection of large LPG storages
The total destruction of the facility occurred because there was a failure of the overall system of protection, which includes layout, emergency isolation and water spray systems.

Gas detection and emergency isolation
One feature which might have averted the disaster is more effective gas detection and emergency isolation. The plant had no gas detector system and, probably as a consequence, emergency isolation was too late.

Planning for emergencies
One particularly unsatisfactory aspect of the emergency was the traffic chaos which built up as residents sought to flee the area and the emergency services tried to get in.

Another was risk run by the large number of rescuers who came on site from a BLEVE of one of the larger spheres.

Fire fighting in BLEVE hazard situations
The fire services appear to have taken a considerable a risk in trying to fight the fire on the two larger spheres. The potential death toll if a BLEVE had occurred was high.

Boiling liquid expanding vapour explosions
After Flixborough the problem of vapour cloud explosions received much attention. Mexico City demonstrates that boiling liquid expanding vapour explosions are an equally important hazard.

Mexico City represents the largest series of major BLEVEs which has occurred and provides much information on these.

Appendix

Bhopal

5

Contents

Early in the morning of 3 December 1984, a relief valve lifted on a storage tank containing highly toxic methyl isocyanate (MIC) at the Union Carbide India Ltd (UCIL) works at Bhopal, India. A cloud of MIC gas was released onto housing, including shantytowns, adjoining the site.

Close on 2000 people died within a short period and tens of thousands were injured. The casualty figures are discussed further in Section A5.9.

The accident at Bhopal is by far the worst disaster which has ever occurred in the chemical industry. Its impact has been felt worldwide, but particularly in India and the United States.

Following the accident the Government of India (GoI) set up an inquiry which reported at the end of 1985 (Varadarajan, 1985). An investigation was conducted by the US parent company the Union Carbide Corporation (UCC), which issued its own report (the *Union Carbide Report*) (Union Carbide, 1975). Another investigation was carried out by the ICFTU-ICEF (1985). In its initial investigation, Union Carbide had limited access to documents and personnel, and it subsequently caused to be published further findings (Kalelkar, 1988). These investigations are described in Section A5.8.

Besides the investigations mentioned, other accounts have been given in Bhopal. *Anatomy of a Crisis* by Shrivastava (1987) and by Badhwar and Trehan (1984), Bhushan and Subramanian (1985), Bowonder (1985, 1987, 1988), Bowonder and Miyake (1988), Kalelkar (1988), Kletz (1988h), van Mynen (1990) and Bowonder, Arvind and Miyake (1991).

Selected references on Bhopal and MIC are given in Table A5.1.

A5.1 The Company and the Management

Union Carbide began operations in India in 1904 and by 1983 had 14 plants operating in the country. Its Indian interests were held by UCIL. UCIL was owned 50.9% by the American parent company UCC and 49.1% by Indian investors. UCC thus retained a majority share, having persuaded the Indian government to waive its usual requirement for Indian majority shareholding, on the basis of the technological sophistication of the plant and the export potential.

Table A5.1   *Selected references on Bhopal and MIC*

Anon. (1984h); Badhwar and Trehan (1984); Anon. (1985b–e,k,ff,gg,ii); Anon. (1985 LPB 63, p. l); Basta (1985); Basta *et al.* (1985); Bhushan and Subramanian (1985); Bowonder (1985, 1987); Bowonder, Kasperson and Kasperson (1985); Chowdhury *et al.* (1985); IBC (1985/59); ICFTU-ICEF (1985); Kharbanda (1985, 1988); Kletz (1985j,p, 1988h,n, 1990g, 1993b); Lepkowski (1985, 1986); Lihou (1985); Resen (1985a); Union Carbide Corp. (1985); Varadarajan (1985); Webber (1985, 1986); Anon. (1986b,l); Bellamy (1986); Bisset (1986); Shrivastava (1987); M.P. Singh and Ghosh (1987); Anon. (1988b,o); Bowonder and Miyake (1988); J. Cox (1988a); Emsley (1988); Kalelkar (1988a,b); Kharbanda (1988); V.C. Marshall (1988a); Sriram and Sahasrabuhde (1988, 1989); Tachakra (1988); van Mynen (1990); M.P. Singh (1990); Bowonder, Arvind and Miyake (1991); Narsimhan (1993)

Methyl isocyanate
Kimmerle and Eben (1964); ten Berge (1985)

UCIL began operations at Bhopal in 1969. Initially, the plant formulated carbamate pesticides from concentrates imported from the United States. In 1975, UCIL was licensed to manufacture its own carbaryl with the trade name Sevin. The process selected was the same as at the UCC plant at W. Virginia, but initially the MIC intermediate was imported from the latter source. Production began in 1979. The plant had a capacity of 5250 te/y, but the market was less than expected. Production peaked at 2704 te in 1981 and fell to 1657 te in 1983. At these levels of sales the plant had problems of profitability.

Prior to the accident, the management structure of UCIL changed and the Bhopal pesticides plant was put under the direction of the Union Carbide battery division in India.

A5.2 The Site and the Works

The location of the UCIL works at Bhopal is shown in Figure A5.1. The works was in a heavily populated area. Much of the housing development closest to the works had occurred since the site began operations in 1969, including the growth of the J.P. Nagar shantytown. Although these settlements were originally illegal, in 1984 the government gave the squatters rights of ownership on the land to avoid having to evict them. Other residential areas which were affected by the gas cloud had been inhabited for over 100 years.

A5.3 The Process and the Plant

In the process used at Bhopal, MIC was made using the reaction scheme shown in Figure A5.2. The process itself is shown in Figure A5.3. Monomethylamine (MMA) is reacted with excess phosgene in the vapour phase to produce methylcarbamoyl chloride (MCC) and hydrogen chloride and the reaction products are quenched in chloroform. The unreacted phosgene is separated by distillation from the quench liquid and recycled to the reactor. The liquid from the still is fed to the pyrolysis section where MIC is formed. The stream from the pyrolyser condenser passes as feed to the MIC refining still (MRS). MIC is obtained as the top product from the still. The MIC is then run to storage.

Phosgene was produced on site by reacting chlorine and carbon monoxide. The carbon monoxide was also produced on site.

The MIC storage system (MSS) consisted of three storage tanks, two for normal use (Tanks 610 and 611) and one for emergency use (Tank 619). The tanks were 8 ft diameter × 40 ft long with a nominal capacity of 15,000 USgal. They were made of 304 stainless steel with a design pressure of 40 psig at 121°C and with hydrostatic test pressure of 60 psig. A diagram of the storage tank system is shown in Figure A5.4.

A 30 ton refrigeration system was provided to keep the tank contents at 0°C by circulating the liquid through an external heat exchanger.

There was on each storage tank a pressure controller which controlled the pressure in the tank by manipulating two diaphragm motor valves (DMVs), a make-up valve to admit nitrogen and a blowdown valve to vent vapour. Each tank had a safety relief valve (SRV) protected by a bursting disc. It also had a high temperature alarm and low and high level alarms.

A vent gas scrubber (VGS) and a flare were provided to handle vented gases. The VGS was a packed column 5 ft 6 in.

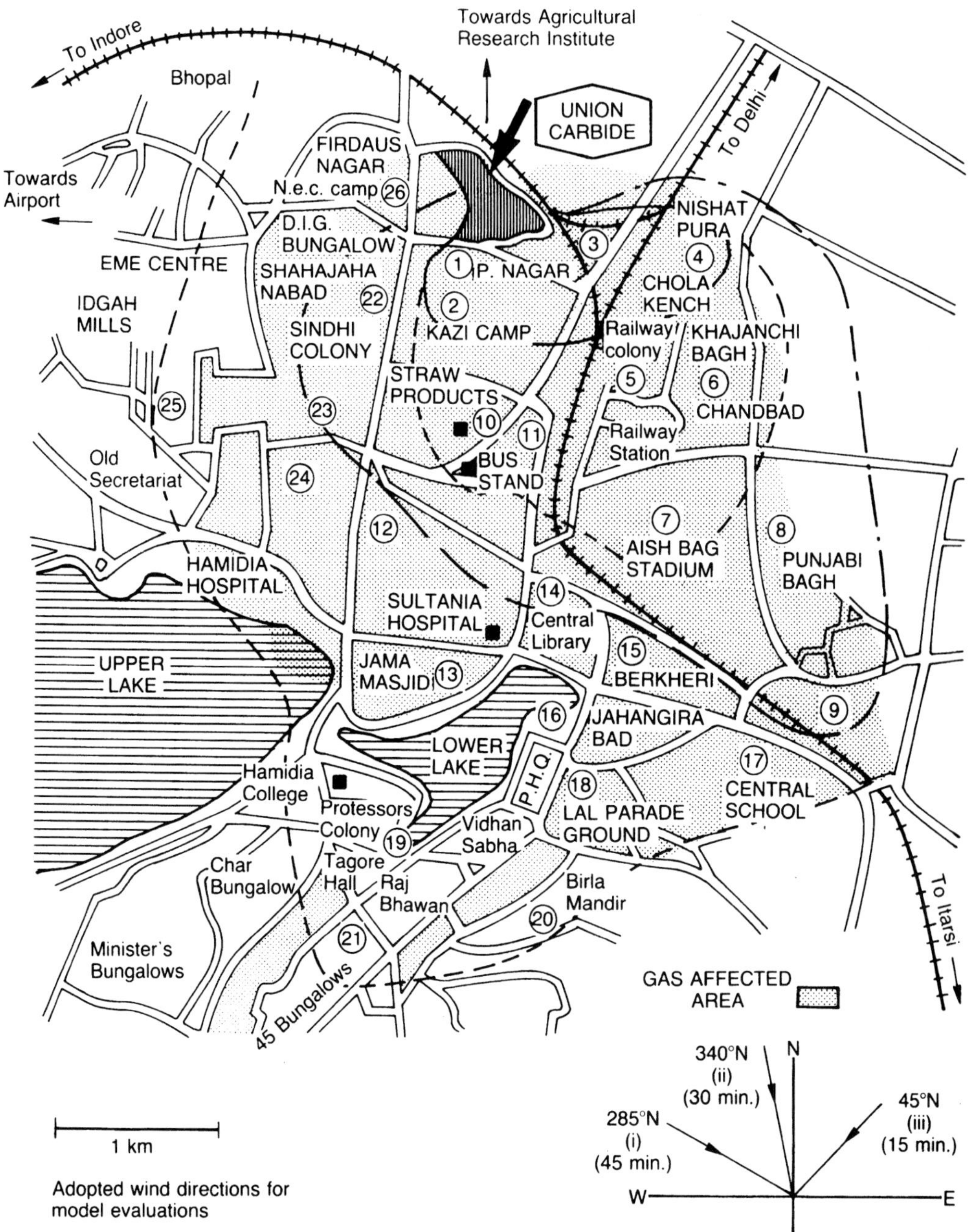

Figure A5.1 *Simplified plan of the area near the Union Carbide India Ltd works at Bhopal (M.P. Singh and Ghosh, 1987). The diagram also shows the estimated dimensions of the gas cloud (Courtesy of Elsevier Science Publishers)*

diameter in which the vent gases were scrubbed with caustic soda. There were two vent headers going into the column: the process vent header (PVH), which collected the MIC system vents, and the relief valve vent header (RVVH), which collected the safety valve discharges. Each vent header was connected both to the VGS and the flare and could be routed to either. The vent stack after the VGS was 100 ft (33 m) high. A diagram of the VGS is shown in Figure A5.5.

(1) $COCl_2$ + CH_3NH_2 ──────────▶ $CH_3NHCOCl$ +   HCl   +   Heat
     Phosgene   Monomethylamine   Methylcarbamoyl   Hydrogen
                        (MMA)   chloride   chloride
                                                   (MCC)

                              Heat CH_3NCo
(2) $CH_3NHCOCl$ ──────────────────▶   Methyl isocyanate   +   HCl
     (MCC)   (MIC)

Figure A5.2   *Reaction scheme (Union Carbide, 1985. Reproduced by permission.)*

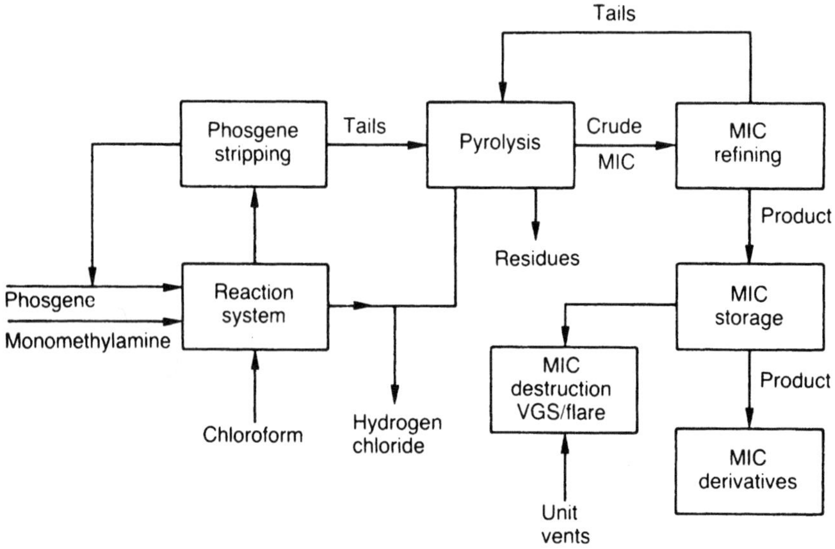

Figure A5.3   *Process for production of MIC (Union Carbide, 1985. Reproduced by permission.)*

The VGS had the function of handling process vents from the PVH and of receiving contaminated MIC, in either vapour or liquid form, and destroying it in a controlled manner.

The function of the flare was to handle vent gases from the carbon monoxide unit and the MMA vaporizer safety valve and also vent gas from the MIC storage tanks, the MRS and the VGS.

In the two years preceding, the number of personnel on site were reduced, 300 temporary workers being laid off and 150 permanent workers pooled and assigned as needed to jobs, some of which they said when interviewed they felt unqualified to do. The production team on the MIC facility was cut from 12 to 6.

A5.4 MIC and Its Properties

MIC is a colourless liquid with a normal boiling point of 39°C. It has a low solubility in water. It is relatively stable when dry, but is highly reactive and in particular can polymerize and will react with water. It is flammable and has a flashpoint of −18°C and a lower flammability limit of 6% v/v. It is biologically active and highly toxic.

The high toxicity of MIC is indicated by the fact that its TLV at the time was 0.02 ppm. This is very low relative to most typical compounds handled in industry.

MIC is an irritant gas and can cause lung oedema, but it also breaks down in the body to form cyanide. The cyanide suppresses the cytochrome oxidase necessary for oxygenation of the cells and causes cellular asphyxiation.

Information on the inhalation toxicity of MIC is given by Kimmerle and Eben (1964) and ten Berge (1985).

MIC can undergo exothermic polymerization to the trimer, the reaction being catalysed by hydrochloric acid and inhibited by phosgene. It also reacts with water, iron being a catalyst for this reaction. This reaction is strongly exothermic.

A5.5 Events Prior to the Release

In 1982, a UCC safety team visited the Bhopal plant. Their report gave a generally favourable summary of the visit, but listed ten safety concerns, including

3. Potentials for release of toxic materials in the phosgene/MIC unit areas and storage areas, either due to equipment failure, operating problems, or maintenance problems.
4. Lack of fixed water spray protection in several areas of the plant.
7. Deficiencies in safety valve and instrument maintenance program.

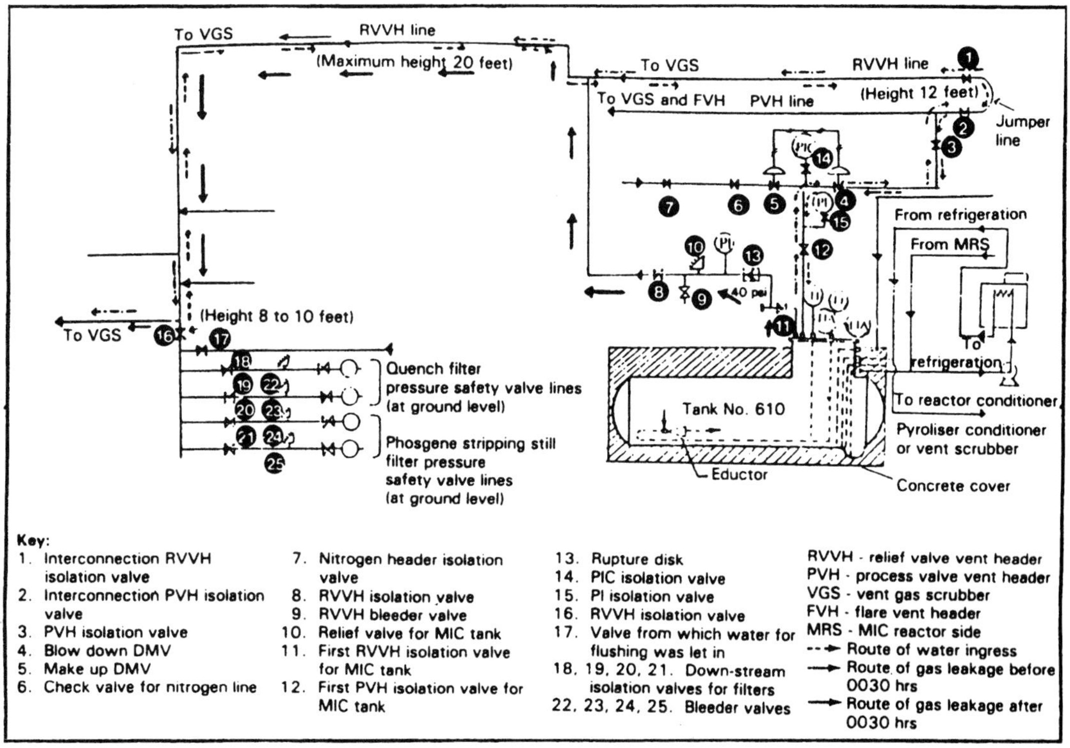

Key:

1. Interconnection RVVH isolation valve	7. Nitrogen header isolation valve	13. Rupture disk
2. Interconnection PVH isolation valve	8. RVVH isolation valve	14. PIC isolation valve
3. PVH isolation valve	9. RVVH bleeder valve	15. PI isolation valve
4. Blow down DMV	10. Relief valve for MIC tank	16. RVVH isolation valve
5. Make up DMV	11. First RVVH isolation valve for MIC tank	17. Valve from which water for flushing was let in
6. Check valve for nitrogen line	12. First PVH isolation valve for MIC tank	18, 19, 20, 21. Down-stream isolation valves for filters
		22, 23, 24, 25. Bleeder valves

RVVH - relief valve vent header
PVH - process valve vent header
VGS - vent gas scrubber
FVH - flare vent header
MRS - MIC reactor side
--→ Route of water ingress
——→ Route of gas leakage before 0030 hrs
➤ Route of gas leakage after 0030 hrs

Figure A5.4 *Flow diagram of Tank 610 system (Bhushan and Subramanian, 1985) (Courtesy of Business India.)*

8. Deficiencies in Master Tag/Lockout procedure application.
10. Problems created by high personnel turnover at the plant, particularly in operations.

Following this visit valves on the MIC plant were replaced, but degraded again. At the time of the accident, the instruments on Tank 610 had been malfunctioning for over a year.

Between 1981 and 1984, there were several serious accidents on the plant. In December 1981, three workers were gassed by phosgene and one died. Two weeks later 24 workers were overcome by another phosgene leak. In February 1982, 18 people were affected by an MIC leak. In October 1982, three workers were injured and nearby residents affected by a leak of hydrochloric acid and chloroform.

Following this latter accident workers from the plant posted a notice in Hindi which read: 'Beware of fatal accidents . . . lives of thousands of workers and citizens in danger because of poison gas . . . Spurt of accidents in the factory, safety measures deficient'. These posters were also distributed in the community.

About a year before the accident, a 'jumper line' was connected between the PVH and the RVVH. Figure A5.4 shows the MIC storage tank and pipework arrangements. The jumper line is between Valves 1 and 2. The object of the modification was to allow gas to be routed to the VGS if repairs had to be done on one of the vent headers.

In June 1984 the 30 ton refrigeration unit cooling the MIC storage tanks was shut down. The charge of Freon refrigerant was drained from the system.

In October the VGS was turned off, apparently because it was thought unnecessary when MIC was only being stored not manufactured. In the same month, the flare tower was taken out of service, a section of corroded pipe leading to it being removed so that it could be replaced.

Another feature was that difficulty was being experienced in pressurizing MIC storage Tank 610. It appeared that since nitrogen was passing through the make-up valve satisfactorily, the blowdown valve was leaking and preventing pressurization.

According to plant workers there were other instrumentation faults. The high temperature alarm had long been faulty. There were also faults on the pressure controller and the level indicator.

The plant had a toxic gas alarm system. This consisted of a loud siren to warn the public and a muted siren to warn the plant. These two sirens were linked and could be activated from a plant toxic alarm box. The loud siren could be stopped from the control room by delinking the two. A procedure had been introduced according to which after delinking, the loud siren could be turned on only by the plant superintendent.

Plant workers stated that on the morning of 2 December washing operations were undertaken. Orders were given to flush out the downstream sections of four filter pressure safety valves lines. These lines are shown in Figure A5.4. In order to carry out this operation Valve 16 on the diagram

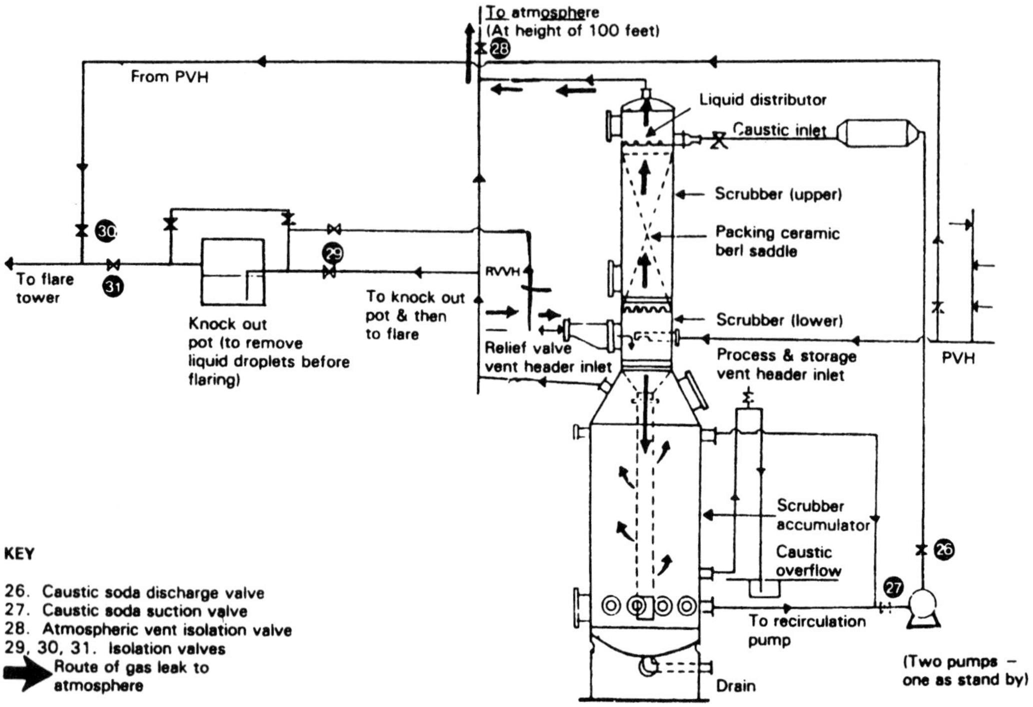

Figure A5.5   *Flow diagram of the vent gas scrubber system (Bhushan and Subramanian, 1985) (Courtesy of Business India)*

was shut, Valves 18–21 and 22–25 opened and then Valve 17 opened to admit water.

It was suggested water might have entered MIC storage Tank 610 as a result of this operation – the water washing theory. On this hypothesis, water evidently leaked through Valve 16, into the RVVH and passed through the jumper line into the PVH and thence into Tank 610. This would require that Valves 3 and 12 were open to connect the tank to the PVH and Valves 1 and 2 open to connect the RVVH to the PVH via the jumper line.

A5.6 The Release

On the evening of 2 December, a shift change took place on the plant at 22.45. At 23.00, the control room operator noticed that the pressure in Tank 610 was 10 psig. This was higher than normal but within the 2–25 psig operating pressure of the tank. At the same time the field operator reported a leak of MIC near the VGS. At 00.15, the field operator reported an MIC release in the process area and the control room operator saw that the pressure on Tank 610 was now 30 psig and rising rapidly. He called the supervisor and ran outside to the tank. He heard rumbling sounds coming from the tank and a screeching noise from the safety valve and felt heat from the tank. He returned to the control room and turned the switch to activate the VGS, but this was not in operational mode, the circulating pump not being on.

At 00.20, the production supervisor informed the plant superintendent of the release. At 00.45 operations in the derivative unit were suspended due to the high concentration of MIC.

At 01.00, an operator in this unit turned on the toxic gas alarm siren. After 5 min the loud siren was switched off leaving the muted siren on.

At about the same time the plant superintendent and control room operator verified that MIC was being emitted from the VGS stack to atmosphere and turned on and directed at the stack fixed fire water monitors to knock down the vapour.

Water was also directed at the MIC tank mound and at the vent header to the VGS. Steam issued from the cracks in the concrete showing that the tank was hot.

One plant supervisor tried to climb the structure to plug the gas leak but was overcome, falling and breaking both legs.

Some time between 01.30 and 02.30, the safety valve on Tank 610 reseated and the release of MIC ceased.

At about 02.30, the loud siren was switched on again.

The cloud of MIC gas spread from the plant towards the populated areas to the south. There was a light wind and inversion conditions.

People in the housing around the plant felt the irritant effect of the gas. Many ran out of their houses, some towards the plant. Within a short period, animals and people began to die.

At Railway Colony some 2 km from the plant, where nearly 10,000 people lived, it was reported that within 4 min 150 died, 200 were paralysed and 600 rendered unconscious and that 5000 were severely affected.

People tried to telephone the plant but were unable to get through. At 01.45, a magistrate contacted the plant superintendent.

The cloud of toxic gas hung around the area for the whole of 3 December. During the day, it stopped moving towards the city, but resumed its movement in that direction during the night.

A5.7 The Emergency and the Immediate Aftermath

Large numbers of people were affected by the toxic gas and very large numbers fled their homes.

The two hospitals principally concerned, the Hamidia and the Javaprakas Hospitals, were overwhelmed with casualties.

The difficulties were compounded by the fact that it was not known what the gas was or what its effects were. Speculation about the gas, including suggestions that it was phosgene, continued in the world press for some days.

The company provided little advice. Initially, it stated that MIC causes eye irritation but is not lethal.

As early as noon on 3 December, doctors at the Gandhi Memorial College carried out post-mortems which gave strong evidence of cyanide poisoning. Victims had died of respiratory arrest, but there was no evidence of the cyanosis due to the deoxygenation of the blood which normally accompanies pulmonary asphyxiation and there were cases where there was no evidence of pulmonary oedema.

There developed a conflict of views on the appropriate treatment. The standard treatment for cyanide poisoning is sodium thiosulfate. One group took the view that this should not be given until cyanide poisoning was established by analyses, another argued that it was well known that in cyanide poisoning the cyanide may be metabolized, leaving little trace. There followed a period in which the advice given was not clear. It was not until 3 February that an authoritative and unambiguous recommendation that sodium thiosulfate be used was issued by the Indian Council for Medical Research.

The Indian Central Bureau of Investigation (CBI) took control of the site and began a criminal investigation.

A5.8 The Investigations

A5.8.1 Government of India investigation
An investigation of the incident was undertaken by the GoI. It issued in December 1985, the *Report on Scientific Studies on the Factors Related to Bhopal Toxic Gas Leakage* by a team chaired by Varadarajan (1985).

The report refers to the fact that it was reported that about 21.30 on 2 December an operator was clearing a possible choke in the RVVH lines downstream of the phosgene stripping still filters by water flushing, without inserting a blind. The 6 in. isolation valve on the RVVH would presumably be closed but if it had not been leaktight, water could enter the RVVH. This water could have found its way into the Tank 610 via the blowdown DMV or through the SRV and bursting disc.

A5.8.2 Union Carbide investigation — 1
A team from UCC arrived in Bhopal on 6 December charged with the tasks of assisting in the safe disposal of the remaining MIC and investigating the accident. The first task was completed on 22 December and the team returned home on 2 January.

The investigation was severely constrained by the CBI control of the site and by the criminal investigation. The team were allowed only limited access to plant records and personnel. They were permitted to talk to certain persons, but not to interview staff directly involved in the incident. They were allowed to take samples from Tank 610, but not to open and inspect the tank and its piping or to take samples from elsewhere on the plant. An account of the situation is given by Kalelkar (1988b).

On their return home, the investigators carried out a programme of some 500 experiments to establish what had occurred. The most abundant component in the residues was MIC trimer, others present in significant quantities being the components conveniently referred to as DMI, DMU, TMU, TMB and TRMB. There were also iron, chromium and nickel in approximately their proportions in 304 stainless steel and some 5% chloride.

Trimerization had obviously occurred, but it was unclear what other reactions had taken place. The investigators carried out experiments in which the principal materials believed to have been in the tank, namely MIC, chloroform, water and iron, were heated at 200°C and developed a reaction scheme which accounted for the production, starting from these materials, of the components found in the residue with the exception of TRMB, which they considered they could account for.

From plant records, Tank 610 contained prior to the incident 41 te (90,400 lb; 11,290 USgal) of liquid. The team estimated that Tank 610 had originally contained 1000–2000 lb (120–240 USgal) water and 1500–3000 lb chloroform. The source of the water was uncertain. The chloroform could be accounted for by the fact that the MRS had been operated at a temperature higher than normal and in preparation for shutdown MIC with a high chloroform content has been sent to Tank 610 rather than Tank 619. The iron could have come from corrosion, given high chloroform and water contents and high temperature.

The scenario which the investigators invoked to explain the events is as follows. The contents of Tank 610 were initially at 15–20°C. Some 1000–2000 lb water entered the tank in a manner unknown. The exothermic reaction between MIC and water led to an increase in temperature and also in pressure due to evolution of carbon dioxide. The higher temperature and presence of chloroform caused accelerated corrosion. The iron thus produced catalysed the exothermic trimerization of MIC.

Calculations showed that reaction of some 40% of the MIC would generate enough heat to vaporize the rest. This would give some 36,000 lb of solids in the tank, but only an estimated 10,000 lb were found. There may have been appreciable loss of solids and liquid through the relief vent.

The period during which the relief valve was open was reported to have been about 2 h. It was calculated that in order to vent most of the tank contents in this time the discharge rate would have had to be 40,000 lb/h, of which 29,000 lb/h were vapour and 11,000 lb/h solids/liquid mixture and that this would have required a pressure averaging 180 psig. The temperature reached was estimated as in excess of 200°C. These conditions would have been attained during the course of the venting. With the safety valve lifting at 40 psig the initial discharge would have been 10,000 lb/h.

The report puts forward the hypothesis that the water was directly introduced into Tank 610, either inadvertently or deliberately through the process vent line, nitrogen line or other piping. It refers to the washing operation on the filter pressure safety valve lines and states that this section of line had not been isolated using a blind, but that passage of water to Tank 610 through several reportedly closed valves is unlikely.

The report draws attention to a number of factors which contributed to the accident. These include the facts that refrigeration had been discontinued, that a blind was not used to isolate the lines being washed out, that the MRS was operated at a higher than normal temperature, that the VSG did not work and that the flare was out of commission.

At the press conference held to present the report, the UCC spokesman suggested that the water may have come from a nearby utility station which supplied water and nitrogen to the area: 'If someone had connected tubing to the water line instead of the nitrogen line, either deliberately or intending to introduce nitrogen into the tank, this could account for the presence of the water. . .'. The press interpreted this as a suggestion by UCC that the cause of the accident was sabotage. Subsequently UCC agreed that there was no direct evidence for this hypothesis.

A5.8.3 Union Carbide investigation – 2

Following the involvement of the Indian Government in litigation in the US courts, UCC was allowed access to personnel who had been involved on the plant and was able to gather evidence in support of its original hypothesis. An account of this extension of the UCC investigation is given by Kalelkar (1988b).

Against the water washing hypothesis he makes the following points. First, the water was introduced through a $1/2$ in. inlet (Valve 17). The difference in head between this point and the inlet to Tank 610 was some 10.4 ft. There were three open bleeder valves close to the inlet point which would limit the backpressure of water to 0.7 ft. Second, there were a number of valves between the inlet point and the tank and for water to pass these would have had to be open or not leak-tight. One of these valves, close to the inlet point, had been shut since 29 November 1984. It was given a one-hour test during which no water leaked through it. Third, for water to reach Tank 610 it would have had to fill the 6 in. diameter connecting pipe, 65 ft of 8 in. RVVH, with numerous branches running off, and then some 340 ft of 4 in. RVVH. The amount of water to do this was estimated as 4500 lb. On 8 February 1985, the CBI ordered a hole to be drilled in the lowest point of the PVH. For the hypothesis to hold this section, which had no bleeders or flanged joints, should have been full of water; it was completely dry.

In support of the hypothesis of direct entry of water into Tank 610 the arguments presented by Kalelkar may be summarized as follows. First, an instrument supervisor, not on duty that night, stated that he had found the local pressure indicator on the tank missing; this was one of the few points to which a water hose could be connected. Second, a water hose was found nearby. Third, there was evidence that the operators had become aware earlier in the evening that water had entered Tank 610 and had taken steps to deal with the situation. Fourth, the plant logs showed evidence of extensive tampering and alteration.

The plant could be supplied with MIC from Tank 610 or Tank 611, the MIC being passed to a one ton tank, the Sevin charge pot. There were difficulties with the pressure in Tank 610 and the transfers had been made from Tank 611. However, investigation showed that the MIC in the Sevin charge pot contained water. It had evidently come from Tank 610. Water is heavier than MIC and Tank 610 had a bottom offtake. The hypothesis is that the operators, aware that water had got into Tank 610 and wishing to remove it,

on this occasion drew the MIC from that tank rather than from Tank 611.

Kalelkar suggests that the addition of water to Tank 610 may have been the act of a disgruntled employee.

A5.9 The Late Aftermath

The precise numbers of the dead and injured at Bhopal are uncertain. The scale of the accident was such that it led to much confusion. People have continued to die of the effects over a period of years. The official Indian Government estimate of the death toll about 2 years after the event was 1754. By 1989, this had risen to 3150 and by 1994 to 4000. Other figures given are 30,000 permanently or totally disabled; 20,000 temporary cases; and 50,000 with minor injury.

The ICFTU-ICEF report in 1985 states that the number of people treated in the state hospitals had been given by Dr Nagu, the Director of Health Services for Madhya Pradesh, as approximately 170,000. Some 130,000 were treated in Bhopal hospitals, mainly for lung and eye injuries, and some 40,000 in 22 other districts. Some 12,000 of the 170,000 were in a critical condition and 484 died. He estimated the total number of dead as 2000.

The disaster led to various sets of court proceedings. The GoI instituted criminal proceedings against UCC, which at the time of writing remain extant. The GoI also became a party to proceedings in the US courts. In 1987, UCC made a 'final settlement' with the GoI of $470 million. Victims tried to challenge this in the US courts, but the US Supreme Court ruled that they lacked legal standing to do so.

A5.10 Some Lessons of Bhopal

A5.10.1 Some lessons

The lessons to be learnt from Bhopal are numerous. A list of some of them is given in Table A5.2. They combine many of the lessons of Flixborough and Seveso.

Some of these lessons are now considered:

Public control of major hazard installations

The disaster at Bhopal received intense publicity for an extended period and put major hazards on the public agenda worldwide, but particularly in India and the United States, which had not reacted so strongly to Flixborough and Seveso, whose impact had been felt most in Europe.

Table A5.2   *Some lessons of Bhopal*

Public control of major hazard installations
Siting of and development control at major hazard
   installations
Management of major hazard installations
Highly toxic substances
Runaway reactions in storage
Water hazard in plants
Relative hazards of materials in process and in storage
Relative priority of safety and production
Limitation of inventory in the plant
Set pressure of relief valves
Disabling of protective systems
Maintenance of plant equipment and instrumentation
Isolation procedures for maintenance
Control of plant and process modifications
Information for authorities and public
Planning for emergencies

Siting of and development control at major hazard installations

Very large numbers of people were at risk from the plant at Bhopal. This situation was due in large part to the encroachment of the shantytowns, which came up to the site boundary. Although these settlements were illegal, the Indian authorities had acquiesced in them.

In this instance, however, this was not the whole story. The accident showed that site was close enough to areas populated before the plant was built to present a hazard when used for the production of a chemical as toxic as MIC. If the manufacture of such a chemical was envisaged from the start, the problem may be regarded as one of siting. If not, it may be viewed as one of intensification of the hazard on the site.

Management of major hazard installations

The plant at Bhopal was by any standards a major hazard and needed to be operated by a suitable competent management. The standards of operation and maintenance do not give confidence that this was so.

There had been recent changes in the responsibility for the plant which suggest that the new management may not have been familiar with the exigencies of major hazards operation. However, many of the problems on the plant appear to have antedated these changes.

Highly toxic substances

MIC is a highly toxic substance, much more toxic than substances such as chlorine which are routinely handled in the chemical industry. The hazard from such highly toxic substances has perhaps been insufficiently appreciated.

This hazard will only be realized if there is a mechanism for dispersion. At Bhopal, this mechanism was the occurrence of exothermic reactions in the storage tank.

Runaway reaction in storage

The hazard of a runaway reaction in a chemical reactor is well understood, but that of such a reaction in a storage tank had received little very attention. At Bhopal, this occurred due to ingress of water. Where such a reaction could act as the mechanism of dispersion for a large inventory of a hazardous substance, the possibility of its occurrence should be carefully reviewed.

Water hazard in plants

In general terms, the hazard of water ingress into plants is well known. In particular, water may contact hot oil and vaporize with explosive force or may cause a frothover, it may corrode the equipment and it may cause a blockage by freezing. Bhopal illustrates the hazard of an exothermic reaction between a process fluid and water.

Relative hazard of materials in process and in storage

There has been a tendency to argue that the risks from materials in storage are less than from materials in process, since, although usually the inventories in storage are larger, the probability of a release is much less. The release at Bhopal was from a storage tank, albeit from one associated with a process.

The relative hazard of materials in process and in storage is discussed Chapter 22.

Relative priority of safety and production

The features which led to the accident have been described above. As indicated, the *Union Carbide Report* itself refers to a number of these.

The ICFTU-ICEF report states that at the time of the accident the plant was losing money and lists a number measures which had been taken, apparently to cut costs. These include the manning cuts and the cessation of refrigeration.

Limitation of inventory in the plant

The hazard at Bhopal was the large inventory of highly toxic MIC. The process was the same as that used at UCC's W. Virginia plant. UCIL had stated that it regarded this inventory as undesirable, but was overruled by the parent company, which wished to operate the same process at both plants.

Processes are available for the manufacture of MIC which require only small inventories of the material. Moreover, carbaryl can be made by a route which does not involve MIC. The alternatives to the use of MIC are discussed by Kletz (1988h).

Set pressure of relief devices

It is desirable from the operational viewpoint for the set pressure of a relief valve to be such that the valve opens when the pressure rise threatens the integrity of the vessel but not when normal minor operating pressure deviations occur. Where the cause of potential pressure rise is a runaway reaction, however, there is a penalty in setting a high set pressure in that this may allow the reaction to reach a higher temperature and to proceed more rapidly before venting starts, so that there is a need to balance these two factors.

Disabling of protective systems

It was evidently not appreciated that the flare system was a critical component for the protection of the plant, since it was allowed to remain out of commission for the 3 months prior to the accident. It is essential that there be strict procedures for the disabling of any item which is critical for protection and that the time for which the item is out of action be kept to a minimum.

Maintenance of plant equipment and instrumentation

The 1982 UCC safety team drew attention to the problems in the maintenance of the plant. The *Union Carbide Report* gives several examples of poor maintenance of plant equipment and instrumentation and the ICFTU-ICEF report gives further details. Workers stated that leaking valves and malfunctioning instruments were common throughout the plant.

Maintenance was also very slow. The flare system, which was a critical protective system, had been out of commission for three months before the accident.

Isolation procedures for maintenance

A particular deficiency in the maintenance procedures was the failure to isolate properly the section of plant being flushed out by positive isolation using a slip plate or equivalent means. The fact that the water may not have entered in this way does not detract from this lesson.

Control of plant and process modifications

A principal hypothesis to explain the entry of water into Tank 610 is that the water passed through the jumper line. The installation of this jumper line was a plant modification. Company procedures called for plant modifications

Event	Recommendations for prevention/mitigation
	1st layer: Immediate technical recommendations *2nd layer: Avoiding the hazard* **3rd layer: Improving the management system**
Public concern compelled other companies to improve standards	**Provide information that will help public keep risks in perspective.**
Emergency not handled well	Provide and practise emergency plans.
About 2000 people killed	**Control building near major hazards.**
Scrubber not in full working order Flare stack out of use Both may have been too small	Keep protective equipment in working order. Size for foreseeable conditions.
Discharge from relief valve	
Refrigeration system out of use	Keep protective equipment in use even though plant is shut down.
Runaway reaction	
Rise in temperature	**Train operators not to ignore unusual readings.**
Water entered MIC tank	*Carry out hazops on new designs.* *Do not allow water near MIC.*
Decision to store over 100 tonnes MIC	*Minimize stocks of hazardous materials.*
Decision to use MIC route	*Avoid use of hazardous materials.*
Joint venture established	**Agree who is responsible for safety.**
	To achieve the above: **Train chemical engineers in loss prevention**

Figure A5.6   *An accident model for Bhopal, showing critical events and recommendations (Kletz, 1988h) (Courtesy of Butterworths)*

to be checked by the main office engineers, but were evidently disregarded.

There was also a process modification which more certainly contributed to the accident. This is the decommissioning of the refrigeration system, so that the temperature in Tank 610 was higher than the 0°C for which the system was designed.

Information for authorities and public

UCIL had not provided full information on the substances on site to the authorities, emergency services, workers or members of the public exposed to the hazards. Many workers interviewed said they had had no information or training about the chemicals.

Planning for emergencies

The response of the company and the authorities to the emergency suggests that there was no effective emergency plan.

Within the works, defects revealed by the emergency include the hesitation about the use of the siren system and the lack of escape routes.

The preliminary condition for emergency planning to protect the public outside the works is provision to the authorities of full information about the hazards. This was not done. In consequence the people exposed did not know what the siren meant or what action to take, the hospitals did not know what they might be called on to handle, and so on.

Likewise, the essential action in an actual emergency is to inform the authorities what has happened and what the hazards are. On the morning of the accident, the hospitals were in the dark about the nature and effects of the toxic chemical whose victims they were trying to treat.

A5.10.2 An accident model

An accident model for Bhopal given by Kletz (1988h) is shown in Figure A5.6.

Appendix

6

Pasadena

Contents

Shortly after 1.00 p.m. on 23 October 1989, a release occurred on a polyethylene plant at the Phillips 66 Company's chemical complex at Pasadena, near Houston, Texas. A vapour cloud formed and ignited, giving rise to a massive vapour cloud explosion. There followed a series of further explosions and a fire. Twenty-two people on the site were killed and one later died from injuries, making a death toll of 23. The number injured are variously given as 130 and 300.

A report on the investigation of the accident has been issued by OSHA (1990a). Other accounts include those of Mahoney (1990), T. Richardson (1991) and J.N. Scott (1992).

Selected references on Pasadena are given in Table A6.1.

A6.1 The Site and the Plant

The Phillips works was sited in the Houston Chemical Complex along the Ship Channel, the location of a number of process companies.

The plant on which the release occurred was Plant V, one of the two active polyethylene plants in the complex. The plant operated at high pressure (700 psi) and high temperature. The process involved the polymerization of ethylene in isobutane, the catalyst carrier. Particles of polyethylene settled out and were removed from settling legs.

A6.2 Events Prior to the Explosion

On the previous day work began to clear three of the six settling legs on Reactor No. 6, which were plugged. The three legs were prepared by a company operator and were handed over to the specialist maintenance contractors, Fish Engineering. The configuration of a typical leg is shown in Figure A6.1.

At 8.00 a.m. on Monday, 23 October, work began on the second of the three blocked legs, Leg No. 4. The isolation procedure was to close the DEMCO ball valve and disconnect the air lines to it.

The maintenance team partially disassembled the leg and were able to remove part of the plug, but part remained lodged in the pipe 12–18 in. below the ball valve. One of the team was sent to the control room to seek assistance. Shortly after, at 1.00 p.m., the release occurred.

Although both industry practice and Phillips corporate safety procedures require isolation by means of a double block system or a blind flange, at local plant level a procedure had been adopted which did not conform to this.

It was subsequently established that the DEMCO ball valve was open at the time of the release. The air hoses to the valve had been cross-connected so that the supply which should have closed it actually opened it. The hose connectors for the 'open' and 'close' sides of the valve were identical, thus allowing this cross-connection to be made. Although procedures laid down that the air hoses should not be connected during maintenance, there was no physical barrier to the making of such a connection. The ball valve had a lockout system but it was inadequate to prevent the valve being inadvertently or intentionally opened during maintenance.

Table A6.1   *Selected references on Pasadena*

Anon. (1989 LPB 90, p. 0); Anon. (1990 LPB 94, p. 30); Mahoney (1990); OSHA (1990a); Redmond (1990); Vervalin (1990b); Bond (1991 LPB 97); Kletz (1991J); T. Richardson (1991); J.N. Scott (1992)

A6.3 The Explosion

The mass of gas released was estimated as some 85,200 lb of a mixture of ethylene, isobutane, hexene and hydrogen, which escaped within seconds. The release was observed by five eyewitnesses. A massive vapour cloud formed and moved rapidly downwind.

Within 90–120 s the vapour cloud found a source of ignition. Possible ignition sources were a gas-fired catalyst activator with an open flame; welding and cutting operations; an operating forklift truck; electrical gear in the control building and the finishing building; 11 vehicles parked near the polyethylene plant office; and a small diesel crane, although this was not operating.

The TNT equivalent of the explosion was estimated in the OSHA report as 2.4 tons. An alternative estimate from seismograph records is 10 tons.

There followed two other major explosions, one when two 20,000 USgal isobutane storage tanks exploded and the other when another polyethylene plant reactor failed catastrophically, the timings being some 10–15 min and some 25–45 min, respectively, after the initial explosion. One witness reported hearing 10 separate explosions over a 2-h period.

Debris from the explosion was found 6 miles from the site.

All 22 of those who died at the scene were within 250 ft of the point of release and 15 of them were within 150 ft.

Injuries which occurred outside the site were mainly due to debris from the explosion.

The explosion resulted in the destruction of two HOPE plants.

A6.4 The Emergency and the Aftermath

People in the immediate area of the release began running away as soon as they realized that gas was escaping. The alarm siren was activated, but the level of noise in the finishing building was such that there was a question whether some employees there failed to hear it.

The immediate response to the emergency was provided by the site fire brigade, which undertook rescue and care of the injured and began fighting the fire. Twenty-three persons were unaccounted for, but for an extended period the area of the explosion remained dangerous to enter.

Severe difficulties were experienced in fighting the fires resulting from the explosion. There was no dedicated fire water system, water for firefighting being drawn from the process water system. The latter suffered severe rupture in the explosion so that water pressure was too low for firefighting purposes. Fire hydrants were sheared off by the blast. Fire water had to be brought by hose from remote sources such as settling ponds, a cooling tower, a water treatment plant and a water main on a neighbouring plant. These difficulties were compounded by failures of the fire pumps. The electrical cables supplying power to the regular fire pumps were damaged by the fire so that these pumps were put out of action. Further, of the three backup diesel fire pumps one was down for maintenance and one quickly ran out of fuel. Despite these problems the fire was brought under control within some 10 h.

The handling of the emergency was handicapped by the facts that the intended command centre had been damaged and that telephone communications were disrupted. Telephone lines were jammed for some hours following the accident.

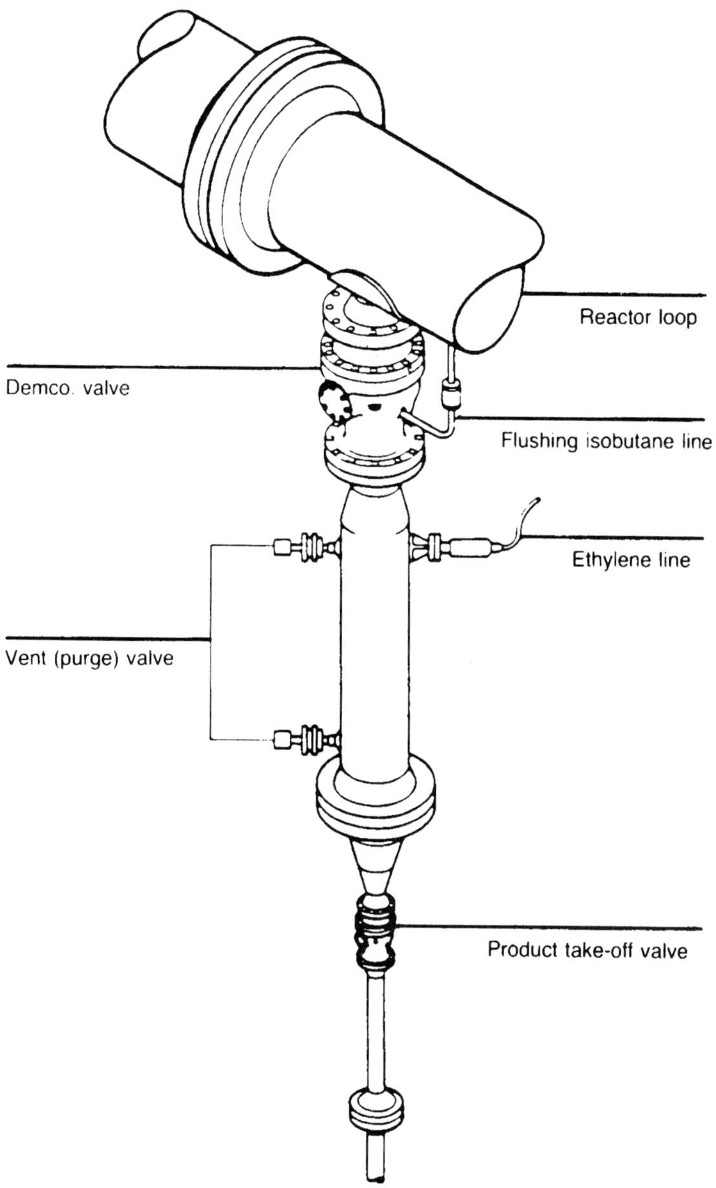

Figure A6.1 *Typical piping settling leg arrangement on reactor (Occupational Safety and Health Administration, 1990a)*

The emergency response was co-ordinated by the site chief fire officer and involved the local Channel Industries Mutual Aid (CIMA) organization, a co-operative of some 106 members in the Houston area.

More than 100 people were evacuated from the administration building across the Houston Ship Channel by the US Coast Guard and by Houston fireboats; they would otherwise have had to cross the area of the explosion to reach safety.

The media were quickly aware of the explosion and within an hour there were on site 150 media personnel from 40 different organizations.

The financial loss in this accident is comparable with, and may exceed, that of the Piper Alpha disaster. Redmond (1990) has quoted a figure of $1400 million, divided almost equally between property damage and business interruption losses.

On the basis of a review of company reports and of the defects found during the investigation of the disaster, OSHA issued a citation to the company for wilful violations of the 'general duty' clause. The citation covered the lack of hazard analysis; plant layout and separation distances; flammable gas detection; ignition sources; building ventilation intakes; and the fire water system; the permit system; isolation for maintenance.

A6.5 Some Lessons of Pasadena

Some of the lessons to be learned from Pasadena are listed in Table A6.2.

Management of major hazard installations
The OSHA report details numerous defects in the management of the installation. Some of these are described below.

Hazard assessment of major hazard installations
According to the report, the company had made no use of hazard analysis or an equivalent method to identify and assess the hazards of the installation.

Plant layout and separation distances
The report was critical of the separation distances in the plant in several respects. It stated that the separation distances between process equipment plant did not accord with accepted engineering practice and did not allow time for personnel to leave the polyethyelene plant safely during the initial vapour release; and that the separation distance between the control room and the reactors was insufficient to allow emergency shut-down procedures to be carried out.

Location of control room
As just mentioned, the control room was too close to the plant. It was destroyed in the initial explosion.

Building ventilation intakes
The ventilation intakes of buildings close to or down-wind of the hydrocarbon processing plants were not arranged so as to prevent intake of gas in the event of a release.

Minimization of exposure of personnel
Closely related to this, there was a failure to minimize the exposure of personnel. Not only the control room but the finishing building had relatively high occupancy.

Table A6.2   Some lessons of Pasadena

Management of major hazard installations
Hazard assessment of major hazard installations
Plant layout and separation distances
Location of control room
Building ventilation intakes
Minimization of exposure of personnel
Escape and escape routes
Gas detection system
Control of ignition sources
Permit-to-work systems
Isolation procedures for maintenance
Integrity of fire water system
Dependability of fire pumps
Audibility of emergency alarm
Follow-up of audits
Planning for emergencies

Escape and escape routes
As already stated, the separation distances were not such as to allow personnel on the polyethylene plant to escape safely. Further, the only escape route available to people in the administration block (other than across the ship channel) was across the area of the explosion.

Gas detection system
Despite the fact that the plant had a large inventory of flammable materials held at high pressure and temperature, there was no fixed flammable gas detection system.

Control of ignition sources
The control of ignition sources around the plant was another feature criticized in the OSHA report.

Permit-to-work systems
The OSHA report stated that an effective permit system was not enforced for the control of the maintenance activities either of the company's employees or of contractors.

Isolation procedures for maintenance
In this incident the sole isolation was a ball valve, which was meant to be closed but was in fact open. There was no double block system or blind flange.

The practice of not providing positive isolation was a local one and violated corporate procedures. The implication is that it had not been brought to light by any safety audits conducted.

Integrity of fire water system
The practice of relying for fire water on the process water system and the failure to provide a dedicated fire water system meant that the fire water system was vulnerable to an explosion.

Dependability of fire pumps
The electrical cables to the regular fire pumps were not laid underground and were therefore vulnerable to damage by explosion and fire. One of the back-up diesel pumps had insufficient fuel and one had been taken out for maintenance without informing the chief fire officer.

Audibility of emergency alarm
As described, the level of noise in some areas was such that the employees might not have been able to hear the siren.

Follow-up of audits
The OSHA report criticized the company's failure to act upon reports issued previously by the company's own safety personnel and by external consultants, which drew attention to unsafe conditions.

Planning for emergencies
The disaster highlighted a number of features of emergency planning. The company had put a good deal of effort into planning and creating personal relationships with the emergency services, by means such as joint exercises, and these paid off. The value of planning, training and personal relations was one of the most positive lessons drawn.

Another area in which a proactive approach proved beneficial was in relations with the media. Senior personnel made themselves available, and the company evidently felt it received fair treatment.

One weakness of the emergency planning identified was that it had not envisaged a disaster of the scale which actually occurred.

The incident brought out the need to be able to respond clearly to calls from those liable to be affected about the toxicity of the fumes and smoke generated in such an event.

The behaviour of rescue helicopters posed a problem. Personnel on the ground had no means of communication with them and the craft tended to come in low, creating the danger of blowing flames or toxic fume onto those below. A need was identified for altitude and distance guidelines for helicopters.

Appendix

7

Canvey Reports

Contents

The most comprehensive hazard assessment of non-nuclear installations in the United Kingdom is the Canvey study carried out for the HSE by SRD.

The first phase of the work is described in *Canvey: An Investigation of Potential Hazards from Operations in the Canvey Island/Thurrock Area* (the First *Canvey Report*) (HSE, 1978b). The report is in two parts: Part 1 is an introduction by the HSE and Part 2 the SRD study.

The origin of the investigation was a proposal to withdraw planning permission for the construction of an additional refinery in the area. Two oil companies, Occidental Refineries Ltd and United Refineries Ltd, had been granted planning permission for the construction of oil refineries. The construction of the Occidental refinery was begun in 1972, but was halted in 1973 pending a major design study review. United Refineries had valid planning consents, but had not started construction. It was a public inquiry into the possible revocation of the planning permission for the United Refineries development which gave rise to the investigation.

Responses to the First *Canvey Report* centred mainly on two aspects: the methodology used and the magnitude of the assessed risks. The HSE commissioned further work, leading to the Second *Canvey Report* (HSE, 1981a), in which the methodology used is revised and the assessed risks are rather lower.

An account of the first report is given in Sections A7.1–A7.6, of the response to that report in Section A7.7 and of the second report in Sections A7.8–A7.10.

A map of the Canvey/Thurrock area showing the principal installations and populated areas is given in Figure A7.1.

Selected references on the *Canvey Reports* are given in Table A7.1.

A7.1 First *Canvey Report*

The terms of reference of the investigation were

In the light of the proposal by United Refineries Limited to construct an additional refinery on Canvey Island, to investigate and determine the overall risks to health and safety arising from any possible major interactions between existing or proposed installations in the area, where significant quantities of dangerous substances are manufactured, stored, handled, processed and transported or used, including the loading and unloading of such substances to and from vessels moored at jetties; to assess the risk; and to report to the Commission.

The members of the investigating team were appointed as inspectors of the HSE under the provisions of the Health

and Safety at Work etc. Act 1974 and were given specified powers to enable them to make the necessary inquiries. The overall approach taken in the investigation was

(1) To identify any potentially hazardous materials, their location and the quantities stored and in process.
(2) To obtain and review the relevant material properties such as flammability and toxicity.
(3) To identify the possible ways in which failure of plants might present a hazard to the community.
(4) To identify possible routes leading to selected failures. Typically, the factors examined included operator errors, fatigue or aging of plant, corrosion, loss of process control, overfilling, impurities, fire, explosion, missiles and flooding.
(5) To quantify the probability of the selected failures occurring and their consequences.

The investigation involved the identification of the principal hazards of the installations and activities in the area, the assessment of the associated risks to society and to individuals, and the proposal of modifications intended to reduce these risks.

Some 30 engineers were engaged in the investigation, which cost about £400,000.

A7.2 First *Canvey Report*: Installations and Activities

The principal hazardous installations and activities identified in the investigation are summarized in Table A7.2. These installations are now briefly described.

A7.2.1 British Gas Corporation

The British Gas Corporation operates on Canvey Island, a methane terminal that is used for the importation and storage of LNG from Algeria. There are some 50 shipments of LNG a year made by two specially designed ships each with a cargo capacity of about 12,000 t. The LNG is pumped ashore from a single jetty into above-ground and in-ground fully refrigerated storage tanks with a combined total capacity of approximately 100,000 t. The natural gas is sent out as vapour by pipeline except for a small amount that goes out by road.

The terminal is also used for the fully refrigerated storage of liquid butane. A pipeline from the butane tanks crosses the area.

In addition, new LNG ships are commissioned at the terminal by using the LNG from the storage tanks to cool down the ships' cargo tanks.

A7.2.2 Texaco Ltd

Texaco Ltd operates on Canvey Island storage for petroleum products, but not LPG, with a total capacity of more than 80,000 t. The petroleum products are brought in by sea via two jetties and are sent out by sea, road and pipeline.

A7.2.3 London and Coastal Wharves Ltd

London and Coastal Wharves Ltd also operates on Canvey Island storage for a wide range of substances, including petroleum products and other flammable and toxic materials, with a total capacity of more than 300,000 t. Part of the site is leased to Texaco. The materials are brought in mainly by sea via a single jetty and also by road and are sent out not only by road but also by sea. Oil is sent out by Texaco by pipeline.

Table A7.1   *Selected references on Canvey Reports*

Canvey Reports
HSE (1978b, 1981a)

Reviews and critiques
N. Turner (1975); Anon. (1980,l,p,s); Cremer and Warner (1980); Hansard (1980); V.C. Marshall (1980a); Anon. (1981r,w); R. Ward (1981); A.F. Grant (1982); IChemE (1982a,b); Rasbash (1982a); Anon. (1983c); Petts (1985a)

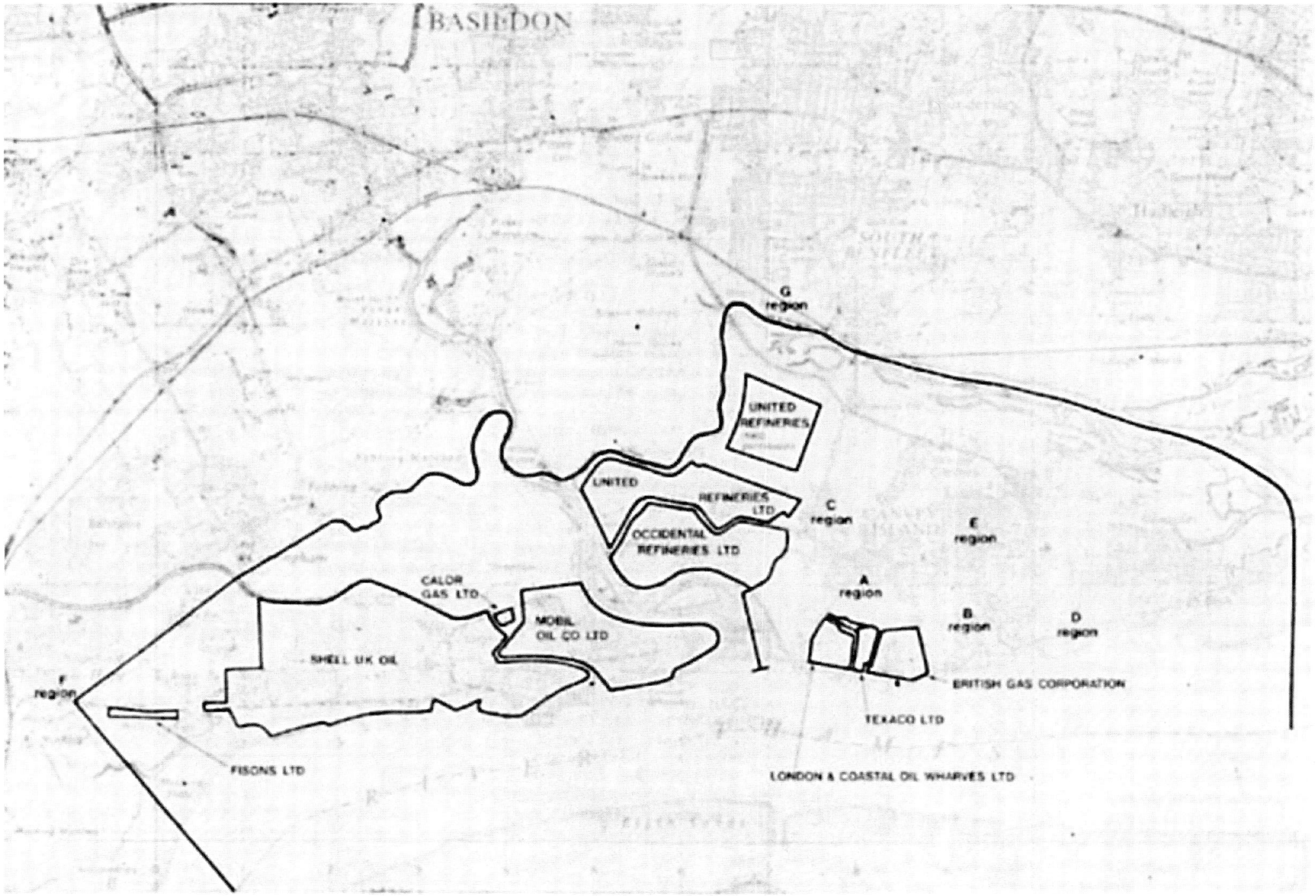

Figure A7.1 First Canvey Report: map of the Canvey/Thurrock area, showing principal installations and populated areas (Health and Safety Executive, 1978b) (Courtesy of HM Stationery Office)

Table A7.2 First Canvey Report: principal hazardous installations and activities at Canvey

Location	Company	Installation or activity	Storage	Employees	Transport in	Transport out
Canvey Island	British Gas Corporation	LNG terminal	Fully refrigerated storage of LNG (atmospheric pressure, <−162°C) 6 × 4000 t above-ground tanks 2 × 1000 t above-ground tanks 4 × 20,000 t in-ground tanks. Fully refrigerated storage of butane (atmospheric pressure, <10°C) 1 × 10,000 t tank 2 × 5000 t tanks	200	Sea	Mainly pipeline (as vapour) but some road
	Texaco Ltd	Petroleum products storage	Atmospheric storage of petroleum products >80,000 t total capacity	130	Sea	Pipeline, road, sea
	London and Coastal Wharves Ltd	Flammable and toxic liquids storage	Atmospheric storage of liquids >300,000 t total capacity	50	Mainly sea but some road	Pipeline (Texaco oil). Rest mainly road but some sea
	Occidental Refineries Ltd	Oil refinery (proposed)	Pressure storage of LPG (atmospheric temperature) 2 × 750 t propane spheres 2 × 400 t butane spheres			
	United Refineries Ltd	Oil refinery (proposed)	Pressure storage of LPG (atmospheric temperature) 4 × 200 t propane spheres 3 × 900 t butane spheres. Process and storage containing hydrogen fluoride			
Coryton	Mobil Oil Co. Ltd	Oil refinery	Pressure storage of LPG (atmospheric temperature) 1 × 1000 t + 14 other vessels, giving 4000 t total capacity	800	LPG produced on site	Pipeline, road, rail, sea
		Oil refinery (extension)	Pressure storage of LPG (atmospheric temperature) 4 × 1000 t LPG spheres. Fully refrigerated storage of LPG (atmospheric pressure) 1 × 5000 t tank. Process and storage containing hydrogen fluoride		LPG produced on site	Pipeline, road, rail, sea

Location	Company	Activity	Description		Method of receipt	Method of dispatch
Shell Haven	Calor Gas Ltd	LPG terminal	Pressure storage of LPG (atmospheric temperature) 3 × 60 t propane vessels 2 × 60 t butane vessels + cylinders, giving 500 t total inventory	100	Pipeline (from oil refineries)	Road (cylinders, bulk tankers)
	Shell Oil U.K. Ltd	Oil refinery	Pressure storage of LPG (atmospheric temperature) 1 × 1700 t butane sphere + 3 other butane spheres, giving 3200 t total butane capacity 4 × 400 t propane spheres 3 × 135t LPG horizontal vessels	1900	LPG produced on site	Pipeline, road, rail, sea
			Fully refrigerated storage of liquid anhydrous ammonia (atmospheric pressure, −33°C) 1 × 14,000 t tank. Process and storage containing hydrogen fluoride 2 × 40 t vessels		Sea	Sea, road, occasionally rail
Stanford-le-Hope	Fisons Ltd	Ammonium nitrate plant	Storage of 92% aqueous ammonium nitrate solution 1 × 5000 t tank 1 × 2000 t tank	80	Ammonium nitrate produced on site	Road
		Ammonia storage	Semi-refrigerated pressure storage of liquid anhydrous ammonia (pressure above atmospheric, 6°C) 1 × 2000 t sphere		Rail	Ammonia used on site
Canvey/ Thurrock area	General	Transport of hazardous Materials by river, road, rail and pipeline				

A7.2.4 Mobil Oil Co. Ltd

Mobil Oil Co. Ltd operates at Coryton, an oil refinery as well as a bulk distribution depot and a large research and technical services laboratory. Oil is brought in by sea into storage tanks and is drawn off from these as required by the refinery. LPG is produced on site. The total storage capacity is more than 1,500,000 t. Products are sent out by sea, road, rail and pipeline.

In addition, at the time of the investigation Mobil was undertaking a major extension of the refinery.

A7.2.5 Calor Gas Ltd

Calor Gas Ltd operates at Coryton, an LPG terminal for the filling of gas cylinders. The gas is supplied by pipeline from both the nearby refineries and is held in storage vessels with a total capacity of 350 t. The gas cylinders bring the total inventory on site to 500 t. LPG is sent out by road in cylinders and in bulk tankers.

A7.2.6. Shell Oil UK Ltd

Shell Oil UK Ltd operates at Shell Haven, an oil refinery. Oil is brought in by sea into storage tanks and is drawn off from these as required to the refinery. LPG is produced on site. The total storage capacity is more than 3,500,000 t. Products are sent out by sea, road, rail and pipeline.

There is also on site a large fully refrigerated storage tank for liquid anhydrous ammonia with a capacity of 14,000 t. Ammonia is brought in by sea and sent out by sea and road, and occasionally by rail.

A7.2.7 Fisons Ltd

Fisons Ltd operates at Stanford-le-Hope, an ammonium nitrate plant in which strong ammonium nitrate solution is produced by reacting together ammonia and nitric acid, the latter itself being made on site from ammonia. The ammonium nitrate is stored in two large heated tanks with a total capacity of 7000 t and is sent out by road. Liquefied anhydrous ammonia for the process is brought in by rail and is stored in a large semi-refrigerated pressure storage sphere with a capacity of 2000 t.

A7.2.8 Other activities

Explosives are trans-shipped at Chapmans Anchorage at the eastern end of Canvey Island, while there is a specified anchorage for explosives ships at the other end of the area.

Ships passing up and down the river are obliged to travel fairly close to the sea walls and jetties within the area.

A7.3 First *Canvey Report:* Identified Hazards

The investigation identified several principal hazards in the area. These are;

(1)   oil spillage over bund;
(2)   LNG vapour cloud release (1000 t);
(3)   LPG vapour cloud release (1000 t);
(4)   ammonium nitrate explosion (4500 t of 92% solution);
(5)   ammonia vapour cloud release (1000 t);
(6)   hydrogen fluoride cloud release (1000 t).

The figures in parentheses indicate standard cases considered in the study.

A severe fire might occur if there is an escape of flammable liquids from storage so that large quantities flow down into the residential area. This hazard is presented by the large storages of flammable liquids at London and Coastal Wharves and Texaco and at the proposed refineries of Occidental and United Refineries, and also by the large butane storage at British Gas.

A severe vapour cloud fire and/or explosion might occur if there is a spillage of LNG so that a vapour cloud forms and ignites. This hazard is presented by the large LNG terminal and storage at British Gas. The spillage might occur at sea or on land.

Similarly, a severe vapour cloud fire and/or explosion might occur if there is a spillage of LPG so that a vapour cloud forms and ignites. This hazard is presented by the LPG terminals and storages at British Gas, Shell and Mobil and by those at the proposed refineries of Occidental and United Refineries.

A severe explosion might occur if there is a rupture of an ammonium nitrate storage tank. This hazard is presented by the storage of ammonium nitrate at Fisons.

A severe toxic release might occur if there is a spillage of ammonia. This hazard is presented by the large terminals and storages at Fisons and at Shell. The spillage might occur at sea or on land.

A severe toxic release might also occur if there is a rupture of storage or process plant containing hydrogen fluoride. This hazard is presented by the alkylation facilities at Shell, at the Mobil extension and at the proposed Occidental refinery.

The investigators also identified and assessed other hazards, but these are not considered here.

A7.4 First *Canvey Report:* Failure and Event Data

The investigation required the estimation of the probabilities of various occurrences and of their consequences.

Some of the sources of information on such probabilities used in the study were

(1)   UK industries, including oil, chemical and other process industries and transport;
(2)   government organizations such as those concerned with fire, and road, rail and sea and air transport;
(3)   professional institutions, for example, Institution of Chemical Engineers, Institution of Civil Engineers, American Institute of Chemical Engineers;
(4)   international safety conference proceedings, for example, loss prevention in the process industries, ammonia plant safety and hazardous materials spills;
(5)   industry-based associations, for example, Chemical Industries Association, Institute of Petroleum, American Petroleum Institute, Liquefied Petroleum Gas Industry Technical Association;
(6)   international insurance interests, for example, Lloyds, Det Norsk Veritas, and industrial risk insurers, Fire Protection Association;
(7)   overseas government and international agencies, for example, US Coast Guard, US Department of Transportation, OECD and EEC;
(8)   specialized research laboratories;
(9)   individual subject specialists known or recommended to the investigating team.

Selected references used in the *Canvey Report* are given in Table A7.3.

Table A7.3 First Canvey Report: Selected references used

Min. of Defence (n.d.); Wardle (n.d.); Brodie (1956); B.R. Morton, Taylor and Turner (1956); Cremer (1959, 1961); Minorsky (1959); Cremer and Callaway (1961); Pasquill (1961, 1965); Patty (1962); R.K. Davis *et al.* (1963); Blokker (1964); Bryant (UKAEA 1964 AHSB(RP) R42); Glasstone (1964); Yih (1965); van Dolah *et al.* (1966 BM RI 6773); Resplandy (1967); Brasie and Simpson (1968); Slade (1968); G.A. Briggs (1969): Clough and Garland (1970b); Kiwan (1970b); Bryce-Smith (1971); Finney (1972); MacArthur (1972); Brobst (1972); Feldbauer *et al.* (1972); Humbert-Basset and Montet (1972); Oppenheim, Kuhl and Kamel (1972); Burgess and Zabetakis (1973 BM RI 7752); Kuhl, Kamel and Oppenheim (1973); May, McQueen and Whipp (1973); Strehlow (1973b); BatteUe Columbus Lab. (1974); L.E. Brown, Wesson and Welker (1974b); Dicken (1974, 1975); Geiger (1974); Hosker (1974b); Kneebone and Prew (1974); Munday and Cave (1974); Murata *et al.* (1974); NTSB (1974 PAR-74-06); Reed (1974); Simmons, Erdmann and Naft (1974); van Ulden (1974); CIA (1975/8); Germeles and Drake (1975); Gething (1975); J. Hall, Barrett and Ralph (1975); Lonsdale (1975); Raj *et al.* (1975); DoE (1976b); Gifford (1976b); MacMullen (1976); Munday (1976a); Neff, Meroney and Cermak (1976); Carver *et al.* (1977); R.A. Cox and Roe (1977); Davenport (1977b); Fitzpatrick and Goddard (1977); Havens (1977)

See also Appendix 28 (HSE and SRD; UKAEA, SRD)

The degree of uncertainty associated with the probability estimates is indicated by the following code:

(a) assessed statistically from historical data – this method is analogous to the use of aggregate estimates in economic forecasting;
(b) based on statistics as far as possible but with some missing figures supplied by judgement;
(c) estimated by comparison with previous cases for which fault tree assessments have been made;
(d) 'dummy' figures – likely always to be uncertain, a subjective judgement must be made;
(e) not used;
(f) fault tree synthesis, an analytically based figure which can be independently arrived at by others.

The coding (a) denotes a value based on historical failure or event data. The coding (b) indicates that the value is again based on historical data as far as possible, although with some exercise of judgement, but also that given more effort a firmer value would probably be obtained. The coding (d) indicates that the value is based on judgement and the prospects of ever reducing the uncertainty are poor. Category (d) factors are effectively dummy values. Conclusions should not be drawn from these without testing for sensitivity. The value of a (d) category factor is generally not very small: it is typically 0.3 and exceptionally 0.1. The coding (f) denotes a value synthesized by fault tree methods. This category is in fact very little used in the study. Instead use is made of (c) category factors. The coding (c) indicates that the value is based on a value previously obtained for a similar situation using fault tree methods. The *Canvey Report* contains much useful information on

failure and event data. Some of these data are given in Section A7.5 and in Table A7.4. It is emphasized that these data are given here for illustrative purposes and that the report itself should be consulted for the description of the background to and the application of these data.

A7.5 First *Canvey Report:* Hazard Models and Risk Estimates

The investigation involved the study of a wide range of hazards and scenarios.

The projects that were initiated as part of the investigation were

(1) consideration of known history of identified storage tanks and their possibility of failure;
(2) probability of particular storage tanks or process vessels being hit by missiles caused by fires or explosions on site or adjacent sites, by fragmentation of rotating machines or pressure vessels, or transport accidents;
(3) effects of vapour cloud explosions on people, houses, engineering structures, etc.;
(4) evaporation of LNG from within a containment area on land or from a spill on water;
(5) special problems of frozen earth storage tanks for LNG and the effect of flooding;
(6) study of possible failures in handling operations;
(7) consideration of the possible benefits and practicality of evacuation;
(8) civil engineering aspects of the sea-wall – the chance of it being breached by subsidence, explosion or impact of ships, consideration of the timing of improved defences, consideration of the time for floods to rise;
(9) statistics of ship collisions and their severity, groundings, etc., applying extensive world experience to the Canvey Island area;
(10) reliability and analysis of fluid handling practices, ship to shore, and store to road vehicles and pipelines;
(11) toxicology of identified hazardous substances;
(12) studies to determine the lethal ranges for various releases of toxic or explosive materials leading to a number of special studies such as
    (a) the behaviour of ammonia spilt on water or land, and
    (b) an assessment of the relative importance of explosion or conflagration from a cloud of methane or liquefied petroleum gas.

The subjects that are considered in appendices to the report are

(1) a review of current information on the causes and effects of explosions of unconfined vapour clouds (F. Briscoe);
(2) fires in bunds – calculations of plume rise and position of downwind concentration maximum (R. Griffiths);
(3) a quantitative study of factors tending to reduce the hazards from airborne toxic clouds (Q.R. Beattie);
(4) the dispersal of ammonia vapour in the atmosphere with particular reference to the dependence on the conditions of emission (F. Abbey, R.F. Griffiths, S.R. Haddock, G.D. Kaiser, R.J. Williams and B.C. Walker);

Table A7.4 First Canvey Report: some failure and event data used (after Health and safety Executive, 1978b)

Installation or activity[a]		Canvey Report page(s)
Pressure vessels (LPG, ammonia, HF)	Frequency of spontaneous failure of pressure vessel	10^{-5}/year–10^{-4}/year	57, 59, 69
Pressure circuit (HF)	Frequency of spontaneous failure of pressure circuit	10^{-4}/year	59
	Frequency of release due to operational fault	10^{-4}/year	59
	Frequency of penetration of pressure circuit by missile	10^{-4}/year	59
High speed rotating machine	Frequency of disintegration of rotor	10^{-4}/year–10^{-3}/year	58
Pipework (LPG)	Frequency of failure of pipework (whole refinery installation)	5×10^{-3}/year	56
Pump(LPG)	Frequency of catastrophic failure of pump	10^{-4}/year	56
LPG filling point	Frequency of large vapour release	5×10^{-3}/year	58, 81
LNG tank (above ground)	Frequency of serious fatigue failure	2×10^{-4}/year	63
	Frequency of overpressurization by overfilling	10^{-5}/year–10^{-4}/year	63, 95
	Frequency of rollover involving structural damage	10^{-5}/year–10^{-4}/year	63, 95
Jetty pipework (LNG)	Frequency of catastrophic failure of jetty pipework	10^{-4}/year–10^{-3}/year	63
Fire	Frequency of major fire in a refinery	0.1/year	55, 130
Explosion	Probability of refinery explosion, given major refinery fire	0.5	130
Missiles	Probability of missile generation, given refinery explosion	0.1	130
	Frequency of missile-generating explosion in refinery	5×10^{-3}/year	130
	Average number of missiles generated per explosion	c. 6	130
	Probability of missile hitting large storage sphere at 300 m	10^{-3}/year	55

Unconfined vapour cloud explosion	Frequency of unconfined vapour cloud explosion in a refinery	10^{-3}/year	69, 130
Pipeline (butane)	Frequency of failure of pipeline (15/20 cm diameter)	3×10^{-4}/km y	85
Rail transport	Frequency of derailment of rail tank car	1×10^{-6}/train km travelled	79
	Probability of overturning given derailment	0.2	79
Road transport	Frequency of accident of road tanker involving spillage	1.6×10^{-8}/km travelled	84
Sea transport (Port of London)	Frequency of ship–ship collision of moderate severity due to harbour movements	0.5×10^{-4}/harbour movement	146
	Frequency of berthing contact	1.5×10^{-4}/harbour movement	146
	Frequency of grounding	0.3×10^{-4}/harbour movement	146
	Frequency of spillage due to harbour movements: ammonia carrier	3.1×10^{-5}/harbour movement	149
	Frequency of fire	0.5×10^{-4}/harbour movement	146
	Frequency of ship–ship collision of moderate severity in transit in estuary	2.3×10^{-5}/movement	147
	Frequency of spillage due to ship–ship collision intransit: ammonia carrier	5×10^{-6}/movement	150
Jetty incidents	Frequency of shipboard explosion at jetty: ammonia ship, Shell jetty	10^{-4}/harbour movement	96
	Frequency of fire or explosion at jetty: LPG, Occidental, Mobil jetty	4×10^{-5}/harbour movement	96
Aircraft movements	*See* Section A.5.5 of this Appendix		139–140
Helicopter movements	Frequency of helicopter accident – total accidents	3×10^{-7}/km flown	141
	– Fatal accidents	8×10^{-8}/km flown	141

[a] The data given in this table are in most cases heavily qualified and the original report should be consulted for the description of the background to and the application of them.

(5) statistics on fires and explosions at refineries Q.H. Bowen);
(6) missiles – penetration capability (EA White);
(7) discussion of data base for pressure vessel failure rate (TA Smith);
(8) risk of aircraft impacts on industrial installations in the vicinity of Canvey Island (LS. Fryer);
(9) the dispersion of gases that are denser than air, with LNG vapour as a particular example (G.D. Kaiser);
(10) not used;
(11) the risk of a liquefied gas spill to the estuary (D.F. Norsworthy);
(12) the toxic and airborne dispersal characteristics of hydrogen fluoride (Q.R. Beattie, F. Abbey, S.R. Haddock, G.D. Kaiser);
(13) transient variation of the wall temperature of an LNG above-ground storage tank during exposure to an LNG fire in an adjacent bund (I.R. Fothergill);
(14) the escape of 1000 t of anhydrous ammonia from a pressurized storage tank (L.S. Fryer, G.D. Kaiser and B.C. Walker);
(15) graphical calculation of toxic ranges for a release of 1000 t of ammonia vapour (J.H. Bowen);
(16) effect of unbunded spill of hydrocarbon liquid from refinery at Canvey Island (A.N. Kinkead);
(17) estimated risk of missile damage causing a vapour cloud release from existing and proposed LPG storage vessels at the Mobil refinery, Coryton (D.F. Norsworthy);
(18) risks of accidents involving road tankers carrying hazardous materials (L.S. Fryer);
(19) toxicology of lead additives (S.R. Haddock);
(20) compatibility of materials stored at London and Coastal Wharves Ltd (S.R. Haddock);
(21) blast loading on a spherical storage vessel (Q. Wall);
(22) statistical comment on data on distribution of cracks found on inspection of steel vessel (Q.C. Moore);
(23) calculation of resistance of ship hull to collision (A.N. Kinkead);
(24) reduction of apparent risk by shared experience (Q.H. Bowen).

The treatment of some of the topics which is given in the report is now described. These topics are

(1) failure of pressure vessels;
(2) failure of pressure piping;
(3) failure of pipelines;
(4) generation of and rupture by missiles;
(5) crash of and rupture by aircraft;
(6) ship collision and other accidents;
(7) flow of a large release of oil;
(8) temperature of the wall of an LNG tank exposed to an LNG fire in an adjacent bund;
(9) evaporation of LNG and ammonia on water;
(10) dispersion of an LNG vapour cloud;
(11) unconfined vapour cloud fire and explosion;
(12) ammonium nitrate explosion;
(13) toxicity of chlorine, ammonia, hydrogen fluoride and lead additives;
(14) dispersion of ammonia and hydrogen fluoride vapour clouds;
(15) factors mitigating casualties from a toxic release;
(16) road tanker hazards;
(17) evacuation.

It is emphasized that the following summary of the methods used in the report is necessarily highly compressed and that the report itself should be consulted for a fuller discussion of the methods themselves and of the background to and application of the methods.

A7.5.1 Failure of pressure vessels
Spontaneous failure of pressure vessels is considered as a possible initiating event for releases of LPG, ammonia and hydrogen fluoride. The report discusses spontaneous failure of pressure vessels in Part 2, Section 5.3.2. and in Appendix 7 and the sensitivity of the results for LPG, ammonia and hydrogen fluoride releases in Part 2, Sections 17.14–17.16.

It reviews the surveys of pressure vessel failures available at that date, as described in Chapter 12, and concludes that the UK survey data are broadly applicable to LPG storage vessels.

For vessels covered by the surveys the frequency of a fault requiring repair or withdrawal from service was about 3×10^{-4}/y. If it can be assumed that the critical defect length is less than the wall thickness, so that detection is easier, and that there is an effective inspection system, a reasonable estimate of the probability of detection is 90%. The frequency of an undetected fault would then be about 3×10^{-5}/y. A catastrophic failure rate of pressure vessels of 10^{-5}/y is used in the study.

In addition, however, the sensitivity of the results to errors in the assumed failure rate was calculated taking a failure rate of 10^{-4}/y instead of 10^{-5}/y. The effect of this change is almost to double the risk from LPG operations for the existing installations after the proposed modifications. Thus for an incident involving >1500 casualties the frequency is 264×10^{-6}/y with a contribution of $20–10^{-6}$/y for spontaneous failure of pressure vessels, but the latter rises to 200×10^{-6}/y if the spontaneous failure rate of pressure vessels is increased from 10^{-5}/y to 10^{-4}/y.

It is concluded that a high standard of inspection of the LPG pressure vessels is necessary.

External threats to pressure vessels were also considered. An event tree for the derailment involving overturning of a rail tank car with possible threat to a sphere containing 2000 t of anhydrous ammonia is shown in Figure A7.2.

Missile threats are described in Section A7.5.4.

A7.5.2 Failure of pressure piping
Failure of pressure piping is considered as a possible initiating event for releases of LPG and LNG. The report discusses the failure of LPG pressure piping in Part 2, Section 5.3.1.

LPG is piped around the refineries in pressure piping. The pipes mainly contain liquid, but there are some vapour return lines. It was considered that probably if a large pipe were to fail, there would be a vapour plume, the plume would find a source of ignition and would burn back to the pipe and there form a burning jet. In burning back there would be the possibility of semi-confined vapour explosions. If on a 25 cm diameter LPG pipe the fluid burned as a jet, the latter could be over 50 m long, while if, due to obstruction, it burned rather as a hemisphere, the radius of the latter could be over 10 m.

Various effects of such fires and explosions from LPG pipe failure are reviewed. An LPG sphere should withstand a semi-confined explosion beneath it; a horizontal cylinder

Figure A7.2 *First Canvey Report: event tree for derailment involving overturning of a rail tank car with possible threat to a sphere containing 2000 te of anhydrous ammonia (Health and Safety Executive, 1978b) (Courtesy of HM Stationery Office)*

could be lifted up some tens of centimetres. The principal hazard foreseen, however, is burn back under an LPG storage sphere.

Three cases are treated: (1) failure of a pipe, (2) failure of a pump and (3) failure of a suction line within about 10 m of a storage sphere. The smallest pipe size considered was 15 cm diameter.

It was estimated that the area of a typical refinery is about 3 km^2 and that the length of LPG pipe of diameter $\geq$15 cm in such a refinery is about 3 km, of which about 10% is 25 cm diameter pipe. The total frequency of pipe fracture for this length of pipe was assessed from data in the SRS data bank as 5×10^{-4}/y. For a pipe to affect the storage it must be sufficiently near. The hazard ranges of 15 and 25 cm diameter pipe were taken as 150 and 500 m, respectively. The width of a plume would be of the order of 20°, so that the possibility of the plume lying in any particular direction is 5%. Then the frequencies of involvement of an LPG storage vessel due to pipe failures are

Frequency for 15 cm pipe
$$= 2 \times 10^{-2} \times 5 \times 10^{-3} \times 5 \times 10^{-2} = 5 \times 10^{-6}/\text{y}$$

Frequency for 25 cm pipe
$$= 1 \times 0^{-1} \times 2 \times 10^{-1} \times 5 \times 10^{-3} \times 5 \times 10^{-2} = 5 \times 10^{-6}/\text{y}$$

Total frequency for all large pipes $= 10^{-5}/\text{y}$.

The number of LPG pumps in a typical refinery was estimated at about 20. The frequency of catastrophic rupture of such pumps was assessed from data in the SRS data bank as 10^{-4}/y. As before, the probability of the plume in any particular direction is 5%. Then the frequency of involvement of an LPG storage vessel due to a pump rupture is

Frequency for pumps $= 20 \times 10^{-4} \times 5 \times 10^{-2} = 10^{-4}/\text{y}$

In addition, there is the possibility that if the pumps are close together (<20 m spacing), a fire at one pump would involve other pumps and would create a large fire regardless of plume direction. Then the frequency for this event is 2×10^{-3}/y.

The proportion of 25 cm diameter LPG pipe within 10 m of a storage vessel was estimated as about 10%. Then the frequency of involvement of an LPG storage vessel due to failure of such pipe is

Frequency for 25 cm pipe beneath vessels
$$= 10^{-1} \times 10^{-6} \times 5 \times 10^{-3} = 5 \times 10^{-5}/\text{y}$$

Thus the frequency of failure of an LPG vessel due to failure of a large LPG pipe or a pump is 1.6×10^{-4}/y. If interaction between pumps can occur as described above, this figure rises to 2.1×10^{-3}/y.

The report also discusses the failure of the pipes that carry LNG from the jetty head to the storage tanks in Part 2, Section 6.5.2.

The frequency of failure of these pipes was assessed as $10^{-4} - 10^{-3}$/y. The lower rate is more likely during actual transfer pumping and the higher rate during warm-up and cool-down when the pumps are not in use.

A7.5.3 Failure of pipelines

Failure of a pipeline is considered as a possible initiating event for LNG and LPG. The report discusses failure of pipelines in Part 2, Section 15.

There are four methane pipelines leaving the British Gas terminal. One of these is a 35 cm diameter line passing close to the built-up area. A serious failure of this line would lead to a large release of methane, but the gas would be highly buoyant and would undergo turbulent mixing with air, so that there is no chance of the formation of a large vapour cloud heavier than air.

The frequency of a serious failure of the 35 cm pipeline in the built-up area was estimated as not greater than 10^{-4}/y.

There is a liquid butane pipeline from the British Gas terminal to a point near the Shell refinery. The line is 20 cm diameter for 4 km leaving the terminal and then reduces to 15 cm diameter. The line is not used, but contains some 585 t of liquid butane at a pressure of 4–6 atm. A serious failure of this line could give an initial liquid release rate of 20–30 t/min. If the pressure were sustained by flashing off of propane some 50–100 t could be released in a few minutes. Unless there is a limit on the amount of propane or other volatile component present, the flashing off of the discharging mixture could form a vapour cloud.

The frequency of a serious failure was estimated from data for similar pipelines in the SRS data bank as 3×10^{-4}/km y. The distance travelled by the pipeline through populated terrain is 1.5 km. Thus the frequency of a serious failure causing casualties was estimated as 4.5×10^{-4}/y.

A7.5.4 Generation of and rupture by missiles

Rupture by a missile is considered as a possible initiating event for releases of LPG, ammonia and hydrogen fluoride. The report discusses the generation of missiles in Appendix 5 and the penetration capability of missiles in Appendix 6.

Some scenarios considered are shown in Table A7.5.

The frequency of a large fire in an oil refinery is, for the United States of America, 0.2/y obtained from API data and, for the United Kingdom, 0.25/y obtained from FRS data. The API figures indicate that the proportion of these fires that are very large is about one-third. Thus the frequency of a very large fire in a refinery was taken as 0.1/y.

The proportion of such very large fires that involves explosion was estimated from IRI data as 0.5. The proportion of such explosions that generate missiles was estimated as 0.1.

The frequency of a very large fire that involves an explosion and generates missiles was thus estimated as 5×10^{-3} y.

It was estimated also that on average such a missile-generating explosion generates some half-dozen missiles.

The frequency of disintegration of the rotor of a rotating machine was estimated as 5×10^{-4}/y from data in the SRS data bank.

The report gives in Appendix 6 methods for the estimation of the probability of a missile landing in a given area at a known distance from the source and methods for determining whether the missile will penetrate a vessel.

Missiles particularly considered are fragments from process plant and gas cylinders. With regard to the former a case is quoted in the United Kingdom in which a fractured vessel produced at least seven missiles of 1 t capacity. Reference is also made to the explosion at Whiting, Indiana, where an exploding hydroformer generated some tens of missiles, one of 60 t, with a velocity of 600 ft/s.

Various empirical formulae for penetration by missiles are given. There is considerable variation in the results obtained from these formulae. When used to determine the penetration thickness for a missile with a velocity of 300 ft/s,

the penetration thicknesses given by the formulae ranged from 0.25 to 2.75 in. One reason for the variation is that some of the formulae are applicable to clamped flat plates rather than spherical or cylindrical vessels.

It is concluded that

(1) LPG spheres in excess of $1\frac{1}{4}$ in. thick will not be penetrated by flying gas cylinders even at velocities of 300 ft/s, which is a pessimistic upper limit;

(2) spheres and cyinders 0.75 in. thick could be penetrated by flying gas cylinders;

(3) process missiles can be considerably larger than a gas cylinder and can impact with substantially greater energy – these could penetrate LPG spheres and cylinders.

The report also discusses missile threat to the LNG tanks in Part 2, Section 6.5.4.

A7.5.5 Crash of and rupture by aircraft

Rupture by aircraft crash is considered a possible initiating event for release of all the hazardous materials. The report discusses aircraft crash and vessel rupture in Appendix 8.

The frequency of a fixed wing aircraft crash at Canvey was assumed to be that for the United Kingdom in general and not that for an area near the airport. Canvey is within the Special Rules Zone of Southend airport, but is neither close to the main runway nor beneath the airspace of an aircraft waiting to land.

The frequency of an aircraft crash was obtained from UK aircraft accident data. Two alternative equations are given for the frequency of an aircraft crash on a target. They are:

Method 1 $\qquad F = A_H F_H + A'_V F_H$ [A7.5.1a]

Method 2 $\qquad F = A_H F_H + A_V F_H$ [A7.5.1b]

where A_H is the horizontal area of the target, A'_V is the vertical area of the target, A_V is the horizontal projection of the vertical area of target; F the frequency of aircraft crash on the target area; F_H is the frequency of aircraft crash per unit of ground area and F_V is the frequency of aircraft crash per unit of vertical area.

In Equation A7.5.1a the value of A'_V was calculated assuming that the impact angle is 15°. In Equation A7.5.1b the value of F_V was calculated from data on the frequency with which aircraft crash into the National Grid system. The latter frequency was taken as 0.1/y.

Results obtained for the frequency of crashes of aircraft of all types on the total target area at Canvey for each type of hazard were identical by both methods except that, for the refrigerated LPG at British Gas, Method 1 gave

Table A7.5 First Canvey Report: some potential missile accidents at Canvey (after Health and Safety Executive, 1978b)

Site	Installation	Missile
Shell	LPG vessel	Missile from storage area
Mobil (including extension)	LPG vessel	Missile from process area
Fisons	Ammonia sphere	Missile from rotating machine
Shell	Hydrogen fluoride plant	Missile from neighbouring plant
Mobil (extension)	Hydrogen fluoride plant	Missile from neighbouring plant

Table A7.6 *First Canvey Report: some potential aircraft crash accidents at Canvey (after Health and Safety Executive, 1978b)*

Site	Installation	Frequency of impact (impacts/y)
Shell, Mobil, UR, Occidental	LPG	1×10^{-6}
British Gas	Methane	1×10^{-6}
Calor Gas	LPG	2×10^{-6}
London and Coastal Wharves	Oil	7×10^{-6}
Fisons	Ammonia	1×10^{-7}
Shell	Ammonia	1×10^{-7}
Shell, Mobil, UR, Occidental	Process area	2×10^{-5}

a frequency of impact of 6×10^{-7}/y and Method 2 one of 4×10^{-7}/y. Other results were as given in Table A7.6.

The frequency of a helicopter crash was also considered. Helicopters are used to survey pipelines in the area. It was assumed that data on the frequency of helicopter crashes obtained from NTSB publications are applicable to these helicopters. The frequency of crash obtained from these data is 3×10^{-7}/km flown. In such crashes descent is almost vertical.

Frequencies of crash estimated were 2.4×10^{-5}/y, 1.4×10^{-5}/y and 3.6×10^{-6}/y for the Occidental, Shell and Texaco sites. These are not, however, the frequencies of impact on vulnerable targets. These latter frequencies should be very much less, provided the helicopter routes are chosen to avoid such targets.

Helicopters are used by British Gas to survey its gas pipelines. There are about 26 inspection flights per year. The pilots are instructed to avoid flying over industrial complexes and to skirt such locations by about 500 yd.

A7.5.6 Ship collision and other accidents

Ship collision is considered as a possible initiating event for release of LNG, LPG and ammonia. The report discusses ship collision in Appendices 11 and 23.

Ship collision may occur while the ship is at sea, is stationary in mid-channel or is moored at a jetty. At the time of the survey the Port of London Authority (PLA) had just imposed on ships a new speed restriction of 8 knot.

Data on the frequency of ship collisions were obtained from the PLA. These data show that over the 12-year period 1965–76, there were within the scheduled area 592,250 harbour movements and 121 accidents, of which 91 were classed as minor. Thus this gives for collisions of at least moderate severity a frequency of 0.5×10^{-4}/harbour movement.

Other PLA data were analysed to obtain the frequencies of berthing contact, grounding and fire. Thus the frequencies for collision and for these other accidents are

Frequency of ship–ship collision (moderate severity)
$= 0.5 \times 10^{-4}$/harbour movement

Frequency of berthing contact
$= 1.5 \times 10^{-4}$/harbour movement

Frequency of grounding
$= 0.3 \times 10^{-4}$/harbour movement

Frequency of fire
$= 0.5 \times 10^{-4}$/harbour movement

Further PLA data were analysed to obtain the frequency of ship–ship collision for ships in transit in the estuary. Thus the frequency of collision in transit is

Frequency of ship–ship collision (moderate severity) in transit $= 2.3 \times 10^{-5}$/movement

The vulnerability of ships to collision damage varies. The LNG carriers are double-hulled ships, whereas the LPG and ammonia carriers were not considered as of comparable strength. The former have much greater cargo protection.

An investigation of the resistance of ships to impact was carried out in 1959 by Minorsky, who correlated his results in terms of the critical impact speed vs the loaded displacement of the striking ship, and the Minorsky curve method is widely used as a means of assessing impact resistance of ships.

For a typical LPG carrier, data on the relation of the critical impact speed vs loaded displacement of the striking ship were obtained from Lloyds Register. These data were interpreted as meaning that a striking ship with a loaded displacement of about 4000 t (2000 GRT) could penetrate an LPG carrier and cause spillage if there is no speed restriction, but that a loaded displacement of about 15,000 t (7500 GRT) would be necessary if the speed of the striking ship is limited to 8 knot.

Data for another typical LPG carrier were obtained from Norsk Veritas. These data differ from the Lloyds data and indicate that this is an area of uncertainty.

The frequencies of spillage for an ammonia carrier, which in this context is broadly similar to an LPG carrier, were estimated by using the historical data in conjunction with the following assumed probabilities:

Probability of spillage from ship–ship collision = 0.2
Probability of spillage from berthing contact = 0.1
Probability of spillage from grounding = 0.2

Thus for an ammonia carrier the frequencies of spillage due to harbour movements are

Frequency of spillage due to ship–ship collision
$= 1 \times 10^{-5}$/harbour movement

Frequency of spillage due to berthing contact
$= 1.5 \times 10^{-5}$/harbour movement

Frequency of spillage due to grounding
$= 0.6 \times 10^{-5}$/harbour movement

Total frequency of spillage due to harbour movements
$= 3.1 \times 10^{-5}$/harbour movement

An alternative calculation of the frequency of spillage due to harbour movement based on world data gave a value of 1.45×10^{-5}/harbour movement.

It is concluded that for an ammonia carrier a reasonable estimate of frequency of spillage due to harbour movements is unlikely to exceed.

Frequency of spillage due to harbour movements
$= 2 \times 10^{-5}$/harbour movement

but that this estimate might be significantly reduced if more information were available on cargo protection.

The frequency of spillage for an ammonia carrier due to collision in transit in the estuary was estimated using an assumed probability of spillage due to ship–ship collision of 0.2. Thus for an ammonia carrier frequency of spillage in transit is unlikely to exceed

Frequency of spillage due to ship–ship collision in transit
$= 5 \times 10^{-6}$/movement

This estimate might be significantly reduced if more information were available on cargo protection.

The frequency of spillage for an LPG carrier was taken as similar to that for an ammonia carrier. Thus for an LPG carrier the frequency of spillage is unlikely to exceed

Frequency of spillage $= 2 \times 10^{-5}$/movement

Again this estimate might be significantly reduced given more information on cargo protection.

For an LNG carrier, the situation is different because the vessel is double-hulled. This case was the subject of a special study. From this study a relation was obtained for an LNG carrier between the critical impact speed and the loaded displacement of the striking ship. This relation for an LNG carrier differs considerably from that for an LPG carrier, since the former, being double-hulled, has much greater protection. The relation shows that for an LNG ship stationary in mid-channel a striking ship with a loaded displacement of 100,000 t (50,000 GRT) would need to have a speed of 9 knot to effect penetration at a 90° impact angle, and that for an LNG ship moored at a jetting striking ships with loaded displacements of 100,000 t (50,000 GRT) and 20,000 t (10,000 GRT) would need to have speeds of 7 and 13.5 knot, respectively, to effect penetration at a 45° impact angle.

It was assumed, therefore, that only ships with a loaded displacement greater than about 20,000 t (10,000 GRT) have the potential to cause a spillage from an LNG carrier. This does not include allowance for the new PLA speed restrictions.

It was also concluded that for an LNG carrier spillage due to berthing contact could be disregarded.

The frequency of spillage for an LNG carrier was estimated from the historical data in conjunction with the following assumed probabilities:

Probability of striking ship $> 20,000$ t (10,000 GRT) $= 0.1$

Probability of strike in vulnerable section $= 0.5$

Then for an LNG carrier the frequency of spillage is

Frequency of spillage $= 2.5 \times 10^{-6}$/movement

This assessment is based on only four incidents over the 12-year period.

An alternative calculation of the frequency of spillage for LNG carriers based on world data gave a value of 1×10^{-6}/movement.

It is concluded that for an LNG carrier a reasonable estimate for frequency of spillage is

Frequency of spillage $= 2 \times 10^{-6}$/movement

A7.5.7 Flow of a large release of oil
A large release of oil from storage and flow of this oil towards housing is one of the hazard situations considered. The report discusses the methods of calculating the flow of a large release of oil from the catastrophic failure of a storage tank in Appendix 16.

Equations are given for the flow of a slumping fluid derived from those of van Ulden (1974) assuming that his constants c and A have the value of unity. It is assumed that the liquid is initially held in a vertical cylindrical source. The equations are

$$r = [r_s^2 + 2t(gV_s/\pi)^{1/2}]^{1/2} \qquad [\text{A7.5.2}]$$

$$h_f = \frac{V_s}{\pi r^2} \qquad [\text{A7.5.3}]$$

$$u_f = (gh_f)^{1/2} \qquad [\text{A7.5.4}]$$

where g is the acceleration due to gravity (m/s^2); h_f the slumped liquid depth (m); r the radius of the flooded zone (m); r_s the radius of the source cylinder (m); t the time after initiation of slumping (s); u_f the slumped liquid velocity (m/s); and V_s the volume of the source cylinder (m^3).

A7.5.8 Temperature of the wall of an LNG tank exposed to an LNG fire in an adjacent bund
The rupture of an LNG tank due to an LNG fire in an adjacent bund is one of the hazard situations considered. The report discusses the effect of an LNG fire in an adjacent bund on the temperature of an LNG tank wall in Appendix 13.

The LNG tanks considered are cylindrical in shape and are 30 m diameter and 20 m high. They stand in a bund of 60 m diameter.

The temperature of the tank wall is determined by the heat balance on the wall. It was assumed that the wall gains heat by radiation and loses it both by radiation and by convection.

For the heat gained by the tank wall use was made of the experimental work of L.E. Brown, Wesson and Welker (1974b). It was assumed that the heat flux from the fire is 44,000 BTU/h ft^2, which is the maximum average value obtained by these workers for liquid pools of up to 30 m diameter. This value is considered reasonably conservative for bund fires with an effective diameter up to 60 m. The flame height calculated from the empirical formula given by these workers is 75 m. The fraction of the radiated heat received by the tank was then calculated by dividing both the flame and tank surfaces into smaller subsurfaces and determining view factors. The total radiated heat Q_i received by the tank wall was calculated as 3.11×10^7 W.

For the heat loss by the tank wall the equation used for radiation is

$$Q_r = \varepsilon\sigma(A_V + A_H)\theta_T^4 \qquad [\text{A7.5.5}]$$

where A_H is the area of the tank roof (m^2); A_V the exposed area of the tank vertical sides (m^2); Q_r the heat loss by radiation (W); ε the emissivity of the tank wall; θ_T the absolute temperature of the tank wall (K); and σ the Stefan–Boltzmann constant (W/m^2 K^4). The equation used for convection is

$$Q_n = (1.76A_V + 2.25A_H)(\Delta\theta)^{1.25} \qquad [\text{A7.5.6}]$$

with

$$\Delta\theta = \theta_T - \theta_A \qquad [\text{A7.5.7}]$$

where Q_n is the heat loss by convection (W); $\Delta\theta$ the temperature difference between the tank wall and ambient air (°C); and θ_A the absolute temperature of ambient air (K).

The areas A_V and A_H of tank surface are the exposed areas and since the whole of the roof but only part of the walls is exposed, A_H is based on the whole of the roof surface, but A_V is based on only part of the vertical wall surface.

The equilibrium tank wall temperature θ_T was found to be 647 K assuming an emissivity ε of 0.5 and did not differ by more than 40°C even assuming an emissivity of unity. The tank wall temperature is therefore relatively insensitive to estimates of the emissivity.

The heat loss by radiation is over three times greater than that lost by convection at a tank wall temperature of 650 K.

The time for the temperature of the tank wall to reach steady state was calculated from the unsteady-state equation

$$mc_p \frac{d\theta_T}{dt} = Q_i - (Q_n + Q_r) \qquad [A7.5.8]$$

where c_p is the specific heat of the tank wall (J/kg °C); m the mass of the exposed tank wall (kg); and t the time (s).

The thermal capacity mc_v of the exposed tank wall was taken to include half that of the associated perlite insulation.

Equation A7.5.8 is non-linear and is solved numerically. It was calculated that the time for the temperature of the wall to reach 95% of its equilibrium value is about 42 min.

A7.5.9 Evaporation of LNG and ammonia on water
The spillage and evaporation of LNG and ammonia on water is one of the hazard situations considered. The report discusses the evaporation rates of LNG and of ammonia spilled on water in Part 2, Sections 6 and 7, respectively.

It is stated that for a rapid spill of LNG on water the evaporation time is short and a puff dispersion model is applicable. But for a continuous spill the evaporation rate rapidly reaches the spill rate and this evaporation rate then determines the pool diameter reached. The evaporation rate has been determined by a number of investigators as about 0.19 kg/m²s, which corresponds to a liquid regression rate of 4.7×10^{-4} m/s.

Similarly, it is stated that for a rapid spill of ammonia on water the evaporation time is short and that all the ammonia will evaporate, except for about 20% which dissolves in the water.

A7.5.10 Dispersion of an LNG vapour cloud
The dispersion of an LNG vapour cloud is one of the hazard situations considered. The report discusses the dispersion of an LNG vapour cloud in Appendix 9.

A cloud of cold LNG vapour tends to behave as a heavy gas cloud. The following elements are identified as particularly important in describing the behaviour of such a cloud: (1) specification of the source, (2) description of the gravitational slumping, (3) description of the air entrainment, and (4) description of the thermal effects.

The model of the source that is assumed can have a significant effect on the results. The release may be instantaneous or continuous. The source may be of fixed size or of increasing size, as with LNG spilled on to water. The process of release may or may not involve the entrainment of large amounts of air.

The differences between the various models currently available for the gravitational slumping phase are discussed and it is suggested that these may be due in large part to the extent to which air entrainment is taken into account. The model of R.A. Cox and Roe (1977) is instanced as representative of the new generation of models that take into account air entrainment.

Information is presented on the distance travelled by an LNG cloud before it is diluted to the lower flammability limit, that is, the hazard range. Experimental work described includes

Source	Release	Weather conditions
API	1/3–10 t on sea	Stable (Pasquill stability category E)
Shell	5–50 t on sea	Relatively stable
AGA	Spillages in bunds giving vapour releases up to 25 kg/s	

Theoretical work mentioned is that of Cox and Roe of Cremer and Warner.

Hazard ranges derived from these sources are shown in Figures A7.3(a) and A7.3(b) for continuous and instantaneous releases, respectively. Models for both instantaneous and continuous releases may be derived from the experimental work described by estimating values of the dispersion coefficients in the Pasquill–Gifford model for an instantaneous release and assuming that these coefficients are the same for a continuous release. This is the basis for the curves shown in Figures A7.3(a) and A7.3(b), except for the API 2.5% case, which is based on the work of Feldbauer *et al.* (1972). The lower flammability limit of methane is approximately 5% and most of the data shown in the figures are for this concentration, but some data are shown for a concentration of 2.5%, which allows for localized short-term concentrations up to the lower flammability limit (LFL).

The weather conditions on which the graphs are based are neutral conditions (Pasquill stability category D) or stable conditions (Pasquill stability category E). For less stable conditions the hazard range is less or at worst no greater. For Canvey more stable conditions (Pasquill stability categories F and G) were not considered, because Canvey Island is an urban site and it has been shown that over such an area the dispersion of a plume in very stable conditions is unlikely to produce airborne concentrations greater than those obtained over open country for neutral and stable conditions (Pasquill stability categories D and E).

The quantities of LNG involved in the releases in Figures A7.3(a) and A7.3(b) are relatively small. The warning is given, therefore, that the use of these graphs for a continuous release of, say, 10 t/s or an instantaneous release of, say, 1000 t involves scaling up by two orders of magnitude with all the uncertainties attendant on such a procedure.

Mention is also made of methods for the prediction of LNG vapour dispersion based on wind tunnel simulations, but it is concluded that the results obtained are subject to too great a range of uncertainty to be useful.

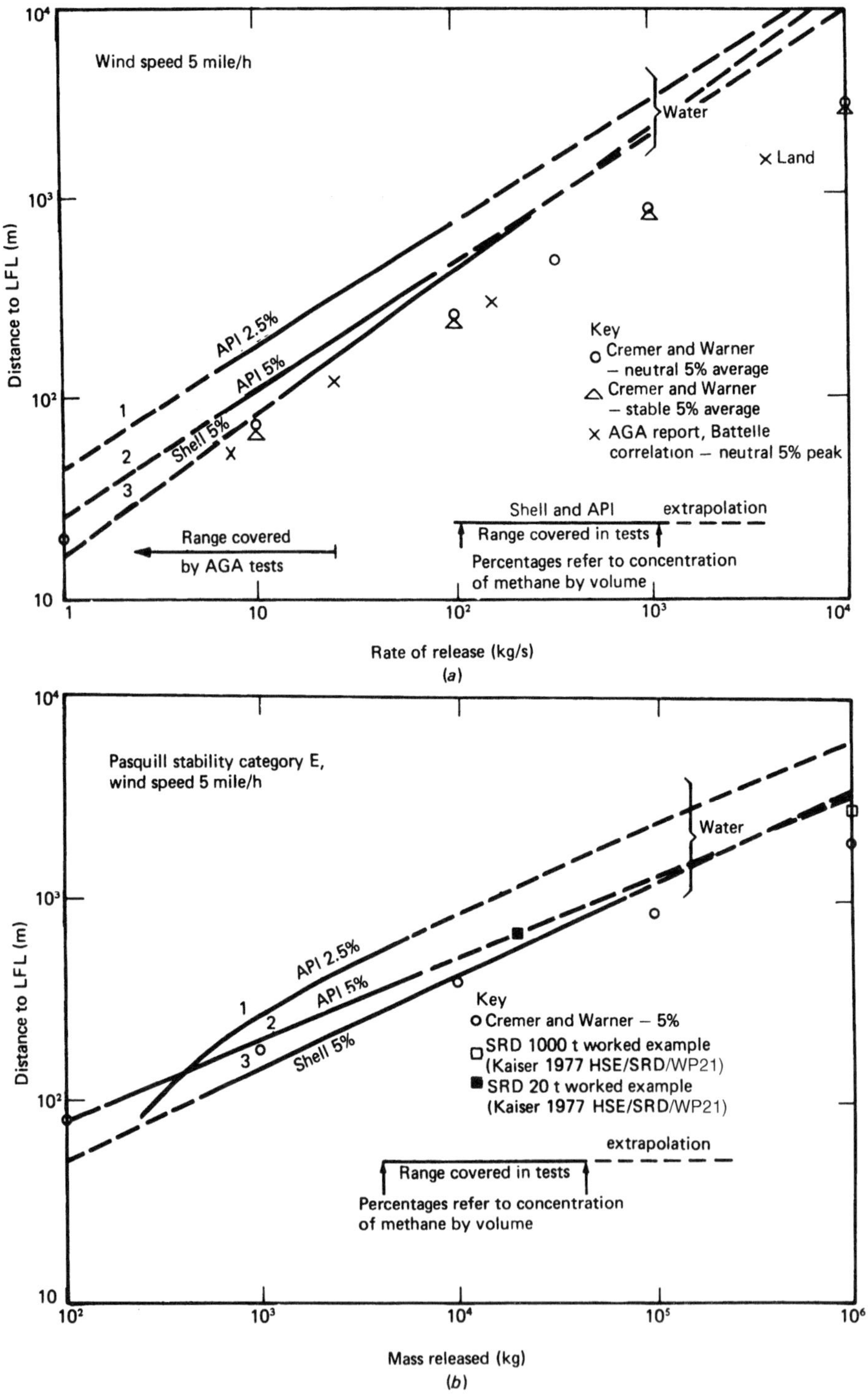

Figure A7.3 First Canvey Report: hazard ranges for releases of LNG vapour in neutral and stable weather conditions (Health and Safety Executive, 1978b): (a) continuous releases — curve 1 API tests model, 2.5% average, stable condition; curve 2 API tests model, 5% average, stable condition; curve 3 Shell tests model, 5% average, stable condition; (b) instantaneous releases — curve 1 API tests model, 2.5% average, stable condition (Feldbauer et al., 1972); curve 2 API tests model, 5% average, stable condition; curve 3 Shell tests model, 5% average, stable condition (Courtesy of HM Stationery Office)

For the purpose of the hazard assessment, however, it was necessary to make estimates of the possible range of vapour clouds from marine spillages of LNG. The distances obtained by scaling up the small spillage data were considered too conservative and the following estimates were used:

Spillage (t)	*Maximim range to LFL* (km)	
	average	peak
10	0.54	0.7
100	1.4	1.8
1,000	3.8	5.2
10,000	10	10

The average concentration corresponds to 5% and the peak concentration to 2.5% that, as explained above, allows for localized short-term concentrations of double the latter value.

A7.5.11 Unconfined vapour cloud fire and explosion

An unconfined vapour cloud fire or explosion is considered as one of the possible accidents that could be the cause of fatalities. The report discusses unconfined vapour cloud fire and explosion in Part 2, Sections 4 and 6, and in Appendix 1. The treatment given of unconfined vapour cloud fire and explosion is broadly as follows.

Two standard vapour clouds are considered. The first contains 100 t and the second 1000 t of hydrocarbons.

Ignition of an unconfined vapour cloud may result in fire or in explosion. The fire and explosion characteristics are given for the two standard vapour clouds just mentioned.

Both complex explosion models and a modified TNT equivalent model were used to estimate the damage effects from an unconfined vapour cloud explosion.

Only the modified TNT equivalent model is described here. This is similar to the usual model, except that the relation between peak overpressure and scaled distance is modified to allow for the fact that the explosion is that of a vapour cloud rather than that of TNT.

For the energy of explosion an equivalence factor α is defined such that the explosion of a flammable gas cloud of W t of hydrocarbons with a combustion energy release H kcal produces the same damage effects as a TNT explosion with an energy release of αH kcal. The value of α used is 0.1. In other words, it is assumed that 10% of the hydrocarbons takes part in the explosion.

For the peak overpressure Δp_m and the peak dynamic pressure q_m at radial distance r and scaled distance $r/H^{1/3}$ the relations used are shown in Figure A7.4. Separate curves of Δp_m are given for TNT explosion, for hydrocarbon detonation and for hydrocarbon deflagration with a deflagration velocity of 170 m/s. Figure A7.4 is for spherical symmetry. Results for explosions with hemispherical symmetry may be obtained by doubling the energy release.

Thus, for example, for a hydrocarbon vapour cloud explosion with an energy release αH where

$$H = 11.1 \times 10^6\,W \text{ kcal}$$

$$\alpha = 0.1$$

Figure A7.4 gives for hemispherical symmetry

Δp_m		r (m)
(psi)	(bar)	
3	0.2	$69.1\,W^{1/3}$
1	0.07	$196\,W^{1/3}$

For the two standard hydrocarbon vapour clouds considered the fire and explosion characteristics used are shown in Section A of Table A7.7. The ranges quoted for dilution to the LFL are for weather conditions of Pasquill stability category D. The overpressures quoted apply only to the cases where an explosion occurs.

The foregoing indicates for the standard hydrocarbon vapour clouds considered the estimated hazard ranges of fire and of explosion, that is, overpressure effects, where these effects occur.

The probability of ignition of a vapour cloud was taken as 0.1 or 1, depending on whether ignition was regarded as improbable or probable, taking into account the ignition sources between the point of release and the population group.

The probability of deflagration with overpressure in a vapour cloud, given ignition, was taken as 1 for large clouds (>100 t) of flammable gas, as 0.1 for smaller clouds of flammable gas other than methane, and as 0.01 for smaller clouds of methane, reflecting the significantly lower probability of explosion of a cloud of methane compared with one of other hydrocarbons.

For the 1000 t hydrocarbon vapour cloud it was assumed that the whole cloud burns with a fireball approximately equal in volume to that of the cloud formed by the stoichiometric gas−air mixture and further that, where overpressure occurs, the proportion of hydrocarbon taking part in the explosion is 10%.

If the vapour cloud is ignited *in situ*, the hazard to the public for the situations considered is that of explosion. In accordance with the data given in Table A7.7 it was assumed that the overpressure at the site boundary would not exceed 0.2 atm.

An incidence of 1½% is quoted for casualties among the population in the areas subject to an overpressure between 0.2 and 0.05 atm, which is approximately the level of overpressure corresponding to the Explosives Acts recommendations.

If the vapour cloud drifts and is ignited off-site, the hazards to the public for the situations considered are those of fireball and explosion. For this case, where the cloud is ignited and burns fully over the populated region, it was assumed that the whole population engulfed in the fireball is killed. For the 1000 t hydrocarbon cloud burning fully over a populated region with the average population density of the area, the number of fatalities would be 2000. In addition, it was assumed that in this situation there would be 100 further fatalities and 2000 serious injuries caused by explosion effects out to a radius where the overpressure is 0.075 atm.

The more probable case, however, is ignition when the leading edge of the cloud is just beginning to encroach on the populated region.

It was assumed that if for ignition of the cloud on the edge of the populated region, the number of casualties is N and

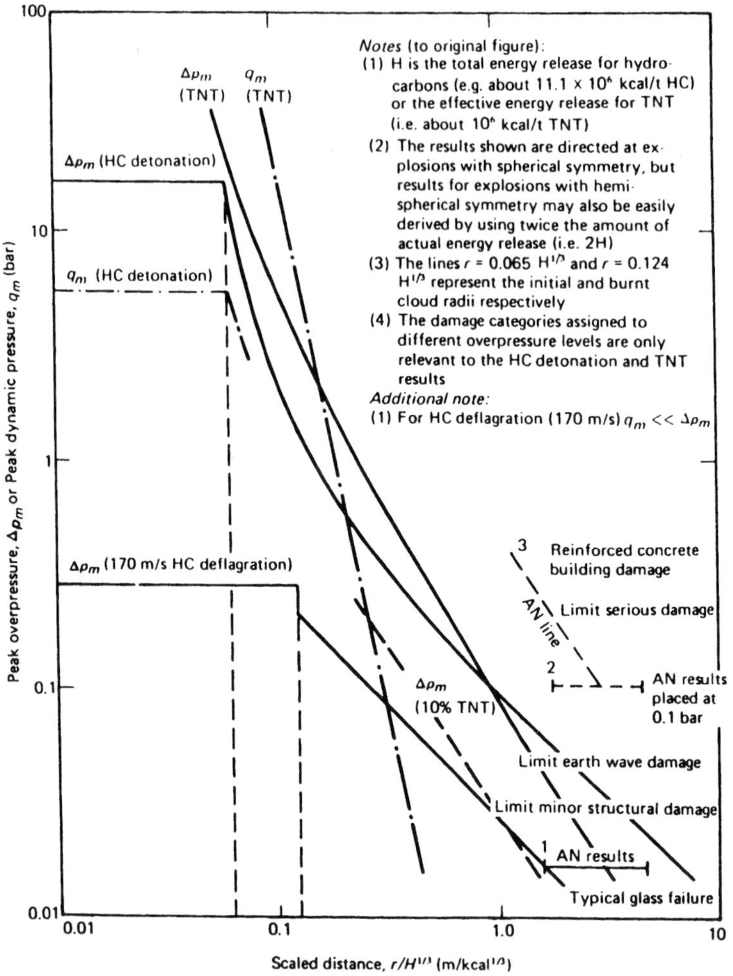

Figure A7.4 First Canvey Report: peak overpressure and peak dynamic pressure vs scaled distance for various types of explosion (Health and Safety Executive, 1978b) (Courtesy of HM Stationery Office)

the probability is P, then for ignition right over the populated region the number of casualties is $4N$ and the probability is $0.25P$. It was also assumed that the range of probabilities may be represented by a linear probability distribution function. The corresponding relationships between the numbers of casualties and the probability of these numbers are given in Section B of Table A7.7. The number of fatalities is taken as half the number of casualties.

For a hydrocarbon vapour cloud greater than 1000 t it was assumed that the number of casualties would vary with the mass of vapour according to a 2/3 power law.

The report discusses unconfined LNG vapour cloud fire and explosion in Part 2, Section 6. The following scenarios are among those considered:

Case 1a Release from storage tank (25 cm diameter hole) – vapour cloud of up to 100 t
    1b Release from ship's cargo tank at jetty (25 cm diameter hole) – vapour cloud of 100 t

2 Failure of storage tank – vapour cloud of 200 t
3 Failure of ship's cargo tank – vapour cloud of 1000 t

The fire and explosion characteristics of the vapour clouds for these cases are given in Section C of Table A7.7.

The probability of explosion, that is, overpressure in the vapour cloud, was taken as unity only for the larger clouds. The explosion effects were again calculated assuming that 10% of the hydrocarbons take part in the explosion.

A7.5.12 Ammonium nitrate explosion
An ammonium nitrate (AN) explosion is considered as one of the possible accidents that could be the cause of fatalities. The report discusses ammonium nitrate explosions in Part 2, Sections 4.1.2 and 11, and in Appendix 1.

The ammonium nitrate considered is the storage of 92.5% aqueous solution in two tanks of 5000 and 2000 t capacity, respectively. A large spillage from this storage would probably give a mushy heap of ammonium nitrate

Table A7.7 *First Canvey Report: vapour cloud fire, explosion and casualty characteristics used (after Health and Safety Executive, 1978b)*

A   Fire an explosion characteristics of hydrocarbon vapour clouds

Mass of hydrocarbon (t)	*Diameter of stoichiometric hemispherical cloud* (m)		*Range of stated overpressure* (km)		*Range for dilution to lower flammability limit* (km)
	Before burning	*After burning*	*0.2 atm* (3 psi)	*0.07 atm* (1 psi)	
100	169	323	0.321	0.910	2
1000	364	696	0.691	1.960	5

B   Casualty characteristics of a 1000 t hydrocarbon vapour cloud fire and explosion

Upper limit				
No. of casualty	0	1500	3000	4500
No. of fatalities	0	750	1500	2250
Cumulative probability	1	0.64	0.35	0.14

C   Fire and explosion characteristics of an LNG vapour cloud

Case	*Mass of hydrocarbon* (t)	*Burnt cloud radius* (m)	*Radius of stated overpressure* (m)		*Probability of overpressure*
			0.2 atm (3 psi)	*0.07 atm* (1 psi)	
1	100	162	321	910	0.01
2	200	204	404	1147	1
3	1000	348	691	1960	1

crystals near the point of release. An escape of 4500 t of 92.5% solution was considered as the standard case.

If ammonium nitrate is ignited, the result is normally a fire rather than explosion. The latter is likely to occur only if conditions are favourable for transition from a fire to an explosion.

These conditions have been investigated theoretically and experimentally by van Dolah *et al.* (BM 1966 RI 6773) and this work is discussed. It is concluded that the work indicates that transition to explosion would be impossible for quantities of ammonium nitrate of tens of tonnes and probably hundreds of tonnes, but that it might occur at the thousands of tonnes level.

The historical record of ammonium nitrate explosions is reviewed and explosions of prilled ammonium nitrate in transport are quoted, one in rail tank cars at Traskwood, Arkansas, in 1960 and one in road tankers in Queensland in 1972.

The only significant possibility foreseen for the occurrence of an ammonium nitrate explosion was the derailment of a train carrying petroleum products that could cause simultaneously the rupture of an ammonium nitrate tank and a liquid petroleum fire.

The frequency of a derailment and overturning was calculated as follows:

Frequency of derailment of rail tank car
$= 1 \times 10^{-6}$/train km

Number of trains with petroleum products
$= 3000$/y

Length of track opposite tanks ≈ 300 m

Probability of derailment leading to overturning $= 0.2$

Probability of overturning on tank side of track $= 0.5$

Frequency of derailment and overturning on tank side of track $= 8.5 \times 10^{-5}$/y

In view of the many uncertainties and of the omission of other possible accident modes this latter value was also taken as the frequency of an ammonium nitrate explosion. Thus

Frequency of ammonium nitrate explosion
$= 8.5 \times 10^{-5}$/y

The effect of an ammonium nitrate explosion was assessed by reference to the data on such explosions given by M.A. Cook (1958) and shown in Table A7.8.

These results were interpreted as follows. It was assumed that the predominant reaction in the explosion of ammonium nitrate is

$$2NH_4NO_3 = 2N_2 + 4H_2O + O_2$$

for which the heat of reaction is 0.35×10^6 k cal/t. Then using this heat of reaction the values obtained for the group $r/H^{1/3}$ (m/(kcal)$^{1/3}$) corrected for spherical symmetry for the distance for major damage in the four explosions listed lie in the range 1.72–4.8. This spread indicates a lack of agreement with the 1/3 power scaling relation, which is reduced but not eliminated if a 1/2 power scaling relation

Table A7.8   First Canvey Report: some major ammonium nitrate explosions (after Health and Safety Executive, 1978b)

Date	Location	Mass of ammonium nitrate (t)	Distance for major damage (m)	Fatalities
1918	Morgan, NJ	500	1600	64
1923	Oppau, Germany	4500	7000	1100[a]
1947	Brest, France	3000	5000	21
1947	Texas City, TX	3500	2300	560

[a] This value differs from that given by Commentz *et al.* (1921).

Figure A7.5   First Canvey Report: lethal exposure time vs concentration for chlorine (Health and Safety Executive, 1978b) (Courtesy of HM Stationery Office)

is used. The disagreement may be due in part to the relative lack of precision in the term 'serious damage'. If these results are located on the curve for overpressure Δp_m (TNT) in Figure A7.4, then, as shown by the full line (line 1) in this figure, they correspond to an overpressure of 0.02 bar. This is an extremely low overpressure to cause serious damage. In view of this dilemma, it was decided to locate the ammonium nitrate results at an overpressure of 0.1 bar, as shown by the dotted line (line 2) in Figure A7.4, although they then lie above the TNT curve. It is the further dotted line (line 3) which is used to calculate the damage ranges for explosions of 2250 and 4500 t of ammonium nitrate in the report.

A7.5.13 Toxicity of chlorine, ammonia, hydrogen fluoride and lead additives

The toxicity of chlorine, ammonia, hydrogen fluoride and lead additives is considered. The report discusses the toxicity of chlorine in Appendices 3 and 12, that of ammonia in Appendix 15, that of hydrogen fluoride in Appendix 12 and that of lead additives in Appendix 19.

There is considerable uncertainty about the lethal dosages for all these chemicals and it was necessary, therefore, to make approximate estimates.

Chlorine was not one of the chemicals considered in the study to constitute a hazard at Canvey, but its toxicity is discussed in relation to factors that mitigate casualties from a toxic release and in relation to the toxicity of hydrogen fluoride.

For chlorine the relation between the lethal concentration and the time of exposure shown in Figure A7.5 was used. This graph is derived from data given by Dicken (1974, 1975).

For ammonia a probit equation was used

$$\Pr = a + b \ln D \qquad [A7.5.9]$$

with

$$D = \int_0^t C^{2.75}\, \mathrm{d}t \qquad [A7.5.10]$$

where C is the concentration (g/m^3); D the dosage $((\mathrm{g/m}^3)^{2.75}\mathrm{min})$; Pr the probit; t the time (min); and a,b are constants.

Precise information on the lethal dosage of ammonia is lacking. Two cases were considered. For case A it was assumed that a concentration of 1.75 g/m^3 (2500 ppm) is lethal to 50% of the exposed population in 30 min, while for case B it was assumed that the same concentration is 50% lethal in 60 min. Then the values of the constants are

	a	b
Case A	1.14	0.782
B	−7.41	2.205

For hydrogen fluoride it is necessary to take care with the units in which concentration is expressed on account of the high degree of association in the vapour phase. Concentrations are therefore expressed in units of mg/m^3 rather than of ppm.

Precise information on the lethal dosage of hydrogen fluoride is again lacking. Insofar as it is an irritant gas it appears to have effects similar to chlorine at similar concentrations. The toxicity of hydrogen fluoride was therefore assumed for the purpose of the study to be similar to that of chlorine at a similar concentration, where the concentration units are mg/m^3.

In addition, however, hydrogen fluoride in large but sublethal dosages may also have other injurious long-term effects.

The lead additives are the alkyl lead compounds, particularly the methyl and ethyl derivatives.

For lead additives precise information on the lethal dosage is again lacking. The medical evidence is reviewed in some detail. It is concluded that there is a significant probability of fatality at a concentration of $500 \; mg/m^3$ for an exposure of 1 h. It is also considered that total dosage is more important than concentration so that the concentration can be scaled inversely with the exposure time for exposure times in the range 15–60 min.

A7.5.14 Dispersion of ammonia and hydrogen fluoride vapour clouds

The dispersion of anhydrous ammonia and hydrogen fluoride gas clouds is one of the hazard situations considered. The report discusses the dispersion of an ammonia gas cloud in Part 2, Section 7, and in Appendices 4 and 14 and that of a hydrogen fluoride gas cloud in Part 2, Section 5.5, and in Appendix 12.

The dispersion of an anhydrous ammonia cloud depends on whether it is lighter or heavier than air. The following conclusions are reached. If an initial fraction of the ammonia released is in the form of suspended droplets, the cloud may become denser than air as it is diluted. There will be a critical value for the initial fraction of droplets such that if the fraction is below this value the cloud will be less dense than air regardless of the humidity of the air. There will be a second critical value such that if the fraction is above this value the cloud will be more dense than air regardless of the humidity of the air. These two critical values are given, respectively, as 4–8% and 16–20% of the total mass of ammonia. Between these critical values the density of the cloud is determined by the degree of dilution and humidity of the air. High values of the air humidity favour low cloud densities and vice versa, since water in the atmosphere will tend to freeze and so liberate latent heat.

The possible chemical reactions of anhydrous ammonia with the water and the carbon dioxide in the air to give ammonium hydroxide and ammonium bicarbonate and carbonate are also reviewed, but it is concluded that it seems unlikely that any such reaction will be a significant effect, although in foggy or rainy conditions it may convert some ammonia to a less toxic form and so effect some reduction in the hazard.

The historical record of anhydrous ammonia releases is reviewed and those at Blair, Conway, Potchefstroo and Houston are quoted. In all these cases the ammonia cloud appears to have behaved as denser than air.

The emission and dispersion of a release of 1000 t of anhydrous ammonia from pressurized storage at 6°C in weather conditions of Pasquill stability category D and wind speed of 3 m/s were calculated.

For emission it was assumed that the whole mass of 1000 t of ammonia is released virtually instantaneously, that 20% of the ammonia forms vapour and that the remaining 80% forms liquid droplets, so that the whole amount spilled is airborne.

The escaping ammonia will entrain air. There is little information available on this aspect, although the Potchefstroom accident provides some evidence. There some 40 t of ammonia were released and eyewitness accounts stated that 'the *immediate* resulting gas cloud was about 150 m in diameter and nearly 20 m in depth'. This corresponds approximately to an air–ammonia mass ratio of about 10 if all the ammonia was airborne and visible. If it is assumed that in the scenario considered enough dry air is mixed with the ammonia to vaporize all the liquid ammonia, leaving an air–ammonia mixture at the boiling point of ammonia, the corresponding air–ammonia mass ratio is about 20.

The following conditions were taken, therefore, as the source for the dispersion model. The release is 1000 t of anhydrous ammonia. This forms a vapour cloud with an air–ammonia mass ratio of 20 at a temperature of −33°C and with a density of $1.42 \; kg/m^3$. This density is greater than that of air at 20°C by a factor of 1.18 so that the cloud is denser than air. Both the radius and height of the cloud are 167 m.

For dispersion it was assumed that the behaviour of the vapour cloud is described initially by a heavy gas slumping model and then by a neutral gas dispersion model. This follows the approach developed by van Ulden (1974).

For slumping, van Ulden's model is used

$$\frac{dr}{dt} = \left(\frac{(\rho_0 - \rho_a)gh}{\rho_0} \right)^{1/2} \qquad [A7.5.11]$$

with

$$h = V_0/\pi r^2 \qquad [A7.5.12]$$

and

$$r^2 - r_0^2 = 2\left[\left(\frac{(\rho_0 - \rho_a)gV_0}{\pi\rho_0} \right)t \right]^{1/2} \qquad [A7.5.13]$$

where g is the acceleration due to gravity; h the height of the cloud; r the radius of the cloud; r_a the initial radius of the cylinder; t the time; V_0 the initial volume of the cylinder; ρ_a the density of air; and ρ_0 the initial density of the cloud.

The termination of slumping depends in this model on the turbulent energy density and hence on the surface roughness length. For a roughness length z_0 of 10 cm the predicted height h at termination of slumping is 0.15 m. For the purposes of this approximate calculation the height h at termination of slumping was taken as 1 m.

The state of the cloud at termination of slumping is then as follows. The radius is 2170 m, the height is 1 m, the density is $1.42 \; kg/m^3$. The time taken is about 12 min and the distance travelled with the wind speed of 3m/s is about 2.6 km.

This model predicts, therefore, that the cloud radius becomes very large and that the cloud travels far downwind. The cloud travel distance is regarded as pessimistic, because it is to be expected that at a height of 1 m the cloud will be slowed down by various obstacles.

For further dispersion the Pasquill–Gifford model for a neutral buoyancy gas was used

$$\chi(x;y,t) = \frac{2Q^*}{(2\pi)^{3/2}\sigma_x\sigma_y\sigma_z} \times \exp\left[-\frac{1}{2}\left(\frac{(x-ut)^2}{\sigma_x^2} + \frac{y^2}{\sigma_y^2}\right)\right] \qquad [A7.5.14]$$

with

$$\sigma_x = \sigma_y \qquad [A7.5.15]$$

where x y and z are the distances in the downwind, crosswind and vertical directions (m); Q^* the mass released instantaneously (kg); t the time (s); u the wind speed (m/s); σ_x, σ_y and σ_z the standard deviations, or dispersion coefficients, in the downwind, crosswind and vertical *(x,y,z)* directions (m); and χ the concentration (kg/m³). The downwind distance x is measured from the source. At the termination of slumping it was assumed that the cloud has a Gaussian concentration distribution with the 10% edges of the cloud equal to the edges of the slumped cloud. Then at transition

$$r = 2.14\sigma_y \qquad [A7.5.16]$$

$$h = 2.14\sigma_z \qquad [A7.5.17]$$

The maximum concentration on the axis at a downwind distance x occurs when the centre of the cloud is at that point so that

$$x = ut \qquad [A7.5.18]$$

and hence

$$\chi_{\max}(x) = \frac{2Q^*}{(2\pi)^{3/2}\sigma_x\sigma_y\sigma_z} \qquad [A7.5.19]$$

The mean concentration $\chi_{av}(x)$ along the axis is related to the maximum concentration $\chi_{\max}(x)$ as follows:

$$\chi_{av}(x) = 0.585\,\chi_{\max}(x) \qquad [A7.5.20]$$

The mean concentrations calculated for the release considered are given in Figure A7.6. Figure A7.6(a) shows the mean concentration as a function of downwind distance and Figure A7.6(b) the mean concentration as a function of exposure time.

The distance at which the ammonia cloud is potentially lethal is obtained from Figure A7.6. If it is assumed that the ammonia is potentially lethal at a concentration of 0.002 kg/m³ for a period of 30–50 min, the range of potential lethality of the cloud is 5 km from the point at which slumping is terminated. This point is itself 2.6 km from the source.

Cloud behaviour was also investigated for other weather conditions.

The approximate area within which people are at risk from an instantaneous release of 1000 t of anhydrous ammonia in weather conditions of Pasquill stability category D and wind speed of 6 m/s is shown in Figure A7.7.

The model just described is a relatively crude one. Further, more sophisticated models were derived which took into account (1) the heating of the cloud and (2) the entrainment of air during the slumping phase. These models indicated that the overall behaviour of the cloud is not greatly altered by these effects and that the original model gave a sufficient representation for approximate calculations.

The dispersion of a hydrogen fluoride gas cloud also depends on whether it is lighter or heavier than air. Hydrogen fluoride monomer has a molecular weight of 20, which suggests that the gas should be lighter than air. But hydrogen fluoride gas is highly associated, forming in particular the hexamer. Although dissociation takes place as the gas is diluted, the process is endothermic so that cooling occurs. Thus the behaviour of a hydrogen fluoride gas cloud is in some ways analogous to that of an ammonia gas cloud.

The approach adopted was to treat the hydrogen fluoride cloud as denser than air.

A7.5.15 Factors mitigating casualties from a toxic release

Large toxic releases are considered as one of the possible accidents which could be the cause of fatalities. Theoretical estimates of large toxic releases tend, however, to give rather large numbers of fatalities and to appear pessimistic compared with the historical record. Consideration was therefore given to the factors mitigating casualties from a toxic release.

The report discusses mitigating factors for a toxic release in Appendix 3.

The historical record of chlorine releases is reviewed. The releases listed by Simmons, Erdmann and Naft (1974) are quoted, as are the releases at Zarnesti, Roumania, in 1939 and at Baton Rouge, Louisiana, in 1976.

The following features are adduced as factors that may mitigate the effects of a chlorine release:

(1)   low population density downwind of the source, particularly within the first kilometre or so;
(2)   favourable weather conditions;
(3)   low rate of release;
(4)   escape;
(5)   shelter.

It is possible to make a quantitative estimate of the effect of these factors. There are other factors such as

(6)   topography

which may also be important, but which in the present state of knowledge cannot be readily assessed.

A series of calculations was done for a 20 t chlorine release over periods of 6 s, 10 min and 6 h in weather conditions of Pasquill stability category D and with a wind speed of 5 m/s.

The calculations were based on the Pasquill model for short continuous releases, as given in Equation 15.16.31,

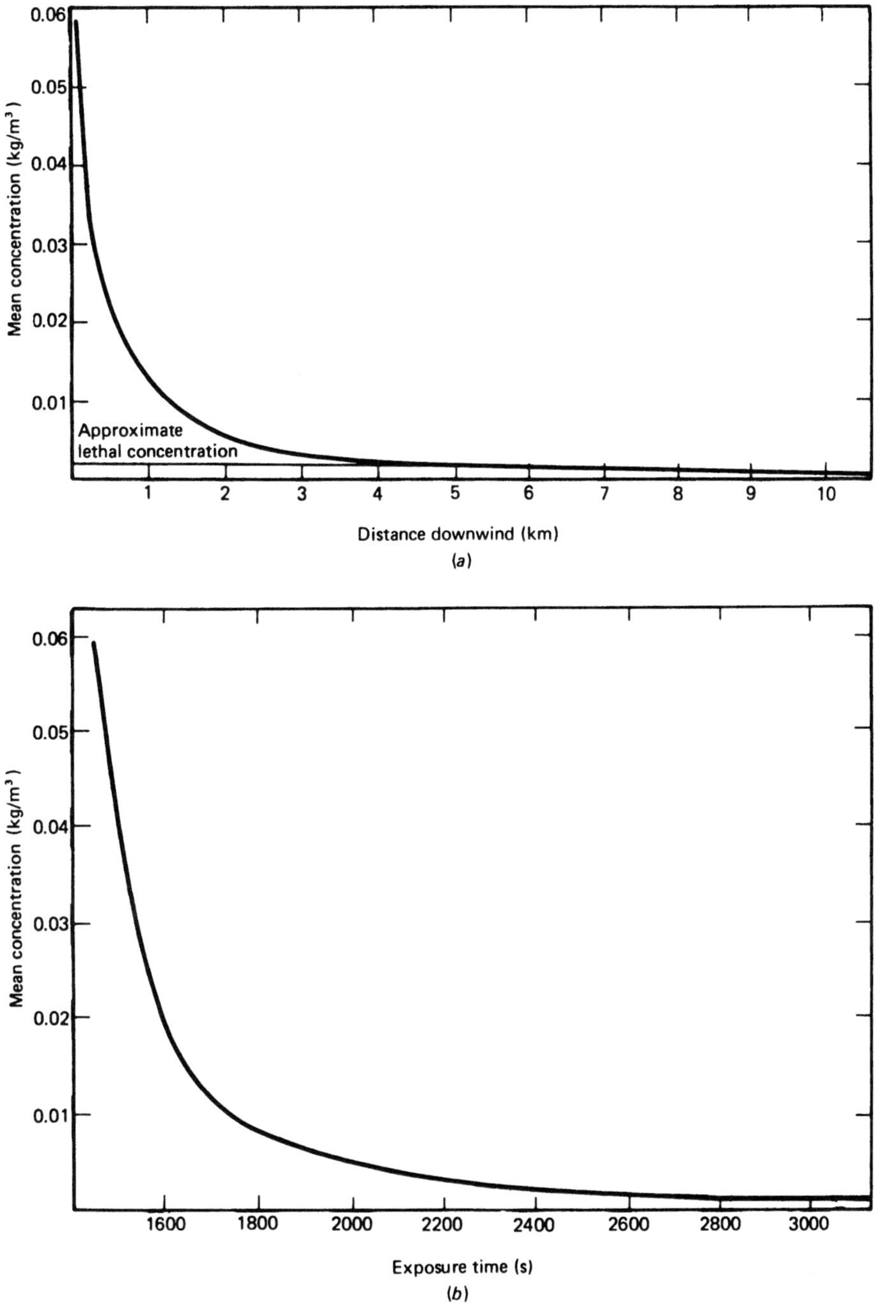

Figure A7.6 First Canvey Report: mean concentration vs distance and exposure time for an instantaneous release of WOO te of anhydrous ammonia (Health and Safety Executive, 1978b): (a) mean concentration vs distance and (b) mean concentration vs exposure time (Courtesy of HM Stationery Office)

using the graphs presented by Bryant (1964 UKAEA AHSB(RP) R42 Figure 4). This model may be adapted to the determination of the time of passage and the dosage from a nearly instantaneous release, as follows. The time of passage for an instantaneous release is obtained as the times between the concentration rising to, and subsequently falling back to, one-tenth of its maximum value. This time of passage is corrected for a nearly instantaneous release

Figure A7.7   *First Canvey Report: approximate area within which people are at risk from an instantaneous release of 1000 te of anhydrous ammonia (Health and Safety Executive, 1978b) (Courtesy of HM Stationery Office)*

by adding to it the time of release. For the dosage use is made of the fact that Equation 15.16.31 gives the total dosage for an instantaneous release, as described in Chapter 15. The model applies strictly to short releases of approximately 3 min duration. No account was taken, therefore, of plume broadening with duration of release, but it was believed that this effect would be small for the stability condition considered.

For various points on the downwind axis of the cloud, calculations were made of the dosage and of the time of passage, and by dividing the former by the latter, of the mean concentration. The chlorine toxicity data given in Figure A7.5 were used to determine for each mean concentration the corresponding lethal exposure time. If the lethal exposure time was less than the time of passage, then the cloud was assumed to be fatal to a person remaining in the open air at that point.

The results of the calculations are summarized in Figures A7.8(a–c), which show the mean concentration, the time of passage and the lethal exposure time vs distance for the three releases studied.

The distances at which the time of passage and lethal exposure time cross over are the lethal distances, or hazard ranges, of the clouds. These distances are 2.7, 1.9 and 0.6 km for release times of 6 s, 10 min and 6 h, respectively. The variation of the lethal range is due entirely to the slope of the line of lethal concentration vs exposure time in Figure A7.5, which is a log–log plot. If the slope were −1 instead of approximately −2, there would be no variation of distance.

The effects of the mitigating factors are discussed in the light of these results.

For population density the effect of low population density is as follows. In a UK urban area with a population density of about 5000/km^2 the numbers in a 15°

sector would be about 650, 2650 and 6000, while in a UK rural area with a population density of about 100/km^2 the corresponding numbers would be 13, 53 and 120 within distances of 1, 2 and 3 km, respectively. Thus it is estimated that a rapid release of 20 t of chlorine gas could kill up to 6000 in an urban area and up to 120 in a rural area.

The effect of more favourable weather conditions was not calculated. It was considered that, in view of the uncertainties in gas dispersion estimates, calculations for other weather conditions were not justified.

The effect of a lower rate of release is to reduce the estimated number of fatalities. The values may be calculated from the data given.

The effect of escape was estimated by calculating the time for a man to walk out of the cloud. It was assumed that if the axial concentration is lethal for the time of passage of the cloud, the concentration halfway out of the cloud may be considered only injurious and that at the edge of the cloud, defined as 10% of that of the centre, relatively harmless. It was assumed that the man would walk out of the cloud across wind at a speed of 3 mile/h (1.34 m/s). Escape was assumed if the time to walk from the cloud centre to the cloud edge was less than the lethal exposure time at the axial concentration. The times to walk out are shown in Figures A7.8(a–c). The distances at which the time to walk out and the lethal exposure time curves cross are the lethal distances of the clouds allowing for escape. For the release time of 6 s the lethal range remains 2.7 km, but for the release times of 10 min and 6 h it is reduced to 1.1 km and less than 100 m, respectively.

The effect of shelter was estimated using the equation

$$C = C_0[1 - \exp(-\lambda t)] \qquad [A7.5.21]$$

where C is the concentration indoors; C_0 the concentration outdoors; t the time; and λ the number of air changes per unit time.

The value of λ is about one change per hour for a modern centrally heated building, but in general can be two or three changes per hour, particularly if doors or windows are open. Assuming that shelter with one air change per hour is used, for a 6 s release the cloud is lethal at 100 m, but not at 200 m, and for a 10 min release it is lethal at 300 m, but not at 500 m. For a 6 h release the concentration in the building rises to that outside it after 1 h and the cloud is lethal at 0.6 km as before. Thus shelter is very effective for the release time of 6 s and quite effective for the release time of 10 min, but is ineffective for the release time of 6 h. In the latter case, however, escape is relatively easy.

A7.5.16 Road tanker hazards

A road tanker accident is considered as a possible initiating event for release of flammable and toxic materials. The report discusses road tanker accidents in Part 2, Section 14, and in Appendix 18.

The nature of the hazard from a road tanker accident is not discussed in as much detail as that from some of the other hazards. Mention is made of the possibility of a 10 t spillage of flammable material which could give rise to a fire (p. 96) and of spillage of LPG which could give rise to a 14 t cloud which could travel 200 m and still remain flammable (p. 178).

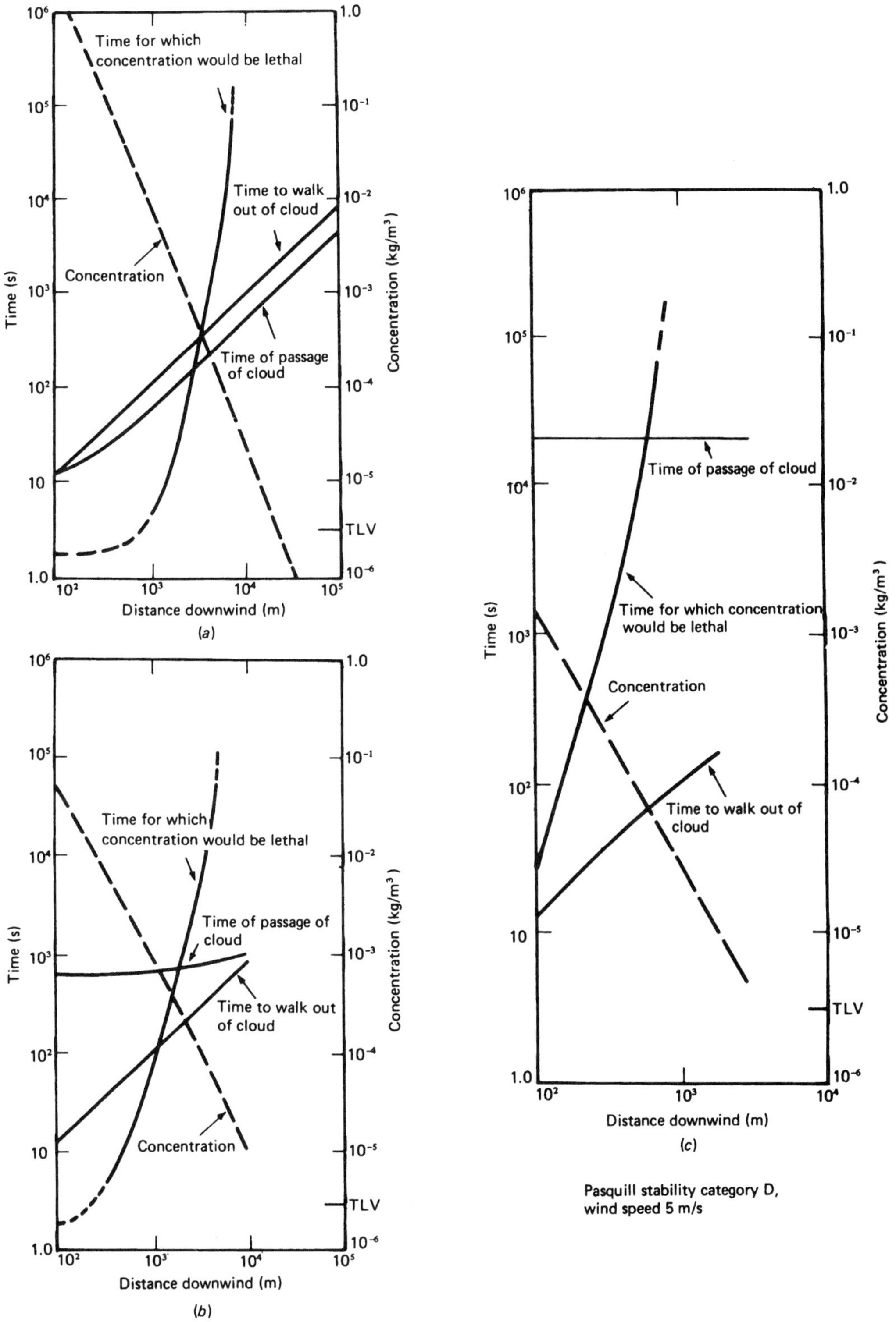

Figure A7.8 *First Canvey Report: mean concentration, lethal exposure time, time of passage and time to walk out vs distance for a 201 release of chlorine with various release times (Health and Safety Executive, 1978b): (a) release time of 6 s; (b) release time of Wmin and (c) release time of 6 h (Courtesy of HM Stationery Office)*

The frequency of spillage is estimated from data recorded by HM Inspector of Explosives on accidents resulting in loss of life and personal injury and involving the contents of petrol tankers. These records indicated that over the past 5 years there were on average five accidents per year that led to a spillage. But from statistics published by the DoE (1976) the quantity of petroleum products moved by road is 4.3×10^9 t km. Then since the average tanker load is 14 t

Frequency of petrol tanker accident involving spillage
$= 1.6 \times 10^{-8}$/km

Tankers carrying LPG are stronger than petrol tankers, but the conservative assumption was made that the frequency of accidents involving spillage is the same for LPG tankers as for petrol tankers.

The frequency of spillage from all hazardous materials transported by road from the British Gas Corporation, London and Coastal Wharves, and Texaco was then estimated as 1.4×10^{-3}. The contribution of flammables to this figure is the largest, that of toxics being relatively small.

A7.5.17 Evacuation

The mitigating effect of evacuation was also considered. The report discusses evacuation in Part 1, Section 10, and in Part 2, Section 16.

There are two respects in which the evacuation situation at Canvey Island is unusual. There are only two roads leading off the island, and these converge at a single roundabout, and there exist already plans for evacuation in the event of flooding by the sea.

The need for an additional road off the island has been the subject of debate in the area.

The evacuation process was modelled. An exponential model was used in which it was assumed that half the population is moved to safety in the first 2 h, half of the remainder in the next 2 h and so on.

Evacuation was found to be possibly beneficial in one or two types of emergency. In particular, consideration was given to evacuation in the event of a large release of ammonia from a ship collision or jetty incident. For the majority of emergencies, however, it is suggested that it would be better to remain indoors rather than to attempt evacuation.

Such evacuation would require prior planning. The existing arrangements for evacuation for flooding would not necessarily suffice, since the advance warning from a chemical emergency is likely to be less.

Similarly, the provision of an additional road was seen as beneficial for the same types of incident, but it was considered that the construction of such a road could not be recommended solely for evacuation without further discussion.

A7.6 First *Canvey Report*: Assessed Risks and Actions

The hazards described in Section A7.3 were assessed using appropriate failure and event data as described in Section A7.4 and hazard models and risk estimates, including those described in Section A7.5. The accidents that might occur, the nature and frequency of the initiating events and the frequency−number relations for casualties, or societal risks, are shown in Table A7.9.

The results of the risk assessments are presented by the investigators as risks of causing casualties, that is, severe hospitalized casualties or worse. This is in accordance with established practice (e.g. Department of Defense, n.d.;

Glasstone, 1964). It was considered misleading to attempt to distinguish between severe injury and death.

These results have several interesting features. The hazards may be ranked for societal risks in order of descending frequency for accidents of different magnitude as shown in Table A7.9.

The hazard arising from the very large quantities of LNG stored is a serious one, but is no worse than that from the considerably smaller quantities of LPG.

The obvious hazards of LNG, LPG and ammonia are equalled by others, such as oil and hydrogen fluoride, which are perhaps less well appreciated.

The relative importance of the hazards changes with the scale of the accident. For the smaller scale accidents oil spillage, flammable vapour clouds and toxic gas clouds are all important. As the scale increases, it is the toxic gas clouds that dominate.

There are a number of interactions identified both within sites and between sites. These include the threat to LPG storage at Mobil from the Calor Gas site, to the oil storage at Texaco from explosives barges, to the ammonia sphere at Fisons from rotating machinery and from the ammonium nitrate plant at Fisons, to various installations from process and jetty explosions, and possibly to the ammonia storage tank at Shell from explosion in the Shell refinery.

The relative hazard of the pressure storage of anhydrous ammonia at Fisons is much greater than that of the refrigerated storage of the same chemical at Shell. The risk for the latter was assessed as negligible with the possible exception of rupture by an explosion.

The assessed societal risks are shown in Table A7.9 and in Figure A7.9. Figure A7.9(a) gives the societal risks for all the existing installations. It shows that the risk of an accident causing more than 10 casualties is 31×10^{-4}/y. Figure A7.9(b) gives the societal risks for all the proposed developments. It shows that the risk of an accident causing more than 10 casualties is 16×10^{-4}/y.

The assessed individual risks are given for all the existing installations. These range from 13×10^{-4}/y in region A to less than 1×10^{-4}/y in region G of the area. The individual risks are also given for all the existing installations and proposed developments. These range from 26.3×10^{-4}/y in region A to less than 2×10^{-4}/y in region G (Table A7.10).

The investigators made a number of recommendations for the reduction of the hazards. These included

(1)   oil spillage − construction of a simple containing wall around the London and Coastal Wharves and Texaco sites and the proposed Occidental and UR refinery sites;

(2)   LNG tank flooding − construction of a dike;

(3)   spontaneous failure of LPG spheres − high standard of inspection;

(4)   LPG vessel rupture by missile (Mobil) − measures including fitting of pressure relief valves on cylinders at Calor Gas depot;

(5)   LPG tank failure at BG jetty − improvement of bund;

(6)   LPG pipeline failure (BG) − removal of pipeline;

(7)   spontaneous failure of ammonia sphere − high standard of inspection, control of ammonia purity;

(8)   ammonia release at Shell jetty − provision of water sprays at jetty;

Table A7.9 First Canvey Report: principal hazards identified at Canvey and associated societal risks (after Health and Safety Executive, 1978b)

Hazardous material	Company	Initiating event	Type of accident	Frequency of initiating event	Frequencies in units of 10^{-6}/year — Frequencies for number of offsite casualties exceeding						
					10	1,500	3,000	4,500	6,000	12,000	18,000
Oil	L & C, Texaco	Explosion at BG jetty / Explosion at Occidental jetty	Oil spillage over bund	1,800 } 3,600	1,230	1,000	500	130			
	Texaco	Explosion of stranded explosives barge		2,500	50	35	17	7			
		Explosion of stranded explosives barge and breaching of sea wall		6	6	4	3	2	1		
	Occidental	Explosion in process area		1,000	25	18	8	4			
	Occidental, UR	LPG explosion due to spontaneous failure of sphere		150	30	20	10	4			
		Refinery fire		100,000	25	18	8	3	1		
		Subtotal			1,366	1,095	546	150	1		
LNG	British Gas	Incident at BG jetty (spillage, ship collision, fire, explosion)	LNG vapour cloud	2,000	168	118	83	56	37	16	7
		LNG ship collision		50	5	3	1	0.5	0.3	0.2	0.1
		Flooding of tanks		500	106	70	37	14			
		Spillage over bund of above-ground tanks		180	56	33	17	10			
		Spillage into bund of above-ground tanks		540	162						
		Subtotal		3,270	497	224	138	80	37	16	7
LPG	British Gas	Explosion at BG jetty	LPG vapour cloud	—	520	320	160	80			
		Pipeline failure		450	450	320	160	80			
		Subtotal			970	320	160	80	37	16	7
LNG and LPG	British Gas	Subtotal		3,720	1467	544	298	160	37	16	7
LPG	Shell	Spontaneous failure of vessel / Missile from storage area / Shock wave from filling point or process area	LPG vapour cloud	140 } 100 1,235	184	110	60	25			
	Mobil	Spontaneous failure of vessel / Missile from process area		230 240							

Table A7.9 (continued)

Hazardous material	Company	Initiating event	Type of accident	Frequency of initiating event	Frequencies in units of 10^{-6}/year — Frequencies for number of offsite casualties exceeding						
					10	1,500	3,000	4,500	6,000	12,000	18,000
		Interaction from Calor Gas; sphere failure		4,100 ⎫	87	50	24	12			
		Interaction from Calor Gas; pipework failure		1,200 ⎭							
	Shell, Mobil	Fire or explosion at jetty		5,700	69	43	30	15			
	Occidental	Spontaneous failure of vessel		90	3	2	1				
		Fire or explosion at jetty		4,000	80	50	28	11			
	UR	Spontaneous failure of vessel		70	18	8	4	2			
	—	LPG ship collision		6,640	196	124	64	31			
		Subtotal		23,745	637	387	211	96			
Ammonium nitrate	Fisons	Rail accident	Ammonium nitrate explosion	85	85	85	85	17	17		
Anhydrous ammonia	Fisons	Spontaneous failure of sphere	Ammonia vapour cloud	100 ⎫	68	40	28	21	15	7	3
		Missile from rotating machine		30							
		Explosion in process area		100 ⎭							
		Rail accident		135	204	120	84	73	47	21	9
	Shell	Fire or explosion at jetty		400	228	132	96	68	40	32	20
		Ammonia ship collision		375	235	153	122	96	72	54	41
		Subtotal		1,140	735	445	330	258	174	114	73
Hydrogen fluoride	Shell	Intrinsic failure of pressure circuit	Hydrogen fluoride vapour cloud	200 ⎫							
		Missiles from neighbouring plant		200 ⎬	125	100	80	75	70	35	15
		Adjacent vapour cloud explosion		100 ⎭							
	Mobil (inc. extension)	Intrinsic failure of pressure circuit		100 ⎫							
		Missiles from neighbouring plant		100 ⎬	71	99	72	51	30	24	15
		Adjacent vapour cloud explosion		100 ⎭							
	Occidental	Intrinsic failure of pressure circuit		100 ⎫							
		Missiles from neighbouring plant		100 ⎬	168	144	132	120	114	80	70
		Subtotal			464	343	284	246	214	139	100

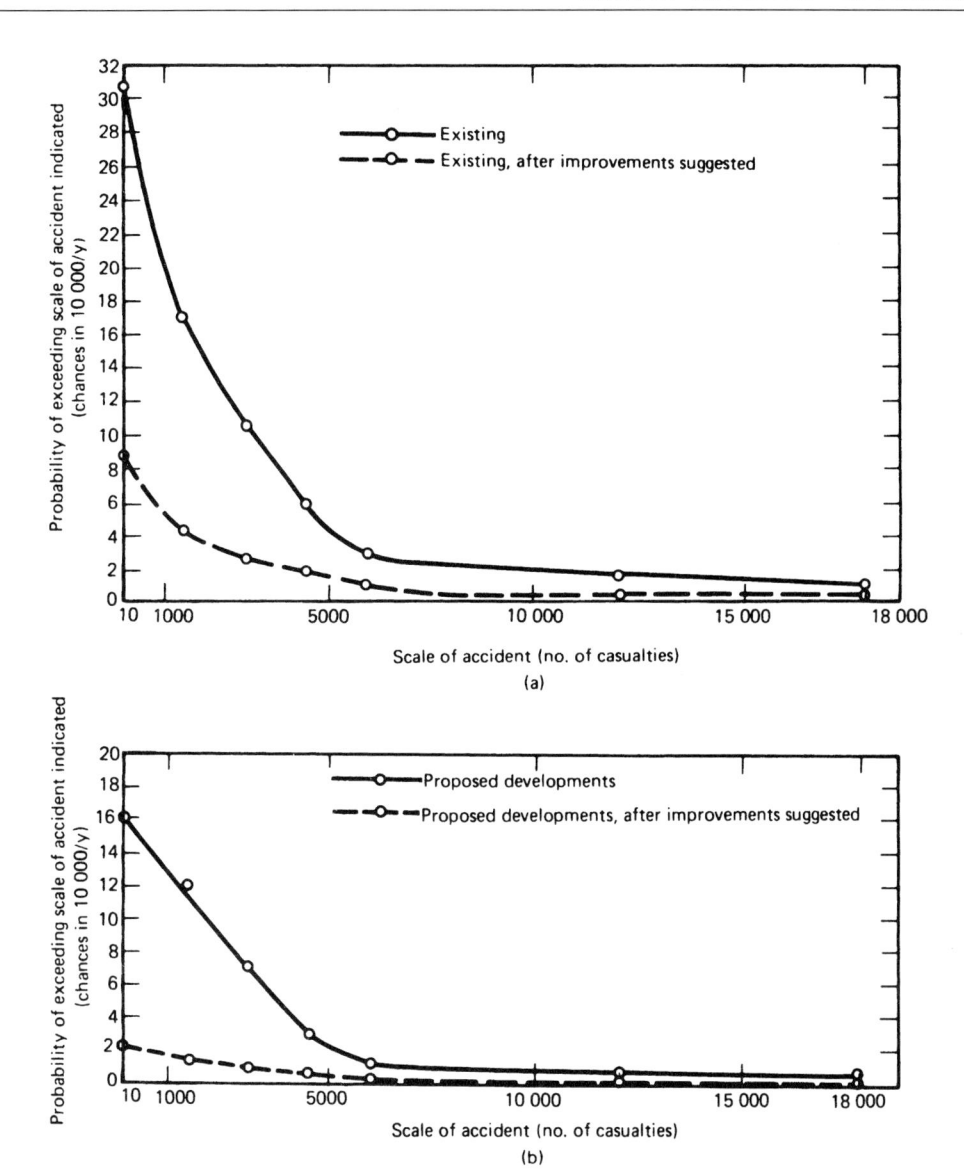

Figure A7.9 *First Canvey Report: societal risks for existing installations and proposed developments (Health and Safety Executive, 1978b): (a) all existing installations and (b) all proposed developments (Courtesy of HM Stationery Office)*

(9) HF plant rupture (Mobil, Occidental) – provision of water sprays;
(10) ship collision – strict enforcement of the speed restriction of 8 knot;
(11) road tanker hazards – road tanker traffic restriction to new road only (if road built).

The assessed effect of the proposed modifications is to eliminate the hazards from oil spillage, from LPG vapour cloud due to explosion at the jetty and pipeline failure at British Gas, and from vessel rupture at Mobil due to missiles from Calor Gas, and to reduce greatly many of the other hazards.

The rank order for societal risk of the hazards assuming that the proposed modifications are carried out is then as shown in Table A7.11. Oil spillage and LPG vapour cloud (BG) are eliminated.

There are some hazards, however, which remain relatively difficult to eliminate. In particular, spontaneous failure of storage vessels, jetty incidents and ship collisions make large contributions to the residual risks.

The effect of the proposed modifications on the societal risks for the existing and proposed installations is given in Figure A7.9. This shows that the risk of an accident causing more than 10 casualties is 9×10^{-4}/y for the existing installations and 2×10^{-4}/y for the proposed developments.

Table A7.10   *First Canvey Report: rank order of societal risks for some principal hazards of existing and proposed installations at Canvey (after Health and Safety Executive, 1978b)*

Frequencies in units of 10^{-6}/year								
>10 casualties			>4,500 casualties			>18,000 casualties		
1	Oil spillage	1366	1	Ammonia vapour cloud	258	1	HF vapour cloud	100
2	LPG vapour cloud (BG)	970	2	HF vapour cloud	246	2	ammonia vapour cloud	73
3	Ammonia vapour cloud	735	3	Oil spillage	150	3	LNG vapour cloud	7
4	LPG vapour cloud (others)	637	4	LPG vapour cloud (other)	96			
5	LNG vapour cloud	497	=5	LNG vapour cloud	80			
6	HF vapour cloud	464	=5	LPG vapour cloud (BG)	80			
7	AN explosion	85	7	AN explosion	17			

Table A7.11   *First Canvey Report: rank order of societal risks for some principal hazards of existing and proposed installations at Canvey, after improvements suggested (after Health and Safety Executive, 1978b)*

Frequencies in units of 10^{-6}/year								
>10 casualties			>4,500 casualties			>18,000 casualties		
1	LPG vapour cloud (other)	421	1	Ammonia vapour cloud	71	1	HF vapour cloud	24
2	LNG vapour cloud	396	2	HF vapour cloud	67	2	Ammonia vapour cloud	18
3	Ammonia vapour cloud	240	3	LNG vapour cloud	66	3	LNG vapour cloud	7
4	HF vapour cloud	115	4	LPG vapour cloud (other)	63			
5	AN explosion	8	5	AN explosion	2			

The assessed individual risks are given for all the existing installations after the proposed modifications. These range from 6×10^{-4}/y in region A to less than 1×10^{-4}/y in region G. The individual risks are also given for all the existing installations and proposed developments, after the proposed modifications. These range from 9×10^{-4}/y in region C to less than 1×10^{-4}/y in Region G.

The risk to an individual member of the public at Canvey may be compared with the average risk to workers in British industry. For the latter the fatal accident rate (FAR) per 10^8 exposed hours, is 4. If it is assumed that a man works 2000 h per year, then the FAR is equivalent to an individual risk of 0.8×10^{-4}/y.

A7.7 First *Canvey Report*: Responses to Report

A7.7.1 Response of HSE
The HSE accepted the report as a significant step forward in quantitative risk assessment, while acknowledging that there were some deficiencies.

It concluded that provided certain improvements were carried out, none of the existing installations need cease operations and the proposed developments could take place.

A7.7.2 Response of other parties
An account of some of the other responses to the report is given by the HSE in the Second *Canvey Report*, described below.

Publication of the report led to a surge of interest both locally and at national and international levels. HSE attended a public meeting at Canvey and explained the report. While the determination of the risks was welcomed, there was criticism of HSE's decision to allow continued operation of the existing installations, subject to the modifications.

The HSE gave a summary of some criticisms made of the report by various parties. One was that the report overestimates the risks. The assessors were required by HSE not to err on the side of optimism. Many in the industry, however, considered that it is preferable to adopt a 'best estimate' approach and that a conservative method which involves making a pessimistic estimate at each stage where there is doubt is liable to give a gross overestimate of the risks.

The HSE was also criticized in appearing to acquiesce in risks, some of which had been assessed as relatively high. On this the HSE draws attention to its requirement that improvements be made. It goes on.

HSE strongly believes that it has a duty, as the body charged with enforcing health and safety legislation, to express opinions about what may be required by that legislation. In HSE's view it is not sufficient merely to identify the risk to health and safety without also recommending a course of action.

A critique of the First *Canvey Report* was published by Cremer and Warner (1980).

A7.7.3 General comments
An investigation of the kind described has perhaps an inevitable tendency to suggest that the hazards revealed were previously unappreciated. In fact companies involved in such a survey are generally very well aware of the

hazards and have taken the measures that they consider appropriate.

The report itself expresses some reservations on the risk estimates and states: 'Practical people dealing with industrial hazards tend to feel in their bones that something is wrong with risk estimates as developed in the body of the report'.

At the time, the industry in the United Kingdom was conscious particularly of the Flixborough explosion. This has been dwarfed by subsequent disasters such as those at Mexico City, Bhopal, Ufa and Piper Alpha.

A7.8 Second *Canvey Report*

In 1981 there appeared the Second *Canvey Report* entitled *Canvey: A Review of Potential Hazards from Operations in the Canvey Island/Thurrock Area Three Years after Publication of the Canvey Report* (HSE, 1981a). This again was in two parts, with Part 1 an introduction by the HSE and Part 2 the SRD study.

In Part 1 HSE outlines the criticism of the first report, as just described, and gives its conclusions concerning the reviewed assessment.

In the next two sections, accounts are given of the reassessed risks and actions and of some technical aspects of the second report.

A7.9 Second *Canvey Report*: Reassessed Risks and Actions

The risks assessed in the Second *Canvey Report* are generally less than those in the first report. Thus, for example, at Stanford-le-Hope the individual fatality risk is given as 0.6×10^{-4}/y, as opposed to 5×10^{-4}/y in the first report, a reduction by a factor of 8.

The HSE in Part 1 adduce the following reasons for this: (1) physical improvements already carried out; (2) changes in operation; (3) detailed studies by companies; (4) changes in assessment techniques; (5) correction of errors and (6) further changes firmly agreed but yet to be made.

One major change in operation was the cessation of ammonia storage at Shell.

In a number of cases in the first report, estimates of the risks had to be made without benefit of detailed studies. Subsequent studies by the companies involved showed that in some cases the risks had been overestimated. A case in point was the limited ammonia spill at Fisons.

In a large proportion of cases the effect of the use of improved models for heavy gas dispersion was to reduce the travel distance of the cloud and hence the risks.

The HSE describe a number of further actions taken or pending to reduce the risks.

A7.10 Second *Canvey Report*: Technical Aspects

The Second *Canvey Report* revisits some of the frequency estimates and gives a revised set of hazard models and injury relations, covering the following aspects: (1) emission; (2) gas dispersion; (3) ignition; (4) fire events (5) vapour cloud explosions and (6) toxic release.

A7.10.1 Pressure vessel failure
The report reviews the frequency of spontaneous failure of pressure vessels, in the light of a further survey of the failure rate of such vessels by T.A. Smith and Warwick (1978) (see also Smith and Warwick 1981 SRD R203). The use of a failure rate of 10^5/y is reaffirmed.

A7.10.2 Emission models
The set of emission models given covers the following cases:

(1) rupture of a pressure vessel;
(2) flow from a pipe:
    (a) non-flashing liquid;
    (b) flashing liquid;
    (c) gas.

The model for emission following rupture of a pressure vessel containing liquefied gas is the adiabatic flash.

A7.10.3 Vaporization model
Vaporization of a pool of liquefied gas is treated using the SRD code SPILL, which can be applied to instantaneous or continuous releases.

A7.10.4 Gas dispersion models
Most of the liquefied gases studied exhibit heavy gas behaviour on release. For these use is made of the SRD codes DENZ and CRUNCH for instantaneous and continuous releases, respectively.

The report gives some results from these models in the form of tables of distance to the LFL and ½LFL for releases of propane and butane. As described in Chapter 15, a correlation model for heavy gas dispersion based on such tables has been developed by Considine and Grint (1985).

A7.10.5 Ignition models
The report gives several sets of ignition probabilities. It distinguishes three distinct cases of gas cloud

Delay before ignition	Cloud composition	Likely consequence
Seconds	Droplets entrained, cloud likely to be rich	Fire burning around edge of cloud (possibly leading to a fireball)
Minutes	Gas well mixed with air, likely to be in flammable range	Flash fire or explosion
An hour	Gas well diluted with air, likely to be lean	Hazard unlikely, though some flammable gas pockets possible

Table A7.12 *Second Canvey Report: some ignition probabilities for a flammable vapour cloud (after Health and Safety Executive, 1981a)*

A Ignition on site

Sources of ignition	Ignition probability
None	0.1
Very few	0.2
Few	0.5
Many	0.9

B Ignition at jetty

Delay before ignition	Probability of ignition following	
	Fire/explosion	Collision
Immediate ignition (within 30 seconds)	0.6	0.33
Delayed ignition (delay ½ to several minutes)	0.3	0.33
No ignition (within a few minutes)	0.1	0.33

C Ignition in transit

Terrain	Ignition probability
Open land	0
Industrial site	0.9

D Ignition over populated area

Type of ignition	Ignition probability
Edge/edge: edge of unignited cloud just reaches edge of populated area when ignition occurs	0.7
Central: unignited cloud right over population when ignition occurs	0.2
No ignition	0.1

dispersion: ignition may occur (1) at source, which may be (a) on site or (b) at a jetty; (2) in transit or (3) over a populated area. Table A7.12 shows some of the ignition probabilities given for a flammable gas cloud in these situations. The values are based partly on data on the proportion of releases ignited given by Science Applications Inc. (SAI) (1974) and D.C. Wilson (1980) and partly on judgement.

A7.10.6 Fire models and injury relations
The report gives a set of models for the following fire events: (1) pool fires; (2) fireballs; (3) BLEVEs; (4) flash fires.

Delay before ignition	Liquefied gas release	
	Refrigerated	Pressurized
Immediate ignition (within seconds)	Pool fire	Fireball
Slight delay (up to 30 s)	Diffusion flash fire	Premixed flash fire/explosion
Longer delay (min)	Premixed flash fire/explosion	Premixed flash fire/explosion

The event that occurs will depend on the delay before ignition. The behaviour assumed is as follows: For the various fire models the report refers to work by Considine (1981), evidently an early version of the paper by Considine and Grint (1985), which gives a similar set of models, as described in Chapter 16.

A7.10.7 Vapour cloud explosion model and injury relations
The vapour cloud explosion model covers both the probability that an explosion occurs given ignition and the effects of such an explosion. For the former the probabilities used are

Size of cloud (te)	Explosion probability
<10	0
100–100	0.1 (LNG = 0.01)
>100	1

For the effects of such an explosion use is made of the following relations based on the work of the ACHM:

$$p_o = kM^{1/3} \qquad [A7.10.1]$$

$$R = 30M^{1/3} \qquad [A7.10.2]$$

hence

$$p_a = k'R \qquad [A7.10.3]$$

where M is the mass released (te), p_a the peak side-on overpressure (kPa), R the radius of the cloud (m) and k, k' are constants. Relationships are given as follows:

Overpressure (kPa)	Hazard range (m)	Casualty probability
0<7	>4.25R	0
7–21	(2–4.25)R	0.1
21–34	(1.35–2)R	0.25
34–48	(1.1–1.35)R	0.70
>48	<1.1R	0.95

Further the authors quote the estimates of the ACMH that the overpressure at the centre of the cloud is likely to be about 103 kPa and at the edge about 69 kPa.

A7.10.8 Toxic injury relations

Injury relations are given for ammonia and for hydrogen fluoride. For these gases the authors state that the best information available to them on the LC_{50}, the concentration lethal at the 50% level, is as follows:

Time t (min)	*Lethal concentration* LC_{50}	
	Ammonia (kg/m^3)	*Hydrogen fluoride* (kg/m^3)
1		1.5×10^{-3}
5	7×10^{-3}	4×10^{-4}
10	3.5×10^{-3}	
30	1.2×10^{-3}	
60	5×10^{-4}	4×10^{-5}

They further state that these data may be fitted to the relation

$$LCt_{50} = kCt^{0.9} \qquad [A7.10.4]$$

where C is the concentration, LCt_{50} the dose lethal at the 50% level and t time.

A7.11 Notation

Section A7.5

Subsection A7.5.5

A_H	horizontal area of target
A_V	vertical area of target
A_V'	horizontal projection of vertical area of target
F	frequency of aircraft crash on target area
F_H	frequency of aircraft crash per unit of ground area
F_V	frequency of aircraft crash per unit of vertical area

Subsection A7.5.7

g	acceleration due to gravity (m/s^2)
h_f	slumped liquid depth (m)
r	radius of flooded zone (m)
r_s	radius of source cylinder (m)
t	time after initiation of slumping (s)
u_f	slumped liquid velocity (m/s)
V_s	volume of source cylinder (m^3)

Subsection A7.5.8

A_H	area of tank roof (m^2)
A_V	exposed area of tank vertical sides (m^2)
c_p	specific heat of tank wall ($J/kg°C$)
m	mass of exposed tank wall (kg)
Q_i	heat input to tank (W)

Q_n	heat loss by convection (W)
Q_r	heat loss by radiation (W)
t	time (s)
ε	emissivity of tank wall
θ_A	absolute temperature of ambient air (K)
θ_T	absolute temperature of tank wall (K)
$\Delta\theta$	temperature difference between tank and ambient air (°C)
α	Stefan–Boltzmann constant ($W/m^2 K^4$)

Subsection A7.5.11

H	heat of combustion of hydrocarbons (kcal)
Δp_m	peak overpressure (bar)
q	peak dynamic pressure (bar)
r	distance (m)
W	mass of hydrocarbons (t)
α	equivalence factor

Subsection A7.5.13

a, b	constants
C	concentration (g/m^3)
D	dosage ($((g/m^3)^{2.75}\, min)$)
P_r	probit
t	time (min)

Subsection A7.5.14

Equations A7.5.11–A7.5.13

g	acceleration due to gravity
h	height of cloud
r	radius of cloud
r_0	initial radius of cylinder
t	time
V_0	initial volume of cylinder
ρ_a	density of air
ρ_0	initial density of cloud

Equations A7.5.14–A7.5.20

h	height of cloud (m)
Q^*	mass released instantaneously (kg)
r	radius of cloud (m)
t	time (s)
u	wind speed (m/s)
x, y	distances in downwind, cross-wind directions (m)
$\sigma_x, \sigma_y, \sigma_z$	dispersion coefficients in downwind, crosswind and vertical directions (m)
χ	concentration (kg/m^3)
χ_{av}	mean concentration along axis (kg/m^3)
χ_{max}	maximum concentration along axis (kg/m^3)

Subsection A7.5.15

C	concentration indoors
C_0	concentration outdoors
t	time
λ	number of air changes per unit time

Section A7.10

Subsection A7.10.7

k, k'	constants
K	constant
M	mass released (te)
p_O	peak side-on overpressure (kPa)
R	radius of cloud (m)

Subsection A7.10.8

C	concentration
k	constant
LCt_{50}	dose lethal at 50% level
t	time

Appendix

Rijnmond Report

8

Contents

Another comprehensive hazard assessment is the Rijnmond study carried out for the Rijnmond Public Authority by Cremer and Warner.

This work is described in *Risk Analysis of Six Potentially Hazardous Industrial Objects in the Rijnmond Area*, a Pilot Study issued by the Rijnmond Public Authority (1982) (the Rijnmond Report).

The Rijnmond is the part of the Rhine delta between Rotterdam and the North Sea. Within the Rijnmond area of $15 \times 40 \text{ km}^2$ are located some one million people and a vast complex of oil, petrochemical and chemical industries as well as the largest harbour in the world. A description of the industrial complex at Rijnmond has been given by Molle and Wever (1984). A map of the Rijnmond showing the principal plant areas is given in Figure A8.1 and a diagram of the petrochemical complex in Figure A8.2. There are five oil refineries with a refining capacity of some 81 Mt/year, including the large refineries of Shell at Pernis and BP near Rozenburg, and major petrochemicals complexes operated by Shell, Gulf and Esso. Some of the other plants are described below.

A first note on the industrial hazards issued in 1976 by the Rijnmond Authority was strongly criticized by industry. Accordingly a commission, COVO, was set up with representatives of the Labour Inspectorate of the Ministry of Social Affairs, the Rijnmond Authority and industry. COVO decided to carry out a pilot study of six industrial installations in the Rijnmond in order to evaluate the applicability of risk assessment to decision-making on safety and commissioned Cremer and Warner to carry out

the study. A steering committee was formed and the work was done to the requirements of this committee.

A8.1 The Investigation

The aim of the work was to evaluate the methods of risk assessment for industrial installations and to obtain experience in the practical application of these methods. Such evaluation was considered to be essential before any decision could be made on the role of such methods in the formulation of safety policy.

The objectives of the study were formulated so as to give an answer to the following questions:

(1)   What is the reliability of the assessment of the consequences and probabilities of possible accidents with industrial installations when the procedures and methodology of risk analysis are carried out to their full extent?
(2)   What problems and gaps in knowledge exist in the field of risk analysis?
(3)   How can the results of a risk analysis be presented conveniently, without losing important details, so that it may be used for safety policy decisions?
(4)   How well can the influence of risk reducing measures on the consequences and probabilities be calculated?
(5)   What resources are required, in time and money, to assess the risk with sufficient accuracy to be useful for safety policy decisions?

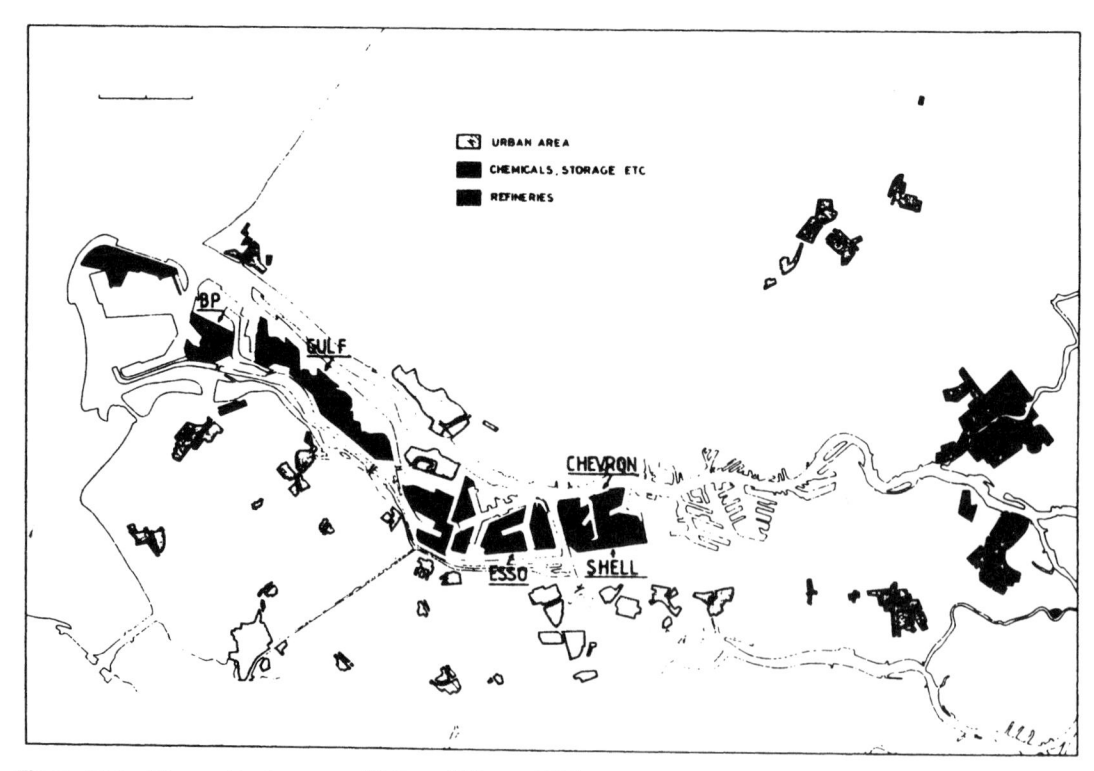

Figure A8.1   *Rijnmond harbour area (Molle and Wever, 1984)*

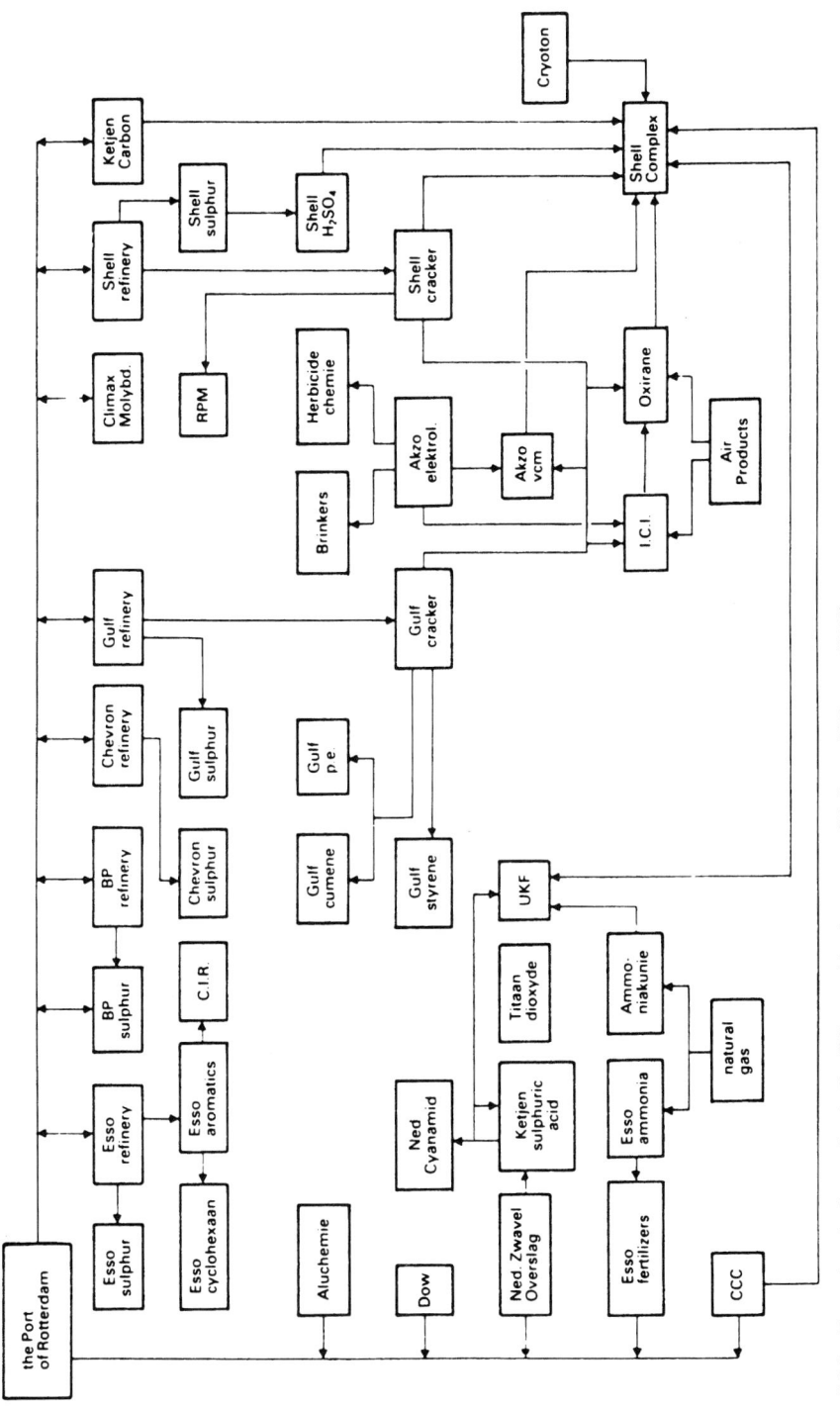

Figure A8.2 *Petrochemical complex in the Rijnmond harbour area (Molle and Wever, 1984)*

The COVO steering committee was reinforced with experts from industry. The committee also included representatives of the Battelle Institute who were charged with the task of scrutinizing the methodology and models used. The overall approach taken in the investigation was:

(1) to collect basic data and define the boundary limits of the installation;
(2) to identify the potential failure scenarios;
(3) to select and apply the best available calculation models for physical phenomena;
(4) to collect and apply the best available basic data and models to calculate the probabilities of such events;
(5) to choose and develop different forms of presentation of the final results;
(6) to investigate the sensitivity of the results for variations in the assumptions used and to estimate the accuracy and reliability of these results;
(7) to investigate the influence of risk reducing measures.

The investigation involved the identification of the principal hazards of the installations in the area, the assessment of the associated individual and societal risks and the proposal of remedial measures to reduce the risks.

The project was planned to last one year but in fact took two-and-a-half years and it cost some 2.5 m guilders. A supplementary contract was placed with the consultants to study methods of presenting the results. The report was published in 1982. It is in five parts. Part 1 is the report of the steering committee, Part 2 is the main Cremer and Warner study, Part 3 the supplementary Cremer and Warner study, Part 4 is the Battelle review and Part 5 contains the industrial and other comments.

A8.2 Installations and Activities

The six installations studied were as follows:

(1) Acrylonitrile storage (Paktank);
(2) Ammonia storage (UKF);
(3) Chlorine storage (AKZO);
(4) LNG storage (Gasunie);
(5) Propylene storage (Oxirane);
(6) Hydrodesulfurizer (Shell).

The storages of acrylonitrile and LNG were at atmospheric pressure and those of ammonia, chlorine and propylene were all liquids under pressure. The acrylonitrile, ammonia and chlorine represented toxic materials and the LNG and propylene flammable materials. The hydrodesulfurizer represented part of a process plant, the section studied being the diethanolamine stripper.

The context of these installations is described by Molle and Wever and accounts of the individual installations are given below.

The report is organized partly by installation and partly by themes and material on a particular installation is therefore given dispersed. Table A8.1 gives a summary of the entries for each installation.

A8.3 Event Data

The event data used in the study were a mixture of generic data, plant data and estimates. Some failure and event data and some external event data used are given

in Tables A8.2 and A8.3, respectively. Other data are given in the sections on particular installations.

A8.4 Hazard Models

The principal hazard models used are given in Table A8.4 and their application is described in the sections on particular installations.

A8.5 Injury Relations

The injury relations used are given in Table A8.5 and their application is described in the sections on particular installations.

A8.6 Population Characteristics

The study covered the risks both to employees and to the public. For the off-site population a grid was used showing the number of people over an area of some 75 km^2, which covered the whole of the Rijnmond. Populations were estimated for each 500 m^2 square using data from the 1971 census, updated in 1975.

Both day and night time populations were determined, day being defined as the working week of 45 h and night as the remaining 123 h.

Outside the population grid average population densities were used. These were required only in cases where an exceptionally large cloud occurred.

For the particular installations studied data were obtained on the numbers of employees present during day and night and on their locations. The on-site population N_e was estimated from the relation

$$N_e = \frac{(5N_d + 16N_n) \times 52}{((5 \times (52 - n_h))}$$ [A8.6.1]

where n_h is the number of weeks off for holidays, sickness, etc., N_d the numbers present during the working day and N_n the numbers present at all other times.

For toxic gas hazard it was assumed that 1% of the population are outdoors. This estimate includes an allowance for people taking shelter. For explosion hazard it was assumed that 10% are outdoors.

A8.7 Mitigation of Exposure

The principal mitigations of exposure considered were evacuation and shelter. For evacuation the estimate of the proportion evacuated is taken as a function of the warning time. In some cases the proportion is estimated as 90–100%.

For the effect of shelter from toxic gas the single-exponential stage model of ventilation is used.

A8.8 Individual Assessments

A8.8.1 Acrylonitrile storage (Paktank)
The acrylonitrile (AN) storage is one of a number of storages at the Botlek terminal of Paktank, a tank storage company providing bulk storage and associated services to the oil and chemical industries.

There are normally two or three tanks containing AN. The study was confined to a single tank and associated transfer systems which had been dedicated to handling AN for some years. This AN tank is one of a number of fixed roof tanks located in a tank bund which is at the centre of the site. The tank has a capacity of 3700 m³. It is designed to withstand small variations of internal pressure, the breather valves being designed to open at an overpressure of 200 mmHg and a vacuum of 50 mmHg.

Operation at the site is flexible, but in the preceding two years only six modes of transfer had been recorded, these being loading of coastal barges, loading of road tankers, loading of rail tank cars, filling from coastal barges and chemical tankers, filling from sample tanks, and transfer between this tank and others at the site.

A summary of the risk assessment for the AN storage is given in Table A8.6. Section A of the table lists the undesired events considered, section B the hazard scenarios and section C the main fault trees.

An escape from the tank into the bund, though instantaneous, gives rise to a constant evaporation rate and hence acts as a continuous source. Releases from the pipework are also continuous sources.

The frequency of the releases is estimated using the seven fault trees listed as shown in Figure A8.3. The table giving the frequency and event data for these trees contains 81 entries. Of these some 30 are derived from the generic data given in Tables A8.2 and A8.3 and the rest from other sources, mainly company data, meteorological data or judgement.

For fire hazard it was assumed that 10% of all major spillages or resulting vapour clouds and 5% of significant spillages ignite and that an operator present at the release

or ignition source receives fatal burns in 80% of cases. The hazard distances obtained are shown in the table as <10 m.

For toxic gas hazard the exposure is continuous. In most cases only employees were at risk. A 10-min exposure was assumed for employees, who are alert to the hazard.

It was found, however, that serious toxic effects were possible to the public from a major spill (PI) if the exposure time was 30 min. For the public credit was given for evacuation, usually 90–100%, depending on the warning time.

Typical hazard distances to the LTL_{50} of 76 m are given in Table A8.6. The distances are greater for adverse weather conditions, being on the centreline for 2.0/F and 1.5/D conditions 575 and 260 m to the LTL_{50} and 879 and 365 m to the LTL_{05}, respectively.

A8.8.2 Ammonia storage (UKF)

The ammonia storage is part of a complex of plants producing ammonia, nitric acid, ammonium nitrate, urea and other fertilizers at the Pernis site of UKF.

The ammonia is stored in a sphere under pressure at atmospheric temperature. The capacity of the sphere is 1000 m³, but the average inventory is only 40% of this. The ambient temperature was taken in the study as 15°C that corresponds to a pressure of 7.1 bar.

Ammonia is transferred to the sphere from rail tank cars or from the ammonia plant. Transfer of the liquid ammonia from the rail tank car is effected by pressurizing the tank

Table A8.1 *Rijnmond Report: summary of entries for the six installations (after Rijnmond Public Authority, 1982)*

Entry	Page					
	Acrylonitrile	*Ammonia*	*Chlorine*	*LNG*	*Propylene*	*Hydrodesulfurizer*
Main entry	81	111	151	183	202	225
Historical review	109, 350	132, 351	181, 355	200, 361	223, 362	240, 370
Release scenarios	91	113	154	187	203	227
Undesired events						
Events key	89	133	157	188	203	229
Event frequency	101	146	171	196	210	237
Failure data	97, 446	122, 462	170, 480	194, 497	209, 513	235
Other data	438			191		235
Fault trees	451	468	491	507	527	
Hazard models	96	115	156	189	205	228
Injury relations						
Fire/explosion	319, 331			319, 331	319, 331	
Toxic effects	93, 340	145, 340	340			231, 340
Consequence results						
Summary table	102	149	174	198	214	239
Fatalities table	440					
Grid printouts						
Fire/explosion				419	422	
Toxic effects	93	135, 397	161, 404			413
Hazard ranges						
Fire/explosion	96, 102					
Toxic effects	95, 102	135	161			
Risk results						
Tables	102	135, 578	161			
Risk contours		560				
FN tables	588	589	596	603	604	608
FN curves	562	563	564	565	566	567
Evaluation	250	250	251	251	252	252
Remedial measures	107	130	180	199	213	238

Table A8.2   *Rijnmond Report: some failure and event data used (after Rijnmond Public Authority, 1982)*

A Data sources

1.	AEC (1975)	10.	W. Marshall *et al.* (1976)
2.	Recht (1966c)	11.	Welker *et al.* (1976)
3.	SAI (1975)	12.	Bush (1975)
4.	Lees (1976b)	13.	SRS data bank
5.	Lees (1976c)	14.	D.L.E. Jacobs (1971)
6.	Skala (1974)	15.	Phillips and Warwick (1969 AHSB(S) R162)
7.	A.E. Green and Bourne (1976)	16.	Kletz (1972)
8.	Upfold (1971)	17.	Lawley and Kletz (1975)
9.	Smith and Warwick (1974 SRD R30)		

B Event data

Equipment	Failure rate[a,b]		Reference
	Rate	Range	
Tanks and vessels			
Pressure vessels			
Serious leakage	1×10^{-5}/year	6×10^{-6}–2.6×10^{-3}/year	7, 9, 10
Catastrophic rupture	1×10^{-6}/year	4.6×10^{-5}–6.3×10^{-7}/year	12, 15
Atmospheric storage tanks			
Serious leakage	1×10^{-4}		
Catastrophic rupture	6×10^{-6}/year		
Refrigerated storage			
Tanks (double wall,			
high integrity)			
Serious leakage from	2×10^{-5}/year		
inner tank			
Catastrophic rupture from	1×10^{-6}/year		
both containments			
Pumps, pipework, etc.			
Pumps			
Failure to start	1×10^{-3}/d	5×10^{-5}–5×10^{-3}/d	1
Failure to run	3×10^{-5}/h	1×10^{-7}–1×10^{-4}/h	1
normally			
Failure to stop	1×10^{-4}/d	4×10^{-5}–1×10^{-3}/d	1, 11
Catastrophic failure	1×10^{-4}/year		3, 13
Pipework[c,d]			
<50 mm			
Significant leakage	1×10^{-8}/sect. h		1, 9, 13, 15
Catastrophic rupture	1×10^{-9}/sect. h		
>50 mm, <150 mm			
Significant leakage	6×10^{-9}/sect. h		
Catastrophic rupture	3×10^{-10}/sect. h		
>150 mm			
Significant leakage	3×10^{-9}/sect. h		
Catastrophic rupture	1×10^{-10}/sect. h		
Hoses			
Heavily stressed	4×10^{-5}/h		11, 14
Lightly stressed	4×10^{-6}/h		
Loading arms			
Leakage	3×10^{-6}/h	1×10^{-7}–1×10^{-4}/h	1, 11
Catastrophic rupture	3×10^{-8}/h	1×10^{-8}–1×10^{-5}/h	
Valves			
Pressure relief valves			
Fails dangerously			
Blocked	0.001/year	0.001–0.01/year	5, 7, 13
Lifts heavy	0.004/year	$(1.4 \times 10^{-5}$–3.6×10^{-5}/d)	1
Lifts light/leakage	0.06/year	0.02–0.09/year	1, 5, 7
Vacuum relief valves	0.005/year	$(1 \times 10^{-5}$–1×10^{-4}/d)	1
Other data taken from references 1, 5, 6, 7			

Table A8.2   *(continued)*
Instrumentation
Data taken from
references 1, 5, 6, 7, 13, 16, 17

Electrical equipment			
Battery supply			
Failure to provide proper output (in standby mode)	1×10^{-6}/h	1×10^{-7}–6×10^{-6}/h	1, 7
Electric motors			
Failure to start	3×10^{-4}/day	7×10^{-5}–3×10^{-3}/d	1
Failure to run	7×10^{-6}/h	5×10^{-7}–1×10^{-4}/h	1, 7
Emergency diesel system (complete)			
Failure to start	3×10^{-2}/day	1×10^{-3}–1×10^{-1}/d	1
Failure to run	3×10^{-3}/h	1×10^{-4}–1×10^{-3}/h	1

[a] The unit d in this table stands for demand.
[b] The failure rates taken from the *Rasmussen Report* correspond to the median values quoted there.
[c] Values per metre run are also given and are one-tenth of those per section.
[d] Following data are given for range:

Industry		Nuclear	
<75 mm	2×10^{-9}–5×10^{-6}/sect. h		3×10^{-11}–3×10^{-8}/sect. h
>75 mm	1×10^{-10}–5×10^{-6}/sect. h		3×10^{-12}–1×10^{-10}/sect. h

Table A8.3   Rijnmond Report: some external events data used (after Rijnmond Public Authority, 1982)

Event	Frequency[a] (events/year)					
	Generic	Acrylonitrile	Ammonia	Chlorine	LNG	Propylene
Earthquake	10^{-8}	X	X		X	X
Subsidence	2×10^{-9}	2×10^{-7}	2×10^{-7}	2×10^{-7}	X	X
Flooding	10^{-7}	X	X	X	10^{-9}	
Vehicular intrusion	2×10^{-7}	X	X	X	2×10^{-9}	X
Mechanical impact	10^{-7}	3×10^{-6}	X	X		
Major fire	10^{-6}	X	10^{-9}	X	10^{-12b}	10^{-5c}

[a] Frequencies for these events are not given for the hydrodesulfurizer.
[b] Estimate based on fact that major fire would cause immediate ignition, not a vapour cloud.
[c] Estimate towards higher end of range due to flammable nature of products stored.

Table A8.4   Rijnmond Report: some hazard models used

Event	Model
A Emission	
Flow from	
A. Vessel containing liquid at atmospheric pressure	
B. Vessel containing liquid above atmospheric boiling point	
C. Vessel containing gas only	
Two-phase flow	Fauske–Cude (Fauske, 1964; Cude, 1975)
B Vaporization	
Spreading of liquid spill	Shaw and Briscoe (1978 SRD R100)
Vaporization of cryogenic liquid	
On water	Shaw and Briscoe (1978 SRD R100)
On land	Shaw and Briscoe (1978 SRD R100); AGA (1974)
Combined spreading and vaporization of cryogenic liquid on land	
Vaporization from spill into complex bunds	
Evaporation of volatile liquid	Pasquill (1943), Opschoor (1978)

Table A8.4   *(continued)*

Event	Model
C Gas dispersion	
Neutral density gas dispersion	Pasquill–Gifford
Dense gas dispersion	R.A. Cox and Roe (1977)
Dispersion of jet of dense gas	Ooms, Mahieu and Zelis (1974)
D Vapour cloud combustion	
Initial dilution with air	
Continuous Release	Ooms, Mahieu and Zelis (1974)
Instantaneous release	Empirical relation
Ignition sources	
Fire vs explosion	
E Vapour cloud explosion	
Vapour cloud explosion	TNO (Wiekema, 1980)
F Pool fire	
Flame emissivity	
Liquid burning rate	Burgess, Strasser and Grumer (1961)
Flame length	P.H. Thomas (1963)
Flame tilt	P.H. Thomas (1965b)
Fraction of heat radiated	
View factor	Rein, Sliepcevich and Welker (1970)
G Jet flame	
Flame length	API (1969 RP 521)
Deflection effect	API (1969 RP 521)
Buoyancy and liquid effects	MITI (1976)
View factor	Rein, Sliepcevich and Welker (1970)
H Fireball	
Fireball dimensions	
Radius	R.W. High (1968)
Duration	R.W. High (1968)
View factor	McGuire (1953)

Table A8.5   *Rijnmond Report: some injury relations used*

A   Fire: Steady state

Incident thermal radiation (kW/m^2)	Effect
1.6	Level insufficient to cause discomfort for long exposures
4.5	Level sufficient to cause pain if subject does not reach cover in 20 s; blistering of skin likely (first degree burns)
12.5	Minimum level for piloted ignition of wood, melting of plastic tubing, etc.
25	Minimum level for unpiloted ignition of wood at infinitely long exposures
37.5	Level sufficient to damage process equipment

B   Fire: Unsteady state

Incident thermal radiation dose threshold (kJ/m^2)	Effect
65	Threshold of pain: no reddening or blistering of skin
125	First degree burns
250	Second degree burns
375	Third degree burns

Table A8.5   *(continued)*

C   Explosion

Blast overpressure (bar)	Effect
0.03	Window breakage, possibly causing some injuries
0.1	Repairable damage; light structures collapse; pressure vessels remain intact
0.3	Major structural damage (assumed fatal to people inside buildings or other structures)

D   Toxic effects

Data given in terms of constants in probit equation for fatalities
$Y = k_1 + k_2 \ln \Sigma C^n T$

Gas	Constants n	k_1	k_2	LTL_{50} ($\mathrm{ppm}^n \cdot \mathrm{min}$)
Acrylonitrile	1			8.15×10^3
Ammonia	2.5	-30.57	1.385	1.22×10^{10}
Chlorine	2.75	-17.1	1.69	5.46×10^4
Hydrogen sulfide	2.75			1.6×10^8

Table A8.6   *Rijnmond Report: summary of assessment of acrylonitrile storage (Paktank) (after Rijnmond public Authority, 1982)*

A   Undesired events

Event[a]		Frequency (events/year)	Mass flow or mass (kg/s or kg)	Duration (s)	Average no. of fatalities to public[b,c] (deaths/year)
P1	Major rupture of tank	4.3×10^{-5}	1.85×10^6		7.6×10^{-6}
P2	Split in tank bottom/shell	2.0×10^{-4}	1.85×10^6		
P3	Rupture of tank roof	1.8×10^{-4}			
P4	Open drain valve on tank	1.0×10^{-4}		1800	
P5	Overfilling of tank	1.5×10^{-3}	$200\,\mathrm{m}^3/\mathrm{h}^d$		
P6	Internal explosion	5×10^{-6}			
P7	Unauthorized entry into tank	2×10^{-5}			
P8	Transfer line rupture	9×10^{-7}	1.85×10^6		3.2×10^{-7}
P9	Transfer line leakage	5.2×10^{-5}		1800	
P10	Hose rupture	2.6×10^{-3}		900	
P11	Catastrophic rupture of transfer line	8.8×10^{-6}		900	
P13	Transfer hose rupture	4×10^{-3}		600	
P14	Transfer hose rupture	4×10^{-3}		600	7.9×10^{-6}
Total					

B   Hazard scenarios

Spillage into bund
Evaporation from bund
Gas dispersion (from continuous source)
Pool fire in bund
Flash fire in vapour cloud

C   Fault trees

P1 Catastrophic rupture of tank
    P1.1 Severe overpressurization of tank
        P1.1.1 Pressurization of tank with nitrogen
        P1.1.2 Severe polymerization within tank
    P1.2 Severe tank depressurization

Table A8.6   (continued)

P1.4 Severe polymerization in tank
P5 Tank overfilled up to breather valve level
   P5.1 High level detected, incorrect action taken

[a] There are 15 further events:

P12 Transfer line leakage	P22 Line leakage
P15 Transfer hose leakage	P23 Sampling hose/line rupture
P16 Transfer line rupture	P24 Transfer line/hose rupture
P17 Transfer hose rupture	P25 Sample tank leakage
P18 Transfer line/hose rupture	P26 Explosion with pre-pump sample
P19 Transfer line/hose rupture	P27 Hose rupture
P20 Transfer line/hose leakage	P28 Flash fire or explosion within AN system during maintenance
P21 AN transfer line/hose leakage	

None of these events causes an average number of fatalities to the public $>10^{-5}$/year.
[b] For both cases P1 and P8 the typical hazard distances to the LTL_{50} are 76 m *and to the* LTL_{05} 126 m. These are the greatest hazard distances quoted.
[c] For all cases the typical hazard distances for fatal effects from fire are <10 m. No hazard distances for fatal effects from explosion are quoted.
[d] Maximum rate (pump rate).

car with ammonia vapour from the sphere and from the ammonia plant by pumps. Ammonia is transferred from the sphere to the downstream plants by pumps.

An escape from a sphere may constitute an instantaneous or a continuous source. Catastrophic rupture of a sphere was taken as equivalent to an instantaneous release, other escapes to continuous releases.

A summary of the risk assessment for the ammonia storage is given in Table A8.7. Sections A–C list the undesired events, the hazard scenarios and the main fault trees.

Ammonia constitutes both a flammable and a toxic hazard, but since the toxic effects can occur at concentrations 100 fold less than the flammable effect, only the former were considered.

A8.8.3 Chlorine storage (AKZO)

The chlorine storage consists of five spheres associated with chlorine cellrooms at the Botlek site of AKZO.

The chlorine is stored in the spheres under pressure at atmospheric temperature. Each sphere has a capacity of 90 m^3 and can therefore hold 100 te, but at any given time one of the spheres is used as a dump tank, so that the capacity of the storage is 400 te. At ambient temperature the sphere pressure is 6.5 bar, but pressure is taken up to 9 bar when the sphere is discharging to consumer plants. The spheres are located in a bund.

Chlorine is received only from the associated production plant, but is sent out to a variety of consumers. There is a chlorine gas disposal plant consisting of caustic soda scrubbers.

An escape from a sphere may constitute an instantaneous or a continuous source. Catastrophic rupture of a sphere was taken as equivalent to an instantaneous release, other escapes to continuous releases or mixed releases.

A summary of the risk assessment for the chlorine storage is given in Table A8.8. Sections A–C list the undesired events, the hazard scenarios and the main fault trees.

A8.8.4 LNG storage (Gasunie)

The LNG storage consists of two tanks and associated equipment at the peak shaving plant of Gasunie at Maasvlakte.

The LNG is contained in an inner tank of 9% nickel steel and an outer tank of carbon steel. This whole container is surrounded by a further containment consisting of a reinforced concrete floor and a pre-stressed concrete wall. The capacity of each tank is 57,000 m^3.

An escape from a tank may constitute an instantaneous or a continuous source. Catastrophic failure of a tank was taken as equivalent to an instantaneous release, other escapes to continuous releases or mixed releases.

A summary of the risk assessment for the LNG storage is given in Table A8.9. Sections A–C list the undesired events, the hazard scenarios and the main fault trees.

A8.8.5 Propylene storage (Oxirane)

The propylene storage is associated with the propylene oxide plant of Oxirane at Seinehaven. There are two spheres each with a capacity of 600 te. The ambient temperature was taken in the study as 15°C, that corresponds to a gauge pressure of 8 bar.

Propylene is transferred to the spheres by pipeline, ship and rail tank cars. It is transferred from the spheres to the downstream plants.

An escape from a sphere may constitute an instantaneous or a continuous source. Catastrophic rupture of a sphere was taken as equivalent to an instantaneous release, other escapes to continuous releases.

A summary of the risk assessment for the propylene storage is given in Table A8.10. Sections A–C list the undesired events, the hazard scenarios and the main fault trees.

A8.8.6 Hydrodesulfurizer (Shell)

The section of the hydrodesulfurizer studied is the di-ethanolamine (DEA) regenerator. The latter is part of an absorber–regenerator unit which removes hydrogen sulfide from sour gas. The hydrogen sulfide is absorbed into DEA in the absorber and the fat DEA passes to the regenerator where a reversible reaction occurs releasing the hydrogen sulfide again.

The escapes that can occur from this unit are of two types. There may be a direct release of hydrogen sulfide gas. Alternatively, the release may be by evolution of the gas from the fat DEA liquid.

Hydrogen sulfide presents both a flammable and a toxic gas hazard. On review the flammable hazard was discounted. However, the flammability of the gas remains relevant in that in a proportion of cases the gas cloud would burn and would then not constitute a toxic hazard.

A summary of the risk assessment for the regenerator section of the hydrosulfurizer is shown in Table A8.11. Section A lists the undesired events and Section B the hazard scenarios. As indicated in section C there are no fault trees.

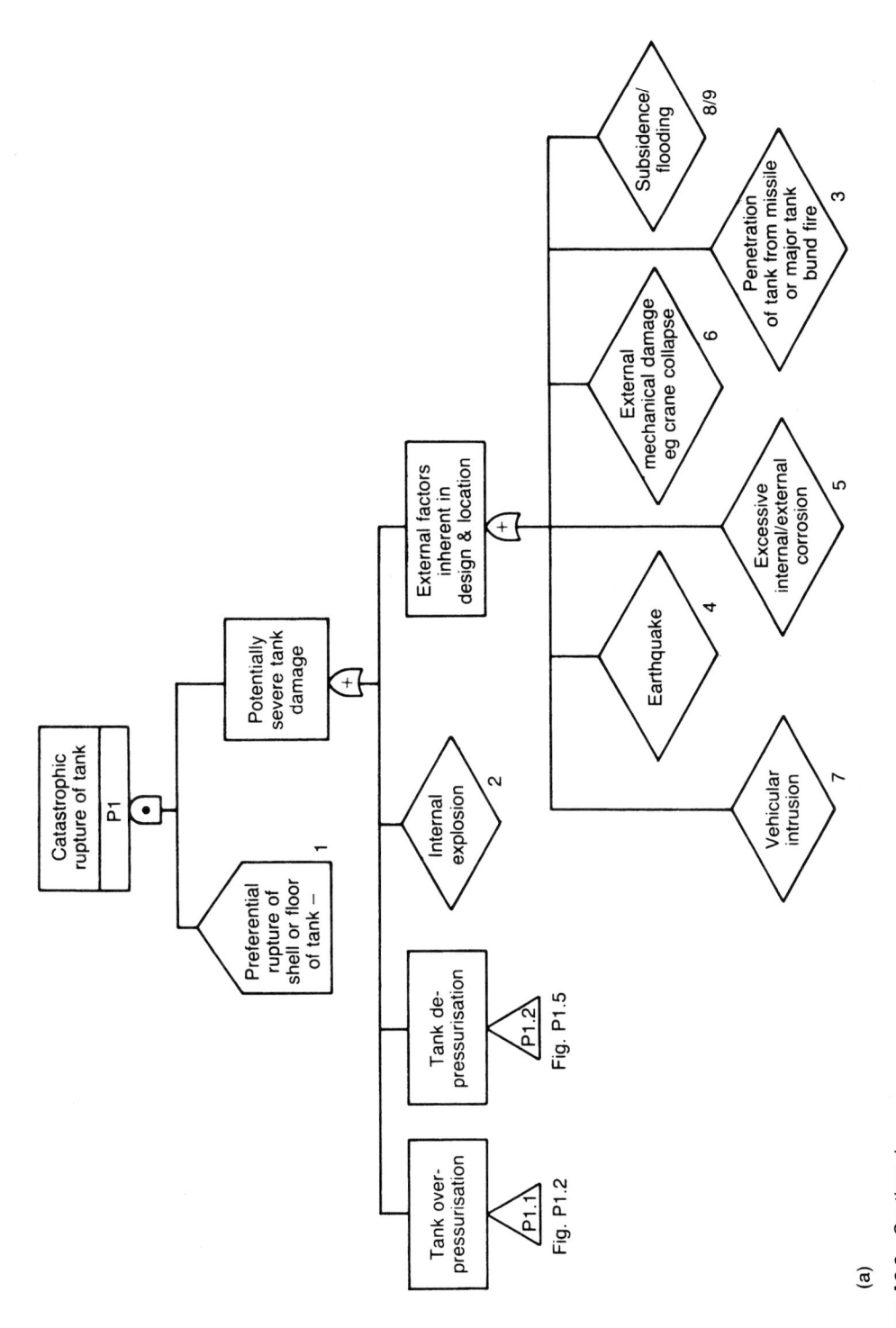

Figure A8.3 *Continued*

(a)

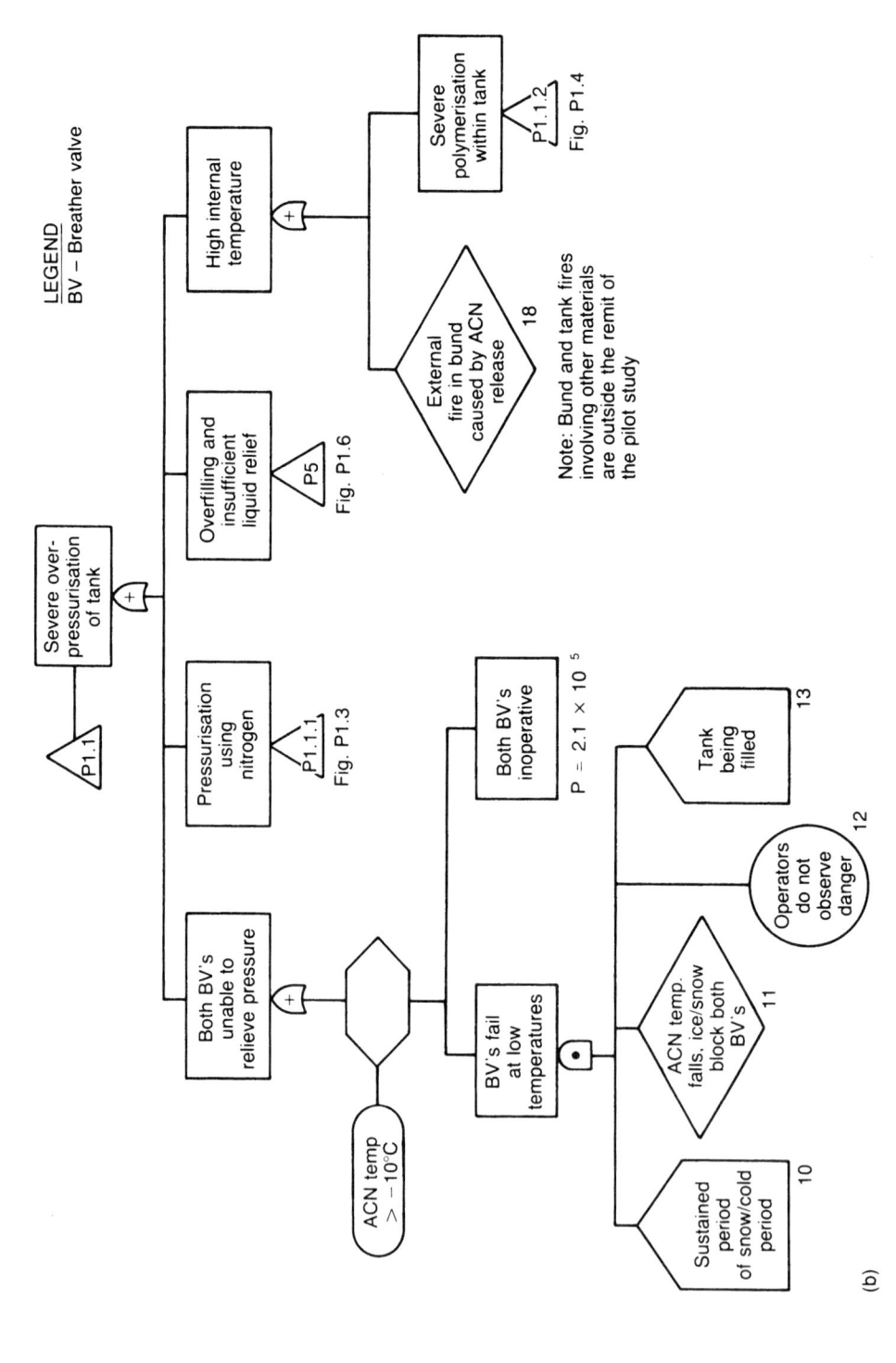

Figure A8.3 *Rijnmond Report: fault trees for release from the acrylonitrile storage (Rijnmond Public Authority, 1982. Reproduced by permission); (a) head of tree for top event 'catastrophic rupture or tank'; (b) sub-tree for 'tank overpressurization' and (c) sub-tree for 'tank depressurization'*

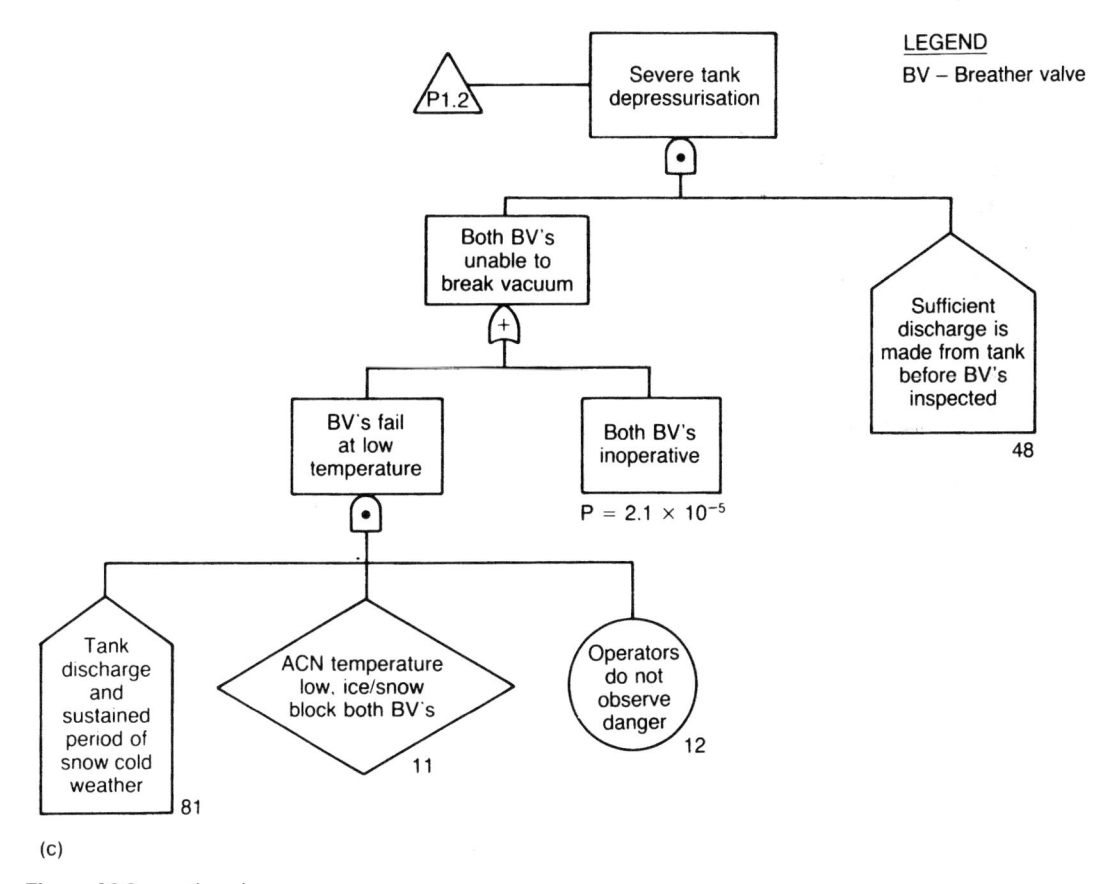

(c)

Figure A8.3   continued

Table A8.7   Rijnmond Report: summary of assessment of ammonia storage (UKF) (after Rijnmond
Public Authority, 1982)

A   Undesired events

Event		Frequency (events/year)	Mass flow or mass (kg/s or kg)	Duration (s)	Average no. of fatalities to public[a] (deaths/year)
U0	Catastrophic failure of sphere when full	2.3×10^{-7}	682,000	I	54×10^{-6}
U1	Catastrophic failure of sphere with average inventory	1.8×10^{-6}	250,000	I	67×10^{-6}
U2.1	Full bore fracture of liquid line during transfer of ammonia	5.6×10^{-7}	166	1,200	27×10^{-6}
U2.3	Full bore fracture of liquid line during transfer from another point	2.1×10^{-7}	83	3,000	11×10^{-6}
U2.4	Full bore fracture of liquid line while line is isolated	1.9×10^{-4}	83	86	0
U3	Full bore fracture of bottom connection on sphere	4.2×10^{-6}	36	6,950	38×10^{-6}
U4	Full bore fracture of top connection on sphere	4.4×10^{-6}	39	6,400	9×10^{-6}
U5	Major crack in sphere shell[b]	1.0×10^{-5}	9.2	>10,000	40×10^{-6}
U7	Full bore fracture of vapour line	1.4×10^{-5}	22	300	56×10^{-6}
Total					197×10^{-6}

Table A8.7   (continued)

B   Hazard scenarios

Sphere catastrophic failure: instantaneous release
Pipeline rupture: continuous release

C   Fault trees

U1   Sphere rupture
    U1.1 Support failure
    U1.2 Excess external heat
    U1.3 Excess of pressure
        U1.3.1 Overfilling with liquid
        U1.3.2 Chemically incompatible material introduced
    U1.4 Mechanical defect
        U1.4.1 Stress corrosion cracking
        U1.4.2 Corrosive material introduced from rail cars

[a] For these nine cases the typical hazard distances to the LTL_{50} are 2308, 1497, 671, 575, 241, 374, 23, 63, 87 m and to the LTL_{05} 2775, 1819, 893, 755, 327, 501, 100, 103 and 142 m respectively.
[b] Equivalent to a hypothetical 50 mm hole.

Table A8.8   Rijnmond Report: summary of assessment of chlorine storage (AKZO) (after Rijnmond Public Authority, 1982)

A   Undesired events

Event[a]	Frequency (events/year)	Mass flow or mass (kg/s or kg)	Duration (s)	Average no. of fatalities to public[b] (deaths/year)
A1.1 Catastrophic burst, full	0.93×10^{-7}	100,000	I	2.59×10^{-5}
A1.2 Catastrophic burst, half full	0.74×10^{-6}	50,000	I	7.47×10^{-5}
A2.1 Split below liquid, $\Delta p = 6.5$ bar	0.83×10^{-6}	63	790	1.03×10^{-4}
A2.2 Split below liquid, $\Delta p = 9$ bar	0.83×10^{-6}	76	660	1.03×10^{-4}
A3.1 Split above liquid level, $\Delta p = 6.5$ bar	0.83×10^{-6}	14	3500	0.82×10^{-4}
A3.2 Split above liquid level, $\Delta p = 9$ bar	0.83×10^{-6}	26	1900	0.82×10^{-4}
A4.1h Fracture of connection, full bore, $\Delta p = 6.5$ bar, angled horizontally	3×10^{-7}	253	200	3.7×10^{-5}
A4.1d Fracture of connection, full bore, $\Delta p = 6.5$ bar, angled down	3×10^{-7}	94	530	3.3×10^{-5}
A4.2h Fracture of connection, full bore, $\Delta p = 9$ bar, angled horizontally	3×10^{-7}	300	166	3.7×10^{-5}
A4.2d Fracture of connection, full bore, $\Delta p = 9$ bar, angled down	3×10^{-7}	110	450	3.3×10^{-5}
A4.3h Fracture of connection, small leak $\Delta p = 6.5$ bar, angled horizontally	6×10^{-6}	5.8	3600	9.1×10^{-5}
A4.3d Fracture of connection, small leak, $\Delta p = 9.5$ bar, angled down	6×10^{-6}	2.2	3600	4.3×10^{-5}
A4.4h Fracture of connection, small leak, $\Delta p = 9$ bar, angled horizontally	6×10^{-6}	6.9	3600	9.1×10^{-5}
A4.4d Fracture of connection, small leak, $\Delta p = 6.5$ bar, angled down	6×10^{-6}	2.6	3600	4.3×10^{-5}
A7.3 Full bore fracture of liquid line, valve open	2.4×10^{-6}	64/32	56/1500	9×10^{-5}
A7.4 Full bore fracture of liquid line, valve open, later shut	7.4×10^{-6}	64/32	56/44	5.8×10^{-5}
A11.5 Full bore fracture of liquid line, at end, valve open, later shut	5.9×10^{-5}	160/100	60/240	4.8×10^{-4}
A11.6 Full bore fracture of liquid line, small leak	1.2×10^{-3}	3.6/1.3	300/2300	1.8×10^{-3}
A15.1 Full bore fracture of vapour line, valves open, later shut	6.6×10^{-6}	15	300	1.8×10^{-4}
A17.2 Full bore fracture of liquid line, at end, valves open, later shut	8×10^{-6}	64/32	23/280	6.4×10^{-5}
Total (37 events)				3.57×10^{-3}

Table A8.8   (continued)

B   Hazard scenarios

Tank catastrophic failure: instantaneous release
Tank leak: mixed and continuous release
Pipework leak: mixed and continuous release

C   Fault trees

A1 Tank rupture
    A1.2 Pressure of water in tank
    A1.3 Internal explosion in tank
    A1.4a Bursting disc does not relieve
    A1.4b Source of overpressure potential
    A1.5 Overpressure due to liquid overfilling

[a] There are 17 further events. None of these events causes an average number of fatalities to the public greater than 10^{-5}/year.
[b] For these 20 cases the typical hazard distances to the LTL_{50} are 6500, 4900, 3400 ($\times$2), 2200 ($\times$2), 4700 ($\times$ 4), 1100 ($\times$4), 2100, 1600 ($\times$2), 1100 and 1600 m and to the LTL_{05} 8600, 6400, 4200 ($\times$2), 2800 ($\times$2), 6000 ($\times$2), 1400 ($\times$4), 2600, 2000 ($\times$2), 1400 and 2000 m, respectively.

Table A8.9   Rijnmond Report: summary of assessment of LNG storage (Gasunie) (after Rijnmond Public Authority, 1982)

A   Undesired events

Event	Frequency[a] (events/year)	Mass flow or mass (kg/s or kg)	Duration (s)	Average no. of fatalities to public[b] (deaths/year)
G1 Catastrophic failure of tank	0.8×10^{-6} 0.126	19×10^6	I	6.3×10^{-10}
G2 Failure of tank into earth bund	1×10^{-9} 0.144	19×10^6	I	3.5×10^{-12}
G3.1 Fracture of line (no pumping)	2.5×10^{-7} 0.051	60	35	4.5×10^{-11}
G3.2 Fracture of line, other place (no pumping)	8.1×10^{-7} 0.027	60	35	0
G4.1 Fracture of line (pumping)	1.5×10^{-8} 0.06	100	600	4.1×10^{-12}
G4.2 Fracture of line, other place (pumping)	4.8×10^{-8} 0.0585	100	600	0
Total				6.8×10^{-10}

B   Hazard scenarios

Inner tank failure: instantaneous release
Concrete containment failure: instantaneous release
Pipework rupture: continuous release (short)
Spreading of liquid
Vaporization of liquid
Pool fire
Flash fire

C   Fault trees

G1 Tank rupture
    G1.1 Other sources of overpressurization
    G1.2 Excessive underpressurization
    G1.3 Internal explosion

[a] For each case the probability of immediate ignition is 0.1 and that of ignition when dispersed is given by the second entry in the column.
[b] For these six cases the typical hazard distances for fatal effects from fire are 211, 174, 152, 210, 151 and 199 m, respectively. No hazard distances are quoted for fatal effects from explosion.

Table A8.10 *Rijnmond Report: summary of assessment of propylene storage (Oxirane) (after Rijnmond Public Authority, 1982)*

A Undesired events

Event[a]	Frequency[b] (events/year)	Mass flow or mass (kg/s or kg)	Duration (s)	Average no. of fatalities to public[c] (deaths/year)
0.01 Failure of full sphere	0.028×10^{-6} 0.96	600,000	I	1.43×10^{-7}
0.02 Failure of sphere containing 300 te	0.233×10^{-6} 0.94	300,000	I	1.52×10^{-6}
0.03 Full rate from sphere relief valve	2.0×10^{-5} 0.001	20	C	0
0.04 Failure of connection	3.5×10^{-7} 0.3	455	1300	2.2×10^{-7}
0.05 Failure of liquid line	1.3×10^{-6} 0.33	233	2570	5.3×10^{-7}
0.08 Failure of connection	1.1×10^{-6} 0.3	202	C	3.6×10^{-7}
0.09 Failure of liquid line	5.8×10^{-6} 0.15	104	C	1.5×10^{-6}
0.20 Failure of liquid line	2.4×10^{-5} 0.33	230	C	13.2×10^{-6}
0.21 Failure of liquid line	2.4×10^{-5} 0.13	230	C	9.0×10^{-6}
0.25 Failure of liquid line	3.6×10^{-6} 0.104	280	C	1.04×10^{-6}
0.26 Failure of liquid line	2.0×10^{-5} 0.17	104	C	2.1×10^{-6}
0.30 Failure of liquid line	1.1×10^{-5} 0.31	230	C	6.2×10^{-6}
Total (31 cases)				3.66×10^{-5}

B Hazard scenarios

Sphere failure: instantaneous release
Vaporizer shell failure: instantaneous release
Pipework failure: continuous release
Gas dispersion
Ignition probability
Flash fire
Explosion

C Fault trees

01 Sphere rupture
    01.1 Gross overstressing of shell
        01.1.1 Gas phase pressurization of sphere
            01.1.1.1 High temperature and pressure from jetty vaporizer
            01.1.1.2 Manual and power valves in circuit round spheres/vaporizer liquid and vapour lines all open
        01.1.2 Internal explosion in sphere
        01.1.3 Wrong material loaded
        01.1.4 Liquid overfilling
            01.1.4.1 Overfilling from ship
            01.1.4.2 Overfilling from rail car
    01.2 Mechanical defect in sphere shell
        01.2.1 Wrong material loaded (see tree for 01.1.3)

[a] There are 19 further cases. None of these events causes an average number if fatalities to the public >10^{-6}/year.

[b] For each case the probability of immediate ignition is 0.1 and that of ignition when dispersed is given by the second entry in the column.

[c] For these 12 cases the typical hazard distances for fatal effects from fire are 353, 338, 47. 125, 154, 105, 110, 138, 218, 268, 118 and 147 m, respectively. The hazard distances for explosion are given for overpressures of 0.3, 0.1 and 0.03 bar. For the former the distances are 207, 164, 28, 62, 74, 51, 56, 68, 88 and 102 m, respectively.

Table A8.11 *Rijnmond Report: summary of assessment of hydrodesulfurizer (Shell) (after Rijnmond Public Authority, 1982)*

A   Undesired events

Event	Frequency (events/year)	Mass flow or mass (kg/s or kg)	Duration (s)	Average no. of fatalities to public[a] (deaths/year)
S1.1 Major failure H_2S line, elevated, 1 min duration	3.2×10^{-6}	1.21	60	0
S1.2 Major failure H_2S line, elevated, 10 min duration	3.2×10^{-6}	1.21	600	0
S1.3 Major failure H_2S line, elevated, extended duration	7.2×10^{-7}	1.21	Long	0
S2.1 Major failure H_2S line, low level, 1 min duration	3.2×10^{-6}	1.21	60	0
S2.2 Major failure H_2S line, low level, 10 min duration	3.2×10^{-6}	1.21	600	0
S2.3 Major failure H_2S line, low level, extended duration	7.2×10^{-7}	1.21	Long	0
S5 Major failure at DEA line	1.6×10^{-6}	0.49	Long	0
Total				0

B   Hazard scenarios

Pipework failure:
  continuous release
Gas dispersion

C   Fault trees

None

[a] For these 7 cases the typical hazard distances to the LTL_{50} are 0, 19, 59, 81, 144, 153, and 64 m and to the LTL_{05} 0, 85, 96, 99, 173, 181 and 81 m, respectively.

Table A8.12 *Rijnmond Report: individual and average annual risk for six installations (Rijnmond Public Authority, 1982)*

Installation	Average annual fatalities		Individual risk employees
	Employees	Public	
Acrylonitrile	2.1×10^{-3}	7.9×10^{-6}	6.6×10^{-6}
Ammonia	2.1×10^{-3}	2.0×10^{-4}	20×10^{-6}
Chlorine	1.1×10^{-2}	3.6×10^{-3}	5.1×10^{-5}
LNG	1.5×10^{-7}	6.8×10^{-10}	5.7×10^{-9}
Propylene	1.1×10^{-4}	3.7×10^{-5}	7.7×10^{-7}
Hydrodesulfurizer	1.0×10^{-6}	0	2.1×10^{-9}

For cases S1–S2, which involve direct release of hydrogen sulfide, it was found that the inventory of gas, which is at a gauge pressure 0.5 bar, was too small to sustain a release of any significant duration and that for these cases the limiting flow was determined by the rate of evolution of the gas, which was 4370 kg/h. This is equivalent to discharge from hole of a 75 mm diameter.

All the emissions are continuous sources.

An ignition probability of 0.2 on release was assumed.

For toxic gas hazard, therefore, the exposure is continuous. In all cases only employees were at risk.

Typical hazard distances to the LTL_{50} are given in Table A8.11, the greatest distance there quoted being 153 m. The distances are greater for adverse weather conditions, being for case S2.3 on the centreline for 2.0/F conditions 395 m.

A8.9 Assessed Risks

The results of the assessment are presented in a number of ways. These include

Risk contours
Risk to employees:
  Individual risk
Risk to public:
  Individual risk
  Societal risk
  FN tables
  FN curves
Average annual fatalities

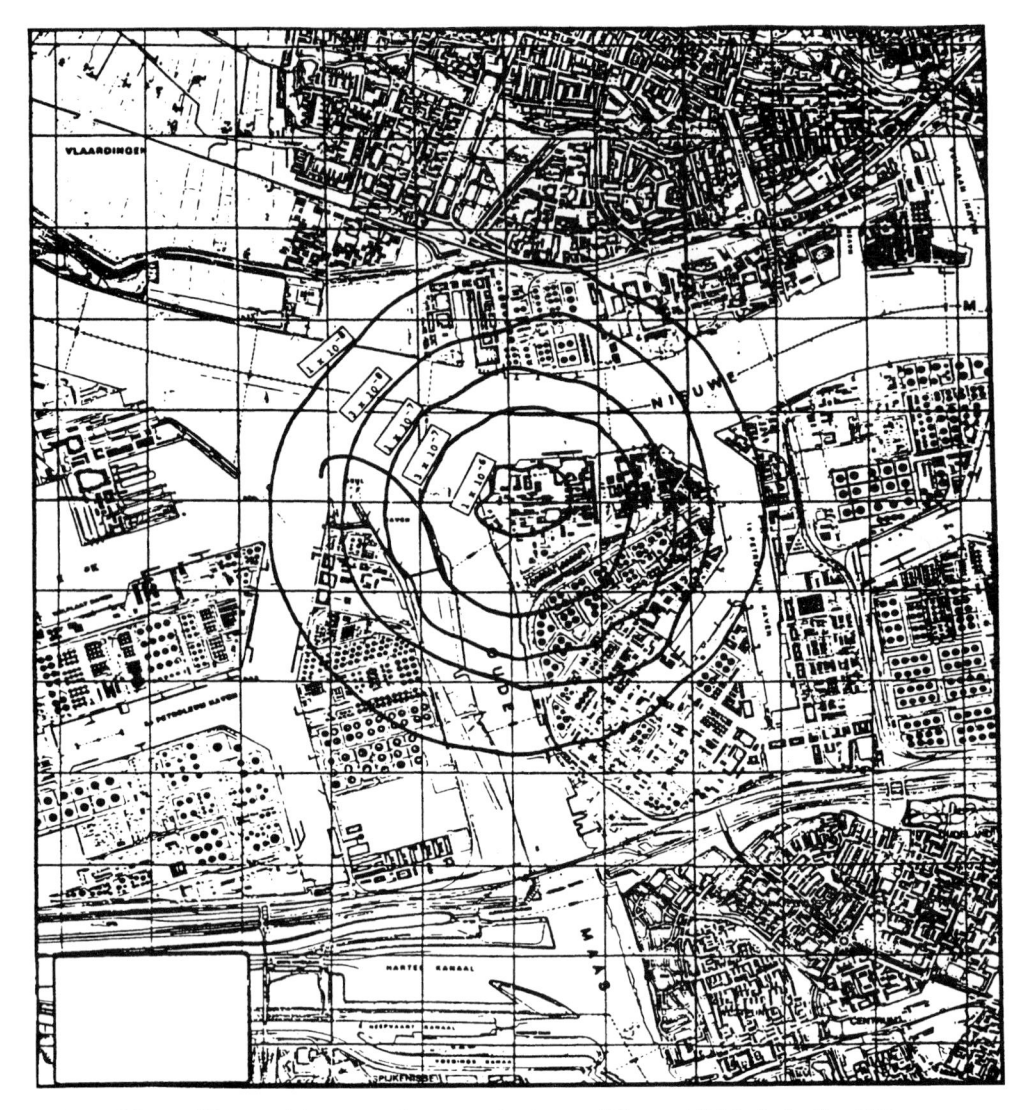

Figure A8.4   *Rijnmond Report: risk contours for ammonia storage (Rijnmond Public Authority, 1982. Reproduced by permission)*

Table A8.13   *FN table for public for chlorine storage in Rijnmond Report (after Rijnmond Public Authority, 1982)*

No. of deaths	Cumulative frequency	No. of deaths	Cumulative frequency
1	4.00×10^{-4}	125	2.09×10^{-6}
10	1.07	151	1.62
20	3.70×10^{-5}	200	9.86×10^{-7}
30	1.50	304	3.06
40	9.46×10^{-6}	404	2.34
50	7.32	518	1.26
60	6.34	738	4.85×10^{-8}
70	5.23	1027	3.40
80	4.31	1697	7.50×10^{-10}
90	3.23		
100	2.87		

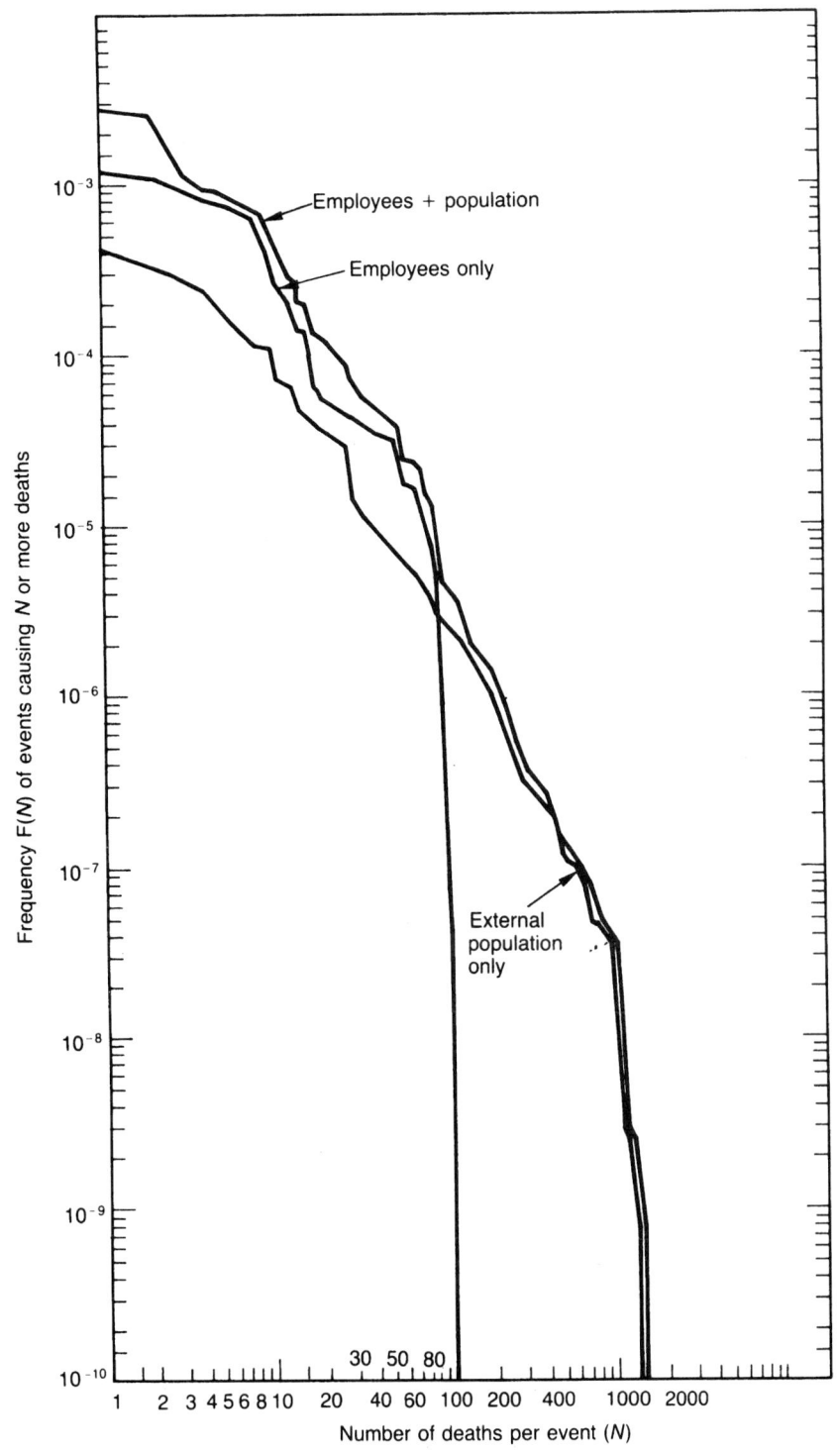

Figure A8.5 *Rijnmond Report: FN curve for chlorine storage (Rijnmond Public Authority, 1982.
Reproduced by permission)*

The individual risks and average annual fatalities are shown in Table A8.12.

The use of risk contours is illustrated here by one of those for the ammonia storage shown in Figure A8.4, while that of FN tables and of FN curves is illustrated by those for the chlorine storage shown in Table A8.13 and Figure A8.5, respectively.

A8.10 Remedial Measures

For each installation recommendations were made for remedial measures to reduce the risks. In accordance with the remit of the study the proposals made were illustrative rather than exhaustive and are suggestions for further evaluation.

For the AN storage the main measure proposed is nitrogen blanketing to prevent internal explosion. It is recognized, however, that there is a contrary view that lack of oxygen may make spontaneous reaction more likely. It was estimated that inciting would reduce the frequencies of events P1 and P3 to 0.1×10^5 and 0.4×10^5/year, respectively. Some 17 other measures are also listed, including hose support, tanker earthing and changes to sampling methods.

For the ammonia storage the principal measure suggested is the provision of a bund. This would do little to retain material released in a catastrophic failure of the sphere, but it might well mitigate the effect of failure of a bottom connection.

Another proposal made is the provision of an emergency air supply at the control room to allow the operators time to close emergency isolation valves.

For the chlorine storage the main recommendations are directed to reduction of the size of escapes from pipework by the use of excess flow valves.

For the LNG storage the main measure proposed is directed towards the prevention of escalation due to brittle fracture of carbon steel from a small release of cold liquefied gas. It is suggested that baffles on the tank roof would reduce this risk. It was estimated that this would reduce the total risk by 5.6%.

For the propylene storage the installation of remote isolation valves is the principal measure proposed. It was estimated that this would reduce the total risk by half. Proposals are also made for modifications to the vapour return system.

For the hydrodesulfurizer the measures proposed are minor. It is suggested that there should be monitoring of hydrogen sulfide in the control room and that the hazard of spread of DEA through the drain system be investigated.

A8.11 Critiques

As indicated, the *Rijnmond Report* itself contains its own critique in the form of a review by outside consultants and industrial comment. Much of this relates to the failure and event data, the hazard models and the injury relations and particularly to the models.

A8.12 Notation

n_h number of weeks off for holidays, sickness, etc.
N_d number of persons on site during day
N_e number of persons on site
N_n number of persons on site at other times

Appendix

Laboratories

9

Laboratories are an integral part of the activities of the process industries. They include not only analytical laboratories and laboratories carrying out research and development work in industry but also university teaching laboratories where future industrial managers are trained. Pilot plants are considered in Appendix 10.

Accounts of laboratories and of laboratory safety are given in *Safety in the Chemical Laboratory* by Pieters and Creighton (1957), *Safety in University Laboratories* by the AUT (1967), *Hazards in the Chemical Laboratory* by Muir (1971) and later Bretherick (1999, 2000), *Handbook of Laboratory Safety* by Steere (1971), *A Guide to Laboratory Design* by Everett and Hughes (1975), *Prudent Practices for Handling Hazardous Chemicals in Laboratories* by NRC Staff (1981), *Introduction to Safety in the Chemical Laboratory* by N.T. Freeman and Whitehead (1982), *Handbook of Laboratory Waste Disposal* by Pitt and Pitt (1985), *Improving Safety in the Chemical Laboratory* by J.A. Young (1987a) and *Safety in Chemical Engineering Research and Development* by V.C. Marshall and Townsend (1991) (the IChemE *Laboratories Guide*).

A *Code of Practice for Chemical Laboratories* (the RIC *Laboratories Code*) has been published by the Royal Institute of Chemistry (RIC) (1976).

Activities in a chemical laboratory are distinguished from those on an industrial plant, above all, by their small scale. The smaller quantities of species in the laboratory can reduce the overall consequence of an event. In addition, heat losses are relatively greater at the laboratory scale. This certainly reduces the hazards. There are, however, factors which may increase the risk of a hazardous initial event, such as the use of inexperienced and untrained personnel, and of escalation of that event, such as the enclosed space and the close proximity of personnel. Further, if the small scale of the hazard leads to general complacency and poor practice, the risk that it will be realized is augmented.

Despite the lesser scale of the hazard, the approach to hazard control used for full-scale plants is applicable in large part to activities in the chemical laboratory. An accident in the laboratory can give rise to considerable personal injury or loss of life and direct damage loss, because the density of personnel and of expensive equipment which are at risk is relatively high. There may also be serious consequential loss due to interruption of the facilities which the laboratory provides for production plants or of research and development activities.

Moreover, where a laboratory, whether in the university, government, or industry, has a training function, it is essential to adhere to, and instill, good practice in hazard control.

Many of these topics touched on here have already been discussed, albeit not specifically in relation to laboratories, in previous chapters, in particular those on legislation (Chapter 3), management and management systems (Chapter 6), hazard identification (Chapter 8), reactive hazards (Chapter 33), process design (Chapter 11), fire, explosion and toxic release (Chapters 16–18), personal safety (Chapter 25) and safety systems (Chapter 28).

Selected references on laboratory safety are given in Table A9.1.

A9.1 Legal Requirements

In the United Kingdom the Factories Act 1961 covers industrial workplaces, including laboratories, while the

Table A9.1 *Selected references on laboratory safety*

BSC (n.d./4); Kintoff (1939); Pieters and Creighton (1957); Quam (1963); H.H. Fawcett (1965e, 1982e); C.H. Gray (1966); AUT (1967); Everett (1969, 1979, 1980 LPB 32); A.F.H.Ward (1969); BDH (1970, 1977a–c); MCA (1970/19, 1972/21); Imperial College Safety Committee (1971); Muir (1971); Steere (1971); Stephenson (1972 MCA/21); Ciba-Geigy Ltd (1974); Fairbairn (1974 SRD R8); Weekman (1974); Everett and Hughes (1975); Bremner (1976); Cooke (1976); RIC (1976); Seidenberger and Barnard (1976); Weissbach (1976); Anon. (1977 LPB 13, p. 14); Anon. (1977 LPB 18, p. 17); V.C. Marshall (1977c, 1979b, 1980 LPB 32, 1981 LPB 37); Anon. (1978 LPB 20); Curry (1978); Dept of Health and Social Security (1978, 1979); Pitt (1978); J.M. Reynolds (1978); Buttolph (1979, 1980 LPB 32); Cohn, linnecar and Ryall (1979); Fowler (1979, 1980 LPB 32); G.F. Martin (1979, 1980 LPB 32); Napier (1979b, 1980 LPB 32); Pogany (1979); Spencer (1979); Penninger and Okazaki (1980); Anon. (1981 LPB 39, p. 11); Bretherick (2000); G. Marshall (1981 LPB 39); NRC Staff (1981); N.T. Freeman and Whitehead (1982); Pitt and Pitt (1985); B. Miller *et al.* (1986); Armour (1987); Baum and Diberardinis (1987); Bulloff (1987); Gerlach (1987); Hoyle and Stricoff (1987); Prokopetz and Walters (1987); Reale and Young (1987); Redden (1987); Saunders (1987); Springer (1987); Varnerin (1987); Williamson and Kingsley (1987); J.A Young (1987a–c); ACGIH (1988/20, 1990/40, 43, 44, 1991/61, 69, 1992/77); Ciolek (1988); Le, Santay and Zabrenski (1988); Lunn and Sansone (1990); Stricoff and Walters (1990); Kalis (1991); V.C. Marshall and Townsend (1991); NFPA (1991 NFPA 45); Woodhead (1992); Bartlett, Hall and Gudde (1993); Senkbeil (1994) BS (Appendix 27 *Laboratory Equipment*)

Fume cupboards, enclosures

Walls and Metzner (1962); BOHS (1980 Monogr. 4, 1987); J. Grant and Rimmer (1980); ASHRAE (1985 ASHRAE 110); OSHA (1990/6); V.C. Marshall and Townsend (1991); AIHA (1993 Z95) BS 7258: 1990–

HSWA 1974 applies to all places of work and thus extends coverage to university as well as industrial laboratories.

Chapter 3 gives an account of these acts and also of other legislation relevant to laboratories such as that on the workplace, including management, work equipment and incident reporting; pressure systems; toxic substances and occupational hygiene; flammable and explosive materials; fire; ionizing radiations; electricity; and personal safety, including safety signs, machine guarding, compressed air and compressed gases, manual handling, protective equipment, eye protection and first aid.

A review of legislation applicable to chemical engineering laboratories is given in the IChemE *Laboratories Guide*.

A9.2 Laboratory Management Systems

The management approach used for the control of hazards in industry is in large part applicable in chemical laboratories. In particular, use should be made of formal systems and procedures as described in Chapter 6.

The laboratory should have a management system with suitable organization and competent people, systems and procedures, standards and codes of practice, and documentation. There should be a clear chain of command and separation of executive from advisory, or staff functions. There should be a safety officer and a safety committee, and safety audits should be conducted.

Some principal systems are described below, and others in later sections.

A9.2.1 Hazard reviews
There should be a system of hazard reviews for experimental work. A review should give a statement of the hazards involved and of the means by which they will be controlled.

A9.2.2 Permit systems
Hazardous activities should be covered by a permit system. Permits may be required for general maintenance activities, for hot work and for vessel entry. The IChemE *Guide* gives example permit forms.

A9.2.3 Incident reporting
There should be a system for the reporting of incidents. As a minimum this should meet the legal requirements given in RIDDOR (1985), but in addition there should be a general requirement for the reporting of unintended events.

A9.2.4 Safety audits
The system of safety audits should cover the management systems for the laboratory, the equipment, the services, storage and waste disposal, and the operations.

The conduct of safety audits is discussed in the IChemE *Guide* and by Bretherick (2000) and J.A. Young (1987b).

A9.3 Laboratory Personnel

There tend to be wide variations in the capability of the personnel who work in laboratories, ranging from well-trained and experienced permanent staff to relatively inexperienced research workers and students. A discussion of people in laboratories is given by Buttolph (1980 LPB 32).

A9.3.1 Training
It follows from the above that for laboratory personnel training assumes particular importance. This training needs to cover the hazards, the equipment, the procedures and the systems. As with all training, it should aim to motivate as well as to inform.

Personnel who use the laboratory should be trained in good laboratory practice. The training should include such aspects as action to take in case of fire and other accidents, methods of eye irrigation and use of fire extinguishers.

The laboratory should have the appropriate number of persons trained in first aid.

Where the personnel concerned are prospective managers of industrial plants, experience in a well-conducted laboratory provides a good foundation.

Accounts of the content of, and techniques for, laboratory training are given by Bretherick (2000), Armour (1987) and Redden (1987).

A9.4 Laboratory Codes

A9.4.1 RIC *Laboratories Code*
The RIC *Laboratories Code* covers the following aspects:

(1) Laboratory services
   (a) electricity
   (b) gas for heating
   (c) water
   (d) steam
   (e) ventilation
   (f) compressed air
   (g) brine circulation
   (h) hot plates, ovens and furnaces
   (i) drainage
(2) Hazardous chemicals
   (a) flammable substances
   (b) explosive substances and mixtures
   (c) corrosive chemicals
   (d) irritant chemicals, dusts, gases and vapours
   (e) toxic chemicals
   (f) radiation and radioactive materials
   (g) cylindered gases
   (h) cryogenic fluids and systems
(3) Hazardous techniques
   (a) pressure reactions
   (b) vacuum techniques
   (c) glass handling
   (d) heating of chemical apparatus
   (e) centrifugation
(4) Organization
   (a) safety officers and safety committees
   (b) appointment and functions of a safety committee
   (c) fire prevention
   (d) personal protection
   (e) rescue
   (f) first aid and medical services
   (g) housekeeping and safety audits
   (h) labelling
   (i) storage
   (j) waste disposal
   (k) noise
   (l) unattended experiments and working alone
   (m) documentation of procedures
(5) Legal responsibility
   (a) legal responsibility
   (b) legislation

A9.4.2 NFPA 45
Guidance on fire protection is available in NFPA 45: 1986 *Fire Protection for Laboratories using Chemicals*.

The standard covers (1) general topics, (2) laboratory unit hazard classification, (3) laboratory unit design and construction, (4) fire protection, (5) explosion hazard protection, (6) laboratory ventilating systems and hood requirements, (7) chemical storage, handling and waste disposal, (8) compressed and liquefied gases, (9) laboratory operations and apparatus and (10) hazard identification.

The classification divides laboratories into high, intermediate and low hazard according to the maximum densities of materials held. For flammable or combustible liquids Class I, II or IIIA, these densities are

Laboratory unit class	Flammable or combustible liquid class	Maximum quantity/ 100ft^2 of unit (US gal)	
		Excluding storage	Including storage [format]
A High hazard	I	10	20
	I, II and IIIA	20	40
B Intermediate hazard	I	5	10
	I, II and IIIA	10	20
C Low hazard	I	2	4
	I, II and IIIA	4	8

The storage referred to is storage in cabinets and cans. The standard states that Class A laboratory units should not be used as instructional laboratories and that for Class B and C instructional laboratories the maximum quantities should be half those given above.

An account of the use of NFPA 45 in laboratory design is given by Le, Santay and Zabrenski (1988).

A9.4.3 IChemE *Laboratories Guide*
The IChemE *Laboratories Guide* has the following coverage: (1) introduction, (2) safety management of laboratories, (3) the law and process laboratories, (4) hazards in chemical engineering laboratories, (5) designing process research and development facilities, (6) services, (7) storage of materials and disposal of waste, (8) design of experiments, (9) operating procedures and safety, (10) emergency procedures, (11) maintenance and modifications and (12) relevant publications.

A9.5 Laboratory Hazards

Reviews of hazards in chemical laboratories are given in the RIC *Code* and the IChemE *Guide* and in most texts of laboratory safety such as those of Steere (1971), Bretherick (2000) and J.A. Young (1987a). The list quoted from the RIC *Code* in Section A9.4 is typical.

In addition to the specific treatments mentioned below, a general account of the handling of hazardous research chemicals is given by Prokopetz and Walters (1987) and one of hazards other than those from chemicals by Bulloff (1987).

Further treatments of the hazards of flammable substances are given in Chapter 16, of toxic substances in Chapter 18 and of others such as radiation, electrical and mechanical hazards in Chapter 25.

A review of the causes of incidents in laboratories is given by J.A. Young (1987c), who utilizes the following headings: (1) management responsibilities inadequate, (2) lack of communication, (3) lack of personal responsibility, (4) lack of proper ventilation, (5) lack of protective equipment, (6) personal hygiene problems, (7) electrical hazards, (8) storage problems and (9) inadequate emergency procedures and equipment.

A9.5.1 Reactive substances
Where reactive substances are handled or processed, every effort should be made to obtain the fullest information on their behaviour. Texts such as *Handbook of Reactive Chemical Hazards* by Bretherick (2000) , Sax's *Dangerous*

Properties of Industrial Materials (2000), and *Hazardous Chemicals Handbook* by Carson and Mumford (1994) are an essential point of departure.

Bretherick (2000) has also provided a review of chemical reactivity in the laboratory context.

A9.5.2 Flammable substances
Many of the liquids and gases handled in laboratories are flammable. Each phase presents its characteristic hazards.

Accounts of the hazards of flammable and combustible materials are given in the various NFPA codes, including for laboratories NFPA 45 and by Steere (1971), Bretherick (2000) and Gerlach (1987).

Where flammable liquids are handled, one hazard is release followed by ignition. Factors which need to be considered are the integrity of the containment, the flow of liquid if it does escape and any sources of ignition.

Another hazard is overheating of the liquid, leading to its escape, either as a liquid following failure of the containment or as vapour from an open vent.

A frequent cause of bench fires is loss, or under-circulation, of cooling water.

Flammable liquids should not be handled near, or heated on, naked flames and should also be kept away from arcing/sparking devices.

Where flammable gases or vapours are handled, the most common hazard is a leak which allows a flammable atmosphere to build up in the enclosed laboratory space. Unless there is a high level of ventilation, this can occur even with a relatively small leak rate.

Another cause of bench fires is contact between water and reactive metals such as sodium.

Oxygen should not be used in such a way as to create the hazard of an oxygen-enriched atmosphere. Oxygen is liable to oxidize and then rupture rubber tubing. There is also the possibility of an 'oxygen fire' in which the metal can be ignited (arising, for example, from adiabatic compression, particularly in non oil-free lines as described in NFPA 53).

A9.5.3 Toxic substances
Where toxic substances are handled, it is necessary to consider all three modes of entry into the body and both short- and long-term toxic effects.

For entry, inhalation may well be the principal mode, but in laboratory work in which chemicals are 'handled' to a much greater extent the ingestion and skin modes may be more significant than is usually the case. High standards of hygiene therefore assume greater importance.

With regard to inhalation, there may be a hazardous short-term effect if the substance is highly toxic and circumstances can be foreseen in which a concentration sufficiently high to cause acute poisoning can occur. This is rather more likely in the enclosed space of a laboratory.

More commonly, however, the hazard arises from exposure to low concentrations over a relatively long period.

Where a toxic hazard exists, the requirements of the Control of Substance Hazardous to Health (COSHH) Regulations 1988 apply.

The design countermeasures are primarily a high standard of containment. Other measures are described in the next section.

Accounts of the effects of toxic chemicals in the laboratory context are given by Steere (1971), Bretherick (2000) and Springer (1987).

A9.5.4 Radiation hazards

There are a variety of radiation hazards, most of which may be found on occasion in laboratories. They include both ionizing and non-ionizing radiations as described in Chapter 25.

With regard to ionizing radiation, some devices containing radioactive sources instanced in the IChemE *Guide* are certain liquid level gauges, gas chromatograph detectors, leakage detectors, anti-static devices on balances, and fire detectors.

Certain apparatus producing voltages above 5 kV may be a source of X-rays.

Non-ionizing radiation sources include lasers, microwaves and UV and IR devices.

A9.5.5 Electrical hazards

The electrical hazards are little different from those in industrial situations generally, but unless good practice is observed a laboratory tends to be especially vulnerable to them.

Personnel may be at risk of electrocution from temporary lashups and repairs, defective and damaged cabling, unearthed components and poor earthing. This may be aggravated by handling equipment with wet hands.

Short circuits due to worn cabling, wrong wiring or over-rated fuses can lead to a fire.

A9.5.6 Mechanical hazards

The mechanical hazards likewise are such as may be found in other industrial situations, but again laboratory personnel may be especially vulnerable unless strict controls are observed.

These mechanical hazards include those associated with rig equipment, workshop machinery, hand and power tools, lifting equipment, rotating equipment, and pinching equipment.

Accidents are liable to occur where laboratory personnel use equipment with which they are unfamiliar in order to progress the job.

A9.5.7 Operating condition hazards

Equipment operating at high or low temperature constitutes a contact hazard. Contact with a very cold surface can cause a 'cold burn' not unlike the burn from a very hot surface.

A cryogenic fluid also involves the hazard that it will vaporize by heat transfer from the atmosphere unless this is positively prevented. Depending on the nature of the fluid, such vaporization may give rise to an atmosphere enriched with a flammable or toxic material or oxygen.

A laboratory usually contains a number of sources of high pressure such as compressed air, steam, compressed gas in cylinders and so on. Water also may act as a high pressure source. Rupture of equipment may occur either because the pressure to which it is subjected is above the design conditions, or because the equipment is weaker than designed. Overpressure can occur if a valve is open or fails, weakening if metal equipment is exposed to extremes of heat, cold or thermal cycling.

Failure may also be caused by vacuum, which may occur not just by connection to a vacuum supply but by rapid cooling of a vapour in the apparatus.

Rupture by a pressurized vapour/gas or by a superheated liquid, however, would have more severe consequences than by a subcooled liquid.

A9.5.8 Water release hazards

Another hazard is the escape of water, particularly in the form of a jet. Some potential effects of such a jet are short circuiting, thermal shock, extinction of gas jets and reaction with water-reactive chemicals.

A9.6 Laboratory Design

Laboratory design and layout is discussed in the IChemE *Guide* and by Steere (1971), Everett and Hughes (1975) and Baum and Diberardinis (1987).

A checklist for laboratory design is given by Everett (1980 LPB 32) and a case study in such design by Martin (1980 LPB 32).

A9.6.1 Laboratory layout

The design and layout of a laboratory should proceed on principles broadly similar to those described in Chapter 10. This is the approach taken in the IChemE *Guide*.

The method described in the guide is to analyse the needs of the experimental activities using a flow diagram showing the flow of materials between the experimental rigs, the workshops, stores, analytical services, waste disposal facilities, etc., and to develop layout diagrams in which the low and high hazard features are separated, with minimization of exposure near the latter.

A9.6.2 Toxic chemicals

The hazard from toxic chemicals has been described in the previous section.

Where toxic chemicals are handled, the design should aim to keep the concentrations in the laboratory environment below the relevant exposure limits. The prime means of ensuring this is to prevent leaks through a high standard of containment.

For toxic chemicals the COSHH Regulations 1988 apply as described in Chapter 25. These regulations include provisions for monitoring the workplace atmosphere.

Means of keeping down the concentration of contaminants which do escape include use of ventilation and of fume hoods.

A9.6.3 Ventilation

The most common method of controlling the concentration of contaminants in the workplace is ventilation. The basic methods are local exhaust ventilation and general ventilation. Ventilation of the workplace is described in Chapter 25.

The exhaust from the ventilation should pass to a safe place (e.g. away from air intakes).

A9.6.4 Fume hoods

For some toxic or noxious chemicals use is made of a fume cupboard. An account of fume hoods is given in Chapter 25.

The resort to fume hoods has been criticized, however, by V.C. Marshall (1981 LPB 37) as outmoded. He argues that in the laboratory the release of noxious chemicals has become too readily accepted. The proper approach is higher standards of containment which avoid such emissions.

A9.6.5 Laboratory support

There are a number of facilities which may be needed to support the laboratory. They include (1) workshop, (2) stores, (3) receipt bays, (4) analytical services and (5) staff facilities. These are essentially self-explanatory.

The storage needs should be examined to identify the extent to which segregation is required. It is normal to devote one store to the materials required for the fabrication of rigs and another to process materials such as flammable and toxic liquids, but further segregation may be required.

The IChemE *Guide* suggests, for example, that separate stores are normally required for solvents, different classes of chemicals, explosives, gas cylinders and cryogenic materials.

Eating and drinking should not be permitted in laboratories. There need to be adequate staff facilities provided.

A9.7 Laboratory Equipment

A9.7.1 Glassware
Much laboratory apparatus is constructed in glass. Glass has a number of advantages such as resistance to corrosion, flexibility in use and visibility of contents. It can be used with a high degree of safety. For this it is necessary that its limitations be respected, that it be used only in suitable applications and that manufacturers guidance be consulted.

A9.7.2 Hotplates, ovens and furnaces
Hotplates, ovens and furnaces involve the hazard that an item being heated, or the heating element itself, becomes overheated.

The use of ovens is frequently not subject to the same degree of formal control as other apparatus, and different people may put in items which are incompatible or may alter the controls, with undesirable results.

A9.7.3 Centrifuges
Laboratory centrifuges are a recognized hazard. They are the subject of BS 4402: 1982 *Specification for Safety Requirements for Laboratory Centrifuges* and are treated in Appendix 6 of the IChemE *Guide*. This is in addition to BS 767 and a further IChemE guide both of which apply to industrial centrifuges.

A laboratory centrifuge should be properly balanced. The buckets should be well fitting and with rubber pads in place. The machine should be cleaned and dried after use to prevent corrosion which is a common cause of mechanical failure.

In use the lid should be closed and locking devices secured before start-up. The machine should be brought up to speed gradually. It should be stopped using the controls, not simply switched off. The lid should not be opened until the machine has come to rest.

Centrifuges are discussed further in Chapter 11.

A9.8 Laboratory Services

Laboratories use a wide range of services. The most common are (1) water, (2) steam, (3) compressed air, (4) fuel gas and (5) electrical power. Others include refrigerated coolants, vacuum, oxygen and other piped gases.

A9.8.1 Water
Water is used as process water and as coolant.

Where the cooling load is larger enough to justify it, an external cooling water system may be provided which is a smaller-scale version of an industrial system with the water pumped through cooling towers.

The use of water in laboratories involves several hazards. The first is the potential for reverse flow of liquid from a laboratory process into a water supply main. Where such a possibility exists, some safeguard is required. A break tank provides a barrier more dependable than a non-return valve.

Another hazard is loss of cooling water. Generally this occurs on a single rig due to a leak at that point.

A third hazard is the escape of water into the laboratory. A typical situation might be failure of a cooling water connection on apparatus in a fume hood where the drains are blocked, so that the water flows into the laboratory. If the equipment is running unattended, a large quantity of water can escape, perhaps damaging equipment such as computers on the floor below.

There is often a requirement for demineralized water, which is supplied by a stand-alone unit.

A9.8.2 Steam
Steam is required in the laboratory for heating on larger rigs. It is typically provided by an external boiler system, but for small quantities may be generated by a small electrically heated, or less commonly, gas-fired, unit.

The pressure of steam for laboratories is typically in the range 4–10 barg. Measures should be taken to ensure that equipment supplied with such steam cannot be over-pressured and that, where loss of coolant with steam still on could be hazardous, steam is shut off.

The use of steam implies the use of steam traps. These devices need regular maintenance.

A9.8.3 Compressed air
Compressed air for laboratories is generally provided by a fixed piped system supplied from an external air receiver charged by a compressor.

The air should be free of oil, which may be achieved either by use of an oil-free compressor or by oil-removal equipment. Regular maintenance is required for the latter.

Compressed air is potentially hazardous and the supply pressure should be limited to that required for the purposes of the laboratory.

Like steam, compressed air is a potential source of over-pressure and measures are needed to safeguard against this.

A9.8.4 Fuel gas
Fuel gas is another service which is usually provided in a laboratory by a fixed pipe system.

A9.8.5 Refrigerated coolants
Cooling to lower temperature that can be achieved by cooling water may be provided using refrigeration, either by circulation of a refrigerant or by the use of an intermediate fluid such as brine or ethylene glycol.

The normal practice is to house the refrigeration set in a separate room.

A9.8.6 Vacuum
In some laboratories a piped system is provided to supply vacuum. Collapse of equipment under vacuum is not unlike an explosion and may generate missiles. Measures are required to ensure that equipment connected to vacuum is not under pressure. Also, backflow prevention should be installed.

A9.8.7 Oxygen
Another gas which may be provided as a fixed piped supply is oxygen, but where this is done special care is needed.

Hazards arising from the use of oxygen are attack and rupture of rubber tubing and creation of an oxygen-enriched atmosphere, which can result in a dramatic enhancement of the flammability of clothing.

A9.8.8 Other piped gases

Laboratory requirements for gases other than air vary greatly. In principle, where the usage is high, a fixed piped system has the very considerable advantage of eliminating the use of temporary facilities. However, frequently what is required is moderate quantities of a relatively large number of gases, so that a piped supply for any one is difficult to justify. In some instances pipes are used for different gases, with suitable precautions.

Where a piped supply is installed, the gas is usually piped from gas cylinders located at a protected shelter on the outside of the building.

Piped systems for hydrogen require special care.

Lines should be carefully labelled and flow restrictors should be installed.

A9.8.9 Compressed gas cylinders

For the reasons just given, supply of many of the gases used is often taken from gas cylinders brought into the laboratory. The use of gas cylinders requires a number of precautions.

The contents of the cylinder should be identified by labelling and colour coding. In use the cylinder should be fitted with a regulator and both valve and regulator should be kept clean and free of grease and oil. The valve should be closed when the cylinder is not in use. The cylinder should be transported in a wheeled carrier and when in use supported in a vertical position.

A more detailed treatment of precautions in the use of gas cylinders is given in Appendix 4 of the IChemE *Guide*.

Cylinders should be secured (e.g. with chains) when in use or transported.

A9.9 Laboratory Storage and Waste Disposal

A9.9.1 Materials storage

The storage of materials for the laboratory is discussed in the IChemE *Guide* and by Steere (1971), Bretherick (2000) and Williamson and Kingsley (1987).

The main inventory of hazardous chemicals is, or should be, in the chemical storage locations. General principles governing such storage concern (1) segregation of incompatible materials, (2) types of containers, (3) receipt points, (4) acquisition, stock-taking and disposal, (5) minimization of inventory and (6) identification, ownership, safety and health information and labelling.

With regard to segregation, some relevant categories of chemical are (1) flammables, (2) toxics, (3) oxidizing agents, (4) corrosive substances, including strong acids and alkalis, (5) cryogenic substances and (6) pyrophoric substances. Toxics should be segregated from flammables and oxidizing agents from flammables, organic chemicals, reducing or dehydrating agents. Categories which should be held in separate stores include corrosive substances, cryogenic substances and pyrophoric substances.

Types of container used are small containers, large containers, bulk containers and gas cylinders. Aspects of good practice with small containers include the use of bottle carriers, non-spill containers and controlled dispensing. Good practice in the storage of large containers, essentially drums, and bulk containers is discussed in Chapter 22.

The receipt points for chemicals should be designed to facilitate handling of incoming loads. Vehicles carrying gas cylinders generally have their own lifting gear, but those carrying drums often do not.

The control of materials in storage is exercised through the processes of acquisition, stock-taking and disposal. It is common for an experimenter to order an appreciable quantity of a chemical, use only a portion of it and then forget about it. Unless close control is exercised, there can build up in the stores, a large quantity of chemicals, much without 'ownership'. It is necessary to keep the situation under control by taking stock periodically and by disposal of unwanted material.

Closely connected with this is the minimization of inventory. The inventory held in the laboratory itself and in storage should be the minimum necessary.

The number of chemicals used in a laboratory can be large and the conditions of use very varied. It is essential to ensure that all materials are identified and labelled. There should be a materials safety data sheet for each chemical. Reale and Young (1987) give an account of labelling and MSDSs in the laboratory context.

The Poison Rules 1982 give in Schedule 1 a list of substances which must be stored in a locked cupboard under the supervision of a registered keeper and the usage of which must be entered in a register.

A9.9.2 Waste disposal

A laboratory usually generates gaseous, liquid and solid wastes, all of which need to be disposed of. The problems of waste disposal are not trivial and have become increasingly severe as environmental standards have risen.

Accounts of this aspect are given in *Handbook of Laboratory Waste Disposal* by Pitt and Pitt (1985), in the IChemE *Guide* and by Steere (1971), Bretherick (2000) and Varnerin (1987).

A distinction is generally made between domestic and chemical wastes. The former category includes paper and broken glass. The latter should be segregated both from paper and from unbroken bottles.

Flammable solvents generally constitute a large proportion of the chemical wastes. Other liquids which are incompatible with such solvents should be collected separately. Separate collection is also required for oil-soaked rags. On no account should chemical wastes be put down the drains.

Separate, clearly labelled receptacles are required for each category of waste. For some categories, such as for flammable solvents, receptacles with appropriate safety features should be used.

Options for the disposal of laboratory wastes are listed by Pitt and Pitt, who give some 13 choices, although many are applicable only to small quantities.

The disposal of wastes has to be formally managed by identifying disposals and obtaining the necessary licenses.

Specialist contractors are used by most laboratories for disposal of at least some of their wastes. This is one of the options discussed by Pitt and Pitt.

A9.10 Laboratory Operation

A9.10.1 Information requirements

It is essential to have the fullest information on the chemicals being handled and on the associated hazards. This applies particularly to chemicals which are highly reactive or which have toxic effects at low concentrations over long periods. Sources of such information are discussed in Chapter 33.

A9.10.2 Design of experiments

Laboratory experiments are undertaken for different purposes and these necessarily influence their design.

In all cases inherently safer design should be a prime aim, but the extent to which it can be achieved will vary. Facets of inherent safety include scale, operating conditions and materials. Where the objective is the pedagogical one of demonstrating a principle, it may be possible to operate on a small scale, at atmospheric pressure and temperature and using less hazardous materials such as air, water and inert solids.

The purpose of the experiment should be well defined. The experiment should not be too ambitious, and if necessary should progress in stages. Information on the materials handled and their potential hazards should be as complete as practical. The basis of safety should be defined. The design should cover the equipment and its operation, and should meet applicable standards and codes.

The design should be subject to a hazard review, utilizing hazard identification techniques such as checklists and hazop studies (covered in A9.10.3.).

Laboratory equipment should be fit for purpose. The standard of construction of the apparatus should match the degree of hazard and its expected life.

It is not uncommon to build a 'temporary' rig with connections made to certain services by hose rather than by fixed piping. If this is accepted in certain applications, care is still needed to ensure that the practice does not extend to unsuitable cases or to operation over extended periods.

Another common practice is the use in a new experiment of a rig 'inherited' from an earlier one. The suitability of the equipment for the new purpose needs careful scrutiny.

For critical features consideration should be given to the provision of automatic protective devices. Trips may be required to guard against loss of cooling, excessive heating, and deviations of pressure, temperature, flow, level and composition.

Full instructions should be prepared for operating procedures and emergency procedures. Consideration should be given to the requirements of unattended operation if this is planned.

Certain activities require special enclosures. These include the use of a fume hood where the materials are toxic and of explosion cubicles where the process involves hazardous reactions or high pressures.

The planning of safe experiments is discussed by Fowler (1989 LPB 32).

A9.10.3 Hazard assessment

If an experiment has the potential to cause serious injury, it is important to carry out a hazard assessment. Case studies of the application of hazard assessment in the laboratory have been described by Le, Santay and Zabrenski (1988).

One example given is a review of the situations involving release of toxic gas sufficient to pose an acute toxic threat. The gases in question were arsine, diborane and phosphine, all in hydrogen. A tabulation was made of the various release scenarios, showing the expected toxic loads and candidate scenarios were then investigated. Fault tree analysis was used to determine the failure paths.

Another case study described by these authors involved fault tree analysis where the top event considered was the formation of a detonable hydrogen–air mixture from a gas phase reactor system.

A9.10.4 COSHH assessment

A COSHH assessment should be made for any laboratory activity which attracts such an assessment under the COSHH Regulations 1988, as described in Chapter 18.

A9.10.5 Operating procedures

Operating procedures may be classified in two ways. One distinction is between general laboratory procedures and procedures specific to a particular rig. Another is between those which concern safety and those which do not.

For the laboratory as a whole, general operating instructions relating to safety are usually set down in a laboratory safety manual.

For the particular rig, it is advisable to separate the procedures which have safety implications from those which do not. Where operating instructions are imported, as with equipment bought from a manufacturer, it may be necessary to rewrite them, abstracting the safety-related parts.

The operating instructions for a rig should cover start-up, running and shut-down, and should indicate the hazards and the countermeasures to be taken. They should not duplicate the general instructions. They should be clear and simple, and should be as concise as possible.

A9.10.6 Emergency procedures

The operating procedures for each rig should cover action to be taken in the event of an emergency (1) on that rig and (2) in the laboratory as a whole.

It is an accepted principle that a laboratory fire should be fought if it can be done without undue risk. The same principle may be extended to other actions in an emergency which threatens to escalate.

Emergency planning is discussed in Section A9.12.

A9.10.7 Equipment maintenance

The general approach to maintenance of equipment in a laboratory is similar to that in a plant. It involves identification of the equipment to be worked on; planning of the work and shut-down of the equipment; electrical and mechanical isolation of the equipment and preparation for the work by removing fluids and solids and by cleaning. The principles are as outlined in Chapter 21.

A9.10.8 Equipment and procedure modification

Similarly, control should be exercised over modifications to the equipment or the activities, analogous to those over equipment and process modification in a full-scale plant described in Chapter 21.

The IChemE *Laboratories Guide* gives detailed guidance on what, in a laboratory context, constitutes a modification and contains in Appendix 14 a specimen form for control of modifications.

A9.10.9 Permit systems

The scope of the permit system, defining which activities need to be so covered and which do not, should generally be agreed between the laboratory superintendent and the rig supervisor.

The activities which require permits and the parties responsible for issuing and receiving them should be listed in the general laboratory instructions or the rig instructions.

Hazardous activities such as maintenance and vessel entry should be covered by a permit system. The principles are similar to those for an industrial plant as described in Chapter 21.

One of these principles is formal handover from operations to maintenance personnel and then handback in the reverse direction. In a laboratory these responsibilities may sometimes be less clear than in a plant and, as just stated, they need therefore to be clearly defined. The effectiveness of the permit system depends on formal observance of such handovers.

A9.10.10 Housekeeping

There should be a high standard of housekeeping. Facilities should be provided for disposal of broken glass and of waste materials. Equipment not in use should be removed or arranged tidily. Spillages should be cleaned up at once. Orderly sample storage with minimal retention should be maintained; glassware/equipment cleaning should be carried out.

A9.10.11 Out-of-hours working

Personnel working out of normal work hours should be required to sign in and sign out.

Working alone in the laboratory should be permitted only where the hazard is low and is best kept to a minimum. Anyone who does work alone should be in contact with, or should be visited by, the security guard at regular intervals.

A9.10.12 Unattended operation

Unattended operation of equipment may present a hazard, especially if flammable or toxic materials are involved. There will be some experiments where such operation is not permissible. When unattended operation is practiced, arrangements should be made for the apparatus to be visited by a trained security guard, who should be given full instructions on the action to be taken in case of accident or doubt and should be able to make telephone contact with the experimenter.

A9.10.13 Access

Access to the laboratory should be subject to control and unauthorized persons should be excluded. The extent of the measures necessary depends on the nature of the hazards and the confidentiality of the work.

A9.11 Laboratory Fire and Explosion Protection

A9.11.1 Fire protection

The laboratory should be designed for fire protection, in accordance with building and fire protection codes and with advice from the fire authorities.

Some basic elements of design for fire protection include the fire resistance of the laboratory envelope, including doors; internal layout; hazardous area classification; mechanical ventilation and a fire alarm system.

It is desirable that there be constructional features with a minimum fire resistance between the laboratory and any adjoining section of the building and that the doors meet a fire resistant standard. Measures should be taken to prevent fire spread through openings in walls and voids in the roof.

It should be a principal aim of the laboratory layout to minimize both the occurrence and escalation of fire.

Hazardous area classification should be applied to effect control of ignition sources. It is suggested in the IChemE *Guide* that for a well-ventilated laboratory classification as Zone 2 should be sufficient. The guide states that ideally the whole space should be so classified, but if this is not practical, the Zone 2 should extend at least 2 m around any potential leak source.

The risk of buildup of a flammable atmosphere may be minimized by the use of mechanical ventilation. The guide recommends that where flammable liquids are in use, the ventilation fan should be operated for the period of use and for 5 min thereafter. (Use of a flammable storage cabinet may be appropriate. Also, labs with fire risk may need to require workers/visitors to use Flame Retardant Clothing.)

There should be a fire alarm system with alarm points inside the exits and at other strategic points.

A9.11.2 Fire fighting

The laboratory should be equipped with suitable fire extinguishers, fire blankets, etc. An account of fire extinguishers in general is given in Chapter 16 and in the laboratory context in the IChemE *Guide*.

A9.11.3 Explosion protection

Where there is to be operation of equipment such as an experimental reactor with a significant risk of vessel rupture, a separate cubicle should be provided in which the work can be conducted. Such arrangements are perhaps more commonly found in pilot plants, but may find place in a laboratory.

The cubicle should be designed to contain both blast and missiles. Accounts of such explosion-proof cells are given in Chapter 17 and Appendix 10.

A9.12 Emergency Planning

Emergency planning should be undertaken to identify the potential causes and types of emergencies and their consequences, and the countermeasures required. The principles are similar to those outlined in Chapter 24.

Planning should cover not only the laboratory and the rigs but also storages and services. The plan should be formulated by conducting audits during design and at commissioning and should be updated by audits at regular intervals thereafter.

The purposes of emergency planning are to protect life and property. They are achieved by early recognition of the problem, rapid raising of the alarm and prompt action to bring the incident under control and to separate the people from the hazard.

Functions which need to be performed in an emergency are to deal with the emergency, to evacuate people and to liaise with the emergency services. The personnel who will perform these functions should be identified and their tasks defined.

The range of situations in laboratories tends to be particularly wide and the emergency plan should aim for flexibility.

Laboratory personnel responsible for emergency response should understand the statutory duties and likely priorities and needs of the external services.

There are certain typical problems faced by the external fire services coming into laboratory situations. Steps should be taken to find out what information the fire brigade is likely to need and to ensure as far as is practical that it can be provided. This means having the information available and appointing someone to communicate it in the actual event. Frequently fire brigades are hampered by hazards from gas cylinders and radioactive materials and by lack of information on the chemicals present and their hazards.

There should be regular fire drills, including joint exercises with the external services.

A detailed account of laboratory emergency planning is given in the IChemE *Guide*.

Appendix

10

Pilot Plants

Contents

Pilot plants are intermediate in scale between the laboratory bench and the full-scale plant. They are used for a variety of purposes, essentially to obtain information on process design and operation and on products, including information relevant to safety. Some hazards are liable to show up first at the pilot plant stage and the pilot plant is therefore an important tool for hazard identification, as described in Chapter 8. Here consideration is given to the more general aspects of pilot plants and particularly to their characteristic features and hazards, and to safety in pilot plants.

Accounts of pilot plants are given in *Pilot Plants, Models and Scale-Up Methods in Chemical Engineering* by Johnstone and Thring (1957), *The Chemical Plant from Process Selection to Commercial Operation* by Landau (1967), *Chemical Process Development* by D.G. Jordan (1968) and *Pilot Plant Design, Construction and Operation* by Palluzzi (1992) and by Constan (1984), Carr (1988), Dore (1988), Palluzzi (1989, 1990, 1991), Lesins and Moritz (1991), J. Jones *et al.* (1993a,b) and Lo and Oakes (1993). Safety is treated in the *Pilot Plant Safety Manual* by the AIChE (1972/33) and by Kohlbrand (1985), Siminski (1987), Carr (1988), Dore (1988), Capraro and Strickland (1989), Carson and Mumford (1989 LPB 89, 90), Constan, Herman and Moeller (1992) and Palluzzi (1992).

A number of authors on pilot plants give checklists, often as tables. These checklists include the following: justification for use (J. Jones *et al.*, 1993b); overall programme (Lesins and Moritz, 1991; J. Jones *et al.*, 1993a); chemical and process information (Dore, 1988); scale-up issues (Lo and Oakes, 1993); project review (J. Jones *et al.*, 1993a); flowsheet review (Lesins and Moritz, 1991); hazards (Dore, 1988); hazard identification—what-if review (Lesins and Moritz, 1991); plant construction (Dore, 1988); staffing requirements (J. Jones *et al.*, 1993a); operating and safety aspects (Dore, 1988); operating manual (J. Jones *et al.*, 1993a); operator training manual (Chaty, 1985) and computer control systems (J. Jones *et al.*, 1993a). Many of the entries in the AIChE *Pilot Plant Safety Manual* may be used as checklists for pilot plant safety.

In general, the principles of laboratory safety, discussed in Appendix 9, are applicable to pilot plants also.

Selected references on pilot plants are given in Table A10.1.

A10.1 Pilot Plant Uses, Types and Strategies

A10.1.1 Uses of pilot plants
The purposes for which pilot plants are built are the development of (1) a new product, (2) a new process or (3) an existing process. For a new product small quantities may be required for testing and for market development. For a new process information is required to prove feasibility, specify operating conditions, resolve scale-up issues, provide design data, identify problems, develop operating procedures and provide experience and training. For an existing process work may be required to check the suitability of different raw materials, improve product quality, explore modified operating conditions, improve the treatment of the effluents and effect cost reductions and other optimizations.

Frequently it is information on reaction kinetics and yield which is lacking and which the pilot plant is required to provide. Other information typically obtained from a pilot plant is data on heat and mass transfer and pressure

Table A10.1   *Selected references on pilot plants*

Michel, Beattie and Goodgame (1954); Johnstone and Thring (1957); E.L. Clark (1958); Lengeman and Hardison (1964); Gernand (1965); Hudson (1967); Landau (1967); Ellis (1968); D.G. Jordan (1968); Anon. (1969a); IChemE (1969/47); Conn (1971); AIChE (1972/68); Katell (1973); Roche *et al.* (1977); Shields and Vander Klay (1977); Braun (1978); Berty (1979); Randhava, Lo and Kronseder (1982); Kunesh and Hollenack (1983); Constan (1984); J.L. Jones (1984); Margolin (1984); Pfeffer, Bhalla and Dope (1984); Chaty (1985); Lowenstein (1985); P. Dawson (1986); Palluzzi (1987, 1990, 1991, 1992); S.G. Phillips (1990); Ginkel and Olander (1991); Lesins and Moritz (1991); Pollack (1991); Uitenham and Munjal (1991); Jones *et al.* (1993a,b); Lo and Oakes (1993)

Safety in pilot plants
A.R. Jones and Hopkins (1966); Deviny (1967); Graf (1967); Marinak (1967); Prugh (1967); Klaassen (1971); Ligi (1964, 1973); Bait and Berg (1974); Weissbach (1976); Roche *et al.* (1977); Lowenstein (1979); Franko-Filipasic and Michaelson (1984); Kohlbrand (1985); Basta (1986b); Siminski (1987); Carr (1988); Dore (1988); Capraro and Strickland (1989); Carson and Mumford (1989 LPB 89, 90); Bond, Carson and Mumford (1990 LPB 93); Constan, Herman and Moeller (1992)

High-pressure pilot plants
E.L. Clark (1963); W.G. High (1967); AIChE (1972/65); B.C. Cox and Saville (1975); Jercinovic (1984); Mehne (1984)

drops; on mixing effects; on impurities, foams and emulsions; and on fouling and corrosion.

The justification for pilot plants is discussed by J. Jones *et al.* (1993a). They discuss the following situations, in increasing order of difficulty:

Process	Product	Customer
New	Existing	Existing
New	New	Existing
New	New	New

A10.1.2 Types of pilot plant
Pilot plants differ in both type and scale. There are three principal types: (1) the general-purpose pilot plant; (2) the specific-purpose pilot plant; and (3) the multi-purpose pilot plant. The merits of these different types of pilot plant are discussed by Palluzzi (1991, 1992).

Representative capacities for different scales of operation are given by J. Jones *et al.* (1993b) as follows:

Scale	Capacity	
	Batch	*Continuous*
Laboratory bench	0.1–1000 g	10–100 g/h
Mini-pilot plant	1–100 kg	0.1–10 kg/h
Pilot plant	100–1000 kg	10–100 kg/h
Semi-works	1000–10,000 kg	100–1000 kg/h

Some pilot plants currently in use are quite old and over the years have been modified and adapted.

Accounts of particular pilot plants include those of Chaty (1985), P. Dawson (1986), Siminski (1987), Carr (1988), Capraro and Strickland (1989) and Lesin and Moritz (1991).

A10.1.3 Strategies for development utilizing pilot plants

It need not be assumed that a pilot plant necessarily has to be designed and located in-house; there are other options. The main options are (1) design and location in-house; (2) location in-house but with design by a design contractor; (3) hiring of outside pilot plant facilities and (4) contracting out of the pilot plant work. The merits of these options are discussed by Palluzzi (1991).

A10.2 Pilot Plant Features and Hazards

While the hazards encountered on pilot plants are essentially similar to those on full-scale plants, the characteristics of pilot plants are such that the hazard profile is somewhat different. The specific features and hazards of pilot plants are discussed by a number of authors, including Carr (1988) and Capraro and Strickland (1989).

Some features characteristic of pilot plants include (1) gaps in knowledge; (2) novelty of chemicals, process, equipment and operations; (3) scale effects; (4) extent of manual activities; (5) frequency of modification; (6) multiplicity of tasks; (7) materials storage and transfer; (8) flow features; (9) recycles; (10) utilities features; (11) frangible elements; (12) plant layout features; (13) location in a building and (14) research staff involvement.

The process reaction may involve a number of potential hazards. Most frequently mentioned is the unidentified reaction exotherm, which could lead to a reaction runaway. There should be a formal system of screening to identify any exotherms. However, this is by no means the only hazard which may be present. The raw materials may contain impurities. The ratio and flow rate of the reactants may vary. There may be inhomogeneities due to poor mixing which affect reaction rate and temperature measurement. Features of the reaction such as induction time may assume greater significance. The handling of the catalyst may pose problems.

There are a number of equipment problems which can occur on pilot plants. One is the use of unfamiliar equipment. Another is the reactivation of idled equipment. When this is in prospect, a check on the integrity of the equipment is advisable; it may have been 'cannibalized'. Another problem can be the use of equipment which differs from that used on the full scale and for which, in consequence, operating procedures may be deficient. An example is an oven used by a number of different users with no formal control over its contents and settings.

A principal reason for building a pilot plant is to fill the gaps in the information necessary for the design and operation of the full-scale plant. This implies that there will be gaps also in the information which ideally would be available for the design of the pilot plant.

A pilot plant is on a scale intermediate between those of the laboratory and of the full-size plant. Relative to the laboratory, the larger scale of the pilot plant means that hazards may become apparent which were obscured at the laboratory scale. Operation on the laboratory scale has some of the features of an inherently safer design. The most obvious is the limited inventory, but there are also others

such as good ventilation. In some cases the problem may be that the change in scale is accompanied by other changes, such as the use of less pure raw materials. In others it may simply be that features always present become more obvious with scale such as the need to dispose of noxious effluents. Moreover, the pilot plant is the stage at which the process is first carried out in process plant as opposed to laboratory equipment and thus the stage at which difficulties of processing, of equipment or of measurement will show up.

Relative to the full-scale plant, the smaller scale of the pilot plant is beneficial in reducing the scale of the hazard, but can lead to problems of flow control, blockage and so on.

The balance between manual and automatic functions tends to be different in a pilot plant from that on the full scale. There are several reasons for this. Foremost reasons are economics and flexibility; others include factors such as measurement problems.

It is a normal, and expected, feature of pilot plant operation that both the process and the plant are subject to frequent modification. A system for the control and documentation of modifications of both types is required, both as a record of the learning process and to assure safety.

Work in a pilot plant is liable to involve tasks which are more numerous, varied and novel features, which can increase the probability of error.

The problems of containers for materials and the identification of the materials and containers can be a significant problem in pilot plants. The number of materials used may be appreciable, creating much greater scope for error than on a regular plant. It is necessary, therefore, that raw materials, intermediates, products and effluents be fully identified. Where common containers are used, as sometimes with effluents, controls need to be established to prevent mixing of incompatible materials.

Flow problems associated with features such as highly viscous fluids, gas locks, lutes, siphons, water hammer, cavitation and so on are features commonly encountered on process plants, but it is on the pilot plant that they may be first experienced and will have to be solved.

Another feature which may well be novel at the pilot stage is recycling with its characteristic problems of positive feedback and build-up of impurities.

Some pilot plants, particularly the larger ones, are on a works site and are served by the works utilities. It is normal practice to give advance warning to the main production plants of an expected change in the status of such utilities, but the Cinderella status of the pilot plant may result in its being overlooked, and suffering unexpected loss of a utility.

Frangible pipework and vessels made of glass or plastic and other features, such as rotameters and sightglasses are common in pilot plants, and make the plant more vulnerable to damage and leaks.

Plant layout in pilot plants tends to be more congested. Such congestion can increase the risk of damage to plant or inadvertent movement of controls. An example of the latter is movement of a control such as a valve as an operator brushes against it. This danger is increased by the fact that the valve is much lighter and more readily moved than its equivalent on the full scale. Similar points apply to damage to plant, which is more vulnerable on the small scale.

In contrast to most large-scale plants, which are in the open, many pilot plants are in buildings, and this has a number of implications. There is often a more congested layout. There is an increased hazard from accumulation of gases or vapours. These may be flammable or toxic vapours, whether from open vessels or from leaks and spillages. Or they may be asphyxiant gas, such as nitrogen, or flammability-enhancing gas, such as oxygen. Likewise, there may be occupational hygiene problems because fugitive emissions disperse less readily.

Research staff involved in pilot plant work are likely to be less attuned than those familiar with plant operation to the disciplines necessary, and specific steps may need to be taken to rectify this.

A10.3 Pilot Plant Scale-up

Historically, a basic reason for the use of a pilot plant has been the difficulty of scaling up from the laboratory bench to the full scale. To a considerable extent, such scale-up is what chemical engineering is all about, and progress in the discipline has made this reason seem rather less cogent.

Despite this, operation on the pilot scale may still be desirable even in respect of fundamental features such as reactor heat transfer, mixing, runaway and venting.

However, even in terms of scale-up, there is more to pilot plant operation than the scale effects typically expressed in terms of dimensionless numbers and engineering models. As already indicated, there are other effects of scaling up such as the use of less pure raw materials; the occurrence of poor mixing; the build-up of impurities due to recycling; the need to dispose of noxious effluents; and so on.

The process of scale-up is frequently envisaged as a linear one, progressing to successively larger scales. An alternative model, however, is one in which the process is iterative, passing to and fro between the smaller and large scales.

The pilot plant stage is the first at which the process is operated in the form of a plant. It therefore provides the first opportunity to examine all those features which are new in passing from laboratory bench to plant. This is so whether the pilot plant is small or large.

A10.4 Pilot Plant Design

There are a number of features which are characteristic of, or tend to bulk large in, the design of pilot plants. Some of these are now considered.

A10.4.1 Some design objectives
The pilot plant should be utilized to design the full-scale plant using recognized engineering methods of scale-up.

The design should accommodate a sufficiently wide range of operating variables, such as pressure and temperature, to allow the effect of these variables to be determined with some confidence.

The purpose of the pilot plant is to provide information for the design of the full-scale plant. The plant should be designed and operated so as to yield information in which confidence can be placed and which is conformable to established design methods. As far as practicable, the process should be modelled and the experimental results compared with the model as they are obtained. Important features of such a model are the mass and heat balance and

pressure drop relations, the reaction kinetics and the unit operations models.

A10.4.2 Design standards and codes
A particular problem in pilot plant design is the application of standards and codes. In general, these are formulated with full-scale plant in mind, and can sometimes pose difficulties for the pilot plant designer.

The standards to be applied should be declared at an early stage and any potential conflict identified, whether between different standards or between a standard and the design.

The application of standards to pilot plants is discussed by Siminski (1987), who describes the particular case of the standards covering an engine test facility. The approach which he describes is to assess the impact of standards on the safety, industrial hygiene and environmental aspects of the plant, to assess the risks and develop alternative strategies, to seek independent review and to test design concepts early on, to examine the specification to check whether features causing conflict with standards are strictly necessary, and to negotiate with the parties to achieve solutions.

A10.4.3 Flexibility and multiple use
Since the purpose of a pilot plant is to extend knowledge and deal with novel issues, it needs to be flexible. Methods of providing such flexibility are discussed by P. Dawson (1986). They include the provision of inherently flexible equipment, a range of equipments and materials of construction and a layout which groups together items which most frequently need to be connected, thus avoiding long pipe runs.

A10.4.4 Plant layout
The basic principles of plant layout apply to pilot plants also, but some features assume particular importance.

One feature just mentioned is layout to minimize the length of pipework and to ensure a convenient arrangement of principal items of equipment.

Another feature is to ensure good access, which includes both access to items of equipment but also minimization of the likelihood of inadvertent damage to plant or operation of controls as personnel move about congested passages. Closely related is minimization of damage by dropped objects.

Another feature is clear identification of equipment. One aspect is the labelling of tanks and vessels holding process materials, since misidentification of materials is an error characteristic of pilot plant operation. Another aspect is labelling and coding of equipment which may need to be worked on and which may contain noxious materials.

Another feature is provision for collecting and disposing of liquid leaks.

A fifth feature is the provision of protective barricades against missiles. In some cases this leads to a design in which the pilot plant space is divided up into separate cells.

A10.4.5 Protective barricades
If the process carried out involves high pressures, and particularly high-pressure reactions, protective barricades may be provided.

Some practical aspects of barricades are discussed by Carr, who describes methods of protection against missiles such as valve stems and reactor fragments, as well as high-pressure steam jets, which can be almost as damaging.

Missiles from an exploding reactor may be stopped by a barricade in the form of a sand bath protected by a pressure relief valve. This reduces the need to house the whole plant in a barricade.

For barricade design in general, design case missiles are specified and the barricade is designed to withstand these missiles, using methods such as those outlined in Chapter 17. It is on pilot plants that the provision of such barricades is most common, and thus where most of the expertise resides.

A10.4.6 Pressure systems

The combination of the facts that a pilot plant is usually in a building and that it is subject to continuous experimentation and modification means that it is necessary to give particular attention to the integrity of the pressure system. Procedures should be laid down for the proof testing and leak testing of the plant.

A10.4.7 Pressure relief

It is common practice in pilot plants to put a safety valve on a vessel. The lines to and from such valves tend to be small diameter and liable to fill with liquid which then solidifies. Countermeasures include design of lines so that the liquid drains away and use of heated lines.

A10.4.8 Frangible elements

Pilot plants are frequently constructed in part or in whole in glass or plastic. They also tend to contain vulnerable features such as rotameters and sightglasses.

The protection of the latter devices is described by Carr. Use is made of protective shields with adequate pressure-relief space between the shield and the device. Specific designs have been developed and tested for most types of device used.

A10.4.9 Gas cylinders

Another hazard discussed by Carr is gas cylinders. Rupture of a line supplied by a high pressure cylinder can result in a large jet of gas. He describes arrangements involving the use of a limiting orifice just downstream of the regulator.

A10.4.10 Ignition source control

Another aspect of the design is the control of ignition sources. The importance of this is enhanced in a pilot plant by the facts that the probability of a leak tends to be higher, that in a building gas or vapour accumulates more readily and that the probability of personnel being present is relatively high.

Since it is necessary to exclude all ignition sources, the proper approach is to make a formal examination of the plant to identify possible sources of release. These sources include activities, both operations and maintenance, as well as fixed plant. Such ignition hazard identification provides a basis for the design both of layout and equipment and of operating controls.

The application of the methods of hazardous area classification is one necessary aspect of this. The principal options are classification as Zone 1 or as Zone 2. Classification as Zone 1 puts constraints on the equipment, particularly instrumentation, which can be used, while

classification as Zone 2 requires the elimination of leaks and a good level of ventilation.

An account of practical measures for an area classified as Class 1, Division 2, in the US system is given by Carr.

Other features typically include liquid catchment and disposal arrangements, gas detectors and mechanical ventilation.

Multipurpose plant places particularly severe requirements on ignition control.

These hardware measures need to be complemented by a system of operational controls to eliminate ignition sources. Failure to do this largely negates the value of the hardware.

A10.4.11 Plant classification

A systematic approach to the design of pilot plants is to use a formal method of classification.

Carr describes the method which is used in one company for classifying pilot plants for the purposes of design and operation and of personnel protection and which utilizes four operating categories:

Category	Hazard potential
1. Detonation reactions	High potential for detonation or massive rupture and large, extremely rapid release
2. Rupture and fires	High potential for large rupture and large fire
3. Leaks and small fires	Low potential for leaks and, at worst, only leaks and fires of limited size
4. Low hazard operations	Near ambient temperatures and low pressures with very low potential for either ruptures or fires

The corresponding protection requirements are given in Table A10.2.

A10.4.12 Extent of automation and protective systems

The operations to be performed on a pilot plant are on a small scale and are required for a limited period, relative to the equivalent operations on the full-scale plant. Further, their automation may well require the development of measuring instrumentation. Hence it will generally not be appropriate to seek to automate these operations to the extent that will pertain on the full-scale plant.

Somewhat similar considerations apply to the provision of protective systems such as trips and interlocks. These definitely have their place in pilot as in full-scale plants, but the particular mix appropriate to the full scale may not suit the pilot scale.

A10.4.13 Control systems

The pilot plant should be provided with instrumentation sufficient not only to obtain design data but to ensure safety. The same principles should be applied as to the design of full-scale plant, such as the use of trips and interlocks, subject to the comments made above on the extent of automation and of protective systems.

The use of computers, mainly PCs, for data acquisition and control of experiments is routine in pilot plants. In some cases the flexibility of such computers has also been exploited to provide alarms and trips.

Table A10.2 Protection requirements scheme for pilot plants (after Carr, 1988) (Courtesy of the American Institute of Chemical Engineers)

Typical hazard	Category 1 Detonating reactions		Category 2 Rupture and fire hazards		Category 3 Leaks and small fires		Category 4 Low hazards	
	Hazard potential	Protection device	Hazard potential	Protection device	Hazard potential	Protection device	Hazard potential	Protection device
Detonation (internal)	High	Barricade Blast relief	Negligible		Negligible		Negligible	
Shrapnel	High	Barricade Shrapnel stop	Negligible		Negligible		Negligible	
Explosive atm. (external)	High	Barricade Blast relief Ventilation Min. cubicle size	Moderate	Baffle Ventilation Cubicle size	Nelgligible		Negligible	
Fire	High	Baffle	High	Baffle	Moderate	Baffle	Negligible	
Pressure leaks	Possible	Baffle (barricade)	Possible	Baffle	Possible	Baffle	Negligible	
Skin toxic	Possible	Baffle (barricade)	Possible	Baffle	Possible	Baffle	Possible	Baffle
Lung toxic	Possible	Ventilation Exculded	Possible	Ventilation Limited	Possible	Ventilation Limited	Possible	Ventilation Unlimited
Permitted access to operating areas								
Summary of protective devices		Barricade Blast relief Shrapnel stop Ventilation Min. cubicle size		Baffle Ventilation Cubicle size		Baffle Ventilation Required if: lung toxic chemicals		Baffle required if: skin toxic chemicals or adjacent operations (see baffle design) Ventilation required if: lung toxic chemicals

Control schemes for pilot plants are discussed by Uitenham and Munjal (1991) and Palluzzi (1992).

The effectiveness of any protective systems depends on the training of the operators.

A10.4.14 High-pressure plants

Some pilot plants involve high-pressure equipment. The AIChE *Pilot Plant Safety Manual* contains sections on the design of pilot plants for high-pressure and for high-temperature operation. For the former further guidance is given in the *High Pressure Safety Code* by E.G. Cox and Savffle (1975) and by E.L. Clark (1963), W.G. High (1967), Jercinovic (1984) and Mehne (1984).

At the pilot plant scale it is quite common to provide blast-proof cubicles and barricades against missiles, which tend to be impractical on the full-scale plant.

A10.5 Pilot Plant Operation

A10.5.1 Suitability of plant

If the pilot plant is a multipurpose one and is thus used to investigate a number of different processes, a particularly careful check should be made that it is fully suitable for the proposed process.

A10.5.2 Personnel and training

The magnitude of the hazard on a pilot plant is less than on the full scale, but in other respects the operations tend to be more demanding. The materials, the process, the equipment, the plant and the procedures are all relatively unfamiliar. The operating team needs strong leadership and experienced personnel.

As with plant operation generally, training is critical. Many of the topics mentioned in this appendix imply the need for training.

Training is required for management and research personnel also. The latter may well be used to laboratory rather than plant situations, and need to become familiar with plant disciplines.

A10.5.3 Operating and emergency procedures

The operations to be performed should be identified and for each a suitable operating procedure should be developed, and examined from the safety viewpoint. Operations which are likely to bulk large in pilot plant work include manual operations, reactor operations, sampling and measurement activities. There also needs to be suitable emergency procedures.

A10.5.4 Specifications and documentation

Despite the small scale, it is desirable to adhere to a certain formality in pilot plant operation. Some aspects of this are discussed by J. Jones *et al.* (1993a).

There should be formal specification for the raw materials, intermediate and products, for yield and throughput, for cost targets and for the completion date.

Records should be kept of the progress of the project during the pilot plant stage, of problems and hazards encountered, and of the steps taken to solve the problems and to eliminate or control the hazards.

The data needed for the design of the full-scale plant should be documented and the information gathered and recorded.

The operating procedures evolved, including the emergency procedures should be properly documented. Again, this should include any problems and hazards encountered and the response made.

A10.5.5 Mothballing

By its nature a pilot plant is liable to be subject to intermittent operation, with periods when it is not in use. Often a pilot plant is shut-down and 'mothballed' for an extended period. If this is a possibility, it should be taken into account in the design. At the start of such a shut-down measures should be taken to prevent deterioration, including cleaning and flushing of the plant. On recommissioning care should be taken to identify hazards which may arise from the prolonged shut-down.

A10.6 Pilot Plant Safety

An accident in a pilot plant, like one in a laboratory, is generally on a much smaller scale than one on a full-scale plant, but again it can give rise to considerable direct and consequential loss.

Safety in pilot plants is treated in the AIChE *Pilot Plant Safety Manual*. This deals with procedures, including transfer of information from the chemist, hazard identification, process design, mechanical design and design review; scale-up from pilot to full scale; engineering standards for flammable, explosive, corrosive and toxic materials and for radiation protection; maintenance procedures; and high-pressure and high-temperature processes.

In general, the same approach should be taken to the design and operation of a pilot plant as is adopted for a full-scale plant. Many of the topics already discussed are equally relevant to pilot plants, including management and management systems (Chapter 6), hazard identification (Chapter 8), reactive hazards (Chapter 3), process, pressure system and control system design (Chapters 11–13), fire, explosion and toxic release (Chapters 16–18), plant operation and maintenance (Chapters 20 and 21), personal safety (Chapter 25) and safety systems (Chapter 28). On the other hand there are certain features which are characteristic of pilot plants.

A10.6.1 Project safety reviews

There should be a system of project safety reviews adapted to pilot plant design and operation. These should be the subject of a formal requirement and should be documented.

The general methods of hazard identification should be used to discover potential hazards in plant design and operation. In addition, there are hazard identification procedures which are particularly relevant to pilot plants.

The information on the chemicals handled, the reactions involved, and the materials of construction for the plant should be as complete and as well documented as practical.

The transfer of information from the chemist to the engineer should be regulated by formal procedures. The chemist should give a full description of the process, including the reaction kinetics and heats of reaction, limits of operating parameters such as pressure and temperature, and procedures and precautions adopted. The engineer should study the research reports to envision problems which may arise in scale-up to the pilot plant.

The process reaction(s) should be screened to identify any reaction exotherm which might lead to a runaway reaction as described in Chapter 33.

A10.7 Pilot Plant Programmes

Accounts of pilot plant programmes emphasize a number of themes. They include keeping attention focused on the basic chemistry; keeping the programme simple, first establishing feasibility, before getting involved in matters such as yield, effluents, etc., applying lateral thinking and seeking alternative approaches; being prepared to take the calculated risk that the process may not work, although taking no risks with safety; and making use of expert advice from other sources.

A10.7.1 Handover and decommissioning

The pilot plant operation must recognize the point at which the baton should be handed over to the next stage.

However, it is advisable not to be in too much of a hurry to decommission a pilot plant. The purpose of the plant is to provide information and/or products to other users. Once they have received these, they too will have to undergo a learning process, and are likely to come back with queries, some of which may require further plant trials. This is true particularly with new products, where the queries may be directed to improving quality or reducing cost and meeting regulatory requirements and environmental restraints.

Appendix

11

Safety, Health and the Environment

Contents

Although the problems of environmental protection (EP) are basically beyond the scope of this book, they cannot be completely neglected, because there are a number of ways in which they are closely linked to safety and loss prevention (SLP).

Selected references on the environment and pollution are given in Table A11.1.

Safety, Health and the Environment

These links are recognized in the common practice of coupling health, safety and environment (HSE) or safety, health and environment (SHE). In the following the latter acronym is preferred, the former already having an established meaning.

A11.1 Common Responsibilities

A11.1.1 Professional responsibilities
The professional engineer has responsibilities both for safety and for environmental matters, which are very similar in nature.

These responsibilities are outlined in the Engineering Council codes *A Code of Professional Practice on Risk Issues (1992) and A Code of Professional Practice on Engineers and the Environment* (1993) described in Chapter 3.

A11.1.2 Legislation
The detailed legislative requirements for safety and environment naturally vary between countries but the worldwide trend is clear. It is towards higher standards both inside and outside the factory; more formal management systems for identification, assessment and control; inherently safer/cleaner designs; and greater legislative and public pressures.

Legislation is described in Sections A11.4 – A11.6.

A11.1.3 Professional education
The education of professional engineers should provide a firm grounding in both safety and environmental matters, creating awareness, instilling responsibility and introducing the appropriate engineering approaches.

A11.1.4 Job descriptions
Safety and environmental issues have both grown progressively in importance for industry and government at all levels. In industry there has been an increasing tendency for job titles and descriptions to assign responsibilities which cover both safety and environment.

A11.2 Common Elements

A11.2.1 Economic factors
Both SLP and EP have a fundamental influence on the design of the plant. Each imposes constraints which may well be decisive in the economics of the design.

A poor design is more likely to require the addition of numerous protective devices and the installation of much pollution control equipment and to involve more last-minute modifications to meet legislative and other safety and environmental standards.

A11.2.2 Inherently safer/cleaner design
The concept of inherently safer design, discussed in Chapter 11 and Chapter 32, has its counterpart in that of inherently cleaner design, or, more broadly, waste minimization.

Table A11.1   *Selected references on pollution, effluents and waste disposal*

Chementator (Table A1.4); EPA (n.d./1, 1988/2); MCA (SG-9, n.d./ll); RCEP (Appendix 28); Mohlman (1950); Brady (1959); Cross (1962); Carson (1963); Byrd (1968); Chem. Engng (1968–); Chieffo and McLean (1968a,b); Marshal (1968); AIChE (1969–/120–126); Bond and Straub (1972, 1973); Cecil (1972); Levine (1972b); Nilsen (1972b); Prober (1972–); Racine (1972); Ripley (1972); S.S. Ross and White (1972); G.F. Bennett (1973); DoE (1973 Poll. Pap. 4, 1974 Poll. Pap. 1, 1976 Poll. Pap. 9); R.D. Fox (1973); Vervalin (1973b,c, 1979); Mabey (1974); McKnight, Marstrand and Sinclair (1974); Napier (1974a); Rudolph (1974); Sax (1974); Strahler and Strahler (1974); Jaffe and Walters (IChemE 1975/60); Roulier, Landreth and Carnes (1975); Singer (1975); Yamaguchi (1975); Boyd and Leotta (1976); CONCAWE (1976 6/76, 1977 2/77, 1979 5/79); R.W. James (1976); Warner (1976, 1979); Anon. (1977f); Allaby (1977); Ashby (1977, 1979); E.B. Harrison (1977); Hillman (1977); Holum (1977); P. Sutton (1977b); Boyd (1978); G. Parkinson (1978, 1980, 1987); Pier *et al.* (1978); Pocock and Docherty (1978); Ricci (1978c); Schneiderman (1978); Bridgwater and Mumford (1979); Golden *et al.* (1979); Harwell (1979); Baasel, Mcallister and Kingsbury (1980); Bakshi and Naveh (1980); Dix (1980); Eggington (1980); Moss (1980); Anon. (1981i); Kletz (1981J, 1993e, 1994 LPB 110); Portman and Norton (1981); Bennett, Feates and Wilder (1982); Trewhitt (1982); API (1983/7); ASME (1983/194); R.M. Harrison (1983); Macrory (1983); Train (1983); Barnes, Forster and Hrudey (1984–); D. Clarke (1984); Jalees (1985); Neely and Blau (1985); Pitt and Pitt (1985); D. Williams (1985); APCA (1986); Hollowood (1986); Vervalin (1986a); Anon. (1987i); Anon. (1988e); C. Butcher (1989c); CIA (1989 CE23, 1991 BT23); Hoover *et al.* (1989); Penkett (1989); Veselind, Peirce and Weiner (1990); IBC (1991/82–84); Jackman and Powell (1991); Shillito (1991); Anon. (1992 LPB 108); Diepolder (1992); Goldsmith (1992a); Parfitt and Andreasen (1992); Reid (1992); Wainwright (1992); Anon. (1993a,f); API (1993 Publ. 311, 317); Doerr (1993); Englehardt (1993); Hydrocarbon Processing (1993b); Karrh (1993); Anon. (1994 LPB 120, p. 13); Agra Europe (1994a); Debeil and Myren (1994 LPB 119); Shillito (1994 IChemE/110)
BS 7750: 1992

Waste minimization, inherently cleaner design
Mencher (1967); Kohn (1978b); Duffy (1983); Tavlarides (1985); Goodfellow and Berry (1986); Koenigsberger (1986); EPA (1987b, 1988, 1989a,b); Gardner and Huisingh (1987); ACGIH (1989/34); CIA (1989 CE1, 1990 CE2); Higgins (1989); Redman (1989c); H.M. Freeman (1990); Hethcoat (1990); Hunter and Benforado (1990); API (1991 Publ. 302); Berglund and Lawson (1991); R. Smith and Petela (1991–); Crittenden and Kolaczkowski (1992); Curran (1992); IBC (1992/88); Chadha and Parmele (1993); IChemE (1993/106, 1994/111); Rossiter, Spriggs and Klee (1993)

AIChE Center for Waste Reduction Technologies
L.L. Ross (1991)

Pilot plants
Gundzik (1983); d'Aco, Campbell and Stone (1984); Pitt and Pitt (1985); Gurvitch and Lowenstein (1990)

Chemicals in the environment, inc. physical and chemical properties

ASTM (Appendix 28); Sax (1957); MCA (1966–/13, 14); CONCAWE (1970 9/70, 1971 9/71, 1985 5/85); Heck, Daines and Hindawi (1970); AIChE (1971/132, 1975/138, 1980/147); Walker (1971); A Tucker (1972); NAS (1975); Moein, Smith and Stewart (1976); Neeley (1976, 1980); Curtis, Copeland and Ward (1978); Mackenthun and Guarria (1978); SCI (1978); S.D. Lee and Mudd (1979); Tinsley (1979); Afghan and Mackay (1980); Alexander (1980); Blair (1980); Bungay, Dedrick and Matthews (1980); Freed (1980); Hanson (1980); Haque (1980a,b); Haque et al. (1980); Keith (1980); Khan (1980); Kimerle (1980); Lijinsky (1980); Mackay, Shiu and Sutherland (1980); Mill (1980); Murphy (1980); Riordan (1980); M.E. Stephenson (1980); Stern (1980); Weber et al. (1980); Zepp (1980); Jorgensen and Johnson (1981); N.J. King and Hinchcliffe (1981); Neely (1981); API (1982 Publ. 4434); P.R. Edwards, Campbell and Milne (1982); Kullenberg (1982); Versino and Ott (1982); Rowe (1983); Dragun, Kuffner and Schneiter (1984); Iyengar (1984); Jorgensen (1984); Okouchi and Sasaki (1984); Piver (1984); Greenland (1985); Neely and Blau (1985); CEFIC Workshop (1986); Donkin and Widdows (1986); Moriarty (1986); Coughtrey, Martin and Unsworth (1987); Ashworth et al. (1988); Valsaraj and Thibodeaux (1988); Pagna and Ottar (1989); Barton, Clark and Seeker (1990); Lyman, Reehl and Rosenblatt (1990); Manahan (1991, 1993); Wakoh and Hirona (1991); Brusseau (1992); Thoma et al. (1992); Gentile et al. (1993); Kollig, Kitchens and Hamrig (1993); Marple and Throop (1993); Larson and Weber (1994)
Environmental limit values, indices: Ott (1978); Sittig (1994)

Particular chemicals

Lead: WHO (EHC3); Bryce-Smith (1971, 1982); DoE (1974 Poll. Paper); Collingridge (1979); RCEP (1983 Rep. 9); *Mercury*: Hartung and Dinman (1972); W.E. Smith and Smith (1975); *PCBs*: WHO (EHC2); EPA (1975a); Higuchi (1976); Willman, Blazevich and Snyder (1976); Stroud, Wilkerson and Smith (1978); Berry (1981); Anon. (1984a); Kokozska and Flood (1985); Motter (1991)
Dioxin, TCDD: Ayres (1981); Rice (1982); R.E. Tucker, Young and Gray (1983); Anon. (1984p); J.L Fox (1984b, 1985); Short et al. (1984); D. Williams (1984); Arienti et al. (1988)
Pesticides: WHO (EHC5, EHC9); Carson (1963); F. Graham (1970); C.A. Edwards (1974)
Nitrates: Nicolson (1979)

Pollution and human health

Howe (1970); Waldbott (1973); Chementator (1975 May 12, 40; Aug. 4, 38; 1977 Jan. 3, 36); Trevethick (1976); Lave and Siskin (1977a,b); Doll and McLean (1979); Moghissi et al. (1980); Bolten (1985); BMA (1991)

Environmental management, environmental risk management

Cecil (1969); Vervalin (1979); IBC (1981/10); APCA (1986); J.J. Stevens (1988); Kolluru (1991); Harwell, Cooper and Flaak (1992); Kraft (1992); Ormon and Isherwood (1992); Welsh (1992, 1993); S. Wilson (1992); Callahan and McCaw (1993); Turney (1993); Petts and Eduljee (1994)
BS 7750: 1992
Environmental planning
D.C. Wilson (1981); Barclay (1987); DoE (1988 Circ. 15/88); Hassan (1993); Petts and Eduljee (1994)
Environmental hazard identification, assessment
Fuquay (1968); Friedl, Hiltz and Marshall (1973); EPA

(1984d); Onishi et al. (1985); Keller and Lamb (1987); B.P. Smith (1987); NSW Govt (1989b); Paustenbach (1989); Cassidy (1990); Picciolo and Metzger (1992); Rouhianen (1992); IBC (1993/98); Petts and Eduljee (1994)
Safety versus environment
Anon. (1992 LPB 108, p. 0); Kletz (1994 LPB 116)
Environmental modelling
W.J. Farmer, Yang and Letey (1980); Burns, Cline and Lassiter (1982); Thibodeaux and Hwang (1982); EPA (1987); Batchelor (1990); Chhiber, Apostolakis and Okrent (1991); McBean (1993)
Environmental auditing
C. Butcher (1972b); Reiter, Sobel and Sullivan (1980); Baumer (1982); Keene (1982); Greeno (1984); D.L Russell (1985); Kingsbury (1994 LPB 118)
Environmental emergency planning
Cashman (1988); Cassidy (1990)

Environmental impact statements

Chementator (1972 May 1, 22; 1974 Apr. 1, 20); Science Applications Inc. (1974); Mueller (1974); Gratt and McGrath (1975a,b, 1976); Jimeson (1975); Dept of Interior (1976); Willis and Henebry (1976, 1982); Cremer and Warner (1977); Jain, Urban and Stacey (1980); Jain and Hutchings (1978); Prater (1978); ICI Australia Ltd (1979); Munn (1979); Atkins Res. and Dev. (1980); Passow (1980); Porter et al. (1980); Pizzi (1982); van Rest (1982); Petts and Edjuljee (1994)

EPA compliance, inspections

Hyland and Moore (1992); Hanisch, Hoffnagle and Walata (1993); Hohman, Sullivan and McNamee (1993)

Cost of pollution, pollution control

RCEP (Appendix 28); Eckenfelder (1969); Popper (1971); Chementator (1975 May 26, 58; Sep. 1, 52; 1977 Jan. 31, 55); E.P. Austin (1977); Isaac (1978); Marion (1978c); Martindale (1979); Vatavuk and Neveril (1980); Anon. (1991h)

Environmental analysis, monitoring

Ottmers et al. (1972); Nemerow (1974); Sittig (1974); McKee (1976); R.G. Thompson (1976); R.M. Harrison and Perry (1977); Cornish (1978a); R. Briggs (1980); Hinchcliffe (1980); Simpson (1980); Hwang and Koerner (1983); Weston (1984); McMorris and Gravley (1993)

Airborne effluents

HSE (Appendix 28 Best Practicable Means Lftls, Emission Test Methods); ASTM (STP 281); HM Chief Alkali Inspector (annual report); MCA (n.d./6–10); WHO (EHC4, EHC7); J.E. Pearson, Nonhebel and Ulander (1935); Hewson (1951); Magill, Holden and Ackley (1956); Dept of Health, Education and Welfare (1960); Lapple (1962); Stern (1962–); Yocom and Wheeler (1963); Hughson (1966); Squires (1967); Brewer (1968); Carlton-Jones and Schneider (1968); Chieffo and McLean (1968a,b); Imperato (1968); Meetham (1968); Munson (1968); Rossano and Cooper (1968); Scorer (1968, 1973); Sickles (1968); Fay and Hoult (1969); Constance (1970, 1972a); Ermenc (1970); Weismantel (1970); Anon. (1971a); AIChE (1971/132, 1972/133, 1974/136, 1975/139, 140, 1976/142, 1979/145, 1980/146, 147, 1981/149); Strauss (1971–, 1975); Anon. (1972a,e); CONCAWE (1972 14/72); Crowley (1972); Elkin and Constable (1972); Iya (1972); Morrow, Brief and Bartrand (1972a,b); Nonhebel (1972); S.S. Ross (1972a); R.S. Smith (1972); Swithenbank (1972); Teller (1972); ASME (1973/32); Cross and Schiff (1973); DoE (1973/2, 3, 1974/4); EPA (1973, 1978a);

IHVE (1973); Lazorko (1973); Peters (1973); Stern et al. (1973); Vandegrift, Shannon and Gorman (1973); Wiley (1973); Dorman (1974); Pfeifer (1974); Ring and Fox (1974); Driscoll (1975); Ecke and Dreyhaupt (1975); Marchello and Kelly (1975); Open Univ. (1975a); Parekh (1975); Rymarz and Klipstein (1975); Schneider et al. (1975a,b); Seinfeld (1975); Crawford (1976); Gammell (1976); O'Connell (1976); RCEP (1976 Rep. 5); Theodore and Buonicore (1976, 1982); API (1977 Waste Manual, vol. 2); M.L. Barker (1977); E. Briggs (1977); Bump (1977); S.K. Friedlander (1977); N. Kaplan and Maxwell (1977); Kohn (1977b); Lasater and Hopkins (1977); Lave and Siskin (1977a); Perry and Young (1977); Preussner and Broz (1977); R.D. Reed (1977); Ricci (1977b); Rosebrook (1977); Straitz (1977); Dunlap and Deland (1978); Freitag and Packbier (1978); A.F. Friedlander (1978); McCarthy (1978); NSCA (1978); A. Parker (1978); Parrish and Seidel (1978); Bochinski, Schoultz and Gideon (1979); Downey and Ni Uid (1979); Hesketh (1979); Parmele, O'Connell and Basdekis (1979); Stockham and Fochtman (1979); Storch (1979); D.P. Wallace (1979); Buonicore (1980); England, Heap and Pershing (1980); Kenson and Holland (1980); Licht (1980); Marzo and Fernandez (1980); Niess (1980); Parungo, Pueschel and Wellman (1980); Theodore, Sosa and Fajardo (1980); Wheeler (1980); Ireland (1981, 1984, 1987); Keene (1981); Lallande (1981, 1982); SCI (1981b); de Wispelaere (1981, 1983, 1984, 1985); Holland and Fitzsimmons (1982); Keith (1982); Record, Bubenick and Kindya (1982); Theodore and Buonicore (1982); E. Weber (1982); Elliot (1983); J.D. Daniels (1984); R.W. Lee (1984); Ogawa (1984); D.D. Adams and Page (1986); APCA (1986); Bretschneider and Kurfurst (1986); Farquhar (1986); Henstock (1986); Perriman (1986); De Wispelaere, Schiermeier and Gillani (1986); Redman (1989a); Veselind, Peirce and Weiner (1990); DoE (1991); ACGIH (1992/75); Benitez (1993); Speight (1993); IBC (1994/109)

Odours
ASTM (STP 164, DS 48); Henderson and Haggard (1922); Summer (1963, 1971); Chementator (1969 Jun. 30, 51); Turk (1969); Yocom and Duffee (1970); Oelert and Florian (1972); H.B.H. Cooper (1973); Dravnieks (1974); Turk, Johnston and Moulton (1974); Charlton, Sarteur and Sharkey (1975); CONCAWE (1975 8/75); Klooster, Vogt and Bernhart (1976); Harrell, Sewell and Walsh (1979); Valentin and North (1980); Kenson (1981); Amoore and Hautala (1983); Artis (1984); D.W. Jones (1984); Keddie (1984); Lynskey (1984 LPB 60); Ferryman (1984); P.J. Young and Parker (1984); Ruth (1986); van Langenhove, Lootens and Schamp (1988); Miedema and Ham (1988); Devis (1990); Valentin (1990); Nagy (1991); A.M. Martin et al. (1992); Sober and Paul (1992)

Liquid effluents
ASTM (STP 130, 148, 1481, 207, 337, 442, 573); CONCAWE (Appendix 28, 1970 3/70, 1975 3/75, 1979 3/79); IP (Appendix 28 Oil Loss Con/s); MCA (n.d./12, 1966–/13, 14); Eckenfelder and O'Connor (1961); Gurnham (1963, 1965); Busch (1965); Eckenfelder (1966, 1969); Fair, Geyer and Okun (1966); AIChE (1967–71/127–129, 1968/130, 1969/ 131, 1972/135, 1974/137, 1975/141, 1977/143, 1981/148); Beychock (1967, 1971); Jaeschke and Trobisch (1967); D.R. Montgomery (1967); SCI (1967); Cross (1968); Dahlstrom (1968); Dykman and Michel (1968); Eliassen and Tchobanoglous (1968); Geinopolos and Katz (1968); Kemmer and Odland (1968); Lesperance (1968); Sebastian and Cardinal (1968); A.R. Thompson (1968);

API (1969 Waste Manual, vol. 1); Cecil (1969); Chopey (1970); Hiser (1970); Min. of Housing and Local Govt (1970); Anon. (1971h); Nemerow (1971, 1974); Characklis and Busch (1972); Moores (1972); Newsom and Sherratt (1972); Nogaj (1972); McLoughlin (1973); Reiterand Sobel (1973); Helliwell and Bossanyi (1975); Huber (1975); IChemE (1975/58); Lash and Kominek (1975); Open Univ. (1975b); H.W. Parker (1975); Bush (1976); Mulligan and Fox (1976); Thompson (1976); Cope (1977); Finelt and Crump (1977); Ford and Tischler (1977); Lederman (1977); Lewin (1977); McDowell and O'Connor (1977); Moss (1977); Paulson (1977); B.A. Bell, Whitmore and Cardenas (1978); Culp, Webster and Culp (1978); IBC (1978/3, 1993/100); Nathan (1978); Ramalho (1978, 1979); C.R. Fox (1979); Freeze and Cherry (1979); Martindale (1979); Sundstrom and Klei (1979); Afghan and Mackay (1980); Brewin and Hellawell (1980); R.A. Freeman (1980); Glaubinger (1980); J.H. Robertson, Cowen and Longfield (1980); D.L Russell (1980); G.K. Anderson and Duarte (1981); Arceivala (1981); Ayling and Castrantas (1981); Chalmers (1981); Howe, Howe and Howe (1981); Iddleden (1981); Sidwick (1981); Walters and Wint (1981); Askew (1982); Olthoff and Oleskiewicz (1982); Selby (1982); Vernick and Walker (1982); Wheatley (1982); Barrs (1983); Langer (1983); Cushie (1984); A.E. James (1984); R.A. Mills (1984); Eckenfelder, Patoczka and Watkin (1985); Crossland (1986); J. Lawrence (1986); NFPA (1990 NFPA 820); Veselind, Peirce and Weiner (1990); Metcalf and Eddy (1991); G. Parkinson and Basta (1991); RCEP (1992 Rep. 16). Deep well disposal: Selm and Hulse (1960); Talbot and Beardon (1964); D.L. Warner (1965); Talbot (1968); Sheldrick (1969); M.E. Smith (1979)
Hydraulic dispersal: Abbott (1961); Fisher et al. (1979); Lloyd, O'Donnell and Wilkinson (1979); Cunge, Holly and Verwey (1980); Kobus (1981); R.E. Lewis (1981); Novak and Cabelka (1981); Komar (1983); ASCE (1986/28)
Oil-water separators: API (1990 Publ. 421)
Stream dispersal, pollution, renewal: Streeter and Phelps (1925); H.A. Thomas (1948); Churchill and Buckingham (1956); Hwang (1980); Nusser (1982); Velz (1984); *Wet air oxidation*: Flynn (1979); Wilhelmi and Knopp (1979); Laughlin, Gallo and Robey (1983); Baillod, Lamparter and Barna (1985); Heimbuch and Wilhelmi (1985)

Spills, spill control
BDH (1970); Anon. (1972b); Attaway (1972); Bock and Sullivan (1972); L.S. Brown (1972); Cairns, Dickson and Grossman (1972); J.B. Cox (1972); G.W. Dawson, Shuckrow and Mercer (1972); Donaldson (1972); Goodson and Jacobs (1972); R.H. Hiltz, Marshall and Friel (1972); Jennings (1972); Laroche (1972); Ludwig and Johnson (1972); D.G. Mason, Gupta and Scholz (1972); R.G. Sanders, Pantezolos and Rich (1972); Schuckrow, Mercer and Dawson (1972); Welch, Marmelstein and Maughan (1972); Wilder and Brugger (1972); Wye (1972); Ziegler and Lafornara (1972); May, McQueen and Whipp (1973); AIChE (1974/99, 1980/100, 144, 1988/100, 1989/70); May and Perumal (1974); Small and Snyder (1974); Univ. Engrs Inc. (1974); Weismantel (1974); Welker, Wesson and Brown (1974); Brugger and Wilder (1975a,b); G.W. Dawson (1975); Lindsey (1975); Otterman (1975); Wirth (1975); Armstrong, Gloyna and Wyss (1976); L.E. Brown et al. (1976); Bunner (1976); F.E. Clark (1976); W.D. Clark (1976b); Conner (1976); Courchaine and Dennis (1976); Dalton (1976); G.W. Dawson, Mercer and Parkhurst (1976); Diefenbach (1976); Eagon et al. (1976); Freestone (1976); Freestone, Lafornara and

McCracken (1976); Fullner and Crump-Wiesner (1976); Gupta 1976); Hooper, Warner and Galen (1976); Imbrie *et al.* (1976); Jennings and Gentry (1976); Jensen (1976); Jouhola and Bougger (1976); Kaufmann (1976); Kirsch *et al.* (1976); Michalovic *et al.* (1976); Moser, Hargens and Wilder (1976); Overfield and Richard (1976); Rice *et al.* (1976); K.C. Roberts (1976); Rodgers (1976); Roehlich, Hiltz and Brugger (1976); Seidenberger and Barnard (1976); S. Shaw and Gower (1976); Silvestri *et al.* (1976); E.A. Simons (1976); Sinclair and Bower (1976); T.J. Thomas, Srinavasan and Gower (1976); B.C. Weber (1976); Whitebloom (1976); Wiener and Heard (1976); Huibregste *et al.* (1977); Martinsen and Muhlenkamp (1977); Badley and Heaton (1978); D.P. Brown (1978); Charlton *et al.* (1978); Corbett (1978); Darnell (1978); G.W. Dawson and McNeese (1978); G.W. Dawson, McNeese and Christensen (1978); DOT (1978); Feates (1978); Freestone and Zaccor (1978); Gilman and Freestone (1978); Glattley (1978); Griwatz and Brugger (1978); Halvorsen (1978); Hand and Ford (1978); Harsch (1978a,b); Hubbs (1978); Huibregste, Lafornara and Kastman (1978); Huigbregste, Moser and Freestone (1978); Imbrie, Karwan and Stone (1978); Khattak and Whitmore (1978); Kirsch *et al.* (1978); Kiygielski and Bennett (1978); Ladick (1978); Lafornara, Frank and Wilder (1978); Lafornara, Lampl *et al.* (1978); Lafornara, Marshall *et al.* (1978); Lampl, Massey and Freestone (1978); Lathrop and Larson (1978); Mercer *et al.* (1978); Michalovic *et al.* (1978); Moien (1978); Neely and Mesler (1978); Norman and Dowell (1978, 1980); T.W. Pearson (1978); Rinka and Schulz (1978); Roth *et al.* (1978); Ryckman, Wiese and Ryckman (1978); Schultz and Shah (1978); Seeman *et al.* (1978); Silvestri *et al.* (1978); J.T. Smith (1978); Soden and Johnson (1978); Stonebreaker, Freestone and Peltier (1978); Thuma *et al.* (1978); Tsahalis (1978); Urban and Losche (1978); Whiting and Shafer (1978); Willmoth (1978); Tracy and Zitrides (1979); Gunderloy and Stone (1980); Weidenbaum (1980); Huibregste and Kastman (1981); Schwartzman (1981); Shooter, Lyman and Sinclair (1981); Wentzel *et al.* (1981); Armstrong (1982); G.F. Bennett, Feates and Wilder (1982); Brugger (1982); Eckenfelder (1982); R.H. Hiltz (1982); Kletz (1982); Lafornara (1982); Ludwigson (1982); Nadeau (1982); Nusser (1982); Scholz (1982); ASTM (1983 STP 825, 1990/1); Jeulink (1983); G.W. Dawson and Mercer (1986); R.H. Hiltz (1988); Martins (1988); Melvoid and Gibson (1988); Abbott and Leeman (1990); ACGIH (1992/87); Browne (1993)

Oil spills

Anon. (1969d); Pattison (1969); Fay (1970); CONCAWE (1974 4/74, 1980 9/80, 1981 7/81, 1981 9/81, 1983 10/83, 1987 1/87, 1988 2/88); d'Alessandro and Cobb (1976b); Fingas and Ross (1977); C.J. Clark (1978); IP (1978 PUB 11, 1983 PUB 18, 1985/3); Middleditch (1981); API (1982 Publ. 2022); ASTM (1984 STP 840)

Marine
RCEP (1972 Rep. 2, 1981 Rep. 8); WSL (1972); Cttee on the Environment (1974); F.W.G. Smith (1974); DoE (1976 Poll. Pap. 8); Johnston (1977); Cole (1979); Jaafe *et al.* (1981); Norcross (1981); Kullenberg (1982); IP (1993 TP 27); Lawrence and Cross (1994)

Torrey Canyon
Cowan (1969)

Hazardous and solid wastes
MCA (SG-9, 1974 SW1-SW3); Mohlman (1950); SCI (1957); Rickles (1965); FPA (S3, 1971/15); Key (1970); Novak (1970, 1972); Anon. (1971f); IChemE (1971/51); Witt (1971b, 1972); AIChE (1972/134); Baum and Parker (1973); Santolieri (1973); IMechE (1975/44, 1977/49); Mantell (1975); Open Univ. (1975d); Patrick (1975); Roulier, Landreth and Carnes (1975); Saxton and Narkus-Kramer (1975); Stephenson *et al.* (ASME 1975/186); Besselievre and Schwartz (1976); Boily (1976); D.R. Davies and McKay (1976); Frisbie (1976, 1978); Lazar (1976a,b); E.P. Austin (1977); Feates (1977); Hillman (1977); Keen (1977); Paulson (1977); S.W. Pearce (1977); G. Jones (1978); Morrison and Ross (1978); Ricci (1978b); Huddlestone (1979); Okey, Digregorio and Kominek (1979); Pojasek (1979a,b); J. Smith *et al.* (1979); API (1980 *Waste Manual*); Conway and Ross (1980); Gradet and Short (1980); Anon. (1981n); Basta (1981b,e); Jamieson (1981); Shen (1981); ASTM (1982 STP 760, 1983 STP 805, 1984 STP 851); Bentley (1982); Farquhar (1982); Kiang and Metry (1982); R.D. Ross (1982); Skitt (1982); Thibodeaux, Springer and Riley (1982); A.N. Clarke and Clarke (1983); Chivers (1983, 1984); Cope, Fuller and Willetts (1983); G.W. Davidson (1983); Dyer and Mignone (1983); Finnecy (1983); Jennings (1983); A. Lawrence (1983); D.A. Mills (1983); Cook (1984 LPB 55); Hawkins (1984); Hillman (1984); Kennard (1984); Luck (1984); Mackie and Niesen (1984); E.F. Wood (1984); Zirschky and Gilbert (1984); Anon. (1985o); Basta, Hughson and Mascone (1985); Pitt and Pitt (1985); Porteous (1985); RCEP (1985 Rep. 11); Anon. (1986k); Arthurs (1986); Hollowood (1986); Payne (1986); W.F. Robinson (1986); Barclay (1987); M. Bradford (1987); CIA (1987 RC15, RC16, 1989 RC17); Colen (1987a,b); Crittenden and Kolaczowski (1987); Crumpler and Martin (1987); EPA (1987); Loehr (1987); E.J. Martin and Johnson (1987); Stone (1987); Wiles (1987); Rendell (1988a); Hammitt and Reuter (1989); Salcedo, Cross and Chrismon (1989); ACGIH (1990/42, 1992/83, 84); Earth (1990); Batchelor (1990); L.W. Jones (1990); Pekelney (1990); Soudarajan (1990); Tedder and Pohland (1990); Veselind, Peirce and Weiner (1990); Weitzman (1990); HMIP (1991); Nemerow and Dasgupta (1991); IBC (1992/87); Paluzzi (1992); IMechE (1994/170); Petts and Eduljee (1994); Sara (1994)
Incineration
ASME (Appendix 28 Solid Waste); Frankel (1966); Monroe (1968, 1983b); Cross (1972); CONCAWE (1975/2); Dunn (1975, 1979); C.R. Lewis, Edwards and Santoro (1976); R. Harrison and Coulson (1977); Bartsch, Gilley and Steele (1978); J. Jones (1978); Dunn (1979); Fabian, Reher and Schon (1979); Hitchcock (1979); Ready and Schwab (1980 LPB 36, 1981); HSE (1981 BPM 11); Deneau (1983); Frankel, Sanders and Vogel (1983); Monroe (1983b); Anon. (1984aa); Feeley (1984); H.M. Freeman (1984); Brunner (1985, 1986, 1989, 1993); Rickman, Holder and Young (1985); Basta (1986a); Zurer (1986); Baker-Counsel (1987a); Beychok (1987); H.M. Freeman *et al.* (1987); Wiley (1987); C. Butcher (1990a); IBC (1990/78, 1992/93); Vervalin (1990a); Veselind, Peirce and Weiner (1990); P.T. Williams (1990); Ondrey and Fouhy (1991); Zeng and Okrent (1991); Altorfer (1992); S.E. Anderson, Dowell and Mynaugh (1992); Haseltine (1992); RCEP (1993 Rep. 17)
Landfill
Perket (1978); Duvael (1979); Basta (1981b); Corbin and Lederman (1982); IBC (1982/22); A. Parker (1982); G.W. Dawson (1983); J.D. Cook (1984); Hoppe (1984a); Crawford

and Smith (1985); Shuckrow, Touhill and Pajak (1987);
Anon. (1988k); Shacke and Sawai (1990); Veselind, Peirce
and Weiner (1990); Kemp and Gerrard (1991)
Transport
R.J. Buchanan (1982); Bromley and Finnecy (1985)

Ocean disposal, ocean incineration
Chementator (1970 Aug.24, 32; 1974 Oct.28, 33; 1975 Dec.8,
82); Zapatka, Hann and Zapatka (1976); K. Pearce (1977);
Ketchum, Kester and Park (1981); Finnecy (1982); Novak
and Pfrommer (1982); Remirez (1982); Anon. (1984d);
NRC (1984); Ditz (1988)

Waste sites, site clean-up, contaminated land
Dennis (1978); Kohn (1978d); Overcash and Paul (1980); SCI
(1980); IBC (1981/18, 1992/89); D.C. Wilson, Smith and
Pearce (1981); Bowders, Koerner and Lord (1982); Long and
Schweitzer (1982); Lord, Koerner and Freestone (1982); D.C.
Wilson (1982); R.E. Edwards, Speed and Verwoert (1983);
Muller, Brodd and Leo (1983); Rogoshenski, Boyson and
Wagner (1983); Tyagi, Lord and Koerner (1983a–c); Block,
Dragun and Kalinowski (1984); Block and Kalinowski
(1984); Ehrenfelder and Bass (1984); Husak, Kissenpfennig
and Gradet (1985); MA Smith (1985); Wagner *et al.* (1986);
Lord and Koerner (1988); des Rosiers (1987); Arenti *et al.*
(1988); Hopper (1989); Porter (1989); Ahlert and Kosson
(1990); Bleicher (1990); IP (1991 PUB 58, 1993/2);
Raghavan, Coles and Dietz (1991); AGA (1992/86);
N. Morgan (1992a); Vandegrift, Reed and Tasker (1992);
Cairnie (1993); Pratt (1993)
Superfund
Ember (1984); Hoppe (1984b); Casler (1985); Sidley and
Austin (1987); Redman (1989b); Bisio (1991b, 1992a)

Love Canal
Glaubinger, Kohn and Remirez (1979); J.L. Fox (1980a);
Gage (1980); Holden (1980); Kolata (1980); Picciano (1980);
M.W. Shaw (1980); M. Brown (1981); Kohn (1982); Anon.
(1984hh)

Rechem
Anon. (1984a,i,o,aa,dd); Anon. (1985y,z); Anon. (1986d,x);
Anon. (1987v); J. Cox (1988b)

The philosophy of inherently cleaner design is to avoid
the generation of noxious effluents. As with inherently
safer design, this is easier said than done, but progress has
been made and the pressures to operate cleaner processes
continue to grow.

A11.2.3 Intermediate storage
A prime theme in inherently safer design is the elimination
of intermediate storage. It transpires that the elimination of
storage tanks is also a significant feature of waste minimization programmes. In the first case the motivation is to
reduce the inventory of hazardous materials, while in the
second it is to reduce the quantities of tank residues and
wash liquids.

A11.2.4 Hazard identification
The general approach to hazard identification developed in
SLP is also applicable to EP.

The hazop method, for example, is increasingly being
adapted or enlarged to cover the environment.

A11.2.5 Hazard assessment
Likewise, hazard assessment is finding application in EP
as well as in SLP.

Although much of the methodology of hazard assessment used in SLP is applicable to environmental problems,
there are also some significant differences. The treatment
of the events leading up to a release uses an essentially
common methodology, but the modelling of the consequences of a release naturally differs more significantly.

Thus whereas for SLP hazard assessment involves estimating and evaluating the consequences in terms of injury
to people and damage to property, for EP the consequences
to be considered are the fate of chemicals in the environment, and their ultimate effects on plant and animal life.

A11.2.6 Hazard models
Many of the models used in hazard assessment for SLP are
applicable also to EP, and indeed it was for the latter that
some were first developed.

Thus the treatment of emission, vaporization and dispersion largely shares a common methodology, although
there are some differences of emphasis. In gas dispersion,
for example, heavy gas dispersion is of particular significance in SLP, while for EP passive dispersion, including dispersion over much greater distances, plays a more
significant role.

A11.2.7 Fugitive emissions
A problem common to SHE and EP is that of low level, or
fugitive, emissions from a plant. Such emissions have an
impact on the health of workers in the plant and on the
population and environment outside.

In SLP the need for stricter controls on emissions within
the plant was evidenced by the vinyl chloride problem in
1975, while in EP pressure for emission controls arose from
studies of the causes of air pollution in cities such as Los
Angeles in the later 1970s, which implicated emissions from
nearby refineries.

A11.2.8 Environmental impact assessments
There has been a general trend to require that an industrial
activity make an environmental impact assessment (EIA)
or environmental impact statement (EIS), also referred to,
respectively, as an environmental assessment (EA) and
an environmental statement (ES). Such environmental
impact assessments and statements include safety issues,
although mainly in respect of hazards to the public.

A11.2.9 Safety cases
Closely related is the requirement in European legislation
for preparation of a safety case. The requirement to address
hazards to the environment as well as humans has received
increasing emphasis.

A11.2.10 Communication with public
Process plants and hazardous waste facilities are both
liable to generate opposition, prior to creation or in the
course of operation. The need to communicate with the
public, and the approach required, are very similar in both
cases.

A11.3 Some Conflicts
Another link between SLP and EP is that in some cases
a degree of conflict arises. In some cases the preferred

solution to a safety problem may be ruled out, or at least discouraged, on environmental grounds; in others stricter environmental controls may push the designer towards solutions which would not otherwise be adopted for safety reasons.

Accounts of such conflicts have been given by Elphick (1972), R.Y. Levine (1972b), Bodurtha (1976) and Kletz (1993e).

Some of the practices which are so affected are now considered.

A11.3.1 Reactor venting
The normal method of protection for a chemical reactor has been the provision of a bursting disc with a short, straight pipe discharging to atmosphere at a safe place.

If such discharge is not permitted but this basic method of protection is retained, it becomes necessary to pipe each bursting disc into a vent header leading to a disposal system such as a scrubber or flare. Some of the problems which this creates have been described by Levine.

It is largely for this reason that the alternative approach of using instrumented protective systems on reactors has become increasingly attractive, as described in Chapter 11.

A11.3.2 Pressure relief valve discharge
An essentially similar conflict and similar problems arise over the practice of fitting pressure relief valves which discharge directly to atmosphere.

Some of the problems of vent collection systems are described by Kletz.

Avoidance of discharge to atmosphere is one of the main arguments for the use of trip systems to protect against overpressure, as described in Chapter 13.

A11.3.3 Flaring
A flare radiates intense heat and light and may be smoky and noisy. These are all aspects of pollution which have become increasingly unacceptable.

In contrast to the oil industry, which has continuous flaring, the chemical industry often makes use of an intermittent flare. The flare may be used to burn small quantities of smelly materials. But since very small quantities are sometimes not ignited by the flare pilot burner, it is sometimes necessary to operate the flare continuously, using fuel gas which could otherwise be used more productively.

A11.3.4 Bleeding
The build-up of impurities in plants is avoided by taking off a bleed. If this practice is inhibited by environmental considerations, the impurity can accumulate and may create a hazard. Levine describes a multimillion dollar loss which occurred due to the build-up of the explosive impurity azomethane.

A11.3.5 Leakages
The degree of leakage which has been tolerated in the past is now frequently unacceptable. The problems which can then arise are illustrated by the leakage of hydrogen from mercury cell chlorine plants. The hydrogen stream is one of the carriers of mercury vapour from the cells. Therefore mercury escapes in the hydrogen leakages. One possible solution is to run with the hydrogen under a slight negative pressure, but this creates the risk of air ingress leading to an explosion.

A11.3.6 Incineration
Burning in open pits has been widely used as method of disposing of waste solvents on chemical plants. Closed incinerators eliminate this pollution, but present hazards of their own.

A11.3.7 Ventilation
The hazard from flammable leaks in a compressor house can be reduced if the building is of open construction such that ventilation is much enhanced. This solution runs into the difficulty, however, that the noise of the machinery carries further.

A11.3.8 Combustion processes
The attempt to reduce the emission of nitrogen oxides from combustion processes can increase the risk of explosion. A marked decrease in nitrogen oxides formation occurs as excess air is reduced. But reduction of excess air increases the probability of an explosion due to unburned fuel in the gas leaving the combustion chamber.

A11.3.9 Pollution control equipment
Equipment for the removal of flammable dust from gases can present another hazard. Although this can be minimized by the use of wet scrubbers, these run into difficulties from water pollution controls.

A11.3.10 Halons
Halons have found a distinctive role in the process industries as fire extinguishants and explosion suppressants. The banning of chlorofluorohydrocarbons (CFCs) creates a gap which for fire extinguishants is not yet fully filled.

Pollution of the Environment

Awareness of the seriousness of environmental pollution grew rapidly in the 1960s. The publication of *Silent Spring* by Carson (1963) was a landmark. Other warnings have followed such as *Bitter Harvest* by Egginton (1980).

Accounts of pollution and environment include *Handbook of Environmental Control* by Bond and Straub (1973–), *The Pollution Handbook* by Mabey (1974), Introduction to Environmental Science by Strahler and Strahler (1974), *A Dictionary of the Environment* by Allaby (1977), *Topics and Terms in Environmental Problems* by Holum (1977), *The Protection Handbook of Pollution Control* by P. Sutton (1977b), *Environmental Education* by Bakshi and Naveh (1980), Britain, Europe and the Environment by MacRory (1983) and *Environmental Pollution and Control* by Veselind, Peirce and Weiner (1990).

Also relevant are the reports of the Royal Commission on Environmental Pollution.

A11.4 Legislation

Although there are significant differences between gaseous effluents, liquid effluents and solid wastes, some of the legislation addresses more than one of these, and it is convenient therefore to take all three together.

A11.4.1 Air pollution
The Alkali etc. Works Regulation Act 1906 was for many years the principal act governing air pollution.

The philosophy of enforcement of the act has been the use of best practicable means (BPM) to minimize emissions.

The HSWA 1974 states in Section 1(1) (d) that its provisions are made with a view to controlling emissions of noxious substances and it contains in Section 5 a general duty to use best practicable means to prevent such emission.

The Health and Safety (Emissions into the Atmosphere) Regulations 1983, amended 1989, create in Section 3 prescribed classes of premises for the purposes of Section 1(1) (d) of the HSWA, listed in Schedule 1, and specify under Section 4 substances prescribed as noxious. These premises and substances are listed in Schedules 1 and 2, respectively.

A11.4.2 Control of Pollution Act 1974

The Control of Pollution Act 1974 (COPA) has four main parts: Part I, Waste on Land; Part II, Pollution of Water; Part III, Noise; and Part IV, Pollution of the Atmosphere.

Part I of the act states requirements for the disposal authority to control waste disposal by a system of licensing. It prohibits the unlicensed disposal of controlled waste.

In Part II the act lays down the duties of the Water Authorities to operate a system of consents for discharges and prohibits noxious discharges.

The provisions of Part IV on atmospheric pollution are limited and deal mainly with pollution by certain fuels and with information about air pollution.

A11.4.3 Environmental Protection Act 1990

The Environmental Protection Act 1990 (EPA) has nine parts, those of principal relevance here being Part I, Integrated Pollution Control and Air Pollution Control by Local Authorities; Part II, Waste on Land; and Part III, Statutory Nuisances and Clean Air.

Part I of the act establishes a system of integrated pollution control (IPC) enforced by HM Inspectorate of Pollution (HMIP) and one of air pollution control (APC) administered by the local authorities (LAAPC).

Section 1 of Part I gives definitions of environment, pollution and harm which read as follows:

(2)   The 'environment' consists of all, or any, of the following media, namely the air, water and land; and the medium of air includes the air within buildings and the air within other natural or man-made structures above or below ground.

(3)   'Pollution of the environment' means pollution of the environment due to the release (into any environmental medium) from a process of substances which are capable of causing harm to man or any other living organisms supported by the environment.

(4)   'Harm' means harm to the health of living organisms or other interference with the ecological systems of which they form part and, in the case of man, includes offence caused to any of his senses or harm to his property; and 'harmless' has a corresponding meaning.

Section 2 creates a system of prescribed processes and substances and Section 3 allows regulations to be made for their control. Subsection 2 of Section 3 allows the prescription of standard limits for the concentration, the amount or the amount in any period to be released; of standard requirements for measurement; and of standards

or requirements for the process. Subsection 5 allows the making of plans for establishing limits for the total amount or total amount in any period to be released in any area; for quotas for operators; and for establishing or reducing limits so as progressively to reduce pollution.

Sections 6–12 create a system of authorizations to operate a prescribed process. Section 7 deals with the conditions of such an authorization. It requires that in carrying on such a process use should be made of the best available technology not entailing excessive cost (BATNEEC) to prevent release of prescribed substances or to render harmless any other potentially harmful substances.

Section 4 establishes the enforcing authorities and Sections 13–19 deal with enforcement. Section 13 gives powers to make an enforcement notice and Section 14 a prohibition notice.

Section 23 gives a list of offences and penalties. Section 27 gives powers such that where an offence causes any harm which it is possible to remedy the inspectorate may arrange for any reasonable steps to be taken to remedy the harm and may recover the cost from any person convicted.

The EPA covers discharges into all environmental media. Some processes have the potential to make significant discharges into more than one medium, in some cases as a result of transfer from one medium to another. An example is removal of substances from a gas stream which then ends up in a liquid effluent. In such situations the EPA requires the selection of the best practical environmental option (BPEO) and the operator is required to demonstrate that its process meets the BPEO criterion.

Part II of the act creates a duty of care in respect of wastes and a system of waste management licences.

Part III defines a number of statutory nuisances which are 'prejudicial to health or a nuisance' and which include emissions such as smoke; fumes and gases; dust, steam, smell and other effluvia; accumulations or deposits; and noise. It lays on a local authority a duty to inspect its area to detect such nuisances and to act on them.

The prescribed processes and substances are given in the Environmental Protection (Prescribed Processes and Substances) Regulations 1991. Schedule 1 contains a number of chapters, each of which deals with a particular industry. The chemical industry is covered in Chapter 4 of the schedule.

A distinction is made between those processes which are subject to control by local authorities in respect of air pollution only and those which have the potential for pollution of more than one environmental medium, which require an IPC authorization from the HMIP – the Part B and Part A processes, respectively.

A11.4.4 Clean Air Act 1993

The Clean Air Act 1993 covers pollution by smoke, grit, dust and fumes from industrial processes not required to be registered under the HSWA 1974 and the Health and Safety (Emissions into the Atmosphere) Regulations 1983.

The act prohibits the emission of 'dark' smoke from any chimney or industrial premises.

A11.4.5 Water pollution

Until 1973 local authorities had the duty of providing sewerage and sewerage treatment not only to domestic but also as far as possible to industrial users. Controls on effluents from industrial premises to the public sewerage system were established by the Public Health (Drainage of Trade Premises) Act 1937.

The Water Act 1973 transferred the responsibilities for industrial effluents to the public Water Authorities (WAs). The operation and maintenance of the sewerage system continued to be done by the local authorities on an agency basis.

The Control of Pollution Act 1974 strengthened the controls on water pollution.

The development of controls on water pollution has been influenced by the apparent deterioration of water quality in the North Sea and by a series of North Sea Conferences of the states concerned, the first being held in 1984.

Substances identified as particularly harmful to the aquatic environment by reason of their toxicity, persistence and bioaccumulation are identified in the 'Red list'.

Control of water pollution is now governed by the Water Act 1989, described below, and the EPA 1990.

The system being developed under these acts is one of integrated pollution control based on environmental quality objectives (EQOs) and environmental quality standards (EQSs) utilizing the BATNEEC criterion.

A11.4.6 Water Act 1989
The Water Act 1989 defines the waters which are subject to control, establishes the powers of control and creates the National Rivers Authority (NRA).

Prior to privatization the WAs were responsible both for the supply of water and for control of pollution of water. In other words, they were both poacher and gamekeeper. The need to separate these functions led to the creation of the NRA, which is now responsible for protecting the aquatic environment.

Section 103 of the Act defines the various waters subject to controls, which are essentially inland waters, including rivers and watercourses; groundwaters; coastal waters; and territorial waters out to a 3 mile limit.

Part III of the Act gives the NRA powers to improve water quality and to control effluent discharges.

Discharge of liquid effluent is governed by a system of discharge consents. The discharge consent procedures are given in Schedule 12 of the Act.

Public participation is allowed by arrangements for advertising a discharge application in the local press and by a duty on the authority to consider representations made.

A11.4.7 Hazardous wastes
Control of hazardous wastes is exercised under the Deposit of Poisonous Waste Act 1972 and the Deposit of Poisonous Waste (Notification of Removal or Deposit) Regulations 1972.

Hazardous waste controls are also major elements of the Control of Pollution Act 1974, Part I and the EPA 1990, Part II, as already described. The EPA 1990 creates a system of waste management licences.

A11.4.8 Environmental impact assessment
EC Directive 85/337/EEC on the assessment of the effect of certain public and private projects on the environment creates for certain projects requirements for an environmental assessment.

In the United Kingdom, implementation of these requirements is largely covered by the Town and Country Planning (Assessment of Environmental Effects) Regulations 1988 (the EA Regulations).

The regulations specify that a developer provide information on the impact of the project in the form of an environmental statement.

A11.4.9 Advisory bodies
There is a standing Royal Commission on Environmental Pollution (RCEP). The reports of the commission are listed in Appendix 28.

Among the topics addressed are coastal water pollution (1972, *Third Report*), integrated air pollution control (1976, *Fifth Report*), sea pollution (1981, *Eighth Report*), waste management (1985, *Eleventh Report*), best practical environmental option (1988, *Twelfth Report*), freshwater quality (1992, *Sixteenth Report*) and waste incineration (1993, *Seventeenth Report*).

A11.4.10 Safety cases
EC Directive 82/501/EEC on major accident hazards creates controls on certain installations which include the submission of a safety report. There have been two subsequent modifying directives, the second 88/610/EEC containing requirements in response to the Rhine pollution incident at Schweizerhalle in 1986.

The initial directive is implemented in the United Kingdom by the CIMAH Regulations 1984 and the two later directives by amendments of these regulations. In the second set of amendments in 1990 the requirement for the safety case is contained in Regulation 7 and its contents are specified in Schedule 6.

Schedule 6 refers to information on the consequences of a major accident without specifically mentioning those to the environment, but guidance given in HS(R) 21 by the HSE (1990) states that in relation to the environment information should include the routes by which harm may be brought about and the effect on the exposed environment including persistence of the substance in the environment.

A11.5 EC Directives

There are a considerable number of EC Directives concerned with the environment. Some of the principal directives are listed in Chapter 3.

An information service on EC legislation is given in *European Environment Law for Industry* by Agra Europe (1994a).

Directive 76/464/EEC gives a list of substances dangerous to the environment. 80/68/EEC addresses groundwater protection. Discharge of certain substances listed in 76/464/EEC is covered by 86/280/EEC. There are directives on discharge of metals such as mercury and cadmium. Directive 84/360/EEC deals with air pollution from industrial plants and 88/609/EEC with large combustion plants. There are directives on ozone, sulfur dioxide, nitrogen dioxide and carbon dioxide. Other specific pollutants covered by directives include lead, asbestos and PCBs. Toxic and dangerous wastes are the subject of Directives 78/319/EEC and 91/689/ EEC, while 84/631/EEC deals with transfrontier shipment of wastes.

Directive 85/337/EEC requires for certain activities the production of an environmental impact assessment.

A11.6 US Legislation

Accounts of US legislation on the environment and on pollution control are given by Lipton and Lynch (1987),

E.J. Martin and Johnson (1987), Nemerow and Dasgupta (1991) and Veselind, Peirce and Weiner (1991).

The account given below complements that given in Chapter 3.

A11.6.1 Air pollution
In the United States, air pollution legislation includes the Air Pollution Control Act 1975, the Clean Air Act 1963 (CAA) and the Clean Air Act Amendments 1970, 1977 and 1990 (CAAA).

The CAAA 1970 required the EPA to establish National Ambient Air Quality Standards (NAAQSs). The CAAA 1977 brought in requirements for the NAAQSs to be regularly up-dated and for the 'prevention of significant deterioration' (PSD) in regions with air cleaner than the NAAQS. A system of New Source Reviews (NSRs) was instituted with New Source Performance Standards (NSPSs). Best available control technology (BACT) is required for compliance.

Title III of the CAAA 1990 gives a list of 189 air toxics and requires the EPA to publish a list of source categories emitting 10 ton/year of any one toxic or 25 ton/year of a combination and to issue for such toxics maximum achievable control technology (MACT) standards.

A11.6.2 Water pollution
Water quality in the United States is protected by legislation on effluent discharges to streams and on drinking water quality. The Federal Water Pollution Control Act 1972 envisaged a nationwide policy of zero discharge of pollutants by 1985. The EPA introduced a discharge permit system in the form of the National Pollutant Discharge Elimination System (NPDES). The Clean Water Act Amendments 1977 backed away from zero discharge and introduced a system requiring instead that in due course discharges be treated with the best conventional pollutant control technology. In addition, the EPA is now charged with setting limits for some 100 toxic substances in effluents.

Advisory standards for drinking water in the United states have long been set by the US Public Health Service (USPHS). The Safe Drinking Water Act 1974 authorizes the EPA to set minimum national drinking water standards.

A11.6.3 Hazardous wastes
Hazardous wastes in the United states, governed previously by the Solid Waste Disposal Act 1965, are now are subject to the Resource Conservation and Recovery Act 1976 (RCRA). The act seeks to ensure proper land disposal of defined hazardous wastes and to fill certain loopholes in control on air and water pollution. Under it the Office of Solid Waste (OSW) of the EPA promulgates regulations defining hazardous wastes and standards applicable to it. In implementation of the Act, the EPA classified waste disposal sites as landfills, lagoons or landspreading operations, defined eight categories of impact which such sites might have and stated the operational and performance standards to be met to minimize such impacts.

Hazardous wastes attract further controls. For these the EPA has set up what is in effect a 'cradle-to-grave' system covering hazardous waste generation, transport and treatment, storage and disposal facilities, imposing duties on the generator, the transporter and the facility operator and involving tracking by a system of identification numbers and control by a system of permits.

The problem of the large number of hazardous waste sites in unsatisfactory condition, or worse, is addressed by the Comprehensive Environmental Resource Conservation and Liability Act 1980 (CERCLA), or Superfund. The act creates controls on such facilities and provides for the EPA to supervise the clean-up of existing and abandoned sites. The EPA seeks to identify Potentially Responsible Persons (PRPs) and, where such parties are not found, it initiates a clean-up paid for from the Superfund.

The Superfund Amendments and Reauthorization Act 1986 (SARA) renews the Superfund. The part known as SARA Title III extends its application to accidental releases.

The numerous other items of legislation which have some bearing on hazardous waste are reviewed by Edelman (1987).

A11.6.4 Fugitive emissions
Standards governing fugitive emissions have been promulgated by the EPA under the CAA. These are the Standards of Performance of New Stationary Sources Equipment Leaks of VOC, Petroleum Refineries and Synthetic Organic Chemical Manufacturing Industry 1983, amended 1984, also referred to as the New Source Performance Standards (NSPSs), the NSPS Regulations or the VOC (SOCMI) Regulations. These set standards for emissions permissible from items of equipment.

In addition to these controls on VOCs, the EPA has also issued National Exposure Standards for Hazardous Air Pollutants (NESHAPs) for certain specific substances. These too give equipment emission standards.

The NSPSs apply to new equipment but also have application to reconstructed and modified equipment, and guidance has been given on determining how far they affect existing equipment. In addition, existing equipment is affected by the assimilation into EPA controls of state requirements on monitoring and maintenance (M&M).

A11.6.5 Environmental impact assessment
The National Environmental Policy Act 1969 (NEPA) contains in Section 102 the far-reaching provision that where the action of a federal agency may have significant consequences for the human environment, an environmental impact statement should be given. The practice is to prepare a draft statement, which is put out for consultation, and then to issue the final statement.

A11.7 Environmental Management

11.7.1 Environmental management systems
The environment needs to be protected by an approach to management and management systems similar to that developed for safety and loss prevention, which has been described in Chapter 6. There should be an environmental management system (EMS) which parallels the safety management system.

A11.7.2 BS7750
Environmental management is the subject of BS 7750: 1992 Specification for Environmental Management Systems. The standard may be regarded as applying in this field the principles of quality systems given in BS 5750. An overview is given by Shillito (1991).

Requirements of BS 7750 include (1) documentation of the environmental management and management systems; (2) collation of regulatory requirements on the environment; (3) inventory of raw materials and energy usage and of wastes and releases; (4) formulation of environmental objectives; (5) an environmental management plan; (6) a system of environmental audits; (7) a system of environmental controls with verification and testing; and (8) personnel qualified and trained in environmental matters.

A11.7.3 Process environmental reviews
The EMS should require a system of process environmental reviews akin to the process safety reviews described in Chapters 6 and 8.

This review system should include formal systems for the identification and assessment of environmental hazards.

The hazop study method is now applied to identify potential problems with the environment as well as with safety and operability. Accounts are given by Isalski et al. (1992) and Ormond and Isherwood (1992).

A11.7.4 Environmental audits
Another essential feature of an EMS is a system of environmental audits. Environmental auditing stands in essentially the same relation to EP as safety auditing does to SLP.

Accounts of environmental audits are given by Baumer (1982), Keene (1982), D.L Russell (1985) and Petts and Eduljee (1994).

A11.7.5 Environmental planning
The essential concern of planning is with the environment as broadly defined, including amenity as well as hazards and pollution.

The planning arrangements vary from country to country. Accounts of planning in the United Kingdom in relation to major hazard installations have been given by Petts (1988b, 1989, 1992).

Accounts of planning in relation to pollution are given by D.C. Wilson (1981) and Petts and Eduljee (1994), both dealing with hazardous wastes.

To a considerable extent planning centres around the EIS, discussed in Section A11.9.

A11.7.6 Environmental emergency planning
Environmental emergency planning is needed to cover both fixed installations and transport. The treatment in Chapter 24, which covers both aspects as far as safety is concerned, is in large part applicable to planning for environmental emergencies.

Further treatments relevant to the environment specifically are given in Hazardous Materials Emergency Response and Control by Cashman (1988) and by G.F. Bennett, Feates and Wilder (1982).

As stated earlier, the CIMAH safety case is required to cover environmental as well as safety aspects of emergency planning.

A11.8 Environmental Hazard Assessment

There is now an increasing activity in environmental hazard assessment. In large part this has been driven by the threat to the environment from hazardous waste sites.

In the United states, the pollution arising from the hazardous wastes dumped at Love Canal, described in Section A11.22, and the resultant arrangements for clearing up waste sites under Superfund have stimulated assessment of such sites to determine the extent of the risks and to set priorities for clean-up.

Hazard assessment is undertaken both for existing or proposed hazardous waste treatment, storage and disposal facilities (TSDFs) and for abandoned hazardous waste dumps.

Accounts of environmental hazard assessment are given in Risk Assessment of Hazardous Waste Sites by Long and Schweitzer (1982), Hazardous Waste Risk Assessment by Asante-Duah (1993) and Standard Handbook for Solid and Hazardous Waste Facility Assessments by Sara (1994) and by H. Bradford (1986a), B.P. Smith (1987), Kemp and Gerrard (1991), Kolluru (1991), Welsh (1992, 1993) and Rouhiainen (1993).

A11.8.1 EPA guidance
Guidance on hazard assessment of hazardous waste sites, in the context of Superfund, is given in Risk Assessment Guidance for Superfund: I, Human Health Evaluation Manual and in Risk Assessment Guidance for Superfund: II, Environmental Evaluation Manual both by the EPA (1989a,b). EPA philosophy is also described by B.P. Smith (1987).

The process outlined by the EPA involves the following steps: (1) hazard identification, (2) toxicity assessment, (3) exposure assessment and (4) risk characterization.

A brief description of the assessment process is given by Kolluru (1991) and a more detailed account by Asante-Duah (1993).

These stages in the assessment are now outlined.

A11.8.2 Hazard identification
Hazard identification is the first stage and, in this context, means the identification of those chemicals which pose the greatest risks, the so-called indicator chemicals.

Since this identification involves the inventory, mobility, persistence and toxicity of the chemicals, some degree of iteration back from the subsequent stages may be necessary.

A11.8.3 Toxicity assessment
Toxicity assessment requires consideration of toxicity both for carcinogens and for non-carcinogens.

For carcinogens the assumption made by the EPA is that there is no threshold below which a dose is harmless. The probability of harm is then obtained from the slope of the dose-response curve, the so-called cancer slope factor (CSF) or potency factor. The EPA also recognizes a weight of evidence factor. Carcinogens are divided into four categories: (1) human carcinogens, (2) probable human carcinogens, (3) probable human carcinogens but with inadequate human data and (4) possible human carcinogens.

For non-carcinogens a threshold is admitted. For the dose of such a chemicals there is therefore a no observable adverse effect level (NOAEL). A reference dose (RfD) is defined which is one hundredth of the NOAEL and which thus incorporates a substantial safety factor. The reference dose is also termed the acceptable daily intake (ADI).

A11.8.4 Exposure assessment
Exposure assessment involves determining the persons exposed and their intake of toxic chemicals.

The EPA has issued standard assumptions for exposure. For adults these include the following:

Air inhaled (m^3/d)   20
Water ingested (1/d)   2

Soil ingested (mg/d)   100
Fish consumption (g/d)   6.5
Lifetime exposure (year)   70

The assessment required is for a reasonable maximum exposure (RME) under present and future land use conditions.

A11.8.5 Risk characterization

Risk characterization involves combining the toxicity and exposure estimates to obtain an assessment of risk to human health and making an assessment of the risks to the environment.

Cancer risks are expressed as the probability of contracting cancer from exposure to the chemicals over a lifetime. Non-cancer risks are expressed for a single substance in terms of a hazard quotient (HQ) or for a set of substances in terms of a hazard index (HI), the sum of the hazard quotients. The hazard quotient is the ratio of the daily intake to the RfD. If the HQ is less than unity, the chemical is not regarded as a threat to public health.

A11.8.6 Source terms and transport models

A full hazard assessment of a hazardous waste TSDF requires the modelling of source terms and of the transport and transformation of chemicals in the atmosphere and the ground. These aspects are discussed in Section A11.12.

A11.9 Environmental Impact Assessment

A11.9.1 Environmental assessments

An EA presented in an ES is for some projects a regulatory requirement under the Environmental Assessment Regulations 1988.

Accounts of ES are given in *Environmental Impact Assessment of Waste Treatment and Disposal Facilities* by Petts and Eduljee (1994) and by E.J. Martin and Johnson (1987) and Veselind, Peirce and Weiner (1990).

The treatment by Petts and Eduljee is comprehensive and although concerned with hazardous waste facilities its general approach is of much wider application.

The authors describe the regulatory requirements for, the purposes of and practice in respect of an EA and they relate the EA to the requirement to select the best practicable environmental option (BPEO) with its implication that alternatives must be considered.

They outline the process of scoping. Following Elkin and Smith (1988), identification of the crucial concerns should cover (1) geographic boundaries, (2) administrative boundaries, (3) project timing and duration, (4) key stakeholders, (5) key resources and land uses, (6) key activities, (7) key policies and (8) interactions. They describe the scoping techniques of consultation and identification of impacts and their significance and indicate other techniques such as checklists, matrix and network methods, and cause—effect diagrams. They then deal specifically with the scoping of landfill and incineration.

For each feature subject to impact, Petts and Eduljee describe the feature itself, the scoping, the baseline conditions and survey, the prediction and evaluation of impacts, and the mitigation of impacts. They apply this treatment to (1) flora and fauna, (2) geology and soils, (3) ground and surface water, (4) air quality and climate, (5) public health, (6) landscape and visual amenity, (7) transport, (8) social and economic features, (9) land use and heritage

and (10) residuals. In addition to the usual impact factors they consider acute events, noise and vibration, and landfill gas.

As a case study, the authors consider the Seal Sands incinerator.

A US perspective is given by Veselind, Peirce and Weiner (1990). They describe an EIS as consisting of three distinct parts: (1) inventory, (2) assessment and (3) evaluation. The relevant factors are broadly similar to those just mentioned. They also outline a scoring scheme in which impacts are rated using scales of importance and magnitude.

In some instances, particularly in the early days, the EIS has constituted a major exercise. For the trans-Alaska pipeline the draft EIS, the comments and the final EIS ran to 30 volumes.

A11.9.2 Safety cases

For major hazard installations the CIMAH Regulations 1984 require the submission of a safety case which covers the consequences of a major accident not only for human but for the environment. This aspect of the safety case is discussed by Singleton (1989) and Cassidy (1990).

The safety case is concerned with acute events and it may be regarded as complementing the regulatory controls dealing with long-term pollution and environmental impact.

Cassidy advises that the safety case should give information on the land uses around the site, the environmental hazards of the substances stored or processed and the potential impact on flora and fauna and on the balance of nature. Where there is potential to contaminate water, it should give comprehensive information on the local watercourses and their uses. Particular attention should be paid to the potential for harm to targets which are rare or unique, to persistence of noxious substances in the environment and to long-term damage. The author gives decision trees for assessing land and water impacts.

He also draws attention to the need to address the environmental aspects in the emergency planning required under the regulations.

Singleton deals with the assessment of the consequences to the environment of an accident involving an ultra-toxic substance. He considers the cases of copper—chrome—arsenate (CCA) wood preservatives and of pesticides.

A11.9.3 Waste facility assessments

A hazardous waste facility needs to be assessed to determine the hazard which it poses to the environment and to humans.

Petts and Eduljee (1994) describe two approaches to such assessment. One is to undertake a full environmental hazard assessment, as described in the previous section.

The other approach is the use of a structured checklist. This is the method used in the project on hazard assessment of landfill (HALO) described by Kemp and Gerard (1991).

A somewhat similar approach is taken in the Hazard Ranking Scheme (HRS) of the EPA (1982a), which utilizes a scoring system.

A11.9.4 Spill impact assessment

A more specific form of impact assessment is the assessment of the effects of a spill of hazardous material.

Typically this might be a liquid spill occurring in road or rail transport or from a pipeline.

Methods of assessing the impact of a spill which has already occurred are given in the literature as one aspect of spill control, as described in Section A11.20, but assessment of the impact of potential spills in advance receives little mention.

A11.10 Environmental Economics

The costs to the process industries of environmental protection are high, but experience indicates that, as with safety, good practice is also good business.

Costs and savings may be computed at national or company level. At the national level costs in the United states have been given in *Environmental Investments: The Costs of a Clean Environment* by the EPA (1991) (Chementator 1991 Mar., 25). In 1972, expenditure on pollution control was some $30 billion ($\approx$1% of GDP). By 1987, it had risen to $98 billion ($\approx$2% GDP) and by 1990 to $115 billion ($\approx$2.1% GDP); the projection for the year 2000 is $140–160 billion ($\approx$3% GDP); the values are all in 1990 dollars. The 1987, figures include $28.9 billion for air and $42.9 billion for water pollution control. The process industries account for a large proportion of this expenditure.

Savings at the national level are inevitably less easy to obtain.

However, as described in Section A11.13, the efforts now devoted to waste minimization are testimony to the fact that there are economic benefits.

A11.11 Environmentally Noxious Chemicals

There are a number of substances which figure particularly prominently as pollutants. Some of the principal ones are mentioned briefly in this section.

Many of these substances appear in gaseous or liquid effluent streams. To this extent, their release is controlled one. The hazard from others may arise from an accident. A pesticide is a case in point, as described below.

A11.11.1 Sulfur dioxide and acid rain

A major pollutant from combustion processes is sulfur dioxide, which is responsible for acid rain.

Acid rain pollution is described in *Sulphur Emissions and the Environment* by the SCI (1981b), *Energy and the Environment: Acid Rain* by Keith (1982), *Acid Rain Information Book* by Record, Bubenick and Kindya (1982) and *Acid Deposition* by Adams and Page (1986) and by Ireland (1987). Control technologies are detailed in *Add Emissions Abatement Techniques* by the DoE (1991–).

A11.11.2 Nitrogen oxides

Nitrogen oxides NO_x arise from combustion processes. Both industrial combustion and cars are major contributors.

The NO_x problem and control technologies are discussed by Marzo and Fernandez (1980), Niess (1980), Redman (1989a) and R. Smith and Petala (1991–).

A11.11.3 Pesticides

Pesticides are highly active chemicals. The expose in *Silent Spring* by Carson (1963) rendered their use controversial.

The production, storage and handling of pesticides creates the potential for release in an event such as a fire, or with the fire water used.

The pesticide problem is discussed in *Persistent Pesticides in the Environment* by C.A. Edwards (1974) and by the SCI (1978).

A11.11.4 Polychlorinated biphenyls

For many years, until their extreme toxicity came to be appreciated, polychlorinated biphenyls (PCBs) were used as electrical transformer fluids. The main concern now is prevention of escape from such equipment and safe methods of disposal. This is discussed by Berry (1981) and Motter (1991). A guide to EPA-approved methods of disposal is given by Kokozska and Flood (1985).

A11.11.5 Toxic metals

Toxic metals which occur in trace quantities in effluents, including those from incinerators, can be very harmful to the environment.

Treatments of toxic trace metals are given in *Control and Fate of Atmospheric Trace Metals* by Pagna and Ottar (1989) and by Barton, Clark and Seeker (1990).

A11.11.6 Mercury

Certain processes, notably traditional processes for chloralkali production, discharge mercury, albeit in very low concentrations. Accumulation of mercury, especially in the form of methyl mercury, is very noxious to the environment.

A11.11.7 Dioxins

TCDD and other dioxins exhibit toxic effects in minute concentrations, and rank as ultratoxic substances. The toxicity of TCDD is discussed in Appendix 3 in the context of Seveso.

TCDD is produced in very low concentrations in combustion processes and can be produced in a chemical reactor.

The dioxin problem and methods of dealing with it are discussed in *Human and Environmental Risks of Chlorinated Dioxins and Related Compounds* by R.E. Tucker, Young and Gray (1983) and *Dioxin Containing Wastes: Treatment Technologies* by Arienti *et al.* (1988) and by Ayres (1981), Rice (1982), Short *et al.* (1984) and Beychok (1987a).

A11.11.8 Nitrates

Nitrates are another significant pollutant, but arise mainly from the use of fertilizers, and are outside the present scope. An account is given by Nicolson (1979).

A11.11.9 Health effects

Threats to health may arise from the use, or more often abuse, of chemicals or from their release to the environment.

Health effects of chemicals in the environment are treated in *Health Effects of Environmental Pollutants* by Waldbott (1973), *Environmental Industrial Health Hazards* by Trevethick (1976), *Air Pollution and Human Health* by Lave and Seskin (1977a), *Long-Term Hazards of Environmental Chemicals* by Doll and McLean (1979) and *Hazardous Waste and Human Health* by the BMA (1991) and by McLean (1981) and Asante-Duah (1993), as well as in texts on environmental hazard assessment.

A11.12 Chemicals Transport, Transformation, Fate and Loading

A fundamental approach to hazard assessment in EP involves tracing the dispersion and determining the fate of chemicals in the environment.

In principle, a full treatment therefore involves study of the dispersion of gaseous effluents in the atmosphere and of the deposition of particulate matter; the transmission of chemicals through waters and soils of all types; and their passage through the food chain.

A great deal of work has been and continues to be done in all these areas. In some cases the prime concern is with potential airborne releases from nuclear power plants or leaks from nuclear waste stores into watercourses, in others it is with the effects of pesticides and other chemicals already in use.

Accounts of the transport, transformation and fate of chemicals in the environment include *Environmental Pollution* by Chemicals Walker (1971), *Persistent Pesticides in the Environment* by C.A. Edwards (1974), *Pollution Criteria for Estuaries* by Helliwell and Bossanyi (1975), *Principles for Evaluating Chemicals in the Environment* by the NAS (1975), *Safety of Chemicals in the Environment* by Harwell (1979), *Chemical Concepts in Pollutant Behaviour* by Tinsley (1979), *Hydrocarbons and Halogenated Hydrocarbons in the Aquatic Environment* by Afghan and Mackay (1980), *Dynamics, Exposure and Hazard Assessment of Toxic Chemicals* by Haque (1980b), *Chemicals in the Environment* by Neely (1981), *Physico-Chemical Behaviour of Atmospheric Pollutants* by Versino and Ott (1982), *Environmental Exposure to Chemicals* by Neely and Blau (1985), *Pollutant Transport and Fate in Ecosystems* by Coughtrey, Martin and Unsworth (1987), *Control and Fate of Atmospheric Trace Metals* by Pagna and Ottar (1989), *Reaction Mechanisms in Environmental Organic Chemistry* by Larson and Weber (1994), *Hazardous Waste Risk Assessment* by Asante-Duah (1993) and *Standard Handbook for Solid and Hazardous Waste Facility Assessment* by Sara (1994). Worldwide limit values are given in *World-wide Limits for Toxic and Hazardous Chemicals in Air, Water and Soil* by Sittig (1994).

A11.12.1 Source terms

The starting point for the modelling of transport through environmental media is the definition of the source terms.

Three principal types of source terms are (1) gaseous emissions from a hazardous waste TSDF, (2) leachate seepage from a TSDF and (3) liquid spill into a watercourse in a transport accident.

A review of gaseous emission source terms at a TSDF is given by B.P. Smith (1987). Models quoted by this author are for surface impoundments, that of Thibodeau, Parker and Heck; for aerated tanks, that of R.A. Freeman (1980); for storage tanks, the usual relations for 'breathing' losses; for landfill, that of W.J. Farmer, Yang and Letey (1980), as modified by Hwang (1980) and Shen (1981); and for land treatment, that of Thibodeaux and Hwang (1982).

For liquid seepage the source term is highly site specific. Seepage from landfill is treated in accounts of landfill and of leachate.

A liquid spill from a transport vehicle is a relatively well-defined source. Another form of transport spills is a pipeline leak, where the issues are the location and the flow rate.

A11.12.2 Transport of chemicals

Chemicals are transported through the environment by atmospheric dispersion and by movement through waters.

A review of air, surface water, groundwater, soil and multimedia transport models applicable in this context is given by McBean (1993).

For transport of gaseous emissions use is made of atmospheric dispersion models. These include the passive gas dispersion models familiar in hazard assessment, but also in some cases models for transport over much greater distances and periods and in more complex meteorological conditions.

The ISC model of the EPA is widely used, both in its short-term version ISCST and its long-term version ISCLT.

For transport through water the review by McBean may be supplemented by the exhaustive treatment of transport through soil and groundwater, based on fundamental geology and hydrology, given by Sara (1994).

Transport through water is typically modelled using compartment, or control volume, models. For movement in surface water a widely used model is EXAMS of Burns, Cline and Lassiter (1982).

The modelling of the dispersion of a chemical spilled into a stream is discussed by Kontaxis and Nusser (1982).

There is no obvious preferred model for movement through groundwater.

The TOX-SCREEN model of the EPA (1984c), which assesses the potential fate of chemicals released to air, surface water or soil, may be used to screen for substances which even on conservative assumptions are unlikely to pose a threat to the environment.

In the United kingdom, as described by Welsh (1993), the HSE is developing a risk assessment tool for pollution risk from accidental influxes into rivers and estuaries (PRAIRIE), which incorporates a family of river dispersion models DYNUT.

A11.12.3 Transformation of chemicals

Chemicals moving through the environment may undergo a number of transformations, including reactions, among which are (1) hydrolysis, (2) chemical oxidation, (3) photolysis and photo-oxidation, (4) biodegradation, (5) absorption, (6) adsorption, (7) volatilization and (8) sedimentation.

Some of these transformation processes are described by Welsh (1993).

A11.12.4 Fate of chemicals

The determination of the fate of chemicals in the environment is an extremely complex matter. In part it has to do with the transport and transformation processes just outlined and in part with the assimilative capacity of environmental media and with the effects of the take-up of chemicals into food chains. The concept of an ultimate fate appears to imply some eventual steady state, although the associated timescales may well be rather long.

A11.12.5 Capacity of the environment

To a greater or lesser degree, environmental media have the capacity to tolerate a certain chemical burden. The assumption that this is so is frequently implicit in industry practices.

In certain instances this assimilative capacity is estimated and the loading is adjusted accordingly. Two such cases are the limiting chemical load in land treatment and the oxygen demand load on watercourses.

An example of calculated loading occurs in land treatment, where the quantity of liquid waste applied to an area of land may be estimated by identifying the soil contaminant control and removal mechanisms and estimating the assimilative capacity.

A11.12.6 Loading of watercourses
Another example of loading occurs when a chemical is spilled into a watercourse, either by accidental discharge from a works or as a result of a transport accident.

The health of a watercourse such as a stream or river depends on the maintenance of sufficient dissolved oxygen (DO).

Industrial wastewater discharged into a watercourse tends to use up the oxygen in the latter, its propensity to do so being measured by its oxygen demand (OD), or more specifically its biological oxygen demand (BOD), described below. If the wastewater is hot, it also tends to raise the temperature of the watercourse. Discharge of wastewater therefore tends to deplete the oxygen in the watercourse both by virtue of its OD demand and its temperature. Reaeration processes operate to make up the dissolved oxygen deficit caused by these deoxygenation processes.

There are a number of models, described by Nemerow and Dasgupta (1991), of the effect of wastewater discharge. They include the formulation of Streeter and Phelps (1925), the method of H.A. Thomas (1948) and the correlation of Churchill and Buckingham (1956). The latter found that they were able to correlate the DO as a function of the BOD, the temperature and the flow.

A11.12.7 Biological oxygen demand
A standard method of expressing the oxygen demand is as the BOD measured under specified conditions over 5 days and at $20°C$.

The solubility of oxygen in water at $20°C$ is 9.2 mg/l. The BOD of domestic sewage is about 200 mg/l, whilst that of some industrial wastes can be as high as 30,000 mg/l.

Another measure is the chemical oxygen demand (COD). The test for COD is quicker than that for BOD, but since in this case all the organics are oxidized, this test gives a higher reading.

A11.13 Waste Minimization
The development of loss prevention is paralleled by that of waste minimization. A large part of the latter is by inherently cleaner design (ICD), which stands in much the same relationship to the environment as inherently safer design does to safety. But in so far as waste minimization involves management commitment and management systems to deliver a systematic approach, the closer parallel is with loss prevention. Another term commonly used is clean technology.

Accounts of waste minimization are given in *Hazardous Waste Minimization Handbook* by Higgins (1989), *Hazardous Waste Minimization* by R.A Freeman (1990), *Waste Minimization Guide* by Crittenden and Kolaczkowski (1992) and by Redman (1989c), Hethcoat (1990), R. Smith and Petala (1991–), Curran (1992), Chadha and Parmele (1993) and Rossiter, Spriggs and Klee (1993).

In the United States, the waste minimization approach is actively promoted by the EPA, relevant publications being *Waste Minimization, Environmental Quality with Economic Benefits* (EPA, 1987b) and *Waste Minimization Opportunity Assessment Manual* (EPA, 1988).

A11.13.1 Waste minimization management systems
The waste minimization philosophy has been widely adopted in the process industries. As with loss prevention, success has been achieved as management has become convinced and committed and has developed effective strategies, put in place formal systems, allocated the necessary resources, including people, and created a self-reinforcing waste minimization culture.

An account of developments in the United States and elsewhere, is given by Redman (1989c), together with numerous examples of case studies and cost savings.

Basic approaches to waste minimization are to modify (1) the chemicals, (2) the process, (3) the equipment and (4) the effluent treatment. Chemicals which tend to be troublesome include toxic metals and halogenated hydrocarbons. Process modifications which may prove fruitful are replacement of (1) chemical by mechanical processes and (2) single pass processes, including rinse operations, with closed processes. Effluent treatment may be modified to recycle or recover materials.

Approaches to waste minimization are outlined in Figure A11.1. On an existing plant the first step is a waste audit.

The ways in which waste minimization may be effected are very varied. This comes across clearly from the examples quoted by Redman. They include minimization of plant washdowns, squeezing of sludges to remove water and reduce mass, design of drums which leave less residue, and so on.

Redman cites a case, described by Huisingh, in which installation of equipment to recover wastes from film development at a cost of $120,000 made annual savings of $2.6 million on silver and on developing, fixing and bleach solutions, a payback time of one month.

An account of a waste minimization programme in one company is given by Koenigsberger (1986).

A11.13.2 Process plant wastes
A review of the application of waste minimization in the process industries covering (1) the basic problem, (2) reactors, (3) separation and recycle systems, (4) process operations and (5) utilities is given by R. Smith and Petala (1991–).

In a reactor five main sources of waste are (1) low conversion, (2) unwanted by-products from the primary reaction, (3) unwanted by-products from the secondary reaction, (4) unwanted by-products from feed impurities and (5) degraded catalyst. There is often scope for reduction of such waste by the application of reaction engineering principles.

In separation processes there may be scope for waste minimization by (1) elimination of feed impurities, (2) elimination of extraneous materials used for separation, (3) recycling of waste streams and (4) additional separation of such streams.

As an illustration of the benefits of eliminating a feed impurity the authors instance the elimination of nitrogen by the use of oxygen rather than air in the oxychlorination stage of ethylene dichloride production. They describe a design of ethylene dichloride reactor which avoids carryover of the ferric chloride catalyst and thus eliminates the need for washing and neutralization stages.

Under process operations they cover (1) start-up and shut-down, (2) product changeover, (3) equipment maintenance, (4) tank filling, (5) accidental spillages and (6) fugitive emissions. During transient operations such as start-up and shut-down product may be off specification, additional by-products may be made and recycles may not be possible. Product changeover and equipment cleaning give rise to cleaning wastes. Tank filling causes losses as vapours are forced out.

In the utilities wastes are associated particularly with combustion processes, steam generation and cooling water. All these wastes can be reduced by a design which promotes energy efficiency. In combustion methods are discussed of minimizing or dealing with oxides of sulfur and nitrogen. Other wastes to be minimized are boiler and cooling tower blowdowns.

A11.13.3 Refinery wastes

The application of waste minimization to refinery operations has been described by Hethcoat (1990) and Curran (1992).

Whereas originally waste minimization was applied mainly to the conservation of hydrocarbons, it now finds much wider application. The overall strategy involves (1) source reduction and (2) recycling as well as (3) treatment.

Some principal wastes are (1) oily materials, (2) catalyst residues, (3) spent caustic and (4) wastewaters.

Oily residues are oil-coated solids, mainly from oil–water separators, dissolved air filtration (DAF) units, heat exchanger cleanings and tank bottoms. Residues also arise in sewers from road dust materials carried in with runoff water. Waste catalysts arise from processes such as catalytic cracking, conversion of heavier products to lighter ones, reforming of gases to liquid products, and sulfur removal. Spent caustics occur as phenolic or sulfide caustics.

The incidence of wastes is highly variable from one refinery to another. Curran states that quantities of oil residues and usage of water can vary by almost two orders of magnitude.

Measures to reduce refinery wastes include use of mixers in crude oil tanks to keep solids suspended, elimination of intermediate storage tanks between units, elimination of blending tanks by use of in-line blending, minimization of surfactants which cause emulsions and promote sludges, preskimming of oil at the separators, use of pressurized air at the dissolved air flotation unit to give a more concentrated sludge, reprocessing of oily residues in suitable refinery units such as a coker, paving of process areas to reduce dust and sweeping of roads for the same purpose.

At the utilities measures taken include maximization of air cooling, use of a closed cooling water system and reduction of solids in the boiler blowdown water.

Segregation also has a part to play. Curran describes segregation of phenolic and sulfide spent caustics, Hethcoat provision of three separate sewer systems for non-oily liquids and one for oily ones.

Hethcoat discusses the identification of options for waste minimization, under the source reduction, recycling and treatment heads, giving as illustrations oily wastewater from separator bottoms and wastes from empty drums.

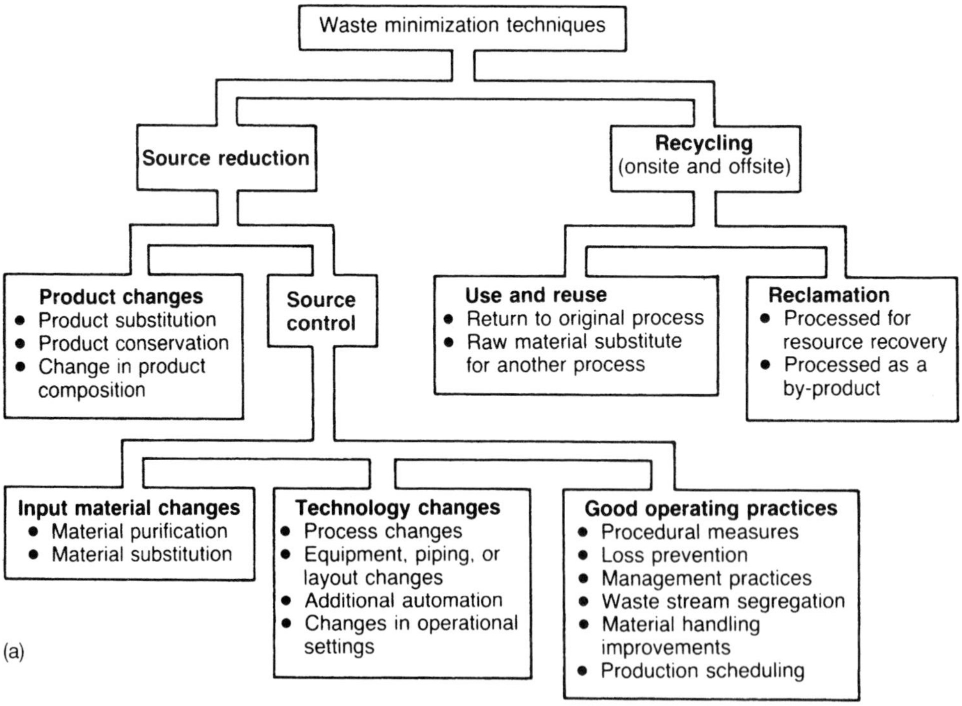

Figure A11.1 *Waste minimization: (a) techniques; and (b) assessment procedure (Environmental Protection Agency; reproduced by permission of Gower Press)*

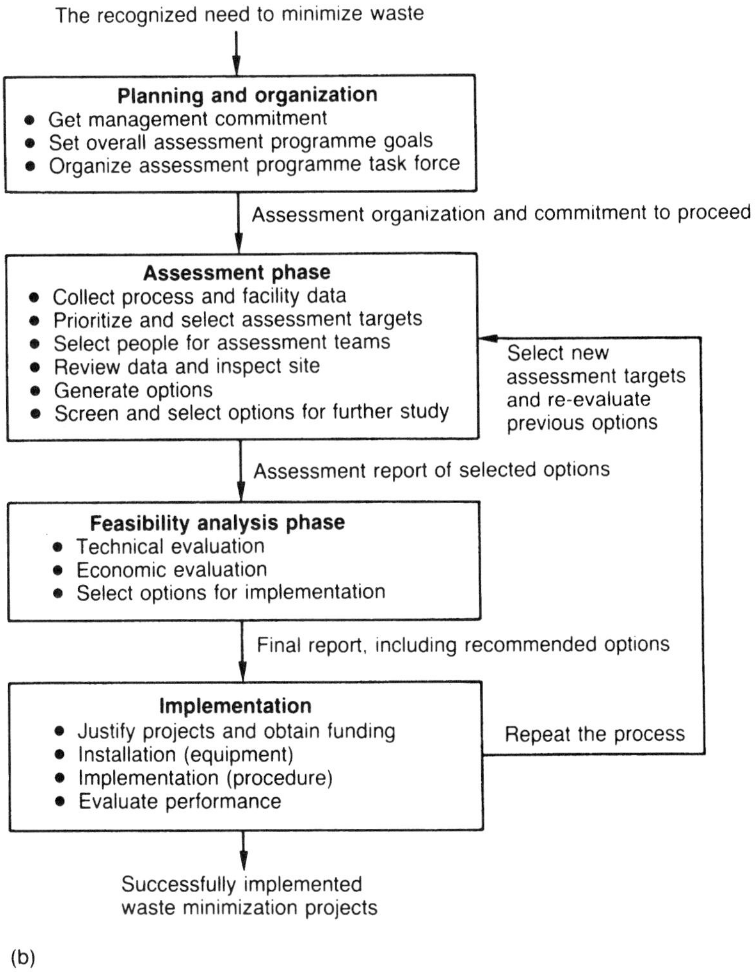

The recognized need to minimize waste

↓

Planning and organization
- Get management commitment
- Set overall assessment programme goals
- Organize assessment programme task force

Assessment organization and commitment to proceed

Assessment phase
- Collect process and facility data
- Prioritize and select assessment targets
- Select people for assessment teams
- Review data and inspect site
- Generate options
- Screen and select options for further study

Select new
assessment targets
and re-evaluate
previous options

Assessment report of selected options

Feasibility analysis phase
- Technical evaluation
- Economic evaluation
- Select options for implementation

Final report, including recommended options

Implementation
- Justify projects and obtain funding
- Installation (equipment)
- Implementation (procedure)
- Evaluate performance

Repeat the process

↓

Successfully implemented
waste minimization projects

(b)

Figure A11.1 *Continued*

A11.13.4 Process integration

Waste minimization may be addressed in terms of process integration. The treatment by R. Smith and Petala (1991), just described, approaches the problem from this viewpoint. Another account is that of Rossiter, Spriggs and Kell (1993).

A11.13.5 Center for Waste Reduction Technology

The AlChE has set up the Center for Waste Reduction Technology (CWRT), a sister centre to the CCPS. It is described by L.L Ross (1991).

A11.14 Gaseous Effluents

Processes tend to generate a number of gaseous effluents which include impurities entering in the feeds, notably nitrogen in process air and air dissolved in liquid feeds; unwanted gaseous products from reactions; gases from purges such as that in the flare stack; and gaseous products of combustion.

Accounts of gaseous effluents, and the technologies of dealing with them, are given in *Air Pollution Handbook* by Magill, Holden and Ackley (1956), *Atmospheric Pollution* by Meetham (1968), *Air Pollution* by Scorer (1968), *Air Pollution Control* by Strauss (1971–), *Gas Purification Processes for Air Pollution Control* by Nonhebel (1972), *Pollution in the Air* by Scorer (1973), *Compilation of Air Pollutant Emission Factors* by the EPA (1978a), *Fundamentals of Air Pollution* by Stern *et al.* (1973), *Dust Control and Air Cleaning* by Dorman (1974), *Gas Cleaning for Air Quality* by Marchello and Kelly (1975), *Air Pollution* by the Open University (1975a), *Air Pollution* by Seinfeld (1976), *Air Pollution Control Equipment* by Theodore and Buonicore (1976), *Approaches to Controlling Air Pollution* by Friedlander (1978), *Reference Book* by the National Society for Clean Air (NSCA) (1978), *Industrial Air Pollution Handbook* by A. Parker (1978), *Air Pollution Impacts and Control* by Downey and Ni Uid (1979), *Air Pollution Control* by Hesketh (1979), *Air Pollution Control Engineering* by Licht (1980), *Paniculate Air Pollution* by Theodore, Sosa and Fajardo (1980), *Air Pollution* by E. Weber (1982), *Hazardous Air*

Emissions from Incineration by Brunner (1985), *Environmental Pollution and Control* by Veselind, Peirce and Weiner (1990), *Process Engineering and Design for Air Pollution Control* by Benitez (1993) and *Gas Processing: Environmental Aspects and Methods* by Speight (1993).

A11.14.1 Types of gaseous effluent

A gas effluent stream may consist of a gas which is itself noxious or of a non-noxious carrier gas which contains noxious impurities. The impurities may be gases or vapours, liquid droplets or solid particles.

Some principal gaseous effluents which require to be treated are (1) hydrogen sulfide, (2) carbon dioxide, (3) sulfur-containing gases SO_x and (4) nitrogen-containing gases NO_x. Hydrogen sulfide and carbon dioxide are commonly referred to as acid gases.

Hydrogen sulfide figures prominently in oil handling and refining. Carbon dioxide is the main acid gas product of combustion and also arises from oil refining and chemical operations. Sulfur oxides SO_x, essentially SO_2, are generated in combustion of fuel containing sulfur and in oil refining and chemical operations. Nitrogen oxides NO_x, essentially NO and NO_2, arise as so-called fuel-bound NO_x and thermal NO_x from combustion processes.

A11.14.2 Gaseous effluent control

Waste minimization may be applied to reduce the quantities of the gaseous effluents to be treated and the concentration of noxious components in these effluents, and to facilitate any treatment necessary.

A principal application of this approach which has long been practised is desulfurization of coal prior to combustion, as described below.

A11.14.3 Gas cleaning

The main methods of removing noxious gases or vapours from a gas stream are absorption, adsorption, reaction to a solid product and reaction with solids.

Absorption processes for removal of hydrogen sulfide and carbon dioxide are the amine family of processes, including the Girbitol process using ethanolamine, and the carbonate family, including the Benfield process using potassium carbonate, and for removal of hydrogen sulfide the Stretford process using sodium carbonate.

Hydrogen sulfide is removed in the Claus process by thermal and then catalytic reaction to sulfur. Hydrogen sulfide is also removed by reaction with iron oxide, a traditional gasworks process.

Processes for removal by absorption of sulfur dioxide include the Battersea process. Prior to combustion sulfur dioxide may be removed by fuel desulfurization. In its application to coal, this approach is also known as clean coal technology. In the process sulfur dioxide may be removed by reaction with limestone.

Absorption in oil is a traditional method of removing acid gas and hydrocarbons.

Flue gas desulfurization (FGD) comprises a group of techniques. One option is desulfurization of the fuel prior to combustion. A traditional post-combustion method is the removal of sulfur dioxide by the use of lime or limestone in 'wet scrubbers'. Advanced post-combustion methods involve contacting the gas with a sorbent such as lime either by in-stack injection or by injection into separate in-line vessels.

There are also combined processes for removal of both sulfur dioxide and nitrogen oxides, in which FGD is combined with a denitrification step. One method of removing nitrogen oxides is catalytic reaction with injected ammonia.

There are a large number of other gas cleaning processes. Outlines of about 80 are given by Speight (1993). Process flow diagrams are given periodically in *Hydrocarbon Processing*.

Removal of gaseous components may also be effected by the use of membrane processes.

Principal methods for the removal of particles include cyclones, fabric filters and electrostatic precipitators.

A review of technologies for acid gas removal in support of the UK programme is given in *Acid Emissions Abatement Technologies* by the DoE (1991).

A11.14.4 Smog reduction

A number of industrialized countries have suffered from serious air pollution in the form of smog. An account of the 'killer smog' in London in 1952 is given below.

In the United Kingdom, considerable progress has been made in reducing this form of air pollution. Thus in the City of London between 1955 and 1974, for example, the annual average concentrations of smoke and sulfur dioxide near the ground were reduced as follows (Warner, 1976):

	Concentration of smoke ($\mu g/m^3$)	Concentration of SO_2 ($\mu g/m^3$)
1955	$\approx$195	230
1974	$\approx$40	110

A11.14.5 Tall chimneys

The use of tall chimneys for avoidance of pollution in the immediate vicinity of a plant is a traditional practice. The drawback is that noxious effluents may be spread, albeit in dilute form, further afield.

An account of the practice of the Alkali Inspectorate in relation to tall chimneys has been given by Ireland (1984).

The topic has also been discussed by Perriman (1986).

A11.14.6 Emission bubble

An approach to air pollution control which in principle allows a company to adopt the most economic trade-offs is that of the 'bubble', in which all emissions from sources within the bubble are treated as if they were from a single controllable source. A formal statement of such a policy is that alternative methods of control are allowed provided it can be demonstrated that they are substantially equivalent, the criterion being the equivalence of the emissions leaving the bubble. The bubble concept was adopted by the EPA in 1979. It had a mixed reception, being welcomed in some cases by industry, but not always by the states. An account is given by Weismantel and Parkinson (1980).

A11.15 Liquid Effluents

The liquid effluents from a process plant include not just those shown as such on the flowsheet for the basic process, but a variety of other liquid flows ranging from washwater to rainwater runoff.

Accounts of liquid effluents, and the technologies of dealing with them, are given in *Biological Waste Treatment* by Eckenfelder and O'Connor (1961), *Industrial Wastewater Control* by Gurnham (1965), *Industrial Water Pollution* by Eckenfelder (1966), *Water and Waste Water Engineering* by Fair, Geyer and Okun (1966), *Aqueous Wastes from Petroleum and Petrochemical Plants* by Beychok (1967), *Liquid Waste of Industry* by Nemerow (1971), *Water Pollution* by Newsom and Sherratt (1972), *Waste Water Systems Engineering* by H.W. Parker (1972), *Clean and Dirty Water* by the Open University (1975b), *Handbook of Advanced Wastewater Treatment* by Gulp, Webster and Gulp (1978), *Wastewater Treatment* by Sundstrom and Klei (1979), *Wastewater Treatment and Disposal* by Arceivala (1981), *Biological Treatment of Waste Water* by Winkler (1981), *Handbook of Wasterwater Treatment Processes* by Vernick and Walker (1982), *Removal of Metals from Wastewater* by Cushie (1984), *Hazardous Waste Management Engineering* by E.J. Martin and Johnson (1987), *Environmental Pollution and Control* by Vesilind, Peirce and Weiner (1990), *Wastewater Engineering* by Metcalf and Eddy Inc. (1991) and *Industrial and Hazardous Waste Treatment* by Nemerow and Dasgupta (1991).

The modelling of the hydraulics of, and dispersion in, rivers, estuaries, etc. is described in *Hydraulic Behaviour of Estuaries* by McDowell and O'Connor (1977); *Groundwater* by Freeze and Cherry (1979); *The Mathematics of Hydrology and Water Resources* by Lloyd, O'Donnell and Wilkinson (1979); *Practical Aspects of Computational River Hydraulics* by Cunge, Holly and Verwey (1980); *Hydraulic Modelling* by Kobus (1981); *Models in Hydraulic Engineering* by Noak and Cabelka (1981); and also in certain texts of computational fluid dynamics.

A11.15.1 Types of liquid effluent

Liquid effluents from process plants are fluids from (1) the process, (2) the cooling system, (3) the process ancillary operations, (4) the sanitary system and (5) the rainwater runoff. The process ancillary operations are those such as cleaning and washing.

A proportion of the fluids from the process are liquids which are either non-aqueous or high concentration aqueous wastes. The rest are wastewaters containing oils and chemicals at lower concentrations.

A detailed review of liquid wastes from different industries, including the oil and chemical industries, is given by Nemerow and Dasgupta (1991). They cover organic chemicals; toxic chemicals; acid and chloralkali wastes; formaldehyde wastes; pesticides wastes; and wastes from the energy, fertilizer, phosphates, plastics and rubber, soap and detergent and explosives industries.

A11.15.2 Wastewater control

Wastewater requires a degree of treatment before it is discharged to receivers such as municipal systems or watercourses. There are a number of prior measures which may be taken to minimize the extent of the treatment required. These measures include (1) volume reduction, (2) strength reduction and (3) flow equalization and proportioning.

Waste minimization may be applied to reduce the volume of wastewater either by generating less in the first place or by reusing it. It may also be applied to keep down the concentration of contaminants in the wastewater. Segregation of wastewaters also has a role to play in reducing the volume to be treated or facilitating treatment.

Flow equalization involves the control of flows discharged to smooth out plant flow fluctuations and flow proportioning the control of discharge flows to match those in the municipal system or the watercourse.

A11.15.3 Wastewater treatment

Complete treatment of wastewater is typically classified as follows: (1) wastewater pretreatment, or conditioning, (2) primary treatment, (3), secondary treatment, (4) tertiary treatment, (5) sludge treatment and (6) liquid and sludge disposal.

Wastewater pretreatment processes are screening and grit removal and neutralization.

Primary treatment of wastewater involves removal of suspended solids by (1) sedimentation and (2) flotation. Either process may be operated with chemical addition. Removal of oil droplets can be effected by gravity separation.

Solids also occur in colloidal form, and the methods of dealing with them are often termed intermediate treatment. They include (1) chemical coagulation, (2) charge neutralization coagulation and (3) adsorption. The second of these involves coagulation by neutralization of the electric charge.

Secondary treatment processes deal with dissolved organic materials and comprise a range of mainly biological treatments. These processes are (1) aerobic or (2) anaerobic.

There is a wide variety of aerobic processes including (1) lagooning, (2) activated sludge, (3) modified aeration, (4) dispersed growth aeration, (5) high rate aerobic treatment, (6) contact stabilization, (7) trickling filtration, (8) biological discs, (9) mechanical aeration, (10) wet combustion and (11) spray irrigation. Holding in lagoons is used to effect biological degradation. Activated sludge involves oxidation by floes from suspended and colloidal solids, the sludge concentration being controlled by a recycle. Modified aeration is a modification of the activated sludge method. Dispersed growth aeration is a variant, which does not involve the use of floes. Contact stablization, or biosorption, is another modification of the activated sludge method. High rate aerobic treatment, or total oxidation, is another sludge process but with a higher sludge return rate. Trickle filtration and biological discs involve oxidation by slimes, the substrate being in the one case a packed bed and in the other rotating discs. Mechanical aeration involves sparging air into the liquid. Wet combustion, or wet air oxidation, is described below. In spray irrigation the liquid is sprayed onto land with aerobic conditions maintained to a certain depth, and is thus both a treatment and a disposal.

Anaerobic digestion takes place in closed vessels in the absence of air. Loadings are low and residence times long. It is most suited to small throughputs of liquids with high concentrations of readily oxidized organics and is used mainly for sludges rather than liquid wastes.

Other methods of removing organics include steam stripping and solvent extraction.

Tertiary treatment processes remove dissolved inorganic materials. They involve both chemical treatments and physical separation using a variety of unit operations. Chemical methods include (1) coagulation and precipitation and (2) oxidation and reduction reactions. Unit operations utilized are (1) evaporation, (2) adsorption,

(3) ion exchange, (4) reverse osmosis and (5) electrodialysis. Adsorbents used are activated carbon and polymers.

Sludge treatment methods include (1) digestion, (2) conditioning, (3) thickening, (4) dewatering and (5) drying. Digestion may be aerobic or anaerobic, utilizing some of the methods already described. Conditioning is effected using chemical and thermal methods. Thickening may be done in (1) gravity, (2) flotation or (3) centrifugal thickeners. Methods of dewatering are (1) vacuum filtration, (2) filter press dewatering, (3) belt filter dewatering and (4) centrifugal dewatering.

A11.15.4 Wet air oxidation

For water contaminated by small amounts of organic compounds a method of treatment which is often attractive is the wet air oxidation (WAO) process, also referred to as wet combustion. Accounts are given by Flynn (1979), Wilhelmi and Knopp (1979), Laughlin, Gallo and Robey (1983), Baillod, Lamparter and Barna (1985), Heimbuch and Wilhelmi (1985), E.J. Martin and Johnson (1987) and Nemerow and Dasgupta (1991).

The aqueous effluent to be treated is pumped to a high pressure, passed through a heater to heat it part way to the reaction temperature, then into a reactor vessel where compressed air is blown in and where the resultant exothermic reaction raises the liquid to its reaction temperature, finally out through a cooler to a separator where the oxidized liquid is taken off the bottom and nitrogen, carbon dioxide, steam, etc., off the top. Typical reaction temperatures are in the range 275–320°C. There are numerous variations on this basic process, including the use of a catalyst.

Wet air oxidation can degrade most industrial organic compounds. The reduction, or dilution, factors obtainable are generally good, though higher in some cases than others. Wilhelmi and Knopp list factors in the range 12–4200.

A11.15.5 Wastewater receivers

Wastewater treated to a suitable quality may be discharged to the municipal sewage system, to watercourses such as streams or rivers, or to land.

Some options available in principle are discharge of (1) raw industrial and sanitary wastes to the municipal system, (2) partially treated industrial and sanitary wastes to the municipal system, (3) completely treated industrial and sanitary wastes to the municipal system, partially treated industrial wastes to a watercourse, completely treated industrial wastes to a watercourse, and (6) completely treated industrial wastes to land. These options are discussed in detail by Nemerow and Dasgupta (1991).

A11.15.6 Deep well disposal

A quite different option for dealing with liquid wastes is deep well injection. Accounts are given by M.E. Smith (1979) and Nemerow and Dasgupta (1991).

This form of disposal has been used particularly in the United States, subject to strict controls and permitted only if no better method is available.

The formation selected to receive the waste is one below the levels holding potable water. The basic technology used derives from oil and gas production. The receiving formation may first be given a conditioning treatment. In order to eliminate solids, the waste is pretreated to remove suspended solids. In addition, a buffer solution such as dilute salt water may first be injected to prevent precipitation from the waste liquid.

A11.15.7 Sludge disposal

Methods of disposal of sludge from wastewater treatment include lagooning, landfill, incineration and sludge barging.

Lagooning involves digestion and is in effect a treatment process which shades into storage or disposal. Landfill may take the form of area landfill or trench landfill. Incineration is considered below.

Sludge barging has been used but tends to involve problems of sludge rising to the surface or being washed ashore, and in many cases has been discontinued.

A11.16 Hazardous and Solid Wastes

In addition to gaseous and liquid effluents, it is also necessary to deal with hazardous and solid wastes, including sludges and residues.

Accounts of hazardous wastes and solid wastes, and the methods of waste disposal, are given in *Disposal of Solid Toxic Wastes* by Key (1970), *Solid Wastes* by Mantell (1975), *Waste Recycling and Pollution Control* by Bridgewater and Mumford (1979), *Toxic and Hazardous Waste Disposal* by Pojasek (1979), *Handbook of Industrial Waste Disposal* by Conway and Ross (1980), *Design of Land Treatment Systems for Industrial Waste* by Overcash and Pal (1980), *Waste Management* by Woolfe (1981), *Hazardous Material Spills Handbook* by G.F. Bennett, Feates and Wilder (1982), *Hazardous Waste Processing Technology* by Kiang and Metry (1982), *Risk Assessment of Hazardous Waste Sites* by Long and Schweitzer (1982), *The Scientific Management of Hazardous Wastes* by Cope, Fuller and Willetts (1983), *Handbook of Industrial Residues* by Dyer and Mignone (1984), *Hazardous Air Emissions from Incineration* by Brunner (1985), *Handbook of Laboratory Waste Disposal* by Pitt and Pitt (1985), *Hazardous Waste Management* by Dawson and Mercer (1986), *The Solid Waste Handbook* by W.F. Robinson (1986). *Hazardous Waste Management Engineering* by E.J. Martin and Johnson (1987), *Handbook of Hazardous Waste Incineration* by Brunner (1989), *Environmental Impacts of Hazardous Waste Treatment, Storage and Disposal Facilities* by Salcedo, Cross and Chrismon (1989), *Emerging Technologies in Hazardous Waste Management* by Tedder and Pohland (1990), *Environmental Pollution and Control* by Veselind, Peirce and Weiner (1990), *Hazardous Waste Treatment Technologies* by Jackman and Powell (1991), *Industrial and Hazardous Waste Treatment* by Nemerow and Dasgupta (1991), *Environmental Impact Assessment for Waste Treatment and Disposal Facilities* by Petts and Eduljee (1994) and *Standard Handbook for Solid and Hazardous Waste Facility Assessment* by Sara (1994).

A11.16.1 Types of hazardous and solid waste

Hazardous and solid wastes fall into three categories (1) hazardous liquid wastes, (2) hazardous solid wastes and (3) non-hazardous solid wastes. The main concern here is with the first two of these.

A11.16.2 Hazardous waste control

The principles of waste minimization should be applied to minimize the quantity of hazardous waste and to ease the problems of handling it in treatment, storage and disposal facilities.

A11.16.3 Hazardous waste treatment, storage and disposal

Hazardous wastes may be stored or sent to ultimate disposal, either with or without prior treatment.

Methods of dealing with solid hazardous wastes include the use of (1) incineration, (2) landfill, (3) waste piles and (4) storage. Incineration, landfill and land treatment, which are usually treated as methods of ultimate disposal, are discussed further below.

A waste pile, or tip, may be regarded as storage or disposal, depending on the acceptability of its being permanent. Storage proper is typically in drums.

There are a number of methods of stabilization, fixation, solidification and encapsulation available for the treatment of residues and sludges, many developed originally for radioactive wastes. An account is given in E.J. Martin and Johnson (1987).

For liquid hazardous wastes methods include the use of (1) incineration, (2) surface impoundment, (3) land treatment and storage. Incineration and land treatment are both methods of ultimate disposal.

Surface impoundment (SI) or lagooning, has been discussed above. The liquid may be aerated or non-aerated. As stated, a lagoon may be regarded as a form intermediate between treatment, storage and ultimate disposal.

Storage is mainly in storage tanks, which may be open, open with external floating roof, closed with fixed roof and closed with internal floating roof. The vapour space may be inerted.

A11.16.4 Incineration

A prime method of dealing with hazardous waste is incineration. It is used for solid, liquid and sludge or residue wastes.

Treatments of incineration are given in *Handbook on Incineration* by Cross (1972), Incineration by Brunner (1987) and *Handbook of Hazardous Waste Incineration* by Brunner (1989) and by Novak and Pfrommer (1982), Grumpier and Martin (1987), Wiley (1987), Vervalin (1990), Veselind, Peirce and Weiner (1990) and Nemerow and Dasgupta (1991). Particular aspects of incineration discussed are the decision whether or not to use incineration by Haseltine (1992), testing of incinerators by Monroe (1983b), off-normal emissions from incinerators by Zeng and Okrent (1991), problems with incinerators by Ready and Schwab (1980), role of incineration in remedial actions by H.M. Freeman (1984) and dioxin problems by Beychok (1987). Plasma arc incineration is described by Ondrey and Fouhy (1991).

Incineration is a preferred method of the EPA.

Incineration is commonly carried out in dedicated equipment. Types of incineration equipment are described by Novak and Pfrommer (1982) and Brunner (1989). Alternatively, in some instances items of process plant may be used as incinerators. Feeley (1984) describes the use of boilers for this purpose.

Incineration of solid wastes is done in open pits, rotary kilns and multiple hearth furnaces.

Ideally, the products of combustion from an incinerator are carbon dioxide and water vapour. In practice, the gaseous effluents may contain more noxious gases due to (1) incomplete combustion and (2) components in the waste. Some effluents which are troublesome in incinerators include sulfur and nitrogen oxides, metals and dioxins.

There are three main factors which determine the completeness of combustion: time, turbulence and temperature – the three Ts. The first two are set by the design of the system, leaving only the temperature as an operating variable. The temperature is controlled by adjusting the air/fuel ratio.

If the effluent gas falls below the concentration at which it can sustain combustion before destruction is sufficiently complete, one option may be to use an afterburn by injecting additional fuel gas. Alternatively, it may be necessary to pass the effluent gas through a scrubber. Thus an incinerator produces both gaseous effluent and ash, and may produce a liquid effluent also.

Combustion products which have proved particularly troublesome are the dioxins, notably TCDD. TCDD is a product in virtually all combustion processes, including barbecues. The production of TCDD at incinerators has been the subject of much concern. Controls are such that operators are required to keep the concentration of TCDD measured in the adjacent area vanishingly small.

Incineration has the advantages that it is an established process, can handle a wide range of wastes, can accept high throughputs and is economical in its requirement for land, especially relative to landfill. Disadvantages are the comparatively high cost and the effluents generated.

Hazardous wastes in sludge form are incinerated in rotary kilns, fluidized beds and multiple hearth furnaces.

Incineration is also practised for liquid wastes, provided they are at least partially combustible. A liquid waste may be categorized as combustible, partially combustible or non-combustible. A waste with a calorific value below about 18,500 kJ/kg is regarded as only partially combustible. Where the properties allow, liquid wastes are generally injected through an atomizer.

A11.16.5 Landfill

Another principal method of disposing of hazardous waste is landfill (LF). This involves depositing the hazardous waste on land which is then covered over. Both area and trench systems are used.

Accounts of the landfill option are given in *Landfill Technology* by Crawford and Smith (1985) and by A. Parker (1982), G.W. Dawson (1983), J.B. Cook (1984), Loehr (1987), Veselind, Peirce and Weiner (1990) and Nemerow and Dasgupta (1991).

The main problem with landfill is the leaching out of hazardous substances and transport of these into watercourses. Leachate management is therefore essential if landfill is used.

This aspect is discussed in *Groundwater Contamination from Hazardous Wastes* by E.F. Wood (1984) and by Hoppe (1984a), Shuckrow, Touhill and Pajak (1987), Ahlert and Kosson (1990) and Batchelor (1990).

There are a number of options for dealing with leachates. The first should be the application of waste minimization to eliminate or reduce potential leachates. The site can be designed to minimize the generation and escape of leachates, which includes diversion of surface waters and provision of barriers such a liners and trenches. Leaching may be reduced and/or leachate treatment simplified by segregation of wastes. Stabilization methods may be used to minimize leaching. The leachates may be collected and treated.

Economic estimates generally show landfill to be appreciably cheaper than incineration but the potential for migration of leachates into watercourses can be a serious problem which may well alter the balance of advantage.

A description of an landfill project associated with the chemical industry is given by Anon. (1988k).

A11.16.6 Land treatment
Land treatment involves spraying the liquid on land, exploiting the assimilative capacity of the soil.

Accounts are given in *Design of Land Treatment Systems for Industrial Wastes* by Overcash and Pal (1980) and by Loehr (1987) and Nemerow and Dasgupta (1991).

A11.16.7 Contract disposal
The producer of hazardous wastes often has the option of having them removed by a specialist contractor. Given the number and severity of problems often arising with such wastes, this can be an attractive option.

The main point to be made here is that regulatory controls are now such that there is a responsibility of the producer to ensure that the wastes are properly disposed of. The producer who hands over the wastes to a 'cowboy' contractor can expect trouble.

A11.16.8 Hazardous waste facility siting
A hazardous waste treatment, storage and disposal facility is liable to pose a more serious siting problem than the generality of process activities. In the United States more than half the applications for such developments fail. An account of the siting problem of TSDFs, with emphasis on the public consultation process, is given by Barclay (1987).

A11.16.9 Hazardous waste facility management
A hazardous waste facility needs to be managed. Incineration, landfill and storage all involve their characteristic management problems.

Active management is required of the basic activity and of the monitoring of effluents from the treatment processes, of spills and other escapes, and of the concentrations of chemicals in the immediate environment.

For landfill, leachate management assumes particular importance. An account is given by Shuckrow, Touhill and Pajak (1987).

A11.16.10 Hazardous waste facility reclamation
In the past hazardous waste was frequently deposited in an uncontrolled manner, with the consequence that many countries have a large number of waste disposal sites, which present major problems in establishing ownership and responsibility, reasserting control and effecting reclamation.

Accounts of land decontamination and reclamation are given in *Reclamation of Contaminated Land* by the SCI (1980), *Remedial Action Technology for Waste Disposal Sites* by Rogoshenski, Boyson and Wagner (1983) and by Wagner *et al.* (1986), *Evaluation of Remedial Action Unit Operations at Hazardous Waste Disposal Sites* by Ehrenfeld and Bass (1984), *Contaminated Land: Reclamation and Treatment* by M.A. Smith (1985), *Dioxin Containing Wastes: Treatment Technologies* by Arienti *et al.* (1988), *Environmental Remediation* by Vandegrift, *Reed and Tasker* (1992) and *Remedial Processes for Contaminated Land by Pratt* (1993).

As described earlier, under CERCLA 1980, or Superfund, the EPA is charged with supervision of site clean-up and federal funds are available for this. An account is given in *Superfund Handbook* by Sidley and Austin (1987).

A11.16.11 Ocean dumping and incineration
A quite different solution to the waste disposal problem is to dump or incinerate the waste at sea.

Accounts of these options are given in *Ocean Dumping of Industrial Wastes* by Ketchum, Kester and Park (1981) and *Ocean Disposal Systems for Sewage Sludge and Effluent* by the NRC (1984) and by Finnecy (1982), Remirez (1982), Crumpler and Martin (1987), Ditz (1988) and Nemerow and Dasgupta (1991).

Essential safeguards in any system of ocean dumping are those substances which are highly toxic and persistent and do not disperse sufficiently are not dumped and that for those materials which are the assimilative capacity of the area should not be overloaded.

Vessels used for ocean incineration include *Matthias I, II* and *III*, Vesta and Vulcanus. Of these the MT Vulcanus, a cargo ship converted in 1972 into a chemical tanker with two large incinerators, is perhaps the best known and documented. These incinerator vessels are described by Finnecy (1982).

Both ocean dumping and ocean incineration are governed by the marine pollution conventions. The Convention for the Prevention of Marine Pollution by Dumping of Wastes and Other Matter 1972 (the London Dumping Convention (LDC)) governs ocean dumping, and provisions on ocean incineration have been added to this.

A review of the operation of this convention is given by the IMO (1991 IMO-532). The unregulated dumping which used to take place has largely stopped, the dumping of certain wastes has been eliminated altogether or is being phased out and stringent programmes are in place to assess the need for and impact of ocean disposal.

A11.16.12 Underground storage tanks
It is convenient at this point to mention briefly the problem of leaking underground storage tanks (USTs) and pipework. The use of underground storage is widespread but in recent years there has been growing concern, particularly in the United States, over leaks from such installations. Accounts are given by G. Parkinson (1987) and Stone (1987).

Underground storage tanks are considered in Chapter 22.

A11.17 Fugitive Emissions

Fugitive emissions occur on process plants from diffuse continuous sources such as valves, seals and flanges; from equipment which operates intermittently such as pressure relief valves; from the 'breathing' of storage tanks; and from activities such as draining and sampling and the opening up of equipment during operations or for maintenance.

Accounts of fugitive emissions are given in *Fugitive Emissions of Vapour from Process Plant Equipment* by the BOHS (1984 TG3) and *Health Hazard Control in the Process Industries* by Lipton and Lynch (1987).

The topic of fugitive emissions can be considered under the following heads: (1) emission sources, (2) emission control, (3) occupational health effects and (4) environmental effects.

Fugitive emission sources, emission control and occupational health aspects are discussed in Chapters 15, 12 and 18, respectively.

The environmental effect which initiated concern over fugitive emissions was smog, originally in Los Angeles.

The EPA has been active to reduce fugitive emissions, as instanced by *Compilation of Air Pollutant Emission Factors* (EPA, 1973), *VOC Fugitive Emissions in Synthetic Organic Chemical Manufacturing Industry – Background Information for Promulgated Standards* (EPA, 1982) and *Standards of Performance for New Stationery Sources Equipment Leaks of VOC, Petroleum Refineries and Synthetic Organic Chemical Manufacturing Industry* (EPA, 1984).

A11.18 Odours

Another form of pollution is noxious odours. The concentration at which a substance can cause an odour can be very low, and certainly well below the levels of the air quality standards set to prevent air pollution.

Accounts of odours and their control are given in *Methods of Air Deodorisation* by Summer (1963), *Odour Pollution of Air* by Summer (1971), *Human Response to Environmental Odors* by Turk, Johnston and Moulton (1974) and *Odour Control* by Valentin and North (1980) and by Oelert and Florian (1972), Amoore and Hautala (1983), Ruth (1986), Langenhove, Lootens and Schamp (1988), Miedema and Ham (1988), Valentin (1990), Nagy (1991), A.M. Martin *et al.* (1992) and Sober and Paul (1992).

A11.18.1 Regulatory controls
Regulatory controls on odours are at present fairly rudimentary. In the UK, odours are statutory nuisances under the EPA 1990 and subject to control by the local authorities.

In the USA the Clean Air Amendments 1990 provide a framework within which controls on industrial odours may develop.

In the Netherlands, the authorities have taken initiatives, described below, aimed at putting odour control on a more scientific basis.

A11.18.2 Odour-generating activities
Many of the activities which give rise to noxious odours lie outside the mainstream process industries, being such as agricultural activities; animal waste rendering and leather tanning; fishmeal processing; sewage and effluent treatment and sludge disposal; brickmaking; and so on. However, petroleum refining and chemical manufacture do generate odours as do coke ovens and tar distillation; pulp and paper processing; plastics and rubber processing; soap and detergent manufacture; and fermentation processes associated with brewing or pharmaceutical manufacture.

A11.18.3 Odour measurement
A fundamental approach to odour measurement would appear to require that (1) the components responsible for the annoyance are identified, (2) their concentrations are measured and (3) their contributions, severally and possibly in combination, are quantified. This is a tall order. It is not surprising therefore that the most practical way of measuring an odour is still the human nose.

The method commonly used is to assemble an odour panel of suitably selected individuals and to use the technique of dynamic dilution. This involves diluting the original sample containing the odour with progressively larger quantities of clean air until 50% of the panel can no longer detect the odour. The results obtained are sensitive to the experimental conditions and can vary by a factor of 10 or more.

The concentration at which the odour is no longer detectable to 50% of the panel is designated as one odour unit (ou). The number D of dilutions required to get down to this concentration is a measure of the odour concentration of the original sample. Thus if dilution by a factor of 2000 is required ($D = 2000$), the odour concentration of the sample is 2000 ou.

A11.18.4 Odour thresholds
There are available a number of compilations of odour thresholds. They include the collections in Standardised Human Olfactory Thresholds by Devis (1990) and by Leonardos, Kendall and Barnard (1969), Amoore and Hautala (1983) and Ruth (1986).

A11.18.5 Odour sources
Odour sources may take the form of point sources, area sources, line sources or fugitive emissions. An area source is exemplified by an open tank containing liquid and a line source by a works transport route. A discussion of such sources is given by Sober and Paul (1992).

Fugitive emission sources, with typical emission rates, are listed in EPA AP-42 Compilation of Air Pollutant Emission Factors (1985).

A11.18.6 Odour dispersion
The dispersion of an odour may be modelled using passive gas dispersion models, which are described in Chapter 15.

The use of such models is discussed by Sober and Paul (1992). They refer in particular to the various EPA models such as ISC and Inpuff.

Models specific to odours, developed at Warren Spring Laboratory (WSL), have been described by Valentin and North (1980) and Valentin (1990). For short-range dispersion from a ground level source, the following simplified relation provides a rough guide:

$$Dr = 7E/UX^2 \qquad [A11.18.1]$$

with

$$E = D_e F \qquad [A11.18.2]$$

where D_e is the odour concentration at the emission point (ou), D_r is the odour concentration at the receptor point (ou), E the contaminant, or odour, emission flow (m^3/s), F the total emission flow (m^3/s), U the wind speed (m/s) and X the distance between the emission and receptor points (m).

Other WSL models relate to the distance at which complaints can be expected. For ground level emissions

$$d_{max} = (\phi E)^{0.6} \qquad [A11.18.3]$$

with

$$E = DF \qquad [A11.18.4]$$

where d_{max} is the maximum distance at which complaints can be expected (m), D the dilution factor (ou) and ϕ a factor. The value of ϕ is 2.2, which is the geometric mean of the uncertainty range 0.7–7. The equation can be used for emissions for chimneys provided that d_{max} is 40–50 times the effective chimney height.

An equation is also given for the effective chimney height for nuisance-free dispersion:

$$H_e = (\psi E)^{9.5} \qquad\qquad [A11.18.5]$$

where H_e is the effective chimney height (m) and ψ a factor. The value of ψ is 0.1, which is close to the geometric mean of the uncertainty range 0.05–0.15. The actual chimney height is obtained by subtracting the plume rise from the effective chimney height.

The last two equations encompass not just dispersion but also human response. The ability to correlate the complaints distance, albeit with a wide uncertainty range, by extrapolation from panel experiments to field assessments on the basis of the dilution factor D is encouraging for the development of a generalized approach.

A11.18.7 Odour control strategy

The outline of an odour control strategy is given by Valentin (1990). A necessary preliminary is to establish that it really is your plant which is the source of the problem. The odour sources should be identified. If practical, an odour source should be eliminated. For those which are not eliminated, ventilation, local where possible, should be used to remove and collect the contaminated air. If the air stream contains mist or dust, these should be removed. This air should then be treated using one of the methods described below.

A11.18.8 Odour control techniques

There are a number of methods available for treating air containing air contaminated with odours. Accounts are given by Valentin (1990) and A.M. Martin et al. (1992).

Martin et al. describe a wide range of techniques, including (1) incineration, (2) adsorption, (3) absorption, (4) condensation, (5) advanced oxidation and (6) biological treatment. Incineration may be carried out in a thermal incinerator, a catalytic incinerator, a flare, a boiler or a process heater. Adsorption is commonly by carbon. Advanced oxidation involves treatment with an oxidizing agent such as ozone using UV radiation to effect photochemical stimulation of the reaction. Biological treatment may be by a biological scrubber or a biological filter. The authors give examples of the efficiencies obtained with the different methods and guidance on selection of a method.

The options available are also discussed by Valentin, who deals in particular with thermal and catalytic incineration, adsorption, absorption and biological filters.

A11.18.9 Odour impact on community

Studies of the impact of odours on the neighbouring community have been described by van Langenhove, Lootens and Schamp (1988) and Miedema and Ham (1988).

The former study involved the investigation of odour from an animal rendering plant. A questionnaire was used to assess the proportion of persons annoyed by the odour at different distances. The effective limit of the nuisance was found at about 600 m. The authors then used a passive gas dispersion model to work back to the effective strength of the source, which they assessed as 68×10^6 ou/h.

Miedema and Ham describe field studies conducted by TNO in support of Dutch government plans for odour control. The authors emphasize the importance of the method used to assess the dilution factor D. They describe and advocate the use of the TNO 'sniffing car' method.

These authors too made use of a questionnaire to determine the relation between the distance and the proportion of persons annoyed by the odour. They correlate their results in terms of the C_{98} value, or the 1 hour average concentration which is exceeded only 2% of the year, and of the probability of being annoyed, expressed as a Z score, from the normal distribution, and obtain the relation

$$Z = 1.19 \log C_{98} - 2.16 \qquad\qquad [A11.18.6]$$

They suggest that a correlation of this type has potential as the basis for standards for odour control.

A11.19 Transport

A proportion of the chemicals produced are consumed on site, but a large proportion are transported from the site.

The transport of chemicals is discussed in Chapter 23 and emergency planning for such transport in Chapter 24.

Such transport, whether by vehicle or a pipeline, has the potential for an accident resulting in a spill. Spills are discussed in the following section.

A11.19.1 Transport of hazardous wastes

There is also the separate question of the transport of hazardous wastes. This has been a severe problem area due to the prevalence of illicit practices in transport and disposal, described in *Laying Waste* by M. Brown (1981). The transport of hazardous wastes, and of material from spill incidents, is discussed by R.J. Buchanan (1982), Bromley and Finnecy (1985) and Colen (1987b).

A11.20 Spills

Another way in which pollution can occur is by a liquid spill, either at a fixed site or during transport, including by pipeline.

Spills are treated in *Hazardous Chemicals – Spills and Waterborne Transportation* by Weidenbaum (1980) and *Hazardous Material Spills Handbook* by G.F. Bennett, Feates and Wilder (1982) and by Lindsey (1975) and Feates (1978).

The variety of situations which can occur and of the responses are illustrated in the numerous case histories described in the conference series *National Conference on Hazardous Materials Spills* in the United States.

In the United States, hazardous materials spills involve several agencies, including the DOT, the EPA and the FEMA, as reflected in the guidance published for the EPA in *Manual for the Control of Hazardous Material Spills* by Huibregtse et al. (1977), in *Emergency Guide for Selected Hazardous Materials* by the DOT (1978) and for FEMA in *Planning Guide and Checklist for Hazardous Materials Contingency Planning* by Gunderloy and Stone (1980).

A11.20.1 Types of spill

Some principal types of spill are those from (1) process plant or storage, (2) pipeline, (3) road tanker, (3) rail tank car, (5) barge, (6) ship and (7) drilling rig or fixed production platform.

A spill within the works occurs in a relatively controlled environment, but this is less true of one in transport.

A liquid spillage on land poses a threat to land and to watercourses. It is characterized by the facts that the release is transient and generally relatively limited in volume but often noxious.

Liquid spilled into a stream can cause a degree of pollution which kills off much of the aquatic life so that a long period elapses before recovery can occur.

A11.20.2 Response to spills onto land
If a spill occurs on land, there are a variety of measures which can be taken. A review is given by Scholz (1982), who considers the problem under the heads of (1) termination of discharge, (2) containment to prevent entry into watercourses and (3) recovery and treatment.

Chemical and physical measures are discussed by Eckenfelder (1982) who covers (1) neutralization, (2) sedimentation, (3) coagulation, (4) precipitation, (5) carbon adsorprtion, (6) ion exchange, (7) oxidation–reduction reactions and (8) biodegradation. A fuller account of biological measures is given by Armstrong (1982).

The EPA has developed a number of items of equipment for dealing with spills, including a portable foam diking system, a mobile physical–chemical treatment unit, a portable sedimentation tank, a mobile incineration system and a mobile soil decontamination system. The foam diking system can be used to restrict the flow of a liquid over asphalt or concrete or to plug drains to prevent liquid from entering. These and other devices are described Freestone and Brugger (1982), in an account of EPA development work on spill control technology, and also by Scholz.

Accounts of specific techniques and devices for dealing with hazardous material spills include those on physical barriers by Friel, Hiltz and Marshall (1973), sorbents by Melvoid and Gibson (1988), a mobile treatment system by M.K. Gupta (1976), a mobile incineration system by Tenzer *et al.* (1979), groundwater protection by Huibregtse and Kastman (1981), *in situ* detoxification of soil by Huibregtse, Lafornara and Kastman (1978) and soil reclamation by Wentzel *et al.* (1981).

A11.20.3 Response to spills into streams
If a spill occurs which threatens a stream may be possible to prevent it from entering the stream, but once it does the measures which can be taken are generally limited.

Strategies for dealing with spills into streams have been described by G.W. Dawson and co-workers (Dawson, Shuckrow and Mercer, 1972; Dawson, 1975; Dawson and Parkhurst, 1976; Dawson and McNeese, 1978).

One approach is stream diversion. The EPA has developed a portable stream diversion system capable of bypassing $0.35 \text{ m}^3/\text{s}$ a distance of 300 m.

Accounts are given by Lafornara (1982) of stream sampling and analysis, by Nadeau (1982) of bioassay and by Kontaxis and Nusser (1982) of dispersion modelling.

A11.20.4 Response to oil spills
Methods for the control of oil spills have been published by CONCAWE. They include a manual on inland oil spill cleanup (CONCAWE 1981 7/81), disposal techniques for spilt oil (CONCAWE 1980 9/80), protection of ground-water from oil pollution (CONCAWE 1979 3/79), strategies for the assessment of biological impacts of large oil spills on European coasts (CONCAWE 1985 5/85), and field guides to inland oil clean-up (CONCAWE 1983 10/83), coastal waters clean-up (CONCAWE 1981 9/81) and oil spill dispersal (CONCAWE 1988 2/88).

A11.20.5 Ultimate disposal of spilled material
Ultimate disposal of spilled material is considered by Lindsey (1975), R.J. Buchanan (1982), Novak and Pfrommer

(1982) and A. Parker (1982). The options described by these authors are primarily incineration and landfill.

A11.21 Marine Pollution

Pollution of the seas occurs by discharges made following cleaning of tanks of oil tankers and as a result of accidental spillages from oil tankers and chemical carriers.

Pollution may occur as the result of illicit discharges or as the result of an accident. Most of the incidents which have hit the headlines have occurred after the ship has run aground.

Accounts of sea pollution are given in *Oil Pollution of the Sea and Shore* by WSL (1972), *Accidental Oil Pollution of the Sea* by the DoE (1976), *Marine Pollution* by R. Johnston (1977) and *Pollutant Transfer and Transport in the Sea* by Kullenberg (1982), *Poisoners of the Sea* by Goudray (1988) and *Marine Pollution* by R.B. Clark (1993) and of pollution criteria in *Pollution Criteria for Estuaries* by Helliwell and Bossanyi (1975).

The WSL publication records experience gained as a result of the pollution mitigation activities following the Torrey Canyon disaster.

A11.22 Pollution Incidents

Incidents involving pollution occur both from fixed installations and in transport and over varying timescales.

A11.22.1 Incident databases
The SRD operates the Environmental Data Service EnvI-DAS which complements its other databases MHIDAS for major hazard incidents and EIDAS for explosion incidents.

In addition, the account of an incident in a hazards database frequently includes some information on environmental effects.

A11.22.2 Air pollution
Air pollution incidents tend to take the form of (1) a sudden large release, (2) a regular lower level release which is aggravated by an unusual weather condition or (3) a regular practice of lower level release which is found to be damaging the environment.

An example of a sudden large release is that at Seveso, described in Appendix 3.

A11.22.3 Smog incidents
At the end of October 1948, a heavy smog developed in inversion conditions in the industrial town of Donora, Pennsylvania. After several days conditions became such that 17 died within one day; a further 4 succumbed before the smog abated.

The worst smog incident recorded was the 'killer smog' in London in 1952. Again a heavy smog developed in inversion conditions. The concentration of sulfur dioxide rose to some seven times its normal level. The maximum mean daily concentration of the gas in Central London reached $3500\mu\text{g}/\text{m}^3$. Two days into the incident the death rate began to soar. The number of excess deaths was estimated at about 4000.

A11.22.4 Bonnybridge
An example of the problems attendant on incineration is provided by the travails of Rechem, a company involved in the incineration of hazardous wastes. During the 1980s, the

incinerators which it operated included one at Bonnybridge in Scotland and one at Pontypool in Wales. At Bonnybridge there were complaints from the public and cattle nearby suffered from an unexplained sickness. Tests did not confirm excessive levels of contamination, but the plant was closed. The company had further troubles in its relations at Pontypool, though again numerous independent surveys failed to find concentrations of contaminants above the background levels.

A11.22.5 Watercourse pollution
Watercourse pollution incidents tend to take the form of (1) a sudden large release, (2) a temporary but appreciable increase in the quantity or concentration of a regular release or (3) a regular practice of lower level release which is found to be damaging the environment.

A11.22.6 Schweizerhalle
On 1 November 1986, a fire in a agrochemical warehouse at Schweizerhalle in Basel, Switzerland, broke out and was extinguished with large quantities of fire water, which became contaminated and then found its way into the Rhine, causing serious pollution. The incident is described in Case History A113.

A11.22.7 Minimata
Discharge of mercury compounds in Minimata Bay in Japan led to severe pollution and to the Minimata mercury poisoning disaster. For some 15 years prior to 1953, the Chisso Chemical Company had been discharging mercury catalyst into the bay. People fell ill, at first in small and then in increasing numbers. In 1957, a ban was placed on fishing in the bay. Of 116 officially recorded cases in 1958, 43 had died. By 1972, the official list of victims was almost 300.

A11.22.8 Hazardous waste pollution
Hazardous waste tends to give rise to developing problems at hazardous waste facilities rather than to incidents as such. However, incidents can arise such as those due to serious malpractice in transport of such wastes.

Accounts of hazardous waste problems and incidents are given in *Laying Waste* by M. Brown (1981) and *World of Waste* by Goudray (1992).

A11.22.9 Love Canal
The scale of the environmental disaster which may result from waste disposal is illustrated by the case of Love Canal. In 1953, this site was sold by the Hooker Chemical Company to the Niagara Falls education authority for a nominal sum. For over a decade the site had been used to dump chemical wastes. In 1976, local residents began to complain of odours and other nuisance. Concern escalated, leading to the relocation of 238 families and causing the President to declare the site a national environmental emergency, making it thereby eligible for federal clean-up funds.

There were numerous damage suits. In support of its suit against the company, the Justice Department commissioned a chromosome study of 36 residents; the study proved controversial.

The engineering measures taken to render the site safe are described by Glaubinger, Kohn and Remirez (1979).

A11.22.10 Land spill pollution
Incidents involving PCBs are described by A.J. Smith (1982), others involving pesticides by Frisbie (1982) and one involving phenol by A. Shepherd (1982a).

The Seveso incident, in which a large area of land was contaminated with TCDD, is described in Appendix 3.

A11.22.11 Sea pollution
Sea pollution incidents which tend to have particularly serious consequences are groundings of tankers on coasts. The three cases given below are all of this kind. Pollution of the sea also occurs from oil production activities, as described in Appendix 18.

A11.22.12 Torrey Canyon
On 18 March 1967, the tanker Torrey Canyon, carrying 114,000 tons of crude oil, ran aground on the coast of Cornwall, releasing 30,000 tons of oil. Over the next 7 days a further 20,000 tons escaped. On 26 March the ship broke its back with release of another 50,000 tons. Following this, the ship was bombed to burn the remaining 20,000 tons and the oil slick was sprayed with detergent to disperse it.

A11.22.13 Amoco Cadiz
On 16 March 1978, the oil tanker Amoco Cadiz lost its steering and ran aground at Portsall on the Brittany coast. The resulting oil spill polluted some 200 km of coastline, causing an incident which at the time was the worst to have occurred.

A11.22.14 Exxon Valdez
On 24 March 1989, the tanker Exxon Valdez ran aground near Valdez, Alaska, rupturing eight cargo tanks and releasing some 258,000 bbl of oil. The result was a major environmental disaster.

A11.23 Notation

Section A11.18
C_{98}	1-h average concentration exceeded only 2% of the year
d_{max}	maximum distance at which complaints can be expected (m)
D	dilution factor (ou)
D_e	odour concentration at emission point (ou)
D_r	odour concentration at reception point (ou)
E	contaminant, or odour, emission flow (m³/s)
F	total emission flow (m³/s)
H_e	effective chimney height (m)
U	wind speed (m/s)
X	distance between emission point and reception point (m)
Z	score value
ϕ	factor
ψ	factor

Appendix

12

Noise

Contents

The environment may be degraded not only by chemical pollution but also by noise. Process plant contains a variety of noise sources covering rotating equipment, piping, combustion equipment and vents to atmosphere. Noise from process plant sources may affect workers from the perspective of noise induced hearing loss and the general public through environmental or community noise impacts. The control of noise from process plants needs to be controlled and noise control design needs to be an integral part of the engineering and planning for any process facility.

Accounts of noise, noise effects on humans and noise control are given in the following references:

- *Handbook of Noise Control* by C.M. Harris (1957),
- *Handbook of Noise Measurement* by A. Peterson and Gross (1967),
- *Noise and Vibration Control* by Beranek (1971),
- *Noise and Hearing Conservation Manual* by Berger *et al.* (1986 AIHA/11),
- *Engineering Noise Control* by Bies and Hansen (1988),
- *Woods Practical Guide to Noise Control* by Sharland (1990),
- *Noise and Health* by Fay (1991 ACGIH/70) and
- *Noise and Vibration Control Engineering* by Beranek and Ver (1992).
- *Code of Practice for Reducing the Exposure of Employed Persons to Noise* by the DoE (1972/1) (the DoE Noise Code).

A12.1 Regulatory Controls

Regulations applicable to noise levels within an industrial facility are intended to protect employees from noise-induced hearing loss. The regulations are usually based on an employee noise exposure limits (e.g. 85 dB(A) averaged over 8 h). The regulations normally allow for exposure to higher noise levels for reduced time periods and mandate the use of hearing protection if necessary. This allows the employees' averaged noise exposure level to remain below the safe limits.

Applicable regulations for the UK/EEC and USA would include the following:

- EEC Machinery Directive – 'Noise at Work Directive', 86/188/EEC and 'Machinery Directive', 89/392/EEC.
- UK Noise at Work Regulations 1989.
- USA OSHA regulation – USA Code of Federal Regulations Title 29 Part 1910.95

Typically these regulations set a series of action levels based on employee noise exposure levels as follows:

(1) First Action Level – daily personal noise exposure exceeds 85 dB(A)
  - Employer is responsible for initiating a hearing conservation programme and having noise assessment made by a competent person.
  - Employer is required to reduce the risk of hearing damage from exposure to noise to the lowest level reasonably practicable.
  - Appropriate hearing protection should be available to the employee on request.
(2) Second Action Level – daily personal noise exposure of 90 dB(A)
  - Appropriate hearing protection must be made available to all persons exposed.
  - Employer must reduce noise as far as reasonably practicable.

- Other requirements include assessment records and audiograms, designated ear protection zones, training on maintenance and use of equipment and provision of information to employees and information to employees.

The following International Institute of Noise Control Engineering (I-INCE) Publication summarizes and compares employee noise exposure regulations for a range of countries worldwide: 'Technical Assessment of Upper Limits on Noise in the Workplace' I-INCE Publication 94-1.

Community or environmental noise impact must also be considered in plant noise control design. Noise limits set to control noise emissions outside of the plant are primarily intended to protect plant neighbours from noise-related annoyance or interference. In setting community noise limits, local regulations need to be considered together with assessments of likely community reaction to the increased noise levels. These issues also need to be considered in the context of the Environmental Impact Assessment (EIA) for the project.

Government regulations to control environmental noise vary greatly from country to country. Typically regulations apply to particularly zoning areas or land use classifications with different noise limits defined for specified daytime and night-time noise levels (e.g. residential, mixed residential/commercial, industrial etc.). These limits are forced to be generic and may not meet the specific requirements of a plant location. An important point, which is often overlooked, is that compliance with community noise regulations does not ensure against community complaints. People will complain because they find the noise disturbing or annoying, not because it exceeds a government regulation.

An EIA for a project would require the evaluation of the existing ambient noise levels in the proposed plant location and the predicted operational noise levels from the facility to assess the noise impact of the new plant.

Applicable standards and references for environmental noise would include the following:

- World Bank – *'Pollution Prevention and Abatement Handbook'* of 1998.
- US Export Import Bank – *'Environmental Procedures'* 1998.
- ISO 1996-1:1982 *'Acoustics Description and Measurement of Environmental Noise – Part 1: Basic Quantities and Procedures'.*
- ISO 1996-2:1987 *'Acoustics – Description and Measurement of Environmental Noise – Part 2: Acquisition of Data Pertinent to Land Use'.*
- ISO 1996-3:1987 *'Acoustics – Description and Measurement of Environmental Noise – Part 3: Application to Noise Limits'.*
- UK Control of Pollution Act 1974.
- UK Environmental Protection Act 1990.

A12.2 Process Plant Noise

The following organizations have a series of useful standards and guidelines relevant to Process Plant Noise:

- Engineering Equipment Material Users Association (EEMUA), 14-15 Belgrave Square, London, England SW1X 8PS UK; Phone: 071-235-5316

- CONCAWE, Madouplein 1, Brussels B-1030, Belgium; Phone: 32-2-220-3111
- American Petroleum Institute, 1220 L Street, NW, Washington, DC 20005-4070; Phone 202-682-8000

A12.3 Noise Control Terminology

A summary of widely used noise control terminology is given below:

- Audible Frequency Range
  The range of sound frequencies normally heard by the human ear. The audible range spans from 20 Hz to 20 kHz, but for most engineering investigations only frequencies between 40 Hz and 11 kHz (i.e. 63 Hz to 8 kHz octave bands) are considered.
- Background Noise
  The sound pressure levels in a given environment from all sources excluding a specific sound source being investigated or measured.
- Decibels (dB)
  Ten times the logarithm (to the base 10) of the ratio of two mean square values of sound pressure, voltage or current.
- Sound Pressure
  A dynamic variation in atmospheric pressure. The pressure at a point minus the static atmospheric pressure at that point (N/m^2).
- Sound Pressure Level (L_p)
  The ratio, expressed in decibels, of the mean square pressure to a reference mean square pressure which by convention has been selected to be equal to the assumed threshold of hearing.

$$L_p = 10 \log_{10} \left(\frac{P}{P_{\text{ref}}} \right)^2$$

where P = RMS sound pressure (N/m^2); $P_{\text{ref}} = 2 \times 10^{-5}\,N/m^2$.

- Sound Intensity
  The rate of sound energy transmission per unit area in a specified direction (W/m^2).
- Sound Intensity Level (L_I)
  The rate of sound energy transmission per unit area in a specified direction related to a reference intensity level.

$$L_I = 10 \log_{10} \left(\frac{I}{I_{\text{ref}}} \right)$$

where I = RMS value sound intensity (W/m^2); $I_{\text{ref}} = 1 \times 10^{-12}$ watts/m^2.

- Sound Power
  The total sound energy radiated by a source per unit time (W).
- Sound Power Level (L_w)
  The acoustic power radiated from a given sound source as related to a reference power level, expressed in decibels.

$$L_w = 10 \log_{10} \left(\frac{W}{W_{\text{ref}}} \right)$$

where W = RMS value sound power (W); $W_{\text{ref}} = 1 \times 10^{-12}$ watts.

- Octave Bands
  Frequency ranges in which the upper limit of each band is twice the lower limit. Octave bands are identified by their geometric mean frequency (center frequency). The table below shows the octave band upper/lower frequency limits.

Octave band cut-off frequencies (Hz)		
Octave band center frequency	Lower band limit	Upper band limit
31.5	22	44
63	44	88
125	88	177
250	177	355
500	355	710
1,000	710	1,420
2,000	1,420	2,840
4,000	2,840	5,680
8,000	5,680	11,360
16,000	11,360	22,720

- 1/3 Octave Bands
  Frequency ranges where each octave is divided into one-third octaves with the upper frequency limit being 1.26 times the lower frequency. Identified by the center frequency of each band.
- A-Weighted Sound Pressure Level (dBA)
  The sound pressure level measured using an electronic filter that approximates the frequency response of the human ear. The table below shows octave band A-weighting correction factors.

Octave band A-weighting correction factors	
Octave band center freq (Hz)	A-weighting correction (dB)
63	−26.2
125	−16.1
250	−8.6
500	−3.2
1000	0
2000	+1.2
4000	+1.0
8000	−1.1

- L_{Aeq} – Equivalent Continuous Sound Level
  The A-weighted energy mean of the noise level averaged over the measurement period. It can be considered as the continuous steady noise level which would have the same total A-weighted acoustic energy as the real fluctuating noise measured over the same time period.

$$L_{\text{Aeq}} = 10 \log_{10} \frac{1}{T} \int_0^T \left(\frac{p_A(t)}{p_0} \right)^2 dt$$

where T = total measurement time; $P_A(t)$ = A-weighted instantaneous sound pressure; $P_{\text{ref}} = 2 \times 10^{-5}\,N/m^2$

L_{Aeq} is, therefore, an important number for the evaluation of a fluctuating noise level, because it reflects the actual energy content of the time varying noise. L_{Aeq} is used in its own right in many national and international standards for rating some forms of community noise.

- L_N – Percentile Level
  The A-weighted sound level equalled or exceeded by a fluctuating sound level × percent of the measurement time period. For example:c

  L_{A10} – the A-weighted sound pressure level exceeded for 10% of the measurement period
  L_{A50} – the A-weighted sound pressure level exceeded for 50% of the measurement period
  L_{A90} – the A-weighted sound pressure level exceeded for 90% of the measurement period

  The various L_N levels over a measurement period are important in the evaluation of the temporal nature of the sound. A widely varying sound level over a measurement period will tend to have large differences between the various L_N levels than a more constant steady noise.

- Daily Personal Exposure
  The daily personal exposure of an employee is defined in the schedule to the HSE *Noise Guide 1* as follows:

$$L_{EP,d} = 10 \log_{10} \left[\frac{1}{T_o} \int_0^{T_e} \left(\frac{pA(t)}{p_o} \right)^2 dt \right]$$

where $L_{EP,d}$ is the daily personal exposure (dB(A)); $p_A(t)$ is the time-varying value of the A-weighted instantaneous sound pressure in the undisturbed field (Pa); t is the time (h); p_o is the base value of the sound pressure (Pa); T_e is the duration of the exposure (h); T_o a period of 8 h. As before, the value of p_o is 2×10^{-5} Pa.

A12.4 Noise Control

Noise control is generally considered under four heads: (1) at source, (2) in transmission, (3) on emission and (4) on receipt.

Noise can be generated only if there is a vibrating source, which may take the form of a vibrating surface or of fluctuation in fluid flow. The topic of noise is therefore inextricably bound up with that of vibration, as attested by the titles quoted earlier.

- Control at Source
  Control of noise at source should be the first option considered. The approach is analogous to inherently safer design. The approach to noise control at source is to identify potential sources and to either eliminate them altogether or to reduce their level of vibration by one or more of the following: the surface, the amplitude, the frequency and, in the case of a fluid, the flow velocity. A number of specific methods for control of noise, such as impact reduction, clamping, damping, acoustic panels and silencers can be considered.
- Control in Transmission
  In principle, noise associated with transmission through pipes and ducts may be structure-borne or fluid-borne. For gas or vapour pipes the sound is predominantly fluid borne as evidenced by the effectiveness of control by the use of in-line silencers. Such silencers are used, for example, on the intakes and exhausts of internal combustion engines and on the intake/exhaust of fans.
- Control on Emission
  Control on emission involves modification of the route by which the noise reaches the worker. Devices used include sound-absorbing materials on the workroom walls, sound-absorbing screens, total or partial enclosure of machines and cabins serving as 'noise refuges'. Another approach is to increase the distance between the worker and the noise source. Distance may be increased by segregation of noisy machines, use of remote controls and siting of exhausts.
- Control at Receiver
  Control at the receiver involves the use of ear protection. Types and selection of ear protectors are discussed in HSE *Noise Guide 5*.
  Limiting the time for which the person is exposed to the noise may effect some reduction in the noise burden, but to be effective the proportion of time exposed has to be cut drastically. Complete removal for half a shift, for example, reduces the daily exposure by only 3 dB(A).

The importance of addressing noise control issues as early as possible in a process plant engineering project cannot be over emphasized. The earlier noise control design issues are identified and resolved on a project, the more effective is the noise control design. That is effectiveness as measured in terms of overall project costs, impact on project schedule and compliance with the project noise requirements.

Effective noise control design for refining or petrochemical projects requires careful planning during the initial stages of the project, with equally careful and prudent execution through the various phases of the project. Before addressing specific project noise control design issues it will be useful to place some perspective on the noise control challenges faced on process plant projects. Any metric, from the capital cost, which can run in hundreds of millions, if not billions, of US dollars to the thousands of individual equipment items and bulk quantities of piping demonstrates the scale and complexity of these projects. The projects can be completely new 'grass-roots' plants or expansions to existing facilities. Often times a large project is further complicated by the need to divide the engineering in to manageable packages, with different engineering contractors responsible for different process units.

The sections below highlight the key noise control design activities from the initial front end engineering design (FEED) phase of the project through Detailed Design to Commissioning and Start-up:

- Front End Engineering Design (FEED)
  Noise control must be executed as early as possible in a project to minimize the effect on project costs layout and schedule. The major noise control activities on a project, therefore, tend to focus towards the FEED phase of the project. During these initial stages of a project, baseline noise surveys may be required to establish existing noise levels at the plant site. For a 'grass-roots' project the primary focus of the baseline survey will be on community noise levels. For an expansion project at

an existing facility the survey must consider both in-plant and community noise levels.

The key noise control task during the early stages of a project is the FEED Noise Study. This study evaluates the potential noise sources within the plant and establishes, through plant noise modelling, the predicted noise levels (in-plant and community) for the new facility. The initial modelling will be based on standard equipment with no special provision for noise control. The modelling will establish a baseline noise control position for the overall plant. This will identify potentially high noise equipment and high noise areas of the plant and will allow noise control design options to be readily included and evaluated in the plant noise model. This 'what if' analysis will allow the noise control design to be optimized to meet the plant owner's overall project noise control requirements.

Effective and accurate plant noise modelling is critical to this evaluation process. The plant noise modelling must be based on international and industry accepted standards (e.g. EEMUA, CONCAWE, ISO 9613). Noise model input data must be based on reliable and verifiable equipment and field noise data. Only with a solid background of field experience on comparable equipment in similar service can the noise control engineer speak with authority on predicted plant noise levels.

The findings of the FEED Noise Study will provide the data to define definite and unambiguous noise limits and requirements for the Detailed Design phase of the project. This will apply to the designation of in-plant noise limits and the definition of 'Restricted Areas' (i.e. hearing protection areas) together with property line noise requirements and overall sound power level limits. The goal must be to define and specify noise control requirements so that they meet the project requirements and can be effectively managed and enforced during the Detailed Design phase of the project.

• Detailed Design.
Noise control for the detailed design phase of a project will involve the engineering contractor, or contractors, executing the noise control design for the project in compliance with the noise limits defined during the FEED phase of the project. Primary areas of noise control activity during the Detailed Design phase of the project would include:

− Equipment requisition and design/data review process
− Piping noise and control valve noise control
    Acoustic insulation definition
    Low noise control valve selection
    Vent and in-line silencers
− Acoustic enclosures
− Acoustic insulation for rotating equipment
− Equipment shop noise testing (only in appropriate situations when good measurements are possible)

• Commissioning and Start-up.
The main noise control activities during commissioning and start-up will be noise surveys to demonstrate/verify compliance with the project noise limits. Depending on the results of the noise surveys corrective action may need to be initiated to reduce operation noise levels. Depending on the project type, long-term environmental noise monitoring in the community or on the plant property may be worthwhile during commissioning/start-up. This type of monitoring would provide data on upset and startup venting operating conditions (e.g. flaring). Obtaining reliable data on these difficult to reproduce operating conditions may prove valuable.

Noise control issues associated with typical classes of process plant equipment area discussed below:

• Rotating Machinery
A major source of noise in process plant is rotating machinery. Centrifugal compressors can produce noise levels of up to 100 dB(A) at 1 m from the equipment surface, often with dominant tonal content at the blade passing frequency. The casing is usually sufficiently massive to prevent transmission of aerodynamic vibration. There are also other noise sources such as motors, turbines, gear units, lube oil pumps and control valves. Full acoustic enclosure is usually unnecessary and it is more common to use a combination of acoustic insulation and spring support for the pipework together with low noise valves. Another source of noise on rotating machinery is electric motors. Large medium voltage motors can produce a noise level above 90 dB(A), due principally to the cooling air fan and also sometimes to magnetically induced vibration. The problem is generally soluble using smaller, more efficient cooling air fans but on occasion increased frame size and/or cooling air silencers may be required.

• Fluid Flow in Pipes
Another main source of plant noise is the flow of fluids in pipes. Noise control of individual plant items may not be sufficient to attain the noise target and it may be necessary to address the noise in the pipework also.

Noise from gas pipes is the subject of studies by CONCAWE described by Marsh *et al.* (1985), Pinder (1985) and van de Loo (1988). The CONCAWE piping model is an essentially empirical one and is somewhat similar to that in VDI 3733. With regard to noise control, Pinder outlines four methods, which relate to: pipewall thickness, plant layout, in-line silencers and acoustic insulation.

Noise due to flow through valves is discussed in ISA S75.17-1989 *Control Valve Aerodynamic Noise Prediction.*

• Combustion Processes
Combustion processes are also a significant source of plant noise. On furnaces with induced or naturally induced air flow the air inlet and combustion noise can exceed 90 dB(A) @ 1 m per burner unless adequate acoustic absorption is included in the design of the burner plenums. With forced draft systems the sound levels are much less, noise in this case coming mainly from the fan intake, casing and motor. Noise can be transmitted by furnace wall and by duct vibration, the solution in this case being adequate stiffness.

• Flares
A particular type of combustion process that is liable to generate a high level of noise, including noise affecting the public, is a flare.

The noise generated by a flare is a function of the energy release in the flame. Swithenbank (1972) has defined a noise generation efficiency equal to the fraction of the net heat release which is given off as noise power, and has correlated it with mass velocity at the flare tip. Ohiwa, Tanaka and Yamaguchi (1993) describe fundamental studies on the noise from a turbulent diffusion flame.

As he points out, proprietary flare tips are available to overcome the noise problem. Effective designs are based on the Coanda effect plus other utilizing mufflers or baffles for steam assist flares. A ground flare design also offers benefits for reduced noise due to flaring.

- Pressure Relief
An alternative method of disposal that also generates noise is venting to atmosphere. A method of estimating the noise from a high pressure relief discharging to atmosphere through a vent stack is given in API RP 521: 1990.

A12.5 Notation

f	fractional exposure
f_{tot}	total fractional exposure
H	sound power level (dB)
L	sound pressure level (dB(A))
L_{eq}	equivalent continuous sound level (dB(A))
$L_{\text{EP,d}}$	daily personal exposure (dB(A))
L_{T}	sound pressure level at distance r (dB)
N	number of decibels (dB)
$p_{\text{A}}(t)$	time-varying value of A-weighted instantaneous sound pressure in the undisturbed field (Pa)
p_{o}	base value of sound pressure (Pa)
P	pressure disturbance (Pa)
r	distance (m)
t	time (h)
T_{e}	duration of exposure (h)
T_{o}	8-h period
W	source power (W)

Subscript

o   reference level

Appendix 13

Safety Factors for Simple Relief Systems

Contents

A13.1 Comments on Safety Factors to be Applied when Sizing a Simple Relief System

It is recommended that a safety factor be applied to take account of both uncertainty in the data and inaccuracy in the calculation methods.

It is assumed in the following discussion that (1) the designer has chosen a method appropriate to the application, intended to tend towards oversizing the relief system rather than the reverse, and (2) the relief system consists at most of a safety valve or bursting disc preceded by a constant diameter pipe and followed by a constant diameter pipe to atmosphere. It is also assumed that the flow through any safety valve has been checked against the manufacturer's data; under no circumstances should the mass flow be assumed to exceed the value so calculated.

The safe case is that which corresponds with the highest estimate of the required relief rate(s) and the lowest estimate of the actual flow(s).

The overall safety factor can be applied in one of two ways:

(1)   by sizing the relief system to pass a flow equal to the required mass flow multiplied by the safety factor; or
(2)   in cases where the vent capacity is determined by one particular cross-sectional area, by multiplying the calculated required vent area by the safety factor. However, it is important then to check that the capacity remains determined by this area.

The overall safety factor F_0 may be represented as the product of a number of sub-factors, each taking account of particular features:

$$F_o = \prod_{i=1}^{n} F_i \qquad \text{[A13.1]}$$

where F_i is the sub-factor for feature i. Some of the sub-factors which may be appropriate are

(1)   A sub-factor ($\geqslant 1$) to reflect any features of the method, unless included in other specific sub-factors such as those suggested below, which might lead to undersizing of the relief system for the application concerned. These might be features of the methods used to calculate the required relief rate(s) or vent area, although in most cases the methods will have been chosen so as to tend to overestimate those. They may also be features of the approach to assessment of the relief system flow capacity not covered below, for example the method may have neglected the presence of solids in the flow. The designer must himself establish the required value for this sub-factor.
(2)   A sub-factor ($\geqslant 1$) to take account of uncertainty in the data. The magnitude and direction of any errors arising from the data assumptions should be assessed. For reactors particular attention should be paid to any errors affecting the rate of reaction including its dependence on temperature.
(3)   A sub-factor ($\geqslant 1$) to allow for the inaccuracy in the fluid flow calculation methods which is present even for incompressible liquid and/or ideal gas. In this discussion it is assumed that the largest realistic pipe roughness has been assumed, otherwise a larger sub-factor will be necessary.

Some recognized wide-applicability two-phase gas/vapour-liquid flow methods, when used for lines in which friction and/or static head are significant, may be accurate only to a factor of ± 2 (i.e. the actual flow is in the range half to two times the calculated value), and would thus correspond to a sub-factor of 2. For poorer methods, a larger sub-factor may be needed, while for methods more targeted at the specific application a lower sub-factor may suffice; expert advice should be sought. For two-phase gas/vapour-liquid flow with negligible friction and no changes in static head, for example through a safety valve alone, lower sub-factor values in the range 1.2–1.5 may sometimes be sufficient; expert advice should be sought.

For vapour-only flow when friction is significant a reasonable value for this sub-factor will usually lie in the range 1.2–1.5, while if only momentum changes are significant, a lower sub-factor may be allowable; expert advice should be sought.

(4)   A subfactor to allow for the effects of gas/vapour-phase nonideality if these have not been taken into account in the mass flow capacity calculation.

For critical (choked) frictionless gas/vapour-only flow (e.g. through a safety valve alone) the factor may be estimated from the work of Leung and Epstein (1988) using their figures 2, 4–6. These figures show the ratio of real-gas critical mass flow to ideal ($Z = 1$) critical mass flow (our sub-factor is the inverse of this ratio) as a function of the ratio of stagnation (inlet reservoir) pressure/thermodynamic critical pressure, for various values of the ideal (zero pressure) ratio of specific heats. The figures are based on the applicability of the Redlich–Kwong equation of state, but give a useful indication of the magnitude of the effects of non-ideality on critical flow. It can be seen that a sub-factor is needed for pressures exceeding about 20 times the critical pressure. At lower pressures the real critical flow is seen to be greater than the ideal critical flow, but caution should be exercised before adopting a sub-factor <1. It should be noted that non-ideality cannot be fully compensated for by assuming a constant Z value or using an adjusted fictitious molecular weight; this is discussed below.

For gas/vapour-only flow with friction, and for nonchoked flow with or without friction, the author is unaware of any treatment analogous to that of Leung and Epstein. For given inlet reservoir conditions and back pressure and assumed constant Z, the adiabatic or isothermal mass flow of a gas through a perfect nozzle, and from a reservoir via a perfect rounded inlet into and through a straight pipe, is proportional to $1/Z^{1/2}$ (Lapple, 1943). This suggests an approximate value for this sub-factor given by

$$F_4 = \frac{\text{Real gas mass flow}}{\text{Ideal gas } (Z = 1) \text{ massflow}} = \frac{1}{Z_{\max}^{1/2}} \qquad \text{[A13.2]}$$

where $Z_{\max}$ is the maximum value of Z occurring in the flow. Clearly this sub-factor will be inaccurate when Z varies significantly (e.g. near the thermodynamic critical point, and near choking conditions); this can be seen by comparing this sub-factor (which if Z were constant would apply equally to critical frictionless flow) with the sub-factor obtained from the work of Leung and Epstein for critical frictionless flow (and also from their figure 9). It would seem prudent to adopt a sub-factor at least as large as the larger of these two estimated sub-factors (but again, care should be exercised before adopting any value <1). If the non-ideality is severe and such as to reduce the flow, then ideally the relief system should be sized using a method which takes full account of non-ideality; the other sub-factors would still be required. If the non-ideality is severe but (as is more likely) is such as to increase the flow, then a sub-factor of 1 may be prudent.

It should be noted that the above sub-factors based on the work of Leung and Epstein and of Lapple are for application to the mass flow calculated on the basis of fully ideal ($Z = 1$) flow. If the flow has been calculated as semi-ideal with an assumed constant Z, or with an adjusted fictitious molecular weight, to give partial compensation for the departure from ideality, then this corresponds to the prior application of a sub-factor of $Z_{assumed}^{1/2}$ or $(MW_{actual}/MW_{assumed})^{1/2}$.

It seems likely that the effects of gas/vapour-phase non-ideality in two-phase gas/vapour-liquid flow, will be less, but the author is not aware of any studies to validate this. It is suggested that this sub-factor be taken as the same as for gas/vapour-only flow.

(5) A sub-factor (≥ 1) to take account of errors in the flow calculation for two-phase gas/vapour-liquid flow resulting from neglecting the variation in liquid density, liquid specific heat and latent heat along the pipe. The variation in these quantities can be judged from the calculated conditions along the pipe and thermodynamic charts. Then either a value can be assigned to this sub-factor or safe values can be assigned to those properties and the sub-factor set to 1. For calculation of mass flow capacity, safe values (giving a low mass flow) are low density, high specific heat and low latent heat. For gas-only or vapour-only flow, this sub-factor is irrelevant.

Finally the designer must check that there are no further sub-factors to be included, and decide whether to accept the calculated overall safety factor F_0. There is little published advice on typical factors. It would be surprising if the overall factor were to be less than 1.2 in any event, and significantly larger factors would be likely for reactors. For polymerization reactors Boyle (1967) said that the specified relief area should be two to three times the area indicated by design calculation (but note that this was before the preparation of the DIERS methods). Boyle's recommendation is consistent with typical overall factors obtained using the approach described above for two-phase vapour/liquid venting.

Caution

It is emphasized that the above comments are offered only to assist the designer in his consideration of what might be an appropriate overall safety factor. The designer must decide whether the advice is relevant to his application (bearing in mind the restrictions defined in the second paragraph of this note) and what sub-factors and overall factor are adequate.

H.A. Duxbury
Industrial Professor, Department of Chemical
Engineering, Loughborough University of Technology
May 1994

Appendix

14

Failure and Event Data

Contents

Many of the methods described in this volume require the use of data for failure and other events. This appendix gives a brief account of such data and a selection of references, which have been published in the open literature.

The data are given in summary form and primarily for illustrative purposes only. It is emphasized that there are many factors that determine the failure rates of equipment and the range of failure rates observed can be quite wide. It should be borne in mind that the same data are often quoted by a number of references. The existence of several sets of data that are in agreement is not necessarily evidence of the validity of the data. For industrial reliability work, therefore, it is necessary to consult the original literature and to make appropriate use of data banks and plant data.

Selected references on event and failure data are:

Event and failure

US Navy, Bur. of Naval Weapons (n.d.); NRC (Appendix 28 *Failure Data*); Dummer and Griffin (1960, 1966); Edison Elec. Inst. (1963, 1967); Timmermann (1968); R.L. Browning (1969a–c, 1970); F.R. Farmer (1971); Jacobs (1971); A.E. Green and Bourne (1972); AEC (1975); Gangadharan and Brown (1977); Cremer and Warner (1978); HSE (1978b); Ericsson and Bjore (1983); Johanson and Fragola (1983); Sherwin (1983); IEEE (1984); OREDA (1984, 1989, 1997, 2002); Rackwitz (1984); Bendixen, Dale and O'Neill (1985); Reliability Analysis Center (1985); Rossi (1985); IAEA (1988); Boykin and Levery (1989); Hauptmans and Homke (1989); EUROSTAT (1991)

Data collection and exchange

Pollocks and Richards (1964); Boesebeck and Homke (1977); Frankel and Dapkunas (1977); Haueter (1977); Richards (1980); B.K. Daniels (1982); Bockholts (1983); Borkowski, Drago and Fragola (1983); Carlesso, Bastia and Borelli (1983); Himanen (1983); Melis *et al.* (1983); Games *et al.* (1985); Lamerse and Bosnian (1985); Walls and Bendell (1985); Blokker and Goos (1986); Turpin and Kamath (1986); Wingender (1986, 1991); Bendell (1987), Davidson (1994); Ireson, Coombs and Moss (1996); Lannoy and Procaccia (1996); Dhillon (1999); EsReDA (1999)

Data problems and validity

Kletz(1973b); J.H. Bowen (1977); R.L. Browning (1977); Cannon (1977); Devereux (1977); Lees (1977c); Parsons (1977); Rex (1977); Vesely (1977b); HSE (1978B); Apostolakis *et al.* (1980); Pitts *et al.* (1980); Apostolakis (1982, 1985a); Martz (1984); OREDA (1984, 1992, 1997, 2002); Mosleh and Apostolakis (1985); Andow (1989), EsReDA (1999), Kletz (1999)

Data banks

Barnes (1967 UKAEA AHSB(S) R138); Naresky (1967); AEC (1975); J.H. Bowen (1977); Collacott (1977b); B.K. Daniels (1982); Bobbio and Saracco (1983); Capobianchi (1983, 1991); K.R. Davies (1983); OREDA (1984, 1992, 1997, 2002); Bendell and Cannon (1985); Scarrone, Piccinini and Massobrio (1989); Bendell (1991a–c); Cannon (1991b); Cannon and Bendell (1991); Cross and Stevens (1991); Mizuta *et al.* (1991); MIL-HDBK 217F; CCPS (1991); Fragola (1996); Ireson, Coombs, and Moss (1996); EsReDA (1999); Cooke and Bedford (2002) SRD: Barnes (1967 UKAEA AHSB(S) R138); Ablitt (1973 UKAEA SRS/GR/14, SRD R16,1975); Fothergffl (1973 UKAEA SRS/GR/22); Barnes

et al. (1976); J.H. Bowen (1977); Cannon (1991a,b); EdF: Silberberg and Meclot (1985); Procaccia (1991); ENI: Avogadri, Bello and Colombari (1983); ERDS: Capobianchi (1991); FACTS: Koehorst (1989); Hoehorst and Bockholts (1991); IAEA: Tomic and Lederman (1991); Offshore: Gaboriaud, Grollier-Baron and Leroy (1983); Gjestad (1983); Tveit, Ostby and Moss (1983); OREDA (1984, 1989, 1992, 1997, 2002); Bruce (1994); Groenendijk (1995); Sandtorv, Hokstad and Thompson (1996); Rausand and Øien (1996); Haugen, Hokstad and Sandtorv (1997); Langseth, Haugen and Sandtorv (1998)

Populations (of plants, equipment, etc.) at risk

Plants: AGA (n.d./lOl); IP (Oil Data Sht 15); IPE (1967–); SRI Int. (1988a–c); Anon. (1985dd); HSE (1986 Major Hazards 8); IPE (2002); Welsh (1993)
Equipment: Hooper (1982)
Transport: ACDS (1991)

Mechanical equipment failure modes (see also Table 7.1)

IMechE (1970/3); Pilborough (1971, 1989); Collins and Monack (1973); B.W. Wilson (1974); Lancaster (1975); OREDA (1984, 1992, 1997, 2002); H.C.D. Phillips (1990), Davidson (1994)

Pressure vessels

Kellerman (1966); Kellerman and Seipel (1967); Phillips and Warwick (1968 UKAEA AHSB(S) R162); Slopianka and Mieze (1968); Butler (1974); Engel (1974); T.A. Smith and Warwick (1974); AEC (1975); Boesbeck (1975); Bush (1975); HSE (1978b); Arulanantham and Lees (1981); Hurst, Hankin *et al.* (1992); Columns: H.C.D. Phillips (1990)

Storage tanks

H.C.D. Phillips (1990)

Pipework

S.A Wilson (1972, 1976); AEC (1975); Boesebeck and Homke (1977); Bush (1977); HSE (1978b,d); Arulanantham and Lees (1981); Cannon (1983); R.E. Wright, Stevenson and Zuroff (1987); Blything and Barry (1988 SRD R441); CIA, Chlorine Sector Gp (1989); de la Mare, Bakouros and Tagaras (1993); Hurst, Davies *et al.* (1994); Strutt, Allsop and Ouchet (1994); G. Thompson (1994)

Heat exchangers

A.E. Green and Bourne (1972); C.F. King and Rudd (1972); H.C.D. Phillips (1990); OREDA (1992, 1997, 2002), Lannoy and Procaccia (1996)

Rotating machinery

Turbomachinery: R.L. Browning (1970); Sohre (1970); OREDA (1992, 1997, 2002)
Compressors: Ufford (1972); Avogadri, Bello and Colombari (1983); H.C.D. Phillips (1990); OREDA (1992, 1997, 2002); Lannoy and Procaccia (1996), Hale and Arno (2001)
Pumps: Anon. (19721); F.R. Farmer (1971); Emelyanov *et al.* (1972); A.E. Green and Bourne (1972); C.F. King and Rudd (1972); Ufford (1972); AEC (1975); R. James (1976); Sherwin and Lees (1980); Dorey (1981); Aupied le Coguiec and Procaccia (1983); Avogadri, Bello and Colombari (1983); Sherwin (1983); Anon. (1985x); Bloch and Johnson (1985); N.M. Wallace and David (1985); Flitney (1987); H.C.D. Phillips (1990); OREDA (1992, 1997, 2002), Lannoy and Procaccia (1996); Sawaryn, Grames and Whelehan (2002)

Valve
Rijnmond Report; Moss (1977 NCSR R11); Aupied, le Coguiec and procaccia (1983); Vivian (1985); OREDA (1992, 1997, 2002)
Pressure relief valves: A.E. Green and Bourne (1972); Kletz (1972a, 1974a); Lawley and Kletz (1975); Aird (1982, 1983); Aupied, Le Coguiec and Procaccia (1983); Oberender and Bung (1984); OREDA (1984, 1992, 1997, 2002); Maher *et al.* (1988); Hanks (1994); D.W. Thompson (1994), Alfares (1999)
Non-return valves: AEC (1975); Aupied, Le Coguiec and Procaccia (1983); McElhaney (2000)
Emergency isolation valves: Haugen, Hokstad and Sandtorv (1997)

Bursting discs
Lawley (1974b)

Instruments, including valves
Barnes (1965 UKAEA AHSB(S) R99, 1966 UKAEA AHSB(S) R119, 1967 UKAEA AHSB(S) R122, R131, 1966); A.E. Green and Bourne (1965 UKAEA AHSB(S) R91, 1965, 1966 UKAEA AHSB(S) R117, 1972); Bourne (1966 UKAEA AHSB(S) R110, 1967); A.E. Green (1966 UKAEA AHSB(S) R113, 1969 R164, 172, 1971 UKAEA SRS/GR/2, 1970, 1972); J.C. Moore (1966); Hensley (1967 UKAEA AHSB(S) R138, 1968, 1969 UKAEA AHSB(S) R178, 1971); R.L. Browning (1969c, 1972, 1977); Mercer (1969); Anyakora (1971); Anyakora, Engel and Lees (1971); Stewart (1971); Upfold (1971); Lees (1973b, 1976b,c); Anon. (1974); Bulloch (1974); Lawley (1974b); Skala (1974); AEC (1975); Lawley and Kletz (1975); Moss (1977 NCSR R12); Bott and Haas (1978); H.S. Wilson (1978); Huyten (1979); de la Mare (1980); Piccinini *et al.* (1982); Aupied, Le Coguiec and Procaccia (1983); Rooney (1983); Unger (1983); Oberender and Bung (1984); OREDA (1984, 1992, 1997, 2002); Vivian (1985); Kumar, Chidambaram and Gopalan (1989); R. Green (1993); Paula (1993); Strutt, Allsop and Ouchet (1994)

Process computers
Barton *et al.* (1970); Hubbe (1970); E. Johnson (1988)

Fire protection, fire and gas detection, sprinkler systems
Rasbash (1975a); M.J. Miller (1974, 1977); P. Nash and Young (1976); Kamath, Keller and Wooliscroft (1981); Peacock, Kamath and Keller (1982); Peacock and Watson (1982); OREDA (1984, 1992, 1997, 2002); Gupta (1985); Finucane and Pinkney (1988 SRD R431); Rose-Pehrsson *et al.* (2000); Rowekamp and Berg (2000); Budnick (2001)

Utilities
R.L. Browning (1969c); Kletz (1975); Coltharp *et al.* (1978); OREDA (1984, 1992, 1997, 2002)

Power supplies
Dickinson (1962); R.L. Browning (1969c); AEC (1975); Jarrett (1983); B. Stevens (1983); Ketron (1980)
Diesel generators, diesel-driven equipment: F.R. Farmer (1971); AE. Green and Bourne (1972); AEC (1975); OREDA (1984, 1992, 1997, 2002); B. Steven (1983); Vesely, Uryasev and Samanta (1994)

Electronic equipment
Dummer and Griffin (1960, 1966); US Army, Dept of Defense (1965); A.E. Green and Bourne (1972)

Electrical equipment
F.R. Farmer (1971); A.E. Green and Bourne (1972); AEC (1975); OREDA (1984, 1992, 1997, 2002)

Electric motors
Dickinson (1962); Benjaminsen and van Wiechen (1968); R.L. Browning (1969c); AEC (1975); Avogadri, Bello and Colombari (1983)

Boilers
F.R. Farmer (1971); H.C.D. Phillips (1990)

Mills
Notman, Gerard and de la Mare (1981)

Belt conveyors
Notman, Gerard and de la Mare (1981)

Cranes
F.R. Farmer (1971); OREDA (1992, 1997, 2002)

Reactor overpressure
Marrs and Lees (1989); Marrs *et al.* (1989)

Leaks and spillages
Davenport (1977b, 1983); Kletz (1977J); HSE (1978b,d); Baldock (1980); Forsth (1981b); Hawksley (1984); A.W. Cox, Lees and Ang (1990); ACDS (1991); Berglund (1992); Taback, Siegell, Martino and Ritter (2000)

Ignition of leaks
R.L. Browning (1969c); A.W. Cox, Lees and Ang (1990); ACDS (1991)

Fire and explosions
Process plant fires: Rutstei and Clarke (1979); G. Wiiliam (1999)
Refinery fires: API (1983, 1984, 1985)
Pump fires: Kletz (1971); Wallace and David (1985); HSE (1978b)
Tank fires: Kletz (1971); API (1983, 1984, 1985)
Furnace explosions: Kletz (1972c); Ostroot (1972)
Warehouse fires: Hymes and Flynn (1992 SRD R578)

Vapour cloud explosions
Wiekema (1983a,b, 1984); Davenport (1977, 1983)
Probability of ignition: Kletz (1977J); HSE (1978b); Moussa *et al.* (1982); A.W. Cox, Lees and Ang (1990); ACDS (1991)
Delay before ignition: Kletz (1977J)
Drift before ignition: Kletz (1977J)

BLEVE
Blything (1986); Hurst, Hankin *et al.* (1992)

Transport
Westbrook (1974); HSE (1978b); Appleton (1988 SRD R474); ACDS (1991); P.A. Davies and Lees (1992); Montiel *et al.* (1996)

Pipelines – *see* Table 23.1

Ships
HSE/SRD (HSE/SRD/WP10, 29); Sandtory and Edwards (1980); Poten and Partners (1982); Blything and Lewis (1985 SRD R340)
Fire and explosion: Blything and Edmondson (1983, 1984 SRD R292); ACDS (1991)

LNG plants

Welker *et al.* (1976); Welker and Schorr (1979); AGA (1981/33); D.W. Johnson and Welker (1981); Moussa *et al.* (1982)

Offshore

Goodwin and Kemp (1980); Sofyanos (1981); Anon. (1983h); Dahl *et al.* (1983); OREDA (1984, 1992, 1997, 2002)

Missiles

HSE/SRD (HSE/SRD/WP7); HSE (1978b); Holden and Reeves (1985); Holden (1988 SRD 477); ACDS (1991)

Aircraft crashes

AEC (1975); HSE (1978b); Phillips (1981 SRD R198); Marriott (1985); Roberts (1987 SRD R388)

A14.1 Type of Data

There are several types of data that may be required. The first type is accident and incident data including the recording of system malfunctions, accident reports and databanks established to store accident information. The second type is event and failure rates data that involve all the abnormal events and the performance of equipment. For equipment, these data include:

(1) Failure frequency, probability
(2) Repair time
(3) Unavailability

and for events:

(4) Event frequency.

Failure rate data are usually given as frequencies or mean time to failure (MTTF). Two types of failure rate data are typically used: time-related and demand-related. Time-related failure rates are established for equipment that is continuously in use while demand-related failure rates are for equipment that is normally dormant and required to operate on demand.

A14.2 Definition and Regimes of Failure

The failure rate recorded for equipment necessarily depends on the definition of failure. The importance of this may vary. Failure of a pump to start on demand would appear relatively unambiguous, but failure of a pressure relief valve may refer either to failure of the set pressure to stay within prescribed limits when removed from the plant and tested in the workshop or to failure to relieve pressure during plant operation. An equipment failure can be catastrophic, degraded or incipient. It is essential to have a uniform definition and classification of failure for data collection, selection and verification.

It is usually assumed that the failure rate of a component, or strictly the hazard rate z, is constant, but this may not necessarily be the case, especially at the burn-in and wearout stages of its life cycle. Consideration may be given to the effect of time-dependent failure characteristics for the equipment and processes perceived critical to plant operation. Bathtub and Weibull distribution have been used to

represent a wide range of failure data. The hazard rate may be decreasing, constant or increasing, corresponding to the regimes of early failure, constant failure or wearout failure and characterized by values of the Weibull shape parameter $\beta < 1, 1, > 1$, respectively. A number of workers have found an early failure regime, corresponding to $1 < \beta$. Some typical results are

Equipment	β	Reference
Pump,		
type A	0.74–1.07	de la Mare (1976)
		Berg (1977)
type B	0.69–0.87	de la Mare (1976)
Valves	0.70–1.02	de la Mare (1976)
Pumps	as low as 0.5	Aird (1977b)

A further account of the Weibull method is given in Chapter 7.

A14.3 Influence Factors

The failure rate of equipment is influenced by a large number of factors, including design, specification, manufacture, application, operating conditions, maintenance, and environment. Process equipment is used in a wide variety of operating conditions and environments, and it is desirable to account for these influencing factors.

For some types of device, the effect of certain influencing or environmental factors can be quite well defined. An environment factor K, which is a multiplier to the nominal failure rate, is widely used for the effect of temperature on items of electrical equipment such as resistors. Some values of environment factors are given by A.E. Green and Bourne (1972). Base failure rate, quality factor, learning factor, and environment factor are defined in MIL-HDBK-217F (1991) to predict failure rates for some electronic equipment. It is more difficult to define the effect of such factors on mechanical equipment, but a few examples may be mentioned. Several authors have given correlations for the effect of particular influencing factors on equipment. Moss (1977 NCSR R11) has given data on the failure rates of mechanical valves in two nuclear power stations. The overall failure rates λ of the valve were

Steam valves	$\lambda = 0.1$ faults/year
Water valves	$\lambda = 0.01$ faults/year

The variation of the failure rate with the severity of the operating conditions pressure and temperature was correlated by an equation of the form

$$\lambda = \lambda_0 \exp\left[\frac{D}{D_0} - 1\right] \qquad [A14.3.1]$$

with

$$D = p + t \qquad [A14.3.2]$$

where D is the severity parameter, p is pressure (psi), t is temperature ($^\circ$C) and subscript 0 indicates base case.

C.F. King and Rudd (1972) have expressed the hazard rate for pumps as function of the form

$$z = At^{-1/2} + B + Ct^3 \qquad [A14.3.3]$$

where t is time and A, B and C are functions of temperature, motor power and pump utilization. An environment factor for instruments has been given by Anyakora, Engel and Lees (1971), as described in Chapter 13. Similarly, a severity index for instruments has been given by Barbin (1973).

A14.4 Collection of Data

Any reliability work would not be successful without an efficient data collection and analysis system, and collection of data is receiving more and more attention of engineers. Failure rate data are characterized by complex interactions of components, specifications, maintenance procedures, operating environments, etc. Each company should examine the sources of data carefully and take full advantages of the available information, internal and external.

The most desirable information for reliability and risk analysis is to have sufficient plant specific data. In general, failure of plant equipment needs to be recorded and investigated, both in order to identify types of failure so that failure rates can be reduced by better engineering, and to obtain failure data for reliability calculations.

It is normal practice to record for production management the downtime of the plant together with its cause. Frequently the cause assigned is the failure of the particular equipment. This system may yield useful data, particularly on plant availability. Likewise, it is normal practice to record for maintenance management failures of plant equipment. Useful data on failure rates may be generated by this system.

Most chemical plants have a wealth of data in their operating and maintenance records. Frequently, the data yielded by the existing production and maintenance management systems are not directly usable for reliability work. It is then necessary to make certain adjustment the records to obtain the desired data. Usually this involves some additional work by the operating and maintenance personnel. It is essential, therefore, to design the data collection system appropriately. Realizing this, CCPS (1991) devoted a whole chapter discussing locating raw plant data and converting them into reliability data.

If data collection is to be instituted for reliability work, it is necessary to define the system carefully. This includes (1) data requirement, (2) data capture, (3) data classification (4) data retrieval, and (5) data utilization. The system should be appropriate in scale and in duration. In some cases the aim is to monitor continuously features such as equipment failure and/or unit availability. In this case continuous collection is clearly necessary. On the other hand if the aim is simply to obtain data for reliability engineering or hazard assessment purposes, it may be sufficient to collect certain data over a limited period on a 'campaign' basis.

One pitfall is to embark upon too ambitious a scheme that is then not fully exploited and soon falls into disuse. A better policy is to start with a more modest system and to put effort at an early stage into utilizing the data and demonstrating their usefulness to those involved in collecting them. At the other extreme, the desire to avoid this error can lead to the converse mistake of seeking so little modification to the existing system that usable data are not obtained. It is desirable, therefore, both to design the system so that it matches the use to be made of it and, once the system is operating, to begin as soon as possible to utilize the data which flow so that those participating can see that use their efforts are not being wasted.

Particular attention should be paid to the point of generation of the data. In some cases the quality of information on documents such as job tickets is sufficiently good for the purpose. But in other cases some form of specific debriefing may be preferable in order to ensure that the data obtained are meaningful.

Further discussions of data collection are given by D.J. Smith and Babb (1973), Wingender (1991), Davidson (1994), Ireson, Coombs and Moss (1996) and Dhillon (1999).

A14.5 Sources of Data

Failure data may be obtained from external sources such as the literature and data banks. Alternatively, they may be collected within the works. The sources available depend on the user. A company operating plant has access to data from its own works that are not available to other parties, unless supplied to an accessible database. The possible sources may include warranty claims, previous experience with similar or identical equipment, repair facility records, factory acceptance testing, records generated during the development phase, customers' failure reporting systems, test, and inspection records generated by quality control/manufacturing group.

Failure rates depend on many factors, including the function of the equipment in the system and the definition of failure; the process environment and the maintenance practices; and the type of equipment and its manufacturer. The ideal situation is to have sufficient in-house data from identical equipment in the same application. On the other hand, internal collection may not be appropriate, or for new plant designs there is apparently no in-house failure data.

Generic data are often used in reliability work, especially at design stage. However, there are available in the literature a number of collections of data, it requires the knowledge of the background and data origin to choose the appropriate generic failure rate data. Table A14.1 is the cross reference between principal data sources prepared by the Data Acquisition Working Party of the Mechanical Reliability Committee of the Institution of Mechanical Engineering (Davidson, 1994), and Table A14.2 is the equipment category associated with Table A14.1.

A selection of data on equipment failure rates obtained by the UKAEA and given by A.E. Green and Bourne are shown in Table A14.3 and Figure A14.1. These data derive from the work of the UKAEA initially in the nuclear field but subsequently in non-nuclear applications also.

Figure A14.2 and Figure A14.3 gives the operating ranges of critical failure rates of some equipment from nuclear plants and EIReDA data.

Further selected failure rate data given in the *Rasmussen Report* are shown in Table A14.4. The failure data in this table include both failure rate per unit time and failure probability on demand and are quoted as a median value together with a range. These data were obtained from both nuclear and non-nuclear sources, but were collected for use in nuclear hazard assessment, in particular on the critical loss of coolant accident (LOCA). The report, which is described more fully in Appendix 23, gives extensive documentation qualifying the data presented.

Table A14.1 *Summary of principal reliability data sources (reproduced from The Reliability of Mechanical Systems edited by J. Davidson (technical editor C. Hunsley), 1994, with permission from Professional Engineering Publishing Ltd)*

Reference	Application	Equipment					
		Rotation	Static	Instrumentation	Safety	Process	Electrical
CCPS	I	Y	Y	Y	Y	Y	Y
Davenport and Warwick	N, I	–	Y	–	–	–	–
DEF STAN 00-41 pt 3 Issue 1	M	Y	Y	Y	Y	Y	Y
De la Mare	O	–	Y	–	–	–	–
Dexter and Perkins DP-1633	N, I	Y	Y	Y	Y	Y	Y
EIReDA	N, I	Y	Y	Y	Y	Y	Y
ENI	I	Y	Y	Y	Y	Y	Y
Green and Bourne	N, I	Y	Y	Y	Y	Y	Y
IAEA-TECDOC	N	Y	Y	Y	Y	Y	Y
IEEE 500	I	Y	–	Y	Y	–	Y
Lees	N, I	Y	Y	Y	Y	Y	Y
MIL-HDBK-217F	M	Y	Y	–	Y	–	Y
NPRD-91	N, I	Y	Y	Y	Y	–	Y
OREDA 84	O	Y	Y	Y	Y	Y	Y
OREDA 92	O	Y	Y	Y	Y	Y	Y
RKS/SKI 85-25	N, I	Y	Y	Y	Y	–	Y
Rothbart	I	Y	Y	Y	Y	–	Y
Smith	N, I	Y	Y	Y	Y	Y	Y
Smith and Warwick	N, I	–	Y	–	–	–	–
Wash 1400	N	Y	Y	Y	Y	–	Y
RMC HARIS	O, N, I	Y	Y	Y	Y	Y	Y
SRD Data Centre	O, N, I, M	Y	Y	Y	Y	Y	Y

O = offshore; N = nuclear; I = industrial; M = military.

Table A14.2 *Equipment category code for Table A14.1 (reproduced from The Reliability of Mechanical Systems edited by J. Davidson (technical editor C. Hunsley), 1994, with the permission from Professional Engineering Publishing Ltd)*

Equipment category	Typical component
1 Rotating equipment	Pumps, compressors, turbines, motors
2 Static equipment	Pipes, pipelines, flowlines, valves, vessels
3 Instrumentation	Sensor (temperature, pressure, level etc.) and controllers
4 Safety	Fire pumps, safety valves, fire/gas/smoke detectors
5 Process	Pumps, compressors, valves vessels, piping
6 Electrical	Cables, motors, circuit boards lamps

[a] This table should be used in conjunction with the above table.
[b] Note that the categories are not intended to be mutually exclusive. For example, pumps can be categorized as bother 'rotating' and 'process'.

Another set of failure rate data are those given by D.J. Smith (1985) and a selection is shown in Table A14.5. In this case the failure rate is given as a single value, a range or a single value and a range.

A synthesis table generated by EIReDA is shown in Table A14.6. The data set are originally from nuclear plants.

Failure rate data for offshore have been given in the collections by OREDA (1984, 1992, 1997, 2002). These data include not only overall failure rates but also failure rates in some failure modes. The number of failures, which determines the confidence bounds on the values, is also given.

In the cases there is no applicable internal or external failure data, careful engineering estimation has to be made. Usually, the final data set to be used is a judicious mixture of data from all these sources.

A14.6 Status of Data

In hazard assessment it is a good practice to indicate the status of the data. Some relevant distinctions are:

(1) Value based on historical data:
  (a) value based on large number of events (narrow confidence limits);
  (b) value based on small number of events (wide confidence limits);
  (c) value based on number of event-free years or occasions;
(2) Value based on judgment of a number of experts.
(3) Value synthesized using fault tree methods.
(4) Value based partly on data and partly on judgment of analyst.
(5) Value entirely on judgement of analyst.

A similar classification is given in the *First Canvey Report* described in Appendix 7.

The status of data is discussed by Andow (1989) and Holloway (1989).

A14.7 Processing of Data

Using failure rate data requires thorough understanding and judicious judgements. The quality of the data from different sources will be varying. For a particular

Table A14.3 *Some data on equipment failure rates published by the UKAEA, (data from Reliability Technology by A.E. Green and J.R. Bourne, Copyright © , 1972, reproduced with permission of John Wiley and Sons, Inc.)*

	Failure rate (failures/10^6 h)
Electric motors (general)	10.0
Transformers (< 15 kV)	0.6
(132–400 kV)	7.0
Circuit breakers (general, < 33 kV)	2.0
(400 kV)	10.0
Pressure vessels (general)	3.0
(high standard)	0.3
Pipes	0.2
Pipe joints	0.5
Ducts	1.0
Gaskets	0.5
Bellows	5.0
Diaphragms (metal)	5.0
(rubber)	8.0
Unions and junctions	0.4
Hoses (heavily stressed)	40.0
(lightly stressed)	4.0
Ball bearings (heavy duty)	20.0
(light duty)	10.0
Roller bearings	5.0
Sleeve bearings	5.0
Shafts (heavily stressed)	0.2
(lightly stressed)	0.02
Relief valves: leakage	2.0
blockage	0.5
Hand-operated valves	15.0
Control valves	30.0
Ball valves	0.5
Solenoid valves	30.0
Rotating seals	7.0
Sliding seals	3.0
'O' ring seals	0.2
Couplings	5.0
Belt drives	40.0
Spur gears	10.0
Helical gears	1.0
Friction clutches	3.0
Magnetic clutches	6.0
Fixed orifices	1.0
Variable orifices	5.0
Nozzle and flapper assemblies: blockage	6.0
breakage	0.2
Filters: blockage	1.0
leakage	1.0
Rack-and-pinion assemblies	2.0
Knife-edge fulcrum: wear	10.0
Springs (heavily stressed)	1.0
(lightly stressed)	0.2
Hair springs	1.0
Calibration springs: creep	2.0
breakage	0.2
Vibration mounts	9.0
Mechanical joints	0.2
Grub screws	0.5
Pins	15.0
Pivots	1.0
Nuts	0.02
Bolts	0.02
Boilers (all types)	1.1
Boiler feed pumps	1012.5
Cranes	7.8

Sources: F.R. Farmer (1971); A.E. Green and Bourne (1972).
Note: Further failure data on electronic, mechanical, pneumatic and hydraulic components are given by A.E. Green and Bourne (1972).

reliability study, it is essential to consult the original data sources and identify appropriate data sources to reach a better estimation of the equipment failure rates. EsReDA (1999) offers a summary of the collection, analysis and quality assurance of reliability data.

Mean failure rate is appropriate for failure rates obtained from testing a large population of samples under controlled environment, typically for electronic components. On the other hand, for mechanical equipment, we may want to look into the plant operation to select the appropriate data and apply some adjustment.

A14.8 Uncertainty of Data

Reliability data can deviate by a factor of three or four, and a factor of 10 is not unusual, as suggested by Trevor Kletz (1999). However, not much information exists in the open literature regarding the uncertainty of data. Since the uncertainty is quite large, it is inappropriate to report failure rate data or calculations using failure rate data to more than two significant figures. A serious implication of precision is suggested when more than two significant figures are reported.

A14.9 Databases

Database and data bank are sometimes referred to in the literature exchangeably. However, a database is a collection of data managed through computer programs for specified applications while a data bank is a collection of data in a specific area of knowledge, such as a handbook. An account of data banks and databases is given in *Reliability Data Banks* by Cannon and Bendell (1991). Fragola (1996) reviewed reliability databases from a historical aspect and suggested the possible improvements of reliability database development.

The two main kinds of database are the incident database and the reliability database. An incident database does not have an inventory of items at risk and concentrates on the attributes and development of the incidents. The information available is often limited to whatever was recorded at the time. A variety of databases have been established to store and share this type of information. Among them are FACTS, WOAD (Worldwide Offshore Accident Database), MARS (the European Union Major Accident Reporting System), etc. A reliability database may well record incidents, but treats them primarily as events from which statistical information on reliability, availability and maintainability are to be derived. Reliability data can be overall failure rates or values separated into failure modes and causes.

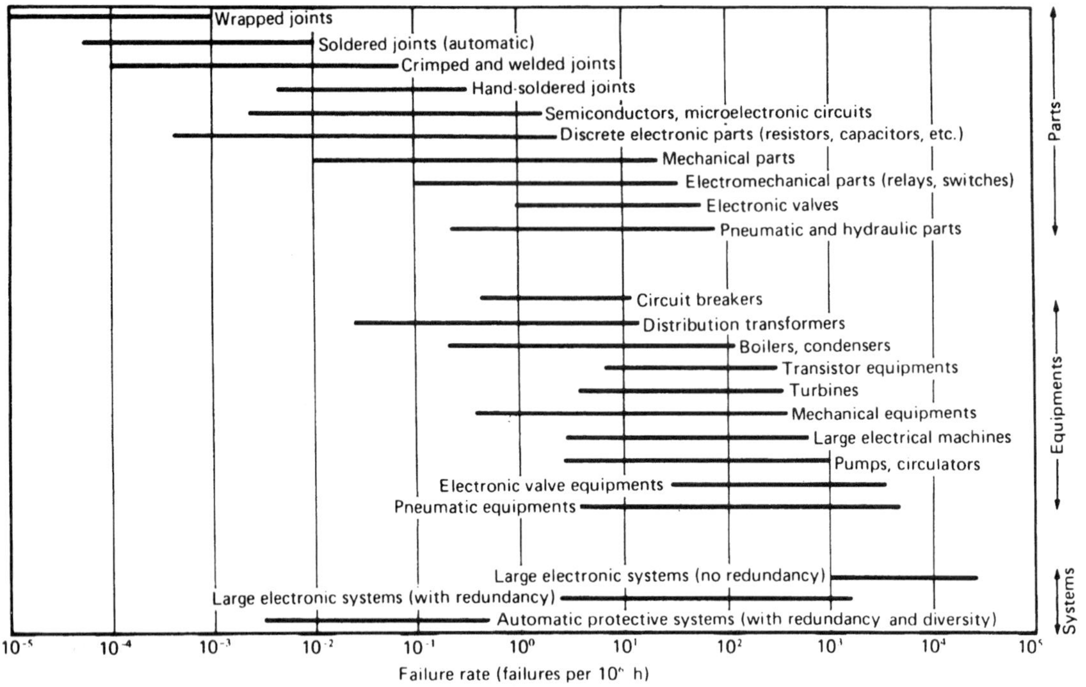

Figure A14.1   *Typical ranges of failure rates for parts, equipments and systems (A.E. Green and Bourne, 1972) (reproduced with permission from Reliability Technology by A.E. Green and J.R. Bourne, Copyright ©, 1972, John Wiley and Sons, Inc.)*

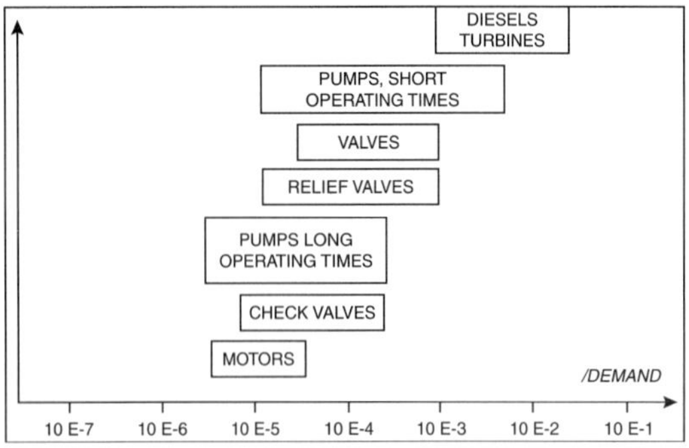

Figure A14.2   *Demand critical failure rate ranges of some equipment at French nuclear plants (sample = 80–170 unit-years depending on equipment) (Lannoy and Procaccia, 1996; reproduced with permission from Elsevier)*

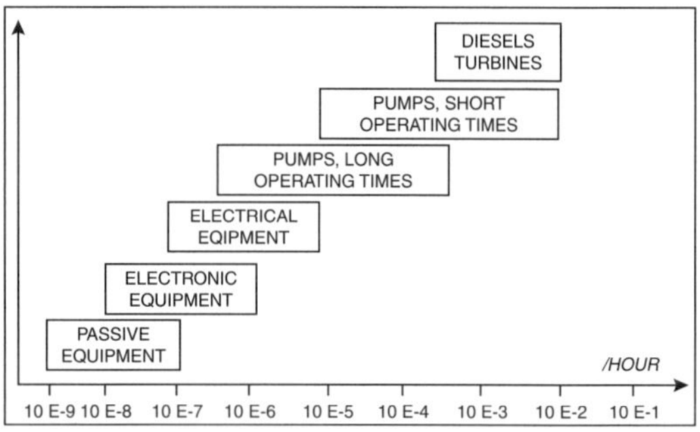

Figure A14.3   *Operating critical failure rate ranges of some equipment at French nuclear plants (sample = 80–170 unit-years depending on equipment) (Lannoy and Procaccia, 1996; reproduced with permission from Elsevier)*

Reliability databases are created by different users for somewhat different purposes. They range from the database created by a single individual through those at company level to those operated by specialist organizations. High quality reliability data are expensive to collect and often require a long time or a large sample to be statistically viable. The investment of effort in creating and operating a reliability database is appreciable and the exercise is not one to be undertaken lightly. It is essential to define clearly the data to be held, the means by which they will be acquired, and the uses to which they will be put.

Most serious reliability databases are part of the activities of an organization that is involved in other aspects of reliability work also. The organization thus has an interest in the database as a user, which is likely to make it more friendly to other users. There are a variety of reliability databases, such as the reliability databank operated by AEA Technology, ERDS, OREDA, etc. CCPS (1991) provides a review of the available databases. CCPS is also conducting a reliability project with selected companies, but the results of this effort are not publically available.

Table A14.4 *Some failure data used in the Rasmussen Report (AEC, 1975)*

		Failure rate	
		Median	Range
Diesels (complete plant)	Failure to start, Q_d[a]	3×10^{-2}/d	1×10^{-2}–1×10^{-1}/d
	Failure to run, given start, in emergency conditions, λ_0	3×10^{-3}/h	3×10^{-4}–3×10^{-2}/h
Diesels (engine only)	Failure to run, given start, in emergency conditions, λ_0	3×10^{-4}/h	3×10^{-5}–3×10^{-3}/h
Battery power systems (wet cell)	Failure to provide proper output λ_s	3×10^{-6}/h	1×10^{-6}–1×10^{-5}/h
Electric motors	Failure to start, Q_d[b]	3×10^{-4}/d	1×10^{-4}–1×10^{-3}/d
	Failure to run, given start, in normal environment, λ_0	1×10^{-5}/h	3×10^{-6}–3×10^{-5}/h
	Failure to run, given start, in extreme environment, λ_0	1×10^{-3}/h	1×10^{-4}–1×10^{-2}/h
Transformers	Open circuit, primary or secondary, λ_0	1×10^{-6}/h	3×10^{-7}–3×10^{-6}/h
	Short, primary to secondary λ_0	1×10^{-6}/h	3×10^{-7}–3×10^{-6}/h
Solid state devices (low power application)	Failure to function, λ_0	1×10^{-6}/h	1×10^{-7}–1×10^{-5}/h
	Failure by short, λ_0	1×10^{-7}/h	1×10^{-8}–1×10^{-6}/h
Circuit breakers	Failure to transfer, Q_d[b]	1×10^{-3}/d	3×10^{-4}–3×10^{-3}/d
	Premature transfer, λ_0	1×10^{-6}/h	3×10^{-7}–3×10^{-6}/h
Fuses	Failure to open, Q_d	1×10^{-5}/d	3×10^{-6}–3×10^{-5}/d
	Premature open, λ_0	1×10^{-6}/h	3×10^{-7}–3×10^{-6}/h
Relays	Failure to energize, Q_d[b]	1×10^{-4}/d	3×10^{-5}–5×10^{-4}/d
	Failure of NO contacts to close, given energization, λ_0	3×10^{-7}/h	1×10^{-7}–1×10^{-6}/h
	Failure of NC contacts by opening, given no energization, λ_0	1×10^{-7}/h	3×10^{-8}–3×10^{-7}/h
	Short across NO/NC contacts, λ_0	1×10^{-8}/h	1×10^{-9}–1×10^{-7}/h
	Coil open, λ_0	1×10^{-7}/h	1×10^{-8}–1×10^{-6}/h
	Coil short to power, λ_0	1×10^{-8}/h	1×10^{-9}–1×10^{-7}/h
Terminal boards	Open connection, λ_0	1×10^{-7}/h	1×10^{-8}–1×10^{-6}/h
	Short to adjacent circuit, λ_0	1×10^{-8}/h	1×10^{-9}–1×10^{-7}/h
Wires (typical circuit, several joints)	Open circuit, λ_0	3×10^{-6}/h	1×10^{-6}–1×10^{-5}/h
	Short to ground, λ_0	3×10^{-7}/h	3×10^{-8}–3×10^{-6}/h
	Short to power, λ_0	1×10^{-8}/h	1×10^{-9}–1×10^{-7}/h
Instrumentation (amplification annunciators transducers, combination)	Failure to operate, λ_0	1×10^{-6}/h	1×10^{-7}–1×10^{-5}/h
	Shift in calibration, λ_0	3×10^{-5}/h	3×10^{-6}–3×10^{-4}/h
Pressure switches	Failure to operate, Q_d	1×10^{-4}/d	3×10^{-5}–3×10^{-4}/d
Limit switches	Failure to operate, Q_d	3×10^{-4}/d	1×10^{-4}–1×10^{-3}/d
Valves (manual)	Failure to remain open (plug), Q_d	1×10^{-4}/d	3×10^{-5}–3×10^{-4}/d
	Rupture, λ_s	1×10^{-8}/h	1×10^{-9}–1×10^{-7}/h
Valves (air-fluid operated)	Failure to operate, Q_d[b]	3×10^{-4}/d	1×10^{-4}–1×10^{-3}/d
	Failure to remain open (plug)[c]		
	Q_d	1×10^{-4}/d	3×10^{-5}–3×10^{-4}/d
	λ_s	3×10^{-7}/h	1×10^{-7}–1×10^{-6}/h
	Rupture, λ_s	1×10^{-8}/h	1×10^{-9}–1×10^{-7}/h
Valves (motor operated, includes driver)	Failure to operate, Q_d[b]	1×10^{-3}/d	3×10^{-4}–3×10^{-3}/d
	Failure to remain open (plug)[c]		
	Q_d	1×10^{-4}/d	3×10^{-5}–3×10^{-4}/d
	λ_s	3×10^{-7}/h	1×10^{-7}–1×10^{-6}/h
	Rupture, λ_s	1×10^{-8}/h	1×10^{-9}–1×10^{-7}/h
Valves (solenoid operated)	Failure to operate, Q_d[b]	1×10^{-3}/d	3×10^{-4}–3×10^{-3}/d
	Failure to remain open (plug), Q_d	1×10^{-4}/d	3×10^{-5}–3×10^{-4}/d
	Rupture, λ_s	1×10^{-8}/h	1×10^{-9}–1×10^{-7}/h
Relief valves	Failure to open, Q_d	1×10^{-5}/d	3×10^{-6}–3×10^{-5}/d
	Premature open, λ_0	1×10^{-5}/h	3×10^{-6}–3×10^{-5}/h

Table A14.4   *(continued)*

		Failure rate	
		Median	*Range*
Check valves	Failure to open, Q_d	1×10^{-4}/d	$3 \times 10^{-5} - 3 \times 10^{-4}$/d
	Internal leak (severe), λ_0	3×10^{-7}/h	$1 \times 10^{-7} - 1 \times 10^{-6}$/h
	Rupture, λ_s	1×10^{-8}/h	$1 \times 10^{-9} - 1 \times 10^{-7}$/h
Pumps (including driver)[d]	Failure to start, Q_d	1×10^{-3}/d	$3 \times 10^{-4} - 3 \times 10^{-3}$/d
	Failure to run, given start, in normal environment, λ_0	3×10^{-5}/h	$3 \times 10^{-6} - 3 \times 10^{-4}$/h
	Failure to run, given start, in extreme post-accident environment inside containment, λ_0	1×10^{-3}/h	$1 \times 10^{-2} - 1 \times 10^{-4}$/h
Pipes ≤ 3 in. (per section)	Rupture/plug, λ_s, λ_0	1×10^{-9}/h	$3 \times 10^{-11} - 3 \times 10^{-8}$/h
> 3 in. (per section)			
	Rupture/plug, λ_s, λ_0[b]	1×10^{-10}/h	$3 \times 10^{-12} - 3 \times 10^{-9}$/h
Gaskets (containment quality)	Leak (serious) in post-accident situation, λ_0	3×10^{-6}/h	$1 \times 10^{-7} - 1 \times 10^{-4}$/h
Elbows, flanges, expansion joints (containment quality)	Leak (serious) in post-accident situation, λ_0	3×10^{-7}/h	$1 \times 10^{-8} - 1 \times 10^{-5}$/h
Welds (containment quality)	Leak (serious) in post-accident situation, λ_0	3×10^{-9}/h	$1 \times 10^{-10} - 1 \times 10^{-7}$/h
Clutches (mechanical)	Failure to operate, Q_d[b]	3×10^{-4}/d	$1 \times 10^{-4} - 1 \times 10^{-3}$/d

[a] Q_d = failure/demand; λ_0 = failure/h in operational mode; λ_s = failure/h in standby mode. The unit d in this table stands for demand.
[b] Demand probabilities are based on the presence of proper input control signals.
[c] Plug probabilities are given as probabilities per demand and as rates per hour, since the plug is generally time-dependent but may only be detected upon a demand on the system.
[d] Turbine-driven pump system may have significantly higher failure rates.

In the following, brief descriptions are given of six principal databases: the NCSR, ERDS, FACTS EIREDA, OREDA, and GIDEP databases. Other databases include the EOF database (Procaccia, 1991); the IAEA database (Tomic and Lederman, 1991); the CREDO database at ORNL (Knee, 1991); the Dante database (Mizuta *et al.*, 1991), the SRDF database (AUPIED and PROCACCIA, 1984), the SKI-PIPE database (Lydell 2000), and the database operated for SRD Association by AEA Technology etc.

A14.9.1 Design of database
As to its definition, a database is first of all a computer system, which may have billions of data collections. The construction of a database involves the translation of the logical data model, giving the relationships between the data field into the physical database embodied in the computer. This is ensured by the database management system (DBMS). Many software companies have developed very effective database management programs that can be tailored for any specific needs. Four main architecture models of the database management system are the hierarchical, network, relational and object models. The characteristics and merits of these different approaches for reliability databases are discussed in EsReDA (1999).

Information about components held in the database comes under two main heads: the component inventory and the component history. The component inventory contains a description of each component on which information is held. The component history gives the detailed failure and maintenance history of each component. Both for component inventory and component history the design of the data set to be held is not a trivial question. For inventory it is first necessary to define the component boundary. A valve actuator, for example, may not remain permanently on the same valve.

It is then necessary to specify the attributes of the components. This involves deciding the degree of subdivision

or, alternatively, the level of aggregation. Excessive subdivision results in samples which are too small, while excessive aggregation lumps together items that differ significantly.

For component history, information is typically recorded on failure mode and failure cause. Here the problem is the tendency toward excessive subdivision by creation of additional failure mode and cause categories. This problem is aggravated if numbers of persons are authorized to extend these categories.

It is also desirable to have a plain language fault descriptor. Experience suggests that a set of pre-specified descriptors is on some occasions too detailed and on others not detailed enough, and that there is value in a facility to use free text.

In general, it is desirable to store the data in their 'original' form rather than in an abstracted form, which involves loss of information. For this reason, the data in a reliability data bank may be stored as a set of events from which statistical data may then be obtained.

Early reliability databases tried to account for environments and other conditions via K or π factors, such as MIL-HDBK 217F and Green and Bourne (1972). Most of them give point estimates of failure rates without indicating the bound and confidence level. Modern reliability databases provide failure rates on categorized failure modes, like IEEE-Std-500 (1984), OREDA (1984, 1992, 1997, 2002), EIREDA (1991) and CCPS (1991). IEEE-Std-500, OREDA (1984, 1992, 1997, 2002) and CCPS (1991) provide upper and lower values on failure rates and EIREDA (1991) provides confidence level with error factors.

Many efforts have been carried out to standardize database design. Further discussion on database design can be found in Ireson, Coombs, and Moss (1996), Cooke and Bedford (2002) and the vol. 51, num. 2 of *Reliability Engineering and System Safety* which is specially devoted to reliability database design.

Table A14.5 *Some data on equipment failure rates (D.J. Smith, 1985; reproduced by permission of Macmillan)*

Equipment	Failure rate[a] (failure/10^6 h)		
Compressor			
Centrifugal, turbine driven	150		
Reciprocating, turbine driven	500		
Electric motor driven	100	300	
Diesel generator	125	4000	
	(0.97 start)		
Electricity supply	110		
Gaskets	0.02	1	
Heat exchanger	1	40	
Pipe joint	0.5		
Pumps			
Centrifugal	10	**30**	80
Boiler	100	500	
Fire[b]	100	150	
Fuel	6	50	
Oil lubrication	10	30	100
Vacuum	20		
Turbine, steam	30	80	
Valves			
Ball	1	3.5	
Butterfly	1	**20**	30
Gate	1.5	15	
Relief	4	9	
Non-return	2	5	
Slam shut	10	30	
Solenoid	1.5	**10**	30
Valve actuators[c]			
Fail open	0.1	4	
Spurious close	5	40	

[a] Entries are given in three formats: a single value, where the various references are in good agreement; two values, indicating a range; and three values with one in bold, indicating a range with the value in bold predominating. Bold is also used for one end of a range, where that value predominates.
[b] Approximately 800 for a complete fire pump and priming system.
[c] Depends on the complexity of the pneumatic circuit; requieres FMEA.

A14.9.2 Operation of database

The operator of a reliability database has to maintain a sufficient flow of data from the data suppliers, synthesize the data from different sources, and satisfy the needs of the data users in a manner which is cost effective. This is no easy task.

Fresh data are the lifeblood of a reliability data bank. There must be a sufficient incentive for an organization to supply them, either by collecting them itself or by allowing access for their collection. Typically it will be one that is also a user of the database.

The data user will have a set of requirements, which often can be met only in part. Ideally the user would like data from which to obtain failure and repair time distributions, and so on. Often these will simply not be available, the best on offer being perhaps failure or repair rates based on a constant rate assumption. There may also be problems of sample size. The data bank operator has to balance the often-conflicting requirements of user demand, data supply and economy in handling.

A14.9.3 Data acquisition

Information on equipment for the component inventory is obtained from the plant documentation, which includes equipment records and equipment and plant drawings. Acquisition of the inventory data for a plant may not be straightforward. Records and drawings may be incomplete from the outset, and it is even more common that modifications made are not entered. There can be confusing identifications, with the same item allocated several quite different code numbers.

There can be a corruption of the component design parameters, with differences between the item as designed and as purchased and installed being quite common. As Cross and Stevens comment: 'It is incredible how poor the average plant inventory records are'.

For information on equipment failure and repair a basic source of information is the job card or ticket. Typically this gives as a minimum the identity of the equipment, the failure notified or diagnosed and the repair work done. Other documentation such as logbooks, permits-to-work and stores requisitions are generally useful as cross-checks rather than prime sources. Sometimes, vendors' test data can also be used as a supplement source.

To the extent practical, the failure data should be such as to permit analysis to obtain failure distributions as well as average values. This implies that the data should give the times to failure for individual equipment rather than just the total number of failures. This greatly enhances the value of the data to the data analyst.

Some data require a degree of reinterpretation or filtering. One common problem is the failure that presents as a cluster of events arising over a short time interval. This may take several forms, each of which involves events that are in some sense dependent. One is the repair that is unsatisfactory so that failure recurs almost immediately and the work has to be done again. Somewhat similar is the repair that is made in several passes such as a leaking valve which is first tightened, then repacked and finally replaced. A third situation is the nearly simultaneous failure of a number of items due to a common cause.

A14.9.4 NCSR database

The NCSR Reliability Data Bank is a major database system which has been in operation for over twenty years serving the nuclear, aerospace, electronics, oil, chemical and other industries. Accounts of the early database and SYREL data bank have been given by Barnes (1967 UKAEA AHSB(S) R138), Ablitt (1973 SRS/GR/14) and Fothergffl (1973 SS/GR/22). The mature database has been described by Cannon (1991a,b). An account of further developments in the database is given by Cross and Stevens (1991).

As described by Cannon (1991a) the data bank about the mid-1980s consisted of data stores for (1) generic reliability, (2) events, (3) accidents, (4) human reliability, and (5) mal-operation. Much of the data collection was carried out by placement of students in cooperating companies. Cannon gives examples of the use of the data bank, including cases where application of the data resulted in reductions in the observed failure rates.

Cross and Stevens (1991), writing from the perspective of the user as well as the operator of a data bank, describe the transition of the NCSR system from a system which was based originally on 1960s technology, using mainframe

Table A14.6 *Example of best-estimated reliability parameters from some equipment (Lannoy and Procaccia, 1996; reproduced with permission from Elsevier)*

Equipment	Main Character	Mean failure rate $(10^{-6}/h)$	Range $(10^{-6}/h)$	Failure on demand $(10^{-3}/d)$	Range $(10^{-3}/d)$	MMTR (h)
Accumulator, nitrogen driven						
Pressurized	$P < 50$ bar; T $< 50°$C	0.52	0.2–0.8	0.03	–	2.6
Compressor, motor driven						
Centrifugal	P: 6–40 bar; kW: 360	81	60–104	0.18	0.09–0.25	2.6
Reciprocating	$P \cong 10$ bar; kW: 5–85	782	506–1108	0.3	0.2–0.38	5
Water filter (paper)	P: 10–190 bar	2.3	1.5–3.1	–	–	6
Heat exchanger						
Plate	P: 7–155 bar; T: 30–220°C	20.7	19–22	0.01	–	9
Shell/tube		0.55	0.08–0.9	–	–	25
Shell/tube	P: 155 bar; T: 200–300°C	0.31	0.005–0.5	0.02	–	25
Shell/tube	P: 7–50 bar; T: 50–220°C	0.83	0.2–1.3	–	–	25
Pump, centrifugal multistage						
Motor/turbine driven[a]	kW: 300/400 head: 900 m flow: 80 m³/h	174	125–228	0.2	0.1–0.3	34
Turbine driven	kW: 2200/5300 head: 570 m flow $\leq$3300 m³/h	10	7–14	0.15	0.11–0.27	13

[a] Standby pump.

computers, large in size and not readily modified, to one utilizing modern database methods and implemented on PCs.

Of the two basic formats for storage of information, summary and complete record, the NCSR system utilizes the latter, so that complete details of each event are held. From these complete records statistical data can then be abstracted as required.

A14.9.5 ERDS database
The European Reliability Data System (ERDS) is an EC database operated by the JRC at Ispra. It is described by Capobianchi (1991).

ERDS acquires its data from European and other databases, and is therefore in effect a database of databases. It is oriented particularly to serve the nuclear industry.

The data are rendered homogeneous by conversion to a uniform format. This format is specified in a detailed system of classification – the Common European Reference Classifications, which is thus by way of being a European standard.

ERDS has four main data subsystems: (1) the Component Event Data Bank (CEDB), (2) the Abnormal Occurrences Reporting System, (3) the Operating Unit Status Report and (4) the Reliability Parameter Data Bank. Among them, the CEDB stores the component characteristics, historical information, and failure information and is the central subsystem, which serves and is connected to the other three subsystems.

A14.9.6 FACTS database
The Failure and Accident Technical Information System (FACTS) is an incident data bank operated by TNO in the Netherlands. It is described by Koehorst and Bockholts (1991).

FACTS gathered the information about near-misses, minor accidents and major accidents associated with hazardous materials. Nuclear materials and military activity are excluded for its scope. The information is derived from the literature; from companies; from inspectorates, fire services, etc.; and from FACTS agencies in other countries, respecting anonymity. Press reports serve as triggers to acquire information.

Features of the database are a schedule of accident attributes and values and a hierarchical keyword structure. Another structure is the cause classification in which the course of the accident is translated into a sequence of occurrences.

This makes it possible to trace back from an event down the causal chain. The original plain text accounts are held on microfiche.

Applications of the system described by the authors include (1) analysis of the role of instrumentation in accidents; (2) analysis of incorrect human response and (3) compilation of a reference book to trace incident causes (the Cause Book), giving a survey of incident causes which can occur in a large number of systems and operations.

Further analysis is given in *An Analysis of Accidents with Casualties in the Chemical Industry Based on Historical Facts* by Koehorst (1989).

Developments of the system include an expert system front end.

A14.9.7 OREDA database
The Offshore Reliability Data (OREDA) project is an offshore reliability database described in the handbook by OREDA (1984, 1992, 1997, 2002). It gives failure rates, failure mode distribution and repair data for offshore installations, mainly in the UK and Norwegian sectors of the North Sea.

A substantial effort was devoted to the design of the system to ensure as far as possible that the components, their boundaries and environment, are well defined and that the data obtained are high quality and statistically meaningful. The database is fully explained in the OREDA handbook, which describes the system, gives the data and outlines the statistical treatment. Some discussion about the data collection and application of this project can be found in Groenendijk (1995), Sandtorv, Hokstad and Thompson (1996), Rausand and Øien (1996), Haugen, Hokstad and Sandtorv (1997), Langseth, Haugen and Sandtorv (1998).

The project therefore serves as a good illustration of a reliability database and is widely recognized as a standard information source for risk analysis.

The project is supported by ten major oil and gas companies: AGIP, BP, Elf, Esso, Norsk Hydro, SAGA, Shell, Statoil and Total. The data cover (1) process systems, (2) control and safety systems, (3) electrical systems, (4) utility systems, (5) crane systems and (6) subsea equipment and drilling systems.

The component inventory gives the following information: (1) brief description, (2) application, (3) operational mode (continuous, standby, protective, etc.), (4) internal environment (fluids handled), (5) external environment and (6) boundary specification, including a sketch. The operating data include (1) population at risk, (2) number of installations supplying data, (3) total calendar and operational times and (4) number of demands (where applicable). The failures are broken down into four different categories of severity: (1) critical, (2) degraded, (3) incipient and (4) unknown severity; there may be a number of modes in each category. The failure and repair data comprise (1) number of failures, (2) failure modes, (3) failure rates for each failure mode, (4) active repair time and (5) repair time. Mean and lower and upper bound values are given for the failure rates and mean, minimum and maximum values for the repair times. The active repair time is the average time required to analyse and repair the item and return it to service, excluding time to detect the fault and isolate the equipment and any delay due to tools or spares. The repair time is the number of manhours required for the repair. The failure and repair data are given for each mode and for the equipment overall.

Some of the events recorded in the OREDA as failures are those such as failure to start and spurious trip.

The number of installations participating, the number of components at risk and the number of failures experienced are very variable. In many cases the population at risk is 10 or less.

The handbook states that data in the form of times between failures are collected, but that confidence in the statistical data varies between the different generic groups and that for most purposes time-independent failure rates are a relevant approximation. It does not give data on times between failures.

A14.9.8 EIReDA database
The EIReDA (European Industry Reliability Data Bank) is operated by ESReDA, which is established in 1992 by EuReDATA (European Reliability Data Bank Association) and ESReDA (European Safety and Reliability Research and Development Association) jointly. The original data were collected from 50 nuclear power plants operated by Electricite de France from 1978 to 1995 and analysed using Bayesian methodology. The data bank contains failure rates (for critical failure modes only) and mean repair time data for mechanical components, electrical components and instrumentation and control equipment. A full description of the data bank is provided in EsReDA (1998), which describes the scope, the choice of components, the definition of boundaries, and the statistical treatment.

Engineering and operational characteristics are provided in detail for each item of equipment, as well as a color photograph, a general drawing and the boundaries. Background information such as inventory and service year is also provided. Comparable data are listed in a separate table, compiled from other published sources (mainly WASH 1400, IEEE 500, RKS 85-25; EuReDATA and ESReDA project reports and publications, and other power plant, modern petrochemical and processing plants). The methodological application to other industries is included in the appendix.

A14.9.9 GIDEP database
GIDEP (Government Industry Data Exchange Program) was established in 1959 as the Inter-service Data Exchange Program (IDEP) by the US government to reduce expenditures of resources in the life cycle of systems facilities and equipment through share of data among government agencies and industry organizations.

To make use of GIDEP database, an agency or company has to be a full or partial participant of GIDEP program. GIDEP is currently participated by thousands of agencies and industries.

Information on parts, processes and materials they use or manufacture is submitted to the GIDEP Operations Center by government and industry participants.

The program maintains six major types of data: (1) engineering data; (2) failure experience data; (3) metrology data; (4) product information data; (5) reliability – maintainability data and (6) urgent data requests. GIDEP is DOD's centralized database for Diminishing Manufacturing Sources and Material Shortages (DMSMS).

A14.10 Inventory

A14.10.1 Inventory of plants
Information is sometimes required on the number of plants at risk, either nationally or world-widely. This is needed, for example, in order to convert data on the number of a particular type of event, for example, vapour cloud explosions, into a frequency per plant.

Data on the number and capacity of chemical and petrochemical plants in various countries are given in the publications by SRI International which include Directory of Chemical Producers – United States 1988 and Directory of Chemical Producers – Western Europe 1988 (SRI Int., 1988b,c).

From the SRI data for the United States (US) and Western Europe (WE):

	US	WE	UK
No. of chemical/petrochemical plants	1420	1455	185

The following ratios apply for the number of plants and for the plant capacity:

	US/WE	UK/WE
No. of plants	0.92	0.12
Plant capacity	1.08	0.12

For petroleum refineries similar data are given in the *International Petroleum Encyclopaedia 2002* (IPE, 2002). From IPE data

Region	Africa	Asia	East Europe	Middle East	North America	South America	West Europe
No. of refineries	46	203	94	46	170	68	105

The countries having over 10 refineries from IPE (2002) data are listed below.

	No. of refineries
United States	143
China	95
Russia	42
Japan	35
Canada	21
Germany	17
India	17
Italy	17
France	13
Brazil	13
United Kingdom	11
Argentina	10
Australia	10
Romania	10
World total	732

A refinery contains a number of major units and this number varies. Information given in periodic surveys (e.g. Anon., 1985dd) suggests an average value of about five such units per refinery.

Information is also available on the number of installations that attract certain regulatory controls. In the United Kingdom, the number of installations notifiable under the NIHHS Regulations 1982 has been given by Pape and Nussey (1985) and by Pape (1989) as

LPG	600
Natural gas	400
Chlorine	120
Total	1600

Of the LPG installations 450 are storage installations and 130 gas cylinder storages.

In a discussion of the CIMAH Regulations 1984 (later superseded by COMAH) Welsh (1993) states that over 400 safety reports have been submitted. In interpreting this figure, it should be borne in mind that the number of reports is necessarily be equated to the number of CIMAH installations. CIMAH Regulation 1984 was later replaced by COMAH Regulation in the United Kingdom, which is stricter and came into force in the February of 1999. A total of 3–5000 top-tier major hazard plants in Europe are expected to submit safety, of which the United Kingdom has 438 with a further 843 lower tier installations.

A14.10.2 Ammonia plants

Estimates of the number of ammonia installations and carriers have been given by Baldock (1980) and are shown in Table A14.7.

A14.11 Inventory of equipment in plants

Another estimate that may be required is the number of potential leak sources on a plant. Information on the number of fittings and valves on pipework of different diameters has been given by Hooper (1982) in the context of cost estimation and is shown in Table A14.8.

Numbers of potential leak sources on various types of plant have been given, mainly in the context of fugitive emissions. Table A14.9 shows the numbers given by T.W. Hughes, Tierney and Khan (1979) for some potential leak sources in four petrochemical plants and Table A14.10 those given by D.P. Wallace (1979) for some sources in a single medium sized plant. Table A14.11 shows data given by Lipton and Lynch (1987) on the number of some sources in a large refinery. Table A14.12 after Wetherold (1983) gives the number of valves in a refinery and in two other plants.

Using such data, it is possible to construct a profile of the potential leak sources on a typical plant.

In a study of fugitive emissions from the ethylene oxide production industry, equipment containing more than 1 wt % ethylene oxide is surveyed in 13 major EO plants by Berglund, Romano, and Randall (1990).

A14.12 Vessel and Tanks

The failure rates of pressure vessels are discussed in Chapter 12.

Table A14.7   *Estimated number of ammonia installations and carriers*

	Estimated no. world-wide
Plants	2,000
Storage areas	1,000
Vessels	10,000
Refrigeration plants	100,000
Transfer points	1,000
Road tankers	1,000
Rail tankers	5,000
Ammonia ships	20
Pipeline miles	2,000

Source: Baldock (1980), table 2.

Table A14.8 *Pipework fittings and valves*

Nominal pipe diameter (in.)	Total length of pipe (ft)	Number of fittings	
		Flanges	*Valves*
$^1/_2$	33,990	1,818	11,589
$^3/_4$	33,123	2,973	7,551
1	124,513	12,552	10,363
$1^1/_2$	121,212	7,299	3,313
2	142,891	11,727	4,199
3	125,550	10,427	2,441
4	84,705	6,608	1,346
6	77,717	4,578	898
8	67,667	3,592	466
10	39,225	1,613	301
12	16,445	762	162
14	3,997	342	72
16	10,292	506	90
18	3,530	362	41
20	5,698	804	34
24	5,983	357	40
30	3,121	255	13
36	1,608	66	12

Source: Hooper (1982), table on p. 128.

Table A14.9 *Inventory of potential leak sources: number of sources on four petrochemical plants*

Leak source	No. of items Monochloro-benzene plant	Butadiene plant	Ethylene oxide/glycol plant	Dimethyl tereph-thalate plant
Flanges	1500	26,000	NA[a]	NA
Valves	640	6,700	NA	NA
Pumps	25	174	69	67

Source: T.W. Hughes, Tierney and Khan (1979), table 3.
[a] NA: not available.

Table A14.10 *Inventory of potential leak sources: estimates of number of leak sources on a medium-sized plant*

Leak source	No. of items
Flanges	2410
Valves	
In-line, gas	365
In-line, liquid	670
Open-ended	415
Pump seals	
Packed	6
Mechanical, single	43
Mechanical, double	10
Compressor seals	2
Safety relief valves	50

Source: D.P. Wallace (1979), p. 92.

Table A14.11 *Inventory of potential leak sources: estimated number of sources in a large refinery*

Leak source	No. of items
Flanges	46,500
Valve	11,500
Pump seals	350
Compressors	70
Relief valves	100
Drains	650

Source: Lipton and Lynch (1987), table 7.2.

Table A14.12 *Inventory of potential leak sources: number of valves in a refinery and two other plants*

A   No. of valves in three plants

	No. of valves
Large integrated refinery	21, 800
Large olefins plant	15,000
Cumene process unit	1,179

B   No. of valves on different duties in large refinery

Leak source	No. of items
Valves	
Gas and light liquid only	13,334
All	21,776

Source: Wetherold (1983), tables 1 and 3.

Estimated failure rates of storage tanks quoted by Batstone and Tomi (1980) are shown in Table A14.13.

Data on the failure rates of storage tanks in ammonia and LNG are given in Sections A14.25.2 and A14.24, respectively.

Data on storage tank fires are given in Section A14.28.5.

A14.13 Pipework

The failure rates of pipes are discussed in Chapter 12.

De la mare and his coworkers developed pipeline inventory and pipeline failure and repair history databanks for the North Sea area. Some of their statistic results are shown in Tables A14.14–A14.18.

Data on the failure rates of gaskets are given in Tables A14.3–14.5 and by Pape and Nussey (1985).

Comparison of these values gives for the failure frequency of gaskets:

	Failure frequency (failures/year)
Smith: Lower limit	0.00018
Upper limit	0.0088
UKAEA	0.0044
Rasmussen Report	0.026
Pape and Nussey	4×10^{-5} (average)

A14.13.1 Bellows
There is a potential ambiguity in the term bellows, which may refer to bellows used in instrumentation or to those used in pipework.

The UKAEA data on the failure rate of bellows in Table A14.3 gives a failure frequency of 5×10^{-6} failures/year. D.J. Smith (1985) gives a failure rate of 4×10^{-6} failures/year.

A14.14 Heat Exchangers

Data on the failure rates of heat exchangers are given in Table A14.5.

Some information on the failure rate of heat exchangers has been given by C.F. King and Rudd (1972) in a reliability study of a heavy water plant. With some 21 heat exchangers, the MTTF ranged from 677 to 7865 h.

For offshore shell and tube heat exchangers, the data given by OREDA (1997) include the following (all failure modes included):

To be statically justifiable, minimums of 3,500,000 calendar hours were established for each component. They reported 0.07601 failures/year for electric air compressor and 0.01033 failures/year for fuel air compressor (Hale and Arno, 2001).

A14.15.2 Fans
Data on the failure rates of fans are given in Table A14.45.

A14.15.3 Pumps
There are wide variations in the type, duty and environment, and hence in the failure rate, of pumps. However, many pumps have a failure rate of some 1–5 failures/year. Data on the failure rates of pumps are given in Tables A14.3–A14.5.

D.J. Smith (1985) gives the failure modes of pumps as about 50% leakage and 50% no transmission.

Some information on the failure rate of pumps has been given by C.F. King and Rudd (1972) in a reliability study of a heavy water plant. MTTFs for four auxiliary

	Population	No. of failures	Failure rate (per 10^6 h)	Mean repair time (manhours)
Crude oil → > sea water (1000–3000 kW)	2	0	35.14 (upper limit)	
Crude oil → > sea water (1000–10,000 kW)	10	29	94.42	145.0
Gas → > gas (1000–10,000 kW)	4	5	75.02	48.2
Gas → > gas (10,000–30,000 kW)	1	5	142.60	3.4
Gas → > sea water (100–1000 kW)	4	3	31.16	14.7
Gas → > sea water (1000–10,000 kW)	4	12	85.62	53.2
Gas → > water/glycol (1000–3000 kW)	3	5	79.61	47.4
Gas → > water/glycol (1000–10,000 kW)	5	13	120.24	45.0
Gas → > water/glycol (10,000–30,000 kW)	12	38	121.64	64.4

A14.15 Rotating Machinery

A14.15.1 Compressors
Data on the failure rates of compressors are given in Table A14.5.

For offshore compressors the data given by OREDA (1997) include the following (all failure modes included):

	Population	No. of failures	Failure rate (per 10^6 h)			Mean repair time (manhours)
			Lower	Mean	Upper	
Centrifugal, electric motor driven						
100–1000 kW	5	58	2.64	550.66	2106.50	20.6
1000–3000 kW	14	204	174.54	880.21	2033.05	25.1
3000–10,000 kW	9	398	1157.79	2433.02	4088.47	48.2
Centrifugal, turbine driven						
	9	586	122.28	2449.88	7341.37	29.6
Reciprocating, electrical motor driven						
1000–3000 kW	4	352	2293.78	2509.70	2741.16	9.6
3000–10,000 kW	4	317	4029.38	5388.42	6909.82	98.6

The Power Reliability Enhancement Program (PREP) of the US Army corps of Engineers conducted a study of the operational and maintenance data on 204 power generation, power distribution and HVAC components (Hale and Arno, 2001). Efforts focused on the equipment installed after 1971.

pumps ranged from 51 to 398 days according to maintenance data, but from 12.5 to 439 days according to production data.

An account has been given by R. James (1976) of a pump maintenance program on 880 major process pumps, mainly centrifugal pumps, aimed at eliminating premature wearout failures, in which the MTBF of the pumps was raised from 8.7 to 12.2 months.

In accounts of pump reliability improvements by a particular manufacturer (Anon., 1985x; Bloch and Johnson, 1985) it is stated that for pumps the industry average MTBF is some 6 months, corresponding to a failure rate of 2 failures/year. Bloch and Johnson state that ANSI standard pumps have an MTBF of 13 months, or failure rate of 0.9 failures/year. It is claimed, however, by Anon. (1985x) that specified improvements have resulted in the achievement of an MTBF of some 25 months, or failure rate of 0.48 failures/year. For individual features improvements are said to have extended the lives of the ball bearings, the mechanical seal system, the shaft and the coupling to 5, 2.5, 15 and 7 years, respectively.

A study of 85 ethylene plant pumps has been reported by Sherwin (1983). There were 243 failures and the failure rate was 1.8 failures/year. The failure modes are shown in Table A14.19

Data on the failure rates of pumps in LNG plants are given in Table A14.44.

A study of feedwater pumps in French nuclear power stations has been made by Aupied, Le Coguiec and Procaccia (1983). The overall failure rate was Failure rate of feedwater pumps $= 5.6 \times 10^{-4}$ failures/h $= 4.9$ failures/year

The failure modes of the pumps are shown in Table A14.20.

A survey of pump mechanical seals by BHRA has been reported by Flitney (1987) in which pumps were surveyed in three refineries and five chemical plants, a sample of some 200–300 pumps being taken at each site. The two principal reasons for seal removal were

Leakage	66%
Bearing replacement	12%

Mechanical seals are also the subject of a study at Esso reported by N.M. Wallace and David (1985). They give a table of seal lives for some 17 cases, the lives varying from 2 to 12 months. For catastrophic failure of a pump the Rijnmond Report uses a value of 1×10^{-4} failures/year.

Operational data of 505 primary coolant pumps of steam supply systems in 146 nuclear power plants were analysed and an average failure rate of 1×10^{-5} was suggested (Milivojevic and Riznic, 1989).

Data from 320 electric submersible pumps gathered over a 13-year period in Milne Point Field, Alaska were analysed and the mean time to failure is estimated to be 330 days (Sawaryn, Grames, and Whelehan, 2002).

Some data on pumps given out by EIReDA is shown in Table A14.21.

Table A14.13   Estimated failure rates of storage tanks

	Failure rate (failures/10^6 year)
Atmospheric tank	30
Refrigerated tank	
Single wall	10
Double wall	1

Source: Batstone and Tomi (1980), table Al.

Table A14.14   Failure frequency varying with pipeline status and location (de la Mare, Bakouros and Tagaras, 1993; reproduced with permission from Elsevier)

	Near platform	Open Sea	Near Shore
Installation	4	19	0
Hydrotest	5	5	0
Operation	15	7	2

Table A14.15   Failure frequency varying with cause of failure and location (de la Mare, Bakouros and Tagaras, 1993; reproduced with permission from Elsevier)

	Near platform	Open sea	Near shore
Material	1	3	0
Construction	3	10	0
Corrosion	4	0	0
Anchors	5	7	0
Fishing activity	0	1	0
Scour-vortex	4	4	2
Manufacturer	1	0	0
Constructor's equipment	2	2	0

Table A14.16   Failure frequency and rate varying with diameter (de la Mare, Bakouros and Tagaras, 1993; reproduced with permission from Elsevier)

Diameter range (inches)	Percent by number of pipelines	Percent by total operational experience[a]	Failure frequency	Failure rate (failures/1000 km-year)
0.0–6.0	4.8	0.9	0	0.00
6.1–12.0	13.6	2.7	2	1.71
12.1–18.0	20.8	10.2	21	4.71
18.1–24.0	24.8	10.2	11	2.47
24.1–30.0	16.8	23.5	7	0.68
>30.0	19.2	52.5	16	0.70
Total	100.0	100.0	57	1.31

Table A14.17   *Failure frequency and rate varying with length (de la Mare, Bakouros and Tagaras, 1993; reproduced with permission from Elsevier)*

Length range (km)	Percent by number of pipelines	Percent by total length	Percent by total operational experience[a]	Failure frequency	Failure rate (failures/1000 km-year)
0.0–2.0	4.95	0.18	0.21	1	11.19
2.1–10.0	24.79	2.37	2.25	8	8.13
10.1–20.0	23.98	6.57	5.03	13	5.92
20.1–50.0	16.53	10.84	9.64	3	0.71
50.1–100.0	11.57	15.96	17.77	4	0.52
> 100.0	18.18	64.08	65.10	25	0.84
Total	100.00	100.00	100.00	54	1.21

[a] Operational experience is defined as the product of length and year of service.

Table A14.18   *Failure frequency (F) and rate (r) varying with commissioning dates and years of operation (de la Mare, Bakouros and Tagaras, 1993; reproduced with permission from Elsevier)*

Years of operation	Commissioning date							
	Year 7–8		Year 9–10		Year 11–12		Year 13–14	
	F	r	F	r	F	r	F	R
1	2	1.8	17	7.2	14	8.2	3	7.5
2	5	2.3	21	4.5	16	4.7	3	3.7
3	8	2.2	21	3.0	17	3.3	3	2.5
4	10	2.0	22	2.4	18	2.7	3	1.9
5	10	1.6	23	2.0	18	2.6	–	–
6	10	1.4	23	1.6	18	1.8	–	–
7	10	1.2	23	1.4	–	–	–	–
8	10	1.0	23	1.2	–	–	–	–
9	10	0.9	–	–	–	–	–	–
10	10	0.8	–	–	–	–	–	–

For offshore pumps failure data are given by OREDA (1992, 1997, 2002), covering motor and turbine driven oil pumps, motor and diesel driven fire pumps, chemical injection pumps, and sea water injection and lift pumps. The following data were extracted from OREDA (1997).

	Population	No. of failures	Mean failure rate (per 10^6 h)	Mean repair time (manhours)
Centrifugal, chemical injection, processing				
51–100 kW	2	10	142.69	57.1
100–1000 kW	2	40	570.78	117.1
Oil handling				
100–1000 kW	16	321	533.28	38.4
1000–2500 kW	5	132	1136.97	27.8
2500–3600 kW	3	98	1357.04	28.7
Diaphragm, chemical injection, processing				
	8	35	124.86	31.6
Reciprocating, oil handling				
	2	6	105.63	15.6

Table A14.19   *Failure modes of ethylene plant pumps*

Failure mode	No. of failures	Proportion (%)
Seals/glands	119	49.0
Overhauls[a]	62	25.5
Cleaning	14	5.8
Repeat overhauls	7	2.9
Leaks[b]	7	2.9
Motor failures	5	2.0
Couplings	5	2.0
Bearings	2	0.8
Other	22	9.1
Total	243	100.0

Source: Sherwin (1983).
[a] Overhauls were not at regular intervals but as a result of conditions found on opening up following a failure.
[b] Leaks other than those due to seals/glands.

A14.15.4 Turbines

Data on the failure rates of turbines are given in Table A14.5.

A study of steam turbine failure with particular reference to catastrophic failure and the generation of missiles has been described by Bush (1973). He quotes data from six major suppliers of steam turbines. The accumulated

Table A14.20 Failure modes of feedwater pumps in some French nuclear power stations

Failure mode	Proportion (%)
Body, shaft	5
Packing	29
Overspeed	10
Contactors	13
Control	16
Lubrication	12
Human error	6

Source: Aupied, Le Coguiec and Procaccia (1983), figure, p. 149.

operating experience is 12,330 years prior to 1950 and 57,950 years in the period 1950–72. There were no failures in the earlier period and 10 in the later period. All four failures after 1959 were due to over-speed. Of the ten incidents seven generated missiles. His analysis indicates that the failure frequency was slowly decreasing. His estimates of the then current failure frequencies are

Frequency of failure $= 9 \times 10^{-5}$/year

Frequency failure resulting in missile generation
    $= 8 \times 10^{-5}$/year

For offshore gas turbines failure data are given by OREDA (1992, 1997, 2002).

A14.16 Valves

A14.16.1 General
A study of valves in French nuclear power stations has been carried out by Aupied, Le Coguiec and Procaccia (1983). Some of the more critical valves are classified as primary valves and the others as secondary valves. Some of the failure rates obtained in this work are shown in Table A14.22, while failure modes are given in Table A14.23.

A survey designed to identify significant and common valve problems such as leakage and jamming and to relate them to valve type, service and manufacturer has been described by Vivian (1985). The survey was conducted in a major oil company and covered 10 businesses involving 17 facilities, of which a significant proportion are in the North Sea. Some quarter of a million valves were covered, of which almost 10% were reported as giving significant problems. The numbers of each type of valve are shown in Figure A14.4(a) and the associated problems in Figure A14.4(b).

The UKAEA data in Table A14.3 gives for control valves a failure rate of 0.25 failures/year and for manual valves 0.13 failures/year. The values used in the *Rijnmond Report* are 0.3 and 0.1 failures/year for control and manual valves, respectively.

Further values for valve failure rates obtained by Moss (1977 NCSR Rll) are discussed in Section A14.25. These values are 0.1 and 0.01 failures/year for valves on steam and water service, respectively. The latter value in particular is a lower than the others quoted here.

For valve rupture the *Rasmussen Report* gives a failure frequency of 1×10^{-8}/h (0.9×10^{-4}/year). For valves in offshore operation, failure data are given by OREDA (1992, 1997, 2002).

A14.16.2 Control valves
Data on the failure rates of control valves are given in Tables A14.3–A14.5. Further data are given in various parts of this appendix.

A14.16.3 Pressure relief valves
For pressure relief valves (PRVs) the definition of failure is particularly important. Definitions of failure that may be used include failure of any kind, failure to lift within a certain proportion of the set pressure and failure to open on demand. The failure rates for these different types of failure are quite different.

Data on the failure rates of PRVs are given in Tables A14.3–A14.5, A14.22 and A14.23.

The following data are for the failure rates of PRVs are given by Kletz (1972a, 1974a) and by Lawley and Kletz (1975):

	Failure rate (faults/year)
Valve fails shut	0.001
Valve lifts heavy	0.004
Total fail danger	0.005 (later modified to 0.01)
Valve fails open or lifts light	0.02
Total fail safe	0.02

A survey of PRV inspections in a large chemical company has been reported by Aird (1983). Failure was defined as lifting 10% outside the set pressure when put under test. The number of useful tests was 866. The proportion of failures was 44.5%.

The failure rate showed no discernible trend with operating time and it was concluded that PRVs may be subject to changes that occur relatively rapidly. One cause often quoted is spring relaxation.

A study of controlled safety valves in power stations has been made by Oberender and Bung (1984). The valves were either intrinsically controlled by the vented fluid or externally controlled by a control fluid (pneumatic or hydraulic). Two basic events were considered: faultless functioning and failure to open on demand. Some 1378 tests were conducted. The proportion of valves giving faultless functioning was approximately:

Intrinsic control	40%
External control	
Pneumatic	80%
Hydraulic	85%

The proportion giving failure on demand was approximately:

Intrinsic control	
Load principle	2%
Relief principle	4.5%
External control	
Pneumatic	0.3%
Hydraulic	0.8%

Table A14.21 Failure rate data of motor driven pump from EIReDA (Lannoy and Procaccia, 1996; reproduced with permission from Elsevier)

Type	Application	Medium	Main character	Mean failure rate (10^{-6}/h)	Range (10^{-6}/h)	Failure on demand (10^{-3}/d)	Range (10^{-3}/d)	MMTR (h)
Centrifugal	Circulating water	Fresh/salt water	kW ≤4500 head: 25 m flow ~8000 m³/h	4.7	3–6.6	0.20	0.04–0.3	41
Centrifugal	Containment drain system	Polluted water	kW ≤200 head: 30 m	200	100–400	1	0.3–3	20
Centrifugal[a]	Refuelling pool	Borated water (200 ppm)	kW ≤75 head: 50 m	10.6	7–14.5	0.01	<0.015	27
Helicoidal	Reactor cooling	Borated water	kW ≤6100 head: 90 m flow: 22,400 m³/h	2.6	1.6–8.8	0.1	0.04–0.15	37
Centrifugal[b]	Chemical and volumetric control	Borated water	kW: 450 head: 1770 m flow: 34 m³/h	59	47–72	0.12	0.1–0.16	22
Centrifugal	Makeup	Demin. water	kW ≤40 head<190 m flow: 36 m³/h	13.4	9.6–17.6	0.02	<0.03	16
Centrifugal[c]	Safety injection	Borated water (2000 ppm)	kW ≤260 head≤82 m flow: ≤750 m³/h	24	11.5–48.1	0.9	0.55–1.25	19

[a] Intermitent operation.
[b] Two out of three systems.
[c] Standby pump operating hour <1 h (test).

Table A14.22 *Failure rates of valves in some French nuclear power stations*

						Failure rate	
						In operation (failures/year)	*On demand*
Primary valves							
Pressurizer safety relief valve	18	175,500	382	9[a]	4	51	0.01
Heat removal loop safety relief valve	12	105,000	12	3	2	29	0.17[b]
Secondary valves Condensate and drain pumps non-return valves	12	88,700	1375	2	0	23	–[c]

Source: Aupied, Le Coguiec and Procaccia (1983), tables 4 and 5.
[a] Failures mainly detected during annual tests.
[b] Original table leaves this space blank.
[c] Original table gives value of 0.0034.

Table A14.23 *Failure modes of valves in some French nuclear power stations*

Type of valve	*Type of fault*				
	Outlet leakage	*Corrosion*	*Untightness*	*Mechanical*	*Non-operation*
Gate valve	23	11	9	26	31
Globe valve	22	20	25	16	17
Plug valve	55	17	22	6	
Safety relief valve	30	17	28	5	20
Non-return valve	42	21	13	9	15
Overall value	30	15	20	15	20

Source: Aupied, Le Coguiec and Procaccia (1983), figure, p. 139.

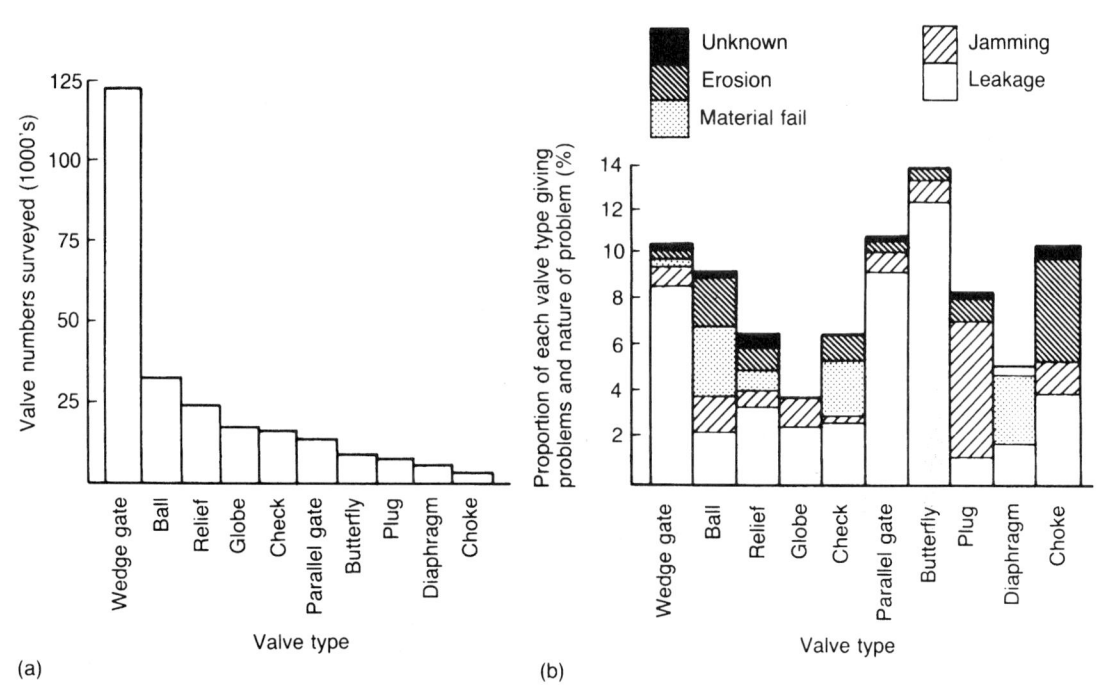

(a)

(b)

Figure A14.4 *Number and failure modes of valves in a large oil company survey (after Vivian, 1985): (a) number of valves; and (b) failure modes of valves (Reproduced by permission of Gower Press)*

For the intrinsically controlled valves, however, there was an improvement over time, the failure rate falling from some 10% in the initial period to <1% in the final period.

Estimates of the success rate of PRVs in particular applications have been given by Prugh (1981) as follows:

Venting of vapour/air explosion	1%
Venting of runaway reaction	95%
Venting of excessive nitrogen purge	99%

Historical data on a 10% random sample (240 valves) out of 2416 relief valves in a petrochemical plant were collected and analysed (Alfares 1999). Figure A14.5 shows the composition of the sample and valves in the plant, and Table A14.24 shows the average time between failure by valve pressure, service, size, temperature and type.

Failure data of relief valves for offshore operations are given by OREDA (1992, 1997, 2002). The following data are extracted from OREDA (1997):

Type	Population	No. of failures	Mean failure rate (per 10^6 h)	Mean repair time (manhours)
Globe (0.1–1.0 in.)	7	6	22.08	1.0
Globe (1.1–5.0 in.)	21	9	12.30	3.4
Globe (5.1–10.0 in.)	8	4	15.99	3.5

A14.16.4 Non-return valves

ORNL (Oak Ridge National Laboratory) and NIC (Nuclear Industry Check Valve Group) jointly conducted a study of check valve failures recorded between 1991 and 1992 in the INPO (Institute of Nuclear Power Operations) Nuclear Plant Reliability Data System (NPRDS) (McElhaney 2000). This study characterized check valve failures by valve type, and correlations between failure

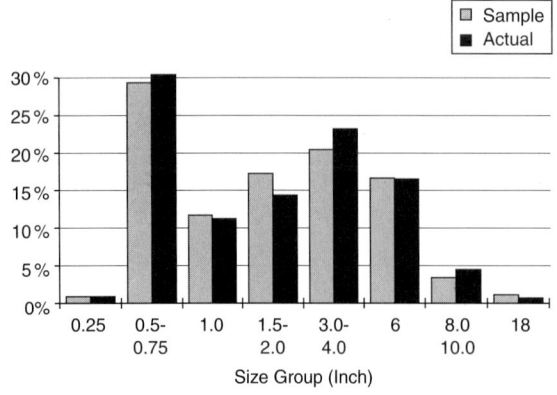

Figure A14.5 *The composition of the sample and relief valves in a petrochemical plant (Alfares, 1999; with permission from Elsevier)*

modes, failure causes, failure area, failure discovery method, extent of degradation with valve types were developed. Figures A14.6(a) and (b) are the distribution of 1991–92 check valve failures and 1991 check valve population by valve type respectively. Tables A14.25 and A14.26 are number of failures by valve type and failure mode and by valve type and failure cause specifically. Figure A14.7 is the relative failure rate by valve type and different system.

Additional data on the failure rates of non-return, or check, valves are given in Tables A14.4, A14.5, A14.22 and A14.23.

A14.16.5 Emergency isolation valves

Failure data for safety shut-down valves from OREDA (1992) has been analysed by Haugen, Hokstad and Sandtorv (1997). The safety shut-down valves included are mainly gate valves of approximately 6 in. handling oil, gas or water in offshore installations. The data are divided into two groups based on the test intervals (6 months or 12 months) as shown in Tables A14.27–A14.30.

OREDA (1997) gives failure rate data for shutoff valves in offshore operations as follows:

Type	Population	No. of critical failures	Mean critical failure rate (critical failures per 10^6 h)	Mean repair time (manhours)
Gate (1.1–5.0 in.)	104	13	2.10	0.3
Gate (20.1–30.0 in.)	1	1	28.54	3.0
Globe	8	1	5.19	4.0

Data on the failure rates of emergency isolation valves, also called slam shut or shut-off valves, are given in Table A14.5. Data on the failure rates of manual isolation valves are given in Tables A14.3 and A14.4.

A14.17 Instruments

For instruments the definition of failure is particularly important. This was discussed in detail in Chapter 13.

The importance of the definition of failure has been studied by Kortland (1983) for differential pressure transmitters. For transmitters required to maintain their calibration within 2% the failure rate observed was 0.1 failures/year. For transmitters with a required calibration within 5%, a less severe specification, the observed failure rate was a function of the calibration interval, being about 0.01 failures/year with a calibration interval of 1 year, but increasing for longer calibration intervals.

Some data on the failure rates of instruments are given in Chapter 13. The data given there refer mainly to instrumentation in the early and mid-1970s. The data given in this section supplement those given in Chapter 13 and include more recent data.

Data on the failure rates of instruments are given in Tables A14.3 and A14.4.

A study of control system failure sequences in ammonia plants has been described by Prijatel (1984). A comparison is given between predicted and actual failure sequences. Information on actual failure sequences was obtained from the work of G.P. Williams and Hoehing (1983). Some 95% of

Table A14.24 *Average time between failures of relief valves in a petrochemical plant (Alfares, 1999; with permission from Elsevier)*

Pressure (psi)	≤0	0–100	101–200	201–300	301–400	401–500	501–600	601–700	≥700
ATBF (m)	22.25	22.00	26.83	23.25	16.79	26.38	19.95	18.36	28.23
Service	Water	Steam	Chemical	Air	HC[a] gas	HC liquid			
ATBF (m)	21.02	20.60	20.26	41.11	23.13	22.88			
Size (in.)	0.25	0.5–0.75	1.0	1.5–2.0	3.0–4.0	6.0	8.0–10.0	18.0	
ATBT (m)	41.00	21.99	24.67	20.68	22.58	29.41	24.33	25.20	
Temp. (°C)	1–99	100–199	200–299	300–399	400–499	600–699	900–999		
ATBF (m)	31.89	22.97	21.94	24.13	19.73	27.80	24.29		
Type[b]	CV	BE	TK	BB	BA	PI			
ATBF	22.76	23.67	32.27	22.92	18.80	13.82			

[a] HC is hydrocarbon.
[b] CV is conventional; BB balanced bellows; BE bellows; BA balanced; TK tank and PI pilot-operated.

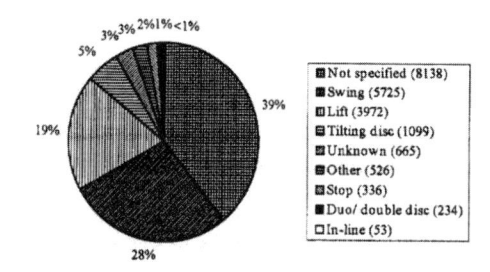

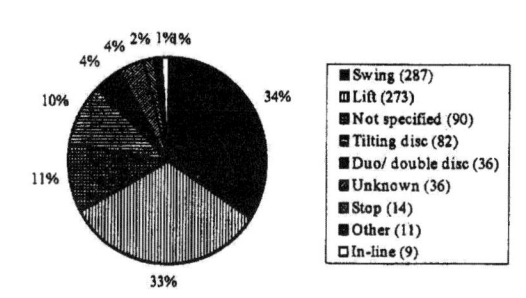

Fig A14.6 *(a) Distribution of 1991–92 check valve failure by valve type. (b) Distribution of 1991 check valve population by valve type (McElhaney, 2000; reproduced with permission from Elsevier)*

the actual failure sequences were single event failures. Data on instrument failure were obtained from plant records and from the literature. Some of the data used in the study are given in Table A14.31.

For control failure sequences the actual shut-down frequencies of particular units such as compressors were some two to four times as high as the predicted frequencies. The ranking of the instruments as a cause of shut-down was in descending order of importance controllers, switches, solenoid valves, control valves and uninterruptible power supply (UPS) system, the contribution of these items to shut-downs being 41.2, 36.7, 7.5, 4.6 and 4.3%, respectively. Thus two items, controllers and switches, accounted for 78% of shut-downs.

Data on the failure rates of process chromatographs and other analytical instruments have been given by Huyten

(1979). Some 870 instruments are considered in the survey. He gives the following availabilities:

Analysers	93.8%
Gas chromatographs	91.0%

He also gives for analysers failure modes that include the following:

Sampling system	39%
Analyser	18%
Non-availability of spares	16%
Plant upsets, start-up and shut-down	11%

A comparative study of the MTBF and maintenance time for controllers from five manufacturers has been described by H.S. Wilson (1978). He gives the following data:

Manufacturer type	MTBF (year)	Maintenance time (h)
A	Electronic 1	0.5
B	Electronic 1	1.5
C	Pneumatic 5	3
D	Electronic 3	4
E	Pneumatic 15	5

He comments that the controller from manufacturer E failed so rarely that the maintenance time was greater due to unfamiliarity. From this comment the maintenance times are evidently times per failure.

Paula (1993) presented failure rates for programmable logic controllers (PLC) in three different processes: a US phenol plant, French nuclear plants and a Canadian nuclear power plant with data input from operation personnel and Electricite de France (EDF). His result is presented in Table A14.32. Rate of coverage, the probability of a PLC successfully switching to the redundant processor upon a single processor failure, is also estimated.

A good deal of the literature data for instrument failure is now quite old. In general, it is to be expected that the failure rates have fallen. The following is a comparison of the failures rates given for certain measuring instruments

Table A14.25 *Number of 1991–92 failures by valve design and failure modes (McElhaney, 2000; reproduced with permission from Elsevier)*

	Disc or other part, off or broken	*Stuck closed*	*Loose or detached part*	*Stuck open*	*Restricted motion of flow*	*Improper seating*	*Unknown/ miscellaneous*
Not specified	9	2	1	12	14	49	3
Duo/double disc	4	0	0	3	1	27	1
In-line	0	0	0	4	0	5	0
Lift	4	24	1	29	12	194	9
Other	0	1	0	1	1	5	3
Stop	0	0	1	2	2	8	1
Swing	14	3	12	51	31	143	34
Tilting disc	5	4	2	8	11	48	4
Unknown	3	0	9	3	3	18	0

Table A14.26 *Distribution of 1991–92 failures by valve design and failure cause (McElhaney, 2000; reproduced with permission from Elsevier)*

	Not specified	*Duo/double disc*	*In-line*	*Lift*	*Other*	*Stop*	*Swing*	*Tilting disc*	*Unknown*
Abnormal wear	6	8	14	4	23	13	8	10	2
Normal wear	31	21	29	25	23	19	33	24	37
Design problem	16	12	14	10	0	38	15	11	11
Human error	1	0	0	0	0	0	1	2	0
Maintenance error	7	0	7	2	8	0	5	8	20
Corrosion	7	25	14	19	8	0	5	4	7
Foreign material	13	6	14	35	15	13	14	8	11
Procedure problem	4	8	0	0	0	0	2	3	0
Stress corrosion cracking	0	0	0	0	0	0	1	0	0
Erosion/erosion-corrosion	1	15	0	0	0	13	2	2	0
Installation problems	3	2	0	0	0	0	2	5	0
Manufacturing defect	2	2	0	1	0	0	3	4	2
Unknown	8	2	7	3	23	6	9	20	11

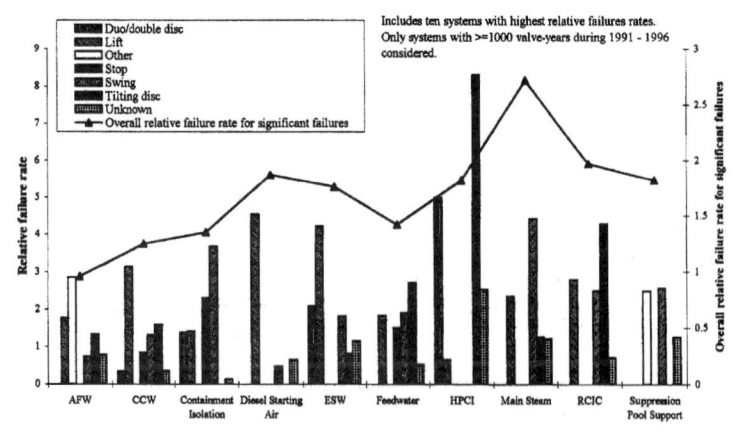

Figure A14.7 *Relative failure rate by valve design and system (McElhaney, 2000; reproduced with permission from Elsevier)*

Table A14.27 *Failure causes for maintainable items – critical (C) and non-critical (NC) failures, $\tau = 6$ months (Haugen, Hokstad and Sandtorv, 1997; reproduced with permission from Elsevier)*

Maintainable item	Failure cause/descriptor									
	Mechanical defects		Material failures (defect/ deterior.)		Instrument/ Electrical failures		Other		Total	
	C	NC	C	NC	C	NC	C	NC	C	NC
Actuating device	0	0	0	2	3	0	0	3	3	5
Actuator	3	0	3	10	0	0	1	0	7	10
Body + internals	2	3	16	34	0	0	7	4	25	41
Control unit	0	0	0	0	5	1	1	0	6	1
Monitor/display	0	2	0	0	0	3	0	0	0	5
Seals	0	0	1	7	0	0	0	0	1	7
Signal transm.	0	0	0	0	0	1	0	0	0	1
Valve seat	0	0	0	3	0	0	0	0	0	3
Subsystems	0	0	5	0	0	0	0	0	5	0
Total	5	5	25	56	8	5	9	7	47	73

Table A14.28 *Failure causes for maintainable items – critical (C) and non-critical (NC) failures, $\tau = 12$ months (Haugen, Hokstad and Sandtorv, 1997; reproduced with permission from Elsevier)*

Maintainable item	Failure cause/descriptor									
	Leakage		Mechanical defects		Material failures (defect/ deterior.)		Instrument/ Electrical failures		Other	
	C	NC	C	NC	C	NC	C	NC	C	NC
Actuating device	0	6	1	1	0	12	1	1	2	9
Actuator	3	11	1	1	25	27	0	0	3	4
Body + internals	0	0	1	6	16	10	0	0	0	1
Control unit	0	0	0	1	0	0	12	2	0	0
Monitor/display	0	0	0	0	0	0	0	3	0	0
Seals	0	0	0	0	1	4	0	0	0	0
Sensor	0	0	0	0	0	0	2	3	0	0
Signal transm.	0	0	0	0	0	0	0	1	0	0
Subsystems	0	0	0	2	4	4	0	0	0	0
Unknown	0	1	0	1	0	3	1	0	1	1
Valve seat	0	0	0	0	1	3	0	0	0	0
Total	3	18	3	12	47	63	16	10	6	15

Table A14.29 *Number of critical failures for various failure modes, $\tau = 6$ months (Haugen, Hokstad and Sandtorv, 1997; reproduced with permission from Elsevier)*

Code	DOP	EXL	FTC	FTO	INL	LCP	PLU	SPO	⋯	Total
1.0			1							1
1.4			5							5
2.0					1	3				4
2.1		2	1			16				19
2.2			2							2
3.0			3	3						6
3.1				1				1		2
6.1	1					1	1			3
6.2				1						1
8.2		2	1			1				4
Total	1	4	13	5	1	21	1	1	0	47

by Anyakora, Engel and Lees (1971) and for sensors in process alarm systems by OREDA (1992):

| | Failure rate (failures/year) | |
	Anyakora, Engel and Lees	OREDA
Pressure	1.41	
(p) (< 1500 psig)		0.019
(e)		0.37
Flow	1.73	
(p) (gas)		0.25
(e) (gas)		0.75
Level	1.71	
(e)		0.096
Temperature	0.88	
(e)		0.1

The first authors refer for pressure to 'pressure measurement', for flow and level to differential pressure transducers and for temperature transducers, but do not distinguish between transducers with pneumatic (p) or electronic (e) output. The instruments selected from the OREDA collection are all transducers with specified type of output. In comparing the figures, allowance needs to be made for the effects on failure rate of environment and calibration requirements. Nevertheless, the data do suggest an improvement in reliability.

Table A14.30 *Number of critical failures for various failure modes, τ = 12 months (Haugen, Hokstad and Sandtorv, 1997; reproduced with permission from Elsevier)*

Code	EXL	FTC	FTO	INL	LCP	SPO	⋯	Total
1.0			1					1
1.2		1	2					3
1.4		1	1		1			3
2.0	1			3	2			6
2.1	16		1	15	8			40
2.2		1						1
3.0		8						8
3.1		6				1		7
3.2		1						1
7.0			1					1
8.2	1	1		2				4
Total	18	19	6	20	11	1	0	75

The codes system used for the failure modes are shown as follows:

DOP	Delayed operation
EXL	External leak
FTC	Fail to close
FTO	Fail to open
INL	Internal leak
LCP	Leakage in closed position
PLU	Plugged
SPO	Spurious operation

For process sensors used in offshore operations, OREDA (1997) gives the following data:

Application	Population	No. of failures	Mean failure rate ($/10^6$ h)	Mean repair time (manhours)
Flow	72	92	26.41	0.4
Level	73	57	32.93	10.4
Pressure	282	137	18.06	6.4
Temperature	8	7	56.36	0.5

A14.18 Process Computers

The reliability of process computer systems is discussed in Chapter 13.

Data on the failure rates of 12 process computer systems in the paper industry given by Hubbe (1970) are shown in Table A14.33.

Data by E. Johnson (1983) on the failure rates of several different process computer configurations are summarized in Table A14.34. The data refer to total system failure, rather than failure of individual items such as printers. Johnson gives full details of the fault and downtime incidents. He states that for the total system an MTBF of 5000 h with a combined availability of about 99.9%, corresponding to no more than 8 h downtime per year, is about the level which may be found acceptable.

A14.19 Relief Systems

Some information is available on the failure rates of the individual elements of a pressure relief system, such as pressure relief valves, bursting discs and vent systems. Data on the failure of these items is given in this section. For chemical reactors, information on the failure rate of pressure relief systems has been given by Marrs and Lees (1989) as described in Chapter 11.

A14.19.1 Pressure relief valves
Data on the failure rate of PRVs are given in Section A14.16.3.

A14.19.2 Bursting discs
Data for the failure rate of bursting discs are given by Lawley (1974b) as follows:

Frequency of disc rupture at normal pressure
= 0.2 failures/year.

Table A14.31 *Failure of instruments in ammonia plants (after Prijatel, 1984)*

	Failure rate (failures/year)
Control valves	0.028
Controllers[a]	0.25
Switches[b]	0.22
Redundant switch systems[c]	1.68×10^{-4}
Solenoid valves	0.046
UPS system	0.026

[a] For flow, pressure, level.
[b] For flow, pressure, temperature, level.
[c] An alternative figure of 2.9×10^{-3} failures/year is also given.

Table A14.32 *Failure and coverage rates of PLC in an US phenol plant, French-design nuclear power plants, and a Canadian nuclear power plant (Paula, 1993; reproduced with permission from Elsevier)*

PLC application	Single processor failure rate (per PLC year)	Total PLC failure rate (per PLC year)		Coverage	
US phenol plant	2	0.0082[b]	0.016[c]	0.9918–0.9959[b]	0.9837–0.9918[c]
French-design nuclear power plants	–[a]	0.011[d]	0.021[e]		
Canadian nuclear power plant	–[a]	0.025			

[a] Information not available.
[b] Excluding early failures.
[c] Including early failures.
[d] Excluding power supply failures.
[e] Including power supply failures.

Table A14.33 *Failure data for 12 process computer systems in the paper industry (Hubbe, 1970)*

	MTBF		Availability	
	Hardware faults (h)	Software faults (h)	Hardware faults (%)	Software faults (%)
Composite average	550	1365	99.1	99.80
Best case	1633	8163	99.94	99.97

Table A14.34 *Failure data for some process computer systems in the chemical industry (after E. Johnson, 1983; reproduced by permission of Elsevier Science Publishers)*

System	Operating time (h)	No. of failures	Downtime		Availability (%)
			Failure (h)	Planned (h)	
Single computer system with analogue standby	66,528	13	65	300	99.9
Twin computer system with analogue standby[a]	35,040	8	30.5	38	99.91
Twin computer system with shared critical loops – 1	78,888	21	172	48	99.78
Twin computer system with shared critical loops – 2	78,888	37	388	73	99.5
Twin computer system with analogue standby[a]	13,848	6	54	17	99.61

[a] Different configurations.

For chemical reactors the failure rate of bursting discs in a single company has been investigated by Marrs and Lees (1989). The estimates obtained refer to the bursting disc itself, including blockage before the disc, but not to the vent pipework after the disc. There were during the period of investigation 11 successful ventings and no failures. The estimated probability of failure from these data is 0.083 failures/demand. An alternative, and less pessimistic, estimate was obtained from the fact that there were four unrevealed fail-to-danger failures in 164 reactor-years with an inspection interval of 1 year, giving an estimated probability of failure of 0.012. The authors' best estimate of the probability of failure is 0.01 failures/demand, but the confidence bounds are relatively wide.

A14.19.3 Vent systems
For chemical reactors and vented vessels, the failure rate of vent systems, excluding bursting discs, in a single company has been investigated by Marrs and Lees (1989). There were 28 successful ventings and no failures during the period of investigation.

The estimated probability of failure from these data is 0.034 failures/demand. An alternative, and less pessimistic, estimate was obtained from the fact that there was one unrevealed fail-to-danger failure in 262 vessel-years with an inspection interval of one year, giving an estimated probability of failure of 0.0019. The authors' best estimate of the probability of failure is 0.002 failures/demand, but the confidence bounds are relatively wide.

A14.20 Fire and Gas Detection Systems

A study of fire detection systems with particular reference to false alarms has been described by Peacock, Kamath and Keller (1982). Some data from this study are shown in Table A14.35. Section A of the table shows the event rates for fire detection systems as assessed for chemical plant by safety officers from a single company. Section B gives the event rates for each type of detector.

Hanks (1983) has described a study of the fire and gas detection system at the gas terminal at St Fergus. The results reported for failure rates are confined to those for the gas detectors, which were mainly in compressor cabs

Table A14.35 *Some event rates in fire detection systems (after Peacock, Kamath and Keller, 1982; reproduced by the permission of the American Institute of Chemical Engineers)*

A   Events by location

	Event rate (events/10^3 detector-year)			
	Real alarms	False alarms	Failure to operate	Real alarms/false alarm
Plant in buildings	4.7	3.4	0.15	1 : 0.7
Plant in open	60	273	27	1 : 4.6
All locations	2.1	7.2	0.16	1 : 3.4

B   Events by type of detector

Smoke	5.9	40	0.5	1 : 6.8
Heat	1.5	5.3	0.3	1 : 3.5
Smoke and heat	16	38	0	1 : 2.4
Flame (UV, IR)	108	622	108	1 : 5.8

and emergency generator rooms. Hanks gives the following failure rates:

	Failure rate (failures/10^6 h)	
	Compressor cab	Other installation
Gas detector	48	8
Gas detector and module	75	35

Y.P. Gupta (1985) describes a survey of automatic fire detection systems at six sites. His account includes estimates of the failure rates of ionization-type smoke detectors synthesized from data on the reliability of electronic components. For detectors in a first class, or non-adverse, environment, the overall failure rate was assessed as 0.057 faults/year, of which 0.04 faults/year were classed as fail-safe and 0.017 faults/year as dangerous. Most adverse environments result in a much higher failure rate. For such environments the failure rate was assessed as 0.46 faults/year. For control units the failure rate was assessed for false alarms as 0.044 faults/year and for unrevealed dangerous faults as 0.06 faults/year.

The US Navy Program, Damage Control Automation for Reduced Manning (DC-ARM) tested the performance of 20 sensors to 120 scenarios (82 fire, 38 nuisance sources and 120 non-fire events) (Rose-Pehrsson *et al.*, 2000). These sensors include electrochemical cell sensors (most of the gas detectors), solid state metal sensor (general hydrocarbon sensor), non-dispersive infrared sensor (carbon dioxide detector), photoelectric smoke detectors and ionization smoke detectors, which is shown in Table A14.36. The performance of smoke detectors is shown in Table A14.37. The data set were analysed with probabilistic neural network and par of the results is shown in Table A14.38.

For offshore fire and gas detection systems failure data are given by OREDA (1992, 1997, 2002). OREDA (1997) include failure rate data for fire and gas detectors as following (all failure modes included):

Type	Population	No. of failures	Mean failure rate (Per 10^6 h)	Mean repair time (man hours)
Flame				
Infrared	520	111	11.94	8.6
Combination IR/UV	132	31	7.36	0.4
Ultraviolet	358	150	48.84	6.1
Heat				
Rate compensated	666	46	2.30	17.4
Rate-of-rise	199	33	7.06	3.8
Smoke/combustion				
Ionization	1214	171	4.60	3.9
Photo electric	683	47	2.13	3.3
Hydrocarbon gas				
Catalytic	2044	1730	31.08	4.5
Infrared	2	19	309.97	10.5
H_2S gas	10	1	3.26	–

A14.21 Fire Protection Systems

Data on the failure rates of sprinkler systems are available from sources such as the Factory Mutual Research Corporation and the NFPA.

A review of failure rates of sprinkler systems that draws on these sources has been given by Rasbash (1975a). For fires in buildings in the United Kingdom in the period 1966–71, the breakdown of incidents given is

Type of incident	No. of incidents
Small fire, extinguished by other means	5,229
Sprinkler system installed and operated	
Fire controlled by other means	275
Fire controlled by sprinkler	3,180
Fire extinguished by sprinkler	651
Effect of sprinkler unknown	73
Sprinkler system installed but did not operate	676
Total	10,084

Thus, excluding the small fires, sprinkler performance was unsatisfactory in some 14% of cases. NPFA statistics for the period 1925–64 quoted by Rasbash indicate that in some 75,290 fires sprinkler performance was unsatisfactory in some 3.8% of cases.

Rasbash also quotes statistics from the Australian Fire Protection Association (AFPA). For some 5734 fires in the period 1886–1968 the proportion of sprinklers which gave unsatisfactory performance was 0.2%. For some 1250 fires in the period 1968–73 the proportion was 0.64%. These very low AFPA values are discussed by Rasbash, who refers to differences in the criteria for satisfactory operation.

Table A14.36 *Sensors used in the multicriteria detection test program (Rose-Pehrsson et al., 2000; reproduced with permission from Elsevier)*

Species	Sensor range	Resolution	Instrument model number	Manufacturer
Oxygen (O_2)	0–25%	0.1% O_2	6C	City Technology
Carbon monoxide with H_2 compensation (CO 4000 ppm)	0–4000 ppm	1 ppm	A3ME/F	City Technology
Carbon monoxide (CO 50 ppm)	0–50 ppm	0.5ppm	TB7E-1A	City Technology
Carbon dioxide (CO_2)	0–5000 ppm	Accuracy = greater than ±5% of reading or ±100 ppm	2001V	Telaire/Englehard
C_1–C_6 hydrocarbons (Ethyl)	0–50 ppm ethylene (C_2H_4)	±2.5 ppm	SM95-S2 with general hydrocarbons solid state sensor	International sensor technology
Hydrogen (H_2)	0–200 ppm	2 ppm	TE1G-1A	City Technology
Hydrogen chloride (HCl)	0–10 ppm	0.5 ppm	TL1B-1A	City Technology
Hydrogen cyanide (HCN)	0–25 ppm	0.1 ppm ±2%F.S. accuracy	4664-40-1-1-1	EIT
Hydrogen sulfide (H_2S)	0–5 ppm	0.1 ppm	TC4A-1A	City Technology
Sulfur dioxide (SO_2)	0–10 ppm	0.5 ppm	TD2B-1A	City Technology
Nitric oxide (NO)	0–20 ppm	0.5 ppm	TF3C-1A	City Technology
Temperature (thermocouple or TC)	−200 to 1250°C	1°C or 0.75%	Type K, 0.127 mm bare bead TC	Omega
Temperature (Temp)	−20°C to 75°C	±0.6°C accuracy	HX93 transmitter (RTD)	Omega
Relative humidity (RH)	3–95%	±2%RH accuracy	HX93 transmitter	Omega
Photoelectric smoke detector (PHOTO)	0–19% obs./m		4098–9701	Simplex
Ionization smoke detector (ION)	1.6–10% obs./m		4098–9716	Simplex
Residential ionization smoke detector (RION)			83R	First Alert
Optical density meter (ODM)			VDM-2 670 nm, 2 mW laser	Meredith
Laser and photodiode with 0.965 m spacing			MRD 500 PIN silicon	Motorola
Measuring ionization chamber (MICX, MICY, MICY20)			EC-912	Photodiode Delta Electronics Testing

Table A14.37 *Summary of smoke detector performance (Rose-Pehrsson et al., 2000; reproduced with permission from Elsevier)*

Alarm level (%obs./m)	Photoelectric			Ionization		
	Fire source (%)	Nuisance source (%)	Overall[a] (%)	Fire source (%)	Nuisance source (%)	Overall (%)
Typical: 11 (Photo) 4.2 (Ion)	38	82	52 (76)	68	71	69 (85)
UL 268 minimum: 1.63	70	53	65 (83)	80	61	74 (87)
Half the UL 268 minimum 0.82	73	53	67 (83)	84	58	76 (88)

[a] Overall percent correct is based on 82 fires and 38 nuisance sources (120 for events). The parenthetical value is the percent correct based on 82 fires, 38 nuisance sources and 120 nonfire (background) events. The parenthetical value is the overall classification parameter that can be compared to the multivariate analysis results.

For the United Kingdom fires the two principal causes of unsatisfactory performance were that the heads were inaccessible to the fire and that the water was shut off, these occurring in 9.6 and 4.3% of the incidents, respectively.

The figures for unsatisfactory performance of sprinklers given by P. Nash and Young (1976) for the NPFA and Australian data are similar, but they give for the United Kingdom for the period 1965–69 a figure of 8.3% and quote a figure of 15% given by the Factory Mutual Corporation for the period 1970–72.

From the UK data, the causes of unsatisfactory performance are given as system shut off, defective system,

Table A14.38   *Probabilistic neural network classification results (Rose-Pehrsson et al., 2000; reproduced with permission from Elsevier)*

Sensor sets	Number wrong	No. of real fires correct	No. of false alarms	Overall percent correct
0.82%				
CO_{50}, CO_{4000}, MICX, ODM, RION	16	73	7	93
CO_{4000}, MICX, ODM, RION	14	75	7	94
CO_{50}, MICX, ODM, RION	19	73	10	92
CO_2, CO_{50}, CO_{4000}, NO, Ethyl, MICX, ODM, RION	15	73	6	94
MICX, ODM, RION	24	66	8	90
CO_{4000}, ODM, RION	19	71	8	92
CO_{4000}, MICX, RION	28	63	9	88
CO_{4000}, MICX, ODM	27	66	11	89
Temp, RH, ODM, ION, PHOTO	14	73	5	94
O_2, H_2, Ethyl, ION, PHOTO	13	74	5	95
O_2, CO_{4000}, H_2, ION, PHOTO	12	75	5	95
O_2, CO_{4000}, Ethyl, ION, PHOTO	13	75	6	95
O_2, H_2S, RH, ION, PHOTO	13	74	5	95
1.63%				
MICX, ODM, RION	22	70	10	91
CO_2, CO_{50}, CO_{4000}, NO, Ethyl, MICX, ODM, RION, PHOTO	13	74	5	95
CO_{4000}, MICX, ODM, RION	18	72	8	93
CO_{50}, MICX, ODM, RION	18	69	5	93
O_2, CO_{4000}, NO_2, ODM, PHOTO	15	74	5	95
O_2, H_2S, RH, ION, PHOTO	6	78	2	98
11%				
MICX, ODM, PHOTO	21	69	8	91
CO_{50}, CO_{4000}, NO, Ethyl, MICX, ODM, RION, PHOTO	10	76	4	96
CO_{4000}, MICX, ODM, RION	13	73	4	95
O_2, CO_2, H_2, MICY2, ODM	10	73	1	96
O_2, CO_{4000}, RH, ION, PHOTO	7	77	2	97
O_2, H_2S, RH, ION, PHOTO	6	78	2	98

system frozen and unknown, these occurring in 4.6, 0.60, 0.07 and 3.0% of the incidents, respectively.

From the NFPA data some principal causes of unsatisfactory performance were valve shut, inadequate water supply and obstruction to distribution, these occurring in 36.0, 9.6 and 8.4% of the incidents, respectively. The causes for the valve being closed were closed for no known reason, closed too early in the fire, closed for system repair or modification and closed to prevent freezing, these occurring in 22, 21 and 20% of cases with a closed valve.

Nash and Young also give failure rates for the components of sprinkler systems as follows:

Component	Failure rate (failures/year)	
	Safe	Dangerous
Wet alarm valve	15×10^{-2}	0.4×10^{-4}
Alternate alarm valve	15×10^{-2}	0.8×10^{-4}
Alarm motor and gong	6×10^{-2}	1.6×10^{-2}
Accelerator	13×10^{-2}	7.9×10^{-3}
Main sprinkler stop valve	–	2×10^{-3}
Non-return valves	–	10×10^{-3}

Budnick (2001) reported an analysis of 16 published reliability data for automatic sprinkler systems and the mean value of reliability was estimated between 93 and 96% in his results. Further data on the effectiveness of fire protection systems have been given by M.J. Miller (1977), based on Factory Mutual experience in the period 1970–75. In presenting data, the author warns that an unknown number of system operations are not reported. He also points out that there is no standard definition of effectiveness so that there is an element of subjectivity. He states as rough guidance that the performance would be considered acceptable if the system is designed to FM/NFPA standards and is operated within the design conditions and if any loss does not exceed the estimated Normal Loss Expectancy. For sprinkler systems, the number of reported events by type of system, with the proportion effective in brackets, was as follows: wet 2102 (91%); dry 650 (86%); deluge 62 (76%) and non-freeze 16.

For special protection systems, the number of events with the proportion of successes, was as follows: water spray 49 (53%); dry chemical 22 (27%); carbon dioxide 100 (51%) and steam 30 (63%).

Information on failure rates of special protection systems is limited. For carbon dioxide systems in Germany, Miller quotes a success rate of some 76–78% over a 14-year period.

For Halon systems, he states that data are even scantier, but refers to a series of 300 tests of such systems by Dupont in which in some 23% of installations significant problems were identified and corrected.

Miller also gives data on the effectiveness of other active protection systems, namely: gas analyser; smoke detector; rate-of-rise detector; spray nozzles; foam water sprinkler; high expansion foam; low expansion foam; explosion suppression; halon (fire protection); halon (explosion suppression); vaporizing liquid; inert gas; in-rack sprinklers; standpipes. For only three of these is the number of reported events 10 or more, the numbers, with the number of successes, being as follows: gas analyser 13 (2); high expansion foam 10 (3) and low expansion foam 12 (5).

Reliability data of fire protection features from one boiling water reaction (BWR) sites and two pressurized water reaction (PWR) sites in German has been statistically analysed (Rowekamp and Berg, 2000). The study covers 7-year operation of BWR and 7 and 4 years of operation for the two PWRs. Failure rates and unavailability for the fire protection equipment were estimated and presented in Table A14.39.

A14.22 Emergency Shut-down Systems

Data on emergency isolation valves are given in Section A14.16.

For offshore emergency shutdown systems, failure data are given by OREDA (1997). It includes emergency shutdown valves as follows (all failure modes included):

	Population	No. of failures	Mean failure rate (failures/10^6 h)
Ball			
5.1–10.0 in.	14	29 (1)	61.61
10.1–20.0 in	13	15 (3)	32.93
20.1–30.0 in	2	0 (0)	6.04
Butterfly	1	1 (1)	28.54
Gate			
1.1–5.0 in.	39	24 (11)	20.13
5.1–10.0 in.	481	245 (105)	26.50
Plug	2	7 (1)	99.89

The values for number of failures in brackets refer to failure to close or to failure involving some degree of internal leakage.

Table A14.39 *Plant specific reliability data estimated for active fire protection features in a German BWR reference plant (Rowekamp and Berg, 2000; reproduced with permission from Carl Hanser Verlag)*

	BWR plant			PWR plant 1			PWR plant 2		
	Inspection period	Failure rate [1/h]	Unavailability	Inspection period	Failure rate [1/h]	Unavailability	Inspection period	Failure rate [1/h]	Unavailability
Fire alarm boards									
Detection drawers	3m, 1a	6.7×10^{-8}	1.2×10^{-4}	3m	3.4×10^{-8}	7.4×10^{-5}	1a	1.3×10^{-6}	1.1×10^{-2}
Detection lines	3m, 1a	2.3×10^{-8}	4.0×10^{-5}	3m	2.0×10^{-8}	4.3×10^{-5}	1a	1.6×10^{-8}	1.4×10^{-4}
Fire detectors									
Automatic	1a	1.4×10^{-7}	1.3×10^{-3}	1a	4.8×10^{-8}	4.2×10^{-4}	1a	4.0×10^{-9}	3.5×10^{-5}
Press button	1a	1.1×10^{-7}	9.4×10^{-4}	1a	3.8×10^{-8}	3.3×10^{-4}	1a	3.8×10^{-8}	3.3×10^{-4}
Fire dampers	3m, 1a, 3a	2.3×10^{-6}	6.7×10^{-3}	6m	6.6×10^{-7}	2.9×10^{-3}	1a	1.4×10^{-7}	1.2×10^{-4}
Fire doors	1a	4.9×10^{-7}	4.3×10^{-3}	1a	7.1×10^{-7}	6.3×10^{-3}	3m, 1a	1.1×10^{-6}	2.5×10^{-3}
Inergen gas extinguishing system	–	–	–	–	–	–	6m	1.5×10^{-5}	5.8×10^{-2}
Dry sprinkler extinguishing system (total failure)	6m, 1a, 5a	2.3×10^{-7}	9.9×10^{-4}	3m, 1a	1.5×10^{-7}	3.2×10^{-4}	6m, 5a	1.6×10^{-7}	6.5×10^{-4}
Dry sprinkler extinguishing system (automatic actuation failure only)	6m, 1a, 5a	4.1×10^{-6}	1.8×10^{-2}	3m, 1a	1.3×10^{-7}	2.9×10^{-2}	6m	5.8×10^{-7}	2.5×10^{-3}
Wet sprinkler extinguishing system	6m, 1a, 5a	7.2×10^{-8}	3.2×10^{-4}	–	–	–	–	–	–
Gas extinguishing system (CO_2)	6m	2.1×10^{-6}	9.2×10^{-3}	–	–	–	–	–	–
Stationary fire pumps	1m, 1a	2.1×10^{-6}	1.4×10^{-3}	1m, 1a	1.4×10^{-6}	8.5×10^{-4}	6m	3.8×10^{-6}	1.6×10^{-5}
Wall hydrants	6m, 1a	4.4×10^{-8}	1.9×10^{-4}	1a	8.5×10^{-7}	7.4×10^{-3}	1a	1.1×10^{-7}	9.5×10^{-4}

For offshore platforms in the Gulf of Mexico, Forsth (1983) has reported that in 12 cases where the emergency shutdown system was mentioned in the fire or explosion incident report, the system operated properly in 11 but failed in one.

For programmable logic controllers used in the emergency shutdown systems, Mitchell and Williams (1993) analysed failure data from natural gas compressor stations. The failure rate of PLC (microprocessor failures only) was estimated to be 0.072 failures/PLC year while the overall PLC failure (all causes) was estimated 0.32 failures/PLC year.

A14.23 Utility Systems

A14.23.1 Electrical power

The failure rate of the outside power supply varies with the country concerned. In the United Kingdom, the National Grid system gives a high reliability supply and the failure rate is relatively low. In addition, chemical works often have their own power station. The failure rate of the power supply to a plant should normally be determined for the particular works.

For the United Status, the *Rasmussen Report* (figure III 6–5) gives data for outage times following a transmission line failure, as shown in Table A14.40.

A detailed study of the power supply to a plant making RDX explosive, on which it is critical that agitation should not stop, has been described by Ketron (1980). His analysis of the distribution of power failures by duration may be summarized as shown in Table A14.41. The dates quoted in the table run from December 1968 to January 1974 .

A study of probability of off-site power outage versus duration at three nuclear plants is published by Arul *et al.* (2003) with the probability of 1-h off-site power outage between 0.1 and 1.0.

Table A14.40   *Outage times of electrical power supply following a transmission line failure (after Atomic Energy Commission, 1975)*

Outage time (%)	Proportion of outages (%)
< 0.01	1.1
0.01–0.032	6.1
0.032–0.1	18.7
0.1–0.32	37.9
0.32–1.0	12.6
1.0–3.2	11.7
3.2–10	8.4

Table A14.41   *Duration of power failures in an explosives plant (after Ketron, 1980; reproduced by permission of the American Institute of Chemical Engineers)*

Duration of power failure t (min)	No. of incidents
Momentary	13
t < 1	3
1 < t < 10	2
10 < t < 60	4
t > 60	6
Total	28

Literature failure rates of equipment used in power supply systems given by Ketron are shown in Table A14.42. Weather-related features of power supplies have been considered by Jarrett (1983).

A14.23.2 Diesel generators

Information on the reliability of diesel generators is given in the *Raosmussen Report* (AEC, 1975). The study gives a probability of failing to start on demand of 3×10^{-2}.

Thus, if there are two diesel generators, but only one is required to provide the full emergency load, it might be calculated that the probability of both generators failing to start on demand is 9×10^{-4}. In fact, however, startup is treated in the study (p. III-72) as a single event that may trip both units. The probability of both generators failing to start on demand is assessed as 10^{-2}. The repair time for a diesel generator is given in the study (p. III-55/56) as a mean time of 21 h with a range of times of 2–300 h.

A survey of emergency generating equipment over the period 1977–82 has been reported by R. Stevens (1983). Much of the equipment was found to be in appalling condition. The failure modes are shown in Table A14.43.

Twenty-six years' operational data of 11 diesel generators of 125–250 kV A at Chalk River Nuclear Laboratory have been analysed presented by Winfield (1988), which gave a mean failure to start probability of 4.4×10^{-2} and failure to run probability between 5.6×10^{-3} and 1.2×10^{-2}.

Table A14.42   *Failure rates of electrical power supply equipment (after Ketron, 1980; reproduced by permission of the American Institute of Chemical Engineers)*

	Failure rate[a] (failures/d)
Diesel engine	0.008
Electric generator	0.0014
Electric motor	0.0014
Steam turbine	0.000057
Solenoid valves	0.0000063
Pneumatic valve	0.00011
Globe valve	0.0034

[a] Literature values.

Table A14.43   *Failure modes of emergency engines/ generators (after B. Stevens, 1983; reproduced by permission of the American Institute of Chemical Engineers)*

	No. of cases	Proportion (%)
Cracking/overheating	36	26.1
Cracking/freezing	13	9.4
Mechanical breakage	6	4.4
Bearings and journals/scoring	18	13.0
Engine block/breakage	14	10.1
Pistons/breakage	9	6.5
Valves/breakage	7	5.1
General mechanical	20	14.5
Piston rings/breakage	7	5.1
Crankshaft/cracking–breaking	8	5.8
Total	138	100.0

Table A14.44 *Mean probability of failure of emergency diesel generators at 63 nuclear plants from 1988–91 (Vesely, Uryasev, and Samanta,1994; reproduced with permission from Elsevier)*

	For individual diesels	Over station	Diesels or stations
Failure to start	5.0×10^{-3}	5.2×10^{-5}	5.0×10^{-3}
Failure to load-run	9.6×10^{-3}	9.5×10^{-5}	9.3×10^{-3}

A total of 195 emergency diesel generators at 63 operating nuclear plants have been analysed from 1988 to 1991 (Vesely, Uryasev and Samanta, 1994). The mean probability of failure to start and failure to load-run is given in Table A14.44. Load-run failure is defined as diesel generator starts but does not run successfully. Estimation of standard deviation and variance is also provided in their paper.

For offshore diesel-driven pumps and emergency power generators, failure data are given by OREDA (1997) as follows.

	Population	No. of failures	Mean failure rate (per 10^6 h)	Failure to start on demand
Main power				
1–100 kW	2	82	731.62	–
100–1000 kW	8	174	479.80	–
Emergency power	7	142	685.44	0.018

A14.23.3 Instrument air
An estimate of the failure rate of the instrument air supply has been given Lawley and Kletz (1975). They give: failure frequency of instrument air supply = 0.05 failures/year.

This evidently refers to the instrument air supply rather than to the connections from the air supply to the instrument, for which separate failure rates are quoted.

A14.23.4 Cooling water
The arrangements for the supply of cooling water in a works vary somewhat. Generally, there is a works cooling water supply system, but use may also be made of other sources of supply such as wells.

A typical estimate of the failure rate of the cooling water supply of about 0.1 failures/year. This is for the supply itself and does not include failure of equipment such as cooling water pumps supplying a particular plant.

A14.23.5 Steam
Information on the failure rates of components of steam supply systems has been given by Coltharp *et al.* (1978), who carried out a study on the steam system of an automobile factory. Some of the data given by these authors is shown in Table A14.45.

A14.24 LNG Plants

A survey of events on LNG plants has been reported by Welker and Schorr (1979). The data were obtained on 25

Table A14.45 *Failure rates of some components of a steam supply system (after Coltharp et al., 1979; reproduced by permission of the American Institute of Chemical Engineers)*

Equipment	Failure rate[a] (failures/10^6 h)
Drives for spreader stoker:	
Electric drive	0.3
Steam drive	50
Boiler feedwater pump	0.9
Condensate collection and return	10
Waterwall tubes	57
Steam generating tubes	0.3
Superheater	0.4
Air preheater	1.1
Fans	
Overfire air	57
Induced draft	1.1
Drives for fans	
(forced draft, overfire air,	
induced draft)	
Electric drive	2
Steam drive	3
Ash conveyor	10

[a] Sources of data are Edison Electric Institute (EEI), Hartford Steam Boiler (HSE) and automobile company.

LNG peak-shaving plants in the United States ranging in age from 10 years to a few months. The survey covered nearly 35,000 h of vaporization experience, more than 400,000 h of liquefaction experience and more than 1.5 million hours of tank storage, as well as nearly 1.5 billion ft-h of pipe use and more than 1.3 million manhours of operating time.

The authors define a minor failure as one which results in an unscheduled shutdown of operating equipment where the repair period is less than 24 h, a major failure as one where the shutdown is more than 24 h and a safety-related failure as one which results in a fire or a large leak of gas or liquid.

Major leaks of gas are defined as those which could result in a release of the order of 100,000 ft^3 of gas. Major leaks of liquid involved primarily vaporizer tubes or pumps, with one major leak due to flange gasket failure.

The fires reported are confined to those involving LNG as gas or liquid. About half the fires were minor and half major. Vaporizers appeared to have a high risk of fire, because they have a relatively high failure rate and are located near fired equipment.

Some failure and event rates given in this survey are shown in Table A14.46.

A14.25 Leaks

A14.25.1 Leak sources
Information on the distribution of leak sources is available in several different classifications. The distribution of the place of origin for large fires in Great Britain 1971–73 has been given in Table 2.11.

The distribution of leak sources for process industry accidents reported to the HSE in the year 1987–88 has

Table A14.46 *Failure and event rates on LNG plants (after Welker and Schorr, 1979; reproduced by permission of the American Gas Association)*

A Failure rates

Plant section or equipment	MTBF (h)			
	Major failure	*Minor failure*	*Safety-related failure*	*Total failures*
Gas pretreatment	20,000	3000	$> 350,000^{a}$	3000
Liquefaction	6500	2500	$> 420,000^{a}$	1800
LNG vaporizers	8000	700	15,000	700
Compressors	3000	900	$> 2 \times 10^{6a}$	700
LNG pumps	$> 35,000$	3500	$> 35,000$	3500
Cryogenic valves	$> 4 \times 10^{7a}$	2×10^{7}	$> 4 \times 10^{7a}$	2×10^{7}
Controls	15×10^{6}	700,000		

B Other events

	MTBF (h)
LNG tanks:	
Gas leaks	$> 1.5 \times 10^{6}$ (no leaks)
Liquid leaks	$> 1.5 \times 10^{6}$ (no leaks)
Cold spots	100,000
Pipelines (per ft):	
LNG pipelines	$> 1.5 \times 10^{9}$ (no failures)[b]
Pipe insulation	$> 1.5 \times 10^{9}$ (no failures)
Fire water mains	$> 1.6 \times 10^{9}$
Hazard detection sensors:	
Gas	$100,000^{c}$
Radiation	$350,000^{c}$
High temperature	$> 4 \times 10^{6}$ (no failures)
Smoke	1.4×10^{6}
Emergency systems:	
Water hydrants, monitors	4×10^{6}
Halon systems	100,000
Dry chemical systems	3×10^{6}
Human errors, leaks and fires:	
Human error incidents	50,000
Major gas leaks	300,000
Major liquid leaks	150,000
Major fires	$200,000^{d}$

[a] No failures.
[b] Small leaks from gaskets not included.
[c] Excluding false alarms.
[d] Leaks only.

been analysed by A.W. Cox, Ang and Lees (1990). Separate analyses are given for normal, closed process plant and for plant and activities with open surfaces, etc., as shown in Tables A14.47 and A14.48, respectively.

For vapour cloud explosions the distribution of leak sources has been obtained by A.W. Cox, Ang and Lees (1990) from the case histories listed by Davenport (1977, 1983) as shown in Table A14.49.

For fires and explosions on offshore installations in the Gulf of Mexico and in the Norwegian North Sea the distribution of leak sources has been obtained by A.W. Cox, Ang and Lees (1990) from data given by Forsth (1981a,b) as shown in Tables A14.50 and A14.51.

Based on the API analysis of the SCAQMD refinery screening data, 90% of reducible fugitive emissions come from only about 0.13% of the piping components (API, 1997; Taback, Siegell, Martino, and Ritter, 2000).

A14.25.2 Leak frequency
Estimates of the frequency of leaks on ammonia installations and carriers have been made by Baldock (1980) and are given in Table A14.52.

In late 1987, 39 chemical plants and EPA jointly conduct a fugitive emission study including all domestic producers of ethylene oxide, phosgene and acrolein. Around 100,000 components were examined between 1988–1990, and the result is presented in Berglund (1992), and leak rate is also found to be a function of monitoring frequency and maintenance procedure.

Table A14.47   Leak sources on closed process plant
(after A.W. Cox, Ang and Lees, 1990; reproduced by
permission of the Institution of Chemical Engineers)

	No. of incidents	Proportion of incidents (%)
Plant	19	22.1
Reactor	8	9.3
Vessel	10	11.6
Tank	7	8.1
Heat exchanger	4	4.7
Vaporizer	2	2.3
Pump	13	15.1
Pipework	17	19.8
Hose	6	7.0
Total	86	100.0

Table A14.48   Leak sources for plants and activities
with open surfaces, etc. (after A.W. Cox, Ang and Lees,
1990; reproduced by permission of the Institution of
Chemical Engineers)

	No. of incidents	Proportion of incidents (%)
Solvent evaporating		
Oven	4	2.9
Spray booth	15	10.8
Small container	31	22.3
Cleaning/degreasing		
Process	15	10.8
Tanker/mobile plant	33	23.7
Other	41	29.5
Total	139	100.0

Table A14.49   Leak sources in vapour cloud explosion
incidents for 1962–1982 (after A.W. Cox, Ang and Lees,
1990; reproduced by permission of the Institution of
Chemical Engineers)

Leak source	No. of incidents	Proportion of incidents (%)
Reactor – reaction	1	2.85
Vessel – explosion	1	2.85
Vessel – rupture	1	2.85
Tanks – reaction	1	2.85
Tanks – overfilling, frothover	1	2.85
Tanks – refrigerated storage, failure	2	5.7
Pipe	9	25.8
Flange	2	5.7
Other fittings	7	20.0
Hose	1	2.85
Valves	3	8.6
Sight glass	1	2.85
Pumps	1	2.85
Flare	2	5.7
Valve opened	1	2.85
Venting	1	2.85
Total	35	100.0

Table A14.50   Leak sources for fire and explosions in
the Gulf of Mexico (after Forsth, 1981b; reproduced by
permission of Det Norske Veritas)

Leak source	No. of incidents	Proportion of incidents (%)
Tanks, vessels, drains, sumps, pans, pits	18	13.7
Holes, cracks	6	4.6
Flanges, unions	26	19.8
Hoses	4	3.1
Nipples	3	2.3
Valves	16	12.2
Exhaust	8	6.1
Other	47	35.9
Unknown	3	2.3
Total	131	100.0

Table A14.51   Leak sources for fires and explosions
in the Norwegian North Sea (after Forsth 1981a;
reproduced by permission of Det Norske Veritas)

Leak source	No. of incidents	Proportion of incidents (%)
Tanks, vessels, drains, sumps, etc.	10	7
Ruptures, holes, cracks	4	3
Flanges, unions	7	5
Valves, vents	8	6
Other	30	21
Unknown	48	36

A14.26 Ignition

A14.26.1 Ignition sources

There is a small amount of information on the distribution
of ignition sources, much of it for offshore installations.

A study by the Fire Protection Association (1974) of ignition sources for large fires in the chemical and petroleum
industries in Great Britain in 1971–73 gave the data shown
in Table 2.11. There were 79 fires of which 23 were solid and
10 unknown, the other 46 being gas, vapour or liquid.

Ignition sources for process industry fire and explosions
reported to the HSE in the year 1987–88 have been analyzed by A.W. Cox, Ang and Lees (1990). Separate analyses
are given for normal, closed process plants and for plants
and activities with open surfaces, etc. The former sets
have been given in Table 16.46 and the latter are shown in
Table A14.53.

Offshore ignition sources have been given for fires and
explosions in the Gulf of Mexico (GoM) and in the Norwegian North Sea (NNS) in reports by workers at Det Norske
Veritas (Sofyanos, 1981; Forsth, 1981a,b, 1983).

Forsth (1983) has given the data shown in Table 16.47 for
ignition sources in these two locations. The number of
accidents considered was for the GoM was 326 over the
period 1956–81 and for the NNS 133 over an unspecified
period.

Table A14.52 *Estimated frequency of releases on ammonia installations and carriers (after Baldock, 1980; reproduced by permission of the American Institution of Chemical Engineers)*

Incident	No. of incidents	Estimated frequency
Major failure of storage vessel	2	1 in 6×10^4 vessel-year
Major release from storage vessel	1	1 in 10^4 storage area-year
Serious release on plant	12	1 in 2000 plant-year
Release on refrigerator plant	15	1 in 10^5 plant-year
Release at transfer point		
Flexible hose failure	11	1 in 1000 transfer point-year
Movement while still connected		
Major release	3	1 in 4000 transfer point-year
Other release	8	1 in 1500 transfer point-year
Major release in transport		
Road	6	1 in 2000 tanker-year
Rail	18	1 in 3000 tanker-year
Pipeline	8	1 in 3000 mile-year
Sea	1	1 in 200 ship-year
Total	79	

Table A14.53 *Ignition sources for plants and activities with open surfaces, etc. (after A.W. Cox, Ang and Lees, 1990; reproduced by permission of the Institution of Chemical Engineers)*

	No. of incidents	Proportion of incidents (%)
Flames: general	27	19.4
LPG fired equipment	2	1.4
Hot surfaces	20	14.4
Friction	11	7.9
Electrical	29	21.0
Hot particles	–	–
Static electricity	10	7.2
Smoking	17	12.2
Auto-ignition	2	1.4
Unknown	21	15.1
Total	139	100.0

Table A14.54 *Ignition of blowouts in the Norwegian North Sea (after Dahl et al. 1983; reproduced by permission of Det Norske Veritas)*

	No. of incidents	Proportion of incidents (%)
Gas:		
No ignition	81	70
Fire	23	20
Explosion	12	10
Subtotal	116	100
Oil:		
No ignition	11	92
Fire	1	8
Explosion	0	0
Subtotal	12	100
Oil and gas:		
No ignition	13	57
Fire	10	43
Explosion	0	0
Subtotal	23	100
Fire	4	
Explosion	1	
Other	15	
Total	171	

Table A14.55 *Number of fires in US petroleum industry 1982–1985 (API, 1983, 1984, and 1985)*

Installations	No. of fires			
	1982	1983	1984	1985
Exploration, production, drilling			68	107
Gas processing			28	19
Exploration, production, drilling and gas processing, not separated			25	0
Total of above	163	182	121	126
Offshore portion only			13	6
Refining	201	173	142	104
Chemical operations	61	70		
Petrochemicals			33	20

Table A14.56 *Number of fires in storage tanks in US refineries 1982–85 (API, 1983, 1984 and 1985)*

No. of fires	Type of tank			
	1982	1983	1984	1985
Floating roof	6	6	10	3
Cone roof	9	6	13	1
Dome roof	0	1	0	0

A14.26.2 Ignition probability
A discussion of the probability of ignition of gas and liquid releases is given in Chapter 16. The account there utilizes data on blowouts on offshore installations given by Dahl et c¹. (1983). A fuller tabulation of these data is shown in Table A14.54.

A14.27 Explosion following ignition

A14.27.1 Explosion probability
The discussion of ignition in Chapter 16 also covers for gas releases the probability of explosion, given ignition.

Table A14.57 *Specific origin of 131 accidents in the transportation of natural gas (Montiel et al., 1996; 2reproduced with permission from Elsevier)*

Origin	No. of accidents	Percentage
Piping	127	96.9
Pumps/compressors	2	1.5
Rail tank	1	0.8
Substation	1	0.8

Table A14.58 *Causes of 131 accidents in the transportation of natural gas (Montiel et al., 1996; reproduced with permission from Elsevier)*

Cause	No. of accidents	Percentage of known[a]
Mechanical failure	39	43.3
Impact failure	37	41.1
Human error	32	35.6
External events	14	15.6
Service failure	1	1.1

[a] Of the 131 accidents, causes of 90 were known.

A14.28 Fires

A14.28.1 Process plant fires
A survey of the frequency of fires in industry in Britain has been reported by Rutstein and Clarke (1979). For the chemical and allied industries (Standard Industrial Classification 5) they correlate the probability of fire per year with the floor space of the building using a function of the form

$$P = aB^c \qquad [A14.28.1]$$

where B is the floor space (m^2), P the probability of fire over 1 year, a is a constant and c an index. They give the following data:

	Probability of fire over 1 year	
	All buildings	Process buildings
Probability function	$P = 0.017B^{0.27}$	$P = 0.069B^{0.46}$
Probability for 1500 m^2 building	0.12	0.21

A survey of the major safety incidents has been conducted by Williams, G.P. (1999). Eighty-nine ammonia plants during the period 1994–96 have one fire every 14.2 months on average, which is slightly better then previously reported value of 14.6 months.

A14.28.2 Petroleum industry fires
Information on the frequency of fires in the petroleum industry in the United States is given in the annual series Reported Fire Losses in the Petroleum Industry by the API. The fires reported are those involving losses greater than $2500. The fires losses for the period 1982–85 are shown in Table A14.55.

A14.28.3 Refinery fires
Data on the frequency of refinery fires are given in Table A14.57. The API fire loss reports also give a breakdown of the fire losses. For 1985 the sizes of fire by loss were

Size of loss (1000$)	No. of fires
2.5–100	65
100–1000	37
>1000	7

The frequency of refinery fires may be estimated from these data and from the number of refineries given in Section A14.11. The *First Canvey Report* gives frequency of fire in a refinery = 0.1 fires/year

A14.28.4 Pump fires
Estimates of the frequency of pump fires have been given by Kletz (1971) and are shown in Table A16.55.

N.M. Wallace and David (1985) have described a study in Esso on mechanical seals, in which losses due to fires from pump failures were apportioned as follows:

Seal failure	54%
Bearing or shaft failure	36%
Unknown	10%

A14.28.5 Storage tank fires
Information on the frequency of fires on storage tanks in the petroleum industry in the United States is given in the annual series *Reported Fire Losses in the Petroleum Industry* by the API. The fires reported are those involving losses greater than $2500. The storage tank fire losses for the period 1982–85 are shown in Table A14.56.

For fires in fixed roof storage tanks for hydrocarbons Kletz (1971) states that based on data from more than over 500 tanks over a period of 20 years the frequency of a tank fire or explosion is once in 883 tank-year.

His estimate of the factor by which the frequency of fire or explosion may be reduced by the use of inerting is 10.

A14.29 Explosions

A14.29.1 Furnace explosions
Data on the distribution of causes of furnace explosions has been given by Ostroot (1972). The numbers of incidents listed are

Cause of explosion	Furnace firing	
	Gas	Oil
Inadequate purge	55	12
Delayed ignition	42	81
Incorrect fuel–air ratio	19	5

In the discussion to the paper Kletz (1972c) said that his company reckoned on a frequency of explosion of 1 in 25 furnace-year. Ostroot quoted for his company a figure of about 1 in 100, or even 1 in 1000 furnace-year.

A14.29.2 Vapour cloud explosions

Data on vapour cloud explosions have been given by Davenport (1977, 1983).

For vapour cloud explosions at fixed installations Davenport (1983) records some 35 cases over the period 1962–82 inclusive. The location of these explosions is

United States	19
Western Europe	10
Other	6

If it is assumed that there were at risk in the United States and Western Europe over this period some 10,000 chemical/petrochemical plants, major refinery units, LPG storages and natural gas plants, then

Frequency of vapour cloud explosion $= 29/(21 \times 10,000)$
$= 1.4 \times 10^{-4}$ explosions/year

Information on the distribution of leak sources in vapor cloud explosions is given in Table A14.49.

Further data on vapour cloud explosions are given in Chapter 17.

A14.30 Transport

Failure data for transport are given in Chapter 23 and Appendix 17.

A total of 185 accidents involving natural gas (gaseous) were surveyed with 179 from the Major Hazard Incident Data Service database (Montiel et al., 1996), of which 131 (70.8%) is transportation related. These 131 accidents during transportation are further analysed, and Tables A14.57 and A14.58 give a summary of their findings.

A14.31 External Events

A14.31.1 Aircraft crash

Information on the probability of a potentially damaging accident due to an aircraft crash at various reactor sites is given in the *Rasmussen Report* (table III 6–4). The highest probability quoted is 1×10^{-6}/year for a crash at a site located 5 miles from an airport with 40,000 air carrier and 40,000 naval flight movements per year. The assumed target area is 0.01 mile^2 for larger aircraft and 0.005 mile^2 for smaller ones.

The impact of the aircraft would not necessarily cause damage within the containment. In fact the estimated probability that such damage would occur given impact is less than 1 in 100.

Other accounts of the risk of aircraft crash are those of D.W. Phillips (1981 SRD R198) and Marriott (1987).

A14.32 Notation

Section A14.23

a	constant
B	floor space (m^2)
c	index
P	probability of fire in 1 year

Section A14.27

A, B, C	constants
D	severity parameter
p	pressure (psi)
t	temperature (°C)
z	hazard rate
λ	failure rate

Subscript

o	base case

Appendix

Earthquakes

15

Contents

An earthquake is one of the principal natural hazards from which process plants world-wide are at risk. Some account is therefore necessary of the seismic design of plants and the assessment of seismic hazard to plants.

Accounts of earthquakes are given in *Seismicity of the Earth and Associated Phenomena* by Gutenberg and Richter (1954), *Elementary Seismology* by Richter (1958), *Introduction to Seismology* by Bath (1979), *Earthquakes and the Urban Environment* by Berlin (1980), *Earthquakes* by Eiby (1980), *An Introduction to the Theory of Seismology* by Bullen and Bolt (1985) and *Earthquakes* by Bolt (1988). Treatments of earthquake engineering, most of which give some coverage of seismology, include *Dynamics of Bases and Foundations* by Barkan (1962), *Earthquake Engineering* by Wiegel (1970), *Fundamentals of Earthquake Engineering* by Newmark and Rosenbleuth (1971), *Earthquake Resistant Design* by Dowrick (1977, 1987), *Ground Motion and Engineering Seismology* by Cakmak (1987) and *Manual of Seismic Design* by Stratta (1987). UK conditions are dealt with in *Earthquake Engineering in Britain* by the Institution of Civil Engineers (1985) and by Lilwall (1976) and Alderson (1982 SRD R246, 1985).

It is proper to recognize also the large amount of work done on seismicity and seismic engineering in Japan.

Earthquake engineering is a specialist discipline. The account given here is limited to an elementary introduction.

Selected references on earthquakes, earthquake-resistant design and earthquake hazard assessment are given in Table A15.1

A15.1 Earthquake Geophysics

A15.1.1 Earth structure
The structure of the earth is illustrated in Figure A15.1 and is approximately as follows: a crust 30 km thick, a mantle 2900 km thick and a core of 3470 km radius, giving a total radius of 6370 km. The core has an inner core of 1400 km radius.

The crust exhibits two layers. In the lower of these layers the velocity of seismic waves is rather higher. The interface between the two is known as the Conrad discontinuity. Likewise, the mantle is divided into the upper and lower mantles. The crust and the upper mantle are termed the lithosphere. Other discontinuities are the Mohorovicic discontinuity at the boundary of the crust and the mantle, and the Gutenberg discontinuity at the boundary of the mantle and the core.

A15.1.2 Crustal strain and elastic rebound
The earth's crust has a degree of elasticity and when subject to stress due to the earth's forces it undergoes crustal strain. This property is the basis of the elastic rebound theory of Reid, which provides one explanation of the way in which earthquakes occur. Reid suggests that 'the crust,

Table A15.1 *Selected references on earthquakes, earthquake-resistant design and earthquake hazard assessment (see also Tables 9.1 and 10.2)*

California Inst. Technol. (n.d.); NRC (Appendix 28 Seismic Events); Mercalli (1902); Wrinch and Jeffreys (1923); Milne (1939); Nordquist (1945); Housner (1947); Dix (1952); EERI (1952); Gutenberg and Richter (1954,

1956, 1965); Neumann (1954); Jenkins and Oakshott (1955); Gutenberg (1956); Richter (1958); Bullen (1963); Evison (1963); Esteva and Rosenblueth (1964); Matuzawa (1964); C.R. Allen (1967); Esteva (1967, 1974); Jennings, Housner and Tsai (1968); Karnik (1968, 1991); Rothe (1969); Stacey (1969); P.O. Marshall (1970); NEIC (1970); Tucker, Cook *et al.* (1970); Anon. (1971e); Greensfelder (1971); Nat. Ocean Atmos. Admin (1972); Bath (1973, 1979); Donovan (1973); Tank (1973); Barbreau, Ferrieux and Mohammadioun (1974); Lomnitz (1974); Hsieh, Okrent and Apostolakis (1975a,b); Okrent (1975); US Geol. Survey (1975); H.D. Foster (1976); H.D. Foster and Carey (1976); Hsieh and Okrent (1976); Lomnitz and Rosenblueth (1976); McGuire (1976); Bolt (1978, 1980, 1982, 1988); Lee, Okrent and Apostolakis (1979); Verney (1979); Berlin (1980); Eiby (1980); BRE (1983 BR31); Veneziano, Cornell and O'Hara (1984); Bullen and Bolt (1985); Muir Wood (1985); EPRI (1986); Mosleh and Apostolakis (1986); ASCE (1987/33, 1988/ 36); Bernreuter, Savy and Mensing (1987); Cassaro and Martinez-Romero (1987); Cermak (1987); J. Evans *et al.* (1987); Grandori, Perotti and Tagliani (1987); Kropp (1987); Nuttli and Herrman (1987); Savy, Bernreuter and Chen (1987); Stratta (1987); Van Gils (1988); EEFLT (1991).

Earthquakes in Britain: Davison (1924); Lilwall (1976); Alderson (1982 SRD R246, 1985); Instn Civil Engrs (1985); Irving (1985); Liam Finn and Atkinson (1985)

Earthquake-resistant design, earthquake hazard assessment
ASCE (Appendix 27, 28, 1979/8, 1981/12, 1983/13, 14, 16, 1984/19, 1985/24, 1986/30, 1987/32, 1990/39, 40, 1991/42, 43); ASME (Appendix 28 Pressure Vessels and Piping, 1982/12, 1984/16); ASTM (STP 450); NRC (Appendix 28 Seismic Fragility, Seismic and Dynamic Qualification, 1973, 1979a,c); EERI (1952); Barkan (1962); NTIS (1963); Benjamin (1968); Benjamin and Cornell (1970); Wiegel (1970); Newmark and Rosenbleuth (1971); BRE (1972 BR6); AEC (1973); Dept of Commerce (1973); Newmark and Hall (1973, 1978); Okamoto (1973); Struct. Engrs Association of California, Seismology Committee (1975); Blume (1976); Int. Conf. Building Officials (1976, 1991); Lomnitz and Rosenblueth (1976); McGuire (1976); Aziz (1977); Dowrick (1977, 1987); P. Arnold (1978); Sachs (1978); Alderson (1979 SRD R135, 1982 SRD R246, 1985); Eberhart (1979); T.Y. Lee, Okrent and Apostolakis (1979); Anon. (1981c); R. Davies (1982); M. Schwartz (1982a); ASCE (1984); R.P. Kennedy and Ravindra (1984); Kunar (1985); Lee and Trifunac (1985); C.R. Smith (1985); Dimming (1986); Stratta (1987); R.P. Kennedy *et al.* (1989); Goschy (1990); Casselli, Masoni and Foraboschi (1992); Ravindra (1992)

Structures: Timoshenko (1936); Flugge (1960); Warburton (1976); Pilkey and Chang (1978)

Storage tanks, vessels: Veletsos and Yang (1976); Eberhart (1979); R.P. Kennedy (1979, 1982a); Haroun and Housner (1981); R. Davies (1982); D.W. Phillips (1982); Veletsos (1984); Priestly *et al.* (1986); Veletsos and Yu Tang (1986a,b)

Fluid–structure interactions, liquid sloshing: ASME (PVP 128, 1988 PVP 145, 1989 PVP 157); ASCE (1984/18)

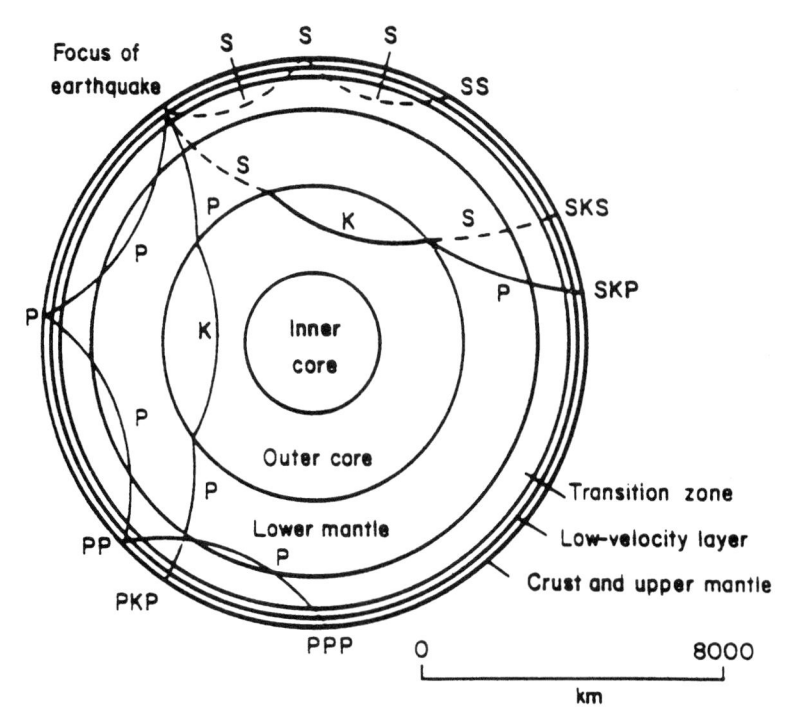

Figure A15.1 *The structure of the earth and passage of seismic waves through it (US Geological Survey, 1975)*

in many parts of the earth, is being slowly displaced, and the difference between displacements in neighbouring regions sets up elastic strains, which may become greater than the rock can endure. A rupture then takes place, and the strained rock rebounds under its own elastic stresses, until the strain is largely or wholly relieved'.

A15.1.3 Plate tectonics, faults and fault creep
In some regions earthquakes are associated with faulting of the earth, while in others there appears to be little apparent correlation between earthquakes and structural features.

Earthquakes are frequently associated with faulting of the earth. There has been debate, however, as to whether the faults have caused the earthquakes or the earthquakes the faults. The latter viewpoint has been argued by Evison (1963).

A fault in the crust may be caused by tension, compression or shear. A normal fault, in which slip occurs between one block and another, is usually considered to be the result of tension and a reverse fault, in which one block is thrust below another, the result of compression or thrust. These two types of fault are also termed dip-slip faults. The slip plane is characterized by the angle of dip, or angle of the plane to the horizontal; an angle of dip close to 90° corresponds to a nearly vertical slip plane. In a shear fault the relative movement of the two blocks is predominantly horizontal. This type is also termed a strike-slip, or transcurrent, fault. There are also oblique slip faults in which there is a combination of vertical and horizontal movement.

As a result of displacement, the ground in a zone extending out from a fault can become permanently deformed. The disturbed zone, or shatter zone, can be up to 1 km wide. The magnitude of displacements is discussed by Berlin, who mentions cases of a 14.4 m vertical displacement, or uplift, in Alaska and of a 4.9 m horizontal displacement in California; there is a 6.1 m horizontal displacement reported for the 1906 San Francisco earthquake but this is subject to some doubt.

The build-up of strain in the crust may be relaxed by a much more gradual movement of the crust, or fault creep. This process, which may be continuous or episodic, does not generate earthquake waves. An account is given by Berlin.

In some regions there are faults that are associated with recent seismic activity, or active faults. Perhaps the best known is the San Andreas fault in California. This fault is a strike-slip fault, or, viewed on a finer scale, a system of strike-slip faults. There are often lesser faults associated with the main fault as branch faults or as subsidiary faults, joining the main fault at oblique angles and sometimes in echelon along it.

There are some regions which are very much more prone to earthquakes than others. One explanation for this is given by the theory of continental drift, and of plate tectonics, according to which the earth's lithosphere is divided into a number of rigid plates that move relative to each other. Earthquakes are much more frequent at these interfaces. The Pacific rim is the most striking example, but there are a number of others.

There are a number of definitions of what constitutes an active fault. For the siting of nuclear power plants in the United States, the NRC considers a fault to be active if there has been movement at least once in last 35,000 years. Other definitions are discussed by Berlin (1980).

As stated earlier, however, earthquakes do also occur in regions where there is no apparent correlation with active fault structures.

A15.1.4 Focus and epicentre
The origin of an earthquake is termed the focus, or hypocentre, and the point on the earth's surface directly above the focus the epicentre.

Earthquakes are classified as shallow focus (focus depth <70 km), intermediate focus (depth 70–300 km) and deep focus (depth >300 km).

A15.1.5 Seismic waves
During an earthquake waves pass through the earth and impart motion to the ground. These waves are elastic in that the rocks through which they pass then return to their original shapes and positions.

There are two broad types of wave: body waves and surface waves. Body waves are classified as primary, or P, waves and secondary, or S, waves. A P wave is a longitudinal or compression wave, is analogous to a sound wave and has a velocity of some 8 km/s. An S wave is a transverse or shear wave, is analogous to electromagnetic waves and has a velocity of some 4.5 km/s. The passage of P and S waves is illustrated in Figure A15.1. Both P and S waves can undergo reflection. Reflection can occur at the earth's surface or at a boundary such as that between the mantle and the core. Refraction can occur through the core. On reflection both types of wave can be transformed to the other type, that is, a P wave to an S wave, or vice versa.

The notation used for P and S waves is illustrated in Figure A15.1. A P wave travelling directly from the focus to the point of measurement without reflection is denoted as P, one undergoing one reflection at the surface PP and one undergoing two reflections at the surface PPP. A P wave undergoing one reflection at the core is denoted as PcP. The corresponding notation for S waves is S, SS and SSS for a wave with zero, one and two surface reflections and ScS for a wave with one core reflection. There are also hybrid waves such as PS, PSS, PPS, PSP; SP, SPP, SSP, SPS; PcS and ScP.

Both P and S waves can undergo refraction through the earth's core. In the core itself there is a difference between P and S waves: an S wave cannot pass through the core. However, the wave leaving the core can be either a P wave or an S wave; in the latter case this is due to transformation at the core boundary. Hence, an S wave can enter the core, undergoing transformation to a P wave on entry and further transformation to an S wave on exit, and then travel on

as an S wave again. The principal refractions through the core are denoted as PKP, PKS, SKP and SKS (K for Kern in German). There are also more complex waves such as PKKP (refraction–reflection–refraction in the core) and PKPPKP (refraction in core–reflection at surface–refraction in core).

The distance between the focus and the point of measurement is generally expressed in degrees rather than kilometres. A degree corresponds to about 111 km.

There is a limiting angle of about 103° at which a P wave is reflected from the core. At any angle greater than this the wave is refracted through the core. The effect is to create between an angle of about 103° and one of about 142° a 'shadow zone' to which waves do not penetrate.

The other main type of wave is surface waves. Surface waves, long waves or L waves, travel round the surface of the earth rather than through the interior. They are classified as Love, or L_Q, waves and Rayleigh, or L_R, waves. Love waves have a horizontal but no vertical component, Rayleigh waves have both horizontal and vertical components. A Rayleigh wave is generally described as having a retrograde elliptical motion. Both types of wave travel more slowly than S waves, with the Rayleigh wave slower than the Love wave.

L waves may travel round the earth several times and generate ground movement with amplitudes of about 1 mm. They have a typical period of 10–20 s. They contain large amounts of energy but do relatively little damage on account of the long period. Often an earthquake is 'heard before felt'. Sounds like a low rumble can be generated by the faster P waves, while the ground shaking is due to the larger amplitude but slower S waves.

A15.1.6 Seismic measurements
An instrument for the recording of the ground motion caused by an earthquake is known as a seismometer and the record that it produces a seismogram.

Measurements may be made of any of the three main time-domain parameters: displacement, velocity and acceleration.

The measurement of ground motion is well developed and records have been obtained for the amplitude and for the acceleration of a large number of earthquakes.

Figure A15.2 gives a seismogram showing the arrival of the P, S and L waves.

There are a number of seismographic networks. A region of high seismic activity may have a number of such networks. Thus, in California, Berlin describes networks operated by the US Geological Survey, the Nuclear Regulatory Commission, the University of California and several other bodies. At the international level there is the World-wide Standard Seismographic Network (WSSN).

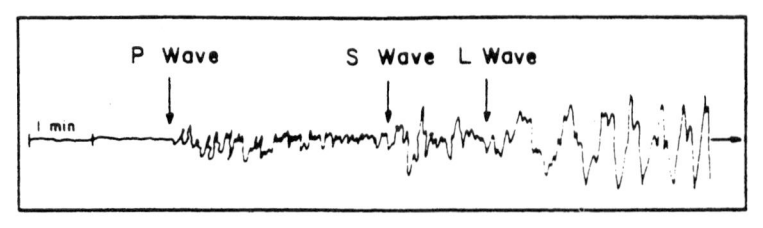

Figure A15.2   *A seismogram showing P, S and L waves (National Oceanic and Atmospheric Administration, 1972)*

A15.2 Earthquake Characterization

The quantitative characterization of earthquakes is largely in terms of magnitude and intensity scales and of empirical correlations. Earthquakes are a geographical phenomenon and hence in many cases the original correlations have been derived for a specific region, often California. This should be borne in mind in respect of the relationships quoted below.

A15.2.1 Focus and epicentre

For a particular location the distance r to the epicentre is the epicentral distance and the distance to the focus is the focal distance so that

$$R^2 = h^2 + r^2 \qquad [A15.2.1]$$

where R is the focal distance, or slant distance (km), h the distance between the focus and the epicentre, or focal depth (km) and r the epicentral distance (km). Use is often made of a modified focal distance defined as

$$R^2 = h^2 + r^2 + k^2 \qquad [A15.2.2]$$

where k is a modifying factor (km). Esteva (1967) has given an estimate of $k = 20$.

A15.2.2 Magnitude and magnitude scales

The severity of an earthquake is defined in terms of its magnitude and intensity. The more objective is the magnitude, which is a measure of the total energy in the seismic waves.

A scale for the representation of the magnitude of an earthquake was devised by Richter and magnitude is commonly quoted in terms of the value on the Richter scale. The value is a measure of the ratio of the maximum amplitude recorded for the earthquake in question to the maximum amplitude for a standard earthquake, with both measurements made on a standard seismograph located at a standard distance from the earthquake, the instrument being a Wood–Anderson seismograph of defined characteristics.

The magnitude M of an earthquake may be measured locally, in which case it is denoted M_L, or at a distant point through surface waves, when it is denoted M_S. The two differ and in order to overcome this difficulty Gutenberg introduced the concept of unified magnitude m, or m_b, which depends on the body waves. The magnitudes quoted in the literature are frequently not fully defined.

The relation of Richter for the magnitude of a local earthquake is

$$M_L = \log_{10}(A/A_O) \qquad [A15.2.3]$$

where A is the maximum amplitude (mm), A_O the maximum amplitude of the standard earthquake (mm) and M_L the local magnitude, under the standard conditions described. The value of A_O assigned to the standard earthquake is $A_O = 0.001$ mm.

The Richter scale is thus a logarithmic one and an earthquake that is one unit higher on the scale has an amplitude 10 times as great as that below it.

The Richter magnitude scale is open-ended with no lower or upper limit. The scale point 0 is an arbitrary one.

A number of equations have been developed for the surface wave magnitude M_S and the body magnitude m_b, relating these quantities to the characteristics of the seismographic record.

It is often necessary to convert one type of magnitude to another. Two widely used approximate relations are those given by Richter (1958)

$$M_S = 1.59m_b - 3.97 \qquad [A15.2.4]$$

and

$$m_b = 2.5 + 0.63M_S \qquad [A15.2.5]$$

These two magnitudes agree at a value of about 6.8; below this m_b is larger and above it M_S is larger. There is also the empirical equation of Gutenberg (1956)

$$m_b = 1.7 + 0.8M_L - 0.01M_L^2 \qquad [A15.2.6]$$

A15.2.3 Energy

The relationship between the total seismic wave energy and the surface wave magnitude was the subject of a series of studies by Gutenberg and Richter, who produced between 1936 and 1956 a number of correlations. The 1956 Gutenberg–Richter equation for energy, quoted by Gutenberg (1956), is

$$\log_{10} E = 11.8 + 1.5M_S \qquad [A15.2.7]$$

where E is the total energy (erg).

Gutenberg has also given the following relations between the total energy and the other magnitudes:

$$\log_{10} E = 9.9 + 1.9M_L - 0.024M_L^2 \qquad [A15.2.8]$$
$$\log_{10} E = 5.8 + 2.4m_b \qquad [A15.2.9]$$

An earthquake that is one unit higher on the Richter magnitude scale has an energy some 27 times as great as that below it.

A15.2.4 Frequency and return period

A correlation between the frequency and magnitude of earthquakes was obtained by Gutenberg and Richter. Their equation is given by Richter (1958) as

$$\log_{10} N = a - bM \qquad [A15.2.10]$$

where N is the frequency of earthquakes exceeding that magnitude (per year), and a and b are constants. This equation is generally referred to as the Gutenberg–Richter equation for frequency.

Various workers have used the Gutenberg–Richter equation to correlate the frequency of earthquakes for different regions and periods.

Estimates for the relationship between the magnitude and the frequency of earthquakes world-wide by Gutenberg and Richter (1940) are given below together with later estimates by the National Earthquake Information Center (NEIC) (1970).

M_L	Gutenberg and Richter description	Annual frequency	NEIC description	Annual frequency
>8.0	Great earthquakes	1	Great	1.1
7.0–7.9	Major earthquakes	10	Major	18
6.0–6.9	Destructive shocks	100	Large (destructive)	120
5.0–5.9	Damaging shocks	1,000	Moderate (damaging)	1,000
4.0–4.9	Minor strong shocks	10,000	Minor (damage slight)	6,000
3.0–3.9	Generally felt	100,000	Generally felt	49,000

A15.2.5 Intensity and intensity scales

The other main measure of the severity of an earthquake is the intensity. Intensity is a measure of the severity of an earthquake as experienced at a particular location.

Thus, an earthquake of a given magnitude may have different intensities at different locations. However, in so far as interest centres on the effects at a given location, such as a town, an intensity may be assigned for the earthquake at that location. Thus, an intensity is often quoted for an earthquake identified by town and year.

There are several scales for the representation of the intensity of an earthquake. One of the earliest was the Rossi–Forelli (RF) scale. This was modified by Mercalli to give the Mercalli scale and subsequently modified again in 1931 to give the Modified Mercalli (MM) scale. The MM scale has since undergone further revisions. The MM scale is the most widely used intensity scale. It is shown in Table A15.2

One way of characterizing a particular earthquake is by an isoseismal map showing the contours of the MM scale intensities.

The MM scale gives the intensity as a Roman numeral. For quantitative work the intensity is required as a numerical value. The practice is to use the Roman numeral directly. Thus, an earthquake of intensity between MM scale I and II has an intensity of 1.5. This is the approach described by Richter (1958, p. 140).

There are numerous empirical relationships that have been developed between intensity and other earthquake parameters. One is the equation of Gutenberg and Richter (1956) between local magnitude, intensity and epicentral intensity

$$M_L = I + I_O/3 \qquad [A15.2.11]$$

where I is the intensity and I_O the epicentral intensity.

Another is the relation of Esteva and Rosenblueth (1964) between intensity, local magnitude and focal depth

$$I = 8.16 + 1.45M_L - 2.46 \log_{10} h \qquad [A15.2.12]$$

The following relation is given by Richter (1958) between peak ground acceleration (PGA) and intensity:

$$\log_{10} a = \frac{I}{3} - \frac{1}{2} \qquad [A15.2.13]$$

where a is the PGA (cm^2/s).

A15.3 Earthquake Effects

The effects of an earthquake may be classified as direct or indirect. The direct effects include

(1)   ground shaking;
(2)   ground lateral displacement;
(3)   ground up-lift and subsidence;

and the indirect effects

(4)   ground settlement;
(5)   soil liquefaction;
(6)   slope failure:
    (a)   avalanches, landslides,
    (b)   mud slides;
(7)   floods;
(8)   tsunamis and seiches;
(9)   fires.

Of the direct effects, ground shaking usually occurs over a wide area. At faults, fault displacement results in lateral movement of the ground and surface breaks. The distance over which the surface breaks occur is variable. There may be tectonic up-lift and subsidence.

Of the indirect effects, ground settlement may occur due to compaction of unconsolidated deposits. There may be soil liquefaction, giving a 'quick condition' failure. Slope failures may occur in the form of avalanches, landslides and mud slides.

The earthquake may cause floods due to the failure of dams or levees. In certain regions there may occur tsunamis. A tsunami is a wave occurring in relatively shallow water and is caused by vertical displacement of the earth beneath the water. Seiches may also occur. A seich is a standing wave on the surface of water such as a lake.

The destruction of buildings and the rupture of gas mains tends to give rise to widespread fires.

A15.4 Earthquake Incidents

Tabulations of earthquakes are given in a number of texts. Eiby (1980) gives a chronological list; this includes the values of the magnitudes. Berlin gives tables showing earthquakes that have caused major fatalities world-wide, major fatalities in the United States and major property damage in the United States. Bolt (1988) gives tables of earthquakes world-wide, in the United States and Canada and in Central and South America; the North American list gives MM intensities. Berlin (1980) also gives lists of earthquakes with special features such as shattered earth, vertical displacement and bad ground and soil liquefaction effects.

Table A15.3 gives some earthquakes in this century that have caused major loss of life world-wide or property damage in the United States. Table A15.4 lists some earthquakes that are frequently discussed in texts on earthquakes or earthquake-resistant design as having features

Table A15.2   *The Modified Mercalli intensity scale*

Intensity	Description[a]
I	Not felt. Marginal and long-period effects of large earthquakes
II	Felt by persons at rest, on upper floors, or favourably placed
III	Felt indoors. Hanging objects swing. Vibration like passing of light trucks. Duration estimated. May not be recognized as an earthquake
IV	Hanging objects swing. Vibration like passing of heavy trucks; or sensation of a jolt like a heavy ball striking the walls. Standing motor cars rock. Windows, dishes, doors rattle. Glasses clink. Crockery clashes. In the upper range of IV, wooden walls and frame creak
V	Felt outdoors; direction estimated. Sleepers wakened. Liquids disturbed, some spilled. Small unstable objects displaced or upset. Doors swing, close, open. Shutters, pictures move. Pendulum clocks stop, start, change rate
VI	Felt by all. Many frightened and run outdoors. Persons walk unsteadily. Windows, dishes, glassware broken. Knickknacks, books, etc., off shelves. Pictures off walls. Furniture moved or overturned. Weak plaster and masonry D cracked. Small bells ring (church, school). Trees, bushes shaken (visibly, or heard to rustle)
VII	Difficult to stand. Noticed by drivers of motor cars. Hanging objects quiver. Furniture broken. Damage to masonry D, including cracks. Weak chimneys broken at roof line. Fall of plaster, loose bricks, stones, tiles, cornices (also unbraced parapets and architectural ornaments – CFR). Some cracks in masonry C. Waves on ponds; water turbid with mud. Small slides and caving in along sand or gravel banks. Large bells ring. Concrete irrigation ditches damaged
VIII	Steering of motor cars affected. Damage to masonry C; partial collapse. Some damage to masonry B; none to masonry A. Fall of stucco and some masonry walls. Twisting, fall of chimneys, factory stacks, monuments, towers, elevated tanks. Frame houses moved on foundations if not bolted down; loose panel walls thrown out. Decayed piling broken off. Branches broken from trees. Changes in flow or temperature of springs and wells. Cracks in wet ground and on steep slopes
IX	General panic. Masonry D destroyed; masonry C heavily damaged, sometimes with complete collapse; masonry B seriously damaged. (General damage to foundations – CFR.) Frame structures, if not bolted, shifted off foundations. Frames cracked. Serious damage to reservoirs. Underground pipes broken. Conspicuous cracks in ground. In alluviated areas sand and mud ejected, earthquake fountains, sand craters
X	Most masonry and frame structures destroyed with their foundations. Some well-built wooden structures and bridges destroyed. Serious damage to dams, dikes, embankments. Large landslides. Water thrown on banks of canals, rivers, lakes, etc. Sand and mud shifted horizontally on beaches and flat land. Rails bent slightly
XI	Rails bent greatly. Underground pipelines completely out of service
XII	Damage nearly total. Large rock masses displaced. Lines of sight and level distorted. Objects thrown into the air

[a] CFR, Charles F. Richter additions to the 1931 scale. Masonry A, good workmanship, mortar, and design; reinforced, especially laterally, and bound together by using steel, concrete, etc.; designed to resist lateral forces. Masonry B, good workmanship and mortar; reinforced, but not designed in detail to resist lateral forces. Masonry C, ordinary workmanship and mortar; no extreme weaknesses like failing to tie in at corners, but neither reinforced nor designed against horizontal forces. Masonry D, weak materials, such as adobe; poor mortar; low standards of workmanship; weak horizontally.
From *Elementary Seismology* by Charles F. Richter, W. H. Freeman and Company, Copyright © 1958. Reproduced by permission.

of particular interest; accounts of these earthquakes are given in the references shown in the table.

Table A15.3 gives only earthquakes that have occurred since 1920. It nevertheless includes some that have caused very large loss of life, in particular those at Kwanto in Japan in 1923 and Tangshan in China in 1976, where the death tolls were 143,000 and 250,000, respectively.

The San Francisco earthquake in 1906 was one of large magnitude (8.3) and high intensity (XI). The ground shook for some 40–60 s. Notable features were the difference in damage between buildings on good ground and those on bad ground and the outbreak of large numbers of fires. Some 700 people were killed.

The Tokyo–Yokohama earthquake in Japan in 1923, also known as the Kwanto earthquake after the province most affected, was another with large magnitude (8.3). Tokyo and Yokohama, both very densely populated, were almost completely destroyed. Fire broke out on a large scale and it was this rather than building collapse that was responsible for the high death toll of some 143,000.

These two disasters gave impetus to the development of seismological studies and seismic engineering in the United States and Japan.

The earthquake in Imperial Valley, California, in 1940, also known as El Centre, which had a magnitude of 6.9, is of particular interest to seismologists. A seismograph in El Centre recorded a PGA of some one third that of gravity. For many years this was the only record available to engineers showing clearly the ground shaking caused by a large earthquake close to its source.

Another earthquake that has been of particular interest to seismological engineers is that at Parkfield in 1966 with a magnitude of 5.6. The Parkfield earthquake was less destructive than that at El Centre in 1940, but it gave a PGA of 0.5 g compared with that of 0.33 g for El Centre.

The interest in these two earthquakes has been enhanced by a further earthquake at Imperial Valley in 1979, which was of similar size to the 1940 event, and by the fact that Parkfield has been the subject of one of the principal examples of the prediction of future earthquakes.

Table A15.3 *Some earthquakes causing major loss of life or property damage*

A   Earthquakes causing major loss of life[a]

Year	Location	Number of deaths
1920	Kansu, China	180,000
1923	Tokyo-Yokohama, Japan	143,000
1932	Kansu, China	70,000
1935	Quetta, India	60,000
1939	Chilian, Chile	30,000
1939	Erzincan, Turkey	23,000
1960	Agadir, Morocco	14,000
1962	North-western Iran	14,000
1968	North-eastern Iran	11,600
1970	Peru	66,000
1972	Managua, Nicaragua	12,000[b]
1976	Guatemala	22,000
1976	Tangshan, China	$\approx 250,000$[c]
1976	Hopei, China	655,000

B   Earthquakes causing major property damage in the United States[d]

Year	Location	Property damage ($ million)
1906	San Francisco, CA	24
		500 (fire loss)
1933	Long Beach, CA	40
1952	Kern County, CA	60
1964	Alaska and US West Coast	500
1971	San Fernando, CA	553

[a] Earthquakes since 1920 causing 10,000 or more deaths.
[b] This is Berlin's figure; Bolt gives 5000.
[c] For China in 1976 this is Bolt's entry; Berlin gives Hopei with 655,000 deaths.
[d] Earthquakes since 1900 causing $40 million or more property damage.
Sources: Berlin (1980); Bolt (1985).

In 1954 the oil fields of Kern County, California, were subject to an earthquake. The damage caused by this event to process installations is described in Section A15.5.

Several earthquakes have occurred in areas that were not regarded as major seismic risk zones. One of these occurred at Agadir in 1960. The epicentre of the earthquake was close to the town and although of relatively low magnitude (5.8) it was very destructive. Much of the town consisted of old buildings and this was a further major contributor to the destruction. The death toll is quoted as 14,000.

The earthquake at Skopje in 1963 had similar features. The area was not regarded as one of particular seismic activity. Its epicentre was close and it was highly destructive despite being of relatively low magnitude (6.0). About 1200 people were killed.

Both the Agadir and Skopje earthquakes were practically a single shock.

The earthquake at Mexico City in 1957, which had a magnitude of 7.5, was notable for the damage to buildings standing on bad ground. In locations where the soil was alluvium, there was widespread soil liquefaction. The old lake bed area of the city experienced intensities of VII, whereas at other locations the intensities were V or less.

Another earthquake in which there was extensive damage due to soil liquefaction was that at Niigata in 1964, which had a magnitude of 7.5. The Niigata earthquake gave impetus to the study of soil effects, and particularly soil liquefaction.

A large earthquake, of magnitude 8.4, occurred in Alaska in 1964. This event was notable for its long duration and wide reach, and for its destructiveness. The earth shook for about 3 min. Ground fissures occurred 725 km from the epicentre. Quite large areas suffered up-lift or subsidence. The docks at the port of Anchorage suffered large displacements. The damage caused to process installations by this event is described in Section A15.5.

The San Fernando earthquake of 1971 has several features of interest. It is a prime instance of the shattered earth effect described in Section A15.7. In localized areas the intensity was assessed as high as XII. Ground accelerations were the highest recorded at 1.25 and 0.7 g for horizontal and vertical accelerations, respectively. Some of the damage effects, particularly to lifelines, are described in Section A15.5.

As far as concerns the United Kingdom, mention should be made of the Great British Earthquake of April 22, 1884, near Colchester, to which a magnitude of 5.3 has been assigned. The PGA has been estimated as between 0.07 and 0.2 g.

A15.5 Earthquake Damage

There are available numerous accounts of earthquake damage. Most texts on seismology contain descriptions and illustrations of the damage caused by particular earthquakes. These accounts deal largely, often exclusively, with damage to buildings, which is not the prime concern here. There is rather less information available on damage to plant.

A review of seismic damage to process plant and utilities has been given by Alderson (1982 SRD R246). Information is also given by Berlin (1980).

The earthquake in Kern County near Bakersfield in 1952 (Case History A20) occurred in an oil production area. The production rate of several oilfields was affected. Many storage tanks skidded but only a few collapsed.

The most spectacular effect occurred at the Paloma Cycling Plant, where the earthquake caused two of the five large butane spheres to collapse, rupturing pipework and giving a major release. A large vapour cloud formed and was ignited at electrical transformers almost three blocks away. There was a large vapour cloud explosion, followed by a major fire.

On vessels, columns and tall heat exchangers steel foundation bolts were stretched. The top of one vessel standing 60 ft high and weighing some 100 te was estimated from the 1 in. stretch of the bolts at its base to have moved over an arc of 3 ft. There were no major failures, however, of horizontal vessels or pressure storage tanks. High pressure pipework withstood the earthquake well, although there were thermal failures in the subsequent fire; only one cold failure was reported.

At other installations damage was minimal. At one plant oil sloshing in storage tanks broke through the roof. At one pipeline station pipe displacements of 5 in. were found. Elevated water tanks with no lateral support suffered badly. Of the 12 in the area, two collapsed and seven had suffered damage to their supports.

Table A15.4 *Some earthquakes of particular interest*

Date		Magnitude	Reference
18 April 1906	San Francisco, CA	8.3	
1 Sept. 1923	Kwanto, Japan	8.3	
18 May 1940	Imperial Valley (El Centre)	6.9	Dowrick (1977)
21 July 1952	Kern County, CA	7.7	Berlin (1980); Alderson (1982)[a]
21 Dec. 1954	Eureka, CA	6.6	Alderson (1982)
22 March 1957	San Francisco, CA	5.3	Alderson (1982)
28 July 1957	Mexico City	7.5	Berlin (1980)
29 Feb. 1960	Agadir, Morocco	5.8	Eiby (1980)
22 May 1960	Arauco, Chile	8.4	Eiby (1980)
26 July 1963	Skopje	6.0	Alderson (1982)
27 March 1964	Alaska	8.4	Berlin (1980); Alderson (1982)
16 June 1964	Niigata, Japan	7.5	Dowrick (1977); Berlin (1980);
27 June 1966	Parkfleld	5.6	Dowrick (1977); V.W. Lee and Trifunac (1987); Bolt (1988)
19 Aug. 1966	Varto, Turkey	7	Berlin (1980)
9 Feb. 1967	Columbia	6.7	Alderson (1982)
31 May 1970	Chimbote, Peru	7.7	Eiby (1980)
9 Feb. 1971	San Fernando, CA	6.8	Berlin (1980)
4 Feb. 1975	Liaoning Province	7.4	Bolt (1988)
6 Sept. 1975	Lice, Turkey	6.7	Berlin (1980)
6 May 1976	Friuli, Italy	6.5	Alderson (1982)
27 July 1976	Tangshan	8.0	Bolt (1988)
13 Aug. 1978	Santa Barbara, CA	5.8	Alderson (1982)
15 Oct. 1979	Imperial Valley (El Centre)	—[b]	Bolt (1988)
2 May 1983	Coalinga, CA	6.7 (M_L)	Bolt (1988)
3 March 1985	Algarrobo, Chile	7.8	Bolt (1988)
19 Sept. 1985	Ixtapa, Mexico	8.1	Bolt (1988)

[a] Alderson (1982) = Alderson (1982 SDR R246).
[b] This earthquake is described by Bolt (1988) as 'similar' to that in Imperial Valley in 1940.

There was damage to a number of electrical power plants and electrical distribution systems.

Overall plant management in this area were earthquake-conscious. Generally, tall vessels were well anchored and pipework provided with flexibility.

Another instructive case is the earthquake at San Fernando in 1971. This was the first occasion on which a major effort was made to investigate the effects on lifeline systems.

The earthquake was particularly severe over an area of some 30 km² at San Fernando and Sylmar. Violent ground movement in this area broke gas mains and valves, but this was the only area where the gas distribution system suffered serious damage. The water supply system to San Fernando was devastated.

The earthquake disrupted power supplies to large parts of Los Angeles. There was damage to Sylmar Converter Station (40% equipment loss), Olive Switching Station (80% loss) and Sylmar Switching Station (90% loss). The restoration time for the first of these was estimated at 1.5–2 years.

At the Sylmar Central Office of the local telephone company 91 te of automatic switching equipment were destroyed, putting the telephone service out of action.

The transport system was also severely affected. Some 62 bridges were damaged, of which 25% collapsed or were severely damaged.

Other earthquakes for which Alderson gives some information on damage to plant include Eureka in 1954, San Francisco in 1957, Alaska in 1964 and Santa Barbara in 1978. It is not uncommon for there to be some rupture of water mains, often due to a combination of corrosion and pressure surge. Some rupture of gas pipes also occurs.

In the 1964 Alaska earthquake there were many failures of tanks and pipework. Tanks of all kinds moved and, if not well anchored, toppled over. Pipes failed at screwed joints and poorly guided thermal loops, as did long unsupported cast iron pipe runs.

A15.6 Ground Motion Characterization

The ground motion caused by an earthquake may be characterized in terms of the peak values of the acceleration, velocity and displacement; of the full time-domain responses of these quantities, particularly the accelerogram, or of the response spectrum. The latter is described below.

The information required for design depends on the method used. Some methods utilize the peak acceleration. The trend is, however, to use the full response in the form of an accelerogram or a response spectrum.

For a single parameter, such as the PGA, methods of prediction are available. For a full response, such as an accelerogram or response spectrum, it is necessary to select one that is judged appropriate for the site.

A15.6.1 Strong ground motion

The typical pattern of ground motion is a sudden transition from zero to maximum shaking, followed first by a period of more or less uniformly intense vibration and then by a rather gradual attenuation of this vibration.

There are, however, appreciable differences in the ground motion from different earthquakes. Newmark and Rosenblueth (1971) distinguish four types of strong ground motion: (1) practically a single shock; (2) a moderately long, extremely irregular motion; (3) a long ground motion exhibiting pronounced prevailing periods of vibration; and (4) a ground motion involving large-scale, permanent deformations of the ground.

They give the earthquakes at Agadir in 1960 and Skopje in 1963 in the first category; that at El Centre in 1940 in the second; that at Mexico City in 1964 in the third and those in Alaska and at Niigata, both in 1964, in the fourth.

A15.6.2 Peak ground motion

Methods are available for the estimation of the peak ground motion parameters. For ground acceleration one widely quoted relation is that given by Donovan (1973), derived from data for several seismic regions:

$$a = \frac{1080 \exp(0.5M)}{(R + 25)^{1.32}} \qquad [A15.6.1]$$

where a is the PGA (cm/s^2).

Esteva (1974) has used data for earthquakes in California to obtain equations for both PGA and peak ground velocity:

$$a = \frac{5600 \exp(0.8M)}{(R + 40)^2}, \quad R > 15 \qquad [A15.6.2]$$

$$v = \frac{32 \exp M}{(R + 25)^{1.7}}, \quad R > 15 \qquad [A15.6.3]$$

where v is peak ground velocity (cm/s).

For peak ground displacement Newmark and Rosen-blueth (1971) give the following approximate empirical relation

$$5 \leq \frac{ad}{v^2} \leq 15 \qquad [A15.6.4]$$

where d is peak ground displacement (cm). The lower limit of 5 and the upper limit of 15 apply to large epicentral distances, say 100 km, and to small ones, respectively.

A15.6.3 Vibration parameters

The vibration characteristics of the earthquake are also important. Correlations are available for the fundamental frequency, or period, of earthquakes. These are described by Dowrick (1977).

A15.6.4 Accelerograms

The basic seismographic measurement is the accelerogram. Profiles of velocity and displacement may be obtained from this by integration. A set of such profiles is illustrated in Figure A15.3.

There are now available accelerograms for quite a large number of real earthquakes. Collections are maintained by major seismographic institutions such as the California Institute of Technology.

Accelerograms are also available for simulated earthquakes. A set of eight such accelerograms have been

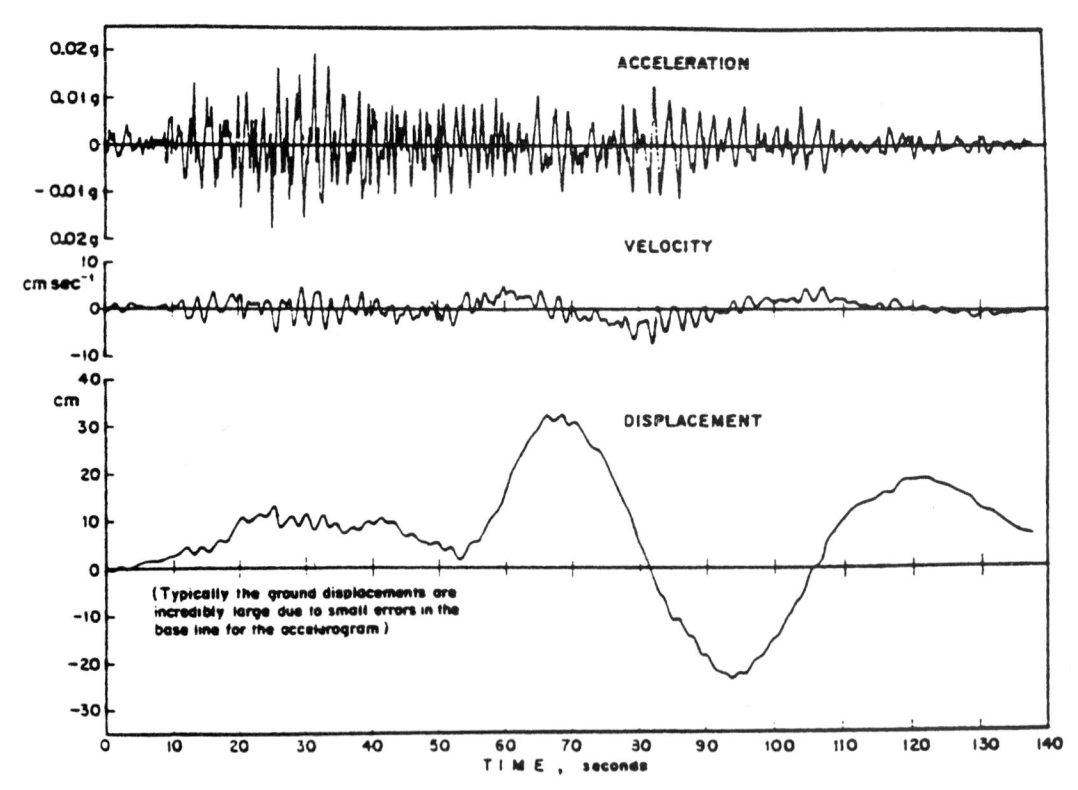

Figure A15.3   *Time histories of acceleration, velocity and displacement of the earthquake at Mexico City, 6 July 1964 (after Rosenblueth, 1996)*

described by Jennings, Housner and Tsai (1968). There exist computer codes for the generation of simulated earthquakes.

The design of earthquake-resistant structures is increasingly based on dynamic analysis. The structure is modelled using the appropriate equations of motion and this model is then perturbed by a suitable forcing function. The natural forcing function to use is an accelerogram.

A15.6.5 Response spectra

The main interest in design, however, is in the maximum values of certain responses of the structure. Many of these depend only on the natural period and the degree of damping. A plot which shows the response of one of the parameters acceleration, velocity or displacement for a given degree of damping is known as a response spectrum and that for a set of values of the damping as the response spectra.

Response spectra may be obtained from accelerograms by Fourier analysis and there are computer codes to do this.

There are response spectra available for both real and simulated earthquakes. A regulatory body may specify a standard set of response spectra for use in design.

Figure A15.4 shows a set of response spectra issued by the NRC.

A15.7 Ground, Soils and Foundations

The characteristics of the ground have profound effects on virtually all aspects of earthquake assessment and earthquake-resistant design.

A15.7.1 Ground effects

Experience of earthquakes shows that differences in the ground can result in large differences in the damage done.

According to Eiby (1980), in extreme cases this can amount to up to four degrees of intensity on the MM scale. It is therefore important to avoid siting structures on bad ground.

Bedrock is generally regarded as a sound foundation and alluvium as less satisfactory. Alluvium tends to absorb the effects of small earthquakes but to amplify those of large ones, giving rise to the effect of soil amplification.

A particular effect of bad ground is soil liquefaction, which can result in very severe damage. This is considered below.

An important class of locations where alluvium is present is old river terraces, which are often the site of ports.

As just stated, structures on the same site may suffer quite different degrees of damage due to differences in the quality of the ground. In the earthquake in 1957 in Mexico MM intensities of VII were determined for the old lake bed area of Mexico City, while other parts had intensities of V or less. The earthquake at Varto, Turkey, in 1966 destroyed some buildings in a school complex but not others, the large difference in damage being attributed to the effect of very localized differences in the ground

Where a structure straddles two types of soil, damage tends to be enhanced, and this is to be avoided.

It is not universally the case that ground motion is worse in bad ground than in bedrock. Where the rock is on a ridge there can occur a phenomenon variously known as shattered earth, exploded outcrop or churned ground. The 1971 San Fernando earthquake produced numerous instances.

The susceptibility of steep slopes to avalanches is affected by the nature of the soil. A major slope failure occurred in the earthquake at Chimbote in Northern Peru in 1970 when an avalanche with an estimated velocity of 200−400 km/h travelled 18 km, buried whole towns and killed some 20,000. However, slope failures can also occur on relatively gentle slopes. In the Alaska earthquake in 1964 slope failures occurred due to liquefaction of lenses of sand contained in clay deposits.

A15.7.2 Soil settlement and liquefaction

Some of the worst earthquake damage has been due to a quick condition failure in which liquefaction of the soil has occurred. It is important, therefore, to be able to assess susceptibility to soil liquefaction. Damage may also occur due to soil compaction and settlement short of liquefaction.

The behaviour of a soil under earthquake loading is a function of the strain magnitude, strain rate and number of cycles of loading. Some soils increase in strength under rapid cyclic loading, while others, such as saturated sands, tend to lose strength.

Broadly, gravel and clay soils are not susceptible to liquefaction and dense sands are less likely to liquefy than loose ones.

Two important characteristics of a soil bearing on liquefaction are its cohesion and its water content. Saturated cohesionless soils are prone to liquefaction. A full discussion of the effect of cohesion and saturation on soil liquefaction is given by Newmark and Rosenblueth (1971).

The susceptibility of a cohesionless soil to liquefaction depends on its water content. For a dry cohesionless soil grain size is another important factor. For a medium grain size soil experiencing vibration a critical void ratio exists below which it tends to dilate and above which it tends to reduce in volume. This effect has been modelled by Barkan (1962). Other effects which can occur are breakage in large grain soils and air entrapment and soil liquefaction in fine grain soils.

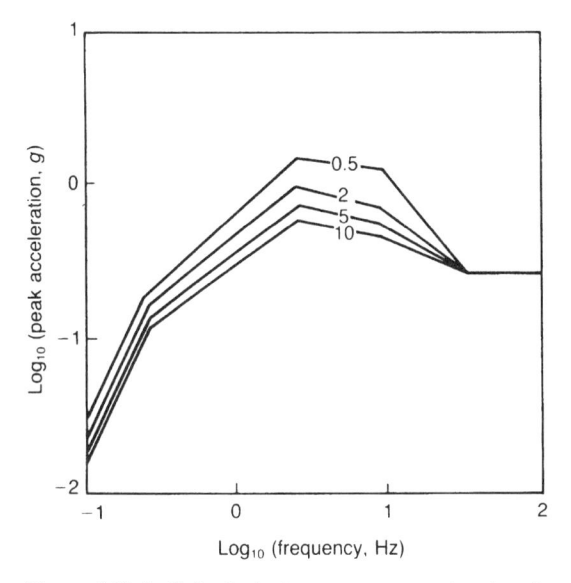

Figure A15.4 *Seismic design response spectra given in Regulatory Guide 1.60 of the Nuclear Regulatory Commission (Nuclear Regulatory Commission, 1973). The four damping levels shown are for viscous damping of 0, 5, 2, 5 and 10% of critical damping)*

A saturated cohesionless soil is most prone to liquefaction. The effect is not well understood, although it is known that relative density, confining pressure and drainage conditions are three important variables. Loose sands often liquefy after only a few loading cycles, whereas dense ones may require several hundred cycles. Conditions which prevent water from escaping promote liquefaction.

Extensive soil liquefaction occurred in the earthquake at Niigata in 1964. There was loose sand and water was not able to escape, partly due to the extent of the foundations.

With a saturated cohesive soil the strength tends to increase due to strain rate but to decrease due to cyclic loading. The net effect may be to increase or decrease the strength, depending on the nature of the soil. In a very sensitive soil a decrease in strength may set in after only a few cycles, whereas with a relatively insensitive one the strength may still be increasing after some 50 cycles.

A15.7.3 Site investigation

It is practice to carry out a site investigation to determine the characteristics of the ground. An account of the laboratory and field tests and of the information which they yield is given by Dowrick (1977). Laboratory tests on relative density and particle size distribution and field tests on groundwater conditions and penetration resistance bear on the problems of soil liquefaction and settlement, while field tests on soil distribution, layer depth and depth to bedrock, groundwater conditions and the natural period of the soil provide data for response calculations. Other tests may be performed to obtain the shear modulus and damping characteristics.

In addition to the use of drill holes to make the field tests just described, use may also be made of other techniques such as inducing and measuring vibrations.

A15.7.4 Design basis earthquake

Where dynamic analysis is to be performed, it may be necessary to modify the accelerogram to take account of the nature of the ground. The case of bedrock would seem at first sight to be the most straightforward, but this is not necessarily so, since most accelerograms are for softer soils, and there are very few for bedrock. The modification of accelerograms to take account of ground conditions is a specialist matter. An account is given by Dowrick (1977).

The ground motion is also affected by any structure placed on it. The accelerogram relevant to the site as a whole is therefore the free field accelerogram.

A15.7.5 Soil–structure interaction

It has been found by experience that there is an appreciable interaction between the soil and the structure and that it is necessary to take this into account.

The effects of bad ground have already been described. One such effect is soil amplification of the earthquake vibrations.

The effects of soil–structure interaction (SSI) are to modify the dynamics of the structure and to dissipate its vibrational energy.

In general, damage is greater if the natural period of a structure is similar to that of the soil on which it is built, for example a low, short-period building on a short-period soil deposit or a tall, long-period building on a long-period deposit. In the earthquake in 1957 at Mexico City extensive damage was done to long-period structures standing on deep (>100 m) alluvium. Another case where the natural

period of structures suffering damage appeared closely correlated with the depth of subsoil was in the earthquake at Caracas, Venezuela, in 1967.

The Uniform Building Code (UBC) in the United States, described below, originally contained a soil amplification factor that was then removed due to lack of knowledge as to what the factor should be. The 1976 version contains a site–structure resonance factor S. This factor is a function of the ratio T/T_s where T and T_s are the natural periods of the structure and soil, respectively.

A soil deposit may be characterized by the width and depth of soil overlying the bedrock. The natural period increases with the depth of the deposit. Relations are available for the determination of the natural period of horizontal vibration of a soil deposit. Dowrick (1977) gives the equation

$$T = 4H/v_s \qquad [A15.7.1]$$

where H is the depth of the soil layer (m), T the natural period of the soil layer (s) and v_s the velocity of the shear wave (m/s).

In determining soil–structure interaction use is made of soil–structure models. In the simpler models the soil layer is represented using models of the lumped mass with spring and damping forces type not dissimilar to those used for the structure itself.

A full dynamic analysis of soil–structure interaction is perhaps the most difficult task in earthquake engineering.

A15.8 Earthquake-resistant Design

The design of an earthquake-resistant structure includes the following principal steps:

(1) seismic assessment;
(2) ground assessment;
(3) selection of structural form;
(4) selection of the materials;
(5) overall design of the structure.

The first and second of these are considered in Sections A15.11 and A15.7, respectively.

A15.8.1 Structural form

It has been found by experience that certain structural forms resist earthquakes well while others do not. An account of the features of structural form that give earthquake resistance is given by Dowrick (1977). In broad terms the structure should be simple and symmetrical and not excessively elongated in plan or elevation. It should have a uniform and continuous distribution of strength. It should be designed so that horizontal members fail before vertical members.

The desirable degree of stiffness has been a matter of debate. The point is discussed by Dowrick, who states that either a stiff or a flexible structure can be made to work.

A15.8.2 Materials

The properties of materials of construction that confer good resistance to earthquakes include homogeneity, ductility and high strength/weight ratio.

Materials generally quoted as being suitable for earthquake-resistant structures include steel, *in situ* reinforced concrete and prestressed concrete.

A15.8.3 Static analysis

The traditional method of earthquake-resistant design is static force analysis, also referred to as the static, or equivalent static, design method.

The static method is based on the following relation for the effective horizontal force, or base shear, on the structure:

$$V = ma \qquad [A15.8.1]$$

where a is the acceleration (m/s^2), m mass (kg) and V horizontal force (N).

The value of the acceleration a used in static design is typically in the range $0.05-0.20$ g.

Methods for the design of structures to withstand a horizontal force in the form of a wind load are well established in civil engineering, and the approach required for the static method for earthquake resistance is not dissimilar.

Equation A15.8.1 is the basic relation traditionally used in earthquake design codes. These codes are described in Section A15.9.

A15.8.4 Dynamic analysis

It has long been recognized that the static design method is oversimplified, and a variety of dynamic design methods have been developed.

In dynamic analysis an unsteady-state model of the structure is produced and the forcing function of the design earthquake is defined. The model is then excited using this forcing function. Methods of dynamic analysis differ in the type of model used, the use of model decomposition and the domain (time, frequency) in which the forcing is applied.

The forces estimated using the more rigorous methods of dynamic analysis can be an order of magnitude greater than those obtained from the static method.

An account of dynamic analysis is given in Section A15.10.

A15.8.5 Detailed structural design

The design methods described give information on the forces to be expected at different points in the structure. The detailed design of the various elements of the structure can then be performed.

Discussions of the behaviour of structural elements and of detailed design using different materials are given by Newmark and Rosenblueth (1971) and Dowrick (1977).

A15.8.6 Detailed non-structural design

It is also necessary to design the non-structural features for seismic resistance. Much of the damage done by earthquakes is non-structural.

These features include in-fill panels and partitions; cladding and doors and windows. An account of their design is given by Dowrick (1977).

A15.8.7 Services

The seismic resistance of services is also relevant. Guidance on this aspect is given by Dowrick (1977).

For services, the specification of the seismic resistance required is of particular importance. There may be some key services whose survival in full working order is regarded as vital.

For such essential services Dowrick suggests the use of earthquake accelerations higher than those utilized in building design; the use of dynamic rather than static design methods and of the response spectrum technique and the design for elastic rather than plastic deformation.

He states that a high degree of seismic resistance can usually be obtained at relatively little extra cost.

Some services which he considers are electrical equipment, such as transformers mechanical equipment, such as boilers and pipework.

A15.8.8 Specific structures

The bulk of guidance on earthquake-resistant design relates to buildings and similar structures, but some is available for other types of structure. The types of structure relevant to process plant include columns, storage tanks and pipes.

Newmark and Rosenblueth (1971) discuss the seismic design of towers, stacks and stack-like structures; tanks and hydraulic structures and pipes. Columns are also considered by Dowrick (1977).

Accounts of particular earthquakes frequently give information on the extent of survival of, or damage to, process plant items.

A15.9 Earthquake Design Codes

Guidance on the design of earthquake-resistant structures is available in earthquake design codes. These codes have been developed particularly in the United States and Japan, and many countries have adopted or adapted the US codes.

Some earthquake design codes which have been developed in the United States include the Uniform Building Code (UBC) of the International Conference of Building Officials (ICBO); the Basic Building Code of the Building Official and Code Administration (BOCA); the Standard Building Code (SBC) of the Southern Building Code Congress International; the recommendations of the Structural Engineers Association of California (SEAOC) and the ANSI code.

The UBC appears the most widely quoted and influential. An account of the development of the code is given by Berlin (1980).

The method of design used in the code is the equivalent static method, also known as the seismic coefficient method. The account given here is based on the 1976 version of the code.

In its simplest form the basic equation of the code is

$$V = CW \qquad [A15.9.1]$$

where V is the lateral or shear force (N), W the load (N) and C a coefficient. The lateral shear force V is commonly known as the base shear coefficient.

The working equation of the code has passed through various versions. That given in the 1976 code is

$$V = ZIKCSW \qquad [A15.9.2]$$

where C is the flexibility factor; I the occupancy importance factor K the framing factor; S the site factor and Z the seismic risk zone factor.

The values to be used for the various factors are treated in the code. A discussion of these factors is given by Berlin (1980).

The flexibility factor C is a function of the natural period of the structure. There are a number of equations for the estimation of C. One principal equation is

$$C = 1/15T^{1/2} \qquad [A15.9.3]$$

where T is the fundamental period of vibration of the structure in the direction under consideration(s).

The occupancy importance factor I is 1.5 for essential facilities, 1.25 for a building where the primary occupancy is more than 300 people and 1.0 for other buildings.

The framing factor K allows for the types of structure. It has values ranging from 0.67 to 1.33. The lower value is for a structure with ductile, moment resisting frame and the higher for a structure with a box system, in which lateral forces are resisted by shear walls or braced frames. A tabulation of values is given for different types of structure.

For the seismic zone risk factor Z the United States is divided into five zones, which are defined in the code and may be characterized as follows:

Zone 0   No damage
Zone 1   Minor damage (MM intensity V and VI)
Zone 2   Moderate damage (intensity VII)
Zone 3   Major damage (intensity VIII)
Zone 4   Areas within Zone 3 close to certain major fault systems.

The values of Z for use in these zones are 3/16, 3/8, 3/4 and 1 for Zones 1, 2, 3 and 4, respectively.

An account of codes in various countries is given by Alderson (1982 SRD R246).

A15.10 Dynamic Analysis of Structures

There are a variety of methods of dynamic analysis. Features of these methods are the unsteady-state model of the structure, the use of decomposition techniques, the forcing function and the domain in which the analysis is conducted.

In the following account, dynamic analysis is described mainly in terms of a single design earthquake. It is generally recommended that the analysis be repeated for further earthquakes.

A15.10.1 Unsteady-state models of structures

Unsteady-state models of the structure are distinguished by the number of degrees of freedom (DoF). Both single degree-of-freedom (SDF) and multiple degree-of-freedom (MDF) models are used. Basic SDF and MDF models are described by Dowrick (1977).

The basic SDF model is the standard second-order system. This may be written as

$$F_I + F_D + F_S = F(t) \qquad [A15.10.1]$$

where F is the applied force, or forcing function (N), F_D the damping force (N), F_I the inertia force (N) and F_S the elastic force (N).

Substituting for the various force terms, Equation A15.10.1 may be written as

$$m\ddot{x} + c\dot{x} + kx = F(t) \qquad [A15.10.2]$$

where c is the damping coefficient (kg/s), k the resistance (kg/s^2), m the mass of the system (kg), t the time (s) and x the displacement (m). In standard form the equation becomes

$$\frac{1}{\omega_n^2}\ddot{x} + \frac{2\zeta}{\omega_n}\dot{x} + x = \frac{1}{k}F(t) \qquad [A15.10.3]$$

with

$$\zeta = (c^2/4mk)^{1/2} \qquad [A15.10.4]$$

$$\omega_n = (k/m)^{1/2} \qquad [A15.10.5]$$

where ζ is the damping factor and ω_n the natural frequency (s^{-1}). The damped frequency is

$$\omega = \omega n(1 - \zeta^2)^{1/2} \qquad [A15.10.6]$$

where ω is the frequency of damped oscillation (s^{-1}).

The SDF model is a lumped mass model. The distribution of mass may be taken into account by the use of an MDF model. Thus an n-storey structure might be represented by an MDF model in which the mass of each storey is assumed to be located at the floor of that storey. The model obtained then consists of a set of n equations similar to Equation A15.10.2.

In the simpler case, the model is formulated as a linear one. For some systems a linear model represents an unacceptable oversimplification and a non-linear model is required.

A15.10.2 Time-domain solution

The dynamic model is then subjected to the forcing function of the design basis earthquake. In time-domain analysis, the unsteady-state model is excited by the time forcing function, which is the accelerogram. The equations are integrated numerically and the time response of the structure is obtained. These responses include the forces, or stresses, on, and displacements of, the structural elements.

A15.10.3 Normal modal analysis

In some cases it is permissible to simplify the solution of an MDF model by a method known as normal modal, or simply modal, analysis. In this technique the vibrations in mutually perpendicular directions are treated as uncoupled and the responses are obtained for the separate modes and are then combined by superposition to give the overall responses. The method is applicable only to linear systems.

A15.10.4 Response spectra

Another approach is the response spectrum method. As described above, a response spectrum represents for a series of SDF systems with a given degree of damping the peak acceleration, velocity or displacement obtained when forced by the acceleration–time profile, or accelerogram, of the design earthquake. The family of curves for different degrees of damping constitutes the response spectra. The response spectra most commonly given are those for acceleration.

The determination of the response spectra from the forcing function $F(t)$ of the design earthquake is described by Dowrick. The displacement $x(t)$ is given by

$$x(t) = \int_0^t \frac{F(\tau)}{m\omega} \exp[-\zeta\omega(t - \tau)]\sin[\omega(t - \tau)]\,d\tau$$

$$[A15.10.7]$$

Then, taking $\omega \approx \omega_n$, a response function is defined as

$$V(t) = \omega_n x(t) \qquad [A15.10.8]$$

where $V(t)$ is the response function (m/s).

It is the maximum values of the responses that are of interest. The maximum value of the response function is termed the spectral velocity, or strictly the pseudo-spectral velocity:

$$S_v = V_{max} \qquad [A15.10.9]$$

where S_v is the spectral velocity (m/s) and V_{max} the maximum value of the response function (m/s). Further

$$S_d = S_v/\omega_n \qquad [A15.10.10]$$
$$S_a = \omega_n S_v \qquad [A15.10.11]$$

where S_a is the spectral acceleration (m/s^2) and S_d the spectral displacement (m).

The application of the response spectra is as follows. The natural frequency and damping of the structure is determined. The response spectra are then used to obtain the peak acceleration. This peak acceleration is then applied to find the forces on, and displacements of, the structural elements.

The basic response spectra technique is described by Dowrick as essentially a special case of modal analysis, in that the method treats the system as a linear one and uses the principle of superposition. He also describes, however, adaptations of the method to non-linear analysis.

A15.10.5 Frequency domain solution

Alternatively, the analysis may be performed in the frequency domain. The method is described by Dowrick. The acceleration–time profile is converted by Fourier transform methods into an acceleration–frequency profile. This is then used to excite the model of the system. The response in the frequency domain is then converted back using the inverse transform to give the response–time profile.

R. Davies (1982) describes the use of a version of this approach. The acceleration–time profile of the design earthquake, the accelerogram, is converted by Fourier analysis into an acceleration–frequency profile. The analysis then proceeds in two stages. In the first stage, using a series of frequencies from the acceleration–frequency profile, the unsteady-state model is excited at each frequency by a sinusoidal force of that frequency and with an acceleration amplitude set at an arbitrary value. The outputs are the forces on, and displacements of, the structural elements as a function of frequency. In the second stage these outputs are converted back to time profiles of forces and displacements.

A15.10.6 Selection of method

The selection of a method of dynamic analysis is discussed by Dowrick (1977). The structures to be analysed range from small, simple structures to large, complex ones. Broadly, more complex structures demand more sophisticated methods. He gives in ascending order of complexity the methods of static analysis, response spectra, modal analysis and non-linear modelling.

A15.11 Seismicity Assessment and Earthquake Prediction

The assessment of the seismic hazard to which an installation may be exposed is normally based on information about the seismicity of the area in question. In principle, this may be complemented by information from the monitoring of ground motions and prediction of any impending earthquake. However, such prediction is not sufficiently precise to be of much use for this purpose.

A15.11.1 Seismic data and maps

There are available a number of compilations of seismic maps and other information on seismicity. They include *Seismicity of the Earth and Associated Phenomena* by Gutenberg and Richter (1954); for Europe *Seismicity of the European Area* by Karnik (1968, 1971) and for Britain *A History of British Earthquakes* by Davison (1924) and *Seismicity and Seismic Hazard in Britain* by Lilwall (1976).

A15.11.2 Seismicity assessment

Seismic design or assessment normally takes as its starting point information on the relationship between the magnitude and frequency of earthquakes at the location of interest. In default of more specific information, use is generally made of the historical record.

The normal method is to use a relation such as the Gutenberg–Richter equation, Equation A15.2.10, from which the frequency of earthquakes of various magnitudes may be determined. Several return periods and magnitudes are selected for study.

For an earthquake of a particular magnitude and frequency the probability of occurrence may be estimated making some assumption about the probability distribution. A common assumption is that the incidence is random.

Then, using a relation such as that of Donovan or Esteva, Equations A15.6.1 or A15.6.2, respectively, the PGA can be obtained.

The approach taken is therefore statistical rather than geological. The potential benefits of a more geological approach are discussed by V. W. Lee and Trifunac (1987). They argue that estimates for the occurrence rates of earthquakes are over a time interval of some 10^3–10^6 years for those based on geological structure, or 'geological seismicity', while those based on 'historical seismicity' are for intervals of the order of 10^2–10^3 years.

The assessment of seismicity in Britain described below serves to illustrate the approach generally taken.

A15.11.3 Probability distributions

The frequency of earthquakes of different magnitudes is commonly obtained from the Gutenberg–Richter equation, as just described.

Use may also be made of the extreme value distribution of Gumbel (1958).

The occurrence of an earthquake of given size is commonly treated as a randomly occurring event for which the probability distribution is the Poisson distribution:

$$P = 1 - \exp(-\lambda) \qquad [A15.11.1]$$

where for an earthquake of a specified size, P is the annual probability of exceedance and λ the expected annual number of earthquakes.

The possibility of complementing this by utilizing information obtained from monitoring seismological indicators points to the possible use of the Bayesian approach.

A discussion of appropriate distributions is given by Newmark and Rosenblueth (1971).

A15.11.4 Seismicity in Britain

A study of the seismicity of Britain which is frequently quoted is that of Lilwall (1976). Figure A15.5 is one of his plots showing intensities of earthquakes with a return period of 200 years.

Lilwall divides the country into the following zones: Great Glen, Comrie and Memstrie, South Wales and Herefordshire and the remainder. The first two are areas of a seismicity relatively high for Britain. The third may be regarded as typical of a relatively active area of the country.

For the frequency of earthquakes Lilwall recasts Equation A15.2.10 as

$$\log_{10} N = a' - b' m_b \qquad [A15.11.2]$$

where a' and b' are constants. He obtains the following values for a' and b':

Area	a'	b'	Surface area (km^2)
Overall	4.13	1.09	2.3×10^5
South Wales and Herefordshire	1.1	0.55	15,000
Remainder	4.0	1.15	2.3×10^5

The return periods for earthquakes of different magnitudes are then as follows:

Magnitude m_b	Return period (year)
4.0	1.5
4.5	6.0
5.0	20.0
5.5	70.0
6.0	260.0

The last two values are extrapolations beyond the observed range.

From analysis of the data, Lilwall proposes for Britain an upper limit of 5.72 for the body wave magnitude and gives the relation

$$P = 1 - \exp\left[-\frac{1}{5}\left(\frac{5.72 - m_b}{5.72 - 4.43}\right)^{2.22}\right] \qquad [A15.11.3]$$

where P is the annual probability of exceedance.

For the relation between M_S and m_b, he uses the equation of P.D. Marshall (1970):

$$M_S = 2.08 m_b - 5.65 \qquad [A15.11.4]$$

For the intensity in Britain he derives, following Esteva and Rosenblueth (1964), the relation

$$I = 6.0 + 1.5 M_s - 1.9 \ln R \qquad [A15.11.5]$$

or, using Equation [A15.11.4]

$$I = -2.5 + 3.1 m_b - 1.9 \ln R \qquad [A15.11.6]$$

He uses for the estimation of R Equation A15.2.2 with $k = 20$.

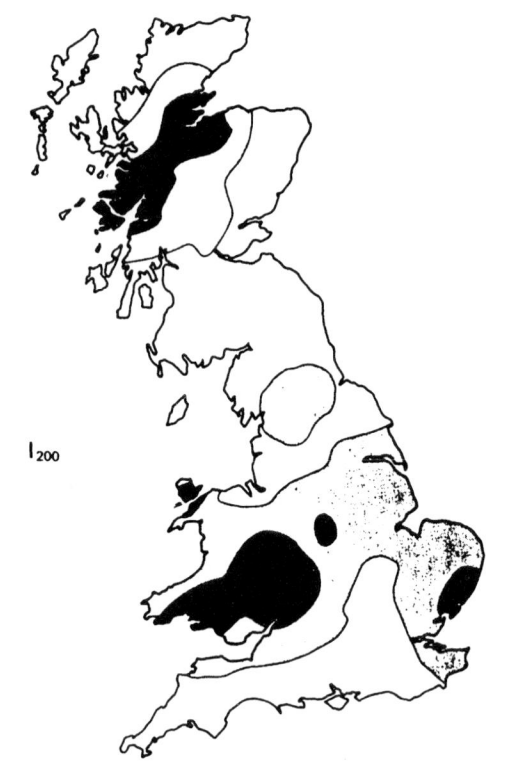

Figure A15.5   *Seismicity map of Great Britain showing intensities of earthquakes with a return period of 200 years (Lilwall, 1976). Despite of shading give intensities of up to 5, 5–6, 6–7 and over 7 (courtesy of HM Stationery Office)*

A15.11.5 Earthquake prediction

There are a number of methods for the prediction of earthquakes. Such methods are applicable mainly to active regions. Accounts are given by Berlin (1980), Eiby (1980) and Bolt (1988).

One important concept is that of the seismic gap. Essentially, a seismic gap is an area that has not had an earthquake for some time and in which strain is accumulating. The argument is that due to this strain accumulation the probability of an earthquake increases with time.

The seismic gap concept has been applied to large earthquakes. Areas at the boundary of tectonic plates that have not had an earthquake for a long period are regarded as more likely to experience one than those that have had a recent event. The seismic gap approach has also been applied to lesser earthquakes.

It has been found that in some cases large, shallow-focus earthquakes tend to migrate along a fault zone.

There are a number of phenomena that are regarded as precursors of earthquakes. They range from fore-shocks, seismic wave anomalies and fault creep to behaviour of animals.

There are also available several earthquake precursor models. Three principal models are (1) the dilatancy model; (2) the fault creep model; and (3) the propagating deformation front model.

Attempts to predict earthquakes have met with mixed success. There is in China an established earthquake prediction programme. One much quoted case is that of the earthquake in Liaoning Province on 4 February 1975, which had a magnitude of 7.4. Officials decided that the evidence justified the issue of a warning that a strong earthquake was probable within 24 h, people left their houses and the earthquake in fact struck that evening. On the other hand no warning was issued for the earthquake at Tangshan in 1976, which had a magnitude of 8.0.

In an account of the prediction of earthquakes in California, Bolt (1988) stated that there was one definitive prediction. This was that a moderate earthquake was likely to occur near Parkfield, California, between 1987 and 1993. This was based on a simple linear prediction of the trend of earthquakes, which have occurred in 1901, 1922, 1934 and 1966.

A15.12 Design Basis Earthquake

In earthquake-resistant design, an installation is designed to withstand seismic events so that the frequencies at which various degrees of damage are incurred do not exceed the specified values. The choice of the design basis earthquake is governed by these frequency specifications.

The characterization of the ground motion of an earthquake has been described in Section A15.6. Essentially, there are three main approaches. These are characterization in terms of peak acceleration; of the response spectra or of the acceleration–time profile. These three methods of characterization accord with the three main methods of analysis described in Section A15.10: static analysis, response spectra and dynamic analysis.

A15.12.1 Seismic hazard curve
There are available empirical correlations between the frequency of exceedance and the magnitude of earthquakes and between their peak acceleration and magnitude, as described in Section A15.2. From these it is possible to obtain the relationship between the frequency of exceedance and peak acceleration. A plot of these two quantities is often referred to as a seismic hazard curve. A seismic hazard curve may be constructed for a region or for a particular site.

Figure A15.6 shows seismic hazard curves for the United Kingdom given by Alderson (1982 SRD R246). Curve A is based on the Gutenberg–Richter equation, Equation A15.2.10, and curve B by a method given by Hsieh, Okrent and Apostolakis (1975).

A15.12.2 Response spectra
As described in Section A15.6, there are available a number of response spectra both for real earthquakes and for simulated earthquakes.

One or more of these response spectra may be selected to define the design basis earthquake. With this method also the choice is governed by the earthquake severity, and hence the return period, for which the installation is to be designed.

Two earthquakes that figure prominently in the literature are those at El Centre in 1940 and Parkfield in 1966. A number of response spectra for the El Centre earthquake and a comparison with that at Parkfield are given by Dowrick (1977). Further accounts of the Parkfield response

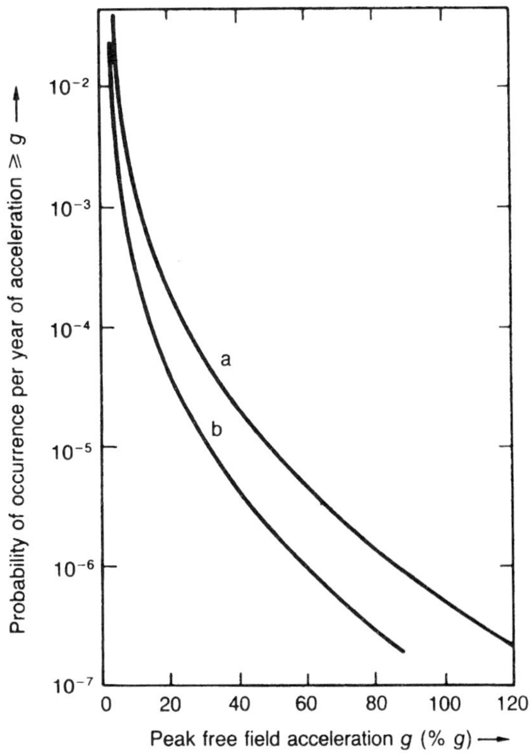

Figure A15.6 *Seismic hazard curves for the United Kingdom (Alderson, 1982 SRD R 246): (a) curve based on Gutenberg–Richter equation and (b) curve based on method of Hsieh, Okrent and Apostolakis (1975) (courtesy of the UKAEA Systems Reliability Directorate)*

spectra are given by D.W. Phillips (1982) and V.W. Lee and Trifunac (1987).

The seismic design response spectra given in *Regulatory Guide 1.60* of the NRC have already been mentioned and are given in Figure A15.4. They are discussed further below.

Another set of response spectra are those developed in New Zealand, as described by Dowrick (1977).

A15.12.3 Accelerograms
For accelerograms the approach is broadly similar to that for response spectra just described. One or more accelerograms may be selected with the choice governed by earthquake severity and return period.

For accelerograms also, the 1940 El Centre and 1966 Parkfield earthquakes figure prominently in the literature.

Newmark and Rosenblueth (1971) give a set of accelerograms which illustrate the distinctions which they make between different types of ground motion.

A15.13 Nuclear Installations

The seismic hazard is of particular importance for nuclear power plants and a large amount of work has been done both on seismic design and on seismic hazard assessment.

The NRC has set earthquake-resistant design criteria and has undertaken a large amount of work on seismic engineering of nuclear plants.

A15.13.1 NRC seismic design response spectra

For the purpose of nuclear plant siting, the NRC considers a fault to be active if there has been movement at least once in 35,000 years. For such a fault, it is to be assumed that the largest earthquake supported by the fault system will occur within the next 50 years.

In 1973 the NRC issued *Regulatory Guide 1.60* (*RG 1.60*) containing its seismic design response spectra, which have already been mentioned and are shown in Figure A15.4.

Background to *RG 1.60* is given by V.W. Lee and Trifunac (1987). The *RG 1.60* response spectra were developed from accelerograms of some 33 earthquakes and model the shapes of spectra that can be expected at intermediate distances from medium to large earthquakes. They state that experience of the *RG 1.60* response spectra since that date shows that they are useful for straightforward cases, but meet with difficulties for responses in the near and far fields.

These authors also describe the scaling of the *RG 1.60* response spectra for other distances, particularly for locations in the near field. The method generally used is based on scaling by the PGA, although this presents some difficulties.

Lee and Trifunac discuss approaches to the improved characterization of near-source ground motion.

A15.13.2 Seismic Safety Margins Research Program

The NRC has for some years had a major programme of work on seismic safety margins. Accounts of this work include those of Cummings (1986) and R.P. Kennedy *et al.* (1989).

In simple terms the seismic safety margin is the earthquake level at which with high confidence failure is extremely unlikely. An earthquake of this level is described as a seismic margin earthquake (SME). The SME is generally characterized by the PGA.

The main body of work is that done under the Seismic Safety Margins Research Program (SSMRP). Among the reasons for the programme were uncertainty about the size of earthquakes to be expected; a lack of experimental data to guide plant design; the potential for common mode failure due to ground shaking and the possibility that 'strengthening' might increase risk.

The SSMRP methodology involves a five-step procedure: (1) characterization of the seismic hazard; (2) determination of the response of the structure and subsystems to seismic excitation; (3) determination of the fragility functions; (4) identification of accident scenarios; and (5) calculation of the frequency of failure and release.

The seismic hazard is characterized by seismic hazard curves, response spectra and acceleration time histories based on a combination of real earthquakes, estimates from magnitude correlations, expert judgement and other methods. The necessary computations are performed by the computer code HAZARD. The soil–structure interaction and the response of structure and subsystems to seismic excitation are treated simultaneously and are handled by the SMACS code. Seismic excitation includes forcing by an ensemble of acceleration time histories in three orthogonal directions. A third code SEISIM is used to handle the radioactive release aspects.

Much of the work in the SSMRP has been concerned with the seismic fragility of items of equipment. The fragility is a function of the seismic load and is expressed as the cumulative probability of failure with load.

A15.13.3 Seismic margin estimation

A review of methods for the calculation of the seismic margin of items of equipment is given by R.P. Kennedy *et al.* (1989).

The seismic margin review of a nuclear power plant requires the determination for certain components of the High Confidence of Low Probability of Failure (HCLPF) capacity. The study describes two candidate methods for doing this, the Conservative Deterministic Failure Margin (CDFM) method and the Fragility Analysis (FA) method.

The CDFM method involves a design calculation which is essentially conventional in method but which utilizes more conservative assumptions.

In the FA method, the fragility of the equipment is modelled as a double log–normal distribution with three parameters: A_m the median ground acceleration and β_R and β_U, the logarithmic standard deviations representing, respectively, the randomness and the uncertainty in the median value. The two latter parameters may be subsumed in the combined parameter $\beta_c (= (\beta_R^2 + \beta_U^2)^{1/2})$. The report describes the estimates made by four experts of the HCLPF by the CDFM and FA methods for a representative set of items, including a storage tank, a heat exchanger and an air tank. The bulk of the report consists of detailed calculations. The results are expressed as values of the PGA.

A15.14 Process Installations

There is rather less work published on the seismic design and seismic assessment of process plants, but there are some accounts available. Moreover, the methodologies developed in the nuclear industry are in large part applicable to the process industries.

The guide TID-7024 *Nuclear Reactors and Earthquakes* (NTIS, 1963) is quoted in the NPFA codes such as NFPA 59A: 1985 *Production, Storage and Handling of Liquefied Natural Gas* (*LNG*) for guidance on seismic loading.

A review of seismic design and assessment for process plants with special reference to major hazards is given in *Seismic Design Criteria and Their Applicability to Major Hazard Plant within the United Kingdom* by Alderson (1982 SRD R 246). Further accounts of seismic assessment of process plants include those of Ravindra (1992), and, for the United Kingdom, of R. Davies (1982), Alderson (1985) and Kunar (1983).

Methods for seismic design are given by the NRC (1979c). There is also a good deal available for particular items. For example, for liquid storage tanks accounts have been published by Veletsos and co-workers (Veletsos and Yang, 1976; Veletsos, 1984; Veletsos and Yu Tang, 1986a,b), Haroun and Housner (1981), R.P. Kennedy (1979, 1989) and Priestly *et al.* (1986). Guidance on analysis of individual failure modes is given in standard texts on structures such as Timoshenko (1936), Warburton (1976), Pilkey and Chang (1978) and, for wind loads, Sachs (1978).

A15.14.1 Seismic assessment of generic plants

R. Davies (1982) has reviewed the seismic hazard for process plants in the United Kingdom. He identifies liquefied gas storage as a matter of principal concern and describes a study to assess the risk from such storage.

The study was in three parts: the assessment of seismicity in the United Kingdom; the selection of representative containment and the identification of principal failure modes and the modelling of the seismic response of these containments.

Three levels of earthquake were defined, Levels 1–3. A Level 1 earthquake was taken as one with a return period of 150 years which would have a 0.25 probability of occurring over a plant life of 40 years. The work of Lilwall suggests that this return period corresponds to an earthquake of body wave magnitude about 4.5, which would typically yield at 15–20 km a site intensity of around VI on the MM scale. The Level 2 earthquake was taken as one with a return period of 2000 years, magnitude of 5.25 and site intensity of VII. For the Level 3 earthquake a frequency of 10^{-4} events/year, or a return period of 10^{-4} years, was taken, following NFPA guidance in NFPA 59A: 1980 on the design earthquake for LNG storage.

Earthquakes corresponding to each of these three levels were identified and the accelerograms of these earthquakes were obtained. The records used were those for the earthquake at New Madrid, Missouri, on 13 June 1975 (body wave magnitude 4.3), those for that at Forgaria-Cornino, Italy, on 11 May 1976 (magnitude 5.2) and the Temblor records of the earthquake at Parkfield on 27 June 1966 (magnitude 5.6).

The method of dynamic analysis used by Davies was described earlier.

Three generic types of liquefied gas storage were considered: a double skin cryogenic tank with a suspended roof, an inner shell of 9% nickel steel and an outer shell of mild steel to hold 20,000 te of LNG; a 1250 te capacity spherical pressure storage vessel to hold 1250 te of propane and a cylindrical pressure storage vessel, or torpedo, to hold 250 te of liquid carbon dioxide. For all three containments a foundation of hard rock was considered and the low temperature tank was also considered with a piled foundation on soft soil.

The failure modes considered were as follows. For the storage tank they were (1a) rupture of the connection between the tank wall and the floor; (1b) ripping of the tie-down strap; and (1c) excessive displacement of the base of the shell and rupture of the connection pipes. For the sphere they were (2a) tear at the connection of the shell and supporting leg; (2b) buckling of the support columns; and (2c) excessive displacement at the connecting pipe takeoff and rupture of the pipe. For the torpedo they were (3a) tear at the connection of the shell and support; (3b) rupture of straps or supports; and (3c) sliding on or rolling off support and rupture of the connecting pipes.

For the Level 1 earthquake the ratio of the seismic response stress to the seismic capacity stress was very small in each of these failure modes; the highest value of the ratio was 0.03 for the cryogenic tank tie-down strap on hard rock (mode 1b). For the Level 2 earthquake the ratio was 1.4 for the cryogenic tank tie-down strap on hard rock (1b), 0.48 for the inner shell base of the tank on hard rock (1a) and 0.21 for the inner shell base on piles (1a again). For the Level 3 earthquake these three latter values were 1.9, 0.72 and 0.29, respectively. Data on displacements are also given.

The conclusion of the study is that for liquefied gas storages of the type considered an earthquake which occurs sufficiently frequently to be likely to be felt during the life of the plant poses a negligible risk of loss of containment, but that one which approaches the maximum likely to occur in the United Kingdom could produce appreciable horizontal displacements in all types of plant and stresses in certain critical parts of some containments which exceed the dynamic design stresses allowed in current design codes.

A15.14.2 Seismic assessment of particular plants

An account of the seismic assessment of an existing process plant as part of the controls on acutely hazardous material (AHM) under the California Risk Management and Prevention Program (RMPP) under has been given by Ravindra (1992). The review covered both the site-specific geology and the engineering facilities.

He describes the investigation of the seismic characteristics of the site, including the impact of active faults; the potential for surface faulting and ground breakage, soil settlement and liquefaction, and slope instability; and local soil amplification. A seismic hazard curve was established.

The seismic capacity of the facilities was assessed by review of the design drawings and design codes used and by a walkaround of the plant.

It was found that some components had been designed using the static design method with an acceleration of 0.2 g; this acceleration is similar to that in the UBC. Detailed evaluation established that some of these items were vulnerable even though they had been designed to a building code.

The author gives examples of the seismic capacity assessed by a deterministic method given in RMPP guidelines and by the FA method and quotes values of the median ground acceleration A_m and the logarithmic standard deviation of the uncertainty in this acceleration. The failure mode determining seismic capacity in most cases is failure of the anchor bolts or anchorage. The author comments that generally this can be upgraded at minimal cost.

A15.14.3 Spherical pressure vessel

An assessment of the seismic capacity of a spherical pressure storage vessel has been described by D.W. Phillips (1982).

The vessel considered is a 15 m diameter vessel supported on 12 columns, or legs. The nominal contents of the vessel are 1200 te, the mass of the shell is 166 te and that of the legs 15.6 te. The active length of the supporting columns is 7.94 m. Each column has a flat steel plate welded to the bottom. The plate is attached by four foundation bolts to the concrete pedestal. The spacing between the bolts is 0.58 m and the inset of the bolts is 0.08 m. The bolts are 0.032 m in diameter. The effective dynamic mass of the vessel itself, with a contribution from the legs, is taken as 171 te ($166 + \frac{1}{3} \times 15.6$), giving a total effective dynamic mass of 1371 te for the full vessel.

From a separate study, it is known that the natural period of the vessel for horizontal vibration is as follows:

State	Mass of vessel and shell (te)	Natural period (s)
Full	1371	0.75
$\frac{1}{2}$ full	711	0.58
$\frac{1}{4}$ full	471	0.25

The vessel is located on a typical UK site. For such a site the seismic characteristics in terms of the free field horizontal acceleration are taken as

Return period (year)	Horizontal acceleration (g)
100	0.05
1000	0.12
10,000	0.25

The method of calculation used is essentially a static analysis, but first a dynamic analysis is conducted to determine a dynamic, or resonance, correction factor. For this use is made of the data given above on the natural period and the horizontal acceleration together with suitable response spectra. Two sets of response spectra are used, the NRC *RG 1.60* spectra and the Temblor–Parkfield spectra. A damping of 5% is taken as typical of welded steel at high strain and of bolted steel. For each degree of fill, the response spectra are used to obtain the acceleration from the natural frequency and damping. This value is scaled by the ratio of the maximum acceleration of the site to the maximum acceleration implied in the response spectra.

The results for the horizontal response with intact bracing are

The vertical excitation is taken as $\frac{2}{3}$ of the horizontal. The overall structural response is assumed to be given by

$$W_O = \left[\sum_{i=1}^{3} W_i^2 \right] \qquad \text{[A15.14.2]}$$

where W_i is the load in the ith direction (N) and W_O the overall load (N).

Then for the 100 year return period with horizontal acceleration of 0.05 g and incorporating the two orthogonal accelerations and the dynamic correction factor of 2, the peak horizontal seismic load of the full tank is

$$W_s = 2^{1/2}(1371 \times 10^3) \times 2 \times 0.05 \times 9.81$$
$$\approx 1.9 \times 10^6 \text{N}$$

Thus, the seismic load is nearly 20 times the wind load.

Phillips gives detailed calculations for the various modes of failure and for the corresponding seismic capacities. These show that the seismic response for which the structure is weakest is the overturning couple on the column supports. The corresponding failure mode is tensile failure of the pedestal bolts. For this failure mode the safety factor has the values of 0.5, 0.2 and 0.1 for the 100, 1000 and 10,000 year return periods, respectively.

Return period (year)	Horizontal response (g)					
	NRC response spectra			Temblor–Parkfield reponse spectra		
	Sphere full	½ full	¼ full	Sphere full	½ full	¼ full
100	0.07	0.11	0.14	0.11	0.16	0.21
1000	0.17	0.25	0.33	0.26	0.37	0.50
10,000	0.36	0.53	0.70	0.54	0.79	1.03

The accelerations are rather greater for the Temblor–Parkfleld than for the NRC response spectra. The greatest acceleration is experienced by the ¼ full tank, but for this case the mass is much less. The dynamic correction factor chosen is 2, which corresponds approximately to the ratio of the PGAs for the site to those for the case of the full tank with the Temblor–Parkfield spectra.

The sphere did not appear to have been designed for earthquake resistance. Probably, the principal horizontal force considered was wind loading. The relation for the wind load is

$$W_w = C_D \rho_a \pi r_s^2 u_w^2 \qquad \text{[A15.14.1]}$$

where C_D is the drag coefficient, r_s the radius of the sphere (m), u_w the wind speed (m/s), W_w the wind load (N) and ρ_a the density of the air (kg/m^3). At the estimated Reynolds number of 10^7 the drag coefficient is 0.5. The wind speed with a 50 year return period would be between 18 and 32 m/s. Taking the higher value for the wind speed and $r_s = 7.6$ m and $\rho_a = 1.25$ kg/m^3

$$W_w \approx 1.1 \times 10^5$$

The seismic loading is assumed to consist of accelerations in three orthogonal directions, two horizontal and one vertical.

The calculation of this failure mode may be summarized as follows. Figure A15.7 shows the overturning couple for which the relation is

$$F_{sb} = \frac{1}{n} F_s \frac{l_1}{l_2} \qquad \text{[A15.14.3]}$$

where F_s is the seismic load (N), F_{sb} the seismic tensile force on the bolts (N), l_1 and l_2 are the distances shown in Figure A15.7 (m) and n is the number of bolts resisting overturning. Taking $l_1 = 7.94$ m and $l_2 = 0.66$ m and using the seismic load for the 100 year return period

$$F_{sb} = \frac{1}{24}(1.9 \times 10^6) \times \frac{7.94}{0.66}$$
$$= 9.5 \times 10^5 \text{ N}$$

Taking a bolt area of 8.0×10^{-4} m^2 per bolt gives a bolt tensile stress of

$$F_{sb} = \frac{9.5 \times 10^5}{8.0 \times 10^{-4}}$$
$$= 1.2 \times 10^9 \text{ Pa}$$

Figure A15.7 *Overturning couple on the support column of a pressure storage sphere (Phillips, 1982) (courtesy of the ukaea Systems Reliability Directorate)*

where F_{sb} is the seismic tensile stress on each bolt (Pa). For the purpose of this assessment, tensile failure is assumed to occur at the yield stress, which for high tensile steel is some $0.6–0.7 \times 10^9$ Pa. Taking the lower value of the yield stress, gives for the safety factor at the 100 year return period a value of 0.5 ($= 0.6 \times 10^9/1.2 \times 10^9$).

For none of the other failure modes considered was the safety factor less than unity at the 100 year return period, though for two others it fell to unity at the 1000 year return period and below it at the 10,000 return period.

A15.14.4 Storage tank
An assessment of the seismic capacity of a flat-bottomed storage tank has been described by R.P. Kennedy (1989). The tank considered is 40 ft in diameter with a shell 37 ft high and an overall height to the tank top of 43.4 ft.

The author describes in detail the use of the CDFM method. This consists of two parts, the determination of the seismic responses and the determination of the seismic capacities. There is not necessarily a one-to-one correspondence between a seismic response and seismic capacity. For a given response there may be several features resisting failure and all these must be considered.

The following responses are considered: (1) overturning moment in the tank shell immediately above the base plate of the tank; (2) overturning moment applied to the tank foundation through a combination of the tank shell and base plate; (3) base shear beneath the tank base plate; (4) combination of the hydrostatic plus hydrodynamic pressures on the tank side wall; (5) average hydrostatic minus hydrodynamic pressure on the base plate of the tank; and (6) fluid slosh height. These responses are all expressed as forces except the last; for this what matters is whether the sloshing fluid hits the roof.

The author comments that the first of these responses is usually the governing one. The third is seldom governing

for tanks with a radius greater than 5 m. The second, fourth and sixth seldom govern.

In investigating these responses at least two horizontal modes, the impulse and the convective, or sloshing, modes and one vertical mode should be considered.

Kennedy gives detailed calculations for the various seismic responses and for the corresponding seismic capacities. The governing response is usually the first, the overturning moment in the tank shell.

The seismic responses are determined using a value of 0.27 g for the peak ground acceleration. The overturning moment response is then estimated as 19,600 kip-ft. The corresponding capacity is determined by considering (1) the compressive buckling capacity of the shell; (2) the tensile hold-down capacity of the anchor bolts, including their anchorage and attachment to the tank; and (3) the hold-down capacity of the fluid pressure acting on the tank base plate. The combined resistance of these gives an overturning moment capacity of 20,800 kip-ft. The SME for this response is then calculated as 0.29 g (0.27 x (20,800/19,600)).

This response gives the lowest value of the SME and the overall SME for the tank is thus determined as 0.29 g.

Kennedy also describes the determination of the seismic capacity in terms of the alternative FA method. Using this method the SME is found to be 0.31 g.

A15.15 Notation

a	peak ground acceleration (fraction of g); (cm/s^2) (Sections A15.2 and A15.6 only); (m/s^2) (Sections A15.8 and A15.10 only)
a, a'	constants in Equations A15.2.10 and A15.11.2, respectively
A	maximum amplitude of earthquake (mm)
A_m	median ground acceleration (fraction of g)
A_O	maximum amplitude of standard earthquake (mm)
b, b'	constants in Equations A15.2.10 and A15.11.2, respectively
c	damping coefficient (kg/s)
C	coefficient (Equation A15.9.1)
C	flexibility factor
C_D	drag coefficient
d	peak ground displacement (cm)
E	energy of earthquake (erg)
F	force (N)
F_s	seismic force (N)
F_{sb}	seismic tensile force on bolts (N)
F_{sb}	seismic tensile stress on bolts (Pa)
g	acceleration due to gravity (m/s^2)
h	focal depth (km)
H	depth of soil layer (m)
I	intensity of earthquake (Equation A15.11.5)
I	occupancy importance factor (Equation A15.9.2)
I_O	epicentral intensity of earthquake
k	modifying factor for focal distance (km) (Equation A15.2.2)
k	resistance (kg/s^2) (Equations A15.10.2–5)
K	framing factor
l_1	active distance of support column (m)
l_2	spacing + inset of bolts on support column base plate (m)
m	mass (kg)
m_b	body wave magnitude of earthquake
M	magnitude of earthquake

M_L	local magnitude of earthquake
M_S	surface wave magnitude of earthquake
n	number of bolts resisting overturning
N	frequency of earthquake (year^{-1})
P	annual probability of exceedance
r	epicentral distance (km)
R	focal distance (km)
S	site factor
S_a	spectral acceleration (m/s^2)
S_d	spectral displacement (m)
S_v	spectral velocity (m/s)
t	time (s)
T	natural period (s)
T	natural period of soil layer (s) (Equation A15.7.1)
u_w	wind speed (m/s)
v_s	velocity of shear wave (m/s)
V	horizontal force, base shear (N)
V	response function (m/s) (Section A15.10 only)
W	load (N)
W_i	load in ith direction (N)
W_O	overall load (N)

W_s	seismic load (N)
W_w	wind load (N)
x	displacement (m)
Z	seismic risk zone factor
β	composite variability (fraction of g)
β_R	log-normal standard deviation of randomness of median ground acceleration (fraction of g)
β_U	log-normal standard deviation of uncertainty in median ground acceleration (fraction of g)
ζ	damping factor
λ	average annual number of events
ρ_a	density of air (kg/m^3)
ω	damped frequency (s^{-1})
ω_n	natural frequency (s^{-1})

Subscripts

D	damping
I	inertia
max	maximum value
S	elastic

Appendix

16

San Carlos de la Rapita

At about 14.30 in the afternoon of 11 July 1978 fire and explosions from a road tanker carrying propylene occurred at a camp site at Los Afaques (The Sand Dunes) at San Carlos de la Rapita between Barcelona and Valencia in Spain. The eventual death toll was 215.

A Court of Inquiry on the disaster was held at Tarragona (Court of Inquiry, Tarragona 1982). A report was submitted to the inquiry by Carrasco (1978). Independent investigations for the HSE were been carried out by Scilly (1978) and Hymes (1983 SRD R275), for the Hampshire Fire Brigade by Stinton (1978, 1979, 1983) and for the Dutch Directorate General of Labour by Ens (1986 LPB 68). Analyses of the accident are given by V.C. Marshall (1985, 1986 LPB 72, 1987).

Selected references on San Carlos are given in Table A16.1.

A16.1 The Camp Site

The camp site was a triangular shaped piece of land between the coast road and the sea some 3 km from the nearest township San Carlos de la Rapita to the north. It was 200 m in length and 10,000 m^2 in area, tapering from about 100 m wide at the north to 30 m at the south end, and 60 m wide at the point of the accident. There was a brick wall on a concrete foundation between the road and the camp.

On the day of the accident, the camp site was fully booked with some 800 people, but not all these were there at the time and in the area affected. Estimates of the number on site and of the number in the area most affected are 500 and 300−400, respectively.

A16.2 The Road Tanker

The road tanker was manufactured in 1973. It had no pressure relief valve and no pressure test certificate. It had nominal and actual capacities of 45 and 44.4 m^3.

It has been alleged that the tanker had been used on previous occasions to carry ammonia, which can lead to cracking of the tank, but this has been denied.

At 12.05 the tanker took on a load of propylene at the ENPETROL Tarragona refinery. The system was that the driver would learn how much he was carrying when the vehicle was weighed at the weighbridge. There was no metering or other device to prevent overfilling. If he wished to reduce the load, he could burn some off using a device like a flamethrower. The maximum load for the vehicle was 19 te, but it was plated as suitable for a maximum load of 22 te and it was actually weighed out as a 23.5 te load. At 12.35 the vehicle left the refinery.

The driver had a recommended route, which was the motorway, and was provided with the toll money, but actually took the N340 coast road, which went past the Los Alfaques camp.

Table A16.1   *Selected references on San Carlos*

Carrasco (1978); Stinton (1978, 1979a,b, 1983); UKAEA (1979); J. Harris (1979); Court of Inquiry, Tarragona (1982); Moodie (1982); Scilly (1982); Hymes (1983 SRD R275, 1986 LBP 61); Manas (1984); V.C. Marshall (1985, 1986 LPB 72, 1987); Ens (1986 LPB 68, 72); Foxcroft (1986 LPB 71); Bond (1987 LPB 73)

A16.3 The Fire and Explosions − 1

The sequence of events at the camp site is not entirely clear, but several eyewitness accounts are available. V.C. Marshall (1987) quotes the following. A young man serving a customer in the camp shop off the main site heard a bang. He got into his car to investigate and 2 min after the first sound heard a severe explosion. A 'fireball' appeared on the site. A tourist in the camp restaurant heard a 'pop' from a tanker on the main road and saw a milky cloud drifting towards him. He ran to move his car and seconds later the cloud ignited. In one newspaper account a motor cyclist following the vehicle said that smoke and fire were issuing from the tank. In another such account a tourist walking his dog at the camp site entrance saw the tanker start to snake and then tip over; there followed almost immediately an explosion and fire.

There were conflicting reports on the last movements of the road tanker, summarized by Hymes (1983 SRD R275). They include reports to the effect that the driver stopped at the camp with the tanker already leaking; that he stopped after it sprang a leak with a sharp report at the camp; that the vehicle veered across the road and broke through the camp site wall, suffering either a small leak or complete disintegration; and that the tanker veered across the road and struck the wall a glancing blow.

Hymes gives two photographs, the first captioned as showing the unignited gas cloud and the second the cloud on fire. The latter photograph is shown in Figure A16.1. The photographer stated that the time between taking his first picture and going to the camp entrance to take further shots was 1−2 min and that the tanker ruptured violently after a few minutes.

It is reported that until the main 'fireball' event, people stood around watching the pall of smoke from the fire. When it did occur large numbers, many scantily clad, were burned, some running into the sea to escape or douse the flames.

Figure A16.1   *The fire preceding the 'fireball' at San Carlos (Hymes, 1983 SRD R275) (Courtesy of the Safety and Reliability Directorate)*

The area of damage and injury, including injury outside the flame envelope, is quoted by Marshall as covering an area equivalent to that of a 125 m radius circle. Over 90% of the camp site was gutted by fire.

The explosion effects were variable. The discotheque about 75 m from the tanker was completely demolished, probably by an explosion of the gas inside it, killing four people, but in the opposite direction at a distance of only 20 m a motor cycle still stood on its footrest. Many other objects in the camp liable to be blown over by blast still stood.

The site after the event is shown in Figure A16.2.

The vehicle was torn into four main pieces. Some two-thirds of the tank, including the rear portion, flew to the north-west, landed on the ground some 150 m away and then slid and came to rest in the outer wall of a restaurant about 300 m from its starting point. The middle section travelled about 100 m to the north-east into the camp. The cab unit travelled some 60 m along the line of the road to the south. The front end cap fetched up about 100 m beyond the cab. The location of some of the fragments is shown in Figure A16.3. The front end cap is the white hemispherical object at the bottom of Figure A16.3.

A16.4 The Emergency and the Aftermath

For about half an hour after the fire there were some 200–300 people milling round the camp, many seriously burned and calling for help. Private cars and taxis began to ferry the injured to a hospital at Amposta 13 km away. The first ambulance came at 14.45 from the Shell oil drilling site at San Carlos. The municipal ambulances came at 15.05 and the fire brigade about 15.30.

The road was still blocked by the burning cab so that victims had to be rescued from both sides of the camp and taken north to Barcelona or south to Valencia as circumstances permitted. Those taken north to Barcelona received primary medical care at points en route, while those taken

south to Valencia 165km away did not. In the first week the death rate at Valencia was twice that at Barcelona, but over 2 months the death rates evened up, since most victims were too badly burned to survive.

Over 100 people died outright. Others died later from burns, so that the final figures were 215 dead and 67 injured.

Most of those treated had very serious burns. Out of 148 cases 122 had third degree burns to 50% or more of the body. Either there were fewer people only slightly burned or they received treatment elsewhere.

A16.5 The Fire and Explosions – 2

The Court of Inquiry held that there had been a rupture of the road tanker due to hydraulic overfill. This was the explanation given in a report submitted to it by Carrasco (1978).

Other authors have put forward alternative explanations. There appear to be three principal hypotheses concerning the initial event: (1) hydraulic rupture of the tank, (2) a small leak on the tank and (3) a road accident.

A version of the second hypothesis has been described by Hymes (1983 SRD R275). This is that a leak occurred on the tanker, that a gas cloud formed and found a source of ignition and that the flames burned back to the tanker. There followed an engulfing fire which caused a BLEVE. The vessel was in a weakened condition and the time for the BLEVE to occur was shorter than usual. Hymes does not state how the tanker came to stop at the camp, but on this scenario it could have been to attend to the leak.

A somewhat similar account is given by Stinton (1983).

The starting point in the argument in favour of the third hypothesis, as developed by Ens (1986 LPB 68), is the need to explain how the tanker came to stop and how the tank itself failed. Relevant here is the evidence of the one eye-witness who saw smoke and flames coming from the tanker and of the other who saw it snaking. The hypothesis is as follows. The tanker was grossly overfilled and under stress

Figure A16.2   *The site after the fire and explosion (Hymes 1983 SfiD R275). The camp site is in the right background behind the breached wall (Courtesy of the Safety and Reliability Directorate)*

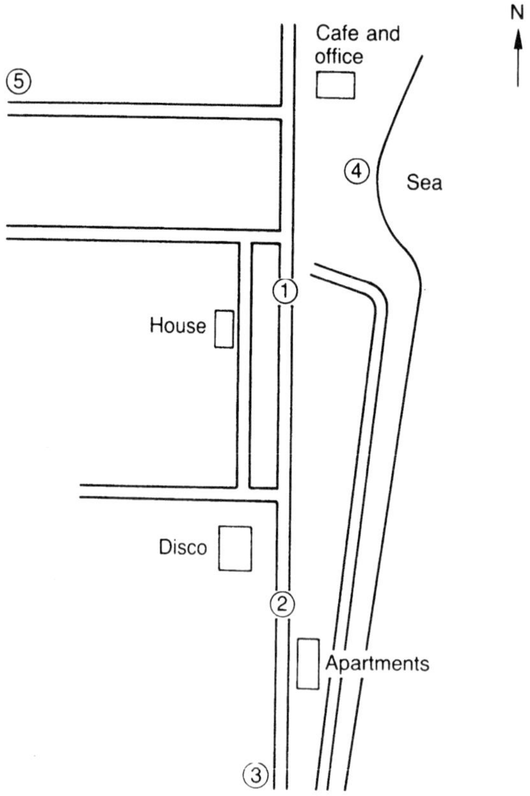

1 Apparent point of disintegration
2 Driver's cab
3 Front end of tank
4 Middle of tank
5 Rear end of tank

Figure A16.3 *Flight of some missiles generated by the explosion (after Stinton, 1983; V.C. Marshall, 1987)*

so that it was liable to burst if subjected to impact. It crashed due to a driver failure or tyre defect, causing it to break through the boundary wall into the camp site. The tank ruptured and released its contents. A gas cloud and liquid pool formed, the gas being at first too rich to burn rapidly. In due course, however, the heat vaporized the liquid pool and there was a fireball. Minor explosions may

have occurred at the edges of the cloud and stronger explosions in the buildings.

With regard to the first hypothesis, that the tank suffered hydraulic rupture, attention has focused on the overfilling of the tanker and on the possible effects of this. The calculations on this aspect are very dependent on the assumptions made about the volume and temperature of the liquid when the tanker left the refinery.

Evidence was presented by Carrasco (1978) that the tank contents could have warmed up sufficiently in 2½ h for the tank to become hydraulically full. He estimates a warm-up rate of 5.6°C/h.

Carrasco's work has been criticized by Marshall and Ens. Marshall draws attention to the work of Moodie (1982) showing the need, in considering hydraulic rupture, to allow for the expansion of the tank metal, the expansion being equivalent to between 3 and 6°C, depending on the steel used. Both authors, however, consider that it is possible that the tank had become hydraulically full. Ens suggests that this becomes credible if allowance is made for a difference between nominal and actual tanker capacity, weighing errors and warming up of the liquid in the terminal pipework. Under such biaxial stress the material would no longer be ductile but would become brittle.

V.C. Marshall (1987) gives an extensive discussion of the possible warm-up of the tank, the hydraulic overfill and physical rupture, the behaviour of the fragments and the fire. Despite his critique of Carrrasco, he concludes that hydraulic rupture of the tank is the most likely cause of the incident.

He suggests that the initial bang heard was probably the tank rupture, that the later explosion was of gas which had entered a building and then exploded inside it and that the fire was a severe flash fire rather than a vapour cloud explosion or a BLEVE. The pattern of injuries, with a large proportion of dead to injured, is consistent with a flash fire.

Hymes, Ens and Marshall all give detailed accounts of the site and of the physical evidence such as the fragments, the explosion damage and the burned area.

A16.6 Lessons of San Carlos

Among the lessons of San Carlos are the importance of

(1) Equipment, procedures, supervision and training to prevent overfilling of vehicles carrying hazardous materials.
(2) Provision of pressure relief on vehicles carrying flammable materials.
(3) Routing of vehicles carrying hazardous materials away from populated areas and vulnerable targets.
(4) Prompt treatment of burn victims.

Appendix

17

ACDS Transport Hazards Report

Contents

A17.1 The Investigation

The Advisory Committee on Major Hazards in its *Third Report* (Harvey, 1984) recommended that although its terms of reference were restricted to fixed installations, the major hazard potential from the transport of hazardous materials should also receive attention.

The HSC accordingly asked the relevant permanent committee, the Advisory Committee on Dangerous Substances (ACDS) to examine the matter. Its findings are presented in *Major Hazard Aspects of the Transport of Dangerous Substances* (ACDS, 1991) (the ACDS *Transport Hazards Report*).

Selected references on the ACDS *Transport Hazards Report* are given in Table A17.1. Its principal contents are shown in Table A17.2.

Table A17.1 *Selected references on the ACDS Transport Hazards Report*

ACSD (1991); Turner (1992 LPB 101); Anon. (1992 LPB 107, p. 29)

Table A17.2 *Principal contents of the ACDS Transport Hazards Report*

1. Introduction
2. Scope
3. Method of Study
4. Quantified Risk Assessment
5. Risk Criteria
6. QRA Results for Rail, Road and Ports
7. The Peterborough Incident
8. Risk Reduction and Mitigation
9. Emergency Planning and Response
10. Comparison of Rail and Road Transport
11. Overview
12. Recommendations

Appendices, including
Appendix 3 Summary of World-wide Major Hazard Transport Accidents
Appendix 4 Summary of Movements and Regulatory Arrangements
Appendix 5 Modes and Consequences of Failure of Road and Rail Tankers Carrying Liquefied Gases and Other Hazardous Liquids
Appendix 6 Criteria for Evaluating Individual and Societal Risks in Transport
Appendix 7 Port Risks
Appendix 8 Transport by Rail
Appendix 9 Transport by Road
Appendix 10 Transport of Explosives by Rail and Road
Appendix 11 Comparison of Risks for Transport by Rail and Road
Appendix 12 Tolerability of Risks of Rail and Road Transport and Risk Reduction and Mitigation
Appendix 13 Emergency Planning
Appendix 14 Management of Safety in Transport

Some of these entries are paraphrased.

A17.2 Substances and Activities

The study deals with four main transport activities: (1) rail transport, (2) road transport, (3) marine transport (ports only) and (4) explosives.

The substances investigated varied as between these different modes of transport. For road and rail four substances were studied: (1) motor spirit, (2) LPG, (3) chlorine and (4) ammonia. These substances are referred to in the report, and therefore here, as the non-explosive materials.

For ports the substances considered were (1) flammable liquids, (2) flammable liquefied gas, (3) toxic liquefied gas, (4) ammonia and (5) ammonium nitrate.

The transport of explosives by rail and by road was the subject of a separate study within the report.

Excluded from the study were air transport, pipelines and radioactive substances.

A17.3 Event Data

The report contains a wealth of data on the movement of hazardous materials by rail, road and sea, on the nature and composition of loads and cargos and on event frequencies and probabilities.

The event frequencies cover not only land transport accidents such as vehicle and wagon impacts and fires and marine accidents such as collisions, strikings, groundings and engine room fires but also data used in the various fault trees.

The probability data include release size distributions, ignition probabilities and so on.

A17.4 Hazard Models

The study makes use of hazard models for fire, explosion and toxic release.

The fire models are for (1) a torch fire, (2) a flash fire and (3) a pool fire and the explosion models are for (1) a vapour cloud explosion (VCE), (2) a BLEVE and (3) a condensed phase explosion. The models relevant to toxic release are those for gas dispersion.

The pool fire model is required for motor spirit, the other fire models and the VCE and BLEVE models for LPG or LFG.

The hazard models fall into two sets. The first set comprises HSE models, which are used for rail and road transport, the second set models incorporated in the SAFETI code, which are used for ports. In the following account the HSE models are described first, followed by those used in SAFETI, where applicable.

A17.4.1 Gas dispersion
The gas dispersion models used are DENZ and CRUNCH for instantaneous and continuous releases, respectively. The main direct application of these models is to toxic release.

Use is also made of the parameterizations by Considine and Grint (1985) of results for flammable gas clouds obtained from DENZ and CRUNCH.

The gas dispersion model used in SAFETI is the dense gas dispersion model of R.A. Cox and Carpenter (1980).

A17.4.2 Torch fire
The torch, or jet, fire model used is that of Considine and Grint (1985), modified to give length and width and representing the flame as a cone.

Atmospheric transmissivity is determined by the method of Simpson (1984 SRD R304).

A17.4.3 Flash fire

For a flash fire use is made of the model of Considine and Grint (1985). The formation of the gas cloud is modelled using the gas dispersion models described in Section A17.4.1 and its combustion by assuming that it burns within the contour of the lower flammability limit.

A17.4.4 Pool fire

For a pool fire use is made of the SPREAD model, a sister model of SPILL (Prince, 1981 SRD R210), which differs from the latter in treating simultaneously the spreading and burning of the pool. The regression rate used is that of Mitzner and Eyre (1982). The other features of the model are not described but are evidently conventional.

A17.4.5 Vapour cloud explosion

The treatment of a VCE involves the determination of the mass of fuel in the cloud and of the overpressure resulting from explosion of this fuel.

For the mass of fuel in the cloud, for an instantaneous release the method is to take this mass as twice the flash fraction. It is assumed that no explosion occurs if the mass is less than 10 te. For a continuous release the mass of fuel in the cloud is determined by the method of Considine and Grint (1985).

The overpressure from explosion of this mass of fuel is determined from the correlation of Kingery and Pannell (1964) between mass of explosive and peak overpressure for TNT. The implication is that a TNT equivalent model is used, but further details are not given.

The VCE model used in SAFETI is the TNO correlation model.

A17.4.6 BLEVE

The BLEVE model used is that of A.F. Roberts (1981/82).

A17.4.7 LPG fire model

The hazard models described are used in conjunction with the release scenarios for rail and road transport to produce an overall LPG fire model.

In this overall model the number of fatalities is obtained from the relation

$$N = D[aP_{od} + bP_{id}P_f|_{id} + cP_{id}(1 - P_f|_{id}) + d]\qquad[\text{A17.4.1}]$$

where D is the population density (persons/m^2), N the number of fatalities, $P_f|_{id}$ the probability of fatal injury given that the person is within the cloud, P_{id} the probability of being indoors, P_{od} the probability of being outdoors and $a-d$ are coefficients. The coefficients c and d are associated with VCEs.

The model is shown in Table A17.3 in the form of the coefficients $a-d$ to be used in Equation A17.4.1.

A17.4.8 LPG fire scenarios

The LPG fire models just described are applied in the report to yield the hazard ranges and areas for the set of scenarios relevant to rail and road transport. These are summarized in Table A17.4.

A17.4.9 Condensed phase explosion

The condensed phase explosion model applicable depends on the type of explosive. This aspect is discussed in Section A17.10. An HD1.1 explosive constitutes a mass explosion hazard, an HD1.2 explosive a projection, or fragmentation, hazard and an HD1.3 explosive a fire hazard.

For the estimation of the overpressure from an HD1.1 explosive use was again made of the correlation of Kingery and Pannell (1964).

For the fragments from an HD1.2 explosive data were supplied by the MoD from explosives trials, which allowed an estimate to be made of the density of fragments possessing a kinetic energy in excess of 80 J, taken as the lethal value. From these data two graphs were constructed, one for persons outdoors and one for those indoors. The data are classified as confidential but the report states that the average probability of fatal injury so obtained for persons in the open at 200 m is 0.02, while that for those indoors at the same distance is an order of magnitude less.

The model used for the fire on an HD1.3 explosive is a vertical flame. Further details are not given.

Table A17.3 Coefficients for LPG hazard model fatal injury equation (Advisory Committee on Dangerous Substances, 1991) (Courtesy of HM Stationery Office)

	a	b	c	d
2 kgs^{-1} torch flame	332	3.3	0	0
36 kgs^{-1} torch flame	5,353	53.4	0	0
20 te BLEVE	47,596	182	0	0
40 te BLEVE	87,013	290	0	0
2 kgs^{-1} flash fire F/2 weather	7,800	78	0	0
36 kgs^{-1} flash fire D/5 weather	7,800	78.8	0	0
36 kgs^{-1} flash fire F/2 weather	185,000	1,850	0	0
Flash fire/torch flame/BLEVE (20 te) D/5	48,416	214.8	0	0
Flash fire/torch flame/BLEVE (40 te) D/5	87,225	298	0	0
Flash fire/torch flame/BLEVE (20 te) F/2	213,587	1,951	0	0
Flash fire/torch flame/BLEVE (20 te) F/2	238,145	2,016	0	0
20 te flash fire D/5	64,808	211	0	0
20 te flash fire F/2	57,969	222	0	0
40 te flash fire D/5	97,644	340	0	0
40 te flash fire F/2	88,180	346	0	0
36 kgs^{-1} VCE	185,000	1,850	925	0
20 te VCE	22,167	222	111	25,874
40 te VCE	34,636	346	173	52,916

Table A17.4   *Hazard ranges and areas for rail and road transport scenarios obtained from the LPG hazard model (after Advisory Committee on Dangerous Substances, 1991)*

A   Torch flames

Release rate (kg/s)	Side-on			End on		
	Length (m)	Distance to 50% lethality[a] (m)	Distance to 1% lethality (m)	Diameter (m)	Distance to 50% lethality[a] (m)	Distance to 1% lethality (m)
2	12.9	8.5	12.3	4.3	6.7	9.4
36	54.6	33	48	18.2	26.3	36.5

B   Flash fires: instantaneous releases (rail tank wagons)

Vessel capacity (te)	Distance to 50% lethality		Distance to 1% lethality	
	D5 (m)	F2 (m)	D5 (m)	F2
20	70	50	90	75
40	80	70	110	95

C   Flash fires: instantaneous releases (road tankers)

Vessel capacity[b] (te)	Distance to 50% lethality (m)	Distance to 1% lethality (m)
15	140	150

D   Flash fires: continuous releases

Release rate (kg/s)	Area of fire	
	D5 (m^2)	F2 (m^2)
2	310	7,800
36	7,800	185,000

E   BLEVEs (rail tank wagons)

Vessel capacity	Fireball radius (m)	Fireball duration (s)	Distance to 50% lethality (m)	Distance to 1% lethality (m)
20	76	12	110	175
40	96	15	160	245

F   BLEVEs (road tankers)

Vessel capacity	Fireball radius (m)	Fireball duration (s)	Distance to 50% lethality (m)	Distance to 1% lethality (m)
15	69	11	95	150

[a] This is also the distance to spontaneous ignition.
[b] Cloud radius 75 m.

A17.4.10 Condensed phase explosion: hazard range

It is convenient to give at this point the estimate quoted in the report of the hazard range of a condensed phase explosion.

The ranges for fatal injury by thermal radiation or overpressure are in practice about 80–85 m, being limited by the maximum permitted loads of explosive of 16 te for a lorry and 20 te for a rail wagon. These are the distances at which the proportion of fatal injuries would be about 5%.

A17.4.11 Toxic release

For toxic release use is made of the appropriate gas dispersion model to determine toxic concentration and hence, for a particular exposure period, the corresponding toxic load.

The conventional infiltration model is used to determine the indoor toxic load.

For the number of persons affected by the release, the general approach is that used in the impact model described in the next section. The concentration contours are calculated at which the probabilities of fatal injury are 1.0, 0.9, 0.5 and 0.1. This gives between these isopleths three zones with concentrations denoted C_1, C_2 and C_3 and with average probabilities of fatality taken as 0.95, 0.3 and 0.1.

For each zone there is a corresponding probability of escape indoors, the values used being 0 for C_1, 0.2 for C_2 and 0.8 for C_3.

An analysis is given of the possible responses of an individual affected by a toxic gas cloud. The results are shown in Figure A17.1.

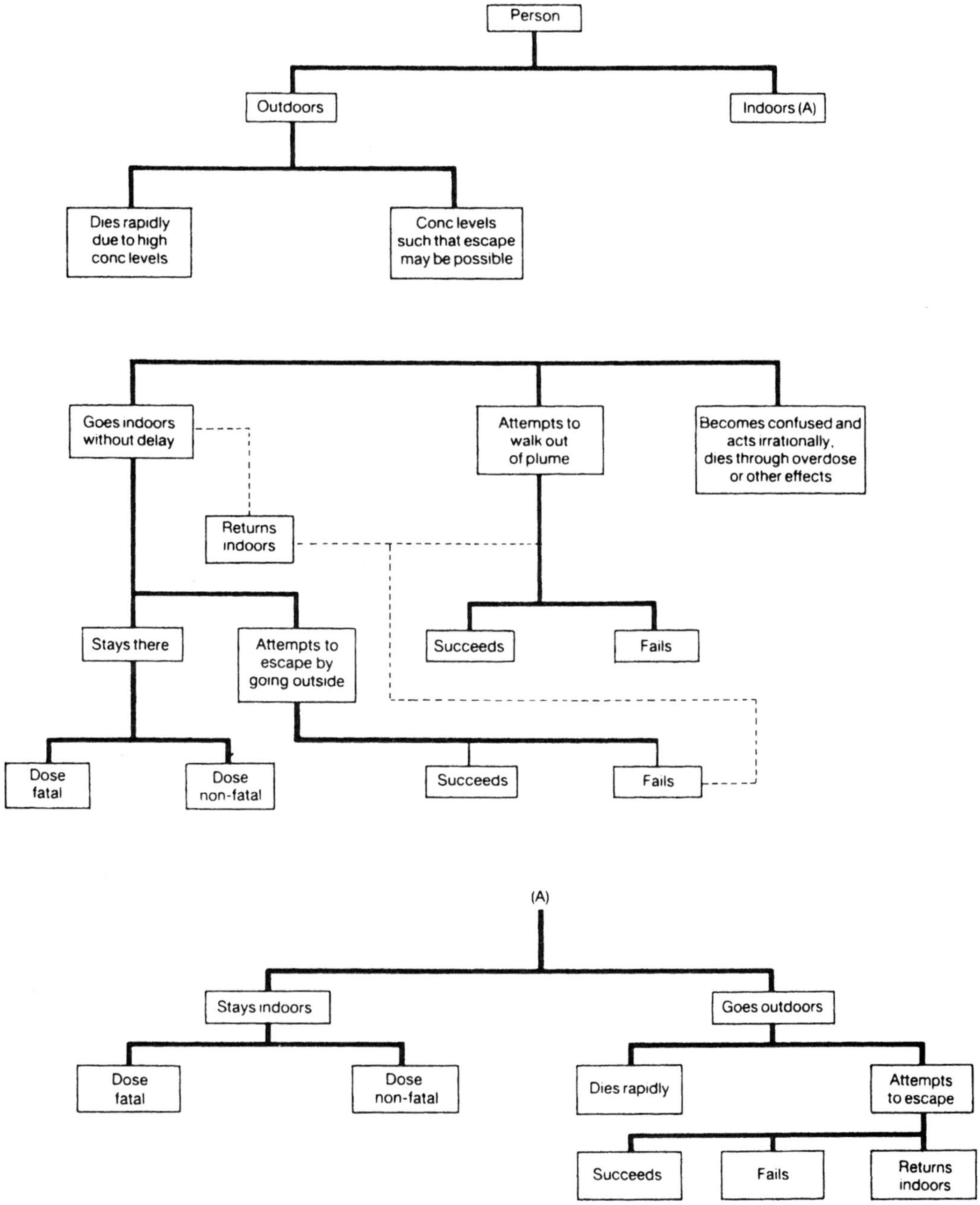

Figure A17.1 *Possible responses of an individual affected by a toxic gas cloud (Advisory Committee on Dangerous Substances, 1991) (Courtesy of HM Stationery Office)*

A17.5 Injury Relations

Injury models are used in the study to relate the intensity of the physical effects to the probability of injury. These are mainly probit equations, but HSE dangerous doses are also utilized. Some use is also made of rules-of-thumb.

The injury relations fall into two sets. The first set comprises HSE models, which are used for rail and road transport, the second set relations incorporated in the SAFETI code, which are used for ports. In the following account the HSE relations are described first, followed by those used in SAFETI, where applicable. In some cases no details are given for the latter; where this is the case, it should not be assumed that they are the same as the HSE relations.

A17.5.1 Thermal radiation

For fatal injury from thermal radiation use is made of the probit equation of Eisenberg, Lynch and Breeding (1975)

$$Y = -14.9 + 2.56 \ln(I^{4/3}t) \qquad [A17.5.1]$$

where I is the thermal radiation (kW/m^2), t the time (s) and Y the probit.

In the SAFETI work on fireballs from hot rupture of large cargos in port, a modification of the Eisenberg equation was used. In the region of thermal radiation above 500 kJ/m^2, it was assumed that 75% of persons outdoors and 25% of those indoors are killed, either by radiation or by secondary ignition effects.

A17.5.2 Fire engulfment

Injury can also occur due to engulfment in a fire. The assumptions made are that a person outdoors is killed, but that for a person indoors there is a certain probability of survival with the complementary probability $P_{f|id}$ of becoming a fatality.

In the SAFETI work the assumptions made were that for a flash fire all persons outside within the LFL contour and 30% of those indoors within the contour are killed but that no fatalities occur beyond this contour.

A17.5.3 Explosive: explosion overpressure

An HD1.1 explosive constitutes a blast hazard. For fatal injury two equations are given, the first being used in the road and rail transport study and the second, and more up-to-date, in the port study, reflecting developments as the work progressed.

The relationship given is for fatalities amongst persons indoors and is based on data on fatal injuries from V-2 rockets in the Second World War, with allowance for the effect of bomb shelters. In this situation building collapse is a more significant cause of injury than overpressure. The data are shown in Figure A17.2(a). The relation is applied also to persons outdoors in the built-up area and therefore near to buildings.

The first version of the equation is

$$Y = 1.47 + 1.37 \ln p^o \qquad [A17.5.2]$$

where p^o is the peak side-on overpressure (psi) and Y the probit. The second version is

$$Y = 2.47 + 1.37 \log_{10} p^o \qquad [A17.5.3]$$

A17.5.4 Explosive: explosion fragments

The hazard from an HD1.2 explosive is fragments. As stated in the previous section, the lethality of a fragment is taken as a function of its kinetic energy, and thus of mass and velocity, the lethal value being taken as one in excess of 80 J.

A17.5.5 Explosive: thermal radiation

The hazard from an HD1.3 explosive is thermal radiation. Figure A17.2(b) shows the relationship given in the report for this.

A17.5.6 Chlorine toxicity

For fatal injury from exposure to chlorine use is made of the following probit equation:

$$Y = -4.4 + 0.52 \ln(C^{2.75}t) \qquad [A17.5.4]$$

where C is the concentration of chlorine, t the exposure time and Y the probit.

A17.5.7 Ammonia toxicity

For fatal injury from ammonia the probit equation used is

$$Y = -12.2 + 0.8 \ln(C^2t) \qquad [A17.5.5]$$

where C is the concentration of ammonia, t the exposure time and Y the probit.

The SAFETI code uses for fatal injury from ammonia the probit equation.

$$Y = -9.82 + 0.71 \ln(C^2t) \qquad [A17.5.6]$$

where C is the concentration of ammonia (ppm), t the exposure time (s) and Y the probit. The group (C^2t) is actually computed as the time integral of the square of the concentration.

A17.5.8 Hazard impact

The method principally used in the report to estimate the impact of the hazard is as follows. The ranges are determined at which the physical effect is lethal at the 0.90, 0.50 and 0.10 probability levels. The areas within these contours are denoted A_{90}, A_{50} and A_{10} respectively. The average lethality within each area is taken as the average of the lethalities at the bounding contours.

Thus, the lethality with the circle A_{90} is 0.95 ($= (1.0 - 0.90)/2$), that within the annulus ($A_{50} - A_{90}$) is 0.7 and that within the annulus ($A_{10} - A_{50}$) is 0.3. This yields the hazard impact relation

$$N = D[0.95A_{10} + 0.7(A_{50} - A_{90}) + 0.3(A_{10} - A_{50})]$$
$$[A17.5.7]$$

where A is an area (m^2), D the population density (persons/m^2) and N the number of fatalities. Use is also made of variations on this basic equation.

A17.6 Population Characteristics

In transport scenarios the characterization of the population exposed tends to be much more complex than for fixed installation scenarios. A significant proportion of the modelling described in the report is addressed to this aspect.

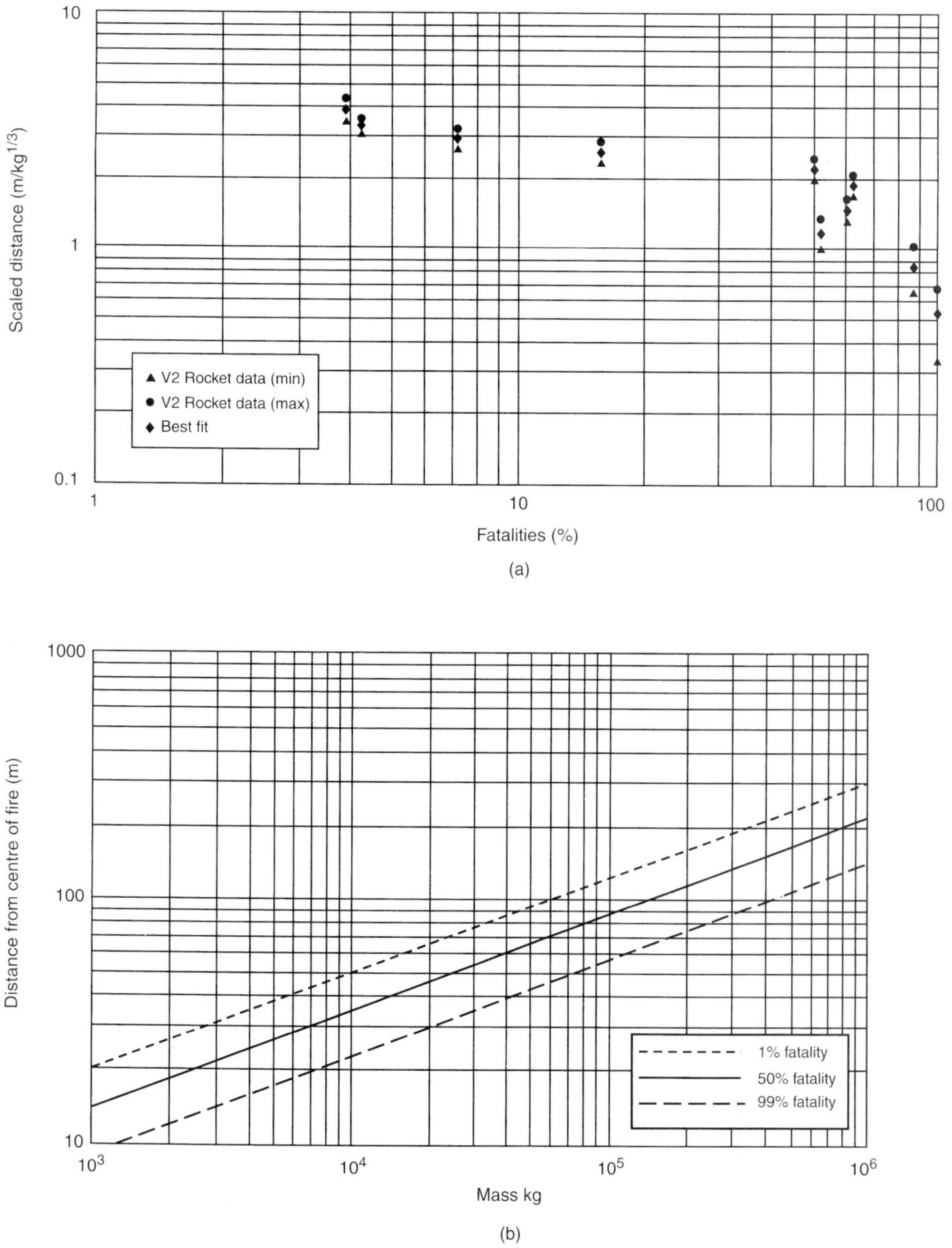

(a)

(b)

Figure A17.2 *Hazard ranges for explosives (Advisory Committee on Dangerous Substances, 1991): (a) HD1.1 explosive (mass explosion hazard) and (b) HD1.3 explosive (fire hazard) (Courtesy of HM Stationery Office)*

For the general population the report utilizes four categories of population density. These are 4210, 1310, 2120 and 20 persons/km^2.

The proportion of the general population outdoors is taken for the purposes of the assessment as a function of the meteorological conditions, particularly conditions D5 and F2.

This characterization of the general population essentially suffices for the assessment of fixed sites such as marshalling yards and ports but needs to be supplemented by additional models for rail and road transport. These models, which are described below, are relatively complex but this complexity is necessary to obtain a realistic estimate of the risks.

A17.7 Rail Transport

The hazard of rail transport of non-explosive materials is considered in Appendix 8 of the report.

An account has been given in Chapter 23 of the rail transport environment in the United Kingdom. It includes data given in the ACDS report on tank wagon capacities and movements and on release frequencies and probabilities.

A17.7.1 Wagons and movements
For the four hazardous materials studied, the rail tank wagon capacities and movements are given in Table 23.23. The wagon capacities are for motor spirit 32 and 75 te, for LPG 20 and 40 te, for chlorine 29 te and for ammonia 53 te.

Rail movements along the major routes in the United Kingdom in 1985 are shown in Table A17.5, which also gives the population densities.

A17.7.2 Exposed population
The population exposed to an en route rail accident considered in the report are the personnel on the trains involved, the passengers in passenger trains (PTs) and the general off-track population.

For the density of the general population along the track the basic method used was to use a map to assign each square kilometre of the trackside area to one of the four population densities described in Section A17.6.

A check was made on the adequacy of using these generalized values by taking one of the rail routes and

determining the population densities with circles of 300 m radius along the track from the 1981 Census enumeration districts. The process was repeated with circles of smaller and larger diameter. The results showed that in general the map and census methods gave similar results, except that the map method underestimated the urban population in a few places.

For the exposure of passengers in a PT to hazard from a heavy goods trains (HGT) five basic scenarios are considered. One is the involvement of a PT waiting at a stop signal in the hazard zone of a release from an HGT, the obedient train case. The other four are (1) collision of the PT and HGT, causing a release; (2) collision of a PT and an HGT, where a release has already occurred; (3) a PT entering the hazard zone of a release from an HGT and (4) a PT entering the hazard zone of a release from an HGT and colliding, causing a further release. Two wind conditions are considered: (1) wind along the track and (2) wind perpendicular to the track.

The proportion of carriages in the PT involved is the ratio of the length of the hazard-affected zone to the length of the train.

With regard to air infiltration into a PT, the ventilation on most modern trains is controlled from the driver's cab and is set at about 13 air changes/h. The driver experiences a ventilation rate about six times this and is thus effectively out of doors.

A17.7.3 Initiating events and event frequencies
The report considers two main types of initiating event.

These are puncture of a tank wagon by collision or derailment and failure or maloperation of the tank wagon equipment.

The treatment given for the frequency of these events for each of the four substances and for the probability of

Table A17.5 *Rail traffic and population densities along some major routes in the United Kingdom in 1985 (Advisory Committee on Dangerous Substances, 1991) (Courtesy of HM Stationery Office)*

	Motor spirit	*LPG*	*Ammonia*	*Chlorine*
All movements (wagon km/year)	10,199,095	1,390,590	1,348,080	313,668
Total wagon journeys/year	55,814	4,334	4,500	2,342
Major route	Merseyside–Leeds–Humberside	Hampshire–Midlands	Teeside–Scotland	Merseyside–Anglesey
Route length (km)	223	329	262	156
Number of wagons on route/year	5300 × 32 te	1,323	2,204	1,216
	3900 × 75 te			
Wagon km/year on route	2,051,600	435,267	530,288	189,696
% Traffic in that substance	20	31	39	60
Population density (people/km²)	*Aggregate length of route with that population density*			
Urban (4210)	15	30	12.5	0
Both sides[a]				
Suburban (1310)				
Both sides[a]	36.5	48	40.5	19
One side only[a]	1.5	20	12.5	24
Built-up rural (210)				
Both sides[a]	17.2	14	15	24
One side only[a]	0	0	0	1.5
Rural (20)				
Both sides[a]	143.2	215	181.5	87.5
One side only[a]	1.5	22	12.5	25.5
Tunnel	9.5			

[a] Indicates whether the population is on both sides of the railway or on one side only, the other side being rural.

ignition, immediate or delayed, of flammables is described in Chapter 23.

A17.7.4 Motor spirit events

For motor spirit, the scenarios considered are instantaneous releases of 32 and 75 te and continuous releases of 25 kg/s (equivalent to a 100 mm hole).

A release of motor spirit, if ignited, results in a pool fire. The pool spread and fire model used is described in Section A17.4.

For a continuous release the times for the 32 te and 75 te wagons to empty are approximately 20 and 50 min, respectively. The size of pool, and pool fire, formed depends on whether ignition occurs immediately. If ignition is delayed the radius of the pool from a 32 te wagon would be some 24 m and that for a 75 te wagon some 37 m. With immediate ignition, the pool radius in both cases would be 12 m. The pool radii for instantaneous releases would be similar.

A17.7.5 LPG events

For LPG, the scenarios are instantaneous releases of 20 and 40 te and continuous releases of 2 and 32 kg/s.

If the release is instantaneous and is ignited immediately, a fireball occurs. Otherwise, a flammable cloud forms. If this cloud is ignited, there is either a flash fire or a VCE.

If the release is continuous and is ignited immediately, a torch fire or a BLEVE occurs. Otherwise a flammable cloud forms. If this cloud is ignited, there is either a flash fire or a VCE. The flash fire may be accompanied by a torch fire or a BLEVE.

The report gives event trees for these scenarios.

A17.7.6 Chlorine and ammonia events

For chlorine and ammonia, the scenarios considered are as follows:

	Chlorine	Ammonia
Two-phase release from valve (kg/s)	1.3	2.1
Single phase release from puncture (kg/s)	45.1	33.5
Instantaneous release from catastrophic failure (te)	29	53

The development of these events is determined using the hazard models described.

A17.7.7 Marshalling yards

In addition to these en route events, the report also addresses releases occurring in marshalling yards. The two materials of interest in this regard are LPG and chlorine. The frequency estimates for these events for the two materials are given in Chapter 23.

The release scenarios considered are the same as for the en route case.

A17.8 Road Transport

The hazard of road transport of non-explosive materials is considered in Appendix 9 of the report.

An account has been given in Chapter 23 of the road transport environment in the United Kingdom. It includes data given in the ACDS report on tanker capacities and movements and on release frequencies and probabilities.

A17.8.1 Vehicles and movements

For the four hazardous materials studied, the road tanker capacities and movements are given in Table 23.6. The tanker capacities are for motor spirit 20–25 te, for LPG 15 te, for chlorine 17 te and for ammonia 15 te.

A17.8.2 Exposed population

The population exposed to an en route road accident considered in the report are the personnel of the vehicles involved, the other road users and the general off-road population.

The road transport situation is rather different from that of rail. In particular, it is necessary to take into account the wide variety of types of road and road usage.

Another basic difference is that the probability of other persons becoming involved in an incident is much higher for road than for rail.

For the density of the general population along the road the method is essentially similar to that used for rail transport.

The exposure of other road users to hazard from a road tanker is treated in terms of two zones, one for users behind the tanker and one for users on the other side of the road. For the former it is assumed that traffic backs up behind the tanker and that the section of the zone ahead of it is clear. In the second zone, on the other carriageway, the population density depends on whether the traffic stops or continues to move. It is assumed that the density in this zone is half that in the first zone.

The density of the road user population backed up behind the tanker is computed as follows. The traffic is assumed to consist of 10% heavy goods vehicles (HGVs). The length of road occupied by an HGV is taken as 20 m and that occupied by other vehicles as 4 m. There are assumed to be 1.5 persons/vehicle. This yields for motorways and for other roads population densities of 0.056 and 0.05 persons/m^2, respectively.

The full set of eight zones defined in the report comprises four on the same side of the road as the tanker and four on the other side. Zone a is one of standard population density, Zone b one of high density, Zone c a clear zone and Zone d the zone of road users behind the tanker. Zones h, g, f and e are, respectively, the corresponding zones on the other carriageway.

With regard to air infiltration into vehicles, work by M. Cooke (1988) has shown that for a car travelling at 40 miles/h the ram effect is sufficient to give one air change per minute and air ingress will be even greater if the ventilation fan is on. For a stationary car use of the ventilation fan will again give about one air change per minute. Even if the fan is off, the car is not expected to provide significant protection, the volume of the air space being small.

A17.8.3 Initiating events and event frequencies

The report considers two main types of initiating event. These are puncture of a tanker by collision and failure or maloperation of the tanker equipment.

The treatment given for the frequency of these events for each of the four substances and for the probability of ignition, immediate or delayed, of flammables is described in Chapter 23.

A17.8.4 Motor spirit events

For motor spirit, the scenarios considered are instantaneous releases of 4, 8 and 12 te and a continuous release of 25 kg/s (equivalent to a 100 mm hole). The tankers contain six compartments and the instantaneous releases correspond to spills from one, two or three of these.

A release of motor spirit, if ignited, results in a pool fire. The pool spread and fire model used is described in Section A17.4.

The pool areas obtained are

Spill	Pool area (m^2)	
	Immediate ignition	Delayed ignition
25 kg/s	314	908
4 te	707	1018
8 te	1385	1964
12 te	2124	3019

A17.8.5 LPG events

For LPG, the scenarios are an instantaneous release of 15 te and continuous releases of 2 and 32 kg/s.

If the release is instantaneous and is ignited immediately, a fireball occurs. Otherwise, a flammable cloud forms. If this cloud is ignited, there is either a flash fire or a VCE.

If the release is continuous and is ignited immediately, a torch fire or a BLEVE occurs. Otherwise a flammable cloud forms. If this cloud is ignited, there is either a flash fire or a VCE. The flash fire may be accompanied by a torch fire or a BLEVE. The report gives event trees for these scenarios.

A17.8.6 Chlorine and ammonia events

For chlorine and ammonia, the scenarios considered are as follows:

	Chlorine	Ammonia
Two-phase release from valve (kg/s)	1.3	2.1
Single phase release from puncture (kg/s)	45.1	33.5
Instantaneous release from catastrophic failure (te)	17.5	15

The development of these events is determined using the hazard models described.

A17.8.7 Lorry stopover points

In addition to these en route events, the report also addresses releases occurring at lorry stopover points. The three materials of interest in this regard are LPG, chlorine and ammonia. The frequency estimates for these events for the three materials are given in Chapter 23. The release scenarios considered are the same as for the en route case.

A17.9 Marine Transport: Ports

The treatment of marine transport in appendix 7 of the ACDS report is confined to the hazard at ports, which constitute fixed installations.

The study of ports was performed using the SAFETI code.

A17.9.1 Ports

There are in the United Kingdom some 42 ports with handling profiles exhibiting wide variety. The approach adopted was to study a set of three ports which had a representative set of major hazards. These ports were River Tees, the largest British chemical port complex; Felixstowe, a busy port with moderate hazardous trades; and Shoreham, a small port with petroleum product trades.

A17.9.2 Materials and movements

The principal hazardous materials and the movements of these materials through British ports are shown in Table A17.6.

A17.9.3 Exposed population and ignition sources

The model for the exposed population is that incorporated in the SAFETI code. The population is characterized in terms of the numbers of persons within 100 m grid squares. Separate distributions are used for night and day, the night-time figures being based on census data and the day-time ones on adjustments to these data. The density of ignition sources is also modelled using the grid square method, with values of ignition probability assigned on the basis of judgement.

A17.9.4 Initiating events and event frequencies

The report first considers the following types of marine event: (1) collision, (2) grounding, (3) striking and (4) impact. A collision occurs where two ships run into each other, a striking where a moored vessel is struck by a passing vessel and an impact where a vessel runs into a dock wall or jetty.

The starting point for the estimation of the frequency of these events at British ports was a study by the National Ports Council (1976), which was somewhat old and has known defects. The data were therefore re-analysed and supplemented by three additional data sets. The results of this analysis are given in Table A17.7(A).

The effects of these events were assessed separately for tankers and for gas carriers. For tankers the effects were taken to be (1) collision or striking below water, (2) collision or striking above water, (3) grounding damage or (4) impact damage. For each case a representative hole size was assigned.

For gas carriers all four marine events were assumed to have one of two effects: (1) cold leak through a hole or (2) cold rupture of the tank contents. For a cold leak the hole size was taken as that of the loading pipe connection, located at the bottom of a prismatic tank or at the mid-height of a spherical or cylindrical tank. For a cold rupture the release was taken as instantaneous from a pressurized tank or as occurring over 5 min from a refrigerated tank. It was assumed that 90% of the releases were leaks and 10% ruptures.

Four other events were also considered: (1) transfer spills, (2) tanker explosion, (3) gas carrier fires and (4) ammonium nitrate ship explosions.

The frequency of transfer spills was estimated from incidents reported to the HSE under NADOR in a 5.25 year period 1981–86. Transfer spills tend to be small and possibly, unreported. There were only eight spills reported, and just one for liquefied gas.

There are significant differences between cargos and transfer arrangements which will affect the frequency of transfer spills. Differences identified included (1) cargo

Table A17.6 *Hazardous cargos handled at British ports (after Advisory Committee on Dangerous Substances, 1991)*

A Overall

Material group	Terminals	Visits (ships/year)	Amount per terminal (te/year)	Amount per movement (te/ship)	Total amount (te/year)
Crude oil	17	1,957	9,244,347	80,291	157,153,891
Low flashpoint products	43	5,163	478,261	3,983	20,565,228
High flashpoint products	41	3,126	340,233	4,463	13,949,554
Flammable liquefied gas	26	2,388	374,180	4,008	9,674,268
Toxic liquefied gas	4	87	113,826	5,233	455,303
Low flash chemicals	21	1,244	105,689	1,771	2,205,975
High flash chemicals	23	1,399	119,689	1,945	2,739,746
Explosive chemicals	19	258	14,283	1,052	271,377
Total		15,662			207,015,342

B Liquefied gas

Liquefied gas	Visits (ships/year)	Total amount (te/year)
Ammonia	87	455,303
Butadiene	14	20,990
Butane	548	2,247,856
C_4 mix	227	236,151
Ethane	76	250,848
Ethylene	255	623,886
LPG[a]	555	2,339,388
Propane	414	3,470,246
Propylene	222	354,824
VCM	78	130,079
Total	2476	10,129,571

[a] LPG comprises propane–butane mixes in varying proportions.

Table 17.7 *Frequency of some marine accidents at British ports (after Advisory Committee on Dangerous Substances, 1991)*

A Accidents by port type

	Collisions (per encounter)	Grounding (per km)	Striking (per passing)	Impact (per visit)
Open sea port	5.0×10^{-4}	6.5×10^{-5}	4.0×10^{-6}	2.2×10^{3}
Wide estuary	4.0×10^{-5}	8.0×10^{-6}	4.0×10^{-6}	2.2×10^{3}
Wide river	1.2×10^{-4}	1.6×10^{-5}	9.0×10^{-6}	2.1×10^{3}
Narrow river	5.0×10^{-4}	6.5×10^{-5}	4.2×10^{-5}	6.5×10^{3}

B Accidents by cargo group

	Transfer accident (per cargo transferred)	Explosion (per visit)
Crude oil	1.9×10^{-4}	1.3×10^{-5}
Low flash products	1.8×10^{-4}	3.5×10^{-6}
High flash products	1.8×10^{-4}	1.5×10^{-6}
Flammable liquefied gas	7.6×10^{-5}	—
Toxic liquefied gas	7.6×10^{-5}	0
Low flash chemicals	1.5×10^{-4}	1.3×10^{-5}
High flash chemicals	1.5×10^{-4}	0
Ammonium nitrate	0	7.0×10^{-6}

(liquid, pressurized liquefied gas or refrigerated liquefied gas), (2) transfer equipment (articulated hard arm or flexible hose), (3) proportion of cargos unloaded, (4) use of vapour recovery, (5) use of ranging alarms, (6) emergency shut-down system (none, basic or advanced), (7) quick release coupling (none, pre-1987 or post-1987) and (8) environment (berth tidal or non-tidal, number of passing vessels).

The various transfer spills were condensed into four: (1) 3 min full bore (LG only), (2) 5 min full bore (liquid cargos only), (3) 15 min full bore and (4) 10 min at 10% full bore.

Event trees were constructed to assist in estimating the proportion of each of these different outcomes. The branches in the event trees were determined by the following three factors (1) immediate operator reaction (within 1 min), (2) operation of BSD within 10 min and (3) effectiveness of BSD when operated.

The results of this work are the transfer spill frequencies shown in Table A17.7(B).

Tanker explosion frequencies were obtained from analysis of cases world-wide in Lloyds Casualty Returns for the period 1977–86. Table A17.7(B) shows the results.

The frequency of gas carrier fires was obtained from the study by Blything and Edmondson (1984 SRD R292). The data were processed to give incident rates per visit and per kilometre of approach.

It was estimated that there have been nearly 100 fire incidents but without any release of a cargo. Applying the statistics for the case of no event over a period of time, the estimate obtained for the probability of cargo release given a fire was 0.007. It was then assumed that 90% of such cases are hot leaks and 10% hot ruptures. Only the latter was considered further, the former being treated as leaks through safety valves which are consumed by the fire on the vessel.

For ammonium nitrate (AN) explosions the use of historical data was eschewed, since such incidents occurred many years ago and involved organic-coated material which is no longer used. Instead use was made of fault trees to predict the explosion frequency, the main path being a fire, probably from the engine room, entering a hold and melting part of the AN. Explosion could then be initiated by confinement, organic contamination or impact shock. The estimated frequency of AN ship explosions so obtained is 6.2×10^{-8} per cargo.

A17.9.5 Releases of flammable liquefied gas
Releases of flammable liquefied gas were modelled using the emission and dispersion models in SAFETI. The detailed treatment of the emission scenarios for cold releases, namely cold leaks from and cold rupture of pressurized and refrigerated tanks, is described in the report. The dispersion of the vapour was then obtained using the dense gas dispersion model.

The ignition model in the code is then used to determine the proportion of cases in which the cloud is ignited. The proportion of ignitions resulting in a VCE is taken as 10%, the other 90% as giving a flash fire.

Hot rupture of a pressurized tank is assumed to give a fireball. The size of tank is typically several hundred tonnes and the size of the resultant fireball is well outside the range for which correlations have been validated.

Hot rupture of a refrigerated tank was initially also assumed to give a fireball, but this outcome is uncertain. A pool fire may well be more likely. Modelling of these two

alternatives for this case indicated that they present comparable hazards.

A17.9.6 Releases of toxic liquefied gas
Releases of liquefied ammonia were modelled using the emission and dispersion models in SAFETI.

A17.9.7 Explosions of flammable liquid
For flammable liquid explosions modelling of representative events in respect of overpressure and fragments indicated that they are not in general significant beyond the vessel.

Such explosions do, however, cause casualties amongst the crew. From analysis of tanker losses in port in the period 1977–86 the following estimates were derived:

Proportion of crew killed	Probability given cargo explosion
0.15	0.37
0.40	0.11
0.65	0.04
1.0	0.04

A17.9.8 Ammonium nitrate explosions
Ammonium nitrate explosions are modelled using the TNT model with an appropriate value for the TNT equivalent of AN. For bagged AN this was taken as 13% of the shipment mass and for bulk AN as 33%. These values include allowance for the TNT equivalent of AN, the probable mass aboard when the explosion occurs and the efficiency of explosion for the different types of storage.

A17.9.9 Representative ports
The report gives the results of the detailed studies for the three representative ports of River Tees, Felixstowe and Shoreham.

A17.9.10 Extension to all ports
The risk results obtained for the three representative ports were applied using a 'simplified method', details of which are given in the report, to obtain an estimate of the national societal risk.

A17.10 Transport of Explosives

The hazard from the transport of explosives by rail and road is considered in Appendix 10 of the report.

A17.10.1 Categorization of explosives
The hazards presented by explosives include blast, fragments and fireball. The hazard which predominates in a particular case depends on the class of explosive.

The classification generally used in transport is the UN scheme, described in Chapter 23. The classes of explosive in this scheme which are of prime interest here are

HD1.1   Mass explosion hazard;
HD1.2   Projection hazard;
HD1.3   Fire hazard;
HD1.4   No significant hazard.

For the purpose of hazard assessment, the ACDS found it necessary to develop a categorization more adapted to this purpose, which reflects particularly the heat sensitivity of the substance or articles:

UN class	ACDS category	
	M	Heat sensitive substances in flammable packaging
HD1.1	N	Heat sensitive articles − not readily ignitable
	P	Heat insensitive substances
HD1.2	Q	Heat sensitive articles
HD1.3	R	Heat sensitive substances in flammable packaging
	T	Heat sensitive articles and substances in non-flammable packaging
HD1.4	W	Heat sensitive articles which present no great hazard

A17.10.2 Limits on loads
In the United Kingdom there are limits set to the load of explosive which can be carried. These are 16 te for a lorry and 20 te for a rail wagon, with a further limit of 40 te for a train.

A17.10.3 Rail transport of explosives: explosives and movements
The explosives moved in rail transport are predominantly (>97%) military explosives. In the year 1988/89 BR data showed that there were 8132 movements of wagons containing explosives. The average explosives train journey is 320 km and the explosives movements are therefore 2.6×10^6 wagon km/year.

A special investigation was made of the composition of this traffic, which was found to be by Hazard Division as follows: HD1.1 21%; HD1.2 12%; HD1.3 10% and HD1.4 44%. The explosives traffic was also analysed by the HSE categorization scheme, which yielded by proportion of wagons the following dominant categories: N 10%; P 11%; Q 12%; T 10%; W 44%; N/T 3% and Q/W 4%. Some 27% of explosives wagons contained HD1.1 explosives, while 44% contained only HD1.4.

An analysis was also conducted to obtain the average values of the NEQ, effectively TNT equivalent, of the loads.

A17.10.4 Rail transport of explosives: exposed population
The population exposed to the rail transport of explosives was determined in the study using a method essentially similar to that used for the four non-explosive substances.

In the event, the assessment showed that for rail the risk is predominantly to the off-site population.

A17.10.5 Rail transport of explosives: initiating events and event frequencies
The main mechanisms identified in the report for the initiation of explosives are (1) fire, (2) impact and (3) unsafe explosives.

An analysis was made for rail and road transport of data from various sources including the minutes of the ESTC of the MoD, the annual reports of HM Inspector of Explosives and the SRD explosives database EIDAS. There was for the United Kingdom over a 40 year period only one accident involving fatality, that at Peterborough in 1989 (HSE, 1990c). However, for both rail and road there were a number of dangerous occurrences, listed in Annex 2 of Appendix 10.

The report discusses in detail the methods by which it obtains its estimates of the explosive events in both forms of transport.

For rail transport it gives the following frequencies:

Initiating mechanism	Frequency (events/wagon km)
Unsafe explosives	1×10^{-9}
Fire	6×10^{-10}
Impact	1×10^{-10}

The frequencies of events from unsafe explosives and from fire are thus assessed as broadly comparable, that assessed for events from impact being much lower.

A17.10.6 Rail transport of explosives: explosions en route
The explosion event depends, as stated earlier, on the class of explosive carried, being for HD1.1 a mass explosion, for HD1.2 fragments and for HD1.3 a fireball. The models described in Section A17.4 were applied using the hazard impact method given in that section.

A17.10.7 Rail transport of explosives: explosions in marshalling yards
The study identified one marshalling yard which handled some 50% of explosives traffic and in which explosives were present almost continuously. An outline assessment was made, although no formal estimates of individual or societal risk were produced.

Application of the hazard model for the different classes of explosive showed that none of the events had a range to 10% lethality of more than 50 m. The population at this yard was beyond 110 m.

Analysis resulted in the following estimates for the frequency of initiation of an explosive load in a marshalling yard

Initiating mechanism	Frequency (events/year)
Unsafe explosives	1×10^{-3}
Fire	2×10^{-4}
Impact	2×10^{-4}

A17.10.8 Road transport of explosives: explosives and movements
An investigation was made of the composition of the road explosive shipments broadly similar to that made for rail. It was found to be by Hazard Division as follows: HD1.1 94%; HD1.2 2%; HD1.3 40%. The explosives traffic was also analysed using the ACSD categorization scheme, which yielded by mileage covered the following categories: M 51%; N 23%; P 20%; Q 2%; R 3%; T 1%.

An analysis was also conducted to obtain the average values of the NEQ of the loads. The loads were classified into three bands with the load being distributed according to the following probabilities

Band	Mean NEQ (kg)	Proportion of total mileage (%)
1	316	18
2	1,778	68
3	14,125	14

A17.10.9 Road transport of explosives: exposed population

The population exposed to the road transport of explosives was determined in the study using a method essentially similar to that used for the four non-explosive substances.

In the event, the assessment showed that for road the risk is predominantly to the road users.

A17.10.10 Road transport of explosives: initiating events and event frequencies

The initiating events in road transport are the same as those for rail transport. The analysis made to obtain the frequency of the road events has been described above. The detailed methods are given in the report.

For road transport it gives the following frequencies:

Initiating mechanism	Frequency (events/vehicle km)
Unsafe explosives	1×10^{-9}
Fire	2×10^{-9}
Impact	2×10^{-10}

The figures are somewhat similar to those for rail transport, the frequencies of events from unsafe explosives and from fire being assessed as broadly comparable, that assessed for events from impact being much lower.

A17.10.11 Road transport of explosives: explosions en route

The explosion event was modelled in a manner essentially similar to that used for rail transport. The models described in Section A17.4 appropriate to the class of explosive – HD1.1, HD1.2 or HD1.3 – were applied using the hazard impact method given in that section.

A17.10.12 Road transport of explosives: explosions at lorry stopover points

For explosives the situation at lorry stopover points is quite different from that at marshalling yards. Such points are always on premises licensed for the purpose. Lorries also stop briefly en route, but such stops are brief and are always attended. This risk was not pursued.

A17.11 Risk Criteria

The ACDS report gives in Appendix 6 a review of the principles underlying the setting of risk criteria and of the application of these to individual and societal risk.

The basic principle is that the risks should be as low as reasonably practicable (ALARP).

A17.11.1 Individual risk

With regard to individual risk to members of the public, the report refers to proposals made by the HSE. At the Hinkley Point Inquiry the HSE proposed that a fatality risk of 10^3/year is 'intolerable', though subsequent discussion suggested a lower level of intolerability. The HSE has also proposed a fatality risk of 10^{-6}/year as 'broadly acceptable'. The report adopts these criteria.

A17.11.2 Societal risk

In respect of societal risk, the ACDS criteria are formulated in terms of FN curves. A distinction is made between an FN curve for risk at national level and one for risk at local level. Both types of risk occur in the study, the en route risks of rail, road and explosives transport being treated mainly as national and the risks at fixed sites such as ports, marshalling yards, lorry stopover points and unloading points as local.

The basic principle is illustrated in Figure 9.39, which shows an FN criterion plot for a local risk. The FN space is divided into four parts by three lines. The upper line is the tolerability line. Risks above this line are regarded as intolerable. The middle line is the scrutiny line. Risks between this line and the tolerability line may be unjustifiable and require further study. The bottom line is the negligible risk line. Risks below this line are regarded as negligible. Those between the negligible risk line and the scrutiny line should be reduced applying the ALARP principle.

For the local risk criterion FN plot the ACDS proceeds as follows. The reference point is the risks assessed at Canvey. These risks were regarded as on the borderline of tolerability. The values used by the ACDS for these risks are one third those predicted in the Second *Canvey Report* (HSE 1981a). This Canvey FN curve is then used as a guide both to the setting of the absolute value of the tolerability line and to its slope. This is illustrated in Figure A17.3.

The negligible risk line for the local FN plot is set three decades below the tolerability line. The arguments for this are based essentially on examination of the implications of setting it either two or four decades below. If the line were two decades down, the expectation value would be six fatalities per 1000 years. Applying a value of a life of £0.5 million gives a potential expenditure of £3000 per annum, which might justify a QRA done every 10 years. It is argued that such a risk could hardly be regarded as negligible. On the other hand, if the negligible risk line were to be set four decades below the tolerability line, this would justify a potential expenditure of £30 per annum or £300 every 10 years. On the basis of the cost of staff time this would barely permit even a cursory discussion, which suggests that this setting of the line is well into the negligible risk region. The slope of the tolerability line, and of the other lines, is –1, again reflecting the assessed risk at Canvey.

The treatment in the report for the national risk criterion FN plot is less firm. A tentative scrutiny line is developed for ports which is derived not from any argument based on direct combination of the number of ports and of the local criteria but rather on the implication of transferring trade from any port where the risks are already at the tolerability limit. This line is entered on the FN plot for national societal risk for ports and also on other national societal risk plots as shown in the following paragraphs.

A17.11.3 Value of a life

As already mentioned, the report refers to the figure of £0.5 million for the avoidance of a statistical fatality, or the value of a life, used by the Department of Transport. It gives in appendix 7 on ports a discussion of cost benefit analysis of proposed remedial measures in which a factor of 4 is applied to this figure, yielding a value of a life of £2 million (1991 value) to allow for 'gross disproportion' and uses this latter in the subsequent analysis.

A17.12 Assessed Risks

The assessed risks quoted in the study are now described. The report is careful to quote confidence bounds and to

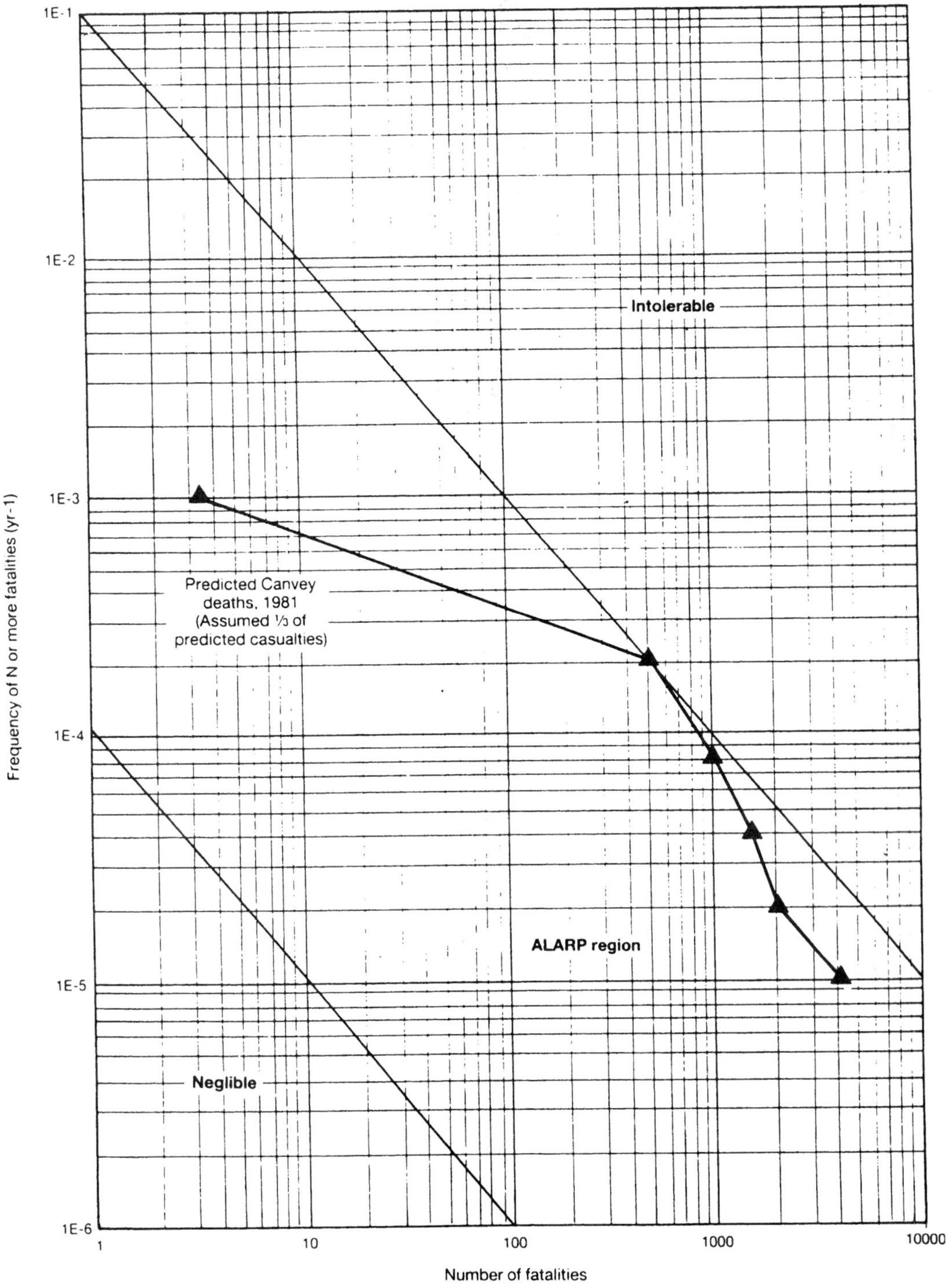

Figure A17.3 *FN curve risk criterion proposed by the ACDS for an identifiable community, $E-n = 10^{-n}$ (Advisory Committee on Dangerous Substances, 1991) (Courtesy of HM Stationery Office). Five concentrations are calculated at which the probabilities of fatal injury are close to 1.0 (concentrations C_1, C_2) and 0.9, 0.5 and 0.1 (three concentrations C_3). For each concentration there is a corresponding probability of escape indoors, the values used being 0 for C_1, 0.2 for C_2 and 0.8 for C_3*

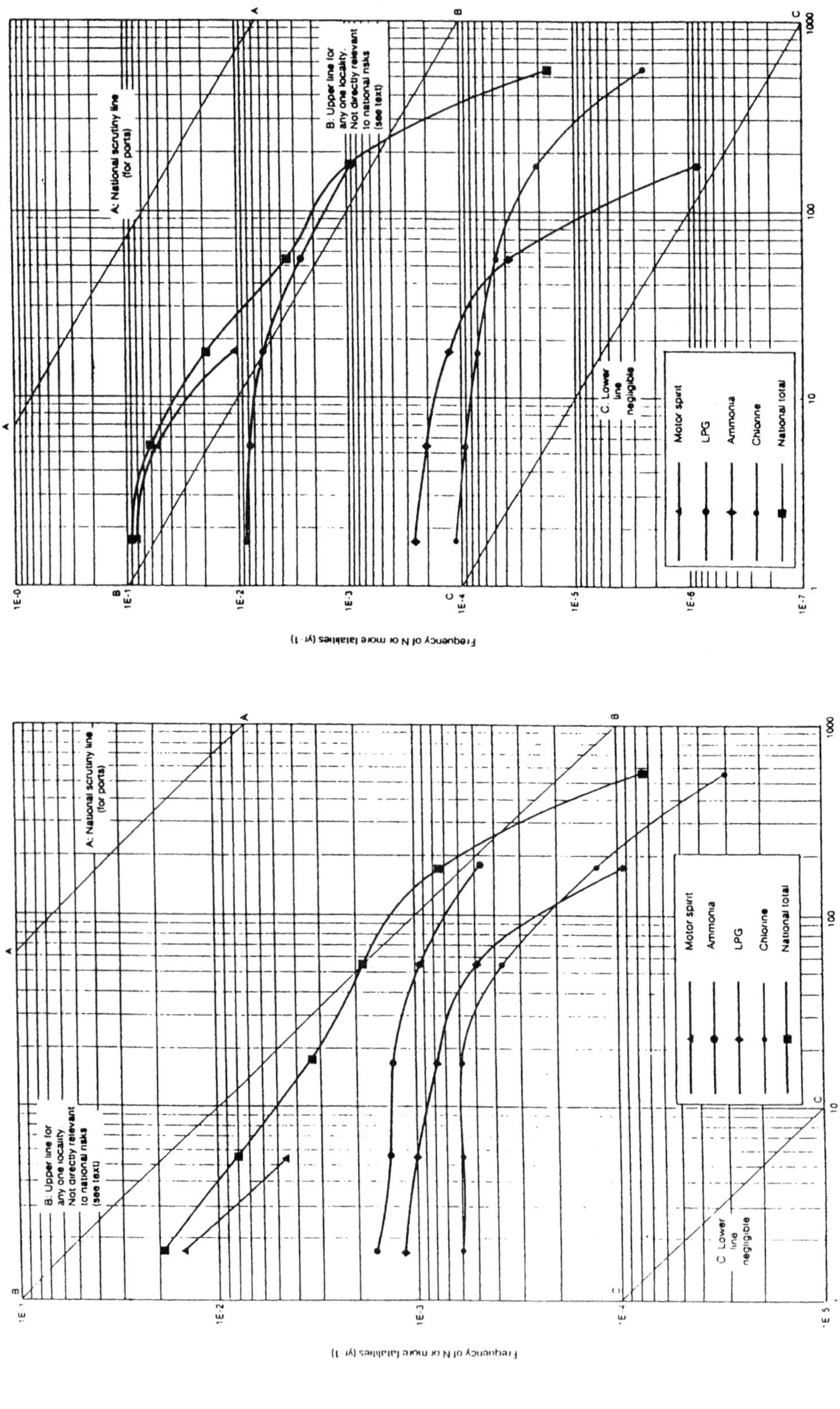

Figure A17.4 FN curves for national societal risks from rail and road transport of four hazardous materials (Advisory Committee on Dangerous Substances, 1991): (a) rail transport and (b) road transport. $E-n = 10^{-n}$ (Courtesy of HM Stationery Office)

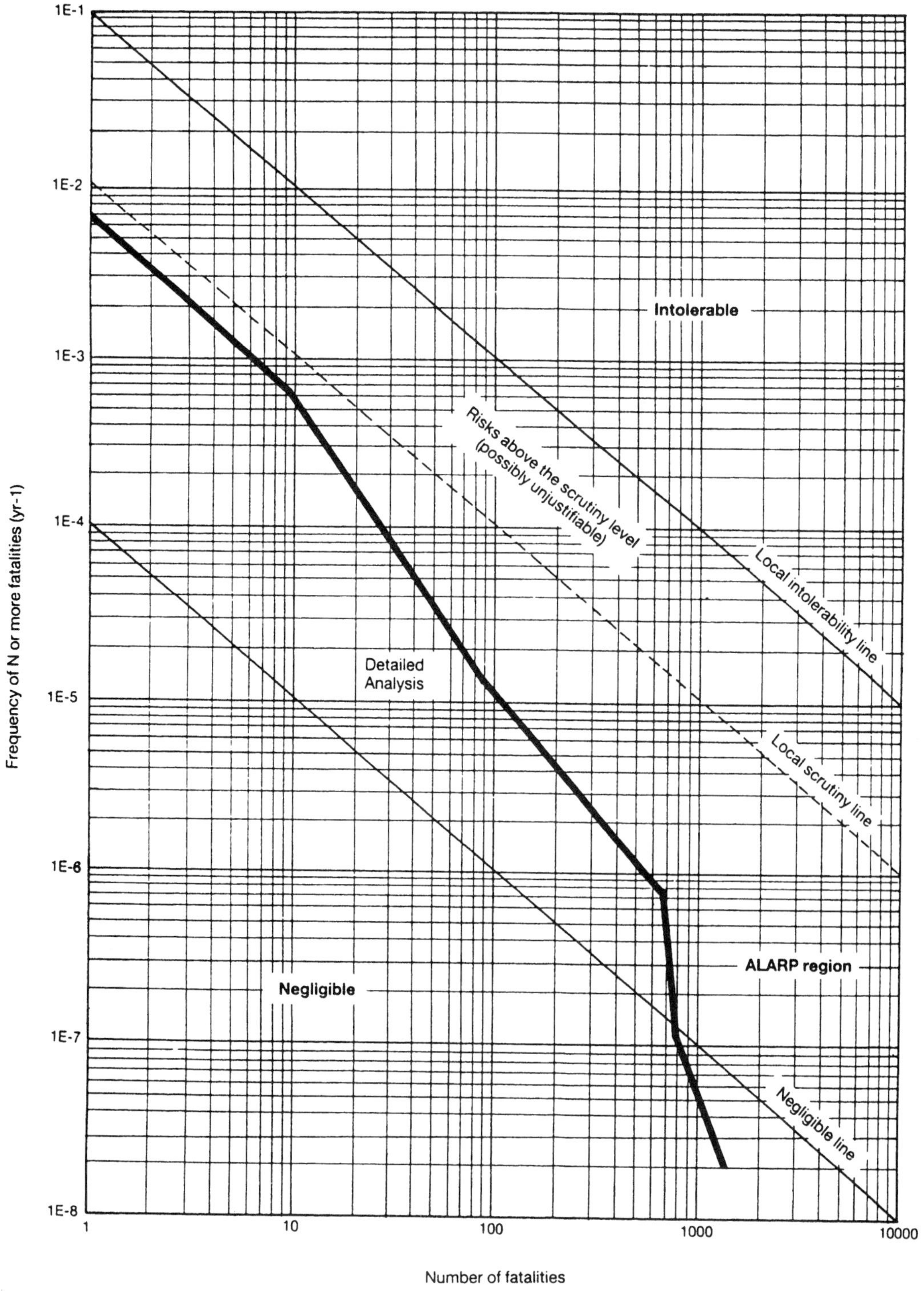

Figure A17.5   *Continued*

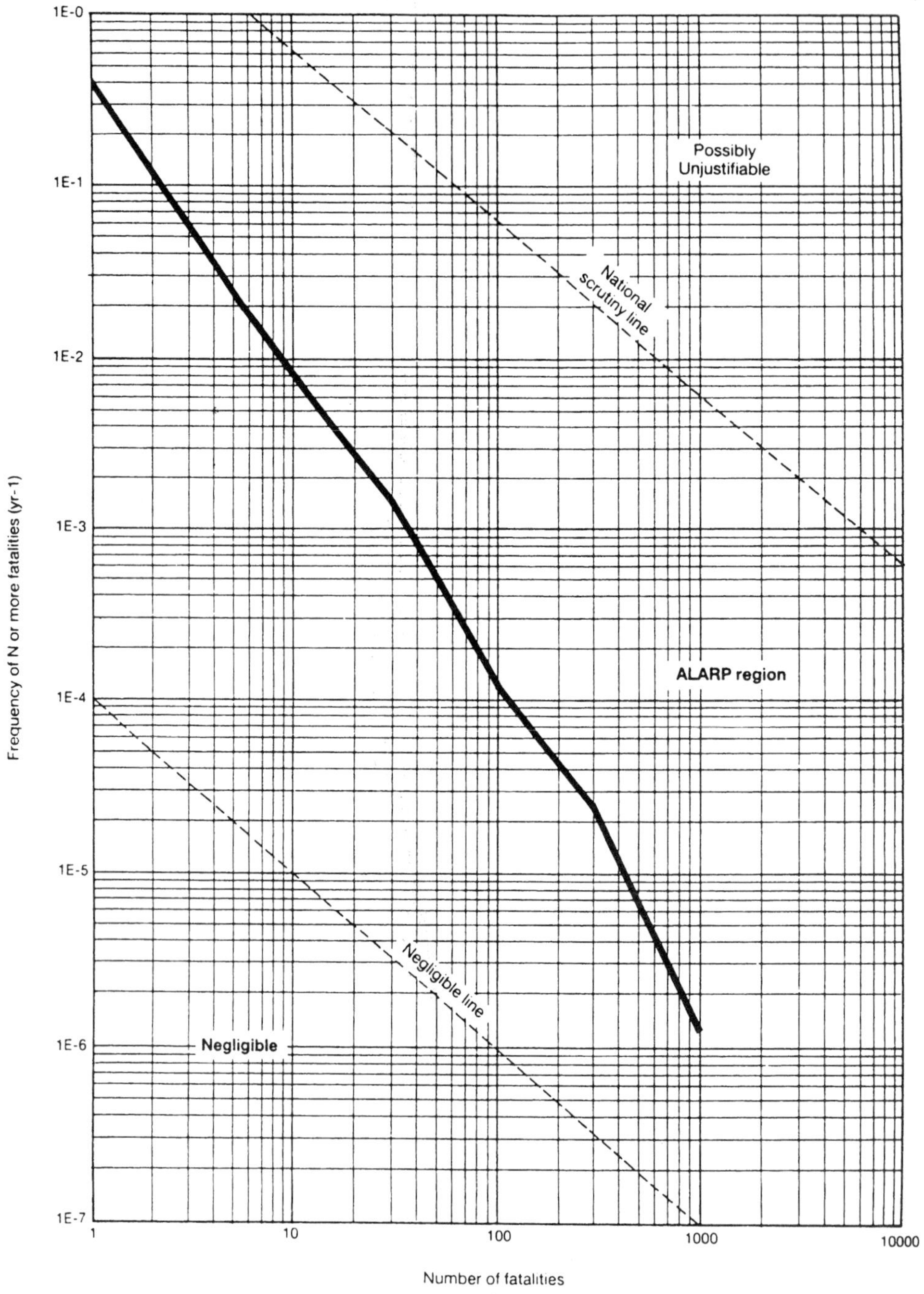

Figure A17.5   *FN curves for societal risks at British ports (Advisory Committee on Dangerous Substances, 1991): (a) local societal risk at Felixstowe and (b) national societal risk (simplified method). $E-n = 10^{-n}$ (Courtesy of HM Stationery Office)*

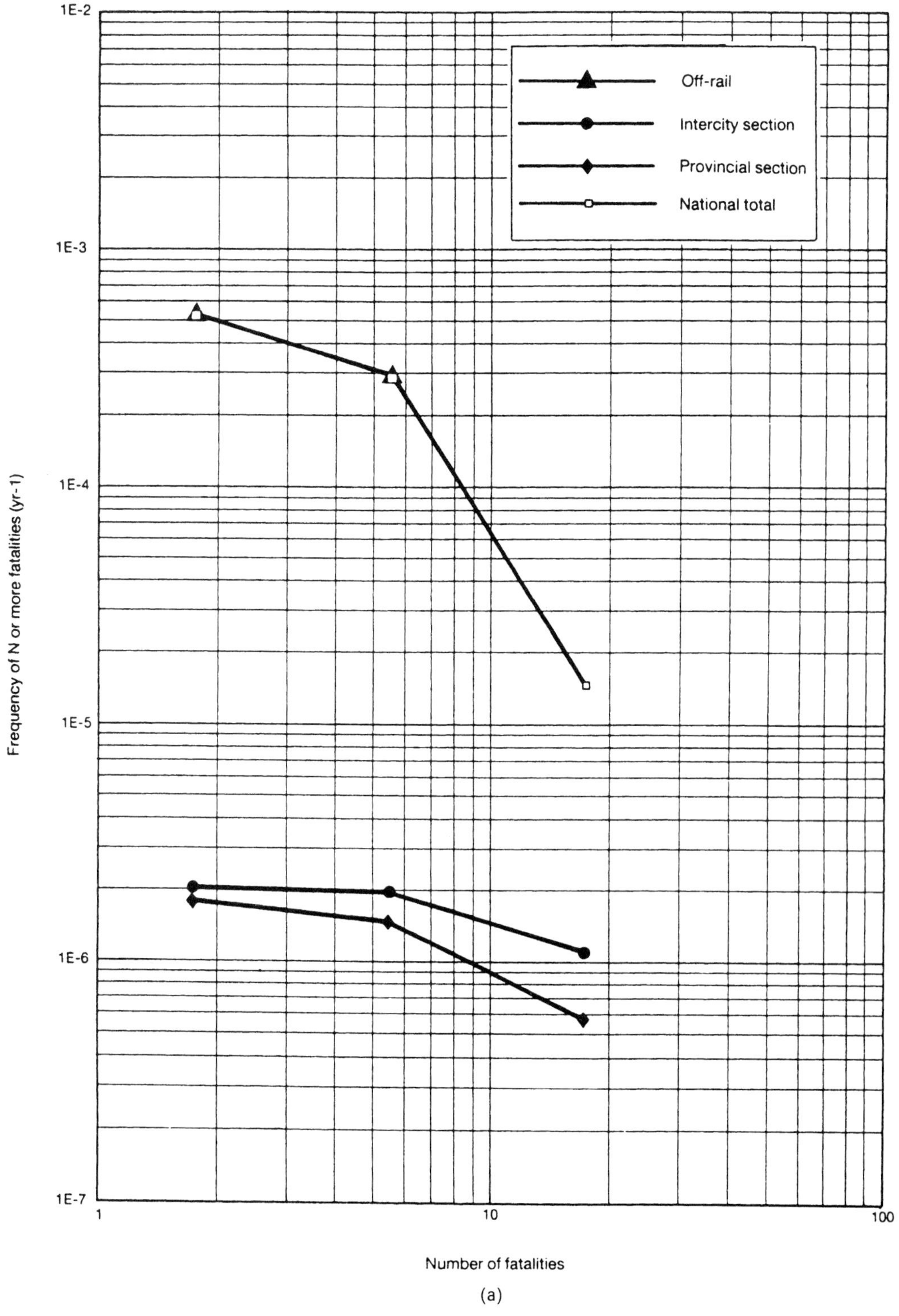

Figure A17.6   *Continued*

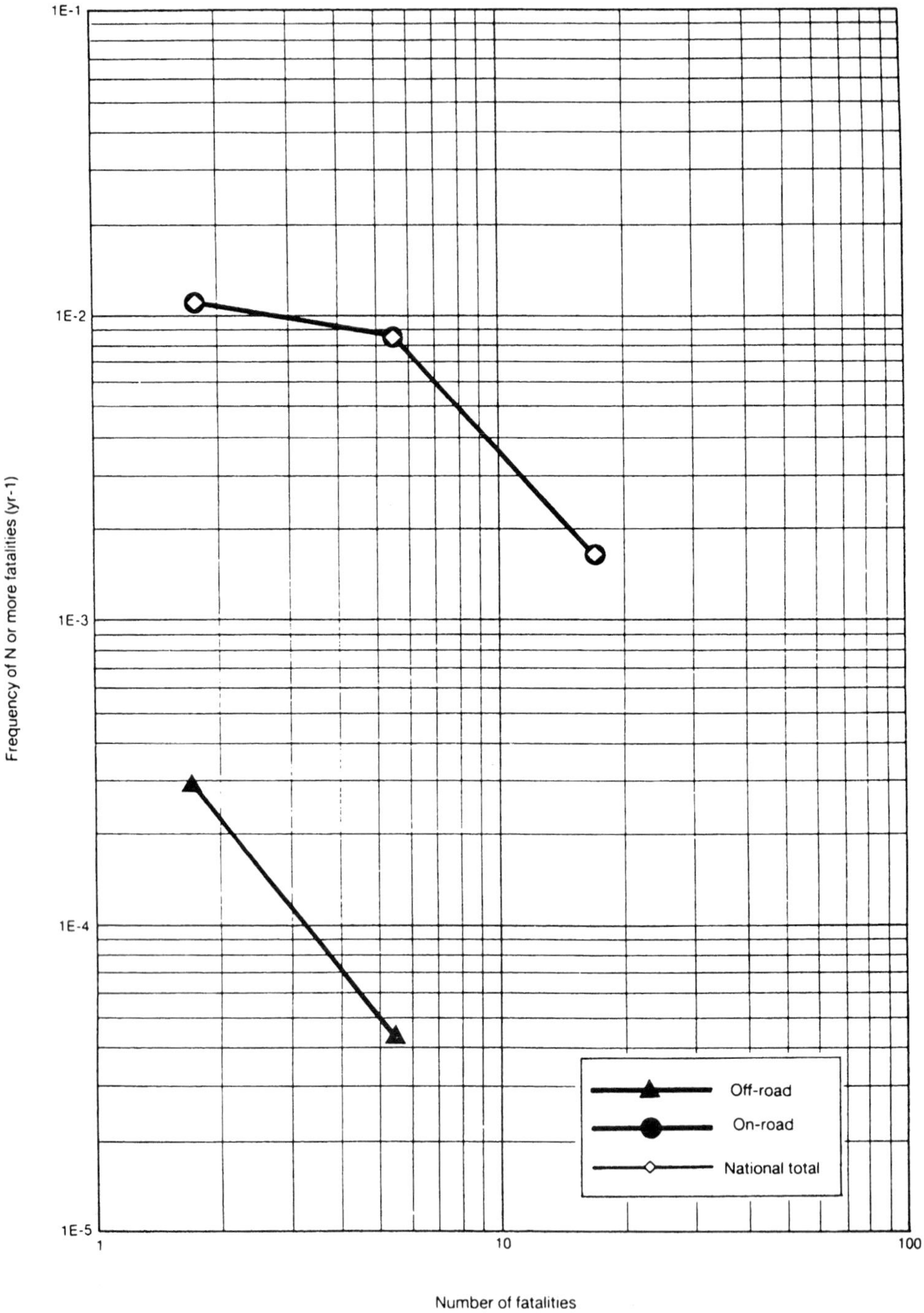

Figure A17.6 *FN curves for national societal risks from rail and road transport of explosives (Advisory Committee on Dangerous Substances, 1991): (a) rail transport and (b) road transport, $E-n = 10^{-n}$ (Courtesy of HM Stationery Office)*

note uncertainties in, and qualifications to, the figures given. The account given here is in outline only.

A17.12.1 Rail and road transport

The assessed risks of rail and road transport for the four non-explosive materials are presented as individual risks, as societal risks in the form of FN curves and as expectation values.

The individual risk to the public for rail transport of the four materials is assessed as negligible. For example, for a person living 50 m from the track the fatality risk from chlorine transport is predicted to be 2.4×10^{-7}/year. For road transport, the situation is rather more complex in that there may be certain locations such as a sharp corner which have a higher than average risk. But in general the risk is assessed as negligible and as only a small fraction of the overall risk faced by road users.

The national societal risks assessed for the transport of the four materials, and for the combination of these, are shown in Figures A17.4(a) and (b) for rail and road, respectively. For rail transport the predominant risks are from motor spirit and ammonia, for road transport from motor spirit and LPG. For both modes the risk lies between the tolerability and negligible risk lines.

The expectation values of the annual number of fatalities for the rail transport of motor spirit and ammonia are 0.074/year and 0.3/year and those for the road transport of motor spirit and LPG are 1.1/year and 0.8/year, respectively.

The report gives in Appendix 11 a comparison of the relative risks from rail and road transport. It makes the point that the choice of mode is not unrestricted. For example, motor spirit deliveries must at some stage be by road. A comparison of the rail and road risks shown in Figures A17.4(a) and (b) suggests that the risks of road transport are higher, but these graphs are for risks from actual traffic flows. Fair comparison needs to be based on a pattern of shipments which is the same for both modes. The report gives a detailed discussion but the overall conclusion is that the assessment made provides little support for a general preference for either mode on the grounds of safety.

A17.12.2 Ports

As far as regards ports, the study of the three ports found that in some limited areas near the ports individual risk was such that on the basis of the risk criteria adopted advice would be against new housing development.

The assessed societal risks are shown in Figure A17.5(a) and (b). The former shows the local societal risk for one of the ports, Felixstowe, obtained from the individual assessment for that port, and the latter the national societal risk for British ports, obtained from the simplified method described above. At none of the three ports was the local societal risk within the scrutiny zone. The national societal risk is also outside that zone.

A17.12.3 Transport of explosives

Turning to the risk from the transport of explosives, individual risk from en route transport by rail or road was not judged significant, but there was one marshalling yard where it might be.

The national societal risks from rail and road transport of explosives are shown in Figures A17.6(a) and (b), respectively. The risks from road transport are higher by an order of magnitude. This is accounted for only in part by the greater volume moved; the proximity of other road

users is also a factor. As Figure A17.6 shows, for explosives the principal contributor to the assessed societal risks in rail transport is the off-site population, whereas for road transport it is the road user population. For rail the fraction of the total risk contributed by the rail users is negligible, while for road the fraction contributed by the off-site population is 2–3%. For both modes the risk lies between the tolerability and negligible risk lines. The report indicates that further scrutiny would be appropriate.

A17.13 Risk Evaluation and Remedial Measures

The ACDS report gives an assessment of individual and of local and national societal risks, identifies certain activities or situations where the risks may need to be reduced or at any rate should be subject to further scrutiny and in Appendices 7, 10 and 12 makes proposals for a number of remedial measures.

One measure of general applicability is adherence to, and enforcement of, good working practice.

There are also various measures appropriate to the particular activity. For rail, it is suggested that the design and construction of tanker wagons for motor spirit and for ammonia, the improvement of communication by provision of radio telephones and the training of train drivers should receive attention.

For road, measures suggested relate to the design and construction of road tankers, use of designated routes, movement by night, provision of improved communication and availability of information in a chemical emergency and the training of drivers. Technical measures referred to for tankers are devices to monitor tyre pressure and so reduce the risk of a tyre fire; additives for middle distillates to reduce the risks from static electricity during switch loading and cut-off devices to prevent engine overrun leading to possible ignition of vapour from spills following a tanker puncture. There may be scope for reduction of risk by the use of designated routes. Communications could be improved by provision of radio telephones. Information on the characteristics and emergency handling of chemicals carried should be available from chemical emergency information centres 24 hours a day.

For ports, the proposals made in the report include measures related to hard arm loading equipment and ESD systems, communications within the port, control of pleasure vessels and land use planning.

While many of these proposals might be made without benefit of a hazard assessment, the study provides guidance on priority areas. It identifies a number of features as meriting further scrutiny. These include the dominant contribution of motor spirit and ammonia to the en route national societal risks of rail transport and of motor spirit and LPG to those of road transport; the existence of individual risk with land use planning implications within 100 m of lorry stopover points and the national societal risk for road transport of explosives.

At least as important as the assessed levels of risk is the improved understanding of the hazards involved. An example is the relative contributions to the national societal risk in the transport of explosives of the off-site and rail or road user populations.

Also of value are the negative results from the study. For the four non-explosive materials the report states that the risks from marshalling yards are not such as to inhibit further housing development in the zone where houses currently exist.

A17.14 Notation

$a-d$	constants
A	area (m^2)
C	concentration
D	population density (persons/m^2)
I	thermal radiation (kW/m^2)
N	number of fatalities
p^o	peak side-on overpressure

$P_{f\mid id}$	probability of fatal injury given that a person is indoors and within the cloud
P_{id}	probability that a person is indoors
$P_{od}\,A$	probability that a person is outdoors
t	exposure time
Y	probit

Equation A17.5.2
p^o	peak side-on overpressure (psi)

Equation A17.5.6
C	concentration of ammonia (ppm)
t	time (s)

Appendix 18

Offshore Process Safety

Contents

Any account of loss prevention needs to include some mention of offshore oil and gas activities, even if this is necessarily brief. There is a continuous interaction between developments onshore and offshore. The treatment given here is confined to an outline of offshore safety activities.

Offshore installations operate in a difficult and often hostile environment. The problems, not only of structures but also of processing, are challenging. The solution of these problems often involves technological innovation. The significance of this for safety and loss prevention is clear.

The report into the Piper Alpha disaster (Cullen, 1990), known as the *Cullen Report*, provides a wealth of detail both on the design and operation of an actual platform, albeit one of the older ones, in the North Sea and on the impact on it of a major accident event.

Accounts of offshore oil and gas are given in *Offshore Oil Technology* by Ranney (1979), *Onshore Impacts of Offshore Oil* by Cairns and Rogers (1981), *Introduction to Petroleum Development* by Skinner (1982), *Technological Guidelines for Offshore Oil and Gas Development* by Gilbert (1983), *Gas Production Operations* by Begg (1984), *Behaviour of Offshore Structures* by Battjes (1985) and *Safety in the Offshore Petroleum Industry* by Barrett, Howells and Hindley (1987) and by R.J.S. Harris (1978), Holmes and Mead (1978), Timar (1978), Shyvers (1981) and Waldie (1986).

The series of booklets by BP Petroleum Development Ltd (1990a–g) constitutes a useful starting point.

Much useful information is also given in *Offshore Installations: Guidance on Design, Construction and Certification* by the HSE (1990b), also called the *HSE Onshore Design Guide*.

Selected references on offshore safety are given in Table A18.1.

A18.1 North Sea Offshore Regulatory Administration

An outline of the legislation governing oil and gas activities in the UK sector of the North Sea has been given in Lees Edition 3, Chapter 3, with a summary of the principal legislation given in Table A18.2.

Two principal elements of the administration are the systems of internal control and of risk assessment. The *Guidelines for the Licencee's Internal Control 1979* describe in effect an SMS. The *Regulations Related to the Licencee's Internal Control 1985* make this a regulatory requirement.

With regard to risk assessment, the *Regulations Concerning Safety Related to Production and Installation 1976* contained a requirement that if the living quarters were to be located on a platform where drilling, production or processing of petroleum was taking place, a risk evaluation should be carried out. At this date the evaluation was largely qualitative. The move to a more quantitative approach came with the *Guidelines for Safety Evaluation of Platform Conceptual Design 1981*. These had as a central feature, the provision of a sheltered area, required the conduct of a concept safety evaluation (CSE) and specified numerical acceptance criteria.

The Guidelines defined a design accidental event as one that does not violate any of the following three criteria:

- at least one escape way from central positions which may be subjected to an accident, shall normally be intact for at least an hour during a design accidental event;

Table A18.1   *Selected references on offshore safety issues*

Offshore regulatory administration
Burgoyne (1980); Barrett, Howells and Hindley (1987); Lyons (1989); Cullen (1990); Higgs (1990); IP (1990 PUB 51, 1992 PUB 64); Petrie (1990); Priddle (1990); Barrell (1992); J.W. Griffiths (1992); HSE (1992 OTI 588); H. Hughes (1992); Lees (1992); Leiser (1992); Bibbings (1992); Heiberg-Andersen (1990); Ognedal (1990, 1994); Tveit (1990)

Offshore hazards
Leblanc (1981); Anon. (1983p); CuUen (1990); Ognedal (1990). **Wind and waves:** DoEn (1977a, 1984); HSE (1990b) *Earthquakes:* Cornell and Vanmarcke (1975); Wiggins, Hasselman and Chrostowski (1976); M.W. Mitchell (1983) *Vessel-platform collisions:* Hathway and Rowe (1981); DoEn (1984); Technica Ltd (1985); HSE (1990b); **Blowouts**: Fischer (1982); Anon. (1983b); Dahl *et al.* (1983); Goins and Sheffield (1983); Podio, Fosdick and Mills (1983); Milgram and Erb (1984); P.G. Mills (1984); Holand and Rausand (1987); Ostebo *et al.* (1989) *Flammable gas clouds:* HSE (1992 OTI 591); Cleaver, Humphreys and Robinson (1994) *Ignition sources* (see also Table 16.23): Forsth (1981a,b, 1983); Dahl *et al.* (1983); Sokolov (1989); J.G. Marshall (1989); Cullen (1990); Eckhoff and Thomassen (1994) *Fires and explosions:* Forsth (1981a,b, 1983); Solberg (1982a); Anon. (1983h,j,t); Dahl *et al.* (1983); Tompkins and Riffle (1983); Anon. (1985s,u); Brandie (1989b); Cullen (1990); Hjertager (1991a,b); HSE (1992 OTI 585, 586, 592, 593, 595–599); Samuels (1992); Gardner *et al.* (1994); Wighus (1994); K. van Wingerden (1994) *Missiles:* R.A. Cox (1989); A.C. Palmer (1989); Cullen (1990); HSE (1992 OTI 603) *Structural loading, response:* DoEn (1984); HSE (1990b, 1992 OTI 594, 601, 602, 608, 610); Haaverstad (1994) *Escalation:* D. Drysdale (1989); Cullen (1990); Four Elements Ltd (1991); M. Morris, Miles and Cooper (1994)

Offshore safety management
Brading (1989); Denton (1989,1991); Ellice (1989); Fotland and Funnemark (1989); Grogan (1989); Littlejohn (1989); Macallan (1989); McKee (1989); McReynolds (1989); G. Richards (1989); R.A. Sheppard (1989); API (1990 RP 750); Cullen (1990); J. King (1992); McKeever and Lawrenson (1992); S. Lewis and Donegani (1993); Jacobson (1994). **Command and control**: Baxendine (1989); Cullen (1990); Larken (1992)

Offshore emergency response, planning
Fischer (1982); Tompkins (1984); Baxendine (1989); Matheson (1989); Cullen (1990); Fitzgerald *et al.* (1990); R. Wilson (1992)

Offshore evacuation, escape and rescue
Booth (1989); Clayson (1989); J.D. Evans (1989); Heiberg-Andersen (1989); Jefferey (1989); Keueher (1989); Lien (1989); McNeill (1989); de la Pena (1989); Perrott (1989); Petrie (1989); Rudd (1989); Side (1989); I.G. Wallace (1989,1992); Cullen (1990); Owen and Spouge (1991); Forland (1992); Forster and Wong (1992)

Offshore safety
DoEn (1977, 1984); Burgoyne (1980); HSE (1980 HS(G) 12); Nat. Res. Coun., Cttee on Assessment of Safety of Outer

Continental Shelf Activities (1981); Dick (1982);
Petrie (1983); Kaarstad and Wulff (1984); Barrett, Howells
and Hindley (1987); Cullen (1990); Knott (1990); Antomakis
and Barnes (1991); W.S. Atkins (1991); Bill (1991); IMechE
(1991/134, 1993/157); Venn (1991); Lees (1992); J. Morgan
(1992); Renwick and Tolloczko (1992); Ronald (1992);
Salter (1992); Eckhoff (1994); Lode (1994); Pappas (1994);
Rushton *et al.* (1994).
Safety system: Bodie (1989); Gordon (1989);
McGeough (1989); Cullen (1990)
Costs of safety: HSC (1992); Potter (1994)
Injuries offshore: Wright (1986); Cullen (1990);
H. Hughes (1992)

Formal safety assessment
Fjeld, Andersen and Myklatun (1978); Borse (1979); Slater,
Ramsay and Cox (1981); Pyman and Gjerstad (1983);
Vinnem (1983); Deaves (1986); Schrader and Mowinckel
(1986); Haugen and Vinnem (1987); R.A. Cox (1989d,
1990, 1993); Ellis (1989); Ferrow (1989); Fleishman (1989);
Gorse (1989); van der Graaf and Visser (1989); Hogh (1989);
Pape (1989); Cullen (1990); Tveit (1990); Burns, Grant and
Fitzgerald (1991); Rock and Butcher (1991); Diaz Correa
(1992); Pape (1992); S.J. Shaw (1992); Sherrard (1992);
Potts (1993); S.J. Shaw and Kristofferson (1993); Gardner
et al. (1994); K Miller (1994); Pitblado (1994); Ramsay *et al.*
(1994); Trbojevic *et al.* (1994)

Offshore safety cases
V.C. Marshall (1989); Sefton (1989); Cullen (1990);
Mansfield (1992); Pape (1992); Wendes (1992); Barrell
(1994); Bellamy (1994); Clegg (1994); Durnin (1994);
H. Hughes (1994); Jacobson (1994); Salter (1994);
Spence (1994); Spiller (1994); Hawker (1995); D. Scott
(1995); D.J. Wilson (1995)

Offshore incidents
Le Blanc (1981); WOAD (1988); Due, McFarlane and
Crowther (1994); H. Hughes (1994)
Sea Gem: R.J. Adams (1967)
Ekofisk Bravo: Gjerde (1977–78); Ognedal (1990)
Alexander Kielland: Naesheim (1981); Ognedal (1990)
Ocean Ranger: NTSB (1983 MAR-83-02); Hickman (1984)

Table A18.2   *Selected legislation of offshore*

A   Acts

1934	Petroleum (Production) Act
1962	Pipelines Act
1964	Continental Shelf Act
1971	Mineral Workings (Offshore Installations) Act
1971	Prevention of Oil Pollution Act
1975	Petroleum and Submarine Pipe-lines Act
1992	Offshore Safety Act
	Offshore Safety (Protection against Victimization) Act

B   Statutory Instruments

1972	SI 703	Offshore Installations (Managers) Regulations
1973	SI 1842	Offshore Installations (Inspectors and Casualties) Regulations
1974	SI 289	Offshore Installations (Construction and Survey) Regulations
1976	SI 1019	Offshore Installations (Operational Safety, Health and Welfare) Regulations
	SI 1542	Offshore Installations (Emergency Procedures) Regulations
	SI 923	Submarine Pipe-lines (Diving Operations) Regulations
1977	SI 486	Offshore Installations (Life-saving Appliances) Regulations
1978	SI 611	Offshore Installations (Fire Fighting Equipment) Regulations
	SI 1759	Offshore Installations (Well Control) Regulations
1981	SI 399	Diving Operations at Work Regulations
1989	SI 1029	Offshore Installations (Emergency Pipe-line Valve) Regulations
	SI 971	Offshore Installations (Safety Representative and Safety Committees) Regulations
1992	SI 2885	Offshore Installations (Safety Case) Regulations

- the shelter area shall be intact during a calculated accidental event until safe evacuation is possible;
- depending on the platform type, function and location, when exposed to the design accidental event, the main support structure must maintain its load carrying capacity for a specified time.

The categories of potential accident event to be evaluated were specified as

(1) blowouts,
(2) fire,
(3) explosion and similar incidents,
(4) falling objects,
(5) ship and helicopter collisions,
(6) earthquakes,
(7) other possible relevant types of accident,
(8) extreme weather conditions and
(9) relevant combinations of these accidents.

The Guidelines gave explicit numerical acceptability criteria. In practical terms, it may be considered necessary to exclude the most improbable accidental events from the analysis. However, the total probability of occurrence of each type of excluded situation should not by best available estimate exceed 1 CT[4] per year for any of the main functions specified. This estimate is meant to indicate the order of magnitude to aim for, as detailed calculations of probabilities in many cases will be impossible due to lack of relevant data.

Risk assessment is now the subject of the *Regulations Relating to the Implementation and Use of Risk Assessment in the Petroleum Activities 1990.* Ognedal emphasized that the Norwegian attitude is flexible in its approach to risk assessment and tries to avoid its degenerating into a 'numbers game'.

The *Cullen Report* recommended far-reaching changes in the regulatory administration in the British sector of the North Sea. The recommendations flow from the evidence given on the Piper Alpha disaster and on the regulatory administration up to that date. The administration envisaged is one of goal-setting rather than prescriptive regulations and of the use of QRA to demonstrate compliance. The report also recommended that an operator should submit a safety case and that this should be given structure by a requirement to demonstrate by QRA the integrity of a temporary safe refuge (TSR). Another major recommendation is that the operator demonstrate, as part of the safety case, an appropriate Safety management system (SMS). A further account of the report's recommendations in relation to the evidence on the Piper Alpha disaster is given in Appendix 19.

The recommendations of the *Cullen Report* were accepted in toto by the government and the new administration with the HSE as the regulatory body was put in place in 1991. Initial legislation under this administration includes the *Offshore Safety Act 1992* and the *Offshore Installations (Safety Case) Regulations 1992*. Development of the new administration has been described by Barrell (1992a,b).

A18.2 Gulf of Mexico Offshore Regulatory Administration

The offshore petroleum industry began in the US Gulf of Mexico in the 1950s. The Coast Guard was the original regulatory agency, but the responsibility for offshore safety was eventually transferred to the Minerals Management Service (MMS). The safety regulations are found in the *UIS Code of Federal Regulations, Title 250*.

MMS directly references many of the American Petroleum standards and recommended practice documents. The API 14 series, contain standards for subsurface safety valves, platform safety systems, piping systems, electrical systems, and fire safety systems. The API t series recommended practices for offshore staff training is also codified by the MMS.

API 14C, the recommended Practice for analysis, design and installation of basic surface safety systems on offshore production platforms was originally promulgated in 1972. API 14C embodies two very important process safety concepts;

- Every identifiable failure mode shall require two functionally independent safeguards.
- Recommended safeguards for each type of offshore process unit (pressure vessel, pump, heater, etc) were developed using a generic failure mode and effects analysis.

The documentation showing the results of a 14C approach are presented in a Safety Analysis and Functional Evaluation diagram, known as a SAFE chart. The SAFE chart is a special type of cause and effect diagram.

A18.3 Offshore Process Safety Management

In the 1990s, MMS requested voluntary adherence to the API 75 and API 14G recommended practice documents, which present process safety concepts to the offshore environment. These standards are very similar to the OSHA process safety management regulation, containing virtually identical wording in many sections.

A18.4 Offshore Safety Management

An offshore platform has both marine and process characteristics. It shares with marine vessels the hazards of severe weather and of collision and *in extremis* those of escape to the sea. At the same time it contains high pressure plant processing oil and gas. The platform is also a self-contained community with its own power plant, accommodation and other facilities. The senior manager on the platform is required in large part to combine the skills of a ship's captain with those of a refinery manager and to exercise command and control in an emergency.

A18.5 Inherently Safer Offshore Design

The processing of hydrocarbons is the *raison d'etre* of an offshore platform. This clearly sets certain limitations on the practice of inherently safer design. Nevertheless, this design principle is just as applicable offshore.

Three examples will suffice. One is reduction in the oil inventory in the production separators. The separators constitute a major oil inventory on the platform and it was the separators that fed the oil pool fire on Piper Alpha. Separators are now in use which have much reduced inventories. It may be noted in passing that evidence on the hydrocarbon inventory on Piper Alpha was given to the Inquiry by Clark (1989).

A second example is adoption of a layout that minimizes the possibility of an oil pool forming near a gas riser.

A third example concerns the potential for a jet flame from a gas riser to impinge on the accommodation. Following Piper Alpha, Shell reviewed their platforms to establish whether in each case this was a possibility (Chamberlain, 1989). In those cases where it was, action was taken.

A18.5.1 Friendly plant
Closely related is the design of plant that is friendly to the process operator. In oil and gas extraction there are strong pressures to maintain production. In these circumstances, it is highly desirable that there be fallback states of plant operation to which the operator can resort, without facing the all-or-nothing choice of continuing normal production or effecting total shut-down.

A case in point is the confidence that the operator has that shutting down the gas plant will not lead to loss of power on the main generators. Evidence at the *Piper Alpha Inquiry* appeared to indicate that in some systems fuel changeover on the generators, from gas to diesel, could not be fully relied on. If this situation exists, it puts a greater pressure on the operator to keep going.

A18.5.2 Plant layout
Offshore production platforms carry large amounts of equipment held in a small space. Space and weight are both at a premium, since they are difficult and expensive to provide. Spacing between equipment in a module has to be less generous than on onshore plants. Where a 15 m distance is widely used in the latter, the distance offshore is often half that. It is also necessary to ensure that the layout is such that the centre of gravity of the total mass of equipment is at the centre of the supporting structure. CAD packages for plant layout provide a powerful tool for achieving a layout that meets this requirement.

The decks of the platform are divided into modules. The modules are separated by fire walls which inevitably reduce the ventilation.

Layout design seeks to separate sources of hazard from vulnerable targets. The application of this principle is seen on many platforms by the separation of the accommodation from the wellhead.

A18.5.3 Platform systems
There are a number of basic systems that are critical for safe operation of the platform. The account given here of these systems is limited to a brief overview. It is convenient to cast it as a description of the systems on Piper Alpha. The *Cullen Report* gives a detailed description of these systems and of their behaviour on the night of the disaster.

Electrical power systems
The first of the systems is the electrical power supply system. Basic power is supplied by a pair of main turbine-driven generators with dual fuel firing, gas being the normal fuel with diesel fuel as standby. As backup, there is an emergency generator, turbine-driven and diesel fuelled. Changeover of fuel on the main generator and start up of the emergency generator are automatic. Drilling is served by a separate power supply from a diesel-driven generator with its own emergency generators.

The emergency generator for the main supply is designed to provide supply to critical services, which include HVAC, instrumentation and valves, and emergency lighting. Further backup is provided by an uninterruptible power supply (UPS) drawn from batteries designed to provide power during the momentary interruption while the emergency generator starts up and, if necessary, for a period in the event of total failure of the main supply.

Electrical systems for offshore production platforms are the subject of API RP 14F: 1991.

Fire protection system
The fire protection system comprises a number of complementary elements. These are

- hazardous area classification,
- fire walls,
- fire and gas detection system,
- fire water deluge system,
- fire pumps,
- foam system,
- Halon or clean agent systems and
- fire fighting arrangements.

Fire prevention and control on offshore production platforms is covered in API RP 14G: 1986.

The first line of defense against fire is the use of hazardous area classification to reduce the risk of ignition of any flammable leak that may occur. This was formerly covered by API RP 500B: 1973, specific to offshore, which is now replaced by the general onshore and offshore code RP 500: 1991.

The traditional form of active fire protection is the water deluge system. This is covered by the NFPA Fire codes, *Construction and Use Regulations 1974* and the associated guidance in the *Offshore Design Guide*. The plant is divided into reference areas with a specified quantity of water to be delivered over each area. In order to limit the size of the reference areas use is made of firewalls. There are firewalls between the main production modules, namely the wellhead, separation and gas compression modules. These provide basic passive fire protection. They are not necessarily designed as blast walls.

There is an extensive fire and gas (F&G) detection system, with sensors in the main production modules and

elsewhere, utilizing both combustible gas detectors and fire detectors.

A fixed water deluge system with distribution throughout the main modules and at the risers furnishes a basic level of active fire protection. Water for the deluge system is drawn from the sea by fire pumps operating off the main power supply but with diesel-driven pumps as standby.

At locations where an oil pool fire may occur, such as the production separators, foam injection is provided.

Certain closed volumes may be provided with a halon total flooding system. Enclosures which may be protected in this way include the centrifugal compressor enclosures, the generator area, the electrical switchgear room and the control room. Where, as in the latter case, operators may be present, a non-toxic agent is used and there is prior alarm. These fixed systems are supplemented by the fire fighting teams.

As just indicated, the traditional means of fire protection has been passive protection by firewalls and active protection by a uniform water deluge. Alternative approaches have tended to be inhibited by the need to comply with the two separate sets of regulations covering fire, one requiring passive and the other active fire protection measures.

In addition to the trade-off between passive and active fire protection, there may also be a trade-off between ventilation and active fire protection. If a module is well ventilated, gas from a leak is less likely to accumulate in the first place.

A plea for greater flexibility was made to the *Piper Alpha Inquiry* by Brandie (1989b), who advocated a scenario-based approach in which specific fire hazards are identified and water deluge arrangements directed more specifically to these scenarios. The *Cullen Report* recommended this approach within the context of *Safety Case*.

There are a number of issues related to fire protection. One is the availability of standard fire tests for hydrocarbon, as opposed to building, fires, such tests being needed for design of fire walls. Another is the use of passive fire protection on risers, which might involve a risk of corrosion beneath the lagging.

Another long-standing issue, which relates to active fire protection, is the availability of the fire pumps.

There is generally a strong argument for the use of passive fire protection in that once installed it appears less subject to human failings. The matter is not, however, straightforward. Thus, for example, in some applications corrosion may occur under fireproofing, which may both promote the corrosion and conceal it.

There is now considerable activity in the investigation of fire events on platforms, covering oil pool fires and jet flames; impingement of flames on vulnerable targets such as risers and passive protection of these targets. Further details are given in Section A18.7.

A18.5.4 Emergency shut-down system
The platform is provided with an emergency shut-down (ESD) system, the main functions of which are (1) to shut down the flow from the reservoir, (2) to shut off the flow through the pipelines entering and leaving the platform, (3) to shut down the main items of equipment and (4) to initiate blowdown of the inventories to flare. There are various levels of ESD, some involving only individual items of equipment and others a full platform ESD, or PESD. Physically, PESD may be activated from the control room or from any of a number of emergency pushbuttons around the platform.

As important as the hardware are the procedures that govern operation of the PESD system. It bears emphasis that the protection apparently afforded can be negated if there are cultural, organizational or human factors which inhibit initiation of the system in a real emergency.

A18.5.5 Evacuation, escape and rescue system

The evacuation, escape and rescue (EER) system is built around the three main means of leaving the platform, which are (1) helicopter, (2) lifeboat and (3) liferaft. The emergency procedure is for personnel to collect at designated muster points. In the vast majority of cases evacuation is by helicopter, personnel being summoned to the helideck from their muster points. There are a number of reasons, however, why evacuation by helicopter may be impractical. They include high winds and smoke from the platform, which can prevent landing. Another is the time taken to reach the platform. A helicopter already in or near the field may arrive relatively quickly but in many locations it takes about an hour for a helicopter to travel from an onshore base.

If helicopter evacuation is not practical, the other main means of getting off the platform is by lifeboat. The use of lifeboats is also subject to limitations. Lifeboats may sometimes be difficult to reach and have limitations in high seas. Even if the conditions do not actually prevent use, they may increase the risk of injuries.

If the lifeboats cannot be used, resort is to escape to the sea by launching liferafts and climbing down knotted ropes. The use of knotted ropes requires a certain degree of fitness and is not without risk. One area of development has been improved lifeboat systems. Prominent among these is the freefall lifeboat, which is launched with its complement straight into the water, as opposed to being lowered from davits. These have been in use on Norwegian platforms and are the method used on the replacement for Piper Alpha, Piper Bravo.

A variety of devices have become available for escape, ranging from individual packages which can be hooked onto a guard rail to chutes and slides.

Another aspect is the integrity of the escape routes from locations where personnel are likely to be to the means of escape. Escape routes need therefore to be designed against a variety of scenarios. Of particular importance are the routes to the lifeboats from the accommodation, where at any given time a large proportion of the personnel will be. One approach is to locate lifeboats in a protected area integral with the accommodation.

The *Piper Alpha Inquiry* heard a considerable amount of evidence on evacuation, escape and rescue. Contributors included Petrie (1989) (life-saving appliances), Rudd (1989) and Heiberg-Andersen (1990) (evacuation), Ginn (1989) (evacuation by helicopter), Kelleher (1989) (lifeboats), Side (1989) (evacuation and rescue), I.G. Wallace (1989) (evacuation and escape), Lien (1989) and Perrott (1989) (escape systems), Booth (1989) (escape routes) and J.D. Evans (1989) and de la Pena (1989) (smoke hoods). Work in this area has been described by Bellamy and Harrison (1988), Forland (1992), Forster and Wong (1992) and I.G. Wallace (1992).

A18.5.6 Explosion protection

In many earlier platform designs, there was little protection against explosion over and above that against fire, namely partition walls being designed as firewalls rather than as blast walls. This situation no longer pertains and much effort is being devoted to explosion protection. CFD simulation is being used to study the overpressures generated by explosions in modules, with particular reference to the enhancing effect of obstacles and to the mitigating effects of venting and of water spray systems.

A large amount of work has been done on the development and venting of explosions in modules and other obstructed spaces. Much of the work using CFD explosion simulation codes has been directed to the offshore module situation. Representative work using such codes is that of Hjertager (1982a, 1986, 1991), Bakke *et al.* (1989), Catlin (1990) and Hjertager *et al.* (1994). Examples of work on module explosions includes that described by Solberg, Pappas and Skramstad (1980, 1981), A.J. Harrison and Eyre (1987b), Catlin (1991), Brenton, Thomas and Al-Hassan (1992), Samuels (1992, 1993), Catlin, Manos and Tite (1993) and Catlin *et al.* (1993). A fuller account is given in Chapter 17.

Blast walls

An increasingly common method of explosion protection is the use of blast walls. The design of a blast wall is partly a matter of formulating suitable accident scenarios and predicting by simulation the resultant overpressure and partly one of engineering the wall so that it fulfils its protective function even if it deforms somewhat.

An account of the design of the blast walls on the Kittiwake platform was given to the *Piper Alpha Inquiry* by Doble (1989). The engineering of blast walls was described by van Beek (1989).

Smoke minimization and protection

Another area of work is the investigation of the hazard presented by smoke, particularly that from an oil pool fire. Aspects of this are the generation and movement of smoke and the exclusion of smoke from the accommodation.

A18.5.7 Design basis accidents

The design of plant to cope with accidental events requires that there be defined a set of design basis accidents. The concept is that the plant is then designed to withstand the design basis accidents in each category but does not have to be designed for events more severe than this. The selection of the design basis accidents is therefore closely linked with the estimates for the frequency of such events, so that overall risk in the design is an appropriate one.

Drilling

Drilling is a major activity on the platform that has no counterpart onshore. One aspect of this activity is the ever-present hazard of a well blowout.

Another aspect that can impinge on the operation of the platform as a whole is the need to avoid loss of power such that the drill becomes stuck.

Diving

Diving is another activity specific to an offshore platform. For the most part it does not impinge to any major extent on other activities except in so far as precautions are necessary to ensure the safety of the divers. At the time of the explosion on Piper Alpha the fire pumps had been turned to manual start to prevent sudden start-up of the pumps with consequent danger of a diver being drawn into the pump water inlet. This was not the policy on the sister platform Claymore, run by the same operator and in the same field.

Contractors

The proportion of contractors on a platform may well be of the order of 70% or more. Typically specialist teams such as drillers and divers are contractors with one person from the operating company as liaison. Some contractors may remain on a platform for long periods, others come and go. The operating company attempts to ensure the quality of contractors admitted to the platform by means of a quality assurance system. Personnel from the company visit the contractor company onshore and make the usual quality assurance audit.

It is the responsibility of a contractor to ensure that its personnel are properly trained in offshore emergency procedures as well as in safe systems of work and PTW systems. However, there may well be features of the systems operated on a platform that are particular to that platform. The operator needs, therefore, to ensure that contractors within its system are familiar with them.

A18.6 Offshore Emergency Planning

Offshore emergency planning has the same broad features as that for onshore. In principle, they include the detection of the incident, the assessment of its nature and seriousness and if necessary, the declaration of the emergency, the assumption of command and control, and the implementation of the emergency plan.

However, the emergency may not present in this ideally structured form. It may well take the form of an emergency shut-down initiated automatically or by personnel in the control room or at one of the shut-down buttons scattered throughout the platform.

A18.6.1 Emergency scenarios

The first step in emergency planning is the definition of the set of scenarios on which the plan is to be based. A wide range of scenarios need to be considered if the plan is to be robust. It is not uncommon for an incident to require the evacuation of the platform, but in the vast majority of cases this will be by helicopter and management may well feel it has not done as well as it might if even one person is injured.

At the other extreme is the sort of situation that arose on Piper Alpha where there was no prospect of an evacuation by any of the conventional means and where escape to the sea was the only option. This implies that the person in command explicitly instruct personnel to make their own escape.

A18.6.2 Goal-setting regulations

The basic framework operated in the new offshore safety administration is that of regulations which are goal-setting rather than prescriptive.

The first set of regulations of the new offshore administration are the *Offshore Installations (Safety Case) Regulations 1992*.

Before considering these, it is interesting to note that the first regulations issued, by the DoEn, in the wake of the Piper Alpha disaster were the *Offshore Installations (Pipe-line Valve) Regulations 1989*, which require the operator to fit EIVs to the pipelines connected to the platform and are thus prescriptive. The *Cullen Report* explicitly

endorsed this, in effect considering that exceptionally this case met the conditions for adopting a prescriptive regulation.

A18.6.3 Safety case

The *Offshore Installations (Safety Case) Regulations 1992* implement Recommendation 1 of the Cullen Report that the operator should submit to the regulator a safety case.

For major hazard installations onshore a requirement for a safety case already existed under the *CIMAH Regulations 1984*. Evidence on the onshore safety case was given to the *Piper Alpha Inquiry* by Sefton (1989) and V.C. Marshall (1989d).

The offshore safety case is largely modelled on the onshore case, but has three features not explicitly contained in the latter.

Safety management system

One of these is the requirement on the operator to demonstrate that it has an SMS, which will assure the safety of the project through the design and operation. The SMS is thus part of the safety case rather than a free-standing requirement.

Temporary safe refuge

The second distinguishing feature of the offshore safety case is the requirement that the platform has a TSR, normally the accommodation, which should provide protection for a defined period. Design of a TSR involves the definition of the events against which it is to provide protection and specification both of the endurance period and the conditions which constitute failure.

Quantitative risk assessment

The third distinguishing feature of the offshore safety case is the requirement to use QRA. Essentially the QRA is a matter for the operator. The *Cullen Report* does, however, propose that one aspect of it be specified by the regulator.

The *Safety Case* involves a demonstration that the frequency of events that threaten the endurance of the accommodation, or TSR, will not exceed a certain value. In order to provide at least one fixed point in the administration, both the minimum endurance and the frequency with which there is a failure of such endurance should be specified by the regulatory body, at least in the first instance.

Accounts of offshore QRA include those of Pape (1992) (regulator), Sherrard (1992) (drillling units), R.F. Evans (1994) (model validation) and Gardner *et al*. (1994) (methodology).

A18.6.4 Costs of offshore safety

In the wake of Piper Alpha the offshore industry has incurred major expenditure to enhance safety. Much of this was incurred in response to the disaster itself and before the Cullen Report appeared.

The HSE itself has made a comparison of the quantifiable costs and benefit of safety measures in the UK North Sea expressed as Net Present Value over 15 years (HSC, 1992). It attributes the bulk of both costs and benefits to the *Safety Case Regulations*, the costs being estimated at £1200–2500 million and the benefits at £2600–4600 million. A commentary on these estimates is given by Potter (1994).

A18.7 Offshore Event Data

As for onshore installations, information on offshore events is of two main types: (1) incident data and (2) equipment failure data. The *World-wide Offshore Accident Databank (WOAD, 1988)* enumerates the principal accidents involving offshore structures involved in oil and gas activities. There is also a need, particularly in hazard assessment, for more detailed information on events such a leaks, fires and explosions, dropped loads and so on. The HSE has created the HCR database, as described by Bruce (1994).

Equipment failure data collection is the subject of a major co-operative exercise carried out under the aegis of OREDA and initiated some years before Piper Alpha. This is described in Appendix 14.

Appendix

19

Piper Alpha

Contents

At 10.00 p.m. on 6 July 1988 an explosion occurred in the gas compression module of the Piper Alpha oil production platform in the North Sea. A large pool fire took hold in the adjacent oil separation module, and a massive plume of black smoke enveloped the platform at and above the production deck, including the accommodation. The pool fire extended to the deck below, where after 20 min it burned through a gas riser from the pipeline connection between the Piper and Tartan platforms. The gas from the riser burned as a huge jet flame. Most of those on board were trapped in the accommodation. The lifeboats were inaccessible due to the smoke. Some 62 men escaped, mainly by climbing down knotted ropes or by jumping from a height, but 167 died, the majority in the quarters.

The Piper Alpha explosion and fire was the worst accident which has occurred on an offshore platform.

Following the disaster a Public Inquiry was set up under the Public Inquiries Regulations – Offshore Installations Regulations 1974 presided over by Lord Cullen to establish the circumstances of the disaster and its cause and to make recommendations to avoid similar accidents in the future. The Inquiry's report *The Public Inquiry into the Piper Alpha Disaster* (the *Piper Alpha Report,* or *Cullen Report*) (Cullen, 1990) is the most comprehensive inquiry conducted in the United Kingdom into an offshore platform disaster, or indeed into any process industry disaster, onshore or offshore.

The Piper Alpha Inquiry has been of crucial importance in the development of the offshore safety regime in the UK sector of the North Sea. Whereas Flixborough was followed first by a Court of Inquiry and then by the Advisory Committee on Major Hazards, the Piper Alpha Inquiry not only discharged the function of an inquiry into the specific disaster, but made recommendations for fundamental changes to the offshore safety regime which were accepted by the government.

The description of the Piper Alpha disaster given below is necessarily a relatively brief one. Nevertheless, it is somewhat fuller than that of the case histories in the other appendices, for several reasons. It provides a good illustration of the work of an accident inquiry. It is replete with lessons on design and operation of hazardous installations. And it has had far-reaching consequences for the offshore safety regime. A fuller account is given in the *Piper Alpha Report.* The daily transcripts, available in copyright libraries, also repay study. An account from the viewpoint of one of the consultants to the Inquiry has been given by Sylvester-Evans (1991).

Selected references on Piper Alpha are given in Table A19.1.

A19.1 The Company, the Management and the Personnel

The Piper Alpha oil platform was owned by a consortium consisting of Occidental Petroleum (Caledonia) Ltd, Texaco Britain Ltd, International Thomson plc and Texas Petroleum Ltd and was operated by Occidental.

The management concerned with the Piper platform included

Offshore Installation Manager	Mr C.D. Seaton
Offshore Superintendent	Mr T.J. Scanlon
Senior Maintenance Superintendent	Mr R.H. Seddon
Maintenance Superintendent	Mr K.D. White
	Mr A.C.B. Todd

Table A19.1   *Selected references on Piper Alpha*

Report of the Public Inquiry
Cullen (1990)

Part 1 Evidence
F.H. Atkinson (1989) (Lloyds Register); Bakke (1989) (explosion simulation); Balfour (1989) (gas detectors); Bett (1989) (reciprocating compressors); Bodie (1989) (Offshore Safety Superintendent); Bollands (1989) (Control Room Operator); Brading (1989) (Chairman, Occidental Petroleum (Caledonia) Ltd); Burns (1989) (Shift Supervisor, MCP01); A.G. Clark (1989) (Maintenance Lead Hand); M.R. Clark (1989) (hydrocarbon inventory); Clayson (1989) (evacuation, escape and rescue); R.A. Cox (1989a) (damage to firewall, explosion simulation); R.A. Cox (1989b) (damage by projectiles); R.A. Cox (1989c) (damage to electrical systems, ESD, F&G, fire protection systems); R.A. Cox (1989d) (damage to risers); Cubbage (1989) (explosion effects analysis); J. Davidson (1989) (Operations Superintendent, Claymore); M.E. Davies (1989) (wind tunnel modelling of gas dispersion); D.D. Drysdale (1989) (escalation of fire); J. Drysdale (1989) (hydrates, methanol injection); Gordon (1989) (Manager, Loss Prevention Dept); P.M. Grant (1989) (contractors); Grieve (1989) (Process Operator); Grogan (1989) (Vice-President Engineering, Occidental); Guiomar (1989) (OIM, MCP01); Henderson (1989) (Lead Operator); Jefferey (1989) (liferafts); Jenkins (1989) (DoEn Inspector); Johnsen (1989) (hydrates, methanol injection); Leeming (1989) (OIM, Tartan); P. Lloyd (1989) (electrical systems); Lockwood (1989) (Lead Production Operator, permits-to-work); Macallan (1989) (Production and Pipeline Manager); A.G. McDonald (1989) (telecommunications); McGeough (1989) (safety training); McLaren (1989) (Lloyds Register, electrical); McNeill (1989) (rescue); McReynolds (1989) (Vice-President Operations, Occidental); J.G. Marshall (1989) (ignition sources); Moreton (1989) (Production Supervisor, Tartan); J. Murray (1989) (gas detectors); A.C. Palmer (1989) (damage by projectiles); Paterson (1989) (hydrates, methanol injection); Petrie (1989) (Director of Safety, DoEn); Pillans (1989) (Lloyds Register, electrical); Rankin (1989) (supervisor, Score PSV certification team); G. Richards (1989) (OIM (back-to-back)); S.M. Richardson (1989a) (hydrates, methanol injection); S.M. Richardson (1989b) (gas pipelines); S.M. Richardson (1989c) (autoignition); S.M. Richardson (1989d) (leaks); Ritchie (1989) (Managing Director, Score UK Ltd); K. Roberts (1989) (Facilities Engineer, Tartan); G.G. Robertson (1989) (Safety Supervisor); J.B. Russell (1989) (hydrates, methanol injection); Saborn (1989) (standby vessel); Sandlin (1989) (OIM, Claymore); Saville (1989a) (condensate admission to PSV 504 system); Saville (1989b) (leaks); Saville (1989c) (hydrates, methanol injection); Saville (1989d) (pipe failure); Scanlon (1989) (Offshore Superintendent, maintenance); Scilly (1989) (explosion effects analysis); Scothern (1989) (gas detectors); Seddon (1989) (Senior Maintenance Superintendent); Smyllie (1989) (flare); Tea (1989) (gas detectors); Thomson (1989) (Lloyds Register); Todd (1989) (Maintenance Superintendent); Tucker (1989) (accommodation); P.C.A. Watts (1989) (flare); Whalley (1989) (PSV recertification); W.P. Wood (1989) (DoTp surveys); Wottge (1989) (platform, facilities and systems)

Part 2 Evidence

A.J. Adams (1989) (pipeline isolation, inc. subsea isolation valves); C.S. Allen (1989) (PTWs); Ashworth(1989) (process control and ESD); Banks (1990) (maintenance supervisors); Baxendine (1989) (emergency command); van Beek (1989) (blast walls); Booth (1989) (escape routes); Brandie (1989a) (safe havens); Brandie (1989b) (alternatives to standard fire water systems); Broadribb (1989) (subsea isolation valves); Chamberlain (1989) (mitigation of vapour cloud explosions); R.A. Cox (1989c) (QRA); Cunningham (safety representatives); Dalzell (1989) (smoke ingress into accommodation); Daniel (1989) (standby vessels); G.H. Davies (1989) (PTWs); Day (1989) (emergency power); Denton (1989) (quality management systems); Doble (1989) (explosion prevention and mitigation – Kittiwake); Drew (1989) (standby vessels); Ellice (1989) (training of OIMs); Ellis (1989) (HSE view of QRA); J.D. Evans (1989) (smoke hoods); Ferrow (1989) (FSA); Fleishman (1989) (Gyda safety evaluation); Gilbert (1989) (subsea isolation valves); Ginn (1989) (evacuation by helicopter); Gorse (1989) (FSA); Heiberg-Andersen (1990) (evacuation, Norwegian sector); Higgs (1990) (offshore safety regime); Hodgkins (1989) (HSC – DoEn Agency Agreement); Hogh (1989) (QRA); M.J. Jones (1990) (training); Keenan (1989) (standby vessels); Kelleher (1989) (lifeboats); Kinloch (1989a) (PTW); Kyle (1989) (PTWs); Lien (1989) (escape systems); Littlejohn (1990) (offshore supervisors); Lyons (1989) (offshore safety regime); McIntosh (1989) (fire and explosion protection); McKee (1990) (safety management); Macey (1989) (standby vessels); Matheson (1989) (offshore emergency medical team); V.C. Marshall (1989d) (safety cases); Middleton (1989) (standby vessels); Nordgard (1990) (accommodation in Norwegian sector); Ognedal (1990) (Norwegian offshore safety regime); Pape (1989) (HSE view on QRA); de la Pena (1990) (smoke hoods); Perrott (1989) (escape systems); Petrie (1989, 1990) (life-saving appliances, offshore safety regime); Priddle (1990) (offshore safety regime); Rimington (1990) (onshore safety regime); Rudd (1989) (evacuation); Scanlon (1989) (PTWs); Sefton (1989) (CIMAH, safety cases); R.A. Sheppard (1989) (safety management); Side (1989) (rescue and evacuation); Spouge (1989) (options for accommodation); B.G. Taylor (1989) (offshore industry developments); Tveit (1990) (Norwegian offshore safety regime, QRA); Vasey (1989) (mitigation of module explosions); I.G. Wallace (1989) (evacuation and escape); Willatt (1989) (offshore pipeline connections)

Further accounts

Anon. (1988g); Johnsen (1989, 1990); Boniface (1990a,c–e); Redmond (1990, 1991 LPB 102); S.M. Richardson, Saville and Griffiths (1990); Sylvester-Evans (1990a,b, 1991); Tombs (1990); Lees (1991,1992a, 1994b)

Other personnel on duty on the evening of July 6 and referred to below include

Lead maintenance hand	Mr A.G. Clark
Lead production operator	Mr R.A. Vernon
Phase 1 operator	Mr R.M. Richard
Phase 2 operator	Mr E.C. Grieve
Control room operator	Mr G. Bollands
Instrument technician	Mr W.H. Young

A19.2 The Field and the Platform

The Piper Alpha platform was located in the Piper field some 110 miles north-east of Aberdeen. The platforms in the field and the pipeline connections between them are shown in Figure A19.1. The Piper platform separated the fluid produced by the wells into oil, gas and condensate. The oil was pumped by pipeline to the Flotta oil terminal in the Orkneys, the condensate being injected back into the oil for transport to shore. The gas was transmitted by pipeline to the manifold compression platform MCP-01, where it joined the major gas pipeline from the Frigg Field to St Fergus.

There were two other platforms connected to Piper Alpha. Oil from the Claymore platform, also operated by Occidental, was piped to join the Piper oil line at the 'Claymore T'. Claymore was short of gas and was therefore connected to Piper Alpha by a gas pipeline so that it could import Piper gas. Oil from Tartan was piped to Claymore and then to Flotta and gas from Tartan was piped to Piper and thence to MCP-01.

Elevation views of the Piper Alpha platform, the layout of the production deck at the 84 ft level and the layout of the deck below, the 68 ft level, are shown in Figures A19.2, A19.3 and A19.4 respectively. The production deck level consisted of four modules, Modules A–D. A Module was the wellhead, B Module the oil separation module, C Module the gas compression module and D Module the power generation and utilities module.

A Module was about 150 ft long east to west, 50 ft wide north to south and 24 ft high. The other modules were of approximately similar size. There were firewalls between A and B Modules, between B and C Modules and between C and D Modules (the A/B, B/C and C/D firewalls, respectively); these firewalls were not designed to resist blast.

The pig traps for the three gas risers from Tartan and to MCP-01 and Claymore were on the 68 ft level. Also on this level were the dive complex and the JT flash drum, the condensate suction vessel and the condensate injection pumps.

There were four accommodation modules: the East Replacement Quarters (ERQ), the main quarters module; the Additional Accommodation East (AAE); the Living Quarters West (LQW) and the Additional Accommodation West (AAW).

The control room was in a mezzanine level in the upper part of D Module. It was located about one quarter of the way along the C/D firewall from the west face.

There were two flares on the south end of the platform, the east and west flares, and there was a heat shield around A Module to provide protection against the heat from the flares.

Platform systems included the electrical supply system, the fire and gas detection system, the fire water deluge system, the emergency shut-down system, the communications system and the evacuation and escape system.

Electrical power was supplied by two main generators which normally ran off the gas supply but could be fired by diesel. There was a diesel-fired emergency generator and also a drilling generator and an emergency drilling generator. In addition, there were uninterrupted power supplies for emergency services.

The main production areas were equipped with a fire and gas detection system. In C Module the gas detection system was divided into five zones: C1 and C2 in the west and east halves of the module and C3, C4 and C5 at the three compressors, respectively.

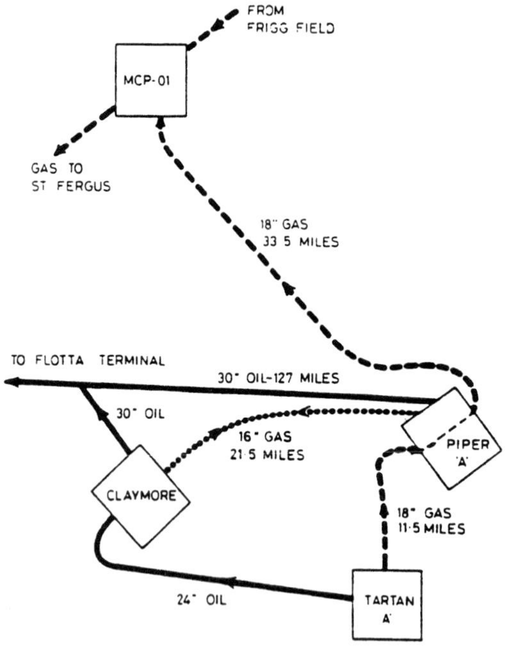

Figure A19.1 *Pipeline connections of the Piper field (Sylvester-Evans, 1991)(Courtesy of the Institution of Chemical Engineers)*

The fire water deluge system consisted of ring mains which delivered foam in A–C Modules and part of D module and at the Tartan and MCP-01 pig traps and water at the condensate injection pumps. The fire pumps were supplied from the main electrical supply but there were backup diesel-driven pumps.

The hydrocarbon inventory in the pipelines was approximately as follows. The main oil line was 30 in. in diameter and 30 miles long and held some 70,000 te of oil. The gas line from Tartan was 18 in. in diameter and 11.5 miles long and held some 450 te; the gas line to MCP-01 was 18 in. in diameter and 33.5 miles long and held 1280 te; the gas line to Claymore was 16 in. in diameter and 21.5 miles long and held 260 te.

A19.3 The Process and the Plant

The fluid from the wellhead, containing oil, gas, condensate and water, passed through the wellhead 'Christmas trees' to the two separators where the gas was separated from the oil and water. The oil was then pumped into the main oil line. The gas was compressed first in three centrifugal compressors to 675 psia, with some gas being taken off at this point as fuel gas for the main generators, and then boosted in the first stage of two reciprocating compressors to 1465 psia. Condensate was removed and the gas was then further compressed in the second stage of the reciprocating compressors to 1735 psia. The gas then went three ways: to serve as lift gas at the wells, to MCP-01 as export gas or to flare. The plant could be operated in two modes, which affected the method of removing condensate. In the normal, or phase 2, mode, the gas passed from the

first stage of the reciprocating compressors to the Gas Conservation Module (GCM), where it was dried. The gas was then cooled by reducing the pressure across a turbo-expander so that condensate was knocked out by the expansion and returned to the outlet of the JT flash drum, which was also the inlet of the second stage of the reciprocating compressors. Condensate from the GCM was passed to the JT flash drum. The process could also revert to the original, or phase 1 mode, dating from a period before the GCM was installed to produce export quality gas, in which the GCM was isolated and gas from the first stage of the reciprocating compressors was let down in pressure across the JT valve into the JT flash drum so that condensate was knocked out by the Joule–Thomson (JT) effect and then passed as before into the second stage of the reciprocating compressors. Condensate from the JT flash drum passed first to two parallel centrifugal condensate booster pumps and then to two reciprocating condensate injection pumps which pumped the condensate into the main oil line. There was normally one condensate injection pump line operating and one on standby.

Each condensate injection pump was protected from overpressure on the delivery side by a single pressure safety valve (PSV). The PSV was on a separate relief line from the delivery head of the pump rather than on the delivery line itself. The valve on A pump was PSV 504 and that on B pump PSV 505. These valves were located in C Module, the relief line running up from the 68 ft level, where the pumps were located, to the PSVs in C Module and back down to the condensate suction vessel on the 68 ft level.

In accordance with standard practice, methanol was injected into the process at various points to prevent formation of hydrates which would tend to cause blockages.

A19.4 Events Prior to the Explosion

On 6 July there was a major work programme on the platform. This included the installation of a new riser for the Chanter field and work on a prover and metering loop.

The extra accommodation for the workforce was provided on the Tharos, a large floating fire fighting vessel anchored near the platform. Also near the platform were the standby vessel, the Silver Pit, a pipeline vessel, the Lowland Cavalier, and Maersk anchor handling vessels for the Tharos.

The GCM was also out of service for changeout of the molecular sieve driers. In consequence, the plant operation had reverted to the phase 1 mode so that the gas was relatively wet.

The resulting increased potential for hydrate formation was recognized by management onshore. The increased methanol injection rates required were calculated and communicated to the platform together with suggestions for the configuration of the methanol pumps. The methanol injection rates needed were some 12 times greater than for normal phase 2 operation.

However, there was an interruption of the methanol supply to the most critical point, at the JT valve, between 4.00 and 8.00 p.m. that evening.

The operating condensate injection pump was B pump. The A pump was down for maintenance. There were three maintenance jobs to be done on this pump: (1) a full 24 month preventive maintenance (PM), (2) repair of the pump coupling and (3) recertification of PSV 504. In order to carry out the 24 month PM, the pump had been isolated

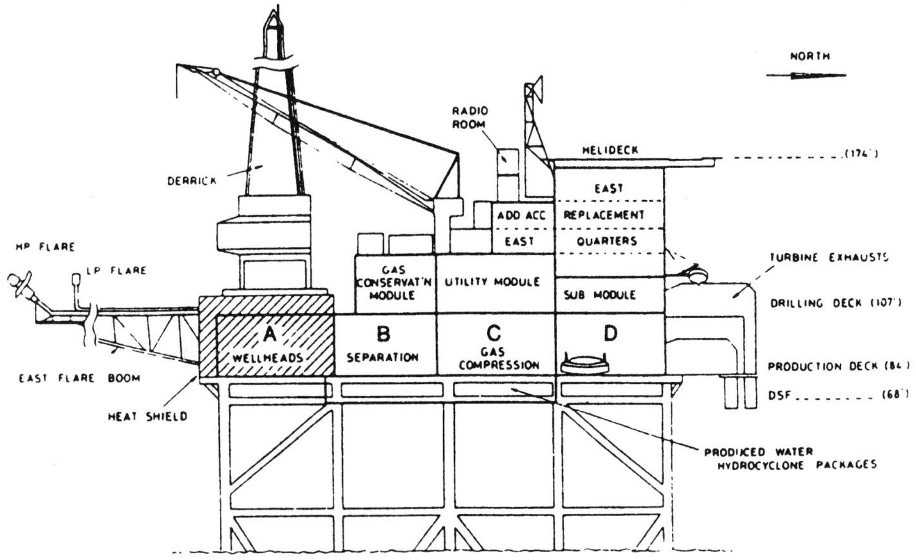

Figure A19.2   *The Pipe Alpha platform: (a) east elevation and (b) west elevation (Sylvester-Evans,1991) (Courtesy of the Institution of Chemical Engineers)*

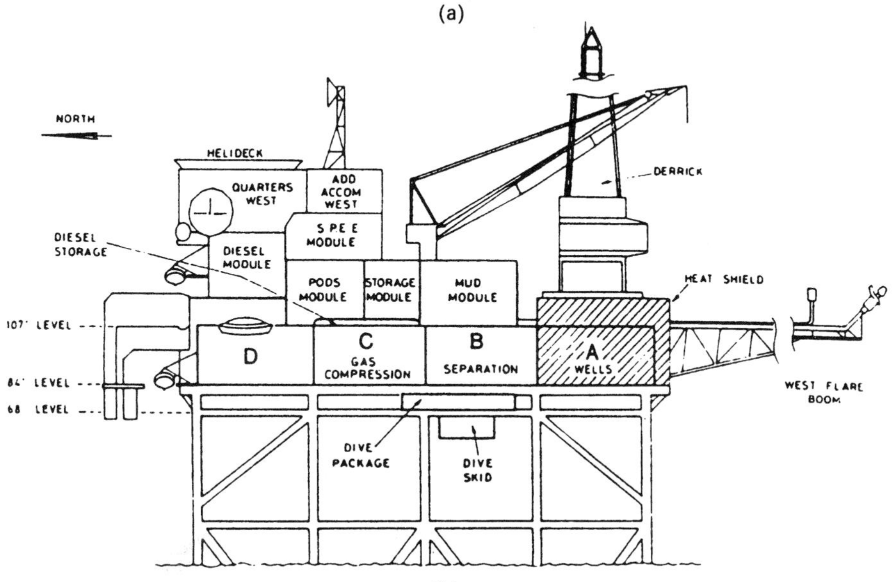

by closing the gas operated valves (GOVs) on the suction and delivery lines but slip plates had not been inserted. Work on the coupling, which was suffering from a vibration problem, would not involve breaking into the pump.

With the pump in this state, with the GOVs closed but without slip plate isolation, access was given to remove PSV 504 for testing. It was taken off in the morning of July 6 by a two-man team from the specialist contractor Score UK Ltd. They were not able to restore the PSV that evening. The supervisor in this team came back to the control room some time before 6.00 p.m. to suspend the permit-to-work (PTW)

and the team then went off duty, intending to put the PSV back the next day.

At about 4.50 p.m. that day, just at shift changeover, the maintenance status of the pump underwent a change. The maintenance superintendent decided that the 24 month PM would not be carried out and that work on the pump should be restricted to the repair of the pump coupling.

About 9.50 p.m. that evening B pump on the 68 ft level tripped out. The lead production operator and the phase 1 operator attempted to restart it but without success. The loss of this pump meant that with A pump also down

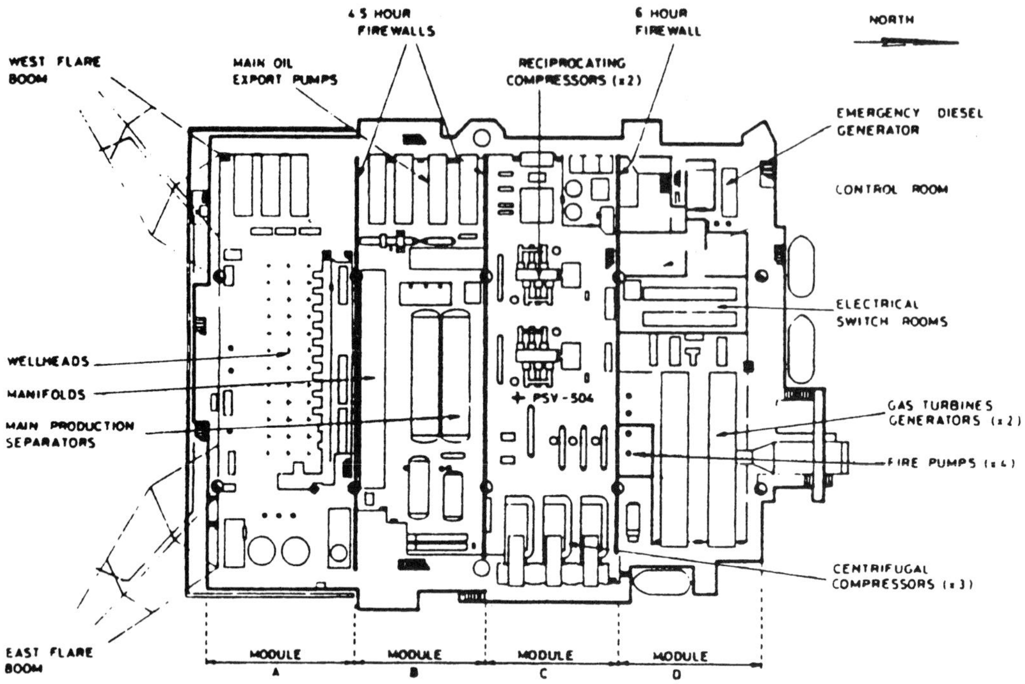

Figure A19.3   *The Piper Alpha platform: the production deck on the 84 ft level (Sylvester-Evans, 1991) (Courtesy of the Institution of Chemical Engineers)*

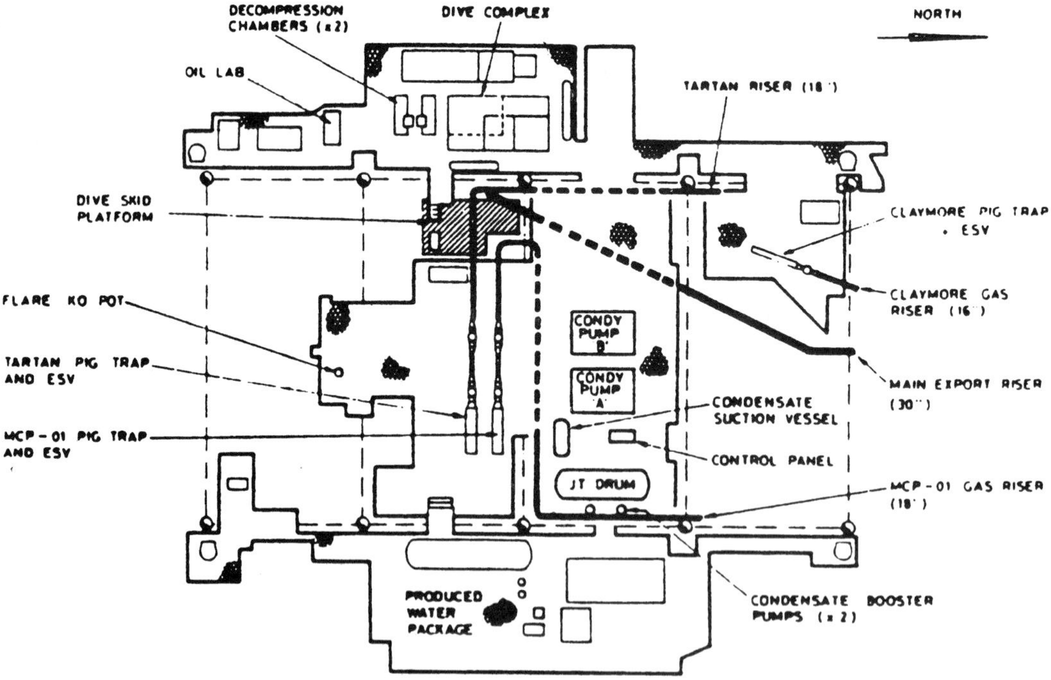

Figure A19.4   *The Piper Alpha platform: the 68 ft level (Sylvester-Evans, 1991) (Courtesy of the Institution of Chemical Engineers)*

condensate would back up in the JT flash drum and within some 30 min would force a shut-down of the gas plant. There was a possibility that if the gas supply to the main generator was lost and if the changeover to the alternative diesel fuel failed, the wells also would have to be shut-down. It would then be necessary to undertake a 'black start'.

The lead operator came up to the control room. He talked on the telephone with the lead maintenance hand and it was agreed to attempt to start A pump. The lead operator signed off the permit for A pump so that it could be electrically deisolated and restarted, and went back down to the pumps. The lead maintenance hand came down to the control room to organize the electricians to deisolate the pump. It is uncertain precisely what action the lead operator and the phase 1 operator took. They were observed at the pumps by the phase 2 operator and an instrument fitter, but the evidence of these witnesses was inconclusive. However, there was no doubt that the lead operator intended to start A pump.

About 9.55 p.m. the signals for the tripping of two of the centrifugal compressors in C Module came up in the control room. This was followed by a low gas alarm in C3 zone on C centrifugal compressor. Then, the third centrifugal compressor tripped. Before the control room operator could take any action a further group of alarms came up: three low gas alarms in zones C2, C4 and C5 and a high gas alarm. The operator had his hand out to cancel the alarms when he was blown across the room by the explosion.

Just prior to the explosion personnel in workshops in D Module heard a loud screeching sound which lasted for about 30 s.

A19.5 The Explosion, the Escalation and the Rescue

The initial explosion occurred at 10.00 p.m. It destroyed most of the B/C and C/D firewalls and blew across the room the two occupants of the control room, the control room operator and the lead maintenance hand.

The emergency shut-down (ESD) system operated, closing the emergency shut-off valve (ESV) on the main oil line and starting blowdown of the gas inventories to flare. The ESVs on the gas pipelines were not designed to close on platform ESD; this would impose an undesirable forced shut-down on the other platforms connected to Piper. Instead there were three separate shut-down buttons for these ESVs in the control room.

The explosion was followed almost immediately by a large fireball which issued from the west side of B Module and a large oil pool fire at the west end of that module. The explosion and fire were witnessed by personnel on the vessels lying off the platform. It so happened that one witness on the Tharos was standing with camera at the ready. He took a sequence of shots of the development of the fireball.

The large oil pool fire gave rise to a massive smoke plume which enveloped the platform from the production deck at the 84 ft level up.

The offshore installation manager (OIM) made his way to the radio room and had a Mayday signal sent.

The Tharos effectively took on the role of On-Scene Commander. The Coast Guard station and Occidental headquarters onshore were informed. Rescue helicopters and a Nimrod aircraft for aerial on-scene command were dispatched. The flight time for the helicopters was about an hour.

Most of the personnel on the platform were in the accommodation, the majority in the ERQ. Within the first minute flames appeared on the north face of the module also and the module was enveloped in the smoke plume coming from the south. The escape routes from the module to the lifeboats were impassable.

At the 68 ft level divers were working with one man under water. They followed procedure, got the man up and briefly through the decompression chamber. They were unable to reach the lifeboats, which were inaccessible due to the smoke. They therefore launched life rafts and climbed down by knotted rope to the lowest level, the 20 ft level.

The drill crew also followed procedure and secured the wellhead.

The oil pool in B Module began to spill over onto the 68 ft level where a further fire took hold. There were drums of rigwash stored on that level which may have fed the fire.

The fire water drench system did not operate. There was only a trickle of water from the sprinkler heads.

The explosion disabled the main communications system which was centred on Piper. The other platforms were unable to communicate with Piper. They became aware that there was a fire on Piper, but did not appreciate its scale. They continued for some time in production and pumping oil. This pumping would have caused some additional oil flow from the leak at Piper.

After some 20 min from the initial explosion the fire on the 68 ft level led to the rupture of the Tartan riser on the side outboard of the ESV. This resulted in a massive jet flame which enveloped much of the platform.

The emergency procedure was for personnel to report to their lifeboat, but in practice most evacuations would be by helicopter and personnel would be directed from the lifeboats to the dining area on the upper deck of the ERQ and then to the helideck. Personnel in the ERQ found the escape routes to the lifeboats blocked and waited in the dining area. The OIM told them that a Mayday signal had been sent and that he expected helicopters to be sent to effect the evacuation. In fact the helideck was already inaccessible to helicopters.

Some 33 min into the incident the Tharos picked up the signal 'People majority in galley. Tharos come. Gangway. Hoses. Getting bad'.

No escape from the ERQ to the sea was organized by the senior management. However, as the quarters began to fill with smoke individuals filtered out by various routes and tried to make their escape.

Some men climbed down knotted ropes to the sea. Others jumped from various levels, including the helideck at 174 ft. One man who had arrived only that afternoon on his first tour jumped from a high level. One standing on pipes protruding from the pipe deck was pushed off by another behind him who could no longer stand the heat of the pipes.

The vessels around the platform launched their fast rescue craft (FRCs). The first man rescued, by the FRC of the Silver Pit, was the oil laboratory chemist, who, on experiencing the explosion, simply walked down to the 20 ft level and was picked up without getting his feet wet. Most survivors, however, were rescued from the sea. Much of the rescue operation took place after the rupture of the Tartan riser.

The FRC of another vessel, the Sandhaven, was destroyed with only one survivor. The FRC of the Silver Pit made repeated runs to the platform; eventually it was blown out of the water, and began to sink, but returned to the platform and then finally sank, its crew being themselves rescued by helicopter.

At about 10.50 p.m. the MCP-01 riser ruptured and about 11.18 p.m. the Claymore riser ruptured. The pipe deck collapsed and the ERQ tipped. By 12.15 a.m. on 7 July the north end of the platform had disappeared. By the morning only A Module, the wellhead, remained standing.

A19.6 The Investigation

An investigation of the disaster was immediately undertaken by the Department of Energy (DoEn) headed by Mr Petrie. Two reports were issued, an interim report (the *Petrie Interim Report*, or simply, *the Petrie Report*) and a final report (*the Petrie Final Report*); the latter included appendices on various technical studies commissioned.

The *Petrie Report* put forward two scenarios for the hydrocarbon leak which led to the explosion: a leak from the site of PSV 504 (Scenario A) and a leak due to ingestion of liquid into the reciprocating compressors (Scenario B).

The Inquiry was presided over by a Scottish judge, Lord Cullen, assisted by three technical assessors. There was a legal counsel to the Inquiry assisted by technical consultants to the Inquiry. Parties to the Inquiry included Occidental, the DoEn, groups representing survivors and the trade unions, the contractors, the specialist contractor Score, several equipment manufacturers and for the second part, the UK Oil Operators Association (UKOOA). Part 1 of the Inquiry dealt with the disaster and its background, Part 2 with the future. The Inquiry heard some 280 witnesses in 180 days of evidence and received some 840 productions, or documents.

It began by considering whether to advise that the debris of the platform could be raised from the sea bed. It was clear at an early stage that the size of leak sought was of the order of 10 mm^2. The evidence was that the operation presented a number of problems and hazards, would involve considerable delay and might well not provide much useful information. The Inquiry decided not to pursue the matter.

In seeking to find the cause of the leak, the *Piper Alpha Report* begins with the evidence on the explosion itself. It concludes that the explosion was at 10.00 p.m., that it was in C Module and in the south-east quadrant of that module, that the fuel involved was condensate, that the leak gave rise to a gas cloud filling less than 25% of the module, that the mass of fuel within the flammable region was some 30–80 kg, that the explosion was a deflagration rather than a detonation, that the maximum peak overpressure was in the range 0.2–0.4 bar and that the ignition source could not be identified. Evidence for these findings included the gas alarms in C Module; the screeching noise heard just prior to the explosion; testimony of and photographs taken by observers on the surrounding vessels of the fires just after the explosion; the effects of the explosion, including the damage to the two firewalls in C Module; the effects of the explosion on the control room and its occupants; the lack of damage to the heat shield on A Module and estimates of overpressure based on some of the explosion effects, such as firewall damage and bodily translation of persons.

It was not initially clear how a gas cloud of sufficient size could develop without setting off certain gas alarms which according to the evidence had not been triggered. In particular there was a gas detector in the roof above the site of PSV 504 or PSV 505 (the two valves were close together)

and another some 2–3 ft above floor level among the heat exchangers between the reciprocating and centrifugal compressors; both these detectors were in C2 zone. However, the first gas alarm was in C3 zone at C centrifugal compressor. Accordingly, wind tunnel tests were commissioned from BMT Fluid Mechanics to explore the pattern of gas alarms for different types of leak. Scenarios investigated included leaks of natural gas and of condensate, the one a buoyant and the other a heavy gas; leaks from various points in the modules and leaks from various types of source, including a leaking flange and an open pipe. It was concluded by the investigator that of the scenarios studied only a leak from the site of PSV 504 or PSV 505 fitted the pattern of gas alarms and, further, that this leak was a two-stage leak, the first stage being small and the second relatively large. It would have been this second stage which gave rise to the gas cloud sought.

The experimental run of main interest simulated a leak of 100 kg/min from PSV 504. Figure A19.5 shows the contours of the lower flammability limit of the gas cloud formed after 30 s from such a leak. The cloud would not set off either of the gas detectors mentioned. Figure A19.6 shows the mass of gas within the flammable limit as a function of the leak flow rate after 30 s and at infinite time. These results were subject to a number of reservations but indicated that a gas cloud of sufficient size could be formed.

The next question considered was whether the explosion of such a gas cloud could cause the damage observed. It was estimated that the B/C firewall would fail at an overpressure of 0.1 bar and the C/D firewall at an overpressure of 0.12 bar.

Simulations of the explosion of flammable mixtures in the module had been commissioned prior to the Inquiry at the Christian Michelsen Institute (CMI) by the DoEn and other parties. The simulations were performed using the FLACS computer code. Following the wind tunnel work, the Inquiry commissioned a single further run for a gas cloud in the south-east quadrant of C Module and containing some 45 kg of propane within the flammable limit. The simulation was subject to a number of reservations but indicated that such an explosion could cause the firewall damage observed.

The simulation also provided an explanation of a point which had seemed puzzling. The two occupants of the control room were thrown across it by the explosion and experienced a rush of cold air, not hot gas. The simulation showed that in the early stages of the explosion the control room wall would be subject to a positive overpressure and inrush of air, but that by the time the hot combustion products approached the control room, the negative phase of the pressure pulse had set in, the velocity vectors had reversed and the direction of air flow was out of the control room into C Module.

The two-stage nature of the leak also presented another point of difficulty. Isolation of A pump was by the closure of the GOVs on the suction and the delivery lines. The suction GOV was electrically isolated and it was uncertain whether power to it had been restored by the time of the initial explosion. In any event restoration of the valve would involve reconnecting a pneumatic line to the valve, which could quickly be done by an operator. It was concluded that this connection was made and that probably the operator gave it a tweak to make sure the valve movement was restored. This would have had the effect of admitting

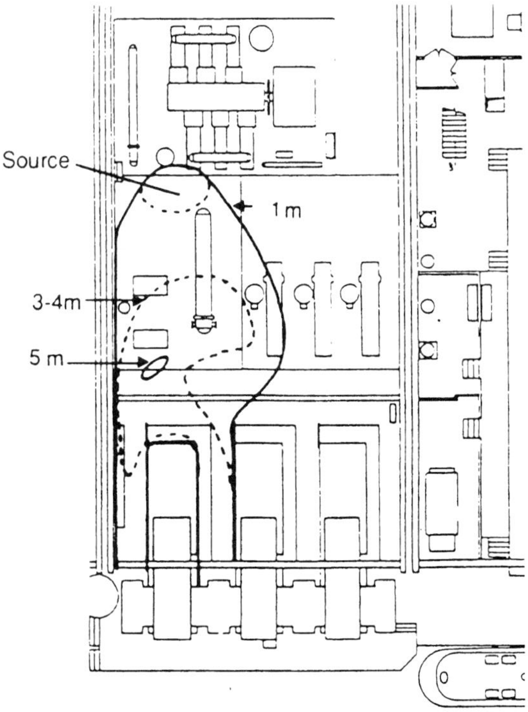

Figure A19.5   *The flammable gas cloud for a leak of 100 kg/min in the BMT wind tunnel tests: LEL contours at 30 s (Cullen, 1990)*

condensate to the relief line to PSV 504, but not of filling it with condensate liquid, thus giving rise to the early, smaller leak. Subsequent opening of the valve and filling of the relief line with condensate liquid could then have caused the later, larger leak.

Evidence was also heard on tests on leaks from blind flanges. The flange at PSV 504 was a ring-type joint (RTJ) flange. Three methods of tightening up were investigated: flogging up with a flogging spanner and hammer; hand tightening with a combination spanner and finger tightening. The results showed that a flange in good condition which had been flogged up or hand tightened did not leak. Even deterioration of the flange would be unlikely to give the leak sought unless the deterioration was gross. However, a finger-tightened flange could give a leak which was directionally downwards and was of the flow rate sought.

The Inquiry concluded that the explosion had been caused by ignition of a gas cloud containing some 45 kg of hydrocarbon within the flammable range, arising from a two-stage leak, in the first stage perhaps some 4 kg/min and in the second stage some 110 kg/min lasting some 30 s, coming from an orifice of equivalent diameter some $8 \, \text{mm}^2$.

There was no obvious explanation why the blind flange was not leak-tight. Much evidence was led to the effect that an experienced and competent fitter would not make up a blind flange which was not leak-tight. The Inquiry noted, however, that the decision not to proceed with the full 24 month PM on A pump was taken just as shift handovers on the platform were starting so that some personnel may have been ignorant of this change in intent and that the lack of leak-tightness of the blind flange may have been connected with the status of A pump.

The lead production operator had clearly had the intention to start A pump. It was difficult to explain this given fact that its sole PSV was off. The Inquiry concluded that the lead operator was indeed ignorant of this, even though

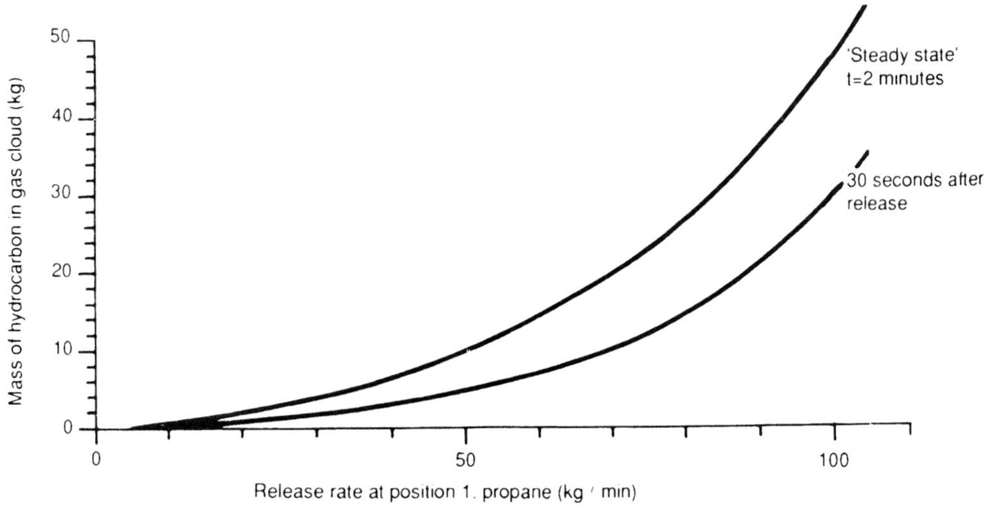

Figure A19.6   *Mass of fuel in the flammable range in the BMT wind tunnel tests: (a) variation with time (b) variation with leak rate (Cullen, 1990)*

this meant a serious breakdown of communications about the work. It implied that the fact that the PSV was off was not communicated in the handovers of the lead maintenance hand, the phase 1 operator and the lead production operator and that the lead operator did not learn of it through the PTW system.

When he found that he was unable to put the PSV back that evening, the Score supervisor came up to the control room to suspend the permit. He was on his first tour as a supervisor and had had no training in the operation of the PTW system in use on the platform. Whom he spoke to and what transactions took place were obscure. It was unclear how he knew that the procedure in filling out the permit for suspension was to write 'SUSP' in the gas test column.

In any event the Score supervisor did not make a final inspection of the job site before going off work and evidently the lead production operator did not inspect the job site either, although in both cases good practice required that this be done.

Further, the leak would not have occurred if there had been a more positive isolation of the pump by means such as the use of a slip plate.

The explanation just described is that adopted in the *Piper Alpha Report* but several other scenarios for the leak were also explored. One group of scenarios was concerned with explanations of the leak from the blind flange following admission of condensate into A pump other than lack of leak-tightness. They include the possibilities of auto-ignition, shock loading, low temperature brittle fracture, and overpressurization by methanol injection. All of these were quickly ruled out except auto-ignition by compression of a flammable mixture formed in the relief line. The line had been left open for an hour before the blind flange was put back on. It was not possible to calculate whether auto-ignition would have occurred due to lack of data on auto-ignition properties of the multi-component mixture at the high pressures involved, some 300 bar. Moreover, company documentation on the rating of the pipework was inconsistent so that it was uncertain whether the flange was a 900 or 1500 lb one. The expert evidence was that if auto-ignition had occurred and the lower rating applied, a leak was possible, although whether it would have had the required characteristics was another matter. An account of this work on auto-ignition has been given by S.M. Richardson, Saville and Griffiths (1990).

The scenario was considered that condensate liquid had backed up in the JT flash drum and thence into the reciprocating compressors. There was evidence that on loss of B pump steps had been taken to reduce the condensate make by unloading and recycling these compressors. The report concludes that there had been insufficient time for backup to occur before the initial explosion and that in addition both the conditions around the compressors and the expected action of protective instrumentation were against this scenario.

A further scenario which emerged in the Inquiry was that the leak occurred from PSV 505 and that it was caused by hydrate blockage. The interruption to the methanol supply to the JT valve lent credibility to this scenario. Experimental work commissioned showed that hydrates would form under the conditions pertaining at the JT valve during the partial loss of methanol supply if the temperature at the valve fell below a critical value; it had in fact been below this temperature on 5 July. The expert evidence was that hydrates could pass through to the condensate

injection pumps and cause blockage there some 2 h after restitution of the full methanol supply. There were several versions of the scenario all leading to blockage of hydrate at PSV 505 and overrunning of the pump so that the delivery pressure rose to a value high enough to cause rupture of the valve, which was the weakest point in the line. The report does not rule out this scenario, but regards it as less likely than the preferred one.

Finally, the consultants to the Inquiry reviewed a large number of other scenarios which were not purely theoretical but had some link with the information available at the time, which included a hazop study, past equipment failures and process conditions that evening. None was found convincing by the Inquiry.

Turning to the escalation, the causes of the oil pool fire and the fireball which occurred in B Module within seconds of the explosion in C Module were unclear. The Inquiry heard evidence on the type, number, velocity and impact effects of the projectiles which would have resulted from the destruction of the B/C firewall. The condensate injection line ran from C Module through into B Module where it joined the main oil line. The report concludes that probably the fireball was caused by a missile rupturing this line near the main oil pumps at the west face of B Module.

Estimates of the size of the oil pool fire indicated that the supply of oil to the fire probably exceeded the oil inventory of the separators and that there was a leak of oil from the main oil line through the main oil line EVS which was not fully closed. This leak would be aggravated by continued pumping of oil by the other platforms.

The fire water deluge system did not work. The initial explosion knocked out the main power supplies. It may also have damaged the water pumps and the water mains. In any event it was the practice on Piper to put the pumps on manual start when divers were in the water and thus in possible danger of being sucked into the pump intakes and they were on manual start that evening. The start controls were at the pumps themselves. After the explosion occurred an attempt was made to get through to the pumps to start them by personnel wearing breathing apparatus, but to no avail. Further evidence was given of quite extensive blockage caused by corrosion products in the fire water deluge system, which operated on sea water, a problem which had persisted for some years.

The initial explosion caused the operation of the platform ESD. This could have occurred through loss of the main power supply and/or rupture of a pneumatic ring main. Also, although dazed by the explosion the control room operator pressed the platform ESD button. He did not, however, press the buttons to close the ESVs on the three gas pipelines. Evidently, these did close but their closure was due rather to the effects of the initial explosion on power supplies to the valves.

Following the initial explosion a period of extended flaring occurred which greatly exceeded that to be expected from the flaring of the gas inventory on the platform. The report accepted that the most probable explanation was a failure of the Claymore ESV to close fully.

The main communications for the Piper field were centred on Piper. The system was knocked out by the initial explosion, so that the other platforms were unable to communicate with Piper and had difficulty communicating with the shore.

The report details a number of management weaknesses. There were severe and numerous defects in the PTW system. For example, it violated more than half of the main points in the code of practice on PTWs issued by the Oil Industry Advisory Committee (OIAC). The system was operated rather casually. The training of the specialist contractors supervisor in the permit system operated on the platform was found to be inadequate.

With regard to handovers, the company had been prosecuted only a year before for a fatality and had pleaded guilty. The report takes the view that a failure in handover procedures was a factor in that accident.

A number of different types of audit were performed by the company, by its partners, by loss prevention specialists and so on. None of these had revealed the defects in the PTW which became apparent very quickly at the Inquiry.

The report is critical of the handling of the emergency by the senior management on the platform and in particular of the failure to recognize that helicopter evacuation was not possible and to take command of the situation and organize escape from the ERQ.

The decision to keep the platform operating despite the large workload is another matter of which the report is critical.

The report states that the company had no system to ensure that all projects were subject to formal safety assessment. Certain techniques such as hazop were used on some projects and quantitative estimates had been made in some studies, but the approach was unsystematic. The report takes the view that as a consequence the hazards presented by the hydrocarbon inventory on the platform and particularly in the pipelines had not been systematically addressed.

Part 2 of the Inquiry was concerned with the future offshore safety regime. The context was not only the Piper Alpha disaster but also the changes taking place in the North Sea oil province. The exploitable oil and gas fields were becoming smaller and the technology to develop them was becoming more varied.

The evidence in Part 1 revealed serious weaknesses in the company management. It was an issue why the DoEn had not discovered these weaknesses. The report is critical of the relative lack of emphasis placed by the Department on the assessment of management and management systems.

In contrast to the British onshore and Norwegian offshore regimes, which had both moved increasingly towards goal-setting regulations, the British offshore regime relied excessively on prescriptive regulations, and associated guidance.

The deficiencies of such a regime were illustrated in the regulations concerning fire protection, which had a number of defects. Passive and active fire protection were covered by two separate sets of regulations. The regulations, and associated guidance, for active fire protection led in practice to systems based on delivery of a uniform quantity of water over large areas of a platform, deluge systems prone to blockage and massive water pumps. Fire protection was not integrated with explosion protection.

The report states that the approach taken by the DoEn to the control of the major hazards from hydrocarbons at high pressure did not impress as an effective one. Further, the inspectorate had relatively little expertise in this area.

Following the Piper Alpha disaster the DoEn brought in regulations to require ESVs to be placed nearer to sea level and for the valves to be of full ESV standard. The report notes that of the 400 risers covered by the regulations, some 70 required modification in the latter respect.

The regime made little use of formal safety assessment (FSA). This was in contrast to the regulatory use of FSA onshore, and in particular the onshore safety case. The DoEn had in fact explicitly rejected the concept of an offshore safety case. This policy also contrasted with the situation in the Norwegian sector, where a concept safety evaluation (CSE) was required. Quantitative risk assessment is required in a CSE and is often necessary to fulfil the requirements of a safety case.

Considerable evidence was heard on quantitative risk assessment (QRA). The burden of this evidence was that QRA is in regular use in many companies as an aid to decision-making both for onshore and offshore installations and that there was no serious impediment to this from any problems of overall methodology, frequency estimation, consequence modelling or risk criteria.

In contrast to the HSE, the DoEn was not well equipped to operate a regime based on goal-setting regulations and FSA. It had no experts in FSA or fire protection.

A number of recommendations which would have met some of the points on which the DoEn was criticized had been made in the *Burgoyne Report* (Burgoyne, 1980), but had not been implemented.

A19.7 Some Lessons of Piper Alpha

A19.7.1 Some lessons

Lessons from the Piper Alpha disaster are considered in this section with the exception of the recommendations of the report on the offshore safety regime which are considered in Section A19.8. A list of some of the lessons is given in Table A19.2. Many apply particularly to offshore installations, but others are of more general applicability.

Regulatory control of offshore installations
The Piper Alpha disaster exposed weaknesses in the offshore regulatory regime which have already been described. The lessons drawn are seen in the recommendations given in Section A19.8.

Quality of safety management
The *Piper Alpha Report* is critical of the quality of management, and particularly safety management, in the company.

It was not that the company did not put effort into safety. On the contrary, there were numerous meetings and much training on safety. The problem was the quality of these activities.

Many managers had come up through the ranks and had minimal qualifications. The culture tended to be somewhat in-grown and insufficiently self-critical.

These defects manifested themselves in various ways such as in the toleration of poor practices in plant isolation and operation of the PTW system; in the failure to appreciate the ineffectiveness of the audits done; in the failure to address the major hazard problem and to use FSA. The report comments:

Senior management were too easily satisfied that the PTW system was being operated correctly, relying on the absence of any feedback of problems as indicating that all was well. They failed to provide the training necessary to ensure that an effective PTW system was operated in practice. In the face of a known problem with the deluge system they did not personally become involved in probing the extent of the problem and what should be done to resolve it as soon as possible. They adopted a superficial

Table A19.2 *Some lessons of Piper Alpha*

Regulatory control of offshore installations
Quality of safety management
Safety management system
Documentation of plant
Fallback states in plant operation
Permit-to-work systems
Isolation of plant for maintenance
Training of contractors personnel
Disabling of protective equipment by explosion itself
Offshore installations: control of pressure systems for
    hydrocarbons at high pressure
Offshore installations: limitation of inventory on
    installation and in its pipelines
Offshore installations: emergency shut-down system
Offshore installations: fire and explosion protection
Offshore installations: temporary safe refuge
Offshore installations: limitation of exposure of
    personnel
Offshore installations: formal safety assessment
Offshore installations: safety case
Offshore installations: use of wind tunnel tests and
    explosion simulations in design
The explosion and fire phenomena
    Explosions in semi-confined modules
    Pool fires
    Jet flames
Publication of reports of accident investigations

response when issues of safety were raised by others, . . .
They failed to ensure that emergency training was being
provided as they intended. Platform personnel and management were not prepared for a major emergency as they
should have been. (para 14.52)

A crucial weakness was failure to appreciate that
absence of feedback to management about problems is
almost certainly an indicator not that there are no problems
but that there are, and they could be serious. Of one OIM the
report states: 'His approach seemed to be, in his own words,
"surely that is all you are concerned with about the permit
system . . . If the system is working and no problems are
identified . . . then you should be reasonably happy with
it, surely?" . . . He had been surprised by the number of
deficiencies in the operation of the permit system which
had been revealed in the Inquiry. He had checked this out
and found it to be true.' Of another manager it states that he
said 'he knew that the system was monitored on a daily
basis by safety personnel. By the lack of feedback he "knew
that things were going all right and there was no indication
that we had any significant permit to work problems".'
(para 14.26)

Safety management system
Onshore the quality of the management and the management system are of prime concern to the HSE in its inspections in general and in the safety case in particular. In
submitting a safety case a company will often give extensive documentation on its systems. Nevertheless, in the
regulations the formal requirements on management are
fairly minimal.
    The Inquiry heard evidence in favour of the assurance of
safety through the use of quality assurance to standards

such as BS 5750 and ISO 9000. It also heard evidence on the
need for better qualified management, including a proposed requirement for all OIMs to be graduates.
    The concept of a safety management system goes part
way towards these insofar as the system itself is based on
principles similar to those of quality assurance and covers
the question of management quality.

Documentation of plant
The discrepancies in the documentation concerning the
rating of the flange on PSV 504 have already been mentioned. The Inquiry in fact heard of a number of other
defects in the documentation of the plant. Failure to maintain correct records can have serious consequences.

Fallback states in plant operation
The loss of the working condensate pump on Piper created
a situation where operating personnel were under some
pressure to start the other pump and avoid a gas plant shut-
down with its possible escalation to a total shut-down, loss
of power and the need for a 'black start'. In this case the
pressure was created partly by the view which an individual took of the probability that the changeover of the main
generators from gas to diesel would fail. This illustrates the
desirability of ensuring that plants have fallback states
short of total shut-down. In this case the problem was in the
reliability of changeover, a type of problem which may lie
with design or with maintenance.

Permit-to-work systems
The defects in the PTW system have already been described. These defects led directly to a situation where condensate was admitted to a pump from which the PSV had
been removed and hence to the disaster. The *Piper Alpha
Report* devotes considerable attention to the need for an
effective system.

Isolation of plant for maintenance
The *Piper Alpha Report* states that the disaster would not
have occurred if A pump had been positively isolated so
that condensate could not be admitted. Positive isolation is
not achieved by shutting a valve but requires means such as
insertion of a slip plate or removal of a pipe section.

Training of contractors' personnel
The proportion of contractors' personnel on an offshore
installation can be as high as 70%. The offshore scene
therefore exemplifies in extreme form a problem which
applies to onshore plants also. This is the need to train
contractors' personnel in the company's operating and
emergency systems and procedures. Failure to train a contractor's supervisor in the operation of the PTW system on
Piper meant that he was unfamiliar with a feature of the
system which turned out to be a critical one.

Disabling of protective equipment by explosion itself
The initial explosion on Piper disabled large parts of the
protective systems, including power supplies and fire
water supplies. It illustrates the importance of taking this
factor into account in design and in FSA.

Offshore installations: control of pressure systems for hydrocarbons at high pressure
An offshore production platform contains a large amount of
plant containing hydrocarbons at high pressure. The feed

to this plant is from the wells, which can sometimes behave in an unpredictable way. The pipelines connected to the platform contain large quantities of hydrocarbon, the high pressure gas pipelines constituting a particularly serious hazard.

There needs therefore to be a comprehensive system for the control of the total pressure system, covering design, fabrication, installation, operation, inspection, maintenance and modification and including control of such features as materials of construction, lifting of loads and so on and personnel need to be trained in the purposes and operation of system.

Offshore installations: limitation of inventory on installation and in its pipelines

The scale of the Piper disaster was due primarily to the large inventory of the three high pressure gas pipelines connected to the platform. The Inquiry heard evidence on the practicalities of reducing the number of gas pipelines connected to a platform. There are many technical problems involved, but the point has been made that such reduction should be a design objective.

The main inventory of hydrocarbons in process on a platform is in the separators. The massive oil pool fire on Piper was fed from the separators. The *Piper Alpha Report* recommends that methods of dumping this inventory be explored.

Evidence was also heard that in some cases the main inventory of hydrocarbons on a platform might be the diesel fuel.

The alternative method of preventing the hydrocarbon inventory from feeding a fire is emergency isolation, which is considered next.

Offshore installations: emergency shut-down system

The ESD system on Piper operated, shut-down ESVs and blew the gas inventory on the platform down to flare.

Nevertheless, the accident drew attention to a number of problems in effecting isolation, some specific to offshore platforms and some more generally applicable.

The Tartan riser ruptured on the outboard side of the ESV so that closure of this valve was of no avail. It is clear that an ESV needs to be located as close to the sea level as practical in order to minimize this risk.

It is possible to go one step further and install a subsea isolation valve, but this is for consideration on a case-by-case basis.

Both types of isolation valve received considerable attention in the Inquiry. However, neither will be effective unless it achieves tight shut-off. The evidence that the main oil line and the Claymore gas line ESVs did not shut-off tightly emphasizes the importance of this feature.

Moreover, in order to be effective the ESD system has to be activated. The fact that on Piper closure of the three gas line ESVs was not part of the platform ESD but had to be effected for each line separately by manual pushbutton, that these buttons were not pushed and that closure only occurred due to loss of power shows that this problem also is not a trivial one.

Offshore installations: fire and explosion protection

An offshore installation is not in general able to call on outside assistance comparable with that available from the fire brigade to an onshore plant. It must be self-reliant.

This implies that both protection against, and mitigation of, fire and explosion on the one hand and fire fighting on the other are of particular importance and that both the hazard assessment and the design and operation of the plant must be of high quality.

Offshore installations: temporary safe refuge

It is clear from the Piper disaster that there needs to be a temporary safe refuge (TSR) where personnel can shelter in an emergency and where the emergency can be controlled and evacuation organized.

This TSR will normally be the accommodation. In most cases it will be on the production platform itself, but it may be on a separate accommodation platform.

The protection of the TSR from ingress of smoke and fumes from outside and from generation of fumes by fires playing on the outside needs careful attention. Measures require to be taken to prevent smoke ingress through doors and through the ventilation system.

Offshore installations: limitation of exposure of personnel

The concept of a TSR is a particular application of the more general principle of limitation of exposure of personnel. The Inquiry also heard evidence of the application of the principle to other aspects such as the location of workshops.

Offshore installations: formal safety assessment

The evidence showed that many companies which operate installations onshore and offshore have formal systems for safety assessment and practise FSA routinely, that FSA has considerable benefits in the design and operation of plant and that it provides a suitable basis for dialogue between the company and the regulatory body.

Offshore installations: safety case

A safety case is a particular form of FSA. The evidence indicated that a safety case is as applicable offshore as onshore and that it is a suitable means for the company to demonstrate to the regulatory body that it has identified the major hazards of its installation and has them under control.

Offshore installations: use of wind tunnel tests and explosion simulations in design

Wind tunnel tests and explosion simulations were used in the Inquiry to investigate the cause of the explosion, but evidence was also heard of their value in platform design.

Wind tunnels may be used to assess the effectiveness of ventilation and of the gas detection system in a module, the wind conditions at the helideck and the movement of smoke from oil pool fires. Explosion simulations may be used to investigate the effect of different module layouts on explosion overpressures and to assess the effectiveness of blast walls.

The explosion and fire phenomena

The Piper disaster drew attention to several important aspects of explosion and fire on offshore installations. These include explosions in semi-confined modules, oil pool fires and jet flames.

Explosions in semi-confined and congested modules are a hazard which assumes particular significance offshore. Although major progress has been made in the last decade in simulating such explosions and developing design methods, this remains an area where further work is needed.

Oil pool fires onshore are relatively well understood, but this does not apply to the behaviour of such a pool fire on an offshore platform. Aspects of some importance are design to prevent accumulation of an oil pool in the first place and the massive smoke plume from such a fire.

Jet flames, including jet flames from risers, are particularly important for offshore platforms. In this case there are available a number of models developed for flares and flames on onshore plant, including pipelines, which can be applied offshore.

Evidence given indicated that in considering the hazard of a jet flame from a riser, the worst case was not necessarily a full bore rupture but a partial rupture, since the latter is sustained for a longer period.

Publication of reports in accident investigation

The Inquiry heard that the company had a policy of severely restricting circulation of accident investigation reports.

Likewise, the DoEn did not make public reports on major offshore accidents. This contrasts markedly with the HSE policy of issuing reports on major accidents, many of which are referred to in this book.

A19.7.2 An accident model

The following outline of a model of the accident highlights the role played by some of the features just mentioned:

	Deficiencies in
Initial event: gas explosion	Operational control
Escalation 1: explosion damage	Hazard identification, assessment and management Explosion mitigation
Escalation 2: oil pool fire	Hazard identification, assessment and management Fire mitigation and fire fighting Inter-platform emergency planning
Escalation 3: riser rupture	Hazard identification, assessment and management Fire protection of risers
Escalation 4: accommodation failure	Hazard identification, assessment and management TSR fire and smoke protection Emergency command and control

A19.8 Recommendations on the Offshore Safety Regime

The *Piper Alpha Report* makes recommendations for fundamental changes in the offshore safety regime.

The basis of the recommendations is that the responsibility for safety should lie with the operator of the installation and that nothing in the regime should detract from this.

The offshore regime envisaged in the recommendations is one in which the emphasis is on the operator demonstrating to the regulatory authority the safe design and operation of its installation rather than demonstrating mere compliance with regulations. In this regime the preferred form of regulations is goal-setting rather than prescriptive.

Table A19.3 *Some elements of the safety management system*

Organizational structure
Management personnel standards
Training, for operations and emergencies
Safety assessment
Design procedures
Procedures, for operations, maintenance, modifications and emergencies
Management of safety by contractors in respect of their work
Involvement of the workforce (operator's and contractors') in safety
Accident and incident reporting, investigation and follow-up
Monitoring and auditing of the operation of the system
Systematic re-appraisal of the system in the light of experience of the operator and industry

The recommendations envisage that FSA will play a major role. It may be used to demonstrate compliance with a goal-setting regulation or with the general requirements of the HSWA.

A central feature of the regime proposed is the safety case for the installation. This safety case is broadly similar to that required for onshore installations but there are some important differences. In the offshore safety case it is required that the operator should demonstrate that the installation has a TSR in which the personnel on the installation may shelter while the emergency is brought under control and evacuation organized.

Further, it is recommended that this demonstration should be by QRA. This means that there must be criteria which define the failure of the TSR and criteria for its endurance time and its failure frequency. The criteria may then be met by reducing the frequency of accidental events, by increasing the durability of the TSR or by some combination of these.

The recommendation on the safety case includes a requirement that the operator should demonstrate that it has a safety management system (SMS) to ensure the safe design and operation of the installation. This SMS should draw on quality assurance principles similar to those of BS 5750 and ISO 9000. The elements of the SMS should include those listed in Table A19.3. They include management personnel standards.

Various measures related to hardware were urged at the Inquiry. These included the provision of separate accommodation platforms, the installation of subsea isolation valves and blast walls, the use of freefall lifeboats and purpose-built standby vessels. The report takes the view, however, that in accordance with its basic philosophy such matters should be dealt with as part of the demonstration of safe design and operation.

The report considers that the then current regulatory body, the DoEn, is unsuitable as the body to be charged with implementing the new regime and recommends the transfer of responsibility for offshore safety to the HSE.

These recommendations were accepted immediately by the government and the new regime under the HSE began in April 1991.

Appendix

20

Nuclear Energy

Contents

The energy industries in general have much in common with the process industries in terms of the management, the plant and the hazards. This is true even of the nuclear industry, although this also has a number of unique features. Historically, the nuclear industry has often had to face particular problems before they have impinged on the process industries and has had to devise solutions for them. In consequence, there are a number of areas where work in the nuclear industry is relevant to the process industries.

In this appendix a general account is given of nuclear energy, essentially with reference to those features which are common to, or have impact on, the process industries. Accounts of the accidents at Three Mile Island and at Chernobyl and of the Rasmussen Report are given in Appendices 21, 22 and 23, spectively.

Some topics common to both industries have been dealt with already in other chapters. The account of major hazard control in Chapter 4 covers nuclear hazards. The treatment of management and management systems in Chapter 6 is applicable to nuclear as well as process systems.

General information on nuclear phenomena and on radioactivity is given in *Source Book on Atomic Energy* by Glasstone (1968), *The Effects of Nuclear Weapons* by Glasstone and Dolan (1979), *Nuclear Energy Technology* by Knief (1981) and *A Dictionary of Nuclear Power and Waste Management* by Foo-Sun Lau (1987).

Frequent reference is made in this appendix to two reports issued following the accident at Three Mile Island. These are *The Report of the Presidents Commission on the Accident at Three Mile Island (the Kemeny Report)* (Kemeny, 1979) and *TMI-2 Lessons Learned Task Force Final Report (the Task Force Report)* (NRC, 1979d).

Selected references on the nuclear industry are given in Table A20.1.

Table A20.1 *Selected references on nuclear energy. See also Appendix 28, Section B (Nuclear Regulatory Commission)*

Nuclear energy and technology
ASME (Appendix 28 *Nuclear Safety*); HSE (Appendix 28 *Nuclear Installations,* 1979 HA 4); AIChE (1954–67/ 101–118, 1972/119); McCullough (1957); J.H. Bowen and Masters (1959); Kopelman (1959); Tipton (1960); El-Wakil (1962); Fozard (1963); Gill (1964); T. Thompson and Beckerley (1964); Lamarsh (1965); Zudans, Yen and Steigelman (1965); Shaw (1967); Glasstone (1968); F.R. Farmer (1969a); Glasstone and Sesonske (1969); ENEA (1970); Ramey (1970); Wechsler (1970); Berry (1971); CEGB (1971); T.J. Thompson (1971); A.E. Green and Bourne (1972); R.V. Moore (1972); Holdren and Herrera (1973); Jaeger (1973, 1975); Inglis (1973); ACRS (1974); Allardice and Trapnell (1974); Bupp and Derian (1974); Cochran (1974); Willrich and Taylor (1974); Flood (1976); Haferle *et al.* (1976); S.E. Hunt (1976); RCEP (1976 Rep. 6); ASTM (1977 STP 616); S. Fawcett (1977); IMechE (1977/38, 1982/64, 1984/82, 1988/99, 1992/ 143); Pocock (1977); Pedersen (1978); Duderstadt (1979); J. Hill (1979a); Kaplan (1979); Cottrell (1981); Knief (1981); IBC (1982/24); IAEA (1983); W. Marshall (1984); Rahn *et al.* (1984); Stahlkopf and Steele (1984); Bennett (1985); Wilkie and Berkovitch (1985); Wilkie (1986); Kletz (1986e, 1987g, 1991c); Inhaber (1987); UKAEA (1987); Varey (1987); Bindon (1988); Lewins (1988); G. Lewis (1988); V.C. Marshall (1988b); Gittus (1988);

Bennett and Thomson (1989); NFPA (1991 NFPA 801, 1993 NFPA 803); Nowlen (1992); Hicks (1993) ANSI N series

Inherently safer design
Kletz (1988h); Forsberg *et al.* (1989)

Sizewell B
CEGB (1982); HSE (Appendix 28 *Nuclear Installations, Sizewell B*, 1982 HA 3, 1982b,c, 1983a); NRPB (1982 R137, R142, R145, R146, 1983 R160, R163); NRC (1983 NUREG-0999); (IMechE (1989/113, 1994/173);
Software reliability: CEC (1987 EUR 11147 EN); Hunns and Wainwright (1991)

Nuclear waste, fuel reprocessing
I.K.G. Williams (1979); Bennett (1981); R. Smith and Hartley (1982); J.A Williams (1982); Bennett (1985); IBC (1985/64, 1986/70, 1987/73); Heafield (1986); N.J. James, Rutherford and Sheppard (1986); Elliot and King (1987); N. James, Sheppard and Williams (1987)

Radioactivity and radioactive contamination
NRPB (Appendix 28); Moelwyn-Hughes (1961); W.J. Moore (1962); Yen Wang (1969); Coggle and Noakes (1972); McDonald, Darley and Clarke (1973); Sternglass (1973); AEC (1975); Brodine (1975); MRC (1975b); G.N. Kelly, Jones and Hunt (1977); R.H. Clarke (1980); Whicker and Schultz (1982); Foo-Sun Lau (1987); NRPB (1989, 1993) BS (Appendix 27 *Radioactivity*)

Nuclear hazard assessment and control
HSE (Appendix 28 *Nuclear Installations*, 1977c, 1979d,f, 1979 HA 5,1983/8, 1992/16); AEC (1957, 1975); Sinclair (1963); F.R. Farmer (1967a,b, 1969a,b, 1970, 1971); Bourgeois (1971); Dale and Harrison (1971); F.R. Farmer and Gilby (1971); Hanauer and Morris (1971); Kirk and Taylor (1971); Ybarrando, Solbrig and Isbin (AlChE 1972/119); Hosemann, Schikarski and Wild (1973); EPA (1974); Karam and Morgan (1974); N.C. Rasmussen (1974, 1981); Chicken (1975, 1982, 1986); Bridenhaugh, Hubbard and Minor (1976); Flood (1976); Freudenthal (1976); Rust and Weaver (1976); Ford (1977); Fussell and Burdick (1977); Comey (1977); HM Chief Inspector of Nuclear Installations (1977); G.N. Kelly, Jones and Hunt (1977); NRC (1977a–c); UCS (1977); Higson (1978); R.J. Parker (1978); E.E. Lewis (1979); Windscale Local Liaison Comm. (1979); Alvares and Hasegawa (1979/80); Joksimovic and Vesely (1980); Strong and Menzies (1980); Charlesworth, Gronow and Kenny (1981); Fleming *et al.* (1981); Sagan (1981); Allan, Adraktas and Campbell (1982); NRC (1982); Openshaw (1982); I.B. Wall (1982); Amendola (1983a,b); D.E. Bennett (1983); Fremlin (1983); Heudorfer and Hartwig (1983); Brooks (1984); CEC (1985/2); Lois and Modares (1985); Frank and Moieni (1986); C.D. Henry and Brauer (1986); Ronen (1986); Anon. (1987p); Ashworth and Western (1987); Durant and Perkins (1987); Gittus (1987); Mullins and Clough (1990 SRD R509); J. Singh, Cave and McBride (1990); Ang (1992); IAEA (1992); Wu and Apostolakis (1992); Hoffmeister (1993); J.R. Thomson (1994)

NRC studies
NRC (Appendix 28, *Loss of Coolant Accident, Loss of Fluid Test, Operational Safety Reliability Program, Reactor Safety Study, Safety Assessment, Safety Goals, Severe Accident Risk Reduction Program*)
Benchmark studies TUV (1979); NEA (1984)

Detailed studies
Coxon (1971); Kazarians and Apostolakis (1978);
Apostolakis and Mosleh (1979); Rasmuson, Wilson and
Burdick (1979); Apostolakis and Kazarians (1980);
Aupied, Le Coguiec and Procaccia (1983); M.G.R. Evans
and Parry (1983); Hendrickx and Lannoy (1983);
Coudray and Mattei (1984); NRC Steam Explosion Rev.
Gp (1985); Bley and Reuland (1987); Ashurst and Davidson
(1993 IPB 112)

Emergency planning
Eppler (1970); HSE (1982/5, 1986b, 1991/15)

Incidents
Chamberlain (1959); Anon. (1962c); Norman (1975); Anon.
(1976g); HM Chief Inspector of Nuclear Installations (1977);
V.C. Marshall (1979c); Bertini *et al.* (1980); Curtis, Hogan
and Horowitz (1980); J.G. Fuller (1984); May (1989); Mosey
(1990); Ashurst and Davidson (1993 LPB 112); Anon. (1994
LPB 119, p. 13)

Windscale
Atom. Energy Office (1957); Fleck (1958); Fremlin (1978);
R.J. Parker (1978); HSE (1981/4, 1986/9); NRPB (1981 R133,
1982 R135); May (1989); Arnold (1992)

Browns Ferry
AEC (1975); Ford, Kendall and Tye (1976); NRC (1976);
Anon. (1982 LPB 47, p. 13); May (1989)

Severe accident risks
Pickard, Lowe and Garrick (1983); NRC (1987a,b, 1989);
Wheeler *et al.* (1989)
Precursor events
Minarick and Kukielka (1982); Cottrell *et al.* (1984);
Minarick *et al.* (1985)

A20.1 Radioactivity

Accounts of radioactivity and of ionizing radiations are
given by Glasstone (1968), Glasstone and Dolan (1979),
Knief (1981), Foo-Sun Lau (1987) and in many texts on
physical chemistry such as Moelwyn-Hughes (1961) and
W.J. Moore (1963).
    Units of radioactivity are defined in Worley and Lewins
(1988) (the *First Watt Committee Report*), based on Foo-
Sung Lau (1987).

A20.1.1 Types of radiation
There are three main types of radioactive particle or ray:
(1) α-particles, (2) β-rays or particles and (3) γ- and X-rays.
    α-particles are helium nuclei, or protons. Their mass is
relatively large and their range correspondingly short. An
air gap of about 8 cm or a sheet of paper may suffice to stop
them. Generally, they present little external radiation
hazard, but they are dangerous if they get inside the body.
    β-rays, or particles, are electrons. They have a much
smaller mass and greater penetrating power. Their range
in air is about 10 m, but they are usually stopped by a few
millimetres of solid material.
    γ- and X-rays are electromagnetic radiations, or photons,
which differ from each other only in respect of source and
wavelength. They have a range of several hundred metres,
but can be stopped by a few millimetres of lead or 0.5–1 m
of concrete.

Both β-rays and γ- and X-rays constitute an external
radiation hazard.
    Ionizing radiation may be classified as direct or indirect.
The former consists of particles and thus includes α-parti-
cles (protons) and β-particles (negatively charged electrons)
and positrons (positively charged electrons), while the latter
includes γ- and X-rays (both photons) and also neutrons.

A20.1.2 Units of radiation
The units of activity of a radioactive source are bequerels
(Bq) and curies (Ci)

1 Bq = 1 disintegration/s

$1 \text{ Ci} = 3.7 \times 10^{10} \text{ Bq}$

Another unit of source strength is the roentgen (R),
which defines the ionization effect of γ- and X-rays.
    1 R produces in 1 cm^3 air ions carrying a charge of 1
electrostatic unit (esu)
    For γ- and X-ray radiation 1 R is equivalent to the absorp-
tion of 87 erg/g of air. However, for other types of radiation
and targets the equivalences are significantly different.
    Units of absorbed dose are rads and grays (Gy), the latter
being the modern unit.

1 rad = 100 erg/g = 0.01 J/kg

1 Gy = 1 J/kg = 100 rad

The effect on man of a given absorbed dose D is a function of the type of radiation, for which a quality factor Q is
defined. The value of Q is unity for γ- and X-rays and electrons, but has other values for other types of radiation. The
value of Q for α-particles is 20. The quality factor is also
referred to as the relative biological effectiveness (RBE).
    The dose equivalent H is the product of the absorbed
dose and the quality factor:

$$H = D \times Q \qquad [A20.1.1]$$

Units of effective dose, or dose equivalent, are rems and
sieverts (Sv), the latter being the modern unit

1 rem = 1 rad × Q

1 Sv = 1 Gy × Q

1 Sv = 100 rem

Thus the roentgen is the unit of total dose, the rad or
gray that of absorbed dose and the rem or sievert that of
effective dose.
    In the context of discussion of biological effects effective
dose is frequently referred to simply as dose.

A20.1.3 Radioactive decay
Specific Activity and Half Life
The basic equation for radioactive decay is

$$-\frac{dn}{dt} = \lambda n \qquad [A20.1.2]$$

where n is the number of radioactive atoms, t time
and λ the radioactive decay constant (s^{-1}). Then from
Equation A20.1.2

$$\frac{n}{n_0} = \exp(-\lambda t) \qquad [A20.1.3]$$

where n_0 is the number of radioactive atoms at arbitrary time zero. The half-life τ is defined by the relations

$$\frac{n}{n_0} = \frac{1}{2} = \exp(-\lambda\tau) \qquad [A20.1.4]$$

$$\tau = \ln 2/\lambda = 0.693/\lambda \qquad [A20.1.5]$$

But

$$n_0 = N_0/M_a \qquad [A20.1.6]$$

where M_a is the atomic weight and N_0 is Avogadro's constant ($= 6.024 \times 10^{23}$). Then from Equations A20.1.2, A20.1.3 and A20.1.6 at arbitrary time $t = 0$ ($n = n_0$)

$$-\frac{dn}{dt} = \lambda\frac{N_0}{M_a} = \frac{\ln 2}{\tau}\frac{N_0}{M_a} \qquad [A20.1.7]$$

Energy release

The release of energy during radioactive decay is governed by the relation of Einstein

$$E = mc^2 \qquad [A20.1.8]$$

where c is the velocity of light, E the energy release and m the mass.

The energy may be calculated from the decay equation by determining the difference in mass of the two sides. An atomic mass unit (amu) is equivalent to 931 MeV.

For example, for the decay of polonium 210 to lead 206 and helium nucleus, or α-particle, the decay equation is

$$_{84}Po^{210} \rightarrow {}_{82}Pb^{206} + {}_2He^4$$

and the atomic mass balance is

		Atomic mass
LHS	$_{84}Po^{210}$	210.049
RHS	$_{82}Pb^{206}$	206.034
	$_2He^4$	4.004
Total RHS		210.038
Mass loss		0.011

Energy $= 931 \times 0.011 = 10.2$ MeV

A20.1.4 Radioactivity health effects

An account of health effects from radiation is given in *Living with Radiation* by the National Radiation Protection Board (NRPB) (1989). An account in the specific context of Chernobyl is given by the Watt Committee on Energy (1991).

Potential health effects of radiation at relatively low doses are short term: (1) birth defects from radiation received by the foetus; (2) thyroid cancers from radioactive iodine; from 3 years onwards: (3) leukaemia; over a period 5–40 years: (4) cancers; and over a generation: (5) genetic effects.

Recommendations for radiation protection standards are given in *Radiological Protection Bulletin 141 1993* by the NRPB (1993). Details are given in Chapter 25.

A20.2 Nuclear Industry

A20.2.1 Nuclear industry regulation

Public control of the nuclear industry is discussed in Chapter 4. In the United Kingdom the regulatory body is the Nuclear Installations Inspectorate (NII), which is part of the HSE. In the United States of America it changed in the mid-1970s from the Atomic Energy Commission (AEC) to the Nuclear Regulatory Commission (NRC).

A20.2.2 Nuclear industry advisory committees

In the United Kingdom there is a standing committee, the Advisory Committee on the Safety of Nuclear Installations (ACSNI) which advises on nuclear safety. In the United States of America there is an Advisory Committee on Reactor Safety (ACRS).

Advice on radiological protection in the United Kingdom is the responsibility of the National Radiological Protection Board (NRPB).

A20.2.3 International bodies

Bodies operating at the international level include the International Atomic Energy Agency (IAEA) in Vienna and the International Committee on Radiological Protection (ICRP). The IAEA was set up in 1957 as an intergovernmental organization linked to the UN.

A20.2.4 Nuclear industry research

Work for the nuclear industry is a main source of support for a number of major research laboratories and organizations whose output is also of interest to the process industries. In the United Kingdom these include the UK Atomic Energy Authority (UKAEA), which operates laboratories at Harwell. The Systems Reliability Directorate (SRD) is part of the UKAEA.

In the United States of America major laboratories with a nuclear interest include the Argonne National Laboratory (ANL), Brookhaven National Laboratory (BNL), Electric Power Research Institute (EPRI), Lawrence Livermore National Laboratory (LLNL), Oak Ridge National Laboratory (ORNL) and Sandia Laboratories (SL).

A20.2.5 Nuclear industry accidents

Early perceptions of the hazard from the nuclear industry were coloured by the image of an explosion from a nuclear weapon. A major accident in the nuclear industry, a core meltdown, is quite different. There would not be the initial intense blast and radiation effects, but rather a dispersion of radioactive material at the site and from the site by the wind. Such an accident occurred at Chernobyl.

A20.2.6 Nuclear industry debate

Virtually from its inception the nuclear industry has been the subject of debate. Some principal critiques are *Nuclear Power* by Patterson (1976), *Soft Energy Paths* by Lovins (1977), *The Risks of Nuclear Power Reactors* by the Union of Concerned Scientists (UCS) (1977), *Nuclear or Not?* by Foley and van Buren (1978), *Cover-up* by Grossman (1980), *The Fast Breeder Reactor* by Sweet (1980), *The PWR Decision* by Cannell and Chudleigh (1984), *Going Critical* by Patterson (1985) and *Nuclear Politics* by T. Hall (1986).

A20.3 Nuclear Reactors

Accounts of nuclear reactor systems include *Nuclear Reactor Control and Instrumentation* by J.H. Bowen and Masters

(1959), *Nuclear Reactor Engineering* by Glasstone and Sesonke (1962), *Nuclear Energy Technology* by Knief (1981), *Nuclear Power Technology* by W. Marshall (1984), *The Elements of Nuclear Power* by D.J. Bennett and Thompson (1989) and *The Technology of Nuclear Reactor Safety* by T. Thompson and Beckerley (1989).

Treatments by the HSE include PWR (HSE, 1979d), *Safety Assessment Principles for Nuclear Reactors* (HSE, 1979f) and *Sizewell B* (HSE, 1982b).

A20.3.1 Nuclear reactor types
Some principal types of nuclear reactor are

(1) Gas-cooled reactors:
   (a) Magnox reactor;
   (b) Advanced gas-cooled reactor (AGR);
   (c) High temperature gas-cooled reactor (HTGR);
(2) Light water reactors (LWRs):
   (a) Pressurized water reactor (PWR);
   (b) Boiling water reactor (BWR);
(3) Heavy water reactors:
   (a) CANDU reactor;
   (b) Steam generating heavy water reactor (SGHWR);
(4) Fast breeder reactors.

Accounts of reactor types are given in *Nuclear Power Reactors* by the UKAEA (1987) and by Patterson (1976) and Lewins (1988).

In the United Kingdom the first reactors were the gas-cooled Magnox reactors. The next generation were the AGRs reactors. The first pressurized water reactor came much later, at Sizewell B.

Elsewhere, notably the United States of America and France, the vast majority of reactors installed have been PWRs.

A20.3.2 Reactor design features
In a nuclear reactor energy is released by a chain reaction involving neutrons. The reaction is sustained by those neutrons which are absorbed. Slower neutrons are absorbed in larger proportion than faster ones. Use is made, therefore, of a 'moderator' to slow down the neutrons. In some designs the moderator is graphite, for example, the Magnox reactor, while in others it is water, for example, the PWR.

The power output of the reactor is controlled by movement of control rods, containing materials which are strong absorbers of neutrons, and the reactor is shut down by their full insertion. Emergency shut down, or scram, is effected by dropping in the rods.

Even when shut down the reactor has a certain output of heat, due to radioactivity in the core. This decay heat needs to be removed if the core is not to overheat. The removal of the decay heat is one of the main problems which have to be handled in reactor design.

The reactor core is cooled by a coolant which in the AGR is carbon dioxide and in the PWR is water.

A20.3.3 Advanced gas-cooled reactors
The underlying concept of the AGR is that if forced convection circulation of the gas coolant is lost, natural convection will still effect a degree of cooling. To this extent, the reactor is an application of inherently safer design.

A20.3.4 Pressurized water reactors
The majority of nuclear reactors installed world-wide are PWRs. In the PWR design the core is cooled by water which also doubles as the moderator.

A PWR is provided with multiple systems to inject cooling water into the reactor to keep the core cool and with numerous protective systems to ensure the operation of this backup cooling. The reactor is vulnerable in the event of the failure of these instrument and cooling systems. On a PWR an emergency can escalate within a relatively short time scale.

In a PWR the fuel is inside three sets of containment. The first is the fuel rods. The second is the reactor pressure vessel (RPV) in which the fuel rods are held. The third is the containment building which encloses the reactor vessel.

The reactor is vulnerable to loss of coolant and much attention has been focused on the large loss of coolant (LOCA) accident.

A20.3.5 Inherently safer design
In view of the nature of the hazard from a nuclear reactor, there is interest in inherently safer design. Mention has already been made of the inherent safety aspects of the AGR.

A review of approaches to inherently safer design of nuclear reactors is given in *Proposed and Existing Passive and Safety-related Structures, Systems and Components (Building Blocks) for Advanced Light Water Reactors* by Forsberg *et al.* (1989).

Some examples are discussed by Kletz (1988h). The HTGR is a small reactor cooled by high pressure helium. The reactor is designed to withstand a high temperature and to have an adequate heat loss by radiation and natural convection if the coolant should fail. The Swedish process-inherent ultimate safety (PIUS) reactor is a water-cooled reactor immersed in boric acid solution, so that if the coolant pumps fail this solution is drawn through the core by natural convection.

Another aspect of inherent safety is the reliability of the overall system for the control and protection of the reactor, which includes both the automatic control and protection and the process operator. As already mentioned, a PWR is vulnerable to equipment failure in the instrument and coolant systems which protect it. It is therefore vulnerable also to operator actions which render this protection ineffective.

This feature is discussed by Franklin (1986) in these terms:

> When operators are subject to conditions of extreme emergency arising from a combination of unforeseen circumstances they will react in ways which lead to a high risk of promoting accidents rather than diminishing them. This is materially increased if operators are aware of the very small time margins that are available to them.

A20.4 Nuclear System Reliability

A20.4.1 Pressure systems
The integrity of the pressure system is critical to the safety of a nuclear reactor such as the PWR. Much attention has been therefore focused on the potential for pressure vessel failure.

Pressure vessel failure is considered in the *Rasmussen Report*. It is also discussed by Bridenbaugh, Hubbard and Minor (1976) and Cottrell (1977).

It was a principal issue at the Sizewell B inquiry, and an account is given in Appendix 26.

The treatment given in Chapter 12 of failure rates of pressure vessels draws heavily on estimates made with nuclear applications in mind.

A20.4.2 Protective systems

From the start the nuclear industry made extensive use of instrumentation, and particularly of instrumented protective systems. Accounts of this work are given in *Reliability Assessment of Protective Systems for Nuclear Installations* by Eames (1965 AHSB(S) R99) and *Safety Assessment with Reference to Automatic Protective Systems for Nuclear Reactors* by A.E. Green and Bourne (1966 AHSB(S) R136).

This work led to interest in reliability technologies, including fault trees and dependent failure. Since it was necessary to have a target to aim for in design of such protective systems, this also prompted the development of risk criteria.

A20.4.3 Reliability engineering

The nuclear industry joined the defence and aerospace industries in developing reliability techniques. A treatment of reliability engineering from the viewpoint of the nuclear industry and with special reference to protective systems, is *Reliability Technology* by A.E. Green and Bourne (1972).

A20.4.4 Failure and event databases

The practice of reliability engineering created the need for failure and event databases. This is illustrated by the development of the database of the UKAEA Systems Reliability Service (SRS) (Ablitt, 1973 SRS/GR/14). The development of this database is described in Appendix 14.

A20.4.5 Fault trees

The need to identify failure paths and to estimate failure frequencies for systems with multiple layers of protection has led in the nuclear industry to extensive use of fault tree analysis. It is the basic method used for frequency estimation in the *Rasmussen Report*. The technique is described in *Reliability and Fault Tree Analysis* by Barlow, Fussel and Singpurwalla (1975) and in *Fault Tree Handbook* by Vesely *et al.* (1981).

Associated with this work are the fault tree analysis computer codes PREP and KITT (Vesely and Narum, 1970).

A20.4.6 Event trees

Likewise, the nuclear industry has made much use of event trees, both to identify and quantify the effects of certain failures, such as those of the power supply, and the outcomes of a release to atmosphere.

A20.4.7 Dependent failures

Nuclear reactors rely on extensive protective systems to give a very high degree of reliability, and it is characteristic of such systems that they are liable to be defeated by dependent failure. In some cases different assumptions about dependent failure can result in a dramatic increase in the estimated frequency of an accident. The nuclear industry has therefore put much effort into the study of this problem.

The problem of dependent failure bulks large in the analyses in the Rasmussen Report and in the critiques of that report.

An account is given in *Defences against Common-mode Failures in Redundant Systems* by Bourne *et al.* (1981 SRD R196).

A20.5 Nuclear Hazard Assessment

A20.5.1 Probabilistic risk assessment

The nuclear industry has for long made use of probabilistic risk assessment (PRA), or probabilistic safety assessment (PSA), as a means of ensuring that its hazards are identified and under control and as a basis for dialogue on this with the regulatory authority. Guidance is given in *PRA Procedures Guide* by the NRC (1982). A more recent treatment is *Procedures for Conducting Probabilistic Safety Assessment of Nuclear Power Plant* by the IAEA (1992).

The *Rasmussen Report*, described in Appendix 23, is one well known generic PRA. Another is the *Deutsche Risikostudie Kernkraftwerke* by the Technische Uberwachungsverein (TUV) (1979).

A20.5.2 Accident sequences and scenarios

In a nuclear PRA considerable effort is devoted to the treatment of the accident sequences, or scenarios.

Thus a large proportion of the IAEA document just referred to is devoted to procedures for the identification of accident initiators and of accident sequences, for the classification of accident sequences into plant damage states and for the determination of the frequency of the accident sequences.

A20.5.3 Rare events

There are a number of natural and man-made threats which may hazard a nuclear plant. Examples of these two types of event are, respectively, an earthquake and an aircraft crash. Such events, their consequences and frequency, have therefore been the subject of much study by the nuclear industry.

A20.5.4 Explosions

One type of the event which may put a nuclear plant at risk is an explosion. The industry has examined various kinds of explosion, including vapour cloud explosions, and certain types of particular interest to it such as steam explosions and hydrogen explosions.

A20.5.5 Hydrogen explosions

The fuel elements in the reactor core are held in metal cladding tubes. At high temperatures the zirconium metal used for this cladding can react with steam to produce hydrogen, with the risk of a hydrogen explosion.

A20.5.6 Missiles

Studies of some relevance to process plants are those on missiles such as the work reported by Baum (1984, 1991).

A20.5.7 Earthquakes

Of rare natural events, the earthquake hazard bulks large both in plant design and in risk assessment. The nuclear regulator has put much effort into defining the type of earthquake which the plant should be required to withstand. Guidance is given in *Design Response Spectra for Seismic Design of Nuclear Power Plants* by the NRC (1973).

Then, given such a design earthquake, it is possible to specify equipment to withstand it, as described below.

A20.5.8 Expert judgement
Where hard information is lacking, resort may be had to the use of expert judgement. Thus expert judgement may be used to obtain estimates of failure and event rates. It may also be applied to other aspects such as the formulation of accident sequences. Guidance on this approach is given in *Eliciting and Analysing Expert Judgement: a Practical Guide* by M.A. Meyer and Booker (1990).

A20.5.9 Computer error
A modern nuclear power station operates under computer control. The control computer, both hardware and software, is therefore a safety critical system.

The quality assurance of the computer software is a particularly intractable problem. An account of this aspect of the Sizewell B computer system is given by Hunns and Wainwright (1991).

A20.5.10 Human error
The conduct of PRA for nuclear plants has led naturally to a requirement for methods of handling the human aspects, particularly in fault trees and in event trees. An overall methodology for this problem is described in *Handbook of Human Reliability Analysis with Emphasis on Nuclear Power Plant Applications* by Swain and Guttman (1983). A method of estimating error probabilities is addressed in *SLIM-MAUD: an Approach to Assessing Human Error Probabilities using Structured Expert Judgement* by Embrey et al. (1984).

Also relevant is the ACSNI Study Group on Human Factors *Second Report: Human Reliability Assessment – a Critical Overview* (ACSNI, 1991).

A20.5.11 Source terms
Each accident scenario needs to be associated with a defined source term, so that the dispersion of radioactive materials may be modelled.

A20.5.12 Emission models
The nuclear industry was one of the first to develop a sustained interest in two-phase flow, both flow within the plant and flow of leaks from it. An example of work from a nuclear background is *Two-Phase Flow and Heat Transfer* by Butterworth and Hewitt (1977).

A20.5.13 Gas dispersion models
Much of the work on gas dispersion modelling has been in support of work on the consequences of a nuclear accident. This is illustrated by one of the early texts, *Meteorology and Atomic Energy 1968* by Slade (1968).

A20.5.14 Severe accident risk
A focus to work on PRA has latterly been provided by the Severe Accident Risk study, for which the basic document is NUREG-1150 *Reactor Risk Reference Document* of the NRC (1987).

A20.5.15 Risk criteria
The quantification of risks from nuclear reactors has necessarily led to the need for risk criteria by which to judge these risks. Some of the first risk comparisons were those given in the *Rasmussen Report*.

Since then, there has been a wide-ranging debate of risk criteria for man-made hazards of all kinds. Nuclear risk criteria emerging from this are given in *The Tolerability of Risk from Nuclear Power Stations* by the HSE (1988b), and associated comments (HSE, 1988a).

A20.6 Nuclear Pressure Systems

A20.6.1 Quality assurance
All aspects of the design and operation of nuclear power plant need to be carried out to high standards. There has to be, therefore, a system of quality assurance to ensure that these standards are met.

This aspect is described in NUREG-1055 *Improving Quality and the Assurance of Quality in the Design and Construction of Commercial Nuclear Power Plants* (NRC, 1984).

A20.6.2 Inspection
High-quality inspection is an essential part of the nuclear industry's armoury. It has a special interest in non-invasive techniques of monitoring and inspection. These include acoustic emission monitoring, ultrasonics and leak detection methods.

Methods of leak detection are described in NUREG/CR-4813 *Assessment of Leak Detection Systems for LWRs* (NRC, 1987).

The industry has also developed risk-based inspection. An account is given in NUREG/CR-5371 *Development and Use of Risk-based Inspection Guides* (NRC, 1989).

A20.6.3 Fracture mechanics
One of the most powerful inspection techniques for pressure systems is fracture mechanics and the nuclear industry has been in the van in developing such methods. They include the CEGB R6 system described in R/H/R6 *Assessment of the Integrity of Structures Containing Defects* by Milne et al. (1986).

A20.6.4 Seismic qualification
One of the principal external hazards to which a nuclear plant is vulnerable is an earthquake. There is therefore a requirement that equipment crucial to safe operation be able to withstand an earthquake of specified intensity. This is implemented by a system of seismic qualification of equipment. An account is given in NUREG-1030 *Seismic Qualification of Equipment in Operating Nuclear Power Plants* by the NRC (1987).

A20.6.5 Ageing
Another aspect which the nuclear industry has explored in some detail is ageing of equipment, and residual, or remanent, life assessment.

An account of basic principles is given in NUREG/CR-5314 Life Assessment Procedure for Major LWR Components (NRC, 1990). There are studies on specific equipment such as electric motors, diesel generators, instrument air systems, motor operated valves, etc.

A20.7 Nuclear Reactor Operation

The *Kemeny Report* and, even more, the *Task Force Report* on TMI contain numerous recommendations on nuclear reactor operation, particularly on the design of display and alarm systems and of control rooms generally, operating and emergency procedures, and operator training.

A20.7.1 Human factors
A considerable effort is made to apply human factors on nuclear power plants, a practice reinforced by the *Kemeny Report's* emphasis on people-related problems.

Some typical titles which illustrate the application of human factors on nuclear power plants are as follows: NUREG/CR-3331 *A Methodology for Allocating Nuclear Power Plant Control Functions to Human or Automatic Control* (NRC, 1983); NUREG/CR-3371 *Task Analysis of Nuclear Power Plant Control Room Crews* (NRC, 1984); and NUREG-1122 *Knowledges and Abilities Catalog for Nuclear Power Plant Operators* (NRC, 1987).

Three reports on human factors in nuclear power plant operation have been issued by the ACSNI Study Group on Human Factors. Reference has already been made to the second report, on human reliability assessment (ACSNI, 1991). The other two are First Report on Training and Related Matters (ACSNI, 1990) and Third Report: Organising for Safety (ACSNI, 1993).

A20.7.2 Process operator
The problems, and modelling, of the process operator have been the subject of extensive studies, notably the work described in *Information Processing and Man–Machine Interaction* by Rasmussen (1986).

A20.7.3 Display and alarm systems
Early developments in computer aids for the process operator were the alarm analysis systems developed at Wylfa and Oldbury nuclear power stations, described by Welbourne (1965) and Patterson (1968), respectively. The purpose of these systems is to assist the operator in interpreting the large number of alarms which tend to come up in a plant emergency.

One of the perceived problems at TMI was the difficulty experienced by operators in assessing the safety status of the plant. A recommendation of the Task Force Report was that every reactor should have a plant safety status display.

TMI also illustrated the alarm interpretation problem just mentioned and the report recommended that the feasibility be explored of developing a suitable aid, which it termed a disturbance analysis system (DAS).

A20.7.4 Control room design
A large proportion of the studies conducted on control rooms have been in the nuclear industry, and these have contributed to the development of understanding of human factors and human error.

In the aftermath of TMI control room design in general was also an issue. This illustrated in NUREG-0700 *Guidelines for Control Room Design Reviews* (NRC, 1981).

A set of studies of the Sizewell B control room are described by Ainsworth and Pendlebury (1994), Umbers (1994) and Whitfield (1994).

A20.7.5 Operating procedures
In the operation of nuclear power plants particular emphasis is placed on written operating procedures and on the training of operators in these. An account is given in NUREG/CR-3968 *Study of Operating Procedures in Nuclear Power Plants* (NRC, 1987).

A20.7.6 Emergency procedures
Operating procedures for dealing with an emergency assume particular importance on nuclear power plants. The *Kemeny Report* and the *Task Force Report* make recommendations on the emergency procedures.

Such procedures are the subject of NUREG/CR-0899 *Guidelines for the Preparation of Emergency Operating Procedures* (NRC, 1982) and NUREG/CR-3177 *Methods for Review and Evaluation of Emergency Procedure Guidelines* (NRC, 1983).

A20.7.7 Operator training
Process operators on nuclear power plants undergo extensive training, for which the industry makes appreciable use of simulators.

The need for such training is a principal theme of both the *Kemeny Report* and of the *Task Force Report* on TMI. It is also the subject of the first report of the Study Group on Human Factors of the ACSNI (1990). Simulator training is treated in NUREG/CR-3725 *Nuclear Power Plant Simulators for Operator Licensing and Training* (NRC, 1984) and numerous other reports.

A20.8 Nuclear Emergency Planning

An emergency plan, covering both on-site and off-site aspects, is an integral part of the arrangements at a nuclear power plant.

A20.8.1 Accident management plans
There has been growing emphasis on the need to plan for and to manage an emergency.

Relevant work is that described in NUREG/CR *Management of Severe Accidents* by the (NRC, 1985) and NUREG/CR-5543 *A Systematic Process for Developing and Assessing Accident Management Plans* by the NRC (1991).

A further account is given by Ang (1992).

A20.8.2 Off-site emergency plans
Off-site emergency planning has been a feature of the nuclear industry from its earliest days. An account is given in *Emergency Plans for Civil Nuclear Installations* by the HSE (1986b).

A20.9 Nuclear Incident Reporting

In most regulatory regimes the nuclear industry is required to report to the regulator the occurrence of specified types of incidents.

A20.9.1 Licensee Event Reports
In the United States of America the system is that described in the NUREG/CR-2000 *Licensee Event Report* (LER) compilation.

A20.9.2 Precursors to severe accidents
The LERs record some events which are of such a nature that they could well have escalated into more severe events, even though in the particular case this has not occurred. These events are precursors of more serious incidents and serve as warnings. Periodic analyses are made of such precursors such as NUREG/CR-4674 *Precursors to Potential Severe Core Damage Accidents 1990: Status Report* by the NRC (1991).

A20.10 Nuclear Incidents

There have been a number of incidents on nuclear plants which are instructive for the process industries also. In addition to TMI and Chernobyl, they include Windscale and Browns Ferry.

Accounts of incidents include those in *Nuclear Power* by Patterson (1976), *The Nuggett File* by the UCS (1979) and *Nuclear Lessons* by Curtis *et al.* (1980); later incidents are covered by May (1989).

A20.10.1 Radioactive sources

Although the incidents of prime interest here are those involving nuclear plant, other nuclear sources should not be neglected. The following case, believed to be the worst of it kind to date, illustrates the hazard from equipment containing a radioactive source.

In 1985 in Goiania, Brazil, a radiotherapy institute moved premises, leaving behind a unit containing a caesium-137 source in a stainless steel container and without notifying the licensing authorities. Subsequently the building was partially demolished. Some 2 years after the institute's departure, on 13 September 1987, two men entered and tried to dismantle the machine for scrap, one taking home the steel cylinder; both fell ill. However, some 5 days later one broke open the steel cylinder and the caesium spilled out. A train of events ensued which resulted in 249 people being exposed to radiation, of whom four died.

A20.10.2 Military reactors and weapons plants
Hanford, the Green Run, 1949

On 2 December 1949 at Hanford, Washington State, there occurred one of a series of releases from an installation of eight nuclear reactors producing plutonium for the Manhattan Project. This was the Green Run, an experiment conducted to investigate monitoring methods which would be useful in intelligence work on the nuclear capabilities of other countries. The release formed a plume 200 miles × 40 miles in dead calm conditions and involved 20,000 Ci of xenon and 7780 Ci of iodine-131.

Rocky Flats, Colorado, 1957

On 11 September 1957 at Rocky Flats, Colorado, a fire occurred in a glove box at an atomic weapons plant. Plutonium shavings ignited spontaneously, workers tried to put the fire out and explosions occurred which blew out the ventilation filters. For half a day there was a release of plutonium contaminated smoke. On 11 May 1969 there occurred another glove box fire. All new weapons production was shut down for 6 months.

Idaho Falls, Idaho, 1961

On 3 January 1961 at Idaho Falls, Idaho, an incident occurred on a prototype military nuclear power plant, the Stationary Low Power Reactor SL-1, one of 17 reactors at the AECs National Reactor Test Station (NRTS). The reactor was shut down and control rods had been disconnected to install additional instrumentation. The emergency started with the sounding of alarms. As it developed, three men working on the control rods were found to be missing. One was discovered dead, pinned to the ceiling by a control rod, another was also dead and the third died soon after. The precise sequence of events is uncertain. One explanation given is that the central control rod was partially withdrawn and there was a power surge, generating steam which lifted the lid of the pressure vessel, causing it to rise 9 ft and then drop back.

A20.10.3 Windscale

Accounts of the Windscale incident are given in *Windscale Fallout* by Breach (1978) and *Windscale 1957* by L. Arnold (1992) and by May (1989).

On 10 October 1957 at Windscale, United Kingdom, an incident occurred which led to a serious radioactive release. The installation consisted of two simple air-cooled, graphite-moderated, atomic piles used for the production of plutonium. The graphite was subject to deformation and build-up of energy due to neutron bombardment, the so-called Wigner effect. A procedure had been established to release the Wigner energy by turning off the fans, making the pile critical and letting heat build up. On 7 October 1957, this procedure had been followed, but it appeared that not all the Wigner energy had been released, so the pile was heated up again. The sensors then showed an abnormal temperature rise and the power was reduced. By 9 October conditions seemed normal except that there was a hot spot. However, on 10 October a rise in radioactivity was detected at the filters in the chimney. An attempt was made to inspect the core using a remote scanner, which jammed. Staff removed a charge plug and looked in. All the fuel channels which they could see were on fire and a fuel cartridge had burst. The air movement caused by the fans was now fanning the blaze. Despite fears that it might lead to a hydrogen–oxygen explosion, the decision was taken to use water to put out the fire, which it did. As a result of the incident, workers on the site were exposed to radiation and milk from cattle in the area had to be discarded. The release was mitigated by the presence of the filters, installed at the insistence of Sir John Cockcroft, prominent at the top of the chimney, and known locally as 'Cockcrofts Folly'. The piles were decommissioned and set in concrete.

A20.10.4 Civil nuclear reactors and reprocessing plants
Chalk River, Ontario, 1957

On 12 December 1957 at Chalk River, Ontario, there was a loss of control on a nuclear reactor leading to damage of the core so that it had to be removed and buried. In outline the initial sequence of events was as follows. Outside the control room, a supervisor found that an operator was opening valves which caused the control rods to withdraw. He immediately set about shutting the valves, but some of the rods did not drop back. He telephoned the operator in the control room, intending to ask him to push a button which would drop control rods in, but by a slip of the tongue actually referred to a button which would withdraw rods. In order to comply, the operator moved away from the phone so that the supervisor temporarily lost touch with him. The operator in the control room soon recognized from the rising reactor temperature that there was something wrong and pressed the scram button, but events had been set in train which led to core damage.

Detroit, Michigan, 1966

On 5 October 1966 near Detroit, Michigan, the Enrico Fermi reactor experienced an incident. The reactor was the first commercial fast breeder reactor in the United States of America. It was being started up for a series of tests when abnormalities were observed and some of the control rods were found not to be in their expected positions. Tests on the sodium coolant showed it to be contaminated, suggesting that part of the fuel core had melted, but the cause was unknown. The authorities were alerted and went on standby for evacuation. Subsequent investigation found that separation had occurred of parts from a steel cone at the bottom of the vessel, installed so that in a meltdown the fuel would spread out and not reach critical mass. In a lastminute modification, zirconium plates had been put onto the cone. One plate had worked loose, had been forced into the core by the sodium coolant and had blocked coolant flow to two of the fuel assemblies.

An account of this incident is given in *We Almost Lost Detroit* by Fuller (1984).

Browns Ferry, Alabama, 1975

Prior to TMI one of the US incidents which received particular attention was that on 22 March 1975 at Browns Ferry. Accounts are given in the *Rasmussen Report* and in *Browns Ferry: the Regulatory Failure* by Ford, Kendall and Tye (1976). The incident is described in Appendix 23.

Beloyarsk, USSR, 1978

On 30–31 December 1978 at Beloyarsk, USSR, there occurred an incident which until Chernobyl was the most serious in the nuclear industry. The complex was 50 km from Sverdlovsk and housed two RMBK reactors and a BM600 fast breeder reactor. A serious fire broke out in the machine hall, causing steel girders and the concrete roof to collapse, opening up a huge hole above No. 2 Generator and disabling the fire protection system. Emergency procedures called for both RMBK reactors to be shut down. But as it was close to −50°C outside, it was feared the reactor cooling systems would freeze and the cores would overheat. Attempts were made to keep No. 1 Reactor and its turbine running, but despite the fire the turbine froze. There followed an extended emergency which eventually, with the arrival of fire fighters from outside, was brought under control.

Cap la Hague, France, 1981

On 6 January 1981 at Cap la Hague, Normandy, France, in a nuclear fuel reprocessing facility there occurred the most serious of a series of incidents. Fire broke out in a spent fuel dry waste silo. Ignition occurred in cotton waste, soaked in solvent and in contact with uranium and magnesium. The cotton waste had been used in a decontamination operation there several weeks earlier. The fire was attacked first by water, which formed steam, and then by liquid nitrogen. Radioactivity was released inside and outside the plant.

New York State, 1982

On 25 January 1982 in New York State the Ginna reactor experienced tube rupture, due to corrosion, in the steam generator, causing contamination of the secondary, clean steam circuit by the radioactive water from the primary water circuit cooling the core. High pressure in the steam circuit caused a pressure relief valve to open, initially in 5 min bursts and then by sticking open for 50 min. Another PRV on the primary circuit was deliberately opened to reduce the pressure in that circuit but also stuck open. A steam bubble formed, but it proved possible to correct it by pumping in more water.

Frankfurt, FRG, 1987

On 16–17 December 1987 at Frankfurt, FRG, the Biblis A nuclear reactor experienced an incident involving the low pressure injection (LPI) system. It was a requirement that this system remain isolated to prevent (1) leakage of primary coolant out through it and (2) overpressure of the system itself by the primary coolant. During start up the main valve TH22 S006 on the pipe between the LPI system and the primary circuit had been left open; there were also two secondary valves on the line. The status of the main valve was shown by a red light on the control panel, but the operators assumed that the fault was on the light itself. There were two changes of shift before it was realized that valve TH22 S006 was open. The operators took 2 h to decide on a plant shut down and then 10 min later changed their minds and tried an alternative, and hazardous, procedure. They decided to initiate flow through valve TH22 S006, expecting that the valve would then be shut by its trip system. They controlled the flow by cracking open one of the secondary valves and primary coolant started to escape, but valve TH22 S0006 did not shut, and after a few seconds they desisted. The operators then shut the reactor down.

A20.11 Notation

c	velocity of light
D	absorbed dose (Sv)
E	energy
H	dose equivalent (Sv)
m	mass
M_a	atomic weight
n	number of radioactive atoms
n_0	number of radioactive atoms at time zero
N_0	Avogadro's number
Q	quality factor
t	time
λ	radioactive decay constant (s^{-1})
τ	half-life (s)

Appendix

21

Three Mile Island

Contents

At 4.00 on 28 March 1979, a transient occurred on Reactor No. 2 at Three Mile Island, near Harrisburg, Pennsylvania. A turbine tripped and caused a plant upset. The operators tried to restore conditions, but, misinterpreting the instrument signals, misjudged the situation and took actions which resulted in the loss of much of the water in the reactor and the partial uncovering of the core. Radioactivity escaped into the containment building. Site and general emergencies were declared.

The accident at Three Mile Island (TMI) (also referred to initially as Harrisburg) was the most serious accident which had occurred in the US nuclear industry.

The President set up a commission to investigate the accident. Their work is described in *The Report of the Presidents Commission on the Accident at Three Mile Island* (the Kemeny Report) (Kemeny, 1979).

The NRC carried out an investigation, reported in *Investigation into the March 28, 1979 Three Mile Island Accident by Office of Inspection and Enforcement* (the NRC Report) (NRC, 1979b). It also set up a task force, the findings of which are given in *TMI-2 Lessons Learned Task Force Final Report* (the Task Force Report) (NRC, 1979d).

Another report is *Analysis of Three Mile Island – Unit 2 accident by the Nuclear Safety Analysis Center* (NSAC) of EPRI (1979a).

Hearings were also held by the US Congress (US Congress, 1979) at which evidence was taken from managers, engineers, process operators and others.

The Accident at Three Mile Island gives the initial response of the HSE (1979a).

The TMI accident was a watershed in the development of nuclear power in the US and worldwide. After 1979, the ordering of new nuclear power stations in the US virtually dried up. Elsewhere, TMI has at the least heightened awareness of nuclear hazards.

In addition to the official reports mentioned, accounts of the TMI accident have been given in *Three Mile Island* by Stephens (1980) and by J.F. Mason (1979), Rubinstein (1979a, b), Collier and Davies (1980), Dunster (1980), H.W. Lewis (1980), Rogovin (1980), Schneider (1980) and Kletz (1982l, 1988h). A detailed timetable of the events during the accident is given by the NRC (1979b) and by J.F. Mason (1979).

Selected references on TMI are given in Table A21.1.

A21.1 The Company and the Management

The company which operated TMI was Metropolitan Edison (Met Ed).

Table A21.1   *Selected references on Three Mile Island*

NRC (Appendix 28 Three Mile Island, 1979b,d); Douglas (1979); EPRI (1979a); Ferrara (1979); HSE (1979a); Kemeny (1979); J.F. Mason (1979); Rubinstein (1979a,b); Savage (1979); US Congress (1979); Brookes and Siddall (1980); Cobb and Marshall (1980); Collier and Davies (1980); Dunster (1980); Lanouette (1980); Levy (1980); H.W. Lewis (1980); Rogovin (1981); Stephens (1980); Torrey (1980); Jaffe (1981); Moss and Sills (1981); Schneider (1981); Starr (1981); Straus (1981); Wolf (1981); Kletz (1982l, 1983j, 1988h); Sills, Wolf and Shelansky (1982); Slear, Long and Jones (1983); Anon. (1984kk); Ural and Zalosh (1985); Ballard (1988); Philley (1992)

Some of the TMI staff who involved in dealing with the incidents were

Station manager	Mr G. Miller
Station superintendent	Mr J. Logan
Superintendent of technical support	Mr G. Kunder
Shift supervisor	Mr W. Zewe
Shift foreman	Mr F. Scheimann
Control room operators	Mr E. Frederick; Mr C. Foust
Others involved included Met Ed vice-president for generation	Mr J. Herbein
Babcock and Wilcox site representative	Mr L. Rogers

A21.2 The Site and the Works

TMI is situated on the Susquehanna River, some 10 miles south-east of Harrisburg, Pennsylvania.

A21.3 The Process and the Plant

The plant consisted of two PWRs of approximately 900 MWe each. Unit 1 had been in commercial service since 1974 and Unit 2 since December 1978.

The contractors for the reactor plant were Babcock and Wilcox.

The nuclear reactor system is shown in Figure A21.1. The reactor itself is a pressure vessel containing the reactor core. The fuel in the core is enriched uranium dioxide pellets encased in zirconium alloy tubing. There are some 37,000 of these fuel pins arranged in clusters in the core. The core is cooled by the primary cooling system containing water at a high pressure ($\ll 15.2$ MPa) to prevent its boiling. The water acts not only as a coolant but also as a moderator.

Heat is removed from the reactor in this primary cooling circuit, the reactor cooling system (RCS), and used in the steam generators to generate steam to drive a pair of turbogenerators. The primary cooling system consists of two independent loops containing water at high pressure.

Since the volume of the cooling water changes with temperature, it is necessary to have some free space or 'bubble' in the primary circuit to control the system pressure. This is provided by the pressurizer, which is equipped with electrical heating to raise steam and thus increase the pressure, and cold water sprays to condense steam and thus lower the pressure. Steam can be blown off through a pilot-operated relief valve (PORV). The tail pipe from the PORV is run to the reactor coolant drain tank.

An emergency core cooling system (ECCS) is provided to give backup cooling. This consists of three separate systems. One is the high pressure injection (HPI) system which pumps borated water into the system at a pressure greater than the normal coolant pressure. Another is the core flooding system which automatically injects cold borated water into the system when the coolant pressure falls below about 4.1 MPa. The third is the low pressure injection system which operates at 2.8 MPa. This takes water either from a large tank of borated water or from the dump of the containment building and is thus capable of keeping up recirculation indefinitely.

The reactor, the primary cooling circuit and the steam generators are housed in a containment building which is

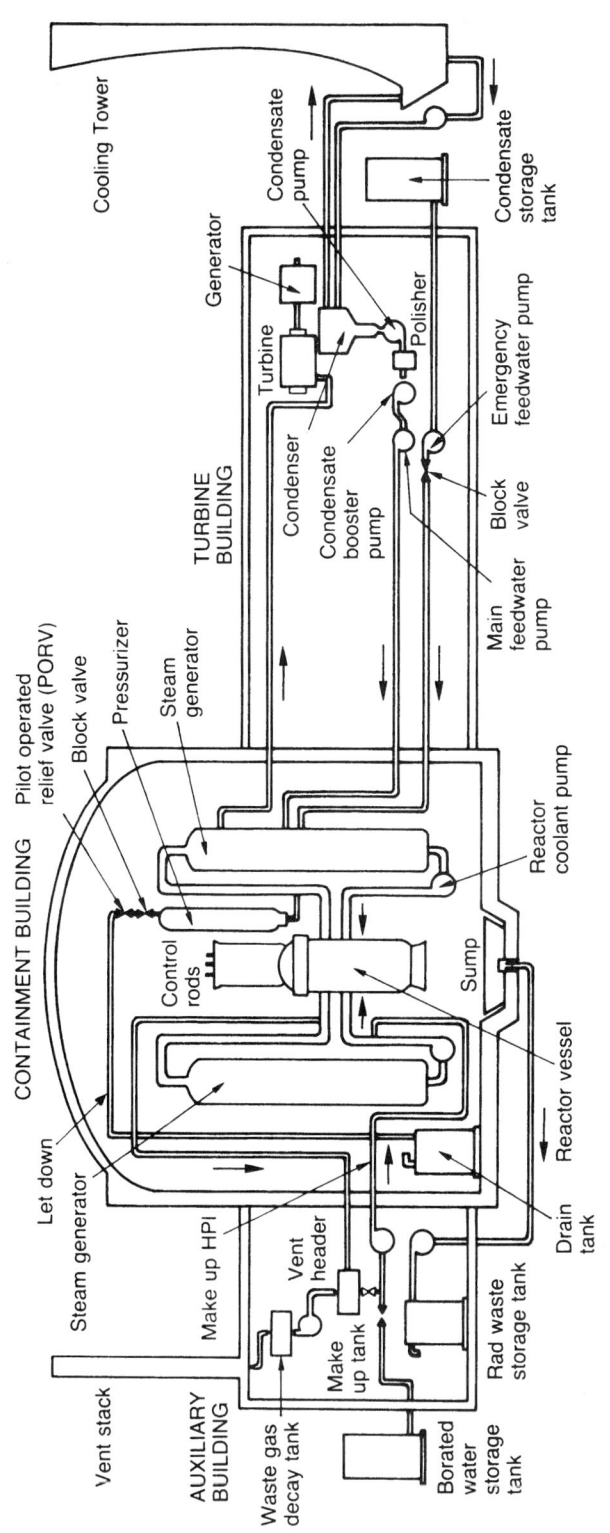

Figure A21.1 *Simplified flow diagram of TMI 2 nuclear power plant (Kemeny, 1979)*

leak-tight. The containment is designed to retain the contents of the primary circuit in the event of a major leak.

The containment building has a spray system containing dilute sodium hydroxide. The spray is triggered by high containment pressure and is intended in the event of a loss of coolant accident to cool down the contents of the containment and to scrub out iodine.

The turbogenerator set is housed in a separate building which also contains the condensate and water treatment plants. Feedwater for the secondary circuit is from the condenser. The condensate water is purified by ion exchange in a condensate polishing system, a package unit.

The water converted to steam on the secondary side in the steam generators is supplied to them by the main feedwater pumps. These are backed up by auxiliary, or emergency, feedwater pumps. Both sets of feedwater pumps are housed in the turbine building.

The control room is a typical one and is shown in Figure A21.2.

A21.4 Events Prior to the Excursion

Some 2 days prior to the incident, routine testing had been done on the valves on the auxiliary feedwater pumps. Two of the valves, the 'twelve valves', had inadvertently been left shut.

For some 11 hours before the incident, the operators had been trying to clear a blockage in the condensate polishing system. In order to do this use was made of compressed air. The service air line used for this purpose was cross-connected to the instrument air line. The pressure of the air in the instrument line was less than that of the water in the units and some water got into the instrument air line, despite the presence of a non-return valve on that line. This water found its way to some of the plant instruments.

Another feature was a persistent slight leak from either the regular safety valves or the PORV.

A21.5 The Excursion – 1

An account is now given of the events which constituted the excursion. A timetable of the events is given in Table A21.2 Sections A and B showing the events and phases, respectively. The trend records of the RCS parameters are shown in Figure A21.3.

At the time of the incident on 28 March, the reactor was operating under automatic control at 97% of its rated output of 961 MWe. The incident began when water which had entered the instrument air line caused the isolation valves on the condensate polishing system to drift shut and the condensate booster pump to lose suction pressure and trip out. The main feedwater pumps in the secondary circuit then tripped and almost immediately the main turbine tripped. The time was 4.00.37.

Valves allowing steam to be dumped to the condenser opened, and the auxiliary feedwater pumps started up. At the steam generators, the removal of heat from the RCS fell, and the pressure of the RCS rose. Within 3–6 s the PORV set pressure of 15.5 MPa was reached and the valve opened, but this was insufficient to relieve the pressure, and at 8 s in, pressure rose to 16.6 MPa, the set point for reactor trip. The control rods were driven into the core to stop the reaction.

The RCS pressure now fell below the PORV set pressure, but the valve failed to shut. However, the valve position indicator in the control room, which actually indicated not the position of the valve but the signal sent to it, showed the valve as closed. The RCS continued to lose water through the open PORV.

In the secondary circuit condensate was no longer being returned to the steam generators, because the valves on the condensate polishing system had closed. The auxiliary feedwater pumps were running, but the valves between the pumps and the steam generators were also, inadvertently, closed. At 1 min 45 s into the incident the steam generators boiled dry.

Figure A21.2   Control room TMI 2 (Kemeny, 1979)

Table A21.2 *Timetable of events at Three Mile Island*

A Timetable (first hours) (after Dunster, 1980)

Absolute time[a]	Time into incident	
4.01	zero	Feed water pump fails
		Pressure rise occurs in primary circuit
		Relief valves open
	8 s	Reactor shut-down
	13 s	PORV fails to close but indicates closure in the control room
		Extra cooling water (normal) injected by the operator
4.02	1 min 45 s	Steam generators dry out
4.03	2 min	Pressure in primary circuit falls and emergency cooling water injection started
4.04	3 min 30 s	Indication of high level of water in the pressurizer causes operator to shut off one emergency cooling pump
4.06	5 min 30 s	Stem generated in the core
		Operators drain off small volumes of primary water in response to rising level in pressurizer
4.11	11 min	Alarm indication of a high level of water in the containment building sump
		Sump water being pumped to auxiliary building
4.39	39 min	Sump pumps stopped by operator
5.00	60 min	Primary coolant pumps vibrating due to pressure of steam
5.14	1 h 14 min	Two primary pumps stopped by operator
5.45	1 h 45 min	Two remaining primary pumps stopped by oprator
5.50	1 h 50 min	Temperature in outlet ducts high enough to show presence of superheated steam – sigificance not recognized
6.22	2 h 22 min	Block valves on the relief valve lines closed by operators
7.00	3 h	Site emergency declared
7.20	3 h 20 min	Indication of 800 rem/h in containment building
7.24	3 h 24 min	General emergency declared
8.00	4 h	Containment building isolated automatically but operators continue to transfer primary let-down water to tanks in the auxiliary building
8.25	4 h 25 min	Local radio carried item concerning general emergency

B Phases (Collier and Davies, 1980)

Phase		
	1	Turbine trip 0–6 min
	2	Loss of coolant
	3	Continued depressurization 20 min–2 h
	4	Heat-up transient 2–6 h
	5	Extended depressurization 6–11 h
	6	Repressurization and ultimate establishment of stable cooling mode 13–16 h
	7	Removal of hydrogen bubble 1–8 d

[a] Absolute times are approximate; the turbine trip was at 4.00.37.

Meanwhile the RCS pressure was dropping. At 2 min the pressure fell to 11 MPa at which point the first of the ECCS systems, the HPI system, came on and injected high pressure water. The liquid level in the pressurizer, however, was rising.

The display of liquid level was an indicator of the level of fluid in the pressurizer, but no longer of the mass of water; the liquid now contained steam bubbles and had thus been rendered less dense. To this extent the level display had become misleading.

The operators became concerned that the HPI system was increasing the inventory of water in the RCS, that the steam bubble in the pressurizer would be lost and that the circuit would become full of water and thus go 'solid'. At 4 min 38 s they therefore shut down one of the HPI pumps and throttled back the other.

With water passing through the PORV the pressure in the reactor coolant drain tank built up and at 7 min 45 s the reactor building sump pump came on, transferring water to the waste tanks in the auxiliary building.

At 8 min the operators realized that the steam generators were dry. They found that the auxiliary feedwater system valves were closed and opened them, thus restoring feedwater to the steam generators.

At 60 min the operators found that the RCS pumps were vibrating. This was due to the presence of steam, though they did not realize this. At 1 h 14 min they shut off the two pumps in Loop B to prevent damage to the pumps and pipework, with possible loss of coolant, and at 1 h 40 min shut off the two pumps in Loop A.

At 2 h 18 min the PORV block valve was at last shut off, thus stopping the loss of water from the RCS. The RCS pressure then began to rise.

In the two hours since the turbine trip periodic alarms had warned of low level radiation in the containment building. At about 2 h in, there was a marked increase in the

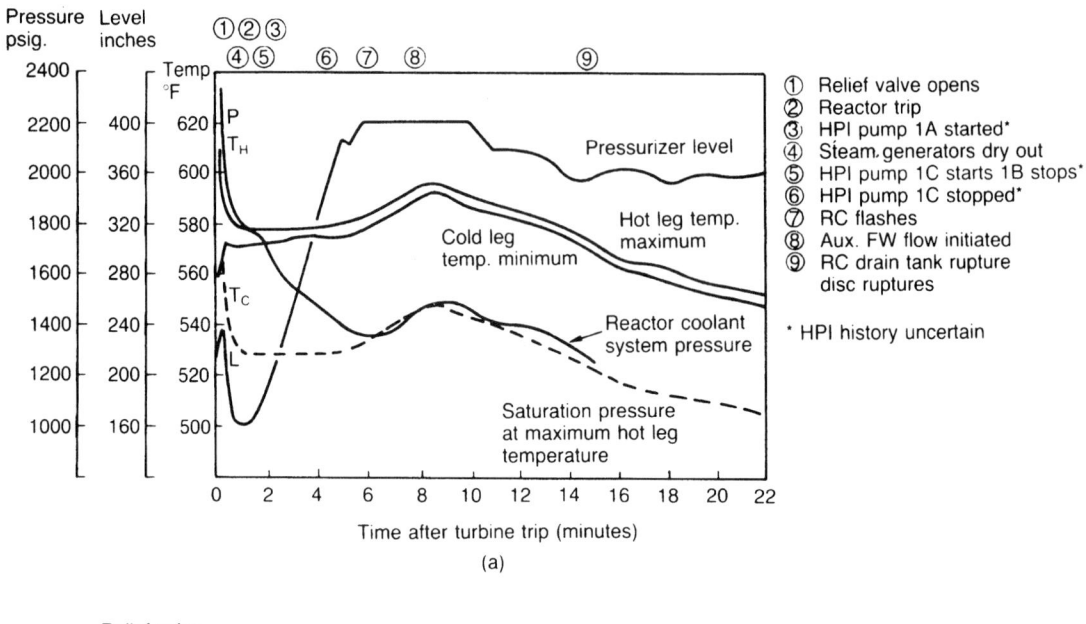

(a)

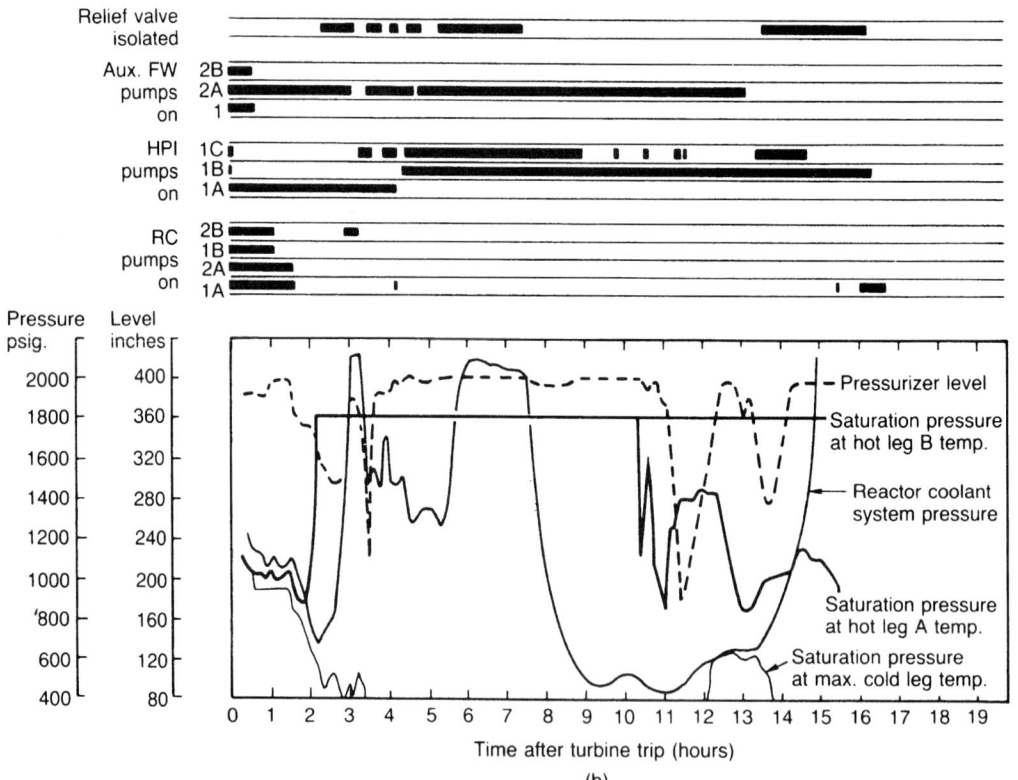

(b)

Figure A21.3 *Trend records of reactor cooling system parameters at TMI 2 (Collier and Davies, 1980): (a) parameters in minutes following turbine trip; and (b) parameters in hours following turbine trip*

radiation readings. By 2 h 48 min in, high radiation levels existed in several areas of the plant. At 2 h 55 min a site emergency was declared.

Attempts were made to restart the RCS pumps. One pump in Loop B operated for some 19 min but was shut down again by vibration trips at 3 h 13 min.

Meanwhile, high radiation levels had been measured in the containment building. At 3 h 30 min a general emergency was declared.

At some point, perhaps about 4 h in, the station manager appears to have realized that the reactor core had suffered damage.

From 4 h 30 min attempts were made to collapse the steam bubbles in the RCS loops, but without success and at 7 h these efforts were abandoned.

The operators then tried to bring in the lower pressure cooling system by reducing pressure in the RCS. They began at 7 h 38 min by opening the PORV block valve. At 8 h 41 min the pressure fell to 4.1 MPa at which point the core flooding system operated.

One consequence of the exposure of the core was that a steam-zirconium reaction occurred, generating hydrogen. During the depressurization, hydrogen from the RCS was vented into the containment building. At 9 h 50 min, there occurred in the containment building what was apparently a hydrogen explosion. There was a thud and a pressure spike of 0.19 MPa occurred. The sprays came on for some 6 min. However, the nature of the event was not immediately understood.

By this time the reactor had been stabilized, although it had suffered core damage.

When the significance of the event in the containment building became appreciated, concern grew about a possible explosion of a hydrogen bubble there.

A21.6 The Emergency and the Aftermath

As already described, a site emergency was declared at 7.00 on 28 March. At 7.24 a general emergency was declared. The emergency which then unfolded over the next week or so is described in the Kemeny Report.

The plant was contacted by the local WKBO radio station. The radio station asked what kind of emergency it was. They were told by the company manager of communications services that it was a general emergency, a 'red-tape' sort of thing required by the NRC when certain conditions exist, but that there was no danger to the public.

The emergency did not end with the establishment of stable operation. The following day, the core damage was found to be more serious than first thought. The hydrogen bubble problem persisted over the next 8 days and caused particular concern. There were also a number of other problems, mainly associated with the large quantities of water contaminated with low level radioactivity. The levels of radioactivity in the atmosphere around the plant were monitored continuously.

The handling of the emergency in respect of evacuation developed into a protracted saga, of which only highlights need be mentioned. In addition to the company and the NRC, the state governor and the President became involved. On 29 March, meetings were held to decide whether to recommend evacuation. On 30 March, a radiation reading outside the plant, later found to be erroneous, led NRC officials to recommend evacuation. The local director of emergency preparedness was notified to stand by for an evacuation; he notified fire departments and had a warning broadcast that an evacuation might be called. Then the NRC chairman assured the Governor that an evacuation was not necessary. Later the Governor decided that pregnant women and pre-school children should be evacuated. Meanwhile, people around the plant had started to make their own arrangements, and many left. By 1 April, the NRC had concluded that the hydrogen bubble posed no threat but failed to announce this properly.

Operations to restore the situation at TMI-2 have lasted over a period of years (Masters, 1984).

A21.7 The Excursion – 2

The operators in the TMI-2 control room made a number of errors. Some of these were failures to make a correct diagnosis of the situation, others were undesirable acts of intervention.

The first was the failure to realize that the PORV had stuck open. The operators had an indication that the PORV had shut again, in the form of a status light. However, this light showed only the shut signal sent to the valve, not the valve position itself. They were also misled by the reading of high water level in the pressurizer.

There were several centra-indications, however, which pointed to the fact that the PORV was open. One was that the PORV outlet pipe was at a higher temperature than usual, 140°C instead of 90°C. Another was that the temperature and pressure in the pressurizer were lower than usual. The third was that there was a high level in the containment building sump. Later, the vibration of the primary coolant pumps was another indication.

Closely connected with this first diagnostic failure was the failure to recognize that the mass of primary coolant water had fallen. Here the operators were misled by the level measurement.

Turning to the interventions, at 2 min into the incident, the HPI pumps tripped in automatically, because the mass of water in the primary circuit was falling. Thirty seconds later, the operators, fearing that the system would fill right up with water and go 'solid', shut the HPI pumps off. This was the wrong action, because the primary circuit then continued to lose water.

Another counterproductive action was the shutting down of the primary coolant pumps. About 60 min into the incident, the pumps began vibrating. This was another clue that the primary circuit fluid was now a two-phase mixture of steam and water. The operators had been trained not to allow violent vibration of the pumps, which could damage them or the piping, and they shut the pumps down. The forced flow of cooling water through the reactor core ceased.

These failures of diagnosis and the fear of the system going solid have frequently been characterized as illustrations of mindset.

The operators were poorly served by the displays and alarms in the control room. Mention has already been made of the fact that some of the indicators did not actually display the variable of direct interest. There were a large number of alarms which were difficult to interpret.

A21.8 The Investigations

Principal investigations into the accident are the NRC Report, the Kemeny Report and the US Congressional hearings. These are described in this section. Recommendations given in the Kemeny Report and the Task Force Report are outlined in Section A21.9.

A21.8.1 NRC report

The NRC Report gives a detailed account of the TMI accident with a full chronology for the first 16 h containing 676 entries. There was little dispute about the basic facts. The investigations are significant mainly for their probing of the deeper causes.

A21.8.2 Congressional hearings

Hearings on the incident which are replete with information were held by the US Congress; they repay study.

Of particular interest here are the sessions dealing with the display and alarm systems. Extracts from the record on these two topics are given, respectively, in Tables A21.3 and A21.4 and are self-explanatory.

Table A21.3   *The display problem at Three Mile Island: evidence given to Congressional committee (after Lees, 1980f, based on US Congress, 1979)*

Weaver: I just want to say that is an extremely important point because one of the things we wanted to find out was how well they were in control. So you are raising the issue, I think, that because these gauges were off at one side they did not see them. Is that a possibility?
Eisenhut: It is certainly a possibility. In fact, that is why I said that is an area – the reason I highlighted it, it is a question area as to what happened to the pilot-operated relief valve, what kind of indications they had that the valve was still open.
Weaver: But they could have seen them?
. . .
Eisenhut: There were at least a couple of other instruments they could have used if those instruments were working. I do not know whether they were.
Weaver: The indications are that they did not see them for several hours; is that correct?
Eisenhut: At least a couple of hours. [p. 27]
Carr: We were told the other day that there was no direct level indicator in the reactor at the level of the coolant, that that had to be inferred from looking at other data, say the pressurizer pressure. Now we are saying that there is no direct indication of this valve at the top of the pressurizer, that we only got an indication, a direct indication on its mechanism rather than on its functioning, and that there were some backup inferences you could draw from other instruments which would lead someone to assemble some data. I suppose jumping way, way ahead of ourselves here I am wondering, is this a special feature of Babcock & Wilcox plants that on their gaging and instrumentation they do not use direct data, and then use the other stuff for backup? Or is this something throughout the nuclear industry, where there is a lot of data points and it leaves it to the operator to sort of assemble all of this, what is happening inside that building, in his head.
Stella: Valve positions normally are indicated by a mechanism which looks at the travel of the stem, which is to some degree on indirect reference to what the valve is actually doing, but a more direct method than you would get from looking at the solenoid which starts the event. [p. 30]
. . .

Stella: But as I tried to understand the accident, I reduced the number of program meters that I really wanted to know about to very few to try to make the judgement of what was going on. I think the operators can do that . . .
   You can reduce the number that you really need to focus on to try to understand the event that you have had to relatively few. For example if you had a major LOCA, he would know very quickly what program meters that he ought to focus his attention on. He would have very quickly a response in the reactor building that says I have a discharge of a lot of high-energy fluid. I either have a break in my secondary system in the reactor building or the primary system. I immediately look at my primary system meters. Then I can tell if that was a primary system rupture. If it is a primary system rupture, I then know what program meters to track. So there is a hierarchy of logic built into the very many alarms that you see that point him at the right direction . . . . But this whole family of alarms and gages that you see, there is a hierarchy of logic built into it that, as you are trained to be an operator, you know what critical program meters to look for very quickly and eliminate or focus your attention in the right direction.
Carr: In the location of the alarms and gages and buzzes and switches and all that stuff, does the location on the panel follow that logic?
Stella: Yes. [p. 31]
. . .
Vento: One other thing before we go on to the other problem in the secondary system, which you did next. This relief valve, of course, is a key part of this, apparently. But there were some tailpipe temperature and some other indicators there that were also available. Who was responsible for watching those at this particular point? The tailpipe temperature?
Frederick: What you are talking about is the outlet temperature of the relief valve.
Vento: There is also a tank that this goes to that is not shown here, and that has an indicator on it and a temperature and so forth. There is a thermocouple on here, and who was responsible for watching those in this particular process? It was just a generally shared thing?
Frederick: The points that you are talking about are not displayed. They are not on the console. They are in the computer. You have to manually call them up.
Vento: You have to call those things up. But were they not called up during this procedure at all?
Frederick: No.
Weaver: How many minutes are we in now? Where are we?
Faust: It is not a normal thing.
Frederick: About an hour.
Faust: We did not know a problem like that existed at the time.
Frederick: If the valve indicated shut, there is no reason to look at the downstream temperature. [p. 135]
. . .
Reis: It is a broad question, but as well as you can answer it, do you think that the control room provided you, in a timely fashion, in an easy-to-read fashion, the information you wanted and needed? And alternatively, in the course of the accident, were the communications in the control room such so that when someone at the panel needed conceptual information, in other words, this instrument, that instrument, and this gage are reading things that do not jibe together, that you had that sort of immediate access to an engineer? When did you get it? Did it work well, so that you were able to figure out what was really going on?
Frederick: The answer to the first question, the instruments, in my opinion, were not adequate. I can come up with a million questions on how to change them and the availability of additional input from the engineering staff.

It was not available to me, because I did not seek it out. I would give whatever question I had to the supervisor, and he would refer them to whoever he could as far as more operators, more engineers, to get more information. And he would bring back whatever suggestions there were if there were any. If things were that confusing, that I had to turn and ask somebody a question, I would wait for their response and their instructions on what to do next and then take the prescribed action according to what they suggested rather than what was on the panel.
Faust: I would way — well, I cannot change it much from what he said — we would continue to feed changes that we were seeing in the plant, to the supervisor; as well as if we thought we knew something that should be done, we would be saying that, too.
Reis: In terms of the layout of the plant, in general, do you think there could be some significant changes that would make your job a lot easier?
Frederick: Yes.
Faust: Definitely. [p. 155]

. . .

Cheney: It seems to me that is a key point. Has this ever been questioned before, whether or not the pressurizer level was an accurate reflection of the reactor coolant level, in a general sense, in terms of the basic design of the reactor?
McMillan: I think a lot of people have asked that question: What does the pressurizer level indicate to you?
Cheney: I would just like to come back to that basic question: Whether or not those operators at Three Mile Island on that day were trained to read that pressurizer level as anything other than indicating reactor level, coolant level in the reactor?
McMillan: Let us trace that through and see if we can find what evidence.
Cheney: Do you think that is some flaw in the design of the reactor?
McMillan: No, sir.
Cheney: Why not have a coolant level indicator from within the reactor itself?
McMillan: That has been suggested by a number of people, post Three Mile Island.
Weaver: Mr McMillan, how do you measure the level of water in the reactor vessel?
McMillan: You do not measure the level of water in the reactor vessel. . . .
Weaver: But is that not an important thing to know? Why do you not have something that tells us the level of the water?
McMillan: Well, I think -
Cheney: I guess I would like to follow up on this. [p. 232]

. . .

Weaver: Are you opposed, Mr McMillan, to water level measurement in the reactor vessel?
McMillan: It is a difficult problem, yes. But I am not at this point saying it is insurmountable . . .
Weaver: I would say the first thing I was going to do, if I were going to design a nuclear plant, is to make sure we knew what the water level was. [p. 237]

[a] In addition to the personnel identified in Section A21.1, the transcript extracts given above refer to the following: Mr Cheney, Mr Vento, Mr Weaver, all Congressmen; Mr Reis, Congressional staff; Mr D.Eisenhut, Deputy Director, Division of Operating Reactors, Office of Nuclear Reactor Regulation, NRC; Mr MacMillan, Vice-President, Nuclear Power Generation Division, Babcock and Wilcox Co.

Table A21.4   *The alarm problem at Three Mile Island: evidence given to Congressional committee (after Lees, 1980e, based on US Congress, 1979)*

Weaver: The plant manager told me he thought 100 audio alarms were going off at one time.
Stella: Is that what he told you?
Weaver: Yes
Stella: I would not be surprised.

. . .

Weaver: All occurring in the control room?
Mattson: Yes.
Carr: And there is no instrumentation that helps the operator sort out the order of importance of these alarms, or the order of importance of gages or lights or switches; right?
Mattson: Only the training.
Carr: Only the training to assemble all of these instruments and react in split seconds. [p. 31]

. . .

Weaver: How are people able to differentiate between these various alarms?
Michelsen: I am not a human factors engineer. I have never studied control rooms enough to understand that area at all. It is a very good question. I have asked it sometimes myself. [p. 47]

. . .

Higgins: There were a lot of short-term problems that were being addressed also; that is, equipment problems, loss of power, additional alarms that continued throughout the day, and we did not mention the number of alarms; there were a large number of equipment starts and stops, large number of individual problems . . . [p. 97]

. . .

Weaver: Would you describe the alarms that were going off? Were they going just constantly or was it -
Higgins: One would occur; the operators would address that. And then another one, perhaps a little bit later another one, or if you have a particular event, a specific event, for example the hydrogen burning in the containment might generate several; or a particular release of radioactivity from an area in the auxiliary building might generate several related alarms. Like generally there are a lot of indicators of a particular event as Mr Michelson described, on relief valve. You have got a relief valve tailpipe temperature alarm; you have got an alarm in the reactor drainpipe, I believe; level indications, and perhaps alarms associated with those system pressure alarms. All this has to do with one relief valve lifting, so when a particular event occurs, you may get several related alarms, all of which have to be addressed. [p. 99]

. . .

Higgins: If you have a scram, many alarms do come in as a result of that. Some are normal; some indicate things not being normal. And it is up to the operator by his training to distinguish which ones he expects, which ones not to expect, that type of thing. [p. 100]

. . .

Higgins: There was a tremendous amount of activity going on in the control room. A lot of people were involved with a lot of the different problems, and that was one of many things. Operators were — alarms were going off; pumps were being started and stopped; valves were being cycled, and this* was just one of a myriad of those things that was occurring throughout the entire day. And I was thoroughly not able to follow what was going on. And I did not pick it up at all. [p. 111] [*A spike on a recorder probably indicating a hydrogen burn — FPL].

. . .

Frederick: The alarms — this is a big problem. There is only one audible alarm in the control room for the 1600 alarm windows that we have, in other words, the ones that are displayed on the front of the console along with the ones on the reverse panel. So that during the emergency, I made a point of announcing that I did not want anybody to acknowledge the alarm, that is, push the acknowledgement to silence the alarm, because that would make all the windows stop flashing, and I wanted to read them all to see what was happening. As we began to run out of ideas, I wanted to review all of the alarms that we received to see if anything was happening that we could not see.

. . .

Frederick: There are several steps in the alarm process. As the alarm comes in, it sounds an alarm and a flashing light. As long as the alarm stays in − and you push the button, and the light will go solid. If in the meantime the alarming condition clears itself or goes away and you push the button, the light will go out and you will not be able to tell that it ever came in. If you have three or four alarms at the same time, only one may stay lit out of the three or four. We had probably 100 or 200 alarms flashing within the first few minutes.

. . .

Frederick: I would not be able to establish the chronology. I just wanted to see what systems were affected by the transient and if we could see something. There are some alarms you expect to get. If you read over them, you just discount them as being normal. But there may be a few that come in that you had not expected, and those are the ones I was looking for. [p. 136]

. . .

Frederick: No. The sequence of alarms that comes out of the computer that they were able to read later on was backlogged. The alarms came in 100 at a time. The computer, the IBM typewriter just types them out one at a time. So as they were coming in rapidly, probably 10 or 15 per second, it just could not keep up. So there was a backlog of maybe 2 or 3 h.

Faust: I do not know what time this fits into, but we had problems with the alarm typewriter, too, at this time. I do not know what it was, but the easiest way for me to say it is, it sort of started eating the paper. In other words, it got off the track. [p. 137]

. . .

Vento: Obviously the proper procedure here would have said that 300° or 280° in the tailpipe should have been an indication.

Frederick: That is the point he is talking about. In other words, what came out of the computer on the typewriter was an alarm that said: Temperature hot. But we could not see that, because the alarm typewriter was backlogged several hours. [p. 139]

. . .

Weaver: Now I am troubled by the newspaper reports. One report, early on, had it that there was just page after page of computer printout of question marks.

Miller: Looking back − and we can come back to you with further information, but there was a period of time in the first 2 to 3 h where the computer failed, I believe.

Zewe: Yes, for over an hour period, it did not give us any information.

Cheney: You mean you would query it and nothing would happen?

Zewe: No. The question marks, periods, and so forth, the computer had hung up, where it really did not scan the parameters and print on in a fashion where we could interpret it.

Cheney: It was unusable.

Miller: I thing Bill would back this up. He was not using that totally. He was just using his console to operate the plant, so when that went out, he just had to go away from it.

Weaver: What caused this? What was going on? What happened? We have it here in the chronology that 1 hour 13 to 2:37 − in other words -

Miller: I think, in fairness, the machine has got a tape, disc, and that sort of thing. It has experienced problems. It is not a vital piece of equipment to Bill's use of the emergency procedures or operations of the plant, so I would guess Bill would just simply have told an electrical engineer to call someone in and go on with the console.

Weaver: You do not have to have it in any way?

Zewe: No, it is just an operator's tool, but it is not anything that is really that critical to the operation. [p. 175]

. . .

[a] In addition to the personnel identified in Section A21.1 and Table A21.3, the transcript extracts given above refer to the following: Mr Carr, Congressman; Mr Stello, Director, Division of Operating Reactors, Office of Nuclear Reactor Regulation, NRC; Mr Mattson, Division of System Safety, Office of Nuclear Reactor Regulation, NRC; Mr Michelson, Nuclear Engineer, Tennessee Valley Authority; and Mr Higgins, Region I, Office of Inspection and Enforcement, NRC.

A21.8.3 Kemeny Report

The Kemeny Report gives an account of the accident, from 29 March to 2 April, presents findings and makes recommendations. Some findings are described in this section.

The reports starts with what it calls 'people-related problems'. It states

Popular discussions of nuclear power plants tend to concentrate on questions of equipment. Equipment can and should be improved to add further safety to nuclear power plants, and some of our recommendations deal with this subject. But as the evidence accumulated, it became clear that the fundamental problems are people-related problems and not equipment problems.

When we say that the basic problems are people-related, we do not mean to limit this term to shortcomings of individual human beings − although these do exist. We mean more generally that our investigation has revealed problems with the 'system' that manufactures, operates and regulates nuclear power plants. There are structural problems in various organizations, there are deficiencies in various processes, and there is a lack of communication among key individuals.

We are convinced that if the only problems were equipment problems, this Presidential Commission would never have been created. The equipment was sufficiently good that except for human failures, the major accident at Three Mile Island would have been a minor incident. But, wherever we looked, we found problems with the human beings who operate the plant, with the management that runs the key organization, and with the agency that is charged with assuring the safety of nuclear power plants. (p. 8)

The form frequently taken by this problem is described as follows: 'In the testimony we received, one word occurred over and over again. That word is "mindset".' (p. 8)

The report is critical of the approach taken to control of the industry, stating:

We note a preoccupation with regulations. It is, of course, the responsibility of the Nuclear Regulatory Commission to issue regulations to assure the safety of nuclear power plants. However, we are convinced that regulations alone cannot assure safety. Indeed, once regulations become as voluminous and complex as those regulations now in place, they can serve as a negative factor in nuclear safety. The regulations are so complex that immense efforts are required by the utility, by it suppliers and by the NRC to assure that regulations are complied with. The satisfaction of regulatory requirements is equated with safety. This Commission believes that it is an absorbing concern with safety that will bring about safety – not just the meeting of narrowly prescribed and complex regulations. (p. 9)

The report makes various criticisms of the NRC, of which a sample, changed in order, are:

We found serious managerial problems within the organization. These problems start at the very top . . . It is not clear to us what the precise role of the five NRC commissioners is, and we have evidence that they themselves are not clear what their role should be . . . NRC's primary focus is on licensing and insufficient attention has been paid to the ongoing process of assuring safety . . . NRC is vulnerable to the charge that it is heavily equipment-oriented, rather than people oriented . . . We are extremely critical of the role the organization played in the response to the accident . . . NRC has sometimes erred on the side of the industry's convenience rather than carrying out its primary mission of assuring safety . . . (pp. 19 – 21)

The Commission identifies an over-concentration on large accident hazards:

We find a fundamental fault even with the existing body of regulations. While scientists and engineers have worried for decades about the safety of nuclear equipment, we find that the approach to nuclear safety has a major flaw. It was natural for the regulators and the industry to ask: 'What is the worst kind of equipment failure that can occur?' Some potentially serious scenarios, such as the break of a huge pipe that carries the water cooling to the nuclear reactor, were studied extensively and diligently, and were used as the basis for the design of plants. A preoccupation developed with such large-break accidents as did the attitude that if they could be controlled, we need not worry about the analysis of 'less important' accidents. (p. 9)
A potentially insignificant incident grew into the TMI accident, with severe damage to the reactor. Since such combinations of minor equipment failures are likely to occur much more often than the huge accidents, they deserve extensive and thorough study. (p. 9)

The report is critical of organizational failure to follow through safety issues:

We find that there is a lack of 'closure' in the system -that is, important safety issues are frequently raised and may be studied to some degree of depth, but are not carried through to resolution; and the lessons learned from these studies do not reach those individuals and agencies that most need to know about them. (p. 11)

The report reopens the issue of the influence of the operator in a developing reactor emergency. The conventional approach has been that any action taken by the operator is likely to be beneficial. It gives this view as an example of designers' mindset: 'They concentrated on equipment, assuming that the presence of operators could only improve the situation – they would not be part of the problem', (pp. 8–9).

On the handling of the emergency the Commission comments:

We are disturbed both by the highly uneven quality of emergency plans and by the problems created by multiple jurisdictions in case of a radiation emergency. Most emergency plans rely on prompt action at local level to initiate a needed evacuation or to take other protective action. We found an almost total lack of detailed plans in the local communities around Three Mile Island. It is one of the many ironies of this event that the most relevant planning by local authorities took place during the accident. In an accident in which prompt defensive steps are necessary within a matter of hours insufficient advance planning could prove extremely dangerous. (p. 15)

The report contains a number of supplemental views by individual members of the commission, one of which is by Professor T.H. Pigford. He is critical of the NRC's approach to safety, which he characterizes as lacking in experienced personnel; failing to take a comprehensive approach with quantified safety objectives and rational priorities; and prone to introduce arbitrary requirements. He calls it 'A stifling approach. The existing process inhibits the interchange of technical information between the NRC and industry. It discourages innovative engineering solutions'. On human factors aspects he states

The emphasis in this report upon equipment versus people obscures the fact that the equipment itself is only one product of the defense-in-depth or multiple-barrier design approach, which also encompasses the analysis of how components must perform and how systems of equipment must operate. The accident demonstrated that this system of equipment performed better than expected.
The nature of the people-related problem needs clarification. One such problem – and a most serious one – was the errors made by the operators and the operator-supervisors, whose training was insufficient in scope and understanding. Another was failure of many individuals to respond adequately to earlier experience from other reactors and to other advance information that might have alerted the operators and avoided the accident.

A21.9 Some Lessons of Three Mile Island

A21.9.1 Kemeny Report recommendations
The Kemeny Report and the Task Force Report both give extensive recommendations. The heads of these recommendations are shown in Table A21.5, in Sections A and B, respectively.
The Kemeny Report recommends that the NRC should be restructured as a new independent agency in the executive branch and that the ACRS should be strengthened.
The report states: 'To the extent that the industrial institutions we have examined are representative of the nuclear industry, the nuclear industry must dramatically change its attitude towards safety and regulations'. It goes on The industry should establish a program that specifies appropriate safety standards including those for management, quality assurance, and operating procedures and practices, and that conducts independent evaluations'.
Improvements in operator training are called for, with particular emphasis on the use of simulators.

Table A21.5 *Heads of some principal recommendations on Three Mile Island*

A   Presidents Commission (Kemeny, 1979)

Nuclear Regulatory Commission
Utility and it suppliers
Training of operating personnel
Technical assessment
Worker and public health and safety
Emergency planning and response
Public's right to information

B   'Lessons Learned' Task Force (Nuclear Regulatory Commission, 1979b)

Personnel qualifications and training
Staffing of control room
Working hours
Emergency procedures
Verification of correct performance of operating activities
Evaluation of operating experience
Man–machine interface
Reliability assessments of final designs
Review of safety classifications and qualifications
Design features of core-damage and core-melt accidents
Safety goal for reactor regulation
Staff review objectives
NRR Emergency Response Team

Under the heading of technical assessment, the report deals with equipment for information display in the control room and for mitigating the consequences of accidents and recommends a formal safety assurance programme involving hazard identification and assessment.

The report recommends that there should be a requirement for an emergency response plan.

It also makes recommendations on worker and public health in relation to nuclear hazards and on the public's right to information.

A21.9.2 Task Force Report recommendations

The recommendations of the Task Force Report are summarized in Section B of Table 21.5. They deal particularly with operator training and with the man-machine interface.

The report calls for a thorough review of control room design and for development of aids to assist the operator in assessing the status of the plant and in analysing disturbances and alarms. It proposes the development both of a plant safety status display system and of a disturbance analysis system (DAS).

On the former the report states

> Each licensee should be required to define and adequately display in the control room a minimum set of plant parameters (in control terminology, a state vector) that defines the safety status of the nuclear power plant. The minimum set of plant parameters should be annotated for sensor limits, process limits, and sensor status. The annotated set of parameters should be presented to the operator in real time by a reliable, single-failure-proof system located in the control room. (p.A-12)

Table A21.6 *Some lessons of Three Mile Island*

Regulatory control of major hazard installations
People-related versus equipment-related problems
Formal safety assurance
Relative importance of large and small failures
Influence of operator in an emergency
Fault diagnosis by the process operator
Accident management for major hazard installations
Follow-up of safety issues
Learning from precursor events
Package and other ancillary units
Display of variable of interest
Plant status display and disturbance analysis systems
Abuse of instrument air
Limitations of non-return valves

A21.9.3 Some lessons

The Three Mile Island incident yields numerous lessons. Some of these are listed in Table A21.6. Many of these lessons are drawn in the reports just described.

Regulatory control of major hazard installations
The comments made in the Kemeny Report on the regulatory regime for US nuclear power plants under the NRC both echo those made in the Robens Report in the UK in 1972 and themselves find an echo in those of the Cullen Report on UK offshore installations in 1990. In all three cases the reports are critical of an approach which they find excessively prescriptive.

People-related versus equipment-related problems
The theme which runs through the Kemeny Report is that equipment-related problems have received attention at the expense of people-related problems. The account of these problems given in the report indicates clearly that the term refers not just to process operators, although operator problems were a central issue, but to management and management systems. Here too, the report may be said to look back to Robens and forward to Cullen.

Formal safety assurance
The Kemeny Report calls for a formal safety assurance programme for nuclear power plants involving hazard identification and assessment. In UK terms, its recommendations parallel those of the ACMH for onshore plant and of the Cullen Report for offshore installations, which uses the term 'formal safety assessment'.

Relative importance of large and small failures
Engineers in the nuclear industry had tended to concentrate their attention on major failures, notably the large loss of coolant accident, to the relative neglect of lesser events, such as a small, or at least slower, loss of coolant, such as actually occurred at TMI. Yet, as the incident showed, such lesser events can escalate. Moreover, lesser events occur much more frequently. There needs to be a more balanced approach which addresses small as well are large failures.

Influence of operator in an emergency
It has been a basic assumption in the industry that in a nuclear emergency operator action would be beneficial, and that the longer the time available for such action the higher

the probability that it would be taken. TMI showed that this is not necessarily so, and that operator action can be detrimental.

Fault diagnosis by the process operator
The incident highlighted the problem of fault diagnosis by the process operator under stress. It showed the limitations of fixed operating rules in dealing with abnormal situations and the dangers of tunnel vision, or mindset. In other words, it showed the inadequacy of rule-based behaviour and the need for knowledge-based behaviour and for training based on development of the latter.

Accident management for major hazard installations
TMI shows that a major accident poses a number of difficult problems and that it needs to be managed. This implies that accident management must be recognized as a specific requirement and planned in advance.

The handling of an emergency in the control room is one aspect of the problem, but there are many others. They include the identification, mobilization, direction and coordination of the necessary resources; planning for an emergency, both on-site and off-site; and handling of the media.

Follow-up of safety issues
The Kemeny Report is critical of the follow-up of safety issues, referring to a lack of closure. Safety issues which are raised and agreed to be valid need to be followed through to resolution. Personnel who most need to know should be informed.

Learning from precursor events
There had been previous incidents in which a PORV had been stuck in the open position, but the management of the TMI plant had no systematic approach to learning from such incidents.

Package and other ancillary units
Turning to some more detailed aspects, the incident started with a fault in the condensate polishing system, a package unit. It illustrates the point that the ancillary equipment and services should receive their due share of attention, starting with specification and continuing right through to operation.

Display of variable of interest
If a process variable is critical, it is highly desirable that the quantity which is displayed should be the variable of direct interest. At TMI there were two cases where this was not done. One is the water level in the pressure vessel. The other is the status of the PORV. The display showed the signal to the valve rather than the actual valve position.

Plant status display and disturbance analysis systems
TMI was a dramatic illustration of the difficulties faced regularly by process operators in interpreting displays and alarms, and of way in which these can be compounded by information overload. It highlighted the scope for the development of improved display and alarm systems and for the exploitation of computers to this end. The principal aids proposed were plant safety status display systems and disturbance analysis systems.

Abuse of instrument air
The air used to clear the blockage on the condensate polishing system was instrument air. It is bad practice to connect instrument air to the process, which can cause it to become contaminated.

Limitations of non-return valves
The entry of water into the instrument air system illustrates the limitations of a non-return valve. Such a valve is intended to stop bulk flow of fluid; it will not necessarily prevent the passage of trace quantities.

Appendix   Chernobyl

22

Contents

On Monday 28 April 1986 a worker at the Forsmark nuclear power station in Sweden put his foot in a radiation detector for a routine check and registered a high reading. The station staff thought they had had a radioactive release from their plant and the alarm was raised. However, as reports came in of high radioactivity in Stockholm and Helsinki, the source of the release was identified as the Soviet Union. In fact, an accident had occurred on Unit 4 at Chernobyl in the Ukraine some 800 miles from Sweden on Saturday 26 April.

At 1.24 on that day an experiment to check the use of the turbine during rundown as an emergency power supply for the reactor went catastrophically wrong. There was a power surge in the reactor, the coolant tubes burst and a series of explosions rent the concrete containment. The graphite caught fire and burnt, sending out a plume of radioactive material. Emergency measures to put out the fire and stop the release were not effective until 6 May.

The Soviet Government set up an investigatory team headed by N.I. Ryzhkov, President of the USSR Council of Ministers, to investigate. A report *The Accident at the Chernobyl Nuclear Power Plant and its Consequences* by the USSR State Committee on the Utilisation of Atomic Energy (1986) (the *USSR State Committee Report*) was presented in August 1986 to a meeting of experts at the IAEA in Vienna by V. Legasov.

Other accounts of the accident include *Summary Report on the Post-Accident Review Meeting on the Chernobyl Accident* by the IAEA (1986), *The Chernobyl Accident and its Consequences* by Gittus *et al.* (1987, 1988) (the *UKAEA Report), Chernobyl and the Safety of Nuclear Reactors in OECD Countries* by the NEA (1987), *Report on the Accident at Chernobyl Nuclear Power Station* by the NRC (1987b), *The Chernobyl Accident and its Consequences for the United Kingdom* by Worley and Lewins (1988) (the First *Watt Committee Report*), Chernobyl. *A Policy Response Study* by Segerstahl (1991) and *Five Years after Chernobyl: 1986–1991* by the Watt Committee (1991) (the Second *Watt Committee Report*) and by Franklin (1986) and Kletz (1988h).

The story of Chernobyl is told in *The Worst Accident in the World* by Haynes *et al.* (1986), *Mayday at Chernobyl* by Hamman and Parrott (1987), *The Chernobyl Disaster* by Haynes and Bojcum (1988), *Chernobyl. The Real Story* by Mould (1988) and *Ablaze* by Read (1993) and by Franklin (1986) and Kletz (1988h).

Selected references on Chernobyl are given in Table 22.1.

Table A22.1   *Selected references on Chernobyl*

NRC (Appendix 28 *Chernobyl*); BNES (1986); Franklin (1986); Hawkes *et al.* (1986); IAEA (1986); IBC (1986/69); USSR State Committee on the Utilization of Atomic Energy (1986); Watt Cttee on Energy (1986, 1991); Collier and Davies (1987); Gittus *et al.* (1987); Hall, Hall and Nixon (1987 LPB 76); Hamman and Parrot (1987); NEA (1987); NRC (1987b); F. Allen (1988); Ballard (1988); Haynes and Bojcum (1988); Kletz (1988h); Marples (1988, 1991); V.C. Marshall (1988 LPB 81); Mould (1988); Wheeler (1988); Worley and Lewins (1988); Gudiksen, Harvey and Lange (1989); Konstantinov and Gonzalez (1989); Parmentier and Nenot (1989); Voznyak, Kovalenko and Troitsky (1989); Hass *et al.* (1990); Ryin *et al.* (1990); Ottewell (1991); Segerstahl (1991); Read (1993)

A22.1 The Operating Organization and the Management

Little information has become available concerning the organization operating the Chernobyl plant or about its management.

A22.2 The Site and the Works

The Chernobyl nuclear power station is situated in a region which at the time was relatively sparsely populated. There were some 135,000 people within a 30 km radius. Of these 49,000 lived in Pripyat to the west of the plant's 3 km safety zone and 12,500 in Chernobyl 15 km to the south-east of the plant.

There were four nuclear reactors on the site. The first pair had been built in the period 1970–77 and the second pair were completed in 1983.

A22.3 The Process and the Plant

Unit 4 at Chernobyl was designed to supply steam to two turbines each with an output of 500 MWe. The reactor was therefore rated at 1000 MWe or 3200 MWt.

Unit 4 was an RBMK-1000 reactor, which is a standard design widely used in the Soviet Union. The RBMK-1000, shown in Figure A22.1, is a boiling water pressure tube, graphite moderated reactor. The reactor is cooled by boiling water, but the water does not double as the moderator, which is graphite.

The design of the reactor avoided the use of a large pressure vessel. Instead, use was made of much smaller individual pressurized fuel channels. The reactor had 1661 fuel channels. There was in each one a fuel assembly, divided into two subassemblies, each containing 18 fuel elements.

The reactor contained 192 te of uranium enriched to 2%. The subdivision of the uranium into a large number of separately cooled fuel channels greatly reduced the risk of total core meltdown.

The reactor was cooled by water passing through the pressure tubes. The water was heated to boiling point and partially vaporized. The steam–water mixture with a mass steam quality of 14% passed to two separators where steam was flashed off and sent to the turbines, while water was mixed with the steam condensate and fed through downcomers to pumps which pumped it back to the reactor. There were two separate loops each with four pumps, three operating and one standby. This system constituted the multiple forced circulation circuit (MFCC).

The moderator used was graphite. The reactor contained a stack of 2488 graphite blocks with a total mass of 1700 te.

An emergency core cooling system (ECCS) was provided to remove residual heat from the core in the event of loss of coolant from the MFCC.

The reactor was provided with a control and protection system (CPS). The control system incorporated a control computer. There was a control system which controlled the power output of the reactor by moving the control rods, in and out. Automatic protective systems included a trip to bring in the ECCS and reactor shut-down trips. Reactor shut-down was effected by the insertion of control rods, and conditions which would trigger shut-down included loss of steam to the turbines and loss of level in the separators.

The reactor was housed in a containment built to withstand a pressure of 0.45 MPa. There was a 'bubble'

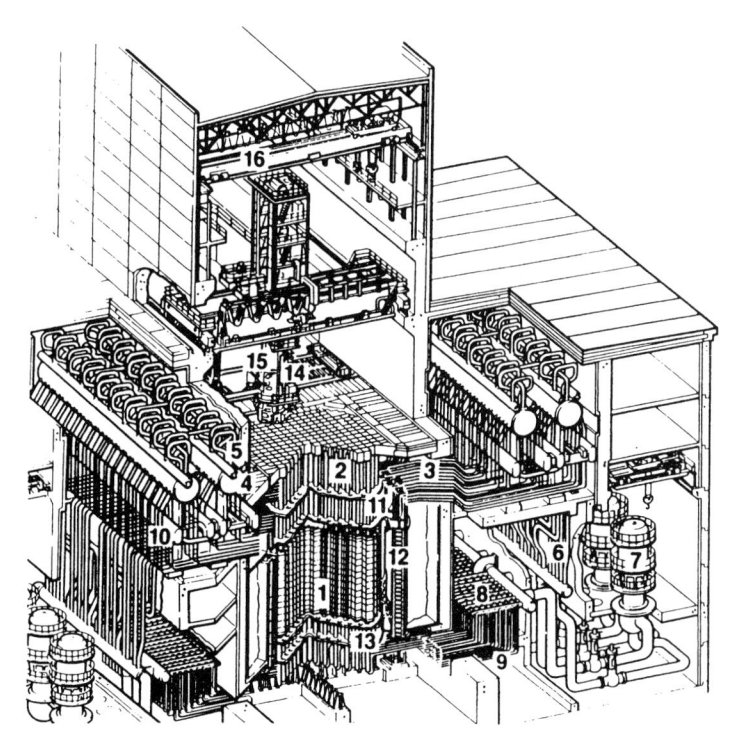

Key

1 **Reactor**
2 **Fuel-channel standpipes**
3 **Steam/water riser pipes**
4 **Steam drums**
5 **Steam headers**
6 **Downcomers**
7 **Main circulating pumps (MCP)**
8 **Group distribution headers**
9 **Reactor inlet water pipes**
10 **Burst-can detection system**
11 **Upper biological shield**
12 **Side biological shield**
13 **Lower biological shield**
14 **Irradiated fuel storage pond**
15 **Fuelling machine**
16 **Bridge crane**

Figure A22.1   *Simplified flow diagram of Chernobyl Unit 4 (Nuclear Regulatory Commission, 1987b)*

condenser designed to condense steam entering the containment from relief valves or from rupture of the MFCC.

An important operating feature of the reactor was its reactivity margin, or excess reactivity. Below a certain power level the reaction would be insufficient to avoid xenon poisoning. It was therefore necessary to operate with a certain excess reactivity. This reactivity could be expressed in terms of the number of CPS control rods which would need to be inserted to counter it. The operating instructions stated that a certain excess reactivity was to be maintained; the equivalent number of control rods was 30.

Another important operating characteristic was the reactor's 'positive void coefficient'. An increase in heat from the fuel elements would cause increased vaporization of water in the fuel channels and this in turn would cause increased reaction and heat output. There was therefore an inherent instability, a positive feedback, which could be controlled only by manipulation of the control rods. Control of the reactor at normal power output presented little problem, but this feature made the reactor highly sensitive at low outputs.

A22.4 Events Prior to the Release

The origin of the accident was the decision to carry out a test on the reactor. Electrical power for the water pumps and other auxiliary equipment on the reactor was supplied by the grid with diesel generators as back-up. However, in an emergency there would be a delay of about a minute

before power became available from these generators. The objective of the test was to determine whether during this period the turbine could be used as it ran down to provide emergency power to the reactor. A test had already been carried out without success, so modifications had been made to the system, and the fresh test was to check whether these had had the desired effect.

A programme for the test was drawn up. It included provision to switch off the ECCS during the test, apparently to prevent its being triggered during the test.

At 1.00 on 25 April the reduction of power began. At 13.05 Turbogenerator 7 was switched off the reactor. At 14.00 the ECCS was disconnected. However, a request was received to delay shutting down since the power output was needed. It was not until 23.10 that power reduction was resumed.

At some stage the trip causing reactor shut-down on loss of steam to Turbogenerator 8 was disarmed, apparently so that if the test did not work the first time, it could be repeated. This action was not in the experimental programme.

The test programme specified that rundown of the turbine and provision of unit power requirements was to be carried out at a power output of 700–1000 MWt, or 20% of rated output. The operator switched off the local automatic control, but was unable to eliminate the resultant imbalance in the measurement function of the overall automatic controls and had difficulty in controlling the power output, which fell to 30 MWt. Only at 1.00 on 26 April was the reactor stabilized at 200 MWt, or 6% of rated output.

Meanwhile the excess reactivity available had been reduced as a result of xenon poisoning.

It was nevertheless decided to continue with the test. At 1.03 the fourth, standby pump in one of the loops of the MFCC was switched on and at 1.07 the standby pump in the other loop.

A22.5 The Release – 1

With the reactor operating at low power, the hydraulic resistance of the core was less and this combined with the use of additional pumps resulted in a high flow of water through the core. This condition was forbidden by the operating instructions, because of the danger of cavitation and vibration. The steam pressure and water level in the separators fell and other process parameters changed. In order to avoid triggering the trips which would shut-down the reactor on these parameters, the trips were disarmed.

The reactivity continued to fall and at 1.22.30 the operator saw from a printout of the reactivity evaluation program that the available excess reactivity had fallen below a level requiring immediate reactor shut-down. Despite this the test was continued.

At 1.23.04 the emergency valves on Turbogenerator 8 were closed. The reactor continued to operate at about 200 MWt, but the power soon began to rise. At 1.23.40 the unit shift foreman gave the order to press the scram button. The rods went down into the core, but within a few seconds shocks were felt and the operator saw that the rods had not gone fully in. He cut off the current to the servo drives to allow the rods to fall in under their own weight.

Within 4 s, by 1.23.44, the reactor power had risen, according to Soviet estimates, to 100 times the nominal value.

At about 1.24, according to observers outside Unit 4, there were two explosions, the second within some 3 s of the first, and debris and sparks shot into the air above the reactor.

It is thought that the fuel fragmented, causing a rapid rise in steam pressure as the water quenched the fuel elements, so that there was extensive failure of the pressure tubes. The explosive release of steam lifted the reactor top shield, exposing the core. Conditions were created for reaction between the zirconium and steam, producing hydrogen. An explosion, involving hydrogen, occurred in and ruptured the containment building, and ejecting debris and sparks were seen.

Some 30 fires broke out.

The accident was aggravated by the fact that the 200 te loading crane fell onto the core and caused further bursts of the pressure tubes.

The stricken reactor is shown in Figure A22.2.

The timetable of these and the following events is shown in Table A22.2.

A22.6 The Emergency and the Immediate Aftermath

The fire brigades from Pripyat and Chernobyl set out at 1.30. The fires in the machine hall over Turbogenerator 7 were particularly serious, because they threatened Unit 3 also. These therefore received priority. By 5.00 the fires in the machine hall roof and in the reactor roof had been extinguished.

However, the heating up of the core and its exposure to the air caused the graphite to burn. The residual activity of the radioactive fuel provided another source of heat. The core therefore became very hot and the site of raging fire.

During the next few days the fire raged and a radioactive plume rose from the reactor. The accident had

Figure A22.2 *Reactor of Chernobyl Unit 4 after the explosions (ITAR-TASS)*

Table A22.2   Timetable of events at Chernobyl

A   Events prior to initial explosion

25 April	1.00	Reduction of power started
	13.05	Turbogenerator 7 disconnected
	14.00	ECCS disconnected
	23.10	Power reduction resumed
26 April	00.28	Operator switches off local automatic control, but is unable to eliminate resultant imbalance in measurement function of the overall automatic controls
		Power output falls to 30 MWt
	1.00	Reactor stabilized at 200 MWt (6% rated output)
	1.03	Standby pump switched on in one loop of MFCC
	1.07	Standby pump switched on in other loop of MFCC
	before 1.19	Trips on low steam pressure and low water level disabled
	1.22.30	Printout confirming fall in excess reactivity
	1.23.04	To begin test, stop valve on steam to Turbogenerator 8 closed
		Trip on closure of both turbogenerator steam stop valves disabled
	1.23.40	Scram button pressed to drop control rods
	1.23.40	Operator hears banging noises and sees rods stopping before they reach bottom; disengages servo drives to allow rods to fall under own weight
		Rapid increase in power
	1.23.48	First explosion – extensive failure of pressure tubes, explosive release of steam, top lifted off reactor
	1.24	Second explosion – explosion within reactor space, possibly involving hydrogen, pressure increase to several MP, rupture of containment building
		Numerous fires

B   Events after initial explosion

26 April	2.54	Fire fighting units from Pripyat and Chernobyl arrive
	3.34	Most of fires in turbine room roof out
	3.54	Fire on reactor building roof out
	5.00	All fires, other than those in core, out
		Unit 3 shut-down
27 April	1.13	Units 1 and 2 shut-down
27 April		Start of smothering operation using helicopters
6 May		Discharge of radioactivity drops to several hundred Ci/h

become a major disaster. It was necessary to evacuate the population from a 30km radius round the plant and deal with the casualties and to take a whole range of measures to dampen and extinguish the fire, to cover and enclose the core, and to deal with the radioactivity in the surrounding area.

Three evacuation zones were established: a special zone, a 10km zone and 30km zone. The latter zone is shown in Figure A22.3. A total of 135,000 people were evacuated.

Accounts of the evacuation are incomplete and sometimes contradictory, but what appears to have happened is as follows. Within a few hours of the accident an emergency headquarters was set up in Pripyat. About 14.00 on 26 April there was an evacuation of some 1000 people from within a 1.6 km zone around the plant. A much larger evacuation took place about 14.00 next day, when about 1000 buses were brought in and within 2 h had evacuated some 40,000 people. However, it was not until some 9 days later that the authorities in Moscow ordered complete evacuation from a 30 km zone around the site.

First reports from Kiev spoke of no apparent concern, but as the nature of the disaster became known, parents hastened to send their children out of the city to relatives elsewhere.

A medical team was set up within 4 h of the accident and within 24 h triage of the 100 most serious cases had been effected.

A system was set up to control movements between the zones. People crossing the zone boundaries had to change clothing and vehicles were decontaminated.

In the immediate aftermath of the accident steps were taken to obtain the measurements essential to the control of the emergency. These included the measurement not only of the radioactivity in and around the plant, but also of conditions in the reactor relevant to bringing it under control. The latter was made more difficult by the fact that the regular measurement system had been disabled.

A decision had to be made whether to let the fire burn itself out or to smother it. In view of the hazard to the surrounding area, the latter course was chosen. A specialist team was assembled who began to cover the damaged reactor with compounds of boron, dolomite, sand, chalk and lead. Between 27 April and 10 May some 5000 te of material were dropped on the reactor by helicopter. By May 6 the release of radioactivity had decreased to a few hundred Ci/day and had ceased to be a major factor.

At the same time the core temperature was tackled by pumping nitrogen under pressure from the compressor

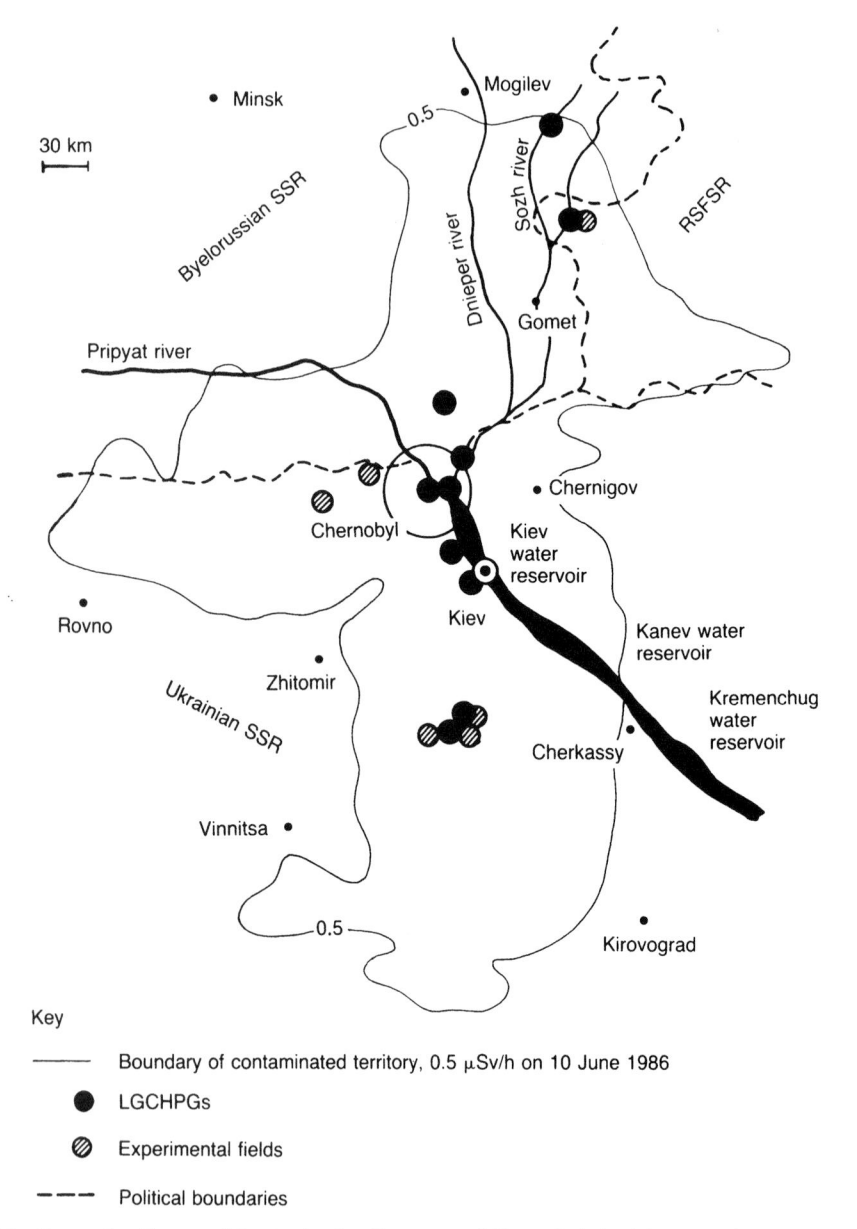

Figure A22.3 *Evacuation Zone and Contamination Zone around Chernobyl site (after Watt Committee, 1991)*

station into the space beneath the reactor vault. This reduced the oxygen concentration and the temperature. By 6 May the temperature rise had been halted and the temperature began to decline.

Meanwhile Soviet experts were concerned about the hazard of the large amount of water directly beneath the reactor. If the core were to come into contact with this water, there could be massive and explosive vaporization into steam. Engineers went down through dark passages flooded with radioactive water and succeeded in opening two large valves to allow the water to drain out.

Next an attempt was made to put a large concrete slab fitted with cooling coils beneath the core to prevent material from the core contaminating the ground and watercourses beneath it. Tunnelling in the soil at this point was difficult and it was necessary to resort to freezing the soil using liquid nitrogen, a method used in building the Leningrad Metro. The construction of this concrete bed was complete by the end of June.

Subsequently the reactor was entombed in concrete.

There was also intense activity to counter the effects of radioactivity in the surrounding area. 7000 wells were sealed. The water supply to Kiev could no longer be taken from water near the plant and a new water supply for the city had to be constructed. A system of dikes was constructed to catch radioactive rain water from the area around the plant. A system of bore holes and barriers was created to prevent contamination of water courses.

Decontamination measures were also set in train. The site itself was divided into zones. The measures taken were to remove debris and contaminated equipment, decontaminate roofs and outer surfaces of buildings, remove a 5–10 cm layer of soil into containers, lay where necessary concrete or fresh earth on the ground, and coat certain surfaces with film-forming compounds.

Decontamination of the area outside the site was complicated by the fact that the distribution of the radionuclides changed with time according to the terrain, particularly during the first 3–4 months. Hence the full value of measures to decontaminate a particular area is generally obtained only temporarily. Buildings were hosed down and the earth around them then removed. Measures were developed to allow the contaminated land to be used for agriculture: these included changes to methods of cultivation and harvesting and the use of dust-suppression materials.

The accident was the subject of intense press interest world-wide, although hard facts were difficult to come by. Interest centred on satellite photographs of the burning core, evacuation measures taken or not taken, the assistance given by an American bone marrow transplant surgeon, potential fallout in other countries and implications for nuclear power generally.

The fire in the reactor was readily seen from above and satellite photographs appeared on the world's television screens. These were soon supplemented by Soviet pictures of the stricken reactor.

The death toll from Chernobyl cannot be known with certainty, since most deaths will be excess cancers. Two men died dealing with the accident itself. By the end of September the toll of dead in the USSR was 31 with 203 injured. The casualties are discussed further in Section A22.8.

Other European countries were also affected. Radioactive contamination was measured across Europe and many countries banned particular products for a period.

A22.7 The Investigations

A22.7.1 USSR State Committee Report
In the West the main source of information on the accident was initially the investigation described in the *USSR State Committee Report*.

Part I of the report gives a brief description of the reactor and describes the accident itself. The seven appendices of Part II describe the reactor system, the detailed design of the reactor, the measures to deal with the consequences, the quantity of radioactivity released, the dispersion of the radioactive plume and the radioactive deposition, the radio-ecological contamination caused, and the medical and biological consequences.

An important feature of the investigation was a simulation of the accident using a mathematical model. This was assisted by the fact that the control computer had a program which logged several hundred critical parameters with a minimum cycle time of 1 s.

According to the simulation a power surge occurred in the reactor which took the output to 530 MWt within 3 s. There was intense steam formation and then nucleate boiling in the fuel channels and abrupt pressure increase, causing bursting of the channels.

The primary explosion was therefore bursting of the fuel channels by high pressure steam. The formation of steam and the rapid rise in core temperature created conditions for the steam–zirconium and other reactions. These reactions gave rise to hydrogen and carbon monoxide. After the reactor space had been vented and destroyed these mixed with air and burnt causing an explosion.

The report criticises both the experimental programme and the conduct of the test. It states

> The quality of the programme was poor and the section on safety measures was drafted in a purely formal way. . . . Apart from the fact that the programme made essentially no provision for additional safety measures, it called for shutting off the reactor's emergency core cooling system.
>
> Because the question of safety in these experiments had not received the necessary attention, the staff involved were not adequately prepared for the tests and were not aware of the possible dangers. Moreover, as we shall see in what follows, the staff departed from the programme and thereby created the conditions for the emergency situation.

The report highlights the positive void coefficient as a problem feature of the reactor design. It lists six violations of the operating instructions by the operators and gives the motivation for and consequences of each.

The test programme specified that the ECCS should be disconnected. In addition, the reactor shut-down trip based on turbogenerator shut-down was disarmed. During the actual test the reactor shut-down trip based on coolant parameters was disarmed. These three actions were all taken to facilitate the test. The three other violations were the reduction in the reactivity margin, the switch from local automatic power control to manual control and the switching on of the standby pumps. The first was connected with attempts to deal with the poisoning problem and it rendered the protective system ineffective. The second was an operator error and it made the reactor much more difficult to control. The third was done to make doubly sure of adequate cooling and it resulted in loss of level in the separator and the decision to disable the reactor shut-down trip based on coolant parameters.

The report also points out that at low power levels the measurement of power density in the core is less accurate.

The report gives an account of the release from the damaged reactor and of the emergency measures taken.

The report records a number of measures being taken to prevent a repetition. They include changes to the organization responsible for the reactors and to the inspection organization and intensification of research on safety aspects.

An important design modification mentioned is the use of uranium fuel enriched to 2.4% rather than 2% in order to try to avoid the positive void coefficient problem.

In presenting the report to the IAEA Legasov is reported as saying that the plant was one of the best in the country with good operators who were so convinced of its safety that they 'had lost all sense of danger'.

A22.7.2 Other reports

As described at the start of the section, there have been a number of other reports on Chernobyl. The coverage of these reports is broadly as follows:

	IAEA	NRC	NEA	Watt Cttee 1 2	UKAEA
Design of RBMK reactors		X	X	X	X
General account of events	X	X		X	X
Chronology of events	X	X	X		X
Radioactive release, source terms	X	X	X	X X	
Radioactive dispersion	X	X	X	X	
Health effects	X	X		X X	
Environmental effects, decontamination	X	X	X	X X	
Emergency response	X	X			
Safety issues	X	X	X	X	

A22.8 The Release – 2

Estimates of the amount of radiation released are given in the *USSR State Committee Report* and several of the other reports mentioned. Table A22.3 Section A, based on figures given by Konstantinov and Gonzalez (1989), gives one such estimate of the total release, while Section B, based on the *USSR State Committee Report*, gives an estimate of the rate of release, showing the fall-off after 6 May.

According to the *USSR State Committee Report*, on 26 April the area around the site was in a low gradient pressure field with a slight wind varying in direction. At an altitude of about 1 km the wind speed was 5–10 m/s in a north-westerly direction.

The most intense plume observed was during the first 2–3 days in a northerly direction. Radiation levels in this stream at 5–10 km from the reactor and at 200m altitude were 1000 mR/h on 27 April and 500 mR/h on 28 April. According to aircraft monitoring data on 27 April the height of the stream exceeded 1200m in a north-westerly direction and the radiation level at this height was 1 mR/h. During the following days the height of the stream did not exceed 200–400 m.

Information on the radiation levels experienced at and around the site is patchy. The report states that those at the site exceeded 100 mR/h. The report also gives the following additional activities:

15 days after the accident at 30–40 km north: 35–40 mR/h

15 days after the accidents at 50–60 km west: 5 mR/h

At beginning of May in Kiev: 0.5–0.8 mR/h

Maps of the radioactive fallout in the Chernobyl region are given in the Second Watt Committee Report and by

Table A22.3 *Estimated radioactive release from Chernobyl Unit 4*

A   Core inventory and estimated fraction released (after Konstantinov and Gonzalez, 1989)

Radionuclide	Core inventory[a] (EBq)	Estimated fraction released[b] (%)	Half-life
Kr-85	0.033	≈100	10.72 year
Xe-133	1.7	≈100	5.25 d
I-131	1.3	20	8.04 d
Te-132	0.32	15	3.26 d
Cs-137	0.29	13	30.0 year
Cs-134	0.19	10	2.06 year
Sr-89	2.0	4	50.5 d
Sr-90	0.2	4	29.12 year
Zr-95	4.4	3	64.0 d
Mo-99	4.8	2	2.75 d
Ru-103	4.1	3	39.3 d
Ru-106	2.1	3	368 d
Ba-140	2.9	6	12.7 d
Ce-141	4.4	2	32.5 d
Ce-144	3.2	3	284 d
Np-239	0.14	3	2.36 d
Pu-238	0.001	3	87.74 year
Pu-239	0.0008	3	24065 year
Pu-240	0.001	3	6537 year
Pu-241	0.17	3	14.4 year
Cm-242	0.026	3	163 d

B   Estimated daily release rates (after USSR State Committee, 1986)

Date	Time after accident (d)	Radioactivity released[a] (MCi)
Apr. 26	0	12
27	1	4.0
28	2	3.4
29	3	2.6
30	4	2.0
May 1	5	2.0
2	6	4.0
3	7	5.0
4	8	7.0
5	9	8.0
6	10	0.1
9	14	≈0.01
23	28	20×10^{-6}

[a] Decay corrected to 6 May 1986. 1 Ebq $= 10^{18}$ Bq.
[b] Stated accuracy ±50% expect for noble gases.

Read. Figure A22.3 from the former shows the boundary of the zone contaminated to the level of 0.5 μ Sv/h on 10 May 1986.

Accounts and maps of the radiation dispersion patterns over Europe and over the United Kingdom are given in the *UKAEA Report* and in the two *Watt Committee Reports* and by Apsimon and Davison (1986), Wheeler *et al.* (1989) and Hass *et al.* (1990).

Figure A22.4 from Gittus *et al.* (1987) shows the estimated radiation dispersion pattern across Europe on 3 May 1986.

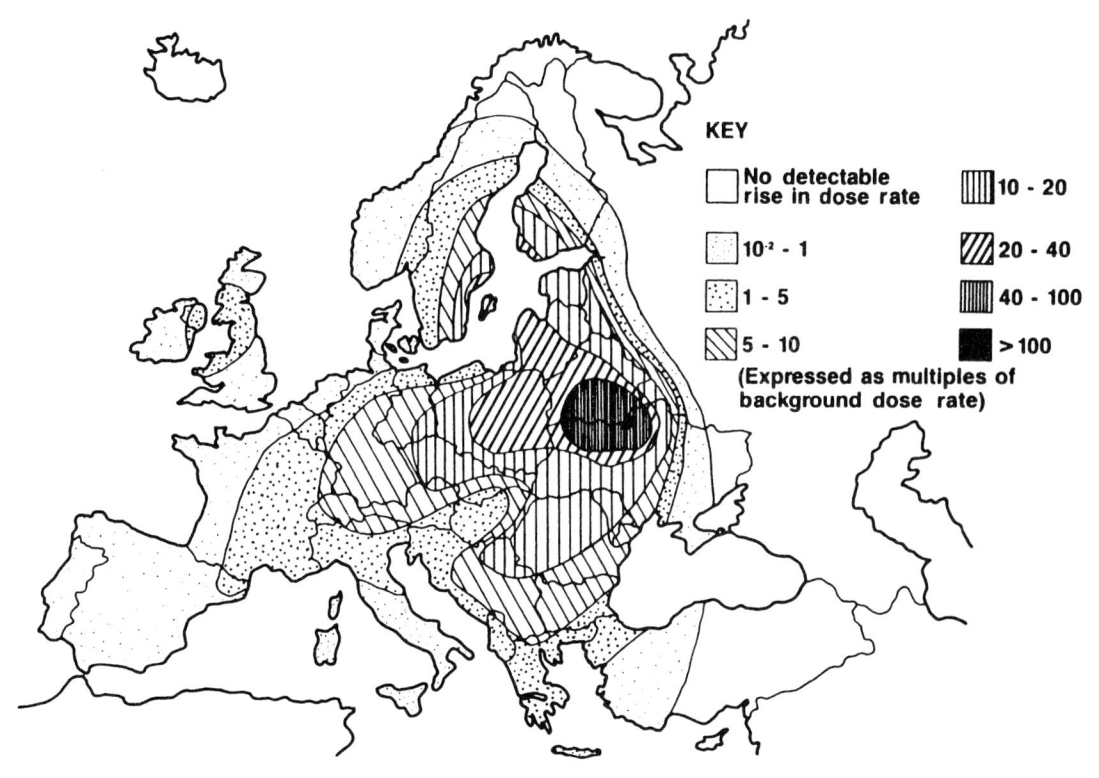

Figure A22.4   *Radiation dispersion pattern across Europe 3 May 1986 (Gittus et al., 1987)*

According to the Second *Watt Committee Report*, of the 4% of the core inventory dispersed, the distribution was approximately 0.3–0.5% on the reactor site, 1.5–2.0% within 20 km and 1.0–15% beyond 20 km. The most significant isotopes outside the USSR were I-131, Cs-134 and Cs-137.

In Europe the highest dose equivalents in the year after the accident were 0.76, 0.67 and 0.59 mSv in Bulgaria, Austria and Greece, respectively, while that in the United Kingdom was 0.03 mSv.

A22.9 The Later Aftermath

Of those on site at the time, two died on site and a further 29 in hospital over the next few weeks. A further 17 are permanent invalids and 57 returned to work but with seriously affected capacity. The others on site, some 200, were affected to varying degrees. There is apparently no detailed information on the effects on the military and civilian personnel brought in to deal with the accident.

Potential health effects at relatively low doses are discussed in Appendix 20. A detailed discussion of such effects from Chernobyl is given in the Second *Watt Committee Report*. For EC countries the estimates of the NRPB in 1986 were 100 fatal thyroid cancers and 2000 fatal general cancers, or 2100 fatal cancers. For the USSR the situation is much more complex, but the report quotes an estimate by Ryin *et al.* (1990) of 1240 fatal leukaemia cases and 38,000 fatal general cancers.

Table A22.4   *Some lessons of Chernobyl*

Management of, and safety culture in, major hazard installations
Adherence to safety-related instructions
Inherent safer design of plants
Sensitivity and operability of plants
Design of plant to minimize effect of violations
Disarming of protective systems
Planning and conduct of experimental work on plants
Accidents involving human error and their assessment
Emergency planning for large accidents
Mitigating features of accidents

Measures to reduce the hazard from Unit 4 reactor have continued long after the events described, involving devoted and heroic work to limit the harm to their fellows, not only in the USSR but elsewhere.

A22.10 Some Lessons of Chernobyl

Some of the lessons of Chernobyl apply specifically to the nuclear industry and these are not considered here, but the more important ones are equally applicable to the process industries. A list of some of these is given in Table A22.4 These lessons are now considered.

A22.10.1 Management of, and safety culture in, major hazard installations

The management of the organization at the Chernobyl plant were clearly inadequate for the operation of a major hazard installation.

The defects highlighted particularly in the foregoing account are a weak safety culture and overconfidence, a potentially lethal combination.

A22.10.2 Adherence to safety-related instructions

Closely linked to this is the lesson which most commentators have highlighted as the principal one, the need to adhere to safety-related instructions. At Chernobyl a number of such instructions were violated by the operators. These violations included disconnecting the ECCS and disabling two sets of trip systems.

A22.10.3 Inherently safer design of plants

The Chernobyl reactor had a low degree of inherent safety, due particularly to the positive temperature coefficient.

The reactor did have some features, however, which might be claimed to be inherently safer. Thus by avoiding the use of a single large pressure vessel the design eliminated the hazard of catastrophic rupture of such a vessel and by subdividing the fuel into individually cooled channels it reduced the risk of total core meltdown.

On the other hand in addition to the sensitivity inherent in the positive temperature coefficient, the design also involved massive graphite blocks which could burn.

Chernobyl thus also illustrates the fact that inherently safer design does not have a single dimension, but is multi-dimensional.

A22.10.4 Sensitivity and operability of plants

Closely related to inherently safer design is the sensitivity and operability of plants. The nuclear reactor at Chernobyl had a regime, that of low excess reactivity, at which it was close to instability and difficult to control. This is an undesirable characteristic in any plant. This fact is recognized in the recommendation made that the uranium fuel enrichment be increased to 2.4%.

A22.10.5 Design of plant to minimize effect of violations

At Chernobyl it was necessary to avoid operating the plant with a power output less than 20% of the rated value. Where there is a feature which is so critical, it is arguable that the plant should be designed to ensure that it cannot be violated. This is another aspect of inherently safer design.

The practicality of such an approach can only be decided in the light of the particular case. As Franklin argues, it is not practical to design against all conceivable violations.

A22.10.6 Disarming of protective systems

The disarming of protective systems is a permissible practice in certain cases, but these need to be well defined and there must be proper procedures for doing so. At Chernobyl, protective systems were disarmed which should not have been.

A22.10.7 Planning and conduct of experimental work on plants

When work is to be done on a plant, whether engineering work or experimental testing, it is necessary, particularly if the work involves modification to safety-related features such as the disarming of trips, to review the potential hazards and to have specific authorized arrangements for the safe operation of the plant.

Once the test is under way, unauthorized modifications should not be made to the equipment or to the test procedure itself which have potential safety implications.

A22.10.8 Accidents involving human error and their assessment

The Chernobyl disaster was caused by a series of actions by the operators of the plant. It appears to be a case of human error which is virtually impossible to foresee and prevent. No doubt the probability of any one of the events would have been assessed as low and that of their combination is virtually incredible. But there was a common factor, namely the determination to carry out the test.

There is a need for the development of techniques, both for design and for hazard assessment of plant, for identifying potentially hazardous operator intervention sequences.

A22.10.9 Emergency planning for large accidents

The scale of the accident at Chernobyl was such that the resources required to deal with the emergency exceeded those available locally. In the event, the authorities responded by mobilizing resources on a military scale. Large numbers of military and civilian personnel were drafted in to work on the reactor itself, on evacuation and on decontamination. The evacuation on 27 April required some 1000 buses.

A22.10.10 Mitigating features of accidents

In terms of the effect of the accident on the population near the site there were several factors which mitigated the effect of the accident at Chernobyl, although the full value of these was partially lost due to hesitation in carrying out the evacuation. These mitigating features are discussed by Hawkes *et al.* (1986).

The accident occurred at night so that fewer people were exposed on site and the population outside were mostly indoors. The intense fire created the conditions for the radioactive plume to rise high and carry the release away. There was little wind so that the plume was not prevented from rising and such wind as existed was away from the large centres of population to the south. The weather was dry so that the radioactivity was not washed down on the nearby population.

Appendix 23

Rasmussen Report

Contents

A comprehensive hazard assessment of nuclear power plants in general is the *Reactor Safety Study: An Assessment of Accident Risks in US Commercial Nuclear Power Plants* by the Nuclear Regulatory Commission (NRC) (1975). The work was done by a team led by Professor N.C. Rasmussen and is often referred to as the *Rasmussen Report*. It is also known as the *Reactor Safety Study* (RSS) and also as *WASH 1400*. It is referred to in this appendix as the *Reactor Safety Study* but elsewhere as the *Rasmussen Report*.This study was a major exercise involving some 70 man years of work and costing $4 million. The report is a document of nine volumes some 15 cm thick.

The RSS constituted a watershed in probabilistic risk assessment. It not only brought the PRA approach to the fore as an aid in decision-making, but created a framework and brought together the various techniques needed to carry out such a study.

The work is of interest in respect of its methodology, its treatment of particular problems in risk assessment, its compilations of failure data and its presentation and evaluation of risks. Also of interest are the critiques of the report.

The methodology is based on the extensive use of fault trees and event trees and addresses the problem of uncertainty in the results obtained.

The report has been followed by further reports which have developed the methodology. In particular, a treatment of overall PRA methodology is given in the *PRA Procedures Guide* of the NRC (1982) and of human factors methodology in the *Handbook of Human Reliability Analysis with Emphasis on Nuclear Power Plant Applications. Final Report* by Swain and Guttmann (1983).

Selected references on the *Reactor Safety Study* are given in Table A23.1.

A23.1 Earlier Studies

The *RSS*, or *WASH 1400*, was not the first study of this subject. There were several earlier reports on nuclear reactor safety.

The first of these was the report *Theoretical Possibilities and Consequences of Major Accidents in Large Nuclear Power Plants* by the Atomic Energy Commission (AEC) (1957). The work was done by the Brookhaven National Laboratory and is often referred to as the Brookhaven Report and also as *WASH 740*.

WASH 740 considered as a pessimistic accident scenario, or maximum credible accident, a situation where a reactor was nearing time for refuelling and thus contained

Table A23.1   *Selected references on the Reactor Safety Study*

Earlier studies
AEC (1957); Ford (1977)

Reactor safety study
AEC (1975)

Reviews and critiques
AEC (1975) (Appendix XI); EPRI (1975); Hocevar (1975); Leverenz and Erdmann (1975, 1979); Boffey (1976); EPA (1976); Yellin (1976); Erdmann *et al.* (1977); Ford Foundation and Mitre Corp. (1977); UCS (1977); H.W. Lewis (1978); S. Levine (1978, 1979); ANS (1980); Hanks (1986)

its largest inventory of fission products and suffered an accident involving loss from containment of half the core inventory in weather conditions which would carry the cloud towards a town of one million people 50 km away. It was estimated that this accident could result in 3,400 deaths, 43,000 injured and $7 billion property damage.

The report did not give an estimate of the frequency of such an event and stated in effect that the methodology to do so did not exist.

A study to update *WASH 740* was carried out by Brookhaven National Laboratory. The study was essentially complete by 1965 but was published only in 1973. It considered an even more pessimistic scenario of a large accident in a city and gave estimates of 45,000 deaths and 70,000 injuries.

Another report on reactor safety was *The Safety of Nuclear Power Reactors (Light Water Cooled) and Related Facilities* by the AEC (1973), or *WASH 1250*. This report included both accounts of policy and collections of data and contained some of the early results of WASH 1400.

A23.2 Risk Assessment Methodology

The report provides a framework for probabilistic risk assessment. The overall methodology described is based on selection of a set of release scenarios; estimation of the frequency of each scenario using fault trees and of the frequency of each associated set of outcomes using event trees; estimation of the consequences using hazard models for the physical phenomena and models of population exposure, mitigation of exposure by shelter and by escape and evacuation and injury from radioactivity and evaluation of these results using risk criteria. Table A23.2(A) indicates the subjects covered by the various volumes of the report and Table 23.2(B) some of the specific topics. Reference to the report is frequently made here, in particular with respect to (1) failure and event data; (2) lognormal distribution; (3) fault trees; (4) event trees; (5) common mode failure; (6) human error; (7) evacuation; (8) shelter; (9) FN curves and (10) risk evaluation.

The *RSS* gives risk assessments of both PWRs and BWRs, but the account given here is largely confined to the former.

A23.3 Event Data

The *RSS* contains a compilation of a large amount of failure and event data. These are based primarily on nuclear industry experience supplemented by other sources.

Most of the data refer to 2 years' experience at 17 nuclear plants which were operational in the United States in 1972. The average plant in the sample was some 4 years old.

Data for equipment are given as failure rates or probabilities of failure on demand.

It is assumed that the failures are random and that the exponential failure distribution is applicable. The possibility of failure due to ageing was considered but excluded. The report states:

It should be recognized that the study did not include extreme ageing consideration since the applicability of its results is limited to only the next five years.

The data on failure rates and probabilities are given not as point values but as ranges. Most of these data are presented in terms of the 90% confidence range and are

correlated using the log–normal distribution with values quoted for the lower and upper bounds, median and error factor. This use of the lognormal distribution reflects the large amount of uncertainty in the data so that for many items it was possible to give only order of magnitude values for the failure rate or probability.

It is convenient to use as one of the parameters which describe the log–normal distribution the median. The report therefore quotes median values for much of the data.

In addition to data on failures, data for test and repair times are also given. These times are also correlated using the log–normal distribution.

It is worth emphasizing that in the report the lognormal distribution is not used to correlate times to failure. These are assumed to be exponentially distributed so that they are described by a constant failure rate. The lognormal distribution is used to describe the spread of these failure rates. On the other hand the log–normal distribution is indeed used to characterize test and repair times.

There are special studies of pipework failure and power supply failure. Reactor pressure vessel failure is also considered.

Some of the failure and event data in the *RSS* are given in Appendix 14.

A23.4 Fault Trees

The *RSS* makes extensive use of fault trees. The need to do this arises because the accident events considered can occur only if there are failures of a number of protective systems and because such failures are rare and insufficient historical data for them are available.

The volume of the report which deals with the fault trees contains some 150 figures, most of them fault tree diagrams.

The computer codes PREP and KITT were used to calculate the frequencies of the top events from those of the base events.

The PWR function unavailabilities obtained from the fault tree analyses are summarized in Table A23.3.

A23.5 Event Trees

The basic accident sequence considered is an initiating event followed by system failure and then containment failure.

The *RSS* makes extensive use of event trees for the accident sequences.

A summary of the event trees is given in Table A23.4.

A23.6 Common Mode Failure

Common mode failures are recognized in the *RSS* as a factor which may greatly increase the frequency of the hazard.

Several techniques are used to handle such common mode failures. For similar events which may in some way be coupled, such as human actions to calibrate an instrument or open a valve, lower and upper bound probabilities of failure are defined. The lower bound probability p_l is the probability calculated assuming complete independence of the events. The upper bound probability p_u is the probability calculated assuming complete coupling between the events. The actual probability p is then taken as the geometric mean, or log–normal median

$$p = (p_l p_u)^{1/2} \qquad [\text{A23.6.1}]$$

The example given is the reclosing of two valves. The probability of failing to close one valve is taken as 10^{-2}. Then, the lower bound is 10^{-4} ($10^{-2} \times 10^{-2}$) and the upper bound 10^{-2} ($10^{-2} \times 1$), while the probability given by Equation A23.6.1 is 10^{-3} ($(10^{-4} \times 10^{-2})^{1/2}$).

The method of handling coupled events in a fault tree is described below.

Other common mode failures may occur as the result of a common event such as a fire or earthquake. The treatment of such external threats is considered below.

The *RSS* concludes that common mode failures do not make a large contribution to the overall frequency of core melt failure.

A23.7 Human Error

The report gives an extensive treatment of human error and estimates for error rates for a number of types of error. It also deals with problems such as non-independence of, or coupling between, errors.

The treatment of human error in the *RSS* and the *Handbook of Human Reliability Analysis* are described in Chapters 9 and 14 and some of their error rate estimates given.

A23.8 Rare Events

An important rare event considered in the *RSS* is failure of the reactor pressure vessel. The report states

Potentially large ruptures in the vessel were considered that could prevent effective cooling of the core by the ECCS. Since certain of these ruptures appeared to be capable of causing missiles (such as the reactor vessel head) with sufficient momentum to rupture the containment, this area was explored with some care. There is some small probability that a large vessel missile could in fact impact directly on the containment and penetrate through the wall. This type of rupture could involve a core meltdown in a non-intact containment.

A study by the Advisory Committee on Reactor Safeguards (ACRS) concluded that 'the disruptive failure probability of reactor vessels designed, constructed and operated according to code is even lower than 1×10^{-6} per vessel year.' The figure used in the *RSS* is 1×10^{-7}/vessel year.

The effect of this low estimate of reactor pressure vessel failure rate is to take this event out of consideration as a cause of release. The *RSS* states

Gross vessel rupture would have to be at least about 100 times more likely than the value estimated in order to contribute to the PWR core melt probability.

A23.9 External Threats

The *RSS* considers external threats such as fire, earthquakes and sabotage.

The prediction of earthquakes is highly uncertain and the report emphasizes this. It nevertheless states that a 'reasonable estimate (of core melt due to earthquake) is 10^{-7} per reactor year'.

The *RSS* acknowledges the threat posed by sabotage, but does not take this into account in the risk estimates made. These estimates apply to bona fide accidents and exclude those due to sabotage.

Table A23.3 *Reactor Safety Study: PWR function unavailabilities from trees (Nuclear Regulatory Commission, 1975)*

System	Unavailability (Q)			Major contributors (point estimates)			
	Q_{upper}	Q_{median}	Q_{lower}	Hardware	Test and maintenance	Human	Common mode
PWR systems							
Electric power	1.0×10^{-4}	1.0×10^{-5}	1.0×10^{-5}				
Reactor protection	1.0×10^{-4}	3.6×10^{-5}	1.3×10^{-5}	2.2×10^{-5}	1×10^{-5}		
Auxilliary feedwater							
0–8 h after small LOCA	3.0×10^{-4}	3.7×10^{-5}	7.0×10^{-6}	2.0×10^{-6}	3.2×10^{-6}		3×10^{-5}
8–24 h after small LOCA	2.6×10^{-3}	1.2×10^{-3}	5.4×10^{-4}	1.1×10^{-3}			
0–8 h without net	2.8×10^{-3}	4.5×10^{-4}	1.2×10^{-4}	2.0×10^{-6}	1.4×10^{-4}		1.1×10^{-4a}
Containment spray injection	7.8×10^{-3}	2.4×10^{-3}	1.0×10^{-3}	3.2×10^{-4}	1.5×10^{-4}		1.9×10^{-3a}
Consequence limiting control							
Hi-single train	7.2×10^{-2}	2.8×10^{-2}	1.4×10^{-2}	2.0×10^{-2}	2.1×10^{-3}	3×10^{-3}	1.0×10^{-3a}
Hi-both trains	4.8×10^{-3}	1.6×10^{-3}	4.8×10^{-4}	4.0×10^{-4}	8.4×10^{-5}		1.0×10^{-3a}
Hi-Hi single train	2.6×10^{-2}	1.2×10^{-2}	7.2×10^{-3}	4.9×10^{-3}	2.1×10^{-3}		1.0×10^{-3a}
Hi-Hi both trains	3.2×10^{-3}	1.1×10^{-3}	4.1×10^{-4}	6.5×10^{-5}	2.0×10^{-5}		1×10^{-3a}
Emergency coolant injection							
Accumulators	1.4×10^{-3}	9.5×10^{-4}	6.2×10^{-4}	4.9×10^{-4}	3.4×10^{-4}		
Low pressure injection	7.4×10^{-3}	4.7×10^{-3}	3.1×10^{-3}	3.2×10^{-4}	9.6×10^{-4}	2.5×10^{-3}	4.5×10^{-5a}
High pressure injection	2.7×10^{-2}	8.6×10^{-3}	4.4×10^{-3}	1.1×10^{-3}		7.0×10^{-4}	4.5×10^{-5a}
Safety injection control							
Single train	1.2×10^{-2}	5.8×10^{-3}	3.3×10^{-3}	3.0×10^{-3}	2.2×10^{-3}		4.5×10^{-5a}
Both trains	2.7×10^{-4}	9.9×10^{-5}	5.4×10^{-5}	9.0×10^{-6}	1.3×10^{-5}		4.5×10^{-5a}
Containment spray recirculation	9.0×10^{-4}	1.0×10^{-4}	2.5×10^{-5}	5.2×10^{-6}	4.3×10^{-5}		2.8×10^{-5a}
Containment heat removal	3.0×10^{-4}	8.5×10^{-5}	3.0×10^{-5}	6.4×10^{-5}			1.0×10^{-5a}
Low pressure recirculation	3.1×10^{-2}	1.3×10^{-2}	4.4×10^{-3}	2.7×10^{-3}	1.0×10^{-4}	1×10^{-5}	6.0×10^{-3a}
High pressure recirculation	2.2×10^{-2}	9.0×10^{-3}	4.3×10^{-3}	2.0×10^{-3}			6.0×10^{-3a}
Containment leakage							
Pressure reduced to $<\frac{1}{2}$psi[b]	7.5×10^{-4}	2.0×10^{-4}	8.4×10^{-5}	1.3×10^{-4}			
Pressure not reduced to $<\frac{1}{2}$psi	1.0×10^{-2}	2.0×10^{-3}	6.0×10^{-4}	8.8×10^{-4}			
Sodium hydroxide addition	1.1×10^{-2}	5.9×10^{-3}	3.6×10^{-3}	2.1×10^{-4}	4.8×10^{-3}		1.2×10^{-3a}
PWR functions							
Electrical power	1.0×10^{-4}	1.0×10^{-5}	1.0×10^{-6}				
Reactor protection	1.0×10^{-4}	3.6×10^{-5}	1.3×10^{-5}	2.7×10^{-5}	1.2×10^{-5}		
Auxilliary feedwater							
0–8 h after small LOCA	3.0×10^{-4}	3.7×10^{-5}	7.0×10^{-6}	2.0×10^{-6}	3.2×10^{-6}		3×10^{-5c}
8–24 h after small LOCA	2.6×10^{-3}	1.2×10^{-3}	5.3×10^{-4}	1.1×10^{-3}			
0–8 h without net	2.8×10^{-3}	4.5×10^{-4}	1.2×10^{-4}	2.0×10^{-6}	1.4×10^{-4}		1.1×10^{-4c}
Containment spray injection	7.8×10^{-3}	2.4×10^{-3}	1.0×10^{-3}	3.2×10^{-4}	1.5×10^{-4}		1.9×10^{-3c}
Emergency coolant injection							
Large LOCA	9.0×10^{-3}	5.6×10^{-3}	3.0×10^{-3}	NA	NA	NA	NA
Small LOCA (2–6 in. diam.)	3.0×10^{-2}	9.5×10^{-3}	4.0×10^{-3}	NA	NA	NA	NA
Small-small LOCA($\frac{1}{2}$–2 in. diam.)	2.7×10^{-2}	8.6×10^{-3}	4.4×10^{-3}	NA	NA	NA	NA
Containment spray recirculation	9.0×10^{-4}	1.0×10^{-4}	2.5×10^{-5}	5.2×10^{-6}	4.3×10^{-5}		2.8×10^{-5c}
Containment head removal	3.0×10^{-4}	8.5×10^{-5}	3.0×10^{-5}	6.4×10^{-5}			1.0×10^{-5c}
Emergency coolant recirculation							
Large LOCA	3.1×10^{-2}	1.3×10^{-2}	4.4×10^{-3}	2.7×10^{-3}	1.0×10^{-4}	1.0×10^{-5}	6.0×10^{-3c}
Small LOCA	2.2×10^{-2}	9.0×10^{-3}	4.3×10^{-3}	2.0×10^{-3}			6.0×10^{-3c}
Small-small LOCA		6.0×10^{-3d}		2.0×10^{-3}			3.0×10^{-3c}
Containment leakage							
Pressure reduced to $<\frac{1}{2}$psi	7.5×10^{-4}	2.0×10^{-4}	8.4×10^{-5}	1.3×10^{-4}			
Pressure not reduced to $<\frac{1}{2}$psi	1.0×10^{-2}	2.0×10^{-3}	6.0×10^{-4}	8.8×10^{-4}			
Sodium hydroxide addition	1.1×10^{-2}	5.9×10^{-3}	3.6×10^{-3}	2.1×10^{-4}	4.8×10^{-3}		1.2×10^{-3c}

[a] Also human error.

[b] Containment pressure must be reduced to $<\frac{1}{2}$ psi to isolate containment spray injection system.

[c] Common mode resulting from human error.

[d] Point estimate value obtained only.

NA, not applicable to combined systems.

Table A23.4 Reactor Safety Study: PWR accident sequences from event trees (Nuclear Regulatory Commission, 1975)

Symbol	Meaning
A	Intermediate to large LOCA
B	Failure of electric power to ESFs
B′	Failure to recover either on-site or off-site electric power within about 1 to 3 h following an initiating transient which is a loss of off-site a.c. power
C	Failure of the containment spray injection system
D	Failure of the emergency core cooling injection system
F	Failure of the containment spray recirculation system
G	Failure of the containment heat removal system
H	Failure of the emergency core cooling recirculation system
K	Failure of the reactor protection system
L	Failure of the secondary system steam relief valves and the auxiliary valves and the auxiliary feedwater system
M	Failure of the secondary system steam relief valves and the power conversion system
Q	Failure of the primary system safety relief valves to reclose after opening
R	Massive rupture of the reactor vessel
S_1	A small LOCA with an equivalent diameter of about 5–15 cm
S_{12}	A small LOCA with an equivalent diameter of about 1–5 cm
T	Transient event
V	Failure of low-pressure injection system check valve
α	Containment rupture due to a reactor vessel steam explosion
β	Containment failure resulting from inadequate isolation of containment openings and penetrations
γ	Containment failure due to hydrogen burning
δ	Containment failure due to overpressure
ε	Containment vessel melt-through

Sequence designation	Event tree failure(s)	Containment event tree failure
PWR sequence		
1. A	Large rupture only	None
2. A-β	Large rupture only	Containment leakage
3. AH-α	ECCS recirculation	Vessel steam explosion
4. AH-β	ECCS recirculation	Containment leakage
5. AH-ε	ECCS recirculation	Melt-through
6. AHI-α	ECCS recirculation plus sodium hydroxide	Vessel steam explosion
7. AHI-β	ECCS recirculation plus sodium hydroxide	Containment leakage
8. AHI-ε	ECCS recirculation plus sodium hydroxide	Melt-through
9. AG-δ	Containment heat removal	Overpressure
10. AHG-δ	ECCS recirculation plus containment heat removal	Overpressure
11. AHG-ε	ECCS recirculation plus containment heat removal	Melt-through
12. AHF-α	ECCS recirculation plus containment spray recirculation	Vessel steam explosion
13. AHF-β	ECCS recirculation plus containment spray recirculation	Containment leakage
14. AHF-δ	ECCS recirculation plus containment spray recirculation	Overpressure
15. AHF-ε	ECCS recirculation plus containment spray recirculation	Melt-through
16. ADI-α	ECCS injection	Vessel steam explosion
17. AD-β	ECCS injection	Containment leakage
18. AD-ε	ECCS injection	Melt-through
19. ADI-α	ECCS injection plus sodium hydroxide	Vessel steam explosion
20. ADI-ε	ECCS injection plus sodium hydroxide	Melt-through
21. ADG-ε	ECCS injection plus containment heat removal	Melt-through
22. ADGI-ε	ECCS injection plus containment heat removal plus sodium hydroxide	Melt-through
23. ADF-β	ECCS injection plus containment spray recirculation	Containment leakage
24. ADF-ε	ECCS injection plus containment spray recirculation	Melt-through
25. ACD-β	Containment spray injection plus ECCS injection	Containment leakage
26. ACD-ε	Containment spray injection plus ECCS injection	Melt-through
27. ACDGI-α	All except electric power and containment spray recirculation	Vessel steam explosion
28. ACDGI-β	All except electric power and containment spray recirculation	Containment leakage
29. ACDGI-δ	All except electric power and containment spray recirculation	Overpressure
30. ACDGI-ε	All except electric power and containment spray recirculation	Melt-through
31. AB-α	Electric power	Vessel steam explosion
32. AB-γ	Electric power	Hydrogen combustion
33. AB-ε	Electric power	Melt-through

Table A23.4   *(continued)*

34. S_2C-α	Small LOCA plus containment spray injection	Vessel steam explosion
35. S_2C-δ	Small LOCA plus containment spray injection	Overpressure
36. $TMLB'$-γ	Transient plus feedwater plus electric power	Hydrogen combustion
37. $TMLB'$-δ	Transient plus feedwater plus electric power	Overpressure
38. $TMLB'$-α	Transient plus feedwater plus electric power	Vessel steam explosion

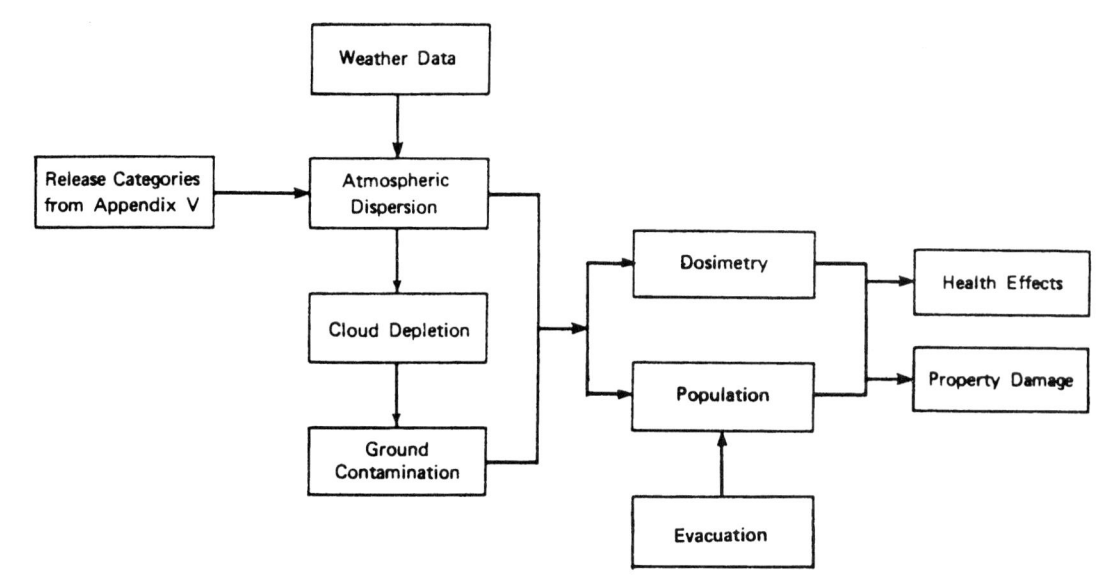

Figure A23.1   *Reactor Safety Study: Schematic outline of consequence models (Nuclear Regulatory Commission 1975)*

A23.10 Release Scenarios

The consequences of a core degradation or meltdown depend on (1) core inventory; (2) fraction released; (3) dispersion and deposition; (4) population exposed; (5) mitigating features (evacuation, shelter); and (6) injury relations.

A schematic outline of the consequence models is shown in Figure A23.1.

The initial radioactivity of the radionuclides in the reactor core is given in Table A23.5 and the reactor core inventory in Table A23.6.

The release scenarios are shown in Tables A23.7 and A23.8.

The release from the reactor is used in the report as the source term for a plume gas dispersion model. This model takes into account the deposition of solid radionuclides.

A23.11 Population Characteristics

The population at risk was characterized by a composite model based on the 68 actual sites holding the first 100 reactors. Six composite sites of different types were defined.

The population density was determined as follows. The first type of site was an Atlantic coastal site and 14 of the reactors were assigned to this composite site. The actual population around each of these reactors was determined from census data for a distance of 50 miles. For each actual

site 16 sectors were defined, making 224 sectors in total. These 224 sectors were then ranked in order of population density and 16 representative sectors were determined for the composite site.

Data from a time use study by J.P. Robinson and Converse (1966) were used to determine the fraction of time spent at different locations and hence the probability that a person was inside shelter.

A23.12 Mitigation of Exposure

Factors which mitigate the exposure of the population include shelter and evacuation.

The *RSS* uses a single exponential stage evacuation model. This model is based on a study of actual evacuations by Hans and Sell (1974).

Measures to mitigate long-term exposure include interdiction of the contaminated zone and decontamination of this zone.

The effect of shelter on concentration is modelled using a single exponential stage model for ventilation.

A23.13 Injury Relations

The effects of radioactivity on people are complex and are both short-term, or prompt, and long-term.

Table A23.5   *Reactor Safety Study: initial activity of radionuclides in reactor core (Nuclear Regulatory Commission, 1975)*

No.	Radionuclide	Radio inventory source (curies $\times 10^{-8}$)	Half-life (days)
1	Cobalt-58	0.0078	71.0
2	Cobalt-60	0.0029	1920
3	Krypton-85	0.0056	3950
4	Krypton-85m	0.24	0.183
5	Krypton-87	0.47	0.0528
6	Krypton-88	0.68	0.117
7	Rubidium-86	0.00026	18.7
8	Strontium-89	0.94	52.1
9	Strontium-90	0.037	11,030
10	Strontium-91	1.1	0.403
11	Yttrium-90	0.039	2.67
12	Yttrium-91	1.2	59.0
13	Zirconium-95	1.5	65.2
14	Zirconium-97	1.5	0.71
15	Niobium-95	1.5	35.0
16	Molybdenum-99	1.6	2.8
17	Technetium-99m	1.4	0.25
18	Ruthenium-103	1.1	39.5
19	Ruthenium-105	0.72	0.185
20	Ruthenium-106	0.25	366
21	Rhodium-105	0.49	1.5
22	Tellurium-127	0.059	0.391
23	Tellurium-127m	0.011	109
24	Tellurium-129	0.31	0.048
25	Tellurium-129m	0.053	0.340
26	Tellurium-131m	0.13	1.25
27	Tellurium-132	1.2	3.25
28	Antimony-127	0.061	3.88
29	Antimony-129	0.33	0.179
30	Iodine-131	0.85	8.05
31	Iodine-132	1.2	0.0958
32	Iodine-133	1.7	0.875
33	Iodine-134	1.9	0.0366
34	Iodine-135	1.5	0.280
35	Xenon-133	1.7	5.28
36	Xenon-135	0.34	0.384
37	Cesium-134	0.075	750
38	Cesium-136	0.030	13.0
39	Cesium-137	0.047	11,000
40	Barium-140	1.6	12.8
41	Lanthanum-140	1.6	1.67
42	Cerium-141	1.5	32.3
43	Cerium-143	1.3	1.38
44	Cerium-144	0.85	284
45	Praseodymium-143	1.3	13.7
46	Neodymium-147	0.60	11.1
47	Neptunium-239	16.4	2.35
48	Plutonium-238	0.00057	32,500
49	Plutonium-239	0.00021	8.9×10^6
50	Plutonium-240	0.00021	2.4×10^6
51	Plutonium-241	0.034	5350
52	Americium-241	0.000017	1.5×10^5
53	Curium-242	0.0050	163
54	Curium-244	0.00023	6630

Table A23.6   *Reactor Safety Study: reactor core inventory (Nuclear Regulatory Commission, 1975)*

Location	Total inventory (curies)			Fraction of core inventory		
	Fuel	*Gap*	*Total*	*Fuel*	*Gap*	*Total*
Core[a]	8.0×10^9	1.4×10^8	8.1×10^9	9.8×10^{-1}	1.8×10^{-2}	1
Spent fuel storage pool (max.)[b]	1.3×10^9	1.3×10^7	1.3×10^9	1.6×10^{-1}	1.6×10^{-3}	1.6×10^{-1}
Spent fuel storage pool (av.)[c]	3.6×10^8	3.8×10^6	3.6×10^8	4.5×10^{-2}	4.8×10^{-4}	4.5×10^{-2}
Shipping cask[d]	2.2×10^7	3.1×10^5	2.2×10^7	2.7×10^{-3}	3.8×10^{-5}	2.7×10^{-3}
Refuelling[e]	2.2×10^7	2×10^5	2.2×10^7	2.7×10^{-3}	2.5×10^{-5}	2.7×10^{-3}
Waste gas storage tank	–	–	9.3×10^{-4}	–	–	1.2×10^{-5}
Liquid waste storage tank	–	–	9.5×10^1	–	–	1.2×10^{-8}

[a] Core inventory based on activity 1/2 h after shutdown.
[b] Inventory of 2/3 core loading; 1/3 core with 3 day decay and 1/3 core with 150 day decay.
[c] Inventory of 1/2 core loading; 1/6 core with 150 day decay and 1/3 core with 60 day decay.
[d] Inventory based on 7 PWR or 17 BWR fuel assemblies with 150 day decay.
[e] Inventory for one fuel assembly with 3 day decay.

In order to determine these effects it is first necessary to estimate the doses of radioactivity received and then to apply a dose−response relation.

The *RSS* gives a detailed treatment of all these aspects and the report contains a large number of tables and graphs showing early and late fatalities and injuries due to different modes of injury and the contribution to these of individual radionuclides.

It also deals with the contamination of land.

A23.14 Uncertainty in Results

An attempt is made in the report to put bounds on the errors of the estimates.

A number of methods are used. Failure rates and probabilities are taken not as single values but as a range of values with a lognormal distribution, as described above.

The inputs to the fault trees are therefore not single values of frequency or probability but distributions of these. The frequency or probability of the top event is then computed, also as a distribution, by sampling from the distributions of these individual events using Monte Carlo simulation.

Such simulation may also be used to take account of common mode failures by arranging coupling between the failure rates or probabilities in question.

A23.15 Presentation of Results

The risks from reactor accidents are presented in the main report principally in the form of tables of frequency−consequence results.

The results for 100 reactors are for the mix of both PWR and BRW reactors considered in the study. The risks for an individual PWR are higher than for a BWR and those for a composite single reactor based on the mix are close to those for a PWR.

Table A23.9(A) shows the societal risk of early effects from one reactor and Table A23.9(B) of those from late effects. The report gives another table showing the societal risk from 100 reactors, the frequencies simply being greater by a factor of 100. Table A23.10 shows the individual and average societal risk from 100 reactors.

The results are also given in graphical form. Figure A23.2 gives a histogram of PWR release categories. Figure A23.3

is the frequency−consequence curve, or FN curve, for early fatalities from a single reactor. The graph shows curves for a PWR, a BWR and the composite reactor.

Figure A23.4 is the frequency−consequence curve for areas in which relocation of people and decontamination would be required.

A23.16 Evaluation of Results

The *RSS* evaluates the risk from the 100 nuclear reactors by making comparisons with those from other hazards.

The main report gives numerous tables showing risks from various natural and man-made hazards such as earthquakes, hurricanes, tornadoes, meteorites, fires, dams, explosions, toxic gas releases and aircraft crashes.

One of the principle comparisons made is the FN curve for the 100 reactors and for other hazards. Figures A23.5 and Figure A23.6 show, respectively, the FN curve for early deaths for natural hazards and for man-made hazards together with the curve for 100 reactors.

The overall conclusion of the report is that the risks to the public from 100 nuclear reactors are very much less than those from other natural and man-made hazards.

The report has a slim Executive Summary which describes the report as a whole and gives a presentation and evaluation of the results. This summary does not include the societal risk estimates given in Table A23.9 and in general tends both to play down the risks and gloss over many qualifying assumptions.

A23.17 Browns Ferry Incident

The incident at Browns Ferry is widely regarded as a near miss and the *RSS* could hardly avoid reference to this incident.

On 22 March 1975 a fire started in the electrical control cables at the plant. Following a plant modification a candle was used to detect air leaks at a point where a set of electrical cables passed through a wall. The fire burned out of control for $7\frac{1}{2}$ hours and badly damaged 1600 cables of which 618 were cables related to the safety systems.

The power interruption caused disturbances on the reactors and the operators, hindered by dense smoke, had to struggle to bring them back under control. Unit 2 was shut-down and stabilized without great difficulty but Unit 1

Table A23.7 Reactor Safety Study: PWR release scenarios ±1 (Nuclear Regulatory Commission, 1975)

Accident category and frequency f_j	Description of accident category	Duration of release (h)	Elevation of release (m)	Energy release rate (Btu/h)	Fraction of core inventory released						
					$Xe-Kr$	I	$Cs=Rb$	$Te-Sb$	$Ba-Sr$	Ru, Rh, Co, Mo, Tc	$Y, La, Zr, Nb, Ce, Pr, Nd, Np, Pu, Am, Cm$
PWR1a and 1b 9×10^{-7} per year	Core melt-down and failure of containment spray and heat removal systems 1a – steam explosion after containment failure due to overpressure; 1b – steam explosion ruptures containment	0.5	25	20×10^6(1a) 520×10^6(1b)	0.9	0.7	0.4	0.4	0.05	0.4	3×10^{-3}
PWR2 8×10^{-6} per year	Core meltdown, failure of containment spray and heat removal systems. Containment fails through overpressure after commencement of core meltdown	0.5	0	170×10^6	0.9	0.7	0.5	0.3	0.06	0.02	4×10^{-3}
PWR3 4×10^{-6} per year	Failure of containment due to overpressure before core meltdown	1.5	0	6×10^6	0.8	0.2	0.2	0.3	0.02	0.03	3×10^{-3}

	Description										
PWR4 5×10^{-7} per year	Core meltdown and failure of containment system properly to isolate; failure of containment spray system	3.0	0	1×10^6	0.6	0.09	0.04	0.03	5×10^{-3}	3×10^{-3}	4×10^{-4}
PWR5 7×10^{-7} per year	As PWR4, but containment spray systems operate to reduce release to atmosphere	4	0	3×10^{-5}	0.3	0.03	9×10^{-3}	5×10^{-3}	10^{-3}	6×10^{-4}	7×10^{-5}
PWR6 6×10^{-6} per year	Core melt down, failure of containment sprays. Containment maintains integrity but core melts through base	10.0	0	0	0.03	8×10^{-4}	8×10^{-4}	9×10^{-8}	9×10^{-5}	7×10^{-5}	10^{-5}
PWR7 4×10^{-5} per year	As PWR6, but containment sprays operate	10.0	0	0	6×10^{-3}	2×10^{-5}	10^{-5}	2×10^{-5}	10^{-6}	10^{-6}	2×10^{-7}
PWR8 4×10^{-5} per year	Large pipe break, containment fails properly to isolate, all other engineered safeguards work, no core meltdown	0.5	0	0	2×10^{-3}	10^{-4}	5×10^{-4}	10^{-6}	10^{-8}	0	0
PWR9 4×10^{-4} per year	As PWR8, but all engineered safeguards function as designed. Essentially PWR design basis accident	0.5	0	0	3×10^{-6}	10^{-7}	6×10^{-7}	10^{-9}	10^{-11}	0	0

Table A23.8 Reactor Safety Study: PWR release scenarios – 2 (Nuclear Regulatory Commission, 1975)

Fission product	Gap release fraction	Meltdown release fraction	Vaporization release fraction[a]	Steam explosion fraction[b]
Xe, Kr	0.030	0.870	0.100	(X)(Y) 0.90
I, Br	0.017	0.883	0.100	(X)(Y) 0.90
Ca, Rb	0.050	0.760	0.190	–
Te[c]	0.0001	0.150	0.850	(X)(Y)(0.60)
Sr, Ba	0.000001	0.100	0.010	–
Ru[d]	–	0.030	0.050	(X)(Y)(0.90)
La[e]	–	0.003	0.010	–

[a] Exponential loss over 2 h with half-time of 30 min. If a steam explosion occurs prior to this, only the core fraction not involved in the steam explosion can experience vaporization.
[b] X = fraction of core involved in the steam explosion. Y = fraction of inventory remaining for release by oxidation.
[c] Includes Se, Sb.
[d] Includes Mo, Pd, Rh, Tc.
[e] Includes Nd, Eu, Y, Ce, Pr, Pm, Sm, Np, Pu, Zr, Nb.

Table A23.9 Reactor Safety Study: societal risk from one reactor (Nuclear Regulatory Commission, 1975)

A Early effects

Chance per reactor-year	Consequences				
	Early fatalities	Early illness	Total property damage (10^9)	Decontamination area (≈square miles)	Relocation area (square miles)
One in 20,000[a]	<1.0	<1.0	<0.1	<0.1	<0.1
One in 1,000,000	<1.0	300	0.9	2,000	130
One in 10,000,000	110	3,000	3	3,200	250
One in 100,000,000	900	14,000	8	–	290
One in 1,000,000,000	3300	45,000	14	–	–

B Cancer and genetic effects

Chance per reactor-year	Consequences		
	Latent cancer[b] fatalities (per year)	Thyroid nodules[b] (per year)	Genetic effects[c] (per year)
One in 20,000[d]	<1.0	<1.0	<1.0
One in 1,000,000	170	1,400	25
One in 10,000,000	460	3,500	60
One in 100,000,000	860	6,000	110
One in 1000,000,000	1,500	8,000	170
Nuclear incidence	17,000	8,000	8,000

[a] This is the predicted chance of core melt per reactor year.
[b] This rate would occur approximately in the 10–40 year period following a potential accident.
[c] This rate would apply to the first generation born after a potential accident. Subsequent generations would experience effects at a lower rate.
[d] This is the predicted chance of core melt per reactor year.

was out of control for some hours and was stabilized only after the fire was put out. Control of Unit 1 was only achieved by the use of equipment which was not part of the safety system.

The fire disabled all five emergency core cooling systems in Unit 1.

At the point when control was lost 7 out of 11 of relief valves were failed. The fire disabled the other four. The *RSS* gives an analysis of the probability that these four valves would in fact be disabled in a fire without any of the valves, from the set of seven or the set of four, being repaired over the duration of the fire. The probability estimated was 0.003.

A23.18 Critical Assumptions

The results obtained in the *RSS* depend on the validity of a number of assumptions.

Some of the more critical assumptions are

plant standards are not seriously below average;
failures are random rather than wearout;
reactor pressure vessel operating conditions are closely controlled;
site is not subject to unusually severe earthquakes;
population density around site is not unusually high;
emergency plans exist and are exercised;

Table A23.10 *Reactor Safety Study: individual and average societal risk from 100 reactors (Nuclear Regulatory Commission, 1975)*[a]

Consequence	Societal	Individual
Early fatalities[b]	3×10^{-3}	2×10^{-10}
Early illness[b]	2×10^{-1}	1×10^{-8}
Latent cancer fatalities[c]	7×10^{-2} per year	3×10^{-10} per year
Thyroid nodules[c]	7×10^{-1} per year	3×10^{-9} per year
Genetic effects[d]	1×10^{-2} per year	7×10^{-11} per year
Property damage ($)	2×10^{6}	–

[a] Based on 100 reactors at 68 current sites.
[b] The individual risk value is based on the 15 million people living in the general vicinity of the first 100 nuclear power plants.
[c] This value is the rate of occurrence per year for about a 30 year period following a potential accident. The individual rate is based on the total US population.
[d] This value is the rate of occurrence per year for the first generation born after a potential accident: subsequent generations would experience effects at a lower rate. The individual rate is based on the total US population.

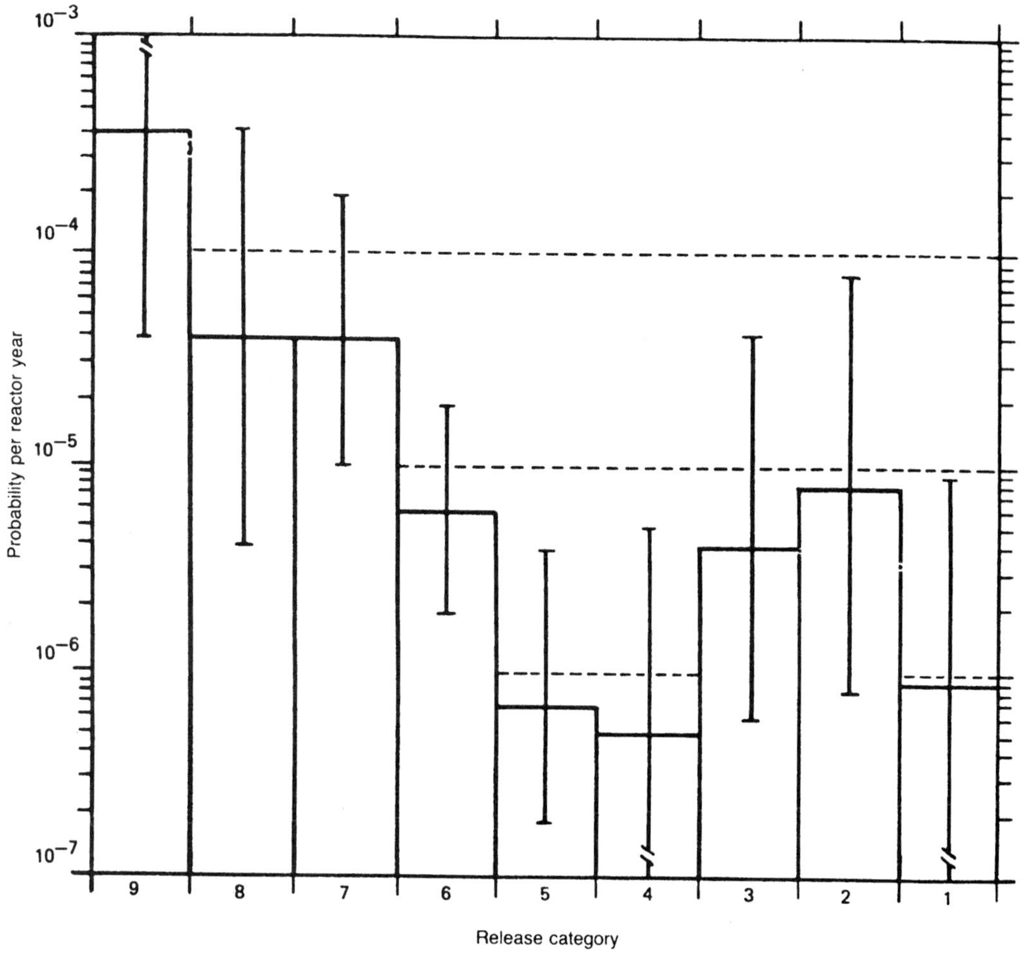

Figure A23.2 *Reactor Safety Study: histogram for frequency of PWR release categories (Nuclear Regulatory Commission, 1975)*

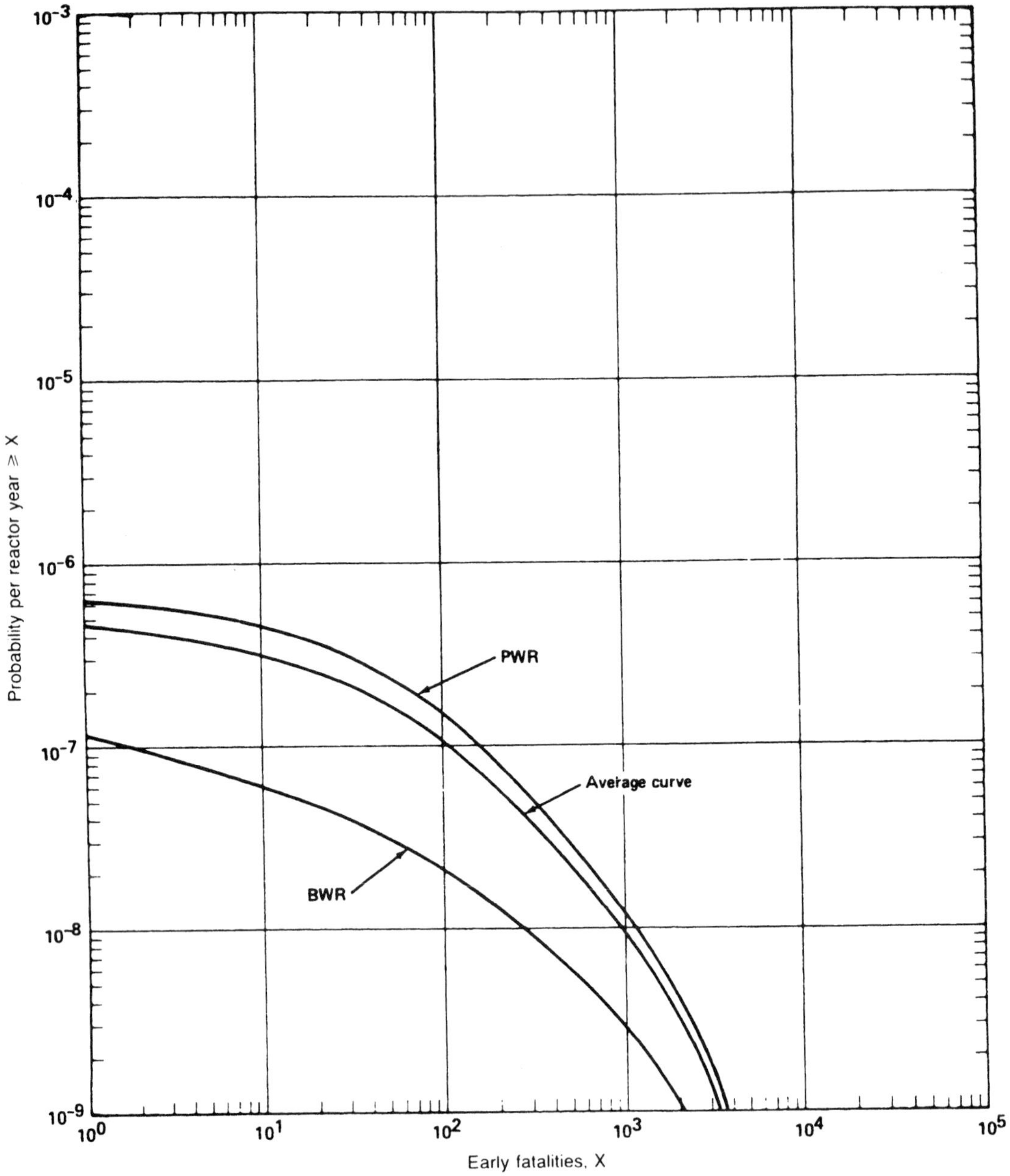

Figure A23.3   *Reactor Safety Study: FN curve for early fatalities for one reactor (Nuclear Regulatory Commission, 1975)*

medical facilities exist to handle large numbers of injured; and also

sabotage is not considered.

A23.19 Critiques

The *RSS* has been the subject of much comment and criticism. Three principal critiques are the comments included as Appendix XI of the report itself, the *Risk Assessment Review Group Report* (H.K. Lewis, 1978) and *The Risks of Nuclear Power Reactors* by the Union of Concerned Scientists (UCS) (1977).

There are also reviews by the EPA (1976) and the American Nuclear Society (1980) and a series of reports by the Electric Power Research Institute (EPRI) (EPRI, 1975; Leverenz and Erdmann, 1975, 1979; Erdmann *et al.*, 1977).

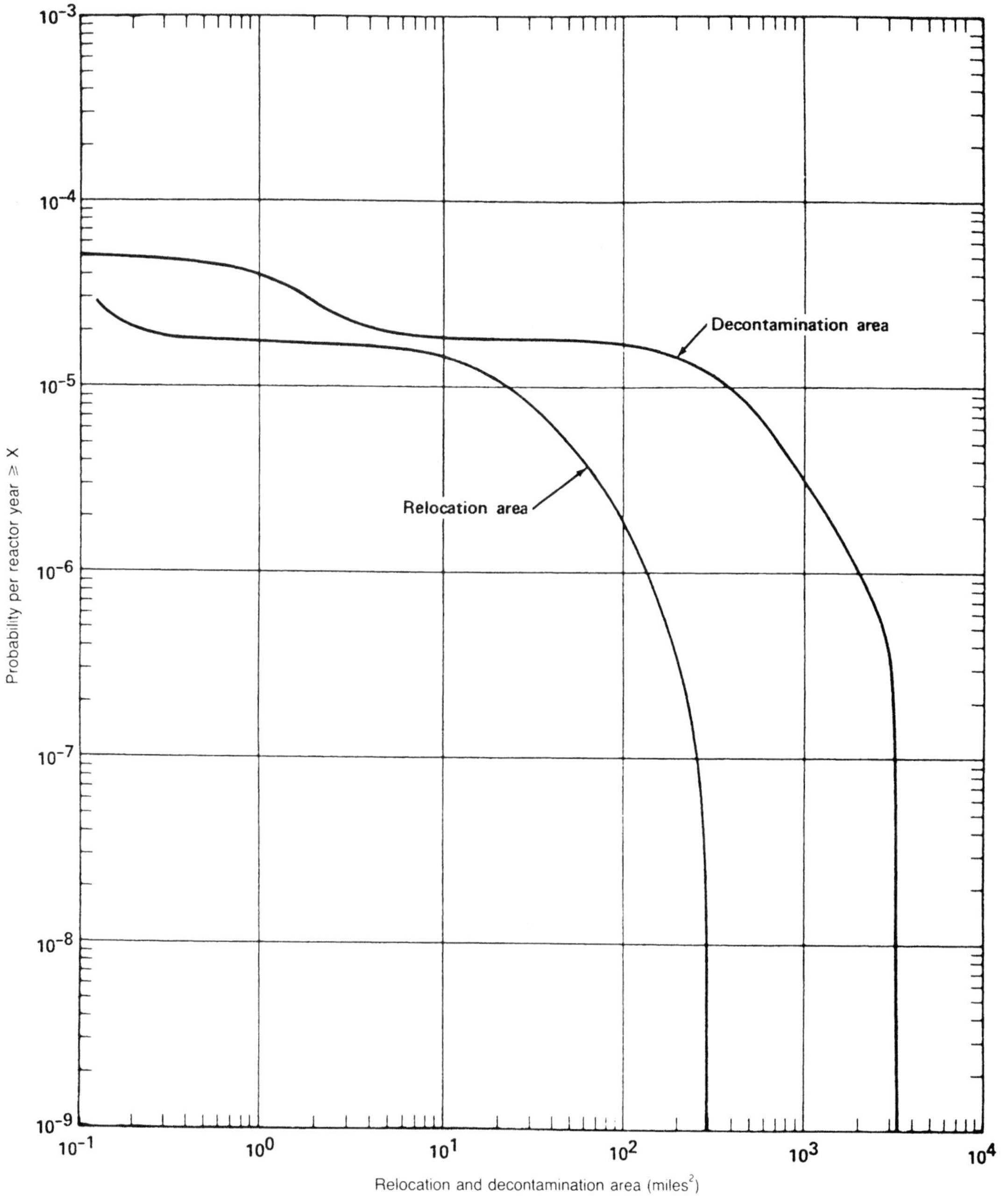

Figure A23.4 *Reactor Safety Study: frequency–consequence curve for areas requiring relocation and decontamination for one reactor (Nuclear Regulatory Commission, 1975)*

Some of the principal topics addressed in the critiques are indicated in Table A23.11.

Some of the comments made in the *Review Group Report* and the UCS review are now described.

A23.19.1 Review Group Report

The review found the *RSS* 'inscrutable'. It was difficult to follow the detailed thread of any calculation and difficult therefore to carry out a peer review.

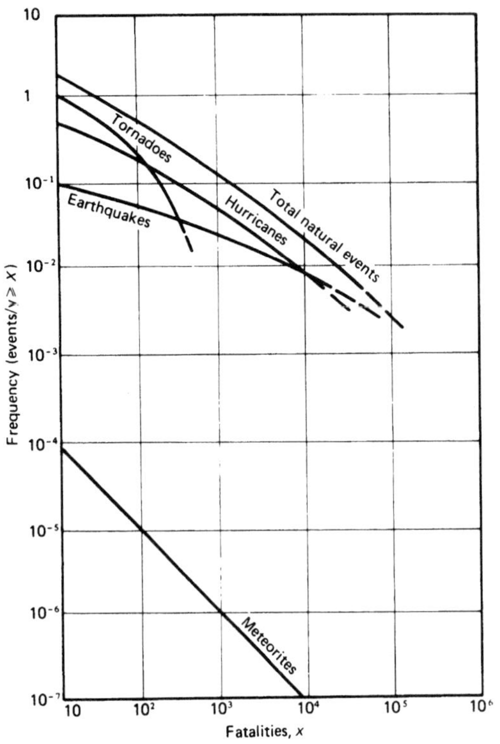

Figure A23.5 *Reactor Safety Study: FN curves for early fatalities for 100 reactors compared with natural hazards (Nuclear Regulatory Commission, 1975)*

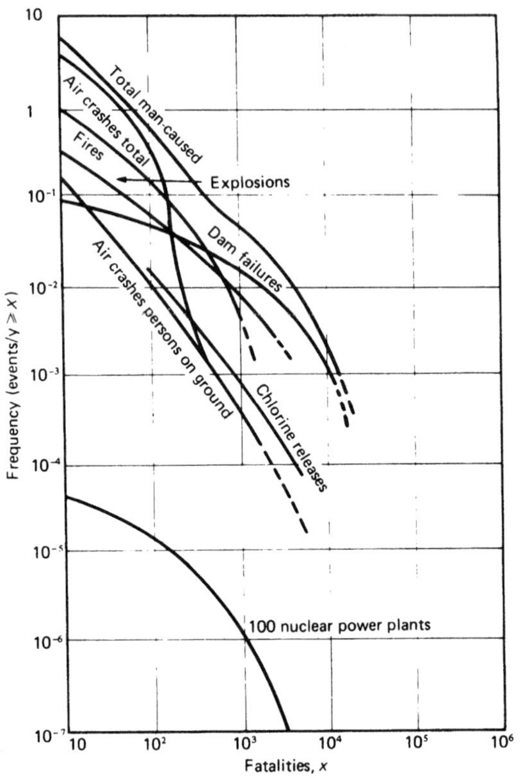

Figure A23.6 *Reactor Safety Study. FN curves for early fatalities for 100 reactors compared with man-made hazards (Nuclear Regulatory Commission, 1975). Note: car accidents caused at the time of the report about 50,000 deaths per year in the United states. These fatalities are not shown because data for large multiple-fatality accidents are not available*

The group was also critical of the Executive Summary which they felt did not give a good summary of the findings of the Main Report and was thus misleading.

The review is nevertheless sympathetic in principle to the attempt to quantify the frequency of events using fault tree and event tree methods and the consequences using hazard and injury models.

It states that it is incorrect to say that the use of fault trees and event trees is fundamentally flawed: it is just an implementation of logic.

On the other hand the practical application of the methodology involves various difficulties, including the availability of data, the statistical treatment of the data and the selection of appropriate models.

The group draws attention to the fact that the methodology is based on converting situations which are essentially continuous into discrete events and that this involves quite drastic simplification.

It is suggested that the problem of data is such that in some areas it is not appropriate to give absolute risk values but only bounding values.

The report criticizes the study for not always distinguishing clearly between different sources of values quoted: historical data, subjective judgements and models.

The review is sympathetic in principle to the use of values determined from expert judgement. Such judgements can be rather accurate as studies of betting demonstrate. It is necessary, however, to stay within the domain of

experience of the expert and not require him to make estimates of events of which he has no experience.

The log–normal distribution is used to fit much of the data. The group are somewhat critical of some of the fits obtained, particularly for sparse data, but they accept the use of this distribution. With regard to the use of the median rather than the mean of this distribution, they state that this is more a matter of presentation than of substance and caused people to infer incorrectly that the study was predicting a lower failure rate than it actually did.

They are much more critical of the use made of the geometric mean, or square root bounding, and quote as an example the estimation of the probability p of failure of three adjacent BWR control rods. This probability is calculated as described above using Equation A23.6.1, where the values of the bound probabilities are the following functions of the failure probability p_x ($= 10^{-4}$) of a single rod

$$p_i = p_x^3 = 10^{-12}$$

$$p_u = 0.01 p_x = 10^{-6}$$

so that

$$p = 10^{-9}$$

An alternative model is the arithmetic mean which gives

$$p = 10^{-6}$$

There is thus a large difference.

The review devotes considerable attention to the problem of common mode failures. Such failures can cause individual equipment failures, can compromise redundant systems and can activate low probability accident sequences. The principal causes of such failure listed by the group are power supply failure, human error and external threats such as fire, explosion, earthquake and sabotage. They were unconvinced that the problem had been fully dealt with in the study and suggest that the best way to handle such events is to mount a defence against them at the start and so eliminate their effects.

It is suggested in the report that the treatment of earthquakes is inadequate. Attention is drawn to the work of Hsieh and Okrent (1976) showing that design and construction errors can greatly reduce the safety factors thought to be present in a design.

Another problem discussed at some length in the review is human factors and human error. The report expresses some disappointment that there were few comments on the human factors aspects of the study. Human factors affect the study in that human error can initiate an accident, can disable protective systems and can cause escalation, while on the other hand human action can mitigate an accident situation. The report recognizes that the human factors aspects of the study represent pioneering work and were well executed, but draws attention to the lack of data. The main thrust of the group's comments, however, is to the effect that the study probably gives too little credit for the effectiveness of human action in averting an accident.

The group identified various sources of uncertainty in the work, some biased towards conservatism and some to nonconservatism. Among the former factors were pervasive regulatory influence towards the use of certain parameter values and reluctance to claim full credit for effective human adaptability during accidents. Among the latter factors were questions of completeness and common-cause failure. They were therefore unable to determine whether the accident frequencies estimated were high or low, but were convinced that the error bounds had been greatly underestimated.

The report discusses the problem of the completeness of the identification of the hazards. The probability of core melt predicted in the study is two orders of magnitude less than the upper bound value which may be estimated from the number of melt-free reactor years accumulated. The report suggests that if there are unidentified accident sequences which can lead to core melt, some precursor events in these sequences should probably be occurring, although they may remain unrecognized. They suggest it is important to subject potentially significant sequences and precursors, as they occur, to the kind of analysis given in the study.

The report suggests that the results of the study cannot be applied directly to particular reactor sites. Application to a specific site would require more detailed study of the site characteristics such as topology and meteorology.

The report gives an account of the Brown's Ferry incident.

A23.19.2 Union of Concerned Scientists report

The review of the RSS by the UCS is more critical.

In the view of the UCS publicly available documents on the study suggest that it was not sufficiently independent of the nuclear industry, that its results were largely predetermined, that critical reviews were suppressed and that certain sensitive issues were avoided.

The report starts with a fundamental critique of the basic methodology. It reviews experience of fault tree analysis in the aerospace industry. It quotes as an example the use of the technique to estimate the reliability of the Apollo Service Propulsion System or SPS engine. One of the authors of the UCS report, W.M. Bryan, was in charge of reliability assessment during the testing of this engine. The estimated failure probability of the engine based on fault tree analysis was 10^{-4} while that estimated after testing was 4×10^{-3} so that the theoretical analysis gave an underestimate by a factor of 40.

The authors state that fault tree analysis for Apollo also failed to assure completeness of hazard identification. Many failures in the programme resulted from events which had not been identified as 'credible' and came as complete surprises. Some 20% of ground test failures and more than 35% of in-flight failures were not identified as credible prior to their occurrence.

The report states

Fault trees, mathematical models and event trees were developed over 20 years ago and were [later] discarded . . . by NASA and . . . by the AEC as viable tools for estimating reliability/safety quantitative values.

It lists as techniques which are used probabilistic design analysis, propagation of error techniques, reliability estimation from small samples, malfunction simulation models and others.

The methodology used in the study is to synthesize accident sequences from failure data for single items. The UCS suggest that this approach fails to utilize information which is available on multiple failure accident sequences which have actually happened.

The review draws attention to the database used in the RSS, which was described above. The average age of the plants used in the study was 4 years. Two difficulties which arise from this are that often data collection was not well established and that failures due to ageing would scarcely have begun to appear.

The UCS criticize the assumption of random failure and suggest that failure rates may be significantly affected by ageing.

Attention is drawn to the danger of using historical failure data. The example is cited of the Skybolt rocket motor. A minor modification was made to the welding procedure for this motor. This resulted in flaws which were not revealed by quality assurance procedures but which caused the unit to explode under pressure and led to mission failures of the first two operationally launched vehicles.

An example is given where the study may have underestimated failure probabilities. For the High Pressure Coolant System (HPCS) the study uses a failure probability of 7.8×10^{-3} per demand. The report quotes data for four reactors in which there were 10 failures in 47 tests, a failure probability of 0.21.

Such instances also suggest that at some plants standards may be generally lower and the probability of an accident correspondingly higher, perhaps by an appreciable factor.

The report is critical of the use made in the study of the lognormal distribution to express the uncertainty in failure rate data. It suggests that it is not at all clear that this is the correct choice.

The report quotes the following comments by Yellin (1976):

The basic fault tree model therefore begins from estimates of the logarithms of the component failure probabilities, rather than from the probabilities themselves. In view of this use of the log−normal input distributions, one should also evaluate the accuracy of roughly 10−20% in their estimates of these output logarithms of failure probabilities, while it is suggested here that the evidence presented in *WASH-1400* is in fact consistent with uncertainties of order 50%. This difference of opinion is crucial, since for a typical failure probability of 10^{-6}, an accuracy of 10−20% results in an overall uncertainty factor of 10 in probability, while an accuracy of 50% results in an overall uncertainty factor of 1000.

The use of the median instead of the mean of the lognormal distribution is also criticized as incorrect. The former is some 2.5 times less than the latter. The authors believe that its use results in the study in underestimation of the risks by this factor.

The assumption of independence of events is criticized by the report, which devotes considerable space to the problem of common mode failures.

The report suggests that the *RSS* underestimates the importance of failures which occur in accident sequences as the result of previous failures and also of failures due to incorrect design.

The *RSS* found that common-cause failures do not make a significant contribution to the overall risk, but the UCS suggest that this contribution may have been underestimated. The example given is the miscalibration of four instruments in turn. The probability p of miscalibration of a single instrument is given in the study as 3×10^{-3}. For dependent failures of this kind the overall probability p of failure would be calculated using the study methodology as the geometric mean of the case for four independent failures and for tight coupling between failures. For the latter case, which gives the upper bound, the probability p_u is

$$p_u = 3 \times 10^{-3} \times 10^{-1} \times 1 \times 1 = 3 \times 10^{-4}$$

The last three terms are based on tight coupling. Then, the estimated probability p is

$$p = [(3 \times 10^{-3})^4 \times 3 \times 10^{-4}]^{1/2} \approx 2 \times 10^{-7}$$

Thus, the estimated ratio of the probability of the common-cause failure (2×10^{-7}) to the probability of the single failure (3×10^{-3}) is approximately 10^{-4}. The UCS compare this figure with their own estimate obtained by examining the common-cause failures in the *RSS* database. Out of 303 failures 33 are identified as common-cause. They take three of these failures as common-cause multiple miscalibration failures and thus obtain an estimate of 10^{-2} for the probability ratio referred to. This differs from the *RSS* estimate by a factor of 100.

The review gives the following breakdown of the 33 common-cause failures in the *RSS*

Failure	No. of failures
Drift of instruments, etc.	4
Miscalibration of instruments, etc.	5
Design error	6
Component failure	9
Environmental effects	6
Human error	2
Unknown	1
Total	33

The review refers to an incident on the Oak Ridge Research Reactor where the reactor had to be operated for 5 h without emergency cooling protection due to seven sequential common mode failures, each involving three parallel elements, making a total of 21 failures in all.

The review also considers what it calls common event failure which may lead to common mode failures. These events include fire, turbine missiles, earthquakes and sabotage. The UCS give an alternative analysis of the probability of core meltdown in the Brown's Ferry fire based on the relief valve failures and obtains a value of 0.03 instead of the *RSS* value of 0.003.

The review draws attention to the problem of design adequacy. The equipment may be reliable in the sense that it functions, but it may not be capable of fulfilling its task. For the Emergency Core Cooling System (ECCS) the *RSS* assigned a probability of failure on demand of less than 0.1. The UCS argue that ECCS systems have not been tested under accident conditions and may well lack the basic capability and thus have a higher failure probability.

Also, in the context of design adequacy the UCS take issue with the value used in the *RSS* for the failure frequency of the reactor pressure vessel. The *RSS* used a figure of 10^{-7} per year. The UCS argue that for non-nuclear pressure vessels the failure frequency is approximately 10^{-5} per year and that there is no justification for using a value 100 times less. They base their argument partly on the fact that the failure rates of components in nuclear plants are not greatly different from those in non-nuclear plants. They quote Professor Rasmussen

Probably one of the most serious issues that the [nuclear power plant] intervenors can raise today, with good statistics to back their case, is that nuclear power plants have not performed with the degree of reliability we would expect from machines built with the care and attention to safety and reliability that we have so often claimed.

The UCS also refer to comments made by Sir Alan Cottrell, formerly Chief Scientific Adviser to the British Government, emphasizing the problem of dealing with deterioration of the vessel in a situation where, being radioactive, it is extremely difficult to replace. The UCS review includes an appendix by Cottrell on reactor pressure vessel integrity.

They also quote the following qualifications given in the *Marshall Report* with respect to such reactor pressure vessels:

That it is essential to confine the operational transients to unusually narrow limits in order to avoid excessive crack growth by metal fatigue.

That emergency core cooling water should be injected at an unusually high temperature in order to minimize the risk of fracture by thermal shock.

The UCS list a number of safety issues which are addressed inadequately or not at all in the *RSS*. These include the hazards of loss of containment through electrical penetration seals, of earthquakes and of sabotage as well as the ageing problem already mentioned.

The UCS consider that the *RSS* dismisses too readily the risk from earthquakes. They give a number of quotations on the inability of seismologists to predict earthquakes. Thus, C.R. Allen (1967) states:

Almost every large earthquake that has occurred in California has proved to be surprising in terms of what would have been expected by geologists, seismologists, and engineers at the time.

and Greensfelder (1971) states:

Thus the San Fernando earthquake occurred in an area which has had relatively low seismicity and was not caused by movement on a major, historically active fault, such as the San Andreas or one of its branches. In fact, portions of the fault associated with this shock were not previously mapped ...

The UCS argue that at the least the earthquake hazard introduces a much larger uncertainty than the *RSS* allows. The review includes an appendix on the earthquake hazards at the Diabolo Canyon nuclear reactor site.

On sabotage the UCS argue that it is indeed difficult to quantify but that it is a relevant factor in considering nuclear safety. They draw attention to the fact that the saboteur may be able to deliberately defeat many of the factors which normally operate to mitigate an accident.

The review makes a number of criticisms of the gas dispersion model, of the models of population exposure and of the relations for injury from radioactivity.

The UCS are doubtful about credit claimed for evacuation. Among their reasons are lack of emergency plans and failures in specific emergencies. They quote the failure to evacuate at Browns Ferry.

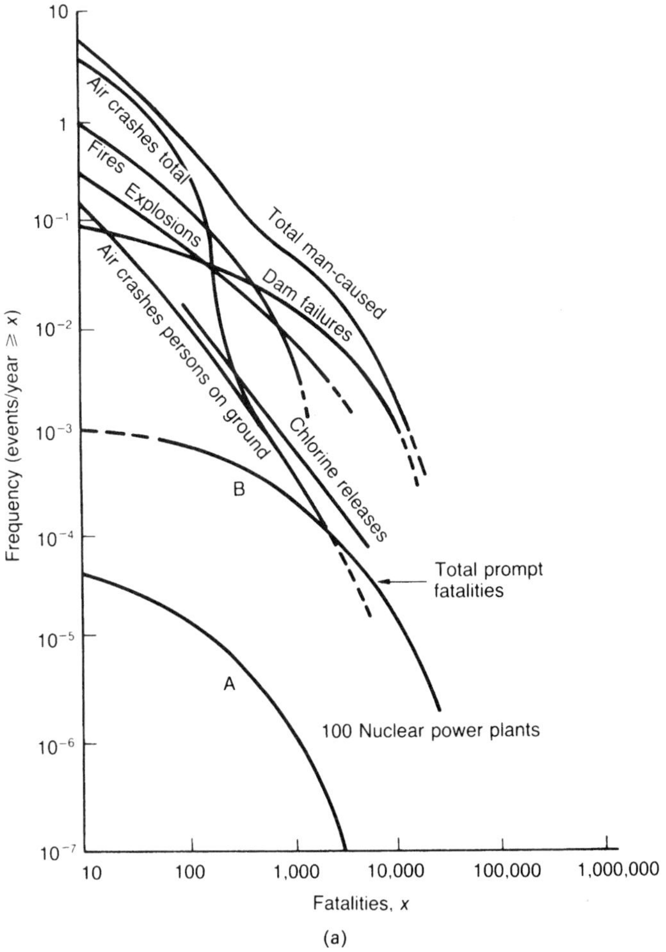

(a)

Figure A23.7   *Reactor Safety Study: Alternative FN curves for fatalities for 100 reactors proposed by the Union of Concerned Scientists (UCS, 1977): (a) early deaths and (b) total deaths. Curve A is that for early deaths given in the original study, curve B the proposed alternative*

Figure A23.7 *continued*

A number of criticisms are made by the UCS of the handling of the results from the *RSS*. One criticism centres on the tendency to present the results simply as showing the risks of nuclear power to be very low and to play down the qualification and the uncertainties. Another is that the way the study was handled did not allow an effective process of peer review.

The large changes made between the draft and final reports also attract criticism. In some cases these changes exceeded the uncertainty bounds given in the draft report.

The Executive Summary is criticized as overoptimistic and misleading.

The UCS conclude that the *RSS* is not usable for policy decisions.

The review gives an alternative statement of the risks, the differences being partly of substance and partly of presentation. The UCS estimate of the frequency of a major release is 1 in 10,000 per reactor year, which is 20 times that of the *RSS*. The upper bound based on the number of incident-free reactor years was 1 in 300. The estimate of the consequences of such a release is that the early fatalities and injuries would be 10 times greater than in the *RSS*. In individual cases the consequences might be 100 or 1000 fold worse than the *RSS* predicts. The UCS estimate of the FN curve for early deaths is shown in Figure 23.7(a).   The UCS also estimate that the late deaths are underestimated in the*RSS* by a factor of 2–5 and further argue that the FN curve should show the total deaths. Figure A23.7(b) shows the curve which they give for this.

A23.20 Notation

p	probability of failure

Subscripts

l	lower bound
u	upper bound

Appendix
24

ACMH Model Licence Conditions

Contents

The Second Report of the Advisory Committee on Major Hazards (Harvey, 1979b) presents in an appendix a set of model controls for a major hazard installation in the form of 'Model Conditions for a Possible Licensing Scheme for Selected High Hazard Notifiable Installations', although the report leaves open the question of whether the use of licensing or regulations is the best method of implementing such controls. The appendix is prefaced by the statement:

'The conditions outlined in this appendix have been written to give guidance as to the range and scope of the requirements that might be required if a licensing scheme of control were adopted for major hazard installations with the highest hazard potential. . . . The Committee is not yet in a position to say if certain installations should be regulated in this way until more information is available, particularly from the proposed Hazardous Installations (Notification and Survey) Regulations. It is, however, clear that there is a gradation of hazard with size and complexity of plant, and as this is increased there should be a greater degree of control and surveillance by the organisation in control of such activities. Thus this appendix has a wider application than the possible requirements for a licensing system of control. It is recommended that any organisation operating a major hazard plant, particularly one at the highest level of hazard potential, should review and satisfy itself that it could demonstrate that the requirements given below are adequately met.'

The full text, which is reproduced by courtesy of the Health and Safety Executive (HSE) and HM Stationery Office is given below.

A24.1 Model Conditions for a Possible Licensing Scheme for Selected High Hazard Notifiable Installations

With regard to a possible licensing system it is recognized that a licence is the most stringent form of control under the Major Hazards arrangements and would be applied only to those Notifiable Installations which present the greatest hazard potential. The licence would be granted to the organization which operates the installation and would be valid only for the specified location.

The approach adopted to licensing would be that foreshadowed by the Robens Committee and embodied in the Health and Safety at Work etc. Act 1974, namely that safety is the responsibility of the organization which should demonstrate that it is taking appropriate measures to ensure effective control of the hazards.

The licensing procedure might be in two stages. At the first stage, the organization would provide the HSE with a statement of intent covering the nature of the proposed installation, the hazards and the features of the design and operation intended to control the hazards. At the second stage, the organization would provide HSE with design and operating information to show how the statement of intent is implemented.

The documentation required might be in two parts corresponding to these two stages and consist of:

Part 1
(1) The Systems Documentation;
(2) The Preliminary Design Document.

Part 2
(1) The Background Documentation;
(2) The Design Document;
(3) The Operating Document.

The Systems Documentation, which is relevant to Licence Conditions 1–9, would be concerned with the general systems which the organization has set up to ensure safety and could be used in support of more than one licence application provided that it is up-to-date.

Details of the Preliminary Design Document are given in Licence Condition 10.

The purpose of the Preliminary Design Document would be to show the general nature of the installation and of any associated hazards. The onus would be on the organization to draw attention at this stage to any special features which might have an important bearing on the granting of a licence.

The Background Document would be documentation on the implementation of Licence Conditions 1–9, in relation to the particular installation.

Details of the main Design Document and the main Operating Document are given in Licence Condition 10.

The main Design Document would be essentially a more up-to-date and detailed version of the Preliminary Design Document, including additional details, such as materials of construction for the main plant items.

The main Operating Document, which incorporates the Operating Manual, would give details of the personnel structure for the operation of the installation.

A licence would be granted for a particular installation on a given site. The licence would not itself deal with questions of siting, but the issue of a licence is an indication that an installation meets certain standards and this is relevant to siting considerations.

The licensee would be required to inform HSE of significant changes which are proposed in any of the matters within the scope of the licence.

A24.1.1 Conditions for a licence

The conditions for the issue of a licence are that the licensee shall demonstrate to the HSE that the design, construction, operation, maintenance and modification of his installation are or will be to the standard appropriate to the installation and in particular that the following features of his organization are to such a standard:

(1) The management system;
(2) The safety system;
(3) The responsible persons;
(4) The arrangement for the identification of hazards;
(5) The arrangements for the assessment of hazards;
(6) The arrangements for the design and operation of pressure systems;
(7) The arrangements for the minimization of exposure of personnel;
(8) The arrangements for the administration of emergencies;
(9) The arrangements for reporting of and learning from incidents;
(10) The design and operating documentation.

A24.1.1.1 Licence Condition: The management system

The organization should show that it has and supports a management system and staff structure which combine to ensure continuing effective control of the installation and its hazards.

The staff concerned include contractors and consultants. This aspect is considered further in Licence Condition 3.

The organization should show the management structure, making clear the distinction between executive and advisory functions, and should give a brief job description for each post.

The management system should give full support to the personnel who are responsible for the design and operation of the installation. Important elements include a suitable and well understood management structure; adequate human resources including coverage of absences, vacancies and emergency situations; recruitment, training and career planning; and effective communications.

The management system should in particular provide satisfactory arrangements in the areas which are the subject of Licence Conditions 2–9. There should be comprehensive, formal and documented set of systems and procedures.

The management should define the objectives of its system of documentation in respect of immediate a communication and of record-keeping and should specify the extent of the documentation required and the procedures for producing it.

The management system should provide for thorough initial and continuing training of personnel in their work generally and in safety in particular.

Full use should be made of appropriate standards and codes of practice. Where there are standards or codes which have statutory backing or which are commonly recognized within the United Kingdom as constituting sound practice, these should be applied as a minimum. Where there are no approved or accepted standards or codes the situation should be covered by the adoption of sound practice and the use of in-house codes.

The management system should include formal procedures for the control of modifications made to the plant or to the process, whether during design or during operation.

The management system should require the independent assessment of features which are critical to the safe operation of the installation. This independent check is essential for inspection of pressure systems and for reliability assessment of instrument trip systems. The guiding principle is that the feature is critical to safe operation of the installation. It is acceptable that the check be done by an in-house authority provided that this is genuinely independent of the interested party. Thus pressure system inspection, for example, must be done by an authority independent of the operating authority, as described by Licence Condition 6.

The management system should contain a variety of arrangements for the periodic audit both of the continuing appropriateness of systems and of the continuing effectiveness of their implementation.

Background It is considered that in the case of major hazard installations the control of the plant and its hazards requires a considerable degree of formalization of communications through written systems and procedures, standards and codes of practice. This is essentially to encourage, and where appropriate enforce, collective and personal discipline by the use of operating methods which have been carefully thought out and which contain an appropriate level of checks and counterchecks to obviate problems and reduce errors.

It is recognized that there is always the problem of over-administration through paperwork and what at times

seems like 'going through the motions' without apparently contributing anything useful. However, it is considered that a careful review of incidents will, time and again, illustrate that there was a loss in discipline because appropriate procedures either did not exist or were not observed.

There should be an interlocking set of systems and procedures to ensure safety through sound engineering and management practices. There is no upper limit to the number of procedures which can be formalized, but there is nothing to be gained by deliberately trying to maximize the number. The optimum number is that which leaves no obvious gaps but avoids creating confusion by overlapping.

Many of the required procedures are implicit in the various Licence Conditions which follow. It is evident that key procedures include those for the identification of hazards, the assessment of hazards, the control of maintenance through permits-to-work, the control of modifications to process or plant, the inspection of equipment, the operation of the process (normal and emergency), the control of access, the conduct of safety audits and the reporting of incidents.

Self-auditing features should be built into the management system in the form of formal instructions for periodic checks on those parts of the system which may become degraded or unnoticed. The operation of a permit-to-work system, for example, should be subjected to regular audit by some means such as an instruction, not merely an exhortation, to the plant manager to sample a proportion of permits each week.

A24.1.1.2 Licence Condition: The safety system

The organization should show that within the management system there is a safety system which is appropriate to the level of hazard inherent in the installation.

The safety system should in particular provide for satisfactory arrangements in the areas of the safety organization, safety objectives and assessment, safety consultative committees, and safety training.

The organization should show that it has people competent to operate the safety system.

Background Most of the aspects of the safety system mentioned are already legal requirements, but it is the object of this section to review their adequacy in relation to the major hazard installation.

A distinction can be drawn between the technological and human sides of safety. On a major hazard plant the technological features are obviously particularly important. The responsibility for these aspects rests primarily with the qualified technical staff in the design and operations areas. There should be no neglect, however, of the human side. On the contrary, on a major hazard plant it is more important than ever to run a 'tight ship' as far as safety is concerned.

The authority of the safety staff should be made clear, particularly in relation to the more technical aspects of the installation. Attention should be paid to the means of ensuring that the safety officers/advisers are effective and are seen to be so.

The safety objectives set for management and the assessment of the performance in meeting these objectives should be indicated. This may not be a simple matter of accident statistics. For major hazard installations there is an additional problem of avoiding rare but catastrophic events. This makes it important to monitor both the occurrence of 'near misses' and the degree of adherence to procedures and rules.

The programme of safety training provided for employees at all levels should be outlined.

Attention is also drawn to Licence Condition 9 which is concerned with the system of reporting of and learning from incidents.

A24.1.3 Licence Condition: The Responsible Persons

The organization should nominate Responsible Persons who are in charge of the design and operation of the installation.

The term 'Responsible Person' has a specific meaning in the context and is explained below.

The organization should show the management structure down to the lowest level of executive technical management, making clear the distinction between executive and advisory functions. There should be a job description for each post; the job descriptions for the posts held by Responsible Persons are particularly important.

The level of seniority held by the Responsible Person should normally be either the lowest or second lowest level of technical executive management.

It is envisaged that persons immediately senior to the Responsible Persons will themselves normally have been or be qualified to be a Responsible Person on major hazard installations, though not necessarily on those processes of which they are now in charge.

The organization should show that a person nominated as a Responsible Person is qualified to hold the post by reasons of his academic qualifications, practical training and recent relevant experience.

On the operations side, this experience should be experience in the operation of the actual process or of a similar process. Experience limited to design of the process or of similar processes and/or to operation of dissimilar processes is not acceptable. On the design side, the experience should be in the design of similar processes.

Where the process incorporates features of considerable technical novelty as a result of which no person has first-hand experience of operation of relevant full-scale plant even greater regard must be paid to the level of competence of the individual and normally experience of pilot plant operation would be required.

Where the installation is to be designed in part or in whole by an outside contractor it is the responsibility of the organization which will operate the installation to satisfy itself and to demonstrate to HSE that the design is to an appropriate standard. The minimum requirement is that there be a nominated Responsible Person in the operating company who has the duty of liaison with the contractor. Where practical it is also desirable to have nominated Responsible Persons in charge of design in the contracting company.

For convenience, reference is made here only to design and operation. The organization should enumerate all the project activities such as fabrication, construction, inspection, commissioning and satisfy itself and HSE that there are nominated Responsible Persons responsible for these activities.

It is emphasized that the ultimate responsibility for the safety of the installation lies with the organization which operates it.

Background   The concepts of 'Responsible' and 'Authorized' Persons occur in various management systems. Our usage of these words is that 'Responsible' relates to a job, for example, plant manager, and 'Authorized' to a task, for example, signing a permit-to-work.

We have considered various models for Responsible Persons. These include those in the Mines and Quarries Act, the Merchant Shipping Act, the Factories Act (radioactive substances), the Explosives Act, the Medicines Act.

We have also considered the arrangement whereby the Department of the Environment advised by the Institution of Civil Engineers recommends individuals for the design of reservoirs.

Competence to do a job must depend on the definition of the job and its relation to other jobs. We began, therefore, with a consideration of management structures and reviewed possible general models for a large and a small firm. It became apparent, however, that this was not a particularly helpful approach and it was not pursued. However, in a concrete situation we do consider the presentation of such a management structure and job descriptions desirable.

With regard to the level of seniority we started with the proposition that the Responsible Person should be at the lowest level of the executive technical management. However, this could give rise to problems in some areas. For example, it is necessary to train for succession and by definition trainees must be at a lower level. Such additional lower levels are not normal in existing practice, would be wasteful and would not offer job satisfaction. We therefore prefer to leave a degree of flexibility. But we do attach importance to ensure that the responsibility is real rather than nominal and is at the lowest practical level.

Thus we envisage that the Responsible Person on the operations side would usually be capable of 'stepping down' and carrying out the job at the next level down.

With regard to selection of Responsible Persons we consider academic qualifications, practical training and recent relevant experience essential.

We think it is desirable that the person chosen should have the sort of broad scientific and technological education which a first degree in science or engineering usually gives. We consider that such a degree should normally be a necessary qualification. Exceptionally, however, people without a degree may be considered. We also attach importance to the ability of the person to recognize problems outside his sphere of competence and to his willingness to seek the advice of other experts.

We consider that although the Flixborough disaster has emphasized the importance of the integrity of the plant, there may be other major hazards where the integrity of the process is of at least equal importance. Thus, whilst it is probable that the Responsible Person will normally be an engineer there will be cases where he will be a scientist, for example, a chemist.

We attach particular importance to the selection of Responsible Persons on the operations side.

We lay particular emphasis on recent relevant experience. There is obviously room for discussion as to the 'relevance' of experience or the 'similarity' of processes. It is up to the organization to convince HSE on these points.

We stress, however, that we would not want this emphasis on recent relevant experience in any way to inhibit technological innovation or normal career development.

We accept that technological progress requires specialization. The professional institutions already recognize this and are considering a register of persons considered competent to design and operate major hazard installations. We are maintaining contact, although, whilst welcoming their interest, we see a number of difficulties in this approach.

The requirement to nominate Responsible Persons should not be seen as detracting in any way from the necessity for a team effort by management to achieve high standards of safety in plants which have major hazards.

A24.1.1.4 Licence Condition: The arrangements for the identification of hazards

The organization should show that it uses appropriate methods of hazard identification at all stages of the project.

The terms 'hazard identification' and 'safety audit' refer to mainly qualitative techniques which review the existence of a hazard.

The application of the techniques should be matched to the stages of the project, starting with coarse scale investigations and progressing to fine scale studies to discover detailed faults.

The management system should contain a formal requirement for the use of such methods, should specify the documentation required arising from this use and should monitor this use.

The organization should show that it has people competent to implement these methods of hazard identification.

Background   The first objective of hazard identification is to reveal the substances or processes which have a hazard potential. The second objective is to identify all conceivable threats to the installation or its processes which might lead to loss of containment.

As technology has progressed identifying hazards has become in some ways more difficult. In particular, there are many hazards which are not revealed by traditional visual inspection. It has become necessary, therefore, to develop additional methods of hazard identification.

An illustrative list of methods is given in Table A24.1

Since every human enterprise involves the possibility of error it follows that the soundness of the management of the potentially hazardous installation is the predominating factor and that the first essential in all cases is an audit of the management system as a whole.

It is recognized that there is a wide variety of methods of hazard identification in use in industry and that different techniques are applicable to different situations. There is no intention of imposing any particular method.

It is necessary that the use of these methods be a requirement of the management system, which also should specify the degree of recording and documentation required and should contain a mechanism for auditing the application of the techniques to ensure that they are used.

The people who have to implement these techniques must be competent to do so. HSE should be able to advise on opportunities for training in this area.

Hazard identification covers much the same ground as safety audits in the broadest sense. The Chemical Industries Association has published two guides entitled *Safety Audits: A Guide for the Chemical Industry* (1973) and *A Guide to Hazard and Operability Studies* (1977).

A24.1.1.5 Licence Condition: The arrangements for the assessment of hazards

The organization should show that the hazards identified by the means described in the preceding section have been removed or that the associated risks have been reduced to a minimal level.

In this context 'minimal' means the probability that an employee or member of the public will be killed or injured or that property will be damaged is at least as low as in good modern industrial practice.

The method of demonstrating that the risks are at a minimal level should be comprehensive and logical.

Table A24.1   *Some methods of hazard identification*

Project stage	Hazard identification method
–	Management and safety system audits
All stages	Checklists
	Feedback for workforce
Research and development	Screening and testing for Chemicals (toxicity, instability, explosibility)
	Reactions (explosibility)
	Impurities
	Pilot plant
Predesign	Hazard indices
	Insurance assessments
	Hazard studies (coarse scale)
Design	Process design checks
	Unit processes
	Unit operations
	Plant equipment
	Hazard and operability studies (fine scale)
	Failure modes and effects analysis
	Fault tree and event trees
	Hazard analysis
	Reability assessments
	Operator task analysis and operating instructions
Commissioning	Checks against design, inspection, examination, testing
	Non-destructive testing, condition monitoring
	Plant safety audits
	Emergency planning
Operation	Inspection testing
	Non-destructive testing, condition monitoring
	Plant safety audits

The method may consist of:

(1) the use of codes of practice generally recognized in the industry;
(2) the use of special testing;
(3) the use of calculations based on appropriate data.

In many cases it will be sufficient to show for all or at least some aspects of the hazard that a generally recognized and accepted code of practice is applicable and has been followed.

Where there is any aspect of the hazard, the risk of which cannot be reduced to a minimal level by following a recognized code of practice or by special testing, then, whenever meaningful quantitative methods should be used to demonstrate that the risk has been reduced to a minimal level. These quantitative methods will normally consist of three steps:

(1) An estimate of the consequence to employees and the public.
(2) An estimate of the frequencies with which hazardous situations will occur.

(3) Comparison of (1) and (2) with the other risks to which people are normally exposed in order to show that the risk under consideration is relatively small.

The management system should contain a formal requirement that such methods of hazard assessment should be applied.

The organization should show that it has access to people competent to implement these methods.

Background Having identified hazards, as described in the previous sections, it is necessary to know that the associated risks have been reduced to a minimal level.

It is envisaged that the method of demonstrating that the risks are at a minimal level will normally be based on a fault tree approach, but that a detailed development of all parts of the tree will not generally be required.

Sometimes it is possible to remove a hazard completely, for example, by replacing a flammable or toxic raw material by a non-flammable or non-toxic one.

More often, the hazard cannot be eliminated completely, though the risk can be reduced to any desired level by the use of protective equipment. For example, the risk that a particular vessel will burst because of overpressure can be reduced by fitting a relief valve suitable for the duty, adequately sized and properly maintained. This does not eliminate the hazard completely as there is a small probability that the relief valve will fail to lift when required. Even if two relief valves are fitted, there is still a very small probability of coincident failure.

In many cases, codes of practice provide generally recognized and accepted methods of reducing a hazard to a minimum level. For example, in the case just considered, it would normally be sufficient to show that the vessel is fitted with a relief valve, adequately sized and properly maintained, as relief valves have been generally recognized for many years as an accepted way of reducing to a minimal level the probability that a vessel will burst.

Similarly, if fracture of pipework has been identified as a hazard, it would be sufficient to show that the pipework has been designed and constructed and will be operated and maintained in accordance with a recognized and relevant code of practice.

Codes of practice should not be used outside their area of applicability. Codes of practice for pipework, for example, do not cover fracture by projectiles and if it is necessary to take the latter into account, a separate study is necessary.

Moreover, codes of practice imply acceptance of some level of probability of the hazard materializing. This level will be unacceptably high in relation to some major hazards.

In some cases, it may be necessary to carry out special tests to quantify aspects of particular hazards.

Where there is no generally accepted code of practice and where the problem cannot be resolved by testing, but where it is reasonably practicable to show quantitatively that the risk has been reduced to a minimal level, this should be done. In some cases, this cannot be done because of lack of data or of a suitable model to describe the system. In these cases, judgement will have to be used.

Examples of hazards which may not be covered by recognized codes of practice and which, if they produce major effects, would have to be individually assessed, are:

(1) runaway reactions (e.g. decomposition and polymerization reactions);

(2) impact of moving objects (e.g. cranes, vehicles, missiles from explosions);

(3) failure of instrumental protective systems;

(4) failure of services (e.g. electricity, water, compressed air).

These can be described as events leading to possible loss of containment.

The hazard quantification will normally consist of three stages in which probability and consequence must both be considered; sometimes it will be necessary to consider a number of possible outcomes differing in probability and consequence. An event which has the potential to kill many people may not cause great concern if the probability of it occurring is sufficiently small. We do not prohibit football matches because there is a small chance that an aeroplane may crash on the crowd. On the other hand, we would not build a new football ground at the end of a busy runway.

(1) The first stage is the estimation of the probable consequence of the hazard. This may be based on past experience or it may be estimated from a theoretical study of the problem. The consequences may be expressed as the probability that an employee or a member of the public will be killed or injured or as the probability that extensive damage will be caused to the property of others or both.

(2) The second stage is the estimation of the probability that the hazard will occur. Again this estimate may be based on past experience, or it may be synthesized from data on the failure rates of individual components or pieces of equipment.

In estimating the probability that the hazard will occur it is necessary to assume that certain standards are followed in the operation of the equipment, for example, that relief valves are tested regularly. If these standards are not followed the conclusions of the hazard quantification are no longer valid.

(3) The third stage is comparison of (1) and (2) with the other hazards to which people are exposed in order to ensure that the risk under consideration is minimal.

This implies the use of a criterion against which risks can be judged. It is not intended that any single criterion should cover all cases.

If it is not possible to carry out a complete study as indicated (i.e. stages 1–3), it may be possible to carry out a partial study and, if so, this should be done, as it helps to identify those aspects of the problem which have most effect on the probability and consequences.

Where a new feature is used in place of one of proven reliability then it should be shown that the new feature is at least as safe as, and preferably safer than, the original. For example, if an instrumented protective system is used in place of a relief valve it should be shown to fail no more often, and preferably less often, than a relief valve.

In other cases it may be appropriate to show that the risk to an employee is no greater than that for employees in the industry as a whole or to show that the risk to a member of the public is comparable with the other risks to which the public are exposed without their consent.

It may be noted that it is difficult to find examples of instances in which members of the public have been killed as a result of accidents on major hazard installations. In most cases, if the risk to employees is minimal the risk to the public will also be minimal; although a lower level of risk is required, the public are usually further away.

It is necessary that the use of quantitative methods, whenever meaningful, should be a requirement of the management system. This should specify the recording and documentation required.

In some cases a detailed study may be required taking many days or even weeks, but in other cases relatively simple calculations may be sufficient.

The organization should show that it has access to, and uses as necessary, people competent to assess hazards, but these people need not be in its full-time employment; they may be consultants.

Quantitative methods provide a means to assist management to choose those measures which are 'reasonably practicable' for providing a safe plant and system of work.

A24.1.1.6 Licence Condition: The arrangements for the design and operation of pressure systems

The organization should show that it has a formal and well-understood management system for controlling and monitoring the design, fabrication, commissioning, operation, inspection and testing of pipework, vessels and other equipment together forming the constituents of a pressure system which may give rise to a serious hazard.

The term 'pressure system' refers to a linked series of equipment items operating at a pressure either above atmospheric or under vacuum, together with all the interconnecting pipework. Such systems commonly form processing units, but the definition includes also storage and handling installations.

The management system of control should be in two parts, and should be effected through two recognized channels of authorization, related to Design and to Operation.

The management design authority should identify and state the design parameters within which the pressure system is to operate and the conditions for which each component part of the pressure system shall be designed. It should also define the code under which the individual components shall be designed, or where no design code exists should cause sufficient work to be undertaken to satisfy themselves, either by experiment or by the use of specialist advisers, that the design is at least as safe under all foreseeable circumstances as the standard demanded by recognized codes.

The design authority should also define the standards to be used for the pressure system during the fabrication and construction stages. Whenever possible the standards specified should be those quoted in recognized codes applicable to the particular pressure system.

Management should ensure that an appropriate inspection system is operated during fabrication and construction work to check that the standards set by the management design authority are being met. This inspection system may be in the same management organization as the design authority, or it may be appropriate to use one of the engineering insurance companies or other approved inspection agency. The inspection system should not be part of the operating authority. The design authority should also specify the written evidence required to demonstrate that the plant has been fabricated, constructed and proof-tested, in accordance with the design requirements. Copies of documentation forming this written evidence should be verified, preferably by the inspection agency before the pressure system enters service.

Copies of all documentation relating to the design parameters, and to the verification by the inspection system, should be retained by the design authority and should also be available on the works on which the pressure system is operated.

The operating authority should prepare a comprehensive set of instructions based on information issued by the design authority and upon an analysis of hazards involved. These instructions should set out clearly the way in which the pressure system shall be operated in both normal and abnormal circumstances, and the way in which the pressure system is to be protected from the effect of conditions more extreme than those permitted by the design authority. These instructions, which should be readily available to all those responsible for operation, inspection and repair of the pressure system, should set the limits within which the system is to be operated. Any variation in these limits outwith the parameters set by the design authority should be referred to the design authority or other independent qualified body for approval before fresh instructions are issued. Such submission and approval should be made in writing and copies retained in the works for future reference.

The operating authority is responsible for ensuring that any repairs and modifications are designed, fabricated and tested to a standard not less than that used by the design authority for the original system. The operating authority should have a system of documentation which controls the repair and modification procedures, so that modification to the pressure system cannot be made without written authority from an authorized person.

Thus, instructions prepared by the operating authority should include formal procedures for identifying and making process modifications, for identifying and making plant modifications and for restarting the plant after discovery of a serious defect.

The operating authority should also provide and enforce the use of a code which ensures the continuing safety of the pressure system by a regular inspection of the equipment and the safety devices which are provided to protect that equipment. Such an inspection code should meet the following criteria. First, that a register is held on each works in which each item of equipment is given a unique designation and an engineering description which adequately describes the design and fabrication details and also details of operating conditions, both normal and maximum. Secondly, that the code should specify the frequency of inspection for the various classes of equipment and should also specify rules concerning the selection and training of inspectors. There should also be rules by which inspection frequencies may be increased or reduced, decisions being based upon the result of inspection records which should be held in the register. The code should specify rules which guarantee the independence of the inspecting system from the operating authority, either by appointing external inspection authorities, or by making satisfactory arrangements for the in-house inspection authority to be responsible to a senior member of the organization who has the design authority within his charge.

The organization should show that it has people competent to execute the control and monitoring functions described above in both the design authority and operating authority areas. It should also show that the inspection service is truly independent of the operating authority.

Background   The control of design and operation of pressure systems is a cardinal feature for ensuring safe operation of major hazard plants.

The separation of management control into the areas of a design authority and an operating authority has been made in order to bring out clearly the separate responsibilities of these two parts of the total organization which designs, builds and operates a major hazard plant. The areas of authority interlock, and in small organizations there may be sharing of some specialist personnel. It is important, however, that the design concept and subsequent modifications to that concept should be controlled by an authority separate from the operating authority. In cases of contractor design, the organization will have to show that the contractor can carry out the duties of a design authority, particularly in the matter of documentation concerning the design parameters for the pressure system and the fabrication details, and in the provision of competent people for authorization.

The use of a regular inspection is an essential feature of the safe continued operation of a pressure system. The register of equipment and inspection reports is the linchpin around which this inspection system is built. Decisions on the frequency of inspection and the nature of satisfactory inspection procedures should be taken by the inspecting authority with advice from the design authority. An essential feature of the integrity of the system is that the inspectors and their management are not under the technical control of the operating authority. In small organizations, considerable use can be made of external inspection agencies and other specialist help. In large organizations, such facilities are likely to be provided in-house. In those cases the organization should take particular care to show the independence of the inspection agency.

It is important that safety devices such as relief valves, non-return valves and vents are included in the register of pressure systems and are subject to the same type of inspection and testing arrangements as are specified for the pressure equipment. Details of testing and frequency of examination may well be different from arrangements made for pressure equipment, but the principle of inclusion in a register, together with test and inspection notes, is essential.

A system for registering and inspecting pressure vessels and other equipment and for the keeping of verification documentation which satisfies many of these criteria is described in BS 5500: 1976 Unfired fusion welded pressure vessels, in the Pressure Vessel Inspection Code and in the draft Piping Systems Inspection Code of the Institute of Petroleum.

A24.1.1.7 Licence Condition: The arrangement for minimization of exposure of personnel

The organization should show that it has assessed the hazards to personnel involved in the installation and that where necessary it has taken steps to reduce these hazards to a minimal level. In particular, measures should be taken to limit the number of people at any one time in areas of high hazard to a minimum consistent with safe and efficient operation and to afford protection to exposed personnel, who may include operating personnel, maintenance personnel, investigation teams, construction workers and visitors.

Background   This matter is fully dealt with in chapter 7 of this report.

A24.1.1.8 Licence Condition: The arrangements for the administration of emergencies

The organization should show that it has assessed the hazards involved in the installation in relation to major emergencies and has maintained emergency plans.

Emergency planning should include identification and assessment of possible major emergencies; nomination of persons responsible for administering an emergency; development of procedures for declaring, communicating and controlling the emergency and for evacuation; provision of buildings such as a control centre or refuge rooms and of equipment such as an alarm system; designation of works emergency teams and definition of duties of other workers; liaison with external services including police, fire and medical services; training and exercises for emergencies.

Background   A guide *Recommended Procedures for Handling Major Emergencies* is published by the Chemical Industries Association (second edition, 1976). This covers primarily the administration of emergencies within the works.

The guide also deals briefly with the planning of action, such as evacuation, which may be required outside the works. This is particularly important for certain major hazards. It should be emphasized in this connection that for toxic gas hazards the value of the time bought by any distance separating the factory and the public may be wasted if there is no evacuation plan.

A24.1.1.9 Licence Condition: The arrangements for reporting of and learning from incidents

The organization should show that it has prepared a schedule of signals to be recorded and that the relevant instruments are housed in such a way that an accident on the plant is unlikely to prevent recovery of the records.

The organization should show that it has a system for the reporting of incidents which might endanger the installation or lead to loss of containment and that it uses the information obtained from this reporting system to learn how to reduce these hazards.

These requirements for the reporting of hazardous incidents are additional to the existing statutory requirements. They have two main objectives. One is to ensure the reporting within the company of the 'near miss' type of incidents which often precede an accident. The other is to obtain data at national level on incidents related to serious hazards. HSE may request that certain types of incident be included.

Background   There already exist certain statutory requirements for the reporting of incidents. These include:

(1)   The Petroleum (Consolidation) Act 1928;
(2)   The Factories Act 1961;
(3)   The Dangerous Occurrences (Notification) Regulations 1947;

as well as those specific to explosives factories, mines and quarries, and nuclear installations.

These requirements are mainly concerned with accidents causing plant downtime, lost-time accidents and fatalities.

It is well-known, however, that for every serious accident there are numerous incidents, including 'near misses'. Since they are more numerous, these incidents offer greater scope for learning and improvement.

It is therefore proposed that the organization should have its own reporting system covering the type of incident significant in relation to its particular hazards.

We have deliberately not specified the incidents which should be reported but we have in mind incidents relating to such matters as:

(1) incidents involving serious operator error;
(2) incidents involving trip system malfunction or disarming;
(3) malfunctions of valves, for example, pressure relief, non-return;
(4) leaks of flammable materials;
(5) leaks of toxic materials;
(6) leaks and fires at pumps;
(7) fires and explosions in furnaces;
(8) storage tank collapse.

The important point is that the reporting system should be tailored to the needs of the organization.

There may also be certain types of incident on which HSE wish to collect information nationally. In this case it may request their inclusion in the reporting scheme. Again we have not specified these incidents but an obvious one is the unconfined vapour cloud explosion.

These reporting requirements should be seen in relation to the fact that we have not recommended that all major hazard installations have a 'black box' recording instrument system. We consider that the arrangements proposed here are a more efficient way of learning both at company and national level.

The reporting system is intended primarily to assist the organization to learn from incidents and it should show that it has formal arrangements to do this.

A24.1.1.10 Licence Condition: the design and operating documentation

The organization should provide full documentation on the design and operation of the installation. The documentation required is as follows:

Part 1 Documentation
(For Part 1(a) The System Documentation see Introductory Section of this appendix.)

Part 2(b) Preliminary Design Document   This should contain:

(1) A brief description of the process and should include the nature of any chemical reaction involved and the various operations to which the material in process is subjected. In addition, any other exothermic reactions which may arise if operating conditions fall outside the design values, should be specified.
(2) A comprehensive description of the nature of the hazards in the materials handled (toxic, flammable, explosive materials), of the objectives to be achieved in order to limit these hazards and of the methods in plant design and operation necessary to achieve these objectives.
(3) A statement of any less hazardous process which could have been used and the reasons for selecting the particular process in question. This might include outstanding economic advantages, factors relating to the availability of raw materials, the avoidance of particularly difficult engineering operations or the necessity of making a product of a particular purity.

(4) A process flow sheet, indicating quantities, the temperatures and pressures of materials at each stage and the vessel inventories, and the flow-rates in each of the principal flowlines. The reasons why such pressures and temperatures must be used should be given. Mass and heat balance diagrams should be given where appropriate.
(5) A list of the main plant items, specifying the capacity, design pressure, temperature limits for safe operation (upper and lower), and any special features of construction, together with the actual operating conditions. Details of services should be given.
(6) Details of the principal standards and codes to be used in the design.
(7) A statement of the inventory of all hazardous materials in process and of the steps taken to keep this at the lowest level consistent with safe and efficient operation.
(8) A statement of the method whereby the process will be controlled. Abnormal features, with particular reference to hazards, should be highlighted and references made to any special features, including trip systems.
(9) A list of all hazardous materials in bulk storage which may be endangered by a process incident and the steps to be taken to minimize the risk of their involvement.
(10) A statement of all materials and services needed to maintain safe operation of the plant and of the steps taken to ensure their continuous availability.
(11) A statement of any noxious effluents and their methods of treatment.
(12) A site layout showing the proposed plant and control room, and their position relative to other installations and buildings in the works, to loading bays at tanker terminals, to plants in neighbouring works and to the public area.
(13) A statement of the location and construction of the control room.
(14) The routing for all vehicles needing access to the plant, whether for the supply of materials and removal of products, or for maintenance or emergency purposes.
(15) An account of the procedures for maintaining effective liaison between the company and any outside organizations involved in the design or construction of the plant.
(16) Details of actual experience of the company, or of availability of experience from external sources, in operation of pilot and production scale plants, for the same or a similar process.
(17) Manning levels on the plant.
(18) Proposals for dealing with emergency situations.

Part 2 Documentation
(For Part 2(a) The Background Documentation see Introductory Section of this appendix.)

Part 2(b) The Design Document   This should contain:

(1) An updated and detailed statement of all materials submitted in the Preliminary Design Document, including the process flowsheet, quantities and flowrates of all materials in process and storage areas, heat and material balances, instrument diagrams and plant layout diagrams.

(2) Details of all principal plant items, as in item 5 of Part 1(b), giving codes used, materials of construction and any special features.

(3) Installation drawings and pipe layouts.

(4) Documentation on hazard assessments and reliability studies.

Part 2(c) The Operating Document   This should contain:

(1) A statement of the technical staff structure, including names of the Responsible Persons, together with details of standby arrangements for absence such as sickness or holidays, and of call-in arrangements.

(2) A statement of the numbers of operating personnel and of their training with particular reference to emergency procedures.

(3) The Operating Manual. This should include details of the start-up, normal shut-down and emergency shut-down procedures.

Appendix

25

HSE Guidelines on Developments near Major Hazards

Contents

The Third Report of the Advisory Committee on Major Hazards (Harvey, 1984) contains an appendix on 'The Siting of Developments in the Vicinities of Major Hazards: HSE's Draft Guidelines to Planning Authorities. The full text, which is reproduced by courtesy of the HM Stationery Office, is given below.

A25.1 The Siting of Developments in the Vicinities of Major Hazards: HSE's Draft Guidelines to Planning Authorities (by the Health and Safety Executive — Major Hazards Assessment Unit)

A25.1.1 Introduction
The control of risks from major hazards may be achieved in various ways, for example:

(1) limit the amount of hazardous material present;
(2) reduce the likelihood of an accidental release of material;
(3) plan for emergencies;
(4) reduce the number of people exposed to risk.

Some of these measures may be taken within the installation, particularly at the project planning stage.

In many cases the situation in the installations is already fixed and the scope for improvement under the HSW Act is fully utilized. Even so, there may remain some residual risk. It is unlikely that the residual risk could justify the removal of existing populations by the termination of viable land uses. However, it is possible to inhibit an increase in the population at risk by control of new developments under planning legislation.

It is the purpose of HSE's advice to facilitate planning control where appropriate. Planning authorities are expected to include the factor of risk in their consideration of a planning decision. Thus, the types of development near major hazards should be deliberately chosen to minimize the population at risk.

HSE takes as a starting point the assumption that the major hazard installation complies with the requirements of HSW Act. It is then necessary to assess the level of residual risk to a proposed development in the vicinity, and to judge whether this might be a significant factor in the planning decision. HSE's assessment takes account of the consequences and likelihoods of various accident scenarios, and the implications of variations in vulnerability between different types of development. In some cases, the assumption of compliance with HSW Act implies a precise standard.

In other cases the standards are not precisely defined, or it may be possible to comply in a variety of ways. The individual characteristics of installations are taken into account where possible, but there are limits to the amount of detail which can be considered. It may be necessary to assume average performance in the absence of obvious individual variations, or for cases which are being assessed on paper.

This note deals only with the provision of advice on proposals for new developments in the vicinity of existing major hazard. The assessment techniques are used in a similar way when considering the siting of new major hazards in relation to existing buildings, but the judgement of the significance of the risks may be different. For example, the siting of a new major hazard may be constrained by the need to integrate it with existing plant. The application of a simple 'reciprocal rule' for the siting of new major

hazards would neglect the differences in availability of sites for the two situations. There is also the possibility of seeking special safety standards on a new plant to facilitate its siting.

A25.1.2 Assessment: likelihood and consequences of an accident
The first part of the assessment is objective in the sense that it does not include social or political judgements, but it does include estimates of various factors where data is more or less unavailable. Thus it includes technical judgement by professional risk assessors. There is inevitably some uncertainty introduced by this judgement process and by lack of precision in basic data. This does not prevent the derivation of reasonable conclusions but it does imply that some caution is necessary. The technical judgements may tend to lead to an over-estimate of the risk level. If a planning decision is finely balanced, a more detailed review of assumptions may be worthwhile.

Assessments of risk from different materials differ in detail but they have many features in common. The basic sequence of events for LPG and chlorine are:

(1) LPG:
    plant failure;
    release;
    vaporization;
    mixture with air;
    ignition;
    explosion and/or fire;
    blast and/or heat propagation;
    impact (either directly on people, or on buildings which collapse or burn).
(2) Chlorine:
    plant failure;
    release;
    vaporization;
    mixture with air;
    travel downwind;
    dilution;
    inhalation;
    injury.

The risks from such sequences can be evaluated in two main ways, namely historical and analytical. In the historical approach, information from past incidents is used to predict the risks. *The Second Report of the Advisory Committee* contains information on some past incidents, and it derives a 'Mortality Index' to indicate the average number of fatalities per tonne of material released. This approach may be useful to devise a hazard rating for a material, but it is noticeable that the average index conceals a wide variation between cases and it is thus of questionable value for predicting individual situations. Other problems with the historical approach are the small number of cases and the incompleteness of information on cases, so it may therefore provide a rather poor statistical basis. This approach might be difficult to adapt for prediction in situations different from those already included by the accidents.

The alternative, analytical approach attempts to synthesize the risk on the basis of sequences outlined above. This approach is complex and requires assumptions to be made in the absence of data. Also, it may be difficult to select appropriate scenarios to include. It is, of course, possible to combine the approaches, and in practice this is what is done. The historical approach is used to provide information on the likelihoods of releases, while the analytical approach is used to predict consequences. To illustrate the approach, the example of bulk pressurized LPG is used. Installations of this type comprise the majority of the major hazards in Britain, and the risks from them are relatively straightforward to assess. The assessment may begin by listing the various possible releases, either by quantity (short release, e.g. tank burst) or by rate (long release, e.g. uncontrolled pipe-split). There is a continuous range of possible release, but this can usually be simplified by identifying obvious sub-divisions. For the LPG case we might have:

damaging but they are so unlikely as to present negligible extra risk. For the LPG example, separation distance guidelines for the siting of the first category of development near existing installations tend to be based on the provision of substantial protection from BLEVE and UVCE; this automatically gives a very high degree of protection from pipe-break and lesser events.

A25.1.3 Assessment: types of development
It was indicated earlier that different types of building may have associated with them populations whose vulnerabilities differ. This may be of great importance in a siting

Event	Probability	Consequences
Minor flange or gland weeping	Quite likely	Not felt off-site
Pipework break	Rather unlikely, especially for major cases	Localized, but could extend off-site from major cases
Tank burst: BLEVE	Very unlikely	May be harmful at considerable distances off-site
Tank burst: UVCE (delayed ignition)	Extremely unlikely	May be harmful at considerable distances off-site
Tank burst: drifting gas clouds	Extremely unlikely	May be harmful at considerable distances off-site

It is now necessary to determine more precisely what are the probabilities and consequences of these events. Obviously there could be wide variations between individual plants and between events in each sub-division. For example, the effects from pipe-break might depend on: size of hole; duration before shut-off; whether liquid gaseous or two-phase; wind-speed; delay before ignition; dispersion to below flammable limits; type of combustion, and specific site features. To simplify the problem, assumptions are made to give an indication of the realistic outcome from the most severe event within a sub-division.

The consequences in terms of likelihood of injury are derived by first calculating the impact intensity of the heat or blast at a particular distance, then relating this to the vulnerability of the population at that distance. The possibility is noted that different types of building, having different populations, might give different degrees of protection from any given intensity.

The result of this process is a list of events and risks. A decision can then be made on which events are trivial in that no injury could result. For remaining events, the consequences are considered in conjunction with the likelihood, to determine the combined risk of release and injury. This completes the objective assessment.

It should be noted that this approach does not automatically rule out from consideration any type of source event, however unlikely. Even the largest, most unlikely events are included, so there is not necessarily any cut-off corresponding to a 'maximum credible accident'. However, attention is paid to the combined risks, which may be very small from an unlikely event, so that the largest events are not necessarily significant determinants of the overall risk. In practice, assessments may be based on one or two events which may thus appear to correspond to 'maximum credible accidents;' this implies that lesser events may be more

judgement. HSE feels that it would be inappropriate to advise against all forms of new development near major hazards nor need new major hazards to be sited away from all existing buildings. Types of developments may be distinguished by various factors which relate to the risks, for example

(1) whether residential, workplace, recreational, shopping, etc.;
(2) length of time any individual may be present;
(3) number of people who may be present and duration of occupancy;
(4) ease of evacuation;
(5) inherent vulnerability of people, for example, elderly, disabled, children;
(6) physical factors, for example building height, materials of construction, etc.

On the basis of these factors, HSE identifies three main categories of development. The first consists of developments where users would be present for most of the time, or where large numbers could be present at any one time, or where people are particularly vulnerable. Examples include housing, shopping centres or very large retail outlets, multi-storey office blocks, etc. The second category consists of less sensitive developments and includes industrial developments with low employment density, warehouses, etc. HSE also identifies a third category containing special cases such as hospitals, schools and other particularly vulnerable developments, to which particular attention should be paid in siting judgements.

It is normally advised that planning control is only necessary in respect of the first and third categories of developments near major hazard installations. There are various mitigating factors associated with the second category, such that it would be difficult to argue that the risks to users justify inhibition of this type of land use. These factors could include ease of evacuation, low occupancy-time, fitness of population (e.g. adult workforce), small number of people, building protection or building not hazardous to occupants, etc.

A25.1.4 Summary of MHAU's advice

For LPG and similar flammable liquid gases, HSE recommends separation distance limits such that the consequences of minor releases at the limit would not be injurious, while the consequences of major releases (BLEVE, UVCE) would be unlikely to seriously injure more than a small fraction of the exposed population. In principle, unignited vapour clouds could travel to the limit, or even beyond it in unfavourable weather conditions, and people enveloped in a cloud would probably be seriously injured, but the risk to an individual from such a scenario is very low.

For chlorine, HSE recommends separation distance limits such that the consequences of major pipework failures would be unlikely to seriously injure more than a small fraction of the population downwind, in typical prevailing weather conditions. In unfavourable weather, or with larger releases from tank failure, people at greater distances could be injured, but the risk to an individual from such a scenario is very low.

A25.1.5 Presentation of advice

In presenting its advice, HSE attempts to give planning authorities clear guidance on the weight to be attached to the factor of safety. HSE tends to use one of three graded responses as follows:

Situation 1 (negligible risk): No significant reasons for refusing planning permission.

Situation 2 (marginal risk): The risks are very small and are not in themselves sufficient grounds for refusing planning permission, but they could be a factor against the development to add to any other adverse factors. This would apply near the outer limit of separation distance, or with small developments.

Situation 3 (substantial risk): While the probability of a serious incident is low, the consequences at the proposed development site could be serious. The safety issue should be a major factor and the planning authority is advised not to grant planning permission. If there were other factors which weighed strongly in favour of the application, HSE would seek to explain the technical assessment and level of risk in more detail before the planning authority's decision was made. (In exceptional cases of situation 3, where the planning authority was minded to grant permission contrary to the HSE's advice, HSE might consider whether to request a planning inquiry where the issues could be examined in detail in a public forum.)

It often happens that an initial unfavourable response from HSE leads to discussions of the scope for modifications to reduce the risk. Significant reduction may enable HSE to advise favourably. In other cases, HSE may give a conditional response, with advice on the use of planning conditions to ensure that the risk is reduced.

Appendix

26

Public Planning Inquiries

Contents

In a limited number of cases, a planning proposal involving a major hazard goes to a public inquiry. The inquiry will then normally hear technical evidence, which is likely to centre around the hazard assessments carried out by the parties. Experience shows that there are a number of problems associated with expert testimony at public inquiries.

A study of large public inquiries is given in *The Big Public Inquiry* by Sieghart (1979).

Accounts have been given of the public inquiries at Mossmorran by McGill (1981, 1982, 1983), Pheasant Wood by Petts (1985b) and at Canvey by Petts (1985a). A comparative review of the problems of expert evidence at public inquiries and proposals for their mitigation has been given by Petts, Withers and Lees (1986).

Public accident inquiries are discussed in Chapter 27.

A26.1 Mossmorran

It is usual in Scotland for planning applications for major oil-related developments to be called in for review by the Secretary of State. In the case of the application of Shell Expro to build a gas plant and of Esso Chem to build an ethylene cracker at Mossmorran together with associated facilities at Braefoot Bay, the minister decided on a public inquiry, which was held in 1977. The local planning authority (LPA) retained the consultants Cremer and Warner to produce an assessment of the hazards. The Health and Safety Executive (HSE) also gave its own assessment. These were the two main sources of technical evidence and neither opposed the development. Opposition came from local interest groups at Aberdour, Braefoot Bay and Dalgety Bay, who produced their own assessment. There was a delay of some 18 months after the inquiry closed until 1979 before the minister gave his approval. The delay was apparently caused in part at least by the issue of ignition of 'radio sparks', which was raised after the inquiry had ended.

The Mossmorran inquiry, which has been studied by McGill (1981, 1982, 1983), illustrated the difficulties of public inquiries on hazardous installations, particularly with regard to expert testimony. The amount of information on the plant limited the contribution of the HSE and severely restricted that of the objectors. The latter considered that the statutory authorities had not made available a sufficiently full hazard assessment and they were not satisfied that the objections made to the assessments presented received sufficient weight or that either the LPA or the HSE were sufficiently independent of government and industry. The LPA were clearly in favour of the proposals before they commissioned the consultant's report.

A planning application is concerned with the development of land. The extent to which a planning inquiry should cover marine activities is perhaps a grey area. The hazards from marine activities at Braefoot Bay were briefly considered in the consultant's report and the inspector did take some evidence on shipping movements in the estuary, but marine hazards were not thoroughly revealed. The objectors were not unnaturally more concerned with the level of risk than with its source and considered that insufficient attention was given to the marine hazards, which might exceed those from land-based activities.

The inquiry exemplified the problem of assessing the risk from a plant for which the completed design is not yet available. The first stage of planning permission is outline planning permission. At this point, the design of the plant is generally available as a feasibility study only. It is outline planning permission, however, which is the important stage and it is difficult thereafter to revoke permission. In this case since the detailed design of the plant had not been done when planning permission was applied for, the assessments which could be made were that much less detailed and the uncertainties in the risks assessed that much greater. The objectors evidently thought that an adequate estimate of the risks could not be made without a hazop study, although this cannot be done until a detailed design is available.

The inquiry did demonstrate, however, the effectiveness of a planning condition. Condition 24 of the planning permission eventually given required the companies to carry out hazop studies and major studies were carried out.

The consultant's report reviewed the various hazard scenarios and graded their likelihood in qualitative terms such as 'low', 'very low' and 'extremely low', but did not give a full assessment of individual and societal risks. The HSE evidence was qualitative. The lack of any full quantitative assessment was criticized by the interest groups, who emphasized that the Canvey study, a quantitative assessment and public report, was regarded by the HSE as a major step forward in hazard control, yet such an exercise was not done for Mossmorran. These groups were also critical of various apparent inconsistencies in the HSE approach. It was alleged that the HSE estimate of the maximum credible spill at Braefoot Bay was not consistent with that given for Canvey and that the HSE was recommending a cordon sanitaire at the gas plant but not at the marine terminal, where many of the hazards were. Some of the HSE's assessments of physical phenomena were challenged and the opposition was able to call expert witnesses of some academic standing. The safety distances acceptable to the HSE were said to be less than those required in certain other countries. The opposition groups produced some risk estimates which indicated that the likelihood of a multiple fatality accident at Aberdour/Dalgety Bay marine terminal was $10^{-4}/$ year as opposed to the estimate of $10^{-6}/$year given by the consultants.

The consultant's report, despite the absence of a quantitative assessment, gave advice on risk criteria. This raised the question whether it is the expert's job not only to do the risk estimates but also to evaluate them. On the other hand there did appear to be some opposition groups would not have been too unhappy with an annual risk of 'one in a million' ($10^{-6}/$year).

After the inquiry had ended, the question was raised of the possible hazard of ignition of a release of flammables by 'radio spark' transmissions from the nearby naval transmitter at Crimond. The HSE undertook extensive work on this 'radio spark' hazard.

The inquiry also illustrates another very basic problem, that if the company makes no modifications to its proposals as a result of objections posed, it appears to ride roughshod over the opposition, while if it does, it appears to be incompetent.

The inquiry and its aftermath give an impression of imbalance in the treatment of the hazards. In particular, the question of hazards associated with radio sparks received a disproportionate amount of attention relative to the hazards at the marine terminal. Despite this concession to the opposition groups, they were clearly dissatisfied with the inquiry.

A26.2 Pheasant Wood

The Pheasant Wood inquiry, which has been studied by Petts (1985b), on proposed housing development near to ICI chlorine and phosgene storages arose when the HSE asked the Secretary of State to call in the proposal for review after the LPA had indicated its intention to grant planning permission. The HSE gave its hazard assessment. The developer retained Cremer and Warner to present an alternative assessment, while ICI, although not directly involved, chose to offer its own assessment. All these parties presented technical evidence. The HSE's assessment included estimates of both frequencies and consequences of events, but did not combine these together to give individual and societal risks. They were criticized by the inquiry assessor and by ICI for this limitation. On the other hand, the other parties, although they did estimate these risks, did not always give the full basis of their estimates. Another point at issue between the HSE and the ICI evidence was some of the data used for the frequency of release. Whereas the HSE had to rely on generic data, the company was able to use data from its own installations which, depending on the viewpoint, were less pessimistic or more optimistic. The appropriateness of the gas dispersion models used was also an issue between the parties. The inspector found against HSE's case and in favour of granting planning permission. A feature of the Pheasant Wood inquiry which distinguishes it from those at Mossmorran and at Canvey was the low level of public interest.

A26.3 Canvey

The Canvey complex of oil refineries, chemical works, etc., has been the subject of several inquiries and hazard assessments, which have been described by Petts (1985a). The period 1965–73 saw a spate of planning applications and three public inquiries for oil refinery development. The economic pressures were clearly strong. The inquiry inspector's recommendations were twice overruled by the minister in favour of development. At the same time, the level of local concern over the hazards and environmental impact of the Canvey complex was rising and led to the opening, in 1975, of an inquiry into the possibility of revoking the planning permission granted to United Refineries Ltd (URL) in 1971. The technical assessor's report discussed the hazards in qualitative terms. No quantitative hazard assessments were presented.

One of the assessor's recommendations was a study of the possible hazardous interaction between the various installations at Canvey. Acting on this, the minister asked the HSE to look into the matter and the HSE commissioned a study from SRD, the First *Canvey Report*.

With the publication of this report, the minister was in a position to reopen the inquiry on the revocation of planning permission for URL. The inquiry was held in 1980 and considered both the bearing of the *Canvey Report* on the URL project and also the validity of the report itself, the parties being essentially the HSE and the objectors. The only full hazard assessment presented was that of the HSE in the form of the report. The report had not been commissioned specifically for the inquiry, but nevertheless there was for the first time available to an inquiry in Britain a full assessment of individual and societal risks.

One of the main hazards considered was a marine spill of LNG or LPG at the British Gas methane terminal.

Disagreements arose over the estimates both of the frequency and the magnitude of this hazard. While the Port of London Authority (PLA) considered that the report overestimated the frequency of ship collision, the objectors presented graphic qualitative evidence of the problems of tanker handling. Estimates of the distance travelled by the flammable gas cloud before it fell to its lower flammability limit varied widely from 1 km to 28 miles.

Doubt was cast on some of the HSE's proposals for mitigating the hazards. The practicality and effectiveness of the water spray barrier against an ammonia cloud was challenged. PLA witnesses stated that the 8 knot speed limit proposed in the Scheduled Area was not practical.

The estimated risks were also matter for debate. The risk values were relatively high. For example, the calculated fatality risk attributed to the URL extension was comparable with the national fatality risk at work. The inspector considered the risks in some areas unacceptable.

Although the *Canvey Report* received praise and its overall approach was not challenged, it had certain deficiencies which detracted from it. The text was very disjointed, there were discrepancies between text and tables and there were some inaccuracies.

The inspector recommended that the URL planning permission should not be revoked, but only if the risk from the British Gas methane terminal were reduced by installing a ring of igniters at the periphery or by closure.

In 1981, the Second *Canvey Report* appeared. In this reassessment, which used revised failure data, more refined hazard models and more accurate population profiles and which took into account changes made on the installations, the assessed risks were less, some by an order of magnitude.

In 1982, another public inquiry was held, this time on the possible discontinuance of the British Gas methane terminal. British Gas produced a full hazard assessment, while the two *convey Reports* formed the basis of the HSE's evidence. The British Gas assessment used by intent broadly realistic estimates, the HSE conservative ones. The risks estimated by British Gas were lower, in some cases by two orders of magnitude. Realism rather than conservatism was not the only reason for this. The British Gas estimates of failure rates were lower and its gas dispersion estimates gave lower travel distances. The inspector stated that he found HSE's evidence much easier to understand and that he preferred their conservative approach, and he put some value on the independence of the HSE.

A number of expert witnesses were called by the objectors, notably the Castle Point District Council and the Castle Point Refineries Resistance Group. In many cases, their evidence failed to stand up to cross-examination due to weak hypotheses on accident scenarios, oversimplified assumptions, incorrect hazard models, misunderstanding of the work of others, especially the Second *Canvey Report*, and numerical errors. The technical assessor played an active role in this inquiry both in advising the inspector and in producing the report.

The inspector recommended against closing the British Gas terminal or requiring igniters.

A26.4 Sizewell

The inquiry on the proposal to construct a PWR at Sizewell B began in January 1983 and ended in February 1985. This is by any standards a 'big' public inquiry and the hazards of a nuclear reactor are one of the topics on which much expert testimony has been heard. In the present context, the inquiry is of interest for its influence on the development of the form of such inquiries; for its influence on the formulation of risk criteria; and for the evidence given on some of the technical safety issues, particularly on rare events.

An important feature of the inquiry was the use of counsel to the inquiry. The inquiry also saw the increased use of informal 'side' meetings between experts at which issues can be clarified and misunderstandings removed.

Another feature of the inquiry is that a preconstruction safety report was available before the inquiry from the CEGB, although many design aspects remained unresolved.

With regard to risk, the Inspector suggested that where risks of the kind considered are accepted they are better described as 'tolerable' rather than 'acceptable' and encouraged the HSE to develop risk criteria. Such criteria have subsequently been published by the HSE (1988b, 1989d).

Among the many issues on hazards which the inquiry illustrates is that of the very rare event. A crucial accident scenario for a PWR is failure of the reactor pressure vessel (RPV). Such a failure is generally agreed to be a very rare event, but it may have very serious consequences. Estimation of the frequency of an event which has a high hazard potential but is very infrequent, perhaps scarcely credible, is one of the most intractable problems in hazard assessment. The inquiry is instructive not only because there were differences between experts in the methodology of estimation of the RPV failure frequency but also because there was scope for differences of perception of what was eventually established.

Evidence on the estimation of the RPV failure frequency was given for the CEGB by Dr B. Edmundson, for the Greater London Council by Technica Ltd and for the Suffolk Local Authorities by Professor Kussmaul from Germany. Another expert, Professor T.A. Kletz, gave evidence at the invitation of the inspector. On the basis of historical pressure vessel failure data Technica estimated the RPV failure rate at 1×10^{-6}/year. This figure was derived from the crude figure of 3.2×10^{-6}/year obtained from the historical data, and was intended to give credit for improved technology embodied in the RPV, but to be nevertheless a realistic rather than a conservative estimate, as the witnesses were at pains to point out. The CEGB derived their estimate of RPV failure rate using engineering principles, such as fracture mechanics, to determine the frequency of the failure modes which they considered credible, and obtained for catastrophic vessel rupture with generation of missiles an estimate of 2.4×10^{-9}/year. Professor Kussmaul preferred the CEGB's approach to that of Technica but advocated an alternative method of pressure vessel construction so as to minimize the risk of weld failure. This construction was not accepted by the CEGB and after further discussion between the parties Professor Kussmaul accepted the CEGB's approach subject to certain procedures which he specified and to which they agreed. As a result, the Suffolk Local Authorities withdrew their objections on grounds of safety, since they accepted Professor Kussmaul's advice that the risks were 'so low as to be incredible'.

There were nearly two orders of magnitude between the CEGB and Technica estimates of the RPV failure rate. The Technica estimate was in fact based on a single historical pressure vessel failure, which the CEGB regarded as so unusual as to be inconceivable on their vessel. However, even if no failure has occurred historically, it is still possible to make a statistical estimate by using the number of failure-free years and assuming that a failure is just about to occur. It therefore makes little difference to the number obtained whether the number of failures is one or zero. More important was the difference between the parties on whether the estimate should be based on historical statistics or engineering principles. Technica argued that even if the single historical failure was a bizarre one, such events cannot be completely discounted, that this stood in for the whole set of bizarre events which might occur, and that the method based on the historical data was the appropriate one to use.

In his cross-examination, counsel for the CEGB sought to identify assumptions in the Technica estimate. For Technica Dr. R.A. Cox stated clearly that the figure of 1×10^6/year was a 'best estimate', intended as realistic rather than conservative. However, counsel drew attention to the fact that in its written evidence Technica had at one point referred to this figure as an 'upper bound'. Dr P.J. Kayes for Technica then explained that what was meant by this was that the figure was an 'upper bound on the type of studies that would claim RPV failure to be incredible'. Counsel then asked: 'Is it intended to provide an upper bound to the likelihood of RPV failure?' and the witness replied: Yes, it is, as a best estimate'. Thus Technica were liable to believe that their figure was accepted as a realistic estimate, the CEGB that it had been conceded to be conservative.

In cross-examination, Professor Kletz commented that the two methods should not be seen as necessarily in conflict. A difference between the two by orders of magnitude was understandable. He did say, however, that while a large improvement in the reliability of the CEGB vessel over historic vessels could be expected, he had some personal scepticism whether any man-made artefact could provide such a low figure as the CEGB was claiming. In the case of the RPV the failure frequency was so low that he felt there was some justification for making allowance as Technica proposed for bizarre events and referred to possible failure due to previously unknown metallurgical phenomena. When cross-examined by one of the assessors, he agreed that his scepticism regarding very low numbers applied to a very low figure for a single event rather than one obtained as the product of the figures for a chain of events.

Thus, while the differences of opinion over points raised by Professor Kussmaul were largely settled by discussion outside the formal proceedings and the main findings reported back to the inquiry, the differences arising from the Technica evidence were not resolved at the side meetings and the parties were liable to interpret the evidence given in the inquiry cross-examination as supporting their particular positions.

A26.5 Expert Evidence

The foregoing account of some of the principal inquiries which have been held on hazardous installations indicates that there are a number of problem areas. These problems have been reviewed and proposals made for mitigating them by Petts, Withers and Lees (1986).

A26.5.1 Problems of adversarial forum
A view commonly expressed by technical experts is that the adversarial situation of a public inquiry is not an

appropriate means of examining technical issues. Many experts find the adversarial process unfamiliar and intimidating and regard it as an inappropriate means of arriving at the truth on technical matters. The adversarial process tends to lead to adoption of rigid, defensive positions.

For interest groups opposed to a development, the adversarial forum of an inquiry has the advantage that they can examine the developer's case, but can be intimidating to them also and it presents considerable difficulties for groups with few resources.

A26.5.2 Problems with hazard assessments

Many of the problems of expert testimony at public inquiries on process installations centre round hazard assessment. Most of these difficulties can be grouped under one or other of the following heads:

(1) obscurity;
(2) incompleteness;
(3) inconsistency.

Obscurity

There may be inherent difficulties in the subject matter or in its presentation.

Thus at the Sizewell inquiry the debate on engineering features of the pressure vessel was difficult to follow without engineering drawings. At the Canvey inquiry, the inspector himself had difficulties with the hazard assessment by British Gas and felt that the mode of presentation was more appropriate to a learned society than an inquiry.

Fundamentally difficult matters are typified by the arguments at Sizewell over the frequency of rare events and over probabilistic risk analyses.

Incompleteness

Information may be incomplete for many reasons. The engineering design may not be complete by the time of the inquiry, there may be deliberate selectivity in the hazards examined and there may be simple failure to identify all the hazards.

Thus at the Mossmorran and Sizewell inquiries much plant-specific information was unavailable because the projects were still in the feasibility stage. Estimates were therefore conceptual and based on generic data. This is likely to be a common situation.

Selectivity is exemplified by Mossmorran, where the objectors expressed concern about hazards at the port which seemed to have received less attention than the gas plant itself. Again at Mossmorran, there were objections to the exclusion of aircraft crashes. Later the possible hazard of ignition from radio sparks was raised and led to a separate investigation.

Inconsistencies

Some inconsistencies appear at an inquiry simply because of the lack of an agreed structure and methodology for the hazard assessment. There are also differences in the particular models used and in the basic approach to probability estimation.

Examples of differing structure and methodologies appeared at Pheasant Wood, where the assessment by the HSE did not provide any estimates of individual or societal risk, while those of the consultants and of ICI did. The HSE case was based on generic data, while the ICI case was based on plant-specific data.

At Canvey there were wide differences in the results obtained from different gas dispersion models. There have also been wide differences in the relations describing injury and damage. This is illustrated by the *Rijnmond Report*, which, although not a document presented to an inquiry, represents the type of hazard assessment which might be submitted. The report revealed ten-fold differences in estimates of the lethal concentrations of chlorine.

Such differences and uncertainties usually result in judgements being made about the likely margin of error. These judgements may occur at a number of points in the chain of calculations which make up the assessment. They may be optimistic or pessimistic.

In general, industry is anxious that the risks should not be exaggerated and presses for realistic assessments wherever possible. It may seek to avoid the use of numerical estimates of risk altogether where the confidence bounds are very wide. The HSE, on the other hand, is bound in the public interest to take a fairly cautious view and may be expected to incline towards the use of values which are to some degree conservative. In the risks assessed for the methane terminal at Canvey, the HSE and British Gas sometimes differed by a factor of a hundred, partly due to the difference between realism and conservatism.

Examples of the use of differences in approach to probability estimation were demonstrated at Sizewell, where the objectors based their estimate of pressure vessel failure rate on historical data and the CEGB its estimate on engineering considerations, the latter thereby obtaining for the event in question a frequency lower by some two orders of magnitude.

It should be recognized that public opposition to an industrial development may in some cases be so strong as to influence the character of the presentation and examination of the hazard assessments. At Pheasant Wood there was little public interest, whereas at Mossmorran this interest was strong and organized.

A26.5.3 Availability of information

An essential requirement of the parties to an inquiry is the availability of information. This information needs to be of the appropriate kind and to be provided at the right time. This requirement would in large part be met if the company makes available at the outset of the inquiry a form of open conceptual safety case.

The LPA may wish to commission an independent report from a consultant, as well as seeking advice from the HSE.

The use of open safety cases presented in a uniform style would greatly assist the LPAs to become familiar with the problems posed by hazardous installations and to instruct consultants.

The HSE may also provide for the inquiry inspector an independent hazard assessment by which he can evaluate the evidence of the other parties.

Twenty-eight days before the inquiry the LPA is required to produce a written statement. It is only the LPA which has this obligation.

A26.5.4 Parties to inquiry

The LPA will always be a party to a planning inquiry. There is no general requirement that staff from other government bodies (such as the HSE) should appear in person. In practice they often do.

The parties will often be represented by legal counsel at the inquiry. An innovation at Sizewell was the appointment of counsel to the inquiry itself. This appears to have been of great assistance to the inspector and to have helped other parties to clarify issues. In an adversarial process, the investigatory role of such counsel appears of great value, particularly for technically complex projects.

The part played by the HSE at an inquiry needs to be seen in the context of its wider and ongoing role in the control of hazards. In order to safeguard this role, it is desirable that as far as possible the HSE should appear as an adviser rather than a protagonist and the situation where the HSE is the instigator of an inquiry is to be avoided as far as possible.

It is appropriate that at the outset of an inquiry the HSE be invited to state its role, the legislation under which it operates and the limits of its powers, that it define the set of principles by which it judges the nature of the risk and the relevant safety criteria required by the HSWA as enforced by the HSE. It may also indicate its expectations of the ability of the developer to meet these standards, while leaving it to the inquiry to balance these expectations with those of others, and with other relevant factors.

Many experts are called by objectors. The problems experienced at past inquiries have often centred around disagreements between these experts and others. It is desirable that these experts play their full part in the formal and informal meetings described below. In this way, issues can be clarified and difficulties which have arisen in the past avoided. Another class of witness is those called by the inspector, but these appear to present fewer problems.

A26.5.5 Arrangements at inquiry

A large inquiry usually has a formal preliminary meeting presided over by the inspector at which a detailed set of objectives, procedures and timetable are developed from the terms of reference. Such a meeting, which largely determines the subsequent course of the inquiry, is essential for large, technically complex projects.

The normal procedure in a public inquiry is to hear evidence by party rather than by issue, with the project's proposer having the final right of reply. This results in safety evidence being interspersed with other evidence on other topics, and not taken in its logical order. An exception tends to be made in big inquiries such as Sizewell, where hearing may be by issue. In general, hearing by issue suits the large organizations, who have little difficulty in making experts available on each issue as it arises. It may, however, create problems for smaller bodies with limited resources.

As described above, there has been some criticism of the adversarial approach inherent in public inquiries. In eyes of the public, however, it remains a safeguard, albeit imperfect, against concealment and is an important factor in legitimizing the inquiry process.

The problem of dealing with expert evidence in an adversarial forum may be mitigated by the use of non-adversarial side meetings between experts. Some of these may be formal meetings with minutes kept which are part of the inquiry record.

Appendix

27

Standards and Codes

Contents

Standards and codes of practice are one of the principal means by which experience of problems and their respective solutions is transmitted and fundamental to recognized and generally accepted good industrial practice. The production of a code is normally undertaken by a group of experienced practitioners under the aegis of an appropriate organization. The code represents the consensus of this group on what constitutes good practice. There is usually some degree of consultation with the parties who will use the code.

There are a vast number of codes produced by a variety of bodies and the status of a code depends on the body which promulgates it. Inevitably a code becomes out-of-date as technology changes, so that a process of continuous revision is necessary. A 5-year update is most commonly-suggested.

The philosophy commonly adopted in writing a new code is to define limited objectives and to get the code finished, recognizing that further revision will be necessary in due course, rather than to seek to produce a definitive work. Standards and codes of practice normally use a specialized and defined terminology.

Mandatory requirements, that is, those that it is essential to satisfy, are expressed by the use of the word 'shall'. The word 'must' is used when an absolute imperative arises, for example when safety considerations apply.

Non-mandatory requirements, that is, those that are recommended or desirable are usually expressed by the use of the auxiliary 'should'. Should implies a recommendation based upon the judgement of experienced people but recognizes that some discretion is appropriate in special circumstances.

A27.1 Globalization of Standards

The ever increasing world of international trade has created a need for internationally accepted standards. When the large majority of products or services in a particular business or industry sector conform to International Standards, a state of industry-wide standardization can be said to exist. This is achieved through consensus agreements between national delegations representing all the economic stakeholders concerned – suppliers, users, government regulators and other interest groups, such as consumers. They agree on specifications and criteria to be applied consistently in the classification of materials, in the manufacture and supply of products, in testing and analysis, in terminology and in the provision of services. Hence, International Standards provide a reference framework, or a common technological language, between suppliers and their customers, thereby facilitating trade and the transfer of technology.

International Organization for Standardization (ISO) together with the International Electrotechnical Commission (IEC) and International Telecommunication Union (ITU) has built a strategic partnership with the World Trade Organization (WTO) with the common goal of promoting a free and fair global trading system. The political agreements reached within the framework of the WTO require underpinning by technical agreements. ISO, IEC and ITU, as the three principal organizations in international standardization, have the complementary scopes, the framework, the expertise and the experience to provide this technical support for the growth of the global market.

ISO is the world's largest developer of standards. ISO is a network of the national standards institutes of

148 countries, on the basis of one member per country, with a Central Secretariat in Geneva, Switzerland, that coordinates the system. ISO is a non-governmental organization, its members are not delegations of national governments. ISO occupies a position between the public and private sectors. This is because many of its member institutes are part of the governmental structure of their countries, while other members have their roots uniquely in the private sector, such as industry associations and technical societies.

Therefore, ISO is able to act as a bridging organization in which a consensus can be reached on solutions that meet both the requirements of business and the broader needs of society, such as the needs of stakeholder groups like consumers and users. The vast majority of ISO standards are highly specific to a particular product, material or process. However, the ISO 9000 and ISO 14000 standards have earned a worldwide reputation are known as 'generic management system standards'. 'Generic' means that the same standards can be applied to any organization, large or small, product or service, and whether it is a business enterprise, a public administration or a government department. 'Management system' refers to what the organization does to manage its processes, or activities. 'Generic' also signifies that no matter what the organization is or does, if it wants to establish a quality management system or an environmental management system, then such a system has a number of essential features that are spelled out in the relevant standards of the ISO 9000 or ISO 14000 standards. The US OSHA process safety management regulation is another example of the performance-based philosophy defined in the generic management system standards.

A27.1.1 Conformity assessment

Conformity assessment means checking that products, materials, services, systems or people measure up to the specifications of a relevant standard. Many products require testing for conformance with specifications or compliance with safety, or other regulations before they can be put in the stream of commerce. Even simpler products may require supporting technical documentation that includes test data. ISO has developed many of the standards against which products are assessed for conformity, as well as the standardized test methods that allow the meaningful comparison of test results so necessary for international trade.

ISO itself does not carry out conformity assessment. ISO develops ISO/IEC guides and standards to be used by organizations that carry out conformity assessment activities. The voluntary criteria contained in these guides and standards represent an international consensus on what constitutes best practice. Their use contributes to the consistency and coherence of conformity assessment worldwide.

A27.2 Where to Find Information on Standards

There are several hundred thousand standards and technical regulations in the world containing special requirements for a particular country or region. Finding information about these, or about related conformity assessment activities, can be a heavy task. ISONET, the SO Information Network, can ease the problem. This is a worldwide network of national standards information centres, which have cooperatively developed a system to provide rapid access to information about standards, technical regulations and testing and conformity assessment activities in operation around the world. The World Trade Organization's Agreement on Technical Barriers to Trade (WTO/TBT) typically establishes a national enquiry point to answer questions on regulations and standards in each country, the ISONET and WTO enquiry points are the same. WSSN is a network of publicly accessible Web servers of standards organizations around the world. It contains links to international, regional and national standardization bodies, and also to other international and regional organizations that develop standards in their specialized subject area; available at http://www.wssn.net/WSSN/index.html.

Leading national standards institutions, which co-operate with ISO are:

- British Standards Institution (BSI);
- American National Standards Institute (ANSI);
- Deutsches Institut fur Normung (DIN);
- Verein Deutscher Ingenieure (VDI);
- Association Francaise de Normalisation (AFNOR);
- Comite de Normalisation Europeen (CEN);
- Comite de Normalisation Europeen Electrotechnique (CENELEC).

Some technical associations that generate standards and codes of practice relevant to safety and loss prevention are listed below.

- American National Standards Institute (New York);
- American Industrial Hygiene Association (Fairfax, VA);
- American Petroleum Institute (Washington, DC);
- American Society of Civil Engineers (New York);
- American Society of Heating, Refrigerating and Air-Conditioning Engineers (Atlanta, GA);
- American Society of Mechanical Engineers (New York);
- American Society for Quality Control (Milwaukee, MI);
- American Society for Testing and Materials (Philadelphia, PA);
- American Welding Society (Miami, FL);
- British Chemical Industry Safety Council;
- British Compressed Gases Association (London);
- Chemical Industry Safety and Health Council (London);
- Det Norske Veritas (Stavanger, Norway);
- Imperial Chemical Industries Pic;

- Institute of Electrical and Electronic Engineers (New York);
- Institute of Petroleum (London);
- Instrument Society of America (Research Triangle Park, NC);
- International Chamber of Shipping (London);
- International Maritime Organization (London);
- Liquefied Petroleum Gas Industry Technical Association (London);
- National Fire Protection Association (Quincy, MA);
- Oil Companies International Marine Forum (London);
- Society of International Gas Tanker and Terminal Operators (London);
- Tubular Exchangers Manufacturers Association;
- Underwriters Laboratories (Northbrook, IL);
- Verein Deutscher Ingenieure.

A27.2.1 Among the first safety codes

Modern fire safety codes and standards, and in particular those developed by the National Fire Protection Association (NFPA), trace their origins to the nineteenth-century development of automatic sprinklers. From the beginning, sprinklers performed properly as extinguishing devices; however, they were originally installed in so many different ways that their reliability was uncertain.

From the beginning, the mission of the NFPA is to reduce the burden of fire and related hazards on the quality of life. The NFPA mission today is accomplished by advocating scientifically based consensus codes and standards, research and education for fire and related safety issues. NFPA's National Fire Codes are developed by technical committees staffed by over 6000 volunteers, and are adopted and enforced throughout the world. NFPA functions as a non-profit membership organization with more than 68,000 members worldwide.

Development of NFPA standards follow the original principles still followed today by NFPA Technical Committees:

'To bring together the experience of different sections and different bodies of underwriters, to come to a mutual understanding, and, if possible, an agreement on general principles governing fire protection, to harmonize and adjust our differences so that we may go before the public with uniform rules and conditions which may appeal to their judgment is the object of this Association.'

Despite the term 'National' in NFPA, the name sometimes seems to imply the lack of international involvement.

Appendix 28

Institutional Publications

A General

American Conference of Government Industrial
Hygienists
(Cincinnati, OH)

American Gas Association
(Arlington, VA)

American Industrial Hygiene ssociation
(Fairfax, VA)

American Institute of Chemical Engineers
(New york)

1989   Experimental Results from the Design
Institute for Physical Property Data,1:
Phase Equilibria and Pure Component
Properties – R II ed. by T.T. Shih and
D.K. Jones   159

American Oil Company (AMOCO)
(Chicago, IL)

Refinery	Process Safety Booklets	Item
1	Hazard of Water	1
2	Hazard of Air	2
3	Safe Furnace Firing	3
4	Safe Ups and Downs	4
5	Hazard of Electricity	5
6	Hazard of Steam	6
7	Safe Handling of light Ends	7
8	Engineering for Safe Operation	8
9	Safe Operation of Air, Ammonia and Ammonium Nitrate Plants	9
10	Safe Operation of Refinery Steam Generators and Water Treating Facilities	10

American Petroleum Institute
(New York)

EA 7301   Guidelines on noise
EA 7102   The chronic toxicity of lead by
D.C. Jessup *et al.*
EA 7402   A mortality study of petroleum refinery
workers by Tabershaw–Cooper
Associates
69–7   A review of the toxicology of lead by
T.J. Haley
75–25   Review of inhalation toxicology of
sulphuric acid and sulphates by
M.C. Battigelli and J.F. Gamble
Bull. 2516   Evaporation loss from low pressure
tanks, 1962
Bull. 2521   Use of pressure–vacuum vent valves for
atmospheric pressure tanks to reduce
evaporation loss, 1966
Bull. 6F2   Bulletin on fire resistance improvements
for API flanges, 2nd ed., 1994
Publ. 301   Above ground storage tank survey:
1989, 1991
Publ. 302   Waste minimization in the petroleum
industry: a compendium of practices, 1991
Publ. 306   An engineering assessment of volumetric
methods of leak detection in above
ground storage tanks, 1991
Publ. 307   An engineering assessment of acoustic
methods of leak detection in above
ground storage tanks, 1991
Publ. 311   Environmental design considerations for
petroleum refining processing units, 1993
Publ. 315   Assessment of tankfield dike lining
materials and methods
Publ. 317   Industry experience with pollution
prevention programs, 1993
Publ. 421   Management of water discharges:
design and operation of oil-water
separators, 1990
Publ. 910   Digest of state boiler and pressure vessel
rules and regulations, 6th ed., 1992

Publ. 920   Prevention of brittle fracture of pressure
vessels, 1990
Publ. 941   Steels for hydrogen service at elevated
temperatures and pressures in petroleum
refineries and petrochemical plants,
4th ed., 1990
Publ. 999   Technical data book – petroleum
refining, 4th ed., 1983
Publ. 1108   API petroleum pipeline maps,
12th ed., 1985
Publ. 2009   Safe practices in gas and electric cutting
and welding in refineries, gasoline
plants, cycling plants and petrochemical
plants, 5th ed., 1988
Publ. 2013   Cleaning mobile tanks in flammable or
combustible liquid service, 6th ed., 1991
Publ. 2015   Cleaning petroleum storage tanks,
3rd ed., 1985
Publ. 2021   Guide for fighting fires in and
around petroleum storage tanks,
3rd ed., 1991
Publ. 2023   Guide for safe storage and handling of
heated petroleum-derived asphalt
products and crude oil residue,
2nd ed., 1988
Publ. 2027   Ignition hazards involved in abrasive
blasting of atmospheric storage tanks in
hydrocarbon service, 2nd ed., 1988
Publ. 2028   Flame arresters in piping systems,
2nd ed., 1991
Publ. 2030   Guidelines for application of water
spray systems for fire protection in
the petroleum industry, 1987
Publ. 2031   Combustible gas detector systems and
environmental/operational factors
influencing their performance, 1991
Publ. 2200   Repairing crude oil, liquefied petroleum
gas, and product pipelines, 2nd ed.,
1983
Publ. 2202   Dismantling and disposing of steel from
tanks which have contained leaded
gasoline, 2nd ed., 1982
Publ. 2203   Fire precautions for fueling fixed,
portable and self-propelled engine-driven
equipment, 3rd ed., 1987
Publ. 2207   Preparing tank bottoms for hot work,
4th ed., 1991
Publ. 2214   Spark ignition properties of hand tools,
3rd ed., 1989
Publ. 2216   Ignition risk of hydrocarbon vapors by
hot surfaces in the open air, 2nd ed., 1991
Publ. 2217A   Guidelines for work in inert confined
spaces in the petroleum industry, 1987
Publ. 2218   Fireproofing practices in petroleum and
petrochemical processing plants, 1988
Publ. 2219   Safe operating guidelines for
vacuum trucks in petroleum service, 1986
Publ. 2510   Design and construction of liquefied
petroleum gas installations (LPG),
6th ed., 1989
Publ. 2510A   Fire protection considerations for
the design and operation of liquefied
petroleum gas (LPG) storage
facilities, 1989

Publ. 2514A Atmospheric hydrocarbon emissions from marine vessel transfer operations, 2nd ed., 1981

Publ. 2517 Evaporation loss from external floating-roof tanks, 2nd ed., 1980

Publ. 2519 Evaporation loss from internal floating-roof tanks, 3rd ed., 1983

Publ. 4322 Fugitive hydrocarbon emissions from petroleum production operations, vols 1–2, 1980

Publ. 4434 Review of groundwater models, 1982

Publ. 4456 Literature survey: aerosol formation, and subsequent transformation and dispersion, during accidental release of chemicals, 1986

Publ. 4457 API evaluation of urban dispersion models with the St Louis RAPS data base, 1987

Publ. 4459 Surface roughness effects on heavier-than-air gas diffusion, 1987

Publ. 4461 Development and evaluation of the OCF/API model, 1988

Publ. 4487 Evaluation of the biodegradation predictive equations in EPA's CHEMDATA6 model, 1989

Publ. 4491 Effect of homogeneous and heterogeneous surface roughness on heavier-than-air gas dispersion, vol. 1, 1989

Publ. 4492 Effect of homogeneous and heterogeneous surface roughness on heavier-than-air gas dispersion, vol. 2, 1989

Publ. 4519 API exposure classification scheme for collection of industrial hygiene monitoring data, 1990

Publ. 4522 Results of hazard response model evaluation using Desert Tortoise (NH_3) and Goldfish (HF) data bases, vol. 1: summary report, 1990

Publ. 4523 Results of hazard response model evaluation using Desert Tortoise (NH_3) and Goldfish (HF) data bases, vol. 1: appendices, 1990

Publ. 4539 Evaluation of area and volume source dispersion models for petroleum and chemical industry facilities Phase I final report, 1992

Publ. 4540 Area and volume source air quality model performance evaluation Phase II final report, 1992

Publ. 4545 Hazard response modeling uncertainty (a quantitative method): user's guide for software for evaluating hazardous gas dispersion models, vol. 1, 1992

Publ. 4546 Hazard response modeling uncertainty (a quantitative method): evaluation of commonly-used hazardous gas dispersion models, vol. 2, 1992

Publ. 4547 Hazard response modeling uncertainty (a quantitative method): components of uncertainty in hazardous gas dispersion models, vol. 3, 1992

Publ. 4559 Results of toxicological studies conducted for the American Petroleum Institute Health and Environmental Sciences Department, 1993

Publ. 4571 A fundamental evaluation of CHEMDATA7 air emissions model, 1993

DR 229 Review of revisions to the offshore and coastal dispersion (OCD) model, 1989

Safety Digest of Lessons Learned

Section 2	Safety in Unit Operations, 1979	1
Section 3	Safe Operation of Auxiliaries, 1980	2
Section 4	Safety in Maintenance, 1981	3
Section 5	Safe Operation of Utilities, 1981	4
Section 6	Safe Operation of Storage Facilities, 1982	5
Section 7	Safe Handling of Petroleum Products, 1983	6
Section 8	Environmental Controls, 1983	7
Section 9	Precautions against Severe Winter Conditions, 1983	8

Annual Summaries

Review of fatal injuries in the petroleum industry 9

Reported fire losses in the petroleum industry 10

Summary of occupational injuries and illnesses in the petroleum industry 11

Other publications

TRC Thermodynamic tables – hydrocarbons 12

TRC Thermodynamic tables – non-hydrocarbons 13

Introduction to the oil pipeline industry, 1984 14

Fracture toughness of steels for API Standard 620 Appendix R Tanks, 1985 15

Validation of heavy gas dispersion models with experimental results of the Thorney Island trials, vols 1–2, 1986 16

Quantified hazards evaluation of marine vapor recovery systems, 1989 17

Safety guidelines for people who live and work near petroleum pipelines, 1990 18

Investigation and prediction of cooling rates during pipeline maintenance welding, and user's manual for Battelle's hot-tap thermal analysis models, 1991 19

Special report on deterioration of arrester safety 'Mitigation of Explosion Hazards of Marine Vapor Recovery Systems' by R.E. White and C.J. Oswald, 1992 20

The following publications are not longer listed

4030 An evaluation of dispersion formulas by G.E. Anderson, R.R. Hipler and G.D. Robinson, 1969

4131 Experimental human exposure to high concentration of carbon monoxide, 1972

4262 Refinery odor control and ambient levels of emissions from refinery operation: final report, 1975

4360 Atmospheric dispersion over water and in the shoreline transition zone, vols 1–2, 1982

Bull.1003 Precautions against electrostatic ignition during loading of tank truck motor vehicles, 2nd ed., 1975

Bull. 2513   Evaporation loss in the petroleum industry –
             causes and control, 1959 (reaffirmed 1973)
Bull. 2518   Evaporation loss from fixed roof tanks, 1962
Bull. 2522   Comparative test methods for evaluating
             conservation mechanisms for evaporation
             loss, 1967
Bull. 2523   Petrochemical evaporation loss from storage
             tanks, 1969
Publ. 700   Checklist for plant completion,
             1981
Publ. 756   Recommended guidelines for documentation
             of training, 1977
Publ. 757   Training and materials catalog, 1979
Publ. 940   Steel deterioration in hydrogen, 1967
Publ. 942   Recommended practice for welded
             plain carbon steel refinery equipment
             for environmental cracking
             service, 1971
Publ. 945   A study of the effects of high temperature,
             high pressure hydrogen on low-alloy
             steels, 1975
Publ. 956   Hydrogen assisted crack growth in $2\frac{1}{4}$ Cr-1
             Mo steel, 1978
Publ. 2002   Inspection for accident prevention, 1984
Publ. 2004   Inspection for fire protection, 1984
Publ. 2007   Safe maintenance practices in refineries,
             2nd ed., 1983
Publ. 2008   Safe operation of inland bulk plants,
             4th ed., 1984
Publ. 2015   A guide for controlling the lead hazard
             associated with tank entry and cleaning,
             2nd ed., 1982
Publ. 2015B Cleaning open-top and covered floating roof
             tanks, 1981
Publ. 2017   First aid training guide, 7th ed., 1979
Publ. 2022   Fire hazards of oil spills on waterways,
             2nd ed., 1982
Publ. 2025   Emergency planning and mutual aid for
             products terminals and bulk plants, 1978
Publ. 2201   Procedures for welding or hot tapping
             on equipment containing flammables,
             3rd ed., 1985
Publ. 2206   Identification of compressed gases
             in cylinders, 1970 (reaffirmed 1981)
Publ. 2209   Pipe plugging practices, 1978
Publ. 2210   Flame arresters for vents of tanks storing
             petroleum products, 2nd ed., 1982
Publ. 2212   Ignition risks of ordinary flashlights,
             2nd ed., 1983
Publ. 2217   Guidelines for confined space work in the
             petroleum industry, 1984
PSD 2213   Ignition risks of ordinary telephones, 1974
PSD 2214   Spark ignition properties of
             hand tools, 1980
PSD 2216   Ignition risk of hot surfaces
             in air, 1980

American Society of Civil Engineers
(New York)

Year		Item
1965	Report on Pipeline Location	1
1974	Pipelines in the Ocean	2
1974	Study of Damage to a Residential Structure from Blast Vibrations by J.F. Wiss and H.R. Nicholls	3
1975	Fire Resistance of Concrete Structures	4
1975	Pipeline Design for Hydrocarbon Gases and Liquids	5
1976	Source Book on Environmental and Safety Considerations for Planning and Design of LNG Marine Terminals	6
1978	Management of Engineering of Control Systems for Water Pipelines	7
1978	Overview of the Alaska Highway Gas Pipeline: The World's Largest Project	8
1979	Pipelines in Adverse Environments	9
1979	Pipelines in Adverse Environments, vols 1–2	10
1979	Predicting and Designing for Natural and Man-Made Hazards	11
1980	Dynamic Response of Structures: Experimentation, Observation, Prediction and Control ed. by G. Hart	12
1980	Ports '80 ed. by J. Mascenik	13
1981	Lifeline Earthquake Engineering: The Current State of Knowledge ed. by D.J. Smith	14
1983	Advisory Notes on Lifeline Earthquake Engineering	15
1983	Pipelines in Adverse Environments II ed. by M.B. Pickell	16
1983	Ports '83 ed. by Kong Wong	17
1983	Seismic Response of Buried Pipes and Structural Components	18
1984	Design and Operation of Pipeline Control Systems ed. by L. Fletcher	19
1984	Fluid/Structure Interaction during Seismic Excitation	20
1984	Lifeline Earthquake Engineering: Performance, Design and Construction ed. by J.D. Cooper	21
1984	Pipeline Materials and Design ed. by B.J. Schrock	22
1985	Design of Structures to Resist Nuclear Weapons Effects, rev. ed.	23
1985	Quality in the Constructed Project ed. by A.J. Fox and H.A. Cornell	24
1985	Reducing Failures of Engineered Facilities	25
1985	Seismic Experience Data – Nuclear and Other Plants ed. by Y.N. Anand	26
1986	Building Motion in Wind ed. by N. Isyumov and T. Schanz	27
1986	Dynamic Response of Structures	28
1986	Modeling Human Error in Structural Design and Construction ed. by A.S. Nowak	29
1986	Physics-Based Modeling of Lakes, Reservoirs and Impoundments ed. by W.G. Gray	30
1986	Ports '86 ed. by P.M. Sorensen	31
1986	Techniques for Rapid Assessment of Seismic Vulnerability ed. by C. Scawthorn	32

American Society of Heating, Refrigerating and Air-Conditioning Engineers
(Atlanta, GA)

American Society of Mechanical Engineers
(New York)

1984   Symposium on Flow Induced
        Vibrations, vols 1–6, ed.
        M.P. Paidoussis *et al.*   182

1984   Two-Phase Flow and Waterhammer
        Loads in Vessels, Piping and
        Structural Systems ed. F.J. Moody and
        Y.W. Shin   183

1985   Pressure Vessel and Piping: Design and
        Analysis – A Decade of Progress ed.
        C. Sundararajian   184

1985   Proceeding of the 1985 Pressure Vessels
        and Piping Conference   185

PVP 35   Fracture Mechanics, Creep and Fatigue
          Analysis ed. by C. Becht, S.K. Bhander
          and B. Tompkins, 1988

PVP 50   Application of Modal Analysis to
          Extreme Loads ed. by C. Lin, 1988

PVP 53   Current Topics in Piping and Pipe
          Support Design ed. by E. van Stijgeren,
          1981

PVP 54   Computer-Aided Design of Pipe and Pipe
          Supports ed. by E. van Stijgeren, 1981

PVP 69   Practical Considerations in Piping
          Analysis ed. by E. van Stijgeren *et al.*,
          1982

PVP 103   Fatigue and Fracture Assessment by
          Analysis and Testing ed. by
          S.K. Bhandari *et al.*, 1986

PVP 104   Flow-Induced Vibration – 1986 ed. by
          S.S. Chen, J.C. Simonis and Y.S. Shin

PVP 105   Design and Analysis of Plate and Shells
          ed. by G.E.O. Widera *et al.*, 1986.

PVP 106   Advances in Impact, Blast, Ballistics, and
          Dynamic Analysis of Structures ed. by
          H.H. Chung and D.W. Nicholson, 1986

PVP 107   Pressure Vessels, Piping, and
          Components – Design and Analysis ed.
          by W.E. Short *et al.*, 1986

PVP 108   Seismic Engineering for Piping Systems,
          Tanks, and Power Plant Equipment ed.
          by T.H. Liu *et al.*

PVP 109   Symposium on ASME Codes and Recent
          Advances in PVP and Valve Technology
          Including a Survey of Operations
          Research Methods in Engineering ed. by
          J.T. Fong, 1986

PVP 110   High Pressure Technology, Design,
          Analysis and Safety of High Pressure
          Equipment ed. by D.P. Kendall, 1986

PVP 111   Materials Property Data: Applications
          and Access ed. by J.G. Kaufman, 1986

PVP 112   Design and Analysis Methods for Plant
          life Assessment ed. by T.V. Narayanan and
          T.J. Kozik, 1986

PVP 114   Properties of High Strength Steels for
          High-Pressure Containments ed. by
          E.G. Nisbett, 1986

PVP 115   Nonlinear Analysis and NDE of
          Composite Material Vessels and
          Components ed. by D. Hui, J.C. Duke
          and H. Chung

PVP 116   New Concepts for Verification of
          Seismic Adequacy of Equipment in
          Operating Plants, 1986

PVP 120   Design and Analysis of Piping,
          Pressure Vessels, and Components
          ed. by W.E. Short *et al.*, 1987

PVP 121   Design and Analysis of Composite
          Material Vessels ed. by D. Hui and
          T.J. Kozik, 1987

PVP 122   Flow-Induced Vibration – 1987 ed. by
          M.K. Au-Yang and S.S. Chen

PVP 125   High Pressure Engineering
          and Technology ed. by
          J.A. Knapp, 1987

PVP 127   Seismic Engineering – Recent Advances
          in Design, Analysis, Testing, and
          Qualification Methods ed. by T.H. Liu
          and A. Marr

PVP 128   Fluid Structure Vibration and Liquid
          Sloshing ed. by D.C. Ma and T.C. Su

PVP 129   Advances in Piping Analysis and Life
          Assessment for Pressure Vessels and
          Piping ed. by S.J. Chang, R.C. Gwaltney
          and T.Q. McCawley, 1987

PVP 130   Advances in Design and Analysis in
          Pressure Vessel Technology ed. by
          H. Chung *et al.*

PVP 133   Damping ed. by F. Kara *et al.*, 1988

PVP 134   Shock and Wave Propagation ed. by
          Y.S. Shin and K.S. Kim, 1988

PVP 136   Codes and Standards and Applications
          for Design and Analysis of Pressure
          Vessel and Piping Components ed. by
          R. Seshadri *et al.*, 1988

PVP 137   Dynamics and Seismic Issues in
          Primary and Secondary Systems ed.
          by K. Karim- Panahi and
          A. Der Kiureghian, 1988

PVP 138   Life Extension and Assessment: Nuclear
          and Fossil Power-Plant Components ed.
          by C.E. Jaske *et al.*, 1988

PVP 139   Design and Analysis of Piping, Pressure
          Vessels, and Components – 1988 ed. by
          Q.N. Truong *et al.*

PVP 140   Unsteady Flows and Design
          Considerations in Vessel and Piping
          Systems ed. by Y.W. Shin and F.J. Moody,
          1988

PVP 141   Elastic–Plastic Failure Modelling of
          Structures with Applications ed. by
          D. Hui and T.J. Kozik, 1988

PVP 143   Advanced Topics in Finite Element
          Analysis ed. by J.F. Cory and J.L. Gordon,
          1988

PVP 144   Seismic Engineering – 1988 ed. by
          T.H. Liu *et al.*

PVP 145   Sloshing, Vibration, and Seismic
          Response of Fluid–Structure Systems
          ed. by D.C. Ma and S.S. Chen, 1988

PVP 147   Seismic, Shock and Vibration Isolation –
          1988 ed. by H.H. Chung and N. Mostaghel

PVP 148   High Pressure Technology – Material
          Design, Stress Analysis, and
          Applications ed. by A.K. Khare, 1988

PVP 149   Advances in Dynamic Analysis
          of Plates and Shells ed. by
          H.H. Chung, 1988

1988	Flow-Induced Vibration in Heat Exchange Equipment ed. by M.P. Paidoussis *et al.*	198
1988	Flow-Induced Vibration due to Internal and Annular Flows, and Special Topics in Fluidelasticity ed. by M.P. Paidoussis, M.K. Au-Yang and S.S. Chen	199
1988	Flow-Induced Vibration and Noise in Cylinder Arrays ed. by M.P. Paidoussis, S.S. Chen and M.D. Bernstein	200
1988	1988 International Design Criteria for Boilers and Pressure Vessels	201
1989	Availability Analysis: A Guide to Efficient Energy Use, rev. ed., ed. by M.J. Moran	202
1991	Computers in Engineering – 1991	203
1992	Computers in Engineering – 1992 ed. by G.A. Gabriele	204
1993	Taguchi on Robust Technology Development: Bringing Quality Engineering Upstream by G. Taguchi	205

American Society for Testing and Materials
(Philadelphia, Pa.)

Special Technical Publications

STP 11	Welding
STP 12	Effect of Temperature on the Properties of Metals
STP 15C	Manual on Quality Control of Materials, rev. ed., 1960
STP 26	Data on Nominal Creep Strength of Steels at Elevated Temperatures
STP 28A	Radiography
STP 30	Wear of Metals
STP 37	Compilation of Available High-Temperature Creep Characteristics
STP 47	Impact Resistance and Tensile Properties of Metals at Subzero Temperatures
STP 62	Magnetic Particle Testing
STP 63	Report on Behavior of Ferritic Steels at Low Temperature
STP 64	Stress Corrosion Cracking of Metals
STP 70	Testing of Bearings
STP 77	Synthetic Lubricants
STP 78	Effects of Low Temperatures on the Properties of Metals
STP 84	Functional Tests for Ball Bearing Greases
STP 85	Magnetic Testing
STP 88	Industrial Gear Lubricants
STP 91	Manual on Fatigue Testing
STP 100	Report on the Strength of Wrought Steels at Elevated Temperatures
STP 101	Ultrasonic Testing
STP 107	Plasticity and Creep of Metals
STP 108	Corrosion of Metals at Elevated Temperatures
STP 112	The Role of Nondestructive Testing in the Economics of Production
STP 130	Continuous Analysis of Industrial Water and Industrial Waste Water
STP 145	Nondestructive Testing
STP 148	Manual on Industrial Water

STP 148I	Manual on Industrial Water and Industrial Waste Water
STP 158	Effect of Temperature on the Brittle Behavior of Metals with Particular Reference to Low Temperatures
STP 164	Odor
STP 165	Effect of Cyclic Heating and Stressing Metals at Elevated Temperatures
STP 171	Basic Effects of Environment on Strengths, Scaling and Embrittlement of Metals at High Temperatures
STP 174	Metallic Materials for Service at Temperatures above 1600°F
STP 176	Impact Testing
STP 179	High-Purity Water Corrosion
STP 187	Relaxation Properties of Steels and Superstrength Alloys at Elevated Temperatures
STP 207	Industrial Water and Industrial Waste Water
STP 213	Nondestructive Testing
STP 231	Brittle Fracture of Rotor Forgings
STP 237	Basic Mechanisms of Fatigue
STP 260	Literature Surveys on: Influence of Stress Concentrations at Elevated Temperatures; and The Effects of Nonsteady Load and Temperature Conditions on the Creep of Metals
STP 264	Report on Stress-Corrosion Cracking of Austenitic Chromium–Nickel Stainless Steel
STP 281	Air Pollution Control
STP 289	Shear and Torsion Testing
STP 307	Erosion and Captation
STP 325	Stress–Strain–Time–Temperature Relationships on Materials
STP 337	Industrial Water and Industrial Waste Water
STP 369	Advances in the Technology of Stainless Steels and Related Alloys
STP 371	Magnetic Testing – Theory and Nomenclature, 1965
STP 371-S1	Direct Current Magnetic Measurements for Soft Magnetic Materials, 1970
STP 374	Irreversible Effects of High Pressure and Temperature on Materials
STP 381	Fracture Toughness Testing and Its Application, 1965
STP 391	Literature Survey on Creep Damage in Metals
STP 394	Fire and Explosion Hazards of Peroxy Compounds
STP 397	Stress Corrosion Cracking of Titanium
STP 408	Erosion by Cavitation or Impingement
STP 410	Plain Strain Crack Toughness Testing of High Strength Materials, 1966
STP 415	Fatigue Crack Propagation
STP 425	Stress Corrosion Testing, 1967
STP 437	Turbine Lubrication Problems
STP 442	Manual on Water, 1969
STP 446	Evaluation of Wear Testing, 1969
STP 450	Vibration Effects of Earthquakes on Soils and Foundations

STP 463	Review of Developments in Plain Strain Fracture Toughness Testing, 1970
STP 466	Impact Testing of Metals, 1970
STP 470A	A Manual on the Use of Thermocouples in Temperature Measurement, 1974
STP 494	Welding of Hot Steels
STP 496	Fracture Toughness Testing at Cryogenic Temperatures, 1971
STP 505	Acoustic Emission, 1972
STP 513	Stress Analysis and Growth of Cracks, 1972
STP 514	Fracture Toughness, 1971
STP 516	Localised Corrosion – Cause of Metal Failure, 1972
STP 518	Stress Corrosion Cracking of Metals – A State of the Art Review, 1972
STP 519	Cyclic Stress-Strain Behaviour – Analysis, Experimentation and Failure Prediction, 1973
STP 520	Fatigue at Elevated Temperatures, 1973
STP 527	Fracture Toughness Evaluation by R-Curve Methods, 1973
STP 534	Manual of Industrial Corrosion Standards and Control, 1974
STP 543	Hydrogen Embrittlement Testing, 1974
STP 550	Non-destructive Rapid Identification of Metals and Alloys by Spot Test, 1973
STP 556	Fatigue and Fracture Toughness – Cryogenic Behavior, 1974
STP 558	Corrosion in Natural Environments, 1974
STP 559	Fracture Toughness and Slow-Stable Cracking, 1974
STP 560	Fracture Analysis, 1974
STP 567	Erosion, Wear, and Interfaces with Corrosion
STP 571	Monitoring Structural Integrity by Acoustic Emissions, 1975
STP 573	Water Quality Parameters
STP 579	Properties of Materials for Liquefied Natural Gas Tankage, 1975
STP 580	Composite Reliability, 1975
STP 581	Thermal Insulations in the Petrochemical Industry, 1975
STP 590	Mechanics of Crack Growth, 1976
STP 593	Fracture Mechanics of Composites, 1976
STP 601	Cracks and Fracture (Ninth Conf.)
STP 605	Properties Related to Fracture Toughness, 1976
STP 610	Stress Corrosion – New Approaches, 1976
STP 614	Fire Standards and Safety, 1977
STP 616	Quality Systems in the Nuclear Industry, 1977
STP 620	Gaskets, 1977
STP 624	Nondestructive Testing Standards – A Review, 1977
STP 627	Fast Fracture and Crack Arrest, 1977
STP 629	Chloride Corrosion of Steel in Concrete, 1977
STP 631	Flow Growth and Fracture (Tenth Conf.), 1977
STP 634	Aquatic Toxicology and Hazard Evaluation (First Symp.), 1977
STP 642	Corrosion – Fatigue Technology, 1978

STP 646	Atmospheric Factors Affecting the Corrosion of Engineering Materials, 1978
STP 649	Walkway Surfaces: Measurement of Slip Resistance, 1978
STP 656	Intergranular Corrosion of Stainless Alloys, 1978
STP 657	Estimating the Hazard of Chemical Substances to Aquatic life, 1978
STP 664	Erosion: Prevention and Useful Application, 1979
STP 665	Stress Corrosion Cracking – the Slow Strain Rate Technique, 1979
STP 667	Aquatic Toxicology (Second Conf), 1979
STP 668	Elastic–Plastic Fracture, 1978
STP 671	Service Fatigue Loads Monitoring, Simulation and Analysis, 1979
STP 675	Fatigue Mechanisms, 1979
STP 677	Fracture Mechanics (Eleventh Conf.), 1979
STP 678	Fracture Mechanics Applied to Brittle Materials (Eleventh Conf), 1979
STP 685	Design of Buildings for Fire Safety, 1979
STP 687	Part-Through Crack Fatigue Life Prediction, 1979
STP 697	Acoustic Emission Monitoring of Pressurized Systems, 1979
STP 700	Fracture Mechanics (Twelfth Conf.), 1980
STP 706	Toughness of Ferritic Stainless Steels, 1980
STP 707	Aquatic Toxicology (Third Conf.)
STP 711	Crack Arrest Methodology and Applications, 1980
STP 713	Corrosion of Reinforcing Steel in Concrete, 1980
STP 714	Effect of Load Variables on Fatigue Crack Initiation and Propagation, 1980
STP 718	Thermal Insulation Performance, 1980
STP 719	Building Air Change Rate and Infiltration Measurements, 1980
STP 721	Sampling and Analysis of Toxic Organics in the Atmosphere, 1981
STP 722	Eddy-Current Characterization of Materials and Structures, 1981
STP 727	Electrochemical Corrosion Testing, 1981
STP 731	Tables for Estimating Median Fatigue Limits, 1981
STP 736	Physical Testing of Plastics, 1981
STP 737	Aquatic Toxicology and Hazard Assessment (Fourth Conf.), 1981
STP 738	Fatigue Crack Growth Measurement and Data Analysis, 1981
STP 741	Underground Corrosion, 1981
STP 743	Fracture Mechanics (Thirteenth Conf.), 1981
STP 744	Statistical Analysis of Fatigue Data, 1981
STP 748	Methods and Models for Predicting Crack Growth under Random Loading, 1981
STP 755	Application of $2\frac{1}{4}$ Cr-1 Mo Steel for Thick-Wall Pressure Vessels, 1982
STP 756	Stainless Steel Castings, 1982
STP 760	Hazardous Solid Waste Testing (First Symp.), 1982

STP 761	Design of Fatigue and Fracture Resistant Structures, 1982
STP 762	Fire Risk Assessment, 1982
STP 766	Aquatic Toxicology and Hazard Assessment (Fifth Conf.), 1982
STP 767	Atmospheric Corrosion of Metals, 1982
STP 770	Low-Cycle Fatigue and life Prediction, 1982
STP 786	Toxic Materials in the Atmosphere, Sampling and Analysis, 1982
STP 789	Thermal Insulations, Materials, and Systems for Energy Conservation in the '80s, 1983
STP 791	Fracture Mechanics (Fourteenth Symp.), 1983
STP 798	Probabilistic Fracture Mechanics and Fatigue Methods: Applications for Structural Design and Maintenance, 1983
STP 801	Corrosion Fatigue: Mechanics, Metallurgy, Electrochemistry and Engineering, 1983
STP 802	Aquatic Toxicology and Hazard Assessment (Sixth Conf.), 1983
STP 803	Elastic–Plastic Fracture (Second Symp.)
STP 805	Hazardous and Industrial Waste Testing (Second Symp.), 1983
STP 806	Standardization of Technical Terminology: Principles and Practice, 1983
STP 811	Fatigue Mechanisms: Advances in Quantitative Measurement of Physical Damage, 1983
STP 812	Flammability and Sensitivity of Oxygen-Enriched Atmospheres, 1983
STP 814	Evaluation and Accreditation of Inspection and Test Activities, 1983
STP 818	Corrosion of Metals in Association with Concrete, 1984
STP 825	A Guide to the Safe Handling of Hazardous Materials Accidents, 1983
STP 826	Fire Resistive Coatings, 1984
STP 833	Fracture Mechanics (Fifteenth Symp.)
STP 834	Definitions for Asbestos and Other Health-Related Silicates, 1984
STP 840	Oil Spill Chemical Dispersants: Research, Experience and Recommendations, 1984
STP 844	Methods for Assessing the Structural Reliability of Brittle Materials, 1984
STP 851	Hazardous and Industrial Waste Management and Testing (Third Symp.), 1984
STP 854	Aquatic Toxicology and Hazard Assessment (Seventh Symp.), 1985
STP 857	Fatigue at Low Temperatures, 1985
STP 868	Fracture Mechanics (Sixteenth Symp.) 1985
STP 872	Inhalation Toxicology of Air Pollutants: Clinical Research Considerations, c. 1985
STP 880	Corrosion of Metals under Thermal Insulation, 1985
STP 882	Fire Safety: Science and Engineering, 1985

Data Series

DS 3	Report on Strength of Wrought Steels at Elevated Temperatures
DS 5	Report on Elevated Temperature Properties of Stainless Steels, 1952
DS 5-S1	Report on Elevated Temperature Properties of Stainless Steels – Supplement, 1965
DS 5-S2	An Evaluation of the Yield, Tensile, Creep and Rupture Strengths of Wrought 304, 316, 321 and 347 Stainless Steels at Elevated Temperatures, 1969
DS 6	Elevated Temperature Properties of Chromium–Molybdenum Steels
DS 11	Elevated Temperature Properties of Carbon Steels
DS 15	Report on Elevated Temperature Properties of Wrought Medium-Carbon Alloy Steels
DS 16	The Elevated-Temperature Properties of Weld-Deposited Metal and Weldments
DS 22	Physical Properties of Metals and Alloys from Cryogenic to Elevated Temperatures
DS 40	Elevated Temperature Properties of Basic-Oxygen Steel, 1952
DS 48	Compilation of Odor and Taste Threshold Values Data, 1973

Other publications
Safe Handling of Hazardous Materials Accidents, 2nd ed., 1990

American Welding Society
(Miami, FA)

Welding Handbook, 1st–8th ed., 1938–1999
Welding Handbook, 9th ed.

A3.0-01	Standard Welding Terms and Definitions, 2001
AWS	Arc Welding and Cutting Safely, 2000
B2.1-00	Specification for Welding Procedure and Performance Qualification, 2000
B4.0/4.0M	Standard Methods for Mechanical Testing of Welds, 1998/2000M
B1.10-86	Guide for the Nondestructive Inspection of Welds, 1986
B1.11-88	Guide for the Nondestructive Inspection of Welds, 1988
B1.11-99	Guide for the Nondestructive Inspection of Welds, 1999
BRS	Braze Safely, 1992
C7.1	Recommended Practices for Electron Beam Welding, 1999
CAWF	Characterization of Arc Welding Fumes, 1983
D16.3M/D16.3	Risk Assessment Guide for Robotic Arc Welding, 2001
QC-1-96	Standard for AWS Certification of Welding Inspectors, 1996
SHF	Safety and Health Fact Sheets, 2nd Edition, 1998
SP	Safe Practices
ULR	Ultraviolet Reflectance of Paint, 1976

Z49.1   Safety in Welding, Cutting, and Allied
             Processes (ANSI Approved), 1999

Effects of Welding on Health
EWH-1   1940–1977
EWH-2   1978–1979
EWH-3   1979–1980
EWH-4   1980–1982
EWH-5   1982–1984
EWH-6   1984–1985
EWH-7   1986–1987
EWH-8   1988–1989
EWH-9   1990–1991
EWH-10   1992–1994
EWH-11   1995–1996
EWH-I   Book Index
WHB 1.9   Welding Science and Technology, 2001
WHB 2.9   Welding Processes, Part 1 (*to be published*)
WHB 3.9   Welding Processes, Part 2 (*to be published*)
WHB 4.9   Materials and Processes, Part 1 (*to be
             published*)
WHB 5.9   Materials and Processes, Part 2 (*to be
             published*)
WI-2000   Welding Inspection Handbook,
             3rd Edition

Associated Octel Company
(Ellesmere Port, Cheshire)

Handbook
Octel Handbook

Bulletins
13   The use of explosimeters in atmospheres
             containing lead alkyl compounds, 1975
18   The Telmatic lead-in-air analyser, 1974
20   How to deal with damaged and defective drums,
             1972
21   Air supplied suit, 1974
24   Lead-in-air threshold limit values, 1972
28   Taking Octel tanks out of service, 1975

Booklets
20/74   Emergencies and safety involving Octel
             antiknock compounds
21/74   Laboratory practice relating to Octel antiknock
             compounds
22/75   Transport of Octel antiknock compounds
23/72   Transport of Octel antiknock compounds by sea

Association of British Chemical Manufacturers
(London)

	Item
Model Safety Rules for Chemical Works	1
1934–60 Quarterly Safety Summary	2
1964 Safety and Management	3

Atomic Energy Research Establishment
Harwell

Harwell Environment Seminars

1   Major Chemical Hazards, 1978
2   Safety of Chemicals in the Environment,
             1979

3   Environmental Quality, 1980
4   Risk, Cost and Pollution, 1982

British Approvals Service for Electrical Equipment in Flammable Atmospheres
(Buxton, Derbyshire)

	Item
BASEEFA Guide, 3rd ed.	1

SFA 3004   Shunt Diode Safety Barriers
SFA 3006   Battery Operated Vehicles
SFA 3007   Instruments for Measuring Gas
             Concentration
SFA 3009   Special Protection
SFA 3012   Intrinsic Safety

British Chemical Industry Safety Council
(London)

Year		Item
1959	Safety in Inspection and Maintenance of Chemical Plant	3
1965–75	Quarterly Safety Summary	4
1965	Safety Training – a Guide for the Chemical Industry	5
1967	The Road Transport of Chemicals – Principal Safety Requirements	6
1968	Guide to Fire Prevention in the Chemical Industry	7
1968	Is it Toxic?	8
1969	Safe and Sound	9
1969	Safe and Sound. Summary for Managements	10
1972	Major Hazards. Memorandum of Guidance on Extensions to Existing Chemical Plant Introducing a Major Hazard	11
1973	Safety Audits	12
1973	Safety Audits. Summary for Managements	13
1974	Vinyl Chloride Monomer. Advisory Notes for PVC Processors	
	British Safety Council Guides: Monthly Publications	14

See also Appendix 27

British Compressed Gases Association
(Crawley, Surrey)

Guidance Notes
GN1   Guidance Notes on the Preparation of
             Major Emergency Procedures, 1977
GN2   Guidance for the Storage of
             Transportable Gas Cylinders for
             Industrial Use, Revision 2, 1997
GN3   Application of the Manual Handling
             Operations Regulations to Gas Cylinders
GN5   The Safe Application of Oxygen Enriched
             Atmospheres when Packaging Food, 1998
GN6   Avoidance and Detection of Internal
             Corrosion of Gas Cylinders

GN7 The Safe Use of Individual Portable or Mobile Cylinder Gas Supply Equipment, 2000
GN8 Catalogue of BCGA Members' Cylinder Test Marks, Revision 4, 2000
GN9 The Application of the Confined Spaces Regulations to the Drinks Dispense Industry, 2000
GN10 Implementation of EIGA Carbon Dioxide Standards, 2000

Technical Reports
TR1 A Method for Estimating the Offsite Risks from Bulk Storage of Liquefied Oxygen R1 (LOX), 1984
TR2 The Probability of Fatality in Oxygen Enriched Atmospheres due to Spillage of Liquid Oxygen R2 (LOX), 1999
TR3 Replacement Substances for the Cleaning of Oxygen System Components, 1999
CP4 Industrial Gas Cylinder Manifolds & Distribution Pipework/Pipelines (excluding acetylene), Revision 2, 1998
CP5 The Design & Construction of Manifolds using Acetylene Gas from 1.5 bar to a Maximum Working Pressure of 25 bar (362 lbf/in^2), Revision 1, 1998
CP6 The Safe Distribution of Acetylene in the Pressure Range 0–1.5 bar (0–22 lbf/in^2), Revision 1, 1998
CP7 The Safe Use of Oxy-Fuel Gas Equipment (Individual Portable or Mobile Cylinder Supply) Revision 2, 1996
CP15-1 The Safe Re-rating of Existing BS 5045, Part 1, 1982: Containers – Amendment AMD 5145, 1986, Part 1: Containers for Permanent Gases, Revision 1, 1996
CP15-2 The Safe Re-rating of Existing BS 5045, Part 1, 1982: Containers – Amendment AMD 5145, 1986, Part 2: Containers for Hydrogen Trailer Service, Revision 1, 1996
CP16 Movement of Static Storage Tanks for Non-Flammable, Non-Toxic Gases by Road, Revision 2, 2002
CP17 The Repair of Hand-held Blowpipes & Gas Regulators used with Compressed Gases for Welding, Cutting & Related Processes, Revision 1, 1998
CP18 The Safe Storage, Handling & Use of Special Gases in the Micro-Electronics Industry, Revision 1, 1995
CP19 Bulk Liquid Oxygen Storage at Users' Premises, Revision 3, 2002
CP20 Bulk Liquid Oxygen Storage at Production Sites, Revision 2, 2002
CP21 Bulk Liquid Argon or Nitrogen Storage at Users' Premises, Revision 1, 1998
CP22 Bulk Liquid Argon or Nitrogen Storage at Production Sites, Revision 1, 2002

CP23 Application of the Pressure Systems and Transportable Gas Containers Regulations 1989 to Industrial and Medical Pressure Systems Installed at Consumer Premises, Revision 1, 2002
CP24 Application of the Pressure Systems and Transportable Gas Containers Regulations 1989 to Operational Process Plants, 1992
CP25 Revalidation of Bulk Liquid Oxygen, Nitrogen, Argon and Hydrogen Cryogenic Storage Tanks, Revision 1, 1998
CP26 Bulk Liquid Carbon Dioxide Storage at Users' Premises, Revision 1, 1999
CP27 Transportable Vacuum Insulated Containers of not more than 1000 litres volume, 1994
CP28 Vacuum Insulated Tanks of not more than 1000 litres volume which are Static Installations at User Premises, 1997
CP29 The Design and Operation of Cylinder and Tube Trailers for the Safe Transport of Compressed Gases by Road, 1999
CP30 The Safe Use of Liquid Nitrogen Dewars up to 50 litres, 2000

Leaflets
L1 Carriage of Gas Cylinders by Road in Cars, Vans and Other Vehicles – Guidance for Drivers at Work
L2 The Safe Handling of Gas Containers at Waste Facilities
L3 Pressure Systems Legislation – A Summary: 2001

Technical Information Sheet
TIS3 Dimensions for Three-seat Cutting Nozzles
TIS4 BS EN ISO 14114 Acetylene Manifold Systems for Welding, Cutting and Allied Processes: General Requirements
TIS6 Cylinder Identification Color Coding and Labeling Requirements, 2001
TIS7 Guidelines for the Safe Transportation, Storage, Use and Disposal of Dry Ice Products, 2001
TIS8 Information for Customers Collecting Gas Cylinders (Flammable, Inert and Oxidizing Gases), 2002
TIS9 Gas Safety in Cellars – Checklist, 2002

British Gas
(London)

Communications
Comm. 1044 The background and implication of IGE/TD/1, ed. 2 by A.E. Knowles and F. Tweedle, 1977
Comm. 1046 Progress in noise control technology – applied to transmission and distribution by D. Headon, 1977

Comm. 1048   Noise generation and suppression in combustion equipment by B.D. Mugridge, C. Hughes and C.A. Roberts, 1977

Comm. 1052   Conservation of natural gas in pipeline operation: system integrity and efficiency by G.C. Britton, G.M. Hugh and R.D. Walker, 1977

Comm. 1053   Hazard prevention – a North American experience by L. James, A. McDiarmid and C.W.P. Maynard, 1977

Comm. 1103   Why pipes fail by D. Needham and M. Howe, 1978

Comm. 1104   A review of construction and commissioning procedures for new pipelines by S. Gorton, D.J. Platts, K.T. Jones, A.E. Hawker and C.W. Osborn, 1978

Comm. 1110   The development of coal liquefaction processes in the United Kingdom by J.M. Topper, 1978

Comm. 1123   Hydrocarbon production from sea bed installations in deep water by A. Dorr, 1978

Comm. 1135   The analysis of engineering projects for the national gas transmission system by D.O. Pugh, 1978

Comm. 1143   North East Frigg, a satellite of the Frigg giant gas field by C.E. Duvet, 1981

Comm. 1149   The northern North Sea gas gathering system by W.J. Walters, R.H. Willmott and I.J. Harthill, 1981

Comm. 1155   Experience with on-line inspection by R.W.E. Shannon, D.W. Hamlyn and D. White, 1981

Comm. 1224   The application of computer control to process plant by A. Brightwell, P. Spittle, T.P. Williams and V.C. Miles, 1983

Comm. 1233   Safety in industry by B.F. Street, 1984

Comm. 1241   Combustion of large scale jet-releases of pressurised liquid propane by W.J.S. Hirst, 1984

Comm. 1242   Assessment of safety systems using fault tree analysis by J.M. Morgan and J.D. Andrews, c.1985

Comm. 1255   BP engineering offshore by M. Woolveridge and A.C. Wake

Comm. 1260   Buried modules – a new concept for district pressure control by D. Needham and C.A. Spearman, 1985

Comm. 1277   Improved methods of PE pipe jointing by L. Maine and T.G. Stafford, 1985

Comm. 1281   Corrosion control of steel pipelines by P. Corcoran and C.J. Argent, 1985

Comm. 1296   Computer-based control and alarm systems for process plant by E.G. Brennan, T.P. Williams and G. Twizell, 1986

Comm. 1297   Building practical expert systems in research and development by C. Ringrose, 1986

Comm. 1298   Smoke, fire and gas detection at British Gas installations by R.S. Davidson and S.J. Peacock, 1986

Comm. 1304   British Gas–a commitment to safety by R. Herbert, J.M. Kitchen and P.M. Mulholland, 1986

Comm. 1305   Standards, certification and testing by W.J. Bennett and R.A. Hancock, 1986

Comm. 1306   Performance testing and monitoring of compressor units by G.W. Fairbairn, J.R. Nisbet and I.C. Robertson, 1986

Comm. 1334   Offshore structural steels – a decade of progress by P.R. Kirkwood, 1987

Comm. 1345   Radar detection of buried pipes and cables by H.F. Scott and D.J. Gunton, 1987

Comm. 1346   Recent developments in on-destructive testing by G.A. Raine and A.D. Batte, 1987

Comm. 1349   Natural gas venting and flaring by D.R. Brown and M. Fairweather, 1987

Comm. 1355   Ventilation in traditional and modern housing by D.W. Etheridge, D.J. Nevrala and R.J. Stanway, 1987

Comm. 1380   The development and application of CAD techniques by T. Nixon, D.W. Hamlyn and J.E. Sharp, c.1988

Comm. 1381   Hazard assessment at full scale–the Spadeadam story by B.J. Flood and R.J. Harris, c.1988

Comm. 1404   Advances in coal hydrogenation for the production of SNG and coal liquids by PA Borrill and F. Noguchi, 1989

Comm. 1408   Understanding vapour cloud explosions by R.J. Harris and M.J. Wickens, 1989

Comm. 1409   The development and exploitation of British Gas pipeline inspection technology by L. Jackson and R. Wilkins, 1989

Comm. 1410   Plant and pipe location using radar methods by D.J. Gunton, 1989

Comm. 1438   Fitness for purpose revalidation of pipelines by C.J. Argent, J.C. Braithwaite and W.P. Jones, 1990

Comm. 1439   Pipe rehabilitation technology by J. Thomas and B.E. McGuire, 1990

Comm. 1456   Programmable electronic systems applied to British Gas plant by R.M. Turner and T.P. Williams, 1991

Comm. 1493   The development of CHAOS: a computer package for the consequence and hazard assessment of offshore structures by K. Greening, M.J.A. Mihsein, D. Piper and M.W. Vasey

Engineering Standards   Item

Code of practice for large gas and dual fuel burners   1

IM/9   Code of practice for the use of gas in atmosphere gas generators and associated plants, 1977

IM/12   Code of practice for the use of gas in high temperature plant, 1980

IM/18   Code of practice for the use of gas in low temperature plant, 1982

PS/SHA1   Code of practice for hazardous area classification. Part 1 – Natural gas (draft), 1981

MRS Reports

MRS E 378   Gaseous and dust explosion venting: determination of explosion relief requirements by M.R. Marshall, 1980

MRS E 374   Ignition probabilities in turbulent mixing flow by A.D. Birch, D.R. Brown and M.G. Dodson, 1980

Recommendations on Transmission and Distribution Practice

IGE/TD/1   Steel pipelines for high pressure gas transmission, ed. 2, with amendments

IGE/TD/5   Transport, handling and storage of plastic pipe. R 1, Polyethylene pipe

IGE/TD/10   Pressure regulating installations for inlet pressures between 75 mbar and 7 bar, with amendments

IGE/SR/10   Procedures for dealing with escapes of gas into underground plant, 1987

IGE/SR/11   Liquefied natural gas: R 1, Temporary storages for periods not exceeding six months, 1971 R 2, Use of road tankers and tank trailers as temporary storages, 1971

IGE/SR/12   Handling of methanol, 1989

IGE/SR/13   Use of breathing apparatus in gas transmission and distribution, 1986

IGE/SR/14   High pressure gas storage: Pt 1, Above ground storage vessels, 1986 R 2, Buried pipe arrays, 1986

IGE/SR/15   Use of programmable electronic systems in safetyrelated applications in the gas industry, 1989

IGE/SR/18   Safe working in the vicinity of gas pipelines and associated installations. R 1, Operating at pressures in excess of 7 bar, 1990

British Occupational Hygiene Society
(London)

Monographs

1   Hazards of occupational exposure to ultraviolet radiation by D. Hughes, 1978

2   Electrical safety interlock systems by D. Hughes, rev ed., 1985

3   The toxicity of ozone by D. Hughes, 1979

4   A literature survey and design study of fume cupboards and fume-dispersion systems by D. Hughes, 1980

5   Notes on ionizing radiations: quantities, units, biological effects and permissible doses by D. Hughes, 1982

10   Education and training in occupation hygiene by P.J. Hewitt, 1983

11   The disposal of hazardous wastes by G.E. Olivers, 1983

12   Allergy to chemicals and organic substances in the workplace by G.W. Cambridge and B.F.J. Goodwin, 1984

Technical Guides

GG1   The Manager's Guide to Control of Hazardous Substances (with 21 case studies)

TG3   Fugitive emissions of vapours from process plant equipment by A.L. Jones, M. Devine, P.R. Janes, D. Oakes and N.J. Western, 1984

TG4   Dustiness estimation methods for dry materials, by C.M. Hammond, N.R. Heriot, R.W. Higman, A.M. Spivey, J.H. Vincent and A.B. Wells, 1985

TG6   Sampling & Analysis of Compressed Air for Breathing Purposes

TG7   Controlling airborne contaminants in the workplace by M. Piney, R.J. Alesbury, B. Fletcher, J. Folwell, F.S. Gill, G.L. Lee, R.J. Sherwood and J.A. Tickner, 1987

TG9   Biological Monitoring Reference Data

TH1   Some applications of statistics in occupational hygiene, 2001

British Safety Council
(London)

	Item
Communicating the Safety Message	1
Compressed air safety code	2
Construction Safety	3
Emergency planning and disaster control	4
Environmental Management	5
Fire Safety	6
Health and Safety in the Office	7
Industrial dermatitis – causes and prevention	8
Laboratory safety	9
Machine guarding	10
Management introduction to total loss control	11
Managing Stress	12
Manual Handling	13
New techniques in loss control management	14
Occupational Health	15
Providing Personal Protection	16
Respiratory Protective Equipment	17
Risk Assessment	18
Safety and Induction Training	19
Safety policies – how to write and enforce them	20
Safety Signs	21
Skin disease from oils – causes and prevention	22
Slips, Trips and Falls	23
Update on Health and Safety Law	24
Waste Management	25
Working with Electricity	26
Working with Substances Hazardous to Health	27
Workplace Noise and Hearing Protection	28
Workplace Transport	29

Building Research Establishment
(Borehamwood)

Current Papers

CP26/71   The breakage of glass windows by gas explosions by R.J. Mainstone, 1971

CPU/73   The hazard of internal blast in buildings by R.J. Mainstone, 1973

CP5/74   Smoke and toxic gases from burning plastic by G.W.V. Stark, 1974

CPU/74   Decomposition products of pvc for studies of fires by W.D. Woolley, 1974

CP2/74   Nitrogen containing products from the thermal decomposition of flexible polyurethane foams by W.D. Woolley, 1974

CP29/74   Effects of fuel geometry in fires by P.M. Thomas, 1974

CP30/74   Fires in enclosures P.M. Thomas, 1974

CP32/74   Fires in model rooms: CIB research programmes P.M. Thomas, 1974

CP35/74   Fires in oil soaked lagging by P.C. Bowes, 1974

CP36/74   Ventilation in relation to toxic and flammable gases in buildings by S.J. Leach and D.P. Bloomfield, 1974

CP43/74   Experimental techniques for wind test tunnels on model buildings by P.F. Grigg and D.E. Sexton, 1974

CP45/74   Structural damage in buildings caused by gaseous explosions and other accidental loadings by N. Taylor and S.J. Alexander, 1974

CP68/74   The rapid extinction of fires in high-racked storages by P. Nash, N.W. Bridge and R.A. Young, 1974

CP72/74   Determination of flame spread and fire resistance by H.L. Malhotra, 1974

CP88/74   Methods of determining the optimum level of safety expenditure by S.J. Melinek, 1974

CP91/74   Polymeric materials in fires by Fire Research Station and Building Regulations Professional Division, 1974

CP16/75   Tornado damage in buildings 26 June 1973 by K.J. Eaton and C.J. Judge, 1975

CP25/75   Problems with the use of fire-fighting foams by J.G. Corrie, 1975

CP29/75   The performance of the sprinkler in detecting fire by P. Nash and R.A. Young, 1975

CP32/75   The behaviour of automatic fire detection systems by J.F. Fry and C. Eveleigh, 1975

CP36/75   The explosion risk of stored foam rubber by W.D. Woolley and S.A. Ames, 1975

CP37/75   Fire resistance – present and future by H.L. Malhotra, 1975

CP42/75   Fire protection of flammable liquid storages by water spray and foam by P. Nash and R.A. Young, 1975

CP67/75   The performance of the sprinkler in the extinction of fire by P. Nash and R.A. Young, 1975

CP77/75   The testing of sprinkler installations by R.A. Young and P. Nash, 1975

CP78/75   The testing of sprinklers by P. Nash and R.A. Young, 1975

CP79/75   The essentials of sprinkler and other water spray fire protection systems by P. Nash, 1975

CP82/75   Some design aspects of fire extinguishers by P. Nash, 1975

CP22/76   Toxic combustion products of plastics by W.D. Woolley and M. Raftery, 1976

CP24/76   The response of buildings to accidental explosions by R.J. Mainstone, 1976

CP43/76   Noise levels at the boundaries of factories and commercial premises by M.P. Jenkins, A.C. Salvidge and W.A. Utley, 1976

CP50/76   The use of automatic sprinklers as fire sensors in chemical plant by P. Nash and C.R. Theobald, 1976

CP52/76   Sprinkler systems for special risks by P. Nash and R.A. Young, 1976

CP72/76   Background noise levels in the United Kingdom by K. Attenborough, S. Clark and W.A. Utley, 1976

CP74/76   Experimental methods for the study of fire-fighting foams by J.G. Corrie, 1976

CP9/77   Fire losses and the effect of sprinkler protection of buildings in a variety of industries and trades by F.E. Rogers, 1977

CP8/78   The estimation of parameters of normal distributions from data restricted to large values, 1978

CP12/78   Thermal theory of ignition burning and extinction of materials in a 'stagnant' gaseous boundary layer, 1978

CP14/78   Soil–structure interaction and deformation with large oil tanks, 1978

CP15/78   Quality control procedures for fire-fighting foams, 1978

CP25/78   On design procedures for wind loading, 1978

CP36/78   The control of noise from fixed premises – some case histories, 1978

CP39/78   Full-scale and model-scale studies of the dynamic behaviour of an offshore structure, 1978

CP40/78   Growth and development of fire in industrial buildings, 1978

CP52/78   life safety: what is it and how much is it worth, 1978

CP59/78   Maintenance and operating costs of modern boilers by R.M. Milner and R.C. Wordsworth

CP71/78   On simulating the atmospheric boundary layer in wind tunnels by N.J. Cook, 1978

CP5/82   Isothermal methods for assessing combustible powders: Pt I Theoretical and experimental approach; Pt II Practical case histories by P.F. Beever and P.F. Thorne, 1982

CP2/83   The incidence of accidental loadings in buildings, 1971–1981 by J.F.A. Moore, 1983

(series discontinued in 1983)

Information Papers

IP6/82   Spontaneous combustion, 1982

IP22/82   Fire and glazing, 1982

IP8/83   The incidence of accidental loads in buildings, 1971 – 1981, 1983

IP15/84   A hydrocarbon fire standard for offshore structures, 1984

IP21/84   Fire spread in buildings, 1984

IP22/84   Smoke spread within a building, 1984

IP20/85   The behaviour of people in fires, 1985

IP5/88   Selection of sprinklers for high rack storage in warehouses, 1988

IP3/89   Subterranean fires in the UK – the
              problem, 1989
IP4/89   Expert systems, 1989
IP21/89   Passive stack ventilation in buildings,
              1989

Notes
N81/81   Dust explosion – the respective roles of
              the Hartmann bomb and 20-litre sphere
              in prescribing the size of explosion relief
              vents: a preliminary study by P. Field and
              A.R. Abrahamsen, 1981
N101/81   Dust explosion protection – a
              comparative study of methods for sizing
              explosion relief vents by P. Field, 1981

Technical Information
TIL 63   Dust explosions, 1979

BRE Reports   *Item*
BR6   Small buildings in earthquake areas, 1972
BR9   Buildings and the hazard of explosion by
              R.J. Mainstone, 1974 Wind loading
              handbook by C.W. Newbury and
              K.J. Eaton, 1974   1
              Structural damage in buildings caused
              by gaseous explosions and other
              accidental loadings 1971–1977 by
              R.J. Mainstone, H.G. Nicholson and
              S.J. Alexander, 1978   2
BR31   Earthquakes and seismic zones in the
              Middle East, 1983
BR60   Directory of UK Fire Research, 1985
BR61   Studies of human behaviour in fire:
              empirical results and their implications
              for education and design, 1985
BR65   A hydrocarbon fire standard – an
              assessment of existing information, 1985
BR70   Assessment of informative fire warning
              systems: a simulation study, 1985
BR80   P.M. Thomas. Fire Research Station
              1951–1986 – selected papers, 1986
BR85   Selections of Statutes relating to fire
              precautions in Great Britain, 1986

United Kingdom Fire and Loss Statistics
United Kingdom Fire and Loss Statistics, 1970
United Kingdom Fire and Loss Statistics, 1971
United Kingdom Fire and Loss Statistics, 1972
(Publication: Department of the Environment)

SO8   Wind environment around buildings,
              1975
SO37   Explosibility assessment of industrial
              dusts and powders, 1983
SO41   Self-heating: evaluating and controlling
              the hazards, 1984
(Publication: HM Stationery Office)

Other publications   *Item*
Loading charts for the design of rigid buried   3
pipelines by N.W.B. Clarke, 1966
Results of fire resistance tests on elements of
building construction   4

Results of fire propagation tests on building
products   5
Results of surface spread of flame tests on
building products   6
The explosion and fire at Nypro UK Ltd
Flixborough on June 1974 by D.M. Tucker, 1976   7
Manual for Testing Flame Arresters by
Z.W. Rogowski, 1978   8
(Publication: HM Stationery Office)

EP series
EP1   The designers guide to wind loading of
              building structures. Part 1: Background,
              damage survey, wind data and structural
              classification, 1986
(Publication: Butterworths, London)

Bureau of Mines
(Washington, DC)

Bulletins
Bull. 231   Investigation of toxic gases from Mexican
              and other high sulphur petroleums and
              products by R.R. Sayers *et al.*, 1925
Bull. 368   Static electricity in nature and industry
              by P.G. Guest, 1938
Bull. 503   Limits of flammability of gases and
              vapours by H.F. Coward and G.W. Jones,
              1952
Bull. 588   Safety at gas processing plants by
              G.M. Kintz and F.C. Hill, 1960
Bull. 604   Structures and propagation of turbulent
              bunsen flames by D.S. Burgess, 1962
Bull. 627   Flammability characteristics of
              combustible gases and vapours by
              M.G. Zabetakis, 1965

Information Circulars
1C 7148   Dust explosion hazards in plant
              producing aluminum, magnesium and
              zinc powders by H. Brown, 1941
IC 8757   Laboratory studies of self-heating of
              coal by A.G. Kerns, 1977

Report Investigations
RI 2491   Hydrogen sulphide as an industrial poison by
              R.R. Sayers, C.W. Mitchell and W.P. Yant, 1923
RI 3274   Accuracy of manometry of explosions by
              H.F. Coward and M.D. Hersey, 1935
RI 3278   Limits of inflammability of diethyl ether and
              ethylene in air and oxygen by G.W. Jones *et al.*,
              1935
RI 3722   Inflammability and explosibility of metal
              powders by I. Hartmann, J. Nagy and
              H.R. Brown, 1943
RI 3731   Inflammability of powders used in the
              metals RI 5671 industry by I. Hartmann and
              J. Nagy, 1944
RI 3751   Inflammability and explosibility of powders
              used in the plastics industry by I. Hartmann
              R.I. 5707 and J. Nagy, 1944
RI 3826   Effect of pressure on the explosibility of
              acetylene–water vapor, acetylene–air and
              acethyphen; ylene–hydrocarbon mixtures by
              G.W. Jones, R.F. Kennedy, I. Spolan and
              W.J. Huff, 1945

RI 3867   Report on the investigation of the fire at the storage and regasification plant of the East Ohio Gas Co., Cleveland, Ohio, October 20, 1944 by M.A. Elliott, C.W. Subel, F.N. Brown, R.T. Artz and L.B. Berger, 1946

RI 3924   Effect of relief vents on reduction of pressures developed by dust explosions by I. Hartmann and J. Nagy, 1946

RI 4191   Investigation of February 10, 1973 explosion, Staten Island liquefied gas storage tank by M.G. Zabetakis and D.S. Burgess, 1973

RI 4245   Explosions of ammonium nitrate fertiliser on board the SS *Grandcamp* and the SS *Highflyer* at Texas City, Tex., April 16–17, 1947 by G.M. Kintz, G.W. Jones and C.B. Carpenter, 1948

RI 4636   Pressure relieving capacities of diaphragms and other devices for venting dust explosions by J. Nagy, J.E. Zeilinger and I. Hartmann, 1950

RI 4725   Recent studies of the explosibility of corn starch by I. Hartmann, A.R. Cooper and M. Jacobsen, 1950

RI 4835   Explosive characteristics of titanium, zirconium, thorium, uranium and their hydrides by I. Hartmann, J. Nagy and M. Jacobsen, 1951

RI 4904   A study to determine factors causing pressure piling in testing explosion-proof enclosures by E.J. Gleim and J.F. March, 1952 RI 6516

RI 5052   Laboratory explosibility of American coals by Hartmann, M. Jacobsen and Williams, 1954

RI 5198   Fire and explosion hazards in thermal coal-drying plants by H.R. Brown, C.J. Dalzell, I. Hartmann, G.J.R. Toothman and C.H. Schwartz, 1956

RI 5225   Fundamental flashback, blowoff and yellow-tip limits of fuel gas–air mixtures by J. Grummer, M.E. Harris and V.R. Rowe, 1956

RI 5457   Communication of flame through cylindrical channels by H.G. Wolfhard and A.E. Bruszak, 1959

RI 5461   Recent developments in spark ignition by E.I. Litchfield and M.V. Blanc, c.1959

RI 5476   Investigations on the explosibility of ammonium nitrate by J. Burns *et al.*, 1953

RI 5610   Flammability limits of methane and ethane in chlorine at ambient and elevated temperatures and pressures by A. Bartkowiak and M.G. Zabetakis, 1960

RI 5624   Laboratory equipment and test procedures for evaluating explosibility of dusts by H.G. Dorsett *et al.*, 1960

RI 5645   Explosion at dephlegmator at Cities Service Oil Company Refinery, Ponca City, Oklahoma, 1959 by M.G. Zabetakis, 1960

RI 5671   Minimum ignition energy concept and its application to safety engineering by E.L. Litchfield, 1960

RI 5707   Research on the hazards associated with the production and handling of liquid hydrogen by M.G. Zabetakis and D.S. Burgess, 1961

RI 5753   Explosibility of agricultural dusts by M. Jacobsen, J. Nagy, A.R. Cooper and F.J. Ball, 1961

RI 5815   Explosibility of coal dust in an atmosphere containing a low percentage of methane by J. Nagy and Portman, 1961

RI 5877   Flammability and detonability studies of hydrogen peroxide systems containing organic substances by J.M. Kuchta, G.H. Martindill, M.G. Zabetakis and G.H. Damon, 1961

RI 5971   Explosibility of dusts used in the plastics industry by M. Jacobsen, J. Nagy and A.R. Cooper, 1962

RI 5992   Flammability and autoignition of hydrocarbon fuels under static and dynamic conditions by J.M. Kuchta, S. Lambiris and M.G. Zabetakis, 1962

RI 6099   Fire and explosion hazards associated with liquefied natural gas by D.S. Burgess and M.G. Zabetakis, 1962

RI 6112   Autoignition of lubricants at elevated pressures by M.G. Zabetakis, G.S. Scott and R.E. Kennedy, 1962

RI 6293   The ignition of combustible mixtures by laminar jets of hot gases by M. Vanpee and A.F. Bruszak, 1963

RI 6309   Hazards in using liquid hydrogen in bubble chambers by M.G. Zabetakis, A.L. Furno and H.E. Perlee, 1963

RI 6369   Ignition of coal dust–methane–air mixtures by hot turbulent gas jets by J.M. Singer, 1964

RI 6516   Explosibility of metal powders by M. Jacobsen, A.R. Cooper and J. Nagy, 1964

RI 6543   Preventing ignition of dust dispersions by inerting by J. Nagy, H.G. Dorsett and M. Jacobson, 1964

RI 6561   Pressure development in laboratory dust explosions by J. Nagy, A.R. Cooper and J.M. Stupar, 1964

RI 6597   Explosibility of carbonaceous dusts by J. Nagy, H.G. Dorsett and A.R. Cooper, 1965

RI 6651   Some fundamental aspects of dust flames by H.M. Cassel, 1964

RI 6654   Autoignition of hydrocarbon jet fuel by J.M. Kuchta, A. Bartkowiak and M.G. Zabetakis, 1965

RI 6659   Flammability characteristics of ethylene by G.S. Scott, R.E. Kennedy, I. Spolan and M.G. Zabetakis, 1965

RI 6747   Sympathetic detonation of ammonium nitrate and ammonium nitrate-fuel oil by R.W. van Dolah *et al.*, 1966

RI 6773   Explosibility of ammonium nitrate under fire exposure by R.W. van Dolah *et al.*

RI 6811   Thermal phenomena during ignition of a heated dust explosion by J. Nagy and D.J. Surincik, 1966

RI 6840   Detonation initiation in alkane–oxygen mixtures by E.L. Litchfield and M.H. Hay, 1966

RI 6851   Explosibility of metal powders by M. Jacobsen, M. Cooper and J. Nagy, 1965

RI 6903   Further studies on sympathetic detonation by R.W. van Dolah, 1966

RI 6931   Equivalence and lower ignition limits of coal dust and methane mixtures by J.M. Singer, E.B. Cook and J. Grumer, 1967

RI 7061   Initiation of spherical detonation in acetylene–oxygen mixtures by E.L. Litchfield and M.H. Hay, 1967

RI 7103   Limits of flame propagation of coal dust – methane–air mixtures by J.M. Singer, A.E. Bruzsak and J. Grumer, 1968

RI 7132   Dust explosibility of chemicals, drugs, dyes and pesticides by H.G. Dorsett and J. Nagy, 1968

RI 7196   Large-scale studies of gas detonations by D.S. Burgess, J.N. Murphy, N.E. Hanna and R.W. van Dolah, 1968

RI 7208   Explosibility of miscellaneous dusts by J. Nagy, A.R. Cooper and H.G. Dorsett, 1968

RI 7279   Explosion development in a spherical vessel by J. Nagy, J.W. Conn and H.C. Verakis, 1969

RI 7448   Hazards associated with the spillage of liquefied natural gas on water by D.S. Burgess, J.N. Murphy and M.G. Zabetakis, 1970

RI 7457   Hydrogen flare stack diffusion phenomena: low and high flow instabilities, burning rates, dilution limits, temperatures and wind effects by J. Grumer *et al.,* 1970

RI 7507   Explosion development in closed vessels by J. Nagy, B.C. Seller, J.W. Conn and H.C. Verakis, 1971

RI 7552   Inhibition of coal–dust air flames by J. Grumer, 1971

RI 7752   Detonation of a flammable cloud following a propane pipeline break: the December 9, 1970, explosion in Port Hudson, Mo. by D.S. Burgess and M.G. Zabetakis, 1973

RI 7839   Constant volume flame propagation by H.E. Perlee, F.N. Fuller and C.H. Saul, 1974

RI 8127   The theory of flammability limits. Natural convection by M. Hertzberg, 1976

RI 8176   Flame and pressure development of large-scale CH_4–air–N_2 explosions. Buoyancy effects and venting requirements by M.J. Sapko, A.L. Furno and J.M. Kuchta, 1976

RI 8187   Compressor and related explosion by H.E. Perlee and M.G. Zabetakis, *c.*1963

RI 8201   Ignition in mixtures of coal–dust, air and methane from abrasive impacts of hard materials with pneumatic pipeline steel by J.S. Kelley and B.L. Forkner, 1976

RI 8360   The flammability of coal dust–air mixtures by M. Hertzberg, K.L. Cashdollar and J.J. Opferman, 1979

Technical Papers

Tech. Pap.   Ignition of natural gas–air mixtures by
475   heated surfaces by P.G. Guest, 1930

Tech. Pap.   Protection of equipment containing explosive
553   acetone–air mixtures by the use of diaphragms by G.W. Jones, E.S. Harris and B.B. Beattie, 1933

PMSRC series

PMSRC   Hazards of spillage of LNG onto water by D.S.
4177   Burgess, J. Biordi and J.N. Murphy, 1972

S series

S 4105   Hazards of LNG spillage in marine transportation by D.S. Burgess, J.N. Murphy and M.G. Zabetakis, 1970

Center for Chemical Process Safety – see American Institute of Chemical Engineers, Center for Chemical Process Safety

Chemical and Allied Products Industry Training Board
(Staines, Middlesex) (now disbanded)

Information Papers

6   A Guide to Analysis of Plant Training Needs, 1971
7   A Guide to Operator Job Analysis, 1971
8   A Guide to Fault Analysis, 1971
9   A Guide to the Preparation of a Training Programme, 1971
10   Safety Training, 1971
13   A Basic Course for Adult Entrants with Chemical Process Operations and an Example Training Unit – Distillation, 1972
16   Health and Safety at Work – Everybody's Concern, 1975
16A   Health and Safety at Work – Everybody's Concern. Supplement A: Permit-to-Work Systems, 1977

Training Recommendations 12, 1971, contains Information Papers 6–9

Other publications	*Item*
Safety for the Smaller Firm, 1973	1

Chemical Industries Association
(London)

		Item
	British Chemicals and Their Manufacturers	1
	Transport Emergency Card (TREMCARD), Manuals, Vols 1–2	2
1970	The Rising Cost of Fire	3
1973	Chemical Industries Association and Chamber of Shipping of the UK: Code of Practice for the Safe Carriage of Dangerous Goods in Freight Containers	4
1973	Road Transport of Hazardous Chemicals: Manual of Safety Requirements	5
1973	CHEMSAFE: Manual of the Chemical Industry Scheme for Assistance in Freight Emergencies	6
1974	Chemical Industries Association and Confederation of British Industry: Company Safety and Health Practices – CBI Guide for Employers	7
1975	Code of Practice for the Large Scale Storage of Fully Refrigerated Ammonia in the United Kingdom	8
1975	Code of Practice for the Safe Handling and Transport of Anhydrous Ammonia by Road in the United Kingdom	9
1975	Determination of Vinyl Chloride	10

1978	Guide to Storage and Use of Highly Flammable Liquids	11
1978	Precautions against Fire and Explosion: Vinyl Chloride	12
1980	Code of Practice for the Storage of Anhydrous Ammonia under Pressure in the United Kingdom	13
1980	Guidelines for Bulk Handling of Chlorine at Consumer Installations	14
PA1	Anhydrous Ammonia – Code of Practice for Transport by Rail, 1975	
PA2	Animal Testing – Why the Chemical Industry Says it is Necessary, 1986	
PA3	Benzene – Health Precaution Guidelines, 1980	
PA8	Chlorine – Safety Advice for Bulk Installations (HS/G 28)	
PA9	CIMAH Safety Case – Ammonia, 1988	
PA11	Codes of Practice for Chemicals with Major Hazards: Acrylonitrile, 1978	
PA12	Codes of Practice for Chemicals with Major Hazards: Anhydrous Hydrogen Chloride, 1975	
PA13	Codes of Practice for Chemicals with Major Hazards: Ethylene Dichloride, 1975	
PA14	Codes of Practice for Chemicals with Major Hazards: Hydrogen Fluoride, 1987	
PA15	Codes of Practice for Chemicals with Major Hazards: Vinyl Chloride, 1978	
PA16	Distribution of Chemicals in the UK – questions and Answers, 1988	
PA18	Guide to Fire Prevention in the Chemical Industry, 1986	
PA21	Formaldehyde. Exposure in Industry, 1983	
PA23	Occupational Exposure Limits, 1985	
PA48	Chlorine – Safety Advice for Bulk Installations, 1986	
RC3	Anhydrous Ammonia – Guidance on Transfer Connections for the Safe Handling in the UK, rev., 1990	
RC4	Bromine in Bulk Quantities – Guidelines for Safe Storage and Handling, 1989	
RC5	Carcinogens in the Workplace – Guidance Document, 1987	
RC6	Chemical Exposure Treatment Cards, rev. ed., 1991	
RC9	COSHH – Guidance on Assessments (Reg 6), 1989	
RC10	COSHH – Guidance on Setting In-House Occupational Exposure Limits (Reg 7), 1990	
RC11	COSHH – Guidance on Health Surveillance, 1989	
RC12	COSHH – Guidance on Information, Instruction and Training (Reg 12), 1990	
RC13	Employment and Reproductive Health, 1984	
RC14	Guidelines for Bulk Handling of Ethylene Oxide, 1979	
RC15	Guidelines on Waste Management – 1, The Use of Waste Disposal Contractors, 1987	
RC16	Guidelines on Waste Management – 2, The Responsible Use of Landfill, 1987	

RC17	Guidelines on Waste Management – 3, Management of Wastes at Production Sites, 1989
RC18	Hazard and Operability Studies, 1977
RC19	Liability Insurance. Models Forms and Guidance, 2nd ed., 1989
RC20	Recommended Procedures for Handling Major Emergencies, 1976
RC20a	Offsite Aspects of Emergency Procedures, 1984
RC21	An Approach to the Categorisation of Process Plant Hazard and Control Building Design, 1979
RC22	Protection of the Eyes, rev., 1990
RC23	Responsible Care. Guidance for Chief Executives, 1990
RC26	Road Transport of Dangerous Substances in Bulk, 1990
RC27	Safe Exportation of Chemicals – An Industry Approach to the Issue, 1989
RC28	Guidelines for Safe Warehousing, 1983
RC29	Toluene Diisocyanate (TDI) and Diphenyl Methane Diisisocyanate (MDI). Guidelines for Reception and Bulk Storage, 1989
RC31	Be Prepared for an Emergency – Guidance for mergency Planning, 1991
RC32	CHEMSAFE, rev. ed., 1991
RC33	Guidance on Safety, Occupational Health and Environmental Protection Auditing, 1991
RC35	Plastic Containers for the Carriage of Hazardous Materials, 1991
RC37	Steel Containers for the Carriage of Hazardous Materials, 1991
RC38	COSHH – Guidance on Record Keeping, 1991
RC39	COSHH – Guidance on Collection and Evaluation of Hazard Information (Reg 2), 1991
RC40	(Ammonia Storage)
RC45	Safe Handling of Potentially Carcinogenic Aromatic Amines and Nitro Compounds, 1992
RC46	Product Stewardship, 1992
RC47	Be Prepared for an Emergency – Training and Exercises, 1991
RC48	Chlorine – Intercompany Collaboration for Emergencies
RC50	COSHH in Distribution, 1992
RC51	Responsible Care Management Systems. Guidelines for Certification to ISO 9001
RC52	Making it Work – A Guide to Implementing Responsible Care
RC53	Responsible Care, 1992
RC54	Risk Assessment Guidelines. Business Interruption, 1992
RC55	Risk Assessment Guidelines. Employer's Liability and Public Liability, 1992
RC56	Risk Assessment Guidelines. Material Damage, 1992
RC58	Volatile Organic Compounds (VOCs) – Guidance on the Management of VOC Emissions, 1992

RC59   Pallets to be Used for the Carriage of
        Hazardous Substances, 1992
CE1   Industrial Waste Management: A CEFIC
        Approach to the Issue, rev., 1989
CE2   CEFIC Guidelines on Waste
        Minimization, 1990
CE3   CEFIC Guidelines on Communication of
        Environmental Information to the Public,
        1989
CE4   CEFIC Recommendations on the Safety of
        Road Hauliers, 1988
CE5   CEFIC Facts and Figures, 1991
CE6   CEFIC Guidelines to ISO 9001: EN29001,
        1991
BT23   UK Chemical Industry Environmental
        Protection Survey: 1991

Chemical Industry Safety and Health Council
(London)

Year		Item
1975–	Quarterly Safety Summary	5
1976	Recommended Procedures for	
	Handling Major Emergencies, 2nd ed.	6
1977	Hazard and Operability Studies	7

Chlorine Institute
(New York)

Pamphlets
001   Chlorine Manual 6th ed., 1997
005   Non-Refrigerated Liquid Chlorine
       Storage 6th ed., 1998
006   Piping Systems for Dry Chlorine 14th ed.,
       1998
008   Chlorine packaging manual, 5th ed., 1985
009   Chlorine Vaporizing Equipment 6th ed.,
       1997
017   Packaging Plant Safety And Operational
       Guidelines – Revision 1, 3rd ed., 2002
018   A bibliography of corrosion by chlorine,
       1956
039   Maintenance Instructions for Chlorine
       Institute Standard Pressure Relief
       Devices, 11th ed., 2001
040   Maintenance Instructions For Chlorine
       Institute Standard Angle Valves, 7th ed.,
       2002
041   Maintenance Instructions for Chlorine
       Institute Standard Pressure Release
       Device, 5th ed., 2001
042   Maintenance Instructions for Chlorine
       Institute Standard Excess Flow Valves,
       5th ed., 2002
049   Recommended Practices for Handling
       Bulk Highway Transports, 8th ed., 2001
057   Emergency Shut-Off Systems for Bulk
       Transfer of Chlorine, 3rd ed., 1997
060   Chlorine Pipelines, 5th ed., 2001
063   First Aid and Medical Management of
       Chlorine Exposures, 6th ed., 1998
064   Emergency Response Plans for Chlorine
       Facilities, 5th ed., 2000

065   Personal Protective Equipment for Chlor-
       Alkali Chemicals, 4th ed., 2001
066   Recommended Practices for Handling
       Chlorine Tank Cars, 3rd ed., 2001
067   Safety guidelines for the manufacture of
       chlorine, 1981
072   Properties of Chlorine in SI Units,
       2nd ed., 1986
073   Atmospheric Monitoring Equipment for
       Chlorine, 6th ed., 1997
074   Estimating the Area Affected by a
       Chlorine Release, 3rd ed., 2001
075   Respiratory protection guidelines for
       chlor-alkali operations, 1982
076   Guidelines for the Safe Vehicular
       Transportation of Chlorine Containers,
       3rd ed., 2001
077   Sampling Liquid Chlorine, 3rd ed., 1997
078   Refrigerated Liquid Chlorine Storage,
       3rd ed., 2000
079   Recommended Practices for Handling
       Chlorine Barges. 3rd ed., 2001
080   Recommended Practices for Handling
       Sodium Hydroxide Solution and
       Potassium Hydroxide Solution Caustic
       Barges, 3rd ed., 2001
085   Recommendations for Prevention of
       Personnel Injuries for Chlorine Producer
       and User Facilities, 3rd ed., 1999
086   Recommendations to Chlor-Alkali
       Manufacturing Facilities for the
       Prevention of Chlorine Releases,
       4th ed., 2001
089   Chlorine Scrubbing Systems,
       2nd ed., 1998
090   Molecular chlorine: Health and
       Environmental Effects, 2nd ed., 1998
098   Recommended Practices for Handling
       Hydrochloric Acid in Tank Cars, 2nd ed.,
       2001
099   Hydrogen Chloride, Anhydrous – Use,
       Handling, & Transportation of Cylinders
       or Tube Trailers, 2nd ed., 2001
121   Explosive Properties of Gaseous
       Mixtures Containing Hydrogen and
       Chlorine, 2nd ed., 1992
125   Guidelines: Medical Surveillance and
       Hygiene Monitoring Practices for Control
       of Worker Exp. to Mercury in the Chlor-
       Alkali Industry, 3rd ed., 1994
126   Guidelines – Medical Surveillance and
       Hygiene Monitoring Practices for Control
       of Worker Exp. To Chlorine in the Chlor-
       Alkali Industry, 3rd ed., 1998
152   Safe Handling of Chlorine Containing
       Nitrogen Trichloride, 1998
160   Estimating the Area Affected by a
       Hydrogen Chloride Release, 1998
161   Guidance Document: Risk Management
       for Hydrogen Chloride, 1998
162   Generic Risk Management Plan for
       Chlorine Packaging Plants & Sodium
       Hypochlorite Production Facilities, 1996
163   Hydrochloric Acid Storage and Piping
       Systems, 2001

| 164 | Reactivity and Compatibility of Chlorine and Sodium Hydroxide with Various Materials, 2001 |
| 167 | Learning From Experience, 2002 |

Other publications

BIBCL	Chlorine: an annotated bibliography, 1971
C-Guide	Communicating Your Chlorine Risk Management Plan, 1998
CPP	Chlorine production processes ed. by J.S. Robinson, 1981
FIRE	Dealing with chlorine emergencies, 1977
IB/A	Instruction Booklet: Chlorine Institute Emergency Kit 'A' for 100-lb. & 150-lb. Chlorine Cylinders, 9th ed., 1996
IB/B	Instruction Booklet: Chlorine Institute Emergency Kit 'B' for Chlorine Ton Containers, 8th ed., 1996
IB/C	Instruction Booklet: Chlorine Institute Emergency Kit 'C' for Chlorine Tank Cars and Tank Trucks, 8th ed., 1996
MIR-121	Explosive properties of gaseous mixtures containing hydrogen and chlorine, 1977
MIR-21	Nitrogen trichloride – a collection of reports and papers, 2nd ed., 1975
MIR-71	Estimating area affected by a chlorine release by A.E. Howerton, 1975
OPFLOW	Developing a chlorine emergency control plan, 1985
PUBLIC	Chlorine Institute Public Information Handbook for Emergency Planning and Community Right-to-Know, 1997
SG/E	Student Guide to 'Handling Packaged Chlorine Safely' Training Program
VPC	The vapour pressure of chlorine by D. Ambrose *et al.*, 1979
WHO 21	Environmental health criteria 21 – chlorine and hydrogen chloride by WHO

Christian Michelsen Institute
(Bergen, Norway)

Reports

CMI 72001-12	An electric spark generator for determination of minimum ignition energies of dust clouds by B. Alvestad, 1975
CMI 77005-2	A new laboratory scale method for determining the minimum explosible concentration of dust clouds by R.K. Eckhoff, 1978
CMI 790750-1	Some recent large scale gas explosion experiments in Norway by R.K. Eckhoff, K. Fuhre, O. Krest, C.M. Guirao and J.J. Lee, 1980
CMI 803301-1	Towards a uniform electric spark ignitability of combustible mixtures by R.K. Eckhoff, 1980
CMI 813307-1	Dust explosion experiments in a vented $500\,m^3$ silo cell by R.K. Eckhoff, K. Fuhre, M. Henery, S. Parker and H. Thorsen, 1982

Commission of the European Communities
(Brussels)

Reports

EUR 6371 EN	Plate inspection programme PISC, vols 1–5, 1979; vol. 6, 1980
EUR 8482 EN	Characterization of the pressure wave originating in the explosion of an extended heavy gas cloud: critical analysis of the treatment of its propagation in air and interaction with obstacles by J.A. Essers, 1983
EUR 8955 EN	Gas cloud explosions and their effect on nuclear power plants. Phase I: Basic development of explosion codes, by S.F. Hall, D. Martin and J. Mackenzie, 1984
EUR 9593 EN	Experimental investigation of the acceleration of deflagration in wake flow, by G. Adomeit and H. Willms, 1984
EUR 9541 EN/I	On the adequacy of numerical codes for the simulation of vapour explosions by C.J.M. van Wingerden and A.C. van den Berg, 1984
EUR 9541 EN/II	Experimental study of the influence of obstacles and partial confinement of flame propagation by C.J.M. van Wingerden, 1984
EUR 9592 EN	Aspects of the dispersion of denser-than-air vapours relevant to gas cloud explosions by C.J. Wheatley and D.M. Webber, 1984
EUR 9679 EN	Caracterisation du champ de pression induit par l'explosion aerienne d'un melange air-hydrocarbure, deflagration lente, deflagration rapide by J. Brossard, D. Desbordes, N. de Fabio, J.L Gamier, A. Lannoy, J.C. Leyer, J. Perrot and J.P. Saint-Cloud, 1985
EUR 9933 EN	Computer processing of visual records from the Thorney Island large scale gas trials by M.L. Rietmuller, J.M. Buchlin and S. Rickaert, 1985
EUR 10029 EN	Large scale field trials on dense vapour EN dispersion by J. McQuaid and B. Roebuck, 1985
EUR 11147 EN	Licensing issues associated with the use of computers in the nuclear industry, by R.E. Bloomfield and W.D. Ehrenburger, 1987
EUR 12602 EN	Fifteen benchmark exercises on vessel depressurisation with non-reacting fluids: hydrodynamic considerations by A.N. Skouloudis, 1990
EUR 13699 EN	Community Documentation Centre on Industrial Risk. Comparison of LPG-related regulations by S. Harris, J.E. Arreger, V.M. Trbojevic and R.A. Cox
EUR 13386 EN	Benchmark exercise on major hazard analysis, vol. 1, Designation of the project, discussion of the results and conclusions
EUR 13597 EN	Benchmark exercise on major hazard analysis, vol. 2, Summary contributions of participants

Other publications   Item
Large scale catastrophic releases of   1
flammable liquids, by D.M. Johnson,
M.J. Pritchard and M.J. Wickens
Res. Contract EV4T.0014 UK(H)
Safety of Thermal Water Reactors by
E. Skupinski, B. Tolley and J. Vilain, 1985
(Publication: Graham and Trottman,   2
London)

Compressed Gas Association
(Arlington, VA)

C-1   Methods for Hydrostatic Testing of
           Compressed Gas Cylinders, 1996
C-4   Fuel Containers and Their Installations
           American National Standard Method of
           Marking Portable Compressed Gas
           Containers to Identify the Material
           Contained, 4th ed., 1990
C-5   Cylinder Service Life – Seamless Steel High
           Pressure Cylinders, 1991
C-6   Standards for Visual Inspection of Steel
           Compressed Gas Cylinders, 2001
C-6.1   Standards for Visual Inspection of High
           Pressure Aluminum Compressed Gas
           Cylinders, 2002
C-6.2   Guidelines for Visual Inspection and
           Requalification of Fiber Reinforced High
           Pressure Cylinders, 1996
C-6.3   Guidelines for Visual Inspection and
           Requalification of Low Pressure Aluminum
           Compressed Gas Cylinders, 1999
C-6.4   Methods for External Visual Inspection of
           Natural Gas Vehicle (NGV), 1998
C-10   Recommended Procedures for Changes of
           Gas Service for Compressed Gas Cylinders,
           1993
C-11   Recommended Practices for Inspection of
           Compressed Gas Cylinders at Time of
           Manufacture, 2001
C-14   Procedures for Fire Testing of DOT Cylinder
           Pressure Relief Device Systems, 1999
C-15   Procedures for Cylinder Design Proof and
           Service Performance Tests, 2002
CGA-341   Standard for Insulated Tank Specification for
           Cryogenic Liquids, 3rd ed., 1987
G-1   Acetylene, 2001
G-1.1   Commodity Specification for Acetylene, 1998
G-1.2   Acetylene Metering and Pipeline
           Transmission, 2000
G-1.3   Acetylene Transmission for Chemical
           Synthesis, 1970, 1984
G-2   Anhydrous Ammonia, 1995
G-2.1   American National Standard Safety
           Requirements for the Storage and Handling
           of Anhydrous Ammonia (ANSI K61.1), 1999
G-2.2   Guideline Method for Determining Minimum
           of 0.2% Water in Anhydrous Ammonia, 1985
G-3   Sulfur Dioxide, 1995
G-4   Oxygen, 1996
G-4.1   Cleaning Equipment for Oxygen Service,
           1996
G-4.3   Commodity Specification for Oxygen, 2000

G-4.4   Industrial Practices for Gaseous Oxygen
           Transmission and Distribution Piping
           Systems, 1993
G-4.8   Safe Use of Aluminum-Structured Packing
           for Oxygen Distillation, 2000
G-4.9   Safe Use of Brazed Aluminum Heat
           Exchangers for Producing Pressurized
           Oxygen, 1996
G-5   Hydrogen, 2002
G-5.3   Commodity Specification for Hydrogen, 1997
G-6   Carbon Dioxide, 1997
G-6.2   Commodity Specification for Carbon
           Dioxide, 2000
G-6.7   Safe Handling of Liquid Carbon Dioxide
           Containers That Have Lost Pressure, 1996
G-6.9   Dry Ice, 1998
G-7   Compressed Air for Human Respiration,
           1990
G-7.1   Commodity Specification for Air, 1997
G-10.1   Commodity Specification for Nitrogen, 1997
G-12   Hydrogen Sulfide, 1996
HB   Handbook of Compressed Gases, 1999
P-1   Safe Handling of Compressed Gases in
           Containers, 2000
P-6   Standard Density Data, Atmospheric Gases
           and Hydrogen, 2000
P-7   Standard for Requalification of Cargo Tank
           Hose Used in the Transfer of Carbon Dioxide
           Refrigerated Liquid, 1996
P-8   Safe Practices Guide for Air Separation
           Plants, 1989
P-8.1   Safe Installation and Operation of PSA and
           Membrane Oxygen and Nitrogen Generators,
           1995
P-8.4   Safe Operation of Reboilers/Condensers in
           Air Separation Units (EIGA DOC 65/99/E),
           2000
P-9   The Inert Gases: Argon, Nitrogen, and
           Helium, 2001
P-11   Metric Practice Guide for the Compressed
           Gas Industry, 1998
P-12   Safe Handling of Cryogenic Liquids, 1993
P-13   Safe Handling of Liquid Carbon Monoxide,
           2nd ed., 1989
P-14   Accident Prevention in Oxygen-Rich and
           Oxygen-Deficient Atmospheres, 1983
P-16   Recommended Procedures for Nitrogen
           Purging of Tank Cars, 1992
P-19   Hazard Ratings for Compressed Gases, 1992
P-20   Standard for the Classification of Toxic Gas
           Mixtures, 1995
P-23   Standard for Categorizing Gas Mixtures
           Containing Flammable and Nonflammable
           Components, 1995
P-24   Guide to the Preparation of Material Safety
           Data Sheets, 1995
P-28   Risk Management Plan Guidance Document
           for Bulk Liquid, 1998
           Hydrogen Systems, 1998
P-29   Application of OSHA PSM and EPA RMP to
           the Compressed Gas Industry, 2000
P-31   Tanker Loading System Guide, 2000
P-32   Safe Storage and Handling of Silane and
           Silane Mixtures, 2000

11/87	Emergency planning guidance notes – Selecting the incident scenarios for off-site emergency planning/Responsibilities of petroleum industry and regulatory authorities in off-site emergency planning/Information to the public
2/88	A field guide to the application of dispersants to oil spills
5/88	Emergency planning guidance notes – Content of emergency plans/Planning for mutual aid
6/88	Emergency planning guidance notes – Refinery emergency planning
2/89	Emergency planning guidance notes – Training, exercises and rehearsal of emergency plans/Communications during emergencies
4/89	Managing safety
7/89	Emergency planning guidance note – Marketing installation emergency planning
5/90	A field guide on reduction and disposal of waste from oil refineries and marketing installations
4/92	Performance of oil industry cross-country pipelines in Western Europe – statistical summary of reported spillages 1991
7/94	Review of European oil industry benzene exposure data (1986–1992)
1/95	Oil refinery waste disposal methods, quantities and costs – 1993 survey
5/96	Collection and disposal of used lubricating oil
2/97	European oil industry guideline for risk-based assessment of contaminated sites
3/97	Task risk assessment
1/98	Methods of prevention, detection and control of spillages in European oil pipelines
4/98	European downstream oil industry safety performance – statistical summary of reported incidents – 1997 and overview 1993 to 1997
7/98	The SEVESO II directive and the oil industry
8/98	Trends in oil discharged with aqueous effluents from oil refineries in Europe – 1997 survey
1/99	European downstream oil industry safety performance – statistical summary of reported incidents – 1998
2/99	Environmental exposure to benzene
1/00	European downstream oil industry safety performance – statistical summary of reported incidents – 1999
2/00	A review of European gasoline exposure data for the period 1993 – 1998
3/01	European downstream oil industry safety performance – statistical summary of reported incidents – 2000
4/01	Performance of cross-country oil pipelines in Western Europe statistical summary of reported spillages – 2000
1/02	Western European cross-country oil pipelines – 30-year performance statistics

7/02	Assessment of personal inhalation exposure to bitumen fume guidance for monitoring benzene-soluble inhalable particulate matter
8/02	Method for monitoring exposure to gasoline vapour in air – revision 2002
9/02	A survey of European gasoline exposures for the period 1999–2001
2/03	Catalogue of CONCAWE reports

Special Interest Reports (White covers)

84/52	Recommended practice for investigating and dealing with
84/54	Petroleum hydrocarbons: their absorption through and effect on skin
84/55	The study of noise from steel pipes
84/64	Experimental verification of the ISVR pipe noise model
86/69	Health aspects of worker exposures to oil mists
87/52	Cost-effectiveness of hydrocarbon emission controls in refineries from crude oil receipt to product dispatch
87/57	Review of strategies for the evaluation of employee exposures to substances hazardous to health
87/59	The prediction of noise radiating from pipe systems – an engineering procedure for plant design
88/56	Quantified risk assessment
89/53	Health risks from working in petroleum refining and from gasoline – a critique and recommendations in response to IARC monograph volume 45
89/55	Review of EC health and safety directives
90/53	An assessment of occupational exposure to noise in Western European oil refineries
91/53	The role of epidemiology in the health assessment of petroleum industry employees
91/55	The European Community legislative process
92/55	The EC Safety Data Sheet Directive
93/52	Industrial hygiene seminar
95/60	Determination of air quality standards bands for ozone
95/62	Air quality standard for particulate matter
96/54	Development of a health surveillance programme for workers in the downstream petroleum industry
96/63	Scientific basis for an air quality standard on benzene
97/51	Scientific basis for an air quality standard for carbon monoxide complaints about noise
00/52	Management of occupational health risks during refinery turnarounds
01/52	A noise exposure threshold value for hearing conservation
01/56	An assessment of occupational exposure to noise in the European oil industry (1989–1999)
02/57	Catalogue of special interest reports

Department of Employment (HM Factory Inspectorate)
(London: HM Stationery Office)

Methods for the Detection of Toxic Substances in Air Booklets

Superseded by HSE MSDS series *Methods for the Determination of Hazardous Substances*

Other publications		Item
1971	Manhole Access in Enclosed Plant	1
1972	The Factories Act 1961. A Short Guide	2
1973	Lead: Code of Practice for Health Precautions	3
1974	Precautions in the Use of Asbestos in the Construction Industry	4
1974	Safety of Scaffolding	5

Department of the Environment
(London: HM Stationery Office)

Circulars

1/72	Development Involving the Use or Storage in Bulk of Hazardous Materials, 1972
22/80	Development Control – Policy and Practice, 1980
9/84	Planning Controls over Hazardous Development, 1984
22/84	Memorandum on Structure and Local Plans, 1984
15/88	Environmental Assessment, 1988

Pollution Papers

1	The Monitoring of the Environment in the United Kingdom, 1974
2	Lead in the Environment and its Significance to Man, 1974
4	Controlling Pollution, 1973
8	Accidental Oil Pollution of the Sea, 1976
9	Pollution Control in Great Britain: How it Works, 1976
10	Environmental Mercury and Man, 1976

Other publications	*Item*
Code of Practice for Reducing the Exposure of Employed Persons to Noise, 1972	1
Towards Cleaner Air, 1973	2
Information about Industrial Emissions to the Atmosphere, 1973	3
Clean Air Today, 1974	4
Code of Practice for the Carriage of Radioactive Materials by Road, 1975	5
Railway accident. Report on the accident that occurred on February 18, 1975, at Moorgate Station on the Northern line, London Transport Railways, 1976	6

Department of Trade and Industry (DTI)
(London)

05/1998	UK chemicals regulatory atlas: an overview of how to guide your chemical through to regulatory compliance.

07/2000	BioGuide: regulations, information and support for biotechnology in the UK (2nd edition).
03/2003	Biotechnology regulatory atlas: an overview of modern biotechnology regulation and how to achieve compliance.

Det Norske Veritas
(Stavanger, Norway)

Research Division Technical Reports

79-0483	Experimental investigations of partly confined gas explosions by D.M. Solberg, J.A. Pappas and E. Skramstad, 1979
83-1317	BLOW-DOWN – a computer program for dynamic analysis of hydrocarbon processes by J. Nylund, 1983

Electrical Equipment Certification Service
(closed 09/2002)
(Sheffield)

EECG2	Electrical Equipment Certification Guide, 1992
SEA 3009	Special Protection
SEA 3012	Intrinsic Safety

Electrical Power Research Institute
(Palo Alto, CA)

EPRI NP-613	On-line power plant alarm and disturbance analysis system. Interim report by B. Frogner and C.H. Meijer, 1978
EPRI NP-1379	On-line power plant alarm and disturbance analysis system. Final report by C.H. Meijer, 1980
EPRI NP-1684	Summary and evaluation of scoping and feasibility studies for disturbance analysis and surveillance systems (DASS) by A.B. Long *et al.* 1980

Engineering Equipment Users Association
(London) (predecessor to Engineering Equipment and Materials Users Association, q.v.)

Handbooks

7	Factory Stairways, Ladders and Handrails, rev. ed., 1973
8	Steam Trapping and Condensate Handling, rev. ed., 1962
9	Agitator Selection and Design, rev. ed., 1962
10	Ball and Roller Bearings (Rolling Bearings), rev. ed., 1965
11	Vacuum Producing Equipment, 1961
12	Thermal Insulation of Pipes and Vessels, 1963
15	Pneumatic Handling of Powdered Materials (including Fluidization), 1963
18	Supporting Mild Steel and Cast Iron Pipe, 1963
19	Separation of Dust from Gases, 1968
23	Pipe Joining Methods, 1968
25	Measurement and Control of Noise, 1968

Environmental Protection Agency
(New York)

	Item
EPA Training Manuals	1
Standard Operating Safety Guides, 1988	2

See also Reference List

Factory Mutual Research Corporation

Reports		*Item*
16215	Evaluation of protection from explosion overpressure in AEC gloveboxes by C. Yao, J. de Ris and S.N. Bajolai, 1969	1
	Explosion venting of low strength equipment and structure by C. Yao, 1973	2

Fire Protection Association
(London)

Compendium of Fire Safety Data (CFSD)

Vol. 1

Organization of fire safety
OR 01   Fire Safety Structure in the United Kingdom, 1990
OR 1   Fire Protection Law, 1989
OR 2   The Fire Service, 1988
OR 3   The Health and Safety Commission and the Health and Safety Executive, 1989
OR 4   Government Departments, 1978
OR 5   Fire Research of fire risks
OR 6   Fire Test Facilities, 1991
OR 7   Security Organizations, 1989

Management of fire risks
MR 2   Fire Safety and Security Planning in Industry and Commerce, 1989
MR4   Fire Safety Patrols, 1989
MR5   Security against Fire-Raisers
MR6   Fire Precautions during Stoppages of Work, 1989
MR7   Procedure in Event of Fire, 1990
MR8   Fire Damage Control: Emergency Planning, 1987
MR9   First Day Induction Training, 1989
MR10   Fire Safety Training,
MR11   Occupational Fire Brigades and Fire Teams: Organization, Role and Functions
MR12   Fire Safety Education and Training of People at Work, 1987
MR14   Occupational Fire Brigades: Reports and Report Writing, 1989

Vol. 2

Industrial and process fire safety
FS 6011   Flammable Liquids and Gases: Explosion Hazards, 1989
FS 6012   Flammable Liquids and Gases: Explosion Control, 1989
FS 6013   Flammable Liquids and Gases: Ventilation, 1979
FS 6014   Flammable Liquids and Gases: Electrical Equipment, 1990

FS 6015   Explosible Dusts, Flammable Liquids and Gases: Explosion Suppression, 1974
FS 6016   Hydraulic Oil Systems Fire Safety, 1976
FS 6017   Piped Services, 1982
FS 6018   Corrosive Smoke: Planning to Minimize the Acid Damage Hazard, 1989
FS 6021   Explosible Dusts: The Hazards, c.1989
FS 6022   Explosible Dusts: Control of Explosions, 1989
FS 6023   Explosible Dusts: Elimination of Ignition Sources, 1989
FS 6024   Explosible Dusts: Extraction, 1989
FS 6027   Bund Walls for Flammable Liquid Storage Tanks, 1990
FS 6028   Safety Containers for Flammable Liquids, 1987
FS 6029   Reduced Sparking Tools (Non-Sparking Tools), 1989
FS 6033   Crushing and Grinding
FS 6034   Reciprocating Compressors
FS 6041   Flammable Liquids and Gases: Application of the HAZCHEM Code to These and Other Substances, 1989
FS 6042   Oil-Fired Installations, 1989

(also Fire safety in various processes)

Occupancy fire safety
(Fire safety in other industries, occupancies and activities)

Vol. 3

Housekeeping and general fire precautions
GP 1   Guide to Fire Prevention with Outside Contractors (withdrawn)
GP 2   Cutting and Welding, 1990
GP 3   Hot Work Permits, 1989
GP 4   Blowlamps
GP 5   Safe Practice in Production Areas, 1988
GP 6   Safe Practice in Storage Areas, 1988
GP 7   Outdoor Storage, 1986
GP 8   Fire Dangers of Smoking, 1989
GP 9   Waste Incinerators, 1987
GP 10   Oil Soaked Floors, 1983
GP 11   Fire Doors and Shutters: Use Care and Maintenance, 1990

Nature and behaviour of fire
NB 1   The Physics and Chemistry of Fire, 1989
NB 2   Ignition, Growth and Development of Fire, 1988
NB 3   The Physiological Effects of Fire, 1989
NB 4   The Combustion Process, 1985
NB 5   Autoignition Temperature (AIT), 1984
NB 6   Ignition of Flammable Gases and Vapours by Sparks from Electrical Equipment, 1984

Vol. 4

Information sheets on hazardous materials
H1   Acetone, 1986
. . .

H174   Dimethyl carbonate, 1992

Vol. 5

Fire protection equipment and systems

PE 1	Automatic fire detection and alarm systems, 1989
PE 2	Fire points, 1988
PE 3	Fire blankets, 1988
PE 4	Portable fire extinguishers, 1990
PE 5	First-aid fire fighting: training, 1989
PE 6	The choice of fixed fire extinguishing systems, 1990
PE 7	Hose reels, 1988
PE 8	Hydrant systems, 1989
PE 9	Automatic sprinklers: an introduction, 1989
PE 10	Automatic sprinklers: components of a system (withdrawn)
PE 11	Automatic sprinklers: design and installation, 1989
PE 12	Automatic sprinklers: care and maintenance, 1989
PE 13	Automatic sprinklers: safety during sprinkler shutdown, 1985
PE 17	Automatic fire ventilation systems, 1989

Security precautions

SEC 1	Security equipment and systems, 1990
SEC 2	Site layout, 1990
SEC 3	Fences, gates and barriers, 1990
SEC 4	External security lighting, 1985
SEC 5	External closed circuit television systems, 1990
SEC 6	Doors, 1985
SEC 7	Windows and rooflights, 1985
SEC 8	Access control, 1991
SEC 9	Locks (withdrawn)
SEC 10	Hardware for fire and escape doors, 1991

Arson

AR 1	Management guide to fire
AR 2	Prevention and control of arson in warehouse and storage buildings, 1992
AR 4	Incendiary devices: information and guidance, 1990
AR 5	Investigation of vehicle fires

Vol. 6

Building and fire

Design guides

FPDG1	Fire prevention design guide: fire and the law, 1989
FPDG2	Fire prevention design guide: site requirements – space separation (withdrawn)
FPDG3	Fire prevention design guide: control of fire and smoke within a building, 1990
FPDG4	Fire prevention design guide: planning means of escape
FPDG5	Fire prevention design guide: fire fighting facilities for the fire brigade (withdrawn)
FPDG6	Fire prevention design guide: equipment for detection and warning of fire and fighting fire, 1991
FPDG7	Fire prevention design guide: fire hazard assessment, 1990
FPDG8	Fire prevention design guide: fire insurance (withdrawn)

FPDG9	Fire prevention design guide: building services, 1990
FPDG10	Fire prevention design guide: cavities and voids in building construction
FPDG11	Fire prevention design guide: suspended ceilings (withdrawn)
FPDG12	Fire prevention design guide: construction of roofs
FPDG13	Fire prevention design guide: flame retardant treatment for wood and its derivatives, 1990
FPDG14	Fire prevention design guide: fire doors and shutters
FPDG15	Fire prevention design guide: upgrading factory buildings to control fire spread
FPDG16	Air conditioning and ventilating systems, 1989

Building products

Fire properties of building products, including

B 15	Fire resistant glazing – wired glass, 1984
B 18	Fire resistant glazing – laminated wired glass, 1985

Fire Safety Sheets

S3		Waste incinerators
S4	1973	Oil-soaked floors
S5	1974	Safety containers for flammable liquids
S6	1974	Fire points
S7	1974	Protection of records, plans and documents
S8	1974	Heat treatment baths
S9		Safe practice in storage areas
S10	1975	Safer practice in production areas
S11	1975	Non-sparking tools

Other publications		*Item*
1964	Storage and handling of liquefied petroleum gas	1
1964	Planning fire safety in industry	2
1965	Checklist: fire safety in industry	3
1965	What is fire?	4
1968	Guide to fire prevention in the chemical industry	5
1968	Planning guide to fire safety with electricity	6
1969	Fire prevention manual for the iron and steel industry	7
1969	Planning guide to procedures in case of fire	8
1970	Fire protection for high bay warehouses	9
1970	Planning guide to fire safety when you have outside contractors in	10
1970	Planning guide to safety during sprinkler shutdown	11
1970	Planning guide to the safe handling of flammable gases	12
1970	Study report on fire risks and fire protection in automated high bay warehouses	13
1971	Background notes on fire and its control	14
1971	The fire hazards and safe disposal of waste	15
1972	Fire precautions during stoppages of work at factories	16
1972	Guide to safety in storage	17

Fire Research Station
(London: HM Stationery Office)

Fire Notes

Fire Research Technical Papers

Fire Research Notes

953   The behaviour of people in fires by P.G. Wood, 1972
957   The performance of some portable gas detectors with aviation fuel vapours at elevated temperatures. Pt 2 – Tests with 'AVCAT', 'KERO B' and 'AVTUR' vapours by P.J. Fardell, 1973
959   Preliminary experiments on the use of water additives for friction reduction in fire hoses by P.F. Thorne and K. Jordan, 1973
961   Dust explosion venting – consideration of further data by K.N. Palmer, 1973
963   Passive and active fire protection – the optimum combination by R. Baldwin and P.M. Thomas, 1973
964   Evaluation of road tanker dip tubes as flame barriers by Z.W. Rogowski and C.P. Finch, 1973
966   The production of oxides of carbon from the thermal and thermal-oxidative decomposition of flexible polyurethane foams by W.D. Woolley and A.I. Wadley, 1973
973   The effect of explosion pressure relief on the maximum experimental safe gap of propane–air and ethylene–air mixtures by Z.W. Rogowski and S.A Ames, 1973
974   Explosion protection of equipment with flame arresters – specification of methods of test and construction, 1973
978   Loss of life expectancy due to fires by S.K. Melinek, 1973
980   Storage properties of four foam liquids (Final report) by S.P. Benson, D.J. Griffiths, D.M. Tucker and J.G. Corrie, 1973
981   The risk of dying by fire by M.A. North, 1973
982   The economics of accident prevention – explosions by S.K.D. Coward, 1973
984   Gas explosions in buildings. Pt 1, Experimental explosion chamber by P.S. Tonkin and C.F.J. Berlemont, 1974
985   Gas explosions in buildings. Pt 2, The measurement of gas explosion pressure by S.A. Ames, 1974
990   Effect of pressure relief on the maximum experimental safe gaps of hydrogen/air mixtures by Z.W. Rogowski and C.P. Finch, 1973
991   Factors affecting fire loss – multiple regression model with extreme values by G. Ramachandran, 1973
992   Dust explosion hazards in pneumatic transport by K.N. Palmer, 1973
993   The compatibility of fluorochemical and fluoro-protein foams by S.P. Benson, D.J. Griffiths and J.G. Corrie, 1973
1003   Experimental appraisal of an American sprinkler system for the protection of goods in high-racked storage by N.W. Bridge and R.A. Young, 1974
1007   A laboratory fire test for foam liquids by S.P. Benson, P.R. Bevan and J.G. Corrie, 1974
1011   Some statistics of fire in road vehicles by N.A. North, 1974
1018   The transmission of explosion of flammable gas/air mixtures through a flame gap on a vented vessel by Z.W. Rogowski, 1974
1019   Pressures produced by gas explosion in a vented compartment by R.N. Butlin and P.S. Tonkin, 1974
1022   Flash points of mixtures of flammable and nonflammable liquids by P.F. Thorne

1025   Casualties attributed to toxic gas and smoke at fires – a survey of statistics by P.C. Bowes, 1975
1026   A review of information on experiments concerning the venting of gas explosions in buildings by R.N. Butlin, 1975
1028   Maximum permissible oxygen concentrations to prevent dust explosions in a large scale vertical explosion tube by P.S. Tonkin, 1975
1029   The early stages of fire growth in a compartment: a co-operative research programme of the CIB (Commission W14). First phase by A.J.M. Heselden and S.J. Melinek, 1975
1037   Performance of a two element crimped ribbon flame arrester by Z.W. Rogowski and A. Pitt
1044   The evaluation of an improved method of gas-freeing an aviation fuel storage tank by P.J. Fardell and B.W. Houghton, 1975
1048   The inhalation toxicity of polyvinyl chloride pyrolysis products by P.C. Bowes, J.A.C. Edginton and R.D. Lynch
1054   Estimation of maximum explosion pressure from damage to surrounding buildings. Explosion at Mersey House, Bootle, 28 August 1975 by R.N. Butlin, 1976
1063   Test performance of automatic fire alarm equipment by G. Cowie, 1977
1068   The protection of high-racked storages by commercial zoned sprinkler systems by S.P. Rogers and R.A. Young, 1977
1073   The toxicity of halon extinguishing agents by P.F. Thorne, 1977
1074   Use of a nitrogen filled high expansion foam to protect a 500 tonne fuel tank by A.A. Briggs, 1977
1078   The construction of flammability diagrams from flash-point measurements by P.F. Thorne and A.A. Briggs, 1978

(series discontinued in 1978)

Fire Research Special Report
2   Heat transfer by radiation by J.H. McGuire

Fire Research Symposia
6   Automatic Fire Detection, 1974

Gas Council
(London)

Communications
GC23   Explosion reliefs for industrial drying ovens by P.A. Cubbage and W.A. Simmonds, 1955
GC43   Explosion reliefs for industrial drying ovens by P.A. Cubbage and W.A. Simmonds, 1957
GC166   Ensuring safety and reliability in industrial gas equipment by P.F. Aris, R.A. Hancock and D.J. Moppett, 1969
GC171   Spark ignition of natural gas, theory and practice by J.F. Sayers, G.P. Tewari, J.R. Wilson and P.F. Jessen, 1970

GE Employers Reinsurance Corporation
(Overland Park, KS)

Health and Safety Commission
(London)

Health and Safety Executive
(HM Stationery Office, London; HSE Books, Sudbury,
Suffolk)

Accident/Incident Reports
See Reference list

Approved Codes of Practice
See Table 3.5

Construction Sheets
15   Confined spaces, 1991

Contract Research Reports
CRR 1/1987   Noise exposure and hearing: a new look at experimental data
CRR 5/1988   Hazard evaluation by thermal analysis
CRR 15/1989   Evaluation of the human contribution to pipework and in-line equipment failure frequencies
CRR 17/1988   Workbook on the dispersion of dense gases
CRR 20/1990   Statutory employee involvement in health and safety at the workplace: a report of the implementation and effectiveness of the Safety Representatives and Safety Committees Regulations 1977
CRR 26/1991   Guidance on HAZOP procedures for computer-controlled plants
CRR 28/1991   Investigations of machinery noise reduction at source
CRR 31/1992   Shiftwork, health and safety: an overview of the scientific literature
CRR 33/1992   Organization, management and human factors in quantified risk assessment. Report 1
CRR 34/1992   Organization, management and human factors in quantified risk assessment. Report 2
CRR 37/1992   Evaluation of the safety of composite transportable gas containers
CRR 38/1992   The development of a model to incorporate management and organisational influences in quantified risk assessment
CRR 39/1992   Review of equations of state of fluids valid to high density
CRR 51/1993   Improving safety on construction sites by changing personnel behaviour
CRR 54/1993   Attitudes towards noise as an occupational hazard, vols 1–2
CRR 57/1993   Occupational health provision at work
CRR 58/1993   A review of the state of knowledge and of current practice in selection techniques for process operators
CRR 60/1993   Safety management of process faults: human factors in control design
CRR 61/1993   Stress research and stress management: putting theory to work
CRR 62/1994   Development and testing of a procedure to evaluate dustiness of powders and dusts in industrial use

Dangerous Substances   *Item*
Major hazard aspects of the transport of dangerous   1
substances, 1991

Emission Test Methods (not in current list)
ETM 1   An introduction: sampling and analysis of emissions to the air from scheduled works, 1981

ETM 2   The sampling of gaseous emissions: sampling and analysis of emissions to the air from scheduled works, 1981

Guidance Notes

Chemical safety
CS 1   Industrial use of flammable gas detectors, 1987
CS 2   The storage of highly flammable liquids, 1977 (not in current list)
CS 3   Storage and use of sodium chlorate, 1985
CS 4   The keeping of LPG in cylinders and similar containers, 1986
CS 5   Storage of LPG at fixed installations, 1981 (not in current list)
CS 6   The storage and use of LPG on construction sites, 1981
CS 9   Bulk storage and use of liquid carbon dioxide: hazards and procedures, 1985
CS 15   Cleaning and gas freeing of tanks containing flammable residues, 1985
CS 16   Chlorine vaporisers, 1985
CS 17   Packed dangerous substances, 1986 (not in current list)
CS 18   Storage and handling of ammonium nitrate, 1986
CS 21   The storage and handling of organic peroxides, 1991

Environmental hygiene
EH 1   Cadmium – health and safety precautions, 1986
EH 2   Chromium – health and safety precautions, rev., 1991
EH 3   Prevention of industrial lead poisoning (not in current list)
EH 4   Aniline – health and safety precautions, 1979
EH 5   Trichloroethylene – health and safety precautions, 1985 (not in current list)
EH 6   Chromic acid concentrations in air, rev., 1990
EH 8   Arsenic – health and safety precautions, rev., 1990
EH 10   Asbestos – exposure limits and measurement of airborne dust concentrations, rev., 1990
EH 11   Arsine – health and safety precautions, rev., 1990
EH 12   Stibine – health and safety precautions, 1977 (not in current list)
EH 13   Beryllium – health and safety precautions, 1977 (not in current list)
EH 14   Level of training for technicians making noise surveys, 1977 (not in current list)
EH 16   Isocyanates – toxic hazards and precautions, 1984
EH 17   Mercury – health and safety precautions, 1977
EH 18   Toxic substances: a precautionary policy, 1977 (not in current list)
EH 19   Antimony – health and safety precautions, 1978
EH 20   Phosphine – health and safety precautions, 1979

EH 21 Carbon dust – health and safety precautions, 1979 (not in current list)

EH 22 Ventilation of the workplace, 1988

EH 26 Occupational skin diseases: health and safety precautions, 1981 (not in current list)

EH 27 Acrylonitrile: personal protective equipment, 1981

EH 28 Control of lead: air sampling techniques and strategies, 1986

EH 34 Benzidine based dyes – health and safety precautions, 1982

EH 35 Probable asbestos dust concentrations at construction processes, rev., 1989

EH 36 Work with asbestos cement, rev., 1990

EH 37 Work with asbestos insulating board, rev., 1989

EH 38 Ozone: health hazards and precautionary measures, 1983

EH 40 Occupational exposure limits 1994, 1994

EH 41 Respiratory protective equipment for use against asbestos, 1985 (not in current list)

EH 42 Monitoring strategies for toxic substances, rev., 1989

EH 43 Carbon monoxide, 1984

EH 44 Dust in the workplace: general principles of protection, rev., 1991

EH 46 Man-made mineral fibres, rev., 1990

EH 47 Provision use and maintenance of hygiene facilities for work with asbestos insulation and coatings, rev., 1990

EH 50 Training operatives and supervisors for work with asbestos insulation, coatings and insulating boards, 1988

EH 51 Enclosures provided for work asbestos insulation, coatings and insulating board, 1989

EH 52 Removal techniques and associated waste handling for asbestos insulation, coatings and insulating board, 1989

EH 53 Respiratory protective equipment for use against airborne radioactivity, 1990

EH 54 Assessment of exposure to fume from welding and allied processes, 1990

EH 55 The control of exposure to fume from welding, brazing and similar processes, 1990

EH 56 Biological monitoring for chemical exposure in the workplace, 1992

EH 57 The problems of asbestos removal at high temperatures, 1993

EH 58 The carcinogenicity of mineral oils, 1990

EH 59 Crystalline silica, 1992

EH 60 Nickel and its inorganic compounds: health and safety precautions, 1991

EH 62 Metalworking fluids – health precautions, 1992

EH 63 Vinyl chloride: toxic hazards and precautions, 1992

EH 64 Occupational exposure limits: criteria document summaries. Synopses of the data used in setting occupational exposure limits, 1993

EH 65/ *Criteria document series, including*

EH 65/4 1,2-dichloroethane: criteria document for an occupational exposure limit, 1993

EH 65/6 Epichlorohydrin: criteria document for an occupational exposure limit, 1993

EH 66 Grain dust, 1993

General

GS 4 Safety in pressure testing, rev., 1992

GS 5 Entry into confined spaces, 1977 (not in current list)

GS 6 Avoidance of danger from overhead electrical lines, rev., 1991

GS 8 Articles and substances for use at work – guidance for designers, manufacturers, importers, suppliers, erectors and installers, 1977 (not in current list)

GS 9 Road transport in factories and similar workplaces, 1992

GS 15 General access scaffolds, 1982

GS 16 Gaseous fire extinguishing systems: precautions for toxic and asphyxiating hazards, 1984

GS 21 Assessment of the radio frequency ignition hazard to process plants where flammable atmospheres may occur, 1983 (not in current list)

GS 24 Electricity on construction sites, 1983 (not in current list)

GS 26 Access to road tankers, 1983 (not in current list)

GS 27 Protection against electric shock, 1984

GS 28 Safe erection of structures, 1984–

GS 29 Health and safety in demolition work, 1984–

GS 31 Safe use of ladders, step ladders and trestles, 1984

GS 33 Avoiding danger from buried electricity cables, 1985 (not in current list)

GS 37 Flexible leads, plugs, sockets, etc., 1985 (not in current list)

GS 38 Electrical test equipment for use by electricians, 1986

GS 39 Training of crane drivers and slingers, 1986

GS 40 Loading and unloading of bulk flammable liquids and gases at harbours and inland waterways, 1986

GS 42 Tower scaffolds, 1987

GS 47 Safety of electricaldistribution systems on factory premises, 1991

GS 49 Pre-stressed concrete, 1991

Medical

MS 4 Organic dust surveys, 1977

MS 5 Lung function, 1977 (not in current list)

MS 7 Colour vision, rev., 1987

MS 8 Isocyanates: medical surveillance, 1983 (not in current list)

MS 12 Mercury: medical surveillance, 1978

MS 13 Asbestos,rev., 1988

MS 15 Welding, 1978

MS 18 Health surveillance by routine procedures (not in current list)

MS 24 Health surveillance of occupational skin disease, 1991

MS 25 Medical aspects of occupational asthma, 1991

Plant and machinery

PM 1 Guarding of portable pipe-threading machines,1984

PM 3 Erection and dismantling of tower cranes, 1976 (not in current list)

PM 5   Automatically controlled steam and hot
         water boilers, rev., 1989
PM 9   Access to tower cranes, 1979 (not in current
         list)
PM 13   Zinc embrittlement of austenitic stainless
         steel, 1977
PM 14   Safety in the use of cartridge operated tools,
         1978 (not in current list)
PM 15   Safety in the use of timber pallets, 1993
PM 16   Eyebolts, 1978
PM 19   Use of lasers for display purposes, 1980
PM 21   Safety in use of woodworking machines, 1981
PM 22   Training advice on the use of abrasive
         wheels, 1983
PM 23   Photo-electric safety systems, 1981
PM 24   Safety at rack and pinion hoists, 1981
PM 27   Construction hoists, 1981 (not in current list)
PM 28   Working platforms on forklift trucks, 1981
PM 29   Electrical hazards from steam/water
         pressure cleaners, rev., 1988
PM 30   Suspended access equipment, 1983
PM 32   Safe use of portable electrical apparatus, 1990
PM 38   Selection and use of electric handlamps, 1992
PM 41   Application of photo-electric safety systems
         to machinery, 1984
PM 53   Emergency private generation: electrical
         safety, 1985
PM 54   Lifting gear standards, 1985
PM 56   Noise from pneumatic systems, 1985
PM 58   Diesel engined lift trucks in hazardous
         areas, 1986
PM 60   Steam boiler blowdown systems, 1987
PM 64   Electrical safety in arc welding, 1986
PM 65   Worker protection at crocodile (alligator)
         shears, 1986
PM 73   Safety at autoclaves, 1990
PM 75   Glass reinforced plastic vessels and tanks:
         advice to users, 1991

Guidance Booklets
HS(G) 3   Highly flammable materials on
           construction sites, 1978
HS(G) 5   Hot work: welding and cutting on plant
           containing flammable materials, 1979
HS(G) 6   Safety in working with lift trucks, 1979
HS(G) 7   Container terminals: safe working
           practice, 1980
HS(G) 11   Flame arresters and explosion reliefs, 1980
HS(G) 12   Offshore construction: health, safety and
           welfare, 1980
HS(G) 13   Electrical testing: safety in electrical
           testing, 1980 (not in current list)
HS(G) 15   Storage of liquefied petroleum gas at
           factories, 1981 (not in current list)
HS(G) 16   Evaporating and other ovens, 1981
HS(G) 17   Safety in the use of abrasive wheels, 1984
HS(G) 19   Safety in working with power operated
           work platforms, 1982
HS(G) 20   Guidelines for occupational health
           services, 1982
HS(G) 22   Electrical apparatus for use in potentially
           explosive atmospheres, 1984
HS(G) 25   Control of Industrial Major Accident
           Hazards Regulations 1984 (CIMAH):
           further guidance on emergency plans, 1985

HS(G) 26   Transport of dangerous substances in
           tank containers, 1986
HS(G) 27   Substances for use to work: provision of
           information, 1985
HS(G) 28   Safety advice for bulk chlorine
           installations, 1986
HS(G) 30   Storage of anhydrous ammonia under
           pressure in the United Kingdom:
           spherical and cylindrical vessels
HS(G) 34   Storage of LPG at fixed installations,1987
HS(G) 37   Introduction to local exhaust
           ventilation, 1987
HS(G) 38   Lighting at work, 1987
HS(G) 39   Compressed air safety, 1990
HS(G) 40   Chlorine from drums and cylinders, 1987
HS(G) 43   Industrial robot safety, 1988
HS(G) 48   Human factors in industrial safety, 1989
HS(G) 51   The storage of flammable liquids in
           containers, 1990
HS(G) 52   The storage of flammable liquids in fixed
           tanks (exceeding 10,000 m^3 total
           capacity), 1991
HS(G) 53   Respiratory protective equipment: a
           practical guide for users, 1990
HS(G) 54   The maintenance, examination and
           testing of local exhaust ventilation, 1990
HS(G) 56   Noise at work. Noise assessment,
           information and control, 1990
HS(G) 61   Surveillance of people exposed to health
           risk at work, 1990
HS(G) 64   Assessment of fire hazards from solid
           materials and the precautions required
           for their safe storage and use, 1991
HS(G) 65   Successful health and safety
           management, 1991
HS(G) 71   Storage of packed dangerous
           substances, 1992
HS(G) 78   Container packing, 1992
HS(G) 96   The costs of accidents at work, 1993

Regulations Booklets
HS(R) 1   Packaging and Labelling of Dangerous
           Substances – Regulations and Guidance
           Notes, 1978
HS(R) 3   Guide to tanker marking regulations,
           1979
HS(R) 6   Guide to the HSW Act, 1983
HS(R) 7   Guide to the Safety Signs Regulations
           1980, 1981
HS(R) 8   Guide to the Diving Operations at Work
           Regulations 1981, 1981
HS(R) 11   First-aid at work, 1981
HS(R) 13   Guide to the Dangerous Substances
           (Conveyance by Road in Road Tankers)
           Regulations 1981, 1981
HS(R) 14   Guide to the Notification of New
           Substances Regulations 1982, 1982
HS(R) 15   Administrative guidance on the
           implementation of the European
           Community 'Explosive Atmospheres'
           Directive (76/117/EEC and 79/196/
           EEC), 1983
HS(R) 16   Guide to the Notification of Installations
           Handling Hazardous Substances
           Regulations 1982, 1983

HS(R) 17   Guide to the Classification and Labelling
            of Explosives Regulations 1983, 1983
HS(R) 18   Administrative guidance on the
            application of the European Community
            'Low Voltage' Directive (73/23/EEC) to
            electrical equipment for use at work in the
            United Kingdom, 1984
HS(R) 19   Guide to the Asbestos (Licensing)
            Regulations 1983, 1984
HS(R) 21   Guide to the Control of Industrial
            Major Accident Hazards Regulations
            1984, 1985
HS(R) 22   Guide to the Classification, Packaging
            and Labelling of Dangerous Substances
            Regulations 1984, 1985
HS(R) 23   Guide to the Reporting of Injuries,
            Diseases and Dangerous Occurrences
            Regulations 1985, 1986
HS(R) 24   Guide to the Road Transport (Carriage of
            Dangerous Substances in Packages, etc.)
            Regulations 1986, 1987
HS(R) 25   Memorandum of guidance on the
            Electricity at Work Regulations 1989, 1989
HS(R) 27   A Guide to Dangerous Substances in
            Harbour Regulations 1987, 1988
HS(R) 29   Notification and marking of sites. The
            Dangerous Substances (Notification and
            Marking of Sites) Regulations, 1990
HS(R) 30   A Guide to the Pressure Systems and
            Transportable Gas Containers
            Regulations 1989, 1990

Hazard Analysis Reports
HA 1   Canvey: A Second Report – A Summary of a
        Review of Potential Hazards from Operations
        in the Canvey Island/Thurrock Area Three
        Years after Publication of the Canvey
        Report, 1981
HA 2   Canvey: A Second Report – A Review of
        Potential Hazards from Operations in the
        Canvey Island/Thurrock Area Three Years
        after Publication of the Canvey Report, 1981
HA 3   Sizewell B: A Review by HM Nuclear
        Installations Inspectorate of the
        Preconstruction Safety Report, 1982
HA 4   PWR: A Report by the Health and Safety
        Executive to the Secretary of State for
        Energy on a Review of the Generic Safety
        Issues of Pressurised Water Reactors, 1979
HA 5   Nuclear Safety: Safety Assessment
        Principles for Nuclear Power Reactors, 1979

Health and Safety at Work Booklets (not in current list)
1   Lifting and Carrying
3   Safety Devices for Hand and Foot Operated
     Presses, 1971
4   Safety in the Use of Abrasive Wheels
6A   Safety in Construction Work: General Site Safety
     Practice
6B   Safety in Construction Work: Roofing
6C   Safety in Construction Work: Excavations
6D   Safety in Construction Work: Scaffolding
6E   Safety in Construction Work: Demolition
6F   Safety in Construction Work: System Building

10   Fire Fighting in Factories, 1970
13   Ionizing Radiations: Precautions for Industrial
      Users, 1970
14   Safety in the Use of Mechanical Power Presses
18   Industrial Dermatitis: Precautionary Measures
20   Drilling Machines: Guarding of Spindles and
      Attachments
22   Dust Explosions in Factories, 1970
24   Electrical Limit Switches and Their
      Applications, 1970
25   Noise and the Worker, 1971
27   Precautions in the Use of Nitrate Salt Baths, 1971
28   Plant and Machinery Maintenance, 1970
29   Carbon Monoxide Poisoning: Causes and
      Prevention
30   Storage of Liquefied Petroleum Gas at
      Factories, 1973
31   Safety in Electrical Testing
32   Repair of Drums and Small Tanks
33   Safety in the Use of Guillotines and Shears
34   Guide to the Use of Flame Arresters and
      Explosion Reliefs, 1965
35   Basic Rules for Safety and Health at Work, 1975
36   First Aid in Factories
37   Liquid Chlorine
38   Electric Arc Welding, 1970
40   Means of Escape in Case of Fire in Offices, Shops
      and Railway Premises
41   Safety in the Use of Woodworking Machines, 1970
42   Guarding of Cutters of Horizontal Milling
      Machines
44   Asbestos: Health Precautions in Industry, 1970
46   Evaporating and Other Ovens, rev., 1974
47   Safety in the Stacking of Materials, 1971

HSE Leaflets
HSE 1   After Flixborough?, 1976
HSE 2   Flixborough, the lessons to be learned, 1976
HSE 5   Introduction to the Employment Medical
          Advisory Service, 1985; rev., 1992
HSE 8   Fires and explosion due to the misuse of
          oxygen, 1984
HSE 9   Area office information services, 1986
HSE 11   Reporting of an injury or a dangerous
          occurrence, rev., 1986
HSE 16   Law on health and safety at work: essential facts
          for small business and the self-employed, 1986
HSE 17   Reporting a case of disease: a brief guide to the
          Reporting of Injuries, Diseases and Dangerous
          Occurrences Regulations 1985, 1986
HSE 21   Report that accident, 1988
HSE 22   The specialists, 1988; rev., 1992
HSEL 1   HSELINE: the new computerised source of health
          and safety at work references, 1983; rev., 1990

Industrial Advisory Committee Leaflets
IAC/L1 Guidance on the implementation of safety
policies, 1982

IND(G) Leaflets
IND(G) 1 (L)   Obligations on importers of
                 machinery, plant and substances in
                 Great Britain, 1982
IND(G) 17 (L)   Asbestos and you, 1984

IND(G) 22 (L)   Danger! transport at work, 1985
IND(G) 30 (L)   Buried cables – beware!, 1985
IND(G) 35 (L)   Hot work on tanks and drums, 1985
IND(G) 39 (L)   Permits to work and you, 1986
IND(G) 64 (L)   Hazard and risk explained, 1989
IND(G) 65 (L)   Introducing COSHH, 1989
IND(G) 67 (L)   Introducing assessment, 1989
                   (The last three constitute the COSHH
                   leaflet 'package')
IND(G) 78 (L)   Transport of LPG cylinders by
                   road, 1990
IND(G) 96 (L)   Are you involved in the transport of
                   dangerous substances by road?, 1990
IND(G) 98 (L)   Permit-to-work systems, 1992
IND(G) 115 (L)   New explosives controls, 1991
IND(G) 117 (L)   Policy statement on publication of
                   reports of accidents, 1991
IND(G) 118 (L)   Policy statement on standards, 1991
IND(G) 119 (L)   Safer representatives and safety
                   committees on offshore
                   installations, 1991

L Series
See Table 3.5
This series is gradually superseding the HS(R) and COP
series

Major Hazards	*Item*
Advisory Committee on Major Hazards. First Report, 1976	1
Assessment of the hazard from radio frequency ignition at the Shell and Esso sites at Braefoot Bay and Mossmorran, Fife, 1978	2
Report of the Steering Committee on radio ignition hazards at St Fergus, Scotland, 1979	3
Safety evaluation of the proposed St Fergus to Mossmorran natural gas liquids and St Fergus to Boddam gas pipelines, 1978	4
A reappraisal of the HSE safety evaluation of the proposed St Fergus to Mossmorran NGL pipeline, 1980	5
Advisory Committee on Major Hazards. Second Report, 1979	6
Advisory Committee on Major Hazards. Third Report, 1984	7
Names and addresses of sites sent to the Health and Safety Executive under Regulation 7 of the Control of Industrial Major Accident Hazards Regulations 1984 (CIMAH) or deemed to have been sent under Regulation 7(3) by means of a notification already made in accordance with Regulations 3(1) of the Notification of Installations Handling Hazardous Substances Regulations 1982 (NIHHS), 1986	8
Quantified risk assessment: its input to decision-making, 1989	9
Risk criteria for land use planning in the vicinity of major hazards, 1989	10

Medical Advice (not in current list)
MA 3   Control of lead at work: medical surveillance
       information and advice from the Health and
       Safety Executive's Medical Division
MA 4   Asbestos (Licensing) Regulations 1983: medical
       surveillance

Medical Series Leaflets (not in current list)
MS(A) 1   Lead and you, 1986
MS(A) 7   Cadmium and you, 1986
MS(A) 12   Hydrogen fluoride and you, 1990
MS(A) 13   Benzene and you, 1992
MS(A) 15   Silica dust and you, 1992
MS(A) 16   Chromium and you, 1991
MS(B) 1   Ulceration of the skin and inside the nose
            caused by chrome, 1974
MS(B) 5   Skin cancer caused by oil, 1983
MS(B) 6   Dermatitis and work (occupational
            dermatitis), 1983

Methods for the Determination of Hazardous Substances
MDHS 70   General methods for sampling airborne
            gases and vapours, 1990
MDHS 71   Analytical quality in workplace air
            monitoring, 1991
MDHS 72   Volatile organic compounds in air, 1992

Nuclear Installations	*Item*
Accident at Three Mile Island, 1979	1
Nuclear safety, 1979	2
PWR, 1979	3
Windscale. The management of safety, 1981	4
Emergency plans for civil nuclear installations, 1982	5
Work of HM Nuclear Installations Inspectorate, 1982	6
HM Nuclear Installations Inspectorate: a bibliography of published work 1962–1982, 1983	7
Safety assessment principles for nuclear chemical plant, 1983	8
Safety audit of BNFLs Sellafield, vols 1–2, 1986	9
Comments received on the tolerability of risk from nuclear power stations, 1988	10
The tolerability of risks from nuclear power stations, 1988 (not in current list)	11
An examination of the CEGB's R6 procedure for the assessment of the integrity of structures containing defects, 1989	12
ACSNI Study Group on Human Factors. First report on training and related matters, 1990	13
ACSNI Study Group on Human Factors. Second report: Human Reliability Assessment – a critical overview, 1991	14
A review of American nuclear emergency planning, 1991	15
Safety assessment principles for nuclear plants, 1992	16
The tolerability of risks from nuclear power stations, 1992	17
ACSNI Study Group on Human Factors. Third report: Organizing for safety, 1993	18

Sizewell B	*Item*
Sizewell B, 1982	19
Appraisal of the current situation on large LOCA calculations in the context of Sizewell B, 1984	20

Sizewell B public inquiry
NII/P/1   Sizewell B public inquiry: proof of
            evidence on the work and responsibilities
            of HM Nuclear Installations Inspectorate,
            1983 (plus additional material denoted
            ADD1, ADD2, etc.)

NII/P/2   Sizewell B public inquiry: proof of evidence on HM Nuclear Installations Inspectorate's view of the Central Electricity Generating Board's safety case, 1983 (plus additional material denoted ADD1, ADD2, etc.)

NII/S/2   Sizewell B inquiry: the Inspectorate's approach to the assessment of code validation submissions, 1983

NII/S/4   Sizewell B inquiry: schedule of NII reservations and issues all of which require to be resolved to the Inspectorate's satisfaction prior to licensing, 1983

NII/S/5   Sizewell B power station public inquiry: safety case computer codes – categorisation and validation assessment, 1983

NII/S/83   Sizewell B power station inquiry: relationship between the NII's assessment principles and levels of risk, 1984

NII 01   Sizewell B: a review by HM Nuclear Installation Inspectorate
             SUPP 1 Degraded core analysis, 1982
             SUPP 2 Safety analysis, 1982
             SUPP 3 External hazards – aircraft crash, 1983
             SUPP 6 Reactor protection system, 1983
             SUPP 7 External hazards – earthquake, 1983
             SUPP 8 External hazards – fire, 1983
             SUPP 9 Code validation for LOCA, 1983
             SUPP 10 Reactor pressure vessel, 1983
             SUPP 11 Pressure circuit components, 1983
             SUPP 12 ALARP strategy for dose reduction, 1983
             SUPP 13 Human factors, 1983
             SUPP 14 Quality assurance, 1983

HA 3   Sizewell B: a review by HM Nuclear Installations Inspectorate of the preconstruction safety report, 1982

Occasional Papers
OP 1   Problem drinker at work, 1981
OP 2   Microprocessors in industry: safety implications of the use of programmable electronic systems in factories, 1981
OP 3   Managing safety, 1981
OP 5   Electrostatic ignition: hazards of insulating materials, 1982
OP 9   Monitoring safety
OP 10   Safety of electrical distribution systems in factories, 1985
OP 11   Measuring the effectiveness of HSE's field activities, 1985

Offshore Technology Reports
OTH 91 337   Pipeline and riser loss of containment study, 1990 (PARLOC 90)
OTH 92 389   Human factors, shift work and alertness in the offshore industry, 1993

OTI 92 585   Generic foundation data to be used in the assessment of blast and fire scenarios and typical structural details for primary, secondary and supporting structures and components, 1992
OTI 92 586   Representative range of blast and fire scenarios, 1992
OTI 92 587   The prediction of single and two-phase release rates, 1992
OTI 92 588   Legislation, codes of practice and certification requirements, 1992
OTI 92 589   Experimental facilities suitable for use in studies of fire and explosion hazards in offshore structures, 1992
OTI 92 590   The use of alternative materials in the design and construction of blast and fire resistant structures, 1992
OTI 92 591   Gas/vapour build up on offshore structures, 1992
OTI 92 592   Confined vented explosions, 1992
OTI 92 593   Explosions in highly congested volumes, 1992
OTI 92 594   The prediction of the pressure loading on structures resulting from an explosion, 1992
OTI 92 595   Possible ways of mitigating explosions on offshore structures, 1992
OTI 92 596   Oil and gas fires: characteristics and impact, 1992
OTI 92 597   Behaviour of oil and gas fires in the presence of confinement and obstacles, 1992
OTI 92 598   Current fire research: experimental, theoretical and productive modelling resources, 1993
OTI 92 599   The effects of simplifications of the explosion pressure – time history, 1988
OTI 92 601   Computerised analysis tools for assessing the response of structures subjected to blast loading, 1992
OTI 92 602   The effects of high strain rate on material properties, 1992
OTI 92 603   Analysis of projectiles, 1992
OTI 92 604   Experimental data relating to the performance of steel components at elevated temperatures, 1992
OTI 92 605   Methodologies and available tools for the design/analysis of steel components at elevated temperatures, 1992
OTI 92 606   Passive fire protection: performance requirements and test methods, 1992
OTI 92 607   Availability and properties of passive and active fire protection systems, 1992
OTI 92 608   Existing fire design criteria for secondary, support and system steelwork, 1992
OTI 92 610   Thermal response of vessels and pipework exposed to fire, 1992

Oil Industry   *Item*
A guide to the principles and operation of permit-to-work procedures as applied to the petroleum industry, 1986   1

Guidelines for occupational health services in
the oil industry, 1987   2
Guidance on health and safety monitoring in the
petroleum industry, 1992   3
Guidance on multiskilling in the petroleum
industry, 1992   4
Guidance on permit-to-work systems in the
petroleum industry, 1991   5
Offshore installations guidance notes:
consolidated edition, 4th ed., 1990–   6

Promotional Leaflets
PML 6   New regulations on packaging and labels for
         dangerous substances: Classification,
         Packaging and Labelling of Dangerous
         Substances Regulations 1984, 1985
PML 7   Deadly maintenance: a study of fatal
         accidents at work, 1985
PML 10   Monitoring safety, 1986

Research Papers
RP 1   Examination of entrained air-flow rate data
         for large-scale water spray installations by
         J. McQuaid, 1978
RP 7   Decay of spherical detonations and shocks by
         H. Phillips, 1980
RP 8   Dispersion of heavier-than-air gases in the
         atmosphere: review of research, and progress
         report on HSE activities by J. McQuaid, 1978
RP 11   Comparative risks of electricity production
         systems: a critical survey of the literature by
         A.V. Cohen and D.K. Pritchard, 1980
RP 18   Reproducibility of asbestos counts by
         T.L. Ogden, 1982
RP 24   Dispersion of heavier-than-air gases in the
         atmosphere: a review of the organization and
         management of the HSE field trials 1982/84
         by D.G. Wilde, 1984
RP 32   First aid retention of knowledge survey by
         M.C. Cullen, 1992

Research Reviews   Item
Shift work and health by J.M. Harrington, 1978   1
Manual handing and lifting by J.D.G. Troup and
F.C. Edwards, 1985   2
Alternatives to asbestos products: a review, 1986   3

Safety Health and Welfare Leaflets
SHW 29   Phenol poisoning (cautionary notice),
            1984
SHW 367   Dermatitis (cautionary notice), 1975
SHW 385   Cyanide poisoning (cautionary
            notice), 1983
SHW 395   Gassing and chemical burns
            (cautionarynotice), 1984
SHW 397   Effects of mineral oil on the skin
            (cautionary notice), 1982
SHW 830   Dust explosions in factories, 1975
SHW 849   Nitrate salt baths (cautionary notice),
            1975
SHW 932   Dangers from carbon monoxide
            (cautionary notice), 1981

SHW 2125   Danger and explosion in oil storage
            tanks fitted with immersed heaters
            (cautionary notice), 1981

Sector Guidance
See Asbestos, Dangerous Substances, Major Hazards,
Nuclear Installations, Oil Industry

Specialist Inspectors Reports
SIR 5   Avoiding water hammer in steam systems
          by S. Mortimer and D.C. Edwards, 1988
SIR 8   The National Exposure Data Base – a
          computer record system for occupational
          hygiene data by P.L. Beaumont, D.K.
          Burns and T.E. Taylor
SIR 9   The fire and explosion hazards of
          hydraulic accumulators by D.B. Pratt
SIR 16   Low volume, high velocity extraction
          systems by B. Fletcher
SIR 19   Fire and explosion hazards associated
          with the storage and handling of
          hydrogen peroxide by R. Merrifield
SIR 21   Assessment of the toxicity of major
          hazard substances by R.M. Turner and
          S. Fairhurst

Technical Data Notes (not in current list)
1   Dust control – the low volume, high velocity
        system, 2nd rev., 1976
2/73   Threshold limit values for 1975, 1976
3   Occupational cancer of the renal tract, rev., 1973
4   Stibine – health and safety precautions, 1976
5   Antimony – health and safety precautions, 1975
6   Arsine – health and safety precautions, 1975
7   Phosphine – health and safety precautions, 1975
8   Chromium – health and safety precautions, 1973
9   Arsenic – health and safety precautions
10   Aniline – health and safety precautions
11   Cadmium – health and safety precautions,
        rev., 1975
12   Notes for the guidance of designers on the
        reduction of machinery noise, 1975
13   Standards for asbestos dust concentration for
        use with Asbestos Regulations 1969, rev., 1974
14   Health – dust in industry
15   Methyl bromide – health and safety precautions
16   Prevention of industrial lead poisoning
17   Trichloroethylene, rev., 1973
18   The safe cleaning repair and demolition of
        large tanks for storing flammable liquids,
        rev., 1975
19   Ventilation of buildings: fresh air requirements,
        rev., 1975
20   Anthrax
21   Mercury, rev., 1975
24   Asbestos Regulations 1969: respiratory
        protective equipment, 2nd rev., 1975
25   Safe operation of automatically controlled steam
        and hot water boilers, 1975
26   Erection and dismantling of tower cranes, 1973
27   Access to tower cranes, 1974
29   Fire risk in the storage and industrial use of
        cellular plastics, rev., 1975

32   Guarding of portable pipe threading machines
33   Power press mechanisms: overrun and fall-back devices
35   Control of asbestos dust, 1975
36   Safety in the use of cartridge operated fixing tools
38   Tripping device for radial and heavy vertical drilling machines
40   Standards for chromic acid concentration in air for use with the chromium plating regulations
41   Isocyanates – toxic hazard and precautions, 1975
42   Probable asbestos dust concentration at construction processes
43   Technical level training: code of practice for reducing the exposure of employed persons to noise, 1976
44   Road transport in factories, 1973
45   Industrial use of flammable gas detectors, 1973
46   Safety at quick opening and other doors of autoclaves, 1974
47   Entry into confined spaces: hazards and precautions, 1975
48   Permissible opening in fixed guards
53/1 Zinc embrittlement of austenitic stainless steel, 1976
53/2 Nitrate stress corrosion of mild steel, 1976
53/3 Creep of metals at elevated temperatures, 1977

Technical Information Leaflets

Engineering and metallurgy
EM 2   Metallurgical examination of broken and defective equipment, 1980
EM 3   Mechanical testing facilities at Sheffield, 1980
EM 4   Fracture mechanics of low and medium-strength steel, 1980
EM 5   Fatigue crack initiation and propagation in steels, 1980
EM 7   Design considerations for mechanical components, 1980

Environmental contaminants
EC 3   Catalytic gas sensing elements: their use in gas monitoring, 1980
EC 4   Diffusive sampler for organic gases and vapours in air, 1980

Frictional ignition
FI 2   Frictional sparking risk with light metals and their alloys, 1980

General
G 1   Health and Safety Laboratories – an introduction, 1980

Hazard analysis
HA 1   Safety aspects of industrial maintenance, 1982

Machine safety
MS 1   Measurement of the dynamic performance of machines and their safety systems, 1980

Respiratory apparatus
RA 1   Filter self-rescuers, 1979

Also series on Electrical hazards, Explosions, Protective equipment, Special instruments and techniques

Technical Papers
TP 1   Examination of entrained air flow rate data for large scale water spray installations by J. McQuaid, 1978 (also referenced as RP 1)
TP 7   Decay of spherical detonations and shocks by H. Phillips, 1980 (also referenced as RP 7)

Toxicity Reviews
TR 1   Styrene, 1981
TR 2   Formaldehyde, 1981
TR 3   Carbon disulphide, 1981
TR 4   Benzene, 1982
TR 20   Toluene, 1989
TR 26   Xylenes, 1992
TR 30   Phosphorous compounds, 1994

Toxicology of Substances in Relation to Major Hazards

		Item
1989	Acrylonitrile by R.M. Turner and S. Fairhurst	1
1990	Ammonia by M.P. Payne, J. Delic and R.M. Turner	2
1990	Chlorine by R.M. Turner and S. Fairhurst	3
1990	Hydrogen Fluoride by R.M. Turner and S. Fairhurst	4
1990	Hydrogen Sulphide by R.M. Turner and S. Fairhurst	5
1992	Sulphuric Acid Mist by R.M. Turner and S. Fairhurst	6
1993	Hydrogen Fluoride, rev., by M. Meldrum	7

Translations
including documents on Seveso, German Risk Study (Risikostudie) and German Order on Hazardous Substances (Störfallverordnung)

Miscellaneous Publications	*Item*
Guide to the Explosives Acts 1875 and 1923, 1941	1
Manhole Access to Enclosed Plant, 1971	2
Code of Practice for Reducing the Exposure of Employed Persons to Noise, 1972	3
Cost Effective Approach to Industrial Safety, 1972	4
Principles of Local Exhaust Ventilation, 1975	5
Fire Certificate (Special Premises) Regulations 1976: 20 questions, 1976	6
Success and Failure in Accident Prevention, 1976	7
A Study of the Causes of Molten Metal and Water Explosions, 1977	8
Brief History of HM Medical Inspectorate, 1979	9
Effective Policies for Health and Safety, 1980	10
Management's Responsibilities in the Safe Operation of Mobile Cranes: Report on Three Crane Accidents, 1980	11
100 Fatal Accidents in Construction, 1981	12
Her Majesty's Inspectors of Factories 1833–1983, 1983	13

Health and Safety Executive and Systems Reliability Directorate

Canvey Island Project Working Papers (assumed date 1977)

HSE/SRD/WP7	The probability of a major hazard being initiated by missiles either from process plant failure or other sources by E.A. White
HSE/SRD/WP10	Risk of a liquefied gas spill due to collision in the estuary
HSE/SRD/WP13	Accidental release of ammonia from a pressurised storage tank by G.D. Kaiser
HSE/SRD/WP14	Properties of ammonia–air mixtures by S.R. Haddock and R.J. Williams
HSE/SRD/WP16	Observations of non-buoyant behaviour of ammonia following accidental discharges to the atmosphere by R.F. Griffiths
HSE/SRD/WP17	Atmospheric release of anti-knock compounds containing tetra-ethyl lead by S.R. Haddock and G.D. Kaiser

HSE/SRD/WP21	Atmospheric dispersion of LNG vapour by G.D. Kaiser
HSE/SRD/WP22	Accidental release of ammonia from a pressurised storage tank. II – thermal effects during dispersion by G.D. Kaiser and B.C. Walker
HSE/SRD/WP28	Characteristics of pressure waves generated by unconfined vapour cloud explosions by F. Briscoe
HSE/SRD/WP29	The risk of a liquefied gas spill to the estuary
HSE/SRD/WP34	Radiation from an LNG fire in a bund to an LNG tank in an adjacent bund by I.R. Fothergill
Unnumbered	Release of anhydrous ammonia from pressurised containers – the importance of denser than air mixtures by G.D. Kaiser and B.C. Walker (given in 1978 as to be published)

Home Office
(London)

Manual of Firemanship

Book 1	Elements of Combustion and Extinction, 1974
Book 2	Fire Brigade Equipment, 1974
Book 3	Hand Pumps, Extinguishers and Foam Equipment, 1988
Book 4	Incidents Involving Aircraft, Shipping and Railways, 1985
Book 5	Ladders and Appliances, 1984
Book 6	Breathing Apparatus and Resuscitation, 1974
Book 7	Hydraulics and Water Supplies, 1975
Book 8	Building Construction and Structural Fire Protection, 1975
Book 9	Fire Protection in Buildings, 1977
Book 10	Fire Brigade Communications, 1978
Book 11	Practical Firemanship I, 1981
Book 12	Practical Firemanship II, 1983

Other publications	*Item*
HAZCHEM Scale Cards	6
Memorandum for the Guidance of Applicants for a License for an Explosives Factory or Explosives Magazine, 1958	7
Civil Defense. Handbook 7: Rescue, 1964	8
Dangerous Substances. Guidance on Dealing with Fires and Spillages, 1972	9
Fire Fighting and Fire Precautions in Automated Warehouses, 1974	10

IBC Technical Services Ltd
(London) (including predecessor Oyez Scientific and Technical Services Ltd)

Intelligence Reports
Design and Siting of Buildings to Resist Explosions and Fire
Hazards in the Process Industries: Identification and Control
Engineering Hazards: Assessment, Frequency and Control

Major Hazards Summer Schools
Major Hazards Summer School – annual schools started
1985

Industrial Risk Insurers
(Hartford, CT)

IR Information: Guidelines for Loss Prevention and Control

Institute of Materials
(London) (successor to Institute of Metals *inter alias*)

440   Corrosion by R.C. Newman, 1991
B135   Refractories for Iron- and Steel-Making by
       J.H. Chesters, 1974
B295   Corrosion Fatigue ed. by R.N. Parkins and
       Y.M. Kolotyrkin, 1983
B303   Microbial Corrosion, 1983
B304   Creep of Metals and Alloys ed. by R.W. Evans
       and B. Wilshire, 1985
B307   The Role of Crack Growth in Metal Fatigue by
       L.P. Pook

B313   Encyclopaedia of Metallurgy and Materials
       by C.R. Tottle, 1984
B316   Metals Databook by C. Robb, 1988
B340   Worked Examples in Fracture Mechanics by
       J.F. Knott and D. Elliott, 1979
B345   Worked Examples in Mass and Heat Transfer
       in Materials Engineering by R.G. Faulkner,
       D.J. Fray and R.D. Jones
B349   The Joining of Metals: Practice and
       Performance, vols 1–2, 1981
B366   Engineering Composite Materials by
       B. Harris, 1986
B367   Worked Examples in Heat Transfer,
       Fuels and Refractories, Fluid Flow
       and Furnace Technology by
       D.N. Gwyther, 1985
B390   Fracture at Stress Concentrators, 1986
B391   Aluminium Technology '86, 1986
B392   Materials at Their Limits, 1986
B393   Designing with Titanium, 1986
B396   Magnesium Technology, 1987
B403   An Introduction to the Modern Theory of
       Metals by Sir A. Cottrell, 1988
B407   Creep and Fracture of Engineering Materials
       and Structures ed. by B. Wilshire and
       R.W. Evans, 1987
B415   Creep and Fracture of Engineering Materials
       and Structures ed. by B. Wilshire and
       R.W. Evans, 1981
B416   Creep and Fracture of Engineering Materials
       and Structures ed. by B. Wilshire and
       R.W. Evans, 1984
B417   Recent Advances in Creep and Fracture of
       Engineering Materials and Structures ed. by
       B. Wilshire and R.W. Evans, 1982
B418   Engineering Approaches to High
       Temperature Design ed. by B. Wilshire and
       D.R.J. Owen, 1983
B423   The Structure of Metals and Alloys by
       W. Hume-Rothery, R.E. Smallman and
       C.W. Haworth, 1988
B426   Stainless Steels '87, 1988
B429   Introduction to Creep by R.W. Evans and
       B. Wilshire
B442   Materials and Engineering Design ed. by
       B.F. Dyson and D.R. Hayhurst, 1989
B445   Mechanical Testing, 1988
B447   Non-Destructive Testing, 1989
B460   Steelmaking by C. Moore and
       R.I. Marshall (2nd ed. of Modern
       Steelmaking Methods)
B463   Creep and Fracture of Engineering Materials
       and Structures ed. by B. Wilshire and
       R.W. Evans, 1991
B464   An Introduction to Grain Boundary Fracture
       in Metals by S.F. Pugh, 1991
B476   Practical Corrosion Principles, 1989
B513   Surface Engineering and Heat Treatment ed.
       By P.H. Morton, 1992
B522   Rupture Ductility of Creep Resistant Steels
       ed. by A. Strang, 1991
B526   Microbial Corrosion, 1992
B527   Corrosion Standards. European and
       International Development, 1991

B528   Building in Steel, 1991
B546   Marine Corrosion of Stainless Steels, 1993
B550   Fracture Mechanics – Worked Examples by
       J. Knott and P. Withey, 1993
B559   Corrosion Inhibitors, 1993
B576   Creep and Fracture of Engineering Materials
       and Structures ed. by B. Wilshire and
       R.W. Evans, 1993
PR1001 Managing Improvement – Where to Start ed.
        By S.H. Coulson, 1989
PR1002 Managing Improvement – A Case Study by
        S.H. Coulson, 1992
PR1004 Total Quality Management ed. by J.M. Buist
        and S.H. Coulson, 1990
PR1005 Engineering Design in Plastics by
        G.H. West, 1986

Institute of Petroleum

(publication – Barking, Essex: Applied Science
Publishers)

Oil Data Sheets
1   Leading oil companies in the United Kingdom
2   Some trade associations relevant to the
     petroleum industry
3   A glossary of terms used in the oil industry
4   Main products and their principal uses
5   Some useful conversion factors used in the oil
     industry
7/8 North Sea oil and gas: information sources
9   UK offshore fields in production and under
     development
10   UK refining distillation capacity
15   World refining capacity

Oil Loss Conferences
1   Oil loss control in the petroleum industry,
     1984
2   Oil loss control, 1987
3   Oil loss control, 1988
4   Oil loss control, 1991 (PUB 65)
5   Oil measurement – new developments in
     tank calibration and meter proving, 1993
     (PUB 73)

TP series
TP11   Guidance notes on refinery and
       distribution terminals work permit
       systems, 1993
TP 20 Recommendation for the safe use of
       telephone installations in cabs of petroleum
       carrying vehicles, 1993
TP 25 Guidelines for health surveillance and
       biological monitoring for occupational
       exposure to benzene, 1993
TP 26 The human health effects of benzene by
       D. Coggan, c. 1993
TP 27 Fate and effects of marine oil pollution in UK
       waters, 1993

Other publications
PUB 11   Inland oil spills – emergency procedures
         and action, 1978

PUB 18   Prevention of water pollution by oil, 1983
PUB 51   Offshore safety – the way ahead ed. by
            A.E. Lodge, 1990
PUB 55   Safety/standby vessels – the new
            requirements ed. by P. Ellis Jones, 1991
PUB 58   Remediation of industrial sites ed. by
            R.J. Watkinson, 1991
PUB 60   The Institute of Petroleum
            epidemiological study: refinery study
            principal results 1951–1989 by
            L. Rushton, 1991
PUB 61   The Institute of Petroleum
            epidemiological study: distribution
            centre study principal results 1951–1989
            by L. Rushton, 1991
PUB 64   Offshore safety – response to
            Cullen, 1992
PUB 69   Offshore supply vessels – regulatory,
            commercial and operational issues ed. by
            P. Ellis Jones, 1993

		Item
Petroleum Measurement Manual, 1968–		1
Modern Petroleum Technology, 5th ed., by G.D. Hobson, 1984		2
Characterization of spilled oil samples – purpose, sampling, analysis and interpretation ed. By J. Butt, 1985		3

Institution of Chemical Engineers
(Rugby)

Safety and Loss Prevention Conferences   *Item*
1960–   Chemical Process Hazards with Special
        Reference to Plant Design, vol.1, 1960;
        vol.2, 1963; vol.3, 1967; vol.4, 1972,; vol.5,
        1974; vol.6, 1977; Chemical Process
        Hazards VII, 1980; (1983 event combined
        with Symposium on Loss Prevention and
        Safety Promotion in the Process
        Industries); Hazards K, 1986; Hazards X,
        1989; Hazards XI, 1991; Hazards XII,
        1994   1–12
1971   Major Loss Prevention in the Process
        Industries   13
1976   Process Industry Hazards – Accidental
        Release, Assessment, Containment and
        Control   14
1981   Reliable Production in the Process
        Industries   15
1981   Runaway Reactions, Unstable Products
        and Combustible Powders   16
1984   The Protection of Exothermic
        Reactors and Pressurised Storage
        Vessels   17
1982   The Assessment of Major Hazards   18
1985   The Assessment and Control of Major
        Hazards   19
1987   Hazards from Pressure: Unstable
        Substances, Pressure Relief and
        Accidental Discharge   20
1991   Piper Alpha: Lessons for Life Cycle
        Management   21
1992   Major Hazards Onshore and Offshore   22

Joint Safety and Loss Prevention Conferences
1987   Internal Conference on Vapor Cloud
        Modelling   23
1988   Preventing Major Chemical and Related
        Process Accidents   24
1990   Safety and Loss Prevention in the
        Chemical and Oil Processing Industries   25

Dust Handling Guides
1981   A User Guide to Dust and Fume Control
        ed. by C.R. Smith   26
1984   Venting Gas and Dust Explosions – A
        Review by G.A. Lunn   27
1985   A User Guide to Dust and Fume Control,
        2nd ed., ed. by D.M. Muir Guide to Dust
        Explosion Prevention and Protection   28
1984   Pt 1 Venting by C. Schofield   29
1988   Pt 2 Ignition Prevention, Containment,
        Inerting, Suppression and Isolation by
        C. Schofield and J.A. Abott   30
1988   Pt 3 Venting of Weak Explosions and the
        Effect of Vent Ducts by G.A. Lunn   31
1992   Dust and Fume Control. A User Guide,
        rev. 2nd ed. Guide to Dust Explosion
        Prevention and Protection   32
1992   Pt 1 Venting, 2nd ed., by G.A. Lunn   33

Major Hazard Monographs
Chlorine Toxicity Monograph, 1987   34
Ammonia Toxicity Monograph, 1988   35
Overpressure Monograph, 1989; rev. as Explosions
in the Process Industries, 1994   36
Thermal Radiation Monograph, 1989   37
Phosgene Toxicity, 1993   38

Other publications
1961   A Problem in Chemical Engineering
        Design: The Manufacture of Acetic
        Anhydride by G.V. Jeffreys   39
1965   Chemical Engineering under Extreme
        Conditions   40
1965   Reaction Kinetics in Product and Process
        Design   41
1966   Advances in Materials   42
1966   Control and Improvement of Plants in
        Operation   43
1966   Movement and Storage of Raw Materials
        and Products   44
1967   The Application of Automation in the
        Process Industries ed. by J.M. Pirie   45
1968   Model Form of Conditions of Contract for
        Process Plant   46
1969   Dropping the Pilot   47
1969   A Guide to Capital Cost Estimation   48
1970   Cryogenic Safety Manual   49
1971   High Temperature Chemical Reaction
        Engineering ed. by F. Roberts, R.F. Taylor
        and T.R. Jenkins   50
1971   Industrial Wastes: A Provisional Code of
        Practice for Disposal of Wastes   51
1972   A Guide for the Economic Evaluation of
        Projects by D.H. Allen   52
1972   Vacuum Insulated Cryogenic Pipe   53

Institution of Electrical Engineers
(London)

Institution of Gas Engineers
(London)

1048 1977 Noise Generation and Suppression in Combustion Equipment by B.D. Mugridge, C. Hughes and C.A. Roberts

1052 1977 Conservation of Natural Gas in Pipeline Operation: System, Integrity and Efficiency by G.C. Britton, G.M. Hugh and R.D. Walker

1053 1977 Hazard Prevention — A North American Experience by L. James, A. McDiarmid and C.W.P. Maynard

1103 1978 Why Pipes Fail by D. Needham and M. Howe

1104 1978 A Review of Construction and Commissioning Procedures for New Pipelines by S. Gorton, D.J. Platts, K.T. Jones, A.E. Hawker and C.W. Osborn

1110 1979 The Development of Coal Liquefaction Processes in the United Kingdom by J.M. Topper

1123 1980 Hydrocarbon Production from Sea Bed Installations in Deep Water by A. Dorr

1135 1980 The Analysis of Engineering Projects for the National Transmission System by G.W. Pickard and B.D. Williams

1143 1981 North East Frigg, A Satellite of the Frigg Giant Gas Field by C.E. Duvet

1149 1981 The Northern North Sea Gas Gathering System by W.J. Walters, R.H. Willmott and I.J. Hartill

1155 1981 Experience with On-Line Inspection by R.W.E. Shannon, D.W. Hamlyn and D. White

1224 1983 The Application of Computer Control to Process Plant by A. Brightwell, P. Spittle, T.P. Williams and V.C. Miles

1233 1984 Safety in Industry by B.F. Street

1241 1984 Combustion of Large Scale Jet Releases of Pressurised Liquid Propane by W.J.S. Hirst

1242 1984 Assessment of Safety Systems Using Fault Tree Analysis by J.M. Morgan and J.D. Andrews

1255 1985 BP Engineering Offshore by M. Woolveridge and A.C. Wake

1260 1985 Buried Modules — A New Concept for District Pressure Control by D. Needham and C.A. Spearman

1277 1985 Improved Methods of PE Pipe Jointing by L. Maine and T.G. Stafford

1281 1985 Corrosion Control of Steel Pipelines by P. Corcoran and C.J. Argent

1296 1986 Computer-Based Control and Alarm Systems for Process Plant by E.G. Brennan, T.P. Williams and G. Twizell

1297 1986 Building Practical Expert Systems in Research and Development by C. Ringrose

1298 1986 Smoke, Fire and Gas Detection at British Gas Installations by R.S. Davidson and S.J. Peacock

1304 1986 British Gas — A Commitment to Safety by R. Herbert, J.M. Kitchen and P.M. Mulholland

1305 1986 Standards, Testing and Certification by W.J. Bennett and R.A. Hancock

1306 1986 Performance Testing and Monitoring of Compressor Units by G.W. Fairburn, J.R. Nisbet and I.C. Robertson

1334 1987 Offshore Structural Steels — A Decade of Progress by P.R. Kirkwood

1345 1987 Radar Detection of Buried Pipes and Cables by H.F. Scott and D.J. Gunton

1346 1987 Recent Developments in Non-Destructive Testing by G.A. Raine and A.D. Batte

1349 1987 Natural Gas Venting and Flaring by D.R. Brown and M. Fairweather

1355 1987 Ventilation in Traditional and Modern Housing by D.W. Etheridge, D.J. Nevrala and R.J. Stanway

1380 1988 The Development and Application of CAD Techniques by T. Nixon, D.W. Hamlyn and J.E. Sharp

1381 1988 Hazard Assessment at Full Scale — The Spadeadam Story by B.J. Flood and R.J. Harris

1404 1989 Advances in Coal Hydrogenation for the Production of SNG and Coal Liquids by P.A. Borrill and F. Noguchi

1408 1989 Understanding Vapour Cloud Explosions — An Experimental Study by R.J. Harris and M.J. Wickens

1409 1989 The Development and Exploitation of British Gas Pipeline Inspection Technology by L. Jackson and R. Wilkins

1410 1989 Plant and Pipeline Location Using Radar Methods by D.J. Gunton

1438 1990 Fitness for Purpose Revalidation of Pipelines by C.J. Argent, J.C. Braithwaite and W.P. Jones

1439 1990 Pipe Rehabilitation Technology by J. Thomas and B.E. McGuire

1456 1991 Programmable Electronic Systems Applied to British Gas Plant by R.M. Turner and T.P. Williams

Technical Recommendations

IGE/SR/5 1987 Entry into and work associated with confined spaces

IGE/SR/7 1989 Bulk storage and handling of highly flammable liquids used within the gas industry

IGE/SR/10 1987 Procedures for dealing with escapes of gas into underground plant

IGE/SR/11 1971 Liquefied natural gas

IGE/SR/12 1989 Handling of methanol

IGE/SR/13 1986 Use of breathing apparatus in gas transmission and distribution

IGE/SR/14 1986 High pressure gas storage

IGE/SR/15 1989 Use of programmable electronic systems in safety related applications in the gas industry

IGE/SR/18 1990   Safe working in the vicinity of gas pipelines and associated installations

IGE/TD/1 1984   Steel pipelines for high pressure gas transmission, 2nd ed., with amendments

IGE/TD/9 1986   Offtakes and pressure regulating installations for inlet pressures between 7 and 100 bar, with amendments

IGE/TD/11 1987   Operation and maintenance of gas pressure raising plant for outlet pressures not exceeding 7 bar

IGE/TD/12 1985   Pipework stress analysis for gas industry plant, with amendments

IGE/ER/1 1990   Occupational hygiene practice in gas industry operations

IGE/ER/2 1990   Assessment and control of environmental noise levels

IGE/TM/2 1978   International index of safety standards and codes relating to gas utilization in industry and commerce

IGE/TM/2A 1985   Supplement to IGE/TM/2

Other publications	*Item*
Gasholders	1
Liquefied Natural Gas	
Pt 1 Temporary Storages for Periods not Exceeding Six Months	2
Pt 2 Use of Road Tankers and Rail Tankers as Temporary Storages	3
Pt 3 Permanent Storages of Capacity Exceeding 1000 tonnes	4
Liquefied Petroleum Gases	5
Storage and Use of light Petroleum Feedstock in Gas Manufacture	6
Liquid Feedstock Pipelines	7
Steel Pipes for High Pressure Gas Transmission, rev., 1967	8
Use of Portable Electrical Apparatus, 1970	9
Mainlaying (Ductile Iron and Steel), 1976	10

Institution of Mechanical Engineers
(London)

		Item
1958	Engine Noise and Noise Suppression	1
1969	Maintenance Developments in the Process Industries	2
1970	Safety and Failure of Components	3
1971	Engineering Materials and Methods	4
1972	Cost Effective Security	5
1973	Terotechnology in the Process Industries	6
1974	Creep Behaviour of Piping	7
1974	Fan Technology and Practice	8
1974	Heat and Fluid Flow in Steam and Gas Turbine Plant	9
1974	Improvement of Reliability in Engineering	10
1974	Pumps for Nuclear Power Plant	11
1974	Steam Plant Operation	12
1974	Wet Steam 4	13
1975	Cavitation and Related Phenomena in Lubrication ed. by D. Dowson, M. Godet and C.M. Taylor	14

1975	Creep and Fatigue at Elevated Temperatures, vols 1–2	15
1975	Engineering Progress through Trouble	16
1975	First European Tribology Congress	17
1975	Periodic Inspection of Pressurised Components 1975	18
1975	Repairs in situ – Maintenance Strategy	19
1975	Retarders for Commercial Vehicles	20
1975	Terotechnology – Does it Work in the Process Industries?	21
1975	Tribology Practical Reviews	22
1975	Vehicle Safety Legislation – Its Engineering and Social Implications	23
1975	Vibration and Noise in Pump, Fan and Compressor Installations	24
1976	Cavitation	25
1976	Corrosion of Motor Vehicles	26
1976	Detail Design	27
1976	Energy Recovery in Process Plants	28
1976	Gas Turbines – Status and Prospects	29
1976	International Vehicle Legislation – Order or Chaos?	30
1976	Interpretation of Complex Signals from Mechanical Systems	31
1977	Braking of Motor Vehicles	32
1977	Brief Cases by H.A.V. Bulleid	33
1977	Design and Operation of Pumps, Fans and Compressors to Meet Current and Projected Noise Legislation	34
1977	Designing with Fibre Reinforced Materials	35
1977	Failure of Components in the Creep Range	36
1977	The Health and Safety at Work Act and its Effect in Industry	37
1977	Heat and Fluid Flow in Water Reactor Saftey	38
1977	High Pressure Engineering	39
1977	The Influence of Environment on Fatigue	40
1977	Periodic Inspection of Pressurised Components – 1997	41
1977	Superlaminar Flow in Bearings ed. by D. Dowson, M. Godet and C.M. Taylor	42
1977	Vibrations in Rotating Machinery	43
1977	Waste in the Process Industries	44
1977	The Wear of Non-Metallic Materials ed. by D. Dowson, M. Godet and C.M. Taylor	45
1978	The Crashworthiness of Vehicles – Land, Sea and Air	46
1978	Creep of Engineering Materials ed. by C.D. Pomeroy	47
1978	Surface Roughness Effects in Lubrication	48
1979	Dynamic Vibration Absorbers by J.B. Hunt	49
1979	Engineering Progress through Development ed. by R.R. Whyte	50
1979	Valve Design by G.H. Pearson	51
1980	The Engineer's Conscience by M.W. Thring	52
1980	Thermal Effects in Tribology ed. by D. Dowson, M. Godet and C.M. Taylor	53
1980	Valve User's Manual	54
1981	Developments in Variable Speed Drives for Fluid Machinery	55

Institution of Metallurgists
See Institute of Materials

International Chamber of Shipping
See Appendix 27 for all publications

International Labour Organisation
(Geneva)

International Maritime Organization
(London)

IMO-532.82.03   Intergovernmental Conference on the
Convention on Dumping Wastes at Sea
1972, 1982 ed.

IMO-013   Convention on the International Maritime
Organization, 1984 ed.

IMO-001   Basic Documents, vol.1, 1986 ed.

IMO-007   Basic Documents, vol.2, 1985 ed.

IMO-970   Global Maritime Distress and Safety
System (GMDSS), 1986 ed.

IMO-110   International Convention for the
Safety of life at Sea 1974 (SOLAS),
1986 ed.

IMO-963   Merchant Ship Search and Rescue
Manual (MERSAR), 1986 ed.

IMO-858   Guidelines on Surveys Required by the
1978 SOLAS Protocol, the IBC Code and
the IGC Code Fire Test Procedures,
1987 ed.

IMO-775   Index of Dangerous Chemicals Carried in
Bulk, 1990 ed.

IMO-904   International Conference on Revision of
the International Regulations for
Preventing Collisions at Sea 1972, 1990 ed.

IMO-520   International Conference on Marine
Pollution 1973 (MARPOL), 1991 ed.

IMO-532   The London Dumping Convention: the
First Decade and Beyond, 1991 ed.

IMO-927   Ships' Routeing, 1991 ed.

IMO-110   SOLAS, 1992 ed.

(The SOLAS documents for 1986 and 1992 both carry the
number IMO-110)

International Occupational Health and Safety Information Centre
(Geneva: Int. Labour Org.)

Information Sheets
15   Human Factors and Safety
17   Noise in Industry

Liquefied Petroleum Gas Industry Technical Association
(London)

Guides
GN1   A Guide to the Writing of LPG Safety Reports,
1989

Other publications   *Item*
An Introduction to Liquefied Petroleum Gas, 1974   1

Lovelace Foundation for Medical Education and Research
(Albuquerque, NM)

See particularly entries in Reference list under: Betz, P.A.;
Bowen, I.G.; Damon, E.G.; Fletcher, E.R.; Goldizen, V.C.;
Hirsch, F.G.; Jones, R.K.; Richmond, D.R.; Taborelli, V.; and
White, C.S.

Manufacturing Chemists Association
(Washington, DC) (predecessor to Chemical Manufacturers
Association q.v.)

Chemical Safety Data Sheets
SD-1-SD99   Chemical Safety Data Sheets, 1952–1975

Safety Guides
SG-1   Health Factors in the Safe Handling of
Chemicals
SG-2   Housekeeping in the Chemical Industry
SG-3   Flammable Liquids – Storage and Handling of
Drum Lots and Smaller Quantities
SG-4   Emergency Organization for the Chemical
Industry
SG-5   Plastic Foams – Storage, Handling and
Fabrication
SG-6   Forklift Operations
SG-7   Guide for Storage and Handling of Shock and
Impact Sensitive Materials
SG-8   Electrical Switch Lockout Procedure
SG-9   Disposal of Hazardous Waste
SG-10   Entering Tanks and Other Enclosed Spaces
SG-11   Off-the-Job Safety
SG-12   Public Relations in Emergencies
SG-13   Maintenance and Inspection of Fire Equipment
SG-14   Safety in the Scale-Up and Transfer of Chemical
Processes
SG-15   Training of Process Operators
SG-16   Liquid Chemicals – Sampling of Tank Car and
Tank Truck Shipments
SG-17   Fire Protection in the Chemical Industry
SG-18   Identification of Materials
SG-19   Electrical Equipment in Hazardous Areas
SG-20   Safety Inspection Committee for
a Small Plant

Cargo Information Cards
CIC-1–CIC-86 Cargo Information Cards

Case Histories	*Item*
Case Histories of Accidents in the Chemical Industry, vol.1, 1962; vol.2, 1966; vol.3, 1979; vol.4, 1975	1–4
Index of Case Histories of Accidents in the Chemical Industry, 1975–	5

Pollution, Effluent and Waste Disposal publications	
Air Pollution – Causes and cures	6
Atmospheric Emissions from Chlor-Alkali Manufacture	7
Atmospheric Emissions from Hydrochloric Acid Manufacturing Processes	8
Atmospheric Emissions from Thermal Process Phosphoric Acid Manufacturing	9
Atmospheric Emissions from Wet Process Phosphoric Acid Manufacturing	10
Pollution Control Primer, vols 1–2	11
Water Pollution – Causes and Cures	12
Behavior of Organic Chemicals in the Aquatic Environment, Pt 1, 1966; Pt 2, 1969	13–14
Odor Threshold for 53 Commercial Chemicals	15
Critical Mathematical Evaluation of Air Pollution Concentration Modeling, 1970	16

Solid Waste Technical Guides
SW-1   A Guide for Loadfill Disposal of Solids Wastes,
1974
SW-2   A Guide for Contract Disposal of Solids Wastes,
1974

SW-3   A Guide for Incineration of Chemical Plant
        Wastes, 1974

Transportation Emergency Guides
CC-1-CC-87   Transportation Emergency Guides

Transportation and Packaging Bulletins
TC-7   Tank Car-Approach Platforms, 1974
TC-8   Recommended Practices for Bulk
         Loading and Unloading Flammable
         Liquid Chemicals to and from Tank
         Trucks, 1975
TC-20   Recommendations for Intransit Loss and
         Damage Report, 1974
TC-27   Loading and Unloading Corrosive
         Liquids – Tank Cars, 1975
TC-28   Loading and Unloading Liquid Caustic
         Tank Cars, 1975
TC-29   Loading and Unloading
         Flammable Liquid Chemical –
         Tank Cars, 1975

Other publications	*Item*
Grassroots – A Community Relations Guide for Chemical Plant Management	17
Guidelines for Risk Evaluation and Loss Prevention in Chemical Plants, 1970	18
Laboratory Waste Disposal Manual, rev. ed., 1970	19
Chemical Statistics Handbook, 7th ed., 1971 –	20
Guide for Safety in the Chemical Laboratory, 2nd ed., by F.G. Stephenson, 1972	21
Guidelines for Chemical Plants in the Prevention, Control and Reporting of Spills, 1973	22
Statistical Summary (Supplement to Handbook), 1975–	23
Guide to the Precautionary Labeling of Hazardous Chemicals, 7th ed., 1976	24

Ministry of Home Security
(London)

Reports
RC 6   (The design of buildings against air
          attack)
RC 216   (Blast pressures from German bombs and
          parachute mines)
RC 270   (Field survey of air raid casualties)
RC 289   (A collection of curves for the estimation
          of the magnitude of blast effects)
RC 420   (The pressure wave produced by an
          under-water explosion)
RC 425   (Measurements of blast pressures from
          bare charges of 60/40 RDX/TNT)
RC 450   Structural defence 1945, by
          D.G. Christopherson, 1946
REN 120   (Derivation of theoretical radius-of-
          damage curves)
REN 144   (Analysis of damage from various
          charges)
REN 188   (Notes on damage to gasholders)

REN 190   The damage caused to houses by the
          Silvertown explosion, 19th January, 1917
          (n.d.),
REN 214   Home damage by HE weapons acting by
          blast
REN 316   (Radius of damage from 12,000 lb HC
          bombs filled with Amatex 9)
REN 317   Casualty effects of high explosive
          bombs in low level daylight raids
          compared with high level night
          raids, by P.M. Blake and
          J.W.B. Douglas
REN 367   (Prediction of blast damage)
REN 368   Observations of blast damage in Great
          Britain from German weapons with
          aluminised and non-aluminised loadings
REN 394   The blast effect of air burst bombs, by
          D.G. Christopherson, 1944
REN 404   ('Fly-bombs': suggested first
          approximation for curves on rate of decay
          of pressure and impulse with distance)
REN 434   (Weapons analysis. British 4000 and
          8000 lb HC bombs and industrial
          buildings)
REN 450   (Clearance of trees)
REN 454   The damage caused by the German flying
          bomb (Fly), by C.G. Lyons and
          P.C. Gardiner (n.d.)
REN 460   (The effect of charge weight ratio on
          equivalent charge weight)
REN 478   (Damage to house property due to a flying
          bomb)
REN 490   Blast impulses associated with specified
          categories of house damage, by
          D.G. Christopherson, 1949
REN 497   (Further notes on equivalent static
          pressure of a blast wave acting on a
          structure of given frequency)
REN 500   Damage caused to multi-storey
          buildings with load-bearing
          walls by German weapons
          acting by blast
REN 523   The screening of blast by buildings, by
          C.G. Lyons and P.C. Gardiner (n.d.)
REN 530   The damage caused by the German long
          range rocket bomb, by C.G. Lyons and
          P.C. Gardiner (n.d.)
REN 549   Distinction between blast and earth
          shock damage
REN 552   (Distance-damage curves. Comparison of
          damage from HE damage with that from
          atom bombs)
REN 558   Empirical formulae for the average circle
          radii of A and C damage, by E.B. Philip
REN 560   (Vulnerability of chimneys – 1)
REN 561   (Vulnerability of chimneys –2)
REN 578   (Vulnerability of chimneys – 3)
REN 580   (Blast impulse on ground due to air-burst
          bomb)
REN 585   Blast and the Ministry of Home
          Security, Research and
          Experiments Department,
          1939–1945, by E.B. Philip

Public Record Office Files
File AIR 20/3439, 1945 File
PREM 3/111, 1945
File HO 191/198, 1945 (a) S101, Summary of information on
the flying bomb to 24/6/44 (b) S104, 'Big Ben' (c) S105,
Casualty rates of flying bombs by day (d) S107, Presumed
V2 projectiles 8th September 1944 (e) S109, Summary of
information on the V2 rocket projectiles to 12 noon 15/9/44
(f) S110 (g) S111, The damage caused by the German long
range rocket bomb (h) S112, Summary of the methods used
for assessing the effectiveness of flying bombs and rocket
bombs (i) S117, Future attacks on London (j) S118, A
comparison of the standardised casualty rates for people in
unprotected parts of dwellings exposed to rocket bombs
(V2), flying bombs (VI) and parachute mines

National Bureau of Standards
(Washington, DC)

Reports
5482   Form characteristics of arrested saline wedges
          by G.H. Keulegan, 1957
5831   The motion of saline fronts in still water by G.H.
          Keulegan, 1958

National Centre for Systems Reliability-see UK
Atomic Energy Authority

National Fire Protection Association	*Item*
Report of Important Dust Explosions, 1957	1
Radiation Control for Fire and Other Emergency Forces by A.A. Keil, 1960	2
Fire Behaviour and Sprinklers by N.J. Thompson, 1964	3
Fire Attack by W.Y. Kimball, vol.1, 1966; vol.2, 1968	4
Industrial Fire Brigades Training Manual, 1968	5
Fire Officers Guide to Extinguishing Systems by C.W. Bahme, 1970	6
The Fire Fighters Responsibility in Arson Detection, 1971	7
Fire Service Communications for Fire Attack by W.Y. Kimball, 1972	8
The Fire Officers Guide to Dangerous Chemicals by C.W. Bahme, 1972	9
Transportation Fire Hazards, 1973	10
The Extinguishment of Fire by W.M. Haessler, 1974	11
A Hazard Study – Natural Gas and Fire Explosions, 1974	12
Fire Protection by Halons, 1975	13
Detecting Fires, 1975	14
Fire Protection by Sprinklers, 1975	15
Guides for Fighting Fires in and around Petroleum Storage Tanks, 1975	16
The Systems Approach to Fire Protection, 1976	17
Fire Command Textbook, 1985	18
Fire Protection Handbook, 16th ed., 1986	19
Electrical Installations in Hazardous Locations, 1988	20
Principles of Fire Protection, 1988	21
SPE Handbook of Fire Protection Engineering, 1988	22
Management in the Fire Service, 1989	23
Principles of Fire Protection Chemistry, 1989	24
Automatic Sprinkler and Standpipe Systems, 1990	25
Industrial Fire Hazards Handbook, 1990	26
Fire Protection Guide to Hazardous Materials, 1991	27
Fire Protection Handbook, 17th ed., 1991	28
Building Construction for the Fire Service, 1992	29
Hazardous Materials Response Handbook, 1992	30
Liquefied Petroleum Gases Handbook, 1992	31
National Fuel Gas Code Handbook, 1992	32
Fire Protection Systems: Inspection, Test and Maintenance, 1993	33
Flammable and Combustible Liquid Code Handbook, 1993	34
National Electrical Code Handbook, 1993	35
National Fire Alarm Code, 1993	36
Automatic Sprinkler Systems Handbook, 1994	37
Fire Alarm Signaling Systems Handbook, 1994	38
Installation of Sprinkler Systems, 1994	39
Life Safety Code Handbook, 1994	40
NFPA Inspection Manual, 1994	41

National Institute for Occupational Safety and Health
(Cincinnati, OH)

Criteria Documents
Publ. 76-195	Acetylene, 1976
Publ. 78-116	Acrylonitrile, 1978
Publ. 74-136	Ammonia, 1974
Publ. 77-169	Asbestos, 1977, rev.
Publ. 74-137	Benzene, 1974; 1976, rev.
Publ. 77-156	Carbon disulphide, 1977
Publ. 73-11000	Carbon monoxide, 1972
Publ. 76-170	Chlorine, 1976
Publ. 80-106	Confined spaces, 1979
Publ. 77-108	Cyanide, hydrogen and cyanide salts, 1976
Publ. 78-215	Diisocyanates, 1978
Publ. 77-200	Ethylene oxide (SHR), 1978
Publ. 77-126	Formaldehyde, 1976
Publ. 72-10269	Hot environments, 1972
Publ. 86-113	Hot environments, 1986, rev.
Publ. 76-143	Hydrogen fluoride, 1976
Publ. 77-158	Hydrogen sulphide, 1977
Publ. 75-126	Identification system for occupationally hazardous materials, 1974
Publ. 78-158	Lead, inorganic, 1978, rev.
Publ. 76-149	Nitrogen, oxides of, 1976
Publ. 73-11001	Noise, 1972
Publ. 76-137	Phosgene, 1976
Publ. 74-;111	Sulfur dioxide, 1974; 1977, rev.
Publ. 78-205	Vinyl acetate, 1978; Vinyl chloride, 1974; Vinyl halides, 1979
Publ. 88-110	Welding, brazing and thermal cutting, 1988

Sampling and Analysis		*Item*
Publs 77-159, 787-189	Manual of Sampling Data Sheets, 1978	1
Publs 77-157A, 77-157B, 77-157C, 78-175, 79-141, 80-125, 82-100	NIOSH Manual of Analytical Methods, 1977–1981	2

R93   c.1979   The Radiological Consequences of
              Notional Accidental Releases of
              Radioactivity from Fast Breeder
              Reactors: Sensitivity of the Incidence
              of Early Effects to the Duration of
              the Release

R98   1980   The Consequences of a Reduction in the
              Administratively Applied Maximum
              Annual Dose Equivalent Level for an
              Individual in a Group of Occupationally
              Exposed Workers by N.T. Harrison

R100 1980   The Radiological Consequences of
              Notional Accidental Releases of
              Radioactivity from Fast Breeder
              Reactors: The Influence of the
              Meteorological Conditions by
              C.R. Hemming *et al.*

R101 1980   ESCLOUD: A Computer Program to
              Calculate the Air Concentration,
              Deposition Rate and External Dose Rate
              from a Continuous Discharge of
              Radioactive Material to Atmosphere
              by J.A. Jones

R109 1980   A Model to Calculate Exposure from
              Radioactive Discharges into the Coastal
              Waters of Northern Europe by
              M.J. Clark *et al.*

R122 1981   The Second Report of a Working Group on
              Atmospheric Dispersion: A Procedure to
              Include Dispersion in the Model for Short
              and Medium Range Atmospheric
              Dispersion of Radionuclides –
              J.A. Jones, group secretary

R123 1981   The Third Report of a Working Group on
              Atmospheric Dispersion: The Estimation
              of Long Range Dispersion and Deposition
              of Continuous Releases of Radionuclides
              to Atmosphere – J.A. Jones, group
              secretary

R124 1981   The Fourth Report of a Working Group on
              Atmospheric Dispersion: A Model for
              Long Range Atmospheric Dispersion of
              Radionuclides Released over a Short
              Period – J.A. Jones, group secretary

R127 1981   MARC – The NRPB Methodology for
              Assessing Radiological Consequences of
              Accidental Releases of Activity by
              R.H. Clarke and G.N. Kelly

R133 c.1981 Accidental Releases of Radionuclides: A
              Preliminary Study of the Consequences
              of Land Contamination by
              J.R. Simmonds *et al.*

R135 1982   An Assessment of the Radiological
              Impact of the Windscale Reactor Fire,
              October 1957 by M.J. Crick and
              G.S. Linsey

R137 1982   An Assessment of the Radiological
              Consequences of Releases from Degraded
              Core Accidents for Sizewell PWR by
              G.N. Kelly and R.H. Clarke

R142 1982   Degraded Core Accidents for the Sizewell
              PWR: A Sensitivity Analysis of the
              Radiological Consequences by
              G.N. Kelly *et al.*

R145 1983   The Influence of Including Consequences
              beyond the UK Boundary on the
              Radiological Impact of Hypothetical
              Accidental Releases by J.A. Jones *et al.*

R146 1983   The Radiological Impact on the Greater
              London Population of Postulated
              Accidental Releases from the Sizewell
              PWR by G.N. Kelly *et al.*

R147 1983   The Radiological Impact of Postulated
              Accidental Releases during the
              Transportation of Irradiated PWR Fuel
              through Greater London by K.B. Shaw *et al.*

R149 1983   Comparison of the MARC and CRAC2
              Programs for Assessing the Radiological
              Consequences for Accidental Releases of
              Radioactive Material by C.R. Hemming
              and D. Charles

R157 1983   The Fifth Report of a Working Group on
              Atmospheric Dispersion: Models to Allow
              for the Effects of Coastal Sites, Plume
              Rise and Buildings on Dispersion of
              Radionuclides and Guidance on the Value
              of Deposition Velocity and Washout
              Coefficients – J.A Jones, group secretary

R160 1983   The Radiological Consequences of
              Degraded Core Accidents for the Sizewell
              PWR: The Impact of Adopting Revised
              Frequencies of Occurrence by G.N. Kelly
              and C.R. Hemming

R163 1983   The Influence of Countermeasures on the
              Predicted Consequences of Degraded
              Core Accidents for the Sizewell PWR by
              G.N. Kelly *et al.*

R173 1984   The Radiation Exposure of the UK
              Population – 1984 Review by J.S. Hughes
              and G.C. Roberts

R175 1985   The Significance of Small Doses of
              Radiation to Members of the Public by
              A.B. Fleishman

R180 1985   Procedures to Relate the NII Safety
              Assessment Principles for Nuclear
              Reactors to Risk by G.N. Kelly *et al.*

National Safety Council
(Chicago, IL)

Safety Data Sheets
Safety Data Sheets NSC1–, 1952

Safe Practice Pamphlets
68   Pressure Vessels – Fired and Unflred
70   Maintenance and Repair Men
105   Welding

Other publications	*Item*
Accident Facts (annual)	1
Accident Prevention Manual for Industrial Operations, 7th ed.	2
Fundamentals of Industrial Hygiene	3
Handbook of Accident Prevention for Business and Industry, 4th ed.	4
Motor Fleet Safety Manual	5
Supervisors Guide to Human Relations	6
Supervisors Safety Manual, 4th ed.	7

National Transportation Safety Board
(Washington, DC)

Highway Accident Reports

HAR-71-06   Liquid Oxygen Tank Truck Explosion
Followed by Fires in Brooklyn, New York,
May 30, 1970

HAR-71-07   Accidental Mixing of Incompatible
Chemicals, Followed by Multiple
Fatalities, During a Bulk Delivery,
Berwick, Maine, April 2, 1971

HAR-72-03   Truck–Automobile Collision Involving
Spilled Methyl Bromide on US 90 near
Gretna, Florida, August 8, 1971

HAR-72-05   Automobile–Truck Collision Followed by
Fire and Explosion of Dynamite Cargo on
US Highway 78, near Waco, Georgia,
June 4, 1971

HAR-73-03   Propane Tractor Semitrailer Overturn
and Fire, US Route 501, Lynchburg,
Virginia, March 9, 1972

HAR-73-04   Multiple Vehicle Collision, Followed by
Propylene Cargo Tank Explosion, New
Jersey Turnpike, Exit 8, September 21,
1972

HAR-76-04   Surtigas SA, Tank-Semitrailer Overturn,
Explosion and Fire, near Eagle Pass,
Texas, April 29, 1975

HAR-76-07   Union Oil Company of California, Tank
Truck and Full Trailer Overturn and Fire,
Seattle, Washington, December 4, 1975

HAR-77-01   Transport Company of Texas, Tractor
Semitrailer (Tank) Collision with Bridge
Column and Sudden Dispersal of
Anhydrous Ammonia Cargo, Houston,
Texas, May 11, 1976

HAR-78-04   Usher Transport Inc.,
Tractor–Cargo–Tank–Semi-Trailer
Overturn and Fire, Beattyville, Kentucky,
September 24, 1977

HAR-80-05   Multiple Vehicle Collision and Fire, US
Route 101, Los Angeles, California,
March 3, 1980

HAR-83-01   Multiple Vehicle Collisions and
Fire, Caldecott Tunnel, Oakland,
California, April 7, 1982

Marine Accident Reports

MAR-74-01   Tank Barge IOS-3301, Port Jefferson, New
York, January 10, 1972

MAR-75-03   Tank Barge *Ocean 89*, Fire and Explosions,
Carteret, New Jersey,
October 25, 1972

MAR-75-05   Tankship *Texaco North Dakota*, Pumproom
Explosion, Gulf of Mexico, October 3, 1973

MAR-78-01   US Tankship SS *Marine Floridan* Collision
with Benjamin Harrison Memorial Bridge,
Hopewell, Virginia, February 24, 1977

MAR-78-02   Tank Barge B-924 Fire and Explosion with
Loss of Life at Greenville, Mississippi,
November 13, 1975

MAR-78-06   Liberian Tankship SS *Sansinena* Explosion
and Fire, Union Oil Terminal Berth 46, Los
Angeles, California, December 17, 1976

MAR-78-09   Collision of Liberian Tankship M/V *Stolt
Viking* and US Crewboat *Candy Bar* in Gulf
of Mexico, January 7, 1978

MAR-78-10   French Tankship SS *Sitala* Collision with
Moored Vessels, New Orleans, Louisiana,
July 28, 1977

MAR-79-04   Luling Destrehan Ferry M/V *George Prince*
Collision with the Tanker SS Frosta on the
Mississippi River, October 20, 1976

MAR-79-05   Capsizing and Sinking of the Self-
Elevating Mobile Offshore Drilling Unit
Ocean Express near Port O'Conner, Texas,
April 15, 1976

MAR-79-06   USS *LY. Spear* (AS-36) Collision with
Liberian Motor Tankship *Zephyros*, Lower
Mississippi River, February 22, 1978

MAR-79-13   US Motor Tankship *Sealift China Sea*
Ramming of the Italian Motor Cargo Vessel
Lorenzo D'Amico, Los Angeles, California,
January 15, 1978

MAR-79-15   Spanish Motor Tankship *Ribaforada*
Ramming of Barge MB-5, Three Wharves
and Cargo Ship M/V *Tiaret*, New Orleans,
Louisiana, December 4, 1977

MAR-79-16   Collision of American Containership SS
SeaLand Venture and Danish Tanker M/T
Netty Maersk, Inner Bar Channel,
Galveston, Texas, August 27, 1978

MAR-80-07   Collision of Peruvian Freighter M/V *Inca
Tupac Yupanqui* and US Butane Barge
Panama City, Good Hope, Louisiana,
August 30, 1979

MAR-80-11   Collision of US Tankship SS *Exxon Chester*
and Liberian Freighter M/V
Regal Sword in the Atlantic Ocean near
Cape Cod, Massachusetts,
June 18, 1979

MAR-80-12   Liberian Tank Vessel M/V *Seatiger*
Explosion and Fire, Sun Oil Terminal,
Nederland, Texas, April 19, 1979

MAR-80-14   Collision of US Coast Guard
Cutter *Blackhorn* and US Tankship
Capricorn, Tampa Bay, Florida,
January 28, 1980

MAR-80-17   Collision of Liberian Tankship M/V *Pina*
and the Towboat *Mr Pete* and its Tow, Mile
99.3 Lower Mississippi River,
December 19, 1979

MAR-80-18   Explosion and Fire on board the SS *Chevron
Hawaii* with Damages to Barges and Deer
Park Shell Oil Company Terminal Houston
Ship Channel, September 1, 1979

MAR-81-04   Tankship SS *Texaco North Dakota* and
Artificial Island EI-361-A, Collision and
Fire, Gulf of Mexico, August 21, 1980

MAR-81-05   Collision of US Mississippi River Steamer *Natchez* and US Tankship SS *Exxon Baltimore*, New Orleans, Louisiana, March 29, 1980

MAR-81-11   Grounding of US Tankship SS *Concha Constable Hook*, Reach of Kill Van Kull, Upper New York Harbor, January 19, 1981

MAR-81-12   Liberian Chemical Tankship M/V *Coastal Transport* Collision with US Offshore Supply Vessel M/V *Sallee P*, Lower Mississippi River, near Venice, Louisiana, November 24, 1980

MAR-81-14   Explosion and Fire on board the US *Tankship Monticello Victory*, Port Arthur, Texas, May 31, 1981

MAR-82-02   Collision of the US Tankship *Pisces* with the Greek Bulk Carrier *Trade Master*, Mile 124, Lower Mississippi River, December 27, 1980

MAR-83-02   Capsizing and Sinking of the US Mobile Offshore Drilling Unit *Ocean Ranger* off the East Coast of Canada, 166 N Miles E of St John's, Newfoundland, February 15, 1983

MAR-83-03   Explosion and Fire on board US Coastal Tankship *Poling Bros* No.9 East River, New York Harbor, February 26, 1982

MAR-83-04   Collision of US Towboat *Creole Genii* and Liberian Tank Vessel Arkas near Mile 130, Mississippi River, March 31, 1982

MAR-83-06   Engine Room Flooding and Near Foundering of US Tankship *Ogden Williamette*, Caribbean Sea, June 16, 1982

MAR-83-07   Explosions and Fire on board the US Tankship SS *Golden Dolphin* in the Atlantic Ocean, March 6, 1982

MAR-84-08   See MAR-87-02

MAR-84-09   Grounding of the US Tankship SS *Mobil Oil* in the Columbia River, near Saint Helens, Oregon, March 19, 1984

MAR-85-04   Collision of the US Towboat *Ann Brent* and Tow with the Greel Tankship *Maninia*, Mile 150, Lower Mississippi River, June 11, 1984

MAR-85-06   Explosion and Sinking of the United States Tankship SS *American Eagle*, Gulf of Mexico, February 26 and 27, 1984

MAR-85-11   Explosion and Fire aboard the US Mobile Offshore Drilling Unit *Zapata Lexington*, Gulf of Mexico, September 14, 1984

MAR-86-03   Explosion and Fire on board the US Mobile Offshore Drilling Unit *Glomar Arctic II* in the North Sea, 130 Nautical Miles E-SE Aberdeen, Scotland, January 15, 1985

MAR-86-05   Explosion and Fire on board US Chemical Tankship *Puerto Rican* in the Pacific Ocean near San Francisco, California, October 31, 1984

MAR-86-10   Collapse of the US Mobile Offshore Drilling Unit *Penrod 61* in the Gulf of Mexico, October 27, 1985

MAR-87-02   Capsizing and Sinking of the US Drillship *Glomar Java Sea* in the South China Sea 65 Nautical Miles South-Southwest of Hainan Island, People's Republic of China, October 25, 1983 (supersedes MAR-84-08)

MAR-87-05   Explosion aboard the US Tank Barge TTT 103, Pascagoula, Mississippi, July 31, 1986

MAR-87-06   Explosions and Fire aboard the US Tankship *Omi Yukon* in the Pacific Ocean about 1000 Miles West of Honolulu, Hawaii, October 28, 1986

MAR-87-07   Engineroom Flooding of the US Tankship *Prince William Sound*, near Puerto Vallarta, Mexico, May 9, 1986

MAR-87-08   Fires on board the Panamanian Tankship *Shoun Vanguard* and the US Tank Barge *Hollywood* 3013, Deer Park, Texas, October 7, 1986

MAR-87-09   Explosion and Fire aboard the US Tank Barge STC 410 at the Steuart Petroleum Company Facility, Piney Point, Maryland, December 20, 1986

MAR-87-10   Grounding of the Panamanian Tankship *Grand Eagle* in the Delaware River, near Marcus Hook, Pennsylvania, September 28, 1985

MAR-88-09   Ramming of the Maltese Bulk Carrier *Mont Fort* by the British Tankship *Maersk Neptune* in Upper New York Bay, February 15, 1988

MAR-89-02   Striking of a Submerged Object by Bahamian Tankship *Esso Puerto Rico*, Mississippi River, Kenner, Louisiana, September 3, 1988

MAR-89-03   Explosion aboard the Maltese Tank Vessel *Fiona* in the Long Island Sound near Northport, New York, August 31, 1988

MAR-89-06   Capsizing and Sinking of the Mobile Offshore Drilling Unit *Rowan Gorilla I* in the North Atlantic Ocean, December 15, 1988

MAR-89-07   Collision between the Swedish Auto Carrier *Figaro* and the French Tankship *Camarque*, Galveston Bay Entrance, November 10, 1988

MAR-90-04   Grounding of the US Tankship *Exxon Valdez* on Bligh Reef, Prince William Sound near Valdez, Alaska, March 24, 1989

MAR-90-06   Engineroom Fire aboard the US Tankship *Charleston* in the Atlantic Ocean about 35 miles off the South Coast, March 7, 1989

MAR-91-01   Grounding of the Greek Tankship *World Prodigy* off the Coast of Rhode Island, June 23, 1989

MAR-91-03   Collision between the Greek Tankship *Shinoussa* and the US Towboat *Chandy N* and Tow near Red Fish Island, Galveston Bay, Texas

MAR-91-04   Explosion and Fire aboard the US Tankship *Jupiter*, Bay City, Michigan, September 16, 1990

MAR-92-01   Grounding of the US Tankship *Star Connecticut*, Pacific Ocean, near Barbers Point, Hawaii, November 6, 1990

MAR-92-02   Explosion and Fire on the US Tankship *Surf City*, Persian Gulf, February 22, 1990

MAR-92-04   Collision of the Hong Kong-Registered Motor Tankship *Mandan* with the US Army Corps of Engineers Barge Flotilla at Mile 10.5 above Head of Passes in the Lower Mississippi River near Venice, Louisiana, August 15, 1990

Pipeline Accident Reports

PAR-71-01   Mobil Oil Corporation High-Pressure Natural Gas Pipeline, near Houston, Texas, September 9, 1969

PAR-71-02   Colonial Pipeline Co., Petroleum Products Pipeline, Jacksonville, Maryland, September 3, 1970

PAR-72-01   Phillips Pipeline Company, Propane Gas Explosion, Franklin County, Missouri, December 9, 1970

PAR-73-02   Exxon Pipeline Company, Crude Oil Explosion at Hearne, Texas, May 14, 1972

PAR-73-04   Phillips Pipeline Company, Natural Gas Liquids Fire, Austin, Texas, February 22, 1973

PAR-74-06   Mid-America Pipeline System, Anhydrous Ammonia Leak, Conway, Kansas, December 6, 1973

PAR-75-01   Michigan-Wisconsin Pipeline Company, Natural Gas Transmission Line Failure, Monroe, Louisiana, March 2, 1974

PAR-75-02   Transcontinental Gas Pipeline Corporation, 30 inch Transmission line Failure, near Bealeton, Virginia, June 9, 1974

PAR-76-01   Mid Valley Pipeline Company, Crude Oil Terminal Fire, near Lima, Ohio, January 17, 1975

PAR-76-05   Dow Chemical Company USA, Natural Gas Liquids Explosion and Fire, near Devers, Texas, May 12, 1975

PAR-76-07   Sun Pipeline Company, Rupture of 8 inch Pipeline, Romulus, Michigan, August 2, 1975

PAR-77-03   Exxon Gas System Inc., Natural Gas Explosion and Fire, Robstown, Texas, December 7, 1976

PAR-78-01   Consolidated Gas Supply Corporation, Propane Pipeline Rupture and Fire, Ruff Creek, Pennsylvania, July 20, 1977

PAR-78-02   Alyeska Pipeline Service Company, Explosion and Fire, Pump Station 8, near Fairbanks, Alaska, July 8, 1977

PAR-79-01   Mid-America Pipeline System, Liquefied Petroleum Gas Pipeline Rupture and Fire, Donnellson, Iowa, August 4, 1978

PAR-80-01   Southern Natural Gas Company, Rupture and Fire of a 14 inch Gas Transmission Pipeline South-East of New Orleans, Louisiana, July 15, 1980

PAR-80-02   Columbia liquefied Natural Gas Corporation, Explosion and Fire, Cove Point, Maryland, October 6, 1979

PAR-80-06   The Pipelines of Puerto Rico Inc., Petroleum Products Pipeline Rupture and Fire, Bayamon, Puerto Rico, January 30, 1980

PAR-83-01   The Gas Company of New Mexico, Natural Gas Explosion and Fire, Portales, New Mexico, June 28, 1982

PAR-83-02   Northern

PAR-83-02   Northern Natural Gas Company, Pipeline Puncture, Explosion and Fire, Hudson, Iowa, November 4, 1982

PAR-83-03   Mississippi River Transmission Corporation, Natural Gas Flash Fire, Pine Bluff, Arkansas, October 1, 1982

PAR-83-04   El Paso Natural Gas Company, Compressor Station Explosion and Fire, Bloomfield, New Mexico, May 26, 1983

PAR-84-01   Mid-America Pipeline System, liquefied Petroleum Gas Pipeline Rupture, West Odessa, Texas, March 15, 1983

PAR-85-01   Arizona Public Service Company, Natural Gas Explosion and Fire, Phoenix, Arizona, September 25, 1984

PAR-85-02   National Fuel Gas Company, Natural Gas Explosion and Fire, Sharpsville, Pennsylvania, February 23, 1985

PAR-86-01   Continental Pipe line Company, Pipeline Rupture and Fire, Kaycee, Wyoming, July 23, 1985

PAR-87-01   Texas Eastern Gas Pipeline Company Ruptures and Fires at Beaumont, Kentucky, on April 27, 1985, and Lancaster, Kentucky, on February 21, 1986

PAR-90-02   Fire on Board the F/V *Northumberland* and Rupture of a Natural Gas Transmission Pipeline in the Gulf of Mexico near Sabine Pass, Texas, October 3, 1989

PAR-91-01   Liquid Propane Pipeline Rupture and Fire, Texas Eastern Products Pipeline Company, North Blenheim, New York, March 3, 1990

Railroad Accident Reports

RAR   Southern Railway Company, Train 154, Derailment with Fire and Explosion, Laurel, Mississippi, January 25, 1969

RAR-70-02   Illinois Central Railroad Company, Derailment at Glendora, Mississippi, September 11, 1969

RAR-71-02   Chicago, Burlington and Quincy RR Company, Train 64 and Train 824, Derailment and Collision with Tank Car Explosion, Crete, Nebraska, February 18, 1969

RAR-72-02   Derailment of Toledo, Peoria and Western Railroad Company's Train No. 20 with Resultant Fire and Tank Car Ruptures, Crescent City, Illinois, June 21, 1970

RAR-72-06   Derailment of Missouri Pacific Railroad Company's Train 94 at Houston, Texas, October 19, 1971

RAR-73-01   Hazardous Materials Railroad Accident in the Alton and Southern Gateway Yard in East St Louis, Illinois, January 22, 1972

RAR-74-04   Derailment and Subsequent Burning of Delaware and Hudson Railway Freight Train at Oneonta, New York, February 12, 1974

RAR-75-02   Southern Pacific Transportation Company, Freight Train 2nd BSM 22 Munitions Explosion, Benson, Arizona, May 24, 1973

RAR-75-04   Hazardous Materials Accident in the Railroad Yard of the Norfolk and Western Railway at Decatur, Illinois, July 19, 1974

RAR-75-07   Hazardous Materials Accident at the
Southern Pacific Transportation
Company's Englewood Yard in Houston,
Texas, September 21, 1974

RAR-76-01   Burlington Northern Inc.,
Monomethylamine Nitrite Explosion,
Wenatchee, Washington, August 6, 1974

RAR-76-08   Derailment of Tank Cars with Subsequent
Fire and Explosion on Chicago, Rock Island
and Pacific Rail Company, near Des
Moines, Iowa, September 1, 1975

RAR-77-07   Derailment of a Burlington Northern
Freight Train at Belt, Montana,
November 26, 1976

RAR-78-04   Louisville and Nashville Railroad
Company, Freight Train Derailment and
Puncture of Anhydrous Ammonia Tank
Cars at Pensacola, Florida, November 9,
1977

RAR-78-07   Derailment of Atlanta and St Andrews Bay
Railroad Company Freight Train,
Youngstown, Florida, February 26, 1978

RAR-78-08   St Louis Southwestern Railway Company
Freight Train Derailment and Rupture of
Vinyl Chloride Tank Car, Lewisville,
Arkansas, March 29, 1978

RAR-79-01   Derailment of L&N Railroad Company's
Train No. 584 and Subsequent Rupture of
Tank Car Containing Liquefied Petroleum
Gas, Waverly, Tennessee, February 22, 1978

RAR-79-11   Louisville and Nashville Railroad
Company, Freight Train Derailment and
Puncture of Hazardous Materials Tank
Cars, Crestview, Florida, April 8, 1979

RAR-81-01   Illinois Central Gulf Railroad Company
Freight Train Derailment, Hazardous
Material Release and Evacuation,
Muldraugh, Kentucky, July 26, 1980

RAR-83-05   Derailment of Illinois Central Gulf
Railroad Freight Train Extra 9629 East
(GS-2-28) and Release of Hazardous
Materials at Livingston, Louisiana,
September 28, 1982

RAR-85-08   Vinyl Chloride Monomer Release from a
Railroad Tank Car and Fire, Formosa
Plastics Corporation Plant, Baton Rouge,
Louisiana, July 30, 1983

RAR-85-10   Denver and Rio Grande Western Railroad
Company Train Yard Accident Involving
Punctured Tank Car, Nitric Acid and Vapor
Cloud, Denver, Colorado, April 3, 1983

RAR-85-12   Derailment of Seaboard System Railroad
Train No. F-690 with Hazardous Material
Release, Jackson, South Carolina,
February 23, 1985

RAR-86-04   Derailment of St Louis Southwestern
Railway Company (Cotton Belt) Freight
Train Extra 4835 North and Release of
Hazardous Materials near Pine Bluff,
Arkansas, June 9, 1985

RAR-89-04   Head-On Collision between Iowa Interstate
Railroad Extra 406 East with Release of
Hazardous Materials near Altoona, Iowa,
July 30, 1988

RAR-89-05   Collision and Derailment of Montana Rail
link Freight Train with Locomotive Units
and Hazardous Materials Release, Helena,
Montana, February 2, 1989

RAR-90-02   Derailment of Southern Pacific
Transportation Company Freight Train on
May 12, 1989, and Subsequent Rupture of
Calney Petroleum Pipeline on May 25, San
Bernadino, California

RAR-91-04   Derailment of CSX Transportation Inc.
Freight Train and Hazardous Materials
Release near Freeland, Michigan, July 22, 1989

Nordic Liaison Committee for Atomic Energy
(available Riso Nat. Lab., Roskilde, Denmark)

NKA/LIT(85)1   The human component in the safety of
complex systems by B. Wahlstrom, 1986

NKA/LIT(85)3   Organization for safety by E. Edsberg,
1985

NKA/LIT(85)4   The design process and the use of
computerized tools in control room design
by B. Wahlstrom, R. Heinonen, J. Ranta
and J. Haarla, 1985

NKA/LIT(85)5   Computer aided operation of complex
systems ed. by LP. Goodstein, 1985

NKA/LIT(85)6   Training diagnostic skills for nuclear
power plants by LP. Goodstein, 1986

Occupational Safety and Health Administration
(Washington, DC)

OSHA series

OSHA	2001	The Occupational Safety and Health Act, 1970
OSHA	2098	OSHA Inspections
OSHA	2220	Carcinogens
OSHA	2224	Carbon Monoxide
OSHA	2225	Vinyl Chloride
OSHA	2230	Lead
OSHA	2234	Mercury
OSHA	2237	Handling Hazardous Materials
OSHA	2248	Toluene Diisocyanate
OSHA	2253	Workers Rights under OSHA
OSHA	2254	Training Requirements for OSHA Standards

Other publications	*Item*
New Hazards for Fire Fighters	1
Occupational Cancer	2
Essentials of Materials Handling	3
Hazard Communication: A Compliance Kit, 1988	4
OSHA Analytical Methods Manual, 1990	5
OSHA Laboratory Standard – Chemical Hygiene Plan, 1990	6
Chemical Information Manual, 1991	7

Oil Companies International Marine Forum
See Appendix 27 for all publications

Oil Insurance Association
(merged with Industrial Risk Insurers, q.v.)

Publications

Publ. 101   Hydrorefining Process Units. Loss Causes and
Guidelines for Loss Prevention, 1974

301 Gasoline Plants. Recommendations and
     Guidelines
302 Our Plant is on Fire. Case Histories of Gasoline
     Plant Losses, 1972
501 Fired Heaters. Loss Causes and Guidelines for
     Safe Design and Operation, 1971
502 Boiler Safety, 1971
631 General Recommendations for Spacing in
     Refineries, Petrochemical Plants, Gasoline
     Plants, Terminals, Oil Pump Stations and
     Offshore Properties, 1972
651 Loss Cause Study of Fires and Explosions in
     Engines, Compressors and Gas Turbines
711 Truck Loading Rack Safety

Loss Information Bulletins
1   Hydrocrackers
2   Alkylation Units, 1972
400-1 Tank Fires, 1974

Loss Control Bulletins
400-2 'Jumbo' Oil Storage Tanks, 1975

Oyez Scientific and Technical Services Ltd
See IBC Technical Services Ltd

Petroleum Industry Training Board
(London)

PITB Safety Training Guides		*Guide*
1975	Personal Safety	1
1975	Static Electricity	2
1975	Safety Legislation for Vehicle and Plant Operators	3

Other publications		*Item*
1971	Selection of Process Operators	1
1971	The Training of Process Operators – Draft Recommendations	2
1972	Technical Education for Process Operators	3
1973	A Guide to Emergency Planning in Oil Installations	4
1973	Oil Storage and Movement Driver and Operator Training – Draft Recommendations	5
1975	Fire Training Manual	6

Riso National Laboratory
(Riso, Denmark) (formerly
Danish Atomic Energy Commission)

Riso 256	Experiments on data presentation to process operators in diagnostic tasks by J. Rasmussen and L.P. Goodstein, 1972
Riso-M-686	On the communication between operators and instrumentation in automatic process plants by J. Rasmussen, 1968
Riso-M-706	On the reliability of process plants and instrumentation systems by J. Rasmussen, 1968
Riso-M-808	Characteristics of operator, automatic equipment and designer in plant automation by J. Rasmussen, 1968
Riso-M-1374	The cause cconsequence diagram method as a basis for quantitative accident analysis by D.S. Nielsen, 1971
Riso-M-1582	A study of mental procedures in electronic trouble shooting by J. Rasmussen and A Jensen, 1973
Riso-M-1654	A formalisation of failure mode analysis of control systems by J.R. Taylor, 1973
Riso-M-1673	The role of the man-machine interface in systems reliability by J. Rasmussen, 1973
Riso-M-1707	A semiautomatic method for qualitative failure mode analysis by J.R. Taylor, 1974
Riso-M-1708	Proving correctness for a real time operating system by J.R. Taylor, 1974
Riso-M-1722	The human data processor as a system component: bits and pieces of a model by J. Rasmussen, 1974
Riso-M-1729	The operator's diagnosis task under abnormal operating conditions in industrial process plant by LP. Goodstein, O.M. Pedersen, J. Rasmussen and P.Z. Skanborg, 1974
Riso-M-1736	Computer-aided diagnosis – results of a literature survey of medical applications by LP. Goodstein, 1974
Riso-1740	Sequential effects in failure mode analysis by J.R. Taylor, 1975
Riso-M-1742	Design errors in nuclear power plant by J.R. Taylor, 1974
Riso-M-1743	Use of cause – consequence charts in practical systems analysis by D.S. Nielsen, 1974
Riso-M-1794	Some assumptions underlying reliability prediction by J.R. Taylor, 1975
Riso-M-1826	Common mode and coupled failure by J.R. Taylor, 1975
Riso-M-1837	A study of abnormal occurrence reports by J.R. Taylor, 1975
Riso-M-1878	Experience with a safety analysis procedure by J.R. Taylor, 1976
Riso-M-1890	Interlock design using fault tree and cause consequence analysis by J.R. Taylor, 1976
Riso-M-1894	Notes on human factors problems in process plant reliability and safety prediction by J. Rasmussen and J.R. Taylor, 1976
Riso-M-1903	Reliability analysis of proposed instrument air system by D.S. Nielsen, O. Platz and H.E. Kongso, 1977
Riso-M-1907	Algorithms and programs for consequence diagram and fault tree construction by E. Hollo and J.R. Taylor, 1976
Riso-M-1908	Some points of advanced alarm system design by E. Hollo, 1977
Riso-M-1983	Notes on diagnostic strategies in process plant environment by J. Rasmussen, 1978
Riso-M-2139	Notes on human error analysis and prediction by J. Rasmussen, 1978
Riso-M-2162	Generality of component models used in automatic fault tree synthesis by J.R. Taylor, 1979

Riso-M-2206	Cause-consequence reporting for accident reduction by O. Bruun, J.R. Taylor and A. Rasmussen, 1979
Rose-M-2285	Notes on human performance analysis by E. Hollnagel, O.M. Pedersen and J. Rasmussen, 1981
Riso-M-2293	Coping with complexity by J. Rasmussen and M. lind, 1981
Riso-M-2304	Human errors. A taxonomy for describing human malfunction in industrial installations by J. Rasmussen, 1981
Riso-M-2311	Automatic fault tree construction with RIKKE – a compendium of examples, Volume 1 Basic models by J.R. Taylor, 1981
Rios-M-2311	Automatic fault tree construction with RIKKE – a compendium of examples, Volume 2 Control and safety loops by J.R. Taylor, 1981
Riso-M-2319	Risk analysis of a distillation unit by J.R. Taylor, O. Hansen, C. Jensen, O.F. Jacobsen, M. Justesen and S. Kjaergaard, 1982
Riso-M-2341	The use of flow models for design of plant operating procedures by M. Lind, 1982
Riso-M-2349	A model of human decision making in complex systems and its use for design of system control strategies by J. Rasmussen and M. Lind, 1982
Riso-M-2351	Formalized search strategies for human risk contributions: a framework for further development by J. Rasmussen and O.M. Pedersen, 1982
Riso-M-2357	Multilevel flow modelling of process plant for diagnosis and control by M. Lind, 1982

Other publications   *Item*
A background to risk analysis, vols I–III by   1
J.R. Taylor, 1979

See also individual authors in Reference List

Royal Commission on Environmental Pollution
(London)
The reports are entitled Royal Commission on Environmental Pollution: First Report, etc.

Reports
1	First Report, 1971
2	Three Issues in Industrial Pollution, 1972
3	Pollution in Some British Estuaries and Coastal Waters, 1972
4	Pollution Control, Progress and Problems, 1974
5	Air Pollution Control: an Integrated Approach, 1976
6	Nuclear Power and the Environment, 1976
7	Agriculture and Pollution, 1979
8	Oil Pollution of the Sea, 1981
9	Lead in the Environment, 1983
10	Tackling Pollution: Experience and Problems, 1984
11	Managing Waste – the Duty of Care, 1985
12	Best Practical Environmental Option, 1988
13	The Release of Genetically Engineered Organisms to the Environment, 1989

14	GENHAZ: A System for the Critical Appraisal of Proposals to Release Genetically Modified Organisms to the Environment, 1991
15	Emissions from Heavy Duty Diesel Engined Vehicles, 1992
16	Freshwater Quality, 1992
17	Incineration of Waste, 1993
18	Transport and the Environment, 1994

Royal Society for the Prevention of Accidents
(Birmingham)
IS/13	Construction Regulations Handbook, 10th ed., 1975
IS/29	Safety Handbook for Mobile Cranes, rev. ed., 1974
IS/32	Occupational Safety Committees Permit-to-Work Systems
IS/72	Factory Accidents: Their Causes and Prevention
IS/73	The Safe Use of Electricity
IS/77	Site Register
IS/103	RoSPA Guide to the Health and Safety at Work Act 1974
IS/106	Safety at Work
IS/108	The History of Work Safety Legislation, 1976

See also Appendix 27 (Imperial Chemical Industries Ltd – RoSPA)

Safety and Reliability Directorate – see UK Atomic Energy Authority, Safety and Reliability Directorate

Safety in Mines Research Board
(Sheffield)

Papers
9	The lag on ignition of firedamp by Naylor and Wheeler, 1925
36	The ignition of gases by hot wires by W.C.F. Shepherd and R.V. Wheeler, 1927
46	The ignition of firedamp by the heat impact of rocks by M.M.J. Burgess and R.V. Wheeler, 1928
54	The ignition of firedamp by the heat impact of metal against rock by M.M.J. Burgess and R.V. Wheeler, 1929
62	The ignition of firedamp by the heat impact of hand picks against rocks by M.M.J. Burgess and R.V. Wheeler, 1930
70	The ignition of firedamp by the heat of coal cutter picks against rocks by M.M.J. Burgess and R.V. Wheeler, 1931
74	International Conf. on Safety in Mines at Buxton
93	The ignition of firedamp by compression by Dixon and Harwood, 1935

Safety in Mines Research Establishment
(Sheffield)

Electrical hazards
1	Testing of intrinsically safe electrical equipment
2	Research on flameproof enclosures for electrical equipment
3	On-load testing of flameproof electrical equipment
4	Arc physics
5	Ignition of flammable gases by arc discharges

Society of International Gas Tanker and Terminal Operators (SIGTTO) (London)

See also Appendix 27
Safety and Reliability Directorate – see UK Atomic
Energy Authority

Technical Research Centre of Finland (VTT) (Espoo, Finland)

516   Evaluation of the validity of four hazard
       identification methods with event descriptions by
       J. Suokas and Pyy, 1988

Other publications   *Item*
25   On the reliability and validity of safety
       analysis by J. Suokas, 1985   1

TNO
(Rijswijk, The Netherlands)

Reports
81-07020   Wind tunnel modeling of the dispersion of
            LNG vapour clouds in the atmospheric
            boundary layer by L.A. Janssen, 1981

See also individual authors in Reference List

**UK Atomic Energy Authority, Health and Safety
Branch and Systems Reliability Service**
(Risley, near Warrington)

Reports in AHSB, AHSB(RP) and SRS/GR series are
prefaced in text by UKAEA, other series are not

AHSB(RP)
AHSB(RP) R42   1964   Methods of estimation of the
                       dispersion of windborne
                       material and data to assist in
                       their application by P.M. Brynt

AHSB series
AHSB(S) R64   1963   An assessment of
                      environmental hazards from
                      fission product releases by
                      J.R. Beattie
AHSB(S) R91   1965   Reliability considerations for
                      automatic protective systems
                      by A.E. Green and A.J. Bourne
AHSB(S) R99   1965   Reliability assessment of
                      protective systems for nuclear
                      installations by AR. Eames
AHSB(S) R110   1966   A criterion for the reliability
                       assessment of protective
                       systems by A.J. Bourne
AHSB(S) R113   1966   Assessment of sensing
                       channels for high integrity
                       protective systems by
                       A.E. Green
AHSB(S) R117   1966   Safety assessment with
                       reference to automatic
                       protective systems for nuclear
                       reactors, Pts 1-3, by A.E. Green
                       and AJ. Bourne
AHSB(S) R119   1966   The use of detailed fault
                       analysis in evaluating fail-safe
                       equipment by A.R. Eames
AHSB(S) R122   1967   The response times of
                       protective systems for nuclear
                       and chemical plant by
                       A.R. Eames
AHSB(S) R131   1967   A method of measuring
                       equipment response time by
                       A.R. Eames

AHSB(S) R136   1967   Safety assessment − a method
                       for determining the
                       performance of alarm and
                       shutdown systems for
                       chemical plants by G. Hensley
AHSB(S) R138   1967   Data store requirements
                       arising out of reliability
                       analyses by A.R. Eames
AHSB(S) R160   1969   A quantitative approach to the
                       evaluation of the safety
                       function of operators on
                       nuclear reactors by J.F. Ablitt
AHSB(S) R162   1968   A survey of defects in pressure
                       vessels built to high standards
                       of construction and its
                       relevance to nuclear primary
                       circuit envelopes by
                       C.A.G. Phillips and
                       R.G. Warwick
AHSB(S) R164   1969   Reliability prediction by
                       AE. Green
AHSB(S) R172   1969   Safety assessment of
                       automatic and manual
                       protective systems for reactors
                       by A.E. Green
AHSB(S) R178   1969   The reliability prediction of
                       mechanical instrumentation
                       equipment for process control
                       by G. Hensley
AHSB(S) R179   1970   Confidence limits for numbers
                       from 0 to 1200 based on the
                       Poisson distribution by
                       E.R. Woodcock and A.R. Eames

SRS/GR series
SRS/GR/1   1973   Plant and process reliability by
                   G. Hensley
SRS/GR/2   1971   Quantitative assessments of
                   system reliability by A.E. Green
SRS/GR/4   1971   Considerations in the use of
                   system reliability models by
                   A.J. Bourne
SRS/GR/9   1971   Human factors by B. Sayers
SRS/GR/12   1973   The effect of equipment
                    reliability upon plant and
                    system performance by
                    A.R. Eames
SRS/GR/13   1973   Some reliability characteristics
                    for operating plant by
                    A.R. Eames and
                    C.D.H. Fothergffl
SRS/GR/14   1973   An introduction to the 'Syrel'
                    data bank by J.F. Ablitt
SRS/GR/22   1973   The analysis and presentation
                    of derived reliability data from
                    a computerised data store by
                    C. Fothergiffl

NCSR series
NCSR R6   1975   Application of diakoptical
                  analysis in system reliability
                  assessment
NCSR R7   1975   Reliability analysis − a
                  systematic approach based on
                  limited data by A.J. Bourne

NCSR R8	1976	Report on the work and activities of the plant availability study group
NCSR RIO	1976	Human reliability in complex systems: an overview by D.E. Embrey
NCSR R11	1977	The effect of operational loading on the failure characteristics of mechanical valves by T.R. Moss
NCSR R12	1976	Advances in Reliability Technology, ed. by T.R. Moss
NCSR R14	1977	The reliability analysis of logical networks by the computer program ALMONA by P. Brock
NCSR 16	1979	A literature survey of computer software reliability by B.K. Daniels and M. Hughes
NCSR 17	1979	Reliability and protection against failure in computer systems by B.K Daniels
NCSR 21	1979	Optimal equipment replacement policies by R.F. de la Mare
NCSR/GR/65	1987	The achievement and assessment of safety in nuclear power reactor systems containing software by C.J. Dale, A. Ball and M.H. Butterfield
NCSR/GR/68	1987	System unavailability for a continuous process having batched input or output by K. Evans
NCSR/GR/71	1987	Performance of pipework in the British sector of the North Sea by A.G. Cannon and R.C.E. Lewis

UK Atomic Energy Authority, Safety and Reliability Directorate
(Risley, near Warrington)

SRD Rl	1971	Principles of reliability of nuclear reactor control and instrumentation systems by A.R. Eames
SRD R8	1974	Chemical plants and laboratories safety assessment by A. Fairbairn, E.S. London, L.E. Lockett and A Cook, 1974
SRD R16	1973	Accident probability reduction by the application of service reliability data by J.F. Ablitt
SRD R21	1973	Stress concentration theory of fracture and its relationship to linear elastic fracture mechanics by W.H. Irvine
SRD R26	1974	A stress concentration theory of fracture and its application to various crack geometries by W.H. Irvine

SRD R30	1974	The second survey of defects in pressure vessels built to high standards of construction and its relevance to nuclear primary circuits by T.A Smith and R.G. Warwick
SRD R48	1977	An elasto-plastic theory of fracture and its application to the analysis of experimental data by W.H. Irvine
SRD R57	1976	An appraisal of the PREP KITT and SAMPLE computer codes for the evaluation of the reliability characteristics of engineered systems by P. Shaw and R.F. White
SRD R63	1976	A description of the mathematical and physical models incorporated in TIRION 2 – a computer program that calculates the consequences of a release of radioactive material to the atmosphere and an example of its use by G.D. Kaiser
SRD R67	1977	Critical review of the USCG report by Raj *et al.* (1974) on spills of liquid anhydrous ammonia on to water, with an alternative assessment of the experimental results by R.F. Griffiths
SRD R68	1976	Accident rates from plant failure at high demand frequencies by D.H. Worledge
SRD R80	1977	The use and interpretation of confidence and tolerance intervals in safety analysis by G.W. Parry, P. Shaw and D.H. Worledge
SRD R85	1977	Effect of duration of release on the dispersion of effluent released to the atmosphere by R.F. Griffiths
SRD R91	1978	A method for analysing cargo protection afforded by ship structures in collision and its application to an LNG carrier by A.N. Kinkead
SRD R95	1977	The use of regeneration diagrams to solve component reliability problems by G.W. Parry and D.H. Worledge
SRD R97	1978	HUBBLE-BUBBLE 1: a computer program for the analysis of non-equilibrium flows of water by D.J. Mather
SRD R100	1978	Evaporation from spills of hazardous liquids on land and water by P. Shaw and F. Briscoe
SRD R103	1978	The density of an ammonia cloud in the early stages of dispersion by S.R. Haddock and R.J. Williams

SRD R110	1978	The incidence of multiple fatality accidents in the UK by R.F. Griffiths and L.S. Fryer
SRD R113	1978	The downtime distribution in reliability engineering by G.W. Parry and D.H. Worledge
SRD R118	1978	HUBBLE-BUBBLE II: a computer program to describe non-equilibrium flow of water in simple pipe systems by D. Martin
SRD R125	1978	Individual risk – a compilation of recent British data by D.R. Grist
SRD R128	1979	The role of event trees in reactor accident analysis by D.H. Worledge
SRD R129	1979	The tolerance–confidence relationship and safety analysis by G.W. Parry, P. Shaw and D.H. Worledge
SRD R131	1978	LNG/water vapour explosions – estimates of pressure and yields by F. Briscoe and G.J. Vaughan
SRD R134	1978	TIRION 4: a computer program for use in nuclear safety studies by L.S. Fryer and G.D. Kaiser
SRD R135	1979	A method for the estimation of the probability of damage due to earthquakes by M.A.H.G. Alderson
SRD R143	1979	Regeneration diagrams by G.W. Parry
SRD R146	1979	A study of common-mode failures by G.T. Edwards and I.A. Watson, 1979
SRD R149	1979	World-wide data on the incidence of multiple fatality accidents by L.S. Fryer and R.F. Griffiths
SRD R150	1979	Examples of the successful application of a simple model for the atmospheric dispersion of dense, cold vapours to the accidental release of anhydrous ammonia from pressurised containers by G.D. Kaiser
SRD R152	1979	DENZ – a computer program for the calculation of the dispersion of dense toxic or explosive gases in the atmosphere by L.S. Fryer and G.D. Kaiser
SRD R154	1979	The accidental release of anhydrous ammonia to the atmosphere – a systematic study of factors influencing cloud density and dispersion by R.F. Griffiths and G.D. Kaiser
SRD R155	c.1979	Validation of BK WAVE: a computer program for the calculation of one-dimensional shock wave propagation from explosions by F. Briscoe
SRD R173	1983	GASEX2: a two dimensional hydrodynamic code for use in gas cloud explosion calculations by D. Martin
SRD R177	1980	The metal/water explosion phenomenon – a review of present understanding by G.J. Vaughan
SRD R190	1980	The characterisation and evaluation of uncertainty in probabilistic risk analysis by G.W. Parry and P.W. Winter
SRD R196	1981	Defences against common-mode failures in redundant systems. A guide for management, designers and operators by A.J. Bourne, G.T. Edwards, D.M. Hunns, D.R. Poulter and I.A. Watson
SRD R198	1981	UK aircraft crash statistics – 1981 revision by D.W. Phillips
SRD R203	1981	A survey of defects in pressure vessels in the UK for the period 1962–1978 and its relevance to nuclear primary circuits by T.A. Smith and R.G. Warwick
SRD R210	1981	A users manual to SPILL by A.J. Prince, 1981
SRD R217	1982	Matching the results of a theoretical model with failure rates obtained from a population of non-nuclear pressure vessels by L.P. Harrop
SRD R221	1982	A survey of turbulence models with particular reference to dense gas dispersion by C.L. Farmer
SRD R229	1983	Development of CRUNCH: a dispersion model for continuous releases of a denser-than-air vapour into the atmosphere by S.F. Jagger
SRD R238	1982	The transition from pressure-energy to gravity-driven slumping during the catastrophic release of a denser-than-air gas by S.F. Jagger
SRD R240	1983	The thermal response of tanks engulfed in fire by D.L.M. Hunt, P. Ramskill, C.G. Shepherd and R. Young, 1983
SRD R241	1983	A guide to the use of a computer program to model the thermal response of a tank engulfed in fire by P.K. Ramskill, 1983
SRD R243	1983	The physics of heavy gas cloud dispersal by D.M. Webber

SRD R245	1983	Noncoherent structure theory: a review and its role in fault tree analysis by B.D. Johnston and R.H. Matthews
SRD R246	1982	Seismic design criteria and their applicability to major hazard plant within the United Kingdom by M.A.H.G. Alderson
SRD R251	1983	GASEX1 — a general one-dimensional code for gas cloud explosions by J. MacKenzie and D. Martin, 1983
SRD R252	1983	Validation calculations for the code GASEX2 by D. Martin
SRD R254	1984	A suggested method for the treatment of human error in the assessment of major hazards by R.F. White
SRD R262	1983	A computational survey of the factors influencing the pressure fields produced by unconfined vapour cloud by J. MacKenzie
SRD R263	1983	The fire hazards and counter measures for the protection of pressurized liquid petroleum gas storage on industrial sites by KW. Blything
SRD R273	1983	A method of risk assessment for a multi-plant site by R.F. White
SRD R275	1983	The physiological and pathological effects of thermal radiation by I. Hymes
SRD R276	1984	A literature review of fire modelling techniques and related subjects by I.C. Simpson and N.A. Taylor
SRD R277	1985	Application of the computer code DENZ by S.F. Jagger
SRD R292	1984	Fire/explosion probabilities on liquid gas ships by K.W. Blything and J.N. Edmondson
SRD R295	1984	Preliminary studies for the analysis of blast loaded cantilever tests by J. Jowett
SRD R297	1984	Thermal radiation hazard ranges from large hydrocarbon pool fires by M. Considine
SRD R299	1984	FIRENET — a program for performing probabilistic risk assessment of fires in buildings by S.F. Hall and W.M. Scattergood
SRD R300	c. 1984	MULTIPLET — a program for calculating large event trees by S.F. Hall
SRD R301	1984	An introduction to fuzzy set theory with a view to the quantification and propagation of vagueness in probabilistic risk and reliability assessment by S.D. Unwin
SRD R302	1985	A study of axisymmetric under-expanded gas jets by P.K. Ramskffl
SRD R303	1984	A study of axi-symmetric two-phase flashing jets by P.R. Appleton
SRD R304	1984	Atmospheric transmissivity — the effects of atmospheric attenuation on thermal radiation by I.C. Simpson
SRD R307	c.1984	A modelling study of elevated plumes at a coastal location during onshore flow by P.J. Cooper and W. Nixon
SRD R310	1984	The assessment of individual and societal risks by M. Considine
SRD R314	1986	Analysis of 100 te propane storage vessel by T.A. Smith
SRD R315	1984	A simple parametric model for the dispersion of dense flammable gases by G.C. Grint
SRD R318	1985	Thorney Island heavy gas dispersion trials — determination of path and area of cloud from photographs by A.J. Prince, D.M. Webber and P.W.M. Brighton
SRD R319	c.1985	Using concentration data to track clouds in the Thorney Island experiments by P.W.M. Brighton
SRD R324	1985	Details and results of spill experiments of cryogenic liquids on to land and water by A.J. Prince
SRD R325	c.1985	Factors influencing pool fires on porous ground by D.M. Webber
SRD R326	1984	In-service reliability data for underground cross-country oil pipelines by K.W. Blything
SRD R327	1985	A model of aerosol deposition in the lung for use in inhalation dose experiments by W. Nixon
SRD R338	1987	The assessment of site-specific aircraft crash hazards by T.M. Roberts
SRD R340	1985	Incident probabilities in liquid gas ships by K.W. Blything and R.C.E. Lewis
SRD R341	c.1985	The estimation of missile velocities following the failure of a vessel containing high temperature pressurised water by D.LM. Hunt and AJ. Wood
SRD R342	1985	Cargo protection afforded by hull structure of LPG carriers during grounding at low or drifting speeds by A.N. Kinkead
SRD R347	1985	Guide to reducing human error in process operation (short version) by Human Factors in Reliability Group

SRD R352	1986	Discharge rate calculation methods for use in plant safety assessment by P.K. Ramskill
SRD R353	1987	A review of stainless steel component failures by T.A. Smith
SRD R354	1987	A description of ENGULF, a computer code to model thermal response of a tank engulfed in fire by P.K. Ramskill and D.L.M. Hunt
SRD R355	1985	Comparison between data from Thorney Island heavy gas trials and predictions of simple dispersion models by C.J. Wheatley, A.J. Prince and P.W.M. Brighton
SRD R357	1986	A theory of heterogeneous equilibrium between vapour and liquid phases of binary systems and formulae for the partial pressure of $HF + H_2O$ vapour by C.J. Wheatley
SRD R371	1986	Similarity solutions for the spreading of liquid pools by D.M. Webber and P.W.M. Brighton
SRD R372	1986	A description of the two dimensional combustion code FLARE by D. Martin
SRD R373	1986	Some calculations using the two dimensional combustion code FLARE by D. Martin
SRD R375	1987	Evaporation from a plane liquid surface into a turbulent boundary layer by P.W.M. Brighton
SRD R376	1986	Development of a model to predict the blast response of cantilevers by J. Jowett and J.P. Byrne
SRD R377	1986	Further development of a cantilever blast response model by J. Jowett, J.P. Byrne and T. Roberts
SRD R378	1986	The effects of missile impact on thin metal structures by J. Jowett
SRD R379	1986	A review of hazard identification techniques and their application to major accident hazards by S.T. Parry
SRD R382	1986	AEROSIM-S user manual by S.A. Ramsdale
SRD R383	1985	Common cause failure reliability benchmark exercise by B.D. Johnston and J. Crackett
SRD R390	1987	An integral model for spreading vaporising pools derived from shallow-layer equations by D.M. Webber and P.W.M. Brighton
SRD R393	1987	A theoretical study of NH_3 concentrations in moist air arising from accidental releases of liquefied NH_3 using the computer code TRAUMA by C.J. Wheatley
SRD R394	1987	A users guide to TRAUMA, a computer code for assessing the consequences of accidental two-phase releases of NH_3 into moist air by C.J. Wheatley
SRD R396	1988	Thermodynamics of mixing and final state of a mixture formed by the dilution of anhydrous hydrogen fluoride with moist air by P.N. Clough and C.J. Wheatley
SRD R401	1987	AEROSIM-N user manual by S.A. Ramsdale
SRD R404	1988	A model for the evaporation and boiling of liquid pools by D.M. Webber
SRD R407	1986	Experiments on some effects of obstacles on dense gas dispersion by R.E. Britter
SRD R410	1987	Discharge of liquid ammonia into moist atmospheres − a survey of experimental data and model for estimating initial conditions of dispersion by C.J. Wheatley
SRD R411	1986	Common cause failure reliability benchmark exercise. Phase II by J. Crackett
SRD R412	1987	Feasibility study of the assessment of risks involved in the conveyance of explosives by road by H. Chapman and P.L. Holden
SRD R413	1987	A users guide to ENGULF, a computer code to model the thermal response of a tank engulfed in fire by P.K. Ramskill
SRD R414	1987	The development of ENGULF to model a multi-component Liquid in a fire engulfed tank by P.K. Ramskill
SRD R418	1987	SRD Dependent Failures Procedures Guide by P. Humphreys and B.D. Johnston
SRD R421	1987	Heat conduction under a spreading pool by D.M. Webber
SRD R423	1987	Dry deposition to an urban complex by B.Y. Underwood
SRD R428	1988	FPAs Guide on Fire Precautions by G. Burke and M. Finucane
SRD R431	1988	Reliability of fire protection and detection systems by M. Finucane and D. Pinkney

SRD R435	1987	Criteria for the rapid assessment of aircraft crash onto major hazard installations by D.W. Phillips	
SRD R437	1987	An integral dynamic turbulence model for heavy gas cloud dispersal close to the source by D.M. Webber and C.J. Wheatley	
SRD R438	1987	Comparison and test of models for atmospheric dispersion of continuous releases of chlorine by C.J. Wheatley, B. Crabol, R.J. Carpenter, S.F. Jagger, C. Nussey, R.P. Cleaver, R.D. Fitzpatrick and A.J. Prince	
SRD R 439	1987	Guidelines for the design and assessment of concrete structures subject to impact by P. Barr	
SRD R441	1988	Pipework failures — a review of the historical incidents by K.W. Blything and S.T. Parry	
SRD R442	1998	Dry deposition to vegetated surfaces parametric dependencies by B.Y. Underwood	
SRD R 443	1987	Likelihood as a measure of uncertainty in system reliability and probabilistic safety assessment by A.R. Garlick	
SRD R444	1990	Similarity solutions for two dimensional steady gravity currents by P.W.M. Brighton	
SRD R455	1987	The application of quality assurance within the UKAEA by F. Allen and T.H. Nixon	
SRD R467	1989	Pressures produced by instantaneous chlorine releases inside buildings by P.W.M. Brighton	
SRD R468	1989	Continuous chlorine releases inside buildings: concentrations by P.W.M. Brighton	
SRD R474	1988	Transport accident frequency data, their sources and their application in risk assessment by P.R. Appleton	
SRD R477	1988	Assessment of missile hazards: review of incident experience relevant to major hazard plant by P.L. Holden	
SRD R480	1989	ENGULF II: a computer code to model the thermal response of a tank partially or totally engulfed in fire by P.K. Ramskffl	
SRD R481	1989	A user's guide to ENGULF II: a computer code to model the thermal response of a tank partially or totally engulfed in fire by P.K. Ramskill and C. Marriott	

SRD R483	1989	On the interpretation of puff-trajectory modelling by B.Y. Underwood
SRD R488	1988	An initial prediction of the BLEVE frequency of a 100 te butane storage vessel by K.W. Blything and A.B. Reeves
SRD R492	1988	The predicted BLEVE frequency of a selected $2000\,m^3$ butane — sphere on a refinery site by M. Selway
SRD R494	1989	Generalisation of the two-phase jet model TRAUMA to substances other than ammonia by D.N. Webber and P.W.M. Brighton
SRD R495	1989	TRAUMA issue 3.5, a users update by S.C. Rutherford and D.M. Webber
SRD R500	1990	The design of bunds by D.S. Barnes
SRD R504	1989	Risk assessment of majorhazards. A brief overview of methods and information sources by P.L. Holden
SRD R507	1990	A model for pool spreading and vaporisation and its implementation in the computer code GASP by D.M. Webber
SRD R509	1990	The long term contribution to severe accident source terms and their mitigation by J.R. Mullins and P.N. Clough
SRD R510	1990	Human error in risk assessment by J. Brazendale
SRD R521	1989	A users guide to G*A*S*P* on microcomputers by D.M. Webber and S.J. Jones
SRD R530	1991	Bund overtopping — the consequences following catastrophic failure of large volume liquid storage vessels by A. Wilkinson
SRD R536	1990	Investigation into the ignition and melting characteristics of apparel fabrics subjected to a radiative heat flux by W. Boydell
SRD R552	1993	A differential phase equilibrium model for clouds, plumes and jets by D.M. Webber and T. Wren
SRD R553	1992	Description of ambient atmospheric conditions for the computer code DRIFT by A.J. Byrne, S.J. Jones, S.C. Rutherford, C.A. Tickle and D.M. Webber
SRD R563	1992	Review of experimental data for validation of the code DRIFT by P.L. Edwards

Welding Institute
(Cambridge)

World Health Organisation
(Geneva)

Environmental Health Criteria

EHC1   Mercury
EHC2   polychlorinated Biphenyls and Terphenyls
EHC3   Lead
EHC4   Oxides of Nitrogen
EHC5   Nitrates, Nitrites and N-Nitroso Compounds
EHC6   Principles and Methods for Evaluating the
        Toxicity of Chemicals, Pt 1
EHC7   Photochemical Oxidants
EHC8   Sulphur Oxides and Suspended Particulate
        Matter
EHC9   DDT and its Derivatives
EHC10   Carbon Disulphide
EHC12   Noise
EHC13   Carbon Monoxide
EHC14   Ultraviolet Radiation
EHC19   Hydrogen Sufflde, 1983
EHC20   Selected Petroleum Products
EHC21   Chlorine and Hydrogen Chloride
EHC28   Acrylonitrile
EHC53   Asbestos
ECH54   Ammonia

Note:
Some of the series in this appendix which are shown with a
closed range of dates are continuing series, e.g. the AlChE
series on Ammonia Plant Safety

B   Nuclear Regulatory Commission

Periodic Reports

NUREG-0020   Licensed operators reactor status
              summary
NUREG-0040   Licensee contractor and vendor
              inspection status report
NUREG-0090   Report to Congress on abnormal
              occurrences
NUREG-0304   Regulatory and technical reports
              (abstract index journal)
NUREG-0525   Safeguards summary event list
NUREG-0540   Title list of documents made publicly
              available
NUREG-0750   Indexes to nuclear regulatory
              commission issuances
NUREG-0933   A prioritization of generic safety issues
NUREG-0936   NRC regulatory agenda
NUREG-0940   Enforcement actions: significant actions
              resolved
NUREG-1125   A compilation of reports of the Advisory
              Committee on Reactor Safeguards
NUREG-1145   US Nuclear Regulatory Commission
              annual report
NUREG-1214   Historical data summary of the
              systematic assessment of licensee
              performance
NUREG-1415   Office of the Inspector General.
              Semi-annual Report

Acoustic Emission Monitoring

NUREG/CR-3825:1985   Acoustic emission/flaw
                      relationship for in-service
                      monitoring of nuclear pressure
                      vessels

NUREG/CR-4300:1985–   Acoustic emission/flaw
                      relationship for in-service
                      monitoring of nuclear pressure
                      vessels. Also 1986, 1987
NUREG/CR-5645:1991   Acoustic emission/flaw
                      relationships for in-service
                      monitoring of LWRs

Aerosols

NUREG/CR-1367: 1981   AEROSOL user's manual
NUREG/CR-2139: 1981   Aerosols generated by free fall
                      of powders and solutions in
                      static air
NUREG/CR-2299: 1982   Aerosol release and transport
                      program
NUREG/CR-3422:1984   Aerosol release and transport
                      program
NUREG/CR-3830:1985   Aerosol release and transport
                      program
NUREG/CR-4342:1985   Uncertainty and sensitivity
                      analysis of a model for
                      multicomponent aerosol
                      dynamics
NUREG/CR-4501:1986   Modeling of vapor generation
                      in flashing flow
NUREG/CR-4658:1986   Aerosols generated by spills of
                      viscous solutions and slurries
NUREG/CR-4779:1987   New data for aerosols
                      generated by releases of
                      pressurized powders and
                      solutions in static air
NUREG/CR-4997:1988   Methods for describing air-
                      borne fractions of free fall
                      spills of powders and liquids
NUREG/CR-5162:1988   CHARM: a model for aerosol
                      behaviour in time varying
                      thermal–hydraulic conditions

Aging

NUREG/CR-3156:1983   A survey of state-of-the-art in
                      aging of electronics with
                      application to nuclear power
                      plant instrumentation
NUREG/CR-3808:1984   Aging–seismic correlation
                      study on class IE equipment
NUREG-1144:1985   Nuclear plant aging research
                      (NPAR) program plan
NUREG/CR-3819:1985   Survey of aged power plant
                      facilities
NUREG/CR-4144:1985   Importance ranking based on
                      aging considerations of
                      components included in
                      probabilistic risk assessments
NUREG/CR-4156:1985   Operating experience and
                      aging–seismic assessment of
                      electric motors
NUREG/CR-4234:1985   Aging and service wear of
                      electric motor-operated valves
                      used in engineered safety-
                      feature systems of nuclear
                      power plants
NUREG/CR-4279:1986   Aging and service wear of
                      hydraulic and mechanical
                      snubbers used in safety-related
                      piping and components of
                      nuclear power plants

NUREG/CR-4548:1986	Correlation of electrical reactor cable failure with materials degradation
NUREG/CR-4597:1986	Aging and service wear of auxiliary feedwater pumps for PWR nuclear power plants
NUREG-1144:1987	Nuclear plant aging research (NPAR) program plan
NUREG/CR-4590:1987	Aging of nuclear station diesel generators
NUREG/CR-4715:1987	An aging assessment of relays and circuit breakers and system, interactions
NUREG/CR-4731:1987	Residual life assessment of major light water reactor components
NUREG/CR-4769:1987	Risk evaluations of aging phenomena
NUREG/CR-4928:1987	Degradation of nuclear plant temperature sensors
NUREG/CR-4747:1988	An aging failure survey of light water reactor safety systems and components
NUREG/CR-4992:1988	Aging and service wear of multistage switches used in safety systems of nuclear power plants
NUREG/CR-5057:1989	Aging mitigation and improved programs for nuclear service diesel generators
NUREG/CR-5314:1990	Life assessment procedures for major LWR components
NUREG/CR-5419:1990	Aging assessment of instrument air systems in nuclear power plants
NUREG/CP-4302: 1991	Aging and service wear of check valves used in engineered safety feature systems of nuclear power plants

Alcohol, Drug Abuse

NUREG-0903: 1982	Survey of industry and government programs to combat drug and alcohol abuse
NUREG/CR-3196:1983	Drug and alcohol abuse: the bases for employee assistance programs in the nuclear utility industry
NUREG/CR-5784:1991	Fitness for duty in the nuclear power industry

Auxiliary Feedwater System

NUREG/CR-5404:1990	Auxiliary feedwater system aging

Bolts

NUREG-1339:1990	Resolution of generic safety issue 29: bolting degradation or failure in nuclear power plants

Chernobyl

NUREG-1250:1987	Report on the accident at the Chernobyl nuclear power station
NUREG-1251:1987	Implications of the accident at Chernobyl for safety regulation of commercial nuclear power plants in the United States (draft)
NUREG-1251:1989	Implications of the accident at Chernobyl for safety regulation of commercial nuclear power plants in the United Sates

Combustion

NUREG/CR-2658:1983	Characteristics of combustion products: a review of the literature
NUREG/CR-4855:1987	Development and application of a computer model for large-scale flame acceleration experiments
NUREG/CR-5275:1989	Flame facility: the effect of obstacles and transverse venting on flame acceleration and transition for hydrogen-air mixtures

Common Cause Failure

NUREG/CR-2099:1982	Common cause fault rates for diesel generators: estimates based on Licensee Event Reports at US commercial nuclear power plants
NUREG/CR-2098:1983	Common cause fault rates for pumps: estimates based on Licensee Event Reports at US commercial nuclear power plants
NUREG/CR-2770:1983	Common cause fault rates for valves
NUREG/CR-2771:1983	Common cause fault rates for instrumentation and control assemblies
NUREG/CR-3289:1983	Common cause fault rates for instrumentation and control assemblies
NUREG/CR-3437:1983	Data analysis using binomial failure rate common cause model
NUREG/CR-4314:1985	Brief survey and comparison of common cause failure analysis
NUREG/CR-4780:1988	Procedures for treating common cause failures in safety and reliability studies
NUREG/CR-5044:1988	Estimation techniques for common cause failure events
NUREG/CR-4780:1989	Procedures for treating common cause failures in safety and reliability studies
NUREG/CR-5460:1990	A cause-defense approach to the understanding and analysis of common cause failures

Computer Reliability

NUREG/CR-2118:1981 — Reactor safety system design using hardened computers

NUREG/CR-2288:1982 — Computer surety

NUREG/CR-3360:1984 — Computer program CDCID, an automated quality control program using CDC update

Computer Software Reliability

NUREG/CR-2186:1982 — Quantitative software reliability analysis of computer codes relevant to nuclear safety

NUREG/CR-4369:1986 — Quality assurance (QA) plan for computer software supporting the US Nuclear Regulatory Commission's high-level waste program

NUREG/CR-4640:1987 — Handbook of software quality assurance techniques applicable to the nuclear industry

Containment

NUREG/CP-0033:1982 — Proceedings of the workshop on Containment Integrity

NUREG/CR-3855:1985 — Characterization of nuclear reactor containment penetration

NUREG/CR-4119:1985 — Integrity of containment penetrations under severe accident conditions

NUREG/CR-4149:1985 — Ultimate pressure capacity of reinforced and prestressed concrete containment

NUREG/CR-4220:1985 — Reliability analysis of containment isolation systems

NUREG/CR-5096:1988 — Evaluation of seals for mechanical penetrations of containment buildings

NUREG/CR-5394:1989 — Evaluation of the leakage behavior of inflatable seals subject to severe accident conditions

NUREG/CR-5602:1990 — Simplified containment event tree analysis for the Sequoyah ice condenser containment

Control System, Including Displays, Alarms, Computers, Computer Aids

NUREG/CR-2776:1982 — Alarms within advanced display systems: alternatives and performance measures

NUREG/CR-2988:1982 — An approach to modeling supervisory control of a nuclear power plant

NUREG/CR-3461:1984 — Response tree evaluations – implications for the use of artificial intelligence in process control rooms

NUREG/CR-3621:1984 — Safety system status monitoring

NUREG/CR-3631:1984 — Response trees and expert systems for nuclear reactor operators

NUREG/CR-3655:1984 — A method for analytical evaluation of computer-based decision aids

NUREG/CR-4017:1984 — Interim criteria for the use of programmable digital devices in safety and control systems

NUREG/CR-3987:1985 — Computerized annunciator systems

NUREG/CR-4227:1985 — Human engineering guidelines for the evaluation and assessment of video display units

NUREG/CR-4272:1985 — Response tree evaluation: experimental assessment of an expert system for nuclear reactor operators

NUREG/CR-3958:1986 — Effects of control system failures on transients, accidents and core-melt frequencies at Combustion Engineering pressurized water reactor

NUREG/CR-4463:1986 — Human factors in annunciator/alarm systems: annunciator experiment plan I

NUREG/CR-4577:1987 — Automated long-term surveillance of a commercial nuclear power plant

NUREG/CR-4797:1987 — Progress reviews of six safety parameter display systems

NUREG/CR-5011:1987 — Application of the adaptive-predictive controllers to plant safety surveillance utilizing online plant analyzer

NUREG-1217:1988 — Evaluation of safety implications of control systems in LWR nuclear power plants (draft)

NUREG-1217:1989 — Evaluation of safety implications of control systems in LWR nuclear power plants (draft)

NUREG-1342:1989 — A status report regarding industry implementation of safety parameter display

NUREG-1361:1989 — Lessons learned in process control at the Halden reactor project

Control Room Design

NUREG-0700:1981 — Guidelines for control room design reviews

NUREG-0801:1981 — Evaluation criteria for detailed control room design review

NUREG/CR-3696:1984 — Potential human factors deficiencies in the design of local control stations and operator interfaces in nuclear power plants

NUREG/CR-3975:1984 — Identification and assessment of anticipated major changes in control rooms

NUREG/CR-4191:1985 — Survey of licensee control room habitability practices

NUREG/CR-5659:1990   Control room habitability system review models

Corrosion

NUREG/CR-2541:1982   Environmentally assisted cracking in light water reactors: critical issues and recommended research

NUREG/CR-3220:1983   Mechanistic aspects of stress-corrosion cracking of type 304 stainless steel in LWR systems

NUREG/CR-3806:1984   Environmentally assisted cracking in light water reactors

NUREG/CR-4221:1985   An evaluation of stress corrosion crack growth in BWR piping systems

NUREG/CR-4287:1985   Environmentally assisted cracking in light water reactors

NUREG/CR-4464:1986   Performance demonstration tests for detection of intergra-nular stress corrosion cracking

NUREG/CR-4667:1986   Environmentally assisted cracking in light water reactors

NUREG/CR-4723:1986   Application of a two-mechanism model for environmentally-assisted crack growth

NUREG/CR-5007:1987   Prediction and mitigation of erosive−corrosive wear in secondary piping systems of nuclear power plants

NUREG/CR-5156:1988   Review of erosion−corrosion in single-phase flows

NUREG-1343:1989   Erosion/corrosion-induced pipe wall thinning in US nuclear power plants

Cost Estimation

NUREG/CR-3971:1984   A handbook for cost estimating

NUREG/CR-4568:1986   A handbook for quick estimates

NUREG/CR-4627:1986   Generic cost estimates

Crack Growth, Fatigue, Fracture Mechanics

NUREG/CR-2301:1982   Fracture mechanics models developed for piping reliability assessment in light water reactors

NUREG/CR-2467:1982   Lower bound KI_{scc} values for bolting materials − a literature study

NUREG/CP-0051:1984   Proceedings of the CSNI specialist meeting on Leak-before-Break in Nuclear Reactor Piping

NUREG/CR-3644:1984   Review of proposed failure criteria for ductile materials

NUREG/CR-3982:1984   Case study of the propagation of a small flaw under PWR loading conditions and comparison with the ASME code design life

NUREG/CR-3723:1985   Stress−intensity-factor influence coefficients for surface flaws in pressure vessels

NUREG/CR-3945:1985   Fatigue crack growth rates of low-carbon and stainless steel piping steels in PWR environments

NUREG/CR-4305:1985   Comments on the leak-before-break concept for nuclear power plant piping systems

NUREG/CR-4325:1985   A parametric study of PWR pressure vessel integrity during overcooling incidents, considering both 2-D and 3-D flaws

NUREG/CR-4422:1985   A review of the models and mechanisms for environmentally-assisted crack growth of pressure vessel and piping steels in PWR environments

NUREG/CP-0067:1986   Proceedings of the second IAEA specialists meeting on Subcritical Crack Growth

NUREG/CP-0077:1986   Proceedings of the seminar on Leak-before-Break

NUREG/CR-4538:1986   Fracture analysis of welded type 304 stainless steel pipe

NUREG/CR-4806:1986   Analysis of cracks in stainless steel TIG welds

NUREG/CR-4724:1987   Fatigue crack growth rates in pressure vessel and piping steels in LWR environments

NUREG/CR-4841:1987   Fracture evaluation of surface cracks embedded in reactor vessel cladding

NUREG/CR-4853:1987   Approximate methods for fracture analysis of through-wall cracked pipe

NUREG/CP-0092:1988   Proceedings of the seminar on Leak-before-Break

NUREG/CR-4828:1988   Fatigue crack growth of part-through cracks in pressure vessel and piping steels

NUREG/CR-4667:1989   Environmentally assisted cracking in light water reactors

NUREG/CP-0109:1990   Proceedings of the seminar on Leak-before-Break

NUREG/CP-0037   Proceedings of the seminar on Assessment of Fracture Prediction Technology: Piping and Pressure Vessels

Data Fitting

NUREG-1378:1989   PLOTNFIT: a basic program for data plotting and curve fitting

Diesel Generators

NUREG-0873:1982   A Bayesian analysis of diesel generator failure data

NUREG/CR-4347:1985   Emergency diesel generator operating experience, 1981−1983

NUREG/CR-4557:1986   A review of issues related to improving nuclear power plant diesel generator reliability

NUREG/CR-4810:1987   Evaluation of diesel unavailability and risk effective surveillance test intervals

NUREG/CR-5078:1988   A reliability program for emergency diesel generators at nuclear power plants

Dispersion in Water Courses

NUREG-0868:1982   A collection of mathematical models for dispersion in surface water and groundwater

Eddy Current Inspection

NUREG/CR-2305:1982   Eddy current inspection for steam generator tubing program

NUREG/CR-2824:1983   Eddy-current inspection for steam generator tubing program

NUREG/CR-3200:1983   Eddy current inspection for steam generator tubing program: quarterly progress report

NUREG/CR-2824:1984   Eddy-current inspection for steam generator tubing program (progress report)

NUREG/CR-3949:1985   Eddy current inspection for steam generator tubing program

Electric Motors

NUREG/CR-4939:1988   Improving motor reliability in nuclear power plants

Electrical Systems

NUREG/CR-2559:1982   Results of phase one of plant electrical system (PES) study

NUREG/CR-3063:1983   A summary of the plant electrical systems (PES) study

NUREG/CR-3122:1983   Potential damaging failure modes of high- and medium-voltage electrical equipment

NUREG/CR-3623:1984   Status report: correlation of electrical cable failure with mechanical degradation

NUREG/CR-4257:1986   Inspection, surveillance and monitoring of electrical equipment in nuclear power plants

Emergency Procedures

NUREG-0696:1981   Functional criteria for emergency response facilities

NUREG-0799:1981   Criteria for preparation of emergency operating procedures (draft)

NUREG/CR-1970:1981   Development of a checklist for evaluating emergency procedures used in nuclear power plants

NUREG/CR-1977:1981   Guidelines for preparing emergency procedures for nuclear power plants

NUREG/CR-1999:1981   Human engineering guidelines for use in preparing emergency operating procedures for nuclear power plants

NUREG/CR-2005:1981   Checklist for evaluating emergency operating procedures used in nuclear power plants

NUREG-0899:1982   Guidelines for the preparation of emergency operating procedures

NUREG/CR-2005:1983   Checklist for evaluating emergency operating procedures used in nuclear power plants

NUREG/CR-3177:1983   Methods for review and evaluation of emergency procedure guidelines

NUREG/CR-3632:1984   Methods for implementing revisions to emergency operating procedures

NUREG-1210:1987   Pilot program: NRC severe reactor accident incident response training manual

NUREG-1358:1989   Lessons learned from the special inspection program for emergency operating procedures

NUREG/CR-5228:1989   Techniques for preparing flow chart-format emergency operating procedures

Emergency Response, Emergency Planning

NUREG-0755:1981   Report to Congress on status of emergency response planning for nuclear power plants

NUREG-0814:1981   Methodology for evaluation of emergency response facilities

NUREG/CP-0032:1982   Workshop on Meteorological Aspects of Emergency Response Plans for Nuclear Power Plants

NUREG/CR-2629:1982   Interim source term assumptions for emergency planning and equipment qualification

NUREG/CR-2654:1982   Procedures for analyzing the effectiveness of siren systems for alerting the public

NUREG/CR-2655:1982   Evaluation of the prompt alerting systems at four nuclear power stations

NUREG/CR-2925:1983   In-plant considerations for optimal offsite response to reactor accidents

NUREG/CR-3524:1984   Organizational interface in reactor emergency planning and response

NUREG/CR-4093:1985 Safety/safeguards interactions during safety-related emergencies at nuclear power reactor facilities

NUREG/CR-4151:1985 Integration of emergency action levels with Combustion Engineering operating procedures

NUREG/CR-4177:1985 Management of severe accidents

NUREG/CR-3365:1986 Report to the NRC on guidance for preparing scenarios for emergency preparedness exercises at nuclear generating stations (draft)

NUREG/CR-5474:1990 Assessment of candidate accident management strategies

NUREG/CR-5513:1990 Accident management information needs

NUREG/CR-5543:1991 A systematic process for developing and assessing accident management plans

Equipment Qualification, Specification
NUREG/CR-4301:1987 Status report on equipment qualification issues research and resolution

NUREG/CR-5012:1988 Similarity principles for equipment qualification by experience

NUREG/CR-5227:1989 Fitness for duty in the nuclear power industry: a review of technical issues

NUREG/CR-5742:1991 Feasibility assessment of a risk-based approach to technical specifications

Error Propagation
NUREG/CR-2839:1983 Survey of error propagation in systems

Evacuation
NUREG/CR-1856:1981 An analysis of evacuation time estimates around 52 nuclear power plant sites

NUREG/CR-2504:1982 CLEAR (calculates logical evacuation and response): a generic transportation network model for the calculation of evacuation time estimates

NUREG/CR-4726:1987 Evaluation of protective action risks

NUREG/CR-4878:1988 The sensitivity of evacuation time estimates to changes in input parameters for the I-DYNEV computer code

Event Trees
NUREG/CR-5174:1989 A reference manual for the event progression analysis code (EVNTRE)

Expert Judgement
NUREG/CR-3115:1983 Expert opinion and ranking methods

NUREG-1290:1987 Differing professional opinions

NUREG/CR-4814:1987 Sources of correlation between experts

NUREG/CR-4962:1987 Methods for the elicitation and use of expert opinion in risk assessment

NUREG-1290:1988 Differing professional opinions

NUREG/CR-5424:1990 Eliciting and analyzing expert judgment

External Hazards
NUREG/CR-1748:1981 Hazards to nuclear power plants from nearby accidents involving hazardous material – preliminary assessment

NUREG/CR-2014:1981 Kinematics of translating tornado wind fields

NUREG/CR-2490:1982 Hazards to nuclear power plants from large liquefied natural gas (LNG) spills on water

NUREG/CR-2555:1982 Data requirements for the evaluation of storm surge models

NUREG/CR-2557:1982 Simulated tornado wind fields and damage patterns

NUREG/CR-2639:1982 Historical extreme winds for the United States-Atlantic and Gulf of Mexico coastlines

NUREG/CR-2678:1982 Flood risk analysis methodology development project

NUREG/CR-2840:1982 Annotated tsunami bibliography

NUREG/CR-2859:1982 Evaluation of aircraft crash hazards for nuclear power plants

NUREG-0965:1983 NRC inventory of dams

NUREG/CR-2462:1983 Capacity of nuclear power plant structures to resist blast loading

NUREG/CR-2879:1983 Feasibility for quantitative assessment of available margins inherent in flood protection of nuclear power plants

NUREG/CR-2944:1983 Tornado damage risk assessment

NUREG/CR-3058:1983 A methodology for tornado hazard probability assessment

NUREG/CR-3060:1983 A critical review of flooding literature

NUREG/CR-3330:1983 Vulnerability of nuclear power plant structures to large external fires

NUREG/CR-3548:1983 Hazardous materials accidents near nuclear power plants: an evaluation of analyses and approaches

NUREG/CR-3670:1984 — Violent tornado climatography, 1880–1982

NUREG/CR-3759:1984 — Lightning strike density for the contiguous United Sates from thunderstorm duration records

NUREG/CR-3848:1984 — Experimental investigation of unsteady tornadic wind loads on structures

NUREG/CR-3874:1984 — Near-ground tornado wind fields

NUREG/CR-4250:1985 — Vehicle barriers: emphasis on natural features

NUREG/CR-4461:1986 — Tornado climatology of the contiguous United States

NUREG/CR-4492:1986 — Methodology for estimating extreme winds for probabilistic risk assessments

NUREG/CR-4496:1986 — A system for generating long streamflow records for study of floods of long return period

NUREG/CR-5042:1987 — Evaluation of external hazards to nuclear power plants in the United States

NUREG/CR-5042:1989 — Evaluation of external hazards to nuclear power plants in the United States

NUREG-1407 — Procedural and submittal guidance for individual plant examination of external events (PEEE) for severe accident vulnerabilities. Final Report

Failures
NUREG/CR-5611:1991 — Issues and approaches for using equipment reliability alert levels

NUREG/CR-5665:1991 — A systematic approach to repetitive failures

Failure Data
NUREG/CR-2232:1981 — Nuclear plant reliability data system 1980 annual report

NUREG/CR-2641:1982 — The in-plant reliability data base for nuclear power plant components

NUREG/CR-3831:1985 — The in-plant reliability data base for nuclear plant components

NUREG/CR-4639:1990 — Nuclear computerized library for assessing nuclear reactor reliability (NUCLARR)

NUREG/CR-5691:1991 — Instrumentation availability for a pressurized water reactor with a large dry containment during severe accidents

Failure Data Analysis
NUREG/CR-2374:1981 — Use of non-conjugate prior distributions in compound failure models

NUREG/CR-2375:1981 — SAFE-R and SAFE-D: computer codes for the analysis of failure data

NUREG/CR-2434:1982 — FRAC (failure rate analysis code): a computer program for analysis of variance of failure rates

NUREG/CR-2464:1982 — Methods for classifying mixtures of exponential distributions based on either exponential or Poisson data

NUREG/CR-2465:1982 — An error and uncertainty analysis of classical and Bayesian failure rate estimates from a Poisson distribution

NUREG/CR-2714:1982 — Analysis of failure rate data with compound models

NUREG/CR-2729:1983 — User's guide to BFR: a computer code based on the binomial failure rate common cause model

NUREG/CR-4217:1985 — A statistical analysis of nuclear power plant valve failure-rate variability – some preliminary results

NUREG/CR-4616:1987 — Root causes of component failures program

NUREG/CR-4672:1987 — Analysis of instrument tube ruptures in Westinghouse 4-loop pressurized water reactors

Fans and Blowers
NUREG/CR-4931:1987 — Response of centrifugal and axi-vane blowers to large pressure transients

Fault Trees, Fault Tree Analysis
NUREG-0492:1981 — Fault tree handbook
NUREG/CR-1965:1981 — The use of the computer code IMPORTANCE with SETS input
NUREG/CR-2327:1981 — Computation of probabilistic importance measures
NUREG/CR-1935:1982 — Quantitative fault tree analysis using the set evaluation program (SEP)
NUREG/CR-3263:1983 — A comparison of methods for uncertainty analysis of nuclear power plant safety system fault tree models
NUREG/CR-3268:1983 — Modular fault tree analysis procedures guide
NUREG/CR-3134:1984 — A SETS user's manual for vital area analysis
NUREG/CR-3547:1984 — A SETS user's manual for accident sequence analysis
NUREG/CR-4598:1986 — A user's guide for the top event matrix analysis code (TEMAC)
NUREG/CR-4800:1987 — SIGPI user's manual for fast computation of probabilistic performance of complex systems
NUREG/CR-5242:1988 — A fast bottom-up algorithm for computing the cut sets of noncoherent fault trees

Fire, Fire Protection

NUREG/CR-2258:1981	Fire risk analysis for nuclear power plants
NUREG/CR-2269:1981	Probabilistic models for the behavior of compartment fires
NUREG/CR-2321:1981	Investigation of fire stop test parameters. Final report
NUREG/CR-2377:1982	Tests and criteria for fire protection of cable penetrations
NUREG/CR-2413:1982	Burn mode analysis of horizontal cable tray fires
NUREG/CR-2607:1983	Fire protection research program for the US Nuclear Regulatory Commission 1975–1981
NUREG/CR-3239:1983	COMPBRN – a computer code for modeling compartment fires
NUREG/CR-3385:1983	Measures of risk importance and their applications
NUREG-1148:1985	Nuclear power plant fire protection research program
NUREG/CR-4112:1985	Investigation of cable and cable system fire test parameters
NUREG/CR-4230:1985	Probability-based evaluation of selected fire protection features in nuclear power plants
NUREG/CR-4321:1985	Full scale measurements of smoke transport and deposition in ventilation system ductwork
NUREG/CR-3656:1986	Evaluation of suppression methods for electrical cable fires
NUREG/CR-4310:1986	Investigation of potential fire-related damage to safety-related equipment in nuclear power plants
NUREG/CR-4479:1986	The use of a field model to assess fire behavior in complex nuclear plant
NUREG/CR-4566:1986	COMPBRN II – a computer code for modeling compartment fires
NUREG/CR-4586:1986	User's guide for a personal computer-based nuclear power plant fire data base
NUREG/CR-4596:1986	Screening test of representative nuclear power plant components exposed to secondary environments created by fires
NUREG/CR-4638:1986	Transient fire environment cable damageability test results
NUREG/CR-4527:1987	An experimental investigation of internally ignited fires in nuclear power plant control cabinets
NUREG/CR-4679:1987	Quantitative data on the fire behaviour of combustible materials found in nuclear power plants
NUREG/CR-5088:1989	Fire risk scoping study: investigation of nuclear power plant fire risk, including previously unaddressed issues
NUREG/CR-5384:1989	A summary of nuclear power plant fire safety research at Sandia National Laboratories, 1975–1987

Fluid Flow

NUREG/CR-4766:1987	User's manual for the NEFTRAN computer code
NUREG/CR-5128:1991	Evaluation and refinement of leak-rate estimation models

Gas Dispersion

NUREG/CR-1395:1981	EOCR building wake effects on atmospheric diffusion
NUREG/CR-1473:1981	The wake and diffusion structure behind a model industrial complex
NUREG/CR-2260:1981	Technical basis for Regulatory Guide 1.145 'Atmospheric dispersion models for potential accident consequence assessment at nuclear power plants'
NUREG/CR-2395:1982	Wind tunnel study of gas dispersion near a cubical model building
NUREG/CR-2521:1982	Method for estimating wake flow and effluent dispersion near simple block-like buildings
NUREG/CR-2584:1982	Meteorological considerations in the development of a real-time atmospheric dispersion model for reactor effluent exposure pathway
NUREG/CR-2754:1982	Critical review of studies on atmospheric dispersion in coastal regions
NUREG/CR-2858:1982	PAVAN: an atmospheric dispersion program for evaluating design basis accidental releases of radioactive materials from nuclear power stations
NUREG/CR-3149:1983	Dispersion coefficients for coastal regions
NUREG/CR-3250:1983	Atmospheric transport and diffusion mechanisms in coastal circulation systems
NUREG/CR-3344:1983	MESO I Version 2.0: an interactive mesoscale Lagrangian puff dispersion model with deposition and decay
NUREG/CR-3352:1983	Fumigation potential at inland and coastal power plant sites
NUREG/CR-3368:1983	Diffusion rates for elevated releases
NUREG/CR-3456:1983	Hanford atmospheric dispersion data: 1960 through June 1967

NUREG/CR-3488:1985 Idaho field experiment
NUREG/CR-4072:1985 The estimation of atmospheric dispersion at nuclear power plants utilizing real-time anemometer statistics
NUREG/CR-4157:1985 A scientific critique of available models for real-time simulations of dispersion
NUREG/CR-4158:1985 A compilation of information on uncertainties involved in deposition modeling
NUREG/CR-4159:1985 Comparison of the 1981 INEL dispersion data with results from a number of different models
NUREG/CR-4113:1986 Flow and dispersion near clusters of buildings
NUREG/CR-4222:1986 Analyses of plume formation, aerosol agglomeration and rainout following containment failure
NUREG/CR-4603:1986 Appraising atmospheric transport and diffusion models for emergency response facilities
NUREG/CR-4950:1987 The shoreline environment atmospheric dispersion experiment (SEADEX)
NUREG/CR-4950:1989 The shoreline environment atmospheric dispersion experiment (SEADEX)

Gas Mixing
NUREG/CR-4020:1985 HMS: a computer program for transient, three-dimensional mixing gases

Glossaries
NUREG-0770:1981 Glossary of terms

Human Error
NUREG-CR-1880:1981 Initial quantification of human errors associated with reactor safety system components in licensed nuclear power plants
NUREG/CR-2416:1982 Initial quantification of human error associated with specific instrumentation and control system components in licensed nuclear power plants
NUREG/CR-2417:1982 Identification and analysis of human errors underlying pump and valve related events reported by nuclear power plant licensees
NUREG/CR-2987:1983 Identification and analysis of human errors underlying electrical/electronic component related events reported by nuclear power plant licensees
NUREG/CR-3309:1984 A simulator-based study of human errors in nuclear power plant control room tasks

NUREG/CR-3518:1984 SLIM-MAUD: an approach to assessing human error probabilities using structured expert judgment
NUREG/CR-4436:1986 Human reliability impact on in-service inspection
NUREG/CR-5319:1989 Risk sensitivity to human error

Human Factors
NUREG/CR-1908:1981 Criteria for safety-related nuclear power plant operator actions
NUREG/CR-2075:1981 Standards for psychological assessment of nuclear facility personnel
NUREG/CR-2076:1981 Behavioral reliability program for the nuclear industry
NUREG-0872:1982 Study of using Licensee Event Reports for a statistical assessment of the effect of overtime and shift work on operator error
NUREG/CR-2586:1982 A survey of methods for improving operator acceptance of computerized aids
NUREG/CR-2833:1982 Critical human factors issues in nuclear power regulation and recommended comprehensive human factors long-range plan
NUREG-0985:1983 US Nuclear Regulatory Commission human factors program plan
NUREG-0985:1984 US Nuclear Regulatory Commission human factors program plan
NUREG/CR-3520:1984 Long term research plan for human factors affecting safeguards at nuclear power plants
NUREG/CR-3517:1986 Recommendations to the NRC on human engineering guidelines for nuclear power plant
NUREG/CR-4862:1987 Cognitive environment simulation – an artificial intelligence system for human performance assessment
NUREG-1384:1989 Human factors regulatory research program plan
NUREG/CR-5348:1989 Man–machine interface issues in nuclear power plants
NUREG/CR-5213:1990 The cognitive environment simulation as a tool for modeling human performance and reliability

Human Reliability Assessment (HRA)
NUREG/CR-2255:1982 Expert estimation of human error probabilities in nuclear power plant operations: review of probability assessment and scaling

NUREG/CR-1278:1983	Handbook of human reliability analysis with emphasis on nuclear power plant applications	
NUREG/CR-2254:1983	A procedure for conducting a human reliability analysis for nuclear power plants	
NUREG/CR-2743:1983	Procedures for using expert judgment to estimate human error probabilities in nuclear power plant operations	
NUREG/CR-2744:1983	Human error data bank for nuclear power plant operations	
NUREG/CR-2986:1983	The use of performance shaping factors and quantified expert judgement in the evaluation of human reliability	
NUREG/CR-3010:1983	Post event human decision errors: operator action tree/time reliability correlation	
NUREG/CR-3519:1985	Human error probability estimation using Licensee Event Reports	
NUREG/CR-3688:1985	Generating human reliability estimates using expert judgment	
NUREG/CR-3837:1985	Multiple-sequential failure model	
NUREG/CR-4009:1985	Human reliability data bank	
NUREG/CR-4010:1985	Specification of a human reliability data bank for conducting HRA segments of PRAs for nuclear power plants	
NUREG/CR-4103:1985	Uses of human reliability analysis probabilistic risk assessment results to resolve personnel performance issues that could affect safety	
NUREG/CR-4772:1987	Accident sequence evaluation program human reliability analysis procedure	
NUREG/CR-4835:1989	Comparison and application of quantitative human reliability analysis methods for the risk methods integration and evaluation program (RMIEP)	
NUREG/CR-5438:1990	Basic considerations in predicting error probabilities in human task performance	

Hydrogen Combustion

NUREG/CR-2475:1983	Hydrogen combustion characteristics related to reactor accidents	
NUREG/CR-3719:1984	Detonation calculations using a modified version of CSQ II: examples for hydrogen-air mixtures	
NUREG/CR-4138:1985	Data analyses for Nevada Test Site (NTS) premixed combustion tests	

Incidents

NUREG/CR-1884:1981	Observations and comments on the turbine failure at Yankee Atomic Electric Company, Rowe, Massachusetts	
NUREG/CR-2993:1983	Examination of failed studs from No. 2 steam generator at the Main Yankee nuclear power station	
NUREG-1154:1985	Loss of main and auxiliary feedwater event at Davis–Besse plant on June 9 1985	
NUREG-1190:1986	Loss of power and water hammer event at San Onofre, Unit 1, on November 21 1985	
NUREG-1195:1986	Loss of integrated control system power and overcooling transient at Rancho Seco on December 16 1985	
NUREG-1410:1990	Loss of vital AC power and the residual heat removal system during mid-loop operations at Vogtle Unit 1 on March 20 1990	
NUREG/CR-5622:1990	Analysis of reactor trips originating in balance of plant systems	
NUREG-1455	Transformer failure and common mode loss of instrument power at Nine Mile Point Unit 2 on August 13 1991	

Incident Investigation

NUREG-1303:1988	Incident investigation manual	
NUREG-1303:1991	Incident investigation manual	

Incident Reporting

NUREG/CR-3119:1983	Nuclear power safety reporting system	
NUREG-1291:1987	BWR and PWR off-normal event descriptions	

Inspection

NUREG/CR-5151:1988	Performance-based inspections	
NUREG/CR-5371:1989	Development and use of risk-based inspection guides	
NUREG/CR-5467:1991	Risk-based inspection guide for Crystal River Unit 3 nuclear power plant	

Leak Detection

NUREG/CR-4813:1987	Assessment of leak detection systems for LWRs	
NUREG/CR-5134:1988	Application of acoustic leak detection technology for the detection and location of leaks in light water reactors	

Licensee Event Reports

NUREG/CR-3824:1984	CONTING program guide	
NUREG-1022:1985	Licensee event report system nuclear reactor accident	

NUREG/CR-3905:1985 Sequence coding and search system for Licensee Event Reports: user's guide

NUREG/CR-4071:1985 Exploratory trend and pattern analysis for 1981 Licensee Event Report data

NUREG/CR-4129:1986 Exploratory trend and pattern analysis of 1981 through 1983 Licensee Event Report data

NUREG/CR-4132:1986 Nuclear power safety reporting system final evaluation results

NUREG/CR-2000 Licensee Event Report (LER) compilation

Light Water Reactors
NUREG/CR-2278:1981 Light water reactor engineered safety features status monitoring. Final report

NUREG-1107:1984 RCSLK9 reactor coolant system leak rate determination for PWRs: user's guide

NUREG/CP-0048:1984 Proceedings of the eleventh Water Reactor Safety Research Information meeting

NUREG/CR-3593:1984 Systems interaction results from the digraph matrix analysis of a nuclear power plant's high pressure safety injection system

NUREG/CR-3943:1985 The BWR plant analyzer

NUREG/CR-4041:1985 System analysis handbook

NUREG/CR-5637:1990 Generic risk insights for Westinghouse and Combustion Engineering pressurized water reactors

NUREG/CR-5640:1990 Overview and comparison of US commercial nuclear power plants

NUREG-1430 Standard technical specifications Babcock and Wilcox plants (draft)

NUREG-1431 Standard technical specifications Westinghouse plants (draft)

NUREG-1432 Standard technical specifications Combustion Engineering plants (draft)

NUREG-1433 Standard technical specifications General Electric plants BWR (draft)

Loss of Coolant Accident (LOCA)
NUREG/CR-4393:1986 Summary of semiscale small break loss-of-coolant accident experiments (1979 to 1985)

NUREG-1230:1987 Compendium of ECCS research for realistic LOCA analysis (draft)

Loss of Fluid Test (LOFT)
NUREG/CR-3005:1985 Summary of the Nuclear Regulatory Commission's LOFT program research findings

NUREG/CR-4046:1985 Determining critical flow valve characteristics using extrapolation techniques

Loose Parts Monitoring
NUREG/CR-3008:1983 Auditory perception in loose-parts monitoring

NUREG/CR-3687:1984 Loose-part monitoring programs and recent operational experience in selected US and Western European commercial nuclear power stations

Maintenance
NUREG/CR-1369:1982 Procedures evaluation checklist for maintenance, test and calibration procedures used in nuclear power plants

NUREG/CR-1786:1982 Maintainability analysis procedure (MAP)

NUREG/CR-3541:1983 Measures of the risk impacts of testing and maintenance activities

NUREG/CR-3817:1985 Development, use and control of maintenance procedures in nuclear power plants

NUREG/CR-3883:1985 Analysis of Japanese–US nuclear power plant maintenance

NUREG/CR-4281:1985 An empirical analysis of selected nuclear power plant maintenance factors and plant safety

NUREG-1212:1986 Status of maintenance in the US nuclear power industry

NUREG/CR-4611:1986 Trends and patterns in maintenance performance in the US nuclear power industry

NUREG/CR-5666:1991 Programmatic root cause analysis of maintenance personnel performance probabilities

NUREG/CR-5695:1991 A process for risk-focused maintenance

Maintenance Personnel Reliability Model
NUREG/CR-2670:1982 Job analysis of maintenance mechanic position for the nuclear power plant maintenance personnel reliability model

NUREG/CR-2668:1983 Job analysis of the maintenance supervisor and instrument and control supervisor positions for the nuclear power plant maintenance personnel reliability model

NUREG/CR-2669:1983 Front-end analysis for nuclear power plant maintenance personnel reliability model

NUREG/CR-3274:1983 — Job analysis of the instrument and control technician position for the nuclear power plant maintenance personnel reliability model

NUREG/CR-3626:1985 — Maintenance personnel performance simulation (MAPPS) model: description of model content, structure and sensitivity testing

NUREG/CR-3634:1985 — Maintenance personnel performance simulation (MAPPS) model: user's manual

NUREG/CR-4104:1985 — Maintenance personnel performance simulation (MAPPS) model

Management and Management Systems

NUREG/CR-3601:1984 — Management and organizational assessments: a review of selected organizations

NUREG/CR-3645:1984 — A guide to literature relevant to the organization and administration of nuclear power plants

NUREG/CR-3737:1984 — An initial empirical analysis of nuclear power plant organization and its effect on safety performance

NUREG/CR-4378:1986 — Objective indicators of organizational performance at nuclear power plants

NUREG/CR-5437:1990 — Organization and safety in nuclear power plants

NUREG/CR-5538:1991 — Influence of organizational factors on performance reliability

Materials of Construction

NUREG/CR-3613:1985 — Evaluation of welded and repair-welded stainless steel for LWR service

NUREG/CR-3911:1985 — Evaluation of welded and repair-welded stainless steel for LWR service

NUREG/CR-3613:1986 — Evaluation of welded and repair-welded stainless steel for LWR service

Meteorology

NUREG-0917:1982 — Nuclear Regulatory Commission staff computer programs for use with meteorological data

NUREG/CR-3773:1984 — Variation of planetary boundary layer dispersion properties with height in unstable conditions

NUREG/CR-3838:1984 — An initial review of several meteorological models suitable for low-level waste disposal facilities

NUREG/CR-3882:1986 — A method to characterize local meteorology at nuclear facilities for application to emergency response needs

Mitigation Systems

NUREG/CR-3980:1988 — Survey of the start of the art in mitigation systems

NUREG/CR-5079:1989 — Experimental results pertaining to thermal igniters

National Reliability Evaluation Program

NUREG/CR-2453:1982 — National reliability evaluation program (NREP) options study

Non-destructive Examination

NUREG/CR-1696:1981 — Integration of NDE reliability and fracture mechanics

NUREG/CR-3110:1983 — Reliability of non-destructive examination

NUREG/CR-3743:1984 — The impact of NDE unreliability on pressure vessel fracture predictions

NUREG/CR-4124:1985 — NDE of stainless steel and on-line leak monitoring of LWRs

NUREG/CR-4368:1985 — NDE of stainless steel and online leak monitoring of LWRs

NUREG/CR-4469:1986 — Integration of non-destructive examination reliability and fracture mechanics

NUREG/CR-4469:1986 (sic) — Non-destructive examination (NDE) reliability for in-service inspection of light water reactors. Semi-annual report

NUREG/CR-4600:1986 — Human factors study conducted in conjunction with a mini-round robin assessment of ultrasonic technician performance

NUREG-1194:1987 — Construction appraisal team inspection results on welding and non-destructive examination activities

NUREG/CR-4469:1987 — Non-destructive examination (NDE) reliability for in-service inspection of light water reactors

NUREG/CR-5075:1988 — The SAF-UT real time inspection system: operational principles and implementation

NUREG/CR-4908:1990 — Ultrasonic inspection reliability for intergranular stress corrosion cracks

NUREG/CR-5553:1990 — Computer programs for eddy current defect studies

Nuclear Accidents

NUREG/CR-2591:1982 — Estimating the potential impacts of a nuclear reactor accident

NUREG/CR-3566:1984 — Socio-economic consequences of nuclear reactor accidents

NUREG/CR-3673:1984 — Economic risks of nuclear power reactor accidents

Nuclear Accident Assessment

NUREG/CR-2326:1983	Calculations of reactor accident consequences, version 2: CRAC2 computer code
NUREG/CR-2901:1983	CRAC calculations for accident sections of environmental statements
NUREG/CR-2552:1984	CRAC2 model description
NUREG/CR-3310:1984	Testing of the CONTAIN code
NUREG/CR-4038:1985	Sensitivity and uncertainty studies of the CRAC2 computer code
NUREG/CR-4085:1985	User's manual for CONTAIN 1.0
NUREG/CR-4199:1985	A demonstration of uncertainty/sensitivity analysis using the health and economic consequence model CRAC2
NUREG-1115:1985	Categorization of reactor safety issues from a risk perspective

Nuclear Power Plant Operation

NUREG/CR-3430:1985	Nuclear power plant operating experience
NUREG-1275:1987	Operating experience feedback report: new plants
NUREG-1272:1988	Report to the US Nuclear Regulatory Commission on analysis and evaluation of operational data
NUREG/CR-4690:1991	Generic communications index

Operating Personnel, Simulators, Training

NUREG/CP-0031:1982	Proceedings of the CSNI specialist meeting on Operator Training and Qualifications
NUREG/CR-2353:1982	Specification and verification of nuclear power plant training simulator response characteristics
NUREG/CR-2587:1982	Functions and operations of nuclear power plant crews
NUREG/CR-2598:1982	Nuclear power plant control room task analysis: pilot study for pressurized water reactors
NUREG/CR-2623:1982	The allocation of functions in man–machine systems: a perspective and literature review
NUREG/CR-3114:1983	Proceedings of the workshop on Cognitive Modeling of Nuclear Plant Control Room Operators
NUREG/CR-3123:1983	Criteria for safety-relatednuclear power plant operator actions: 1982 pressurized water reactor (PWR) simulator exercises
NUREG/CR-3331:1983	A methodology for allocating nuclear power plant control functions to human or automatic control

NUREG/CR-3371:1983	Task analysis of nuclear power plant control room crews
NUREG/CR-3415:1983	Nuclear power plant control room task analysis: pilot study for boiling water reactors
NUREG/CR-3458:1983	An evaluation of the nuclear power plant operator licensing examination
NUREG/CR-3371:1984	Task analysis of nuclear power plant control room crews
NUREG/CR-3396:1984	Experience with the shift technical adviser position
NUREG/CR-3515:1984	Safety-related operation actions: methodology for developing criteria
NUREG/CR-3606:1984	Nuclear power plant control room crew task analysis database – SEEK system: user's manual
NUREG/CR-3725:1984	Nuclear power plant simulators for operator licensing and training
NUREG/CR-3726:1984	Simulator fidelity and training effectiveness
NUREG/CR-3739:1984	The Operator Feedback workshop: a technique for obtaining feedback from operations personnel
NUREG-1021:1985	Operator licensing examiner standards
NUREG-1122:1985	Knowledges and abilities catalog for nuclear power plant operators
NUREG/CR-3481:1985	Nuclear power plant personnel qualifications and training
NUREG/CR-4040:1985	Operational decision-making and action selection under psychological stress in nuclear power plants
NUREG/CR-4051:1985	Assessment of job-related educational qualifications for nuclear power plant operators
NUREG/CR-4206:1985	A select review of the recent (1979–1983) behavioral literature on training simulators
NUREG/CR-4258:1985	An approach to team skills training of nuclear power plant control room crews
NUREG/CR-4280:1985	The effects of supervisor experience and assistance of a shift technical advisor (STA) on crew performance in control room simulators
NUREG/CR-4344:1985	Instructional skills evaluation in nuclear industry training
NUREG/CR-4439:1985	Training review criteria and procedures
NUREG-1220:1986	Training review criteria and procedures
NUREG/CR-4411:1986	Assessment of specialized educational programs for licensed nuclear reactor operators

NUREG/CR-4532:1986	Models of cognitive behavior in nuclear power plant personnel
NUREG/CR-4188:1986	Nuclear power plant simulation facility evaluation methodology
NUREG-1122:1987	Knowledges and abilities catalog for nuclear power plant operators
NUREG/CR-4613:1987	Evaluation of nuclear power plant operating procedures classification and interfaces
NUREG/CR-3968:1987	Study of operating procedures in nuclear power plants

Operational Safety Reliability Program

NUREG/CR-4505:1986	A scoping study of the potential effectiveness of an operational safety reliability program in addressing generic safety problems
NUREG/CR-4506:1986	An operational safety reliability program approach with recommendations for further development and evaluation

Personal Protective Equipment

NUREG/CR-2652: 1982	Evaluation and performance of closed-circuit breathing apparatus

Physics of Reactor Safety

NUREG/CR-3804:1984	Physics of reactor safety
NUREG/CR-3804:1985	Physics of reactor safety
NUREG/CR-4240:1985	Physics of reactor safety

Pipework

NUREG-0803:1981	Generic safety evaluation report regarding integrity of BWR scram system piping
NUREG/CR-2189:1981	Probability of pipe fracture
NUREG/CR-2801:1983	Piping reliability mode validation and potential use for licensing regulation development
NUREG/CR-3059:1983	Parametric calculations of fatigue crack growth in piping
NUREG/CR-3176:1983	Crack growth evaluation for small cracks in reactor coolant piping
NUREG-1061:1984	Report of the US Nuclear Regulatory Commission Piping Review Committee
NUREG/CR-3483:1984	A study of the regulatory position on postulated pipe rupture location criteria
NUREG/CR-3718:1984	Reliability analysis of stiff versus flexible piping
NUREG/CR-3869:1984	Analysis of the impact of in-service inspection using a piping reliability model
NUREG/CR-3996:1984	Response margins of the dynamic analysis of piping systems

NUREG-1061:1985	Report of the US Nuclear Regulatory Commission Piping Review Committee
NUREG/CR-1677:1985	Piping benchmark problems
NUREG/CR-3660:1985	Probability of pipe failure in the reactor coolant loops of Westinghouse PWR plants
NUREG/CR-3663:1985	Probability of pipe failure in the reactor coolant loops of Combustion Engineering PWR plants
NUREG/CR-4082:1985	Degraded piping program – Phase II
NUREG/CR-4263:1985	Reliability analysis of stiff versus flexible piping. Final project report
NUREG/CR-4291:1985	Conclusion and summary report on physical benchmarking of piping systems
NUREG/CR-4082:1986	Degraded piping program
NUREG/CR-4290:1986	Probability of pipe failure in the reactor coolant loops of Babcock and Wilcox PWR plants
NUREG/CR-4529:1986	Piping damping – experimental results from laboratory tests in the seismic range
NUREG/CR-4562:1986	Pipe damping – results of vibration tests in the 33 to 100 hertz frequency range
NUREG/CR-4574:1986	An experimental and analytical assessment of circumferential through-wall cracked pipes under pure bending
NUREG/CR-4792:1986	Probability of failure in BWR reactor coolant piping
NUREG/CR-4407:1987	Pipe break frequency estimation for nuclear power plants
NUREG/CR-4872:1987	Experimental and analytical assessment of circumferentially surface-cracked pipes under bending
NUREG/CR-4943:1987	Preparation of design specifications and design reports for pumps, valves, piping and piping supports used in safety-related portions of nuclear power plants
NUREG-1222:1988	Piping research program plan
NUREG/CR-4082:1989	Degraded piping program
NUREG/CR-4792:1989	Probability of failure in BWR reactor coolant piping
NUREG/CR-5287:1989	Closeout of NRC Bulletin 87–01: thinning of pipe walls in nuclear power plants
NUREG/CR-5603:1990	Pressure-dependent fragilities for piping components

NUREG/CR-5757:1991   Verification of piping response calculation of SMACS code with data from seismic testing of an in-plant piping system

Pipe Whip
NUREG/CR-1721:1981   Computer code development for pipe whip and impact analysis
NUREG/CR-3686:1984   WIPS – computer code for whip and impact analysis of piping systems
NUREG/CR-3231:1987   Pipe-to-pipe impact program
NUREG/CR-5014:1987   Parametric study of pipe whip analysis

Plant Operation
NUREG/CR-2378: 1982   Nuclear power plant operating experience 1980

Population Distribution
NUREG/CR-3056:1984   Population distribution analysis for nuclear power plant siting

Power Supply
NUREG/CR-2989:1983   Reliability of emergency AC power system at nuclear power plants
NUREG/CR-3226:1983   Station blackout accident analyses
NUREG-1032:1985   Evaluation of station blackout accidents at nuclear power plants (draft)
NUREG-1109:1986   Regulatory analysis for the resolution of unresolved safety issue A-44, station blackout
NUREG/CR-4466:1986   Station blackout transients in the semiscale facility
NUREG/CR-4470:1986   Survey and evaluation of vital instrumentation and control power supply events
NUREG/CR-4533:1986   Program to analyze the failure mode of lead-acid batteries
NUREG/CR-4564:1986   Operating experience and aging–seismic assessment of battery chargers and inverters
NUREG/CR-4942:1987   Equipment operability during station blackout events
NUREG-1032:1988   Evaluation of station blackout accidents at nuclear power plants
NUREG-1109:1988   Regulatory/backflt analysis for the resolution of unresolved safety issue A-44: station blackout
NUREG/CR-5032:1988   Modeling time to recovery and initiating event frequency for loss of offsite power incidents at nuclear power plants

Precursors to Potential Accidents
NUREG/CR-2497: 1982   Precursors to potential severe core damage accidents: 1969–1979

NUREG/CR-3591:1984   Precursors to potential severe core damage accidents
NUREG/CR-4674:1991   Precursors to potential severe core damage accidents 1990: a status report

Pressure Systems
NUREG/CR-3604:1984   Bolting applications
NUREG/CR-3853:1984   Preloading of bolted connections in nuclear reactor component supports
NUREG/CR-3228:1985   Structural integrity of water reactor pressure boundary components

Pressure Vessels
NUREG/CR-2308:1982   Design criteria for the spacing of nozzles and reinforced openings in cylindrical nuclear pressure vessels and pipe
NUREG/CR-4267:1985   Vessel integrity (VISA) simulation code sensitivity study
NUREG/CR-4486:1986   VISA II: a computer code for predicting the probability of reactor pressure vessel failure
NUREG/CR-4614:1986   VISA II sensitivity study of code calculations input and analytical model parameters

Probabilistic Risk Assessment
NUREG/CR-2300:1981   PRA procedures guide (draft)
NUREG/CR-2300:1982   PRA procedures guide (draft)
NUREG/CR-2300:1983   PRA procedures guide
NUREG-1050:1984   Probabilistic risk assessment (PRA) reference document
NUREG/CR-2815:1984   Probabilistic safety analysis procedure
NUREG-1093:1984   Reliability and risk analysis methods research plan
NUREG/CR-3440:1984   Identification of severe accident uncertainties
NUREG/CR-3852:1984   Insight into PRA methodologies
NUREG/CR-2815:1985   Probabilistic safety analysis procedures guide
NUREG/CR-3485:1985   PRA review manual
NUREG/CR-3887:1985   Human factors review for severe accident sequence analysis
NUREG/CR-3904:1985   A comparison of uncertainty and sensitivity analysis techniques for computer models
NUREG/CR-4229:1985   Evaluation of current methodology employed in probabilistic risk assessment (PRA) of fire events at nuclear power plants
NUREG/CR-4231:1985   Evaluation of available data for probabilistic risk assessment (PRA) of fire events at nuclear power plants
NUREG/CR-4350:1985   Probabilistic risk assessment course documentation

NUREG/CR-4377:1985 Evaluations and utilizations of risk importances

NUREG/CR-3969:1986 Independent code assessment: Sandia-proposed accuracy quantification methodology

NUREG/CR-4372:1986 Probabilistic risk assessment (PRA) applications

NUREG/CR-4514:1986 Controlling principles for prior probability assignments in nuclear risk assessment

NUREG/CR-4405:1987 Probabilistic risk assessment (PRA) insights

NUREG/CR-4836:1988 Approaches to uncertainty analysis in probabilistic risk assessment

NUREG/CR-5050:1988 Annotated bibliography of reliability and risk data sources

NUREG/CR-5323:1989 Validation of risk-based performance indicators: safety system function trends

NUREG/CR-5425:1989 Evaluation of allowed outage times (AOTs) from a risk and reliability standpoint

NUREG/CR-5262:1990 PRAMIS: probability risk assessment model integration system

NUREG/CR-5477:1990 An evaluation of the reliability and usefulness of external initiator PRA methodologies

NUREG/CR-5520:1991 Procedures guide for extracting and loading probabilistic risk assessment data into MAR-D using IRRAS 2.5

Pumps

NUREG/CR-1205:1982 Data summaries of Licensee Event Reports of pumps at US commercial nuclear power plants

NUREG/CR-2886:1983 The in-plant reliability data base for nuclear power components: interim data report – the pump component

NUREG/CR-3650:1984 A statistical analysis of nuclear power plant pump failure rate variability

NUREG/CR-3914:1985 Pump and valve qualification review guide

NUREG/CR-4077:1985 Reactor coolant pump shaft seal behavior during station blackout

NUREG/CR-4400:1985 The impact of mechanical- and maintenance-induced failures of main reactor coolant pump seals on plant safety

NUREG/CR-4544:1986 Reactor coolant pump seal related instrumentation and operator response

NUREG/CR-4821:1987 Reactor coolant pump shaft seal stability during station blackout

NUREG/CR-4948:1989 Technical findings related to generic issue 23: reactor coolant pump seal failure

NUREG/CR-5706:1991 Potential safety-related pump loss. An assessment of industry data

Quality Assurance

NUREG-1055:1984 Improving quality and the assurance of quality in the design and construction of commercial nuclear power plants

NUREG/CR-4678:1986 A method for using PRA to establish quality program applicability

NUREG/CR-4921:1987 Engineering and quality assurance cost factors associated with nuclear plant modification

Reactor Safety Research

NUREG/CR-2127:1982 Reactor safety research programs: quarterly report

NUREG/CR-2716:1982 Reactor safety research programs: quarterly report

NUREG/CR-2531:1983 Introductory user's manual for the US Nuclear Regulatory Commission reactor safety research data bank

NUREG/CR-3307:1984 Reactor safety research programs

NUREG/CR-3810:1984 Reactor safety research programs

NUREG-1080:1985 Long-range research plan

NUREG-1155:1985 Research program plan

NUREG/CP-0058:1985 Proceedings of the twelfth Water Reactor Safety Research Information meeting

NUREG/CR-3810:1985 Reactor safety research programs

NUREG/CR-3816:1985 Reactor safety research

NUREG/CR-4318:1985 Reactor safety research programs

NUREG/CR-4340:1985 Reactor safety research semi-annual report

NUREG/CR-4340:1986 Reactor safety research semi-annual report

NUREG/CR-4805:1987 Reactor safety research seminar annual report

NUREG-1260:1988 A report to Congress on nuclear regulatory research

NUREG/CR-5039:1988 Reactor safety research semiannual report

NUREG/CR-2331:1990 Safety research programs sponsored by the Office of Nuclear Regulatory Research

Reactor Safety Study

NUREG/CR-1659:1981 Reactor Safety Study methodology applications program

NUREG/CR-1879:1981 Sensitivity of risk parameters to human errors in Reactor Safety Study for a PWR

NUREG/CR-2906:1983   Sensitivity of risk parameters component unavailability in Reactor Safety Study (PSAP/PSAB computer codes)

NUREG/CR-2531:1985   Introductory user's manual for the US Nuclear Regulatory Commission Reactor Safety Research Data Bank

Regulation

NUREG-0834:1981   NRC licensee assessments

NUREG/CP-0017:1981   Executive seminar on the Future Role of Risk Assessment and Reliability Engineering in Nuclear Regulation

NUREG/CR-1750:1981   Analysis, conclusions and recommendations concerning operator licensing

NUREG/CR-2040:1981   A study of the implications of applying quantitative risk criteria in the licensing of nuclear power plants in the United States

NUREG-0845:1982   Agency procedures for the NRC incident response plan

NUREG/CR-2664:1982   Selected review of foreign licensing practices for nuclear power plants

NUREG-0992:1983   Report of the Committee to Review Safeguards Requirements at Power Reactors

NUREG-1024:1983   Technical specifications – enhancing the safety impact (prescriptive regs)

NUREG/CR-3175:1983   Institutional implications of establishing safety goals for nuclear power plants

NUREG-0748:1984   Operating reactors licensing actions summary

NUREG-1090:1984   US Nuclear Regulatory Commission annual report

NUREG/CR-0200:1984   SCALE, a modular code system for performing standardized computer analyses for licensing evaluation

NUREG/CR-3508:1984   Standard setting standards: a systematic approach to managing public health and safety risks

NUREG/CR-3932:1984   Benchmark description of current regulatory requirements and practices in nuclear safety and reliability assurance

NUREG/CR-3976:1984   Regulatory analyses for severe accident issues: an example

NUREG-0885:1985   US Nuclear Regulatory Commission policy and planning guidance

NUREG-0981:1985   NRC/FEMA operational response procedures for response to a commercial nuclear reactor accident

NUREG-1070:1985   NRC policy of future reactor designs

NUREG/CR-4125:1985   Guidelines and workbook for assessment of organization and administration of utilities seeking operating license for a nuclear power plant

NUREG/CR-4133:1985   Nuclear power safety reporting system implementation and operational specifications

NUREG/CR-4152:1985   An independent safety organization

NUREG/CR-4153:1985   Applications of foreign probabilistic safety assessment experience to the US nuclear regulatory process

NUREG/CR-4248:1985   Recommendations for NRC policy on shift scheduling and overtime at nuclear power plants

NUREG-0748:1986   Operating reactors licensing actions summary

NUREG-0980:1986   Nuclear regulatory legislation

NUREG-1080:1986   Long-range research plan

NUREG-1175:1986   NRC safety research in support of regulation

NUREG-1214:1986   Historical data summary of the systematic assessment of licensee performance

NUREG/CR-4125:1986   Guidelines and workbook for assessment of organization and administration of utilities seeking operating license for a nuclear power plant

NUREG/CR-4446:1986   The nuclear industry and its regulators: a new compact is needed

NUREG-0856:1987   US Nuclear Regulatory Commission policy and planning guidance

NUREG-1395:1990   Industry perceptions of the impact of US Nuclear Regulatory Commission on nuclear power plant activities (draft)

NUREG-1266:1991   NRC safety research in support of regulation

NUREG-1435:1991   Status of safety issues at licensed power plants

Relays

NUREG/CR-4867:1991   Relay test program

Reliability Engineering

NUREG/CR-1924:1981   FRANTIC II – a computer code for time dependent unavailability analysis

NUREG/CR-2332:1982   Time dependent unavailability of a continuously monitored component

NUREG/CR-2542:1982	Sensitivity study using the FRANTIC code for the unavailability of a system to the failure characteristics of the components and the operating conditions	NUREG-1231:1988	Safety evaluation report related to Babcock and Wilcox Owners Group plant reassessment program
NUREG/CR-2915:1983	Initial guidance of digraph-matrix analysis for systems interaction studies	NUREG-1232:1988	Safety evaluation report on Tennessee Valley Authority
NUREG/CR-3624:1984	A Fortran 77 program and user's guide for the generation of Latin hypercube and random samples for use with computer models	NUREG/CR-4834:1988	Recovery actions in PRA for the risk methods integration and evaluation program
		NUREG/CR-5033:1988	Summary description of the SCALE modular code system
NUREG/CR-3627:1984	FRANTIC II applications to standby safety systems	NUREG/CR-5264:1988	Guide for licensing evaluations using CRAC2, a computer program for calculating reactor accident consequences
NUREG/CR-4213:1985	SETS reference manual		
NUREG/CR-3495:1987	Calculation of failure importance measures for basic units and plant systems in nuclear power plants	NUREG/CR-5206:1989	A probabilistic evaluation of the safety of Babcock and Wilcox nuclear reactor power plants with emphasis on historically observed operational events
NUREG/CR-4844:1987	Integrated reliability and risk analysis system (IRRAS) user's guide		
NUREG/CR-4618:1989	Evaluation of reliability technology applicable to LWR operational safety	NUREG-0800:1990	Standard review plan for the review of safety analysis reports for nuclear power plants LWR
NUREG/CR-5324:1989	Detecting component failure potential using proportional hazard model	NUREG/CR-4691:1990	MELCOR accident sequence consequence code system (MACCS)
NUREG/CR-5111:1990	Integrated reliability and risk analysis system (IRRAS). Version 2	NUREG/CR-4550:1990	Analysis of core damage frequency: internal events methodology
		NUREG/CR-5376:1990	Quality assurance and verification of the MACCS code
Retrofitting			
NUREG-1409:1990	Backfitting guidelines	NUREG/CR-5510:1990	Evaluations of core melt frequency effects due to component aging and maintenance
Risk Criteria			
NUREG/CR-1916:1981	A risk comparison	NUREG/CR-5527:1990	Risk sensitivity to human error in the LaSalle PRA
NUREG/CR-1930:1981	Index of risk exposure and risk acceptance criteria	NUREG/CR-5518:1991	Quality assurance procedures for the CONTAIN severe reactor accident computer code
Safety Assessment			
NUREG/CR-2631:1983	A logical approach for identifying equipment important to safety in nuclear power plants	*Safety Goals*	
		NUREG-0764:1981	Toward a safety goal: discussion of preliminary policy
NUREG/CR-2728:1983	Interim reliability evaluation program procedures guide	NUREG/CP-0018:1981	Workshop on Frameworks for Developing a Safety Goal
NUREG/CR-3241:1983	Probabilistic methods for identification of significant accident sequences in loop-type LMFBRs	NUREG/CP-0020:1981	Workshop on a Proposed Safety Goal
NUREG/CR-3762:1984	Identification of equipment and components predicted as significant contributors to severe core damage	NUREG/CR-2065:1981	Comments on the ACRS quantitative safety goals
		NUREG/CR-2226:1981	A survey of safety levels in Federal regulation
NUREG/CR-3301:1985	Catalog of PRA dominant accident sequence information	NUREG-0880:1982	Safety goals for nuclear power plants: a discussion
NUREG/CR-4834:1987	Recovery actions in PRA for the risk methods integration and evaluation program (RMIEP)	NUREG-0880:1983	Safety goals for nuclear power plant operation
		NUREG/CR-3507:1984	An analysis of the NRC safety goals for nuclear power

NUREG-1128:1985 | Trial evaluations in comparison with the 1983 safety goals
NUREG/CR-4197:1985 | Safety goal sensitivity studies
NUREG/CR-4048:1986 | A methodology for allocating reliability and risk

Safety Issues, Prioritization of Safety Issues, Unresolved Safety Issues

NUREG-0705:1981 | Identification of new unresolved safety issues relating to nuclear power plants
NUREG-0606:1985 | Unresolved safety issues summary
NUREG-0844:1985 | NRC integrated program for resolution of unresolved safety issues A-3, A-4 and A-5 regarding steam generator tube integrity (draft)
NUREG/CR-2800:1986 | Guidelines for nuclear power plant safety issue prioritization information development

Safety Measurement

NUREG/CR-5568:1990 | Industry based performance indicators for nuclear power plants

Safety Relief Valves

NUREG-0763:1981 | Guidelines for confirmatory inplant tests of safety relief valve discharges for BWR plants

Safeguards

NUREG/CR-1246:1981 | SAFE user's manual
NUREG/CR-2404:1982 | Analysing safeguards alarms and response decisions
NUREG/CR-1246:1983 | SAFE user's manual
NUREG/CR-4392:1985 | Measures of safeguards risk employing PRA (MOSREP)
NUREG-1304:1988 | Reporting of safeguards events

Security

NUREG-0768:1981 | People-related problems affecting security in the licensed nuclear industry
NUREG/CR-1345:1981 | Nuclear power plant design concepts for sabotage control
NUREG/CR-1589:1981 | Pathfinding simulation (PATHS) user's guide
NUREG/CR-2297:1982 | Security management techniques and evaluative checklists for security force effectiveness
NUREG/CR-2546:1982 | Reactor safeguards against inside sabotage
NUREG/CR-2643:1982 | A review of selected methods for protection against sabotage by an insider

NUREG/CR-3251:1984 | The role of security during safety-related emergencies at nuclear power plants
NUREG/CR-3619:1984 | Survey of commercial nonnuclear security programs
NUREG/CR-4462:1986 | A ranking of sabotage/ tampering avoidance technology alternatives
NUREG-1178:1988 | Vital equipment/area guidelines study
NUREG-1267:1989 | Technical resolution of generic safety issue A-29: nuclear power plant design for reduction of vulnerability to industrial sabotage

Seismic Events, Earthquakes

NUREG/CR-1508:1981 | Evaluation of the cost effects on nuclear power plant construction, resulting from the increase in seismic design level
NUREG/CR-1582:1981 | Seismic hazard analysis
NUREG/CR-1751:1981 | ARMA models for earthquake ground motions
NUREG/CR-1889:1981 | Large LOCA-earthquake combination probability
NUREG/CR-2015:1981 | Seismic safety margins research program. Phase I Final Report — overview
NUREG/CR-1120:1982 | Seismic safety margins research program
NUREG/CR-2405:1982 | Subsystem fragility: seismic safety margins research program
NUREG-0967:1983 | Seismic hazard review for the systematic evaluation program — a use of probability in decision-making
NUREG/CR-1582:1983 | Seismic hazard analysis — overview and executive summary
NUREG/CR-2015:1983 | Seismic safety margins research program
NUREG/CR-2945:1983 | Characterization of earthquake forces for probability-based design of nuclear structures
NUREG/CR-3315:1983 | A consensus estimation study of nuclear power plant structural loads (seismic)
NUREG/CR-3342:1983 | Probabilistic models for operational and accidental loads on seismic category I structures
NUREG/CR-3357:1983 | Identification of seismically risk sensitive systems and components in nuclear power plants
NUREG/CR-3127:1984 | Probabilistic seismic resistance of steel containments

NUREG/CR-3480:1984   Value/impact assessment for seismic design criteria USI A-40

NUREG/CR-3756:1984   Seismic hazard characterization of the eastern United States

NUREG/CR-3811:1984   Alternate procedures for the seismic analysis of multiply supported piping systems

NUREG-1147:1985   Seismic safety research program plan

NUREG/CP-0059:1985   Proceedings of the MITI-NRC Seismic Information Exchange meeting

NUREG/CR-3805:1985   Engineering characterization of ground motion

NUREG/CR-4145:1985   Earthquake recurrence intervals at nuclear power plants: analysis and ranking

NUREG/CR-4331:1985   Simplified seismic probabilistic risk assessment

NUREG/CR-4334:1985   An approach to the quantification of seismic margins in nuclear power plants

NUREG/CR-4430:1985   Current methodologies for assessing the potential for earthquake-induced liquefaction in soils

NUREG/CR-3805:1986   Engineering characterization of ground motion

NUREG/CR-4431:1986   Summary report on the seismic safety margins research program

NUREG/CR-4482:1986   Recommendations to the Nuclear Regulatory Commission on trial guidelines for seismic margin reviews of nuclear power plants (draft)

NUREG/CR-4822:1987   Broad band seismic data analysis

NUREG/CR-4903:1987   Selection of earthquake resistant design criteria for nuclear power plants

NUREG/CR-5073:1988   Quantification of margins in piping system seismic response

NUREG/CR-5076:1988   An approach to the quantification of seismic margins in nuclear power plant

NUREG/CR-3006:1989   Damping in building structures during earthquakes

NUREG/CR-5098:1989   An analytical study of seismic threat to containment integrity

NUREG/CR-5250:1989   Seismic hazard characterization of 69 nuclear power plants east of the Rocky Mountains

NUREG/CR-5270:1989   Assessment of seismic margin calculation methods

Seismic Fragility
NUREG/CP-0070:1985   Proceedings of the workshop on Seismic and Dynamic Fragility of Nuclear Power Plant Components

NUREG/CR-3558:1985   Handbook of nuclear power plant seismic fragilities

NUREG/CR-4123:1985   Seismic fragility of reinforced concrete structures and components for application to nuclear facilities

NUREG/CR-4659:1986   Seismic fragility of nuclear power plant components (Phase 1)

NUREG/CR-4734:1987   Seismic testing of typical containment piping penetration systems

NUREG/CR-4776:1987   Response of seismic category I tanks to earthquake excitation

NUREG/CR-4859:1987   Seismic fragility test of a 6-inch diameter pipe system

NUREG/CR-4899:1987   Component fragility research program

NUREG/CR-4910:1987   Relay chatter and operator response after a large earthquake

NUREG/CR-5588:1990   CARES (computer analysis for rapid evaluation of structures). Version 1

NUREG/CR-4659:1991   Seismic fragility of nuclear power plant components (Phase II). A fragility handbook of eighteen components.

Seismic and Dynamic Qualification
NUREG-1018:1983   Seismic qualification of equipment in operating plants

NUREG/CR-1161   Recommended revisions to Nuclear Regulatory Commissions seismic design criteria

NUREG/CR-3266:1983   Seismic and dynamic qualification of safety-related electrical and mechanical equipment in operating nuclear power plants

NUREG/CR-3630:1984   Equipment qualification methodology research: tests of pressure switches

NUREG/CR-3875:1984   The use of in-situ procedures for seismic qualification of equipment in currently operating plants

NUREG-1030:1985   Seismic qualification of equipment in operating nuclear power plants (draft)

NUREG-1209:1986   Program plan for environmental qualification of mechanical and dynamic (including seismic) qualification of mechanical and electrical equipment program (EDQP)

NUREG/CR-3137:1986   Seismic and dynamic qualification of related electrical and mechanical equipment

NUREG-1030:1987   Seismic qualification of equipment in operating nuclear power plants

NUREG-1211:1987   Regulatory analysis for resolution of unresolved safety issue A-46: seismic qualification of equipment in operating plants

Severe Accident Risk Reduction Program

NUREG-0900:1986   Nuclear power plant severe accident research plan

NUREG/CR-4569:1986   A review of the severe accident risk reduction program (SARRP) containment event trees

NUREG/CR-4781:1987   Study of severe accident mitigation systems

NUREG/CR-5132:1988   Severe accidents insight report

NUREG/CR-4551:1990   Evaluation of severe accident risks: quantification of major input parameters

NUREG/CR-4551:1991   Evaluation of severe accident risks: quantification of major input parameters

NUREG/CR-5809:1991   An integrated structure and scaling methodology for severe accident technical issue resolution (draft)

Severe Accident Risk Report (NUREG-1150)

NUREG-1150:1987   Reactor risk reference document (draft)

NUREG/CR-5000:1988   Methodology for uncertainty estimation in NUREG-1150 (draft)

NUREG/CR-5113:1988   Findings of the peer review panel on the draft reactor risk reference document NUREG 1150

NUREG-1420:1990   Special committee review of the Nuclear Regulatory Commission's severe accident risk report (NUREG-1150)

NUREG-1150:1990   Severe accident risks: an assessment for five US nuclear power plants

NUREG/CR-4840:1990   Procedures for the external event core damage frequency analyses for NUREG-1150

NUREG-1150:1991   Severe accident risks: an assessment for five US nuclear power plants

Sizewell B

NUREG-0999:1983   Sizewell B – analysis of British application of US PWR technology

Siting

NUREG/CR-1684: 1981   A comparison of site evaluation methods

SLIM-MAUD

NUREG/CR-3518:1984   SLIM-MAUD: an approach to assessing human error probabilities using structured expert judgment

NUREG/CR-4016:1987   Application of SLIM-MAUD: a test of an interactive computer-based method for organizing expert assessment of human performance and reliability

Source Terms

NUREG-0856:1985   Reassessment of the technical bases for estimating source terms (draft)

NUREG-0956:1985   Reassessment of the technical bases for estimating source terms (draft)

NUREG-0956:1986   Reassessment of the technical bases for estimating source terms

NUREG/CR-4587:1986   Source term code package: a user's guide (MOD1)

NUREG/CR-4656:1986   Verification tests calculations for the source term code package

NUREG/CR-4688:1986   Quantification and uncertainty analysis of source terms for severe accidents in light water reactors (QUASAR)

NUREG-1265:1987   Uncertainty papers on severe accident source terms

NUREG/CR-4722:1987   Source term estimation using MENU-TACT

NUREG/CR-4883:1987   Review of research on uncertainties in estimates of source terms from severe accidents in nuclear power plants

NUREG-1228:1988   Source term estimation during incident response to severe nuclear power plant accidents

NUREG/CR-5531:1991   MELCOR 1.8: a computer code for nuclear reactor severe accident source term and risk assessment analyses

Steam Generators

NUREG/CR-4079:1985   Analytic studies pertaining to steam generator tube rupture accidents

NUREG/CR-4276:1985   Vibration and wear in steam generating tubes following chemical clearning

NUREG/CR-4276:1986   Vibration and wear in steam generating tubes following chemical clearning

Steam Supply

NUREG-0852:1987   Safety evaluation report related to the final design of the standard nuclear steam supply reference system CESSAR

NUREG-1413   Safety evaluation report related to the preliminary design of the standard steam supply reference system RESAR

Structures
NUREG/CR-2347:1982   Margins to failure – category I structures program: background and experimental program plan
NUREG/CR-2638:1982   Snow loads for the design of nuclear power plant structures
NUREG/CR-3628:1984   Probability based safety checking of nuclear plant structures

System Interactions
NUREG/CR-3922:1985   Survey and evaluation of system interaction events and sources
NUREG/CR-4261:1986   Assessment of system interaction experience in nuclear power plants

Thermal Hydraulic Analysis
NUREG/CR-1157:1981   Thermal/hydraulic analysis research program: quarterly report
NUREG/CR-1998:1981   RELAP4/MOD7: a best estimate computer program to calculate thermal and hydraulic phenomena in a nuclear reactor or related system
NUREG/CR-1826:1982   RELAP5/MOD1 code manual
NUREG/CR-3046:1983   COBRA/TRAC – a thermal hydraulics code for transient analysis of nuclear reactor vessels and primary coolant
NUREG/CR-3257:1983   RELAP5 assessment: LOFT turbine trip L6–7/L9–2
NUREG/CR-2896:1984   COMMIX 1A a three-dimensional transient single-phase computer program for thermal hydraulic analysis of single and multicomponent systems
NUREG/CR-3329:1984   Thermal/hydraulic analysis research program
NUREG/CR-3504:1984   Turbulence modeling in the COMMIX computer code
NUREG/CR-3772:1985   RELAP5 assessment: semiscale small break tests S-UT-1, S-UT-2, S-UT-6, S-UT-7 and S-UT-8
NUREG/CR-3802:1985   RELAP5 assessment: quantitative key parameters and run time statistics
NUREG/CR-3820:1985   Thermal/hydraulic analysis research program (TRAC-PF1/MOD1)
NUREG/CR-3936:1985   RELAP5 assessment: conclusions and user guidelines

NUREG/CR-4278:1985   TRAC-PF1/MOD1 development assessment
NUREG/CR-4292:1985   A comparative analysis of constitutive relations in TRAC-PFL and RELAP5/MOD1
NUREG/CR-3262:1986   COBRA-NC: a thermal hydraulic code for transient analysis of nuclear reactor components
NUREG/CR-4371:1986   COMMIX-2: a three-dimensional transient computer program for thermal – hydraulic analysis of two-phase flows
NUREG/CR-4830:1987   MELCOR validation and verification 1986 papers
NUREG/CR-4312:1988   RELAP5/MOD2 code manual
NUREG/CR-5194:1988   RELAP5/MOD2 models and correlations
NUREG/CR-5273:1989   SCDAP/RELAP5/MOD2 code manual

Three Mile Island
NUREG-0680:1981   TMI-1 restart
NUREG-0752:1981   Control room design report for TMI-1
NUREG-0698:1984   NRC plan for cleanup operations at Three Mile Island Unit 2
NUREG/CR-4978:1987   The cooldown aspects of the TMI-2 accident
NUREG-1355:1989   The status of recommendations of the President's Commission on the Accident at Three Mile Island. A ten-year review
NUREG/CR-5457:1989   A review of the Three Mile Island-1 probabilistic risk assessment

Toxic Gas
NUREG/CR-1741:1981   Models for the estimation of incapacitation times following exposure to toxic gases or vapors
NUREG/CR-3685:1984   Toxic gas accident analysis code user's manual
NUREG/CR-5656:1991   EXTRAN: a computer code for estimating concentrations oftoxic substances at control room air intakes
NUREG/CR-5669:1991   Evaluation of exposure limits to toxic gases for nuclear reactor control room operators

Transport
NUREG/CR-3499:1983   Severe rail and truck accidents:toward a definition of bounding environments for transportation packages
NUREG/CR-5717:1991   Packaging supplier inspection guide

Two-Phase Flow

NUREG/CR-2671:1982 — The Marviken full scale critical flow tests

NUREG/CR-3372:1983 — An analysis of wave dispersion, sonic velocity and critical flow in two-phase mixtures

NUREG/CR-4761:1986 — Small break critical discharge – roles of vapor and liquid entrainment in stratified two-phase region upstream of the break

Utilities

NUREG/CR-2796:1982 — Compressed air and backup nitrogen systems in nuclear power plants

NUREG/CR-2797:1982 — Evaluation of events involving service water systems in nuclear power plants

NUREG/CR-3351:1985 — Safety implications associated with in-plant pressurized gas storage

NUREG/CR-5472:1990 — A risk-based review of instrument air systems at nuclear power plants distribution

Value-Impact Analysis

NUREG/CR-4941:1987 — The application of value-impact analysis to USI A-45

NUREG/CR-5579:1990 — Value/impact assessment of jet impingement loads and pipe-to-pipe impact damage

Valves

NUREG/CR-1363:1982 — Data summaries of Licensee Event Reports of valves at US commercial nuclear power plants

NUREG/CR-3154:1983 — The in-plant reliability data base for nuclear power components: interim data report – the valve component

NUREG/CR-3914:1985 — Pump and valve qualification review guide

NUREG/CR-4141:1985 — Containment purge and vent valve test program. Final report

NUREG/CR-4648:1986 — A study of typical nuclear containment purge valves in an accident environment

NUREG/CR-4692:1987 — Operating experience review of failures of power operated relief valves and block valves in nuclear power plants

NUREG/CR-4999:1988 — Estimation of risk reduction from improved PORV reliability in PWRs

NUREG/CR-5159:1988 — Prediction of check valve performance and degradation in nuclear power plant systems

NUREG-1316:1989 — Technical findings and regulatory analysis related to generic issue 70. Evaluation of power operated relief valve and block valve reliability in PWR nuclear power plants

NUREG-1352:1990 — Action plans for motor-operated valves and check valves

NUREG-1372:1990 — Regulatory analysis for the resolution of generic issue C-8 'Main steam isolation valve leakage and LCS failure'

NUREG/CR-5583:1990 — Prediction of check valve performance and degradation in nuclear power plant systems – wear and impact

Water Hammer

NUREG-0918:1982 — Prevention and mitigation of steam generator water hammer events in PWR plants

NUREG/CR-2059:1982 — Compilation of data concerning known and suspected water hammer events in nuclear powerplants

NUREG/CR-2781:1982 — Evaluation of water hammer events in light water reactors

NUREG-0927:1983 — Evaluation of water hammer experience in nuclear power plants

NUREG-0927:1984 — Evaluation of water hammer occurrence in nuclear power plants

NUREG-0993:1984 — Regulatory analysis for USI A-1: 'water hammer'

NUREG/CR-3939:1984 — Water hammer, flow induced vibration and safety/relief valve loads

NUREG/CR-5220:1988 — Diagnosis of condensation-induced water hammer

Particular Nuclear Power Stations

A   Technical specifications for (name) station
B   Safety evaluation report related to the operation of (name) station

	A – Technical specification	B – Safety evaluation
Beaver Valley	1259, 1279	1057
Braidwood	1223, 1261, 1276	1002
Byron	1097, 1113	0876
Calaway		0830
Catawba	1072, 1099, 1106, 1182, 1191	0954
Clinton	1202, 1235	0853
Comanche Peak		0797
Diablo Canyon	0817, 1132, 1151	0675
Enrico Fermi	1089, 1141	0798
Grand Gulf		0831
Hope Creek	1186, 1202	1048
LaSalle	1013	0519
Limerick	1088, 1149	0974, 0991
Millstone	1161, 1176	1031
Nine Mile Point	1193, 1253	1047

Appendix
29

Information Sources

Contents

A29.1 Selected Organizations Relevant to Safety and Loss Prevention

United Kingdom
Associated Octel Company Limited,
http://www.octel-corp.com/home/home.htm
P.O. Box 17, Oil Sites Road, Ellesmere Port,
    Cheshire, CH65 4HF
Phone: +44 (0) 151 355 3611, Fax: +44 (0) 151 356 2349

Association of British Chemical Manufacturers –
    succeeded by Chemical Industries Association
Association of British Insurers,
http://www.abi.org.uk/
51 Gresham Street, London EC2V 7HQ
Phone: +44 (0) 20 7600 3333, Fax: +44 (0) 20 7696 8999

British Approvals Service for Electrical Equipment in
    Flammable Atmospheres (BASEEFA) – succeeded
    by Electrical Equipment Certification Service

British Chemical Engineering Contractors
    Association (BCECA),
http://www.bceca.org.uk/
1 Regent Street, London SW1Y 4NR
Phone: +44 (0) 20 7839 6514

British Chemical Industries Safety Council – succeeded
    by Chemical Industries Association

British Cryogenic Council (BCC) – succeeded by British
    Cryoengineering Society (BCS)

British Cryoengineering Society (BCS),
http://www.bcryo.org.uk/
P.O. Box 41, Leatherhead, KT22 9YY
Phone: +44 (0) 1372 376544, Fax: +44 (0) 1372 376544

British Compressed Gas Association (BCGA),
http://www.bcga.co.uk/
6 St Mary's St, Wallingford, OX10 0EL
Phone: +44 (0) 1491 825533, Fax: +44 (0) 1491 826689

British Computer Society (BCS),
http://www1.bcs.org.uk/
1 Sanford Street, Swindon, Wiltshire, SN1 1HJ
Phone: +44 (0) 1793 417417, Fax: +44 (0) 1793 480270

British Engine Insurance Company,
39 Windsor Place, Cardiff South Glamorgan, CF10 3BW
Phone: +44 (0) 29 20237199

British Fire Services Association,
http://www.bfsa.org.uk/
86 London Road, Leicester LE2 0QR
Phone: +44 (0) 116 2542879, Fax: +44 (0) 116 2542879

British Geological Survey,
Murchison House, West Mains Road, Edinburgh EH9 3LA
Phone: +44 (0) 131 667 1000, Fax: +44 (0) 131 668 2683

British Hydromechanics Research Association –
    succeeded by BHR Group Limited

BHR GROUP Limited
http://www.bhrgroup.co.uk/
    The Fluid Engineering Centre, Cranfield,
    Bedfordshire MK43 0AJ
Phone: +44 (0) 1234 750422, Fax: +44 (0) 1234 750074

British Insurance Association – succeeded by Association
    of British Insurers

British Nuclear Society (BNES),
http://www.bnes.com/
1-7 Great George Street, London, SW1P 3AA
Phone: +44 (0) 20 7665 2241, Fax: +44 (0) 20 7799 1325

British Occupational Hygiene Society (BoHS),
http://www.bohs.org/
Suite 2, Georgian House, Great Northern Road,
    Derby, DE1 1LT
Phone: +44 (0) 1332 298101, Fax: +44 (0) 1332 298099

British Pump Manufacturers Association (BPMA),
http://www.bpma.org.uk/
The McLaren Building, 35 Dale End, Birmingham, B4 7LM
Phone: +44 (0) 121 200 1299, Fax: +44 (0) 121 200 1306

British Rail – privatized and succeeded by the
    Rail Industry Training Council (RITC Ltd)

British Red Cross Society,
http://www.redcross.org.uk/
9 Grosvenor Crescent, London SW1X 7EJ
Phone: +44 (0) 20 7235 5454, Fax: +44 (0) 20 7245 6315

British Safety Council,
http://ww2.britishsafetycouncil.org/
70 Chancellors Road, London W6 9RS
Phone: +44 (0) 20 8741 1231, Fax: +44 (0) 20 8741 4555

British Standards Institution (BSI),
http://www.bsi-global.com/
389 Chiswick High Road, London, W4 4AL
Phone: +44 (0) 20 8996 9000, Fax: +44 (0) 20 8996 7001

British Waterways,
http://www.britishwaterways.co.uk/
Willow Grange, Church Road, Watford, WD17 4QA
Phone: +44 (0) 1923 201 120

Building Research Establishment Ltd (BRE),
http://www.bre.co.uk/
Garston, Watford WD25 9XX
Phone: +44 (0) 1923 664 000

Chamber of Shipping of the UK,
http://www.british-shipping.org/
Carthusian Court, 12 Carthusian Street,
    London EC1M 6EZ
Phone: +44 (0) 20 7417 2800, Fax: +44 (0) 20 7726 2080

Chemical and Allied Product Industry Training Board –
    now disbanded

Chemical Industries Association (CIA),
http://www.cia.org.uk/
Kings Buildings, Smith Square, London SW1P 3JJ
Phone: +44 (0) 20 7834 3399, Fax: 020 7834 4469

Chemical Industry Safety and Health Council – succeeded
    by Chemical Industries Association

The Chemical Society – see the Royal Society
    for Chemistry

Chief and Assistant Fire Officers Association (CACFOA),
http://www.fire-uk.org/
9-11 Pebble Close, Tamworth, Staffordshire, B77 4RD
Phone: +44 (0) 1827 302300, Fax: +44 (0) 1827 302399

Confederation of British Industry (CBI),
http://www.cbi.org.uk/
103 New Oxford Street, London WC1A 1DU
Phone: +44 (0) 20 7379 7400, Fax: +44 (0) 20 7240 8287

Council for Oil and Gas Extraction, Chemicals
    Manufacturing and Petroleum Industries (COGENT)
http://www.cogent-ssc.com/
Monticello House, 45 Russell Square, London WC1B 4JP
Phone: +44 (0) 20 637 9533, Fax: +44 (0) 20 7291 6129

Electrical Equipment Certification Service (EECS) –
    part of HSE but disbanded in 2002

Engineering Council, (UK),
http://www.engc.org.uk/
10 Maltravers Street, London WC2R 3ER
Phone: +44 (0) 20 7240 7891, Fax: +44 (0) 20 7379 5586

Engineering Equipment Users Association – succeeded
    by Engineering Equipment and Materials Users
    Association

Engineering Equipment and Materials Users Association,
http://www.eemua.co.uk/
3rd Floor, 30 Long Lane, London EC1A 9HL
Phone: +44 (0) 20 7796 1293, Fax: +44 (0) 20 7796 1294

The Fellowship of Engineering – now The Royal Academy
    of Engineering

Fire Officers Committee – see Association of
    British Insurers

Fire Protection Association (FPA),
http://www.thefpa.co.uk/
Bastille Court, 2 Paris Garden, London SE1 8ND
Phone: +44 (0) 20 7902 5300, Fax: +44 (0) 20 7902 5301

Fire Industry Confederation (FIC)
http://www.the-fic.org.uk/
Neville House, 55 Eden Street, Kingston upon Thames,
    Surrey, KT1 1BW
Phone: +44 (0) 20 8549 8839, Fax: +44 (0) 20 8547 1564

Fire Research Station (FRS),
http://www.bre.co.uk/frs/
FRS, the Fire Division of BRE, Garston,
    Watford WD25 9XX
Phone: +44 (0) 1923 664700, Fax: +44 (0) 1923 664910

Fire Service College,
http://www.fireservicecollege.ac.uk/
Friends of the Earth Ltd.,
http://www.foe.co.uk/
26-28 Underwood Street, London N1 7JQ
Phone: +44 (0) 20 7490 1555, Fax: +44 (0) 20 7490 0881

Health and Safety Commission/Executive (HSC/E),
http://www.hse.gov.uk/
2 Rose Court, 2 Southwark Bridge, London SE1 9HS
Phone: +44 (0) 20 7556 2100, Fax: +44 (0) 20 7556 2109

Health and Safety Executive,
Research and Laboratory Division,
http://www.hse.gov.uk/research/
Broad Lane, Sheffield, S3 7HQ
Phone: +44 (0) 114 289 2000, Fax: +44 (0) 114 289 2500

High Pressure Technology Association,
http://www.liv.ac.uk/~js1/hpta.html
c/o Department of Chemistry, University of Liverpool,
    Liverpool L69 7ZD
Phone: +44 (0) 151 794 3530, Fax: +44 (0) 151 794 3588

HM Stationery Office (HMSO),
http://www.hmso.gov.uk/
The Stationery Office Ltd, P.O. Box 29, Norwich, NR3 1GN
Phone: +44 (0) 870 600 5522, Fax: +44 (0) 870 600 5533

Home Office,
http://www.homeoffice.gov.uk/
Room 856, 50 Queen Anne's Gate, London SW1 9AT
Phone: +44 (0) 870 000 1585, Fax: +44 (0) 20 7273 2065

Imperial Chemical Industries plc,
20 Manchester Square, London, W1U 3AN
Phone: +44 (0) 20 7009 5000, Fax: 44 (0) 20 7009 5001

Imperial Chemicals Insurance,
1 Adam Street, London WC2

Industrial Fire Protection Association of Great Britain,
140 Aldersgate Street, London EC1A 4HX
Phone: +44 (0) 20 7606 3757

Industrial Safety (Protective Equipment) Manufacturers
    Association,
69 Cannon Street, London EC4N 5AB

The Industrial Society – now disbanded

Institute of Cancer Research,
http://www.icr.ac.uk/
123 Old Brompton Road, London SW7 3RP
Phone: +44 (0) 20 7352 8133, Fax: +44 (0) 20 7370 5261

Institute of Geological Sciences – see British
    Geological Survey

Institute of Materials (IOM),
http://www.iom3.org/
1 Carlton House Terrace, London SW1Y 5DB
Phone: +44 (0) 20 7451 7300, Fax: +44 (0) 20 7839 1702

Institute of Measurement and Control,
http://www.instmc.org.uk/
87 Gower Street, London WC1E 6AA
Phone: +44 (0) 20 7387 4949, Fax: +44 (0) 20 7388 8431

Institute of Metals – see Institute of Materials

Institute of Oceanographic Sciences (IOSDL),
http://wofiles.nwo.ac.uk/
Deacon Laboratory, Brook Road, Wormley, Godalming,
    Surrey GU8 5UB
Phone: +44 (0) 428 684141, Fax: +44 (0) 428 683066

Institute of Offshore Engineering (IOE),
http://www.civ.hw.ac.uk/
Heriot–Watt University, Research Park, Riccarton,
    Currie, Edinburgh EH14 4AP
Phone: +44 (0) 131 451 3151, Fax: +44 (0) 131 449 6254

Institute of Petroleum,
http://www.petroleum.co.uk/
61 New Cavendish Street, London, W1G 7AR
Phone: +44 (0) 20 7467 7100, Fax: +44 (0) 20 7255 1472

Institute of Physics (IOP),
http://www.iop.org/
76 Portland Place, London W1B 1NT
Phone: +44 (0) 20 7470 4800, Fax: +44 (0) 20 7470 4848

Institution of Chemical Engineers (IChemE),
http://www.icheme.org/
Davis Building, 165–189 Railway Terrace,
    Rugby CV21 3HQ
Phone: +44 (0) 1788 578214, Fax: +44 (0) 1788 560833

Institution of Civil Engineers (ICE),
http://www.ice.org.uk/
One Great George Street, Westminster, London SW1P 3AA
Phone: +44 (0) 20 7222 7722

Institution of Electrical Engineers (IEE),
http://www.iee.org/
Savoy Place, London WC2R OBL
Phone: +44 (0) 20 7240 1871, Fax: +44 (0) 20 7240 7735

Institution of Fire Engineers (IFE),
http://www.ife.org.uk/
148 Upper New Walk, Leicester LEI 7QB
Phone: +44 (0) 116 255 3654, Fax: +44 (0) 116 247 1231

Institution of Gas Engineers,
http://www.igaseng.com/
17 Grosvenor Crescent, London SW1X 7ES 12 York Gate
London NW1 4QG
Phone: +44 (0) 20 7487 0650, Fax: +44 (0) 20 7224 4762

Institution of Industrial Safety Officers – see Institution of
    Occupational Safety and Health

Institution of Marine Engineers (IMarEST),
http://www.imarest.org/

Institution of Mechanical Engineers (IMechE),
http://www.imeche.org.uk/
1 Birdcage Walk, Westminster, London, SW1H 9JJ
Phone: +44 (0) 20 7222 7899, Fax: +44 (0) 20 7222 4557

Institution of Nuclear Engineers (INE),
http://www.inuce.org.uk/
Allan House, 1 Penerley Road, London SE6 2LQ
Phone: +44 (0) 208 698 1500, Fax: +44 (0) 208 695 6409

Institution of Occupational Safety and Health (IOSH),
http://www.iosh.co.uk/
The Grange, Highfield Drive, Wigston,
    Leicestershire LE18 1NN
Phone: +44 (0) 116 257 3100, Fax: +44 (0) 116 257 3101

Institution of Plant Engineers – succeeded by
    The Society of Operation Engineers (SOE)

Insurance Technical Bureau – now disbanded

Joint Fire Research Organisation – see Fire Research
    Station

Liquefied Petroleum Gas Industry Technical Association,
17 Grosvenor Crescent, London SW1X 7EF

Loss Prevention Council – purchased by BRE and now
    is the Fire Division (FRS) of the Building
    Research Establishment Ltd (BRE)

Medical Research Council (MRC),
http://www.mrc.ac.uk/
20 Park Crescent, London W1 B 1AL
Phone: +44 (0) 20 7636 5422, Fax: +44 (0) 20 7436 6179

Meteorological Office (MET),
http://www.met-office.gov.uk/
London Road, Bracknell, Berkshire RG12 2SZ
Phone: +44 (0) 845 300 0300, Fax: +44 (0) 845 300 1300

Ministry of Defence, Defence Science and
    Technology Laboratory (DSTL),
http://www.dstl.gov.uk/
Porton Down, Salisbury, Wiltshire SP4 0JQ
Phone: +44 (0) 1980 613121, Fax: +44 (0) 1980 613085

Ministry of Defence, Directorate of Safety Environment
    and Fire Policy (DSEF),
http://www.mod.uk/dsef/
Old War Office, Whitehall, London SW1A 2EU
Phone: +44 (0) 870 607 4455

Down Ministry of Defence,
Tidal Branch, Hydrographic Department (UKHO)
http://www.hydro.gov.uk/
Admiralty Way, Taunton, Somerset TA1 2DN
Phone: +44 (0) 1823 337900, Fax: +44 (0) 1823 284077

National Engineering Laboratory (NEL),
http://www.nel.uk/
East Kilbride, Glasgow, G75 0QU
Phone: +44 (0) 1355 220222, Fax: +44 (0) 1355 272999

National Society for Clean Air (NSCA),
http://www.nsca.org.uk/
44 Grand Parade Brighton, UK BN2 9QA
Phone: +44 (0) 1273 878770, Fax: +44 (0) 1273 606626

National Radiological Protection Board (NRPB),
http://www.nrpb.org/
Chilton, Didcot, Oxon, OX11 0RQ
Phone: +44 (0) 1235 831600, Fax: +44 (0) 1235 833891

National Vulcan Engineering Insurance Group,
3 St Mary, Parsonage, Manchester M60 9AP
Phone: +44 61 834 8124

Oil and Chemical Plant Manufacturers Association,
    Suites 41–48, 87 Regent Street, London W1R 7HF

Petroleum Industry Training Board – succeeded by
    Petroleum Industry National Training Organisations
    (PINTO), and then by the Council for Oil and
    Gas Extraction, Chemicals Manufacturing and
    Petroleoum Industries (COGENT)

Rail Industry Training Council (RITC)
http://www.ritc.org.uk/
B118 Macmillan House, Paddington Station,
    London W2 1FT
Phone: +44 (0) 870 2202773, Fax: +44 (0) 870 2202774

The Royal Academy of Engineering,
http:/www.raeng.org.uk/
29 Great Peter Street, Westminster, London SW1P 3LW
Phone: +44 (0) 20 7222 2688, Fax: +44 (0) 20 7233 0054

The Royal Institute of Chemistry – incorporated in
    The Royal Society of Chemistry

The Royal Meteorological Society,
http://www.royal-met-soc.org.uk/
104 Oxford Road, Berkshire RG1 7LL
Phone: +44 (0) 118 9568500, Fax: +44 (0) 118 9568571

The Royal Society,
http://www.royalsoc.ac.uk/
6-9 Carlton House Terrace, London SW1Y 5AG
Phone: +44 (0) 20 7839 5561, Fax: +44 (0) 20 7930 2170

The Royal Society of Chemistry (RSC),
http://www.rsc.org/
Burlington House, Piccadilly, London W1J 0BA
Phone: +44 (0) 20 7437 8656, Fax: +44 (0) 20 7437 8883

The Royal Society for the Prevention of Accidents (RoSPA),
http://www.rospa.com/CMS/
Edgbaston Park, 353 Bristol Road, Edgbaston,
    Birmingham B5 7ST
Phone: +44 (0) 121 248 2000, Fax: +44 (0) 121 248 2001

Safety in Mines Research Establishment

Sheffield Safety and Reliability Directorate – see
    UK Atomic Energy Authority

Safety and Reliability Society,
http://www.sars.u-net.com
Clayton House, 59 Piccadilly, Manchester M1 2AQ
Phone: +44 (0) 161 228 7824, Fax: +44 (0) 161 236 6977

Society of Chemical Industry (SCI),
http://www.soci.org/SCI/
14-15 Belgrave Square, London SW1X 8BS
Phone: +44 (0) 20 7598 1500, Fax: +44 (0) 20 7598 1545

Society of Gas Tanker and Terminal Operators (SIGTTO),
http://www.sigtto.org/
Liaison Office, 17 St. Helen's Palace, London, EC3A 6DG
Phone: +44 (0) 20 7628 1134, Fax: +44 (0) 20 7628 3163

Society of Occupational Medicine (SOM),
http://www.som.org.uk/
Royal College of Physicians, 6 St. Andrew's Place,
    Regent's Park, London NW1 4LB
Phone: +44 (0) 20 7486 2641, Fax: +44 (0) 20 7486 0028

The Society of Operation Engineers (SOE),
http://www.soe.org.uk/soe.org/
22 Greencoat place, London, SW1P 1PR,
Phone: +44 (0) 20 7630 1111, Fax: +44 (0) 20 7630 6677

Trades Union Congress (TUC),
http://www.tuc.org.uk/
23-28 Great Russell Street, London WC1B 3LS

Transport and Road Research Laboratory (TRL),
http://www.trl.co.uk/
Old Wokingham Road, Crowthorne
Berkshire, RG45 6AU
Phone: +44 (0) 1344 770007, Fax: +44 (0) 1344 770880

TUC Centenary Institute of Occupational Health,
    London School of Hygiene and Tropical Medicine,
http://www.tuc.org.uk/h_and_s/
http://www.lshtm.ac.uk/
Keppel Street, London WC1E 7HT
Phone: +44 (0) 20 7636 8636, Fax: +44 (0) 20 7436 5389

UK Atomic Energy Authority (UKAEA),
http://www.ukaea.org.uk/
Harwell International Business Centre, Didcot,
    Oxon, OX11 0RA
Phone: +44 (0) 1235 820220

UK Atomic Energy Authority, Safety and Reliability
    Directorate
Wigshaw Lane, Culcheth, Warrington, Lancashire

The Welding Institute (TWI),
http://www.twi.co.uk/
Granta Park, Great Abington, Cambridge CB1 6AL, UK
Phone: +44 (0) 1223 891162, Fax: +44 (0) 1223 892588

United States
Air Pollution Control Association,
Pittsburgh, PA 15230

American Association for Artificial Intelligence (AAAI)
http://www.aaai.org/
445 Burgess Drive, Menlo Park, CA 94025-3442
Phone: +1 (650) 328 3123, Fax: +1 (650) 321 4457

American Chemistry Council (ACC),
http://www.americanchemistry.com/
1300 Wilson Blvd. Arlington, VA 22209
Phone: +1 (703) 741 5000, Fax: +1 (703) 741 6000

American Chemical Society (ACS),
http://www.chemistry.org/
1155 16th Street, NW, Washington, DC 20036
1300 Wilson Blvd. Arlington, VA 22209
Phone: +1 (703) 741 5000, Fax: +1 (703) 741 6000

American Chemical Society (ACS),
http://ww.chemistry.org/
Phone: +1 (800) 227 5558 (US only), +1 (202) 872 4600
    (outside the US), Fax: +1 (202) 872 4615

American Gas Association (AGA),
http://www.aga.org/
400 N. Capitol Street, N.W. Washington, DC 20001
Phone: +1 (202) 824 7000, Fax: +1 (202) 824 7115

American Industrial Hygiene Association (AIHA),
http://www.aiha.org/
2700 Prosperity Avenue, Suite 250, Fairfax, VA 22031
Phone: +1 (703) 849 8888, Fax: +1 (703) 207 3561

American Institute for Aeronautics and
    Astronautics (AIAA),
http://www.aiaa.org/
1801 Alexander Bell Drive, Suite 500
Reston, VA 20191-4344
Phone: +1 (703) 264 7500, Fax: +1 (703) 264 7551

American Institute of Chemical Engineers (AIChE),
http://www.aiche.org/
3 Park Avenue, New York, NY 10016-5991
General Inquiries: +1 (212) 591 7338,
    Fax: +1 (212) 591 8897

American Insurance Association (AIA),
http://www.aiadc.org/
Washington, D.C. Headquarters
1130 Connecticut Ave, NW
Ste. 1000, Washington, DC 20036
Phone: +1 (202) 828 7100, Fax: +1 (202) 293 1219

American Meteorological Society (AMS),
http://www.ametsoc.org/AMS/
Headquarters, 45 Beacon Street, Boston, MA 02108-3693
Phone: +1 (617) 227 2425, Fax: +1 (617) 742 8718

American National Standards Institute (ANSI),
http://www.ansi.org/
New York City Office (Operations), 25 West 43rd Street,
    4th Floor, New York, NY 10036
Phone: +1 (212) 642 4900, Fax: +1 (212) 398 0023

American Nuclear Society (ANS),
http://www.ans.org/
555 N Kensington Avenue, Lagrange Park, IL 60525
Phone: +1 (708) 352 6611, Fax: +1 (708) 352 0499

American Petroleum Institute (API),
http:// www.api.org /
1220 L Street, NW, Washington, DC 20005-4070
Phone: +1 (202) 682 8000

American Society of Civil Engineers (ASCE),
http://www.asce.org/
1801 Alexander Bell Drive, Reston, Virginia 20191-4400
Phone: +1 (703) 295 6300, Fax: +1 (703) 295 6222

American Society of Heating, Refrigerating and
    Ventilating Engineers – succeeded by American
    Society of Heating, Refrigerating and Air-Conditioning
    Engineers (ASHRAE),
http://www.ashrae.org/
1791 Tullie Circle NE, Atlanta, GA 30329
Phone: +1 (404) 636 8400, Fax: +1 (404) 321 5478

American Society of Mechanical Engineers (ASME)
http://www.asme.org/
Headquarters, Three Park Avenue, New York,
    NY 10016-5990
Phone: +1 (212) 591 7722, Fax: +1 (212) 591 7674

American Society for Quality (ASQ),
http://www.asq.org/
600 North Plankinton Avenue, Milwaukee, WI 53203
Phone: +1 (414) 272 8575, Fax: +1 (414) 272 1734

American Society of Safety Engineers (ASSE),
http://www.asse.org/
Customer Service: 1800 E Oakton St., Des Plaines, IL 60018
Phone: +1 (847) 699 2929, Fax: +1 (847) 768 3434

American Society for Testing and Materials (ASTM),
http://www.astm.org/
100 Barr Harbor Drive, P.O. Box C700, West Conshohocken,
    PA 19428-2959
Phone: +1 (610) 832 9585, Fax: +1 (610) 832 9555

American Standards Association – now
    American National Standards Institute (ANSI)

American Trucking Association (ATA),
http://www.truckline.com/
2200 Mill Road, Alexandria, VA 22314-4677
Phone: +1 (703) 838 1700

American Welding Society (AWS),
http://www.aws.org/
550 Lejeune Road, Miami, FL 33126
Phone: +1 (800) 443 9353

Argonne National Laboratory,
http://www.anl.gov/
Main site: 9700 South Cass Avenue, Argonne, IL 60439
Phone: +1 (630) 252 2000
Idaho site: Argonne-West, P.O. Box 2528,
    Idaho Falls, ID 83403-2528.
Phone: +1 (208) 533 7341

Association of American Railroads (AAR),
http://www.aar.org/
50 F Street NW, Washington, DC 20001-1564
Phone: +1 (202) 639 2200, Fax: +1 (202) 639 2439

Atomic Energy Commission – now Nuclear
    Regulatory Commission

Ballistics Research Laboratories,
http://www.apg.army.mil/default.htm
Aberdeen Proving Ground, MD 21005
Phone: +1 (410) 306 1403

Battelle Columbus Laboratory,
http://www.battelle.org/default.stm
505 King Avenue, Columbus, OH 43201
Phone: +1 (614) 424 6424, Fax: (614) 424 5263

Battelle Northwest Laboratory,
http://www.pnl.gov/
092 Battelle Boulevard, Richland, WA 99352
Phone: +1 (509) 375 2121

Brookhaven National Laboratory,
http://www.bnl.gov/world/
P.O. Box 5000, Upton, NY 11973-5000
Phone: +1 (631) 344 8000

Bureau of Mines (USMB) – closed in 1996 see
    Minerals Information Team, USGS Geologic Division

Center for Chemical Process Safety (CCPS),
http://www.aiche.org/ccps/
American Institute of Chemical Engineers, 3 Park Ave,
    New York, N.Y., 10016-5991
Phone: +1 (212) 591 7319, Fax: +1 (212) 591 8895

Chemical Manufacturers Association (CMA),
www.cmahq.com/
Wilson Boulevard, Arlington, Virginia 22209
Phone: +1 (703) 741 5922, Fax: +1 (703) 741 6922

Chemical Transport Emergency Center (CHEMTREC),
http://www.chemtrec.org/
1300 Wilson Blvd, Arlington, VA. 22209
Phone: +1 (703) 741 5523, Fax: +1 (703) 741 6037

Chemical Safety and Hazard Investigation Board (CSB),
http://www.chemsafety.gov/
2175 K. Street N.W.-Suite 400, Washington,
    D.C. 20037-1809
Phone: +1 (202) 261 7600, Fax: +1 (202) 261 7650

Chlorine Institute,
http://www.cl2.com/
1300 Wilson Blvd., Rosslyn, VA 22209
Phone: +1 (703) 741 5760, Fax: +1 (703) 741 6068

Clearinghouse for Occupational Safety and
    Health Information, Public Health Service,
4676 Columbia Parkway, Cincinnati, OH 45226

Combustion Institute,
http://www.combustioninstitute.org/
5001 Baum Road, Suite 635, Pittsburgh,
    PA 15213-1851 USA
Phone: +1 (412) 687 1366
Fax: +1 (412) 687 0340

Compressed Gas Association (CGA),
http://www.cganet.com/
4221 Walney Road, 5th Floor, Chantilly VA 20151-2923
Phone: +1 (703) 788 2700, Fax: +1 (703) 961 1831

Department of Transportation Research and
    Special Programs Administration,
    Office of Hazardous Materials (OHM)
http://hazmat.dot.gov/
DHM-1400 7th Street SW, Washington, DC 20590
Washington, DC 20590-0001
Phone: +1 (202) 366 0656, Fax: +1 (202) 366 5713

Department of Transportation, US Coast Guard
http://www.uscg.mil/hq/g-m/gmhome.htm
2100 Second Street SW, Washington, DC 20593-0001
Phone: +1 (202) 267 2200, Fax: +1 (202) 267 4839

Dow Chemical Company,
Bldg 566, Midland, MI 48640

Earthquake Engineering Research Institute (EERI),
http://www.eeri.org/
499 14th Street Suite 320, Oakland, CA 94612-1934
Phone: +1 (510) 451 0905, Fax: +1 (510) 451 5411

Electric Power Research Institute (EPRI),
http://www.epri.com/
3412 Hillview Avenue, Palo Alto, CA 94304
Phone: +1 (800) 313 3774

Environmental Protection Agency (EPA),
    Office of Research and Development (ORD),
http://www.epa.gov/ord/
EPA Headquarters: Ariel Rios Building,
1200 Pennsylvania Avenue, N.W.
Mail Code 3213A, Washington, DC 20460
Phone: +1 (202) 260 2090

Ethylene Oxide Industry Council (EOIC),
2501 M Street NW, Washington, DC 20037

Factory Insurance Association – see Industrial
    Risk Insurers

Factory Mutual Research Corporation (FMRC),
www.fmglobal.com
1151 Boston-Providence Turnpike, Norwood, MA 02062
Phone: +1 (781) 762 4300

Factory Mutual System,
http://www.fmglobal.com/
1301 Atwood Avenue, P.O. Box 7500 Johnston, R.I. 02919
Phone: +1 (401) 275 3000, Fax: +1 (401) 275 3029

Federal Emergency Management Agency (FEMA),
http://www.fema.gov/
500 C Street, SW Washington, DC 20472
Phone: +1 (202) 566 1600

Federal Energy Regulatory Commission (FERC),
http://www.ferc.gov/
888 First Street, N.E., Washington, DC 20426
Phone: +1 (866) 208 3372

Federal Power Commission – see Federal
    Energy Regulatory Commission

Gas Technology Institute (GTI), (formed through the
    merger of Institute of Gas Technology and Gas
    Research Institute)
http://www.gri.org/
1700 S. Mount Prospect Road, Des Plaines, IL 60018
Phone: +1 (847) 768 0500, Fax: +1 (847) 768 0501

Hartford Steam Boiler Inspection and Insurance Company,
http://www.hsb.com/
One State Street, P.O. Box 5024, Hartford CT 06102-5024
Phone: +1 (860) 722 1866, Fax: +1 (860) 722 5106

Human Factors Society – succeeded by Human Factors
    and Ergonomics Society

Human Factors and Ergonomics Society (HFES),
http://hfes.org/
Box 1369, Santa Monica, CA 90406
Phone: +1 (310) 394 1811, Fax: +1 (310) 394 2410

Idaho National Laboratory,
http://www.inel.gov/
2525 Fremont Avenue, P.O. Box 1625, Idaho Falls, ID 83415
Phone: +1 (800) 708 2680

Industrial Risk Insurers (IRI),
http://www.industrialrisk.com/
85 Woodland Street, Hartford, CT 06102-5010
Phone: +1 (860) 520 7300, Fax: +1 (860) 520 7397

Institute of Electrical and Electronic Engineers (IEEE),
http://www.ieee.org/portal/
IEEE Corporate Office, 3 Park Avenue, 17th Floor,
    New York, New York, 10016-5997
Phone: +1 (212) 419 7900, Fax: +1 (212) 752 4929

Institute of Electrical and Electronic Engineers,
    Reliability Society (IEEE RS),
See Institute of Electrical and Electronic Engineers (IEEE)
    information

Institute of Gas Technology,
See Gas Technology Institute (GTI)

The Instrumentation, Systems, and
    Automation Society (ISA),
http://www.isa.org/
67 Alexander Drive, P.O. Box 12277
Research Triangle Park, NC 27709
Phone: +1 (919) 549 8411, Fax: +1 (919) 549 8288

Lawrence Livermore National Laboratories,
http://www.llnl.gov/
7000 East Ave., Livermore, CA 94550-9234
P.O. Box 808, Livermore, CA 94551-0808
Phone: +1 (925) 422 1100, Fax: +1 (925) 422 1370

Los Alamos National Laboratory,
http://www.lanl.gov/
P.O. Box 1663, Los Alamos, NM 87545
Phone: +1 (505) 664 5265, Fax: +1 (505) 667 9819

Lovelace Biomedical and Environmental Research
    Institute,
http://www.lrri.org/
2425 Ridgecrest Dr. SE, Albuquerque, NM 87108-5127
Phone: +1 (505) 348 9400, Fax: +1 (505) 348 8541

Manufacturers Standardization Society of the Valve
   and Fittings Industry (MSS),
http://www.mss-hq.com/
127 Park Street NE, Vienna, VA, 22180-4602
Phone: +1 (703) 281 6613, Fax: +1 (703) 281 6671

Manufacturing Chemists Association – now
   Chemical Manufacturers Association

Minerals Information Team, USGS Geologic Division
http://minerals.usgs.gov/minerals/
http://minerals.usgs.gov/minerals/pubs/mit/
U.S. Geological Survey, 988 National Center,
   Reston, VA 20192
Phone: +1 (703) 648 6140, Fax: +1 (703) 648 4995

National Academy of Science – National Academy of
   Engineering – National Research Council,
2101 Constitution Avenue NW, Washington, DC 20418

National Academies
http://nationalacademies.org/
2101 Constitution Avenue NW, Washington, DC 20418
Phone: +1 (202) 334 2000

National Academy of Engineering (NAE),
http://www.nae.edu/
(See National Academies)

National Academy of Science,
http://www4.nationalacademies.org/nas/nashome.nsf
(See National Academies)

National Aeronautics and Space Administration (NASA),
http://www.hq.nasa.gov/
Headquarters, Washington, DC 20546-0001
Phone: +1 (202) 358 0000

National Board of Fire Underwriters – defunct

National Bureau of Standards,
Chemical Thermodynamics Data Center
http://www.cstl.nist.gov/
Chemical Science and Technology Laboratory, NIST,
   100 Bureau Drive, Stop 8300, Gaithersburg,
   MD 20899-8300
Phone: +1 (301) 975-NIST (6478)

National Center for Atmospheric Research (NCAR),
http://www.ncar.ucar.edu/
Box 3000, 1850 Table Mesa Drive, Boulder, CO 80305
Phone: +1 (303) 497 1000

National Fire Protection Association (NFPA),
http://www.nfpa.org/
1 Batterymarch Park, P.O. Box 9101, Quincy,
   MA 02269-9101
Phone: +1 (617) 770 3000, Fax: +1 (617) 770 0700

National Institute of Standards and Technology (NIST)
http://www.nist.gov/
Building and Fire Research Laboratory, Gaithersburg,
   MD 20899
Phone: +1 (301) 975 6850, Fax: +1 (301) 975 4032

National Institute for Occupational Safety and
   Health (NIOSH),
http://www.cdc.gov/niosh/
4676 Columbia Parkway, Cincinnati, OH 45226
Phone: +1 (800) 356 4674, Fax: +1 (513) 533 8573

National LP Gas Association (LPGA) – now
   National Propane Gas Association (NPGA)

National Propane Gas Association (NPGA),
http://www.npga.org/
1150 17th Street, NW, Suite 310, Washington, DC 20036
Phone: +1 (202) 466 7200

National Research Council (NRC),
http://nationalacademies.org/nrc/
(See National Academies)

National Response Center (NRC)
http://www.nrc.uscg.mil/
c/o United States Coast Guard (G-OPF)-Room 2611,
   2100 2nd Street, Southwest Washington, DC 20593-0001
Phone: +1 (202) 267 2675, Fax: +1 (202) 267 2165
TDD: +1 (202) 267 4477 (TDD means Telecommunications
   Device for the Deaf)

National Safety Council (NSC),
http://www.nsc.org/
1121 Spring Lake Drive, Itasca, IL 60143-3201
Phone: +1 (630) 285 1121, Fax: +1 (630) 285 1315

National Technical Information Service (NTIS),
http://www.ntis.gov/
U.S. Department of Commerce Technology Administration,
   5285 Port Royal Road, Springfield, VA 22161
Phone: +1 (703) 605 6000
TDD: (703) 487 4639

National Transportation Safety Board (NTSB),
http://www.ntsb.gov/
490 L'Enfant Plaza, SW, Washington, DC 20594
Phone: +1 (202) 314 6000

Naval Surface Weapons Center (NSWC) – now disbanded

Nuclear Regulatory Commission,
http://www.nrc.gov/
Office of Public Affairs (OPA), Washington, DC 20555
Phone: +1 (800) 368 5642

Oak Ridge National Laboratory,
http://www.ornl.gov/
One Bethel Valley Road or P.O. Box 2008, Oak Ridge,
   TN 37831
Phone: +1 (865) 241 7600, Fax: +1 (865) 241 7603

Occupational Safety and Health Administration (OSHA),
http://www.osha.gov/
200 Constitution Avenue NW, Washington, DC 20210
Phone: +1 (800) 321 6742

Oil Insurance Association – merged in Industrial
   Risk Insurers (IRI)

Reliability Division American Society for Quality (ASQ) –
   see information for American Society for Quality (ASQ)

Sandia National Laboratory,
http://www.sandia.gov/
Sandia National Laboratories, New Mexico, P.O. Box 5800,
   Albuquerque, NM 87185-(mail stop)
Corporate Ombudsman Office: Mail Stop 0620
Phone: (505) 844 9763

System Safety Society (SSS),
http://www.system-safety.org/
P.O. Box 70, Unionville, VA 22567-0070
Phone: +1 (540) 854 8630

Society of Reliability Engineers (SRE)
http://www.sre.org/

Synthetic Organic Chemical Manufacturers Association (SOCMA),
http://www.socma.com/
1850 M Street NW, Suite 700, Washington, DC 20036
Phone: +1 (202) 721 4100, Fax: +1 (202) 296 8120

Southwestern Research Institute (SwRI),
http://www.swri.edu/default.htm
6220 Culebra Road, P.O. Drawer 28510, San Antonio, TX 78228-0510
Phone: +1 (210) 684 5111

Underwriters Laboratories Inc. (UL),
http://www.ul.com/
333 Pfingsten Road, Northbrook, IL 60062-2096
Phone: +1 (708) 272 8800, Fax: +1 (708) 272 8129

Union of Concerned Scientists (UCS),
http://www.ucsusa.org/
2 Brattle Square, Cambridge, MA 02238-9105
Phone: +1 (617) 547 5552, Fax: +1 (617) 864 9405

USA Standards Institute – now American National Standards Institute (NIST)

US Department of Transportation (DOT),
http://www.dot.gov/
400 7th Street, S.W., Washington, D.C. 20590
Phone: +1 (202) 366 4000

US Government Printing Office,
http://www.access.gpo.gov/
Office of Congressional and Public Affairs, Room C804, Stop: LP
U.S. Government Printing Office
Washington, DC 20402
Phone: +1 (202) 512 1991, Fax: +1 (202) 512 12

Europe
European Chemical Industry Council (CEFIC)
http://www.cefic.be/
Avenue E. van Nieuwenhuyse, 4, B-1160, Brussels, Belgium
Phone: +32 (2) 676 72 11, Fax: +32 (2) 676 73 00

European Commission,
http://europa.eu.int/comm/
B-1049, Brussels, Belgium
Phone: +32 (2) 299 11 11

European Commission, Joint Research Centre,
http://www.jrc.cec.eu.int/
Via E. Fermi 1, I-21020 Ispra (VA), Italy
Phone: +39 (0332) 789111, Fax: +39 (0332) 789001

European Process Safety Centre,
c/o Institution of Chemical Engineers, q.v.
http://www.epsc.org/
165-189 Railway Terrace, Rugby, CV21 3HQ, UK
Phone: +44 (0) 1788 534409, Fax: +44 (0) 1788 551542

Other Countries
Canada
Canadian Centre for Occupational Health and Safety (CCOHS)
http://www.ccohs.ca/
250 Main Street East, Hamilton, Ontario, L8N 1H6
Phone: 1 (905) 570 8094, Fax: 1 (905) 572 2206

Finland
Ministry of Social Affairs and Health (STM, sosiaali- ja terveysministeriö),
http://www.vn.fi/stm/english/
Postal address: P.O. Box 33, FIN-00023 Government
Visiting address: Meritullinkatu 8, 00170 Helsinki
Phone: +358 (9) 160 01, Fax: +358 (9) 160 74126

VTT, Technical Research Centre of Finland,
http://www.vtt.fi/
Espoo: P.O. Box 1700, FIN-02044 VTT
Street address: Kemistintie 3, Espoo
Phone: +358 (9) 4561, Fax +358 (9) 456 7020

France
Societe de Chimie Industrielle,
http://www.scifrance.org/
28, Rue Saint Dominique, F-75007 Paris
Phone: +33 (0) 1 53 59 02 10, Fax: +33 (0) 1 45 55 40 33

Germany
Berufsgenossenschaft der Chemischen Industrie
Bundesanstalt fur Materialprufung, Berlin
Bundesanstalt für Materialforschung (BAM)
Federal Institute for Materials Research and Testing
http://www.bam.de/english/
Unter den Eichen 87, 12205 Berlin
Phone: +49 (30) 8104 0, Fax: +49 (30) 8112029

German Berufsgenossenschaften (GB)
(institutions for statutory accident insurance and prevention for trade and industry)
http://www.hvbg.de/

Berufsgenossenschaftliches Institut für Arbeitsschutz (BIA)
http://www.bgchemie.de/
Alte Heerstraße 111, 53754 Sankt Augustin, Bundesrepublik Deutschland
Phone: +49 (22) 41 2 31 02, Fax: +49 (22) 41 2 31 22 34

DECHEMA, Gesellschaft für Chemische Technik und Biotechnologie e.V.
http://www.dechema.de/homepage/welcome-e.html
Theodor-Heuss-Allee 25, D-60486 Frankfurt am Main
Phone: +49 (069) 7564 0, Fax: +49 (069) 7564 201

Deutsches Institut fur Normung (DIN),
http://www2.din.de/
Berlin: 10772 Berlin
Phone: +49 (030) 2601 0, Fax: +49 (030) 2601 1260
Cologne Office: Kamekestraße 8, 50672 Köln
Phone: +49 (02 21) 57 13 0

Technische Uberwachungsverein,
Postfach 21 04 20, Westendstrasse 199, W-8000 Munchen

Verein Deutscher Ingenieure (VDI),
  The Association of Engineers
http://www.vdi.de/vdi/english/
Graf-Recke-Str. 84 40239 Düsseldorf
Postfach 10 11 39 40002 Düsseldorf
Phone: +49 (0) 211 62 14 0

Hong Kong
Labour Department, Occupational Safety and Health
http://www.info.gov.hk/labour/eng/osh/index.htm
Service Headquarters, 15/F, Harbour Building,
  38 Pier Road, Central
Phone: 2852 4041

Japan
National Research Institute of Fire and Disaster, (NRIFD),
http://www.fri.go.jp/indexe.html
14-1, Nakahara 3 Chome, Mitaka, Tokyo 181-8633
Phone: +81 422 44 8331, Fax: +81 422 42 7719

National Institute of Industrial Safety,
http://www.anken.go.jp/english/top.html
1-4-6, Umezono, Kiyose, Tokyo 204-0024
Phone: +81 424 91 4512, Fax: +81 424 91 7846

Latin America
Pan American Center for Sanitary Engineering and
  Environmental Sciences (CEPIS),
http://www.cepis.ops-oms.org/indexeng.html
Los Pinos 259, Urb. Camacho, La Molina, Lima 12
P.O. BOX 4337, Lima 100, Peru
Phone: +51 (1) 437 1077, Fax: +51 (1) 437 8289

The Netherlands
Committee for Prevention of Disasters — see Ministry of
  Social Affairs

Ministry of Social Affairs,
P.O. Box 90804, 2509LV, The Hague Stichting CONCAWE,
  22 President Kennedylaan, The Hague

TNO Prins Maurits Laboratory (TNO-PML)
http://www.pml.tno.nl/
Lange Kleiweg 137, P.O. Box 45, 2280 AA Rijswijk
Phone: +31 (15) 284 2842, Fax: +31 (15) 284 3991

TNO Environment, Energy and Process Innovation
  (TNO-MEP)
http://www.mep.tno.nl/
Laan van Westenenk 501, P.O. Box 342, 7300 AH Apeldoorn
Phone: +31 (55) 549 35 44

Norway
Christian Michelsen Research AS,
http://www.cmr.no/
Postal address: P.O. Box 6031 Postterminalen,
  N-5892 Bergen, Norway
Visiting address: Fantoftvegen 38, Fantoft
Phone: +47 (55) 57 40 40, Fax: +47 (55) 57 40 41

DNV Technology Services
http://www.dnv.com/technologyservices/
Corporate Headquarters, Oslo: Veritasveien 1, 1322 Høvik
Phone: +47 (67) 57 99 00, Fax: +47 (67) 57 99 11

Norwegian Fire Research Laboratory (NBL),
http://www.nbl.sintef.no/
Postal address: N-7465 Trondheim,
Visiting address: Tiller Bru, Trondheim
Phone: +47 (73) 59 10 78, Fax: +47 (73) 59 10 44

Norwegian Society of Chartered Engineers,
http://www.nif.no/
P.O. Box 2312, Solli, N-0201 Oslo,
Phone: +47 (22) 94 75 29, Fax: +47 (22) 94 75 59

Oreda,
http://www.sintef.no/static/TL/projects/oreda/
c/o DNV Technology Services,
c/o SINTEF Industrial Management,
SINTEF Industrial Management
http://www.sintef.no/units/indman/index.html
S.P. Andersensv. 5, 7465 Trondheim
Phone: +47 (73) 59 03 00, Fax: +47 (73) 59 03 30

Switzerland
Schweizerische Chemische Gesellschaft,
http://www.swiss-chem-soc.ch/
c/o SANW, Bärenplatz 2, 3011 Bern
Phone: +41 (0) 31 310 40 90, Fax: +41 (0) 31 312 16 78

International
Intergovernmental Maritime Consultative Organisation —
  see International Maritime Organization

International Association of Fire Chiefs (IAFC),
http://www.iafc.org/
4025 Fair Ridge Dr., Suite 300, Fairfax,
  VA 22033-2868, USA
Phone: +1 (703) 273 0911, Fax: +1 (703) 273 9363

International Association of Firefighters (IAFF),
http://www.iaff.org/
1750 New York Ave. NW, Washington, DC 20006-5395, USA
Phone: +1 (202) 737 8484, Fax: +1 (202) 737 8418

International Atomic Energy Agency,
http://www.iaea.or.at/
P.O. Box 100, Wagramer Strasse 5, A-1400 Vienna, Austria
Phone: +431 2600 0, Fax: +431 2600 7

International Council of Chemical Associations (ICCA),
http://www.icca-chem.org/
International Chamber of Shipping,
http://www.marisec.org/
12 Carthusian Street, London EC1M 6EZ, UK
Phone: +44 (20) 7417 8844, Fax: +44 (20) 7417 8877

International Commission on Radiological Protection,
http://www.icrp.org/
SE-171 16 Stockholm Sweden
Fax: +46 (8) 729 729 8

International Federation of Chemical and General
  Workers Unions (ICF) — see International Federation
  of Chemical, Energy, Mine, and General Workers'
  Unions (ICEM)

International Federation of Chemical, Energy, Mine,
  and General Workers' Unions (ICEM)
http://www.icem.org/index.html
Avenue Emile de Béco, 109, B-1050 Brussels, Belgium.
Phone: +32 (2) 6262020
Fax: +32 (2) 6484316

International Gas Union (IGU)
http://www.igu.org/
P.O. Box 550, c/o DONG A/S, Agern Alle 24-26,
    2970 Hoersholm, Denmark
Phone: +45 4517 1200, Fax: +45 4517 1900

International Institute for Applied Systems
    Analysis (IIASA),
http://www.iiasa.ac.at/
A-2361, Laxenburg, Austria
Phone: +43 (2236) 807 0, Fax: +43 (2236) 71 313

International Institute for Refrigeration (IIR),
    Institut International du Froid
http://www.iifiir.org/
177, boulevard Malesherbes, 75017 Paris – France
Phone: +33 (0) 1 42 27 32 35, Fax: +33 (0) 1 47 63 17 98

International Institute of Ammonia Refrigeration (IIAR)
http://www.iiar.org/
1110 North Glebe Road, Suite 250, Arlington,
    VA 22201, USA
Phone: +1 730 312 4200, Fax: +1 703 312 0065

International Maritime Organization,
http://www.imo.org/home.asp
4 Albert Embankment, London, SE1 7SR, United Kingdom
Phone: +44 (0) 20 7735 7611, Fax: +44 (0) 20 7587 3210

International Labour Organisation (ILO),
http://www.ilo.org/public/english/index.htm
4, route des Morillons
CH-1211 Geneva 22, Switzerland
Phone: +41 (22) 799 6111, Fax: +41 (22) 798 8685

International Occupational Safety and Health Information
    Centre, International Labour Office (ILO-CIS),
http://www.ilo.org/public/english/protection/safework/
    cis/index.htm
CH-1211 Geneva 22, Switzerland
Phone: +41 (22) 799 6740, Fax: +41 (22) 799 8516

International Labour Office, International Standards
    Organization (ISO),
http://www.iso.ch/iso/en/ISOOnline.frontpage
1 rue de Varembe, Case postale 56, CH-1211 Geneva 20,
    Switzerland
Phone: +41 (22) 749 01 11, Telefax: +41 (22) 733 34 30

Oil Companies International Maritime Forum (OCIMF),
http://www.ocimf.com/
27 Queen Anne's Gate, London, SW1H 9BU,
    United Kingdom
Phone: +44 (0) 20 7654 1200, Fax: +44 (0) 20 7654 1205
Victoria Street, London SW1E 1BH

Organization for Economic Co-operation and
    Development (OECD),
http://www.oecd.org/
2, rue André Pascal
F-75775 Paris Cedex 16, France
Phone : +33 (1) 45 24 82 00

United Nations Environment Programme (UNEP),
http://www.unep.org/default.asp
United Nations Avenue, Gigiri, P.O. Box 30552,
    Nairobi, Kenya
Phone: +254 (2) 621234, Fax: +254 (2) 624489/90

World LP Gas Association (WLPGA),
http://www.worldlpgas.com/
World LP Gas Association, 9 rue Anatole de la Forge,
    75017 Paris, France
Phone: +33 (0) 1 58 05 28 00, Fax: +33 (0) 1 58 05 28 01

United Nations Environment Programme/Division
    of Technology, Industry and Economics
    (UNEP/DTIE)
http://www.uneptie.org/
39-43, Quai André Citroën, 75739 Paris Cedex 15, France
Phone: +33 (1) 44 37 14 41, Fax: +33 (1) 44 37 14 74

Appendix 30

Units and Unit Conversions

Contents

SI units are used in this book unless otherwise stated, these are described in BS 5555:1981 *SI Units and Recommendations for the Use of their Multiples and of Certain Other Units*.

Table A30.1 gives selected references on units.

A30.1 Absolute and Gauge Pressures

It is common practice to attach to units of pressure an 'a' or a 'g', denoting 'absolute' or 'gauge', respectively (e.g. bara, psig). This is not part of the SI system, where instead the designation 'absolute' or 'gauge' or some other defining term is applied to the pressure described. In this book both practices are used.

A30.2 Other Units and Conversions

Some other units which are used in the book are

Table A30.1 *Selected references on units*

Chappelear (1981); Horvath (1986); Mullin (1986)

A30.2.1 Volumetric flux

$1 \, l/m^2 \, s = 10^{-3} \, m^3/m^2 \, s$
$1 \, l/m^2 \, min = 1.667 \times 10^{-5} \, m^3/m^2 \, s$
$1 \, UK \, gal/ft^2 \, min = 8.149 \times 10^{-4} \, m^3/m^2 \, s$
$1 \, US \, gal/ft^2 \, min = 6.791 \times 10^{-4} \, m^3/m^2 \, s$

A30.2.2 Other quantities

SI unit conversion

Conversions of other units to and from SI units are given in tables compiled by Mullin (1967) and Lees (1968). The conversion table given by the latter is reproduced in Table A30.2.

$1 \, bbl \, (barrel) = 42 \, US \, gal$
$1 \, lusec = 10^{-3} \, torr \, l/s$
$1 \, micron = 10^{-6} \, m$
$1 \, mil = 10^{-3} \, in. = 2.540 \times 10^{-5} \, m$
$1 \, torr = 1 \, mmHg = 1.333 \times 10^2 \, N/m^2$

Table A30.2 *SI unit conversion tables (Lees, 1968)*

Acceleration	$1 \, cm/s^2$	$1.0000 \times 10^{-2} \, m/s^2$
	$1 \, m/h^2$	$7.7160 \times 10^{-8} \, m/s^2$
	$1 \, ft/s^2$	$3.0480 \times 10^{-1} \, m/s^2$
	$1 \, ft/h^2$	$2.3519 \times 10^{-8} \, m/s^2$
	$1 \, g$	$9.8067 \, m/s^2$
Area	$1 \, cm^2$	$1.0000 \times 10^{-4} \, m^2$
	$1 \, ft^2$	$9.2903 \times 10^{-2} \, m^2$
	$1 \, in.^2$	$6.4516 \times 10^{-4} \, m^2$
	$1 \, yd^2$	$8.3613 \times 10^{-1} \, m^2$
	$1 \, acre$	$4.0469 \times 10^3 \, m^2$
	$1 \, mile^2$	$2.5900 \times 10^6 \, m^2$
	$1 \, ha$	$1.0000 \times 10^4 \, m^2$
Calorific value (volumetric)	$1 \, cal/cm^3$	$4.1868 \times 10^6 \, J/m^3$
	$1 \, kcal/m^3$	$4.1868 \times 10^3 \, J/m^3$
	$1 \, Btu/ft^3$	$3.7260 \times 10^4 \, J/m^3$
	$1 \, Chu/ft^3$	$6.7067 \times 10^4 \, J/m^3$
	$1 \, therm/ft^3$	$3.7260 \times 10^9 \, J/m^3$
	$1 \, kcal/ft^3$	$1.4786 \times 10^5 \, J/m^3$
Coefficient of expansion (volumetric)	$1 \, g/cm^3 \, °C$	$1.0000 \times 10^3 \, kg/m^3 \, °C$
	$1 \, lb/ft^3 \, °F$	$2.8833 \times 10 \, kg/m^3 \, °C$
	$1 \, lb/ft^3 \, °C$	$1.6018 \times 10 \, kg/m^3 \, °C$
Density	$1 \, g/cm^3$	$1.0000 \times 10^3 \, kg/m^3$
	$1 \, lb/ft^3$	$1.6018 \times 10 \, kg/m^3$
	$1 \, lb/UK \, gal$	$9.9776 \times 10 \, kg/m^3$
	$1 \, lb/US \, gal$	$1.1983 \times 10^2 \, kg/m^3$
	$1 \, kg/ft^3$	$3.5315 \times 10 \, kg/m^3$
	$1 \, ton/vd^3$	$1.3289 \times 10^3 \, kg/m^3$
	$1 \, slug/ft^3$	$5.1538 \times 10^2 \, kg/m^3$
	$1 \, pdl/in.^3$	$8.6032 \times 10^2 \, kg/m^3$
	$1 \, dyn/cm^3$	$1.0197 \, kg/m^3$
Diffusion coefficient – *see* Viscosity, kinematic		
Energy	$1 \, cal$	$4.1868 \, J$
	$1 \, kcal$	$4.1868 \times 10^3 \, J$
	$1 \, Btu$	$1.0551 \times 10^3 \, J$
	$1 \, erg$	$1.0000 \times 10^{-7} \, J$
	$1 \, hp \, h \, (metric)$	$2.6477 \times 10^6 \, J$
	$1 \, kW \, h$	$3.6000 \times 10^6 \, J$
	$1 \, ft \, pdl$	$4.2139 \times 10^{-2} \, J$

	1 ft lbf	1.3558 J
	1 Chu	1.8991×10^3 J
	1 hp h (British)	2.6845×10^6 J
	1 therm	1.0551×10^8 J
	1 thermic	4.1858×10^6 J
	1 ft kgf	2.9891 J
	1 kgf m	9.8066 J
	1 dyn cm	1.0000×10^{-7} J
Force	1 dyn	1.0000×10^{-5} N
	1 kgf	9.8067 N
	1 pdl	1.3826×10^{-1} N
	1 lbf	4.4482 N
	1 tonf (UK)	9.9640×10^3 N
	1 tonf (US)	8.8964×10^3 N
	1 tonnef	9.8067×10^3 N

Heat − *see* Energy
Heat of combustion, formation, etc. − *see* Specific enthalpy
Heat capacity − *see* Specific heat
Heat flow − *see* Power

Heat flux	1 cal/s cm^2	4.1868×10^4 W/m^2
	1 kcal/h m^2	1.1630 W/m^2
	1 Btu/h ft^2	3.1546 W/m^2
	1 Chu/h ft^2	5.6784 W/m^2
	1 kcal/h ft^2	1.2518×10 W/m^2
	1 ft lb/min ft^2	2.4323×10^{-1} W/m^2
	1 hp/ft^2	7.9168×10^3 W/m^2
	1 W/in.^{2}	1.5500×10^3 W/m^2
Heat release rate (mass)	1 cal/s g	4.1868×10^3 W/kg
	1 kcal/h kg	1.1630 W/kg
	1 Btu/h lb	6.4611×10^{-1} W/kg
Heat release rate (volumetric)	1 cal/s cm^3	4.1868×10^6 W/m^3
	1 kcal/h m^3	1.1630 W/m^3
	1 Btu/h ft^3	1.0350×10 W/m^3
	1 Chu/h ft^3	1.8629×10 W/m^3
	1 kcal/h ft^3	4.1071×10 W/m^3
Heat transfer coefficient	1 cal/s cm^2 °C	4.1868×10^4 W/m^2 °C
	1 kcal/h m^2 °C	1.1630 W/m^2 °C
	1 Btu/h ft^2 °F	5.6783 W/m^2 °C
	1 Chu/h ft^2 °C	
	1 kcal/h ft^2 °C	1.2518×10 W/m^2 °C
Henry's law constant	1 atm/(g/cm^3)	1.0133×10^2 (N/m^2)/(kg/m^3)
	1 atm/(kg/m^3)	1.0133×10^5 (N/m^2)/(kg/m^3)
	1 atm/(Ib/ft^3)	6.3258×10^3 (N/m^2)/(kg/m^3)
	1 atm/(kg/ft^3)	2.8693×10^3 (N/m^2)/(kg/m^3)

Latent heat − *see* Specific enthalpy

Length	1 cm	1.0000×10^{-2} m
	1 ft	3.0480×10^{-1} m
	1 Ångstrom	1.0000×10^{-10} m
	1 micron	1.0000×10^{-6} m
	1 in.	2.5400×10^{-2} m
	1 yd	9.1440×10^{-1} m
	1 mile	1.6093×10^3 m
Mass	1 g	1.0000×10^{-3} kg
	1 lb	4.5359237×10^{-1} kg
	1 t (tonne)	1.0000×10^3 kg
	1 grain	6.4799×10^{-5} kg
	1 oz	2.8350×10^{-2} kg
	1 cwt	5.0802×10 kg
	1 ton (long)	1.0160×10^3 kg
	1 ton (short)	9.0719×10^2 kg
	1 dram	1.7718×10^{-3} kg

Mass per unit area	$1\,g/cm^2$	$1.0000 \times 10\,kg/m^2$
	$1\,dyn/cm^2$	$1.0197 \times 10^{-2}\,kg/m^2$
	$1\,lb/ft^2$	$4.8824\,kg/m^2$
	$1\,oz/ft^2$	$3.0515 \times 10^{-1}\,kg/m^2$
	$1\,lb/in.^2$	$7.0307 \times 10^2\,kg/m^2$
	$1\,lb/yd^2$	$5.4249 \times 10^{-1}\,kg/m^2$
	$1\,long$	$3.9230 \times 10^{-4}\,kg/m^2$
	$1\,kg/ft^2$	$1.0764 \times 10\,kg/m^2$
Mass flow	$1\,g/s$	$1.0000 \times 10^{-3}\,kg/s$
	$1\,kg/h$	$2.7778 \times 10^{-4}\,kg/s$
	$1\,lb/s$	$4.5359 \times 10^{-1}\,kg/s$
	$1\,t/h$	$2.7778 \times 10^{-1}\,kg/s$
	$1\,lb/h$	$1.2600 \times 10^{-4}\,kg/s$
	$1\,ton/h$	$2.8224 \times 10^{-1}\,kg/s$
Mass flux, mass velocity	$1\,g/s\,cm^2$	$1.0000 \times 10\,kg/s\,m^2$
	$1\,kg/h\,m^2$	$2.7778 \times 10^{-4}\,kg/s\,m^2$
	$1\,lb/s\,ft^2$	$4.8824\,kg/s\,m^2$
	$1\,lb/h\,ft^2$	$1.3562 \times 10^{-1}\,kg/s\,m^2$
	$1\,kg/h\,ft^2$	$2.9900 \times 10^{-3}\,kg/s\,m^2$
Mass release rate (volumetric)	$1\,g/s\,cm^3$	$1.0000 \times 10\,kg/s\,m^3$
	$1\,kg/h\,m^3$	$2.7778 \times 10^{-1}\,kg/s\,m^3$
	$1\,lb/s\,ft^3$	$1.6018 \times 10\,kg/s\,m^3$
	$1\,lb/h\,ft^3$	$4.4496 \times 10^{-3}\,kg/s\,m^3$
	$1\,kg/h\,ft^3$	$9.8096 \times 10^{-1}\,kg/s\,m^3$
Mass transfer coefficient, concentration driving force − *see* Velocity		
Mass transfer coefficient, dimensionless driving force − *see* Mass flux		
Mass transfer coefficient, pressure driving force	$1\,g/s\,cm^2\,atm$	$9.8687 \times 10^{-5}\,kg/s\,m^2\,(N/m^2)$
	$1\,kg/h\,m\,atm$	$2.7413 \times 10^{-9}\,kg/s\,m^2\,(N/m^2)$
	$1\,lb/h\,ft^2\,atm$	$1.3384 \times 10^{-8}\,kg/s\,m^2\,(N/m^2)$
	$1\,kg/h\,ft^2\,atm$	$2.9507 \times 10^{-8}\,kg/s\,m^2\,(N/m^2)$
Momentum, angular	$1\,g\,cm^2/s$	$1.0000 \times 10^{-7}\,kg\,m^2/s$
	$1\,lb\,ft^2/s$	$4.2140 \times 10^{-2}\,kg\,m^2/s$
	$1\,lb\,ft^2/h$	$1.1706 \times 10^{-5}\,kg\,m^2/s$
Momentum, linear	$1\,g\,cm/s$	$1.0000 \times 10^{-5}\,kg\,m/s$
	$1\,lb\,ft/s$	$1.3826 \times 10^{-1}\,kg\,m/s$
	$1\,lb\,ft/h$	$3.8404 \times 10^{-5}\,kg\,m/s$
Moment of inertia	$1\,g\,cm^2$	$1.0000 \times 10^{-7}\,kg\,m^2$
	$1\,lb\,ft^2$	$4.2140 \times 10^{-2}\,kg\,m^2$
	$1\,oz\,in.^2$	$1.8290 \times 10^{-5}\,kg\,m^2$
	$1\,lb\,yd^2$	$3.7926 \times 10^{-1}\,kg\,m^2$
Power	$1\,cal/s$	$4.1868\,W$
	$1\,kcal/h$	$1.1630\,W$
	$1\,Btu/s$	$1.0551 \times 10^3\,W$
	$1\,erg/s$	$1.0000 \times 10^{-7}\,W$
	$1\,tonne\,cal/h$	$1.1630 \times 10^3\,W$
	$1\,hp\,(metric)$	$7.3550 \times 10^2\,W$
	$1\,ft\,pdl/s$	$4.2140 \times 10^{-2}\,W$
	$1\,ft\,lbf/s$	$1.3558\,W$
	$1\,Btu/h$	$2.9307 \times 10^{-1}\,W$
	$1\,Chu/h$	$5.2754 \times 10^{-1}\,W$
	$1\,hp\,(British)$	$7.4570 \times 10^2\,W$
	$1\,ton\,refrigeration$	$3.5169 \times 10^3\,W$
	$1\,J/h$	$2.7778 \times 10^{-4}\,W$
	$1\,cm^3\,atm/h$	$2.8146 \times 10^{-5}\,W$
Pressure	$1\,dyn/cm^2$	$1.0000 \times 10^{-1}\,N/m^2$
	$1\,kgf/m^2$	$9.8067\,N/m^2$
	$1\,pdl/ft^2$	$1.4882\,N/m^2$

	1 atm (standard)	1.0133×10^5 N/m^2
	1 at (1 kgf/cm^2)	9.8067×10^4 N/m^2
	1 bar	1.0000×10^5 N/m^2
	1 lbf/ft^2	4.7880×10 N/m^2
	1 psi (lbf/in.^{2})	6.8948×10^3 N/m^2
	1 tonf/in.^{2}	1.5444×10^7 N/m^2
	1 in. water	2.4909×10^2 N/m^2
	1 ft water	2.9891×10^3 N/m^2
	1 mmHg	1.3332×10^2 N/m^2
	1 inHg	3.3864×10^3 N/m^2
	1 torr	1.3332×10^2 N/m^2
	1 ksi	6.8948×10^6 N/m^2
	1 oz/ft^2	2.9926 N/m^2
Reaction rate − *see* Mass release rate		
Shear stress − *see* Pressure		
Specific enthalpy	1 cal/g	4.1868×10^3 J/kg
	1 Btu/lb	2.3260×10^3 J/kg
	1 Chu/lb	4.1868×10^3 J/kg
	1 kWh/lb	7.9366×10^6 J/kg
	1 ft lbf/lb	2.9891 J/kg
Specific heat	1 cal/g °C	4.1868×10^3 J/kg °C
	1 Btu/lb °F	4.1868×10^3 J/kg °C
	1 ft lbf/lb °F	5.3803 J/kg °C
	1 m kgf/kg °C	9.8067 J/kg °C
Specific volume	1 cm^3/g	1.0000×10^{-3} m^3/kg
	1 ft^3/lb	6.2428×10^{-2} m^3/kg
	1 ft^3/kg	2.8317×10^{-2} m^3/kg
Surface per unit mass	1 cm^2/g	1.0000×10^{-1} m^2/kg
	1 ft^2/lb	2.0482×10^{-1} m^2/kg
	1 m^2/g	1.0000×10^3 m^2/kg
	1 ft^2/kg	9.2903×10^{-2} m^2/kg
	1 yd^2/oz	2.9494×10 m^2/kg
	1 ft^2/oz	3.2771 m^2/kg
	1 mm^2/mg	1.0000 m^2/kg
Surface per unit volume	1 cm^2/cm^3	1.0000×10^2 m^2/m^3
	1 ft^2/ft^3	3.2808 m^2/m^3
	1 ft^2/gal	2.0436×10 m^2/m^3
	1 yd^2/gal	1.8392×10^2 m^2/m^3
Surface tension	1 dyn/cm	1.0000×10^{-3} N/m
	1 erg/cm	1.0000×10^{-3} N/m
	1 mg/in.	3.8609×10^{-4} N/m
	1 kgf/m	9.8070 N/m
	1 pdl/in.	5.4431 N/m
Temperature difference	1 deg F (deg R)	5/9 deg C (deg K)
Thermal conductivity	1 cal/s cm^2	4.1868×10^2 W/m^2 (°C/m)
	1 kcal/h m^2	1.1630 W/m^2 (°C/m)
	1 Btu/h ft^2	1.7307 W/m^2 (°C/m)
	1 Btu/h ft^2	1.4423×10^{-1} W/m^2 (°C/m)
	1 kcal/h ft^2	3.8156 W/m^2 (°C/m)
Time	1 h	3.6000×10^3 s
	1 min	6.0000×10 s
	1 d (day)	8.6400×10^4 s
	1 yr	3.1558×10^7 s
Torque − *see* Energy		
Velocity	1 cm/s	1.0000×10^{-2} m/s
	1 m/h	2.7778×10^{-4} m/s
	1 ft/s	3.0480×10^{-1} m/s
	1 ft/h	8.4667×10^{-5} m/s
	1 mile/h	4.4704×10^{-1} m/s
	1 kn	5.1444×10^{-1} m/s

Viscosity, absolute (or dynamic)	1 g/cm s (1 poise)	1.0000×10^{-1} kg/m s
	1 kg/m h	2.7778×10^{-4} kg/m s
	1 lb/ft s	1.4882 kg/m s
	1 lb/ft h	4.1338×10^{-4} kg/m s
	1 kg/ft h	9.1134×10^{-4} kg/m s
	1 lbf s/in.	6.8948×10^{3} kg/m s
	1 slug/ft s	4.7880×10 kg/m s
Viscosity, kinematic	1 cm²/s (1 stokes)	1.0000×10^{-4} m²/s
	1 m²/h	2.7778×10^{-4} m²/s
	1 ft²/s	9.2903×10^{-2} m²/s
	1 ft²/h	2.5806×10^{-5} m²/s
Volume	1 cm³	1.0000×10^{-6} m³
	1 ft³	2.8317×10^{-2} m³
	1 l (litre)	1.0000×10^{-3} m³
	1 in.³	1.6387×10^{-5} m³
	1 yd³	7.6456×10^{-1} m³
	1 UK gal	4.5461×10^{-3} m³
	1 US gal	3.7854×10^{-3} m³
	1 UK barrel	1.6366×10^{-1} m³
	1 US barrel	1.1924×10^{-1} m³
	1 pt	4.7318×10^{-4} m³
Volumetric flow	1 cm³/s	1.0000×10^{-6} m³/s
	1 m³/h	2.7778×10^{-4} m³/s
	1 ft³/s	2.8317×10^{-2} m³/s
	1 cm³/min	1.6667×10^{-8} m³/s
	1 l/min	1.6667×10^{-5} m³/s
	1 ft³/min	4.7195×10^{-4} m³/s
	1 ft³/h	7.8658×10^{-6} m³/s
	1 yd³/min	1.2743×10^{-2} m³/s
	1 UK gal/min	7.5768×10^{-5} m³/s
	1 US gal/min	6.3090×10^{-5} m³/s
	1 UK gal/h	1.2628×10^{-6} m³/s
	1 US gal/h	1.0515×10^{-6} m³/s

Wetting rate (mass) – *see* Velocity, absolute

| Wetting rate (volumetric) | 1 l/h in. | 1.0936×10^{-5} m³/s m |
| | – *see also* Viscosity, kinematic | |

Work – *see* Energy

Appendix

31

Process Safety Management (PSM) Regulation in the United States

Contents

The history of safety regulations in the United States can be traced back to the year before the beginning of the twentieth century. The River and Harbor Act, the first known federal legislation in the United States relevant to safety was promulgated in 1899. This Act prohibited the creation of any obstruction not authorized by the US Congress, to the navigable capacity of any waters of the United States except on plans authorized by the Secretary of the Army. The Act was promulgated expressly to protect the nation's waterways from excessive dumping. Subsequent to the River and Harbor Act, the US Congress has passed numerous laws, which impose environmental or safety regulations on businesses. Since then, the total number of legislations has steadily increased. In addition to the federal government, local entities such as the state, county, and cities have also promulgated regulations and ordnances, which impose safety requirements on process facilities.

In 1936, the United States federal government enacted the Walsh–Healy Act to establish federal safety and health standards for activities relating to federal contracts. The Walsh–Healy Act led to early research into the identification and control of occupational diseases. The ideas behind this Act are the basis of many of today's occupational health and safety regulations.

In 1970, Congress promulgated the Occupational Safety and Health Act (OSHA). As a result of this landmark legislation, OSHA and the National Institute for Occupational Safety and Health (NIOSH) were established within the Department of Labor and the Department of Health and Human Services respectively. OSHA's mission is to, 'Assure so far as possible every working man and woman in the nation safe and healthful working conditions'. The OSHA Act allows OSHA to set and enforce standards that require employers to maintain safe and healthful workplaces. NIOSH, on the other hand, does not have any regulatory or enforcement authority, but is charged with the responsibility of training of professionals and with the research and recommendation of new regulations to the Secretary of Labor.

In 1990, EPA analysed chemical incidents in the early to mid-1980s and compared them to the Bhopal incident. The analysis concluded that 17 incidents released sufficient volumes of chemicals that could have been more severe than Bhopal if the weather conditions and plant location were different. Thus, the Clean Air Act Amendments of 1990 contained specific mandates requiring OSHA and EPA to establish regulations to protect workplace employees and the public and the environment, respectively. OSHA fulfilled its mandate in 1992 by promulgating the process safety management regulation.

Although NIOSH and OSHA were created by the same Act of Congress, they are two distinct agencies with separate responsibilities. OSHA is in the Department of Labor and is responsible for creating and enforcing workplace safety and health regulations. NIOSH is in the Department of Health and Human Services and is a research agency. NIOSH identifies the causes of work-related diseases and injuries and the potential hazards of new work technologies and practices. With this information, NIOSH determines new and effective ways to protect workers from chemicals, machinery, and hazardous working conditions. Creating new ways to prevent workplace hazards is the job of NIOSH.

A31.1 The Process Safety Management Programme

The fourteen elements of the OSHA Process Safety Management (PSM) regulation (*29 CFR 1910.119*) were published in the federal register on 24 February 1992. The objective of the regulation is to prevent or minimize the consequences of catastrophic releases of toxic, reactive, flammable, or explosive chemicals. The regulation requires a comprehensive management programme: a holistic approach that integrates technologies, procedures, and management practices.

The process safety management regulation applies to processes that involve certain specified chemicals at or above threshold quantities, processes that involve flammable liquids or gases on-site in one location, in quantities of 10,000 lb or more (subject to few exceptions), and processes which involve the manufacture of explosives and pyrotechnics. Hydrocarbon fuels, which may be excluded if used solely as a fuel, are included if the fuel is part of a process covered by this regulation. In addition, the regulation does not apply to retail facilities, oil or gas well drilling or servicing operations, or normally unoccupied remote facilities.

The management system required by OSHA's process safety management regulation envisions a holistic program with checks and balances aimed at creating multiple barriers of protection. The performance-based approach does not prescribe specific methods and approaches, thus giving facilities the flexibility for customizing the methods to best meet their needs and organizational structures. The process hazard analysis (PHA) is the heart of the programme and impacts or interfaces with all of the other elements. However, it must also be pointed out that all elements of the programme must be implemented in their entirety to get the maximum benefit and accomplish the ultimate objective, that is, reduce the frequency and severity of chemical plant accidents. Each element of the process safety programme is discussed in more detail here.

A31.1.1 Employee participation
This element of the regulation requires developing a written plan of action regarding employee participation; consulting with employees and their representatives on the conduct and development of other elements of process safety management required under the regulation; and providing to employees and their representatives access to process hazard analyses and to all other information required to be developed under this regulation.

A31.1.2 Process safety information
This element of the PSM regulation requires employers to develop and maintain important information about the different processes involved. This information is intended to provide a foundation for identifying and understanding potential hazards involved in the process.

The process safety information covers three different areas, that is, chemicals, technology and equipment. A complete listing of the process safety information that must be compiled in these three areas is shown in Table A31.1. This information is intended to provide a foundation for identifying and understanding potential hazards involved in the process.

The information in Table A31.1 is essential for developing and implementing an effective process safety management programme. The fundamental concept is that complete, accurate, and up-to-date process knowledge is essential for safe and profitable operations. The information contained in the first column of Table A31.1 should be available from the Material Safety Data Sheets (MSDS) for the hazardous chemicals, which are used as primary or

Table A31.1   *Process safety information*

Chemicals	Technology	Equipment
Toxicity	Block flow diagram or	Design codes employed
Permissible exposure limit	process flow diagram	Materials of construction
Physical data	Process chemistry	Piping and instrumentation
Reactivity data	Maximum intended inventory	diagrams
Thermal and chemical	Safe limits for process parameters	Electrical classification
stability data	Consequence of deviations	Ventilation system design
Effects of mixing		Material and energy balances
		Safety systems
		Relief system design
		and design basis

intermediate feedstocks or are produced as products at the plant.

The information contained in the second column of Table A31.1 pertains to the technology of the process itself. The block flow diagram can be replaced by a process flow diagram. The process chemistry information must contain the basic chemical reactions involved and a brief description of the chemistry involved. The maximum intended inventory refers to the maximum amount of any regulated chemical that may be expected to be present in the whole facility at any time. The safe limits for process parameters refer to the upper and lower bounds for the process parameters outside of which the process would be hazardous. For example, in the case of a distillation process, the upper and lower limits of the process parameters outside which the operation of the process could cause significant damage to the tower or other attached equipment would have to be stated. In this example, the process parameters for which upper and lower bounds are to be specified are temperature, pressure, composition, and flow rate. The consequence of deviation from these stated bounds must also be compiled.

Safe upper and lower limits for process parameters and equipment are also necessary for calibrating instrumentation. It is important to understand the distinction between process parameter limits and equipment limits. Safe upper and lower limits for process parameters can be defined as follows:

(1) For non-reactive processes, process parameter safe limits are defined based on equipment design ratings, relief device setpoints, and upstream and downstream conditions of the process and/or equipment.
(2) For reactive processes, process parameter safe limits are defined based on the process chemistry information or any restricted physical conditions for the reaction as well as the criteria used for non-reactive processes.

In contrast, equipment limits are specified by the manufacturer based on materials of construction and design basis. Processes must be operated and maintained within both the process parameter and equipment limits.

The final type of information that must be compiled pertains to the equipment used in the process. The intent is to assure that all equipment used in the process meets appropriate standards and codes such as those published by the American Society of Mechanical Engineers (ASME), the American Petroleum Institute (API), the American Institute of Chemical Engineers (AIChE), the American National Standards Institute (ANSI), the American Society of Testing

and Materials (ASTM), and the National Fire Protection Association (NFPA). Accepted industry practices can be used to decide which standards apply. In cases where standards do not exist, generally acceptable engineering practices can be used. The materials of construction for each equipment item and the design codes employed can be compiled as separate lists or may be listed in the Piping and Instrumentation Diagram (P&ID). The P&IDs must represent the facility exactly as it exists with flanges, valves, and all other connections shown. The different electrical classifications must also be compiled for different parts of the facility. A simple plot plan showing the different areas of electrical classification would be considered to be in compliance with the regulation. Information must also be compiled on any ventilation system. This information would indicate the areas in the facility that are ventilated and the nature of ventilation. A listing of all safety systems must also be compiled that are available to the workers. This listing should include any and all equipment that is available for protection of the workers from any hazard or emergency. Information should also be available on the location of these safety systems and the procedures to use these systems.

As is apparent from the foregoing discussion, compilation of the process safety information database represents a major challenge. This is complicated even more because the process safety information must also be kept up-to-date and accurate and made accessible to employees. Many plants have therefore implemented or are in the process of implementing electronic data management systems to manage, access, and use these data. The completion and accuracy of process safety information is crucial to the implementation of other PSM elements including PHAs and mechanical integrity.

The PSM regulation requires that process equipment should comply with generally accepted engineering practices. It is therefore not only important to compile all equipment information but also to ensure that it complies with consensus standards. OSHA is recognizing that there are consensus standards for design and fabrication, installation, maintenance procedures, and inspection and testing. Therefore, equipment constructed in accordance with codes, standards, or practices no longer in use should be evaluated to ensure that they are compatible with existing standards. For example, a multi-layered vessel built years ago should be evaluated to ensure that it complies with today's engineering standards.

A31.1.3 Process hazards analysis
This element of the PSM regulation requires facilities to perform a PHA. The PHA must address the hazards of the

process, previous hazardous incidents, engineering and administrative controls, the consequences of the failure of engineering and administrative controls, human factors, and an evaluation of effects of failure of controls on employees. This element requires that the PHA be performed by one or more of the following methods or any other equivalent method:

(1) what-if;
(2) checklist;
(3) what-if/checklist;
(4) hazard and operability (HAZOP) studies;
(5) failure modes and effects analysis (FMEA);
(6) fault tree analysis.

The regulation suggests a performance-oriented requirement with respect to the PHA so that the facility has the flexibility to choose the type of analysis that will best address a particular process. PHAs may not be performed unless complete Process Safety Information is available for the process.

PHAs and the results of PHAs can and will impact the development, implementation, and practice of other elements of the PSM regulation. A PHA would help facilities identify hazards and ways to address them. PHAs also identify situations where process incidents due to control failure (e.g. pressure gauges, overfill alarms) could be prevented by redundant or backup control or by frequent maintenance and inspection practices.

Many other elements of the PSM program should flow from, or at least be revised based on, the results of the PHA. Existing standard operating procedures, training and maintenance programs, and pre-startup safety reviews may need to be revised to reflect changes in either practices or equipment that derive from the PHA. The PHA may help define critical equipment that require preventive maintenance, inspection, and testing programs. It may also help a facility focus its emergency response programs on the most likely and most serious release scenarios.

The PSM regulation also requires that the PHAs be updated and revalidated, based on their completion date. First, the PSM regulation requires that the PHAs shall be updated and revalidated. Thus, the whole PHA should not only be examined to verify that the PHA is consistent with the current process but it also should be analyzed to examine if the original analysis is valid. Second, the PSM regulation requires that the updating and revalidating be done by a team similar in qualifications to the team that conducted the original PHA. The intent of a PHA revalidation is to ensure that the PHA team evaluates the previous PHA, examines the extent of any changes that might have occurred since the PHA was implemented (or last reviewed) and decide what work is needed to make the PHA current.

A31.1.4 Operating procedures
The operating procedures must be in writing and provide clear instructions for safely operating processes; must include steps for each operating phase, operating limits, safety and health considerations and safety systems. Procedures must be readily accessible to employees, must be reviewed as often as necessary to assure they are up-to-date and must cover special circumstances such as lockout/tagout and confined space entry. The employer must certify annually that the operating procedures are current and accurate. The synergism and commonality of operating procedures to maintenance procedures is in safe

work practices. These safe work practices include lockout/tagout, confined space entry, opening of process equipment or piping (i.e. hot tapping), and control of entrance into the battery limits. Even though the operations department is involved with all of these practices, maintenance also plays a very vital role in ensuring that these procedures are followed during all maintenance tasks. Many incidents have resulted from inadequate safe work practices or a failure to follow procedures when they exist.

A31.1.5 Training
The regulation requires that facilities certify that employees responsible for operating the facility have successfully completed (including means to verify understanding) the required training. The training must cover specific safety and health hazards, emergency operations and safe work practices. Initial training must occur before assignment. Refresher training must be provided at least every 3 years.

Even though this element of the PSM regulation pertains to operations staff only, it is important to remember that operations and maintenance training should be coordinated.

A31.1.6 Contractors
The PSM regulation identifies responsibilities of the employer regarding contractors involved in maintenance, repair, turnaround, major renovation or specialty work, on or near covered processes. The host employer is required to: consider safety records in selecting contractors; inform contractors of potential process hazards; explain the facility's emergency action plan; develop safe work practices for contractors in process areas; periodically evaluate contractor safety performance; and maintain an injury/illness log for contractors working in process areas. In addition, the contract employer is required to train their employees in safe work practices and document that training, assure that employees know about potential process hazards and the host employer's emergency action plan, assure that employees follow safety rules of facility and advise host employer of hazards contract work itself poses or hazards identified by contract employees.

A31.1.7 Pre-startup safety review
This element of the PSM regulation requires a pre-startup safety review of all new and modified facilities to confirm integrity of equipment; to assure that appropriate safety, operating, maintenance and emergency procedures are in place; and to verify that a PHA has been performed. Modified facilities for this purpose are defined as those for which the modification required a change in the process safety information.

Usually, changes occur during maintenance and therefore, maintenance personnel should be well versed in pre-startup safety review procedures. Maintenance should ensure that all necessary procedures have been completed prior to start-up.

A31.1.8 Mechanical integrity
This element of the PSM regulation mandates written procedures, training for process maintenance employees and inspection and testing for process equipment including pressure vessels and storage tanks; piping systems; relief and vent systems and devices; emergency shutdown systems; pumps; and controls such as monitoring devices, sensors, alarms and interlocks. The PSM regulation calls for correction of equipment deficiencies and assurance that

new equipment and maintenance materials and spare parts are suitable for the process and properly installed.

A31.1.9 Hot work permit
This element of the PSM regulation mandates a permit system for hot work operations conducted on or near a covered process. The purpose of this element of the regulation is to assure that the employer is aware of the hot work being performed, and that appropriate safety precautions have been taken prior to beginning the work. Since welding shops authorized by the employer are locations specifically designated and suited for hot work operations, the regulation does not require a permit for hot work in these locations. Additionally, hot work permits are not required in cases where the employer or an individual to whom the employer has assigned the authority to grant hot work permits is present while the hot work is performed.

A31.1.10 Management of change
This element of the regulation specifies a written programme to manage changes in chemicals, technology, equipment and procedures which addresses the technical basis for the change, impact of the change on safety and health, modification to operating procedures, time period necessary for the change, and authorization requirements for the change. The regulation requires employers to notify and train affected employees and update process safety information and operating procedures as necessary.

A31.1.11 Incident investigation
This element of the regulation requires employers to investigate as soon as possible (but no later than 48 h) incidents, which did result or could have resulted in catastrophic releases of covered chemicals. The regulation calls for an investigation team, including at least one person knowledgeable in the process (a contractor employee, if

appropriate), to develop a written report of the incident. Employers must address and document their response to report findings and recommendations and review findings with affected employees and contractor employees. Reports must be retained for 5 years.

A31.1.12 Emergency planning and response
This element requires employers to develop and implement an emergency action plan according to the requirements of *29 CFR 1910.38(a)* and *29 CFR 1910.120(a), (p), and (q).*

A31.1.13 Compliance audits
This element of the regulation requires employers to certify that they have evaluated compliance with process safety requirements every 3 years and specifies retention of the audit report findings and the employer's response. Employer must retain the two most recent audits.

A31.1.14 Trade secrets
Similar to the trade secret provisions of the hazard communication regulation, the PSM regulation also requires information to be available to employees from the process hazard analyses and other documents required by the regulation. The regulation permits employers to enter into confidentiality agreements to prevent disclosure of trade secrets.

A31.2 Summary Comparison of OSHA Elements with CCPS Elements

It is important to point out that the programmes required by OSHA's PSM elements as given in 29 CFR 1910.119 are similar to the requirements of other programme requirements. The comparison matrix in Table A31.2 demonstrates the relevance of OSHA PSM elements to other programmes.

Table A31.2   *Comparison matrix showing the relevance of OSHA PSM elements to other programs*

Program element	OSHA PSM[1] element	CAAA Sec. 304[2] Par. C Item(s)	AIChE CCPS[3] Guidelines	ACC PSC[4] Practice(s)
Employee Participation	c	[5]	1	[5]
Process Safety Information	d	1	2,10	7,8
Process Hazards Analysis	e	2,3,4	4,6,7	9
Operating Procedures	f	6	6,7,8	7,18
Training	g	7	8	14,17,19,20
Pre-startup Safety Review	i	12	3,4,8	13
Mechanical Integrity	j	10,11	3,6,10	2,12,14,18
Management of Change	l	14	5	10,11
Incident Investigation	m	13	9	4,5
Audits	o	5	11	3
Safe Work Practices	[5], k	[5]	8	18
Emergency Planning/Control/Response	n	9	8	5,6,11,16
Contractors	h	8	–	12,22
Employee Fitness for Duty	–	–	–	21
Multiple Safeguards	[5]	[5]	–	15
Number of Chemicals	140	100 min.	–	–

[1] OSHA PSM: Occupational Safety and Health Administration Process Safety Management.
[2] CAAA: Clean Air Act Amendments.
[3] AIChE CCPS: American Institute of Chemical Engineers Center for Chemical Process Safety.
[4] ACC PSC: American Chemistry Council Process Safety Code.
[5] Either stated or inherent in other elements.

Appendix

32

Risk Management Program Regulation in the United States

Contents

Environmental issues affecting the public health and the environment also received widespread attention. As a result, the Environmental Protection Agency (EPA) was established in 1970 to protect the nation's public health and environment. The EPA is responsible, 'to find ways to cleanup and prevent pollution, ensure compliance and enforcement of environmental laws, assisting states in environmental protection efforts, and scientific research and education to advance the nation's understanding of environmental issues'. In 1970, the EPA promulgated the Clean Air Act, followed by amendments to the Act in 1977 and 1990.

The Toxic Substances Control Act (TSCA), passed in 1976 gave EPA the ability to track and study the 75,000 industrial chemicals produced or imported to the United States. The TSCA is a federally enforced law and is not delegated to the state. Under this Act, the EPA has the authority to ban the manufacture or distribution in commerce, limit the use, require labelling, or place other restrictions on chemicals that pose unreasonable risk. Asbestos, chlorofluorocarbons, and ploychlorinated biphenyls are some of the chemicals regulated by EPA under TSCA.

In 1977, the International Safe Container Act established uniform structural requirements for international cargo containers designed to be transported interchangeably by sea and land carriers. In 1983, the Surface Transportation Assistance Act established protection from reprisal by employers for truckers and certain other employees in the trucking industry involved in activity related to interstate commercial motor vehicle safety and health.

In 1990, EPA analysed chemical incidents in the early to mid-1980s and compared them to the Bhopal incident. The analysis concluded that 17 incidents released sufficient volumes of chemicals that could have been more severe than Bhopal if the weather conditions and plant location were different. Thus, the Clean Air Act Amendments of 1990 contained specific mandates requiring OSHA and EPA to establish regulations to protect workplace employees and the public and the environment respectively. OSHA fulfilled its mandate in 1992 by promulgating the process safety management regulation. EPA on the other hand promulgated the Risk Management Program regulation in 1996. The Clean Air Act Amendments of 1990 also established the Chemical Safety and Hazard Investigation Board.

A32.1 The Risk Management Program

In 1996, EPA promulgated the regulation for *Risk Management Programs for Chemical Accident Release Prevention* (40 CFR 68). This federal regulation was mandated by section 112(r) of the Clean Air Act Amendments of 1990. The regulation requires regulated facilities to develop and implement appropriate Risk Management Program to minimize the frequency and severity of chemical plant accidents. In keeping with regulatory trends, EPA required a performance-based approach towards compliance with the Risk Management Program regulation. The eligibility criteria and requirements for the three different program levels are given in Table A32.1.

The EPA regulation also requires regulated facilities to develop a Risk Management Plan (RMP). The RMP includes a description of the hazard assessment, prevention programme, and the emergency response programme. Facilities submit the RMP to the EPA and subsequently is made available to governmental agencies, the state

emergency response commission, the local emergency planning committees, and communicated to the public.

The Risk Management Program regulation defines the worst-case release as the release of the largest quantity of a regulated substance from a vessel or process line failure, including administrative controls and passive mitigation that limit the total quantity involved or release rate. For gases, the worst-case release scenario assumes the quantity is released in 10 min. For liquids, the scenario assumes an instantaneous spill and that the release rate to the air is the volatilization rate from a pool 1 cm deep unless passive mitigation systems contain the substance in a smaller area. For flammables, the scenario assumes an instantaneous release and a vapour cloud explosion using a 10% yield factor. For alternative scenarios (note: EPA used the term *alternative scenario* as compared to the term *more-likely scenario* used earlier in the proposed regulation), facilities may take credit for both passive and active mitigation systems.

Appendix A of the final regulation lists endpoints for toxic substances to be used in worst-case and alternative scenario assessment. The toxic endpoints are based on ERPG-2 (Emergency Response Planning Guidelines – Level 2) or level of concern data compiled by EPA. The flammable endpoints represent vapour cloud explosion distances based on overpressure of 1 psi or radiant heat distances based on exposure to 5 kW/m^2 for 40 seconds.

Similar to OSHA's PSM regulation, coverage under the the EPA risk management program regulation is determined by a list of chemicals with associated threshold quantities. However, in contrast to the OSHA PSM regulation, the EPA Risk Management Program regulation prescribes a tiered programme based on the risk and incident history of the facility. In order to determine coverage and compliance requirements, a facility may follow the steps described below in chronological order:

(1) Check the EPA Risk Management Program list of chemicals to determine if any of the chemicals are present on site above the listed threshold quantity at any time. The threshold quantity determination is not cumulative, but the maximum that could be present on site at any given time. Other considerations that may enter into the threshold quantity determination are proximity, interconnectivity, concentration, etc. EPA also provides rules for mixtures and effect of dilution. If threshold quantity is not exceeded, then the facility is not subject to the EPA risk management program, and steps 2–4 need not be completed.

(2) If any of the listed chemicals are present at or above the listed threshold quantity, the facility must perform a worst-case scenario analysis. If the impact zone from worst-case scenario does not contain any public receptors, the facility has had no incidents in the last 5 years, and the facility's emergency response plan is coordinated with local emergency response officials; then the facility belongs in Program 1.

(3) If the facility can not claim coverage under Program 1 and the facility is classified in any of the industrial classification codes listed in Table A32.1 or the facility is covered by the OSHA PSM regulation, then the facility belongs in Program 3 and must comply with Program 3 requirements.

(4) If the facility does not belong in Program 1 or Program 3, then by default the facility is required to comply with Program 2 requirements.

Table A32.1 *Eligibility criteria and compliance requirements for different program levels – EPA's Risk Management Program regulation*

Program 1	Program 2	Program 3
Program Eligibility Criteria		
No offsite accident history No public receptors in worst-case circle Emergency response coordinated with local responders	Process not eligible for program 1 or 3	Process is subject to OSHA PSM (29 CFR 1910.119) Process is SIC code 2611, 2812, 2819, 2821, 2865, 2869, 2873, 2879, or 2911
Program Requirements		
Hazard assessment Worst-case analysis 5-year accident history Certify no additional steps needed	*Hazard assessment* Worst-case analysis Alternative releases 5-year accident history	*Hazard assessment* Worst-case analysis Alternative releases 5-year accident history
	Management program Document management system	*Management program* Document management system
	Prevention Program Safety information Hazard review Operating procedures Training Maintenance Incident investigation Compliance audit	*Prevention Program* Process safety information Process hazard analysis Operating procedures Training Mechanical integrity Incident investigation Compliance audit Management of change Pre-startup safety review Contractors Employee participation Hot work permits
	Emergency response program Develop plan and program	*Emergency response program* Develop plan and program

As stated earlier, EPA used a tiered approach based on risk and the facility's incident history. It is a recognition of the fact that one size does not fit all and the risk management program is based on a three-pronged approach. First, the hazard assessment provides information about potential off-site consequences and the incident history of the facility. The prevention programme and the emergency response programme are then developed based on the hazard assessments. It should be noted that the prevention programme requirements under Program 3 mirror the OSHA PSM programme requirements.

Appendix 33

Incident Databases

Contents

Accident prevention and mitigation of consequences is the focus of a number of industry programmes, regulatory initiatives. As part of these programmes and regulations, accident history data is often collected. There are two basic types of information. One is a database consisting of standardized fields of data usually for a large number of incidents. The second are more detailed reports of individual incidents.

Analysis of these accident history databases can provide insight into accident prevention needs. While the analysis and conclusions obtained from the accident database is often limited by the shortcomings of the databases themselves, the fact remains that accident history databases are very useful and can be a powerful tool in focusing risk reduction efforts. The conclusions can be used to systematically identify the greatest risks to allow prioritization of efforts to improve process safety. At the plant level this might entail identifying certain processes, types of equipment, chemicals, operations and other factors most commonly associated with incidents. Databases that cover a very large number of facilities are likely to reveal trends and patterns that no one company or facility could determine from their own experience.

Statistical knowledge of the likelihood of the release of certain types of chemicals could help emergency responders, state emergency response commissions, and local emergency planning committees determine the most likely and most serious chemical releases in their areas and plan appropriate chemical accident responses.

Incident databases may also help identify technologies and practices to prevent chemical accidents, or the need to develop them. For example, the data could indicate that inspection and preventive maintenance of equipment and instruments should become more thorough or more frequent.

The Mary Kay O'Connor Process Safety Center (MKOPSC) provides an assessment of the usefulness of US Federal databases in *Feasibility of Using Federal Incident Databases to Measure and Improve Chemical Safety*. This study identifies a number of trends and patterns of chemical incidents. It also makes a number of recommendations for improving these databases. The report defines an incident as, "An incident is the sudden unintended release of or exposure to a hazardous substance, which results in or might reasonably have resulted in, deaths, injuries, significant property or environmental damage, evacuation or sheltering in place." This definition will be used as a reference to describe the type of incidents reported in the following section.

A33.1 Incident Databases

A33.1.1 National Response Center (NRC) Incident Reporting Information System (IRIS)

The NRC is the primary initial notification system for the US federal government. It is operated 24 h a day, 7 days a week. It receives the notification of a release and forwards that information to numerous federal, state and local agencies. Approximately 25,000 incidents are reported each year. However, the majority of these incidents involve very small or unknown quantities. While reporting to the NRC is required by a number of statutes, it also receives numerous 'complaints' from the public. Many of the incidents are spills of fuel from motor vehicles, small spills of low hazard materials such as lubricating oil and oil sheens

on water with unknown sources, quantities and effects. Only about 30% of the reported cases constitute an incident as defined in the MKOPSC report.

Because this system contains initial reports the information is preliminary and many times inaccurate or incomplete. There is also duplicate reporting of incidents. The completeness and accuracy of chemical names depends on the knowledge level of the person reporting the incident. It has been found that those reports from chemical manufacturing facilities are more accurate and do provide useful information. For a limited number of incidents, NRC also presents a limited number of detailed incident summaries.

A33.1.2 EPA Risk Management Plan (RMP) – 5-year accident history

The US EPA as part of its Risk Management Program requires certain facilities to report their 5-year accident history. This requirement applies to releases of a listed substance that is stored above a threshold quantity and results in fatalities, injuries, or significant environmental or property damage. The initial 5 years covered the period from mid-1993 to 1998.

Only 1900 releases are reported from about 14,500 facilities for the 5-year period. The combination of the three main reporting criteria appears to exclude a large number of incidents. The reports do address such items as the causes and consequences of the release and steps taken to prevent or mitigate future incidents. Since the reporting is from a defined universe of facilities it allows statistical treatment of the frequency of releases not available from systems that report releases from an undefined universe of facilities.

A33.1.3 Agency for Toxic Substances Disease Registry (ATSDR) – Hazardous Substances Emergency Events Surveillance (HSEES)

The ATSDR administers a programme with approximately 15 states to collect information regarding chemical releases and their health effects on workers, emergency responders and the public. This database now contains over 50,000 events and adds more than 5000 per year. This is probably the largest high quality dataset in existence. Releases from both fixed facilities and transportation systems are included. The reporting excludes releases of petroleum products unless they contain other hazardous materials. This system focuses on the health effects that are gathered from numerous sources such as public records, hospitals, media and interviews. The database is expected to be available to the general public by the time this handbook is published. The agency issues annual reports summarizing the results. These reports generally exclude some details such as the quantity released that would be helpful in the analysis of incidents. However, such information will be available when the database is released.

A33.1.4 EPA Accidental Release Information Program (ARIP)

This discontinued program contains nearly 5,000 incidents collected primarily in a 3-year period in the late 1980s. Prior to the establishment of other programmes in the 1990s this was one of the best sources of data. While there was no strict criteria for reporting to this database it is one of the larger collections of incidents with details concerning, causes, consequences, operating mode, and corrective actions. Facilities reported in response to a request from the

EPA. Requests were sent to facilities that reported frequent or significant incidents. While no trends can be gathered from this system it does provide detailed information not available in other systems. The database generally excludes releases of simple hydrocarbons.

A33.1.5 Accident Investigation System (AIS) –
Occupational Safety and Health Administration,
US Department of Labor

This system contains reports from all OSHA-regulated facilities with accidents resulting in a fatality or three or more injuries. The most useful feature of the system is that it includes a brief description of how the incident occurred. Most systems lack this feature making it very difficult to determine what actually occurred during an incident. Also, there are no arbitrary limits on the type of chemical involved or its quantity, only on the consequences. This database contains many accidents unrelated to chemicals but does include a text search.

A33.1.6 US Department of Transportation (DOT) –
Hazardous Material Incident Reporting System (HMIRS)

This reporting system includes all modes of transportation except pipelines. Reporting is required from all interstate transport companies and results in a total of about 17,000 reports per year. Like the NRC, many of these cases are of little significance and do not meet the definition of an incident in the MKOPSC report.

A33.1.7 US DOT Integrated Pipeline Information
System (IPIS)

Pipelines are covered in three categories, one for natural gas distribution, one for natural gas transmission and one for all other hazardous substances.

A33.1.8 Center for Chemical Process Safety (CCPS)

CCPS maintains a private database. Only members that contribute incidents to the database are allowed to view the data.

A33.1.9 Major Accident Hazards Bureau (MAHB),
Major Accident Reporting System (MARS)

This system is maintained by the European Union's MAHB. It is a hybrid between a database and a report-based system. While the data is structured it contains extensive text descriptions of the incidents. The data is sanitized to remove identifying references. There are currently about 400 cases available to the public.

A33.2 Injury and Fatality Databases
(Not Tied to Specific Incidents)

The following are sources of statistical information regarding injuries and fatalities due to chemicals. They do not reveal the details of specific incidents but do provide statistical information about injuries due to chemicals in the United States.

A33.2.1 Occupational Illness and Injury Statistics (OIIS) –
Occupational Safety and Health Administration,
US Department of Labor

These statistics are based on an annual sampling of about 250,000 companies in the United States. This data has been collected and reported on a consistent basis since 1992. There is good detail for chemicals and industries involved in injuries. There are additional details concerning the type of injury and the body parts involved.

A33.2.2 Census of Fatal Occupational Injuries (CFOI) –
Occupational Safety and Health Administration,
US Department of Labor

Beginning in 1992, OSHA implemented a rigorous system of counting all fatal occupational injuries. This is based on cross checking at least two sources of information. This system also contains the chemical and industry involved. The OSHA web site features both a querying system as well as formatted tables of data.

A33.2.3 Mortality Database, Centers for Disease Control,
United States

This system is based on an analysis of all death certificates in the United States. It therefore includes all commercial, transportation and residential settings. The system also includes intentional acts such as assaults, suicides and substance abuse, which are not normally considered to be chemical incidents.

US death certificates are coded to indicate the cause of death. There is also a set of codes that describe the source of injury if applicable. A variety of codes describe hazardous substances.

A33.2.4 US Coast Guard (USCG)

The USCG maintains a database of oil spills and another for chemical spills to navigable waters. The incidents are typically those observed by the Coast Guard. The vast majority involve small or unknown quantities. The data show clear progress in reducing the quantity of oil spilled over the last 30 years.

A33.2.5 National Electronic Injury Surveillance System –
Consumer Product Safety Commission, USA

This system is based on data from the emergency rooms of approximately 100 US hospitals. These hospitals are chosen to reflect US demographics. Specific events are not reported to maintain patient confidentiality. Statistical information for chemical related injuries are reported.

A33.3 Incident Investigation Reports

A33.3.1 United States
United States Chemical Hazard and Safety Board (CSB)
The CSB is the primary investigatory body for fixed facility chemical accidents. They perform detailed examinations of major accidents. They also perform studies of various classes of accidents such as reactive chemicals.

United States Department of Energy
The DOE has an on-line database of Red Alerts, Yellow Alerts, Lessons and Good Work Practices. These mainly involve the contractors that operate various DOE facilities.

United States Environmental Protection Agency (EPA)
The EPA performed several investigations before the advent of the CSB.

United States Fire Administration (USFA)
The USFA investigates major fires of all types. About a dozen of these reports cover chemical incidents.

*National Institute for Occupational Safety
and Health (NIOSH)*
The National Institute for Occupational Safety and Health (NIOSH) Fatality Assessment and Control Evaluation

(FACE) programme is a research programme designed to identify and study fatal occupational injuries. The goal of the FACE programme is to prevent occupational fatalities across the nation by identifying and investigating work situations at high risk for injury and then formulating and disseminating prevention strategies to those who can intervene in the workplace.

The FACE programme currently has two components:

(1) NIOSH in-house FACE began in 1982. Participating states voluntarily notify NIOSH of traumatic occupational fatalities resulting from specific causes of death that have included *confined spaces*, electrocutions, machine-related, falls from elevation, and logging. The confined space incidents are the ones of interest in terms of process safety. In-house FACE is currently not targeting confined space incidents.

(2) NIOSH State-based FACE began in 1989. Currently, 15 State health or labour departments have cooperative agreements with NIOSH for conducting surveillance, targeting investigations, and recommending prevention activities at the State level using the FACE model.

There are approximately 75 NIOSH reports and 70 state reports.

NIOSH has also published a series of 'Fire Fighter Fatality Investigation Reports'. Several of these are chemical related incidents.

Occupational Safety and Health Administration (OSHA)
OSHA, sometimes in collaboration with the EPA, investigated several incidents before the creation of the CSB.

National Transportation Safety Board (NTSB)
The NTSB has responsibility for investigating transportation related hazardous materials spills on highways, airlines, railroads and pipelines. NTSB classifies incidents as Hazardous Materials Accidents for all modes of transportation except pipelines. Pipeline Accident reports are listed separately. Maritime incidents are excluded, they are handled by the USCG.

The Pipeline Accident reports include about 17 incidents since 1994 which are on-line and 69 reports prior to that time which are available on paper. The Hazardous Materials Accident reports include 18 since 1998 that are on-line and 55 available on paper.

Minerals Management Service (MMS)
The MMS investigates incidents related to offshore drilling activities. In-depth reports are called 'Accident Investigation Panel Reports'. Approximately 25 of these are related to pipeline leaks, fires, explosions or blowouts.

Department of Labour – Mine Safety and Health Administration (MSHA)
The MSHA reports include approximately 20 incidents involving fatalities that meet the criteria of chemical incident. The incidents of interest are generally classified as 'Ignition or Explosion of Gas or Dust' or 'Exploding Vessels Under Pressure'.

National Fire Protection Association (NFPA)
The NFPA has produced 141 detailed fire investigations. These reports are available on-line. They are free of charge to NFPA members. Paper copies are available for purchase. Approximately 15 of the reports are chemical related incidents.

A33.3.2 Australia
Australian Transport Safety Bureau (ATSB)
The ATSB have produced 16 reports of interest. These are primarily related to engine room fires on ships.

Department of Mineral and Petroleum Resources (Western Australia)
The DMPR has produced 10 reports of interest concerning a variety of topics.

Victorian Work Cover Authority
Various reports and guidance are found in 'Alerts & Guidance Notes'.

A33.3.3 Canada
Alberta Energy and Utilities Board
The AEUB has produced several incident reports as well as studies related to the hazards of sour natural gas.

Workplace Health and Safety, Alberta
They have produced about 16 reports concerning various fires, explosions and exposures. The reports under the heading 'Fatality Reports' cover the period from 1997 to 2002.

Canadian Center for Occupational Health and Safety (CCOHS)
The CCOHS has produced approximately 60 reports mostly covering fires and explosions. Fatality Reports contains information about the circumstances surrounding occupationally related fatalities. Information is extracted from reports on inquests and inquiries into occupationally related fatalities across Canada. These reports may be from Coroners, Medical Examiners, Government Ministries, or any other body responsible for inquiry into occupational fatalities.

Database highlights are:

(1) unique safety database identifies unsafe equipment, practices, work conditions;
(2) inquest reports offer practical recommendations to prevent recurrences of fatalities.

Transport Safety Board of Canada (TSB)
The TSB is quite similar to the NTSB in the United States. Their interest is in identifying causes and recommending solutions. They do not have regulatory authority. Approximately 45 rail incidents, 14 pipeline incidents and 7 marine incidents involve hazardous materials are available. The reports are under the headings Pipeline, Marine and Rail on their website.

A33.3.4 Finland
Finland Accident Investigation Board (English summary)
The FAIB has produced 16 reports covering Rail, Maritime and Fixed facilities. Their website contains English summaries that are about one page. The actual reports in Finnish are much more detailed.

A33.3.5 United Kingdom
Health and Safety Executive
MHIDAS (Major Hazard Incident Data Service) is a database of incidents involving hazardous materials that had an off-site impact, or had the potential to have an off-site impact. Such impacts include human casualties or damage to plant, property or the natural environment.

MHIDAS is maintained by, AEA Technology plc on behalf of the UK Health and Safety Executive, and features:

(1) searchable codified and textual information on more than 11,000 incidents, involving the transportation, storage and processing of hazardous materials;
(2) worldwide coverage, with a focus on incidents occurring in the United Kingdom and United States;
(3) comprehensive coverage of incidents that have occurred during the last twenty years, plus significant earlier incidents;
(4) new data made available on a quarterly basis.

Each MHIDAS incident is coded in a standard format to include details such as:

(1) the hazardous material involved in the incident;
(2) the number of fatalities, injuries and evacuations attributable to the incident;
(3) a brief description of the circumstances of the incident.

Information from MHIDAS can be accessed:

(1) through a personalized service, provided on a fee basis through AEA Technology, or
(2) on a subscription basis, via CD-ROM or Internet, or
(3) via an on-line information Retrieval Service, for which charges are made.

Marine Accident Investigation Branch (MAIB)

The MAIB has produced seven relevant detailed reports and nine articles in their monthly "Safety Digest". The fundamental purpose of investigating an accident under these regulations is to determine its circumstances and the causes with the aim of improving the safety of life at sea and the avoidance of accidents in the future. It is not the purpose to apportion liability, nor, except so far as is necessary to achieve the fundamental purpose, to apportion blame.

Appendix

34

Web Links

Contents

A34.1 General Information

Safety Information and the Internet, http://hazard.com/course/

The site provides general information on how to use the internet as a tool to obtain required information.

A34.2 Technical Information

Government Research Center, NTIS search engine, http://grc.ntis.gov/

The National Technical Information Service (NTIS) database offers unparalleled bibliographic coverage of research sponsored by the United States and selected foreign governments.

Risk World – First Internet Chemical Process Safety Business Portal, http://www.riskworld.com

This site has a comprehensive source of risk related information on all aspects of life – chemicals, bioterrorism, business, safety, etc.

Risk Analysis Methodologies, http://home1.pacific.net.sg/~thk/risk.html

The site is maintained by T.H. Keong and provides an overview of risk assessment methods.

Root Cause Analysis, http://www.rootcauselive.com/

A website devoted to root cause analysis practitioners.

A34.2.1 Energetic materials/reactive chemicals

Chemweb – Brethericks Databases, http://chemweb.com/databases/brethericks

Bretherick's database is a comprehensive source of information on energetic or reactive materials.

Explosion Hazards Database – National Institute of Materials and Chemical Research (NIMC), http://www.aist.go.jp/RIODB/db005/index.html

This database includes explosion hazard test data for energetic materials obtained at laboratories of the NIMC in Japan.

UK Chemical Reaction Hazard Forum, http://www.crhf.org.uk/

UK Chemical Reaction Hazards Forum is a group of process safety professionals from the UK chemical industry for mutual exchange of information, expertise and ideas

A34.2.2 Toxicology

Environmental Defense Fund – Chemical Scoreboard, http://www.scorecard.org/

Scorecard is a source for free and easily accessible local environmental information. One can learn about environmental issues in your community by typing in zip code. Scorecard carries health and environmental related information for the chemicals in its database.

National Toxicology Program, http://ntp-server.niehs.nih.gov/

National Toxicology Program (NTP) coordinates toxicological testing programmes within the Department of Health and Human Services, strengthens the science base in toxicology; develops and validates improved testing methods; and provides information about potentially toxic chemicals to health regulatory and research agencies, the scientific and medical communities, and the public.

NIOSH – Registry of Toxic Effects of Chemical Substances (RTECS), http://www.cdc.gov/niosh/rtecs.html

RTECS, maintained by National Institute for Occupational, Health, and Safety (NIOSH) is a compendium of data extracted from the open scientific literature. Six types of toxicity data are included for a chemical: (1) primary irritation; (2) mutagenic effects; (3) reproductive effects; (4) tumorigenic effects; (5) acute toxicity; and (6) other multiple dose toxicity.

A34.3 University Academic Programmes

Fire Protection Engineering Program, University of Maryland, College Park, http://www.enfp.umd.edu/

The programme offers fully accredited undergraduate program and one of two graduate degree programs in the Unitied states.

Hazard Lab Group, Michigan Tech University, http://www.chem.mtu.edu/org/hazards/hazards.html

The research of the group members focuses on generating data for flammable and reactive compounds.

Mary Kay O'Connor Process Safety Center (MKOPSC), Texas A&M University, http://process-safety.tamu.edu

The research of the center focuses on process safety in the chemical industry.

Safety, Environment and Loss Prevention, Chemical Engineering Department, Loughborough University, http://www.lboro.ac.uk/departments/cg/research.html

The research of the group focuses on sustainable process development.

Thermal Hazards Laboratory, Queens University, http://me.queensu.ca/people/birk/research/thermalHazards/

The focus of the research at the Thermal Hazards Laboratory is the safety of the storage and transportation of pressure liquified gases. The laboratory uses large scale experiments and thermodynamic modelling to study the response of a vessel in an accident situation and the possible hazards arising from vessel failure.

A34.4 Government Organizations

Health and Safety Executive (HSE), http://www.hse.gov.uk/

The UK Health and Safety Commission (HSC) and the Health and Safety Executive (HSE) are responsible for the regulation of almost all the risks to health and safety arising from work activity in Britain.

Environmental Protection Agency (EPA), http://www.epa.gov

EPA provides leadership in the nation's environmental science, research, education and assessment efforts. The Agency also works with industries and all levels of government in a wide variety of voluntary pollution prevention programmes and energy conservation efforts.

Occupational, Safety and Heath Administration, http://www.osha.gov

OSHA's mission is to prevent work-related injuries, illnesses, and deaths.

US Chemical Safety Board (CSB), http://www.chemsafety.gov/

The mission of the US Chemical Safety and Hazard Investigation Board is to promote the prevention of major chemical accidents at fixed facilities.

A34.5 Societies, Councils, Institutes

American Industrial Hygiene Association (AIHA), http://www.aiha.org/

AIHA promotes, protects, and enhances industrial hygienists and other occupational health, safety and environmental professionals in their efforts to improve the health and well being of workers, the community, and the environment.

Center for Chemical Process Safety (CCPS), http://www.aiche.org/ccps/

CCPS brings together manufacturers, insurers, government, academia, and expert consultants to lead the way in improving manufacturing process safety.

Institute of Chemical Engineers (IChemE), http://www.icheme.org

IChemE is a leading international body, providing services for and representing the interests of those involved in chemical, biochemical and process engineering worldwide.

A34.6 Security and Vulnerability Assessment

American Chemistry Council (ACC) Site Security Guidelines, http://www.responsiblecaretoolkit.com/

ACC offers Guidelines for the U.S. Chemical Industry, developed by a group of company security professionals and designed specifically for the chemical industry, can help member companies build upon their existing security programmes.

CCPS Security and Vulnerability Analysis (SVA), http://www.aiche.org/ccps/sva/

CCPS recommends a method for industry to analyse their chemical security risks.

Sandia National Laboratory's, http://www.nlectc.org/ccfp/

Center for Civil Force Protection (CCFP) at Sandia National Laboratories (SNL) provides physical security counter terrorism assistance to state and local agencies.

References

Notes

(1) References are given in the text by surname only, except where there would be ambiguity, in which case the initials are given also.

(2) References for which no date has been established are denoted (n.d.) and head their group alphabetically. Where the approximate date is known, this is indicated, e.g. SMITH, J. (*c.* 1988).

(3) Anonymous entries are grouped at the start of the references and are in date order.

(4) In ranking a name alphabetically, elements such as de or von are disregarded, e.g. von Bismark is listed under B rather than V.

(5) Chinese names are given in the form used by the particular author, i.e. either in full or by surname with initials.

(6) The following abbreviated titles are used:

Abbreviated title	Full title
Ammonia Plant Safety, **1**–	*Safety in Air and Ammonia Operations Prog.s*, 1960, Vol. 2– (*Ammonia Plant Safety and Related Facilities*), 1970, Vol. 12– (both New York: Am. Inst. Chem. Engrs)
Carriage of Dangerous Goods, **1**	*Carriage of Dangerous Goods 1*
Chemical Process Hazards **1**–	*Chemical Process Hazards with Special Reference to Plant Design*, 1960, **I**-; then Hazards IX, 1986, IX- (both London, then Rugby: Inst. Chem. Engrs)
Combustion, **1**–	*Symp. on Combustion*, 1928, **1**; 1937, **2** (both Pittsburgh, PA: Combustion Inst.); 1949, **3**; 1953, **4** (both Baltimore, MD: Williams & Wilkins); 1955, **5**; 1957, **6** (both New Tork: Reinhold); 1959, **7** (London: Butterworths); 1960, **8** (Baltimore, MD: Williams & Wilkins); 1963, **9** (New York: Academic Press); 1964, **10**; 1967, **11**; 1968, **12**; 1970, **13**; 1972, **14**; 1975, **15**; 1977, **16**; 1979, **17**; 1981, **18**; 1982, **19**; 1985, **20**, 1988, **21**; 1989, **22**; 1990, **23**; 1991, **24** (all Pittsburgh, PA: Combustion Inst.)
Control of Hazardous Material Spills, **1**–	*Nat. Conf. on Hazardous Material Spills*, 1971, Houston, TX; 1974, Fransisco, CA; 1976, New Orleans, LA; 1978, 4, Miami Beach, FL; *Nat. Conf on Hazardous Material Spills*, 1980, Louisville, KY; 1986; 1988, Chicago, IL
Electrical Safety, **1**–	*Electrical Safety in Hazardous Environments*, 1971, **1**; 1975, **2**; 1982, **3**; 1984, **4** (London: Inst. Elec. Engrs)
Gastech, 6	*Gastech*, 1978, **6**; 1979, **7**; 1982, **8**; 1983, **9**; 1985, **10**; 1986, **11**, 1987, **12**; 1989, **13**; 1991, **14**; 1993, **15** (dates are those of proceedings not of conference itself)
Hazard Identification and Risk Analysis	*Int. Conf. on Hazard Identification and Risk Analysis, Human Factors and Human Reliability in Process Safety* (Rugby: Inst. Chem. Engrs), 1992
Hazards from Pressure	*Hazards from Pressure: Exothermic Reactions, Unstable Substances, Pressure Relief and Accidental Discharge* (Rugby: Inst. Chem. Engrs), 1987
Health, Safety and Loss Prevention	*Health, Safety and Loss Prevention in the Oil, Chemical and Process Industries* (Oxford: Butterworth-Heinemann), 1993
LNG Conf. 1–	*Int. LNG Conf.*, 1968, **1**, Chicago, IL; 1970, **2**, Paris; 1972, **3**, Washington, DC; 1974, **4**, Algiers; 1977, **5**, Dusseldorf (Chicago, Ill: Inst. Gas Technol. and other organizations)
LPB	*Loss Prevention Bulletin* (Institution of Chemical Engineers)
Loss Prevention and Safety Promotion 1–	*Loss Prevention and Safety Promotion in the Process Industries*, 1974, **1** (ed. by C.H. Buschmann) (Amsterdam: Elsevier); 1977, **2** (Frankfurt: DECHEMA); 1980, **3** (Basel: Swiss Soc. Chem. Ind.); 1983, **4** (Rugby: Inst. Chem. Engrs); 1986, **5** (Paris. Soc. Chim. Ind.); 1989, **6** (Oslo: Norwegian Soc. Engrs); 1992, **7** (Taormina, Sicily: Italian Federation of Chemical Industry)
Major Loss Prevention	*Major Loss Prevention in the Process Industries* (London: Inst. Chem. Engrs), 1971
Major Hazards Onshore and Offshore	*Major Hazards Onshore and Offshore* (Rugby: Inst. Chem. Engrs), 1992
Modelling and Mitigating the Consequences of Accidental Releases	*Int. Conf. and Workshop on Modelling and Mitigating the Consequences of Accidental Releases of Hazardous Materials* (New York: Am. Inst. Chem. Engrs), 1991
Preventing Major Chemical Accidents	*Int. Symp. on Preventing Major Chemical Accidents* (ed. J.L. Woodward) (New York: Am. Inst. Chem. Engrs), 1987
Preventing Major Chemical and Related Process Accidents	*Preventing Major Chemical and Related Process Accidents* (Rugby: Inst. Chem. Engrs), 1988
Prevention of Occupational Risks, **1**–	*Int. Symp .on Prevention of Occupational Risks*, 1970, **1**, Frankfurt-am-Main; 1973, **2**, Frankfurt-am-Main; 1976, **3** Frankfurt-am-Main; 1977, **5**, Bucharest (Heidelberg: Berufsgenossenschaft der Chem. Ind.)

Process Industry Hazards	*Process Industry Hazard – Accidental Release, Assessment, Containment and Control* (London: Inst. Chem. Engrs), 1976
Process Safety Management	*Process Safety Management* (New York: Am. Inst. Chem. Engrs), 1993
Protection of Exothermic Reactors	*The Protection of Exothermic Reactors and Pressurised Storage Vessels* (Rugby: Inst. Chem. Engrs), 1984
Runaway Reactions	*Int. Symp. on Runaway Reactions* (New York: Am. Inst. Chem. Engrs), 1989
Safety in Polyethylene Plants	*Safety in High Pressure Polyethylene Plants*, 1973, **1**; 1974, **2**; 1978, **3** (New York: Am. Inst. Chem. Engrs)
Safety and Loss Prevention in the Chemical and Oil Processing Industries	*Safety and Loss Prevention in the Chemical and Oil Processing Industries* (Rugby: Inst. Chem. Engrs), 1990
Static Electrification	*Inst. of Physics, Confs on Static Electrification*, 1954, **1** (Br. J. Appl. Phys., 1954, **4**, Suppl. 2); 1967, **2**; 1971, **3**; 1974, 4; 1979, **5**; 1983, **6**; 1987, **7**; 1991, **8**
Vapour Cloud Modelling	*Int. Conf. on Vapour Cloud Modelling* (New York: Am. Inst. Chem. Engrs), 1987

ANON. (n.d.a). *Memorandum on Explosion and Gassing Risks in the Cleaning, Examination and Repair of Stills, Tanks, etc.* (London: HM Stationery Office)

ANON. (n.d.b). Report of the Committee appointed by the Secretary of State to inquire into the cause of the explosion which occurred on Friday, 19th January 1917, at the Chemical Works of Messrs Brunner Mond and Company, Limited, Crescent Wharf, Silvertown, in the County of Essex (London)

ANON. (1915a). Chlorine poisoning. *The Lancet*, May 15, 1036

ANON. (1915b). French observations on the effects of the irritant gases used by the Germans. *The Lancet*, Jun. 19, 1317

ANON. (1915c). Gas poisoning: physiological symptoms and clinical treatment (report on lecture 4 by L. Hill). *The Lancet*, Dec., 4, 1248

ANON. (1915d). The German use of asphyxiating gases. *Br. Med. J.*, May 1, 774

ANON. (1915e). Medical arrangements of the British Expeditionary Force. *Br. Med. J.*, Jun. 5, 984

ANON. (1915f). Poisonous gases in warfare (review of *The Poison War* by A.A. Roberts). *Br. Med. J.*, July 3, 15

ANON. (1915g). Precautions with asphyxiating gases (Parliamentary report). *The Lancet*, Jun. 19, 1317

ANON. (1918). War gas poisoning. *Br. Med. J.*, Aug. 10, 138

ANON. (1919). Irritant poison gas. *Br. Med. J.*, Dec. 20, 816

ANON. (1920). Gas poisoning. *Br. Med. J.*, Feb. 14, 221

ANON. (1923). *History of the Great War, Medical Services* (London: HM Stationery Office)

ANON. (1926). *The Medical Department of the United States Army in the World War, vol. 4, Medical Aspects of Gas Warfare* (Washington, DC: US Govt Printing Office)

ANON. (1928). Explosion of phosgene at Hamburg. *J. R. United Services Inst.*, **73**, 505

ANON. (1930). *Manual of Treatment of Gas Casualties* (London: HM Stationery Office)

ANON. (1933a). The disaster at Neunkirchen iron works. A preliminary report. *Gas World*, Feb. 18, 1

ANON. (1933b). The disaster at Neunkirchen iron works. Result of German inquiry. *Gas World*, Apr. 22, 3

ANON. (1933c). *Gasschutz und Luftschutz*, **3**, 326

ANON. (1946). The LCC Hospitals in Wartime, *The Medical Officer (new series)*, **76**(8), Aug. 24

ANON. (1947). Missile wounds of the head – review of war experience. *The Lancet*, **2**, 254

ANON. (1948). Ferdinand Flury 1877–1947. *J. Ind. Hyg. Toxicol.*, **30**, 1

ANON. (1949–50). *NFPA Quart.*, **43**, 113

ANON. (1951). Charts of maximum horizontal range for fragments. US Army, Ballistic Res. Lab., Aberdeen Proving Ground, MD, *Tech. Note 496*

ANON. (1952). The Avonmouth oil fire. *FPA J.*, **217**

ANON. (1952–53). LPG gas fires and explosions, Denison, Texas. *NFPA Quart.*, **46**, 63

ANON. (1953). Deaths during the fog. *The Lancet*, **264**, 288

ANON. (1957–58). Flame failure protection. *Br. Chem. Engng*, 1957, **2**, 654 and 1958, **3**, 30

ANON. (1958). Emissions to the Atmosphere from Eight Miscellaneous Sources in Oil Refineries. Joint District, Federal and State Project for the Evaluation of Refinery Emissions, Air Pollution Control District, County of Los Angeles, Calif., Rep. 8

ANON. (1959a). Fire in a Californian oil refinery. *FPA J.*, **46**, 114

ANON. (1959b). Safety in air and ammonia plants. *Chem. Engng Prog.*, **55**(6), 54; **55**(7), 76; **55**(8), 69

ANON. (1960a). Beware detonations. *Chem. Engng*, **67** (Jan. 11), 105

ANON. (1960b). Computer reliability sets record. *Instrum. Control Systems*, **3**(2), 192

ANON. (1960c). Explosion-proof lubricants. *Chem. Engng Prog.*, **56**(1), 130

ANON. (1960d). Safety in air and ammonia plants. *Chem. Engng Prog.*, **56**(6), 73

ANON. (1960e). Safety operating that low temperature plant. *Chem. Engng Prog.*, **56**(5), 37

ANON. (1960f). Safe ups and downs of refinery units. *Am. Petrol. Inst. Proc.*, **40**(3), 220

ANON. (1960–61). Large loss of life fires: transportation – 1959. *NFPA Quart.*, **54**, 35

ANON. (1961a). Operating practices. *Chem. Engng Prog.*, **57**(4), 59

ANON. (1961b). Plan in advance for disaster control. *Chem. Engng*, **68**(Aug. 7), 111

ANON. (1961c). Revised flowsheet cuts maintenance costs. *Chem. Engng*, **68**(Nov. 13), 116

ANON. (1962a). Chemical safety: a session with experts. *Ind. Engng Chem.*, **54**(10), 30; **54**(11), 29

ANON. (1962b). A guide to crushing and grinding practice. *Chem. Engng*, **69**(Dec. 10), 129

ANON. (1962c). Hanford reactor incidents: the 1955 incident. *Nucl. Safety*, **4**, 104

ANON. (1962d). How hydrogen attacks steel. *Chem. Engng*, **69**(Mar. 5), 142

ANON. (1962e). Report on non-sparking hand tools. *J. Inst. Petrol.*, **49**(474), 180

ANON. (1962f). Safe unloading of liquid olefins and diolefins. *Chem. Engng*, **69**(May 28), 146

ANON. (1962g). TV keeps tabs on distant pumps. *Chem. Engng*, **69**(Apr. 2), 56

ANON. (1963). New ratios for estimating plant costs. *Chem. Engng*, **70**(Sep. 30), 120

ANON. (1964a). Compressors for process plants. *Chem. Engng*, **71**(Mar. 16), 163

ANON. (1964b). Picking a plant site by digital computer. *Chem. Engng*, **71**(Jan. 20), 88

ANON. (1964c). Pipeline safety controversy to be aired. *Chem. Engng*, **71**(Jun. 22), 86

ANON. (1964d). Profiting by mechanical seal advances. *Chem. Engng*, **71**(Mar. 2), 83

ANON. (1965a). Dictionary of Organic Compounds, 4th ed. (London: Eyre Spottiswoode)

ANON. (1965b). Instrument maintenance cost trends. *Chem. Engng*, **72**(May 24), 144

ANON. (1966a). Aluminum for process equipment. *Chem. Engng*, **73**(Aug. 29), 116; **73**(Sep. 26), 142

ANON. (1966b). Dry powder in the hydrocarbon industry. *Fire Int.*, **1**(12), 60

ANON. (1966c). Explosion at a refinery in Germany. *Fire Int.*, **1**(12), 35

ANON. (1966d). Fire in a Cyprus oil installation. *Fire Int.*, **1**(120), 82

ANON. (1966e). Heavy casualties at two refinery explosions. *FPA J.*, **70**, 224

ANON. (1966f). Noise suppression in process plant. *Chem. Process Engng*, **47**(8), 54

ANON. (1966g). Protecting liquid petroleum gas vessels from fire. *The Engineer*, Mar. 25, 475

ANON. (1966h). Report on Feyzin. *Fire Int.*, **1**(12), 16

ANON. (1966–67). Material factors. *Chem. Engng Prog.*, 1960, **62**(9), 127; **62**(10), 105; **62**(11), 125; **62**(12), 115; 1967, **63**(1), 127; **63**(2), 89; **63**(3), 99; **63**(4), 91; **63**(5), 97; **63**(6), 117; **63**(7), 125

ANON. (1967a). Chemical technology for sale or license. *Chem. Engng*, **74**(Sep. 25), 147

ANON. (1967b). Fatal chemical plant explosion. *Fire J.*, **61**(5), 8

ANON. (1967c). Five killed in flour dust explosion. *FPA J.*, **74**(Apr.), 45

ANON. (1967d). Kellogg's single train large ammonia plants achieve lowest costs. *Chem. Engng*, **74**(Nov. 20), 112

ANON. (1968a). The challenge of pollution control. State regulations. *Chem. Engng*, **75**(Oct. 14), 25

ANON. (1968b). Electrical fault on motor causes benzene fire. *Accidents*, **66**, 18

ANON. (1969a). Dropping the pilot. *Br. Chem. Engng*, **14**, 1453

ANON. (1969b). Four die in polythene plant explosion. *FPA J.*, **83**, 351

ANON. (1969c). Process piping. *Chem. Engng*, **76**(Apr. 14), 95

ANON. (1969d). Santa Barbara post spill. *Chem. Engng*, **76**(May 19), 100

ANON. (1969e). Setting a value on creep strength. *Chem. Engng*, **76**(May 19), 208

ANON. (1969f). What does Div. 2 of the ASME Vessel Code mean? *Hydrocarbon Process.*, **48**(1), 147

ANON. (1970a). Commendations for two at Dudgeons Wharf. *Fire*, Feb., 467

ANON. (1970b). How reliable is vendor's equipment? *Chem. Engng Prog.*, **66**(10), 29

ANON. (1970c). New stainless steel needs no nickel. *Chem. Engng*, **77**(Nov. 30), 92

ANON. (1970d). Sun closes last of 170 ddc loops on cat reformer: computer process 99.7% available first year. *Instrum. Technol.*, **17**(4), 35

ANON. (1970e). Wilton explosion – diesel Engine explosion. *Chem. Britain*, **6**, 3

ANON. (1971a). Air pollution control. *Chem. Engng*, **78**, Jun. 21, 131

ANON. (1971b). Making sense from alloy compositions. *Chem. Engng*, **78**(Nov. 1), 92

ANON. (1971c). Occupational Safety and Health Act. *Hydrocarbon Process.*, **50**(4), 146

ANON. (1971d). Probe begins after Iranian explosion. *Oil Gas J.*, Jan. 11, 28

ANON. (1971e). The San Fernando earthquake – one of the most important in history. What it did to the earth. *Engng and Sci.*, Mar., 4

ANON. (1971f). Solid waste disposal. *Chem. Engng*, **78** (Jun. 21), 155

ANON. (1971g). State regulations. *Chem. Engng*, **78** (Jun. 21), 19

ANON. (1971h). Water pollution control. *Chem. Engng*, **78** (Jun. 21), 65

ANON. (1972a). ASTM study will evaluate stack-emission test methods. *Chem. Engng*, **79**(May 1), 30

ANON. (1972b). *British Labour Struggles: Contemporary Pamphlets 1727–1850. The Factory Act of 1819* (New York: Arno Press)

ANON. (1972c). Defusing hazardous materials. *Chem. Engng*, **79**, May 15, 60

ANON. (1972d). First NPRA panel on refinery maintenance (and similar titles). *Hydrocarbon Process.*, 1972, **51**(6), 121; 1973, **52**(6), 85; 1974, **53**(7), 159

ANON. (1972e). Industry examines six air-quality tests. *Chem. Engng*, **79**(Apr. 3), 52

ANON. (1972f). LNG Plant explosion. *Fire J.*, **66**(4), 38

ANON. (1972g). Pitting corrosion. *Hydrocarbon Process.*, **51**(8), 70

ANON. (1972h). Reliability in process pumps. *Br. Chem. Engng*, **17**(2), 141

ANON. (1972i). SNG: the process options. *Chem. Engng*, **79**(Apr. 17), 64

ANON. (1972j). Stress corrosion cracking. *Hydrocarbon Process.*, **51**(8), 68

ANON. (1972k). When safety valves aren't enough. *New Scientist*, 54(Jun. 1), 494

ANON. (1972l). *Eur. Chem. News*, Apr. 19, 4

ANON. (1972m). *Eur. Chem. News*, Sep., 10

ANON. (1973a). Board raps proposed HI placard systems for hazardous cargo. *Fire Engng*, **126**, Apr., 84

ANON. (1973b). How people react to fire casts doubt on training. *Fire Engng*, **126**, Nov. 53

ANON. (1973c). Method for measuring flammability of liquids. *Fire Engng*, **126**(May), 52

ANON. (1973d). Occupational Safety and Health Act. *Chem. Engng*, **80**(Feb. 26), 13

ANON. (1973e). The Occupational Safety and Health Act. *Chem. Engng*, **80**(Jun. 18), 3

ANON. (1973f). OSHA standards for chemical carcinogens. *Chem. Engng*, **80**(Sep. 3), 60

ANON. (1973g). Quick reference guide to air pollution codes. *Chem. Engng*, **80**(Jun. 18),11

ANON. (1973h). Relationship of vibration to gear quality. *Hydrocarbon Process.*, **52**(12), 56

ANON. (1973i). Standby power supplies – can process industries afford to be without? *Process Engng*, Jun., 96

ANON. (1973j). *Daily Telegraph*, Feb. 3

ANON. (1974a). Drain orifice outperforms steam traps, other devices. *Chem. Engng*, **81**(Oct. 28), 62

ANON. (1974b). Experts differ on possible cause of Flixborough explosion. *Fire Int.*, **45**, 17

ANON. (1974c). Experts discuss SNG manufacture. *Hydrocarbon Process.*, **53**(1), 151

ANON. (1974d). The HMFI experiment. *Occup. Safety Health*, **4**(7), 17

ANON. (1974e). Increasing number and cost of fires in the chemical process industries. *Fire Prev.*, **106**, 15

ANON. (1974f). Maintenance is costing more. *Chem. Engng*, **81**(Oct. 14), 60

ANON. (1974g). Multiple death fires, 1973. *Fire J.*, **68**(May), 69

ANON. (1974h). OSHA citations – an analysis. *Chem. Engng*, **81**(Nov. 25), 44

ANON. (1974i). Predicting and preventing brittle fracture. *Chem. Engng*, **81**(Aug. 5), 114

ANON. (1974j). VCM exposure standards: too tough too soon? *Chem. Engng*, **81**(Jun. 24), 110

ANON. (1974k). USSR rejects Pitesti sabotage rumours. *Eur. Chem. News*, Aug. 16, 6

ANON. (1974l). Valve malfunctions in nuclear power plants. *Nucl. Safety*, **15**, 316

ANON. (1974m). 1973 large loss fires, United States and Canada. *Fire J.*, **68**(Jul.), 77

ANON. (1974n). 3000 flee butane gas blast. *Herald American*, Aug., **26**

ANON. (1974–). Fixed gas extinguishing systems. *Fire Surveyor*, 1974, **3**(5); 1974,**3**(6); 1975, **4**(1), 49

ANON. (1975a). *Carbon Monoxide, Industry and Performance* (Oxford: Pergamon Press)

ANON. (1975b). Fixed extinguishing systems – dry chemical powder. *Fire Surveyor*, **4**(3), 25

ANON. (1975c). Fixed extinguishing systems – foam. *Fire Surveyor*, **4**(4), 45; **4**(5), 35; **4**(6), 39

ANON. (1975d). The Flixborough explosion: how it happened. *Fire Int.*, **48**, 92

ANON. (1975e). Flixborough goes bomb in Vietnam. *New Scientist*, May 1, 262

ANON. (1975f). Gas pipelines. *Nat. Safety News*, Sep., 345

ANON. (1975g). Guide to ASME Pressure Vessel Code. *Hydrocarbon Process.*, **54**(12), 60

ANON. (1975h). Hirwaun – triumph or tragedy. *Gas World*, Nov., 555

ANON. (1975i). Instrument questions answered. *Hydrocarbon Process.*, **54**(2), 77

ANON. (1975j). Is there an economy of size for tomorrow's refinery. *Hydrocarbon Process.*, **54**(5), 111

ANON. (1975k). Lubrication failure on a 3500 HP Engine. *Ammonia Plant Safety*, **17**, 151

ANON. (1975l). Recent developments in the ASME Pressure Vessel Code. *Hydrocarbon Process.*, **54**(12), 53

ANON. (1975m). Safety in refinery operations. *Hydrocarbon Process.*, **54**(5), 107

ANON. (1975n). US refinery fire kills eight firemen. *Fire Int.*, **50**, 69

ANON. (1975o). VCM decision questioned. *Process News*, Jan. 21

ANON. (1976a). Das plant poses special problems. *Oil Gas J.*, Jul. 12, 72

ANON. (1976b). Extinguishing fires with foam and wetting agents. *Fire Int.*, **53**, 59

ANON. (1976c). Feedback, 18 years of LNG service. *Marine Week*, Nov. 19, 29

ANON. (1976d). Fire protection of offshore platforms. *Fire Surveyor*, **5**(1), 20; **5**(2), 29; **5**(3), 20

ANON. (1976e). Firms propose East Coast LNG plant. *Oil Gas J.*, Aug. 23

ANON. (1976f). History of the safety movement. *Occup. Hazards, Sep.*

ANON. (1976g). Hydrogen explosion injures two. *Nucl. Safety*, **17**, 249

ANON. (1976h). Lightning strikes tank. *Fire J.*, Sep., 46

ANON. (1976i). Reactor shutdown eliminated by pushbutton changer for ruptured pressure disc. *Chem. Process*, Chicago, **39**(1), 79

ANON. (1976j). Risks present in start-up of high pressure vessels. *Oil Gas J.*, Dec. 20

ANON. (1976k). *Sapere*, **6**(Nov./Dec.) (Special issue on Seveso)

ANON. (1976l). UV detectors. *Fire Surveyor*, **5**(4), 34

ANON. (1976m). Valve protects pump against overheat damage. *Chem. Engng, Albany*, **83**(Feb. 16), 67

ANON. (1976n). *The Morning Advocate*, Baton Rouge, Dec.

ANON. (1976o). The Times, May 13

ANON. (1977a). Anticipator – developing a safety system for industrial hazards. *Measmnt Control*, **10**(2), 51

ANON. (1977b). Blast destroys Montedisons' Brindisi chemical complex. *Eur. Chem. News*, Dec. 16/23, 12

ANON. (1977c). Blast, fire damage Qatar plant. *Oil Gas J.*, Apr. 18, 29

ANON. (1977d). Danger from explosion. *OS & H Bull.*, Jun., 1

ANON. (1977e). Designing monitors to deal with fires on offshore oil rigs. *Fire Int.*, **57**, 81

ANON. (1977f). Directory of environmental offices and officials. *Chem. Engng*, **84**(Oct. 17), 10

ANON. (1977g). Fire protection for living quarters on an oil platform. *Fire Int.*, **57**, 41

ANON. (1977h). Gas detection. *Fire Surveyor*, **6**(2), 26; **6**(3), 24

ANON. (1977i). Getting the best out of conveyor and elevators. *Plant Engr*, **21**(10), 20

ANON. (1977j). A method of sealing small leaks of dangerous substances. *Fire Int.*, **57**, 52

ANON. (1977k). Norwegian paint factory. *Fire Prev.*, **119** (Jun.), 32

ANON. (1977l). The operators view of furnaces. *Hydrocarbon Process.*, **56**(2), 81

ANON. (1977m). OSHA: moving to balance. *Chem. Engng*, **84**(Apr. 11), 108

ANON. (1977n). Private communication

ANON. (1977o). *The Guardian*, Dec. 24

ANON. (1977p). *The Guardian*, Dec. 29

ANON. (1977–). Breathing apparatus. *Fire Surveyor*, 1977, **6**(6), 21; 1978, **7**(1), 21; 1978, **7**(2), 21

ANON. (1978a). The dangers of acetylene. *Chem. Engr, Lond.*, **329**, 85

ANON. (1978b). *First National School on Explosibility of Industrial Dusts* (English translation: Dept. of Chem. Eng., Kansas State Univ., Manhattan, KS)

ANON. (1978c). Guide to ASME Sec. VIII, Div. 1. *Hydrocarbon Process.*, **57**(12), 90

ANON. (1978d). Human error led to heavy losses in tank farm. *Fire Int.*, **62**, 51

ANON. (1978e). Policy for occupational health. *Sci. Publ. Policy*, **5**, 39

ANON. (1978f). *Prevention of Dust Explosions in Grain Elevators*. Task force report, (Washington, DC: US Dept of Agriculture, Office of the Special Co-ordinator for Grain Elevator Safety and Security)

ANON. (1978g). Risk analysis makes chemical plants safer. *Chem. Engng News*, **56**(40), 8

ANON. (1978h). Scorecard. *Pipeline Gas J.*, Jun., 19

ANON. (1978i). Selecting hi-temp heat transfer fluids. *Chem. Engng*, **85** Aug. 14, 104

ANON. (1978j). Tanker incidents show importance of proper labelling. *Fire Int.*, **62**, 74

ANON. (1978k). *Daily Telegraph*, Feb. 27

ANON. (1978l). *The Guardian*, Feb. 26

ANON. (1978m). *The Guardian*, Apr. 28

ANON. (1979a). *Best Practicable Means: An Introduction* (London: HM Stationery Office)

ANON. (1979b). Butane tanks at Flixborough opposed by HSE. *Fire Prev.*, **131**, 5

ANON. (1979c). Countering the rise in flammable risk. *Processing*, Apr., 25

ANON. (1979d). 'Defend yourself', Dunster tells chemical industry. *Chem. Ind.*, May 19, 324

ANON. (1979e). *Fire and explosion damage, Roland Mill in Bremen, 6 Feb. 1979* (Bremen: Bremen Fire Brigade and Police)

ANON. (1979f). Flame detectors – their use and application. *Fire Surveyor*, **8**(6), 37

ANON. (1979g). Hazard identification problem at a road tanker incident. *Fire Int.*, **63**, 94

ANON. (1979h). Industry criticises EEC's hazards proposals. *Chem. Ind.*, Dec. 15, 857

ANON. (1979i). Questions on sodium chlorate after container depot fire. *Fire Int.*, **63**, 43

ANON. (1979j). *Chem. Ind.*, Nov. 3, 719

ANON. (1979k). *Fire Prev.*, **139**, 29

ANON. (1979l). *The Guardian*, Jan. 9

ANON. (1980a). The Bantry Bay tanker disaster. *Fire Prev.*, **139**, 13

ANON. (1980b). *Bericht uber den Störfall in Sevesco* (Frankfurt: Dechema-Haus)

ANON. (1980c). Cables: the weak spot in fire defence. *Fire Int.*, **68**, 83

ANON. (1980d). Comparative tests on liquid fuel fires. *Fire Int.*, **68**, 65

ANON. (1980e). *Effective Policies for Health and Safety at Work* (London: HM Stationery Office)

ANON. (1980f). The engineer's liability to prosecution. *Chem. Engr, Lond.*, **359/360**, 548

ANON. (1980g). Fire alarm systems: the new code of practice. *Fire Surveyor*, **9**(2), 26

ANON. (1980h). Fire and explosion hazards with aerosols. *Fire Prev.*, **139**, 22

ANON. (1980i). Flight deck resource management. *Flight Deck*, Nov., 1; Dec., 1

ANON. (1980j). Generation of static electricity during the discharge of portable CO_2 fire extinguishers. *Fire Prev.*, **139**, 27

ANON. (1980k). Heat radiation aspects of liquid fuel fires. *Fire Int.*, **69**, 60

ANON. (1980l). HSE acts on Canvey Island LNG tanks. *Chem. Ind.*, May 17, 391

ANON. (1980m). HSE and UKAEA reassess risk. *Chem. Ind.*, Jun. 7, 430

ANON. (1980n). Individual risk of early fatality. *Sci. Publ. Policy*, **7**, 152

ANON. (1980o). Inert gas systems for tanker safety. *Fire Int.*, **69**, 31

ANON. (1980p). New study pinpoints HSE mistakes on Canvey. *Chem. Ind.*, Feb. 16, 119

ANON. (1980q). *Prevention of Dust Explosions in Grain Elevators, an Achievable Goal, Dept of Agriculture Report* (Washington, DC: US Govt Printing Office)

ANON. (1980r). Report pinpoints sodium chlorate danger. *Chem. Ind.*, Jun. 21, 471

ANON. (1980s). Risk analysis is misused, say experts. *Chem. Ind.*, Oct. 4, 756

ANON. (1980t). Sprinklers: high piled and rack storage. *Fire Surveyor*, **9**(1), 21

ANON. (1980u). Tank farm fire in the world's largest inland harbour. *Fire Int.*, **67**, 72

ANON. (1980v). Tanker explosion costs 50 lives. *Fire Int.*, **69**, 24

ANON. (1980w). Twelfth regulation in implementation of the Federal Pollution Control Law (Accident Regulation) (Störfall-Verordnung). *Bundesgesetzblatt, pt 1*, **32**, 772

ANON. (1980x). Union calls for tighter control on VCM. *Chem. Ind.*, Aug. 2, 581

ANON. (1980y). *Fire Prev.*, **139**, 29

ANON. (1980z). *Fire Prev.*, **139**, 34

ANON. (1980aa). *Fire Prev.*, **139**, 35

ANON. (1981a). Dust clue to Nypro blast. *Chem. Ind.*, Mar.21, 169

ANON. (1981b). Dust eliminated during filling operation. *Chem. Eng.*, **88**(Jun. 29), 55

ANON. (1981c). Earthshaking news for the CPI. *Chem. Eng.*, **88**(Oct. 19), 57

ANON. (1981d). Electronic corrosion monitor saves on equipment wear. *Chem. Eng.*, **88**(Apr. 6), 47

ANON. (1981e). Energy-saving motors offer fast payback. *Chem. Eng.*, **88**(Jan. 26), 69

ANON. (1981f). Fiberglass pumps stand up to harsh chemicals. *Chem. Eng.*, **88**(Feb. 9), 63

ANON. (1981g). Flames leap 100 m into the air. *Fire Int.*, **71**, 75

ANON. (1981h). Flight crew responsibility. *Flight Deck*, Mar., 1

ANON. (1981i). HSE comes under fire from Lords. *Chem. Ind.*, Jun. 6, 372

ANON. (1981j). HSE investigates heavy gases. *Chem. Ind.*, Aug. 1, 512

ANON. (1981k). Industry toxicology group changes leaders. *Chem. Eng.*, **88**(Feb. 23), 34

ANON. (1981l). Lessons learned from Mississauga disaster. *Fire Int.*, **70**, 64

ANON. (1981m). The mystery remains. *Haz. Cargo Bull.*, Jun., 15

ANON. (1981n). New hazardous waste definition in the offing? *Chem. Eng.*, **88** (July 27), 28

ANON. (1981o). New mechanism for Seveso explosion proposed. *Chem. Ind.*, Apr. 4, 197

ANON. (1981p). New Mexico refinery fire kills one. *Chem. Eng.*, **88** (Jun. 29), 31

ANON. (1981q). The NFPA: an international institution. *Fire Int.*, **73**, 80

ANON. (1981r). One more inquiry on Canvey Island. *Chem. Engr, Lond.*, **371/372**, 403

ANON. (1981s). Plastic-lined pipe emphasizes ease of installation. *Chem. Eng.*, **88** (Mar. 9), 61

ANON. (1981t). Regulations on halon fire extinguishing systems relaxed in Britain. *Fire Int.*, **72**, 49

ANON. (1981u). Safety expenditure maintained despite recession. *Chem. Ind.*, May 2, 299

ANON. (1981v). Steam traps – the quiet thief in our plants. *Chem. Eng.*, **88**(Feb. 9), 85

ANON. (1981w). Substantial improvements in safety at Canvey Island says HSE. *Chem. Ind.*, Dec. 19, 846

ANON. (1982a). Animal experiments – no end in sight. *Chem. Ind.*, Feb. 6, 71

ANON. (1982b). Chemicals EDC probes health and safety costs. *Chem. Ind.*, Mar. 6, 142

ANON. (1982c). Cleaning up the Valley of the Drums. *Chem. Eng.*, **89** (Jan. 25), 25

ANON. (1982d). CLEAR moves to stop using lead. *Chem. Ind.*, Feb. 6, 67

ANON. (1982e). Early warning system for toxic waste spills. *Chem. Eng.*, **89** (Jul. 24), 35

ANON. (1982f). Gas spillage incidents on trial. *Dock and Harbour Authority*, Jun., 99

ANON. (1982g). Help for people who deal with disaster. *Business Week*, Aug. 30, 812

ANON. (1982h). How a chemical spillage accident was handled. *Fire Int.*, **75**, 66

ANON. (1982i). Major incident at a chemical Operations Prog. in the south of Germany. *Fire Int.*, **77**, 38

ANON. (1982j). The MFL fire wall concept. *Fire Int.*, **77**, 53

ANON. (1982k). Realistic shipboard test for halon system. *Fire Int.*, **74**, 64

ANON. (1982l). Safety standards May drop, warns Duncan. *Chem. Ind.*, Jan. 2, 5

ANON. (1982m). Salford shaken by chemicals explosions. *Chem. Ind.*, Oct. 2, 730

ANON. (1982n). Unions suspend safety talks with CIA. *Chem. Ind.*, Apr. 3, 210

ANON. (1982–83). Smoke mask controversy: the present situation in Britain. *Fire Int.*, **78**, 58

ANON. (1983a). Blast rips chemical plant in Louisiana. *Chem. Eng.*, **90**(Aug. 8), 29

ANON. (1983b). Blowouts are the most serious offshore hazard. *Oil Gas J.*, **81**(May 30), 105

ANON. (1983c). Canvey Island report accepted by government. *Chem. Engr, Lond.*, **393**, 10

ANON. (1983d). A conceptual approach towards a probability based design guide on structural fire safety. *Fire Safety J.*, **6**, 1

ANON. (1983e). Correct and rapid treatment for burn victims. *Fire Int.*, **80**, 25

ANON. (1983f). Cutting equipment costs. *Chem. Engr, Lond.*, **397**, 7

ANON. (1983g). Equipment survey: seals and gaskets. *Chem. Engr, Lond.*, **398**, 43

ANON. (1983h). Fire study covers causes and cures. *Petrol. Engr Int.*, Feb., 12

ANON. (1983i). Five managers convicted in Seveso case. *Chem. Eng.*, **90**(Oct. 17), 12C

ANON. (1983j). Five recent offshore fires produce new safety alerts. *Oil Gas J.*, **81**(Dec. 26), 154

ANON. (1983k). Foam and its application. *Fire Int.*, **83**, 48

ANON. (1983l). Fresh demands delay Seveso trial. *Chem. Ind.*, May 2, 324

ANON. (1983m). ICI gets pat on the back from fire chief. *Chem. Ind.*, Dec. 19, 904

ANON. (1983n). Idea with flare scoops MacRobert award. *Chem. Ind.*, Jan. 3, 3

ANON. (1983o). Meanwhile Italy is probing the TCDD disappearance. *Chem. Eng., Albany*, **90**(Jun. 27), 12

ANON. (1983p). MMS safety alert pinpoint offshore work hazards. *Oil Gas J.*, **81**(May 2), 160

ANON. (1983q). New ICI process cuts ammonia cost. *Chem. Engr, Lond.*, **390**, 30

ANON. (1983r). Panel discussion on stress corrosion cracking in ammonia storage spheres. *Plant/Operations Prog.*, **2**, 247

ANON. (1983s). Revolutionary distillation. *Chem. Engr, Lond.*, **388**, 7

ANON. (1983t). Safety Alerts probes cause of offshore fires and explosions unrelated to well pressure. *Oil Gas J.*, **81** (May 16), 84

ANON. (1983u). Seveso wastes slated for disposal in Switzerland. *Chem. Eng.*, **90**(Dec. 26), 12C

ANON. (1983v). The Swiss will dispose of Seveso's TCDD-laden wastes. *Chem. Eng.*, **90** (Jun. 27), 12C

ANON. (1984a). Ban burning of PCBs, says Greenpeace. *Chem. Ind.*, Oct. 1, 675

ANON. (1984b). Britain's unique approach. *Hazardous Cargo Bull.*, **5**(12), 3

ANON. (1984c). Burning the midnight oil. *Hazardous Cargo Bull.*, **5**(7), 3

ANON. (1984d). Burning wastes at sea. *Hazardous Cargo Bull.*, **5**(4), 11

ANON. (1984e). CHEMTREC hotline extended into Canada. *Hazardous Cargo Bull.*, **5**(12), 10

ANON. (1984f). Chlorine poisoning. *The Lancet*, Feb. 11, 321

ANON. (1984g). Controversy over new HSE safety package. *Chem. Ind.*, Sep. 3, 594

ANON. (1984h). Creeping death in Bhopal. *Chem. Ind.*, Dec. 17, 854

ANON. (1984i). Dioxin scare sparks inquiry into ReChem plant. *Chem. Ind.*, Jul. 16, 486

ANON. (1984j). The 'do's' and 'don'ts' of portable electric tools. *Health and Safety at Work*, Sep., 32

ANON. (1984k). Early suppression fast response sprinklers. *Fire Int.*, **87**, 44

ANON. (1984l). Emergency backup on tap. *Hazardous Cargo Bull.*, **5**(5), 15

ANON. (1984m). Face to face with the HSE hazards chief. *Chem. Engr, Lond.*, **406**, 26

ANON. (1984n). Focusing on pipes, valves and fittings. *Process Eng.*, May, 73

ANON. (1984o). Further monitoring at ReChem site. *Chem. Ind.*, Sep. 17, 638

ANON. (1984p). Getting rid of dioxins safely. *Chem. Engr, Lond.*, **410**, 21

ANON. (1984q). Heavy gas releases – dispersion and control. *Chem. Engr, Lond.*, **405**, 28

ANON. (1984r). IMDG Code – current status. *Hazardous Cargo Bull.*, **5**(4), 4

ANON. (1984s). Inerting on chemships. *Hazardous Cargo Bull.*, **5**(5), 6

ANON. (1984t). Livingston Engineer convicted. *Hazardous Cargo Bull.*, **5**(6), 34

ANON. (1984u). Looking at national safety policies. *Chem. Engr, Lond.*, **400**, 8

ANON. (1984v). LPG emergency response advisers. *Hazardous Cargo Bull.*, **5**(12), 10

ANON. (1984w). *Major hazard installations: Planning and assessment.* Loughborough Univ. of Technol., Dept of Chem. Eng., Seminar Proc.

ANON. (1984x). New data on hydrogen/air explosions. *Chem. Engr, Lond.*, **408**, 20

ANON. (1984y). New testing facility for liquefied gas spills. *Hazardous Cargo Bull.*, **5**(12), 10

ANON. (1984z). New VCM/cancer link exposed by Norwegians. *Chem. Ind.*, May 21, 359

ANON. (1984aa). Outside monitoring of PCB incineration agreed by ReChem. *Chem. Ind.*, Aug. 20, 558

ANON. (1984bb). Pinch concept helps to evaluate heat recovery networks for improved petrochem operation. *Oil Gas J.*, **82**(May 28), 113

ANON. (1984cc). Plant wide cabling systems: why Ethernet won't do. *Process Eng.*, Jun., 69

ANON. (1984dd). ReChem hastens closure of PCB plant. *Chem. Ind.*, Oct. 15, 718

ANON. (1984ee). Refrigerated peroxides. *Hazardous Cargo Bull.*, **5**(5), 12

ANON. (1984ff). The rise of the Engineer. *Hazardous Cargo Bull.*, **5**(7), 24

ANON. (1984gg). Searching for the seal of approval. *Process Eng.*, Mar., 63

ANON. (1984hh). Talk of new leaks in Love Canal area. *Chem. Eng.*, Albany, 91(Nov. 26), 28

ANON. (1984ii). Temperature control for batch processes. *Chem. Engr, Lond.*, **402**, 29

ANON. (1984jj). Third piece in the jigsaw. *Hazardous Cargo Bull.*, **5**(4), 8

ANON. (1984kk). TMI-2 head is lifted. *Nucl. Eng. Int.*, **29** Sep., 2

ANON. (1984ll). Towards leaner machines. *Hazardous Cargo Bull.*, **5**(12), 13

ANON. (1984mm). Unions dismayed by workplace safety initiative. *Chem. Ind.*, May 21

ANON. (1984nn). USA pointers to Euro cuts. *Chem. Ind.*, Nov. 5, 756

ANON. (1984oo). Waiting to happen. *Hazardous Cargo Bull.*, **5**(6), 34

ANON. (1984pp). What really happened at Abbeystead? *Chem. Engr, Lond.*, **410**, 7

ANON. (1984qq). When is an explosion not an explosion? *Chem. Engr, Lond.*, **407**, 5

ANON. (1984rr). *Workshop on Computer-based Systems in process control. Ergonomic Problems in Process Operations* (Rugby: Inst. Chem. Engrs), p. 151

ANON. (1984ss). *Workshop on Effects of Monotony and Reduced Activity in Monitoring Automated Plant. Ergonomic Problems in Process Operations* (Rugby: Inst. Chem. Engrs), p. 33

ANON. (1984tt). *Workshop on Expert Systems in Process Control. Ergonomic Problems in Process Operations* (Rugby: Inst. Chem. Engrs), p. 147

ANON. (1984uu). *Workshop on Techniques for Human Reliability Assessment. Ergonomic Problems in Process Operations* (Rugby: Inst. Chem. Engrs), p. 39

ANON. (1984vv). Workstations. *Process Eng.*, Dec., 46

ANON. (1985a). Abbeystead – would hazop have revealed the causes beforehand? *Chem. Engr, Lond.*, **413**, 10

ANON. (1985b). Bhopal – Carbide gets to the core of the tragedy. *Chem. Ind.*, July 15, 450

ANON. (1985c). Bhopal – was it sabotage? *Chem. Ind.*, Feb. 18, 94

ANON. (1985d). Carbide blames fatal leak on water. *Chem. Ind.*, Apr. 1, 202

ANON. (1985e). Carbide springs another leak. *Chem. Ind.*, Aug. 19, 530

ANON. (1985f). Chemical plant safety, better engineering education discussed at Seattle meeting. *Chem. Eng. Prog.*, **81**(11), 27

ANON. (1985g). Chemical plants everywhere 'May be unsafe'. *Chem. Ind.*, Nov. 18, 738

ANON. (1985h). CIMAH Regulations 'inadequate' say MPs. *Chem. Ind.*, Mar. 4, 136

ANON. (1985i). Concrete inhibits LNG boil-off. *Chem. Engr, Lond.*, **417**, 35

ANON. (1985j). Cost indices. *Process Eng.*, Feb., 9

ANON. (1985k). Delhi leak marks Bhopal anniversary. *Chem. Ind.*, Dec. 16, 802

ANON. (1985l). Expert system can supervise many processes. *Process Eng.*, Nov., 24

ANON. (1985m). Fouling May now be predicted. *Chem. Engr, Lond.*, **411**, 24

ANON. (1985n). Fuming war over phosgene at Widnes. *Chem. Ind.*, Sep. 2, 566

ANON. (1985o). Government dithers over hazardous waste issue. *Chem. Ind.*, May 20, 319

ANON. (1985p). *The Hazardous Chemical Data Manual* (Washington, DC: US Govt Printing Office)

ANON. (1985q). Insulating ethylene tanks. *Process Eng.*, Feb., 13

ANON. (1985r). Leak sealed. *Process Eng.*, Oct., 21

ANON. (1985s). Leaks cause two rig fires. *Oil Gas J.*, **83** (Aug. 5), 100

ANON. (1985t). New fluidised bed drier shows Higee-style intensification. *Chem. Engr, Lond.*, **416**, 25

ANON. (1985u). Offshore platform fires and explosions analysed. *Oil Gas J.*, **83**(Apr. 5), 72

ANON. (1985v). Pemex – the forgotten disaster. *Chem. Engr, Lond.*, **418**, 16

ANON. (1985w). Program to protect safety complainers. *Chem. Eng.*, **92**(Nov. 11), 41

ANON. (1985x). Pump lasts 3½ times longer than standard ANSI pumps. *Chem. Eng.*, **92**(Oct. 14), 43

ANON. (1985y). ReChem operations given clean bill of health. *Chem. Ind.*, Mar. 4, 139

ANON. (1985z). ReChem rebuffs negligence claims. *Chem. Ind.*, Jan. 21, 35

ANON. (1985aa). Regulations from HSC on getting explosives from A to B. *Chem. Engr, Lond.*, **417**, 10

ANON. (1985bb). Safety and reliability at Culcheth. *Chem. Ind.*, Mar. 4, 141

ANON. (1985cc). Staveley prompts call for safer storage. *Chem. Ind.*, Sep. 16, 606

ANON. (1985dd). Survey of operating refineries worldwide. *Oil Gas J.*, **83**(Dec. 30), 100

ANON. (1985ee). Survivor lives on. *Haz. Cargo Bull.*, Nov., 57

ANON. (1985ff). Uncharted pipe could explain Bhopal disaster. *New Scientist*, Apr. 11, 4

ANON. (1985gg). Union Carbide suffers set-back of further gas leaks in US plant. *Chem. Engr, Lond.*, **417**, 9

ANON. (1985hh). Worst accident in industrial history. *Process Eng.*, **66**(1) Jan., 5

ANON. (1985ii). *Chem. Ind.*, Feb. 18, 100

ANON. (1986a). Animal experiments 'here to stay'. *Chem. Ind.*, Apr. 7, 220

ANON. (1986b). Carbide prepares for Bhopal payouts. *Chem. Ind.*, Feb. 3, 79

ANON. (1986c). CCPS poised for the future. *Chem. Eng. Prog.*, **82**(8), 48

ANON. (1986d). A clean bill of health for ReChem. *Chem. Engr, Lond.*, **426**, 9

ANON. (1986e). Combined method improves design for creep. *Process Eng.*, Mar., 25

ANON. (1986f). Corrosion in sour media. *Process Eng.,* Feb., 65

ANON. (1986g). 3–D CAD: a user's opinion. *Process Eng.,* Mar., 36

ANON. (1986h). Expert systems for industry: SIRA club mounts five projects. *SIRA Newsletter,* **60**, Apr.–Jun., 133

ANON. (1986i). Experts thrash out thresholds. *Chem. Ind.,* Oct. 20, 677

ANON. (1986j). Fire risks from electrostatics. *Process Eng.,* Mar., 27

ANON. (1986k). Improved hazardous wastes management needs. *Chem. Eng. Prog.,* **82**(9), 29

ANON. (1986l). Indian anger at $350m Carbide offer. *Chem. Ind.,* Apr. 7, 219

ANON. (1986m). Industrial thermograph. *Chem. Engr, Lond.,* **423**, 49

ANON. (1986n). Leak sealed. *Process Eng.,* Oct., 21

ANON. (1986o). Mossmorran – technical wonder or costly blunder. *Chem. Ind.,* Jun. 16, 407

ANON. (1986p). Need data on a chemical's toxicity in a hurry? *Chem. Eng.,* **93**, Sep. 1, 12C

ANON. (1986q). A new, safer route for making phosgene. *Chem. Eng.,* **93**(May 12), 20C

ANON. (1986r). On-site chlorination is a water industry first. *Process Eng.,* May, 25

ANON. (1986s). One-step method determines sour water H_2S hazard. *Oil Gas J.,* **84** (Feb. 24), 55

ANON. (1986t). Pilot-operated safety relief valves for gas separation plant. *Process Eng.,* Dec., 17

ANON. (1986u). Proposed law ensnares warehouse owners. *Chem. Ind.,* Feb. 17, 111

ANON. (1986v). The Rhine disaster exacts its toll. *Chem. Eng.,* **93**(Dec. 8), 12C

ANON. (1986w). Toxic spills take toll on Rhine. *Chem. Ind.,* Nov. 17, 755

ANON. (1986x). UEA study exonerates ReChem. *Chem. Ind.,* May 5, 295

ANON. (1986y). Warehouse blaze at Sandoz causes Europe's worst environmental disaster. *Process Eng.,* Dec.

ANON. (1986z). Editorial. *Chem. Engr, Lond.,* **426**, 1

ANON. (1986aa). Editorial. *Chem. Engr, Lond.,* **429**, 1

ANON. (1987a). Batch processing. *Process Eng.,* Sep., 32

ANON. (1987b). Check valve selection and performance. *Reliab. Eng.,* **17**(1), 81

ANON. (1987c). Computerised maintenance. *Plant Engr,* **31**(3), 19

ANON. (1987d). Court decisions put safety before cost. *Chem. Eng.,* **94**(12), 33

ANON. (1987e). A dry mechanical seal for compressors. *Chem. Engr, Lond.,* **417**, 28

ANON. (1987f). DTI launches £150,000 study of maintenance practices. *Chem. Engr, Lond.,* **441**, 7

ANON. (1987g). Dual-column reduces heat loss. *Chem. Engr, Lond.,* **443**, 21

ANON. (1987h). Earthquake simulator inaugurated. *Chem. Engr, Lond.,* **436**, 12

ANON. (1987i). EPA pollutant study: surprising results. *Chem. Eng.,* **94**(17), 27

ANON. (1987j). EPA writes new rule for toxics control. *Chem. Eng.,* **94**(16), 17

ANON. (1987k). Expert systems developed to support Hazop analysis. *Process Eng.,* Oct., 28

ANON. (1987l). Explosions kill three at BP. *Process Eng.,* Apr., 5

ANON. (1987m). Flammable gas detector uses infra-red light. *Chem. Engr, Lond.,* **442**, 21

ANON. (1987n). Fourteen ways to avoid bogus valves. *Chem. Eng.,* **94**(10), 19

ANON. (1987o). Fugitive emissions are focus of study. *Chem. Eng.,* **94**(14), 30

ANON. (1987p). HSE safety audit of BNFL Sellafield minces no words. *Chem. Engr, Lond.,* **433**, 9

ANON. (1987q). Lightning is a problem for refineries in 'JR' territory. *Process Eng.,* June, 25

ANON. (1987r). LNG world overview. *Hydrocarbon Processing,* **66**(3), 124–B

ANON. (1987s). PC-based expert systems are developed in only a few weeks. *Process Eng.,* Feb., 25

ANON. (1987t). Pinpointing leaks. *Chem. Eng. Prog.,* **83**(12), 13

ANON. (1987u). Programmable electronic systems – safety guidelines launched. *Chem. Engr, Lond.,* **438**, 7

ANON. (1987v). Re-Chem. acquitted. *Chem. Engr, Lond.,* **442**, 7

ANON. (1987w). Storage-tank rules are set in motion. *Chem. Eng.,* **94**(8), 22

ANON. (1987x). Three gas detectors from the Ministry of Defence. *Chem. Engr, Lond.,* **433**, 24

ANON. (1987y). When the six-tenths rule doesn't apply. *Process Eng.,* Sep., 15; Oct., 13

ANON. (1987z). Why are de-aerators cracking up? *Process Eng.,* Jan., 37

ANON. (1988a). AIChE's design institutes help reduce industrial risks. *Chem. Eng. Prog.,* **84**(9), 22

ANON. (1988b). Bhopal caused by disgruntled operator says Arthur D. Little. *Chem. Engr, Lond.,* **449**, 6

ANON. (1988c). BP fined £750,000 for fatal Grangemouth accidents. *Process Eng.,* Apr., 4

ANON. (1988d). Call for reassessment of maintenance after record fines of £750,000 for BP. *Chem. Engr, Lond.,* **448**, 7

ANON. (1988e). CEFIC reiterates environmental guide. *Hydrocarbon Process.,* **67**(10), 15

ANON. (1988f). The design of inherently safer plants. *Chem. Eng. Prog.,* **84**(9), 21

ANON. (1988g). Did a blank flange fail? – Piper Alpha interim report. *Chem. Engrs, Lond.,* **454**, 9

ANON. (1988h). Explosion wrecks Wendstone plant. *Process Engng,* Apr., 7

ANON. (1988i). ICE breaks into safety training. *Audio Visual,* May, 44

ANON. (1988j). In the hot seat. *Haz. Cargo Bull.,* Apr., 73

ANON. (1988k). New precautions in landfill safety. *Chem. Engr, Lond.,* **444**, 18

ANON. (1988l). Pipe freezing and dust suppression at Southampton. *Chem. Engr, Lond.,* **448**, 29

ANON. (1988m). Steel oil storage tanks should be reevaluated, new NBS study notes. (1988). *Hydrocarbon Process.,* **67**(8), 17

ANON. (1988n). Strong industry support for expert system research club. *Process Engng,* Apr., 27

ANON. (1988o). Union Carbide's cheap trick. *Business India,* Jun. 13–26, 39

ANON. (1988p). Users discuss piping and valve problems. *Hydrocarbon Process.,* **67**(8), 43

ANON. (1988q). What would you do if . . . *Haz. Cargo Bull.,* Mar., 48

ANON. (1989a). Drives. *Plant Engr,* Nov./Dec., 23

ANON. (1989b). Human error a common element in BP incidents. *Chem. Engr, Lond.,* **461**, 11

ANON. (1989c). Underground storage tanks. *Chem. Engng Prog.*, **84**(11), 15

ANON. (1989d). A wealth of information on-line. *Chem. Engng*, **96**(6), 112

ANON. (1990a). Combustion gases. *Chem. Engr, Lond.*, **481**, 22

ANON. (1990b). Computerised maintenance. *Plant Engr*, Mar./Apr., 15

ANON. (1990c). Dangers of transporting LPG cylinders by road. *Chem. Engr, Lond.*, **469**, 12

ANON. (1990d). Houston explosion. *Chem. Engr, Lond.*, **477**, 11

ANON. (1990e). Investigation begins into Stanlow explosion and fire. *Chem. Engr, Lond.*, **471**, 20

ANON. (1990f). Phoenics. *Chem. Engr, Lond.*, Sep. 26, 40

ANON. (1990g). Refinery fire. *Chem. Engr, Lond.*, **487**, 12

ANON. (1990h). Risk management. *Chem. Engng Prog.*, **86**(3), 21

ANON. (1990i). Romanian plant explosion. *Chem. Engr, Lond.*, **483**, 9

ANON. (1990j). Safe pressure systems. *Plant Engr*, May/June, 31

ANON. (1990k). SARA goes public. *Chem. Engng*, **97**(3), 30

ANON. (1991a). Arguments at SRD. *Chem. Engr, Lond.*, **506**, 7

ANON. (1991b). Blast rocks chemical store. *Chem. Engr, Lond.*, **503**, 14

ANON. (1991c). Breathing Apparatus (London: HM Stationery Office)

ANON. (1991d). Citgo explosion explained. *Chem. Engr, Lond.*, **502**, 12

ANON. (1991e). Fugitive emissions rule. *Chem. Engr, Lond.*, **496**, 11

ANON. (1991f). Fulmar explosion. *Chem. Engr, Lond.*, **502**, 8

ANON. (1991g). Halons: the search for alternatives. *Fire Surveyor*, **20**(1), 5

ANON. (1991h). How to comply with the Clean Air Act. *Chem. Engng*, **98**(7), 121

ANON. (1991i). Keeping pump emissions under control. *Chem. Engng*, **98**(2), 93

ANON. (1991j). Safety confusion fears. *Chem. Engr, Lond.*, **506**, 7

ANON. (1991k). Shell's near-disaster lands a fine of £100,000. *Chem. Engr, Lond.*, **509/510**

ANON. (1991l). Six die on Albright blast. *Chem. Engr, Lond.*, **499**, 10

ANON. (1992a). Boilers. *Plant Engr*, **36**(1), 33

ANON. (1992b). Exothermic runaway caused Hickson deaths. *Chem. Engr, Lond.*, **532/533**, 7

ANON. (1992c). Fire fighters in the dark as chemical tank complex blazes. *Fire Int.*, **134**, Jun./Jul., 32

ANON. (1992d). The new Building Regulations. *Fire Surveyor*, **21**(1), 24

ANON. (1992e). OSHA issues final rule for chemical PSM. *Hydrocarbon Process.*, **71**(4), 33

ANON. (1992f). Progress and politics. *Chem. Engr, Lond.*, **529**, s3

ANON. (1993a). BP Chemicals gets virtuous. *Chem. Engr, Lond.*, **546/547**, 5

ANON. (1993b). EC environmental approach unsustainable, says CIA. *Chem. Engr, Lond.*, **549**, 6

ANON. (1993c). EO blast mechanism found. *Chem. Engr, Lond.*, **540**, 8

ANON. (1993d). Foam can cause fires, clams BP. *Chem. Engr, Lond.*, **539**, 6

ANON. (1993e). Hoechst's costly error. *Chem. Engr, Lond.*, **538**, 8

ANON. (1993f). IPPC neutered at final hurdle. *Chem. Engr, Lond.*, **550**, 4

ANON. (1994a). Disk specification. *J. Loss Prev. Process Ind.*, **7**, 202

ANON. (1994b). Elimination not limitation. *Chem. Engr, Lond.*, **561**

ANON. (1995). *The Guardian*, Apr. 28

AALBERSBERG, H.G. (1991). Health, safety, environment: a preconstruction focus. *Hydrocarbon Process.*, **70**(6), 34B

AAREBROT, E. and SVENES, S. (1992). Increased level of safety by use, of high integrity pressure protection systems in process plant. *Loss Prevention and Safety Promotion*, **7**, paper 66

AARSTAD, I.E. (1990). *The Norwegian regulations concerning risk assessment and their implications in emergency planning. Piper Alpha: Lessons for Life Cycle Management* (Rugby: Instn Chem. Engrs), p. 133

AARTS, J.J. and MORRISON, O.M. (1981). Refrigerated storage tank retainment walls. *Ammonia Plant Safety*, **23**, 124

ABADIE, V.H. (1966). Making, handling and storing liquefied natural gas. *Chem. Engng*, **73**(Jan. 17), 151

ABBEY, D.R. (1964). Electrical codes vs process hazards. *Chem. Engng*, **71**(Aug. 17), 133

ABBOTT, C. (1974). The measurement and use of oxygen index. *Combust. Flame*, **23**, 1

ABBOTT, J.A. (1990). *Prevention of Fires and Explosions in Dryers*, 2nd ed. (Rugby: Instn Chem. Engrs)

ABBOTT, K.W. (1993). TQM can improve project control. *Hydrocarbon Process.*, **72**(12), 44

ABBOTT, M.A. (1961). On the spreading of one fluid over another. *La Houille Blanche*, **5–6**, 827

ABBOTT, M.B. and BASCO, D.K. (1989). Computational Fluid Dynamics (London: Longmans)

ABBOTT, M.B. and CUNGE, J.A. (eds) (1982). *Engineering Applications of Computational Hydraulics*, vols 1 and 2 (London: Pitman)

ABBOTT, N.J. and SCHULMAN, S. (1976). Protection from fire: nonflammable clothing – a review. *Fire Technol.*, **12**, 204

ABBOTT, R.E. and LEEMANN, J.E. (1990). *Spill Reporting Procedures Guide* (Edison, NJ: BNA Books)

ABBOTT, W.N. (1933). The incidence of pulmonary disease following exposure to vesicant and asphyxiating gases. *Br. Med. J.*, Nov. 11, 862

ABDEL-GAYED, R.G. and BRADLEY, D. (1977). Dependence of turbulent burning velocity on turbulent Reynolds number and ratio of laminar burning velocity to r.m.s. turbulent velocity. *Combustion*, **16**, 1725

ABDEL-GAYED, R.G. and BRADLEY, D. (1981). A two-eddy theory of premixed turbulent flame propagation. *Phil. Trans. R. Soc., Ser. A*, **A301**, 1457

ABDEL-GAYED, R.G. and BRADLEY, D. (1982). The influence of turbulence upon the rate of turbulent burning. In Lee, J.H.S. and Guirao, C.M. (1982), *op. cit.*, p. 51

ABDEL-GAYED, R.G., BRADLEY, D. and LAWES, M. (1987). Turbulent burning velocities: a general correlation in terms of straining rates. *Proc. R. Soc., Ser. A*, **414**, 389

ABDOLLAHIAN, D. *et al.* (1984). Analytical prediction of single-phase and two-phase flow through cracks in pipes and tubes. *Am. Inst. Chem. Engrs, Symp. Ser.*, **80**, 19

ABDULLAH, A.H. and YACCOB, K.A.R. (1990). Strategies in practising occupational health and industrial hygiene in chemical processing industries. *Safety and Loss Prevention in the Chemical and Oil Processing Industries*, p. 293

ABE, K. (1978). Levels of trust and reaction to various sources of information in catastrophic situations. In Quanterelli, E.L. (1978), *op. cit.*, p. 147

ABEDIAN, B. and SONIN, A.A. (1982). Theory of electric charging in turbulent pipe flow. *J. Fluid Mech.*, **120**, 199

ABELSON, P.H. (1973). Prevention of cancer. *Science*, **182**, 973

ABERNATHY, M.W. (1977). Design horizontal gravity settlers. *Hydrocarbon Process.*, **56**(9), 199

ABKOWITZ, M. and CHENG, P.D. (1989). Hazardous materials risk estimation under conditions of limited data availability. *Nat. Res. Coun., Transport. Res. Board, Transport. Res. Record*, **1245**, 14

ABLITT, I. (1987). Gas compressor selection. *Plant Engr*, **31**(6), 28

ABLITT, J.F. (1975). Accident probability reduction by the application of service reliability data. In Hendry, W.D. (1975), *op. cit.*, paper C2

ABOU-ARAB, T.W., ENAYET, M.M. and KAMEL, M.M. (1991). Recent measurements of flame acceleration in semi-confined geometries. *J. Loss Prev. Process Ind.*, **4**, 202

ABOUSEIF, G. and TOONG, T.Y. (1982a). On direct initiation of gaseous detonation. *Combust. Flame*, **45**, 39

ABOUSEIF, G. and TOONG, T.Y. (1982b). Theory of unstable one-dimensional detonations. *Combust. Flame*, **45**, 67

ABOUSEIF, G. and TOONG, T.Y. (1986). Theory of unstable two-dimensional detonations: genesis of the transverse waves. *Combust. Flame*, **63**, 191

ABRAHAM, G. (n.d.). *Jet Diffusion in Stagnant Ambient Fluid. Rep. 29.* Delft Hydraulics Lab., The Netherlands

ABRAHAM, G. (1970). Round buoyant jet in cross flow. *Fifth Int. Conf. on Water Pollution Res.*, San Francisco

ABRAHAM, R.W. (1973). Reliability of rotating equipment. *Chem. Engng*, **80**(Oct. 15), 96

ABRAMOVICH, G.N. (1963). *Theory of Turbulent Jets* (Cambridge, MA: MIT)

ABRAMOVICH, G.N. (ed.) (1969). *Turbulent Jets of Air, Plasma and Real Gas* (Consultants Bureau)

ABRAMOWITZ, M. and STEGUN, I.A. (1964). *Handbook of Mathematical Functions with Formulas, Graphs and Mathematical Tables* (Washington, DC: US Govt Printing Office)

ABRAMS, P. and OLSON, R. (1988). Get the right seal. *Hydrocarbon Process.*, **67**(11), 98

ABRAMSON, L.R. (1981a). A rejoinder to Kaplan and Garrick (letter). *Risk Analysis*, **1**, 235

ABRAMSON, L.R. (1981b). Some misconceptions about the foundations of risk analysis (letter). *Risk Analysis*, **1**, 229

ABUAF, N., JONES, O.C. and WU, B.J.C. (1983). Critical flashing flow in nozzles with subcooled inlet conditions. *J. Heat Transfer*, **105**, 379

ACHARD, C. (1919). The effects of poisoning by war gases. *Rev. Sci.*, **57**, 265. See also *Bull. Med.*, **1919**, 59

ACHESON, E.D. (1967). *Medical Record Linkage* (Oxford: Oxford Univ. Press)

ACHESON, E.D. (1979). Record linkage and the identification of long-term environmental hazards. *Proc. R. Soc., Ser. B*, **B205**, 165. See also Doll, Sir R. and McLean, A.E.M. (1979), *op. cit.*, p. 165

ACHESON, E.D. and GARDNER, M.J. (1981). Dose–response relations from epidemiological studies. *Proc. R. Soc., Ser. A*, **A376**, 79. See also Warner, Sir F. and Slater, D.H. (1981), *op. cit.*, p. 79

ACKERMANN, D. and HUTMACHER, W. (1987). Experiences and results of gas tanker operation with the CATO online computer. *Gastech 86*, p. 213

ACKERMANN, D.M. (1986). The status of safety techniques for storage, loading, and unloading of LPG. In Hartwig, S. (1986), *op. cit.*, p. 365

ACKETTS, A.G. (1987). The commercial implications in the premium gas market of Calor's underground storage at Humberside. *Gastech 86*, p. 347

ACKROYD, G.C. (1974). Insurance aspects of loss prevention. *Loss Prevention and Safety Promotion*, **1**, 139

D'ACO, V.J., CAMPBELL, G.R. and STONE, R.W. (1984). Industrial waste treatment pilot plants. *Chem. Engng Prog.*, **80**(9), 45

ACTON, M.R., SUTTON, P. and WICKENS, M.J. (1990). An investigation of the mitigation of gas cloud explosions by water sprays. *Piper Alpha: Lessons for Life Cycle Management* (Rugby: Instn Chem. Engrs), p. 61

AD HOC COMMITTEE ON THE EVALUATION OF LOW LEVELS OF ENVIRONMENTAL CARCINOGENS (1972). *Evaluation of Environmental Carcinogens – Report to the Surgeon General, USPHS, Congressional Record E952–958*, Feb. 9

ADAM, L. (1982). Acoustic emission testing of ammonia storage spheres. *Am. Inst. Chem. Engrs Mtg* (New York: Am. Inst. Chem. Engrs)

ADAMCZYK, A.A. (1976). *An Investigation into Waves Generated from Non-ideal Energy Sources. Rep. UILUENG 76 0506.* Univ. of Illinois, Urbana, IL

ADAMS, A.J. (1989). *Piper Alpha Public Inquiry. Transcript, Day 136*, 137

ADAMS, D.D. and PAGE, W.P. (1986). *Acid Deposition* (New York: Plenum Press)

ADAMS, E.M. (1968). Procedures for the evaluation of chemical reactivity. *Carriage of Dangerous Goods*, **1**, 41

ADAMS, G.K. and PACK, D.C. (1959). Some observations on the problem of transition between deflagration and detonation. *Combustion*, **7**, 812

ADAMS, G.K. and SCRIVENER, J. (1955). Flame propagation in methyl and ethyl nitrate vapors. *Combustion*, **5**, 656

ADAMS, J.A. and ROGERS, D.F. (1973). *Computer Aided Heat Transfer Analysis* (New York: McGraw-Hill)

ADAMS, J.A. (1982). Issues in human reliability. *Human Factors*, **24**, 1

ADAMS, J.G.U. (1985). *Risk and Freedom* (Cardiff: Transport Publ. Projects)

ADAMS, J.G.U. (1988). Risk homeostasis and the purpose of safety regulation. *Ergonomics*, **31**, 407

ADAMS, M. (1984). Don't overspecify control valves. *Chem. Engng*, **91**(Oct. 29), 121

ADAMS, R.J. (1967). *Report of the Inquiry into the Causes of the Accident to the Drilling Rig* Sea Gem, *Cmnd 3409* (London: HM Stationery Office)

ADAMS, W.B. (1967). Reforming system failure. *Ammonia Plant Safety* **9**, 34

ADAMS, W.V. (1982). Advancements in sealing ethylene and other LPG pumps. *Hydrocarbon Process.*, **61**(1), 126

ADAMS, W.V. (1983). Troubleshooting mechanical seals. *Chem. Engng*, **90**(Feb. 7), 48

ADAMS, W.V. (1987). Better high temperature sealing. *Hydrocarbon Process.*, **66**(1), 53

ADAMS, W.V. (1991). Control fugitive emissions from mechanical seals. *Chem. Engng Prog.*, **87**(8), 36

ADAMS, W.V. (1992). Controlling fugitive emissions from mechanical seals. *Hydrocarbon Process.*, **71**(3), 99

ADAMSON, M.S. and ROBERGE, P.R. (1991). The development of a deep knowledge diagnostic expert system using fault tree analysis information. *Can. J. Chem. Engng*, **69**, 76

ADCOCK, C.T. and WELDON, J.D. (1967a). To minimize the loss in a chemical plant explosion. *Safety Mainten.*, **134**, 45

ADCOCK, C.T. and WELDON, J.D. (1967b). Vapor release and explosion in a light hydrocarbon plant. *Loss Prevention*, **1**, 70. See also *Chem. Engng Prog.*, 1967, **63**(8), 54

ADCOCK, C.T. and WELDON, J.D. (1973). Light hydrocarbon plant explosion. In Vervalin, C.H. (1973a), *op. cit.*, p. 8

ADCOCK, H.W. (1978). The Engineer's obligations as related to man-made hazards. *Joint Conf. on Predicting and Designing for Natural and Man-Made Hazards* (New York: Am. Soc. Civil Engrs), p. 39

ADDERTON, D.V. (1974). Gas inferno rages. *The News Gazette*, Champaign, IL, Sep. 15

ADDIS, T.R. (1982). Expert systems: an evolution in information retrieval. *Inform. Technol. Res. Dev.*, **1**, 301

ADDISON, H. (1966). *Centrifugal and Other Rotodynamic Pumps*, 3rd ed. (London: Chapman & Hall)

ADELMAN, H.G. (1981). A time dependent theory of spark ignition. *Combustion*, **18**, 1333

ADELMAN, L., COHEN, M.S., BRESNICK, T.A., CHINNIS, J.O. and LASKEY, K.B. (1993). Real-time expert system interfaces, cognitive processes, and task performance: an empirical assessment. *Human Factors*, **35**, 243

ADELMAN, M.A. (1972). *The World Petroleum Market* (Baltimore, MD: John Hopkins Press)

ADELSON, L. and KAUFMANN, J. (1971). Fatal chlorine poisoning: report of two cases with clinicopathological correlation. *Am. J. Clin. Pathol.*, **56**, 430

ADELSON, L. and SUNSHINE, I. (1966). Fatal hydrogen sulphide intoxication. Report of three cases occurring in a sewer. *Arch. Path.*, **81**, 375

ADHIKARI, J.P., SEN, G.K., TEWARY, B.K., BANERJEE, A., MUKHERJEE, R.N., RAJWAR, D.P. and SINGH, B. (1990). Modification and testing of a Gaussian dispersion model for particulate matter in the respirable size range. *Atmos. Environ.* **24A**, 1647

ADIGA, K.C., RAMAKER, D.E., TALEM, P.A. and WILLIAMS, F.W. (1989). Modeling pool-like gas flames of propane. *Fire Safety J.*, **14**(3), 241

ADIGA, K.C., RAMAKER, D.E., TATEM, P.A. and WILLIAMS, F.W. (1990). Numerical predictions for a simulated methane fire. *Fire Safety J.*, **16**, 443

ADLER, J. (1978). Asymmetric self-heating of a circular cylinder. *Combust. Flame*, **32**, 163

ADLER, J. (1987). Thermal explosion theory for a slab with partial insulation: a correction to d(e). *Combust. Flame*, **69**, 367

ADLER, J. (1983). Thermal explosion theory for a slab with partial insulation. *Combust. Flame*, **50**, 1

ADLER, J. and BOWES, P.C. (1983). Criticality in the hollow cylinder: a correction to Anisimov's result. *Combust. Flame*, **51**, 113

ADLER, J. and ENIG, J.W. (1964). The critical conditions in thermal explosion theory with reactant consumption. *Combust. Flame*, **8**, 97

ADLER, J. and ZATURSKA, M.B. (1966). The dependence of flame ignition on transport properties and chemical kinetic parameters. *Combust. Flame*, **10**, 273

ADLER, S.B. and LIN, T.C.T. (1985). K-constants: water in hydrocarbons. *Hydrocarbon Process.*, **64**(4), 99

ADOMEIT, G. (1965). Ignition of gases at hot surfaces under unsteady-state conditions. *Combustion* **10**, 237

ADORJAN, A.S., CRAWFORD, D.B. and HANDMAN, S.E. (1982). *Damping effect of perlite/fiberglass insulation on outer tank dynamic loads in double-walled cryogenic tanks* (Washington DC: Am. Chem. Soc.)

ADVISORY COMMITTEE ON THE CARRIAGE OF DANGEROUS GOODS BY SEA (1971). *Report of the Advisory Committee on the Carriage of Dangerous Goods by Sea* (London: HM Stationery Office)

ADVISORY COMMITTEE ON DANGEROUS SUBSTANCES, HEALTH AND SAFETY COMMISSION (1991). *Major Hazard Aspects of the Transport of Dangerous Substances* (London: HM Stationery Office)

ADVISORY COMMITTEE ON REACTOR SAFEGUARDS (1974). *Integrity of Reactor Vessels for Light Water Power Reactors*

ADVISORY COMMITTEE ON THE SAFETY OF NUCLEAR INSTALLATIONS (1990). *Study Group on Human Factors. First Report on training and related matters* (London: HM Stationery Office)

ADVISORY COMMITTEE ON THE SAFETY OF NUCLEAR INSTALLATIONS (1991). *Study Group on Human Factors. Second report: Human Reliability Assessment: a Critical Overview* (London: HM Stationery Office)

ADVISORY COMMITTEE ON THE SAFETY OF NUCLEAR INSTALLATIONS (1993). *Study Group on Human Factors. Third report: Organising for safety* (London: HM Stationery Office)

AELION, V. and POWERS, G.J. (1991). A unified strategy for the retrofit synthesis of flowsheet structures for attaining or improving operating procedures. *Comput. Chem. Engng*, **15**, 349

VAN AERDE, M., STEWART, A. and SACCOMANNO, F. (1988). Estimating the impacts of LPG spills during transportation accidents. *J. Haz. Mats*, **20**, 375

AERSTIN, F. and STREET, G. (1978). *Applied Chemical Process Design* (New York: Plenum Press)

AFFENS, W.A. (1968). Flammability properties of hydro-carbon fuels. *J. Chem. Engng Data*, **11**, 197

AFFENS, W.A. and CARHART, H.W. (1974). *Ignition studies. Pt VII, The determination of autoignition temperatures of hydrocarbon fuels. Rep. 7665*. Naval Res. Lab.

AFFENS, W.A., CARHART, H.W. and MCLAREN, G.W. (1977). Determination of flammability index of hydrocarbon fuels by means of hydrogen flame ionization detector. *J. Fire Flammability*, **8**, 141

AFFENS, W.A., JOHNSON, J.E. and CARHART, H.W. (1961). Effect of chemical structure on spontaneous ignition of hydrocarbons. *J. Chem. Engng Data*, **6**, 613

AFFENS, W.A. and SHEINSON, R.S. (1980). Autoignition: the importance of the cool flame in the two-stage process. *Loss Prevention*, **13**, 83

AFGAN, N.H. and BEER, J.M. (eds) (1974). *Heat Transfer in Flames* (New York: Wiley)

AFGHAN, B.K. and MACKAY, D. (eds) (1980). *Hydrocarbons and Halogenated Hydrocarbons in the Aquatic Environment* (New York: Plenum Press)

AFZAL, S.M.M. and LIVINGSTON, E.H. (1974). *Safety in Polyethylene Plant Components. Safety in Polyethylene Plants 2* (New York: Am. Inst. Chem. Engrs), p. 25

AGAM, G. (1994). *Industrial Chemicals* (Amsterdam: Elsevier)

AGAR, J. (1978). Smokeless flare control. *API Proc., Refining Dept, 43rd Midyear Mtg*, **57**, 42

AGGARWAL, S.K. (1989). Ignition behaviour of a multi-component fuel spray. *Combust. Flame*, **76**, 5

AGGARWAL, S.K. and CHA, S. (1988). A parametric study on spray ignition and comparison with experiments. *Combust. Sci. Technol.*, **59**, 213

AGNEW, J.T. and GRAIFF, L.B. (1961). Pressure dependence of laminar burning velocity by the spherical bomb method. *Combust. Flame*, **5**, 209

AGNEW, W.G. and AGNEW, J.T. (1965). Composition profiles of the diethyl ether–air two-stage reaction stabilized in a flat-flame burner. *Combustion*, **10**, 123

AGOSTON, G.A. (1962). Controlling mechanisms in the burning of pools of liquid fuels. *Combust. Flame*, **6**, 212

AGOSTON, G.A., WISE, H. and ROSSER, W.A. (1957). Dynamic factors affecting the combustion of liquid spheres. *Combustion*, **6**, 708

AGRA EUROPE (1994a). *European Community Environment Legislation for Industry* (London)

AGRA EUROPE (1994b). *Eurochem Monitor* (London)

AGRAWAL, D.D. (1981). Experimental determination of burning velocity of methane-air mixtures in a constant volume vessel. *Combust. Flame*, **42**, 243

AGREDA, V.H., PARTIN, L.R. and HEISE, W.H. (1990). High-Purity methyl acetate via reactive distillation. *Chemi. Engi. Prog.*, (February), 40–46.

AGUILO, A. and HORLENKO, T. (1980). Formic acid. *Hydrocarbon Process.*, **59**(11), 120

AHAMMED, M. and MELCHERS, R.E. (1993). Probabilistic assessment of pipeline service life based on pitting corrosion leaks. In Melchers, R.E. and Stewart, M.G. (1993), *op. cit.*, p. 131

AHEARNE, S. (1982). Petrochemicals: is there a future? *Chem. Ind.*, Jan. 16, 42

AHERN, A.A. (1991). Properly use explosion-proof electrical equipment. *Chem. Engng Prog.*, **87**(7), 65

AHERN, W. (1980). California meets the LNG terminal. *Coastal Zone Mngmt J.*, **7**, 185

AHLBORG, G. (1951). Hydrogen sulphide poisoning in shale oil industry. *Am. Med. Ass. Arch. Ind. Hyg. Occup. Med.*, **3**, 247

AHLERT, R.C. and KOSSON, D.S. (1990). Treatment of hazardous landfill leachates and contaminated groundwater. *J. Haz. Materials*, **23**(3), 331

AHLMANN (1905). Dissertation, Univ. of Wurzburg

AHMED, M., FISHER, H.G. and JANESHEK, A.M. (1989). Reaction kinetics from self-heat data – correction for the depletion of sample. *Runaway Reactions*, 331

AHMED, M. and LAVIN, M. (1991). Exothermic reactions of DCPD and protection against them. *Plant/Operations Prog.*, **10**, 143

AHMED, N. and KHAN, A.A. (1992). Common telltales can identify safety hazards. *Chem. Engng Prog.*, **88**(7), 73

AHMED, S. (1984). Tank farm construction. *Oil Gas J.*, **82**, Jan. 23, 55; Jan. 30, 153; Feb. 6, 127

AHMED, S. and METCALF, D.R. (1983). On the use of point estimates in risk assessment of nuclear power plants. *Reliab. Engng*, **6**, 211

AIGUEPERSE, J. *et al.* (1988). Fluorine compounds, inorganic: hydrofluoric acid. *Ullmanns Encyclopaedia of Chemical Technology*, rev. ed., vol. A11 (Weinheim: VCH), p. 307

AIKEN, W.S. (1958). Building reliability into digital process control systems. *Control Engng*, **5**(10), 76

AILOR, W.H. (ed.) (1971). *Handbook on Corrosion Testing and Evaluation* (New York: Wiley)

AILOR, W.H. (ed.) (1982). Atmospheric Corrosion (New York: Wiley)

AINSWORTH, L. and PENDLEBURY, G., (1995). Task-based contributions to the design and assessment of the man/machine interfaces for a pressurised water reactor. *Ergonomics*, **38**, 463

AINSWORTH, R.A. and GOODALL, I.W. (1983). Defect assessments at elevated temperature. *J. Press. Ves. Technol.*, **105**, 263

L'AIR LIQUIDE (1976). *Gas Encyclopaedia* (Amsterdam: Elsevier)

AIR POLLUTION CONTROL ASSOCIATION (1986). *Avoiding and Managing Environmental Damage from Major Chemical Accidents* (Pittsburgh, PA)

AIR WEATHER SERVICE (1978). *Calculation of Toxic Corridors. Pamphlet, 105–57*

AIRD, R.J. (1977a). Interactive computation for maintenance data analysis. *First Nat. Conf. on Reliability* (Culcheth, Lancashire: UK Atom. Energy Auth.), paper NRC5/4

AIRD, R.J. (1977b). Private communication

AIRD, R.J. (1978). The bath tube curve myth (letter). *Quality Assurance*, **4**(1)

AIRD, R.J. (1980). The application of reliability engineering to process plant maintenance. *Chem. Engr, Lond.*, **356**, 301

AIRD, R.J. (1981). *The application of reliability engineering to process plant maintenance. In Reliable Production in the Process Industries* (Rugby: Instn Chem. Engrs), p. 51

AIRD, R.J. (1982). Reliability assessment of safety/relief valves. *Trans. Instn Chem. Engrs*, **60**, 314

AIRD, R.J. (1983). Failure criteria for pressure relief valves. *Euredata 4*, paper 4.5

AIRD, R.J. (1984). Practical estimation of reliability parameters for process equipment. *Reliab. Engng*, **7**, 77

AIT LAOUSSINE, N. (1980). LNG grows as world energy source. *Hydrocarbon Process.*, **59**(4), 130

AITCHESON, J. and BROWN, J.A.C. (1969). *The Lognormal Distribution* (Cambridge: Cambridge Univ. Press)

AITKEN, A. (1982). Fault analysis. In Green, A.E. (1982b), *op. cit.*, p. 67

AITKEN, R. (1987). Safety in handling. *Plant Engr*, **31**(2), 18

VAN AKEN, J.A., SCHMIDT, A.C.G., VAN DER VET, R.P. and WOLTERS, W.K. (1984). A reliability-based method for the exploitation of maintenance opportunities. *Eighth Symp. on Advances in Reliability*

AKERSTEDT, T. and TORSVALL, L. (1981). Shift work. Shift-dependent well being and individual differences. *Ergonomics*, **24**, 265

AKITA, K. (1959). Studies on the mechanism of ignition of wood. Report of the Fire Research Institute of Japan 1959, **9**(1–2), 99 (English summary)

AKITA, K. (1973). Some problems of flame spread across a liquid surface. *Combustion*, **14**, 1075

AKITA, K. and FUJIWARA, O. (1971). Pulsating flame spread across the surface of liquid fuels (letter). *Combust. Flame*, **17**, 268

AKITA, K.A. and YUMOTO, T. (1965). Heat transfer in small pools and rates of burning of liquid methanol. *Combustion*, **10**, 943

AKITSUNE, K., TAKAHASHI, H. and JOJIMA, T. (1978). Start-up of ammonia-urea fertilizers plant. *Ammonia Plant Safety*, **20**, 51

AKIYAMA, S., HISAMOTO, T. and MOCHIZUKI, S. (1981). Chloromethanes from methanol. *Hydrocarbon Process.*, **60**(3), 76

VAN DEN AKKER, H.E.A. (1984). Discharges of saturated and superheated liquids from pressure vessels. Prediction of homogeneous choked two-phase flow through pipes. In *The Protection of Exothermic Reactors and Pressurised Storage Vessels* (Rugby: Instn Chem. Engrs), p. 91

VAN DEN AKKER, H.E.A., SNOEY, H. and SPOELSTRA, H. (1983). Discharges of pressurised liquefied gases through apertures and pipes. *Loss Prevention and Safety Promotion 4*, vol. 1, p. E23

ALAMARO, M. (1994). Bigger is not always better. *Chemtech* 34, 9 (September), 49–50

AL-ABDULALLY, F., AL-SHUWAIB, S. and GUPTA, B.L. (1987). Hazard analysis and safety considerations in refrigerated ammonia storage tanks. *Ammonia Plant Safety* **27**, 153. See also *Plant/Operations Prog.*, 1987, **6**, 84

AL-ASADI, H. (1980). Computer aided layout of chemical plant. Ph.D. Thesis, University College of Swansea

AL-ATTAR, A. and HASSAN, T. (1994). *Rule Induction in Knowledge Engineering* (Chichester: Ellis Horwood)

ALAGY, J. *et al.* (1968). Selectively oxidise cyclohexane. *Hydrocarbon Process.*, **47**(12), 131

ALAGY, J., TRAMBOUZE, P. and VAN LANDEGHEM, H. (1974). Designing a cyclohexane reactor. *Ind. Engng Chem. Process Des. Dev.*, **13**, 317

ALARIE, Y. (1980). Toxicological evaluation of airborne chemical irritants and allergens using respiratory reflex reactions. *Proc., Symp. on Inhalation Toxicity and Technology* (Michigan: Ann Arbor Science), p. 207

ALBA, C. (1970). Size rupture discs by nomograph. *Hydrocarbon Process.*, **49**(9), 297

ALBAUGH, L.R., WILLIAMS, G.M., KATZ, M.J. and PRATT, T.H. (1979). How hazard analysis can help. *Loss Prevention*, **12**, p. 30

ALBERT, A. (1985). *Selective Toxicity*, 7th ed. (London: Chapman & Hall)

ALBERTSON, M.L., DAI, Y.B., JENSEN, R.A. and ROUSE, H. (1950). Diffusion of submerged jets. *Trans. Am. Soc. Civil Engrs*, **115**, 639

ALBINI, F.A. (1981). A model for the wind blown flame from a line fire. *Combust. Flame*, **43**, 151

ALBISSER, R.H. and SILVER, L.H. (1960). Safety evaluation of new processes. *Ind. Engng Chem.*, **52**(11), 61A

ALBISSER, R.H. and SILVER, L.H. (1964). Safety evaluation of new processes. In Vervalin, C.H. (1964), *op. cit.*, p. 72

ALBRECHT, A.R. and SEIFERT, W.F. (1970). Accident prevention in high temperature heat transfer fluid systems. *Loss Prevention*, **4**, p. 67

ALBRIGHT, H.E. (1979). Apply large high-speed synchronous motors. *Hydrocarbon Process.*, **58**(1), 181; **58**(2), 129

ALBRIGHT, L.F. (1966a). Alkylation processes using hydrogen fluoride as catalyst. *Chem. Engng*, **73**(Sep. 12), 205

ALBRIGHT, L.F. (1966b). Chemistry of aromatic nitration. *Chem. Engng*, **73**, Apr. 25, 169

ALBRIGHT, L.F. (1966c). Comparisons of alkylation processes. *Chem. Engng*, **73**(Oct. 10), 209

ALBRIGHT, L.F. (1966d). High-pressure processes for polymerising ethylene. *Chem. Engng*, **73**(Dec. 19), 113

ALBRIGHT, L.F. (1966e). Nitration of paraffins. *Chem. Engng*, **73**(Jun. 6), 149

ALBRIGHT, L.F. (1966f). Polymerization of ethylene. *Chem. Engng*, **73**(Nov. 21), 127

ALBRIGHT, L.F. (1966g). Processes for nitration of aromatic hydrocarbons. *Chem. Engng*, **73**(May 9), 161

ALBRIGHT, L.F. (1967a). Commercial low-pressure processes for polymerization of ethylene. *Chem. Engng*, **74**(Feb. 13), 159

ALBRIGHT, L.F. (1967b). Commercial vapor phase processes for partial oxidation of light paraffins. *Chem. Engng*, **74**(Aug. 14), 165

ALBRIGHT, L.F. (1967c). Low pressure polymerization of ethylene with solid catalysts. *Chem. Engng*, **74**(Jan. 16), 169

ALBRIGHT, L.F. (1967d). Manufacturers of vinyl chloride. *Chem. Engng*, **74**(Apr. 10), 219

ALBRIGHT, L.F. (1967e). Partial oxidation of light paraffins. *Chem. Engng*, **74**(Jul. 17), 197

ALBRIGHT, L.F. (1967f). The properties, chemistry and synthesis of aluminum alkyls. *Chem. Engng*, **74**(Dec. 4), 179

ALBRIGHT, L.F. (1967g). Vinyl chloride processes. *Chem. Engng*, **74**(Mar. 27), 123

ALBRIGHT, L.F., CRYNES, B.L. and CORCORAN, W.H. (1983). *Pyrolysis: Theory and Practice* (Orlando, FL: Academic Press)

ALBRIGHT, L.F. and HANSON, C. (1969). Kinetics and mechanism of aromatic nitrations. *Loss Prevention*, **3**, 26

ALBRIGHT, L.F. and HANSON, C. (1976). *Industrial and Laboratory Nitrations* (Washington, DC: Am. Chem. Soc.)

ALBRIGHT, L.F., VAN MUNSTER, F.H. and FORMAN, J.C. (1968). High pressure: principles and process trends. *Chem. Engng*, **75**(Sep. 23), 194

ALCAZAR, C. and AMILLO, M.S. (1984). Get fuel from the flare. *Hydrocarbon Process.*, **63**(7), 63

ALCOCK, P. (1985). Dynamic simulation as an Engineering support tool. *Process Engng*, Sep., 40

ALDER, W.A.T. and ASHURST, J.A.S. (1992). The development of risk criteria for application to new industries. *Hazard Identification and Risk Analysis, Human Factors and Human Reliability in Process Safety*, p. 229

ALDERSEY, M.L., LEES, F.P. and RUSHTON, A.G. (1991). Knowledge elicitation and representation for diagnostic tanks in the process industries: forms of representation and translation between these forms. *Process Safety Environ.*, **69B**, 187

ALDERSON, M. (1976). *An Introduction to Epidemiology* (London: Macmillan)

ALDERSON, M.A.H.G. (1985). *Seismic Engineering requirements in British industry. Earthquake Engineering in Britain* (London: Instn Civil Engrs), p. 243

ALDIS, D.F. and GIDASPOW, D. (1990). Two-dimensional analysis of a dust explosion. *AIChE J.*, **36**, 1087

ALDIS, D.F. and LAI, F.S. (1979). *Review of Literature Related to Engineering Aspects of Grain Dust Explosions*. Misc. Publ. 1375. Sci. and Educ. Admin., Agricult. Res., Dept of Agricult. and Kansas State Univ., Manhattan, KS

ALDRED, J.W. and WILLIAMS, A. (1966). The burning rates of drops of n-alkanes. *Combust. Flame*, **10**, 396

ALDRICH, C.K. (1960a). Design tips reduce tank failures. *Chem. Engng*, **67**(Aug. 8), 154

ALDRICH, C.K. (1960b). How to extend storage tank life. *Chem. Engng*, **67**(Jul. 25), 148

ALDRICH, C.K. (1960c). Selecting materials for process piping. *Chem. Engng*, **67**(Nov. 14), 183

ALDWINKLE, D.S. and MCLEAN, D. (1985). A safety review of ships for liquefied gases and future legislative needs. *Gastech* **84**, 120

ALDWINCKLE, D.S. and SLATER, D.H. (1983). Risk and reliability methods used in the analysis of an offshore LNG liquefaction and storage ship. *Loss Prevention and Safety Promotion* 4, vol. 2, p. B78

ALE, B.J.M. (1980). Analysis of the operating stability of a caprolactam reactor. *Loss Prevention and Safety Promotion* 3, vol. 4, p. 1541

ALE, B.J.M. (1991). Risk analysis and risk policy in the Netherlands and the EEC. *J. Loss Prev. Process Ind.*, **4**(1), 58

ALE, B.M.J. (1992). The implementation of an external safety policy in the Netherlands. *Hazard Identification and Risk Analysis, Human Factors and Human Reliability in Process Safety*, p. 173

ALE, B.J.M. and BRUNING, F. (1980). Unconfined vapour cloud explosions (letter). *Chem. Engr, Lond.*, **352**, 47 and **358**, 494

ALE, B.J.M., BRUNING, F. and KOENDERS, H.A.A. (1980). The limits of flammability of mixtures of ammonia, hydrogen, nitrogen and oxygen at elevated temperatures and pressures. In *Loss Prevention and Safety Promotion 3*, vol. 3, p. 1405

ALE, B.J.M. and WHITEHOUSE, J.J. (1986). Computer based system for risk analysis of process plants. In Hartwig, S. (1986), *op. cit.*, p. 343

D'ALESSANDRO, P.L. and COBB, C.B. (1976a). Hazardous material control for bulk storage facilities. In *Control of Hazardous Materials Spills 3*, p. 39

D'ALESSANDRO, P.L. and COBB, C.B. (1976b). Oil spill control. *Hydrocarbon Process.*, **55**(2), 121; **55**(3), 145

ALESSO, H.P. (1983). Some algebraic aspects of decomposed non-coherent structure functions. *Reliab. Engng*, **5**, 129

ALESSO, H.P. (1985). On the relationship of digraph matrix analysis to Petri Net theory and fault trees. *Reliab. Engng*, **10**(2), 93

ALESSO, H.P., PRASSINOS, P. and SMITH, C.F. (1985). Beyond fault trees to fault graphs. *Reliab. Engng*, **12**(2), 79

ALEXANDER, D.S. and FINIGAN, C.M. (1964). Inert gas + poor piping = explosion. In Vervalin, C.H. (1964), *op. cit.*, p. 335

ALEXANDER, J.M. (1974). The hazard of gas phase oxidation in liquid phase air oxidation processes. *Chemical Process Hazards V*, p. 157

ALEXANDER, J.M. (1990a). Chemical plant underwriting and risk assessment – a personal view. *Safety and Loss Prevention in the Chemical and Oil Processing Industries*, p. 75

ALEXANDER, J.M. (1990b). Gas phase ignition in liquid phase air oxidation processes: a recipe for disaster. *Process Safety Environ.*, **68B**, 17

ALEXANDER, J.M. (1990c). Gas phase ignition in liquid phase air oxidation processes – a recipe for disaster. *Safety and Loss Prevention in the Chemical and Oil Processing Industries*, p. 195

ALEXANDER, J. (1992a). Coode hot spots. *Chem. Engr, Lond.*, **514**, 4

ALEXANDER, J.M. (1992b). Fire risk management for bulk atmospheric storages. *CHEMECA 92*, 3.2

ALEXANDER, M. (1980). Biodegradation of toxic chemicals in water and soil. *In Haque, R. (1980), op. cit.*, p. 179

ALEXANDER, S.J. and HAMBLY, E.C. (1970). The design of structures to withstand gaseous explosions. *Concrete*, **4**, 62, 107

ALEXANDER, S.J. and TAYLOR, N. (1973a). Field surveys of the damage caused by gaseous explosions. *Proc. Br. Ceram. Soc.*, **21**, 221

ALEXANDER, S.J. and TAYLOR, N. (1973b). *The Frequency and Severity of Gaseous Explosions* (London: Instn Gas Engrs)

ALEXEEFF, G.V., LEE, Y.C., WHITE, J.A. and PUTAS, N.D. (1986). Use of an approximate lethal exposure method for examining the acute inhalation toxicity of combustion products. *J. Fire Sci.*, **4**(2), 100

ALEXEEFF, G., LEWIS, D. and LIPSETT, M. (1992). Use of toxicity information in risk assessment for accidental releases of toxic gases. *J. Haz. Materials*, **29**, 387

ALFORD, P.B. (1965). 'Design maintenance' optimizes on-stream time. *Chem. Engng*, **72**(Feb. 1), 96

ALGER, R.S. and CAPENER, E.L. (1972). *Aircraft Ground Fire Suppression and Rescue Systems – Basic Relationships for Military Fires – Phases I and II. Rep. AGFSRS-72-1.* US Naval Ordinance Lab, MD

ALGER, R.S., CORLETT, R.C., GORDON, A.S. and WILLIAMS, F.A. (1979). Some aspects of structures of turbulent pool fires. *Fire Technol.*, **15**, 142

ALI, Y. and NARASIMHAN, S. (1993). Sensor network design for maximizing reliability of linear processes. *AIChE J.*, **39**, 820

ALIMONTI, C., FRITTE, A. and GIOT, M. (1990). *Critical Two-Phase-Flow: Models and Data. Eur.* Two-Phase Flow Group Mtg, EDF, Paris

ALKIDAS, A. and DURBETAKI, P. (1973). Ignition of a gaseous mixture by a heated surface. Combust. *Sci. Technol.*, **7**, 135

ALLABY, M. (1977). *A Dictionary of the Environment* (New York: van Nostrand Reinhold)

ALLAN, D. (1986). Compressed air. *Plant Engr*, **30**(5), 13

ALLAN, D.S. and ATHENS, P. (1968). Influence of explosions on design. *Loss Prevention*, **2**, 103

ALLAN, D.S., BROWN, A.A. and ATHENS, P. (1975). Risks associated with an LNG shipping operation. *Transport of Hazardous Cargoes*, **4**

ALLAN, D.S. and SCHIFF, D.S. (1975). A cryogenic gas leak detector. *Fire Technol.*, **11**, 270

ALLAN, F.J., CAMERON, D.E. and LAMBIE, D.A. (1966). The calorific values of partially decomposed wood samples (letter). *Combust. Flame*, **10**, 394

ALLAN, R.N., ADRAKTAS, A. and CAMPBELL, J.F. (1982). Interactive safety assessment of nuclear reactor systems. *Reliab. Engng*, **3**(5), 393

ALLAN, R.N., SHAMAN, A.M. and OUCH, JR (1986). Cost benefit and reliability assessment of electrical generating systems. *Reliab. Engng*, **14**(4), 1

ALLARDICE, C. and TRAPNELL, E. (1974). *The Atomic Energy Commission* (New York: Praeger)

ALLEN, A. (ed.) (1988). *Chemical Safety Data Sheets*, vols 2–5 (London: Royal Society of Chemistry). See also Walsh, D. (1988)

ALLEN, A.G., HARRISON, R.M. and NICHOLSON, K.W. (1991). Dry deposition of fine aerosol to a short grass surface. *Atmos. Environ.*, **25A**, 2671

ALLEN, C.B. and JANZ, G.J. (1980). Molten salts safety and hazards: an annotated bibliography. *J. Haz. Materials*, **4**(2), 145

ALLEN, C.R. (1967). Earthquakes, faulting and nuclear reactors, *Engng and Sci.*, Nov., 10

ALLEN, C.S. (1989). *Piper Alpha Public Inquiry. Transcript, Day* 154

ALLEN, D.F. (1976). Colour coding of piping systems – identification for safety. *BSI News*, Sep. 7

ALLEN, D.H. (1968). Credibility forecasts and their application to the economic assessment of novel research and development projects. *Oper. Res. Quart.*, **19**(1), 25

ALLEN, D.H. (1973). Economic aspects of plant reliability. *Chem. Engr, Lond.*, **278**, 467

ALLEN, D.H. (1975). Investment decisions – evaluation techniques in perspective. *Chem. Engng, Lond.*, **293**, 42

ALLEN, D.H. (1988). *Economic Evaluation of Projects*, 3rd ed. (Rugby: *Instn Chem. Engrs*)

ALLEN, D.H. and COKER, F. (1977). A case study in system availability analysis – a beer filtering and kegging plant. *First Nat. Conf. on Reliability. NRC4/13* (Culcheth, Lancashire: UK Atom. Energy Auth.)

ALLEN, D.H., GARNER, F.H., LONG, R. and TODD, J.R. (1959). The low-temperature vapour-phase oxidation of cyclohexane. *Combust. Flame*, **3**, 75

ALLEN, D.H. and PEARSON, G.D.M. (1976). *Techniques for the availability evaluation of chemical processing systems. Advances in Technology* (Culcheth, Lancashire: UK Atom. Energy Auth.), ART4

ALLEN, D.J. (1981). Fault tree analysis of a proposed ethylene vaporisation unit. *Ind. Engng Chem. Fund.*, **20**, 304

ALLEN, D.J. (1984). Digraphs and fault trees. *Ind. Engng Chem. Fund.*, **23**, 175

ALLEN, D.J. and RAO, M.S.M. (1980). New algorithms for the synthesis and analysis of fault trees. *Ind. Engng Chem. Fund.*, **19**, 79

ALLEN, D.W. (1969). Safety considerations for ammonia plant catalysts. *Safety in Air and Ammonia plants*, **11**, 51

ALLEN, E.E. (1972). Valves can be quiet. *Hydrocarbon Process.*, **51**(10), 137

ALLEN, F. (1988). Description of the Chernobyl accident. In Worley, N.G. and Lewins, J. (1988), *op. cit.*, p. 19

ALLEN, SIR G. (1984). Impact of industrial and academic collaboration on new technology. *Proc. R. Soc. Lond., Ser. A*, **393**, 1

ALLEN, G. (1984). Do cone roof tanks court disaster? *Hydrocarbon Process.*, **63**(8), 109

ALLEN, G.D., WEY, R.E. and CHAN, H.H. (1983). Flare gas recovery success at Canadian refineries. *Oil Gas J.*, **81**(Jun. 27), 79

ALLEN, J. (1978). Anatomy of LISP (New York: McGraw-Hill)

ALLEN, J. (1979). An overview of LISP. *Byte*, **4**(8)

ALLEN, J. (1987). *Understanding Natural Language* (Menlo Park, CA: Benjamin Cummings)

ALLEN, J., HENDLER, J. and TATE, A. (eds) (1990). Readings in Planning (San Mateo, CA: Morgan Kaufmann)

ALLEN, L.J. (1975). *The Trans-Alaska Pipeline* (Seattle, WA: Scribe Publ. Corp.)

ALLEN, O.M. and HANNA, O.J. (1971). Arrow diagrams improve operational safety. *Chem. Engng*, **78**(Sep. 20), 166

ALLEN, R.L. (1977a). Ammunition and explosives – an approach to risk management in a hazardous field. In Council for Science and Society (1977), *op. cit.*, p. 84

ALLEN, R.L. (1977b). Some comments on the report of the Court of Inquiry into the Flixborough disaster. In Council for Science and Society (1977), *op. cit.*, p. 75

ALLEN, T. (1975). *Particle Size Measurement* (London: Chapman & Hall)

ALLEN, W.F. (1952–). How to size piping and valves for flashing mixtures. *Power Engng*, 1952: May 1960; Jun. 1983; Aug. 1985; 1953; Oct. 1994

ALLGOOD, JR and SWIHART, G.R. (1970). Design of flexural members for static and blast loading. *ACI Monograph 5* (Detroit: Am. Concrete Inst.)

ALLIANZ VERSICHERUNG AG (1978). *Handbook of Loss Prevention* (Berlin: Springer)

ALLINSON, J.P. (1966). Protecting liquid petroleum vessels from fire (letter). *The Engineer*, July

VON ALLMEN, E. (1960). Economics are basic for plant location. *Chem. Engng Prog.*, **56**(11), 37

ALLSOP, P. (ed.) (1962). *Encyclopaedia of Factories Shops and Offices: Law and Practices* (London: Sweet & Maxwell)

ALLSOP, R.E., BANNISTER, D.J., HOLDEN, D.J. BIRD, J. and DOWNE, H.E. (1986). A methodology for assessing social considerations in transport of low and intermediate level radioactive waste, Dept of the Environment, Rep. DoE/RW/86–020

ALLUM, S. and WELLS, G.L. (1993). Short-cut risk assessment. *Process Safety Environ.*, **71B**, 161

ALMON, W.S. (1991). Pressure vessel management by computer. *Hydrocarbon Process.*, **70**(6), 97

ALONSO, J.R.F. (1971). Estimating the costs of gas-cleaning plant. *Chem. Engng*, **78**(Dec. 13), 86

ALP, E. (1985). *COBRA: an LNG model. Heavy Gas Workshop* (Downsview, Ont.: Concord Sci. Corp.)

ALP, E., ALP, S., BROWN, P.M. and PORTELLI, R.V. (1986). Heavy gas dispersion modelling. In de Wispelaere, C., Schiermeier, F.A. and Gillani, N.V. (1986), *op. cit.*, p. 653

ALP, E. and MATTHIAS, C.S. (1991). COBRA. A heavy gas/liquid spill and dispersion modelling system. *J. Loss Prev. Process Ind.*, **4**, 139

ALPATOV, I.M. (1964). A study of gaseous ammonia toxicity. *Gig. Tr. Prof. Zabol.*, **2**, 14

ALPATOV, I.M. and MICHAILOV, V.I. (1963). Inquiries into gaseous ammonia toxicity. *Gig. Tr. Prof. Zabol.*, **7**, Dec., **51**

ALPERT, M. and RAIFFA, H. (1969). A progress report on the training of probability assessors. Unpublished paper

ALPERT, R.L. (1975). Turbulent ceiling-jet induced by large-scale fires. *Combust. Sci. Technol.*, **11**, 197

ALPERT, R.L. and WARD, E.J. (1984). Evaluation of unsprinklered fire hazards. *Fire Safety J.*, **7**, 127

ALRAMADHAN, M.A., ARPACI, V.S. and SELAMET, A. (1991). Radiation affected liquid pool burning on water. *Combust. Sci. Technol.*, **72**, 233

ALSPACH, J. and BIANCHI, R.J. (1984). Safe handling of phosgene in chemical processing. *Plant/Operations Prog.*, **3**, 40

ALTMAN, J.W. (1964). A central store of human performance data. *Symp. on Quantification of Human Performance*, Sandia Labs, Albuquerque, NM

ALTMAN, P.L. and DITTMER, D.S. (1961). *Blood and Other Body Fluids* (Bethesda, MD: Fed. Am Soc. Exp. Biol.)

ALTMAN, P.L. and DITTMER, D.S. (1968). *Metabolism* (Bethesda, MD: Fed. Am Soc. Exp. Biol.)

ALTMAN, P.L. and DITTMER, D.S. (1971). *Respiration and Circulation* (Bethesda, MD: Fed. Am Soc. Exp. Biol.)

ALTMAN, P.L. and DITTMER, D.S. (ed.) (1974). *Biology Data Book*, 2nd ed., vol. 3 (Bethesda, Md: Fed. of Am. Soc. Exp. Biol.)

ALTMAN, P.L. and KATZ, D.D. (eds) (1979). *Inbred and Genetically Defined Strains of Laboratory Animals. Pt 1, Mouse and Rat* (Bethesda, MD: Fed. Am. Soc. Exp. Biol.)

ALTORFER, F. (1992). Explosion protection for waste gas combustion. *Loss Prevention and Safety Promotion 7*, vol. 3, 105.1

ALTY, J.L. and BERGAN, M. (1992). Guidelines for multimedia design in a process control application. *Int. Conf. on Design and Safety of Nuclear Reactors ANP'92* (ed. Oka and Koshizuka) (Atomic Energy Soc. of Japan), vol. 4, p. 43 IV-1

ALTY, J.L. and COOMBS, M.J. (1984). *Expert Systems. Concepts and Examples* (Manchester: NCC)

ALTY, J.L. and MCCARTNEY, M.D. (1991). Managing multimedia resources in process control: problems and solutions. In *Multimedia: Systems, Interactions and Applications* (ed. L. Kjelldahl) (Berlin: Springer), p. 293

ALVARES, N.J. (1975). Some experiments to delineate the conditions of flashover in enclosure fires. *Int. Symp. on Fire Safety of Combustible Materials* (Edinburgh: Univ. of Edinburgh), p. 375

ALVARES, N.J. and HASEGAWA, H.K. (1979/80). Fire hazard analysis for fusion experiments. *Fire Safety J.*, **2**(3), 191

ALVARES, N.J., FOOTE, K.L. and PAGNI, P.J. (1984). Forced ventilated enclosure fires. *Combust. Sci. Technol.*, **39**, 55

ALVARES, N.J. and MARTIN, S.B. (1971). Mechanisms of ignition of thermally irradiated cellulose. *Combustion*, **13**, 905

VON ALVEN, W.H. (ed.) (1964). *Reliability Engineering* (New York: Wiley)

ALVEY, P.L., MYERS, C.D. and GREAVES, M.F. (1984). *An analysis of the problems of augmenting a small expert system. Research and Development in Expert Systems* (ed. R.A. Bramer) (Cambridge: Cambridge Univ. Press), p. 61

AMAL LTD (1980). *Flame Arresters* (Birmingham)

AMBLE, T. (1987). *Logic Programming and Knowledge Engineering* (Wokingham: Addison-Wesley)

AMDUR, M.O. (1958). The respiratory response of guinea pigs to sulphuric acid mist. *Arch. Ind. Health*, **18**, 407

AMDUR, M.O., DUBRIEL, M. and CREASIA, D.A. (1978). Respiratory response of guinea pigs to low levels of sulphuric acid. *Environ. Res.*, **15**, 418

AMDUR, M.O., SCHULZ, R.Z. and DRINKER, P. (1952). Toxicity of sulphuric acid mist to guinea pigs. *Arch. Ind. Hyg. Occup. Med.*, **5**, 318

AMDUR, M.O., SILVERMAN, L. and DRINKER, P. (1952). Inhalation of sulphuric acid mist by human subjects. *Arch. Ind. Hyg. Occup. Med.*, **6**, 305

AMENDOLA, A. (1983a). An international benchmark exercise on system reliability. *Euredata* 4, paper 19.4

AMENDOLA, A. (1983b). Presentation of the reliability benchmark exercise. *Tech. Note 1.05.01.83.36* Comm. Eur. Commun., Joint Res. Centre, Ispra, Italy

AMENDOLA, A., CONTINI, S. and NICHELE, P. (1988). MARS: the major accident reporting system. *Preventing Major Chemical and Related Process Accidents*, p. 445

AMENDOLA, A., CONTINI, S. and ZIOMAS, I. (1992a). Procedures and uncertainties in chemical risk assessment: results of a European benchmark exercise. *Loss Prevention and Safety Promotion 7*, vol. 1, 33.1

AMENDOLA, A., CONTINI, S. and ZIOMAS, I. (1992b). Uncertainties in chemical risk assessment: results of a European benchmark exercise. *J. Haz. Materials*, **29**, 347

AMENDOLA, A. and MELIS, M. (1983). Reliability component models and data banks. *Euredata 4*, paper 5.4

AMENDOLA, A.M., SILVESTRI, E., CONTINI, B. and SQUELLATI, E. (1983). Algorithms for automated fault tree construction. *Euredata 4*, paper 3.1

AMENDOLA, S.C., SHARP-GOLDMAN, S.L., JANJUA, M.S., KELLY, M.T., PETILLO, P.J. and BINDER, M. (2000). *An ultrasafe hydrogen generator: aqueous, alkaline borohydride solutions and ru catalyst. J. Power Sources* **85**, 186–9.

AMERICAN CHEMISTRY COUNCIL RESPONSIBLE CARE® *Code for Process Safety, (Management Practice 4)* (Washington DC: ACC)

AMERICAN CHEMICAL SOCIETY (1956). Industrial foams – their characteristics and applications. *Ind. Engng Chem.*, **48**

AMERICAN CONFERENCE OF GOVERNMENT INDUSTRIAL HYGIENISTS (n.d.). *A Guide to the Control of Laser Hazards*

AMERICAN CONFERENCE OF GOVERNMENT INDUSTRIAL HYGIENISTS (1968). *Industrial Ventilation*, 10th ed. (Lansing, MI)

AMERICAN CONFERENCE OF GOVERNMENT INDUSTRIAL HYGIENISTS (1972). *Air Sampling Instruments for Evaluation of Atmospheric Contaminants*, 4th ed. (Lansing, MI)

AMERICAN CONFERENCE OF GOVERNMENT INDUSTRIAL HYGIENISTS – see also Appendix 28

AMERICAN CONFERENCE OF GOVERNMENT INDUSTRIAL HYGIENISTS and AMERICAN INDUSTRIAL HYGIENE ASSOCIATION (1963). *Respiratory Protective Devices Manual* (Ann Arbor, MI: Braun-Brumfield)

AMERICAN GAS ASSOCIATION (1954). *Purging Principles and Practice* (New York)

AMERICAN GAS ASSOCIATION (1965). *Gas Engineers Handbook* (New York: Industrial Press)

AMERICAN GAS ASSOCIATION (1973). *LNG Information Book* (Arlington, VA: Nat. Tech. Inf. Service)

AMERICAN GAS ASSOCIATION (1974). *LNG Safety Program: Phase II, Proj. IS–3–1* (Arlington, VA: Nat. Tech. Inf. Service)

AMERICAN GAS ASSOCIATION (1975). *Purging Principles and Practice*, 2nd ed. (Arlington, VA)

AMERICAN GAS ASSOCIATION – see also Appendix 28

AMERICAN INDUSTRIAL HYGIENE ASSOCIATION – see Appendices 27 and 28

AMERICAN INSTITUTE OF CHEMICAL ENGINEERS (1994). *1994 CEP Software Directory* (New York)

AMERICAN INSTITUTE OF CHEMICAL ENGINEERS, CENTER FOR CHEMICAL PROCESS SAFETY (1985). *Guidelines for Hazard Evaluation Procedures* (New York)

AMERICAN INSTITUTE OF CHEMICAL ENGINEERS – see also Appendix 28

AMERICAN INSURANCE ASSOCIATION (1962). *Handbook of Industrial Safety Standards* (New York)

AMERICAN INSURANCE ASSOCIATION (1967). *National Building Code* (New York)

AMERICAN INSURANCE ASSOCIATION (1979). *Hazard Survey of the Chemical and Allied Industries* (New York)

AMERICAN INSURANCE SERVICES GROUP (1986). *Fire Resistance Ratings*

AMERICAN IRON AND STEEL INSTITUTE (1980). *Designing Fire Protection for Steel Columns*, 3rd ed.

AMERICAN METEOROLOGICAL SOCIETY (1977). Stability classification schemes and sigma curves—a summary of recommendations. *Bull. Am. Met. Soc.*, **58**, 1305

AMERICAN METEOROLOGICAL SOCIETY (1978). Accuracy of dispersion models. *Bull. Am. Met. Soc.*, **59**, 1025

AMERICAN NATIONAL STANDARDS – see Appendix 27

AMERICAN NUCLEAR SOCIETY (1980). An evaluation of the Rasmussen and Lewis reports and of the January 1979 NRC policy statement on risk assessment. *Nucl. News*, June, 177

AMERICAN OIL COMPANY – see Appendix 28

AMERICAN PETROLEUM INSTITUTE (1956). *Sparks from Hand Tools* (New York)

AMERICAN PETROLEUM INSTITUTE (1966). Training for refinery fire fighting. *Fire Int.*, **1**(12), 47

AMERICAN PETROLEUM INSTITUTE (1966). *API Exposure Classification Scheme for Collection of Industrial Hygiene Monitoring Data* (New York)

AMERICAN PETROLEUM INSTITUTE – see also Appendices 27 and 28

AMERICAN SOCIETY OF CIVIL ENGINEERS (1963). Design of structures to resist nuclear weapons effects. *Manual of Engineering Practice 42*, Suppl. (New York)

AMERICAN SOCIETY OF CIVIL ENGINEERS (1984). *Guidelines for the Seismic Design of Oil and Gas Pipeline Systems* (New York)

AMERICAN SOCIETY OF CIVIL ENGINEERS – see also Appendices 27 and 28

AMERICAN SOCIETY OF HEATING, REFRIGERATING AND AIR CONDITIONING ENGINEERS (1972). *ASHRAE Handbook of Fundamentals* (New York)

AMERICAN SOCIETY OF HEATING, REFRIGERATING AND AIR-CONDITIONING ENGINEERS (1973). *Standards for Natural and Mechanical Ventilation* (New York)

AMERICAN SOCIETY OF HEATING, REFRIGERATING AND AIR-CONDITIONING ENGINEERS – see also Appendices 27 and 28

AMERICAN SOCIETY OF MECHANICAL ENGINEERS (1982). Use of acoustic emission examination in lieu of radiography.

ASME Boiler and Pressure Vessel Code, Section VIII, Division 1, Code Case 1968 (New York)

AMERICAN SOCIETY OF MECHANICAL ENGINEERS (1983). Acoustic emission examination of fiber-reinforced plastic vessels. *ASME Boiler and Pressure Vessel Code*, Section V, Subsection A, Article 11 (New York)

AMERICAN SOCIETY OF MECHANICAL ENGINEERS (1989). Acoustic emission examination of metallic pressure vessels during pressure testing. *ASME Boiler and Pressure Vessel Code*, Section V, Article 12 (New York)

AMERICAN SOCIETY OF MECHANICAL ENGINEERS — see *also Appendices 27 and 28*

AMERICAN SOCIETY OF PERSONNEL ADMINISTRATION (1970). *A Handy Reference Guide to the OSHA—1970* (Berea, OH)

AMERICAN SOCIETY FOR QUALITY CONTROL — see Appendix 27

AMERICAN SOCIETY OF SAFETY ENGINEERS (1966). *Scope and function of the professional safety position*

AMERICAN SOCIETY FOR TESTING AND MATERIALS (1967). *ASTM Standards in Building Codes* (Philadelphia, PA)

AMERICAN SOCIETY FOR TESTING AND MATERIALS (1975). *CHE-TAH Program* (Philadelphia, PA)

AMERICAN SOCIETY FOR TESTING AND MATERIALS (1980). *The Prevention of Occupational Cancer* (Philadelphia, PA)

AMERICAN SOCIETY FOR TESTING AND MATERIALS — see also Appendices 27 and 28

AMERICAN WELDING SOCIETY — see Appendices 27 and 28

VAN AMERONGEN, L. (1968). A guide to rigging for operations and maintenance supervisors. *Chem. Engng*, **75**(Apr. 22), 202

VAN AMERONGEN, L. (1970). Safe maintenance rigging. *Chem. Engng*, **77**(Sep. 21), 193

AMES, B.N. (1971). *The Detection of Chemical Mutagens with Enteric Bacteria. Chemical Mutagens: Principles and Methods for Their Detection*, vol. 1 (ed. A. Hollaender) (New York: Plenum Press), p. 267

AMES, B.N., DURSTON, W.E., YAMASAKI, E. and LEE, F.D. (1973). Carcinogens and mutagens: a simple test system combining liver homogenates for activation and bacteria for detection. *Proc. Nat. Acad. Sci.*, **70**, 2281

AMES, J., MYERS, T.C., REID, L.E., WHITNEY, D.C., GOLDING, S.H., HAYES, S.R. and REYNOLDS, S.D. (1985). *SAI Airshed Model Operations Manuals*, vol. 1, *Users manual. Rep. EPA-600/8—85—007a* Environ. Prot. Agency, Research Triangle Park, NC

AMESZ, J., FRANCOCCI, G., PRIMAVERA, R. and VAN DER PAS, T. (1983). The European abnormal occurrences reporting system. *Euredata* 4, paper 6.2

AMMANN, H.M. (1986). A new look at physiological respiratory response to H$_2$S poisoning. *J. Haz. Materials*, **13**(3), 369

AMOORE, J.E. and HAUTALA, E. (1983). Odor as an aid to chemical safety: odor thresholds compared with Threshold Limit Values and volatilities for 214 industrial chemicals in air and water dilution. *J. Appl. Toxicol.*, **3**, 272

AMOS, C.N. *et al.* (1983). *Critical Discharge of Initially Subcooled Water Through Slits. Rep. NUREG/CR-3475.* Nucl. Regul. Comm., Washington, DC

AMREHN, H. (1969). Operating experiences with process computers in chemical industry. *Regelungstechnische Praxis*, **11**, 19

AMROGOWICZ, J. and KORDYLEWSKI, W. (1991). Effectiveness of dust explosion suppression by carbonates and phosphates. *Combust. Flame*, **85**, 520

AMSON, J.E. (1976). An analysis of the economic impact of hazardous substances the *Hazardous Materials Spills*, **3**, 49

AMSON, J.E. and GOODIER, J.L. (1978). An analysis of the economic impact of spill prevention costs on the chemical industry. *Control of Hazardous Materials Spills* 4, p. 39

AMSTADTER, R.L. (1972). *Reliability Mathematics* (New York: McGraw-Hill)

AMYOT, D.E. (1980). Report on Mississauga Train Derailment and Evacuation Emergency Planning Canada, Toronto, Ont., Rep.

AMYOTTE, P.R., BAXTER, B.K. and PEGG, M.J. (1990). Influence of initial pressure on spark-ignited dust explosions. *J. Loss Prev. Process Ind.*, **3**(2), 261

AMYOTTE, P.R., MINTZ, K.J. and PEGG, M.J. (1992). Effectiveness of various rock dusts as agents of coal dust inerting. *J. Loss Prev. Process Ind.*, **5**, 196

AMYOTTE, P.R., KHAN, F.I. and DASTIDAR, A.G. (2003). Reduce dust explosions the inherently safer way. *Chemical Eng. Prog.* **99**(Oct. 10), 36—43.

AMYOTTE, P.R., MINTZ, K.J., PEGG, M.J., SUN, Y.H. and WILKIE, K.I. (1991). Laboratory investigation of the dust explosibility characteristics of three Nova Scotia coals. *J. Loss Prev. Process Ind.*, **4**(2), 102

AMYOTTE, P.R. and PEGG, M.J. (1989). Lycopodium dust explosions in a Hartmann bomb: effects of turbulence. *J. Loss Prev. Process Ind.*, **2**(2), 87

AMYOTTE, P.R. and PEGG, M.J. (1992). Dust explosion prevention by addition of thermal inhibitors. *Plant/ Operations Prog.*, **11**, 166

ANAGNOSTOU, E. and POTTER, A.E. (1959). Quenching diameters of some fast flames at low pressures. *Combust. Flame*, **3**, 453

ANASTAS, P.T. (1994). Benign by design chemistry: pollution prevention through synthetic design. *Preprints of Papers Presented at the 208th ACS National Meeting*, Aug. 21—25, 1994, Washington, D.C., 204—205. Center for Great Lakes Studies, University of Wisconsin-Milwaukee (Milwaukee, WI: Division of Environmental Chemistry, American Chemical Society)

ANASTAS, P.T. and BREEN, J.J. (1997). Design for the environment and green chemistry: The heart and soul of industrial ecology. *J. Cleaner Prod.* **5**, 1—2; 97—102

ANCHOR, R.D., MALHOTRA, H.L. and PARKISS, J.A. (eds) (1986). *Design of Structures against Fire* (New York: Elsevier)

ANDERBERG, Y. (1988). Modelling steel behaviour. *Fire Safety J.*, **13**, 17

ANDERSEN, P.K. (1982). Testing goodness of fit of Cox's regression life model. *Biometrics*, **38**, 67

ANDERSEN, W.H. (1979). Critical energy relation for projectile impact ignition. *Combust. Sci. Technol.*, **19**, 259

ANDERSEN, BJORN and FAGERHAUD, (2000). *Root Cause Analysis — Simplified Tools and Techniques* (Milwaukee. ASQ Quality Press) ISBN 0—87389—466—9

ANDERSON, C. and NORRIS, E.B. (1974). *Fragmentation and Metallurgical Analysis of Tank Car RAX 201. Rep.*

ANDERSON, C., TOWNSEND, W., MARKLAND, R. and ZOOK, J. (1975). *Comparison of Various Thermal Systems for the Protection of Railroad Tank Cars Tested at the FRA/ BRL Torching Facility. Interim Memo Rep. 459.* Ballistic Res. Lab.

ANDERSON, C., TOWNSEND, W., ZOOK, J. and COWGILL, G. (1974). *The effects of a fire environment on a rail tank car filled with LPG. Rep. FRA/OR&D 75—31.* Dept of Transportation, Washington, DC

ANDERSON, C.E., PREDEBON, W.W. and KARPP, R.R. (1985). Computational modeling of explosive filled cylinders. *Int. J. Engng Sci.*, **23**, 1317

ANDERSON, C.M. and PAKULAK, J.M. (1978). The prediction of the reaction of an explosive system in a fire environment. Coated RDX systems for pressed explosives. *J. Haz. Materials*, **2**(2), 143

ANDERSON, C.P. (1960). Unattended oxygen plants. *Safety in Air and Ammonia Plants*, **2**, 22. See also *Chem. Engng Prog.*, 1960, **56**(5), 58

ANDERSON, C.W. (1976). Extreme value theory and its approximations. *Advances in Reliability Technology* (Culcheth, Lancashire: UK Atom. Energy Auth.), ART11

ANDERSON, D., ISTANCE, H. and SPENCER, J. (eds). (1977). *Human Factors in the Design of Ships* (Stockholm: Ergonomilaboratiet AB)

ANDERSON, E.H. (1966). How zinc controls corrosion. *Chem. Engng*, **73**(Oct. 24), 154

ANDERSON, F.E. (1976). Pressure-relieving devices. *Chem. Engng*, **83**(May 24), 128

ANDERSON, F.V. (1982). Plant layout. In Kirk, R.E. and Othmer, D.F. (1982), *op. cit.*, vol. 18, p. 23

ANDERSON, G.E., HIPPLER, R.R. and ROBINSON, G.D., (1969). *An Evaluation of Dispersion Formulas* (New York: Am. Petrol Inst.)

ANDERSON, G.K. and DUARTE, A.C. (1981). Applications of biological treatment methods to industrial effluents. Basic principles. *Chem. Ind.*, July 4, 446

ANDERSON, J. (1976). *Language, Memory and Thought* (Hillsdale, NJ; Erlbaum Associates)

ANDERSON, J. and BOWER, G. (1973). *Human Associative Memory* (Washington, D.C.: Winston)

ANDERSON, J.E. (1967). Explosions and safety problems in handling urea-ammonium nitrate solutions. *Ammonia Plant Safety*, **9**, 70

ANDERSON, J.E. (1974). Long-range ammonia future bright. *Ammonia Plant Safety*, **16**, 1

ANDERSON, JR and BOUDART, M. (eds) (1981). *Catalysis – Science and Technology* (Berlin: Springer)

ANDERSON, J.L., PARKER, F.L. and BENEDICT, B.A. (1973). *Negatively Buoyant Jets in a Crossflow. Rep. EPA 660/2–73–012.* Environ. Prot. Agency

ANDERSON, M.M. (1971). Transportation emergency procedures. *Loss Prevention*, **5**, p. 116

ANDERSON, M.W., HOEL, D.G. and KAPLAN, N.L. (1980). A general scheme for the incorporation of pharmacokinetics in low-dose risk estimation for chemical carcinogenesis. Example – vinyl chloride. *Toxicol. Appl. Pharmacol.*, **55**, 154

ANDERSON, N. (1974). Cognitive algebra: integration theory applied to social attribution. *Exp. Soc. Psychol.*, **7**, 1

ANDERSON, N. (1991). Tailor-made training for international students. *Fire Int.*, **131**(Oct./Nov.), 53

ANDERSON, O., BELL, R., MEFFERT, K. and VANTRIN, J.P. (1987). European Collaborative Project on the Assessment of Programmable Systems. *J. Occup. Accid.*, **9**, 123

ANDERSON, P.J. and BODLE, W.W. (1974). Safety considerations in the design and operation of LNG terminals. *LNG Conf.*, 4

ANDERSON, P.J. and DANIELS, E.J. (1978). LNG uses – now and later. *Hydrocarbon Process.*, **57**(5), 191

ANDERSON, P.N. (1977). Coding of chemicals for road transport. *Symp. Hazard 77 – Advances in the Safe Handling and Movement of Hazardous Substances, Teesside Polytechnic, Middlesborough*

ANDERSON, R.B. (1979). *Proving Programs Correct* (New York: Wiley)

ANDERSON, R.J. and RUSSELL, T.W.F. (1965–). Designing for two-phase flow. *Chem. Engng*, 1965, **72**(Dec. 6), 139; **72**(Dec. 20), 99; 1966, **73**(Jan. 3), 87

ANDERSON, R.P. and ARMSTRONG, D.R. (1972). Experimental study of vapor explosions. *LNG Conf.*, **3**

ANDERSON, R.P. and ARMSTRONG, D.R. (1974). *Comparison between vapor explosion models and recent experiment results. Heat Transfer Research and Development Symp. Ser. 70* (New York: Am. Inst. Chem. Engrs)

ANDERSON, R.T. and NERI, L. (1990). *Reliability-Centred Maintenance* (London: Elsevier)

ANDERSON, S.E., DOWELL, A.M. and MYNAUGH, J.B. (1992). Flashback from waste gas incinerator into air supply piping. *Plant/Operations Prog.*, **11**, 85

ANDERSON, S.E. and SKLOSS, R.E. (1991). More bang for the buck: getting the most from an accident investigation. *Plant/Operations Prog.*, **11**, 85. See also CCPS (1992/1), p. 296

ANDERSON, T. (1987). Design fault tolerance in practical systems. In Littlewood, B. (1987b), *op. cit.*, p. 44

ANDERSSON, B.O. (1991a). An accident at a Lithuanian fertilizer plant. *Plant/Operations Prog.*, **10**, 221

ANDERSSON, B.O. (1991b). Lithuanian ammonia accident, Mar. 20th, 1989. *Hazards*, **XI**, p. 15

ANDERSSON, M. (1989). An object-oriented language for model representation. Paper presented at CACSD 89, Tampa, FL

ANDO, M. and KAWAHARA, H. (1976). How skin effect tracing works. *Hydrocarbon Process.*, **55**(12), 90

ANDO, M. and OTHMER, D.F. (1970). Heating pipelines with electrical skin current. *Chem. Engng*, **77**(Mar. 9), 154

ANDO, T., FUJIMOTO, Y. and MORISAKI, S. (1991). Analysis of differential scanning calorimetric data for reactive chemicals. *J. Haz. Materials*, **28**, 251

ANDOW, P.K. (1973). A method for process computer alarm analysis. Ph.D. Thesis, Loughborough Univ. of Technol., Loughborough, Leicestershire

ANDOW, P.K. (1976). Event trees and fault trees. *Course on Loss Prevention in the Process Industries* Dept of Chem. Engng, Loughborough Univ. of Technol., Loughborough, Leicestershire

ANDOW, P.K. (1980). Difficulties in fault tree synthesis for process plant. *IEEE Trans. Reliab.*, **R-29**, 2

ANDOW, P.K. (1981). Fault trees and failure analysis: discrete state representation problems. *Trans. Instn Chem. Engrs*, **59**, 125

ANDOW, P.K. (1982). Improving process safety with better monitoring and control systems. *Chem. Engr, Lond.*, **383/4**, 325

ANDOW, P.K. (1984). *Expert Systems in Process plant Fault Diagnosis. Ergonomic Problems in Process Operations* (Rugby: Instn Chem. Engrs), p. 49

ANDOW, P.K. (1985a). Alarm systems and alarm analysis. *Plant/Operations Prog.*, **4**, 116

ANDOW, P.K. (1985b). Fault diagnosis using intelligent knowledge based systems. *Chem. Engng Res. Des.*, **63**, 368

ANDOW, P.K. (1985c). Improvement of operator reliability using expert systems. *Reliability 85*, p. 4B/4/1

ANDOW, P.K. (1986). Improvement of operator reliability using expert systems. *Reliab. Engng*, **14**(4), 309

ANDOW, P.K. (1989). Estimation of event frequencies: system reliability, component reliability data, fault tree analysis. In Cox, R.A. (1989a), *op. cit.*, p. 59

ANDOW, P.K. (1991). *Guidance on HAZOP Procedures for Computer-Controlled plants* (London: HM Stationery Office)

ANDOW, P.K. and FERGUSON, G. (1987). Applications of knowledge-based systems in chemical process safety. *Preventing Major Chemical Accidents*, p. 1.129

ANDOW, P.K. and LEES, F.P. (1974). Process plant alarm systems: general considerations. *Loss Prevention and Safety Promotion* 1, p. 299

ANDOW, P.K. and LEES, F.P. (1975). Process computer alarm analysis: outline of a method based on list processing. *Trans Inst. Chem. Engrs*, **54**, 195

ANDOW, P.K., LEES, F.P. and MURPHY, C.P. (1980). The propagation of faults in process plants: a state of the art review. *Chemical Process Hazards*, **7**, 225

ANDRADE, R.C., GATES, J.A. and MCCARTHY, J.W. (1983). Controlling boiler carryover. *Chem. Eng*, **90**(Dec. 26), 51

ANDRE, P.D. (1977). Development of maintainability prediction methods based upon checklists and multiple regression techniques. *First Nat. Conf. on Reliability, NRC4/6* (Culcheth, Lancashire: UK Atom. Energy Auth.)

ANDREASEN, P. and RASMUSSEN, B. (1990). Comparison of methods of hazard identification at plant level. *J. Loss Prev. Process Ind.*, **3**(4), 339

ANDREASSEN, T.E. and GJERSTAD, T. (1989). Risk analysis of petrol and propane transport through Oslo city. *Loss Prevention and Safety Promotion 6*, p. 16.1

ANDREIEV, G., NEFF, D.E. and MERONEY, R.H. (1983). *Heat Transfer Effects During Cold Dense Gas Dispersion. Rep. CER-83–84–GA-DEN-RNM-3.* Dept of Civil Engineering, Colorado State Univ., Fort Collins, CO

ANDREN, A. (1987). A combined first-order closure/Gaussian dispersion model. *Atmos. Environ.*, **21**, 1045

ANDREOPOULOS, A.G., KARAYANNIS, A.N. and MARKATOS, N.C. (1992). Experimental and computational investigation of ventilation effectiveness in an industrial building. *Process Safety Environ.* **70B**, 75

ANDREOZZI, R., AQUILA, T., INSOLA, A. and CAPRIO, V. (1992). Runaway reactions in nitration of amidic compounds. *Loss Prevention and Safety Promotion 7*, vol. 3, 122.1

ANDREW, S.P.S. (1981). Fuels and the future. *Chem. Ind.*, Jun. 6, 383

ANDREW, W.G. (1971). Startup made easier. Instrumentation. *Hydrocarbon Process.*, **50**(12), 87

ANDREW, W.G. (1975). *Applied Instrumentation in the Process Industries* (Houston, TX: Gulf Publ. Co.)

ANDREW, W.G. and WILLIAMS, H.B. (1980). *Applied Instrumentation in the Process Industries*, vol. 2, *Practical Guidelines* (Houston, TX: Gulf Publ. Co.)

ANDREWS, D. (1988). Nuclear power (letter). *Chem. Engr Lond.*, **450**, 6

ANDREWS, D.C. (1993). Industry views on chemical process safety. *Process Safety Prog.*, **12**, 104

ANDREWS, D.G.R. and GRAY, P. (1964). Combustion of ammonia supported by oxygen, nitrous oxide or nitric oxide: laminar flame propagation at low pressures in binary mixtures. *Combust. Flame*, **8**, 113

ANDREWS, G.E. and BRADLEY, D. (1972a). The burning velocity of methane-air mixtures. *Combust. Flame*, **19**, 275

ANDREWS, G.E. and BRADLEY, D. (1972b). Determination of burning velocities: a critical review. *Combust. Flame*, **18**, 133

ANDREWS, G.E. and BRADLEY, D. (1973a). Determination of burning velocity by double ignition in a closed vessel. *Combust. Flame*, **20**, 77

ANDREWS, G.E. and BRADLEY, D. (1973b). Limits of flammability and natural convection for methane-air mixtures. *Combustion*, **14**, 1119

ANDREWS, G.E., BRADLEY, D. and LWAKABAMBA, S.B. (1975a). Measurement of turbulent burning velocity for large turbulent Reynolds numbers. *Combustion*, **15**, 655

ANDREWS, G.E., BRADLEY, D. and LWAKABAMBA, S.B. (1975b). Turbulence and turbulent flame propagation. *Combust. Flame*, **24**, 285

ANDREWS, G.E., HERATH, P. and PHYLAKTOU, H.N. (1990). The influence of flow blockage on the rate of pressure rise in large L/D cylindrical closed vessel explosions. *J. Loss Prev. Process Ind.*, **3**(3), 291

ANDREWS, H.C. (1970). *Computer Techniques in Image Processing* (New York: Academic Press)

ANDREWS, J.D. and MORGAN, J.M. (1985). *Application of the Digraph Method of Fault Tree Construction to Process Plant., Rep. MRS E 452.* Midlands Res. Station, Br Gas, Solihull

ANDREWS, J.D. and MORGAN, J.M. (1986). Application of the digraph method of fault tree construction to process plant. *Reliab. Engng*, **14**, 85

ANDREWS, J.D. and MOSS, T.R. (1993) *Reliability and Risk Assessment* (London: Longman)

ANDREWS, R.K. and WEBER, P.R. (1993). High pressure operations: upkeep of equipment. *Chem Engng*, **100**(9), 98

ANDREWS, T.A. (1970). How to repair leaks and control burning gases from below-ground LPG pipelines without interrupting supplies. *Fire*, Feb., 464

ANDRONOPOULOS, S., BARTZIS, J.G., WÜRTZ, J. and ASIMAKOPOULOS, D. (1993). Simulation of the Thorney Island Dense Gas Trial 8 using the code ADREA-HF. *Process Safety Prog.*, **12**, 61

ANDRUS, E.C. *et al.* (1948). *Advances in Military Medicine* (Boston, MA.: Little, Brown & Co.)

ANFENGER, M.B. and JOHNSON, C.W. (1941). Proc. Am. Petrol. Inst., **22**, 54

ANFOSSI, D. (1982). Plume rise measurements at Turbigo. *Atmos. Environ.*, **16**, 2565

ANFOSSI, D. (1989). A discussion of nocturnal temperature profiles. *Atmos. Environ.*, **23**, 1177

ANFOSSI, D., FERRERO, E., BRUSASCA, G., MARZORATI, A. and TINARELLI, G. (1993). A simple way of computing buoyant plume rise in Lagrangian stochastic dispersion models. *Atmos. Environ.*, **27**A, 1443

ANG, A.H.S. and TANG, W.H. (1974–75). *Probability and Decision: Concepts in Engineering, Planning and Design*, vol. 1, 1974; vol. 2, 1975 (New York: Wiley)

ANG, M.L. (1992). *Severe Accident and Accident Management Modelling in a PSA for a PWR. PSA/ PRA – Safety and Risk Assessment* (London: IBC)

ANG, M.L. and LEES, F.P. (1989). Annotated bibliography. In Lees, F.P. and Ang, M.L. (1989), *op. cit.*, p. 256

ANGAS, G. (1982). Simulation and its role in liquefied gas carrier personnel training. *Gastech 81*, p. 313

ANGAS, G.B. (1985). Onboard operations and safety training for liquefied gas tanker personnel. *Gastech*, p. 152

ANGEL, R.F. (1975). The classification of maximum surface temperatures of electric surface heaters used in explosive atmospheres. *Electrical Safety*, **2**, 75

ANGELIN, L., DELLA VALLE, G., LOVATO, A., MONTE, A. and RIENZI, S.A. (1992). Safety simulation and control for a reactor in presence of highly exothermic reaction. *Loss Prevention and Safety Promotion 7*, vol. 3, 129-1

ANGELINI, E. (1984). Protect plants with water spray. *Hydrocarbon Process.*, **63**(12), 89

ANGLE, R.P., BRENNAND, M. and SANDHU, H.S. (1992). Surface albedo measurements at 53°N latitude. *Atmos. Environ.*, **26**A, 1545

ANGUS, C.J. and WINTER, P. (1981). Information management and draughting for the project Engineer. *Second World Cong. of Chemical Engineering, Montreal*

ANGUS, H.C. (1984). Storage, distribution and compression of hydrogen. *Chem. Ind.*, Jan. 16, 68

ANKE, K., KALTENECKER, H. and OETKER, R. (eds) (1970). *Prozessrechner Wirkungsweise und Einsatz* (Munich: Oldenbourg)

ANNAMALAI, K. and DURBETAKI, P. (1975). Characteristics of an open diffusion flame. *Combust. Flame*, **25**, 137

ANNAMALAI, K. and SIBULKIN, M. (1979a). Flame spread over combustible surfaces for laminar flow systems – part I: excess fuel and heat flux. *Combust. Sci. Technol.*, **19**, 167

ANNAMALAI, K. and SIBULKIN, M. (1979b). Flame spread over combustible surfaces for laminar flow systems – part II: flame heights and flame spread rates. *Combust. Sci. Technol.*, **19**, 185

ANNAND, W.J.D. (1970). Turbulent flame propagation: comments on Sanematsu's correlation (letter). *Combust. Flame*, **15**, 81

ANNARUMMA, M.O., MOST, J.M. and JOULAIN, P. (1991). On the numerical modeling of buoyancy-dominated turbulent vertical diffusion flames. *Combust. Flame*, **85**, 403

ANNEMAIER, D. and GRAF, R. (1988). Curbing the threat of electrical cables. *Fire. Int.*, **112**, 22

ANNETT, J., DUNCAN, K.D., STAMMERS, R.B. and GRAY, M.J. (1971). *Task Analysis. Dept of Employment, Training Inf. Paper 6* (London: HM Stationery Office)

ANSPACH, G., BASELER, R. and GLASFELD, R. (1979). A floating LNG receiving terminal. *Chem. Engng Prog.*, **75**(10), 86

ANTHONY, E.J. (1977a). Some aspects of unconfined gas and vapour explosions. *J. Haz. Materials*, **1**(4), 289

ANTHONY, E.J. (1977b). The use of venting formulae in the design and protection of buildings and industrial plant from damage by gas or vapour explosions. *J. Haz. Materials*, **2**(1), 23

ANTHONY, E.J. (1977/78). The use of venting formulae in the design and protection of building and industrial plant from damage by gas or vapour explosions. *J. Haz. Materials*, **2**, 23

ANTHONY, E.J. and GREANEY, D. (1979). The safety of hot self-heating materials. *Combust. Sci. Technol.*, **21**, 79

ANTHONY, G. (1985). Technical Reviews: seeking that essential edge. *Process Engng*, Dec., 65

ANTHONY, L.J. (1985). *Information Sources in Engineering*, 2nd ed. (London: Butterworths)

ANTHROP, D.F. (1973). *Noise Pollution* (Lexington, MA: Lexington Books)

ANTONAKIS, C.J. and BARNES, R.H. (1991). *Safety is good Engineering. Safety Developments in the Offshore Oil and Gas Industry* (London: Instn Mech. Engrs), p. 1

ANTONELLO, F. and BUZZI, G. (1992). The release of heavy gases – hydrogen fluoride. *Loss Prevention and Safety Promotion 7*, vol. 2, 93-1

ANTONY, S.M. and GAJJAR, D.B. (1982). Preventing motor burnout. *Chem. Engng*, **89**(Dec. 13), 107

ANTRIM, D.S. (1963). Better ball valves spark wider use. *Chem. Engng*, **70**(May 13), 185

ANYAKORA, S.N. (1971). Malfunction of process instruments and its detection using a process computer. Ph.D. Thesis, Loughborough Univ. of Technol.

ANYAKORA, S.N. and LEES, F.P. (1972a). Detection of instrument malfunction by the process operator. *Chem. Engr, Lond.*, **264**, 304

ANYAKORA, S.N. and LEES, F.P. (1972b). *Principles of the detection of malfunction using a process computer. Decision, Design and the Computer* (London: Instn Chem. Engrs), p. 6:7

ANYAKORA, S.N. and LEES, F.P. (1973). *The detection of malfunction using a process control computer: simple noise power techniques for instrument malfunction. The Use of Digital Computers in Measurement* (London: Instn Elec. Engrs), p. 35

ANYAKORA, S.N., ENGEL, G.F.M. and LEES, F.P. (1971). Some data on the reliability of instruments in the chemical plant environment. *Chem. Engr, Lond.*, **255**, 396

API 9100. *A and B Model Environmental, Health & Safety Management System* (Washington D.C.: American Petroleum Institute)

APOLLO – Incident Investigation and Problem Solving Techniques, (Richland, WA: Apollonian Publications), ISBN 1–883677–00–9

APOSTOLAKIS, G.E. (1974). *Mathematical Methods of Probabilistic Safety Analysis. Rep. UCLA-ENG-7464.* Univ. of California, Los Angeles, CA

APOSTOLAKIS, G.E. (1975). *The effect of a certain class of potential common mode failures on the reliability of redundant systems. Rep. UCLA-ENG-7528.* Univ. of California, Los Angeles, CA

APOSTOLAKIS, G.E. (1976). The effect of a certain class of potential common mode failures on the reliability of redundant systems. *Nucl. Engng Des.*, **36**, 123

APOSTOLAKIS, G.E. (1978). Probability and risk assessment: the subjectivistic viewpoint and some suggestions. *Nucl. Safety*, **19**, 305

APOSTOLAKIS, G. (1981). *Bayesian methods in risk assessment. Advances in Nuclear Science and Technology*, vol. 13 (ed. J. Lewins and M. Becker) (New York: Plenum Press)

APOSTOLAKIS, G. (1982). Data analysis in risk assessment. *Nucl. Engng Des.*, **71**, 375

APOSTOLAKIS, G. (1985a). The broadening of failure distributions in risk analysis: how good are the experts? (letter) *Risk Analysis*, **5**, 89

APOSTOLAKIS, G. (1985b). On the assessment of human error rates using operational experience. *Reliab. Engng*, **12**(2), 93

APOSTOLAKIS, G. (1986). On the use of judgment in probabilistic risk assessment. *Nucl. Engng Des.*, **93**, 161

APOSTOLAKIS, G. (1988). Expert judgement in probabilistic safety assessment. In Clarotti, C.A. and Lindley, D.V. (1988), *op. cit.*

APOSTOLAKIS, G. (1990). The concept of probability in safety assessments of technological systems. *Science*, **250**, 1359

APOSTOLAKIS, G. (ed.) (1991). *Probabilistic Safety Assessment and Management*, vols 1–2 (New York: Elsevier)

APOSTOLAKIS, G.E. and BANSAL, P.P. (1976). *The Effect of Human Error on the Availability of Periodically Inspected Redundant Systems. Rep. UCLA-ENG-7650.* Univ. of California, Los Angeles, CA

APOSTOLAKIS, G.E. and BANSAL, P.P. (1977). Effect of human error on the availability of periodically inspected redundant systems. *IEEE Trans. Reliab.*, **R-26**, 220

APOSTOLAKIS, G.E. and CHU, T.L. (1980). The unavailability of systems with periodic test and maintenance. *Nucl. Technol.*, **50**, 5

APOSTOLAKIS, G. and CHU, T.L. (1984). Time-dependent accident sequences including human actions. *Nucl. Technol.*, **64**, Feb., 115

APOSTOLAKIS, G., GARRIBA, S. and VOLTA, G. (eds) (1980). *Synthesis and Analysis Methods for Safety and Reliability Studies* (New York: Plenum Press)

APOSTOLAKIS, G.E. and KAPLAN, S. (1981). Pitfalls in risk calculations. *Reliab. Engng*, **2**(2), 135

APOSTOLAKIS, G., KAPLAN, S., GARRICK, B.J. and DUPHILY, R.J. (1980). Data specialization for plant specific risk studies. *Nucl. Engng Des.*, **56**, 321

APOSTOLAKIS, G. and KAZARIANS, M. (1980). The frequency of fires in light water reactor components. *ANS/ENS Topical Mtg on Thermal Reactor Safety, Knoxville, TN*

APOSTOLAKIS, G.E. and LEE, YUM TONG (1977). Methods for the estimation of confidence bounds for the top event unavailability of fault trees. *Nucl. Engng Des.*, **41**, 411

APOSTOLAKIS, G. and MOIENI, P. (1987). The foundations of models of dependence in probabilistic safety assessment. *Reliab. Engng*, **18**, 177

APOSTOLAKIS, G. and MOSLEH, A. (1979). Expert opinion and statistical evidence: an application to reactor core melt frequency. *Nucl. Sci. Engng*, **70**, 135

APOSTOLAKIS, G.E., SALEM, S.L. and WU, J.S. (1978). *CAT: a computer code for the automated construction of fault trees. Rep. EPRI NP-705.* Elec. Power Res. Inst., Palo Alto, CA

APPEL, K.E., PETER, H. and BOLT, H.M. (1981). Effect of potential antidotes on the acute toxicity of acrylonitrile. *Int. Arch. Occup. Environ. Health*, **49**, 157

APPELMAN, L.M., TEN BERGE, W.F. and REUZEL, P.G.J. (1982). Acute inhalation toxicity studies of ammonia in rats with variable exposure periods. *Am. Ind. Hyg. Assoc. J.*, **43**, 662

APPL, M. (1975). Tube crack and vibrations in a CO_2 removal section. *Ammonia Plant Safety*, **17**, 150

APPL, M., FÄSSLER, K., FROMM, D., GEBHARD, H. and PORTL, H. (1990). New cases of stress corrosion cracking in large atmospheric storage tanks. *Ammonia Plant Safety*, **30**, 22

APPL, M. and FEIND, K. (1976). Solving corrosion problems in steam reforming. *Ammonia Plant Safety*, **18**, 112

APPL, M., FEIND, K. and LIEBE, W. (1976). An ammonia shift converter failure. *Ammonia Plant Safety*, **19**, 40

APPLEGATE, F.L. (1965). Ammonia is easy to handle. *Chem. Engng Prog.*, **61**(1), 66

APPLETON, A. (1986). Latent damage. *Chem. Engr, Lond.*, **429**, 15

APPLETON, A. (1988). The Consumer Protection Act 1987 and product liability. *Chem. Engr, Lond.*, **444**, 11

APPLEYARD, C. (1977). Marking of storage places. *Symp. Hazard 77 – Advances in the Safe Handling and Movement of Hazardous Substances, Teesside Polytechnic, Middlesborough*

APPLEYARD, R.D. (1980). *Testing and Evaluation of EXPLO-SAFE System as a Method of Controlling the Boiling Liquid Expanding Vapour Explosion. Rep. TP 2740.* Transport Canada

APREA, G. (1983). World LPG sea transport requirements to 1990. *Gastech 82*, 470

APSIMON, H.M. and DAVISON, A.C. (1986). A statistical model for deriving probability distributions of contamination for accidental releases. *Atmos. Environ.*, **20**, 1249

AQUILO, A., ALDER, J.S., FREEMAN, D.N. and VOORHOEVE, R.J.H. (1983). Focus on chemistry. *Hydrocarbon Process.*, **62**(3), 57

ARABITO, G., CAPRIO, V., CRESCITELLI, S., RUSSO, G. and TUFANO, V. (1983). Runaway reactions in a polyacrylonitrile wet-spinning plant. *Loss Prevention and Safety Promotion 4*, vol. 3, B10

ARAI, M., SAITO, K. and ALTENKIRCH, R.A. (1990). A study of boilover in liquid pool fires supported on water. Pt. 1, Effects of a water sublayer on pool fires. *Combust. Sci. Technol.*, **71**, 25

ARANT, J.B. (1972). Special control valves reduce noise and vibration. *Chem. Engng*, **79**(Mar. 6), 92

ARANT, J.B. (1981). What is a 'fire-safe' valve? *Hydrocarbon Process.*, **60**(5), 269

ARAVAMUDAN, K.S. and DESGROSEILLERS, G.J. (1981). *Thermal Radiation Hazards from Hydrocarbon Fireballs. Health and Safety Executive Seminar on Thermal Radiation and Overpressure Calculations* (London: Health and Safety Executive)

ARBOUS, A.G. and KERRICH, J.E. (1953). The phenomenon of accident proneness. *Ind. Med. Surg.*, **22**(5), 141

ARCEIVALA, S.J. (1981). *Wastewater Treatment and Disposal* (Basel: Dekker)

ARCHAMBAULT, J.P., BALDINI, R. and TRAIN, P. (1980). Ammonia-based process units with integral environmental protection. *Ammonia Plant Safety*, **22**, p. 19

ARCHARD, T.F. (1952). The temperature of rubbing surfaces. *Wear*, **2**, 438

ARCHBOLD, T., LAIDLAW, J.C. and MCKECHNIE, J. (1984). *Engineering Research Centres* (London: Longmans)

ARCHIBALD, R.G. (1977). Process sour gas safely. *Hydrocarbon Process.*, **56**(3), 219

ARCURI, K. (1973). Quiet steamline design to comply with OSHA rules. *Chem. Engng*, **80**(Jan. 8), 134

ARDEN, T.V. (1968). *Water Purification by Ion Exchange* (London: Butterworths)

ARDILLON, E. and BOUCHACOURT, M. (1994). Probabilistic methods for evaluation of erosion-corrosion wall thinning in French pressurized water reactors. *Conf. on Valve and Pipeline Reliability*. Engng Dept, Univ of Manchester

ARDRAN, G.M. (1964). Phosgene poisoning (letter). *Br. Med. J.*, **1**, 375

ARDRON, K.H. (1978). A two-fluid model for critical vapour-liquid flow. *Int. J. Multiphase Flow*, **4**, 323

ARDRON, K.H., BAUM, M.R. and LEE, M.H. (1977). On the evaluation of dynamic forces and missile energies arising in a water reactor loss-of-coolant accident. *J. Br. Nucl. Energy Soc.*, **16**, 81

ARENDT, H. (1970). *On Violence* (New York: Harcourt, Brace & World)

ARENDT, J.A. and MARRA, R.P. (1989). QRA of runaway reactions in a peroxy acid CSTR. *Runaway Reactions*, 547

ARENDT, J.S. (1986). Determining heater retrofits through risk assessment. *Plant/Operations Prog.*, **5**, 228

ARENDT, J.S. (1990). Management of quantitative risk assessment in the chemical process industry. *Plant/Operations Prog.*, **9**, 262

ARENDT, J.S. and FUSSELL, J.B. (1981). System reliability Engineering methodology for industrial application. *Loss Prevention*, **14**, 18

ARENDT, J.S. and LORENZO, D.K. (1991). Investigation of a filter explosion. *J. Loss Prev. Process Ind.*, **4**, 338

ARENDT, J.S., CASADA, M.L. and ROONEY, J.J. (1986). Reliability and hazards analysis of a cumene hydroperoxide plant. *Plant/Operations Prog.*, **5**, 97

ARENDT, J.S., LORENZO, D.K., MONTAGUE, D.F. and DYCUS, F.M. (1987). Ensuring operator reliability during off-normal conditions using an expert system. *Preventing Major Chemical Accidents*, p. 6.69

ARENDT, J.S., CAMPBELL, D.J., CASADA, M.L. and LORENZO, D.K. (1984). Risk analysis of a petroleum refinery. *Chem. Engng Prog.*, **80**(8), 58

ARENDT, J.S., MITCHELL, C.M. and REMSON, A.C. (1993). The life-cycle approach to nurturing process safety management systems. *Process Safety Management*, p. 339

ARENDT, S. (1994). PSM implementation. *Process Safety Prog.* **13**, 20

ARGENT, R.P., COOK, P. and GOLDSTONE, P. (1992). Life cycle risk assessment and implementation on an expansion project for a hazardous facility. *Major Hazards Onshore and Offshore*, p. 439

ARGYRIS, C. (1957). *Personality and Organization* (New York: Harper & Row)

ARIENTI, M., WILK, L., JASINSKI, M. and PROMINSKI, N. (1988). *Dioxin Containing Wastes: Treatment Technologies* (Park Ridge, NJ: Noyes Data Corp.)

ARIES, R.S. and NEWTON, R.D. (1955). *Chemical Engineering Cost Estimation* (New York: McGraw-Hill)

ARIMOCHI, K., NAKANISHI, M., TOYODA, M. and SATO, H.K. (1984). *Fracture Safety Assessment of 9% Ni Steel Weldments. Transport and Storage of LPG and LNG* (Antwerp: Koninklijke Vlaamse Ingenieursvereniging), p. 71

ARIS, R. (1990). *Elementary Chemical Reactor Analysis* (Oxford: Butterworth-Heinemann)

ARIS, R.S. and AMUNDSON, N.R. (1957). Stability of some chemical systems under control. *Chem. Engng Prog.*, **53**, 227

ARIS, R.S. and AMUNDSON, N.R. (1958). An analysis of chemical reactor stability and control. *Chem. Engng Sci.*, **7**, 121, 132

ARKIN, H. and COLTON, R.R. (1950). *Tables for Statisticians* (New York: Barnes & Noble)

ARLOING, F., BERTHET, E. and VIALLIER, J. (1940). Action of chronic intoxication of chlorine fumes on experimental guinea pigs. *Presse Med.*, **48**, 361

ARMAND, J.L. (1982). Response of spherical cargo tanks for liquefied gas to large support deformation. *Gastech 81*, 209

ARMISTEAD, G. (1959). *Safety in Petroleum Refinery and Related Industries*, 2nd ed. (St Michaels, MD: Security Publ. Corp.)

ARMISTEAD, J.G. (1966). Du Pont's Louisville Works explosion. *Fire J.*, **60**(1)

ARMISTEAD, J.G. (1973). Polymerization explosion at Dupont. In Vervalin, C.H. (1973a), *op. cit.*, p. 38

ARMITAGE, J.B. and STRAUSS, H.W. (1964). Safety considerations in industrial use of organic peroxides. *Ind. Engng Chem.*, **56**(12), 28

ARMITAGE, J.W. and GRAY, P. (1965). Flame speeds and flammability limits in the combustion of ammonia: ternary mixtures with hydrogen, nitric oxide, nitrous oxide or oxygen. *Combust. Flame*, **9**, 173

ARMITAGE, P.M., ELKINS, K.J., KITCHEN, D. and PINTO, A. (1992). Leading concept ammonia process: first two years. *Ammonia Plant Safety*, **32**, 111

ARMITAGE, R.N. (1983). Suppression of transient surge pressures during ship/shore cargo transfer. *Loss Prevention and Safety Promotion 4*, vol. 2, D30

ARMITAGE, W. (1970). Maintenance effectiveness. In Jardine, A.K.S. (1970b), *op. cit.*, p. 196

ARMOUR, G.C., BUFFA, E.S. and VOLLMAN, T.E. (1964). Allocating facilities with CRAFT. *Harvard Business Rev.*, **42**(2), 130

ARMOUR, M.A. (1987). Doing it right. In Young, J.A, (1987a), *op. cit.*, p. 23

ARMSTRONG, N.E. (1982). Biological measures. In Bennett, G.F., Feates and Wilder (1982), *op. cit.*, pp. 9–40

ARMSTRONG, N.E., GLOYNA, E.F. and WYSS, O. (1976). Use of biological countermeasures for removal of hazardous material spills. *Control of Hazardous Material Spills*, **3**, 396

ARMSTRONG, R. (1972). Better ways to build process plants. *Chem. Engng*, **79**(Apr. 17), 86

ARMSTRONG, W.S. (1983). Multi-product batch reactor control. *Chem. Engng Prog.*, **79**(1), 56

ARMSTRONG, W.S. and COE, B.F. (1983). Multi-product batch reactor control. *Chem. Engng Prog.*, **79**(1), 56

ARNAUD, F.C. (1971). 'Risk point' determination as an alternative to hazardous area classification: Part 1. *Electrical Safety*, **1**, 145

ARNAUD, F.C. and JORDAN, V.C. (1975). Implications of the Health and Safety at Work, etc., Act 1974. *Electrical Safety 2*, 6

ARNAUD, R. *et al.* (1992). The overpressure generated by the ignition of a large scale free natural gas jet. *Loss Prevention and Safety Promotion 7*, vol. 3, 113-1

ARNBERG, B.T. (1962). Review of critical flowmeters for gas flow measurements. *J. Basic Engng*, **84**, 447

ARNDT, J.H. and KIDDOO, D.B. (1975). Special purpose gearing – how to avoid start-up problems. *Hydrocarbon Process.*, **54**(3), 113

ARNE, F. (1967). Vinyl process: petrochemical pacesetter. *Chem. Engng*, **74**(Mar. 27), 48

ARNER, L.D., JOHNSON, G.R. and SKOVRONEK, H.S. (1986). Delineating toxic areas by canine olfaction. *J. Haz. Materials*, **13**(3), 375

ARNEY, H.E. (1977a). Clean tanks and vessels safely. *Hydrocarbon Process.*, **56**(5), 141

ARNEY, H.E. (1977b). Safe entry and cleaning of storage tanks and vessels. *API Proc., Refining Dept, 42nd Midyear Mtg*, **56**, 341

ARNOLD *et al.* (1973). *Hazards from Burning Garments. Final Report. Rep. NTIS COM-73-10957*. Gillette Res. Inst.

ARNOLD, B., REDFEARN, C. and MAAREN, R. (1992). Safety – a catalyst to improve organizational performance. *Loss Prevention and Safety Promotion 7*, vol. 1, 45-1

ARNOLD, C.G. (1978). Using real time corrosion monitors in chemical plants. *Chem. Engng Prog.*, **74**(3), 43

ARNOLD, C.G., MUELLER, W.H. and ROSS, B.W. (1986). Reducing risks through a pressure vessel management program. *Am. Inst. Chem. Engrs, Symp on Loss Prevention* (New York)

ARNOLD, L. (1992). *Windscale 1957* (London: Macmillan)

ARNOLD, M.W. and DARIUS, I.H. (1988). Alarm management in batch process control. *Paper 88-1473* (Pittsburgh, PA: Instrum. Soc. Am.)

ARNOLD, P. (1978). Seismic risk analysis combined with random and nonrandom earthquake occurrence. *Tenth Ann. Conf. on Offshore Technology*, p. 513

ARNOLD, R.M. (1993). Catastrophes the underlying causes (poster item). *Process Safety management*, p. 507

ARNOLD, W. (1959). *Der Apparatebau* (Munich: Carl Hanser)

ARNONI, Y.G. (1979). The marriage of LNG and offshore facilities. *Chem. Engng Prog.*, **75**(10), 60

ARNOT, W. and HIRONS, C. (1974). Commissioning of a 260,000 T/A VCM complex. *Instn Chem. Engrs Symp. on Commissioning of Chemical and Allied Plant* (London)

ARON, R.B. (1964). *Antistatic agents. Modern Plastics Encyclopaedia* (New York: McGraw-Hill), p. 443

ARPACI, V. and SELAMET, A. (1991). Buoyancy driven turbulent diffusion flames. *Combust. Flame*, **86**, 203

ARSCOTT, J.E., STREET, P.J. and TWAMLEY, C.S. (1973). The application of water sprays to control fire in coal milling plant at power stations. *Fire Prev. Sci. Technol.*, **7**, 4

ARSCOTT, P.E. (1975). Accident and injury data as a guide to safety performance. In Hendry, W.D. (1975), *op. cit.*, paper E5

ARSCOTT, P. and ARMSTRONG, M. (1977). *An Employers Guide to Health and Safety Management* (London: Engng Employers Fed.)

ARTHUR, J.R. and NAPIER, D.H. (1955). Formation of carbon and related materials in diffusion flames. *Combustion*, **5**, 303

ARTHURS, J. (1986). Disposal of industrial wastes: major elements of environmental control – an introduction. *Chem. Ind.*, Jun. 16, 410

ARTINGSTALL, G. (1972). Displacement of gas for a ruptured container and its dispersal in the atmosphere. *Chemical Process Hazards IV*, p. 3

ARTINGSTALL, G. (1975). Appraisal of the damage to the reactor vessels in Section 25A. *Rep. to Flixborough Court of Inquiry* (Safety in Mines Res. Estab.)

ARTIS, D. (1984). Legal controls over odour nuisance. *Chem. Ind.*, May 7, 320

ARTUS, C.H. (1976). All about chemical hose. *Chem. Engng*, **83**(Apr. 26), 121

ARTZ, C.P., MONCRIEF, J.A. and PRUIT, B.A. (1979). *Burns: A Team Approach* (Philadelphia, PA: W.B. Saunders)

ARULANANTHAM, D.C. and LEES, F.P. (1981). Some data on the reliability of pressure equipment in the chemical plant environment. *Int. J. Pres. Ves. Piping*, **9**, 327

ARUP, H. (1977). Stress corrosion cracking in ammonia storage. *Ammonia Plant Safety*, **19**, 73

ARWOOD, R., HAMMOND, J. and WARD, G.G. (1985). Ammonia inhalation. *J. Trauma*, **25**, 444

ARYA, S.C., DREWYER, R.P. and PINCUS, G. (1975). Foundation design for vibrating machines. *Hydrocarbon Process.*, **54**(11), 273

ARYA, S.C., DREWYER, R.P. and PINCUS, G. (1977). Foundation design for reciprocating compressors. *Hydrocarbon Process.*, **56**(5), 223

ARYA, S.P.S., CAPUANO, M.E. and FAGEN, L.C. (1987). Some fluid modeling studies of flow and dispersion over two-dimensional low hills. *Atmos. Environ.*, **21**, 753

ARYA, S.P.S. and GADIYARAM, P.S. (1986). An experimental study of flow and dispersion in the wakes of three-dimensional low hills. *Atmos. Environ.*, **20**, 729

ARYA, S.P.S. and LAPE, J.F. (1990). A comparative study of the different criteria for the physical modeling of buoyant plume rise in a neutral atmosphere. *Atmos. Environ.* **24A**, 289

ARYA, S.P.S. and SHIPMAN, M.S. (1981). An experimental investigation of flow and diffusion in the disturbed boundary layer over a ridge – I, Mean flow and turbulence structure. *Atmos. Environ.*, **15**, 1173

ARYA, S.P.S., SHIPMAN, M.S. and COURTNEY, L.Y. (1981). An experimental investigation of flow and diffusion in the disturbed boundary layer over a ridge – II, Diffusion from a continuous point source. *Atmos. Environ.*, **15**, 1185

ASADI, G.V. and VAUGHAN, H. (1990). Dynamics and structural damage of tanker ships running aground. *J. Haz. Materials*, **25**(1/2), 61

ASANO, K., KRAMER, H. and SCHON, G. (1977). Electrostatic fields around a sphere in a cylindrical tank partly filled with charged liquid. *Proc. Inst. Electrostatics, Japan*, 114

ASANTE-DUAH, D.K. (1993). Hazardous Waste Risk Assessment (Boca Raton, FL: Lewis Publs)

ASBJORNSEN, O.A. (1976). *Private communication*

ASBJORNSEN, O.A. (1986). Error in the propagation of error formula. *AIChE J.*, **32**, 332

ASBJORNSEN, O.A. (1989). Control and operability of process plants. *Comput. Chem. Engng*, **13**, 351

ASH, P.J.N., BAVERSTOCK, K.F. and VENNART, J. (1976). Risks of everyday life. *Community Health*, **8**, 29

ASHBAUGH, W.G. (1965a). Corrosion in heat exchangers using brackish water. *Chem. Engng*, **72**(Jul. 5), 146

ASHBAUGH, W.G. (1965b). External stress corrosion cracking of stainless steel under thermal insulation. *Mater. Protection*, May, 18

ASHBAUGH, W.G. (1970). Stress corrosion cracking of process equipment. *Chem. Engng Prog.*, **66**(10), 44

ASHBAUGH, W.G. (1976). How plant operating decisions affect ethylene furnace tube life. *Hydrocarbon Process.*, **55**(8), 162

ASHBAUGH, J.M. (1987). Equipment and technology for the subsea frontier. *Gastech 86*, 180

ASHBY, LORD (1976). Protection of the environment: the human dimension. *J. R. Soc. Med.*, **69**, 721

ASHBY, LORD (1977). The risk equations: the subjective side of assessing risks. *New Scient.*, **10**(May), 398

ASHBY, LORD. (1979). Environmental pollution: modern issues. *Chem. Ind.*, Apr. 7, 235

ASHBY, J. and PATON, D. (1993). The influence of chemical structure on the extent and sites of carcinogenesis for 522 rodent carcinogens and 55 different human carcinogen exposures. *Mutat. Res.* **286**, 3–74.

ASHBY, J. and STYLES, J.J. (1978). Does carcinogenic potency correlate with mutagenic potency in the Ames assay? *Nature*, **271**, 452

ASHBY, M.F. and BROWN, L.M. (eds) (1983). *Perspectives in Creep Fracture* (Oxford: Pergamon Press)

ASHBY, M.F. and JONES, D.R.H. (1980). *Engineering Materials* (Oxford: Pergamon Press)

ASHE, J.B. (1971). The fire insurance surveyor's view. *Major Loss Prevention*, p. 9

ASHER, J., CONLON, T.W., TOFIELD, B.C. and WILKINS, N.J.M. (1983). Thin layer activation – a new plant corrosion-monitoring technique. In Butcher, D.W. (1983), *op. cit.*, p. 95

ASHER, R.C. (1982). Ultrasonic techniques for non-invasive instrumentation on chemical and process plant. *Chem. Engr, Lond.*, **383/384**, 317

ASHFIELD, A. (1979). Calculation and presentation of sprinkler hydraulic calculations. *Fire Surveyor*, **8**(2), 32; **8**(3), 21

ASHFIELD, A. (1993). Sprinkler calculations: computer programs. *Fire Surveyor*, **22**(1), 14

ASHFORD, N.A. (1993). *The Encouragement of Technological Change for Preventing Chemical Accidents: Moving Firms From Secondary Prevention and Mitigation to Primary Prevention* (Cambridge, MA: Center for Technology, Policy and Industrial Development, Massachusetts Institute of Technology)

ASHFORD, N.A. (1997a). Industrial safety: the neglected issue in industrial ecology. *J. Cleaner Prod.* **5**, 1–2, 115–21

ASHFORD, N.A. (1997b). Policies for the promotion of inherent safety. *New Solutions* (Summer), 46–52

ASHFORD, N.A. (1997c). Policies for the promotion of inherent safety: The United States experience. In *Environmentally-Conscious Design and Manufacturing: European, Japanese and North American Perspectives* (ed. D. Wallace and T. Tomiyama) (Cambridge, MA: MIT Press)

ASHFORD, N.A. and ZWETSLOOT, G. (2000). Encouraging inherently safer production in european firms: a report from the field. *J. of Hazard. Mater.* **78**, 123–44

ASHILL, E.R. (1966a). Fire incident at Fawley refinery. *Fire Int.*, **1**(12), 43

ASHILL, E.R. (1966b). When to cool and when to foam. *Fire Int.*, **1**(12), 45

ASHMAN, L.E. and BÜCHLER, A. (1961). The ignition of gases by electrically heated wires. *Combust. Flame*, **5**, 113

ASHMORE, F.S. (1987a). Area classification for electrical equipment. *Fire Surveyor*, **16**(3), 9

ASHMORE, F.S. (1987b). Understanding area classification. *Fire Surveyor*, **16**(4), 5

ASHMORE, F.S. and SHAMA, Y. (1988). Decisions on risk. *Fire Safety J.*, **13**, 125

ASHMORE, P.G. (1967). The sensitization and inhibition of ignitions. In Ashmore, P.G., Dainton, F.S. and Sugden, T.M. (1967), *op. cit.*, p. 287

ASHMORE, P.G., DAINTON, F.S. and SUGDEN, T.M. (eds) (1967). *Photochemistry and Reaction Kinetics* (Cambridge: Cambridge Univ. Press)

ASHMORE, P.G. and LEVITT, B.P. (1959). Further studies of induction periods in mixtures of H_2, O_2 and NO_2. *Combustion*, **7**, 45

ASHMORE, P.G. and PRESTON, K.F. (1967). Sensitised ignitions of methane and oxygen. *Combust. Flame*, **11**, 125

ASHTON, L.K. (1961). *Fire Protection* (London: Newnes)

ASHTON, W.G. (1977). Translating UKHIS from a voluntary to a mandatory scheme. *Symp. Hazard 77 – Advances in the Safe Handling and Movement of Hazardous Substances, Teesside Polytechnic, Middlesborough*

ASHTON, W.G. and BUTCHER, D.W. (1976). Emergency procedures for the transport of dangerous chemicals in Great Britain. *Prevention of Occupational Risks*, **3**, 473

ASHWORTH, F.P.O. and WESTERN, D.J. (1987). Degraded core analysis. *Nucl. Energy*, **26**, 233

ASHWORTH, M. (1989). *Piper Alpha Public Inquiry. Transcript, Day 140*

ASHWORTH, R.A., HOWE, G.B., MULLINS, M.E. and ROGERS, T.N. (1988). Air-water partitioning coefficients of organics in dilute aqueous solutions, *J. Haz. Materials*, **18**(1), 25

ASIMOW, M. (1962). *Introduction to Design* (Englewood Cliffs, NJ: Prentice-Hall)

ASKARI, A., BULLMAN, S.J., FAIRWEATHER, M. and SWAFFIELD, F. (1990). The concentration field of a turbulent jet in a cross-wind. *Combust. Sci. Technol.*, **73**, 463

ASKEW, M.W. (1981). High-rate biological treatment. *Chem. Ind.*, July 4, 448

ASKREN, W.B. and REGULINKSI, T.L. (1969). Quantifying human performance for reliability analysis of systems. *Human Factors*, **11**, 393

ASLAM, M. (1970). Minimum length of process piping. *Hydrocarbon Process.*, **49**(11), 203

ASLANOV, S.K. and GOLINSKY, O.S. (1989). Energy of an asymptotic equivalent point detonation for the detonation of a charge of finite volume in an ideal gas. *Combustion, Explosion and Shock Waves*, 801

ASPHAHANI, A.I. (1986). Overview of advanced materials technology. *Chem. Engng Prog.*, **82**(6), 33

ASQUITH, J.P. (1974). Plant commissioning: one contractor's view on involvement. *Instn. Chem. Engrs Symp. on Commissioning of Chemical and Allied Plant* (London)

ASSELINEAU, P., HUTELLIER, R., LASNERET, J. and MEUNIER, C. (1972). The second methane terminal of Gaz de France. *LNG Conf. 3*

ASSHETON, R. (1930). *History of Explosions* (Wilmington, DL: Charles L. Storey Co. for Inst. of Makers of Explosives)

ASSINI, J. (1974). Choosing welding fittings and flanges. *Chem. Engng*, **81**(Sep. 2), 90

ASSOCIATED OCTEL COMPANY – see Appendix 28

ASSOCIATED OFFICES TECHNICAL COMMITTEE (n.d.). *Rules for the Construction, Testing and Scantlings of Metal Arc Welded Steel Boilers and Pressure Vessels* (London)

ASSOCIATION OF AMERICAN RAILROADS (1972). Report on summary of ruptured rail tank cars involved in past accidents. *Railroad Tank Car Safety Research and Test Proj. RA-01-2-7.* Chicago Res. Center, Ass. Am. Railroads, Chicago, IL.

ASSOCIATION OF AMERICAN RAILROADS (1991). Procedures for acoustic emission evaluation of tank cars and IM-101 tanks

ASSOCIATION OF THE BRITISH CHEMICAL MANUFACTURERS – see Appendix 28

ASSOCIATION OF BRITISH INSURERS (1992). *Insurance Statistics 1987–91* (London)

ASSOCIATION OF THE BRITISH PHARMACEUTICAL INDUSTRY (1981). *Guidance Notes on Chemical Reaction Hazard Analysis* (London)

ASSOCIATION OF THE BRITISH PHARMACEUTICAL INDUSTRY. (1989). *Guidelines for Chemical Reaction Hazard Evaluation*, 2nd ed. (London)

ASSOCIATION NATIONAL POUR LA PROTECTION CONTRE L'INCENDIE (1972). Studies of automatic techniques for the control of fires. *Fire Prev. Sci. Technol.*, **2**, 4

ASSOCIATION OF UNIVERSITY TEACHERS (1967). *Safety in University Laboratories* (London)

ASTARITA, G., SAVAGE, D.W. and BISIO, A. (1983). *Gas Treating with Chemical Solvents* (New York: Wiley-Interscience)

ASTBURY, G.R. (1991). Water jet warning. *Chem. Engr, Lond*, **503**, 4

ASTBURY, N.F. and VAUGHAN, G.N. (1972). Motion of a brickwork structure under certain assumed conditions. *Tech. Note 191*. Br. Ceramics Res. Ass., Stoke-on-Trent

ASTBURY, N.F., WEST, W.H.W. and HODGKINSON, H.R. (1972). Experimental gas explosion: report of further tests at Potters Marsden. *Special Publ. 74*. Br. Ceramics Res. Ass., Stoke-on-Trent

ASTBURY, N.F., WEST, H.W.H. and HODGKINSON, H.R. (1973). Experimental gas explosions – report of further tests at Potters Marson. *Proc. Br. Ceram. Soc.*, **21**, 195

ASTBURY, N.F., WEST, W.H.W., HODGKINSON, H.R., CUBBAGE, P.A. and CLARE, R. (1970). Gas explosions in load-bearing structures. *Special Publ. 68*. Br. Ceramics Res. Ass., Stoke-on-Trent,

ASTLEFORD, W.J., BASS, R.L. and COLONNA, G.R. (1985). Modeling of shiptank ventilation and occupational exposures to chemical vapors during tank entry. *Plant/Operations Prog.*, **4**, 90

ASTLEFORD, W.J., MORROW, T.B. and BUCKINGHAM, J.C. (1983). *Hazardous Chemical Vapor Handbook for Marine Vessels. Final Report. Rep. CG-D-12-83.* US Coast Guard. Washington, DC

ASTOLFI, M. and ELBAZ, J. (1977). COVAL, a computer code for random variables combination and reliability – how to use. *Rep. EUR-5804.* Comm. Eur. Commun.

ASTOLFI, M., CONTINI, S. and VAN DEN MUYZENBERG, C.L. (1978). A technique suitable for the analysis of large fault trees on minicomputers. *Int. Conf. on Reliability and Maintainability, Paris*

ASTOLFI, M. *et al.* (1978). SALP-3 – a computer program for fault tree analysis. *Rep. EUR 6183*

ATALLA, R.H. and WOHL, K. (1965). The role of inerts in hydrocarbon flames. *Combustion*, **10**, 259

ATALLAH, S. (1965). Flame heights and burning rates of liquid fuels in open tanks (letter). *Combust. Flame*, **9**, 203

ATALLAH, S. (1977). Security for CPI plants. *Chem. Engng*, **84**(Oct. 10), 139

ATALLAH, S. (1979). Fumigants and grain dust explosions. *Fire Technol.*, **15**, 5

ATALLAH, S. (1980). Assessing and managing industrial risk. *Chem. Engng*, **87**(Sep. 8), 94

ATALLAH, S. (1982). LNG safety research overview. *Gastech 81*, 307

ATALLAH, S. (1983). LNG safety research in the USA *J. Haz. Materials*, **8**(1), 25

ATALLAH, S. and ALLAN, D.S. (1971). Safe separation distances from liquid fuel fires. *Fire Technol.*, **7**(1), 47

ATALLAH, S. and GUZMAN, E. (1987). Remote sensing of fire and hazardous gases. *Preventing Major Chemical Accidents*, p. 5.63

ATALLAH, S. and GUZMAN, E. (1988). Safety audits in developing countries. *Preventing Major Chemical and Related Process Accidents*, p. 35

ATALLAH, S. and RAJ, P.P.K. (c. 1973). Radiation from LNG fires. *Proj. IS-3-1, paper G-30-34*. Am. Gas Ass.

ATHEARN, F.H. (1979). Selection guide for steam turbines. *Hydrocarbon Process.*, **58**(8), 87

ATHERLEY, G.R.C. (1977a). Noise. In Handley, W. (1977), *op. cit.*, p. 332

ATHERLEY, G.R.C. (1977b). *People and Safety – Stress and its Role in Serious Accidents. Chemical Engineering in a Hostile World* (London: Clapp & Poliak)

ATHERLEY, G.R.C. (1977c). The scope of occupational safety. In Handley, W. (1977), *op. cit.*, p. 1

ATHERLEY, G.R.C. (1978). *Occupational Health and Safety Concepts* (London: Appl. Sci. Publs)

ATHERLEY, G.R.C. and PURNELL, G.V. (1969). Noise control. In Handley, W. (1977), *op. cit.*, p. 330

ATHERTON, G.K. (1971). Safety training by involvement. *Major Loss Prevention*, p. 217

ATHEY, T.W., TELL, R.A. and JANES, O.E. (1973). *The Use of an Automated Population Data Base in Population Exposure Calculations*. Environ. Protect. Agency

ATIYAH, P.S. (1970). *Accidents, Compensation and the Law* (London: Weidenfeld and Nicolson)

ATKINS, K.T.G., FYFE, D. and RANKIN, J.D. (1974). Corrosion in CO_2 removal plant towers. *Ammonia Plant Safety* **16**, 39

ATKINS, R.S. (1970). Which process for xylene? *Hydrocarbon Process.*, **49**(11), 127

ATKINS, W.S. (1991). Some thoughts on Lord Cullen's 'Four forthwiths'. *Safety Reliab.*, **11**(3), 6

ATKINS RESEARCH AND DEVELOPMENT (1980). Assessment of industrial environmental impact. *Sci. Public Policy*, **7**, 145

ATKINSON, A. (1988). Reducing waterside corrosion in boiler plant. *Process Engng*, Apr., 59

ATKINSON, F.H. (1970). A classification society's approach to vessels designed for the carriage of liquefied gases. LNG Conf. 2

ATKINSON, F.H. (1989). *Piper Alpha Public Inquiry. Transcript, Day 99*

ATKINSON, G.T. and DRYSDALE, D.D. (1992). Convective heat transfer from fire gases. *Fire Safety J.*, **19**, 217

ATKINSON, G.T. and JAGGER, S.F. (1994). Assessment of hazards from warehouse fires involving toxic materials. *Fire Safety J.*, **22**, 107

ATKINSON, G.T., JAGGER, S.F. and KIRK, P.G. (1992). Assessment of individual risks from fires in warehouses containing chemicals. *Hazard Identification and Risk Analysis, Human Factors and Human Reliability in Process Safety*, p. 141

ATKINSON, M.J. (1985). Liquid levels. *Plant Engr*, **29**(1), 25

ATKINSON, N. (1987a). The desktop gurus. *Process Engng*, Oct., 33

ATKINSON, N. (1987b). Taking 3D modelling into conceptual design. *Process Engng*, Nov., 53

ATKINSON, N. (1988a). BS 5750: it's no guarantee. *Process Engng*, Jan., 34

ATKINSON, N. (1988b). Contracting CAD. *Process Engng*, Apr., 53

ATKINSON, N. (1988c). When it comes to safety, are you keeping up with the Joneses? *Process Engng*, Mar., 47

ATKINSON, P.G., GRIMSEY, R.M. and HANCOCK, R.A. (1967). *Control Units for Automatic Burners. Res. Comm. GC139* (London: Gas Council)

ATKINSON, R. (1987). Planning and implementing plant revamps. *Chem. Engng*, **94**(3), 51

ATKINSON, R., BULL, D.C. and SHUFF, P.J. (1980). Initiation of spherical detonation in hydrogen air. *Combust. Flame*, **39**, 287

ATOMIC ENERGY ESTABLISHMENT, HARWELL – see Appendix 28

ATOMIC ENERGY COMMISSION (1955). *Meteorology and Atomic Energy*, 1st ed.; 1968, 2nd ed. (Washington, DC)

ATOMIC ENERGY COMMISSION (1957). *Theoretical Possibilities and Consequences of Major Accidents in Large Nuclear Power Plants. Rep. WASH-740.* (Washington, DC)

ATOMIC ENERGY COMMISSION (1973). *Design Response Spectra for Seismic Design of Nuclear Power Plants. Regul. Guide 1.60, rev. 1.* (Washington, DC)

ATOMIC ENERGY COMMISSION (1975). *Reactor Safety Study: An Assessment of Accident Risks in US Commercial Nuclear Reactors. Rep. WASH-1400.* (Washington DC)

ATOMIC ENERGY OFFICE (1957). *Accident at Windscale 1 Pile on 10th Oct. 1957. Cmnd 302* (London: HM Stationery Office)

VAN ATTA, C.M. (1967). *Vacuum Science and Engineering* (New York: McGraw-Hill)

VAN ATTA, F.A. (1982). Federal standards on occupational safety and health. In Fawcett, H.H. and Wood, W.S. (1982), *op. cit.*, p. 29

ATTAWAY, L.D. (1972). Hazardous material spills – the national problem. *Control of Hazardous Material Spills*, **1**, 1

ATTEBERY, JR (1970). Failure of a primary waste heat boiler. *Ammonia Plant Safety*, **12**, 34

ATTEBERY, JR and THOMPSON, L.E. (1972). Primary reformer transfer header failure. *Ammonia Plant Safety*, **14**, 37

ATTWOOD, C.L. and SUITT, W.J. (1983). *Users Guide to BFR – Computer Code on Binomial Failure Rate Common Cause Model. Rep. NUREG/CR-2729*. Nucl. Regul. Comm., Washington, DC

ATWATER, M.A. and LONDERGAN, R.J. (1985). Differences caused by stability class on dispersion in tracer experiments. *Atmos. Environ.*, **19**, 1045

ATWOOD, C.L. (1991). *Estimation for the BFR common cause model. Draft Rep. NUREG/CR-1401*. Nucl. Regul. Comm., Washington, DC

ATTWOOD, D. (1970). The interaction between human and automatic control. In *Paper-Making Systems and Their Control* (ed. F. Bolam) (London: Br. Paper and Board Makers Ass.), p. 69. See also Edwards, E. and Lees, F.P. (1974), *op. cit.*, p. 120

ATWOOD, J.D. (1976). How hot is too hot? *Ammonia Plant Safety*, **18**, 109

ATWOOD, J.D. (1977). Public communications in an emergency. *Ammonia Plant Safety*, **19**, p. 79

ATWOOD, J.D. (1984). Energy pressures on ammonia. *Plant/Operations Prog.*, **3**, 95

ATWOOD, K.F., LOMBARD, J.F. and MERRIAM, J.S. (1978). More efficient methane usage in ammonia plants. *Ammonia Plant Safety*, **20**, 171

AUBRECHT, D.A., LINDEMANN, F.O. and SCHREIBER, S. (1968). Reducing fire and explosion hazards. *Loss Prevention*, **2**, 123

AUBREY, C. (1991). *Meltdown* (London: Collins & Brown)

AUFFRET, A., BOULVERT, L. and THIBAULT, E. (1983). A new approach to automation of process loss prevention and safety actions in refining operations. *Loss Prevention and Safety Promotion 4*, vol. 1, p. 146

AUGER, J.E. (1993). Compliance strategies that work. *Process Safety Prog.*, **12**, 195

AUGUSTINE, R.L. (1965). *Catalytic Hydrogenation* (London: Arnold)

AUGUSTINE, R.L. (ed.) (1985). *Catalysis of Organic Reactions* (New York: Dekker)

AUPIED, JR, LE COGUIEC, A. and PROCACCIA, H. (1983). Valves and pumps operating experience in French nuclear Plants. *Reliab. Engng*, **6**, 133

AUSTIN, D.G. (1979). *Chemical Engineering Drawing Symbols* (London: Godwin)

AUSTIN, D.G. and JEFFREYS, G.V. (1979). *The Manufacture of Methyl-Ethyl-Ketone from 2–Butanol: A Worked Solution to a Problem in Chemical Engineering Design* (London: Instn Chem. Engrs)

AUSTIN, E.P. (1977). *Economics of Waste Disposal – A Contractor's View. Chemical Engineering in a Hostile World* (London: Clapp & Poliak)

AUSTIN, G.T. (1965a). Hazards of commercial chemical operations. In Fawcett, H.H. and Wood, W.S. (1965), *op. cit.*, p. 92

AUSTIN, G.T. (1965b). Hazards of commercial chemical reactions. In Fawcett, H.H. and Wood, W.S. (1965), *op. cit.*, p. 80

AUSTIN, G.T. (1982a). Hazards of commercial chemical operations. In Fawcett, H.H. and Wood, W.S. (1982), *op. cit.*, p. 73

AUSTIN, G.T. (1982b). Hazards of commercial chemical reactions. In Fawcett, H.H. and Wood, W.S. (1982), *op. cit.*, p. 61

AUSTIN, G.T. (1984). *Shreve's Chemical Process Industries* (New York: McGraw-Hill)

AUSTIN, K. (1984). Pressure surges. *Process Engng*, Mar., 35

AUSTIN, R.M. and NAU, B.S. (1974). *The Seal Users Handbook* (Cranfield: Br. Hydromech. Res. Ass.)

AUSTIN, W.N. (1979). *Fire Fighting Techniques in Petrochemical Installations. A Bibliography* (Billingham, Cleveland: Cleveland County Libraries)

AUTENRIET, JR (1962). Money-saving methods for maintaining electrical equipment. *Chem. Engng*, **69**(Apr. 2), 128

AUTHEN, T.K., SKRAMSTAD, E. and NYLUND, J. (1976). Gas carriers – effects of fire on the cargo containment system. *Gastech 76*, p. 99

AUTON, T.R. and PICKLES, J.H. (1978). *The Calculation of Blast Waves from the Explosion of Pancake-shaped Vapour Clouds. Note Rd/l/N 210/78*. Central Elec. Generating Board

AUTON, T.R. and PICKLES, J.H. (1980). Deflagration in heavy flammable vapours. *Bull. Inst. Math. Appl.*, **16**, 126

AVEN, T. (1985). Reliability/availability evaluations of coherent systems based on minimal cut sets. *Reliab. Engng*, **13**(2), 93

AVEN, T. (1986). Two new component importance measures for a flow network system. *Reliab. Engng*, **14**(1), 75

AVERY, A.W. (1961). High vacuum closures. *Chem. Engr, Lond.*, **153**, A61

AVERY, H.S. (1972a). Abrasive wear. *Hydrocarbon Process.*, **51**(8), 79

AVERY, H.S. (1972b). Thermal fatigue. *Hydrocarbon Process.*, **51**(8), 78

AVERY, R.E. (1991). Pay attention to dissimilar-metal welds. *Chem. Engng Prog.*, **87**(5), 70

AVERY, R.E. (1991b). Resist chlorides, retain strength with Duplex stainless steel. *Chem. Engng Prog.*, **87**(3), 78

AVERY, R.E. and VALENTINE, H.L. (1968). High-temperature piping systems for petrochemical processing. *Safety in Air and Ammonia Plants*, **10**, 20. See also *Chem. Engng Prog.*, 1968, **64**(1), 89

AVES, C.M. (1929). Hydrogen sulfide poisoning in Texas *Texas State J. Med.*, **24**, 761

AVGERINOS, G.F. (1987). The LPG supply overhang after 1986. *Gastech 86*, p. 85

AVISSE, P. (1989). OPERFORM – formation des operateurs. *Loss Prevention and Safety Promotion 6*, p. 18.1

AVNI, E., KANDEL, A. and GUPTA, M. (1985). Expert systems in chemical process design. In Gupta, M., Bandler, W. and Kiszka, J.B. (1985), *op. cit.*

AVOGADRI, S., BELLO, G.C. and COLOMBARI, V. (1983). The ENI reliability data bank scope, organization, an example of reports: pumps, compressors, electric motors. *Euredata 4*, paper 7.3

AXELROD, L.C. and FINNERAN, J.A. (1965). Design of fail-safe control systems for steam reforming plants. *Ammonia Plant Safety*, **7**, 1

AXELROD, L.C. and FINNERAN, J.A. (1966). Make steam reformer controls safe. *Hydrocarbon Process.*, **45**(10), 140

AXELROD, L.C., DAZE, R.E. and WICKHAM, H.P. (1968). The large plant concept. *Chem. Engng Prog.*, **64**(7), 17

AYDEMIR, N.U., MAGAPU, UK, SOUSA, A.C.M. and VENART, J.E.S. (1988). Thermal response analysis of LPG tanks exposed to fire. *J. Haz. Materials*, **20**, 239

AYERS, E.D. and RHODES, A.W. (1963). Materials handling and bulk packaging. *Chem. Engng*, **70**(Sep. 16), 157

AYERS, J.A. (ed.) (1964). *Decontamination of Nuclear Reactors and Equipment* (New York: Ronald Press)

AYLING, G.W. and CASTRANTAS, H.M. (1981). Waste treatment with hydrogen peroxide. *Chem. Engng*, **88**(Nov. 30), 79

AYOUB, M.A. (1982). The manual lifting problem: the illusive solution. *J. Occup. Accid.*, **4**(1), 1

AYOUB, M.M., MITAL, A., ASFOUR, S.S. and BETHEA, N.J. (1980). Review, evaluation and comparison of models for predicting lifting capacity. *Hum. Factors*, **22**, 257

AYOUB, M.M., MITAL, A., BAKKEN, G.M., ASFOUR, S.S. and BETHEA, N.J. (1980). Development of STRENgtH and capacity norms for manual materials handling activities: the state of the art. *Hum. Factors*, **22**, 271

AYOUB, M.M., SELAN, J.L. and LILES, D.H. (1983). An ergonomics approach for the design of manual materials-handling tasks. *Hum. Factors*, **25**, 507

AYRAL, T.E. and MELVILLE, S.J. (1992). Conduct a process control technology audit. *Hydrocarbon Process.*, **71**(6), 43

AYRAULT, M., BALINT, J.-L. and MOREL, R. (1991). An experimental study on the evolution and dispersion of a cloud of gas heavier than air. *J. Haz. Materials*, **26**(1), 1

AYRES, D.C. (1981). Destruction of polychlorodibenzo-p-dioxins. *Nature*, **290**(Mar. 26), 323

AZBEL. D. (1981). *Two Phase Flows in Chemical Engineering* (Cambridge: Cambridge Univ. Press)

AZBEL, D.S. and CHEREMISINOFF, N.P. (1981). *Chemical and Process Equipment Design: Vessel Design and Selection* (Woburn, Ma.: Ann Arbor Sci. Publs)

AZIZ, T.S. (1977). Seismic risk analysis for Canadian nuclear power plants. Fourth Int. *Conf. on Structural Mechanics in Reactor Technology*, vol. K(b), paper K9/3

AZZOPARDI, B.J. and GIBBONS, D.B. (1983). Annular two phase flow in large diameter tube. *Chem. Engr, Lond.*, **398**, 19

BAADER, E.W. (1952). Chlorine anhydride poisoning. *Med. Deporte. Trab.*, **17**, 5252

BAAS, J. and WARNER, R. (1992). How much 'life' is left in your olefin unit? *Hydrocarbon Process.*, **71**(12), 81

BAASEL, W.D., MCALLISTER, R.A. and KINGSBURY, G.L. (1980). Multimedia environmental goals. *Chem. Engng Prog.*, **76**(10), 37

BAATZ, H. (1977). Protection of structures. In Golde, R.H. (1977a), *op. cit.*, p. 599

BABA, T.B. and KENNEDY, J.R. (1976). Ethylene and its co-products: the new economics. *Chem. Engng*, **83**(Jan. 5), 116

BABBIDGE, L.G., PARTRIDGE, C.C. and D'ANGELO, J.A. (1977). Fire-test valves for reliability. *Hydrocarbon Process.*, **56**(12), 179

BABCOCK and WILCOX (1993). *Steam, Its Generation and Use* (London)

BABER, R.L. (1987). *The Spine of Software: Designing Probably Correct Software – Theory and Practice* (New York: Wiley)

BABKIN, V.S., KORZHAVIN, A.A. and BUNEV, V.A. (1991). Propagation of premixed gaseous explosion flames in porous media. *Combust. Flame*, **87**, 182

BABRAUSKAS, V. (1981). A closed form approximation for post-flashover compartment fire temperatures. *Fire Safety J.*, **4**, 63

BABRAUSKAS, V. (1983). Estimating large pool fire burning rates. *Fire Technol.*, **19**, 251

BABRAUSKAS, V. (1986a). Free burning fires. *Fire Safety J.*, **11**(1), 33

BABRAUSKAS, V. (1986b). Pool fires: burning rates and heat fluxes. In *Fire Protection Handbook*, 16th ed. (edited by A.E. Cote and J.M. Linville), pp. 21–36

BABRAUSKAS, V. (1986c). Room fire temperature computations. In *Fire Protection Handbook*, 16th ed. (edited by A.E. Cote and J.M. Linville), pp. 21–31

BABRAUSKAS, V. (1993). Toxic hazards from fires: a simple assessment method. *Fire Safety J.*, **20**, 1

BABRAUSKAS, V., LEVIN, B.C., GANN, R.C., PAABO, M., HARRIS, R.H., PEACOCK, R.D. and YUSA, S. (1992). Toxic potency measurement for fire hazard analysis. *Fire Technol.*, **28**, 163

BABRAUSKAS, V. and PEACOCK, R.D. (1992). Heat release rate: the single most important variable in fire hazard. *Fire Safety J.*, **18**, 255

BABRAUSKAS, V. and WILLIAMSON, R.B. (1978). Post-flashover compartment fires: basis of a theoretical model, *Fire and Materials*, **2**, 239

BABRAUSKAS, V. and WILLIAMSON, R.B. (1979). Post-flashover compartment fires: application of a theoretical model, *Fire and Materials*, **3**, 1

BABUSHOK, V., GOLDSHTEIN, V.M. and SOBOLEV, V.A. (1990). Critical conditions for thermal explosion with reactant consumption. *Combust. Sci. Technol.*, **70**, 81

BACCARO, G.P., GAITONDE, N.Y. and DOUGLAS, J.M. (1970). An experimental study of oscillating reactors. *AIChE J.*, **16**, 249

BACCHELLI, B. and BIANCO, E. (1971). Startup made easier. Turbine driven compressors. *Hydrocarbon Process.*, **50**(12), 90

BACCHETTI, J.A. (1967). Unusual start-up problems. *Chem. Engng Prog.*, **63**(12), 53

BACH, G.G., KNYSTAUTAS, R. and LEE, J.H. (1969). Direct initiation of spherical detonations in gaseous explosives. *Combustion*, **12**, 853

BACH, G.G., KNYSTAUTAS, R. and LEE, J.H. (1971). Initiation criteria for diverging gaseous detonations. *Combustion*, **13**, 1097

BACH, N.G. (1968). Plant communications: a radio for everyone who needs it. *Chem. Engng*, **75**(Apr. 22), 189

BACHANT, J. and MACDERMOTT, J. (1984). R1 revisited: four years in the trenches. *AI Magazine*, **5**(3)

BACHE, D.H., LAWSON, T.J. and UK, S. (1988). Development of a criterion for defining spray drift. *Atmos. Environ.*, **22**, 131

BACHLIN, W. and PLATE, E.J. (1986). The dispersion of accidentally released gases in a built up area. *Loss Prevention and Safety Promotion*, **5**, p. 27-1

BACHLIN, W. and PLATE, E.J. (1987). *Wind Tunnel; Simulation of Dispersion Processes in Built-up Areas. Fortschritte der Sicherheitstechnik 1* (Frankfurt-am-Main: DECHEMA), p. 313

BACHLIN, W., PLATE, E.J. and THEURER, W. (1989). The influence of city canyons on the dispersion of accidentally released gases. *Loss Prevention and Safety Promotion 6*, p. 70.1

BACK, N.C., THOMAS, A.A. and MACEWEN, J.D. (1972). *Reclassification of Materials Listed as Transportation Hazards. Rep. TSA-20-72-3* US Air Force, Aerospace Med. Res. Lab., Wright Patterson Air Force Base, OH

BACKENSTO, E.B., PRIOR, J.E. and PORTER, J.E. (1966). Monitor skin temperature to reduce heater tube carburization. *Hydrocarbon Process.*, **45**(9), 291

BACKERT, W. and RIPPIN, D.W.T. (1985). The determination of maintenance strategies for plants subject to breakdown. *Comput. Chem. Engng*, **9**, 113

BACKHAUS, H. (1982). Carrying liquid gas on inland waterways. *Fire Int.*, **77**, 58

BACKHAUS, H. (1983). Subsea pipelines for low-temperature liquids to be used for gas tanker loading/unloading in foreshore areas. *Gastech 82*, p. 197

BACKHAUS, H.W. and JANNSEN, R. (1974). LNG inland transportation with railway tanks cars and river-going tankers. *LNG Conf. 4*

BACKHAUS, H. and ÖLSCHLAGER, R. (1985). A 30 000 m³ semi-pressurised ethylene carrier. *Gastech 84*, p. 256

BACKHOUSE, R. (1986). *Program Construction and Verification* (Englewood Cliffs, NJ: Prentice-Hall)

BACKHURST, J.R. and HARKER, J.H. (1973). *Process plant Design* (London: Heinemann)

BACON, D.H. and STEPHEN, R.C. (1977). *Mechanical Technology* (London: Newnes-Butterworths)

BACOW, L.S. and WHEELER, M. (1984). *Environmental Dispute Resolution* (New York: Plenum Press)

BADAME, P.J. (1986). A 15,000 ton atmospheric ammonia storage tank roof rupture. *Ammonia Plant Safety*, **26**, p. 158

BADAMI, V.N. (1983). Safety valve protects vacuum lines. *Chem. Engng*, **90**(Jan. 10), 144

BADER, B.E., DONALDSON, A.B. and HARDEE, H.C. (1971). Liquid propellant rocket abort fireball model, *J. Spacecraft Rockets*, **8**, 1216

BADER, F.P. (1979). Interface computers to existing plants. *Hydrocarbon Process.*, **58**(6), 83

BADGER, F.W. (1968). The alternatives to financial losses connected with accidents at the new ammonia plants. *Ammonia Plant Safety*, **10**, 71

BADHWAR, I. and TREHAN, M. (1984). Bhopal: city of death. *India Today*, Dec. 31, 4

BADISCHE ANILIN UND SODAFABRIK (1921). *Werkzeitung der BASF*, **9**(10), 22

BADISCHE ANILIN UND SODAFABRIK (1948). *Explosions Katastrophe in Ludwigshafen-am-Rhein* (Ludwigshafen: BASF)

BADISCHE ANILIN UND SODAFABRIK (1958). *Firmengeschichte* (Ludwigshafen; BASF)

BADISCHE ANILIN UND SODAFABRIK (1965). *In the Realm of Chemistry* (Dusseldorf: Econ Verlag)

BADISCHE ANILIN UND SODAFABRIK (1979). *Sicherheit in der Chemie* (Ludwigshafen)

BADLEY, J.H. and HEATON, R.C. (1978). A portable system for determination of trace volatiles pollutants in water. In *Control of Hazardous Material Spills*, vol. 4, p. 115

BADOUX, R.A.J. (1983). Some experiences of a consulting statistician in industrial safety and reliability. Third *Nat. Reliability Conf*. See also *Reliab. Engng*, 1985, **10**, 219

BADR, A. (1984). Temperature measurements in a negatively buoyant round vertical jet issued in a horizontal crossflow. In Ooms, G. and Tennekes, H. (1984), *op. cit*., p. 167

BAGGIO, J.L. and SAINTHERANT, P. (1980). NGL refrigerant improves recovery. *Hydrocarbon Process*., **59**(4), 138

BAGIN, Y.P. *et al.* (1985). Optical observations of an explosion during the MASSA experiment. Izvestia. *Akad. Sci. USSR, Phys. Solid Earth*, **21**(11), 826

BAGNARD, G.M. (1960). Use swivel joints for flexibility. *Chem. Engng*, **67**(Sept. 9), 175

BAGSTER, D. (1993). A comparison of plume models in estimation of individual risk contours. *Health, Safety and Loss Prevention*, p. 179

BAGSTER, D.F. and DAVIS, B.S. (1990). Computational aspects of radiation view factors of pool and jet fires. *Safety and Loss Prevention in the Chemical and Oil Processing Industries*, p. 209

BAGSTER, D.F. and PITBLADO, R.M. (1989). Thermal hazards in the process industry. *Chem. Engng Prog*., **85**(7), 69

BAGSTER, D.F. and PITBLADO, R.M. (1991). The estimation of domino incident frequencies – an approach. *Process Safety Environ*., **69B**, 196

BAGULEY, P. (1986). Condition-based maintenance. *Chem. Engr, Lond*., **427**, 38

BAGULEY, P. (1989). Total quality management. *Chem. Engr, Lond*., **463**, 50

BAGULEY, P. (1990). BS 5750 (ISO 9000) and all that jazz. *Chem. Engr, Lond*., **470**, 71

BAHME, C.W. (1961). *Fire Protection for Chemicals* (Boston, MA: Nat. Fire Prot. Ass.)

BAHME, C.W. (1972). *Fire Officers Guide to Dangerous Chemicals* (Boston, MA: Nat. Fire Prot. Ass.)

BAIER, R.E., MICHALOVIC, J.G., DEPALMA, V.A. and PILIE, R.J. (1975). Universal gelling agent for the control of hazardous material spills. *J. Haz. Materials*, **1**(1), 21

BAILEY, A. (1984). Carriage of dangerous substances (letter). *Chem. Engr, Lond*., **396**, 3

BAILEY, A.G., SMALLWOOD, J.M. and TOMITA, H. (1991). Electrical discharges from the human body. *Electrostatics*, **8**, 101

BAILEY, J.D. (1983). Is your plant safe from terrorists? *Hydrocarbon Process*., **62**(5), 168

BAILEY, J.D. (1992). Safer pesticide packaging and formulations for agricultural and residential applications. In *Reviews of Environmental Contamination and Toxicology*, (New York: Springer-Verlag) pp. 17–27

BAILLOD, C.R., LAMPARTER, R.A. and BARNA, B.A. (1985). Wet oxidation for industrial waste treatment. *Chem. Engng Prog*., **81**(3), 52

BAILLOU, F., LISBET, R., DUPRE, G., PAILLARD, C. and GUSTIN, J.L. (1992). Gas phase explosion of nitrogen trichloride: application to the safety of chlorine plants and chlorination process. *Loss Prevention and Safety Promotion 7*, vol. 3, p. 106–1

BAILLY, P., BROSSARD, J., DESROSIER, C., RENARD, J., BRANDT, B. and LAZARE, B. (1989). Prediction of Ariane 4 explosion effects on mechanical structures on the launch pad. *Loss Prevention and Safety Promotion 6*, p. 81.1

BAINBRIDGE, B.L. and KELTNER, N.R. (1988). Heat transfer to large objects in large pool fires. *J. Haz. Materials*, **20**, 21

BAINBRIDGE, L. (1971). *The Influence of Display Type on Decision-making Strategy. Displays* (London: Instn Elec. Engrs)

BAINBRIDGE, L. (1972). An analysis of a verbal protocol from a process control task. Ph.D. Thesis, Univ. of Bristol

BAINBRIDGE, L. (1974). Analysis of verbal protocol from a process control task. In Edwards, E. and Lees, F.P. (1974), *op. cit*., p. 146

BAINBRIDGE, L. (1978). Forgotten alternatives to skill and work-load. *Ergonomics*, **21**, 169

BAINBRIDGE, L. (1981). Mathematical equations or processing routines? In Rasmussen, J. and Rouse, W.B. (1981), *op. cit*., p. 259

BAINBRIDGE, L. (1983). Ironies of automation. *Automatica*, **19**, 775

BAINBRIDGE, L. (1993a). Planning the training of a complex skill. *Le Travail Humain*, **56**, 211

BAINBRIDGE, L. (1993b). Types of hierarchy imply types of model. *Ergonomics*, **36**, 1399

BAINBRIDGE, L., BEISHON, J., HEMMING, J.H. and SPLAINE, M. (1968). A study of real-time human decision-making using a plant simulator. *Opl. Res. Quart.* **19** (Special Conf. Issue), p. 91

BAINBRIDGE, L., LENOIR, T.M.J. and VAN DER SCHAAF, T.W. (1993). Cognitive processes in complex tasks: introduction and discussion. *Ergonomics*, **36**, 1273

BAINES, D. (1984). GRP in the process industries. *Chem. Engr, Lond*., **405**, 24

BAINES, W.D. and HOPFINGER, E.J. (1984). Thermals with large density difference. *Atmos. Environ*., **18**, 1051

BAINES, W.D. and TURNER, J.S. (1969). Turbulent buoyant convection from a source in a confined region. *J. Fluid Mech*., **37**, 51

BAIS, A.F., ZEREFOS, C.S. and ZIOMAS, I.C. (1989). Design of a system for real-time modelling of the dispersion of hazardous gas releases in industrial plants: 1, Emissions from industrial stacks. *J. Loss Prev. Process Ind*., **2**(3), 155

BAJPAI, S.N. (1980). Flammability of ethylene oxide in sterilizer operations. In *Loss Prevention*, **13**, p. 119

BAKEMEIER, H. *et al.* (1985). Ammonia. *Ullmanns Encyclopaedia of Chemical Technology*, rev. ed., vol. A2 (Weinheim: VCH), p. 143

BAKER, LORD (1978). *Enterprise versus Bureaucracy. The Development of Structural Air-Raid Precautions during the 2nd World War* (Oxford: Pergamon)

BAKER, A. (1979), private communication to A.F. Roberts

BAKER, A.C. and ENTRESS, J.H. (1992). The VASPS subsea separation and pumping system. *Trans. Inst. Chem. Engrs*, **70A**, 9

BAKER, A.R. (1976). *Accident investigation*. Dept of Chem. Engng, Loughborough Univ. of Technol., Course on Loss Prevention in the Process Industries

BAKER, A.R. (1982). The role of physical simulation experiments in the investigation of accidents. *J. Occup. Accid.*, **4**(1), 33

BAKER, D.F. (1982). Surge control for multistage centrifugal compressors. *Chem. Engng*, **89**(May 31), 117

BAKER, D.P. and SALAS, E. (1992). Principles for measuring teamwork skills. *Hum. Factors*, **34**, 469

BAKER, D.S. (1981). Wood in fire, flame spread and flame retardant treatments. *Chemy Ind.*, July 18, 485

BAKER, F.E. and SHEPPERD, R.E. (1971). Fire testing of electric cables and the benefits of fire-retardant paints. *Fire Technol.*, **7**(4), 285

BAKER, G. (1989). Condition monitoring. *Plant Engr*, May/June, 26

BAKER, J.A. (1963). Productivity rating and reliability index. *ISA J.*, **10**(9), 61

BAKER, J.D. (1974). Pre start-up safety review. *Loss Prevention*, **8**, p. 38

BAKER, J.E. and BURT, R.D. (1967). Startup and operating problems in chlorine plants. *Chem. Engng Prog.*, **63**(12), 47

BAKER, J.F., WILLIAMS, E.L. and LAX, D. (1948). The design of frames buildings against high explosive bombs. In *The Civil Engineer at War*, vol. 3 (London: Inst. Civil Engrs), p. 80

BAKER, J.M. (1976). *Vibration Isolation* (Oxford: Oxford Univ. Press)

BAKER, L.S. *et al.* (1978). *TREDRA – A Computer Program to Draft Fault Trees. Rep. JBFA-128-78*, JBF Associates, Knoxville, TN

BAKER, O. (1954). Designing for simultaneous flow of oil and gas. *Oil Gas J.*, Jul. 26, 185

BAKER, O. (1958). Multiphase flow in pipelines. *Oil Gas J.*, Nov. 10, 156

BAKER, P.S. (1968). Radioisotopes in chemical processes. *Chem. Engng*, **75**(Mar. 11), 179

BAKER, Q.A. and BAKER, W.E. (1991). Pros and cons of TNT equivalence for industrial explosion accidents. *Modeling and Mitigating the Consequences of Accidental Releases*, p. 585

BAKER, Q.A., DOOLITTLE, C.M., FITZGERALD, G.A. and TANG, M.J. (1997). Recent developments in the Baker-Strehlow VCE analysis methodology, AIChE 31st Loss Prev. Symp. Houston, Paper 42f, Mar. 9–13

BAKER, Q.A., TANG, M.J., SCHEIER, E. and SILVA, G.J. (1994). Vapor cloud explosion analysis. AIChE 28th Annu. Loss Prev. Symp. Houston, Apr. 17–21

BAKER, S.B., TRIPOD, J. and JACOB, J. (1970). The problem of species difference and statistics. *Proc. Eur. Symp. for the Study of Drug Toxicity, XI*, (Amsterdam: Excerpta Medica)

BAKER, T.C. and PRESTON, F.W. (1946). Fatigue of glass under static loads. *J. Appl. Phys.*, **17**, 170

BAKER, W.E. (1973). Explosions in Air (Austin, TX: Univ. of Austin Press)

BAKER, W.E. (1982). Explosion accident investigation. *Plant/Operations Prog.*, **1**, 144

BAKER, W.E., COX, P.A., WESTINE, P.S., KULESZ, J.J. and STREHLOW, R.A. (1978). *A short course on explosion hazards evaluation*, South Western Res. Inst., San Antonio, TX

BAKER, W.E., COX, P.A., WESTINE, P.S., KULESZ, J.J. and STREHLOW, R.A. (1983). *Explosion Hazards and Evaluation* (Amsterdam: Elsevier)

BAKER, W.E., ESPARZA, E.D. and KULESZ, J.J. (1977). Venting of chemical explosions and reactions. *Loss Prevention and Safety Promotion 2*, p. 241

BAKER, W.E., WESTINE, P.S. and COX, P.A. (1977). Methods for prediction of damage to structures from accidental explosions. *Loss Prevention and Safety Promotion 2*, p. 339

BAKER, W.E., WESTINE, P.S. and DODGE, F.T. (1973). *Similarity Methods in Engineering Dynamics* (Rochelle Park, N.J.: Spartan Hayden)

BAKER, W.E., HOKANSON, J.C. and KULESZ, J.J. (1980). A model analysis for vented dust explosions. In *Loss Prevention and Safety Promotion 3*, vol. 3, p. 1339

BAKER, W.E., WHITE, R.E. and HOKANSON, J.C. (1991). *Gas, Dust and Hybrid Explosions* (Amsterdam: Elsevier)

BAKER, W.E. *et al.* (1975). *Workbook for Predicting Pressure Waves and Fragment Effects of Exploding Propellant Tanks and Gas Storage Vessels NASA Contract 134906, Rep.* NASA Lewis Res. Center

BAKER, W.E. *et al.* (1977). *Workbook for Predicting Pressure Waves and Fragment Effects of Exploding Propellant Tanks and Gas Storage Vessels. NASA Contract 134906, Rep.* NASA Lewis Res. Center

BAKER, W.E. *et al.* (1978). *Handbook for Estimating the Effects of Accidental Explosions in Propellant Handling Systems. NASA Contractor Rep. 3023.* NASA, Lewis Res. Center

BAKER, W.E. *et al.* (1979). *A short course on Explosion Hazards Evaluation*, South Western Res. Inst., San Antonio, TX

BAKER, W.J. (1980). Selecting and specifying air-cooled heat exchangers. *Hydrocarbon Process.*, **59**(5), 173

BAKER-COUNSELL, J. (1984a). Speciality steels. *Process Engng*, Nov., 57

BAKER-COUNSELL, J. (1984b). Testing time for pumps left out in the cold. *Process Engng*, Dec., 25

BAKER-COUNSELL, J. (1985a). Coating systems: are you getting the best from them. *Process Engng*, Nov., 65

BAKER-COUNSELL, J. (1985b). Corrosion problems? – ask the expert. *Process Engng*, Apr., 61

BAKER-COUNSELL, J. (1985c). Increased demands on offshore processing are coming. *Process Engng*, **66**(1), 18

BAKER-COUNSELL, J. (1985d). On-site materials management goes on-line. *Process Engng*, Mar., 25

BAKER-COUNSELL, J. (1985e). Planning for that successful Operations Prog. revamp. *Process Engng*, Apr., 21

BAKER-COUNSELL, J. (1985f). PVQAB: who can help? *Process Engng*, Feb., 43

BAKER-COUNSELL, J. (1985g). Radioactivation: the latest aid to corrosion monitoring. *Process Engng*, Feb., 47

BAKER-COUNSELL, J. (1985h). Seal developments improve safety and service life. *Process Engng*, Aug., 33

BAKER-COUNSELL, J. (1986a). Database/CAD link-up organises corrosion data gathering. *Process Engng*, June, 33

BAKER-COUNSELL, J. (1986b). Exit the safety inspector. *Process Engng*, Apr., 29

BAKER-COUNSELL, J. (1986c). Feeling the heat. *Process Engng*, Jul., 55

BAKER-COUNSELL, J. (1986d). Miniature modeller. *Process Engng*, Sept., 25

BAKER-COUNSELL, J. (1987a). Hazardous wastes: the future for incineration. *Process Engng*, Apr., 25

BAKER-COUNSELL, J. (1987b). How long will your plant run safely. *Process Engng*, May, 61

BAKHMETEFF, B.A. (1941). *The Mechanics of Turbulent Flow* (Princeton, NJ: Princeton Univ. Press)

BAKIR, F. *et al.* (1973). *Methylmercury poisoning in Iraq. Science*, **181**, 239

BAKKE, J. (1986a). The LNG/LIN scheme as a transport alternative for Northern Norwegian gas. *Gastech 85*, p. 219

BAKKE, J.R. (1986b). Numerical simulation of gas explosions. Ph.D. Thesis, Univ. of Bergen

BAKKE, J. (1989). *Piper Alpha Inquiry Transcript, Days 77*, 123

BAKKE, J. and ANDERSEN, P.G. (1982). Baseload LNG plants with spherical storage tanks, all built as very large modules. *Gastech 81*, p. 131

BAKKE, J.R., BJERKETVEDT, D. and BJØRKHAUG, M. (1990). FLACS as a tool for safe design against accidental gas explosions. *Piper Alpha: Lessons for Life Cycle Management* (Rugby: Instn Chem. Engrs), p. 141

BAKKE, J.R., BJERKETVEDT, D., BJØRKHAUG, M. and SORHEIM, H.R. (1989). Practical applications of advanced gas explosion research. *Loss Prevention and Safety Promotion 6*, p. 24.1

BAKKE, J.R. and HJERTAGER, B.H. (1987). The effect of explosion venting in empty vessels. *Int. J. Num. Meth. Engng*, **24**, 129

BAKSHI, T.S. and NAVEH, Z. (eds) (1980). *Environmental Education* (New York: Plenum Press)

BALAAM, E. (1977). Dynamic predictive maintenance for refinery equipment. *Hydrocarbon Process.*, **56**(5), 131

BALCERZAK, M.H., JOHNSON, M.R. and KURZ, F.R. (1966). *Nuclear blast simulation, Pt I: Detonable gas explosion. Final Rep. DASA 1972-1*. Gen. Am. Res. Div., Niles, IL

BALDEWICZ, W. *et al.* (1974). *Historical perspectives on risk for large scale technological systems. Rep. UCLA-ENG-7485*. Univ. of California, Los Angeles, CA

BALDINI, R. and KOMOSINSKY, P. (1988). Consequence analysis of toxic substance clouds. *J. Loss Prev. Process Ind.*, **1**(3), 147

BALDOCK, P.J. (1980). Accidental releases of ammonia: an analysis of reported incidents. In *Loss Prevention* **13**, p. 35

BALDOVINETTI, G.B. (1983). Warehouse fire protection and the special problem of high shelving. *Fire Int.*, **81**, 38

BALDWIN, R. (1968). Flame merging in multiple fires. *Combust. Flame*, **12**, 318

BALEMANS, A.W.M. (1975). Public emergency limits. *Transport of Hazardous Cargoes*, 4

BALEMANS, A.W.M. and VAN DE PUTTE, T. (1977). Guidelines for explosion-resistant control buildings in the chemical process industries. *Loss Prevention and Safety Promotion 2*, p. 213

BALEMANS, A.W.M. *et al.* (1974). Check-list. *Loss Prevention and Safety Promotion 1*, p. 7

BALENTINE, H.W. and ELGROTH, M.W. (1985). Validation of a hazardous spill model using N_2O_4 and LNG spill data. *Am. Poll. Control Assoc. Ann. Mtg*, paper 85-25B.1

BALFOUR, D. (1989). *Piper Alpha Public Inquiry Transcript, Day 76*

BALL, D.F. and PEARSON, A.W. (1976). The future size of process plant. *Long Range Planning*, Aug., 18

BALL, D.F. and STEWARD, H.F. (1974). *Caprolactam – after Flixborough?* Res. and Dev. Unit, Manchester Business School, Manchester

BALL, H.C. and HOWARTH, H.M. (1984). Improvements in valve quality and reliability. *API Proc, Refining Dept, 49th Midyear Mtg*, **63**, 304

BALL, J.G. (1975a). *The Flixborough disaster – a metallurgical assessment. Rep. to Flixborough Court of Inquiry.* Imperial College, London

BALL, J.G. (1975b). The metallurgy of the 8 inch pipe in the context of the Flixborough disaster. *Instn Chem. Engrs, Symp. on Technical Lessons of Flixborough* (London)

BALL, J.G. (1976). Technical lessons of Flixborough. The metallurgy of the 8 inch pipe in the context of Flixborough. *Chem. Engr, Lond.*, **308**, 275

BALL, W.L. (1962). Hazard level of hydrocarbon films in oxygen systems. *Ammonia Plant Safety*, **4**, 16

BALL, W.L. (1963). Discussion of energy release in a liquid oxygen pump. *Ammonia Plant Safety*, **5**

BALL, W.L. (1964). Explosion in an aftercooler of a compressor using synthetic lubricant. *Ammonia Plant Safety*, **6**, 11

BALL, W.L. (1965). Safe level of oxygen in nitrogen compression systems. *Ammonia Plant Safety*, **7**, 19

BALL, W.L. (1966a). Explosion in expander discharge piping. *Ammonia Plant Safety*, **8**, 1

BALL, W.L. (1966b). Review of new solubility data for certain hydrocarbons in liquid oxygen. *Ammonia Plant Safety*, **8**, 12

BALL, W.L. (1968a). Air separation plant incident. *Ammonia Plant Safety*, **10**, 41

BALL, W.L. (1968b). Status report of the Ammonia Storage Committee. *Ammonia Plant Safety*, **10**, 38

BALLAL, D.R. and LEFEBVRE, A.H. (1973). Turbulence effects on enclosed flames. Fourth Int. Symp. on Gas Dynamics and Reactive Systems, San Diego, CA

BALLAL, D.R. and LEFEBVRE, A.H. (1975a). The influence of flow parameters on minimum ignition energy and quenching distance. *Combustion*, **15**, 1473

BALLAL, D.R. and LEFEBVRE, A.H. (1975b). The influence of spark discharge characteristics on minimum ignition energy in flowing gases. *Combust. Flame*, **24**, 99

BALLAL, D.R. and LEFEBVRE, A.H. (1975c). The structure and propagation of turbulent flames. *Proc. R. Soc., Ser. A.*, **344**, 217

BALLAL, D.R. and LEFEBVRE, A.H. (1977). Flame quenching in turbulent flowing gaseous mixtures. *Combustion*, **16**, 1689

BALLAL, D.R. and LEFEBVRE, A.H. (1978). Ignition and flame quenching of quiescent fuel mists. *Proc. R. Soc., Ser. A.*, **364**, 277

BALLAL, D.R. and LEFEBVRE, A.H. (1979). Ignition and flame quenching of flow heterogeneous fuel-air mixtures. *Combust. Flame*, **35**, 155

BALLAL, D.R. and LEFEBVRE, A.H. (1981a). Flame propagation in heterogeneous mixtures of fuel droplets. *Combustion*, **18**, 321

BALLAL, D.R. and LEFEBVRE, A.H. (1981b). A general model for spark ignition for gaseous and liquid fuel-air mixtures. *Combustion*, **18**, 1737

BALLAL, D.R. (1983a). Flame propagation through dust clouds of carbon, coal, aluminium and magnesium in an environment of zero gravity. *Proc. R. Soc.*, Ser. A., **385**, 21

BALLAL, D.R. (1983b). Further studies on the ignition and flame quenching of dust clouds. *Proc. R. Soc.*, Ser. A., **385**, 1

BALLANTYNE, A.W. (1972). The reformer information network. *Ammonia Operations Prog. Safety*, **14**, 40

BALLANTYNE, B. and SCHWABE, P. (eds) (1981). *Respiratory Protection* (London: Chapman & Hall)

BALLARD, D.H. and BROWN, C. (1982). *Computer Vision* (Englewood Cliffs, NJ: Prentice-Hall)

BALLARD, G.M. (1987). Reliability analysis – present capability and future developments. *SRS Quart. Digest*, Oct., **12**

BALLARD, G.M. (1988a). Dependent failure analysis in probabilistic safety assessment. *SRS Quart. Dig.*, Apr., 3

BALLARD, G.M. (1988b). *Nuclear Safety after Three Mile Island and Chernobyl* (London: Elsevier)

BALLER, H. (1983). Rig winterisation to allow year-round drilling of northern Norway. *Oil Gas J.*, **81** (Aug. 1), 71

BALLMER, R.W. (1964). Safe start-up and shutdown of vacuum and crude units. In Vervalin, C.H. (1964a), *op. cit.*, p. 247

BALLOU, D.F., LYONS, T.A. and TACQUARD, J.R. (1967). Mechanical refrigeration. *Hydrocarbon Process.*, **46**(6), 118

BALT, L. and VAN DEN BERG, H. (1974). The lay-out of a chlorination laboratory. *Loss Prevention and Safety Promotion 1*, p. 363

BALTZELL, H.J. (1978). Estimate refinery cost better. *Hydrocarbon Process.*, **57**(1), 129

BAMFORD, C.H., CRANK, J. and MALAN, D.H. (1946). The combustion of wood. *Proc. Cambridge Phil. Soc.*, **42** (2), 166

BAMFORD, W.D. (1961). *Control of Airborne Dust* (Br. Cast Iron Res. Ass.)

BANARES-ALCANTARA, R., KO, E.I., WESTERBERG, A.W., and RYCHENER, M.D. (1988). DECADE – a hybrid expert system for catalyst selection. II, Final architecture and results. *Comput. Chem. Engng*, **12**, 923

BANARES-ALCANTARA, R., SRIRAM, D., VENKATASUBRAMANIAN, V., WESTERBERG, A. and RYCHENER, M. (1985). Knowledge-based expert systems for CAD. *Chem. Engng Prog.*, **81**(9), 25

BANARES-ALCANTARA, R., WESTERBERG, A.W. and RYCHENER, M.D. (1985). Development of an expert system for physical property predictions. *Compu. and Chem. Engng*, **9**, 127

BANARES-ALCANTARA, R., WESTERBERG, A.W., KO, E.I. and RYCHENER, M.D. (1987). DECADE – a hybrid expert system for catalyst selection. I, Expert system consideration. *Comput. Chem. Engng*, **11**, 3

BANERJEE, K., CHEREMISINOFF, N.P. and CHEREMISINOFF, P.N. (1985). *Flare Gas Systems Pocket Handbook* (Houston, TX: Gulf Publ. Co.)

BANERJEE, S. (1988). Multiphase flow aspects of chemical plant safety. *Short Course on Modelling of Multiphase Systems for Industrial Applications*, Swiss Fed. Inst. of Technol. (ETH), Geneva

BANERJEE, S., PRANDINI, F. and PATRONCINI, G. (1990). Aspects of vent sizing for homopolymerization reactors for polypropylene plants based on pilot plant studies. *J. Loss Prev. Process Ind.*, **3**(1), 8

BANG, A. and BIRKELAND, S. (1973). Purging and gas-freeing of cargo tanks on liquefied gas tankers. Safety aspects in comparison with similar operations on crude oil tankers. *Transport of Hazardous Cargoes, 3*

BANGE, J.B. and OSMAN, R.H. (1993). Select the right centrifuge drives for batch processes. *Chem. Engng Prog.*, **89**(8), 34

BANKS, B. (1973). Pressure vessel failure during hydrotest. *Weld. Metal Fabrication*, Jan. 4

BANKS, J. and RARDIN, R.L. (1982). International comparison of fire loss. *Fire Technol.*, **18**, 268

BANKS, Q.H. (1972). Governmental considerations on methods for prevention of spills in the rail transportation of hazardous materials. *Control of Hazardous Material Spills*, **1**, 45

BANKS, R. (1978). Ammonia output increased by hydrogen recovery. *Ammonia Plant Safety*, **20**, 79

BANKS, R.P. (1990). *Piper Alpha Public Inquiry. Transcript, Day 168*

BANKS, W.W. and CERVEN, F. (1984). Predictor displays: the application of human engineering in process control system. *Plant/Operations Prog.*, **3**, 215

BANKS, W.W. and WELLS, J.E. (1992). A probabilistic risk assessment using human reliability analysis methods. *Hazard Identification and Risk Analysis, Human Factors and Human Reliability in Process Safety*, p. 315

BANNISTER, J. (1989). *Managing Risk* (London: Insurance and Reinsurance Res. Group)

BAO-CHUN FAN and SICHEL, M. (1988). A comprehensive model for the structure of dust detonations. *Combustion*, **22**, 1741

BAR-OR, R., SICHEL, M. and NICHOLLS, J.A. (1981). The propagation of cylindrical detonations in monodisperse sprays. *Combustion*, **18**, 1599

BAR-OR, R., SICHEL, M. and NICHOLLS, J.A. (1982). The reaction zone structure of cylindrical detonations in monodisperse sprays. *Combustion*, **19**, 665

BARA, B.M., WILSON, D.J. and ZELT, B.W. (1992). Concentration fluctuation profiles from a water channel simulation of a ground level release. *Atmos. Environ.*, **26A**, 1053

BARAD, M.L. (1951). Diffusion of stack gases in very stable atmospheres. *Met. Monogr.*, **1**, 9

BARAD, M.L. (ed.) (1958). *Project Prairie Gas, a field program in diffusion*, vols 1–3. *Geophysical Res. Paper 59* (ASTUA Doc. AD-151572). US Air Force, Cambridge Res. Center

BARAD, M.L. and FUQUAY, J.J. (eds) (1962). *The Green Glow diffusion program*, vols 1–2. *Geophysical Res. Paper 73*. US Air Force, Cambridge Res. Center

BARBER, D.H. (1975). *Safety in petroleum refinery operations*. *Proc. Ninth World Petrol. Cong.*, vol. 6 (London: Applied Science), p. 133

BARBER, D.H. and TIBBETTS, A.M. (1976). Developing safe workers in refineries. *Chem. Engng*, **83**(Feb. 2), 117

BARBIERI, A. *et al.* (1977). Review paper of industrial loss and prevention history. *Loss Prevention and Safety Promotion 2*, p. 461

BARBIERI, S. *et al.* (1992). COMPARE: a computer aided risk evaluation package. *Loss Prevention and Safety Promotion 7*, vol. 3, 136.1

BARBIN, F.M. (1973). Different PM intervals for identical instruments. *Instrum. Technol.*, **20**(11), 44

BARBOUR, H.G. (1919). The effects of chlorine on the body temperature. *J. Pharmacol. Exp. Therapeut.*, **14**, 65

BARBOUR, H.G. and WILLIAMS, H.W. (1919). The effects of chlorine gas upon isolated bronchi and pulmonary vessels. *J. Pharmacol. Exp. Therapeut.*, **14**, 47

BARBOZA, M.J., MILITANA, L.M. and HAYMES, N.M. (1986). *Consequence modeling of accidental releases during transportation, handling and storage of liquid hazardous wastes. Fifth Joint Conf. on Applications of Air Pollution Modeling* (Boston, MA: Am. Met. Soc.), p. 311

BARBREAU, A., FERRIEUX, H. and MOHAMMADIOUN, B. (1974). Seismological studies carried out by the CEA in connection with the safety of nuclear sites. *Symp. on Siting of Nuclear Facilities*, p. 517

BARCLAY, D.A. (1988). Protecting process safety interlocks. *Chem. Engng Prog.*, **84**(6), 22

BARCLAY, M. (1987). Hazardous-waste-management facility siting. In Martin, E., and Johnson, J.H. (1987), *op. cit.*, p. 469

BARCROFT, J. (1918). Report on the condition of the blood in cases of gas poisoning. *Chem. Warfare Med. Cttee, Med. Res. Coun.*, London, p. 6

BARCROFT, J. (1920). Some problems of the circulation during gas poison. *J. Army Med. Corps*, **34**, 155

BARCROFT, J. (1925). *The Respiratory Function of the Blood*, 2nd ed. (Cambridge: Cambridge Univ. Press)

BARCROFT, J. (1931). The toxicity of atmospheres containing hydrocyanic acid gas. *J. Hyg. (Camb.)*, **31**, 1

BARDWELL, J. (1955). Cool flames in butane oxidation. *Combustion*, **5**, 529

BARIN, I. (1989). *Thermochemical data of pure substances* (Weinheim: VCH)

BARKAN, D.D. (1962). *Dynamics of Bases and Foundations* (New York: McGraw-Hill)

BARKER, A. (1984). Planning inquiries: a role for Parliament. *R. Soc. Arts J.*, **132**, 619

BARKER C.D., (1982). A virtual source model for building wake dispersion in nuclear safety calculations. Central Electricity Generating Board, Rep. TPRD/B/0072/N82.

BARKER, D.D., WHITNEY, M.G. and WACLAWCZYK, J.H. (1993). Failure and design limit criteria for blast loaded structures. *Process Safety Prog.*, **12**, 216

BARKER, G. (1969). Eye protection. In Handley, W. (1969), *op. cit.*, p. 371

BARKER, G. (1977). Eye protection. In Handley, W. (1977), *op. cit.*, p. 351

BARKER, G.F., KLETZ, T.A. and KNIGHT, H.A. (1977). Olefin plant safety during the last 15 years. *Loss Prevention*, **11**, p. 4. See also *Chem. Engng Prog.*, 1977, **73**(9), 64

BARKER, J. (1981). Pneumatic pressure conveying. *Plant Engr*, **25**(2), 23

BARKER, M.L. (1977). *Specialists and Air Pollution: Occupations and Preoccupations* (Univ. of Victoria)

BARKLEY, N.E. (1986). Ammonia plant operating difficulties related to oxygen in natural gas feedstock. *Ammonia Plant Safety*, **26**, 115

BARKS, R.A. (1972). *Distribution Theory* (London: Hutchinson)

BARLOW, D.H. (1963). Intrinsic safety, its growing status in Europe. *Control Engng*, **10**(3), 85

BARLOW, J.A. (1975). Energy survey in a petrochemical plant. *Chem. Engng*, **82**(Jul. 7), 93

BARLOW, R. and PROSCHAN, F. (1965). *Mathematical Theory of Reliability* (New York: Wiley)

BARLOW, R. and PROSCHAN, F. (1975). *Statistical Theory of Reliability and Life Testing* (New York: Holt, Rinehart & Winston)

BARLOW, R.E. and CHATTERJEE, P. (1973). *Introduction to fault tree analysis. Rep AD 774072* Nat. Tech. Inf. Service, Springfield, VA.,

BARLOW, R.E., FUSSELL, J.B. and SINGPURWALLA, N.D. (eds) (1975). *Reliability and Fault Tree Analysis* (Philadelphia, PA: Soc. for Ind. and Appl. Maths)

BARNABY, K.C. (1969). *Basic Naval Architecture*, 6th ed. (London: Hutchinson)

BARNARD, J.A. and BRADLEY, J.N. (1985). *Flame and Combustion* (London: Chapman & Hall)

BARNARD, J.A. and HARWOOD, B.A. (1973a). Slow combustion and cool flame behaviour of iso-octane. *Combust. Flame* **21**, 345

BARNARD, J. and HARWOOD, B.A. (1973b). The spontaneous combustion of *n*-heptane. *Combust. Flame*, **21**, 141

BARNARD, J.A. and HARWOOD, B.A. (1974). Physical factors in the study of the spontaneous ignition of hydrocarbons in static systems. *Combust. Flame*, **22**, 35

BARNARD, J.A. and HAWTIN, P. (1961). The gaseous oxidation of paraxylene. *Combust. Flame*, **5**, 249

BARNARD, J.A. and IBBERSON, V.J. (1965). The gaseous oxidation of toluene. *Combust. Flame*, **9**, 81 and 149

BARNARD, J.A. and KIRSCHNER, E. (1967). The slow combustion of ketene. *Combust. Flame*, **11**, 496

BARNARD, J.A. and SANKEY, B.M. (1968). The slow combustion of the isomeric xylenes. *Combust. Flame*, **12**, 345 and 353

BARNARD, J. and WATTS, A. (1969). The cool flame oxidation of ketones. *Combustion*, **12**, 365

BARNARD, J.A. and WATTS, A. (1972). The relationship between slow reaction and cool flames in combustion processes. *Combust. Sci. Technol.*, **6**, 125

BARNARD, R.C. (1984). Science, policy and the law: a developing partnership in reducing the risk of cancer. In Deisler, P.F. (1984), *op. cit.*, p. 215

BARNER, H.E. and SCHEUERMAN, R.V. (1977). *Handbook of Thermochemical Data for Compounds and Aqueous Species* (New York: Wiley-Interscience)

BARNES, D., FORSTER, C.F. and HRUDEY, S.E. (eds) (1984–). *Environmental and Public Health Engineering*, vols 1–3 (London: Pitman)

BARNES, E.F. (1976). New materials solve reciprocating compressor problems. *Hydrocarbon Process.*, **55**(7), 147

BARNES, H. (1993). Rheology for the chemical engineer. *Chem. Engr, Lond.*, **545**, 17

BARNES, M. (1990). *Hinkley Point Public Inquiries* (London: HM Stationery Office)

BARNES, M.H. and FLETCHER, E.A. (1974). Thermal diffusion and flame stoichiometry. *Combust. Flame*, **23**, 399

BARNES, R.E. (1974). Commissioning large rotating machinery. *Instn Chem. Engrs Symp. on Commissioning of Chemical and Allied plant* (London)

BARNES, R.W. and DOAK, R.L. (1990). Troubleshooting control valves: learn the tricks, avoid the pitfalls. *Chem. Engng*, **97**(5), 96

BARNETT, E.P. (1973). Early planning for start-up. In Vervalin, C.H. (1973a), *op. cit.*, p. 401

BARNETT, H.C. and HIBBARD, R.R. (eds) (1957). *Basic considerations in the combustion of hydrocarbon fuels with air. Rep. NACA 1300.* Nat. Assn Corrosion Engrs

BARNETT, J.A. (1982). Some issues of control in expert systems. *Proc. Int. Conf. on Cybernetics and Society, Seattle.*

BARNETT, J.W. (1980). Better ways to clear crude storage tanks and desalters. *Hydrocarbon Process.*, **59**(1), 82

BARNEY, G.C. (1985). Intelligent Instrumentation (Englewood Cliffs, NJ: Prentice-Hall)

BARNFIELD, J. (1986). The UK procedure for the appraisal of fire protection of structural steelwork. *CTICM, Proc. EGOLF Symp.* (Brussels)

BARNFIELD, J.R. and HAY, A. (1990). A quantitative approach to fire safety. *Safety and Loss Prevention in the Chemical and Oil Processing Industries*, p. 449

BARNHART, D.A. (1963). Sensing side effects for supplementary checkout data. *Int. Conf. on Aerospace Ground Support Equipment*

BARNWELL, J. (1979). Designing safety into an ethylene plant. *Loss Prevention*, **12**, 13. See also *Chem. Engng Prog.*, 1978, **74**(10), 66

BARNWELL, J. and DERBYSHIRE, J. (1979). Opportunities for energy saving in process plant. *Chemy Ind.*, Sep. 1, 574

BARNWELL, J. and ERTL, B. (1987). Expert systems and the chemical engineer. *Chem. Engr, Lond.*, **440**, 41

BARNEY, G.C. (1985). *Intelligent Instrumentation* (Englewood Cliffs, NJ: Prentice-Hall)

BAROCZY, C.J. (1966). A systematic correlation for two-phase pressure drop. In *Heat Transfer – Los Angeles* (New York: Am. Inst. Chem. Engrs), p. 232

BARON, S. *et al.* (1982). An approach to modeling supervisory control of a nuclear power plant. *Rep. NUREG/CR-2988.* Nucl. Regul. Comm., Washington, DC, also *Rep. ORNL/ sub/81–70523/1,* Oak Ridge Nat. Lab., Oak Ridge, TN

BARON, T. (1954). Reactions in turbulent free jets – the turbulent diffusion flame. *Chem. Engng Prog.,* **50**, 73

BARONA, N. and PRENGLE, H.W. (1973). Design reactors this way for liquid phase processes. *Hydrocarbon Process.,* **52**(3), 63; **52**(12), 73

BARR, A. and FEIGENBAUM, E.A. (1981–). *The Handbook of Artificial Intelligence,* vols 1–2 (Los Altos, CA: Kaufmann)

BARR, J. (1953). Diffusion flames. *Combustion,* 4, 765

BARR, J.T. (1986). Safety and environmental concerns in resin manufacture. In Nass, L.I. and Heiberger, C.A. (1986), *op. cit.*

BARR, P., BROWN, M.L., CARTER, P.G., HOWE, W.D., JOWETT, J., NEILSON, A.J. and YOUNG, R.L.D. (1980). Studies of missile impact with reinforced concrete structures. *Nucl. Engng,* 19, 179

BARR, S., KYLE, T.G., CLEMENTS, W.E. and SEDLACEK, W. (1983). Plume dispersion in a nocturnal drainage wind. *Atmos. Environ.,* 17, 1423

BARRATT, M. (1986). Machinery monitoring. *Plant Engr,* **30**(4), 21

BARRELL, A.C. (1971). The safe cleaning, maintenance and demolition of large tanks for storing flammable liquids. *Major Loss Prevention,* p. 184

BARRELL, A.C. (1980). Assessment of chemical plant for the proposed Hazardous Installations (Notification and Survey) Regulations. *Chemical Process Hazards VII,* p. 219

BARRELL, A.C. (1984). Face to face with the hazards chief. *Chem. Engr, Lond.,* **406**, 26

BARRELL, A.C. (1985). *Developments in the control of major hazards. The Assessment and Control of Major Hazards* (Rugby: Instn Chem. Engrs), p. 1

BARRELL, A. (1988a). Inherent safety – only by design? (editorial). *Chem. Engr, Lond.,* **451**, 3

BARRELL, A.C. (1988b). Mossmorran – planning and safety issues in the construction of a large natural gas liquids fractionation plant in Scotland. *Preventing Major Chemical and Related Process Accidents,* p. 621

BARRELL, A.C. (1990). Major hazard controls in Europe. *Safety and Loss Prevention in the Chemical and Oil Processing Industries,* p. 1

BARRELL, A.C. (1992a). Control of major hazards offshore – implementing Lord Cullen's recommendations. *Major Hazards Onshore and Offshore,* p. 1

BARRELL, (1992b). *Risk analysis and its contribution to health and safety. Health and Safety in the Offshore Oil and Gas Industries* (Rugby: Instn Chem. Engrs)

BARRELL, A.C., DANIELS, J.T. and HAGON, D.O. (1983). Quantified risk analysis in the process industries (letter). *Chem. Engr, Lond.,* **393**, 4

BARRELL, A.C. and MCQUAID, J. (1985). *The HSE programme of research and model development on heavy gas dispersion. The Assessment and Control of Major Hazards* (Rugby: Instn Chem. Engrs), p. 165

BARRELL, A.C. and SCOTT, P.J.C. (1982). The regulatory position – United Kingdom and elsewhere. In Green, A.E. (1982b), *op. cit.,* p. 15

BARRELL, A.C. and THOMAS, D.C. (1982). Assessing management in major hazard operations. In *Developments 82* (Rugby: Instn Chem. Engrs)

BARRELL, T. (1994). *The HSE position and the safety case regime. Offshore Safety Case Management* (Rugby: Instn Chem. Engrs)

BARRETO, E. (1975). The hazard of static electricity with hydrocarbons. *Transport of Hazardous Cargoes,* 4

BARRETT, B., HOWELLS, R. and HINDLEY, B. (1987). *Safety in the Offshore Petroleum Industry. The Law and Practice for Management* (London: Kogan Page)

BARRETT, O.H. (1981). Installed cost of corrosion-resistant piping. *Chem. Engng,* **88**(Nov. 2), 97

BARRINGTON, E.A. (1973). Acoustic vibrations in tubular heat exchangers. *Chem. Engng Prog.,* **69**(7), 62

BARRINGTON, E.A. (1978). Cure exchanger acoustic vibration. *Hydrocarbon Process.,* **57**(7), 193

BARRIS, J.P. and CONTE-RUSSIAN, D.L. (1980). Sprinkler trade-offs: are they justified? *Fire J.,* May, 63

BARRISH, R.A. (1993). Process safety management in Delaware. *Process Safety Prog.,* **12**, 115

BARRON, M.J. (1971). Railroads and transportation safety. *Loss Prevention,* **5**, 7

BARRON, R.F. (1970). LNG plants use these systems. *Hydrocarbon Process.,* **49**(11), 192

BARRON, R.F. (1966–). Cryogenic Systems; 1988, 2nd ed. (New York: McGraw-Hill)

BARROW, C.S., ALARIE, Y., WARRICK, J.C. and STOCK, M.F. (1977). Comparison of the sensory irritation response in mice to chlorine and hydrogen chloride. *Arch. Environ. Health,* **31**, 68

BARROW, C.S., and DODD, D.E. (1979). Ammonia production in inhalation chambers and its relevance to chlorine inhalation studies. *Toxicol. Appl. Pharmacol.,* **49**, 89

BARROW, C.S., KOCIBA, R.J., RAMPY, L.W., KEYES, D.G. and ALBEE, R.R. (1979). An inhalation toxicity study of chlorine in Fischer 344 rats following 30 days exposure. *Toxicol. Appl. Pharmacol.,* **49**, 77

BARROW, C.S., LUCIA, H. and ALARIE, Y.C. (1979). A comparison of the acute inhalation toxicity of hydrogen chloride versus the thermal decomposition products of polyvinyl chloride. *J. Combust. Toxicol.,* **6**, 3

BARROW, D.B., CEFAI, J. and TAYLOR, S. (1999). Shrinking to fit. *Chem. Ind.,* 15(Aug. 2), 591–94.

BARROW, R.E. and SMITH, R.G. (1975). Chlorine-induced pulmonary function changes in rabbits. *Am. Ind. Hyg. Ass. J.,* **36**, 398

BARRS, T.W. (1983). Monitoring refinery sewers for amines. *Chem. Engng Prog.,* **79**(3), 56

BARRY, A. (1978). *Errors in Practical Measurement in Science, Engineering and Technology* (New York; Wiley)

BARRY, D.M. and HUDSON, M.W. (1983). A probabilistic information system for plant maintenance. *Euredata* 4, paper 12.4

BARRY, I. (1991). Vermiculite cement and passive fire protection in the hydrocarbon and chemical industries. *Fire Surveyor,* **20**(5), 7

BARRY, P.J. (1964). *Estimation of downwind concentration of airborne effluents discharged in the neighbourhood of buildings. Canadian Rep. AECL-4043*

BARRY, P.J. (1971). A note on peak-to-mean concentration ratios. *Boundary Layer Met.,* **2**, 122

BARRY, P.J. (1972). Relationship between concentrations of atmospheric pollutants and averaging time. *Atmos. Environ.,* **6**, 581

BARRY, T.F. (1987). Computer room risk assessment. *Plant/ Operations Prog.,* **6**, 68

BARRY, T.I. (1980). *The Industrial Use of Thermochemical Data* (London: R. Soc. Chem.)

BARSNESS, D. (1982). Adhesives and sealants make maintenance easier. *Chem. Engng*, **89**(Jan. 25), 129

BARTELS, A.L. (1970). *The use of light metals and their alloys in hazardous areas. Rep. 70-32* Elec. Res. Ass., Leatherhead

BARTELS, A.L. (1971a). Ignition by hot surfaces in electrical equipment. *Electrical Safety*, **1**, 109

BARTELS, A.L. (1971b). Low power arcing in flameproof enclosures. *Electrical Safety* **1**, 104

BARTELS, A.L. (1975a). Ignition energies of methane/air mixtures as a function of electrical discharge-type. *Electrical Safety*, **2**, 148

BARTELS, A.L. (1975b). The relationship between stress and strain in flameproof enclosures. *Electrical Safety*, **1**, 153

BARTELS, A.L. (1975c). The revised United Kingdom Code of Practice for electrical apparatus in potentially explosive atmospheres. *Electrical Safety*, **1**, 164

BARTELS, A.L. (1977–). Electrical equipment in hazardous atmospheres. *Fire Surveyor*, 1977, **6**(6), 35; 1978, **7**(1), 28

BARTELS, A.L. (1983). Intrinsic safety. *Electron. Power*, Apr, 301

BARTELS, A.L. (1990). *Explosions in Air Compressor Electric Motor Drives* (London: HM Stationery Office)

BARTELS, A.L. and DAY, A.G. (1975). Insulation requirements for explosion-protected equipment. *Electrical Safety*, **1**, 158

BARTELS, A.L. and HOWES, J.A. (1971). The variation of minimum igniting currents with pressure, temperature and oxygen enrichment. *Electrical Safety*, **1**, 120

BARTH, E.F. (1990). An overview of the history, present status, and future direction of solidification/stabilization technologies for hazardous waste treatment. *J. Haz. Materials*, **24**(2/3), 103

BARTH, J. and MAARLEVELD, A. (1967). *Operational aspects of a DDC system. The Application of Automation in the Process Industries* (ed. by J.M. Pirie) (London: Instn *Chem. Engrs*)

BARTH, U. (1992). Steam curtains – mitigation of heavy gas clouds on industrial terrain. *Loss Prevention and Safety Promotion*, **7**, vol. 2, 96-1

BARTHELEMY, B. and LAMBOLEY, G. (1978). Some lessons from warehouse fires. *Fire Int.*, **59**, 42

BARTHOLOME, G. and VASOUKIS, G. (1975). Analysis of defects in welds. *J. Pres. Ves. Technol.*, **97**, 315

BARTHOLOMEW, C.H. (1984). Catalyst deactivation. *Chem. Engng*, **91**(Nov. 12), 96

BARTHOLOMEW, D.J. (1963). Two stage replacement strategies. *Opl Res. Q.*, **14**, 71

BARTZIS, J.G. (1983). Simple approach to deposition effects on air concentration of radioactive material. *Atmos. Environ.*, **17**, 2097

BARTZIS, J.G. (1989). Turbulent diffusion modelling for wind flow and dispersion analysis. *Atmos. Environ.*, **23**, 1963

BARTKNECHT, W. (1968). Precautions against explosions in pipelines. *Stahl u. Eisen*, **88**(16), 857

BARTKNECHT, W. (1971). Suppression of dust explosions in containers. *Staub Reinhaltung der Luft*, **31**(3), 28 (English ed.)

BARTKNECHT, W. (1972a). Characteristics of combustible types of dust in relation to explosions in pipes with narrow cross-sections. *Int. Symp. on Dust and Explosion Risks in Mines and Industry* (Heidelberg: Berufsgenossenschaft der Chem. Ind.), p. 84

BARTKNECHT, W. (1972b). Fuel gas and dust explosions in containers and pipelines – progress and safety measures. *Stahl u. Eisen*, **92**, 245

BARTKNECHT, W. (1973). Suppression of explosions of gases and dusts in containers. *Prevention of Occupational Risks*, **2**, 341

BARTKNECHT, W. (1974a). The course of gas and dust explosions and their control. *Loss Prevention and Safety Promotion*, **1**, p. 159

BARTKNECHT, (1974b). Report on investigations on the problem of pressure relief of explosions of combustible dusts in vessels. *Staub Reinhaltung der Luft*, **34**(11), 289; **34**(12), 358 (English ed.)

BARTKNECHT, W. (1975). Nature and prevention of dust and gas explosions in containers and pipes. *Chem. Ing. Tech.*, **47**(6), 236

BARTKNECHT, W. (1977a). Explosion pressure relief. *Loss Prevention*, **11**, 93. See also *Chem. Engng Prog.*, 1977, **73**(9), 45

BARTKNECHT, W. (1977b). Explosion is suppressed in milliseconds. *Chem. Ind.*, **29**(7), 393

BARTKNECHT, W. (1980). *Explosionen – Ablauf und Schutzmassnahmen*, 2nd ed. (Berlin: Springer)

BARTKNECHT, W. (1981a). *Explosions: Course, Prevention, Protection* (Berlin Springer)

BARTKNECHT, W. (1981b). *Use of protective measures against the threat of dust explosion in process equipment. The Hazards of Industrial Explosion from Dusts* (London: Oyez/IBC), paper 9

BARTKNECHT, W. (1985). Effectiveness of explosion venting as a protective measure for silos. *Plant/Operations Prog.*, **4**, 4

BARTKNECHT, W. (1986). Pressure venting of dust explosions in large vessels. *Loss Prevention and Safety Promotion 5*, p. 59-1. See also *Plant/Operations Prog.*, **5**, 196

BARTKNECHT, W. (1988). Ignition capabilities of hot surfaces and mechanically generated sparks in flammable gas and dust/air mixtures. *Plant/Operations Prog.*, **7**, 114

BARTKNECHT, W., (1989). *Dust Explosions: Course, Prevention, Protection* (Berlin: Springer)

BARTKNECHT, W., HURLIMANN, H., SIWEK, R. and PELLMONT, G. (1980). Recent results of the department 'Explosion Research' of the Central Safety Service of Ciba-Geigy Ltd. *Loss Prevention and Safety Promotion 3*, vol. 2, p. 823

BARTKOWIAK, A., LAMBIRIS, S. and ZABETAKIS, M.G. (1959). Flame propagation through kerosine foams. *Combust. Flame*, **3**, 347

BARTLETT, C.J.S., HALL, D.E. and GUDDE, N.J. (1993). The development of flammable atmospheres above residual fuels in laboratory storage. *Health, Safety and Loss Prevention*, p. 234

BARTLETT, F.C. (1943). Fatigue following highly skilled work. *Proc. R. Soc, Ser. B.*, **131**, 247

BARTOK, W. and SAROFIM, A.F. (eds) (1991). *Fossil Fuel Combustion – A Source Book* (New York: Wiley Interscience)

BARTON, A.D., JENKINS, D.G., LEES, F.P. and MURTAGH, R.W. (1970). Commissioning and operation of a direct digital control computer on a cement plant. *Measmnt Control*, **3**, T77

BARTON, J.A. and NOLAN, P.F. (1984). *Runaway reactions in batch reactors. The Protection of Exothermic Reactors and Pressurised Storage Vessels* (Rugby: Instn Chem. Engrs), p. 13

BARTON, J.C., FINLEY, C.R. and FAULKNER, J.R. (1986). CAD system used for major offshore project. *Oil Gas J.*, **84**(Mar. 3), 96

BARTON, J. and ROGERS, R. (1993). *Chemical Reaction Hazards* (Rugby: Instn Chem. Engrs)

BARTON, J.A. and ROGERS, R.L. (1993). Techniques for evaluating chemical reaction hazards. In *Chemical Reaction Hazards — A Guide* (Rugby, Warwickshire, UK: Institution of Chemical Engineers), pp. 19–26.

BARTON, R.G., CLARK, W.D. and SEEKER, W.R. (1990). Fate of metals in waste combustion systems. *Combust. Sci. Technol.*, **74**, 327

BARTON, R.R., CARVER, F.W.S. and ROBERTS, T.A. (1974). The use of reticulated metal foam as flash-back arrestor element. *Chemical Process Hazards V*, p. 223

BARTSCH, E.H., GILLEY, W.F. and STEELE, J.D. (1978). Virginia's organisation and management structure for mitigating Kepone incineration testing. *Control of Hazardous Materials Spills*, **4**, 250

BARWELL, F.T. (1979). Bearing Systems (Oxford: Oxford Univ. Press)

BASCOMBE, K.N. (1967). Calculation of ignition delays in the hydrogen-air system. *Combust. Flame*, **11**, 1

BASDEN, A. and HINES, J.G. (1986). Implications of relations between information and knowledge in use of computers to handle corrosion knowledge. *Br. Corrosion J.*, **21**(3), 157

BASHIR, S. *et al.* (1992). Thermal runaway limit of tubular reactors, defined at the inflection point of the temperature profile. *Ind. Engng Chem. Res.*, **31**, 2164

BASHIR, S., CHOVAN, T., MASRI, B.J., MUKHERJEE, A., PANT, A., SEN, S. and VIJAYARAGHAVAN, P. (1992). Thermal runaway limit of tubular reactors, defined at the inflection point of the temperature profile. *Ind. Eng. Chem. Res.* **31** (9), 2164–71

BASILA, M.R., STEFANEK, G. and CINAR, A. (1990). A model-object based supervisory expert system for fault tolerant chemical reactor control. *Comput. Chem. Engng*, **14**, 551

BASKERVILLE, C. (1919). Certain war gases and health., *Science*, **50**, 50

BASKETT, P.J.F. and ZORAB, J.S.M. (1977). Essentials of emergency management. In Easton, K.D. (1977), *op. cit.*, p. 1

BASS, A., HOFFNAGLE, G.F. and EGAN, B.A. (1977). Predictive capabilities of Gaussian air quality models as applied to refinery emission sources. *API Proc., Refining Dept, 42nd Midyear Mtg*, **56**, 172

BASS, C.D. (1982). Safe governing of turbine speed. *Plant/Operations Prog.*, **1**, 101

BASS, H.G. (1984). *Intrinsic Safety — Instrumentation for Flammable Atmospheres* (Quartermaine House Ltd)

BASSETT, J.D. (1975). Study of the charging mechanisms during the production of water aerosols. *J. Electrostatics*, 1, 47

BASSETT, L.C. and NATARAJAN, R.S. (1980). Hydrogen — buy it or make it? *Chem. Engng Prog.*, **76**(3), 93

BASSO, D. (1992). Electric motors power up the CPI. *Chem. Engng*, **99**(7), 65

BAST (1970). Explosion of a stirring vessel caused by catalytic decomposition of benzoyl peroxide. *Prevention of Occupational Risks*, 1

BASTA, N. (1980). Motor controller spell savings for the CPI. *Chem. Engng*, **87**(Dec. 29), 19

BASTA, N. (1981a). Asbestos: enter new substitutes. *Chem. Engng*, **88**(Feb. 9), 47

BASTA, N. (1981b). It's back to a rewrite for land-disposal rules. *Chem. Engng*, **88**(Oct. 5), 79

BASTA, N. (1981c). The RCRA law — when the states take over. *Chem. Engng*, **88**(June 1), 24

BASTA, N. (1981d). Standards are needed for process-control systems. *Chem. Engng*, **88**(Aug. 24), 51

BASTA, N. (1981e). Waste exchanges grow in scope and number. *Chem. Engng*, **88**(July 13), 68

BASTA, N. (1985). Union Carbide engineers speak out. *Chem. Engng*, **92**(Dec. 9/23), 21

BASTA, N. (1986a). Hazardous waste incineration — a burning CPI concern. *Chem. Engng*, **93**(Mar. 3), 21

BASTA, N. (1986b). Safety becomes the big issue at pilot plant level. *Chem. Engng*, **93**(May 12), 29

BASTA, N. (1989). PCs invade new frontiers. *Chem. Engng*, **96**(4), 43

BASTA, N. (1992). Experts ponder new ways to assess risk. *Chem. Engng*, **99**(3), 35

BASTA, N., FARRELL, P., PRICE, W. and DWYER, P. (1985). US CPI start to feel effects of Bhopal tragedy. *Chem. Engng*, **92**(Mar. 18), 27

BASTA, N., HUGHSON, R.V. and MASCONE, C.F. (1985). Chemical engineers speak out on hazardous waste management. *Chem. Engng*, **92**(Sept. 16), 58

BASTL, W. and FELKEL, L. (1981). Disturbance analysis systems. In Rasmussen, J. and Rouse, W.B. (1981). *op. cit.*, p. 451

BASU, P. and BHADURI, D. (1972). Structure of premixed turbulent flame (letter). *Combust. Flame*, **18**, 303

BATCHELOR, B. (1990). Leach models: theory and application. *J. Haz. Materials*, **24**(2/3), 255

BATCHELOR, G.K. (1952). Diffusion in a field of homogeneous turbulence. II, The relative motion of particles. *Proc. Cambridge Phil. Soc.*, **48**, 345

BATCHELOR, G.K. (1953). *The Theory of Homogeneous Turbulence* (London: Cambridge Univ. Press)

BATCHELOR, G.K. (1956). *Turbulent diffusion. Surveys in Mechanics* (London: Cambridge Univ. Press), p. 475

BATCHELOR, G.K. (1964). Diffusion from sources in a turbulent boundary layer. *Archiv. Mechaniki Stosowanej*, **3**, 661

BATCHELOR, G.K. (1967). *An Introduction to Fluid Dynamics* (London: Cambridge Univ. Press)

BATEL, W. (1976). *Dust Extraction Technology* (Stonehouse, Gloucestershire: Technicopy Ltd)

BATEMAN, J.W. (1963). Fire protection handbook. *Hydrocarbon Process.*, **42**(4), 179

BATEMAN, T.L., SMALL, F.H. and SNYDER, G.E. (1974). Dinitrotoluene pipeline explosion. *Loss Prevention*, **8**, p. 117

BATEMAN, W.J. (1964). An alkylation unit detonates. In Vervalin, C.H. (1964a), *op. cit.*, p. 352

BATES, J.C. (1992). Lift ropes. *Plant Engr*, **36**(3), 26

BATES, J.F. (1966). Effect of stress on corrosion. *Ind. Engng Chem.*, **58**(2), 18

BATES, J.M. (1963). Why so mainy stainless steel. *Chem. Engng Prog.*, **59**(6), 92

BATH, M. (1979). Introduction to Seismology (Basel: Birkhauser Verlag)

BATRA, B., BORCHERT, C.A., LEWIS, K.R. and SMITH, A.R. (1993). Design process equipment for corrosion control. *Chem. Engng Prog.*, **89**(5), 68

BATSTONE, R.J. and TOMI, D.T. (1980). Hazard analysis in planning industrial developments. *Loss Prevention*, **13**, 7

BATTARRA, V., MARIANI, O., GENTILI, M. and GIACHETTA, G. (1985). Condensate-line correlations for calculating holdup, friction compared to field data. *Oil Gas J.*, **83**(Dec. 30), 148

BATTELLE COLUMBUS LABORATORIES (1972). *Damage Tolerant Design Handbook.* (Battelle Columbus, OH: Metals and Ceramic Information Center, Battelle Columbus Laboratories)

BATTELLE COLUMBUS LABORATORIES (1974). LNG safety program. Interim report on Phase II work. *Am. Gas Ass., Proj. IS-3–1*

BATTERHAM, R.J. (1985). External corrosion – how long before pipework inspection? *Ammonia Plant Safety*, **25**, 47. See also *Plant/Operations Prog.*, **4**, 154

BATTILANA, R.E. (1971). Fretting corrosion under mechanical seals. *Chem. Engng*, **78**(Mar. 8), 130

BATTILANA, R.E. (1989). Better seals will boost pump performance. *Chem. Engng*, **96**(7), 106

BATTIN-LECLERC, F., MARQUAIRE, P.M., COME, G.M., BARONNET, F. and GUSTIN, J.L. (1992). Auto-ignitions of gas-phase mixtures of 1, 4 dioxane and chlorine. *Loss Prevention and Safety Promotion 7*, vol. 2, 82-1

BATTISON, G. (1974). Impressions of LNG-4. *Gas World*, Aug.

BATTJES, J.A. (ed.) (1985). *Behaviour of Offshore Structures* (Amsterdam: Elsevier)

BATTOCLETTI, J.H. (1976). *Electromagnetism, Man and the Environment* (London: Paul Elek)

BATTS, J.R. (1971). A computer program for approximating system reliability – Part 2. *IEEE Trans. Reliab.*, **R-20**, 88

BATURIN, V.V. (1972). *Fundamentals of Industrial Ventilation* (Oxford: Pergamon Press)

BAUER, D. (1980). Exposure to dimethylformamide, why and how to monitor. *Loss Prevention and Safety Promotion 3*, vol. 2, p. 695

BAUER, F. (1982). ISL shock wave facilities. In Nellis, W.J., Seaman, L. and Graham, A.R. (1982), *op. cit.*, p. 674

BAUER, F.L. (1975a). *Software Engineering* (Berlin: Springer)

BAUER, F.L. (1975b). Software engineering. In Bauer, F.L. (1975a), *op. cit.*, p. 522

BAUER, H., SCHULZ, K.H. and SPIEGELBERG, U. (1961). Occupational poisoning in the manufacture of chlorophenol compounds. *Arch. Gewerbepath. Gewerbehyg.*, **18**, 538

BAUER, P., BROCHET, C. and PRESLES, H.N. (1984). The influence of initial pressure on critical diameters of gaseous explosive mixtures. In Bowen, J.R. *et al.*, (1984b), *op, cit.*, p. 118

BAUER, P., PRESLES, H.N. and HEUZE, D. (1986). Influence of initial pressure and ignition energy on the deflagration behaviour of fuel/air mixtures. *Loss Prevention and Safety Promotion 5*, p. 64-1

BAUER, P., PRESLES, H.N., HEUZE', O., DUNAND, M. and DESBORDES, D. (1989). Flame propagation in hydrocarbon air mixtures confined in a tube: experimental safety assessment related to elevated initial pressures. *Loss Prevention and Safety Promotion 6*, p. 82.1

BAUERMEISTER, K.J. (1975). Turbocompressors for air and oxygen service. *Hydrocarbon Process.*, **54**(6), 63

BAULCH, D.L., DRYSDALE, D.D., HORNE, D.G. and LLOYD, A.C. (1973). *Evaluated Kinetic Data for High Temperature Reactions*, vol. 1 (London: Butterworths)

BAUM, B. and PARKER, C.H. (1973–). *Solid Waste Disposal*, vol. 1, 1973; vol. 2, 1974 (Ann Arbor, MI: Ann Arbor Sci. Publs)

BAUM, J. (1991). Noise at work. *Plant Engr*, **35**(Mar./ Apr.), 30

BAUM, J. and DIBERARDINIS, L. (1987). Designing safety into the laboratory. In Young, J.A, (1987a), *op. cit.*, p. 275

BAUM, M.M. (1981). On predicting spontaneous combustion. *Combust. Flame*, **41**, 187

BAUM, M.R. (1979). Blast waves generated by the rupture of gas pressurised ductile pipes. *Trans. Instn Chem. Engrs*, 57, 15

BAUM, M.R. (1984). The velocity of missiles generated by the disintegration of gas-pressurised vessels and pipes. *J. Pressure Ves. Technol.*, **106**, 362

BAUM, M.R. (1987). Disruptive failure of pressure vessels: preliminary design guidelines for fragment velocity and the extent of the hazard zone. *Advances in Impact, Blast Ballistics and Dynamic Analysis of Structures* (New York: Am. Soc. Mech. Engrs)

BAUM, M.R. (1988). Disruptive failure of pressure vessels: preliminary design guidelines for fragment velocity and the extent of the hazard zone. *J. Pres. Ves. Technol.*, **110**, 168

BAUM, M.R. (1989). Rupture of gas-pressurised cylindrical pressure vessel: the dynamics of end-cap missiles. *Chem. Engng Res. Des.*, **67**, 353

BAUM, M.R. (1991). Rupture of a gas-pressurised cylindrical pressure vessel. The velocity of rocketing fragments. *J.Loss Prev. Process Ind.*, **4**(2), 73

BAUM, M.R. (1993). Velocity of a single missile ejected from a vessel containing high pressure gas. *J. Loss Prev. Process Ind.*, **6**, 251

BAUMAN, D.E. (1982). Segregate shutdowns to avoid serious problems. *Hydrocarbon Process.*, **61**(1), 101

BAUMANN, H.C. (1964). *Fundamentals of Cost Engineering in the Chemical Industry* (New York: Reinhold)

BAUMANN, H.D. (1971). Control-valve noise: cause and cure. *Chem. Engng*, **78**(May 17), 120

BAUMANN, H.D. (1981). Control valve vs. variable-speed pump. *Chem. Engng*, **88**(Jun. 29), 81

BAUMANN, H.D. (1992). Reduce pressures quietly and safely. *Chem. Engng*, **99**(9), 138

BAUMANN, J. and ORUM, P. (1999). *Accidents Waiting to Happen: Hazardous Chemicals in the U.S. Fifteen Years After Bhopal* (Washington, D.C.: U.S. PRIG Education Fund)

BAUMANN, H.D. and VILLIER, R. (1974). Avoiding control valve failure. *Chem. Process.*, **20**(9), 83

BAUMEISTER, T. (ed.) (1978). *Mark's Standard Handbook for Mechanical Engineers*, 8th ed. (New York: McGraw-Hill)

BAUMER, A.R. (1982). Making environmental audits. *Chem. Engng*, **89**(Nov. 1), 101

BAUSBACHER, E. and HUNT, R. (1993). *Process Plant Layout and Piping Design* (Englewood Cliffs, NJ: PTR Prentice Hall)

BAWN, C.E.H. and SKIRROW, G. (1955). The oxidation of olefins. *Combustion*, **5**, 521

BAXENDINE, M.R. (1989). *Piper Alpha Public Inquiry. Transcript, Day 147*

BAXTER, P.J. (1985). Medical implications of the release of toxic substances into the community. *Chemy Ind.*, Jul. 15, 465

BAXTER, P.J., ANTHONY, P.P., MACSWEEN, R.N. and SCHEUER, P.J. (1977). Angiosarcoma of the liver in Great Britain, 1963–73. *Br. Med. J.*, **2**(6092), Oct. 8, 919

BAYBUTT, P. (1983). Prcedures for the use of risk analysis in decision-making. In Hartwig, S. (1983), *op. cit.* p. 261

BAYBUTT, P. (1986). Decision support systems for risk and safety analysis. In *Hazards IX*, p. 227

BAYBUTT, P. (1989). Uncertainty in risk analysis. In Cox, R.A. (1989a), *op. cit.*, p. 247

BAYBUTT, P. (2003). Inherent security: protecting process plants against threats. *Chemical Eng. Prog.* **99**(Dec. 12), 35–38

BAYER, R. and KHANDROS, I. (1985). Model aids alloy life prediction. *Hydrocarbon Process.*, **64**(1), 72

BAYLISS, M., SHORT, D. and BAX, M. (1988). Underwater Inspection (London: Spon)

BAYLISS, R.S. (1987). Dynamic modeling of gas pipeline and processing plant network under relief valve discharge conditions. *Vapor Cloud Modeling*, p. 274

BAYLISS, R.S. (1991). Overpressure protection on an oil-producing platform: an evolving situation. *Safety*

Developments in the Offshore Oil and Gas Industry (London: Instn Mech. Engrs), p. 27

BAZOVSKY, I. (1960). Electron. Equipment Engr, **8**(Mar.), 113

BAZOVSKY, I. (1961). *Reliability Theory and Practice* (Englewood Cliffs, NJ: Prentice-Hall)

BDM CORP. (1975). *Oil Spill Risk Assessment for OCS Oil and Gas Processing Facilities and Deepwater Ports*, vol. II

BEACH, F.X.M., JONES, E.S. and SCARROW, G.D. (1969). Respiratory effects of chlorine gas. *Br. J. Ind. Med.*, **26**, 231

BEACH, R. (1964–). Preventing static electricity fires. *Chem. Engng*, 1964, **71**(Dec. 21), 73; 1965, **72**(Jan. 4), 63; **72**(Feb. 1), 85

BEADLE, R.W. and SEMONIN, R.G. (eds) (1974). Precipitation scavenging. *ERDA Symp. Ser. CONF 741003*

BEAK, M., COLWELL, S.A., CROWHURST, D. and ELLIS, B.R. (1994). The behaviour of masonry and concrete panels under explosion and static loading. *Hazards XII*, p. 227

BEALE, R. and JACKSON, T. (1990). Neural Computing (Bristol: Adam Hilger)

BEALS, G.A. (1971). *Guide to local diffusion of air pollutants. Tech. Rep. 214*. Air Weather Service, Scott Air Force Base, IL, Tech. Rep. 214.

BEAM, B.H. and GIOVANNETTI, A. (1975). Underground storage systems for high-pressure air and gases. *J. Pres. Ves. Technol.*, **97**, 101

BEAN, D.W. (1976a). Health and safety trends in the UK chemical industry. *Prevention of Occupational Risks*, **3**, 411 Dept of Chem. Engng, Loughborough Univ. of Technol.

BEAN, N.W. (1976b). Sources of information and advice on loss prevention. *Course on Loss Prevention in the Process Industries*. Dept. of Chem. Engng, Loughborough Univ. of Technol

BEARD, A.N. (1979). Towards a rational approach to fire safety. *Fire Prevention Sci. Technol.*, **22**, 16

BEARD, A.N. (1981/82). A stochastic model for the number of deaths in a fire. *Fire Safety J.*, **4**(3), 169

BEARD, A. (1992a). Evaluation of deterministic fire models. *Fire Safety J.*, **19**, 295

BEARD, A. (1992b). Limitations of computer models. *Fire Safety J.*, **18**, 375

BEARD, A., DRYSDALE, D., HOLBORN, P. and BISHOP, S. (1993). Configuration factor for radiation in a tunnel or partial cylinder. *Fire Technol.*, **29**, 281

BEARD, M.W. (1973). High speed gear units. Shipping and installation specs. *Hydrocarbon Process.*, **52**(12), 66

BEARDALL, L.R. (1983). Developments in offshore process plant. *Chem. Engr, Lond.*, **396**, 28

BEARE, A.N., DORRIS, R.E., BOVELL, C.R., CROWE, D.S. and KOZINSKY, E.J. (1983). *A simulator-based study of human errors in nuclear power plant control room tasks. Rep. NUREG/CR-3309*. Nucl. Regul. Comm., Washington, DC

BEARMAN, D. (1983). Intelligent fire alarm systems outdate BS code of practice. Fire Surveyor, **12**(2), 5

BEASELY, M.E. (1992). Look hard at ball valve materials for abrasive chemical processes. *Chem. Engng Prog.*, **88**(12), 73

BEATSON, C. (1978). Turning a computer into a maintenance man's friend. *The Engineer*, **246**(Apr. 6), 42

BEATTIE, D. (1989). Should dangerous goods be moved by rail rather than road? In Guy, G.B. (1989), *op. cit.*, p. 149

BEATTY, R.E. and KRUEGER, R.G. (1978). How the steam trap saves plant energy. *Chem. Engng Prog.*, **74**(9), 94

BEAUBOIS, H. and DEMAND, C. (1974). *Airships* (London: McDonald & Janes)

BEAUMONT, S. (1981). Fire precautions in chemical plant. *Chemy Ind.*, July 18, 481

BEAUSOLEIL, R.W., PHILLIPS, C.W. and SNELL, J.E. (1974). Field investigation of natural gas pipeline accident. *Fire J.*, **68**(May), 77

BEAZLEY, W.G. (1986). Prevention of chemical leaks using expert systems. *API Proc. Refining Dept, 51st Midyear Mtg*, **65**, 515

BECK, E. (1986). Fire at warehouse 956. *Chemy Ind.*, Dec. 1, 801

BECK, H. (1959). Experimental determination of the olfactory thresholds of some important gases (chlorine, sulphur dioxide, ozone, nitrous gases) and symptoms induced in humans by low concentrations. Dissertation, Univ. of Wurzburg

BECK, J.D. and RAIDL, J.H. (1973). Relief valve reliability is upgraded. *Oil Gas J.*, Nov. 5, 74

BECK, J.F., CORMIER, F. and DONINI, J.C. (1979). The combined toxicity of ethanol and hydrogen sulphide. *Tox. Letts*, **3**, 311

BECK, J.V. and ARNOLD, K.J. (1977). *Parameter Estimation in Engineering and Science* (New York: Wiley)

BECK, M.S., HEWITT, P.J., HOBSON, J.H. and PYE, D.N. (1971). Electrostatic charge measurement in liquid conveyors with special reference to explosion risk. *Electrical Safety*, **1**, 72

BECKER, H. (1983). *Information Integrity: A Structure for its Definition and Management* (Maidenhead: McGraw-Hill)

BECKER, H.A., HOTTEL, H.C. and WILLIAMS, G.C. (1965). Concentration intermittency in jets. *Combustion*, **10**, 1253

BECKER, H.A., HOTTEL, H.C. and WILLIAMS, G.C. (1967). The nozzle-fluid concentration field of the round, turbulent, free jet. *J. Fluid Mech.*, **30**, 285

BECKER, H.A. and LIANG, D. (1978). Visible length of vertical free turbulent diffusion flames. *Combust. Flame*, **32**, 115

BECKER, H.A. and LIANG, D. (1982). Total emission of soot and thermal radiation by free turbulent diffusion flames. *Combust. Flame*, **44**, 305

BECKER, H.A. and YAMAZAKI, S. (1977). Soot concentration field of turbulent propane/air diffusion flames. *Combustion*, **16**, 681

BECKER, H.A. and YAMAZAKI, S. (1978). Entrainment, momentum flux and temperature in vertical free turbulent diffusion flames. *Combust. Flame*, **33**, 123

BECKER, J.V. (1979). Designing safe interlock systems. *Chem. Engng*, **86**(Oct. 15), 103

BECKER, J.V. and HILL, R. (1979). Fundamentals of interlock systems. *Chem. Engng*, **86**(Oct. 8), 101

BECKER, R. (1922). *Z. Physik*, **8**, 321

BECKER, R. (1980a). Mathematical model of luminous flame radiation for the determination of safety zones. *Ger. Chem. Engng*, **3**, 229

BECKER, R. (1980b). Mathematical model of luminous flame radiation to determine safety zones and protective devices. *Loss Prevention and Safety Promotion 3*, p. 229

BECKER, W. and HERTEL, H. (1991). Fire safety concept. *Fire Safety J.*, **17**, 103

BECKERDITE, J.M., POWELL, D.R. and ADAMS, E.J. (1983). Self-association of gases: 2, The assocation of hydrogen fluoride. *J. Chem. Engng Data*, **28**, 287

BECKERS, H.L. (1993). Industrial R&D and competition. *Chem. Engng Res. Des.*, **71A**, 523

BECKETT, H.E. (1937). *The nature and properties of glass. Glass in Architecture and Decoration* (ed. by R. McGrath and A.C. Frost) (London: Architect. Press), p. 413

BECKETT, H.E. (1958). Determination of glass thickness for windows. *Glass Age*, p. 40

BECKMAN, J.R. (1984). Breathing losses from fixed roof tanks by heat and mass transfer diffusion. *Ind. Engng Chem. Process Des. Dev.*, **23**, 472

BECKMAN, J.R. and GILLMER, J.R. (1981). Model for prediction of emission from fixed roof storage tanks. *Ind. Engng Chem. Process Des. Dev.*, **20**, 646

BECKMAN, J.R. and HOLCOMB, J.A. (1986). Experimental and theoretical investigation of working emissions from fixed tank roofs. *Ind. Engng Chem. Process Des. Dev.*, **25**, 293

BECKMAN, L.V. (1992a). Optimum proof testing of programmable safety systems. *Hydrocarbon Process.*, **71**(11), 87

BECKMAN, L.V. (1992b). How reliable is your safety system? *Chem. Engng*, **99**(1), 108

BECKMAN, L.V. (1993). Selecting redundant microprocessor based systems. *Hydrocarbon Process.*, **72**(2), 71

BECKWITH, B.M. (1987). Evaluating cladding processes. *Chem. Engng*, **94**(2), 191

BEDDOE, A.A. (1969). The cost of industrial injuries. In Handley, W. (1969), *op. cit.*, p. 449

BEDDOW, K.J. (ed.) (1984). *Particle Characterization in Technology* (Boca Raton, FL: CRC Press)

BEDDOWS, N.A. (1976). Safety and health criteria in plant layout. *Chem. Engng*, **83**(Oct. 18), 133

BEDER, S. (1991). The fallible engineer. *New Scient.*, Nov. 2, 38

BEDFORD, T. (1974). Basic Principles of Heating and Ventilation (London: H.K.Lewis)

BEDNAR, H.H. (1986). *Pressure Vessel Design Handbook*, 2nd ed. (New York: Van Nostrand Reinhold)

BEDSON, J.H. (1980). Elastomeric expansion joints. *Chem. Engng*, **87**(Dec. 29), 34

BEEBEE, D.W. (1976). New techniques in overcoming electrical runout. *Hydrocarbon Process.*, **55**(8), 131

BEEBY, G.H. (1974a). British legislation concerned with health and safety at work. *Loss Prevention and Safety Promotion 1*, p. 49

BEEBY, G.H. (1974b). The Robens Report – Health and Safety at Work. *Course on Process Safety – Theory and Practice.* Dept. of Chem. Engng, Teesside Polytechnic, Middlesbrough

BEECH, J.W. (1985). Planning for emergencies. *Chemy Ind.*, July 1, 436

VAN BEEK, A.W. (1989). *Piper Alpha Public Inquiry. Transcript, Day 163*

BEELEY, P., GRIFFITHS, J.F. and GRAY, P. (1980). Rapid compression studies on spontaneous ignition of isopropyl nitrate. *Combust. Flame*, **39**, 255 and 269

BEER, F.P. and JOHNSTONE, E.R. (1981). *Mechanics of Materials*; 2nd ed., 1992 (New York: McGraw-Hill)

BEER, J.M. and HOWARTH, C.R. (1969). Radiation from flames in furnaces. *Combustion*, **12**, 1205

BEERBOUWER, A. (1974). *ASTM Comm D-2, Tech. Div. J.*

BEEREL, A.C. (1987). *Expert Systems: Strategic Implications and Applications* (Chichester: Ellis Horwood)

BEERS, Y. (1957). *Introduction to the Theory of Error*, 2nd ed. (Reading, MA: Addison-Wesley)

BEESE, J.G., ORGAN, T.M. and WADE, M.V. (1980). The monitoring of safety-rupture diaphragm. *Loss Prevention and Safety Promotion 3*, p. 1394

BEESE, J., TROLLUX, J. and JEAN, P. (1989). A new LNG alternative: the 'floating LNG receiving terminal' concept. *Gastech 88*, paper 5.4

BEEVER, P.F. (1981). Isothermic methods for assessing combustible powders – practical case histories. In *Runaway Reactions, Unstable Products and Combustible Powders* (London: Instn Chem. Engrs), paper 4/X

BEEVER, P.F. and GRIFFITHS, J.F. (1989). Scaling rules for prediction of thermal runaway. *Runaway Reactions*, p. 1

BEEVER, P.F. and THORNE, P.F. (1981). Isothermal methods for assessing combustible powders – theoretical and experimental approach. In *Runaway Reactions, Unstable Products and Combustible Powders* (London: Instn Chem. Engrs), paper 1/B

BEEVERS, A. (1985a). Computerised risk analysis. *Process Engng*, Mar., 29

BEEVERS, A. (1985b). Efficiency drive. *Process Engng*, Feb., 55

BEEVERS, A. (1986). Low power routes to intrinsic safety. *Process Engng*, Sep., 61

BEEVERS, C.L. (1982). The shipping and terminalling of gas liquids in Europe. *Gastech 81*

BEGG, G.A.J. (1963–64). Mechanical engineering design in the chemical industry. *Proc. Instn Mech. Engrs*, **178**, pt 1, 685

BEGG, H.D. (1984). *Gas Production Operations* (Tulsa, OK: Oil and Gas Consultants Inc.)

BEGG, V. (1984). *Developing Expert CAD Systems* (London: Kogan Page)

BEGGS, H.D. and BRILL, J.P. (1973). A study of two-phase flow in inclined pipes. *J. Petrol. Technol.*, May, 607

BEGHIN, P., HOPFINGER, E.J. and BRITTER, R.E. (1981). Gravitational convection from instantaneous sources on inclined boundaries. *J. Fluid Mech.*, **107**, 407

BEGLEY, J.A. and LANDES, J.D. (1972). The J integral as a fracture criterion. *ASTM STP 514*. Am. Soc. Testing and Mats, Philadelphia, PA, p. 24

BEIGLER, S.E. (1983). The siting and construction of buildings. *Loss Prevention and Safety Promotion 4*, vol. 1, p. H10

BEISHON, R.J. (1967). Problems of task description process control. *Ergonomics*, **10**, 177

BEISHON, R.J. (1969). An analysis and simulation of an operator's behaviour in controlling continuous baking ovens. *The Simulation of Human Behaviour* (ed. by F. Bresson and M. de Montmollin)(Paris: Dunod). See also Edwards, E and Lees, F.P. (1974), *op. cit.*, p. 79

BELAND, B. (1981). Arcing phenomenon as related to fire investigation. *Fire Technol*, **17**, 189

BELAND, B. (1982). Considerations of arcing as a fire cause. *Fire Technol*, **18**, 188

BELANGER, J.M. (1990). Responsible care. *Plant/Operations Prog.*, **9**, 150

BELARDO, S., PIPKIN, J. and SEAGLE, J.P. (1985). Information support for control of hazardous materials movement. *J. Haz. Materials*, **10**(1), 13

BELCHER, D.W., SMITH, D.A. and COOK, E.M. (1963). Design and use of spray drier. *Chem. Engng*, **70**(Sep. 30), 83; Oct. 14, 201

BELL, A. (1966). Noise. *An Occupational Hazard and Public Nuisance* (Geneva: World Health Org.)

BELL, A.T., MANZER, L.E., CHEN, N.Y., WEEKMAN, V.W., HEGEDUS, L.L. and PEREIRA, C.J. (1995). Protecting the environment through catalysis. *Chemical Eng. Prog.*, (Feb.), 26–34

BELL, B.A., WHITMORE, F.C. and CARDENAS, R.R. (1978). Anaerobic biodegradation of Kepone in sewage sludge. *Control of Hazardous Material Spills*, **4**, p. 240

BELL, B.J. (1987). Evaluation of the contribution of human errors to accidents. *Preventing Major Chemical Accidents*, p. 6.49

BELL, B.J. and SWAIN, A.D. (1981). *A procedure for conducting human reliability analysis for nuclear power plants* (draft). *Rep. SAND 81–1655*. Sandia Labs, Albuquerque, NM, Also Rep. NUREG/CR-2254. Nucl. Regul. Comm., Washington DC

BELL, B.J. and SWAIN, A.D. (1983). *A Procedure for Conducting a Human Reliability Analysis for Nuclear Power Plants. Final Report.* Rep. NUREG/CR-2254 Nucl. Regul. Comm., Washington DC

BELL, C.L. (1977). *Flixborough and the Health and Safety at Work Act. Design Developments Related to Safety, Costs and Solid Systems* (London: Instn Chem. Engrs), paper A-1–1

BELL, C.L. (1979). Planning at Flixborough. Design 79 (London: Instn Chem. Engrs)

BELL, D.I., SKIRROW, G. and TIPPER, C.F.H. (1968). The low temperature gas phase oxidation of acetaldehyde and propionaldehyde. *Combust. Flame*, **12**, 557

BELL, D.P. and ELMES, P.C. (1965). The effects of chlorine gas on the lungs of rats without spontaneous pulmonary disease. *J. Pathol. Bacteriol*, **89**, 307

BELL, F.G. (1987). *Blasting and vibrations. Ground Engineers Reference Book*, (London: Butterworths), sec. 39.2.5.

BELL, G.D. (n.d.). Unpublished data. UK Atom. Energy Authority, Health and Safety Branch

BELL, G.D. (1970). Siting criteria. *Nucl. Engng Des.*, **13**(2), 187

BELL, G.D. and BEATTIE, J.R. (1974). Assessing risks in the nuclear and non-nuclear industries. *Loss Prevention and Safety Promotion 1*, p. 61

BELL, H.V. (1982). Shutdown and start-up of a cryogenic tank. *Chem. Engng Prog.*, **78**(2), 74

BELL, J. (1984). Indecision is the biggest risk. *New Scientist*, Sept. 13, 17

BELL, K. and MORRIS, S.D. (1992). Flow visualizations during top venting from a small vessel. *J. Loss Prev. Process Ind.*, **5**, 160

BELL, K., MORRIS, S.D. and OSTER, R. (1993). Vent line void fractions and mass flow rates during top venting of high viscosity fluids. *J. Loss Prev. Process Ind.*, **6**, 31

BELL, K.J. (1983). Coping with an improperly vented condenser. *Chem. Engng Prog.*, **79**(7), 54

BELL, N.A.R. (1971). Loss prevention in the manufacture of nitroglycerine. *Major Loss Prevention*, p. 50

BELL, P. (ed.) (1971a). *Industrial Fuels* (London: Macmillan)

BELL, P. (1971b). *Mechanical Prime Movers* (London: Macmillan)

BELL, R. (1986). The assessment, architecture and performance of industrial programmable electronic systems (PES) with particular reference to robotics safety. In Daniels, B.K. (1986), *op. cit.*

BELL, R.C.(1944). An analysis of 259 of the recent flying bomb casualties. *Br. Med. J.*, Nov 25, 689; and related letter Dec.16 (1944) 799

BELL, R.P. (1978). Isopleth calculations for ruptures of sour gas pipelines. *Energy Processing/Canada*, July-Aug., **36**

BELL, Z.G., LAFLEUR, J.C., LYNCH, R.P. and WORK, G.A. (1975). Control methods for vinyl chloride. *Chem. Engng Prog.*, **71**(9), 45

BELLAMY, L.J. (1983). Neglected individual, social and organisational factors, *Reliability 83*, p. 2B/5/1

BELLAMY, L.J. (1984). Not waving but drowning: problems of human communication in the design of safe systems.

Ergonomic Problems in Process Operations (Rugby: Instn Chem. Engrs), p. 167

BELLAMY, L. (1985). How people's behaviour shapes your Operations Prog. operation. *Process Engng*, July, 27

BELLAMY, L.J. (1986). The safety management factor: an analysis of the human error aspects of the Bhopal disaster. *Safety and Reliab. Soc., Symp. on Statutory Safety – Meeting the Requirments.*

BELLAMY, L.J. (1989). Informative fire warning systems: a study of their effectiveness. *Fire Surveyor*, **18**(4), 17

BELLAMY, L.J. (1991). The quantification of human fallibility. *J. Health Safety*, **6**, 13

BELLAMY, L.J. (1994a). The human element. *Offshore Safety Case Management* (Rugby: Instn Chem. Engrs), p. lb1

BELLAMY, L.J. (1994b). The influence of human factors science on safety in the offshore industry. *J. Loss Prev. Process Ind.*, **7**, 370

BELLAMY, L.J. and GEYER, T.A.W. (1988). Addressing human factors issues in the safe design and operation of computer controlled process systems. In Sayers, B.A. (1988), *op. cit.*, p. 189

BELLAMY, L.J. and GEYER, T.A.W. (1991). Incorporating human factors into formal safety assessments: the offshore safety case. *Management and Engineering of Fire Safety and Loss Prevention* (ed. by BHR Group Ltd) (London: Elsevier), p. 55

BELLAMY, L.J. and HARRISON, P.I. (1988). *An evacuation model for major accidents. Disasters and Emergencies. The Need for Planning* (London: IBC Tech. Services Ltd)

BELLAMY, L.J., KIRWAN, B. and COX, R.A. (1986). Incorporating human reliability into probabilistic risk assessment. *Loss Prevention and Safety Promotion 5*, 6.1

BELLAMY, L.J., GEYER, T.A.W. and ASTLEY, J.A. (1989). *Evaluation of the Human Contribution to Pipework and In-line Equipment Failures Frequencies. HSE Contract Res. Rep. 15/1989.* Technica Ltd, London,

BELLAMY, L.J., GEYER, T.A.W. and ASTLEY, J.A. (1991). A review of human factors guidance applicable to the design of offshore platforms. *Safety Developments in the Offshore Oil and Gas Industry* (London: Instn Mech. Engrs), p. 79

BELLAMY, L.J., WRIGHT, M.S. and HURST, N.W. (1993). History and development of a safety management system audit for incorporation into quantitative risk assessment (poster item). *Process Safety Management*, p. 523

BELLANI, A., CANNALIRE, C. and BELTRAME, D. (1992). Safety provisions and optimized design for underground and mounded LPG storages. *Loss Prevention and Safety Promotion 7*, vol. 3, 158–1

BELLES, F.E. (1959). Detonability and chemical kinetics: prediction of limits of detonability of hydrogen. *Combustion*, **7**, p. 745

BELLES, F.E. and O'NEAL, C. (1957). Effects of halogenated extinguishing agents on flame quenching and a chemical interpretation of their action. *Combustion*, **6**, p. 806

BELLEW, E.F. (1978). Comparing chemical precipitation methods for water treatment. *Chem. Engng, Albany*, **85**(Mar. 13), 85

BELLINGHAM, B. (1976). The detection of malfunction using a process computer. PhD Thesis, Loughborough Univ. of Technol.

BELLINGHAM, B. and LEES, F.P. (1976a). The effect of malfunction monitoring of reliability estimation. Diagnostic Techniques (London: Instn Chem Engrs)

BELLINGHAM, B. and LEES, F.P. (1976b). The effect of malfunction monitoring on reliability, maintainability and availability. *Advances in Reliability Technology* (Culcheth, Lancashire: UK Atom. Energy Auth.), ART17

BELLINGHAM, B. and LEES, F.P. (1977a). The detection of malfunction using a process control computer: a Kalman filtering technique for general control loops. *Trans Instn Chem. Engrs*, **55**, 253

BELLINGHAM, B. and LEES, F.P. (1977b). The detection of malfunction using a process control computer: a simple filtering technique for flow control loops. *Trans Instn Chem. Engrs*, **55**, 1

BELLO, G.C. and AVOGADRI, S. (1982). Offshore pipeline system – how to decide the optimal redundancies using reliability engineering. *Reliab. Engng*, **3**(2), 89

BELLO, G.C. and COLOMBARI, V. (1980). The human factors in risk analyses of process plants: the control room operator model 'TESEO'. *Reliab. Engng*, **1**(1), 3

BELLO, G.C. and ROMANO, A. (1983a). Abnormal release of hazardous material and energy. A survey of theory and experiments. *Eurodata 4*, paper 1.4

BELLO, G.C. and ROMANO, A. (1983b). Safety analysis of an LPG coastal storage facility. *Euredata 4*, paper 4.3

BELLO, G.C., ROMANO, A. and DOSI, W. (1983). Probabilistic risk assessment of a large marine terminal of liquefied petroleum gases. *Loss Prevention and Safety Promotion* 4, vol. 2, p. B21

BELLOMO, P.J. (1993). High performance risk management. *Process Safety Management*, p. 323

BELLUS, F., REVEILLARD, Y., BONNAURE, C. and CHEVALIER, L. (1977). Tests at the Fos terminal on LNG behaviour in big tanks. *LNG Conf.*, 5

BELLUS, F., VINCENT, R., COCHARD, H. and MAUGER, J. (1977). Controlling the hazards from LNG spills on ground: LNG fire fighting methods. *LNG Conf.*, 5

BELORE, R. and BUIST, I. (1986). *A computer model for computing leak rates of chemicals from damaged storage and transportation tanks*. Rep. to Environment Canada, Env. Prot. Directorate, Ont.

BELTRAMINI, L. and MOTARD, R.L. (1988). KNOD – a knowledge based approach for process design. *Comput. Chem. Engng*, **12**, 939

BENARIE, M.M. (1977). Test fires for the evaluation of fire detectors. *Fire Int.*, **58**, 72

BENARIE, M.M. (1987). The limits of air pollution modelling. *Atmos. Environ.*, **21**, 1

BENATT, F.G.S. and EISENKLAM, P. (1969). Gaseous entrainment into axisymmetric liquid sprays. *J. Inst. Fuel*, **42**, 309

BENAYOUNE, M. and PREECE, P.E. (1987). Review of information management in computer-aided engineering. *Comput. Chem. Engng*, **11**, 1

BENAZZOUZ, A. and ABBOU, H. (1987). More than 20 years of LNG operations at the GLAZ (ex-CAMEL) plant, Arzew, Algeria. *Gastech 86*, p. 377

BENAZZOUZ, A. and LASNAMI, A. (1989). Changes in LNG cargo composition due to long storage and prevention of roll-over during the unloading operation. *Gastech 88*, paper 3.4

BENDANI, A. (1989). Safety improvements in LNG and LPG plants in Algeria from 1962 until now. *Gastech 88*, paper 4.2

BENDELL, A. (1985). Proportional hazards modelling in reliability assessment. *Reliab. Engng*, **11**, 175

BENDELL, A. (1987). How to collect and use process plant reliability data. *Process Engng*, June, 39

BENDELL, A. (1991a). Analysis methodologies. In Cannon, A.G. and Bendell, A. (1991), *op. cit.*, p. 19

BENDELL, A. (1991b). Developments (data banks). In Cannon, A.G. and Bendell, A. (1991), *op. cit.*, p. 283

BENDELL, A. (1991c). Overview: into the future (data banks). In Cannon, A.G. and Bendell, A. (1991), *op. cit.*, p. 287

BENDELL, A. (1991d). Principles of reliability data bases. In Cannon, A.G. and Bendell, A. (1991), *op. cit.*, p. 11

BENDELL, A. (1991e). Reliability. In Cannon, A.G. and Bendell, A. (1991), *op. cit.*, p. 1

BENDELL, A. and ANSELL, J. (1984). The incoherency process of multistate coherentsystems. *Reliab. Engng*, **8**(3), 165

BENDELL, A. and CANNON, A.G. (1985). The SRS data bank concept. *Reliab. Engng*, 11.131

BENDELL, A. and WALLS, L.A. (1985). Exploring reliability data. *Quality and Reliab. Engng Int.*, **1**, 37

BENDER, M. (1979). Fume hoods, open canopy type – their ability to capture pollutants in various environments. *Am. Ind. Hyg. Ass. J.*, **40**, 118

BENDIXEN, L.M., DALE, S.E. and O'NEILL, J.K. (1985). Improved safety and reliability of chemical and petroleum plants. Am. Inst. *Chem. Engrs, Summer Nat. Mtg*, paper 5d

BENDIXEN, L.M. and O'NEILL, J.K. (1984). Chemical plant risk assessment using HAZOP and fault tree methods. *Plant/Operations Prog.*, **3**, 179

BENDIXEN, L.M. (2002). Integrate EHS for better process design. *Chem. Eng. Prog.* **98**(Feb. 2), 26–32

BENEDETTI, R.P. (1982). NFPA's impact on the chemical industry. *Plant/Operations Prog.*, **1**, 45

BENEDICK, W.B. (1979). High explosive initiation of methane-air detonation. *Combust. Flame*, **35**, 89

BENEDICK, W.B. (1982). Review of large scale fuel-air explosion tests and techniques. In Lee, J.H.S. and Guirao, C.M. (1982), *op. cit.*, p. 507

BENEDICK, W.B., KENNEDY, J.D. and MOROSIN, B. (1970). Detonation limits of unconfined hydrocarbon-air mixtures (letter). *Combust. Flame*, **15**, 83

BENEDICK, W.B., KNYSTAUTAS, R. and LEE, J.H. (1984). Large-scale experiments on the transmission of fuel-air detonations from two-dimensional channels. In Bowen, J.R. et al., (1984b), *op, cit.*, p. 546

BENEDICK, W.B., MOROSIN, B. and KENNEDY, J.D. (1971). Initiation and propagation of detonation in several fuel air explosives. *First Fuel Air Explosives Conf.*, Eglin AFB, FL

BENEDICT, B.A. (1978). Analytical models for toxic spills. Control of Hazardous Material Spills 4, p. 439

BENEDICT, H.M. and BREEN, W.H. (1955). The use of weeds as a means of evaluating vegetation damage caused by air pollution. *Proc. Third Nat. Air Pollution Symp.*, p. 177

BENEDICT, R.P. (1977). *Fundamental of Temperature, Pressure and Flow Measurements* (New York: Wiley)

BENEDICTO, L., MAHER, S.T., MORANDINI, S.J., SLOANE, B.D. and PHILLIPPI, H.I. (1991). Offshore platform safety shutdown system effectiveness. *Safety Developments in the Offshore Oil and Gas Industry* (London: Instn Mech. Engrs), p. 45

BENEL, D.C.R., MCCAFFERTY, D.B., NEAL, V. and MALLORY, K.M. (1981). Issues in the design of annunciator systems. *Proc. Human Factors Soc. Twenty-Fifth Ann. Mtg*, p. 122

BENHAM, P.P. and WARNOCK, F.V. (1973). *Mechanics of Solids and Structures* (London: Pitman)

BENIN, A.I., KOSSOI, A.A. and SHARIKOV, Y.V. (1990). Automated system for study of thermal explosion of chemical technology processes. *Zh. Vses. Khim. O-va im D.I. Mendeleeva*, **35**(4), 428

BENITEZ, J. (1993). *Process Engineering and Design for Air Pollution Control* (Englewood Cliffs, NJ: Prentice-Hall)

BENJAMIN, A.S. *et al.* (1986). *Evaluations of severe accident risks and the potential for risk reduction: Surry Power Station, Unit 1. Rep. NUREG/CR-4551.* Nucl. Regul. Comm., Washington, DC

BENJAMIN, F.B. (1953). Relationships between injury and pain in the human skin. *J. Appl. Physiol,* **5,** 740

BENJAMIN, J.R. (1968). Probabilistic models for seismic force design. *J. Struct. Div. ASCE,* **94**(ST5), paper 5950

BENJAMIN, J.R. and CORNELL, C.A. (1970). *Probability, Statistics and Decision for Civil Engineers* (New York: McGraw-Hill)

BENJAMIN, T.B. (1968). Gravity currents and related phenomena. *J.Fluid Mech.,* **31,** 209

BENJAMINSEN, J.M. and VAN WIECHEN, P.H. (1968). You can calculate mean time to explosions. *Hydrocarbon Process.,* **47**(8), 121

BENJAMINSEN, M.W. and MILLER, J.G. (1941). The flow of saturated water through throttling orifices. *Trans ASME,* **63**(5), 419

BENN PUBLICATIONS (1980). *Chemical Industry Directory* (Tunbridge Wells, Kent)

BENNELLICK, E.J. (1969). Ionizing radiation. In Handley, W., (1969), *op. cit.,* p. 138

BENNELLICK, E.J. (1977). Ionizing radiation. In Handley, W., (1977), *op. cit.,* p. 53

BENNER, L. (1975a). Accident investigation: multilinear events sequencing methods. *J. Safety Res.,* **7**(2), 67

BENNER, L. (1975b). How the NTSB functions. *Loss Prevention,* **9,** p. 73

BENNER, L. (1984). Accident models: how underlying differences affect workplace safety (abstract). *J. Occup. Accidents,* **6**(1/3), 127

BENNER, L. (1985). Rating accident models and investigation methodologies. *J. Safety Res.,* **16**(3), 105

BENNER, L. *Accident Investigation — Multilinear Events Sequence Method, J. Saf. Res.,* Volume 7, #2, Washington D.C.

BENNER, L. and ROTE, R.A. (1981). NTSB Hazardous Materials Spills Maps: a new safety information resource. Loss Prevention **14,** p. 54

BENNET, D.J. (1981). *The Elements of Nuclear Power* (London: Longmans)

BENNET, T.C., COWLEY, L., DAVENPOR, J.N. and ROWSON, J.J. (1991). Large-scale natural gas and LPT jet fires. Thornton Res. Centre, Shell Res. Ltd., Rep. TNER.91.022

BENNET, T.C. (1972). Automate for plant emergencies. *Hydrocarbon Process.* **51**(12), 57

BENNET, B. (1985). Thermal oxide reprocessing at Sellafield. *Chem. Engr, Lond.,* **415,** 35

BENNETT, C.O. and MYERS, J.E. (1962). *Momentum, Heat and Mass Transfer* (New York: McGraw-Hill)

BENNETT, D.E. (1983). *Vulnerability of nuclear power plant structures to large external fires. Rep. SAND83-1178.* Sandia Nat. Labs, Albuquerque, NM. Also *Rep. NUREG/CR-3330.* Nucl. Regul. Comm., Washington, DC

BENNETT, D.J. (1977). The contribution of reliability engineering. *Plant Engr,* **21**(11), 20

BENNETT, D.J. and THOMSON, J.R. (1989). *The Elements of Nuclear Power,* 3rd ed. (Longman)

BENNETT, E. (1988). Policy and control within the European chemical industry. *Preventing Major Chemical and Related Process Accidents,* p. 601

BENNETT, E., DEGAN, J. and SPIEGEL, J. (1963). *Human Factors in Technology* (New York: McGraw-Hill)

BENNETT, G.F. (1973). Information sources. *Chem. Engng,* **80**(Jun. 18), 39

BENNETT, G.F. and WILDER, I. (1981). Evolution of hazardous material spills regulations in the United States. *J. Haz. Materials,* **4**(3), 257

BENNETT, G.F., FEATES, F.S. and WILDER, I. (1982). *Hazardous Materials Spills Handbook* (New York: McGraw-Hill)

BENNETT, M. (1988). A simple physical model of dry deposition to a rough surface. *Atmos. Environ.,* **22,** 2701

BENNETT, N., CAIRNS, F. and COOPER, S.O. (1986). Fabric dust collector explosion venting. *Am. Inst. Chem. Engrs, Symp on Loss Prevention*

BENNETT, P.A. (1984). The safety of industrially based controllers incorporating software. PhD. Thesis, The Open Univ.

BENNETT, P.A. (1991a). Safety. In McDermid, J. (1991), *op. cit.,* p. 60/1

BENNETT, P.A. (1991b). Specification, design and testing of safety critical software. The Fellowship of Engineering (1991b), *op. cit.,* p. 21

BENNETT, P. (ed.) (1993). *Safety Aspects of Computer Control* (Oxford: Butterworth Heinemann)

BENNETTS, R.G. (1972). A realistic approach to detection of test set generation for combinatorial logic circuits. *Comput. J.,* **15,** 238

BENNETTS, R.G. (1975). On the analysis of fault trees. *IEEE Trans. Reliab.,* **R-24,** 175

BENOIT, J. (1982). Classification of LPG carriers: aspects of new and foreseeable IMCO rules. *Gastech 81,* 427

BENSEL, C.K. (1993). The effects of various thicknesses of chemical protective gloves on manual dexterity. *Ergonomics,* **36,** 687

BENSON, P.E. (1979). CALINE 3 — a versatile dispersion model for predicting air pollutant levels near highways and arterial streets. *Rep FHWA/CA/TL-79/23.* Fed. Highway Admin., Washington, DC

BENSON, R.E. (1973). Analyzing piping flexibility. *Chem. Engng,* **80**(Oct. 29), 102

BENSON, R.S. (1987). The state of chemical process control: an industrialist's view. *Chem. Engng Res. Des.,* **65,** 451

BENSON, R.S. (1992). Computer aided process engineering — an industrial perspective. *Comput. Chem. Engng,* **16,** S1

BENSON, R.S. and BURGOYNE, J.H. (1951). *Br. Shipbuilding Res. Ass., Rep. 76*

BENSON, R.S. and PONTON, J.W. (1993). Process miniaturization — a route to total environmental acceptability? *Chem. Engng Res. Des.,* **71A,** 160

BENSON, R.S. and PONTON, J.W. (1993). Process miniaturisation — A route to total environmental acceptability? *Trans. IChemE* **71,** Part A, (Mar), 160–168.

BENSON, T.H. and SINCLAIR, J.R. (1982). US Coast Guard research and development program for hazardous chemical discharge response. *J. Haz. Materials,* **6**(3), 267

BENTLEY, J. (1982). Controlled disposal of special wastes. A view from the centre. Chemy Ind., April 17, 251

BENTLY, D.E. (1970). Need advance warning of thrust bearing failure. *Hydrocarbon Process.,* **49**(1), 111

BENTLY, D.E. (1974). Monitor machinery condition for safe operation. *Hydrocarbon Process.,* **53**(11), 205

BENTLY, D.E. and REID, C.P. (1970). Unreliability factors in chemical plants. *Chem. Engng Prog.,* **66**(12), 50

BENUZZI, A., KOTTOWSKI, H., VERSINO, B. and ZALDIVAR, J.M. (1989). Runaway reactions: Joint Research Centre of the Commission of the European Communities project. *J. Loss Prev. Process Ind.,* **2**(3), 128

BENZER, W.L. (1970). Trends in materials applications. Steels. *Chem. Engng*, **77**, Oct.12, 101

BERANEK, L.L. (1960). *Noise Reduction* (New York: McGraw-Hill)

BERANEK, L. (ed.) (1971). *Noise and Vibration Control* (New York: McGraw-Hill)

BERANEK, L.L. and VER, I.L. (1992). *Noise and Vibration Control Engineering* (New York: Wiley)

BERCANI, U., DEVOOGHT, J. and SMIDTS, C. (1988). Recent progress in human reliability models for nuclear power safety. *Nucl. Europe*, **8**(10), 13

BERECZ, E. and BALLA-ACHS, M. (1983). *Gas Hydrates* (Amsterdam: Elsevier)

BERENBLUT, B.J. and MENASHE, J. (1982). Ranking of risks in commerce and industry. *Developments 82* (Rugby: Instn Chem. Engrs)

BERENBLUT, B.J. and WHITEHOUSE, H.B. (1976). *A Method for Monitoring Process Plant Based on a Decision Table Analysis* (London: Insurance Tech. Bur.)

BERENBLUT, B.J. and WHITEHOUSE, H.B. (1977). A method for monitoring process plant based on a decision table analysis. *Chem. Engr, Lond.*, **318**, 175

BERENBLUT, B.J. and WHITEHOUSE, S.M. (1988). Optimisation of loss prevention measures for chemical process plant. *Preventing Major Chemical and Related Process Accidents*, p. 51

BERENBLUT, B.J., WHITEHOUSE, S.M., PHILLIPS, H.C.D. and SALTER, R.E. (1985). Pemex in retrospect: lessons in risk assessment and loss prevention. *Chem. Engr, Lond.*, **418**, 17

BERENBLUT, I.I. and DOWNES, A.B. (1960). *Tables for Petroleum Gas/Oxygen Flames* (London: Oxford Univ. Press)

BERENDT, R.D. and CORLISS, E.L.R. (1976). *Quieting: A Practical Guide to Noise Control* (Washington, DC: Nat. Bur. Stand.)

BERESFORD, T.C. (ca. 1977). Private communication to J. McQuaid

BERESFORD, T.C. (1981). The use of water spray monitors and fan sprays for dispersing gas leakages. In Greenwood, D.V. (1981), *op. cit.*, paper 6

BERET, S.E., MUHLE, M.E. and VILLAMIL, I.A. (1978). Purging criteria for LPDE make bins. *Safety in Polyethylene Plants*, 3, p. 70. See also *Chem. Engng Prog.*, 1977, **73**(12), 44

BERETS, D.G., GREENE, E.F. and KISTIAKOWSKY, G.B. (1950). Gaseous detonations. I, Stationary waves in hydrogen−oxygen mixtures. *J. Am. Chem. Soc.*, **72**, 1080

BEREZIN, I.V., DENISOV, E.T. and EMMANUEL, N.M. (1966). *The Oxidation of Cyclohexane* (Oxford: Pergamon Press)

VAN DEN BERG, A.C. (1978). *Calculation on the Explosive Contents of Gas Clouds with a Density Greater than Air. Rep. PML 1978−10.* Prinz Mauritz Lab., TNO, The Netherlands

VAN DEN BERG, A.C. (1980). *"BLAST" − A 1−D Variable Flame Speed Blast simulation Code using a Flux-corrected-transport Algorithm. Rep. PML 1980−162.* Prins Maurits Lab., TNO, Rijswijk, The Netherlands

VAN DEN BERG, A.C. (1982). *Computational Investigation of Blast Due to Non-steady Deflagrative Flames using a Low-flame-speed Explosion Simulation Model. Rep. PML 1982−107.* Prins Maurits Lab. TNO, Rijswijk, The Netherlands

VAN DEN BERG, A.C. (1984). Blast effects from vapour cloud explosions. *Ninth Int. Symp. on The Prevention of Occupational Accidents and Diseases in the Chemical Industry, Lucerne*

VAN DEN BERG, A.C. (1985). The multienergy method. A framework for vapour cloud explosion blast prediction. *J. Haz. Materials*, **12**(1), 1

VAN DEN BERG, A.C. (1987). *On the Possibility of Vapour Cloud Detonation. Rep. PML 1987−IN50.* TNO Prins Maurits Lab., Rijswijk,

VAN DEN BERG, A.C. (1989). *REAGAS − A Code for Numerical Simulation of 2−D Reactive Gas Dynamics. Rep. PML-1989−IN-48.* TNO Prins Maurits Lab, Rijswijk, The Netherlands,

VAN DEN BERG, A.C. and LANNOY, A. (1993). Methods for vapour cloud explosion blast modelling. *J. Haz. Materials*, **34**, 151

VAN DEN BERG, A.C., VAN WINGERDEN, C.J.M. and THE, H.G. (1991a). Vapor cloud explosion blast modeling. *Modeling and Mitigating the Consequences of Accidental Releases*, p. 543

VAN DEN BERG, A.C., VAN WINGERDEN, C.M.J. and THE, H.G. (1991b). Vapour cloud explosions − experimental investigations of the key parameters and blast modelling. *Hazards XI*, unbound paper. See also *Process Safety Environ.*, **69B**, 139

VAN DEN BERG, A.C., VAN WINGERDEN, C.J.M. and VERHAGEN, T.L.A. (1989). Numerical simulation of gas explosions. *Europex Newsletter*, **11**, 9

VAN DEN BERG, A.C., VAN WINGERDEN, C.J.M., ZEEUWEN, J.P. and PASMAN, H.J. (1987). Current research at TNO on vapor cloud explosion modeling. *Vapor Cloud Modeling*, p. 687

VAN DEN BERG, C. *et al.* (1980). The effects of modifications on safety in the process industry. *Loss Prevention and Safety Promotion* 3, vol. 2, p. 741

BERG, G.G. and MAILLIE, H.D. (1981). *Measurement of Risks* (New York: Plenum Press)

VAN DEN BERG, P.J. and DEJONG, W.A. (1980). *Introduction to Chemical Process Technology* (Dordrecht: Reidel)

BERG, Ø. (1977). The terotechnological implications of capital breakdown. *Loss Prevention and Safety Promotion* 2, p. 447

BERGE, H. (1970). Liquefied natural gas storage units in a densely populated area. *LNG Conf.*, **2**

BERGE, H. and POLL, J. (1972). *Experience Gained in the Commissioning of the LNG Storage Installation at Stuttgart. Comm. 87* (London: Instn Gas Engrs)

TEN BERGE, W.F. (1985). The toxicity of methylisocyanate for rats. *J. Haz. Materials*, **12**(3), 309

TEN BERGE, W.F. and ZWART, A. (1989). More efficient use of animals in acute inhalation toxicity testing. *J. Haz. Materials*, **21**, 65

TEN BERGE, W.F. and VAN HEEMST, M.V. (1983). Validity and accuracy of a commonly used toxicity assessment model in risk analysis. *Loss Prevention and Safety Promotion* 4, vol. 1, p. 11

TEN BERGE, W.F., ZWART, A. and APPELMAN, L.M. (1986). Concentration−time mortality response relationship of irritant and systemically acting vapours and gases. *J. Haz. Materials*, **13**(3), 301

BERGENTHAL, M., FROMM, D. and LIEBE, W. (1990). Severe damage to the rotor of a syngas compressor. *Ammonia Plant Safety*, **30**, 248

BERGER, D.L. (1979). *Industrial Security* (London: Butterworths)

BERGER, D.M. (1977). How corrosion theory relates to protective coatings. *Chem. Engng*, **84**(Aug. 1), 77; **84** (Aug. 29), 89

BERGER, D.M. (1982). Electrochemical and galvanic corrosion of coated steel surfaces. *Chem. Engng*, **89**(Jun. 28), 109

BERGER, D.M. (1983). How to inspect paints and coatings. *Chem. Engng*, **90**(Mar. 7), 177; **90**(Apr. 4), 105

BERGER, E., (1983). Boil-off reliquefaction facilities for diesel-driven LNG carriers. *Gastech 82*, p. 225

BERGER, K. (1977). Protection of underground blasting operations. In Golde, R.H. (1977), *op. cit.* p. 633

BERGER, S.A. (1994). The pollution prevention hierarchy as an R&D management tool. In *Pollution Prevention Via Process and Product Modifications*, (ed. M. El-Halwagi, and D.P. Petrides), pp. 23–28. AIChE Symposium Series, 303. (New York: American Institute of Chemical Engineers)

BERGER, J.S. and PERLMUTTER, D.D. (1964). The effect of feedback control on chemical reactor stability. *AIChE J.*, **10**, 238

BERGERON, C.A. and HALLETT, W.L.H. (1989a). Autoignition of single droplets of two-component liquid fuels. *Combust. Sci. Technol.*, **65**, 181

BERGERON, C.A. and HALLETT, W.L.H. (1989b). Ignition characteristics of liquid hydrocarbon fuels as single droplets. *Can. J. Chem. Engng*, **67**, 142

BERGLES, A.E. and ISHIGAI, S. (eds) (1981). *Two-Phase Flow Dynamics* (Washington, DC: Hemisphere)

BERGER, S.A. and LANTZY, R.J. (1996). Reducing inherent risk through consequence modeling. *1996 Process Plant Saf. Symp.*, Volume 1, April 1–2, 1996, Houston, TX, (ed. H. Cullingford), 15–23. (Houston, TX: South Texas Section of the American Institute of Chemical Engineers.)

BERGLES, A.E., COLLIER, J.G., DELHAYE, J.M., HEWITT, G.F. and MAYINGER, F. (1981). *Two-Phase Flow and Heat Transfer in the Power and Process Industries* (Washington, DC: Hemisphere)

BERGLUND, I. (1975). Guide to Swedish Pressure Vessel Code. *Hydrocarbon Process.*, **54**(12), 64

BERGLUND, I. (1978). Guide to Swedish Pressure Vessel Code. *Hydrocarbon Process.*, **57**(12), 94

BERGLUND, R.L. and LAWSON, C.T. (1991). Preventing pollution in the CPI. *Chem. Engng*, **98**(9), 120

BERGLUND, R.L. and WHIPPLE, G.M. (1987). Predictive modeling of organic emissions. *Chem. Engng. Prog.*, **83**(11), 46

BERGMAN, A. (1982). Radio homes in on sparks. *Chem. Engng*, **89**(May 31), 149

BERGMANN, D. and MAFI, S. (1979). Selection guide for expansion turbines. *Hydrocarbon Process.*, **58**(8), 83

BERGMANN, E.P. (1993). Approximate risk assessment prioritises remedial decisions. *Hydrocarbon Process.*, **72**(8), 111

BERGMANN, E.P. and RIEGEL, J.P. (1983). Deep water port – fire and explosion hazards assessment. *Plant/ Operations Prog.*, **2**, 89

BERGMAN, E.P., BLAYLOCK, N.W. and KETCHAM, D.E. (1992). Risk management success calls for hard decisions. *Hydrocarbon Process.*, **71**(8), 89

BERGROTH, K. (1992). Explosion in a sulphonation process. *Loss Prevention and Safety Promotion 7*, vol. 1, p. 23–1

BERGSTROM, D.R. and LADD, D.J. (1963). Effects of wall temperatures. *Chem. Engng*, **70**(Jul. 8), 176

BERGTRAUN, E.M. (1978). Safety and planning for construction and alteration projects. *Chem. Engng*, **85**(Feb. 27), 133

BERHINIG, R.M. (1985). Fire resistance test for petrochemical facility structural elements. *Plant/Operations Prog.*, **4**, 230

BERKEL, T.G. (1982). Experience is the most responsible teacher. *Chem. Engr, Lond.*, **382**, 281

BERKEY, B. and WORKMAN, W. (1987). Relating chemical reactivity to process hazards. *Preventing Major Chemical Accidents*, p. 4.17

BERKEY, B.D., PRATT, T.H. and WILLIAMS, G.M. (1988). Review of literature related to human spark scenarios. *Plant/ Operations Prog.*, **7**, 32

BERKEY, B.D., DOWD, W.M. and JONES, P.I. (1993). Integration of quality, process safety, and environmental management systems. *Process Safety Management*, p. 91

BERKOVITCH, I. (1985). How to get the best from your technical manuals. *Process Engng*, Aug. 29

BERKOVITCH, I. and BAKER-COUNSELL, J. (1985). Proving the use of expert systems. *Process Engng*, Dec. 25

BERKOWICZ, R., OLESEN, H.R. and TORP, U. (1986). The Danish Gaussian air pollution model (OML): description, test and sensitivity analysis in view of regulatory applications. In de Wispelaere, C., Schiermeier, F.A. and Gillani, N.V. (1986), *op. cit.*, p. 453

BERKOWITZ, N. (1979). *An Introduction to Coal Technology* (London: Academic Press)

BERKUN, M.M. (1964). Performance decrement under psychological stress. *Hum. Factors*, **6**, 21

BERKUN, M.M., BIALEK, H.M., KERN, R.P. and YAGI, K. (1962). Experimental studies of psychological stress in man. *Psychol. Monographs: General and Applied*, vol. 76

BERLAD, A.L. (1961). Inflammability limits and flame radiation loss rates (letter). *Combust. Flame*, **5**, 389

BERLAD, A.L. (1973). Thermokinetics and combustion phenomena in nonflowing gaseous systems: an invited review. *Combust. Flame*, **21**, 275

BERLAD, A.L. and POTTER, A.E. (1955). Prediction of the quenching effect of various surface geometries. *Combustion*, **5**, p. 728

BERLAD, A.L. and YANG, C.H. (1959). On the existence of steady state flames. *Combust. Flame*, **3**, 447

BERLAD, A.L. and YANG, C.H. (1960). A theory of flame extinction limits. *Combust. Flame*, **4**, 325

BERLIN, G.L. (1980). *Earthquakes and the Urban Environment*, vols 1–3 (Boca Raton, FL: CRC Press)

BERMAN, H.L. (1978). Fired heaters. *Chem. Engng*, **85**(Jun. 19), 99; **85**(Jul. 31), 87; **85**(Aug. 14), 129; **85**(Sep. 11), 165

BERMAN, W. (1982). Hydrogen combustion research at Sandia. In Lee, J.H.S. and Guirao, C.M. (1982), *op. cit.*, p. 861

BERNARD, J.W. and WUJKOWSKI, J.W. (1965). Direct digital control experiences in a chemical process. *ISA J.*, **12**(12), 43

BERNIE, J. (1983). Corrosion – should we care? *Chem. Engr, Lond.*, **397**, 38

BERNOTAT, R.K. and GARTNER, K.P.L. (eds) (1972). *Displays and Controls* (Amsterdam: Swets & Zeitlinger)

BERNREUTER, D.L., SAVY, J.B. and MENSING, R.W. (1987). Comparison of seismic hazard estimates obtained by using alternative seismic hazard methodologies. In Cermak, A.S. (1987), *op. cit.*, p. 219

BERNSTEIN, G., CARLSON, E.C., FELDER, R.M. and BOKENY, R.E. (1993). A simulation-based decision support system for a pharmaceuticals manufacturing plant. *Process Safety Prog.*, **12**, 71

BERNSTEIN, H. and YOUNG, G.C. (1960). Sparking characteristics of metals used in tools. *Material Des. Engng*, **52**(7), 104

BEROGGI, G.E.G. and WALLACE, W.A. (1991). Closing the gap: transit control for hazardous material flow. *J. Haz. Materials*, **27**, 61

BERRIAUD, C. *et al.* (1978). Comportement local des enceintes en beton sous l'impact d'un projectile rigide. *Nucl. Engng Des.*, **45**, 457

BERRY, R. (1969). Setting standards for safe equipment. In Handley, W. (1969), *op. cit.*, p. 272

BERRY, R. (1977). Setting standards for safe equipment. In Handley, W. (1977), *op. cit.*, p. 409

BERRY, R.I. (1978). Fiber optics: guiding light for instruments. *Chem. Engng*, **85**(Nov. 20), 89

BERRY, R.I. (1981). New ways to destroy PCBs. *Chem. Engng*, **88**(Aug. 10), 37

BERRY, W.E. (1971). *Corrosion in Nuclear Applications* (New York: Wiley)

BERRYMAN, R. and DANIELS, L. (1985). Pipeline technology — the challenge facing industry. *Chem. Engr, Lond.*, **419**, 17

BERSIER, P., VALPIANA, L. and ZUBLER, H. (1971). Thermal stability of aromatic amines and their diazonium salts. *Chem. Ing. Tech.*, **43**(24), 1311

BERSINI, V., DEVOOGHT, J. and SMIDTS, C. (1988). Recent progress in human reliability models for nuclear power safety. *Nucl. Eur.*, **8**(1), 13

BERTA, I. and GASTANEK, N. (1979). The energy of electro-static discharges. *Electrostatics*, **5**, p. 67

BERTEIN, H. (1973). Charges on insulators generated by breakdown of gas. *J. Phys. D: Appl. Phys.*, **6**, 1910

BERTEIN, H. (1975). Explosion of hydrocarbons by partial discharges. *Static Electrification*, **4**, p. 290

BERTEIN, H. (1977). The importance of electrostatic laws for understanding partial discharges. *Proc. Third Int. Conf. on Static Electricity* (Paris: Soc. de Chimie Industrielle)

BERTHELOT, M. (1892). *Explosives and Their Power* (London: Murray)

BERTHOLD, W., HECKLE, M., LUDECKE, H.J. and ZIEGER, A. (1975). Simple investigations of the avoidance of thermal explosions in bulk containers. *Chem. Ing. Tech.*, **47**, 368

BERTHOLD, W. and LOFFLER, U. (1976). Testing of thermally unstable products and reaction product mixtures by the pressure heat accumulation method. *Prevention of Occupational Risks*, **3**, p. 211

BERTHOLD, W. and LOFFLER, U. (1980). Thermal stability of hydrogen peroxide. *Loss Prevention and Safety Promotion 3*, vol. 3, p. 1431

BERTHOLD, W. and LOFFLER, U. (1981). *Lexikon sicherheitstechnischer Begriffe in der Chemie* (Weinheim: Verlag Chemie)

BERTHOLD, W. and LOFFLER, U. (1983). Self-accelerating decomposition of ethylene cyanide. *Loss Prevention and Safety Promotion 4*, vol. 3, p. C1

BERTINI, H.W. *et al.* (1980). Nucl. Safety Information Center, Oak Ridge Nat. Lab., Oak Ridge, TN, Rep.

BERTRAM, C.G., DESAI, V.J. and INTERESS, E. (1972). Designing steam tracing. *Chem. Engng*, **79**(Apr. 3), 74

BERTREM, B.E. (1976), Butterfly valves for flow of process fluids. *Chem. Engng*, **83**(Dec. 20), 63

BERTY, J.M. (1979). The changing role of the pilot plant. *Chem. Engng Prog.*, **75**(9), 48

BERTY, J.M. (1984). 20 years of recycle reactors in reaction engineering. *Plant/Operations Prog.*, **3**, 163

BERYLAND, M.E. (1972). Investigations of atmospheric diffusion providing meteorological basis for air pollution control. *Atmos Environ.*, **6**(6), 379

BERYLAND, M.Y. (1975). *Contemporary Problems of Atmospheric Diffusion and Pollution in the Atmosphere* (Leningrad: Gidrometizdat)

BERZ, I. (1961). On inductive sparks of high incendive power. *Combust. Flame*, **3**, 131

BERZINS, V. and JONES, K. (1989). Maximising the potential of process engineering databases: a review. *Computer Integrated Process Engineering* (Rugby: Instn Chem. Engrs), p. 225

BESNARD, J.-C. (1983). The safety engineering approach to industrial hazards. *Fire Int.*, **83**, 53

BESS, F.D. (1972). Operational management considerations for the prevention of hazardous material spills in process plants. In *Control of Hazardous Material Spills*, vol. 1, p. 25

BESSANT, J.R. and DICKSON, K. (1982). Determinism or design: some notes on job content and microelectronics. *Design 82* (Rugby: Instn Chem. Engrs)

BESSELIEVRE, E. and SCHWARZ, M. (1976). *The Treatment of Industrial Wastes*, 2nd ed. (New York: McGraw-Hill)

BESSEY, R.L. (1974). Fragment velocities from exploding liquid propellant tanks. *Shock and Vibration Bull.*, Aug., Bull. 44

BESSEY, R.L. and KULESZ, J.J. (1976). Fragment velocities from bursting cylindrical and spherical pressure vessels. *Shock and Vibration Bull.*, Aug, Bull. 46

BEST, A.C. (1957). Maximum gas concentration at ground level from industrial chimneys. *J. Inst. Fuel*, **30**, 329

BETILLE, M. and LEBRETON, J.M. (1986). The low boil-off Technigaz membrane system: an assessment of safety, reliability and operational economy. *Gastech 85*, p. 237

BETT, K.E. (1989). *Piper Alpha Public Inquiry. Transcript, Day 92*

BETT, K.E. and BURNS, D.J. (1967). Safety and design of high pressure plant. *Chemy Ind.*, May 27, 859

BETTIS, R.G. and JAGGER, S.F. (1992). Some experimental aspects of transient releases of pressurised liquefied gases. *Major Hazards Onshore and Offshore*, p. 211

BETTIS, R.J., MAKHVILADZE, G.M. and NOLAN, P.F. (1987). Expansion and evaluation of heavy gas and particulate clouds. *J. Haz. Materials*, **14**(2), 213

BETTIS, R.J., NOLAN, P.F. and MOODIE, K. (1987). Two-phase flashing releases following rapid depressurisation due to vessel failure. *Hazards from Pressure* (Rugby: Instn Chem. Engrs), p. 247

BETTMAN, O.L. (1974). *The Good Old Days – They Were Terrible* (New York: Random House)

BETTS, P.L. and HAROUTUNIAN, V. (1988). Finite element calculations of transient dense gas dispersion. In Puttock, J.S. (1988b), *op. cit.*, p. 349

BETZ, P.A., CHIFFELLE, T.L., DAMON, E.G. and RICHMOND, D.R. (1965). *The Effects of a 500-ton Explosion on Goats Exposed in the Higher Pressure Regions. Special Rep. DASA-1642–1.* (Santa Barbara, CA: DASA Data Center, Defense Atom. Support Agency, Dept of Defense) p. 400

BETZ LABORATORIES INC. (1976). *Betz Handbook of Industrial Water Conditioning*, 7th ed. (Trevose, PA)

BEVERIDGE, G.S.G. (1979). HSC draft Regulations for the Notification and Survey of Hazardous Installations. *Chem. Engr, Lond.*, **344**, 340

BEVERIDGE, G.S.G. and WAITE, P.J. (1985). Safety and loss prevention – international comparisons. *Multistream 85* (Rugby: Instn Chem. Engrs), p. 247

BEVERIDGE, H. (1985). Improving the efficiency of rupture-disk systems. *Process Engng*, **66**(1), 35

BEVERIDGE, H.J.R. and JONES, C.G. (1984). Shock effects on a bursting disc in a relief manifold. *The Protection of Exothermic Reactors and Pressurised Storage Vessels* (Rugby: Instn Chem. Engrs), p. 207

BEVIRT, J.L. (1979). How much Federal regulations cost Dow. *Chem. Engng Prog.*, **75**(5), 30

BEYCHOK, M.R. (1967). *Aqueous Wastes from Petroleum and Petrochemical Operations Prog.s* (New York: Wiley)

BEYCHOK, M.R. (1971). Wastewater treatment. *Hydrocarbon. Process.*, **50**(12), 109

BEYCHOK, M.R. (1979). How accurate are dispersion predictions? *Hydrocarbon Process.*, **58**(10), 113

BEYCHOK, M.R. (1983). Calculate tank losses easier. *Hydrocarbon Process.*, **62**(3), 71

BEYCHOK, M.R. (1987a). A data base of dioxin and furan emissions from municipal refuse incinerators. *Atmos. Environ.*, **21**, 29

BEYCHOK, M.R. (1987b). Observations and predictions of jet diffusion flame behaviour (letter). *Atmos. Environ.*, **21**, 1868

BEYER, J.C. (1962). *Wound Ballistics.* (Washington, DC: Office of the Surgeon General, Dept of the Army)

BEYER, W.H. (ed.) (1968). *CRC Handbook Tables for Probability and Statistics*, 2nd ed. (Boca Raton, FL.: CRC Press)

BEYER, W.H. (ed.) (1978). *Handbook of Mathematical Sciences*, 5th ed. (Boca Raton, FL: CRC Press)

BEYER, W.H. (ed.) (1984). *CRC Standard Mathematical Tables*, 27th ed. (Boca Raton, FL: CRC Press)

BEYERS, C.K. (1980). Failure analysis: tool for HPI loss prevention. *Hydrocarbon Process.*, **59**(8), 155

BEYLER, C.L. (1977). An evaluation of sprinkler discharge calculation methods. *Fire Technol.*, **13**, 185

BEYLER, C.L. (1984). A design method for flaming fire detection. *Fire Technol.*, **20**(4), 5

BEYLER, C.L. (1986). Fire plumes and ceiling jets. *Fire Safety J.*, **11**(1), 53

BEYLING (1906). Gluckauf, **42, 97, 129**

BEYNON, D. (1982). Rationalisation: an ICI view. *Chemy Ind.*, Jan. 16, 46

BEYNON, G.V., COWLEY, L.T., SMALL, L.M. and WILLIAMS, I. (1988). Fire engulfment of LPG tanks: HEATUP, a predictive model. *J. Haz. Materials*, **20**, 227

BEYNON-DAVIES, P. (1987). Software engineering . . . and knowledge engineering: unhappy bedfellows? *Comput. Bull.*, Dec., 9

BHADURI, D. (1970). Turbulent flame propagation: comments on Sanematsu's correlation (letter). *Combust. Flame*, **15**, 79

BHAGAT, P. (1990). An introduction to neural networks. *Chem. Engng Prog.*, **86**(8), 55

BHASIN, V.C. (1990). Actuator selection. *Chem. Engng*, **97**(11), 140

BHAT, N. and MCAVOY, T.J. (1990). Use of neural nets for dynamic modelling and control of chemical process systems. *Comput. Chem. Engng*, **14**, 573

BHATTACHARYYA, B.C. (1976). *Introduction to Chemical Equipment Design. Mechanical Aspects* (Kharagpur, Madras: Ind. Inst. Technol.)

BHOONCHAISRI (1990). Occupational health and safety curriculum through 'The Distance Teaching System'. *Safety and Loss Prevention in the Chemical and Oil Processing Industries*, p. 311

BHUSHAN, B. and SUBRAMANIAN, A. (1985). Bhopal: what really happened? *Business India*, Feb. 25 – Mar. 10, 102

BIACH, W.L. and WATT, T. (1993). Improve understanding of equipment via visualization. *Chem. Engng Prog.*, **89**(10), 37

BIANCONI, R. and TAMPONI, M. (1993). A mathematical model of diffusion from a steady source of short duration in a finite mixing layer. *Atmos. Environ.*, **27A**, 781

BIARNES, R. (1982). Measurement of acid-dewpoint temperature – a useful tool. *Plant/Operations Prog.*, **1**, 230

BIASUTTI, G.S. (1976). Safety during continuous nitration. *Prevention of Occupational Risks* **3**, p. 227

BIASUTTI, G.S. (1979). *Histoire des Accidents dans l'Industrie des Explosifs* (Montreux: Corbas (printers))

BIASUTTI, G.S. and CAMERA, E. (1974). Safety aspects of continuous nitration. *Loss Prevention* **8**, p. 123

BIBBINGS, R. (1992). Towards a stronger safety culture: a union view. *Health and Safety in the Offshore Oil and Gas Industries* (Rugby: Instn Chem. Engrs)

BICHARD, S.H. and ROGERS, L.M. (1976a). Industrial thermography: how representative of plant condition is the thermal image produced by an infrared camera? *Br. J. NDT*, **18**, 111

BICHARD, S.H. and ROGERS, L.M. (1976b). A review of industrial applications of thermography. *Br. J. NDT*, **18**, 2

BICKELL, M.B. and RUIZ, C. (1967). *Pressure Vessel Design and Analysis* (London: Macmillan)

BICKENBACH, O. (1947). Deposition before French military tribunal, May 6, 1947. *Tribunal Militaire de la Gieme Region, Doc. 1852* (translation by the Office of the Chief of Council for War Crimes, US Army)

BICUM, A.W. and BARNETTE, A.E. (1993). Common documentation systems for ISO 9000 and process safety management. *Process Safety Management*, p. 269

BIEFEL, R. and POLECK, T. (1980). On poisoning from coal volatile matter and town gas. *Z. Biologie*, **16**, 279

BIEGLER, L.T. (1989). Chemical process simulation. *Chem. Engng Prog.*, **85**(10), 50

BIERL, A. (1978). Investigation of leak rates of rubber/asbestos seals in flange connections. Doctoral Thesis, Ruhr Univ. of Bochum

BIERL, A. (1980). Emissions from seals – their detection, causes and reduction by intensified maintenance. *CONCAWE Seminar on Hydrocarbon Emissions from Refineries*, The Hague

BIERL, A. and KREMER, H. (1978). Calculation of gaseous leakages from flange connections with IT seals. *Chem. Ing. Tech.*, **50**(7), 542

BIERL, A., STOECKEL, A., KREMER, H. and SINN, R. (1977). Leak rates of seal elements. *Chem. Ing. Tech.*, **49**(2), 89

BIERLEIN, L.W. (1984). Regulatory considerations of hazardous materials transportation. *Plant/Operations Prog.*, **3**, 112

BIERWERT, D.V. and KRONE, F.A. (1955). How to find the best site for new plant. *Chem. Engng*, **62**(Dec.), 191

BIES, D.A. and HANSEN, C.H. (1988). *Engineering Noise Control* (London: Unwin Hyman)

BIGELOW, C.R. (1970). A system for reporting transportation accidents. *Chem. Engng Prog.*, **66**(4), 71

BIGELOW, C.R. (1971). Chemical industry transportation emergency information systems. *Loss Prevention*, **5**, 21

BIGELOW, C.R. (1972). Prevention of hazardous material spills in the transportation industry – introductory remarks. *Control of Hazardous Material Spills*, **1**, p. 35

BIGELOW, N., HARRISON, I., GOODELL, H. and WOLF, H.G. (1945). Studies on pain: quantitative measurements of two pain sensations of the skin, with reference to the nature of hyperalgesia of peripheral neuritis. *J. Clin. Invest.*, **24**, 503

BIGGER, C.J. and COUPLAND, J.W. (eds) (1983). *Expert Systems: A Bibliography* (London: Instn Elec. Engrs)

BIGGERS, E.W. and SMITH, T.C. (1972). Train operators for safety. *Hydrocarbon Process.*, **51**(12), 54

BIGGS, D.H. (1976). Electrical runout in rotating machinery. *Hydrocarbon Process.*, **55**(8), 133

BIGGS, J.M. (1964). *Introduction to Structural Dynamics* (New York: McGraw-Hill)

BIGGS, N., LLOYD, E.K. and WILSON, R.J. (1976). *Graph Theory: 1736–1936* (Oxford: Oxford Univ. Press)

BIGGS, W.D. (1960). *The Brittle Fracture of Steel* (London: McDonald and Evans)

BIGHAM, J.E. (1958). Check your procedures on rupture disc installation. *Chem. Engng*, **65**(3), 133

BIGNELL, V., PETERS, G. and PYM, C. (1977). *Catastrophic Failures* (Milton Keynes: Open Univ. Press)

BIKERMAN, J.J. (1947). *Surface Chemistry for Industrial Research* (New York: Academic Press)

BIKERMAN, J.J. *et al.* (1953). *Foams* (New York: Reinhold)

BILBY, B.A., COTTRELL, A.H. and SWINDEN, K.H. (1963). The spread of plastic yield from a notch. *Proc. R. Soc., Ser. A.*, **272**, 304

BILGER, R.W. (1975). The structure of diffusion flames. *Combust. Sci. Technol.*, **13**, 155

BILGER, R.W. (1976). Turbulent jet diffusion flames. *Prog. Energy Combust. Sci.*, **1**, 87

BILGER, R.W. (1977). Reaction rates in diffusion flames. *Combust. Flame*, **30**, 277

BILGER, R.W. (1979). Effects of kinetics and mixing in turbulent combustion. *Combust. Sci. Technol.*, **19**, 89

BILGER, R.W. (1989). The structure of turbulent nonpremixed flames. *Combustion*, **22**, 475

BILICKI, Z. and KESTIN, J. (1990). Physical aspects of the relaxation model in two-phase flow. *Proc. R. Soc, Ser. A.*, **428**, 379

BILL, F. (1991). Holistic safety engineering from an authority point of view. *Safety Developments in the Offshore Oil and Gas Industry* (London: Instn Mech. Engrs), p. 9

BILLET, R. (1979). *Distillation Engineering* (London: Heyden)

BILLETER, L. and FANNELOP (1989). Gas concentration over an underwater gas release. *Atmos. Environ.*, **23**, 1683

BILLINGE, K. (1981). The frictional ignition hazard in industry – a survey of reported incidents from 1958 to 1978. *Fire Prev. Sci. Technol.*, **24**, 13

BILLINGE, K., MOODIE, K. and BECKETT, H. (1986). The use of water sprays to protect fire engulfed LPG storage tanks. *Loss Prevention and Safety Promotion* 5, p. 47–1

BILLINGS, J. and COPLAND, D. (1992). *Ashton Munitions Explosion* (Tameside: Tameside Leisure Services)

BILLINTON, R. (1970). *Power System Reliability Evaluation* (New York: Gordon & Breach)

BILLINTON, R. and HOSSAIN, K.L. (1983). Reliability equivalents – basic concepts. *Reliab. Engng*, **5**, 223

BILLINTON, R. and ALLAN, R.N. (1983). *Reliability Evaluation of Power Systems* (London: Pitman)

BILLINTON, R., RINGLEE, R.J. and WOOD, A.J. (1973). *Power System Reliability Calculations* (Cambridge, MA: MIT Press)

BILOUS, O. and AMUNDSON, N.R. (1955). Chemical reactor stability and sensitivity. *AIChE J.*, **1**, 513

BINDON, J. (1988). Reactor operation and operator training in the United Kingdom. In Worley, N.G. and Lewins, J. (1988), *op. cit.*, p. 61

BINES, R.D. (1993). Best practicable means and BATNEEC – what they mean to the process industry. In Instn Mech. Engrs (1993), *op. cit.*, p. 27

BINGHAM, D.A. (1973). *The Law and Administration Relating to Protection of the Environment* (London: Oyez Publ.)

BINGHAM, E.C. (1966). Compact design pays off at new nitric acid plant. *Chem. Engng*, **73**(May 23), 116

BINNS, C.H.B. (1978). Health aspects of petroleum products. *Operations Prog. Engr*, **22**(9), 24

BINNS, G.D., NELSON, J.I.A. and THOMPSON, E. (1973). *Knights Looseleaf Building Regulations* (London: Knight)

BINNS, J.S. and BARRETT, P. (1984). Dispersion from a large bore relief of a batch reactor: a practical application. *The Protection of Exothermic Reactors and Pressurised Storage Vessels* (Rugby: Instn Chem. Engrs), p. 279

BINSTED, D.S. (1960). Work study in project planning of a chemical process. Pt 2, Description of techniques. *Chemy Ind.*, Jan. 16, 59

BIOTECH LIFE SCIENCE DICTIONARY (1995). BioTech Resources and Indiana University, www.biotech.icmb.utexas.edu

BIRBARA, P.J. (1976). A management system for product assurance. *Chem. Engng Prog.*, **72**(5), 33

BIRCH, A.D. and BROWN, D.R. (1989). The use of integral models for predicting jet flows. In Cox, R.A. (1989a), *op. cit.*, p. 95

BIRCH, A.D., BROWN, D.R., COOK, D.K. and HARGRAVE, G.K. (1988). Flame stability in underexpanded natural gas jets. *Combust. Sci. Technol.*, **58**, 267

BIRCH, A.D., BROWN, D.R. and DODSON, M.G. (1980). *Ignition Probabilities in Turbulent Mixing Flow. Rep. MRS E 374.* Midlands Res. Station, Br. Gas, Solihull

BIRCH, A.D., BROWN, D.R. and DODSON, M.G. (1981). Ignition probabilities in turbulent mixing flows. *Combustion*, **18**, p. 1775

BIRCH, A.D., BROWN, D.R., DODSON, M.G. and SWAFFIELD, F. (1984). The structure and concentration decay of high pressure jets of natural gas. *Combust. Sci. Technol.*, **36**, 249

BIRCH, A.D., BROWN, D.R., DODSON, M.G. and THOMAS, J.R. (1978). The turbulent concentration field of a methane jet. *J. Fluid Mech.*, **88**, 431

BIRCH, A.D., BROWN, D.R., FAIRWEATHER, M. and HARGRAVE, G.K. (1989). An experimental study of a turbulent natural gas jet in a cross-flow. *Combust. Sci. Technol.*, **66**, 217

BIRCH, A.D. and HARGRAVE, G.K. (1989). Lift-off heights in underexpanded natural gas jet flames. *Combustion*, **22**, 825

BIRCH, A.D., HUGHES, D.J. and SWAFFIELD, F. (1987). Velocity decay of high pressure jets. *Combust. Sci. Technol.*, **52**, 161

BIRCHALL, H. (1960). Work study in project planning of a chemical process. Pt 1, General approach and description. *Chemy Ind.*, Jan. 16, 56

BIRCHALL, J.D. (1970). On the mechanism of flame inhibition by alkali metal salts. *Combust. Flame*, **14**, 85

BIRCHER, H. (1899). *Die Wirkung der Artilleriegeschütze* (Aarau, Switzerland: Sauerlander)

BIRCHLER, G.J. (1972). Design for fire protection. *Fire Engng*, **125**(Sep.), 34

BIRCHON, D. (1975). *Non-Destructive Testing* (Oxford: Oxford Univ. Press)

BIRCHON, D. (1988). *Non-Destructive Testing for Engineering Design* (Br. Inst. Non-Destructive Testing)

BIRCHON, D., BROMLEY, D.E. and WINGFIELD, P.M. (1967). Some recent developments in non-destructive testing. *The Engineer*, Nov. 3, 590

BIRD, F.E. (1966). Property damage: safety's missing link. *Nat. Safety News*, Sep.

BIRD, F.E. (1974). *Management Guide to Loss Control* (Atlanta, GA: Institute Press)

BIRD, F.E. and GERMAIN, G.L. (1966). *Damage Control* (New York: Am. Mgmt Ass.)

BIRD, F.E. and GERMAIN, G.L. (1985). *Practical Loss Control Leadership* (Atlanta, GA: Int. Loss Control Inst.)

BIRD, R.B., STEWART, W.E. and LIGHTFOOT, E.N. (1960). *Transport Phenomena* (New York: Wiley)

BIRD, R.K. (1980). Fire-fighting techniques on offshore helidecks. *Fire Int.*, **67**, 35

BIRK, A.M. (1988). Modelling the response of tankers exposed to external fire impingement. *J. Haz. Materials*, **20**, 197

BIRK, A.M. (1989). Modelling the effects of a torch type fire impingement on a rail or highway tanker. *Fire Safety J.*, **15**, 277

BIRK, A.M., ANDERSON, R.J. and COPPENS, A.J. (1990). A computer simulation of a detailment accident. *J. Haz. Materials*, **25**(1/2), 121, 149

BIRK, A.M. and CUNNINGHAM, M.H. (1994). The boiling liquid expanding vapour explosion. *J. Loss Prev. Process Ind.*, **7**, 474

BIRK, A.M. and OOSTHUIZE, P.H. (1982). *Model for the Prediction of Radiant Heat Transfer to a Horizontal Cylinder Engulfed in Flames. Paper 82–WA/HI-52* Am. Soc. Mech. Engrs, New York

BIRKBY, C., BROWN, J. and STREET, P.J. (1972). Carbon monoxide: an indicator of spontaneous combustion in coal milling plant at power stations. *Fire Prev. Sci. Technol.*, **2**, 12

BIRKELAND, P.W., LANIER, M.W., MAGURA, D.D. and MAST, R.F. (1979). Pre-stressed concrete hulls for barge-mounted Operations Prog.s. *Chem. Engng Prog.*, **75**(11), 44

BIRKHAHN, W. and WALLIS, A. (1979). Safety in the company. In BASF (1979), *op. cit.*, p. 269

BIRKY, G.J. and MCAVOY, T.J. (1990). A general framework for creating expert systems for control system design. *Comput. Chem. Engng*, **14**, 713

BIRKY, G.J., MCAVOY, T.J. and MODARRES, M. (1988). An expert system for control system design. *Comput. Chem. Engng*, **12**, 1045

BIRNBAUM, Z.W. and SAUNDERS, S.C. (1969). A new family of life distributions. *J. Appl. Probability*, **6**, 319

BISHOP, K. and KNITTEL, T. (1993). Do you have the 'right' flame arrrester in service? *Hydrocarbon Process.*, **72**(2), 99

BISHOP, P.G. (ed.) (1990). *Dependability of Critical Computer Systems – 3* (London: Elsevier Applied Science)

BISHOP, R.C. (1981). Self-heating of materials laboratory testing. *Runaway Reactions, Unstable Products and Combustible Powders* (London: Instn Chem. Engrs), paper 2/H

BISHOP, R.E.D. and JOHNSON, D.C. (1979). *Mechanics of Vibration* (Cambridge: Cambridge Univ. Press)

BISHOP, W.D. and MUDAHAR, M.S. (1979). Fertilizer: solving the food problem for the developing world. *Ammonia Plant Safety*, **21**, 1

BISANTI, L. *et al.* (1979). Experience of the accident at Seveso. *Proc. Sixth Eur. Teratological Soc. Conf.* (Budapest: Akademai Kiado)

BISIO, A. (1989). LA Law – Proposition 65. *Chem. Engr, Lond.*, **458**, 17

BISIO, A. (1991a). Making sites clean again. *Chem. Engr, Lond.*, **497**, 37

BISIO, A. (1991b). US chemical plant safety receives attention. *Chem. Engr, Lond.*, **505**, 37

BISIO, A. (1992a). The front end of Superfund. *Chem. Engr, Lond.*, **529**, s24

BISIO, A. (1992b). The right to know. *Chem. Engr, Lond.*, **529**, s14

BISIO, A. (1993). Safety management: a new beginning. *Chem. Engr, Lond.* **534**, 13

BISIO, A. and KABEL, R.L. (1985). *Scaleup of Chemical Processes* (New York: Wiley-Interscience)

BISSERET, A. (1981). Application of signal detection theory to decision-making in supervisory control. The effect of operators experience. *Ergonomics*, **24**, 81

BISSET, D.W. (chrmn) (1986). *Bhopal Alternatives Review: An Assessment of the Canadian Situation* (Ottawa, Ont.: Environment Canada)

BITRON, M.D. and AHARONSON, E.F. (1978). Delayed mortality of mice following inhalation of acute doses of CH_2O, SO_2, Cl_2 and Br_2. *Am. Ind. Hyg. J.*, **39**(2), 129

BITTERMAN, H.J. (1993). Avoid downtime in centrifugal pumps. *Chem. Engng*, **100**(10), 145

BJERKEVEDT, D. and SONJU, O.K. (1984). Detonation transmission across an inert region. *Prog. Aeronaut. Astronaut.*, **95**

BJERKEVEDT, D., SONJU, O.K. and MOEN, I.O. (1986). The influence of experimental condition on the re-initiation of detonation across an inert region. *Prog. Aeronaut. Astronaut.*, **106**, 109

BJORDAL, E.N. (1980). Are 'Risk analyses' obsolete. In *Loss Prevention and Safety Promotion 3*, vol. 2, p. 440

BJORDALE, E.N. (1993). Norwegian exence of legislation concerning quality assurance of safety in industry. *Process Safety Management*, p. 175

BJORDALE, E.N. and KILPONEN, K.C. (1994). International activities. *Process Safety Prog.*, **13**, 16

BJORKLUND, R.A., KSHIDA, R.O. and FLESSNER, M.F. (1982). Experimental evaluation of flashback flame arrestors. *Plant/Operations Prog.*, **1**, 254

BJORKMAN, A.B. (1982). Views of a shipbuilder towards modern gas carriers. *Gastech 81*, p. 429

BJORKMAN, J., KOKKALA, M.A. and AHOLA, H. (1992). Measurements of the characteristic lengths of smoke detectors. *Fire Technol.*, **28**, 99

BLACK, D. (1958). *The Theory of Committees and Elections* (London: Cambridge Univ. Press)

BLACK, H. (1996). Supercritical carbon dioxide: the greener solvent. *Environ. Sci and Technol.* **30** (3), 124A–7A

BLACK, D.E. and HORN, S.J. (1968). Radioactive materials in pilot plants. *Chem. Engng Prog.*, **64**(10), 43

BLACK, H.F., FRANCE, D., JENSSEN, D.N. and BROWN, R. (1974). Theoretical and experimental investigations relating to centrifugal pump rotor vibration. *Plant Engr*, **18**(7/8), 16

BLACK, J.E., GLENNY, E.T. and MCNEE, J.W. (1915). Observations on 685 cases of poisoning by noxious gases used by the enemy. *Br. Med. J., Jul. 31*, 165

BLACK, J.M. and PONTON, J.W. (1992). A hierarchical method for line-by-line hazard and operability studies. *Workshop on Interactions between Process Design and Process Control, London*, p. 227

BLACK, M.W. (1978). Problems in siting of hazardous waste disposal facilities – the Peerless experience. *Control of Hazardous Material Spills 4*, p. 232

BLACK, W.S. (1989). Non-self-acting safety systems for the process industries. *Hazards X*, p. 291

BLACKADAR, A.K. (1976). Modeling the nocturnal boundary layer. *Third Symp. on Atmospheric Turbulence, Diffusion and Air Quality* (Boston, MA: Am. Met. Soc.), p. 46

BLACKADDER, E. (1976). Toxic releases and their biological effects. *Process Industry Hazards*, p. 1

BLACKBURN, G.M. and KELLARD, B. (1986). Chemical carcinogens. *Chemy Ind.*, Sept. 15, 607; Oct. 20, 687; Nov. 17, 770

BLACKBURN, J. (1977). Cast reformer headers and manifolds. *Ammonia Plant Safety*, **19**, p. 6

BLACKETT, P.M.S. (1962). *Studies of War* (Edinburgh: Oliver & Boyd)

BLACKMAR, G.E. (1973). Find leaks faster with sonic detectors. *Hydrocarbon Process.*, **52**(1), 73

BLACKMORE, D.R., EYRE, J., HOMER, J.B. and MARTIN, J.A. (1981). Refrigerated gas safety research. *Am. Gas Ass., Transmission Conf., Atlanda, GA*

BLACKMORE, D.R., EYRE, J.A. and SUMMERS, G.G. (1982). Dispersion and combustion behaviour of gas clouds resulting from large spillages of LNG and LPG on to the sea. *Trans. Inst. Mar. Engrs*, **94**, paper 29

BLACKMORE, D.R., HERMAN, M.N. and WOODWARD, J.L. (1982). Heavy gas dispersion models. *J. Haz. Materials*, **6**(1/2), 107. See also in Britter, R.E. and Griffiths, R.F. (1982), *op. cit.*, p. 107

BLACKNEY, A.B. (1982). RAF maintenance data system. *Reliab. Engng*, **3**(4), 315

BLACKSHEAR, P.L. (ed.) (1974). *Heat Transfer in Fires* (New York: Wiley)

BLACKSHEAR, P.L. and MURTY, K.A. (1965). Heat and mass transfer to, from, and within cellulosic solids burning in air. *Combustion*, **10**, p. 911

BLACKSHIELD, A. (1969). Grinding operations. In Handley, W. (1969), *op. cit.*, p. 128

BLACKWELL, W.W. (1981). Calculating two-phase pressure drop. *Chem. Engng*, **88**(Sep. 7), 121

BLACKWELL, W.W. (1982a). Estimating heat-tracing requirements for pipelines. *Chem. Engng*, **89**(Sep. 6), 115

BLACKWELL, W.W. (1982b). Sizing condensate-return lines. *Chem. Engng*, **89**(Jul. 12), 105

BLACKWOOD, T.R. (1982). Fugitive particle emissions – an overview. *Plant/Operations Prog.*, **1**, 263

BLACTIN, S.C. and ROBINSON, H. (n.d.). *Spontaneous Electrification of Coal Dust Clouds. Paper 71.* Safety in Mines Res. Estab.

BLAHA, M.R., YAMASHITA, Y. and MOTARD, R.L. (1985). Database management systems for the process engineer. *Chem. Engng Prog.*, **81**(9), 45

BLAIN, P. (1987). Background to COSHH. Toxic responses. *Health and Safety at Work*, Nov. 25

BLAIR, D.A. (1977). Operate your refinery safely. *Hydrocarbon Process.*, **56**(4), 241

BLAIR, E.H. (1980). A role of transport and fate studies in industrial decision-making for toxic chemicals. In Haque, R. (1980), *op. cit.*, p. 5

BLAIR, H. (1993). The environmental pluses of diaphragm pumps. *Chem. Engng*, **100**(5), 118

BLAKE, M.P. (1964). New vibration measurement standards. *Hydrocarbon Process.*, **43**(1), 111

BLAKE, M.P. (1966). New vibration standards for electric motors. *Hydrocarbon Process.*, **45**(3), 126

BLAKE, M.P. (1967). Use phase measuring to balance rotors in place. *Hydrocarbon Process.*, **46**(7), 127

BLAKE, M.P. and MITCHELL, W.S. (1972). *Vibration and Acoustic Measurement Handbook* (New York: Spartan Books)

BLAKE, P.M. and DOUGLAS, J.W.B. (n.d.). *Casualty Effects of High Explosive Bombs in Low Level Daylight Raids Compared with High Level Night Raids. Rep. REN 317.* Research and Experiments Dept, Ministry of Home Security

BLAKE, P.M., DOUGLAS, J.W.B., KROHN, P.L. and ZUCKERMAN, S. (1941). *Rupture of eardrums by blast. Rep 43/179/WS 21.* Med. Res. Coun., Oxford

BLAKE, R.P. (1943). *Industrial Safety* (Englewood Cliffs, NJ: Prentice-Hall)

BLAKE, T.R. and MCDONALD, M. (1993). An examination of flame length data from vertical turbulent diffusion flames. *Combust. Flame*, **94**, 426

BLAKE, W.K. (1986). *Mechanics of Flow-Induced Sound and Vibration*, vols 1–2 (Orlando, FL: Academic Press)

BLAKEY, P. (1981). Cryogenic gases. *Chem. Engng*, **88**(Oct. 5), 113

BLAKEY, P. and ORLANDO, G. (1984). Using inert gas for purging, blanketing and transfer. *Chem. Engng*, **91**(May 28), 97

BLAKEY, W.G. (1974). Statutory requirement for chemical safety. *Course on Process Safety – Theory and Practice.* Dept of Chem. Engng, Teesside Polytechnic, Middlesborough

LE BLANC, L. (1981). Tracing the causes of rig mishaps. *Offshore*, Mar.

BLANC, M.V., GUEST, P.G., VON ELBE, G. and LEWIS, B. (1947). Ignition of explosive gases by electric sparks. I, Minimum ignition energies and quenching distances of mixtures of methane, oxygen and inert gases. *J. Chem. Phys.*, **15**, 798

BLANC, M.V., GUEST, P.G., VON ELBE, G. and LEWIS, B. (1949). Ignition of explosive gas mixtures by electric sparks. III, Minimum ignition energies and quenching distances of hydrocarbons and ether with oxygen and inert gases. *Combustion*, **3**, p. 363

BLANCHARD, B.S. and LOWERY, E.E. (1969). *Maintainability Principles and Practice* (New York: McGraw-Hill)

BLANCHARD, P.C. *et al.* (1966). *Likelihood-of-accomplishment Scale for a Sample of Man–Machine Activities.* Dunlop and Associates Inc., Santa Monica, CA.

BLANCHARD, R. (1980). Coping with catastrophe. *Int. Mngmt*, Apr. 12

BLANCHARD, R.L. (1975). Measurement of density in custody transfer systems. *Gastech 75*

BLANCHARD, R.L. and SHERBURNE, A.E. (1985). Calibrating accurate level gauges in partly filled LNG/LPG tanks. The transfer calibrator. *Gastech 84*, p. 506

BLAND, J.A. and BEDDOW, J.L. (1974). Plants with hazards to spare. *Fire Engng*, **127**(Feb.), 37

BLAND, J.G.V. (1977). Portable electric tools: safe working. In Handley, W. (1977), *op. cit.*, p. 162

BLAND, W.F. and DAVIDSON, R.L. (eds) (1967). *Petroleum Processing Handbook* (New York: McGraw-Hill)

BLANKEN, J.M. (1975). Swaying of a CO_2 stripper. *Ammonia Plant Safety*, **17**, p. 146

BLANKEN, J.M. (1978). Pitfalls in handling ammonia. *Ammonia Plant Safety*, **20**, p. 32

BLANKEN, J.M. (1980). Behaviour of ammonia in the event of a spillage. *Ammonia Plant Safety*, **22**, 25

BLANKEN, J.M. (1982). Panel discussion: stress corrosion cracking of ammonia spheres – results of survey. *Am. Inst. Chem. Engrs, Symp. on Safety in Ammonia Plants and Related Facilities, Los Angeles, CA*

BLANKEN, J.M. (1984). Stress corrosion cracking of ammonia storage spheres: survey and panel discussion. *Ammonia Plant Safety*, **24**, 140

BLANKEN, J.M. (1987). Atmospheric ammonia storage tanks without and with secondary containment (discussion contribution). *Ammonia Plant Safety*, **27**, 175

BLANKEN, J.M. (1990). (Letter on liquid jet disintegration). *Plant/Operations Prog.*, **9**, O6

BLANKEN, J.M. and GROEFSMA, T. (1983). Sudden pressure increase in a vent header. *Plant/Operations Prog.*, **2**, 137

BLANKS, H.S. (1977). The generation and use of component failure data. *Quality Assurance*, **3**(3), 85

BLANSHAN, S.A. (1978). A time model: hospital organizational response to disaster. In Quanterelli, E.L. (1978), *op. cit.*, p. 173

VAN BLARCOM, P.P. (1974). Bypass systems for centrifugal pumps. *Chem. Engng*, 81(Feb. 4), 94

VAN BLARCOM, P.P. (1980). Recirculation systems for cooling centrifugal pumps. *Chem. Engng*, 87(Mar.), 131

BLASS, E. (1985). Methodological development of chemical engineering processes. *Chem. Ing. Tech.*, 57, 201

BLATTE, H.J. (1985). Camera pays for itself after one incident. *Fire Int.*, 92, 79

BLATZ, H. (1959). *Radiation Hygiene Handbook* (New York: McGraw-Hill)

BLAZEK, A. (1973). *Thermal Analysis* (New York: van Nostrand Reinhold)

BLEARS, D.G. and COVENTRY, R.J. (1974). Development of continuous organic lead-in-air monitors. *Chemical Process Hazards*, V, p. 322

BLEEZE, A. (1992). Cracking health and safety management. *Major Hazards Onshore and Offshore*, p. 123

BLEIBERG, J., WALLEN, J., BRODKIN, R. and APPELBAUM, I.L. (1964). Industrially acquired porphyria. *Arch. Derm*, 89, 793

BLEICHER, S.A. (1990). Strategic alternatives for dealing with the regulatory maze governing environmental cleanup. *J. Haz. Materials*, 25(3), 269

BLENKINSOP, R. (1992). Building on a firm foundation. *Chem. Engr, Lond.*, 522, 22

BLESS, S.J. (1982). Impact physics facilities at the University of Dayton Research Institute. In Nellis. W.J., Seaman, L. and Graham, A.R. (1982), *op. cit.*, p. 668

BLEVINS, R.D. (1977). *Flow Induced Vibrations* (New York: van Nostrand Reinhold)

BLEVINS, R.D. (1979). *Formulas for Natural Frequency and Mode Shape* (New York: van Nostrand Reinhold)

BLEWITT, D.N., PETERSEN, R.L., RATCLIFF, M.A. and HESKSTAD, G. (1991). Evaluation of water spray mitigation system for an industrial facility. *Modeling and Mitigating the Consequences of Accidental Releases*, p. 483

BLEWITT, D.N., YOHN, J.F. and ERMAK, D.L. (1987). An evaluation of SLAB and DEGADIS heavy gas dispersion models using the HF spill test data. *Vapor Cloud Modeling*, p. 56

BLEWITT, D.N., YOHN, J.F., KOOPMAN, R.P. and BROWN, T.C. (1987). Conduct of anhydrous hydrofluoric acid spill experiments. *Vapor Cloud Modeling*, p. 1

BLEWITT, D.N., YOHN, J.F., KOOPMAN, R.P., BROWN, T.C. and HAGUE, W.J. (1987). Effectiveness of water spays on mitigating anhydrous hydrofluoric acid releases. *Vapor Cloud Modeling*, p. 155

BLEY, D.C. and REULAND, W.B. (1987). Application of a process flow model to a shut-down nuclear plant probabilistic safety assessment. *Reliab. Engng*, 17(4), 241

BLICK, R.G. (1965). Stress analysis. *Chem. Engng*, 72(Jan. 18), 159; 72(Feb. 15), 169

BLICKENSDERFER, R. (1975). Methane ignition by frictional impact heating. *Combust. Flame*, 25, 143

DE BLIECK, J.L., CIJFER, H.J. and JUNGERHANS, R.R.J. (1971). Which feeds for olefins? *Hydrocarbon Process.*, 50(9), 177

BLIN, A., CARNINO, A., DUCHEMIN, B., KOEN, B.V. and LANORE, J.M. (1977). The state of the art. A computer code for the evaluation of reliability and availability of complex systems. *First Nat. Conf. on Reliability.* NRC3/10 (Culcheth, Lancashire: UK Atom Energy Auth.)

BLINOV, V.I. and KHUDIAKOV, G.N. (1957). Certain laws governing diffusive burning of liquids. *Academia Nauk, SSSR Doklady*, 113, 1094

BLISS, C.I. and CATTELL, M. (1943). Biological assay. *Ann. Rev. Physiol.*, 5, 479

BLISS, C.I. (1934). The method of probits. *Science*, 79(2037), 38

BLISS, C.I. (1935). The calculation of the dosage-mortality curve. *Ann. Appl. Biol.*, 22, 134

BLISS, C.I. (1952). *The Statistics of Bioassay* (New York: Academic Press)

BLISS, C.I. (1957). Some principles of bioassay. *Am. Sci.*, 45, 449

BLOCH, H.P. (1974). Why properly rated gears still fail. *Hydrocarbon Process.*, 52(12), 95

BLOCH, H.P. (1976). How to uprate turbo equipment by optimized coupling selection. *Hydrocarbon Process.*, 55(1), 87

BLOCH, H.P. (1977a). Improve safety and reliability of pumps and drivers. *Hydrocarbon Process.*, 56(1), 97; 56(2), 123; 56(3), 133; 56(4), 181; 56(5), 213

BLOCH, H.P. (1977b). Predict problems with acoustic incipient failure detection systems. *Hydrocarbon Process.*, 56(1), 191

BLOCH, H.P. (1978). How to select a centrifugal pump vendor. *Hydrocarbon Process.*, 57(6), 93

BLOCH, H.P. (1980). Predict pump problems with IFD. *Hydrocarbon Process.*, 59(1), 87

BLOCH, H.P. (1982). How to buy a better pump. *Hydrocarbon Process.*, 61(1), 81

BLOCH, H.P. (1983a). Centrifugal pump failure-reduction program can show quick success. *Oil Gas. J.*, 81(Jan 10), 79

BLOCH, H.P. (1983b). Selection strategy for mechanical shaft seals. *Hydrocarbon Process.*, 62(1), 75

BLOCH, H.P. (1988). *Improving Machinery Reliability*, 2nd ed. (Houston, TX: Gulf Publ Co.).

BLOCH, H.P. (ed.) (1989). *Process Plant Machinery* (Oxford: Butterworth-Heinemann)

BLOCH, H.P. (1991). Use laser optics for on-stream alignment verification. *Hydrocarbon Process.*, 70(1), 71

BLOCH, H.P. (1993). Bearing seals ward off failures. *Chem Engng*, 100(3), 149

BLOCH, H.P. and GEITNER, H.P. (1983). *Machinery Failure Analysis and Troubleshooting* (Houston, TX: Gulf Publ. Co.)

BLOCH, H.P. and GEITNER, H.P. (1990). *An Introduction to Machinery Reliability Assessment* (New York: van Nostrand Reinhold).

BLOCH, H.P. and JOHNSON, D.A. (1985). Downtime prompts upgrading of centrifugal pumps. *Chem. Engng*, 92(Nov. 25), 35

BLOCH, H.P. and NOACK, P.W. (1992). The expanded capabilities of screw compressors. *Chem. Engng*, 99(2), 100

BLOCH, P. (1926). *La Guerre Chimique* (Paris)

BLOCK, B. (1962). Emergency venting for tanks and reactors. *Chem. Engng*, 69(Jan. 22), 111

BLOCK, J.A. (1971). A theoretical and experimental study of nonpropagating free-burning fires. *Combustion*, 13, p. 971

BLOCK, R.M. and KALINOWSKI, T.W. (1983). Disposing of those old drums. *Chem. Engng*, 90(Mar. 21), 77; 90(Apr. 18), 103

BLOCK, R.M., DRAGUN, J. and KALINOWKSI, T.W. (1984). Health and environmental aspects of setting cleanup criteria. *Chem. Engng*, 91(Nov. 26), 70

BLOCKER, T.G., BLOCKER, V., GRAHAM, J.E. and JACOBSON, H. (1969). Follow-up medical survey of the Texas City disaster. *Am. J. Surgery*, 97, 604

BLOCKER, V. (1949). The Texas City Disaster. Pattern of injury in 3,000 casualties. *Texas Rep. Biol. Med.*, 7, 22

BLOCKER, V. and BLOCKER, T.G. (1949). The Texas City disaster. A survey of 3,000 casualties. *Am. J. Surg*, Nov., 756

BLOCKLEY, D.I. (1980). *The Nature of Structural Design and Safety* (Chichester: Ellis Horwood)

BLOEBAUM, W.D. and LEWIS, J.P. (1969). Economics of LNG production and distribution for remote satellite operations. *Proc. Int. Conf. on Liquefied Natural Gas* (London)

BLOGG, E.C. (1987). A safety training programme onboard LNG carriers. *Gastech 86*, p. 143

BLOHM, K.H. (1977). Testing of the operational safety of machines for the chemical industry by the expert committee 'Chemistry'. *Prevention of Occupational Risks*, **5**, p. 146

BLOKKER, E.F. (1986). From risk analysis to decision analysis. In Hartwig, S. (1986), *op. cit.* p. 315

BLOKKER, E.F. (1987). Safety policy for the Rijnmond region. *Preventing Major Chemical Accidents*, p. 2.31

BLOKKER, E.F., COX, R.A., LANS, H.J.D. and MEPPELDER, F.H. (1980). Evaluation of risks associated with process industries in the Rijnmond area − a pilot study. *Loss Prevention and Safety Promotion 3*, vol. 3, p. 1078

BLOKKER, E.F. and GOOS, D. (1986). Event data collection for the Rijnmond process industry. *Loss Prevention and Safety Promotion 5*, p. 66−1

BLOKKER, P.C. (1964). Spreading and evaporation of petroleum products on water. *Fourth Int. Harbour Conf., Antwerp*

BLOMQUIST, D.L. (1988). New LPG loss-control standards. *Hydrocarbon Process.*, **67**(12), 58

BLOOM, B. and PEBWORTH, L.A. (1979). On-stream repairs on reformer bottom manifold insulation. *Ammonia Plant Safety*, **21**, p. 32

BLOOM, S.G. (1980). A mathematical model for reactive, negatively buoyant atmospheric plumes. In Hartwig, S. (1980), *op. cit.*, p. 103

BLOOMER, R.N. and SMALLEY, J. (1975). New procedures for measuring air leakage and conditions in steam condensing plants under vacuum. *Plant Engr*, **19**(5), 14

BLOORE, P.D. (1974). Analysis of a batch process explosion. *Chemical Process Hazards*, **5**, p. 133

BLUBAUGH, R.L. and WATTS, H.J. (1969). Aligning rotating equipment. *Loss Prevention*, **3**, 66. See also *Chem. Engng Prog.*, 1969, **65**(4), p. 44

BLUHM, W.C. (1961). How to operate a flare system safely. *Oil Gas J.*, **59**(Aug. 28), 73

BLUHM, W.C. (1964a). Protective facilities for refinery process units. In Vervalin, C.H. (1964a), *op. cit.*, p. 219

BLUHM, W.C. (1964b). Safe operation of refinery flare systems. In Vervalin, C.H. (1964a), *op. cit.*, p. 256

BLUME, J.A. (1971). Civil structures and earthquake safety. *Interim Rep. of the Special Subcommittee of the San Fernando Earthquake Study*

BLUME, J.A. (1978). Predicting natural hazards − state of the art. *Joint Conf. on Predicting and Designing for Natural and Man-Made Hazards* (New York: Am. Inst. Civil Engrs), p. 1

BLUMEL, P. and SCHULZ-FORBERG, B. (1989). Dangerous goods/Gefahrgut databank. *Loss Prevention and Safety Promotion 6*, p. 83.1

BLUMRICH, W.E., KOENING, H.J. and SCHWEHR, E.W. (1978). The Fisons HDH phosphoric acid process. *Chem. Engng Prog.*, **74**(11), 58

BLYTHE, A.R. and REDDISH, W. (1979). Charges on powders and bulking effects. *Electrostatics*, **5**, p. 107

BLYTHING, K.W. (1984). In-service reliablity data for underground cross-country oil pipelines. Safety Reliablity Directorate, UKAEA, 19 844

BLYTHING, K.W. (1986). 'BLEVE' probability of a 100 te LPG storage vessel. *Gastech 85*, p. 147

BLYTHING, K.W. and EDMONDSON, J.N. (1983). Incident rates on liquid gas carriers with special reference to fire and explosion. *Loss Prevention and Safety Promotion 4*, vol. 2, p. B49

BOARD, S.J., HALL, R.W. and HALL, R.S. (1975). Detonation of fuel coolant explosions. *Nature*, **254**, 319

BOARDMAN, J. *et al.* (1972). 'The UK probabilistic accident consequence code CONDOR'. Seminar on Methods and Codes for Assessing the Off-Site Consequences of Nuclear Accidents, Athens

BOARDMAN, R.M. (1972). A state contingency plan for hazardous spills. *Control of Hazardous Material Spills*, **1**, p. 59

BOBBIO, A. and SARACCO, O. (1983). The reliability data bank on electronic components of the 'Circolo dell'Affidabilita': operating procedures and some results. *Euredata 4*, paper 11.3

BOBERG, I.E. (1970). Choosing tank design. *Chem. Engng*, **77**(Aug. 10), 134

BOBROW, D.G. (1984). Qualitative reasoning about physical systems: an introduction. *Artificial Intelligence*, **24**, 1

BOBROW, D.G. (ed.) (1985). *Qualitative Reasoning about Physical Systems* (Cambridge, MA: MIT Press)

BOBROW, D.G. and COLLINS, A. (eds) (1975). *Representation and Understanding: Studies in Cognitive Science* (New York: Academic Press)

BOBROW, D.G. and WINOGRAD, T. (1977). An overview of KRL, a knowledge representation language. *Cognitive Sci.*, **1**, 3

BOBY, W.M.T. and SOLT, G.S. (1965). *Water Treatment Data* (London: Hutchinson)

BOCCIO, J.L. (1982). *Requirements for Establishing Detector Siting Criteria in Fire involving Electrical Materials*. Rep. *SAND81−7168*. See also *NUREG/CR-2409*. Sandia Nat. Lab., Albuquerque, NM

BOCHINSKI, J.H., SCHOULTZ, K.S. and GIDEON, J.A. (1979). Meeting emission standards for acrylonitrile. *Chem. Engng Prog.*, **75**(8), 53

BOCK, D.H. and SULLIVAN, P.F. (1972). Methods for the detection and marking of spilled hazardous material. *Control of Hazardous Material Spills*, **1**, p. 119

BOCK, E.A. (ed.) (1965). *Government Regulation of Business: A Casebook* (Englewood Cliffs, NJ: Prentice-Hall)

BOCKENHAUER, M. (1980). Some considerations on gas carrier safety. *Gastech 79*, p. 235

BOCKENHAUER, M. (1982). Some notes on the practical application of the IMCO Gas Carrier Code to pressure vessel type cargo tanks. *Gastech 81*, p. 237

BOCKENHAUER, M. (1985). The filling limitations of cargo tankers − review of the IMO Gas Carrier Code requirements. *Gastech 84*, p. 142

BOCKENHAUER, M. (1987). Some design aspects of multigrade liquefied gas/chemical/product carriers. *Gastech 86*, p. 187

BOCKENHAUER, M. (1989). Coping with the phenomenon of ammonia stress corrosion cracking. *Gastech 88*, paper 5.10

BOCKHOLTS, P. (1983). Collection and application of incident data. *Loss Prevention and Safety Promotion 4*, vol. 1, p. K11

BOCKHOLTZ, P. (1983). Collection and application of incident data. *Euredata 4*, paper 6.3

BOCKMANN, T., INGEBRIGTSEN, G.H. and HAKSTAD, T. (1981). Safety design of the ethylene plant. *Loss Prevention*, **14**, p. 127. See also *Chem. Engng. Prog., 1980*, **76**(11), 37

BODDAERT, H.P. (1975). Tank cleaning. *Transport of Hazardous Cargoes*, 4

BODDINGTON, T. (1963). The growth and decay of hot spots and the relation between structure and stability. *Combustion*, **9**, p. 287

BODDINGTON, T., CHANG-GEN FENG and GRAY, P. (1982). Disappearance of criticality in thermal explosion under the Frank–Kamenetski boundary conditions: comment on Zaturska and Gill, Donaldson and Shouman. *Combust. Flame*, **48**, 303

BODDINGTON, T. and GRAY, P. (1970). Temperature profiles in endothermic and exothermic reactions and the interpretation of experimental rate data. *Proc. R. Soc., Ser. A.*, **320**, 71

BODDINGTON, T., GRAY, P. and HARVEY, D.I. (1971a). Thermal explosions and critical size: bounds on criteria of criticality (letter). *Combust. Flame*, **17**, 263

BODDINGTON, T., GRAY, P. and HARVEY, D.I. (1971b). Thermal theory of spontaneous ignition: criticality in bodies of arbitrary shape. *Phil. Trans R. Soc., Ser. A.*, **270**, 467

BODDINGTON, T., GRAY, P. and ROBINSON, C. (1979). Thermal explosions and the disappearance of criticality at small activation energies: exact results for the slab. *Proc. R. Soc., Ser. A.*, **368**, 441

BODDINGTON, T., GRAY, P. and SCOTT, S.K. (1981). Correction of kinetic data in non-isothermal reactors with nonuniform temperatures: analytical treatment for spherical reactant masses. *Proc. R. Soc., Ser. A.*, **378**, 27

BODDINGTON, T., GRAY, P. and WALKER, I.K. (1981). Runaway reactions and thermal explosion theory. In *Runaway Reactions, Unstable Products and Combustible Powders* (London: Instn Chem. Engrs), paper 1/C

BODDINGTON, T., GRIFFITHS, J.F., LAYE, P.G. and NELSON, D.C. (1989). Fundamental aspects of thermal analysis and heat flux calorimetry. *Runaway Reactions*, p. 21

BODEN, M. (1977). *Artificial Intelligence and Natural Man* (Brighton: Harvester Press)

BODIE, A. (1989). *Piper Alpha Public Inquiry. Transcript, Days 121, 122*

BODSBERG, L.A. and INGSTAD, O. (1989). Technical and human implications of automatic safety systems. *Loss Prevention and Safety Promotion* 6, p. 4.1

BODURTHA, F.T. (1955). A technique for the rapid solution of an air-pollution equation. *J. Air Poll. Control Ass.*, **5**, 127

BODURTHA, F.T. (1961). The behaviour of dense stack gases. *J. Air Poll. Control Ass.*, **11**, 431

BODURTHA, F.T. (1976). Explosion hazards in pollution control. *Loss Prevention*, **10**, p. 88

BODURTHA, F.T. (1980). *Industrial Explosion Prevention and Protection* (New York: McGraw-Hill)

BODURTHA, F.T. (1988). Vent heights for emergency releases of heavy gases. *Plant/Operations Prog.*, **7**, 122

BODURTHA, F.T., PALMER, P.A. and WALSH, W.H. (1973). Discharge of heavy gas from relief valves. *Loss Prevention*, **7**, p. 61. See also *Chem. Engng Prog., 1973*, **69**(4), 37

BØE, C. (1978). Some views on risk and risk acceptance. *Fifth Symp on Reliability Technology, Univ. of Bradford*

BØE, C. and FOLEIDE, A. (1975). Criteria for safety at sea. *Transport of Hazardous Cargoes*, **4**

BOEING COMPANY (1965). *Systems Safety Symp.* (Seattle, WA)

BOEL, M. and DANIELLOU, F. (1984). Elements of process control operators reasoning: activity planning and system and process response times. *Ergonomic Problems in Process Operations* (Rugby: Instn Chem. Engrs), p. 1

DE BOER, J. and BAILLIE, T.W. (1980a). A disaster plan for a district hospital. In De Boer, J. and Baillie, T.W. (1980), *op. cit.*, p. 65

DE BOER, J. and BAILLIE, T.W. (eds) (1980b). *Disasters: Medical Organisation* (Oxford: Pergamon)

BOESEBECK, K. (1975). Failure probability of reactor pressure vessels derived from failure statistics for pressure vessels from the conventional domain. *Technische Uberwachung*, **16**(10), 281

BOESEBECK, K. and HOMKE, P. (1977). Failure data collection and analysis in the Federal Republic of Germany. In Gangadharan, A.C. and Brown, S.J. (1977), *op. cit.*, p. 7

BOESINGER, D.D., NIELSEN, R.H. and ALBRIGHT, M.A. (1980). *Hydrocarbons: butane. Kirk–Othmer Encyclopaedia of Chemical Technology*, 3rd ed., vol. 12 (New York: Wiley), p. 910

BOEYE, C.G. (1979). Flare relief systems. In Minors, C. (1979), *op. cit.*, paper 6

BOFFEY, P.M. (1976). Rasmussen issues revised odds on a nuclear catastrophe. *Sci. Publ. Policy*, **3**, 146

BOFFEY, T.B. (1981). *Graph Theory in Operations Research* (London: Macmillan)

BOGER, H.W. (1969). Methods for sizing control valves. *Chem. Engng*, **74**(Nov. 6), 247

BOGNAR, J., PETERS, H. and SCHATZMAYR, G. (1978). Calculation of autofrettage on tubes and fittings. *Safety in Polyethylene Operations Progs.*, **3**, p. 33

BOHLIN, A. (1973). Port safety regulations relating to dangerous cargoes, liquid and packed goods. *Transport of Hazardous Cargoes*, **3**

BOHLKEN, S.F. (1961). Heat exchanger explosion at nitrogen wash unit. *Ammonia Plant Safety*, **3**, p. 13. See also *Chem. Engng Prog., 1961*, **57**(4), 149

BOHLMAN, M.T. (1975). The interrelation between hazard evaluation and tank container design. *Transport of Hazardous Cargoes*, **4**

BOHM, J. (1982). *Electrostatic Precipitation* (Amsterdam: Elsevier)

BOIKO, V.M., PAPYRIN, A.N., WOLINSKI, M. and WOLANSKI, P. (1984). Dynamics of dispersion and ignition of dust layers by a shock wave. In Bowen, J.R. et al. (1984b), *op, cit., p. 293*

BOILY, A. (1976). Central hazardous waste treatment at your disposal. *Control of Hazardous Material Spills*, **3**, p. 441

BOISSIERAS, J. (1983). *Causal Tree. Description of the Method*. Rhone-Poulenc, Princeton, NJ

BOIX, J.A. (1985). Emergency flare system – some practical design features. *Plant/Operations Prog.*, **4**, 222

BOKOR, A. (1980). Partition of impact energy between local deformation and target response. *Nucl. Energy*, **19**, 211

BOLAND, D. and HINDMARSH, E. (1984). Heat exchanger network improvements. *Chem. Engng Prog.*, **80**(7), 47

BOLAND, R.F., CTVRTNICEK, T.E., DELANEY, J.L., EARLEY, D.E. and KHAN, Z.S. (1976). *Screening Study for Miscellaneous Sources of Hydrocarbon Emissions in Petroleum Refineries*. Rep. EPA 450/3–76–041. Environ. Prot. Agency, Res. Triangle Park, NC

BOLDING, A. (1994). Calculating operating weights of offshore fixed platforms. *Offshore*, Jun., 42

BOLLANDS, G. (1989). *Piper Alpha Public Inquiry. Transcript, Days 48, 49*

BOLLEN, R.F. (1960). Hydrocarbon contamination of an air separation Plant. *Ammonia Plant Safety*, **2**, p. 25. See also *Chem. Engng Prog., 1960*, **56**(5), 61

BOLLIGER, D.H. (1982). Assessing heat transfer in process-vessel jackets. *Chem. Engng*, **89**(Sep. 20), 95

BOLLINGER, L.M. and WILLIAMS, D.T. (1949). *The Effect of Reynolds Number in Turbulent-flow Range on Flame Speeds on Bunsen Burners. Rep. 932*. Nat. Advisory Cttee Aeronaut.

BOLLINGER, U. and WRIGHT, J.W.D. (1977). Options for planned maintenance. *Hydrocarbon Process.*, **56**(1), 85

BOLME, D.W. and HORTON, A. (1980). The Humphreys and Glasgow/Bolme nitric acid process. *Ammonia Plant Safety*, **21**, p. 184

BOLOGNA, S. (1993). Independent software verification and validation in practice: methodological and managerial aspects. In Bennett, P. (1993), *op. cit.*, p. 88

BOLT, B.A. (1978). *Earthquakes; 1988*, rev. ed.; 1993, rev. ed. (New York: W.H. Freeman)

BOLT, B.A. (ed.) (1980). *Earthquakes and Volcanoes* (San Francisco, CA: W.H. Freeman)

BOLT, B.A. (1982). *Inside the Earth* (New York: Freeman)

BOLT, J.A. and SAAD, M.A. (1957). Combustion rates of freely falling fuel drops in a hot atmosphere. *Combustion*, **6**, p. 717

BOLT, R. and LOGTENBERG, T. (1989). Pipelines once buried never to be forgotten. In Guy, G.B. (1989), *op. cit.*, p. 195

BOLT, R.M. and ARZYMANOW, G.W. (1982). Pre-assembly of the Sullom Voe process facilities. *Chem. Engng Prog.*, **78**(11), 64

BOLTEN, J.G. (1985). Estimating the chronic health risk from coal-fired power plant toxic emissions. *J. Haz. Materials*, **10**(2/3), 351

BOLTON, L. (1978). What happened in Seveso. *Chem. Engng*, **85**(Sep. 11), 104C

BOLTON, L. (1991). Seveso revisited. *Chem. Engr, Lond.*, **500**, 13

BOLZ, R.E. and TUVE, G.L. (eds) (1973). *Handbook of Tables for Applied Engineering Science*, 2nd ed. (Boca Raton, FL: CRC Press)

BOLZERN, O. and BOURNE, J.R. (1985). Rapid chemical reactions in a centrifugal pump. *Chem. Engng Res. Des.*, **63**, 275

BOMHARD, H. (1987). Concrete pressure vessels – the preventive answer to the Mexico City LPG disaster. *Gastech 86*, p. 415

BOMPAS-SMITH, J.H. (1973). *Mechanical Survival: the Use of Reliability Data* (New York: McGraw-Hill)

BONAR, J.A. (1972). Fuel ash corrosion. *Hydrocarbon Process.*, **51**(8), 76

BOND, A. (1979). Safety standards spur gas detection development. *Process Engng*, Dec., **59**

BOND, B.B. (1965). *Spectrometric Oil Analysis. REP. O&R-P-1.* Naval Air Station, Pensacola, FL

BOND, J. (1987). Case studies on unstable substances. *Hazards from Pressure*, p. 37

BOND, J. (1989a). Loss prevention management. *Loss Prevention and Safety Promotion* 6, p. 11.1

BOND, J. (1989b). Under the iceberg. *Chem. Engr, Lond.*, **463**, 4

BOND, J. (1990). Loss prevention: an important part of total quality management. *Qualilty Forum*, **16**(4), 179

BOND, J. (1991). *Sources of Ignition* (London: Butterworth-Heinemann)

BOND, J. and BRYANS, J.W. (1975). The two safeguard approach for minimising human failure injuries. *Chem. Engr, Lond.*, **296**, 245

BOND, N.A. (1981). Troubleshooting in the commercial computer industry: a success story. In Rasmussen, J. and Rouse, W.B. (1981), *op. cit.*, p. 75

BOND, P.G. (1978). Pesticides containers: design and manufacture. *Chemy Ind.*, Feb. 18, 122

BOND, R.G. and STRAUB, C.P. (1972). *Handbook of Environmental Control* (Cleveland, OH: Chemical Rubber Co.)

BOND, R.G. and STRAUB, C.P. (eds) (1973–). *Handbook of Environmental Control*, vols 1–4 (Boca Raton, FL: CRC Press)

BONE, W.A. and NEWITT, D.M. (1929). *Combustion at High Pressures* (New York)

BONE, W.A. and TOWNEND, D.T.A. (1927). *Flame and Combustion in Gases* (London: Longmans Green)

BONGARD, N. (1970). *Pattern Recognition* (New York: Spartan Books)

BONGERS, A., TEN BRINK, G. and RULKENS, P. (1980). Release of pressurized hydrocarbon from a pilot-plant-scale pressure vessel; experiments and theory. *Loss Prevention and Safety Promotion* 3, vol. 3, p. 1198

BONGERS, P.J.W., DOLS, P.L.M. and LINSSEN, T.J.M. (1978). Fatigue crack initiation in high-pressure tubes. *Safety in Polyethylene Operations Progs.*, **3**, p. 10

BONGIORNO, S.J., CONNOR, J.M. and WALTON, B.C. (1970). The module concept of reformer design. *Ammonia Plant Safety*, **12**, p. 39

BONHAM, E.A. (1977). The design of a flare system for a naphtha cracking complex. *Eleventh Loss Prevention Symp.* (New York: Am. Inst. Chem. Engrs), paper

BONIFACE, A. (1989). Britain's waterways. *Chem. Engr, Lond.*, **464**, 24

BONIFACE, A. (1990a). Action on offshore safety. *Chem. Engr, Lond.*, **486**, 12

BONIFACE, A. (1990b). Getting to grips with COSHH. *Chem. Engr, Lond.*, **479**, 14

BONIFACE, A. (1990c). Hydrates – a factor in Piper Alpha? *Chem. Engr, Lond.*, **480**, 12

BONIFACE, A. (1990d). Piper Alpha inquiry sparks auto-ignition concern. *Chem. Engr, Lond.*, **483**, 16

BONIFACE, A. (1990e). Piper report set to reform. *Chem. Engr, Lond.*, **485**, 15

BONILLA, J.A. (1978). Estimate safe flare-headers quickly. *Chem. Engng*, **85**(Apr. 10), 135

BONNAFOUS, G. and DIVINE, R. (1986). Commissioning of a new LPG storage system at Le Havre. *Gastech 85*, p. 435

BONNE, U., HOMANN, K.H. and WAGNER, H.G. (1965). Carbon formation in premixed flames. *Combustion*, **10**, p. 503

BONNEL, W.S. and BURNS, J.A. (1964). Distillation unit startup and shutdown. In Vervalin, C.H. (1964a), *op. cit.*, p. 240

BONNER, B.H. and TIPPER, C.F.H. (1965a). The cool flame combustion of hydrocarbons. *Combust. Flame*, **9**, 317 and 387

BONNER, B.H. and TIPPER, C.F.H. (1965b). Cool-flame combustion of hydrocarbons. *Combustion*, **10**, p. 145

BONNER, W.A. (1977). Preview of new Nelson curves. *Hydrocarbon Process.*, **56**(5), 165

BONNER, W.W. (1977). Revisions to the Nelson curves, 1977. *API Proc., Refining Dept, 42 Midyear Mtg*, **56**, 3

BONNEY, M.C. and WILLIAMS, R.W. (1977). CAPABLE. A computer program to lay out controls and panels. *Ergonomics*, **20**, 297

DE BONO, E. (1985). Thinking about self-organising systems. *Computing the Magazine*, Apr. 25, 25

BONYUN, M.E. (1935). Frangible discs as protection for pressure vessels. *Trans. Am. Inst. Chem. Engrs*, **31**, 256

BONYUN, M.E. (1945). Protecting pressure vessels with rupture discs. *Chem. Met. Engng*, **42**, 260

BOOCOCK, G. (1978). Some aspects of laboratory testing of pesticides packaging. *Chemy Ind., Feb. 18*, 111

BOOGAARD, J. (1976). The handling of data from wall thickness measurements on tubes to achieve reliable operating conditions. *Diagnostic Techniques* (London: Instn Chem. Engrs)

BOOHER, H.R. (1975). Relative comprehensibility of pictorial information and printed words in procedurized instructions. *Hum. Factors*, **17**, 266

BOOIJ, C.G. (1979). The investigation into the spreading and dispersion of a heavy gas cloud by means of plant damages in the affected area. *Proc. Tenth Int. Tech. Mtg on Air Pollution Modelling and Its Application* (NATO/CCMS), p. 655

BOOK, N.L. and CHALLAGULLA, V.U.B. (2000). Inherently safer analysis of the attainable region process for the adiabatic oixidation of sulfur dioxide. *Comput. Chem. Eng.* **24**, 1421−27

BOON, D. (1983). Selection of electrical apparatus for use in explosive gas atmospheres. *Electron. Power*, Apr., 305

BOORD, C.E. (1949). The oxidation of hydrocarbons. *Combustion*, **3**, p. 416

BOOSE, J.H. (1986). *Expertise Transfer for Expert System Design* (Amsterdam: Elsevier)

BOOSER, E.R. (1983). *CRC Handbook of Lubrication*, vols 1−2 (Boca Raton, FL: CRC Press)

BOOTH, A.D., KARMARKAR, M., KNIGHT, K. and POTTER, R.C.L. (1980). Design of emergency venting system for phenolic resin reactors. *Trans. Instn Chem. Engrs*, **58**, 75 and 80

BOOTH, G. (1976). Process changes can cause accidents. *Loss Prevention*, **10**, p. 76

BOOTH, M.J. (1989). *Piper Alpha Public Inquiry. Transcript, Days 145, 146*

BOOTH, R. (1973). *Industrial Gases* (Oxford: Pergamon Press)

BOOTH, R.L. (1979). Complying with NPDES requirements. *Chem. Engng Prog.*, **75**(3), 82

BOOTMAN, J. (1983). The biology of mutagenesis and carcinogenesis and its relevance to the chemical industry. *Chemy Ind., Jan. 17*, 59

BOOZ-ALLEN APPLIED RESEARCH INC. (1973). *Analysis of LNG Marine Transportation* (Bethesda, MD)

BOPP, H.P. (1967). Startup problems in Mexico. *Chem. Engng Prog.*, **63**(12), 57

BORD, R.J. and O'CONNOR, R.E. (1990). Risk communication, knowledge and attitudes: explaining reactions to a technology perceived as risky. *Risk Analysis*, **10**, 499

BOREHAM, B.W. (1986). Preliminary wind tunnel investigation of the dispersion of pollutant particles in model building wake flows using bipolar space charge. *Atmos. Environ.*, **20**, 1523

BORER, J. (1985). *Instrumentation and Control for the Process Industries* (Amsterdam: Elsevier)

BORER, J. (1991). *Microprocessors in Process Control* (New York: Elsevier)

BORGER, G., HUNING, W., NENTWIG, H.U., SCHELLMANN, E. and SCHWEITZER, M. (1980). Hydraulic flash-back protection. *Loss Prevention and Safety Promotion 3*, vol. 3, p. 1261

BORGHESE, A., DIANA, M., MOCCIA, V. and TAMAI, R. (1991). Early growth of flames, ignited by fast sparks. *Combust. Sci. Technol.*, **76**, 219

BORGMANN, E.H. (1963). Failures in gas compressor equipment. *Hydrocarbon Process.*, **42**(10), 135

BORGNES, O. (1979a). *Effect of Insulation Faults and Uneven Distribution of Spray Water on Process Vessels Metal Temperature during Hydrocarbon Fires. Rep. 79−0759.* Det norske Veritas

BORGNES, O. (1979b). *Temperature Rise of Insulated and Uninsulated Hydrocarbon Process Vessels and Process Piping during Hydrocarbon Fires. Rep. 79−0362.* Det norske Veritas

BORGNES, O. and KARLSEN, R. (1979). *Temperature Rise of Partly Filled Insulated and Uninsulated Propane Vessels and Crude Oil Separators during Hydrocarbon Fires. Rep. 79−0541.* Det norske Veritas

BORHAUG, J.E. and MITCHELL, J.S. (1972). New tool for vibration analysis. *Hydrocarbon Process.*, **51**(11), 147

BORIS, J.P. (1976). *Flux-Corrected Transport Modules for Solving Generalized Continuity Equations. NRL Memo. Rep 3237.* Naval Res. Lab., Washington, DC

BORIS, J.P. and BOOK, D.L. (1976). Solution of continuity equations by the method of Flux-Corrected Transport. *Meth. Comput. Phys., vol. 16* (New York: Academic Press)

BORISOV, A.A., GELFAND, B.E., TIMOFEEV, E.I., TSYGANOV, S.A. and KHOMIC, S.V. (1984). Ignition of dust suspensions behind shock waves. In Bowen, J.R. *et al.*, (1984b), *op. cit.*, p. 332

BORISOV, A.A., KOGARKO, S.M. and LYUBIMOV, A.V. (1968). Ignition of fuel films behind shock waves in air and oxygen. *Combust. Flame*, **12**, 465

BORISOV, A.A. and LOBAN, S.A. (1977). Detonation limits of hydrocarbon/air mixtures in tubes. *Combust. Explosion Shock Waves*, **13**, 618

BORKOWSKI, R.J., DRAGO, J.P. and FRAGOLA, J.R. (1983). The inplant reliability data system. *Euredata 4*, paper 9.1

BORNAT, R. (1979). *Understanding and Writing Compilers* (London: Macmillan)

BORSE, E. (1979). Design basis accidents and accident analysis with particular reference to offshore platforms. *J. Occup. Accid.*, **2**, 227

BORSE, E. (1983a). Historic development and important references to the development of safety and reliability theory. *Course on Risk Analysis.* Division of Machine Design, Norwegian Inst. of Technol., Univ. of Trondheim

BORSE, E. (1983b). Risk analysis. *Course on Risk Analysis.* Division of Machine Design, Norwegian Inst. of Technol., Univ. of Trondheim

BORSE, E. (1983c). Risk analysis, example. *Course on Risk Analysis.* Division of Machine Design, Norwegian Inst. of Technol., Univ. of Trondheim

BORSETH, K., HUSE, P.T. and OLSEN, S. (1980). A deepwater NGL/LPG liquefaction and storage plant. *Gastech 79*, p. 277

BOS, P.G. and WILLIAMS, J.H. (1979). Selecting co-generation system. *Chem. Engng*, **86**(Feb. 26), 104

BOSANQUET, C.H. (1935). Eddy diffusion of smoke and gases in atmosphere. In Pearson, J.E., Nonhebel, G and Ulander, P. H.N. (1935), *op. cit.*, appendix

BOSANQUET, C.H. (1957). The rise of a hot waste gas plume. *J. Inst. Fuel*, **30**, 322

BOSANQUET, C.H., CAREY, W.F. and HALTON, E.M. (1950). Dust deposition from chimney stacks. *Proc. Inst. Mech. Engrs*, **162**, 355

BOSANQUET, C.H. and PEARSON, J.L. (1936). The spread of smoke and gases from chimneys. *Trans. Faraday Soc.*, **32**, 1249

BOSCH, F.M. and DE KEYSER, P.M. (1983). The washout of hydrogen fluoride. In de Wispelaere, C. (1983), *op. cit.*, p. 175

BOSCH, W.W. (1992). Identify, screen and mark toxic chemicals. *Chem. Engng Prog.*, **88**(8), 33

BOSCHUNG, P. and GLOR, M. (1980). Methods for investigating the electrostatic behaviour of powders. *J. Electrostatics*, **8**, 205

VAN BOSKIRK, B.A. (1982). Sensitivity of relief valves to inlet and outlet line lengths. *Chem. Engng*, **89**(Aug. 23), 77

VON BOSKIRK, B.A. (1987). Sizing and selecting conservation vents. *Chem. Engng*, **94**(17), 173

BOSMAN, J. (1980). Disaster organisation and the railways. In De Boer, J. and Baillie, T.W. (1980), *op. cit.*, p. 35

BOSMAN, M. (1985). Availability analysis of a natural gas compressor plant. *Reliab. Engng*, **11**(1), 13

BOSSART, C.J. (1974). Monitoring and control of combustible gas detectors below the lower explosive limit. *Twentieth ISA Analysis Instrumentation Symp.* (Pittsburgh, PA: Instrum. Soc. Am.), p. 71

BOSSCHE, A. (1991a). Computer aided fault tree synthesis I: system modelling and causal trees. *Reliab. Engng System Safety,* **32,** 217

BOSSCHE, A. (1991b). Computer aided fault tree synthesis II: fault tree construction. *Reliab. Engng System Safety,* **33,** 1

BOSSCHE, A. (1991c). Computer aided fault tree synthesis III: real-time fault location. *Reliab. Engng System Safety,* **33,** 161

BOSSELAAR, H. (1976). Inspection and inspection data handling for predictive maintenance. *Mtg of Materials Preservation Gp, London,* paper

BOSSERT, J.S. (1976). Instrument safety in explosive atmospheres. *ISA Trans.,* **14**(2), 122

BOSTON, J.F., BRITT, H.I. and TAYYABKHAN, M.T. (1993). Software: tackling tougher tasks. *Chem. Engng Prog.,* **89**(11), 38

BOSWORTH, M.A. (1991). Dilute phase pneumatic conveying: instrument selection guide. *Chem. Engng,* **98**(9), 166

BOTHE, H. and BRANDES, E. (1992). Testing the combustibility of water reduced liquids. *Loss Prevention and Safety Promotion 7,* vol. 2, 73–1

BOTHE, H., BRANDES, E. and REDEKER, T. (1986). Investigation of the formation of explosive heavy fuel oil vapour/air mixtures. *Loss Prevention and Safety Promotion 5,* p. 57–1

BOTHE, H. and STEEN, H. (1989). The ignition of flammable vapours by hot pipes and plates. *Loss Prevention and Safety Promotion 6,* p. 85.1

BOTKIN, L.A. (1981). Chemical transport over the highway: equipment design and regulations. *Loss Prevention,* **14,** p. 44

BOTT, T.F. and HAAS, P.M. (1978). Initial data collection efforts of CREDO: sodium valve failures. *Fifth Symp. on Reliability Technology, Univ. of Bradford*

BOTT, T.R. (1990). *Fouling Notebook* (Rugby: Instn Chem. Engrs)

BOTTELBERGHS, P.H. (1995). *QRA in the Netherlands. Safety Cases* (London: IBC Technical Services)

BOTTERILL, J.A., WILLILAMS, I. and WOODHEAD, T.J. (1984). The transient release of pressurised LPG from a small diameter pipe. *The Protection of Exothermic Reactors and Pressurised Storage Vessels* (Rugby: Instn Chem. Engrs), p. 265

BOUBLIK, T., FRIED, V. and HALA, E. (1984). *The Vapour Pressures of Pure Substances* (Amsterdam: Elsevier)

BOUCHARD, J.G., PAYNE, P.A. and SZYSZKO, S. (1994). Non-invasive measurement of process states using acoustic emission techniques coupled with advanced signal processing. *Chem. Engng Res. Des.,* **72A,** 19

BOUCHER, D.F. and ALVES, G.E. (1959–). Dimensionless numbers. *Chem. Engng Prog.,* 1959, **55**(9), 55; 1963, **59**(8), 75

BOUCKAERT, M. and CAPPOEN, L. (1984). *Maritime Transport of Liquefied Gases. Transport and Storage of LPG and LNG* (Antwerp: Koninklijke Vlaamse Ingenieursvereniging), p. 209

BOUDART, M. (1968). *Kinetics of Chemical Processes* (Englewood Cliffs, NJ: Prentice-Hall)

BOUGHEN, D.M. (1979). Designing a Halon 1301 system. *Fire Surveyor,* **8**(5), 21

BOUGHEN, D.M. (1980). Extinguishing natural gas fires. *Fire Surveyor,* **9**(1), 29

BOUGHEN, D.M. (1982). Piped dry chemical fire suppression systems. *Fire Surveyor,* **11**(3), 33

BOUGHEN, D.M. (1983). Fixed foam fire suppression systems. *Fire Surveyor,* **12**(3), 5

BOUHAFID, A., VANTELON, J.P., JOULAIN, P. and FERNANDEZ-PELLO, A.C. (1989). On the flame structure at the base of a pool fire. *Combustion,* **22,** 1291

BOUHAFID, A., VANTELON, J.P., SOUIL, J.M., BOSSEBOEUF, G. and RINGERE, F.X. (1989). Characterisation of thermal radiation from freely burning oil pool fires. *Fire Safety J.,* **15,** 367

BOULANGER, A. and LUYTEN, W. (1983a). Pilot project shows liquefied gas is stored successfully at low temperatures in clay. *Oil Gas J.,* **81**(Mar. 7), 118

BOULANGER, A. and LUYTEN, W. (1983b). Underground storage of liquefied gas at low temperature. *Gastech 82,* p. 409

BOULT, D.M., PITBLADO, R.M. and KENNEY, G.D. (2001). Lessons learned from the explosion and fire at the Esso gas processing plant at Longford, Australia, *AIChE Loss Prev. Symp.*

BOULT, M.A. and NAPIER, D.H. (1972). Behaviour of polyurethane foams under controlled heating. *Fire Prev. Sc. Technol.,* **3,** 13

BOULT, M.A., GAMADIA, R.K. and NAPIER, D.H. (1972). Thermal degradation of polyurethane foams. *Chemical Process Hazards,* **4,** p. 56

BOULTON, P.I.P. and KITTLER, M.A.R. (1979). Estimating program reliability. *Comput. J.,* **22,** 328

BOUMANNS, A.A. (1957). *Physica,* **23,** 1007

BOURDEAU, M. (1991). The CD-REEF: the French building technical rules on CR-ROM. In Kahkonen, K. and Bjork (1991), *op. cit.*

BOURDEAU, P. and GREEN, G. (1989). *Methods for Assessing and Reducing Injury from Chemical Accidents* (New York: Wiley)

BOURDELON, T.R. and LAI, Y.S. (1979). Vibration effects on pressure relief valve function. In Haupt, R.W. and Meyer, R.A. (1979), *op. cit.,* p. 179

BOURGEOIS, J., CANDES, P., CLEMENT, B., COSTES, D., DELCHAMBRE, P., OULLION, J. and PETIT, J. (1971). Evolution of French ideas on the safety of reactors. *Proc. Fourth Int. Conf. on Peaceful Uses of Atomic Energy* (New York: United Nations and Vienna: Int. Atom. Energy Agency), p. 163

BOURGUET, J.M. (1968). Selection of refrigerating compressor driving machines for large-capacity natural gas liquefaction installations. *LNG Conf.,* **1**

BOURGUET, J.M. (1972). Experience of Arzew and its effect on the design of the Skikda natural gas liquefaction Operations Prog. *LNG Conf.,* **3**

BOURGUET, J.M. (1981). LNG cold: still a problem. *Hydrocarbon Process.,* **60**(1), 167

BOURGUET, J.M., GARNAUD, R. and GRENIER, M. (1970). Large capacity LNG installations. *LNG Conf.,* **2**

BOURLIOUX, A. and MAJDA, A.J. (1992). Theoretical and numerical structure for unstable two-dimensional detonations. *Combust. Flame,* **90,** 211

BOURNE, A.J. (1967). A criterion of the reliability assessment of protective systems. *Control,* **11**(12), 495

BOURNE, A.J. (1970). Techniques of quantitative reliability analysis. *Nucl. Engng Des.,* **13**(2), 191

BOURNE, A.J. (1972). Measuring reliability. *Electron. Power,* **18,** 181

BOURNE, A.J. (1982). Probability assessment. In Green, A.E. (1982), *op. cit.,* p. 73

BOURNE, J.R., BROGLI, F., HOCH, F. and REGENASS, W. (1987). Heat transfer from exothermically reacting fluid in vertical unstirred vessels. *Hazards from Pressure,* p. 315

BOURNE, J.R. and TOVSTIGA, G. (1988). Micromixing and fast chemical reactions in a turbulent reactor. *Chem. Engng Res. Dev.*, **66**, 26

BOURTON, K. (1967). *Chemical and Process Engineering Unit Operations* (London: MacDonald)

BOVET, D.M. (1973). *Preliminary Analysis of Tanker Grounding and Collisions*. US Coast Guard

BOWDEN, B.V. (ed.) (1953). *Faster than Thought* (London: Pitman)

BOWDEN, F.P. (1949). Initiation of solid explosives by impact and friction. *Proc. R. Soc., Ser. A.* **198**, 337

BOWDEN, F.P. (1953). The development of combustion and explosion in liquids and solids. *Combustion*, **4**, p. 161

BOWDEN, F.P., MULCAHY, M.F.R., VINES, R.G. and YOFFE, A. (1947). The detonation of liquid explosives by gentle impact. The effect of minute gas spaces. *Proc. R. Soc., Ser. A.*, **188**, 291

BOWDEN, F.P. and TABOR, D. (1954). *The Friction and Lubrication of Solids* (Oxford: Clarendon Press)

BOWDEN, F.P. and THOMAS, P.H. (1954). The surface temperature of sliding. *Proc. R. Soc., Ser. A.*, **223**, 29

BOWDEN, F.P., STOKE, M.A. and TUDOR, G.K. (1947). Hot spots on rubbing surfaces and the detonation of explosives by friction. *Proc. R. Soc.*, **188**, 329

BOWDEN, F.P. and YOFFE, A. (1949). Hot spots and the initiation of explosions. *Combustion*, **3**, p. 551

BOWDEN, F.P. and YOFFE, A.D. (1952). *Initiation and Growth of Explosions in Liquids and Solids* (Cambridge: Cambridge Univ. Press)

BOWDEN, F.P. and YOFFE, A.D. (1958). *Fast Reactions in Solids* (London: Butterworths)

BOWDERS, J.J., KOERNER, R.M. and LORD, A.E. (1982). Buried container detection using ground-probing radar. *J. Haz. Materials*, **7**(1), 1

BOWDITCH, M., DRINKER, C.K., DRINKER, P., HAGGARD, H.H. and HAMILTON, A. (1940). Code for safe concentrations of certain common toxic substances used in industry. *J. Ind. Hyg. Toxicol.*, **22**, 251

BOWEN, D.G. (1983). Low-level ethylene decompositions – an inconvenience or a warning? *Loss Prevention and Safety Promotion* 4, vol. 1, p. M1

BOWEN, E.R., ROTONDI, R.H. and REID, J.V. (1980). Ferrography in the process industries: predicting incipient machine failure. *Ammonia Plant Safety*, **22**, p. 213

BOWEN, G. (1984). Gas compressors. *Plant Engr*, **28**(6), 20

BOWEN, H.H. (1967). The 'imp' in the system. In Singleton, W.T., Easterby, R.S. and Whitfield, D.C. (1967), *op. cit.*, p. 12

BOWEN, I.G., ALBRIGHT, R.W., FLETCHER, E.R. and WHITE, C.S. (1961). *A Model Designed to Predict the Motion of Objects Translated by Classical Blast Waves. Rep. CEX-58.9*. Civil Effects Test Operations, Atom. Energy Comm., Office of Tech. Services, Dept of Commerce, Washington, DC

BOWEN, I.G., FLETCHER, E.R. and PERRETT, R.F.D. (1965). *Translation of Goats and Anthropomorphic Dummies by Blast Waves. Rep. DASA-1642–1*, p. 454. DASA Data Center, Defense Atom. Support Agency, Dept of Defense, Santa Barbara, CA

BOWEN, I.G., FLETCHER, E.R. and RICHMOND, D.R. (1968). *Estimate of Man's Tolerance to the Direct Effects of Air Blast. Rep. DASA-2113*. Defense Atom. Support Agency, Dept of Defense, Washington, DC

BOWEN, I.G., FLETCHER, E.R., RICHMOND, D.R., HIRSCH, F.G. and WHITE, C.S. (1968). *Biophysical Mechanisms and Scaling Procedures Applicable in Assessing Responses of the Thorax Energized by Air-blast Overpressures or Non-penetrating Missiles. REP. DASA-1857*. Defense Atom. Support Agency, Dept of Defense, Washington. See also *Ann. N.Y. Acad. Sci*, 1968, **152** art. 1, 122

BOWEN, I.G., FRANKLIN, M.E., FLETCHER, E.R. and ALBRIGHT, R.W. (1963a). *Secondary Missiles Generated by Nuclear-produced Blast Waves. Rep. WT-1468*. Civil Effects Test Group, US Atom. Energy Comm., Office of Tech. Services, Dept of Commerce, Washington, DC

BOWEN, I.G., FRANKLIN, M.E., FLETCHER, E.R. and ALBRIGHT, R.W. (1963b). Secondary missiles generated by nuclear produced blast waves. *Report to Test Director, Operation PLUMBOB, Program 33*

BOWEN, I.G., RICHMOND, D.R., WETHERBE, M.B. and WHITE, C.S. (1956). *Biological Effects of Blast from Bombs. Glass Fragments as Penetrating Missiles and some of the Biological Implications of Glass Fragmented by Atomic Explosions. Rep. AECU-3350*. US Atom. Energy Comm., Office of Tech. Services, Dept of Commerce, Washington, DC

BOWEN, I.G., STREHLER, A.F. AND WETHERBE, M.B. (1956). *Distribution and Density of Missiles from Nuclear Explosions. Rep. WT-1168*. Civil Effects Test Group, US Atom. Energy Comm., Office of Tech. Services, Dept of Commerce, Washington, DC

BOWEN, I.G., WOODWORTH, P.B., FRANKLIN, M.E. and WHITE, C.S. (1962). *Translational Effects from High Explosives. Rep. DASA-1336*. Defense Atom. Support Agency, Dept of Defense, Washington, DC

BOWEN, J. (1980). A computer-based system for the management of operational factors on major hazard plants. *Course on Loss Prevention in the Process Industries*. Dept of Chem. Engng, Loughborough Univ. of Technol.

BOWEN, J.H. (1976). Individual risk vs public risk criteria. *Chem. Engng Prog.*, **72**(2), 63

BOWEN, J.H. (1977). The application of an event data store to safe operation. *Loss Prevention and Safety Promotion* 2, p. 385

BOWEN, J.H. (1978). In Health and Safety Executive (1978b), *op. cit.*, Appendix 24

BOWEN, J.H. (1980). Missile problems in the chemical and nuclear power industries. *Nucl. Engng*, **19**, 149

BOWEN, J.H. and MASTERS, E.F. (1959). *Nuclear Reactor Control and Instrumentation* (London: Temple Press)

BOWEN, J.R., MANSON, N., OPPENHEIM, A.K. and SOLOUKHIN, R.I. (eds) (1984a). *Dynamics of Flames and Reactive Systems. (Proc. Ninth Int. Coll. on Gasdynamics of Explosions and Reactive Systems, Poitiers, 1983; Prog. Astronaut. and Aeronaut., vol. 95)* (New York: Am. Inst. Aeronaut. Astronaut.)

BOWEN, J.R., MANSON, N., OPPENHEIM, A.K. and SOLOUKHIN, R.I. (eds) (1984b). *Dynamics of Shock Waves, Explosions, and Detonations (Proc. Ninth Int. Coll. on Gasdynamics of Explosions and Reactive Systems, Poitiers, 1983; Prog. Astronaut. Aeronaut., vol. 94)* (New York: Am. Inst. Aeronautics and Astronaut.)

BOWEN, J.R., RAGLAND, K.W., STEFFES, F.J. and LOFLIN, T.G. (1971). Heterogenous detonation supported by fuel fogs or films. *Combustion*, **13**, p. 1131

BOWEN, J.M. and CAMPBELL, J.D. (1986). Metallurgical examination of existing piping in an ammonia plant. *Ammonia Plant Safety*, **26**, p. 39. See also *Plant/Operations Prog.*, 1986, **5**, 81

BOWEN, P.J. and SHIRVILL, L.C. (1994a). Combustion hazards posed by the pressurized atomization of high-flash-point liquids. *J. Loss Prev. Process Ind.*, **7**, 233

BOWEN, P.J. and SHIRVILL, L.C. (1994b). Pressurised atomisation of high flashpoint liquids – implications for hazardous area classification. *Hazards*, **XII**, p. 27

BOWES, P.M., OSWALD, C.J., VARGAS, L.M. and BAKER, W.E. (1991). *Building Debris Hazard Prediction Model. Final Report. Proj. SWRI 06–2945.* Southwest Res. Inst., San Antonio, TX

BOWER, J.N. and PETERSON, K.R. (1963). Guide to steam tracing design. *Hydrocarbon Process.*, **42**(3), 149

BOWER, J.S. and SULLIVAN, E.J. (1981). Polar isopleth diagrams: a new way of presenting wind and pollution data. *Atmos. Environ.*, **15**, 537

BOWERS, J.R., MANSON, N., OPPENHEIM, A.K. and SOLOUKIN, R.I. (eds) (1981). Gasdynamics of detonations and explosions. *Prog. Astronaut. Aeronaut.*, **75**

BOWERS, J.R., MANSON, N., OPPENHEIM, A.K. and SOLOUKIN, R.I. (eds) (1983). Shock waves, explosions and detonations. *Prog. Astronaut. Aeronaut.*, **87**

BOWERS, W.J., SHEPPARD, G.T. and WILLIAMS, J.A. (1981). Reliability in irradiated fuel processing. *Reliable Production in the Process Industries* (Rugby: Instn Chem. Engrs), p. 13

BOWES, C. (1985). Academic accidents. *Occup. Safety and Health*, Feb., **32**

BOWES, P.C. (1954). Factors limiting general application of the Mackey test for spontaneous heating and ignition. *J. Appl. Chem.*, **4**, 140

BOWES, P.C. (1962). Spontaneous heating. *Encyclopaedic Dictionary of Physics* (ed. J. Thewlis) (Oxford: Pergamon Press)

BOWES, P.C. (1968). Thermal explosion of benzoyl peroxide. *Combust. Flame*, **12**, 289

BOWES, P.C. (1969). Thermal ignition in two-component systems, theoretical model. *Combust. Flame*, **13**, 521

BOWES, P.C. (1972). Thermal ignition in two component systems. Pt II – Experimental study. *Combust. Flame*, **19**, 55

BOWES, P.C. (1974a). Fires in oil-soaked lagging. *Fire Prev. Sci. Technol.*, **9**, 13

BOWES, P.C. (1974b). The thermal explosion of unstable substances. *Loss Prevention and Safety Promotion* 1, p. 279

BOWES, P.C. (1976). Hazards associated with self-heating of solids. *Process Industry Hazards*, p. 93

BOWES, P.C. (1981). A general approach to the prediction and control of potential runaway reaction. *Runaway Reactions, Unstable Products and Combustible Powders* (London: Intn Chem. Engrs), paper 1/A

BOWES, P.C. (1984). *Self-Heating: Evaluating and Controlling the Hazards* (Amsterdam: Elsevier)

BOWES, P.C., BURGOYNE, J.H. and RASBASH, D.J. (1948). The inflammability in suspension of mixtures of combustible and incombustible dusts. *J. Soc. Chem. Ind.*, **67**, 125

BOWES, P.C. and LANGFORD, B. (1968). Spontaneous ignition of oil-soaked lagging. *Chem. Process Engng*, **49**(5), 108

BOWES, P.C. and THOMAS, P.H. (1966). Ignition and extinction phenomena accompanying oxygen dependent self-heating of porous bodies. *Combust. Flame*, **10**, 221

BOWES, P.C. and TOWNSHEND, S.E. (1962). Ignition of combustible dusts on hot surfaces. *Br. J. Appl. Phys.*, **13**, 105

BOWKER, A.H. and LIEBERMAN, G.J. (1959). *Engineering Statistics* (Englewood Cliffs, NJ: Prentice-Hall)

BOWLES, P.M., OSWALD, C.J., VARGAS, L.M. and BAKER, W.E. (1991). Building debris hazard prediction model. *Final report. Southwest Res. Inst., San Antonio, TX, Rep. SWRI Proj. 06–2945*

BOWLING, S.A. (1985). Modifications necessary to use standard dispersion models at high latitudes. *Atmos. Environ.*, **19**, 93

BOWMAN, C.T. (1970). An experimental and analytical investigation of the high temperature oxidation mechanisms of hydrocarbon fuels. *Combust. Sci. Technol.*, **2**, 161

BOWMAN, J.A. (1989). The consequences of Section 313. *Chem. Engng Prog.*, **85**(6), 48

BOWMAN, J.C.R. and PALMER, K. (1974). Insurance aspects of the oil and chemical industries. *Loss Prevention and Safety Promotion* 1, p. 133

BOWMAN, J.D. (1991). Use column scanning for predictive maintenance. *Chem. Engng Prog.*, **87**(2), 25

BOWMAN, J.W. and PERKINS, R.P. (1990). Preventing fires with high temperature vaporizers. *Plant/Operations Prog.*, **9**, 39

BOWMAN, M.C. (1979). *Carcinogens and Related Substances: Analytical Chemistry for Toxicological Research* (New York: Dekker)

BOWMAN, M.C. (ed.) (1982). *Handbook of Carcinogens and Hazardous Substances* (New York; Dekker)

BOWNE, N.E. (1961). Some measurements of diffusion parameters from smoke plumes. *Bull. Am. Met. Soc.*, **42**, 101

BOWNE, N.E. (1974). Diffusion rates. *J. Air Poll. Control Ass.*, **24**, 832

BOWNE, N.E. and BALL, J.T. (1970). Observational comparison of rural and urban boundary layer turbulence. *J. Appl. Met.*, **9**, 862

BOWNE, N.E., BALL, J.T. and ANDERSON, G.E. (1968). *Some Measurements of the Atmospheric Boundary Layer in an Urban Complex. Rep. TRC-7237–295 (AD-842951L).* Travelers Res. Center

BOWNE, N.E., CHA, S. and MURRAY, D.R. (1978). Use of fluctuating plume puff model for the prediction of the impact of odorous emissions. *Ann. Mtg Air Poll. Control Ass., Houston, Tex., paper 787–68.61*

BOWNE, N.E. and YOCOM, J.E. (1979). A chemical engineer's guide to meteorology. *Chem. Engng*, **86**(Jul. 30), 57

BOWONDER, B. (1985). The Bhopal incident: implications for developing countries. *The Environmentalist*, **5**, 89

BOWONDER, B. (1987). An analysis of the Bhopal accident. *Project Appraisal*, **2**(3), 157

BOWONDER, B., ARVIND, S.S. and MIYAKE, T. (1991). Low probability-high consequence accidents: application of systems theory for preventing hazardous failures. *Systems Res.*, **9**, 5

BOWONDER, B., KASPERSON, J.X. and KASPERSON, R.E. (1985). Avoiding future Bhopals. *Environment*, **27**(7), 6

BOWONDER, B. and MIYAKE, T. (1988). Managing hazardous facilities: lessons from the Bhopal incident. *J. Haz. Materials*, **19**, 237

BOX, G.E.P. and CULLUMBINE, H. (1947). The relationship between survival time and dosage with certain toxic gases. *Br. J. Pharmacol.*, **2**, 27

BOXHORN, L.C. (1979). Remote multiplexing systems for process control and monitoring. *Chem. Engng*, **86** (Oct. 15), 73

BOYCE, M.G. (1990). Halon systems and the COSHH Regulations. *Fire Surveyor*, **19**(4), 9

BOYCE, M.P. (1978). How to achieve on-line availability of centrifugal compressors. *Chem. Engng*, **85**(Jun. 5), 115

BOYD, B.D. (1978). The augmentation of the pollution incident reporting system designed for spillage prevention. *Control of Hazardous Material Spills*, **4**, 46

BOYD, D. and LEOTTA, J. (1976). Prevention: the usage of the United States Coast Guard's pollution incident reporting system in program management. *Control of Hazardous Material Spills*, **3**, p. 94

BOYD, E.M., MACLACHLAN, M.L. and PERRY, W.F. (1944). Experimental ammonia poisoning in rabbits and cats. *J. Ind. Hyg. Toxicol.*, **26**, 29

BOYD, G.M. (1970). Brittle Fracture of Steel Structures (London: Butterworths)

BOYD, M.G., MARSHALL, T.P., MARTIN, F.J. and NOESEN, S.J. (1981). Explosion pressures in enclosures compartmented by porous barriers. *Combustion*, **18**, p. 1683

BOYDELL, T.H. (1970). *A Guide to Job Analysis* (Br. Ass. Commercial and Ind. Educ.)

BOYDELL, T.H. (1976). *A Guide to the Identification of Training Needs* (Br. Ass. Commercial and Ind. Educ.)

BOYEN, J.L. (1975). *Practical Heat Recovery* (New York: Wiley)

BOYEN, V.E., BRANDES, R.L., BURK, A.F. and BURNS, A.W. (1987). Process hazards management: the key elements. *Chemical Manufacturers Assn Conf. on Toxic Hazards* (Washington, DC)

BOYER, C.B. (1984). High pressure codes and standards. *Plant/Operations Prog.*, **3**, 3

BOYER, D.W., BRODE, H.L., GLASS, I.I. and HALL, J.G. (1958). *Blast from a Pressurized Sphere. Rep. UTIA 48.* Univ. of Toronto, Inst. of Aerophysics

BOYES, J.A. (1975). Aids to safety. *Plant Engr*, **19**(10), 26

BOYETT, J.L. (1979). Finding and defusing ethylene plant booby traps. *Loss Prevention*, **12**, p. 5

BOYKIN, R.F. and LEVARY, R.R. (1989). Risk analysis using simulation for a chemical storage problem. *J. Loss Prev. Process Ind.*, **2**(2), 108

BOYKIN, R.F. and KAZARIANS, M. (1987). Quantitative risk assessment for chemical operations. *Preventing Major Chemical Accidents*, p. 1.87

BOYLAND, E. (1977). The importance of occupational cancer. *Chem. Engr, Lond.*, **323**, 569

BOYLAND, E., MCDONALD, F.F. and RUMENS, M.J. (1946). The variation in the toxicity of phosgene for small animals with the duration of exposure. *Br. J. Pharmcol.*, **1**, 81

BOYLE, A.R. and LLEWELLYN, F.J. (1947). The electrostatic ignitability of various solvent vapour-air mixtures. *J. Soc. Chem. Ind.*, **66**, 99

BOYLE, A.R. and LLEWELLYN, F.J. (1950a). The electrification of metal dusts. *J. Chem. Soc.*, **69**, 45

BOYLE, A.R. and LLEWELLYN, F.J. (1950b). The electrostatic ignitability of dust clouds and powders. *J. Soc. Chem. Ind.*, **69**, 173

BOYLE, W.J. (1967). Sizing relief area for polymerization reactors. *Loss Prevention*, **1**, 78. See also *Chem. Engng Prog., 1967*, **63**(8), 61

BOYNE, W.J. (1966). Liquid oxygen disposal vessel explosion. *Ammonia Plant Safety*, **8**, 7

BOYNTON, D.E., NICHOLS, W.B. and SPURLIN, H.M. (1959). How to tame dangerous chemical reactions. *Ind. Engng Chem.*, **51**, 489

BP PETROLEUM DEVELOPMENT LTD (1982a). *The Forties Oil Field* (London)

BP PETROLEUM DEVELOPMENT LTD (1982b). *Sullom Voe Terminal* (London)

BP PETROLEUM DEVELOPMENT LTD (1983a). *The Buchan Oil Field* (London)

BP PETROLEUM DEVELOPMENT LTD (ca. 1983b). *Eighty Years Experience in Exploration and Production* (London)

BP PETROLEUM DEVELOPMENT LTD (ca. 1983c). *The Southeast Forties Development* (London)

BP PETROLEUM DEVELOPMENT LTD (1984a). *The Magnus Oil Field* (London)

BP PETROLEUM DEVELOPMENT LTD (1984b). *The West Sole Gas Field* (London)

BP PETROLEUM DEVELOPMENT LTD (ca. 1985). *Floating Production Systems* (London)

BP PETROLEUM DEVELOPMENT LTD (1988). *Miller Gas System. Proposed St Fergus to Boddam Pipeline Environmental Assessment* (Aberdeen)

BP PETROLEUM DEVELOPMENT LTD (ca. 1990a). *The Beatrice Oilfield* (London)

BP PETROLEUM DEVELOPMENT LTD (ca. 1990b). *The Buchan Oilfield* (London)

BP PETROLEUM DEVELOPMENT LTD (ca. 1990c). *The Clyde Oilfield* (London)

BP PETROLEUM DEVELOPMENT LTD (ca. 1990d). *The Forties Oilfield* (London)

BP PETROLEUM DEVELOPMENT LTD (ca. 1990e). *The Magnus Oilfield* (London)

BP PETROLEUM DEVELOPMENT LTD (ca. 1990f). *The Southern North Sea Fields* (London)

BP PETROLEUM DEVELOPMENT LTD (ca. 1990g). *The Thistle Oilfield* (London)

BP TRADING LTD (1980). *Requirements for the Protection of Diesel Engines Operating in Zone 2 Hazardous Areas, BP Std 200* (London)

BRABY, C.D., HARRIS, D. and MUIR, H.C. (1993). A psychophysiological approach to the assessment of work underload. *Ergonomics*, **36**, 1035

BRADBURY, F. (1980). Democratic accountability and the votes for nuclear energy. *Sci. Publ. Policy*, **7**, 346

BRADBURY, R. (1982). Reliability: its application to maintenance decision-making. *Plant Engr*, **26**(2), 18

BRADFORD, B.W. and FINCH, G.I. (1937). The mechanism of ignition by electric discharges. *Combustion*, **2**, 112

BRADFORD, H. (1986a). EPA's risk-assessment guidelines take shape despite OMB opposition. *Chem. Engng*, **93**(19), Oct. 13, 19

BRADFORD, H. (1986b). OSHA's new program gets tepid reviews. *Chem. Engng*, **93**(17), 24

BRADFORD, M. (1987). Process plants as hazardous-waste facilities. *Chem. Engng*, **94**(3), 69

BRADFORD, J.R. and ELLIOTT, T.R. (1915–16). Cases of gas poisoning among the British troops in Flanders. *Br. J. Surg.*, **3**, 234

BRADFORD, M. and DURRETT, D.G. (1984). Avoiding common mistakes in sizing distillation safety valves. *Chem. Engng*, **91**(Jul. 9), 78

BRADFORD, W.J. (1985). Standards on safety: how helpful are they for practicing engineers? *Chem. Engng Prog.*, **81**(8), 16

BRADFORD, W.J. (1986a). Chemicals. *Fire Protection Handbook*, 16th ed. (ed. by A.E. Cote and J.M. Linville), p. 5–59

BRADFORD, W.J. (1986b). Storage and handling of chemicals. *Fire Protection Handbook*, 16th ed. (ed. A.E. Cote and J.L. Linville), p. 11–49

BRADFORD, W.J. and CULBERTSON, T.L. (1967). Design of control houses to withstand explosive forces. *Loss Prevention*, **1**, 28

BRADING, J.E. (1989). *Piper Alpha Public Inquiry. Transcript, Days 126, 127*

BRADLEY, B.W. (1985). Coalescers offer unique design features. *Oil Gas J.*, **83**(Nov. 18), 136

BRADLEY, C.I. and CARPENTER, R.J. (1983). The simulation of dense vapour cloud dispersion using wind tunnels and water flumes. *Loss Prevention and Safety Promotion 4*, vol. 1, p. F56

BRADLEY, C.I., CARPENTER, R.J., WAITE, P.J., RAMSAY, C.G. and ENGLISH, M.A. (1983). Recent development of a simple box-type model for dense vapour cloud dispersion. In Hartwig, S. (1983), *op. cit.*, p. 77

BRADLEY, C.W.J. (n.d.). *Criteria for Plant Separation Distances and Location*, pts 1–2. Dept of Chem. Engng, Polytechnic of the South Bank, London

BRADLEY, C.W.J. (1985). Criteria for plant separation distances and location. *The Assessment and Control of Major Hazards* (Rugby: Instn Chem. Engrs), p. 247

BRADLEY, D. (1979). Unconfined vapour cloud explosions. *Fire Prev. Sci. Technol.*, **21**, 13

BRADLEY, D. (1982). Turbulent flame. In Lee, J.H.S. and Guirao, C.M. (1982), *op. cit.*, p. 971

BRADLEY, D. (1992). How fast can we burn? *Combustion*, **24**, 247

BRADLEY, D., CHEN, Z. and SWITHENBANK, J.R. (1989). Burning rates in turbulent fine dust–air explosions. *Combustion*, **22**, 1767

BRADLEY, D., GASKELL, P.H. and GU, X.J. (1994). Application of a Reynolds stress, stretched flamelet, mathematical model to computations of turbulent burning velocities and comparison with experiments. *Combust. Flame*, **96**, 221

BRADLEY, D. and HUNDY, G.F. (1971). Burning velocities of methane-air mixtures using hot-wire anemometers in closed vessel explosions. *Combustion*, **13**, 575

BRADLEY, D., LAU, A.K.C. and LAWES, M. (1992). Flame stretch rate as a determinant of burning velocity. *Phil Trans R. Soc. Lond., Ser. A.*, **338**, 359

BRADLEY, D. and MITCHESON, A. (1976). Mathematical solutions for explosions in spherical vessels. *Combust. Flame*, **26**, 201

BRADLEY, D. and MITCHESON, A. (1978). The venting of gaseous explosions in spherical vessels. *Combust. Flame*, **32**, 221 and 237

BRADLEY, J. (1986). The changing patterns of gas seaborne trades. *Gastech 85*, p. 93

BRADLEY, J.N. (1962). *Shock Waves in Chemistry and Physics* (London: Methuen)

BRADLEY, J.N. (1969). *Flame and Combustion Phenomena* (London: Methuen)

BRADLEY, J.R. (1973). Future ports – onshore and offshore. *Conf. on Bulk Transportation of Hazardous Materials by Water in the Future* (Washington, DC: Nat Acad. Sci., Nat. Acad. Engng, Nat. Res. Coun.)

BRADLEY, J.R. and NIMMO, N.M. (1968). Design your large ammonia plant for start-up. *Ammonia Plant Safety*, **10**, 84

BRADSHAW, P. (1971). *Introduction to Turbulence and its Measurement* (Oxford: Pergamon Press)

BRADSHAW, P. (ed.) (1978). *Turbulence* (Berlin: Springer)

BRADY, G.S. and CLAUSER, H.R. (1978). *Materials Handbook* (New York: McGraw-Hill)

BRADY, S.O. (1959). In-plant pollution control in practice. 1, The petroleum industry. *Chem. Engng Prog.*, **55**(3), 45

BRAGDON, C.R. (1970). *Noise Pollution* (Philadelphia, PA: Univ. of Pennsylvania Press)

BRAHMBHATT, S.R. (1984). Are liquid thermal-relief valves needed? *Chem. Engng*, **91**(May 14), 69

BRAITHWAITE, J.C. (1985). Operational pipeline inspection. *The Assessment and Control of Major Hazards* (Rugby: Instn Chem. Engrs), p. 45

BRAKER, W. and MOSSMAN, A.L. (1970). *Effects of Exposure to Toxic Gases. First aid and Medical Treatment*. Matheson Gas Products, East Rutherford, NJ

BRAKESTAD, H. and WIIK, T. (1979). Riser rupture, causes and consequences. Pt 2, *The Influence of Surface Flaws on the Strength of Pressurised Steel Pipes. Rep. 79–0303*. Det Norske Veritas

BRAMBLE, D.M. (1983). Respiratory patterns and control in unrestrained human running. In Whipp, B.J. and Wiberg, D.M. (1983), *op. cit.*, p. 213

BRAMER, M.A. (1983). A survey and critical review of expert systems research. In Hayes, P.J. and Michie, D. (1983), *op. cit.*, p. 3

BRAMER, M.A. (1984). Expert systems: the vision and the reality. *Research and Development in Expert Systems* (ed. by R.A. Bramer) (Cambridge: Cambridge Univ. Press), p. 1

BRAMLEY-HARKER, R. (1969). Working with machinery. In Handley, W. (1969), *op. cit.*, p. 279

BRAMLEY-HARKER, R. (1977). Working with machinery. In Handley, W. (1977), *op. cit.*, p. 156

BRANCH, J. (1986). Interacting with computers – expert systems. *Chemy Ind.*, May **5**, 321

BRANCHEREAU, P. and BONJOUR, E. (1986). The CHAGAL offshore refrigerated LPG terminal – latest developments and applications. *Gastech 85*, p. 409

BRAND, M.R. (1980). An examination of certain Bayesian methods used in reliability analysis. *Reliab. Engng*, **1**(2), 115

BRAND, R.R. and BURGESS, F.L. (1973). Reducing hazards of handling reactive catalysts. *Chem. Engng*, **80**(Nov. 12), 248

BRANDEIS, J. (1985). Effect of obstacles on flames. *Combust. Sci. Technol.*, **44**, 61

BRANDEIS, J. and BERGMANN, D.J. (1983). A numerical study of tunnel fires. *Combust. Sci. Technol.*, **35**, 133

BRANDIE, E.F. (1989a). *Piper Alpha Public Inquiry. Transcript, Day 139*

BRANDIE, E.F. (1989b). *Piper Alpha Public Inquiry. Transcript, Days 143, 144*

BRANDT, D., GEORGE, W., HATHAWAY, C. and MCCLINTOCK, N. (1992). Plant layout: the impact of codes, standards and regulations. *Chem. Engng*, **99**(4), 89

BRANDYBERRY, M.D. and APOSTOLAKIS, G.E. (1991a). Fire risk in buildings: frequency of exposure and physical model. *Fire Safety J.*, **17**, 339

BRANDYBERRY, M.D. and APOSTOLAKIS, G.E. (1991b). Fire risk in buildings: scenario definition and ignition frequency calculations. *Fire Safety J.*, **17**, 363

BRANNEGAN, D.P. (1985). Hazards evaluation in process development. In Hoffman, J.M. and Maser, D.C. (1985), *op. cit.*, p. 17

BRANNON, A.R. (1974). Preplanning for emergencies in the chemical industry. *Fire Engrs J.*, **34**(94), 18

BRANNON, A.R. (1976). Disaster planning for the chemical industry. *Course on Fire and Explosion Safety in the Chemical Industry*. Dept of Fire Safety Engng, Univ. of Edinburgh,

BRANTLEY, J.P. (1964). Tank truck trilogy. *Chem. Engng Prog.*, **60**(3), 32

BRASIE, W.C. (1976a). The hazard potential of chemicals. *Loss Prevention*, **10**, 135

BRASIE, W.C. (1976b). In discussion of paper by Vincent, G.C. *et al.* (1976), *op. cit.*, p. 71

BRASIE, W. (1981). Guidelines for estimating damage from grain dust explosions. *The Hazards of Industrial Explosion from Dusts* (London: Oyez/IBC), paper 8

BRASIE, W.C. and SIMPSON, D.W. (1968). Guidelines for estimating damage explosion. *Loss Prevention*, **2**, 91

BRATKO, I. (1986). *Prolog Programming for Artificial Intelligence* (Wokingham: Addison-Wesley)

BRAUBACH, H.O. (1982). Die Störfallverordnung. *Chem. Ing. Tech.*, **54**(4), A236

BRAUER, F.E., GREEN, N.W. and MESLER, R.B. (1968). Metal/water explosions. *Nucl. Sci. Engng*, **31**, 551

BRAUN, J. (1979). Flour dust explosion kills 12. *Fire Int.*, **65**, 18

BRAUN, R.L. (1978). Low-temperature processing in pilot plants. *Chem. Engng*, **85**(Jan. 16), 129

BRAUNTON, P.N. (1980). Corrosion problems and answers in hot water and steam handling and condensers. *Interflow 80* (Rugby: Instn Chem. Engrs)

BRAUTIGAM, F.C. (1977). Welding practices that minimize corrosion. *Chem. Engng*, **84**(Jan. 17), 145; **84**(Feb. 14), 97

BRAUWEILER, J.R. (1963). Economics of long vs short-life materials. *Chem. Engng*, **70**(Jan 21), 128

BRAVENEC, E.V. (1972). Why steels fracture. *Chem. Engng*, **79**(Jul. 10), 100

BRAVO, F. and BEATTY, B.D. (1993). Decide whether to use thermal relief valves. *Chem. Engng Prog.*, **89**(12), 35

BRAY, G.A. (1964). Fire protection of liquid petroleum gas storage tanks. *Inst. Gas Engrs J.*, **4**, 776. See also *Gas J.*, **319**(5264), 152

BRAY, G.A. (1966). Fire protection of liquid petroleum gas storage tanks. *Fire Int.*, **1**(12), 75

BRAY, G. (1973). Sprinkler systems for computer protection. *Fire Prev. Sci. Technol.*, **6**, 20

BRAY, G. (1980). Sprinkler system design with particular reference to the use of hydraulic calculations. *Fire Prev. Sci. Technol.*, **23**, 8

BRAY, K.N.C., CHAMPION, M. and LIBBY, P.A. (1989). Mean reaction rates in premixed turbulent flames. *Combustion*, **22**, 763

BRAYE, J. (1985). Planning to completion. *Process Engng*, Mar., 45

BRAZDIL, J.F. (1991). *Acrylonitrile. Kirk–Othmer Encyclopaedia of Chemical Technology*, 4th ed., vol. 1 (New York: Wiley), p. 352

BRAZIER, G. (1993). Keeping score in rupture disks. *Chem. Engng*, **100**(9), 155

BRAZIER, A.J. (1994). A summary of incident reporting in the process industry. *J. Loss Prev. Process Ind.*, **7**, 243

BREACH, I. (1978). *Windscale Fallout* (Harmondsworth, Middlesex: Penguin Books)

BREBBIA, C.A. and FERRANTE, A.J. (1983). *Computational Hydraulics* (London: Butterworths)

BREHMER, B. (1981). Models of diagnostic judgments. In Rasmussen, J. and Rouse, W.B. (1981), *op. cit.*, p. 231

BREIPOHL, A.M. (1970). *Probabilistic Systems Analysis* (New York: Wiley)

BREMNER, M. (1976). Hazards of school science. *New Scientist*, **70**, 550

BRENCHLEY, D.L. and WEGENG, R.S. (1998). Status of microchemical systems development in the United States of America. *AIChE 1998 Spring National Meeting*, Mar., 1998, New Orleans, LA

BRENDEN, J.J. (1967). Calorific values of the volatile pyrolysis products of wood (letter). *Combust. Flame*, **11**, 437

BRENNAN, D.J. (1992). Evaluating scale economies in process plants: a review. *Chem. Engng Res. Des.*, **70A**, 516

BRENNAN, D.J. (1993). Some challenges of cleaner production for process design. *Environ. Protect. Bull.* 024, (May), 3–7

BRENNAN, E.G., BROWN, D.G. and DODSON, M.G. (1984). Dispersion of high pressure jets of natural gas in the atmosphere. *The Protection of Exothermic Reactors and Pressurised Storage Vessels* (Rugby: Instn Chem. Engrs), p. 239

BRENNAN, P.J. (1962). Radioisotopes for routine inspection of valves and piping. *Chem. Engng*, **69**(Oct. 29), 140

BRENNAN, P.J. and STOW, A. (1977). Fire services. In Easton, K.D. (1977), *op. cit.*, p. 399

BRENNECKE, H. (1982). Isolating slide valves for explosion protection – system Kammerer. *The Control and Prevention of Dust Explosions* (London: Oyez/IBC), paper 12

BRENTON, J.R., THOMAS, G.O. and AL-HASSAN, T. (1994). Small-scale studies of water spray dynamics during explosion mitigation tests. *Hazards XII*, p. 393

BRESLER, S.A. (1970). Guide to trouble-free compressors. *Chem. Engng*, **77**(Jun. 1), 161

BRESLER, S.A. and HERTZ, M.J. (1965). Negotiating with engineering contractors. *Chem. Engng*, **72**(Oct. 11), 209

BRESLER, S.A. and IRELAND, J.D. (1972). Substitute natural gas: processes, equipment, costs. *Chem. Engng*, **79**(Oct. 16), 94

BRESLER, G.A. and JAMES, G.R. (1965). Questions and answers on todays ammonia plants. *Chem. Engng*, **72**(Jun. 21), 109.

BRESS, D.F. and PACKBIER, M.W. (1977). The start-up of two major urea plants. *Ammonia Plant Safety*, **19**, 122. See also *Chem. Engng Prog., 1977*, **73**(5), 80

BRESSIN, M. (1985). *Road Accidents Involving the Transport of Dangerous Cargo 1982–1984*. Federal Highway Res. Inst., Accident Res. Div. (Bergisch Gladbach, Germany)

BRESSON, E. and PEREZ, P. (1977). Isolated work stations. *Loss Prevention and Safety Promotion 2*, p. 97

BRESSOU (1926). War gases and horses. *Rev. Vet.*, **78**, 611

BRESTAL, R. *et al.* (1991). Minimize fugitive emissions with a new approach to valve packings. *Chem. Engng Prog.*, **87**(8), 42

BRETHERICK, L. (1974). Assessment of reactive chemical hazards – sources of information. *Chemical Process Hazards V*, p. 1

BRETHERICK, L. (1975). *Handbook of Reactive Chemical Hazards* (London: Butterworths)

BRETHERICK, L. (1976). Reactive chemical hazards: review of availability of information. *Prevention of Occupational Risks*, **3**, 263

BRETHERICK, L. (1979). Nitromethane–zeolite ignition. *Chemy Ind.*, **532**

BRETHERICK, L. (1981). *Hazards in the Chemical Laboratory*, 3rd ed. (London: R. Soc. Chem.). See also Muir, G.D. (1971)

BRETHERICK, L. (1985). *Handbook of Reactive Chemical Hazards*, 3rd ed. (London: Butterworths)

BRETHERICK, L. (1987a). Chemical reactivity: instability and incompatible combinations. In Young, J.A, (1987a), *op. cit.*, p. 93

BRETHERICK, L. (1987b). Reactive chemical hazards: an overview. *Preventing Major Chemical Accidents*, p. 4.1

BRETHERICK, L. (1990a). Chemicals, reactions, hazards. *Safety and Loss Prevention in the Chemical and Oil Processing Industries*, p. 247

BRETHERICK, L. (1990b). *Bretherick's Handbook of Reactive Chemical Hazards*, 4th ed. (Oxford: Butterworth-Heinemann)

BRETSCHNEIDER, B. and KURFURST, J. (1987). *Air Pollution Control Technology* (Amsterdam: Elsevier)

BRETT-CROWTHER, M.R. (1980a). The imbroglio of risk. *Sci. Publ. Policy*, **7**, 306

BRETT-CROWTHER, M.R. (1980b). Uncertain decision making on environmental problems. *Sci. Publ. Policy*, **7**, 377

BREWER, G.L. (1968). Air pollution control. Fume incineration. *Chem. Engng*, **75**(Oct. 14), 160

BREWIN, D.J. and HELLAWELL, J.M. (1980). A water authority viewpoint on monitoring. *Chemy Ind.*, Aug. 2, 595

BREYER, S.G. (1982). *Regulation and Its Reform* (Cambridge, MA: Harvard Univ. Press)

BREYER, S.G. and STEWART, R.B. (1985). *Administrative Law and Regulatory Policy*, 2nd ed. (Boston, MA: Little Brown)

BREYSSE, P.A. (1961). Hydrogen sulphide fatality in a poultry feather fertiliser plant. *Am. Ind. Hyg. Ass. J.*, **22**, 220

BRIAN, P.L.T. (1988). Managing safety in the chemical industry. *Preventing Major Chemical and Related Process Accidents*, p. 615

BRIAN, P.L.T., VIVIAN, J.E. and HABIB, A.G. (1962). The effect of the hydrolysis reaction upon the rate of absorption of chlorine into water. *AIChE J.*, **8**, 205

BRIAN, P.L.T., VIVIAN, J.E. and PIAZZA, C. (1966). The effect of temperature on the rate of absorption of chlorine into water. *Chem. Engng Sci.*, **21**, 551

BRIDENBAUGH, D.G., HUBBARD, R.B. and MINOR, G.C. (1976). *Testimony of Dale G. Bridenbaugh, Richard B. Hubbard, Gregory C. Minor before the Joint Committee on Atomic Energy, February 18, 1976* (Cambridge, MA: Union of Concerned Scientists)

BRIDGE, N.W. (1977). Some monitoring circuits for multi-zone fire detector systems. *Fire Prev. Sci. Technol.*, **18**, 21

BRIDGER, G.W. (1976). Catalyst for steam reforming of hydrocarbons. *Ammonia Plant Safety*, **18**, 24

BRIDGES, W.G. (1985). The assessment of toxic hazards. *The Assessment and Control of Major Hazards* (Rugby: Instn Chem. Engrs), p. 413

BRIDGES, G.W. (1985). PSM cost/benefit survey results. *Process Safety Prog.*, **13**, 23

BRIDGES, W.G., KIRKMAN, J.Q. and LORENZO, D.K. (1992). Strategies for integrating human reliability analysis with process hazard evaluations. *Hazard Identification and Risk Analysis, Human Factors and Human Reliability in Process Safety*, p. 361

BRIDGEWATER, A.V. (1979). International construction cost location factors. *Chem. Engng*, **86**(Nov. 5), 119

BRIDGEWATER, A.V. and MUMFORD, C.J. (1979). *Waste Recycling and Pollution Control* (London: Godwin)

BRIDGWATER, J. (1983). Dust formation and suppression. *Chem. Engr, Lond.*, **395**, 32

BRIDGMAN, P.W. (1931). *Dimensional Analysis* (New Haven, CT: Yale Univ. Press)

BRIEGER, H., RIEDERS, F. and HODERS, W.A. (1952). Acrylonitrile: spectrophotometric determination, acute toxicity and mechanism of action. *Ind. Hyg. Occup. Med.*, **6**, 128

BRIFFA, F.E.J. (1981). Transient drag in sprays. *Combustion*, **18**, 307

BRIGGS, A.A. (1979). Protection of large flammable storage tanks during repair by means of nitrogen-filled high expansion foam. *J. Haz. Materials*, **3**(1), 17

BRIGGS, A.A. and WEBB, J.S. (1988). Gasoline fires and foams. *Fire Technol.*, **24**, 48

BRIGGS, E. (1977). Constraints and procedures for discharge to the atmosphere. *Chemical Engineering in a Hostile World* (London: Clapp & Poliack)

BRIGGS, G.A. (1965). A plume rise model compared with observations. *J. Air Poll. Control Ass.*, **15**, 433

BRIGGS, G.A. (1968). Momentum and buoyancy effects. In Slade, D.H. (1968), *op. cit.*, p. 189

BRIGGS, G.A. (1969). *Plume Rise. Div. of Tech. Inf., Atom. Energy Comm.*, Oak Ridge, TN, Rep.

BRIGGS, G.A. (1974). *Diffusion Estimation for small Emissions. Ann. Rep. 1973 Rep. ATDL-106.* Air Resources Atmos. Turbulence and Diffusion Lab., Environ. Res. Labs, Washington, DC

BRIGGS, G.A. (1976). Plume rise predictions. In Haugen, D.A. (1976), *op. cit.*, p. 59

BRIGGS, G.A. (1984). Plume rise buoyancy effects. In *Atmospheric Science and Power Production*, Rep. DOE/TIC-27601, p. 327

BRIGGS, G.A., EBERHARD, W.L., GAYNOR, J.E., MONINGER, W.R. and UTTAL, T. (1986). Convective diffusion field measurement compared with laboratory and numerical experiments. *Fifth Joint Conf. on Applications of Air Pollution Meteorology* (Boston, MA: Am. Met. Soc), p. 340

BRIGGS, G.A., HUBER, A.H., SNYDER, W.H. AND THOMPSON, R.S. (1992). Diffusion in building wakes for ground level releases (discussion). *Atmos. Environ.*, **26B**, 513. See also Ramsdell, J.V. (1990)

BRIGGS, G.A., THOMPSON, R.S. and SNYDER, W.H. (1990). Dense gas removal from a valley by crosswinds. *J. Haz. Materials*, **24**(1), 1

BRIGGS, M. (ed.) (1994) *Decommissioning, Mothballing and Revamping of Process Plant* (Rugby: Instn Chem. Engrs)

BRIGGS, P.P., RICHARDS, J.M. and FIESINGER, E.G. (1986). Inspection of a 30,000 ton atmospheric ammonia storage tank. *Ammonia Plant Safety*, **26**, 89

BRIGGS, R. (1980). Instrumentation for water pollution monitoring. *Chemy Ind.*, Aug. 2, 587

BRIGGS, R. (1988). How will your operators react in an emergency? *Process Engng*, Feb., **43**

BRIGHAM, F.R. and LAIOS, L. (1975). Operator performance in the control of a laboratory process plant. *Ergonomics*, **18**, 53

BRIGHAM, W.E., HOLSTEIN, E.D. and HUNTINGTON, R.L. (1957). How uphill and downhill flow affect pressure drop. *Oil Gas J.*, **55**(Nov. 11), 145

BRIGHT, A.W. (1977). Electrostatic hazards in liquids and powders. *J. Electrostatics*, **4**, 131

BRIGHT, A.W. and HAIG, I.G. (1977). Safety criteria applicable to the loading of large plastic tanks. *IEEE Conf. of Industrial Applications Society* (New York)

BRIGHT, A.W. and HUGHES, J.F. (1975). Research on electrostatic hazards associated with tank washing in very large crude carriers − introduction and experimental modelling. *J. Electrostatics*, **1**, 37

BRIGHT, G.F. (1972). Halting product loss through pressure relief valves. *Loss Prevention*, **6**, 88. See also *Chem. Engng Prog.*, *1972*, **68**(5), 59

BRIGHT, R.G. (1977). A new test method for automatic fire detection systems. *Fire Technol.*, **13**, 105

BRIGHTON, P.W.M. (1978). Strongly stratified flow past three-dimensional obstacles. *Quart. J. R. Met. Soc*, **104**, 289

BRIGHTON, P.W.M. (1985a). Area-averaged concentrations, height-scales and mass balances. *J. Haz. Materials*, **11**(1/3), 189. See also in McQuaid, J. (1985), *op. cit.*, p. 189

BRIGHTON, P.W. (1985b). Evaporation from a plane liquid surface into a turbulent boundary layer. *J. Fluid Mech.*, **159**, 323

BRIGHTON, P.W.M. (1986). Heavy-gas dispersion from sources inside buildings or in their wakes. In *Refinement of Estimates of the Consequences of Heavy Toxic Vapour Release* (Rugby: Instn Chem. Engrs), p. 2.1

BRIGHTON, P.W.M. (1987). A user's critique of the Thorney Island dataset. *J. Haz. Materials*, **16**, 457

BRIGHTON, P.W.M. (1988). Similarity solutions for two-dimensional steady gravity currents. In Puttock, J.S. (1988B), *op. cit.*, p. 115

BRIGHTON, P.W.M. (1990). Further verification of a theory for mass and heat transfer from evaporating pools. *J. Haz. Materials*, **23**(2), 215

BRIGHTON, P.W.M. and PRINCE, A.J. (1987). Overall properties of the heavy gas clouds in the Thorney Island Phase II trials. *J. Haz. Materials*, **16**, 103

BRIGHTON, P.W.M., PRINCE, A.J. and WEBBER, D.M. (1985). Determination of cloud area and path from visual and concentration records. *J. Haz. Materials*, **11**(1/3), 155. See also in McQuaid, J. (1985), *op. cit.*, p. 155

BRIGHTON, P.W.M. *et al.* (1994). Comparison of heavy gas dispersion models for instantaneous releases. *J. Haz. Materials*, **36**, 193

BRILEY, G.G. (1967). Safety on ammonia barges. *Ammonia Operations Prog. Safety*, **9**, 10

VAN DEN BRINK, C. (1983). Changing trends in chemical shipping and storage. *Chemy Ind.*, Nov. 21, 857

BRINK, J.A. (1962). New process applications of fiber mist eliminators. *Ammonia Plant Safety*, **4**, 35

BRINKLEY, S.R. (1969). Determination of explosion yields. *Loss Prevention*, **3**, 79

BRINKLEY, S.R. and KIRKWOOD, J.G. (1947). Theory of the propagation of shock waves. *Phys. Rev.*, **71**, 606

BRINKLEY, S.R. and KIRKWOOD, J.G. (1949). On the condition of stability of the plane detonation wave. *Combustion*, **3**, 586

BRINKLEY, S.R. and KIRKWOOD, J.G. (1961). Theory of the propagation of shock waves. *Phys. Rev.*, **71**, 9

BRINKLEY, S.R. and LEWIS, B. (1959). On the transition from deflagration to detonation. *Combustion*, **7**, p. 807

BRINKLEY, S.R. and SEELY, L.B. (1969). Construction of the Hugoniot curve and calculation of the Chapman–Jouguet points for general equations of state. *Combust. Flame*, **13**, 506

BRINKMAN, J.A. (1993). Verbal protocol accuracy in fault diagnosis. *Ergonomics*, **36**, 1381

BRISCOE, B. (1982). Friction, fact and fiction. *Chemy Ind.*, July 17, 467

BRISCOE, G.J. and NERTNEY, R.J. (1977). *Safety Analysis Guide*. Dept of Energy, Rep. (Washington, D.C.)

BRISCOE, W.G. (1976). Stud tension. *Oil Gas J.*, Sep. 13, 76

BRISMAR, B. and BERGENWALD, L. (1982). The terrorist bomb explosion in Bologna, Italy, 1980: an analysis of the effects and injuries sustained. *J. Trauma*, **22**(3), 216

BRISTOL, E.G. (1985). A design tool kit for batch process control: terminology and a structural model. *InTech*, Oct., 47

BRISTOL, E.H. (1980). After DDC: idiomatic control. *Chem. Engng Prog.*, **76**(11), 84

BRISTOL, J.M. (1981). Diaphragm metering pumps. *Chem. Engng*, **88**(Sep. 21), 124

BRISTOWS ENVIRONMENTAL LAW GROUP (1991). *Environment. Operations Prog. Engr*, **35**(May/Jun.), 33

BRITISH APPROVALS SERVICE FOR ELECTRICAL EQUIPMENT IN FLAMMABLE ATMOSPHERES — see Appendix 28

BRITISH BROADCASTING CORP. (1982). *Transcript of Checkpoint, 2 Jun., 1982* (London)

BRITISH CHEMICAL INDUSTRY SAFETY COUNCIL — see Appendices 27 and 28

BRITISH COMPRESSED GASES ASSOCIATION (1984). *A Method for Estimating the Offsite Risks from Bulk Storage of Liquefied Oxygen (LOX)* (London)

BRITISH COMPRESSED GASES ASSOCIATION — see also Appendices 27 and 28

BRITISH CRYOGENICS COUNCIL (n.d.). *Vacuum Insulated Cryogenic Pipes — Some Recommendations Concerned with Pipe and Coupling Design* (London)

BRITISH CRYOGENICS COUNCIL (1970). *Cryogenic Safety Manual* (London)

BRITISH CRYOGENICS COUNCIL (1975). *Equipment Guide* (London)

BRITISH DRUG HOUSES LTD (1962). *Flash Points* (Poole)

BRITISH DRUG HOUSES LTD (1970). *Dealing with Spillages of Hazardous Chemicals* (wallchart) (Poole)

BRITISH DRUG HOUSES LTD (1977a). *Hazard Warning Symbols and Their Meaning* (wallchart) (Poole)

BRITISH DRUG HOUSES LTD (1977b). *Laboratory First Aid* (wallchart) (Poole)

BRITISH DRUG HOUSES LTD (1977c). *Properties of Gases* (wallchart) (Poole)

BRITISH GAS (1990). *Review of the Applicability of Predictive Methods to Gas Explosions in Offshore Modules* (London: HM Stationery Office)

BRITISH GAS — see Appendices 27 and 28

BRITISH GAS CORPORATION (1978). *Statement by the British Gas Corporation on the Possibility and Consequences of an Unconfined Vapour Cloud Explosion Involving LNG.* Midlands Res. Station, Solihull

BRITISH INSTITUTE OF MANAGEMENT (1990). *Managing Occupational Health and Safety* (OHS) (London)

BRITISH INSURANCE ASSOCIATION (annual). *Annual Report* (London)

BRITISH INSURANCE ASSOCIATION (1975). *Insurance Facts and Figures 1974* (London)

BRITISH INTELLIGENCE OBJECTIVES SUBCOMMITTEE (1946). *Intelligence Report. Japanese Chemical Warfare*, vol. 2, App. A, *Fundamental Research on Effectiveness of Toxic Gases, Rep. BIOS/JAP/PR/394* (London: H. M. Stationery Office)

BRITISH MEDICAL ASSOCIATION (1987). *Living with Risk* (London)

BRITISH MEDICAL ASSOCIATION (1991). *Hazardous Waste and Human Health* (Oxford: Oxford Univ. Press)

BRITISH NUCLEAR ENERGY SOCIETY (1986). *Proc. Conf. on Chernobyl* (London)

BRITISH OCCUPATIONAL HYGIENE SOCIETY (1984). *Fugitive Emissions of Vapours from Process Equipment* (Leeds: Science Reviews Ltd)

BRITISH OCCUPATIONAL HYGIENE SOCIETY (1987). *Controlling Airborne Contaminants in the Workplace* (Northwood, Middlesex: Science Reviews Ltd)

BRITISH OCCUPATIONAL HYGIENE SOCIETY — see Appendix 28

BRITISH PETROLEUM CO. LTD (1973). *Gas Making and Natural Gas* (London)

BRITISH PETROLEUM CO. LTD (1977). *Our Industry Petroleum*, 5th ed. (London)

BRITISH PLASTICS FEDERATION (1979). *Guidelines for the Safer Production of Phenolic Resins* (London)

BRITISH PUMP MANUFACTURERS ASSOCIATION (1974). Pump Users Handbook (London: Trade & Technical Press)

BRITISH RAIL (n.d.a). *Dangerous Goods by Merchandise Trains* (London)

BRITISH RAIL (n.d.b). *List of Dangerous Goods and Conditions of Acceptance. BR 22426* (London)

BRITISH RAIL (n.d.c). *Working Manual for Rail Staff, Pt 3, Handling and Carriage of Dangerous Goods. BR 30054/3* (London)

BRITISH RUBBER MANUFACTURERS ASSOCIATION LTD (1977). *Isocyanates in Industry*, rev. ed. (Birmingham)

BRITISH SAFETY COUNCIL — see Appendix 28

BRITISH SOCIETY FOR SOCIAL RESPONSIBILITY IN SCIENCE (1979). *Asbestos Killer Dust* (London: BSSRS Publs)

BRITISH STANDARDS INSTITUTION — see Appendix 27

BRITISH STANDARDS INSTITUTION (1986). Private communication to the Institution of Chemical Engineers

BRITISH TRANSPORT COMMISSION, BRITISH WATERWAYS (1962). *Petroleum Spirit and Carbide of Calcium Bye Laws* (London)

BRITISH VALVE MANUFACTURERS ASSOCIATION (1972). *Technical Reference Book on Valves for the Control of Fluids* (London)

BRITON, O.J., DECLERCK, D.H. and VORHIS, F.H. (1970). Trends in materials applications. Linings. *Chem. Engng*, 77(Oct. 12), 123

BRITT, C.H. (1975). Hot tapping in cryogenic service. *Loss Prevention*, 9, p. 4

BRITT, F. (1993). Properly design FRP process piping. *Chem. Engng Prog.*, 89(8), 57

BRITTER, R. (1979). The spread of a negatively buoyant plume in a calm environment. *Atmos. Environ.*, 13, 1241

BRITTER, R. (1980). The ground level extent of a negatively buoyant plume in a turbulent boundary layer. *Atmos. Environ.*, 14, 779

BRITTER, R.E. (1982). *Special Topics on Dispersion of Dense Gases. Rep. RPG 1200.* Health and Safety Executive, Sheffield

BRITTER, R.E. (1988a). Fluid modelling of dense gas dispersion over a ramp. *J. Haz. Materials*, 18(1), 37

BRITTER, R.E. (1988b). A review of some mixing experiments relevant to dense gas dispersion. In Puttock, J.S. (1988B), *op. cit.*, p. 1

BRITTER, R.E. and GRIFFITHS, R.F. (1982). The role of dense gases in the assessment of industrial hazards. *J. Haz. Materials*, 6(1/2), 3. See also in Britter, R.E. and Griffiths, R.F. (1982), *op. cit.*, p. 3

BRITTER, R.E. and LINDEN, P.F. (1980). The motion of the front of a gravity current travelling down an incline. *J. Fluid Mech.*, 99, 531

BRITTER, R.E. and MCQUAID, J. (1987). The development of a workbook on the dispersion of dense gases. *Vapor Cloud Modeling*, p. 399

BRITTER, R.E. and MCQUAID, J. (1988). *Workbook on the Dispersion of Dense Gases* (London: Health and Safety Executive)

BRITTER, R.E. and SIMPSON (1978). Experiments on the dynamics of a gravity current head. *J. Fluid Mech.*, 88, 223

BRITTER, R.E. and SNYDER, W.H. (1988). Fluid modelling of dense gas dispersion over a ramp. *J. Haz Materials*, 18(1), 37

BRITTON, C.F. (1981). Preventing unscheduled shutdowns. *Chem. Engng*, 88(Jun. 1), 83

BRITTON, L.G. (1988). Systems for electrostatic evaluation in industrial silos. *Plant/Operations Prog.*, 7, 40

BRITTON, L.G. (1990a). Combustion hazards of silane and its chlorides. *Plant/Operations Prog.*, 9, 16

BRITTON, L.G. (1990b). Thermal stability and deflagration of ethylene oxide. *Plant/Operations Prog.*, 9, 75

BRITTON, L. (1991). Spontaneous fires in insulation. *Plant/Operations Prog.*, 10, 27

BRITTON, L.G. (1992). Using material data in static hazard assessment. *Plant/Operations Prog.*, 11, 56

BRITTON, L.G. (1993). Static hazards using flexible intermediate bulk containers for powder handling. *Process Safety Prog.*, 12, 240

BRITTON, L.G. (1994). Loss case histories in pressurized ethylene systems. *Process Safety Prog.*, 13, 128

BRITTON, L.G. and CHIPPETT, S. (1989). Practical aspects of dust deflagration testing. *J. Loss Prev. Process Ind.*, 2(3), 161

BRITTON, L.G. and CLEM, H.G. (1991). Include safety in insulation system selection. *Chem. Engng Prog.*, 87(11), 87

BRITTON, L.G. and KIRBY, D.C. (1989). Analysis of a dust deflagration. *Plant/Operations Prog.*, 8, 177

BRITTON, L.G. and SMITH, J.A. (1988). Static hazards of drum filling. Part I: Actual incidents and guidelines; Part II Electrostatic models: theory and experiment. *Plant/Operations Prog.*, 7, 53

BRITTON, L.G., TAYLOR, D.A. and WOBSER, D.C. (1986). Thermal stability of ethylene at elevated pressures. *Plant/Operations Prog.*, 5, 238

BRITTON, L.G. and WILLIAMS, T.J. (1982). Some characteristics of liquid-metal discharges involving charged low risk oil. *J. Electrostatics*, 13, 185

BRO, M.I. and PILLSBURY, R.D. (1980). Applications of Teflon and Tefzel fluoroplastics to increase loss prevention and personal safety. *Loss Prevention*, 13, p. 57

BRO, P. and LEVY, S.C. (1990). *Quality and Reliability Methods for Primary Batteries* (New York: Wiley)

BROADBENT, D.E., REASON, J. and BADDELEY, A. (eds) (1990). *Human Factors in Hazardous Situations* (Oxford: Clarendon Press)

BROADBENT, W. (1915). Some results of German gas poisoning. *Br. Med. J.*, Aug. 14, 247

BROADHURST, V.A. (1971). Basic legal requirements. *Major Loss Prevention*, p. 21

BROADRIBB, M.P. (1989). *Piper Alpha Public Inquiry. Transcript, Days 135, 136*

BROBST, W.A. (1972). *The Probability of Transportation Accidents. Minutes Fourteenth Explosives Safety Seminar.* Explosives Safety Board, Dept of Defense

BROCH, J.T. (1971). *Acoustic Noise Measurements* (Hounslow: B&K Labs)

BROCH, J.T. (1980). *Mechanical Vibration and Shock Measurements* (Naeman, Denmark: Brüel & Kjaer)

BROCK, P. (1977). *The Reliability Analysis of Logical Networks by the Computer Program ALMONA. Rep. NCSR R14.* Nat. Centre for System Reliab., Culcheth

BROCK, P. and CROSS, A. (n.d.). *Computer Programs for Reliability Analysis* (Culcheth, Lancashire: Systems Reliability Directorate)

BROCKHOFF, L.H. (1992). A method for assessment of risks of transport of dangerous goods. A micro approach to macro analysis. *Loss Prevention and Safety Promotion 7*, vol. 3, 160–1

BROCKHOFF, L., PETERSEN, H.J.S. and HAASTRUP, P. (1992). A consequence model for chlorine and ammonia based on a fatality index approach. *J. Haz. Materials*, 29, 405

BROCKWELL, J.L. (1990). Prediction of decomposition limits for ethylene oxide–nitrogen vapor mixtures. *Plant/Operations Prog.*, 9, 98

BRODE, H.L. (1955). Numerical solutions of spherical blast waves. *J. Appl. Phys.*, 26(6), 766

BRODE, H. (1957). A calculation of the blast wave from a spherical charge of TNT. *Rand Corp., Santa Monica, CA, Rep.*

BRODE, H.L. (1959). Blast wave from a spherical charge. *Phys. Fluids*, 2, 217

BRODEUR, P. (1974). *The Expendable Americans* (New York: Viking)

BRODGESELL, A. (1971).Valve selection. *Chem. Engng*, **78**(Oct. 11), 119

BRODIE, B.B. (1956). Pathways of drug metabolism. *J. Pharm. Pharmacol.*, **8**, 1

BRODIE, G.W. (1973). Dome bursting discs come under pressure from reverse buckling type. *Process Engng*, **12**, 92

BRODINE, V. (1975). *Radioactive Contamination* (New York: Harcourt Brace Johanovich)

BRODKEY, R.S. (1967). *The Phenomena of Fluid Motion* (London: Addison-Wesley)

BRODSKY, A. (ed.) (1978–). *Handbook of Radiation Measurement and Protection*, vols 1–2 (Boca Raton, FL: CRC Press)

BROEK, D. (1974). *Elementary Engineering Fracture Mechanics* (Leiden: Noordhoff)

BROECKMANN, B. and SCHECKER, H.G. (1992).'Boil over' effects in burning oil tanks. *Loss Prevention and Safety Promotion* 7, vol. 1, 15.1

BROEKMATE, A. (1980). Use of diaphragm couplings in a Syngas compressor. *Ammonia Plant Safety*, **22**, 77

BROGLI, F. (1980). Thermally safe conditions for long term storage of bulk materials. In *Loss Prevention and Safety Promotion* 3, vol. 2, p. 369

BROGLI, F., GRIMM, P., MEYER, M. and ZUBLER, H. (1980). Hazards of self-accelerating reactions. In *Loss Prevention and Safety Promotion* 3, vol. 2, p. 665

BROGLI, F., GIGER, G., RANDEGGER, H. and REGENASS, W. (1981). Assessment of reaction hazards by means of a bench scale heat flow calorimeter. *Runaway Reactions, Unstable Products and Combustible Powders* (London: Instn Chem. Engrs), paper 3/M

BROK, J.F.C. and VAN DER VET, R.P. (1984). *A Method for the Analysis and Processing of Marine Collision Data. Rep. A.MER. 83.042.* Shell Res.

BROK, J.F., VAN DUYVENBODE, E.A. and KILAAS, L. (1985). An enhanced integrated methodology to determine the reliability function of a specific pipeline and riser system. *Reliab. Engng*, **12**, 1

BROKAW, R.S. and GERSTEIN, M. (1957). Correlations of burning velocity, quenching distances, and minimum ignition energies for hydrocarbon–oxygen–nitrogen systems. *Combustion*, **6**, 66

BROKAW, R.S. and JACKSON, J.L. (1955). Effect of temperature, pressure and composition on ignition delays for propane flames. *Combustion*, **5**, 563

BROMLEY, J. and FINNECY, E.E. (1985). The transfrontier movement of hazardous wastes. *Chemy Ind.*, Jan. 21, 44

BROODO, A., GILMORE, J. and ARMSTRONG, A. (1976). Did arson cause that plant fire? *Hydrocarbon Process.*, **55**(2), 163

BROOK, N. (1971). *Mechanics of Bulk Materials Handling* (London: Butterworths)

BROOK, P.H. (1978). Solution to burner ignition problems. *Ammonia Plant Safety*, **20**, p. 156

BROOK, R.W.H. (1974). How safe is safe? *Chart. Mech. Engr*, Nov., 75

BROOKE, J.B. (1981). Tools for debugging computer programs – how much do they help? In Rasmussen, J. and Rouse, W.B. (1981), *op. cit.*, p. 87

BROOKE, J.B., COOK, J.F. and DUNCAN, K.D. (1983). Effects of computer aiding and pretraining on fault location. *Ergonomics*, **26**, 669

BROOKE, J.B. and DUNCAN, K.D. (1980a). An experimental study of flowcharts as an aid to identification of procedural faults. *Ergonomics*, **23**, 387

BROOKE, J.B. and DUNCAN, K.D. (1980b). Experimental studies of flowchart use at different stages of program debugging. *Ergonomics*, **23**, 1057

BROOKE, J.B. and DUNCAN, K.D. (1981). Effects of system display format on performance in a fault location task. *Ergonomics*, **24**, 175

BROOKE, J.B. and DUNCAN, K.D. (1983a). A comparison of hierarchically paged and scrolling displays for fault finding. *Ergonomics*, **26**, 465

BROOKE, J.B. and DUNCAN, K.D. (1983b). Effects of prolonged practice in a fault-location task. *Ergonomics*, **26**, 379

BROOKE, J.M. (1970). Inhibitors: new demands for corrosion control. *Hydrocarbon Process.*, **49**(1), 121; **49**(3), 138; **49**(8), 107; and **49**(9), 299

BROOKE, M. (1962). Corrosion inhibitors checklist. *Chem. Engng*, **69**(Feb. 5), 135

BROOKE, M. (1970). Process plant utilities. Water. *Chem. Engng*, **77**(Dec. 14), 135

BROOKES, G.L. and SIDDALL, E. (1980). An analysis of the Three Mile Island accident. *Canadian Nucl. Soc, First Ann. Conf., Montreal*

BROOKES, V.J. (1948). *Poisons*; 1958, 2nd ed. (New York: van Nostrand Reinhold); 1975, rev. ed. (Huntington, New York: R.E. Krieger)

BROOKLES, L.G. (1977). The health hazards of not going nuclear (review). *Sci. Publ. Policy*, **4**, 291

BROOKS, A.C. (1983). Realistic cold redundant systems. *Reliab. Engng*, **6**, 127

BROOKS, A.C. (1984). The application of availability analysis to nuclear power plants. *Reliab. Engng*, **9**(3), 127

BROOKS, C.E.P. and CARRUTHERS, N. (1953). *Handbook of Statistical Methods in Meteorology* (London: HM Stationery Office)

BROOKS, D.G., MILLER, A.C., HAMM, G.L. and MERKHOFER, M.W. (1993). An approach to prioritizing risk-reducing projects (poster item). *Process Safety Management*, p. 497

BROOKS, G.T. (1974). *Chlorinated Insecticides*, vols 1–2 (Boca Raton, FL: CRC Press)

BROOKS, H. (1972). Science and trans-science. *Minerva*, **10**, 484

BROOKS, J.K., HANEY, R.W., KAISER, G.D., LEYVA, A.J. and MCKELVEY, T.C. (1988). A survey of recent major accident legislation in the USA. *Preventing Major Chemical and Related Process Accidents*, p. 425. See also Lees, F.P. and Ang, M.L. (1989), *op. cit.*, p. 45

BROOKS, K., BALAKOTAIAH, V. and LUSS, D. (1988). Effect of natural convection on spontaneous combustion of coal stockpiles. *AIChE J.*, **34**, 353

BROOKS, R. (1986). A robust layered control system for a mobile robot. *IEEE J. Rob. Aut.*, **RA2**(1), 14

BROOKS, R. (1989). A robot that walks: emergent behaviors from a carefully evolved network. *Neural Computation*, **1**(1)

BROOMFIELD, E.J. and CHUNG, P.W.H. (1994). Hazard identification in programmable systems: a methodology and case study. *ACM Appl. Comput. Rev.*, **2**(1), 7

BROOMHEAD, L.A. (1986). A new mass flowmeter for LPG/LNG. *Gastech 85*, p. 497

BROPHY, C.P., MILES, W.D. and COCHRANE, R.C. (1959). *United States Army in World War II: the Technical Services Chemical Warfare Service: from Laboratory to Field* (Washington, DC)

BROSCHKA, G.L., GINSBURGH, I., MANCINI, R.A. and WILL, R.G. (1983). A study of flame arrestors in piping systems. *Plant/Operations Prog.*, **2**, 5

BROSMER, M.A. and TIEN, C.L. (1986). Thermal radiation properties of propylene. *Combust. Sci. Technol.*, **48**, 163

BROSMER, M.A. and TIEN, C.L. (1987). Radiative energy block-age in large pool fires. *Combust. Sci. Technol.*, **51**, 21

BROSSARD, J., DESBORDES, D., DIFABIO, N., GARNIER, J.L., LANNOY, A., LEYER, J.C., PERROT, J. and SAINT-CLOUD, J.P. (1985). Truly unconfined deflagrations of ethylene–air mixtures. *Tenth Int. Coll. on Dynamics of Explosions and Reactive Systems, Berkley, CA*

BROSSARD, J., LEYER, J.C., DESBORDES, D., GARNIER, J.L., HEN-DRICKX, S., LANNOY, A., PERROT, J. and SAINT CLOUD, J.P. (1983). Experimental analysis of unconfined explosions of air–hydrocarbon mixtures – characterisation of the pressure field. *Loss Prevention and Safety Promotion* 4, vol. 1, p. D10

BROSSARD, J., LEYER, J.C., DESBORDES, D., SAINT-CLOUD, J., HENDRICKX, S., GARNIER, J.L., LANNOY, A. and PERROT, J. (1984). Air blast from unconfined gaseous detonations. In Bowen, J.R. *et al.* (1984b), *op. cit.*, p. 556

BRÖTZ, W. and SCHÖNBUCHER, A. (1978). Heat and mass transfer in tank flames. *Chem. Ing. Tech.*, **50**, 573

BRÖTZ, W., SCHÖNBUCHER, A. and SCHÄBLE, R. (1977). Statistical investigations of pool flames. *Loss Prevention and Safety Promotion* 2, p. 289

BROUGH, C.W. (1981). Dealing with hazard and risk in plan-ning (2). In Griffiths, R.F. (1981), *op. cit.*, p. 89

BROUGHTON, J. (ed.) (1993). *Process Utility Systems* (Rugby: Instn Chem. Engrs)

BROUWERS, A.A.F. (1984). Automation and the human opera-tor: effects in process operations. *Ergonomic Problems in Process Operations* (Rugby: Instn Chem. Engrs), p. 179

BROUWERS, J.J.H. (1986). Probabilistic descriptions of irre-gular system downtime. *Reliab. Engng*, **15**(4), 263

BROUWERS, J.J.H., VERBEEK, P.H.J. and THOMSON, W.R. (1987). Analytical system availability techniques. *Reliab. Engng*, **17**(1), 9

BROWN, A. (1964). Storage and handling on process plants of light hydrocarbon liquid feedstocks. *Chem. Engr, Lond.*, **178**, CE99

BROWN, A.A. (1969). How to design pile-supported founda-tions for flare stacks. *Hydrocarbon Process.*, **48**(12), 139

BROWN, A.A. (1971). An easy way to compute tower founda-tion flexure. *Hydrocarbon Process.*, **50**(1), 121

BROWN, A.A. (1973). New foundation design method for pile supported tanks. *Hydrocarbon Process.*, **52**(9), 175

BROWN, A.A. (1974). A new approach to tank foundation design. *Hydrocarbon Process.*, **53**(10), 153

BROWN, B. (1964). *Improving the Reliability of Estimation Obtained from a Consensus of Experts. Rep. P-2986.* Rand Corp., Santa Monica, CA

BROWN, B. (1969). *General Properties of Matter* (London: Butterworths)

BROWN, B. (1989). Research into evacuation of aircraft pas-sengers. *Fire Int.*, **119**, 29

BROWN, B., COCHRANE, S. and DALKEY, N. (1969). *The Delphi Method II: Structure of Experiments. Rep. RM-5957–PR.* Rand Corp., Santa Monica, CA

BROWN, C.C. (1984). High- to low-dose extrapolation in ani-mals. In Rodricks, J.V. and Tardiff, R.G. (1984a), *op. cit.*, p. 57

BROWN, D., (1978). Medical record-keeping: new trauma for CPI. *Chem. Engng*, **85**(Jan. 2), 28

BROWN, D.B. (1971). A computerised algorithm for determin-ing the reliability of redundant configurations. *IEEE Trans Reliab.*, **R-20**(3), 121

BROWN, D.B. (1976). *System Analysis and Design for Safety* (New York: McGraw-Hill)

BROWN, D.C. and CHANDRASEKARAN, B. (1989). *Design Problem Solving* (London: Pitman)

BROWN, D.M. (ca. 1986) *Chemical Plant Design to Withstand External Effects.* Dept of Chem. Engng, Polytechnic of the South Bank, London

BROWN, D.M. and BALL, P.W. (1980). A simple method for the approximate evaluation of fault trees. *Loss Prevention and Safety Promotion* 3, vol. 2, p. 466

BROWN, D.M. and NOLAN, P.F. (1985). The effect of external blast on cylindrical structures. *The Assessment and Control of Major Hazards* (Rugby: Instn Chem. Engrs), p. 229

BROWN, D.M. and NOLAN, P.F. (1987). The interaction of shock waves on cylindrical structures. *Plant/Operations Prog.*, **6**, 52

BROWN, D.P. (1978). Proposed volatile spill suppression con-cepts. *Control of Hazardous Material Spills*, **4**, 303

BROWN, D.R. and GREGORY, R.J. (1982). The development of an industry code for hazardous area classification of natur-al gas transmission installations. *Flammable Atmo-spheres and Area Classification, IEE Coll. Dig. 1982/26* (London), p. 9/1

BROWN, G.A., CRAMER, J.J. and SAMELA, D. (1988). The impact of the proposed Clean Air Act amendments. *Chem. Engng Prog.*, **84**(12), 41

BROWN, G.H. (1976). Implementation of a Chemical Hazards Response Information System (CHRIS). *Control of Hazardous Material Spills*, **3**, 103

BROWN, G.W. (1974). Valve problems: causes and cures. *Hydrocarbon Process.*, **53**(6), 97

BROWN, H. (1946). Design of explosion pressure vents. *Engng News Record*, Oct. 3

BROWN, I.D. (1990a). Accident reporting and analysis. In Wilson, J.R. and Corlett, E.N. (1990), *op. cit.*, p. 755

BROWN, I.D. (1990b). Drivers' margins of safety considered as a focus for research on error. *Human Factors*, **33**, 1307

BROWN, J. (ed.) (1989). *Environmental Risks: Perception Ana-lysis and Management* (London: Belhaven)

BROWN, J.A. (1973). A study of growing danger of detonation in unconfined gas cloud explosions. *John Brown Associ-ates, Berkeley Heights, NJ*

BROWN, J.D. and SHANNON, G.T. (1963a). Design guide to refinery sewers. *Hydrocarbon Process.*, **42**(5), 141

BROWN, J.D. and SHANNON, G.T. (1963b). Minimizing hazards of refinery sewer systems. *API Proc., Div. of Refining*, **43**, 296

BROWN, J.F.W. (1958). The effect of operating temperature on the bursting pressure of some materials commonly used for bursting discs. *Trans Instn Chem. Engrs*, **36**, 81

BROWN, J.M. (1976). Linearity vs non-linearity of dose–response for radiation carcinogenesis. *Health Phys.*, **31**, 231

BROWN, J.S. and DE KLEER, J. (1981). Towards a theory of qua-litative reasoning about mechanisms and its role in troubleshooting. In Rasmussen, J. and Rouse, W.B. (1981), *op. cit.*, p. 317

BROWN, K. (1982). Ease thermal-stress ills. *Chem. Engng*, **89**(Feb. 8), 131

BROWN, K.A.P. (1985). In pursuit of reliability. *Reliab. Engng*, **10**(3), 141

BROWN, K.C. (1951). *Dust Explosions in Factories: Protection of Elevator Casings by Pressure Relief Vents. SMRE Res. Rep. 22.* Safety in Mines Res. Estab., Sheffield

BROWN, K.C. and CURZON, G.E. (1962). *Dust Explosions in Fac-tories: Explosion Vents in Pulverised Fuel Plants. SMRE Res. Rep. 212.* Safety in Mines Res. Estab., Sheffield

BROWN, K.C. and CURZON, G.E. (1963). *Dust Explosions in Factories – Explosion Vents in Pulverised Fuel Plants* (London: HM Stationery Office)

BROWN, K.C. and JAMES, G.J. (1962). *SMRE Res. Rep. 201.* Safety in Mines Res. Estab., Sheffield

BROWN, K.C. *et al.* (1962). *Trans. Instn Mining Engrs*, **74**, 261

BROWN, K.W. and THOMAS, J.C. (1988). Leak rates into drainage systems underlying lined retention facilities. *J. Haz. Materials*, **18**(2), 179

BROWN, L.E., JOHNSON, D.W. and MARTINSEN, W.E. (1987). Hazard control methods of high volatility chemicals. *Preventing Major Chemical Accidents*, p. 5.41

BROWN, L.E., MARTINSEN, W.E., MUHLENKAMP, S.P. and PUCKETT, G.L. (1976). *Small Scale Tests on Control Methods for some Liquefied Natural Gas Hazards. Rep DOT-CG-42, 355A, Task 1.* MIT, CAMBRIDGE, IL

BROWN, L.E. and ROMINE, L.M. (1979). Liquefied gas fires: which foam? *Hydrocarbon Process.*, **58**(9), 321

BROWN, L.E. AND ROMINE, L.M. (1981). Liquefied gas fires: which foam? *Fire Prev. Sci. Technol.*, **24**, 4

BROWN, L.E., WESSON, H.R. and WELKER, J.R. (1974a). Predict LNG fire radiation. *Hydrocarbon Process.*, **53**(5), 141

BROWN, L.E., WESSON, H.R. and WELKER, J.R. (1974b). Thermal radiation from LNG fires and LNG fire suppression. *Proc. Conf. Nat. Gas Res. Tech.*, **3**, 31

BROWN, L.E., WESSON, H.R. and WELKER, J.R. (1975). Predicting LNG fire radiation. *Fire Prev. Sci. Technol.*, **11**, 26

BROWN, L.S. (1972). A physical barrier system for control of hazardous material spills in waterways. *Control of Hazardous Material Spills*, **1**, 93

BROWN, M. (1981). *Laying Waste. The Poisoning of America by Toxic Chemicals*, rev. ed. (New York: Washington Square Press)

BROWN, N.J., FRISTROM, R.M. and SAWYER, R.F. (1974–). A simple premixed flame model including an application to H_2 + air flames. *Combust. Flame*, 1974, **23**, 269; 1975, **24**, 409

BROWN, N.R. (1981). High reliability process computer systems. *Reliable Production in the Process Industries* (Rugby: Instn Chem. Engrs), p. 195

BROWN, P.J. (1972). Compressor and engine maintenance. *Chem. Engng Prog.*, **68**(6), 77

BROWN, P.M.M. and FRANCE, D.W. (1975). How to protect air-cooled heat exchangers against overpressure. *Hydrocarbon Process.*, **54**(8), 103

BROWN, R. (1978). Design of air-cooled heat exchangers. A procedure for preliminary estimates. *Chem. Engng*, **85**(Mar. 27), 108

BROWN, R. (1985). The electronic expert teams up with human skills. *Computing*, Feb. 21, 10

BROWN, R. and YORK, J.L. (1962). Sprays formed by flashing liquids. *AIChE J.*, **8**, 149

BROWN, R.D. and BENNETT, B.A. (1978). Commercial routes to vinyl acetate. *Chem. Engng Prog.*, **74**(11), 11

BROWN, R.G., VON HERRMAN, J.L. and QUILLIAM, J.F. (1982). *Operator Action Event Trees for Zion 1 Pressurised Water Reactor. Rep. NUREG/CR-2888.* Nucl. Regul. Comm., Washington, DC

BROWN, R.N. (1964). Control systems for centrifugal gas compressors. *Chem. Engng*, **71**(Feb. 17), 135

BROWN, R.N. (1972). Can specifications improve compressor reliability. *Hydrocarbon Process.*, **51**(7), 89

BROWN, R.N. (1974). Test compressor performance – in the shop. *Hydrocarbon Process.*, **53**(4), 133

BROWN, R.S. (1981). Selecting stainless steel for pumps, valves and fittings. *Chem. Engng*, **88**(Mar. 9), 109

BROWN, R.S. (1982). Damage detection, repair and prevention in an ammonia storage sphere. *Plant/Operations Prog.*, **1**, 97

BROWN, S. (1991). Quantitative risk assessment in the refining industries. *Process Engng*, Apr., 43

BROWN, S. and FREETH, A. (1990). Research and development activity at the Petroleum Science and Technology Institute. *Chem. Engng Res. Des.*, **68A**, 384

BROWN, S. and MARTIN, J. (1977). *Human Aspects of Man-Made Systems* (Milton Keynes: Open Univ.)

BROWN, S.J. (1987). Effect of system dynamics on failure: illustrative examples. *Plant/Operations Prog.*, **6**, 20

BROWN, S.T. (1987). Concentration prediction for meandering plumes based on probability theory. *Atmos. Environ.*, **21**, 1321

BROWN, T. and CADICK, J.L. (1979–). Electrical energy. *Chem. Engng*, 1979, **86**(Jan. 1), 72; **86**(Jan. 29), 111; **86**(Mar. 12), 85; **86**(May 7), 89; **86**(Jun. 18), 119; **86**(Jul. 16), 107; **86**(Sep. 10), 137; **86**(Oct. 27), 127; **86**(Dec. 3), 101; 1980, **87**(Jan. 28), 117; **87**(Apr. 21), 145

BROWN, T.A.J. (1959). *Pressure Piling in Compartmented Enclosures. I, Effects of Bore and Length of Connecting Tube in Explosions of Pentane–Air Mixtures. Tech. Rep. D/T 109.* Br. Elec. and Allied Ind. Res Ass.

BROWN, T.C., CEDERWALL, R.T., ERMAK, D.L., KOOPMAN, R.P., MCCLURE, J.W. and MORRIS, L.K. (1989). *Falcon Series Data Report: 1987 LNG Vapor Barrier Verification Field Trials.* Lawrence Livermore Nat. Lab., Livermore, CA

BROWNE, G. (1993). Evaluating external spill response sources. *Process Safety Prog.*, **12**, 206

BROWNE, H.C. *et al.* (1961). Barricades for high pressure research. *Ind. Engng Chem.*, **53**(1), 52A.

BROWNE, W. and CONSTANTINIS, D.A. (1983). Condition monitoring using mechanised ultrasonics. In Butcher, D.W. (1983), *op. cit.*, p. 51

BROWNELL, L.E. (1963). Mechanical design of tall towers. *Hydrocarbon Process.*, **42**(6), 109

BROWNELL, L.E. and YOUNG, E.H. (1959). *Process Equipment Design* (New York: Wiley)

BROWNING, B. and SEARSON, A.H. (1989). The lessons of the Thessaloniki oil terminal fire. *Loss Prevention and Safety Promotion 6*, p. 39.1

BROWNING, E. (n.d.a). *Toxicity of Industrial Metals* (London: HM Stationery Office)

BROWNING, E. (n.d.b). *Toxicity of Industrial Organic Solvents* (London: HM Stationery Office)

BROWNING, E. (1965). *Toxicity and Metabolism of Industrial Solvents* (Amsterdam: Elsevier)

BROWNING, J.A. and KRALL, W.G. (1955). Effect of fuel droplets on flame stability, flame velocity, and inflammability limits. *Combustion*, **5**, 159

BROWNING, J.A., TYLER, T.L. and KRALL, W.G. (1957). Effect of particle size on combustion of uniform suspensions. *Ind. Engng Chem.*, **49**, 142

BROWNING, J.E. (1968). Tougher safety measures foreseen for CPI. *Chem. Engng*, **75**(Jun. 17), 102

BROWNING, J.E. (1969a). Hazardous-chemical cargo: time bomb for catastrophe. *Chem. Engng*, **76**(May 5), 44

BROWNING, J.E. (1969b) Plot safer shipping course. *Chem. Engng*, **76**(Sep. 8), 64

BROWNING, R.L. (1969a). Analyzing industrial risks. *Chem. Engng*, **76**(Oct. 20), 109

BROWNING, R.L. (1969b). Calculating loss exposures. *Chem. Engng*, **76**(Nov. 17), 239

BROWNING, R.L. (1969c). Estimating loss probabilities. *Chem. Engng*, **76**(Dec. 15), 135

BROWNING, R.L. (1970). Finding the critical path to loss. *Chem. Engng*, **77**(Jan. 26), 119

BROWNING, R.L. (1972). Loss analysis improves process protective instrumentation. *Instrum. Technol.*, **19**(10), 27

BROWNING, R.L. (1973). Safety and reliability decision-making by loss rates. *Loss Prevention*, **7**, 1

BROWNING, R.L. (1975). Analyze losses by diagram. *Hydrocarbon Process.*, **54**(9), 253

BROWNING, R.L. (1976). Human factors in the fault tree. *Chem. Engng Prog.*, **72**(6), 72

BROWNING, R.L. (1977). A solution for the failure data problem in the processing industries. In Gangadharan, A.C. and Brown, S.J. (1977), *op. cit.*, p. 51

BROWNING, R.L. (1979). Use a fault tree to check safeguards. *Loss Prevention*, **12**, 20

BROWNING, R.L. (1980). The Loss Rate Concept in Safety Engineering (Basel; Dekker)

BROWNLEE, K.A. (1949). *Industrial Experimentation*, 2nd ed. (London: H.M. Stationery Office)

BROWNSTON, L., FARRELL, R., KANT, E. and MARTIN, N. (1985). *Programming Expert Systems in OPS5* (Reading, MA: Addison-Wesley)

BROYLES, R.K. (1985). Pipe loop or expansion joint. *Chem. Engng*, **92**(Oct. 14), 103

BRUBAKER, L.A., KOERNER, W.R. and MATHURA, R.E. (1972). LNG and regulation. *LNG Conf.*, **3**

BRUCE, D.J. and DIGGLE, W.M. (1974). Major emergencies in an industrial conurbation. *Loss Prevention and Safety Promotion 1*, p. 111

BRUCE, F.E. (ed.) (1985). *Trenchless Construction for Utilities* (London: Instn Pub. Health Engrs)

BRUCE, R.A.P. (1994). The offshore hydrocarbon releases (HCR) database. *Hazards XII*, p. 107

BRUCE, R.D. and WERCHAN, R.E. (1975). What does noise control cost? *Hydrocarbon Process.*, **54**(5), 234

BRUDERER, R.E. (1989a). Design example: explosion protection selected for a spray drying installation. *Plant/Operations Prog.*, **8**, 141

BRUDERER, R.E. (1989b). Ignition properties of mechanical sparks and hot surfaces in dust/air mixtures. *Plant/Operations Prog.*, **8**, 152

BRUDERER, R.E. (1993). Use bulk bags cautiously. *Chem. Engng Prog.*, **89**(5), 28

BRUDERER, R.E. (1994). Advantages of the new version VDI 3673 Part 1 in comparison with the chapter 7 of NFPA 68; *AIChE Meeting*, Denver, CO, Paper 49A, Aug. 14–17

BRUERE, B. and COENEN, P. (1992). Control strategy for emissions of volatile organic compounds: the case of diffuse emissions. *Loss Prevention and Safety Promotion 7*, vol. 3, 167–1

VAN DE BRUG, A., BACHANT, J. and MACDERMOTT, J. (1986). The taming of R1. *IEEE Expert*, **1**(3), 33

BRUGGER, J.E. and WILDER, I. (1975a). Prevention and control of hazardous material spills in the water environment: an update. *Transport of Hazardous Cargoes*, **4**

BRUGGER, J.E. and WILDER, I. (1975b). A review of the US Environmental Protection Agency's research program on the prevention and control of hazardous material spills. *J. Haz. Materials*, **1**(1), 3

BRUGGER, J.E. (1982). Technology development. In Bennett, G.F., Feates and Wilder (1982), *op. cit.*, p. 9–24

BRUGGINK, R.H. and SHADLEY, J.R. (1973). A workable furnace noise control program. *Chem. Engng Prog.*, **69**(10), 56

BRUINZEEL, C. (1963). Electric discharges during a simulated aircraft refuelling. *J. Inst. Petrol.*, **49**(473), 125

BRUMMEL, R.N. (1989). Procedures for preventing pilot plant runaways. *Plant/Operations Prog.*, **8**, 228

BRUMSHAGEN, W. (1982). Unloading of large LPG carriers into salt and rock caverns. *Gastech 81*, p. 395

BRUMSHAGEN, W. (1983). Sea transport of liquefied CO_2 in gas tankers. *Gastech 82*, p. 219

BRUNDRETT, G.W. (1977). Ventilation: a behavioural approach. *Energy Res.*, **1**, 289

BRUNE, R.L., WEINSTEIN, M. and FITZWATER, M.E. (1983). *Peer Review Study of the Draft Handbook for Human Reliability Analysis with Emphasis on Nuclear Power Plant Applications. Rep. NUREG/CR-1278.* Nucl. Regul. Comm., Washington, DC

BRUNER, J.S., GOODNOW, J.J. and AUSTIN, G.A. (1956). *A Study of Thinking* (New York: Wiley)

BRUNING, F. and OHM, A. (1983). *Accident Casuistics and External Safety,* Draft paper for Loss Prevention and Safety promotion 4

BRUNNER, C.R. (ed.) (1985). *Hazardous Air Emissions from Incineration* (London: Chapman & Hall)

BRUNNER, C.R. (1987). Incineration. *Chem. Engng*, **94**(14), 96

BRUNNER, C.R. (1989). *Handbook of Hazardous Waste Incineration* (New York: Wiley)

BRUNNER, C.R. (1993). *Hazardous Waste Incineration*, 2nd ed. (New York: McGraw-Hill)

BRUNT, D. (1934). *Physical and Dynamical Meteorology* (Cambridge: Cambridge Univ. Press)

BRUSSEAU, M.L. (1992). Factors influencing the transport and fate of contaminants in the subsurface. *J. Haz. Materials*, **32**, 137

BRUTSAERT, W.H. (1982). *Evaporation into the Atmosphere* (Dordrecht: Reidel)

BRUUN, O., TAYLOR, J.R. and RASMUSSEN, A. (1979). *Cause–Consequence Reporting for Accident Reduction. Rep. Riso-M-2206.* Riso Nat. Lab., Roskilde, Denmark

BRYAN, A. (1975). *The Evolution of Health and Safety in Mines* (Ashire Publ.)

BRYAN, J.L. (1974). *Fire Suppression and Detection Systems* (Beverly Hills, CA: Glencoe Press)

BRYAN, J.L. (1976). The determination of behavior responses exhibited in fire situations. *J. Fire Flammability*, **7**, 319

BRYAN, J.L. (1978). An analysis of the variations in the behaviour in fire situations between selected British and United States populations. *J. Fire Flammability*, **9**, 249

BRYAN, J.L. (1986). Damageability of buildings, contents and personnel from exposure to fire. *Fire Safety J.*, **11**(1), 15

BRYAN, W. (1976). *Testimony before the Federal Power Commission in the application of Eascogas LNG Inc. and Distrigas Corp. Docket Nos CP73–47 and CP73–132*

BRYAN, W.R. and SHIMKIN, M.B. (1945). Quantitative analysis of dose-response data obtained with three carcinogenic hydrocarbons in strain C3H male mice. *J. Nat. Cancer Inst.*, **3**(5), 503

BRYANT, E.A. (1991). *Natural Hazards* (Cambridge: Cambridge Univ. Press)

BRYANT, J.A. (1976). Design of fail-safe control systems. *Power*, Jan.

BRYANT, J.T. (1971). The combustion of premixed laminary graphite dust flames at atmospheric pressure. *Combust. Sci. Technol.*, **2**, 389

BRYANT, J.T. (1975). Ignition time delay of fuel droplets, an empirical correlation with flash point. *Combust. Sci. Technol.*, **10**, 185

BRYANT, P. (1991). Gas flooding systems: a review of British Standard 7273 Part 1: 1990. *Fire Surveyor*, **20**(1), 26

BRYANT, W.M.D. (1933). Empirical molecular heat equations from spectroscopic data. *Ind. Engng Chem.*, **25**, 820

BRYCE, D.J. (1980). Safety considerations relative to shipping terminals for hazardous bulk material. *Marichem 80*

BRYCE, D.J. and TURNER, M.J. (1979). Safety evaluation of pipelines. *Br. Hydromech. Res. Ass. Third Int. Conf. on Internal and External Protection of Pipes* (Cranfield)

BRYCE, D.J., RAMSEY, K.W. and SEATON, H.D. (1982). Differing industries handling flammable materials and the role of the enforcing authority. *Flammable Atmospheres and Area Classification, IEE Coll. Dig. 1982/26 (London)*, p. 4/1

BRYCE, D.J. and ROBERTSON, S.S.J. (1983). The classification of hazardous areas – legal and economic considerations. *Electron. Power*, Apr., 298

BRYCE, S.G. and FRYER-TAYLOR, R.E.J. (1994). Reducing the purge in hydrocarbon vent stacks. *J. Loss Prev. Process Ind.*, **7**, 249

BRYCE-SMITH, D. (1971). Lead pollution – a growing hazard to public health. *Chem. Britain*, p. 54

BRYCE-SMITH, D. (1982). Lead pollution – causes and control. *Chemy Ind.*, Jan. 2, 29

BRYERTON, G. (ed.) (1970). *Nuclear Dilemma* (New York: Ballantine)

BRYSON, E.J. and DICKERT, E.J. (1977). Ethylene plant turbomachinery. *Hydrocarbon Process.*, **56**(11), 309

BRZUSTOWSKI, T.A. (1973). A new criterion for the length of a gaseous turbulent diffusion flame. *Combust. Sci. Technol.*, **6**, 313

BRZUSTOWSKI, T.A. (1976). Flaring in the energy industry. *Prog. Energy Combust. Sci.*, **2**, 129

BRZUSTOWSKI, T.A. (1977). Flaring: the state of the art. *Loss Prevention*, **11**, 15

BRZUSTOWSKI, T.A., GOLLAHALLI, S.R. and SULLIVAN, H.F. (1975). The turbulent hydrogen diffusion flame in a cross wind. *Combust. Sci. Technol.*, **11**, 29

BRZUSTOWSKI, T.A., GOLLAHALLI, S.R., GUPTA, M.P., KAPTEIN, M. and SULLIVAN, H.F. (1975). *Radiant Heating from Flares*. *ASME paper 75–HT-4*. Am. Soc. Mech. Engrs, New York

BRZUSTOWSKI, T.A. and SOMMER, E.C. (1973). Predicting radiant heating from flares. *Proc. API Div. of Refining*, **53**, 865

BRZUSTOWSKI, T.A. and TWARDUS, E.M. (1982). A study of the burning of a slick of crude oil on water. *Combustion*, **19**, 847

BS 8800, (1996), *Guide to Occupational Health and Safety Management Systems* (London: Health & Safety Executive)

BUCCI, R.J., PARIS, P.C., LANDES, J.C. and RICE, J.D. (1972). J integral estimation procedures. *ASTM STP 514*, p. 40. Am. Soc. Testing and Mats, Philadelphia, PA

BUCH, G. and DIEHL, A. (1984). An investigation of the effectiveness of pilot judgment training. *Hum. Factors*, **26**, 557

BUCH, R.R. and FILSINGER, D.H. (1985). Method for fire hazard assessment of fluid-soaked thermal insulation. *Plant/Operations Prog.*, **4**, 176

BUCHANAN, B.G. (1979). Issues of representation in conveying the scope and limitations of intelligent assistant programs. *Machine Intelligence 9* (ed. by J.E. Hayes, D. Michie and L.I. Mikulich) (Chichester: Ellis Horwood), p. 407

BUCHANAN, B.G. and FEIGENBAUM, E.A. (1978). DENDRAL and Meta-DENDRAL: their applications dimension. *Artificial Intelligence*, **11**, 5

BUCHANAN, B.G., FEIGENBAUM, E.A. and SRIDHARAN, N.S. (1972). Heuristic theory formation: data interpretation and rule formation. *Machine Intelligence 7* (ed. by B. Meltzer and D. Michie) (Edinburgh: Edinburgh Univ. Press), p. 165

BUCHANAN, B.G. and SHORTLIFFE, E.H. (1984). *Rule-Based Expert Systems* (Reading, MA: Addison-Wesley)

BUCHANAN, B.G., SUTHERLAND, G.L. and FEIGENBAUM, E.A. (1969). Heuristic DENDRAL: a program for generating explanatory hypotheses in organic chemistry. *Machine Intelligence 4* (ed. by B. Meltzer and D. Michie) (Edinburgh: Edinburgh Univ. Press)

BUCHANAN, B.G., SUTHERLAND, G.L. and FEIGENBAUM, E.A. (1970). Towards an understanding of information processes of scientific inference in the context of organic chemistry. *Machine Intelligence 5* (ed. by B. Meltzer and D. Michie) (Edinburgh: Edinburgh Univ. Press)

BUCHANAN, B.G., SMITH, D.H., WHITE, W.C., GRITTER, R.J., FEIGENBAUM, E.A., LEDERBERG, J. and DJERASSI, C. (1976). Applications of artificial intelligence for chemical inference XXII: Automatic rule formation in mass spectrometry by means of the meta-DENDRAL program. *J. Am. Chem. Soc.*, **98**, 6168

BUCHANAN, J.R. and HUTTON, F.C. (1968). Analysis and automated handling of technical information at the nuclear safety information centre. *J. Chem. Educ.*, **45**(6), A487

BUCHANAN, P.J. (1979). Responsibilities for the plant engineer under health and safety legislation. *Plant Engr*, **23**(6), 24

BUCHANAN, R.J. (1982). Transportation. In Bennett, G.F., Feates and Wilder (1982), *op. cit.*, p. 14−2

BUCHANAN, R.J. and METRY, A.A. (1978). Closing the gap in hazardous waste management in New Jersey. *Control of Hazardous Material Spills*, 4, p. 196

BUCHER, W. and FRETZ, R. (1986). Safety aspects of computer controlled chemical plants. *Loss Prevention and Safety Promotion 5*, p. 53−1

BUCHLIN, J.M. (1994). Mitigation of problem clouds. *J. Loss Prev. Process Ind.*, **7**, 167

BUCHTER, H.H. (1967). *Apparate und Armaturen der chemischen Hochdrucktechnik. Konstruktion, Berechnung und Herstellung* (Berlin: Springer)

BUCK, E. (1978). Letter symbols for chemical engineering. *Chem. Engng Prog.*, **74**(10), 73

BUCK, E. and FRANKL, E.M. (1984). Gaps in the pure component experimental physical property data base. *Chem. Engng Prog.*, **80**(3), 82

BUCK, G.S. (1993). Implications of the new pump shaft seal standard. *Hydrocarbon Process.*, **72**(6), 53

BUCK, M.E. and BELASON, E.B. (1985). ASTM test for effects of large hydrocarbon pool fires on structural members. *Plant/Operations Prog.*, **4**, 225

BUCK, P.C. and COLEMAN, V.P. (1985). Slipping, tripping and falling accidents at work: a national picture. *Ergonomics*, **28**, 949

BUCKLAND, I.G. (1980). Explosions of gas layers in a room size chamber. *Chemical Process Hazards VII*, p. 289

BUCKLAND, I.G. (1992). The use of fire and gas detection systems as part of the safety control package. *Major Hazards Onshore and Offshore*, p. 283

BUCKLEY, J.L. and WEINER, S.A. (1978). *Documentation and Analysis of Historical Spill Data to determine Hazardous Material Spill Prevention Research Priorities. Rep. PB-281−090/IGA*. Environmental Protection Agency

BUCKLEY, L.A., JIANG, X.Z., JAMES, R.A., MORGAN, K.T. and BARROW, C.S. (1984). Respiratory tract lesions induced by sensory irritants at the RD_{50} concentration. *Toxicol. Appl. Pharmacol.*, **74**, 417

BUCKLEY, D.H. (1981). *Surface Effects in Adhesion, Friction, Wear, and Lubrication* (Amsterdam: Elsevier)

BUCKLEY, J.A. (1966). Vinyl chloride via direct chlorination and oxychlorination *Chem. Engng*, **73**(Nov. 21), 102

BUCKLEY, P.S. (1964). *Techniques of Process Control* (New York: Wiley)

BUCKLEY, P.S. (1970). Protective controls for a chemical reactor. *Chem. Engng*, **77**(Apr. 20), 145

BUCKMASTER, J. (1976). The quenching of deflagration flames. *Combust. Flame*, **26**, 151

BUCKMASTER, J., GESSMAN, R. and RONNEY, P. (1992). The three-dimensional dynamics of flame-balls. *Combustion*, **24**, p. 53

BUCKMASTER, J.D. and JOULIN, G. (1993). Influence of boundary-induced losses on the structure and dynamics of flame-balls. *Combust. Sci. Technol.*, **89**, 57

BUCKMASTER, J.D., JOULIN, G. and RONNEY, P.D. (1991). The structure and stability of nonadiabatic flame balls: II: Effects of far field losses. *Combust. Flame*, **84**, 411

BUCKMASTER, J. and LEE, C.J. (1992). The effects of confinement and heat loss on outwardly propagating spherical flames. *Combustion*, **24**, 45

BUCKMASTER, J.D. and LUDFORD, G.S.S. (1982). *Theory of Laminar Flames* (Cambridge: Cambridge Univ. Press)

BUCKMASTER, J. and MIKOLAITIS, D. (1982–). A flammability-limit method for upward propagation through lean methane/air mixtures in a standard flammability tube. *Combust. Flame*, 1982, **45**, 109; 1985, **59**, 99; 101; **60**, 103

BUCKMASTER, J.D., SMOOKE, M. and GIOVANGIGLI, V. (1993). Analytical and numerical modeling of flame-balls in hydrogen-air mixtures. *Combust. Flame*, **94**, 113

BUCSKO, R.T. (1983). Glass as a material of construction. *Chem. Engng Prog.*, **79**(2), 82

BUDNICK, E.K. (1986). Quantitative fire hazards analysis – overview of needs, methods and limitations. *Fire Safety J.*, **11**(1), 3

BUDNICK, E.K. and EVANS, D.D. (1986). Hand calculations for enclosure fires. *Fire Protection Handbook*, 16th ed. (ed. by A.E. Cote and J.M. Linville), p. 21–19

BUDNICK, E.K. and WALTON, W.D. (1986). Computer fire models. *Fire Protection Handbook*, 16th ed. (ed. by A.E. Cote and J.M. Linville), p. 21–25

BUEHLER, C.A. (1967). Safety showers that work in winter. *Chem. Engng*, **74**(May 22), 204

BUEHLER, J.H., FREEMAN, R.H., KEISTER, R.G., MCREADY, M.P., PESETSKY, B.I. and WATTERS, D.T. (1970). Report on explosion at Union Carbide's Texas City butadiene refining unit. *Chem. Engng*, **77**(Sep. 7), 77

BUEKER, D.J. *et al.* (1993). Fault tree analysis of an HCN pumping system. *Process Safety Environ.*, **71B**, 259

BUETTNER, K. (1950). Conflagration heat. *Survey of German Aviation Medicine, World War II* (Washington, DC: US Govt Printing Office)

BUETTNER, K. (1951a). Effects of extreme heat and cold on human skin. I, Analysis of temperature changes caused by different kinds of heat application. *J. Appl. Physiol.*, **3**, 691

BUETTNER, K. (1951b). Effects of extreme heat and cold on human skin. II, Surface temperature, pain and heat conductivity in experiments with radiant heat. *J. Appl. Physiol.*, **3**, 703

BUETTNER, K. (1952). Effects of extreme heat and cold on human skin. III, Numerical analysis and pilot experiments in penetrating flash radiation effects. *J. Appl. Physiol.*, **5**, 207

BUETTNER, K.J.K. (1957). *Heat Transfer and Safe Exposure Time for Man in Extreme Thermal Environment. ASME paper 57–SA-20*. Am. Soc. Mech. Engrs, New York

BUFFA, E.S. (1980). *Modern Production/Operations Management* (New York: Wiley)

BUFFHAM, B.A. and FRESHWATER, D.C. (1969). Reliability engineering for the process plant industry. *Chem. Engr, Lond.*, **231**, 367

BUFFHAM, B.A., FRESHWATER, D.C. and LEES, F.P. (1971). Reliability engineering – a rational technique for loss prevention. *Major Loss Prevention*, p. 87

BUFFINGTON, M.A. (1973). How to select package boilers. *Chem. Engng*, **80**(Oct. 27), 98

BUGLIARELLO, G., ALEXANDRE, A., BARNES, J. and WAKSTEIN, C. (1976). *The Impact of Noise Pollution* (New York: Pergamon Press)

BUHROW, R.P. (1971). Scheduling piping inspections. *Chem. Engng*, **78**(Jul. 26), 120

BUHROW, R.P. (1980). New API vessel inspection code. *Hydrocarbon Process.*, **59**(5), 135

BUHROW, R.P. (1983). State of the refinery NDTart. *API Proc., Refining Dept, 48th Midyear Mtg*, **62**, 223

BUHROW, R.P. (1985). Areas of vulnerability. *API 50th Midyear Mtg*, **64**, 134

BUHROW, R.P. (1986). Know 'areas of vulnerability'. *Hydrocarbon Process.*, **65**(8), 99

VAN BUIJTENEN, C.J.P. (1980). Calculation of the amount of gas in the explosive region of a vapour cloud released in the atmosphere. *J. Haz. Materials*, **3**(3), 201

BUILDING RESEARCH ESTABLISHMENT (1959). *Principles of Modern Building*, vol. 1, 3rd ed. (London: H.M. Stationery Office)

BUILDING RESEARCH ESTABLISHMENT –see Appendix 28

BUILTJES, P.J. and GULDEMOND, C.P. (1984). Wind tunnel modelling of heavy gas dispersion. *Instn Chem. Engrs, Eur. Branch, Symp. on Heavy Gas Releases – Dispersion and Control*

BUIVID, M.G. and SUSSMAN, M.V. (1978). Superheated liquids containing suspended particles. *Nature*, **275**, 203

BUIVIDAS, L.J. (1981). Coal to ammonia: its status. *Ammonia Plant Safety*, **23**, 67

BULGIN, D. (1945). Electrically conductive rubber. *Trans. Rubber Ind.*, **21**, 188

BULKLEY, W.L. (1967). Technical investigation of major process-industry accidents. *Loss Prevention*, **1**, 23

BULKLEY, W.L. and GINSBURGH, I. (1968). How to load distillates safely. *Hydrocarbon Process. Petrol. Refiner*, **47**(5), 121

BUCKLEY, W.L. and HUSA, H.W. (1962). Combustion properties of ammonia. *Ammonia Plant Safety*, **4**, 81. See also *Chem. Engng Prog.*, **58**(2), 81

BULKLEY, W.L. and JACOBS, R.B. (1966). Hazard of atmospheric releases of large volumes of hydrogen. *Facilities Subcttee, Am. Petrol. Inst., Oct. 4*

BULL, D.C. (1977). Autoignition at hot surfaces. *Loss Prevention and Safety Promotion* 2, p. 165

BULL, D.C. (1979). Concentration limits to the initiation of unconfined detonation in fuel/air mixtures. *Trans. Inst. Chem. Engrs*, **57**, 219

BULL, D. (1982a). Gas phase detonation. In Lee, J.H.S. and Guirao, C.M. (1982), *op. cit.*, p. 975

BULL, D.C. (1982b). Towards an understanding of the detonability of vapour clouds. In Lee, J.H.S. and Guirao, C.M. (1982), *op. cit.*, p. 139

BULL, D.C. (1992). Review of large-scale explosion experiments. *Plant/Operations Prog.*, **11**, 33

BULL, D.C., CAIRNIE, L.R., HARRISON, A.J. and MORGAN, P.A. (1980). The influence of natural convection on the critical conditions for hot surface ignition. *Loss Prevention and Safety Promotion 3*, vol. 2, p. 634

BULL, D.C., ELSWORTH, J.E. and HOOPER, G. (1979a). Concentration limits to unconfined detonation of ethane−air. *Combust. Flame*, **35**, 27

BULL, D.C., ELSWORTH, J.E. and HOOPER, G. (1979b). Susceptibility of methane−ethane mixtures to gaseous detonation in air. *Combust. Flame*, **34**, 327

BULL, D.C., ELSWORTH, J.E., HOOPER, G. and QUINN, C.P. (1976). A study of spherical detonation in mixtures of methane and oxygen diluted by nitrogen. *J. Phys., D Appl. Phys.*, **9**, 1991

BULL, D.C., ELSWORTH, J.E., MCCLEOD, M.A. and HUGHES, D. (1981). Initiation of unconfined gas detonations in hydrocarbon-air mixtures by a sympathetic mechanism. *Prog. Astronaut Aeronaut.*, **75**, 61

BULL, D.C., ELSWORTH, J.E., SHUFF, P.J. and METCALFE, E. (1982). Detonation cell structures in fuel/air mixtures. *Combust. Flame*, **45**, 7

BULL, D.C. and GRANT, A.J. (1975). The influence of natural convection on critical autoignition temperatures. *Second European Symp. on Combustion*, p. 545

BULL, D.C., PYE, D.B. and QUINN, C.P. (1976). Autoignition and cool-flame behaviour in a convective system. *Combust. Flame*, **27**, 399; 28, 207

BULL, D.C. and QUINN, C.P. (1975). *Second European Symp. on Combustion*, p. 551

BULL, D. and STRACHAN, D. (1992). *Liquefied natural gas safety research* (London: Shell Int. Petrol. Co.)

BULL, J.P. (1971). Revised analysis of mortality due to burns. *The Lancet*, Nov. 20, 1133

BULL, J.P. and FISHER, A.J. (1954). A study of mortality in a burns unit: a revised estimate. *Ann. Surgery*, **139**, 269

BULL, J.P. and SQUIRE, J.R. (1949). A study of mortality in a burns unit. *Ann. Surgery*, **130**, 160

BULLEN, K.E. (1963). *Introduction to Theoretical Seismology* (Cambridge: Cambridge Univ. Press)

BULLEN, K.E. and BOLT, B.A. (1985). *An Introduction to the Theory of Seismology* (Cambridge: Cambridge Univ. Press)

BULLEN, M.L. (1978). A combined overall and surface energy balance for fully-developed ventilation-controlled liquid fuel fires in compartments. *Fire Res.*, **1**(3), 171

BULLEN, M.L. and THOMAS, P.H. (1979). Compartment fires with non-cellulosic fuels. *Combustion*, **17**, p. 1139

BULLINGER, H.-J., KERN, P. and MUNTZINGER, W.F. (1987). Design of controls. In Salvendy, G. (1987), *op. cit.*, p. 577

BULLOCH, B.C. (1974). The development and application of quantitative risk criteria for chemical processes. *Chemical Process Hazards V*, p. 289

BULLOFF, J.J. (1987). Other hazards. In Young, J.A, (1987a), *op. cit.*, p. 171

BULLOFF, J.J. and SINCLAIR, J.R. (1976). Agents for amelioration of discharges of hazardous chemicals in water. *Control of Hazardous Chemical Spills*, **3**, p. 277

BULMER, G.H. (1972). Fire safety considerations related to the large scale usage of sodium. *Chemical Process Hazards IV*, p. 79

BULOW, C.L. (1964). Copper alloys for extreme temperatures. *Chem. Engng*, **71**(Jan. 20), 162

BULTZO, C. (1968). Unique compressor problems. *Hydrocarbon Process.*, **47**(3), 115

BULTZO, C. (1972). Problems with rotating and reciprocating equipment. *Chem. Engng Prog.*, **68**(3), 66

BUMP, R.L. (1977). Electrostatic precipitators in industry. *Chem. Engng*, **84**(Jan. 17), 129

BUNDY, A. (1984). *Incidence Calculus: A Mechanism for Probabilistic Reasoning. Res. Paper 216*. Dept of Artificial Intell., Univ. of Edinburgh

BUNDY, F.P. and STRONG, H.M. (1955). Measurement of flame speed, pressure and velocity. *High Speed Aerodynamics and Jet Propulsion*, vol. IX, pt 2 (ed. by B. Lewis, R.N. Pease and H.S. Taylor) (London: Geoffrey Cumberledge and Oxford Univ. Press), p. 343

BUNGAY, P.M., DEDRICK, R.L. and MATTHEWS, H.B. (1980). Pharmacokinetics of environmental contaminants. In Haque, R. (1980), *op. cit.*, p. 369

BUNGE, P.F. and HONIGH, C.G.F. (1969). Absorption of chlorine from an explosive mixture. *Symp. on Inherent Hazards of Manufacturing and Storage in the Process Industry*, The Hague, p. 203

BUNN, A.R. and LEES, F.P. (1988). Expert design of plant handling hazardous materials. Design expertise and computer aided design methods with illustrative examples. *Chem. Engng Res. Des.*, **66**, 419

BUNNER, W.R. (1976). How to develop training for hazardous material spill control. *Control of Hazardous Material Spills*, **3**, p. 161

BUNTON, J.F. and BUCKLAND, W.G.N. (1975). Recovery of low grade heat. *Plant Engr*, **19**(4), 9

BUONICORE, A.J. (1980). Air pollution control. *Chem. Engng*, **87**(Jun. 30), 81

BUPP, I. and DERIAN, J.C. (1974). *Trends in Light Water Reactor Capital Costs in the US: Causes and Consequences*. Center for Policy Alternatives, Massachusetts Inst. Technol, Cambridge, MA

BURCAT, A. and HASSON, A. (1982). Deflagration to detonation transitions in hexane and heptane mixtures with oxygen. *Combustion*, **19**, p. 625

BURCAT, A., SCHELLER, K. and LIFSHITZ, A. (1971). Shock tube investigation of comparative ignition delay time for C1−C5 alkanes. *Combust. Flame*, **16**, 29

BURCH, W.M. (1986). Process modifications and new chemicals. *Chem. Engng Prog.*, **82**(4), 5

BURCHELL, L.R. (1980). Energy and petrochemicals in the Middle East. *Chemy Ind., Jan. 5*, 32

BURCHETT, D.K. (1980). Sizing of emergency vents taking liquid entrainment into account. *Loss Prevention and Safety Promotion 3*, vol. 2, p. 915

BURCK, C.G. (1975). What we can learn from BART's misadventures. *Fortune*, **42**(1), 104

BURDEKIN, F.M. and STONE, D.E.W. (1966). The crack opening displacement approach to fracture mechanics in yielding materials. *J. Strain Analysis*, **1**, 145

BURDEN, F.A. and BURGOYNE, J.H. (1949). The ignition of flame reactions of ethylene oxide. *Proc. R. Soc., Ser. A.*, **199**, 328

BURDICK, G.R., FUSSELL, J.B., RASMUSON, D.M. and WILSON, J.R. (1977). Phased mission analysis: a review of new developments and an application. *IEEE Trans Reliab.*, **R-26**, 43

BURDICK, G.R., MARSHALL, N.H. and WILSON, J.R. (1976). *COMCAN − A Computer Program for Common Cause Analysis. Rep. ANCR-1314*. Aerojet Nucl. Corp., Idaho Falls, ID

BUREAU OF MINES − *see Appendix 28*

BUREAU OF NATIONAL AFFAIRS INC. (1978). *Chemical Regulation Reporter* (Washington, DC)

BURET, J., HERVO, Y. and TESSIER, J. (1985). LPG carriers at Nord Mediterranee − experience and new trends. *Gastech 84*, p. 350

BURFORD, R.R. (1976). The use of AFFF in sprinkler systems. *Fire Technol.*, **12**, 5

BURGE, M. and SCOTT, D. (1992). Assessment of low probability/high consequence risks with emphasis on management systems. *Loss Prevention and Safety Promotion 7*, vol. 1, 48–1

BURGE, S.J. (1990). Control of exposure to hazardous substances. *Safety and Loss Prevention in the Chemical and Oil Processing Industries*, p. 139

BURGE, W.H. (1975). *Recursive Programming Techniques* (Reading, MA: Addison-Wesley)

BURGER, J.R. and HOOGENDAM, C. (1965). The development of a new check valve and some special applications. *Am. Soc. Mech. Engrs, Twentieth Ann. Petrol. Mech. Engng Conf.* (New York)

BURGER, R. (1982). Cooling towers can make money. *Chem. Engng Prog.*, **78**(2), 84

BURGER, R. (1983). Conservation strategies for cooling towers. *Chem. Engng Prog.*, **79**(12), 33

BURGESS, D. (1969). Thermochemical criteria for explosion hazards. *Am. Inst. Chem. Engrs, Sixty Second Ann. Mtg* (New York), paper 10b

BURGESS, A.R. and LAUGHLIN, G.W. (1972). The cool flame oxidation of n-heptane. Part I – The kinetic features of the reaction. *Combust. Flame*, **19**, 315

BURGESS, D.S., BIORDI, J. and MURPHY, J.N. (1970). Hazards of spillage of LNG in marine transportation. *Bureau of Mines, Washington, DC, Rep. AD-705078*

BURGESS, D., BIORDI, J. and MURPHY, J. (1972). Hazards of spillage of LNG into water. *Bureau of Mines, PMSRC Rep. 4177*

BURGESS, D.S. and GRUMER, J. (1962). Comments on 'The burning rate of liquid fuels from open trays by natural convection' by D.B. Spalding. *Fire Res. Abs. Revs*, **4**, 236

BURGESS, D. and HERTZBERG, M. (1974). Radiation from pool flames. In Afgan, N.H. and Beer, J.M. (1974), *op. cit.*, p. 413

BURGESS, D. and HERTZBERG, M. (1975). The flammability limits of lean fuel-air mixtures: thermochemical and kinetic criteria for explosion hazards. *ISA Trans.*, **14**(2), 129

BURGESS, D.S., MURPHY, J.N. and ZABETAKIS, M.G. (1970). Hazards of LNG spillage in marine transportation. *SRC Rep. MIPR Z-70099–9–92317 S-4105*

BURGESS, D., MURPHY, J.N., ZABETAKIS, M.G. and PERLEE, H.E. (1975). Volume of flammable mixture resulting from the atmospheric dispersion of a leak. *Combustion*, **15**, p. 289

BURGESS, D.S., STRASSER, A. and GRUMER, J. (1961). Diffusive burning of liquid fuels in open trays. *Fire Res. Abs. Revs*, **3**, 177

BURGESS, D. and ZABETAKIS, M.G. (1964). Fire and explosion hazards of LNG. In Vervalin, C.H. (1964a), *op. cit.*, p. 49

BURGESS, I.W., OLAWALE, A.O. and PLANK, R.J. (1992). Failure of steel columns in fire. *Fire Safety J.*, **18**, 183

BURGESS, L.H. (1974). System reliability. *Hospital Engng*, Oct., 3

BURGESS, M.J. and WHEELER, R.V. (1911). Lower limit of inflammation of mixtures of paraffin hydrocarbons with air. *J. Chem. Soc.*, **99**, 2013

BURGESS, M.J. and WHEELER, R.V. (1929). *Paper 54*. Safety in Mines Res. Board, Sheffield

BURGESS, W.A. (1981). *Recognition of Health Hazards in Industry* (New York: Wiley)

BURGOYNE, J.H. (1945). Dust explosions. *IFBA*, Mar., 4

BURGOYNE, J.H. (1948). Low limits of inflammability. *Research*, **1**, 528

BURGOYNE, J.H. (1949a). Inflammability of gases. *Research*, **2**, 512

BURGOYNE, J.H. (1949b). The inflammability of oil mists. *Fuel*, **28**(1), 2

BURGOYNE, J.H. (1950). Oil storage tank fires. *Oil*, **1**(9), 12

BURGOYNE, J.H. (1952). Tank fire control by agitation (letter). *Petrol. Times*, **56**(1427), Apr. 18, 315

BURGOYNE, J.H. (1956). Gaseous explosions. *J. R. Inst. Chem.*, **80**, 67

BURGOYNE, J.H. (1957). Mist and spray explosions. *Chem. Engng Prog.*, **53**(3), 121

BURGOYNE, J.H. (1961). Principles of explosion prevention. *Chem. Process Engng*, **42**, 157

BURGOYNE, J.H. (1963). The flammability of mists and sprays. *Chemical Process Hazards II*, p. 1

BURGOYNE, J.H. (1965a). Accidental ignitions and explosions of gases in ships. *Instn Mar. Engrs Trans.*, **77**(5), 129

BURGOYNE, J.H. (1965b). Explosion and fire hazards associated with the use of low-temperature industrial fluids. *Chem. Engr, Lond.*, **185**, CE7

BURGOYNE, J.H. (1965c). Intrinsic safety in the United Kingdom. *ISA Monogr. 110* (Pittsburgh, PA: Instrum. Soc. Am.), p. 95

BURGOYNE, J.H. (1967a). Designing for protection against dust explosions. *Chemy Ind.*, May 27, 854

BURGOYNE, J.H. (1967b). Electrical safety in Great Britain. In *ISA Monogr. 111* (Pittsburgh, PA: Instrum. Soc. Am.), p. 328

BURGOYNE, J.H. (1969). Electrical safety in Great Britain. *Electrical Safety ISA Monogr. 112* (Pittsburgh, PA: Instrum. Soc. Am.), p. 19

BURGOYNE, J.H. (1971). Division 2 as an objective. Electrical safety in Great Britain. *Electrical Safety Practices, ISA Monogr. 133* (Pittsburgh, PA: Instrum. Soc. Am.)

BURGOYNE, J.H. (1975). Risk management and chemical manufacture. *Chem. Engr, Lond.*, **295**, 151

BURGOYNE, J.H. (1977). Dust explosion reliefs. A note on design calculations. *Chemical Process Hazards VI*, unbound paper

BURGOYNE, J.H. (1978). The testing and assessment of materials liable to dust explosion or fire. *Chemy Ind.*, Feb. 4, 81

BURGOYNE, J.H. (chrmn) (1980). *Offshore Safety, Cmnd 7866* (London: H.M. Stationery Office)

BURGOYNE, J.H. (1981/82). The scientific investigation of occurrence of fire. *Fire Safety J.*, **4**(3), 159

BURGOYNE, J.H. (1982). Accident investigation. *J. Occup. Accidents*, **3**(4), 289

BURGOYNE, J.H. (1985a). Abbeystead (letter). *Chem. Engr, Lond.*, **411**, 7

BURGOYNE, J.H. (1985b). Legislative control of industrial operations. *Chem. Engr, Lond.*, **414**, 70

BURGOYNE, J.H. (1985c). Preventing emergencies in the process industries. *Chem. Engr*, **413**, 33

BURGOYNE, J.H. (1986a). The flash-back hazard in stacks discharging flammable gases. *Hazards IX*, p. 37

BURGOYNE, J.H. (1986b). Reflections on process safety. *Hazards IX*, p. 1

BURGOYNE, J.H. (1987). Safe disposal of relief discharges. *Hazards from Pressure*, p. 201

BURGOYNE, J. (1990). Qualifications in process safety. *Chem. Engr, Lond.*, **484**, 50

BURGOYNE, J.H., BETT, K.E. and LEE, R. (1967). The explosive decomposition of ethylene oxide vapour under pressure. Pt II. *Chemical Process Hazards II*, p. 1

BURGOYNE, J.H., BETT, K.E. and MUIR, R. (1960). The explosive decomposition of ethylene oxide vapour under pressure. *Chemical Process Hazards I*, p. 30

BURGOYNE, J.H. and BURDEN, F.A. (1948). Ethylene oxide explosions. *Nature*, **162**(2), 181

BURGOYNE, J.H. and BURDEN, F.A. (1949). Explosive decomposition of ethylene oxide. *Nature*, **163**, 723

BURGOYNE, J.H. and COHEN, L. (1954). The effect of droplet size on flame propagation in liquid aerosols. *Proc. R. Soc., Ser. A.*, **225**, 375

BURGOYNE, J.H. and COX, R.A. (1953). The oxidation of ethylene and propylene in the gas phase. *J. Chem Soc.*, **180**, 876

BURGOYNE, J.H. and CRAVEN, A.D. (1973). Fire and explosion hazards in compressed air systems. *Loss Prevention*, **7**, 79

BURGOYNE, J.H. and FAULS, J. (1963). Dust explosion prevention in the United Kingdom with special reference to starch and allied industries. *Die Starke*, **15**(7), 260

BURGOYNE, J.H. and HIRSCH, H. (1954). The combustion of methane at high temperatures. *Proc. R. Soc., Ser. A.*, **227**, 73

BURGOYNE, J.H. and KAPUR, P.K. (1952). The oxidation of ethylene oxide in the vapour phase. *Trans Faraday Soc.*, no. 351, **48**(3), 234

BURGOYNE, J.H. and KATAN, L.L. (1947). Fires in open tanks of petroleum products: some fundamental aspects. *J. Inst. Petrol.*, **33**, 158

BURGOYNE, J.H., KATAN, L.L. and RICHARDSON, J.F. (1949a). The application of air foams to oils burning in bulk. *J. Inst. Petrol.*, **35**(312), 795

BURGOYNE, J.H., KATAN, L.L. and RICHARDSON, J.F. (1949b). The use of solid carbon dioxide to extinguish the burning of liquids. *J. Inst. Petrol.*, **35**(312), 818

BURGOYNE, J.H. and NEALE, R.F. (1953a). Limits of inflammability and spontaneous ignition of some organic combustibles in air. *Fuel*, **32**(1), 17

BURGOYNE, J.H. and NEALE, R.F. (1953b). Some new measurements of inflammability ranges in air. *Fuel*, **32**(1), 5

BURGOYNE, J.H. and NEWITT, D.M. (1955). Crankcase explosions in marine engines. *Trans Instn Mar. Engrs*, **67**(8), 255

BURGOYNE, J.H., NEWITT, D.M. and THOMAS, A. (1954). Explosion characteristics of lubricating oil mist (in crank cases). *The Engineer, Lond.*, **198**(5140), Jul. 30, 165

BURGOYNE, J.H. and RICHARDSON, J.F. (1948). Fire and explosion risks associated with liquid methane. *Fuel*, **27**(2), 37

BURGOYNE, J.H. and RICHARDSON, J.F. (1949a). Extinguishing burning liquids by the application of non-flammable gases and liquids. *Fuel*, **28**(7), 150

BURGOYNE, J.H. and RICHARDSON, J.F. (1949b). The inflammability of oil mists. *Fuel*, **28**, 2

BURGOYNE, J.H. and ROBERTS, A.F. (1968a). The spread of flame across a liquid surface: II, Steady-state conditions. *Proc. R. Soc., Ser. A.*, **308**, 55

BURGOYNE, J.H. and ROBERTS, A.F. (1968b). The spread of flame across a liquid surface: III, A theoretical model. *Proc. R. Soc., Ser. A.*, **308**, 69

BURGOYNE, J.H., ROBERTS, A.F. and QUINTON, P.G. (1968). The spread of flame across a liquid surface: I, The induction period. *Proc. R. Soc., Ser. A.*, **308**, 39

BURGOYNE, J.H. and STEEL, A.J. (1962). Flammability of methane-air mixtures in water-base foams. *Fire Res. Abs Revs*, 4(1–2), 67

BURGOYNE, J.H. and WEINBERG, F.J. (1956). The mechanism of flame propagation. *Univ. of Sheffield Fuel Soc. J.*, 7, 6

BURGOYNE, J.H. and WILLIAMS-LEIR, G. (1948a). The influence of incombustible vapours on the limits of inflammability of gases and vapours in air. *Proc. R. Soc., Ser. A.*, **193**, 525

BURGOYNE, J.H. and WILLIAMS-LEIR, G. (1948b). The limits of inflammability of gases in the presence of diluents. A critical review. *Fuel*, **27**(4), 118

BURGOYNE, J.H. and WILLIAMS-LEIR, G. (1949). Inflammability of liquids. *Fuel*, **28**(7), 145

BURGOYNE, J.H. and WILSON, M.J.G. (1957). Tests on commercial explosion relief valves. *Br. Shipbuilding Res. Ass., Rep. 220*

BURGOYNE, J.H. and WILSON, M.J.G. (1960). The relief of pentane vapour-air explosions in vessels. *Chemical Process Hazards I*, p. 25

BURHOP, E.H.S. (1977). Acceptable risk related to nuclear radiation. In Council for Science and Society (1977), *op. cit.*, p. 63

BURK, A.F. (1990). Regulatory initiatives and related activities in the United States. *Plant/Operations Prog.*, **9**, 254

BURK, A.F. (1992). Strengthen process hazards reviews. *Chem. Engng Prog.*, **88**(6), 90

BURK, A.F. (1994). International regulations. *Process Safety Prog.*, **13**, 9

BURK, A.F. and SMITH, W.L. (1990). Process safety management within Dupont. *Plant/Operations Prog.*, **9**, 269

BURK, H.S. (1981). Conceptual design of refinery tankage. *Chem. Engng*, **88**(Aug. 24), 107

BURKARDT, F. (1986). The human factor in safety, a review. *Loss Prevention and Safety Promotion*, **5**, 1.1

BURKE, B.G. and MOORE, D.E. (1990). Leak-before-break assessment of two anhydrous ammonia spheres. *Ammonia Plant Safety*, **30**, p. 91

BURKE, J.J. and WEISS, V. (eds) (1980). *Risk and Failure Analysis for Improved Performance and Reliability* (New York: Plenum Press)

BURKE, P.Y. (1982). Compressor intercoolers and after-coolers: predicting off-performance. *Chem. Engng*, **89**(Sep. 20), 107

BURKE, S.P. and SCHUMANN, T.E.W. (1928a). Diffusion flames. *Combustion*, **1**, p. 2

BURKE, S.P. and SCHUMANN, T.E.W. (1928b). Diffusion flames. *Ind. Engng Chem.*, **20**, 998

BURKINSHAW, L.D. (1968). PPO foam for cryogenic applications. *LNG Conf. 1*

BURKITT, J.K. (1965). Reliability performance of an on-line digital computer when controlling a plant without the use of conventional controllers. In: *Automatic Control in the Chemical Process and Allied Industries* (London: Soc. Chem. Ind.)

BURKLIN, C.E., SUGAREK, R.L. and KNAPF, F.C. (1977). *Development of Improved Emission Factors and Process Descriptions for Petroleum Refining. EPA Proj. 68–02–1889.* Radian Corp, Rep. to Environmental Protection Agency

BURKLIN, C.R. (1972). Safety standards, codes and practices for plant design. *Chem. Engng*, **79**(Oct. 2), 56; **79**(Oct. 16), 113; **79**(Nov. 13), 143

BURKLIN, C.R. (1979). *The Process Plant Designer's Handbook of Codes and Standards* (Houston, TX: Gulf Publ. Co.)

BURLESON, D.C., MALPASS, D.B., MUDRY, W.L. and WATSON, S.C. (1975). A simple method for measuring the 'pyrophoric limit' of metal alkyl solutions. *Transport of Hazardous Cargoes 4*

BURMAN, N.M. (ca. 1989). Fragmentation and performance prediction of high explosive munitions. *Int. Symp on Ballistics, Brussels*, p. 1

BURNELL, J.G. (1947). Flow of boiling water through nozzles, orifices and pipes. *Engineering, Lond.*, **164**(4272), 572

BURNETT, W.W., KING, E.G., GRACE, M. and HALL, W.F. (1977). Hydrogen sulphide poisoning: review of 5 years experience. *Can. Med. Ass. J.*, **117**, 1277

BURNHAM, W.D. and HALL, A.R. (1985). *Prolog Programming and Applications* (London: Macmillan)

BURNS (1973). Hazardous materials accidents and their impact on the chemical industry. *Prevention of Occupational Risks 2*, p. 475

BURNS, B.W. (1967). Design control centres to resist explosions. *Hydrocarbon Process.*, **46**(11), 257

BURNS, C.C. (1988). Interactions of risk analysis and contingency planning in risk management. *J. Loss Prev. Process Ind.*, **1**(3), 141

BURNS, D.J. and PITBLADO, R.M. (1993). A modified HAZOP methodology for safety critical system assessment. In Redmill, F. and Anderson, T. (1993), *op. cit.*

BURNS, D.J., GRANT, M.M. and FITZGERALD, B.P. (1991). A structured approach to hazard assessment. *Safety Developments in the Offshore Oil and Gas Industry* (London: Instn Mech. Engrs), p. 115

BURNS, J. (1989). *Piper Alpha Public Inquiry. Transcript, Day 70*

BURNS, L.A., CLINE, D.M. and LASSITER, R.R. (1982). *Exposure Assessment Modeling System (EXAMS). User Manual and System Documentation, REP. EPA-600/3–82–023.* (Athens, GA: EPA Athens Res. Lab.)

BURNS, R. (1974). Blast rips tanker apart in dock. *Fire Engng.*, **127**(Nov.), **32**

BURNS, R. (1990). The Canadian Chemical Producers' Association transportation code of practice. *Plant/ Operations Prog.*, **9**, 242

BURNS, R.C. (1971). How to specify centrifugal compressors. The casing, nozzles and auxiliary piping. *Hydrocarbon Process.*, **50**(10), 79

BURNS, W. (1968). *Noise and Man* (London: John Murray)

BURNS, W. and ROBINSON, D.W. (1970). *Hearing and Noise in Industry* (London: HM Stationery Office)

BURNUP, R.G. (1986). Confidence in probability risk assessment and the Sample computer programme. *SRS Quart. Digest*, Jul., 23

BURR, A.H. (1981). *Mechanical Analysis and Design* (Amsterdam: Elsevier)

BURR, I.W. (1953). *Engineering Statistics and Quality Control* (New York: McGraw-Hill)

BURRAGE, L.J. (1971). Organized safety training in educational establishments as a necessary prerequisite for industrial safety. *Major Loss Prevention*, p. 199

BURRELL, G.A. (1919). Contributions from the Chemical Warfare Service, USA. *Ind. Engng Chem.*, **11**, 93

BURST, J.F. and SPIECKERMAN, J.A. (1967). A guide to selecting modern refractories. *Chem. Engng*, **74**(Jul. 31), 85

BURSTALL, R.M. (ed.) (1971). *Programming in POP-2* (Edinburgh: Edinburgh Univ. Press)

BURSTALL, R.M. and COLLINS, A. (1971). A primer of POP-2 programming. In Burstall, R.M. (1971), *op. cit.*

BURSTALL, R.M. and POPPLESTONE, R.J. (1971). POP-2 Reference Manual. In Burstall, R.M. (1971), *op. cit.*

BURSTALL, R.M., COLLINS, A. and POPPLESTONE, R.J. (1971). *Programming in POP-2* (Edinburgh: Edinburgh Univ. Press)

BURSTEIN, S. (1989). Crushsyndrome. In *Praxis der Intensivdebehandlung* (ed. P. Lawin) (Stuttgart: Thieme)

BURSTOW, D.J., LOVELAND, R.J., TOMLINSON, R. and WIDGINGTON, D.W. (1981). Radio frequency ignition hazards. *Radio Electron. Engr*, **51**, 151

BURTON, A. (1989). Hydraulic drives. *Plant Engr*, Nov./Dec., 28

BURTON, C.S., LIU, M.K., ROTH, P.M., SEIGNEUR, C. and WHITTEN, G.Z. (1983). Chemical transformations in plumes. In de Wispelaere, C. (1983), *op. cit.*, p. 3

BURTON, I., KATES, R.W. and WHITE, G.F. (1978). *The Environment as Hazard* (New York: Oxford Univ. Press)

BURTON, J.E. (1980). Safety analysis of a shaft-winding system. In Moss, T.R. (1980), *op. cit.*, p. 88

BURTON, P.I. (1981). Keeping the operators in the picture. *Chem. Engr, Lond.*, **366**, 103

BURWELL, J.T. (1957–58). Survey of possible wear mechanisms. *Wear*, **1**, 119

BURZSTEIN, S. (1989). Crush syndrome. *Praxis der Intensivbehandlung* (ed. P. Lawin) (New York: George Thierne Verlag)

BUSCARELLO, R.T. (1968). Practical solutions to vibration problems. *Chem. Engng*, **75**(Aug. 12), 157

BUSCH, A.E. (1965). Liquid waste disposal system design. *Chem. Engng*, **72**(Mar. 29), 83

BUSCH, N.E. (1973). On the mechanics of atmospheric turbulence. In Haugen, D.A. (1973), *op. cit.*, p. 1

BUSCHART, R.J. (1975). An analytical approach to electrical-area classification flammable vapors and gases. *Electrical Safety*, **2**, 195

BUSCHART, R.J. (1991). *Electrical and Instrumentation Safety for Chemical Processes* (London: Chapman & Hall)

BUSCHER, H. (1932). *Grün- und Gelbkreuz* (Hamburg)

BUSCHMANN, C.H. (1968). Hazard evaluation of poisonous cargo. *Carriage of Dangerous Goods*, **1**, 174

BUSCHMANN, C.H. (1973). Container transport of dangerous goods. *Transport of Hazardous Cargoes*, **3**

BUSCHMANN, C.H. (ed.) (1974). *Loss Prevention and Safety Promotion in the Process Industries* (Amsterdam: Elsevier)

BUSCHMANN, C.H. (1975). Experiments in the dispersion of heavy gases and abatement of chlorine clouds. *Transport of Hazardous Cargoes*, **4**

BUSE, F. (1975). The effects of dimensional variations on centrifugal pumps. *Chem. Engng*, **84**(Sep. 26), 93

BUSE, F.W. (1985). Power consumption of double mechanical seals. *Chem. Engng*, **92**(Dec. 9/23), 125

BUSE, F.W. (1992). Take many factors into consideration when selecting pump materials. *Chem. Engng Prog.*, **88**(5), 84; **88**(9), 50

BUSH, A.W., LEWIS, B.A. and WARREN, M.D. (eds) (1989). *Flow Modelling in Industrial Processes* (Chichester: Ellis Horwood)

BUSH, K.E. (1976). Refinery wastewater treatment and reuse. *Chem. Engng*, **83**(Apr. 12), 113

BUSH, M.J. and WELLS, G.L. (1971). Unit plot plans for plant layout. *Br. Chem. Engng*, **16**, 325, 514

BUSH, M.J. and WELLS, G.L. (1972). Computer aided production of unit plot plans for chemical plant. *Decision, Design and the Computer* (London: Instn Chem. Engrs), p. 2:15

BUSH, S.H. (1975). Pressure vessel reliability. *J. Press. Ves. Technol.*, **97**, 54

BUSH, S.H. (1977). Review of reliability in piping in light water reactors. In Gangadharan, A.C. and Brown, S.J. (1977), *op. cit.*, p. 37

BUSH, W.D. (1980). When you're a project engineer. *Hydrocarbon Process.*, **59**(11), 263

BUSHAR, T.A. (1973). Bearings – why they fail. *Chem. Engng*, **80**(Feb. 5), 92

BUSHNELL, D.M. and GOODERUM, P.B. (1968). Atomisation of superheated water jets at low ambient pressures. *J. Spacecraft Rockets*, **5**, 231

BUSINGER, J.A. (1973). Turbulent transfer in the atmospheric surface layer. In Haugen, D.A. (1973), *op. cit.*, p. 67

BUSINGER, J.A. (1982). Equations and concepts. In Nieuwstadt, F.T.M. and van Dop, H. (1982), *op. cit.*, p. 1

BUSINGER, J.A. and ARYA, S.P.S. (1974). Height of the mixed layer in the stably stratified planetary boundry layer. *Adv. Geophys.*, **18A**, 73

BUSINGER, J.A. and YAGLOM, A.M. (1971). Introduction to Obukhov's paper on 'Turbulence in an atmosphere with a non-uniform temperature'. *Boundary Layer Met.*, **2**, 3

BUSINGER, J.A., WYNGAARD, J.C., IZUME, Y. and BRADLEY, E.F. (1971). Flux-profile relationships in the atmosphere surface layer. *J. Atmos. Sci.*, **28**, 181

BUSTIN, W.M. (1963). Static electricity safety studies — marine operations. *API Proc., Div. of Refining*, **43**(III), 307

BUSTIN, W.M. (1973–74). *API Statics Res. Program, Pt 1* (Am. Petrol. Inst.)

BUSTIN, W.M. and DUKEK, W.G. (1983). *Electrostatic Hazards in the Petroleum Industry* (New York: Wiley)

BUSTIN, W.M., CULBERTSON, T.L. and SCHLECKSER, L.E. (1957). General consideration of static electricity in petroleum products. *Proc. Am. Petrol. Inst., pt III*, **37**, 24

BUSTIN, W.M., KOSZMAN, I. and TOBYE, I.T. (1964). New theory for static relaxation in high resistivity field. *Hydrocarbon Process.*, **43**(11), 209

BUTCHER, C. (1989a). The dark side of the chemical industry. *Chem. Engr, Lond.*, **458**, 44

BUTCHER, C. (1989b). Polypropylene. *Chem. Engr, Lond.*, **459**, 32

BUTCHER, C. (1989c). Short is beautiful. *Chem. Engr, Lond.*, **462**, 33

BUTCHER, C. (1990a). Incinerating hazardous waste. *Chem. Engr, Lond.*, **471**, 27

BUTCHER, C. (1990b). Load cell basics. *Chem. Engr, Lond.*, **478**, 25

BUTCHER, C. (1991a). CFD comes to the desktop. *Chem. Engr, Lond.*, **498**, 23

BUTCHER, C. (1991b). Emergency shutdown valves. *Chem. Engr, Lond.*, **488**, 21

BUTCHER, C. (1991c). Low flow measurement. *Chem. Engr, Lond.*, **491**, 21

BUTCHER, C. (1991d). Vacuum pumps are looking green. *Chem. Engr, Lond.*, **28**

BUTCHER, C. (1992a). A bright future for PC-CAD. *Chem. Engr, Lond.*, **532/3**, 19

BUTCHER, C. (1992b). Environmental auditing: a sea of standards. *Chem. Engr, Lond.*, **529**, s26

BUTCHER, C. (1992c). More poke, less pounds. *Chem. Engr, Lond.*, **532/3**, 18

BUTCHER, C. (1992d). Secondhand, not second-best. *Chem. Engr, Lond.*, **520**, 21

BUTCHER, C. (1993a). Buy used, be careful. *Chem. Engr, Lond.*, **548**, 17

BUTCHER, C. (1993b). Trouble-free condensate handling. *Chem. Engr, Lond.*, **549**, 24

BUTCHER, C. (1993c). Steady advance in control. *Chem. Engr, Lond.*, **542**, 17

BUTCHER, C. (1994). Safety management: learning from the Hickson fire. *Chem. Technol. Europe*, **1**(3), 30

BUTCHER, C., MASCONE, C.F. and SANTAQUILANI, A.G. (1991). Engineering ethics: how chemical engineers respond. *Chem. Engr, Lond.*, **507**, 25

BUTCHER, D.W. (ed.) (1983). *On-Line Monitoring of Continuous Process Plants* (Chichester: Ellis Horwood)

BUTCHER, D.W. and SHARPE, H.M. (1983). Problems associated with the marine transportation of monomers in bulk. In: *Loss Prevention and Safety Promotion 4*, vol. 2, p. C-15

BUTCHER, E.G. (1985). Analysis of fire behaviour. *Fire Surveyor*, **14**(3), 27

BUTCHER, E.G. and MOSS, C. (1991). Pressurisation systems as a fire safety measure: difficulties associated with their design and installation. *Fire Surveyor*, **20**(3), 28

BUTCHER, E.G. and PARNELL, A.C. (1979). *Smoke Control in Fire Safety Design* (London: Spon)

BUTCHER, G. (1991). Fire resistance and buildings. *Fire Surveyor*, **20**(6), 6

BUTIKOFER, R.E. (1974). Stable utilities and unit operations. *Chem. Engng Prog.*, **70**(10), 75

BUTLER, C.P., MARTIN, S.B. and WIERSMA, S.J. (1973). Measurements of the dynamics of structural fires. *Combustion*, **14**, 1053

BUTLER, J. and BONSALL, D.J. (1993). Permit to work systems after the Cullen Report. *Health, Safety and Loss Prevention*, p. 478

BUTLER, J., BALL, D.F. and PEARSON, A.W. (1978). Improving perceptions of health and safety needs in industry. *Chem. Ind.*, Oct. 21, 798

BUTLER, J.A.V. (ed.) (1951). *Electrical Phenomena at Interfaces* (London: Methuen)

BUTLER, M.E., NAYAR, M.P. and WHEELER, M.E. (1993). How to facilitate start-up. *Chem Engng*, **100**(6), 82

BUTLER, P. (1973). Standby power supplies — can process industries afford to be without? *Process Engng*, Jun., 96

BUTLER, P. (1974). Some reasons for Class I vessel failures. *Process Engng*, Sept., 4

BUTLIN, R.N. (1971). Polyethylene-dust air flames (letter). *Combust. Flame*, **17**, 446

BUTRAGUENO, J.L. and COSTELLO, J.F. (1978). Safe stand-off distances for pressurised hydrocarbons. In *Control of Hazardous Material Spills*, **4**, 411

BUTT, J.M. (1975). Hazard information systems. *Transport of Hazardous Cargoes*, **4**

BUTT, J.R. (1980). *Reaction Kinetics and Reactor Design* (Englewood Cliffs, NJ: Prentice-Hall)

BUTTERFIELD, J.M. (1981). High reliability flow system — an assessment of pump reliability and an optimisation of the number of pumps. *Reliab. Engng*, **2**(1), 7

BUTTERS, J.N. (1971). *Holography and its Technology* (London: Peter Peregrinus)

BUTTERWICK, B. (1976). *User Guide for the Safe Operation of Centrifuges* (Rugby: Instn Chem. Engrs)

BUTTERWORTH, B.E. (1979). *Strategies for Short Term Testing of Mutagens/Carcinogens* (Boca Raton, FL.: CRC Press)

BUTTERWORTH, D. (1977). *Introduction to Heat Transfer* (Oxford: Oxford Univ. Press)

BUTTERWORTH, D. (1980). Unresolved problems in heat exchanger design. *Interflow 80* (Rugby: Instn Chem. Engrs), p. 231

BUTTERWORTH, D. (1992). New twists in heat exchange. *Chem. Engr, Lond.*, **526**, 23

BUTTERWORTH, D. and HEWITT, G.F. (1977). *Two-phase Flow and Heat Transfer* (Oxford: Oxford Univ. Press)

BUTTERWORTH, G.J. (1979a). The detection and characterisation of electrostatic sparks by radio methods. *J. Electrostatics*, **5**, 97

BUTTERWORTH, G.J. (1979b). Electrostatic ignition hazards associated with the preventative release of fire extinguishing fluids. *Electrostatics*, **5**, 161

BUTTERWORTH, G.J. and BROWN, K.P. (1982). Assessment of electrostatic ignition hazards during tank filling with the aid of computer modelling. *J. Electrostatics*, **13**, 9

BUTTERWORTH, G.J. and DOWLING, P.D. (1981). Electrostatic effects with portable CO2 fire extinguishers. *J. Electrostatics*, **11**, 43

BUTTERY, L. (1978). New survey of US maintenance costs. *Hydrocarbon Process.*, **57**(1), 85

BUTTIGIEG, M.A. and SANDERSON, P.M. (1991). Emergent features in visual display design for two types of failure detection tasks. *Hum. Factors*, **33**, 631

BUTTOLPH, M.A. (1979). People in laboratories. *Instn Chem. Engrs, Symp. on Safety in Chemical Engineering Departments* (London)

BUTWELL, K.F., HAWKES, E.N. and MAGO, B.F. (1973). Corrosion control in CO_2 removal systems. *Ammonia Plant Safety*, **15**, p. 50. See also *Chem. Engng Prog., 1973*, **69**(2), 57

BUTWELL, K.F., KUBEK, D.J. and SIGMUND, P.W. (1980). Amine Guard III. *Ammonia Plant Safety*, **21**, p. 156

BUTZ, C.H. (1966). When electricity is cheaper than steam for pipe tracing. *Chem. Engng*, **73**(Oct. 10), 232

BÜTZER, P. and NAEF, H. (1991). MET. *Model for effects of toxic gases.* (Basel: Swiss Fed. Comm. for Chemical Emergencies and Accidents)

BÜTZER, P. and NAEF, H. (1992). Model for effects of toxic gases (MET). *Swiss Chem.*, **14**(1), 7

BUTZERT, H.E. (1976). Ammonia/methanol plant start-up. *Ammonia Plant Safety*, **18**, p. 78

BUXTON, A.G. (1980). Handling of toxic gases. In Hughes, D.J. (1980), *op. cit.*, paper 4

BUYERS, T. (1972). Philosophy for plant commissioning. *Chem. Process.*, **18**(8), 89

BUZACOTT, J.A. (1970a). Markov approach to finding failure times of repairable systems. *IEEE Trans. Reliab.*, **R-19**, 128

BUZACOTT, J.A. (1970b). Network approaches to finding the reliability of repairable systems. *IEEE Trans. Reliab.*, **R-19**, 140

BUZACOTT, J.A. (1976). Automatic transfer lines with buffer stocks. *Int. J. Prod. Res.*, **5**, 183

BUZACOTT, J.A., WEAVING, A.H. and WESOLOWSKI-LOW, T.A. (1967). Quantitative safety. *Trans. Soc. Instrum. Technol.*, **19**, 60

BUZZELLI, D.T. (1990). Restoring public confidence in the safety of chemical plants. *Plant/Operations Prog.*, **9**, 145

BYERS, H.R. (1959). *General Meteorology* (New York: McGraw-Hill)

BYGGSTO/YL, S. and SAETRAN, L.R. (1983). An integral model for gravity spreading of heavy gas clouds. *Atmos. Environ.*, **17**, 1615

BYRD, J.F. (1968). The challenge of pollution control. How to organise a control program. *Chem. Engng*, **75**(Oct. 14), 50

BYRD, W.K. (1970). Transportation of hazardous materials. In National Academy of Engineering (1970), *op. cit.*, p. 29

BYRNE, E.J. (1969). Measuring and controlling. *Chem. Engng*, **76**(Apr. 14), 189

BYRNE, J.F. (1975). Good records control accidents. *Hydrocarbon Process.*, **54**(11), 335

BYRNE, J.R., MOIR, F.E. and WILLIAMS, R.D. (1989). Stress corrosion in a 12 ktonne fully-refrigerated ammonia storage tank. *Ammonia Plant Safety*, **29**, 122

CABINET OFFICE (1967). *The Torrey Canyon* (London: HM Stationery Office)

CABRINI, G. and CUSMAI, G. (1977). Failure in urea plant, carbamate condenser. *Ammonia Operations Prog., Safety*, **19**, p. 129

CABROL, B., ROUX, A. and LHOMME, V. (1987). Interpretation of the Thorney Island Phase I trials with the box model CIGALE2. *J. Haz. Materials*, **16**, 201

CACERES, S. and HENLEY, E.J. (1976). Process failure analysis by block diagrams and fault trees. *Ind Engng Chem. Fund.*, **15**(2), 128

CADEDDU, M., FAORO, S., VAN DEN BERG, A.C., VAN LEEUWEN, D., GUISEPPETTI, G. and MAZZA, G. (1992). Blast prevention inside underground hydro-electric power plants. *Loss Prevention and Safety Promotion 7*, vol. 1, 37–1

CADLE, S.H., DASCH, J.M. and MULAWA, P.A. (1985). Atmospheric concentrations and the deposition velocity to snow of nitric acid, sulfur dioxide and various particulate species. *Atmos. Environ.*, **19**, 1819

CAGNETTI, P. (1975). Downwind concentrations of an airborne tracer in the neighbourhood of a building. *Atmos. Environ.*, **9**, 739

CAGNOLATTI, J.P. (1988). Stress corrosion cracking of condensing turbine rotors in ammonia plants (panel discussion). *Ammonia Plant Safety*, **28**, 187

CAHAN, V. (1980). Benzene verdict: a knell for pending OSHA rules? *Chem. Engng*, **87**(Aug. 11), 73

CAHAN, V. (1981). Cost-benefit analysis is axed by court ruling. *Chem. Engng*, **88**(Jul. 13), 71

CAHAN, V. (1982a). In limbo: OSHA's cancer policy. *Chem. Engng*, **89**(Jul. 24), 27

CAHAN, V. (1982b). OSHA: is it going back instead of forward? *Chem. Engng*, **89**(Jan. 25), 47

CAHAN, V., RECIO, M., TREWHITT, J. and HOPPE, R. (1982). Regulatory reform: moving, but is it enough? *Chem. Engng*, **89**(Sept. 20), 57

CAHILL, L.B. (1973). Defining plant, noise problems. *Chem. Engng Prog.*, **69**(10), 65

CAHNERS EXHIBITIONS LTD (1985). Electronics in Oil and Gas 85. *Proceedings of the Technical Programme* (London)

CAIRNEY, E.M. and CUDE, A.L. (1971). The safe dispersal of large clouds of flammable heavy vapours. *Major Loss Prevention*, p. 163

CAIRNIE, T. (1993). *Contaminated Land* (London: Chapman & Hall)

CAIRNS, A. and GARRETT, M. (1990). Getting the plant right first time. *Chem. Engr, Lond.*, **470**, 53

CAIRNS, F. (1984). Explosion protection for fabric dust collectors. *Chem. Engr, Lond.*, **401**, 27

CAIRNS, J. (1975). The cancer problem. *Sci. American*, **233**(5), 64

CAIRNS, J. (1981). The origin of human cancers. *Nature*, **289**, 353

CAIRNS, J. (ed.) (1986). *Ecoaccidents* (New York: Plenum Press)

CAIRNS, J., DICKSON, K.L. and CROSSMAN, J.S. (1972). The response of aquatic communities to spills of hazardous materials. *Control of Hazardous Material Spills*, **1**, 179

CAIRNS, J.J. and FLEMING, K.N. (1977). *STADIC—A Computer Code for combining Probability Distributions. Rep. GA 14055.* General Atomic

CAIRNS, W.J. and ROGERS, P.M. (eds) (1981). *Onshore Impacts of Offshore Oil* (London: Applied Science)

CAKIR, A., HART, D.J. and STEWART, T.F.M. (1980). *Visual Display Terminals* (Chichester: Wiley)

CAKMAK, A.D. (ed.) (1987). *Ground Motion and Engineering Seismology* (Amsterdam: Elsevier)

CALABRESE, A.M. and KREJCI, L.D. (1974). Safety instrumentation for ammonia plants. *Ammonia Plant Safety*, **16**, 100

CALABRESI, G. (1970). *The Costs of Accidents: A Legal and Economic Analysis* (New Haven: Yale Univ. Press)

CALABRO, S.R. (1962). *Reliability Principles and Practice* (New York: McGraw-Hill)

CALCOTE, H.F. (1981). Mechanisms of soot nucleation in flames – a critical review. Combust. *Flame*, **42**, 215

CALCOTE, H.F. *et al.* (1952a) Spark ignition. *Ind. Engng Chem.*, **44**, 2656

CALCOTE, H.F., GREGORY, C.A., BARNETT, C.M. and GILMER, R.B. (1952b). Spark ignition. The effect of molecular structure. *Ind. Engng Chem.*, **44**, 2656

CALDER, C.A. and GOLDSMITH, W. (1971). Plastic deformation and perforation of thin plates resulting from projectile impact. *Int. J. Solids Struct.*, **7**, 863

CALDER, K.L. (1973). On estimating air pollution concentrations from a highway in an oblique wind. *Atmos. Environ.*, **7**, 863

CALDER, W. (1982). Intrinsic safety: effects on loss prevention. *Plant/Operations Prog.*, **1**, 12

CALDER, W. and MAGISON, E.C. (1983). *Electrical Safety in Hazardous Locations* (Pittsburgh, PA: Instrum. Soc. Am.)

CALDWELL, D., ELFERS, F. and FANKHANEL, M. (1978). The vaporized fuel oil system. *Chem. Engng Prog.*, **74**(11), 66

CALDWELL, L.G. (1976). A pneumatic conveying primer. *Chem. Engng Prog.*, **72**(3), 63

CALEF, R.H. (1964). Diaphragm valves. *Chem. Engng*, **71**(Oct. 26), 125

CALIFORNIA ASSEMBLY (1977). Safety of liquefied natural gas facilities. Regular Session, Jan. 17, *No. AB220 1977–78*

CALIFORNIA CONTROL COMMISSION (1978). *Offshore LNG Terminal Study.* Draft rep.

CALIFORNIA ENERGY RESOURCES CONSERVATION AND DEVELOPMENT COMMISSION (1977). *LNG Study: An Assessment of Aisk.* Rep. (Sacramento, CA)

CALIFORNIA INSTITUTE OF TECHNOLOGY (n.d.). *Strong Motion Earthquake Accelerograms.* Earthquake Engng Res. Lab, California Inst. Technol., Pasadena, CA, Serial Reps

CALIFORNIA STATE DEPARTMENT OF PUBLIC HEALTH (1953–57). *Home Safety Project. Final report*

CALIFORNIA STATE LEGISLATURE, ASSEMBLY SUBCOMMITTEE ON ENERGY (1976). *Liquefied Natural Gas Hearings*, Jul. 22–29, 1976, Sacramento, CA

CALIFORNIA STATE LEGISLATURE, ASSEMBLY SUBCOMMITTEE ON ENERGY (1977). *Final Report on the Natural Gas Shortage*, Feb. 8, 1977, Sacramento, CA

CALISTRAT, M.M. (1975). What causes wear in gear-type couplings. *Hydrocarbon Process.*, **54**(1), 53

CALISTRAT, M. (1978–). Extend gear couping life. *Hydrocarbon Process.*, 1978, **57**(1), 112; 1979, **58**(1), 115

CALLAHAN, B. and MCCAW, P.A. (1993). Risk assessment alleviates environmental liabilities. *Hydrocarbon Process.*, **72**(8), 61

CALLAHAN, H.E. (1968). Practical problems in the designing and siting of explosives facilities. *Ann. NY Acad. Sci*, **152**(art.1), 61

CALLANDER, B.A. (1986). Short range dispersion within a system of regular valleys. In de Wispelaere, C., Schiermeier, F.A. and Gillani, N.V. (1986), *op. cit.*, p. 253

CALLEN, R.B., DIPERNA, C.J. and MILLER, D.E. (1993). A multidimensional crisis management programme. *Health, Safety and Loss Prevention*, p. 427

CALLISTER, W.G. (1976). The role of non-destructive testing in loss prevention. *Soc. Chem. Ind., Symp. on Loss Prevention* (London)

CALMON, C. (1980). Explosion hazards of nitric acid in ion-exchange equipment. *Chem. Engng*, **87**(Nov.17), 271

CALVERT, J.G. and PITTS, J.N. (1966). *Photochemistry* (New York: Wiley)

CALVERT, W.R. (1967). Fundamentals of air purification. *Ammonia Plant Safety*, **9**, 1

CALVERT, C. (1992a). Environmentally-friendly catalysis using non-toxic supported reagents. *Environ. Protect. Bull.* 021 (November 1992), 3–9

CALVERT, C. (1992b). Environmentally-friendly catalysts using non-toxic supported reagents. *IChemE North Western Branch Symposium Papers*, 3, 4.1–4.16

CALZADA, M., ESPASA, M.L. and BARRERA, A. (1992). Expert systems in safety promotion. *Loss Prevention and Safety Promotion 7*, vol. 3, 148–1

CAMARDA, P., CORSI, F. and TRENTADUE, A. (1978). An efficient simple algorithm for fault tree automatic synthesis from the reliability graph. *IEEE Trans. Reliab.*, **R-27**, 215

CAMARINOPOULOS, L. and YLLERA, J. (1985). An improved top-down algorithm combined with modularization as a highly efficient method for fault tree analysis. *Reliab. Engng*, **11**, 93

CAMBRAY, P. and DESHAIES, B. (1978). Ecoulement engendree par un piston spherique: solution analytique approchee. *Acta Astronaut.*, **5**, 611

CAMERON, A. (1976). *Basic Lubrication Theory*, 2nd ed. (Chichester: Ellis Horwood)

CAMERON, A. and MCETTLES, C.M. (1983). *Basic Lubication Theory*, 3rd ed. (Chichester: Wiley)

CAMERON, A.M. (1948). *Chemistry in Relation to Fire Risk and Fire Extinction* (London: Pitman)

CAMERON, G.R., COURTICE, D.C. and FOSS, G.L. (1941). Effect of exposing different animals to a low concentration of phosgene 1:1,000,000 (4 mg/m3) for five hours. *First Report on Phosgene Poisoning*

CAMERON, J.A. and DANOWSKI, F.M. (1974). How to select materials for centrifugal compressors. *Hydrocarbon Process.*, **53**(6), 115

CAMERON, J.D. (1969). First aid and casualty treatment. In Handley, W. (1969), *op. cit.*, p. 349

CAMERON, J.D. (1977). First aid and casualty treatment. In Handley, W. (1977), *op. cit.*, p. 386

CAMERON, R.F. (1993). Incorporating human error in risk assessment. In Melchers, R.E. and Stewart, M.G. (1993), *op. cit.*, p. 187

CAMERON, R.F. and CORRAN, E.R. (1993). Reconciling risk criteria in radiation protection and land use planning. In Melchers, R.E. and Stewart, M.G. (1993), *op. cit.*, p. 59

CAMM, A.W. (1977). The handling and disposal of hazardous spillages. *Symp. Hazard 77 – Advances in the Safe Handling and Movement of Hazardous Substances*, Teesside Polytechnic, Middlesbrough

VAN CAMP, C.E., VAN DAMME, P.S., WILLEMS, P.A., CLYMANS, P.J. and FROMENT, G.F. (1985). Severity in the pyrolysis of petroleum fractions. Fundamentals and industrial applications. *Ind. Engng Chem. Process Des. Dev.*, **24**, 561

CAMP, R.C. (1989). *Bench Marking* (Milwaukee, WI: Quality Press)

CAMPAGNE, W.V.L. (1985). Select HPI gas turbines. *Hydrocarbon Process.*, **64**(3), 77; **64**(4), 105

CAMPBELL, A., HOLLISTER, V., DUDA, R. and HART, P. (1982). Recognition of a hidden mineral deposit by an artificial intelligence program. *Science*, **217**, 927

CAMPBELL, C. (1991–). Positive pressure ventilation – tradition vs technology. *Fire Int.*, **132**(Dec./Jan.), 18

CAMPBELL, C., LITTLER, W.B. and WHITWORTH, C. (1932). The measurement of pressures developed in explosion waves. *Proc. R. Soc., Ser. A.*, **137**, 380

CAMPBELL, D.B. (1965). You don't have to hurt people. *Chem. Engng Prog.*, **61**(9), 35

CAMPBELL, D.F. (1958). *Process Dynamics* (New York: Wiley)

CAMPBELL, G.A. and RUTLEDGE, P.V. (1972). Detonation of hydrogen peroxide vapour. *Chemical Proces Hazards*, IV, p. 37

CAMPBELL, H. (1915). Poisonous gases (letter). *Br. Med. J.*, Jun. 19, 1065

CAMPBELL, I.E. (1956) *High Temperature Technology* (New York: Wiley)

CAMPBELL, I.E. and SHERWOOD, E.M. (1967). *High Temperature Materials and Technology* (New York: Wiley)

CAMPBELL, J.A. (ed.) (1983). *Issues in PROLOG Implementation* (New York: Wiley)

CAMPBELL, J.A. (ed.) (1984). *Implementations of Prolog* (Chichester: Ellis Horwood)

CAMPBELL, J.A. and CUENA, J. (1989). *Perspectives in Artificial Intelligence*. vol. 1, *Expert Systems: Applications and Technical Foundations* (Chichester: Ellis Horwood)

CAMPBELL, J.D. (1973). Dynamic plasticity: macroscopic and micrscopic aspects. *Mats Sci. Engng*, **12**, 3

CAMPBELL, J.M. (1978). *Gas Conditioning and Processing* (Norman, OK: Campbell Petroleum Services)

CAMPBELL, J.R. and ALONSO, J.R.F. (1978). To get curves from data points. *Hydrocarbon Process.*, **57**(1), 123

CAMPBELL, J.R. and GADDY, J.L. (1976). Methodology for simultaneous optimization with reliability: nuclear PWR example. *AIChE J.*, **22**, 1050

CAMPBELL, J.W. and MEDES, J.S. (1985). A comparative analysis of the flow of flashing liquids through pipe systems. *Paper C19/85*, p. 217. Instn Mech. Engrs, London

CAMPBELL, P.C. (1981). Inspection and repair of urea reactor and ammonia receiver. *Ammonia Plant Safety*, **23**, 4

CAMPBELL, P.G. (1990). Management systems for control of hazards. *Safety and Loss Prevention in the Chemical and Oil Processing Industries*, p. 165

CAMPBELL, R.W. (1967). Materials of construction for cryogenic processes. *Chem. Engng*, **74**(Oct. 23), 188

CANADINE, I.C. (1991). The management of safety in transport. In Advisory Committee on Dangerous Substances (1991), *op. cit.*, p. 367

CANADINE, I.C. and PURDY, G. (1989). The transportation of chlorine by road and rail in Britain – a consideration of the risks. *Loss Prevention and Safety Promotion 6*, vol. 1, p. 15–1

CANAVAN, H.M. (1967). Design, operation and corrosion trends in pressure vessels. *Loss Prevention*, **1**, 21

CANAWAY, R.T. (1989). Business interruption insurance for North Sea oil and gas production. *Soc. Petrol. Engrs, Conf. on Offshore Europe 89, paper 19220*

CANAWAY, R.T. (1991). Offshore risk reduction measures to improve insurability. *Offshore Operations Post Piper Alpha* (Marine Mgmt Holdings Ltd)

CANAWAY, R.T. (1992). Loss estimation for refineries and chemical plant, and risk improvement. *Major Hazards Onshore and Offshore*, p. 459

CANEILL, J.Y., HAUGUEL, R., DE LA BASTIDE, SOUFFLAND, D. (1985). *Electricite de France, Rep. EDF HE/32.85.15*

CANFIELD, F.B. (1985). Computer-aided engineering in the process industry. *Chem. Engng Prog.*, **81**(9), 31

CANFIELD, F.B. and NAIR, P.K. (1992). The key to computer integrated processing. Comput. *Chem. Engng*, **16**, S473

CANFIELD, J.A. and RUSSELL, L.H. (1969). Measurement of the heat flux within a luminous aviation fuel flame. *Combustion Inst., Eastern Section Mtg* (Pittsburgh, PA)

CANGI, J.W. (1978). Characteristics and corrosion properties of cast alloys. *Chem. Engng Prog.*, **74**(3), 61

CANHAM, W.G. (1966). Piping codes and the chemicals plant, *Chem. Engng*, **73**(Oct. 10), 199

CANHAM, W.G. (1967). New directions in company standards. *Chem. Engng*, **74**(Nov. 20), 127

CANHAM, W.G. (1981). The ANSI/ASME B31.3 piping code: an update. *API Proc., Refining Dept, 46th Midyear Mtg*, **60**, 299

CANHAM, W.G. and HAGERMAN, J.R. (1970). Reduce piping connection costs. *Hydrocarbon Process.*, **49**(5), 131

CANJAR, L.N., MANNING, F.S. *et al.* (1962–). Thermo properties of hydrocarbons. *Hydrocarbon Process.*, **41**(8), 121; **41**(9), 253; **41**(10), 149; **41**(11), 203; **41**(12), 115; 1963, **42**(1), 129; **42**(8), 127; 1964, **43**(6), 177; 1965, **44**(9), 219; **44**(10), 137; **44**(10), 141; **44**(11), 291; **44**(11), 293; **44**(12), 154

CANJAR, L.N., MANNING, F.S. *et al.* (1966). Thermo properties of non-hydrocarbons. *Hydrocarbon Process.*, **45**(1), 135; **45**(1), 139; **45**(2), 158; **45**(2), 161; **45**(3), 137; **45**(3), 141; **45**(4), 161; **45**(4), 165

CANJAR, L.N. and MANNING, F.S. (1967). *Thermodynamic Properties and Reduced Correlations for Gases* (Houston, TX: Gulf Publ. Corp.)

CANNALIRE, C., CLARKE, R., MAZZAROTTA, B. and NICOLETTI, R. (1993). Optimizing safety of control rooms in case of flammable releases. *J. Loss Prev. Process Ind.*, **6**, 313

CANNELL, W. and CHUDLEIGH, R. (1984). *The PWR Decision* (London: Friends of the Earth)

CANNON, A.G. (1977). The collection and assembly of reliability data for components in chemical plants. *First Nat. Conf. on Reliability* (Culcheth, Lancashire: UK Atom. Energy Auth.), NRC 5/8

CANNON, A.G. (1983). Reliability data for pipework. *Euredata 4*, paper 8.2

CANNON, A.G. (1983). Transport of hazardous materials. *Euredata 4*, paper 2.1

CANNON, A.G. (1991a). Development of a large data bank. In Cannon, A.G. and Bendell, A. (1991), *op. cit.*, p. 255

CANNON, A.G. (1991b). Some achievements due to the development of data banks. In Cannon, A.G. and Bendell, A. (1991), *op. cit.*, p. 43

CANNON, A.G. and BENDELL, A. (eds) (1991). *Reliability Data Banks* (London: Elsevier Appl. Sci.)

CANNON, K.J. and DENBIGH, K.G. (1957). Studies on gas-solid reactions – causes of thermal instability. *Chem. Engn Sci.*, **6**, 155

CANNON, N.T. (1976). Plant, maintenance and modification. *Course on Loss Prevention in the Process Industries*. Dept of Chem. Engng, Loughborough Univ. of Technol.

CANNON-BOWERS, J.A., TANNENBAUM, S.I., SALAS, E. and CONVERSE, S.A. (1991). Toward an integration of training theory and technique. *Hum. Factors*, **33**, 281

CANON, J. (1969). Guidelines for selecting process-control valves. *Chem. Engng*, **76**(Apr. 21), 109; **76**(May 5), 105

CANT, R.S. and BRAY, K.N.C. (1989). A theoretical model of premixed turbulent combustion in closed vessels. *Combust. Flame*, **76**, 243

CANTER, D. (ed.) (1980). *Fires and Human Behaviour* (London: Wiley); 1990, 2nd ed. (London: Fulton)

CANTER, D. (1980/81). Fires and human behaviour: emerging issues. *Fire Safety J.*, **3**(1), 41

CANTILLI, E.J. *et al.* (1976). *Transportation Systems Survey, A Literature Survey and Annotated Bibliography* (Washginton, DC: Dept of Transportation)

CANTRELL, C.J. (1982). Vapor recovery for refineries and petrochemical plants. *Chem. Engng Prog.*, **78**(10), 56

CANTWELL, J.R. and BRYANT, R.E. (1973). How to avoid alloy failures in (1) piping by liquid metal attack and (2) flare tips by severe cracking. *Hydrocarbon Process.*, **52**(5), 114

CANU, P., ROTA, R.A., CARRA, S. and MORBIDELLI, M. (1990). Vented gas deflagrations. A detailed mathematical model tuned to a large set of experimental data. *Combust. Flame*, **80**, 49

CAPDEVIELLE, J.P. and GOY, A. (1985). Ultrasonic testing of the wall to bottom weld in a 9% nickel steel storage tank. *Gastech 84*, p. 418

CAPEL, L. and VAN DER HORST, J.M.A. (1970). Stainless steel corrosion in urea synthesis. *Chem. Engng Prog.*, **66**(10), 38

CAPIZZANI, R.E. (1985). An overview of flammable liquid drum storage and protection. *Plant/Operations Prog.*, **4**, 135

CAPLAN, K.J. (1982). Ventilation basics. *Plant/Operations Prog.*, **1**(3), 194

CAPLEN, R.H. (1972). *A Practical Approach to Reliability* (London: Business Books)

CAPLIN, M. (1941). Ammonia-gas poisoning. Forty-seven cases in a London shelter. *The Lancet*, **241**(Jul. 26), 95

CAPLOW, S.D. (1967). Economics of gas turbine drives. *Chem. Engng*, **75**(Mar. 27), 103

CAPOBIANCHI, S. (1983). General informatics structure and software facilities developed for the European Reliability Data System. *Euredata 4*, paper 14.1

CAPOBIANCHI, S. (1991). The European Reliability Data System – ERDS: status and future developments. In Cannon, A.G. and Bendell, A. (1991), *op. cit.*, p. 213

CAPODAGLIO, E., PEZZAGNO, G., BOBBIO, G.C. and CAZZOLI, F. (1969). A study of the respiratory function in workers engaged in the electrolytic production of chlorine and soda. *Med. Lav.*, **60**, 192

CAPORALI, G. (1972). How Montedison makes acrylo. *Hydrocarbon Process.*, **51**(11), 145

CAPP, B. (1988). The maximum experimental safe gap for a spray of higher flash point liquid. *J. Haz. Materials*, **18**(1), 91

CAPP, B. (1992). Temperature rise of a rigid element flame arrester in endurance burning with propane. *J. Loss Prev. Process Ind.*, **5**, 215

CAPP, B. (1994). Flame arresters endurance burning. *Hazards XII*, p. 405

CAPPER, R. (1983). Halon extinguishing agents and their use. *Fire Surveyor*, **12**(1), 31

CAPPER, R. (1985). Controlling oil and LPG fires. *Fire Surveyor*, **14**(2), 33

CAPPS, R.W. and THOMPSON, J.R. (1993). Statistical safety factors reduce overdesign. *Hydrocarbon Process.*, **72**(11), 77

CAPRARO, M.A. and STRICKLAND, J.H. (1989). Preventing fires and explosions in pilot. *Plant/Operations Prog.*, **8**, 189

CAPRIO, V., INSOLA, A. and LIGNOLA, P.G. (1977). Isobutane cool flame investigation in a continuous stirred tank reactor. *Combustion*, **16**, p. 1155

CAPUTO, R.J. (1987). Engineering for safer plants. *Preventing Major Chemical Accidents*, p. 3.43, 3.57

CAPUTO, R., DENMARK, E., HEITNER, I. and MAYO, H.C. (1967). Noise levels in high capacity ammonia plants. *Ammonia Operations Prog. Safety*, **9**, p. 56

CARBERRY, J. (1976). *Chemical and Catalytic Reaction Engineering* (New York: McGraw-Hill)

CARBONELL, J.R. (1966). A queueing model of many- instrument visual sampling. *IEEE Trans. Hum. Fact. Electron.*, *HFE-4, 157*

CARDILLO, P. (1988). Thermal decomposition of 3−amino-5- methyl isoxazole. *J. Loss Prevention Process Ind.*, **1**(1), 46

CARDILLO, P. (1994). Calorimetric data for hazard process assessment: alkene epoxidation with peracids. *J. Loss Prev. Process Ind.*, **7**, 33

CARDILLO, P. and ANTHONY, E.J. (1978). Hybrid gas and dust systems. *La Rivista dei Combustibili*, **32**, 390

CARDILLO, P. and ANTHONY, E.J. (1979). *Dust Explosion and Fires. Guide to the Literature.* Stazione Sperimentale per i Combustibili, S. Donato Milanese

CARDILLO, P. and CATTANEO, M. (1991). Application of DSC with atmospheric control in thermal hazard evaluation: a discussion of sources of error. *J. Loss Prev. Process Ind.*, **4**, 283

CARDILLO, P., CATTANEO, M., BERATTO, S. and CANAVESI, L. (1993). Data-oriented hazard assessment as process optimization data: the isomerization of an 'oxime'. *J. Loss Prev. Process Ind.*, **6**, 69

CARDILLO, P. and GIRELLI, A. (1981). The Seveso runaway reaction: a thermoanalytical study. *Runaway Reactions, Unstable Products and Combustible Powders* (London: Instn Chem. Engrs), paper 3/N

CARDILLO, P., GIRELLI, A. and FERRAIOLO, G. (1984). The Seveso case and the safety problem in the production of 2,4,5−trichlorophenol. *J. Haz. Materials*, **9**(2), 221

CARDILLO, P. and NEBULONI, M. (1991). Theoretical and calorimetric evaluation of thermal stability of glycidol. *J. Loss Prev. Process Ind.*, **4**, 242

CARDILLO, P. and NEBULONI, M. (1992). Reactivity limits of aluminium and halohydrocarbon mixtures: determination by the ASTM CHETAH program. *J. Loss Prev. Process Ind.*, **5**, 81

CARDINALE, A., GRILLO, P. and MESSINA, S. (1983). Risk management in project execution. *Euredata 4*, paper 4.4

CARDOZO, R.L. (1991). Enthalpies of combustion, formation, vaporization and sublimation of organics. *AIChE J.*, **37**, 290

CARDWELL, D.S.L. (1957). *The Organisation of Science in England* (London: Heinemann)

CARELL, R. (1992). Plant design. *Plant Engr*, **36**(6), 33

CARHART, H.W. (1968). Fire hazard classification of chemical vapors relative to explosion-proof electrical equipment. *Carriage of Dangerous Goods*, **1**, p. 121

CARHART, H.W. (1994). Impact of low O2 on fires. *J. Haz. Materials*, **36**, 133

CARHART, R.E. (1977). Reprogramming DENDRAL. *AISB Quart.*, **28**, 20

CARHART, R.R. (1953). *A Survey of the Current Status of the Reliability Problem. Res. Memo RM-1131.* Rand Corp, Santa Monica, CA

CARISSIMO, B., RIOU, Y., CANEILL, J.Y., GIRARD, A., BREFORT, B. and NEDELKA, D. (1989). Simulations of the Thorney Island continuous and instantaneous releases with obstacles using a three dimensional model. *Loss Prevention and Safety Promotion*, **6**, 86.1

CARLESSO, S., BASTIN, F. and BORELLA, A. (1983). Principal achievements in the development of the component event data bank. *Euredata 4*, paper 14.2

CARLSON, G.A. (1973). Spherical detonations in gas−oxygen mixtures. *Combust. Flame*, **21**, 383

CARLSON, R.A., SULLIVAN, M.A. and SCHNEIDER, W. (1989). Component fluency in a problem-solving context. *Hum. Factors*, **31**, 489

CARLTON-JONES, D. and SCHNEIDER, H.B. (1968). Air pollution control. Tall chimneys. *Chem. Engng*, **75**(Oct. 14), 166

CARMACK, B.J. (1976). Human behaviour in fires. *J. Fire Flammability*, **7**, 559

CARMICHAEL, R.L. *et al.* (1966-). Nonferrous metals and the chemical engineer. *Chem. Engng*, 1966, **73**, Nov. 21, 109; Dec. 5, 139; Dec. 19, 105; 1967, **74**, Jan. 16, 159; Jan. 30, 139

CARMICHAEL, R.S. (ed.) (1982). *Handbook of Physical Properties of Rocks*, vols 1–3 (Boca Raton, FL: CRC Press)

CARMODY, D.B. (1983). The contractor's marketing effectiveness: an owner's viewpoint. *Chem. Engng Prog.*, **79**(12), 15

CARMODY, E.O. and STAFFIN, B.E. (1970). How to blow up a plant, without damaging it. *Chem. Engng*, **77**(Jul. 27), 166

CARMODY, T.W. (1988). Organization and projects of the Center for Chemical Process Safety. *Plant/Operations Prog.*, **7**, J5

CARMODY, T.W. (1989). Organisation and projects of the Center for Chemical Process Safety. *Loss Prevention and Safety Promotion* 6, p. 0.1

CARMODY, T.W. (1990a). Current research developments and future needs. *Plant/Operations Prog.*, **9**, 204

CARMODY, T.W. (1990b). Update on AIChE's Center for Chemical Process Safety. *Plant/Operations Prog.*, **9**, 141

CARN, K.K. (1987). Estimates of the mean concentration and variance for the Thorney Island Phase I dense gas dispersion experiments. *J. Haz. Materials*, **16**, 75

CARN, K.K. and CHATWIN, P.C. (1985). Variability and heavy gas dispersion. *J. Haz. Materials*, **11**(1/3), 281. See also in McQuaid, J. (1985), *op. cit.*, p. 281

CARN, K.K., SHERRELL, S.J. and CHATWIN, P.C. (1988b). Analysis of Thorney Island data: variability and box models. In Puttock, J.S. (1988b), *op. cit.*, p. 205

CARNASCIALI, F., LEE, J.H.S., KNYSTAUTAS, R. and FINSECHI, F. (1991). Turbulent jet initiation of detonation. *Combust. Flame*, **84**, 170

CARNES, S.A. (1986). Institutional issues affecting the transport of hazardous materials in the United States: anticipating strategic management needs. *J. Haz. Materials*, **13**(2), 257

CARNINO, A. (1976). Reliability techniques in assessment of nuclear reactor safety. In Henley, E.J. and Lynn, J.W. (1976), *op. cit.*, p. 83

CARNINO, A. (1986). Human reliability. *Loss Prevention and Safety Promotion* 5, 3–1

CARNINO, A. and GRIFFON, M. (1982). Causes of human error. In Green, A.E. (1982), *op. cit.*, p. 171

CARNINO, A., NICOLET, J.-L. and WANNER, J.-C. (1990). *Man and Risks, Technological and Human Risk Prevention* (Basel: Dekker)

CARO, D. and IRVING, M. (1973). The Old Bailey bomb explosion. *The Lancet*, June 23, 1433

CARPENTER, C.P., SMYTH, H.F. and POZZANI, U.C. (1949). The assay of acute vapor toxicity, and the grading and interpretation of the results of 96 chemical compounds. *J. Ind. Hyg. Toxicol.*, **31**, 343

CARPENTER, D.L. (1957). The explosibility characteristics of coal dust clouds using electric spark ignition. *Combust. Flame*, **1**, 63

CARPENTER, D.L. and DAVIES, D.R. (1958). The variation with temperature of the explosibility characteristics of coal

dust clouds using electric spark ignition. *Combust. Flame*, **2**, 35

CARPENTER, K.G. (1964). Anticipating hazards, the key to safe operations. *Chem. Engng Prog.*, **69**(4), 37

CARPENTER, L.G. (1970). *Vacuum Technology* (London: Adam Hilger)

CARPENTER, R.J., CLEAVER, R.P., WAITE, P.J. and ENGLISH, M.A. (1987). The calibration of simple model for dense gas dispersion using the Thorney Island Phase I trials data. *J. Haz. Materials*, **16**, 293

CARPENTER, S.B. *et al.* (1971). Principal plume dispersion models: TVA power plants. *J. Air Poll. Control Ass.*, **21**, 491

CARPER, K. (1989). *Forensic Engineering* (New York: Elsevier)

CARR, B. (1986). Chlor-aid – the intercompany collaboration for chlorine engineers. In Wall, K. (1986), *op. cit.*, p. 113

CARR, C.F. (1975). Industrial accident data: how much scope is there for improvement? In Hendry, W.D. (1975), *op. cit.*, paper D5

CARR, F.H. (1919). Gas warfare. *Br. Med. J.*, Aug. 2, 140

CARR, J.W. (1988). Taking the guesswork out of pilot plant safety. *Chem. Engng Prog.*, **84**(9), 52

CARR-BRION, K. (1986). Ways to measure water content online. *Process Engng*, March, 55

CARRARA (1973). Explosion during distillation of mother-liquors from oxidation of p-nitro-ethylbenzene to p-nitro-acetophenone. *Prevention of Occupational Risks*, **2**, p. 491

CARRAS, J.N.A. and WILLIAMS, D.J. (1984-). Experimental studies of plume dispersion in convective conditions. *Atmos. Environ.*, **18**, 135; 1986, **20**, 2307

CARRASCO (1978). *Report to Tarragona Court of Inquiry on San Carlos Disaster*

CARRE, B. (1979). *Graphs and Networks* (Oxford: Oxford Univ. Press)

CARRE, C. and BRE, P. (1991). Mounded storage – Gaz de France's first experience. *Gastech 90*, p. 7.7

CARRIER, A.G.E. (1971). UK and German rotating equipment standards. *Hydrocarbon Process.*, **50**(6), 101

CARRIER, G., FENDELL, F., CHEN, K. and COOK, S. (1991). Evaluating a simple model for laminar flame propagation rates. *Combust. Sci. Technol.*, **79**, 207, 229

CARRIER, G.F., FENDELL, F.E. and FELDMANN, P.S. (1980). Wind-aided flame spread along a horizontal fuel slab. *Combust. Sci. Technol.*, **23**, 41

CARRIER, G., FENDELL, F. and FELDMAN, P. (1984). Big fires. *Combust. Sci. Technol.*, **39**, 135

CARRIER, G.F., FENDELL, F.E. and WOLFF, M.F. (1991). Wind-aided firespread across arrays of discrete fuel elements, Pt 1. *Combust. Sci. Technol.*, **75**, 31. See also Wolff, M.F., Carrier, G.F. and Fendell, F.E. (1991)

CARRIKER, D. (1979). Steam hose couplings. *Chem. Engng*, **86**(Jan. 29), 121

CARROLL, M.M. (1985). Polyvinylchloride (PVC) pipe reliability and failure modes. *Reliab. Engng*, **13**(1), 11

CARRITHERS, G.W., DOWELL, A.M. and HENDERSHOT, D.C. (1996). It's Never Too Late for Inherent Safety. *Int. Conf. and Workshop on Process Saf. Manag. and Inherently Safer Processes*, Oct. 8–11, 1996, Orlando, FL, 227–41. (New York: American Institute of Chemical Engineers)

CARRUTHERS, D.J. and HUNT, J.C.R. (1988). Turbulence, waves and entrainment near density inversion layers. In Puttock, J.S. (1988b), *op. cit.*, p. 77

CARRUTHERS, J.A. and MARSH, K.J. (1962). Charge relaxation in hydrocarbon liquids flowing through conducting and non-conducting pipes. *J. Inst. Petrol.*, **48**, 169

CARRUTHERS, J.A. and WIGLEY, K.J. (1962). The estimation of electrostatic potentials, fields, and energies in a rectangular metal tank containing charged fuel. *J. Inst. Petrol.*, **48**(462), 180

CARSLAW, H.S. and JAEGER, J.C. (1959). *Conduction of Heat in Solids*, 2nd ed. (Oxford: Clarendon Press)

CARSON, D.J. and RICHARDS, P.J.R. (1978). Modeling surface turbulent fluxes in stable conditions. *Boundary Layer Met.*, **14**, 67

CARSON, J. (1986). A report on the Symposium on Uncertainty in Modelling Atmospheric Dispersion. *Atmos. Environ.*, **20**, 1047

CARSON, J.S. (1981). *Treatment of Burns* (London: Chapman & Hall)

CARSON, P.A. and JONES, K. (1984). Effective management of safety and hygiene information to promote safe handling of chemicals. *J. Haz. Materials*, **9**(3), 305

CARSON, P.A. and MUMFORD, C.J. (1979). An analysis of incidents involving major hazards in the chemical industry. *J. Haz. Materials*, **3**(2), 149

CARSON, P.A. and MUMFORD, C.J. (1989). *The Safe Handling of Chemicals in Industry* (London: Longmans)

CARSON, P.A. and MUMFORD, C.J. (1994). *Hazards Chemicals Handbook* (Oxford: Butterworth-Heinemann)

CARSON, R. (1963). *Silent Spring* (London: Hamish Hamilton)

CARSON, W.G. (1981). *The Other Price of British Oil* (Oxford: Martin Robertson)

CARTER, A.D.S. (1972). *Mechanical Reliability* (London: Macmillan)

CARTER, A.D.S. (1973). The bath tub curve for mechanical components – fact or fiction? *Terotechnology in the Process Industries* (London: Instn Mech. Engrs)

CARTER, A.D.S. (1976). Loading roughness and reliability. *Advances in Reliability Technology. ART6* (Culcheth, Lancashire: UK Atom. Energy Auth.)

CARTER, A.D.S (1979). Reliability reviewed. *Proc. Instn Mech. Engrs*, **193**, 81

CARTER, A.D.S. (1980). A view of mechanical reliability. In Moss, T.R. (1980), *op. cit.*, p. 1

CARTER, A.D.S (1983). Early life failures. *Reliab. Engng*, **4**, 41

CARTER, A.D.S. (1986). *Mechanical Reliability*, 2nd ed. (London: Macmillan)

CARTER, C.J. (1980). Why contract maintenance? *Hydrocarbon Process.*, **59**(1), 103

CARTER, D.A. (1988). Single and two-phase flow calculations. *Course on Hazard Control and Major Hazards*. Dept of Chem. Engng, Loughborough Univ. of Technol.

CARTER, D.A. (1989). Methods for estimating the dispersion of toxic combustion products from large fires. *Chem. Engng Res. Des.*, **67**, 348

CARTER, D.A. (1991). Aspects of risk assessment for hazardous pipelines containing flammable substances. *J. Loss Prev. Process Ind.*, **4**, 68

CARTER, D.A. (1992). Dispersion of toxic combustion products from large fires. *Fire Technol.*, **28**, 168

CARTER, H.R. (1966). Management information systems – tomorrow's management tool. *Hydrocarbon Process.*, **45**(1), 123

CARTER, J.A. (1986). Salvage of cargo from the war damaged *Gaz Fountain. Gastech 85*, p. 141

CARTER, J.T. (1976). The biological basis for chemical environmental standards. *Process Industry Hazards*, p. 11

CARTER, J.T. (1980). The medical contribution to plant, design. *Chemical Process Hazards*, **VII**, p. 131

CARTER, K.A., LONSDALE, H., MARTENS, G.J. and KLOPPERS, J.J. (1977). Ammonia manufacture from refinery gas. *Ammonia Plant Safety*, **19**, 34

CARTER, N. and MENCKEL, E. (1985). Near-accident reporting: a review of Swedish research. *J. Occup. Accid.*, **7**(1), 41

CARTER, R.J. (1964). Control systems for process plant. *Control*, **8**(76), 507

CARTER, R.J. (1982). Automation of design and process operations. The technology is easy; using it is hard. *Chem. Ind.*, Oct. 2, 759

CARTER, R.L. and ROE, F.J.C. (1975). Chemical carcinogens in industry. *J. Soc. Occup. Med.*, **25**, 86

CARTON, B. and KAUFFER, E. (1980). The metrology of asbestos. *Atmos. Environ.*, **14**, 1181

CARTWRIGHT, D.J. and ROOKE, D.P. (1975). Evaluation of stress intensity factors. *J. Strain Analysis*, **10**(4), 217

CARTWRIGHT, J. and WIDGINTON, D.W. (1971). Safety barriers for use in intrinsically safe circuits. *Electrical Safety*, **1**, 1

CARTWRIGHT, P. (1987). Test your plant, before your process does. *Process Engng*, Dec., 45

CARTWRIGHT, P. (1988). Earthing electrostatic charge. *Chem. Engr, Lond.*, **454**, 49

CARTWRIGHT, P. (1990). What price a dust explosion? *Chem. Engr, Lond.*, **482**, 22

CARTWRIGHT, P. (1992). Look at electrostatics closely. *Chem. Engng Prog.*, **88**(9), 76

CARTWRIGHT, P.A. (1975). Safety at Work Act (letter). *Chem. Engr, Lond.*, **302**, 615

CARUANA, C.M. (1986). Hazardous waste management – a top priority with EPA. *Chem. Engng Prog.*, **82**(7), 50

CARVER, F.W.S., HARDS, D.J., FREEDER, B.G.P. and TAYLOR, A.J. (1977). The prevention of explosion in liquefied petroleum gas cylinders in fire conditions. *Chemical Process Hazards*, **VI**, unbound paper

CARVER, F.W.S., HARDS, D.J. and HUNT, K. (1974). The explosibility of some unsaturated C4 hydrocarbon fractions. *Chemical Process Hazards*, **V**, p. 99

CARVER, F.W.S., SMITH, C.M. and WEBSTER, G.A. (1972). The explosive decomposition of acetylene in pipelines. *Chemical Process Hazards*, **IV**, p. 44

CARVER, J. (1976). *Report on Explosion at Houghton Main Colliery, Yorkshire* (London: HM Stationery Office)

CARY, H.B. (1979). *Modern Welding Technology* (Englewood Cliffs, NJ: Prentice-Hall)

CASADA, M.L., KIRKMAN, J.Q. and PAULA, H.M. (1990). Facility risk review as an approach to prioritizing loss prevention efforts. *Plant/Operations Prog.*, **9**, 213

CASARETT, L.J. and DOULL, J. (1975). *Toxicology: the Basis Science of Poisons* (New York: Macmillan)

CASE, R.C. (1970). A method to determine exchanger relief valve requirements. *API Proc., Div. of Refining*, **50**, 1082

CASERTA, L.V. (1972a). Failure of an ammonia loading hose. *Ammonia Plant Safety*, **14**, p. 31

CASERTA, L.V. (1972b). Noise abatement in ammonia Plants. *Ammonia Plant Safety*, **14**, p. 6

CASEY, G.B. (1976). Rupture in an ammonia shift conversion unit. *Ammonia Plant Safety*, **18**, p. 103

CASHDOLLAR, K.L. (1994). Flammability of metals and other elemental dust clouds. *Process Safety Prog.*, **13**, 139

CASHDOLLAR, K.L. and CHATRATHI, K. (1992). Minimum explosible dust concentrations measured in 20 l and 1 m^3 chambers. *Combust. Sci. Technol.*, **87**, 157

CASHDOLLAR, K.L., HERTZBERG, M. and ZLOCHOWER, I.A. (1989). Effect of volatility on dust flammability limits for coals, gilsonite and polyethylene. *Combustion*, **22**, 1757

CASHDOLLAR, K.L., WEISS, E.S., GRENINGER, N.B. and CHATRATHI, K. (1992). Laboratory and large-scale dust explosion research. *Plant/Operations Prog.*, **11**, 247

CASHMAN, J.R. (1988). *Hazardous Materials Emergencies Response and Control* (Lancaster, PA: Technomic Publ. Co.)

CASLER, J. (1985). *Superfund Handbook* (Chicago, IL: Sidley & Austin and Concorde, MA: ERT Inc.). See also Sidley and Austin (1987)

CASON, R.L. (1971). Will PM help or hurt? *Hydrocarbon Process.*, **50**(1), 95

CASON, R.L. (1972). Estimate the downtime your improvement will save. *Hydrocarbon Process.*, **51**(1), 73

CASSARO, M.A. and MARTINEZ-RAMIREZ, E. (eds) (1987). *The Mexico Earthquakes 1985* (New York: Am. Soc. Civil Engrs)

CASSATA, J.R., DASGUPTA, S. and GANDHI, S.L. (1993). Modeling of tower relief dynamics. *Hydrocarbon Process.*, **72**(10), 71; **72**(11), 69

CASSEL, H.M., DAS GUPTA, A.K. and GURUSWAMY, S. (1949). Factors affecting flames propagation through dust clouds. *Combustion*, **3**, p. 185

CASSEL, H.M. and LIEBMAN, I. (1959). The cooperative mechanism in the ignition of dust dispersions. *Combust. Flame*, **3**, 467

CASSEL, H.M., LIEBMAN, I. and MOCK, W.K. (1957). Radiative transfer in dust flames. *Combustion*, **6**, p. 602

CASSELLI, A., MASONI, P. and FORABOSCHI, F.P. (1992). A methodology for evaluating the seismic safety margin of a chemical plant. *Loss Prevention and Safety Promotion 7*, vol. 2, 71-1

CASSELLS, W. (1986-). Exhaust ventilating. *Plant Engr*, 1986, **30**(3), 23; 1987, **31**(1), 21; **31**(3), 17

CASSIDY, K. (1987). CIMAH safety cases: the HSE approach. *Chem. Engr, Lond.*, **439**, 6

CASSIDY, K. (1988a). CIMAH safety cases. *Course on Hazard Control and Major Hazards*. Dept of Chem. Engng, Loughborough Univ. of Technol.

CASSIDY, K. (1988b). CIMAH – where we are now. *J. Loss Prev. Process Ind.*, **1**(4), 221

CASSIDY, K. (1988c). The national and international framework for legislation for industrial major accident hazards. *Preventing Major Chemical and Related Process Accidents*, p. 401

CASSIDY, K. (1989). CIMAH safety cases: 1, overview. In Lees, F.P. and Ang, M.L. (1989), *op. cit.*, p. 220

CASSIDY, K. (1990). CIMAH and the environment. *Process Safety Environ.*, **68B**, 195

CASSIDY, K. (1993). Risk management: a business risk or a risky business? *Process Safety Management*, p. 29

CASSIDY, K. (1994). European regulations. *Process Safety Prog.*, **13**, 10

CASSIDY, K. and PANTONY, M.F. (1988). Major industrial risks: a technical and predictive basis for on and off site emergency planning in the context of UK legislation. *Preventing Major Chemical and Related Process Accidents*, p. 75

CASSON, A. (*c.* 1992). Intelligent support for authors and readers of technical documents. *AIRING*, **15**, 11. Artificial Intelligence Applications Inst., Edinburgh

CASSON, A. and STONE, D. (1992). An expertext system for building standards. *Computers and Building Standards Worshop*, Univ. of Montreal

CASSUT, L.H., MADOCKS, F.E. and SAWYER, W.A. (1964). *A study of the hazards in the storage and handling of liquid hydrogen* (New York: AD Little Inc.)

CASTANTRAS, H.M., BANERJEE, D.K. and NOLLER, D.C. (1965). *Fire and Explosion Hazards of Peroxy Compounds* (Philadelphia, PA: Am. Soc Testing Materials)

CASTLE, G.K. (1974). Fire protection of structural steel. *Loss Prevention*, **8**, p. 57

CASTLE, G.K. (1979). A contractor views fireproofing. *Hydrocarbon Process.*, **58**(8), 179

CASTLE, G.K. (1984). Improve control system fire protection. *Hydrocarbon Process.*, **63**(3), 113

CASTLE, G.K. and CASTLE, G.G. (1987). Effect of fireproofing design on thermal exposure of horizontal members with top flange exposed. *Plant/Operations Prog.*, **6**, 193

CASTLE, G.S.P. and INCULET, I.I. (1991). Induction charge limits of small water droplets. *Electrostatics*, **8**, p. 141

CASTLE, G.S.P., INCULET, I.I. and LITTLEWOOD, R. (1983). Ionic space charges generated by evaporating charged liquid sprays. *Electrostatics*, **6**, 59

CASTLEMAN, J.R. (1990). Planning and use of a risk assessment system for multiple applications in the processing industry. *Safety and Loss Prevention in the Chemical and Oil Processing Industries*, p. 499

CASTO, W.R. (1973). Safety-related occurrences reported in October and November 1972. *Nucl. Safety*, **14**(2), 120

CASTRO, G.O. (1982). Selecting and installing plastic-lined pipe. *Chem. Engng*, **89**(Mar. 22), 112

CASTRO, I.P., KUMAR, A., SNYDER, W.H. and ARYA, S.P.S. (1993). Removal of slightly heavy gases from a valley by cross-winds. *J. Haz. Materials*, **34**, 271

CASTRO, I.P. and SNYDER, W.H. (1982). A wind tunnel study of dispersion from sources downwind of three-dimensional hills. *Atmos. Environ.*, **16**, 1869

CASTRO, I.P., SNYDER, W.H. and LAWSON, R.E. (1988). Wind direction effects on dispersion from sources downwind of steep hills. *Atmos. Environ.*, **22**, 2229

CATCHPOLE, J.O., KELLY, M.J. and MUSGRAVE, C. (1984). Reliability growth of gas turbine powered compressor units. *Reliab. Engng*, **8**(4), 235

CATCHPOLE, J.P. and FULFORD, G. (1966). Dimensionless groups. *Ind. Engng Chem.*, **58**(3)46

CATE, C.L. and FUSSELL, J.B. (1977). *BACFIRE – A Computer Program for Common Cause Failure Analysis. Rep. NERS-77-02*. Nucl Engng Dept, Univ. of Tennessee

CATES, A. and SAMUELS, B. (1991). A simple assessment methodology for vented explosions. *J. Loss Prev. Process Ind.*, **4**, 287

CATES, A.T. (1992). Shell Stanlow fluoroaromatics explosion – 20 March 1990: assessment of the explosion and of blast damage. *J. Haz. Materials*, **32**, 1

CATES, J.H. (1972). Electric motors: principles and applications in the chemical process plant., *Chem. Engng*, **79**(Dec. 11), 82

CATHERALL, J.A. (1977). *Fibre Reinforcement* (London: Mills & Boon)

CATINO, C.A., GRANTHAM, S.D. and UNGAR, L.H. (1991). Automatic generation of qualitative models of chemical process units. *Comput. Chem. Engng*, **15**, 583

CATLIN, C.A. (1985). An acoustic model for predicting overpressures caused by the deflagration of a ground lying vapour cloud. *The Assessment and Control of Major Hazards* (Rugby: Instn Chem. Engrs), p. 193

CATLIN, C.A. (1990). CLICHE – a generally applicable and practicable offshore explosion model. *Piper Alpha: Lessons for Life Cycle Management* (Rugby: Instn Chem. Engrs), p. 25. See also *Process Safety Environ.*, **68B**, 245

CATLIN, C.A. (1991). Scale effects on the external combustion caused by venting of a confined explosion. *Combust. Flame*, **83**, 399

CATLIN, C.A., GREGORY, C.A.J., JOHNSON, D.M. and WALKER, D.G. (1993). Explosion mitigation in offshore modules by general area deluge. *Process Safety Environ.*, **71B**, 101

CATLIN, C.A. and JOHNSON, D.M. (1992). Experimental scaling of the flame acceleration phase of an explosion by changing fuel gas reactivity. *Combust. Flame*, **88**, 15

CATLIN, C.A. and LINDSTEDT, R.P. (1991). Premixed turbulent burning velocities derived from mixing controlled reaction models with cold front quenching. *Combust. Flame*, **85**, 427

CATLIN, C.A., MANOS, A. and TITE, J.P. (1993). Mathematical modeling of confined explosions in empty cube and duct shaped enclosures: effect of scale and geometry. *Process Safety Environ.*, **71B**, 89

CATO, G.A. and DOBBS, W.F. (1971). An ammonia tank car emergency. *Ammonia Plant Safety*, **13**, 1

CATTABENI, F., CAVALLARO, A. and GALLI, G. (eds) (1978). *Dioxin, Toxicological and Chemical Aspects* (New York: Halsted Press)

CAUDWELL, R., BOLTON, L., JAFFE, M. and SMITH, J. (1981). Europe to tighten controls on hazardous solid wastes. *Chem. Engng*, **88**(Dec. 14), 20E

CAUGHEY, S.J. (1982). Observed characteristics of the atmospheric boundary layer. In Nieuwstadt, F.T.M. and van Dop, H. (1982), *op. cit.*, p. 107

CAUGHEY, S.J. and KAIMAL, J.C. (1977). Vertical heat flux in the convective boundary layer. *Quart. J. R. Met. Soc.*, **103**, 811

CAUGHEY, S.J., WYNGAARD, J.C. and KAIMAL, J.C. (1979). Turbulence in the evolving boundary layer. *J. Atmos. Sci.*, **36**, 1041

CAULFIELD, M., CLEAVER, R.P., COOK, D.K. and FAIRWEATHER, M. (1993). A integral model of turbulent jets in a cross-flow. Pt 1 – Gas dispersion. *Process Safety Environ.*, **71B**, 235

CAUSE, J.L. *et al.* (1980). Ethylene oxide. *Kirk-Othmer Encyclopaedia of Chemical Technology*, 3rd ed., vol. 9 (New York: Wiley), p. 432

CAVALLARO, A., TEBALDI, G. and GUALDI, R. (1979). Analysis by means of meteorological and physical considerations about the transport and ground depositon of the TCDD emitted the 10/7/76 from the Icmesa factory. *Proc. Tenth Int. Tech. Mtg on Air Pollution Modelling and Its Application* (NATO/CCMS), p. 175

CAVALLARO, A., TEBALDI, G. and GUALDI, R. (1982). Analysis of transport and ground deposition of TCDD emitted on 10 July 1976 from the ICMESA factory (Seveso, Italy). *Atmos. Environ.*, **16**, 731

CAVALLA, J.F. (1979). Notification scheme – too much safety? *Chem. Ind.*, Feb. 17, 123

CAVASENO, V. (1978a). EPA rule to curb hazardous chemicals. *Chem. Engng*, **85**(Apr. 24), 50

CAVASENO, V. (1978b). Flowmeter choice widens. *Chem. Engng*, **85**(Jan. 30), 55

CAVASENO, V. (1978c). Latest word on asbestos won't be the last. *Chem. Engng*, **85**(Nov. 6), 76

CAVASENO, V. and CHEMICAL ENGINEERING STAFF (1979). *Process Heat exchangers* (New York: McGraw-Hill)

CAVE, L. (1988). Guidelines for the use of vapour cloud dispersion models (book review). *Chem. Engr, Lond.*, **444**, 41

CAVE, S.R. and EDWARDS, D.W. (1997a). Chemical process route selection based on assessment of inherent environmental hazard. *Comp. Chem. Eng.* **21**, S965–S970

CAVE, S.R. and EDWARDS, D.W. (1997b). A methodology for chemical process route selection based on assessment of inherent environmental hazard. *The 1997 IChemE Jubilee Research Event*, 49–52.

CAWKWELL, M.G. and CHARLES, M.E. (1985). Pressures, temperatures predicted for two-phase pipelines. *Oil Gas J.*, **83**(May 27), 101

CAZAMIAN, P. (1983). Accident proneness. In Parmegiani, L. (1983), *op. cit.*, p. 20

CEAGLSKE, N.H. (1956). *Automatic Process Control for Chemical Engineers* (New York: Wiley)

CECIL, L.K. (1969). Water reuse and disposal. *Chem. Engng*, **76**(May 5), 92

CECIL, L.K. (1972). Manage the environment with imagination. *Hydrocarbon Process.*, **51**(10), 79

CEFIC WORKSHOP (1986). The occurrence of chlorinated solvents in the environment. *Chem. Ind.*, Dec. 15, 861

CEKALIN, E.K. (1962). Propagation of flame in turbulent flow of two-phase fuel-air mixture. *Combustion*, **8**, 1125

CELLIER, J.M. and EYROLLE, H. (1992). Interference between switched tasks. *Ergonomics*, **35**, 25

CENDROWSKA, J. and BRAMER, M. (1984). Inside an expert system: a rational reconstruction of the MYCIN consultation system. In T.O'Shea and M. Eisenstadt (1984), *op. cit.*, p. 453

CENTER FOR CHEMICAL PROCESS SAFETY (*c.* 1985). *A Challenge to Commitment* (New York: Am. Inst Chem. Engrs)

CCPS (1989). *Guidelines For Chemical Process Quantitative Risk Analysis* (New York) ISBN 0-8169-0402-2, p. 585.

CENTER FOR CHEMICAL PROCESS SAFETY (1990). *Safety, Health and Loss Prevention: Problems for Undergraduate Engineering Curricula – Instructor's Guide* (New York)

CCPS (1992). *Guidelines For Hazard Evaluation Procedures, 2nd Edition With Worked Examples.* (New York: AICHE), ISBN 0-8169-0491-X, p. 461

CCPS (1992). *Guidelines for Investigating Chemical Process Incidents* (New York), ISBN 0-8169-0555-X, p. 347

CCPS (1994). *Guidelines For Evaluating the Characteristics of Vapor Cloud Explosions, Flash Fires*, and *Bleves* (New York), ISBN 0-8169-0474-X, pp. 387

CCPS (1995). *Guidelines for Chemical Transportation Risk Analysis* (New York), ISBN 0-8169-0629-7, p. 382

CCPS, (1995). *Understanding Atmospheric Dispersion of Accidental Releases* (New York), ISBN 0-8169-0681-5, pp. 64

CCPS (1996). *Guidelines for Evaluating Process Plant Buildings for External Explosions and Fires* (New York), ISBN 0-8169-0646-7, p. 208

CCPS (1996). *Guidelines for Use of Vapor Cloud Dispersion Models*, 2nd Ed. (New York), ISBN 0-8169-0702-1, p. 304

CCPS (1997). *Guidelines for Post-Release Mitigation Technology* (New York) ISBN 0-8169-0588-6, p. 175

CCPS (1998). *RELEASE: A Model with Data to Predict Aerosol Rainout in Accidental Releases* (New York), ISBN 0-8169-0745-5, p. 150

CCPS (1992). *Guidelines for Hazard Evaluation Procedures.* 2nd Edition (New York: American Institute of Chemical Engineers)

CCPS (1993a). *Guidelines for Safe Automation of Chemical Processes* (New York: American Institute of Chemical Engineers)

CCPS (1993b). *Guidelines for Engineering Design for Process Safety* (New York: American Institute of Chemical Engineers)

CCPS (1994). *Guidelines for Preventing Human Error in Process Safety* (New York: American Institute of Chemical Engineers)

CCPS (1995). *Tools for Making Acute Risk Decisions With Chemical Process Safety Applications* (New York: American Institute of Chemical Engineers)

CCPS (1996). *Inherently Safer Chemical Processes: A Life Cycle Approach* (ed. D. A. Crowl) (New York: American Institute of Chemical Engineers)

CCPS (1998). *Guidelines for Design Solutions for Process Equipment Failures* (New York: American Institute of Chemical Engineers)

CCPS (2000). *Guidelines for Chemical Process Quantitative Risk Analysis.* 2nd Edition. (New York: American Institute of Chemical Engineers)

CCPS (2001). *Layer of Protection Analysis* (New York: American Institute of Chemical Engineers)

CCPS and Center for Waste Reduction Technologies (CWRT) (2001). *Making EHS an Integral Part of Process Design.* (New York: American Institute of Chemical Engineers)

CCPS (2003). *Guidelines for Investigating Chemical Process Incidents.* 2nd Edition (New York: American Institute of Chemical Engineers)

CENTER FOR CRITICAL THINKING GLOSSARY. *Glossary Of Critical Thinking Terms* (Wye Mills, MD: Center For Critical Thinking,) 21679, www.criticalthinking.org

CENTRAL ELECTRICITY GENERATING BOARD (1966). Study of operators information requirements at Trawsfynydd (Cheltenham, Gloucestershire). See also Edwards, E. and Lees, F.P. (1974), *op. cit.*, p. 348

CENTRAL ELECTRICITY GENERATING BOARD (1971). *Modern Power Station Practice*, vol. 1, *Nuclear Power Generation* (Oxford: Pergamon Press)

CENTRAL ELECTRICITY BOARD (1982). *Sizewell B Inquiry Pre-Construction Safety Report* (London)

CENTRAL FIRE BRIGADES ADVISORY COUNCIL (1978). *Major Disaster Planning.* Home Office Fire Dept, London

CENTRAL STATISTICAL OFFICE (1985). *Annual Abstract of Statistics*, 1985 ed. (London: H.M. Stationery Office)

CENTRAL STATISTICAL OFFICE (1986). *Annual Abstract of Statistics* (London: H.M. Stationery Office)

LA CERDA, D.J. (1968). Better instrument and plant, air systems. *Hydrocarbon. Process.*, **47**(6), 107

CERDA, H.W. and NAPIER, D.H. (1977). Reliability and hazard analysis in the context of sugar dust explosions. *First Nat. Conf. on Reliability. NRC 4/8* (Culcheth, Lancashire: UK Atom. Energy Auth.)

CERMAK, J.E. (ed.) (1980). *Wind Engineering*, vols 1–2 (Oxford: Pergamon Press)

CESARI, F. and HELLEN, T.K. (1979). Evaluation of stress intensity factors for internally pressurised cylinders with surface flaws. *Int. J. Pres. Ves. Piping*, **7**, 199

CESS, R.D. (1974). Infrared gaseous radiation. In Afgan, N. and Beer, J.M. (1974), *op. cit.*, p. 231

CETEGEN, B.M. and AHMED, T.A. (1993). Experiments on the periodic instability of buoyant plumes and pool fires. *Combust. Flame*, **93**, 157

CHACHRA, V., GHARE, P.M. and MOORE, J.M. (1979). *Applications of Graph Theory Algorithms* (Amsterdam: Elsevier)

CHADHA, N. and PARMELE, C.S. (1993). Minimize emissions of air toxics via process changes. *Chem. Engng Prog.*, **89**(1), 37

CHADWICK, D. and JAMIE, M. (1977). How important is corrosion as a cause of accidents? (abstract). *Chemical Process Hazards*, **VI**, p. 168

CHADWICK, G.A. (1972). *Mettellography of Phase Transformations* (London: Butterworths)

CHAFFEE, C.C. (1970). Transfer line failure. *Ammonia Plant Safety*, **12**, p. 30. See also *Chem. Engng Prog.*, 1970, **66**(1), 54

CHAILLOT, H. (1966). The cooling of liquid and liquefied hydrocarbon storage tanks. *Fire Int.*, **1**(12), 22

CHAINEAUX, J. and DANNIN, E. (1989). Experimental determination of the risk of auto-ignition explosion or detonation in industrial chemical reactors. *Loss Prevention and Safety Promotion* 6, p. 87.1

CHAINEAUX, J. and DANNIN, E. (1992). Sizing of explosion vents for the protection of vessels in which gases are processed at P and T higher than ambient. *Loss Prevention and Safety Promotion* 7, vol. 2, 60–1

CHAINEAUX, J. and MAVROTHALASSITIS, G. (1990). Discharge in air of a vessel pressurized by a flammable gas, and the volume of the resulting flammable mixture generated. *J. Loss Prev. Process Ind.*, **3**(1), 88

CHAKIR, A., CATHONNET, M., BOETTNER, J.C. and GAILLARD, F. (1989). Kinetic study of butane oxidation. *Combust. Sci. Technol.*, **65**, 207

CHAKRABARTI, A., STEINER, E.C., WERLING, C.L. and YOSHIMINE, M. (1985). Kinetic and reactor modeling: hazard evaluation and scale-up of a complex reaction. In Hoffman, J.M. and Maser, D.C. (1985), *op. cit.*, p. 91

CHAKRAVARTY, T. (1993). Use cannonical correlation analysis instead of regression. *Chem. Engng Prog.*, **89**(10), 76

CHAKRAVARTY, T., FISHER, H.G. and VOYT, L.A. (1993). Reactive monomer tank – a thermal stability analysis. *J. Haz. Materials*, **33**, 203

CHAKRAVARTY, A., LOCKWOOD, F.C. and SINICROPI, G. (1984). The prediction of burner stability limits. *Combust. Sci. Technol.*, **42**, 67

CHALK, R. (1973). Citizen participation in technical planning. *Pub. Sci. Newsletter*, **4**(6), 4

CHALLIS, E.J. (1979). Management of major hazard plant. *J. Haz. Materials*, **3**(1), 49. See also Lees, F.P. and Ang, M.L. (1989), *op. cit.*, p. 109

CHALLONER, A. (1986). Sensing the pressure. *Process Engng*, Aug. 36

CHALMERS, R. (1981). Amounts and effects of toxic materials discharged to sewers. *Chem. Ind.*, Apr. 18, 271

CHAMANY, B.F., MURTY, L.G.K. and RAY, R.N. (1981). A methodology for choosing protective signals and settings for a nuclear power plant. *Int. Atom. Energy Agency, Conf. on Current Nuclear Power Plant, Safety Issues, IAEA -CN-39/56* (Vienna)

CHAMBARD, J.L. (1978). Selecting materials for high pressure fittings. *Safety in Polyethylene Plants*, **3**, 50. See also *Chem. Engng Prog.*, 1977, **73**(12), 33

CHAMBARD, J.L. (1980). Designing process relieving systems for safe performance. In *Loss Prevention and Safety Promotion*, vol. 3, p. 1230

CHAMBERLAIN, A.C. (1953). Aspects of travel and deposition of aerosol and vapour clouds. *UK Atom. Energy Res. Estab., Rep. HP/R 1261* (London: HM Stationery Office)

CHAMBERLAIN, A.C. (1959). Deposition of iodine-131 in Northern England in October 1957. *Quart. J. R. Met. Soc.,* **85**, 350

CHAMBERLAIN, A.C. (1961). Aspects of the deposition of radioactive and other gases and particles. *Int. J. Air Poll*, **3**, 63

CHAMBERLAIN, A.C. (1966a). Transport of gases to and from grass-like surfaces. *Proc. R. Soc., Ser. A*, **290**, 236

CHAMBERLAIN, A.C. (1966b). Transport of lycopodium spores and other small particles to rough surfaces. *Proc. R. Soc., Ser. B*, **296**, 45

CHAMBERLAIN, A.C. (1975). The movement of particles in plant communities. *Vegetation and the Atmosphere*, vol. 1 (ed. by J.L. Monteith) (London: Academic Press), p. 155

CHAMBERLAIN, A.C. (1983). Effect of airborne lead on blood lead. *Atmos. Environ.*, **17**, 677

CHAMBERLAIN, A.C. and CHADWICK, R.C. (1953). Deposition of airborne radioiodine vapour. *Nucleonics*, **II**, 22

CHAMBERLAIN, A.C. and CHADWICK, R.C. (1966). Transport of iodine from atmosphere to ground. *Tellus*, **18**, 226

CHAMBERLAIN, G. (1980). Fault diagnosis. *Plant Engr*, **24**(2), 18

CHAMBERLAIN, G.A. (1987). Developments in design methods for predicting thermal radiation from flares. *Chem. Engng Res. Des.*, **65**, 299

CHAMBERLAIN, G.A. (1989). *Piper Alpha Public Inquiry. Transcript, Days 134 and 138*

CHAMBERLAIN, G.A. (1994). An experimental study of large-scale compartment fires. *Hazards*, **XII**, p. 155

CHAMBERS, A.D. and FISHER, L.R. (1989). Sizing and selection of safety/relief valves. *Selection and Use of Pressure Relief Systems for Process Plant* (Manchester: Instn. Chem. Engrs, North-West Branch), p. 2.1

CHAMBERS, D.B. (1978). Oil-water separation using fibrous materials. *Chem. Ind.*, Nov. 4, 834

CHAMBERS, J.L. *et al.* (n.d.). *Graphical Methods for Data Analysis* (California: Wadsworth Inc. Gp)

CHAMBERS, L.E. and POTTER, W.S. (1974). Design ethylene furnaces. *Hydrocarbon Process.*, **53**(1), 121; **53**(3), 95; **53**(8), 99

CHAMBRE', P.L. (1952). On the solution of the Poisson- Boltzmann equation with application to the theory of thermal explosions. *J. Chem. Phys.*, **20**, 1795

CHAMOW, M.F. (1978). Direct graph techniques for the analysis of fault trees. *IEEE Trans. Reliab.*, **R-27**, 7

CHAMPION, M., DESHAIES, B., JOULIN, G. and KINOSHITA, K. (1986). Spherical flame initiation: theory versus experiments for lean propane–air mixtures. *Combust. Flame*, **65**, 319

CHAMPNESS, M. and JENKINS, G. (1985). *Oil Tanker Data Book 1985* (Amsterdam: Elsevier)

CHAN, C., LEE, J.H.S., MOEN, I.O. and THIBAULT, P. (1980). Turbulent flame acceleration and pressure development in tubes. *Combust. Inst., First Specialists Mtg*, Bordeaux

CHAN, C., MOEN, I.O. and LEE, J.H. (1983). Influence of confinement on flame acceleration due to repeated obstacles. *Combust. Flame*, **49**, 27

CHAN, C.K. and GREIG, D.R. (1989). The structures of fast deflagrations and quasi-detonations. *Combustion*, **22**, 1733

CHAN, C.K., LAU, D. and RADFORD, D. (1991). Transition to detonation resulting from burning in a confined vortex. *Combustion*, **23**, 1797

CHAN, S.T. (1983). *FEM3 – A Finite Element Model for the Simulation of Heavy Gas Dispersion and Incompressible Flow. Users Manual. Rep. UCRL-53397*. Lawrence Livermore Nat. Lab., Livermore, CA

CHAN, S.T. (1992). Numerical simulations of LNG vapor dispersion from a fenced area. *J. Haz. Materials*, **30**, 195

CHAN, S.T. and ERMAK, D.L. (1984). Recent results in simulating LNG vapor dispersion over variable terrain. In Ooms, G. and Tennekes, H. (1984), *op. cit.*, p. 105

CHAN, S.T., ERMAK, D.L. and MORRIS, L.K. (1987). FEM3 model simulations of selected Thorney Island Phase I trials. *J. Haz. Materials*, **16**, 267

CHAN, S.T., GRESHO, P.M. and ERMAK, D.L. (1981). *A Three-dimensional Conservation Model for Simulating LNG Vapor Dispersion in the Atmosphere. Rep. UCID-19210*. Lawrence Livermore Nat. Lab., Livermore, CA

CHAN, S.T., RODEAN, H.C. and BLEWITT, D.N. (1987). FEM3 modeling of ammonia and hydrogen fluoride dispersion. *Vapor Cloud Modeling*, p. 116

CHAN, S.T., RODEAN, H.C. and ERMAK, D.L. (1984). Numerical simulations of atmospheric releases of heavy gases over variable terrain. In de Wispelaere, C. (1984), *op. cit.*, p. 295

CHANDLER, F.J. and BROOKS, J.B. (1981). Preventive maintenance and inspection. *Ammonia Plant Safety*, **23**, 157

CHANDLER, S.E. (1969). Multiple death fires. *Fire Res. Tech. Paper 22* (London: H.M. Stationery Office)

CHANDNANI, M.K. (1981). Design HPI plants for safety. *Hydrocarbon Process.*, **60**(11), 324

CHANDRA, J. and DAVIS, P.W. (1980). Rigorous bounds and relations among spatial and temporal approximations in the theory of combustion. *Combust. Sci. Technol.*, **23**, 153

CHANDRASEKHAR, S. (1960). *Radiative Transfer* (New York: Dover)

CHANDRASEKHAR, S. (1961). *Hydrodynamic and Hydromagnetic Stability* (London: Oxford Univ. Press)

CHANEY, J.F., RAMDAS, V., RODRIGUEZ, C.R. and WU, M.H. (eds) (1982). *Thermophysical Properties Research Literature Retrieval Guide 1900–1980* (New York: Plenum Press)

CHANG, C.D. (1983). *Hydrocarbons from Methanol* (New York: Dekker)

CHANG, C.T., HWANG, H.C. and HWANG, D.M. (1993). Fault-free synthesis techniques for process systems with coupled feedforward and feedback loops. *Health, Safety and Loss Prevention*, p. 167

CHANG, H.R. and REID, R.C. (1982). Spreading-boiling model for instantaneous spills of liquified petroleum gas (LPG) on water. *J. Haz. Materials*, **7**(1), 19

CHANG, S.L. and RHEE, K.T. (1983). Adiabatic flame temperature estimates of lean fuel/air mixtures. *Combust. Sci. Technol.*, **35**, 203

CHANG, T.Y. and WEINSTOCK, B. (1974). Rollback modeling of urban air pollution. *First Symp. on Atmospheric Diffusion and Air Pollution* (Boston, MA: Am. Met. Soc.), p. 184

CHAO, B.H., LAW, C.K. and TIEN, J.S. (1991). Structure and extinction of diffusion flames with flame radiation. *Combustion*, **23**, p. 523

CHAO, J. (1980). Properties of elemental sulphur. *Hydrocarbon Processing*, **59**(11), 217

CHAPANIS, A. (1965). *Man–Machine Engineering* (Belmont, CA: Wadsworth)

CHAPANIS, A., GARNER, W.R. and MORGAN, C.T. (1949). *Applied Experimental Psychology: Human Factors in Engineering Design* (New York: Wiley)

CHAPMAN, D. (1985). *Planning for Conjuctive Goals Rep. AI-TR-802*. AI Lab, Massachusetts Inst. Technol.

CHAPMAN, D.J.V. and LLOYD, R.B. (1992). Towards risk-based structural integrity methods for PWRs. *Nucl. Energy,* **31**, 363

CHAPMAN, D.L. (1989). On the rate of explosions in gases. *Phil. Mag.,* **47**, 90

CHAPMAN, D.L. (1913). Theory of electrocapillarity. *Phil. Mag.,* **25**, 475

CHAPMAN, F.S. and HOLLAND, F.A. (1965a). Heat transfer correlations for agitated liquids in process vessels. *Chem. Engng,* **72**(Jan. 18), 153

CHAPMAN, F.S. and HOLLAND, F.A. (1965b). Heat transfer correlations for jacketed vessels. *Chem. Engng,* **72**(Feb. 15), 175

CHAPMAN, F.S. and HOLLAND, F.A. (1965−). Keeping piping hot. *Chem. Engng,* 1965, **72**(Dec. 20), 79; 1966, **73**(Jan. 17), 133

CHAPMAN, F.S. and HOLLAND, F.A. (1966). Positive displacement pumps. *Chem. Engng,* **73**(Feb. 14), 129

CHAPMAN, K.R. (1975). Tankers − the ship design aspects of the bulk transport of hazardous liquid cargoes by sea. *Instn Chem Engrs, Symp. on Transport of Hazardous Materials,* Birkenhead Coll. of Technol.

CHAPMAN, M.J. and FIELD, D.L. (1979). Lessons from carbon bed adsorption losses. *Loss Prevention,* **12**, 136

CHAPMAN, W.R. (1921). Propagation of flame in mixtures of ethylene and air. *J. Chem. Soc.,* 1677

CHAPMAN, W.R. and WHEELER, R.V. (1926). The propagation of flame in mixtures of methane and air. Part IV, The effect of restrictions on the path of the flame. *J. Chem. Soc.,* 2139

CHAPMAN, W.R. and WHEELER, R.V. (1927). The propagation of flame in mixtures of methane and air. Pt V, The movement of the medium in which the flame travels. *J. Chem. Soc.,* 38

CHAPPELEAR, P.S.L. (1981). Industry initiates SI conversion. *Hydrocarbon Process.,* **60**(4), 150

CHAPPELL, W.G. (1977). Pressure/time diagram for explosion vented space. *Loss Prevention,* **11**, p. 76

CHAR, C.V. (1975a). Check pipe support orientation. *Hydrocarbon Process.,* **54**(9), 207

CHAR, C.V. (1975b). New approach to heat exchange foundations design. *Hydrocarbon Process.,* **54**(3), 121

CHAR, C.V. (1979). Guide to pipe support design. *Hydrocarbon Process.,* **58**(3), 133; **58**(9), 241

CHARACKLIS, W.G. and BUSCH, A.W. (1972). Industrial wastewater treatment. *Chem. Engng,* **79**(May 8), 61

CHARBONNEAU, A. (1985). Test system spots MOV degradation. *Oil Gas J.,* **83**(Dec.16), 102

CHARLES, A. (1974). *The Effects of a Fire Environment on a Rail Tank Car Filled with LPG. Rep. PB-241358.* Dept of Transportation, London

CHARLES, A. and NORRIS, E.B. (1974). *Fragmentation and Metallurgical Analysis of Tank Car Rax 201. Rep. FRA/R8D 7530.* Dept of Transportation, London

CHARLES, R.J. (1958). Static fatigue of glass. *J. Appl. Phys,* **29**, 1549, 1554

CHARLESWORTH, F.R., GRONOW, W.S. and KENNY, A.W. (1981). *Windscale: The Management of Safety* (London: Health and Safety Executive)

CHARLEUX, J. and HUTHER, M. (1983). Selection of steel qualities for low temperature use. *Gastech 82,* p. 426

CHARLTON, D. (1965). Select safe shutdown system for fired heaters. *Hydrocarbon Process.,* **44**(5), 255

CHARLTON, D. (1973). Selection of safe shutdown system for fired heaters. In Vervalin, C.H. (1973a), *op. cit.,* p. 410

CHARLTON, F.E. (1963). Why paint? *Chem. Engng,* **70**(Oct. 28), 158

CHARLTON, H.E. (1960). The fundamental factors influencing tightness in the design of stop valves. *Chem. Engr, Lond.,* **152**, A59

CHARLTON, I. and HUITSON, P. (1983). Teroman. *Plant Engr,* **27**(2), 33

CHARLTON, J.S. (1976). Radioactive tracer techniques in plant investigation. *Diagnostic Techniques* (London: Instn Chem. Engrs)

CHARLTON, J., SARTEUR, R. and SHARKEY, J.M. (1975). Identify refinery odors. *Hydrocarbon Process.,* **54**(5), 97

CHARLTON, T.J., ALLISON, R.C., BREYETTE, J.L., HOLDER, L.K. and WARD, C.H. (1978). Relating handling of hazardous materials to spill prevention. *Control of Hazardous Material Spills,* **4**, p. 32

CHARNEY, M. (1967). Explosive venting vs explosion venting. *Loss Prevention,* **1**, 35

CHARNEY, M. (1969). Flame inhibition of vapor-air mixtures. *Loss Prevention,* **3**, 74

CHARNEY, M. and LAWLER, F.K. (1967). Stop explosions after they start. *Food Engng,* **39**(10), 82

CHARNIAK, E. and MCDERMOTT, D. (1985). *Introduction to Artificial Intelligence* (Reading, MA: Addison-Wesley)

CHARNIAK, E., RIESBECK, C.K. and MCDERMOTT, D.V. (1980). *Artificial Intelligence Programming* (Hillsdale, NJ: Erlbaum Associates)

CHARPENTIER, P. (1979). Barge-mounted Plants. *Chem. Engng Prog.,* **75**(11), 35

CHARSLEY, P. and BROWN, B. (1993). HAZOP studies under ISO 9000. *Process Safety Management,* p. 281

CHARTERED INSTITUTE OF BUILDING SERVICES (1976). *CIBS Guide, Pt 1, A4: Air Infiltration* (London)

CHARTERED INSTITUTE OF BUILDING SERVICES (1977a). *Building Energy Code, pt 1* (London)

CHARTERED INSTITUTE OF BUILDING SERVICES (1977b). *Code of Practice for the Commissioning of Ventilation Systems* (London)

CHARTERED INSTITUTE OF BUILDING SERVICES ENGINEERS (1977). *CIBSE Guide* (London)

CHARTERED INSTITUTE OF BUILDING SERVICES ENGINEERS (1978−). *Commissioning Code. Air Distribution Systems* (London)

CHARTERS, D.A., GRAY, W.A. and MCINTOSH, A.C. (1994). A computer model to assess the hazards in tunnels (FASIT). *Fire Technol.,* **30**, 134

CHASE, G. (1979). The corrosive connection. *Chem. Engng,* **86**(Sep. 10), 149

CHASE, J.D. (1966). Searching for physical property data. *Chem. Engng,* **73**(Sep. 12), 190

CHASE, J.D. (1970). Plant cost vs capacity: new way to use exponents. *Chem. Engng,* **77**(Apr. 6), 113

CHASE, J.D. (1984). The qualification of pure component physical property data. *Chem. Engng Prog.,* **80**(4), 63

CHASE, W.P. (1965). Measurement and prediction of human performance as a quantitative factor in system effectiveness. *Ann. Reliab. Maintainab.,* **4**, 355

CHASIS, D.A. (1976). *Plastic Piping Systems* (New York: Industrial Press)

CHASIS, H. *et al.* (1947). Chlorine accident at Brooklyn. *Occup. Med.,* **4**, 152

LE CHATELIER, H. (1891). Estimation of fire damp by flammability limits. *Ann. Mines,* **19**(Sep. 8), 388

LE CHATELIER (1937). Referenced in Frank-Kamenetsky, D.A. (1955), *op. cit.*, p. 205, 235, quoting Jouguet, E. (1917), *op. cit.*

LE CHATELIER, H. and BOUDOUARD, O. (1898). Limits of flammability of gaseous mixtures. *Bull. Soc. Chim. (Paris)*, **19**, 483

CHATHAM-SAVANNAH DEFENSE COUNCIL (1971). *Disaster Control Directive*. Savannah-Chatham County, GA

CHATRATHI, K. (1991). Explosion isolation systems. *Chem. Engng*, **98**(10), 134

CHATRATHI, K. (1992a). Analysis of vent area estimation methods for non-nomograph gases. *Plant/Operations Prog.*, **11**, 174

CHATRATHI, K. (1992b). Deflagration protection of pipes. *Loss Prevention and Safety Promotion 7*, vol. 3, 107−1

CHATRATHI, K. (1992c). Deflagration protection of pipes. *Plant/Operations Prog.*, **11**, 116

CHATRATHI, K. (1994). Dust and hybrid explosibility in a 1 m^3 spherical chamber. *Process Safety Prog.*, **13**, 183

CHATRATHI, K. and DEGOOD, R. (1991). Explosion isolation systems used in conjunction with explosion vents. *Plant/Operations Prog.*, **10**, 159

CHATTERJEE, N. and GEIST, J.M. (1972). The effects of stratification on boil-off rates in LNG tanks. *Pipeline Gas J.*, **199**(Sep.), 40

CHATWIN, P. (1982a). The use of statistics in describing the dispersion of toxic and flammable gas clouds, and associated hazards. *Seminar on Toxic and Flammable Clouds − Hazards and Protection* (London: Instn Mech. Engrs)

CHATWIN, P.C. (1982b). The use of statistics in describing and predicting the effects of dispersing gas clouds. *J. Haz. Materials*, **6**(1/2), 213. See also in Britter, R.E. and Griffiths, R.F. (1982), *op. cit.*, p. 213

CHATWIN, P.C. (1983). The incorporation of wind shear effects into box models of heavy gas dispersion. In Ooms, G. and Tennekes, H. (1984), *op. cit.*, p. 63

CHATWIN, P.C. (1991). New research on the role of concentration fluctuations in useful models of the consequences of accidental releases of dangerous gases. *Modeling and Mitigating the Consequences of Accidental Releases*, p. 327

CHATWIN, P.C. and SULLIVAN, P.J. (1980). Some turbulent diffusion invariants. *J. Fluid Mech.*, **97**, 405

CHATY, J.C. (1985). A general purpose pilot plant facility. *Plant/Operations Prog.*, **4**, 105

CHAUDHRI, M.M. (1976). Stab initiation of explosives, *Nature*, **263**, 121

CHAUVIN (1973). The mechanism of dust ignition. *Prevention of Occupational Risks*, **2**, p. 327

CHAUVIN, J.M. and BONJOUR, E. (1985). Refrigerated LPG loading/unloading system using a CALM buoy. *Gastech 84*, p. 341

CHEAH, S.C., CHUA, S.K., CLEAVER, J.W. and MILLWARD, A. (1984). Modelling of heavy gas plumes in a water channel. Ooms, G. and Tennekes, H. (1984), *op. cit.*, p. 199

CHEAH, S.C., CLEAVER, J.W. and MILWARD, A. (1983a). The physical modelling of heavy gas plumes in a water channel. *Loss Prevention and Safety Promotion 4*, vol. 1, p. F35

CHEAH, S.C., CLEAVER, J.W. and MILLWARD, A. (1983b). Water channel simulation of the atmospheric boundary layer. *Atmos. Environ.*, **17**, 1439

CHECKEL, M.D. and THOMAS, A. (1994). Turbulent combustion of premixed flames in closed vessels. *Combust. Flame*, **96**, 351

CHEFFERS, S.J. (1969). Respiratory equipment. In Handley, W. (1969), *op. cit.*, p. 388

CHEFFERS, S.J. (1977). Respiratory equipment. In Handley, W. (1969), *op. cit.*, p. 375

CHELAPATI, C.V., KENNEDY, R.P. and WALL, I.B. (1972). Probabilistic assessment of aircraft hazard for nuclear power plants. *Nucl. Engng Des.*, **19**, 333

CHELIUS, J. (1962). Understanding the refractory metals. *Chem. Engng*, **69**(Dec. 10), 178

CHELL, G.G. (1978). A procedure for incorporating thermal and residual stresses into the concept of a failure assessment diagram. *ASTM STP 668*. Am. Soc. Testing and Mats, Philadelphia, PA, p. 581

CHEMENTATOR (1949−). *Chem. Engng*, 1949, **56**, and subsequent issues

CHEMICAL AND ALLIED PRODUCTS INDUSTRY TRAINING BOARD (1979). *Product Liability and the Manager* (draft) (Staines, Middlesex)

CHEMICAL AND ALLIED PRODUCTS INDUSTRY TRAINING BOARD − see also Appendix 28

CHEMICAL ENGINEERING (1958−). Chemical Engineering cost file. *Chem. Engng*, 1958, **65**, and subsequent issues

CHEMICAL ENGINEERING (1962−). Materials of construction. Twentieth biennial report. *Chem. Engng*, 1962, **69**, and then biennially

CHEMICAL ENGINEERING (1963). Cost file index for 1958 to 1963. *Chem. Engng*, **70**(Dec. 23), 104

CHEMICAL ENGINEERING (1968−). Pollution control (later Environmental engineering). *Chem. Engng*, 1968, **75**(Oct. 14) (Deskbook issue); 1970, **77**(Apr. 27); 1971, **78**(Jun. 21); 1972, **79**(May 8); 1973, **80**(Jun. 18); 1974, **81**(Oct. 2); 1975, **82**(Oct. 6); 1976, **83**(Oct. 18); 1977, **84**(Oct. 17)

CHEMICAL ENGINEERING (1969a). Liquids handling. *Chem. Engng*, **76**(Apr. 14) (Deskbook issue)

CHEMICAL ENGINEERING (1969b). Solids handling. *Chem. Engng*, **76**(Oct. 13) (Deskbook issue)

CHEMICAL ENGINEERING (1969c). Special issue on Process control including Guide to process instrument elements. *Chem. Engng*, **76**(Jun. 2)

CHEMICAL ENGINEERING (1970−). Engineering materials. *Chem. Engng*, 1970, **77**(Oct. 12) (Deskbook issue); 1972, **79**(Dec. 4)

CHEMICAL ENGINEERING (1971). Pump and valve selector. *Chem. Engng*, **78**(Oct. 11) (Deskbook issue)

CHEMICAL ENGINEERING (1973a). Chemical products/manufacturing distribution. *Chem. Engng*, **80**(Oct. 18) (Deskbook issue)

CHEMICAL ENGINEERING (1973b). Plant maintenance and engineering. *Chem. Engng*, **80**(Feb. 26) (Deskbook issue)

CHEMICAL ENGINEERING (1974a). *Physical Properties* (New York: McGraw-Hill)

CHEMICAL ENGINEERING (1974b). *Supplementary Readings in Engineering* (New York: McGraw-Hill)

CHEMICAL ENGINEERING (1978a). Liquids handling. *Chem. Engng*, **85**(Apr. 3), 9 (Deskbook issue)

CHEMICAL ENGINEERING (1978b). Materials handling. *Chem. Engng*, **85**(Oct. 30), 9 (Deskbook issue)

CHEMICAL ENGINEERING (1978c). *Modern Cost Engineering: Methods and Data* (New York: McGraw-Hill)

CHEMICAL ENGINEERING (1979a). *Modern Cost Engineering* (New York: McGraw-Hill)

CHEMICAL ENGINEERING (1979b). *Process Technology and Flowsheets* (New York: McGraw-Hill)

CHEMICAL ENGINEERING (1980). Twenty-ninth Materials of Construction Report. *Chem. Engng*, **87**(Nov.3), 86

CHEMICAL ENGINEERING (1984). Thirty-first Materials of Construction Report. *Chem. Engng*, **91**(Oct. 29), 68

CHEMICAL ENGINEERING PROGRESS (1989). First Annual Software Directory. *Chem. Engng Prog.*, **85**(10) (supplement)

CHEMICAL ENGINEERING PROGRESS (1994). *1994 CEP Software Directory* (New York: Am. Inst. Chem. Engrs)

CHEMICAL ENGINEERING STAFF (1960). Acetylene transmission systems. *Chem. Engng*, **67**(Sep. 5), 131

CHEMICAL ENGINEERING STAFF (1980). *Practical Process Instrumentation and Control* (New York: McGraw-Hill)

CHEMICAL ENGINEERING STAFF (1980). *Safe and Efficient plant Operation and Maintenance* (New York)

CHEMICAL INDUSTRIES ASSOCIATION (1975). *Hazard Identification – A Voluntary Scheme for the Marking of Tank Vehicles and Dangerous Substances* (London)

CHEMICAL INDUSTRIES ASSOCIATION (1978). *Guide to the Safe Practice in the Use and Handling of Hydrogen Fluoride* (London)

CHEMICAL INDUSTRIES ASSOCIATION (1979). *An Approach to the Categorisation of Process plant Hazard and Control Building Design* (London)

CHEMICAL INDUSTRIES ASSOCIATION (1980). *Emergency – dealing with Press, Radio and Television* (London)

CHEMICAL INDUSTRIES ASSOCIATION (1981). *The Control of Occupational Carcinogens* (London)

CHEMICAL INDUSTRIES ASSOCIATION (1982). *Inter-Company Collaboration for Ammonia Transport Emergencies* (London)

CHEMICAL INDUSTRIES ASSOCIATION (1983). *Inter-Company Collaboration for Chlorine Emergencies* (London)

CHEMICAL INDUSTRIES ASSOCIATION (1982). *Inter-Company Collaboration for Chlorine Emergencies* (London)

CHEMICAL INDUSTRIES ASSOCIATION, CHLORINE SECTOR GROUP (1987). General Guidance on Emergency Planning within the CIMAH Regulations for Chlorine Installations (London)

CHEMICAL INDUSTRIES ASSOCIATION, CHLORINE SECTOR GROUP (1989). Industrial guidance on the safety case: 1, Chlorine. In Lees, F.P. and Ang, M.L. (1989), *op. cit.*, p. 135

CHEMICAL INDUSTRIES ASSOCIATION – see also Appendix 28.

CHEMICAL INDUSTRY SAFETY AND HEALTH COUNCIL – see Appendices 27 and 28

CHEMICAL MANUFACTURERS ASSOCIATION (1990). *A Manager's Guide to Reducing Human Error* (Washington, DC)

CHEMICAL SOCIETY (1975). *Handling Toxic Chemicals – Environmental Considerations* (London)

CHEMICAL WARFARE MEDICAL COMMITTEE (1918–). *Reports 1–20* (London: H.M. Stationery Office)

CHEN, B., SHEN, J., SUN, Q. and HU, S. (1989). Development of an expert system for synthesis of heat exchanger networks. *Comput. Chem. Engng*, **13**, 1221

CHEN, C.J. and RODI, W. (1980). *Vertical Turbulent Buoyant Jets* (Oxford: Pergamon Press)

CHEN, E. (1975). Analysis of ship collision probabilities. *Transport of Hazardous Cargoes*, 4

CHEN, G.K. (1987). Troubleshooting distribution problems in distillation columns. *Chem. Engr, Lond.*, **440**(Distillation Suppl.), 10

CHEN, J. and MADDOCK, M.J. (1973). How much spare heater for ethylene plants. *Hydrocarbon Process*, **52**(5), 147

CHEN, J.R., RICHARDSON, S.M. and SAVILLE, G. (1992). Numerical simulation of full-bore ruptures of pipelines containing perfect gases. *Process Safety Environ.*, **70B**, 59

CHEN, L.D. and FAETH, G.M. (1981a). Ignition of a combustible gas near heated vertical surfaces. *Combust. Flame*, **42**, 77

CHEN, L.D. and FAETH, G.M. (1981b). Initiation and properties of decomposition waves in liquid ethylene oxide. *Combust. Flame*, **40**, 13

CHEN, L.W. and MODARRES, M. (1992). Hierarchical decision process for fault administration. *Comput. Chem. Engng*, **16**, 425

CHEN, N.H. (1979). An explicit equation for friction factor in pipe. *Ind. Engng. Chem. Fund.*, 1979, **18**, 296; 1980, **19**, 228

CHEN, N.H. (1980). An explicit equation for friction factor in pipe (letter). *Ind. Engng. Chem. Fund.*, **19**, 228

CHEN, Y.N. (1968). Flow induced vibration and noise in tube bank heat exchangers due to von Karman streets. *J. Engng Ind.*, **90**, 134

CHEN, Z., GOVIND, R. and WEISMAN, J. (1983). Discharge of a saturated liquid through a relief valve. In Veziroglu, T.N. (1983), *op. cit.*, p. 58

CHEN-HWA CHUI (1982). Depressuring analysis applied to cryogenic plant safety. *Sixteenth Loss Prevention Symp.* (New York: Am. Inst. Chem. Engrs)

CHEN-HWA CHIU and MURRAY, L.K. (1992). Retrofit baseload LNG plants. *Hydrocarbon Process.*, **71**(1), 69

CHEN-SHAN KAO, TAMHANE, A.C. and MAH, R.H.S. (1990). Gross error detection in serially correlated process data. *Ind. Engng Chem. Res.*, **29**, 1004

CHENG-CHING YU and CHYUAN LEE (1991). Fault diagnosis based on qualitative/quantitative process knowledge. *AIChE J.*, **37**, 617

CHENIER, P.J. (1986). *Survey of Industrial Chemistry* (New York; Wiley)

CHER, M. (1985). A sulfur hexafluoride tracer study at the Tracy power plant – I. Preliminary studies. *Atmos. Environ.*, **19**, 1417

CHEREMISINOFF, N.P. (1979). *Applied Fluid Flow Measurement* (Basel: Dekker)

CHEREMISINOFF, N.P. (1981). *Process Level Instrumentation and Control* (Basel: Dekker)

CHEREMISINOFF, N.P. (1986). *Complex Fluid Flows* (Lancaster, PA: Technomic Publishing Co.)

CHEREMISINOFF, N.P. and CHEREMISINOFF, P.N. (1978). *Fibreglass-Reinforced Plastics Deskbook* (Ann Arbor, MI: Ann Arbor Sci. Publs)

CHEREMISINOFF, P.N. and MORRESI, A.C. (1979). *Benzene. Basic and Hazardous Properties* (New York: Dekker)

CHEREMISINOFF, P.N. and YOUNG, R.A. (1975). *Industrial Odor Technology and Assessment* (Ann Arbor, MI: Ann Arbor Sci. Publs)

CHERRINGTON, D.C. and CIUFFREDA, A.R. (1967). Piping design for hydrogen service. *Hydrocarbon Process.*, **46**(5), 148

CHERRY, D.H., GROGAN, J.C., HOLMES, W.A. and PERRIS, F.A. (1977). Availability analysis for chemical plants. *Eleventh Loss Prevention Symp.* (New York: Am. Inst. Chem. Engrs). See also *Chem. Engng Prog.*, 1978, **74**(1), 55

CHERRY, D.H., GROGAN, J.C., KNAPP, G.L. and PERRIS, F.A. (1982). Use of data bases in engineering design. *Chem. Engng Prog.*, **78**(5), 59

CHERRY, D.H., PRESTON, M.L. and FRANK, G.W.H. (1985). *Batch Process Schedule Generation – ICI's BatchMASTER. PSE 95* (Rugby: Instn Chem. Engrs), p. 91

CHERRY, K.F. (1980). Prevent fluid mixups with a modified check valve. *Chem. Engng*, **87**(Nov. 17), 275

CHERUBIN, L.J. (1960–). Minimizing radiation hazards. *Chem. Engng*, 1960, **67**(Jun. 27), 105; 1961, **68**(Apr. 3), 163

CHERVINSKY, A. and MANHEIMER-TIMNAT, Y. (1969). A study of axisymmetrical turbulent jet diffusion flames. *Combust. Flame*, **13**, 157

CHESLER, S. and JESSER, B.W. (1952). Some aspects of the design and economic problems involved in safe disposal of flammable vapours from safety relief valves. *Trans. ASME*, **74**, 229

CHESTER, E.H., GILLESPIE, D.G. and KRAUSE, F.D. (1969). The prevalence of chronic obstructive pulmonary disease in chlorine gas workers. *Am. Rev. Respir. Dis.*, **99**, 365

CHESTER, E.H., KAIMAL, P.J., PAYNE, C.B. and KOHN, P.M. (1977). Pulmonary injury following exposure to chlorine gas. *Chest*, **72**, 247

CHESTER, W. (1954). The quasi-cylindrical shock tube. *Phil. Mag.*, **45**, 1293

CHESTER, J.R. and PHILLIPS, R.D. (1985). The best pressure relief valves: simple or intelligent? *InTech*, Jan., 51

CHESTERS, J.H. (1961). *Steel Plant Refractories* (Pittsburgh, PA: United Steel Co.)

CHESTERTON INT. (1987). Pump monitoring. *Plant Engr*, **30**(2), 21

CHETTY, V.J. and VOINOV, A.N. (1972). An analysis of the chemico-kinetic nature of ignition of liquid fuels and the effect of physical factors. *Combust. Flame*, **18**, 417

CHEUNG, J.T.Y. and STEPHANOPOLOUS, G. (1990). Representation of process trends – Part I: a formal representation framework. *Comput. Chem. Engng*, **14**, 495

CHEVALIER, A.B.H. (1991). Passive safety features – a considered view of UK safety arguments. *Nuc. Energy* **30** (2), 79–83

CHEYREZY, M. (1983). Cryogenic storage tanks – design of the outer concrete containment. *Gastech 82*, p. 377

CHHIBBER, S., APOSTOLAKIS, G. and OKRENT, D. (1991). A probabilistic framework for the analysis of model uncertainty. *Process Safety Environ.*, **69B**, 67

CHICAGO BRIDGE AND IRON COMPANY (1974). *Horton Sphere Pressure Vessels. Bull. 4300* (Chicago, IL)

CHICAGO BRIDGE AND IRON COMPANY (1975a). *Evaporation Loss. Bull. 533* (Chicago, IL)

CHICAGO BRIDGE AND IRON COMPANY (1975b). *Oil Storage Tanks with Fixed Roofs. Bull. 3310* (Chicago, IL)

CHICAGO BRIDGE AND IRON COMPANY (1976a). *Horton Floating Roof Tanks. Bull. 3200* (Chicago, IL)

CHICAGO BRIDGE AND IRON COMPANY (1976b). *SOHIO/CBI Floating Roof Emission Test Program: Final Report* (Chicago, IL.)

CHICHELLI, M.T. and BONILLA, C.F. (1945–). Heat transfer to liquids boiling under pressure. *Trans. Am. Inst. Chem. Engrs*, 1945, **41**, 755; 1946, **42**, 411

CHICKEN, J.C. (1975). *Hazard Control Policy in Britain* (Oxford: Pergamon Press)

CHICKEN, J.C. (1982). *Nuclear Power Hazard Control Policy* (Oxford: Pergamon Press)

CHICKEN, J.C. (1986). *Risk Assessment for Hazardous Installations* (Oxford: Pergamon Press)

CHIDAMBARIAH, V., TRAVIS, C.C., TRABALKA, J.R. and THOMAS, J.K. (1991). A relative risk index for prioritization of inactive underground storage tanks. *J. Haz. Materials*, **27**, 327

CHIEFFO, A.B. and MCLEAN, R.H. (1968). Reduce pollution using early leak detection. *Hydrocarbon Process.*, **47**(5), 149

CHIEFFO, A.B. and MCLEAN, R.H. (1968a). Fast detection of leaks cuts hydrocarbon losses and pollution. *Chem. Engng*, **75**(Jul. 15), 144

CHIESA, P.J. (1980). Severe laboratory fire test for fire fighting foams. *Fire Technol.*, **16**, 12

CHIEU, J. and FOSTER, S. (1993). Improve storm water management for refineries. *Hydrocarbon Process.*, **72**(12), 73

CHIGIER, N.A. and STROKIN, V. (1974). Mixing processes in a free turbulent diffusion flame. *Combust. Sci. Technol.*, **9**, 111

CHIGNELL, M.H. and PETERSON, J.G. (1988). Strategic issues in knowledge engineering. *Hum. Factors*, **30**, 381

CHIKHLIWALA, E.D. and GELINAS, C.G. (1988). Early detection and monitoring system for identification of leak location and quantification of spill rate. *Preventing Major Chemical and Related Process Accidents*, p. 255

CHIKHLIWALA, E.D. and HAGUE, W.J. (1987). Specialized techniques for modeling the unique phenomena exhibited in HF releases. *Vapor Cloud Modeling*, p. 955

CHIKHLIWALA, E.D., OLIVER, M., GELINAS, C.G. and SCHIPHOLT, B.L. (1989). Comparison of transient modeling dynamics between different hazardous chemical tank ruptures. *Loss Prevention and Safety Promotion 6*, p. 68.1

CHILDS, P. (1992). Asbestos-free gaskets. *Chem. Engr, Lond.*, **522**, 19

CHILES, JAMES R. (2001) *Inviting Disaster, Lessons From the Edge of Technology* (New York: HarperCollins Publishing) ISBN 0-06-662081-5

CHILTON BOOK CO. (1994). *Instrument Engineer's Handbook*, 3rd ed. (Radnor, PA)

CHILTON, A.B., SHULTIS, J.K. and FAW, R.E. (1984). *Principles of Radiation Shielding* (Englewood Cliffs, NJ: Prentice-Hall)

CHILTON, C.H. (1966). How readers view pipeline safety. *Chem. Engng*, **73**(May 23), 145

CHILTON, T.H. and COLBURN, A.P. (1934). Mass transfer (absorption) coefficients – production from data on heat transfer and fluid friction. *Ind. Engng Chem.*, **26**, 1183

CHILVER, R.C. (1973). The Insurance Technical Bureau. *The Post Magazine*, Aug. 9, 2196

CHIN-SHU WANG and SIBULKIN, M. (1993). Comparison of minimum ignition energy for four alkanes: effects of ignition kernel size and equivalence ratio. *Combust. Sci. Technol.*, **91**, 163

CHING-MING SHEIH (1981). Pasquill–Taylor dispersion parameters over water near shore. *Atmos. Environ.*, **15**, 101

CHIPPERFIELD, P.N.J. (1979). EEC 'Dangerous Substances' Directive. The UK industry view. *Chem. Ind.*, May 5, 313

CHIPPETT, S. (1984). Modeling of vented deflagrations. *Combust. Flame*, **55**, 127

CHIRONNA, R.J. and VOEPEL, L.P. (1983). Variable duty scrubber for TDI fumes. *Plant/Operations Prog.*, **2**, 243

CHISHOLM, D. (1981). Flow of compressible two-phase mixtures through sharp-edged orifices. *J. Mech. Engng Sci.*, **23**, 45

CHISHOLM, D. (1983). *Two-phase Flow in Pipelines and Heat Exchangers* (London: Godwin)

CHISHOLM, D. and LAIRD, A.D.K. (1958). Two-phase flow in rough tubes. *Trans ASME*, **80**, 276

CHISHOLM, D. and SUTHERLAND, L.A. (1969–70). Prediction of pressure gradients in pipeline systems during two-phase flow. *Proc. Instn Mech. Engrs*, **184**(3c), 24

CHISHOLM, D. and WATSON, G.G. (1966). *The Flow of Steam/Water Mixtures through Sharp Edged Orifices. Rep. NEL 213*. Nat. Engng Lab, East Kilbride

CHISOLM, L.H. and SLEEMAN, D.H. (1979). An aide for theory formation. In Michie, D. (1979), *op. cit.*, p. 202

CHISNELL, B. (1978). Toxic gas in fires. *Fire Prev. Sci. Technol.*, **19**, 16

CHISNELL, R.F. (1957). The motion of a shock wave in a channel with applications to cylindrical and spherical shock waves. *J. Fluid Mech.*, **2**, 286

CHISSICK, S.S., BRIGGS, K., DUNBAR, C. and DOOHER, T. (1984). *Hazardous Substances in the Workplace* (Chichester: Wiley) (educational package, including films and slides)

CHISSICK, S.S. and DERRICOTT, R. (eds) (1981). *Occupational Health and Safety Management* (Chichester: Wiley)

CHITRA, S.P. (1993). Use neural networks for problem solving. *Chem. Engng Prog.*, **89**(4), 44

CHIU, C. (1982). Apply depressuring analysis to cryogenic plant safety. *Hydrocarbon Process.*, **61**(11), 255

CHIU, K.W. and LEE, J.H. (1976). A simplified version of the Barthel model for transverse wave spacings in gaseous detonation. *Combust. Flame*, **26**, 353

CHIU, K.W., LEE, J.H. and KNYSTAUTAS, R. (1977). The blast waves from asymmetrical explosions. *J. Fluid Mech.*, **82**, 193

CHIVERS, G.E. (1983). *The Disposal of Hazardous Wastes* (Leeds: H&H Consultants Ltd)

CHIVERS, G.E. (1984). The petrochemical industry. In Barnes, D., Forster, C.F. and Hrudey, S.E. (1984), *op. cit.*

CHIVES, J., MITCHELL, L. and ROWE, G. (1974). The variation of real contact area between surfaces with contact pressure and material hardness. *Wear*, **28**, 171

CHLOPIN (1927–). Military and sanitary principles of gas protection. *Z. Gesamt. Schiess-Sprengstoffwesen*, **22**, 192, 227, 261, 264, 297, 333, 369; 1928, **23, 29**

CHLORINE INSTITUTE — see Appendix 28

CHLUMSY, V. (1966). *Reciprocating and Rotating Compressors* (London: Spon)

CHO, C.-K. (1980). *An Introduction to Software Quality Control* (New York: Wiley)

CHODNOWSKY, N.M. (1968). Centrifugal compressors for high pressure service. *Chem. Engng*, **75**(Dec. 2), 110

CHOE, W.G. and WEISMAN, J. (1974). *Flow Pattern and Pressure Drop in Cocurrent Vapour–Liquid Flow.* State of the art report. Univ. of Cincinnati, OH

CHOI, K.K. (1987). Fire stops for plastic pipe. *Fire Technol.*, **23**, 267

CHOLIN, J. (1989a). The current state of the art in optical fire detection. *Plant/Operations Prog.*, **8**, 12

CHOLIN, J. (1989b). Optical fire detection. *Chem. Engng Prog.*, **85**(7), 62

CHOMIAK, J. (1973). Flame turbulence interaction. *Combust. Flame*, **20**, 143

CHOMIAK, J. and JAROSINSKI, J. (1982). Flame quenching by turbulence. *Combust. Flame*, **48**, 241

CHOMSKY, N. (1957). *Syntactic Structures* (The Hague: Mouton)

CHONG, O.Y.M.K., LAM, K.Y., ALWIS, W.A.M., CHIN, J.C. and TAN, S.T. (1993). PC based software system for blast propagation. *Health, Safety and Loss Prevention*, p. 226

CHOPEY, N.P. (1967a). Blackouts put power supply under scrutiny. *Chem. Engng*, **74**(Sep. 11), 102

CHOPEY, N.P. (1967b). Fearful citizens scare off new CPI plant. *Chem. Engng*, **74**(Jul. 31), 38

CHOPEY, N.P. (1970). Mercury's turn as villain. *Chem. Engng*, **77**(Jul. 27), 84

CHOPEY, N.P. (1972). Hydrogen: tomorrow's fuel. *Chem. Engng*, **79**(Dec. 25), 24

CHOPPIN, J. (1991). Quality. *Plant Engr*, **35**, Jan./Feb., 36

CHOPPIN, J. (1993). TQM. *Plant Engr*, **37**(2), 34

CHOQUETTE, A.E. (1984). Evaluating fire-tested valves: a critical look. *Hydrocarbon Process.*, **63**(7), 85

CHOROFAS, D. (1960). *Statistical Processes and Reliability Engineering* (Princeton, NJ: van Nostrand)

CHOUDHARY, G. (ed.) (1981). *Chemical Hazards in the Workplace* (Washington, DC: Am. Chem. Soc.)

CHOW, W.K. and FONG, N.K. (1991). Numerical simulation on cooling of the fire induced air flow by water sprays. *Fire Safety J.*, **17**, 263

CHOWDHURY, J. (1982). New rules seek to place limits on plant, radiation. *Chem. Engng*, **89**(Nov. 15), 71

CHOWDHURY, J. (1984). Process trio selectively produce olefins from LPG. *Chem. Engng*, **91**(Jun. 25), 40

CHOWDHURY, J. (1985a). CPI flutter over flawed valves. *Chem. Engng*, **92**(Sep. 2), 22

CHOWDHURY, J. (1985b). Expert systems gear up for process synthesis jobs. *Chem. Engng*, **92**(Aug. 19), 17

CHOWDHURY, J. (1986). Troubleshooting comes online in the CPI. *Chem. Engng*, **93**(Oct. 13), 14

CHOWDHURY, J. and CAUDWELL, R. (1982). Thinking machines and the CPI. *Chem. Engng*, **89**(Sep. 20), 45

CHOWDHURY, J. and PARKINSON, G. (1992). OSHA tightens its hold. *Chem. Engng*, **99**(5), 37

CHOWDHURY, J., SHORT, H. and GIBB, R. (1985). Process controllers don 'expert' guises. *Chem. Engng*, **92**(Jun. 24), 14

CHOWDHURY, J. *et al.* (1985). Bhopal, a year later: learning from a tragedy. *Chem. Engng*, **92**(Dec. 9/23), 14

CHRISTEN, P., BOHNENBLUST, H. and SEITZ, S. (1994). A methodology for assessing catastrophic damage to the population and environment: a quantitative multi-attribute approach for risk analysis based on fuzzy set theory. *Process Safety Prog.*, **13**, 234

CHRISTEN, R. (1980). Possibilities of controlling thermic dangers. *Loss Prevention and Safety Promotion 3*, vol. 2, p. 379

CHRISTENSEN, J.J., HANKS, R.W. and IZATT, R.M. (1982). *Handbook of Heats of Mixing* (New York: Wiley)

CHRISTENSEN, J.M. (1976). Ergonomics: where have we been and where are we going: II. *Ergonomics*, **19**, 287

CHRISTENSEN, J.M. and HOWARD, J.M. (1981). Field experience in maintenance. In Rasmussen, J. and Rouse, W.B. (1981), *op. cit.*, p. 111

CHRISTIAN, A.H. (1960). Guidebook for technical management. *Ind. Engng Chem.*, **52**(1), 93A

CHRISTIAN, A.H. (1965). The future of chemical safety. In Fawcett, H.H. and Wood, W.S. (1965), *op. cit.*, p. 554

CHRISTIAN MICHELSEN INSTITUTE (1992). *From Ideas to Practical Results* (Bergen)

CHRISTIAN MICHELSEN INSTITUTE — see Appendix 28

CHRISTIANSEN, J.H. and JORGENSEN, L.E. (1983). Propylene discharge — a case study. *Loss Prevention and Safety Promotion 4*, vol. 1, p. L9

CHRISTIANSEN, O.B. (1978). Closed loop spray drying systems. *Chem. Engng Prog.*, **74**(1), 83

CHRISTIANSEN, V. (1994). Combustion of some pesticides and evaluation of the environmental impact. *J. Loss Prev. Process Ind.*, **7**, 39

CHRISTIANSEN, V., KAKKO, R. and KOIVISTO, R. (1993). Environmental impact of a warehouse fire containing ammonium nitrate. *J. Loss Prev. Process Ind.*, **6**, 233

CHRISTOPHER, D.H. (1978). Toxicological testing of industrial chemicals. *Chem. Ind.*, Apr. 15, 256

CHRISTOPHERSON, D.G. (1944). *The Blast Effect of Air Burst Bombs. Rep. REN 394.* Ministry of Home Security, London

CHRISTOPHERSON, D.G. (1946). *Structural Defence. Rep. RC 450.* Ministry of Home Security, London

CHRISTOPHERSON, D.G. (1949). *Blast Impulse Associated with Specific Categories of Housing Damage. Rep. REN 490.* Ministry of Home Security, London

CHRISWELL, L.I. (1983). Modern processes for the manufacture of polyethylene and polypropylene. *Chem. Engng Prog.*, **79**(4), 84

CHU, B.B. (1976). *A Computer Oriented Approach to Fault Tree Construction. Rep. NP-288.* Elec. Power Res. Inst., Palo Alot, CA

CHU, V.H. (1975). Turbulent dense plumes in laminar crossflow. *J. Hydraulic Res.,* **13,** 263

CHUBB, J.N., POLLARD, I.E. and BUTTERWORTH, G.J. (1975). Techniques for assessment of electrostatic ignition hazards. *Electrical Safety,* **2,** 21

CHUDLEIGH, M.F. and CATMUR, J.R. (1992). Implications for new standards for safety-critical software on computer control of process plants. *Loss Prevention and Safety Promotion* 7, vol. 1, 40–1

CHUEI-TIN CHANG and HER-CHUAN HWANG (1992). New developments of the digraph-based techniques for fault tree synthesis. *Ind. Engng Chem. Res.,* **31,** 1490

CHUEI-TIN CHANG, KAI-NAN MAH and CHII SHIANG TSAI (1993). A simple design strategy for fault monitoring systems. *AIChE J.,* **39,** 1146

CHUNG, B.T.F. and NARAGHI, M.H.N. (1981). Some exact solutions for radiation view factors from spheres. *AIAA J.,* **19,** 1077

CHUNG, B.T.F. and SUMITRA, P.S. (1972). Radiation shape factors from plane point sources. *ASME J. Heat Transfer,* **94,** 328

CHUNG, D.T., MODARRES, M. and HUNT, R.N.M. (1989). GOTRES: an expert system for fault detection and analysis. *Comput. Chem. Engng,* **24,** 113

CHUNG, P.W.H. (1993). Qualitive analysis of process plant behaviour. *Proc. Sixth Int. Conf. on Industrial and Engineering Applications of Artificial Intelligence and Expert Systems* (Gordon & Breach), p. 277

CHUNG, P.W.H., ABBAS, S. and ROBERTSON, D. (1993). Representing design information and safety constraints. *Proc. Sixth Int. Conf. on Industrial and Engineering Applications of Artificial Intelligence and Expert Systems* (Gordon & Breach), p. 201

CHUNG, P.W.H. and BROOMFIELD, E.J. (1994). Computer control and human error in the process industries. Draft book chapter

CHUNG, P.W.H. and GOODWIN, R. (1994). Representing design history. *Articial Intelligence in Design '94* (ed. by J.Gero) (Amsterdam: Kluwer)

CHUNG, P.W.H. and STONE, D. (1994). Approaches to representing and reasoning with technical regulatory information. *Knowledge Engng Rev.,* **9**(2), 147

CHUNG-CHEN CHANG and CHENG-CHING YU (1990). On-line fault diagnosis using the signed directed graph. *Ind. Engng Chem. Res.,* **29,** 1290

CHUNG-CHIANG PENG and SHEUE-LING HWANG (1994). The design of an emergency operating procedure in process control systems: a case study of a refrigeration system in an ammonia plant. *Ergonomics,* **37,** 689

CHUNG-YOU WU (1983). Are your flare systems adequate? *Chem. Eng,* **90**(Oct. 31), 41

CHUNGHU TSAI, T. (1985). Flare system design by microcomputer. *Chem. Engng,* **92**(Aug. 19), 55

CHURCH, A.H. (1944). *Centrifugal Pumps and Blowers* (New York: Wiley)

CHURCH, A.H. (1963). *Mechanical Vibrations,* 2nd ed. (New York: Wiley)

CHURCH, B.J. (1978). Gas pipes in ducts. *Fire Prev. Sci. Technol.,* **19,** 11

CHURCH, D.M. (1962). How to pick centrifugal compressors. *Chem. Engng,* **69**(Apr. 2), 117

CHURCH, E.H. and PRENTICE, J.S. (1978). Safe purging of an LPDE train. *Safety in Polyethylene Plants,* **3,** p. 1

CHURCH, J. (1986). Effective monitoring of health and safety. *Chem. Ind., Sep. 15,* 605

CHURCH, P.E. (1949). Dilution of waste gases in the atmosphere. *Ind. Engng. Chem.,* **41,** 2753

CHURCH, H.W. (1976). The atmospheric dispersion model as used in the Reactor Safety Study, WASH-1400. *Third Symp. on Atmospheric Turbulence, Diffusion and Air Quality* (Boston, MA: Am. Met. Soc.), p. 188

CHURCHILL, M.A. and BUCKINGHAM, R.A. (1956). Statistical method for analysis of stream purification capacity. *Sewage Ind. Wastes,* **28,** 517

CHURCHILL, S.W. (1973). Empirical expressions for the shear stress in turbulent flow in commercial pipe. *AIChE J.,* **19,** 375

CHURCHILL, S.W. (1974). *The Interpretation and Use of Rate Data: The Rate Concept* (New York: McGraw-Hill)

CHURCHILL, S.W. (1977). Friction-factor equation spans all fluid-flow regimes. *Chem. Engng,* **84**(Nov. 7), 91

CHURCHILL, S.W. (1980). Comment on 'An explicit equation for friction factor in pipe'. *Ind Engng Fund.,* **18,** 228

CHURCHILL, S.W. (1989). *Viscous Flows* (Oxford: Butterworth-Heinemann)

CHURCHLEY, A.R. (1987). A rationale for the reliability assessment of high integrity mechanical systems. *Reliab. Engng,* **19**(1), 59

CHUSE, R. (1954–). *Pressure Vessels;* 1956, 2nd ed.; 1958, 3rd ed.; 1960, 4th ed.; 1977, 5th ed: (New York: McGraw-Hill)

CHUSE, R. and EBER, S.M. (1984). *Pressure Vessels. The ASME Code Simplified,* 6th ed. (New York: McGraw-Hill)

CHUSE, R. and CARSON, B.E. (1993). *Pressure Vessels. The ASME Code Simplified,* 7th ed. (New York: McGraw-Hill)

CHUSHKIN, P.I. and SHURSHALOV, L.V. (1982). Numerical computations of explosions in gases. *Proc Eighth Int. Conf. on Numerical Methods in Fluid Dynamics* (Berlin: Springer), p. 21

CHYNOWETH, E. (1986). Trouble-shooting with gamma rays on your plant. *Process Engng,* Aug., 44

CHYNOWETH, E. (1987). Seal of approval – graphite foil does the job safely. *Process Engng,* Jul., 63

CHYUAN-CHUNG CHEN (1984). Predicting the performance of a heat-exchanger train. *Chem. Engng,* **91**(Mar. 19), 155

CHYUAN-CHUNG CHEN (1993). Cope with dissolved gases in pump calculations. *Chem Engng,* **100**(10), 106

CIANCIA, J. and STEYMANN, E.H. (1965). Valves in the chemical process industries. *Chem. Engng,* **72**(Aug. 30), 97

CIARAMBINO, I., MERLA, C. and MESSINA, S. (1983). Interactive computer codes concerning safety shutdown systems design starting from operability analysis. *Euredata 4,* paper 3.3

CIBA-GEIGY (1974). *Safety in Laboratories* (London)

CIBOROWSKI, S. and KRZYSZTOFORSKI, A. (1981). Risks in the cyclohexane oxidation. *Loss Prevention,* **14,** 140

CIHAL, V. (1984). *Intergranular Corrosion of Steels and Alloys* (Amsterdam: Elsevier)

CINDRIC, D.T. (1985a). Design a safe front-end vent for an ammonia plant. *Plant/Operations Prog.,* **4,** 242

CINDRIC, D.T. (1985b). Safe design of a front-end vent. *Ammonia Plant Safety,* **25,** p. 167.

CINDRIC, D.T., GANDHI, S.L. and WILLIAMS, R.A. (1987). Designing piping systems for two-phase flow. *Chem. Engng. Prog.,* **83**(3), 51

CINDRIC, D.T. and HASSETT, M.J. (1989). Noise in ammonia plants: a twenty-year perspective. *Ammonia Plant Safety,* **29,** p. 41

CINKOTAI, K.I., JONES, C.C., TOPHAM, J.C. and WATKINS, P.A. (1976). Experience with mutagenic tests as indicators of carcinogenic potential. *Toxicology Club Mtg*, New York

CINNAMON, S.J. and MYERS, J.E. (1965). Reducing liquid storage losses. *Chem. Engng Prog.*, **61**(12), 69

CIOLEK, W.H. (1988). Laboratory shielding for projectiles. *Plant/Operations Prog.*, **7**, 79

CIRILLO, M.C. and POLI, A.A. (1992). An intercomparison of semiempirical diffusion models under low wind speed, stable conditions. *Atmos. Environ.*, **26A**, 765

CITY AND GUILDS INSTITUTE (1972). *086 Chemical Technicians Certificate* (London). See also Edwards, E. and Lees, F.P. (1974), *op. cit.*, p. 327

CIUFFREDA, A.R. and GREENE, B.N. (1972). Hydrogen plants shutdown reduced. *Hydrocarbon Process.*, **51**(5), 113

CIUFFREDA, A.R. and HOPKINSON, B.E. (1968). API surveys hydrogen plant corrosion. *Hydrocarbon Process.*, **47**(5), 114

CIVIL AVIATION AUTHORITY (1977). 1977 first nine month statistics. *Flight Int.*, Oct. 22, 1215

CIZMAR, L.E. (1983). Mechanical design of exchangers. *Chem. Engng Prog.*, **79**(7), 47

CLAGETT, D.C., LIZZI, J., VOS, A. and MUDAN, K.S. (1993). Chemical process quantitative risk assessment (CPQRA) and risk management at GE Plastics. *Health, Safety and Loss Prevention*, p. 155

CLAGUE, T.E. (1972). Dust explosion mechanisms. *Birmingham Univ. Chem. Engr*, **23**

CLANCEY, V.J. (n.d.). *Tests Applied to Industrial Explosives. RARDE Memo 18/63*

CLANCEY, V.J. (1972a). Assessment of explosion hazards of unstable substances. *Chemical Process Hazards*, **IV**, p. 50

CLANCEY, V.J. (1972b). Diagnostic features of explosion damage. *Sixth Int. Mtg of Forensic Sciences, Edinburgh*

CLANCEY, V.J. (1974a). The evaporation and dispersion of flammable liquid spillages. *Chemical Process Hazards* V, p. 80

CLANCEY, V.J. (1974b). The interpretation of explosibility test results. *Loss Prevention and Safety Promotion 1*, p. 253

CLANCEY, V.J. (1975a). *Appraisal of the Approach Adopted in the Cremer and Warner Report No. 9*. J.H. Burgoyne and Partners, London, Rep. to Flixborough Court of Inquiry

CLANCEY, V.J. (1975b). Fire hazards of calcium hypochlorite. *J. Haz. Materials*, **1**(1), 83

CLANCEY, V.J. (1975c). Fire hazards of calcium hypochlorite. *Transport of Hazardous Cargoes*, **4**

CLANCEY, V.J. (1976a). Liquid and vapour emission and dispersion. *Course on Loss Prevention in the Process Industries*. Dept of Chem. Engng, Loughborough Univ. of Technol.

CLANCEY, V.J. (1976b). Private communication

CLANCEY, V.J. (1977a). Dangerous clouds, their growth and explosive properties. Chemical Process Hazards, **VI**, p. 111

CLANCEY, V.J. (1977b). Deposition from particulate clouds. *Chemical Process Hazards*, **VI**, p. 21

CLANCEY, V.J. (1977c). The development of hazardous clouds. *Loss Prevention and Safety Promotion 2*, p. 323

CLANCEY, V.J. (1977d). Private communication

CLANCEY, V.J. (1981). The investigation of explosions with special reference to aircraft and ships. *J. Occup. Accidents*, **3**(3), 195

CLANCEY, V.J. (1982). The effects of explosions. *The Assessment of Major Hazards* (Rugby: Instn Chem. Engrs), p. 87

CLANCEY, V.J. (1983). Emergency actions – two case histories. *Loss Prevention and Safety Promotion* 4, vol. **1**, p. M28

CLANCEY, V.J. (1987). Calcium hypochlorite. A fire and explosion hazard. *Hazards from Pressure*, p. 11

CLANCEY, W.J. (1979). Tutoring rules for guiding a case method dialogue. *Int. J. Man–Machine Studies*, **11**, 25

CLANCEY, W.J. (1983). The epistemology of a rule-based expert system: a framework for explanation. *Artificial Intelligence*, **20**, 215

CLANCEY, W.J. and LETSINGER, R. (1981). NEOMYCIN: reconfiguring a rule-based expert system for application to teaching. *Proc. Seventh Int. Joint Conf. on Artificial Intelligence* (Vancouver, BC: Int. Joint Conf. on Artificial Intelligence), p. 215

CLANCEY, W.J., SHORTLIFFE, E.H. and BUCHANAN, B.G. (1979). Intelligent computer-aided instruction for medical diagnosis. *Proc. Third Ann. Symp. on Computer Applications in Medical Care*, p. 175

CLANCY, L.J. (1975). *Aerodynamics* (Harlow: Longmans)

CLARK, A.G. (1989) *Piper Alpha Inquiry. Transcript, Days 46, 47*

CLARK, B.D., CHAPMAN, K., BISSET, R., WATHERN, P. and BARRETT, M. (1981). *A Manual for the Assessment of Major Development Proposals* (London: H.M. Stationery Office)

CLARK, B.D., GILAD, A., BISSET, R. and TOMLINSON, P. (1984). *Perspectives on Environmental Impact Assessment* (Dordrecht: Reidel)

CLARK, C. (1987). Metal storage tanks – safety in standards. *Preventing Major Chemical Accidents*, p. 2.83

CLARK, C.C. (1978). High-temperature hydrogen: threat to alloy steels. *Chem. Engng*, **85**(Jul. 3), 87

CLARK, C.J. (1978). Oil pollution – local effluent problems. *Chemy Ind.*, Nov. 4, 821

CLARK, C.R.N. (1970). Bursting discs in process plant protection. *Chem. Process Engng*, **51**, 43

CLARK, D. (1984). Enforcement of nuisance legislation – its successes and failures. *Chemy Ind.*, May 7, 345

CLARK, D.A.R. (1956). *Materials and Structures* (London: Blackie)

CLARK, D.K. (1959). *Effectiveness of Chemical Weapons in World War 1*. Operations Res. Office, John Hopkins Univ., Bethesda, MD, Rep.

CLARK, E.L. (1958). Equipment and safety: keys to pilot plant success. *Chem. Engng*, **65**, (Jul. 28), 129

CLARK, E.L. (1963). A hard look at safety in high pressure research. *Chem. Engng*, **70**(Mar. 18), 183

CLARK, F.D. and LORENZONI, A.B. (1985). *Applied Cost Engineering*, 2nd ed. (New York: Dekker)

CLARK, F.E. (1976). State of New York, Region 8 – hazardous spill response team – government policies – programs – case history. *Control of Hazardous Material Spills*, 3, p. 118

CLARK, F.R.S. (1981). Fire spread tests – a critique. *Fire Technol.*, **17**, 131

CLARK, J.K. and HELMICK, N.E. (1980). How to optimize the design of steam systems. *Chem. Engng*, **87**(Mar. 10), 116

CLARK, K.L. (1983). An introduction to logic programming. In Hayes, P.J. and Michie, D. (1983), *op. cit.*, p. 93

CLARK, K.L. and MCCABE, F.G. (1982). PROLOG: a language for implementing expert systems. *Machine Intelligence*, **10**, 455

CLARK, K.L. and TARNLUND, S.A. (eds) (1982). *Logic Programming* (New York: Academic Press)

CLARK, M.A. (1992). Fluoropolymer linings: the low cost advantages. *Chem. Engng,* **99**(8), 106

CLARK, M.R. (1989). *Piper Alpha Public Inquiry. Transcript, Day 80*

CLARK, P.A. and COCKS, A.T. (1988). Mixing models for the simulation of plume interaction with ambient air. *Atmos. Environ.,* **22,** 1097

CLARK, R. (1987). CIMAH safety cases: the BP Group strategy. *Chem. Engr, Lond.,* **439,** 10

CLARK, R., DYSON, R., EBERLEIN, A.J. and HAWKSLEY, J.L. (1989). Industrial approaches to the safety case. In Lees, F.P. and Ang, M.L. (1989), *op. cit.,* p. 122

CLARK, R.B. (1992). *Marine Pollution, 3rd ed.* (Oxford: Clarendon Press)

CLARK, R.L. (1978). How to prepare inspection reports. *Hydrocarbon Process.,* **57**(2), 159; **57**(3), 159

CLARK, R.N., FOSTH, D.C. and WALTON, V.M. (1975). Detecting instrument malfunctions in control systems. *IEEE Trans. Aerospace Electron. Systems,* **AES-11,** 465

CLARK, R.W. and CONNAUGHTON, G.E. (1984). Failure of a 1500 psig steam line. *Ammonia Plant Safety,* **24,** 177

CLARK, S. (1981). Preventing dust explosions. *Chem. Engng,* **88**(Oct. 5), 153

CLARK, S.G.M. and CAMIRAND, G.D. (1971). Acrylonitrile – how, where, who – future. *Hydrocarbon Process.,* **50**(1), 109

CLARK, W. and FROMM, B.S. (1987). *Burn Mortality* (Stockholm: Almquist & Wiksell)

CLARK, W.D. (1976a). Hydrodesulfurization incidents teach lessons. *Ammonia Plant Safety,* **18,** p. 91

CLARK, W.D. (1976b). Using fire foam on ammonia spills. *Ammonia Plant Safety,* **18,** 40

CLARK, W.D. (1977). Cracking in high temperature shift converter. *Ammonia Plant Safety,* **19,** p. 51

CLARK, W.D. and CRACKNELL, A. (1977). Avoiding stress corrosion cracking. *Ammonia Plant Safety,* **19,** p. 69

CLARK, W.D. and DUNMORE, O.J. (1976). Repairing titanium lined urea reactors. *Ammonia Plant Safety,* **18,** 98

CLARK, W.D. and GEORGE, E.A. (1973). Cracking and blistering in a shift converter. *Ammonia Plant Safety,* **15,** 58

CLARK, W.D. and MANTLE, K.G. (1967). Brittle fracture of ammonia converter. *Ammonia Plant Safety,* **9,** 26

CLARK, W.D. and SUTTON, L.J. (1974). Avoidance of losses caused by confusion of materials in piping systems. *Weld. Metal Fabrication,* **42**(Jan. 21)

CLARK, W.E. (1974). Fire fighting strategy. *Fire Engng,* **127**(Dec. **31**)

CLARKE, A.J., LUCAS, D.H. and ROSS, F.F. (1970). Tall stacks – how effective are they? *Second Int. Clean Air Conf., Washington, DC*

CLARKE, A.N. and CLARKE, J.H. (1983). Personnel training – hazardous materials and wastes. *J. Haz. Materials,* **8**(2), 129

CLARKE, A.W. (1971a). Harmonising codes on the carriage of dangerous goods. *Transport of Hazardous Cargoes,* **2,** paper 15

CLARKE, A.W. (1971b). Safe containment of dangerous goods during carriage. *Major Loss Prevention,* p. 175

CLARKE, C. (1989). Sea change in the North Sea. *Energy Economist,* **93** (Jul. 2)

CLARKE, D. (1988a). An introduction to water hammer. *Chem. Engr, Lond.,* **449,** 44

CLARKE, D. (1988b). Water hammer. *Chem. Engr, Lond.,* 1988, **452,** 34; 1989, **457,** 14

CLARKE, F.B. (1986a). Fire hazard assessment. *In Fire Protection Handbook,* 16th ed. (ed. A.E. Cote and J.M. Linville), p. 21–2

CLARKE, F.B. (1986b). Fire hazards of materials: an overview. *In Fire Protection Handbook,* 16th ed. (ed. A.E. Cote and J.M. Linville), p. 5–2

CLARKE, G. (1978). The strategic use of water for fire fighting. *Fire Surveyor,* **7**(4), 41

CLARKE, H. (1980). Safety problems arising from the lack of application of existing technology; specific problems of the small firm. *Loss Prevention and Safety Promotion 3,* vol. 3, p. 1374

CLARKE, J.F. and KASSOY, D.R. (1984). High-speed deflagration with compressibility effects. In Bowen, J.R. *et al.* (1984b), *op, cit.,* p. 175

CLARKE, J.F. and MCCHESNEY, M. (1975). *Dynamics of Real Gases* (London: Butterworths)

CLARKE, J.R.P. (1985). Monitoring gaseous emissions. *Chem. Engr, Lond.,* **416,** 36

CLARKE, J.S. (1970). Why not just build reactors to the code. *Hydrocarbon Process.,* **49**(6), 79

CLARKE, J.S. (1971). New rules to prevent tank failures. *Hydrocarbon Process.,* **50**(5), 92

CLARKE, L. and DAVIDSON, R.L. (1962). *Manual for Process Engineering Calculations,* 2nd ed. (New York: McGraw-Hill)

CLARKE, M. (1992). Fire fighting foams. *Fire Surveyor,* **21**(6), 11

CLARKE, N.W.B. (1968). *Buried Pipelines* (London: Maclaren & Sons)

CLARKE, R.H. (1980). Radiological protection criteria for controlling doses to the public in the event of unplanned releases of radioactivity. *Proc. IIASA Workshop on Procedural and Organisational Measures for Accident Management* (Vienna)

CLARKE, R. (1989–1990). Who is liable for software errors? *Comput. Law Security Rev.,* **28**

CLARKE, R.J. and NICHOLSON, P.T.H. (1992). Environmental and safety risk assessment of major oil and gas developments in the USSR. *Loss Prevention and Safety Promotion 7,* vol. 1, p. 7-1

CLARKE, R.W. and CONNAUGHTON, G.E. (1984). Failure of a 1500 psig steam line. *Plant/Operations Prog.,* **3,** 141

CLAROTTI, C.A. and LINDLEY, D.V. (eds) (1988). *Accelerated Life Testing and Experts' Opinions in Reliability* (Amsterdam: North Holland)

CLARY, J.J., GIBSON, J.E. and WARITZ, R.S. (eds) (1983). *Formaldehyde: Toxicology–Epidemiology Mechanisms* (New York: Dekker)

CLATWORTHY, S. (1988). PES guidelines – ally or adversary? *Chem. Engr Lond.,* **451,** 16

CLAUDE, J. and ETIENNE, M. (1991). Repairs to the LNG carrier 'Tellier': longevity of the Technigaz tank technology. *Gastech 90,* p. 4.10

CLAUDE, R.E. (1959). Control of axial compressors in variable capacity service. *Chem. Engng,* **66**(May 18), 175

CLAUSER, F.H. (1954). Turbulent boundary layers in adverse pressure gradients. *J. Aeronaut. Sci.,* **21,** 91

CLAUSER, F.H. (1956). The turbulent boundary layer. *Advances in Applied Mechanics,* **4,** 1

CLAUSS, F.J. (1969). *Engineers Guide to High Temperature Materials* (Reading, MA: Addison-Wesley)

CLAVEL, R.L. and ESTRABAUD, P. (1980). Application of multivariate analysis for studying the explosion of dust clouds. *Loss Prevention and Safety Promotion 3,* vol. 3, p. 1183

CLAVIN, P. (1984). Introduction to modern laminar flame theory. In Bowen, J.R. *et al.* (1984a), *op, cit.,* p. 1

CLAY, G.A., FITZPATICK, R.R.D., HURST, N.W., CARTER, D.A. and CROSSTHWAITE, P.J. (1988). Risk assessment for installation where liquefied petroleum gas (LPG) is stored in bulk vessels above ground. *J. Haz. Materials*, **20**, 357

CLAYDON, D.A. (1967). The influence of safety considerations on the design of an acetic acid plant. *Chemical Process Hazards*, **III**, p. 36

CLAYSON, D.B., KREWSKI, D. and MUNRO, I. (eds) (1985). *Toxicological Risk Assessment* (Boca Raton, FL: CRC Press)

CLAYSON, P.G. (1989). *Piper Alpha Public Inquiry. Transcript, Day 103*

CLAYTON, A. (1976). Fire fighting in chemical plant. *Course on Fire and Explosive Safety in the Chemical Industry.* Dept of Fire Safety Engng, Univ. of Edinburgh

CLAYTON, B.R. and BISHOP, R.E.D. (1982). *Mechanics of Marine Vehicles* (London: Spon)

CLAYTON, C.H. and JOHNSON, T.E. (1978). Engineering materials for pumps and valves. *Chem. Engng Prog.*, **74**(9), 54

CLAYTON, G.D. and CLAYTON, F.E. (eds) (1978). *Patty's Industrial Hygiene and Toxicology*, 3rd ed. (New York: Wiley-Interscience)

CLAYTON, G.D. and CLAYTON, F.E. (1981). *Patty's Industrial Hygiene and Toxicology*, vol. 2B, *Toxicology* (New York: Wiley)

CLAYTON, G.D. and CLAYTON, F.E. (1982). *Patty's Industrial Hygiene and Toxicology*, vol. 2C, *Toxicology* (New York: Wiley)

CLAYTON, H.M. (1977). Toxicity and vinyl chloride. *Dev. PVC Prod. Process.*, **1**, 43

CLAYTON, W.E. (1977). Transport hazard information systems. *Symp. Hazards 77 – Advances in the Safe Handling and Movement of Hazardous Substances*, Teesside Polytechnic, Middlesborough

CLAYTON, W.E. and GRIFFIN, M.L. (1994). Catastrophic failure of a liquid carbon dioxide storage vessel, *Process Safety Prog.*, **13**, 202

CLEAL, D.M. and HEATON, N.O. (1988). *Knowledge Based Systems. Implications for Human–Computer Interfaces* (Chichester: Ellis Horwood)

CLEARE, I.M. (1991). *Certification of Explosion Protected Electrical Equipment. Safety Developments in the Offshore Oil and Gas Industry* (London: Instn Mech. Engrs), p. 53

CLEARY, J.A. (1993). Overview of the new sealless pump standard. *Hydrocarbon Process.*, **72**(6), 49

CLEARY, J.A. (1993). Overview of the new sealless pump standard. *Hydrocarbon Proc.* (June), 49–51.

CLEAVER, P., HUMPHREYS, C.E. and ROBINSON, C.G. (1994). Accidental generation of gas clouds on offshore process installations. *J. Loss Prev. Process Ind.*, **7**, 273

CLEAVER, R.P. and EDWARDS, P.D. (1990). Comparison of an integral model for predicting the dispersion of a turbulent jet in a crossflow with experimental data. *J. Loss Prev. Process Ind.*, **3**(1), 91

CLEAVER, R.P., MARSHALL, M.R. and LINDEN, P.F. (1994). The build-up of concentration within a single enclosed volume following a release of natural gas. *J. Haz. Materials*, **36**, 209

CLEAVER, T.N. (1983). How experimental factors influence mechanical seal design. *Chem. Engr, Lond.*, **398**, 32

CLEGG, J. (1994). The next stage – an HSE perspective. *Offshore Safety Case Management* (Rugby: Instn Chem. Engrs), p. jc1

CLELAND, D.I. and KING, W.P. (ed.) (1983). *Project Management Handbook* (New York: van Nostrand Reinhold)

CLEMEDSON, C.J. (1956). Blast injury. *Phys. Rev.*, **36**, 336

CLEMEDSON, C.J., HELLSTROM, G. and LINDGREN, S. (1968). The relative tolerance of the head, thorax and abdomen to blunt trauma. *Ann N.Y. Acad. Sci*, **152**(Art. 1), 187

CLEMENT, R. (1989). Deciding when to use modular construction. *Chem. Engng*, **96**(8), 169

CLEMENTSON, A.T. (1963). Statistical analysis of plant yields. *Br. Chem. Engng*, **8**(8), 564

CLEMMOW, D.M. and HUFFINGTON, J.D. (1965). An extension of the theory of thermal explosion and its application to oscillatory burning of explosives. *Trans. Faraday Soc.*, **52**, 385

CLEREHUGH, G. (1991). *British Gas Research in the Field of Safety and Environment.* (Vernon Clancey Memorial Lecture). (London: Br. Gas). See also *Process Safety Environ.*, **69B**, 123

CLEREHUGH, G., SHANNON, E. and JACKSON, L. (1983). On-line gas pipeline inspection. *Oil Gas J.*, **81**(Jan. 31), 117; (Feb. 7), 62

CLERINX, J. (chrmn) (1986). *CEFIC Views on the Quantitative Assessment of Risks from Installations in the Chemical Industry* (Brussels: Conseil Europeen des Federations de l'Industrie Chimique)

VAN CLEVE, G. (1973). Disaster plan for high pressure process plant. *Safety in Polyethylene plants*, **1**, 60

CLEVELAND, A. and HEADON, D. (1972). The Attenuation of Noise Produced by Compressor plant. *Comm. GC 200.* Instn Gas Engrs, Lond., See also *Instn Gas Engr J.*, **12**, 277

CLEVELAND, A. and YOUNG, D.A. (1972). *Some Design Considerations and Operational Experience with UK Gas Compressor Stations. Comm. 873.* Instn Gas Engrs, London. See also *Instn Gas Engr J.*, **12**, 72

CLEVETT, K.J. (1986). *Process Analyser Technology* (New York: Wiley)

CLIFT, R., GRACE, J.R. and WEBER, M.E. (1978). *Bubbles, Drops and Particles* (New York: Academic Press)

CLIFTON, P.V., VINCI, G.F. and GANS, M. (1971). The engineering constructor's point of view. *Chem. Engng, Lond.*, **254**, 430

CLIFTON, R.H. (1974). *Principles of Planned Maintenance* (London: Arnold)

CLINE, D.D. and KOENIG, L.N. (1983). The transient growth of an unconfined pool fire. *Fire Technol.*, **19**, 149

CLOCKSIN, W.F. (1984). An introduction to PROLOG. In O'Shea, T. and Eisenstadt, M. (1984), *op. cit.*, p. 1

CLOCKSIN, W.F. and MELLISH, C.S. (1981). *Programming in PROLOG* (Berlin: Springer)

CLOKE, M. (1988). *A Guide to Plant Management* (Rugby: Instn Chem. Engrs)

CLOPPER, C.J. and PEARSON, E.S. (1934). The use of confidence or fiducial limits illustrated in the case of the binomial. *Biometrika*, **26**, 404

CLOSNER, J.J. (1968). LNG storage with prestressed concrete. *LNG Conf. 1*

CLOSNER, J.J. (1970). Very large prestressed concrete tanks for LNG storage. *LNG Conf. 2*

CLOSNER, J.J. (1987a). Concrete in chemical plants. *Preventing Major Chemical Accidents*, p. 2.47

CLOSNER, J.J. (1987b). Synopsis of research needs in improving plant safety. *Preventing Major Chemical Accidents*, p. 2.155

CLOSNER, J.J. and WESSON, H.R. (1983). Refrigerated liquefied gases (RLG) storage safety – what does it cost? *Gastech 82*, p. 313

CLOUDT, W.O. (1978). Tensioning bolts in the field. *Chem. Engng*, **85**(Mar. 27), 147

CLOUGH, P.N., GRIST, D.R. and WHEATLEY, C.J. (1987). Mixing of anhydrous hydrogen fluoride with moist air. *Vapor Cloud Modeling*, p. 39

CLOUGH, R.B. (1987). The energetics of AE source characterization. *Mats Evaluation*, **45**(5), 556

CLOUGH, R.H. and SEARS, G.A. (1979). *Construction Project Management* (New York: Wiley-Interscience)

CLOUGH, W.S. and GARLAND, J.A. (1970). *The Behaviour in the Atmosphere of the Aerosol from a Sodium Fire. AERER6460*

CLOUGHTON, D.W. (1981). Aspects of interruption surveys. *Fire Surveyor*, **10**(3), 35

CLOUTIER, S.E. (1985). Vapor flow equation for determining the capacity of a pressure relief valve. *API 50th Midyear Mtg*, **64**, 183

CLOYD, D.R. and MURPHY, W.J. (1965). *Handling Hazardous Materials* (Washington, DC: Nat. Aeronaut. Space Admin.)

CLULEY, J.C. (1993). *Reliability in Instrumentation and Control* (Oxford: Butterworth-Heinemann)

CLYMER, A.B. (1985). Simulate your way to safety. *Hydrocarbon Process.*, **64**(12), 79

COATES, C.F. (1984a). The ARC in chemical hazard evaluation. *Chemy Ind.*, Mar. 19, 212

COATES, C.F. (1984b). The ARC in chemical hazard evaluation. *Thermochimica Acta*, **85**, 369

COATES, C.F. and RIDDELL, W. (1981a). Assessment of thermal hazards in batch processing. *Chemy Ind.*, Feb. 7, 84

COATES, C.F. and RIDDELL, W. (1981b). A system of hazard evaluation for a medium size manufacture of bulk organic chemicals. *Runaway Reactions, Unstable Products and Combustible Powders* (London: Instn Chem. Engrs), paper 4/Y

COATES, J.F. (1973). Institutional and technical problems in risk analysis. *Conf. on Bulk Transportation of Hazardous Materials by Water in the Future* (Washington, DC: Nat. Acad. Sci., Nat Acad. Engng and Nat. Res. Coun.), p. 249

COATES, J.J. (1960). Safety in process operations. *Am. Petrol. Inst. Proc.*, **39**(3), 21

COATES, T. (1966). Occupational health hazards in the chemical industry. *Br. J. Ind. Safety*, **6**(20), 237

COATES, T. (1969). The effects of noise on human being working or living near industrial concerns. *Chem. Engr, Lond.*, **227**, CE112

COATS, H.V. (1967). Tube and piping failure. *Ammonia Plant Safety*, **9**, p. 35

COBB, A.J. and MONIER-WILLIAMS, S. (1988). CIL's experience with a computerized ammonia plant trip system. *Ammonia Plant Safety*, **28**, 150. See also *Plant/Operations Prog.*, **7**, 243

COBB, E.C. and MARSHALL, V.C. (1980). Harrisburg: the report of the President's Commission. *Chem. Engr, Lond.*, **353**, 107

COCHEO, S. (1981). How to evaluate distributed computer systems. *Hydrocarbon Process.*, **60**(6), 95

COCHEO. S. (1983). Avoid vulnerable control system architecture. *Hydrocarbon Process.*, **62**(6), 59

COCHRAN, T. (1974). *The Liquid Metal Fast Breeder Reactor: An Environmental and Economic Critique* (Baltimore, MD: Resources for the Future)

COCKRAM, M.D. (1977). Graphite bursting discs give better pressure relief. *Process Engng*, Oct.

COCKRAM, R.W.J. (1971). The generation of static electricity during discharge of liquefied CO_2 gas. *Electrical Safety*, **1**, 155

COCKS, R.E. (1979). Dust explosions: prevention and control. *Chem. Engng*, **86**(Nov. 5), 94

COCKS, R.E. and MEYER, R.C. (1981). Fabrication and use of a 20-1 spherical dust testing apparatus. *Loss Prevention*, **14**, 154

COCKS, R.E. and ROGERSON, J.E. (1978). Organising a process safety program. *Chem. Engng*, **85** (Oct. 23), 138

COCO, J.C., ed. (2001). *Large Property Damage Losses in The Hydrocarbon-Chemical Industries: A Thirty-year Review – Nineteenth Edition* (New York)

CODIER, E.O. (1968). Reliability growth in real life. *Proc. 1968 Ann. Symp. on Reliability*, p. 458

CODLIN, E.M. (1968). *Cryogenics and Refrigeration – A Bibliographic Guide* (London: Macdonald)

CODY, D.J., VANDELL, C.A. and SPRATT, D. (1985). Selecting positive displacement pumps. *Chem. Engng*, **92**(Jul. 22), 38

COEHN, A. (1898). *Ann. Phys.*, **64**, 217

COELHO, H. *et al.* (1980). *How to Solve it with Prolog*, 2nd ed., (Portugal: Ministerio da Habitacao e Obras Publicas)

COEVERT, K., GROOTHUIZEN, T.M., PASMAN, H.J. and TRENSE, R.W. (1974). Explosions of unconfined vapour clouds. *Loss Prevention and Safety Promotion*, **1**, 145

LE COFF, J. (1970). The new ASME pressure vessel code. *Loss Prevention*, **4**, 114

COFFEE, R.D. (1968). Dust explosions: an approach to protection design. *Fire Technol.*, **4**(2), 81

COFFEE, R.D. (1969). Hazard evaluation testing. *Loss Prevention*, **3**, 18

COFFEE, R.D. (1972). Evaluation of chemical stability. *J. Chem. Educ.*, **49**(6), A343

COFFEE, R.D. (1973). Hazard evaluation: the basis for chemical plant design. *Loss Prevention*, **7**, 58

COFFEE, R.D. (1976). Design your plant for survival. *Hydrocarbon Process.*, **55**(5), 301

COFFEE, R.D. (1980). Cool flames and autoignitions: two oxidation processes. *Loss Prevention*, **13**, 74

COFFEE, R.D. (1982a). Chemical stability. In Fawcett, H.H. and Wood, W.S. (1982), *op. cit.* p. 305

COFFEE, R.D. (1982b). Cool flames. In Fawcett, H.H. and Wood, W.S. (1982), *op. cit.* p. 323

COFFEE, R.D. and WHEELER, J.J. (1967). Explosibility and stabilization of propargyl bromide. *Loss Prevention*, **1**, 6

COFFEE, R.D., RAYMOND, C.L. and CROUCH, H.W. (1950). *The Testing of Materials for Pressure Relief Vents. Eastman Kodak Co.*, Kodak Park Works

COFFMAN, J.T. and BERNSTEIN, M.D. (1979). Failure of safety valves due to flow induced vibrations. In Haupt, R.W. and Meyer, R.A. (1979), *op. cit.*, p. 191

COFIELD, W.W. (1978). Regulatory problems in siting and constructing SNG and LNG plants. *API Proc., Refining Dept, 43rd Midyear Mtg*, **57**, 3

COGAN, J.L. (1985). Monte Carlo simulation of buoyant dispersion. *Atmos. Environ.*, **19**, 867

COGGLE, J.E. and NOAKES, G.R. (1972). *Biological Effects of Radiation* (London: Wykeham Publs)

COHEN, A.V. (1981). The nature of the decisions in risk management. In Griffiths, R.F. (1981), *op. cit.*, p. 21

COHEN, A.V. and PRITCHARD, D.K. (1980). *Comparative Risks of Electricity Production Systems: A Critical Survey of the Literature* (London: H.M. Stationery Office)

COHEN, B.L. (1983). *Before Its Too Late* (New York: Plenum Press)

COHEN, B.L. (1989). Journalism's failure on nuclear power information. *J. Haz. Materials*, **20**, 255

COHEN, E. (ed.) (1968). Prevention and Protection against Accidental Explosion of Fuels and Other Hazardous Mixtures. *Ann N.Y. Acad. Sci.*, **152**(Art. 1)

COHEN, E. and DOBBS, N. (1968a). Design procedures and details for reinforced concrete structures utilised in explosive storage and transport facilities. *Ann. N.Y. Acad. Sci.*, **152**(Art. 1), 452

COHEN, E. and DOBBS, N. (1968b). Models for determining the response of reinforced concrete structures to blast loads. *Ann. N.Y. Acad. Sci.*, **152**(Art. 1), 810

COHEN, E.M. (1985). Fault tolerant processes. *Chem. Engng*, **92**(Sep. 16) 73

COHEN, E.M. and FEHERVARI, W. (1985). Sequential control. *Chem. Engng*, **92**(Apr. 29) 61

COHEN, E.R. and TAYLOR, B.N. (1987). The 1986 CODATA recommended values of the fundamental physical constants. *J. Res. Nat. Bur. Stand.*, **92**(2), 85

COHEN, H., ROGERS, G.F.C. and SARAVANAMUTTOO, H.I.H. (1972). *Gas Turbine Theory*, 2nd ed. (London: Longmans)

COHEN, J. (1985). Describing Prolog by its interpretation and compilation. *Comm. ACM*, **25**, 1311

COHEN, J. and HANSEL, M. (1956). *Risk and Gambling* (New York: Philosophical Library)

COHEN, J., DEARNALEY, E.J. and HANSEL, C.E.M. (1956). Risk and hazard. *Opl. Res. Quart.*, **7**(3), 67

COHEN, J., DEARNALEY, E.J. and HANSEL, C.E.M. (1958). The risk taken in driving under the influence of alcohol. *Br. Med. J.*, Jun. 21, 1438

COHEN, P.R. (1985). *Heuristic Reasoning about Uncertainty: an Artificial Intelligence Approach* (Boston, MA: Pitman)

COHEN, P.R. and FEIGENBAUM, E.A. (eds) (1982). *The Handbook of Artificial Intelligence*, vol. 3 (Los Altos, CA: Kaufmann)

COHN, D. (1983). Chain reaction power system failure in a fertiliser complex. *Plant/Operations Prog.*, **2**, 168

COHN, P. and LINNECAR, D.F.C. (1979). Toxic effects of metals. *Chemy Ind.*, May 19, 326

COHN, P., LINNECAR, D. and RYALL, C. (1979). Health hazards in chemical and biological research laboratories. *Chemy Ind.*, Nov. 3, 734

COIT, D.W., DEY, K.A. and TURKOWSKI, W.E. (1986). Practical reliability data and analysis. *Reliab. Engng*, **14**(1), 1

COKE, M. (1988). *A Guide to plant Management* (Rugby: Instn Chem. Engrs)

COKER, A.K. (1991a). Determine process-pipe sizes. *Chem. Engng Prog.*, **87**(3), 33

COKER, A.K. (1991b). Sizing process piping for single-phase fluids. *Chem. Engr, Lond.*, **505**, 19

COKER, A.K. (1992). Size relief valves sensibly. *Chem. Engng Prog.*, **88**(8), 20; **88**(11), 94

COKER, O. (1978). Development of availability analysis techniques for process plants. Ph.D. Thesis, Univ. of Nottingham

COLANGELO, V.J. and HEISER, F.A. (1974). *Analysis of Metallurgical Failures* (New York: Wiley)

COLANNINO, J. (1993). Prevent boiler tube failures. *Chem. Engng Prog.*, **89**(10), 33; **89**(11), 73

COLARDYN, F., VAN DER STRAETEN, M., TASSON, J. and VAN EGMOND, J. (1976). Acute chlorine gas intoxication. *Acta Clin. Belg.* **31**, 70

COLBERT, R.W. (1982). Industrial heat exchange networks. *Chem. Engng Prog.*, **78**(7), 47

COLBY, J.H., WHITE, G.A. and NOTWICK, P.N. (1980). Safety, simplicity and saving with Selectoxo catalyst process in a 600 t/d Kellogg ammonia plant. *Ammonia Plant Safety*, **21**, p. 138

COLE, B.N., BAUM, M.R. and MOBBS, F.R. (1969–70). An investigation of electrostatic charging effects in high- speed gas-solids pipe flows. *Proc. Inst. Mech. Engrs*, **184**, 3C77

COLE, H.A. (1979). Pollution of the sea and its effects. *Proc. R. Soc., Ser. B*, **205**, 17. See also in Doll, Sir R. and McLean, A.E.M. (1979), *op. cit.*, p. 17

COLE, P.T. (1983). On-line monitoring using an acoustic emission testing service – what's really involved? In Butcher, D.W. (1983), *op. cit.*, p. 27

COLE, P.T. (1988). *The capabilities and limitations of NDT. Part 7: Acoustic emission* (Br. Inst. of NDT)

COLE, P.T. (1990). Structural integrity testing of pressure vessels and storage tanks using Monsanto 'MONPAC' technology. *Engineering Integrity through Testing – EIS 90* (Engng Integrity Soc.)

COLE, P.T. (1992). Acoustic methods for the evaluation of tank integrity and floor condition. *Proc. First Conf. on Environmental Management and Maintenance of Hydrocarbon Storage Tanks*

COLE, P.T. and HUNTER, M. (1991). An acoustic emission technique for detection and quantification of gas through-leakage to reduce gas losses from process pipe. *Proc. Fourth Oil Loss Conf.* (London: Inst. Petrol.)

COLE, S.T. and WICKS, P.J. (1994). European Community research in major industrial hazards. *J. Loss Prev. Process Ind.*, **7**, 68

COLEBROOK, C.F. (1939). Turbulent flow in pipes, with particular reference to the transition region between the smooth and rough pipe laws. *J. Instn Civil Engrs*, **1**, 133

COLEBROOK, C.F. and WHITE, C.M. (1937). Experiments with fluid-friction in roughened pipes. *Proc. R. Soc.*, **161**, 367

COLEMAN, L.W. (1961). Automatic hydrocarbon analyzer. *Ammonia Plant Safety*, **3**, 19

COLEMAN, L.W. (1962). Individual hydrocarbon paths through air separation plant. *Ammonia Plant Safety*, **4**, 26

COLEMAN, R.J. (1978). *Management of Fire Service Operations* (N. Scituate, MA: Duxbury Press)

COLEMAN, T. (1984). Electrostatic precipitators. *Chemy Ind.*, Feb. 20, 129

COLEMAN, T. (1989). Safety of liquid carbon dioxide storage vessels. *Chem. Engr, Lond.*, **461**, 29

COLEN, M.S. (1987a). Causation and the defense of toxic tort litigation. *J. Haz. Materials*, **15**(1/2), 57

COLEN, M.S. (1987b). Regulation of hazardous waste transportation: Federal, state, local or all of the above. *J. Haz. Materials*, **15**(1/2), 7

COLEN, M.S. (1988). Litigating toxic shock syndrome: lessons to be learned. *J. Haz. Materials*, **18**(3), 285

COLENBRANDER, G.W. (1980). A mathematical model for the transient behaviour of dense vapour clouds. *Loss Prevention and Safety Promotion 3*, p. 1104

COLENBRANDER, G.W.L. and PUTTOCK, J.S. (1983). Dense gas dispersion behaviour: experimental observations and model developments. *Loss Prevention and Safety Promotion 4*, vol. 1, p. F66

COLENBRANDER, G.W. and PUTTOCK, J.S. (1984). Maplin Sands experiments 1980: interpretation and modelling of liquified gas spills onto the sea. In Ooms, G. and Tennekes, H. (1984), *op. cit.*

COLES, R.R.A. and RICE, C.G. (1970). Towards a criterion of impulse noise in industry. *Ann. Occup. Hyg.*, **13**, 43

COLEY, J.W. (1985). Inspection equipment and techniques – past, present and future. *API 50th Midyear Mtg*, **64**, 137

COLLACOTT, R.A. (1947). *Mechanical Vibrations* (London: Pitman)

COLLACOTT, R.A. (1975). Condition monitoring techniques in maintenance. *Course on Maintenance Management*. Dept of Chem. Engng, Loughborough Univ. of Technol.

COLLACOTT, R.A. (1976a). *Contamination analysis. Diagnostic Techniques* (London: Instn Chem. Engrs)

COLLACOTT, R.A. (1976b). *Performance trend monitoring. Diagnostic Techniques* (London: Instn Chem. Engrs)

COLLACOTT, R.A. (1977a). Loss prevention through fault diagnosis. *Chem. Engr, Lond.*, **323**, 582

COLLACOTT, R.A. (1977b). *Mechanical Fault Diagnosis* (London: Chapman & Hall)

COLLACOTT, R.A. (1985). *Structural Integrity Monitoring* (London: Chapman & Hall)

COLLARD, R.S. (1976). Developing a catalyst regeneration procedure. *Ammonia Plant Safety*, **18**, 32

COLLIER, B. (1957). *The Defence of the United Kingdom* (London: HM Stationery Office)

COLLIER, J.A. (1964). Chemical plant safety: the insurance company viewpoint. *Chem. Engng*, **71**(Oct. 12), 207

COLLIER, J.G. (1972). *Convective Boiling and Condensation*, 1st edn. 1981, 2nd ed. (New York: McGraw-Hill); 1994, 3rd ed. (Oxford: Oxford Univ. Press)

COLLIER, J.G. (1988). Prevention of major process accidents within the UK and overseas. *Preventing Major Chemical and Related Process Accidents*, p. 515

COLLIER, J.G. and DAVIES, L.M. (1980). The accident at Three Mile Island. *Heat Transfer Engng*, **1**(3), 56

COLLIER, J.G. and DAVIES, L.N. (1987). *Chernobyl*. Central Elec. Generating Board, London

COLLIER, J.G. and WALLIS, G.B. (1968). *Two-Phase Flow and Heat Transfer, vols 1–4* (Hanover, NH: Thayer School of Engng, Dartmouth College)

COLLIER, J., DAVIES, M. and GARNE, L. (1982). Design for safety. *Atom*, **308**(Jun.), 127

COLLIER, R.P. (1984). *Two-phase Flow through Intergranular Stress Corrosion Cracks. Rep. EPRI-NP-3540-LD*. Elec. Power Res. Inst., Palo Alto, CA

COLLIER, S.L. (1983). How valve hammer causes knocking in triplex pumps. *Oil Gas J.*, **81**(Mar. 28), 95

COLLINGRIDGE, D. (1980). *The Social Control of Technology* (Milton Keynes: Open Univ.)

COLLINRIDGE, D.G. (1979). The entrenchment of technology: the case of lead petrol additives. *Sci. Publ. Policy*, **6**, 332

COLLINGS, A.J. and LUXON, S.G. (eds) (1982). *Safe Use of Solvents* (New York: Academic Press)

COLLINGTON, D. (1989). Computer controlled maintenance. *plant Engr*, Jan./Feb., 26

COLLINS, C., VENTHAM, W.M., CRAWFORD, D.B. and HANDMAN, S.E. (1983). LNG storage – a review of current design considerations. *Gastech 82*, p. 352

COLLINS, C., CHEN, R.J.J. and ELLIOT, D.G. (1985). Trends in NGL recovery from natural and associated gases. *Gastech 84*, p. 287

COLLINS, C.H. and BEALE, A.J. (1992). *Safety in Industrial Microbiology and Biotechnology* (Oxford: Butterworth-Heinemann)

COLLINS, J. (1982). Industrial robotics. *plant Engr*, **26**(5), 22

COLLINS, J.A. and MONACK, M.L. (1973). Stress corrosion cracking in the chemical process industry. *Material. Prot. Performance*, **12**, 11

COLLINS, P.M. (1974). Detonation initiation in unconfined fuel-air mixtures. *Acta Astronautica*, **1**, 259

COLLOCOTT, S.J., MORGAN, V.T. and MORROW, R. (1980). The electrification of operating powder fire extinguishers. *J. Electrostatics*, **9**, 191

COLMERAUER, A. (1985). Prolog in 10 figures. *Comm. ACM*, **28**, 1296

COLMERAUER, A., KANOUL, H. and VAN CANEGHEM, M. (1983). Prolog, theoretical principles and current trends. *Technol. and Sci. of Informatics*, **2**, 255

COLOMBARI, V., CAZZOLA, F., SALVATORE, A., FUREGATO, E. and SCOPESI, L. (1992). Industrial pipelines and environmental protection: methodology for environmental evaluation. *Loss Prevention and Safety Promotion 7*, vol. 1, 6-1

COLOMBO, A.G. (1973). *CADI, a Computer Code for System Availability and Reliability Evaluation. Rep. EUR 4940E*. Comm. Eur. Commun., Brussels

COLOMBO, A.G. and JAARSMA, R.J. (1983). Combination of reliability parameters from different data sources. *Euredata 4*, paper 10.4

COLOMBO, A.G. and SARACCO, O. (1983). Bayesian estimation of the time-independent failure rate of an item taking into account its quality and operational conditions. *Euredata 4*, paper 12.1

COLOMBO, A.G., JAARSMA, R.J. and OLIVI, L. (1978). On the statistical data processing for a safety data system. *Proc. ANS Conf. on Probabilistic Analysis of Nuclear Reactor Safety*, Los Angeles, CA

COLONNA, G.R. (1984). Coast Guard confined spaces safety procedures. *Plant/Operations Prog.*, **3**, 188

COLONNA, G.R. (1986). Evaluating hazardous chemicals in the marine workplace. *Chem. Engng Prog.*, **82**(4), 10

COLONNA, J.L., LECOMTE, B. and CAUDRON, S. (1986). Regasification optimisation at Montoir-de-Bretagne LNG terminal. *Gastech 85*, p. 381

COLT, W.J. (1984). Assessing plant damage. *Hydrocarbon Process.*, **63**(10), 85

COLTHARP, W.M., DELLENEY, R.D., RIESE, C.E. and STRANGE, T.I. (1979). Determining availability of steam supply systems. *Loss Prevention 12*, p. 53

COLUMBIA SCIENTIFIC INDUSTRIES (1990). *CHARM* (Milton Keynes)

COLUMBO, U. and THRING, M.W. (1972). Soot in flames: the technological significance of the problems of soot and other particles formed in flames. *Combust. Sci. Technol.*, **5**, 189

COLYER, R.S. and MEYER, J. (1991). Understand the regulations governing equipment leaks. *Chem. Engng Prog.*, **87**(8), 22

COMBLEY, R.C. (1987). Flammable gases: hazards and detection. *Fire Surveyor*, **16**(1), 22

COMBUSTION INSTITUTE (1928–). International Symposia on Combustion (Pittsburgh, PA)

COMEAU, E.T. (1972). Precautions for ammonia storage tanks. *Ammonia Plant Safety*, **14**, 69

COMEAU, E.T. and WEBER, M.L. (1977). Problems with tank foundation heaters. *Ammonia Plant Safety*, **19**, 63

COMER, P.J. (1977). The dispersion of large-scale accidental releases such as Seveso. *R. Met. Soc. Mtg on Atmospheric–Surface Exchanges of Pollution*, London

COMER, P.J. and KIRWAN, B. (1985). A reliability study of a platform blowdown system. *Symp. on Automation and Safety in Shipping and Offshore Petroleum Operation*, Trondheim

COMER, P.J., COX, R.A. and SYLVESTER-EVANS, R. (1980). A formalised procedure for the analysis of risks posed by complex petrochemical developments. *Bull. Inst. Math. Appl.*, **16**(May/Jun.)

COMER, P.J., SLATER, D.H., MERCER, A. and PERRIMAN, R.J. (1983). An application of a multiple point source atmospheric dispersion model. *Atmos. Environ.*, **17**, 43

COMEY, D. (1977). A review of the Wechsler Report. In Union of Concerned Scientists (1977), *op. cit.*, p. 193

COMFORT, L.K. (1988). *Managing Disaster* (Durham: Duke Univ. Press)

COMINGS, E.W. (1956). *High Pressure Technology* (New York: McGraw-Hill)

COMITIS, S.C., GLASSER, D. and YOUNG, B.D. (1994). An experimental and modeling study of fires in ventilated ducts. Part I: Liquid fuels. *Combust. Flame*, **96**, 428

COMLEY, E.A. (1968). Catalyst experiences in the new large ammonia plants. *Ammonia Plant Safety*, **10**, 105

COMMENTZ, C. *et al.* (1921). The explosion of the nitrate plant at Oppau. *Chem. Metall. Engng*, **25**(18), 818

COMMISSION OF THE EUROPEAN COMMUNITIES − see also Appendix 28

COMMISSION OF THE EUROPEAN COMMUNITIES (1989). *Environmental Research Programme. Major Technological Hazards. Progress Report June−December 1989* (Brussels)

COMMISSION OF THE EUROPEAN COMMUNITIES (1994). *Proposal for a Council Directive on the Control of Major-Accident Hazards involving Dangerous Substances (COMAH)* (Brussels)

COMMISSION OF THE EUROPEAN COMMUNITIES AND ITALIAN MINISTRY OF HEALTH (1976). *Proc. of the Expert Meeting on the Problems Raised by TCDD Pollution* (Milan)

COMMISSIONER OF POLICE (1945). *Report of Commissioner of Police of the Metropolis for the year 1944* (London: HM Stationery Office)

COMMITTEE ON THE ENVIRONMENT (1974). *North Sea Oil and the Environment* (London: HM Stationery Office)

COMMITTEE FOR EXPLOSION RESEARCH (1958). *Final Report on Bromma Tests* (Stockholm)

COMMITTEE FOR PREVENTION OF DISASTERS. (1974). Public Emergency Limits. *Staub Reinhaltung der Luft*, **34**, 73 (English ed.) (p. 85 in German ed.)

COMMITTEE FOR THE PREVENTION OF DISASTERS (1988). *Methods for Determining and Processing Probabilities* (Voorburg) (the 'Red Book')

COMMITTEE FOR THE PREVENTION OF DISASTERS (1992a). *Methods for the Calculation of Physical Effects Resulting from Releases of Hazardous Materials (Liquid and Gases)*, 2nd ed., *Rep. CPR 14E* (Voorburg) (the 'Yellow Book')

COMMITTEE FOR THE PREVENTION OF DISASTERS (1992b). *Methods for the Determination of Possible Damage to People and Objects Resulting from Releases of Hazardous Materials. Rep. CPR 16E* (Voorburg) (the 'Green Book')

COMMITTEE OF SCIENTISTS (1967). *The Torrey Canyon* (London: HM Stationery Office)

COMMITTEE OF VICE-CHANCELLORS AND PRINCIPALS OF THE UNIVERSITIES OF THE UNITED KINGDOM (1966). *Radiological Protection in Universities* (London)

COMMITTEE OF VICE-CHANCELLORS AND PRINCIPALS OF THE UNIVERSITIES OF THE UNITED KINGDOM (1992). *Safety in Universities − Notes of Guidance, Pt 2:1 Lasers,* 3rd ed. (London)

COMMONER, B. (1966). *Science and Survival* (London: Gollancz)

COMMONER, B. (1973). Workplace burden. *Environment*, **15**(6), 15

COMMONWEALTH DEPARTMENT OF PRODUCTIVITY (1979). *Safety and Health at Work. Guide to Sources of Information* (Canberra: Aust. Govt Publ. Service)

COMPRESSED GAS ASSOCIATION − see Appendix 28

COMPTROLLER GENERAL OF THE UNITED STATES (1978). *Liquefied Energy Gases Safety*, vols 1−3 (Washington, DC: US Govt Printing Office)

COMPUTER DATA SYTEMS INC. (1985). *VAX ONDEK 84 System Documentation. Rep. to US Coast Guard, Contract W6903−07G, (Rovkville, MD)*

COMPUTER SCIENCES CORP. (1973). *Vessel Traffic Systems Issue Study*, vol. 3. *Rep. to US Coast Guard*

COMROE, J.H. (1974). *Physiology of Respiration*, 2nd ed. (Chicago, IL: Year Book Med. Publs)

COMSTOCK, J.P. (ed.) (1967). *Principles of Naval Architecture* (New York: Soc. Nav. Architects Mar. Engnrs)

COMSTOCK, J.P. and ROBERTSON, J.B. (1969). Survival of collision damage versus the 1960 Convention on Safety of Life at Sea. *Soc. Nav. Architects Mar. Engrs Trans.*, **69**

CONAWAY, C.W., DAVENPORT, J.A. and NORSTROM, G.P. (1980). Analysis of losses in the chemical industry as it applies to the polyvinyl chloride industry. *Mtg of Vinyl Chloride Safety Ass.*, Bermuda

CONCAWE (1984). Methodologies for hazard analysis and risk assessment in the petroleum refining and storage industry. *Fire Technol.*, **20**(3), 23; **20**(4), 43

CONCAWE − see also Appendix 28

CONFEDERATION OF BRITISH INDUSTRY (1968). *Industrial Safety Organization* (London)

CONFEDERATION OF BRITISH INDUSTRY (1974). *Company Safety and Health Policies. CBI Guide for Employers* (London)

CONFEDERATION OF BRITISH INDUSTRY (1983). *The Regulation of Occupational Carcinogens − A CBI Policy Statement* (London)

CONFEDERATION OF BRITISH INDUSTRY (1990). *Developing a Safety Culture* (London)

CONGER, H.C. (1979). New approach to cooling water treatment. *Hydrocarbon Process.*, **58**(1), 119

CONISON, J. (1960). How to design a pressure relief system. *Chem. Engng*, **67**(15), 109

CONISON, J. (1963). Factors in sizing a safe and economical vapor relief system. *API Proc., Div. of Refining*, **43**, 434

CONKLIN, J. and BEGEMAN, L. (1988). gIBIS: a hypertext tool for exploratory policy discussion. *ACM Trans. Office Inf. Sys.*, **6**, 303

CONKLIN, J.H. (1961). Nitrogen wash incident. *Ammonia Plant Safety*, **3**, p. 16. See also *Chem. Engng Prog.*, **57**(4), 52

CONLEY, M.J., ANGELSEN, S. and WILLIAMS, D. (1991). Ammonia vessel integrity problem: a modern approach. *Plant/Operations Prog.*, **10**, 201

CONLEY, R.C. and CLERRICO, A.T. (1992). Select and implement a plantwide information system. *Hydrocarbon Process.*, **71**(5), 47

CONN, A.L. (1971). How much experimentation before commercialisation? *Chem. Engng Prog.*, **67**(6), 22

CONNAUGHTON, G.E. (1981). Failure of a superheated steam header. *Ammonia Plant Safety*, **23**, 26

CONNAUGHTON, G.E. and CLARK, R.W. (1984). Secondary reformer catalyst support dome: failure and repair. *Ammonia Plant Safety*, **24**, 73

CONNELL, J.R. and CHURCH, H.W. (1980). *Recommended Methods for Estimating Atmospheric Concentrations of Hazardous Vapours after Accidental Release near Nuclear Reactor Sites. Rep. NUREG/CR-1152.* Nucl. Regul. Comm., Washington, DC

CONNELL, M.J.G. (1983). *ESTC Flying Glass Hazard Trials.* Property Services Agency, Dept of the Environ.

CONNELL, M.J.G. (1992). *Thamesport Office Building. Lethality of Glazing Fragments. UB 822/12/IMC1.* Explosion Effects Branch, Property Services Agency, Dept of the Environ.

CONNELLY, C.S. (1992). Emergency procedure writing for oil refinery process units. *Hazard Identification and Risk Analysis, Human Factors and Human Reliability in Process Safety*, p. 309

CONNELLY, E.M., HAAS, P. and MYERS, K. (1993a). Method for building performance measures for process safety management. *Process Safety Management*, p. 293

CONNELLY, E.M., HAAS, P. and MYERS, K. (1993b). Performance measures developed for CCPS's elements of process safety management. *Process Safety Management*, p. 307

CONNER, J.R. (1976). Chemical fixation of hazardous material spill residues. *Control of Hazardous Material Spills*, 3, p. 416

CONNOR, J. (1974). Explosion risk of unstable substances. Test methods employed by EM2 (Home Office) Branch, Royal Armament Research and Development Establishment. *Loss Prevention and Safety Promotion* 1, p. 261

CONNOR, S. (1984). Carbon monoxide – the unknown gas. *Fire Int.*, 87, 41

CONNOR, W.K. (1973). Noise control. *Chem. Engng*, 80 (Jun. 18), 105

CONNORS, D.C. (1984). The variability of system failure probability. *Reliab. Engng*, 9(2), 117

CONRAD (1973). Deactivating explosible gas systems as a primary explosion protection measure. *Prevention of Occupational Risks*, 2, 131

CONRAD, D., DIETLEN, S. and SCHENDLER, T. (1992). Preventing explosion hazards of decomposable gases. *Loss Prevention and Safety Promotion* 7, vol. 3, 111-1

CONRAD, D., GREWER, T., HENTSCHEL, B. and REDEKER, T. (1992). Development of a standard method for flammability limits of vapour-air mixtures. *Loss Prevention and Safety Promotion* 7, vol. 2, 79-1

CONRAD, J. (ed.) (1980). *Society, Technology and Risk Assessment* (London: Academic Press)

CONRAD, J. (1984). Total plant-safety audit. *Chem. Engng*, 91(May 14), 83

CONROY, J.A. (1987). A computer-based inspection system. *Chem. Engng Prog.*, 83(5), 39

CONSIDINE, D.M. (1968). Process instrumentation. *Chem. Engng*, 75(Jan. 29), 84; (Feb. 12), 137

CONSIDINE, D.M. (ed.) (1971). *Encyclopaedia of Instrumentation and Control* (New York: McGraw-Hill)

CONSIDINE, D.M. (ed.) (1974). *Chemical and Process Technology Encyclopaedia* (New York: McGraw-Hill)

CONSIDINE, D.M. (1977). *Energy Technology Handbook* (New York: McGraw-Hill)

CONSIDINE, D.M. (1985). *Process Instruments and Controls Handbook*, 3rd ed. (New York: McGraw-Hill)

CONSIDINE, M. (1981). *The Hazards from Catastrophic Releases of Flammable Liquids. Draft Working Paper HSE/SRD/PD024/WP7.* UK Atom. Energy Auth., Systems Reliab. Directorate, Culcheth, Warrington

CONSIDINE, M. (1983). A hazard assessment of the Morecambe Bay gas pipeline and coastal reception terminal. *Loss Prevention and Safety Promotion* 4, vol. 1, p. I23

CONSIDINE, M. (1986). *Risk Assessment of the Transportation of Hazardous Substances through Road Tunnels. State of the Art Report.* Transportation Safety Board, Nat. Res. Coun., Washington, DC

CONSIDINE, M. and GRINT, G.C. (1985). Rapid assessment of the consequences of LPG releases. *Gastech 84*, p. 187

CONSIDINE, M., GRINT, G.C. and HOLDEN, P.L. (1982). Bulk storage of LPG–factors affecting offsite risk. *The Assess-

ment of Major Hazards* (Rugby: Instn Chem. Engrs), p. 291

CONSTABLE, G.E.P. and PARKES, D. (1979). *Design for maintainability. Design 79* (London: Instn Chem. Engrs)

CONSTAN, G.L. (1984). Pilot plants for medium-sized companies. *Chem. Engng Prog.*, 80(2), 56

CONSTAN, G., HERMAN, J. and MOELLER, P. (1992). Ensuring safe design and operation of pilot plants. *J. Loss Prev. Process Ind.*, 5, 42

CONSTANCE, J.D. (1970). Estimating exhaust-air requirements for processes. *Chem. Engng*, 77(Aug. 10), 116

CONSTANCE, J.D. (1971a). Control of explosive or toxic air–gas mixtures. *Chem. Engng*, 78(Apr. 19), 121

CONSTANCE, J.D. (1971b). Pressure ventilate for explosion protection. *Chem. Engng*, 78(Sep. 20), 150

CONSTANCE, J.D. (1972a). Calculate effective stack height quickly. *Chem. Engng*, 79(Sep. 4), 81

CONSTANCE, J.D. (1972b). Calculating the masking effects of noise. *Chem. Engng*, 79(Apr. 17), 124

CONSTANCE, J.D. (1972c). Estimating fan noise from tip speeds. *Chem. Engng*, 79(Oct. 2), 92

CONSTANCE, J.D. (1978). How to pressure-ventilate large motors for corrosion, explosion and moisture protection. *Chem. Engng*, 85(Feb. 27), 113

CONSTANCE, J.D. (1979). Piping around control valve. *Chem. Engng*, 86(Nov. 5), 129

CONSTANCE, J.D. (1980). Solve gas purging problems graphically. *Chem. Engng*, 87(Dec. 29), 65

CONSTANCE, J.D. (1983). *Controlling In-plant Airborne Contaminants* (Basel: Dekker)

CONSTANCE, J.D. (1985). Ventilating CPI buildings. *Chem. Engng*, 92(Dec. 9/23), 89

CONSTRADO (1983). *Fire Protection for Structural Steel in Buildings* (Croydon)

CONSUMERS ASSOCIATION (1972). *Electricity Supply and Safety* (London)

CONTI, R.S. and HERTZBERG, M. (1989). Thermal autoignition temperatures for hydrogen–air and methane–air mixtures. *J. Fire Sciences*, 6(5), 348

CONTINI, S., SILVESTRI, E. and DE COLA, G. (1983). Development in fault tree analysis: the NEWSALP code. *Euredata 4*, paper 3.2

CONTINILLO, G., CRESCITELLI, S., FUMO, E., NAPOLITANO, F. and RUSSO, G. (1991). Coal dust explosions in a spherical bomb. *J. Loss Prev. Process Ind.*, 4, 223

CONTRA COSTA COUNTY HEALTH SERVICES (1997). *Inherently Safer Practices* (Martinez, CA: Contra Costa County)

CONTROL ENGINEERING (1966). *Digital Computers in Industry.* Reprint 931

CONVEYOR EQUIPMENT MANUFACTURERS ASSOCIATION (1979). *Belt Conveyors for Bulk Materials,* 2nd ed. (Boston, MA: CBI Publ. Co.)

CONWAY, R.A. (1982). *Environment Risk Analysis for Chemicals* (New York: van Nostrand Reinhold)

CONWAY, R.A. and ROSS, R.D. (1980). *Handbook of Industrial Waste Disposal* (New York: van Nostrand Reinhold)

CONWAY, R.W., MAXWELL, W.L. and MILLER, L.W. (1967). *Theory of Scheduling* (Reading, MA: Addison-Wesley)

COOK, B.E. (1979). Instrumentation and plant integrity. *Measmnt Control*, 12, 4

COOK, D.K. (1991a). An integral model of turbulent non-premixed jet flames in a cross-flow. *Combustion*, 23, 653

COOK, D.K. (1991b). A one-dimensional model of turbulent jet diffusion. *Combust. Flame*, 85, 143

COOK, D.K., FAIRWEATHER, M., HAMMONDS, J. and HUGHES, D.J. (1987). Size and radiative characteristics of natural gas flares. *Chem. Engng Res. Des.*, **65**, 310, 318

COOK, D.K., FAIRWEATHER, M., HANKINSON, G. and O'BRIEN, K.O. (1987). Flaring of natural gas from inclined vent stacks. *Hazards from Pressure*, p. 289

COOK, D.T. (1982). Selecting hand-operated valves for process plants. *Chem. Engng*, **89**(Jun. 14), 126

COOK, E.M. (1964). Air cooled heat exchangers. *Chem. Engng*, **71**(May 25), 137; (Jul. 6), 131; (Aug. 3), 97

COOK, G.B. (1959). Some developments in the theory of thermal explosions. *Combustion*, **6**, 626

COOK, I. and UNWIN, S.D. (1989). Controlling principles for the assignment of probability densities for uncertainty propagation. In Cox, R.A. (1989a), *op. cit.*, p. 233

COOK, J. and WOODWARD, J. (1993a). A new integrated model for pool spreading, evaporation and solution on land and water. *Health, Safety and Loss Prevention*, p. 359

COOK, J. and WOODWARD, J. (1993b). A new unified model for jet, dense, passive and buoyant dispersion including droplet evaporation and pool modelling. *Health, Safety and Loss Prevention*, p. 371

COOK, J., BAHRAMI, Z. and WHITEHOUSE, R.J. (1990). A comprehensive program for calculation of flame radiation levels. *J. Loss Prev. Process Ind.*, **3**(1), 150

COOK, J.B. (1991). The employer's approach to warnings. In The Fellowship of Engineering (1991b), *op. cit.*, p. 89

COOK, J.D. (1984). Landfill as a disposal route for difficult wastes. *Chemy Ind.*, Sep. 3, 615

COOK, J.D. and HUGHES, D. (1986). *Fume Cupboards Revisited* (Leeds: H&H Sci. Consultants Ltd)

COOK, J.L. (1972). Nitriding in a synthesis loop steam generator. *Ammonia Plant Safety*, **14**, 39

COOK, M.A. (1958). *The Science of High Explosives* (New York: Reinhold)

COOK, M.A. (1960). Modern blasting agents. *Science, 132*, 1105

COOK, M.A., PACK, D.H. and GEY, W.A. (1959). Deflagration to detonation transition. *Combustion*, **7**, 820

COOK, R. and GUHA, P. (1981). Principles governing the production of sound stainless steel castings. *Chemy Ind.*, Oct. 17, 733

COOK, R.J. (1981). Thermal expansion in pipes. *plant Engr*, **25**(1), 29

COOK, S. (1987). Object-oriented programming in the UK. *Comput. Bull.*, **3**(4), 22

COOK, S.J., CULLIS, C.F. and GOOD, A.J. (1977). The measurement of flammability limits of mists. *Combust. Flame*, **30**, 304

COOKE, A.J.D. (1976). *A Guide to Laboratory Law* (London: Butterworths)

COOKE, E.F. (1971). A guide for selecting adjustable speed drives. *Chem. Engng*, **78**(Sep. 6), 70

COOKE, E.F. (1973). Power transmission. *Chem. Engng*, **80**(Feb. 26), 53

COOKE, G.M.E. (1974). Fire protection methods for external steelwork. *Fire Prev. Sci. Technol.*, **9**, 4

COOKE, G.M.E. (1975). Problems in the development and application of new technologies for fire grading of buildings. *Fire Prev. Sci. Technol.*, **12**, 4

COOKE, G.M.E. (1988). An introduction to the mechanical properties of structural steel at elevated temperatures. *Fire Safety J.*, **13**, 45

COOKE, M. (1988). Risk assessment safety: a case study of two chlorine plants. M.Sc. Thesis, Univ of Manchester Inst. Sci. Technol.

COOKE, N.E. and KHANDHADIA, P.S. (1987). Unconfined vapor clouds I: kinetics of dispersed clouds of liquid. *Vapor Cloud Modeling*, p. 597

COOKE, N.E. and KHANDHADIA, P.S. (1987). Unconfined vapor clouds II: kinematics of explosively dispersed clouds of liquid. *Vapor Cloud Modeling*, p. 625

COOMBE, R.A. (1968). *Radioactivity for Engineers* (London: Macmillan)

COON, R.A., JONES, R.A., JENKINS, L.J. and SIEGEL, J. (1970). Animal inhalation studies on ammonia, ethylene glycol, formaldehyde, dimethylamine and ethanol. *Toxicol. Appl. Pharmacol.*, **16**, 646

COONEY, W.D.C. (1985). Chemical transport emergency – the aftermath. *Chemy Ind.*, Jan. 21, 53

COONEY, W.D.C. (1991). Emergency planning: consideration for a major transport accident involving dangerous substances. In Advisory Committee on Dangerous Substances (1991), *op. cit.*, p. 360

COOPER, A.R. and JEFFREYS, G.V. (1973). *Chemical Kinetics and Reactor Design* (Englewood Cliffs, NJ: Prentice-Hall)

COOPER, A.S. and MCCONOMY, H.F. (1966). Cracking failures of austenitic steel. *Hydrocarbon Process.*, **45**(5), 181

COOPER, B.D. (1982). National codes – a contracting view. Flammable Atmospheres and Area Classification, *IEE Coll. Dig. 1982/26 (London)*, p. 3/1

COOPER, C.M. (1965). Stop weld failures in hot hydrogen service. *Hydrocarbon Process.*, **44**(1), 101

COOPER, C.M. (1972a). Hydrogen attack. *Hydrocarbon Process.*, **51**(8), 69

COOPER, C.M. (1972b). Naphthenic acid corrosion. *Hydrocarbon Process.*, **51**(8), 75

COOPER, D.F. and CHAPMAN, C.B. (1987). *Risk Analysis for Large Projects* (Chichester: Wiley)

COOPER, D.J. and LALONDE, A.M. (1990). Process behaviour diagnostics and adaptive process control. *Comput. Chem. Engng*, **14**, 541

COOPER, D.O. and DAVIDSON, L.B. (1976). Parameter method for risk analysis. *Chem. Engng Prog.*, **72**(11), 73

COOPER, G.J., MAYNARD, R.L., CROSS, N.L. and HILL, J.F. (1983). Casualties from terrorist bombings. *J. Trauma*, **23**(11), 955

COOPER, H.B.H. (1973). Can you measure odor? *Hydrocarbon Process.*, **52**(10), 97

COOPER, J. (1961). Human initiated failures and malfunction reporting. *IRE Trans. Hum. Factors Electron.*, **HFE-2**, 104

COOPER, M.G. (1985). *Risk: Man-Made Hazards to Man* (Oxford: Clarendon Press)

COOPER, M.G., FAIRWEATHER, M. and TITE, J.P. (1986). On the mechanisms of pressure generation in vented explosions. *Combust. Flame*, **65**, 1

COOPER, P. (1974). *Poisoning by Drugs and Chemicals,* 3rd ed. (London: Alchemist Publs)

COOPER, P. (1986–). Condition monitoring. *Plant Engr*, 1986, **30**(6), 19; 1987, **31**(4), 24

COOPER, R.K. (1980). On the probability of trains overturning in high winds. In Cermak, J.E. (1980), *op. cit.*, vol. 2, p. 1185

COOPER, S.P. and MOORE, P.E. (1993). Active detonation protection. *Health, Safety and Loss Prevention*, p. 279

COOPER, S.P., MOORE, P.E., CAPP, B. and SENECA, J. (1992). Investigation into the use of active detonation arresters for solvent and waste gas recovery systems. *Major Hazards Onshore and Offshore*, p. 385

COOPER, W.F. (1953a). The electrification of fluids in motion. *Br. J. Appl. Phys.*, **4**(Suppl. 2), S11

COOPER, W.F. (1953b). The practical estimation of electrostatic hazards. *Br. J. Appl. Phys.*, **4**(Suppl. 2), S71

COOPER, W.F. (1971). Electrical hazards in danger areas (classification of risks and certification of equipment). *Electrical Safety*, **1**, 177

COOPEY, W. (1961). Water-spray keeps compressors clean. *Chem. Engng*, **68**(May 1), 106

COOPEY, W. (1965). New seal for supershaft pressures. *Chem. Engng*, **72**(Sep. 27), 153

COOPEY, W. (1967). A fresh look at spring loaded packing. *Chem. Engng*, **74**(Nov. 6), 278

COOPEY, W. (1969). Pumps leak? Try spring-loaded packing. *Hydrocarbon Process.*, **48**(9), 215

COPE, B. (1977). *The Chemical Treatment of Toxic Waste. Chemical Engineering in a Hostile World* (London: Clapp & Poliak)

COPE, C.B., FULLER, W.H. and WILLETTS, S.L. (1983). *The Scientific Management Hazardous Wastes* (Cambridge: Cambridge Univ. Press)

COPIGNEAUX, P. (1980). Compare codes for relief valve calculations. *Hydrocarbon Process.*, **59**(8), 69

COPIGNEAUX, P. (1982). Calculating theoretical flow rates of safety relief valves. *Hydrocarbon Process.*, **61**(11), 209

COPLIEN, J.O. (1992). *Advanced C++* (Reading, MA: Addison Wesley)

COPP, D.L. (1971). *Report on the Public Safety Aspects of the Proposed Liquid Natural Gas Storage and Liquefaction plant at Hirwaun, Aberdare*

COPPACK, M.E. (1980). A survey of flowmeters commonly used in the chemical industry. *Interflow 80* (Rugby: Instn Chem. Engrs), p. 63

COPPALLE, A. and JOYEUX, D. (1993). Experimental and theoretical studies of soot formation in an ethylene jet flame. *Combust. Sci. Technol.*, **93**, 375

COPPALLE, A. and VERVISCH, P. (1983). The total emissivities of high-temperature flames. *Combust. Flame*, **49**, 101

COPPALLE, A., NEDELKA, D. and BAUER, B. (1993). Fire protection water curtains. *Fire Safety J.*, **20**, 241

COPPELL, D.L. (1976). Blast injuries of the lungs. *Br. J. Surgery*, **63**, 735

COPSON, H.R. and CHENG, C.F. (1957). Some case histories of stress corrosion cracking of austenitic stainless steels associated with chlorides. *Corrosion*, **13**, 397

CORBETT, C.R. (1978). A dynamic regional response team. *Control of Hazardous Material Spills*, **4**, 4

CORBETT, E. (1963). Safety aspects of handling isocyanates in urethane foam production. *Chemical Process Hazards*, **II**, p. 41

CORBETT, H.J. (1988). Chemical risk: living up to public expectations. *Preventing Major Chemical and Related Process Accidents*, p. 607

CORBETT, J.O. (1981). The validity of source-depletion and alternative approximation methods for a Gaussian plume subject to dry deposition. *Atmos. Environ.*, **15**, 1207

CORBETT, R.A. (1985). Flare-system operating costs can be reduced. *Oil Gas J.*, **83**(Sep. 23), 136

CORBIN, M.H. and LEDERMAN, P.B. (1982). Hazardous landfill management, control options. *J. Haz. Materials*, **5**(3), 235

CORDEA, J.N., FRISBY, D.L. and KAMPSCHAEFER, G.E. (1972). Steels for storage and transportation of liquid natural gas. *LNG Conf. 3*

CORDER, A.S. (1976). *Maintenance Management Techniques* (London: McGraw-Hill)

CORDES, R.J. (1977). Use compressors safety. *Hydrocarbon Process.*, **56**(10), 227; **56**(11), 469

COREA, R., THAM, M.T. and MORRIS, A.J. (1992). Modelling for intelligent fault diagnosis — integrating numerical and qualitative techniques. *Inst. Elec. Engrs, Colloquium on Intelligent Fault Diagnosis, Dig. 1992/045* (London), p. 2/1

CORKHILL, M. (1975). *LNG Carriers: The Ships and Their Market* (London: Fairplay Publs)

CORLETT, R.C. (1968). Gas fires with pool like boundary conditions. *Combust. Flame*, **12**, 19

CORLETT, R.C. (1970). Gas fires with pool-like boundary conditions: further results and interpretation. *Combust. Flame*, **14**, 351

CORLETT, R.C. and FU, T.M. (1966). Some recent experiments with pool fires. *Pyrodynamics*, **1**, 253

CORLETT, R.C. and WILLIAMS, F.A. (1979). Modeling direct suppression of open fires. *Fire Res.*, **1**(6), 323

CORLETT, R.C., STONE, J.P. and WILLIAMS, F.W. (1980). Scale modeling of inert pressurant distribution. *Fire Technol.*, **16**, 259

CORLEY, J.W. (1972). Process plant design for control of hazardous material spills. *Control of Hazardous Material Spills*, **1**, 15

CORN, J.K. and CORN, M. (1984). The history and accomplishments of the Occupational Safety and Health Administration in reducing cancer risks. In Deisler, P.F. (1984), *op. cit.*, p. 175

CORN, M. (1976). OSHA's Morton Corn talks to CE. *Chem. Engng*, **83**(Aug. 30), 57

CORNELL, C.A. and VANMARCKE, E.H. (1975). Seismic risk analysis for offshore structures. *Seventh Ann. Conf. on Offshore Technol.*, p. 145

CORNELL, C.E. (1968). Minimizing human errors. *Space Aeronaut.*, **49**(Mar.), 72

CORNETT, C.L. and JONES, J.L. (1970). Reliability revisited. *Chem. Engng Prog.*, **66**(12), 29

CORNISH, D.C. (1978a). The evaluation of instruments for pollution and environmental monitoring. *Chemy Ind.*, Apr. 15, 264

CORNISH, D.C. (1978b). Evaluation statistics: their significance for the improvement of instrumentation. *Sci. Instrum. Res. Ass., Symp. on Instrumentation Reliability — Can it be Improved?* (Chislehurst, Kent)

CORNISH, D.C. (1978c). The value of instrument evaluation. *Int. Ass. Water Poll., Int. Workshop on Instrumentation and Control for Water and Wastewater Treatment and Transport Systems* (London: Pergamon Press)

CORNISH, D., JEPSON, G. and SMURTHWAITE, M. (1981). *Sampling Systems for Process Analysis* (London: Butterworths)

CORNISH, E.H. (1987). *Materials and the Designer* (Cambridge: Cambridge Univ. Press)

CORNWELL, J.B. and PFENNING, D.B. (1987). Comparison of Thorney Island data with heavy gas dispersion models. *J. Haz. Materials*, **16**, 315

CORONA, A. (1984). *Fireproofing in Refineries and Petrochemical plants and Offshore Platforms*. Am. Petrol. Inst. Symp.

CORONER (1945). *Coroner's Report on East Ohio Gas Company Disaster on October 20, 1944*, Cuyahoya County, Cleveland

CORPSTEIN, R.R., DOVE, R.A. and DICKEY, D.S. (1979). Stirred tank reactor design. *Chem. Engng Prog.*, **75**(2), 66

CORRAN, E.R. and WITT, H.T. (1982). Reliability analysis techniques for the design engineer. *Reliab. Engng*, **3**(1), 47

CORRIE, J.G. (1974). The fire protection of flammable liquid risks by foams. *Chemical Process Hazards,* V, p. 68

CORRIE, J.G. (1977). Quality control procedures for fire-fighting foams. *Chemical Process Hazards,* VI, p. 133

CORRIGAN, M.K. (1990). Which instrumentation for hazardous-area plants. *Chem. Engng Prog.,* 86(9), 55

CORRIPIO, A.B., CHRIEN, K.S. and EVANS, L.B. (1982a). Estimate costs of centrifugal pumps and electric motors. *Chem. Engng,* 89(Feb. 22), 115

CORRIPIO, A.B., CHRIEN, K.S. and EVANS, L.B. (1982b). Estimate costs of heat exchangers and storage tanks via correlations. *Chem. Engng,* 89(Jan. 25), 125

CORTI, M. and SANSONE, A. (1992). Safety in drilling operations: real time expert systems help operators in preventing accidents. *Loss Prevention and Safety Promotion* 7, vol. 3, 153.1

CORY, J.M. and RICCIOLI, F.D. (1985). What are fire-safe valves? *Chem. Engng,* 92(May 27), 147

COSTA, G., LIEVORE, F., CASALETTE, G., GAFFURI, E. and FOLKARD, S. (1989). Circadian characteristics influencing inter-individual differences in tolerance to shiftwork. *Ergonomics,* 32, 373

COSTA, R., RECASENS, F. and VELO, E. (1995). Inherent thermal safety of stirred-tank batch reactors: a prognosis tool based on pattern recognition of hazardous states. *Int. Symp. on Runaway React. and Pressure Relief Design,* August 2–4, 1995, Boston, MA, ed. G. A. Melhem and H. G. Fisher, 690–709 (New York: American Institute of Chemical Engineers)

COSTANZA, P.A., HAGEN, G.L., KALAGNAMAN, R.S. and KOMISINSKY, P.J. (1987). Estimation of catastrophic quantities of toxic chemicals by atmospheric dispersion modeling. *Plant/Operations Prog.,* 6, 215

COSTE, J.O. and PECHERY, P. (1977). Contribution to the study of charges generated by friction between metals and polymers. *Proc. Third Int. Conf. on Static Electricity* (Paris: Soc. de Chimie Industrielle)

COSTELLO, THE HONOURABLE MR JUSTICE D. (1979). *Disaster at Whiddy Island, Bantry Bay, Co. Cork. Report of Tribunal of Inquiry.* (Dublin: Stationery Office)

COSTELLO, E.J., RHODES, R.D. and YOVINO, J. (1965). Reinforced plastics for corrosion resistance. *Chem. Engng. Prog.,* 61(4), 49

COTE, A. (1990). The computer and fire protection engineering – 25 years of progress. *Fire Technol.,* 26, 195

COTE, A.E. (1986). Assessing the magnitude of the fire problem. *Fire Protection Handbook,* 16th ed. (ed. by A.E. Cote and J.M. Linville), pp. 1–2

COTE, A.E. and LINVILLE, J.M. (eds) (1986). *Fire Protection Handbook* (Quincy, MA: Nat. Fire Prot. Ass.)

COTE, R.P. and WELLS, P.G. (1991). Controlling Chemical Hazards. *Fundamentals of the Management of Toxic Chemicals* (London: Unwin Hyman)

COTES, J.E. (1979). Lung Function, 4th ed. (Oxford: Blackwell)

COTGROVE, S. (1981). Risk, value, conflict and political legitimacy. In Griffiths, R.F. (1981), *op. cit.,* p. 122

COTGROVE, S. (1982). *Catastrophe or Cornucopia* (Chichester: Wiley)

COTT, B.J. and MACCHIETTO, S. (1989). An integrated approach to computer-aided operation of batch chemical plants. *Comput. Chem. Engng,* 13, 1263

COTT, B.J., DURHAM, R.G., LEE, P.L. and SULLIVAN, G.R. (1989). Process model-based engineering. *Comput. Chem. Engng,* 13, 973

COTTAM, A.M. (1991). Risk assessment and control in biotechnology. *Hazards,* XI, p. 341

COTTON, H.C. and DENHAM, J.B. (1968). British view on storage tank design. *Hydrocarbon Process.,* 47(5), 135

COTTON, J.B. (1970). Using titanium in the chemical plant. *Chem. Engng Prog.,* 66(10), 57

COTTRELL, A. (1977). Reactor pressure vessel failure. In *Union of Concerned Scientists* (1977), *op. cit.,* p. 197

COTTRELL, A.H. (1967). *An Introduction to Metallurgy* (London: Arnold)

COTTRELL, A.H. and SWANN, P.R. (1975a). Metallurgical examination of the failure of the 8 inch diameter stainless steel pipe. *Rep. to Flixborough Court of Inquiry* (Cambridge Univ. and Imperial Coll.)

COTTRELL, A.H. and SWANN, P.R. (1975b). A metallurgical examination of the failures of the 8″ diameter stainless steel pipe. *Instn Chem. Engrs Symp. on Technical Lessons of Flixborough* (London)

COTTRELL, A.H. and SWANN, P.R. (1976). Technical lessons of Flixborough. A metallurgical examination of the 8–inch line. *Chem. Engr, Lond.,* 308, 266

COTTRELL, S.R.A. (1981). *How Safe is Nuclear Energy?* (London: Heinemann)

COTTRELL, W.B. and SAVOLAINEN, J. (eds) (1965). *US Reactor Containment Technology – A Compilation of Current Practice in Analysis, Design, Construction, Test and Operation. Rep. ORNL-NSIC-5O.* Oak Ridge Nat. Lab., Oak Ridge, TN

COTTRELL, W.B., MINARICK, J.W., AUSTIN, P.N., HAGEN, E.W. and HARRIS, J.D. (1984). *Precursors to Potential Severe Core Damage Accidents: 1981–1981: A Status Report. Rep. NUREG/CR-3591.* Nucl. Regul. Comm., Washington, DC

COUCH, R.O. (1991). Guide to multiple pipe containment. *Chem. Engng,* 98, 158

COUDRAY, R. and MATTEI, J.M. (1984). System reliability; an example of nuclear reactor analysis. *Reliab. Engng,* 7(2), 89

COUGHANOWR, D.R. and KOPPELL, L.B. (1965). *Process Systems Analysis and Control* (New York: McGraw-Hill)

COUGHTREY, P.J., MARTIN, M.H. and UNSWORTH, M.H. (1987). *Pollutant Transport and Fate in Ecosystems* (Oxford: Blackwells)

COULSEY, H. (1986). How expert was my system? *Chem. Engr, Lond.,* 425, 55

COULSON, J.M. and RICHARDSON, J.F. (1955–). *Chemical Engineering,* vol. 1, 1955; vol. 2, 1955; vol. 3, 1971 (Oxford: Pergamon Press)

COULSON, J.M. and RICHARDSON, J.F. (1977–). *Chemical Engineering* (SI ed.) (Oxford: Pergamon Press)

COULSON, K.L. (1975). *Solar and Terrestrial Radiation* (New York: Academic Press)

COULSTON, K. (1993). Valve verification procedure works with or without pressure. *Offshore Int.,* Nov., 88

COULTER, B.M. (1984). *Compressible Flow Manual* (Houston, TX Gulf Publishing Co.)

COULTER, J.B. (1972). A state's approach to the protection of the environment from hazardous material spills. *Control of Hazardous Material Spills,* 1, 5

COULTER, K.E. (1965). Depth analysis in plant safety. *Chem. Engng Prog.,* 61(2), 47

COULTER, K.E. and MORELLO, V.S. (1971). *Process and equipment reliability in large chemical plants* (Midland, MI: Dow Chem. Co.)

COULTER, K.E. and MORELLO, V.S. (1972). Improving on-stream time in process plants. *Chem. Engng Prog.,* 68(3), 56

COULTER, K.E. and TUTTLE, F.C. (1976). Safe furnace design and operation. *Loss Prevention 10*, p. 1. See also *Chem. Engng Prog.*, **72**(11), 41

COULTER, M.O. (ed.) (1986). *Modern Chlor-Alkali Technology*, vol. 1 (Chichester: Ellis Horwood)

COULTHARD, L. (1984). High velocity flares offer several advantages. *Oil Gas J.*, **82**(Jun. 11), 117

COUNCIL OF THE EUROPEAN COMMISSION (1986). *Assessment Architecture and performance of Industrial Programmable Electronic Systems with Particular Reference to Robotics safety. Collaborative Proj. AES/ TAPDS/002/A-12/02.86/ SAH.* Comm. Eur. Commun.

COUNCIL FOR SCIENCE AND SOCIETY (1976). *Superstar Technologies* (London: Barry Rose)

COUNCIL FOR SCIENCE AND SOCIETY (1977). *The Acceptability of Risks* (London: Barry Rose)

COUNCIL, J. (1993). A review of the role of cost—benefit analysis as an input to the management of safety. *Process Safety Management,* p. 79

COUNIHAN, J. (1975). Adiabatic atmospheric surface layers: a review and analysis of data from the period 1880— 1972. *Atmos. Environ.*, **9**, 871

COUPLAND, M. (1980). Belt drives. *plant Engr*, **23**(6), 19

COURANT, R. and FRIEDRICHS, K.O., (1948). *Supersonic Flow and Shock Waves* (New York: Wiley)

COURCHAINE, R.J. and DENNIS, D.M. (1976). Michigan's program for prevention and control of hazardous material spills. *Control of Hazardous Material Spills*, **3**, 132

COURT, E.T. (1981). To specify a reliability requirement does not ensure its achievement. *Reliab. Engng*, **2**(4), 243

COURT OF INQUIRY, TARRAGONA (1982). *Judgement on San Carlos disaster*

COURTENEY, W.G. (1962). Review of 'Studies on the mechanism of ignition of wood' by K. Akita. *Fire Res. Abstracts Rev.*, 4(1—2), 109. See also Akita, K. (1959)

COURTICE, F.C. and FOSS, G.L. (1946). Acute phosgene poisoning. *The Lancet*, **251**(Nov. 9), 670

COURTY, P., ARLIE, J.P., CONVERS, A., MIKITENKO, P. and SUGIER, A. (1984). C_1—C_6 alcohols from syngas. *Hydrocarbon Process.*, **63**(11), 105

COUSINS, E.W. and COTTON, P.E. (1951a). Design closed vessels to withstand internal explosions. *Chem. Engng*, **58**(8), 133

COUSINS, E.W. and COTTON, P.E. (1951b). *Protection of Closed Vessels Against Internal Explosions. Paper 51—PRI-2.* Am. Soc. Mech. Engrs, New York

COVELLO, V.T. (1984). Social and behavioral research on risk: uses in risk management decision-making. *Environ. Int.*, **10**, 541

COVELLO, V.T. (1991). Risk comparisons and risk communication: issues and problems in comparing health and environmental risks. In Kasperson, R.E. and Stallen, P.J.M. (1991), *op. cit.*

COVELLO, V.T. and ALLEN, F. (1988). *Seven cardinal rules of risk communication* (Washington, DC : Environ. Prot. Agency)

COVELLO, V.T., MENKES, J. and NEHNEVAJSA, J. (1982). Risk analysis, philosophy, and the social and behavioral sciences: reflections on the scope of risk analysis research. *Risk Analysis*, **2**, 53

COVELLO, V.T., FLAMM, W.G., RODRICKS, J.V. and TARDIFF, R.G. (1983). *The Analysis of Actual versus Perceived Risks* (New York: Plenum Press)

COVELLO, V.T., SANDMAN, P. and SLOVIC, P. (1988). *Risk Communication, Risk Statistics and Risk Comparisons: a Manual for plant Managers* (Washington, DC: Chem. Manufacturers Ass.)

COVINGTON, L.C., SHUTZ, R.W. and FRANSON, I.A. (1978). Titanium alloys for corrosion resistance. *Chem. Engng Prog.*, **74**(3), 67

COWAN, E. (1969). *Oil and Water. The Torrey Canyon Disaster* (London: William Kimber)

COWAN, H.J. (1982). *Design of Reinforced Concrete Structures* (Englewood Cliffs, NJ: Prentice-Hall)

COWAN, J.L. (1975). Techniques for the chose of redundancy in chemical process plants. Ph.D Thesis, Univ. of Cambridge

COWAN, J.L., SHEARER, C.J. and PERKINS, J.D. (1976). A technique for the optimal choice of redundancy for chemical processes. *Advances in Reliability Technology. ART5* (Culcheth, Lancashire: UK Atom. Energy Auth.)

COWARD, H.F. and BRINSLEY, F. (1914). The dilution-limits of inflammability of gaseous mixtures. J. *Chem. Soc.*, 1859, 1865

COWARD, H.F. and GUEST, P.G. (1927). Ignition of natural gas—air mixtures with heated metal bars. *J. Am. Chem. Soc.*, **49**, 2479

COWARD, H.F. and PAYMAN, W. (1937). Problems in flame propagation. *Combustion*, **2**, 189

COWARD, H.F. and WHEELER, R.V. (1934). The movement of flame in firedamp explosions. *Safety in Mines Res. Board, Paper 82* (London: HM Stationery Office)

COWARD, H.F., COOPER, C. and JACOBS, J. (1914). The ignition of some gaseous mixtures by electrical discharge. *J. Chem. Soc.*, **105**, 1069

COWEN, P.A., GODSMARK, P.H. and GOODWIN, R.C. (1971). Safety procedures necessary for the installation and maintenance of electrical apparatus in petrochemical works. *Electrical Safety*, **1**, 38

COWFER, J.A. and MAGISTRO, A.J. (1983). Vinyl chloride. *Kirk—Othmer Encyclopaedia of Chemical Technology*, 3rd ed., vol. 23 (New York: Wiley), p. 865

COWIE, C.T.Y. (1976). Hazard and operability studies — a new safety technique for chemical plants. *Prevention of Occupational Risks*, **3**, 341

COWLES, S.W. (1966). Oxygen fires. *Ammonia Plant Safety*, **8**, p. 19

COWLEY, L.T., (1989). *Proc. LPGITA Seminar* (Reigate: LPGITA)

COWLEY, L.T. and PRITCHARD, M.J. (1991). Large-scale natural gas and LPG jet fires and thermal impact on structures. *Gastech 90*, p. 3.6

COWLEY, W.E. and WYLDE, L.E. (1978). The behaviour of butt fusion-welded polyethylene pipelines under fatigue loading conditions. *Chemy Ind.,* Jun. 3, 371

COWLEY, W.E., DENT, N.P. and MORRIS, R.H. (1978). The brittle failure of UPVC lined glass reinforced plastics pipelines. *Chemy Ind.,* Jun. 3, 365

COX, A.P. and PASMAN, H.J. (1983). Risk analysis in the process industries. *Loss Prevention and Safety Promotion 4*, vol. 1, p. G2

COX, A.W., LEES, F.P. and ANG, M.L. (1990). *Classification of Hazardous Locations* (Rugby: Instn Chem. Engrs)

COX, B.G. and SAVILLE, G. (eds) (1975). *High Pressure Safety Code.* High Pressure Technol. Ass., Imperial Coll., London

COX, B.J. (1986). *Object Oriented Programming* (Reading, MA: Addison-Welsey)

COX, D.C. and BAYBUTT, P. (1982). Limit lines for risk. *Nucl. Technol.*, **57**, 320

COX, D.R. (1962). *Renewal Theory* (New York: McGraw-Hill)

COX, D.R. (1972). Regression models and life tables. *J.R. Stat. Soc., Ser. B*, **34**, 187

COX, D.R. and MILLER, H.D. (1972). *The Theory of Stochastic Processes* (London: Methuen)

COX, G. (1977). On radiant heat transfer from turbulent flames. *Combust. Sci. Technol.*, **17**, 75

COX, G. and CHITTY, R. (1980). A study of the deterministic properties of unbounded fire plumes. *Combust Flame*, **39**, 191

COX, J. (1971). Ship design aspects of the safe containment of dangerous chemicals. *Transport of Hazardous Cargoes*, **2**, paper 2

COX, J. (1988a). Bhopal (letter). *Chem. Engr, Lond.*, **455**, 6

COX, J. (1988b). Re-Chem dispute (letter). *Chem. Engr, Lond.*, **444**, 4

COX, J.B. (1972). The spill prevention, containment and countermeasure (SPCC) plan of the Refuse Act permit program. *Control of Hazardous Material Spills*, **1**, 27

COX, J.B. (1989). Chronic hazards and risk perception – an overview. *J. Haz. Materials*, **21**, 197

COX, J.I. (1975a). Flixborough – some additional lessons. *Instn Chem. Engrs Symp. on Technical Lessons of Flixborough*, Nottingham

COX, J.I. (1975b). *Report into the causes of the Flixborough disaster*. Rep. to Flixborough Court of Inquiry (L.H. Mandelstam and Partners (UK) Ltd)

COX, J.I. (1976a). Engineering lessons of the Flixborough disaster. *Course on Loss Prevention in the Process Industries*. Dept of Chem. Engng, Loughborough Univ. of Technol.

COX, J.I. (1976b). Flixborough – some additional lessons. *Chem. Engr, Lond.*, **309**, 353

COX, J.L. (ed.) (1983). *Natural Gas Hydrates* (Boston, MA: Butterworths)

COX, K.E. and WILLIAMSON, K.D. (eds) (1977–). *Hydrogen: Its Technology and Applications*, vols 1–5 (Boca Raton, FL.: CRC Press)

COX, L.A. (1982). Artifactual uncertainty in risk analysis. *Risk Analysis*, **2**, 121

COX, N.D. (1973). Applications of reliability theory to process design. *Loss Prevention 7*, p. 12

COX, N.D. (1976). Evaluating the profitability of standby components. *Chem. Engng Prog.*, **72**(7), 76

COX, O.J. and WEIRICK, M.L. (1981). Sizing safety valve inlet lines. *Loss Prevention 14*, p. 173. See also *Chem. Engng Prog.*, 1980, **76**(11), 51

COX, P. (1984). How we built Micro Expert. In Forsyth, R. (1984), *op. cit.*, p. 112

COX, R.A. (1977). Methods for predicting the atmospheric dispersion of massive releases of flammable vapour. *Prog. Energy Combust. Sci.*, **6**, 141

COX, R.A. (1981). Some considerations relating to the mass of fuel in an unconfined vapour cloud explosion. *Health and Safety Executive Seminar on Thermal Radiation and Overpressure Calculations* (London: Health and Safety Executive)

COX, R.A. (1982a). Improving risk assessment methods for process plant. *J. Haz. Materials*, **6**(3), 249

COX, R.A. (1982b). *Uses and limitations of analytical methods in hazard assessment and loss prevention. Developments 82* (Rugby: Instn Chem. Engrs)

COX, R.A. (1986). Appraisal of the utility of risk analysis in the process industry. *Loss Prevention and Safety Promotion 5*, vol. I, p. 14-1; vol. III, p. 206

COX, R.A. (1987). An overview of hazard analysis. *Preventing Major Chemical Accidents*, p. 1.37

COX, R.A. (ed.) (1989a). *Mathematics in Major Accident Risk Assessment* (Oxford: Clarendon Press)

COX, R.A. (1989b). *Piper Alpha Inquiry. Transcript, Days 79, 80, 94*

COX, R.A. (1989c). *Piper Alpha Public Inquiry. Transcript, Days 131, 132*

COX, R.A. (1990). The application of risk assessment to existing installations. *Piper Alpha: Lessons for Life Cycle Management* (Rugby: Instn Chem. Engrs), p. 153

COX, R.A. (1993). Risk analysis of offshore installations – past, present and future. *Health, Safety and Loss Prevention*, p. 57

COX, R.A. and CARPENTER, R.J. (1980). Further development of a dense vapour cloud dispersion model for hazard analysis. In Hartwig, S. (1980), *op. cit.*, p. 55

COX, R.A. and COMER, P.J. (1980). Containment failure discharge and vapour cloud formation. *Safety Promotion and Loss Prevention in the Process Industries* (London: Oyez)

COX, R.A. and COMER, P.J. (1982). Development of low-cost risk analysis methods for process plant. *The Assessment of Major Hazards* (Rugby: Instn Chem. Engrs), p. 353

COX, R.A. and ROE, D.R. (1977). A model of the dispersion of dense vapour clouds. *Loss Prevention and Safety Promotion 2*, p. 359

COX, R.A., COMER, P.J., PYMAN, M.A.F. and SLATER, D.H. (1980). Safety issues in the siting of modern LNG terminals. *Gastech 79*, p. 183

COX, R.A., PYMAN, M.A., SHEPPARD, P.A. and COOPER, J.P.S. (1980). Models for evaluating the probabilities and consequences of possible outcomes following releases of toxic or flammable gases. In Hartwig, S. (1980)., *op. cit.*, p. 185

COX, S. and COX, T. (1984). Women at work: summary and review. *Ergonomics*, **27**, 597

COX, S.J. (1992). *Risk Assessment Toolkit*. Centre for Extension Studies, Loughborough Univ. of Technol., Loughborough

COX, S.J. and TAIT, N.R.S. (1991). *Reliability, Safety and Risk Management* (Oxford: Butterworth-Heinemann)

COX, W.E. and WALKER, W.R. (1978). Common law liability: a supplemental control over hazardous material spills. *Control of Hazardous Material Spills*, **4**, 72

COX, W.M. and TIKVART, J.A. (1986). Assessing the performance level of quality models. In de Wispelaere, C., Schiermeier, F.A. and Gillani, N.V. (1986), *op. cit.*, p. 425

COXHEAD, P. (1987). *Starting LISP for AI* (Oxford: Blackwell)

COXON, B.A. and GILBERT, I. (1983). The incorporation of fault patterns associated with weather into reliability models. *Euredata 4*, paper 15.5

COXON, T. (1971). The design of nuclear plant with integral coolant circuits as related to the maximum credible accident approach. *Proc. Fourth Int. Conf. on Peaceful Uses of Atomic Energy* (New York: United Nations and Vienna: Int. Atom. Energy Agency), p. 227

CRACKNELL, A. (1971). Explosion in the feed-gas section of an ammonia plant. *Ammonia Plant Safety*, **13**, 57

CRACKNELL, A. (1976). The effects of hydrogen on steel. *Chem. Engr, Lond.*, **306**, 92

CRACKNELL, A. (1980). Stress corrosion cracking in ammonia storage. *Ammonia Plant Safety*, **22**, 63

CRAFT, J. (1985). Designing a database for the process engineer. *Process Engng*, May, 47

CRAIG, A. (1978). Is the vigilance decrement simply a response adjustment towards probability matching? *Hum. Factors*, **20**, 441

CRAIG, A. (1979). Nonparametric measures of sensory efficiency for sustained monitoring tasks. *Hum. Factors*, **21**, 69

CRAIG, A. (1980). Effect of prior knowledge of signal probabilities on vigilance performance at a two-signal task. *Hum. Factors*, **22**, 361

CRAIG, A. (1981). Monitoring for one kind of signal in the presence of another: the effects of signal mix on detectability. *Hum. Factors*, **23**, 191

CRAIG, A. (1987). Signal detection theory and probability matching apply to vigilance. *Hum. Factors*, **29**, 645

CRAIG, B. (1985). Critical velocity examined for effects of erosion–corrosion. *Oil Gas J.*, **83**(May 27), 99

CRAIG, L.W. (1989). Control structure for a batch reactor. *Plant/Operations Prog.*, **8**, 35

CRAIG, R.G. (1970). Route gives volume output of cyclohexane. *Chem. Engng*, **77**(Jun. 1), 108

CRAIG, R.G. and WHITE, E.A. (1980). Refine LPG for process flexibility. *Hydrocarbon Process.*, **59**(12), 111

CRAIG, R.M. (1978). Onshore gas terminals. Institute of Petroleum (1978), *op. cit.*, p. 129

CRAIK, K. (1943). *The Nature of Explanation* (Cambridge: Cambridge Univ. Press)

CRAIN, S. (1981). Another tank farm fire . . . *Hydrocarbon Process.*, **60**(1), 262

CRAKER, J.M., SCOTT, R.M. and DUTTON, R.S. (1986). Design and project management aspects of the LPG facilities at Port Bonython, South Australia. *Gastech 85*, p. 478

CRAKER, W.E. (1981). A fast low cost methodology for reliability modelling of large production processes. *Reliable Production in the Process Industries* (Rugby: Instn Chem. Engrs), p. 85

CRAKER, W.E. and MOBBS, D.B. (1977). Simple modelling aids for process design and evaluation. *First Nat. Conf. on Reliability*. *NRC3/8* (Culcheth, Lancashire: UK Atom. Energy Auth.)

CRALLEY, L.V. (1942). The effects of irritant gases upon the rate of ciliary activity. *J. Ind. Hyg. Toxicol.*, **24**, 193

CRALLEY, L.V. and CRALLEY, L.J. (eds) (1978). *Patty's Industrial Hygiene and Toxicology*, vol. 3, *Theory and Rationale of Industrial Hygiene Practice*, 3rd rev. ed. (New York: Wiley-Interscience)

CRALLEY, L.V. and CRALLEY, L.J. (eds) (1983). *Industrial Hygiene Aspects of plant Operations*, vols 1–3 (New York: Macmillan)

CRALLEY, L.V. and CRALLEY, J. (1987). *Industrial Hygiene Aspects of plant Operations*, vols 1–3 (Chichester: Wiley)

CRAMER, H. (1946). *Mathematical Methods of Statistics* (Princeton, NJ: Pergamon Press)

CRAMER, H.E. (1957). A practical method for estimating the dispersion of atmospheric contaminants. *Proc. First Nat. Conf. on Applied Meteorology* (Am. Met. Soc.), p. C-33

CRAMER, H.E. (1959a). A brief survey of the meteorological aspects of atmospheric pollution. *Bull. Am. Met. Soc.*, **40**(4), 165

CRAMER, H.E. (1959b). Engineering estimates of atmospheric dispersal capacity. *Am Ind. Hyg. Ass. J.*, **20**(3), 183

CRAMER, H.E. (1976). Improved techniques for modeling the dispersion of tall stack plumes. *Proc. Seventh Int. Tech.*

Mtg Air Pollution Modeling and its Applications (Environ. Prot. Agency), p. 731

CRAMER, H.E., RECORD, F.A. and VAUGHAN, H.C. (1958). *The Study of the Diffusion of Gases or Aerosols in the Lower Atmosphere. Final Rep. Contract AF19(604)-1058*. Dept of Meteorology, Massachusetts Inst. Technol., Cambridge, MA

CRAMER, H.E., DESANTO, G.M., DUMBAULD, R.K., MORGENSTERN, P. and SWANSON, R.N. (1964). *Meteorological Prediction Techniques and Data System. Rep. GCA-64-3-G* Geophysical Corp. of America, Bedford, MA

CRAMP, J.H.W. (1982). Laser monitoring of process plant. *Chemy Ind.*, Feb. 20, 124

CRAN, J. (1981). Improved factored method gives better preliminary cost estimates. *Chem. Engng*, **88**(Apr. 6), 65

CRANE COMPANY (1986). *Flow of Fluids through Valves, Fittings and Pipes*, (New York)

CRANE, F.A.A. and CHARLES, J.A. (1984). *Selection and Use of Engineering Materials* (London: Butterworths)

CRANE, R. and GREGG, R. (1983). Program for evaluating shell-and-tube heat exchangers. *Chem. Engng*, **90**(Jul. 25), 76

CRANK, J. (1956). *The Mathematics of Diffusion;* 1979, 2nd ed. (Oxford: Clarendon Press)

CRANZ, C. (1926). *Lehrbuch der Ballistik* (Berlin: Springer)

CRANZ, C. and BECKER, K. (1921). *Handbook of Ballistics*, vol. 1, *Exterior Ballistics* (London: HM Stationery Office)

CRAPPER, G.D. (1984). *Introduction to Water Waves* (Chichester: Ellis Horwood)

CRAVEN, A.D. (1972). Thermal radiation hazards from the ignition of emergency vents. *Chemical Process Hazards*, **IV**, p. 7

CRAVEN, A.D. (1975). Atmospheric control in process equipment. *Loss Prevention* **9**, 60

CRAVEN, A.D. (1976). Fire and explosion hazards associated with the ignition of small-scale spillages. *Process Industry Hazards*, p. 39

CRAVEN, A.D. (1981). Fire and explosion investigation. *J. Occup. Accid.*, **3**(3), 207

CRAVEN, A.D. (1982). Fire and explosion investigations on chemical plants and oil refineries. In Fawcett, H.H. and Wood, W.S. (1982), *op. cit.*, p. 659

CRAVEN, A.D. (1987). A simple method of estimating exothermicity by average bond energy summation. *Hazards from Pressure*, p. 97

CRAVEN, A.D. and FOSTER, M.G. (1966). The limits of flammability of ethylene in oxygen, air and air-nitrogen mixtures at elevated temperatures and pressures. *Combust. Flame*, **10**, 95

CRAVEN, A.D. and FOSTER, M.G. (1967). Dust explosion prevention – determination of critical oxygen concentration by the vertical tube method. *Combust. Flame*, **11**, 408

CRAVEN, A.D. and GREIG, T.R. (1967). The development of detonation overpressures in pipelines. *Chemical Process Hazards*, **III**, p. 41

CRAWFORD, A.M. and CRAWFORD, K.S. (1978). Simulation of operational equipment with a computer-based instructional system: a low cost training technology. *Hum. Factors*, **20**, 215

CRAWFORD, D.B., DURR, C.A. and HANDMAN, S.E. (1986). Inside an LNG storage tank. *Gastech 85*, p. 449

CRAWFORD, J.F. and SMITH, P.G. (1985). *Landfill Technology* (London: Butterworths)

CRAWFORD, M. (1976). *Air Pollution Control Theory* (New York: McGraw-Hill)

CRAWFORD, M. and SANTOS, L. (1986). Overcoming steam hammer in high pressure pipework. *Process Engng,* Nov., 57

CRAWFORD, T.V. (1966). A computer program for calculating atmospheric dispersion of large clouds. Lawrence Radiation Lab., Univ. of California, CA, Rep. UCRL-50179

CRAWLEY, F.K. (1982). The effects of the ignition of a major fuel spillage. *The Assessment of Major Hazards* (Rugby: Instn Chem. Engrs), p. 125

CRAWLEY, F.K. and ERSKIN, J.B. (1981). Monitoring of rolling bearings. *Reliable Production in the Process Industries* (Rugby: Inst. Chem. Engrs), p. 137

CRAWLEY, F.K. and SCOTT, D.S. (1992). The development of safety teaching in UK chemical engineering colleges. *Loss Prevention and Safety Promotion 7,* vol. 2, 54.1

CRAWLEY, F.K. and SCOTT, D.S. (1984). The design and operation of offshore relief systems. *The Protection of Exothermic Reactors and Pressurised Storage Vessels* (Rugby: Instn Chem. Engrs), p. 291

CRAWLEY, F.K., GRANT, M.M. and GREEN, M.D. (1992). Can we identify potential major hazards? *Major Hazards Onshore and Offshore,* p. 427

CRAWLEY, J.E. (1968). The present and future contribution of the human operator to the control of LD steelmaking. *Proc. BISRA Symp. on Steelmaking,* p. 101. See also Edwards, E. and Lees, F.P. (1974), *op. cit.,* p. 249

CREAMER, E.L. (1974). Stress corrosion cracking of reformer tubes. *Chem. Engng Prog.,* **70**(8), 105

CREANEY, P.R. (1981). If downtime is the problem – FSD is the answer. *plant Engr,* **25**(5), 15

CREBER, F.L. (n.d.). *Safety for Industry* (London: R. Soc Prev. Accid.)

CREDE, C.E. (1965). *Shock and Vibration Concepts in Engineering Design* (Englewood Cliffs, NJ: Prentice-Hall)

CREECH, J.L. and JOHNSON, M.N. (1974). Angiosarcoma of the liver in the manufacture of polyvinyl chloride. *J. Occup. Med.,* **16**, 150

CREECH, J.L. and MAKK, L. (1975). Liver disease among polyvinyl chloride production workers. *Ann. N.Y. Acad. Sci.,* **246**, 88

CREECH, M.D. (1940). *Study of combustion explosion in pressure vessels* (Kansas City, MO: Black, Syvals and Bryson)

CREECH, M.D. (1941). Combustion explosions in pressure vessel protected with rupture disks. *Trans. ASME,* **63**(7), 583

CREED, M.J. and FAUSKE, H.K. (1990). An easy, inexpensive approach to the DIERS procedure. *Chem. Engng Prog.,* **86**(3), 45

CREEDY, G.D. (1990). The CCPA manufacturing code of practice. *Plant/Operations Prog.,* **9**, 238

CREEDY, G.D. (1993). Planning guidelines for acute risk management: the Canadian chemical industry experience. *Process Safety Management,* p. 15

CRELLIN, G.L. and SMITH, A.M. (1982). Reliability measurement and confirmation. In Green, A.E. (1982), *op. cit.,* p. 293

CRELLIN, G.L., JACOBS, I.M., SMITH, A.M. and WORLEDGE, D.H. (1986). Organizing dependent event data – a classification and analysis of multiple component fault reports. *Reliab. Engng,* **15**(2), 145

CREMER, H.W. (1945). The siting and layout of industrial works. *Trans Instn Chem. Engrs,* **23**, 52

CREMER, H. (ed.) (1956). *Chemical Engineering Practice,* vols 1–12 (London: Butterworths)

CREMER, J. (1959). Biochemical studies on the toxicity of tetraethyl lead and other organo-lead compounds. *Br. J. Ind. Med.,* **16**, 191

CREMER, J. (1961). The toxicity of tetraethyl lead and related alkyl metallic compounds. *Ann. Occup. Hyg.,* **3**, 226

CREMER, J. and CALLAWAY, S. (1961). Further studies on the toxicity of some tetra and trialkyl lead compounds. *Br. J. Ind. Med.,* **18**, 277

CREMER, L. and HECKL, M. (1973). *Structure-borne Sound* (Berlin: Springer)

CREMER AND WARNER (1977). *The Hazard and Environmental Impact of the Proposed Shell NGL plant and Esso Ethylene plant at Mossmorran and Export Facilities at Braefoot Bay. Report* (London)

CREMER AND WARNER (1978). *Guidelines for the Layout and Safety Zones in Petrochemical Developments. Rep. C-2056* (London)

CREMER and WARNER (1980). *An analysis of the Canvey Report* (London: Oyez)

CRESCITELLI, S., RUSSO, G., TUFANO, V., NAPOLITANO, F. and TRANCHINO, L. (1977). Flame propagation in closed vessels and flammability limits. *Combust. Sci. Technol.,* **15**, 201

CRESCITELLI, S., RUSSO, G. and TUFANO, V. (1979a). Analysis and design of venting systems: a simplified approach. *J. Occup. Accid.,* **2**, 125

CRESCITELLI, S., RUSSO, G. and TUFANO, V. (1979b). The influence of different diluents on the flammability limits of ethylene at high temperatures and pressures. *J. Haz. Materials,* **3**(2), 177

CRESCITELLI, S., RUSSO, G. and TUFANO, V. (1980). Mathematical modelling of relief venting of gas explosions: theory and experiments. *Loss Prevention and Safety Promotion 3,* vol. 3, p. 1187

CRESCITELLI, S., DE STEFANO, G., PISTONE, L. RUSSO, G. and TUFANO, V. (1982). An experimental study on the conditions of safe oxidation of toluene with oxygen-carbon dioxide mixture. *J. Haz. Materials,* **5**(3), 189

CRESCITELLI, S., DE STEFANO, G., RUSSO, G. and TUFANO, V. (1982). On the flammability limits of saturated vapours of hydrocarbons. *J. Haz. Materials,* **5**(3), 177

CRESPO, A. *et al.* (1991). *One-dimensional Model of Torch Fires. Rep. UPM-FIRE.* Univ. Politecnica de Madrid

CRESSWELL, D.C. (1976). Temperature runaway and thermal stability of reactors. *Course on Fire and Explosion Safety in the Chemical Industry.* Dept of Fire Engng, Univ. of Edinburgh,

CRESSWELL, F.L. (1980). Background to sprinkler design and the use of pumps. *Fire Surveyor,* **9**(6), 39

CRESSWELL, W.L. and FROGGATT, P. (1963). *The Causation of Bus Driver Accidents: An Epidemiological Study* (Oxford: Oxford Univ. Press)

CRIBB, G.S. and HILDREW, R.G. (1970). The strategy for LNG in the United Kingdom. *LNG Conf. 2*

CRICK, M.J., MORREY, M.E. and HILL, M.D. (1986). Uncertainty due to lack of knowledge of parameter values in the predictions of a model for calculating the health and economic impact of accidental releases of radioactive materials to the atmosphere. *Inst. of Mathematics and its Applications, Mtg on Mathematics in Major Accident Risk Assessment*

CRISI, G.S. (1992). Buying heat exchangers. *Chem. Engng,* **99**(2), 94

CRITCHFIELD, C.E.M. (1974). Engineering for human safety. *Loss Prevention and Safety Promotion 1,* p. 1

CRITCHLEY, O.H. (1976). Risk prediction, safety analysis and quantitative probability methods – a *caveat. J. Br. Nucl. Energy Soc.,* **15**(1), 18

CRITCHLEY, O.H. (1978). Aspects of the historical, philosophical and mathematical background to the statutory management of nuclear plant risks in the United Kingdom. *Radiation Protection in Nuclear Power Plants and the Fuel Cycle* (London: Br. Nucl. Energy Soc.), p. 11

CRITCHLEY, O.H. (1981). Technological progress, safety and the guardian role of inspection. *Sci. Public Policy,* **8**, 291

CRITTENDEN, B. and KOLACZKOWSKI, S. (1987). Management of hazardous and toxic wastes. *Process Engng,* May, 81

CRITTENDEN, B.D. and KOLACZKOWSKI, S.T. (1992). *Waste Minimization Guide* (Rugby: Instn. Chem. Engrs)

CRITTENDEN, B.D., KOLACZKOWSKI, S.T. and HOUT, S.A. (1987). Modelling hydrocarbon fouling. *Chem. Engng Res. Des.,* **65**, 171

CROCE, P.A. (ed.) (1975). *A Study of Room Fire Development: the Second Full Scale Bedroom Fire Test of the Home Fire Project. Rep. RC-75–T-31.* Factory Mutual Res. Corp.

CROCE, P.A. and MUDAN, K.S. (1986). Calculating impacts for large hydrocarbon fires. *Fire Safety J.,* **11**(1), 99

CROCE, P.A., MUDAN, K.S. and WIERSMA, S.J. (1986). Thermal radiation from LNG trench fires. *Gastech 85,* p. 158

CROCE, P.A., DRAKE, E.M., WADE, D.E., SMITH, R.A. and MUNSON, R.E. (1988). Guidelines for the safe storage and handling of highly toxic materials. *Preventing Major Chemical and Related Process Accidents,* p. 95

CROCKER, B.B. (1970). Preventing hazardous pollution during plant catastrophes. *Chem. Engng,* **77**, May 4, 97

CROCKER, M.J. and KESSLER, F.M. (1982). *Noise and Noise Control,* vol. 2 (Boca Raton, FL: CRC Press)

CROCKER, M.J. and PRICE, A.J. (1975). *Noise and Noise Control* (Cleveland, OH: CRC Press)

CROCKER, R. (1985). Ultrasound for diagnosis and measurement. *Chem. Engr, Lond.,* **420**, 27

CROCKER, W.P. and NAPIER, D.H. (1986). Thermal radiation hazards of liquid pool fires and tank fires. In *Hazards, IX,* p. 159

CROCKER, W.P. and NAPIER, D.H. (1988a). Assessment of mathematical models for fire and explosion hazards of liquefied petroleum gases. *J. Haz. Materials,* **20**, 109

CROCKER, W.P. and NAPIER, D.H. (1988b). Mathematical models for the prediction of thermal radiation from jet fires. *Preventing Major Chemical and Related Process Accidents,* p. 331

CROCKFORD, G.W. (1980). Protective clothing and respiratory protection. In Waldron, H.A. and Harrington, J.M. (1980), *op. cit.,* p. 314

CROFT, W.M. (1980/81). Fires involving explosions – a literature review. *Fire Safety J.,* **3**(1), 3

CROM, R.C.W. (1977). Safeguarding against shock hazards. *Chem. Engng,* **84**(Mar. 28), 90

CROMARTY, B.J. (1992). Reformer tools: failure mechanisms, inspection methods and repair techniques. *Ammonia Plant Safety,* **32**, 197

CROMBIE, R. and GREEN, V.R. (1994). Some aspects of through-life integrity assessment – a safety assessor's view. *Br. Nucl. Energy Soc. Symp. on Thermal Reactor Safety Assessment* (London)

CROMEANS, J.S. and FLEMING, H.W. (1972). A review of catalysts used in ammonia production. *Ammonia Plant Safety,* **14**, 89

CROMEANS, J.S. and FLEMING, H.W. (1976). Improved ammonia plant catalysts. *Ammonia Plant Safety,* **18**, 19

CROMEANS, J.S. and KNIGHT, B.C. (1969). Safety problems and practices for catalytic units. *Ammonia Plant Safety,* **11**, 56

CRONAN, C.S. (1960a). New details roundout liquid H_2 story. *Chem. Engng,* **67**(Feb. 22), 61

CRONAN, C.S. (1960b). New process gets ultimate safeguards. *Chem. Engng,* **67**(Jul. 25), 68

CRONAN, C.S. (1960c). Safety lags new technology. *Chem. Engng,* **67**(Feb. 22), 60

CRONENBERG, A.W. and BENZ, R. (1980). Vapour explosion phenomena with respect to nuclear reactor safety assessment. *Adv. Nucl. Sci. Technol.,* **12**, 247

CRONENWETT, R.H. (1976). Low cost automation of steam tracing. *Chem. Engng,* **83**(May 10), 141

CRONER PUBLISHING LTD (1993). *Transportation Operations* (Kingston, Surrey)

CRONIN, C.H. (1978). The impact of international pesticide regulations on pesticide labelling, transport and storage. *Chemy Ind.,* Feb. 18, 115

CRONIN, J.L. and NOLAN, P.F. (1987a). The comparative sensitivity of test methods for determining initial exotherm temperatures in thermal decompositions of single substances. *J. Haz. Materials,* **14**(3), 293

CRONIN, J.L. and NOLAN, P.F. (1987b). Laboratory techniques for the quantitative study of thermal decompositions. *Plant/Operations Prog.,* **6**, 89

CRONIN, J.L., NOLAN, P.F. and BARTON, J.A. (1987). A strategy for thermal hazards assessment in batch chemical manufacturing. *Hazards from Pressure,* p. 113

CRONIN, J.L., NOLAN, P.F. and BARTON, J.A. (1989). Strategy for the thermal hazard evaluation of chemical reactions, illustrated by an analysis of the nitration of toluene. *Runaway Reactions,* p. 633

CRONJE, J.S., BISHNOI, P.R. and SVRCEK, W.Y. (1980). The application of the characteristic method to shock tube data that simulate a gas pipeline rupture. *Can. J. Chem. Engng* **58**, 289

CROOK, J. (1992). Protection of platforms from gas blasts. *Fire Int.,* **133**(Feb./Mar.), 26

CROOK, P.A. and ASPHAHANI, A. (1983). Alloys to protect against corrosion and wear. *Chem. Engng,* **90**(Jan. 10), 127

CROOKS, C.A., KURIYAN, K. and MACCHIETTO, S. (1992). Integration of batch plant design, automation and operation software tools. *Comput. Chem. Engng,* **16**, Suppl., S289

CROOKS, E. (1989). Legislation and the use of pressure relief devices. *Selection and Use of Pressure Relief Systems for Process plant* (Manchester: Instn. Chem. Engrs, North-West Branch), p. 1.1

CROOKS, E. (1994). Pipelines and pressure systems regulations. *Conf. on Valve and Pipeline Reliability.* Engng Dept, Univ of Manchester

CROOME-GALE, D.J. and ROBERTS, B.M. (1975). *Air Conditioning and Ventilation of Buildings* (Oxford: Pergamon)

CROSBY, H.L. (1979). Suppressing centrifuge chatter. *Chem. Engng Prog.,* **75**(10), 92

CROSBY, P. (1979). *Quality is Free* (New York: McGraw-Hill)

CROSBY, P. (1984). *Quality without Tears* (New York: McGraw-Hill)

CROSBY, P. (1986). *Running Things* (New York: McGraw-Hill)

CROSBY VALVE AND GAGE COMPANY (1994). *CROSBY-SIZE* (Wrentham, MA)

CROSETTI, P.A. (1971). Fault tree analysis with probability evaluation. *IEEE Trans Nucl. Sci.,* **NS-18**, 465

CROSETTI, P.A. and BRUCE, R.A. (1970). Commercial application of fault tree analysis. *Proc. Reliab. and Maintainab. Conf.,* **9**, 230

CROSS, A. (1982). Fault trees and event trees. In Green, A.E. (1982), *op. cit.*, p. 49

CROSS, A. and STEVENS, B. (1991). Reliability data banks – friend, foe or a waste of time? In Cannon, A.G. and Bendell, A. (1991), *op. cit.*, p. 265

CROSS, B. (1962). Aerial photos: new weapons against pollution. *Chem. Engng*, **69**(Apr. 2), 42

CROSS, E.F. (1968). Water pollution control. Sampling and analysis. *Chem. Engng*, **75**(Oct. 14), 75

CROSS, F.L. (1972). *Handbook on Incineration* (Westport, CN: Technomic Publ. Co.)

CROSS, F.L. and SCHIFF, H.F. (1973). Continuous source monitoring. *Chem. Engng*, **80**(Jun. 18), 125

CROSS, F.L. and SIMONS, J.L. (eds) (1975). *Industrial plant Siting* (Westport, CT: Technomic)

CROSS, J.A. (1980). Electrostatic charging and ignition of dusts. *Chemical Process Hazards*, VII, p. 1

CROSS, J.A. (1981). Hazards in powder processing. *Particulate Technology* (Rugby: Instn. *Chem. Engrs*), p. 1

CROSS, J.A. (1987). *Electrostatics Principles, Problems and Applications* (Bristol: Adam Hilger)

CROSS, J. and FARRER, D. (1982). *Dust Explosions* (New York: Plenum Press)

CROSS, J.A., HAIG, I. and CETRONIO, A. (1977). Electrostatic hazards from pumping insulating liquids in glass pipes. *Proc. Third Int. Conf. on Static Electricity* (Paris: Soc. de Chimie Industrielle), p. 2a

CROSS, M., RICHARDS, C.W., IEROTEOU, C. and LEGGETT, P. (1989). Software engineering in CFD future trends. In Bush, A.W. Lewis, B.A. and Warren, M.D. (1989), *op. cit.*, p. 54

CROSSLAND, SIR B. (1991). Thoughts of an assessor in a disaster inquiry. In The Fellowship of Engineering (1991b), *op. cit.*, p. 117

CROSSLAND, N.O. (1986). Predicting the hazards of chemicals to aquatic environments. *Chemy Ind.*, Nov. 3, 740

CROSSLAND, S. (1972). Process liquids to SNG. *Hydrocarbon Process.*, **51**(4), 89

CROSSLEY, P.B., and RIPLING, E.J. (1975). Plane strain crack arrest characterisation of steels. *J. Pres. Ves. Technol.*, **97**, 291

CROSSLEY, R.W., DORKO, E.A., SCHALLER, K. and BURCAT, A. (1972). The effect of higher alkanes on the ignition of methane–oxygen–argon mixtures in shock waves. *Combust. Flame*, **19**, 373

CROSSMAN, E.R.F.W. (1956). Perception study – a complement to motion study. *Manager*, **24**, 141

CROSSMAN, E.R.F.W. (1960). *Automation and Skill* (London: H.M. Stationery Office). See also Edwards, E. and Lees, F.P. (1974), *op. cit.*, p. 1

CROSSMAN, E.R.F.W. and COOKE, J.E. (1962). Manual control of slow-response systems. *Int. Cong. on Human Factors in Electronics*, Long Beach, CA. See also Edwards, E. and Lees, F.P. (1974), *op. cit.*, p. 51

CROSSMAN, E.R.F.W., COOKE, J.E. and BEISHON, R.J. (1964). *Visual Attention and the Sampling of Displayed Information in Process Control. Rep. HFT 64–11–7.* Hum. Factors in Technol. Res. Gp, Univ. of California, Berkeley, CA. See also Edwards, E. and Lees, F.P. (1974), *op. cit.*, p. 25

CROSSTHWAITE, P.J. (1984). HSE's approach to the control of developments near to notifiable LPG installations. In Petts, J.I. (1984), *op. cit.*

CROSSTHWAITE, P.J. (1986). Development control in the vicinity of certain LPG installations. *Gastech 85*, p. 128

CROSSTHWAITE, P.J. and CROWTHER, J.H. (1992). Use of risk analysis for the location of buildings in the vicinity of chemical processing plant. *Hazard Identification and Risk Analysis, Human Factors and Human Reliability in Process Safety*, p. 211

CROSSTHWAITE, P.J., FITZPATRICK, R.D. and HURST, N.W. (1988). Risk assessment for the siting of developments near liquefied petroleum gas installations. *Preventing Major Chemical and Related Process Accidents*, p. 373

CROUCH, E.A. and WILSON, R. (1984). Inter-risk comparisons. In Rodricks, R.V. and Tardiff, R.G. (1984a), *op. cit.*, p. 97

CROUCH, H.W., CHAPMAN, C.H., RAYMOND, C.L. and WISCHMEYER, F.W. (1952). *Maximum Pressures and Rates of Pressure Rise Due to Explosions of Various Solvents.* Eastman Kodak Co., Kodak Park Div.

CROUCH, W.W. (1968). Evaluating hazards of chemicals in bulk water transportation. *Carriage of Dangerous Goods*, **1**, p. 159

CROUCH, W.W. and HILLYER, W.C. (1972). What happens when LNG spills? *Chemtech*, Apr.

CROWE, B.L. (1969). The tragedy of the commons revisited. *Science*, **166**, 1103

CROWE, C., DOUGLAS, P.L., GLASGOW, J. and MALLICK, S.K. (1992). Development of an expert system for process synthesis. *Chem. Engng Res. Des.*, **70A**, 110

CROWE, C.M. (1988). Recursive identification of gross errors in linear data reconciliation. *AIChE J.*, **34**, 541

CROWE, W.H. and MARYSIUK, A.M. (1965). How to work safely with alkylation. *Hydrocarbon Process.*, **44**(5), 192

CROWHURST, D. (1989). Dust explosion hazards in dryers. *Fire Surveyor*, **18**(3), 17

CROWL, D.A. (1991). Using thermodynamic availability to determine energy of explosion. *Plant/Operations Prog.*, **10**, 136

CROWL, D.A. (1992a). Calculating the energy of explosion using thermodynamic availability. *J. Loss Prev. Process Ind.*, **5**, 109

CROWL, D.A. (1992b). Liquid discharge from process and storage vessels. *J. Loss Prev. Process Ind.*, **5**, 73

CROWL, D.A. (1992c). Using thermodynamic availability to determine the energy of explosion of compressed gases. *Plant/Operations Prog.*, **11**, 47

CROWL, D.A. and LOUVAR, J.F. (1989). Instructional videotapes on chemical process safety. *Plant/Operations Prog.*, **8**, 225

CROWL, D.A. and LOUVAR, J.F. (1990). *Chemical Process Safety: Fundamentals with Applications* (Englewood Cliffs, NJ: Prentice Hall)

CROWLEY, C.J. and BLOCK, J.A. (1989). Safety relief system design and performance of a hydrogen peroxide system. *Runaway Reactions*, p. 395

CROWLEY, E.D. (1993). Stemming leaks with spiral wound gaskets. *Chem. Engng*, **100**(7), 139

CROWLEY, E.D. and HART, D.G. (1992). Minimize fugitive emissions. *Hydrocarbon Process.*, **71**(7), 93

CROWLEY, J.B. (1972). Good sampling saves money. *Hydrocarbon Process.*, **51**(10), 164

CROWLEY, J.E. (1982). *Hydraulic Design and Estimation of Fire Sprinkler Systems.* Crowley Design Gp, King of Prussia, PA

CROWLEY, J.M. (1986). *Fundamentals of Applied Electrostatics* (New York: Wiley)

CROWLEY, M.S. (1965). Tips on insulating tanks. *Ammonia Plant Safety*, **7**, 30. See also *Chem. Engng Prog.*, 1965, **61**(1), 88

CROWLEY, M.S. (1968). New data on hydrogen–silica reactions in refractories. *Ammonia Plant Safety*, **10**, 25

CROYSDALE, L.G., SAMUELS, W.E. and WAGNER, J.A. (1965). Urea—ammonia nitrate solutions: are they safe? *Ammonia Plant Safety*, **7**, 49. See also *Chem. Engng Prog.*, 1965, **61**(1), 72

CROZIER, R.A. (1980). Pressure relief to prevent heat exchanger tube failure. *Chem. Engng*, **87**(Dec. 15), 79

CROZIER, R.A. (1985). Sizing relief valves for fire emergencies. *Chem. Engng*, **92**(Oct. 28), 49

CRUICE, W.J. (1986). Explosions. *Fire Protection Handbook*, 16th ed. (ed. by A.E. Cote and J.M. Linville), pp. 4–17

CRUISE, D.R. (1964). Notes on the rapid computation of chemical equilibrium. *J. Phys. Chem.*, **68**, 3797

CRUMP, K.S. and CROCKETT, P.W. (1985). Improved confidence limits for low-dose carcinogenic risk assessment from animal data. *J. Haz. Materials*, **10**(2/3), 419

CRUMPLER, E. and MARTIN, E.J. (1987). Incineration. In Martin, E.J. and Johnson, J.H. (1987), *op. cit.*, p. 227

CRUVER, J.E. and MOULTON, R.W. (1967). Critical flow of liquid—vapour mixtures. *AIChE J.*, **13**, 52

CSANADY, G.T. (1955). Dispersal of dust particles from elevated sources. *Aust. J. Phys.*, **8**, 545

CSANADY, G.T. (1969). Dosage probabilities and area coverage from instantaneous point sources at ground level. *Atmos. Environ.*, **3**, 25

CSANADY, G.T. (1973). *Turbulent Diffusion in the Environment* (Dordrecht: Reidel)

CSATHY, D. (1967). Evaluating boiler designs for process heat recovery. *Chem. Engng*, **74**(Jun. 5), 117

CSATHY, D. (1981). Design of waste-heat boilers. *Chem. Engr, Lond.*, **367**, 159

CUBBAGE, P.A. (1959). *Flame Traps for Use with Town Gas/Air Mixtures. Res. Comm. GC 63.* Gas Coun., London

CUBBAGE, P.A. (1963). The protection by flame traps of pipes conveying combustible mixtures. *Chemical Process Hazards*, **II**, p. 29

CUBBAGE, P.A. (1989). *Piper Alpha Public Inquiry. Transcript, Days 92, 93*

CUBBAGE, P.A. and MARSHALL, M.R. (1972). Pressures generated in combustion chambers by the ignition of air—gas mixtures. *Chemical Process Hazards*, **IV**, p. 24

CUBBAGE, P.A. and MARSHALL, M.R. (1973). *Pressures Generated by Gas—Air Mixtures in Vented Enclosures* (London: Instn Gas Engrs)

CUBBAGE, P.A. and MARSHALL, M.R. (1974). Explosion relief protection for industrial plant of intermediate strength. *Chemical Process Hazards*, **V**, p. 196

CUBBAGE, P.A. and MOPPETT, D.J. (1970). Explosion generation and recording. Appendix A in Astbury, N.F. *et al.* (1970), *op. cit.*

CUBBAGE, P.A. and SIMMONDS, W.A. (1955a). *An Investigation of Explosion Reliefs for Industrial Drying Ovens. Res. Comm. GC23.* Gas Coun., London

CUBBAGE, P.A. and SIMMONDS, W.A. (1955b). Investigation of explosion reliefs for industrial drying ovens. *Gas World*, **142**(3719), 1380. See also *Gas J.*, 1955, **284**(4828), 725; 1957, **292**(4926), 574

CUBBAGE, P.A. and SIMMONDS, W.A. (1955c). An investigation of explosion reliefs for industrial drying ovens – I, Top reliefs in box ovens. *Trans. Inst. Gas Engrs*, **105**, 470

CUBBAGE, P.A. and SIMMONDS, W.A. (1957a). *An Investigation of Explosion Reliefs for Industrial Drying Ovens. Res. Comm. GC 43.* Gas Coun., London

CUBBAGE, P.A. and SIMMONDS, W.A. (1957b). An investigation of explosion reliefs for industrial drying ovens – II, Back

reliefs in box ovens: relief in conveyor ovens. *Trans. Inst. Gas Engrs*, **107**, 503

CUBBAGE, P.A. and SIMMONDS, W.A. (1957c). Top reliefs in box ovens. *Gas World* **145** (3779, Suppl.), 10

CUDAHY, J.J. and TROXLER, W.L. (1983). Autoignition temperature as an indicator of thermal ignition stability. *J. Haz. Materials*, **8**(1), 59

CUDE, A.L. (1974a). Dispersion of gases vented to atmosphere from relief valves. *Course on Process Safety – Theory and Practice.* Dept of Chem. Engng, Teesside Polytechnic, Middlesborough

CUDE, A.L. (1974b). Dispersion of gases vented to atmosphere from relief valves. *Chem. Engr, Lond.*, **290**, 629

CUDE, A.L. (1975). The generation, spread and decay of flammable vapour clouds. *Course on Process Safety – Theory and Practice.* Dept of Chem. Engng, Teesside Polytechnic, Middlesborough.

CUDE, A.L. (1977). Private communication

CULBERTSON, W.L. and HORN, J. (1968). The Phillips— Marathon Alaska to Japan LNG project. *LNG Conf.* 1

CULICOVER, P.W. (1982). *Syntax*, 2nd ed. (New York: Academic Press)

CULLEN, THE HONOURABLE LORD (1990). *The Public Inquiry into the Piper Alpha Disaster* (London: HM Stationery Office)

CULLEN, E.J. (1991). Responsibility of the individual. In The Fellowship of Engineering (1991b), *op. cit.*, p. 81

CULLEN, J. (1984). Flixborough. HSC on the wider issues. *Chem. Engr, Lond.*, **405**, 25

CULLEN, J. (1985). What can we learn from Pemex? *Chem. Engr, Lond.*, **418**, 21

CULLEN, SIR J. (1994). Health and safety: a burden or business? *Process Safety Environ.*, **72B**, 3

CULLINGFORD, M.C., SHAH, S.M. and GITTUS, J.H. (1985). *Implications of PRA* (London: Elsevier Appl. Sci.)

CULLINGFORD, R.E. (1986). *Natural Language Processing* (Totowa, NJ: Rowman & Littlefield)

CULLINGTON, R.F. (1987). Fire stopping systems. *Fire Surveyor*, **13**(4), 9

CULLIS, C.F. and FOSTER, C.D. (1973). Study of the spontaneous ignition of hydrocarbons and the application of computer techniques. *Combustion*, **14**, 423

CULLIS, C.F. and FOSTER, C.D. (1974). Studies of the spontaneous ignition in air of binary hydrocarbon mixtures. *Combust. Flame*, **23**, 347

CULLIS, J.F. and FIRTH, J.G. (eds) (1981). *Detection and Measurement of Hazardous Materials* (London: Heinemann)

CULLIS, C.F., HIRSCHLER, M.M., TOWNSEND, R.P. and VISANUVIMOL, V. (1983a). The combustion of cellulose under conditions of rapid heating. *Combust. Flame*, **49**, 249

CULLIS, C.F., HIRSCHLER, M.M., TOWNSEND, R.P. and VISANUVIMOL, V. (1983b). The pyrolysis of cellulose under conditions of rapid heating. *Combust. Flame*, **49**, 235

CULP, R.L., WEBSTER, G.M. and CULP, G.L. (1978). *Handbook of Advanced Wastewater Treatment*, 2nd ed. (New York: van Nostrand Reinhold)

CULSHAW, G.W. and GARSIDE, J.E. (1949). A study of burning velocity. *Combustion*, **3**, 204

CUMBER, P.S. and CLEAVER, R.P. (1994). Methods of predicting the dispersal of natural gas releases in ventilated ducts. *Hazards*, **XII**, p. 213

CUMBERLAND, R.F. (1978). Role of the National Chemical Emergency Centre at Harwell, UK. *Control of Hazardous Material Spills*, 4, 60

CUMBERLAND, R.F. (1982). The control of hazardous chemical spills in the United Kingdom. *J. Haz. Materials*, **6**(3), 277

CUMBERLAND, R.F. and HEBDEN, M.D. (1975). A scheme for recognising chemicals and their hazards in an emergency. *J. Haz. Materials*, **1**(1), 35

CUMMING, R.H. and BROWN, D. (1991). Process validation and safety in biotechnology. *Hazards,* **XI**, p. 363

CUMMINGS, D.K. and BRADLEY, C.W.J. (1988). Consequence analysis for risk assessment of major hazards in port installations. *Preventing Major Chemical and Related Process Accidents,* p. 699

CUMMINGS, D.L., LAPP, S.A. and POWERS, G.J. (1983). Fault tree synthesis from a directed graph model for a power distribution network. *IEEE Trans. Reliab.,* **R-32**, 140

CUMMINGS, G.E. (1986). *Summary Report on the Seismic Safety Margins Research Program. Rep. NUREG/CR-4431.* Nuclear Regulatory Comm., Washington, DC

CUMMINGS, J.C., LEE, J.H.S., CAMP, A.L. and MARX, K.D. (1984). Analysis of combustion in closed or vented rooms or vessels. *Plant/Operations Prog.,* **3**, 239

CUMO, M. and NAVIGLIO, A. (eds)(1989). *Safety Design Criteria for Industrial plants* (Boca Raton, FL: CRC Press)

CUNGE, J.A., HOLLY, F.M. and VERWEY, A. (1980). *Practical Aspects of Computational River Hydraulics* (Boston, MA: Pitman)

CUNNINGHAM, A.W.T. (1990). *Piper Alpha Public Inquiry. Transcript, Day 168*

CUNNINGHAM, C.E. and COX, W. (1972). *Applied Maintainability Engineering* (New York: Wiley)

CUNNINGHAM, E.R. (1985). Keeping fluid handling systems safe with overpressure protection. *plant Engng,* **39**(Feb. 14), 34

CUNNINGHAM, P.F. (1962). Mechanical drive selection. *Chem. Engng Prog.,* **58**(1), 106

CUNNINGHAM, R.G. (1970). Electrification of insulating belts passing over grounded rollers. *J. Colloid Interface Sci.,* **32**, 373

CUNY, F.C. (1983). *Disasters and Development* (Oxford: Oxford Univ. Press)

CUNY, X. and BOY, M. (1981). Analyse semiologique et apprentissage des outilssignes: l'apprentissage du schema d'electricite. *Communications,* **33**, 103

CUPERUS, N.J. (1979). Cryogenic storage facilities for LNG and NGL. *Tenth World Petrol. Cong.,* Bucharest, panel discussion 17, paper 3

CUPERUS, N.J. (1980). Developments in cryogenic storage tanks. *LNG Conf. 6,* session 2, paper 13

CURLIN, L.V., BOMMARAJU, T.V. and HANSSON, C.B. (1991). Alkali and chlorine products. In: *Kirk–Othmer Encyclopaedia of Chemical Technology,* 4th ed., vol. 1 (New York: Wiley), p. 938

CURRAN, L.M. (1992). Waste minimization practices in the petroleum refining industry. *J. Haz. Materials,* **29**, 189

CURRAN, L.M. and KIZIOR, G.J. (1992). A computerized method for reporting SARA Title III Section 313 emissions from a petroleum refinery. *J. Haz. Materials,* **31**, 255

CURRIEO, R.A. (1993). How to choose process equipment made from reinforced plastic. *Chem Engng,* **100**(8), 92

CURRIE, I.M. (1960). *Work Study* (London: Pitman)

CURRY, N. (1978). *Health and Safety Executive Pilot Study. Working Conditions in Universities* (London: Health and Safety Executive)

CURRY, R.E. (1981). A model of human fault detection for computer dynamic processes. In Rasmussen, J. and Rouse, W.B. (1981), *op. cit.,* p. 171

CURRY, R.E., KLEINMAN, D.L. and HOOFMAN, W.C. (1977). A design procedure for control/display systems. *Hum. Factors,* **19**, 421

CURTELIN, R. (1991). Decompression d'un reservoir de propane liquefie. Ph.D. Thesis, L'Ecole Centrale, Paris

CURTIS, J.S. (1990). Safe storage of dilute ethylene oxide mixtures in water. *Plant/Operations Prog.,* **9**, 91

CURTIS, M.W., COPELAND, T.L. and WARD, C.H. (1978). Aquatic toxicity of substances proposed for spill prevention regulation. *Control of Hazardous Material Spills,* 4, p. 99

CURTIS, R., HOGAN, E.R. AND HOROWITZ, S. (1980). *Nuclear Lessons* (Wellingborough: Turnstone Press)

CUSHIE, G.C. (ed.) (1984). *Removal of Metals from Wastewater* (Park Ridge, NJ: Noyes Pubs)

CUSSELL, F. (1979). EEC 'Dangerous Substances' Directive. Fixed emission standards with reference to the EEC Directive. *Chemy Ind.,* May 5, 306

CUSTER, R.L.P. (1972). Oxidation is only the beginning. *Fire Engng,* **125**(Jul. 30)

CUSWORTH, G.R.N. (ed.) (1975). *Health and Safety at Work etc. Act* (London: Butterworths)

CUTLER, D.P. (1974). The ignition of gases by rapidly heated surfaces. *Combust. Flame,* **22**, 105

CUTLER, D.P. (1978). Further studies of the ignition of gases by transiently heated surfaces. *Combust. Flame,* **33**, 85

CUTLER, D.P. (1986). Current techniques for the assessment of unstable substances. *Hazards,* **IX**, p. 133

CUTMORE, W.H. (1972). *Shaws Commentary on the Building Regulations 1972* (London: Shaw & Son)

CYBULSKI, A., STANKIEWICZ, A.I., EDVINSSON ALBERS, R.K. and MOULIJN, J.A. (1999). Monolithic reactors for fine chemicals industries: a comparative analysis of a monolithic reactor and a mechanically agitated slurry reactor. *Chem. Eng. Sci.* **54**, 2351–8

CZAGA, S.J. and DRURY, C.G. (1981). Training programs for inspection. *Hum. Factors,* **23**, 473

CZAIKOWSKI, M.P. and BAYNE, A.R. (1980). Make formic acid from CO. *Hydrocarbon Process.,* **59**(11), 103

CZERNIAK, E. (1969). Foundation design guide for stacks and towers. *Hydrocarbon Process.,* **48**(6), 95

CZICHOS, H. (1978). *Tribology* (Amsterdam: Elsevier)

CZUPPON, T.A. and BUIVIDAS, L.J. (1979). Which feedstock for ammonia? *Hydrocarbon Process.,* **58**(9), 197

CZUPPON, T.A., KNEZ, S.A. and ROUNES, J.M. (1992). Ammonia. *Kirk-Othmer Encyclopaedia of Chemical Technology,* vol. 2, 4th ed. (New York: Wiley)

DABBERDT, W.F. and DAVIS, P.A. (1974). Determination of energetic characteristics of urban–rural surfaces in the Greater St Louis area. *First Symp. on Atmospheric Diffusion and Air Pollution* (Boston, MA: Am. Met. Soc.), p. 133

DABBERDT, W.F., BRODZINSKY, R., CANTREL, L.B.C., RUFF, R.E., DIETZ, R. and SETHURAMAN, S. (1982). *Atmospheric Dispersion over Water and in the Shoreline Transition Zone,* vol. 1, *Analysis and Results. Rep. SRI 3450.* Stanford Res. Inst., Menlo Park, CA

DABBERDT, W.F., JOHNSON, W.B., BRODZINSKY, R. and RUFF, R.E. (1983). *Central California Coastal Air Quality Model Validation Study: Data Analysis and Model Evaluation. Rep. SRI 3868.* Standford Res. Inst., Menlo Park, CA

DABORA, E.K. (1982). Fundamental mechanisms of liquid spray detonations. In Lee, J.H.S. and Guirao, C.M. (1982), *op. cit.,* p. 245

DABORA, E.K., NICHOLLS, J.A. and MORRISON, R.B. (1965). The influence of a compressible boundary on the propagation of gaseous detonations. *Combustion,* **10**, p. 817

DAHL, E., BERN, T.I., GOLAN, M. and ENGEN, G. (1983). *Risk of Oil and Gas Blow-out on the Norwegian Continental Shelf. Rep. STF 88A82062.* SINTRF, Trondheim

DAHLE, T.G. (1994). The Norwegian approach to safety in offshore petroleum activity. *J. Loss Prev. Process Ind.*, **7**, 379

DAHLL, G. (1976). A system for on-line analysis of abnormal situations. *Automation in Offshore Field Operation* (eds. Galtung, Rosenhaug and Williams) (Amsterdam: North Holland)

DAHLL *et al.* (1976). On-line analysis of abnormal plant situations. Unpublished paper.

DAHLSTROM, D. (1968). Water pollution control. *Chem. Engng*, **75**(Oct. 14), 103

DAHMS, R.H. (1964). How to prevent plastics fires with retardants. In Vervalin, C.H. (1964a), *op. cit.*, p. 93

DAHN, C.J. (1980). Chemical compatibility and safe storage considerations for process systems hazard analysis. *J. Haz. Materials*, **4**(1), 121

DAHN, C.J. (1981). *Electrostatic Hazards of Dusts. The Hazards of Industrial Explosion from Dusts* (London: Oyez/IBC), paper 4

DAHN, C.J. (1992). Electrostatic hazards of pneumatic conveying of powders. *Plant/Operations Prog.*, **11**, 201

DAHN, C.J. (1993). How safe are your pneumatic conveyors? *Chem. Engng Prog.*, **89**(5), 17

DAHN, C.J., ASHUM, M. and WILLIAMS, K. (1986). Contribution of low-level flammable concentrations to dust explosion output. *Plant/Operations Prog.*, **5**, 57

DAIGRE, L.C. and NIEMAN, G.R. (1974). Computer control of ammonia plants. *Ammonia Plant Safety*, **16**, 45

DAKIN, T.W. *et al.* (1974). *Electra*, **32**, 61

DAILEY, W.L. (1970). Maximising on-stream time for large plants. *Chem. Engng Prog.*, **66**(12), 34

DAILEY, V.W. (1976). Area monitoring for flammable and toxic hazards. *Loss Prevention*, **10**, 8

DAILEY, W.V. (1993). Improve process analyzer availability. *Hydrocarbon Process.*, **72**(1), 69

DALAL, S.S. (1980). Hazards from neighbouring plants. *Loss Prevention and Safety Promotion 3*, vol. 4, p. 1528

DALDERUP, L.M. (1974). T. Soc. *Geneesk*, **52**, 582

DALE, C. (1987). Data requirements for software reliability prediction. In Littlewood, B. (1987b), *op. cit.*, p. 144

DALE, C.J. (1985). Application of the proportional hazard model in the reliability field. *Reliab. Engng*, **10**(1), 1

DALE, G.C. and HARRISON, J.R. (1971). Safety of nuclear power plants. *Proc. Fourth Int. Conf. on Peaceful Uses of Atomic Energy* (New York/Vienna: United Nation Int. Atom. Energy Agency)

DALE, H.C.A. (1958). Fault finding in electronic equipment. *Ergonomics*, **1**, 356

DALE, H.C.A. (1964). Factors affecting the choice of strategy in searching. *Ergonomics*, **7**, 73

DALE, H.C.A. (1968). Weighing evidence: an attempt to assess the efficiency of the human operator. *Ergonomics*, **11**, 215

DALE, S. (1984). The identification of risk. *The Management of Technological Risk* (London: A.D. Little)

DALE, S.E. (1987). Cost effective design considerations for safer chemical plants. *Preventing Major Chemical Accidents*, p. 3.79

DALE, S.E. and CROCE, P.A. (1985). Operation and maintenance safety audit for an existing liquefied natural gas. *Gastech 84*, p. 171

DALIN, M.A., KOLCHIN, I.K. and SEREBRYAKOV, B.R. (1971). Acrylonitrile (Westport, CT: Technomic)

DALKEY, N. (1969). An experimental study of group opinion. The Delphi method. *Futures*, **1**(5), 408

DALLAVALLE, J.M. (1952). *Exhaust Hoods*, 2nd ed. (New York: The Industrial Press)

DALLE MOLLE, D.T. and HIMMELBLAU, D.M. (1987). Fault detection in a single-stage evaporator via parameter estimation using the Kalman filter. *Ind. Engng Chem. Res.*, **26**, 2482

DALLE MOLLE, D.T., KUIPERS, B.J. and EDGAR, T.F. (1988). Qualitative modeling and simulation of dynamic systems. Comput. *Chem. Engng*, **12**, 853

DALLIMONTI, R. (1972). Future operator consoles for improved decision-making and safety. *Instrum. Technol.*, **19**(8), 23

DALLIMONTI, R. (1973). New designs for process control consoles. *Instrum. Technol.*, **20**(11), 48

DALL'ORA, F. (1971). European pressure vessel codes. *Hydrocarbon Process.*, **50**(6), 93

DALMAI, G. (1958). Investigations on the inflammability of methane-air and butane-air mixtures ignited by electric sparks. *Combust. Flame*, **2**, 171

DALTON, M.L. and BRICKER, D.L. (1978). Anhydrous ammonia burns of the respiratory tract. *Tox. Med.*, **74**, 51

DALTON, T.F. (1976). Hazardous materials cleanup in industrial areas. *Control of Hazardous Material Spills*, **3**, 407

DALY, E.F. and SUTHERLAND, G.B.B.M. (1949). The emission spectrum of carbon dioxide at 4.3 m. *Combustion*, **3**, 530

DALZELL, G.A. (1989). *Piper Alpha Public Inquiry. Transcript, Days 150, 151*

DALZELL, G. and MELVILLE, G.S. (1992). A method for preliminary design of fire protection requirements on an offshore oil production platform. *Major Hazards Onshore and Offshore*, p. 559

DALZELL, G.A. and WILLING, P.R. (2000). Safer design – an attitude. *IChemE Symp. Ser. 147*, 213–20

DALZELL, W.H. and SAROFIM, A.F. (1969). Optical constants of soot and their applications to heat flux calculations. *J. Heat Transfer*, **91**, 100

DALZELL, W.H., WILLIAMS, G.C. and HOTTEL, H.C. (1970). A light-scattering method for soot concentration measurements. *Combust. Flame*, **14**, 161

DAMAMME, G. (1982). The divergent quasistationary detonation wave. In Nellis, W.J., Seaman, L. and Graham, A.R. (1982), *op. cit.*, p. 603

DAMIANO, E.R. (1990). Observations and predictions of jet diffusion flame behaviour. *Atmos. Environ.*, **24A**, 2527

DAMKOHLER, G. (1940). The effect of turbulence on the flame velocity in gas mixtures. *Z. Elektrochem.*, **46**, 601

DAMMERS, W.R. (1969). Thermal instabilities in stirred flow reactors. *Symp. on Inherent Hazards of Manufacturing and Storage in the Process Industry*, The Hague, p. 105

DAMON, E.G., GAYLORD, C.S., HICKS, W., YELVERTON, J.T., RICHMOND, D.R. and WHITE, C.S. (1966). *The Effects of Ambient Pressure on Tolerance of Mammals to Air Blast. Rep. DASA-1852.* Defense Atom. Support Agency, Dept of Defense, Washington, DC

DAMON, E.G., GAYLORD, C.S., YELVERTON, J.T., RICHMOND, D.R., BOWEN, I.G. and WHITE, C.S. (1968). Effects of ambient pressure on tolerance of mammals to air blast. *Aerospace Med.*, **39**, 1039

DANA, C. (1993). Consider quarter-turn gates for gate valve applications. *Hydrocarbon Process.*, **72**(6), 78

DANA, M.T. and HALES, J.M. (1976). Statistical aspects of the washout of polydisperse aerosols. *Atmos. Environ.*, **10**, 45

DANAHER, J.W. (1980). Human error in ATC system operations. *Hum. Factors*, **22**, 535

DANCER, D. (1990). Pressure and temperature response of liquefied gases in containers and pressure vessels which are subjected to accidental heat input. *J. Haz. Materials*, **25**(1/2), 3

DANEKIND, W.E. (1976). Steam management in a refinery. *Hydrocarbon Process.*, **55**(12), 71

DANEKIND, W.E. (1979). Steam management in a petroleum refinery. *Chem. Engng Prog.*, **75**(2), 51

DANGREAUX, J. (1986). Destructive explosion effects. *Loss Prevention and Safety Promotion 5*, 9.1

DANICIC, I. (1983). *Lisp Programming* (Oxford: Blackwell)

DANIEL, J., PUFFLER, M. and STRIZENEC, M. (1971). Analysis of operator's work at different levels of automated production. *Studia Psychologica*, 13, 326. See also Edwards, E. and Lees, F.P. (1974), *op. cit.*, p. 159

DANIEL, J.S.S. (1989). *Piper Alpha Public Inquiry. Transcript, Day 149*

DANIEL, R.L. (1973). Chlorination hazards. *Loss Prevention*, **7**, 74

DANIELLOU, F. and BOEL, M. (1984). Automated process control: the roles of computer-available information and field information. *Ergonomic Problems in Process Operations* (Rugby: Instn Chem. Engrs), p. 59

DANIELS, B.K. (1982). Data banks for events, incidents and reliability. In Green, A.E. (1982), *op. cit.*, p. 259

DANIELS, B.K. (1983a). Error, fault and failure in software. *Instn Chem. Engrs, Conf. on Human Reliability in the Process Control Centre* (Rugby)

DANIELS, B.K. (1983b). Software reliability. *Reliab. Engng*, 4, 199

DANIELS, B.K. (ed.) (1986). *Safety and Reliability of Programmable Electronic Systems* (London: Elsevier Appl. Sci.)

DANIELS, B.K. (ed.) (1987). *Achieving Safety and Reliability with Computer Systems* (London: Elsevier Appl. Sci.)

DANIELS, E.J. and ANDERSON, P.J. (1977). International LNG projects continue to progress as new plans evolve. *Pipeline and Gas J.*, Jun. 29

DANIELS, J.D. (1984). Removal of particulate matter from air by the use of wet scrubbers. *Chemy Ind.*, Feb. 20, 126

DANIELS, J.T. and HOLDEN, P.L. (1983). Quantification of risk. *Loss Prevention and Safety Promotion 4*, vol. 1, p. G33

DANIELS, T. (1973). *Thermal Analysis* (New York: Halstead Press)

DANIELSEN, L. (1993). The problems of an industrial supplier — and how to achieve quality. In Bennett, P. (1993), *op. cit.*, p. 205

DANIELSON, G.L. (1964). Handling chlorine — 1. Tank car quantities. *Chem. Engng Prog.*, **60**(9), 86

DANILOV, B. (1981). Examples of corrosion control. Hydrocarbon Process., **60**(2), 95; **60**(3), 115; **60**(4), 192

DANN, J.N. and HENSON, T.L. (1978). Computer simulation of a chemical plant. *Chem. Engng Prog.*, **74**(6), 64

DANSKIN, N. (1969). Commercial vehicles. In Handley, W. (1969), *op. cit.*, p. 51

VAN DANTZIG, D. (1960). The economic decision problems concerning the security of the Netherlands against storm surges. *Report of the Delta Committee*, vol. 3 (The Hague: Netherlands Govt Printing Office)

DARBY, M.S. and MOBBS, S.D. (1988). Observations of internal gravity waves in the stably stratified atmospheric boundary layer. In Puttock, J.S. (1988b), *op. cit.*, p. 55

DARILEK, G.T. and PARRA, J.O. (1989). The electrical leak location method for geomembrane liners. *J. Haz. Materials*, **20**, 177

DARLING, H.L. (1967). Insulation fire on a storage sphere. *Ammonia Plant Safety*, **9**, 16

DARLING, J. (1991). Stresswaves. *Plant Engr*, **35**(Jan./Feb.), 28

DARLING, J.A. (1974). The encouragement of the use of low-hazard explosives. *Loss Prevention and Safety Promotion. 1*, p. 287

DARMER, K.I., HAUN, C.C. and MACEWAN, J.D. (1972). The acute inhalation toxicology of chlorine pentafluoride. *Am. Ind. Hyg. Ass. J.*, **33**, 661

DARMER, K.I., KINKEAD, E.R. and DIPASQUALE, L.C. (1972). *Acute Toxicity in Rats and Mice Resulting from Exposure to HCl Gas and HCl Aerosol for 5 and 30 Minutes. Rep. AMRL-TR-72–21.* US Air Force, Aerospace Med. Div., Aerospace Med. Res. Lab, Wright Patterson Air Force Base

DARMER, K.I., KINKEAD, E.R. and DIPASQUALE, L.C. (1974). Acute toxicity in rats and mice exposed to hydrogen chloride. *Am. Ind. Hyg. Assoc. J.*, **35**, 623

DARNELL, A.J. (1978). Disposal of spilled hazardous materials by the bromination process. *Control of Hazardous Material Spills*, 4, 221

DARTNELL, R.C. and VENTRONE, T.A. (1971). Explosion of a *para*-nitrometacresol unit. *Loss Prevention*, 5, 53. See also *Chem. Engng Prog.*, *1971*, **67**(6), 58

DARTON, R. (1986). Letting it flow. *Chem. Engr, Lond.*, **424**, 28

DARTT, C.B. and DAVIS, M.E. (1994). Catalysis for environmentally benign processing. *Ind. Eng. Chem. Res.* **33**, 2887–99

DARTT, S.R. (1982). Distributed digital control — its pros and cons. *Chem. Engng*, **89**(Sep. 6), 111

DASARATHY, B. and YANG, C. (1980). A transformation on ordered trees. *Computer J.*, **23**, 161

DASGUPTA, P.K. and SHEN DONG (1986). Solubility of ammonia in liquid water and generation of trace levels of standard gaseous ammonia. *Atmos. Environ.*, **20**, 565

DASSAU, W.R. and WOLFGANG, G.H. (1964). Reactor design considers stability and control. *Hydrocarbon Process.*, **43**(12), 76

DASSONVILLE, C. and LECHAT, J.F. (1989). Decommissioning and entry of two 35 000 m³ LNG tanks at Fos-sur-Mer after 15 years in service. *Gastech 88*, paper 3.3

DATE, C.J. (1977). *An Introduction to Database Systems, 3rd ed.* (Reading, MA: Addison-Wesley)

DAUBERT, T.E. (1980). State-of-the-art: property predictions. *Hydrocarbon Process.*, **59**(3), 107

DAUBERT, T.E. and DANNER, R.P. (eds) (1985–). *Data Compilation Tables of Properties of Pure Compounds* (New York: Am. Inst. Chem. Engrs)

DAUERMAN, L. (1982). Where does liability begin and end? *Chem. Engng Prog.*, **78**(10), 11

DAUGHERTY, M.L., WATSON, A.P. and VO-DINH, T. (1992). Currently available permeability and breakthrough data characterizing chemical warfare agents and their simulants in civilian protective clothing materials. *J. Haz. Materials*, **30**, 243

DAUGHERTY, V. (1983). The identification and safe life prediction of hazardous materials. *J. Haz. Materials*, **7**(3), 247

DAUTREBANDE, L. (1933). *Les Gaz Toxiques* (Paris: Masson)

DAVE, N.D. (1987). Loss-control representatives. What they do. *Chem. Engng*, **94**(7), 49

DAVENPORT, G.B. (1992a). The ABCs of hazardous waste legislation. *Chem. Engng Prog.*, **88**(5), 45

DAVENPORT, G.B. (1992b). Understand the air pollution laws that affect CPI plants. *Chem. Engng Prog.*, **88**(4), 30

DAVENPORT, G.B. (1992c). Understand the water pollution laws governing CPI plants. *Chem. Engng Prog.*, **88**(9), 30

DAVENPORT, J.A. (1977a). Prevent vapor cloud explosions. *Hydrocarbon Process.*, **86**(3), 205

DAVENPORT, J.A. (1977b). A survey of vapor cloud incidents. *Loss Prevention*, **11**, 39

DAVENPORT, J.A. (1977c). Private communication

DAVENPORT, J.A. (1981a). Examine your pre-emergency plan. *Hydrocarbon Process.*, **60**(5), 287

DAVENPORT, J.A. (1981b). Explosion losses in industry. *Fire J.*, Jan., 52

DAVENPORT, J.A. (1983). A study of vapor cloud incidents – an update. *Loss Prevention and Safety Promotion 4*, vol. 1, p. C1

DAVENPORT, J.A. (1986). Hazards and protection of pressure storage of liquefied petroleum gases. *Loss Prevention and Safety Promotion 5*, p. 22.1

DAVENPORT, J.A. (1987). Gas plant and fuel handling facilities: an insurer's view. *Plant/Operations Prog.*, **6**, 199

DAVENPORT, J.A. (1988). Hazards and protection of pressure storage and transport of LP-gas. *J. Haz. Materials*, **20**, 3

DAVENPORT, R. and BRAMELLER, A. (1972). *Optimum Pressure Control in Gas Networks for the Reduction of Losses. Commun. 882. Instn Gas Engrs, London.* See also *Instn Gas Engrs J.*, **12**, 272

DAVENPORT, T.J. (1991). A further survey of pressure vessel failures in the UK. *Reliability '91* (Culcheth, Warrington: Safety and Reliability Directorate)

DAVID, P. (1990). Thermodynamics and process design. *Chem. Engr, Lond.*, **473**, 31

DAVID, P. (1991). Looking after steam traps. *Chem. Engr, Lond.*, **489**, 28

DAVID, W. (1933a). Chlorine accident in Tilsit. *Med. Klin.*, **6**, 184

DAVID, W. (1933b). Industrial chlorine gas poisonings. *Vergiftungsfälle*, **4**, 145

DAVIDSON, A.T.G. and MACDONALD, I.F. (1977). *Evaluation of Reliability and Availability of Complex Systems. Rep. NCSR R12.* Nat. Centre for Systems Reliab., Culcheth

DAVIDSON, G.A. (1986). Gaussian versus top-hat profile assumptions in integral plume models. *Atmos. Environ.*, **20**, 471

DAVIDSON, G.A. and SLAWSON, P.R. (1982). Effective source flux parameters for use in analytical plume rise models. *Atmos. Environ.*, **16**, 223; 1983, **17**, 197

DAVIDSON, J. (ed.) (1988). *The Reliability of Mechanical Systems* (London: Instn Mech. Engrs)

DAVIDSON, J. (1989). *Piper Alpha Public Inquiry. Transcript, Day 63*

DAVIDSON, J.F. (1984). On public inquiries. *Chem. Engr, Lond.*, **404**, 31

DAVIDSON, J., CLIFT, R. and HARRISON, D. (eds) (1985). *Fluidization*, 2nd ed. (London: Academic Press)

DAVIDSON, R. (1970). *Peril on the Job – A Study of Hazards in the Chemical Industry* (Washington, DC: Public Affairs Press)

DAVIDSON, R. (1974). A union view of HPI safety. *Hydrocarbon Process.*, **53**(1), 159

DAVIDSON, R.L. (1967). Trend is to big plants. *Chem. Engng*, **74**(Jan. 2), 68

DAVIE, F.M., MORES, S., NOLAN, P.F. and HOBAN, T.W.S. (1993). Evidence of the oxidation of deposits in heated bitumen storage tanks. *J. Loss Prev. Process Ind.*, **6**, 145

DAVIE, F.M., NOLAN, P.F. and HOBAN, T.W.S. (1993). Study of iron sulfide as a possible ignition source in the storage of heated bitumen. *J. Loss Prev. Process Ind.*, **6**, 139

DAVIE, F.M., NOLAN, P.F. and HOBAN, T.W.S. (1994). Case histories of incidents in heated bitumen storage tanks. *J. Loss Prev. Process Ind.*, **7**, 217

DAVIE, F.M., NOLAN, P.F. and TUCKER, R.F. (1993). Mathematical model for the self-heating of deposits found in heated bitumen storage tanks. *J. Loss Prev. Process Ind.*, **6**, 203

DAVIE, R.A. (1989). *Piper Alpha Public Inquiry. Transcript, Day 114*

DAVIES, A. (1978). Assessing maintenance performance by cusum monitoring. *Plant Engr*, **22**(4), 15

DAVIES, A.C. (1984). *The Science and Practice of Welding*, 8th ed. (Cambridge: Cambridge Univ. Press)

DAVIES, A.D. (1987). Some tools for fire model validation. *Fire Technol.*, **23**, 95

DAVIES, A.E. (1972). Principles and practice of aircraft power plant maintenance. *Trans Inst. Mar. Engrs*, **84**, 441

DAVIES, A.G. (1972). *Organic Peroxides* (London: Butterworths)

DAVIES, B. (1993). The great Euro-transport fiasco. *Chem. Engr, Lond.*, **537**, 16

DAVIES, B.O. and ROYCE, A.T. (1961). Bulk handling of liquid sulphur trioxide. *Chem. Engr, Lond.*, **155**, A47

DAVIES, C. (1970). *Calculations in Furnace Technology* (Oxford: Pergamon Press)

DAVIES, C. (1979). Computer-aided design. *Chemy Ind.*, Aug. 4, 505

DAVIES, C.A., KIPNIS, I.M., CHASE, M.W. and TREWEEK, D.N. (1985). The thermochemical and hazard data of chemicals: estimation using the ASTM CHETAH program. In Hoffman, J.M. and Maser, D.C. (1985), *op. cit.*, p. 81

DAVIES, C.N. (1954). *Dust is Dangerous* (London: Faber & Faber)

DAVIES, C.N. (ed.) (1961). *Inhaled Particles and Vapours*, vol. 1 (Oxford: Pergamon Press)

DAVIES, C.N. (1966). *Aerosol Science* (New York: Academic Press)

DAVIES, C.N. (ed.) (1967). *Inhaled Particles and Vapours*, vol. 2 (Oxford: Pergamon Press)

DAVIES, C.N., DAVIS, P.R. and TYRER, F.H. (eds) (1967). *The Effects of Abnormal Physical Conditions at Work* (Edinburgh: E & S Livingstone)

DAVIES, D.G. (1967). A psycho-physiological investigation of process control skill. *MSc Thesis, Univ. of Aston in Birmingham*

DAVIES, D.K. (1967). The generation and dissipation of static charge on dielectrics in a vacuum. *Static Electrification*, **2**, 29

DAVIES, D.K. (1970). Charge generation on solids. *Advances in Static Electricity*, **1**, 10

DAVIES, D.L. (1975). Labelling for user protection. *Distribution Conf.* (London: Chem. Ind. Ass.), p. 27

DAVIES, D.R. and MACKAY, G.A. (1976). Recent developments in the transport of liquid wastes. *J. Haz. Materials*, **1**(3), 199

DAVIES, D.S. (1962). *Nomography and Empirical Equations* (New York: Reinhold)

DAVIES, D.S. (1989). *Piper Alpha Public Inquiry. Transcript, Days 153, 154*

DAVIES, G.O. (1977). How to make the correct economic decision on spare equipment. *Chem. Engng*, **84**(Nov. 21), 187

DAVIES, I.L. (1980). Damage effects from the impact of missiles against reinforced concrete structures. *Nucl. Engng*, **19**, 199

DAVIES, J. (1977). A simulator to train process operators of substitute for natural gas. *Measmt Control*, **10**, 350

DAVIES, J.K.W. (1987). A comparison between the variability exhibited in small scale experiments and in the Thorney Island Phase I trials. *J. Haz. Materials*, **16**, 339

DAVIES, J.K.W. (1989a). The application of box models in the analysis of toxic hazards by using the probit dose–response relationship. *J. Haz. Materials*, **22**(3), 319

DAVIES, J.K.W. (1989b). Probabilistic methods in the development of mathematical models of system failure. In Cox, R.A. (1989), *op. cit.*, p. 47

DAVIES, J.K.W. (1992). The CEC Major Technological Project BA: research on continuous and instantaneous heavy gas clouds. *J. Haz. Materials*, **31**, 177

DAVIES, J.P., PALMER, K.N. and ROGOWSKI, Z.W. (1973). Performance of metallic foams as flame arresters. *Fire Prev. Sci. Technol.*, **5**, 14

DAVIES, J.T. (1972). *Turbulence Phenomena* (New York: Academic Press)

DAVIES, J.W.L. (1982). *Physiological Responses to Burn Injury* (London: Academic Press)

DAVIES, K.R. (1983). Computerised data base for the storage and retrieval of abnormal occurrence reports. *Loss Prevention and Safety Promotion 4*, vol. 1, p. K1

DAVIES, K.R. (1985). Techniques for the identification and assessment of major accident hazards. *The Assessment and Control of Major Hazards* (Rugby: Instn Chem. Engrs), p. 289

DAVIES, M. (1949). *The Physical Principles of Gas Liquefaction and Low Temperature Rectification* (London: Longmans Green)

DAVIES, M.E. (1985). *The Validity of Densimetric Froude Number Scaling in Heavy Gas Clouds. Rep. TPRD/L/2945/N85.* Central Elec. Generating Board, Leatherhead

DAVIES, M.E. (1989). *Piper Alpha Inquiry Transcript, Days 116, 117*

DAVIES, M.E. and INMAN, P.N. (1987). A statistical examination of wind tunnel modelling of the Thorney Island trials. *J. Haz. Materials*, **16**, 149

DAVIES, M.E. and SINGH, S. (1985a). The Phase II trials: a data set on the effect of obstructions. *J. Haz. Materials*, **11**(1/3), 301. See also in McQuaid, J. (1985), *op. cit.*, p. 301

DAVIES, M.E. and SINGH, S. (1985b). Thorney Island: its geography and meteorology. *J. Haz. Materials*, **11**(1/3), 91. See also in McQuaid, J. (1985), *op. cit.*, p. 91

DAVIES, N.V. and TEASDALE, P. (1994). *The Costs to the British Economy of Work Accidents and Work-related Ill Health* (London: Health and Safety Executive)

DAVIES, P.A. (1990). A methodology for the quantitative risk assessment of the road and rail transport of explosives. Ph.D. Thesis, Loughborough Univ. of Technol.

DAVIES, P.A. (1993). A guide to the evaluation of condensed phase explosions. *J. Haz. Materials*, **33**, 1

DAVIES, P.A. and LEES, F.P. (1991a). Accident speed of freight trains. *J. Haz. Materials*, **28**, 367

DAVIES, P.A. and LEES, F.P. (1991b). Impact speed of goods vehicles. *J. Haz. Materials*, **26**, 213

DAVIES, P.A. and LEES, F.P. (1992). The assessment of major hazards: the road transport environment for conveyance of hazardous materials in Great Britain. *J. Haz. Materials*, **32**, 41

DAVIES, P. and HYMES, I. (1985). Chlorine toxicity criteria for hazard assessment. *Chem. Engr, Lond.*, **415**, 30; **416**, 7

DAVIES, P.C. and PURDY, G. (1986). Toxic gas risk assessments – the effects of being indoors. In *Refinement of Estimates of the Consequences of Heavy Toxic Vapour Release* (Rugby: Instn Chem. Engrs), p. 1.1

DAVIES, P.C. and PURDY, G. (1987). The assessment of major hazards: the lethal toxicity of bromine (letter). *J. Haz. Materials*, **14**, 273

DAVIES, R. (1979). Air oxidation process hazards – an overview. *Loss Prevention*, **12**, 103

DAVIES, R. (1982). Seismic risk to liquefied gas storage plant in the United Kingdom. *The Assessment of Major Hazards* (Rugby: Instn Chem. Engrs), p. 245

DAVIES, R. (1983). ICI Paint's new £20 resin plant. *Chem. Engr, Lond.*, **392**, 85

DAVIES, R.M. and TAYLOR, SIR G. (1950). The mechanisms of large bubbles rising through extended liquids and through liquids in tubes. *Proc. R. Soc., Ser. A*, **200**, 375

DAVIS, A.G. and GODFREY, P.R. (1991). The regulatory implications of the use of glass reinforced plastics as a structural material on platform topsides. *Safety Developments in the Offshore Oil and Gas Industry* (London: Instn Mech. Engrs), p. 61

DAVIS, B. (1985). Upgrading measurement systems: one key to company success. *Process Engng*, **66**(1), 25

DAVIS, B.C. (1983). Flare efficiency studies. *Plant/Operations Prog.*, **2**, 191

DAVIS, B.C. (1985). US EPA's flare policy: update and review. *Chem. Engng Prog.*, **81**(4), 7

DAVIS, B.C. and BAGSTER, D.F. (1989–). The computation of view factors of fire models. *J. Loss Prev. Process Ind.*, 1989, **2**(4), 224; 1990, **3**(3), 327

DAVIS, D. (1993). Legal liability. In Bennett, P. (1993), *op. cit.*, p. 35

DAVIS, D.R. (1958). Human errors and transport accidents. *Ergonomics*, **2**, 24

DAVIS, D.S. (1961). Emergency vent size for tanks containing flammable liquids (nomogram). *Br. Chem. Engng*, **6**, 644

DAVIS, E.J. and AKE, J.A. (1973). Equilibrium thermochemistry computer programs as predictors of energy hazard potential. *Loss Prevention*, **7**, p. 28

DAVIS, E.J., HAMPSON, B.H. and YATES, B. (1981). Analysis of the development of runaway reactions in powders. *Runaway Reactions, Unstable Products and Combustible Powders* (London: Instn Chem. Engrs), paper 1/D

DAVIS, F.E. (1976). *New Techniques in Loss Control Management* (Hawthorn, Vic.: Hawthorn Press)

DAVIS, G.A., KINCAID, L., MENKE, D., GRIFFITH, B., JONES, S., BROWN, K. and GOERGEN, M. (1994). *The Product Side of Pollution Prevention: Evaluating the Potential for Safe Substitutes.* (Cincinnati, Ohio: Risk Reduction Engineering Laboratory, Office of Research and Development, U. S. Environmental Protection Agency)

DAVIS, G.D. (1987). Investigation and repair of an auxiliary boiler explosion. *Ammonia Plant Safety*, **27**, 95. See also *Plant/Operations Prog.*, **6**, 42

DAVIS, H. (1983). Evaluating multistage centrifugal compressors. *Chem. Eng*, **90**(Dec. 26), 35

DAVIS, H. (1990). Pump selection. *Plant Engr*, Nov./Dec., 41

DAVIS, H.M. (1971). How to specify centrifugal compressors. Inspection, shipping and erection. *Hydrocarbon Process.*, **50**(10), 86

DAVIS, H.M. (1974a). How to improve compressor operation and maintenance. *Hydrocarbon Process.*, **53**(1), 93

DAVIS, H.M. (1974b). Test compressor performance – in the field. *Hydrocarbon Process.*, **53**(4), 138

DAVIS, J. (1973). LNG: growth or safety? *Chem Engng*, **80**(May 28), 50

DAVIS, J.C. (1978). Are bigger plants better? No small debate. *Chem. Engng*, **85**(May 22), 49

DAVIS, J.C. (1979). Alaskan refinery challenges winter's fury. *Chem. Engng*, **86**(Feb. 26), 82

DAVIS, K.G. and MANCHANDA, K.D. (1974). Unit operations for drying fluids. *Chem. Engng*, **81**(Sep. 16), 102

DAVIS, L.E. and WACKER, G.J. (1987). Job design. In Salvendy, G. (1987), *op. cit.*, p. 431

DAVIS, L.N. (1979). *Frozen Fire* (San Francisco, CA: Friends of the Earth)

DAVIS, L.N. (1984). *The Corporate Alchemists* (Hounslow, Middlesex: Maurice Temple Smith)

DAVIS, M.R. (1991). Compressible gas–liquid mixture flow at abrupt pipe enlargements. *Exptl Thermal Fluid Sci.*, **4**, 684

DAVIS, P.A., REIMER, A., SAKIYAMA, S.K. and SLAWSON, P.R. (1986). Short-range atmospheric dispersion over a heterogenous surface – I. *Atmos. Environ.*, **20**, 41

DAVIS, P.O. and DOTSON, C.O. (1987). Physiological aspects of fire fighting. *Fire Technol.*, **23**, 280

DAVIS, P.R. (1983). Human factors contributing to slips, trips and falls. *Ergonomics*, **26**, 51

DAVIS, R. (1977). Interactive transfer of expertise: acquisition of new inference rules. *Proc. Int. Joint Conf. on Artificial Intelligence*, **5**, 321

DAVIS, R. (1978). Knowledge acquisition in rule-based systems: knowledge about representation as a basis for system construction and maintenance. In D.A. Waterman and F. Hayes-Roth (1978), *op. cit.*, p. 99

DAVIS, R. (1980a). Application of meta-knowledge to the construction, maintenance and use of large knowledge bases. In Davis, R. and Lenat, D. (1980), *op. cit.*, p. 229

DAVIS, R. (1980b). Meta-rules: reasoning about control. *Artificial Intelligence*, **15**, 179

DAVIS, R. (1982). Expert systems. Where are we? And where do we go from here? *AI Magazine*, **3**(2), 3

DAVIS, R. (1984). Diagnostic reasoning based on structure and behaviour. *Artificial Intelligence*, **24**, 347

DAVIS, R., BUCHANAN, B.G. and SHORTLIFFE, E.H. (1977). Production rules as a representation for a knowledge-based consultation program. *Artificial Intelligence*, **8**, 15

DAVIS, R. and KING, J. (1976). An overview of production systems. In *Machine Intelligence 8* (ed. E.W. Elcock and D. Michie) (Chichester: Ellis Horwood)

DAVIS, R. and LENAT, D.B. (1982). *Knowledge-Based Systems in Artificial Intelligence* (New York: McGraw-Hill)

DAVIS, R.K., HORTON, A.W., LARSON, E.E. and STEMMER, K.L. (1963). Inhalation of tetramethyl lead and tetraethyl lead. *Arch. Environ. Health*, **6**, 473

DAVIS, R.W., MOORE, E.F., SANTORO, R.J. and NESS, J.R. (1990). Isolation of buoyancy effects in jet diffusion flame experiments. *Combust. Sci. Technol.*, **73**, 625

DAVIS, T.L. (1943). *Chemistry of Powder and Explosives* (New York: Wiley)

DAVISON, C. (1924). *A History of British Earthquakes* (Cambridge: Cambridge Univ. Press)

DAVISON, E. (1977). Road and sea interface situations. *Symp. Hazard 77 – Advances in the Safe Handling and Movement of* Polytechnic, Middlesborough

DAVISON, L. and LORD, H. (1974). Some aspects of fire and explosion hazards in enclosure-protected electrical apparatus. *Fire Prev. Sci. Technol.*, 1013

DAVISON, R.M. and MISKA, K.H. (1979). Stainless steel heat exchangers. *Chem. Engng*, **86**(Feb. 12), 129; (Mar. 12), 111

DAVY, J.R. (1951). *Industrial High Vacuum* (London: Pitman)

DAWES, J.G. and DAVIES, J.K.W. (1977). Significant numbers of accidents. *J. Occup. Accid.*, **1**(2), 123

DAWSON, G.W. (1975). Treatment of hazardous materials spills in flowing streams with floating mass transfer agents. *J. Haz. Materials*, **1**(1), 65

DAWSON, G.W. (1983). Risk management and the landfill in hazardous waste disposal. *J. Haz. Materials,* **9**(1), 43

DAWSON, G.W. and MCNEESE, J.A. (1978). Advanced system for the application of buoyant media to spill contaminated waters. *Control of Hazardous Material Spills*, 4, p. 325

DAWSON, G.W., MCNEESE, J.A. and CHRISTENSEN, D.C. (1978). An evaluation of alternatives for the removal/destruction of Kepone residuals in the environment. *Control of Hazardous Material Spills*, 4, p. 244

DAWSON, G.W. and MERCER, B.W. (1986). *Hazardous Waste Management* (New York: Wiley)

DAWSON, G.W., MERCER, B.W. and PARKHURST, R.G. (1976). Comparative evaluation of *in situ* approaches to the treatment of flowing streams. *Control of Hazardous Material Spills*, 3, p. 266

DAWSON, G.W., SHUCKROW, W.J. and MERCER, B.W. (1972). Strategy for treatment of waters contaminated by hazardous materials. Control of Hazardous Material Spills, 1, p. 141

DAWSON, J.D. (1979). Designing high temperature piping. *Chem. Engng*, **86**(Apr. 9), 127

DAWSON, J.E. (1960a). Keys to corrosion-wise reactor design. *Chem. Engng*, **67**(May 16), 170

DAWSON, J.E. (1960b). Often overlooked reactor design factors. *Chem. Engng*, 67(May 2), 150

DAWSON, J.L., EDEN, D.A. and HLADKY, H. (1983). The use of electrochemical noise for corrosion monitoring. In Butcher, D.W. (1983), *op. cit.*, p. 119

DAWSON, P. (1986). Pilot plant design and operational concepts. *Chemy Ind.*, Feb. 3, 99

DAWSON, P., LAMB, B. and STOCK, D. (1986). A numerical model to simulate the atmospheric transport and diffusion of pollutants over complex terrain. *Fifth Joint Conf. on Applications of Air Pollution Meteorology* (Boston, MA: Am. Met. Soc.), p. 223

DAWSON, R.F.F. (1967). *Costing Road Accidents in Great Britain. Rep. LR79.* Road Res. Lab, Crowthorne

DAWSON, R.F.F. (1971). *Current Costs of Road Accidents in Great Britain. Rep. LR396.* Road Res. Lab., Crowthorne

DAWSON, S., POYNTER, P. and STEVENS, D. (1982). Activities and outcomes: monitoring safety performance. *Chemy Ind.*, Oct. 16, 790

DAWSON, S., POYNTER, P. and STEVENS, D. (1985). Safety specialists in industry: roles, constraints and opportunities. *J. Occup. Behaviour*, **5**, 253

DAWSON, S., WILLMAN, P., CLINTON, A. and BAMFORD, M. (1988). *Safety at Work: The Limits of Self-Regulation* (Cambridge: Cambridge Univ. Press)

DAWSON, T.H. (1983). *Offshore Structural Engineering* (Englewood Cliffs, NJ: Prentice-Hall)

DAWSON, T.J. (1956). Behaviour of welded pressure vessels in agricultural ammonia service. *Welding J.*, **35**, 568

DAY, A.T. (1978). Water treatment. *Plant Engr*, **22**(8), 18

DAY, B.B. (1949). Application of statistical methods to research and development in engineering. *Rev. Inst. Int. Statist.*, **17**, 129

DAY, B.F. (1978). Ammonia transportation accident report. *Ammonia Plant Safety*, **20**, 30

DAY, B.F., FORD, R.D. and LORD, P. (1969). Building Acoustics (Amsterdam: Elsevier)

DAY, J. (1989). *Piper Alpha Public Inquiry. Transcript, Day 141*

DAY, M.J. and MOOREHOUSE, J. (1973). The effectiveness of natural ventilation in preventing gas accumulations in heating plant. *Instn Gas Engrs, London, Commun. 915.* See also *Instn Gas Engrs J.*, **13**, 270

DAYAN, A. and TIEN, C.L. (1974). Radiant heating from a cylindrical fire column. *Combust. Sci. Technol.*, **9**, 41

DAZE, R.E. (1965). How to instrument centrifugal compressors. *Hydrocarbon Process.*, **44**(10), 125

DEACON, A.R. (1988). Deadly maintenance. *Plant Engr*, **32**(2), 24

DEACON, E.L. (1949). Vertical diffusion in the lowest layers of the atmosphere. *Quart. J. R. Met. Soc.*, **75**, 89

DEACON, E.L. (1957). Wind profiles and the shearing stress – an anomaly resolved. *Quart. J. R. Met. Soc.*, **83**, 537

DEACON, E.L. (1977). Gas transfer to and across an air – water interface. *Tellus*, **29**, 363

DEAN, B.V. (1963). Replacement theory. *Progress in Operations Research*, **1**, 327

DEAN, J.A. (ed.) (1978). *Lange's Handbook of Chemistry*, 12th ed. (New York: McGraw-Hill)

DEAN, J.C., HULBURT, D.A., MATTHEWS, A.L. and WILLIAMS, G.M. (1992). Quantitative evaluation of electrostatic hazards. *Hazard Identification and Risk Analysis, Human Factors and Human Reliability in Process Safety*, p. 85

DEAN, M.M. (1987). *Process Modelling* (London: Longman)

DEAN, S.W. (1989). Assuring plant reliability through optimum materials selection. *Chem. Engng Prog.*, **85**(6), 36

DEAR, T. (1974). Noise control in large gas handling plants. *Ammonia Plant Safety*, **16**, p. 108

DEARDORFF, J.W. (1970). A three dimensional numerical investigation of the idealized planetary boundary layer. *J. Geophys. Fluid Dyn.*, **1**, 377

DEARDORFF, J.W. (1973). Three-dimensional numerical modeling of the planetary boundary layer. In Haugen, D.A. (1973), *op. cit.*, p. 217

DEARDORFF, J.W. (1974). Three-dimensional numerical study of the height and mean structure of a heated planetary boundary layer. *Boundary Layer Met.*, **7**, 81

DEARDORFF, J.W. and WILLIS, G.E. (1974). Physical modeling of diffusion in the mixed layer. *First Symp. on Atmospheric Diffusion and Air Pollution* (Boston, MA: Am. Met. Soc.), p. 387

DEARDORFF, J.W. and WILLIS, G.E. (1975). A parameterization of diffusion into the mixed layer. *J. Appl. Met.*, **14**, 1451

DEARDORFF, J.W. and WILLIS, G.E. (1984). Ground-level concentration fluctuations from a buoyant and a non-buoyant source within a laboratory convectively mixed layer. *Atmos. Environ.*, **18**, 1297

DEASON, W.R., KOERNER, W.E. and MUNCH, R.H. (1959). Determining maximum heat load in equipment design: determination of heat evolution rates; effects of ring substituents and added contaminants on nitrobenzenes. *Ind. Engng Chem.*, **51**, 1001

DEASY, K. (1979). Detailed study of the causes of dust explosions. *Fire Int.*, **66**, 18

DEAVES, D.M. (1983a). Application of a turbulence flow model to heavy gas dispersion in complex situations. In Hartwig, S. (1983), *op. cit.*, p. 91

DEAVES, D.M. (1983b). Experimental and computational assessment of full-scale water spray barriers for dispersing dense gases. *Loss Prevention and Safety Promotion 4*, vol. 1, p. F45

DEAVES, D.M. (1984). Application of advanced turbulence models in determining the structure and dispersion of heavy gas clouds. In Ooms, G. and Tennekes, H. (1984), *op. cit.*

DEAVES, D.M. (1985). 3-Dimensional model predictions for the upwind building trial of Thorney Island Phase II. *J. Haz. Materials*, **11**(1/3), 341. See also in McQuaid, J. (1985), *op. cit.*, p. 341

DEAVES, D.M. (1986). Some implications of gas dispersion and wind flows on equipment siting on offshore process installations. *Loss Prevention and Safety Promotion 5*, p. 31–1

DEAVES, D.M. (1987a). Development and application of heavy gas dispersion models of varying complexity. *J. Haz. Materials*, **16**, 427

DEAVES, D.M. (1987b). Incorporation of the effects of buildings and obstructions on gas cloud consequence analysis. *Vapor Cloud Modeling*, p. 844

DEAVES, D.M. (1989). Gas dispersion: applications and new approaches. *J. Loss Prev. Process Ind.*, **2**(1), 39

DEAVES, D.M. (1992). Dense gas dispersion modelling. *J. Loss Prev. Process Ind.*, **5**, 219

DEAVES, D.M. and HALL, R.C. (1990). The effects of sloping terrain on dense gas dispersion. *J. Loss Prev. Process Ind.*, **3**(1), 142

DEB, S.K., EVANS, B.L. and YOFFE, A.D. (1962). Ignition and sensitized reactions in explosive inorganic azides. *Combustion*, **8**, p. 829

DEBOLD, T.A. (1991). Select the right stainless steel. *Chem. Engng Prog.*, **87**(11), 38

DECHAUX, J.C. and DELFOSSE, L. (1979). The negative temperature coefficient in the C_2 to C_{13} hydrocarbon oxidation. *Combust. Flame*, **34**, 161, 169

DECHAUX, J.C., FLAMENT, J.L. and LUCQUIN, M. (1971). Negative temperature coefficient in the oxidation of butane and other hydrocarbons. *Combust. Flame*, **17**, 205

DECHEMA (1953–). *Werkstoff-Tabelle*, 3rd ed. (Frankfurt)

DECHEMA (1987–). *DECHEMA Corrosion Handbook* (New York: VCH)

DECKER, D.A. (1971). Explosion venting guide. *Fire Technol.*, **7**(3), 219

DECKER, D.A. (1974). An analytical method for estimating overpressure from theoretical atmospheric explosions. *Nat. Fire Prot. Ass., Ann Mtg*

DECKERT, A.H. (1986). Emergency responses to spills of sulfuric acid during transit. *Chem. Engng Prog.*, **82**(7), 11

DECLERCK, D.H. and PATARCITY, A.J. (1986). Guidelines for selecting corrosion-resistant materials. *Chem. Engng*, **93**(22), 46

DECORTIS, F. (1993). Operator strategies in a dynamic environment in relation to an operator model. *Ergonomics*, **36**, 1291

DEES, J.P. and TAYLOR, R. (1990). Health care management – a tool for the future? *AAOHN J.*, **38**(2), 52

DEFENSE TECHNICAL INFORMATION CENTER (1980). *Methodology for Chemical Hazard Prediction.* Rep. DDESB TP 10, Change 3. Explosives Safety Board, Defense Logistics Agency, Dept of Defense, Alexandra, VA

DEGANCE, A.E. and ATHERTON, R.W. (1970). Chemical engineering aspects of two-phase flow. *Chem. Engng*, **77**(Mar. 23), 135; (Apr. 20), 151; (May 4), 113; (Jul. 13), 95; (Aug. 10), 119; (Oct. 5), 87; (Nov. 2), 101

DEGHETTO, K. and LONG, W. (1966). A new design method to check to towers for dynamic stability. *Hydrocarbon Process.*, **45**(2), 143

DEGOOD, R. (1986). Explosion venting (letter). *Chem. Engng*, **93**(Jan. 6), 5

DEGOOD, R. and CHATRATHI, K. (1991). Comparative analysis of test work studying factors influencing pressure developed in vented deflagrations. *J. Loss Prev. Process Ind.*, **4**, 297

DE GREENE, K.B. (1970). Systems analysis techniques. *Systems Psychology* (ed. K.B. De Greene) (New York: McGraw-Hill)

DEGUINGAND, B. and GALANT, S. (1981). Upper flammability limits of coal dust–air mixtures. *Combustion*, **18**, p. 705

DEHAVEN, E.S. (1979). Using kinetics to evaluate reactivity hazards. *Loss Prevention*, **12**, p. 41

DEHN, J. (1972). *Thermal Effects which may Result in the Unexpected Ignition of Gaseous Mixtures. Memo Rep. 2179.* US Army, Ballistics Res. Lab., Aberdeen Proving Ground, MD, Memo

DEHN, J.T. (1973). Modeling the ignition of fuel-oxidiser systems. *Loss Prevention*, **7**, p. 38

DEHN, J.T. (1975). Flammable limits over liquid surfaces. *Combust. Flame*, **24**, 231

DEICHMAN, W.B. and GERARDE, H.W. (1969). *Toxicology of Drugs and Chemicals*, 4th ed. (New York: Academic Press)

DEIESO, D.A., MULVEY, N.P. and KELLY, J. (1988). Accidental release prevention: regulator's perspective. *Preventing Major Chemical and Related Process Accidents*, p. 435

VON DEIMLING (1931). *Errinerungen aus meinem Leben* (Paris: Montaigne)

DEISLER, P.F. (1984a). A methodology for reducing industrially related cancer risk. In Deisler, P.F. (1984), *op. cit.*, p. 135

DEISLER, P.F. (ed.) (1984b). *Reducing the Carcinogenic Risks in Industry* (New York: Dekker)

DEJA, D. (1989). Programmable controllers. *Plant Engr*, Jul./Aug., 18

DEKKER, C.M. and SLIGGERS, C.J. (1992). Good manufacturing practice for modelling air pollution: quality criteria for computer models to calculate air pollution. *Atmos. Environ.*, **26A**, 1019

DEKTAR, C. (1974). River of blazing gasoline on freeway. *Fire Engng*, **127**(Mar.), 42

DELBOURGO, R. and LAFFITTE, P. (1957). Some recent experimental results on the flammability regions of various gases and combustible vapours at low pressures. *Combustion*, **6**, p. 288

DELBRIDGE, P. (1990). Public participation in risk assessment. *Plant/Operations Prog.*, **9**, 183

DELBOY, W.J., DUBNANSKY, R.F. and LAPP, S.A. (1991). Sensitivity of process risk to human error in an ammonia plant. *Plant/Operations Prog.*, **10**, 207

DELICHATSIOS, M.A. (1976). Fire growth rates in wood cribs. *Combust. Flame*, **27**, 267

DELICHATSIOS, M.A. (1980). Strong turbulent buoyant plumes. 1, Similarity. *Combust. Sci. Technol.*, **24**, 191

DELICHATSIOS, M.A. (1982). Exposure of steel drums to an external spill fire. *Plant/Operations Prog.*, **1**, 37

DELICHATSIOS, M.A. (1985). Exact solution for the rate of creeping flame spread over thermally thin materials. *Combust. Sci. Technol.*, **44**, 257

DELICHATSIOS, M.A. (1987) Air entrainment into buoyant jet flames and pool fires. *Combust. Flame*, **70**, 33

DELICHATSIOS, M.A. (1988). On the similarity of velocity and temperature profiles in strong (variable density) turbulent buoyant plumes. *Combust. Sci. Technol.*, **60**, 253

DELICHATSIOS, M.A. (1993a). Smoke yields from turbulent buoyant jet flames. *Fire Safety J.*, **20**, 299

DELICHATSIOS, M.A. (1993b). Transition from momentum to buoyancy controlled turbulent jet diffusion flames and flame height relationships. *Combust. Flame*, **92**, 349

DELICHATSIOS, M.A. and ORLOFF, L. (1985). Entrainment measurements in turbulent buoyant jet flames and implications for modeling. *Combustion* **20**, p. 367

DELICHATSIOS, M.A. and ORLOFF, L. (1989). Effects of turbulence on flame radiation from diffusion flames. *Combustion* **22**, 1271

DELIQUET, A. (1979). Computer program derives equations. *Hydrocarbon Process.*, **58**(3), 140

DELLNER, W.J. (1981). The user's role in automated fault detection and system recovery. In Rasmussen, J. and Rouse, W.B. (1981), *op. cit.*, p. 487

DELONEY, H.C. (1972). An evaluation of intrinsically safe instrumentation. *Chem. Engng*, **79**(May 29), 67

DELONG, D.F. (1971). Removal and replacement of catalyst in an ammonia converter. *Ammonia Plant Safety*, **13**, 47

DELPLACE, M. and VOS, E. (1983). Electric short circuits help the investigators determine where the fire started. *Fire Technol.*, **19**, 185

DELSETH, R. (1998). Production industrielle avec le phosgene. *Chemia* **52**, (Dec. 12), 698–701

DELUMYEA, R.D., MOORE, K.P., MILLER, G.I., DUKES, S.A., MCKINNEY, K.J. and MORGAN, M.A. (1986). Determination of rates of product combustion and soot formation during combustion of solvents commonly carried by rail. *J. Haz. Materials*, **13**(2), 169

DELUMYEA, R.D., MOORE, K.P. and MORGAN, M.A. (1986). Determination of products of combustion from fires involving solvents commonly carried by rail. *J. Haz. Materials*, **13**(3), 337

DEMAIO, J., PARKINSON, S.R. and CROSBY, J.V. (1978). A reaction time analysis of instrument scanning. *Hum. Factors*, **20**, 467

DEMAREST, K.D. (1971). Current status of reactor furnaces. *Chem. Engng Prog.*, **67**(1), 57

DEMAREST, K.D. (1975). Converting reformer furnaces to fuel oil firing. *Ammonia Plant Safety*, **17**, p. 24

DEMARS, K.R. *et al.* (1977). *Pipeline Failure: A Need for Improved Analysis and Site Surveys. OTC paper 2966*

DEMENA, H.F., LITWILLER, J.A., SIMPSON, M.P. and UNRUH, W. (1984). Nitrogen complex: ammonia plant operator training simulator. *Ammonia Plant Safety*, **24**, 171

DEMERJIAN, K.L., KERR, J.A. and CALVERT, J.G. (1974). The mechanism of photochemical smog formation. *Adv. Engng Sci. Technol.*, **4**, 382

DEMERY, W.P. (1977). OSHA – where it stands, where it's going. *Chem. Engng*, **84**(Apr. 11), 110

DEMING, W.E. (1982). *Quality, Productivity and Competitive Position* (Cambridge, MA: MIT Center for Advanced Engineering Study)

DEMING, W.E. (1986). *Out of the Crisis* (Cambridge, MA: MIT Center for Advanced Engineering Study)

DEMINSKI, R.M. (1964a). Development and application of non-lubricated reciprocating compressors. *Ammonia Plant Safety*, **6**, p. 33

DEMINSKI, R.M. (1964b). Development of non-interchangeable suction and discharge valves. *Ammonia Plant Safety*, **6**, p. 4

DEMINSKI, R.M. and HUNTER, F.J. (1962). Safety in compressor cylinder design and application. *Ammonia Plant Safety*, **4**, p. 44

DEMOREST, W.J. (1985). Pressure measurement. *Chem. Engng*, **92**(Sep. 30), 56

DEMPSTER, M.A.H., FISHER, M.L., JANSEN, L., LAGEWEG, B.J., LENSTRA, J.K. and RINNOOY KAN, A.H.G. (1981). Analytical evaluation of hierarchical planning systems. *Opers Res.*, **29**, 707

DENBIGH, K.G. (1965). *Chemical Reactor Theory* (Cambridge: Cambridge Univ. Press)

DENBIGH, K.G. and TURNER, J.C.R. (1971). *Chemical Reactor Theory*, 2nd ed. (Cambridge: Cambridge Univ. Press)

DENBIGH, K.G. and TURNER, J.C.R. (1984). *Chemical Reactor Theory*, 3rd ed. (Cambridge: Cambridge Univ. Press)

DENBOW, N. and BRIGHT, A.W. (1979). The design and performance of novel on-line charge-density monitors, injectors and neutralisers for use in fuel systems. *Electrostatics*, **5**, 171

DENEAU, K.S. (1983). Pyrolytic destruction of hazardous waste. *Plant/Operations Prog.*, **2**, 34

DENHAM, J.B., RUSSELL, J. and WILLS, C.M.R. (1968). How to design a 600 000 bbl tank. *Hydrocarbon Process.*, **47**(5), 137

DENISON, D.M. and TONKINS, W.J. (1967). *Further Studies upon Human Aspects of Fire in Artificial Gas Environments. Report.* RAF Inst. Aviation Medicine

DENISON, D.M. *et al.* (1968). Problem of fire in oxygen-rich surroundings. *Nature*, **218**, 1110

DENN, M.M. (1975). *Stability of Reaction and Transport Processes* (Englewood Cliffs, NJ: Prentice-Hall)

DENNETT, M.F. (1980). *Fire Investigation* (Oxford: Pergamon Press)

DENNING, P.J. (1986). Towards a science of expert systems. *IEEE Expert*, **1**(2), 60

DENNIS, D.M. (1978). A bankrupt disposal industry – who is responsible for cleanup. *Control of Hazardous Material Spills*, 4, p. 166

DENNIS, J.A. (1982). Ionizing radiation. In Green, A.E. (1982), *op. cit.*, p. 537

DENNIS, N.T.M. and HEPPEL, T.A. (1968). *Vacuum System Design* (London: Chapman & Hall)

DENOUX, R.A. (ed.) (1975). *Instrument Maintenance Managers Source Book* (New York: Wiley)

DENT, B. (1988). Switched on? *Process Engng*, Mar., 59

DENTON, A.A. (1989). *Piper Alpha Public Inquiry. Transcript, Day 161*

DENTON, A.A. (1991). Quality management offshore. *Safety Developments in the Offshore Oil and Gas Industry* (London: Instn Mech. Engrs), p. 73

DEPAOLA, T.J. and MESSINA, C.A. (1984). Nitrogen blanketing. *Plant/Operations Prog.*, **3**, 203

DEPARTMENT OF AGRICULTURE (ca. 1980). See Cross, J. and Farrer, D. (1982), *op. cit.*, p. 7

DEPARTMENT OF THE ARMY (1967). *Military Explosives. Tech. Manual TM 9–1300–214* (Washington, DC)

DEPARTMENT OF THE ARMY (1969). *Nuclear Handbook for Medical Service Personnel. Tech. Manual TM8–215* (Washington, DC)

DEPARTMENT OF THE ARMY, THE NAVY AND THE AIR FORCE (1969). *Structures to Resist the Effects of Accidental Explosions. Rep. TM 5–1300* (Washington, DC)

DEPARTMENT OF COMMERCE (1973). *Scenarios of Buildings in Given Earthquake Damage States, Rev. 1. Rep. PB 80/106156.* Nat. Tech. Inf. Service, Springfield, VA

DEPARTMENT OF COMMERCE, MARITIME ADMINISTRATION (1977). *Status of LNG Vessels* (Washington, DC)

DEPARTMENT OF EMPLOYMENT – see Appendix 28

DEPARTMENT OF EMPLOYMENT AND PRODUCTIVITY (1968). *Code of Practice for the Protection of Persons Exposed to Radiation in Research and Teaching* (London: HM Stationery Office)

DEPARTMENT OF EMPLOYMENT AND PRODUCTIVITY (1970). *Safety in Mechanical Handling* (London: HM Stationery Office)

DEPARTMENT OF ENERGY (1977a). *Offshore Installations. Guidance on Design and Construction* (London: HM Stationery Office)

DEPARTMENT OF ENERGY (1977b). *Report of the Inquiry into Serious Gas Explosions* (London: HM Stationery Office)

DEPARTMENT OF ENERGY (1978). *An Approach to Liquefied Natural Gas (LNG) Safety and Environmental Control Research. Rep.* DOE/EV-0002 (Washington, DC)

DEPARTMENT OF ENERGY (1980). *A Manual for Prediction of Blast and Fragment Loadings on Structures. Rep. DOE/TIC-11268* (Washington, DC)

DEPARTMENT OF ENERGY (1980). *Offshore Installations: Guidance on Fire Fighting Equipment* (London: HM Stationery Office)

DEPARTMENT OF ENERGY (1984). *Offshore Installations: Guidance on Design and Construction*, 3rd ed. (London: HM Stationery Office)

DEPARTMENT OF THE ENVIRONMENT (n.d.). *Code of Practice – Safety of Loads on Vehicles* (London: HM Stationery Office)

DEPARTMENT OF THE ENVIRONMENT (1974). *Distribution Pattern of Dangerous Goods Transported in Bulk by Road.* Paper of Working Party on the Transport of Dangerous Goods

DEPARTMENT OF THE ENVIRONMENT (1976a). *Accidental Oil Pollution of the Sea* (London: HM Stationery Office)

DEPARTMENT OF THE ENVIRONMENT (1976b). *Transport Statistics Great Britain 1974* (London: HM Stationery Office)

DEPARTMENT OF THE ENVIRONMENT (1983). *Planning Appeals* (London: HM Stationery Office)

DEPARTMENT OF THE ENVIRONMENT (1984a). *Local Plans: Public Local Inquiries* (London: HM Stationery Office)

DEPARTMENT OF THE ENVIRONMENT (1984b). *Planning Permission. A Guide for Industry* (London: HM Stationery Office)

DEPARTMENT OF THE ENVIRONMENT (1984c). *Structure Plans: The Examination in Public* (London: HM Stationery Office)

DEPARTMENT OF THE ENVIRONMENT (1991). *Acid Emissions Abatement Techniques*, vols 1–4 (London: HM Stationery Office)

DEPARTMENT OF THE ENVIRONMENT – see also Appendix 28

DEPARTMENT OF THE ENVIRONMENT and CHEMICAL SOCIETY (1976). *Data Base DESCNET (Network of Data on Environmentally Significant Chemicals)* (London)

DEPARTMENT OF HEALTH, EDUCATION AND WELFARE (1960). *Atmospheric Emissions from Petroleum Refineries* (Washington, DC: US Govt Printing Office)

DEPARTMENT OF HEALTH, EDUCATION AND WELFARE and DEPARTMENT OF LABOR, DIVISION OF OCCUPATIONAL HEALTH (n.d.). *Occupational Skin Diseases* (Washington, DC: US Govt Printing Office)

DEPARTMENT OF HEALTH AND SOCIAL SECURITY (1978). *Code of Practice for the Prevention of Infection in Chemical Laboratories and Post-Mortem Rooms* (London)

DEPARTMENT OF HEALTH AND SOCIAL SECURITY (1979). *Report of the Investigation into the Cause of the 1978 Birmingham Smallpox Occurrence* (London)

DEPARTMENT OF INDUSTRY (1975). *Corrosion Directory* (London: HM Stationery Office)

DEPARTMENT OF INDUSTRY (1981). *The Handling and Storage of Coated Wrapped Steel Pipes.* (London: HM Stationery Office)

DEPARTMENT OF INDUSTRY – see also Appendix 28

DEPARTMENT OF THE INTERIOR (1976). *Alaska National Gas Transportation System. Final Environmental Impact Statement* (Washington, DC)

DEPARTMENT OF LABOR (1963). *Static Electricity. Bull. 256* (Washington, DC)

DEPARTMENT OF SCIENTIFIC AND INDUSTRIAL RESEARCH (1963). Report on 'non-sparking' hand tools. *J. Inst. Petrol.*, **49**(474), 180

DEPARTMENT OF TRADE (1977). *Code for Portable Tanks and Road Tank Vehicles for the Carriage of Liquid Dangerous Goods in Ships* (London: HM Stationery Office)

DEPARTMENT OF TRADE AND INDUSTRY (1971). *Carriage of Dangerous Goods in Ships*, 2nd ed. (London)

DEPARTMENT OF TRADE AND INDUSTRY (1974). *Materials Handling* (London: HM Stationery Office)

DEPARTMENT OF TRADE AND INDUSTRY (1986). *Register of Quality Assessed United Kingdom Companies* (London: HM Stationery Office)

DEPARTMENT OF TRADE AND INDUSTRY (1989). *Instructions for Consumer Products* (London: HM Stationery Office)

DEPARTMENT OF TRADE AND INDUSTRY (1990). *SafeIT – A Government Consultation Document on the Safety of Computer-controlled Systems* (London)

DEPARTMENT OF TRADE AND INDUSTRY (1993). *Guidance Notes for Applications and Notifications of Onshore Pipelines under the Pipelines Act 1962* (London: HM Stationery Office)

DEPARTMENT OF TRANSPORT (1978). *The Highway Code* (London: HM Stationery Office)

DEPARTMENT OF TRANSPORT (n.d.). *Driver and Vehicle Licensing Agency, What New Licence?* (London)

DEPARTMENT OF TRANSPORT (1982). *Road Accidents Great Britain 1981* (London: HM Stationery Office)

DEPARTMENT OF TRANSPORT (1985). *Road Accidents Great Britain 1984* (London: HM Stationery Office)

DEPARTMENT OF TRANSPORT (1987a). *Road Accidents in Great Britain 1986 – the casualty report* (London: HM Stationery Office)

DEPARTMENT OF TRANSPORT (1987b). *Transport Statistics Great Britain 1976–86* (London: HM Stationery Office)

DEPARTMENT OF TRANSPORT (1992a). Survey of chemical tankers. *Instructions for the Guidance of Surveyors* (London: HM Stationery Office)

DEPARTMENT OF TRANSPORT (1992b). *Transport Statistics Great Britain 1991* (London)

DEPARTMENT OF TRANSPORT (1993a). *European Agreement concerning the International Carriage of Dangerous Goods by Road (ADR)* (London: HM Stationery Office)

DEPARTMENT OF TRANSPORT (1993b). *Regulations concerning the International Carriage of Dangerous Goods by Rail (RID)* (London: HM Stationery Office)

DEPARTMENT OF TRANSPORTATION (1974). *Summary of Liquid Pipeline Accidents Reported on DOT Form 7000–1 January 1965–December 1973* (Washington, DC)

DEPARTMENT OF TRANSPORTATION (1978). *Emergency Guide for Selected Hazardous Materials* (Washington, DC)

DEPARTMENT OF TRANSPORTATION, COAST GUARD (1974a). *CHRIS. A Condensed Guide to Chemical Hazards. Rep. CG446–1* (Washington, DC)

DEPARTMENT OF TRANSPORTATION, COAST GUARD (1974b). *CHRIS. Hazardous Chemical Data. Rep. CG446–2* (Washington, DC)

DEPARTMENT OF TRANSPORTATION, COAST GUARD (1974c). *CHRIS. Hazard Assessment Handbook. Rep. CG446–3* (Washington, DC)

DEPARTMENT OF TRANSPORTATION, COAST GUARD (1974d). *CHRIS. Response Methods Handbook. Rep. CG446–4* (Washington, DC)

DEPARTMENT OF TRANSPORTATION, COAST GUARD (1976). *Liquefied Natural Gas, Views and Practices, Policy and Safety. Rep. CG-478* (Washington, DC)

DEPARTMENT OF TRANSPORTATION, COAST GUARD (1977a). *The Port of Boston: LNG–LPG Operation/Emergency Plan* (Washington, DC)

DEPARTMENT OF TRANSPORTATION, COAST GUARD (1977b). *Predictability of LNG Vapor Dispersion from Catastrophic Spills in Water* (Washington, DC)

DEPARTMENT OF TRANSPORTATION, OFFICE OF PIPELINE SAFETY (1975). *Seventh Annual Report on the Natural Gas Pipeline Safety Act* (Washington, DC)

DEPARTMENT OF TRANSPORTATION, OFFICE OF PIPELINE SAFETY OPERATIONS (1977a). *Liquefied Natural Gas Facilities (LNG)*: Federal Safety Standards. Federal Register 42, No. 77, Apr. 21, 20776 (Washington, DC)

DEPARTMENT OF TRANSPORTATION, OFFICE OF PIPELINE SAFETY OPERATIONS (1977b). *Summary of Liquid Pipeline Accidents Reported on DOT Form 7000–1 from January 1, 1976, through December 31, 1976* (Washington, DC)

DEPAUL, F.T. and SHEIH, C.M. (1985). A tracer study of dispersion in an urban street canyon. *Atmos. Environ.*, **19**, 555

DEPAUL, F.T. and SHEIH, C.M. (1986). Measurements of wind velocities in a street canyon. *Atmos. Environ.*, **20**, 455

DEPUE, C. (1991). Carefully plan equipment monitoring. *Chem. Engng Prog.*, **87**(9), 75

DERBYSHIRE, D.H. (1960). Explosion hazards in the manufacture of terephthalic acid. *Chemical Process Hazards I*, p. 37

DERKS, R. (1991). Be safe – eliminate PCBs from transformers. *Chem. Engng*, **98**(1), 133

DERKSEN, R.W. and SULLIVAN, P.J. (1987). Deriving contaminant concentration fluctuations from time-averaged records. *Atmos. Environ.*, **21**, 789

DEROBERT, L. (1964) Anatomical study of four cases of acute intoxication by ammonia gas. *Ann. Med. Leg.*, **44**, 362

DERRINGTON, J.A. (1991). Hazard and the civil engineer. In The Fellowship of Engineering (1991b), *op. cit.*, p. 53

DESAEDELEER, G., SCHENE, R., MAHER, S.T., SHARP, D.R. and RODIBAUGH, R.K. (1989). Relief valve testing interval optimization program for the cost-effective control of major hazards. *Loss Prevention and Safety Promotion 6*, p. 9.1

DESAI, M.B. (1981). Preliminary cost estimating for process plants. *Chem. Engng*, **88**(Jul. 27), 65

DESAI, P.R. and BUONICORE, A.J. (1990). Engineering controls to minimize worker exposure to ethylene oxide at sterilization facilities. *Plant/Operations Prog.*, **9**, 103

DESCH, H.E. (1981). *Timber. Its Structure and Properties* (London: Macmillan)

DESCHNER, W.W., GIECK, J. and POTTS, P. (1968). Can small ammonia plants compete? *Hydrocarbon Process.*, **47**(9), 261

DESENBY, M.G.J. (1970). Materials for sea water cooling. *Chem. Engng*, **77**, (Jun. 15), 182

DESHAIES, B. and LEYER, J.D. (1981). Flow field induced by unconfined spherical accelerating flames. *Combust. Flame*, **40**, 141

DESHOTELS, R. and GOYAL, R. (1992). Hazop methods applied to process design interfaces (poster item). *Hazard Identification and Risk Analysis, Human Factors and Human Reliability in Process Safety*, p. 443

DESIGN INSTITUTE FOR EMERGENCY RELIEF SYSTEMS (1987). *Safety Valve Stability and Capacity Test Results: Pressure*

Relief Valve Performance Study. Final Report (New York: Am. Inst. Chem. Engrs)

DESIGN INSTITUTE FOR EMERGENCY RELIEF SYSTEMS (1992). *Emergency Relief Design Using DIERS Methodology* (New York: Am. Inst. Chem. Engrs)

DESIMONE, J.M., MAURY, E.E., GUAN, Z., COMBES, J.R., MENCELOGLU, Y.Z., CLARK, M.R., MCCLAIN, J.B., ROMACK, T.J. and MISTELE, C.D. (1994). Homogeneous and heterogeneous polymerizations in environmentally responsible carbon dioxide. *Preprints of Papers Presented at the 208th ACS National Meeting*, Aug. 21–25, 1994, Washington, DC, 212–14. Center for Great Lakes Studies, University of Wisconsin-Milwaukee (Milwaukee, WI: Division of Environmental Chemistry, American Chemical Society)

DESJARDIN, M., GUSTAFSON, D., HELLING, R., LADD, M.J., STEELE, K. and VANDERHOOF, D. (1991). Dow Chemical Western Division Research pre-operational review for new projects. *Plant/Operations Prog.*, **10**, 69

DESTEESE, J.G. and RHOADS, R.E. (1983). Research and development need and opportunities in LFG safety and environmental control. In Hartwig, S. (1983), *op. cit.*, p. 127

DET NORSKE VERITAS – see Appendices 27 and 28

DET NORSKE VERITAS (DNV) (1996) *Practical Loss Control Leadership*, Chapter 4, Revised Edition (Atlanta: DNV), ISBN 0-88061-054-9

DETLOR, L.T. (1975). Code section update: non-destructive tests. *Hydrocarbon Process.*, **54**(12), 56

DETMAN, R. (1977). Economics of six coal-to-SNG processes. *Hydrocarbon Process.*, **56**(3), 115

DETZ, C.M. (1976). Threshold conditions for the ignition of acetylene gas by a heated wire. *Combust. Flame*, **26**, 45

DEUSCHLE, R. and TIFFANY, F. (1973). Barrier intrinsic safety. *Hydrocarbon Process.*, **52**(12), 111

DEUTSCH, D.J. (1979). Electronic devices aim at motor protection. *Chem. Engng*, **86**(Jun. 18), 76

DEUTSCH, D.J. (1981). No health risks are seen yet for synfuel workers. *Chem. Engng*, **88**(Feb. 9), 51

DEUTSCHE FORSCHUNGSGEMEINSCHAFT (1982). *Maximum Concentrations at the Workplace and Biological Tolerance Values for Working Materials* (Weinheim: VCH)

DEVAUCHELLE, D. and NEY, J. (1962). *Prevention des Accidents dans les Industries Chimque* (Paris: Dunod)

DEVERALL, L.I. and LAI, W. (1969). A criterion for thermal ignition of cellulosic materials. *Combust. Flame*, **13**, 8

DEVEREUX, I.F. (1977). The role of incident reporting in project reliability programme management. *First Nat. Conf on Reliability* (Culcheth, Lancashire: UK Atom. Energy Auth.), NRC5/10

DEVIA, N. and LUYBEN, W.L. (1978). Reactors: size versus stability. *Hydrocarbon Process.*, **57**(6), 119. See also in Kane, L.A. (1980), *op. cit.*, p. 223

DEVINE, M.D. *et al.* (1974). *Analysis and Management of a Pipeline Safety System. Rep. PB-238 828*. Nat. Tech. Inf. Service, Springfield, VA

DEVINY, W.M. (1967). Pilot plant procedures. Disposal of hazardous chemicals. *Chem. Engng Prog.*, **63**(11), 56

DEVIS, M. (1990). *Standardized Human Olfactory Thresholds* (Oxford: Oxford Univ. Press)

DEVORE, A., VAGO, G.J. and PICOZZI, G.J. (1980). Heat exchangers. Specifying and selecting. *Chem. Engng*, **87** (Oct. 6), 133

DEVRIES, E.A. (1983). Avoid programmable controller problems. *Hydrocarbon Process.*, **62**(6), 65

DEWEY, M. (1978). Contract vs in house research. *Chemy Ind.*, **Apr. 15**, 262

DEWHIRST, F. (1981a). Chlorine in schools. *School Sci. Rev.*, Sep., 177

DEWHIRST, F. (1981b). Chlorine toxicity. *Chemy Britain*, **17**(4), 154

DEWHIRST, F. (1981c). Voluntary chlorine inhalation. *Br. Med. J.*, **282**, 565

DEWHURST, P. (1986). Communicating with the public. *Chemy Ind.*, Jan. 6, 22

DEWIS, M. (1985a). Control of industrial major accidents hazards – the new regulations. *Chemy Ind.*, Mar. 18, 187

DEWIS, M. (1985b). Deregulation of health and safety at work. *Chemy Ind.*, May 20, 321

DEWIS, M. (1985c). Ionising radiations – the new regulations. *Chemy Ind.*, Dec. 16, 812

DEWIS, M. (1986). Product liability – the proposed new law. *Chemy Ind.*, Sep. 1, 578

DEWITT, J.D. *et al.* (1974). Olefins outlook. *Hydrocarbon Process.*, **53**(11), 121

DEWITT, S.H. (1999). Microreactors for chemical synthesis. *Curr. Opin. in Chem. Biol.* **3**, 350–6

DEWITTE, M., VREBOSCH, J. and VAN TIGGELEN, A. (1964). Inhibition and extinction of premixed flames by dust particles. *Combust. Flame*, **8**, 257

DHAN, C.J., KASHANI, A. and REYES, B. (1994). Static electricity hazards of ÂÊexible intermediate bulk container (FIBC). *Process Safety Prog.*, **13**, 123

DHARMADHIKARI, S. and HECK, G. (1991). Liquefied gas storage. *Chem. Engr, Lond.*, **499**, 17

DHILLON, B.S. (1986). *Human Reliability with Human Factors* (Oxford: Pergamon Press)

DHILLON, B.S. and MISRA, R.B. (1985). Effect of critical human error on system reliability. *Reliab. Engng*, **12**(1), 17

DHILLON, B.S. and RAYAPATI, S.N. (1986). A complex system reliability evaluation method. *Reliab. Engng*, **16**(2), 163

DHILLON, B.S. and SINGH, C. (1981). *Engineering Reliability. New Techniques and Applications* (New York: Wiley-Interscience)

DIAB, S. and MADDOX, R.N. (1982). Absorption. *Chem. Engng*, **89**(Dec. 27), 38

DIAB, S.Y. (1983). Drying gas pipelines. *Oil Gas J.*, **81**(Feb. 28), 81; (Mar. 7), 113

DIACCI, E., RIZZI, R. and RONCHETTI, C. (1983). On-line monitoring of the true corrosion rate in power plants. Butcher, D.W. (1983), *op. cit.*, p. 217

DIACONICOLAOU, G.J., MUMFORD, C.J. and LIHOU, D.A. (1980). Prediction of acceptable spacing distance for plant involving an unconfined vapour cloud explosion hazard. *Symp. on Explosions, Fire Hazards and Relief Venting*, Sheffield Univ.

DIAL, R.E. (1968). High alumina refractory materials for gas reforming. *Ammonia Plant Safety*, **10**, 29

DIAMAUI, R.M.E. (1971). *The Prevention of Corrosion* (London: Business Books)

DIAPER, D. (1989). *Knowledge Elicitation* (Chichester: Ellis Horwood)

DIAZ CORREA, P.A. (1992). Maximum credible accident analysis for a gas compression plant. *Loss Prevention and Safety Promotion 7*, vol. 3, 133.1

DIBBO, A. (1984). Plastic process plant. *Chem. Engr, Lond.*, **408**, 38

DICK, D. (1982). Safety technology for offshore oil platforms. In Green, A.E. (1982), *op. cit.*, p. 407

DICK, J.B. (1949). Experimental studies in natural ventilation of houses. *J. Inst. Heat. Ventil. Engrs*, **17**(173), 420

DICK, J.B. (1950a). The fundamentals of natural ventilation of houses. *J. Inst. Heat. Ventil. Engrs*, **18**(Jun.), 133

DICK, J.B. (1950b). Measurement of ventilation using tracer gas technique. *Heat. Piping Air Cond.*, **22**(5), 131

DICK, J.B. and THOMAS, B.A. (1951). Ventilation research in occupied houses. *J. Inst. Heat. Ventil. Engrs*, **19**(Oct.), 306

DICKEN, A.N.A. (1974). The quantitative assessment of chlorine emission hazards. *Chlorine Bicentennial Symp.* (New York: Chlorine Inst.)

DICKEN, A.N.A. (1975). The quantitative assessment of chlorine emission hazards. *London Chemical Engineering Cong.* (London: Instn Chem Engrs)

DICKENSON, S.W. (1976). Pressure systems: general considerations. *Course on Loss Prevention in the Process Industries.* Dept of Chem. Engng, Loughborough Univ. of Technol.

DICKER, D.W.G. and RAMSEY, K.W. (1983). Precautions against fire and explosion hazards at jetties. *Loss Prevention and Safety Promotion 4*, vol. 2, p. D48

DICKEY, D.S. (1991). Succeed at stirred-tank-reactor design. *Chem. Engng Prog.*, **87**(12), 22

DICKIE, P. and LUNN, G.A. (1984). The maximum experimental safe gap: comparison of results from 8-litre and 20-cm³ explosion vessels. *J. Haz. Materials*, **8**(3), 271

DICKINSON, H.A. (1969). The port transport industry. In Handley, W. (1969), *op. cit.*, p. 266

DICKINSON, W.H. (1962). Report on reliability of electrical equipment in industrial plants. *AIEE Trans. Appl. Ind.*, **81**, 132

DICKSON, C.R., START, G.E. and MARKEE, E.H. (1969). Aerodynamics effects of the EBR-II reactor complex on effluent concentration. *Nucl. Safety*, **10**, 228

DICKSON, G.T. and TEATHER, F.W. (1982). Safety technology in industry. Pharmaceutical. In Green, A.E. (1982), *op. cit.*, p. 425

DICKSON, L.J. (1978). Pipelines — oil and gas. In Institute of Petroleum (1978), *op. cit.*, p. 104

DICRISTOFARO, D.C. and EGAN, B.A. (1986). A comparison of complex terrain models using the full scale plume study (ESPS) database: Complex I, RTDM, and CTDM. *Fifth Joint Conf. on Applications of Air Pollution Meteorology* (Boston, MA: Am. Met. Soc.), p. 219

DIDION, GENERAL (1860). *Traite' de Balistique*, 2nd ed. (Paris: Dunod)

DIEDRONKS, A.G.M. and TENNEKES, H. (1984). Entrainment effects in the well-mixed atmospheric boundary layer. *Boundary Layer Met.*, **30**, 75

DIEFENBACH, R.E. (1976). On-scene decisions — the resultant effects—impacts. *Control of Hazardous Material Spills*, **3**, p. 356

DIEHL, G.M. (1974). Machinery vibration analysis. *Chem. Engng Prog.*, **66**(5), 52

DIEHL, G.M. (1975). *Machinery Acoustics* (New York: Wiley)

DIENER, R. (1991). Mitigation of HF releases. *J. Loss Prev. Process Ind.*, **4**(1), 35

VAN DIENST, J., BASLER, J.C., BENOCCI, C., OLIVARI, D., RIETHMULLER, M.L. and VERGISON, E. (1986). *Atmospheric Dispersion of Heavy Gases in a Complex Environment. Contract CEC ENV 713–B CRS, Final Rep.* Comm. Eur. Commun.

DIENST CENTRAAL MILIEUBEHEER RIJNMOND (1985). *Study into Risk Transportation of Liquid Chloride and Ammonia in the Rijnmond Area* (Rotterdam)

DIEPOLDER, P. (1992). Is 'zero discharge' realistic? *Hydrocarbon Process.*, **71**(10), 129

DIERKES, M., EDWARDS, S. and COPPOCK, R. (eds) (1980). *Technological Risk* (Cambridge, MA: Oelgeschlager, Gunn & Hain)

DIETER, G.E. (1961). *Mechanical Metallurgy* (New York: McGraw-Hill)

DIETTERICH, T.C. and MICHALSKI, R.S. (1981). Inductive learning of structural descriptions: evaluation criteria and comparative review of selected methods. *Artificial Intelligence*, **16**, 257

DIGGLE, W.M. (1976). Major emergencies in a petrochemicals complex: planning action by emergency services. *Process Industry Hazards*, p. 217

DILLIBERTO, M.C. and GRATZOL, O.K. (1985). Fire proofing materials: what's the best test? *Hydrocarbon Process.*, **64**(4), 121

DILLER, W.F. (1985). Pathogenesis of phosgene poisoning. *Tox. Ind. Health*, **1**(2), 7

DILLER, W.F. and ZANTE, R. (1982). Dose—effect relationship of the effects of phosgene on man and animals (literature study). *Zbl. Arbeitsmed. Arbeitschutz Prophyl.*, **32**, 360

DILLON, A. (1992). Reading from paper versus screens: a critical review of the empirical literature. *Ergonomics*, **35**, 1297

DILLON, C.P. (1986). *Corrosion Control in the Chemical Process Industries* (New York: McGraw-Hill)

DILWALI, K.M. and MUDAN, K.S. (ca. 1986) *Dike Design Alternatives to Reduce Toxic Vapor Dispersion Hazards.* (Cambridge, MA: A.D. Little)

DILWALI, K.M. and MUDAN, K.S. (1987). Dike design alternatives to reduce toxic vapor dispersion hazards. *Vapor Cloud Modeling*, p. 539

DILWORTH, P.J. (1979). The design of pipework for chemical plants. *Design 79* (London: Instn *Chem. Engrs*), p. K4.1

DIMAIO, L.R. and LANGE, R.F. (1984). Effect of water quality on fire fighting foams. *Plant/Operations Prog.*, **3**, 42

DIMAIO, L.R., LANGE, R.F. and CONE, F.J. (1984). Aspirating vs non-aspirating nozzles for making fire fighting foams — evaluation of a non-aspirating nozzle. *Fire Technol.*, **20**(1), 5

DIMAIO, L.R. and NORMAN, E.C. (1988). Performance of aqueous hazmat foams on selected hazardous materials. *Plant/Operations Prog.*, **7**, 195

DIMAIO, L.R. and NORMAN, E.C. (1990). Continuing studies of hazardous material vapor mitigation using aqueous foam. *Plant/Operations Prog.*, **9**, 135

DIMENTBERG, M. (1972). The design and operation of an LNG pipeline from the Arctic. *LNG Conf.* **3**

DIMEO, M. (1975). Explosion at Flixborough. *Fire J.*, Nov.

DIMMER, P.R. (1986). Fire fighting and damage control in warships. *Fire Int.*, **97**, 33

DIMOPLON, W. (1974). How to determine the geometry of pressure vessel heads. *Hydrocarbon Process.*, **53**(8), 71

DIMOPLON, W. (1978). What process engineers need to know about compressor design. *Hydrocarbon Process.*, **57**(5), 221

DIN, M.B. (1975). Waste heat boiler failure. *Ammonia Plant Safety*, **17**, p. 149

DINAPOLI, R.N. (1978). Costs are estimated for LNG terminals. *Oil Gas J.*, Mar. 13, 83

DINCER, A.K., DRAKE, E.M. and REID, R.C. (1977). Boiling of liquid nitrogen and methane on water. The effect of initial water temperature. *Int. J. Heat Mass Transfer*, **20**, 176

DINENNO, P.J. (1982). Simplified radiation heat transfer calculations from large open hydrocarbon fires. *Soc. Fire Protection Engrs, Ann. Mtg.*

DINGER, R.H. (1981). Improve project control. *Hydrocarbon Process.*, **60**(12), 88

DINGMAN, J.C. (1967). Effect of variables on corrosion. *Ammonia Plant Safety*, **9**, p. 48

LE DIRAISON, J.C. and BAILLEUL, P. (1986). The Nantes cryogenic testing centre. *Gastech 85*, p. 488

DIRECTORATE GENERAL FOR ENVIRONMENTAL PROTECTION (1988–89). *Premises for Risk Management – Risk Limits in the Context of Environmental Policy. National Environmental Policy Plan, Annex* (The Hague)

DIRKMAAT, J.J. (1981). Experimental studies on the dispersion of heavy gases in the vicinity of structures and buildings. *Proc. Symp. on Designing with the Wind*, Nantes.

DISERENS, N.J., SMITH, J.R. and BRIGHT, A.W. (1975). A preliminary study of electric field problems in plastic tanks and their theoretical modelling by means of a finite difference computer program. *J. Electrostatics*, **5**, 169

DISHART, P.C. and BRUCE, S.M. (1975). *X-ray Methods of Analysis and their Value in the Gas Industry. Commun. 972*. Instn Gas Engrs, London. See also *Instn Gas Engrs J.*, **15**, 262

DISS, E., KARAM, H. and JONES, C. (1961). Practical way to design safety disks. *Chem. Engng*, **68**(Sep. 18), 187

DITALI, S., ROVATI, A. and RUBINO, F. (1992). Experimental model to assess thermal radiation from hydrocarbon pool fires. *Loss Prevention and Safety Promotion 7*, vol. 1, 13–1

DITTMER, D.C. (1997). 'No Solvent' Organic Synthesis. *Chem. Ind.*, (Oct. 6), 779–84

DITZ, D.W. (1988). The risk of hazardous waste spills from incineration at sea. *J. Haz. Materials*, **17**(2), 149

DIX, C.H. (1952). *Seismic Prospecting for Oil* (New York: Harper)

DIX, H.M. (1980). *Environmental Pollution* (Chichester: Wiley)

DIXON, H.B. (1910). The union of hydrogen and oxygen in flame. *J. Chem. Soc.*, **97**, 661

DIXON, H.B., BRADSHAW, W.L. and CAMPBELL, C. (1914). The firing of gases by adiabatic compression – 1. *J. Chem. Soc.*, **105**, 2027

DIXON, H.B. and CROFTS, J.M. (1914). The firing of gases by adiabatic compression. – 2. *J. Chem. Soc.*, **105**, 2036

DIXON, H.B., HARWOOD, J. and HIGGINS, W.F. (1926). On the ignition-point of gases. *Trans. Faraday Soc.*, **22**, 267

DIXON, J.R. (1966). *Design Engineering* (New York: McGraw-Hill)

DIXON, J.S. (1967). Distance piece explosion. *Ammonia Plant Safety*, **9**, p. 36

DIXON, W.M. and DREW, D. (1968). Fatal chlorine poisoning. *J. Occup. Med.*, **10**, 249

DIXON-JACKSON, K. (1989). The use and interpretation of DSC and HFC data in assessing the need for venting. *Selection and Use of Pressure Relief Systems for Process Plant* (Manchester: Instn. Chem. Engrs, North-West Branch), p. 5.1

DIXON-JACKSON, K. (1994). Incident during nitration in a batch reactor. *Hazards*, **XII**, p. 5.4

DIXON-LEWIS, G. (1955). Second limits in mixtures of hydrogen, carbon monoxide and oxygen. *Combustion*, **5**, p. 603

DIXON-LEWIS, G. (1991). Structure of laminar flames. *Combustion*, **23**, p. 305

DIXON-LEWIS, G. and ISLAM, S.M. (1982). Flame modeling and burning velocity measurement. *Combustion*, **19**, p. 283

DIXON-LEWIS, G. and ISLES, G.L. (1959). Limits of inflammability. *Combustion*, **7**, p. 475

DIXON-LEWIS, G. and ISLES, G. (1961). Inflammability limits (letter). *Combust. Flame*, **5**, 111

DIXON-LEWIS, G. and SHEPHERD, I.G. (1975). Some aspects of ignition by localized sources, and of cylindrical and spherical flames. *Combustion*, **15**, p. 1483

DIXON-LEWIS, G. and SIMPSON, R.J. (1977). Aspects of flame inhibition by halogen compounds. *Combustion*, **16**, 1111

DJANG, T.K. (1942). *Factory Inspection in Great Britain* (London: Allen & Unwin)

DOBBINS, R.A. (1979). *Atmospheric Motion and Air Pollution* (New York: Wiley)

DOBBS, N., COHEN, E. and WEISSMAN, S. (1968). Blast pressures and impulse loads for use in the design and analysis of explosive storage and manufacturing facilities. *Ann. N.Y. Acad. Sci.*, **152**(Art. 1), 137

DOBIE, W.C., CUNNINGHAM, N. and REID, M.K. (1976). Electric surface heating for explosive and non-explosive atmospheres. *Br. Nat. Cttee for Electroheat, Conf. on Health and Safety Aspects of the Design and Operation of Electroheat Equipment* (London)

DOBLE, P.A.C. (1989). *Piper Alpha Public Inquiry. Transcript, Days 137, 138*

DOBROWOLSKI, Z.C. (1984). Mechanical vacuum pumps in the process industry. *Chem. Engng Prog.*, **80**(7), 75

DOBRZYNSKI, J. (1979). BAT and BPT may spell zero discharge for many. *Chem. Engng*, **86**(Jan. 1), 52

DOCKENDORFF, R.L. (1980). Pilot-operated relief valves in crude desalter service. *API Proc., Refining Dept, 41st Midyear Mtg*, **59**, 248

DODD, K.T. and GROSS, D.R. (1980). Ammonia inhalation toxicity in cats: a study of acute and chronic respiratory dysfunction. *Arch. Environ. Health*, **35**, 543

DODD, V.R. (1984). Condition monitoring of turbo-machinery cuts costs over 4–year period. *Oil Gas J.*, **82**(Mar. 12), 96

DODDING, R.A., SIMMONS, R.F. and STEPHENS, A. (1970). The extinction of methane-air diffusion flames by sodium bicarbonate powders (letter). *Combust. Flame*, **15**, 313

DODDS, R.L. (1963). Stream lining maintenance paperwork. *Chem. Engng*, **70**(Sep. 16), 200

DODGE, B.F. (1944). *Chemical Engineering Thermodynamics* (New York: McGraw-Hill)

DODGE, F.T., PARK, J.T., BUCKINGHAM, J.C. and MAGOTT, R.J. (1983). *Revision and Experimental Verification of the Hazard Assessment Computer System Models for Spreading Movement, Dissolution and Dissipation of Insoluble Chemicals Spilled onto Water. Rep. CG-D-35 – 83*. US Coast Guard.

DOE, J.E., CURRY, A.M., ROBINSON, M. and MELIA, B.J. (1986). The acute toxicity of the breakdown products of the fire extinguishant bromochlorodifluoromethane (BCF, Halon 1211). *J. Fire Sci.*, **4**(2), 113

DOE, J.E. and MILBURN, G.M. (1983). The relationship between exposure concentration, duration of exposure and inhalation LC50 values. Private communication

DOEBELIN, E.O. (1975). *Measurement Systems*, rev. ed. (Tokyo: McGraw-Hill)

DOELP, L.C. and BRIAN, P.L.T. (1982). Reliability of pressure protective systems. A Markov analysis. *Ind. Engng Chem. Fund.*, **21**, 101

DOELP, L.C., LEE, G.K., LINNEY, R.E. and ORMSBY, R.M. (1984). Quantitative fault tree analysis: gate-by-gate method. *Plant/Operations Prog.*, **3**, 227

DOERING, E.J. and GADDY, J.L. (1977). Optimal process reliability with flowsheet simulation. *Eleventh Loss Prevention Symp.* (New York: Am. Inst. Chem. Engrs)

DOERING, W. and BURKHARDT, G. (1949). *Contributions to the Theory of Detonation. Rep. TR F-TS-1227-1A*. US Air Force, Wright Patterson Air Force Base, OH

DOERR, W.W. (1993). Plan for the future with pollution prevention. *Chem. Engng Prog.*, **89**(1), 26

DOERR, W.W. and HESSIAN, R.T. (1991). Control toxic emissions from batch operations. *Chem. Engng Prog.*, **87**(9), 57

DOHERTY, M. and BUZAD, G. (1992). Reactive distillation by design. *The Chem. Eng.*, (Aug. 27), s17-s19

DOHERTY, R., SHORT, J.M., AKST, I.B. and KAMLET, M.J. (1989). Estimation of the efficiency of non-ideal explosives. *Combust. Flame*, **76**, 369

DOHNAL, M., VYKYDAL, J., KVAPILIK, M. and BURES, P. (1992). Practical uncertainty assessment of reasoning paths (fault trees) under total uncertainty ignorance. *J. Loss Prev. Process Ind.*, **5**, 125

DOI, M. and OSAWA, H. (1983). A storage process with non-constant inflow and outflow. *Fourth Int. Conf. on Mathematical Modelling in Science and Technology* (ed. by J. R. Avula *et al.)* (New York: Pergamon Press), p. 305

DOIG, I.D. (1977). Training of process plant malfunction analysts. *CHEMECA 77*

DOIG, I.D. and REINTEN, R.H. (1979). Sensitivity of a process to component faults. *Comput. Chem. Engng*, **3**, 497

DOKTER, T. (1985a). Fire and explosion hazards of chlorine-containing systems. *J. Haz. Materials*, **10**, 73

DOKTER, T. (1985b). Formation of NCl_3 and N_2O in the reaction of NaOCl and nitrogen compounds. *J. Haz. Materials*, **12**(3), 207

DOKTER, T., BRUNINK, J.P., EGGELTE, H. and KRAMER, T. (1985). Chlorine container experiment. *J. Haz. Materials*, **12**(1), 11

VAN DOLAH, R.W. (1961). Evaluating the explosive character of chemicals. *Ind. Engng. Chem.*, **53**(7), 59A

VAN DOLAH, R.W. (1965). Stability and its evaluation. In Fawcett, H.H. and Wood, W.S. (1965), *op. cit.*, p. 319

VAN DOLAH, R.W. (1969a). Detonation potential of nitric acid systems. *Loss Prevention*, **3**, p. 32

VAN DOLAH, R.W. (1969b). Phenomenology of explosions. *Symp. on Inherent Hazards of Manufacturing and Storage in the Process Industry* (The Hague), p. 1

VAN DOLAH, R.W. and BURGESS, D.S. (1968). *Explosion Problems in the Chemical Industry*. Am. Chem. Soc., Course Notes

VAN DOLAH, R.W., ZABETAKIS, M.G., BURGESS, D.S. and SCOTT, G.S. (1963). *Review of Fire and Explosion Hazards of Flight Vehicle Combustibles*. Inf. Circ. 8137. Bur. Mines, Washington, DC

DOLAN, J.E. (1957). The suppression of methane/air ignitions by fine powders. *Combustion*, **6**, p.787

DOLD, J.W. and CLARKE, J.F. (1985). Ignition and criticality of a finite quantity of gas released in the atmosphere. *Combustion*, **20**, p. 1861

DOLL, R. (1955). Mortality from lung cancer in asbestos workers. *Br. J. Ind. Med.*, **12**(2), 81

DOLL, SIR. R. (1977). Strategy for detection of cancer hazards to man. *Nature*, **265**, 589

DOLL, SIR R. (1979). The pattern of disease in the post- infection era: national trends. *Proc. Roy. Soc., Ser. B*, **205**, 47. In Doll, Sir R. and McLean, A.E.M. (1979), *op. cit.*, p. 47

DOLL, SIR R. and MCLEAN, A.E.M. (eds) (1979). *Long-term Hazards from Environmental Chemicals* (London: The Royal Society)

DOLL, SIR R. and PETO, R. (1981). *The Causes of Cancer* (Oxford: Oxford Univ. Press)

DOLL, T.R. (1982). Making the proper choice of adjustable-speed drives. *Chem. Engng*, **89**(Aug. 9), 46

DOLLE, J. and GILBOURNE, D. (1976). Start-up of the Skikda LNG plant. *Chem. Engng Prog.*, **72**(1), 39

DOLLIMORE, D. (1981). Heated argument (letter). *Chemy Ind.*, Apr. 4, 196

DOMALSKI, E.S. and TSANG, W. (1975). Assessing the hazard potential of chemical substances. *Transport of Hazardous Cargoes*, **4**

DOMBROWSKI, N., FOUMENY, E.A. and RIZA, A. (1993). Know the CFD codes. *Chem. Engng Prog.*, **89**(9), 46

DOMBROWSKI, S.L.S. (1988). Of mice and men: the effect of science on regulations. *Chem. Engng Prog.*, **84**(12), 35

DOMINGUEZ, G.S. (1977). *Guidebook: Toxic Substances Control Act* (Boca Raton, FL: CRC Press)

DOMINGUEZ, G.S. (1978). What TOSCA means for the CPI. *Chem. Engng*, **85**(Apr. 24), 77

DOMINGUEZ, R. (1933). Studies of renal excretion of creatinine. II, Volume of distribution. *Proc. Soc. Exp. Biol. Med.*, **31**, 1146

DONAHUE, M.L. (1994). CHEMTREC's role in hazardous materials emergency response. *J. Haz. Materials*, **36**, 177

DONAKOWSKI, T.D. (1981). Is liquid hydrogen safer than liquid methane? *Fire Technol.*, **17**, 183

DONALD, I. and CANTER, D. (1994). Employee attitudes and safety in the chemical industry. *J. Loss Prev. Process Ind.*, **7**, 203

DONALDSON, LORD (chrmn) (1994). *Safer Ships, Cleaner Seas* (London: HM Stationery Office)

DONALDSON, C.R. (1973). Handling and storage of organic peroxides. *Safety in Polyethylene Plants*, **1**, p. 31

DONALDSON, H.H. (ed.) (1924). *The Rat — Data and Reference Tables for the Albino Rat*, 2nd ed. (Philadelphia, PA: Wistar)

DONALDSON, M. (1986). Could water-jetting solve your plant cleaning problems? *Process Engng*, Feb., 49

DONALDSON, W.T. (1972). Identification of spilled hazardous materials. *Control of Hazardous Material Spills*, **1**, p. 109

DONAT, C. (1971a). Selection and dimensioning of pressure relief devices for dust explosions. *Staub Reinhaltung der Luft*, **31**(4), 17 (English ed.)

DONAT, C. (1971b). *VDI Berichte 165*, p. 58

DONAT, C. (1973a). Discharge of explosion pressure with rupture discs and bleeder valve explosion doors. *Prevention of Occupational Risks*, **2**, p. 289

DONAT, C. (1973b). Release of the pressure of an explosion with rupture discs and explosion valves. *ACHEMA 73* (Frankfurt-am-Main)

DONAT, C. (1973c). Use of bursting safety devices for slow and rapid increases in pressure. *Chem. Ing. Tech.*, **45**, 790

DONAT, C. (1977a). *Pressure release of dust explosions. VDI 3673 (draft)*. Verein Deutscher Ing., Düsseldorf

DONAT, C. (1977b). Pressure relief as used in explosion protection. *Loss Prevention*, **11**, 87

DONAT, C. (1982). Explosion-proof type of construction. *The Control and Prevention of Dust Explosions* (London: Oyez/IBC), paper 6

DONEGANI, A. and JONES, D.W. (1993). Some recent global trends in the regulatory control of major hazards in the oil, chemical and process industries. *Health, Safety and Loss Prevention, p. 38*

DONIGER, D.D. (1978). *The Law and Policy of Toxic Substances Control: A Case Study of Vinyl Chloride* (Baltimore, MD: John Hopkins Univ. Press)

DONKIN, P. and WIDDOWS, J. (1986). Scope for growth as a measurement of environmental pollution and its interpretation using structure–activity relationships. *Chemy Ind.*, Nov. 3, 732

DONNELLY, R.E. An overview of OSHA's Process Safety Management Standards (USA). *Process Safety Prog.*, **13**, 53

DONOHUE, M.D. and GEIGER, J.L. (1994). Reduction of VOC emission during spray painting operations: a new process using supercritical carbon dioxide. *Preprints of Papers Presented at the 208th ACS National Meeting*, Aug. 21–25, 1994, Washington, D.C. 218–19. Center for Great Lakes Studies, University of Wisconsin-Milwaukee (Milwaukee, WI: Division of Environmental Chemistry, American Chemical Society)

DONOVAN, F.X. (1973). Streams limit tank collapse in Jersey oil terminal fire. *Fire Engng*, **126**(Sep.), 128

DONOVAN, J.R. (1962). Safe handling of molten sulfur. *Chem. Engng Prog.*, **58**(1), 92

DONOVAN, N.C. (1973). A statistical evaluation of strong motion data including the February 9, 1971, San Fernando earthquake. *Proc Fifth World Conf. on Earthquake Engineering*, **1**, 1252

DOOLEY, D.A. (1957). Heat Transfer Fluid Mechanics Inst., Stanford Univ., p. 321

DOOLIN, J.H. (1961). Centrifugal pump lubrication. *Chem. Engng*, **68**(Oct. 16), 210

DOOLIN, J.H. (1963). Centrifugal pumps and entrained air problems. *Chem. Engng*, **70**(Jan. 7), 103

DOOLIN, J.H. (1978). Install pumps for minimum stress. *Hydrocarbon Process.*, **57**(6), 96

DOOLIN, J.H. (1984). How reliable are pumps with induced impellers? *Hydrocarbon Process.*, **63**(1), 69

DOOLIN, J.H. (1990). Pump base plate design and installation. *Hydrocarbon Process.*, **69**(7), 59

DOOLIN, J.H. and TEASDALE, L.M. (1990). Sealless magnetic-driven pumps: stronger than ever. *Chem. Engng*, **97**(8), 112

DOOLIN, J., KAWOHL, R., MCGUIRE, J.T., MCCAUL, C., PRANG, A. and DODSON, J. (1991). Pumping difficult fluids. *Chem. Engng*, **98**(12), 67

DOONER, R. (1986). Looking for trouble. *Process Engng, Mar.* 51

VAN DOORN, M. and SMITH, J.M. (1980). Gas cloud containment with water sprays. *Loss Prevention and Safety Promotion* **3**, vol. 3, p. 1068

DOOYEWEERD, E. (1975). Explosion in an East Asian urea plant. *Ammonia Plant Safety*, **17**, 152

VAN DOP, H. (1992). Buoyant plume rise in a Lagrangian framework. *Atmos. Environ.*, **26A**, **1335**

DORAN, D. (1980). Ionizing radiation: physics, measurement and control. In Waldron, H.A. and Harrington, J.M. (1980), *op. cit.*, p. 280

DORAN, J.C. and HORST, T.W. (1985). An evaluation of Gaussian models with dual-tracer field measurements. *Atmos. Environ.*, **19**, 939

DORAN, J.C., HORST, T.W. and NICKOLA, P.W. (1978). Variations in measured values of lateral diffusion parameters. *J. Appl. Met.*, **17**, 825

DORAN, J.E. (1968). New developments of the Graph Traverser. *Machine Intelligence 2* (ed. by E. Dale and D. Michie) (Edinburgh Oliver and Boyd), p. 119

DORAN, J.E. and MICHIE, D. (1966). Experiments with the graph traverser program. *Proc. R. Soc., Ser. A*, **294**, 235

DORE, B. (1991). ESA research and development strategy on sneak analysis. *Proc ESA Symp. on Space Product Assurance for Europe in the 1990s. Rep. ESA SP-316.* Eur. Space Agency, Noordwijk, The Netherlands

DORE, J.C. (1988). Process safety – an integral part of pilot plants. *Plant/Operations Prog.*, **7**, 223

DOREY, J. (1981). Consideration of the reliability of pumps, derived from the first year of experience of the SRDF, the reliability data collection system of Electricité de France. *Reliab. Engng*, **2**(3), 179

DORF, R.C. (1980). *Modern Control System*, 3rd ed. (Reading, MA: Addison-Wesley)

DÖRGE, K.J., PANGRITZ, D. and WAGNER, H.G. (1976). Experiments on velocity augmentation of spherical flames by grids. *Acta Astronaut.*, **3**, 1067

DÖRGE, K.J., PANGRITZ, D. and WAGNER, H.G. (1979). On the effect of obstacles on the spread of flame. *ICI Jahrestagung*, p. 441

DÖRGE, K.J., PANGRITZ, D. and WAGNER, H.G. (1981). On the influence of several apertures on the spread of flame: an extension of Wheeler's experiments. *Z. Phys. Chem., Neue Folge*, **127**, 61

DORING, W. (1943). On detonation processes in gases. *Ann. Physik*, **43**, 421

DORMAN, R.G. (1974). *Dust Control and Air Cleaning* (Oxford: Pergamon Press)

DÖRNER, D. and PFEIFER, E. (1993). Strategic thinking and stress. *Ergonomics*, **36**, 1345

DOROFEEV, S.B., EFIMENKO, A.A., KOCHURKO, A.S. and KUZNETSOV, M.S. (1993). Industrial fuel ignition conditions by flame radiation. *J. Loss Prev. Process Ind.*, **6**, 159

DORON, E. and ASCULAI, E. (1983). A gaussian trajectory model for hazard evaluation for prolonged releases from nuclear reactors. In de Wispelaere, C. (1983), *op. cit.*, p. 231

DÖRRE, P. (1987). An event-based multiple malfunction model. *Reliab. Engng*, **17**(1), 73

DORSEY, J.J. (1967). Methanol plant explosion. *Ammonia Plant Safety*, **9**, p. 80

DORSEY, J.S. (1976). Static sparks – how to exorcise 'go devils'. *Chem. Engng*, **83**(Sep. 13), 203

DORSEY, J.S. (1984). Controlling external underground corrosion. *Chem. Engng*, **91**(Mar. 5), 77

DOSHER, J.R. (1970). Trends in petroleum refining. *Chem. Engng*, **77**(Aug. 10), 96

DOTIWALLA, K.K. (1991). Process heating systems: steam or thermal fluid. *Hydrocarbon Process.*, **70**(10), 67

DOTSON, R.L. (1978). Modern electrochemical technology. *Chem. Engng*, **85**(Jul. 17), 106

DOUGAN, K.W. and REILLY, M.C. (1993). Quantitative reliability methods improve plant uptime. *Hydrocarbon Process.*, **72**(8), 131

DOUGHERTY, E.M. and FRAGOLA, J.R. (1988). *Human Reliability Analysis* (New York: Wiley)

DOUGHERTY, R.J. (1965). Co-ordinate safety for plant construction. *Hydrocarbon Process.*, **44**(3), 163

DOUGLAS, J. (1979). California report pinpoints hazards in layout of Three Mile Island control room. *Nature*, **279** (Jun. 7), 464

DOUGLAS, J.M. (1988). *The Conceptual Design of Chemical Processes* (New York: McGraw-Hill)

DOUGLAS, M. (1966). *Purity and Danger* (London: Routledge & Kegan Paul)

DOUGLAS, M. (1982). *Cultural bias. The Active Voice* (London: Routledge & Kegan Paul), Ch. 9

DOUGLAS, M. (1985). *Risk Acceptability According to the Social Sciences* (London: Routledge & Kegan Paul)

DOUGLAS, M. (1987). *How Institutions Think* (London: Routledge)

DOUGLAS, M. (1990). Risk as a forensic resource. *Daedalus*, **119**(4), 1

DOUGLAS, M. (1992). *Risk and Blame* (London: Routledge)

DOUGLAS, M. and WILDAVSKY, A. (1982). How do we know the risks we face? Why risk selection is a social process. *Risk Analysis*, **2**(2), 45

DOUGLAS, M. and WILDAVSKY, A. (1982b). *Risk and Culture* (Berkeley, CA: Univ. of California Press)

DOUGLAS, T.L., VON BRAMER, P.T. and JENKINS, W.L. (1982). Screen and select vinyl monomer inhibitors. *Hydrocarbon Process.*, **61**(6), 109

DOULL, J. (1987). Strategies for the prevention of acute human toxicity effects. *Preventing Major Chemical Accidents*, p. 1.55

DOURY, A. (1981). *Le Vademecum des Transferts Atmospheriques. CEA-DSN Rep. 440*

DOURY, A. (1988). A design basis for the operational modelling of atmospheric dispersion. *J. Loss Prev. Process Ind.*, **1**(3), 156

DOUWES, C. and VAN DER WAARDEN, M. (1967). Current decay during measurement of DC electrical conductivity in solutions of hydrocarbons. *J. Inst. Petrol.*, **53**, 237

DOW CHEMICAL COMPANY (1964). *Dow's Process Safety Guide*, 1st ed. (Midland, MI)

DOW CHEMICAL COMPANY (1966a). *Dow's Process Safety Guide*, 2nd ed. (Midland, MI)

DOW CHEMICAL COMPANY (1966b). Process safety manual. *Chem. Engng Prog.*, **62**(8), 93

DOW CHEMICAL COMPANY (1976). *Fire and Explosion Index. Hazard Classification Guide*, 4th ed. (Midland, MI)

DOW CHEMICAL COMPANY (1980). *Fire and Explosion Index. Hazard Classification Guide*, 5th ed. (Midland, MI)

DOW CHEMICAL COMPANY (1987). *Dow's Fire and Explosion Index Hazard Classification Guide* (Midland, MI)

DOW CHEMICAL COMPANY (1994). *Dow's Fire and Explosion Index Hazard Classification Guide – Seventh Edition.* (New York: American Institute of Chemical Engineers.) ISBN 0-8169-0623-8.

DOW CHEMICAL COMPANY. (1994a). *Dow's Chemical Exposure Index* (New York: Am. Inst. Chem. Engrs)

DOW CHEMICAL COMPANY (1994b). *Dow's Fire and Explosion Index Hazard Classification Guide* (Midland, MI)

DOWDING, C.H. (1985). *Blast Vibration Monitoring and Control* (Englewood Cliffs, NJ: Prentice-Hall)

DOWELL, A.M. (1990). *Guidelines for Systems Oriented Multiple Cause Incident Investigation. Report.* Risk Analysis Dept, Rohm and Haas Texas Inc., Deer Park, TX

DOWELL, A.M. (1994a). Low cooling tower level causes evacuation. *Process Safety Prog.*, **13**, 114

DOWELL, A.M. (1994b). Managing the PHA team. *Process Safety Prog.*, **13**, 30

DOWELL, A.M. (1999). Layer of protection analysis and inherently safer processes. *Process Saf. Prog.* **18** (4) (Winter), 214–20

DOWELL, ARTHUR. M., III (1997). Layer of protection analysis: A New PHA Tool, After HAZOP, Before Fault Tree. *Int. Conf. and Workshop on Risk Anal. in Process Saf.*, Oct. 21–24, 1997, Atlanta, GA, 13–28, (New York: American Institute of Chemical Engineers)

DOWELL, ARTHUR.M., III (1998). Layer of protection analysis for determining safety integrity level, *ISA Trans.* **37**, 155–166

DOWELL, ARTHUR. M., III (1999). Layer of protection analysis and inherently safer processes, **33rd** *Annu. Loss Prev. Symp.*, American Institute of Chemical Engineers

DOWELL, D.L. (1971). Handling vinyl chloride emergencies. *Loss Prevention*, **5**, p. 29

DOWIE, J. and LEFRERE, P. (1980). *Risk and Chance* (Milton Keynes: Open Univ. Press)

DOWLING, T.E. (1960). Containers for chemicals. *Chem. Engng*, **67**(Oct. 3), 85

DOWNEY, W.K. and NI UID, G. (1979). *Air Pollution Impacts and Control* (Dublin: Nat. Board for Sci. and Technol.)

DOWNING, A.G. (1964). Frictional sparking of cast magnesium, aluminium and zinc. *NFPA Quart.*, **57**(3), 235

DOWNING, J.V. and SANDERS, M.S. (1987). The effects of panel arrangement and locus of attention on performance. *Hum. Factors*, **29**, 551

DOWRICK, D.J. (1977). *Earthquake Resistant Design* (Chichester: Wiley)

DOWRICK, D.J. (1987). *Earthquake Resistant Design*, 2nd ed. (Chichester: Wiley)

DOYAL, L. and EPSTEIN, S.S. (1983). *Cancer in Britain. The Politics of Prevention* (London: Pluto Press)

DOYLE, D. (1971). Monitoring hazardous environments. *Measmt Control*, **10**(10), 359

DOYLE, H.E. (1972). Highlights of API 610 pump standard. *Hydrocarbon Process.*, **51**(6), 85

DOYLE, P.T. and BENKLY, G.L. (1973). Use fanless air coolers. *Hydrocarbon Process.*, **52**(7), 81

DOYLE, W.E. (1965). Piping tierod design made simple. *Hydrocarbon Process.*, **44**(8), 118

DOYLE, W.H. (1964). Explosion venting of buildings and tanks. *Nat. Safety Cong., Chem. Ind. Trans.*

DOYLE, W.H. (1969). Industrial explosions and insurance. *Loss Prevention*, **3**, p. 11

DOYLE, W.H. (1972a). Instrument connected losses in the CPI. *Instrum. Technol.*, **19**(10), 38

DOYLE, W.H. (1972b). Some major instrument-connected CPI losses. *Instrumentation in the Chemical and Process Industries 8* (Pittsburgh, PA: Instrum. Soc. Am.)

DOYLE, W.H. (1980). Private communication

DOYLE, W.H. (1981). Case histories in duct fire incident. *Loss Prevention*, **14**, p. 181

DRABEK, T.E. (1969). *Laboratory Simulation of a Police Communications System Under Stress. Report.* (Ohio State Univ.)

DRAGER, K.H., SOMA, H.S. and FALMYR, O. (1988). PROF, a computer code for prediction of operator failure rate. Sayers, B.A. (1988), *op. cit.*, p. 158

DRAGOSAVIC, M. (1972a). Research on gas explosions in the Netherlands. *Building Research Estab. Symp. on Buildings and the Hazard of Explosion*, paper 3

DRAGOSAVIC, M. (1972b). *Structural Measures to Prevent Explosions of Natural Gas in Multi-storey Domestic Structures. Rep. B1–72–6/04.3.02.520.* TNO, Rijswijk, The Netherlands.

DRAGOSAVIC, M. (1973). Structural measures against gas explosion in high rise blocks of flats. *Heron*, **19**, 5

DRAGOSET, W.H. (1976). Cyclohexane oxidation process corrosion, piping failure and corrective action. *Prevention of Occupational Risks*, **3**, 323

DRAGUN, J., KUFFNER, A.C. and SCHNEITER, R.W. (1984). Transport and transformations of organic chemicals. *Chem. Engng*, **91** (Nov. 26), 65

DRAKE, A. (1989). Detecting incipient fires: a new approach. *Fire Surveyor*, **18**(5), 5

DRAKE, C.E. and CARP, J.R. (1960). Shell-and-tube heat exchangers. *Chem. Engng*, **67**(Apr. 18), 165

DRAKE, E.M. (1976). LNG rollover – update. *Hydrocarbon Process.*, **55**(1), 119

DRAKE, E.M. and CROCE, P.A. (1988). Guidelines for Safe Storage and Handling of Highly Toxic Hazard Materials: a new CCPS publication. *Plant/Operations Prog.*, **7**, 259

DRAKE, E.M. and PUTNAM, A.A. (1975). Vapor dispersion from spills of LNG on land. *Advances in Cryogenic Engineering*, **20**, 134

DRAKE, E.M. and REID, R.C. (1975). How LNG boils on soils. *Hydrocarbon Process.*, **54**(5), 191

DRAKE, E.M. and REID, R.C. (1977). The importation of liquefied natural gas. *Sci. American*, **236**(4), 29

DRAKE, E.M. and THURSTON, C.W. (1993). A safety evaluation framework for process hazards management in chemical facilities with PES-based controls. *Process Safety Prog.*, **12**, 92

DRAKE, E.M., GEIST, J.M. and SMITH, K.A. (1973). Prevent LNG 'rollover'. *Hydrocarbon Process.*, **52**(3), 87

DRAKE, E.M., HARRIS, S.H. and REID, R.C. (ca. 1973). *Analysis of Vapor Dispersion Experiments. Proj. IS-3-1*, paper D-76–80. (Am. Gas Ass)

DRAKE, E.M., JEJE, A.A. and REID, R.C. (1975). Transient boiling of liquefied cryogens on a water surface. 1, Nitrogen, methane and ethane. *Int. J. Heat Mass Transfer*, **18**, 1361

DRALEY, J.E. (1962). Corrosion of film-forming metals. *Chem. Engng*, **69**(Nov. 12), 256; **69**(Nov. 26), 152

DRANGEID, G.A. (1989). Safety is profitable business. *Noroil*, Feb., 23

DRANSFIELD, P.B. and GREIG, T.R. (1982). Decontaminating vessels containing hazardous materials. *Chem. Engng Prog.*, **78**(1), 35

DRANSFIELD, P.B. and LOWE, D.R.T. (1981). The problems involved in designing reliability into a process at its early research or development stages. *Reliable Production in the Process Industries* (Rugby: Instn *Chem. Engrs*), p. 67

DRAPER, E. and WATSON, R.R. (1970). *Collated Data on Fragments from Stacks of High Explosive Projectiles. Tech. Memo 2/70*. Directorate of Safety, Ministry of Defence.

DRAVNIEKS, A. (1974). Measuring industrial odors. *Chem. Engng*, **81**(Oct. 21), 91

DRAXLER, R.R. (1976). Determination of atmospheric diffusion parameters. *Atmos. Environ.*, **10**, 99

DRAXLER, R.R. (1979). Some observations on the along-wind dispersion parameter. *Fourth Symp. on Turbulence, Diffusion and Air Pollution* (Boston, MA: Am. Met. Soc.), p. 5

DRAXLER, R.R. (1987). Accuracy of various diffusion and stability schemes over Washington, DC. *Atmos. Environ.*, **21**, 491

DREIER, H. and ENGEL, U. (1971). Measures for protection of and practical experience with type 'e' explosion protected electrical machines. *Electrical Safety*, **1**, 22

DREISBACH, R. (1955–). *Physical Properties of Chemical Compounds*, vol. 1, 1955; vol. 2, 1959; vol. 3, 1961 (Washington, DC: Am. Chem. Soc.)

DREMIN, A.N. (1969). Critical phenomena in the detonation of liquid explosives. *Combustion*, **12**, 691

DRENICK, R.F. (1960). The failure law of complex equipment. *J. Soc. Ind. Appl. Math.*, **8**(4), 680

DRESCHER (1920). On the uptake of carbon monoxide and hydrogen cyanide through the skin. Dissertation, Univ. of Wurzburg

DRESSEL, M., HEINKE, G. and STEINHOFF, U. (1991). Inspect for maintenance with NDT methods. *Chem. Engng Prog.*, **87**(4), 30

DRESSLER, H. (1977). Problems of technical safety involved in the pulverisation of products which may cause dust explosions. *Prevention of Occupational Risks*, **5**, p. 175

DREW, B.C. (1989). *Piper Alpha Public Inquiry. Transcript, Day 150*

DREWITT, D.F. (1975). The insurance of chemical plants. *Course on Process Safety – Theory and Practice*. Dept of Chem. Engng, Teesside Polytechnic, Middlesborough

DREYFUS, H. and DREYFUS, S. (1986). Why expert systems do not exhibit expertise. *IEEE Expert*, **1**(2), 86

DREYFUS, H.L. (1972). *What Computers Can't Do* (New York: Harper & Row)

DRIEDONKS, A.G.M. and TENNEKES, H. (1984). Entrainment effects in the well-mixed atmospheric boundary layer. *Boundary Layer Met.*, **30**, 75

DRINKENBURG, B. (1988). Distillation – sophisticate or dinosaur? *Chem. Engr, Lond.*, **450**, 26

DRINKER, C.K. (1945). *Pulmonary Edema and Inflammation* (Cambridge, MA: Harvard Univ. Press)

DRINKER, P. and HATCH, T. (1954). *Industrial Dust*, 2nd ed. (New York: McGraw-Hill)

DRISCOLL, J.N. (1975). *Flue Gas Monitoring Techniques* (New York: Wiley)

DRISKELL, J.E. and SALAS, E. (1992). Collective behavior and team performance. *Hum. Factors*, **34**, 277

DRISKELL, L. (1976). Coping with high-pressure let down. *Chem. Engng*, **83**(Oct. 25), 113

DRISKELL, L. (1983). Predicting flow through control valves. *Chem. Eng*, **90**(Sep. 5), 94

DRISKELL, L. (1987). Select the right control valve. *Chem. Engng*, **94**(11), 123

DRISKELL, L.R. (1964). Piping of pressure relief devices. In Vervalin, C.H. (1964a), *op. cit.*, p. 130

DRISKELL, L.R. (1969). New approach to control valve sizing. *Hydrocarbon Process.*, **48**(7), 131

DRIVAS, P.J. (1982). Calculation of evaporative emissions from multi-component liquid spills. *Environ. Sci. Tech.*, **16**, 726

DRIVAS, P.J., SABNIS, J.S. and TEUSCHER, L.H. (1983). Model simulates pipeline, storage tank failures. *Oil Gas J.*, Sep. 12, 162

DROGARIS, G. (1991). *Major Accident Reporting System: Lessons Learned from Accidents Notified*. Comm. Europ Commun., Community Documentation Centre on Industrial Risk

DROGARIS, G. (1993). *Major Accident Reporting System* (Amsterdam: Elsevier)

DROPPO, J.G. and WATSON, E.C. (1975). Meteorological aspects of hazardous cargo risk assessments of hazardous materials transport. *Transport of Hazardous Cargoes*, **4**

DROSTE, B. (1992). Fire protection of LPG tanks with thin sublimation and intumescent coating. *Fire Technol.*, **28**, 257

DROSTE, B. and MALLON, M. (1989). A statistical review of the reasons leading to LPG accidents. *Loss Prevention and Safety Promotion 6*, p. 79.1

DROSTE, B. and SCHOEN, W. (1988). Full scale fire tests with unprotected and thermal insulation LPG storage tanks. *J. Haz. Materials*, **20**, 41

DROSTE, B. *et al.* (1984). Failure mechanisms of propane tanks under thermal stresses including fire engulfment. *Int. Conf. on Transport and Storage of LPG and LNG*, Brugge

DROZDOV, N.P. and ZELDOVICH, Y.B. (1943). *Zhur. Fiz. Khim.*, **17**, 134

DRUIJFF, H.J. (1977). Transport regulations for dangerous goods in the Netherlands. *Symp Hazard 77 – Advances in the Safe Handling and Movement of Hazardous Substances*, Teesside Polytechnic, Middlesborough

DRURY, C.G. (1983). Task analysis methods in industry. *Appl. Ergonomics*, **14**(1), 19

DRURY, C.G. and ADDISON, J.L. (1973). An industrial study of the effects of feedback and fault density on inspection performance. *Ergonomics*, **16**, 159

DRURY, C.G. and FOX, J.G. (eds) (1975). *Human Reliability in Quality Control* (London: Taylor & Francis)

DRURY, C.G. and SINCLAIR, M.A. (1983). Human and machine performance in an inspection task. *Hum. Factors*, **25**, 391

DRURY, C.G., PARAMORE, B., VAN COTT, H.P., GREY, S.M. and CORLETT, E.N. (1987). Task analysis. In Salvendy, G. (1987), *op. cit.*, p. 370

DRUST, J.G. (1972). Preventive maintenance through oil analysis. *Chem. Engng Prog.*, **68**(8), 84

DRYDEN, H.L. (1939). Turbulence and diffusion. *Ind. Engng Chem.*, **31**, 416

DRYDEN, I.G.C. (ed.) (1975). *The Efficient Use of Energy* (London: IPC and Sci & Tech. Press)

DRYDEN, P.S. and GAWECKI, L.J. (1987). The transport of hazardous goods: an approach to identifying and apportioning costs. *Twelfth Australian Transport Research Forum*, Brisbane

DRYDEN, R.J., RHODE, R.V. and KUHN, P. (1952). *Fatigue and Fracture of Metals* (New York: Wiley)

DRYER, F.L. and SCHUG, K.P. (1982). A comprehensive mechanism for the pyrolysis and oxidation of ethylene. *Combustion*, **19**, p. 153

DRYSDALE, D.D. (1971). The induction period in low-temperature hydrocarbon oxidation (letter). *Combust. Flame*, **17**, 261

DRYSDALE, D.D. (1976a). Dispersion of vapours. *Course on Fire and Explosion Safety in the Chemical Industry.* Dept of Fire Safety Engng, Univ. of Edinburgh

DRYSDALE, D.D. (1976b). Properties of open fires. *Course on Fire and Explosion Safety in the Chemical Industry.* Dept of Fire Safety Engng, Univ. of Edinburgh

DRYSDALE, D.D. (1980). Aspects of smouldering combustion. *Fire Prev. Sci. Technol.*, **23**, 18

DRYSDALE, D.D. (1985). *An Introduction to Fire Dynamics* (Chichester: Wiley)

DRYSDALE, D.D. (1989). *Piper Alphs Public Inquiry. Transcript, Day 111*

DRYSDALE, D.D. and DAVID, G.J. (1979/80). Hazard analysis for a storage sphere of pressurised liquefied flammable gas. *Fire Safety J.*, **2**(2), 91

DRYSDALE, D.D. and MACMILLAN, A.J.R. (1992). Flame spread on inclined surfaces. *Fire Safety J.*, **18**, 245

DRYSDALE, J. (1989). *Piper Alpha Public Inquiry. Transcript, Day 70*

DSCHANG (1928). Dissertation, Univ. of Wurzburg

DUA, P., MACFARLANE, J.W. and KRAUSE, B.D. (1994). Maintenance related incidents in topside systems. *Hazards*, **XII**, p. 367

DUANE, J.T. (1962). *Tech. Inf. Series Rep. DF 62 MD300*. General Electric Co.

DUBAR, J. and CALZIA, J. (1968). Explosive properties of mixtures of nitric acid, acetic anhydride and water. *Compt. Rend., Ser. C*, **266**, 1114

DUBE, J.C., ECKHARDT, L.L. and SMALLEY, A.J. (1991). Reciprocating compressor bottle sizing. *Hydrocarbon Process.*, **70**(7), 79

DUBNER, R.M. (1975). High speed gear failures – user experiences. *Hydrocarbon Process.*, **52**(12), 54

DUBOIS, D. and PRADE, H. (1980). *Fuzzy Sets and Systems* (New York: Academic Press)

DUCH, M.W., MARCALI, K., GORDON, M.D., HENSLER, C.J. and O'BRIEN, G.J. (1982). Thermal stability evaluation using differential scanning calorimetry and accelerating rate calorimetry. *Plant/Operations Prog.*, **1**, 19

DUCHSCHERER, H. and MOLZAHN, M. (1986). A practical approach to reviewing the design of chemical plants for safety. *Loss Prevention and Safety Promotion 5*, p. 17–1

DUCKWORTH, T. and MCGREGOR, A.A. (1989). The maintenance of relief systems. *Selection and Use of Pressure Relief Systems for Process Plant* (Manchester: Instn. Chem. Engrs, North-West Branch), p. 4.1

DUCLOS, J.M., VEYNANTE, D. and POINSOT, T. (1993). A comparison of flamelet models for premixed turbulent combustion. *Combust. Flame*, **95**, 101

DUCLOS, P. and BINDER, S. (1990). Public health consequences of acute chemical releases, Louisiana, 1986. *J. Haz. Materials*, **23**(1), 109

DUCLOS, P., BINDER, S. and RIETER, R. (1989). Community evacuation following the Spencer metal processing plant fire, Nanticoke, Pennsylvania. *J. Haz. Materials*, **22**(1), 1

DUCOMMUN, J.C. (1962). What status process safety. *Hydrocarbon Process.*, **41**(11), 246

DUCOMMUN, J.C. (1964a). How to use sample bombs safely. In Vervalin, C.H. (1964a), *op. cit.*, p. 137

DUCOMMUN, J.C. (1964b). This sample bomb exploded. In Vervalin, C.H. (1964a), *op. cit.*, p. 345

DUCROS, M. and SANNER, H. (1988). Application du programme CHETAH a l'e'tude de la sensibilisation de composes oxyge'ne's a l'estimation des limites infe'rieures d'inflammabilite'. *J. Haz. Materials*, **19**(1), 33

DUDA, R.O. and GASCHNIG, J.G. (1981). Knowledge-based expert systems come of age. *Byte*, Sep., 238

DUDA, R.O. and SHORTLIFFE, E. (1983). Expert systems research. *Science*, **220**, 261

DUDA, R.O., HART, P.E. and NILSSON, N.J. (1976). Subjective Bayesian methods for rule-based inference systems. *Proc. AFIP 1976 Nat. Computer Conf.*, **45**, 1075

DUDA, R.O., HART, P.E., NILSSON, N.J. and SUTHERLAND, G. (1978). Semantic network representations in rule-based inference systems. In D.A. Waterman and F. Hayes-Roth (1978), *op. cit.*, p. 203

DUDA, R.O., HART, P.E., BARRETT, P., GASCHNIG, J., KONOLIGE, J., REBOH, R. and SLOCUM, J. (1978). *Development of the PROSPECTOR Consultation System for Mineral Exploration. Tech. Rep.* Standford Res. Inst., Standford, CT

DUDA, R.O., GASCHNIG, J. and HART, P. (1979). Model design in the PROSPECTOR consultant system for mineral exploitation. In Michie, D. (1979), *op. cit.*, p. 153

DUDERSTADT, J.J. (1979). *Nuclear Power* (New York: Dekker)

DUDLEY, H.C. and NEAL, P.A. (1942). Toxicology of acrylonitrile (vinyl cyanide). *J. Ind. Hyg. Toxicol.*, **24**, 27

DUDLEY, H.C., SWEENEY, T.R. and MILLER, J.W. (1942). Toxicology of acrylonitrile (vinyl cyanide). II, Studies of effects of daily inhalation. *J. Ind. Hyg.*, **24**, 255

DUFAU, F.A., ESCHAR, D.F., HADDAD, A.C. and THONON, C.H. (1964). High purity cyclohexane. *Chem. Engng Prog.*, **60**(9), 43

DUFF, G.M.S. and HUSBAND, P. (1974). Emergency planning. *Loss Prevention and Safety Promotion 1*, p. 121

DUFF, J.C. (1976). Identifying metals on site. *The Metallurgist and Materials Technologist*, June

DUFF, W.R. (1975). Statistical analysis of fire and explosions in a petroleum company. In Hendry, W.D. (1975), *op. cit.*, paper E3

DUFFAUT, F. (1983). Progress in the manufacture and application of cryogenic invar. *Gastech 82*, p. 433

DUFFIELD, J.S. and NIJSING, R. (1994). Computer simulations describing the emergency pressure relief of hybrid and gassy systems. *Hazards*, **XII**, p. 505

DUFFY, A.R. and DAINORA, J. (1968). Considerations for LNG pipe material selection. *LNG Conf.*, **1**

DUFFY, A.R., GIDEON, D.N. and PUTTNAM, A.A. (1974). Dispersion and radiation experiments. *LNG Safety Program. Interim Report on Phase II Work. AGA Project IS-3–1.* Battelle Columbus Labs, Columbus, OH

DUFFY, J.J. (1983). Source control: modification of a flame retardant chemical. *Plant/Operations Prog.*, **2**, 241

DUFFY, T.M., CURRAN, T.E. and SASS, D. (1983). Document design for technical job tasks: an evaluation. *Hum. Factors*, **25**, 143

DUFOUR, J.W. (1989). Revised API pump standard. *Chem. Engng*, **96**(7), 101

DUGDALE, D.S. (1960). Yielding of steel sheets containing slits. *J. Mech. Phys Solids*, **8**, 100

DUGDALE, D.S. (1968). *Elements of Elasticity* (Oxford: Pergamon Press)

DUGGAN, J.J. (1964a). Basics of plant layout and design. In Vervalin, C.H. (1964a), *op. cit.*, p. 17

DUGGAN, J.J. (1964b). Five safer methods of chemical storage. In Vervalin, C.H. (1964a), *op. cit.*, p. 20

DUGGAN, P.C. (1978). Shore terminals – oil. In Institute of Petroleum (1978), *op. cit.*, p. 137

DUGGAN, P.J., JOHNSON, A.A. and ROGERS, R.L. (1994). Chemical reaction hazards associated with the use of sodium borohydride. *Hazards*, **XII**, p. 553

DUGGAN, T.V. and BYRNE, J. (1977). *Fatigue as a Design Criterion* (London: Macmillan)

DUGGER, G.L. (1962). *Effect of Initial Mixture Temperature on Flame Speed of Methane–air, Propane–Air and Ethylene–Air Mixtures. Rep. 1061.* Nat. Advisory Cttee Aeronaut

DUGGER, G.L. and SIMON, D.M. (1953). Prediction of flame velocities of hydrocarbon flames. *Combustion*, **4**, 336

DUGGER, G.L., WEAST, R.C. and HEIMEL, S. (1955). Effect of preflame reaction on flame velocity of propane–air mixtures. *Combustion*, **5**, 589

DUGLEUX, P. and FRELING, E. (1955). On the combustion of paraffin hydrocarbons. *Combustion*, **5**, 491

DUGSTAD, I. and VENKATRAM, A. (1989). An interpretation of Taylor's statistical analysis of particle dispersion. *Atmos. Environ.*, **23**, 1163

DUGUID, H.Q. (1965). How to prevent air compressor fires. *Hydrocarbon Process.*, **44**(4), 177

DUGUID, I.M. (1987). Analysing past incidents (letter). *Chem. Engr. Lond.*, **441**, 3

DUHL, R.W. (1967). Zirconium as a reactor lining. *Ammonia Plant Safety*, **9**, 78

DUHNE, C.R. (1979). Viscosity-temperature correlations for liquids. *Chem. Engng*, **86**(Jul. 16), 83

DUIJM, N.J. (1994). Jet flame attack on vessels. *J. Loss Prev. Process Ind.*, **7**, 160

DUIJM, N.J. and WEBBER, D.M. (1994). Dispersion in the presence of buildings. *J. Loss Prev. Process Ind.*, **7**, 118

DUIJM, N.J., VAN ULDEN, A.P., VAN HEUGTEN, W.H.H. and BUILTJES, P.J.H. (1986). Physical and mathematical modelling of heavy gas dispersion. In de Wispelaere, C., Schiermeier, F.A. and Gillani, N.V. (1986), *op. cit.*, p. 655

DUISER, J.A. (1984). Escape of liquefied gases from broken pipes. *The Protection of Exothermic Reactors and Pressurised Storage Vessels* (Rugby: Instn Chem. Engrs), p. 163

DUKES, J.A. (1975). Total damage control: practical experience on the reporting of the dangerous situation. In Hendry, W.D. (1975), *op. cit.*, paper B3

DUKES, R.R. and SCHWARTING, C.H. (1968). Choosing materials for making chlorine and caustic. *Chem. Engng*, **75**(Mar. 11), 206; (Apr. 8), 172

DUKLER, A.E and HUBBARD, M.G. (1975). A model for gas–liquid slug flow in horizontal and near-horizontal tubes. *Ind. Engng Fund.*, **14**, 337

DUKLER, A.E. and TAITEL, Y. (1984). *Flow Patterns for Steady Upward Gas-Liquid Flow in Vertical Tubes. Des. Manual VUFP-1.* Am. Inst. Chem. Engrs; Design Inst. for Multiphase Flow, New York

DUKLER, A.E., WICKS, M. and CLEVELAND, R.G. (1964). Frictional pressure drop in two phase flow. *AIChE J.*, **10**, 38 and 44

DUMAS, T. and BULANI, W. (1974). *Oxidation of Petrochemicals: Chemistry and Technology* (London: Applied Sciences)

DUMAY, J.M. (1982). Cryogenic flexible pipes for offshore LNG/LPG production. *Gastech 81*, p. 145

DUMITRIU, C. (1977). Questions coming up when ergonomy is applied in the design of machines and units of the chemical industry. *Prevention of Occupational Risks*, **5**, 51

DUMM, K. and FORTMANN, M. (1975). An initial approach to establishing an experimentally proven 'leak before rupture criterion' for SNR-300 components. *Int. J. Pres. Ves. Piping*, **3**, 43

DUMMER, G.W.A. and GRIFFIN, N. (1960). *Electronic Equipment Reliability* (London: Pitman)

DUMMER, G.W.A. and GRIFFIN, N. (1966). *Electronics Reliability – Calculation and Design* (Oxford: Pergamon Press)

DUNBAR, I.H., HIORNS, D.N., MATHER, D.J. and TICKEL, G.A. (1994). Atomisation and dispersion of toxic liquids resulting from accidental pressurised releases. *Hazards*, **XII**, p. 143

DUNBAVIN, P.R. (1993). Noise control at source. In Institution of Medical Engineers *op. cit.*, p. 31

DUNCAN, K.D. (1974). Analytical techniques in training design. In Edwards, E. and Lees, F.P. (1974), *op. cit.*, p. 283

DUNCAN, K.D. (1981). Training for fault diagnosis in industrial process plant. In Rasmussen, J. and Rouse, W.B. (1981), *op. cit.*, p. 553

DUNCAN, K.D. and GRAY, M.J. (1975a). An evaluation of a fault-finding course for refinery process operators. *J. Occup. Psychol.*, **48**, 199

DUNCAN, K.D. and GRAY, M.J. (1975b). Scoring methods for verification and diagnostic performance in industrial fault-finding problems. *J. Occup. Psychol.*, **48**, 93

DUNCAN, K.D. and SHEPHERD, A. (1975a). *Analysis and Training of Fault Location Tasks in the Chemical Industry. Report.* Chemical and Allied Products Ind. Training Board, Staines, Middlesex

DUNCAN, K.D. and SHEPHERD, A. (1975b). A simulator and training technique for diagnosing plant failures from control panels. *Ergonomics*, **18**, 627

DUNCAN, K.D., GRUNEBERG, M.M. and WILLIS, D. (eds) (1980). *Changes in Working Life* (Chichester: Wiley)

DUNCAN, K.P. (1977). The detection of long-term environmental hazards to health: occupational experience. *R. Soc. Mtg on Long-Term Hazards to Man from Man-made Chemicals in the Environment* (London)

DUNCAN, K.P. (1979). Occupational experience. *Proc. R. Soc., Ser. B*, **205**, 157. See also Doll, Sir R. and McLean, A.E.M. (1979), *op. cit.*, p. 157

DUNCAN, N. (1987). Private communication to P.A. Davies

DUNCAN, W.J., THOM, A.S. and YOUNG, A.D. (1960). *The Mechanics of Fluids* (London: Arnold)

DUNDEE DEPARTMENT OF TOWN AND REGIONAL PLANNING (1979). *The Implications of Large-scale Petrochemical Developments* (Dundee)

DUNHILL, S. (1984a). Integration bridges the gap between engineering programs. *Process Engng, Jun., 45*

DUNHILL, S. (1984b). Seals and packings keep the product in the process. *Process Engng, Aug., 42*

DUNKER, A.M. (1981). Efficient calculation of sensitivity coefficients for complex atmospheric models. *Atmos. Environ.*, **15**, 1155

DUNKLE, R.V. (1963). Configuration factors for radiant heat-transfer calculations involving people. *J. Heat Transfer*, **85**(1), 71

DUNLAP, R.W. and DELAND, M.R. (1978). Latest clean air requirements. *Hydrocarbon Process.*, **57**(10), 91

DUNLOP, A.K. (1970). Using corrosion inhibitors. *Chem. Engng*, **77**(Oct. 5), 108

DUNLOP, C.L. (1990). *Practical Guide to Maintenance Engineering* (Oxford: Butterworth-Heinemann)

DUNN (1918). *Report on Pathological Changes in the Lungs of Goats Resulting from the Inhalation of Phosgene. Rep. 9.* (London: Chemical Warfare Med. Cttee, Med. Res. Coun.)

DUNN, C.L. (ed.) (1952). *The Emergency Medical Services*, vol. 1 (London: HM Stationery Office)

DUNN, D.J. and STERN, T.E. (1952). *Hand Grenades for Rapid Incapacitation. Rep. BRL R-806.* Ballistic Res. Lab., Aberdeen Proving Ground, MD

DUNN, D.W., JOHNSON, M.L., HEDLEY, W.H., PATE, J.B., BARRETT, G.J. and MCKINNERY, W.N. (1983). Control practices at formaldehyde plants. *Chem. Engng Prog.*, **79**(3), 35

DUNN, K.S. (1975). Incineration's role in ultimate disposal of process wastes. *Chem. Engng*, **82**(Oct. 6), 141

DUNN, K.S. (1979). Problems and practicalities of incinerating chemical wastes. *Chem. Engr, Lond.*, **350**, 779

DUNN, R. (1984). *Software Defect Removal* (Maidenhead: McGraw-Hill)

DUNN, R. and ULLMAN, R. (1982). *Quality Assurance for Computer Software* (Maidenhead: McGraw-Hill)

DUNN, W.A. (1974). *Spill Risk Analysis Program. Phase II, Methodology Development and Demonstration.* Operations Res. Inc.

DUNN, W. and TULLIER, P.M. (1974). *Spill Risk Analysis Program. Phase II, Methodology and Demonstration.* Operations Res. Inc., Rep. to US Coast Guard

DUNN-RANKIN, D., BARR, P.K. and SAWYER, R.F. (1986). Numerical and experimental study of "tulip" flame formation in a closed vessel. *Combustion*, **21**, 1291

DUNNE, J.J.W. and HIGGINS, M.J. (1989). Downstream – safely. *Gastech 88*, paper 4.5

DUNBOBBIN, B.R., WERLEY, B.L. and HANSEL, J.G. (1991). Structured packings for use in oxygen service: flammability considerations. *Plant/Operations Prog.*, **10**, 45

DUNSTER, H.J. (1977). Reliability as a contribution to safety. *First Nat. Conf. on Reliability. NRC1/3* (Culcheth, Lancashire: UK Atom. Energy Auth.)

DUNSTER, H.J. (1980). Some reactions to the accident at Three Mile Island. *Nucl. Energy*, **19**, 139

DUNSTER, H.J. (1981). Regulation of risk. *Proc. R. Soc., Ser. A.*, **376**, 199. See also Warner, Sir F. and Slater, D.H. (1981), *op. cit.*, p. 199

DUNSTER, H.J. (1982). The assessment of the risks of energy; an iconoclastic view. *Energy World*, Jan., 2

DUNSTER, H.J. and VINCK, W. (1979). The assessment of risk – its value and limitations. *Nucl. Engng Int.*, Aug.

DUNSTER, J. (1973). Costs and benefits of nuclear power. *New Scientist*, Oct. 18, 192

DUNT, G.G.A. (1976). Classification and inspection frequencies for pressure vessels according to current in-service inspection codes. *Diagnostic Techniques* (London: Instn *Chem. Engrs*)

DUONG, D.Q. (1991). Modeling and hazard analysis of Sandia National Laboratories/Underwriters Laboratories' experiments. *Proc. Third Int. Symp on Fire Safety Science* (ed. by G. Cox and B. Langford) (London: Elsevier), p. 515

DU PONT COMPANY (1980). *Blaster's Handbook*, 16th ed. (Wilmington, DC)

DU PONT, R., THEODORE, L. and REYNOLDS, J. (1991). *Accident and Emergency Management* (New York: VCH)

DUPRE', G., JOANNON, J., KNYSTAUTAS, R. and LEE, J. (1991). Unstable detonations in the near-limit regime in tubes. *Combustion*, **23**, p. 1813

DURAND, E. (1966). *Electrostatique*, vols 1–3 (Paris: Masson)

DURANT, W.S. and PERKINS, W.C. (1987). A methodology for chemical hazards analysis at nuclear fuels reprocessing plants. *Preventing Major Chemical Accidents*, p. 6.85

DURHAM, A. (1984a). Keeping the machine on a tight leash at all times. *Computing*, Dec. 13, 26

DURHAM, A. (1984b). Programming an apprentice. *Computing*, Dec. 6, 24

DURHAM, A. (1985a). The expert systems helper. *Comput. Mag.*, May 30, 5

DURHAM, A. (1985b). A model approach to problem-solving. *Comput. Mag.* Feb. 21, 4

DURHAM, A. (1985c). Natural language. The PROLOG way. *Comput. Mag.*, Apr. 11, 4

DURHAM, A. (1985d). The tree of knowledge. *Comput. Mag.*, Apr. 25, 5

DURHAM, A. (1985e). Trying to mimic the mind. *Comput. Mag.*, Apr. 18, 7

DURHAM, A. (1985f). What expert systems need to become streetwise. *Computing*, Nov. 28, 2

DURHAM, A. (1986). How to serve industry by the appliance of science. *Computing*, Apr. 24, 42

DURHAM, A. (1987). Moving experts forward one small step at a time. *Computing*, Sep. 17

DURNIN, D. (1994). The contractor's response. *Offshore Safety Case Management* (Rugby: Instn *Chem. Engrs*), p. dd1

DURR, A.C. (1982). Accurate reliability prediction. *Reliab. Engng*, **3**(6), 475

DURR, C.A. and VAN LAERHOVEN, F.H.L. (1985). Vapour recovery from liquid hydrocarbon storage tanks. *Gastech 84*, p. 521

DURRANT, O.W. and LANSING, E.G. (1976). Furnace implosions and explosions. *Combustion*, **48**(3), 12

DUSWALT, A.A. (1991). Experimental errors in predictions of thermal hazards. *Plant/Operations Prog.*, **10**, 55

DUTT, S. (1996). Safe design of direct steam injection heaters. *The 5th World Cong. of Chem. Eng.*, July 14–18, (1996), San Diego, CA, Vol. II, pp. 1102–07. (New York: American Institute of Chemical Engineers)

DUTY, J.M., FISHER, D. and LEWIS, W.W. (1993). New realities of modular construction. *Hydrocarbon Process.*, **72**(12), 57

DUVAEL, W.A. (1979). Solid-waste disposal: land filling. *Chem. Engng*, **86**(Jul. 2), 77

DUVAL, R.C. (1985). Thermochemical hazard evaluation. In Hoffman, J.M. and Maser, D.C. (1985), *op. cit.*, p. 57

DUWEZ, P.E., CLARK, D.S. and BOHNENBLUST, H.F. (1950). The behavior of long beams under impact loading. *J. Appl. Mech.*, **17**, 27

DUXBURY, H.A. (1976). Gas vent sizing methods. *Loss Prevention*, **10**, 147

DUXBURY, H.A. (1979a). Relief line sizing for gases. *Chem. Engr, Lond.*, **350**, 783; **351**, 851

DUXBURY, H.A. (1979b). Relief system sizing for polymerisation reactions. In Minors, C. (1979), *op. cit.*, paper 3

DUXBURY, H.A. (1980). The sizing of relief systems for polymerisation reactors. *Chem. Engr, Lond.*, **352**, 31

DUXBURY, H.A. (1993). Private communication

DUXBURY, H.A. and WILDAY, A.J. (1987). Calculation methods for reactor relief: a perspective based on ICI experience. *Hazards from Pressure*, p. 175

DUXBURY, H.A. and WILDAY, A.J. (1989). Efficient design of reactor relief systems. *Runaway Reactions*, p. 372

DUXBURY, H.A. and WILDAY, A.J. (1990). The design of reactor relief systems. *Process Safety Environ.*, **68B**, 24

DVORE, D.S. and VAGLIO-LAURIN, R. (1982). Atmospheric diffusion of small instantaneous point releases near the ground. *Atmos. Environ.*, **16**, 2791

DWYER, J.J. (1971a). How to specify centrifugal compressors. Air filters and intercoolers. *Hydrocarbon Process.*, **50**(10), 80

DWYER, J.J. (1971b). How to specify centrifugal compressors. Operating checks and data. *Hydrocarbon Process.*, **50**(10), 87

DWYER, J.J. (1973). Compressor problems – causes and cures. *Hydrocarbon Process.*, **52**(1), 77

DWYER, J.J. (1974). Solve centrifugal compressor problems before start-up. *Chem. Engng Prog.*, **70**(10), 71

DWYER, P. (1985). Coming up: regulation of underground tanks. *Chem. Engng*, **92**(Apr. 29), 17

DYBDAL, E.C. (1966). Subject index of CE cost files: 1958 to 1966. *Chem. Engng*, **73**(Dec. 5), 166

DYBKJAER, I., BOHLBRO, H., ALDRIDGE, C.L. and RILEY, K.L. (1980). Sulphur tolerant shift process and catalyst. *Ammonia Plant Safety*, **21**, p. 145

DYE, M.S. (1979). The Anticipator hazard warning system. *Design 79* (London: Instn Chem. Engrs)

DYE, R. (1977). Major field repair on ammonia converter. *Ammonia Plant Safety*, **19**, 90

DYER, A.F. (1969). LNG from Alaska to Japan. *Loss Prevention*, **3**, p. 92. See also *Chem. Engng Prog.*, 1969, **65**(4), 53

DYER, A.J. (1974). A review of flux–profile relationships. *Boundary Layer Met.*, **7**, 363

DYER, A.J. and HICKS, B.B. (1970). Flux–gradient relationships in the constant flux layer. *Quart. J. R. Met. Soc.*, **96**, 715

DYER, J.C. and MIGNONE, N.A. (1983). *Handbook of Industrial Residues* (Park Ridge, NJ: Noyes)

DYESS, C.E. (1973). Putting a 70,000 bbl/day crude still on stream. *Chem. Engng Prog.*, **69**(8), 98

DYFED COUNTY FIRE BRIGADE (1983). *Report of the Investigation into the Fire at Amoco Refinery – 30th August, 1983* (Carmarthen, Dyfed)

VAN DYKE, B.H. and KAWALLER, S.I. (1980). Chemically reactive coatings for passive fire protection in LNG and LPG storage and transportation. *Gastech 79*, p. 201

DYKES, G.J. (1992). Flexible regulation of the chemical industry. *Loss Prevention and Safety Promotion 7*, vol. 1, lecture

DYKMAN, M. and MICHEL, B.J. (1968). Comparing mechanical aerator designs. *Chem. Engng*, **76**(Mar. 10), 117

DYSON, R. (1987). CIMAH safety cases: a view from British Gas. *Chem. Engr, Lond.*, **439**, 7

DYSON, R. (1989). Industrial approaches to the safety case. In Lees, F.P. and Ang, M.L. (1989b), *op. cit.*, p. 122

DZIEWULSKI, T.A. (1973). Factors affecting reliability of large rotary and reciprocating equipment. *Chem. Engr, Lond.*, **278**, 477

EADES, M. (1991). Assessing and minimizing safety risks. *Process Ind. J.*, Sep., 44

EAGLEMAN, J.R. (1983). *Severe and Unusual Weather* (New York: van Nostrand)

EAGLEMAN, J.R., MUIRHEAD, V.U. and WILLIAMS, N. (1975). *Thunderstorms, Tornadoes and Building Damage* (Lexington, MA: Lexington Books)

EAGON, H.B., HOWARD, J.F., SMITH, A.C. and DIEFENBACH, R.E. (1976). Removal of hazardous fluid from the ground water in a congested area – a case history. *Control of Hazardous Material Spills*, **3**, 373

EALES, L.C. (1977a). Earthing welding currents. *Plant Engr*, **21**(7), Jul./Aug., 16

EALES, L.C. (1977b). Electric shock. *Plant Engr*, **22**(1), 21

EALES, L.C. (1977c). Other methods of reducing shock to earth hazards. *Plant Engr*, **21**(4), 20

EAMES, A.R. (1966). Assessment of protective systems. *Nucl. Engng*, **11**, 189

EAMES, A.R., FOTHERGILL, C.D.H., ROUGHLEY, R. and WOODCOCK, E.R. (1976). Operation of a reliability data bank. In Henley, E.J. and Lynn, J.W. (1976), *op. cit.*, p. 99

EAMES, J.P., STURGIS, N.H. and WEEKS, C.F. (1959). The contractor's role in plant start-up *Chem. Engng Prog.*, **55**(8), 47

EARLE (1975). Insurance. HASAWA's effect on the plant engineer. *Plant Engr*, **23**(5), 23

EARLY, W.F. (1991). Process safety management: an emerging consensus. *Hydrocarbon Process.*, **70**(2), 105

EARLY, W.F. and EIDSON, M.A. (1990). Design for zero releases. *Hydrocarbon Process.*, **69**(8), 47

EARLY, W.F., LIVINGSTON, M.C. and NEWSOM, D.E. (1990). ARCHIE evaluates hazard incidents. *Hydrocarbon Process.*, **69**(11), 69

EARTHQUAKE ENGINEERING FIELD INVESTIGATION TEAM (1991). *The Luzon, Philippines, Earthquake of 16 July 1990* (London)

EARTHQUAKE ENGINEERING RESEARCH INSTITUTE (1952). *Earthquake and Blast Effects on Structures* (Univ. of California)

EASON, N.W. and WHITE, E.N. (1977). Computers as aids to maintenance. *Chart. Mech. Engr*, Oct., 78

EASTERBROOK, J. and GAGLIARDI, D.V. (1984). Sewers can pass on problems. *Plant/Operations Prog.*, **3**, 29

EASTMAN, C. (1978). The representation of design problems and maintenance of their structure. In J.C. Latombe (1978), *op. cit.*, p. 335

EASTON, K. (1977a). Immediate care. In Easton, K.D. (1977), *op. cit.*, p. 174

EASTON, K. (ed.) (1977b). *Rescue Emergency Care* (London: Heinemann)

EASTOP, T.D. and McCONKEY, A. (1963). *Applied Thermodynamics for Engineering Technologists; 1970*, 2nd ed.; 1978, 3rd ed.; 1986, 4th ed. (London: Longmans); 1993, 5th ed. (Harlow: Longmans)

EASTWOOD, D.G. (1971). Application and design of flameproof transformers and switchgear. *Electrical Safety*, **1**, 92

EASTWOOD, T.W. (1994). Analyze hazards for existing plant. *Chem. Engng Prog.*, **90**(1), 72

EBADAT, V. (1991). Influence of powder streaming current on surface potential of bulked high density polyethylene. *Electrostatics*, **8**, 77

EBADAT, V. (1993). Keep dust-cloud explosions at bay. *Chem Engng*, **100**(12), 123

EBADAT, V. and CARTWRIGHT, P. (1991a). Electrostatic hazards associated with the use of flexible intermediate bulk containers in flammable atmospheres. *Electrostatics*, **8**, 91

EBADAT, V. and CARTWRIGHT, P. (1991b). Electrostatic hazards in the use of flexible intermediate bulk containers. *Hazards*, **XI**, 105

EBBIN, S. and KASPER, R. (1974). *Citizen Groups and the Nuclear Power Controversy* (Cambridge, MA: MIT Press)

EBERHART, T.M. (1979). Designing fiberglass tanks for earthquake conditions. *Chem. Engng*, **86**(Jan. 15), 147

EBERLEIN, A.J. (1974). Loss prevention in relation to process plant design. *Course on Process Safety – Theory and Practice*. Dept of Chem. Engng, Teesside Polytechnic, Middlesborough

EBERLEIN, J. (1987). CIMAH safety cases: safety case modelling at Shell UK. *Chem. Engr, Lond.*, **439**, 11

EBERT, F. (1982–83). Threefold explosion protection. *Fire Int.*, **78**, 27

EBERT, F. and BECKER, T. (1982). Model calculations on the explosion of free gas clouds containing mixtures of hydrocarbons and air. *Chem. Ing. Tech.*, **54**, 736

EBERTS, R.E. (1985). Cognitive skills and process control. *Chem. Engng Prog.*, **81**(12), 30

ECHIGO, R., NISHIWAKI, N. and HIRATA, M. (1967). A study on the radiation of luminous flames. *Combustion*, **11**, 381

ECHOLS, W.T. (1976). Estimating the hazardous radius of exposure from accidental releases of hydrogen sulfide gas. *Control of Hazardous Material Spills*, **3**, 201

ECK, B. (1974). *Fans* (Oxford: Pergamon Press)

ECKE, H.O.G. and DREYHAUPT, F.J. (1975). Environmental protection policy and measures to reduce hydrocarbon emissions from oil refineries. *Proc. Ninth World Petrol. Cong.*, vol. 6 (London: Applied. Science), p. 229

ECKENFELDER, W. (1966). *Industrial Water Pollution Control* (New York: McGraw-Hill)

ECKENFELDER, W. (1969). Economics of wastewater treatment. *Chem. Engng*, **76**(Aug. 25), 109

ECKENFELDER, W.W. (1982). Chemical and physical measures. In Bennett, G.F., Feates and Wilder (1982), *op. cit.*, pp. 9–50

ECKENFELDER, W. and O'CONNOR, D.J. (1961). *Biological Waste Treatment* (Oxford: Pergamon Press)

ECKENFELDER, W.W., PATOCZKA, J. and WATKIN, A.T. (1985). Wastewater treatment. *Chem. Engng*, **92**(Sep. 2), 60

ECKER, H.W., JAMES, B.A. and TOENSING, R.H. (1974). Electrical safety: designing purged enclosures. *Chem. Engng*, **81**(May 13), 93

ECKERT, E.R.G. and DRAKE, R.M. (1959). *Heat and Mass Transfer* (New York: McGraw-Hill)

ECKERT, H.K. (1965). Insurance management and loss prevention. *Chem. Engng. Prog.*, **61**(8), 24

ECKHART, R.A. (1969). Dynamic simulation of an LPG vaporizer. *Ind. Engng Chem. Proc. Des. Dev.*, **8**, 491

ECKHOFF, R.K. (1975). Towards absolute minimum ignition energies for dust clouds? *Combust. Flame*, **24**, 53

ECKHOFF, R.K. (1976a). *A Study of Selected Problems related to the Assessment of Ignitability and Explosibility of Dust Clouds. Report* (Bergen: Christian Michelsen Inst.)

ECKHOFF, R.K. (1976b). *A Study of Selected Problems Related to the Assessment of Ignitability and Explosibility of Dust Clouds* (Bergen: A.S. John Griegs Boktrykkeri)

ECKHOFF, R.K. (1977). Pressure development during explosion in clouds of dusts from grain feedstuffs and other natural organic materials. *Fire Res.*, **1**(2), 71

ECKHOFF, R.K. (1982). Necessary dust explosion vent areas for large storage silos in the grain, feed and flour industries. *The Control and Prevention of Dust Explosions* (London: Oyez/IBC), paper 16

ECKHOFF, R.K. (1984). Dust explosion experiments in a 500 m^3 silo cell. *J. Occup Accid.*, **6**(4), 229

ECKHOFF, R.K. (1985). Use of (dP/dt)$_{max}$ from closed bomb tests for predicting violence of accidental dust explosions in industrial plants. *Fire Safety J.*, **8**, 159

ECKHOFF, R.K. (1989). Sizing of dust explosion vents in the process industries. *Loss Prevention and Safety Promotion* 6, p. 21.1

ECKHOFF, R.K. (1990). Sizing of dust explosion vents in the process industries. Advances made during the 1980s. *J. Loss Prev. Process Ind.*, **3**(3), 268

ECKHOFF, R.K. (1991). *Dust Explosions in the Process Industries* (Oxford: Butterworth-Heinemann)

ECKHOFF, R.K. (1993). Dust explosion research. State-of-the-art and outstanding problems. *J. Haz. Materials*, **35**, 103

ECKHOFF, R.K. (ed.) (1994). *Safety on Offshore Process Installations* (Oxford: Butterworth-Heinemann)

ECKHOFF, R.K., ALFERT, F., FUHRE, K. and PEDERSEN, G.H. (1988). Maize starch explosions in a 236 m^3 experimental silo with vents in the silo wall. *J. Loss Prev. Process Ind.*, **1**(1), 16

ECKHOFF, R.K. and ENSTAD, G. (1976). Why are "long" electric sparks more effective dust explosion initiators then 'short' ones? *Combust. Flame*, **27**, 129

ECKHOFF, R.K. and FUHRE, K. (1983). Dust explosion experiments in a vented 500 m^3 silo cell. In *Loss Prevention and Safety Promotion 4*, vol. 3, F11

ECKHOFF, R.K. and MATHISEN, K.P. (1978). A critical examination of the effect of dust moisture on the rate of pressure rise in Hartmann bomb tests. *Fire Res.*, **1**(4/5), 273

ECKHOFF, R.K. and THOMASSEN, O. (1994). Possible sources of ignition of potential explosive gas atmospheres on offshore process installations. *Loss Prev. Process Ind.*, **7**, 281

ECKHOFF, R.K., FUHRE, K., HENERY, M., PARKER, S. and THORSEN, H. (1982). *Dust Explosion Experiments in a 500 m^3 Vented Silo*. (Bergen: Christian Michelsens Inst)

ECKHOFF, R.K., FUHRE, K., GUIRAO, C.M. and LEE, J.H.S. (1984). Venting of turbulent gas explosions in a 50 m^3 chamber. *Fire Safety J.*, **7**, 191

ECKHOFF, R.K., FUHRE, K. and PEDERSEN, G.H. (1986). Dust explosion experiments in a vented 236 m^3 silo cell. *Loss Prevention and Safety Promotion 5*, 62–1

ECKHOFF, R.J., FUHRE, K. and PEDERSEN, G.H. (1987). Dust explosion experiments in a vented 236 m^3 silo cell. *J. Occu Accid.*, **9**(3), 161

ECKLES, A.J. and BENZ, P.H. (1992). The basics of vacuum processing. *Chem. Engng*, **99**(1), 78

ECKMAN, R.M. (1994). Re-examination of empirically derived formulas for horizontal diffusion from surface sources. *Atmos. Environ.*, **28A**, 265

ECKSTROM, C.M. (1973). In-plant health hazards. *Chem. Engng*, **80**(Jun. 18), 173

ECOLOGY AND ENVIRONMENT INC. (1985). *Toxic Substance Storage Tank Containment* (New York: Noyes)

ECONOMOPOULOS, A.P. (1978). A fast computer method for distillation calculations. *Chem. Engng*, **85**(April 24), 91

EDA, H. (1976). *Vessel Manoeuvring Simulation*. Stevens Inst. of Technol., Re to US Coast Guard

EDDY, D.E., SCHROEDTER, P.M. and STRAUCH, W.G. (1966). Safety features of refrigeration systems for atmospheric storage of ammonia. *Ammonia Plant Safety*, **8**, 21.

EDDY, P., POTTER, E. and PAGE, B. (1976). *Destination Disaster* (London: Hart-Davis)

EDELEANU, C.B. (1976). A general case for on-line equipment monitoring. *Diagnostic Techniques* (London: Instn Chem. Engrs)

EDELEANU, C. (1978). Safety in process plant in relation to corrosion. *Chemy Ind.*, Sep. 2, 649

EDELEANU, C. (1981). Equipment on line surveillance: an aid to reliability. *Reliable Production in the Process Industries* (Rugby: Instn Chem. Engrs), p. 129

EDEN, H.F. (1971). Aspects of electrostatic charging by powders: analysis of potential hazards. *Am. Inst. Chem. Engrs, Proc. Sixty-Fourth Ann. Mtg* (New York)

EDEN, H.F. (1972). Potential hazards of electrostatic charging by powders. *Loss Prevention* **6**, 27

EDESKUTY, F.J. and WILLIAMSON, K.D. (eds) (1983). *Liquid Cryogens* (Boca Raton, FL: CRC Press)

EDGAR, M.D., JOHNSON, A.D., PISTORIUS, J.T. and VARADI, T. (1984). Troubleshooting made easy. *Hydrocarbon Process.*, **63**(5), 65

EDGE, R.J. (1976). Asbestos-related disease in Barrow-in-Furness. *Environ. Res.*, **11**, 244

EDHOLM, O.G. (1967). *The Biology of Work* (London: World Univ. Library)

EDIGER, D. (1978). Washington's LNG role: no more in deep freeze. *Chem. Engng*, **85**(Jun. 19), 70

EDINBERG, D.L., VAN MATER, P.R., ORSERO, and FINIFTER, D. (1982). A comparison of the collision resistance of membrane tank-type and spherical tank-type LNG tankers. *Gastech* **81**, 217

EDISON ELECTRIC INSTITUTE (1963). *Analytical Report of Equipment Availability for the Seven Year Period 1955–61. Publ. 63–42*

EDISON ELECTRIC INSTITUTE (1966). *Report on Co-ordination of Electric Power Supply. Publ. 66–30*

EDISON ELECTRIC INSTITUTE (1967). *Report on Equipment Availability for the Seven Year Period 1960–66. Publ. 67–23*

EDKINS, J.S. and TWEEDY, N. (1918). *The Early Changes produced in the Lungs by Gas Poisoning and their Significance*. Rep. 2. Chemical Warfare Med. Cttee, Med. Res. Coun., London.

EDMISTER, W.C. (1947–). Application of thermodynamics to hydrocarbon processing. *Petrol. Refiner, 1947*, **26**(7), 77; **26**(8), 83; **26**(9), 105; **26**(10), 99; **26**(11), 110; **26**(12), 112; 1948, **27**(1), 74; **27**(2), 100; **27**(3), 83; **27**(4), 134; **27**(5), 93; **27**(6), 104

EDMISTER, W.C. and LEE, B.I. (1984). *Applied Hydrocarbon Thermodynamics, 2nd ed.* (Houston, TX: Gulf Publ. Co.)

EDMONSON, H. and HEAP, M.P. (1969). A precise test of the flame-stretch theory of blow-off. *Combustion*, **12**, 1007

EDMONDSON, H. and HEAP, M.P. (1970). The correlation of burning velocity and blowoff data by the flame stretch principle. *Combust. Flame*, **15**, 179

EDMONDSON, H. and HEAP, M.P. (1971). The burning velocity of hydrogen–air flames. *Combust. Flame*, **16**, 161

EDMONDSON, I.R. (1977). Removal of flue gas tunnels solves problems. *Ammonia Plant Safety*, **19**, 138

EDMONDSON, J.N. (1990). The current state of risk assessment. *Safety and Loss Prevention in the Chemical and Oil Processing Industries*, p. 509

EDMONDS, J.E. and WYNNE, G.C. (1927). *History of the Great War: Military Operations: France and Belgium, 1915*, vol. 1 (London)

EDSALL, J.T. (1975). Scientific freedom and responsibility. *Science*, **188**, 687

EDSTROM, J.O. and LJUNGBERG, L. (1964). Low-carbon stainless steels for process uses. *Chem. Engng*, **71**(Dec. 21), 114

EDSTROM, J.O. and LJUNGBERG, L. (1965). Low-carbon stainless. *Chem. Engng*, **72**(Jan. 18), 184

EDWARDS, A.R. (1978). Water reactor containment. *Course on Reactor Safety* UKAEA, Safety and Reliab. Directorate, Harwell

EDWARDS, C.A. (1974). *Persistent Pesticides in the Environment* (Boca Raton, FL: CRC Press)

EDWARDS, D.G., WILLIAMS, G.T. and BREEZE, J.C. (1959). Pressure and velocity measurements in detonation waves in hydrogen-oxygen mixtures. *J. Fluid Mech.*, **6**, 497

EDWARDS, D.H., THOMAS, G.O. and NETTLETON, M.A. (1979). The diffraction of a planar detonation wave at an abrupt area change. *J. Fluid Mech.*, **95**, 79

EDWARDS, D.H., FEARNLEY, P., THOMAS, G.O. and NETTLETON, M.A. (1981). Shocks and detonations in channels with 90 bends. *Combust. Inst., First Specialist Mtg, French Section* (Pittsburgh, PA), 431

EDWARDS, D.H., THOMAS, G.O. and NETTLETON, M.A. (1983). The diffraction of detonation waves in channels with 90° bends. *Arch. Comb.*, **3**, 65

EDWARDS, D.K. (1963). *Studies of Infrared Radiation in Gases. Rep 62–65.* Univ. of Calif., Los Angeles, CA

EDWARDS, D.W. and GUPTA, J.P. (2002). Inherently Safer design – present and future. *Trans. IChemE* **80**, Part B, (May), 115–24

EDWARDS, D.W. and LAWRENCE, D. (1993). Assessing the inherent safety of chemical process routes: is there a relation between plant costs and inherent safety? *Process Safety Environ.*, **71B, 252**

EDWARDS, D.W. and LAWRENCE, D. (1993). Assessing the inherent safety of chemical process routes: is there a relation between plant costs and inherent safety? *Trans. IChemE* 71, Part B, (November), 252–58

EDWARDS, D.W. and LAWRENCE, D. (1995). Inherent safety assessment of chemical process routes. *The 1995 ChemE Research Event, First Eur. Conf. Young Res. Chem. Eng.*, 62–64. (Rugby, UK: Inst. Chem. Eng)

EDWARDS, D.W., LAWRENCE, D. and RUSHTON, A.G. (1996). Quantifying the inherent safety of chemical process routes. *5th World Cong. of Chem. Eng.*, July 14–18, 1996, San Diego, CA, Vol. II, 1113–18 (New York: American Institute of Chemical Engineers)

EDWARDS, E. (1964). *Information Transmission* (London: Chapman & Hall)

EDWARDS, E. (1973). Ergonomics in control. *Proc. IEE*, **120**, 1181

EDWARDS, E. (1979). Flight deck alarm systems. *Aircraft Engng.* Feb., 11

EDWARDS, E. and LEES, F.P. (1971a). The development of the role of the human operator in process control. *Interfaces with the Process Control Computer* (ed. T.J. Williams) (Pittsburgh, PA: Instrum. Soc. Am.), 138

EDWARDS, E. and LEES, F.P. (1971b). The influence of the process characteristics on the role of the human operator in process control. *Data Reduction, Communication and Presentation for the Process Operator* (London: Inst. Measamt Control)

EDWARDS, E. and LEES, F.P. (1973). *Man and Computer in Process Control* (London: Instn. Chem. Engrs)

EDWARDS, E. and LEES, F.P. (eds) (1974). *The Human Operator in Process Control* (London: Taylor & Francis)

EDWARDS, G.T. (1987a). Defences against dependent failure. *SRS Quart. Digest*, Jul. 4

EDWARDS, G.T. (1987b). Dependent failure modelling. *SRS Quart. Digest*, Oct. 12

EDWARDS, G.T. (1987c). Options for maintenance strategy. *SRS Quart. Digest*, Jul. 19

EDWARDS, G.T. (1988). A dependent failures data study. *SRS Quart. Digest*, Jan. 9

EDWARDS, H.R. (1983). Electrostatic hazards associated with marine tanker operations, water washing and product handling. *Loss Prevention and Safety Promotion 4*, vol. 2, C1

EDWARDS, H.R. and UNDERWOOD, M.C. (1984). The ignition of powder air mixtures by discharge of static electricity. *J. Electrostatics*, **15**, 123

EDWARDS, J.C. and BUJAC, P.D.B. (1990). Design for hygiene. *Process Safety Environ.*, **68B**, 203

EDWARDS, J.D. (1980). Management of the chemical plant maintenance function. *Chem. Engr, Lond.*, **356**, 316

EDWARDS, K.W., FARIDANY, E.K., SLOGGETT, J.E. and ARREGGER, J.E. (1982). The OLASCO offshore liquefaction and shipping system for marginal gas fields. *Gastech 81*, 108

EDWARDS, P.R. and PUMPHREY, N.W.J. (1982). Ingestion and retention of mercury by sheep grazing near a chlor-alkali plant. *J. Sci. Food Agric.*, **23**, 237

EDWARDS, P.R., CAMPBELL, I. and MILNE, G.S. (1982). The impact of chloromethanes on the environment. *Chemy Ind.*, Aug. 21, 574; Sep. 4, 619; Sep. 18, 714

EDWARDS, R.M. (1987). Improve your acceptance criteria for high energy pumps. *Process Engng*, Apr. 33

EDWARDS, R.E., SPEED, N.A. and VERWOERT, D.E. (1983). Clean-up of chemically contaminated sites. *Chem. Engng*, **90**(Feb. 21), 73

EDWARDS, W. (1965). Probabilistic information processing systems for action selection. *Informatiom System Sciences* (ed. J. Spiegel and D. Walker) (Washington, DC: Spartan)

EDWARDS, W. and TVERSKY, A. (1976). *Decision Making* (Baltimore, MD: Penguin)

EDWORTHY, J., LOXLEY, S. and DENNIS, I. (1991). Improving auditory warning design: relationships between warning sound parameters and perceived urgency. *Hum. Factors*, **33**, 205

EFTIS, J. and LIEBOWITZ, H. (1972). On the modified Westergaard equations for certain plane crack problems. *Int. J. Frac. Mech.*, **8**, 383

EFTIS, J., SUBRAMANIAN, N. and LIEBOWTIZ, H. (1977). Crack border stress and displacement equations revisited. *Engng Frac. Mech.*, **9**, 189

EGAN, B. (1975). *Safety Policies – How to Write and Enforce Them* (London: Commercial)

EGAN, B. (1978). *The Manual of Safety Representation*. Vol. 2, *Safety Inspections* (London: New Commercial)

EGAN, B. (1979). *Safety Policies* (London: New Commercial)

EGAN, B.A. (1976). Turbulent diffusion in complex terrain. In Haugen, D.A. (1976), *op. cit.*, p. 112

EGAN, B.A., D'ERRICO, R. and VAUDO, C. (1979). Estimating air quality levels in regions of high terrain under stable atmospheric conditions. *Fourth Sym on Turbulence, Diffusion and Air Pollution* (Boston, MA: Am. Met. Soc.), 177

EGEIBAN, O.M., GRIFFITHS, J.F., MULLINS, J.J. and SCOTT, S.K. (1982). Explosion hazards in exothermic materials: critical conditions and scaling rules for masses of different geometries. Combustion., **19**, 825

EGERTON, SIR A. (1953). Limits of inflammability. *Combustion*, **4**, 4

EGERTON, SIR A., EVERETT, A.J. and MOORE, N.P.W. (1953). Sintered metals as flame traps. *Combustion*, **4**, 689

EGERTON, A. and LEFEBVRE, A.H. (1954). The effect of pressure variations on burning velocities. *Proc. R. Soc., Ser. A*, **222**, 206

EGERTON, SIR A.C. and POWLING, J. (1948). *Proc. R. Soc., Ser. A*, **190**, 172 and 190

EGERTON, SIR A.C. and THABET, S.K. (1952). *Proc. R. Soc., Ser. A*, **211**, 445

EGERTON, H.B. (1969). *Non-Destructive Testing* (Oxford: Oxford Univ. Press)

EGGE, E.D. and BOCKENHAUER, M. (1991). Calculation of the collision resistance of ships and its assessment for classification purposes. *Marine Structures*, **4**, 35

EGGERS, R. and GREEN, V. (1990). Pressure discharge from a pressure vessel filled with CO_2. *J. Loss Prev. Process Ind.*, **3**(1), 59

EGGERSTEDT, P.M., ZIEVERS, J.F. and ZIEVERS, E.C. (1993). Choose the right ceramic for filtering hot gases. *Chem. Engng Prog.*, **89**(1), 62

EGGINTON, J. (1980). *Bitter Harvest. The Poisoning of Michigan* (London: Secker & Warburg)

EGGLESTON, L.A. (1967). Dust and the process engineer. *Loss Prevention*, **1**, 39

EGGLESTON, L.A., HERRERA, W.R. and PISH, M.D. (1976). Water spray to reduce vapor cloud spray. *Loss Prevention*, **10**, 31

EGLI, U.M. and RIPPIN, D.W.T. (1986). Short-term scheduling for multiproduct batch chemical plants. *Comput. Chem. Engng*, **10**, 303

EGOL, L. (1989a). Hazardous materials trigger an explosion. *Chem. Engng*, **96**(4), 179

EGOL, L. (1989b). Simulation and process control. *Chem. Engng*, **96**(2), 161

EGOLF, L.M. and JURS, P.C. (1992). Estimation of autoignition temperatures of hydrocarbons, alcohols and esters from molecular structure. *Ind. Engng Chem. Res.*, **31**, 1798

EHRENBERGER, W. (1987). Systematic versus probabilistic verification efforts—some examples. In Littlewood, B. (1987b), *op. cit.*, p. 125

EHRENBERGER, W.D. (1993). Design and licensing of safety-related software. In Bennett, P. (1993), *op. cit.*, p. 227

EHRENFELD, J. and BASS, J. (1984). *Evaluation of Remedial Action Unit Operations at Hazardous Waste Disposal Sites* (Park Ridge, NJ: Noyes)

EHRLER, J.J. and JURAN, B. (1982). VAM and AA by carbonylation. *Hydrocarbon Process.*, **61**(2), 109

EIBER, R.J. (1981). Review of delayed tank car failure. *Ammonia Plant Safety*, **23**, 146

EIBY, G.A. (1980). *Earthquakes* (London: Heinemann)

EICHEL, F.G. (1967). Electrostatics. *Chem. Engng*, **74**(Mar. 13), 153

EICHELBERGER, W.G., SMURA, B.B. and BERGENN, W.R. (1961). Explosions and detonations in chlorine production. *Chem. Engng., Prog.*, **57**(8), 95

EICHHORN, J. (1964). Mists can cause explosions. In Vervalin, C.H. (1964a), *op. cit.*, p. 106

EICHLER, T.V. and NAPADENSKY, H.S. (1977). *Accidental Vapor Phase Explosions on Transportation Routes near Nuclear Plants. Rep. J6405 ANL-K77-3776-1*. Argonne Nat. Lab., Argonne, IL.

EICHMANN, E. (1971). US structural and foundation standards. *Hydrocarbon Process.*, **50**(6), 112

EIDELMAN, S. (1982). A method of calculation of the critical energy for direct initiation of unconfined detonation. *Combust. Sci. Technol.*, **30**, 133

EIDELMAN, S. and BURCAT, A. (1981). The mechanism of a detonation wave enhancement in a two-phase combustible medium. *Combustion*, **18**, 1661

EIDELMAN, S. and XIAOLONG YANG (1993). Detonation wave propagation in combustible particle/air mixtures with variable particle density distributions. *Combust. Sci. Technol.*, **89**, 201

EIDELMAN, S., OVED, Y., HASSON, A. and BURCAT, A. (1981). Reinforced blast waves in gases containing solid particles. *Combust. Sci. Technol.*, **25**, 21

EIDSVIK, K.J. (1978). *Dispersion of Heavy Gas Clouds in the Atmosphere. Rep. 32/78*. Norwegian Inst. for Air Res., Lillestrom.

EIDSVIK, K.J. (1979). A model for heavy gas dispersion in the atmosphere. *Proc. Tenth Int. Tech. Mtg on Air Pollution Modelling and Its Application (NATO/CCMS)*, p. 675

EIDSVIK, K.J. (1980). A model for heavy gas dispersion in the atmosphere. *Atmos. Environ.*, **14**, 769

EIDSVICK, K.J. (1981a). Heavy gas dispersion model with liquified release. *Atmos. Environ.*, **15**, 1163

EIDSVIK, K.J. (1981b). Prediction of hazard area of an accidental gas release. *Am. Met. Soc., Proc. Seventh Conf. on Probability and Statistics in Atmospheric Science*, p. 86

EIERMAN, R.G. (1995). Improving inherent safety with sealless pumps. *Proc. of the 29th Annu. Loss Prev. Symp.*, Jul. 31 – Aug. 2, 1995, Boston, MA, (ed. E. D. Wixom and R. P. Benedetti) Paper 1e. (New York: American Institute of Chemical Engineers)

EIFEL, P.J. (1968). Railroad and highway transportation of LNG. *LNG Conf.*, 1

EIGENBERGER, G. and SCHULER, H. (1987). Reactor stability and safe reactor *Fortschritte der Sicherheitstechnik 1* (Frankfurt-am-Main: DECHEMA), p. 85

EIGENMANN, K. (1976). Hazard studies applying thermo-analytic micro-methods. *Prevention of Occupational Risks*, **3**, 127

EIGENMANN, K. (1977). A program for testing the thermal safety of chemical processes. *Loss Prevention and Safety Promotion, 2*, p. 124

VAN EIJK, F.P. (1975). Instrument trip system maintenance and improvement program. *Ammonia Plant Safety*, **17**, 69

VAN EIJNATTEN, A.L.M. (1977). Explosion in a naphtha cracking unit. *Loss Prevention*, **11**, 11

EIJNDHOVEN, P.P., NIEUWENHUIZEN, J.K. and WALLY, R. (1974). The cooling of LPG storage tanks in case of fire. *Loss Prevention and Safety Promotion 1*, p. 405

EILERS, J.F. and SMALL, W.M. (1973). Tube vibration in a thermosiphon reboiler. *Chem. Engng Prog.*, **69**(7), 57

EIMUTIS, E.C. and KONICEK, M.G. (1972). Derivations of continuous functions for the lateral and vertical atmospheric dispersion coefficients. *Atmos. Environ.*, **6**, 859

EINHORN, I.N. and GRUNNET, M.L. (1978). Physiological and toxilogical aspects of combustion of natural and synthetic materials: past, present and future. *Fire Res.*, **1**(3), 143

EISEN, C.L., GROSS, R.A. and RIVLIN, T.J. (1960). Theoretical calculations in gaseous detonations. *Combust. Flame*, **4**, 137

EISENBERG, G.M. (1975). Guide to butt joint factors based on radiographic examination in accordance with UW-11 and UW-12 of Section VIII, Division 1. *J. Pressure Vessel Technol.*, **97**, 35

EISENBERG, N.A., LYNCH, C.J. and BREEDING, R.J. (1975). *Vulnerability Model: A Simulation System for Assessing Damage Resulting from Marine Spills. Rep. CG-D-136–75. Enviro Control Inc., Rockville, MD*

EISENBERG, N.A., LYNCH, C.J., GARDENIER, J.S. and PARNAR-OUSKIS, M.C. (1976). The vulnerability model: assessing damage from maritime spills by computer simulation. *Control of Hazardous Material Spills*, **3**, 178

EISENKLAM, and ARUNACHALAM, S.A. (1966). The drag resistance of burning drops. *Combust. Flame*, **10**, 171

EISENKLAM, P., ARUNACHALAM, S.A. and WESTON, J.A. (1967). Evaporation rates and drag resistance of burning drops. *Combustion*, **11**, 715

EISHOUT, R.V. and KILSTROM, M.S. (1988). Tips for better retrofits. *Chem. Engng*, **95**(Aug.) 15), 95

EISNER, H. (1979). Hazard assessment. *Sci. Publ. Policy*, **6**, 146

EISNER, H.S. and LEGER, J.P. (1988). The International Safety Rating System in South African mining. *J. Occup Accid.*, **10**(2), 141

EKE, P.W.A. and GIBSON, G.A. (1977). Experiences in the commissioning of liquid natural gas carriers. *LNG Conf 5*

EKE, P.W.A., GRAHAM, E.B. and MALYN, T.H. (1974). Liquefaction and boil-off reliquefaction facilities at Canvey Island. *LNG Conf. 4*

EKSTRUM, J.D. (1981). Sizing pulsation dampers for reciprocating pumps. *Chem. Engng*, **88**(Jan. 12), 111

EL-SAYED, S.A. (1994). Critical conditions in uniform temperature systems with variable heat transfer coefficients in thermal explosion theory. *J. Loss Prev. Process Ind.*, **7**, 223

EL-SHOBOKSHY, M.S. (1985). The dependence of airborne particulate deposition on atmospheric stability and surface conditions. *Atmos. Environ.*, **19**, 1191

EL-SHOBOKSHY, M.S. and HUSSEIN, F.M. (1988). Correlation between indoor—outdoor inhalable particulate concentrations and meteorological variables. *Atmos. Environ.*, **22**, 2667

ELAND, K.G. (1966). New guide to steam tracing design. *Hydrocarbon Process.*, **45**(11), 218

VON ELBE, G. (1953). The problem of ignition. *Combustion*, **4**, p. 13

VON ELBE, G. (1955). Chemical kinetics of hydrocarbon combustion. *Combustion*, **5**, 79

VON ELBE, G. and LEWIS, B. (1937a). The combustion of paraffin hydrocarbons. *J. Am. Chem. Soc.*, **59**, 976

VON ELBE, G. and LEWIS, B. (1937b). The mechanism of the combustion of hydrocarbons. *Combustion*, **2**, 169

VON ELBE, G. and LEWIS, B. (1937c). Steady state rate of a chain reaction for the case of chain destruction at walls of varying efficiencies. *J. Am. Chem. Soc.*, **59**, 970

VON ELBE, G. and LEWIS, B. (1937d). Theory of flame propagation. *Combustion*, **2**, 183

VON ELBE, G. and LEWIS, B. (1949). Theory of ignition, quenching, and stabilization of flames of nonturbulent gas mixtures. *Combustion*, **3**, 68

ELDER, L.L. and BATTEN, C.H. (1975). Guidelines for safe hot taps on gas pipe. *Loss Prevention*, **9**, 42

ELDER, T. and STRONG, J. (1953). The infra-red transmission of atmospheric windows. *J. Franklin Inst.*, **255**, 189

ELECTRIC POWER RESEARCH INSTITUTE (1975). *Summary of the AEC Reactor Safety Study (WASH 1400)* (Palo Alto, CA)

ELECTRIC POWER RESEARCH INSTITUTE (1979a). *Analysis of Three Mile Island – Unit 2 Accident. Rep. NSAC-1 and NSAC-1 Suppl.* (Nucl. Safety Analysis Center, Electric Power Res. Inst., Palo Alto, CA)

ELECTRIC POWER RESEARCH INSTITUTE (1979b). *Categorisation of Cable Flammability. Rep. EPRI-NP-1200* (Palo Alto, CA)

ELECTRIC POWER RESEARCH INSTITUTE (1981a). Ignitability of High Fire Point Liquids. Rep. EPRI-NP-1731 (Palo Alto, CA)

ELECTRICAL POWER RESEARCH INSTITUTE (1981b). *SPASM, a Computer Code for Monte Carlo System Evaluation. Rep. EPRI NP-1685* (Palo Alto, CA)

ELECTRICAL POWER RESEARCH INSTITUTE (1984). *Human Factors Guide for Nuclear Power Plant Control Room Development. Rep. EPRI NP 3659* (Palo Alto, CA)

ELECTRIC POWER RESEARCH INSTITUTE (1986). *Seismic Hazard Methodology for the Central and Eastern United States*, vol. 1. *Rep. NP-4726* (Palo Alto, CA)

ELECTRIC POWER RESEARCH INSTITUTE — see also Appendix 28

ELECTRICAL RESEARCH ASSOCIATION (1981). Radio Frequency Ignition Thresholds of a Large Mobile Crane. ERA Technol. Rep. 81–57 (Leatherhead)

ELECTRICAL RESEARCH ASSOCIATION (1982). *Ignition at Radio Frequencies using simple Tuned and Equivalent Circuits. ERA Technol. Rep. 82–14* (Leatherhead)

ELECTRICITY COUNCIL and BRITISH LIGHTING COUNCIL (1967). *Lighting in Industry*, 3rd ed. (London)

EL-EMAM, S.H. and MANSOUR, H. (1983). An analytical model for two-phase flow field of liquid sprays, In Veziroglu, T.N. (1983), *op. cit.*, p. 256

ELEY, C. (1992). Compliance audit checklist for hazardous chemicals. *Hydrocarbon Process.*, **71**(8), 97

ELFIMOVA, E.V. (1959). Determination of limit of allowable concentration of hydrochloric acid aerosol (hydrogen chloride) in atmospheric air. *Gig Sanit.*, **24**(1), 144

ELFIMOVA, E.V. (1964). Data for the hygienic evaluation of hydrochloric acid aerosol (hydrogen chloride). *Limits of Allowable Concentrations of Atmospheric Pollutants* (ed. V.A. Rjaboiv) (in Russian)

ELGEE, M. (1965). Investing in major spare parts. *Chem. Engng*, **72**(Sep. 13), 218

ELIAS, J. (1978). An intelligent approach to burner safety. *Loss Prevention*, **12**, 73

ELIASSEN, R. and TCHOBANOGLOUS, G. (1968). Water pollution control. Advanced treatment processes. *Chem. Engng*, **75**(Oct. 14), 95

ELKIN, H.F. and CONSTABLE, R.A. (1972). Source/control of air emissions. *Hydrocarbon Process.*, **51**(10), 113

ELKIN, T.J. and SMITH, P.G.R. (1988). What is a good Environmental Impact Statement? Reviewing screening reports from Canada's National Parks. *J. Environ. Mngmt*, **26**, 71

ELKINS, H.B. (1939). Toxic fumes (in Massachusetts industries). *Ind. Med.*, **8**, 426

ELKINS, H.B. (1959). *The Chemistry of Industrial Toxicology*, 2nd ed. (New York: Wiley)

ELLGEHAUSEN, D. (1980). A gas-chromatographic monitor for toxic substances in the air of a production plant. *Loss Prevention and Safety Promotion 3*, vol. 2, p. 249

ELLICE, K.A.J. (1989). *Piper Alpha Public Inquiry. Transcript, Day 147*

ELLINOR, R.C. (1990). Fire engineering in petrochemical and offshore applications — the spectrum of optical fire detection. *Safety and Loss Prevention in the Chemical and Oil Processing Industries*, p. 255

ELLIOT, L.B. (1986). Analogical problem-solving and expert systems. *IEEE Expert*, **1**(2), 17

ELLIOT, P. (1984). Process plant corrosion: recognising the threat. *Plant Engr*, Nov. 43

ELLIOT, T.A. and KING, R.A. (1987). Nuclear waste — a new industry. *Chem. Engr*, **436**, 45

ELLIOT, W.H. (1983). The Chambers Works generic VOC bubble. *Plant/Operations Prog.*, **2**, 38

ELLIOTT, D.E. and CHIESA, P.J. (1976). Rheological properties of fire fighting foams. *Fire Technol.*, **12**, 141

ELLIOTT, D.M. and OWEN, J.M. (1968). Critical examination in process design. *Chem. Engr. Lond.*, **223**, CE377

ELLIOTT, H.K. (1969). Electrical problems in chlorine manufacture. *Loss Prevention*, **3**, p. 57

ELLIOTT, I. (1986). The art of specification. *Process Engng*, Feb., **76**(Jun.), 69; (Sep.), 85; (Dec.), 55

ELLIOTT, J. (1987). Piping up for flexibility. *Chem. Engr. Lond.*, **435**, 51

ELLIOTT, M. (1976). Remote visual inspection — an undiscovered technique. *Diagnostic Techniques* (London: Instn Chem. Engrs)

ELLIS, A.F. (1989). *Piper Alpha Public Inquiry. Transcript, Day 133*

ELLIS, C. (1989). *Expert Knowledge and Explanation: The Knowledge–Language Interface* (Chichester: Ellis Horwood)

ELLIS, C.E. (1968). Unattended pilot plants. *Chem. Engng Prog.*, **64**(10), 50

ELLIS, DE C.O.C. (1928). Flame movement in gaseous explosive mixtures. *Fuel*, **7**, 5

ELLIS, DE C.O.C. and WHEELER, R.V. (1925). The movement of flame in closed vessels. *Fuel*, **4**, 356

ELLIS, DE C.O.C. and WHEELER, R.V. (1928a). Explosions in closed cylinders. Pt 3, The manner of movement of flame. *J. Chem. Soc.*, 3215

ELLIS, DE C.O.C. and WHEELER, R.V. (1928b). Explosions in closed vessels: the correlation of pressure development with flame movement. *Fuel*, **7**, 169

ELLIS, D.L. (1974). Propane disaster averted after tank car derailment. *Fire Engng*, **127**(Mar.), **35**

ELLIS, D.M. (1990). Micro-organisms. *Plant Engr*, Jan./Feb., 35

ELLIS, J. and MUALLA, W. (1986). Selecting check valves. *Process Engng*, Nov. 47

ELLIS, J.M. (1975). Inspection and maintenance of electrical equipment in hazardous environments. *Electrical Safety*, **2**, 138

ELLIS, W.P. (1969). How to select mastics for thermal insulation. *Hydrocarbon Process.*, **48**(1), 119

ELLISON, T. and TURNER, J.S. (1959). Turbulent entrainment in stratified flows. *J. Fluid Mech.*, **6**, 423

ELLISON, T.H. and TURNER, J.S. (1960). Mixing of dense fluid in a turbulent pipe flow. *J. Fluid Mech.*, **8**, 514

ELLUL, I. (1989). Pipeline leak

ELLUL, I. (1989). Pipeline leak detection. *Chem. Engr, Lond.*, **461**, 39

ELLWOOD, P. (1969a). Acetic acid via methanol and synthesis gas. *Chem. Engng*, **76**(May 19), 149

ELLWOOD, P. (1969b). New life for fixed bed phthalic anhydride route. *Chem. Engng*, **76**(Jun. 2), 80

ELLWOOD and DANATOS, S. (1966). Process furnaces. *Chem. Engng*, **73**(Apr. 11), 151

ELMES, P.C. and BELL, D. (1963). The effects of chlorine gas on the lungs of rats with spontaneous pulmonary disease. *J. Pathol. Bacteriol.*, **86**, 317

ELMES, P.C. and SIMPSON, M.J.A. (1971). Insulation workers in Belfast, 3: Mortality 1940–66. *Br. J. Ind. Med.*, **28**, 226

ELMES, P.C. and SIMPSON, M.J.A. (1977). A further study of mortality due to asbestos exposure. *Br. J. Ind. Med.*, **34**, 174

ELONKA, S.M. (1975). *Standard Plant Operators' Manual* (New York: McGraw-Hill)

ELPHICK, N.C. (1972). Safety and pollution regulations. *Loss Prevention*, **6**, 126

EL-RIFAI, M. (1979). Estimate equipment weights. *Hydrocarbon Process.*, **58**(9), 225

ELSHOUT, R. and WETHERILL, F. (1987). Effective operator training program. *Chem. Engng*, **94**(10), 115

ELSWORTH, J.E. and EYRE, J.A. (1984). The susceptibility of propene–propane/air mixtures to detonation. *Combust. Flame*, **55**, 237

ELSWORTH, J.E. and EYRE, J.A. (1986). Indirect initiation of spherical detonation in fuel/air mixtures by a transition process. In Hartwig, S. (1986), *op. cit.*, p. 233

ELSWORTH, J.E., EYRE, J.A. and WAYNE, F.D. (1983). Combustion of refrigerated liquefied propane in partially confined spaces. *Loss Prevention and Safety Promotion 4*, vol. 2, p. C35

ELSWORTH, J., EYRE, J. and WAYNE, D. (1984). Liquefied gas spillages in partially confined spaces. *Chem. Engr, Lond.*, **400**, 26

ELSWORTH, J.E., SHUFF, P.J. and UNGUT, A. (1984). "Galloping" gas detonations in the spherical mode. In Bowen, J.R. *et al.* (1984b), *op. cit.*, p. 130

ELTON, R.L. (1980). Designing stormwater handling systems. *Chem. Engng*, **87**(Jun. 2), 64

ELVERS, B., HAWKINS, S. and SCHULZ, G. (1992). Principles of chemical reaction engineering and plant design. *Ullmans Encyclopaedia of Chemical Technology*, rev., ed., vol. B4 (Weinheim: VCH)

ELVERS, B., RAVENSCROFT, M., ROUNSAVILLE, J.F.A. and SCHULZ, G. (1988). Unit operations I and II. *Ullmans Encyclopaedia of Chemical Technology*, rev. ed., vols B2 and B3 (Weinheim: VCH)

EL-WAKIL, M.H. (1962). *Nuclear Power Engineering* (New York: McGraw-Hill)

EMANUEL, N.M. (1965). *The Oxidation of Hydrocarbons in the Liquid Phase* (Oxford: Pergamon Press)

EMBER, L. (1984). Superfund: a captive of politics. *Chemy Ind.*, Nov. 19, 785

EMBLEM, K. and FANNELOP, T.K. (1983). Entrainment mechanisms of air in heavy gas clouds. In Hartwig, S. (1983), *op. cit.*, p. 67

EMBLEM, K., KROGSTAD, P.A. and FANNELOP, T.K. (1984). Experimental and theoretical studies in heavy gas dispersion. Part I, Experiments. Ooms, G. and Tennekes, H. (1984), *op. cit.*, p. 307

EMBLEM, K. and MADSEN, O.K. (1986). Full scale test of a water curtain in operation. *Loss Prevention and Safety Promotion 5*, p. 46–1

EMBLING, A.W. (1986). The risk management systems approach explained. *CHEMECA 86*, p. 284

EMBREY, D.E. (1976a). The human operator and system reliability. *Advances in Reliability Technology* (Culcheth, Lancashire: UK Atom. Energy Auth.), ART15

EMBREY, D.E. (1976b). *Human Reliability in Complex Systems: An Overview. Rep. NCSR 10.* National Centre for Systems Reliability, Culcheth, Lancashire

EMBREY, D.E. (1979a). Design approaches to reducing human error in process plant operation. *Design 79* (London: Instn Chem. Engrs)

EMBREY, D.E. (1979b). Human reliability in the process industries. *Terotechnica*, **1**, 109

EMBREY, D.E. (1981). Approaches to the evaluation and reduction of human error in the process industries. *Reliable Production in the Process Industries* (Rugby: Instn Chem. Engrs), p. 181

EMBREY, D.E. (1983a). The quantification of human reliability using expert judgement: current findings and future developments. *Instn Chem. Engrs Conf. on Human Reliability in the Process Control Centre* (London)

EMBREY, D.E. (1983b). *The Use of Performance Shaping Factors and Quantified Expert Judgment in the Evaluation of Human Reliability: an Initial Appraisal. Rep. 2986.* Nucl. Regul. Comm., Washington, DC. Also Rep. 51591, Brookhaven Nat. Lab., Upton, NY

EMBREY, D.E. (1984). Application of human reliability assessment techniques to process plant design. *Ergonomic Problems in Process Operations* (Rugby: Instn Chem. Engrs), p. 65

EMBREY, D.E. (1985). Closing the gap between ergonomics and engineering. *Process Engng*, July, 31

EMBREY, D.E. (1986). Approaches to aiding and training operators' diagnoses in abnormal situations. *Chemy Ind.*, Jul. 7, 454

EMBREY, D.E. (1990). *Top–down Human Error and Task Analysis (THETA) software manual.* Human Reliability Associates, Wigan, Lancashire

EMBREY, D.E. (1992a). Managing human error in the chemical process industries. *Loss Prevention and Safety Promotion 7*, vol. 2, 50–1

EMBREY, D.E. (1992b). Managing human error in the chemical process industry. *Hazard Identification and Risk Analysis, Human Factors and Human Reliability in Process Safety*, p. 399

EMBREY, D.E. (1992c). Quantitative and qualitative prediction of human error in safety assessments. *Major Hazards Onshore and Offshore*, p. 329

EMBREY, D.E. and KIRWAN, B. (1983). A comparative evaluation of three subjective human reliability quantification techniques. *Proc. Ergonomics Soc. Ann. Conf.*

EMBREY, D., HUMPHREYS, P.C., ROSE, E.A., KIRWAN, B. and REA, K. (1984). *SLIM-MAUD: an Approach to Assessing Human Error Probabilities Using Structured Expert Judgment. Rep. NUREG/CR-3518.* Nucl. Regul. Comm., Washington, DC

VAN EMDEN, M.H. (1977). Programming with resolution logic. *Machine Intelligence 8* (ed. E. Elcock and D. Michie (Chichester: Ellis Horwood), p. 266

EMELYANOV, I.Y., SHASHARIN, G.A., KIREEV, G.A., KLEMIN, A.I., POLYAKOV, E.F., STRIGULIN, M.M. and SHIVERSII, E.A. (1972) Evaluating the reliability of the Beloyarsk atomic power stations pumps on the basis of operating data. *Soviet Atom. Energy*, **33**, 637

EMERICK, R.H. (1967). A near-death day in the plant. *Hydrocarbon Process.*, **48**(4), 163

EMERSON, G.B. (1985). Selecting pressure relief valves. *Chem. Engng*, **92**(Mar. 18), 195

EMERSON, G.B. (1988). Pressure relief valve types and applications. *Hydrocarbon Process.*, **67**(5), 71

EMERSON, M.C. (ca. 1986a). *Dense Cloud Behaviour in Momentum Jet Dispersion.* (London: Technica Ltd)

EMERSON, M.C. (1986b). A new 'unbounded' jet dispersion model. *Inst. of Mathematics and its Applications, Mtg on Mathematics in Major Accident Risk Assessment*

EMERSON, M.C. (1987a). Heavy gas dispersion. *Course on Hazard Control and Major Hazards.* Dept of Chem. Engng, Loughborough Univ. of Technol.

EMERSON, M.C. (1987b). A model of pressurized releases with aerosol effects. *Vapor Cloud Modeling*, p. 181

EMERSON, M.C. (1989). Dense cloud behaviour in momentum jet dispersion. In Cox, R.A. (1989a), *op. cit.*, p. 113

EMERSON, M.C., PITBLADO, R.M. and SHARIFI, T. (1988). Quantification of release consequences. *Preventing Major Chemical and Related Process Accidents*, p. 669

EMERSON, S.D. (1990). Management of process safety at Kerr-McGee Corporation. *Plant/Operations Prog.*, **9**, 161

EMERY, S.J. (1983). Protecting against backflow in process lines. *Chem. Eng*, **90**(Nov. 28), 81

EMMONS, H.W. (ed.) (1958). *High Speed Aerodynamics and Jet Propulsion, vol. III, Fundamentals of Gas Dynamics* (London: Oxford Univ. Press)

EMMONS, H.W. (1965). Fundamental problems of the free-burning fire. *Combustion* **10**, 951

EMMONS, H.W. (1971). Fluid mechanics and combustion. *Combustion*, **13**, 1

EMMONS, H.W. (1974). Fire and fire protection. *Sci. American*, **231**, 21

EMMONS, H.W. (1979). The prediction of fire in buildings. *Combustion*, **17**, 1101

EMMONS, H.W. (1990). Firesafety science – the promise of a better future. *Fire Technol.*, **26**(1), 5

EMMRICH, F. (1971). Dry chemical fire extinguishing systems and installations in chemical industries. *Major Loss Prevention*, p. 139

EMMS, H.W. (1988). Why fire model? The MGM fire and toxicity testing. *Fire Safety J.*, **13**, 77

EMORI, R.I. and SAITO, K. (1983). A study of scaling laws in pool and crib fires. *Combust. Sci. Technol.*, **31**, 217

EMSLEY, J. (1988). Riding the chemical tiger (book review). *New Scientist*, Jul. 14, 64

VAN DER ENDE, E.C. (1969). The hazard of compressing oxygen under high pressure. *Sym on Inherent Hazards of Manufacturing and Storage in the Process Industry*, The Hague, p. 169

ENG, N. and BARNES, C.M. (1984). Ethylene from NGL feedstocks, Pt 6. *Hydrocarbon Process.*, **63**(3), 83

ENGEL, J.R. (1974). Pressure vessel failure statistics and probabilities. *Nucl. Safety*, **15**(4), 387

ENGEL, P.A. (1976). *Impact Wear of Materials* (Amsterdam: Elsevier)

ENGEL, R. (1969). A report on two train wrecks. *Ammonia Plant Safety*, **11**, 28

ENGEL, U. and WICKBOLDT, H. (1975). Special types of explosion protected electrical machines of type of protection 'Increased Safety'. *Electrical Safety*, **2**, 118

ENGELHARDT, F.R. and HOLLIDAY, M.G. (1985). *Dose–Lethality Relationships of Acute Exposure to Anhydrous Ammonia.* Res. Rep. INFO-0153. Atom Energy Control Board, Ottawa, Ont.

ENGELMANN, H.D., ERDMANN, H.H., FUNDER, R. and SIMMROCK, K.H. (1989). The solving of complex process synthesis problems using distributed expert systems. *Comput. Chem. Engng*, **13**, 459

ENGELMANN, R.J. (1968). The calculation of precipitation scavenging. In Slade, D.H. (1968), *op. cit.*, p. 208

ENGELMANN, R.J. (1992). Sheltering effectiveness against plutonium provided by buildings. *Atmos. Environ.*, **26A**, 2037

ENGELMORE, R. and MORGAN, T. (eds) (1988). *Blackboard Systems* (Wokingham: Addison-Wesley)

ENGELSTAD, P.H. (1970). A socio-technical approach to problems of process control. *In Paper-Making Systems and their Control* (ed. F. Bolam) (London: Br. Paper and Boardmakers Ass.), 91. See also Edwards, E. and Lees, F.P. (1974), *op. cit.*, p. 367

ENGER, L. (1986). A higher order closure model applied to dispersion in a convective PBL. *Atmos. Environ.*, **20**, 879

ENGER, T. and HARTMAN, D.E. (1972a). Mechanics of the LNG–water interaction. *Am. Gas. Ass., Distribution Conf.*

ENGER, T. and HARTMAN, D. (1972b). Rapid phase transformation during LNG spillage on water. *LNG Conf. 3*

ENGESSER, B., ZEHNDER, H.B., IMAI, T., TANAKA, H., KOYAMA, Y., SUZUKI, Y. and TAKAHASHI, K. (1987). Development of the Sulzer dual fuel diesel engine. *Gastech 86*, p. 261

ENGH, T.A. and LARSEN, K. (1979). Break-up length of turbulent water and steel jets in air. *Scand. J. Metall.*, **8**, 161

ENGINEERING APPLIANCES LTD (1977). Confidence in bellows after Flixborough. *Newsletter 67*, Sep., p. 2

THE ENGINEERING COUNCIL (1983). *Technical Reviews for Manufacturing, Process and Construction Companies* (London).

THE ENGINEERING COUNCIL (1991). Engineers and risk issues. An 'embryo' code of practice. *Safety Reliab.*, **11**(3), 17

THE ENGINEERING COUNCIL (1992). *Engineers and Risk Issues* (London)

THE ENGINEERING COUNCIL (1993). *Engineers and the Environment* (London)

ENGINEERING EQUIPMENT USERS ASSOCIATION – see Appendix 28

ENGINEERING EQUIPMENT AND MATERIALS USERS ASSOCIATION – see Appendix 28

ENGINEERING SCIENCES DATA UNIT (1977). *Examples of the Use of Data Items on Fatigue Crack Propagation Rates. Item 74071*

ENGLAND, G.C., HEAP, M. and PERSHING, D.W. (1980). Control of NO emissions. *Hydrocarbon Process.*, **59**(1), 167

ENGLAND, W.G., TEUSCHER, L.H., HAUSER, L.E. and FREEMAN, B. (1978). *Atmospheric Dispersion of Liquefied Natural Gas Vapor Clouds using SIGMET, a Three-dimensional Time-dependent Hydrodynamic Computer Model.* Heat Transfer and Fluid Mechanics Inst., Washington State Univ., Pullman, WA

ENGLE, J.P. (1971). Cleaning boiler tubes chemically. *Chem. Engng*, **78**(Oct. 18), 154

ENGLEHARDT, J.D. (1993). Pollution prevention techniques: a review and classification. *J. Haz. Materials*, **35**, 119

ENGLEHARDT, J.D. (1993). Pollution Prevention Technologies: A Review and Classification. *J. of Hazard. Mater.* **35**, 119–50

ENGLER, M.R. (1968). LNG utilization: San Diego Gas and Electric opens new fields. *LNG Conf. 1*

ENGLISH, J. (1986). Deepwater challenges to the subsea service industry. *Gastech 85*, p. 168

ENGLISH, M.A. and WAITE, P.J. (1989). The effect of macro modelling on risk analysis results. In Cox, R.A. (1989a), *op. cit.*, p. 217

ENGLISH, P. (1987). Plant condition monitoring. *Plant Engr*, **31**(5), 27

ENGLISH, P.C. and BOSWORTH, C. (1978). How is availability to be obtained in steelworks instrumentation? *Sci. Instrum. Res. Ass. Sym on Instrumentation Reliability – Can it be Improved?* (Chislehurst, Kent)

ENGLUND, S.M. (1990). Opportunities in the design of inherently safer chemical plants. *Adv. in Chem. Eng.* **15**, 69–135

ENGLUND, S.M. (1991). Design and operate plants for inherent safety. *Chem. Engng Prog.*, **87**(3), 85; **87**(5), 79

ENGLUND, S.M. (1991a). Design and operate plants for inherent safety – Part 1. *Chem. Eng. Prog.* (March), 85–91

ENGLUND, S.M. (1991b). Design and operate plants for inherent safety – Part 2. *Chem. Eng. Prog.*, (May), 79–86

ENGLUND, S.M. (1993). Process and design options for inherently safer plants. *Prevention and Control of Accidental Releases of Hazardous Gases* (ed. V. M. Fthenakis) (New York: Van Nostrand Reinhold), pp. 9–62

ENGLUND, S.M. (1994). Inherently safer plants – practical applications. *Am Inst. of Chem Engrs. 1994 Summer National Meeting*, Aug. 14–17, 1994, Denver, CO, Paper No. 47b

ENGLUND, S.M. and GRINWIS, D.J. (1992). Provide the right redundancy for control systems. *Chem. Engng Prog.*, **88**(10), 36

ENGLUND, S.M., MALLORY, J.L. and GRINWIS, D.J. (1992). Prevent backflow. *Chem. Engng Prog.*, **88**(2), 47

ENIG, J.W. (1956). *The Solution of the Steady-state Heat Conduction Equation with Chemical Reaction for the Hollow Cylinder. Rep. NAVORD 4267. Naval Ordnance Lab.*

ENIG, J.W. (1966). Critical parameters of the Poisson–Boltzmann equation of steady-state thermal explosion theory. *Combust. Flame*, **10**, 197

ENIG, J.W., SHANKS, D. and SOUTHWORTH, R.W. (1956). The numerical solution of the heat conduction equation occurring in the theory of thermal explosion. Naval Ordnance Lab., *Rep. NAVORD 4377*

ENNIS, B. and LEBLANC, J.R. (1984). Primary reformer riser liner collapse. *Ammonia Plant Safety*, **24**, 92

ENNIS, R. and LESUR, P.F. (1977). How small NH$_3$ plants compete. *Hydrocarbon Process.*, **56**(12), 121

ENNIS, R.H. (1969). *Logic in Teaching* (Englewood Cliffs, NJ: Prentice-Hall)

ENSMINGER, J.T., LU, P.Y. and OEN, C. (1988). Material safety data sheets: The Martin Marietta Energy Systems Inc. experience. *J. Haz. Materials*, **17**(3), 259

ENVIRO CONTROL INC. (ca. 1976). *A Critical Review of Six Hazard Assessment Models* (Rockville, MD)

ENVIRO CONTROL INC. (1977). *The Vulnerability Model for Damage Assessment of the Effects of Hazardous Spills* (Rockville, MD)

ENVIRON CORPORATION (1986). *Elements of Toxicology and Chemical Risk Assessment* (Washington, DC)

ENVIRON CORPORATION (1989). Elements of toxicology. *Chem. Engng Prog.*, **85**(8), 37

ENVIRONMENTAL MUTAGEN SOCIETY (1975). Environmental mutagenic hazards. *Science*, **187**, 503

ENVIRONMENTAL PROTECTION AGENCY (1971). *Effects of Noise on People* (Washington, DC)

ENVIRONMENTAL PROTECTION AGENCY (1973). *Compilation of Air Pollutant Emission Factors*, 2nd ed. (Research Triangle Park, NC)

ENVIRONMENTAL PROTECTION AGENCY (1974). *Comments by the Environmental Protection Agency on Reactor Safety Study: An Assessment of Accident Risks in US Commercial Nuclear Power Plants* (Washington, DC)

ENVIRONMENTAL PROTECTION AGENCY (1975a). *National Conference on Polychlorinated Biphenyls.* (Washington, DC: Office of Toxic Substances, Environ. Prot. Agency)

ENVIRONMENTAL PROTECTION AGENCY (1975b). *Scientific and Technical Assessment Report on Vinyl Chloride and Polyvinyl Chloride. Rep. EPA-600/6-75-004* (Washington, DC)

ENVIRONMENTAL PROTECTION AGENCY (1975c). *Standard Support and Environmental Impact Statement: Emission Standard for Vinyl Chloride. Rep. EPA-450/2-75-009* (Research Triangle Park, NC)

ENVIRONMENTAL PROTECTION AGENCY (1976). *Reactor Safety Study WASH 1400. A Review of the Final Report. Rep. EPA-520/3-76-009* (Washington, DC)

ENVIRONMENTAL PROTECTION AGENCY (1977). *User's Manual for Single-Source (CRSTER) Model. Re EPA-450/2-77-013* (Research Triangle Park, NC)

ENVIRONMENTAL PROTECTION AGENCY (1978a). *Compilation of Air Pollutant Emission Factors*, 3rd ed., suppl. 8 (Research Triangle Park, NC)

ENVIRONMENTAL PROTECTION AGENCY (1978b). *Control of Volatile Organic Compound Leaks from Petroleum Refinery Equipment. Guideline Doc. EPA-450/2-78-036* (Research Triangle Park, NC)

ENVIRONMENTAL PROTECTION AGENCY (1981). *Peer Review Process for Scientific and Information Documents* (Washington, DC)

ENVIRONMENTAL PROTECTION AGENCY (1982a). *Uncontrolled Hazardous Waste Sites Ranking: A Users Manual. National Oil and Hazardous Substances Pollution Contingency Plan, Federal Register*, vol. 4, (Washington, DC)

ENVIRONMENTAL PROTECTION AGENCY (1982b). *VOC Fugitive Emissions in Synthetic Organic Chemicals Manufacturing Industry – Background Information for Promulgated Standards. Rep. EPA-450/3-80-033b* (Washington, DC)

ENVIRONMENTAL PROTECTION AGENCY (1983). *Health Hazard Assessment Document for Acrylonitrile* (Washington, DC)

ENVIRONMENTAL PROTECTION AGENCY (1984a). *Evaluation and Selection of Models for Estimating Air Emissions from Hazardous Waste Treatment, Storage, and Disposal Facilities. Re EPA-450/3-84-020* (Research Triangle Park, NC)

ENVIRONMENTAL PROTECTION AGENCY (1984b). *Interim Procedures for Evaluating Air Quality Models (rev.). Rep. EPA-450/4-84-023* (Research Triangle Park, NC)

ENVIRONMENTAL PROTECTION AGENCY (1984c). *Standards of Performance for New Stationary Sources Equipment Leaks of VOC, Petroleum Refineries and Synthetic Organic Chemical Manufacturing Industry. Federal Register, 1984*, 49(105), May 30, p. 22, 607

ENVIRONMENTAL PROTECTION AGENCY (1984d). *Users Manual for TOX-SCREEN: A Multimedia Screening-Level Program for Assessing the Potential of Chemicals Released to the Environment. Rep. EPA-560/5-830024* (Office of Toxic Substances, Environ. Prot. Agency, Oak Ridge Nat. Lab., TN)

ENVIRONMENTAL PROTECTION AGENCY (1986). *Industrial Source Complex (ISC) Dispersion Model User's Guide*, 2nd ed., vols 1–2 (Research Triangle Park, NC)

ENVIRONMENTAL PROTECTION AGENCY (1987a). *Technical Guidance for Hazards Analysis* (Washington, DC)

ENVIRONMENTAL PROTECTION AGENCY (ca. 1987b). *Waste Minimization Environmental Quality* (Washington, DC)

ENVIRONMENTAL PROTECTION AGENCY (1988a). *Protocols for Generating Unit-Specific Emission Estimates for Equipment Leaks of VOC and VHA Doc. EPA-450/3-88-010* (Research Triangle Park, NC)

ENVIRONMENTAL PROTECTION AGENCY (ca. 1988b). *Waste Minimization Opportunity Assessment Manual. Re EPA-625/7-88-003* (Washington, DC)

ENVIRONMENTAL PROTECTION AGENCY (1989a). *Hazardous Waste Treatment, Storage and Disposal Facilities (TSDF)*

—*Air Emission Models. Doc. EPA-450/3—87—026* (Research Triangle Park, NC)

ENVIRONMENTAL PROTECTION AGENCY (1989b). *Risk Assessment Guidance for Superfund:* vol. I, *Human Health Evaluation Manual. Rep. 540/1—89/002* (Washington, DC)

ENVIRONMENTAL PROTECTION AGENCY (1989c). *Risk Assessment Guidance for Superfund: II, Environmental Evaluation Manual. Rep. 540/1—89/001* (Washington, DC)

ENVIRONMENTAL PROTECTION AGENCY (1994a). *Estimated Exposure to Dioxin-Like Compounds* (Research Triangle Park, NC)

ENVIRONMENTAL PROTECTION AGENCY (1994b). *Health Assessment Document for 2,3,7,8-Tetrachlorodibenzo-p-dioxin and Related Compounds* (Research Triangle Park, NC)

ENVIRONMENTAL PROTECTION AGENCY, OFFICE OF TOXIC SUBSTANCES (1979). *Toxic Substances Control Act Inventory,* Initial Inventory vols I–IV (Washington, DC)

ENVIRONMENTAL PROTECTION AGENCY, OFFICE OF TOXIC SUBSTANCES (1982). *Toxic Substances Control Act Inventory,* Cumulative suppl. II (Washington, DC)

ENVIRONMENTAL PROTECTION AGENCY — *Informational Documents* (Washington, DC)

ENVIRONMENTAL PROTECTION AGENCY — see also Appendix 28

ENZINNA, R. (1985). Improving safety with longer trip test intervals. *Nucl. Engng Int.*, **30**(374), 29

EPA RMP United States Department Of Environmental Protection, *Risk Management Plan 40,* CFR Part 68

EPHRATH, A.R. and YOUNG, L.R. (1981). Monitoring vs man-inthe-loop detection of aircraft control failures. In Rasmussen, J. and Rouse, W.B. (1981), *op. cit.*, p. 143

EPPE, J., FONTANGES, R. and GROLLIER-BARON, R. (1980). Problems caused by the outcoming of allergic phenomena in a fermentation pilot plant. *Loss Prevention and Safety Promotion 3,* vol. 2, p. 704

EPPLER, E.P. (1969). CMF considerations in the design of systems for protection and control. *Nucl. Safety,* **10**, 38

EPPLER, E.P. (1970). The ORR emergency cooling failure. *Nucl. Safety,* **11**, 323

EPPS, M. (1988). Safety practices and procedures in ammonia production, storage and handling facilities (panel discussion). *Ammonia Plant Safety* **28**, 169

EPSTEIN, B. (1960). Elements of the theory of extreme values. *Technometrics,* **2**(7), 27

EPSTEIN, C.F. (1971). Correlation and prediction of explosive metal-water reaction temperatures. *Nucl. Sci. Engng,* **10**(Jul.), 247

EPSTEIN, H.I. and BUZEK, J.R. (1978). Stresses in ruptured floating roofs. *J. Pres. Ves. Technol.,* **100**, 291

EPSTEIN, J.P. (1967). How to buy or sell used equipment. *Chem. Engng,* **74**(Jan. 2), 104

EPSTEIN, J.P. (1978). How to buy and sell used equipment. *Chem. Engng,* **85**(Nov. 20), 185

EPSTEIN, L.D. (1981). Costs of standard vertical storage tanks and reactors. *Chem. Engng,* **88**(Jul. 13), 141

EPSTEIN, M. (1987). Buoyancy-driven exchange flow through openings in horizontal partitions. *Vapor Cloud Modeling,* p. 756

EPSTEIN, M., HENRY, R.E., MIDVIDY, W. and PAULS, R. (1983). One-dimensional modeling of two-phase jet expansion and impingement. *Second Int. Topical Mtg on Nuclear Reactor Thermal Hydraulics,* Santa Barbara, CA

EPSTEIN, M., SWIFT, I. and FAUSKE, H.K. (1986). Estimation of peak pressure for sonic-vented hydrocarbon explosions in spherical vessels. *Combust. Flame,* **66**, 1

EPSTEIN, M., FAUSKE, H.K. and HAUSER, G.M. (1989). The onset of two-phase venting via entrainment in liquid-filled storage vessels exposed to fire. *J. Loss Prev. Process Ind.,* **2**(1), 45

EPSTEIN, M., FAUSKE, H.K. and HAUSER, G.M. (1990). A model of the dilution of a forced two-phase chemical plume in a horizontal wind. *J. Loss Prev. Process Ind.,* **3**(3), 280

EPSTEIN, M. *et al.* (1990). A computer model for the estimation of peak pressure for sonic-vented tetrafluoroethylene decompositions. *J. Loss Prev. Process Ind.,* **3**(4), 370

EPSTEIN, S.S. (1978). *The Politics of Cancer* (San Francisco, CA: Sierra Club)

EPSTEIN, S.S. and GRUNDY, R.D. (eds) (1974). *The Legislation of Product Safety* (Cambridge, MA: MIT Press)

EPSTEIN, S.S. and SWARTZ, J.B. (1981). Fallacies of life-style cancer theories. *Nature,* **289**(Jan. 15)

ERB, H.A. (1975). Repair techniques for machinery rotor and case damage. *Hydrocarbon Process.,* **54**(1), 60

ERDMANN, R.C. (1977). A method of quantifying logic models for safety analysis. *Nuclear Systems Reliability Engineering and Risk Assessment* (Philadelphia, PA: SIAM), p. 732

ERDMANN, R.C., LEVERENZ, F.L. and KIRCH, H. (1978). *WAMCUT, a computer Code for Fault Tree Evaluation. Rep. EPRI NP-803.* Elec. Power Res. Inst., Palo Alto, CA

ERDMANN, R.C. *et al.* (1975). *Probabilistic Safety Analysis. Rep. EPRI 217—2—4.* Elec. Power Res. Inst., Palo Alto, CA

ERDMANN, R.C. *et al.* (1977). *Probabilistic Safety Analysis. Rep. NP-424.* Elec. Power Res. Inst., Palo Alto, CA

ERDMANN, R.C. *et al.* (1978). *Probabilistic Safety Analysis III. Rep. EPRI NI-749.* Elec. Power Res. Inst., Palo Alto, CA

ERDOGAN, F. (1976). Ductile fracture theories for pressurised pipes and containers. *Int. J. Pres. Ves. Piping,* **4**, 253

ERGONOMICS SOCIETY (1982). *Statement of Case by the Ergonomics Society to the Sizewell B Inquiry. Statement of Case*

ERGONOMICS SOCIETY (1983). *Proof of Evidence to Sizewell B Public Inquiry* (Loughborough)

ERICSSON, G.A. and BJÖRE, S. (1983). Component failure data of Swedish boiling water reactors. *Euredata 4,* paper 9.2

ERICSSON, V. and EDIN, K. (1960). On complete blast scaling. *J. Phys. Fluids,* **3**(5), 893

ERIKSON, L. (1982). Lawrence Livermore National Laboratory single-stage 101 mm gun. In Nellis, W.J., Seaman, L. and Graham, A.R. (1982), *op. cit.*, p. 685

ERMAK, D.L. (1991). Gravity spreading in the dispersion of dense gas plumes. *Modeling and Mitigating the Consequences of Accidental Releases,* p. 341

ERMAK, D.L. and CHAN, S.T. (1985). A study of heavy gas effects on the atmospheric dispersion of dense gases. *NATO/CCMS, Proc. Fifteenth Int. Tech. Mtg on Air Pollution Modelling and its Applications,* paper VI.6

ERMAK, D.L. and CHAN, S.T. (1986). A study of heavy gas effects on the atmospheric dispersion of dense gases. In de Wispelaere, C., Schiermeier, F.A. and Gillani, N.V. (1986), *op. cit.*, p. 723

ERMAK, D.L. and CHAN, S.T. (1988). Recent developments on the FEM3 and SLAB atmospheric dispersion models. In Puttock, J.S. (1988b), *op. cit.*, p. 261

ERMAK, D.L., CHAN, S.T., MORGAN, D.J. and MORRIS, L.K. (1982). A comparison of dense gas dispersion model simulations with Burro series LNG spill test results. *J. Haz. Materials,* 6 (1/2), 129. See also in Britter, R.E. and Griffiths, R.F., (1982), *op. cit.*, p. 129

ERMAK, D.L., GOLDWIRE, H.C., HOGAN, W.J., KOOPMAN, R. and McRAE, T.G. (1983). Results of 40-m^3 LNG spills on water. In Hartwig, S. (1983), *op. cit.*, p. 163

ERMAK, D.L., KOOPMAN, R.P., MCRAE, T.G. and HOGAN, W.J. (1982). LNG spill experiments: dispersion, RPT, and vapor burn analysis. *Am. Gas Ass., Operating Section Proc.*, p. T-203

ERMENC, E.D. (1970). Controlling nitric oxide emission. *Chem. Engng*, **77**(Jun. 1), 193

ERNST, G.W. and NEWELL, A. (1969). *GPS: A Case Study in Generality and Problem-solving* (New York: Academic Press)

ERNST, W. (1983). Fire caused by vinyl chloride release from rail tank car. *Loss Prevention and Safety Promotion 4*, vol. 1, p. N1

ERNST, W. (1989). Severe explosion in a paint factory. *Loss Prevention and Safety Promotion 6*, p. 33.1

ERNST, W.R. (1992). Approximation of the error function (letter). *Comput. Chem. Engng*, **16**, 225

ERSKINE, J.B. (1973). Acoustic design for a petrochemical plant. *Chem. Engr, Lond.*, **274**, 312

ERSKINE, J.B. (1976a). Condition monitoring in the heavy chemical industry using noise and vibration measures. *Diagnostic Techniques* (London: Instn Chem. Engrs)

ERSKINE, J.B. (1976b). Inspection techniques for heavy chemical plant pressure vessels: current restraints and scope for future developments in NDT applications. *Diagnostic Techniques* (London: Instn Chem. Engrs)

ERSKINE, R. (1980). Vessel inspection procedure and philosophy. *Ammonia Plant Safety*, **22**, 98

ERSKINE, R. (1986). Inspection and maintenance problems with ageing ammonia plants. *Ammonia Plant Safety*, **26**, 9

ERVIN, M.A. and LUSS, D. (1970). Effect of fouling on stability of adiabatic packed bed reactors. *AIChE J.*, **16**, 979

ESARY, J.D. and PROSCHAN, F. (1963). Coherent structures of non-identical components. *Technometrics*, **5**, 191

ESARY, J.D. and ZIEHMS, H. (1975). Reliability analysis of phased missions. In Barlow, R.E., Fussell, J.B. and Singpurwalla, N.D. (1975), *op. cit.*

ESCHENBACH, R.C. and AGNEW, J.T. (1958). Use of the constant-volume bomb technique for measuring burning velocity. *Combust. Flame*, **2**, 273

ESCHENBRENNER, G. and WAGNER, G.A. (1972). A new high capacity ammonia converter. *Ammonia Plant Safety*, **14**, 51

ESCHENBRENNER, G.P., HONIGSBERG, C.A. and IMPAGLIAZZO, A.M. (1968). Fabrication problems with heavy pressure equipment for large ammonia plants. *Ammonia Plant Safety*, **10**, 43

ESCHENBRENNER, G.P., THIELSCH, H. and CLARK, W.D. (1968). Air ammonia panel. *Ammonia Plant Safety*, **10**, 115

ESCHENROEDER, A., IRVINE, E., LLOYD, A., TASHIMA, C. and TRAN, K. (1980). Computer simulation models for assessment of toxic substances. In Haque, R. (1980a), *op. cit.*, p. 323

ESCOE, A.K. (1986). *Mechanical Design of Process Systems*, vols 1–2 (Houston, TX: Gulf Publ. Co.)

ESCOE, A.K. (1992). Maximum allowable pressures for nozzle openings. *Hydrocarbon Process.*, **71**(5), 95

ESCUDIER, M.P. (1972). Aerodynamics of a burning turbulent gas jet in a crossflow. *Combust. Sci. Technol.*, **4**, 293

ESCUDIER, M.P. (1975). Analysis and observations of inclined turbulent flame plumes. *Combustion Sci. Technol.*, **10**, 163

ESPARZA, E.D. (1986). Blast measurements and equivalency for spherical charges at small scaled distances. *Int. J. Impact Engng*, **4**, 23

ESPARZA, E.D. and BAKER, W.E. (1977a). *Measurement of blast waves from Bursting Pressurized Frangible Spheres. Rep. NASA CR-2843* (Washington, DC: NASA Sci. and Tech. Inf. Office)

ESPARZA, E.D. and BAKER, W.E. (1977b). *Measurement of Blast Waves from Bursting Pressurized Frangible Spheres with Flash Evaporating Vapour or Liquid. Rep. NASA CR-2811* (Washington, DC: NASA Sci. and Tech. Inf. Office)

ESPENSON, J.H. (1981). *Chemical Kinetics and Reaction Mechanism* (New York: McGraw-Hill)

ESPLUND, D. and SCHILDWACHTER, J. (1962). Wash away your compressor problems. *Ind. Engng Chem.*, **54**(10), 35

ESPOSITO, F.M. and ALARIE, Y. (1989). Inhalation toxicity of carbon monoxide and hydrogen cyanide gases released during thermal decomposition of polymers. *J. Fire Sciences*, **6**(3), 195

ESRIG, M.I., AHMAD, S. and MAYO, C. (1975). Settlement of ammonia storage tanks. *Ammonia Plant Safety*, **17**, 93

ESSENHIGH, R.H. (1967). Combustion phenomena in industrial flames. The Rosin equation. *Ind. Engng Chem.*, **59**(7), 52

ESSENHIGH, R.H. (1968). Dust explosion research. *Loss Prevention*, **2**, 44

ESSENHIGH, R.H. and WOODHEAD, D.W. (1958). Speed of flame in slowly moving clouds of cork dust. *Combust. Flame*, **2**, 365

ESSENHIGH, G.L. (1991). On-site emergency planning and the use of predictive techniques. *J. Loss Prev. Process Ind.*, **4**(1), 44

ESSEX COUNTY COUNCIL (n.d.). *Canvey Island: Advice to Householders* (Castle Point)

ESSINGER, J.N. (1973). A closer look at turbo-machinery alignment. *Hydrocarbon Process.*, **52**(9), 185

ESSINGER, J.N. (1974). Hot alignment too complicated? *Hydrocarbon Process.*, **53**(1), 99

ESTEVA, L. (1967). Criteria for the construction of spectra for seismic design. *Third Pan-American Symp. on Structures* (Caracas, Venezuela)

ESTEVA, L. (1974). Geology and predictability in the assessment of seismic risk. *Proc. Second Int. Conf. Ass. Engng. Geologists*

ESTEVA, L. and ROSENBLUETH, E. (1964). Spectra of earthquakes at moderate and large distances. *Soc. Mex. de Ing. Sismica*, **2**, 1

ETCHELLS, J.C. (1994). The protection of reactors containing exothermic reactions: an HSE view. *Hazards*, **XII**, p. 377

ETHENAKIS, V.M., SCHATZ, K.W. and ZAKKAY, V. (1991). Modeling of water spraying of field releases of hydrogen fluoride. *Modeling and Mitigating the Consequences of Accidental Releases*, p. 403

ETHERIDGE, R. (1985). Control of centrifugal pumps. *Chem. Engr, Lond.*, **411**, 46

ETLING, D. (1990). On plume meandering under stable stratification. *Atmos. Environ.*, **24A**, 1979

ETOWA, C.B., AMYOTTE, P.R., PEGG, M.J. and KHAN, F.I. (2002). Quantification of inherent safety aspects of the dow indices. *J. of Loss Prev. Process Ind* **15**, 477–87

ETTLING, B.V. (1978). Electrical wiring in building fires. *Fire Technol.*, **14**, 317

EUBANK, C.S., RABINOWITZ, M.J., GARDINER, W.C. and ZELLNER, R.E. (1981). Shock-initiated ignition of natural gas-air mixtures. *Combustion*, **18**, 1767

EUGEN, T. (1982). *The Perception of Odors* (New York: Academic Press)

EULENBERG, H. (1865). *Die Lehre von den schädlichen Gasen und Dampfen* (Braunschweig)

EUROPEAN CHEMICAL INDUSTRY ECOLOGY AND TOXICOLOGY CENTRE (1991). *Emergency Exposure Indices for Industrial Chemicals* (Brussels)

EUROPEAN COMMISSION (1977). *Public Health Risks of Exposure to Asbestos* (Oxford: Pergamon Press)

EUROPEAN COMMITTEE OF MANUFACTURERS OF COMPRESSORS, VACUUM PUMPS AND PNEUMATIC TOOLS (n.d.a). *Pneumatic Tools – Safety in operation*

EUROPEAN COMMITTEE OF MANUFACTURERS OF COMPRESSORS, VACUUM PUMPS AND PNEUMATIC TOOLS (n.d.b). *Safety Recommendations for the Use and Operation of Portable Pneumatic Tools*

EUROPEAN COMMITTEE OF MANUFACTURERS OF COMPRESSORS, VACUUM PUMPS AND PNEUMATIC TOOLS (1986). *Recommendations for the Proper Use of Hand Held and Hand Operated Pneumatic Tools*

EUROPEAN CONVENTION FOR CONSTRUCTIONAL STEELWORK (1985). *Design Manual on the European Recommendations for the Fire Safety of Steel Structures* (Brussels)

EUROPEAN INFORMATION CENTRE FOR EXPLOSION PROTECTION (1984). *Explosion Protection in Practice. Pt 2, Dust Explosion Protection* (Hove-Antwerp)

EUROPEAN NUCLEAR ENERGY AGENCY (1970). *Water Cooled Reactor Safety* (Paris)

EUROPEAN PARLIAMENT (1979). *Working Documents 1979–1980, Doc. 117/79*, 23 April 1979 (Strasbourg), p. 246

EUROPEAN PETROLEUM ORGANISATIONS (1974). *European Model Code of Safe Practice for Dealing with Oil Spills at Sea or on Shore* (London: Appl. Sci. Publs)

EUROPEAN PROCESS SAFETY CENTRE (1993). *Annual Report* (Rugby)

EUROPEAN PROCESS SAFETY CENTRE (1994). *Safety Management Systems* (Rugby: Instn Chem. Engrs)

EUROSTAT (1991). *European Industry Reliability Handbook* (Paris)

EUTECTIC + CASTOLIN LTD (1979). Protective maintenance welding. *Plant Engr*, **23**(2), 20

EVANGER, T., FAGERTUN, J.A., HALS, T.E. and MAGNUSSEN, B. (1989). PIA – an advanced computer program system for simulation of blow-down of process equipment under hazardous conditions. *Loss Prevention and Safety Promotion 6*, p. 101.1

EVANGER, T., FAGERTUN, J.A., HALS, T.E. and MAGNUISSEN, B.H. (1990). PIA: a computer program for simulation of blow-down of process equipment under hazardous conditions. *Proc. Eur. Seminar 14 on Heat Transfer and Major Technological Hazards* (ed. L. Bolle), 19/1

EVANS, A. (1994). Food and drink warning for chemical incidents. *Fire Int.*, **142**(Feb.), 11

EVANS, A. and PUTTOCK, J.S. (1986). Experiments on the ignition of dense flammable gas clouds. *Loss Prevention and Safety Promotion 5*, p. 35–1

EVANS, A.A. (1989). Mathematical solution of burnt region temperatures for explosions in spherical vessels. *Proc. Fourth Australian Heat and Mass Transfer Conf., Christchurch, New Zealand*, p. 559

EVANS, A.A. (1990). The determination of the burning velocity from explosion data obtained in spherical vessels. *CHEMECA 90*

EVANS, A.A. (1991). Deflagrations in spherical vessels: a theoretical analysis of the pressure-time relationship. *Process Safety and Environment*, **69B**, 90

EVANS, A.A. (1992a). Deflagration in spherical vessels: laminar burning velocity formulae derived from the pressure ratio approximation. *Process Safety Environ.*, **70B**, 189

EVANS, A.A. (1992b). Deflagrations in spherical vessels: an assessment of 'technically correct' laminar burning velocity formulae. *Process Safety Environ.*, **70B**, 115

EVANS, A.G. (1975). Flixborough disaster – appraisal of 20 inch/28 inch bridging pipe assembly. Rep. to Flixborough Court of Inquiry (HM Factory Inspectorate)

EVANS, C.L. (1967). The toxicity of hydrogen sulphide and other sulphides. *Quart., J. Ex Physiol.*, **52**, 231

EVANS, C.W. (1974). *Hose Technology* (Barking: Applied Science)

EVANS, D. and PFENNING, D. (1985). Water sprays suppress gas-well blow-out fires. *Oil Gas J.*, **83**(Apr. 29), 80

EVANS, D.J. (1970). Non-destructive testing in the field. *Chem. Engng Prog.*, **66**(9), 66

EVANS, E.M. and WHITTLE, J. (1974). Base injection of foam to fight oil-tank fires. *Fire Prev. Sci. Technol.*, **8**, 21

EVANS, F.L. (1962). *Maintenance Supervisor's Handbook* (Houston, TX: Gulf Publ. Co.)

EVANS, F.L. (1967). The new look in HPI maintenance. Hydrocarbon Process., **46**(1), 114

EVANS, F.L. (1969). The best of recent maintenance management ideas. *Hydrocarbon Process.*, **48**(1), 107

EVANS, F.L. (ed.) (1971). *Equipment Design Handbook for Refineries and Chemical Plants*, vols 1 and 2 (Houston, TX: Gulf Publ. Co.)

EVANS, F.L. (ed.) (1974a). Refiners face corrosion facts. *Hydrocarbon Process.*, **53**(4), 109

EVANS, F.L. (ed.) (1974b). Valves for the HPI. *Hydrocarbon Process.*, **53**(6), 87

EVANS, F.L. (1979). *Equipment Design Handbook for Refineries and Chemical Plants*, vol. 1, 2nd ed. (Houston, TX: Gulf Publ. Co.)

EVANS, F.W., MEYER, and OPPLIGER, W. (1977). Safety considerations in the development and design of an industrial nitration plant. *Loss Prevention and Safety Promotion 2*, p. 191

EVANS, H. (1988). A guide to IT for the process engineer. *Process Engng*, Mar., 37

EVANS, H.E. (1984). *Mechanisms of Creep Fracture* (Amsterdam: Elsevier Applied Science)

EVANS, I. (1988). Inverter drives: are we playing with fire? *Chem. Engr, Lond.*, **445**, 55

EVANS, J., HENDRY, S., MILES, J.C., PAPPIN, J. and STUBBS, R. (1987). A non-linear site response analysis for earthquake loading using an explicit time-stepping finite element code. In Cermak, A.S. (1987), *op. cit.*, p. 609

EVANS, J.D. (1989). *Piper Alpha Public Inquiry. Transcript, Days 163, 167*

EVANS, J.L. (1985). Fixed foam extinguishing systems. *Fire Surveyor*, **14**(6), 4

EVANS, J.L. (1988). Foams for flammable liquid fires: choice and evaluation. *Fire Surveyor*, **17**(5), 5

EVANS, J.M. and VERDEN, M.L. (1982). Safety considerations for synfuels plants. *Chem. Engng Prog.*, **78**(4), 69

EVANS, J.R. (1987). Petrochemical gas shipping developments. *Gastech 86*, p. 336

EVANS, L. (1974). *Selecting Engineering Materials for Chemical Process Plant* (London: Business Books)

EVANS, L.B., JOSEPH, B. and SEIDER, W.D. (1977). System structures for process simulation. *AIChE J.*, **23**, 658

EVANS, M.G.K. and PARRY, G.W. (1983). Quantification of the contribution to light water reactor core melt frequency of loss of offsite power. *Reliab. Engng*, **6**, 43

EVANS, M.G.K., PARRY, G.W. and WREATHALL, J. (1984). On the treatment of common-cause failures in system analysis. *Reliab. Engng*, **9**(2), 107

EVANS, M.W. and ABLOW, C.M. (1961). Theories of detonation. *J. Chem. Rev.*, **129**, 129

EVANS, M.W., SCHEER, M.D., SCHOEN, L.J. and MILLER, E.L. (1949). A study of high velocity flames developed by grids in tubes. *Combustion*, **3**, 168

EVANS, R.A. (1974). Fault trees and cause-consequence charts. *IEEE Trans. Reliab.*, **R-23**, 1

EVANS, R.A. (1975). Statistical independence and common-mode failures (editorial). *IEEE Trans. Reliab.*, **R-24**, 289

EVANS, R.F. (1994). Requirements for validation of mathematical models in safety cases. *Hazards*, **XII**, p. 263

EVANS, T.W. (1979). *Nuclear Litigation* (New York: Practising Law Institute)

EVANS, U.R. (1960). *Corrosion and Oxidation of Metals* (London: Arnold); Supplement 1968

EVERETT, K. (1969). *Safety Handbook* (Leeds, Yorkshire: Univ. of Leeds)

EVERETT, K. (1979). The designing of chemical engineering laboratories. *Instn Chem. Engrs, Sym on Safety in Chemical Engineering Departments* (London)

EVERETT, K. and HUGHES, D. (1975). *A Guide to Laboratory Design* (London: Butterworths)

EVERETT, W.S. (1964). Causes and cures of pressure pulsation. *Hydrocarbon Process.*, **43**(8), 117

EVERTON, A., HOLYOAK, J. and ALLEN, D. (1990). Fire Safety and the Law, 2nd ed. (Borehamwood: Paramount)

EVERTON, A.R. (1972). Fire and the Law (London: Butterworths)

EVES, D. (1994). Safety management: are we learning the lessons? *Process Safety Prog.*, **13**, 4

EVISON, F.F. (1963). Earthquakes and faults. *Bull. Seismol. Soc. Am.*, **53**, 873; *Economic Benefits Rep. EPA-530-SW-87-026* (Washington, DC)

EVITT, S.D. and MUKADDAM, W.A. (1988). Expert systems in the HPI. *Hydrocarbon Process.*, **67**(1), 92-5

EWALDS, H. and GIELISSE, J. (1984). Assessment of acceptable weld defect size in an LPG storage tank. In *Transport and Storage of LPG and LNG* (Antwerp: Koninklijke Vlaamse Ingenieursvereniging), p. 277

EWAN, B.C.R. and MOODIE, K. (1986). Structure and velocity measurements in underexpanded jets. *Combust. Sci. Technol.* **45**, 275

EWAN, B.C.R., MOODIE, K. and HARPER, P.J. (1988). A study of two phase critical pipe flow models for superheated liquid releases. *J. Haz. Materials*, **20**, 273

EWART CHAINBELT CO. LTD (1977). Getting the best out of conveyors and elevators. *Plant Engr*, **21**(10), 20

EWELL, T.I. and CROWL, D.A. (2002). Thermodynamic availability analysis applied to the characterization of reactive chemicals. AIChE 36th Loss Prev. Symp., New Orleans, Paper LPS-5d Mar. 13

EWING, C.T., FAITH, F.R., HUGHES, J.T. and CARHART, H.W. (1989a). Evidence for flame extinguishment by thermal mechanisms. *Fire Technol.*, **25**(3), 195

EWING, C.T., FAITH, F.R., HUGHES, J.T. and CARHART, H.W. (1989b). Flame extinguishment properties of dry chemicals. Fire Technol., **25**(32), 134

EXCELL, P.S. and HOWSON, D.P. (1978). Performance of long dipoles as unintended receiving antennas. *IERE Conf. on Electromagnetic Compatibility*

EXCELL, P.S. and MADDOCKS, A.J. (1984). Gain, effective aperture and efficiency of industrial structures acting as electrically-small loop antennas. *IERE Conf. on Electromagnetic Compatibility*

EXPERT COMMISSION FOR SAFETY IN THE SWISS CHEMICAL INDUSTRY (1988). Static electricity: rules for plant safety. *Plant/Operations Prog.*, **7**, 1

EXXON (n.d.). Damage estimates from BLEVEs. Unpublished work, quoted by CCPS

EYCOTT, R. (1989). H&V systems: the importance of chemical cleaning. *Plant Engr, Mar./Apr.*, 30

EYRES, D.J. (1978). *Ship Construction* (London: Heinemann)

EYSTER, L.A.E. and MEEK, W.J. (1920). Experiments on the pathological physiology of acute phosgene poisoning. *Am. J. Physiol*, **51**, 303

EZEKOYE, L.I. (1979). Power operated relief valve actuator selection: a comparative study. In Haupt, R.W. and Meyer, R.A. (1979), *op. cit.*, p. 55

EZRA, A.A. (1958). The plastic response of a simply supported beam to an impact load at its centre. *Proc. Third US Nat. Conf. on Applied Mechanics*, p. 513

FABER, J. and ALSOP, D. (1979). *Reinforced Concrete Simply Explained* (Oxford: Oxford Univ. Press)

FABER, J. and JOHNSON, B. (1979). *Foundation Design Simply Explained* (Oxford: Oxford Univ. Press)

FABER, M. (1982). Reaction forces in explosion pressure relief. *The Control and Prevention of Dust Explosions* (London: Oyez/IBC), paper 8

FABIAN, C.L. (1982a). ASPEN - a new process simulator in debut. *Chem. Engng*, **89**(May 31), 59

FABIAN, C.L. (1982b). EPA's new bubble: not everyone's delighted. *Chem. Engng*, **89**(Jul. 24), 24

FABIAN, P. et al. (1993). Demystifying the selection of mist eliminators. *Chem. Engng*, **100**(11), 148; **100**(12), 106

FABIAN, H.W., REHER, P. and SCHON, M. (1979). How Bayer incinerates waste. *Hydrocarbon Process.*, **58**(4), 183

FABIANI, P. (1986). POSEIDON - a key of the future subsea production. *Gastech 85*, p. 176

FABIC, S. (1967). *Computer Program WHAM for Calculation of Pressure, Velocity and Force Transients in Liquid Filled Piping Networks. Rep. 67-49-R.* Kaiser Engineers

FABRICK, A.J. (1982). CHARM: an operational model for predicting the fate of elevated or surface releases of dense or buoyant gases. *Am. Met. Soc., Third Joint Conf. on Applications of Air Pollution Modeling*

FACER, J.R. and RICH, R.W. (1984). Shutdown winterization of a 1500 ton per day ammonia plant. *Ammonia Plant Safety*, **24**, 56. See also *Plant/Operations Prog.*, 1984, **3**, 12

FACER, R.I. (1980). *High reliability design of fluid flow systems in the nuclear industry. Interflow 80* (Rugby: Instn Chem. Engrs)

FACEY, B. (1991). Full scale system verification through logic modelling. *Safety and Reliability*, **11**(2), 21

FACKRELL, J.E. (1984a). An examination of simple models for building influenced dispersion. *Atmos. Environ.*, **18**, 89

FACKRELL, J.E. (1984b). Parameters characterising dispersion in the near wake of buildings. *J. Wind Engng Ind. Aerodynamics*, **16**, 97

FACKRELL, J.E. and ROBINS, A.G. (1982). Concentration fluctuations and fluxes in plumes from point sources in a turbulent boundary layer. *J. Fluid Mech.*, **117**, 1

FACTORY INSURANCE ASSOCIATION (n.d.). *Fire Prevention and Protection for Chemical Plants* (Hartford, CT)

FACTORY MUTUAL (1983). A large loss for K Mart: a lesson for others. *Record*, **60**(3), 3

FACTORY MUTUAL ENGINEERING CORPORATION (1967). *Handbook of Industrial Loss Prevention* (New York: McGraw-Hill)

FACTORY MUTUAL ENGINEERING CORPORATION (1977). *Hydraulics of Fire Protection Systems* (Norwood, MA)

FACTORY MUTUAL INSURANCE COMPANY (1940). Properties of flammable liquids, gases and solids. *Ind. Engng Chem.*, **32**, 880

FACTORY MUTUAL INT. (1991a). Planned maintenance. *Plant Engr*, **35**(May/Jun.), 29

FACTORY MUTUAL INT. (1991b). Pressure vessels. *Plant Engr*, **35**(Jul./Aug.), 32

FACTORY MUTUAL RESEARCH CORPORATION (1990). Private communication to CCPS

FACTORY MUTUAL RESEARCH CORPORATION (1991). Comparative study on the performance of the US standard spray sprinkler and the European conventional style sprinkler. *Fire Technol.*, **27**, 350

FACTORY MUTUAL RESEARCH CORPORATION − see also Appendix 28

FADEL, T.M. (1987). The safe way to install restriction orifices. *Chem. Engng*, **94**(5), 114

FADER, B. (1982). *Industrial Noise Control* (New York: Wiley)

FAETH, G.M. (1967). Monopropellant droplet burning at low Reynolds numbers. *Combust. Flame*, **11**, 167

FAETH, G.M. (1977). Current status of drop and liquid combustion. *Prog. Energy Combust. Sci.*, **3**, 191

FAETH, G.M. (1979). Spray combustion model − a review. *Seventeenth Aerospace Sciences Mtg*

FAETH, G.M. (1983). Evaporation and combustion of sprays. *Prog. Energy Combust. Sci.*, **9**, 1

FAGAN, F.L. (1976). Design and code inspections to reduce errors in program development. *IBM Syst. J.*, **15**, 182

FAGERLUND, A.C. (1976). Pipewall vibrations reveal . . . valve-generated noise. *Hydrocarbon Process.*, **55**(10), 147

FAGERLUND, A. (1984). Predicting aerodynamic noise from control valves. *Chem. Engng*, **91**(May 14), 65

FAIR, E.W. (1973). Shall we buy that new equipment? *Chem. Engng*, **80**(Oct. 15), 124

FAIR, E.W. (1974). Before you buy, what about maintenance costs? *Chem. Engng*, **81**(Mar. 4), 132

FAIR, G.M., GEYER, J.G. and OKUN, D.A. (1966). *Water and Waste Water Engineering* (New York: Wiley)

FAIR, J.R., CROCKER, B.B. and NULL, H.R. (1972a). Sampling and analyzing trace quantities. *Chem. Engng*, **79**(Sep. 18), 147

FAIR, J.R., CROCKER, B.B. and NULL, H.R. (1972b). Trace-quantity engineering. *Chem. Engng*, **79**(Aug. 7), 60

FAIRCHILD, E.J. (1979). *Suspected Carcinogens* (Tunbridge Wells: Castle House)

FAIRCLOUGH, A.J. (1986). Environmental action in the European Community. *Chemy Ind.*, Jun. 2, 388

FAIRCLOUGH, J.V. (1969). Mechanical handling. In Handley, W. (1969), *op. cit.*, p. 188

FAIRCLOUGH, M.P. (1988). Plant availability modelling techniques. *SRS Quart. Dig.*, Jan. 3

FAIRHALL, L.T. (1957). *Industrial Toxicology*, 2nd ed. (Baltimore, MD: Williams & Wilkins)

FAIRHURST, S. and TURNER, R.M. (1993). Toxicological assessments in relation to major hazards. *J. Haz. Materials*, **33**, 215

FAIRLEY, W.B. (1977). Evaluating the 'small' probability of a catastrophic accident from the marine transportation of liquefied natural gas. In Fairley, W.B. and Mosteller, F. (1977), *op. cit.*, p. 123

FAIRLEY, W.B. and MOSTELLER, F. (eds) (1977). *Statistics and Public Policy* (Reading, MA: Addison-Wesley)

FAIRWEATHER, M. and VASEY, M.W. (1982). A mathematical model for the prediction of overpressures generated in totally confined and vented explosions. *Combustion*, **19**, p. 546

FAIRWEATHER, M., JONES, W.P. and MARQUIS, A.J. (1988). Predictions of the concentration field of a turbulent jet in a cross-flow. *Combust. Sci. Technol.*, **62**, 61

FAIRWEATHER, M., JONES, W.P., LINDSTEDT, R.P. and MARQUIS, A.J. (1991). Predictions of a turbulent reacting jet in a crossflow. *Combust. Flame*, **84**, 361

FAIRWEATHER, M., JONES, W.P., LEDIN, H.S. and LINDSTEDT, R.P. (1992). Predictions soot formation in turbulent, non-premixed propane flames. *Combustion*, **24**, 1067

FAIRWEATHER, M., JONES, W.P. and LINSTEDT, R.P. (1992). Predictions of radiative transfer from a turbulent reacting jet in a cross wind. *Combust. Flame*, **89**, 45

FAITH, W.L., KEYES, D.B. and CLARK, R.L. (1965). *Industrial Chemicals*, 2nd ed. (London: Wiley)

FALCK-MUUS, R. (1967). New process solves nitrate corrosion. *Chem. Engng*, **74**(Jul. 3), 108

FALCKE, F.K. and LORENTZ, G. (eds) (1985). *Handbook of Acid-Proof Construction* (Weinheim: VCH)

FALCONER, P. and DRURY, J. (1975). *Building and Planning for Industrial Storage and Distribution* (London: Architectural Press/Halsted Press)

FALCK, A. (1992). *VT VESSEL* (Hovik, Norway: DnV Technica)

FALICOFF, W. and POPOVICH, J. (1981). Solar-energy method for temperature maintenance of residual oil storage tanks. *API Proc., Refinery Dept, 46 Midyear Mtg*, **60**, 164

FALK, K.G. (1906). The ignition temperature of hydrogen−oxygen mixtures. *J. Am. Chem. Soc.*, **28**, 1517

FALK, K.G. (1907). The ignition temperatures of gaseous mixtures. *J. Am. Chem. Soc.*, **29**, 1536

FALK, L.L. *et al.* (1953). *Savannah River Plant Stack Gas Dispersion and Microclimate Survey. Rep. DP-19.* E.I. Dupont de Nemours

FALLOWS, R.G. (1982). Confusing hazard and risk (letter). *Chem. Engr, Lond.*, **378**, 103

FALTINGS, B. and STRUSS, P. (eds) (1991). *Recent Advances in Qualitative Physics* (Cambridge, MA: MIT Press)

FALTUS, F. (1985). *Joints with Fillet Welds* (Amsterdam: Elsevier)

FAN, B.C. and SICHEL, M. (1989). A comprehensive model for the structure of dust detonations. *Combustion*, **22**, 1741

FAN, J.Y., NIKOLAOU, M. and WHITE, R.E. (1993). An approach to fault diagnosis of chemical processes via neural networks. *AIChE J.*, **39**, 82

FANAKI, F. (1984). Diffusion of heavy gases. In de Wispelaere, C. (1984), *op. cit.*, p. 329

FANARITIS, J.P. and BEVEVINO, J.W. (1976). How to select the optimum shell-and-tube heat exchangers. *Chem. Engng*, **83**(Jul. 5), 62

FANG, J.B. (1975). *Measurement of the Behaviour of Incidental Fire in a Compartment. Rep. NBS IR 75−679* Nat. Bureau Standards, Washington, DC

FANGER, P.O. and CHRISTENSEN, N.K. (1986). Perception of draught in ventilated spaces. *Ergonomics*, **29**, 215

FANNELOP, T.K. and WALDMAN, G.D. (1972). Dynamics of oil slicks. *AIAA J.*, **10**, 506

FANNELOP, T.K. and ZUMSTEG, F. (1986). Special problems in heavy gas dispersion. In Hartwig, S. (1986), *op. cit.*, p. 123

FANTI, M., CHUNG, P.W.H. and RUSHTON, A.G. (1993). Resolving ambiguity in qualitative models (applied to fault propagation) through high level constraints. *Instn Chem. Engrs Res. Event 93* (Rugby)

FARDELL, P.J. and HOUGHTON, B.W. (1976). The evaluation of an improved method of gas-freeing an aviation fuel storage tank. *J. Haz. Materials*, **1**(3), 237

FARIDANY, E.K. (1982). World LNG trade: present and future. *Hydrocarbon Process.*, **61**(1), 68−C

FARIDANY, E.K. and FFOOKS, R.C. (1983). The Ocean Phoenix multi-lobe tank for LPG carriers. *Gastech 82*, p. 175

FARINOLA, G. and LANGANA, V. (1979). Why barge mounted ammonia–urea plants. *Hydrocarbon Process.*, **58**(9), 202

FARKAS, A. (1970). What you should know ... catalytic hydrocarbon oxidation. *Hydrocarbon Process.*, **49**(7), 121

FARLEY, H.E. (1979). Choosing and applying line-blind valves. *Chem. Engng*, **86**(Oct. 8), 121

FARMER, D. (1975). *Health and Safety at Work – An Appraisal for Management* (London: Keith Shipman Developments)

FARMER, D. (1977). *Legal accountability of the chemical producer. Chemical Engineering in a Hostile World* (London: Clapp & Poliak)

FARMER, D. (1982). Hygienists monitor dusts and gases. *Health and Safety at Work*, Aug., 15

FARMER, D.H. (1978). Loss management. *Plant Engr*, **22**(1) Jan./Feb., 19

FARMER, E. (1932). *The Causes of Accidents* (London: Pitman)

FARMER, E. and CHAMBERS, E.G. (1926). *A Psychological Study of Individual Differences in Accident Rates. Ind. Fatigue Res. Board, Rep. 38* (London: HM Stationery Office)

FARMER, E. and CHAMBERS, E.G. (1939). *A Study of Accident Proneness among Motor Drivers. Ind. Health Res. Board, Rep. 84* (London: HM Stationery Office)

FARMER, F.R. (1967a). Siting criteria – a new approach. *Atom*, **128**(Jun.), 152

FARMER, F.R. (1967b). Siting criteria – a new approach. *Br. Nucl. Energy Soc. J.*, **6**(3), 219

FARMER, F.R. (1969a). *Major Safety Aspects for Construction of HTGRs and AGRs in More Densely Populated Areas. Proc. Sci. IAEA STI/PUB/197*

FARMER, F.R. (1969b). Progress towards the identification of safety requirements for advanced reactors. *Br. Nucl. Energy Soc., Symp. on Safety and Siting*, p. 123

FARMER, F.R. (1970). Reactor safety – the carrot or the stick? In National Academy of Engineering (1970), *op. cit.*, p. 106

FARMER, F.R. (1971). Experience in the reduction of risk. *Major Loss Prevention*, p. 82

FARMER, F.R. (1975). Accident probability data. *J. Inst. Nucl. Engrs*, **16**, 1

FARMER, F.R. (1977a). Relationship between safety and reliability in major hazard situations. *First Nat. Conf. on Reliability. NRC6/4* (Culcheth, Lancashire: UK Atom. Energy Auth.)

FARMER, F.R. (1977b). Today's risks: thinking the unthinkable. *Nature*, **267**, 92

FARMER, F.R. (1981). Experience in the quantification of engineering risks. *Proc. R. Soc., Ser. A*, **376**, 103. See also Warner, Sir F. and Slater, D.H. (1981), *op. cit.*, p. 103

FARMER, F.R. (1982). The general scene in the United Kingdom. In Green, A.E. (1982), *op. cit.*, p. 571

FARMER, F.R. and GILBY, E.V. (1971). Reactor safety philosophy and experimental verification. In *Proc. Fourth Int. Conf. on Peaceful Uses of Atomic Energy* (New York: United Nations and Vienna: Int. Atom. Energy Agency), p. 187

FARMER, W.J., YANG, M.S. and LETEY, J. (1980). *Land Disposal of Hexachlorobenzene Wastes – Controlling Vapor Movement in Soil. Rep. EPA-600/2–80–119.* Environ. Prot. Agency, Durham, NC

FARQUHAR, J.T. (1982). Controlled disposal of special wastes. The industrial viewpoint. *Chemy Ind.*, Apr. 17, 259

FARQUHAR, J.T. (1986). Best practicable means. *Chemy Ind.*, Aug. 18, 541

FARRELL, T.R. (1982). The relationship between cargo characteristics and bulk chemical tanker design and construction. *Chem. Engr, Lond.*, **382**, 275

FARRIS, E.J. and GRIFFITHS, J.Q. (1949). *The Rat in Laboratory Investigation* (New York: Hafner)

FARROW, J.F. (1971). User guide to steam turbines. *Hydrocarbon Process.*, **50**(3), 71

FASANO, J.B. and EBERHART, T.M. (1980). Designing fiber-glass-reinforced plastic vessels for agitator service. *Chem. Engng*, **87**(May 5), 115

FATHI, Z., RAMIREZ, W.F. and KORBICZ, J. (1993). Analytical and knowledge-based redundancy fault diagnosis in process plants. *AIChE J.*, **39**, 42

FAUPEL, J.H. (1964). *Engineering Design* (New York: Wiley)

FAUQUIGNON, C. and CHERET, R. (1969). Generation of detonation in solid explosives. *Combustion*, **12**, 745

FAURE, J., SIBILLE, M., FAURE, H., STEPHAN, C., YACOUB, M. and MOTIN, J. (1970). Acute poisoning by chlorine and phosgene. *Le Poumon et le Coeur*, **26**, 913

FAUSKE, H.K. (1961). *Proc. Heat Transfer Fluid Mech. Inst.*, 79

FAUSKE, H.K. (1962). *Contribution to the Theory of Two-phase, One Component Critical Flow. Rep. ANL 6633.* Argonne Nat. Lab., Argonne, IL

FAUSKE, H.K. (1963). A theory for predicting pressure gradients for two-phase critical flow. *Nucl. Sci. Engng*, **17**, 1

FAUSKE, H.K. (1964). The discharge of saturated water through tubes. *Seventh Nat. Heat Transfer Conf.* (New York: Am. Inst. Chem. Engrs), p. 210

FAUSKE, H.K. (1983). Multi-phase flow considerations in sizing emergency relief systems for runaway chemical reactions. In Veziroglu, T.N. (1983), *op. cit.*, p. 203

FAUSKE, H.K. (1984a). Generalized vent sizing monogram for runaway chemical reactions. *Plant/Operations Prog.*, **3**, 213

FAUSKE, H.K. (1984b). A quick approach to reactor vent sizing. *Plant/Operations Prog.*, **3**, 145

FAUSKE, H.K. (1984c). Scale-up for safety relief runaway reactions. *Plant/Operations Prog.*, **3**, 7

FAUSKE, H.K. (1985a). Emergency relief system (ERS) design. *Chem. Engng Prog.*, **81**(8), 53

FAUSKE, H.K. (1985b). Flashing flows – some practical guidelines for emergency releases. *Plant/Operations Prog.*, **4**, 132

FAUSKE, H.K. (1986a). Disposal of two-phase emergency releases. *Proc. Fourth Int. Symp. on Multiphase Transport and Particulate Phenomena* (ed. T.N. Veziroglu)

FAUSKE, H.K. (1986b). Turbulent and laminar flashing flow considerations. *Proc. Fourth Int. Symp. on Multiphase Transport and Particulate Phenomena* (ed. T.N. Veziroglu)

FAUSKE, H.K. (1986c). Two-phase liquid swell considerations in storage vessels subjected to external heat source. *Proc. Fourth Int. Symp. on Multiphase Transport and Particulate Phenomena* (ed. T.N. Veziroglu)

FAUSKE, H.K. (1987a). Pressure relief and venting: some practical considerations related to hazard control. *Hazards from Pressure* (Rugby: Instn Chem. Engrs), p. 133

FAUSKE, H.K. (1987b). Relief system design for runaway chemical reactions. *Preventing Major Chemical Accidents*, p. 3.17

FAUSKE, H.K. (1988a). Emergency relief system design for reactive and non-reactive systems: extension of the DIERS methodology. *Plant/Operations Prog.*, **7**, 153

FAUSKE, H.K. (1988b). The next step in accommodating major chemical accidents. *Preventing Major Chemical and Related Process Accidents*, p. 185

FAUSKE, H.K. (1989a). Emergency relief system design for runaway chemical reaction: extension of the DIERS methodology. *Chem. Engng Res. Des.*, **67**, 199

FAUSKE, H.K. (1989b). Preventing explosions during chemical and materials storage. *Plant/Operations Prog.*, **8**, 181

FAUSKE, H.K. (1990). Practical containment concepts in connection with short duration high rate two-phase discharges. *J. Loss Prev. Process Ind.*, **3**(1), 130

FAUSKE, H.K. and EPSTEIN, M. (1987). Source term considerations in connection with chemical accidents and vapor cloud modeling. *Vapor Cloud Modeling*, p. 251

FAUSKE, H.K. and EPSTEIN, M. (1988). Source term consideration in connection with chemical accidents and vapour cloud modelling. *J. Loss Prev. Process Ind.*, **1**(2), 75

FAUSKE, H.K. and EPSTEIN, M. (1989). Hazardous vapor clouds – release type, aerosol formation and mitigation. *Loss Prevention and Safety Promotion 6*, p. 69.1

FAUSKE, H.K. and GROLMES, M.A. (1992). Mitigation of hazardous emergency release source terms via quench tanks. *Plant/Operations Prog.*, **11**, 121

FAUSKE, H.K. and HENRY, R.E. (1983). Scale considerations and vapor explosions. In Veziroglu, T.N. (1983), *op. cit.*, p. 263

FAUSKE, H.K. and LEUNG, J.C. (1985). New experimental technique for characterizing runaway chemical reactions. *Chem. Engng Prog.*, **81**(8), 39

FAUSKE, H.K. and MIN, T.C. (1963). *A Study of the Flow of Saturated Freon 11 through Apertures and Short Tubes. Rep. ANL 6667.* Argonne Nat. Lab., Argonne, IL

FAUSKE, H.K., GROLMES, M.A., HENRY, R.E. and THEOFANOUS, T.G. (1980). Emergency pressure relief systems associated with flashing liquids. *Swiss Chem.*, **2** (7/8), 73

FAUSKE, H.K., GROLMES, M.A. and HENRY, R.E. (1983). Emergency relief systems – sizing and scaleup. *Plant/Operations Prog.*, **2**(1), 27

FAUSKE, H.K., GROLMES, M.A. and LEUNG, J.C. (1984). Multiphase flow considerations in sizing emergency relief systems for runaway chemical reactions. In *Multi-phase Flow and Heat Transfer III, Pt B: Applications* (ed. T.N. Veziroglu and A.E. Bergles) (Amsterdam: Elsevier), p. 899

FAUSKE, H.K., EPSTEIN, M., GROLMES, M.A. and LEUNG, J.C. (1986). Emergency relief vent sizing for fire emergencies involving liquid filled atmospheric storage vessels. *Plant/Operations Prog.*, **5**, 205

FAUSKE, H.K., GROLMES, M.A. and CLARE, G.H. (1989). Process safety evaluation applying DIERS methodology to existing plant operations. *Plant/Operations Prog.*, **8**, 19

FAUSKE, H.K., CLARE, G.H. and CREED, M.J. (1989). Laboratory tool for characterizing chemical systems. *Runaway Reactions*, p. 364

FAVERGE, J.M. (1970). L'homme, agent d'infiabilite' et de fiabilite' des processus industriels. *Ergonomics*, **13**, 301

DE FAVERI, D.M., HANZEVACK, E.L. and DELANEY, B.T. (1982). Dispersion from safety valves and other momentum emission sources: intermittent. *Atmos. Environ.*, **16**, 1239, 2535

DE FAVERI, D.M., PASTORINO, R., FUMAROLA, G. and FERRAIOLO, G. (1984). Mixing of gas clouds by water barriers. *J. Occup. Accid.*, **5**(4), 257

DE FAVERI, D.M., FUMAROLA, G., ZONATO, C. and FERRAIOLO, G. (1985). Estimate flare radiation intensity. *Hydrocarbon Process.*, **64**(5), 89

DE FAVERI, D.M., VIDILI, A., PASTORINI, R. and FERRAIOLO, G. (1989). Wind effects on diffusion of flames of high source momentum. *J. Haz. Materials*, **22**(1), 85

DE FAVERI, D.M., ZONATO, C., PAGELLA, C., VIDILI, A. and FERRAIOLO, G. (1989). Runaway reaction and safety measures in organic processes. *Runaway Reactions*, p. 155

DE FAVERI, D.M., ZONATO, C., VIDILI, A. and FERRAIOLO, G. (1989). Theoretical and experimental study on the propagation of heat inside deposits of coal. *J. Haz. Materials*, **20**, 135

FAWBERT, J. (1985). Going for open tender on plant equipment. *Process Engng*, Mar., 36

FAWCETT, H.H. (1965a). The literature of chemical safety. *J. Chem. Educ.*, **42**(10), A815; **42**(11), A897

FAWCETT, H.H. (1965b). Other personal protective equipment. In Fawcett, H.H. and Wood, W.S. (1965), *op. cit.*, p. 438

FAWCETT, H.H. (1965c). Radiation – controllable energy. In Fawcett, H.H. and Wood, W.S. (1965), *op. cit.*, p. 347

FAWCETT, H.H. (1965d). Respiratory hazards and protection. In Fawcett, H.H. and Wood, W.S. (1965), *op. cit.*, p. 373

FAWCETT, H.H. (1965e). Safer experimentation: probing the frontiers while respecting the unknown. In Fawcett, H.H. and Wood, W.S. (1965), *op. cit.*, p. 290

FAWCETT, H.H. (1965f). Toxicity as a hazard. In Fawcett, H.H. and Wood, W.S. (1965), *op. cit.*, p. 279

FAWCETT, H.H. (1981). The changing nature of acute chemical hazards: a historical perspective. *J. Haz. Materials*, **4**(4), 313

FAWCETT, H.H. (1982a). Chemical health and safety: is our future secure? In Fawcett, H.H. and Wood, W.S. (1982), *op. cit.*, p. 1

FAWCETT, H.H. (1982b). Chemical wastes: new frontiers for the chemist and engineers – RCRA and Superfund. In Fawcett, H.H. and Wood, W.S. (1982), *op. cit.*, p. 597

FAWCETT, H.H. (1982c). Other personal protective equipment. In Fawcett, H.H. and Wood, W.S. (1982), *op. cit.*, p. 583

FAWCETT, H.H. (1982d). Respiratory hazards and protection. In Fawcett, H.H. and Wood, W.S. (1982), *op. cit.*, p. 529

FAWCETT, H.H. (1982e). Safer experimentation: probing the frontiers while respecting the unknown. In Fawcett, H.H. and Wood, W.S. (1982), *op. cit.*, p. 421

FAWCETT, H.H. (1982f). Toxicity: chemicals react with living systems. In Fawcett, H.H. and Wood, W.S. (1982), *op. cit.*, p. 261

FAWCETT, H.H. (1982g). Toxicity versus hazard. In Fawcett, H.H. and Wood, W.S. (1982), *op. cit.*, p. 245

FAWCETT, H.H. (1985). *Hazardous and Toxic Materials: Safe Handling and Disposal* (New York: Wiley- Interscience)

FAWCETT, H.H. (ed.) (1988). *Hazardous and Toxic Materials*, 2nd ed. (New York: Wiley)

FAWCETT, H.H. and WOOD, W.S. (1965). *Safety and Accident Prevention in Chemical Operations* (New York: Wiley)

FAWCETT, H.H. and WOOD, W.S. (1982). *Safety and Accident Prevention in Chemical Operations*, 2nd ed. (New York: Wiley)

FAWCETT, R.W. and KLETZ, T.A. (1982). One organisation's memory: the use of a computerised system to store and retrieve information on loss prevention. *Plant/Operations Prog.*, **1**, 7

FAWCETT, S. (1977). The design of nuclear power plant for high reliability. *First Nat. Conf. on Reliability. NRC2/5.* (Culcheth, Lancashire: UK Atom. Energy Auth.)

FAWELL, J. (2001). Balancing the chemical and microbial risks in drinking water. *Microl. Pathog. Disinfect. By-Prod Drink. Water*, 1–57881–1, pp. 503–13

FAY, J. (1952). A mechanical theory of spinning detonation. *J. Chem. Phys.*, **20**, 942

FAY, J.A. (1962). The structure of gaseous detonation waves. *Combustion*, **8**, p. 30

FAY, J.A. (1969). The spread of oil on a calm sea. In Hoult, D.P. (1969), *op. cit.*

FAY, J.A. (1970). Oil spills, need for law and science. *Technol. Rev.*, **72**, 33

FAY, J.A. (1973). Unusual fire hazard of LNG tanker spills. *Combust. Sci. Technol.*, **7**, 47

FAY, J.A. (1980). Gravitational spread and dilution of heavy vapour clouds. *Proc. Second Int. Symp. on Stratified Flows* (Trondheim: Norwegian Inst. of Technol.), vol. 1, p. 471

FAY, J.A. (1984). Experimental observations on entrainment rates in dense gas dispersion tests. In Ooms, G. and Tennekes, H. (1984), *op. cit.*, p. 39

FAY, J.A. (1988). Modeling the Lake Nyos gas disaster. *Atmos. Environ.*, **22**, 417

FAY, J.A. and HOULT, D.P. (eds) (1969). *Aerophysics of Air Pollution* (Am. Inst. Aeronaut. and Astronaut.)

FAY, J.A. and LEWIS, D.H. (1975). The inflammability and dispersion of LNG vapor clouds. *Transport of Hazardous Cargoes*, 4

FAY, J.A. and LEWIS, D.H. (1977). Unsteady burning of unconfined fuel vapor clouds. *Combustion*, **16**, 1397

FAY, J.A. and MACKENZIE, J.J. (1972). Cold cargo. *Environment*, **14**(9), 21

FAY, J.A. and RANCK, D. (1981). *Scale Effects in Liquefied Fuel Vapor Dispersion. Rep. DOE/EP-0032.* Dept of Energy, Washington, DC

FAY, J.A. and RANCK, D.A. (1983). Comparison of experiments on dense gas cloud dispersion. *Atmos. Environment*, **17**, 239

FAY, J.A. and ZEMBA, S.G. (1985). Dispersion of initially compact dense gas clouds. *Atmos. Environ.*, **19**, 1257

FAY, J.A. and ZEMBA, S.G. (1986). Integral model of dense gas plume dispersion. *Atmos. Environ.*, **20**, 1347

FAY, J.A. and ZEMBA, S.G. (1987). Modeling the dispersion of initially compact dense gas clouds. *Vapor Cloud Modeling*, p. 439

FAY, J.A., DESGROSEILLIERS, G.J. and LEWIS, D.H. (1979). Radiation from burning hydrocarbon clouds. *Combust. Sci. Technol.*, **20**, 141

FAYED, A.S. and OLTEN, L. (1983). Comparing measured with calculated multiphase flow pressure drop. *Oil Gas J.*, **81**(Aug. 22), 136

FAZZALARI, F.A. (ed.) (1978). *Compilation of Odor and Taste Threshold Values Data* (Philadelphia, PA: Am. Soc. Testing and Materials)

FEARNEHOUGH, G.D. (1974). Fracture propagation control in gas pipelines: a survey of relevant studies. *Int. J. Pres. Ves. Piping*, **2**, 257

FEARNEHOUGH, G.D. (1985). The control of risk in gas transmission pipelines. *The Assessment and Control of Major Hazards* (Rugby: Instn Chem. Engrs), p. 25

FEARNLEY, P. and NETTLETON, M.A. (1983). Pressures generated in blast waves in channels with 90° bends. *Loss Prevention and Safety Promotion 4*, vol. 3, p. E34

FEATES, F. (1977). Is disposal necessary? *Chemical Engineering in a Hostile World* (London: Clapp & Poliak)

FEATES, F.S. (1978). Spills experience – costs and solutions. *Chemy Ind.*, Feb. 18, 118

FEATHERSTONE, D.F. (1962). *War Games* (London: Stanley)

FEAY, B.A. and BOWEN, J.R. (1973). A model of the ignition of cylindrically-confined explosive gas mixtures. *Combust. Flame*, **20**, 111

FEAY, B.A., DAUBERT, T.E., DANNER, R.P., HIGH, M.S. and RHODES, C.L. (1984). Interactive computer for DIPPR data. *Chem. Engng Prog.*, **80**(8), 55

FEDERAL EMERGENCY MANAGEMENT AGENCY (1989). *Handbook of Chemical Hazard Analysis Procedures* (Washington, DC)

FEDERAL ENERGY REGULATORY AGENCY (1978). *Environmental Impact Statement, Western LNG Project. Draft Rep. FERC/EIS-0002* (Washington, DC)

FEDERAL ENERGY REGULATORY COMMISSION (1977). *Initial Decision Upon Applications to Import LNG from Algeria, El Paso Eastern Company et al. (Dockets CP 77–330 et al.) Oct. 25, 1977* (Washington, DC)

FEDERAL POWER COMMISSION (1966). *Safety of Interstate Natural Gas Pipelines* (Washington, DC: US Govt Printing Office)

FEDERAL POWER COMMISSION (1974). *Order Requiring Cessation of Operations and Compliance with the Conditions of Temporary Authorisation, Chattanooga Gas Company (Docket CP 73–329) Mar. 13* (Washington, DC)

FEDERAL POWER COMMISSION, BUREAU OF NATURAL GAS (1973). *Preliminary Staff Report on Investigation of Disaster at Texas Eastern Transmission Corporation LNG Storage Tank on Staten Island, Borough of Richmond, New York City, New York, February 10, 1973 (Docket CP 73–235, Jul. 9)* (Washington, DC)

FEDERAL POWER COMMISSION, BUREAU OF NATURAL GAS (1974). *Final Environmental Impact Statement for the Construction and Operation of an LNG Terminal at Staten Island, New York, July, 1974* (Washington, DC)

FEDERAL POWER COMMISSION, BUREAU OF NATURAL GAS (1976). *Final Environmental Impact Statement for the Construction and Operation of an LNG Import Terminal at Everett, Massachusetts (Port of Boston) (Docket CP 73–135 et al.) Distrigas of Massachusetts Corporation et al.* (Washington, DC)

FEDERAL POWER COMMISSION, BUREAU OF NATURAL GAS (1977). *Matagorda Bay Project Final Environmental Impact Statement* (Washington, DC)

FEDEROFF, B.T. (ed.) (1960). *Encyclopaedia of Explosives and Related Compounds* (Dover, NJ: Piccatinny Arsenal)

FEDOR, O.H., PARSONS, W.N. and DE COUTINHO, S.J. (1975). Risk management technique for design and operation of LNG facilities and equipment. *Transport of Hazardous Cargoes*, 4

FEELEY, F.G. (1979). Fuel conversion problems in industry. *Loss Prevention*, **12**, 45

FEELEY, F.G. (1984). Burning wastes in steam boilers. *Plant/Operations Prog.*, **3**, 31

FEELY, F.J., MARSHALL, B.T., PALMER, A.J. and SOMMER, E.C. (1980). Safety audit of refrigerated liquefied gases facilities. *API Proc., Refining Dept, 45th Midyear Mtg*, **59**, 237

FEGAN, J.P. (1981). Noise legislation. *Plant Engr*, **25**(4), 26

FEGGETTER, A.J. (1982). A method for investigating human factors aspects of aircraft accidents and incidents. *Ergonomics*, **25**, 1065

FEHLING, H.R. and LESER, T. (1949). Determination of the true composition of the products of the theoretical combustion with oxygen and oxygen/nitrogen mixtures at temperatures up to 2500°C at atmospheric pressure. *Combustion*, **3**, 634

FEIGE, N.G. and KANE, R.L. (1970). CPI applications of titanium. *Chem. Engng Prog.*, **66**(10), 53

FEIGEL, F. (1958). *Spot Tests* (Amsterdam: Elsevier)

FEIGENBAUM, A.V. (1983). *Total Quality Control*, 3rd ed. (New York: McGraw-Hill)

FEIGENBAUM, E.A. (1979). Themes and case studies of knowledge engineering. In Michie, D. (1979), *op. cit.*, p. 3

FEIGENBAUM, E.A. (1983). Knowledge engineering: the applied side. In Hayes, P.J. and Michie, D. (1983), *op. cit.*, p. 37

FEIGENBAUM, E.A. and FELDMAN, J. (eds) (1963). *Computers and Thought* (New York: McGraw-Hill)

FEIGENBAUM, E.A. and MCCORDUCK, P. (1983). *The Fifth Generation* (London: Pan Books)

FEIGENBAUM, E.A., BUCHANAN, B.G. and LEDERBERG, J. (1971). *On Generality and Problem-solving: A Case Study Using the DENDRAL Program. Machine Intelligence 6* (ed. B. Meltzer and D. Michie) (Edinburgh: Edinburgh Univ. Press), p. 165

FEINBERG, J. (1974). *Doing and Deserving: Essays in the Theory of Responsibility* (Princeton, NJ: Princeton Univ. Press)

FEIND, K. (1975). Reducing vapor loss in ammonia tank spills. *Ammonia Plant Safety*, **17**, 114

FEIND, K. (1978). Flame-resistant insulation for ammonia tanks. *Ammonia Plant Safety*, **20**, 46

FEINGOLD, A. (1966). Radiant-interchange configuration factors between surfaces. *Proc. R. Soc., Ser. A*, **292**, 51

FELDBAUER, G.F., HEIGL, J.J., MCQUEEN, W., WHIPP, R.H. and MAY, W.G. (1972). *Spills of LNG on Water − Vaporization and Downwind Drift of Combustible Mixtures. Rep. EE61E- 72.* Esso Res. and Engng Co.

FELDMAN, E.J. (1987). Specifying electric motors. *Chem. Engng*, **94**(7), 37

FELDMAN, R. (1974). Subliming coatings for fireproofing of steel. *Loss Prevention*, **8**, 65

FELDMAN, R.J. and GROSSEL, S.S. (1968). Safe design of an ethylene plant. *Loss Prevention*, **2**, 131

FELDMAN, R.P. (1969). Economics of plant start-up. *Chem. Engng*, **76**(Nov. 3), 87

FELICI, N.J. (1979). Bubbles, partial discharges and liquid breakdown. *Electrostatics*, **5**, 181

FELICI, N.J. and LARIGALDIE, S. (1980). Experimental study of a static discharger for aircraft with special reference to helicopters. *J. Electrostatics*, **9**, 59

FELLENSIEK, J., BUCHHOLZ, R., ONKEN, U., OVERHOFF, K.H. and SCHECKER, H.G. (1987). Water trap flame arresters in pipework on chemical plants. *Fortschritte der Sicherheitstechnik 1* (Frankfurt-am-Main: DECHEMA)

FELLER, W. (1951). *An Introduction to Probability Theory and Its Applications*; 1965, 2nd ed.; 1968, 3rd ed. (New York: Wiley)

FELLS, I. (1980). Nuclear power in an age of reason. *J. Inst. Nucl. Engrs*, **21**(4), 102

FELLS, I. (1981). Nuclear energy − the way ahead. *Sci. Public Policy*, **8**, 217

FELLS, I. and RUTHERFORD, A.G. (1969). Burning velocity of methane−air flames. *Combust. Flame*, **13**, 130

THE FELLOWSHIP OF ENGINEERING (1991a). The Fellowship of Engineering Guidelines for People Making or Receiving Warnings of Disaster. In The Fellowship of Engineering (1991b), *op. cit.*, p. 1

THE FELLOWSHIP OF ENGINEERING (1991b). *Preventing Disasters* (London)

FELSKE, J.D. and TIEN, C.L. (1973). Calculation of the emissivity of luminous flames. *Combustion Sci. Technol.*, **7**, 25

FELSKE, J.D. and TIEN, C.L. (1974a). Infrared radiation from nonhomogeneous gas mixtures having overlapping bands. *J. Quant. Spectroscop. Radiat. Transfer*, **14**, 35

FELSKE, J.D. and TIEN, C.L. (1974b). A theoretical closed form expression for the total band absorptance of infrared-radiation gases. *Int. J. Heat Mass Transfer*, **17**, 155

FELSTEAD, D.K., ROGERS, R.L. and YOUNG, D.G. (1983). The effect of electrode characteristics on the measurement of minimum ignition energy of dust clouds. *Electrostatics*, **6**, 105

FENDELL, F.E. and MITCHELL, J.A. (1993). Capping clouds and smoke densities above big fires. *Combust. Sci. Technol.*, **92**, 35

FENG, C.C. and SIRIGNANO, W.A. (1977). Further calculations based on a theory of flame spread across solid fuels. *Combust. Flame*, **29**, 247

FENG, C.C., WILLIAMS, F.A. and KASSOY, D.R. (1975). Flame propagation through layered fuel-air mixtures. *Combustion Sci. Technol.*, **10**, 59

FENIMORE, C.P. and JONES, G.W. (1969). Coagulation of soot to smoke in hydrocarbon flames. *Combust. Flame*, **13**, 303

FENIMORE, C.P. and MARTIN, F.J. (1966). Flammability of polymers. *Combust. Flame*, **10**, 135

FENLON, W.J. (1984). A comparison of ARC and other thermal stability test methods. *Plant/Operations Prog.*, **3**, 197

FENLON, W.J. (1987). Calorimetric methods used in the assessment of thermally unstable or reactive chemicals. *Preventing Major Chemical Accidents*, p. 4.31

FENN, J.B. (1951). Lean flammability limit and minimum spark ignition energy. *Ind. Engng Chem.*, **43**, 2865

FENN, W.O. and RAHN, H. (eds) (1964). *Handbook of Physiology, Sec. 3, Respiration*, vol. 1 (Washington, DC: Am. Physiol. Soc.)

FENNELL, D. (1988). *Investigation into the Kings Cross Underground Fire* (London: HM Stationery Office)

FENNELL, THE HON MR JUSTICE (1991). Disaster inquiries: the view of the judiciary. In The Fellowship of Engineering (1991b), *op. cit.*, p. 112

FENNER, O. (1964). Plastics for process equipment. *Chem. Engng*, **71**(Nov. 9), 170

FENNER, O.H. (1967). Where to use plastic materials. *Loss Prevention*, **1**, 44

FENNER, O.H. (1968). Plastic materials of construction. *Chem. Engng*, **75**(Nov. 4), 126

FENNER, O.H. (1970). Why materials fail. Plastics. *Chem. Engng*, **77**(Oct. 12), 23

FENNING, R.W. (1926). Gaseous combustion at medium pressures. *Phil. Trans. R. Soc., Ser. A*, **225**, 331

FENSOM, D.H. and CLARK, B. (1984). Tantalum − its uses in the chemical industry. *Chem. Engr, Lond.*, **406**, 46

FENSTERMACHER, T.E., WOODARD, K. and ADDERSON, N. (1987). A quick running personal computer model for the evaporation and dispersion of dense gases. *Vapor Cloud Modeling*, p. 342

FENTON, R. (1977). Lubricating grease. *Plant Engr*, **21**(10), 18; **21**(11), 13; 1978, **22**(1), 16

FERCHO, W.W. and RINGER, L.J. (1972). Small sample power of some tests of constant failure rate. *Technometrics*, **14**, 713

FERGE, D.T. (1972). Dynamic pressure testing. *Chem. Engng Prog.*, **68**(9), 85

FERGUS, E., HEDLEY, D., RIDDELL, I. and HENNELL, M. (1991). Software testing tools. In Ince, D. (1991a), *op. cit.*, p. 56

FERGUSON, D.M. (1976). Short term exposure limits. *Ann. Occup. Hyg.*, **19**, 275

FERGUSON, G. and ANDOW, P.K. (1986). Process plant safety and artificial intelligence. *World Cong. of Chemical Engineering, Tokyo*, p. 1092

FERGUSON, J.M., MCLEAN, D. and SAKAI, D. (1989). Structural assessments for a new generation of spherical tank LNG carriers. *Gastech 88*, paper 5.7

FERGUSON, P. (1975). Gas men abandon Canvey earth-pit fight. *New Civil Engr*, Nov. 20, 20

FERGUSON, R.A.D. (1981). *Comparative Risks of Electricity Generating Fuel Systems in the UK* (Stevenage: Peter Peregrinus)

FERGUSON, R.A.D. (1982). Risk assessment and evaluation. *Sci. Public Policy*, **9**, 251

FERGUSON, R.E. and YOKLEY, C.R. (1959). Products of cool flame oxidation of propane-2-^{13}C. *Combustion*, **7**, 113

FERKISS, V.C. (1969). *Technological Man: The Myth and the Reality* (New York: New American Library)

FERLAND, R.H. (1988). Packing options for rotating equipment. *Hydrocarbon Processing*, **67**(4), 39

FERN, A.G. and NAU, B. (1976). *Seals* (Oxford: Oxford Univ. Press)

FERNANDES, R.C., SEMIAT, R. and DUKLER, A.E. (1983). Hydrodynamic model for gas–liquid slug flow in vertical tubes. *AIChE J.*, **29**, 981

FERNANDEZ-PELLO, A.C. (1977a). Downward flame spread under the influence of externally applied thermal radiation. *Combust. Sci. Technol.*, **17**, 1

FERNANDEZ-PELLO, A.C. (1977b). Upward laminar flame spread under the influence of externally applied thermal radiation. *Combust. Sci. Technol.*, **17**, 87

FERNANDEZ-PELLO, A.C. (1984). Flame spread modeling. *Combust. Sci. Technol.*, **39**, 119

FERNANDEZ-PELLO, A.C. and HIRANO, T. (1983). Controlling mechanisms of flame spread. *Combust. Sci. Technol.*, **32**, 1

FERNANDEZ-PELLO, A.C. and MAO, C.P. (1981). A unified analysis of concurrent modes of flame spread. *Combust. Sci. Technol.*, **26**, 147

FERNANDEZ-PELLO, A.C. and WILLIAMS, F.A. (1974). Laminar flame spread over PMMA surfaces. *Combustion*, **15**, 217

FERNANDEZ-PELLO, A. and WILLIAMS, F.A. (1977). A theory of laminar flame spread over flat surfaces of solid combustibles. *Combust. Flame*, **28**, 251

FERNEY, M.J. (1991). Process simulators for safety. *Plant/Operations Prog.*, **10**, 133

FERRADA, J.J. and HOLMES, J.M. (1990). Developing expert systems. *Chem. Engng Prog.*, **86**(4), 34

FERRAIOLO, G. (1979). *La Chimica e l'Industria*, **61**, 108

FERRARA, G.M. (1979). *The Disaster File: the 1970s* (Kingsport, TN: Kingsport Press)

FERRARA, V. and CAGNETTI, P. (1980). A simple model for estimating airborne concentrations downwind of buildings for discharges near ground level. *Seminar on Radioactive Releases and Their Dispersion in the Atmosphere Following a Hypothetical Reactor Accident* (Riso, Denmark: Comm. Eur. Commun.)

FERRIS, B.G., BURGESS, W.A. and WORCESTER, J. (1967). Prevalence of chronic respiratory disease in a pulp mill and a paper mill in the United States. *Br. J. Ind. Med.*, **24**, 26

FERRIS, B.G., PULEO, S. and CHEN, H.Y. (1979). Mortality and morbidity in a pulp and a paper mill in the United States: a ten-year follow-up. *Br. J. Ind. Med.*, **36**, 127

FERRIS, J. (2002). Inherently safer technology and the risk management program. *17th Annu. Int. Conf. and Workshop: Risk, Reliability, and Security*, Oct. 8–11, 2002, Jacksonville, FL, Regulation of Inherent Safety Workshop

FERRIS, T.V. (1974). The explosion of methanol-air mixtures at above atmospheric conditions. *Loss Prevention*, **8**, 15

FERROW, M. (1989). *Piper Alpha Public Inquiry. Transcript, Days 158, 159*

FERRY (1930). What happened on the Yser? *Revue des Vivants*, Jul.

FERRY, TED S. (1988). *Modern Accident Investigation and Analysis*, 2nd Edition (New York) John Wiley & Sons ISBN 0–471–62481–0

FERRY, GENERAL (1935). The first gas attack. Gaz de Combat. Defense Passive. *Feu. Securite. No. 3*, May

FERRY, T.S. (1988). *Modern Accident Investigation and Analysis* (New York: Wiley)

FERTILIO, A. and PRINCIP, B. (1975). Inspect reformer tubes for repair. *Hydrocarbon Process.*, **54**(9), 174

FERTILISER MANUFACTURERS ASSOCIATION (1989). Industrial guidance on the safety case: 2, Ammonium nitrate. In Lees, F.P. and Ang, M.L. (1989b), *op. cit.*, p. 155

FESSLER, GEBELE and PRANDTL (1933). *Gaskampfstoffe- Gasvergiftungen* (Munich: Gmelin)

FEUGIER, A. (1972). Soot oxidation in laminar hydrocarbon flames. *Combust. Flame*, **19**, 249

FEULESS, S.C. and MADHAVEN, S. (1988). 7 steps to shorter turnarounds. *Chem. Engng*, **95**(Mar. 14), 126

FEUTZ, R.J. and WALDECK, T.A. (1965). The application of fault tree analysis to dynamic systems. In Boeing Co. (1965), *op. cit.*

FFOOKS, R.C. (1977). A review of current LNG ship technology – and an attempt to rationalise. *LNG Conf 5*

FICKETT, W. and DAVIS, W.C. (1979). *Detonation* (Berkeley, CA: Univ. of California Press)

FIEDLER, F. and TANGERMANN-DLUGI, G. (1980). Turbulent dispersion of heavy gases in the atmospheric boundary layer after ground level release. In Hartwig, S. (1980), *op. cit.*, p. 157

FIELD, E.M. (1963). The growth structure and breakdown of magnetite films on mild steel. *Proc. Second Int. Metallic Corrosion Conf., New York*, p. 829

FIELD, P. (1982). *Dust Explosions* (Amsterdam: Elsevier)

FIELD, P. (1983). *Explosibility Assessment of Industrial Powders and Dusts* (London: HM Stationery Office)

FIELD, P. (1984). Dust explosion protection – a comparative study of selected methods for sizing explosion relief vents. *J. Haz. Materials*, **8**(3), 223

FIELD, P. (1985). Effective sprinkler protection for high racked storage. *Fire Surveyor*, **14**(5), 9

FIELD, P. and MURRELL, J.W. (1988). The fire hazard and protection of bin storage. *Fire Surveyor*, **17**(6), 5

FIELDER, E. *et al.* (1990). Methanol. *Ullmanns Encyclopaedia of Chemical Technology*, rev. ed., vol. A16 (Weinheim: VCH), p. 465

FIELDHOUSE, K.N. (1970). Techniques for identifying sources of noise and vibration. *Sound Vibration*, **4**(12), 14

FIELDING, L. and MONDY, R.E. (1981). A better way to balance turbo-machinery. *Hydrocarbon Process.*, **60**(1), 97

FIELDING, L.W., PRESTON, M.L. and SINCLAIR, P.A. (1986). Gas dispersion modelling and its application in ICI. The disp2 model. *Loss Prevention and Safety Promotion 5*, p. 34.1

FIELDNER, A.C., KATZ, S.H. and KINNEY, S.P. (1921). *Gas Masks for Gases Met in Fighting Fires. Tech. Paper 248*. Bureau of Mines, Washington, DC

FIERZ, H. (1994). Assessment of thermal safety during distillation of DMSO. *Hazards*, **XII**, p. 563

FIERZ, H. and ZWAHLEN, G. (1989). Deflagration in solids – a special hazard in chemical production. *Loss Prevention and Safety Promotion*, **6**, p. 12.1

FIERZ, H., FINCK, P., GIGER, G. and GYGAX, R. (1983). Thermally stable operating conditions of chemical processes. *Loss Prevention and Safety Promotion 4*, vol. 3, p. A12

FIERZ, H., FINCK, P., GIGER, G. and GYGAX, R. (1984). Thermally stable operation of chemical processes. *Chem. Engr, Lond.*, **400**, 9

FIFE, I. and MACHIN, E. (1963). *The Offices, Shops and Railway Premises Act 1963* (London: Butterworths)

FIFE, I. and MACHIN, E. (1972). *Redgrave's Factories Acts*, 22nd ed. (London: Butterworths)

FIFE, I. and MACHIN, E. (1982). *Redgrave's Health and Safety in Factories*, 2nd ed. (London: Butterworths)

FIFE, I. and MACHIN, E.A. (1988). *Redgrave's Health and Safety in Factories*, 2nd ed. (London: Butterworths)

FIFE, I. and MACHIN, E.A. (1990). *Health and Safety* (London: Butterworths)

FIGG, J. (1979). Corrosion of steel in concrete. *Chem. Ind.*, Jan. 20, 39

FIGG, J. (1983). Chloride and sulphate attack on concrete. *Chem. Ind.*, Oct. 17, 770

FIJAS, D.F. (1989). Getting top performance from heat exchangers. *Chem. Engng*, **96**(12), 141

FIKES, R.E. and NILSSON, N.J. (1971). STRIPS: a new approach to the application of theorem proving to problem solving. *Artificial Intelligence*, **2**, 198

FIKES, R.E., HART, P.E. and NILSSON, N.J. (1972). Learning and executing generalised robot plans. *Artificial Intelligence*, **3**, 1

FIKSEL, J. (1985). Quantitative risk analysis for toxic chemicals in the environment. *J. Haz. Materials*, **10**(2/3), 227

FILE, W.T. (1991). *Cost Effective Maintenance* (Oxford: Butterworth-Heinemann)

FILETTI, E.G. and TRUMPLER, P.R. (1977). Increase plant availability with trend monitoring. *Hydrocarbon Process.*, **56**(9), 233

FILIPCZYNSKI, L. (1960). *Ultrasonic Testing of Materials* (London: Butterworths)

FILLER, W.S. (1974). *The Effect of a Case on Airblast Measurements. Pt 1, Friable Inert Cases. Rep. NOLTR 74-62.* Naval Ordnance Systems Command, Washington, DC

FILLO, J.P. and KEYWORTH, C.J. (1992). SARA Title III – a new era of corporate responsibility and accountability. *J. Haz. Materials*, **31**, 219

FILSTEAD, C.G. (1965). CAMEL LNG plant, world's largest. *Hydrocarbon Process.*, **44**(7), 135

FILSTEAD, C.G. and ROOKE, D.E. (1970). Six years' operational experience with the 'Methane Princess' and 'Methane Progress'. *LNG Conf, 2*

FINANCIAL TIMES (1979). *Oil and Gas International Year Book 1979/80* (London)

FINCH, C.A. (1986). Chemical warfare – then and now? *Chem. Ind.*, Dec. 15, 870

FINCH, F.E. and KRAMER, M.A. (1988). Narrowing diagnostic focus using functional decomposition. *AIChE J.*, **34**, 25

FINCH, F.E., OYELEYE, O.O. and KRAMER, M.A. (1990). A robust event-oriented methodology for diagnosis of dynamic process systems. *Comput. Chem. Engng*, **14**, 1379

FINCH, R.N. and SERTH, R.W. (1990). Model air emission better. *Hydrocarbon Process.*, **69**(1), 75

FINDLATER, A.E. and PREW, L.R. (1977). Operational experience with LNG ships. *LNG Conf. 5*

FINDLAY, J.V. and KUHLMAN, R.L. (1980). *Leadership in Safety* (Loganville: Georgia Inst. Press)

FINDLER, N.V. and MELTZER, B. (eds) (1971). *Artificial Intelligence and Heuristic Programming* (Edinburgh: Edinburgh Univ. Press)

FINE, B. (1958). Effect of pressure on turbulent burning velocity. *Combust. Flame*, **2**, 109

FINE, D.H. and GRAY, P. (1967). Explosive decomposition of solid benzoyl peroxide. *Combust. Flame*, **11**, 71

FINE, D.H., GRAY, P. and MACKINVEN, R. (1970). Thermal effects accompanying spontaneous ignitions in gases. *Proc. R. Soc., Ser. A*, **316**, 223, 241

FINE, W.T. (1971). Mathematical evaluation for controlling hazards. *J. Safety Res.*, **3**(4), 157

FINE, W.T. (1973). Mathematical evaluations for controlling hazards. In Widner, J.T. (1973), *op. cit.*, p. 68

FINELT, S. and CRUMP, J.R. (1977). Pick the right water reuse system. *Hydrocarbon Process.*, **56**(10), 111

DE FINETTI, B. (1974). *Theory of Probability*, vols 1–2 (New York: Wiley)

FINGAS, M.F. and ROSS, C.W. (1977). *An Oil Spill Bibliography, March, 1975 to December, 1976.* Environ. Protection Service, Dept of Fisheries and Environment Canada

FINK, D.G. and BEATY, H.W. (ed.) (1978). *Standard Handbook for Electrical Engineers*, 11th ed. (New York: McGraw-Hill)

FINK, R., OPPERT, S., COLLINSON, P., COOKE, G., DHANJAL, S., LESAN, H. and SHAW, R. (1993). Data management in clinical laboratory information systems. In Redmill, F. and Anderson, T. (1993), *op. cit.*

FINK, Z.J. and VANPEE, M. (1975). Overall kinetics of hot gas ignition. *Combust. Sci. Technol.*, **11**, 229

FINKELSTEIN, E. (1982). Feeding liquefied gas to a gas–liquid reaction. *Chem. Engng*, **89**(Oct. 18), 132

FINLAYSON, K. and GANS, M. (1967). Planning the successful start-up. *Chem. Engng Prog.*, **63**(12), 33

FINLEY, B.H., WEBSTER, R.G. and SWAIN, A.D. (1974). Reduction of human errors in field test programs. *Hum. Factors*, **16**, 215

FINLEY, F. (1972). How cost-effective is your maintenance organisation? *Hydrocarbon Process.*, **51**(1), 81

FINLEY, H.F. (1973). Total life cycle costs of plant and equipment. *Process Technol.*, Apr., 177

FINLEY, H.F. (1978). Maintenance management for today's high technology plants. *Hydrocarbon Process.*, **57**(1), 101

FINLEY, H.F. (1987). Reduce costs with knowledge-based maintenance. *Hydrocarbon Process.*, Jan., 64

FINLEY, R.W. (1980). Incipient failure detection in rotating machinery. *Chem. Engng*, **87**(Jul. 14), 104

FINN, D.P. (1980). Select equipment drive to cut operating energy costs. *Chem. Engng*, **87**(Mar. 24), 121

FINNECY, E.E. (1982). Incineration at sea. In Bennett, G.F., Feates and Wilder (1982), *op. cit.*, p. 14.37

FINNECY, E.E. (1983). Aspects of control of hazardous waste disposal in the European Community. *Chem. Ind.*, Jun. 6, 416

FINNERAN, J., SWEENEY, N.J. and HUTCHINSON, T.G. (1968). Start-up performance of large ammonia plants. *Chem. Engng Prog.*, **64**(8), 72

FINNERAN, J.A., BUIVIDAS, L.J. and WALEN, N. (1972). Advanced ammonia technology. *Hydrocarbon Process.*, **51**(4), 127

FINNEY, D.J. (1941). On the distribution of a variate whose logarithm is normally distributed. *J. R. Stat. Soc.*, **7**(Suppl.), 155

FINNEY, D.J. (1971). *Probit Analysis* (London: Cambridge Univ. Press)

FINNEY, J.C. (1973). What to specify for gear reliability. *Hydrocarbon Process.*, **52**(12), 53

FINNIE, I. (1963). Use graphs to predict furnace tube life. *Hydrocarbon Process.*, **42**(3), 155

FINNIE, I. and HELLER, W.R. (1959). *Creep of Engineering Materials* (New York: McGraw-Hill)

FIOCK, E.F. (1955). Measurement of burning velocity. In *High Speed Aerodynamic and Jet Propulsion*, vol. IX, pt 2 (ed.

B. Lewis, R.N. Pease and H.S. Taylor) (London: Geoffrey Cumberledge/Oxford Univ. Press), p. 409

FIOCK, E.F. and MARVIN, C.F. (1937a). The measurement of flame speeds. *Chem. Rev.*, **21**, 367

FIOCK, E.F. and MARVIN, C.F. (1937b). The measurement of flame speeds. *Combustion*, **2**, p. 194

FIOCK, E.F., MARVIN, C.F., CALDWELL, F.R. and ROEDER, C.H. (1940). *Flame Speeds and Energy Considerations for Explosions in a Spherical Bomb. REP. 682.* Nat. Adv. Comm. Aero.

FIRE CONTROL ENGINEERING COMPANY (n.d.). *Dry Chemical Fire Extinguishing Agents*, 2nd rev. ed. (Fort Worth, TX)

FIRE JOURNAL STAFF (1967). Fatal chemical plant explosion. *Fire J.*, **61**(5), 8

FIRE JOURNAL STAFF (1973a). Distillation, sulfonating process explosion. In Vervalin, C.H. (1973a), *op. cit.*, p. 46

FIRE JOURNAL STAFF (1973b). Gasoline-tank failure and fire. In Vervalin, C.H. (1973a), *op. cit.*, p. 55

FIRE JOURNAL STAFF (1973c). Liquid propane leak in tank truck. In Vervalin, C.H. (1973a), *op. cit.*, p. 53

FIRE JOURNAL STAFF (1973d). Polystyrene explosion at Monsanto, Canada. In Vervalin, C.H. (1973a), *op. cit.*, p. 43

FIRE JOURNAL STAFF (1973e). Unusual tank rupture. In Vervalin, C.H. (1973a), *op. cit.*, p. 48

FIRE METAL PRODUCTS CORPORATION (n.d.). *Rupture Discs, Deluge Valves and Explosion Venting Assemblies*

FIRE OFFICES COMMITTEE (n.d.a). *List of Approved Portable Fire Extinguishers* (London)

FIRE OFFICES COMMITTEE (n.d.b). *Rules for Automatic Fire Alarm Installations* (London)

FIRE OFFICES COMMITTEE (1968). *Rules for Automatic Sprinkler Installations*, 29th ed. (London)

FIRE PROTECTION ASSOCIATION (n.d.). *Fire Resistance of Protected Steel Columns and Beams. Tech. Inf. Sheet 4011* (London)

FIRE PROTECTION ASSOCIATION (Annual). *Annual Report* (London)

FIRE PROTECTION ASSOCIATION (1966). Bulk storage of liquefied petroleum gases. *Fire Int.*, **1**(12), 53

FIRE PROTECTION ASSOCIATION (1974a). Fires in electric cables. *Fire Int.*, **46**, 41

FIRE PROTECTION ASSOCIATION (1974b). Increasing number and cost of fires in chemical process industries. *Fire Prev.*, **106**, 15

FIRE PROTECTION ASSOCIATION (1976). Facts and figures about £15,000 plus fires in 1974. *Fire Prev.*, **112**, 19

FIRE PROTECTION ASSOCIATION (1991). *FPA Large Fire Loss Analysis for 1989* (London)

FIRE PROTECTION ASSOCIATION – see also Appendix 28

FIRE RESEARCH STATION – see Appendix 28

FIRST, K.E. and HUFF, J.E. (1989). Design charts for two-phase flashing flow in emergency pressure relief systems. *Runaway Reactions*, p. 681. See also *Plant/Operations Prog.*, 1989, **8**, 40

FIRST, M. (1983). Engineering control of health hazards. *Am. Ind. Hyg. Ass. J.*, **44**, 621

FIRTH, J.G., JONES, A. and JONES, T.A. (1973). The principles of the detection of flammable atmospheres by catalytic devices. *Combust. Flame*, **20**, 303

FIRTH, J.G., JONES, A. and JONES, T.A. (1974). Flammable gas detectors. *Chemical Process Hazards*, V, p. 307

FISCH, E. (1984). Winterizing process plants. *Chem. Engng*, **91**(Aug. 20), 128

FISCHER, D.W. (ed.) (1982). *Managing Technological Accidents: Two Blowouts in the North Sea* (Oxford: Pergamon)

FISCHER, H.B., LIST, E.J., KOH, R.C., IMBERGER, J. and BROOKS, N. (1979). *Mixing in Inland and Coastal Waters* (London: Academic Press)

FISCHER, S.J. (1988). Study of the radiation properties of liquid pool fires. Ph.D. Thesis, Univ. of Washington

FISCHER, S.J. and GROSSHANDLER, W.L. (1989). Radiance, soot, and temperature interactions in turbulent alcohol fires. *Combustion*, **22**, 1241

FISCHER, S.J., HARDOUIN-DUPARC, B. and GROSSHANDLER, W.L. (1987). The structure and radiation of an ethanol pool fire. *Combust. Flame*, **70**, 291

FISCHER, W. (1965). *Chem. Tech. (Berlin)*, **17**(5), 298

FISCHHOFF, B. and MACGREGOR, D. (1983). Judged lethality: how much people seem to know depends upon how they are asked. *Risk Analysis*, **3**, 229

FISCHHOFF, B., HOHENEMSER, C., KASPERSON, R.E. and KATES, R.W. (1978). Handling hazards. *Environment*, **20**(7), 16

FISCHHOFF, B., SLOVIC, P., LICHTENSTEIN, S., READ, S. and COMBS, B. (1978). How safe is safe enough? A psychometric study of attitudes towards technological risks and benefits. *Policy Sciences*, **9**, 127

FISCHHOFF, B., LICHTENSTEIN, S., SLOVIC, P., DERBY, S.L. and KEENEY, R.L. (1981). *Acceptable Risk* (Cambridge: Cambridge Univ. Press)

FISCHMANN, J. (1924). *Gas Warfare* (Moscow)

FISEROVA-BERGEROVA, V. (ed.) (1983). *Modeling of Inhalation Exposure to Vapors: Uptake, Distribution and Elimination* (Boca Raton, FL: CRC Press)

FISH, A. (1966). The non-isothermal oxidation of 2–methylpentane. I, The properties of cool flames. *Proc. R. Soc., Ser. A*, **293**, 378

FISH, A. (1968). The cool flames of hydrocarbons. *Angew. Chem. Int. Ed. Engl.*, **7**(1), 45

FISH, A., READ, A., AFFLECK, W.S. and HASKELL, W.W. (1969). The controlling role of cool flames in two stage ignition. *Combust. Flame*, **13**, 39

FISHBEIN, L. (1979). *Potential Industrial Carcinogens and Mutagens* (Amsterdam: Elsevier)

FISHBEIN, M. and AJZEN, I. (1975). *Belief, Attitudes, Intention and Behavior* (Reading, MA: Addison-Wesley)

FISHBURN, B. (1976). Some aspects of blast from fuel–air explosives. *AScta Astronaut.*, **3**, 1049

FISHBURN, B., SLAGG, N. and LU, P. (1981). Blast effect from a pancake shaped fuel drop–air cloud detonation (theory and experiment). *J. Haz. Materials*, **5**, 65

FISHBURNE, E.S. and PERGAMENT, H.S. (1979). The dynamics and radiant intensity of large hydrogen flames. *Combustion*, **17**, 1063

FISHEL, F.D. and HOWE, C.D. (1979). Cut energy costs with variable frequency motor drives. *Hydrocarbon Process.*, **58**(9), 231

FISHER, A. (1991). Risk communication challenges. *Risk Analysis*, **11**(2), 173

FISHER, A., CHESTNUT, L.G. and VIOLETTE, D.M. (1989). The value of reducing risks of death: a note of new evidence. *J. Policy Analysis Management*, **8**(1), 88

FISHER, B. (1984). Operating centrifugal compressors in parallel. *Plant Engr*, Jun., 59

FISHER, B.E. (1976). Underground disaster. *Emergency Planning Digest*, **3**(1), 14

FISHER, B.E. (1978a). Mass emergency problems and planning in the UK from the perspective of the police. *Mass Emergencies*, **3**, 41

FISHER, B.E. (1978b). Training in personnel, exercises and studies of contingency planning: practical experiences of a British emergency planning officer. *Mass Emergencies*, **3**, 83

FISHER, B.E. (1980a). The Moorgate Underground disaster. In Frey, R. and Safar, P. (1980), *op. cit.*, p. 26

FISHER, B.E. (1980b). Police command structure at major incidents in Great Britain. In Frey, R. and Safar, P. (1980), *op. cit.*, p. 261

FISHER, C.H. (1982). Estimating liquid densities and critical properties of n-alkanes. *Chem. Engng*, **89**(Jan. 11), 107

FISHER, E. (1950). *Spherical Cast TNT Charges: Air Blast Measurements. Rep. NOLM 10789*. Naval Ordnance Lab., White Oak, MD

FISHER, E.M. (1953a). *Air Blast Resulting from the Detonation of Small TNT Clouds. REP. 2890*. Naval Ordnance Lab., White Oak, MD

FISHER, E.M. (1953b). *The Effect of the Steel Case on the Air Blast from High Explosives. Rep. NAVORD 2753*. Naval Ordnance Lab., White Oak, MD

FISHER, H.G. (1985). DIERS research program on emergency relief systems. *Chem. Engng Prog.*, **81**(8), 33

FISHER, H.G. (1986). Use of test information to improve emergency relief designs. *Am. Inst. Chem. Engrs, Symp on Loss Prevention*

FISHER, H.G. (1989). The DIERS Users Group: a forum for development/dissemination of emergency relief system design technology. *Plant/Operations Prog.*, **8**, 70

FISHER, H.G. (1991). An overview of emergency system design practice. *Plant/Operations Prog.*, **10**, 1

FISHER, H.G. and GOETZ, D.D. (1991). Determination of self-accelerating decomposition temperature using the Accelerating Rate Calorimeter. *J. Loss Prev. Process Ind.*, **4**, 305

FISHER, H.G. and GOETZ, D.D. (1993). Determination of self-accelerating decomposition temperatures of self-reactive substances. *J. Loss Prev. Process Ind.*, **6**, 183

FISHER, H.R. (1963). Obsolescence and optimum replacement timing. *Chem. Engr, Lond.*, **167**, CE86

FISHER, J.T. (1973). Designing emergency shutdown systems. *Chem. Engng*, **80**(Dec. 10), 138

FISHER, L.L. (1981). Vibration monitoring as a predictive maintenance tool. *Ammonia Plant Safety*, **23**, 166

FISHER, L.L. and FEENEY, R.L. (1984). How important is turbine control? *Ammonia Plant Safety*, **24**, 125. See also *Chem. Engng Prog.*, 1984, **80**(3), 47

FISHER, R.A. and TIPPETT, L.H.C. (1928). Limiting forms of the largest or smallest member of sample. *Proc. Cambridge Phil. Soc.*, V(XXIV), pt II, 180

FISHER, R.A. and YATES, F. (1957). *Statistical Tables for Biological, Agricultural and Medical Research*, 5th ed. (London: Oliver & Boyd)

FISHER, T.G. (1990). *Batch Control Systems* (Pittsburgh, PA: Instrum. Soc. Am.)

FISHER, W.R., DOHERTY, M.F. and DOUGLAS, J.M. (1988). The interface between design and control. *Ind. Engng. Chem. Res.*, **27**, 597, 606, and 611

FISHKIN, C.I. and SMITH, R.L. (1964). Handling flammable dusts. *Chem. Engng Prog.*, **60**(4), 45

FISK, A.D. and SCHNEIDER, W. (1981). Control and automatic processing during tasks requiring sustained attention: a new approach. *Hum. Factors*, **23**, 737

FISK, P.R. and HOWES, D.A. (1993). New challenges for physical chemists: laws, chemicals and regulators. *J. Loss Prev. Process Ind.*, **6**, 165

FITCH, R.K. (1986). Rail transportation systems. *Fire Protection Handbook*, 16th ed. (ed. by A.E. Cote and J.L. Linville), pp. 13–31

FITT, J.S. (1974). The process engineering of pressure relief and blowdown systems. *Loss Prevention and Safety Promotion 1*, p. 317

FITT, J.S. (1976a). Design for safety in the contractor's office. *Process Industry Hazards*, p. 175

FITT, J.S. (1976b). Pressure systems: protection, relief and blowdown. *Course on Loss Prevention in the Process Industries*. Dept of Chem. Engng, Loughborough Univ. of Technol.

FITT, J.S. (1981a). Pressure piling: a problem for the process engineer. *Chem. Engr, Lond.*, **368**, 237

FITT, J.S. (1981b). Pressure piling – a supplementary note (letter). *Chem. Engr, Lond.*, **374**, 523

FITT, J.S. (1983). Protection of distillation columns against vacuum caused by loss of heating. In *Loss Prevention and Safety Promotion 4*, vol. 3, p. G1

FITTS, P.M. (1962). Functions of man in complex systems. *Aerospace Engng*, **21**(1), 34

FITTS, P.M. and JONES, R.E. (1947). *Analysis of Factors Contributing to 460 'Pilot Error' Experiences in Operating Aircraft Controls. Memo Rep. TSEAA-694–12*. Air Materiel Command, US Air Force

FITZER, E. *et al.* (1988). Fibers, 5. Synthetic inorganic. *Ullmanns Encyclopaedia of Chemical Technology*, rev. ed., vol. A11 (Weinheim: VCH), p. 1

FITZGERALD, B. (1990). Control valves: give them attention before they fail. *Chem. Engng*, **97**(5), 90

FITZGERALD, B.P., GREEN, M.D., PENINGTON, J. and SMITH, A.J. (1990). A human factors approach to the effective design of evacuation systems. *Piper Alpha: Lessons for Life Cycle Management* (Rugby: Instn Chem. Engrs), p. 167

FITZPATRICK, J. and GODDARD, A.J.H. (1977). *Sector Population Distributions, Based on the 1971 Census in the Vicinity of Thurrock* (paper). Imperial Coll, London

FITZPATRICK, K. (1973). Polyethylene plant safety devices. *Safety in Polyethylene Plants*, **1**, 39

FITZPATRICK, K. (1974). Closures for high pressure piping systems. *Safety in Polyethylene Plants*, **2**, 12

FITZSIMMONS, P.E. and COCKRAM, M.D. (1979). Safety bursting discs. In Minors, C. (1979), paper 10

FITZSIMMONS, W.A. and HANCOCK, R.A. (1973). Safety in industrial gas-fired plant. *Instn Gas Engrs, London, Commun.* **924**. See also *Instn Gas Engrs J.*, **13**, 269

FIUMARA, A. (1977). Explosion pressure as a fundamental parameter for the safety in the process industries. *Loss Prevention and Safety Promotion 2*, p. 173

FIUMARA, A. and AVELLA, F. (1983). Effects of power frequency discharges on flammable mixtures. In *Loss Prevention and Safety Promotion 4*, vol. 1, p. M20

FIUMARA, A. and CARDILLO, P. (1976). Safety rules for the use of ethylene in the chemical industry. *Prevention of Occupational Risks*, **3**, 239

FIUMI, L. and GIORGIO, Z. (1992). An intelligent system to control quality. *Loss Prevention and Safety Promotion 7*, vol. 3, 149.1

FIZER, J.C. (1976). Transporting hazardous materials by rail. *Control of Hazardous Materials Spills*, vol. 3, p. 332

FJELD, S., ANDERSEN, T. and MYKLATUN, B. (1978). Risk analysis of offshore production and drilling platform. *Tenth Ann. Conf. on Offshore Technology*, vol. 2, p. 879

FLAKE, R.E. (1964). Chlorine inhalation (letter). *New England J. Med.*, **217**, 1373

FLAM, F. (1994). Laser Chemistry: the light choice. *Science* **266**, (Oct. 14), 215–17

FLAMBERG, S.A., TORTI, K.S. and MYERS, P.M. (1998). Reduce toxic hazards using passive mitigation. *Chem. Eng.* (July), 133–39

FLAMM, L. and MACHE, H. (1917). Combustion of an explosive gas mixture in a closed vessel. *Wien Ber.*, **126**, 9

FLANAGAN, G. (1984). Selecting a volatile organic chemical detector. *Chem. Engng Prog.*, **80**(9), 37

FLANAGAN, J.C. (1954). The critical incident technique. *Psychol. Bull*, **51**, 327

FLATZ, W. (1981). Sizing equipment for multiproduct plants. *Chem. Engng*, **88**(Jul. 13), 105

FLECK, SIR A. (1958). *Windscale Piles. Final Report of the Committee. Cmnd 417* (London: HM Stationery Office)

FLEISCHER, M.T. (1980). *SPILLS, an Evaporation/Air Dispersion Model for Chemical Spills on Land.* Shell Dev. Center, Westhollow Res. Center, Houston, TX

FLEISCHMANN, M. and OVERSTALL, J. (1983). The status of electrochemical technology. *Chem. Engr, Lond.*, **393**, 24

FLEISHMAN, A.B. (1989). *Piper Alpha Public Inquiry. Transcript, Days 136, 137*

FLEISHMAN, A.B. and HOGH, M.S. (1989). The use of cost benefit analysis in evaluating the acceptability of industrial risks – an illustrative case. *Loss Prevention and Safety Promotion 6*, p. 60.1

FLEISIG, D.J. (1975). The German explosion-proof concept for hazardous atmospheres. *Electrical Safety*, **2**, 174

FLEMING, H.W. and CROMEANS, J.S. (1971). Primary reforming catalyst. *Ammonia Plant Safety*, **13**, 85

FLEMING, J.B., LAMBRIX, J.R. and NIXON, J.R. (1976). Safety in phenol-from-cumene process. *Hydrocarbon Process.*, **55**(1), 185

FLEMING, K. (1984). Safety takes a front seat. *Hazardous Cargo Bull.*, **5**(6), 9

FLEMING, K.N. (1974). *A Reliability Model for Common Mode Failures in Redundant Safety Systems. Rep. GA-13284.* General Atomic Co.,

FLEMING, K.N. and HANNAMAN, G.W. (1976). *Common Cause Failure Considerations in the Prediction of HTGR Cooling System Reliability. Rep. GA-13658.* General Atomic Co.

FLEMING, K.N., HOUGHTON, W.J., HANNAMAN, G.W. and JOKSIMOVIC, V. (1981). Probabilistic risk assessment of HTGRs. *Reliab. Engng*, **2**(1), 17

FLEMING, R. (1958). *Scale-Up in Practice* (New York: Reinhold)

FLESCH, E. and DOURCHE, M. (1986). LNG sampling systems. *Gastech 85*, p. 494

FLESCH, E. and LOOTVOET, R. (1987). Evacuated insulating boxes giving low boil-off for gas-transport methane carriers. *Gastech 86*, p. 225

FLESSNER, M. and BJORKLUND, R. (1980). Control of gas detonations in pipes. *Loss Prevention and Safety Promotion 3*, vol. 3, p. 863

FLESSNER, M.F. and BJORKLUND, R.A. (1981). Control of gas detonations in pipes. *Loss Prevention*, **14**, 113

FLETCHER, B. (1977). Centreline velocity characteristics of rectangular unflanged hoods and slots under suction. *Ann. Occup. Hyg.*, **20**, 141

FLETCHER, B. (1982). Sudden discharge of superheated fluid to atmosphere. *The Assessment of Major Hazards* (Rugby: Instn Chem. Engrs), p. 25

FLETCHER, B. (1984a). Discharge of saturated liquids through pipes. *J. Haz. Materials*, **8**(4), 377

FLETCHER, B. (1984b). Flashing flow through orifices and pipes. *Chem. Engng Prog.*, **80**(3), 76

FLETCHER, B. and JOHNSON, A.E. (1984). The discharge of superheated liquids from pipes. *The Protection of Exo-thermic Reactors and Pressurised Storage Vessels* (Rugby: Instn Chem. Engrs), p. 149

FLETCHER, B. and JOHNSON, A.E. (1986). The capture efficiency of local exhaust ventilation hoods and the role of capture velocity. In Goodfellow, H.D. (1986), *op. cit.*, p. 369

FLETCHER, D.F. and THEOFANOUS, T.G. (1994). Recent progress in the understanding of steam explosions. *J. Loss Prev. Process Ind.*, **7**, 457

FLETCHER, E.R. (1974). Characteristics and biological effects of fragments from glass and acrylic windows broken by airblast. *Proc. DCPA All Effects Research Contractors Conf.* (Washington, DC: Defense Civil Preparedness Agency), p. 57

FLETCHER, E.R. and BOWEN, I.G. (1968). *Blast-induced Translational Effects. Rep DASA-1859.* Defense Atom. Support Agency, Dept of Defence, Washington, DC. See also *Ann. N.Y. Acad. Sci*, 1968, **152**(art. 1), 378

FLETCHER, E.R. and RICHMOND, D.R. (1970). *Blast Displacement of Prone Dummies – Project LN402. Rep. DASA-2377–1.* DASA Data Information and Analysis Center, Defense Atom. Support Agency, Santa Barbara, CA

FLETCHER, E.R. and RICHMOND, D.R. (1971). *A Model to Simulate Thoracic Responses to Air Blast and its Impact. Rep. (AD 740438).* US Air Force, Wright Patterson AFB, OH

FLETCHER, E.R. and RICHMOND, D.R. (1974). Characteristics and biological effects of fragments from glass and acrylic windows broken by air blast. *Explosives Safety Board, Dept of Defense, Proc. Sixteenth Explosives Safety Seminar*, vol. 1, p. 57

FLETCHER, E.R., ALBRIGHT, R.W., GOLDIZEN, V.C. and BOWEN, I.G. (1961). *Determinations Aerodynamic Drag Parameters of Small Irregular Objects by Means of Drop Tests. Civil Effects Test Operations, Rep. CEX-59.14.* Atom. Energy Comm., Office of Tech. Services, Dept of Commerce, Washington, DC

FLETCHER, E.R., BOWEN, I.G. and PERRET, R.F.D. (1965). *Impact Velocities of Steel Spheres Translated by Air Blast. Rep. DASA-1656.* Defense Atom. Support Agency, Dept of Defense, Washington, DC

FLETCHER, E.R., RICHMOND, D.R., BETZ, P.A. and JONES, R.K. (1971). *Blast Displacement of Dummies on the Surface – Project LN402. Rep DASA-2606–1*, vol. 1, p. 659. DASA Information and Analysis Center, Defense Atom. Support Agency, Dept of Defense, Washington, DC

FLETCHER, E.R., RICHMOND, D.R. and JONES, R.K. (1971a). *Blast Displacement of Prone Dummies. Rep. DASA-2710.* Defense Nucl. Agency, Dept of Defense, Washington, DC

FLETCHER, E.R., RICHMOND, D.R. and JONES, R.K. (1971b). Project LN402 – Blast displacement of dummies in open terrain and in field fortifications. *Defence Res. Board of Canada, Symp. Rep. on Event Dial Pack* (Ottawa), vol. 2, p. 607

FLETCHER, E.R., BOWEN, I.G., JONES, R.K., RICHMOND, D.R. and DORN, R.V. (1972). *Personnel casualties. Section 1, Air blast (U). Rep DNA EM-1 (Revision of Technical Manual 23–200).* Defense Nucl. Agency, Dept of Defence, Washington, DC

FLETCHER, E.R., RICHMOND, D.R. and JONES, R.K. (1973). Air blast effects on windows in buildings and automobiles on the Eskimo II event. *Explosives Safety Board, Dept of Defense, Proc. Fifteenth Explosives Safety Seminar*, p. 251

FLETCHER, E.R., RICHMOND, D.R. and RICHMOND, D.W. (1974). Air blast effects on windows in buildings and automobiles on the Eskimo II event. *Explosives Safety Board, Dept of Defense, Proc. Sixteenth Explosives Safety Seminar*, vol. 1, p. 185

FLETCHER, E.R., RICHMOND, D.R. and WHITE, C.S. (1974). Biological hazards. *Explosives Safety Board, Dept of Defense, Proc. Sixteenth Explosives Safety Seminar*, vol. 1, p. 1

FLETCHER, E.R., YELVERTON, J.T., HUTTON, R.A. and RICHMOND, D.R. (1975). *Probability of Injury from Airblast Displacement as a Function of Range and Yield. Rep. DNA 3779T.* Defense Nucl. Agency, Dept of Defense, Washington, DC

FLETCHER, E.R., RICHMOND, D.R. and JONES, R.K. (1976). *Velocities, Masses and Spatial Distributions of Glass Fragments from Windows Broken by Air Blast. Report.* Defense Nucl. Agency, Dept of Defense, Washington, DC

FLETCHER, E.R., RICHMOND, D.R. and YELVERTON, J.T. (1980). *Glass Fragment Hazard from Windows Broken by Air Blast. Rep. DNA 5593T.* Defense Nucl. Agency, Dept of Defence, Washington, DC

FLETCHER, J.A. and DOUGLAS, H.M. (1971). *Total Loss Control* (London: Ass. Business Programmes)

FLETCHER, J.R. (1984). New stainless steels for the process industries. *Chem. Engr, Lond.*, **403**, 39

FLETCHER, R. (1979). Maintain combustion systems. *Hydrocarbon Process.*, **58**(1), 99

FLETCHER, R.F. (1968). Liquid propellant explosions. *J. Spacecraft Rockets*, **5**(10), 1227

FLEXMAN, R.E. and STARK, E.A. (1987). Training simulators. In Salvendy, G. (1987), *op. cit.*, p. 1012

FLINT, A.R. (1981). Risks and their control in civil engineering. *Proc. R. Soc., Ser. A*, **376**, 167. See also in Warner, Sir F. and Slater, D.H. (1981), *op. cit.*, p. 167

FLINTA, J., GERNBORG, G. and ADESSON, H. (1971). Results from the blowdown test for the exercise. *Eur. Two-Phase Flow Group Mtg*, Denmark

FLITNEY, R. (1986). Mechanical seals for the process industries. *Chem. Engr, Lond.*, **428**, 37

FLITNEY, R. (1987). Reliability of seals in centrifugal process pumps. *Process Engng*, Jul., 39

FLOOD, B.J. and STEPHENSON, R. (1988). The management and safe operation of a comprehensive hazard test site for the process industries. *Preventing Major Chemical and Related Process Accidents*, p. 149

FLOOD, M. (1976). Nuclear sabotage. *Bull. Atom. Scientists*, Oct., 29

FLOQUET, P., PIBOLEAU, L. and DOMENECH, S. (1988). Mathematical programming tools for chemical engineering process design synthesis. *Chem. Engng Process.*, **23**(2), 99

FLORENCE, D.M. (1977). A novel fire test for pipe insulation. *Fire Technol.*, **13**, 199

FLORES, A. (1983). Safety in design: an ethical viewpoint. *Chem. Engng Prog.*, **79**(11), 11

FLORES, G.H. (1983). Controlling exposure to alkylene oxides. *Chem. Engng Prog.*, **79**(3), 39

FLORIO, J. (1987). New right-to-know radically changes approach to toxic threats by providing information officials need for emergency planning. *New Jersey's First Congressional District, News Release*, March 13

FLORY, K., PAOLI, P. and MESLER, R. (1969). Molten metal-water explosions. *Chem. Engng Prog.*, **65**(12), 50

FLOTHMANN, D. and MJAAVATTEN, A. (1986). Qualitative methods for hazard identification. In Hartwig, S. (1986), *op. cit.*, p. 329

FLOTHMANN, D. and NIKODEM, H.J. (1980). A heavy-gas dispersion model with continuous transition from gravity to spreading to tracer diffusion. In Hartwig, S. (1980), *op. cit.*, p. 89

FLOTHMANN, D., HEUDORFER, W. and LANGBEIN, D. (1980). Evaporation of volatile liquids and its limitation by heat transfer. *Loss Prevention and Safety Promotion 3*, vol. 3, p. 1133

FLOYD, R.W. (1967). Assigning meanings to programs. *Mathematical Aspects of Computer Science* (ed. J.T. Schwartz) (Providence, RI: Am. Math. Soc.)

FLUGGE, W. (1960). *Stresses in Shells* (Berlin: Springer)

FLUGGEN, E., NUSSBAUMER, M. and REUTER, K. (1985). A concrete storage barge — results of a large scale offshore model test. *Gastech 84*, p. 485

FLUID CONTROLS INSTITUTE (1985). Preventing backflow with lift-disk silent-check valves. *Plant Engng*, **39**(Aug. 22), 52

FLURY, F. (1921a). On war gas poisoning. I, On irritant gases. *Z. Ges. Exp. Med.*, **13**, 1

FLURY, F. (1921b). On war gas poisoning. IX, Locally irritant arsenic compounds. *Z. Ges. Exp. Med.*, **13**, 523

FLURY, F. (1928). Toxic effects in modern industry from the pharmacological–toxicological viewpoint. *Arch. Exp. Pathol. Pharmak.*, **138**, 65

FLURY, F. (1938). Fundamental toxilogical problems of the effect of war gases. *Deutsch. Mil. Artz*, **2**, 56

FLURY, F. and HEUBNER, W. (1919). On effect and detoxification of inhaled hydrogen cyanide. *Biochem. Z.*, **95**, 249

FLURY, F. and ZANGER, H. (1928). *Lehrbuch der Toxikologie* (Berlin: Springer)

FLURY, F. and ZERNICK, F. (1931). *Schädliche Gase* (Berlin: Springer)

FLURY, F. and ZERNIK, F. (1932). The quantitative evaluation of the toxic effects of gases and vapours. *Gasschutz Luftschutz*, **2**, 149

FLURY, K.E., DINES, D.E., RODARTE, J.R. and RODGERS, R. (1983). Airway obstruction due to inhalation of ammonia. *Mayo Clin. Proc.*, **58**, 339

FLYGT PUMPS LTD (1990). Submersible pumps. *Plant Engr*, Nov./Dec., 43

FLYNN, B.L. (1979). Wet air oxidation of waste streams. *Chem. Engng*, **75**(4), 66

FLYNN, J.P. and MORRISSETTE, M.D. (1977). Development of a compatibility guide for the water transport of bulk chemicals. *J. Haz. Materials*, **1**(4), 331

FLYNN, J.P. and ROSSOW, H.E. (1970a). *Classification* of Chemical Reactivity Hazards. Report. US Coast Guard

FLYNN, J.P. and ROSSOW, H.E. (1970b). Development of a compatibility guide for water transport of bulk chemical cargoes. *J. Haz. Materials*, **1**(4), 331

FOA, V., EMMET, E.A., MARONI, M. and COLOMBARI, A. (1987). *Occupational and Environmental Chemical Hazards* (Chichester: Ellis Horwood)

FOCHTMAN, E.G., LANGION, W.M. and HOWARD, D.R. (1963). Continuous corrosion measurements. *Chem. Engng*, **70**(Jan. 21), 140

FOGEL, L.J. (1963). *Biotechnology: Concepts and Applications* (Englewood Cliffs, NJ: Prentice-Hall)

FOGLER, H.S. (ed.) (1981). *Chemical Reactors* (Washington, DC: Am. Chem. Soc.)

FOGLER, H.S. (1986). *Elements of Chemical Reaction Engineering*; 1992, 2nd ed. (Englewood Cliffs, NJ: Prentice-Hall)

FOLEY, G. and VAN BUREN, A. (eds) (1978). *Nuclear or Not?* (London: Heinemann)

FOLKARD, S. (1992). Is there a 'best compromise' shift system? *Ergonomics*, **35**, 1453

FOLKARD, S. and CONDON, R. (1987). Night shift paralysis in air traffic controllers. *Ergonomics*, **30**, 1353

FOLKARD, S. and MONK, T.H. (1979). Shiftwork and performance. *Hum. Factors*, **21**, 483

FONS, W.L. (1961). Rate of combustion from free surfaces of liquid hydrocarbons. *Combust. Flame*, **5**, 283

FONS, W.L., CLEMENTS, H.B. and GEORGE, P.M. (1963). Scale effects on propagation rate of laboratory scale crib fires. *Combustion*, **9**, 860

FONTANA, M.G. (1957). Corrosion and its manifestations. *Chem. Engng Prog.*, **53**(11), 523

FONTANA, M.G. (1970). Selecting construction materials for pumps. *Chem. Engng Prog.*, **66**(5), 65

FONTANA, M.G. (1986). *Corrosion Engineering*, 3rd ed. (New York: McGraw-Hill)

FONTANA, M.G. and GREENE, N.D. (1978). *Corrosion Engineering*, 2nd ed. (New York: McGraw-Hill)

FONTEIN, J. (1968). Disastrous fire at Shell Oil refinery at Rotterdam. *Instn Fire Engrs Quart.*, **28**(Dec.), 408

FOO, S.-C. (1990). Benzene exposure from gasoline vapour. *Safety and Loss Prevention in the Chemical and Oil Processing Industries*, p. 355

FOO-SUN LAU (1987). *A Dictionary of Nuclear Power and Waste Management* (New York: Wiley)

FOOD SAFETY COUNCIL, SCIENTIFIC COMMITTEE (1978). Proposed system for food safety assessment. *Food Cosmetics Technol.*, **16**(Suppl. 2), 1

FOORD, A. (1967–68). Investment appraisal for chemical engineers. *Chem. Process Engng*, 1967, **48**, 71, 95; 1968, **49**, 87

FORBES, D.J. (1982). Design of blast-resistant buildings in petroleum and chemical plant. In Fawcett, H.H. and Wood, W.S. (1982), *op. cit.*, p. 489

FORBES, W.H., SARGENT, F. and ROUGHTON, F.J.W. (1945). The rate of carbon monoxide uptake by normal men. *Am. J. Physiol.*, **143**, 594

FORBUS, K.D. (1981). Qualitative reasoning about physical processes. *Proc. Seventh Int. joint Conf. on Artificial Intelligence*, Vancouver, BC

FORBUS, K.D. (1983). Qualitative reasoning about space and motion. In Gentner, D. and Stevens, A.L. (1983), *op. cit.*, p. 53

FORBUS, K.D. (1984). Qualitative process theory. *Artificial Intelligence*, **24**, 85

FORBUS, K.D. (1990). The qualitative process engine. In Weld, D.S. and de Kleer, J. (1990), *op. cit.*

FORD FOUNDATION and MITRE CORPORATION (1977). *Nuclear Power Issues and Choices* (Cambridge, MA: Ballinger)

FORD, B. (1972). *German Secret Weapons* (London: Pan/Ballantine)

FORD, D.F. (1977). *A History of Federal Nuclear Safety Assessments: from WASH-740 through the Reactor Safety Study* (Cambridge, MA: Union of Concerned Scientists)

FORD, D.F., KENDALL, H.W. and TYE, L.S. (1976). *Browns Ferry: the regulatory failure.* (Cambridge, MA: Union of Concerned Scientists)

FORD, D.L. and TISCHLER, L.F. (1977). Biological system development. *Chem. Engng*, **84**(Aug. 15), 131

FORD, H., BETT, K.E., ROGAN, J. and GARDNER, H.F. (1973). Design criteria for pressure vessels. *Safety in Polyethylene Plants*, **1**, 15. See also *Chem. Engng Prog.*, 1972, **68**(11), 77

FORD, N. (1991). *Expert Systems and Artificial Intelligence: An Information Manager's Guide* (London: Library Assoc. Publ.)

FORDESTROMMEN, N.T. and HAUGSET, K. (1991). Integrating surveillance and control without overwhelming the operator. *Nucl. Engng Int.*, **41**

FORDHAM, S. (1966). *High Explosives and Propellants* (Oxford: Pergamon Press)

FORDHAM, S. (1980). *High Explosives and Propellants*, 2nd ed. (Oxford: Pergamon Press)

FORDHAM COOPER, W. (1986). *Electrical Safety Engineering*, 2nd ed. (London: Butterworths)

FOREST, H.S. (1985). Emergency relief vent sizing for fire exposure when two-phase flow must be considered. *Am. Inst. Chem. Engrs, Nineteenth Loss Prevention Symp.* (New York)

FORESTI, R.J. (1955). The stabilization and temperature measurement of flat cool flames. *Combustion*, **5**, 582

FORGY, C. (1981). *The OPS5 User's Manual. Rep. CMU-CS-81–135.* Dept of Computer Sci., Carnegie Mellon Univ., Pittsburgh, PA

FORGY, C. and MCDERMOTT, J. (1977). OPS, a domain-independent production system language. *Proc. Int. Joint Conf. on Artificial Intelligence*, **5**, 933

FORLAND, A. (1992). Simulation of evacuation, escape and rescue. *Major Hazards Onshore and Offshore*, p. 679

FORMAN, E.R. (1978). How to improve control valve performance. *Chem. Engng*, **85**(Jun. 5), 128

FORMBY, C.L., KIRBY, J.H. and RATCLIFFE, R. (1973). Fracture of a steampipe-analysis in terms of fracture mechanics. *Int. J. Pres. Ves. Piping*, **1**, 245

FORREST, H.S. (1985). Emergency relief vent sizing for fire exposure when two-phase flow must be considered. *Am. Inst. Chem. Engrs, Nineteenth Loss Prevention Symp.*, paper 56e

FORRESTER, J.G. (1982). Halon 1301 liquid discharge – sustained by self-chilling. *Fire Technol.*, **18**, 138

FORRESTER, J.H. (1974). Computer planning Marcus Hook safety valve maintenance. *Oil Gas J.*, Mar. 4, 118

FORRY, K. and SCHRAGE, C. (1966). Trouble-shooting HF alkylation. *Hydrocarbon Process.*, **45**(1), 107

FORSBERG, C.W. (1990). Passive and inherent safety technologies for light-water nuclear reactors. *Am Inst. of Chem. Engrs. 1990 Summer National Meeting*, Aug. 19–22, 1990, San Diego, CA, Session 43

FORSBERG, C.W. *et al.* (1989). *Proposed and Existing Passive and Safety-related Structures, Systems and Components (Building Blocks) for Advanced Light Water Reactors. Rep ORNL-6554.* Oak Ridge Nat. Lab, Oak Ridge, TN

FORSMAN, J.P. (1972). What's new in HDPE processes. *Hydrocarbon Process.*, **51**(11), 131

FORSTER, H., KRAMER, H. and SCHON, G. (1980). Dispersion of gasoline in road tunnels after release of major quantities. In Hartwig, S. (1980), *op. cit.*, p. 141

FORSTER, H., SCHAMPEL, K. and STEEN, H. (1986). Breathing of storage tanks for flammable liquids under atmospheric influence and relevant safety requirements for venting devices and inert gas systems. *Loss Prevention and Safety Promotion 5*, p. 56.1

FORSTER, J.H. and WONG, N. (1992). Analytical methods and working experience with post-Cullen analysis of evacuation, escape and rescue. *Major Hazards Onshore and Offshore*, p. 689

FORSTER, J.M. (1967). *List Processing* (New York: Elsevier)

FORSTER, R.E. (1957). Exchange of gases between alveolar air and pulmonary capillary blood: pulmonary diffusing capacity. *Physiol. Rev.*, **37**, 391

FORSTER, R.E. (1964a). Diffusion of gases. In Fenn, W.O. and Rahn, H. (1964), *op. cit.*, p. 839

FORSTER, R.E. (1964b). Rate of gas uptake by red cells. In Fenn, W.O. and Rahn, H. (1964), *op. cit.*, p. 827

FORSTH, L.R. (1981a). Fires and explosions on fixed platforms in the Norwegian North Sea. *Det norske Veritas, Rep. 81–1175*

FORSTH, L.R. (1981b). Fires and explosions on offshore platforms in the Gulf of Mexico. *Det norske Veritas, Rep. 81–1221*

FORSTH, L.R. (1983). Survey identifies causes of offshore fires and explosions. *Oil Gas J.*, **81**(Jul. 25), 154

FORSYTH, J.S. and GARSIDE, J.E. (1949). The mechanism of flash-back of aerated flames. *Combustion*, **3**, 99

FORSYTH, R. (1984a). The architecture of expert systems. In Forsyth, R. (1984), *op. cit.*, p. 9

FORSYTH, R. (ed.) (1984b). *Expert Systems* (London: Chapman & Hall)

FORSYTH, R. (1984c). Fuzzy reasoning systems. In Forsyth, R. (1984), *op. cit.*, p. 51

FORSYTH, R. (1984d). Machine learning strategies. In Forsyth, R. (1984), *op. cit.*, p. 153

FORSYTH, R.A. (1984). Cyclone separation in natural gas transmission systems. *Chem. Engr, Lond.*, **404**, 37

FORT, J. and JEHL, J. (1988). Magnetic bearings and dry seals improve compressor operation. *Hydrocarbon Process.*, **67**(10), 53

FORT, N. (1989). Radio interference to alarm systems. *Fire Surveyor*, **17**(6), 36

FORTENBERRY, J.C. and SMITH, L.A. (1981). A comparison of risk selections. *Hum. Factors*, **23**, 693

FORTSON, M. *et al.* (1973). *Maritime Accident Spill Risk Analysis Program. Phase I: Methodology and Planning. Report.* Operations Res. Inc.

FOSCANTE, R.E. (1990). Materials of construction: tailoring the optimum system. *Chem. Engng*, **97**(10), 92

FOSS, A.R. (1959). Chemical reactor system dynamics. *Chem. Engng Prog.*, **54**(9), 39

VAN FOSSAN, N.E. (1965). Handling liquefied gas. *Chem. Engng*, **72**(Aug. 2), 103

FOSTER, A. and BURD, J.E. (1993). Hold the line on OSHA compliance costs. *Chem. Engng Prog.*, **89**(6), 71

FOSTER, G. (1993). Noise contour mapping in an oil refinery. *Health, Safety and Loss Prevention*, p. 557

FOSTER, H.D. (ed.) (1976). *Victoria Physical Environment and Development*. Univ. of Victoria, Victoria, BC

FOSTER, H.D. (1980). *Disaster Planning. The Preservation of Life and Property* (New York: Springer)

FOSTER, H.D. and CAREY, R.F. (1976). The simulation of earthquake damage. In Forster, H.D. (1976), *op. cit.*

FOSTER, J.W. (1981). *Reliability, Availability and Maintainability* (Beaverton, OR: M/A Press)

FOSTER, M.D. (1967). The dissipation of electrostatic charges in purified petroleum products. *Static Electrification*, **2**, 78

FOSTER, M.D. and MARSH, K.J. (1964). A portable instrument for measuring the electrical conductivity of aviation fuels. *J. Sci. Instrum.*, **41**, 379

FOSTER, P.J. and MCGRATH, I.A. (1960). Prediction of the emissivity of large hydrocarbon flames from the Mie theory (letter). *Combust. Flame*, **4**, 191

FOSTER, P.M. (1981). Maximum ground level concentrations of washed chimney emissions. *Atmos. Environ*, **15**, 1659

FOSTER, P.M. (1982). Particle fallout during plume rise. *Atmos. Environ.*, **16**, 2777

FOSTER, T.C. (1981). Time required to empty a vessel. *Chem. Engng*, **88**(May 4), 105

FOTHERGILL, J.W. (1984). Smoke movement within a building. *Fire Safety J.*, **7**, 47

FOTLAND, K. and FUNNEMARK, E. (1989). Safety management in offshore projects. *Loss Prevention and Safety Promotion 6*, p. 118.1

FOUHY, K. (1991). Process simulation gains a new dimension. *Chem. Engng*, **98**(10), 47

FOUHY, K., SAMDANI, G. and MOORE, S. (1992). ISO 9000: a new road to quality. *Chem. Engng*, **99**(10), 43

FOULGER, J.H. (1965). Effect of toxic agents. In Fawcett, H.H. and Wood, W.S. (1965), *op. cit.*, p. 250

FOULKES, C.H. (1934). *Gas. The Story of the Special Brigade* (Edinburgh: Blackwood)

FOULKES, N.R., WALTON, M.J., ANDOW, P.K. and GALLUZZO, M. (1988). Computer-aided synthesis of complex pump and valve operations. *Comput. Chem. Engng*, **12**, 1035

FOUMENY, E.A. and BENYAHIA, F. (1993). Can CFD improve the handling of air, gas and gas–liquid mixtures? *Chem. Engng Prog.*, **89**(2), 21

FOUR ELEMENTS LTD (*c.* 1991). *PLATO* (London)

FOUSEK, T. (1983). How ultrasonic instruments aid maintenance. *Hydrocarbon Processing*, **62**(1), 93

FOUSSAT, A. (1981). *Modele de Dispersion Atmos-pherique Non-isotherme d'un Pollutant Gazeux de Densite Quelconque en Presence de Non-uniformites Orographiques.* Tech. Note 135. Van Karman Inst.

FOUST, A.S., WENZEL, L.A., CLUMP, C.W., MANS, L. and ANDERSON, L.B. (1980). *Principles of Unit Operations*, 2nd ed. (New York: Wiley)

FOWELL, D.N. (1949). Exposure to chlorine and heart disease. *J. Am. Med. Ass.*, **140**, 1376

FOWLE, D.J. (1978). The design and use of reinforced plastics piping systems in chemical plant. *Chem. Ind.*, Jun. 3, 361

FOWLE, T.I. (1973). Air compressor fires and explosions. *Fire Prev. Sci. Technol.*, **7**, 20

FOWLER, D.W. (1969). New analysis method for pressure vessel column supports. *Hydrocarbon Process.*, **48**(5), 157

FOWLER, E.W. and SPIEGELMAN, A. (1968). *Hazard Survey of the Chemical and Allied Industries* (New York: Am. Insurance Ass.)

FOWLER, H.W. (1979). Planning safe chemical engineering experiments. *Instn Chem. Engrs Symp. on Safety in Chemical Engineering Departments* (London)

FOWLER, R. (1989). Higee – a status report. *Chem. Engr, Lond.*, **456**, 35

FOWLER, T.J. (1980). Acoustic emission inspection of process equipment. In *Loss Prevention and Safety Promotion 3*, vol. 2, p. 684

FOWLER, T.J. (1986). Acoustic emission testing of process industry vessels and piping. *Loss Prevention and Safety Promotion 5*, p. 51.1

FOWLER, T.J. (1987). Acoustic emission testing of vessels and piping. *Chem. Engng Prog.*, **83**(5), 25

FOWLER, T.J. (1988). Acoustic emission testing of vessels. *Chem. Engng Prog.*, **84**(9), 59

FOWLER, T.J. (1992). Chemical industry applications of acoustic emission. *Mats Evaluation*, **50**, 875

FOWLER, T.J. and SCARPELLINI, R.S. (1980). Acoustic emission testing of FRP equipment. *Chem. Engng*, **87**(Oct. 20), 145; (Nov. 17), 293

FOX, A.J. (1981). Mortality statistics and the assessment of risk. *Proc. Roy. Soc., Ser. A*, **376**, 65. See also in Warner, Sir F. and Slater, D.H. (1981), *op. cit.*, p. 65

FOX, A.J. and COLLIER, P.F. (1976). A survey of occupational cancer in the rubber and cable making industries: analysis of deaths occurring in 1972–74. *Br. J. Ind. Med.*, **33**(4), 249

FOX, C.R. (1979). Removing toxic organics from waste water. *Chem. Engng Prog.*, **75**(8), 70

FOX, C.T. (1967). Blast it clean with water. *Chem. Engng*, **74**(Aug. 14), 180

FOX, J.A. (1977). *Hydraulic Analysis of Unsteady Flow in Pipe Networks* (London: Macmillan)

FOX, J.L. (1980a). Love Canal story epitomises chemical waste disposal problems. *Chem. Ind.*, Jul. 5, 515

FOX, J.L. (1980b). Proposed carcinogen regulations draw quick fire. *Chem. Ind.*, Feb. 16, 125

FOX, J.L. (1981). Toxic waste regulations undergo tough rites of passage in US. *Chem. Ind.*, Jan. 3, 4

FOX, J.L. (1984a). EPA's chemical data system under threat. *Chem. Ind.*, Aug. 20, 565

FOX, J.L. (1984b). Huge settlement avoids dioxin trial. *Chem. Ind.*, Jun. 4, 394

FOX, J.L. (1985). Dioxin trial's mixed message. *Chem. Ind.*, Jul. 1, 425

FOX, J.P., HALL, C.E. and ELVABACK, L.R. (1970). *Epidemiology and Disease* (New York: Macmillan)

FOX, J.S. and SARKAR, F. (1972). The influence of Coanda effect on blow-off limit. *Combust. Sci. Technol.*, **6**, 51

FOX, R.D. (1973). Pollution control at the source. *Chem. Engng*, **80**(Aug. 6), 72

FOXBORO COMPANY (1980). *National and International Codes and Safety Standards for Electrical and Electronic Process Control Equipment. Tech. Inf. Sheet TI 005–100*

FOZARD, B. (1963). *Instrumentation and Control of Nuclear Reactors* (London: Iliffe)

FOZARD, M. (1976). The ambulance experience at Flixborough. In Howard, J. (1976), *op. cit.*

FRAAS, A.P. (1989). *Heat Exchanger Design*, 2nd ed. (Chichester: Wiley)

FRAAS, A.P. and OSOZIK, M.N. (1965). *Heat Exchanger Design* (New York: Wiley-Interscience)

FRANCIS, J.O. and SHACKELTON, W.E. (1985). A calculation of relieving requirements in the critical region. *API 50th Midyear Mtg*, **64**, 179

FRANCIS, M.F. (1964). Failure of primary reformer outlet piping. *Ammonia Plant Safety*, **6**, 1

FRANCOIS, W.F. (1964). When to use and how to select hinged expansion joints. *Hydrocarbon Process.*, **43**(6), 169

FRANK, E.H. and WARDALE, J.K.S. (1970). Failure of liquid ethylene storage tank. *LNG Conf. 2*

FRANK, M.V. and MOIENI, P. (1986). A probabilistic model for flammable pool fire damage in nuclear power plants. *Reliab. Engng*, **16**(2), 129

FRANK, O. (1977). Short-cuts for distillation design. *Chem. Engng*, **84**(Mar. 14), 110

FRANK, P.M. (1987). Advanced fault detection and isolation schemes using nonlinear and robust observers. *IFAC Tenth Triennial World Congress*, Munich, p. 63

FRANK, S.M. and LAMBRIX, J.R. (1966). Ethylene plant capacities increase as co-products value dominates economics. *Hydrocarbon Process.*, **45**(4), 199

FRANK, T.E. (1981). Fire explosion control in bag filter dust collection systems. *The Hazards of Industrial Explosion from Dusts* (London: Oyez/IBC), paper 14

FRANK, W.L. and ZODEH, O.M. (1991). The process safety impact of distributed control systems. *Plant/Operations Prog.*, **10**, 93

FRANK, W.L., GIFFIN, J.E. and HENDERSHOT, D.C. (1993). Team make-up – an essential element of a successful process hazard analysis. *Process Safety Management*, p. 129

FRANKE, F.R., BURNETT, J.C., DANLY, D.E. and HOWARD, W.B. (1975). Explosion protection for process systems. *Loss Prevention*, **9**, 97

FRANKEL, E.G. (1984). *Systems Reliability and Risk Analysis* (Dordrecht: Martinus Nijhoff)

FRANKEL, H.E. and DAPKUNAS, S.J. (1977). Failure analysis and failure data collection in the ERDA coal conversion systems. In Gangadharan, A.C. and Brown, S.J. (1977), *op. cit.*, p. 21

FRANKEL, I., SANDERS, N. and VOGEL, G. (1983). Survey of the incinerator manufacturing industry. *Chem. Engng Prog.*, **79**(3), 44

FRANKEL, J.I. (1966). Incineration of process wastes. *Chem. Engng*, **73**(Aug. 29), 91

FRANKEL, M.L. and SIVASHINSKY, G.I. (1983). On effects due to thermal expansion and Lewis number in spherical flame propagation. *Combust. Sci. Technol.*, **31**, 131

FRANK-KAMENETSKY, D.A. (1939a). *Acta Physicochim. (USSR)*, **10**, 365

FRANK-KAMENETSKY, D.A. (1939b). *Zh. Fiz. Khim.*, **13**, 738

FRANK-KAMENETSKY, D.A. (1942). *Acta Physicochim. (USSR)*, **16**, 357

FRANK-KAMENETSKY, D.A. (1945). *Acta Physicochim. (USSR)*, **20**, 729

FRANK-KAMENETSKY, D.A. (1955). *Diffusion and Heat Exchange in Chemical Kinetics*. (Princeton, NJ: Princeton Univ. Press)

FRANKLAND, P.B. (1978). Relief valves . . . what needs protection? *Hydrocarbon Process.*, **57**(4), 189. See also in Kane, L.A. (1980), *op. cit.*, p. 139

FRANKLAND, P.B. (1981). Solve the wide turndown flow control problem. *Hydrocarbon Process.*, **60**(4), 187

FRANKLIN, D.W. (1991). *Estimating Acute Toxicity Effects on Human Populations from Atypical Releases of Industrial Chemicals. Report.* School of Biological and Biomedical Sci., Univ of Technology, Sydney

FRANKLIN, H.L. (1972). A local contingency plan for hazardous material spills. *Control of Hazardous Material Spills 1*, p. 65

FRANKLIN, N. (1985). The hazards of nuclear power: myth or reality? *Chem. Engr*, **418**, 24

FRANKLIN, N. (1986). The accident at Chernobyl. *Chem. Engr*, **430**, 17

FRANKO-FILIPASIC, B.R. and MICHAELSON, R.C. (1984). Safety in the design of high-pressure laboratories. *Chem. Engng Prog.*, **80**(3), 65

FRANKS, F.C. (1961). Double rupture disks cope with severe conditions. *Chem. Engng*, **68**(May 15), 182

FRANKS, R.G.E. (1967). *Mathematical Modeling in Chemical Engineering* (New York: Wiley)

FRANSSEN, J.M. and BRULS, A. (1986). Resistance au feu des profiles en acier isoles. *CTICM, Proc. EGOLF Symp.* (Brussels)

FRANSSEN, J.M. and DOTREPPE, J.C. (1992). Fire resistance of columns in steel frames. *Fire Safety J.*, **19**, 159

FRANZ, G. and SHELDON, R.A. (1991). Oxidation. *Ullmanns Encyclopaedia of Chemical Technology*, rev. ed., vol. A18 (Weinheim: VCH), p. 261

FRANZEL, H.L. (1970). Guide to trouble-free quality control of equipment engineering. *Chem. Engng*, **77**(Jun. 1), 135

FRANZEN, J.E. and GOEKE, E.K. (1976). Gasify coal for petrochemicals. *Hydrocarbon Process.*, **55**(11), 134

FRASER, R.P. (1957). Liquid fuel atomization. *Combustion*, **6**, 687

FRASER, W.H. (1981). Recirculation in centrifugal pumps. *Am. Soc. Mech. Engrs, Winter Ann. Mtg*

FRASER, W.J., BUSSEY, R.J. and JOHNSTON, J.C. (1990). Semi-quantitative risk assessment method for product management. *Plant/Operations Prog.*, **9**, 247

FRAUMENI, J.F. (ed.) (1975). *Persons at High Risk of Cancer. An Approach to Cancer Aetiology and Control* (New York: Academic Press)

FRAYLINK, J.R. (1984). How to start up major facilities. *Oil Gas J.*, **82**(Dec. 3), 112. Pt 1

FREDERICKSON, T. and BECKMAN, L.V. (1990). *Comparison of Fault Tolerant Controllers used in Safety Applications. ISA paper 90–1315.* (Pittsburgh, PA: Instrum. Soc. Am.)

FREDHOLM, O. (1983). Safety on gas cloud explosions – a Swedish approach. In *Loss Prevention and Safety Promotion 4*, vol. 1, p. H1

FREEBERG, C.R. and AARNI, C.W. (1982). Hydrocarbon emission control in petroleum refineries. *Chem. Engng Prog.*, **78**(6), 35

FREED, V.H. (1980). The university chemist – role in transport and fate of chemical problems. In Haque, R. (1980a), *op. cit.*, p. 27

FREEDER, B.G. and SNEE, T.J. (1988). Alkali-catalysed polymerization of ethylene oxide and propylene oxide – hazard evaluation using accelerating rate calorimetry. *J. Loss Prev. Process Ind.*, **1**(3), 164

FREEDMAN, L. (1983). Boiler treatment for high purity feedwater. *Hydrocarbon Process.*, **62**(1), 101

FREEDY, A. and LUCACCINI, L.F. (1981). Adaptive Computer Training System (ACTS) for fault diagnosis in maintenance tasks. In Rasmussen, J. and Rouse, W.B. (1981), *op. cit.*, p. 637

FREEMAN, G. and LEFEBVRE, A.H. (1984). Spontaneous ignition characteristics of gaseous hydrocarbon–air mixtures. *Combust. Flame*, **58**, 153

FREEMAN, H.M. (1984). The role of incineration in remedial actions. *Plant/Operations Prog.*, **3**, 113

FREEMAN, H.M. (1990). *Hazardous Waste Minimization* (New York: McGraw-Hill)

FREEMAN, H.M., OLEXSEY, R.A., OBERACKER, D.A. and MOURNIGHAN, R.E. (1987). Thermal destruction of hazardous waste – a state-of-the art review. *J. Haz. Materials*, **14**(1), 103

FREEMAN, N.T. (1962). *Protective Clothing and Devices* (London: United Trade Press)

FREEMAN, N.T. (1977). *Appointment with Fire* (London: Alan Osborne)

FREEMAN, N.T. and PICKBOURNE, D. (1971). Loss prevention – today, tomorrow, and the role of the engineer. *Major Loss Prevention*, p. 5

FREEMAN, N.T. and WHITEHEAD, J. (1982). *Introduction to Safety in the Chemical Laboratory* (New York: Academic Press)

FREEMAN, P. and NEWELL, A. (1971). A model for functional reasoning in design. *Proc. Int. Joint Conf. on Artificial Intelligence 2*, p. 621

FREEMAN, R.A. (1980). Laboratory study of biological system kinetics. *Environ. Prot. Agency*, Washington, DC, unpublished rep.

FREEMAN, R.A. (1983). Problems with risk analysis in the chemical industry. *Plant/Operations Prog.*, **2**, 185

FREEMAN, R.A. (1985). The use of risk assessment in the chemical industries. *Plant/Operations Prog.*, **4**, 85

FREEMAN, R.A. (1989). What should you worry about when doing a risk assessment? *Chem. Engng Prog.*, **85**(11), 29

FREEMAN, R.A. (1990). CCPS guidelines for chemical process quantitative risk analysis. *Plant/Operations Prog.*, **9**, 231

FREEMAN, R.A. (1991). Documentation of hazard and operability studies. *Plant/Operations Prog.*, **10**, 155

FREEMAN, R.A. (1991). Reliability of interlocking systems. *Process Safety Prog.*, **13**, 146

FREEMAN, R.A. and GADDY, J.L. (1975). Quantitative overdesign of chemical processes. *AIChE J.*, **21**, 436

FREEMAN, R.A. and SHAW, D.A. (1988). Sizing excess flow valves. *Plant/Operations Prog.*, **7**, 176

FREEMAN, R.A. and SHAW, D.A. (1992). The use of spreadsheets in modeling accidental release of toxic chemicals. *Plant/Operations Prog.*, **11**, 71

FREEMAN, R.A., LEE, R. and MCNAMARA, T.P. (1992). Plan hazop studies with an expert system. *Chem. Engng Prog.*, **88**(8), 28

FREEMAN, R.H. and MCCREADY, M.P. (1971). Butadiene explosion at Texas City – 2. *Loss Prevention*, **5**, 61. See also *Chem. Engng Prog.*, 1971, **67**(6), 45

FREEMAN, S. (1985). Remote measurement of a plant's structural co-ordinates. *Process Engng*, **66**(1) Jan., 45

FREESE, R.E. (1973). Solvent recovery from waste chemical sludge. An explosion case history. *Loss Prevention*, **7**, 108

FREESTON, H.G., ROBERTS, J.P. and THOMAS, A. (1956). Crankcase explosions: an investigation into some factors governing the selection of protective devices. *Proc. Instn Mech. Engrs*, **170**, 811

FREESTONE, F.J. (1976). Shakedown and demonstration of R & D prototype devices and techniques at hazardous material spills. *Control of Hazardous Materials Spills 3*, vol. 2, p. 323

FREESTONE, F.J. and ZACCOR, J. (1978). Design, fabrication, and demonstration of a mobile stream diversion system for hazardous material spill containment. *Control of Hazardous Material Spills 4*, p. 371

FREESTONE, F.J., LAFORNARA, J.P. and MCCRACKEN, W.E. (1976). Performance testing of several oil spill control devices on selected flotable hazardous materials. *Control of Hazardous Material Spills 3*, p. 326

FREEZE, R.A. and CHERRY, J.A. (1979). *Groundwater* (Englewood Cliffs, NJ: Prentice-Hall)

FREGERT, S. (1974). *Manual of Contact Dermatoses* (Copenhagen: Munksgard)

FREIMAN, J.P. and HILL, J. (1992). Use dispersion modeling update. *Hydrocarbon Process.*, **71**(8), 73

FREIREICH, E.J., GEHAN, E.A., RAIL, D.P., SCHMIDT, L.H. and SKIPPER, H.E. (1966). Quantitative comparison of toxicity of anticancer agents in mouse, rat, hamster, dog, monkey and man. *Cancer Chemother. Rep.*, **50**, 219

FREITAG (1940). Dangers of chlorine gas. *Z. Gesamt. Schiess-Sprengstoffwesen*, **35**, 159

FREITAG, W. and PACKBIER, M.W. (1978). Solving nitric acid plant pollution problems. In *Ammonia Plant Safety*, **20**, 11

FREIWALD, H. and KOCH, H.W. (1963). Spherical detonations of acetylene–oxygen–nitrogen mixtures as a function of nature and strength of initiation. *Combustion*, **9**, 275

FREMLIN, J.H. (1978). The Windscale inquiry. *Sci. Publ. Policy*, **5**, 214

FREMLIN, J.H. (1983). Power production at minimum risk. *Nucl. Energy*, **22**, 67

FREMLIN, J.H. (1985). *Power Production: What are the Risks?* (Oxford: Oxford Univ. Press)

FRENCH, D. and RICHARDS, H.A. (1973). Hazardous materials flow in intercoastal waterways – a case study of risk-exposure factors for the future of a specific area. *Conf. on Bulk Transportation of Hazardous Materials by Water in the Future* (Washington, DC: Nat. Acad. Sci., Nat. Acad. Engng, Nat. Res. Coun.), p. 200

FRENCH, H.M. (1979). *The Anatomy of Arson* (New York: ARCO Publishing Co.)

FRENCH, R.W. and CALLENDER, G.R. (1962). Ballistic characteristics of wounding agents. In Beyer, J.C. (ed.) (1962), *op. cit.*, chapter 2

FRENCH, R.W., OLSEN, R.E. and PELOQUIN, G.L. (1990a). Quantified risk as a decision aid. *Process Safety Environ.*, **68B**, 7

FRENCH, R.W., OLSEN, R.E. and PELOQUIN, G.L. (1990b). Quantified risk as a decision aid. *Safety and Loss Prevention in the Chemical and Oil Processing Industries*, p. 461

FRENCH, R.W., WILLIAMS, D.D. and WIXOM, E.D. (1995). Inherent safety, health and environmental (SHE) reviews. *Proc. of the 29th Annu. Loss Prev. Symp.*, July 31–Aug. 2, 1995, Boston, MA, (ed. E.D. Wixom and R.P. Benedetti,) Paper 1c. (New York: AIChE)

FRENCH, W.W. (1992). Tips for troubleshooting pumps. *Chem. Engng Prog.*, **88**(6), 65

FRENDI, A. and SIBULKIN, M. (1990). Dependence of minimum ignition energy on ignition parameters. *Combust. Sci. Technol.*, **73**, 395

FRENKEL, D.I. (1978). Tuning electrostatic precipitators. *Chem. Engng*, **85**(Jun. 19), 105

FRENKEL, L. and HATHAWAY, W. (1976). *Risk Analysis Methods for Deepwater Port Transfer Systems. Report.* US Coast Guard

FRENKIEL, F.N. (1952). Application of the statistical theory of turbulent diffusion to micrometeorology. *J. Met.*, **9**, 252

FRENKLACH, M., TAKI, S. and MATULA, R.A. (1983). A conceptual model for soot formation in pyrolysis of aromatic hydrocarbons. *Combust. Flame*, **49**, 275

FRESHWATER, D.C. and BUFFHAM, B.A. (1969). Reliability engineering for the process plant industries. *Chem. Engr, Lond.*, **232**, 367

FREUDENTHAL, A.M. (1976). Extreme value risk analysis in the structural design of reactor components. *Nucl. Engng Des.*, **37**, 179

FREY, R. and SAFAR, P. (eds) (1980). *Types and Events of Disasters Organization in Various Disaster Situations* (Berlin: Springer)

FREY, R.C. and FINNERAN, J.A. (1969). Process control considerations for large ammonia plants. *Ammonia Plant Safety*, **11**, 1

FREY, R.C. and HANDMAN, S.E. (1987). Safety in storage and transfer from an engineering/construction contractor's viewpoint. *Int. Symp. on Preventing Major Chemical Accidents*, p. 2.71

FREYTAG, H.H. (ed.) (1965). *Handbuch der Raumexplosionen* (Weinheim: Verlag Chemie)

FRICK, T. (ed.) (1962). *Petroleum Production Handbook*, vols 1–2 (New York: McGraw-Hill)

FRICK, W.E. (1984). Non-empirical closure of the plume equations. *Atmos. Environ.*, **18**, 653

FRICKE, M.P. (1975). *Preliminary Civilian Casualty Criteria for Low-yield Nuclear Weapons. Rep. DNA 3547T* (now declassified). Defence Nuclear Agency, Washington, DC

FRIE, J.L., PAGE, G.A., DIAZ, V. and KIVES, J.J. (1992). Evaporation of contained and uncontained spills of multicomponent nonideal solutions. *Loss Prevention and Safety Promotion 7*, vol. 1, p. 16.1

FRIEDEL, L. (1986). Boiling and dissolution delay in two phase systems during vessel top venting. *Loss Prevention and Safety Promotion 5*, vol. 1, p. 43.1 and vol. 3, p. 225

FRIEDEL, L. (1987a). Boiling and dissolution delay in two- phase systems in the sudden depressurisation of pressure vessels. *Fortschritte der Sicherheitstechnik 1* (Frankfurt-am-Main: DECHEMA), p. 219

FRIEDEL, L. (1987b). Leakage through cracks: the applicability of available prediction methods. *Hazards from Pressure*, p. 265

FRIEDEL, L. (1988). Two phase pressure drop in a rupture disc/safety valve unit. *J. Loss Prevention Process Ind.*, **1**(1), 31

FRIEDEL, L. and KISSNER, H.M. (1985). Pressure loss in safety valves during depressurisation. *Int. J. Heat Fluid Flow*, **4**, 229

FRIEDEL, L. and KISSNER, H.M. (1987). Two-phase pressure loss model for safety valves. *Proc. Third Int. Conf. on Multiphase Flow* (Bedford: Br. Hydromech. Res. Ass.), p. 309

FRIEDEL, L. and KISSNER, H.M. (1988). Single-phase and two-phase resistance across rupture discs. *Preventing Major Chemical and Related Process Accidents*, p. 301

FRIEDEL, L. and LOHR, G. (1982). The design of pressure-relief devices for gas–liquid reaction systems. *Int. Chem. Engng*, **22**, 619

FRIEDEL, L. and MOLTER, E. (1989). *Depressurisation Rates during Emergency Vessel Top and Bottom Venting of Two-phase Systems*. Eur. Two-Phase Flow Group, EDF, Paris

FRIEDEL, L. and PURPS, S. (1983). Thermohydraulic processes in pressure vessel and discharge line during emergency relieving. *Loss Prevention and Safety Promotion 4*, vol. 3, p. B18

FRIEDEL, L. and PURPS, S. (1984a). Modelle und Berechnungsverfahren fü die plötzliche Druckentlastung von Gas/Dampf-Flüssigkeits- Reaktionensystemen. *Chem. Ing. Tech.*, **56**, 361

FRIEDEL, L. and PURPS, S. (1984b). Phase distribution in vessels during depressurisation. *Int. J. Heat Fluid Flow*, **5**, 229

FRIEDEL, L. and SCHMIDT, J. (1993). Design of long safety valve vent line systems for gas relief. *J. Loss Prev. Process Ind.*, **6**, 293

FRIEDEL, L. and WEHMEIER, G. (1991). Modelling of the vented methanol/acetic acid anhydride runaway reaction using SAFIRE. *J. Loss Prev. Process Ind.*, **4**(2), 110

FRIEDEL, L. and WESTPHAL, F. (1987). Models for the calculation of leak rates from pressure equipment and pipework in chemical plants. *Fortschritte der Sicherheitstechnik 1* (Frankfurt-am-Main: DECHEMA), p. 261

FRIEDEL, L. and WESTPHAL, F. (1988). Validity of water leak rate prediction methods. *J. Loss Prev. Process Ind.*, **1**(4), 213

FRIEDEL, L. and WESTPHAL, F. (1989). Leak rates through cracks. *Loss Prevention and Safety Promotion 6*, p. 8.1

FRIEDEL, L., SCHMIDT, J. and WEHMEIER, G. (1992). Modelling of transient reactor and catchtank vessel pressure and of mass discharge during emergency venting. *Loss Prevention and Safety Promotion 7*, vol. 1, p. 11.1

FRIEDEL, L., SCHECKER, H.G. and WEHMEIER, G. (1992). Recalculation of the pressure behaviour in the reactor and in the catchtank during emergency depressurization. *J. Loss Prev. Process Ind.*, **5**, 229

FRIEDLANDER, A.F. (1978). *Approaches to Controlling Air Pollution* (Cambridge, MA: MIT Press)

FRIEDLANDER, E. and REED, J.R. (1953). Electrical discharges in air gaps facing solid insulation in high voltage equipment. *Proc. Instn Elec. Engrs*, **100**(pt IIA), 121

FRIEDLANDER, G.D. (1974). The case of the three engineers vs BART. *IEEE Spectrum*, **11**(Oct.), 69

FRIEDLANDER, G.D. (1975). Fixing BART. *IEEE Spectrum*, **12**(Feb.), 43

FRIEDLANDER, S.K. (1977). *Smoke, Dust and Haze: Fundamentals of Aerosol Behaviour* (New York: Wiley)

FRIEDLANDER, S.K. and TOPPER, L. (ed.) (1961). *Turbulence: Classic Papers on Statistical Theory* (New York: Interscience)

FRIEDMAN, D.P. (1974). *The Little LISPer* (Chicago, IL: Science Research Associates)

FRIEDMAN, K.A. (1989a). Can the EPA meet the challenges of the 1990s? *Chem. Engng Prog.*, **85**(8), 47

FRIEDMAN, K.A. (1989b). EPA and SARA Title III. *Chem. Engng Prog.*, **85**(6), 18

FRIEDMAN, K.A. (1990). CCPS: setting the standard for industrial safety. *Chem. Engng Prog.*, **86**(3), 60

FRIEDMAN, L., HOWELL, W.C. and JENSEN, C.R. (1985). Diagnostic judgment as a function of the preprocessing of evidence. *Hum. Factors*, **27**, 665

FRIEDMAN, M.H. (1963a). A correlation of impact sensitivity by means of the hot spot model. *Combustion*, **9**, 294

FRIEDMAN, M.H. (1963b). Size of 'hot spots' in the impact explosion of exothermic materials. *Trans. Faraday Soc.*, **59**, 1865

FRIEDMAN, M.H. (1967). A generalized thermal explosion criterion – exposition and illustrative applications. *Combust. Flame*, **11**, 239

FRIEDMAN, M.H. (1968). On critical hot spot size (letter). *Combust. Flame*, **12**, 281

FRIEDMAN, M.P. (1961). A simplified analysis of spherical and cylindrical blast waves. *J. Fluid Mech.*, **11**, 1

FRIEDMAN, P.G. (1978). Evaluating computer control of ammonia plant. *Ammonia Plant Safety*, **20**, 85

FRIEDMAN, R. (1949). The quenching of laminar oxyhydrogen flames by solid surfaces. *Combustion*, **3**, 110

FRIEDMAN, R. (1953). *Am. Rocket Soc. J.*, **24**, 349

FRIEDMAN, R. (1977). Ignition and burning of solids. *Fire Standards and Safety, ASTM STP 614* (ed. A.F. Robertson), (Philadelphia, PA: Am. Soc. Testing Mats), p. 91

FRIEDMAN, R. and BURKE, E. (1955). A study of ethylene oxide decomposition flame. *Combustion*, **5**, 596

FRIEDMAN, S.B. (1990). *Logical Design of Automation System* (Englewood Cliffs, NJ: Prentice-Hall)

FRIEDMAN, Y.Z. (1992). Avoid advanced control project mistakes. *Hydrocarbon Process.*, **71**(10), 115

FRIEDRICH, E.R. (1974). Floors for process areas. *Chem. Engng*, **81**(Jun. 24), 157

FRIEDRICH, H. and VETTER, G. (1962). *Energie*, **14**(1), 3

FRIEDRICHSEN, F.G. (1983). An ammonia accident in Denmark. *Plant/Operations Prog.*, **2**, 122

FRIEL, J.V., HILTZ, R.H. and MARSHALL, M.D. (1973). *Control of Hazardous Chemical Spills by Physical Barriers. Rep. EPA-600/2–77–162.* Environ. Prot Agency, Washington, DC

FRIEND, L. and ADLER, G.D. (1958). *Transport Properties of Gases* (Evanston, IL: Northwestern Univ. Press)

FRIES, A.A. and WEST, C.J. (1921). *Chemical Warfare* (New York: McGraw-Hill)

FRIESHEL, M.A. and JONES, R.H. (1988). Acoustic emission during intergranular stress corrosion cracking of iron. *J. Acoustic Emission*, **7**, 119

FRIEZE, P.A., MCGREGOR, R.C. and WINKLE, I.E. (eds) (1984). *Marine and Offshore Safety* (Amsterdam: Elsevier)

FRIKKEN, D.R., ROSENBERG, K.S. and STEINMEYER, D.E. (1975). Understanding vapor-phase heat-transfer media. *Chem. Engng*, **82**(Jun. 9), 86

FRINGS, A. and KASTHURI, R. (1986). Fans for the process industries. *Chem. Engr, Lond.*, **421**, 20

FRIS, M. (1972). Present problems regarding the explosibility of flammable dusts as a mixture in air. *Int. Symp on Dust and Explosion Risks in Mines and Industry* (Heidelberg: Berufsgenossenschaft der Chem. Ind.), p. 9

FRISBIE, L.H. (1976). Safe disposal practices for pesticide wastes. *Control of Hazardous Material Spills 3*, p. 434

FRISBIE, L.H. (1978). Safe disposal practices for hazardous wastes. *Control of Hazardous Material Spills 4*, p. 213

FRISBIE, L. (1982). Pesticides. In Bennett, G.F., Feates and Wilder (1982), *op. cit.*, pp. 12–18

FRISTROM, R.M. (1957). The structure of laminar flames. *Combustion*, **6**, 96

FRISTROM, R.M. and WESTENBERG, A.A. (1965). *Flame Structure* (New York: McGraw-Hill)

FRITSCH (1970). Origin and cause of the inflammation of metallic lithium. *Prevention of Occupational Risks*, **1**

FRITZ, E.J. (1969). Anatomy of a nitration explosion. *Loss Prevention*, **3**, 41

FRITZ, R.H. and JACK, G.G. (1983). Water in loss prevention: where do we go from here? *Hydrocarbon Process.*, **62**(8), 77

FRITZ, S. (1954). Scattering of solar energy by clouds of large drops. *J. Met.*, **11**, 291

FROEBE, L.R. (1976). The organisation of a local environmental emergency response team. *Control of Hazardous Material Spills 3*, p. 156

FROEBE, L.R. (1985). State and national resources for community spill disaster preparedness in the United States. *J. Haz. Materials*, **10**(1), 107

FROGNER, B. and MEIJER, C.H. (1978). *On-line Power Plant Alarm and Disturbance Analysis System. Rep. EPRI NP-613.* Elec. Power Res. Inst., Palo Alto, CA

FROGNER, B. and MEIJER, C.H. (1980). *On-line Power Plant Alarm and Disturbance Analysis System. Final Report. Rep. EPRI NP-1379.* Elec. Power Res. Inst., Palo Alto, CA

FROHLICH, B. (1990). Management of safe operations in the process industry. *Plant/Operations Prog.*, **9**, 152

FROHLICH, B. and ROSEN, D. (1992a). Inherent safety – the quality way. *Loss Prevention and Safety Promotion 7*, vol. 1, 39–1

FROHLICH, B. and ROSEN, D. (1992b). Safety and risk management systems in the process industry. *Loss Prevention and Safety Promotion 7*, vol. 1, 46–1

FROHLICH, H. (1949). *Theory of Dielectrics* (Oxford: Clarendon Press)

FROLOV, S.M., GELFAND, B.E. and TSYGANOV, S.A. (1990). A possible mechanism for the onset of pressure oscillation during venting. *J. Loss Prev. Process Ind.*, **3**(1), 64

FROMENT, G.F. (ed.) (1979). *Large Chemical Plants, Efficient Energy Utilisation – Plant Design and Analysis – Processes – Feedstocks* (New York: Elsevier)

FROMENT, G.F. and BISCHOFF, K.B. (1979). *Chemical Reactor Analysis and Design* (New York: Wiley)

FROMENT, G.F. and BISCHOFF, K.B. (1990). *Chemical Reactor Analysis and Design* (New York: Wiley)

FROMM, D. and RAIL, W. (1987a). Damage at syngas compressor by failure of antisurge control. *Ammonia Plant Safety*, **27**, 13

FROMM, D. and RAIL, W. (1987b). Fire at semi-lean pump by reverse motion. *Ammonia Plant Safety*, **27**, 69. See also *Plant/Operations Prog.*, 1987, **6**, 162

FROMM, D., LIEBE, W. and SIEGEL, W. (1988). Pipe rupture caused by vibrations of a safety valve. *Ammonia Plant Safety*, **28**, 39

FROMME, W.A. (1976). Prediction of equipment failures. *Loss Prevention*, **10**, p. 108

FROMMER, W. and KRAMER, P. (1989). Safety aspects in biotechnology. *Loss Prevention and Safety Promotion 6*, p. 5.1

FROST, A.A. and PEARSON, R.G. (1961). Kinetics and Mechanism (New York: Wiley)

FROST, M.J. (1971). *Values for Money – The Techniques of Cost Benefit Analysis* (London: Gower)

FROST, N.E., MARSH, K.J. and POOK, L.P. (1975). *Metal Fatigue* (Oxford: Oxford Univ. Press)

FROST, P. (1983). International fire detection standards. *Fire Surveyor*, **12**(3), 12

FROST, R. (1947). *Met. Mag.*, **76**, 14

FROST, R.A. (1986). *Introduction to Knowledge Based Systems* (London: Collins)

FRUCI, L. (1993). Valves put a plug on emissions. *Chem. Engng*, **100**(6), 139

FRURIP, D.J. (1992). Using the ASTM CHETAH program in chemical process hazard evaluation. *Plant/Operations Prog.*, **11**, 224

FRURIP, D.J., FREEDMAN, E. and HERTEL, G.R. (1989). A new release of the ASTM CHETAH program for hazard evaluation: versions for mainframe and personal computer. *Runaway Reactions*, p. 39. See also *Plant/Operations Prog.*, **8**, 100

FRY, J.E. (1971). Gas explosions attended by fire brigades in buildings. *J. Inst. Fuel*, **44**, 470

FRY, J.F. (1975). The collection and use of fire statistics. In Hendry, W.D. (1975), *op. cit.*, paper F4

FRY, M.A. and BOOK, D.L. (1984). The interaction of explosively produced shock waves with internal discontinuities and external objects. In Bowen, J.R. *et al.* (1984b), *op. cit.*, p. 449

FRY, R.S. and NICHOLLS, J.A. (1975). Blast wave initiation of gaseous and heterogeneous cylindrical detonation waves. *Combustion*, **15**, 43

FRY, T.I.C. (1928). *Probability and its Engineering Uses* (Princeton, NJ: van Nostrand)

FRYCKBERG, E.R. and TEPAS, J.J. (1988). Terrorist bombings: lessons learned from Belfast to Beirut. *Ann. Surgery*, **208**(5), 569

FRYER, D.M. (1978). Solving tube gland corrosion cracking. *Safety in Polyethylene Plants*, **3**, 57

FRYER, D.R.H. (1966). Siting policy and safety evaluation in the UK. *Br. Nucl. Energy Soc., Symp on Safety and Siting* (London)

FRYER, L. and MACKENZIE, J. (1992). SARAH: a system for performing risk analysis of major hazards. *Hazard Identification and Risk Analysis, Human Factors and Human Reliability in Process Safety*, p. 185

FTHENAKIS, V.M. (1993). HGSPRAY: a complete model of spraying unconfined gaseous releases. *J. Loss Prev. Process Ind.*, **6**, 327

FTHENAKIS, V.M. and BLEWITT, D.N. (1993). Mitigation of hydrofluoric acid releases: simulation of the performance of water spraying systems. *J. Loss Prev. Process Ind.*, **6**, 209

FTHENAKIS, V.M. and ZAKKAY, V. (1990). A theoretical study of absorption of toxic gases by spraying. *J. Loss Prev. Process Ind.*, **3**(2), 197

FU, K.S. (ed.) (1981). *Applications of Pattern Recognition* (Boca Raton, FL: CRC Press)

FU, T.T. (1972). Heat radiation from fires of aviation fuels. *Combustion Inst., Eastern Section, Fall Mtg*, Princeton, NJ

FU, T.T. (1974). Heat radiation from fires of aviation fuels. *Fire Technol.*, **10**, 54

FUCHS, H. and BLANKEN, J.M. (1986). Performance of bayonet tube boilers. *Ammonia Plant Safety*, **26**, 187

FUCHS, J. (1974). Solving a steam reformer tube failure. *Ammonia Plant Safety*, **16**, 72

FUCHS, J. and RUBINSTEIN, J. (1975). Secondary reformer air inlet nozzles failure. *Ammonia Plant Safety*, **17**, 148

FUCHS, P. (1984). On the extinguishing effect of various extinguishing agents and extinguishing methods with different fuels. *Fire Safety J.*, **7**, 165

FUGE, C. and SOHNS, D. (1976). Cut costs in olefin plants. *Hydrocarbon Process.*, **55**(11), 165

FUGELSO, L.E., WEINER, L.M. and SCHIFFMAN, T.H. (1972). *Explosion Effects Computation Aids. GARD Prog. 1540*. Gen. Am. Res. Div., Gen. Am Transportation Co., Niles, IL

FUHNER (1919). Hydrogen cyanide poisoning and its treatment. *Deutsch. Med. Wschr.*, **45**, 847

FUHR, J.C. (1992). Prevent fires in thermal oil heat transfer systems. *Chem. Engng Prog.*, **88**(5), 42

FUJITA, A. and RAI, C. (1977). Development and operational experiences of LNG in-ground storage tanks in Japan, Pt II. *LNG Conf 5*

FUJITANI, T., MIURA, M., MOTOZUNA, K., MINODA, K. and MIYANARI, T. (1983). Gas carriers with self-supporting prismatic-type cargo tanks. *Gastech 82*, p. 203

FUJITANI, T., MIYANARI, T., OKUMURA, Y., ANDO, A., MINODA, K., SAKAI, K., IINO, N., USHIROKAWA, O. and NAGANO, T. (1985). IHI SPB LNG carrier – fatigue strength, quality control and recent design development. *Gastech 84*, p. 232

FUJITANI, T., OKUMURA, Y., ANDO, A., NAGA NO, T., AOKI, E., YAMAKAWA, K. and ABE, A. (1987). An overview of the design features of the self-supporting prismatic tank system (SPB) LNG carrier. *Gastech 86*, p. 197

FUJITANI, T., OKUMURA, Y. and ANDO, A. (1989). An experimental LNG carrier using the SPB tank design. *Gastech 88*, paper 5.5

FUJITANI, T., OKUMURA, Y. and ANDO, A. (1991). The IHI SPB LNG carrier – past, present and future. *Gastech 90*, p. 4.4

FULFORD, G. and CATCHPOLE, J.P. (1968). Dimensionless groups. *Ind. Engng. Chem.*, **60**(3), 71

FULKS, B.D. (1982). Planning and organizing for less troublesome plant startups. *Chem. Engng*, **89**(Sep. 6), 96

FULLAM, B.W. (1987). The fire protection of LPG vessels. *Gastech 86*, p. 152

FULLAM, B.W. (1989). CIMAH safety cases: 2, LPG safety cases – failures and omissions. In Lees, F.P. and Ang, M.L. (1989), *op. cit.*, p. 233

FULLAM, B.W. (1995). Safety reports or safety cases: held or hindrance. *Safety Cases* (London: IBC Technical Services)

FULLER, B. (1989). Signal conditioners. *Chem. Engng*, **96**(12), 104

FULLER, D.D. (1984). *Theory and Practice of Lubrication for Engineers* (New York: Wiley)

FULLER, H.T. and BRISTLINE, R.E. (1964). Safe operation of spheres. In Vervalin, C.H. (1964a), *op. cit.*, p. 278

FULLER, J.A. (1984). *The History of Safety Valves Leading to the Development of British Standards for Safety Valves. Paper C365/84.* (London: Instn Mech. Engrs)

FULLER, J.G. (1984). *We Almost Lost Detroit* (New York: Berkley Books)

FULLNER, R.W. and CRUMP-WIESNER, H.J. (1976). Use of EPA's Environmental Response Unit in a pesticide spill. *Control of Hazardous Material Spills*, **3**, 345

FULMER RESEARCH INSTITUTE (1976). *Asbestos – Characteristics, Applications and Alternatives* (Stoke Poges, Slough)

FULTON, A.S. and BARRETT, D.J. (1986). PES – an opportunity for better safety systems. In *Hazards*, **IX**, p. 215

FULTON, G.P. (1960). Control storm and flood water. *Chem. Engng*, **67**(May 2), 129

FUMAROLA, G., PALAZZI, E., PALMIERI, S., RAPPINI, G. and FER-RAIOLO, G. (1980). Vapour clouds release and explosion in naphtha cracking units. *Loss Prevention and Safety Promotion 3*, vol. 2, p. 148

FUMAROLA, G., DE FAVERI, D.M., PALAZZI, E. and FERRAIOLO, G. (1982). Determine plume rise for elevated flares. *Hydrocarbon Process.*, **61**(1), 165

FUMAROLA, G., DE FAVERI, D.M., PASTORINO, R. and FERRAIOLO, G. (1983). Determine safety zones for exposure to flare radiation. In *Loss Prevention and Safety Promotion 4*, vol. 3, p. G23

FUNDERBURG, I.M. (1985). Corrosion-promoted failure mechanisms in nitric acid train cooler/condensers. *Ammonia Plant Safety*, **25**, 146

FUNES-CRAVIOTO, F. *et al.* (1977). Chromosome aberrations and sister-chromatid exchange in workers in chemical laboratories and a roto-printing factory and in children of women laboratory workers. *The Lancet*, **2**(8033), 322

FUNK, E.J. (1963). Inert gas systems: a roundup. *Chem. Engng*, **70**(Oct. 28), 117

FUNK, E.R. (1965). The use of reactive materials in the CPI. *Chem. Engng Prog.*, **61**(5), 121

FUNK, J.W., MURRAY, S.B., WARD, S. and MONE, I.O. (1982). A brief description of the DRES fuel–air explosives testing facility and current research program. In Lee, J.H.S. and Guirao, C.M. (1982), *op. cit.*, p. 565

FUNKE, H. (1976). Explosion in a centrifuge and countermeasures. *Prevention of Occupational Risks*, **3**, 399

FUNTOWICZ, S.O. and RAVETZ, J.R. (1992). *Uncertainty and Quality in Science for Policy* (Dordrecht: Kluwer)

FUQUAY, J. (1958). Meteorological factors in the appraisal and control of acute exposures to stack effluents. *Proc. Second UN Int. Conf. on the Peaceful Uses of Atomic Energy*, **18**, 272

FUQUAY, J.J. (1968). Environmental safety analysis. In Slade, D.H. (1968), *op. cit.*, p. 379

FUQUAY, J.J., SIMPSON, C.L. and HINDS, T.W. (1964). Estimates of ground level exposures resulting from protracted emissions for 70–meter stacks at Hanford. *J. Appl. Met.*, **3**, 761

FUREY, B.T. (1974). 80–hour blaze in tunnel after freight train derails. *Fire Engng*, **127**(Aug.), 24

FUREY, M.J. (1969). Friction wear and lubrication. *Ind. Engng Chem.*, **61**(3), 12

FURLONG, J. (1991). The EC Dangerous Substances Directive: a seventh amendment. *Chem. Ind.*, Jun. 17, 418

FURNO, A.L., COOK, E.B., KUCHTA, J.M. and BURGESS, D.S. (1971). Some observations on near-limit flames. *Combustion*, **13**, p. 593

FUSCO, J.J. and SHARSON, W.S. (1962). Use the fail-safe concept. *Ind. Engng. Chem.*, **54**(3), 40

FUSILLO, R.H. and POWERS, G.J. (1987). A synthesis method for chemical plant operating procedures. *Comput. Chem. Engng*, **11**, 369

FUSILLO, R.H. and POWERS, G.J. (1988a). Computer-aided planning of purge operations. *AIChE J.*, **34**, 558

FUSILLO, R.H. and POWERS, G.J. (1988b). Operating procedure synthesis using local models and distributed goals. *Comput. Chem. Engng*, **12**, 1023

FUSSELL, J.B. (1973a). A formal methodology for fault tree construction. *Nucl. Sci. Engng*, **52**, 421

FUSSELL, J.B. (1973b). *Synthetic Fault Tree Model. A Formal Methodology for Fault Tree Construction. Rep. ANCR 1098*. Nat. Reactor Testing Station, Aerojet Nucl. Co., Idaho Falls, ID

FUSSELL, J.B. (1975). How to hand-calculate system reliability and safety characteristics. *IEEE Trans. Reliab.*, **R-24**, 169

FUSSELL, J.B. (1976). Fault tree analysis: concepts and techniques. In Henley, E.J. and Lynn, J.W. (1976), *op. cit.*, p. 133

FUSSELL, J.B. (1978a). *Common Cause Failure Analysis*. NATO Advanced Study Inst. on Synthesis and Analysis Methods for Safety and Reliability Studies, Urbino, Italy

FUSSELL, J.B. (1978b). *Phased Mission Systems*. NATO Advanced Study Inst. on Synthesis and Analysis Methods for Safety and Reliability Studies, Urbino, Italy

FUSSELL, J.B. and BURDICK, G.R. (1977). *Nuclear Systems Reliability Engineering Risk Assessment* (Philadelphia, PA: Soc. Ind. Appl. Mathematics)

FUSSELL, J.B. and VESELY, W.E. (1972). A new methodology for obtaining cut sets. *Trans Am. Nucl. Soc.*, **15**(1), 262

FUSSELL, J.B., HENRY, E.B. and MARSHALL, N.H. (1971). *MOCUS – a Computer Program to Obtain Minimal Cut Sets from Fault Trees. Rep. ANCR-1156*. Aerojet Nucl. Corp., Idaho Falls, ID

FUSSELL, J.B., POWERS, G.J. and BENNETTS, R.G. (1974). Fault trees – a state of the art discussion. *IEEE Trans Reliab.*, **R-23**, 51

FUSSELL, J.B. *et al.* (1977). *SUPERPOCUS – a Computer Program for Calculating System Reliability and Safety Characteristics. Rep. NERS-77–01*. Dept of Nucl. Engng, Univ. of Tennessee

FUVEL, P. and CLAUDE, J. (1986). GMS 2000 project: building the first LPG membrane tank. *Gastech 85*, p. 427

VAN GAALEN, A.S. (1974). Loss prevention and property protection in the chemical industry. *Loss Prevention and Safety Promotion 1*, p. 35

GABE, D.R. (1972). *Principles of Metal Surface Treatment and Protection* (Oxford: Pergamon Press)

GABOR, T. (1981). Mutual aid systems in the United States for chemical emergencies. *J. Haz. Materials*, **4**(4), 343

GABOR, T. and GRIFFITH, T.K. (1980). The assessment of community vulnerability to acute hazardous materials incidents. *J. Haz. Materials*, **3**(4), 323

GABORIAUD, M., GROLLIER-BARON, R. and LEROY, A. (1983). Data collection in the offshore industry. *Euredata 4*, paper 7.2

GABRIEL, R. (1992). LISP. In Shapiro, S.C. (1992), *op. cit.*, p. 819

GACKENBACH, R.E. (1983). Plastics as a material of construction. *Chem. Engng Prog.*, **59**(10), 97

GADDES, W.M. and BRADY, L. (1981). A method for optimizing human performance in detecting and diagnosing mission avionics faults. In Rasmussen, J. and Rouse, W.B. (1981), *op. cit.*, p. 523

GADDUM, J.H. (1945). Lognormal distributions. *Nature*, **156** (3964) Cot. 20, 463

GADDY, J.L. and CULBERTSON, O.L. (1973). Prediction of variable process performance by stochastic flowsheet simulation. *AIChE J.*, **19**, 1239

GADIAN, T. (1967). Asbestos – a new hazard. *Chemical Process Hazards*, **III**, p. 61

GADIAN, T. (1972). Industrial cancer. *Chemical Process Hazards* **IV**, p. 98

GADIAN, T. (1974). The principle of acceptable risk. *Chemical Process Hazards* **V**, p. 280

GADIAN, T. (1978). Hazards of the chemical industry. *Chemy Ind.*, May 6, 293

GADIAN, T. (1983). Skin problems in the chemical industry. *Chemy Ind.*, Jan. 17, 52

GADSBY, R.E. and LIVINGSTONE, J.G. (1978). Catalysts and some incidents they have survived. *Ammonia Plant Safety*, **20**, 70

GAGE, J.C. (1972). The importance of toxicological information in process design. *Chemical Process Hazards*, IV, p. 95

GAGE, S.J. (1980). Love Canal chromosome study (letter). *Science*, **209**, 752

GAGE, W.L. (1967). *Value Analysis* (London: McGraw-Hill)

GAGLIARDI, D.V. (1965). Prevention in addition to protection. *Chem. Engng Prog.*, **61**(2), 44

GAGLIARDI, D.V. (1985). Common sense approach to loss prevention. *Chem. Engng Prog.*, **81**(8), 26

GAGNE, R.M. (1985). *The Conditions of Learning and the Theory of Instruction*, 4th ed. (New York: CBS Coll. Publ.)

GAI, E.G. and CURRY, R.E. (1976). A model of the human observer in failure detection tasks. *IEEE Trans Syst. Man Cybern.*, **SMC-6**, 85

GAILLARD, A.W.K. (1993). Comparing the concepts of mental load and stress. *Ergonomics*, **36**, 991

GAINES, B.R. (1976). Foundations of fuzzy reasoning. *Int. J. Man–Mach. Stud.*, **8**, 623

GAINES, B.R. and KOHOUT, L.J. (1977). The fuzzy decade: a bibliography of fuzzy systems and closely related topics. *Int. J. Man–Mach. Stud.*, **9**, 1

GAINSBORO, N.R. (1968). Tube vibration in a large bearing water cooler. *Chem. Engng Prog.*, **64**(3), 85

GAIT, A.J. (1967). *Heavy Organic Chemicals* (Oxford: Pergamon Press)

GAITONDE, N.Y. and DOUGLAS, J.M. (1969). The use of positive feedback control systems to improve reactor performance. *AIChE J.*, **15**, 902

GALE, W.E. (1985). Module ventilation rates quantified. *Oil Gas J.*, **83** (Dec. 23), 39

GALAGHER, A.J. and TWEEDALE, H.M. (1990). A development in auditing of major hazard plants. *CHEMECA 91*

GALAGHER, A.J. and TWEEDALE, H.M. (1992). A development in auditing of major hazard plant. *Trans Inst. Chem. Engrs*, **70B**, 18

GALL, J.F. (1980). Fluorine compounds, inorganic: hydrogen. *Kirk–Othmer Encyclopaedia of Chemical Technology*, 3rd ed., vol. 10 (New York: Wiley), p. 733

GALL, W. (1990). The application of two recently developed human reliability techniques to cognitive error analysis. *Eleventh Symp. on Advances in Reliability Technology* (Amsterdam: Elsevier Science)

GALLAGHER, G.A. (1986). Lessons in hazardous materials transportation based on case histories. *Am. Inst. Chem. Engrs, Symp. on Loss Prevention* (New York)

GALLAGHER, G.A. and MCCONE, S.W. (1986). Lessons in hazardous material transportation based on case histories. *Plant/Operations Prog.*, **5**, 186

GALLAGHER, J.T. (1967). Rapid estimation of plant costs. *Chem. Engng*, **74**(Dec. 18), 89

GALLAGHER, R. (1980). Beat corrosion with rubber hose. *Chem. Engng*, **87**(Sep. 8), 105

GALLAGHER, T. (1983). Estimating horizontal-tank volumes. *Chem. Engng*, **90**(Aug. 22), 100

GALLAGHER, W.E. (1980). Minimize stormwater system costs. *Hydrocarbon Process.*, **59**(10), 59

GALLANT, R.W. (1965–). Physical properties of hydrocarbons. *Hydrocarbon Process.*, **44**(7), 95; **44**(8), 127; **44**(9), 255; **44**(10), 151; 1966, **45**(3), 161; **45**(6), 153; **45**(7),

111; **45**(10), 171; **45**(12), 113; 1967, **46**(1), 183; **46**(2), 133; **46**(3), 143; **46**(4), 183; **46**(5), 201; **46**(6), 153; **46**(7), 121; **46**(8), 135; **46**(9), 155; **46**(10), 135; **46**(12), 119; 1968, **47**(1), 135; **47**(2), 113; **47**(3), 89; **47**(4), 128; **47**(5), 151; **47**(6), 139; **47**(7), 141; **47**(8), 127; **47**(9), 269; **47**(10), 115; **47**(11), 223; **47**(12), 89; 1969, **48**(1), 153; **48**(4), 151; **48**(5), 143; **48**(6), 141; **48**(7), 135; **48**(8), 117; **48**(9), 199; **48**(11), 263; **48**(12), 113; 1970, **49**(1), 137; **49**(2), 112; **49**(3), 131; **49**(4), 132

GALLANT, S.I. (1988). Connectionist expert systems. *Comm. ACM*, **31**, 2

GALLIER, P.W. (1968). Operating compressors via computer. *Chem. Engng Prog.*, **64**(6), 71

GALLUN, S.E., LUECKE, R.H., SCOTT, D.E. and MORSHEDI, A.M. (1992). Use open equations for better models. *Hydrocarbon Process.*, **71**(7), 78

GALLUZZO, J.F. (1973). Maintenance aspects of a pharmaceutical plant start-up. *Chem. Engng*, **80**(Apr. 2), 82

GALLWEY, T.J. (1982). Selection tests for visual inspection on a multiple fault type task. *Ergonomics*, **25**, 1077

GALUZZO, M. and ANDOW, P.K. (1984). Failures in control systems. *Reliab. Engng*, **7**(4), 193

GAMBOA, R. (1993). Consider insulation reliability. *Hydrocarbon Process.*, **72**(1), 73

GAMBHIR, S.P., HEIL, T.J. and SCHUELKE, T.F. (1978). Steam use and distribution. *Chem. Engng*, **85**(Dec. 18), 91

GAMBILL, W.R. (1957–). How to estimate engineering properties. *Chem. Engng*, 1957, **64**, Feb., 235; Feb., 237; Mar., 271; Apr., 227; May, 263; Jun., 243; Jul., 263; Aug., 257; Sep., 267; Oct., 283; Dec., 261; 1958, **65**, Jan. 13, 159; Feb. 10, 137; Mar. 10, 147; Apr. 7, 146; May 5, 143; Jun. 2, 125; Jun. 30, 113; Aug. 25, 121; Sept. 22, 169; Oct. 20, 157; Nov. 17, 157; 1959, **66**, Jan. 12, 127; Feb. 9, 123; Mar. 9, 151; Apr. 6, 139; May 18, 240; Jun. 15, 181; Jul. 13, 157; Aug. 10, 129; Oct. 19, 195; Nov. 16, 193; Dec. 14, 169; 1960, **67**, Jan. 11, 109

GAMBRO, A.J., MUENZ, K. and ABRAHAMS, M. (1972). Optimize ethylene complex. *Hydrocarbon Process.*, **51**(3), 73

GAMES, A.M., BREEWOOD, M., MARTKIN, P., K ELLER, A.Z. and AMENDOLA, A. (1985). Common cause failure investigations using the European Reliability Data System. *Reliab. Engng*, **13**(1), 33

GAMMELL, D.M. (1976). Blanket tanks for gas control. *Hydrocarbon Process.*, **55**(12), 101. See also in Kane, L.A. (1980), *op. cit.*, p. 160

EL GANAINY, O. (1985). Failure of dissimilar metals weld in reformer tubes. *Ammonia Plant Safety*, **25**, 10. See also *Plant/Operations Prog.*, **4**, 149

GANAPATHY, V. (1976). Rupture disks for gases and liquids. *Chem. Engng*, **83**(Oct. 25), 152

GANAPATHY, V. *et al.* (1977). Physical properties of selected gas streams. *Chem. Engng*, **84**(Feb. 28), 195

GANAPATHY, V. (1978). Design of air-cooled heat exchangers, process design criteria. *Chem. Engng*, **85**(Mar. 27), 112

GANAPATHY, V. (1981). Estimate nonluminous radiation heat transfer coefficients. *Hydrocarbon Process.*, **60**(4), 235

GANAPATHY, V. (1985). Heat-recovery boiler design for cogeneration process applications covers many parameters. *Oil Gas J.*, **83**(Dec. 2), 116

GANAPATHY, V. (1992). Fouling – the silent heat transfer thief. *Hydrocarbon Process.*, **71**(10), 49

GANAPATHY, V. (1993). Specify packaged steam generators properly. *Chem. Engng Prog.*, **89**(9), 62

GANAPATHY, V. (1994). *Steam Plant Calculations Manual*, 2nd ed. (New York: Marcel Dekker)

GANDHI, P.D. (1988). Correlations of steel column fire test data. *Fire Technol.*, **24**, 20

GANDHI, P.D. and KANURY, A.M. (1988). Thresholds for spontaneous ignition of organic solids exposed to radiant heating. *Combust. Sci. Technol.*, **57**, 113

GANDY, T. and BAILEY, J. (1992). Reduce costs with acoustic emission cool-down monitoring. *Hydrocarbon Process.*, **71**(1), 66

GANESHAN, R., FINGER, S. and GARRETT, J. (1991). Representing and reasoning with design intent. *Artificial Intelligence in Design* (ed. by J. Gero) (Oxford: Butterworth-Heinemann)

GANGADHARAN, A.C., RAO, M.S.M. and SUNDARARAJAN, C. (1977). Computer methods for qualitative fault tree analysis. *Failure Prev. Reliab.*, **251**

GANGADHARAN, A.C. and BROWN, S.J. (eds) (1977). *Failure Data and Failure Analysis: In Power and Processing Industries* (New York: Am. Soc. Mech. Engrs)

GANGER, J.J. and BEARROW, M.E. (1993). How to prioritize process hazard analyses. *Hydrocarbon Process.*, **72**(10), 95

GANI, R. and O'CONNELL, J.P. (1989). A knowledge based system for the selection of thermodynamic models. *Comput. Chem. Engng*, **13**, 397

GANN, R.G. (ed.) (1976). *Halogenated Fire Suppressants* (Washington, DC: Am. Chem. Soc.)

GANN, R.G., STONE, J.P., TATEM, P.A., WILLIAMS, F.W. and CARHART, H.W. (1978). Suppression of fires in confined spaces by nitrogen pressurization: III, Extinction limits for liquid pool fires. *Combust. Sci. Technol.*, **18**, 155

GANS, M. (1976a). The A to Z of plant start-up. *Chem. Engng*, **83**(Mar. 15), 72

GANS, M. (1976b). Which route to aniline? *Hydrocarbon Process.*, **55**(11), 145

GANS, M. (1979). Choosing between air and oxygen for chemical processes. *Chem. Engng Prog.*, **75**(1), 67

GANS, M. and BENGE, J.N. (1974). Scheduling the commissioning operation. *Instn Chem Engrs, Symp. on Commissioning of Chemical and Allied Plant* (London)

GANS, M., KIORPES, S.A. and FITZGERALD, F.A. (1983). Plant start-up – step by step by step. *Chem. Eng*, **90**(Oct. 3), 74

GANS, M., KOHAN, D. and PALMER, B. (1991). Systematize troubleshooting techniques. *Chem. Engng Prog.*, **87**(4), 25

GANS, M. and OZERO, B.J. (1976). For EO: air or oxygen. *Hydrocarbon Process.*, **55**(3), 73

GANONY, W.F. (1983). *Review of Medical Physiology* (Los Altos, CA: Lange)

GANT, J.E. (1978). Hydrogen requirements for hydroprocessing technologies. *Chem. Engng Prog.*, **74**(12), 66

GARBERS, A.W.F. and WASFI, A.K. (1990). Prevent cavitation in high energy centrifugal pumps. *Hydrocarbon Process.*, **69**(7), 48

GARCIA, E.A. (1967). Minimising start-up difficulties. *Chem. Engng Prog.*, **63**(12), 44

GARDENIER, J.S. (1981). Ship navigational failure detection and diagnosis. In Rasmussen, J. and Rouse, W.B. (1981), *op. cit.*, p. 49

GARDENIER, J.S., DWYER, J.A., LYNCH, C.J. and STOEHR, L.A. (1975). Risk management techniques for coast guard regulatory decision-making. *Transport of Hazardous Cargoes*, **4**

GARDNER (1973). Hazard management – systems of control. *Prevention of Occupational Risks*, **2**, 505

GARDNER, A. (1981). Search. In A. Barr and E.A. Feigenbaum (1981), *op. cit.*, p. 19

GARDNER, D.J., HULME, G., HUGHES, D.J., EVANS, R.F. and BRIGHTON, P. (1994). A survey of current predictive methods for explosion hazards in the UK offshore industry. *Hazards*, **XII**, p. 91

GARDNER, G. (1994). Holy air pollutants, BATNEEC! *Chem. Engr, Lond.*, **562**, 28

GARDNER, J.B. (1961). Some factors in the safe operation of air separation units. *Chem. Engr, Lond.*, **156**, A43; **158**, A45

GARDNER, J.D. and MOFFATT, E.A. (eds) (1982). *Highway Truck Collision Analysis* (New York: ASME)

GARDNER, J.F. (1991). Selecting values for reduced emissions. *Hydrocarbon Process.*, **70**(8), 87

GARDNER, J.F. and SPOCK, T.F. (1992). Control emissions from valves. *Hydrocarbon Process.*, **71**(8), 49

GARDNER, L.C. and HUISINGH, D. (1987). Alternative approaches to waste reduction in materials coating processes. *Haz. Waste Haz Materials*, **4**(2), 177

GARDNER, M.B. (1983). Prestressing enhances LNG piping integrity. *Oil Gas J.*, **81**(Nov. 14), 202

GARDNER, W. and TAYLOR, P. (1975). *Health at Work* (London: Ass. Bus. Programmes)

GARDNER, W.A. and ROYLANCE, P.J. (1967). *New Essential First Aid* (London: Pan Books)

GARDON, R. (1953). *Temperature Attained in Wood Exposed to High Intensity Thermal Radiation. Tech. Rep. 3. Fuels Res. Lab, Massachusetts Inst. of Technol., Cambridge, MA*

GARDON, R. (1959). Temperatures attained in wood exposed to high intensity thermal radiation. Ph.D. Thesis, Univ. of London

GARFORTH, A.M. and RALLIS, C.J. (1978). Laminar burning velocity of stoichiometric methane-air: pressure and temperature dependence. *Combust. Flame*, **31**, 53

GARG, A. and AYOUB, M.M. (1980). What criteria exist for determining how much load can be lifted safely? *Hum. Factors*, **22**, 475

GARG, D.R. and STEWARD, F.R. (1971). Pilot ignition of cellulosic materials containing high void spaces. *Combust. Flame*, **17**, 287

GARLAND, J.A. and COX, L.C. (1982). Deposition of small particles to grass. *Atmos. Environ.*, **16**, 2699

GARN, P.D. (1965). *Thermoanalytical Methods of Investigation* (New York: Academic Press)

GARNER, D. (1993). Bolt lubrication. *Plant Engr*, **37**(5), 34

GARNER, F.H., LONG, R., GRAHAM, A.J. and BADAKHSHAN, A. (1957). The effect of certain halogenated methanes on pre-mixed and diffusion flames. *Combustion*, **6**, 802

GARNER, J.F. (1975a). *Control of Pollution Act 1974* (London: Butterworths)

GARNER, J.F. (1975b). *The Law of Sewers and Drains*, 5th ed. (London: Shaw)

GARNER, J.F. and CROW, R.K. (1969). *Clean Air Law and Practice*, 3rd ed. (London: Shaw)

GARNER, P.J. (1976). Toxic vapours (letter). *Petrol. Rev.*, **30**(352), 243

GARNER, R.C. (1979). Carcinogen prediction in the laboratory: a personal view. *Proc. R. Soc., Ser. B*, **205**, 121. See also in Doll, Sir R. and McLean, A.E.M. (1979), *op. cit.*, p. 121

GARNETT, D.I. and PATIENCE, G.S. (1993). Why do scale-up power laws work? *Chem. Engng Prog.*, **89**(8), 76

GARRAD, C. (1973). LP-gas transport truck fire. In Vervalin, C.H. (1973a), *op. cit.*, p. 51

GARRETT, D.E. (1989). *Chemical Engineering Economics* (New York: van Nostrand Reinhold)

GARRETT, L.T. (1979). Multiplexing: justifying, specifying and proposal evaluation. *Hydrocarbon Process.*, **58**(3), 145

GARRIBA, S., MUSSIO, P., NALDI, F., REINA, G. and VOLTA, G. (1977). Efficient construction of minimal cut sets from fault trees. *IEEE Trans Reliab.*, **R-26**, 88

GARRIBA, S. and OVI, A. (1980). Assessment of chemical risk and safety by the use of multi-attribute decision techniques. *Loss Prevention and Safety Promotion 3*, vol. 2, p. 302

GARRICK, B.J. and KAPLAN, S. (1982). Safety technology in industry. Electric power. In Green, A.E. (1982), *op. cit.*, p. 443

GARRISON, C.M. (1981). How to cut agitation costs. *Chem. Engng*, **88**(Nov. 30), 73

GARRISON, P.W. (1975). A method for predicting blast wave overpressure produced by rupture of a gas storage vessel. *AIAA/NASA/ASTM/IES Seventh Joint Space Simulation Conf.* (McDonnel Douglas Astronautics Co.), paper 25

GARRISON, W.G. (1988a). Major fires and explosions analyzed for 30 year period. *Hydrocarbon Process.*, **67**(9), 115

GARRISON, W.G. (1988b). 100 large losses. *A Thirty-Year Review of Property Damage Losses in the Hydrocarbon-chemical Industries*, 11th ed. (New York: M&M Protection Consultants)

GARSIDE, J., DAVEY, R.J. and JONES, A.G. (eds) (1991). *Advances in Industrial Crystallisation* (Oxford: Butterworth-Heinemann)

GARSIDE, R. (1982). *Intrinsically Safe Instrumentation* (Safety Technology Ltd)

GÄRTNER, D., GIESBRECHT, H. and LEUCKEL, W. (1978). Effect of thermodynamic and fluid dynamic processes on the emergency venting of chemical reactors. *Chem. Ing. Tech.*, **50**(7), 503

GÄRTNER, D., GIESBRECHT, H. and LEUCKEL, W. (1980). Effects of two phase flow upon the emergency relief of reactor tanks and upon the subsequent atmospheric jet dispersion. *Loss Prevention and Safety Promotion 3*, vol. 3, p. 954

GARTRELL, F.E., THOMAS, F.W., CARPENTER, S.B., POOLER, F., TURNER, B.B. and LEAVITT, J.M. (1964). *Full Scale Study of Dispersion of Stack Gases. A Summary Report.* Tennessee Valley Authority, Public Health Service, Chattanooga, TN

GARY, J.H. and HANDWERK, G.E. (eds) (1984). *Petroleum Refining Technology and Economics* (New York: Marcel Dekker)

GAS COUNCIL (1960). *Industrial Gas Handbook* (London)

GASCHE, F. (1956). Design of reactors and closures. *Ind. Engng Chem.*, **48**(5), 838

GASPER, K.E. (1978). Non-chromate methods of cooling water treatment. *Chem. Engng Prog.*, **74**(3), 52

GASTON, J.R. (1982). Antisurge control schemes for turbocompressors. *Chem. Engng*, **89**(Apr. 19), 139

GATELEY, W.Y. and WILLIAMS, R.L. (1978). *GO Methodology – System Reliability Assessment and Computer Code Manual. Rep. EPRI NP-766.* Elec. Power Res. Inst., Palo Alto, CA

GATELEY, W.Y., STODDARD, D.W. and WILLIAMS, R.L. (1968). *GO – A Computer Program for Reliability Analysis of Complex Systems.* Rep. KN-67-704. Kaman Sciences, Colorado Springs, CO

GATES, L.E., HICKS, R.W. and DICKEY, D.S. (1976). Application guidelines for turbine agitators. *Chem. Engng*, **83** (Dec. 6), 165

GATZ, D.F. (1991). Urban precipitation chemistry: a review and synthesis. *Atmos. Environ.*, **25B**, 1

GAUERKE, J.R. (1972). Work gloves to meet OSHA rules. *Chem. Engng*, **79**(Apr. 3), 108

GAUGH, R.R. (1972). A new stainless steel for the CPI. *Chem. Engng*, **79**(Oct. 2), 84

GAUGH, R.R. (1976). New stainless for high temperatures. *Hydrocarbon Process.*, **55**(6), 87

GAULTIER, M. (1964). A summary of three cases of acute intoxication of ammonia. *Ann. Med. Leg.*, **44**, 357

GAUMER, L.S., GEIST, J.M., HARTNETT, G.J. and PHANNENSTIEL, L.L. (1972). The design, fabrication, and operation of large cryogenic heat exchangers. *LNG Conf.*, **3**

GAUVIN, S. (n.d.). *Occupational Health – Guide to Sources of Information* (London: Heinemann)

GAVIS, J. (1964). Transport of electric charge in low dielectric constant fluids. *Chem. Engng Sci.*, **19**, 237

GAVIS, J. (1967a). Charge generation in low dielectric constant liquids flowing past surfaces: non-linear theory for turbulent flow in tubes. *Chem. Engng Sci.*, **22**, 365

GAVIS, J. (1967b). Non-linear equation for transport of electric charge in low dielectric constant fluids. *Chem. Engng Sci.*, **22**, 359

GAVIS, J. (1969). The effect of ionic dissociation and recombination on the relaxation of charge in low conductivity dielectric constant liquids. *Chem. Engng Sci.*, 24, 451

GAVIS, J. (1972). The origin of electric charge in flowing hydrocarbons. *Loss Prevention*, **6**, 44

GAVIS, J. and HOELSCHER, H.E. (1964). Phenomenological aspects of hydrocarbons. In Vervalin, C.H. (1964a), *op. cit.*, p. 167

GAVIS, J. and KOSZMAN, I. (1961). Development of charge in low conductivity liquids flowing past surfaces: a theory of the phenomenon in tubes. *J. Colloid Sci.*, **16**, 3

GAVIS, J. and WAGNER, J.P. (1967). Electric charge generation during flow of hydrocarbons through microporous media. *Static Electrification*, **2**, 122

GAVIS, J. and WAGNER, J.P. (1968). Electric charge generation during flow of hydrocarbons through microporous media. *Chem. Engng Sci.*, **23**, 381

GAVRILA, G.N. and SETHI, S. (1991). Best performance characteristic of pressure relief valves and the amount of lost product. *J. Loss Prev. Process Ind.*, **4**, 265

GAY, H.H. (1963). Exposure to chlorine gas. *J. Am. Med. Ass.*, **183**, 806

GAY, N.R., AGNEW, J.T., WITZELL, O.W. and KARABELL, C.E. (1961). Thermochemical equilibrium in hydrocarbon–oxygen reactions involving polyatomic forms of carbon. *Combust. Flame*, **5**, 257

GAY, W.F. (1975). *Summary of National Transportation Statistics* (Washington, DC: US Govt Printing Office)

GAYDON, A.G. (1967). The use of shock tubes for studying fundamental combustion processes. *Combustion*, **11**, 1

GAYDON, A.G. and WOLFHARD, H.G. (1953–). *Flames, Their Structure, Radiation and Temperature*; 1960, 2nd ed.; 1970, 3rd ed.; 1979, 4th ed. (London: Chapman & Hall)

GAYLE, J.B. and BRANSFORD, J.W. (1965). *Size and Duration of Fireballs from Propellant Explosions. Rep. NASA TM X- 53314.* George C. Marshall Space Flight Center, Huntsville, AL

GAZZI, L. and PASERO, R. (1970). Process cooling systems – selection. *Hydrocarbon Process.*, **49**(10), 83

GEARY, K. (1989). Hazard analysis – a standardised approach. In Guy, G.B. (1989), *op. cit.*, p. 1

GEBHARDT, M. (1980). Preplanning for the supertanker threat to ports and coastlines. *Fire Int.*, Special issue, 49

GEBHARDT, M. (1984). Building a relationship with industry. *Fire Int.*, **88**, 37

GEBHARDT, M. (1989). The changing face of fire protection in Hamburg. *Fire Int.*, **120**, 9

GEBHART, D.H. and CALDWELL, M.J. (1986). Techniques for evaluating environmental risks of sour gas development.

Fifth Joint Conf. on Applications of Air Pollution Modeling (Boston, MA: Am. Met. Soc.), p. 303

GEE, D. (1980). Union view of the safety of insulating materials. *Chemy Ind.*, Mar. 1, 180

VAN GEEL, J.L.C. (1966). Safe radius of heat generating substances. *Chem. Tech.*, Jan., **24**

GEETA, B., TRIPATHI, A. and NARASIMHAN, S. (1993). Analytical expressions for estimating damage area in toxic gas releases. *J. Loss Prev. Process Ind.*, **6**, 125

GEHM, K.H. and BITTNER, G. (1971). Electrical installation in explosion hazarded areas, classified as division 1. *Electrical Safety*, **1**, 27

GEHRI, E. (1985). The fire resistance of steel structures. *Fire Technol.*, **21**, 22

GEHRING, P.J., WATANABE, P.G. and PARK, C.N. (1979). Risk of angiosarcoma in workers exposed to vinyl chloride as predicted from studies in rats. *Toxicol. Appl. Pharmacol.*, **49**(1), 15

GEIGER, G.H. (ed.) (1975). *Supplementary Readings in Engineering Design* (New York: McGraw-Hill)

GEIGER, R. (1950). *The Climate near the Ground*; 1965, rev. ed. (Cambridge, MA: Harvard Univ. Press)

GEIGER, W. (1974). Generation and propagation of pressure waves due to unconfined chemical explosions and their impact on nuclear power plant structures. *Nucl. Engng Des.*, **27**(2), 189

GEIGER, W. (1982). Quantification of hazards from vapour cloud explosions – research within the German reactor safety program. In Lee, J.H.S. and Guirao, C.M. (1982), *op. cit.*, p. 841

GEIGER, W. (1983). Present understanding of the explosion properties of flat vapour clouds. In Hartwig, S. (1983), *op. cit.*, p. 225

GEIGER, W. (1986). Pressure-enhancing mechanisms in unconfined vapour cloud explosions. In Hartwig, S. (1986), *op. cit.*, p. 251

GEIGER, W. (1987). Model investigations on the course of gas explosions in free air and implications for layout of chemical plants. *Fortschritte der Sicherheitstechnik 1* (Frankfurt-am-Main: DECHEMA), p. 293

GEIGER, W. and SYNOFZIK, R. (1980a). Influence of dispersion behaviour of dense explosive gases on the possible strength of explosion. In Hartwig, S. (1980), *op. cit.*, p. 245

GEIGER, W. and SYNOFZYK, R. (1980b). A simple model for the explosion of pancake-shaped vapour-clouds. *Loss Prevention and Safety Promotion 3*, vol. 2, p. 505

GEIHSLER, V.G. (1979). Major effects from minor features in ethylene plants. *Loss Prevention*, **12**, 9

GEILING, E.M.K. and MCLEAN, F.C. (1941). *Progress Report on Toxicity of Chlorine Gas for Mice to Nov. 6*, 1941. Nat. Defense Res. Cttee, Office of Sci. Res. and Dev.

GEINOPOLOS, A. and KATZ, W.J. (1968). Water pollution control. Primary treatment. *Chem. Engng*, **75**(Oct. 14), 78

GEIPEL, R. (1982). *Disaster and Reconstruction* (London: Allen & Unwin)

GEIPEL, R. (1991). *Long-Term Consequences of Disasters* (New York: Springer)

GEIRINGER, P.L. (1962). *Handbook of Heat Transfer Media* (New York: Reinhold)

GEIST, J.M. and CHATTERJEE, N. (1972). The effects of stratification on boil-off rates in LNG tanks. *Pipeline Gas J.*, Sep.

GEIST, J.M. (1985). Refrigeration cycles for future base-load LNG plants need a close look. *Oil Gas J.*, **83**(Feb. 4), 56

GELBEIN, A.P. and NISLICK, A.S. (1978). Make phenol from benzoic acid. *Hydrocarbon Process.*, **57**(11), 125

GELBURD, R.M. (1977). A comprehensive approach to occupational safety and health. *Chem. Engng*, Apr. 11, 114

GELDERBLOM (1980). Design of layouts of chemical plants. *IF-PEI III Mtg*

GELFAND, B.E., MEDVEDEV, S.P., POLENOV, A.N., FROLOV, S.M. and TSYGANOV, S.A. (1989). Expanding and loading of stratified dusty systems. *Loss Prevention and Safety Promotion 6*, p. 22.1

GELFAND, B., MEDVEDEV, S., POLENOV, A. and TSYGANOV, S. (1992). Shock waves by sudden expansion of hot liquid. *Loss Prevention and Safety Promotion 7*, vol. 2, 89–1

GELFER, G.A. *et al.* (1964). *Explosion Proof Electrical Equipment : Manual for Oil Refining and Gas Industry* (Israel Program for Scientific Translation)

GELINAS, G. (1985). Computerized system tracks toxic vapor releases. *Nat. Safety News*, Apr., 42

GEMANT, A. (1933). *Liquid Dielectrics* (New York: Wiley)

GEMANT, A. (1962). *Ions in Hydrocarbons* (New York: Interscience)

GEMMILL, A.V. (1961a). Ethylene oxidation: low cost route to acetaldehyde. *Chem. Engng*, **68**(May 15), 66

GEMMILL, A.V. (1961b). Vast safety network covers HCN in transit. *Chem. Engng*, **68**(Sep. 18), 68

GENAIDY, A.M. (1991). A training programme to improve human physical capability for manual handling jobs. *Ergonomics*, **34**, 1

GENERAL ACCOUNTING OFFICE (1978). *Liquefied Energy Gases Safety* (Washington, DC)

GENERAL ELECTRIC COMPANY (1955). *Properties of Combustion Gases*, vol. 1–2 (New York: McGraw-Hill)

GENCO, J.M. and LEMMON, A.W. (1970). Physical explosion in handling molten slags and metals. *Trans Am. Foundrymen's Soc.*, **78**, 317

GENET, S. and PERDRIX, C. (1987). Resistance of hot reduced pipes to hydrogen induced cracking. *Gastech 86*, p. 475

GENGE, C.L. (1989). Enclosure testing of halon: the NFPA method. *Fire Surveyor*, **18**(6), 6

GENSYM CORPORATION (1992). *G2 Reference Manual* (Cambridge, MA)

GENTILE, J.H., HARWELL, M.A., VAN DER SCHALIE, W.H., NORTON, S.B. and RODIER, D.J. (1993). Ecological risk assessment: a scientific perspective. *J. Haz. Materials*, **35**, 241

GENTILE, M. (2003). Personal Communication, June

GENTILE, M., ROGERS, W.J. and MANNAN, M.S. (2001). Inherent safety index for transportation of chemicals. *4th Annu. Symp.: Beyond Regulatory Compliance, Making Safety Second Nature*, 509 (College Station, TX: Mary Kay O'Connor Process Safety Center)

GENTILE, M., ROGERS, W.J. and MANNAN, M.S. (2002). Inherent safety index for transportation of chemicals. *5th Annu. Symp.: Beyond Regulatory Compliance, Making Safety Second Nature*, Oct. 29–30, 2002, College Station, TX, Track I: Inherent Safety Session (College Station, TX: Mary Kay O'Connor Process Safety Center.)

GENTILE, M., ROGERS, W.J. and MANNAN, M.S. (2003). Development of an inherent safety index based on fuzzy logic. *AIChE J.* **49**(4), 959–68

GENTILE, M., ROGERS, W.J. and MANNAN, M.S. (2003a). Integrating inherent safety analysis and environmental protection into a quantitative index. *Proc. of the 37th Annu. Loss Prev. Symp.* March 31–April 3, 2003, New Orleans, LA, (ed. J. F. Murphy and R. W. Johnson), Paper LPS-4f (New York: AIChE)

GENTNER, D. and STEVENS, A.L. (eds) (1983). *Mental Models* (Hillsdale, NJ: Lawrence Erlbaum)

GENTRY, J.W., SPURNY, K.R., SCHORMANN, J., OPIELA, H. and WEISS, G. (1982). Measurement of the diffusion coefficient for chrysotile and crocidolite fibers. *Atmos. Environ.*, **16**, 2947

GEORGAS, C.S. (1983). Ventilation requirements in container ship holds carrying dangerous goods. *Loss Prevention and Safety Promotion 4*, vol. 2, p. D16

GEORGE, A.F., PHILLIPS, D. and SILK, S.J. (1980). Health and safety legislation relating to toxic gases. In Hughes, D.J. (1980), *op. cit.*, paper 1

GEORGE, C.J., BENNETT, G.F., SIMONEAUX, D. and GEORGE, W.J. (1988). Polychlorinated biphenyls. A toxicological review. *J. Haz. Materials*, **18**(2), 113

GEORGE, E.A. (1981a). Case history on transportation. *Loss Prevention*, **14**, 185

GEORGE, E.A. (1981b). Incident involving a loaded anhydrous ammonia road tanker. *Loss Prevention*, **14**, 182

GEORGOPOULOS, P.G. and SEINFELD, J.H. (1986). Instantaneous concentration fluctuations in point source plumes. *AIChE J.*, **32**, 1642

GEORGOPOULOS, P.G. and SEINFELD, J.H. (1988). Estimation of relative dispersion parameters from atmospheric turbulence spectra. *Atmos. Environ.*, **22**, 31

GEPHART, L. and MOSES, S. (1989). An approach to evaluate the acute impacts from simulated accidental releases of chlorine and ammonia. *Plant/Operations Prog.*, **8**, 8

GEPPERT, J. (1889). On the nature of hydrogen cyanide poisoning. *Z. Klin. Med.*, **15**, 307

GERAERDS, W.M.J. (1970). Towards a theory of maintenance. *NATO Conf. on Problems in the Organization and Introduction of Large Logistic Support Systems*, Luxemburg

GERARD, R.W. (1948). Recent research on respiratory agents. In Andrus, E.C. *et al.* (1948), *op. cit.*, p. 565

GERARDU, N.H. (1981). Safe blowoff conditions for storage vessels. *Loss Prevention*, **14**, p. 134. See also *Chem. Engng Prog.*, 1981, **77**(1), 49

GERBERT, K. and KEMMLER, R. (1986). The causes of causes: deterministic and background variables of human factors incidents and accidents. *Ergonomics*, **29**, 1439

GERBIS, H. (1927). Death of a locomotive stoker from the smoulder gas from the self-ignition of lignite briquette. *Zbl. Gewerbehyg. Unfallverhuüutung*, **4**, 330

GERCHIK, M. (1939). Medical experience of Americans with chemical poison gas during the World War. *Protar*, **5**, 173

GERCHOW, F.J. (1975). How to select a pneumatic conveying system. *Chem. Engng*, **82**(Feb. 17), 72

GERHARDT, H.J. (1982). Technical considerations in the design of smoke and heat vents. *Fire Int.*, **77**, 34

GERLACH, H.D. (1980). The German pressure vessel code – philosophy and safety aspects. *Int. J. Pres. Ves. Piping*, **8**, 283

GERLACH, R. (1987). Flammability, combustibility. In Young, J.A. (1987A), *op. cit.*, p. 59

GERMAN COMMISSION FOR THE INVESTIGATION OF HEALTH HAZARDS OF CHEMICAL COMPOUNDS IN THE WORKPLACE (1994). *List of MAK and BAT Values 1993* (New York: VCH)

GERMELES, A.E. (1975a). Forced plumes and mixing of liquids in tanks. *J. Fluid Mech.*, **71**, 601

GERMELES, A.E. (1975b). A model for LNG tank rollover. *Advances in Cryogenic Engineering*, **21**, 326

GERMELES, A.E. and DRAKE, E.M. (1975). Gravity spreading and atmospheric dispersion of LNG vapour clouds. *Transport of Hazardous Cargoes*, 4

GERNAND, M.O. (1965). Streamlined data gathering systems. *Chem. Engng Prog.*, **61**(6), 62

GERRARD, M. (1994). International cost indices. *Chem. Engr, Lond.*, **557**, 18

GERRITSEN, H.G. and VAN'T LAND, C.M. (1988). Intrinsic continuous process safeguarding. *Preventing Major Chemical and Related Process Accidents*, p. 107

GERRITSEN, H.G. and C.M. VAN'T LAND (1988). Intrinsic continuous process safeguarding. *IChemE Symp. Series* No. 110, 107–15

GERSTEIN, M. (1953). The structure of laminar flames. *Combustion*, **4**, 35

GERSTEIN, M., CARLSON, E.R. and HILL, F.U. (1954). Natural gas–air explosions at reduced pressure. Detonation velocities and pressures. *Ind. Engng Chem*, **46**, 2558

GERSTEIN, M., LEVINE, O. and WONG, E.L. (1950). *Fundamental Flame Velocities of Pure Hydrocarbons. Res. Memo RM E50G24.* NACA

GERSTEIN, M. and STINE, W.B. (1973). Analytical criteria for flammability limits. *Combustion*, **14**, p. 1109

GERSUMSKY, W.D. (1977). Technology transfer problems. *Hydrocarbon Process.*, **56**(3), 73

GERTLER, J. and QIANG LUO (1989). Robust isolable models for failure diagnosis. *AIChE J.*, **35**, 1856

GERUNDA, A. (1981). How to size liquid-vapor separators. *Chem. Engng*, **88**(May 4), 81

GERVAIS, P. *et al.* (1965). Poisoning by chlorine vapours. *Presse Med.*, **73**, 725

GESELLSCHAFT FÜR REAKTORSICHERHEIT (1980). *Procedures and Systems for Assisting an Operator during Normal and Anomalous Nuclear Power Plant Operation Situations* (Cologne, Germany)

DI GESSO, J. (1989). Business interruption insurance. *Chem. Engr, Lond.*, **465**, 52

GETHING, J. (1975). Tetramethyl lead absorption: a report of human exposure to a high level of tetramethyl lead. *Br. J. Ind. Med.*, **32**, 329

GETTY, S.M., RICKERT, D.E. and TRAPP, A.L. (1977). Polyborminated biphenyl (PBB) toxicosis: an environment accident. *Critical Revs Environ. Control*, **7**(4), 309

GETZ, R. (1978). How to stress analyze jacketed piping. *Hydrocarbon Process.*, **57**(2), 139

GEVARTER, W.B. (1987). Introduction to artificial intelligence. *Chem. Engng. Prog.*, **83**(9), 21

GEYER, L.H. and PERRY, R.F. (1982). Variation in detectability of multiple flaws with allowed inspection time. *Hum. Factors*, **24**, 361

GEYER, T.A.W. and BELLAMY, L.J. (1991). *Pipework Failure, Failure Causes and the Management Factor.* Paper C425/012 (London: Instn Mech. Engrs)

GEYER, T.A.W., BELLAMY, L.J., ASTLEY, J.A. and HURST, N.W. (1990). Prevent pipe failures due to human errors. *Chem. Engng Prog.*, **86**(11), 66

GEYER, W. (1992). Bring storage tanks to the surface. *Chem. Engng*, **99**(7), 95

GHALI, A. (1979). *Circular Storage Tanks and Silos* (London: Spon)

GHERING, W.L. (1978). *Reference Data for Acoustic Noise Control* (Ann Arbor, MI: Ann Arbor Science)

GHIA, S., REGIS, V., POSSA, G. and TONOLINI, F. (1983). On-line surveillance and monitoring developments for application in power plants. In Butcher, D.W. (1983), *op. cit.*, p. 233

GHONIEM, A.F., ZHANG, X., KNIO, O., BAUM, H.R. and REHM, R.G. (1993). Dispersion and deposition of smoke plumes generated in massive fires. *J. Haz. Materials*, **33**, 275

GHORMLEY, E.L. (1958). Guard against detonation. *Petrol. Refiner*, **37**(1), 185

GHOSH, A. (1980). Checklist for batch process computer control. *Chem. Engng*, **87**(eb. 25), 88

GHOSH, H. (1992). Improve your fired heaters. *Chem. Engng*, **99**(3), 116

GIANETTO, A. and SILVESTON, P.L. (eds) (1986). *Multiphase Chemical Reactors* (Washington, DC: Hemisphere)

GIANNICI, B. and GALLUZO, M. (1983). Socially acceptable levels of risk: some quantitative considerations. *Reliab. Engng*, **5**, 37

GIARINI, O. (1980). Economic development, economic theory and risks. *Sci. Publ. Policy*, **7**, 336

GIBBINGS, J.C. (1967). Non-dimensional groups describing electrostatic charging in moving fluids. *Electrochemica Acta*, **12**, 106

GIBBINGS, J.C. (1979). Interaction of electrostatics and fluid motion. *Electrostatics*, **5**, 145

GIBBINGS, J.C. and HIGNETT, E.T. (1966). Dimensional analysis of Introduction electrostatic streaming current. *Electrochim. Acta*, **11**, 815

GIBBINGS, J.C., SALUJA, G.S. and MACKEY, M.A. (1971). Electrostatic charging in stationary liquids. *Static Electrification*, **3**, 93

GIBBON, H.J., WAINWRIGHT, J. and ROGERS, R.L. (1994). Experimental determination of flammability limits of solvents at elevated temperatures and pressures. *Hazards*, **XII**, p. 1

GIBBON, R.B., KNOWLES, A.E. and MERCER, W.L. (1973). *Pipeline Safety Assurance* (London: Instn Gas Engrs)

GIBBONS, E.J. (1962). Probability, downtime and doomed designs. *Chem. Engng*, **69**(Nov. 12), 250

GIBBONS, W.G., ANDREWS, W.R. and CLARKE, G.A. (1975). Fracture of defects approaching a free surface. *J. Pres. Ves. Technol.*, **97**, 270

GIBBONS, W.S. and HACKNEY, B.D. (1964). *Survey of Piping Failures for the Reactor Primary Coolant Pipe Rupture Study.* Rep. 4574. General Electric Company

GIBBS, C.W. (1960). Compressor explosion. *Ammonia Plant Safety*, **2**, p. 27. See also *Chem. Engng Prog.*, 1960, **56**(6), 64

GIBBS, C.W. and WHITE, K.H. (1961). The importance of maintaining valve records. *Ammonia Plant Safety*, **3**, A13

GIBBS, G.J. and CALCOTE, H.F. (1959). Effect of molecular structure on burning velocity. J. *Chem. Engng Data*, **4**, 226

GIBSON, A. (1980). The changing technology of boiler management. *Plant Engr*, **24**(1), Jan./Feb., 24

GIBSON, A.E. and PITKIN, M. (1972). Developments in LNG transportation in the US. *LNG Conf.*, **3**

GIBSON, B. and LAIOS, L. (1978). The presentation of information to the job-shop scheduler. *Hum. Factors*, **20**, 725

GIBSON, C. (1973). Brunei's LNG blockbuster. *Chem. Engng*, **80**(Nov. 12), 112

GIBSON, C.J. (1968). Feasibility of large super-insulated pipelines for transport of LNG. *LNG Conf.*, 1

GIBSON, G.H. and WALTERS, W.J. (1970). Some aspects of LNG storage. *LNG Conf.*, 2

GIBSON, G.J. (1987). Efficient test runs. *Chem. Engng*, **94**(7), 75

GIBSON, H. and JACKSON, E. (1972). *The Valuation of Plant and Machinery for Rating* (London: Estates Gazette)

GIBSON, K. (1986). Bolt tensioning need not be a strain. *Process Engng*, Jul., 46

GIBSON, M.R. (1978). In discussion of paper by English, P.E. and Bosworth, C. (1978), *op. cit.*

GIBSON, M.R. and KNOWLES, G. (1971). Design and maintenance of instrument trip systems. *First Nat. Conf. on Reliability.* NRC4/11 (Culcheth, Lancashire: UK Atom. Energy Auth.)

GIBSON, N. (1969). Static electricity. In Handley, W. (1969), *op. cit.*, p. 107

GIBSON, N. (1970). The explosion hazard from static electricity in powder systems. *Process Engng*, Sep. 11, 67

GIBSON, N. (1971). Static in fluids. *Static Electrification*, **3**, 71

GIBSON, N. (1972). Characteristics of powders processed in the chemical industry. *Int. Symp. on Dust and Explosion Risks in Mines and Industry* (Heidelberg: Berufsgenossenschaft der Chem. Ind.), p. 158

GIBSON, N. (1973a). Explosion protection of powder processing plant in the chemical industry. *Prevention of Occupational Risks*, **2**, 385

GIBSON, N. (1973b). Safety problems associated with electrostatically charged solids. *Eur. Fed. Chem. Engng, Second Int. Conf. on Static Electricity* (Frankfurt-am-Main: DECHEMA)

GIBSON, N. (1977). Design of explosion reliefs: an informal contribution for discussion. *Chemical Process Hazards VI*, unbound paper

GIBSON, N. (1979). Electrostatic hazards in filters. *Filtr. Sep.*, **16**, 382

GIBSON, N. (1981). Design criteria for equipment handling flammable dusts. *The Hazards of Industrial Explosion from Dusts* (London: Oyez/IBC), paper 10

GIBSON, N. (1983). Electrostatic hazards: a review of modern trends. *Electrostatics*, **6**, 1

GIBSON, N. (1984). Comparison of simple and complex methods for identifying potentially dangerous decompositions in reaction masses and powders. *Chemy Ind.*, Mar. 19, 209

GIBSON, N. (1986a). Electrostatics – a hazard under control. *Loss Prevention and Safety Promotion 5*, 7-1

GIBSON, N. (1986b). Hazard evaluation and process design. *Control and Prevention of Chemical Reaction Hazards* (London: IBC Tech. Services Ltd)

GIBSON, N. (1990a). Chemical reaction hazards in chemical and pharmaceutical industries. *Safety and Loss Prevention in the Chemical and Oil Processing Industries*, p. 541

GIBSON, N. (1990b). Electrostatic hazards in the chemical industry. *Safety and Loss Prevention in the Chemical and Oil Processing Industries*, p. 265

GIBSON, N. and HARPER, D.J. (1981a). Evaluation of electrostatic hazards associated with non-conducting materials. *J. Electrostatics*, **11**, 27

GIBSON, N. and HARPER, D.J. (1981b). Friction and localised heat initiation of powders capable of fast decomposition—an exploratory study. *Runaway Reactions, Unstable Products and Combustible Powders* (London: Instn Chem. Engrs), paper 3/R

GIBSON, N., HARPER, D.J. and ROGERS, R.L. (1985). Evaluation of the fire and explosion risk in drying powders. *Plant/Operations Prog.*, **4**, 181

GIBSON, N., HARPER, D.J. and ROGERS, R.L. (1986). Hazards from exothermic decomposition of powders. *Control and Prevention of Chemical Reaction Hazards* (London: IBC Tech. Services Ltd)

GIBSON, N. and HARRIS, G.F.P. (1976). The calculation of dust explosion vents. *Loss Prevention*, **10**, 141. See also *Chem. Engng Prog.*, 1976, **72**(11), 62

GIBSON, N. and LLOYD, F.C. (1965). Incendivity of discharge from electrostatically charged plastics. *Br. J. Appl. Phys.*, **16**, 1619

GIBSON, N. and LLOYD, F.C. (1967). Electrification of toluene in pipeline flow. *Static Electrification*, **2**, 89

GIBSON, N. and LLOYD, F.C. (1970a). Effect of contamination on the electrification of toluene flowing in metal pipes. *Chem. Engng Sci.*, **25**, 87

GIBSON, N. and LLOYD, F.C. (1970b). Electrification of toluene flowing in large-diameter metal pipes. *J. Phys. D, Appl. Phys.*, **3**, 563

GIBSON, N. and LLOYD, F.C. (1975). Effect of electrification in ball valves on ignition risk in liquid pipeline systems. *J. Electrostatics*, **1**, 339

GIBSON, N., LLOYD, F.C. and PERRY, G.R. (1967). Fire hazards in chemical plant from friction sparks involving the thermite reaction. *Chemical Process Hazards* **III**, p. 26

GIBSON, N., MADDISON, N. and ROGERS, R.L. (1987). Case studies in the application of DIERS venting methods to fine chemical batch and semi-batch reactors. *Hazards from Pressure* (Rugby: Instn Chem. Engrs), p. 157

GIBSON, N., MADDISON, ROUNSLEY, J.S. and STOKES, P.S.N. (1986). Assessment of dust explosion hazards: effect of changes in test methods and criteria. In *Hazards*, **IX**, p. 63

GIBSON, N. and ROGERS, R.L. (1980). Ignition and combustion of dust clouds in hot environments – an exploratory study. *Chemical Process Hazards*, **VII**, p. 209

GIBSON, N., ROGERS, R.L. and WRIGHT, T.K. (1987). Chemical reaction hazards: an integrated approach. *Hazards from Pressure* (Rugby: Instn Chem. Engrs), p. 61

GIBSON, N. and SCHOFIELD, F. (1977). Fire and explosion hazards in spray dryers. *Chemical Process Hazards*, **VI**, p. 53

GIBSON, R. (1992). Warehouse protection. *Plant Engr*, **36**(6), 30

GIBSON, S.B. (1974). Reliability engineering applied to the safety of new projects. *Dept of Chem. Engng, Teesside Polytechnic, Middlesbrough, Course on Process Safety – Theory and Practice*

GIBSON, S.B. (1976a). The design of new chemical plants using hazard analysis. *Process Industry Hazards*, p. 135

GIBSON, S.B. (1976b). How we fixed the flowsheet – safely. *Process Engng*, Jun., 119

GIBSON, S.B. (1976c). Reliability engineering applied to the safety of new projects. *Chem. Engr, Lond.*, **306**, 105

GIBSON, S.B. (1976d). Risk criteria in hazard analysis. *Chem. Engng Prog.*, **72**(2), 59

GIBSON, S.B. (1976e). Safety in the design of new chemical plants (hazard and operability studies). *Course on Loss Prevention in the Process Industries*. Dept of Chem. Engng, Loughborough Univ. of Technol.

GIBSON, S.B. (1976f). The use of quantitative risk criteria in hazard analysis. *J. Occup. Accid*, **1**(1), 85

GIBSON, S.B. (1977a). Hazard analysis in the design of new chemical plants. *Design Developments Related to Safety, Costs and Solid Systems* (London: Instn Chem. Engrs), paper A-4–1

GIBSON, S.B. (1977b). Major hazards – should they be prevented at all costs? *Chemical Engineering in a Hostile World* (London: Clapp & Poliak)

GIBSON, S.B. (1977c). The quantitative assessment of process safety. *Chemical Process Hazards*, **VI**, p. 1

GIBSON, S.B. (1978). Major hazards – should they be prevented at all costs? *Engng Process Economics*, **3**, 25

GIBSON, S.B. (1981). Hazard analysis and risk criteria. *Loss Prevention*, **14**, p. 11. See also *Chem. Engng Prog.*, 1980, **76**(11), 46

GIBSON, S.B. (1987). Investment in human factors pays dividends. *Preventing Major Chemical Accidents*, p. 6.41

GIBSON, S.B. (1992). A comprehensive review of alarm interlock testing. *Hazard Identification and Risk Analysis, Human Factors and Human Reliability in Process Safety*, p. 119

GIBSON, T. (1975). Waste heat boiler possibilities. *Plant Engr*, **19**(3), 9

GIBSON, T.O. (1984). Learning value from recent loss. *Plant/Operations Prog.*, **3**, 18

GIBSON, T.O. (1986). Experience with cost effective loss prevention. *Am. Inst. Chem. Engrs, Symp on Loss Prevention* (New York)

GIBSON, T.O. (1988). Loss prevention features perform in fired heater loss incident. *Plant/Operations Prog.*, **7**, 258

GIBSON, T.O. (1989). Learning value from a blown fuse. *Plant/Operations Prog.*, **8**, 209

GIBSON, W.H. (1976). Disaster planning. *J. Soc. Occup. Med.*, **26**, 136

GIDDINGS, J.C. and HIRSCHFELDER, J.O. (1957). Flame properties and the kinetics of chain-branching reactions. *Combustion*, **6**, p. 199

VON GIERKE, H.E. (1968). Response of the body to mechanical forces – an overview. *Ann. N.Y. Acad. Sci.*, **152**(Art. 1), 172

GIESBRECHT, H. (1988). Evaluation of vapour cloud explosions by damage analysis. *J. Haz. Materials*, **17**, 247

GIESBRECHT, H., BECHTEL, S., GILBERT, N. and SCHLIEPHAKE, V. (1992). Ignition of unconfined vapour clouds by hot pipes. *Loss Prevention and Safety Promotion 7*, vol. 2, 77-1

GIESBRECHT, H., HEMMER, G., HESS, K., LEUCHEL, W. and STOCKEL, A. (1981). Analysis of the potential explosion effect of flammable gases released into the atmosphere over a short time. *Chem. Ing. Tech.*, **53**(1), 256; 271

GIESBRECHT, H., HESS, K., LEUCKEL, W. and MAURER, B. (1980–). Analysis of the potential explosive effect of amounts of combustible gases released into the atmosphere. *Chem. Ing. Tech.*, **52**, 114; 1981, **53**, 1

GIESBRECHT, H., HESS, K., LEUCKEL, W., MAURER, B. and STOECKEL, A. (1980). Analysis of the potential explosion effects of flammable gases during short time release into the atmosphere. In Hartwig, S. (1980), *op. cit.*, p. 207

GIESBRECHT, H., HESS, K., and MAURER, B. (1981a). Analysis of explosion hazards on spontaneous release of inflammable gases into the atmosphere. Pt 1: Propagation and deflagration of vapour clouds on the basis of burning tests on model vessels. Ger. *Chem. Engng*, **4**, 305

GIESBRECHT, H., HESS, K., and MAURER, B. (1981b). Analysis of explosion hazards on spontaneous release of inflammable gases into the atmosphere. Pt 2: Comparison of explosion model from experiments with damage effects of explosion accidents. Ger. *Chem. Engng*, **4**

GIESBRECHT, H. and SEIFERT, H. (1986). Sizing of emergency vents for vessels on the basis of laboratory scale experiments. *Loss Prevention and Safety Promotion 5*, p. 42-1

GIESBRECHT, H., SEIFERT, H. and LEUCKEL, W. (1983). Dispersion of vertical free jets. In Hartwig, S. (1983), *op. cit.*, p. 103

GIFFIN, W.C. and ROCKWELL, T.H. (1984). Computer aided testing of pilot response to critical in-flight events. *Hum. Factors*, **26**, 573

GIFFORD, F.A. (1957). Relative atmospheric diffusion of smoke puffs. *J. Met.*, **14**, 410

GIFFORD, F.A. (1959). Statistical properties of a fluctuating plume dispersion model. *Adv. Geophys.*, **6**

GIFFORD, F.A. (1960a). Atmospheric diffusion calculations using the generalized Gaussian plume model. *Nucl. Safety*, **2**(2), 56

GIFFORD, F. (1960b). Peak to average concentration ratios according to a fluctuating plume dispersion model. *Int. J. Air Poll.*, **3**, 253

GIFFORD, F.A. (1961). Use of routine meteorological observations for estimating atmospheric dispersion. *Nucl. Safety*, **2**(4), 47

GIFFORD, F.A. (1962a). The area within ground level dosage isopleths. *Nucl. Safety*, **4**(2), 91

GIFFORD, F.A. (1962b). Diffusion in the diabatic surface layer. *J. Geophys. Res.*, **67**, 3207

GIFFORD, F.A. (1968). An outline of the theories of diffusion in the lower layers of the atmosphere. In Slade, D. (1968), *op. cit.*

GIFFORD, F.A. (1970). Peak to mean concentration ratios according to a 'top-hat' fluctuating plume model. *Am. Met. Soc., Conf. on Air Pollution Meteorology*

GIFFORD, F.A. (1972). Atmospheric transport and dispersion over cities. *Nucl. Safety*, **13**, 391

GIFFORD, F.A. (1976a). Atmospheric dispersion models for environmental pollution applications. In Haugen, D.A. (1976), *op. cit.*, p. 35

GIFFORD, F.A. (1976b). Turbulent diffusion-typing schemes: a review. *Nucl. Safety*, **17**, 68

GIFFORD, F.A. (1977). Tropospheric relative diffusion observations. *J. Appl. Met.*, **16**, 311

GIFFORD, F.A. (1980). Smoke as a quantitative atmospheric diffusion tracer. *Atmos. Environ.*, **14**, 1119

GIFFORD, F.A. (1986). Atmospheric diffusion in the range 20 to 2000 kilometers. In de Wispelaere, C., Schiermeier, F.A. and Gillani, N.V. (1986), *op. cit.*, p. 247

GIFFORD, F.A. (1987). The time-scale of atmospheric diffusion considered in relation to the universal diffusion function, f1. *Atmos. Environ.*, **21**, 1315

GIFFORD, F.A. and PACK, D.H. (1962). Surface deposition of airborne material. *Nucl Safety*, **3**, 76

GIGER, G., GYGAX, R. and HOCH, F. (1983). Experimental principles for predicting safe conditions for the storage of bulk chemicals. *Loss Prevention and Safety Promotion 4*, vol. 3, p. C14

GILBERT, A. (1964). Designing for safety. Design engineer's viewpoint. *Chem. Engng*, **71**(Jul. 20), 147

GILBERT, B.R., FOX, J.M., BOHANNAN, W.R. and WHITE, R.A. (1985). Cost reduction ideas for an LNG project. *Chem. Engng Prog.*, **81**(4), 17

GILBERT, D.F. (1979). One plant's experience with government regulations. *Hydrocarbon Process.*, **58**(1), 219

GILBERT, H. (1994). Management of risk. In Rawson, K. (1994), *op. cit.*, p. 39

GILBERT, J.T.E. (1983). *Technological Guidelines for Offshore Oil and Gas Development* (Tulsa, OK: PennWell)

GILBERT, M. (1957). The influence of pressure on flame speed. *Combustion*, **6**, p. 74

GILBERT, M. (1958). The hydrazine flame. *Combust. Flame*, **2**, 137

GILBERT, N. and EAGLE, R.S. (1977). Risks in high-pressure vessel start-up. *Ammonia Plant Safety*, **19**, p. 83

GILBERT, R.B. (1989). *Piper Alpha Public Inquiry. Transcript, Day 138*

GILBERT, S.M. (1989). *A model for the effects of a condensed phase explosion in a built-up area*. PhD thesis, Loughborough Univ. of Technology

GILBERT, S.M., LEES, F.P. and SCILLY, N.S. (1994a). A model for the effects of a condensed phase explosion in a built-up area: 1. Outline of explosion effects model and exposure scenarios. Private communication

GILBERT, S.M., LEES, F.P. and SCILLY, N.S. (1994b). A model for the effects of a condensed phase explosion in a built-up area: 2. The explosive power of some weapons causing housing damage in Britain in World War 2. Private communication

GILBERT, S.M., LEES, F.P. and SCILLY, N.S. (1994c). A model for the effects of a condensed phase explosion in a built-up area, 3. Housing damage caused in Britain by some weapons in World War 2 and certain other incidents. Private communication

GILBERT, S.M., LEES, F.P. and SCILLY, N.S. (1994d). A model for the effects of a condensed phase explosion in a built-up area, 4. Housing damage and injury caused in Britain by V-1 flying bombs and V-2 rockets in World War 2. Private communication

GILBERT, S.M., LEES, F.P. and SCILLY, N.S. (1994e). A model for the effects of a condensed phase explosion in a built-up area, 5. A model for the fireball from a condensed phase explosive. Private communication

GILBERT, S.M., LEES, F.P. and SCILLY, N.S. (1994f). A model for the effects of a condensed phase explosion in a built-up area, 6. A model for the fragments and fragment injury from a cased condensed phase explosive. Private communication

GILBERT, S.M., LEES, F.P. and SCILLY, N.S. (1994g). A model for the effects of a condensed phase explosion in a built-up area, 7. A model for the fragments and fragment injury from a cased condensed phase explosive – computer program and numerical results. Private communication

GILBERT, S.M., LEES, F.P. and SCILLY, N.S. (1994h). A model for the effects of a condensed phase explosion in a built-up area, 8. Models for injury to persons indoors and outdoors. Private communication

GILBERT, S.M., LEES, F.P. and SCILLY, N.S. (1994i). A model for the effects of a condensed phase explosion in a built-up area, 9. The model and its application. Private communication

GILCHRIST, H.L. (1924). *Poisoning Warfare Gases* (New York: Appleton)

GILCHRIST, H.L. (1928). *A Comparative Study of World War Casualties from Gas and Other Weapons* (Washington, DC: US Govt Printing Office)

GILCHRIST, H.L. and MATZ, P.B. (1933). *The Residual Effects of Warfare Gases*. 1, Chlorine (Washington, DC: US Govt Printing Office)

GILL, D.A. and PICOU, J.S. (1991). The social physiological impacts of a technological accident: collective stress and perceived health risks. *J. Haz. Materials*, **27**, 77

GILL, F.S. (1980a). Heat. In Waldron, H.A. and Harrington, J.M. (1980), *op. cit.*, p. 225

GILL, F.S. (1980b). Ventilation. In Waldron, H.A. and Harrington, J.M. (1980), *op. cit.*, p. 112

GILL, F.S. and ASHTON, I. (1982). *Monitoring for Health Hazards at Work* (Oxford: Blackwell)

GILL, R.J. and OLSON, D.B. (1984). Estimation of soot thresholds for fuel mixtures. *Combust. Sci. Technol.*, **40**, 307

GILL, S. (1964). *Structures for Nuclear Power* (London: C.R. Brooks)

GILL, S.S. (ed.) (1970). *The Stress Analysis of Pressure Vessels and Pressure Vessel Components* (Oxford: Pergamon Press)

GILL, S.S. (1975). *The Nypro By-Pass Line. Rep. to Flixborough Court of Inquiry* (Manchester: Univ. of Manchester Inst. Sci. Technol.)

GILL, W., DONALDSON, A.B. and SHOUMAN, A.R. (1979-). The Frank–Kamenetski problem revisited. *Combust. Flame*, **36**, 217; 1981, **43**, 219

GILL, W., SHOUMAN, A.R. and DONALDSON, A.B. (1981). The Frank–Kamenetski problem revisited. *Combust. Flame*, **41**, 99

GILLAM, E. (1969). *Materials under Stress* (London: Newnes-Butterworths)

GILLES, A. (1972). 200,000 cubic meter LNG tankers. *LNG Conf.*, 3

GILLES, E. and SCHULER, H. (1982). Early detection of hazardous states in chemical reactors. Ger. *Chem. Engng*, **5**, 69

GILLESPIE, P.J. and DIMAIO, L.R. (1977). How foam can protect HPI plants. *Hydrocarbon Process.*, **56**(8), 111

GILLETT, F.J. (1960). Choosing drives for centrifugal compressors. *Chem. Engng*, **67**(Mar. 7), 143

GILLETT, F.L. (1984). Ethylene from NGL feedstocks, Pt 5. *Hydrocarbon Process.*, **63**(2), 69

GILLETT, J. (1985). Rapid ranking of hazards. *Process Engng*, Feb. 19

GILLETT, J.W. (1980). Terrestrial microcosm technology in assessing fate, transport and effects of toxic chemicals. In Haque, R. (1980), *op. cit.*, p. 231

GILLETTE, R.H. (1980). Comments on "Thermal hazards from LNG fireballs". *Combust. Sci. Technol.*, **22**, 185

GILLINGS, D.W. (1958). The cost of instrumentation and control installation. *Chem. Engr, Lond.*, **140**, A45

GILLIS, J. (1968). Testing for explosiveness of dusts. *Loss Prevention*, **2**, 34

GILLIS, J. and DALE, S. (1980). Explosion venting and explosion suppression in an operating bucket elevator. *Loss Prevention and Safety Promotion 3*, vol. 3, p. 1304

GILLIS, J.P. (1981). Explosion venting and suppression of bucket elevators. *The Hazards of Industrial Explosion from Dusts* (London: Oyez/IBC), paper 16

GILLON, R.D. (1989). Intumescent coatings. *Fire Surveyor*, **18**(2), 13

GILMAN et al. (1948). *Advances in Military Medicine* (Boston, MA: Little, Brown and Co.)

GILMAN, A.Z. and CATTELL, M. (1948). Systemic agents: action and treatment. In Andrus, E.C. et al. (1948), *op. cit.*, p. 546

GILMER, H. and FREESTONE, F.J. (1978). Cleanup of an oil and mixed chemical spill at Dittmer, Missouri, April–May 1977. *Control of Hazardous Material Spills*, 4, 131

GILMORE, C.L. (1965). Loss: your conscience. *Chem. Engng Prog.*, **61**(8), 32

GILMORE, C.L. (1967). A statistical approach to preventing accidents. *Chem. Engng*, **74**(Jul. 17), 224

GILMORE, C.L. (1968). Planning the emergency organization. *Chem. Engng*, **75**(Mar. 11), 179

GILMORE, C.L. (1970). *Accident Prevention and Loss Control* (New York: Am. Mngmt Ass.)

GILMORE, R. (1981). *Catastrophe Theory for Scientists and Engineers* (New York: Wiley)

GILMORE, W.E., GERTMAN, D.I. and BLACKMAN, H.S. (1989). *User-Computer Interface in Process Control* (New York: Academic Press)

GILMOUR, C.H. (1965). No fooling – no fouling. *Chem. Engng Prog.*, **61**(7), 49

GILMOUR, C.H. (1967). Trouble-shooting heat-exchanger design. *Chem. Engng*, **74**(Jun. 19), 221

GILOVICH, THOMAS, *How We Know What Is Not So, The Fallibility of Human Reason In Everyday Life*, (New York: Maxwell Macmillan), ISBN 0-02-911706-2

GILPIN, R. and WRIGHT, C. (1964). *Scientists and National Policy-Making* (New York: Columbia Univ. Press)

VAN GILS, J.M. (1988). *Catalogue of European Earthquakes and Atlas of European Seismic Maps. Rep. EUR 11344 EN*. Comm. of the Eur. Commun., Brussels

GILWOOD, M.E. (1963). Water treatment for plant use. *Chem. Engng*, **70**(Jun. 10), 193

GIMBEL, M.A. (1965a). Personal behaviour – rules and regulations. In Fawcett, H.H. and Wood, W.S. (1965), *op. cit.*, p. 219

GIMBEL, M.A. (1965b). Safety education and training. In Fawcett, H.H. and Wood, W.S. (1965), *op. cit.*, p. 209

GINESTE, M. and LECOMTE, M. (1970). Operation since 1965 of the LNG transport line from Arzew to Le Havre and of the reception terminal. *LNG Conf.*, 2

GINKEL, S.L. and OLANDER, D.L. (1991). Stick to the basics of pilot plant design. *Chem. Engng Prog.*, **87**(1), 27

GINN, M.C. (1989). *Piper Alpha Public Inquiry. Transcript, Day 145*

GINNITY, B. (1988). A monitoring system for accidental releases of gases and vapours. *Preventing Major Chemical and Related Process Accidents*, p. 223

GINSBURGH, I. and BULKLEY, W.L. (1963). Hydrocarbon-air detonations – industrial aspects. *Chem. Engng Prog.*, **59**(2), 82

GINSBURGH, I. and BULKLEY, W.L. (1964). Industrial aspects of hydrocarbon air detonations. In Vervalin, C.H. (1964a), *op. cit.*, p. 116

GIOT, M. (1994). Two-phase releases. *J. Loss Prev. Process Ind.*, **7**, 77

GIRARD, P., HUNEAU, M., RABASSE, C. and LEYER, J.C. (1979). Flame propagation through unconfined and confined hemispherical stratified gaseous mixtures. *Combustion*, **17**, 1247

GIRDHAR, H.Z. and ARORA, A.J. (1977). Effect of pressure on heat of explosion. *Combust. Flame*, **28**, 291

GIRIN, H.A. (1978). Help in the design of high-pressure joints. *Safety in Polyethylene Plants*, **3**, 28

GISELLE, A.M. and KRAUSE, S.R. (1992). Transient heat transfer model in pressure vessels submitted to external heat sources. *Loss Prevention and Safety Promotion 7*, vol. 3, p. 147-1

GITTUS, J. (1988). Safety and nuclear power. *Chem. Engr, Lond.*, **448**, 12

GITTUS, J.H. (1987). Probabilistic safety analysis for the pressurised water reactor. *Science and Public Affairs, No. 2* (London: The Royal Society), p. 121

GITTUS, J.H. et al. (1987). *The Chernobyl Accident and Its Consequences* (London: HM Stationery Office)

GITTUS, J.H. et al. (1988). *The Chernobyl Accident and Its Consequences*, 2nd ed. (London: HM Stationery Office)

GITZLAFF, T.R. and BATTON, F.E. (1979). Two inexpensive sampling systems reduce exposure. *Chem. Engng*, **86**(Mar. 12), 117

GIUGIOIU, G. (1977). The use of 'SEFRO' pneumatic logic elements and devices in chemical systems to prevent fires and explosions. *Prevention of Occupational Risks*, **5**, 111

GJERDE, B. (minister) (1977–78). *The Uncontrolled Blow-Out on the Ekofisk Field (the Bravo Platform) 22 April 1977. Storting Rep. 65*. Ministry of Petroleum and Energy, Oslo

GJESTADT, T. (1983). Practical experience with offshore reliability data collection. *Euredata 4*, paper 7.1

GJESTER, F. (1994). Norway offshore. *Offshore Engr*, Apr., 104

GLADIS, G.P. (1960). Effects of moisture on corrosion in petrochemical environments. *Chem. Engng Prog.*, **56**(10), 43

GLASER, E.M. (1977). Acceptable and unacceptable risks. *Br. Med. J.*, 1028

GLASER, L.B. and KRAMER, J. (1983). Does modularization reduce plant investment? *Chem. Engng Prog.*, **79**(10), 63

GLASER, L.B. and YOUNG, R.M. (1961). Critical path planning and scheduling. *Chem. Engng Prog.*, **57**(11), 60

GLASER, L.B., KRAMER, J. and CAUSEY, E.D. (1979). Practical aspects of modular and barge-mounted plants. *Chem. Engng Prog.*, **75**(10), 49

GLASFELD, R. (1980). Some aspects of LNG ship safety and reliability. *Gastech 79*, p. 137

GLASS, D.S. (1987). Developments in the world-wide trade in ethylene, propylene and butadiene. *Gastech 86*, p. 323

GLASS, I.I. (1960). *Aerodynamics of Blasts. UTIA Rev. 17.* Inst. Aerodynamics., Univ. of Toronto

GLASS, J. (1978). Design of air-cooled heat exchangers. Specifying and rating fans. *Chem. Engng*, **85**(Mar. 27), 120

GLASS, J.A. and RIVARD, J.L. (1987). Exchange of safety information makes dollars and good sense. *Chem. Engng. Prog.*, **83**(5), 18

GLASS, R.L. (1979). *Software Reliability Guide Book* (Englewood Cliffs, NJ: Prentice-Hall)

GLASSBURN, L.E. (1981). Industrial noise: properties, sources and solutions. *Hydrocarbon Process.*, **60**(8), 127

GLASSMAN, I. (1977). *Combustion* (New York: Academic Press)

GLASSMAN, I. (1989). Soot formation in combustion processes. *Combustion*, **22**, 295

GLASSMAN, I. (1993). Comments on "The combustion phase of burning metals" by T.A. Steinberg, D.B. Wilson and F. Benz. *Combust. Flame*, **93**, 338

GLASSMAN, I. and DRYER, F.L. (1980/81). Flame spreading across liquid fuels. *Fire Safety J.*, **3**(2), 123

GLASSMAN, I. and HANSEL, J.G. (1968). Some thoughts and experiments on liquid fuel spreading, steady burning and ignitability in quiescent atmospheres. *Fire Res. Abs. Revs*, **10**, 217

GLASSMAN, I., HANSEL, J.G. and EKLUND, T. (1969). Hydrodynamic effects in the flame spreading, ignitability and steady burning of liquid fuels (letter). *Combust. Flame*, **13**, 99

GLASSMAN, I. and YACCARINO, P. (1981). The temperature effect in sooting diffusion flames. *Combustion*, **18**, 1175

GLASSTONE, S. (1957). *The Effects of Nuclear Weapons* (Washington, DC: Atom. Energy Comm.)

GLASSTONE, S. (1962). *The Effects of Nuclear Weapons, rev. ed.* (Washington, DC: Atom. Energy Comm.)

GLASSTONE, S. (1964). *The Effects of Nuclear Weapons*, rev. ed. (Washington, DC: Atom. Energy Comm.)

GLASSTONE, S. (1968). *Source Book on Atomic Energy*, 3rd ed. (New York: van Nostrand Reinhold)

GLASSTONE, S. and DOLAN, P.J. (1977). *The Effects of Nuclear Weapons*, 3rd ed. (Washington, DC: Dept of Defence)

GLASSTONE, S. and DOLAN, P.J. (1980). *The Effects of Nuclear Weapons*, 3rd ed. (Tonbridge Wells: Castle House Publs)

GLASSTONE, S. and SESONKE, A. (1962). *Nuclear Reactor Engineering* (New York: van Nostrand Reinhold)

GLATT, W. (1977). Explosion protection measures in fluidized bed driers and fluidized bed spray granulators. *Prevention of Occupational Risks*, **5**, p. 161

GLATTLY, J.E. (1978). Spill control and contingency response plan. *Control of Hazardous Material Spills 4*, p. 13

GLAUBINGER, R.S. (1979). A guide to the Resource Conservation and Recovery Act. *Chem. Engng*, **86**(Jan. 29), 79

GLAUBINGER, R.S. (1980). Ground water regulations: trouble from the deep? *Chem. Engng*, **87**(Jul. 28), 27

GLAUBINGER, R.S., KOHN, P.M. and REMIREZ, R. (1979). Love Canal aftermath: learning from a tragedy. *Chem. Engng*, **86**(Oct. 22), 86

GLEEKMAN, L.W. (1963). Corrosion resistant metals. *Chem. Engng*, **70**(Nov. 11), 217

GLEEKMAN, L.W. (1964). Corrosion of aluminum alloys in chlor-alkali environment. *Chem. Engng Prog.*, **60**(5), 49

GLEEKMAN, L.W. (1966). On the job non-destructive testing. *Chem. Engng*, **73**(Apr. 11), 198

GLEEKMAN, L.W. (1970). Trends in materials applications. Non-ferrous metals. *Chem. Engng*, **77**(Oct. 12), 111

GLICKMAN, M. (1971). Positive displacement pumps. *Chem. Engng*, **78**(Oct. 11), 37

GLICKMAN, M. and HEHN, A.H. (1969). Valves. *Chem. Engng*, **76**(Apr. 14), 131

GLICKMAN, T.S. (1988). Benchmark estimates of release accident rates in hazardous materials transportation by rail and truck. *Transport. Res. Record 1193*, (Transport. Res. Board, Nat. Res. Coun.), p. 22

GLICKMAN, T.S. and UJIHARA, A.M. (1990). Deciding between in-place protection and evacuation in toxic vapor cloud emergencies. *J. Haz. Materials*, **23**(1), 57

GLINOS, K. and MYERS, R.D. (1991). Sizing of vacuum relief valves for atmospheric distillation. *J. Loss Prev. Process Ind.*, **4**, 166

GLOAG, H.L. (1961). *Colouring in Factories* (London: HM Stationery Office)

GLOR, M. (1981). Ignition of gas/air mixtures by discharges between electrostatically charged plastic surfaces. *J. Electrostatics*, **10**, 327

GLOR, M. (1984). Conditions for the appearance of discharges from the gravitational compaction of powders. *J. Electrostatics*, **15**, 223

GLOR, M. (1985). Hazards due to the electrostatic charging of powders. *J. Electrostatics*, **16**, 175

GLOR, M. (1987). Discharges and hazards associated with the handling of powders. *Electrostatics*, **7**, 207

GLOR, M. (1988). *Electrostatic Hazards in Powder Handling.* (Letchworth/New York: Research Studies Press/ Wiley)

GLOR, M. and BOSCHUNG, P. (1980). Methods for investigating the electrostatic behaviour of powders. *Loss Prevention and Safety Promotion 3*, vol. 2, p. 623

GLOR, M., LÜTTGENS, G., MAURER, B. and POST, L. (1987). Propagating brush discharges in chemical plant technology. *Fortschritte der Sicherheitstechnik 1* (Frankfurt-am-Main: DECHEMA), p. 363

GLOR, M. and SIWEK, R. (1992). Minimum energy necessary to initiate decomposition of ethylene oxide vapour. *Loss Prevention and Safety Promotion 7*, vol. 2, 78-1

GLUCK, S.E. (1968). Design tips for pneumatic conveyors. *Hydrocarbon Process.*, **47**(1), 88

GMEHLING, J. and ONKEN, U. (1978). *Vapor—Liquid Equilibrium Data Collection: Aqueous-Organic Systems*, vol. 1, pt. 1 (Frankfurtam-Main: DECHEMA)

GNEDENKO, B.V., BELAYEV, Y.K. and SOLOVYEV, A.D. (1969). *Mathematical Methods of Reliability Theory* (New York: Academic Press)

GNOSIS (1984). *Learning Lisp* (Englewood Cliffs, NJ: Prentice-Hall)

GODARD, K.E. (1973). Gas plant start-up problems. *Hydrocarbon Process.*, **52**(9), 151

GODBEY, T. (1991). Brace your plant against dust explosion. *Chem. Engng*, **98**(4), 130

GODFREY, D.E.R. (1959). *Theoretical Elasticity and Plasticity for Engineers* (London: Thames & Hudson)

GODRIDGE, A.M. and HAMMOND, E.G. (1969). Emissivity of a very large residual fuel oil flame. *Combustion*, **12**, 1219

GODSAVE, G.A.E. (1953). Studies of the combustion of drops in a fuel spray – the burning of single drops of fuel. *Combustion*, **4**, 818

GODSE, A.G. (1990). Rotating equipment shop testing. *Hydrocarbon Process.*, **69**(1), 47

GODSE, A.G. (1991). Implement machinery predictive maintenance. *Hydrocarbon Process.*, **70**(9), 163

GOETZ, D.D. and SAWREY, J.S. (1994). Aspects of emergency relief design for reactive systems. *Process Safety Prog.*, **13**, 177

GOFMAN, J. and TAMPLIN, A. (1973). *Poisoned Power: The Case against Nuclear Power Plants* (London: Chatto & Windus)

GOFORTH, C.P. (1970). Functions of a loss control program. *Loss Prevention*, **4**, 1

GOGGIN, D.G. (1987). Vibration monitoring – a programmed approach. *Plant/Operations Prog.*, **6**, 1

GOH, C.B., TAN, R.B.H. and CHING, C.B. (1993). Risk analysis for the road transportation of hazardous chemicals in Singapore. *Health, Safety and Loss Prevention*, p. 191

GOINS, W.C. and SHEFFIELD, R. (1983). *Blowout Prevention* (Houston, TX: Gulf Publ. Co.)

GOKALP, I. (1980). On the correlation of turbulent burning velocities. *Combust. Sci. Technol.*, **23**, 137

GOLAND, L. and CRITELLI, F.X. (1975). The destator: a new method of tanker explosion prevention. *Transport of Hazardous Cargoes*, **4**

GOLAY, M.W. (1982). Numerical modeling of buoyant plumes in a turbulent, stratified atmosphere. *Atmos. Environ.*, **16**, 2373

GOLD, P.I. (1968–). Estimating thermophysical properties of liquids. *Chem. Engng*, 1968, **75**(Oct. 7), 152; (Nov. 4), 185; (Nov. 18), 170; 1969, **76**(Jan. 13), 119; (Feb. 24), 109; (Mar. 10), 122; (Apr. 7), 130; (May 19), 192; (Jun. 30), 129; (Jul. 14), 121; (Aug. 11), 97; (Sep. 8), 141

GOLD, W.D. (1969). Legal aspects of industrial safety. In Handley, W. (1969), *op. cit.*, p. 434

GOLDBERG, A. and ROBSEN, D. (1983). *SMALLTALK-80, The Language and its Implementation* (Reading, MA: Addison-Wesley)

GOLDBERG, F.F. and VESELY, W.E. (1977). Time dependent availability analysis of nuclear safety systems using randomly sampled failure rates. *Fourth Nat. Reliab. Conf., NRC 4/14*

GOLDBERG, L. (ed.) (1974). *Carcinogenesis Testing of Chemicals* (Boca Raton, FL: CRC Press)

GOLDBERGER, W.M. (1966). Trends in high temperature chemical processing. *Chem. Engng*, **73**(Mar. 14), 173; (Mar. 28), 125

GOLDBURG, A. and SIN-I CHENG (1965). A review of the fluid dynamic problem posed by the laminar jet diffusion flame. *Combust. Flame*, **9**, 259

GOLDE, R.H. (1977a). *Lightning*, vols 1–2 (London: Academic Press)

GOLDE, R.H. (1977b). The lightning conductor. In Golde, R.H. (1977a), *op. cit.* p., p. 545

GOLDEN, B.G. (1968). Ways to reduce plant noise. *Hydrocarbon Process.*, **47**(12), 75

GOLDEN, B.G. (1969). What to do about plant noise. *Chem. Engng. Prog.*, **65**(4), 47

GOLDEN, B.G., OUELLETTE, R.P., SAARI, S. and CHEREMISINOFF, P.N. (1979). *Environmental Impact Data Book* (Ann Arbor, MI: Ann Arbor Science)

GOLDEN, D.M. and LARSON, C.W. (1985). Rate constants for use in modeling. *Combustion*, **20**, 595

GOLDENBERG, S.A. and PELEVIN, V.S. (1959). Influence of pressure on rate of flame propagation in turbulent flow. *Combustion*, **7**, 590

GOLDER, D. (1972). Relations among stability parameters in the surface layer. *Boundary Layer Met.*, **3**, 46

GOLDFARB, A.S., GOLDGRABEN, G.R., HERRICK, E.C., OUELLETTE, R.P. and CHEREMISINOFF, P.N. (1981). *Organic Chemicals Manufacturing Hazards* (Ann Arbor, MI: Ann Arbor Science)

GOLDFIELD, J. (1980). Contaminant concentration reduction: general ventilation versus local exhaust ventilation. *Am. Ind. Hyg. Ass. J.*, **41**, 812

GOLDFINGER, P. (1949). The vapor pressure and heat of sublimation of carbon. *Combustion*, **3**, 618

GOLDIN, J.H. (1987). *Plastic Surgery* (Oxford: Blackwell)

GOLDIZEN, V.C., RICHMOND, D.R. and CHIFFELLE, T.L. (1957). *Missile Studies with a Biological Target. Rep. ITR-1470*. Civil Effects Test Group, US Atom. Energy Comm., Office of Tech. Services, Dept of Commerce, Washington, DC

GOLDIZEN, V.C., RICHMOND, D.R., CHIFFELLE, T.L., BOWEN, I.G. and WHITE, C.S. (1961). *Missile Studies with a Biological Target. Rep. WT-1470*. Civil Effects Test Group, US Atom. Energy Comm., Office of Tech. Services, Dept of Commerce, Washington, DC

GOLDMAN, A. and SLATTERY, T. (1964). *Maintainability* (New York: Wiley)

GOLDMAN, M. and GOLDMAN, A. (1978). Corona discharges. *Gaseous Electronics* (ed. M. Hirsch and H.J. Oakham) (New York: Academic Press)

GOLDMAN, R. (1972). Start-up afterthoughts. *Chem. Engng*, **79**(Dec. 25), 72

GOLDMAN, S. (1984). Periodic machinery monitoring: do it right. *Hydrocarbon Process.*, **63**(8), 51

GOLDMAN, S. (1986). Why you need a spectrum analyzer. *Hydrocarbon Process.*, **65**(1), 51

GOLDMANN, F. (1929). *Z. Physik. Chem.*, **B5**, 307

GOLDMANN, S.F. and SARGENT, R.W.H. (1971). Application of linear estimation theory to chemical processes – a feasibility study. *Chem. Engng Sci.*, **26**, 1535

GOLDSACK, S. (ed.) (1985). *Ada for Specification and Design* (Cambridge: Cambridge Univ. Press)

GOLDSCHMIDT, G. and FILSKOV, P. (1990). Substitution – A way to obtain protection against harmful substances at work. *Staub – Reinhaltung Der Luft* **50**, 403–5

GOLDSMITH, J.A. and SCHUBACH, S.A. (1993). Some practical aspects of the application of inhalation toxicity measures in risk analysis. *Health, Safety and Loss Prevention*, p. 531

GOLDSMITH, M. (ed.) (1973). *Science and Social Responsibility* (London: Macmillan)

GOLDSMITH, P. (1992a). A gale of drafts. *Chem. Engr, Lond.*, **525**, 19

GOLDSMITH, P. (1992b). GRP still has far to go. *Chem. Engr, Lond.*, **522**, 28

GOLDSMITH, P. (1992c). Still in the dark. *Chem. Engr, Lond.*, **529**, s15

GOLDSMITH, P. and JOHN, M. (1992). Belgium in focus. *Chem. Engr, Lond.*, **531**, 17

GOLDSMITH, W. (1960). *Impact* (London: Arnold)

GOLDSTEIN, H. and VON NEUMANN, J. (1953). *Blast Wave Calculation, Collected Works of J. von Neumann*, vol. VI (Oxford: Pergamon Press), p. 386

GOLDSTEIN, I.L. (1978). The pursuit of validity in the evaluation of training programs. *Hum. Factors*, **20**, 131

GOLDSTEIN, I.L. (1987). The relationship between training goals and training systems. In Salvendy, G. (1987), *op. cit.*, p. 963

GOLDSTEIN, I.L. and DORFMAN, P.W. (1978). Speed and load stress as determinants of performance in a time sharing task. *Hum. Factors*, **20**, 603

GOLDSTEIN, R.F. and WADDAMS, A.L. (1967). *The Petroleum Chemicals Industry* (London: Spon)

GOLDTHWAITE, W.H., BAYBUTT, P., LEVERENZ, F.L. and KEMP, H.S. (1987). Guidelines for hazard evaluation procedures: preparation and content. *Plant/Operations Prog.*, **6**, 63

GOLDWIRE, H.C. (1985). Status report on the Frenchman Flat ammonia spill study. *Am. Inst. Chem. Engrs, 1985 Ammonia Symp.*

GOLDWIRE, H.C. (1986). Large-scale ammonia spill tests. *Ammonia Plant Safety*, **26**, p. 203. See also *Chem. Engng Prog.*, 1986, **82**(4), 35

GOLDWIRE, H.C., MCRAE, T.G., JOHNSON, G.W., HIPPLE, D.L., KOOPMAN, R.P., MCCLURE, J.W., MORRIS, L.K. and CEDERWALL, R.T. (1983). *Coyote Series Data Report: LLNL/NWC 1981 LNG Spill Tests, Dispersion, Vapour Burn, and Rapid Phase Transition. Rep. UCID-19953*. Lawrence Livermore Lab., Livermore, CA

GOLDWIRE, H.C., MCRAE, T.G., JOHNSON, G.W., HIPPLE, D.L., KOOPMAN, R.P., MCCLURE, J.W., MORRIS, L.K. and CEDERWALL, R.T. (1985). *Desert Tortoise Series Data Report: 1983 Pressurized Ammonia Spills. Rep. UCRL-20562*. Lawrence Livermore Lab., Livermore, CA

GOLEC, R.A. (1985). The Milford Haven tank fire. *Hydrocarbon Process.*, **64**(10), 97

GOLLA, F. and SYMES, W.L. (1915). The immediate effects of the inhalation of chlorine gas (letter). *Br. Med. J.*, Aug. 28, 348

GOLLAHALLI, S.R., KHANNA, T. and PRABHU, N. (1992). Diffusion flames of gas jets issued from circular and elliptical nozzles. *Combust. Sci. Technol.*, **86**, 267

GOLLAHALLI, S.R. and SULLIVAN, H.F. (1974). *Effect of Pool Shape on Burning Rate, Radiation and Flame Height of Liquid Pool Fires. Report.* Thermal Engng Group, Dept of Mech. Engng, Univ. of Waterloo, Ont.

GÖLLER (1970). Safety distances, zones of protection and receiving basins for the storage of liquefied gases. *Prevention of Occupational Risks*, **1**

GÖLLER, O. (1974). Report of three serious accidents in oxygen plants. *Loss Prevention and Safety Promotion 1*, p. 335

GOLOVINA, E.S. and FYODOROV, G.G. (1957). Influence of physical and chemical factors on burning velocity. *Combustion*, **6**, 88

GOLTZ, G. (1984). Fugitive emissions (letter). *Chem. Engr, Lond.*, **407**, 3

GOMES, V.G. (1985). Controlling fired heaters. *Chem. Engng*, **92**(Jan. 7), 63

GOMEZ, A., WAKE, G.C. and GRAY, B.F. (1985). Friction and localized heat initiation of ignition: the asymmetrical slab and cylindrical annulus. *Combust. Flame*, **61**, 177

GOMI, S. (1964). Japan's new vinyl chloride process. *Hydrocarbon Process.*, **43**(11), 165

GONDO, S. and KUSUNOKI, K. (1969). A new look at gravity settlers. *Hydrocarbon Process.*, **48**(9), 209

GONDOUIN, M. and MURAT, F. (1972). Transportation and storage of LNG. *Chem. Engng Prog.*, **68**(9), 71

GONZALEZ, A.J. and DANKELL, D.D. (1993). *Engineering of Knowledge Based Systems* (Englewood Cliffs, NJ: Prentice Hall)

GONZALEZ, M.H., SUBRAMANIAN, T., KAO, R. and LEE, A.L. (1968). Physical properties of natural gas at cryogenic conditions. *LNG Conf. 1*

GOOCH, T.G., HURST, R., KROECKEL, H. and MERZ, M. (eds) (1983). *Behaviour of Joints in High Temperature Materials* (Amsterdam: Elsevier Applied Science)

GOOD, A. (1967). *The Petroleum Officers Handbook* (London: Knight)

GOODALL, A. (1985). *The Guide to Expert Systems* (Oxford: Learned Information)

GOODALL, A. (1988). An introduction to expert systems. *Computer Newsletter*, Oct., 8

GOODALL, D.G. and INGLE, R. (1967). The ignition of flammable liquids by hot surfaces. *Fire Technol.*, **3**, 115

GOODALL, K.L. (1988). Halon fire protection. Are discharge tests necessary? *Fire Surveyor*, **17**(4), 57

GOODCHILD, K. (1985). Control and protection of electric motors. *Plant Engr*, **29**(3), 14

GOODFELLOW, H.D. (1985). *Advanced Design of Ventilation Systems for Contaminant Control* (Amsterdam: Elsevier)

GOODFELLOW, H.D. (ed.) (1986). *Ventilation 85* (Amsterdam: Elsevier)

GOODFELLOW, H.D. and BENDER, M. (1980). Design considerations for fume hoods for process plant. *Am. Ind. Hyg. Ass. J.*, **41**, 473

GOODFELLOW, H.D. and BERRY, J. (1986). Clean-plant design. *Chem. Engng*, **93**(Jan. 6), 55

GOODFELLOW, H.D. and GRAYDON, W.F. (1968a). Dependence of electrostatic charging currents on fluid properties. *Canad. J. Chem. Engng*, **46**, 342

GOODFELLOW, H.D. and GRAYDON, W.F. (1968b). Electrostatic charging currents for different fluid systems. *Chem. Engng Sci.*, **23**, 1267

GOODFELLOW, H.D. and SMITH, J.W. (1982). Industrial ventilation – review and update. *Am. Ind. Hyg. Ass. J.*, **41**, 812

GOODGER, E.M. (1976). *Hydrocarbon Fuels* (London: Macmillan)

GOODGER, E.M. (1977). *Combustion Calculations* (London: Macmillan)

GOODIER, J.L. and CECE, J.M. (1983). Hazardous materials containment via spill prevention and failsafe engineering. *J. Haz. Materials*, **7**(2), 145

GOODIN, R.E. (1980). The ethics of social risks. *Sci. Publ. Policy*, **7**, 318

GOODMAN, H., GRAY, P. and JONES, D.T. (1972). Self-heating during the spontaneous ignition of methyl nitrate vapour. *Combust. Flame*, **19**, 157

GOODMAN, H.J. (1960). *Compiled Free-air Blast Data on Bare Spherical Pentolite. BRL Memo Rep. 1092*. US Army, Ballistic Res. Lab., Aberdeen Proving Ground, MD

GOODNER, H.W. (1992). Risk analysis of human exposure to process systems. *Hazard Identification and Risk Analysis, Human Factors and Human Reliability in Process Safety*, p. 151

GOODNER, H.W. (1993). New way of quantifying risks. *Chem. Engr*, **100**(10), 114; **100**(11), 140

GOODSON, L.H. and JACOBS, W.B. (1972). A rapid detection system for organo-phosphates in water. *Control of Hazardous Material Spills*, **1**, 129

GOODSTEIN, L.P. (1981). Discriminative display support for process operators. In Rasmussen, J. and Rouse, W.B. (1981), *op. cit.*, p. 433

GOODSTEIN, L.P. (1982). Computer-based operating aids. *Design 82* (Rugby: Instn *Chem. Engrs*)

GOODSTEIN, L.P. (1984). *Information Presentation for Decision Making. Rep. M-2461*. Riso Nat. Lab., Riso, Denmark

GOODSTEIN, L.P., HEDEGARD, J., HOJBERG, K.S. and LIND, M. (1983). The GNP testbed for operator support evaluation (paper). *Enlarged Halden Programme Group Mtg*, Loen, Norway

GOODSTEIN, L.P., PETERSEN, O.M., RASMUSSEN, J. and SKANBORG, P.Z. (1974). *The Operator's Diagnosis Task under Abnormal Operating Conditions in Industrial Process Plant. Rep. Riso-M-1929*. Atom. Energy Res. Estab., Riso, Denmark

GOODWIN, C. (1984). Piloting practical uses for on-line information. *Computing*, Dec. 13, 17

GOODWIN, D. (1978). Noise and the plant engineer. *Plant Engr*, **22**(3), 13; **22**(4), 21

GOODWIN, ROBERT. (2000). *Apollo 13 NASA Mission Report* (Washington D.C.: US National Aeronautics and Space Administration), ISBN 1-896522-55-6

GOODWIN, E.M. and KEMP, J.F. (1980). Collision risks for fixed off-shore structures. *J. Nav.*, **33**, 351

GOODWIN, R. and CHUNG, P.W.H. (1994a). *Intelligent Information System for Design. ESCAPE 4*

GOODWIN, R. and CHUNG, P.W.H. (1994b). Representing design issues. *Inst. Chem. Engrs, Research Event '94* (Rugby)

GOOS, G. (1975). Documentation. In Bauer, F.L. (1975a), *op. cit.*, p. 385

GOOSSENS, L.H.J., COOKE, R.M. and VAN STEEN, J.F.J. (1989). *Expert Opinions in Safety Studies: Final Report*. Delft Univ. of Technol.

GOPALAN, M.N. and NATESAN, J. (1982). Cost benefit analysis of a *1*-server *n*-unit system with general repair and different repair policies. *Reliab. Engng*, **3**(4), 253

GOPALAN, M.N. and VENKATESWARLU, P. (1986a). A reliability physics model for one-unit and two-unit standby systems with single repair facility. *Reliab. Engng*, **14**(1), 37

GOPALAN, M.N. and VENKATESWARLU, P. (1986b). Reliability and safety factor of two-unit redundant systems. *Reliab. Engng*, **14**(3), 183

GORBELL, G.L. (1982). Loss prevention evaluations for chemical operations. In Fawcett, H.H. and Wood, W.S. (1982), *op. cit.*, p. 711

GORCZYNSKI, E.W. (1985). Making the most of your plant data. *PSE 85* (Rugby: Instn Chem. Engrs), p. 101

GORDEN, R.L. (1987). *Interviewing: Strategy, Techniques and Tactics*, 4th ed. (Chicago, IL: Dorsey Press)

GORDON, B.A. and POPE, K.S. (1994). Performing due diligence environment audits on US offshore properties. *Offshore*, Jun., 33

GORDON, E., MURPHY, R.E. and DEAN, P.D. (1985). Leak, rupture detection can be improved. *Oil Gas J.*, **83**(Nov. 4), 84

GORDON, H.S. and MASCONE, C.F. (1988). How safe is your plant? *Chem. Engng*, **95**(5), 52

GORDON, J.E. (1949). The epidemiology of accidents. *Am. J. Pub. Health*, **39**, 504

GORDON, M. (1948). Reaction kinetics of adiabatic system. *Trans. Faraday Soc.*, **44**, 196

GORDON, M.D. (1992). On-line modeling of thermally hazardous batch processes. *Plant/Operations Prog.*, **11**, 102

GORDON, M.D., O'BRIEN, G.J., HENSLER, C.J. and MARCALI, K. (1982). Mathematical modeling in thermal hazards evaluation. *Plant/Operations Prog.*, **1**, 27

GORDON, R.M. (1989). *Piper Alpha Public Inquiry. Transcript, Days 120, 121*

GORDON, S. and MCBRIDE, B.J. (1971). *Computer Program for Calculation of Complex Chemical Equilibrium Compositions, Rocket Performance, Incident and Reflected Shocks, and Chapman–Jouguet Detonations. Rep. SP-273*. Nat. Aeronaut. Space Admin.

GORDON, S. and MCBRIDE, B.J. (1976). *Computer Program for Calculation of Complex Chemical Equilibrium Compositions, Rocket Performance, Incident and Reflected Shocks, and Chapman–Jouguet Detonations SP-273. Interim Revision, Rep. N78-17724. Nat. Aeronaut. Space Admin.*

GORDON, W.E. (1965). Loss: the designer's concern. *Chem. Engng Prog.*, **61**(8), 19

GORE, J.P., EVANS, D.D. and MCCAFFREY, B.J. (1991). Temperature and radiation of diffusion flames with suppression. *Combust. Sci. Technol.*, **77**, 189

GORELOV, G.N. and SOBOLEV, V.A. (1991). Mathematical modeling of critical phenomena in thermal explosion theory. *Combust. Flame*, **87**, 203

GOREV, V.A. and BYSTROV, S.A. (1985). Explosion waves generated by deflagration combustion. *Combust. Explosion Shock Waves*, **20**, 614

GORE WILLSE, P.J. and SMITH, A.C. (1993). Fire walls, the last line of defense. *Chem Engng*, **100**(7), 110

GORING, M.H. and SCHECKER, H.G. (1992). HAZEXPERT – an expert system for supporting hazard identification for process plants. *Loss Prevention and Safety Promotion 7*, vol. 3, 150.1

GORMAN, J.J., MORROW, W.E. and ANHORN, V.J. (1965). Scaleup by advanced methods. *Chem. Engng Prog.*, **61**(6), 57

GORMAN, R.V. (1962). Investigation of hydrogen damage to an ammonia converter. *Ammonia Plant Safety*, **14**, 85

GORSE, E.J. (1989). *Piper Alpha Public Inquiry. Transcript, Day 155*

GORTER, J. and KLIJN, A.J. (1986). Vibration measurements provide condition monitoring of rotating equipment. *Oil Gas J.*, **84**(Jan. 13), 64

GORYS, J. (1990). Transportation of dangerous goods in the Province of Ontario. *Transport. Res. Record 1264* (Transport. Res. Board, Nat. Res. Coun.), p. 57

GOSCHY, B. (1990). *The Design of Buildings to Withstand Abnormal Loading* (London: Butterworths)

GOSDEN, D.W.F., SMITH, M. and ELKINGTON, P. (1982). Staying safe and retaining earnings: a team approach to systems integrity on LPG carriers. *Gastech 81*

GOSLINE, C.A. (1952). Dispersion from short stacks. *Chem. Engng Prog.*, **48**(4), 165

GOSLINE, C.A., FALK, L.L. and HELMERS, E.N. (1955). Evaluation of weather effects. In Magill, P.L., Holden, F.R. and Ackley, C. (1955), *op. cit.*

GOSSETT, J.L. (1982). Stop stress-corrosion cracking. *Chem. Engng*, **89**(Nov. 15), 143

GOSTELOW, J.G. (1983). Chemical development in computer control of production plants. *Chemy Ind.*, May 2, 339

GOTAAS, Y. (1985). Heavy gas dispersion and environmental conditions as revealed by the Thorney Island experiments. *J. Haz. Materials*, **11**(1/3), 399. See also McQuaid, J. (1985), *op. cit.*, p. 399

GÖTTGENS, J., MAUSS, F. and PETERS, N. (1992). Analytic approximations of burning velocities and flame thicknesses of lean hydrogen, methane, ethylene, ethane, acetylene, and propane flames. *Combustion*, **24**, 129

GOUDRAY, K.A. (1988). *Poisoners of the Sea* (London: Zed Books)

GOUDRAY, K.A. (1992). *World of Waste* (London: Zed Books)

GOUGH, M. (1984). Federal research aimed at reducing cancer risks. In Deisler, P.F. (1984), *op. cit.*, p. 67

GOULD, J. (1988). *Graph Theory* (Wokingham: Benjamin/ Cummings)

GOULD, L.A. (1969). *Chemical Process Control* (Reading, MA: Addison-Wesley)

GOULD, R.F. (1966). *Advanced Propellant Chemistry* (Washington, DC: Am. *Chem. Soc*)

GOUSE, S.W. (1966). *An Index to the Two Phase Gas-Liquid Flow Literature* (Cambridge, MA: MIT Press)

GOUY, G. (1879). *Ann. Chim. Phys*, **18**, 27

GOUY, M. (1910). On the constitution of the electric charge at the surface of an electrolyte. *J. Phys. Théor. Appl., Ser. 4*, **9**, 457

GOUY, C. (1917). *Ann. Phys*, **7**, 129

GOVARDHAN, C.P. and MARGOLIN, A.L. (1995). Extremozymes for industry – from nature and by design. *Chem. Ind.* (4 Sep.), 689–93

GOVERNALE, L.J., RUHLIN, W.B. and SILVUS, R.F. (1965). How to handle aluminum alkyl compounds. *Chem. Engng Prog.*, **61**(4), 103

GOVIL, K.K. and AGGARWAL, K.K. (1982). Modelling early, chance and wearout failures by a single failure rate equation. *Reliab. Engng*, **3**(5), 339

GOVIL, K.K. and AGARWAL, K.K. (1983a). Mean residual life function for normal, gamma and lognormal densities. *Reliab. Engng*, **5**, 47

GOVIL, K.K. and AGARWALA, R.A. (1983b). A Weibull type rate process equation for physical failure modelling. *Reliab. Engng*, **6**, 125

GOWAR, R.G. (1978). Modern electric fire detection devices. *Plant Engr*, **22**(7), 20

GOWLAND, R.T. (1995). Applying inherently safer concepts to an acquisition which handles phosgene. *Proc. of the 29th Annu. Loss Prev. Symp.* July 31 – August 2, 1995, Boston, MA, (ed. E.D. Wixom and R.P. Benedetti), Paper No. 1d. (New York: AIChE)

GOYAL, O.P. (1980). Guidelines help combustion engineers. *Hydrocarbon Process.*, **59**(11), 205

GOYAL, O.P. (1990). Guidelines on fluid flow systems. *Hydrocarbon Process.*, **69**(4), 47

GOYAL, R.K. (1993). Practical examples of CPQRA from the petrochemical industries. *Process Safety Environ.*, **71B**, 117

GOYAL, R.K. (1993). FMEA, the alternative process hazard method. *Hydrocarbon Process.*, **72**(5), 95

GOYAL, R.K. and AL-JURASHI, N.M. (1991). Gas dispersion models. *J. Loss Prev. Process Ind.*, **4**, 151

GOYAL, S.K. (1967). Better planning for heater maintenance. *Hydrocarbon Process.*, **46**(9), 164

GOYAL, S.K. (1975). Improve turnaround manpower planning. *Hydrocarbon Process.*, **54**(1), 71

GOYER, R.A. and MEHLMAN, M.A. (1977). *Toxicology of Trace Elements* (Chichester: Wiley)

GOZZO and CARRARO (1973). Peroxidic polymer from the auto-oxidation of tetrafluorethylene. *Prevention of Occupational Risks*, **2**, p. 498

DE GRAAF, E.L. (1989). Chemical exposure index. *Loss Prevention and Safety Promotion 6*, p. 43.1

VAN DE GRAAF, G.C. and VISSER, J.P. (1989). Risk assessment in exploration and production. *Loss Prevention and Safety Promotion 6*, p. 50.1

GRABBE, E.M., RAMO, S. and WOOLDRIDGE, D.E. (1958). *Handbook of Automation, Computation and Control* (New York: Wiley)

GRABOWSKI, G.J. (1958). Explosion protection operating experience. *NFPA Quart.*, **52**(2), 109

GRABOWSKI, G.J. (1964). Theoretical and practical aspects of explosion protection. In Vervalin, C.H. (1964a), *op. cit.*, p. 98

GRABOWSKI, G.J. (1965). Suppression by design. *Chem. Engng Prog.*, **61**(9), 38

GRABOWSKI, G.J. (1968). Progress in industrial explosion protection systems. *Fire J.*, **62**(1), 21

GRACZYK, J. and HANNON, B. (1993). Modernize valve selection. *Chem. Engng Prog.*, **89**(12), 20

GRADET, A. and SHORT, W.L. (1980). Managing hazardous wastes under RCRA. *Chem. Engng*, **87** July 28, 60

GRADON, F. (1973). *Maintenance Engineering Organisation and Management* (Barking: Applied Science)

GRADY, D.E. (1982a). Fragment size prediction in dynamic fragmentation. *Am. Inst. Physics, Proc. Conf. on Shock Waves in Condensed Matter 1981*

GRADY, D.E. (1982b). Local inertia effects in dynamic fragmentation. *J. Appl. Phys.*, **53**, 322

GRAF, F.A. (1967). Safe handling of perchloric acid. *Chem. Engng Prog.*, **62**(10), 109

GRAÈFEN, H., GERISCHER, K. and GRAMBERG, U. (1977). Damage prevention in chemical plants through systematic failure analysis. *Loss Prevention and Safety Promotion 2*, p. 145

GRAFTON, R.W. (1983). Ignition/explosion risks in fluidised dryers – avoidance and containment (practical aspects – a user's viewpoint). *Loss Prevention and Safety Promotion 4*, vol. 3, p. E44

GRAHAM, F. (1970). *Since Silent Spring* (Boston, MA: Houghton Mifflin)

GRAHAM, J.F. (1992). Understand ISO 9000's application and requirements. *Hydrocarbon Process.*, **71**(5), 83

GRAHAM, J.J. (1970). The fluidized bed phthalic anhydride process. *Loss Prevention 4*, p. 31. See also *Chem. Engng Prog., 1970*, **66**(9), 54

GRAHAM, J.W. (1964). Gas leakage in sealed systems. *Chem. Engng*, **71**(May 11), 169

GRAHAM, N. (1979). *Artificial Intelligence* (Blue Ridge Summit, PA: Tab Books)

GRAHAM, P.W.W. (1973). Casualty prevention by routeing of ships *Transport of Hazardous Cargoes*, **3**

GRAILLE, M. (1978). Sound proofing of Gaz de France compressor stations. In United Nations (1978), *op. cit.*, p. 131

GRAM, A. (1968). Developing, building and operating an LNG utility system: is it feasible? *LNG Conf. 1*

GRAMBERG, U., GUNTHER, T. and GRAFEN, H. (1980). The role of materials technology in reliability considerations for chemical plants. *Loss Prevention and Safety Promotion 3*, vol. 2, p. 714

GRANDIDGE, F.E. (1969). Atmospheric conditions. In Handley, W. (1969), *op. cit.*, p. 287

GRANDIDGE, F.E. (1977). Atmospheric conditions. In Handley, W. (1977), *op. cit.*, p. 30

GRANDORI, G., PEROTTI, F. and TAGLIANI, A. (1987). On the attenuation of macroseismic intensity with epicentral distance. In Cermak, A.S. (1987), *op. cit.*, p. 581

GRANEK, K. and HECKENKAMP, F.W. (1981). Spiral-wound gaskets – panacea or problem? *Chem. Engng*, **88**(Mar. 23), 231

GRANGE, S.G.M. (1980). *Report of the Mississauga Accident Inquiry* (Ottawa, Ont.: Transport Canada)

GRANGER, J.E. (1981). Plant-site selection. *Chem. Engng*, **88**(Jun. 15), 88

GRANSTROM, S.A. (1972). The Scandinavian approach to structural safety. *Building Research Estab.*, Symp. on Buildings and the Hazard of Explosion

GRANT, A.F. (1982). Canvey Island (letter). *Chem. Engr, Lond.*, **376**, 27

GRANT, C.C. (1985). Computer-aided Halon 1301 piping calculations. *Fire Safety J.*, **9**, 171

GRANT, C.D. (1979). *Energy Conservation in the Chemical and Process Industries* (London: Instn Chem Engrs)

GRANT, E.L. (1964). *Statistical Quality Control*, 3rd ed. (New York: McGraw-Hill)

GRANT, J. (1971). An economic design method for computer supervision of a large plant. *On Line Computer Methods Relevant to Chemical Engineering* (London: Instn. Chem. Engrs)

GRANT, J. and GRANT, C. (eds) (1987). *Hackh's Chemical Dictionary*, 5th ed. (New York: McGraw-Hill)

GRANT, J. and MARSHALL, V.C. (1976). The potential of power fluidics for plant protection. *Measamt Control*, **9**, 145

GRANT, J. and MARSHALL, V.C. (1977). The potential of power fluidics for plant protection. *Measamt Control*, **10**, 183

GRANT, J. and RIMMER, E. (1980). The vortex control of safety enclosures. In Hughes, D.J. (1980), *op. cit.*, paper 8

GRANT, L.B. (1985). Risk analysis of hazardous materials in oil shale. *J. Haz. Materials*, **10**(2/3), 317

GRANT, M.M. (1989). The use of risk analysis in the formulation of policy for the transport of dangerous goods through tunnels. In Guy, G.B. (1989), *op. cit.*, p. 126

GRANT, M.M. and HARVEY, D.W. (1989). A progress review on the development of FIABEX – an intelligent knowledge based system for automated safety, reliability and maintainability analysis. *Reliability 89*, 4C/2/1

GRANT, P.M. (1989). *Piper Alpha Public Inquiry. Transcript, Days 102, 120, 121*

GRANT, W.M. (1962). *Toxicology of the Eye* (Springfield, IL: C. C. Thomas)

GRANTHAM, S.D. and UNGAR, L.H. (1990). A first principles approach to automated troubleshooting of chemical plants. *Comput. Chem. Engng*, **14**, 783

GRANTHAM, S.D. and UNGAR, L.H. (1991). Comparative analysis of qualitative models when the model changes. *AIChE J.*, **37**, 931

GRANTOM, R.L. (1985). Tube wall temperature monitoring technique. *Chem. Engng Prog.*, **81**(7), 41

GRANTON, R.L. and ROGER, D.J. (1987). Ethylene. *Ullmanns Encyclopaedia of Chemical Technology*, rev. ed., **vol. A10** (Weinheim: VCH), p. 45

GRASSMAN, P. (1971). *Physical Principles of Chemical Engineering.* (Oxford: Pergamon Press)

GRASSO, P. (1979). Methodology in assessing carcinogenic risk. *Chemy Ind.*, Feb. 3, 73

GRATT, L. and MCGRATH, E. (1975a). *LNG Terminal Risk Assessment Study for Los Angeles, California. Report.* Science Applications Int.

GRATT, L. and MCGRATH, E. (1975b). *LNG Terminal Risk Assessment Study for Oxnard, California. Report.* Science Applications Int.

GRATT, L. and MCGRATH, E. (1976). *LNG Terminal Risk Assessment Study for Point Conception, California. Report.* Science Applications Int.

GRATTAN, E. and HOBBS, J.A. (1978). *Injuries to Occupants of Heavy Goods Vehicles. Rep. LR 854.* Transport and Road Res. Lab., Crowthorne

GRAVENSTINE, R. (1989). Air operated diaphragm pumps. *Plant Engr*, Jul./Aug., 20

GRAVES, F.E. (1966). How to correctly select and install bolts and nuts. *Chem. Engng*, **73**, Dec. 5, 162

GRAVES, K.W. (1970). *Fire Fighters Exposure Study. Rep. NTIS AD 722 744.* Cornell Aeronaut. Lab., Ithaca, NY

GRAVINER (COLNBROOK) LTD (1966). *Explosion Protection and Suppression for Industry* (Slough, Middlesex)

GRAY, B.F. (1969a). On the critical conditions in a spherical reacting mass. *Combust. Flame*, **13**, 50

GRAY, B.F. (1969b). A note on thermal explosion theory (letter). *Combust. Flame*, **13**, 97

GRAY, B.F. (1970). The dependence of spontaneous ignition temperature on surface to volume ratio in static systems for fuels showing a negative temperature coefficient. *Combust. Flame*, **14**, 113

GRAY, B.F. (1973–). Critical behaviour in chemically reacting systems. *Combust. Flame*, 1973, **20**, 313, 317; 1975, **24**, 43

GRAY, B.F. and JONES, J.C. (1981). Critical conditions in chemically reacting systems. *Combust. Flame*, **40**, 331

GRAY, B.F. and SCOTT, S.K. (1985). The influence of initial temperature excess on critical conditions for thermal explosion. *Combust. Flame*, **61**, 227

GRAY, B.F. and SHERRINGTON, M.E. (1972). Explosive systems with reactant consumption. *Combust. Flame*, **19**, 435, 445

GRAY, B.F. and WAKE, G.C. (1984). Criticality in the infinite slab and cylinder with surface heat sources. *Combust. Flame*, **55**, 23

GRAY, B.F. and WAKE, G.C. (1988). On the determination of critical ambient temperature and critical initial temperature for thermal ignition. *Combust. Flame*, **71**, 101

GRAY, B.F. and YANG, C.H. (1967). The present theoretical position in explosion theory. *Combustion*, **11**, p. 1057

GRAY, B.F. and YANG, C.H. (1969). On the Lotka Frank–Kamenetskii isothermal theory of cool flames. *Combust. Flame*, **13**, 20, 273

GRAY, B.F., DEWYNNE, J., HOOD, M., WAKE, G.C. and WEBER, R. (1990). Effect of deposition of combustible matter on electric power cables. *Fire Safety J.*, **16**, 459

GRAY, B.F., MERKIN, J.H. and GRIFFITHS, J.F. (1991). The prediction of a practical lower bound for ignition time delays and a method of scaling times-to-ignition in large reactant masses from laboratory data. *Combustion*, **23**, p. 1775

GRAY, B.F., LITTLE, S.G. and WAKE, G.C. (1992). The prediction of a practical lower bound for ignition time delays in large reactant masses from laboratory data – II. *Combustion*, **24**, p. 1785

GRAY, B.F., COPPERSTHWAITE, D.P. and GRIFFITHS, J.F. (1993). A novel, thermal instability in a 'semi-batch' reactor. *Process Safety Prog.*, **12**, 49

GRAY, C.H. (1966). *Laboratory Handbook of Toxic Agents*, 2nd ed. (London: R. Inst. Chem.)

GRAY, D. (1988). A career in loss prevention. *Chem. Engr, Lond.*, **451**, 48

GRAY, D.A., DRYSDALE, D.D. and LEWIS, F.A.S. (1983). Identification of electrical sources of ignition in fires (letter). *Fire Safety J.*, **6**, 147

GRAY, H.L. and LEWIS, T.O. (1967). A confidence interval for the availability ratio. *Technometrics*, **9**, 465

GRAY, J. (1981). Characteristic patterns of an variations in community response to acute chemical emergencies. *J. Haz. Materials*, **4**(4), 357

GRAY, J. and QUARANTELLI, E.L. (1981). Social aspects of acute chemical emergencies. An editorial introduction. *J. Haz. Materials*, **4**(4), 309

GRAY, J.A. (1960). Physical aspects of plant site selection. *Chem. Engng Prog.*, **56**(1), 46

GRAY, J.A. (1973). *The Contribution of Research and Development to Safety in the Gas Industry. Commun. 920.* Instn Gas Engrs, London See also *Instn. Gas Engrs J.*, **13**, 267

GRAY, J.A. (1980). Gasification of carbonaceous feed- stocks: a policy for the future. *Chem. Engng Prog.*, **76**(3), 73

GRAY, J.N.P. and MACDONALD, I.F. (1982). Safety study of part of a dynamic positioning system for a diving-support ship. *Reliab. Engng*, **3**(3), 179

GRAY, J.N.P. (1985). Continuous-time Markov models in the solution of practical reliability problems. *Reliab. Engng*, **11**(4), 233

GRAY, P. and GRIFFITHS, J. (1989). Thermokinetic combustion oscillations as an alternative to thermal explosion. *Combust. Flame*, **78**, 87

GRAY, P. and HARPER, M.J. (1959a). Thermal explosions I — Induction periods and temperature changes before thermal ignition. *Trans Faraday Soc.*, **55**, 581

GRAY, P. and HARPER, M.J. (1959b). The thermal theory of induction periods and ignition delays. *Combustion*, **7**, 425

GRAY, P. and LEE, P.R. (1965). Thermal explosions and the effect of reactant consumption on critical conditions (letter). *Combust. Flame*, **9**, 201

GRAY, P. and LEE, P.R. (1967a). Studies on criticality: temperature profiles in explosive systems and criteria for criticality in thermal explosions. *Combustion*, **11**, 1123

GRAY, P. and LEE, P.R. (1967b). Thermal explosion theory. *Oxid. Combust. Revs.*, **2**, 1

GRAY, P. and SHERRINGTON, M.E. (1977). Self-heating chemical kinetics and spontaneously unstable systems. *Chemical Soc., London, Specialist Periodical Reports*, vol. 2, Ch. 8

GRAY, P. and WAKE, G.C. (1990). The ignition of hygroscopic combustible materials by water. *Combust. Flame*, **79**, 2

GRAY, P., LEE, P.R. and MACDONALD, J.A. (1969). Thermal explosion in the sphere and in the hollow spherical shell heated at the inner face. *Combust. Flame*, **13**, 461

GRAY, P.P. (1991). Chemistry and combustion. *Combustion*, **23**, p. 1

GRAY, R.C. and COPPELL, D.L. (1975). Intensive care of patients with bomb blast and gunshot injuries. *Br. Med. J.*, Mar. 1, 502

GRAY, T.G.F. (1977). Convenient closed form stress intensity factors for common crack configurations. *Int. J. Frac.*, **13**(1), 66

GRAY, W.A. (1979). The burning of solid materials. *Fire Prev. Sci. Technol.*, **22**, 9

GRAY, W.A. and MULLER, R. (1974). *Engineering Calculations in Radiative Heat Transfer* (Oxford: Pergamon Press)

GRAY, W.A., KILHAM, J.K. and MULLER, R. (1976). *Heat Transfer from Flames* (London: Paul Elek)

GREASON, W.D. (1979). Test standard for the simulation of discharge of static electricity from human bodies. *Electrostatics Soc. Am., Conf. on Electrostatics*

GREATHOUSE, G.A. and WESSEL, C.J. (1954). *Deterioration of Materials* (New York: Reinhold)

GREEN, A. (1998). Process intensification: The key to survival in global markets? *Chem. Ind.*, **5** (Mar. 2), 168–72

GREEN, A.E. (1968). Assessment of sensing channels for high integrity protective systems. *Instrum. Pract.*, **22**(10), 862

GREEN, A.E. (1969–70). Reliability prediction. *Proc. Instn. Mech. Engrs*, 184(pt 3B), 17

GREEN, A.E. (1970). Safety assessment of automatic and manual protective systems for reactors. *Instrum. Pract.*, **24**(2), 109

GREEN, A.E. (1972). Quantitative assessments of system reliability. *Chemical Process Hazards*, **IV**, p. 103

GREEN, A.E. (1973). Safety assessment of systems. *Int. Atom. Energy Agency, Vienna, Symp. on Principles and Standards of Reactor Safety*

GREEN, A.E. (1974a). The reliability assessment of industrial plant systems. *Loss Prevention and Safety Promotion 1*, p. 81

GREEN, A.E. (1974b). Systems Reliability Service and its generic techniques. *IEEE Trans Reliab.*, **R-23**, 140

GREEN, A.E. (1976). A review of system reliability assessment. In Henley, E.J. and Lynn, J.W. (1976), *op. cit.*, p. 183

GREEN, A.E. (1982a). Glossary of terms. In Green, A.E. (1982), *op. cit.*, p. 621

GREEN, A.E. (ed.) (1982b). *High Risk Safety Technology* (Chichester: Wiley)

GREEN, A.E. (1983). *Safety Systems* (Chichester: Wiley)

GREEN, A.E. and BOURNE, A.J. (1965). Reliability considerations for automatic protective systems. *Nucl. Engng*, **10**, 303

GREEN, A.E. and BOURNE, A.J. (1972). *Reliability Technology* (New York: Wiley)

GREEN, A.J. (1988). Transmission of noise through pipe walls: validation of a computer based model. *Proc. Internoise 88*, p. 1535

GREEN, C.C. (1969). Applications of theorem-proving to problem-solving. *Proc. Int. Joint Conf. on Artificial Intelligence* (ed. D. Walker and L.M. Morton) (Washington, DC), p. 219

GREEN, C.H. (1979). Someone out there is trying to kill me: acceptable risk as a problem definition. *Int. Conf. on Environmental Psychology*, Univ. of Surrey

GREEN, C.H. and BROWN, R.A. (1976/77). Comment on 'The use of quantitative risk criteria in hazard analysis' by S.B. Gibson. *J. Occup. Accid.*, **1**, 265

GREEN, C.H. and BROWN, R.A. (1978). Counting lives. *J. Occup. Accid.*, **2**, 55

GREEN, C.H. and BROWN, R.A. (1980a). The acceptability of risk. *Sci. Publ. Policy*, **7**, 307

GREEN, C.H. and BROWN, R.A. (1980b). Through a glass darkly: perceiving perceived risks to health and safety. *Workshop on Perceived Risks*, Eugene, OR

GREEN, C.H., TUNSTALL, S.M. and FORDHAM, M. (1991). The risks from flooding: which risks and whose perception? *Disasters*, **15**, 227

GREEN, G. and FARBER, R.C. (1974). *Introduction to Security* (London: Butterworths)

GREEN, H.L. and LANE, W.R. (1957–). *Particulate Clouds: Dusts, Smokes and Mists*; 1964, 2nd ed. (London: Spon)

GREEN, J.O. (1978). *Self-checking Safety Level Switches.* Endress and Hauser (UK) Ltd, Manchester

GREEN, M.E. (1978). *Safety in Working with Chemicals* (London: Collier Macmillan)

GREEN, R. (1993). Need for 'moving' instrumentation reliability database (letter). *Chem. Engng*, **100**(3), 8

GREEN, R.W. (ed.) (1987). *The Chemical Engineering Guide to Corrosion Control in the Process Industries* (New York: McGraw-Hill)

GREEN, S.D. (1980). The engineering response to health hazards. *Chemical Process Hazards*, **VII**, p. 137

GREEN, S.R. (1952). How to start up and shut down an integral furnace type boiler. *Power Engng*, **56**(May), 77

GREENBERG, D.S. (1967). *The Politics of American Science* (London: Penguin Books)

GREENBERG, H.R. and CRAMER, J.J. (eds) (1991). *Risk Assessment and Risk Management for the Chemical Process Industry* (New York: van Nostrand Reinhold)

GREENBERG, J.B. (1989). The Burke–Schumann diffusion flame revisited – with fuel spray injection. *Combust. Flame*, **77**, 229

GREENBERG, L. (1972). Does Government enforcement help industrial society? *Engineering*, Sep. 854

GREENE, B.N. (1980). Heat exchangers. Materials of construction. *Chem. Engng*, **87**(Oct. 6), 149

DE GREENE, K.B. (1974). Models of man in system in retrospect and prospect. *Ergonomics*, **17**, 437

GREENE, W. *et al.* (eds) (1985a). *The Chemical Engineering Guide to Compressors* (New York: McGraw-Hill)

GREENE, W. *et al.* (1985b). *The Chemical Engineering Guide to Valves* (New York: McGraw-Hill)

GREENFIELD, L.J., SINGLETON, R.P., MCAFFREE, D.R. and COALSON, J.J. (1969). Pulmonary effects of experimental graded aspiration of hydrochloric acid. *Ann Surg.*, **170**, 74

GREENHUT, M.L. (1956). *Plant Location in Theory and Practise* (Chapel Hill, NC: Univ. of North Carolina Press)

GREENLAND, S. (1985). Elementary models for biological interaction. *J. Haz. Materials*, **10**(2/3), 449

GREENO, L. (1984). Environmental auditing. *The Management of Technological Risks* (London: A.D. Little)

GREENSFELDER, R. (1971). Seismological and crustal movement investigations of the San Fernando earthquake. *Calif. Geology*, **24**(4–5), 62

GREENSPAN, H.P. and JOHANSSON, A.V. (1981). An experimental study of flow over an improvised dike. *Stud. Appl. Math.*, **64**, 211

GREENSPAN, H.P. and YOUNG, R.E. (1978). Flow over a containment dike. *J. Fluid Mech.*, **87**, 179

GREENWAY, P. and SPENCE, A. (1985). Numerical calculation of critical points for a slab with partial insulation. *Combust. Flame*, **62**, 141

GREENWELL, M. (1988). *Knowledge Engineering for Expert Systems* (Chichester: Ellis Horwood)

GREENWOOD, D.V. (ed.) (1981). *The Containment and Dispersion of Gases by Water Sprays* (Rugby: Instn Chem. Engrs)

GREENWOOD, M. and WOODS, H.M. (1919). The incidence of industrial accidents with special reference to multiple accidents. *Ind. Fatigue Res. Board, Rep. 4* (London: HM Stationery Office)

GREGORY, A.H. (1978). Factors affecting the selection of pesticides packaging. *Chemy Ind.*, Feb. 18, 107

GREGORY, C.E. (1977). *Explosives*, 3rd ed. (St Lucia, Queensland: Univ. of Queensland Press)

GREGORY, D. (1993). Lubrication. *Plant Engr*, **36**(6), 36

GREGORY, E.N. (1990). Weld repair. *Plant Engr*, Sep./Oct., **29**

GREGORY, J.G. (ed.) (1970). *Materials Information* (London: Soc. Chem. Ind.)

GREGORY, P.H. (1945). *Br Mycological Soc. Trans*, **28**, 26

GREGORY, R. (1981). Transport of chemicals by sea. *Chemy Ind.*, Nov. 7, 751

GREGORY, R.W. (1985). Computerised corrosion monitoring steps towards preventing unexpected corrosion failures. *The Assessment and Control of Major Hazards* (Rugby: Instn Chem. Engrs), p. 13

GREIN, W. (1973). Explosion-resistant construction of explosion-endangered equipment. *Prevention of Occupational Risks*, **2**, p. 276

GREIN, W. and DONAT, C. (1967). Application, selection and sizing of bursting safety devices. *Z. tech. Ü Ë berwaching*, **8**(6), 185

GREMILLION, J.A. (1979). Computer control without the computer. *Ammonia Plant Safety*, **21**, p. 44

GREINER, M.L. (1984). Emergency response procedures for anhydrous ammonia vapor release. *Ammonia Plant Safety*, **24**, 109. See also *Plant/Operations Prog.*, 1984, **3**, 66

GRESH, M.T. (1992). Troubleshooting compressor performance. *Hydrocarbon Process.*, **71**(1), 57

GRESKOVICH, E.J. and SHRIER, A.L. (1971). Pressure drop and hold-up in horizontal slug flow. *AIChE J.*, **17**, 1214

GREWER, T. (1970). Explosion of a stirring vessel during a nitration reaction. *Prevention of Occupational Risks*, **1**

GREWER, T. (1974). Thermal stability of reaction mixtures. *Loss Prevention and Safety Promotion 1*, p. 271

GREWER, T. (1975). Thermal stability of reaction mixtures. *Chem. Ing. Tech.*, **47**, 230

GREWER, T. (1976). Laboratory tests on runaway reactions. *Prevention of Occupational Risks*, **3**, 73

GREWER, T. (1977). Chemical structure and exothermic decomposition. *Loss Prevention and Safety Promotion 2*, p. 105

GREWER, T. (1979). The hazard of undesired reactions of organic materials in chemical plants. *Chem. Ing. Tech.*, **51**, 928

GREWER, T. (1981). Direct assessment and scaling of runaway reaction and decomposition behaviour using oven and Dewar tests. *Runaway Reactions, Unstable Products and Combustible Powders* (London: Instn Chem. Engrs), *paper 2/E*

GREWER, T. (1987a). Adiabatic methods for the investigation of exothermic reactions. *Thermochimica Acta*, **119**, 1

GREWER, T. (1987b). Chemical self-acceleration of exothermic decomposition *Fortschritte der Sicherheitstechnik 1* (Frankfurt-am-Main: DECHEMA), p. 117

GREWER, T. (1991). The influence of chemical structure on exothermic decomposition. *Thermochimica Acta*, **187**, 133

GREWER, T. and DUCH, E. (1983). Thermochemistry of exothermic decomposition reactions. *Loss Prevention and Safety Promotion 4*, vol. 3, p. A1

GREWER, T. and HESSEMER, W. (1987). The exothermic decomposition of nitrocompounds under the influence of additives. *Chem. Ing. Tech.*, **59**, 796

GREWER, T. and KLAIS, O. (1980). Pressure measurements in exothermic decomposition reactions. *Loss Prevention and Safety Promotion 3*, vol. 2, p. 657

GREWER, T. and KLAIS, O. (1987). Pressure rise during homogeneous decomposition and deflagration. *Hazards from Pressure* (Rugby: Instn Chem. Engrs), p. 1

GREWER, T. and KLAIS, O. (1988). *Exotherme Zersetzung* (DuÈusseldorf: VDI)

GREWER, T., KLUSACEK, H., LÖFFLER, U., ROGERS, R.L. and STEINBACH, J. (1989). Determination and assessment of the characteristics values for the evaluation of the thermal safety of chemical processes. *J. Loss Prev. Process Ind.*, **2**(4), 215

GREWER, T., KLUSACEK, H., LOFFLER, U., ROGERS, R.L. and STEINBACH, J. (1989). Determination and assessment of the characteristic values for the evaluation of the thermal safety of chemical processes. *J. of Loss Prev. in the Process Ind.* **2**, 215–23

GRICE, C.S.W. and WHEELER, R.V. (1929). *Firedamp Explosions within Closed Vessels: 'Pressure Piling'. SMRB Paper 43* (London: HM Stationery Office)

VAN GRIEKEN, K.A. (1976). Stress corrosion cracking in storage spheres. *Ammonia Plant Safety*, **18**, p. 15

VAN GRIEKEN, K.A. (1979). Stress corrosion cracked synthesis gas line in an ammonia plant. *Ammonia Plant Safety*, **21**, p. 18

VAN GRIEKEN, K.A. (1989). Factors controlling the nitriding of stainless steels in ammonia synthesis loops. *Ammonia Plant Safety*, **29**, p. 145

GRIER, N.T. (1969). *Tabulations of Configuration Factors between any Two Spheres and their Parts. Rep. SP-3050.* Nat. Aeronaut. Space Admin.

GRIEVE, E.C. (1989). *Piper Alpha Public Inquiry. Transcript, Days 44, 45*

GRIFFIN, M.L. and GURRY, F.H. (1994). Case histories of some power and control-based process safety incidents. *Process Safety Prog.*, 13, 41

GRIFFITH, A.A. (1920–21). The phenomenon of rupture and flow in solids. *Phil. Trans R. Soc., Ser. A*, **221**, 163 reactions. *Occupational Risks*, **2**, p. 276

GRIFFITH, S. and KEISTER, R.G. (1973). Butadiene unit explodes. In Vervalin, C.H. (1973a), *op. cit.*, p. 32

GRIFFITH, W.C. (1978). Dust explosions. *Ann. Rev. Fluid Mech.*, **10**, 93

GRIFFITHS, D.J. (1993). Safety management systems. In Melchers, R.E. and Stewart, M.G. (1993), *op. cit.*, p. 207

GRIFFITHS, H., PUGSLEY, SIR A. and SAUNDERS, SIR O. (1968). *Report of the Inquiry into the Collapse of Flats at Ronan Point, Canning Town* (London: HM Stationery Office)

GRIFFITHS, J.F. (1993). Kinetic fundamentals of alkane autoignition at low temperatures. *Combust. Flame*, **93**, 202

GRIFFITHS, J.F. and KORDYLEWSKI, W. (1992a). Combustion hazards associated with the hot stacking of processed materials. *Combustion*, **24**, 1777

GRIFFITHS, J.F. and KORDYLEWSKI, W. (1992b). The prediction of spontaneous ignition hazards resulting from the 'hot stacking' of process materials. *Plant/Operations Prog.*, **11**, 77

GRIFFITHS, J.F. and PERCHE, A. (1981). The spontaneous decomposition, oxidation and ignition of ethylene oxide under rapid compression. *Combustion*, **18**, 893

GRIFFITHS, J.F., SKIRROW, G. and TIPPER, C.F.H. (1968–). Studies of stabilised low temperature ignition. *Combust. Flame*, 1968, **12**, 443; 1969, **13**, 195

GRIFFITHS, J.F., GRAY, B.F. and GRAY, P. (1971). Multistage ignition in hydrocarbon combustion: temperature effects and theories of nonisothermal combustion. *Combustion*, **13**, p. 239

GRIFFITHS, J.F., HASKO, S.M. and TONG, A.W. (1985). Thermal ignition of packed particulate solids: criticality, under conditions of variable Biot number. *Combust. Flame*, **59**, 1; 1987, **67**, 265

GRIFFITHS, J.F., COPPERSTHWAITE, D., PHILLIPS, C.H., WESTBROOK, C.K. and PITZ, W.J. (1991). Autoignition temperatures of binary mixtures of alkanes in a closed vessel: comparisons between experimental measurements and numerical predictions. *Combustion*, **23**, 1745

GRIFFITHS, J.F., JIAO, Q., SCHREIBER, M., MEYER, J. and KNOCHE, K.F. (1992). Development of thermokinetic models for autoignition in a CFD code: experimental validation and application of the results to rapid compression studies. *Combustion*, **24**, p. 1809

GRIFFITHS, J.F., HALFORD-MAW, P.A. and ROSE, D.J. (1993). Fundamental features of hydrocarbon autoignition in a rapid compression machine. *Combust. Flame*, **95**, 291

GRIFFITHS, J.W. (1992). The use of risk analysis in design. Towards a goal-setting regime: plans and issues. *Health and Safety in the Offshore Oil and Gas Industries* (Rugby: Instn Chem. Engrs)

GRIFFITHS, M. and LINKLATER, D.R. (1984). *Accidents Involving Road Tankers with Flammable Loads. Res. Rep. 1/84.* Traffic Authority of New South Wales, Traffic Accident Res. Unit

GRIFFITHS, R.F. (1980). Acceptability and estimation in risk management. *Sci. Public Policy*, **7**, 154

GRIFFITHS, R.F. (1981a). Ammonia release hazards. In Oyez Publications (1981), *op. cit.*

GRIFFITHS, R.F. (1981b). A background of risks. In Oyez Publications (1981), *op. cit.*

GRIFFITHS, R.F. (ed.) (1981c). Dealing with Risk. *The Planning, Management and Acceptibility of Technological Risk* (Manchester: Manchester Univ. Press)

GRIFFITHS, R.F. (1981d). The nature of risk assessment. In Griffiths, R.F. (1981), *op. cit.*, p. 1

GRIFFITHS, R.F. (1981e). Problems in the use of risk criteria. In Griffiths, R.F. (1981), *op. cit.*, p. 54

GRIFFITHS, R.F. (1984a). Aspects of dispersion of pollutants into the atmosphere. *Sci. Prog. Oxf.*, **69**, 157

GRIFFITHS, R.F. (1984b). Changes in attitudes towards nuclear safety over the last twenty-five years. *J. Instn Nucl. Engrs*, **25**(5), 144

GRIFFITHS, R.F. (1989). The calculation and evaluation of risk estimates. In Cox, R.A. (1989), *op. cit.*, p. 201

GRIFFITHS, R.F. (1991a). Research requirements in major hazards risk analysis. *Risk Analysis*, **11**, 399

GRIFFITHS, R.F. (1991b). The use of probit expressions in the assessment of acute population impact of toxic releases. *J. Loss Prev. Process Ind.*, **4**(1), 49

GRIFFITHS, R.F. (1994a). Discounting of delayed versus early mortality in societal risk criteria. *J. Loss Prev. Process Ind.*, **7**, 432

GRIFFITHS, R.F. (1994b). Errors in the use of the Briggs parameterisation for atmospheric depression coefficients. *Atoms. Environ.*, **28**, 2861

GRIFFITHS, R.F. and FRYER, L.S. (1988). A comparison between dense-gas and passive tracer dispersion estimates for near-source toxic effects of chlorine releases. *J. Haz. Materials*, **19**(2), 169

GRIFFITHS, R.F. and HARPER, A.S. (1985). A speculation on the importance of concentration fluctuations in the estimation of toxic response to irritant gases. *J. Haz. Materials*, **11**(1/3), 369. See also in McQuaid, J. (1985), *op. cit.*, p. 369

GRIFFITHS, R.F. and KAISER, G.D. (1982). Production of dense gas mixtures from ammonia releases – a review. *J. Haz. Materials*, **6**(1/2), 197. See also in Britter, R.E. and Griffiths, R.F. (1982), *op. cit.*, p. 197

GRIFFITHS, R.F. and MEGSON, L.C. (1982). Uncertainties in modelling the consequences of toxic dense gas releases. *Seminar on Toxic and Flammable Clouds – Hazards and Protection* (London: Instn Mech. Engrs)

GRIFFITHS, R.F. and MEGSON, L.C. (1984). The effect of uncertainties in human toxic response on hazard range estimation of ammonia and chlorine. *Atmos. Environ.*, **18**, 1195

GRIFFITHS, R.F. and SMITH, L.M. (1990). Development of a vegetation-damage indicator as a means of post-accident investigation for chlorine releases. *J. Haz. Materials*, **23**(2), 137

GRIFFITHS, W.A.D. and WILKINSON, D.S. (1985). *Essentials of Industrial Dermatology* (Oxford: Blackwell)

GRIFFON, M. (1980). A method for analysing incidents due to human errors on nuclear installations. *Reliab. Engng*, **1**(2), 83

GRIGORIEV, V.S., POLYAKOV, V.S. and ARTEMEV, A.G. (1990). Some aspects of localization and elimination of accidental emissions of toxic substances. *Zh. Vses. Khim. O-va im D.I. Mendeleeva*, **35**(4), 463

GRIGORIU, M. (ed.) (1984). *Risk, Structural Engineering and Human Error* (Waterloo, Ont.: Univ. of Waterloo Press)

GRILC, V., GOLOB, J. and MODIC, R. (1984). Drop coalescence in liquid/liquid dispersions by flow through glass fibre beds. *Trans. Instn Chem. Engrs*, **62**, 48

GRILLO, P. and QURESHI, A.R. (1988). Aspects of reliability design of hydrocarbon offshore platforms. *Reliab. Engng System Safety*, **20**, 277

GRIMALDI, J.V. (1960). Appraising safety effectiveness. *J. Am. Soc. Safety Engng*, Nov., 57

GRIMALDI, J.V. (1965). Measuring safety effectiveness. In Fawcett, H.H. and Wood, W.S. (1965), *op. cit.*, p. 537

GRIMALDI, J.V. (1970). The measurement of safety engineering performance. *J. Safety Res.*, Sep.

GRIME, G. (1987). *Handbook of Road Safety Research* (London: Butterworths)

GRIMM, W.E.H. (1974). Case studies of fires and explosions in refineries and petrochemical plants. *Loss Prevention and Safety Promotion 1*, p. 355

GRIMMOND, C.S.B., CLEUGH, H.A. and OKE, T.R. (1991). An objective urban heat storage model and its comparison with other schemes. *Atmos. Environ.*, **25B**, 311

GRINT, G.C. (1989). Some fire and explosion hazard ranges. In Lees, F.P. and Ang, M.L. (1989), *op. cit.*, p. 236. (Appendix to paper by Fullam, B.W. (1989), *op. cit.*)

GRINT, G. and PURDY, G. (1990). Sulphur trioxide and oleum hazard assessment. *J. Loss Prev. Process Ind.*, **3**(1), 177

GRIST, J. (1985). Fine chemical plant: standardising on diversity. *Process Engng*, May, 19

GRISWOLD, J. (1946). *Fuels, Combustion and Furnaces* (New York: McGraw-Hill)

GRIWATZ, G.H. and BRUGGER, J.E. (1978). Activated carbon regeneration mobile field-use system. *Control of Hazardous Material Spills 4*, p. 350

GRODZOVSKY, G.L. and KUKANOV, F.A. (1965). Motions of fragments of vessels bursting in a vacuum. *Inzhenemyi Zhumal*, **5**(2), 352

GROFF, E.G. (1982). The cellular nature of confined spherical propane–air flames. *Combust. Flame*, **48**, 51

GROGAN, G.E. (1989). *Piper Alpha Public Inquiry. Transcript, Day 126*

GROGGINS, P.H. (ed.) (1952). *Unit Processes in Organic Synthesis* (New York: McGraw-Hill)

GROHMANN, M. (1979). Extend pump application with inducers. *Hydrocarbon Process.*, **58**(12), 121

GROLLIER-BARON, R. (1986). Working party 'Loss Prevention and Safety' *Loss Prevention and Safety Promotion 5*, vol. III, p. 31

GROLLIER-BARON, R. (1989). Techniques for improving human behaviour with respect to safety and quality. *Loss Prevention and Safety Promotion 6*, p. 30.1

GROLLIER BARON, R. (1992a). Hazards caused by trace substances. *Hazard Identification and Risk Analysis, Human Factors and Human Reliability in Process Safety*, p. 107

GROLLIER-BARON, R. (1992a). Hazards caused by trace substances. *Loss Prevention and Safety Promotion 7*, vol. 2, 63.1

GROLLIER-BARON, R. (1992c). Industrial hazards and external communication (lecture). *Loss Prevention and Safety Promotion 7*, vol. 1

GROLMES, M.A. (1983). A simple approach to transient two-phase level swell. In Veziroglu, T.N. (1983), *op. cit.*, p. 301

GROLMES, M.A. and EPSTEIN, M., (1985). Vapor-liquid disengagement in atmospheric liquid storage vessels subjected to external heat source. *Plant/Operations Prog.*, **4**, 200

GROLMES, M.A. and FAUSKE, H.K. (1983). An evaluation of incomplete vapor phase separation in Freon 12 top vented depressurization experiments. In Veziroglu, T.N. (1983), *op. cit.*, p. 273

GROLMES, M.A. and LEUNG, J.C. (1983). Scaling considerations for two-phase critical flow. In Veziroglu, T.N. (1983), *op. cit.*, p. 259

GROLMES, M.A. and LEUNG, J.C. (1984). Scaling considerations for two-phase critical flow. In *Multiphase Flow and Heat Transfer III A: Fundamentals* (ed. T.N. Veziroglu and A.E. Bergles) (Amsterdam: Elsevier), p. 549

GROLMES, M.A. and LEUNG, J.C. (1985). Code method for evaluating integrated relief phenomena. *Chem. Engng Prog.*, **81**(8), 47

GROLMES, M.A., LEUNG, J.C. and FAUSKE, H.K. (1983). Transient two-phase flow discharge of flashing liquids following a break in a long transmission pipe line. In Veziroglu, T.N. (1983), *op. cit.*, p. 299

GROLMES, M.A., LEUNG, J.C. and FAUSKE, H.K. (1985). Large-scale experiments of emergency relief systems. *Chem. Engng Prog.*, **81**(8), 57

GROLMES, M.A., LEUNG, J.C. and FAUSKE, H.K. (1989). Reactive systems vent sizing evaluations. *Runaway Reactions*, p. 451

GROLMES, M.A. and YUE, M.H. (1993). Relief vent sizing and location for long tubular reactors. *J. Haz. Materials*, **33**, 261

GRONE, O. and PEDERSEN, P.S. (1987). Large diesel engines using high pressure gas injection technology. *Gastech 86*, p. 302

GRONOW, W.S. and GAUSDEN, R. (1973). Licensing and regulatory control of thermal power reactors in the United Kingdom. *Proc. Symp. on Principles and Standards of Reactor Safety* (Vienna: Int. Atom. Energy Agency), p. 527

DE GROOT, J.J. and HEEMSKERK, A.H. (1989). Venting techniques for thermal explosions. *Loss Prevention and Safety Promotion 6*, p. 74.1

DE GROOT, J.J. and HUPKENS VAN DER ELST, F. (1981). Thermal properties of peroxides. *Runaway Reactions, Unstable Products and Combustible Powders* (London: Instn Chem. Engrs), paper 3/V

DE GROOT, J.J., GROOTHUIZEN, T.M. and VERHOEFF, J. (1980). Relief venting of thermal explosions. *Loss Prevention and Safety Promotion 3*, vol. 3, p. 944

GROOTHUIZEN, T.M. (ed.) (1988). *TNO Prins Mauritz Laboratory: Diverse and Dynamic* (Rijswijk, The Netherlands: TNO)

GROOTHUIZEN, T.M. and PASMAN, H.J. (1975). Explosions in liquid and solids. *Loss Prevention*, **9**, 91

GROOTHUIZEN, T.M. and ROMIJN, J. (1976). Heat radiation from fires of organic peroxides as compared with propellant fires. *J. Haz. Materials*, **1**(3), 191

GROOTHUIZEN, T.M., HARTGERINK, J.W. and PASMAN, H.J. (1974). Phenomenology, test methods and case histories of explosions in liquids and solids. *Loss Prevention and Safety Promotion 1*, p. 239

GROOTHUIZEN, T.M., VERHOEFF, J. and DE GROOT, J.J. (1977). Determination of the effect of a thermal explosion of organic peroxides (homogeneous explosion test). *J. Haz. Materials*, **2**(1), 11

GROSE, V.L. (1987). *Managing Risk* (Englewood Cliffs, NJ: Prentice-Hall)

GROSS, D. (1991). Estimating air leakage through doors for smoke control. *Fire Safety J.*, **17**, 171

GROSS, D. and AMSTER, A.B. (1962). Thermal explosions: adiabatic self-heating of explosives and propellants. *Combustion*, **8**, 728

GROSS, H.G. (1968). Safety regulations for the transport of dangerous goods. *Carriage of Dangerous Goods*, **1**, 79

GROSS, S.S. (1978). Evaluation of foams for mitigating air pollution from hazardous material spills. *Control of Hazardous Material Spills*, 4, 394

GROSSEL, S.S. (1986). Design and sizing of knock-out drums/catchtanks for reactor emergency relief systems. *Plant/Operations Prog.*, **5**, 129

GROSSEL, S.S. (1988). Safety considerations in conveying of bulk solids and powders. *J. Loss Prev. Process Ind.*, **1**(2), 62

GROSSEL, S.S. (1990a). Highly toxic liquids: moving them around the plant. *Chem. Engng*, **97**(4), 110

GROSSEL, S.S. (1990b). An overview of equipment for containment and disposal of emergency relief system effluents. *J. Loss Prev. Process Ind.*, **3**(1), 112

GROSSEL, S.S. (1991). DIERS Users Group awareness letter. *Plant/Operations Prog.*, **10**, J6

GROSSER, G.H. *et al.* (eds) (1964). *The Threat of Impending Disaster: Contributions to the Psychology of Stress* (Cambridge, MA: MIT Press)

GROSSHANDLER, S. (1987a). Predictive maintenance of pressure vessels. *Process Engng*, Aug., **27**

GROSSHANDLER, S. (1987b). Safe and cost-effective design of pressure vessels. *Process Engng*, Oct., **51**

GROSSHANDLER, W.L. and MODAK, A.T. (1981). Radiation from nonhomogeneous combustion products. *Combustion*, **18**, 601

GROSSMAN, A.P. (1975). Find heat exchanger leakage accurately. *Hydrocarbon Process.*, **54**(1), 58

GROSSMAN, K. (1980). *Cover Up: What You are Not Supposed to Know about Nuclear Power* (Sagaponack, NY: Permanent Press)

GROSSMAN, M.A. and BAIN, E.C. (1964). *Principles of Heat Treatment* (New York: Am. Soc. Metals)

GROSSMANN, G. and DEJAEGER, J. (1992). Optimization of the start-up of an ammonia plant using a simulator. *Ammonia Plant Safety*, **32**, 164

GROSSMANN, G. and FROMM, D. (1991). HAZOP-proof ammonia plant: a new way of defining a safe and reliable design. *Plant/Operations Prog.*, **10**, 223

GROSSMANN, I.E. (1989). MINLP optimization strategies and algorithms for process synthesis. *Proc. FOCAPD 89*, Snowmass, CO

GROSSMAN, I.E. and FLOUDAS, C.A. (1987). Active constraint strategy for flexibility analysis in chemical processes. *Comput. Chem. Engng*, **11**, 675

GROSSMANN, I.E., WESTERBERG, A.W. and BIEGLER, L.T. (1987). Retrofit design of processes. *Proc. First Int. Conf. Found. Comput. Aided Process Operation*, Park City, UT

GROTE, S.H. (1967). Calculating pressure-release times. *Chem. Engng*, **74**(Jul. 17), 203

GROTZ, B.J., GOSNELL, J.H. and GRISOLIA, L. (1985). Start-up of new 1,500 metric ton/d purifier ammonia plant. *Ammonia Plant Safety*, **25**, 38

GROVE, J.R. (1966). Quenching diameters of moist carbon monoxide air mixtures. *Combust. Flame*, **10**, 308

GROVE, J.R. (1967). The measurement of quenching diameters and their relation to the flameproof grouping of gases and vapours. *Chemical Process Hazards*, **III**, p. 51

GROVE, J.R. and WALSH, A.D. (1953). The propagation of cool flames. *Combustion*, 4, 429

GROVE, J.R., PATEL, J.C. and WEBSTER, P. (1972). Flammability limits of halogenated hydrocarbons. *Chemical Process Hazards*, **IV**, p. 12

GROVER, P. (1990). Infrared can warn of an impending disaster. *Chem. Engng*, **97**(7), 141

GROVER, S.F. and WILSON, B.G. (1960). The storage and transmission of hazardous liquids. *Chemical Process Hazards*, **I**, p. 102

GROVES, M.J. (ed.) (1978). *Particle Size Analysis* (London: Heyden)

GRUHN, P. (1992a). Applying risk assessment information in the specification of safety control systems. *Hazard Identification and Risk Analysis, Human Factors and Human Reliability in Process Safety*, p. 49

GRUHN, P. (1992b). Use the right PLCs for safety control. *Chem. Engng Prog.*, **88**(1), 59

GRUHN, P. (1993). Safety system performance terms: clearing up the confusion. *Hydrocarbon Process.*, **72**(2), 63

GRUMBLES, T.G. (1990). An overview of exposures and exposure control in the ethylene oxide producer and ethoxylation industries. *Plant/Operations Prog.*, **9**, 87

GRUMER, J. (1955). Early oxidation products of acetaldehyde and propane: interpretation. *Combustion*, **5**, 447

GRUMER, J., COOK, E.B. and KUBALA, T.A. (1959). Considerations pertaining to spherical vessel combustion. *Combust. Flame*, **3**, 437

GRUMER, J., STRASSER, A., KUBALA, T.A. and BURGESS, D.S. (1961). Uncontrolled diffusive burning of some new liquid propellants. *Fire Res. Abs. Revs*, **3**, 159

GRUMSTRUP, B. (1992). Shut out leaks with bellows seals valves. *Chem. Engng*, **99**(4), 124

GRUNDFEST, H., KORR, I.M., MCMILLEN, J.H. and BUTLER, E.G. (1945). *Ballistics of the Penetration of Human Skin by Small Spheres. Missile Casualties Rep. 11*. Cttee on Med. Res. Dev.

GRUNDFOS PUMPS LTD (1984). Pump replacement has advantages over repairs. *Plant Engr*, **28**(4), 16

GRUNDY, R.E. and ROTTMAN, J.W. (1985). The approach to self-similarity of the solutions of the shallow-water equations representing gravity-current releases. *J. Fluid Mech.*, **156**, 39

GRUSCHKA, H.D. (1971). *Gas Dynamic Theory of Detonation* (New York: Gordon & Breach)

GRYNING, S.E., HOLTSLAG, A.A.M., IRWIN, J.S. and SIVERTSEN, B. (1987). Applied dispersion modelling based on meteorological scaling parameters. *Atmos. Environ.*, **21**, 79

GRYNING, S.E. and LARSEN, S.E. (1984). Evaluation of a K-model formulated in terms of Monin–Obukhov similarity with the results from the Prairie Grass experiments. In de Wispelaere, C. (1984), *op. cit.*, p. 659

GRYNING, S.E., VAN ULDEN, A.P. and LARSEN, S.E. (1983). Dispersion from a continuous ground-level source investigated by a K-model. *Q. J. Roy. Met. Soc.*, **109**, 355

GUARRO, S. and OKRENT, D. (1984). The logic flowgraph: a new approach to failure modelling and diagnosis for disturbance analysis applications. *Nucl. Technol.*, **67**

GUBIN, S.A. and SICHEL, M. (1977). Calculation of the detonation velocity of a mixture of liquid fuel droplets and a gaseous oxidiser. *Combust. Sci. Technol.*, **17**, 109

GUBITZ, M. and OTT, K.O. (1983). Quantifying software reliability by a probabilistic model. *Reliab. Engng*, **5**, 157

GUCCIONE, E. (1961). Fire extinguisher outstrips itself – and it theory. *Chem. Engng*, **68**(Aug. 7), 78

GUCCIONE, E. (1963). New look in air separation plants. *Chem. Engng*, **70**(Sep. 16), 150

GUCCIONE, E. (1964). On stream liquefaction plant for Sahara natural gas. *Chem. Engng*, **71**(Oct. 12), 182

GUCCIONE, E. (1965a). Acrylonitriles latest synthesis from propylene. *Chem. Engng*, **72**(Mar. 15), 150

GUCCIONE, E. (1965b). The new look in ammonia plants. *Chem. Engng*, **72**(Nov. 22), 124

GUDIKSEN, P.H., HARVEY, T.F. and LANGE, R. (1989). Chernobyl source term – atmospheric dispersion and dose estimation. *Health Phys.*, 697

GUDIVAKA, V. and KUMAR, A. (1990). An evaluation of four box models for instantaneous dense gas release. *J. Haz. Materials*, **25**(1/2), 237

GUELICH, J. (1956). *Chemical Safety Supervision* (New York: Reinhold)

GUELICH, J. (1965). Eye safety in chemical processing. In Fawcett, H.H. and Wood, W.S. (1965), *op. cit.*, p. 427

GUENKEL, A.A., PRIME, H.C. and RAE, J.M. (1981). Nitrobenzene via an adiabatic reaction. *Chem. Engng*, **88**(Aug. 10), 50

GUENOCHE, H. (1964). Flame propagation in tubes and vessels. In Markstein, G.H. (1964), *op. cit.*, p. 107

GUENTHER, D.A., MCGARRY, D.L., SHEARER, R.P. and MACCLEARY, R.C. (1976). Fire analyses from mechanical properties. *Fire Technol.*, **12**, 173

GUENTHER, D.A., GOODWIN, L.G. and BRININGER, F.P. (1978). Temperature determination of a fire from mechanical properties. *Fire Technol.*, **14**, 61

DE GUERIN, D. (1977). Condition monitoring. *Chart. Mech. Engr*, Mar., 88

GUERRERI, G. (1969). Short-cut distillation design – beware! *Hydrocarbon Process.*, **48**(8), 137

GUGAN, K. (1967). The evaluation and control of hazards due to self-heating. *Chemical Process Hazards*, **III**, p. 8

GUGAN, K. (1974a). Lagging fires: the present position. *Chemical Process Hazards*, **V**, p. 28

GUGAN, K. (1974b). A liquid phase oxidation incident. *Chemical Process Hazards* **V**, p. 169

GUGAN, K. (1975a). *The Explosion and Fire at Nypro (UK) Ltd, Flixborough. Report to Flixborough Court of Inquiry* (London: J.H. Burgoyne and Partners)

GUGAN, K. (1975b). Flixborough – a combustion specialist's standpoint. *Instn Chem. Engrs Symp. on Technical Lessons of Flixborough*, Nottingham (London)

GUGAN, K. (1976). Flixborough – a combustion specialist's viewpoint. *Chem. Engr, Lond.*, **309**, 341

GUGAN, K. (1979). *Unconfined Vapour Cloud Explosions* (Rugby: Instn Chem. Engrs)

GUGAN, K. (1980a). Unconfined vapour cloud explosions (letter). *Chem. Engr, Lond.*, **358**, 491

GUGAN, K. (1980b). Unconfined vapour cloud explosions (letter). *Chem. Engr, Lond.*, **359/360**, 567

GUGGER *et al.* (1954). *Sicherheit im Chemiebetrieb* (Dusseldorf: Econ)

GUGUEN, B. and CHERIFI, M.S. (1974). Incidents encountered with the axial turbocompressor at Skikda. *LNG Conf. 4*

GUIBERT, C., DESBORDES, D., LEYER, J.C., LANGUY, J.P. and NICOLAS, N. (1992). Structure loading in the field of a deflagrative unconfined explosion. *Loss Prevention and Safety Promotion 7*, vol. 3, 114.1

GUIDA, M. (1985). On the confidence limits for Weibull reliability and quantiles: the case of maximum likelihood estimation from small size censored samples. *Reliab. Engng*, **12**(4), 217

GUILFORD, J.P. (1954). *Psychometric Methods*, 2nd ed. (New York: McGraw-Hill)

GUILHEM, J. and RICHARD, L. (1970). Lessons learned from the construction and commissioning of the methane vessels 'Polar Alaska' and 'Arctic Tokyo'. *LNG Conf. 2*

GUILL, A.W. (1973). Protective design for reactor, compressor and high pressure piping. *Safety in High Pressure Polyethylene Plants*, **1**, 49. See also *Chem. Engng Prog.*, 1972, **68**(11), 61

GUILLAUME, M., PTACEK, J., WARREN, P.R. and WEBB, B.C. (1978). *Measurements of Ventilation Rates in Houses with Natural and Mechanical Ventilation Systems. Rep. PD 43/78*. Building Res. Estab.

GUINNESS PUBLISHING COMPANY (1991). *The Guinness Book of Records 1992* (London)

GUINNUP, D. (1992). Non-buoyant puff and plume dispersion modeling issues. *Plant/Operations Prog.*, **11**, 12

GUIOMER, D. (1989). *Piper Alpha Public Inquiry. Transcript, Day 70*

GUIRAIO, C.M. and BACH, G.G. (1979). On the scaling of blast waves form fuel–air explosives. *Proc. Sixth Symp. on Blast Simulation*

GUIRAO, C.M., BACH, G.G. and LEE, J.H. (1976). Pressure waves generated by spherical flames. *Combust. Flame*, **27**, 341

GUIRAO, C.M., KNYSTAUTAS, R., LEE, J.H., BENEDICK, W. and BERMAN, M. (1982). Hydrogen–air detonations. *Combustion*, **19**, 583

GUISE, A.B. (1967). Extinguishment of natural gas pressure fires. *Fire Technol.*, **3**(Aug.), 175

GUISE, A.B. (1975). How to fight natural gas fires. *Hydrocarbon Process.*, **54**(8), 76

GUISE, A.B. (1976). How to fight natural gas fires. *Fire Int.*, **53**, 41

GUISE, A.B. and ZERATSKY, E.D. (1965). Fire extinguishing agents and their application. In Fawcett, H.H. and Wood, W.S. (1965), *op. cit.*, p. 485

GUISE, A.B. and ZERATSKY, E.D. (1982). Fire extinguishing agents and their application. In Fawcett, H.H. and Wood, W.S. (1982), *op. cit.*, p. 357

GULBRANDSEN, T. (1991). Expanding the applicability of risk assessments by using expert system technology. *First ESReDA Seminar on The Use of Expert Systems in Safety Assessment and Management*, London

GULDEMOND, C.P. (1986). The behaviour of denser than air ammonia in the presence of obstacles – wind tunnel experiments. *Ammonia Plant Safety*, **26**, 140. See also *Plant/Operations Prog.*, **1986**, 5, 93

GÜLDER, O.L. (1982). Laminar burning velocities of methanol, ethanol and isooctane-air mixtures. *Combustion*, **19**, p. 275

GÜLDER, O.L. (1989). Influence of hydrocarbon fuel structural constitution and flame temperature on soot formation in laminar diffusion flames. *Combust. Flame*, **78**, 179

GULF PUBLISHING CO. (1977). *Map of Oil and Gas Wells/ Fields and Pipelines Offshore Texas and Louisiana* (Houston, TX)

GÜLICH, J.F. and RÖSCH, A. (1989). Cavitation erosion in centrifugal pumps. *Chem. Engng Prog.*, **85**(11), 68

GULYA, J.A. and MARSHALL, E.G. (1970). New property data for hydrocracker steels. *Hydrocarbon Process.*, **49**(6), 91

GUMBEL, E.J. (1942). On the frequency distribution of extreme values in meteorological data. *Bull. Am. Met. Soc.*, **23**, 95

GUMBEL, E.J. (1958). *Statistics of Extremes* (New York: Univ. of Columbia Press)

GUMLEY, J.R., INVERNIZZO, C.G. and KHALED, M. (1977). Lightning protection – a proven system. *Fire Technol.*, **13**, 114

GUMM, W.G. and TURNER, J.E. (1976). The non-destructive testing spectrum. *Chem. Engng*, **83**(Aug. 16), 65

GUNDERLOY, F.C. and STONE, W.L. (ca. 1980). *Planning Guide and Checklist for Hazardous Materials Contingency Planning. Report.* Fed. Emergency Mngmt Agency, Washington, DC

GUNDERSEN, T. and NAESS, L. (1988). The synthesis of cost optimal heat exchanger networks. *Comput. Chem. Engng*, **12**, 503

GUNDZIK, R.M. (1983). Standardized pilot plant procedures. *Chem. Engng Prog.*, **79**(8), 29

GUNN, A.S. and VESILIND, A. (1986). *Environmental Ethics for Engineers* (Chelsea, MI: Lewis)

GUNN, D.C. and HORTON, R. (1989). *Industrial Boilers* (London: Longmans)

GUNN, D.J. (1970). The optimized layout of a chemical plant by digital computer. *Computer Aided Design*, **2**(3), 111

GUNN, D.J. (1982). Spreading and dispersion of dense vapours and gases. *The Assessment of Major Hazards* (Rugby: Instn Chem. Engrs), p. 57

GUNN, D.J. (1984). Dispersion and containment of dense vapours. *The Protection of Exothermic Reactors and Pressurised Storage Vessels* (Rugby: Instn Chem. Engrs), p. 67

GUNN, J.A. (1920). The action of chlorine, etc., on the bronchi. *J. Med.*, **13**, 121

GUNNING, A. (1987). Nuclear fatalities. *Chem. Engr, Lond.*, **435**, 4

GUNTHER, R. (1969). Transfer processes in flames. *Chem. Ing. Tech.*, **41**(5–6), 315

GUOXIANG L.I. and WANG CHANGYING (1981). Comprehensive study of electric spark sensitivity of ignitable gases and explosive powders. *J. Electrostatics*, **11**, 319

GUPTA, A.K. (1949). The explosive polycondensation of ethylene oxide. *J. Soc. Chem. Ind.*, **68**(Jun.), 179

GUPTA, B.P. and JEFFREY, M.F. (1979). Optimize centrifugal compressor performance. *Hydrocarbon Process.*, **58**(6), 88

GUPTA, J.P. (1986). *Fundamentals of Heat Exchanger and Pressure Vessel Technology* (Washington, DC: Hemisphere)

GUPTA, J.P. (1999). Inherently safer design in chemistry and chemical engineering education. *Green Chemistry Network Newsletter*, **2** (July), 9–10

GUPTA, J.P. (2000). A course on inherently safer design. *J. of Loss Prev. in the Process Ind.* **13**, 63–66

GUPTA, J.P. and BABU, B.S. (1999). A new hazardous waste index. *J. of Hazard. Mater.* **A67**, 1–7

GUPTA, J.P. and EDWARDS, D.W. (2002). Some thoughts on measuring inherent safety. *5th Annu. Symp.: Beyond Regulatory Compliance, Making Safety Second Nature*, October 29–30, 2002, College Station, TX, Track I: Inherent Safety Session (College Station, TX: Mary Kay O'Connor Process Safety Center)

GUPTA, J.P. and EDWARDS, D.W. (2003). A Simple Graphical Method for Measuring Inherent Safety. *J. of Hazard. Mater.* 104, 1–3, 15–30

GUPTA, J.P., HENDERSHOT, D.C. and MANNAN, M.S. (2003). Real Cost of Process Safety – A Clear Case for Inherent Safety. *Trans. IChemE* 81, Part B, (Nov.), 406–13

GUPTA, M.K. (1976). *Development of a Mobile Treatment System for Handling Spilled Hazardous Materials. Rep. EPA-600/2-76-109.* Environ. Prot. Agency, Cincinnati, OH

GUPTA, M.M., BANDLER, W. and KISZKA, J.B. (eds) (1985). *Approximate Reasoning in Expert Systems* (Amsterdam: North Holland)

GUPTA, Y. and DHARMADHIKARI, A. (1985). Analysis of false alarms given by automatic fire detection systems. *Reliab. Engng*, **13**(3), 163

GUPTA, Y., DHARMADHIKARI, A. and WING SING CHOW (1986). On the use of maximum entropy concept in two-unit repairable systems. *Reliab. Engng*, **16**(3), 181

GUPTA, Y.P. (1985). Automatic fire detection systems: aspects of reliability, capacity and selection criteria. *Fire Safety J.*, **8**, 105

GUPTON, P.S. and KREISHER, A.S. (1973). Waste heat boiler failures. *Chem. Engng Prog.*, **69**(1), 47

GUPTON, P.S., STAL, J.A. and STOCKWELL, W.R. (1986). Renovating and updating of a large syngas reformer. *Ammonia Plant Safety*, **26**, p. 33

GURDJIAN, E.S., WEBSTER, J.E. and LISSNER, H.R. (1949). Studies on skull fracture with particular reference to engineering factors. *Am. J. Surgery*, **78**, 736

GURNEY, R.W. (1943). *The Initial Velocities of Fragments from Bombs, Shells and Grenades. Rep BRL 405.* US Army, Ballistics Res Lab., Aberdeen Proving Ground, MD

GURNEY, R.W. (1944). *A New Casualty Criterion. Rep BRL 498.* US Army, Ballistics Res Lab., Aberdeen Proving Ground, MD

GURNEY, R.W. (1946). *Velocity and Range of Fragments from a Bursting Gun. Rep. BRL-563.* US Army, , Ballistics Res Lab., Aberdeen Proving Ground, MD

GURNHAM, C.F. (1963). Control of water pollution. *Chem. Engng*, **70**(Jun. 10), 190

GURNHAM, C.F. (1965). *Industrial Wastewater Control* (New York: Academic Press)

GURNHAM, F.C. (ed.) (1965). *Industrial Wastewater Control* (New York: Academic Press)

GURUZ, A.G., GURUZ, H.K., OSUWAN, S. and STEWARD, F.R. (1974). Aerodynamics of a confined burning jet. *Combust. Sci. Technol.*, **9**, 103

GURVITCH, M.M. and LOWENSTEIN, J.G. (1990). Waste disposal in pilot plants. *Chem. Engng*, **87**(12), 108

GUSCIORA, P.H. and FOSS, A.S. (1989). Detecting and avoiding unstable operation of autothermal reactors. *AIChE J.*, **35**, 881

GUSMAN, S., VON MOLTKE, K., IRWIN, F. and WHITEHEAD, C. (1980). *Public Policy for Chemicals* (Washington, DC: The Conservation Foundation)

GUSTAFERRO, A.H. and LIN, T.D. (1986). Rational design of reinforced concrete members for fire resistance. *Fire Safety J.*, **11**(1), 85

GUSTAFSON, R.M. and MUDAN, K. (1987). Ignition potential distribution for heavy gas plumes. *Vapor Cloud Modeling*, p. 816

GUSTIN, J.L., (1989a). Gas-phase explosions of mixtures of organic compounds with chlorine. *Loss Prevention and Safety Promotion 6*, p. 91.1

GUSTIN, J.L. (1989b). Vent sizing for runaway reactions using the DIERS method. *Loss Prevention and Safety Promotion 6*, p. 75.1

GUSTIN, J.L. (1990). Runaway reactions, their courses and methods to establish safe process conditions. *Proc. Eur. Seminar 14 on Heat Transfer and Major Technological Hazards* (ed. L. Bolle), p. 75/1

GUSTIN, J.L. (1993). Thermal stability screening and reaction calorimetry. Application to runaway reaction hazard assessment and process safety management. *J. Loss Prev. Process Ind.*, **6**, 275

GUSTIN, J.L. and VANDERMARLIERE, P. (1992). The runaway bulk polymerisation of vinyl acetate initiated by a peroxide. Case study – venting vs pressure containment. *Loss Prevention and Safety Promotion 7*, vol. 3, 128.1

GUSTIN, J.L. and VIDAL, P. (1992). The hazard of processing nitric solutions of organic diacids. Experimental investigation of the slow reaction – deflagration – detonation. *Loss Prevention and Safety Promotion 7*, vol. 3, 124.1

GUSTIN, J.L., FILLION, J., TRE'AND, G. and EL BIYAALI, K. (1993). The phenol + formaldehyde runaway reaction. Vent sizing for reactor protection. *J. Loss Prev. Process Ind.*, **6**, 103

GUSTIN, R.E. and NOVACEK, D.A. (1979). Ammonia storage vent accident. *Ammonia Plant Safety*, **21**, 80

GUTENBERG, B. (1956). The energy of earthquakes. *Quart. J. Geol. Soc., Lond.*, **112**, 1

GUTENBERG, B. and RICHTER, C.F. (1954). *Seismicity of the Earth and Associated Phenomena* (Princeton NJ: Princeton Univ. Press)

GUTENBERG, B. and RICHTER, C.F. (1956). Earthquake magnitude, intensity, energy and acceleration. *Bull. Seismol. Soc. Am.*, **46**, 105

GUTENBERG, B. and RICHTER, C.F. (1965). *Seismicity of the Earth and Associated Phenomena* (New York: Hafner)

GUTERMUTH, W. (1992). Applications of computer integrated process engineering in the chemical industry. *Comput. Chem. Engng*, **16**, S15

GUTH, D.C. and CLARK, D.A. (1985). Inspection and repair to two ammonia spheres. *Ammonia Plant Safety*, **25**, 60. See also *Plant/Operations Prog.*, **4**, 16

GUTH, R. (ed.) (1986). *Computer Systems for Process Control* (New York: Plenum Press)

GUTHRIE, K.M. (1968). Data and techniques for preliminary capital cost estimating. *Chem. Engng*, **76**(Mar. 24), 114

GUTHRIE, K.M. (1969). Costs. *Chem. Engng*, **76**(Apr. 14), 201

GUTHRIE, K.M. (1970). Capital and operating costs for 54 chemical processes. *Chem. Engng*, **77**(Jun. 15), 140

GUTHRIE, K.M. (1971). Pump and valve costs. *Chem. Engng*, **78**(Oct. 11), 151

GUTHRIE, K.M. (1974). *Process Plant Estimating, Evaluation and Control* (Solana Beach, CA: Craftsman Book Co. of America)

GUTSCHE, B., LUDWIG, C. and VAHRENHOLT, F. (1979). Failure problems and dispersion processes of heavy gases. *Proc. Tenth Int. Tech. Mtg on Air Pollution Modelling and Its Application* (NATO/CCMS), p. 627

GUTTERMAN, G. (1980). Specify the right heat exchanger. *Hydrocarbon Processing*, **59**(4), 161

GUY, A.G. (1959). *Elements of Physical Metallurgy* (New York: Addison-Wesley)

GUY, A.G. (1976). *Essential of Materials Science* (New York: McGraw-Hill)

GUY, G.B. (ed.) (1989). *Reliability on the Move: Safety and Reliability in Transportation* (London: Elsevier)

GUY, J.L. (1982a). Modeling heat transfer systems. *Chem. Engng*, **89**(May 3), 93

GUY, J.L. (1982b). Modeling process systems via digital computers. *Chem. Engng*, **89**(Mar. 8), 97

GUY, J.L. (1983). Dynamic modeling of continuous-distillation and emergency systems. *Chem. Engng*, Mar. 21, 65

GUYMER, P., KAISER, G.D., MCKELVEY, T.C. and HANNAMAN, G.W. (1987). Probabilistic risk assessment in the CPI. *Chem. Engng. Prog.*, **83**(1), 37

GVOZDEVA, L.G. (1961). *Prykl. Mekh.*, **33**, 731

GWALTNEY, R.C. (1968). *Missile Generation and Protection in Light Water Cooled Power Reactor Plants. Rep. ORNL-NSIC-22*. Oak Ridge Nat. Lab., Oak Ridge, MD

GWYNN, J.E. (1988). Positive pressure testing system (PPTS). *Ammonia Plant Safety* **28**, 22

GWYTHER, A.W. (1982). Mechanical problems of containing flammables. The design engineers' viewpoint. *Flammable Atmospheres and Area Classification, IEE Coll. Dig. 1982/26* (London), p. 5/1

GYARMATHY, G. (1976). Behaviour of wet steam in two-phase steam flow in turbines and separators. In Moore, M.J. and Sieverding, C.H. (1976), *op. cit.*

GYGAX, R. (1989). Explicit and implicit use of scale-up principles for the assessment of thermal runaway risks in chemical production. *Runaway Reactions*, p. 52

GYGAX, R.W. (1988). Chemical reaction engineering for safety. *Chem. Eng. Sci.* **43**(8), 1759–71

GYGAX, R.W. (1990). Scaleup principles for assessing thermal runaway risks. *Chem. Engng Prog.*, **86**(2), 53

GYÖRI, I. (1985). Calculator program for compressible flow in pipes. *Chem. Engng*, **92**(Oct. 28), 55

HAAS, J.H. (1990). Steam traps key to process heating. *Chem. Engng*, **97**(1), 151

HAAS, P.M. and BOTT, T.F. (1982). Criteria for safety related nuclear plant operator actions: a preliminary assessment of available data. *Reliab. Engng*, **3**(1), 59

HAASE, H. (1977). *Electrostatic Hazards. Their Evaluation and Control* (New York: VCH)

HAASL, D.F. (1965). Advanced concepts in fault tree analysis. In Boeing Company (1965), *op. cit.*

HAASTRUP, P. (1983). Design error in the chemical industry. *Loss Prevention and Safety Promotion 4*, vol. 1, p. J15

HAASTRUP, P. (1984). Indoor fatal effects of outdoor toxic gas clouds. *J. Occup. Accidents*, **5**(4), 279

HAASTRUP, P. and BROCKHOFF, L. (1990). Severity of accidents with hazardous materials. A comparison between transportation and fixed installations. *J. Loss Prev. Process Ind.*, **3**(4), 395

HAASTRUP, P. and BROCKHOFF, L.H. (1991). Reliability of accident case histories concerning hazardous chemicals. An analysis of uncertainty and quality aspects. *J. Haz. Materials*, **27**, 339

HAASTRUP, P. and STYHR PETERSEN, H.J. (1992). Hazardous materials: a comparison of accidents from fixed installations and transport. *Loss Prevention and Safety Promotion 7*, vol. 2, 97.1

HAAVERSTAD, T.A. (1992). Development and design of blast relief panels. *Loss Prevention and Safety Promotion 7*, vol. 3, 117.1

HAAVERSTAD, T.A. (1994). Structural response to accidental explosions and fires on offshore process installations. *J. Loss Prev. Process Ind.*, **7**, 310

HABER, F. (1924). *Fünf Vortrage aus den Jahren 1920–1923* (Berlin: Springer)

HABER, F. and WOLFF, H. (1923). Mist explosions. *Z. angew. Chem.*, **36**, 373

HABER, L.F. (1975). *Gas warfare 1915–1945. The legend and the facts.* Stevenston Lecture, Bedford College, Univ. of London

HABER, L.F. (1986). *The Poisonous Cloud. Chemical Warfare in the First World War* (Oxford: Clarendon Press)

HÄBERLE, M. (1971). Hazards and loss prevention in cyclohexane–air oxidation process. *Major Loss Prevention*, p. 54

HÄBERLE, M. (1986). Environmental protection – the German experience. *Chemy. Ind.*, Jan. 6, 12

HABICHT, F.H. (1988). The role of the agency for toxic substances and disease registry under the Superfund Amendments and Reauthorization Act of 1986. *J. Haz. Materials*, **18**(3), 219

HADAS, J. (1974). Case studies on accidents in the process industries, with particular respect to the legal expert methods. *Loss Prevention and Safety Promotion in the Process Industries 1*, p. 347

HADDEN, S.G. (1989a). *A Citizen't to Know* (Boulder, CO: Westview Press)

HADDEN, S.G. (1989b). Institutional barriers to risk communication. *Risk Analysis*, **9**, 301

HADDEN, W.A., RUTHERFORD, W.H. and MERRETT, J.D. (1978). The injuries of terrorist bombing: a study of 1532 consecutive patients. *Br J. Surgery*, **65**, 325

HADDENHORST, H.G. and LORENZEN, H. (1977). Studies of the storage of LNG in salt cavities. *LNG Conf. 5*

HADDOCK, S.R. and WILLIAMS, R.J. (1979). The density of an ammonia cloud in the early stages of atmospheric dispersion. *J. Chem. Technol. Biotechnol.*, **29**, 655

HADDON, W. (1973a). Energy damage and the ten countermeasure strategies. *Hum. Factors*, **15**, 355

HADDON, W. (1973b). On the escape of tigers: an ecological note. In Widner, J.T. (1973), *op. cit.*, p. 120

HADDON, W., SUCHMAN, E.A. and KLEIN, D. (1964). *Accident Research* (New York: Harper & Row)

HADJIPAVLOU, S. and CARR-HILL, G. (1986). *A Review of the Blast Casualties Rules Applicable to UK Houses. Publ. 34/86* (London: Sci. Res. Dev. Branch, Home Office)

HADJISOPHOCLEOUS, G.V., SOUSA, A.C.M.A. and VENART, J.E.S. (1990). A study of the effect of the tank diameter on the thermal stratification in LPG tanks subjected to fire engulfment. *J. Haz. Materials*, **25**(1/2), 19

VON HAEFEN, E. (1992). DUSTEXPERT – an expert system for the assessment of dust explosion hazards. *Loss Prevention and Safety Promotion 7*, vol. 3, 157.1

HAENEN, H.T.M. (1976). Experimental investigation of the relationship between generation and decay of charges on dielectrics. *J. Electrostatics*, **2**, 151

HAESSLER, W. (1986). Theory of fire and explosion control. *Fire Protection Handbook*, 16th ed. (ed. by A.E. Cote and J.M. Linville), p. 4-42

HAESSLER, W. (1986). Dry chemical agents and application systems. In Cote, A.E. and Linville, J.L. (1986), *op. cit.*, p. 19–24

HAESSLER, W.M. (1963). Static electricity generation by carbon dioxide extinguishers. *NFPA Quart.*, **56**(3), 221

HAESSLER, W.M. (1973). Fire extinguishing chemicals. *Chem. Engng*, **80**(Feb. 26), 95

HAESSLER, W.M. (1975). System approach vital to design of early alert detector installation. *Fire Engng*, **128**(Jan.), 92

HÄFELE, W. (1974). Hypotheticality and the new challenges: the pathfinder role of nuclear energy. *Minerva*, **12**, 303

HÄFELE, W. *et al.* (1976). *Fusion and Fast Breeder Reactors* (Laxenberg, Austria: Inst. for Appl. Systems Analysis)

HAGE, K.D. (1964). *Particle Fallout and Dispersion Below 30 km in the Atmosphere. Rep. DC-64-1463*. Sandia Corp., Albuquerque, NM

HAGEL, W.C. and MISKA, K.H. (1980). How to select alloy steels for pressure vessels. *Chem. Engng*, **87**(Jul. 28), 89; (Aug. 25), 105

HAGEN, B.C. and MILKE, J.A. (1994). Software review: PCLs analysis of thermal hazards offshore (PATHOS 2). *Fire Technol.*, **30**, 173

HAGENKÖTTER, M. (1979). Humanisation of work – handicap or incentive to industry? In BASF (1979), *op. cit.*, p. 23

HAGER, C.C., LONG, W.D. and HEMPENSTALL, G.T. (1978). Safe, efficient, economic catalyst changes. In *Ammonia Plant Safety*, **20**, p. 111

HAGGARD, H.W. (1924). Action of irritant gases upon the respiratory tract. *J. Ind. Hyg. Toxicol.*, **5**, 390

HAGGARD, H.W. (1925). The toxicology of hydrogen sulphide. *J. Ind. Hyg.*, **7**, 113

HAGGARD, H.W. and HENDERSON, Y. (1922). The influence of hydrogen sulphide on respiration. *Am. J. Physiol.*, **61**, 289

HÄGGLUND, B. and PERSSON, L.E. (1974). *FoU Brand*, **1**, 2

HÄGGLUND, B. and PERSSON, L.E. (1976). The heat radiation from petroleum fires. *Forsvarets Forskningsanstalt, Stockholm, FOA Rep.*

HÄGGLUND, B., JANSSON, R. and ONNERMARK, B. (1974). *Fire Development in Residential Rooms after Ignition from Nuclear Explosions. FOA Rep. C 20016-D6 (A3)*. Forsvarets Forskningsanstalt, Stockholm, Sweden

HAGLER, D.L. (1960). Controlling compressors for chemical processes. *ISA J.*, **7**(9), 104

HAGON, D.O. (1983). (Letter). *Reliab. Engng*, **6**, 193

HAGON, D.O. (1984). Use of frequency – consequence curves to examine the conclusions of published risk analysis and to define broad criteria for major hazard installations. *Chem. Engng Res. Dev.*, **62**, 381

HAGON, D.O. (1986). Pemex (letter). *Chem. Engr, Lond.*, **421**, 3

HAGSTROM, W.O. (1965). *The Scientific Community* (New York: Basic Books)

HAGUE, W.J. (1992). Mitigation systems. *Plant/Operations Prog.*, **11**, 16

HAGUE, W.J. and PEPE, W. (1990). Flow chamber simulations of aerosol formation and liquid jet break-up for pressurized releases of hydrogen fluoride. *Plant/Operations Prog.*, **9**, 125

HAHN, A. and WILLLIAMS, R. (1970). *The Petrochemical Industry* (New York: McGraw-Hill)

HAHN, A.H. *et al.* (1991). Applying sneak analysis to the identification of human errors of commission. *Reliab. Engng System Safety*, **33**, 289

HAHN, F.P. (1975). Assessing the safety in hot tapping operations. *Loss Prevention*, **9**, p. 20

HAHN, F.P., BROWNLEE, A. and THOMPSON, R.H. (1969). Underpressure connections involving welding. *Pipe Welding Conf.*, London

HAHN, G.E. (1994). Getting the most out of steam. *Chem. Engng*, **101**(1), 80

HAHN, G.J. and SHAPIRO, S.S. (1967). *Statistical Models in Engineering* (New York: Wiley)

HAHN, G.T. and ROSENFIELD, A.R. (1965). Local yielding and extension of a crack under plane stress. *Acta Metall.*, **13**, 293

HAHN, G.T. and ROSENFIELD, A.R. (1967). *Sources of Fracture Toughness: the Relationship between K_{IC} and the Ordinary Tensile Properties of Metals. Report*. Battelle Memorial Inst., Columbus, OH

HAHN, G.T. and ROSENFIELD, A.R. (1968). Sources of Fracture Toughness: the Relationship between K_{IC} and the Ordinary Tensile Properties of Materials. ASTM STP 432 (Philadelphia, PA: Am. Soc. Testing Materials), p. 5

HAHNE, E. and GRIGULL, U. (eds) (1977). *Heat Transfer in Boiling* (Washington, DC: Hemisphere)

HAIG, I.G. and BRIGHT, A.W. (1977). Experimental techniques for the study of surface discharges from hydrocarbon surfaces. *Proc. Third Int. Conf. on Static Electricity* (Paris: Soc. de Chimie Industrielle), p. 28a

HAIGH, R.D. (1969). Mechanical safety and electrical control gear. In Handley, W. (1969), *op. cit.*, p. 170

HAIGH, R.D. (1977). Mechanical safety and electrical control gear. In Handley, W. (1977), *op. cit.*, p. 251

HAINER, R.M. (1955). The application of kinetics to the hazardous behavior of ammonium nitrate. *Combustion*, **5**, p. 224

HAINES, H.W. (1962). Cyclohexane. *Ind. Engng Chem.*, **54**(7), 23

HAINES, H.W. (1963). Polypropylene. *Ind Engng Chem.*, **55**(2), 30

HAINES, N.F. (1983). Progress towards an understanding of reliability in non-destructive evaluation. *Nucl. Energy*, **22**, 349

HAI WANG and FRENKLACH, M. (1991). Detailed reduction of reaction mechanisms for flame modeling. *Combust. Flame*, **87**, 365

HAJEK, J.D. and LUDWIG, E.E. (1960). How to design safe flare stacks. *Petro/Chem. Engng*, **XXXII**(6), C31; (7), C44

HAKANSSON, R. (1977). Ammonia loading line rupture. *Ammonia Plant Safety*, **19**, p. 119

HAKE, H.W. (1915). The action of chlorine on the blood. *The Lancet*, Jul. 10, 86

HAKL, J. (1980). Advanced SEDEX (Sensitive Detector of Exothermic Processes). *Loss Prevention and Safety Promotion 3*, vol. 3, p. 1425

HAKL, J. (1981). SEDEX: (Sensitive Detector of Exothermic Processes) – a versatile instrument for investigating thermal stability. *Runaway Reactions, Unstable Products and Combustible Powders* (Rugby: Instn Chem. Engrs), paper 3/L

HALBIG, J. (1966). Steels for the CPI. *Chem. Engng*, **73**(Dec. 19), 130

HALD, A. (1952a). *Statistical Tables and Formulas* (New York: Wiley)

HALD, A. (1952b). *Statistical Theory with Engineering Applications* (New York: Wiley)

HALDANE, J.S. (1896). Poisoning by gas in sewers. *The Lancet*, Jan. 25, 220

HALDANE, J.S. (1917). The therapeutic administration of oxygen. *Br. Med. J.*, Feb. 10, 181

HALDANE, J.S. (1919). Symptoms, causes and prevention of anoxaemia. *Br. Med. J.*, Jul. 12, 65

HALDANE, J.S. (1925). *Callinicus – A Defence of Chemical Warfare* (London: Kegan Paul)

HALDANE, J.S. and PRIESTLEY, J.G. (1935). *Respiration* (Oxford: Oxford Univ. Press)

HALDANE, J.S., MEAKINS, J.C. and PRIESTLEY, J.G. (1918a). *Investigations of Chronic Cases of Gas Poisoning. Rep. II* (London: Chemical Warfare Med. Cttee, Med. Res. Coun.)

HALDANE, J.S., MEAKINS, J.C. and PRIESTLEY, J.G. (1918b). *The Reflex Restriction of Respiration after Gas Poisoning. Report 5* (London: Chemical Warfare Med. Cttee, Med. Res. Coun.)

HALDANE, J.S., MEAKINS, J.C. and PRIESTLEY, J.G. (1918–1919). The effects of shallow breathing. *J. Physiol.*, **52**, 433

HALE, A.R. (1984). Is safety training worthwhile? *J. Occup. Accid.*, **6**, 17

HALE, A.R. and GLENDON, A.I. (1987). *Individual Behaviour and the Control of Danger* (Amsterdam: Elsevier)

HALE, A.R. and HALE, M. (1970). Accidents in perspective. *Occup. Psychol.*, **44**, 115

HALE, A.R. and HALE, M. (1972). *A Review of the Accident Research Literature* (London: HM Stationery Office)

HALE, A.R., KARCZEWSKI, J., KOORNNEEF, F. and OTTO, E. (1991). IDA: an interactive program for the collection and processing of accident data. In Van der Schaaf, T.W., Lucas, D.A. and Hale A.R. (1991), *op. cit.*, p. 65

HALE, C.C. (1974). Ammonia storage design practice. *Ammonia Plant Safety*, **16**, p. 23

HALE, C.C. (1980). Refrigerated ammonia storage: design and practice. *Ammonia Plant Safety* **21**, p. 61

HALE, C.C. (1982). Weather effects on ammonia storage. *Plant/Operations Prog.*, **1**, 107

HALE, C.C. (1984). 1983 survey of refrigerated ammonia storage in the United States and Canada. *Ammonia Plant Safety*, **24**, 181. See also *Plant/Operations Prog.*, 1984, **3**, 147

HALE, C.C. (1987). Application of mechanical refrigeration for ammonia terminal (discussion contribution). *Ammonia Plant Safety*, **27**, p. 178

HALE, C.C. and LICHTENBERG, W.H. (1980). Ammonia storage terminals safety program. *Ammonia Plant Safety*, **22**, p. 35

HALE, C.C. and LICHTENBERG, W.H. (1988). U.S. anhydrous ammonia distribution system in transition. *Ammonia Plant Safety*, **28**, 76

HALE, C.C. and LICHTENBERG, W.H. (1990). Survey of refrigerated storage in the US and Canada. *Ammonia Plant Safety*, **30**, p. 225

HALE, K. (1985). Using optical-fibre sensors to detect cracks. *Process Engng*, Aug., 27

HALE, P.W. (1986). Stress testing and reliability. *Reliab. Engng*, **14**(4), 247

HALE, P.W. (1987). The one-parameter Weibull distribution as a means to reduce sample size – a cautionary tale. *Reliab. Engng*, **17**(2), 89

HALES, J.M. (1976). Atmospheric transformations of pollutants. In Haugen, D.A. (1976), *op. cit.*, p. 228

HALITSKY, J. (1963). Gas diffusion near buildings. *ASHRAE Trans.*, **69**, 464

HALITSKY, J. (1968). Gas diffusion near buildings. In Slade, D.H. (1968), *op. cit.*, p. 221

HALITSKY, J. (1977). Wake and dispersion models for the EBR-II building complex. *Atmos. Environ.*, **11**, 577

HALITSKY, J. and WOODARD, K. (1974). Atmospheric diffusion experiments at a nuclear power plant site under light wind inversion conditions. *First Symp. on Atmospheric Diffusion and Air Pollution* (Boston, MA: Am. Met. Soc.), p. 172

HALL, A. and BURSTOW, D.J. (1980). Risk of ignition of flammable gases and vapours by radio transmission. *Electrotechnology*, **8**, 12

HALL, A. and LOVELAND, R.J. (1980). Radio frequency ignition hazards. *Radio storage and Modulation Techniques, IEE Digest 40* (London: Instn Elec. Engrs), p. 58

HALL, A.R. (ca. 1972). *Pool Burning – A Review. Tech. Rep. 72/11*. Rocket Propulsion Estab., Westcott

HALL, A.R. (1973). Pool burning. *Oxidation Combust. Revs*, **6**, 169

HALL, D.J. (1979a). Comparison of a wind tunnel model of a denser-than-air gas spill with field experiments. *Proc. Tenth Int. Tech. Mtg on Air Pollution Modelling and Its Application* (NATO/CCMS), p. 701

HALL, D.J. (1979b). *Further Experiments on a Model of an Escape of Heavy Gas. Rep. LR 312 (AP)*. Warren Spring Lab., Stevenage

HALL, D.J. and WATERS, R.A. (1985). Wind tunnel model comparisons with the Thorney Island dense gas release field trials. *J. Haz. Materials*, **11**(1/3), 209. See also in McQuaid, J. (1985), *op. cit.*, p. 209

HALL, D.J. and WATERS, R.A. (1989). *Investigation of Two Features of Continuously Released Heavy Gas Plumes. Rep. LR 707(PA)M.* Warren Spring Lab., Stevenage, Hertfordshire

HALL, D.J., BARRETT, C.F. and RALPH, M.O. (1976). *Experiments on a Model of an Escape of Heavy Gas. Rep. LP 217 (AP).* Warren Spring Lab., Stevenage

HALL, D.J., HOLLIS, E.J. and ISHAQ, H. (1982). *A Wind Tunnel Model of the Porton Dense Gas Spill Field Trials. Rep. 394.* Warren Spring Lab., Stevenage

HALL, D.J., HOLLIS, E.J. and ISHAQ, H. (1984). A wind tunnel model of the Porton dense gas spill field trials. In Ooms, G. and Tennekes, H. (1984), *op. cit.*, p. 189

HALL, D.J., KUKADIA, V., UPTON, S.L., MARSLAND, G.A. and EMMOTT, M.A. *Variability in Instantaneously Released Heavy Gas Clouds Dispersing Over Fences – Some Wind Tunnel Model Experiments. Rep. LR 805 (PA).* Warren Spring Lab., Stevenage, Hertfordshire

HALL, D.J., WATERS, R.A., MARSLAND, G.W., UPTON, S.L. and EMMOTT, M.A. (1991). *Repeat Variability in Instantaneously Released Heavy Gas Clouds – Some Wind Tunnel Model Experiments. Rep. LR 804 (PA).* Warren Spring Lab., Stevenage, Hertfordshire, Rep

HALL, J.R. (1985). *Intrinsic Safety in British Coal Mines* (London: Marylebone Press)

HALL, J.R. (1986). Fire risk analysis. *Fire Protection Handbook*, 16th ed. (ed. A.E. Cote and J.M. Linville), p. 21–10

HALL, J.R. and SEKIZAWA, A. (1991). Fire risk analysis: general conceptual framework for describing models. *Fire Technol.*, **27**, 33

HALL, M.H. (1969). Liaison with the hospital emergency department. In Handley, W. (1969), *op. cit.*, p. 361

HALL, M.H. (1977). Liaison with the hospital emergency department. In Handley, W. (1977), *op. cit.*, p. 398

HALL, N. (1994). Chemists clean up synthesis with one-pot reactions. *Science* 266(Oct 7), 32–34

HALL, P.A.V. (1991) Building in quality through the use of software tools. In Ince, D. (1991a), *op. cit.* p. 80

HALL, R.E., FRAGOLA, J.R. and WREATHALL, J. (1982). *Post Event Human Decision Errors: Operator Action Trees/Time Reliability Correlation. Rep. NUREG/CR-3010.* Nucl. Regul. Comm., Washington, DC

HALL, R.H., HAIGH, D.H., MARTIN, R.G. and WAMPFLER, D.J. (1980). Use of absorbent (imbibing) polymers in spill control. *Loss Prevention*, **13**, p. 170

HALL, R.S., MATLEY, J. and MCNAUGHTON, K.J. (1982). Current costs of process equipment. *Chem. Engng*, **89**(Apr. 5), 80

HALL, S.F., MARTIN, D. and MACKENZIE, J. (1982). UKAEA work on the interaction of vapour cloud explosions with structures. In Lee, J.H.S. and Guirao, C.M. (1982), *op. cit.*, p. 475

HALL, S.M. (1993). Size and design relief headers. *Chem. Engng Prog.*, **89**(3), 117

HALL, T. (1986). *Nuclear Politics* (Harmondsworth: Penguin)

HALL, W.J., KIHARA, H., SOETE, W. and WELLS, A.A. (1967). *Brittle Fracture of Welded Plate* (Englewood Cliffs, NJ: Prentice-Hall)

HALLAM, J.L. (1983). Suction speed is important in pump failure rate. *Oil Gas. J.*, **81**(Jan. 17), 114

HALLAM, J.L. (1992). Commissioning compressors with gas seals. *Hydrocarbon Process.*, **71**(7), 41

HALLAM, M.J. (1990). How the insurance industry views risk and some common risk improvement recommendations. *Safety and Loss Prevention in the Chemical and Oil Processing Industries*, p. 87

HALLAN, T. (1994). In-service inspection of storage tanks: a new non-destructive evaluation method. *Process Safety Prog.*, **13**, 101

HALLEY, P.D. (1965). Toxicity in design and production. *Chem. Engng Prog.*, **61**(2), 59

HALLEY, P.D. (1977). OSHA standards impact refining. *Hydrocarbon Process.*, **56**(5), 137

HALLIDAY, B.R. and DEVEREUX, I.F. (1984). Managing reliability growth in practice. *Reliab. Engng*, **9**(2), 81

HALLIDAY, J. (1969). Manual handling. In Handley, W. (1969), *op. cit.*, p. 224

HALLIDAY, J. (1977). Manual handling. In Handley, W. (1969), *op. cit.*, p. 224

HALLING, J. (ed.) (1975). *Principles of Tribology* (London: Macmillan)

HALLOCK, D.C., FARBER, R.M. and DAVIS, C.C. (1972). Compressors and drivers for LNG plants. *Chem. Engng Prog.*, **68**(9), 77

HALM, R.L. (1979). Low ignition energies. *Chem. Engng News*, **57**(33), 13, 2

HALMSHAW, R. (1971). *Industrial Radiology Techniques* (London: Wykeham Publs)

HALPERN, G.S., NYCE, D. and WRENN, C. (1986). Inerting for safety. *Am. Inst. Chem. Engrs, Symp on Loss Prevention*

HALPERN, S. (1978). *The Assurance Sciences* (Englewood Cliffs, NJ: Prentice-Hall)

HALSTEAD, M.P., PROTHERO, A. and QUINN, C.P. (1971). A mathematical model of the cool flame oxidation of acetaldehyde. *Proc. R. Soc., Ser. A*, **322**, 377

HALSTEAD, M.P., PROTHERO, A. and QUINN, C.P. (1973). Modeling the ignition and cool flame limits of acetaldehyde oxidation. *Combust. Flame*, **20**, 211

HALSTEAD, M.P., PYE, D.B. and QUINN, C.P. (1974). Laminar burning velocities and weak flammability limits under engine-like conditions. *Combust. Flame*, **22**, 89

HALSTEAD, M.P., KIRSCH, L.J., PROTHERO, A. and QUINN, C.P. (1975). A mathematical model for hydrocarbon ignition at high pressures. *Proc. R. Soc., Ser. A*, **346**, 515

HALSTEAD, M.P., KIRSCH, L.J. and QUINN, C.P. (1977). The autoignition of hydrocarbon fuels at high temperatures and pressures – fitting a mathematical model. *Combust. Flame*, **30**, 45

HALTINER, G. and MARTIN, F. (1957). *Dynamical and Physical Meteorology* (New York: McGraw-Hill)

HALVORSEN, F.H. (1975). Inerting could prevent some tanker explosions. *Loss Prevention*, **9**, p. 76

HALVORSEN, F.H. (1978). The Coast Guard's new hazardous chemicals training course. *Control of Hazardous Materials Spills 4*, p. 455

HAM, J.M. and GANSEVOORT, J. (1992). Risk analysis and emergency management of ammonia installations. *Ammonia Plant Safety*, **32**, 86

HAMBLY, E.C. (1991). Please don't shoot the messenger, unless . . . The Fellowship of Engineering (1991b), *op. cit.*, p. 7

HAMILTON, A. (1925). *Industrial Poisons in the United States* (New York)

HAMILTON, B.A. (1986). On-line access to chemical standards information. *Chem. Engng Prog.*, **82**(1), 7

HAMILTON, D.C. and MORGAN, W.R. (1952). *Radiant-inter-change Configuration Factors. Tech. Note 2836.* Nat. Aeronaut. and Space Admin.

HAMILTON, P. (1967). *Espionage and Subversion in an Industrial Society* (London: Hutchison)

HAMILTON, T.B. (1984). Safety in combustion of unconventional fuels. *Plant/Operations Prog.*, **3**, 86

HAMINS, A., KLASSEN, M., GORE, J. and KASHIWAGI, T. (1991). Estimate of flame radiance via a single location measurement in liquid pool fires. *Combust. Flame*, **86**, 223

HAMM, G.L. and SCHWARTZ, R.G. (1993). Issues and strategies in risk decision making. *Process Safety Management*, p. 351

HAMM, H.W. (1964). Sealing methods for bolted and screwed connections. *Chem. Engng*, **71**(Aug. 3), 122

HAMM, W. (1984). Safety and loss prevention in the food industry. *Chem. Engr, Lond.*, **406**, 22

HAMMAN, H. and PARROTT, S. (1987). *Mayday at Chernobyl* (London: New English Library)

HAMMER, G. *et al.* (1991). Natural gas. *Ullmanns Encyclopaedia of Chemical Technology*, rev. ed., vol. A17 (Weinheim: VCH), p. 73

HAMMER, W. (1972). *Handbook of System Product Safety* (Englewood Cliffs, NJ: Prentice-Hall)

HAMMERSLEY, J.M. and HANDSCOMB, D.C. (1964). *Monte Carlo Methods* (New York: Wiley)

HAMMETT, J.L. (1980). CRT based systems: what's in it for the operator? *Hydrocarbon Process.*, **59**(7), 135

HAMMINK, M.W.J. and WESTEN, P.C. (1986). Corrosion and erosion of steel in liquid chlorine at different conditions of velocity, water content and temperature. In Wall, K. (1986), *op. cit.*, p. 71

HAMMITT, J.K. and REUTER, P. (1989). Illegal hazardous-water disposal and enforcement in the United States: a preliminary assessment. *J. Haz. Materials*, **22**(1), 101

HAMMOCK, A.A. (1979). Operate large plants safely. *Hydrocarbon Process.*, **58**(4), 263

HAMMOND, C.B. (1961). Explosion suppression: new safety tool. *Chem. Engng*, **68**(Dec. 25), 85

HAMMOND, K.R. and ADELMAN, L. (1976). Science, values and human judgment. *Science*, **194**(Oct. 22), 389

HAMMOND, K.R., MCCLELLAND, G.H. and MUMPOWER, J. (1980). *Human Judgment and Decision Making* (New York: Hemisphere Frederick A. Praeger)

HAMPEL, B. and LUTHER, H. (1957). Effect of impurities on the electrostatic charging of flowing hydrocarbons. *Chem. Ing. Tech.*, **29**, 323

HAMPSON, T.R. (1976). Chemical leak at a bulk terminals tank farm. *Control of Hazardous Material Spills 3*, p. 214

HAMZA, R. and GOLAY, M.W. (1986). Modeling of moist plumes in turbulent, stratified atmospheres. *Atmos. Environ.*, **20**, 9

HANAUER, S.H. and MORRIS, P. (1971). Technical safety issues for large nuclear power plants – risk analysis. *Proc. Fourth Int. Conf. on Peaceful Uses of Atomic Energy*, III, p. 151

HANCOCK, B.M. (1983). Safety and loss prevention – the IchemE's involvement. *Chem. Engr, Lond.*, **392**, 99

HANCOCK, B.M. (1986). Human factors and systems failure: case study of the fire and explosion at Chemstar. *Loss Prevention and Safety Promotion 5*, p. 24–1

HANCOCK, B.M. (1989). Learning from accidents. *Loss Prevention and Safety Promotion 6*, p. 36.1

HANCOCK, B.M. (1990a). Preventing pipe failures. *Safety and Loss Prevention in the Chemical and Oil Processing Industries*, p. 589

HANCOCK, B.M. (1990b). Safer piping: awareness training for the process industries. *Plant/Operations Prog.*, **9**, 114

HANCOCK, B.M. (1992). Learning from accident experience – are the old lessons really being applied? *Loss Prevention and Safety Promotion 7*, vol. 1, 25–1

HANCOCK, E.G. (1973). *Propylene and its Industrial Derivatives* (London: Benn)

HANCOCK, E.G. (ed.) (1975). *Benzene and its Industrial Derivatives* (London: Benn)

HANCOCK, E.G. (ed.) (1982). *Toluene, the Xylenes and their Industrial Derivatives* (Amsterdam: Elsevier)

HANCOCK, P. and CLIFTON, T.E. (1976). A corrosion and deposition monitor for industrial plant. *Diagnostic Techniques* (London: Instn *Chem. Engrs*)

HANCOCK, W.P. (1974). How to control pump vibration. *Hydrocarbon Process.*, **53**(3), 107

HAND, T.D. and FORD, A.W. (1978). The feasibility of dredging for bottom recovery of spills of dense, hazardous chemicals. *Control of Hazardous Material Spills 4*, p. 315

HANDLEY, J.R. (1985). Hazard avoidance in the process of pharmaceuticals. In Hoffman, J.M. and Maser, D.C. (1985), *op. cit.*, p. 41

HANDLEY, T.H. and BARTON, C.J. (1973). *Home Ventilation Rates – A Literature Survey. Rep. ORNL-TM4318*. Oak Ridge Nat. Lab., Oak Ridge, TN

HANDLEY, T.T.J. (1980). Operating experience with a Benfield carbon dioxide removal system. *Ammonia Plant Safety*, **21**, p. 149

HANDLEY, W. (ed.) (1969). *Industrial Safety Handbook* (London: McGraw-Hill).

HANDLEY, W. (ed.) (1977). *Industrial Safety Handbook, 2nd ed.* (London: McGraw-Hill). (Many of the articles in the first edition are rewritten with identical or similar titles in the second edition)

HANDMER, J. and PENNING-ROWSELL, E.C. (1990). *Hazards and the Communication of Risk* (Aldershot: Gower Press)

HANDS, A.H. (1969). Ergonomics. In Handley, W. (1969), *op. cit.*, p. 307

HANES, R.L. (1978). Design alarm systems to favor critical information. *Control Engng*, **25**(10), 49

HANES, R.L. (1980). Understanding annunciators . . . selec-selection and application factors. *Plant Engng*, May, 1

HANISCH, J.L., HOFFNAGLE, G.F. and WALATA, S. (1993). How to prepare a successful operating permit. *Hydrocarbon Process.*, **72**(4), 95

HANKS, B.J. (1986). WASH 1400, a landmark in safety studies. *Safety and Reliab.*, **6**(1), 12

HANKS, B.J. (1994). Reliability of safety relief valves in the petroleum and gas industries. *Conf. on Valve and Pipeline Reliability*, Engng Dept, Univ of Manchester

HANN, R.W. and JENSEN, P.A. (1975). The interrelationship of system properties, spill quantities, material properties, and toxicity in the evolution of water pollution hazard of transported materials. *Transport of Hazardous Cargoes*, 4

HANNA, R.C. (1979). Apply decision theory to hazard evaluation. *Hydrocarbon Process.*, **58**(12), 133

HANNA, S.D., RYDBRINK, K., VLAHOVIC, M. and SHULOCK, C.M. (1992). California SARA Title III Section 313 data for reporting years 1987 and 1988. *J. Haz. Materials*, **31**, 277

HANNA, S.R. (1971). A simple method of calculating dispersion from urban area sources. *J. Air Poll. Control Ass.*, **21**, 774

HANNA, S.R. (1976). Urban diffusion problems. In Haugen, D.A. (1976), *op. cit.*, p. 209

HANNA, S.R. (1979a). Some statistics of Lagrangian and Eulerian wind fluctuations. *J. Appl. Met.*, **18**, 518

HANNA, S.R. (1979b). A statistical diffusion model for use with variable wind fields. *Fourth Symp. on Turbulence, Diffusion and Air Pollution* (Boston, MA: Am. Met. Soc.), p. 15

HANNA, S.R. (1981a). Effects of release height on σ_y and σ_z in daytime conditions. In de Wispelaere, C. (1981), *op. cit.*, p. 337

HANNA, S.R. (1981b). Lagrangian and Eulerian time-scale relations in the daytime boundary layer. *J. Appl. Met.*, **20**, 242

HANNA, S.R. (1982). Applications in air pollution modeling. In Nieuwstadt, F.T.M. and van Dop, H. (1982), *op. cit.*, p. 275

HANNA, S.R. (1984). Concentration fluctuations in a smoke plume. *Atmos. Environ.*, **18**, 1091

HANNA, S.R. (1986). Spectra of concentration fluctuations: the two time scales of a meandering plume. *Atmos. Environ.*, **20**, 1131

HANNA, S.R. (1987a). An empirical formula for the height of the coastal internal boundary layer. *Boundary-Layer Met.*, **40**, 205

HANNA, S.R. (1987b). Review of vapor cloud dispersion models. *Vapor Cloud Modeling*, p. 379

HANNA, S.R. (1994). Hazardous gas model evaluations: is an equitable comparison possible? *J. Loss Prev. Process Ind.*, **7**, 133

HANNA, S.R. and MUNGER, B. (1983). A survey of emergency models and applications, Pt 1. *Am. Poll. Control Ass., Ann Mtg*, paper 83–26.1

HANNA, S.R. and PAINE, R.J. (1987). Convective scaling applied to diffusion of buoyant plumes from tall stacks. *Atmos. Environ.*, **21**, 2153

HANNA, S.R., BRIGGS, G.A., DEARDORFF, J., EAGAN, B.A., GIFFORD, F.A. and PASQUILL, F. (1977). AMS workshop on stability classification schemes and sigma curves – summary of recommendations. *Bull. Am. Met. Soc.*, **58**, 1305

HANNA, S.R., BRIGGS, G.A. and HOSKER, R.P. (1982). *Handbook on Atmospheric Diffusion. Rep. DOE/TIC-11223*. Atmos. Turbulence and Diffusion Lab., Dept of Energy, Washington, DC

HANNA, S.R., PAINE, R.J. and SCHULMAN, L.L. (1984). Overwater dispersion in coastal regions. *Boundary Layer Met.*, **30**, 389

HANNA, S.R., SCHULMAN, L.L., PAINE, R.J. and PLEIM, J.E. (1984). *The Offshore and Coastal Dispersion* (OCD) Model User's Guide, rev. ed. Rep. MMS 84-0069. Environ. Res. and Technol. Inc., Concord, MA

HANNA, S.R., CHANG, J.C. and STRIMAITIS, D.G. (1990). Uncertainties in source emission rate estimates using dispersion models. *Atmos. Environ.*, **24A**, 2971

HANNA, S.R., STRIMAITIS, D.G. and CHANG, J.C. (1991a). Evaluation of fourteen different heavy gas models with ammonia and hydrogen fluoride field data. *J. Haz. Materials*, **26**, 127

HANNA, S.R., STRIMAITIS, D.G. and CHANG, J.C. (1991b). Uncertainties in hazardous gas model predictions. *Modeling and Mitigating the Consequences of Accidental Releases*, p. 345

HANNA, S.R., CHATWIN, P., CHIKHLIWALA, E., LONDERGAN, R., SPICER, T. and WEIL, J. (1992). Results from model evaluation panel. *Plant/Operations Prog.*, **11**, 2

HANNA, S.R., CHANG, J.C. and STRIMAITIS, D.G. (1993). Hazardous gas model evaluation with field observations. *Atmos. Environ.*, **27A**, 2265

HANNAFUSA, T., LEE, C.B. and LO, A.K. (1986). Dependence of the exponent in power law wind profiles on stability and height interval. *Atmos. Environ.*, **20**, 2059

HANNAMAN, G.W. and WORLEDGE, D.H. (1987). Some developments in human reliability analysis approaches and tools. *Int. Post-SMIRT 9 Seminar on Accident Sequence Modelling*, Munich

HANNAMAN, G.W. *et al.* (1983). *Systematic Human Action Reliability Procedure (SHARP). Rep. EPRI-NP-2170-3*. Elec. Power Res. Inst., Palo Alto, CA

HANNAMAN, G.W. *et al.* (1985). A model for assessing human cogitative reliability in PRA studies. *Proc. Third IEEE Conf. on Human Factors and Nuclear Plants* (New York: Inst. Elec. Electron. Engrs)

HANNAMAN, G.W., JOKSIMOVICH, V., WORLEDGE, D.H. and SPURGIN, A.J. (1986). The role of human reliability analysis for enhancing crew performance. *Int. ANS/ENS Topical Mtg on Advances in Human Factors in Nuclear Power Systems*, Knoxville, TN

HANNON, J. (1992). Clean technology through process intensification. *IChemE North Western Branch Symp. Papers*, **3**(7), 1–7.6

HANNUM (ed.) (1984). *Hazards of Chemical Rockets and Propellants*, rev. ed. Chemical Propulsion Information Agency, Appl. Phys. Lab., John Hopkins Univ., Laurel, MD. See also Jensen, A.V. (1972), *op. cit.*

HANRATTY, P.J. and JOSEPH, B. (1992). Decision-making in chemical engineering and expert systems: application of the analytic hierarchy process to reactor selection. *Comput. Chem. Engng*, **16**, 849

HANS, J.M. and SELL, T.C. (1974). *Evacuation Risks – An Evaluation. Rep. EPA-520/6-74-002*. Nat. Environ. Res. Center, Environ. Prot. Agency, Las Vegas

HANSARD (1980). Canvey Island (industrial installations). *Hansard*, May 8, col. 683

HANSCH, C. (1980). The role of the partition coefficient in environmental toxicity. In Haque, R., (1980), *op. cit.*, p. 273

HANSEL, J.G., MITCHELL, J.W. and KLOTZ, H.C. (1992). Predicting and controlling flammability of multiple fuel and multiple inert mixtures. *Plant/Operations Prog.*, **11**, 213

HANSEN, D.E. (1986). Boiler and machinery protection: through control of piping reactions. *Plant/Operations Prog.*, **5**, 183

HANSEN, D.E. (1991). Control mechanical reactions in piping. *Chem. Engng Prog.*, **87**(3), 40

HANSEN, E.W.M. (1981). *A Water Tunnel Technique for Simulation of Dense Gas Dispersion Problems. Rep. NHL 281077*. Norwegian Hydrodynamic Labs, Trondheim

HANSEN, F.V. and SHREVE, J.D. (eds) (1968). *The Theory and Measurement of Atmospheric Turbulence and Diffusion in the Planetary Boundary Layer* (Albuquerque, NM: Sandia Labs)

HANSEN, G. and BERTHOLD, W. (1972). On the sensitivity of ammonium nitrates to detonation impact. *Chem. Zeit.*, **96**, 449

HANSEN, H.R. and VEDELER, B. (1974). *Vessel Design for Sea Transportation of Liquefied Natural Gas. Information 14*. Det Norske Veritas

HANSLIAN, R. (1931). On the history of gas warfare. *Gasschutz Luftschutz*, Oct., **49**

HANSLIAN, R. (1934). The German attack at Ypres on 22.4.1915. *Gasschutz Luftschutz*, **4**, 98; **5**, 123; **6**, 155; **7**, 184; **8**, 207

HANSLIAN, R. (1937). *Der chemische Krieg* (ES Mittler & Sohn)

HANSON, A.R. and RISEMAN, E.M. (1978). *Computer Vision Systems* (New York: Academic Press)

HANSON, D.J.(1986). OSHA issues Interim Report on Special Emphasis Program. *Chem. Engng News*, Aug. 4

HANSON, D.J. (1991). Toxic release inventory growing more contentious. *Chem. Engng News*, Jun. 3, 20

HANSON, D.N. and WILKE, C.R. (1969). Electrostatic precipitator analysis. *Ind. Engng Chem. Proc. Des. Dev.*, **8**, 357

HANSON, F.R. (1980). Arson: its effect and control. *Fire Surveyor*, **9**(3), 33

HANSSEN, A.J. (1961). Accurate valve sizing for flashing liquids. *Control Engng*, **8**(2), 87

HANZEVACK, E.L. (1982). Dispersion from safety valves and other momentum emission sources: continuous. *Atmos. Environ.*, **16**, 1231

HANZEVACK, K.M., ANDERSON, B.D. and RUSSELL, R.L. (1983). Methods to quantify tank losses are improved. *Oil Gas J.*, **81**(Sep. 19), 121

HAPPEL, J. (1958). *Chemical Process Economics* (New York: Wiley)

HAQ, I. (1993). Improve machinery vibration data. *Hydrocarbon Process.*, **72**(3), 94

HAQUE, M.A., RICHARDSON, S.M. and SAVILLE, G. (1992). Blowdown of pressure vessels. I, Computer model. *Process Safety Environ.*, **70B**, 3

HAQUE, M.A., RICHARDSON, S.M., SAVILLE, G. and CHAMBERLAIN, G. (1990). Rapid depressurization of pressure vessels. *J. Loss Prev. Process Ind.*, **3**(1), 4

HAQUE, M.A., RICHARDSON, S.M., SAVILLE, G., CHAMBERLAIN, G. and SHIRVILL, L. (1992). Blowdown of pressure vessels. II, Experimental validation of computer model and case studies. *Process Safety Environ.*, **70B**, 10

HAQUE, R. (1980a). *Dynamics, Exposure and Hazard Assessment of Toxic Chemicals* (Ann Arbor, MI: Ann Arbor Science)

HAQUE, R. (1980b). Dynamics, exposure and hazard assessment of toxic chemicals in the environment: an introduction. In Haque, R. (1980a), *op. cit.*, p. 1

HAQUE, R., FALCO, J., COHEN, S. and RIORDAN, C. (1980). Role of transport and fate studies in the exposure, assessment and screening of toxic chemicals. In Haque, R. (1980), *op. cit.*, p. 47

HARARY, F., NORMAN, R. and CARTWRIGHT, T.D. (1975). *Structural Models: An Introduction of the Theory of Directed Graphs* (New York: Wiley)

HARBERT, F.C. (1983). On-line laser detection of gases. *Plant/ Operations Prog.*, **2**, 58

HARBERT, F.C. (1984). Detection of fugitive emissions using laser beams. *Chem. Engr, Lond.*, **407**, 41

HARBISON and KELLY, G.N. (1985). An interpretation of the Nuclear Inspectorate's Safety Assessment Principles for accidental releases. *Proc. Seminar on Implications of Probabilistic Risk Analysis. IAEA-SR-111/20* (Vienna: Int. Atom. Energy Agency)

HARDEE, H.C. and LEE, D.O. (1973). Thermal hazard from propane fireballs. *Transport Planning Technol*, **2**, 121

HARDEE, H.C. and LEE, D.O. (1974). *Expansions of Clouds from Pressurised Liquids. Rep. SLA-73-7012.* Sandia Lab., Albuquerque, NM

HARDEE, H.C. and LEE, D.O. (1975). Expansion of clouds from pressurised liquids. *Accid. Anal. Prev.*, **7**, 91

HARDEE, H.C. and LEE, D.O. (1977/78). A simple conduction model for skin burns from exposure to chemical fireballs. *Fire Res.*, **1**, 199

HARDEE, H.C., LEE, D.O. and DONALDSON, A.B. (1972). A new method of predicting the critical temperature of explosives for various geometries. *Combust. Flame*, **18**, 403

HARDEE, H.C., DONALDSON, A.B. and LEE, D.O. (1972). Predicting the critical boundary temperature of multi-dimensional explosives. *Combust. Flame*, **19**, 331

HARDEE, H.C., LEE, D.O. and BENEDICK, W.B. (1978). Thermal hazard from LNG fireballs. *Combust. Sci. Technol.*, **17**, 189

HARDEKOPF, F. and MEWES, D. (1988). Critical pressure ratio of two-phase flows. *J. Loss Prev. Process Ind.*, **1**(3), 134

HARDENBURGER, T. (1992). Producing nitrogen at the point of use. *Chem. Engng*, **99**(10), 136

HARDESTY, D.R. and KENNEDY, J.E. (1977). Thermochemical estimation of explosive energy output. *Combust. Flame*, **28**, 45

HARDIE, D.W.F. (1965). *Acetylene Manufacture and Uses* (London: Oxford Univ. Press)

HARDING, A.J. (1959). *Ammonia Manufacture and Uses* (London: Oxford Univ. Press)

HARDING, D.A. (1964). *A Note on the Representation of Pipe Friction in Pressure Surge Analysis. Rep. TN 862.* Br. Hydromechanics Res. Ass., Cranfield, Bedfordshire

HARDING, D.A. (1965–66). A method of programming graphical surge analysis for medium speed computers. *Proc. Instn Mech. Engrs*, **180**(pt 3E), 83

HARDING, E.R. (1977). Safety problems concerning processing machinery in the chemical industry. *Prevention of Occupational Risks*, **5**, 259

HARDING, R.V., PARNAROUSKIS, M.C. and POTTS, R.G. (1978). The development and implementation of the hazard assessment computer system (HACS). *Control of Hazardous Material Spills 4*, p. 51

HARDT, S.L. (1992). Physics, naive. In Shapiro, S.C. (1992), *op. cit.*, p. 1147

HARDY, C.J. and COLLINS, C. (1981). Testing of industrial chemicals for inhalation toxicity. *Chem. Ind.*, Feb. 7, 89

HARDY, E.G. (1982). Phosgene. *Kirk-Othmer Encyclopaedia of Chemical Technology*, 3rd ed., **vol. 17** (New York: Wiley), p. 416

HARDY, J.D., WOLFF, H.G. and GOODELL, H. (1952). *Pain Sensations and Reactions* (Baltimore, MD: Williams & Wilkins)

HARDY, J.D., JACOBS, I. and MEIXNER, M.D. (1953). Thresholds of pain and reflex contraction as related to noxious stimulation. *J. Appl. Physiol.*, **5**, 725

HARDY, S. (1984). A new software environment for list processing and logic programming. In T. O'Shea and M. Eisenstadt (1984), *op. cit.*, p. 110

HARE, J.E. (1982). Ammonia separator failure. *Plant/Operations Prog.*, **1**, 166

HARFORD, T.W. (1969). Chemical handling. In Handley, W. (1969), *op. cit.*, p. 33

HARGIS, A.M. (1968). Heat exchangers that didn't work. *Chem. Engng Prog.*, **64**(3), 80

HARKER, J. (1985). Take another look at your pipe insulation economics. *Process Engng*, **66**(1), 60

HARKER, M. (1990). Foam in sprinkler systems. *Fire Surveyor*, **19**(4), 4

HARKER, R.G. (1977). What can minicomputers do for machinery reliability? *Hydrocarbon Process.*, **56**(8), 137

HARLOW, F.H. and AMSDEN, A.A. (1971); A numerical fluid dynamics calculation method for all flow speeds. *J. Comput. Phys.*, **8**, 197

HARMAN, R.T.C. (1981). *Gas Turbine Engineering* (London: Macmillan)

HARMANNY, A. and OPSCHOOR, G. (1986). Breaking pressure of window panes loaded by explosions. *Loss Prevention and Safety Promotion 5*, p. 58–1

HARMATHY, T.Z. (1972). A new look at compartment fires. *Fire Technol.*, **8**, 196; 326

HARMATHY, T.Z. (1977). Performance of building elements in spreading fires. *Fire Res.*, **1**(2), 119

HARMATHY, T.Z. (1985). Thermal inertia or thermal absorptivity. *Fire Technol.*, **21**, 146

HARMATHY, T.Z. and MEHAFFEY, J.R. (1987). The normalized heat load concept and its use. *Fire Safety J.*, **12**(1), 75

HARMATHY, T.Z. and OLESKIEWICZ, I. (1987). Fire draining system. *Fire Technol.*, **23**, 26

HARMATHY, T.Z. and SULTAN, M.A. (1988). Correlation between the severities of the ASTM E119 and ISO 834 fire exposures. *Fire Safety J.*, **13**, 163

HARMON, G.W. and MARTIN, H.A. (1970). Sizing rupture discs for vessels containing monomers. *Loss Prevention*, **4**, 95

HARMON, G.W. and STUPER, W.W. (1984). Sizing emergency relief systems on vessels containing monomers. *Chem. Engng Prog.*, **80**(3), 53

HARMON, P. and KING, D. (1985). *Expert Systems* (New York: Wiley)

HARMSWORTH, B. (1987). Water down, boiler down. *Plant Engr*, **31**(1), 18

HARNBY, G. (1968). Brittle fracture of centrifugally cast stainless steels. *Ammonia Plant Safety*, **10**, p. 16

HARNBY, N., EDWARDS, M.F. and NIENOW, A.W. (1985). *Mixing in the Process Industries* (London: Butterworths)

HAROLD, M.P. and OGUNNAIKE, B.A. (2000). Process engineering in the evolving chemical industry. *AIChE J.* **46**(11), 2123–27

HAROUN, M.A. and HOUSNER, G.W. (1981). Seismic design of liquid storage tanks. *J. Tech. Councils ASCE*, **107**, 191

HAROUTUNIAN, V. and LAUNDER, B.E. (1988). Second-moment modelling of free buoyant shear flows: a comparison of parabolic and elliptical solutions. In Puttock, J.S. (1988b), *op. cit.*, p. 409

HARPER, E. (1986). Monitoring for compliance. *Chem. Ind.*, Jun. 2, 384

HARPER, W.R. (1967a). *Contact and Frictional Electrification* (Oxford: Clarendon Press)

HARPER, W.R. (1967b). How do solid surfaces become charged? *Static Electrification*, **2**, p. 3

HARPUR, W.W. (1971). A review of recent advances in fire extinguishing chemicals. *Major Loss Prevention*, p. 137

HARR, M.E. (1987). *Reliability-Based Design in Civil Engineering* (New York: McGraw-Hill)

HARRELL, J.B. (1978). Corrosion monitoring in the CPI. *Chem. Engng Prog.*, **74**(3), 57

HARRELL, R.W., SEWELL, J.O. and WALSH, T.J. (1979). Control of malodorous compounds by carbon adsorption. *Loss Prevention*, **12**, p. 124

HARRIES, C.J., WEBSTER, M., SAYLES, R.S. and MACPHERSON, P.B. (1983). Bi-modal failure mechanisms in tribological components. *Reliab. Engng*, **4**(3), 169

HARRIES, P.G. (1976). Experience with asbestos disease and its control in Great Britain's naval dockyards. *Environ. Res.*, **11**, 261

HARRIES, W.L. (1953). The electrical breakdown of air between insulators. *Proc. Instn Elec. Engrs*, **100**(Pt IIA, no. 3, p. 132

HARRINGTON, J.M. (1979). Safety and chemicals in the environment: effects and epidemiology. *Chem. Ind.*, Nov. 3, 722

HARRINGTON, J.M. (1980a). Epidemiology. In Waldron, H.A. and Harrington, J.M. (1980), *op. cit.*, p. 379

HARRINGTON, J.M. (1980b). The health effects of physical agents. In Waldron, H.A. and Harrington, J.M. (1980), *op. cit.*, p. 301

HARRINGTON, J.M. and GILL, F.S. (1983). *Occupational Health*; 1987, 2nd ed.; 1992, 3rd ed. (Oxford: Blackwell)

HARRINGTON, J.M. and WALDRON, H.A. (1980). The structure and function of the lungs. In Waldron, H.A. and Harrington, J.M. (1980), *op. cit.*, p. 61

HARRINGTON, K.H., AINSWORTH, A.M.N., ARENDT, J.S. and GUTHRIE, V.H. (1986). Process unit loss prevention using a risk-based preliminary hazards analysis. *Am. Inst. Chem. Engrs, Symp. on Loss Prevention*

HARRIOTT, P. (1961). Designing temperature stable reactors. *Chem. Engng*, **68**(May 15), 165; (May 29), 81

HARRIOTT, P. (1962). An example of instability in a packed bed reactor. *AIChE J.*, **8**, 562

HARRIOTT, P. (1964). *Process Control* (New York: McGraw-Hill)

HARRIS, C. (1987). Source term – problems with modeling the release. *Vapor Cloud Modeling*, p. 204

HARRIS, C. (1993). Containment enclosures. *Prev. and Control of Accidental Releases of Hazard. Gases*, (ed. V.M. Fthenakis), 404–10 (New York: Van Nostrand Reinhold)

HARRIS, C.J. (ed.) (1979). *Mathematical Modelling of Turbulent Diffusion in the Environment* (New York: Academic Press)

HARRIS, C.M. (ed.) (1957). *Handbook of Noise Control* (New York: McGraw-Hill)

HARRIS, C.M. and CREDE, C.E. (1976). *Shock and Vibration Handbook* (New York: McGraw-Hill)

HARRIS, D.V., KAREL, G. and LUDWIG, A.L. (1961). Electrostatic discharges in aircraft systems. *Am. Pet. Inst., Proc. Ann. Mtg*

HARRIS, F.S. (1980). The practical application of the IMCO Gas Carrier Code for existing ships. *Gastech 79*, p. 169

HARRIS, F.S. (1986). Time for collision-resistant LPG carrier fleet. *Gastech 85*, p. 203

HARRIS, G.F.P. (1967). The effect of vessel size and degree of turbulence on gas phase explosion pressures in closed vessels. *Combust. Flame*, **11**, 17

HARRIS, G.F.P. and BRISCOE, P.G. (1967). The venting of pentane vapour air explosions in a large vessel. *Combust. Flame*, **11**, 329

HARRIS, G.F.P. and MACDERMOTT, P.E. (1977). Flammability and explosibility of ammonia. *Chemical Process Hazards*, **VI**, p. 29

HARRIS, G.F.P. and MACDERMOTT, P.E. (1980). Flammability aspects of the manufacture of phthalic anhydride. *Chemical Process Hazards*, **VII**, p. 67

HARRIS, G.F.P., HARRISON, N. and MACDERMOTT, P.E. (1981). Hazards of the distillation of mononitrotoluenes. *Runaway Reactions, Unstable Products and Combustible Powders* (London: Instn Chem. Engrs), paper 4/W

HARRIS, J. (1979). Inferno at Los Alfraques. *Readers Digest (Canadian ed.)*, **114**(686), 54

HARRIS, J. (1980). What 'Whirlpool case' means for your plant operation. *Hydrocarbon Process.*, **59**(6), 161

HARRIS, J. (1985). Keeping the public informed in a crisis. *Fire Int.*, **94**, 27

HARRIS, L.R. (1983a). Rupture disk technology: when it fails, when it works. *API Proc., Refining Dept, 48th Midyear Mtg*, **62**, 87

HARRIS, L.R. (1983b). Select the right rupture disc. *Hydrocarbon Process.*, **62**(5), 75

HARRIS, M.E., GRUMER, J. VON ELBE, G. and LEWIS, B. (1949). Burning velocities, quenching and stability data on nonturbulent flames of methane and propane with oxygen and nitrogen. *Combustion*, **3**, 80

HARRIS, M.J. and KELLY, A. (1975). The maintenance spares inventory and its control. *Plant Engr*, **19**(10), 22

HARRIS, M.J. and ROWE, A. (1980). The modelling and control of large-unit downtime: observations from the early stages of study. In Moss, T.R. (1980), *op. cit.*, p. 69

HARRIS, N.C. (1975). Risk assessment. *Electrical Safety*, **2**, 132

HARRIS, N.C. (1976). Consideration of criteria of acceptability for risks arising from spills. *Control of Hazardous Material Spills*, **3**, 55

HARRIS, N.C. (1980). The control of vapour emission from liquefied gas spillages. *Loss Prevention and Safety Promotion 3*, vol. 3, p. 1058

HARRIS, N.C. (1981a). The design of effective water sprays – what we need to know. In Greenwood, D.V. (1981), *op. cit.*, paper 7

HARRIS, N.C. (1981b). *Proof of Evidence to Pheasant Wood Inquiry.* (London: Dept of Environ.), document, pp. 327ff.

HARRIS, N.C. (1982). Problems in hazard analysis and risk assessment. *The Assessment of Major Hazards* (Rugby: Instn Chem. Engrs), p. 267

HARRIS, N.C. (1983). Some problems in risk assessment involving gas dispersion. In Hartwig, S. (1983), *op. cit.* p. 293

HARRIS, N.C. (1985). A user's view on the data obtained and those which may still be required. *J. Haz. Materials*, **11**(1/3), 425. See also in McQuaid, J. (1985), *op. cit.*, p. 425

HARRIS, N.C. (1986). The application of heavy gas dispersion in a risk assessment model. In Hartwig, S. (1986), *op. cit.* p. 353

HARRIS, N.C. (1987). Mitigation of accidental toxic gas releases. *Preventing Major Chemical Accidents*, p. 3.139

HARRIS, N.C. (1991). Containment building and scrubber for toxic gas plants. *Modeling and Mitigating the Consequences of Accidental Releases*, p. 443

HARRIS, N.C. and MOSES, A.M. (1983). The use of acute toxicity data in the risk assessment of the effect of accidental releases of toxic gases. *Loss Prevention and Safety Promotion 4*, vol. 1, p. I36

HARRIS, N.C. and ROODBOL, H.G. (1985). A risk assessment model applied to transportation problems. *The Assessment and Control of Major Hazards* (Rugby: Instn Chem. Engrs), p. 389

HARRIS, N.S. (1989). *Modern Vacuum Practice* (New York: McGraw-Hill)

HARRIS, R.J. (1983). *Investigation and Control of Gas Explosions in Buildings and Heating Plant* (London: Spon)

HARRIS, R.J. and WICKENS, M.J. (1989). *Understanding Vapour Cloud Explosions – An Experimental Study. Comm. 1408.* Instn Gas Engrs, London

HARRIS, R.J.S. (1978). Platforms – concrete structures. In Institute of Petroleum (1978), *op. cit.*, p. 66

HARRIS, W.J. (1976). *The Significance of Fatigue* (Oxford: Oxford Univ. Press)

HARRISON, A.J. and CAIRNIE, L.R. (1988). The development and experimental validation of a mathematical model for predicting hot surface autoignition hazards using complex chemistry. *Combust. Flame*, **71**, 1

HARRISON, A.J. and EYRE, J.A. (1986). Vapour cloud explosions – the effect of obstacles and jet ignition on the combustion of gas clouds. *Loss Prevention and Safety Promotion 5*, p. 38–1

HARRISON, A.J. and EYRE, J.A. (1987a). The effect of obstacle arrays on the combustion of large premixed gas/air clouds. *Combust. Sci. Technol.*, **52**, 121

HARRISON, A.J. and EYRE, J.A. (1987b). 'External explosions' as a result of combustion venting. *Combust. Sci. Technol.*, **52**, 91

HARRISON, A.J., FURZELAND, R.M., SUMMERS, R. and CAIRNIE, L.R. (1988). An experimental and theoretical study of auto-ignition on a horizontal pipe. *Combust. Flame*, **72**, 119

HARRISON, D. and HEATH, A. (1983). Valve condition monitoring. In Butcher, D.W. (1983), *op. cit.*, p. 157

HARRISON, D. and WATKINS, R. (1983). The detection of mechanical seal failure due to face vaporisation in laboratory tests. In Butcher, D.W. (1983), *op. cit.*, p. 281

HARRISON, E.B. (1977). Environmental trends. *Hydrocarbon Process.*, **56**(10), 121; **56**(11), 425; **56**(12), 153

HARRISON, E.B. (1978a). Clean Air Act '77. *Hydrocarbon Process.*, **57**(7), 255; **57**(8), 173

HARRISON, E.B. (1978b). What the new Clean Water Act means to HPI plant managers. *Hydrocarbon Process.*, **57**(2), 165

HARRISON, G.A. (1982). Emergency flare tip repair. *Hydrocarbon Process.*, **61**(7), 219

HARRISON, G.W. (1980). On stream repair methods. *Ammonia Plant Safety*, **21**, p. 100

HARRISON, M.E. and FRANCE, J.J. (1989). Trouble shooting distillation columns. *Chem. Engng*, **96**(5), 126

HARRISON, M.R. (1979). Select the right insulation. *Hydrocarbon Process.*, **58**(7), 144

HARRISON, M.R. and PELANNE, C.M. (1977). Cost-effective thermal insulation. *Chem. Engng*, **84**(Dec. 19), 63

HARRISON, P.J. and BELLAMY, L.J. (1989/90). Modelling the evacuation of the public in the event of hazardous release: a decision support tool and aid for emergency planning. *J. Safety Reliab. Soc.*, **9**(4), 14

HARRISON, R. and COULSON, G.S. (1977). The incineration of solid and liquid wastes. *Chemical Engineering in a Hostile World* (London: Clapp & Poliak)

HARRISON, R.M. (ed.) (1983). *Pollution: Causes, Effects and Control* (London: R. Soc. Chem.)

HARRISON, R.M. and MCCARTNEY, H.A. (1980). A comparison of the predictions of a simple Gaussian plume dispersion model with measurements of pollutant concentration at ground-level and aloft. *Atmos. Environ.*, **14**, 589

HARRISON, R.M. and PERRY, R. (1977). *Handbook of Air Pollution Analysis; 1986*, 2nd ed. (London: Chapman & Hall)

HARRISON, R.P. and MILNE, I. (1981). Assessment of defects: the CEGB approach. *Phil. Trans. R. Soc., Ser. A*, **299**, 145

HARRISON, R.P., LOOSEMORE, K. and MILNE, I. (1979). *Rep. R/H/R6, Suppl. 1.* (Barnwell: Central Elec. Generating Board)

HARRISON, S.F. (1984). Care needed in repairing safety valves. *Oil Gas J.*, **82**(Mar. 5), 104

HARRISON, T.J. (ed.) (1972). *Handbook of Industrial Control Computers* (New York: Wiley-Interscience)

HARROD, S.M. (1981). Hazard survey and assessment (letter). *Chem. Engr, Lond.*, **366**, 121

HARRON, J.G. (1986). Loss prevention challenge: how one refinery meets it. *Hydrocarbon Process.*, **65**(9), 167

HARRY, E. and SHIPP, E. (1975). Signature analysis for large rotating machinery benefits plant. *Oil Gas J.*, **73**(15), 59

HARSH, K.M. (1978a). In situ neutralization of an acrylonitrile spill. *Control of Hazardous Material Spills 4*, p. 187

HARSH, K.M. (1978b). Toxicity modification of an anhydrous ammonia spill. *Control of Hazardous Material Spills 4*, p. 148

HART, A. (1984). Experience in the use of an inductive system in knowledge engineering. In Bramer, M.A. (1984), *op. cit.*, p. 117

HART, D. and CROFT, T. (1988). *Modelling with Projectiles* (Chichester: Ellis Horwood)

HART, F.E., BAKER, N.C. and WILLIAMS, I. (1972). CRG route to SNG. *Hydrocarbon Process.*, **51**(4), 94

HART, G.S. (1986). Avoiding and managing environmental damage from major industrial accidents. *J. Air Poll. Control Ass.*, **36**, 127

HART, H.N. (1980). Disaster organisation of Europort harbour (Rotterdam). In De Boer, J. and Baillie, T.W. (1980), *op. cit.*, p. 53

HART, P.E., DUDA, R.O. and EINAUDY, M.T. (1978). PROSPECTOR – a computer-based consultation system for mineral exploration. *Math. Geol.*, **10**(5), 589

HART, P.W. and SOMMERFELD, J.T. (1993). Fluid discharge resulting from puncture of spherical process vessels. *J. Haz. Materials*, **33**, 295

HART, R.L. (1985). Formulas for sizing explosion vents. *Plant/Operations Prog.*, **4**, 1

HARTGERINK, J.W. (1974). Investigation into the cause of an explosion during the drying of porous nitrocellulose powder. *Chemical Process Hazards*, V, p. 125

HARTH, J.M. (1975). The use of inert gas generators for the protection of hazardous cargoes. *Transport of Hazardous Cargoes*, 4

HARTLEY, J. (1985). *Designing Instructional Text* (London: Kogan Page)

HARTLICH, E. (1977). Safety provisions at a German tanker terminal. *Fire Int.*, **57**, 18; **58**, 18

HARTMAN, R.R. (1993). Piping design: one phase at a time. *Chem. Engng*, **100**(10), 82

HARTMANN, I. (1946). Pressure release for dust explosions. *NFPA Quart.*, **40**(1), 47

HARTMANN, I. (1948a). Explosion and fire hazards of combustible dusts. *Industrial Hygiene and Toxicology*, vol. 1 (New York: Interscience)

HARTMANN, I. (1948b). Recent research on the explosibility of dust dispersions. *Ind., Engng Chem.*, **40**(4), 752

HARTMANN, I. (1950). Dust explosions. *Marks Mechanical Engineer's Handbook*, 5th ed. (New York: McGraw-Hill)

HARTMANN, I. (1957). Recent findings on dust explosions. *Chem. Engng Prog.*, **53**(3), 107

HARTMANN, I. (1958). *Industrial Hygiene and Toxicology*, vol. 1, *Explosion and Fire Hazards of Combustible Dusts* (New York: Interscience)

HARTMANN, I. and GREENWALD, H.P. (1945). The explosibility of metal-powder dust clouds. *Mining Metall.*, **26**, 331

HARTMANN, I. and NAGY, J. (1949). The explosibility of starch dust. *Chem. Engng News*, **27**(Jul. 18), 2071

HARTMANN, I. and NAGY, J. (1957). Venting dust explosions. *Ind. Engng Chem.*, **49**(10), 1734

HARTMANN, K. (1971). *Analyse und Steuerung von Prozessen der Stoffwirtschaft* (Leipzig: Veb Deutscher Verlag fur Grundstoff Industrie)

HARTMANN, R.J. (1984). How variable speed and constant speed motors compare for pipeline pump drivers. *Oil Gas J.*, **82**(Mar. 26), 127

DEN HARTOG, J.P. (1952). *Advanced Strength of Materials* (New York: McGraw-Hill)

DEN HARTOG, J.P. (1956). *Mechanical Vibrations* (New York: McGraw-Hill)

HARTON, E.E. (1968). Safety planning for the use of reactive cryogens in large volume. *Ann. N.Y. Acad. Sci.*, **152**(art. 1), 51

HARTUNG, R. and DINMAN, B.D. (eds) (1972). *Environmental Mercury Contamination* (Ann Arbor, MI: Ann Arbor Science)

HARTWIG, S. (ed.) (1980a). *Heavy Gas and Risk Assessment* (Dordrecht: Reidel)

HARTWIG, S. (1980b). Resume: what is still to do and where are we on safe ground. In Hartwig, S. (1980a), *op. cit.*, p. 287

HARTWIG, S. (ed.) (1983a). *Heavy Gas and Risk Assessment II* (Dordrecht: Reidel)

HARTWIG, S. (1983b). Open and controversial topics in heavy gas dispersion and related risk assessment problems. In Hartwig, S. (1983a), *op. cit.* p. 1

HARTWIG, S. (1985). Improved understanding of heavy gas dispersion due to the analysis of the Thorney Island trials data. *J. Haz. Materials*, **11**(1/3), 417. See also in McQuaid, J. (1985), *op. cit.*, p. 417

HARTWIG, S. (ed.) (1986a). *Heavy Gas and Risk Assessment III* (Dordrecht: Reidel)

HARTWIG, S. (1986b). Heavy gases in our industrial and technical society and our needs for further research. In Hartwig, S. (1986a), *op. cit.*, p. 1

HARTWIG, S. and FLOTHMANN, D. (1980). Open and controversial problems in the development of models for the dispersion of heavy-gas. In Hartwig, S. (1980a), *op. cit.*, p. 1

HARTWIG, S., SCHNATZ, G. and HEUDORFER, W. (1984). Improved understanding of heavy gas dispersion and its modelling. In Ooms, G. and Tennekes, H. (1984), *op. cit.*, p. 139

HARTZELL, G.E. (1989). Prediction of the toxic effects of fire effluents. *J. Fire Sciences*, **7**(3), 179

HARTZELL, G.E., PACKHAM, S.C., GRAND, A.F. and SWITZER, W.G. (1985). Modeling of toxicological effects of fire gases. III, Quantification of post-exposure lethality of rats from exposure to HCl atmospheres. *J. Fire Sci.*, **3**(3), 195

HARTZELL, G.E., PRIEST, D.N. and SWITZER, W.G. (1985). Modeling of toxicological effects of fire gases. II, Mathematical modeling of intoxication of rats by carbon monoxide and hydrogen cyanide. *J. Fire Sci.*, **3**(2), 115

HARTZELL, G.E., STACY, H.W., SWITZER, W.G., PRIEST, D.N. and PACKHAM, S.C. (1985). Modeling of toxicological effects of fire gases. IV, Intoxication of rats by carbon monoxide in the presence of an irritant. *J. Fire Sci.*, **3**(4), 263

HARTZELL, G.E., GRAND, A.F. and SWITZER, W.G. (1987). Modeling of toxicological effects of fire gases. *J. Fire Sciences*, **5**(6), 368; **6**(6), 411 (parts 6 and 7)

HARTZELL, G.O. (1987). Criteria and methods for evaluation of toxic hazard. *Fire Safety J.*, **12**(3), 179

HARVEY, B.H. (chrmn) (1976). *First Report of the Advisory Committee on Major Hazards* (London: HM Stationery Office)

HARVEY, B.H. (1979a). Flixborough – five years later. *Chem. Engr, Lond.*, **349**, 697

HARVEY, B.H. (chrmn) (1979b). *Second Report of the Advisory Committee on Major Hazards* (London: HM Stationery Office)

HARVEY, B.H. (chrmn) (1984). *Third Report of the Advisory Committee on Major Hazards* (London: H.M. Stationery Office)

HARVEY, B.H. (ed.) (1980–). *The Handbook of Occupational Hygiene* (Brentford: Kluwer). (Looseleaf with updates)

HARVEY, B.H. and MURRAY, R. (1958). *Industrial Health Technology* (London: Butterworths)

HARVEY, E.N. (1948). Studies in wound ballistics. In Andrus, E.C. *et al.* (1948), *op. cit.*, p. 191

HARVEY, J.F. (1974). *Theory and Design of Modern Pressure Vessels*, 2nd ed. (New York: van Nostrand Reinhold)

HARVEY, J.F. (1980). *Pressure Component Construction* (New York: van Nostrand Reinhold)

HARVEY, P.G. (1981). An industrialist's attitude to risk. *Proc. R. Soc., Ser. A*, **376**, 193. See also in Warner, Sir F. and Slater, D.H. (1981), *op. cit.*, p. 193

HARVEY-JONES, J.H. (1983). New directions for the chemical industry. *Chem. Ind.*, Dec. 19, 923

HARVIN, R.L., LERAY, D.G. and ROUDIER, L.R. (1980). Single pressure or dual pressure nitric acid: an objective comparison. *Ammonia Plant Safety*, **21**, p. 173

HARVISON, C.J. (1972). The exposure of the tank truck carrier to hazardous material spills. *Control of Hazardous Material Spills 1*, p. 53

HARVISON, C.J. (1973). Highway transport of hazardous materials. *Conf. on Bulk Transportation of Hazardous Materials by Water in the Future* (Washington, DC: Nat. Acad. Sci.; Nat Acad. Engng; Nat Res. Coun.), p. 40

HARVISON, C.J. (1981). Safety and loss prevention in the tank truck transportation of chemicals. *Loss Prevention*, **14**, 60

HARWELL (1979). *Safety of Chemicals in the Environment* (Harwell, Oxfordshire)

HARWELL, M.A., COOPER, W. and FLAAK, R. (1992). Prioritizing ecological and human welfare risks from environmental stresses. *Environ. Mngmt*, **15**, 451

HARWOOD, D.W. and RUSSELL, E.R. (1989). *Present Practices of Highway Transportation of Hazardous Materials. Rep. FHWA-RD-89-02.* Res. and Special Programs Admin., Dept of Transportation

HARWOOD, D.W., RUSSELL, E.R. and VINER, J.G. (1989). Characteristics of accidents and incidents in highway transportation of hazardous materials. *Transport. Res. Record 1245*, p. 23. Transport. Res. Board, Nat. Res. Coun.

HARWOOD, D.W., VINER, J.G. and RUSSELL, E.R. (1990). Truck accident rate model for hazardous materials routing. *Transport. Res. Record 1264*, p. 12. Transport. Res. Board, Nat. Res. Coun.

HASABALLAH, E.M. (1981). Two problems causing metal overheating and equipment failure. *Ammonia Plant Safety*, **23**, 10

HASEBA, S., SHIMOSE, T. and KITAGAWA, T. (1966). Nitric oxide explosion. *Ammonia Plant Safety*, **8**, p. 14. See also *Chem. Engng Prog.*, **1966**, 62(4), 92

HASEGAWA, K. (1992). An experimental study on the film detonation of higher hydrocarbons. *Loss Prevention and Safety Promotion 7*, vol. 3, 109.1

HASEGAWA, K. and SATO, K. (1977). Study on the fireball following steam explosion of n-pentane. *Loss Prevention and Safety Promotion 2*, p. 297

HASEGAWA, K. and SATO, K. (1978). *Tech. Memo. 12.* Fire Res. Inst. of Japan, Tokyo

HASEGAWA, K. and SATO, K. (1983). A study of the blast wave from deflagrative explosions. *Fire Safety J.*, **5**, 265

HASEGAWA, K. and SATO, K. (1987). *Experimental Investigation of Unconfined Vapour Cloud Explosions of Hydrocarbons. Tech. Memo 16.* Fire Res. Inst. of Japan, Tokyo

HASEGAWA, K., KONDO, K., HIRATA, Y. and NABA, H. (1989). Characteristics of the conical pile type burning test method for determining the potential hazards of oxidizing materials. *J. Loss Prev. Process Ind.*, **2**(3), 135

HASELBARTH, J.E. (1967). Updated investment costs for 60 types of chemical plants. *Chem. Engng*, **74**(Dec. 4), 214

HASELDEN, G.G. (1970). LNG – the cold fuel. *LNG Conf.*, **2**

HASELDEN, G.G. (1971). *Cryogenic Fundamentals* (London: Academic Press)

HASELDEN, G.G. (1980). Process gas compression. *Interflow 80* (Rugby: Instn Chem. Engrs), p. 13

HASELTINE, D.M. (1992). Wastes: to burn or not to burn. *Chem. Engng Prog.*, **88**(7), 53

HASEMER, T. (1984a). *A Beginner's Guide to LISP* (Wokingham, Berkshire: Addison-Wesley)

HASEMER, T. (1984b). An introduction to LISP. In T. O' Shea and M. Eisenstadt (1984), *op. cit.*, p. 22

HASHEMI, H.T. and WESSON, H.R. (1971). Cut LNG storage costs. *Hydrocarbon Process.*, **50**(8), 117

HASLER, E.F. and MARTIN, R.E. (1971). Hydrastep: a 'fail operative' gauge system to replace visual boiler water level gauges. *Measmt Control*, **4**, 366

HASLER, E.F. and MARTIN, R.E. (1973). Liaproof measurements: the multiple binary philosophy of the 'Hydrastep' boiler water level gauge. *Measmt Control*, **6**, 141

HASLER, E.F. and MARTIN, R.E. (1974). *Hydratect: An Instrument for the Detection of Water in Bled Steam Pipes and Feed Heaters. Rep. RD/L/R 1888.* Cent. Elec. Res. Lab., Cent. Elec. Generating Board, Leatherhead, Surrey

HASS, H., MEMMESHEIMER, M., GEISS, H., JA KOBS, H.J., LAUBE, M. and EBEL, A. (1990). Simulation of the Chernobyl radioactive cloud over Europe using the EURAD model. *Atmos. Environ.*, **24A**, 673

HASSAN, H.M.K. (1992). Code check floating roof tanks. *Hydrocarbon Process.*, **71**(10), 101

HASSAN, M. (1993). Environmental planning for the hazardous chemicals industry. *Health, Safety and Loss Prevention*, p. 420

HASSE, L. and WEBER, H. (1985). On the conversion of Pasquill categories for use over sea. *Boundary Layer Met.*, **31**, 177

HASSELBARTH, J.E. (1977). *Modern Cost Engineering Techniques* (New York: McGraw-Hill)

HASSELBAUM, R. (1992). Keeping tabs from afar. *Chem. Engng*, **99**(10), 148

HASSELMANN (1925a). *Klin. Wschr.*, **4**, 2154

HASSELMANN (1925b). *Arch. Exp. Pathol. Pharmakol.*, **108**, 106

HASSELMAN, D.P.H. and HELLER, R.A. (eds) (1980). *Thermal Stresses in Severe Environments* (New York: Plenum Press)

HASTINGS, M.R. (ed.) (1990). *The Health and Safety Directory 1990/91* (Kingston-on-Thames, Surrey: Croner)

HASTINGS, N.A.J. (1967–68) A stoichastic model for Weibull failures as an aid to forecasting replacement requirements. *Proc. Instn Mech. Engrs*, **182**(pt 1), 715

HASTINGS, N.A.J. (1970). Equipment replacement and the repair limit method. In Jardine, A.K.S. (1970b), *op. cit.*, p. 100

HASTINGS, N.A.J. (1973). Equipment replacement and the repair limit method. In Jardine, A.K.S. (1973b), *op. cit.*, p. 100

HASTINGS, N.A.J. and JARDINE, A.K.S. (1979). The use of RELCODE. *Microelectron. Reliab.*, **19**(1/2)

HASTINGS, N.A.J. and MELLOW, J.M.C. (1978). *Decision Networks* (New York: Wiley)

HASTINGS, N.A.J. and PEACOCK, J.B. (1974). *Statistical Distributions* (London: Butterworths)

HATFIELD, P.E. (1973). Safety around equipment. *Safety in High Pressure Polyethylene Plants*, **1**, 57. See also *Chem. Engng Prog.*, 1972, **68**(11), 65

HATHAWAY, R.S. and ROWE, S.J. (1981). *Collision Velocities between Offshore Supply Vessels and Fixed Platforms. Rep. OT-R-8201*. Nat. Maritime Inst.

HATHI, K.K., SENGUPTA, R. and PURANIK, S.A. (1990). Prevention of hazards through technical alternatives. *Safety and Loss Prevention in the Chemical and Oil Processing Industries*, p. 581

HATTIANGADI, U.S. (1970a). Sizing safety valves for heat sensitive materials. *Chem. Engng*, **77**(Sep. 7), 96

HATTIANGADI, U.S. (1970b). Specifying centrifugal and reciprocating pumps. *Chem. Engng*, **77**(Feb. 23), 101

HATTON, C. (1967). Controlling a gas compressor station electronically. *Instrum. Control Systems*, **40**(5), 145

HATTON, D.V., LEACH, C.S., BEAUDET, A.L., DILLMAN, R.D. and DIFERRANTE, N. (1979). Collagen breakdown and ammonia inhalation. *Arch. Environ. Health*, **34**, 83

HATTWIG, M. (1977). Selected aspects of explosion venting. *Loss Prevention and Safety Promotion 2*, p. 249

HATTWIG, M. (1980). Investigations on the recoil of vented vessels. *Loss Prevention and Safety Promotion 3*, vol. 3, p. 968

HAUETER, R.L. (1977). Current program on power plant availability and reliability data system. In Gangadharan, A.C. and Brown, S.J. (1977), *op. cit.*, p. 1

HAUGEN, D.A. (ed.) (1959). *Project Prairie Grass: A Field Program in Diffusion. Rep. AFCRC-TR-58-235*. US Air Force, Cambridge Res. Center

HAUGEN, D.A. (ed.) (1973). *AMS Workshop on Micrometeorology* (Princeton, NJ: Science Press)

HAUGEN, D.A. (ed.) (1976). *Lectures on Air Pollution and Environmental Impact Analysis* (Boston, MA: Am. Met. Soc.)

HAUGEN, D.A. and FUQUAY, J.J. (1963). *The Ocean Breeze and Dry Gulch Diffusion Programs. Contract AT(45-1)-1350*. US Air Force, Cambridge Res. Labs

HAUGEN, D.A. and TAYLOR, J.H. (eds) (1963). *The Ocean Breeze and Dry Gulch Diffusion Programs. Rep. AFCRL-63-791*, vol. II. Air Force Cambridge Res. Labs

HAUGEN, D.A., BARAD, M.L. and ATHANAITAS, P. (1961). Values of parameters appearing in Sutton's diffusion models. *J. Met.*, **18**, 368

HAUGEN, E.B. (1968). *Probabilistic Approaches to Design* (New York: Wiley)

HAUGEN, S. and VINNEM, J.E. (1987). Assessment of transport regularity. *Reliab. Engng*, **19**(4), 277

HAUIN, K. (1993). Intumescent seals and their use in restricting fire spread. *Fire Surveyor*, **22**(4), 12

HAUPT, R.W. and MEYER, R.A. (eds) (1979). *Safety Relief Valves* (New York: Am. Soc. Mech. Engrs)

HAUPTMANNS, U. (1980). Fault tree analysis of a proposed ethylene vaporisation unit. *Ind. Engng Fundamentals*, **19**, 300. See also Allen, D.J. (1980), *op. cit.*

HAUPTMANNS, U. (1986). Chemical plant safety analysis using fault trees. *Loss Prevention and Safety Promotion 5*, 13.1

HAUPTMANNS, U. and HOÖMKE, P. (1989). Bayesian estimation of failure rate distributions for components in process plants. *Ind. Engng Chem. Res.*, **28**, 1639

HAUPTMANNS, U. and YLLERA, J. (1983). Fault-tree evaluation by Monte-Carlo simulation. *Chem. Engng*, **90**(Jan. 10), 91

HAUSEN, H. (1981). *Heat Transfer in Counter-Current, Co-Current and Cross Current Flow* (New York: McGraw-Hill)

HAUSER, R.L., MCKEEVER, R.B. and STULL, D.P. (1985). Heat flux sensors for insulation economy and process control. *Chem. Engng Prog.*, **81**(2), 24

HÄUSSINGER, P., LOHMÜLLER, R. and WATSON, A.M. (1989). Hydrogen. *Ullmanns Encyclopaedia of Chemical Technology*, rev. ed., vol. A13 (Weinheim: VCH), p. 297

DE HAVEN, E.S. (1983). Approximate hazard ratings and venting requirements from CSI-ARC data. *Plant/Operations Prog.*, **2**, 21

DE HAVEN, E.S. and DIETSCHE, T.J. (1990). Catalyst explosion: a case history. *Plant/Operations Prog.*, **9**, 131

HAVENER, W.H. (1971). *Synopsis of Ophthalmology* (St Louis, MO: Mosby)

HAVENS, J.A. (1977). *Predictability of LNG Vapor Dispersion from Catastrophic Spills onto Water: an Assessment. Rep. CG-M-09-77*. Dept of Chem. Engng, Univ. of Arkansas, AR

HAVENS, J.A. (1978). An assessment of the predictability of LNG vapor dispersion from catastrophic spills onto water. *Fifth Int. Symp. on Transport of Dangerous Goods by Sea and Inland Waterways*, paper C2

HAVENS, J.A. (1979). *A Description and Assessment of the SIGMET LNG Vapour Dispersion Model. Rep. CG-M-3-79*. Dept of Chem. Engng, Univ. of Arkansas, AR

HAVENS, J.A. (1980). An assessment of predictability of LNG vapor dispersion from catastrophic spills onto water. *J. Haz. Materials*, **3**(3), 267

HAVENS, J.A. (1982a). A description and computational assessment of the SIGMET LNG vapor dispersion model. *J. Haz. Materials*, **6**(1/2), 181. See also in Britter, R.E. and Griffiths, R.F. (1982), *op. cit.* p. 181

HAVENS, J.A. (1982b). A review of mathematical models for prediction of heavy gas atmospheric dispersion. *The Assessment of Major Hazards* (Rugby: Instn Chem. Engrs), p. 1

HAVENS, J.A. (1985). The atmospheric dispersion of heavy gases: an update. *The Assessment and Control of Major Hazards* (Rugby: Instn Chem. Engrs), p. 143

HAVENS, J.A. (1986). Mathematical modeling of heavy gas dispersion – an overview. *Loss Prevention and Safety Promotion 5*, p. 32.2

HAVENS, J.A. (1987). Mathematical models for atmospheric dispersion of hazardous chemical gas releases: an overview. *Preventing Major Chemical Accidents*, p. 1.1

HAVENS, J.A. (1992). Review of dense gas dispersion field experiments. *J. Loss Prev. Process Ind.*, **5**, 28

HAVENS, J.A. and SPICER, T.O. (1983). Further analysis of catastrophic LNG spill vapor dispersion. In Hartwig, S. (1983a), *op. cit.* p. 181

HAVENS, J.A. and SPICER, T.O. (1984). Gravity spreading and air entrainment by heavy gases instantaneously released in a calm atmosphere. In Ooms, G. and Tennekes, H. (1984), *op. cit.*, p. 177

HAVENS, J.A. and SPICER, T.O. (1985). *Development of an Atmospheric Dispersion Model for Heavier-than-air Gas Mixtures. Rep. CG-D-22-85*. Dept of Chem. Engng, Univ. of Arkansas, AR. Also *Rep. CG-D-23-85*. US Coast Guard, Washington, DC

HAVENS, J.A., HASHEMI, H.T., BROWN, L.E. and WELKER, J.R. (1972). A mathematical model of the thermal decomposition of wood. *Combust. Sci. Technol.*, **5**, 91

HAVENS, J.A., SPICER, T.O. and SCHREURS, P.J. (1987a). Evaluation of three-dimensional numerical models for atmospheric dispersion of LNG. *Vapor Cloud Modeling*, p. 495

HAVENS, J.A., SPICER, T.O. and SCHREURS, P.J. (1987b). *Evaluation of 3-D Hydrodynamic Computer Models for Prediction of LNG Vapor Dispersion in the Atmosphere. Rep. GRI-87-0173*. Gas Res. Inst., Chicago, IL

HAVENS, J.A., SCHREURS, P.J. and SPICER, T.O. (1987). Analysis and simulation of Thorney Island Trial 34. *J. Haz. Materials*, **16**, 139

HAVENS, J.A., SPICER, T.O. and LAYLAND, D.E. (1987). A dispersion model for elevated dense gas jet releases. *Vapor Cloud Modeling*, p. 568

HAVENS, J.A., SPICER, T.O. and GUINNUP, D.W. (1989). Extension of the DEGADIS atmospheric dispersion model for elevated jet releases. *Loss Prevention and Safety Promotion 6*, p. 72.1

HAVENS, J.A., SPICER, T., KHAJEHNAJAFI, S. and WILLIAMS, T. (1991). Developments in liquefied natural gas dispersion modeling. *Modeling and Mitigating the Consequences of Accidental Releases*, p. 261

HAVENS, J.A., WALKER, H. and SPICER, T. (1994). Wind-tunnel data sets for complex dispersion model evaluation. *J. Loss Prev. Process Ind.*, **7**, 124

HAVERSTAD, T.A. (1989). Explosion and fire protection. *Loss Prevention and Safety Promotion 6*, p. 59.1

HAVILAND, R.P. (1964). *Reliability Engineering and Long Life Design* (Princeton, NJ: van Nostrand)

HAWICKHORST, W. (1975). A general approach to computerised inspection of engineered safety systems. *Symp. on Reliability of Nuclear Power Plant* (Vienna: Int. Atom. Energy Agency)

HAWK, C.W. (1977). Do you understand galvanic corrosion? *Chem. Engng*, **84**(Jun. 6), 193

HAWK, C.W. (1980). Why mechanical seals fail. *Chem. Engng*, **87**(Aug. 11), 171

HAWKER, C.R. (1995). Offshore safety cases – BG E&E's experience. *Safety Cases* (London: IBC Technical Services)

HAWKES, N., LEAN, G., LEIGH, D., MCKIE, R., PRINGLE, P. and WILSON, A. (1986). *The Worst Accident in the World* (London: Pan/Heinemann)

HAWKINS, R. (1984). Misconceptions over waste. *Hazardous Cargo Bull.*, **5**(4), 14

HAWKINS, R.G.P. (1979). Recent legislation in the context of waste management and the role of the chemical engineer. *Chem. Engr, Lond.*, **350**, 773

HAWKINS, R.G.P. (1985). Prevention of losses from the containment of flammable or toxic materials from road and rail tankers. *Chem. Ind.*, Jul. 15, 475

HAWKS, J. (1992). Process improvement and pumps. *Chem. Engng Prog.*, **88**(11), 26

HAWKS, J.L. and MERIAN, J.L. (1991). Create a good PSM system. *Hydrocarbon Process.*, **70**(8), 131

HAWKSLEY, J.L. (1984). Some social, technical and economic aspects of the risks of large plants. *CHEMRAWN III*

HAWKSLEY, J.L. (1987). CIMAH safety cases: the ICI safety case. *Chem. Engr, Lond.*, **439**, 8.

HAWKSLEY, J.L. (1988). Process safety management: a UK approach. *Plant/Operations Prog.*, **7**, 265

HAWKSLEY, J.L. (1989). The selective use of the elements of quantitative risk analysis. *Loss Prevention and Safety Promotion 6*, p. 45.1

HAWKSLEY, J.L. (1993a). International activities: International Process Safety Group session (workshop). *Process Safety Management*, p. 381

HAWKSLEY, J.L. (1993b). A multilevel approach to monitoring the implementation of safety, health, environment (SHE) standards in the ICI Group. *Process Safety Management*, p. 227

HAWKSLEY, J.L. and PRESTON, M.L. (1996). Inherent SHE – 20 years of evolution. *Int. Conf. and Workshop on Process Saf. Manage. and Inherently Safer Processes*, Oct. 8–11, 1996, Orlando, FL, 183–196 (New York: American Institute of Chemical Engineers)

HAWLEY, G.G. (1977). *Condensed Chemical Dictionary*, 9th ed. (New York: van Nostrand Reinhold)

HAWTHORNE, E.P. (1978). *The Management of Technology* (London: McGraw-Hill)

HAWTHORNE, W.R., WEDDELL, D.S. and HOTTEL, H.C. (1949). Mixing and combustion in turbulent gas jets. *Combustion*, **3**, p. 266

HAXO, H.E. (1988). Assessment of techniques for in situ repair of flexible membrane liners: final report. *J. Haz. Materials*, **17**(2), 223

HAY, A. (1976a). Seveso: the aftermath. *Nature*, **263**, 537

HAY, A. (1976b). Seveso: dioxin damage. *Nature*, **266**, 7

HAY, A. (1976c). Seveso: seven months on. *Nature*, **265**, 490

HAY, A. (1976d). Seveso solicitude. *Nature*, **266**, 384

HAY, A. (1976e). Toxic cloud over Seveso. *Nature*, **262**(Aug. 19), 636

HAY, A. (1981). Seveso – the intrigue and the infighting. *Nature*, **290**(Mar. 5), 71

HAY, A. (1982). *The Chemical Scythe. Lessons of 2,4,5-T and Dioxin* (New York: Plenum Press)

HAY, J.S. and PASQUILL, F. (1957). Diffusion from a fixed source at a height of a few hundred feet in the atmosphere. *J. Fluid Mech.*, **2**, 299

HAY, J.S. and PASQUILL, F. (1959). Diffusion from a continuous source in relation to the spectrum and scale of turbulence. *Adv. Geophys.*, **6**, 345

HAY, T.T. (1982). Some maintenance problems of containing flammable materials. *Flammable Atmospheres and Area Classification, IEE Coll. Dig. 1982/26* (London), p. 6/1

HAYASAKA, H., KOSEKI, H. and TASHIRO, Y. (1992). Radiation measurements on large scale kerosene pool flames using high-speed thermography. *Fire Technol.*, **28**, 110

HAYASHI, S. and KUMAGAI, S. (1975). Flame propagation in fuel droplet-vapor-air mixtures. *Combustion*, **15**, p. 445

HAYASHI, S., KUMAGAI, S. and SAKAI, T. (1977). Propagation velocity and structure of flames in droplet-vapor-air mixtures. *Combust. Sci. Technol.*, **15**, 169

HAYASHI, S., OHTANI, T., IINUMA, K. and KUMAGAI, S. (1981). Limiting factor of flame propagation in fuel clouds. *Combustion*, **18**, 361

HAYASHI, T. and RANSFORD, G. (1960). Sudden opening or closing of an outlet valve on a pipeline. *La Houille Blanche*, **15**(6), 657

HAYASHI, T., HOWELL, R.H., SHIBATA, M. and TSUJI, K. (1985). *Industrial Ventilation and Air Conditioning* (Boca Raton, FL: CRC Press)

HAYASHI, Y. (1985). Hazard analysis in chemical complexes in Japan – especially those caused by human error. *Ergonomics*, **28**, 835

HAYES, A.H. and MELAVEN, R.M. (1964). Safe start-up and shutdown of refinery process units. In Vervalin, C.H. (1964a), *op. cit.*, p. 231

HAYES, A.W. (ed.) (1989). *Principles and Methods of Toxicology*, 2nd ed. (New York: Raven)

HAYES, D. (1977). Nuclear terrorism. *Sci. Publ. Policy*, **4**, 82

HAYES, J.E. and MICHIE, D. (1983). *Intelligent Systems* (Chichester: Ellis Horwood)

HAYES, P.J. (1977). In defence of logic. *Int. Joint Conf. on Artificial Intelligence*, **5**, 559

HAYES, P.J. (1979). The naive physics manifesto. In Michie, D. (1979), *op. cit.*, p. 242

HAYES, P.J. (1985). The naive physics manifesto. In Hobbs, J.R. and Moore, R.C. (1985), *op. cit.*, p. 1

HAYES, S.R. and MOORE, G.E. (1986). Air quality model performance: a comparative analysis of 15 model evaluation studies. *Atmos. Environ.*, **20**, 1897

HAYES, T. (1971). The fire service and hazardous materials transport. *Loss Prevention*, **5**, 3

HAYES, T.C. (1972). Houston train derailment and the subsequent hearing. *Fire J.*, **66**(2)

HAYES-ROTH, F. (1985). Rule-based systems. *Comm. ACM*, **28**, 921

HAYES-ROTH, F., WATERMAN, D.A. and LENAT, D.B. (eds) (1983). *Building Expert Systems* (Reading, MA: Addison-Wesley)

HAYNES, V. and BOJCUM, M. (1988). *The Chernobyl Disaster* (London: Hogarth Press)

HAYS, A. and BERGGREN, G. (1976). Control valve packing – some useful guidelines. *Hydrocarbon Process.*, **55**(1), 91. See also in Kane, L.A. (1980), *op. cit.*, p. 176

HAYS, F.L. (1972). Studies of the effects of atmospheric hydrogen sulphide in animals. Thesis, Univ. of Missouri

HAYWARD, A.T. (1977). *Repeatability and Accuracy* (Bury St Edmunds, Suffolk: Mech. Engng Publs)

HAYWARD, A.T.J. (1979). *Flowmeters: A Basic Guide and Source Book for Users* (London: Macmillan)

HAYWARD, J.A. (1969). The construction industry. In Handley, W. (1969), *op. cit.*, p. 256

HAYZELDEN, J.E. (1968). The value of human life. *Pub. Admin.*, **46**

HAZARDS FORUM (1992). *An Engineer's Responsibility for Safety* (London)

HAZARDS PREVENTION MAGAZINE, Second Quarter 2001, System Safety Society, Unionville, VA

HAZEN, F.E. (1991). Concrete advice about containing leaks and spills. *Chem. Engng Prog.*, **87**(8), 75

HE, L. and CLAVIN, P. (1992). Critical conditions for detonation initiation in cold gaseous mixtures by nonuniform hot pockets of reactive gases. *Combustion*, **24**, 1861.

HEAD, J. and RUMLEY, J. (1987). Design of production facilities for floating offshore platforms. *Chem. Engr, Lond.*, **432**, 17

HEAD, J. and RUMLEY, J. (1987). Production design for floating platforms. *Chem. Engr, Lond.*, **438**, 17

HEAFIELD, W. (1986). Nuclear waste disposal. *Chem. Engr, Lond.*, **421**, 14

HEALD, P.T., SPINK, G.M. and WORTHINGTON, P.J. (1972). Post yield fracture mechanics. *Mat. Sci. Engng*, **10**, 129

HEALTH AND SAFETY COMMISSION (1977). *Report 1974–76* (London: HM Stationery Office)

HEALTH AND SAFETY COMMISSION (1978a). *Asbestos. Measurement and Monitoring of Asbestos in Air* (London: HM Stationery Office)

HEALTH AND SAFETY COMMISSION (1978b). *Asbestos. Work on Thermal and Acoustic Insulation and Sprayed Coatings* (London: HM Stationery Office)

HEALTH AND SAFETY COMMISSION (1978c). *The Hazards of Conventional Sources of Energy* (London: HM Stationery Office)

HEALTH AND SAFETY COMMISSION (1983). *Professional Training and Qualification in Occupational Health and Safety* (London: HM Stationery Office)

HEALTH AND SAFETY COMMISSION (1984). *Statement on Future Policy towards Reference to Standards* (London: HM Stationery Office)

HEALTH AND SAFETY COMMISSION (1985). *Plan of Work 1985/86 and onwards* (London: HM Stationery Office)

HEALTH AND SAFETY COMMISSION (1989). Quantitative assessment of major hazard installations: 1, The ACMH reports. In Lees, F.P. and Ang, M.L. (1989b), *op. cit.*, p. 175

HEALTH AND SAFETY COMMISSION (1992). *Draft Offshore Installations (Safety Case) Regulations 199–* (London)

HEALTH AND SAFETY COMMISSION (1993). *Draft Offshore Installations (Prevention of Fire and Explosion and Emergency Response) Regulations and Approved Code of Practice* (London)

HEALTH AND SAFETY COMMISSION – see also Appendix 28

HEALTH AND SAFETY EXECUTIVE (1975a). *Principles of Local Exhaust Ventilation* (London: HM Stationery Office)

HEALTH AND SAFETY EXECUTIVE (1975b). *Vinyl Chloride – Code of Practice for Heath Precautions* (London: HM Stationery Office)

HEALTH AND SAFETY EXECUTIVE (1976a). *The explosion at Appleby-Frodingham steelworks Scunthorpe, 4 November 1975* (London: HM Stationery Office)

HEALTH AND SAFETY EXECUTIVE (1976b). *The Explosion at Laporte Industries Ltd, Ilford 5 April 1975* (London: HM Stationery Office)

HEALTH AND SAFETY EXECUTIVE (1976c). *Safety Officers: Sample Survey of Role and Functions, Discussion Doc.* (London: HM Stationery Office)

HEALTH AND SAFETY EXECUTIVE (1976d). *Success and Failure in Accident Prevention* (London: HM Stationery Office)

HEALTH AND SAFETY EXECUTIVE (1977a). *A Code of Practice for the Bulk Storage of Liquid Oxygen at Production Sites* (London: HM Stationery Office)

HEALTH AND SAFETY EXECUTIVE (1977b). *The Explosion at King's Lynn 27th June 1976* (London: HM Stationery Office)

HEALTH AND SAFETY EXECUTIVE (1977c). *The Generic Safety Issues of Pressurised Water Reactors. An Account of the Study Carried out by the Nuclear Installations Inspectorate of the Health and Safety Executive* (London: HM Stationery Office)

HEALTH AND SAFETY EXECUTIVE (1977d). *Health and Safety Statistics 1975* (London: HM Stationery Office)

HEALTH AND SAFETY EXECUTIVE (1977e). *Occupational Health Services: The Way Ahead* (London: HM Stationery Office)

HEALTH AND SAFETY EXECUTIVE (1977f). *Proposals for New Legislation on Pressurised Systems* (London: HM Stationery Office)

HEALTH AND SAFETY EXECUTIVE (1977g). *Test and Approval of Diesel and Storage Powered Locomotives and Vehicles and Diesel Powered Equipment for Use in Underground Mines. Testing Memo 12* (London: HM Stationery Office)

HEALTH AND SAFETY EXECUTIVE (1978a). *Assessment of the Hazard from Radio Frequency Ignition at the Shell and Esso Sites at Braefoot Bay and Moss Morran, Fife* (London: HM Stationery Office)

HEALTH AND SAFETY EXECUTIVE (1978b). *Canvey: An Investigation of Potential Hazards from Operations in the Canvey Island/Thurrock Area* (London: HM Stationery Office)

HEALTH AND SAFETY EXECUTIVE (1978c). *The Fire on HMS Glasgow, 23 September 1976* (London: HM Stationery Office)

HEALTH AND SAFETY EXECUTIVE (1978d). *A Safety Evaluation of the Proposed St Fergus to Moss Morran Natural Gas Liquids and St Fergus to Boddam Gas Pipelines* (London: HM Stationery Office)

HEALTH AND SAFETY EXECUTIVE (1978e). Private communication

HEALTH AND SAFETY EXECUTIVE (1979a). *The Accident at Three Mile Island* (London: HM Stationery Office)

HEALTH AND SAFETY EXECUTIVE (1979b). *The Fire and Explosion at Braehead Container Depot, Renfrew, 4 January 1977* (London: HM Stationery Office)

HEALTH AND SAFETY EXECUTIVE (1979c). *Hot Work, Welding and Cutting in Plant Containing Flammable Materials* (London: HM Stationery Office)

HEALTH AND SAFETY EXECUTIVE (1979d). *Nuclear Safety. Safety Assessment Principles for Nuclear Reactors* (London: HM Stationery Office)

HEALTH AND SAFETY EXECUTIVE (1979e). *PWR.* (London: HM Stationery Office)

HEALTH AND SAFETY EXECUTIVE (1979f). *Report of the Steering Committee on Radio Frequency Ignition Hazards at St. Fergus, Scotland* (London: HM Stationery Office)

HEALTH AND SAFETY EXECUTIVE (1980a). *The Fire and Explosions at River Road, Barking, Essex, 21 January 1980* (London: HM Stationery Office)

HEALTH AND SAFETY EXECUTIVE (1980b). *A Guide to the Health and Safety at Work Act* (London: HM Stationery Office)

HEALTH AND SAFETY EXECUTIVE (1980c). *A Reappraisal of the HSE Safety Evaluation of the Proposed St. Fergus to Moss Morran NGL Pipeline* (London: HM Stationery Office)

HEALTH AND SAFETY EXECUTIVE (1981a). *Canvey: a Second Report. A Review of the Potential Hazards from Operations in the Canvey Island/Thurrock Area Three Years after Publication of the Canvey Report* (London: HM Stationery Office)

HEALTH AND SAFETY EXECUTIVE (1981b). *The Fire and Explosions at Permaflex Ltd, Trubshaw Cross, Longport, Stoke on Trent, 11 February 1980* (London: HM Stationery Office)

HEALTH AND SAFETY EXECUTIVE (1981c). *Leakage of Propane at Whitefriars Glass, Wealdstone, Middlesex, 20 November 1980* (London: HM Stationery Office)

HEALTH AND SAFETY EXECUTIVE (1981d). *Managing Safety* (London: HM Stationery Office)

HEALTH AND SAFETY EXECUTIVE (1982a). *The Explosion and Fire at Chemstar Limited on 6 September 1981* (London: HM Stationery Office)

HEALTH AND SAFETY EXECUTIVE (1982b). *Sizewell B* (London: HM Stationery Office)

HEALTH AND SAFETY EXECUTIVE (1982c). *Sizewell B Supplement 1: Degraded Core Analysis* (London: HM Stationery Office)

HEALTH AND SAFETY EXECUTIVE (1983a). *Evidence to Sizewell B Inquiry, NII 01*, Suppl. 13 (London: HM Stationery Office)

HEALTH AND SAFETY EXECUTIVE (1983b). *The Fire and Explosions at B&R Hauliers, Salford 25 September 1982* (London: HM Stationery Office)

HEALTH AND SAFETY EXECUTIVE (1983c). *Health and Safety Statistics 1983* (London: HM Stationery Office)

HEALTH AND SAFETY EXECUTIVE (1983d). *Her Majesty's Inspectors of 1933–1983. Essays to Commemorate 150 Years of Health and Safety Inspection* (London: HM Stationery Office)

HEALTH AND SAFETY EXECUTIVE (1983e). *The Shell Expro and Essochem Installations in Fife, Scotland. A Report by the Health and Safety Executive on the Independent Safety Audits of the Land Terminal Operations* (London: HM Stationery Office)

HEALTH AND SAFETY EXECUTIVE (1984). The Factory Inspectorate. *Plant Engr*, **28**(2), 24

HEALTH AND SAFETY EXECUTIVE (1985a). *The Abbeystead Explosion* (London: HM Stationery Office)

HEALTH AND SAFETY EXECUTIVE (1985b). *Deadly Maintenance* (London: HM Stationery Office)

HEALTH AND SAFETY EXECUTIVE (1985c). *Monitoring Safety* (London: HM Stationery Office)

HEALTH AND SAFETY EXECUTIVE (1985d). *The Putney Explosion* (London: HM Stationery Office)

HEALTH AND SAFETY EXECUTIVE (1985e). *The Summit Tunnel Fire* (London: HM Stationery Office)

HEALTH AND SAFETY EXECUTIVE (1986a). *Effective Policies for Health and Safety* (London: HM Stationery Office)

HEALTH AND SAFETY EXECUTIVE (1986b). *Emergency Plans for Civil Nuclear Installations* (HM Stationery Office)

HEALTH AND SAFETY EXECUTIVE (1986c). *Health and Safety Statistics 1985–86* (London: HM Stationery Office)

HEALTH AND SAFETY EXECUTIVE (1987a). *Dangerous Maintenance* (London: HM Stationery Office)

HEALTH AND SAFETY EXECUTIVE (1987b). *Programmable Electronic Systems in Safety Related Applications: Part 1*, An Introductory Guide (London: HM Stationery Office)

HEALTH AND SAFETY EXECUTIVE (1987c). *Programmable Electronic Systems in Safety Related Applications: Part 2*, General Technical Guidelines (London: HM Stationery Office)

HEALTH AND SAFETY EXECUTIVE (1987d). *The Work of a Factory Inspector* (London: HM Stationery Office)

HEALTH AND SAFETY EXECUTIVE (1988a). *Comments Received on 'The Tolerability of Risk from Nuclear Power Stations'* (London: HM Stationery Office)

HEALTH AND SAFETY EXECUTIVE (1988b). *COSHH Assessments* (London: HM Stationery Office)

HEALTH AND SAFETY EXECUTIVE (1988c). *The Tolerability of Risks from Nuclear Power Stations* (London: HM Stationery Office)

HEALTH AND SAFETY EXECUTIVE (1989a). *The Fires and Explosion at BP Oil* (Grangemouth) *Refinery Ltd* (London: HM Stationery Office)

HEALTH AND SAFETY EXECUTIVE (1989b). *Human Factors in Industrial Safety* (London: HM Stationery Office)

HEALTH AND SAFETY EXECUTIVE (1989c). *Noise at Work. Noise Guide Nos 1 and 2* (London: HM Stationery Office)

HEALTH AND SAFETY EXECUTIVE (1989d). *Quantified Risk Assessment: its Input to Decision Making* (London: HM Stationery Office)

HEALTH AND SAFETY EXECUTIVE (1989e). *Risk Criteria for Land Use Planning in the Vicinity of Major Hazard Installations* (London: HM Stationery Office)

HEALTH AND SAFETY EXECUTIVE (1989f). Safety in flame cutting and welding with compressed gases. *Plant Engr*, Mar./Apr., 28

HEALTH AND SAFETY EXECUTIVE (1990a). *A Guide to the Health and Safety at Work etc. Act 1974* (London: HM Stationery Office)

HEALTH AND SAFETY EXECUTIVE (1990b). *Offshore Installations: Guidance on Design, Construction and Certification* (London: HM Stationery Office)

HEALTH AND SAFETY EXECUTIVE (1990c). *The Peterborough Explosion. A Report of the Investigation by the HSE into the Explosion of a Vehicle Carrying Explosives at Fengate Industrial Estate, Peterborough, on 22 March 1989* (London: HM Stationery Office).

HEALTH AND SAFETY EXECUTIVE (1991a). *Control of Exothermic Reactions* (video commentary) (London: HSE)

HEALTH AND SAFETY EXECUTIVE (1991b). *Successful Health and Safety Management* (London: HM Stationery Office)

HEALTH AND SAFETY EXECUTIVE (1992a). *Guide to the Factories Act* (London: HM Stationery Office)

HEALTH AND SAFETY EXECUTIVE (1992b). Health and Safety Statistics 1990/91 (London). *Issue of Employment Gazette*, **110**(9)

HEALTH AND SAFETY EXECUTIVE (1992c). *The Health and Safety System in Great Britain* (London: HM Stationery Office)

HEALTH AND SAFETY EXECUTIVE (1992d). *Safety Assessment Principles for Nuclear Reactors* (London: HM Stationery Office)

HEALTH AND SAFETY EXECUTIVE (1993a). *The Costs of Accidents at Work* (London: HM Stationery Office)

HEALTH AND SAFETY EXECUTIVE (1993b). *The Fire at Allied Colloids Limited* (London: HM Stationery Office)

HEALTH AND SAFETY EXECUTIVE (1994a). *Criteria Document to Review the Effects on Health of Airborne Substances in the Ferrous Foundry Environment* (Sudbury, Suffolk: HSE Books)

HEALTH AND SAFETY EXECUTIVE (1994b). *The Fire at Hickson and Welch Ltd* (Sudbury, Suffolk: HSE Books)

HEALTH AND SAFETY EXECUTIVE – see also Appendix 28

HEALTH AND SAFETY EXECUTIVE, ACCIDENT PREVENTION ADVISORY UNIT(1992). *The Management of Health and Safety* (London: The Industrial Soc.)

HEALY, F. (1959). *Notes on the Basis of Outside Safety Distances for Explosives Involving the Risk of Mass Explosion. Rep. 3/7/EXPLOS/43.* Explosives Storage and Transport Committee, Min. of Defence

HEALY, F. (1965). Private communication (to V.C. Marshall)

HEALY, R.J. (1969). *Emergency and Disaster Planning* (New York: Wiley)

HEARD, D.B. (1986). Fail-safe devices for the prevention of hazardous materials spills. *J. Haz. Materials*, **13**(2), 233

HEARFIELD, F. (1970). *Design for Safety – Fire and the Modern Large-scale Chemical Plant (Course).* Dept of Chem. Engng, Teesside Polytechnic, Middlesbrough

HEARFIELD, F. (1974). Fire and the large-scale chemical plant. *Course on Process Safety – Theory and Practice.* Dept of Chem. Engng, Teesside Polytechnic, Middlesbrough

HEARFIELD, F. (1975). Total loss control. *Course on Process Safety – Theory and Practice.* Dept of Chem. Engng, Teesside Polytechnic, Middlesbrough

HEARFIELD, F. (1976). The philosophy of loss prevention. *Chem. Engr, Lond.*, **308**, 257

HEARFIELD, F. (1979). A simpler plant: trends in plant design and adipic acid reactors. *Design 79* (London: Instn Chem. Engrs)

HEARFIELD, F. (1980a). Adipic acid reactor development – with benefits in energy and safety. *Chem. Engr, Lond.*, **361**, 625

HEARFIELD, F. (1980b). The hazards of pressure testing. *Loss Prevention and Safety Promotion 3*, vol. 3, p. 1359

HEARN, G. (1991). Coping with electrostatic hazards. *Chem. Engng*, **98**(11), 188

HEARN, R.W. (1969). Safety organization in the works. In Handley, W. (1969), *op. cit.*, p. 414

HEATH, C.W., FALK, H. and CREECH, J.L. (1975). Characteristics of cases of angiosarcoma of the liver among vinyl chloride workers in the United States. *Ann. N.Y. Acad. Sci.*, **246**, 231

HEATHERSHAW, R. (1978). Proposed legislation on pressurised systems. *Chem. Ind.*, Jan. 21, 45

HEATON, B. and HANDLEY, R. (1988). Water treatment. *Plant Engr*, **32**(2), 29

HEATON, C.A. (1985). *The Chemical Industry* (Glasgow: Blackie); 1993, 2nd ed. (London: Chapman & Hall)

HECK, W.B. and JOHNSON, R.L. (1962). Aluminum alkyls. Safe handling. *Ind. Engng. Chem.*, **54**(12), 35

HECK, W.W., DAINES, R.H. and HINDAWI, I.J. (1970). Other phytotoxic pollutants. In Jacobson, J.S. and Hill, A.C. (1970), *op. cit.*

HECKLER, N.B. (1969). Finding a profit in failures. *Chem. Engng*, **76**(Nov. 3), 100

HECKLER, N.B. *et al.* (1962). Cast iron valves unsafe for hydrocarbon service. *Hydrocarbon Process.*, **41**(10), 144

HECKMAN, A.T. (1993). Vibration monitoring yields big benefits. *Chem. Engng*, **100**(5), 126

HEDGPATH, J.W. (1965). Bodega Head – a partisan view. *Bull. Atom. Sci.*, 21(Mar.), 2

HEDLEY, B., POWERS, W. and STOBAUGH, R.B. (1970). Methanol – how, where, who – future. *Hydrocarbon Process.*, **49**(7), 131; **49**(8), 117; **49**(9), 275

HEDRICK, L.J. (1993). Improve the sequence descriptions of batch control projects. *Chem. Engng Prog.*, **89**(2), 40

HEEMSKERK, A.H. (1986). Process safety analysis: the batch reaction system. *Loss Prevention and Safety Promotion 5*, p. 55.1

HEEMSKERK, A.H. and FORTUIN, J.M.H. (1980). Aspects of a safety analysis of a liquid phase reaction system. *Loss Prevention and Safety Promotion 3*, vol. 3, p. 1315

HEEMSKERK, A.H. and SCHUURMAN, P. (1989). Investigation into the initiation of a detonation of molten ammonium nitrate by falling objects. *Loss Prevention and Safety Promotion 6*, p. 92.1

HEENAN, W.A. and SERTH, R.W. (1986). Detecting errors in process data. *Chem. Engng*, **93**(21), 99

DE HEER, H.J. (1973). Calculating how much safety is enough. *Chem. Engng*, **80**(Feb. 19), 121

DE HEER, H.J. (1974). A basic theory on the probability of failure of safeguarding systems. *Loss Prevention and Safety Promotion 1*, p. 101

DE HEER, H.J. (1975). Choosing an economical automatic protective system. *Chem. Engng*, **82**(Mar. 17), 73

DE HEER, H.J., KORTLANDT, D. and HANSEN, J. (1980). Comparative risk analysis of processing plant. *Loss Prevention and Safety Promotion 3*, vol. 2, p. 455

VAN HEERDEN, C. (1953). Autothermic processes. *Ind. Engng Chem.*, **45**, 1242

HEERES-SANITATSINSPEKTION DES REICHSKRIEGSMINISTERIUM (1934). *Sanitätsbericht über das deutsche Heer im Weltkrieg 1914–1918*, vol. 3 (Berlin: E.S. Mittler)

HEESTAND, J., SHIPMAN, C.W. and MEADER, J.W. (1983). A predictive model for rollover in stratified LNG tanks. *AIChE*, **29**, 199

HEETVELD, H. (1993). The objectives and scope of plant technical safety reviews. *Health, Safety and Loss Prevention*, p. 457

HEGEDUS, L.L. and MCCABE, R.W. (1984). *Catalyst Poisoning* (New York: Dekker)

HEGLER, C. (1928). On the mass poisoning by phosgene at Hamburg. *Deutsch. Med. Mschr.*, **37**, 1551

HEIBERG-ANDERSEN, G. (1990). *Piper Alpha Public Inquiry. Transcript, Day 165*

HEICKLEN, J. (1976). *Atmospheric Chemistry* (New York: Academic Press)

HEIDELBERG, E. (1959). *Arch. Elektron.*, **44**(2), 116

HEIDELBERG, E. (1967). Generation of igniting brush discharges by charged layers on earthed conductors. *Static Electrification*, **2**, 147

HEIDELBERG, E. (1970a). *Discharges from Electrostatically Charged Non-conductive Layers on Metal Surfaces. Mitteilung 80.* Physikalisch Technische Bundesanstalt

HEIDELBERG, E. (1970b). Electrostatical charges and risk of explosion in the application of synthetics. *Prevention of Occupational Risks*, **1**

HEIDELBERG, E. (1971). Ignition hazards by the use of non-conductive electrostatically chargeable solid materials. *Electrical Safety*, **1**, 68

HEIDELBERG, E. and SCHON, G. (1960). Explosion hazards due to the electrostatic charging of plastic containers. *Die Berufsgenossenschaft/Betriebssicherheit*, **265**

HEIDELBERG, E. and SCHON, G. (1961). Ignition of explosible mixtures by space charge clouds. *Die Berufsgenossenschaft/Betriebssicherheit*, May, 196

HEIDELBERG, E., NABERT, K. and SCHON, G. (1958a). Electrical charging of carbon dioxide. *Arbeitsschutz*, **221**

HEIDELBERG, E., NABERT, K. and SCHON, G. (1958b). Ignition dangers from electrostatically charged carbon dioxide clouds. *Arbeitsschutz*, **242**

HEIDER, F. (1958). *The Psychology of Interpersonal Relations* (New York: Wiley)

HEIDORN, K.C., MURPHY, M.C., IRWIN, P.A., SAHOTA, H., MISRA, P.K. and BLOXAM, R. (1992). Effects of obstacles on the spread of a heavy gas – wind tunnel simulations. *J. Haz. Materials*, **30**, 151

HEIKES, K.E., RANSOHOFF, L.M. and SMALL, R.D. (1990). Numerical simulation of small area fires. *Atmos. Environ.*, **24A**, 297

HEIKKILA, A.-M. (1999). *Inherent Safety in Process Plant Design: An Index-Based Approach* (Espoo, Finland: VTT (Technical Research Centre of Finland))

HEIKKILA, A.-M., HURME, A.M. and JäRVELäINEN, M. (1996). Safety considerations in process synthesis. *Comp. Chem. Engng.* **20**, Suppl. A, S115–S120

HEIKKILA, J. and HEINO, P. (1992). Intelligent tool for screening and preliminary analysis of process hazards. *Loss Prevention and Safety Promotion 7*, vol. 3, 151–1

HEIL, J.A. (1995). *Inherent Safety Characteristics of Innovative Nuclear Reactors* (Petten, Netherlands: Netherlands Energy Res. Foundation)

HEILBRUNNER, S. (ed.) (1988) *ADA in Industry* (Cambridge: Cambridge Univ. Press)

HEIMBUCH, J.A. and WILHELMI, A.R. (1985). Wet air oxidation – a treatment means for aqueous hazardous waste streams. *J. Haz. Materials*, **12**(2), 187

HEINHOLD, D.W., PAINE, R.J., WALKER, K.C. and SMITH, D.G. (1986). Evaluation of the Airtox dispersion algorithms using data from heavy gas field experiments. *Fifth Joint Conf. on Applications of Air Pollution Modeling* (Boston, MA: Am. Met. Soc.), p. 299

HEINHOLD, D.W., WALKER, K.C. and PAINE, R.J. (1987). Effect of sloping terrain on the dispersion and transport of heavier-than-air gaseous releases. *Vapor Cloud Modeling*, p. 584

HEINO, P., SUOKAS, J. and KARVONEN, I. (1989a). Computer aided safety analysis of process systems. *Safety Engng Anniversary Seminar on Automation Safety in Design in Process and Manufacturing Industry*, Espoo, Finland

HEINO, P., SUOKAS, J. and KARVONEN, I. (1989b). Expert system for hazop-studies. *Loss Prevention and Safety Promotion 6*, p. 112.1

HEINO, P., POUCET, A. and SUOKAS, J. (1992). Computer tools for hazard identification, modelling and analysis. *J. Haz. Materials*, **29**, 445

HEINO, P., KARVONEN, I., PETTERSEN, T., WENNERSTEIN, R. and ANDERSEN, T. (1994). Monitoring and analysis of hazards using HAZOP-based plant safety model. *Reliab. Engng System Safety*, **44**, 335

HEINOLD, D.W. (1992). Quantifying potential off-site impacts of SARA Title III releases. *J. Haz. Materials*, **31**, 297

HEINRICH (1973). Report on the calculation of the size of pressure-relieving apertures. *Prevention of Occupational Risks*, **2**, 309

HEINRICH, H.J. (1966a). Determination of relief openings for protection of plants subject to explosion in the chemical industry. *Chem. Ing. Tech.*, **38**, 1125

HEINRICH, H.J. (1966b). Sizing of pressure relief openings for the explosion protection of plants in the chemical industry. *Chem. Ing. Tech.*, **38**(11), 1125

HEINRICH, H.J. (1972). Technical problems of the determination of the safety and technical parameters of flammable dusts. *Int. Symp. on Dust and Explosion Risks in Industry* (Heidelberg: Berufsgenossenschaft der Chem. Ind.), p. 51

HEINRICH, H.J. (1974). Pressure relief of dust explosions. *Arbeitsschutz*, **11**, 314

HEINRICH, H.J. (1975). *Schadenprisma*, **4**, 7

HEINRICH, H.J. and KOWALL, R. (1971). Results of recent pressure relief experiments in connection with dust explosions. *Staub Reinhaltung der Luft*, **31**(4), 10

HEINRICH, H.J. and KOWALL, R. (1972). On the course of pressure-relieved dust explosions with ignition through turbulent flames. *Staub Reinhaltung der Luft*, **32**(7), 22 (English ed.)

HEINRICH, H.W. (1959). *Industrial Accident Prevention*, 4th ed. (New York: McGraw-Hill)

HEINRICH, H.W., PETERSEN, D. and ROOS, N. (1980). *Industrial Accident Prevention*, 5th ed. (New York: McGraw-Hill)

HEINRICH, M., GEROLD, E. and WIETFELDT, P. (1988). Large scale propane release experiments over land at different atmospheric stability classes. *J. Haz. Materials*, **20**, 287; **22**(3) 407

HEINRICHS, R. and HALLENBECK, R.M. (1986). Legal implications of chemical releases. *Chem. Engng Prog.*, **82**(11), 16

HEINZE, A.J. (1979). Pressure vessel design for process engineers. *Hydrocarbon Process.*, **58**(5), 181

HEIPLE, C.R. and CARPENTER (1987). Acoustic emission produced by deformation of metal and alloys. *J. Acoustic Emission*, **6**, 177, 215

HEISING, C.D. (1983). Development of unavailability expressions for one- and two-component systems with periodic testing and common-cause failures. *Reliab. Engng*, **6**, 229

HEISING, C.D. (1987). Risk assessment: controlling hazardous materials. *J. Haz. Materials*, **15**(1/2), 123

HEISING, C.D. and LUCIANI, D.M. (1987). Application of a computerized methodology for performing common cause failure analysis: the Mocus–Bacfire Beta Factor (MOBB) code. *Reliab. Engng*, **17**(3), 193

HEITMAN, H. (1983). Design and testing of ultraviolet flame radiation detectors for automatic fire detection. *Fire Safety J.*, **6**, 183

HEITNER, I. (1968). How to estimate plant noises. *Hydrocarbon Process.*, **47**(12), 67

HEITNER, I. (1970). A critical look at API RP 521. *Hydrocarbon Process.*, **49**(11), 209

HEITNER, I., TRAUTMANIS, T. and MORRISSEY, M. (1983a). Relieving requirements for gas-filled vessels exposed to fire. *API Proc, Refining Dept, 48th Midyear Mtg*, **62**, 112

HEITNER, I., TRAUTMANIS, T. and MORRISSEY, M. (1983b). When gas-filled vessels are exposed to fire ... *Hydrocarbon Process.*, **62**(11), 263

HELANDER, M.G. (1987). Design of visual displays. In Salvendy, G. (1987), *op. cit.*, p. 507

HELD, M. (1968). Calculation of fragment mass distribution of fragmenting munitions. *Explosivstoffe*, **16**, 241

HELD, M. (1979). Consideration of the mass distribution of fragments by natural fragmentation in combination with preformed fragments. *Propellants Explos.*, **1**, 20

HELD, M. (1983a). Blast waves in free air. *Propellants, High Explosives, Pyrotechnics*, **8**, 1

HELD, M. (1983b). TNT equivalent. *Propellants, High Explosives, Pyrotechnics*, **8**, 158

HELD, M. (1990). Fragment mass distribution of HE projectiles. *Propellants, High Explosives, Pyrotechnics*, **15**, 254

HELD, M. (1991). Fragment mass distribution of 'secondary fragments'. *Propellants, High Explosives, Pyrotechnics*, **16**, 21

HELGUERO, V. (1986). *Piping Stress Handbook*, 2nd ed. (Houston, TX: Gulf Publ. Co.)

HÉLIE, M. (1884). *Traité de Balistique Expérimentale*, vol. 1 (Paris: Gauthier-Villars)

HELLER, F.J. (1965). A method for determining filling densities for anhydrous ammonia tank cars. *Ammonia Plant Safety*, **7**, 46

HELLER, F.J. (1981). Ammonia tank car safety. *Ammonia Plant Safety*, **23**, 132

HELLER, F.J. (1983). Safety relief valve sizing: API versus CGA requirements plus a new concept for tank cars. *API Proc., Refining Dept, Forty Eighth Midyear Mtg*, **62**, 123

HELLER, H. (1978). Choosing the right formula for calculator curve fitting. *Chem. Engng*, **85**(Feb. 13), 119

HELLHAKE, F.J. (1981). Setting up a preventive maintenance program. *Chem. Engng*, **88**(Jul. 13), 145

HELLIWELL, P.R. and BOSSANYI, J. (eds) (1975). *Pollution Criteria for Estuaries* (London: Pentech)

HELMERS, E.N. and SCHALLER, L.C. (1982). Calculated process risks and hazards management. *Plant/Operations Prog.*, **1**, 190

HELMERS, S., TOP, F.H. and KNAPP, L.W. (1971). Ammonia injuries in agriculture. *J. Iowa Med. Soc.*, **61**, 271

HELMESTE, R.J. and PHILLIPS, C.R. (1981). A risk evaluation of the chlorine industry in North America. *Risk Analysis*, **1**, 259

HELMHOLTZ, H. (1879). Studies on electrical boundary layers. *Ann. Phys. Chemie.*, **7**(7), 337

VON HELMHOLTZ (1895). *Popular Lectures on Scientific Subjects* (London: Longmans)

HELMREICH, R.L. (1984). Cockpit management attitudes. *Hum. Factors*, **26**, 583

HELMS, H.L. (1983). *McGraw-Hill Computer Handbook* (Maidenhead: McGraw-Hill)

HELMS, J. (1981). Threat perceptions in acute chemical disasters. *J. Haz. Materials*, **4**(4), 321

HELPS, K.A. (1986). Some software verification tools and methods for airborne safety-critical software. *Software Engng J.*, **1**(6), 248

HELSBY, G.H.A. and WHITE, R.F. (1985). Criteria for use in assessment and control of major hazards. *The Assessment and Control of Major Hazards* (Rugby: Instn Chem. Engrs), p. 273

HELSBY, G.H., PUGH, J. and HENSLEY, G. (1982). Application of regulatory control to high risk plant. In Green, A.E. (1982), *op. cit.*, p. 129

HELWIG, C., KLEE, C. and HUBNER, W. (ca. 1989). Mass and shape distributions of secondary fragments of an SAP warhead. *Int. Symp on Ballistics, Brussels*, p. 453

VAN HEMEL, S.B., CONNELLY, E.M. and HAAS, P. (1990). Development of leading indicators of process safety. *Plant/Operations Prog.*, **9**, 181

HEMEON, W.C.L. (1963). *Plant and Process Ventilation*, 2nd ed. (New York: Ind. Press)

HEMPEL, C.G. (1952). *Fundamentals of Concept Formation in Empirical Science* (Chicago, IL: Univ. of Chicago Press)

HEMPSEED, J.W. and ORMSBY, R.W. (1991). Explosion within a helium purifier. *Plant/Operations Prog.*, **10**, 184

HEMS, P. (1979). The increasing use of water spray systems. *Fire Surveyor*, **8**(4), 21

HEMS, P. (1983). Water spray systems. *Fire Surveyor*, **12**(4), 11

HEMSWORTH, B.N. (1974). Evaluation of the potential hazards to health by environmental chemicals. *Course on Process Safety – Theory and Practice*. Dept of Chem. Engng, Teesside Polytechnic, Middlesbrough

HENDERSHOT, D.C. (1987). Safety considerations in the design of batch processing plants. *Preventing Major Chemical Accidents*, p. 3.1

HENDERSHOT, D.C. (1992). Documentation and utilization of the results of hazard evaluation studies. *Plant/Operations Prog.*, **11**, 256

HENDERSHOT, D.C. (1988). Alternatives for reducing the risks of hazardous material storage facilities. *Environ. Progress* **7**(3), (August), 180–4

HENDERSHOT, D.C. (1993). Inherently safer plants. *In Guidelines for Engineering Design for Process Safety* (New York: American Institute of Chemical Engineers, Center for Chemical Process Safety), pp. 5–52

HENDERSHOT, D.C. (1995a). Conflicts and decisions in the search for inherently safer process options. *Process Safety Prog.* **14**(1) (Jan.), 52–6

HENDERSHOT, D.C. (1995b). Some thoughts on the difference between inherent safety and safety. *Process Safety Prog.* **14**(4) (Oct.), 227–8

HENDERSHOT, D.C. (1997a). Measuring inherent safety, health and environmental characteristics early in process development. *Process Safety Prog.* **16**(2) (Summer), 78–9

HENDERSHOT, D.C. (1997b). Safety through design in the chemical process industry: inherently safer process design. *Symp.: Benchmarks for World Class Safety Through Design Programs*, August 19–20, 1997, Bloomingdale, IL (Itasca, IL: National Safety Council, Institute for Safety Through Design)

HENDERSHOT, D.C. (1998). Inherent safety strategies for process chemistry. *Chem. Health Saf.* **5**(4) (July/August), 18–22

HENDERSHOT, D.C. (1999). Safety through design in industry: application in the chemical industry. *In Safety Through Design* (ed. W. C. Christensen and F. A. Manuele), 195–205 (Itasca, IL: National Safety Council, NSC Press)

HENDERSHOT, D.C. (2000). Process minimization: making plants safer. *Chem. Eng. Prog.* **96**(1) (January), 35–40

HENDERSHOT, D.C. (2002). A checklist for inherently safer chemical reaction process design and operation. *17th Annu. Int. Conf. and Workshop: Risk, Reliability, and Security*, Oct. 8–11, 2002, Jacksonville, FL, 211–218. (New York: American Institute of Chemical Engineers

HENDERSHOT, D.C. and POST, R.L. (2000). Inherent safety and reliability in plant design. *2000 Mary Kay O'Connor Process Saf. Cent. Symp. Proc.: Beyond Regulatory Compliance, Making Safety Second Nature*, Oct. 24–25, 2000, 268–281 (College Station, TX: Mary Kay O'Connor Process Safety Center)

HENDERSON, D. (1975). Explosion of ammonia liquor tank. *Ammonia Plant Safety*, **17**, p. 132

HENDERSON, D.K. and TYLER, B.J. (1988). Dual ignition temperatures for dust layers. *J. Haz. Materials*, **19**(2), 155

HENDERSON, E. (1941). Combustible gas mixtures in pipelines. *Proc. Pacific Coast Gas Ass.*, **32**(Sep.), 98

HENDERSON, J.M. and KLETZ, T.A. (1976). Must plant modifications lead to accidents. *Process Industry Hazards*, p. 53

HENDERSON, N.C., SRINIVANSAN, S. and GOWER, C.W. (1976). Techniques to prevent or reduce discharges of hazardous chemicals from marine vessels. *Control of Hazardous Material Spills 3*, p. 238

HENDERSON, S., KING, K. and STONE, G. (1990). Stainless steel – its types and uses. *Chem. Engr, Lond.*, **479**, 22

HENDERSON, T.A. (1989). *Piper Alpha Public Inquiry. Transcript, Days 68, 69, 83*

HENDERSON, Y. and HAGGARD, H.W. (1922). The elimination of industrial odours. *Ind. Engng Chem.*, **14**, 548

HENDERSON, Y. and HAGGARD, H.W. (1943). *Noxious Gases* (New York: Reinhold)

HENDERSON-SELLERS, B. (1986). A theoretical analysis of the plume dispersion parameter sigma z. *Fifth Joint Conf. on Applications of Air Pollution Modeling* (Boston, MA: Am. Met. Soc.), p. 352

HENDRICK, K. and BENNER, L. (1987). *Investigating Accidents with S-T-E-P.* (Dekker: New York)

HENDRICKX, S. and LANNOY, A. (1983). Probabilistic and deterministic safety study of the sea-borne traffic of liquefied gases in the vicinity of a nuclear site. *Loss Prevention and Safety Promotion 4*, vol. 1, p. D1

HENDRIE, G.C. and SONNENFELDT, R.W. (1963). Computer reliability – what's that? *ISA J.*, **10**(1), 51

HENDRIKS, N.A. (1980). Safety wall systems for ammonia storage protection. *Ammonia Plant Safety*, **21**, 69

HENDRY, A.W. and HORSBURGH, E.L. (1915). Some general notes on suffocation by poisonous gases, with detailed notes on one fatal case. *Br. Med. J.*, Jun. 5, 964

HENDRY, A.W. and SINHA, B.P. (1971). Shear tests on full scale, single-storey brickwork structures subjected to precompression. *Civ. Engng Pub. Works Rev.*, **66**(785), 1339

HENDRY, A.W., SINHA, B.P. and MAURENBRECHER, A.H.P. (1971). Full scale tests on the lateral strength of brick cavity walls with precompression. *Proc. Br. Ceramic Soc.*, **21**, 165

HENDRY, J.E., RUDD, D.F. and SEADER, J.D. (1973). Synthesis in the design of chemical processes. *AIChE J.*, **19**, 1

HENDRY, W.D. (ed.) (1975). *The Collection Analysis and Interpretation of Accident Data* (Sheffield: Safety in Mines Res. Estab.)

HENEBRY, M. and BROWN, D. (1978). What every manager should know about environmental regulations. *Hydrocarbon Process.*, **57**(8), 87

HENGLEIN, F.A. (1968). *Grundriss der chemischen Technik*, 12 ed. (Weinheim: Verlag Chemie)

HENGSTENBERG, J., STURM, B. and WINKLER, O. (1964). *Messen und Regeln in der chemischen Technik* (Berlin: Springer)

HENKEL, M.J., HUMMEL, H. and SPAULDING, W.P. (1949). Theory of propagation of flames. Pt 3 – Numerical integrations. *Combustion*, **3**, 135

HENKEL, M.J., SPAULDING, W.P. and HIRSCHFELDER, J.O. (1949). Theory of propagation of flames. Pt 2 – Approximate solutions. *Combustion*, **3**, 127

HENLEY, E.J. (1971). Plant and process reliability. *Instrum. Pract.*, Nov., 625

HENLEY, E.J. (1977). SCEH – A Computer Program that Converts Reliability Block Diagrams to Fault Trees, Univ. of Houston, Houston, TX

HENLEY, E.J. and GANDHI, S.L. (1975). Process reliability analysis. *AIChE J.*, **21**, 677

HENLEY, E.J. and HOSHINO, H. (1977). Storage tanks and plant availability. *Am. Inst. Chem. Engrs, Eleventh Loss Prevention Symp.* (New York)

HENLEY, E.J. and KUMAMOTO, H. (1977). Comment on 'Computer aided synthesis of fault trees'. *IEEE Trans. Reliab.*, **R-26**, 316

HENLEY, E.J. and KUMAMOTO, H. (1981). *Reliability Engineering and Risk Assessment* (Englewood Cliffs, NJ: Prentice-Hall)

HENLEY, E.J. and KUMAMOTO, H. (1985). *Designing for Reliability and Safety Control* (Englewood Cliffs, NJ: Prentice-Hall)

HENLEY, E.J. and KUMAMOTO, H. (1992). *Probabilistic Risk Assessment, Reliability Engineering, Design and Analysis* (Englewood Cliffs, NJ: Prentice Hall) (rev. ed. of Henley, E.J. and Kumamoto, H. (1981), *op. cit.*)

HENLEY, E.J. and LYNN, J.W. (eds) (1976). *Generic Techniques in Systems Reliability Assessment* (Leyden: Noordhoff)

HENLEY, E.J. and WILLIAMS, R.A. (1973). *Graph Theory in Modern Engineering* (New York: Academic Press)

HENN, D.S. and SYKES, R.I. (1992). Large-eddy simulation of dispersion in the convective boundary layer. *Atmos. Environ.*, **26A**, 3145

HENNELL, M.A. (1987). Requirements, specification and testing. In Littlewood, B. (1987b), *op. cit.*, p. 19

HENNY, R.W. and CARLSON, R.H. (1968). Distribution of natural missiles resulting from cratering explosions in hard rock. *Ann N.Y. Acad. Sci*, **152**(art. 1), 404

HENRIKSEN, J.A. (1983). Systems engineering approach to pharmaceutical plant design. *Plant/Operations Prog.*, **2**, 79

HENRIQUES, F.C. (1947). Studies of thermal injury. V, The predictability and the significance of thermally induced rate processes leading to irreversible epidermal injury. *Arch. Path. (Lab. Med.)*, **43**, 489

HENRY, C.D. and BRAUER, D.C. (1986). A safety prediction technique for nuclear power plants. *Reliab. Engng*, **16**(4), 265

HENRY, G.A. (1945). Blast injuries of the ear. *Laryingoscope*, 55

HENRY, I.G. (1967). The Gurney formula and related approximations for high-explosive deployment of fragments. Rep. PUB-198. Hughes Aircraft Co. (unpublished)

HENRY, M. (1985). Expert systems and all that. *Chem. Engr, Lond.*, **420**, 32

HENRY, M., BAILEY, G. and ABOU-LOUKH, S. (1984). Sequences in the control of processes. *Chem. Engr, Lond.*, **405**, 9

HENRY, M.F. (1985). NFPA's consensus standards at work. *Chem. Engng Prog.*, **81**(8), 20

HENRY, M.F. (1986). Flammable and combustible liquids. *Fire Protection Handbook*, 16th ed. (ed. A.E. Cote and J.M. Linville), p. 5.28

HENRY, P.S.H. (1953). Survey of generation and dissipation of static electricity. *Br. J. Appl. Phys*, **4**(Suppl. 2), S6

HENRY, P.S.H. (1971a). Risks of ignition due to static on outer clothing. *Static Electrification*, **3**, 212

HENRY, P.S.H. (1971b). Static in Industry. *Static Electrification*, **3**, 195

HENRY, R.C. (1987). Psychophysics, visibility and perceived atmospheric transparency. *Atmos. Environ.*, **21**, 159

HENRY, R.E. (1968). A Study of One- and Two-component, Two-phase Critical Flows at Low Qualities. Rep. ANL-7430. Argonne Nat. Lab., Argonne, IL

HENRY, R.E. (1970). The two-phase critical discharge of initially saturated or subcooled liquid. *Nucl. Sci. Engng*, **41**, 336

HENRY, R.E. and FAUSKE, H.K. (1971). The two-phase critical flow of one-component mixtures in nozzles, orifices and short tubes. *J. Heat Transfer*, **93**(May), 179

HENRY, R.E., FAUSKE, H.K. and MCCOMAS, S.T. (1970). Two phase critical flow at low qualities – II, Analysis. *Nucl. Sci. Engng*, **41**, 92

HENRY, R.W. (1979). Let's profit from experience. *Loss Prevention*, **12**, 1

HENRY, T.A. (1979). The simple approach to condition monitoring. *Terotechnica*, 1, 131

HENRY, T.A. (1990). Availability centred maintenance. *Plant Engr*, Jul./Aug., 27

HENRY, T.A. (1993a). Availability centred maintenance. *Plant Engr*, **37**(4), 23

HENRY, T.A. (1993b). The management of condition-based maintenance. *Plant Engr*, **37**(1), 27

HENSCHLER, D. (1979). MAK and RRK Values in workplace protection. In BASF (1979), *op. cit.*, p. 81

HENSCHLER, D. and LAUX, W. (1960). On the specificity of a tolerance increase by repeated inhalation of pulmonary oedema producing gases. *Arch. Exp. Path. Pharmacol.*, **239**, 433

HENSEL, W., KRAUSE, U., JOHN, W. and MACHNOW, K. (1994). Critical parameters for the ignition of dust layers at constant heat flux boundary conditions. *Process Safety Prog.*, **13**, 210

HENSHAW, T.L. (1979). Improve power pump suction systems. *Hydrocarbon Process.*, **58**(10), 161

HENSHAW, T.L. (1981). Reciprocating pumps. *Chem. Engng*, **88**(Sep. 21), 105

HENSLEY, G. (1968). Safety assessment – a method for determining the performance of alarm and shutdown systems for chemical plants. *Measmt Control*, 1, T72

HENSLEY, G. (1971). Process plant reliability. *Instrum. Pract.*, **25**, 624

HENSTOCK, W.H. (1986). NO*x* in the cryogenic hydrogen recovery section of an olefins production unit. *Plant/Operations Prog.*, **5**, 232

HENTHORN, G. and LEDNICKY, R. (1992). Sealing tips for plastic-lined pipe. *Chem. Engng*, **99**(11), 142

HENTHORNE, M. (1971–). Fundamentals of corrosion. *Chem. Engng*, 1971, **78**(May 17), 127; (Jun. 14), 120; (Jul. 26), 99; (Aug. 23), 89; (Sep. 20), 159; (Oct. 18), 139; (Nov. 15), 163; (Dec. 27), 73; 1972, **79**(Jan. 10), 103; (Feb. 7), 82; (Mar. 6), 113; (Apr. 3), 97

HEPPEL, L.A. *et al.* (1946). The toxicity of 1, 2 dichloroethane (ethylene dichloride) 5. *J. Ind. Hyg. Toxicol*, **28**, 113

HEPPLE, P. (ed.) (1971). *Lead in the Environment* (Barking: Applied Science)

HEPPLE, P. (ed.) (1973). *Outlook for Natural Gas – A Quality Fuel* (Barking: Applied Science)

HERBERT, K.J. and RAWLINGS, R.L. (1985). Energy conservation in flare purging systems. *API 50th Midyear Mtg*, **64**, 188

HERBERT, M.R. (1984). A review of on-line diagnostic aids for nuclear power plant operators. *Nucl. Energy*, **23**, 259

HERBERTSSON, G. (1992). Some improvements in design of atmospheric ammonia storage tanks of the double integrity type. *Plant/Operations Prog.*, **11**, 126

HERMANCE, C.E. (1973). Ignition analysis of adiabatic homogeneous systems including reactant consumption. *AIAA J.*, **11**, 1728

HERMANN, K., SCHECKER, H.G. and SCHOFT, H. (1992). Direct contact condensation of released vapour–gas mixtures in a jet condenser. *Loss Prevention and Safety Promotion 7*, vol. 2, 98.1

HERNANDEZ, J. and CRESPO, A. (1992). PHOENICS. *J. Comput. Fluid Dynamics Appls*, **2**, 205

HERNANDEZ, L.A. (1968). Controlled-volume pumps. *Chem. Engng*, **75**(Oct. 21, 124

HERON, P.M. (1976). Accidents caused by plant modifications. *Loss Prevention*, **10**, 72

HERPOL, C. and VANDERVELDE, P. (1981/82). Use of toxicity test results and confrontation of some toxicity test methods with fire scenarios. *Fire Safety J.*, **4**(4), 271

HERR, E.A., BAKER, D.W. and FOX, A. (1968). The TEG – a test element for the control of quality and reliability of integrated circuits. *1968 Symp. on Reliability*, p. 201

HERRICK, L.K. (1976). Instrumentation for monitoring toxic and flammable work areas. *Chem. Engng*, **83**(Oct. 18), 147

HERRINGHAM, MAJOR-GENERAL SIR W. (1919). Medicine in the war. *Br. Med. J.*, Jan. 4, 20

HERRINGHAM, W.P. SIR (1920). Gas poisoning. *The Lancet*, **198**(Feb. 21), 423

HERRMANN, H.R. (1972). *Zuverlässigkeitsverfahren für die Prozesstechnik* (Munich: Oldenbourg)

HERSCH, S.M. (1968). *Chemical and Biological Warfare* (London: MacGibbon & Key)

DE HERTOGH, J. and ILLEGHEMS, M. (1974). Welding natural gas-filled pipelines. Metal Constr. *Br. Weld. J.*, **6**(7)

HERTZBERG, M. (1973). The theory of free ambient fires. The convectively mixed combustion of fuel reservoirs. *Combust. Flame*, **21**, 195

HERTZBERG, M. (1982). The flammability limits of gases, vapours and dusts: theory and experiment. In Lee, J.H.S. and Guirao, C.M. (1982), *op. cit.*, p. 3

HERTZBERG, M. (1985). Flame stretch extinction in flow gradients, flammability limits under natural convention. *Combustion*, **20**, 1967

HERTZBERG, M., JOHNSON, A.L., KUCHTA, J.M. and FURNO, A.L. (1977). The spectral radiance growth, flame temperatures and flammability behavior of large-scale, spherical combustion waves. *Combustion*, **16**, 767

HERTZBERG, M., CASHDOLLAR, K.L. and LAZZARA, C.P. (1981). The limits of flammability of pulverized coals and other dusts. *Combustion*, **18**, 717

HERTZBERG, M., CASHDOLLAR, K.L., NG, D.L. and CONTI, R.S. (1982). Domains of flammability and thermal ignitability for pulverized coals and other dusts: particle size dependencies and microscopic residue analyses. *Combustion*, **19**, 1169

HERTZBERG, M., CASHDOLLAR, K.L., ZLOCHOWER, I. and NG, D.L. (1985). Inhibition and extinction of explosions in heterogeneous mixtures. *Combustion*, **20**, 1691

HERTZBERG, M., CONTI, R.S. and CASHDOLLAR, K.L. (1985). Spark ignition energies for dust–air mixtures: temperature and concentration dependencies. *Combustion*, **20**, 1681

HERTZBERG, M., ZLOCHOWER, I.A. and CASDOLLAR, K. (1991). Explosibility of metal dusts. *Combust. Sci. Technol.*, **75**, 161

HERTZBERG, M., CASHDOLLAR, K.L., ZLOCHOWER, I.A. and GREEN, G.M. (1992). Explosives dust cloud combustion. *Combustion*, **24**, 1837

HERTZBERG, M., ZLOCHOWER, I.A. and CASHDOLLAR, K.L. (1992). Metal dust combustion: explosion limits, pressures and temperatures. *Combustion*, **24**, 1827

HERTZBERG, O. and LAMNEVIK, S. (1983). Unconfined gas clouds. *Loss Prevention and Safety Promotion 4*, vol. 1, p. H3

HERXHEIMER, K. (1899). On chloracne. Münch. Med. Wschr., **46**, 278

HERZBERG, F. (1966). *Work and the Nature of Man* (Cleveland, OH: World)

HERZOG, G.R. (1974). Recent major floating roof tank fires and their extinguishment. *Fire J.*, **68**(Jul.), 93

HERZOG, G.R. (1979). When there's a tank farm fire. . . *Hydrocarbon Process.*, **58**(2), 165

HERZOG, R.E. (1967). *Proc. Am. Petrol. Inst.*, Div. Refining, 47, 73

HERZOG, R.E., BALLARD, E.C. and HARTUNG, H.A. (1961). Evaluating electrostatic hazard during the loading of tank trucks. *Proc. Am. Petrol. Inst.*, **41**(VI), 36

HERZOG, R.E., BALLARD, E.C. and HARTUNG, H.A. (1964). Evaluating electrostatic hazard during the loading of tank trucks. In Vervalin, C.H. (1964a), *op. cit.* p. 178

HESEL, D.A. and SCHNADT, H. (1982). Emergency planning for large scale accidents using a standardized evacuation computer code. *Proc. Conf. on Hazardous Materials Spills* (Rockville, MD: Govt Insts), p. 399

HESELDEN, A.J.M. (1982). The interaction of sprinklers and fire venting. *Fire Surveyor*, **11**(5), 13

HESELDEN, A.J.M., THEOBALD, C.R. and BEDFORD, G.K. (1972). Thermal measurements on unprotected steel columns exposed to wood and petrol fires. *Fire Prev. Sci. Technol.*, **2**, 21

HESKESTAD, G. (1965). Hot-wire measurements in a plane turbulent jet. *J. Appl. Mech.*, **32**, 721

HESKESTAD, G. (1966). Hot-wire measurements in a radial turbulent jet. *J. Appl. Mech.*, **33**, 417

HESKESTAD, G. (1980). The role of water in suppression of fire: a review. *J. Fire Flammability*, **11**, 254

HESKESTAD, G. (1981). Peak gas velocities and flame heights of buoyancy-controlled turbulent diffusion flames. *Combustion*, **18**, 951

HESKESTAD, G. (1983a). Luminous heights of turbulent diffusion flames. *Fire Safety J.*, **5**, 103

HESKESTAD, G. (1983b). Virtual origins of fire plumes. *Fire Safety J.*, **5**, 109

HESKESTAD, G. (1984). Engineering relations for fire plumes. *Fire Safety J.*, **7**, 25

HESKESTAD, G. (1986). Smoke movement and venting. *Fire Safety J.*, **11**(1), 77

HESKESTAD, G. (1989). Note on maximum rise of fire plumes in temperature stratified ambients. *Fire Safety J.*, **15**, 271

HESKESTAD, G. (1991). A reduced-scale mass fire experiment. *Combust. Flame*, **83**, 293

HESKESTAD, G. and DELICHATSIOS, M.A. (1989). The initial convective flow in fire. *Fire Safety J.*, **15**, 471

HESKESTAD, G., KUNG, H.C. and TODTENKOPF, N.F. (1976). *Air Entrainment in Water Sprays and Spray Curtains* (paper). *ASME 76–WA/FE-40.* Am. Soc. Mech. Engrs, New York

HESKETH, H.E. (1979). *Air Pollution Control* (Ann Arbor, MI: Ann Arbor Science)

HESKETH, H.E. and CROSS, F.L. (1983). *Fugitive Emissions and Controls* (Ann Arbor, MI: Ann Arbor Science)

HESLER, W.E. (1990). Modular design – where it fits. *Chem. Engng Prog.*, **86**(10), 76

HESLINGER, G. (1985). Analysis of human reliability on performing a specific action. *Reliab. Engng*, **12**(2), 63

HESS, J.L. and KITTLEMAN, T.A. (1983). Predicting volatile emissions. *Chem. Engng Prog.*, **79**(11), 40

HESS, K. (1973). Formation of explosible gas clouds during expansion and measures for their avoidance. *Chem. Ing. Tech.*, **45**, 323

HESS, K. and STICKEL, R. (1970). Überdachentspannung von Gases aus chemsichen Anlagen. *Chem. Ing. Tech.*, **42**, 467

HESS, K., LEUCKEL, W., and STOECKEL, A. (1973). Formation of gas clouds during expansion above roof level and measures for their avoidance. *Chem. Ing. Tech.*, **45**, 323

HESS, K., HOFFMANN, W. and STOECKEL, A. (1974). Propagation processes after the bursting of tanks filled with liquid propane – experiments and mathematical model. *Loss Prevention and Safety Promotion 1*, p. 227

HESS, M. (1976). Ammonia; coal versus gas. *Hydrocarbon Process.*, **55**(11), 97

HESS, R.E. (1972). The national contingency plan – present and future. *Control of Hazardous Material Spills 1*, p. 55

HESS, W. (1911). Experiences with industrial poisonings and their relation to the Swiss factory liability insurance law. Dissertation, Univ. of Zurich

HESSE, D.J. (1991a). A computational procedure for calculating the mass of flammable vapor in a neutrally buoyant cloud. *Modeling and Mitigating the Consequences of Accidental Releases*, p. 511

HESSE, D.J. (1991b). Modeling the time and spatial dependence of the transition from a continuous plume to a detached vapor cloud (poster item). *Modeling and Mitigating the Consequences of Accidental Releases*, p. 643

HESSON, J.C. and PECK, R.C. (1958). Flow of two-phase carbon dioxide through orifices. *AIChE J.*, **4**, 207

HETHCOAT, H.G. (1990). Minimize refinery waste. *Hydrocarbon Process.*, **69**(8), 51

HETSRONI, G. and SOKOLOV, M. (1971). Distribution of mass, velocity and intensity of turbulence in a two-phase turbulent jet. *J. Appl. Mech.*, **38**, 315

HETTIG, S.B. (1966). A project checklist of safety hazards. *Chem. Engng*, **73**(Dec. 19), 102

HETTIG, S.B. (1967). Emergency shutdown system. *Chem. Engng*, **74**(Feb. 27), 141

HEUDORFER, W. (1986). Heavy gas cloud and turbulence. In Hartwig, S. (1986a), *op. cit.* p. 53

HEUDORFER, W. and HARTWIG, S. (1983). A comparison of consequences of different types of UF6 accidents. In Hartwig, S. (1983a), *op. cit.* p. 279

VAN HEUGTEN, W.H.H. and DUIJM, N.J. (1985). Some findings based on wind tunnel simulation and model calculations of Thorney Island trial No. 008. *J. Haz. Materials*, **11**(1/3), 409. See also in McQuaid, J. (1985), *op. cit.*, p. 409

VAN DEN HEUVEL, M.J. (1982). The Notification of New Substances Regulations: the role of the UK competent authority. *Chem. Ind.*, Oct. 16, 798

HEWERDINE, S. (1991). Use of acoustic emission in inspection investigations within ICI. *Hazards*, **XI**, p. 277

HEWERDINE, S. (ed.) (1993). *Plant Integrity Assessment by the Acoustic Emission Testing Method* (Rugby: Instn. Chem. Engrs)

HEWETT, N.H. (1988). Boiler plant. *Plant Engr*, **32**(1), 33

HEWISON, C.H. (1983). *Locomotive Boiler Explosions* (Newton Abbott: David & Charles)

HEWITT, A.G. (1980). The control of toxic emissions in selected batch and continuous processes. In Hughes, D.J. (1980), *op. cit.*, paper 3

HEWITT, C. (1972). *Description and Theoretical Analysis (using Schemata) of PLANNER, a Language for Proving Theorems and Manipulating Models in a Robot. Rep. TR-258.* Massachusetts Inst. of Technol., AI Lab.

HEWITT, D. (1978). In *Report of the Royal Commission on the Health and Safety of Workers in Mines* (Govt of Ontario), p. 78 and Appendix C

HEWITT, F. (1976). Toxic releases. *Course on Loss Prevention in the Process Industries.* Dept of Chem. Engng, Loughborough Univ. of Technol.

HEWITT, G.F. (1982). Applications of two-phase flow. *Chem. Engng Prog.*, **78**(7), 38

HEWITT, G.F. *et al.* (1993). *Process Heat Transfer* (Boca Raton, FL: CRC Press)

HEWITT, P.J. (1991). Statutory records (COSHH Regulations 1988). In Simpson, D. and Simpson, W.G. (1991), *op. cit.* p. 67

HEWKIN, D.J. (1992). Consequences of pressure blast – the probability of fatality inside buildings. *Explosives Safety Board, Dept of Defense, Proc. Twenty-Fifth Explosives Safety Seminar*

HEWSON, E.W. (1951). Atmospheric pollution. *Compendium of Meteorology* (Boston, MA: Am. Met. Soc.)

HEWSON, J.E. (1981). *Process Instrumentation Manifolds: Their Selection and Use*; 1985, rev. ed. (Research Triangle Park, NC: Instrum. Soc. Am.)

HEY, M. (1991). Pressure relief of dust explosions through large diameter ducts and effect of changing the position of the ignition source. *J. Loss Prev. Process Ind.*, **4**, 217

HEYMAN, J. (1977). *Equilibrium of Shell Structures* (Oxford: Oxford Univ. Press)

HICKES, W.F. (1968). Intrinsic safety and the CPI. *Loss Prevention*, **2**, p. 17. See also *Chem. Engng Prog.*, 1968, **64**(4), 67

HICKES, W.F. (1969). Fundamental of intrinsic safety. *Chem. Engng*, **76**(Jun. 30), 139

HICKES, W.F. (1972). Intrinsic safety. *Chem. Engng*, **79**(May 1), 64

HICKES, W.F. and BROWN, K.J. (1971). Assessment of explosion probability for intrinsically safe apparatus. *Electrical Safety*, **1**, 54

HICKMAN, T.A. (chrmn) (1984). *Royal Commission on the Ocean Ranger* Marine Disaster. *Report One: The Loss of the Semisubmersible Drill Rig* Ocean Ranger *and Its Crew* (St John's Newfoundland)

HICKMAN, W.E. and MOORE, W.D. (1986). Managing the maintenance dollar. *Chem. Engng*, **93**(Apr. 14), 68

HICKS, B.L. (1954). Theory of ignition considered as a thermal reaction. *J. Chem. Phys*, **22**, 414

HICKS, D. (1993). Nuclear safety research on thermal reactors. *Process Safety Environ.*, **71B**, 75

HICKS, G.V. *et al.* (1973). *Summary of National Transportation Statistics* (Washington, DC: Dept of Transportation)

HICKS, J.G. (1987). *Welded Joint Design*, 2nd ed. (Oxford: BSP Professional)

HICKS, R.W. and DICKEY, D.S. (1976). Applications analysis for turbine agitators. *Chem. Engng*, **83**(Nov. 8), 127

HICKS, T.G. (1957). *Pump Selection and Application* (New York: McGraw-Hill)

HICKS, T.G. (1958). Pump Operation and Maintenance (New York: McGraw-Hill)

HIDDEN, A. (1989). Investigation into the Clapham Junction Railway Accident (London: HM Stationery Office)

HIDE, D. (1986). Computer control: when does a customised system make sense? *Process Engng*, Apr., 42

HIGGINBOTHAM, G.R. *et al.* (1968). Chemical and toxological evaluations of isolated and synthetic chloro derivatives of dibenzo-*p*-dioxin. *Nature*, **220**, 702

HIGGINS, E.A., FIORCA, V., THOMAS, A.A. and DAVIS, H.V. (1972). Acute toxicity to brief exposures of HF, HCl, NO2 and HCN with and without CO. *Fire Technol.*, **8**, 120

HIGGINS, L.R. and MORROW, L.C. (eds) (1977). *Maintenance Engineering Handbook*, 3rd ed. (New York: McGraw-Hill)

HIGGINS, T.E. (1989). *Hazardous Waste Minimization Handbook* (Chelsea, MI: Lewis)

HIGGINS, T.E. and BYERS, W. (1989). Leaking underground tanks: conventional and innovative cleanup techniques. *Chem. Engng Prog.*, **85**(5), 12

HIGGINSON, J. (1984). Existing risks for cancer. In Deisler, P.F. (1984), *op. cit.*, p. 1

HIGGINSON, J. and MUIR, C.S. (1979). Environmental carcinogenesis: misconceptions and limitations to cancer control. *J. Nat. Cancer Inst.*, **63**, 1291

HIGGS, F. (1990). *Piper Alpha Public Inquiry. Transcript, Day 167*

HIGH, M.S. and DANNER, R.P. (1987). Prediction of upper flammability limit by a group contribution method. *Ind. Engng Chem. Res.*, **26**, 1395

HIGH, R.W. (1968). The Saturn fireball. *Ann. N.Y. Acad. Sci.*, **152**(Art 1), 441

HIGH, W.G. (1967). The design and scale model testing of a cubicle to house oxidation or high pressure equipment. *Chemy Ind.*, Jun. 3, 899

HIGH, W.G. (1974). Explosions, their containment and consequences. *Course on Process Safety – Theory and Practice.* Dept of Chem. Engng, Teesside Polytechnic, Middlesbrough

HIGH, W.G. (1976). Explosions: occurrence, containment and effects. *Course on Loss Prevention in the Process Industries.* Dept of Chem. Engng, Loughborough Univ. of Technol.

HIGH, W.G. (1980). The assessment of the fragment hazard from rupture of HP vessels. In Vodar, B. and Marteau, P. (1980), *op. cit.*, p. 1060

HIGHAM, C.E. (1983). Centralised lubrication. *Plant Engr*, **27**(4), 27

HIGHAM.E. (1985a). Optical fibres as sensors. *Chem. Engr, Lond.*, **415**, 45

HIGHAM, E. (1985b). Optical fibres as transmission cables. *Chem. Engr, Lond.*, **417**, 34

HIGH PRESSURE WATER JETTING CONTRACTORS (1982). *Code of Practice for the use of High Pressure Water Jetting Equipment* (London)

HIGNETT, E.T. and GIBBINGS, J.C. (1968). Electrostatic streaming current developed in turbulent flow through a pipe. *J. Electroanalytical Chem.*, **16**, 239

HIGSON, D.J. (1978). The development of safety criteria for use in the nuclear industry. *Sixth Nat. Chem. Engng Conf.*, Surfer's Paradise, Australia

HIGSON, D.J. (1981). Hazard and risk. *Chem. Engr, Lond.*, **371/2**, 393

HIGSON, K.J., JOHNSON, E., STOTT, A.L. and WILLIAMS, E.W. (1971). Operating experience with an on-line control computer system on the paraquat plant. *Measmnt Control*, **4**, T65

HIGUCHI, T. (ed.) (1976). *PCB Poisoning and Pollution* (Tokyo: Kodansha)

HILADO, C.J. (1970). How to predict if materials will burn. *Chem. Engng*, **77**(Dec.), 14, 174

HILADO, C.J. (ed.) (1973). *Smoke and Products of Combustion* (Westport, CT: Technomic)

HILADO, C.J. and CLARK, S.W. (1972a). Autoignition temperatures of organic chemicals. *Chem. Engng*, **79**(Sept. 4), 75

HILADO, C.J. and CLARK, S.W. (1972b). Discrepancies and correlations of reported autoignition temperatures. *Fire Technol.*, **8**(3), 218

HILADO, C.J. and CUMMING, H.J. (1977). The HC value: a method of estimating the flammability of mixtures of combustible gases. *Fire Technol.*, **13**, 195

HILADO, C.J. and CUMMING, H.J. (1978). Short term LC$_{50}$ values: an update on available information. *Fire Technol.*, **14**, 46

HILADO, C.J. and CUMMING, H.J. (1979). Limits of flammability of organic chemicals. *J. Fire Flammability*, **10**, 252

HILADO, C.J. and FURST, A. (1976). Estimating short term LC$_{50}$ values. *Fire Technol.*, **12**, p. 109

HILADO, C.J. and HUTTLINGER, P.A. (1980). Concentration–time relationships in off-gas toxicity. *Fire Technol.*, **16**, 192

HILADO, C.J. and HUTTLINGER, P.A. (1981a). Government publications relevant *Flammability*, **12**, 301

HILADO, C.J. and HUTTLINGER, P.A. (1981b). A method for describing the potential hazards of flammability, toxicity and other parameters. *J. Fire Flammability*, **12**, 108

HILADO, C.J. and HUTTLINGER, P.A. (1981c). The PH value: a method for describing potential hazards of flammability and toxicity. *Fire Technol.*, **17**, 139

HILADO, C.J. and KOSOLA, K.L. (1977). Bibliography of published information on smoke evolution. *J. Fire Flammability*, **8**, 532

HILADO, C.J., CASEY, C.J. and FURST, A. (1977). Effect of ammonia on Swiss Albino mice. *J. Comb. Toxicol.*, **41**, 385

HILADO, C.J., CUMMING, H.J. and CASEY, C.J. (1978). Toxicity of pyrolysis gases from natural and synthetic materials. *Fire Technol.*, **14**, 136

HILDEBRAND, M. (1985). API aids in loss prevention. *Hydrocarbon Process.*, **64**(12), 84

HILDENBRAND, C.V. (1975). Urethane foam insulation hazards. *Ammonia Plant Safety*, **17**, 109

HILDER, M.H. (1982). The evaporation of hazardous materials from solution. *The Assessment of Major Hazards* (Rugby: Instn Chem. Engrs), p. 45

HILDYARD, N. (1981). *Cover Up* (London: New English Library)

HILE, H.B. (1967). Centrifugal compressors for large ammonia plants. *Ammonia Plant Safety*, **9**, 52

HILKER, R.W. (1962). Which to use . . . ball joints or expansion loops? *Hydrocarbon Process.*, **41**(2), 185

HILL, C.G. (1977). *An Introduction to Chemical Engineering Kinetic and Reactor Design* (New York: Wiley)

HILL, E.J. and BOSE, L.J. (1975). Sneak circuit analysis of military systems. *Proc. Second Int. Systems Safety Conf.*, p. 351

HILL, F.J. (1983). Pump recirculation systems. *Plant Engr*, **27**(3), 29

HILL, F.J. and PETERSON, G.R. (1968). *Introduction to Switching Theory and Logical Design* (New York: Wiley)

HILL, G.B. (1983). Recycling cooling water. *Plant Engr*, **27**(6), 33

HILL, H. (1975). Eyes in industry. *Plant Engr*, 19(1), 25

HILL, H.R., BRUCE, D.J. and DIGGLE, W.M. (1977). The interaction between local authority services and industry to deal with major emergencies. *Loss Prevention and Safety Promotion in the Process Industries 2*, p. 51

HILL, SIR J. (1976). Nuclear power: achievements, problems and myths. *Sci. Publ. Policy*, **3**, 355

HILL, SIR J. (1979a). Future requirements for fast reactor fuel reprocessing. *Chemy Ind.*, Jun. 16, 406

HILL, SIR J. (1979b). Quest for public acceptance of nuclear power. *Nucl. Energy*, **18**, 301

HILL, SIR J. (1981). Nuclear power – the real problems and the problems as perceived by the public. *Petrol. Rev.*, **35**(408), 41

HILL, J.F. (1979). Blast injury with particular reference to recent terrorist bombing incidents. *Ann. R. College Surgeons*, **61**, 4

HILL, J.W. (1980). Man-made mineral fibres and lung disease. *Chemy Ind.*, Mar. 1, 182

HILL, K.M. and KNOTT, H. (1960). The design of plants for handling hydrofluoric acid. *Chemical Process Hazards I*, p. 95

HILL, L. (1915). Gas poisoning. *Br. Med. J.*, Dec. 4, 801

HILL, L. (1920). Lecture on the pathology of war poison gases. *J. R. Army Med. Corps*, **35**, 334

HILL, M. (1971). Concorde electronics. *Electron. Power*, Oct., 402

HILL, P.G. and HUNG, J. (1988). Laminar burning velocities of stoichiometric mixtures of methane with propane and ethane additives. *Combust. Sci. Technol.*, **60**, 7

HILL, R. (1988). The design of alarms and interlocks – process aspects. *J. Loss Prevention Process Ind.*, **1**(1), 25

HILL, R. (1991). The role of instrumentation and process controls in minimizing accidental releases. *Plant/Operations Prog.*, **10**, 129

HILL, R. and KOHAN, D. (1986). The dynamics of safety systems. *Loss Prevention and Safety Promotion 5*, p. 54.1

HILL, R.C. (1974). Collisions and groundings: preventing the preventable. *US Naval Inst. Proc.*, Dec., 42

HILL, R.C.F. (1977). Presenting in-service information to management: use of the Duane method. Paper presented to UK Health Monitoring Group

HILL, R.G. (1968). Drawing effective flowsheet symbols. *Chem. Engng*, **75**(Jan. 2), 84

HILL, R.G. (1977). The sizing of vacuum breakers for distillation equipment. *Loss Prevention and Safety Promotion 2*, p. 257

HILL, R.S. and KIME, D.L. (1976). How to specify drive trains for turbine agitators. *Chem. Engng*, **83**(Aug. 2), 89

HILL, R.T. (1992). Pipeline risk analysis. *Major Hazards Onshore and Offshore*, p. 657

HILLBERG, T. (1982). Control of dynamic bodies in an open seaway. *Gastech 81*, p. 121

HILLER, H. and MARSCHNER, F. (1970). Lurgi makes low pressure methanol. *Hydrocarbon Process.*, **49**(9), 286

HILLMAN, D.J. (1985). Artificial intelligence. *Hum. Factors*, **27**, 21

HILLMAN, T. (1977). Constraints and procedures for waste disposal to land. *Chemical Engineering in a Hostile World* (London: Clapp & Poliak)

HILLMAN, T.C. (1984). The waste disposal authority role. *Chemy Ind.*, Sep. 3, 602

HILLS, P.C., ZHANG, D.K., SAMSON, P.J. and WALL, T.F. (1992). Laser ignition of combustible gases by radiation heating of small particles. *Combust. Flame*, **91**, 399

HILLS, P.D. (1983). Designing piping for gravity flows. *Chem. Eng.*, **90**(Sep. 5), 111

HILLS, P.J. (1981). The Carriage of Hazardous Goods Overland (London: Oyez), p. 27

HILLS, R.E. and NEELY, D.K. (1978). Pick the right seal. *Hydrocarbon Process.*, **57**(10), 119

HILLSTROM, W.W. (1971). Ignition and combustion of unconfined liquid fuels. *Combust. Sci. Technol.*, **3**, 179

HILST, G.R. (1957). The dispersion of stack gases in stable atmospheres. *J. Air Poll. Control Ass.*, **7**, 205

HILST, G.R. and SIMPSON, C.L. (1958). Observations of vertical diffusion rates in stable atmospheres. *J. Met.*, **15**, 125

HILTON, P.J. (ed.) (1976). *Structural Stability, the Theory of Catastrophes and its Applications in the Sciences* (Berlin: Springer)

HILTZ, R.H. (1982). Vapor hazard control. In Bennett, G.F., Feates, F.S. and Wilder, I. (1982), *op. cit.*, p. 10–21

HILTZ, R.H. (1987). The potential of aqueous foams to mitigate the vapor hazard from released volatile chemicals. *Preventing Major Chemical Accidents*, p. 5.87

HILTZ, R.H. (1988). Design and construction of a mobile activated carbon regenerator system. *J. Haz. Materials*, **18**(2), 207

HILTZ, R.H. and BRUGGER, J.E. (1989). Foam scrubbing as a mechanism to mitigate large toxic vapour releases. *Loss Prevention and Safety Promotion 6*, p. 93.1

HILTZ, R.H. and FRIEL, J.V. (1976). Application of foams to the control of hazardous chemical spills. *Control of Hazardous Materials Spills 3*, p. 293

HILTZ, R.H., MARSHALL, M.D. and FRIEL, J.V. (1972). The physical containment of land spills by a foam diking system. *Control of Hazardous Material Spills 1*, p. 85

HILTZ, R.M. (1982). Mitigation of the vapor hazard from silicon tetrachloride using water-based foams. *J. Haz. Materials*, **5**(3), 169

HIMANEN, R. (1983). Experiences of data collection for the availability analysis of a coal fired boiler system. *Euredata 4*, paper 19.2

HIMMELBLAU, D.M. (1970). *Process Analysis by Statistical Methods* (New York: Wiley)

HIMMELBLAU, D.M. (1978). *Fault Detection and Diagnosis in Chemical and Petrochemical Processes* (Amsterdam: Elsevier)

HIMMELBLAU, D.M. (1979). Fault detection in chemical plant equipment via fluid noise analysis. Comput. *Chem. Engng*, **3**, 507

HIMMELBLAU, D.M. (1985). *Material Balance Rectification via Interval Arithmetic. PSE 85* (Rugby: Instn Chem Engrs), p. 121

HIMMELBLAU, D.M. and BISCHOFF, K.G. (1968). *Process Analysis and Simulation* (New York: Wiley)

HINCHCLIFFE, P.R. (1980). The UK approach to monitoring river and tidal water quality. *Chemy Ind.*, Aug. 2, 603

HINCHLEY, J.W. and HIMUS, G.W. (1924). Evaporation in currents of air. *Trans. Instn Chem. Engrs*, **2**, 57

HINCHLEY, P. (1975). How to avoid problems of waste-heat boilers. *Chem. Engng*, **82**(Sep. 1), 94

HINCHLEY, P. (1977). Waste heat boilers: problems and solutions. *Ammonia Plant Safety*, **19**, 8. See also *Chem. Engng Prog.*, 1977, **73**(3), 90

HINCHLEY, P. (1979a). The engineering of reliability into waste heat boiler systems. *Proc. Instn Mech. Engrs*, **193**, 1

HINCHLEY, P. (1979b). Specifying and operating reliable waste-heat boilers. *Chem. Engng*, **86**(Aug. 13), 120

HINDAWI, I.J. (1970). *Air Pollution Injury to Vegetation. Rep. AP-71.* Nat. Air Poll. Control Admin.

HINDE, C. (1986). Fuzzy Prolog. *Int. J. Man-Machine Studies*, **24**, 569

HINDS, F.W. (1975). Safety tests for purged and pressurized machines. *Electrical Safety*, **2**, p. 70

HINDS, P.R., NEWTON, D.W. and JARDINE, A.K.S. (1977). Problems of Weibull parameter estimation from small samples. *First Nat. Conf. on Reliability. NRC5/3.* (Culcheth, Lancashire: UK Atom. Energy Auth.)

HINDS, W.C. (1982). *Aerosol Technology* (New York: Wiley-Interscience)

HINDS, W.T. (1967). On the variation of concentration in plumes and building wakes. *Proc. USAEC Meteorological Information Mtg*, Chalk River, Ont.

HINDS, W.T. (1969). Peak-to-mean concentration ratios from ground level sources in building wakes. *Atmos. Environ.*, **3**, 145

HINES, A.L. and MADDOX, R.N. (1985). *Mass Transfer: Fundamentals and Applications* (Englewood Cliffs, NJ: Prentice-Hall)

HINES, J.G. (chrmn) (1978). *Industrial Corrosion Monitoring* (London: HM Stationery Office)

HINES, J.G. (1986). Corrosion information and computers. *Br. Corrosion J.*, **21**(2), 81

HINES, J.G. and BASDEN, A. (1986). Experience with the use of computers to handle corrosion knowledge. *Br. Corrosion J.*, **21**(3), 151

HINES, W.A. (1966). *Noise Control in Industry* (London: Business Books)

HINGORANEY, R. (1994). Putting expert systems to work. *Chem. Engng*, **101**(1), 121

HINKLEY, P.L. (1984). Standard time/temperature curve: its relevance. *Fire Surveyor*, **13**(3), 5

HINKLEY, P.L. (1986). *Fire venting. Fire Surveyor*, **15**(4), 5

HINKLEY, P.L., HANSELL, G.O., MARSHALL, N.R. and HARRISON, R. (1992). Sprinklers and vents interaction: experiments at Ghent. *Fire Surveyor*, **21**(5), 18

HINRICHSEN, K. (1986). Comparison of four analytical dispersion models for near-surface releases above a grass surface. *Atmos. Environ.*, **20**, 29

HINSHAW, J.R. (1957). *Histological Studies of some Reactions of Skin to Radiant Thermal Energy. ASME Paper 57–SA-21.* Am. Soc. Mech. Engrs, New York

HINSLEY, J.F. (1959). *Non-destructive Testing* (London: Macdonald & Evans)

HINST, M.F. (1963). Controlling corrosion in carbon-steel tubes. *Chem. Engng*, **70**(Jan. 7), 110

HINZE, J.O. (1959). *Turbulence* (New York: McGraw-Hill)

HIRANO, T. (1982). Some problems in the prediction of gas explosions. In Lee, J.H.S. and Guirao, C.M. (1982), *op. cit.*, p. 823

HIRANO, T. (1984). Gas explosion processes in enclosures. *Plant/Operations Prog.*, **3**, 247

HIRANO, T. (1990a). Effectiveness evaluation of expense for loss prevention. *Safety and Loss Prevention in the Chemical and Oil Processing Industries*, p. 93

HIRANO, T. (1990b). A theoretical approach to hazard assessments. *J. Loss Prev. Process Ind.*, **3**(3), 321

HIRANO, T. (1993). A manual proposed by Fire Defence Agency of Japan for effectiveness evaluation of loss prevention systems. *Health, Safety and Loss Prevention*, p. 593

HIRANO, T. and KINOSHITA, M. (1975). Gas velocity and temperature profiles of a diffusion flame stabilized in the stream over liquid fuel. *Combustion*, **15**, 379

HIRANO, T. and SATO, J. (1993). Fire spread along structural metal pieces in oxygen. *J. Loss Prev. Process Ind.*, **6**, 151

HIRANO, T. and SUZUKI, T. (1993). Flame propagation across liquids – a review of gas phase phenomena. *Fire Safety J.*, **21**, 207

HIRANO, T. and TAZAWA, K. (1978). A further study of the effects of external thermal radiation on flame spread over paper. *Combust Flame*, **32**, 95

HIRANO, T., NOREIKIS, S.E. and WATERMAN, T.E. (1974). Postulations of flame spread mechanisms. *Combust. Flame*, **22**, 353

HIRANO, T., SUZUKI, T., MASHIKO, I. and IWAI, K. (1977). Flame propagation through mixtures with concentration gradient. *Combustion*, **16**, p. 1307

HIRATA, I. (1990). Maintenance for loss prevention in the process industries. *Safety and Loss Prevention in the Chemical and Oil Processing Industries*, p. 599

HIRATA, M., OHE, S. and NAGAHAMA, K. (1975). *Computer Aided Data Book of Vapor–Liquid Equilibria* (Amsterdam: Elsevier)

HIRD, D. and FIPPES, D.W. (1960). *The Effect of Various Powdered Materials on the Stability of Protein Foam*. Br. Joint Fire Res. Organ.

HIRD, D., RODRIGUES, A. and SMITH, D. (1970). Foam – its efficiency in tank fires. *Fire Technol.*, **6**(1), 5

HIRSCH, A.E. (1968). The tolerance of man to impact. *Ann. N.Y. Acad. Sci*, 1968, **152**(Art. 1), 168

HIRSCH, E. (1986). Improved Gurney formulas for exploding cylinders and spheres using 'hard core' approximation. *Propellants, Explosives, Pyrotechnics*, **11**, 81

HIRSCH, F.G. (1968). *Effects of Overpressure on the ear – a Review. Rep DASA-1858*. Defense Atom. Support Agency, Dept of Defence, Washington, DC. See also *Ann. N.Y. Acad. Sci*, 1968, **152** (Art. 1), 147

HIRSCHFELDER, J.O. (1963). Some remarks on the theory of flame propagation. *Combustion*, **9**, 553

HIRSCHFELDER, J.O. and CURTISS, C.F. (1949). Theory of propagation of flames. Pt 1 – General equations. *Combustion*, **3**, 121

HIRSCHFELDER, J.O., CURTISS, C.F. and CAMPBELL, D.E. (1953). The theory of flames and detonations. *Combustion*, **4**, p. 190

HIRSCHLER, M.M. (1985). Soot from fires. *J. Fire Sci.*, **3**(5), 343; **3**(6), 380

HIRSCHMANN, H., NICOLESCU, T. and WEBER, R. (1983). Steady state availability of a *k*-out-of-*n* system with total repair. *Reliab. Engng*, **4**(3), 181

HIRSCHORN, J.S. and OLDENBURG, K.U. (1989). Are we cleaned up? An assessment of Superfund. *Chem. Engng Prog.*, **96**, 55

HIRSHLEIFER, J., BERGSTROM, T. and RAPPAPORT, E. (1974). *Applying Cost-Benefit Concept to Projects which alter Human Mortality. Rep.* UCLA-ENG-7478. Univ. of California, Los Angeles, CA

HIRST, R. (1974a). Chemical extinguishants. *Course on Process Safety – Theory and Practice*. Dept of Chem. Engng, Teesside Polytechnic, Middlesbrough

HIRST, R. (1974b). Chemical extinguishants. *Chem. Engr, Lond.*, **290** 627

HIRST, R. (1989). *Underwood's Practical Fire Precautions*, 3rd ed. (Aldershot: Gower)

HIRST, R. (1992). Fire protection offshore – the emphasis has changed for the better. *Fire Int.*, **133**(Feb./Mar.), 20

HIRST, R. and BOOTH, K. (1977). Measurement of flame extinguishing concentrations. *Fire Technol.*, **13**, 296

HIRST, R. and SAVAGE, N. (1981/82). Measurement of inerting concentrations. *Fire Safety J.*, **4**(3), 147

HIRST, W.J.S. (1984). *Combustion of Large Scale Jet-Releases of Pressurised Liquid Propane. Comm. 1241* (London: Instn Gas Engrs)

HIRST, W.J.S. (1986). Combustion of large-scale releases of pressurized liquid propane. In Hartwig, S. (1986a), *op. cit.*, p. 267

HIRST, W.J.S. and EYRE, J.A. (1983). Maplin Sands experiments 1980: combustion of large LNG and refrigerated liquid propane spills on sea. In Hartwig, S. (1983a), *op. cit.* p. 211

HISCOCK, W.G. (1938). Selection tests for chemical process workers. *Occup. Psychol.*, **12**, 178

HISER, L.L. (1970). Selecting a wastewater treatment process. *Chem. Engng*, **77**(Nov. 30), 76

HITCHCOCK, D.A. (1979). Solid waste disposal: incineration. *Chem. Engng*, **86**(May 21), 185

HITCHEN, I.R. (1980). Vibration monitoring for rotating machinery. *Measamnt Control*, **13**, 97

HIX, A.H. (1972). Safety and instrumentation systems. *Loss Prevention*, **6**, 4. See also *Chem. Engng Prog.*, 1972, **68**(5), 42

HJERTAGER, B.H. (1982a). Numerical simulation of turbulent flame and pressure development in gas explosions. In Lee, J.H.S. and Guirao, C.M. (1982), *op. cit.*, p. 407

HJERTAGER, B.H. (1982b). Simulation of transient compressible turbulent flows. *Combust. Sci. Technol.*, **27**, 159

HJERTAGER, B.H. (1984). Influence of turbulence on gas explosions. *J. Haz. Materials*, **9**(3), 315

HJERTAGER, B.H. (1986). Three-dimensional modelling of flow, heat transfer and combustion. In *Handbook of Heat and Mass Transfer* (ed. by N.P. Cheremisinoff) (Houston, TX: Gulf), p. 1303

HJERTAGER, B.H. (1991). Explosions in offshore modules. *Hazards*, **XI**, p. 19. See also *Process Safety Environ.*, 1991, **69B**, 59

HJERTAGER, B.H. (1993). Computer modeling of turbulent gas explosions in complex 2D and 3D geometries. *J. Haz. Materials*, **34**, 173

HJERTAGER, B.H., FUHRE, K., PARKER, S.J. and BAKKE, J.R. (1984). Flame acceleration of propane–air in a large-scale obstructed tube. In Bowen, J.R. *et al.*, (1984b), *op, cit.*, p. 504

HJERTAGER, B.H., FUHRE, K. and BJORKHAUG, M. (1988a). Concentration effects on flame acceleration by obstacles in large-scale methane–air and propane–air vented explosions. *Combust. Sci. Technol.*, **62**, 239

HJERTAGER, B.H., FUHRE, K. and BJORKHAUG, M. (1988b). Gas explosion experiments in 1:33 and 1:5 scale offshore separator and compressor modules using stoichiometric homogeneous fuel/air clouds. *J. Loss Prev. Process Ind.*, **1**(4), 197

HJERTAGER, B.H., BJORKHAUG, M. and FUHRE, K. (1988). Explosion propagation of non-homogeneous methane– air clouds inside an obstructed 50 m^3 vented vessel. *J. Haz. Materials*, **19**, 139

HJERTAGER, B.H., SOLBERG, T. and NYMOEN, K.O. (1992). Computer modelling of gas explosion propagation in offshore modules. *J. Loss Prev. Process Ind.*, **5**, 165

HJERTAGER, B.H., ENGGRAV, S., FO RRISDAHL, J.E. and SOLBERG, T. (1994). A case study of gas explosions in a process plant using a three-dimensional code. In *Guidelines for Evaluating the Characteristics Vapor Cloud Explosions, Flash Fires and BLEVEs* (New York: Am. Inst. Chem. Engrs), Appendix F

HJORT, A.M. and TAYLOR, F.A. (1919). The effect of morphine upon the alkali reserve of blood of dogs gassed with fatal concentration of chlorine. *J. Pharmacol. Exp. Therapeut.*, **13**, 409

HM CHIEF ALKALI INSPECTOR (Annual). *Annual Report on Alkali etc. Works Regulations* (London: HM Stationery Office)

HM CHIEF ALKALI INSPECTOR (1974). *Annual Report on Alkali etc. Works Regulations* (London: HM Stationery Office)

HM CHIEF INSPECTOR OF EXPLOSIVES (Annual). *Annual Report of HM Chief Inspector of Explosives* (London: HM Stationery Office)

HM CHIEF INSPECTOR OF FACTORIES (Annual). *Analysis of Report Accidents and Dangerous Occurrences, HM Chief Inspector of Factories, Dept of Employment* (London: HM Stationary Office)

HM CHIEF INSPECTOR OF FACTORIES (Annual). *Annual Report of HM Chief Inspector of Factories* (London: HM Stationery Office)

HM CHIEF INSPECTOR OF FACTORIES (1968). *Dust Explosions in Factories. Classification List of Dusts that have been Tested in the Form of a Dust Cloud* (London: HM Stationery Office)

HM CHIEF INSPECTOR OF FACTORIES (1974). *Annual Report of HM Chief Inspector of Factories 1974* (London: HM Stationery Office)

HM CHIEF INSPECTOR OF FACTORIES (Annual: 1975–). *Industry and Services Report* (London: HM Stationery Office)

HM CHIEF INSPECTOR OF FACTORIES (1986a). *Health and Safety at Work. Report of HM Chief Inspector of Factories 1985* (London: HM Stationery Office)

HM CHIEF INSPECTOR OF FACTORIES (1986b). Health and safety in the chemical industry: the work of the Chemical National Industry Group. In *Health and Safety at Work. Report of HM Chief Inspector of Factories 1985* (London: HM Stationery Office)

HM CHIEF INSPECTOR OF FACTORIES (1988a). *Health and Safety at Work. Report of HM Chief Inspector of Factories 1986–87* (London: HM Stationery Office)

HM CHIEF INSPECTOR OF FACTORIES (1988b). Major hazard installations. In *Health and Safety at Work. Report of HM Chief Inspector of Factories 1986–87* (London: HM Stationery Office)

HM CHIEF INSPECTOR OF NUCLEAR INSTALLATIONS (1977). *Health and Safety in Nuclear Establishments 1975– 1976* (London: HM Stationery Office)

HM FACTORY INSPECTORATE (1968). *Dust Explosions in Factories. Classified List of Dusts that have been Tested in the Form of a Dust Cloud. SHW 830* (London: HM Stationery Office)

HM INSPECTORATE OF POLLUTION (1991). *Chief Inspector's Guidance to Inspectors. Waste Disposal Industry Sector. IPR5* (London: HM Stationery Office)

HM SENIOR ELECTRICAL INSPECTOR OF FACTORIES (Annual). *Annual Report of HM Senior Electrical Inspector of Factories* (London: HM Stationery Office)

HM STATIONERY OFFICE (Periodic). *Factory Orders* (London: HM Stationary Office)

HO, S.Y. (1992). Impact ignition mechanisms of rocket propellants. *Combust. Flame*, **91**, 131

HO, S.Y. and FONG, C.W. (1989). Relationship between impact ignition sensitivity and kinetics of the thermal decomposition of solid propellants. *Combust. Flame*, **75**, 139

HO, S.Y., FONG, C.W. and HAMSHERE, B.L. (1989). Assessment of the response of rocket propellants to high-velocity projectile impact using small-scale laboratory tests. *Combust. Flame*, **77**, 395

HO, V., SIU, N. and APOSTOLAKIS, G. (1988). COMPBRN III – a fire hazard model for risk analysis. *Fire Safety J.*, **13**, 137

HOAR, T.P. (chrmn) (1971). *Report of the Committee on Corrosion and Protection* (London: HM Stationery Office)

HOARE, D.E. and WALSH, A.D. (1955). The oxidation of methane. *Combustion*, **5**, 467

HOARE, D.E., TING-MAN, L.I. and WALSH, A.D. (1967). Cool flames and molecular structure. *Combustion*, **11**, 879

HOBBS, C.A., GRATTAN, E. and HOBBS, J.A. (1979). *Classification of Injury: Severity by Length of Stay in Hospital. Rep. LR 871*. Transport Road Res. Lab., Crowthorne

HOBBS, C.C. (1980). Hydrocarbon oxidation. *Kirk–Othmer Encyclopaedia of Chemical Technology*, 3rd ed., vol. 12 (New York: Wiley), p. 826

HOBBS, J.R. and MOORE, R.C. (1985). *Formal Theories of the Commonsense World* (Norwood, NJ: Ablex)

HOBSON, G.D. and POHL, W. (eds) (1975). *Modern Petroleum Technology*, 4th ed. (Barking: Applied Science)

HOBSON, L. and DAY, J. (1985). Induction heating of vessels. *Int. J. Elect. Engng Educ.*, **22**, 129

HOC, J.M. (1993). Some dimensions of a cognitive typology of process control situations. *Ergonomics*, **36**, 1445

HOCEVAR, C.J. (1975). *Nuclear Reactor Licensing: A Critique of the Computer Safety Prediction Methods* (Cambridge, MA: Union of Concerned Scientists)

HOCKEY, G.R.J., BRINER, R.B., TATTERSALL, A.J. and WIETHOFF, M. (1989). Assessing the impact of computer workload on operator stress: the role of system reliability. *Ergonomics*, **32**, 1401

HOCKMAN, K.K. and ERDMAN, D.A. (1993). Gearing up for ISO 9000 registration. *Chem Engng*, **100**(4), 128

HODGE, B. and MANTEY, J.P. (1967). How the computer performs arithmetic and logic – 1. *Chem. Engng*, **74**(Nov. 20), 141

HODGE, B.K. (1985). Analysis and design program for series piping systems. *Fire Technol.*, **21**, 310

HODGE, H.C. and STERNER, J.H. (1949). Tabulation of toxicity classes. *Am. Ind. Hyg. Ass. Quart.*, **10**(4), 93

HODGE, R.N. and WHISTON, J. (1989). Quality improvement – a view from the chemical industry. *Phil. Trans. R. Soc., Ser. A*, **327**, 565

HODGES, S. (1994). What is a 'seaworthy' vessel? In Rawson, K. (1994), *op. cit.*, p. 51

HODGES, W.F. (1978). *Logic* (Harmondsworth: Penguin Books)

HODGKINS, D.J. (1989). *Piper Alpha Public Inquiry. Transcript, Day 158*

HODGKINS, P. (1977). Unexpected effects of government policy. *Sci. Publ. Policy*, **4**, 142

HODGSON, E. and GUTHRIE, F.E. (eds) (1980). *Introduction to Biochemical Toxicology* (Oxford: Blackwell)

HODGSON, J.D. (1971). The new Occupational Safety and Health Act. *Chem. Engng*, **78**(Jun. 28), 108

HODGSON, J.P.E. (1991). *Knowledge Representation and Language in AI* (Chichester: Ellis Horwood)

HODGSON, SIR M. (1982). *Making More out of Less*. (George E. Davis Memorial Lecture) (Rugby: Instn Chem. Engrs)

HODNETT, R.M. (1986a). Automatic sprinklers. *Fire Protection Handbook*, 16th ed. (ed. A.E. Cote and J.L. Linville), p. 18.36

HODNETT, R.M. (1986b). Automatic sprinkler systems. *Fire Protection Handbook, 16th ed.* (ed. A.E. Cote and J.L. Linville), p. 18.2

HODNETT, R.M. (1986c). Water spray protection. *Fire Protection Handbook*, 16th ed. (ed. A.E. Cote and J.L. Linville), p. 18.77

HODNETT, R.M. (1986d). Water supplies for sprinkler systems. *Fire Protection Handbook, 16th ed.* (ed. A.E. Cote and J.L. Linville), p. 18.27

HODNETT, R.M. (1986e). Water and water additives for fire fighting. *Fire Protection Handbook*, 16th ed. (ed. A.E. Cote and J.L. Linville), p. 17.2

HODNICK, H.V. (1971). Thwarting maintenance death traps. *Chem. Engng*, **78**(Apr. 5), 142

HODNICK, H.V. (1976). High pressure fog against fire. *Protection*, **13**, 20

HODSON, W.R. (1993). Designing for noise control (or building in silence). In Instn Mech. Engrs (1993), *op. cit.*, 7

HOEHNE, V.O. and LUCE, R.G. (1970). The effect of velocity, temperature and molecular weight on the inflammability limits in wind blown jets of hydrocarbon gases. *Am. Petrol. Inst., Proc. Div Refining, Thirty-Fifth Mid Year Mtg*

HOEHNE, V.O., LUCE, R.G. and MIGA, L.W. (1970). *The Effect of Velocity, Temperature and Gas Molecular Weight on Flammability Limits in Wind-blown Jets of Hydrocarbon Gases. Report.* Battelle Memorial Inst., Columbus, OH

HOEL, D.G., KAPLAN, N.L. and ANDERSON, M.W. (1983). Implications of non-linear kinetics on risk estimation in carcinogenesis. *Science*, **219**, 1032

HOENIG, S.A. (1981). New technology for dust control. *The Hazards of Industrial Explosion from Dusts* (London: Oyez/IBC), paper 11

HOENIG, S. (1989). Reducing the hazard of dust explosions: agglomerate the fines, cool off the hot spots, collect the fines at the point of origin, neutralize local electrostatic charges. *Plant/Operations Prog.*, **8**, 119

HOERNER, S.F. (1958). *Fluid Dynamics Drag* (Midland Park, NJ: published by author)

HOFELICH, T.C. (1989). The use of calorimetry for studying the early part of potential thermal runaway reactions. *Runaway Reactions*, p. 232

HOFELICH, T.C. and THOMAS, R.C. (1989). The use/misuse of the 100 degree rule in the interpretation of thermal hazard tests. *Runaway Reactions*, p. 74

HOFELICHT, T.C., POWERS, J.B. and FRURIP, D.J. (1994). Determination of compatibility via thermal analysis and mathematical modeling. *Process Safety Prog.*, **13**, 227

HOFER, H., REINDLE, E. and HRUBY, E. (1981). Acute inhalation toxicity of ammonia – evaluation of an apparatus. *Chem. Abs.*, **17**, 139864

VAN'T HOFF (1884). *Etudes de Dynamique Chimique* (Amsterdam)

HOFF, A.B.M. (1983). An experimental study of the ignition of natural gas in a simulated pipeline rupture. *Combust. Flame*, **49**, 51

HOFFMAN, A. (1967). Loss of insulation on an ammonia storage tank. *Ammonia Plant Safety*, **9**, 13

HOFFMAN, A.J. and MILLS, S.N. (1956). *Air Blast Measurements about Explosive Charges at Side-on and Normal Incidence. BRL Memo Rep. 988.* US Army, Ballistic Res. Lab., Aberdeen Proving Ground, MD

HOFFMAN, B.M., HANSEN, F.A. and DOELLING, R.J. (1980). Proper usage and savings through rupture discs. *API Proc., Refining Dept, 41st Midyear Mtg*, **59**, 253

HOFFMAN, E.J. (1978). *Coal Conversion* (Laramie, WY: Energon)

HOFFMAN, H.L. (1977). Better furnace operations. *Hydrocarbon Process.*, **56**(2), 79

HOFFMANN, J.M. (1985). Chemical process hazard review. In Hoffmann, J.M. and Maser, D.C. (1985), *op. cit.*, p. 1

HOFFMANN, J.M. and MASER, D.C. (eds) (1985). *Chemical Process Hazard Review* (Washington, DC: Am. Chem. Soc.)

HOFFMAN, P.D. (1974). Hazard evaluation and risk control. *Loss Prevention*, **8**, 80

HOFFMAN, T.R. (1971). *Production Management and Manufacturing Systems* (Belmont, CA: Wadsworth)

HOFFMANN-BERLING, E., GÜNTHER, R. and LEUCKEL, W. (1982). The effect of the flame front structure on flame propagation in premixed gases. *Combustion*, **19**, p. 433

HOFFMEISTER, J.A. (1993). Safety analysis for existing facilities (poster item). *Process Safety Management*, p. 459

HOFHEINZ, P. and KOHAN, J. (1989). Hard lessons and unhappy citizens. *Time*, Jun. 19, 24

HOFMAIER, H.H. (1963). Quick tests for acetylene in oxygen rich liquid streams. *Ammonia Plant Safety*, **5**

HOFMANN (1968). *Zuverlässigkeit von Mess-Steuer-Regel-und Sicherheitssysteme* (Stuttgart Kurt Thieme)

HOFMANN, H.T. (1957). *Exp. Pathol. Pharmakol.*, **232**, 228

HOFMANN, J. (1983). Modelling of fire for risk assessment in petrochemical industries. In Hartwig, S. (1983a), *op. cit.*, p. 249

HOFMANN, M.J. (1985). Process hazard review in a chemical research environment. In Hoffman, J.M. and Maser, D.C. (1985), *op. cit.*, p. 7

HOFMEISTER, M., HALASZ, L. and RIPPIN, D.W.T. (1989). Knowledge-based tools for batch processing systems. *Comput. Chem. Engng*, **13**, 1255

HOFTIJZER, G.W., VAN DER SCHAAF, J.B.R. and LUPKER, H.A. (1989). The risks of shunting-yards due to the transport of LPG in the Netherlands and the possibilities of reducing the risk of a BLEVE of a LPG rail tanker. *Loss Prevention and Safety Promotion 6*, p. 94.1

HOGAN, W.J. (1982). The liquefied gaseous fuel spill effects program. In Lee, J.H.S. and Guirao, C.M. (1982a), *op. cit.*, p. 949

HOGAN, W.J., ERMAK, D.L. and KOOPMAN, R.P. (1981). *Predicting the Hazards from Large Spills of Liquefied Gaseous Fuels. Preprint UCRL-86164.* Lawrence Livermore Nat. Lab., San Francisco, CA

HOGARTH, C.A. and BLITZ, J. (eds) (1960). *Techniques of Non-Destructive Testing* (London: Butterworths)

HOGGER, E.I. (1975). The use of decision tables in control software. *Trends in On-Line Computer Control Systems* (London: Instn Elec. Engrs), p. 238

HOGH, M.S. (1989). *Piper Alpha Public Inquiry. Transcript, Days 132, 133*

HÖGSTROM, U. (1964). An experimental study on atmospheric diffusion. *Tellus*, **16**, 205

HOHENBICHLER, M. and RACKWITZ, R. (1981). On structural reliability of parallel systems. *Reliab. Engng*, **2**(1), 1

HOHENEMSER, C. and KASPERSON, J.X. (eds) (1982). *Risk in the Technological Society* (Boulder, CO: Westview)

HOHMAN, T., SULLLIVAN, C. and MCNAMEE, T.J. (1993). Be prepared for that EPA inspection. *Chem. Engng Prog.*, **89**(11), 67

HOKSTAD, P., ARO, R. and TAYLOR, J.R. (1991). *Integration of SA and RAMS Techniques. Rep. 757302.* SINTEF, Trondheim

HOLAND, P. and RAUSAND, M. (1987). Reliability of subsea BOP systems. *Reliab. Engng*, **19**(4), 263

HOLCOMB, G. and LASCHOBER, S. (1979). Want faster reactor cool-down? *Hydrocarbon Process.*, **58**(9), 163

HOLCOMBE, B.V. (1981). The evaluation of protective clothing. *Fire Safety J.*, **4**(2), 91

HOLDEN, C. (1980). Love Canal residents under stress. *Science*, **208**, 1242

HOLDEN, D.G. and HODGSON, S.S. (1993). Scaling new heights in plant automation. *Chem Engng*, **100**(7), 103

HOLDEN, P.L. (1984). Difficulties in formulating risk criteria. *J. Occup. Accid.*, **6**(4), 241

HOLDEN, P.L. (1985). Developments in risk assessment and its applications in process plant safety studies. *The Assessment and Control of Major Hazards* (Rugby: Instn Chem. Engrs), p. 263

HOLDEN, P.L. and REEVES, A.B. (1985). Fragment hazards from failures of pressurised liquefied gas vessels. *The Assessment and Control of Major Hazards* (Rugby: Instn Chem. Engrs), p. 205

HOLDEN, P.L., LOWE, D.R.T. and OPSCHOOR, G. (1985). Risk analysis in the process industries – an ISGRA update. *Plant/ Operations Prog.*, **4**, 63

HOLDER (1973). When disaster does strike . . . ! *Prevention of Occupational Risks*, **2**, 481

HOLDER, G.D., CORBIN, G. and PAPADOPOULOS, K.D. (1980). Thermodynamic and molecular properties of gas hydrates from mixtures containing methane, argon and krypton. *Ind. Engng Fund.*, **19**, 282

HOLDGATE, M.W. (1981). Risks in the built environment. *Proc. R. Soc., Ser. A*, **376**, 151. See also in Warner, Sir F. and Slater, D.H. (1981), *op. cit.*, p. 151

HOLDING, D.H. (1965). *Principles of Training* (Oxford: Pergamon Press)

HOLDING, D.H. (1987). Concepts of training. In Salvendy, G. (1987), *op. cit.*, p. 939

HOLDREN, J. and HERRERA, P. (1973). *Energy* (New York: Sierra Club)

HOLDSWORTH, M.P. (1982). Review of part year's activities of the Society of International Gas Tanker and Terminal Operators (SIGTTO). *Gastech 81*, p. 101

HOLDSWORTH, M.P. (1983). The year's report on the activities of the Society of International Gas Tanker and Terminal Operators Ltd. *Gastech 82*, p. 191

HOLDSWORTH, M.P. (1985). An up-date on the activities of the Society of International Gas Tanker and Terminal Operators Ltd. *Gastech 84*, p. 205

HOLDSWORTH, M.P. *et al.* (1962). Electrical charging during white oil loading of tankers. *Trans. Inst. Marine Engrs*, **74**

HOLDSWORTH, M.P., VAN DER MINNE, J.L. and VELLENGA, S.J. (1964). Electrostatic charging during the white-oil loading of tanks. In Vervalin, C.H. (1964a), *op. cit.*, p. 185

HOLEN, J., BROSTRØM, M. and MAGNUSSEN, B.F. (1991). Finite difference calculation of pool fires. *Combustion*, **23**, p. 1677

HOLIDAY, A.D. (1982). Conserving and reusing water. *Chem. Engng*, **89**(Apr. 19), 118

HOLLAENDER, A. (ed.) (1973). *Chemical Mutagens* (New York: Plenum Press)

HOLLAN, J., HUTCHINGS, E. and WEITZMAN, L. (1984). STEAMER: an interactive inspectable simulation-based training system. *AI Magazine*, **5**(2), 15

HOLLAND, C.D. and ANTHONY, R.G. (1979). *Fundamentals of Chemical Reaction Engineering* (Englewood Cliffs, NJ: Prentice-Hall)

HOLLAND, D.M. and FITZ-SIMMONS, T. (1982). Fitting statistical distributions to air quality data by the maximum likelihood method. *Atmos. Environ.*, **16**, 1071

HOLLAND, F.A. and BRAGG, K. (1995). *Fluid Flow for Chemical Engineers*, 2nd ed. (London: Arnold)

HOLLAND, F.A. and CHAPMAN, E.S. (1966a). Centrifugal pumps. *Chem. Engng*, **73**(Jul. 4), 91

HOLLAND, F.A. and CHAPMAN, E.S. (1966b). *Pumping of Liquids* (New York: Reinhold)

HOLLAND, F.A. and WATSON, F.A. (1977). Economic penalties of operating a process at reduced capacity. *Chem. Engng*, **84**(Jan. 3), 91

HOLLAND, F.A., MOORES, R.M., WATSON, F.A. and WILKINSON, J.K. (1971). *Heat Transfer* (London: Heinemann)

HOLLAND, F.A., WATSON, F.A. and WILKINSON, J.K. (1974). *Introduction to Process Economics*; 1983, 2d ed. (London: Wiley)

HOLLNAGEL, E. (1984). A conceptual framework for the description and analysis of man–machine system interaction. *Ergonomic Problems in Process Operations* (Rugby: Instn Chem. Engrs), p. 81

HOLLNAGEL, E. (1989). *The Reliability of Expert Systems* (Chichester: Ellis Horwood)

HOLLNAGEL, E. *et al.* (1986). *Intelligent Decision Support in Process Environments* (Berlin: Springer)

HOLLNAGEL, E., CACCIABUE, C. and ROUHET, J.C. (1992). The use of integrated system simulation for risk and reliability assessment. *Loss Prevention and Safety Promotion 7*, vol. 2, 49.1

HOLLOWAY, L.W. (1971). Ammonia reformers and the tube deviation analyzer. *Ammonia Plant Safety*, **13**, 78

HOLLOWAY, L.W. and KELLUM, G.B. (1975). Non-destructive testing saves time and money. *Ammonia Plant Safety*, **17**, 61

HOLLOWAY, L.W., BAUER, R.F. and PITTMAN, T.D. (1981). Eddy current inspection of ferrous tubular materials. *Ammonia Plant Safety*, **23**, 179

HOLLOWAY, N.J. (1988). The significance of human actions for plant safety. In Ballard, G.M. (1988), *op. cit.*, p. 146

HOLLOWAY, N.J. (1989). Pilot study methods based on generic failure rate estimates. In Cox, R.A. (1989a), *op. cit.*, p. 71

HOLLOWOOD, J. (1986). In-house disposal as an option. *Chem Ind.*, Jun. 16, 414

HOLLY, F.M. and GRACE, J.L. (1921). *Proc Am. Soc. Civil Engrs, J. Hydr. Div.*, **98/9365**, p. 1921

HOLM, R. (1952). Temperature development in a heated contact with application to sliding contacts. *J. Appl. Mech.*, **19**, 369

HOLM, W.K. (1984). Use of magnetic inspection pigging provides valuable tool in pipeline maintenance. *Oil Gas J.*, **82**(Aug. 27), 88

HOLMAN, A., ROKSTAD, O.A. and SOLBAKKEN, A. (1976–). High temperature pyrolysis of hydrocarbons. *Ind. Engng Chem. Process Des. Dev.*, 1976, **15**, 439; 1979, **18**, 652

HOLMBERG, E.G. (1960). Valve design: special for corrosives. *Chem. Engng*, **56**(Jun. 13), 238

HOLMES, D.R. (ed.) (1985). *Dewpoint Corrosion* (Chichester: Ellis Horwood)

HOLMES, E. (1973). *Handbook of Industrial Pipework Engineering* (New York: McGraw-Hill)

HOLMES, J.T. and RAMASWAMI, D. (1966). Control valves eliminated from reactor cooling system. *Chem. Engng*, **73**(Nov. 21), 154

HOLMES, J.W. and MEAD, G.R. (1978). Topside facilities and processing. In Institute of Petroleum (1978), *op. cit.*, p. 89

HOLMES, K. (1991). *Implementing BS 5750* (Leatherhead, Surrey: Pira Int.)

HOLMES, T.L. and CHEN, G.K. (1984). Design and selection of spray/mist elimination equipment. *Chem. Engng*, **91**(Oct. 15), 82

HOLMES, W.A. (1976). Availability analysis of chemical plant systems. *Advances in Reliability* (Culcheth, Lancashire: UK Atom. Energy Auth.), ART1

HOLMES, W.S. (1971). An introduction to electrical control circuits. *Chem. Engng*, **78**(Nov. 15), 176

HOLROYD, R. (1967). Ultra large single stream chemical plants: their advantages and disadvantages. (*Chem. Ind.*, **1310**

HOLSCHER, H. and RADER, J. (1986). *TUV Study Group on Microprocessors in Safety Technique* (TUV Bayern/TUV Rheinland)

HOLSTEIN, W.H. (1966). What it costs to steam and electrically trace pipelines. *Chem. Engng Prog.*, **62**(3), 107

HOLT, B.R. *et al.* (1987). CONSYD – integrated software for computer aided control system design and analysis. *Comput. Chem. Engng*, **11**, 187

HOLT, J.C. (1982). Sprinklers and fire venting. *Fire Surveyor*, **11**(6), 13

HOLT, P.F. (1987). *Inhaled Dust and Disease* (Chichester: Wiley)

HOLT, W.H. and RUSSELL, J.L. (1981). Scheduling in the engineering phase. *Hydrocarbon Process.*, **60**(12), 92

HOLTKAMP, R.J. and KEECH, A.E. (1989). SIGTTO guidelines on alleviation of excessive surge pressures in emergency shut-down of cargo transfer. *Gastech 88*, paper 4.4

HOLTON, G.A. and MONTAGUE, D.A. (1988). Application of health effects data to chemical process accidents. *Plant/Operations Prog.*, **7**, 204

HOLTON, G.A., BAKER, L.S. and CASADA, M.L. (1987). Exposure and consequence analysis using Lotus 1–2–3. *Vapor Cloud Modeling*, p. 974

HOLTSLAG, A.A.M. and NIEUWSTADT, F.T.M. (1986). Scaling the atmospheric boundary layer. *Boundary-Layer Met.*, **36**, 201

HOLTSLAG, A.A.M. and VAN ULDEN, A.P. (1983). A simple scheme for daytime estimates of the surface fluxes from routine weather data. *J. Climate Appl. Met.*, **22**, 517

HOLTSLAG, A.A.M., DE BRUIN, H.A.R. and VAN ULDEN, A.P. (1981). Estimation of the sensible heat flux from standard meteorological data for stability calculations during daytime. In de Wispelaere, C. (1981, *op. cit.*, p. 401

HOLUM, J.R. (1977). *Topics and Terms in Environmental Problems* (New York: Wiley)

HOLVE, D.J., HARVILL, T.L., SCHATZ, K.W. and KOOPMAN, R.P. (1990). In-situ optical measurements of hydro-fluoric acid aerosols. *J. Loss Prev. Process Ind.*, **3**(2), 234

HOLZAPFEL, G. and SCHON, G. (1965). On the development of gas detonations in pipes. *Chem. Ing. Tech.*, **37**, 493

HOLZE, K., DECKWER, W.D. and KELM, M. (1980). Application of thermokinetic methods on the denitration with formic acid. *Loss Prevention and Safety Promotion 3*, vol. 3, p. 1413

HOLZWORTH, G.C. (1967). Mixing depths, wind speed and air pollution potential for selected locations in the USA. *J. Appl. Met.*, **6**, 1039

HOLZWORTH, G.C. (1972). *Mixing Depths, Wind Speed, and Potential for Urban Air Pollution throughout the Contiguous United States*. Rep. AP-101. Environ. Prot. Agency, Washington, DC

HOMAN, C., MACCRONE, R.K. and WHALLEY, E. (eds) (1983). *High Pressure in Science and Technology* (Amsterdam: Elsevier)

HOMAN, H.S. (1983). The dependence of soot yield on flame conditions – a unifying picture. *Combust. Sci. Technol.*, **33**, 1

HOMANN, K.H. (1967). Carbon formation in premixed flames. *Combust. Flame*, **11**, 265

HOMANN, K.H. and POSS, R. (1972). The effect of pressure on the inhibition of ethylene flames (letter). *Combust. Flame*, **18**, 300

HOMBERGER, E., REGGIANI, G., SAMBETH, J. and WIPF, H. (1979). *The Seveso Accident: its Nature, Extent and Consequences*. (Hofmann La Roche)

HOME OFFICE (1953). *Basic Studies on the Casualties and Homeless to be Expected from Heavy Air Attacks. Rep. CD/SA 46* (London)

HOME OFFICE (1968). *Report of the Home Office Gas Cylinders and Containers Committee* (London)

HOME OFFICE (1972). *Emergency Services, Circular 1* (London)

HOME OFFICE (1992). *Fire Statistics United Kingdom 1990* (London)

HOME OFFICE – see also Appendices 27 and 28

HONEYCOMBE, R.W.K. (1967). *The Plastic Deformation of Metals* (London: Arnold)

HONEYMAN, G.G. (chrmn) (1960). *Report of the Advisory Committee on the Examination of Steam Boilers, Cmnd 1173* (London: HM Stationery Office)

HONNERY, D.R. and KENT, J.H. (1990). Soot formation in long ethylene diffusion flames. *Combust. Flame*, **82**, 426

HONNERY, D.R. and KENT, J.H. (1992). Soot mass growth modelling in laminar diffusion flames. *Combustion*, **24**, 1041

HONTI, G. (ed.) (1976). *The Nitrogen Industry* (Budapest: Akademiai Kiado)

HOOD, J.E. (1974). Fracture of steel pipelines. *Int. J. Pres. Ves. Piping*, **2**, 165

VAN HOOF, W.C. and OFRENCHUK, J.P. (1982). Foundation failure and its remedy for a liquefied gas storage tank. *Gastech 81*, p. 407

HOOFT, J.P., KEITH, V.F. and PORRICELLI, J.D. (1975). The influence of human behaviour on the controllability of ships. *Transport of Hazardous Cargoes*, 4

HOOPER, E.G. (1975). Measuring safety performance in industry. *Plant Engr*, **19**(6/7) Jun./Jul., 31

HOOPER, M.W., WARNER, J.E. and GALEN, A.M. (1976). Over the road/in-plant tank truck hazardous material spills. One company's management program. *Control of Hazardous Material Spills 3*, p. 135

HOOPER, W.B. (1982). Predict fittings for piping systems. *Chem. Engng*, **89** (May 17), 127

HOORMAN, J.H. (1969). Electrical problems in the chemical industry. *Loss Prevention*, **3**, 54

HOOSE, J. (1984). Using liquid N2 to cool reactors. *Hydrocarbon Process.*, **63**(10), 71

HOOT, T.G. and MERONEY, R.N. (1974). The behaviour of negatively buoyant stack gases. *Sixty-Seventh Ann. Mtg, Air Poll. Control Ass.*, Denver, CO

HOOT, T.G., MERONEY, R.N. and PETERKA, J.A. (1973). *Wind Tunnel Tests of Negatively Buoyant Plumes. Rep. CER73–74TGH-RNM-JAP-13*. Colorado State Univ., Fort Collins, CO

HOOVER, R.C. and FRAUMENI, J.F. (1975). Cancer mortality in US counties with chemical industries. *Environ. Res.*, **9**, 196

HOOVER, R.L., HANCOCK, R.L., HYLTON, K.L., DICKERSON, B. and HARRIS, G.E. (1989). *Health, Safety and Environmental Control* (New York: van Nostrand Reinhold)

HOOVER, T.E. (1970). Technical feasibility and cost of LNG pipelines. *LNG Conf. 2*

HOPE, S. (1976). The management of loss prevention. *Course on Loss Prevention in the Process Industries*. Dept of Chem. Engng, Loughborough Univ. of Technol

HOPE, S. (1983). The CONCAWE report on methodologies for hazard analysis and risk assessment. *Loss Prevention and Safety Promotion 4*, vol. 1, p. B1

HOPF, P.S. (1975). *Designer's Guide to OSHA* (New York: McGraw-Hill)

HOPKIN, D.V. (1977). The reliability of man as a system component. *First Nat. Conf. on Reliability. NRC6/1.* (Culcheth, Lancashire: UK Atom. Energy Auth.)

HOPKIN, P. (1991). Recording COSHH assessments. In Simpson, D. and Simpson, W.G. (1991), *op. cit.*, p. 80

HOPKINS, H.C. (1952). The study of temperature in inflammable solids exposed to high intensity thermal radiation. S.M. Thesis, Massachusetts Inst. Technol., Cambridge, MA

HOPKINS, H.F. (1993). *TRANSPIRE: an Expert System Package for the Assessment of the Risks and Hazards of Gas Transmission Pipelines. Comm. 1536.* (London: Inst Gas Engrs)

HOPKINS, H.F., LEWIS, S.E. and RAMAGE, A.D. (1993). The development and application of the British Gas TRANSPIRE pipeline risk assessment package. *Inst. Gas Engrs, Midland Section Mtg,* Loughborough

HOPKINS, J.M. and CONGELLIERE, R.H. (1968). Noise pollution: a new problem. *Hydrocarbon Process.*, **47**(5), 124

HOPKINSON, B. (1915). *British Ordnance Board Minutes 13565*

HOPKIRK, R.J., LYMPANY, S.D. and MARTI, J. (1980). Three-dimensional numerical predictions of impact effects. *Nucl. Energy*, **19**, 207

HOPPE, R.M. (1982). Widespread dissension over RCRA amendments. *Chem. Engng*, **89**(Aug. 9), 31

HOPPE, R. (1984a). Leaking landfills push EPA into remedial action. *Chem. Engng*, **91**(Oct. 15), 25

HOPPE, R. (1984b). Superfund revamping may raise CPI's cleanup bill. *Chem. Engng*, **91**(Sept. 3), 25

HOPPE, T. (1992). Use reaction calorimetry for safer process design. *Chem. Engng Prog.*, **88**(9), 70

HOPPE, T. and BRUDERER, R. (1989). Thermal investigation into a runaway reaction. *Loss Prevention and Safety Promotion 6*, p. 37.1

HOPPE, T. and GROB, B. (1989). Heat flow calorimetry as a testing method for preventing runaway reactions. *Runaway Reactions*, p. 132

HOPPER, D.R. (1989). Cleaning up contaminated waste sites. *Chem. Engng*, **96**(8), 94

HORAN, E.L. (1970). Planning for emergencies. *Loss Prevention*, 4, 90

HORLOCK, J.H. (1958). *Axial Flow Compressors* (London: Butterworths)

VAN HORN, A.J. and WILSON, R. (1976). *Liquefied Natural Gas Safety Issues, Public Concerns and Decision-Making* (Cambridge, MA: Harvard Univ.)

HORN, B.K.P. (1986). *Robot Vision* (Cambridge, MA: MIT Press)

VAN HORN, D.J. (1985). Risk assessment techniques for experimentalists. In Hoffman, J.M. and Maser, D.C. (1985), *op. cit.*, p. 23

HORN, J. et al. (1974). Alaska to Japan LNG project – Kenai revisited. *LNG Conf. 4*

HORNBOSTEL, C. (1978). *Construction Materials – Types, Uses and Applications* (New York: Wiley-Interscience)

HORNER, G. (1984). Selecting suitable materials for tower packings and internals. *Chem. Engng, Lond.*, **408**, 21

HORNER, N. (1971). Noise control in plants. *Ammonia Plant Safety*, **13**, p. 105

HORNER, R.A. (1989). Direction of process plant safety regulation in the United States. *J. Loss Prev. Process Ind.*, **2**(3), 123

HORNER, T.W. (1974). Probability assessment of LNG accidents on New York barge transits from Staten Island to Newtown Creek and Steinway Creek. In Federal Power Commission (1974), *op. cit.*, vol. II, Attachment 4

HORNER, T.W. and ECOSYSTEMS INC. (1974). Probability assessment of LNG ship accidents in the New York and Providence Harbors. In Federal Power Commission (1974), *op. cit.*, vol. II, Attachment 2

HORNSBY, I. (1984). How to buy a pump. *Chem. Engr, Lond.*, **399**, 30

HORODNICEANU, M. and CANTILLI, E.J. (1979). *Transportation System Safety* (Lexington, MA: Lexington Books)

HORODNICEANU, M. et al. (1976). *Transportion System Safety Methodology* (Washington, DC: Dept of Transportation)

HORODNICEANU, M. et al. (1977). System safety techniques useful to transportation safety. *Transportation Res. Record*, **629**

HOROWITZ, B.A. (1994). Make the most of process design software. *Chem. Engng Prog.*, **90**(1), 54

HOROWITZ, N.C. (1981). Selecting materials for chlorine-gas neutralization. *Chem. Engng*, **88**(Apr. 6), 105

HORRELL, R. (1991). Alignment: the engineer's responsibilities. *Chem. Engr, Lond.*, **488**, 17

HORRELL, W.R., NEAL, M. and NEEDHAM, P. (1991). *Couplings and Shaft Alignments* (Mech. Engng Publs)

HORSLER, A.G. and LUCAS, D.M. (1972). *The Firing of Shell Boilers* (London: Instn Gas Engrs)

HORSLEY, B. (1974). Commissioning a methanol plant. *Instn Chem. Engrs, Symp. on Commissioning of Chemical and Allied Plant, Swansea* (London)

HORSLEY, D.C.M. and PARKINSON, J.S. (1990). *Process Plant Commissioning* (Rugby: Instn Chem Engrs)

VAN DER HORST, J. and VAN DER SCHAAF, J. (1983). Nautical risk of liquefied gas carriers. *Loss Prevention and Safety Promotion 4*, vol. 2, p. B98

VAN DER HORST, J.M.A. (1972). Stress corrosion cracking in ammonia urea synthesis corrosion of austenitic stainless steel. *Ammonia Plant Safety*, **14**, 98

VAN DER HORST, J.M.A. and SLOAN, C.R. (1974). Wrong heat treat causes expansion joint failure. *Ammonia Plant Safety*, **16**, 76

VAN DER HORST, J.M.A. and SLOAN, C.R. (1975). Reformer gas line flange failure. *Loss Prevention*, **9**, 130

HORST, T.W. (1977). A surface depletion model for deposition from a Gaussian plume. *Atmos. Environ.*, **11**, 42

HORST, T.W. (1979). Lagrangian similarity modelling of vertical diffusion from a ground-level source. *J. Appl. Met.*, **18**, 733

HORST, T.W. (1980). A review of Gaussian diffusion– deposition models. *In Atmospheric Sulfur Deposition* (ed. D.S. Shriner, C.R. Richmond and S.E. Lindberg) (Ann Arbor, MI: Ann Arbor Science), p. 275

HORST, T.W. and DORAN, J.C. (1986). Nocturnal drainage flow on simple slopes. *Boundary Layer Met.*, **34**, 263

HORST, T.W., DORAN, J.C. and NICKOLA, P.W. (1979). *Evaluation of Empirical Atmospheric Diffusion Data. Rep. NUREG/CR-0789* (Washington, DC: Nucl. Regul. Comm.)

HORT, E.V. and TAYLOR, P. (1991). Acetylene-derived chemicals. *Kirk–Othmer Encyclopaedia of Chemical Technology*, 4th ed., vol. 1 (New York: Wiley), p. 195

HORVATH, A.C. (1975). *Physical Properties of Organic Compounds* (London: Arnold)

HORVATH, A.L. (1982). *Halogenated Hydrocarbons: Solubility–Miscibility with Water* (Basel: Dekker)

HORVATH, A.L. (1986). *Conversion Tables of Units in Science and Engineering* (London: Macmillan)

HORVATH, H. (1981). Atmospheric visibility. *Atmos. Environ*, **15**, 1785

HORVATH, T. and BERTHA, I. (1975). Mathematical simulation of electrostatic hazards. *Static Electrification*, **4**, p. 256

HORZELLA, T.I. (1978). Selecting, installing and maintaining cyclone dust collectors. *Chem. Engng*, **85**(Jan. 30), 85

HOSEMANN, J.P., SCHIKARSKI, W. and WILD, H. (1973). Radioactive pollutants released in accidents at LWR power plants: review and attempt at classification. In *Physical Behaviour of Radioactive Contaminants in the Atmosphere* (Vienna: Int. Atom. Energy Agency), p. 279

HOSFORD, J.E. (1960). Measures of dependability. *Operations Res.*, **8**, 53

HOSKER, R.P. (1974a). A comparison of estimation procedures for over-water plume dispersion. *First Symp. on Atmospheric Diffusion and Air Pollution* (Boston, MA: Am. Met. Soc.), p. 281

HOSKER, R.P. (1974b). Estimates of dry deposition and plume depletion over forests and grassland. *Physical Behaviour of Radioactive Contaminants in the Atmosphere ST1/POB/354* (Vienna: Int. Atom. Energy Agency), p. 291

HOSKER, R.P. (1979). Empirical estimation of wake cavity size behind block-type structures. *Fourth Symp. on Turbulence, Diffusion and Air Pollution* (Boston, MA: Am. Met. Soc.), p. 603

HOSKER, R.P. (1980). Practical application of air pollution deposition models—current status, data requirements, and research needs. *Proc. Int. Conf. on Air Pollutants and Their Effects on the Terrestrial Ecosystem* (ed. S.V. Krupa and A.H. Legge) (New York: Wiley)

HOSKER, R.P. and LINDBERG, S.E. (1982). Atmospheric deposition and plant assimilation of gases and particles. *Atmos. Environ.*, **16**, 899

HOSKINS, J.C. and HIMMELBLAU, D.M. (1988). Artificial neural network models of knowledge representation in chemical engineering. *Comput. Chem. Engng*, **12**, 881

HOSKINS, J.C., KALIYUR, K.M. and HIMMELBLAU, D.M. (1991). Fault diagnosis in complex chemical plants using artificial neural networks. *AIChE J.*, **37**, 137

HOSKINS, R.K. and WORM, G.H. (1993). Develop an effective management of change manual. *Chem. Engng Prog.*, **89**(11), 77

HOSKINS, S. (1985). Modular halon systems. *Fire Surveyor*, **14**(4), 9

HOSSER, D., DORN, T. and RICHTER, E. (1994). Evaluation of simplified calculation methods for structural fire design. *Fire Safety J.*, **22**, 249

HOTCHKISS, A.G. and WEBER, H.M. (1953). *Protective Atmospheres* (New York: Wiley)

HOTCHKISS, C. (1991). The long arm of valve control. *Chem. Engng*, **98**(8), 104

HOTTEL, H.C. (1930). *Mech. Engng*, **52**, 699

HOTTEL, H.C. (1931). Radiant heat transmission between surfaces separated by non-absorbing media. *Trans ASME*, **53**, 265

HOTTEL, H.C. (1949). *Thermodynamic Charts for Combustion Processes* (New York: Wiley)

HOTTEL, H.C. (1953). Burning in laminar and turbulent fuel jets. *Combustion*, **4**, 97

HOTTEL, H.C. (1954). Radiant heat transmission. In McAdams, W.H. (1954), *op. cit.*, p. 55

HOTTEL, H.C. (1958). Certain laws governing diffusive burning of liquids by V.I. Blinov and G.N. Khudiakov (book review). *Fire Res. Abs. Revs*, **1**, 41

HOTTEL, H.C. (1984). Stimulation of fire research in the United States after 1940 (a historical account). *Combust. Sci. Technol.*, **39**, 1

HOTTEL, H.C. and EGBERT (1941). Radiation of furnace gases. *Trans. ASME*, **63**, 297

HOTTEL, H.C. and EGBERT, R.B. (1942). Radiant heat transmission from water vapours. *Trans. Am. Inst. Chem. Engrs*, **38**, 531

HOTTEL, H.C. and HAWTHORNE, W.R. (1949). Diffusion on laminar flame jets. *Combustion*, **3**, 254

HOTTEL, H.C. and MANGELSDORF, H.G. (1935). Heat transmission by radiation from non-luminous gases. II, Experimental study of carbon dioxide and water vapour. *Trans. Am. Inst. Chem. Engrs*, **31**, 517

HOTTEL, H.C. and SAROFIM, A.F. (1967). *Radiative Transfer* (New York: McGraw-Hill)

HOTTEL, H.C. and SMITH, V.C. (1935). Radiation from non-luminous flames. *Trans. Am. Soc. Mech Engrs*, **57**, 463

HOTTEL, H.C. and WILLIAMS, C.C. (1955). Transient heat flow in organic materials exposed to high intensity thermal radiation. *Ind. Engng Chem.*, **47**, 1136

HOTTEL, H.C., WILLIAMS, G.C. and SATTERFIELD, C.N. (1949). *Thermodynamic Charts for Combustion Processes* (New York: Wiley)

HOTTEL, H.C., WILLIAMS, G.C. and SIMPSON, H.C. (1955). Combustion of droplets of heavy liquid fuels. *Combustion*, **5**, 101

HOUGEN, J.O. (1968). Potential problems and design proposals for turbine/compressor control system. *Ammonia Plant Safety*, **10**, 79

HOUGEN, J.O. (1979). *Measurements and Control Applications*, 2nd ed. (Pittsburg, PA: Instrum. Soc. Am.)

HOUGEN, O.A. and WATSON, K.M. (1947). *Chemical Process Principles*, vol. 3 (New York: Wiley)

HOUGEN, O.A., WATSON, K.M. and RAGATZ, R.A. (1954–). *Chemical Process Principles*, vol. 1, 1954; vol. 2, 1959 (New York: Wiley)

HOUGH, B. (1983). Risk assessment of shipping terminal operations. *Loss Prevention and Safety Promotion 4*, vol. 2, p. B7

HOUGH, E. (1993). Halons use – the big question … but is there a real answer? *Fire Int.*, **138**(Feb.), 47

HOUGHLAND, J.W. (1978). Hazards of fuel oil firing. *Ammonia Plant Safety*, **20**, 148

HOUGHTON, A.J., SIMMONS, J.A. and GONSO, W.E. (1976). A failsafe transfer line for hazardous fluids. *Control of Hazardous Material Spills 3*, p. 29

HOUGHTON, D.D. (ed.) (1985). *Handbook of Applied Meteorology* (New York: Wiley)

HOUGHTON, J.A. (1965). Where insurance companies stand. *Chem. Engng Prog.*, **61**(9), 30

HOUGHTON, J.D. (1979). The right way to overhaul turbo machinery. *Hydrocarbon Process.*, **58**(6), 129

HOUGHTON, P.S. (1976). *Ball and Roller Bearings* (Barking: Applied Science)

HOULDCROFT, P.T. (1975). *Welding Processes* (Oxford: Oxford Univ. Press)

HOULT, D.P. (1969). *Oil on the Sea* (New York: Plenum Press)

HOULT, D.P. (1972a). The fire hazard of LNG spilled on water. *Proc. Conf. on LNG Importation and Safety*, Boston, MA, p. 87

HOULT, D.P. (1972b). Oil spreading on the sea. *Ann. Rev. Fluid Mech.*, **4**, 341

HOULT, D. (1983). Glass-lined steel equipment can solve process problems. *Chem. Engr, Lond.*, **397**, 32

HOULT, D.P., FAY, J.A. and FORNEY, L.J. (1969). A theory of plume rise compared with field observations. *J. Air Poll. Control Ass.*, **19**, 585

HOUSE, F.F. (1967). Winterizing chemical plants. *Chem. Engng*, **74**(Sep. 11), 173

HOUSE, F.F. (1968a). Explosion-proof electrical equipment. *Loss Prevention*, **2**, 10

HOUSE, F.F. (1968b). Pipe tracing and insulation. *Chem. Engng*, **75**(Jun. 17), 243

HOUSE, F.F. (1969). Engineer's guide to process plant layout. *Chem. Engng*, **76**(Jul. 28), 120

HOUSNER, G.W. (1947). Characteristics of strong motion earthquakes. *Bull. Seismol. Soc. Am.*, **37** (1), 19

HOUSTON, D.E.L. (1971). New approaches to the safety problem. *Major Loss Prevention*, p. 210

HOUSTON, D.E.L. (1977). Private communication

HOUSTON, D.E.L. (1980). Why do people take risks? A simple feedback model. *Loss Prevention and Safety Promotion 3*, vol. 2, p. 777

HOUWELING, J. (1980). A generalized shock wave model. *Loss Prevention and Safety Promotion 3*, vol. 2, p. 478

HOVEID, P. (1966). Follow-up study of the chlorine gas accident of January 26, 1940 in Mjondalen. *Norwegian J. Publ. Health*, **37**, 59

VAN DER HOVEN, I. (1968). Deposition of particles and gases. In Slade, D.H. (1968), *op. cit.*, p. 202

VAN DER HOVEN, I. (1976). A survey of field measurements of atmospheric diffusion under low wind speed, inversion conditions. *Nucl. Safety*, **17**, 223

HOWARD, G.W. (1971). V-belt problems and solutions. *Chem. Engng*, **84**(Aug. 15), 167

HOWARD, G.W. (1977). Installing and maintaining V-belts. *Chem. Engng*, **84**(Jul. 18), 117

HOWARD, J. (ed.) (1976). Developments in disaster. *Action for Disaster*, Glasgow

HOWARD, J.B. (1969). On the mechanism of carbon formation in flames. *Combustion*, **12**, 877

HOWARD, J.C. (1958). Static electricity in the petroleum industry. *Elec. Engng*, **77**(7), 610

HOWARD, J.C. (1959). Hazard of static electricity. *Pet. Refin.*, **38**(11), 367

HOWARD, J.C. (1964). The hazards of static electricity. In Vervalin, C.H. (1964a), *op. cit.*, p. 146

HOWARD, J.L. and ANDERSEN, P.G. (1979). A barge-mounted gas liquefaction and storage plant. *Chem. Engng Prog.*, **75**(10), 76

HOWARD, J.R. (1959). Are standard procedures the whole answer? *API Proc*, **39**(3)

HOWARD, R.A. (1971). *Dynamic Probabilistic Systems*, vol. 1, *Markov Models*; vol. 2, *Semi Markov and Decision Processes* (New York: Wiley)

HOWARD, R.A. and MATHESON, J.E. (1980). *Influence diagrams. Report.* Standford Res. Inst., Menlo Park, CA

HOWARD, W.B. (1970a). Flammability limits of fluidized bed mixture. *Loss Prevention*, **4**, 6

HOWARD, W.B. (1970b). Hazards with flammable mixtures. *Loss Prevention*, **4**, 12. See also *Chem. Engng Prog., 1970*, **66**(9), 59

HOWARD, W.B. (1972). Interpretation of a building accident. *Loss Prevention*, **6**, 68

HOWARD, W.B. (1973). *Reactor Relief Systems for Phenolic Resins*, (Monsanto Co.)

HOWARD, W.B. (1975a). Ethylene behavior related to hot tapping. *Loss Prevention 9*, 9

HOWARD, W.B. (1975b). Private communication (to N.A. Eisenberg *et al.*)

HOWARD, W.B. (1980). Tests of explosion venting of buildings. *Loss Prevention and Safety Promotion 3*, 979

HOWARD, W.B. (1982). Flame arresters and flashback arresters. *Plant/Operations Prog.*, **1**, 203

HOWARD, W.B. (1983). Efficient time use to achieve safety of processes. *Loss Prevention and Safety Promotion 4*, vol. 1, p. A11

HOWARD, W.B. (1984). Efficient time use to achieve safety of process or "'We ain't farmin' as good as we know how". *Plant/Operations Prog.*, **3**, 129

HOWARD, W.B. (1985). Seveso: cause: prevention. *Plant/Operations Prog.*, **4**, 103

HOWARD, W.B. (1988). Process safety technology and the responsibility of industry. *Chem. Engng Prog.*, **84**(9), 25

HOWARD, W.B. (1989). On the need for small chemical operators to know and use proper chemical process safety technology. *Plant/Operations Prog.*, **8**, 65

HOWARD, W.B. (1992a). Precautions in selection, installation and use of flame arresters. *Loss Prevention and Safety Promotion 7*, vol. 1, 18.1

HOWARD, W.B. (1992b). Use precaution in selection, installation and operation of flame arresters. *Chem. Engng Prog.*, **88**(4), 69

HOWARD, W.B. and KARABINIS, A.H. (1980). Tests of explosion venting of buildings. *Loss Prevention and Safety Promotion 3*, vol. 3, p. 979

HOWARD, W.B. and KARABINIS, A.H. (1982). Tests of explosion venting of buildings. *Plant/Operations Prog.*, **1**, 51

HOWARD, W.B. and RUSSELL, W.W. (1974). A procedure for designing gas combustion venting systems. *Chemical Process Hazards*, **V**, p. 179

HOWARD, W.B., RODEHORST, C.W. and SMALL, G.E. (1975). Flame arresters for high-hydrogen fuel−air mixtures. *Loss Prevention*, **9**, 46

HOWARD, W.N. (1972). Interpretation of a building explosion accident. *Loss Prevention*, **6**, 68

HOWARD, W.P. (1976). Asbestos: the industry defends itself. *New Civil Engr*, Jun. 17, 32

HOWDEN, D.G. (1975). Welding on pressurized pipe. *Loss Prevention*, **9**, 15

HOWE, D.E. and TING-MAN LI (1968). The combustion of simple ketones. *Combust. Flame*, **12**, 136, 145

HOWE, G.M. (1970). *National Atlas of Disease Mortality in the United Kingdom*, 2nd ed. (London: Nelson)

HOWE, R.H.L., HOWE, R.C. and HOWE, J.M. (1981). The complications of BOD tests. *Chem. Engng*, **88**(Nov. 30), 99

HOWE, S. (1989). Glass and glazing systems: their role in fire protection. *Fire Surveyor*, **18**(4), 12

HOWE, W.H., DRINKER, P.H. and GREEN, R.M. (1961). Progress in plant measurements. *Chem. Engng*, **68**(Jun. 12), 199

HOWELL, G.M. and SCHULLER, M. (1985). Design and construction of the Port Botany, Australia, LPG/butane import terminal. *Gastech 84*, p. 460

HOWELL, J.R. (1982). *A catalog of Radiation Configuration Factors* (New York: McGraw-Hill)

HOWELLS, P. (1993). Electrostatic hazards of foam blanketing operations. *Health, Safety and Loss Prevention*, p. 312

HOWER, N.R. (1961). Liquefied gases: cryogenic or pressure storage? *Chem. Engng*, **68**(May 29), 77

HOWER, T.C. and KISTER, H.Z. (1991). Solve process column problems. *Hydrocarbon Process.*, **70**(5), 89; **70**(6), 83

HOWERTON, A.E. (n.d.). *Estimating Area Affected by a Chlorine Release. Pamphlet R71.* Chlorine Inst., New York

HOWERTON, A.E. (1969). Estimating the area affected by a chlorine release. *Loss Prevention*, **3**, 48

HOWGRAVE-GRAHAM, H.M. (1947). *The Metropolitan Police at War* (London: HM Stationery Office)

HOWLAND, A.W. (1980). Hazard analysis and the human element. *Loss Prevention and Safety Promotion 3*, vol. 2, p. 174

HOWLETT, J. (1982). *Maximum Credible Accident* (London: Arrow)

HOWSON, D.P., EXCELL, P.S. and BUTCHER, G.H. (1981). Ignition of flammable gas/air mixtures by sparks from 2 MHz and 9 MHz sources. *Radio Electron. Engr*, **51**, 170

HOYLE, E.R. and STRICOFF, R.S. (1987). Functional requirements and design criteria for the design of high hazard containment laboratories. *Plant/Operations Prog.*, **6**, 146

HOYLE, R. (1978). How to select and use mechanical packings. *Chem. Engng*, **85**(Oct. 9), 103

HOYLE, W.C. (1982). Bulk Terminals: silicon tetrachloride incident. In Bennett, G.F., Feates, F.S. and Wilder, I. (1982), *op. cit.*, p. 11–11

HOYLE, W.C. and MELVIN, G.L. (1976). A toxic substance leak in retrospect: prevention and response. *Control of Hazardous Material Spills 3*, p. 187

HOYOS, C.G. (1987). Safety through safe operation? *Fortschritte der Sicherheitstechnik 1* (Frankfurt-am-Main: DECHEMA), p. 69

HOYOS, C.G. and ZIMOLONG, B.M. (1988). *Occupational Safety* (Amsterdam: Elsevier)

HØY-PETERSEN, R. (1974). Fire prevention in solvent extraction plants. *Loss Prevention and Safety Promotion 1*, p. 325

HRYCEK, W. (1978). Optimum size for the diked-in areas. *Chem. Engng*, **85**(Mar. 13), 108

HSIANG-CHENG KUNG (1972). A mathematical model of wood pyrolysis. *Combust. Flame*, **18**, 185

HSIANG-CHENG KUNG and HILL, J.P. (1975–). Extinction of wood crib and pallet fires. *Combust. Flame, 1975*, **24**, 305; 1976, **26**, 414

HSIANG-CHENG KUNG and KALELKAR, A.S. (1973). On the heat of reaction in wood pyrolysis. *Combust. Flame*, **20**, 91

HSIANG-CHENG KUNG and STAVRIANIDIS, P. (1982). Buoyant plumes of large-scale pool fires. *Combustion*, **19**, 905

HSIEH, M.S. and TOWNEND, D.T.A. (1939a). An examination of the mechanism by which "cool" flames may give rise to "normal" flames. *J. Chem. Soc.*, p. 337

HSIEH, M.S. and TOWNEND, D.T.A. (1939b). The inflammable ranges of higher paraffin hydrocarbons in air. *J. Chem. Soc.*, p. 341

HSIEH, M.S. and TOWNEND, D.T.A. (1939c). The inflammation of mixtures of air with diethyl ether and with various hydrocarbons at reduced pressures: green flames. *J. Chem. Soc.*, p. 332

HSIEH, T. and OKRENT, D. (1976). *Some Probabilistic Aspects of the Seismic Risk of Nuclear Reactors. Rep. UCLAENG-7613. Univ. of California, Los Angeles, CA*

HSIEH, T., OKRENT, D. and APOSTOLAKIS, G.E. (1975a). On the average probability distribution of peak ground acceleration in the US continent due to strong earthquakes. *Ann. Nucl. Energy*, **2**, 615

HSIEH, T., OKRENT, D. and APOSTOLAKIS, G.E. (1975b). *On the Average Probability Distribution of Peak Ground Accelerations in the US Continent due to Strong Earthquakes. Rep. UCLA-ENG-7516.* Univ. of California, Los Angeles, CA

HSI-JEN CHEN (1987). An exact solution to the Colebrook equation. *Chem. Engng*, **94**(2),

HSU, Y.Y. and GRAHAM, R.W. (1976). *Transport Processes in Boiling and Two-Phase Systems* (New York: McGraw-Hill)

HSU-CHERNG CHIANG and SILL, B.L. (1986). Entrainment models and their application to jets in a turbulent cross flow. *Atmos. Environ.*, **19**, 1425

HTUN, T.L. (1965). Reliability prediction techniques for complex systems. *IEEE Trans Reliab.*, **R-15**(2), 58

HU, D. (1987). *Programmers Reference Guide to Expert Systems* (Indianapolis, IN: Howard W. Sams)

HUANG, L., TAN, R.B.H., and PRESTON, M.L. (2003). An integrated tool for rapid safety, health and environmental risk assessment. *Proc. of the 37th Annu. Loss Prev. Symp.*, March 31–April 3, 2003, New Orleans, LA (ed. J. F. Murphy and R. W. Johnson) Paper LPS-4e (New York: American Institute of Chemical Engineers)

HUANG, Y.W. and FAN, L.T. (1988). Designing an object-relation hybrid database for chemical process engineering. *Comput. Chem. Engng*, **12**, 973

HUANG ZHONG-WEI and XU BIN (1987). Relation between cell size and detonation parameters of overdriven gaseous detonations. *Combust Flame*, **67**, 95

HUB, L. (1975). Development of test methods for increasing safety in the conduct of chemical processes. Doctoral Thesis, Eidgenoss. Tech. Hochschule, Zurich

HUB, L. (1976). Hazard studies applying the SIKAREX calorimeter. *Prevention of Occupational Risks*, **3**, 141

HUB, L. (1977a). Calorimetric methods for investigating dangerous reactions. *Chemical Process Hazards VI*, p. 39

HUB, L. (1977b). Comparison of test methods used for assessment of exothermicity. *Loss Prevention and Safety Promotion 2*, p. 131

HUB, L. (1977c). On-line hazards identification during chemical processes. *Loss Prevention and Safety Promotion 2*, p. 265

HUB, L. (1980). Can thermic hazards remain undetected? Loss Prevention and Safety Promotion 3, vol. 2, p. 356

HUB, L. (1981). Adiabatic calorimetry and the Sikarex technique. *Runaway Reactions, Unstable Products and Combustible Powders* (Rugby: Instn Chem. Engrs), paper 3/K

HUB, L. and JONES, J.D. (1986). Early on-line detection of exothermic reactions. *Plant/Operations Prog.*, **5**, 221

HUB, L., KUPR, T. and STAEHELI, E. (1979). Program development for safety tests. *Comput. Chem. Engng*, **3**, 503

HUBBARD, M. (1967). The comparative costs of oil transport – to and within Europe. *J. Inst. Petrol.*, **53**, 1

HUBBARD, M.G. and DUKLER, A.A. (1968). Flow model for horizontal gas–liquid slug flow. *Am. Inst. Chem. Engrs*, 65th Nat. Mtg

HUBBARD, R.W. (1967). Electrification characteristics of belts of materials running over earthed metal rollers. *Static Electrification*, **2**, 156

HUBBARD, R.W. (1971). Electrification of belts running over earthed metal rollers. *Static Electrification*, **3**, 59

HUBBE, P.D. (1970). Computer control dependability study. *Tappi*, **53**, 1777

HUBBS, S.A. (1978). Case history and contingency plan for toxic spills on the Ohio river. *Control of Hazardous Material Spills 4*, p. 178

HUBER, A.H. (1979). An evaluation of obstacle wake effects on plume dispersion. *Fourth Symp. on Turbulence, Diffusion and Air Pollution* (Boston, MA: Am. Met. Soc.), p. 610

HUBER, A.H. (1984). Evaluation of a method for estimating pollution concentrations downwind of influencing buildings. *Atmos. Environ.*, **18**, 2313

HUBER, A.H. (1988). Performance of a Gaussian model for centerline concentrations in the wake of buildings. *Atmos. Environ.*, **22**, 1039

HUBER, A.H. (1989). The influence of building width and orientation on plume dispersion in the wake of a building. *Atmos. Environ.*, **23**, 2109

HUBER, A.H. and SNYDER, W.H. (1976). Building wake effects on short stack effluents. *Third Symp. on Atmospheric Turbulence, Diffusion and Air Quality* (Boston, MA: Am. Met. Soc.), p. 235

HUBER, A.H. and SNYDER, W.H. (1982). Wind tunnel investigation of the effects of a rectangular-shaped building on dispersion of effluents from short adjacent stacks. *Atmos. Environ.*, **16**, 2837

HUBER, L.J. (1975). State and development in refinery effluent purification. *Proc. Ninth World Petrol. Cong.*, vol. 6 (Barking: Applied Science)

HUBER, P.W. and SONIN, A.A. (1975). A model for the electric charging process in fuel filtration. *Loss Prevention*, **9**, 127

HUBERICH, T. (1982). Safety and environmental protection in ammonia systems. *Plant/Operations Prog.*, **1**, 117

HUBERT, P. and PAGES, P. (1989). Risk management for hazardous materials transportation: a local study in Lyons. *Risk Analysis*, **9**, 445

HUCKINS, H.A. (1978). Conserving energy in chemical plants. *Chem. Engng Prog.*, **74**(5), 85

HUCKNALL, D.J. (1974). *Selective Oxidation of Hydrocarbons* (London: Academic Press)

HUDDLESTONE, R.L. (1979). Solid-waste disposal: land farming. *Chem. Engng*, **86**(Feb. 26), 119

HUDSON, C.M. and SEWARD, S.K. (1978). A compendium of sources of fracture toughness and fatigue-crack growth data for metallic alloys. *Int. J. Frac.*, **14**, R151

HUDSON, K.A. (1977). The implication for industry of the proposed tanker marking regulations 1977. *Symp. Hazard 77 – Advances in the Safe Handling and Movement of Hazardous Substances*, Teesside Polytechnic, Middlesbrough

HUDSON, W.G. (1965). Process design. In Fawcett, H.H. and Wood, W.S. (1965), *op. cit.*, p. 108

HUDSON, W.G. (1967). Pilot plant procedure. Safe scale-up. *Chem. Engng Prog.*, **63**(11), 64

HUEBENER, D.J. and HUGHES, R.J. (1985). Development of push–pull ventilation. *Am. Ind. Hyg. Ass. J.*, 262

HUEBLER, J., EAKIN, B.E. and LEE, A.L. (1970). Physical properties of natural gas at cryogenic conditions. *LNG Conf. 2*

HUERZELER, B. and FANNELOP, T.K. (1991). Small spills of heavy gas from continuous sources. *J. Haz. Materials*, **26**, 187

HUETTNER, D. (1974). *Plant Size. Technological Change and Investment Requirements* (New York: Praeger)

HUEY, D.C. (1973). Electrical controls. *Chem. Engng*, **80**(Feb. 26), 59

HUFF, D. (1954). *How to Lie with Statistics* (New York: Norton)

HUFF, ANDREW M. and MONTGOMERY, RANDAL L. (1997). A risk assessment methodology for evaluating the effectiveness of safeguards and determining safety instrumented system requirements, *Int. Conf. and Workshop on Risk Anal. in Process Saf.*, Oct. 21–24, 1997, Atlanta, GA, 111–126 (New York: American Institute of Chemical Engineers)

HUFF, J.E. (1973). Computer simulation of polymerizer pressure relief. *Loss Prevention*, **7**, 45

HUFF, J.E. (1977a). A general approach to the sizing of emergency pressure relief systems. *Loss Prevention and Safety Promotion 2*, p. 233

HUFF, J.E. (1977b). Supplementary notes to Huff (1977a) issued at the symposium

HUFF, J.E. (1982a). Determination of emergency venting requirements from runaway tests in closed systems. *Am. Inst Chem. Engrs, Spring Mtg* (New York)

HUFF, J.E. (1982b). Emergency venting requirements. *Plant/Operations Prog.*, **1**, 211

HUFF, J.E. (1983). Intrinsic back pressure in safety valves. *API Proc., Refining Dept, 48th Midyear Mtg*, **62**, 105

HUFF, J.E. (1984a). Computer programming for runaway reaction venting. *The Protection of Exothermic Reactors and Pressurised Storage Vessels* (Rugby: Instn Chem. Engrs), p. 109

HUFF, J.E. (1984b). Emergency venting requirements for gassy reactions from closed-system tests. *Plant/Operations Prog.*, **3**, 50

HUFF, J.E. (1985). Multi-phase flashing flow in pressure relief systems. *Plant/Operations Prog.*, **4**, 191

HUFF, J.E. (1987). The role of pressure relief in reactive chemical safety. *Preventing Major Chemical Accidents*, p. 4.43

HUFF, J.E. (1988). Frontiers in pressure relief system design. *Chem. Engng Prog.*, **84**(9), 44

HUFF, J.E. (1990). Flow through emergency relief devices and reaction forces. *J. Loss Prev. Process Ind.*, **3**(1), 43

HUFF, J.E. (1992). Relief system design scope of CCPS effluent handling guidelines. *Plant/Operations Prog.*, **11**, 53

HUFF, J.E. and SHAW, K.R. (1992). Measurement of flow resistance of rupture disk devices. *Plant/Operations Prog.*, **11**, 187

HUFF, V.N., GORDON, S. and MORRELL, V.C. (1951). *Tech. Rep. 1037.* Nat. Adv. Comm. Aero.)

HUFFMAN, K.G., WELKER, J.R. and SLIEPCEVICH, C.M. (1967). *Wind and Interaction Effects on Free-burning Fires. Tech Rep. 1441–3*, Contract OCD-OS-62–89. Univ. of Oklahoma Res. Inst., Norman, OK

HUFFMAN, K.G., WELKER, J.R. and SLIEPCEVICH, C.M. (1969). Interaction effects of multiple pool fires. *Fire Technol.*, **5**, 225

HUG, E. (1932). Poisoning by hydrogen cyanide. Antidote action of sodium hyposulphite. *Comptes Rend. Biol. Soc Paris*, **87**

HUGGETT, C. (1973). Habitable atmospheres which do not support combustion (letter). *Combust. Flame*, **20**, 140

HUGHES, D. (n.d.). *Electrical Safety Interlock Systems* (Leeds: Science Review)

HUGHES, D. (1980). *A Literature Survey and Design Study of Fume Cupboards and Fume-Dispersal Systems* (London: Br Occup Hygiene Soc.)

HUGHES, D.J. (ed.) (1980). *The Control of Toxic Atmospheres* (Manchester: Instn Chem. Engrs, North Western Branch)

HUGHES, E. (1973). Pink was the color of death. *Readers Digest*, Dec., p. 184

HUGHES, H. (1992). Towards a goal-setting regime: plans and issues: the operator's view. *Health and Safety in the Offshore Oil and Gas Industries* (Rugby: Instn Chem. Engrs)

HUGHES, H. (1994). The operator's response. *Offshore Safety Case Management* (Rugby: Instn Chem. Engrs), p. hh1

HUGHES, J.F. and BRIGHT, A.W. (1979). Electrostatic hazards associated with powder handling in silo installations. *IEEE Trans. Ind. Appl.*, **IA-15**, 100

HUGHES, J.F., BRIGHT, A.W., MAKIN, B. and PARKER, I.F. (1973). A study of electrical discharges in a charged water aerosol. *J. Phys. D: Appl. Phys.*, **6**, 966

HUGHES, J.F., CORBETT, R.P., BRIGHT, A.W. and BAILEY, A.G. (1975). Explosion hazards and diagnostic techniques associated with powder handling in large silos. *Static Electrification*, 4, 264

HUGHES, J.R. (1967). Modernize or build new refinery. *Hydrocarbon Process.*, 46(5), 155

HUGHES, J.R. (1970). *Storage and Handling of Petroleum Liquids – Practice and Law, 2nd ed.* (London: Griffin)

HUGHES, M.A. (1991). A selected annotated bibliography of social science research on planning for and responding to hazardous materials disasters. *J. Haz. Materials*, 27, 91

HUGHES, R. (1984). *Deactivation of Catalysts* (London: Academic Press)

HUGHES, R. and PRODHAN, A.S. (1973). The combustion of n-pentenes in the cool flame region. *Combust. Flame*, 20, 297

HUGHES, R. and SIMMONS, R.F. (1970). Cool flame phenomena in the oxidation of n-pentane. *Combust. Flame*, 14, 103

HUGHES, R.N. (1976). Are you a good 'risk manager'? *Hydrocarbon Process.*, 55(12), 163

HUGHES, R.P. (1987a). Fault tree truncation error bounds. *Reliab. Engng*, 17(1), 37

HUGHES, R.P. (1987b). A new approach to common cause failure. *Reliab. Engng*, 17(3), 211

HUGHES, S. (1985). Maximising the reliability of roto-dynamic pumps. *Process Engng*, Aug., 21

HUGHES, S. (1986). Question classification in rule-based systems. In Bramer, M.A. (1986), *op. cit.*, p. 123

HUGHES, S. (1987). A new market for centrifugal pumps. *Chem. Engr, Lond.*, 436, 40

HUGHES, T.W., TIERNEY, D.R. and KHAN, Z.S. (1979). Measuring fugitive emissions from petrochemical plants. *Chem. Engng Prog.*, 75(8), 35

HUGHES, W. and GAYLORD, E. (1964). *Basic Equations of Engineering Science* (New York: Schaum)

HUGHSON, R.V. (1961a). More insight on stress corrosion. *Chem. Engng*, 68(Apr. 17), 207

HUGHSON, R.V. (1961b). Stainless steel protects pipeline insulations *Chem. Engng*, 68(Aug. 21), 146

HUGHSON, R.V. (1961c). Study erosion with supersonic water jets. *Chem. Engng*, 68(Sept. 4), 164

HUGHSON, R.V. (1961d). Where standards stand. *Chem. Engng*, 68(Nov. 13), 201

HUGHSON, R.V. (1963). Combating hot sulphur-bearing gases. *Chem. Engng*, 70(Jun. 24), 138

HUGHSON, R.V. (1965). Insulation covering mastic on stretch-fabric expands with tanks. *Chem. Engng*, 72(Nov. 22), 180

HUGHSON, R.V. (1966). Controlling air pollution. *Chem. Engng*, 73(Aug. 29), 71

HUGHSON, R.V. (1969). Where standards stand now. *Chem. Engng*, 76(Nov. 17), 234

HUGHSON, R.V. (1976). High nickel allows for corrosion resistance. *Chem. Engng*, 83(Nov. 22), 125

HUGHSON, R.V. (1987). Equation-solving programs. *Chem. Engng*, 94(14), 123

HUGHSON, R.V. (1991a). Four programs keep plant emissions in check. *Chem. Engng Prog.*, 87(8), 84

HUGHSON, R.V. (1991b). Keep your plant running smoothly with maintenance software. *Chem. Engng Prog.*, 87(4), 36

HUGHSON, R.V. (1992a). Plumb the power of piping programs. *Chem. Engng Prog.*, 88(6), 95

HUGHSON, R.V. (1992b). Programs predict pump performance. *Chem. Engng Prog.*, 88(11), 43

HUGHSON, R.V. (1992c). Track safety compliance by computer. *Chem. Engng Prog.*, 88(8), 69

HUGHSON, R.V. and KOHN, P.M. (1980). Engineering ethics. *Chem. Engng*, 87(Sep. 22), 132

HUGHSON, R.V. and LABINE, R.A. (1964). Materials of construction. *Chem. Engng*, 71(Nov. 9), 169

HUGO, P. (1980). Start-up and operating behaviour of batch processes. *Chem. Ing. Tech.*, 52, 712

HUGO, P. (1981). Calculation of isothermal semi-batch reactors. *Chem. Ing. Tech.*, 53, 107

HUGO, P. and STEINBACH, J. (1986). A comparison of the limits of safe operation of a SBR and a CSTR. *Chem. Eng. Sci.* 41(4), 1081–87

HUGO, P., KONCZALLA, M. and MAUSER, H. (1980). Approximate solutions for the interpretation of exothermic batch processes with indirect cooling. *Chem. Ing. Tech.*, 52, 761

HUGONIOT, H. (1887–89). Memoire sur la propagation du mouvement dans les corps et specialement dans les gaz parfaits. *J. l'Ecole Polytechnique, Paris*, 1887, 57; 1889, 59

HUIBREGTSE, K.R. and KASTMAN, K.H. (1981). *Development of a System to Protect Groundwater Threatened by Hazardous Spills on Land*. Rep. EPA-600/2–81–085. Environ. Prot. Agency, Cincinnati, OH

HUIBREGTSE, K.R., SCHOLZ, R.C., WULLSCHLEGER, R.E., HANSEN, C.A. and WILDER, I. (1976). User's manual for treating hazardous spills. *Control of Hazardous Material Spills 3*, p. 249

HUIBREGTSE, K.R. *et al.* (1977). *Manual for the Control of Hazardous Material Spills*, vol. 1, *Spill Assessment and Water Treatment Techniques*. Rep. EPA-600/2–77–227. Environ. Prot. Agency, Washington, DC

HUIBREGTSE, K.R., LAFORNARA, J.P. and KASTMAN, K.H. (1978). In place detoxification of hazardous material spills in soil. *Control of Hazardous Material Spills 4*, p. 362

HUIBREGTSE, K.R., MOSER, J.H. and FREESTONE, F. (1978). Control of hazardous spills using improvised treatment processes. *Control of Hazardous Material Spills 4*, p. 338

HUIJGEN, D.G. and PERBAL, G. (1969). Ammonium nitrate-containing compound fertilisers. *Symp. on Inherent Hazards of Manufacturing and Storage in the Process Industry, The Hague*, p. 81

HUKKI, K. and NORROS, L. (1993). Diagnostic orientation in control of disturbance situations. *Ergonomics*, 36, 1317

HUKKOO, R.K., BAPAT, V.N. and SHIRVAIKAR, V.V. (1985). Atmospheric dispersion evaluation under elevated inversion lid. *Atmos. Environ.*, 19, 529

HULIN, K. (1993). Intumescent seals and their use in restricting Äre spread. *Fire Surveyor*, 22(4), 12

HULL, D. (1966). *Introduction to Dislocations* (Oxford: Pergamon Press)

HULLAH, J. (1976). On line trip testing. *Ammonia Plant Safety*. See also *Chem. Engng Prog.*, 1976, 72(4), 72

HULME, D. and LA TROBE-BATEMAN, J. (1983). Take a closer look at modular construction. *Hydrocarbon Process.*, 82(1), 34–C

HULME, J. (1994). Maintenance. *Plant Engr*, 38(1), 25

HUMBERT-BASSET, R. and MONTET, A. (1972). Flammable mixtures penetration in the atmosphere from spillage of LNG. *LNG Conf. 3*

HUMMEL, R.P. (1982). *Clinical Burn Therapy* (Bristol: John Wright)

HUMMER, H.B. (1966). 7–way pump check cuts seal failure. *Hydrocarbon Process.*, 45(1), 131

HUMPHREY, P.A. (1975). Meteorological support and evaluation for hazardous cargo emergencies. *Transport of Hazardous Cargoes*, 4

HUMPHREY, W.S. (1989). *Managing the Software Process* (Reading, MA: Addison-Wesley)

HUMPHREYS, K.K. and KATELL, S. (1981). *Basic Cost Engineering* (Dekker)

HUMPHREYS, P. (1987). Diversity by design – reliability aspects of systems with embedded software. In Littlewood, B. (1987b), *op. cit.*, p. 96

HUMPHREYS, P. (1988a). *Human Reliability Assessors Guide. Rep. RTS 88/95Q.* Safety and Reliab. Directorate, Culcheth

HUMPHREYS, P. (1988b). Human reliability assessors guide – an overview. In Sayers, B.A. (1988), *op. cit.*, p. 71

HUMPHREYS, R.A. (1987). Assigning a numerical value to the beta factor common cause evaluation. *Reliability 87*, p. 2C/5/1

HUNDTOFTE, V.A. (1978). Safe method in reharping reformer furnaces. *Ammonia Plant Safety*, **20**, p. 161

HUNNS, D.M. (1977). Reliability of large systems. *Measmnt Control*, **10**, 297

HUNNS, D.M. (1980). Discussions around a human factors data-base: an interim solution: the method of paired comparisons. *Sixth Advances in Reliability Technology Symp.*, vol. 1 (Culcheth, Warrington: Nat. Centre for Systems Reliab.), p. 33

HUNNS, D.M. (1981). How much automation? *Terotechnica*, **2**, 159

HUNNS, D.M. (1982). Discussion around a human factors data-base. An interim solution: the method of paired comparisons. In Green, A.E. (1982), *op. cit.*, p. 181

HUNNS, D.M. and WAINWRIGHT, N. (1991). Software-based protection for Sizewell B: the regulator's perspective. *Nucl. Engng Int.*, Sep., 38

HUNT and PRICE JONES (1918). *A Note on the Later Effects of Poisoning by Asphyxiating Gas. Rep. 15.* Chemical Warfare Med. Cttee, Med. Res. Coun., London

HUNT, A. (1989). Maintenance – the future. *Plant Engr*, Jan./Feb., 24

HUNT, A., KELLY, B.E., MULLHI, J.S., LEES, F.P. and RUSHTON, A.G. (1992a). The propagation of faults in process plants: overview of, and modelling for, fault tree synthesis. *Reliab. Engng, System Safety*, **39**, 173

HUNT, A., KELLY, B.E., MULLHI, J.S., LEES, F.P. and RUSHTON, A.G. (1992b). The propagation of faults in process plants: dividers and headers in fault tree synthesis. *Reliab. Engng System Safety*, **39**, 195

HUNT, A., KELLY, B.E., MULLHI, J.S., LEES, F.P. and RUSHTON, A.G. (1992c). The propagation of faults in process plants: control systems in fault tree synthesis. *Reliab. Engng, System Safety*, **39**, 211

HUNT, A., KELLY, B.E., MULLHI, J.S., LEES, F.P. and RUSHTON, A.G. (1992d). The propagation of faults in process plants: trip systems in fault tree synthesis. *Reliab. Engng, System Safety*, **39**, 229

HUNT, A., KELLY, B.E., MULLHI, J.S., LEES, F.P. and RUSHTON, A.G. (1993e). The propagation of faults in process plants: fault tree synthesis – 2. *Reliab. Engng System Safety*, **39**, 243

HUNT, D.L.M. and RAMSKILL, P.K. (1985). The behaviour of tanks engulfed in fire – the development of a computer program. *The Assessment and Control of Major Hazards* (Rugby: Instn Chem. Engrs), p. 71

HUNT, E.B. (1975). *Artificial Intelligence* (New York: Academic Press)

HUNT, G.A. (1980). Small scale chemical operations in toxic atmospheres. In Hughes, D.J. (1980), *op. cit.*, paper 6

HUNT, J.C.R. (1971). Effects of buildings on the local wind: the effect of single buildings and structures. *Phil. Trans. R. Soc., Ser. A*, **269**, 457

HUNT, J.C.R. (1982). Diffusion in the stable boundary layer. In Nieuwstadt, F.T.M. and van Dop, H. (1982), *op. cit.*, p. 231

HUNT, J.C.R. (1985). *Turbulence and Diffusion in Stable Environments* (Oxford: Oxford Univ. Press)

HUNT, J.C.R. (1980). Recent trends and developments (in physical modelling). In Cermak, J.E. (1980), *op. cit.*, vol. 2, p. 957

HUNT, J.C.R. and SNYDER, W.H. (1980). Experiments on stably and neutrally stratified flow over a model three- dimensional hill. *J. Fluid Mech.*, **96**, 671

HUNT, J.C.R., ABELL, C.J., PETERKA, J.A. and WOO, H. (1978). Kinematical studies of the flows around free or surface-mounted obstacles; applying topology to flow visualisation. *J. Fluid Mech.*, **86**, 179

HUNT, J.C.R., SNYDER, W.H. and LAWSON, R.E. (1978). *Turbulent Diffusion from a Point Source in Stratified and Neutral Flows around a Three-dimensional Hill – I. Rep. EPA–600/4–78–041.* Environ. Prot. Agency, Res. Triangle Park, NC

HUNT, J.C.R., ROTTMAN, J.W. and BRITTER, R.E. (1984). Some physical processes involved in the dispersion of dense gases. Ooms, G. and Tennekes, H. (1984), *op. cit.*, p. 361

HUNT, J.C.R., STRETCH, D.D. and BRITTER, R.E. (1988). Length scales in stably stratified turbulent flows and their use in turbulence models. In Puttock, J.S. (1988b), *op. cit.*, p. 285

HUNT, L. (1991). Health and safety – the issues. *Process Ind. J.*, Sep., 35

HUNT, R.M. and ROUSE, W.B. (1981). Problem solving skills of maintenance trainees in diagnosing faults in simulated powerplants. *Hum. Factors*, **23**, 317

HUNT, S.E. (1976). Technological problems of nuclear safety. *Chem. Engr, Lond.*, **308**, 265

HUNT, V.R. (1979). *Work and the Health of Women* (Boca Raton, FL: CRC Press)

HUNTER, D. (1959). *Health in Industry* (Harmondsworth: Penguin)

HUNTER, D. (1975). *Diseases of Occupations* (London: English Univ. Press)

HUNTER, D. (1988). New valves promise safer, cheaper chlorine shipping. *Chem. Engng*, **95**(Jan. 18), 27

HUNTER, H.H. (1973). Speak up when disaster strikes. In Vervalin, C.H. (1975a), *op. cit.*, p. 267

HUNTER, J.S. and BENFORADO, D.M. (1990). Avoiding site contamination by waste minimization. *J. Haz. Materials*, **25**(3), 387

HUNTER, L.W. and FAVIN, S. (1981). Fire propagation in a thermally thin fuel-lined duct with a parallel well-mixed forced flow. *Combust. Flame*, **42**, 7

HUPKENS VAN DER ELST, F. (1969). Organic peroxides. Some remarks regarding their properties and handling. *Symp. on Inherent Hazards of Manufacturing and Storage in the Process Industry*, The Hague, p. 51

HUPPERT, H.E. and SIMPSON, J.E. (1980). The slumping of gravity currents. *J. Fluid Mech.*, **99**, 785

HURA, G.S. (1984). On the use of Petri nets for the enumeration of all trees in a graph. *Reliab. Engng*, **7**(4), 229

HURLICH, A. (1963). Low temperature metals. *Chem. Engng*, **70**(Nov. 25), 104

HURON, B.S. (1963). *Elements of Arson Investigation* (New York: Reuben H. Donnelley)

HURRELL, D.J. (1982). Biological hazards. In Green, A.E. (1982), *op. cit.*, p. 549

HURST, A.E. (1969). Colour and environment. In Handley, W. (1969), *op. cit.*, p. 293

HURST, N. (1991). Immediate and underlying causes of vessel failures: implications for including management and organisational factors in quantified risk assessment. *Hazards*, **XI**, p. 155

HURST, N.W. and RATCLIFFE, K. (1994). Development and application of a structured audit technique for the assessment of safety management systems (STATAS). *Hazards*, **XII**, p. 315

HURST, N.W., NUSSEY, C. and PAPE, R.P. (1989). Development and application of a risk assessment tool (RISKAT) in the Health and Safety Executive. *Chem. Engng Res. Des.*, **67**, 362

HURST, N.W., BELLAMY, L.J., GEYER, T.A.W. and ASTLEY, J.A. (1991). A classification scheme for pipework failures to include human and sociotechnical errors and their contribution to pipework failure frequencies. *J. Haz. Materials*, **26**, 159

HURST, N.W., BELLAMY, L.J. and WRIGHT, M.S. (1992). Research models of safety management of onshore hazards and their possible application to offshore. *Major Hazards Onshore and Offshore*, p. 129

HURST, N.W., HANKIN, R.K.S., WILKINSON, J.A., NUSSEY, C. and WILLIAMS, J.C. (1992). Failure rate and incident data bases for major hazards. *Loss Prevention and Safety Promotion 7*, vol. 3, 143.1

HURST, N.W., DAVIES, J.K.W., HANKIN, R. and SIMPSON, G. (1994). Failure rates for pipework – underlying causes. *Engng Dept, Univ. of Manchester, Conf. on Valve and Pipeline Reliability*

HURST, N.W., HANKIN, R., BELLAMY, L.J. and WRIGHT, M.J. (1994). Auditing – a European perspective. *J. Loss Prev. Process Ind.*, **7**, 197

HURST, R.C. (ed.) (1984). *Design of High Temperature Metallic Components* (Amsterdam: Elsevier Applied Science)

HURWITZ, L.J. and TAYLOR, G.I. (1954). Poisoning by sewer gas with unusual sequelae. *Lancet*, **1110**

HUSA, H.W. (1964). How to compute safe purge rates. In Vervalin, C.H. (1964a), *op. cit.*, p. 267. See also *Hydrocarbon Process.*, 1964, **43**(5), 179

HUSA, H.W. and BULKLEY, W.L. (1965). Hazard of liquid ammonia spills from low pressure storage tanks. *Ammonia Plant Safety*, **7**, 41

HUSA, H.W. and RUNES, E. (1963). How hazardous are hot metal surfaces? *Oil Gas J.*, **61**(45), 180

HUSAIN, A. and HAMILEC, A.E. (1978). *J. Appl. Polymer Sci.*, **22**, 1207

HUSAIN, T. (1995). *Kuwait Oil Fires* (Oxford: Pergamon Press)

HUSAK, A.D., KISSENPFENNIG, J.F. and GRADET, A.P. (1985). Waste site clean-up: removal and in-place remediation. *Chem. Engng Prog.*, **81**(10), 25

HUSBAND, T.M. (1974). Trends in maintenance organisation. *Plant Engr*, **18**(5), 23

HUSBAND, T.M. (1976). *Maintenance Management and Terotechnology* (Farnborough: Saxon)

HUSBAND, T. and BASKER, B.A. (1976). Maintenance engineering, the current state of the art. *Production Engr*, Feb., 77

HUSBANDS, B.J. (1985). Chemical emergencies. *Chem Ind.*, Jan. 21, 49

HUSEN, C. and SAMANS, C.H. (1969). Avoiding the problems of stainless steel. *Chem. Engng*, **76**(Jan. 27), 178

HUSEN, C. and SAMANS, C.H. (1970). Why materials fail. Metals. *Chem. Engng*, **77**(Oct. 12), 15

HUSHON, J.M. and GHOVANLOU, A. (1980). *Development of a Methodology to Identify Susceptible Human Population. Rep. MTR 79W79*. Mitre Corp., McLean, VA

HUSKINS, D.J. (1977). *Gas Chromatographs as Industrial Process Analysers* (Bristol: Adam Hilger)

HUSMANN, C.A.W.A. and ENS, H. (1989). The Dutch Occupational Safety Report. In Lees, F.P. and Ang, M.L. (1989), *op. cit.*, p. 21

HUSMANN, C.A.W.A. and VAN DE PUTTE, T. (1982). Process safety analysis: identification of inherent process hazards. *The Assessment of Major Hazards* (Rugby: Instn Chem. Engrs), p. 285

HUSSAIN, M.S. (1993). Improve planetary gear vibration analysis. *Hydrocarbon Process.*, **72**(5), 105

HUSSEINY, A.A., HAGENSON, R.L., DIX, T.E. and SABRI, Z.A. (1982). Unavailability of redundant diesel generators in nuclear power plants. *Reliab. Engng*, **3**(2), 109

HUSTAD, J.E. and SONJU, O.K. (1986). Radiation and size scaling of large gas and gas/oil diffusion flames. *Am. Inst Aeronaut. and Astronaut., Tenth Int. Coll. on Dynamics of Explosions and Reactive Systems* (Berkeley, CA), p. 365

HUSTAD, J.E. and SØNJU, O.K. (1988). Experimental studies of lower flammability limits of gases and mixtures of gases at elevated temperatures. *Combust. Flame*, **71**, 283

HUSTON, W.C. (1983). Easy method for finding leaks in storage tank floors. *Hydrocarbon Process.*, **62**(4), 133

HUTCHEON, D. (1988). GP engineers (letter). *Chem. Engr, Lond.*, **450**, 4

HUTCHEON, I.C. (1981). Intrinsic safety rules OK for process instrumentation. *Second World Instrumentation Symp.*, paper TP1051. See also *Measmnt Control*, **15**, 129, 187

HUTCHERSON, M.N., HENRY, R.E. and WOLLERSHEIM, D.E. (1983). Two-phase vessel blowdown of an initially saturated liquid. *J. Heat Transfer*, **105**, 687, 694

HUTCHINGS, F.R. (1976). Service failure in chemical plant. *Soc. Chem. Ind., Symp. on Loss Prevention* (London)

HUTCHINGS, J., SANDERSON, G., DAVIES, D.G.S. and DEWEY, M.A.P. (1972). Stress corrosion of steels. Anhydrous ammonia. *Ammonia Plant Safety*, **14**, 102

HUTCHINS, B.I. and HARRISON, A. (1966). *A History of Factory Legislation*, 3rd ed. (London: Cass)

HUTCHINSON, G.M. (1983). The specification and achievement of reliability in military equipment. *Reliab. Engng*, **4**(3), 159

HUTCHISON, J.W. (1976). Valve trim. *ISA Handbook of Control Valves* (Pittsburgh, PA)

HUTHER, M. and BENOIT, J. (1983). Resistance analysis of integrated systems. *Gastech 82*, p. 159

HUTHER, M., ZEHRI, M. and ANSLOT, P. (1985). Verification of polyurethane foam insulation reliability. *Gastech 84*, p. 514

HUTHER, M., ANSLOT, P. and ZEHRI, M. (1987). Reliability and safety verification of membrane containments. *Gastech 86*, p. 232

HUTTON, B. (1973). Repair flange leaks – on stream. *Hydrocarbon Process.*, **52**(1), 75

HUTTON, D., PONTON, J.W. and WATERS, A. (1990). AI applications in process design, operation and safety. *Knowledge Engng Rev.*, **5**, 69

HUTZINGER, O. (1974). *Chemistry of PCB's* (Boca Raton, FL: CRC Press)

HUYTEN, F.H. (1979). Process chromatography. *Measamnt Control*, **12**, 422

HWANG, J.C. and KOERNER, R.M. (1983). Groundwater pollution source identification from limited monitoring well data. Pt 1 – theory and feasibility. *J. Haz. Materials*, **8**(2), 105

HWANG, R.R. and CHAING, T.P. (1986). Buoyant jets in a cross-flow of stably stratified fluid. *Atmos. Environ.*, **20**, 1887

HWANG, S.M., GARDINER, W.C., FRENKLACH, M. and HIKADA, Y. (1987). Induction zone exothermicity of acetylene ignition. *Combust. Flame.*, **67**, 65

HWANG, S.T. (1980). Treatability and pathways of priority pollutants in the biological wastewater treatment. *Water — 1980* (New York: Am. Inst. Chem. Engrs), p. 209

HWOSCHINSKY, V. (1962). Cryogenic equipment for lique-faction of natural gas. *Chem. Engng Prog.*, **58**(11), 39

HYDE, J.M. (1985). New developments in CIP practices. *Chem. Engng Prog.*, **81**(1), 39

HYDROCARBON PROCESSING (1962–). Refining process handbook. *Hydrocarbon Process.*, 1962, **41**(Sep.); and then biennially

HYDROCARBON PROCESSING (1963–). Petrochemical handbook. *Hydrocarbon Process.*, 1963, **42**(Nov.); and then biennially

HYDROCARBON PROCESSING (1975). Gas processing handbook. *Hydrocarbon Process.*, **54**(4), 79

HYDROCARBON PROCESSING (1977). Petrochemical handbook. *Hydrocarbon Process.*, **56**(11), 115

HYDROCARBON PROCESSING (1981). Petrochemical handbook. *Hydrocarbon Process.*, **60**(11), 117

HYDROCARBON PROCESSING (1984a). Gas process handbook. *Hydrocarbon Process.*, **63**(4), 51

HYDROCARBON PROCESSING (1984b). Refining process handbook. *Hydrocarbon Process.*, **63**(9), 67

HYDROCARBON PROCESSING (1985). Petrochemical handbook. *Hydrocarbon Process.*, **64**(11), 115

HYDROCARBON PROCESSING (1986a). Advanced process control handbook. *Hydrocarbon Process.*, **65**(2), 37

HYDROCARBON PROCESSING (1986b). Gas process handbook. *Hydrocarbon Process.*, **65**(4), 75

HYDROCARBON PROCESSING (1986c). Refining process handbook. *Hydrocarbon Process.*, **65**(9), 83

HYDROCARBON PROCESSING (1987a). Advanced process control handbook II. *Hydrocarbon Process.*, **66**(3), 55

HYDROCARBON PROCESSING (1987b). Petrochemical handbook. *Hydrocarbon Process.*, **66**(11), 59

HYDROCARBON PROCESSING (1988a). Advanced process control handbook. *Hydrocarbon Process.*, **67**(3), 55

HYDROCARBON PROCESSING (1988b). Gas processing handbook. *Hydrocarbon Process.*, **67**(4), 51

HYDROCARBON PROCESSING (1988c). Refining handbook. *Hydrocarbon Process.*, **67**(9), 61

HYDROCARBON PROCESSING (1990a). Advanced control handbook V. *Hydrocarbon Process.*, **69**(9), 51

HYDROCARBON PROCESSING (1990b). Gas process handbook. *Hydrocarbon Process.*, **69**(4), 69

HYDROCARBON PROCESSING (1990c). Refinery handbook. *Hydrocarbon Process.*, **69**(11), 83

HYDROCARBON PROCESSING (1991a). Advanced Process Control Handbook VI. *Hydrocarbon Process.*, **70**(9), 51

HYDROCARBON PROCESSING (1991b). Petrochemical Handbook 91. *Hydrocarbon Process.*, **70**(3), 123

HYDROCARBON PROCESSING (1992a). Advanced Process Control Handbook VII. *Hydrocarbon Process.*, **71**(9), 99

HYDROCARBON PROCESSING (1992b). Gas Process Handbook 92. *Hydrocarbon Process.*, **71**(4), 83

HYDROCARBON PROCESSING (1992c). Refining Handbook 92. *Hydrocarbon Process.*, **71**(11), 133

HYDROCARBON PROCESSING (1993a). Advanced Process Control Strategies 93. *Hydrocarbon Process.*, **72**(9), 77

HYDROCARBON PROCESSING (1993b). Environmental Processes 93. *Hydrocarbon Process.*, **72**(8), 67

HYDROCARBON PROCESSING (1993c). Petrochemical Handbook 93. *Hydrocarbon Process.*, **72**(3), 153

HYDROCARBON PROCESSING EDITORS (1973). Fundamental of disaster planning. In Vervalin, C.H. (1973a), *op. cit.*, p. 451

HYDROSCIENCE INC. (1979). *Emission Control Options for the Synthetic Organic Chemicals Manufacturing Industry — Fugitive Emissions Report.* Rep. to Emissions Standards and Engineering Div., Office of Air Quality Planning and Standards, Environ. Prot. Agency

HYLAND, R.P. and MOORE, J.F. (1992). Better environmental and safety compliance. *Hydrocarbon Process.*, **71**(9), 82–C

HYMES, I. (1984). The effects on people and structures of explosion, blast, thermal radiation and toxicity. In Petts, J.I. (1984), *op. cit.*

IAMMARTINO, N.R. (1975). New batteries are coming. *Chem. Engng*, **82**(Jan. 20), 48

IANNELLO, V., ROTHE, P.H., WALLIS, G.B., DIENER, R. and SCHREIBER, S. (1989). Aerosol research program: improved source term definition for modeling the ambient impact of accidental release of hazardous liquids. *Loss Prevention and Safety Promotion* 6, p. 58.1

IBC TECHNICAL SERVICES LTD — see Appendix 28

IBRAHIM, M., BARRIE, L.A. and FANAKI, F. (1983). An experimental and theoretical investigation of the dry deposition of particles to snow, pine trees and artificial collectors. *Atmos. Environ.*, **17**, 781

ICERMAN, L. (1977). Perspectives on the world-wide debate and public opinion on nuclear power. *Sci Publ. Policy*, **4**, 465

ICHIKAWA, M. (1983). New formula for upper bound of probability of failure. *Reliab. Engng*, **5**, 173

ICHIKAWA, M. (1984a). General formula for upper bound of probability of failure. *Reliab. Engng*, **8**(1), 57

ICHIKAWA, M. (1984b). Proposal of a new distribution free method for reliability demonstration tests. *Reliab. Engng*, **9**(2), 99

ICHIKAWA, M. (1984c). A theoretical study of flaw size distributions. *Reliab. Engng*, **9**(4), 221

ICHIKAWA, M. (1985). A theoretical study of defect detection probability of non-destructive inspection. *Reliab. Engng*, **10**(3), 175

ICHIKAWA, M. (1986). Some complementary notes to the new distribution-free for reliability demonstration. *Reliab. Engng*, **16**(4), 311

ICHIKAWA, M. (1987). An extension of the reliability-based design method for variable amplitude fatigue. *Reliab. Engng*, **17**(3), 181

ICHNIOWSKI, T. (1984). New measures to bolster safety in transportation. *Chem. Engng*, **91**(Nov. 12), 35

ICHNIOWSKI, T. (1987). Making it safer to move hazardous materials. *Chem. Engng*, **94**(10), 14

ICI AUSTRALIA LTD (1979). *Botany Expansion. Stage II, Environmental Impact Statement* (Matraville, NSW)

ICI SAFETY NEWSLETTER (1979). The man in the middle. *Safety Newsletter, No. 123*

IDA, E.S. (1983). Electronic alarm switch uses fail-safe design concepts. *Control Engng*, **30**(3), 86

IDELCHIK, I.E. (1986). *Handbook of Hydraulic Resistance*, 2nd ed. (New York: Hemisphere)

IGGLEDEN, G.J. (1981). Rotating biological contactors. *Chem Ind.*, Jul. 4, 458

IHNEN, J.L. (1984). Genetic engineering and pollution control: regulatory and legal aspects. *Chem. Engng Prog.*, **80**(12), 19

IIJIMA, T. and TAKENO, T. (1986). Effects of temperature and pressure on burning velocity. *Combust. Flame*, **65**, 35

IIZUKA, Y. and FUJITA, A. (1992). The thermal stability of the Vilsmeier complex. *Loss Prevention and Safety Promotion 7*, vol. 2, 76.1

IJSSELSTEIN, R.R. and VAN DER STEEN, A.C. (1988). Growth of a shock wave into a detonation wave in high explosives. In Groothuizen, T.M. (1988), *op. cit.*, p. 107

IKEDA, K. (1982). Warning of disaster and evacuation behaviour in a Japanese chemical fire. J. *Haz. Materials*, **7**, 51

ILIADIS, A.L. (1990). Explosion in a syngas compressor. *Ammonia Plant Safety* **30**, p. 55

ILLIDGE, J.T. and WOLSTENHOLME, J. (1978). Hazards of oxychlorination. *Loss Prevention*, **12**, 111

ILLUMINATING ENGINEERING SOCIETY (n.d.). *Lighting in Corrosive, Flammable and Explosive Situations. Rep. 1*

IMAI, J., NAKAYA, I., YAJIMA, H., TADA, M., YAMAGATA, S. and DEGUCHI, A. (1984). Safety evaluation of LNG storage tanks of 9% Ni steel based on fracture analysis. *Transport and Storage of LPG and LNG* (Antwerp: Koninklijke Vlaamse Ingenieursvereniging), p. 91

IMBRIE, R.J., KARWAN, K.R. and STONE, C.M. (1978). The spill cleanup equipment inventory system (SKIM). *Control of Hazardous Material Spills 4*, p. 65

IMGRAM, A.G. and MCCANDLESS, J.B. (1972). Check furnace tube condition method in-place metallography. *Hydrocarbon Process.*, **51**(4), 125

IMGRAM, A.G. and MCCANDLESS, J.B. (1982). Use infrared to find hot spots. *Hydrocarbon Process.*, **61**(5), 219

IMHOF, H. (1973). Protecting process plants from power failures. *Chem. Engng*, **80**(Apr. 2), 56

IMOTO, Y., TERADA, S. and MAKI, K. (1982). Predicting creep damage. *Plant/Operations Prog.*, **1**, 127

IMPAGGLIAZZO, A.M. and MURPHY, J.J. (1970). A re-evaluation of external water jackets. *Ammonia Plant Safety*, **12**, 23. See also *Chem. Engng Prog.*, 1970, **66**(3), 66

IMPERATO, N.F. (1968). Air pollution control. Gas scrubbers. *Chem. Engng*, **75**(Oct.,14), 152

IMPERIAL CHEMICAL INDUSTRIES LTD (n.d.a). *Company Engineering Procedures – Hazardous Freight* (London)

IMPERIAL CHEMICAL INDUSTRIES LTD (n.d.b). *Electrical Engineering Codes and Regulations: Group C – Electrical* (London)

IMPERIAL CHEMICAL INDUSTRIES LTD (1961). *Explosives. The Sale, Storage and Conveyance by Road*, 4th ed. (London)

IMPERIAL CHEMICAL INDUSTRIES LTD (1968). *Assessing Projects* (London: Methuen)

IMPERIAL CHEMICAL INDUSTRIES LTD (1974). *Hazard and Operability Studies. Process Safety Rep. 2* (London)

IMPERIAL CHEMICAL INDUSTRIES LTD (1977). *Specification for the Protection of Diesel Powered Mobile Equipment for Use in Zone 2 (Vapour and Gas)* (Wilton)

IMPERIAL CHEMICAL INDUSTRIES (1985). *The Mond Index*, 2nd ed. (Runcorn)

IMPERIAL CHEMICAL INDUSTRIES (1987). *Quality Assurance: A Guide for Staff* (London)

IMI (1993). *The Mond Index, Amended 2nd Edition*. Chesire, U.K.: Mond Index Services

IMPERIAL CHEMICAL INDUSTRIES LTD (1994). Private communication

IMPERIAL CHEMICAL INDUSTRIES LTD – see also Appendix 27

IMPERIAL COLLEGE (1958). *Code of Practice against Radiation Hazards* (London)

IMPERIAL COLLEGE (1975). *Education in Designing for Safety* (London)

IMPERIAL COLLEGE SAFETY COMMITTEE (1971). *Safety in Chemical Laboratories and the Use of Chemicals* (London)

INAMURA, T., SAITO, K. and TAGAVI, K.A. (1992). A study of boilover in liquid pool fires supported on water. Pt 2: Effect of in-depth radiation absorption. *Combust. Sci. Technol.*, **86**, 105. See also Arai, M., Saito, K. and Altenkirch, R.A. (1990)

INBBAR, M. (1965). *Simulation of Social Process: The Disaster Game*. Dept of Social Relations, Johns Hopkins Univ., Baltimore, MD

INCE, D. (1988). Unravelling the evidence for the prosecution. *Computing*, Nov., 32

INCE, D. (ed.) (1991a). *Software Quality and Reliability* (London: Chapman & Hall)

INCE, D. (1991b). Software requirements analysis, formal methods and software prototyping. In Ince, D. (1991a), *op. cit.*, p. 95

INCE, F. and WILLIGES, R.C. (1974). Detecting slow changes in system dynamics. *Hum. Factors, 16*, 278

INDIAN PETROCHEMICALS CORPORATION (1990). *Synthera*, **6**(15)

INDUSTRIAL OIL INSURERS (1985). *The Evaluation of Estimated Maximum Loss from Fire and Explosion in Oil, Gas and Petrochemical Industries with Reference to Percussive Unconfined Vapour Cloud Explosion* (London)

INDUSTRIAL RISK INSURERS (1982). Analyzing IRI's 1981 losses. *The Sentinel*, Fourth Quarter, 10

INDUSTRIAL RISK INSURERS (1990). *Oil and Chemical Properties Loss Potential Estimation Guide* (Hartford, CN)

INDUSTRIAL RISK INSURERS (1991). Plant spacing and layout for oil and chemical plants. *IRI Information Manual* (Hartford, CN)

INDUSTRIAL RISK INSURERS (1992). Plant spacing and layout for oil and chemical plants. *IRI Information Manual, rev. ed.* (Hartford, CN)

INDUSTRIAL RISK INSURERS – see also Appendix 28

THE INDUSTRIAL SOCIETY (1988). *The Management of Health and Safety* (London)

INFANTE, P.F., RINSKY, R.A., WAGONER, J.K. and YOUNG, R.J. (1977a). Benzene and leukaemia. *The Lancet*, **ii**, 868

INFANTE, P.F., RINSKY, R.A., WAGONER, J.K. and YOUNG, R.J. (1977b). Leukaemia in benzene workers. *The Lancet* **ii**, 76

INGBERG, S.H. (1928). Tests of the severity of building fires. *NFPA Quart.*, **22**, 43

INGELS, D.M. and POWERS, J.E. (1964). Analysis of pipeline networks. *Chem. Engng Prog.*, **60**(2), 65

INGLES, O.G. (1980). Safety in civil engineering – its perception and promotion. *Reliab. Engng*, **1**, 15

INGLIS, D. (1973). *Nuclear Energy. Its Physics and its Social Challenge* (New York: Addison-Wesley)

INHABER, H. (1981a). The risk of producing energy. *Proc. R. Soc., Ser. A*, **376**, 121. See also in Warner, Sir F. and Slater, D.H. (1981), *op. cit.*, p. 121

INHABER, H. (1981b). What is the total risk of energy systems. In Berg, G.G. and Maillie, H.D. (1981), *op. cit.*, p. 295

INHABER, H. (1987). The MRS in Oak Ridge. *J. Haz. Materials*, **15**(1/2), 241

INKOFER, W.A. (1969). Ammonia transport via pipeline. *Ammonia Plant Safety*, **11**, p. 40. See also *Chem. Engng Prog.*, 1969, **65**(3), 64

INNES, C.B.T. (1977). How you can reduce maintenance paperwork. *Hydrocarbon Process.*, **56**(1), 77

D'INNOCENZIO, M. (1972). Pump reliability specifications. *Hydrocarbon Process.*, **51**(6), 86

INONE, K., GANDHI, S.L. and HENLEY, E.J. (1974). Optimal reliability design of chemical process systems. *IEEE Trans Reliab.*, **R-23**(2), 29

INSIDE PROJECT TEAM (1997). *The INSET Toolkit*. Chesire, UK: Available from AEA Technology

INSTITUT NATIONAL DE SECURITÉ (1967). Trichloroethylene, Perchloroethylene. *Hygiène et Sécurité*, vols 1 and 2 (Paris)

INSTITUTE OF ELECTRICAL AND ELECTRONIC ENGINEERS (1973). *IEEE Recommended Practice for Grounding of Industrial and Commercial Power Systems* (New York: IEEE Press)

INSTITUTE OF ELECTRICAL AND ELECTRONIC ENGINEERS (1975a). *IEEE Recommended Practice for Engineering and Stand-by Power Systems* (New York: IEEE Press)

INSTITUTE OF ELECTRICAL AND ELECTRONIC ENGINEERS (1975b). *National Electrical Safety Code* (New York: IEEE Press)

INSTITUTE OF ELECTRICAL AND ELECTRONIC ENGINEERS (1984). *IEEE Guide to the Collection and Presentation of Electrical, Electronic, Sensing Component and Mechanical Equipment Reliability Data for Nuclear Power Generating Stations. Std 500* (New York: IEEE Press)

INSTITUTE OF ELECTRICAL AND ELECTRONIC ENGINEERS — see also Appendix 27

INSTITUTE OF FUEL (1963). *Waste-Heat Recovery* (London: Chapman & Hall)

INSTITUTE OF FUEL (1975). *Plant at Risk* (London)

INSTITUTE OF GAS TECHNOLOGY — see material at head of this list of references (LNG Confs)

INSTITUTE OF MATERIALS — see Appendix 28

INSTITUTE OF PERSONNEL MANAGEMENT (1961). *Job Analysis* (London)

INSTITUTE OF PETROLEUM (n.d.). *International Oil Tanker and Terminal Safety Guide* (London)

INSTITUTE OF PETROLEUM (1962). *Modern Petroleum Technology*, 3rd ed. (London)

INSTITUTE OF PETROLEUM (1976). *Refining Safety Code*, 2nd ed. (London)

INSTITUTE OF PETROLEUM (1977). *Health and Safety in the Oil Industry* (London: Heyden)

INSTITUTE OF PETROLEUM (1978). *A Guide to North Sea Oil and Gas Technology* (London: Heyden)

INSTITUTE OF PETROLEUM (1979). *Pressure Piping System Inspection Safety Code* (London)

INSTITUTE OF PETROLEUM (1985). *Guidance on Controls of Organic Lead in Refineries and Bulk Storage and Distribution of Leaded Gasoline* (London)

INSTITUTE OF PETROLEUM — see also Appendices 27 and 28

INSTITUTE FOR SYSTEMS ENGINEERING AND INFORMATICS (1991). *Community Documentation Centre on Industrial Risk*, vol., 3 (containing also vols 1 and 2) (Ispra, Italy: EC Joint Res. Centre)

INSTITUTE FOR SYSTEMS ENGINEERING AND INFORMATICS (1991). *Community Documentation Centre on Industrial Risk. Bull. 4* (Ispra, Italy: EC Joint Res. Centre)

INSTITUTION OF CHEMICAL ENGINEERS (1982a). Canvey Island methane terminal inquiry. *Chem. Engr, Lond.*, **376**, 7

INSTITUTION OF CHEMICAL ENGINEERS (1982b). Canvey Island methane terminal inquiry (evidence to inquiry). *Chem. Engr, Lond.*, **382**, 300

INSTITUTION OF CHEMICAL ENGINEERS (1987). *The Engineer's Responsibility for Computer Based Decisions* (Rugby)

INSTITUTION OF CHEMICAL ENGINEERS (1988). *A Guide to Capital Cost Estimating* (Rugby)

INSTITUTION OF CHEMICAL ENGINEERS (1992a). *Model Form of Conditions of Contract for Process Plants: Subcontracts* (Rugby) (the '*Yellow Book*')

INSTITUTION OF CHEMICAL ENGINEERS (1992b). *Model Form of Conditions of Contract for Process Plants: Suitable for Lump Sum Contracts* (Rugby) (the '*Red Book*')

INSTITUTION OF CHEMICAL ENGINEERS (1992c). *Model Form of Conditions of Contract for Process Plants: suitable for reimbursable contracts* (Rugby) (the '*Green Book*')

INSTITUTION OF CHEMICAL ENGINEERS — see also Appendix 28

INSTITUTION OF CIVIL ENGINEERS (1968). *Prestressed Concrete Pressure Vessels* (London)

INSTITUTION OF CIVIL ENGINEERS (1969). *Safety in Sewers and at Sewage Works* (London)

INSTITUTION OF CIVIL ENGINEERS (1978). *Transport of Hazardous Materials* (London)

INSTITUTION OF CIVIL ENGINEERS (1985). *Earthquake Engineering in Britain* (London)

INSTITUTION OF ELECTRICAL ENGINEERS (1985). *Guidelines for the Documentation of Software in Industrial Computer Systems* (London)

INSTITUTION OF ELECTRICAL ENGINEERS (1989). *Software in Safety-Related Systems* (London)

INSTITUTION OF ELECTRICAL ENGINEERS (1991). *Formal Methods in Safety Critical Systems* (London)

INSTITUTION OF ELECTRICAL ENGINEERS (1992) *Safety Related Systems: A Professional Brief* (London)

INSTITUTION OF ELECTRICAL ENGINEERS — see also Appendix 28

INSTITUTION OF FIRE ENGINEERS (1962). *Chemical Fire and Chemicals at Fires*, 5th ed. (Leicester)

INSTITUTION OF FIRE ENGINEERS (1972). *Hazardous Loads*, 2nd ed. (Leicester)

INSTITUTION OF GAS ENGINEERS — see Appendix 28

INSTITUTION OF HEATING AND VENTILATING ENGINEERS (1970). *IHVE Guide* (London)

INSTITUTION OF HEATING AND VENTILATING ENGINEERS (1973). *Conservation and Clean Air. A Study of Atmospheric Pollution in London* (London)

INSTITUTION OF INDUSTRIAL SAFETY OFFICERS (n.d.). *The Training of Industrial Safety Officers* (Leicester)

INSTITUTION OF MECHANICAL ENGINEERS (1979). *Pressure Relief Devices* (London)

INSTITUTION OF MECHANICAL ENGINEERS (1980). *Modular Design and Construction of On-Shore Process Plant* (London)

INSTITUTION OF MECHANICAL ENGINEEERS (1983). *An Annotated Bibliography on Risk Analysis* (London)

INSTITUTION OF MECHANICAL ENGINEERS (1984). *Safe Pressure Relief* (London)

INSTITUTION OF MECHANICAL ENGINEERS (1987). *Fluid Machinery for the Oil, Petrochemical and Related Industries* (Mech. Engng Publs)

INSTITUTION OF MECHANICAL ENGINEERS (1993). *Noise Control in Industry* (London)

INSTITUTION OF MECHANICAL ENGINEERS — see Appendix 28

INSTITUTION OF METALLURGISTS — see Appendix 28

INSTITUTION OF PRODUCTION ENGINEERS and INSTITUTE OF COST AND WORKS ACCOUNTANTS (1971). *An Engineers' Guide to Costing*, 2nd ed. (both London)

INSTONE, B.T. (1989). Losses in the hydrocarbon process industries. *Loss Prevention and Safety Promotion 6*, p. 119.1

INSTONE, B.T. (1990). The management factor: property insurance hazard assessment in hydrocarbon processing and petrochemical plants. *Safety and Loss Prevention in the Chemical and Oil Processing Industries*, p. 113

INSTRUMENT SOCIETY OF AMERICA (1976). *Binary Logic Diagrams for Process Operations. Std ISA-S5.2* (Research Triangle Park, NC)

INSTRUMENT SOCIETY OF AMERICA (1982). *Graphic Symbols for Distributed Control/Shared Display Instrumentation, Logic and Computer Systems. Std ISA-S5.3* (Research Triangle Park, NC)

INSTRUMENT SOCIETY OF AMERICA (1988). *Batch Control Systems. SP 88* (Pittsburgh, PA)

INSTRUMENT SOCIETY OF AMERICA – see also Appendix 27

INSTRUMENTATION, SYSTEMS, and AUTOMATION SOCIETY (ISA) (1996). ISA-S84.01-1996. *Application of Safety Instrumented Systems to the Process Industries* (Research Triangle Park, NC: Instrumentation, Systems, and Automation Society)

INSURANCE TECHNICAL BUREAU (1981). *IFAL p Factor Workbook* (London)

INTERGOVERNMENTAL MARITIME CONSULTATIVE ORGANISATION (1977). *International Maritime Dangerous Goods Code*, vols 1–4

INTERNATIONAL AGENCY FOR RESEARCH ON CANCER (1972–). *Monographs on the Evaluation of Carcinogenic Risk of Chemicals to Man*, vol. 1

INTERNATIONAL AGENCY FOR RESEARCH ON CANCER (1979a). *Chemical and Industrial Processes Associated with Cancer in Humans. Monogrs Suppl. 1* (Lyons)

INTERNATIONAL AGENCY FOR RESEARCH ON CANCER (1979b). *Evaluation of the Carcinogenic Risk of Chemicals to Humans, vol. 10: Some Monomers, Plastics and Elastomers, and Acrolein* (Lyons)

INTERNATIONAL AGENCY FOR RESEARCH ON CANCER (1980). *The Evaluation of Carcinogenic Risk of Chemicals to Humans, Monogr. 22* (Lyons)

INTERNATIONAL ATOMIC ENERGY AGENCY (1983). *Reliability of Reactor Pressure Components* (Vienna)

INTERNATIONAL ATOMIC ENERGY AGENCY (1984). *Risks and Benefits of Energy Systems* (Vienna)

INTERNATIONAL ATOMIC ENERGY AGENCY (1986). *Summary Report on the Post-Accident Review Meeting on the Chernobyl Accident. INSAG-1* (Vienna)

INTERNATIONAL ATOMIC ENERGY AGENCY (1988). *Component Reliability Data for Use in Probabilistic Safety Assessment. Rep. IAEA-TECDOC-478* (Vienna)

INTERNATIONAL ATOMIC ENERGY AGENCY (1989). *Guidebook on Training to Establish and Maintain the Qualification and Competence of Nuclear Power Plant Operations Personnel. Rep. IAEA-TECDOC-525* (Vienna)

INTERNATIONAL ATOMIC ENERGY AGENCY (1990). *Computer Codes for Level 1 Probabilistic Safety Assessment* (Vienna)

INTERNATIONAL ATOMIC ENERGY AGENCY (1991). *Safety Culture. A Report by the International Nuclear Safety Advisory Group. Rep. 75–INSAG-4* (Vienna)

INTERNATIONAL ATOMIC ENERGY AGENCY (1992). *Procedures for Conducting Probabilistic Safety Assessment of Nuclear Power Plant* (Vienna)

INTERNATIONAL CHAMBER OF SHIPPING (1974). *Tanker Safety Guide* (London)

INTERNATIONAL CHAMBER OF SHIPPING – see also Appendices 27 and 28

INTERNATIONAL CHAMBER OF SHIPPING AND OIL COMPANIES INTERNATIONAL MARINE FORUM (1978). *International Safety Guide for Oil Tankers and Terminals* (London: Witherly)

INTERNATIONAL COMMISSION ON RADIOLOGICAL PROTECTION (1977a). Problems involved in developing an index of harm. *Ann. ICRP*, **1**(4), 1

INTERNATIONAL COMMITTEE ON RADIOLOGICAL PROTECTION (1977b). *Problems Involved in Developing an Index of Harm* (Oxford: Pergamon)

INTERNATIONAL COMMITTEE ON RADIOLOGICAL PROTECTION (1977c). *Recommendations of the International Committee on Radiological Protection* (Oxford: Pergamon)

INTERNATIONAL COMMITTEE ON RADIOLOGICAL PROTECTION (1979). *Limits for Intakes of Radionuclides by Workers* (Oxford: Pergamon)

INTERNATIONAL CONFERENCE OF BUILDING OFFICIALS (1976). *Uniform Building Code* (Whittier, CA)

INTERNATIONAL CONFERENCE OF BUILDING OFFICIALS (1991). *Uniform Building Code*

INTERNATIONAL ELECTROTECHNICAL COMMISSION (n.d.). *Electrical Apparatus for Explosive Gas Atmospheres. Publ. 79*

INTERNATIONAL ELECTROTECHNICAL COMMISSION (1969). *Preliminary List of Basic Terms and Definitions for the Reliability of Electronic Equipment (or Parts) Used Therein. Publ. 271*

INTERNATIONAL ELECTROTECHNICAL COMMISSION (1991). *Software for Computers in the Application of Industrially Related Systems* (Geneva)

INTERNATIONAL ELECTROTECHNICAL COMMISSION (IEC) (2000). IEC 61508, Functional safety of electrical/electronic/programmable electronic safety related systems, Geneva, Switzerland

INTERNATIONAL ELECTROTECHNICAL COMMISSION (IEC) (2003). IEC 61511, Functional safety: safety instrumented systems for the process sector, Geneva, Switzerland

INTERNATIONAL GAS UNION (1968–). *First International Conference on Liquefied Natural Gas*

INTERNATIONAL LABOUR OFFICE (1961). *Accident Prevention* (Geneva)

INTERNATIONAL LABOUR OFFICE (1972). *Accident Prevention: A Worker's Education Manual* (Geneva)

INTERNATIONAL LABOUR OFFICE (1980a). *The Training of Nationals as Technicians and Workers in the Petroleum Industry in Developing Countries* (Geneva)

INTERNATIONAL LABOUR OFFICE (1980b). *Working Conditions and Working Environment in the Petroleum Industry, Including Offshore Activities* (Geneva)

INTERNATIONAL LABOUR OFFICE (1985a). *Working Paper on Control of Major Hazards in Industry and Prevention of Major Accidents. Rep. MHC/1985/1* (Geneva)

INTERNATIONAL LABOUR OFFICE (1985b). *Yearbook of Labour Statistics* (Geneva)

INTERNATIONAL LABOUR OFFICE (1989). *Major Hazard Control* (Geneva)

INTERNATIONAL LABOUR OFFICE (1991). *Prevention of Major Industrial Accidents* (Geneva)

INTERNATIONAL LABOUR OFFICE (1992). *Yearbook of Labour Statistics 1991* (Geneva)

INTERNATIONAL LABOUR ORGANISATION (n.d.). *Model Code of Safety Regulations for Industrial Establishments* (Geneva)

INTERNATIONAL LABOUR ORGANISATION (1971–72). *Encyclopaedia of Occupational Health and Safety, vols 1 and 2* (Geneva)

INTERNATIONAL LABOUR ORGANISATION – see also Appendix 28

INTERNATIONAL LOSS CONTROL INSTITUTE (1990). *SCAT – Systematic Cause Analysis Technique. Report. (Loganville. GA)*

INTERNATIONAL MARITIME ORGANISATION (1984). *MSC Circ. 373*

INTERNATIONAL MARITIME ORGANISATION – see also Appendices 27 and 28

INTERNATIONAL OCCUPATIONAL SAFETY AND HEALTH INFORMATION CENTRE – see Appendix 28

INTERNATIONAL PETROLEUM ENCYCLOPEDIA (1967–) (Tulsa, OK: Pennwalt Publ. Co.) (annual)

INTERNATIONAL PETROLEUM ENCYCLOPEDIA (1988) (Tulsa, OK: Pennwalt)

IPSG/IChemE (1995). *Inherently Safer Process Design* (Rugby, England: The Institution of Chemical Engineers)

INTERNATIONAL SOCIETY FOR TRENCHLESS TECHNOLOGY (1991). *Introduction to Trenchless Technology* (London)

INTERNATIONAL STANDARDS ORGANISATION (1968). *Fire Resistance Tests of Structures. Int. Std. Rec. R834*

INTERNATIONAL STANDARDS ORGANISATION – see also Appendix 27

INTERNATIONAL STUDY GROUP ON HYDROCARBON OXIDATION (1985). *Guidance Notes on the Use of Acoustic Emission Testing in Process Plants* (Rugby: Instn Chem. Engrs)

INTERNATIONAL STUDY GROUP ON RISK ANALYSIS (1982). Quantified risk analysis in the process industries. *Chem. Engr, Lond.*, **385**, 385

INTERNATIONAL STUDY GROUP ON RISK ANALYSIS (1985). *Risk Analysis in the Process Industries* (Rugby: Instn Chem. Engrs)

INTERNATIONAL TECHNICAL INFORMATION INSTITUTE, JAPAN (1975). *Toxic and Hazardous Industrial Chemical Safety Manual* (Japan)

INTEROX CHEMICALS LTD (1975). *Code of Practice for the Storage and Handling of Organic Peroxides* (Luton, Bedfordshire)

IONEL, I.I. (1986). *Pumps and Pumping* (Amsterdam: Elsevier)

IORDACHE, C., MAH, R.S.H. and TAMHANE, A.C. (1985). Performance studies of the measurement test for detection of gross errors in process data. *AIChE J.*, **31**, 1187

IORIO, A.F. and CRESPI, J.C. (1981). Analysis of the failure performance of internally pressurised piping with surface flaws. *Int. J. Pres. Ves. Piping*, **9**,

IRELAND, F.E. (n.d.). *Reflections of an Alkali Inspector* (Third Sir Hugh Beaver Memorial Lecture)

IRELAND, F.E. (1981). Emission to air – standards for control in the United Kingdom. *Chem. Engr, Lond.*, **369**, 281

IRELAND, F.E. (1984). The UK policy on chimneys. *Chem. Engr, Lond.*, **405**, 20

IRELAND, F. (1987). A is for acid rain, B is for bandwagon. *Chem. Engr, Lond.*, **436**, 64

IRELAND, N. (1985). Pipework for sulphuric acid. *Chem. Engr, Lond.*, **419**, 21

IRELAND, W.F. (1926). *The Medical Department of the United States Army in the World War*, vol. 14, *Medical Aspects of Gas Warfare* (Washington, DC: US Govt Printing Office)

IRESON, W.G. (ed.) (1968). *Reliability Handbook* (New York: McGraw-Hill)

IRHAYEM, A. (1985). Prevent plug valves from sticking and jamming. *Chem. Engng*, **92**(Apr. 29), 88

IRI, M., AOKI, K., O'SHIMA, E. and MATSUYAMA, H. (1979). An algorithm for diagnosis of system failures. *Comput. Chem. Engng*, **3**, 489

IRIBANE, J.V. and MASON, B.J. (1967). Electrification accompanying the bursting of bubbles in water and dilute suspensions. *Trans. Faraday Soc.*, **63**, 2234

IRISH, J.L. (1993). The distribution of the present worth of risky outcomes. In Melchers, R.E. and Stewart, M.G. (1993), *op. cit.*, p. 125

IRON AND STEEL INSTITUTE (1969). *Organisation of Maintenance. Publ. 118* (London)

IRVING, D. (1963). *The Destruction of Dresden* (London: Kimber)

IRVING, J. (1985). Earthquake hazard in Britain. *Earthquake Engineering in Britain* (London: Instn Civil Engrs), p. 261

IRWIN, G.R. (1948). *Fracture dynamics. Fracturing of Metals* (Cleveland, OH: Am. Soc. Metals)

IRWIN, G.R. (1957). Analysis of stresses and strains near the end of a crack traversing a plate. *J. Appl. Mech.*, **24**, 361

IRWIN, G.R. (1958). *Fracture. Handbuch der Physik*, vol. VI (Heidelberg: Springer), p. 551

IRWIN, G.R. (1962). The crack extension force for a part through crack in a plate. *J. Appl. Mech.*, **29**, 651

IRWIN, G.R. (1968). Linear fracture mechanics, fracture transition and fracture control. *Engng. Frac. Mech.*, **1**(2), 651

IRWIN, G.R., KRAFFT, J.M., PARIS, P.C. and WELLS, A.A. (1967). *Basic Aspects of Crack Growth and Fracture. Rep. NRL 6598.* Navel Res. Lab., Washington, DC

IRWIN, I.A., LEVITZ, J.J. and FREED, A.M. (1964). *Human Reliability in the Performance of Maintenance. Rep. LRP 317/ TDR-63–218.* Aerojet-General Corp., Sacramento, CA

IRWIN, I.A., LEVITZ, J.J. and FORD, A.A. (1964). *Human Reliability in the Performance of Maintenance.* Electronics Ind. Ass. and Univ. of New Mexico, *Proc. Symp. on Quantification of Human Performance* (ed. by H.R. Leuba)

IRWIN, J.D. and GRAF, E.F. (1979). *Industrial Noise and Vibration Control* (Englewood Cliffs, NJ: Prentice-Hall)

IRWIN, J.S. (1979a). Estimating plume dispersion – a recommended generalized scheme. *Fourth Symp. on Turbulence, Diffusion and Air Pollution* (Boston, MA: Am. Met. Soc.), p. 62

IRWIN, J.S. (1979b). A theoretical variation of the wind profile power law exponent as a function of surface roughness and stability. *Atmos. Environ.*, **13**, 191

IRWIN, J.S. (1983). Estimating plume dispersion – -a comparison of several sigma schemes. *J. Climate Appl. Met.*, **22**, 92

IRWIN, J.S., CHICO, T. and CATALANO, J. (1985). *CDM 2.0 – Climatological Dispersion Model – User's Guide. Report.* Environ. Prot. Agency, Research Triangle Park, NC

IRWIN, J.S. (1984). Site-to-site variation in performance of dispersion parameter estimation schemes. In de Wispelaere, C. (1984), *op. cit.*, p. 605

IRWIN, J.S., RAO, S.T., PETERSEN, W.B. and TURNER, D.B. (1987). Relating error bounds for maximum concentration estimates to diffusion meteorology uncertainty. *Atmos. Environ.*, **21**, 1927

IRWIN, W. (1991a). Avoid common mistakes when insulating piping. *Hydrocarbon Process.*, **70**(10), 64

IRWIN, W. (1991b). Insulate intelligently. *Chem. Engng Prog.*, **87**(5), 51

IRWING, M.H. (1986). Built in test – past mistakes, present problems and future solutions. *Reliab. Engng*, **15**(4), 245

ISAAC, A.H. (1960). Automatic control and instrumentation for hazardous processes. *Chemical Process Hazards*, **I**, p. 13

ISAAC, P.C.G. (1978). Environmental protection: new aims and real costs. *Chem Ind.*, July 15, 497

ISAACS, M. (1968). Selecting efficient, economical insulation. *Chem. Engng*, **76**(Mar. 24), 143

ISAACS, M. (1971). Pressure relief systems. *Chem. Engng*, **78**(Feb. 22), 113

ISAACS, M. and SETTERLUND, R.B. (1971). A guide to pipe joining methods. *Chem. Engng*, **78**(May 3), 74

ISALSKI, H., ANDERSON, T., POTTER, N., ELLIS, J. and WOOD, J. (1992). Hazop for the environment. *Chem. Engr, Lond.*, **529**, s27

ISASZEGI, J. and TIMAR, L. (1979). Reliability tests of a TCB plant and its application to maintenance development. *Hungarian J. Ind. Chem.*, **7**, 323

ISBELL, J.R. (1977). Failure in ammonia plant transfer line. *Ammonia Plant Safety*, **19**, 144

ISBIN, H.S., MOY, J.E. and DA CRUZ, A.J.R. (1957). Two-phase steam water critical flow. *AIChE J.*, **3**, 361

ISBIN, H.S., FAUSKE, H.K., GRACE, T. and GARCIA, I. (1962). Two-phase steam-water pressure drops for critical flows. *Symp. on Two-Phase Fluid Flow* (London: Instn Mech. Engrs)

ISERMANN, R. (1984). Process fault detection based on modeling and estimation methods. *Automatica*, **20**, 387

ISHACK, G. (1992). Reduction of human errors through adequate feedback of operating experience. *Hazard Identification and Risk Analysis, Human Factors and Human Reliability in Process Safety*, p. 429

ISHIDA, H. (1986). Flame spread over fuel soaked ground. *Fire Safety J.*, **10**(3), 163

ISHIDA, H. (1988). Flame spread over ground soaked with highly volatile liquid fuel. *Fire Safety J.*, **13**, 115

ISHIDA, H. (1992). Initiation of fire growth in fuel-soaked ground. *Fire Safety J.*, **18**, 213

ISHIDA, H. and IWAMA, A. (1984). Flame spread along gelled (O/W emulsified) pool of high-volatile hydrocarbon fuel. *Combust. Sci. Technol.*, **37**, 79

ISHIHAMA, W. and ENOMOTO, H. (1973). A new experimental method for studies of dust explosions. *Combust. Flame*, **21**, 177

ISHIHAMA, W. and ENOMOTO, H. (1975). Experimental study of the explosion characteristics of metal dust clouds. *Combustion*, **15**, 479

ISHII, H., ONO, T., YAMAUCHI, Y. and OHTARI, S. (1991). An algorithm for improving reliability of detection with processing of multiple sensors. *Fire Safety J.*, **17**, 469

ISHII, M. (1975). *Thermo-Fluid Dynamics Theory of Two-Phase Flow* (Paris: Eyrolles)

ISHII, M., CHAWLA, T.C. and ZUBER, N. (1976). Constitutive equation for vapor drift velocity in two-phase annular flow. *AIChE J.*, **22**, 283

ISHIKAWA KAORU (1976). *Guide to Quality Control* (Tokyo: Asian Productivity Organisation)

ISHIKAWA KAORU (1985). *What is Total Quality Control? – the Japanese Way* (Englewood Cliffs, NJ: Prentice-Hall)

ISHIKAWA, N. (1982). Combustion of stratified methane-air layers. *Combust. Sci. Technol.*, **30**, 311

ISHIMARU, H. and TAKEGAWA, T. (1980). Preventing stress corrosion cracking in the carbon dioxide absorber of ammonia plants. *Ammonia Plant Safety*, **22**, 170

ISHIMASA, Y. and UMEMURA, J. (1977). Development and operational experiences of LNG in-ground storage tanks in Japan, Pt I. *LNG Conf. 5*

ISHIMOTO, S., SASANO, T. and KAWAMURA, K. (1968). Liquid-phase oxidation of cyclohexanol. *Ind. Engng Chem. Proc. Des. Dev.*, **7**, 469

ISHIZUKA, S. (1991). Determination of flammability limits using a tubular flame geometry. *J. Loss Prev. Process Ind.*, **4**, 185

ISLES, D.L. (1977). Maintaining steam traps for best efficiency. *Hydrocarbon Process.*, **56**(1), 103

ISLITZER, N.F. (1961). Short-range atmospheric dispersion measurements from an elevated source. *J. Met.*, **18**, 443

ISLITZER, N.F. (1965). *Program Review and Summary of Recent Accomplishments at NRTS. Rep. BNL–915*. Atom. Energy Comm., Conf. on AEC Meteorological Activities, Brookhaven Nat. Labs

ISLITZER, N.F. and DUMBAULD, R.K. (1963). Atmospheric diffusion–deposition studies over flat terrain. *Int. J. Air Water Poll.*, **7**, 999

ISLITZER, N.F. and MARKEE, E.H. (1964). Puff diffusion measurements from reactor destructive tests. *Am. Met. Soc., Nat. Conf. on Micrometerology*

ISLITZER, N.F. and SLADE, D.H. (1964). *Diffusion and Transport Experiments, in Meteorology and Atomic Energy – 1968. Rep. TID–24190*. Atom. Energy Comm.

ISLITZER, N.F. and SLADE, D.H. (1968). Diffusion and transport experiments. In Slade, D.H. (1968), *op. cit.*, p. 117

ISMAN, W.E. (1978). Pesticide storage facility fire, Montgomery County, Maryland. *Control of Hazardous Material Spills 4*, p. 417

ISMAN, W.E. (1981). Managing a hazardous materials disaster. *Fire Int.*, **73**, 65

ISMAN, W.E. and CARLSON, G.P. (1980). *Hazardous Materials* (Encino, CA: Glencoe)

ISON, T.G. (1967). *The Forensic Lottery* (London: Staples Press)

ISRAEL, D.J. (1983). The role of logic in knowledge representation. *Computer*, Oct., 37

ISTRE, C.R. (1992). Improved water treatment for high-pressure boilers. *Ammonia Plant Safety*, **32**, 278

ISYUMOV, N. and TANAKA, H. (1979). Wind tunnel modelling of stack gas dispersion – difficulties and approximations. *Proc. Fifth Int. Conf. on Wind Engineering*, Fort Collins, CO

ITAGAKI, H., MIYAKE, A. and OGAWA, T. (1990). Prediction of the relief of gas explosion in a tubular vessel by venting using a mathematical model. *J. Loss Prev. Process Ind.*, **3**(4), 365

ITO, T. and SAWADA, N. (1976). Ground flares aid safety. *Hydrocarbon Process.*, **55**(6), 175

ITOYAMA, N., ENOMOTO, A., KAWAMURA, A., SAKAI, D., MATOBA, M. and TOZAWA, S. (1989). The design of an MRV-type spherical LNG storage tank. *Gastech 88*, paper 3.1

ITOYAMA, N., FUKUSHIMA, S., HATA, S., KOBAYASHI, Y. and KAWABATA, K. (1986). Design and actual service of cargo handling equipment on an LNG carrier. *Gastech 85*, p. 271

ITOYAMA, N. *et al.* (1991). Planning for longevity: a review of R&D, design and construction of three series of large LNG carriers. *Gastech 90*, p. 4.9

IVANTSOV, O.M. (1972). Study of the problem of LNG transmission by pipelines. *LNG Conf. 3*

IVANTSOV, O.M., LIVSHITS, L.S. and ROZHDESTVENSKY, V.V. (1970). Liquefied natural gas mains. *LNG Conf. 2*

IVERSON, J.H. (1968). *Summary of Existing Structures Evaluation, Pt II: Window Glass and Applications. Contract OCD-DAH20–67–C-0136 Final Rep.* Stanford Res. Inst., Menlo Park, CA

IVERSON, W.P. (1968). Microbiological corrosion. *Chem. Engng*, **75**(Sep. 23), 242

IVERSTINE, J.C. and STURROCK, R.H. (1980). Controlling the cost of maintenance. *Chem. Engng*, **87**(Mar. 24), 103

IVES, G. (1991). 'Near miss' reporting pitfalls for nuclear plants. In Van der Schaaf, T.W., Lucas, D.A. and Hale A.R. (1991), *op. cit.*, p. 51

IYA, K.S. (1972). Reduce NO$_x$ in stack gases. *Hydrocarbon Process.*, **51**(11), 163

IYA, K.S., WOLLOWITZ, S. and KASKAN, W.E. (1975). The mechanism of flame inhibition by sodium salts. *Combustion*, **15**, p. 329

IYENGAR, S.S. (ed.) (1984). *Computer Modeling of Complex Biological Systems* (Boca Raton, FL: CRC Press)

JAAFE, M., BOLTON, L., REED, C. and STEWART, J. (1981). A cleanup is in store for the Mediterranean. *Chem. Engng*, **88** (Sep. 7), 20E

JAASUND, S.A. (1987). Electrostatic precipitators: better wet than dry. *Chem. Engng*, **94**(17), 159

JACKMAN, A.P. and POWELL, R.L. (1991). *Hazardous Waste Treatment Technologies* (Park Ridge, NJ: Noyes)

JACKMAN, L.A. and NOLAN, P.F. (1994). Investigation into the effect of wind on fire protection sprayers, using the model SPLASH. *Hazards*, **XII**, p. 415

JACKMAN, P.E. (1993). Fire resistant glazing. *Fire Surveyor*, **22**(1), 19

JACKMAN, P.J. (1980). Natural ventilation principles in design proceedings. *Chart. Inst. of Building Services, Conf. on Natural Ventilation by Design* (London)

JACKS, R.L. (1961). Sound suppression in the chemical process industries. *Chem. Engng*, **68**(Feb. 20), 127

JACKSON, A. (1984). Program designs battery backup. *Hydrocarbon Process.*, **63**(4), 147

JACKSON, C. (1965). Process pumps: 99 plant proven tips. *Hydrocarbon Process.*, **44**(8), 121

JACKSON, C. (1968). Practical guide to mechanical seals. *Hydrocarbon Process.*, **47**(1), 100

JACKSON, C. (1969). New look at vibration measurement. *Hydrocarbon Process.*, **48**(1), 89

JACKSON, C. (1970). An experience with thrust bearing failures. *Hydrocarbon Process.*, **49**(1), 107

JACKSON, C. (1971a). Balance rotors by orbit analysis. *Hydrocarbon Process.*, **50**(1), 73

JACKSON, C. (1971b). How to align barrel-type centrifugal compressors. *Hydrocarbon Process.*, **50**(9), 189

JACKSON, C. (1972a). Specifying pump couplings, bearings, seals and packings. *Hydrocarbon Process.*, **51**(6), 90

JACKSON, C. (1972b). Vibration measurement on turbomachinery. *Ammonia Plant Safety*, **14**, p. 112. See also *Chem. Engng Prog.*, 1972, **68**(3), 60

JACKSON, C. (1973). How to prevent pump cavitation. *Hydrocarbon Process.*, **52**(5), 157

JACKSON, C. (1974). Optimise your vibration analysis procedures. *Hydrocarbon Process.*, **53**(1), 89

JACKSON, C. (1975). How to prevent turbomachinery thrust failures. *Hydrocarbon Process.*, **54**(6), 73

JACKSON, C. (1975−). A practical vibration primer. *Hydrocarbon Processing, 1975*, **54**(4), 161; **54**(6), 109; **54**(8), 109; **54**(11), 251; 1976, **55**(4), 171; **55**(10), 141; 1978, **57**(3), 119; **57**(4), 209

JACKSON, C. (1976a). Alignment of rotating equipment. *Chem. Engng Prog.*, **72**(2), 51

JACKSON, C. (1976b). Techniques for alignment of rotating equipment. *Hydrocarbon Process.*, **55**(1), 81

JACKSON, C. (1978). Install compressors for high availability. *Hydrocarbon Process.*, **57**(11), 243

JACKSON, C. (1981). Guidelines for improving rotating equipment reliability. *Hydrocarbon Process.*, **60**(9), 223

JACKSON, C. and LEADER, M.E. (1979). Turbomachines: how to avoid operating problems. *Hydrocarbon Process.*, **58**(11), 281

JACKSON, C. and KELHAM, S.F. (1984). Developments and trends in the chlor-alkali industry. *Chem Ind.*, June 4, 397

JACKSON, C. (ed.) (1986). *Modern Chlor-Alkali Technology*, vol. 2 (Chichester: Ellis Horwood)

JACKSON, D.H. (1948). Selection and use of ejectors. *Chem. Engng Prog.*, **44**, 347

JACKSON, H.L., MCCORMACK, W.B., RONDESTVEDT, C.S., SMELTZ, K.C. and VIELE, I.E. (1970). Control of peroxidizable compounds. J. *Chem. Educ.*, **47**(3), A175

JACKSON, J. (1975). *Health and Safety − the New Law* (London: EEC Studies)

JACKSON, J. (1979). *Health and Safety − The Law* (London: New Commercial)

JACKSON, J.D. (1961). Corrosion in cryogenic liquids. *Chem. Engng Prog.*, **57**(4), 61

JACKSON, J.D. (1980). *Steam Operation* (Englewood Cliffs, NJ: Prentice-Hall)

JACKSON, J.L. (1951). Spontaneous ignition temperatures. *Ind. Engng Chem.*, **43**, 2869

JACKSON, L., SHANNON, E. and ADLEY, G.A. (1985). British Gas has seamless pipe inspection program. *Oil Gas J.*, **83**(Sep. 9), 147

JACKSON, R. (1977). *Transport in Porous Catalysts* (Amsterdam: Elsevier)

JACKSON, COMMANDER SIR R. (1986). Disasters and the United Nations. *Interdisciplinary Sci. Rev.*, **11**, 326

JACKSON, R.E., KLOMPARENS, A.J. and WESTBROOK, G.T. (1965). Long range planning. *Chem. Engng Prog.*, **61**(1), 83

JACKSON, R.G. (1979). A shipyard-built methanol plant. *Chem. Engng Prog.*, **75**(10), 56

JACKSON, S.M. and DREIBLATT, A. (1990). Application of twin screw extruders for reactive processing. *Plant/Operations Prog.*, **9**, 220

JACOB, A. (1990). The power of water jetting. *Chem. Engr, Lond.*, **500**, 16

JACOB, A. (1991a). Alternative refrigerants. *Chem. Engr, Lond.*, **489**, 14

JACOB, A. (1991b). Cut VOCs to beat new laws. *Chem. Engr, Lond.*, **509/510**, 25

JACOB, A. (1992). Responsible care − is it working? *Chem. Engr, Lond.*, **514**, 25

JACOB, R.J.K., EGETH, H.E. and BEVAN, W. (1976). The face as a data display. *Hum. Factors*, **18**, 189

JACOBOWITZ, J.L. and ZEIS, L.A. (1969). Recent improvements in primary reformer furnaces. *Ammonia Plant Safety*, **11**, p. 6

JACOBS, D.L.E. (1971). How to specify centrifugal compressors. Reliability, performance, specification. *Hydrocarbon Process.*, **50**(10), 69

JACOBS, J.K. (1965). How to select and specify pumps. *Hydrocarbon Process.*, **44**(6), 122

JACOBS, L.W., CHOU, S.F. and TIEDJE, J.M. (1976). Fate of polybrominated biphenyls (PBBs) in soils. Persistence and plant uptake. J. *Agric. Food Chem.*, **24**, 1198

JACOBS, R.B. (1964). The violent nature of detonations. In Vervalin, C.H. (1964a), *op. cit.*, p. 101

JACOBS, R.B. and BULKLEY, W.L. (1967). *Petrol. Ind.*, **19**, 9

JACOBS, R.B., BULKLEY, W.L., RHODES, A.B. and SPEER, T.L. (1957). Destruction of a large refining unit by gaseous detonation. *Chem. Engn Prog.*, **53**, 565

JACOBS, R.B., BLUNK, F.H. and SCHEINEMAN, W. (1964). Beware process explosion hazards. In Vervalin, C.H. (1964a), *op. cit.*, p. 110

JACOBS, R.B. *et al.* (1973). Reformer detonation triggers the great Whiting fire. In Vervalin, C.H. (1973a), *op. cit.*, p. 2

JACOBS, R.M. (1971). Minimizing hazards in design. *Qual Prog.*, Oct., 23

JACOBSEN, C.T. (1986). High temperature polymeric fire barriers. *Plant/Operations Prog.*, **5**, 148

JACOBSEN, L.S. and AYRE, R.S. (1958). *Engineering Vibrations* (New York: McGraw-Hill)

JACOBSEN, M. (1981). Historical background of dust explosion research in the United States 1910–1970. *The Hazards of Industrial Explosion from Dusts* (London: Oyez/IBC), paper 2

JACOBSEN, O. and FANNELOP, T.K. (1984). Experimental and theoretical studies in heavy gas dispersion. Part II: Theory. In Ooms, G. and Tennekes, H. (1984), *op. cit.*, p. 407

JACOBSEN, O. and MAGNUSSEN, B.F. (1987). 3–D numerical simulation of heavy gas dispersion. *J. Haz. Materials*, **16**, 215

JACOBSEN, O.F. (1980). Methodology problems in predicting accidents which have actually occurred. *Loss Prevention and Safety Promotion 3*, p. 409

JACOBSON, J.S. and HILL, A.C. (1970). *Recognition of Air Pollution Injury to Vegetation: a Pictorial Atlas. Informative Rep. 1.* Air Poll. Control Ass.

JACOBSON, K. (1994). Practical issues in safety case management. *Offshore Safety Case Management* (Rugby: Instn Chem. Engrs), p. kj1

JACOBZINER, H. and RAYBIN, H.W. (1962). Hydrochloric acid poisoning. *Arch. Ped.*, **79**, 177

JAEGER, T.A. (ed.) (1973). *Second Int. Conf. on Structural Mechanics in Reactor Technology* (Berlin: Bundesanstalt für Materialprüfung)

JAEGER, T.A. (ed.) (1975). *Third Int. Conf. on Structural Mechanics in Reactor Technology* (Berlin: Bundesanstalt für Materialprüfung)

JAESCHKE, L. and TROBISCH, K. (1967). Treat HPI wastes biologically. *Hydrocarbon Process.*, **46**(7), 111

JAFFEE, R.I. (1964). Ductile materials. *Chem. Engng*, **71** (Jan. 6), 91

JAFFE, L. (1981). Technical aspects and chronology of the Three Mile Island Accident. *Ann. N.Y. Acad. Sci*, **365**, 37

JAGACINSKI, R.J. and MILLER, R.A. (1978). Describing the human operator's internal model of a dynamic system. *Hum. Factors*, **20**, 425

JAGER, P. (1980). Risk analyses for chemical plants. *Loss Prevention and Safety Promotion 3*, p. 561

JAGER, P. (1983). The question of quality of risk analysis for chemical plants: discussing the results for a sulphuric acid plant. *Loss Prevention and Safety Promotion 4*, vol. 1, p. B9

JAGER, P. and HAFERKAMP, K. (1992). A checklist for the systematic examination of storage and production facilities with high risk potential. *Prevention and Safety Promotion 7*, vol. 3, 134.1

JÄGER, P. (1984). Influencing the risk involved with the road transport of propane gas by taking technological measures. In *Transport and Storage of LPG and LNG* (Antwerp: Koninklijke Vlaamse Ingenieursvereniging), p. 307

JÄGER, P., DIEDERSHAGEN, G. and KÜHNREICH, K. (1989). Expert model to assess the hazard potential of technical installations. *Loss Prevention and Safety Promotion 6*, p. 96.1

JAGGER, S.F. (1982). The dispersion of dense toxic and flammable gas clouds in the atmosphere. *Seminar on Toxic and Flammable Clouds – Hazards and Protection* (London: Instn Mech. Engrs)

JAGGER, S.F. (1983). Formulations of the dense gas dispersion problem. In Hartwig, S. (1983), *op. cit.*, p. 27

JAGGER, S.F. and EDMUNDSON, J.N. (1984). Elevated discharge of dense gases: a method for calculating safe dis-charge velocities to allow for different weather conditions. *The Protection of Exothermic Reactors and Pressurised Storage Vessels* (Rugby: Instn Chem. Engrs), p. 215

JAGGER, S.F. and KAISER, G.D. (1981). The accidental release of dense flammable and toxic gases from pressurized containment – transition from pressure driven to gravity driven phase. In de Wispelaere, C. (1981), *op. cit.*, p. 463

JAGGERS, H.C., FRANKLIN, D.P., WARD, D.R. and ROPER, F.G. (1986). Factors controlling burning time for nonpremixed clouds of fuel gas. *Hazards*, **IX**, p. 77

JAGUSCH, G. (ca. 1989). A statistical method of computing the fragment endangering from high explosive shells. *Int. Symp on Ballistics*, Brussels, p. 472

JAHN, G. (1934). *Der Zündvorgang in Gasgemischen* (Berlin: Oldenbourg)

JAIN, A.K. (1976). *Proc ASCE, J. Hyd. Div.*, **102**, 674

JAIN, R.K. and HUTCHINGS, B. (eds) (1978). *Environmental Impact Analysis: Emerging Issues in Planning* (Univ. of Illinois Press)

JAIN, R.K., URBAN, L.V. and STACEY, G.S. (1980). *Environmental Impact Analysis: A New Dimension in Decision Making*, 2nd ed. (New York: van Nostrand Reinhold)

JAIN, S.P. and GOPAL, K. (1985). Reliability of k-to-l-out-of-n systems. *Reliab. Engng*, **12**(3), 175

JAIN, V.K. and EBENEZER, J.S. (1966). Analytical solution of a steady cylindrical flame (letter). *Combust. Flame*, **10**, 391

JAKEMAN, A.J., JUN, B. and TAYLOR, J.A.L. (1988). On the variability of the wind speed component in urban air pollution models. *Atmos. Environ.*, **22**, 2013

JAKEMAN, A.J., TAYLOR, J.A. and SIMPSON, R.W. (1986). Modeling distributions of air pollutant concentrations – II. Estimation of one and two parameter statistical distributions. *Atmos. Environ.*, **20**, 2435

JALEES, K. (1985). India's deteriorating environment. *Chem. Ind.*, Jan. 7, 11

JALURIA, Y. and KAPOOR, K. (1988). Importance of wall flows at the early stages of fire growth in the mathematical modeling of enclosure fires. *Combust. Sci. Technol.*, **59**, 355

JAMBULINGAM, N. and JARDINE, A.K.S. (1986). Life cycle costing considerations in reliability centred maintenance: an application to maritime equipment. *Reliab. Engng*, **15**(4), 307

JAMES, A. (1986). *The Probability of Traffic Accidents Associated with the Transport of Radioactive Wastes, Rep. DoE/RW/85–175.* Dept of the Environ.

JAMES, A.E. (1984). Implementation of COPA II – the industry view. *Chem. Ind.*, Aug. 20, 581

JAMES, G.B. (1947–48). Fire prevention in the chemical industry. *NFPA Quart.*, **41**, 256

JAMES, G.R. (1960). Which process best for producing hydrogen? *Chem. Engng*, **67**(Dec. 12), 161

JAMES, G.R. (1968). A look at high temperature reform piping. *Ammonia Plant Safety*, **10**, 1

JAMES, H., HARRIS, M.J. and HALL, S.F. (1992). Comparison of event tree, fault tree and Markov methods for probabilistic safety assessment and application to accident mitigation. *Major Hazards Onshore and Offshore*, p. 59

JAMES, H.R. (1982). Modular and barge-mounted process plants. *Chem. Engng Prog.*, **78**(11), 59

JAMES, I.A. (1986). *The Probability of Traffic Accidents Associated with the Transport of Radioactive Wastes. Rep. DoE/RW/85–175.* Dept. of the Environment

JAMES, J.L. (1979). Rational evaluation of revamps. *Hydrocarbon Process.*, **58**(11), 261

JAMES, K. (1994). The regulatory framework for the carriage of bulk and packaged chemicals by ships. In Rawson, K. (1994), *op. cit.*, p. 61

JAMES, M. (1985). *Classification Algorithms* (London: Collins)

JAMES, N., SHEPPARD, G. and WILLIAMS, R. (1987). THORP and the safety case. *Chem. Engr, Lond.*, **434**, 17

JAMES, N.J., RUTHERFORD, J. and SHEPPARD, G.T. (1986). Zircaloy hazards in nuclear fuel reprocessing. *Hazards IX*, p. 143

JAMES, P.E. and MENZEL, R.G. (1973). *Research on Removing Radioactive Fallout from Farm Land. Bull. 1464.* Agric. Res. Service, Dept of Agric., Washington, DC

JAMES, P.E. and WILKIN, D.C. (1969). Deep plowing – an engineering appraisal. *Am. Soc. Agric. Engrs*, paper 69–152

JAMES, R. (1976). Pump maintenance. *Chem. Engng Prog.*, **72**(2), 35

JAMES, R. (1988). Intumescent fire seals. Fire Surveyor, **17**(2), 13

JAMES, R. and BLOCH, H.P. (1976). Instrumentation for predictive maintenance monitoring. *Loss Prevention*, **10**, 101

JAMES, R. and WELLS, G. (1994). Safety reviews and their timing. *J. Loss Prev. Process Ind.*, **7**, 11

JAMES, R.A., EXCELL, P.S. and KELLER, A.Z. (1987). Probabilistic factors in radio-frequency ignition and detonation hazards analyses. *Reliab. Engng*, **17**(2), 139

JAMES, R.W. (1976). How safe is your pollution solution? *Ammonia Plant Safety* **18**, 84

JAMES, R.W. (1977). Is that regulation really necessary? *Ammonia Plant Safety* **19**, 108

JAMES, R.W. (1994). Safe use of refrigerants for process applications. *J. Loss Prev. Process Ind.*, **7**, 492

JAMES, R.W. and LAWRENCE, J.A. (1973). Failure repair of an ammonia separator. *Ammonia Plant Safety*, **15**, 65

JAMES, W.H. (1978). Try spring-mounted pumps. *Hydrocarbon Process.*, **57**(9), 247

JAMESON, C.W. and WALTERS, D.B. (eds) (1984). *Chemistry for Toxicity Testing* (London: Butterworths).

JAMIESON, D. (1981). Ultimate disposal to land. *Chem. Ind.*, Apr. 18, 267

JAMIESON, R.M. (1985). Canadian line-inspection procedures detailed. *Oil Gas J.*, **83**(Mar. 25), 107

JANAF International Thermochemical Tables – see Stull, D.R. and Prophet, H. (1971)

JANE'S INFORMATION GROUP (1993a). *Jane's Airports and Handling Agents 1993–94* (Coulsdon, Surrey)

JANE'S INFORMATION GROUP (1993b). *Jane's Containerisation Directory 1993–94* (Coulsdon, Surrey)

JANE'S INFORMATION GROUP (1993c). *Jane's World Railways 1993–94* (Coulsdon, Surrey)

JANIN, R. (1976). Experimental investigation of the thermal behaviour of products and reaction products: quantity and flow measurements of the gases and calories released as a function of temperature. *Prevention of Occupational Risks*, **3**, 89

JANIS, I.L. and MANN, L. (1977). *Decision Making* (New York: Free Press)

JANN, P. (1989). Evaluation of temporary safe havens. *J. Loss Prev. Process Ind.*, **2**(1), 33

JANSEN, J. (1976). Lessons learned from radiation tube cracking. *Ammonia Plant Safety*, **18**, 60

JANSEN, J.P. (1984). The control room of the future. *Chem. Engr, Lond.*, **401**, 22

JANSS, J. and MINNE, R. (1981/82). Buckling of steel columns in fire conditions. *Fire Safety J.*, **4**(4), 221

JANSSEN, A.J.M., SIRAA, N. and BLANKEN, J.M. (1981). Temperature runaway of a methanator. *Ammonia Plant Safety*, **23**, 19

JANSSEN, W. and TENKINK, E. (1988). Risk homeostasis theory and its critics: time for an agreement. *Ergonomics*, **31**, 429

JANSSENS, H. (1984). Dust explosion protection: a practical philosophy. *Process Engng*, Jul., 47

JANSSENS, M. (1991). The rate of heat release in wood products. *Fire Safety J.*, **17**, 217

JANSSON, A., NILSSON, G.T., KJAER, H.G. and VOHTZ, E. (1979). Barge-mounted process plants. *Chem. Engng Prog.*, **75**(10), 66

JANZEN, P. (n.d.). *A Study of Piping Failures in US Nuclear Power Reactors. Pub. AECL-MISC-204.* Atom. Energy of Canada Ltd

JAPAN GAS ASSOCIATION (1976). *A Study of Dispersion of Evaporated Gas and Ignition of LNG Pool Resulted from Continuous Spillage of LNG* (Tokyo)

JAPAN GAS ASSOCIATION (1981). *LNG Terminals of the World* (Tokyo)

JAPAN INSTITUTE FOR SAFETY ENGINEERING (1982). *Report on Burning of Petroleum Fires* (Tokyo) (in Japanese)

JARDAS, Z. (1980). Man monitoring. Sampling efficiency, desorption efficiency and other aspects. *Loss Prevention and Safety Promotion*, **3**, 255

JARDINE, A.K.S. (1970a). Equipment replacement strategies. In Jardine, A.K.S. (1970b), *op. cit.*, p. 20

JARDINE, A.K.S. (ed.) (1970b). *Operational Research in Maintenance* (Manchester: Manchester Univ. Press)

JARDINE, A.K.S. (1973b). *Maintenance, Replacement and Reliability* (London: Pitman)

JARDINE, A.K.S. (1976). The use of mathematical models in industrial maintenance. *Inst. Math Appls*, Aug./Sep., 232

JARDINE, A.K.S. and KIRKHAM, A.C.J. (1973). Maintenance policy for sugar refinery centrifuges. *Proc. Instn Mech. Engrs*, **187**, 679

JARDINE INSURANCE BROKERS LTD (1988). *Risk Management – Practical Techniques to Minimize Exposure to Accidental Losses* (London: Kogan Page)

JAROSINSKI, J. (1983). Flame quenching by a cold wall. *Combust. Flame*, **50**, 167

JAROSINSKI, J. and STREHLOW, R.A. (1978). *The Thermal Structure of a Methane–Air Flame Propagating in a Square Flammability Tube. Rep. AAE 78-6.* Aeronaut. and Astronaut. Engng Dept, Univ. of Illinois

JARRATT, J. (1986). Self regulation in small and large companies. *Chem. Ind.*, Sep. 15, 604

JARRETT, D.E. (1952). *The Flame Area from Propellants Burning in Bulk and its Relation to the Safety Spacing of Category Y Stores. Rep. 5/52.* Armament Res. Estab., Ministry of Supply, Fort Halstead, Appl. Res. Div.

JARRETT, D.E. (1968). Derivation of the British explosives safety distances. *Ann. N.Y. Acad. Sci.*, **152**(art 1), 18

JARRETT, K.R. (1983). The extraction of weather-related features from power transmission reliability data collection schemes. *Euredata 4*, paper 11.5

JARVIS, H.C. (1971). Butadiene explosion at Texas City – 1. *Loss Prevention*, **5**, p. 57. See also *Chem. Engng Prog.*, 1971, **67**(6), 41

JARVIS, R.B. and PANTELIDES, C.C. (1991). Robust dynamic simulation of chemical engineering processes. *Process Safety Environ.*, **69A**, 213

JASKIEWICZ, S.A. (1991). Modify sealless pumps. *Chem. Engng Prog.*, **87**(11), 72

JASTRZEBSKI, Z.D. (1959). *Nature and Properties of Engineering Materials* (New York: Wiley)

JAVADI, S.H.S. and NAPIER, D.H. (1974). Electrification of dielectric belts running on earthed metal rollers. *Chemical Process Hazards*, V, p. 266

JAWAD, M.H. and FARR, J.R. (1984). *Structural Analysis and Design of Process Equipment* (New York: Wiley-Interscience)

J.C. CONSULTANCY LTD. (1986). *Risk Assessment for Hazardous Installations* (Oxford: Pergamon Press)

JEAN, P. and BOURGEOIS, M. (1985). A new generation of LNG carriers based on a proven cargo containment system. *Gastech 84*, p. 332

JEAN, P. and LOOTVOET, R. (1983). The Gaz-Transport membrane, a proven system combining lower operating costs with a high degree of safety. *Gastech 82*, p. 231

JEAN, P., LOOTVOET, R. and BENNETT, H. (1989). Successful service experience and performance of the 130,000 m^3 membrane-type LNG carriers in the MISC fleet. *Gastech 88*, paper 5.1

JEBENS, W.L. (1986). Reduce plant operating risks. *Hydrocarbon Process.*, **65**(12), 65

JEFFEREY, P.G. (1989). *Piper Alpha Public Inquiry. Transcript, Day 98*

JEFFERSON CHEMICAL COMPANY (1963). *Propylene Oxide* (Austin, TX)

JEFFERSON, M., CHUNG, P.W.H. and RUSHTON, A.G. (1995). Automated hazard identification by emulation of hazard and operability studies. *Int. Conf. on Industrial and Engineering Applications of Artificial Intelligence and Expert System*, Australia

JEFFREYS, G.V. (1961). *A Problem in Chemical Engineering Design. The Manufacture of Acetic Acid* (London: Instn Chem. Engrs)

JEFFREYS, SIR H. (1984). *Theory of Probability* (Oxford: Oxford Univ. Press)

JEFRI, M.A. and DAOUS, M.A. (1989). Modelling and simulating a worker's core-temperature in a thermal stress environment. *J. Loss Prev. Process Ind.*, **2**(1), 50

JEGLIC, M.F. and LINDLEY, N.L. (1992). Demystify vacuum capabilities of plastic-lined pipe. *Chem. Engng Prog.*, **88**(4), 56

JEGLIC, M.F. and LINDLEY, N.L. (1994). Reduce flanged connections in plastic lined pipe. *Chem. Engng Prog.*, **90**(3), 30

JELEN, F.C. and COLE, M.S. (1971). Can new equipment cost less? *Hydrocarbon Process.*, **50**(7), 97

JELEN, F.C. and YAWS, C.L. (1977). How energy affects life cycle costs. *Hydrocarbon Process.*, **56**(7), 89

JELINEK, R.V. (1958–). Corrosion. *Chem. Engng, 1958*, **65** (Jul. 28), 114; (Aug. 25), 125; (Sep. 22), 163; (Oct. 20), 163; (Nov. 17), 151; (Dec. 29), 56; 1959, **69**(Jan. 26), 105

JENETT, E. (1963a). Components of pressure-relieving systems. *Chem. Engng*, **70**(Aug. 19), 151

JENETT, E. (1963b). Design considerations for pressure relieving systems. *Chem. Engng*, **70**(Jul. 8), 125

JENETT, E. (1963c). How to calculate back pressure in vent lines. *Chem. Engng*, **70**(Sep. 2), 83

JENETT, E. (1964a). Safety in instrumentation and electrical design. *Chem. Engng*, **71**(Aug. 31), 63

JENETT, E. (1964b). Safety in process and mechanical design. *Chem. Engng*, **71**(Aug. 17), 139

JENETT, E. (1964c). Safety in structural design, layout and piping. *Chem. Engng*, **71**(Sep. 14), 207

JENETT, E. (1969). Experience with and evaluation of critical path methods. *Chem. Engng*, **76**(Feb. 10), 96

JENETT, E. *et al.* (1964). Designing for safety. *Chem. Engng*, **71**(Jul. 20), 135

JENG, S.M., CHEN, L.D. and FEATH, G.M. (1982). The structure of buoyant methane and propane diffusion flames. *Combustion*, **19**, 349

JENKIN, D.B., SINGLETON, P. and WOODWARD, C.C. (1985). The interdependence of plant, port, shipping and customers' facilities in an LNG scheme. *Gastech 84*, p. 215

JENKINS, D.R. and MARTIN, J.A. (1983). Refrigerated LPGs safety research. *Gastech 82*, p. 263

JENKINS, G.M. (1969). A systems study of a petrochemical plant. *J. Syst. Engng*, **1**(1), 90

JENKINS, G.M. and YOULE, P.V. (1971). *Systems Engineering* (London: Watts)

JENKINS, G.M., OTTLEY, D.J. and PACKER, S.H. (1971). A systems study of the effect of unreliability on the profitability of a petrochemical complex. *J. Syst. Engng*, **2**(1), 65

JENKINS, J.E., FRIESEMAN, F. and PREW, L.R. (1974). Early operating experience with the Brunei–Japan LNG project. *LNG Conf.*, 4

JENKINS, J.H., KELLY, P.E. and COBB, C.B. (1977). Design for better safety relief. *Hydrocarbon Process.*, **56**(8), 93. See also in Kane, L.A. (1980), *op. cit.*, p. 142

JENKINS, O.P. and OAKSHOTT, G.B. (1955). *Earthquake in Kern County, California, during 1952. Bull. 171*. Div. of Mines, Dept of Nat. Resources

JENKINS, R.D. (1989). *Piper Alpha Public Inquiry. Transcript, Days 105, 106*

JENKINS, R.H. and CROOKSTON, W.H.K. (1974). Environmental problems in commissioning a major chemical plant. *Instn Chem. Engrs, Symp. on Commissioning of Chemical and Allied Plant* (London)

JENKINS, V. (1994). The reliability of marine propulsion and steering systems. In Rawson, K. (1994), *op. cit.*, p. 1

JENKINS, W. (1962). Safety and low pressure ammonia storage. *Ammonia Plant Safety* 4, 30

JENNINGS, A.A. (1983). Profiling hazardous waste generation for management planning. *J. Haz. Materials*, **8**(1), 69

JENNINGS, A.E. and CHILES, W.D. (1977). An investigation of time-sharing ability as a factor in complex performance. *Hum. Factors*, **19**, 535

JENNINGS, A.J.D. (1974a). The physical chemistry of safety. *Chem. Engr, Lond*, **290**, 637

JENNINGS, A.J.D. (1974b). The physical chemistry of safety. *Course on Process Safety – Theory and Practice*. Dept of Chem. Engng, Teesside Polytechnic

JENNINGS, A.L. and GENTRY, C.R. (1976). The use of spill potential in the designation of hazardous substances. *Control of Hazardous Material Spills*, **3**, 44

JENNINGS, A.L. (1972). Spill damage restoration. *Control of Hazardous Material Spills*, **1**, 221

JENNINGS, P.C., HOUSNER, G.W. and TSAI, N.C. (1968). *Simulated Earthquake Motions. Report*. Earthquake Engng Res. Lab, California Inst. Technol., Pasadena, CA

JENNINGS, S.G. (1975). The electrical conductivity within water droplet clouds. *Static Electrification*, **4**, 34

JENNSEN, T.K. and LARSEN, P. (1980). Risk analysis applied to the transportation of hazardous cargoes: some examples related to public risk. *Marichem 80*

JENSEN, A.V. (1972). *Hazards of Chemical Rockets and Propellants Handbook*, vols 1–3, *CPIA Publ-194*. (Chem. Propulsion Inf. Agency, John Hopkins Univ., Silver Spring, MD). See also Hannum (1984), *op. cit.*

JENSEN, D.E. and JONES, G.A. (1978). Reaction rate coefficients for flame calculations. *Combust. Flame*, **32**, 1

JENSEN, F. (1982). Case studies in system burn-in. *Reliab. Engng*, **3**(1), 13

JENSEN, F. and PETERSEN, N.E. (1982). *Burn-In* (Chichester: Wiley)

JENSEN, J.H. (1978). Combustion safeguards for gas-and-oil fired furnaces. In *Loss Prevention*, **12**, 77. See also *Chem. Engng Prog.*, 1978, **74**(10), 59

JENSEN, K.F. (1999). Microchemical systems: status, challenges, and opportunities. *AIChE J.* **45**(10), (Oct.), 2051–4

JENSEN, N.O. (1980). A simplified diffusion-deposition model. *Atmos. Environ.*, **14**, 953

JENSEN, N.O. (1983). On cryogenic liquid pool evaporation. *J. Haz. Materials*, **8**, 157

JENSEN, N.O. (1981). Entrainment through the top of a heavy gas cloud. In de Wispelaere, C. (1981), *op. cit.*, p. 477

JENSEN, N.O. (1984). On the dilution of a dense gas plume: investigation of the effect of surface mounted obstacles. In Ooms, G. and Tennekes, H. (1984), *op. cit.*, p. 115

JENSEN, N.O. and MIKKELSEN, T. (1984). Entrainment through the top of a heavy gas cloud, numerical treatment. In de Wispelaere, C. (1984), *op. cit.*, p. 343

JENSEN, R.A. (1976). Determining severity of a spill to a waterway. *Control of Hazardous Material Spills 3*, p. 60

JENSEN, R.H. (1964). Compatibility of mechanical foam and dry chemical. *NFPA Quart.*, **57**(3), 296

JENSON, V.G. and JEFFREYS, G.V. (1963). *Mathematical Methods in Chemical Engineering* (London: Academic Press)

JENSSEN, T.K. (1978). Risk analysis of LPG transportation by ship. *Fifth Symp. on Reliability Technology*, Univ. of Bradford

JENSSEN, T.K. (1993). Systems for good management practices in quantified risk analysis. *Process Safety Prog.*, **12**, 137

JERCINOVIC, L.M. (1984). A laboratory pressure safety program. *Plant/Operations Prog.*, **3**, 34

JERVIS, M.W. (1973). Control and instrumentation systems in nuclear power plants. *Int. Atom. Energy Agency, Symp. on Nuclear Power Plant Control and Instrumentation* (Vienna)

JERVIS, M.W. and MADDOCK, P.R. (1965). Electric power station start-up and control. In Miller, W.E. (1965), *op. cit.*, p. 251

JERVIS, M.W. and POPE, R.H. (1977). *Trends in Operator – Process communication. Rep. E/REP/054/77*. Generating Dev. and Constr. Div., Central Elec. Generating Board, Barnwood, Gloucester

JESSOP, P.G., IKARIYA T. and NOYORI, R. (1994). Homogeneous catalytic hydrogenation of supercritical carbon dioxide. *Nature* **368**, (Mar.17), 231–3

JESSOP, T.J. (1980). Periodic inspection for pressurized components. *Int. J. Pres. Ves. Piping*, **8**, 247

JETTA, R.I., BROWN, S.J. and PAMIDI, M.R. (eds) (1981). *Metallic Bellows and Expansion Joints* (New York: Am. Soc. Mech. Engrs)

JEULINK, J. (1983). Mitigation of the evaporation of liquids by fire fighting foams. *Loss Prevention and Safety Promotion 4*, vol. 1, p. E12

JEZZARD, F. (1988). Artificial intelligence in the maintenance field. *Plant Engr*, **32**(1), 39

JIANG MINGXIANG (1985). Theory and algorithm of the quantitative analysis of fault trees. *Reliab. Engng*, **12**(4), 241

JIMESON, R.M. (1975). Elements of an adequate environmental impact statement for the petroleum, petro- chemical and bulk energy industries. *Proc. Ninth World Petrol. Cong.*, vol. 6 (London: Applied Science), p. 223

JIN, T. (1981). Studies of emotional instability in smoke from fires. *J. Fire Flammability*, **12**, 130

JINSI, B. (1985). Soil, seismic studies essential for lines in earthquake areas. *Oil Gas J.*, **83**(Nov. 11), 72

JIRA, R., BLAU, W. and GRIMM, D. (1976). Acetaldehyde via air or oxygen. *Hydrocarbon Process.*, **55**(3), 97

JMP CONSULTANTS (1986). *Results from the M6 (Doxey) Axle Weight Survey (1985), Contractor Rep. 39*. Transport and Road Res. Lab., Crowthorne, Berkshire

JOBACK, K.G. (1994). Solvent substitution for pollution prevention. *In Pollution Prevention Via Process and Product Modifications*, (ed. M. El-Halwagi, and D.P. Petrides) pp. 98–103. AIChE Symposium Series, 303 (New York: American Institute of Chemical Engineers)

JOCHUM, C. (1990). Integrated safety. *J. Occup. Accidents*, **13**(1–2), 139

JOCKEL, H. and TRIEBSKORN, B.E. (1973). GASYNTHAN process for SNG. *Hydrocarbon Process.*, **52**(1), 93

JOFFRE, S.M. (1988). Modelling the dry deposition velocity of highly soluble gases to the sea surface. *Atmos. Environ.*, **22**, 1137

JOHANNESSEN, T. (1987). Nitrogen production by membrane separation. *Gastech 86*, p. 247

JOHANNSEN, G. (1981). Fault management and supervisory control of decentralised systems. In Rasmussen, J. and Rouse, W.B. (1981), *op. cit.*, p. 353

JOHANSEN, T., RAGHURAMAN, K.S. and HACKETT, L.A. (1992). Trends in hydrogen plant design. *Hydrocarbon Process.*, **71**(8), 119

JOHANSON, K.A. (1974). Gas detectors by the acre. *Instrum. Technol.*, **21**(Aug. 33)

JOHANSON, K.A. (1976). Design of a gas monitoring system. *Loss Prevention*, **10**, 15

JOHANSSON, C.H. and PERSSON, P.A. (1970). *Detonics of High Explosives* (London: Academic Press)

JOHANSSON, C.H., PERSSON, A. and SELBERG, H.L. (1957). Ignition of explosives. *Combustion* **6**, 606

JOHANSSON, C.H., LUNDBORG, N. and SJÖLIN, T. (1962). The initiation of solid explosives by shock waves. *Combustion*, **8**, 842

JOHANSSON, G. and FRAGOLA, J.R. (1983). Synthesis of the data base for the Ringhals 2 PRA using Swedish ATV data system. *Euredata 4*, paper 9.3

JOHANNSON, O. (1986). BLEVEs at San Juanico. *Face au Risque*, **222**(4), 35, 55

JOHN, H. *et al.* (1986). Critical two-phase flow through rough slits. *Eur. Two-Phase Flow Group Mtg*, Munich

JOHN, M.R. (1977). Public attitude to the transport of hazardous substances. *Symp. Hazard 77 – Advance in Safe Handling and Movement of Hazardous Substances*, Teesside Polytechnic, Middlesbrough

JOHNS, D.J. (1965). *Thermal Stress Analysis* (Oxford: Pergamon Press)

JOHNS, T.G. and MCCONNILL, D.P. (1984). Assessing fatigue damage when laying pipelines in 1,000–3,000 ft water depths. *Oil Gas J.*, **82**(Aug. 13), 75

JOHNS, W.R. and SADLOWSKI, J. (1977). Optimal maintenance and operation policy for plant with semi-predictable failure characteristics. *Loss Prevention and Safety Promotion 2*, p. 441

JOHNSEN, H. (1989). Hydrate dispute. *Chem. Engr, Lond.*, **462**, 6

JOHNSEN, H.K. (1989). *Piper Alpha Public Inquiry. Transcript, Day 118*

JOHNSEN, H.K. (1990). Hydrates – the start of a chain reaction leading to the Piper Alpha disaster? *Piper Alpha: Lessons for Life Cycle Management* (Rugby: Instn Chem. Engrs), p. 1

JOHNSEN, J.H. and YAHNKE, R.L. (1962). Chlorine–naphtha detonation. *Chem. Engng Prog.*, **58**(6), 71

JOHNSEN, J.J. and YAHNKE, R.L. (1973). Chlorine–naphtha detonation. In Vervalin, C.H. (1973a), *op. cit.*, p. 4

JOHNSEN, K.R. and JENSEN, T. (ca. 1983). Private communication to H.R. Edwards

JOHNSON, A.D. (1992). A model for predicting thermal radiation hazards from large scale LNG pool fires. *Major Hazards Onshore and Offshore*, p. 507

JOHNSON, A.D., BRIGHTWELL, H.M. and CARSLEY, A.J. (1994). A model for predicting the thermal radiation hazards from large-scale horizontally released natural gas jet fires. *Hazards, XII*, p. 123

JOHNSON, B.B. (1992). The 'mental model' meets the planning process: wrestling with risk communication research and practice (unpublished paper). New Jersey Dept of Environmental Protection and Energy

JOHNSON, B.B. and COVELLO, V.T. (eds) (1987). *The Social and Cultural Construction of Risk* (Dordrecht: Reidel)

JOHNSON, C.D. (1977). *Process Control Instrumentation Technology* (New York: Wiley)

JOHNSON, C.M. (1985). How membrane couplings meet API 671. *Hydrocarbon Process.*, **64**(1), 57

JOHNSON, D.M. (1969). The billion gallon tank farm. *Loss Prevention*, **3**, 98. See also *Chem. Engng Prog., 1969*, **65**(4), 41

JOHNSON, D.M. and PRITCHARD, M.J. (1991). Large scale experimental study of boiling liquid expanding vapour explosions (BLEVEs). *Gastech 90*, p. 3.3

JOHNSON, D.M., SUTTON, P. and WICKENS, M.J. (1991). Scaled experiments to study vapour cloud explosions. *Hazards, XI*, p. 67. See also *Process Safety Environ.*, 1991, **69B**, 76

JOHNSON, D.R. (1985). Thorney Island trials: systems development and operational procedures. *J. Haz. Materials*, **11**(1/3), 35. See also in McQuaid, J. (1985), *op. cit.*, p. 35

JOHNSON, D.S. and BORNSTEIN, R.D. (1974). Urban–rural wind velocity differences and their effects on computed pollution concentrations in New York City. *First Symp. on Atmospheric Diffusion and Air Pollution* (Boston, MA: Am. Met. Soc.), p. 267

JOHNSON, D.W. (1991). Prediction of aerosol formation from the release of pressurized, superheated liquids to the atmosphere. *Modeling and Mitigating the Consequences of Accidental Releases*, p. 1

JOHNSON, D.W. and DIENER, R. (1991). Prediction of aerosol formation from the release of pressurized, superheated liquids to the atmosphere. *Hazards, XI*, p. 87

JOHNSON, D.W. and WELKER, J.R. (1978). Diked-in storage areas, revisited. *Chem. Engng*, **85**(Jul. 31), 112

JOHNSON, D.W. and WELKER, J.R. (1981). *Development of an Improved LNG Plant Failure Rate Data Base. Final Report.* Appl. Technol. Corp., Norman, OK

JOHNSON, E. (1966). Five ways to measure maintenance. *Hydrocarbon Process.*, **45**(5), 187

JOHNSON, E. (1983). Redundancy for on-line computer control systems. *Reliab. Engng*, **4**(2), 105

JOHNSON, E. (1988). Reliability assessment of computer systems used in the control of chemical plant. *Preventing Major Chemical and Related Process Accidents*, p. 209

JOHNSON, E. (1989a). Answers surface for Europe's underground woes. *Chem. Engng*, **96**(6), 47

JOHNSON, E. (1989b). Competing for the EC's 1992 pie. *Chem. Engng*, **96**(11), 39

JOHNSON, E. (1989c). The French CPI. *Chem. Engng*, **96**(3), 64CC

JOHNSON, E. *et al.* (1990). Germany in focus. *Chem. Engr, Lond.*, **476**, 18

JOHNSON, E.F. (1967). *Automatic Process Control* (New York: McGraw-Hill)

JOHNSON, E.J. and TEVERSKY, A. (1984). Representations and perceptions of risk. *J. Exp. Psychol.* (General), **113**, 55

JOHNSON, E.R. (1975). Ammonia reformer freeze-up. *Ammonia Plant Safety*, **17**, p. 138

JOHNSON, E.R. (1977). Firing oil on an ammonia reformer. *Ammonia Plant Safety*, **19**, p. 26

JOHNSON, E.W. (1960). Aluminum alloys tough and ductile down to $-423°F$. *Chem. Engng*, **67**(Aug. 8), 133

JOHNSON, F.R. and FISHER, A. (1989). Conventional wisdom on risk communication and evidence from a field experiment. *Risk Analysis*, **9**, 209

JOHNSON, G. and MCFARLAN, T. (1978). Better data with eddy current. *Hydrocarbon Process.*, **57**(1), 117

JOHNSON, G.O. (1980). Probabilistic fracture mechanics (PFM). In Moss, T.R. (1980), *op. cit.*, p. 8

JOHNSON, H.D. (1981). *Sample Survey of Roads of Great Britain. Rep. LR 1005.* Transport and Road Res. Lab., Crowthorne

JOHNSON, H.D. and GARWOOD, F. (1971). *Notes on Road Accident Statistics.* Road Res. Lab., Crowthorne

JOHNSON, J.C. (1979). *Emulsifiers and Emulsifying Techniques* (Park Ridge, NJ: Noyes Data Corp.)

JOHNSON, J.D. (1981). Variable-speed drives can cut pumping costs. *Chem. Engng*, **88**(Aug. 10), 107

JOHNSON, J.E., CRELLIN, J.W. and CARHART, H.W. (1953). Spontaneous ignition properties of fuels and hydrocarbons. *Ind. Engng Chem.*, **45**, 1749

JOHNSON, J.E., CRELLIN, J.W. and CARHART, H.W. (1954). Ignition behaviour of hexane. *Ind. Engng Chem.*, **46**, 1512

JOHNSON, J.H., ERIKSON, E.D., SMITH, S.R., KNIGHT, J., FINE, D.A. and HELLER, C.A. (1988). Products from the detonation of trinitrotoluene and some other navy explosives in air and nitrogen. *J. Haz. Materials*, **18**(2), 145, 161

JOHNSON, J.K. (1977). The ignition of vapour droplets by liquid to metal sparks. *J. Electrostatics*, **4**, 53

JOHNSON, J.P. and JAMISON, L.R. (1975). Pipe line to Japan . . . five years of LNG shipping. *Chem. Engng Prog.*, **71**(7), 97

JOHNSON, L.G. (1964). *The Statistical Treatment of Fatigue Experiments* (Amsterdam: Elsevier)

JOHNSON, M.R., WELCH, R.E., TAKATA, A.N. and BRUCE, R.W. (1974). Fire induced stresses in tanks containing liquefied gas. *Am. Soc. Mech. Engrs, Pressure Vessel and Piping Conf.* (New York)

JOHNSON, N.L. and KATZ, S. (1969). *Distributions in Statistics. Discrete Distributions* (Boston, MA: Houghton Mifflin)

JOHNSON, N.L. and LEONE, F.C. (1964). *Statistics and Experimental Design* (New York: Wiley)

JOHNSON, O.W. (1983). An oil industry viewpoint on flame arresters in pipelines. *Plant/Operations Prog.*, **2**, 75

JOHNSON, P. (1977). Safety considerations in the design and operation of batch reaction systems. *Chem. Ind.*, Apr. 16, 305

JOHNSON, P.C. (1968). Satellite peakshaving systems. *LNG Conf. 1*

JOHNSON, P.F. (1986). Special systems and extinguishing techniques. In Cote, A.E. and Linville, J.L. (1986), *op. cit.*, p. 19–55

JOHNSON, P.L. (1993). *ISO 9000* (New York: McGraw-Hill)

JOHNSON, R.A. (1980). Use of coalescers and settling tubes to remove dispersed liquids. *Ind. Engng Chem. Fund.*, **19**, 71

JOHNSON, R.E. (1965). *Fracture mechanics: a basis for brittle fracture prevention. Rep. WAPD-TM-505*, US-PO Reactor Technology TID-4500. Dept of Commerce

JOHNSON, R.J. (1960). Nickel steel alloys for liquids at −320°F, 9% excels. *Chem. Engng*, **67**(Jul. 25), 115

JOHNSON, R.K. (1985). Legal implications of compliance and noncompliance with standards and internal corporate guidelines. *Chem. Engng Prog.*, **81**(8), 9

JOHNSON, R.T. and MILLER, S.E. (1960). *Occupational Diseases and Industrial Medicine* (Philadelphia, PA: Saunders)

JOHNSON, R.W. (1981). Ignition of flammable vapors by human electrostatic discharges. *Loss Prevention*, **14**, 29

JOHNSON, R.W., UNWIN, S.D. and MCSWEENEY, T.I. (1996). Inherent safety: how to measure it and why we need It. *Int. Conf. and Workshop on Process Saf. Manage. and inherently safer processes*, Oct. 8–11, 1996, Orlando, FL, 118–127 (New York: American Institute of Chemical Engineers)

JOHNSON, R.W., PRUGH, R.W., SCHECHTER, S.J., VIERA, G.A. and BASSLER, E.J. (1989). Guidelines for vapour release mitigation. *Loss Prevention and Safety Promotion*, p. 35.1

JOHNSON, S.C. (1976). US EO/EG – past, present and future. *Hydrocarbon Process.*, **55**(6), 109

JOHNSON, S.L. (1981). Effect of training device on retention of transfer of a procedural task. *Hum. Factors*, **23**, 257

JOHNSON, T.E. (1963). Buying chemical pumps. *Chem. Engng*, **70**(Aug. 5), 138

JOHNSON, W. (1972). *Impact Strength of Materials* (London: Arnold)

JOHNSON, W. and MELLER, P.B. (1962). *Plasticity for Mechanical Engineers* (New York: van Nostrand)

JOHNSON, W.B. and ROUSE, W.B. (1982). Training maintenance technicians for troubleshooting: two experiments with computer simulations. *Hum. Factors*, **24**, 271

JOHNSON, W.B. *et al.* (1971). *Field Study for Initial Evaluation of an Urban Diffusion Model for Carbon Monoxide. Rep. SRI Proj. 8563*. Stanford Res. Inst., Stanford, CA

JOHNSON, W.G. (1973a). *The Management Oversight and Risk Tree MORT. Rep. San 821–UC–41*. Atom Energy Comm., Washington, DC

JOHNSON, W.G. (1973b). Sequences in accident causation. *J. Safety Res.*, Jun., 54

JOHNSON, W.G. (1975). *Accident/Incident Investigation Manual. Report*. Dept of Energy, Washington, DC

JOHNSON, W.G. (1977). *Fundamentals of Safety Management: Systems for Effective Investigation, Evaluation and Control* (New York: Penton Learning Systems)

JOHNSON, W.G. (1980). *MORT Safety Assurance Systems* (Basel: Dekker)

JOHNSON, WILLIAM G. Management Oversight and *Risk Tree (MORT) Safety Assurance Sytems* (Chicago: National Safety Council) ISBN 0-8247-6897-3

JOHNSSON, L., STRID, K.G. and JOHANSSON, S.R. (1972). Measurement of the energy released in nanosecond electric sparks. *Combust. Sci. Technol.*, **5**, 1

JOHNSTON, A.G. (1984). The organisation of a multi-sponsored international research project on dispersion of heavy gas clouds. *Am. Inst. Chem Engrs, Symp. on Loss Prevention* (New York)

JOHNSTON, A.G. (1985). Multinational research project on dispersion of heavy gas clouds. *Chem. Engng Prog.*, **81**(4), 11

JOHNSTON, A.G., MCQUAID, J. and GAMES, G.A.C. (1980). Systematic safety assessment in the mining industry. *Mining Engr*, Mar., 723

JOHNSTON, B.D. (1987a). A structured procedure for dependent failure analysis (DFA). *Reliab. Engng*, **19**(2), 125

JOHNSTON, B.D. (1987b). A structured procedure for dependent failure analysis (DFA). *SRS Quart. Dig.*, Jul., 11

JOHNSTON, C. (1990). Safe air systems. *Plant Engr*, Sep./Oct., 35

JOHNSTON, G.O. (1982). A review of probabilistic fracture mechanics. *Reliab. Engng*, **3**(6), 423

JOHNSTON, H.G. (1968). Designing remote control shelters for personnel safety. *Ann. N.Y. Acad. Sci.*, **152**(art. 1), 585

JOHNSTON, I.H.A. (1993). Fault-tolerant control for safety. In Bennett, P. (1993), *op. cit.*, p. 187

JOHNSTON, J.C. (1974). Estimating flash points for organic aqueous solutions. *Chem. Engng*, **81**(Nov. 25), 122

JOHNSTON, R. (ed.) (1977). *Marine Pollution* (London: Academic Press)

JOHNSTONE, R.E. and THRING, M.W. (1957). *Pilot Plants, Models and Scale Up Methods in Chemical Engineering* (New York: McGraw-Hill)

JOINT FIRE RESEARCH ORGANISATION (1960). *Symposium on Flame Arresters and Relief Vents*

JOJIMA, T. (1980). Urea reactor failure. *Ammonia Plant Safety*, **21**, 111

JOKSIMOVIC, V. and SOLOMON, K.A. (1983). Quantitative safety goals through more adequate risk management and risk assessment. *Reliab. Engng*, **4**(2), 65

JOKSIMOVIC, V. and VESELY, W.E. (1980). Use of PRA in evaluating safety of nuclear power. *Reliab. Engng*, **1**(1), 69

JOLLS, K.R. and RIEDINGER, R.L. (1972–). Basic electrical concepts. *Chem. Engng, 1972*, **79**(May 15), 95; (Jun. 12), 101; (Jul. 24), 137; (Aug. 21), 104; (Sep. 18), 165; (Oct. 2), 67; (Oct. 30), 117; (Nov. 27), 93; (Dec. 25), 67; 1973, **80**(Jan. 8), 121; (Feb. 19), 129; (Apr. 2), 61; (May 14), 140

JOLLES, Z.E. (ed.) (1966). *Bromine and Its Compounds* (London: Benn)

JONAS, P.H. and MASON, B.J. (1968). Systematic charging of water droplets produced by break up of liquid jet and filaments. *Trans. Faraday Soc.*, **64**, 1971

JONAS, P.J. (1982). The practice of energy conservation. *Plant Engr*, **26**(1), 22

JONASSEN, N., HANSSON, I. and NIELSEN, A.R. (1979). On the correlation between decay of charge and resistance parameters of sheet materials. *Electrostatics*, **5**, 215

JONES, A. (1984). Fugitive emissions of volatile hydrocarbons. *Chem. Engr, Lond.*, **406**, 12

JONES, A. (1985). *The Summit Tunnel Fire. Rep. IR/L/FR/ 85/ 26*. Health and Safety Executive, London

JONES, A. and THOMAS, G.O. (1992). The mitigation of small-scale hydrocarbon-air explosions by water sprays. *Chem. Engng Res. Des.*, **70A**, 197

JONES, A. and THOMAS, G.O. (1993). The action of water sprays on fires and explosions: a review of experimental work. *Process Safety Environ.*, **71B**, 41

JONES, A.L. (1984). Accrediting the safety professional (letter). *Chem. Engr, Lond.*, **396**, 3

JONES, A.L. (ed.) (1994). *The Health and Safety Factbook* (London: Professional)

JONES, A.R. and HOPKINS, D.J. (1966). Controlling health hazards in pilot plant operations. *Chem. Engng Prog.*, **62**(12), 59

JONES, A.T. (1952). Noxious gases and fumes. *Proc. R. Soc. Med.*, **45**, 609

JONES, B.G. and DUCKETT, R.C. (1985). Thermographic survey of the integrity of a process plant pressure relief system. *Plant/Operations Prog.*, **4**, 161

JONES, C. and SANDS, R.L. (1981). *Great Balls of Fire* (Coventry: published by authors)

JONES, C.B. (1986). *Systematic Software Development Using VDM* (Hemel Hempstead: Prentice-Hall)

JONES, C.B. and SHAW, R.C.F. (1990). *Case Studies in Systematic Software Development* (Englewood Cliffs, NJ: Prentice-Hall)

JONES, C.D. (1979). Stationarity of the concentration intermittency in short-range atmospheric dispersion. *Proc. Tenth Int. Tech. Mtg on Air Pollution Modelling and Its Application* (NATO/CCMS), p. 187

JONES, C.D. (1983). On the structure of instantaneous plumes in the atmosphere. *J. Haz. Materials*, **7**(2), 87

JONES, C.D. (1988). Heavy gas dispersion. *Course on Hazard Control and Major Hazards*. Dept of Chem. Engng, Loughborough Univ. of Technol.

JONES, C.D. and GRIFFITHS, R.F. (1984). Full scale experiments on dispersion around an isolated building using an ionized air tracer technique with very short averaging time. *Atmos. Environ.*, **18**, 903

JONES, C.J. (1978). The ranking of hazardous materials by means of hazard indices. *J. Haz. Materials*, **2**(4), 363

JONES, D. (ed.) (1992). *Nomenclature of Hazard and Risk Assessment*, 2nd ed. (Rugby: Instn Chem. Engrs)

JONES, D.A. (1983). A method for assessing the offsite risks from bulk liquid oxygen storage. *Loss Prevention and Safety Promotion 4*, vol. 1, p. I13

JONES, D.A. (1985). A review of the developments in LNG storage safety as reflected by risk assessment. *Gastech 84*, p. 133

JONES, D.A. (1987). Can regulations help to prevent accidents? *Preventing Major Chemical Accidents*, p. 2.9

JONES, D.A. (1989). Acceptable risk – the use of HSE risk criteria for the siting of public developments near major hazards. *Loss Prevention and Safety Promotion 6*, p. 65.1

JONES, D.A. and FEARNEHOUGH, G.D. (1986). Natural gas transmission by pipeline. A re-assessment of the hazards and risks, and their influence on UK legislative safety procedures. *Loss Prevention and Safety Promotion 5*, 15–1

JONES, D.L. (1968). Intermediate strength blast waves. *J. Phys. Fluids*, **11**, 1664

JONES, D.L. (1970). Blast waves and scaling laws. *J. Phys. Fluids*, **13**(5), 1398

JONES, D.O. (1996). *Process Intensification of Batch, Exothermic Reactors* (Sudbury, Suffolk, UK: Health and Safety Executive)

JONES, D.W. (1975). Use a safety audit program. *Hydrocarbon Process.*, **54**(8), 67

JONES, D.W. (1984). Control of odour nuisance from oil refineries and downstream processes. *Chem. Ind.*, May 7, 342

JONES, D.W. (1992). Lessons from hazop experiences. *Hydrocarbon Process.*, **71**(4), 77

JONES, D.W. (1994). Implementing process safety management. *Process Safety Prog.*, **13**, 8

JONES, E.O. (1978). Drilling. In Institute of Petroleum (1978), *op. cit.*, p. 33

JONES, F.G. (1971). Prevention of catastrophic failure of reciprocating compressors. *Loss Prevention*, **5**, 100

JONES, F.W.S. (1970). Special considerations for primary reformer tubes. *Ammonia Plant Safety*, **12**, 61

JONES, G. (1978). Solid waste disposal. *Plant Engr*, **22**(7), Sep., 23

JONES, G.E., KENNEDY, J.E. and BERTHOLF, L.D. (1980). Ballistics calculations of R.W. Gurney. *Am. J. Phys.*, **48**(4), 264

JONES, G.G. (1977). Legislation and the role of the Alkali and Clean Air Inspectorate. *Chem. Ind.*, Jan., 15

JONES, G.W. (1928). Flammability of refrigerants. *Ind. Engng Chem.*, **20**, 367

JONES, G.W. (1937). Inflammation limits and their practical application in hazardous industrial operations. *Combustion*, **2**, 248

JONES, G.W. (1938). Inflammation limits and their practical application in hazardous industrial operations. *Chem. Revs*, **22**(1), 1

JONES, G.W., CAPPS, J.H. and KATZ, S.H. (1918). Simple method for determining sulphur dioxide. *Mining Sci. Press*, **117**, 415

JONES, G.W., LEWIS, B., FRIAUF, J.B. and PERROTT, G.S.T.J. (1931). Flame temperatures of hydrocarbon gases. *J. Am. Chem. Soc.*, **53**, 869

JONES, G.W., LEWIS, B. and SEAMAN, H. (1931). The flame temperatures of mixtures of methane-oxygen, methane-hydrogen and methane-acetylene with air. *J. Am. Chem. Soc.*, **53**, 3922

JONES, G.W., LEWIS, B. and SEAMAN, H. (1932). The flame temperatures of mixtures of ammonia and its products of dissociation. *J. Am. Chem. Soc.*, **54**, 2166

JONES, J. (1974). The role of the safety officer in industry. *Chem. Ind.*, Apr. 20, 329

JONES, J. (1977). Safety and loss prevention at a chemical plant site. *Chemical Engineering in a Hostile World* (London: Clapp & Poliak)

JONES, J. (1978). Converting solid wastes and residues to fuel. *Chem. Engng*, **85**(Jan. 2), 87

JONES, J. et al. (1993a). Keep pilot plants on the fast track. *Chem Engng*, **100**(11), 98

JONES, J. et al. (1993b). Tips for justifying pilot plants. *Chem Engng*, **100**(4), 138

JONES, J.E. and OLSON, D.L. (1986). Selecting arc welding processes. *Chem. Engng*, **93**(Mar. 3), 117; (Mar. 31), 131

JONES, J.L. (1984). The make or buy decision for R&D. *Chem. Engng Prog.*, **80**(7), 11

JONES, J.M. and ROSENFELD, J.L.J. (1972). A model for sooting in diffusion flames. *Combust. Flame*, **19**, 427

JONES, L. (1991). Development of safety critical software for the Ministry of Defence. *Safety Reliab.*, **11**(2), 6

JONES, L.W. (1990). Interference mechanisms in waste stabilization/solidification process. *J. Haz. Materials*, **24**(1), 83

JONES, M.C. (1989). Assessing runaway of a Grignard reaction. *Runaway Reactions*, p. 477. See also *Plant/Operations Prog.*, **8**, 200

JONES, M.E. (1990). Materials of construction: lining up to resist corrosion. *Chem. Engng*, **97**(10), 104

JONES, M.H. and SCOTT, D. (eds) (1983). *Industrial Tribology* (Amsterdam: Elsevier)

JONES, M.H. and SEFTON, V.B. (1962). The behaviour of contaminants in air separation plants. *Ammonia Plant Safety*, **4**, 39

JONES, M.J. (1990). *Piper Alpha Public Inquiry. Transcript, Day 170*

JONES, M.R.O. (1984).Vapour explosions resulting from rapid depressurisation of liquids – the importance of initial temperature. *The Protection of Exothermic Reactors and Pressurised Storage Vessels* (Rugby: Instn Chem. Engrs), p. 357

JONES, M.R.O. and BOND, J. (1984). Electrostatic hazards associated with marine chemical tanker operations. Criteria of incendivity in tank cleaning operations. *Chem. Engng Res. Des.*, **62**, 327

JONES, M.R.O. and BOND, J. (1985). Electrostatic hazards associated with marine chemical tanker operations. Criteria of safety in tank cleaning operations. *Chem. Engng Res. Des.*, **63**, 383

JONES, M.R.O. and UNDERWOOD, M.C. (1983). An appraisal of expressions used to calculate the release rate of pressurised liquefied gases. *Chem. Engng J.*, **26**, 251

JONES, M.R.O. and UNDERWOOD, M.C. (1984). The small-scale release rate of pressurised liquefied propane to the atmosphere. *The Protection of Exothermic Reactors and Pressurised Storage Vessels* (Rugby: Instn Chem. Engrs), p. 157

JONES, N. (1976). Plastic behavior of ships. Paper presented at Ann. Mtg of Soc. of Naval Architects and Marine Engrs

JONES, N. (1983). Structural aspects of ship collisions. *Structural Crashworthiness* (ed. N. Jones and T. Wierzbicki) (Oxford: Pergamon Press), p. 308

JONES, P. (1992). Inherent safety: action by HSE. *Chem. Engr, Lond.*, **526**, 29

JONES, P.G. (1991). Computers in chemical plant – a need for safety awareness. *Hazards*, **XI**, p. 289

JONES, P.L. (1981). Design innovations and considerations for a pressure relieving system. *Measmnt Control*, **14**, 211

JONES, P.M., LARRINAGA, M.A.B. and WILSON, C.B. (1971). The urban wind velocity profile. *Atmos. Environ.*, **5**, 89

JONES, P.T. (1980). Resolving seal and bearing failures in process pumps. *Chem. Engng*, **87**(Jun. 16), 145

JONES, R.K. (1971). The biomedical effects of combined injuries. *Probleme des Baulichen Schutzes* (Ernst Mach Inst., Germany), p. 37

JONES, R.K. and RICHMOND, D.R. (1968). The physical and biological aspects of blast injury. *Proc. Workshop on Mass Burns* (Washington, DC: Nat. Acad. Sci.), p. 215

JONES, R.K., RICHMOND, D.R. and FLETCHER, E.R. (1969). A reappraisal of man's tolerance to indirect (tertiary) injury. *Minutes of Eleventh Mtg of Panel P-2*, Technical Co-operation Programme, Pt II. Ministry of Defence, Atom. Weapons Res. Establ., Foulness,

JONES, R.V. (1982). A concurrence in arms and learning. *J. Opl. Res. Soc.*, **33**, 779

JONES, S. (1968). *Design of Instruction* (London: HM Stationery Office)

JONES, S.J. and WEBBER, D.M. (1993). The interaction of a dense gas plume with a fence. *Process Safety Environ.*, **71B**, 15

JONES, T.B. (1975). *Ionospheric Effects of the Flixborough Explosion. Rep. to Flixborough Court of Inquiry* (Univ. of Leicester)

JONES, T.B. (1987). Electroquasistatics of charged powders. *Electrostatics*, **7**, p. 223

JONES, T.B. and SPRACKLEN, C.T. (1974). Ionospheric effects of the Flixborough disaster. *Nature*, **250**(Aug. 30), 719

JONES, W.G., RICHARDSON, J.A., ERSKINE, R. and HEWERDINE, S. (1989). A policy for the safe operation of anhydrous ammonia storage spheres susceptible to stress corrosion cracking. *Loss Prevention and Safety Promotion 6*, p. 10.1

JONES, W.M. (1940a). *Determination of Dust Explosion Possibilities. Special Hazard Study 4.* Factory Insurance Ass., Hartford, CT

JONES, W.M. (1940b). *Prevention and Minimizing the Effects of Dust Explosions in Manufacturing Plants. Special Study 5.* Factory Insurance Ass., Hartford, CT

JONES, W.P. and WHITELAW, J.H. (1982). Calculation methods for reacting turbulent flows: a review. *Combust. Flame*, **48**, 1

JONES, W.T. (1975). *Health and Safety at Work Act – A Practical Handbook* (London: Graham & Trotman)

JONES-LEE, M. (1969). The valuation of reduction in probability of death by road accident. *J. Transport Econ. Policy*, Jan., p. 37

JONES-LEE, M.W. (1976). *The Value of Life* (London: Robertson)

JONES-LEE, M.W. (ed.) (1982). *The Value of Safety* (Amsterdam: North Holland)

JONES-LEE, M.W. (1989). The Economics of Safety and Physical Risk (Oxford: Blackwell)

JONES-LEE, M.W., HAMMERTON, M. and PHILIPS, P.R. (1985). The value of safety: results of a national survey. *Economic J.*, **95**(377), 49

DE JONG, J.J. (1980). Hypothesis about the causes of industrial calamities. *Loss Prevention and Safety Promotion 3*, vol. 2, p. 220

DE JONG, J.J. and KÖSTER, E.P. (1971). The human operator in the computer controlled refinery. *Proc. Eighth World Petrol. Conf.* (London: Inst. Petrol.). See also Edwards, E. and Lees, F.P. (1974), *op. cit.*, p. 196

JONGELEN, E.M., DEN HEIJER, C. and VAN ZEE, G.A. (1988). Detection of gross errors in process data using studentized residuals. *Comput. Chem. Engng*, **12**, 845

JONKER, P.E., PORTER, W.J. and SCOTT, C.B. (1977a). Control floating roof tank emissions. *Hydrocarbon Process.*, **56**(5), 151

JONKER, P.E., PORTER, W.J. and SCOTT, C.B. (1977b). Hydrocarbon emissions from floating roof tanks. *API Proc., Refining Dept, 42nd Midyear Mtg*, **56**, 163

JONSTAD, O. (1983). Risk assessment for an emergency shutdown system. *Euredata 4*, paper 16.1

JOPSON, R.J. and BALE, A.G. (1984). *Dangerous Goods on British Rail. Rep. MR HAZ 1.* Res. and Dev. Div., British Rail

JORDAN, D.G. (1968). *Chemical Process Development*, vols 1 and 2 (New York: Interscience)

JORDAN, E.C. and BALMAIN, K.G. (1968). *Electromagnetic Waves and Radiating Systems* (Englewood Cliffs, NJ: Prentice Hall)

JORDAN, E.W. and ROHLFING, R.L. (1979). Reformer riser pressure shell leaks. *Ammonia Plant Safety*, **21**, p. 37

JORDAN, J.H. (1968). How to evaluate the advantages of contract maintenance. *Chem. Engng*, **75**(Mar. 25), 124

JORDAN, T.E. (1954). *Vapor Pressure of Organic Compounds* (New York: Interscience)

JORDAN, W.E. (1972). Failure modes, effects and criticality analysis. *Proc. Ann. Reliability and Maintainability Symp.* (New York: Am. Soc. Mech. Engrs)

JORGENSEN, S.E. (ed.) (1984). *Modelling the Fate and Effect of Toxic Substances in the Environment* (Amsterdam: Elsevier)

JORGENSEN, S.E.A. and JOHNSEN, I. (1981). *Principles of Environmental Science and Technology* (Amsterdam: Elsevier)

JOSCHEK, H.I. (1983). Risk assessment in the chemical industries. *Plant/Operations Prog.*, **2**, 1

JOSCHEK, H.I., HRLMSTÄDTER, G., DEHLER, J. and LANGENSIEPEN, H.W. (1980). Quantitative risk analysis in the chemical industry. *Chem. Ing. Tech.*, **52**, 123

JOSCHEK, H.I. (1987). Safety and loss prevention in the chemical industry. *Fortschritte der Sicherheitstechnik 1* (Frankfurt-am-Main: DECHEMA), p. 53

JOSCHEK, H.I. (1986). Safety and loss prevention in the process industries. *Loss Prevention and Safety Promotion 5*, vol. III, p. 9

JOSEFSON, A. (1987). Transportation/storage/handling of ammonia (panel discussion). *Ammonia Plant Safety*, **27**, 175

JOSHI, A.A. and PAGAI, P.J. (1994). Fire induced thermal fields in window glass. *Fire Safety J.*, **22**, 25; 45

JOSHI, M.N. (1983). International agreement is needed on hazardous-area classification. *Oil Gas J.*, **81**(Mar. 14), 65

JOSHI, S. and FAIR, J.R. (1991). Adsorptive drying of hydrocarbon liquids. *Ind. Engng Chem. Res.*, **30**, 177

JOSHI, S.K. and DOUGLAS, J.M. (1992). Avoiding accumulation of trace components. *Ind. Engng Chem. Res.*, **31**, 1502

JOST, W. (1939). *Explosions- und Verbrennungsvorgänge in Gasen* (Berlin: Springer)

JOST, W. (1946). *Explosions and Combustion Processes in Gases* (New York: McGraw-Hill)

JOST, W. (1949). Reactions of adiabatically compressed hydrocarbon–air mixtures. *Combustion*, **3**, 424

JOST, W. (1950). *Z. Phys. Chem*, **196**, 4

JOST, W., VON MÜFFLING and ROHRMANN (1936). *Z. Elektrochem.*, **42**, 488

JOUGUET, E. (1905). Sur la propagation des réactions chimiques dans les gaz. *J. Mathématique, 347*

JOUGUET, E. (1917). *Mécanique des Explosives* (Paris: Doin)

JOUHAR, A.J. (1984). *Risk in Society* (Libbey)

JOURDAN, L. (1990). The role of CEFIC in the promotion of process safety. *Plant/Operations Prog.*, **9**, 155

JOURNEE, C. (1907). Relation between energy of bullets and severity of wounds which they can cause. *Rev. Artillerie*, **70**, 81

JOURNET, C. (1993). Computer aids for the safe design and operation of chemical plants. Ph.D. Thesis, Univ. of Bath

JOWETT, C.E. (1976). Control of static electricity. *Fire Prev. Sci. Technol.*, **15**, 4

JOWITT, R. (ed.) *Hygienic Design and Operation of Food Plant* (Chichester: Ellis Horwood)

JOYNER, R.E. (1959). Eye protection in the chemical industry: a report on six years' experience. *Ind. Med. Surg.*, **28**, 174

JOYNER, R.E. and DUREL, E.G. (1962). Accidental liquid chlorine spill in a rural community. *J. Occup. Med.*, **4**, 152

JUDD, S.H. (1970). Noise abatement in existing refineries. *Proc. Am. Petrol. Inst, Div. of Refining*, **50**, 125

JUDD, S.H. (1971). Noise abatement in process plants. *Chem. Engng*, **78**(Jan. 11), 139

JUDD, S.H. and SPENCE, J.H. (1971). Noise control for electric motors. *J. Am. Soc. Safety Environ.*, **16**, 25

JUDSON, R.W. (1966). What information is essential for good piping design? *Hydrocarbon Process.*, **45**(10), 114

JUHOLA, A.J. and BRUGGER, J.E. (1976). Pilot study of activated carbon regeneration. *Control of Hazardous Material Spills 3*, p. 219

JUMPER, C. (1971). New approach to inventory control. *Hydrocarbon Process.*, **50**(8), 137

JUNE, R.K. and DYE, R.F. (1990). Explosive decomposition of ethylene oxide. *Plant/Operations Prog.*, **9**, 67

JUNIERE, P. and SIGWALT, M. (1964). *Aluminium: Its Applications in the Chemical and Food Industries* (London: Crosby Lockwood)

JUNIQUE, J.C. (1988). Flush or blow lines adequately. *Hydrocarbon Process.*, **67**(7), 55

JURAN, J.M. (1964). *Managerial Breakthrough* (New York: McGraw-Hill)

JURAN, J.M. (1979). *Quality Control Handbook*, 3rd ed. (New York: McGraw-Hill)

JURAN, J.M. and GRYNA, F.M. (1980). *Quality Planning and Analysis*, 2nd ed. (New York: McGraw-Hill)

JUSTI, E. (1938). *Spezifische Wärme, Enthalpie, Entropie und Dissoziation* (Berlin: Springer)

JUSTI, E. and LÜDER, H. (1935). Specific heat, entropy and dissociation of industrial gases and vapours. *Forsch. Gebiet. Ingwesen*, **6**(5), 209

JUTLA, T. (1987). Probabilistic fracture mechanics. *Reliab. Engng*, **19**(4), 287

JUVEKAR, V.A. and SHARMA, M.M. (1977). Some aspects of process design of gas/liquid reactors. *Trans. Instn Chem. Engrs*, **55**, 77

KAARSTAD, O. and WULFF, E. (1984). *Safety Offshore* (Stavanger: Universitetsforlaget)

KAC, M.I. and STRIZAK, N.S. (1965). *Safety and Health Engineering in the Chemical Industry* (Moscow: Izdatelstvo 'Khimiya')

KACPRZYK, J. (1983). *Multistage Decision-Making under Fuzziness* (Cologne: TUV Rheinland)

KACPRZYK, J. (1992). Fuzzy sets and fuzzy logic. In Shapiro, S.C. (1992), *op. cit.*, p. 537

KAEFER, D.L. (1977). Tank farm fire security at Valdez. *Hydrocarbon Process.*, **56**(8), 109

KAESS, D. (1970). Guide to trouble-free plant layout. *Chem. Engng*, **77**(Jun. 1), 122

KAFAROV, V.V. *et al.* (1974). Method for quantitative analysis of chemical process system reliability. *Doklady Chem. Technol.*, **215**, 36

KAHANE, S.W., PHINNEY, S.L. and WRIGHT, A.P. (1985). Tightening regulations for hazardous waste management: mandates and compliance strategies. *Chem. Engng Prog.*, **81**(3), 11

KAHKONEN, K. and BJORK, B. (eds) (1991). *Computers and Building Regulations* (Espoo, Finland: VTT)

KAHN, D., TALBOT, T. and WOODWARD, J. (1974). *Vessel Safety Model*. Report to US Coast Guard

KAHN, H. (1957). *Applications of Monte Carlo. Rep. AECU-3259*. Tech. Inf. Service, Atom. Energy Comm., Oak Ridge, TN

KAILASANATH, K. and ORAN, E.S. (1984). Power–energy relations for the direct initiation of gaseous detonations. In Bowen, J.R. *et al.* (1984b), *op. cit.*, p. 38

KAILASANATH, K., ORAN, E.S., BORIS, J.P. and YOUNG, T.R. (1985). Determination of detonation cell size and the role of transverse waves in two dimensional detonations. *Combust. Flame*, **61**, 199

KAIMAL, J.C., HAUGEN, D.A., COTÉ, O.R., IZUMI, Y., CAUGHEY, S.J. and READINGS, C.J. (1976). Turbulence structure in the convective boundary layer. *J. Atmos. Sci.*, **33**, 2152

KAINER, H. (1984). Product safety and producer liability. *Chem. Ind.*, Jun. 4, 407

KAIPAINEN, W.J. (1954). Hydrogen sulphide intoxication. Rapidly transient changes in the electrocardiogram suggestive of myocardial infarction. *Ann. Med. Int. Fenn*, **43**, 97

KAISER, G.D. (1980). The modelling of large scale accidental releases of dense cold vapours; and some examples. *Gastech 79*, p. 227

KAISER, G.D. (1982a). Consequence assessment. In Green, A.E. (1982b), *op. cit.*, p. 93

KAISER, G.D. (1982b). Gas clouds. In Green, A.E. (1982b), *op. cit.*, p. 513

KAISER, G.D. (1982c). Overall assessment. In Green, A.E. (1982), *op. cit.*, p. 101

KAISER, G.D. (1989). A review of models for predicting the dispersion of ammonia in the atmosphere. *Ammonia Plant Safety*, **29**, 100. See also *Plant/Operations Prog.*, 1989, **8**, 58

KAISER, G.D. (1993). Accident prevention and the Clean Air Act Amendments of 1990 with particular reference to anhydrous hydrogen fluoride. *Process Safety Prog.*, **12**, 176

KAISER, G.D. and GRIFFITHS, R.F. (1982). The accidental release of anhydrous ammonia to the atmosphere: a study of factors influencing cloud density and dispersion. *J. Air Poll. Control Ass.*, **32**, 66

KAISER, G.D. and WALKER, B.C. (1978). Releases of anhydrous ammonia from pressurized containers – the importance of denser-than-air mixtures. *Atmos. Environment*, **12**, 2289

KAISER, M.A. (1991). Forestalling a serious accident: a detective story. *Plant/Operations Prog.*, **10**, 100

KAISER, V., SALHI, O. and POCINI, C. (1978). Analyze mixed refrigerant cycles. *Hydrocarbon Process.*, **57**(7), 163

KAKAC, S. (1991). *Boilers, Evaporators and Condensers* (New York: Wiley)

KAKAC, S., BERGLES, A.E. and MAYHINGER, F. (eds) (1981). *Heat Exchangers. Thermal-Hydraulic Fundamentals and Design* (New York: McGraw-Hill)

KAKKO, R. (1989). Vapour cloud modelling in the assessment of major toxic hazards. *J. Loss Prev. Process Ind.*, **2**(2), 102

KAKKO, R. (1991). Computer aided consequence analysis and some future needs. *J. Haz. Materials*, **26**, 105

KAKKO, R. (1992). Effects of the ventilation system and chemical reactions on the concentration and individual risk estimates of toxic materials. *J. Loss Prev. Process Ind.*, **5**, 145

KAKKO, R., VIRTANEN, J. and LAUTKASKI, R. (1992). Experiences in planning process changes on the bases of consequence and risk assessment. *J. Haz. Materials*, **29**, 427

KALBFLEISCH, J.D. and PRENTICE, R.L. (1980). *The Statistical Analysis of Failure Time Data* (New York: Wiley)

KALELKAR, A.S. (1971). Understanding sprinkler performance: modelling of combustion and extinction. *Fire Technol.*, **7**(4), 293

KALELKAR, A.S. (1988a). Bhopal reply (letter). *Chem. Engr, Lond.*, **453**, 8

KALELKAR, A.S. (1988b). Investigation of large magnitude accidents – Bhopal as a case study. *Preventing Major Chemical and Related Process Accidents*, p. 543

KALELKAR, A.S. and CECE, J. (1974). Hazards of liquid ammonia spills. *Ammonia Plant Safety*, **16**, p. 56

KALGHATGI, G.T. (1981–). Blowout stability of gaseous jet diffusion flames. *Combust. Sci. Technol.*, 1981, **26**, 233, 241; 1982, **28**, 241

KALGHATGI, G.T. (1983). The visible shape and size of a turbulent hydrocarbon jet diffusion flame in a cross wind. *Combust. Flame*, **52**, 91. See also *Atmos. Environ.*, **24A**, 2527

KALGHATGI, G.T. (1984). Lift-off heights and visible lengths of vertical turbulent jet diffusion flames in still air. *Combust. Sci. Technol.*, **41**, 17

KALININS, R.V. (1987). Emergency preparedness and response. *Plant/Operations Prog.*, **6**, 6

KALIS, B. (1991). Selecting a laboratory management information system. *Hydrocarbon Process.*, **70**(6), 55

KALKERT, N. and SCHECKER, H.G. (1980). The influence of particle size distribution on the minimum energy of explodable dusts. *Chem. Ing. Tech*, **52**, 515

KALTENECKER, L.V. (1968). Plant location. In Kirk, R.E. and Othmer, D.F. (1963), *op. cit.*

KAM, J.C.P. and RUDLIN, J.R. (1991). Review of the fatigue integrity maintenance of offshore structures. *Safety Developments in the Offshore Oil and Gas Industry* (London: Instn Mech. Engrs), p. 67

KAMATH, A.R.R., KELLER, A.Z. and WOOLISCROFT, M. (1981). Reliability assessment of smoke detectors. *Reliab. Engng*, **2**(4), 283

KAMINS, M. and GROSS, A.J. (1967). *Reliability Assessment in the Presence of Reliability Growth. Rep. RM-5246-PR.* Santa Monica, CA

KAMLET, M.J. and FINGER, M. (1979). An alternative method of calculating Gurney velocities. *Combust. Flame*, **34**, 213

KAMLET, M.J. and JACOBS, S.J. (1968). Chemistry of detonations. I, A simple method for calculating detonation properties of C–H–N–O explosives. *J. Chem. Phys.*, **48**(1), 23

KAMMANN, R. (1975). The comprehensibility of printed instructions and the flow chart alternative. *Hum. Factors*, **17**, 183

KAMPEL, I. (1986). *A Practical Introduction to the New Logic Symbols*, 2nd ed. (London: Butterworths)

KAMPTNER, H.T., KRAUSE, W.R. and SCHILKEN, H.P. (1966a). Acetylene from naphtha pyrolysis. *Chem. Engng*, **73** (Feb. 28), 80

KAMPTNER, H.T., KRAUSE, W.R. and SCHILKEN, H.P. (1966b). High temperature cracking. *Chem. Engng*, **73**(Feb. 28), 93

KAMST, F.H. and LYONS, T.J. (1982). The evaluation of diffusivities for a non-uniform site. *Atmos. Environ.*, **16**, 379

KANAGASABAY, S. (1980). Non-ionizing radiation. In Waldron, H.A. and Harrington, J.M. (1980), *op. cit.*, p. 257

KANAL, L.N. and LEMMER, J.F. (1986). *Uncertainty in Artificial Intelligence* (Amsterdam: North-Holland)

KANDELL, P. (1981). Program sizes pipe and flare manifolds for compressible flow. *Chem. Engng*, **88**(Jun. 29), 89

KANDOLA, B.S. and MORRIS, M. (1994). Smoke hazard assessment using computational fluid dynamics (CFD) modelling. *Hazards*, **XII**, p. 171

KANE, E.R. (1979). An overview of the US chemical industry. *Chem. Ind.*, Jun. 2, 363

KANE, F. (1986). AI and MAP in the processing industries. *Hydrocarbon Process.*, **65**(6), 55

KANE, J.E. and HORST, R.L. (1963). Get better acquainted with aluminum. *Chem. Engng Prog.*, **59**(7), 88

KANE, L.A. (ed.) (1980). *Process Control and Optimisation Handbook for the Hydrocarbon Processing Industries* (Houston, TX: Gulf)

KANE, L.A. (1986). AI and MAP in the processing industries. *Hydrocarbon Process.*, **65**(6), 55

KANE, L.E., BARROW, C.S. and ALARIE, Y. (1979). A short term test to predict acceptable levels of exposure to airborne sensory irritants. *Am. Ind. Hyg. Ass. J.*, **40**, 207

KANE, R.D. (1991). Prevent corrosion of advanced ceramics. *Chem. Engng Prog.*, **87**(6), 77

KANE, R.D. (1992). Know when to use 'advanced' materials. *Chem. Engng Prog.*, **88**(8), 76

KANE, R.D., WILHELM, S.M., ASHBOUGH, W.G. and TARABORELLI, R.G. (1993). Simulate chemical process corrosion in the laboratory. *Chem. Engng Prog.*, **89**(6), 65

KANNAPPAN, S. (1982). Determining centrifugal compressor piping loads. *Hydrocarbon Process.*, **61**(2), 91

KANSA, E.J. and PERLEE, H.E. (1980). A transient dust-flame model: application to coal dust flames. *Combust Flame*, **38**, 17

KANTHA, L.H., PHILLIPS, O.M. and AZAD, R.S. (1977). On turbulent entrainment at a stable density interface. *J. Fluid Mech.*, **79**, 753

KANTOWITZ, B.H. and HANSON, R.H. (1981). Models and experimental results concerning the detection of operator failures in display monitoring. In Rasmussen, J. and Rouse, W.B. (1981), *op. cit.*, p. 301

KANTOWITZ, B.H. and SORKIN, R.D. (1987). Allocation of function. In Salvendy, G. (1987), *op. cit.*, p. 355

KANTYKA, T.A. (1974a). Loss prevention in the process industries. *Course on Process Safety — Theory and Practice*. Dept of Chem. Engng Teesside Polytechnic, Middlesbrough

KANTYKA, T.A. (1974b). Major loss prevention in the process industries. *Chem. Engr, Lond.*, **290**, 620

KANTYKA, T.A. (1979). Exploiting waste heat recovery in the chemical industry. *Chem. Ind.*, Sep. 1, 571

KANURY, A.M. (1972a). Ignition of cellulosic materials: a review. *Fire Res. Abstracts Rev.*, **14**, 24

KANURY, A.M. (1972b). Rate of burning of wood (a simple thermal model). *Combust. Sci. Technol.*, **5**, 135

KANURY, A.M. (1972c). Thermal decomposition kinetics of wood pyrolysis. *Combust. Flame*, **18**, 75

KANURY, A.M. (1975). *Introduction to Combustion Phenomena* (New York: Gordon & Breach)

KANURY, A.M. (1983). A relationship between the flash point, boiling point and the lean limit of flammability of liquid fuels. *Combust. Sci. Technol.*, **31**, 297

KANURY, A.M. (1987). On the craft of modeling in engineering and science. *Fire Safety J.*, **12**(1), 65

KANURY, A.M. and BLACKSHEAR, P.L. (1970). Some considerations pertaining to the problem of wood-burning. *Combust. Sci. Technol.*, **1**, 339

KANURY, A.M. (1981). Limiting case fires arising from fuel tank/pipeline ruptures. *Fire Safety J.*, **3**(3), 215

KAO, J.H.K. (1965). Statistical models in mechanical reliability. *Eleventh Symp. on Reliability and Quality Control* (New York: Instn Elec. Electron. Engrs), p. 240

KAPEGHIAN, J.C., MINCER, H.H., JONES, A.B., VERLANGIERI, A.J. and WATERS, I.W. (1982). Acute inhalation toxicity of ammonia in mice. *Bull. Environ. Contam. Toxicol.*, **29**, 317

VAN KAPEL, K. (1990). Make the most of your boiler inspection. *Chem. Engng Prog.*, **86**(9), 60

KAPILA, A.K., LUDFORD, G.S.S. and BUCKMASTER, J.D. (1975). Ignition and extinction of a monopropellant droplet in an inert atmosphere. *Combust. Flame*, **25**, 361

KAPLAN, C.R. and ORAN, E.D. (1991). Mechanisms of ignition and detonation formation in propane—air mixtures. *Combust. Sci. Technol.*, **80**, 185

KAPLAN, C.R., SUNG WOOK BAEK, ORAN, E.S. and ELLZEY, J. (1994). Dynamics of a strongly radiating unsteady ethylene jet diffusion flame. *Combust. Flame*, **96**, 1

KAPLAN, G. (1979). Nuclear power around the world. *IEEE Spectrum*, **16**(11), 97

KAPLAN, H.L. and HARTZELL, G.O. (1984). Modeling of toxicological effects of fire gases. I, Incapacitating effects of narcotic fire gases. *J. Fire Sci.*, **2**(4), 286

KAPLAN, K., GABRIELSEN, B.L. and VAN HORN, W.H. (1975). Wall failures from explosive forces. *Loss Prevention*, **9**, p. 56

KAPLAN, L.J. (1981a). Fire-safe-valve testing — an explosive issue. *Chem. Engng*, **88**(Sep. 7), 57

KAPLAN, L.J. (1981b). Manufacturers' group enters safety testing arena. *Chem. Engng*, **88**(Oct. 21), 87

KAPLAN, N. (ed.) (1965). *Science and Society* (Chicago, IL: Rand McNally)

KAPLAN, N. and MAXWELL, M.A. (1977). Removal of SO_2 from industrial waste gases. *Chem. Engng*, **84**(Oct. 17), 127

KAPLAN, S. (1982a). Matrix theory formalism for event tree analysis: application to nuclear-risk analysis. *Risk Analysis*, **2**, 9

KAPLAN, S. (1982b). Safety goals and related questions. *Reliab. Engng*, **3**(4), 267

KAPLAN, S. (1983). On a 'two-stage' Bayesian procedure for determining failure rates from experiential data. *IEEE Trans Power Apparatus Systems*, **PAS-102**, 195

KAPLAN, S. and GARRICK, B.J. (1981). Some misconceptions about misconceptions: a response to Abramson (letter). *Risk Analysis*, **1**, 231

KAPTEIN, M. and HERMANCE, C.E. (1977). Horizontal propagation of laminar flames through vertically diffusing mixtures above a ground plane. *Combustion*, **16**, p. 1295

KAPUR, K.C. and LAMBERSON, L.R. (1977). *Reliability Engineering in Design* (New York: Wiley)

KARAM, R.A. and MORGAN, K.Z. (eds) (1974). *Environmental Impact of Nuclear Power Plants* (New York: Pergamon Press)

KARASSIK, I.J. (1959). Increase equipment reliability by studying causes of failure. *Chem. Engng*, Nov. 2, 112

KARASSIK, I.J. (1972). Pump performance characteristics. *Hydrocarbon Process.*, **51**(6), 101

KARASSIK, I.J. (1977). Tomorrow's centrifugal pump. *Hydrocarbon Process.*, **56**(9), 247

KARASSIK, I.J. (1982). Centrifugal pumps and system hydraulics. *Chem. Engng*, **89**(Oct. 4), 84

KARASSIK, I.J. (1993). When to maintain centrifugal pumps. *Hydrocarbon Process.*, **72**(4), 101

KARASSIK, I.J. and KRTOSCH, W.C. (1977). *Pump Handbook* (New York: McGraw-Hill)

KARASSIK, I.J., KRUTZSCH, W.C., FRASER, W.H. and MESSINA, J.P. (1986). *Pump Handbook*, 2nd ed. (New York: McGraw-Hill)

KARCHER, G.G. (1978). New design calculations for high-temperature storage tanks. *Hydrocarbon Process.*, **57**(10), 137

KARCHER, G.G. (1981). Stresses at the shell-to-bottom junction of elevated-temperature tanks. *API Proc., Refinery Dept, 46 Midyear Mtg*, **60**, 154

KARCHER, G.G. and BALL, H.C. (1977). Fabricated valves: a new concept for refinery service. *Hydrocarbon Process.*, **56**(6), 129

KARDOS, J. and VONDRAN, J. (1973). On cost optimal reliability and availability of plants with redundant units. *Chem. Tech.*, **25**(2), 268

KARDOS, J.L. *et al.* (1974–). State progress and trends in reliability work for chemical plants. *Chem. Tech., 1974*, **26**(10), 623; 1976, **28**(9), 520

KAREL, L. and WESTON, R.E. (1947). The biological assay of inhaled substances by the dosimetric method: the retained median lethal dose and respiratory response in unanaesthetised, normal goats exposed to different concentrations of phosgene. *J. Ind. Hyg. Ass.*, **29**, 23

KARIM, R., EMBONG, M.H., RUTTEN, J.F. and BOUTEN, J.W. (1987). Approach and experience in training for a gas processing plant. *Gastech 86*, p. 148

KARINEN, H.K. (1971). Brittle fracture of an ammonia synthesis heat exchanger. *Ammonia Plant Safety*, **13**, 53

KARKHANIS, S. and VAN MORSEL, W.H. (1987). Failure of ammonia converter/feed exchanger. *Ammonia Plant Safety*, **27**, 22

KARL, E. (1973). Strip-wound pressure vessels. *Safety in Polyethylene Plants*, **1**, 26. See also *Chem. Engng Prog.*, 1972, **68**(11), 56

KARLEKAR, B.V. and DESMOND, R.M. (1977). *Engineering Heat Transfer* (St Paul: West)

KARLOVITZ, B. (1953). Open turbulent flames. *Combustion*, **4**, 60

KARLOVITZ, B., DENNISTON, D.W., KNAPSCHAEFER, D.H. and WELLS, F.E. (1953). Studies on turbulent flames. *Combustion*, **4**, 613

KARLOVITZ, B., DENNISTON, D.W. and WELLS, F.E. (1951). Investigation of turbulent flames. *J. Chem. Phys.*, **19**, 541

KARLSSON, H.T. and BJERLE, I. (1980). A simple approximation to the error function. *Comput. Chem. Engng*, **4**, 67

VON KARMAN, T. (1957). The present status of the theory of laminar flame propagation. *Combustion*, **6**, 1

KARMARKAR, M. (1979). Selection and sizing of vents for polymerisation reactors. In Minors, C. (1979), *op. cit.*, paper 4

KARNOPP, D. and ROSENBERG, R.C. (1968). *Analysis and Simulation of Multiport Systems* (Cambridge, MA.: MIT Press)

KARPENKO, G.V. and VASILENKO, I.I. (1979). *Stress Corrosion Cracking of Steels* (Tel Aviv, Israel: Freund)

KARPLUS, H.B. (1961). *Propagation of Pressure Waves in Mixtures of Water and Steam. Rep. ARF-4132–12.* Armour Res. Foundation

KARPP, R.R. and PREDEBON, W.W. (1974). Calculation of fragment velocities from fragmentation munitions. *First Int. Symp on Ballistics*, p. IV-145

KARRH, B.W. (1984). An illustration of voluntary actions to reduce carcinogenic risks in the workplace. In Deisler, P.F. (1984), *op. cit.*, p. 121

KARRH, B.W. (1993). Corporate environmentalism: a global commitment. *Health, Safety and Loss Prevention*, p. 12

KARVONEN, I., HEINO, P. and SUOKAS, J. (1990). *Knowledge Base Approach to Support Hazop Studies. Res. Rep. 704.* VTT, Espoo, Finland

KARVONEN, I., HEINO, P., WENNERSTEN, R., ANDERSEN, T. and PETTERSEN, T. (1992). Experiences from intelligent acquisition and use of safety information. *Loss Prevention and Safety Promotion 7*, vol. 1, 10–1

KARVONEN, I., SUOKAS, J. and HEINO, P. (1987). *Hazopex – Expert System Supporting the Safety Analysis* (paper). Tech. Res. Centre of Finland

KARWAT, E. (1961a). Hydrocarbon control in air separators. *Ammonia Plant Safety*, **3**, 5. See also *Chem. Engng Prog.*, 1961, **57**(4), 41

KARWAT, E. (1961b). Safety measures in low temperature separation of acetylene-containing gas mixtures. *Ammonia Plant Safety*, **3**, A18

KARWAT, E. (1963). Hydrocarbon control in air separators. *Ammonia Plant Safety*, **3**, 5

KASE, D.W. and WIESE, K.J. (1990). *Safety Auditing* (New York: van Nostrand Reinhold)

KASHIWAGI, T. (1974). A radiative ignition model of a solid fuel. *Combust. Sci. Technol.*, **8**, 225

KASHIWAGI, T. (1976). A study of flame spread over a porous material under external radiation fluxes. *Combustion*, **15**, 255

KASHIWAGI, T. (1979a). Effects of attenuation of radiation on surface temperature for radiative ignition. *Combust. Sci. Technol.*, **20**, 225

KASHIWAGI, T. (1979b). Experimental observation of radiative ignition mechanisms. *Combust. Flame*, **34**, 231

KASHIWAGI, T. (1980). Ignition of a liquid fuel under high intensity radiation. *Combust. Sci. Technol.*, **21**, 131

KASHIWAGI, T. (1981). Radiative ignition mechanism of solid fuels. *Fire Safety J.*, **3**(3), 185

KASKAN, W.E. (1957). The dependence of flame temperature on mass burning velocity. *Combustion*, **6**, p. 134

KASPERSON, R.E., RENN, O., SLOVIC, P., BROWN, H.S., EMEL, J., GOBLE, R., KASPERSON, J.X. and RATICK, S. (1988). The social amplification of risk: a conceptual framework. *Risk Analysis*, **8**, 177

KASPERSON, R.E. and STALLEN, P.J.M. (eds) (1991). *Communicating Risks to the Public* (Dordrecht: Kluwer)

KASPRAK, A., VIGEANT, S. and MCBRIDE, G.W. (1986). Determining the impact of an accidental hazardous chemical release. *Fifth Joint Conf. on Applications of Air Pollution Modeling* (Boston, MA: Am. Met. Soc.), p. 286

KASS, I., ZAMEL, N., DOBRY, C.A. and HOLZER, M. (1972). Bronchiectasis following ammonia burns of the respiratory tract. A review of two cases. *Chest*, **62**, 282

KASSATLY, J. (1992). Successfully fast-track process vessels. *Chem. Engng Prog.*, **88**(7), 68

KASSEL, L.S. (1937). The mechanism of the combustion of hydrogen. *Combustion*, **2**, p. 175

KASSOY, D.R. (1975). A theory of adiabatic, homogeneous explosion from initiation to completion. *Combust. Sci. Technol.*, **10**, 27

KASSOY, D.R. and POLAND, J. (1975). The subcritical, spatially homogeneous explosion: initiation to completion. *Combust. Sci. Technol.*, **11**, 147

KASSOY, D.R., KAPILA, A.K. and STEWART, D.S. (1989). A unified formulation for diffusive and nondiffusive explosion theory. *Combust. Sci. Technol.*, **63**, 33

KASTENBERG, W.E., MCKONE, T.E. and OKRENT, D. (1976). *On Risk Assessment in the Absence of Complete Data. Rep. UCLA-ENG-7677.* Univ. of California, Los Angeles, CA

KASTENBERG, W.E. and SOLOMON, K.A. (1985). On the use of confidence levels in risk management. *J. Haz. Materials*, **10**, 263

KASTNER, W., RÖHRICH, E., SCHMITT, W. and STEINBUCH, R. (1981). Critical crack sizes in ductile piping. *Int. J. Press. Ves. Piping*, **9**

KATAN, L.L. (1951). The fire hazard of fuelling aircraft in the open. *Fire Res. Tech. Paper 1* (London: HM Stationery Office)

KATELL, S. (1973). Justifying pilot plant operations. *Chem. Engng Prog.*, **69**(4), 55

KATES, R.W. (1978). *Risk Assessment of Environmental Hazard* (Chichester: Wiley)

KATO, D. (1977). Ultra-low temperature LNG compressors. *LNG Conf. 5*

KATSUNA, M. (ed.) (1968). *Minamata Disease* (Kumamoto Univ.)

KATZ, D.L. (1968). Introduction of the Committee on Hazardous Materials. Advisory to the US Coast Guard. *Carriage of Dangerous Goods*, **1**, 4

KATZ, D.L. (1970). A view of safety in the petroleum industry. In National Academy of Engineering (1970), *op. cit.*, p. 65

KATZ, D.L. (1972). Superheat limit explosions. *Chem. Engng Prog.*, **68**(5), 68

KATZ, D.L. and HASHEMI, H.T. (1971). Pipelining LNG in the dense phase. *Oil Gas J.*, **69**(Jun. 7), 55

KATZ, D.L. and SLIEPCEVICH, C.M. (1971). LNG/water explosions: cause and effect. *Hydrocarbon Process.*, **50**(11), 240

KATZ, M.J. (1976). Hazard and risk evaluation. *Loss Prevention*, **10**, 126

KATZ, W.B. (1976). A new pair of eyes. *Control of Hazardous Material Spills 3*, p. 1

KATZEN, R. (1968). When is the pilot plant necessary? *Chem. Engng*, **75**(Mar. 25), 95

KATZLER, J. (1980). Tank insulation costs up? Use urethane foam. *Hydrocarbon Process.*, **59**(5), 197

KAUBER, G.A. (1973). Formal operating instructions, operations audits – are they worth while? *Chem. Engng*, **80**(Sep. 17), 146

KAUDERS, P. (1984). Design for plant upset conditions. *Chem. Engr, Lond.*, **399**, 9

KAUDERS, P.G. (1980–). Process engineering design. *Chem. Engr, Lond.*, **352**, 51; **358**, 475; **359/360**, 530; **361**, 634; 1981, **367**, 146

KAUFFMAN, C.W. (1981). Agricultural dust explosions in grain handling facilities. *The Hazards of Industrial Explosion from Dusts* (London: Oyez/IBC), paper 18

KAUFFMAN, C.W. (1982). Agricultural dust explosions in grain handling facilities. In Lee, J.H.S. and Guirao, C.M. (1982), *op. cit.*, p. 305

KAUFMAN, C.W. (1986). Recent dust explosion experiences in the US grain industry. *Industrial Dust Explosions* (Philadelphia, PA: Am. Soc. Testing Mats)

KAUFFMAN, C.W., WOLANSKI, P., ARISOY, A., ADAMS, P.R., MAKER, B.N. and NICHOLLS, J.A. (1984). Dust, hybrid and dusty detonations. In Bowen, J.R. *et al.* (1984b), *op. cit.*, p. 221

KAUFFMAN, C.W., SRINATH, S.R., TEZOK, F.I., NICHOLLS, J.A. and SICHEL, M. (1985). Turbulent and accelerating dust flames. *Combustion*, **20**, 1701

KAUFFMAN, D. and CHEN, H.J. (1990). Fault-dynamic modelling of phthalic anhydride reactors. *J. Loss Prev. Process Ind.*, **3**(4), 386

KAUFFMAN, G.J. (1987). *Combustion safety inspections*. *Chem. Engng. Prog.*, **83**(7), 23

KAUFFMANN, W.M. (1969). Safe operation of oxygen compressors. *Chem. Engng*, **76**(Oct. 20), 140

KAUFMAN, D. (1987). Health, safety and loss control topics in the senior design courses. *Plant/Operations Prog.*, **6**, 73

KAUFMAN, F. (1982). Chemical kinetics and combustion: intricate paths and simple steps. *Combustion*, **19**, 1

KAUFMAN, F.X., MAJONE, G. and OSTROM, V. (eds) (1986). *Guidance, Control and Evaluation in the Public Sector* (Berlin: de Gruyter)

KAUFMAN, J. and BURKONS, D. (1971). Clinical, roentgenologic, and physiologic effects of acute chlorine exposure. *Arch. Environ. Health*, **23**, 29

KAUFMAN, J.B. and PERZ, J. (1978). The best choice . . . type MC continuous aluminum sheath cable. *Hydrocarbon Process.*, **57**(8), 123

KAUFMAN, R. (1988). Preparing useful performance indicators. *Training Dev. J.*, Sep., 80

KAUFMAN, R.A., KAUFMAN, P.E., JONES, A.L. and SIEGELL, J.H. (1990). Quantification of emission using dispersion modelling and air monitoring. *Process Safety Environ.*, **68B**, 231

KAUFMANN, S. (1976). Decision making and flammable spill handling. *Control of Hazardous Material Spills 3*, p. 320

KAUPS, T. (1985). Experimental dynamic compaction of perlite insulation. *Gastech 84*, p. 396

KAURA, M.L. (1980a). Aid to firewater design. *Hydrocarbon Process.*, **59**(12), 137

KAURA, M.L. (1980b). Plot plans must include safety. *Hydrocarbon Process.*, **59**(7), 183

KAUTTO, A., NORROS, L., RANTA, J. and HEIMBURGER, H. (1984). The role of visual information in improving the acquisition of adequate work orientation. *Ergonomic Problems in Process Operations* (Rugby: Instn Chem. Engrs), p. 97

KAVASMANECK, P.R. (1990). CMA responsible care process safety code of management practice. *Plant/Operations Prog.*, **9**, 148

KAVIANIAN, H.R., RAO, J.K. and BROWN, G.V. (1992). *Application of Hazard Evaluation Techniques to the Design of Potentially Hazardous Industrial Chemical Processes. Report.* Div. of Training and Manpower Dev., Nat. Inst. Occup. Safety and Health, Cincinnati, OH

KAWAGOE, K. (1958). *Fire Behaviour in Rooms. Rep. 27.* Building Res. Inst., Tokyo

KAWAGOE, K. and SEKINE, T. (1963). *Estimation of Temperature–Time Curves in Rooms. Occasional Rep. 11.* Building Res. Inst., Tokyo

KAWAI, M. (1973). Inhalation toxicity of phosgene and trichlorinitromethane. *Sangyo Igaku*, **15**, 406

KAWAI, T., MOHRI, T., TAKEMURA, K. and SHIBASAKI, T. (1984). Stress analysis for prolonging tube life. *Ammonia Plant Safety*, **24**, 131

KAWALLER, S.I. (1977). Update on magnesium oxychloride fireproofing. *Fire Technol.*, **13**, 139

KAWALLER, S.I. (1980). Chemically reactive coatings for the protection of structural steel. *Fire Surveyor*, **9**(4), 36

KAWAMURA, H. (1980). Specifying valves for the HPI. *Hydrocarbon Process.*, **59**(8), 61

KAWAMURA, P.I. and MACKAY, D. (1987). The evaporation of volatile liquids. *J. Haz. Materials*, **15**(3), 343

KAY, J.M. and NEDDERMAN, R.M. (1985). *Fluid Mechanics and Transfer Processes*, 4th ed. (Cambridge: Cambridge Univ. Press)

KAY, K. (1977). Polybrominated biphenyls (PBB). Environmental contamination in Michigan, 1973–76. *Environ. Res.*, **13**(1), 74

KAY, P.C.M. (1966). On-line computer alarm analysis. *Ind. Electron.*, **4**, 50

KAY, P.C.M. and HEYWOOD, P.W. (1966). Alarm analysis and indication at Oldbury nuclear power station. *Automatic Control in Electricity Supply* (London: Instn. Elec. Engrs), p. 295

KAYE, B.H. (1981). *Direct Characterisation of Fineparticles* (New York: Wiley)

KAYSER, D.S. (1972). Rupture disc selection. *Loss Prevention*, **6**, 82. See also *Chem. Engng Prog.*, 1972, **68**(5), 61

KAYSER, J.N. (1974). Testing fireproofing for structural steel. *Loss Prevention*, **8**, 45

KAZANSKI, A.B. and MONIN, A.S. (1960). A turbulent regime above the surface atmosphere layer. *Izv. Acad. Sci. USSR, Geof. Ser.*, **1**, 110

KAZARIANS, M. and APOSTOLAKIS, G. (1978). On the fire hazard in nuclear plants. *Nucl. Engng Des.*, **47**, 157

KAZARIANS, M., BOYKIN, R.F. and KAPLAN, S. (1986). Transportation risk management – a case study. *Am. Inst. Chem. Engrs, Symp. on Loss Prevention*

KEANE, D.P., STOBAUGH, R.B. and TOWNSEND, P.L. (1973). Vinyl chloride: how, where, who – future. *Hydrocarbon Process.*, **52**(2), 99

KEATING, F.H. (1968). *Chromium Nickel Austenitic Steels* (London: Butterworths)

KEATING, J.P. (1982). The myth of panic. *Fire J.*, May, 57

KEATS, J. (1971). *An Introduction to Quantitative Psychology* (Sydney: Wiley)

KEATTCH, C.J. and DOLLIMORE, D. (1975). *An Introduction to TG*, 2nd ed. (London: Heyden)

KECECIOGLU, D. and DINGJUN LI (1985). Optimum two-sided confidence interval for the location parameter of the exponential distribution. *Reliab. Engng*, **13**(1), 1

KECECIOGLU, D. and JACKS, J.A. (1985). The Weibull stress−life, log−log stress−life, and the overload−stress reliability models in accelerated life testing. *Reliab. Engng*, **11**(2), 109

KEDDIE, A.W.C. (1984). Prediction of odour nuisance. *Chem. Ind.*, May 7, 323

KEEFER, REISLER, GIGLIO-TOS AND KELLNER (1966). *Airblast phenomena. Rep. POR-3001*. Ballistic Res. Lab., Aberdeen Proving Ground, MD

KEELE, C.A., NEIL, E. and JOELS, N. (1982). *Samson Wright's Applied Physiology* (Oxford: Oxford Univ. Press)

KEEN, B. (1977). Safe disposal of toxic waste. *Chemical Engineering in a Hostile World* (London: Clapp & Poliak)

KEENAN, J. (1941). *Thermodynamics* (New York: Wiley)

KEENAN, J.H. (1970). *Thermodynamics* (Cambridge, MA: MIT Press)

KEENAN, J.M. (1989). *Piper Alpha Public Inquiry. Transcript, Day 151*

KEENE, J. (1981). Swooping down on air pollution. *Chem. Engng*, **88**(Jun. 1), 31

KEENE, J. (1982). Environmental auditors see fair weather ahead. *Chem. Engng*, **89**(Dec. 13), 47

KEENEY, R. (1980). *Siting Energy Facilities* (New York: Academic Press)

KEENEY, R.L. and RAIFFA, H. (1976). *Decisions with Multiple Objectives, Preferences and Value Trade-offs* (New York: Wiley)

KEEY, R.B. (1973). *Drying Principles and Practice* (Oxford: Pergamon Press)

KEEY, R.B. (1977). Hazards of heating immiscible liquids. *Chemical Process Hazards*, **VI**, p. 147

KEEY, R.B. (1978). *Introduction to Industrial Drying Operations* (Oxford: Pergamon Press)

KEEY, R.B. (1986a). The analysis of hazard warnings. *Proc. World Cong. III*, Tokyo, vol. II, p. 1084

KEEY, R.B. (1986b). The use of hazard-warning analysis. *Chem. Engng Process.*, **20**, 289

KEEY, R.B. (1987). *Reliability in the Process Industries* (Wellington, New Zealand: IPENZ)

KEEY, R.B. (1990). Some hazards of dispensing automotive fuels at retail service stations. *Safety and Loss Prevention in the Chemical and Oil Processing Industries*, p. 373

KEEY, R.B. (1991). A rapid hazard-assessment method for smaller-scale industries. *Process Safety Environ.*, **69B**, 85

KEEY, R.B. and SMITH, C.H. (1984). The propagation of uncertainties in failure events. *Reliab. Engng*, **10**, 105

KEFER, V. *et al.* (1986). *Leckraten bei untercritischen Rohrleitungsrissen*. Jahrestagung Kerntechnik, Aachen

KEFFER, J.F. and BAINES, W.D. (1963). The round turbulent jet in a cross-wind. *J. Fluid Mech.*, **15**, 481

KEFFER, W.J. (1991). Management of hazardous materials at the local community level. *Plant/Operations Prog.*, **10**, 194

KEHOE, R.A., MACHLE, W.F., KITZMILLER, K. and LEBLANC, T.J. (1932). On the effects of prolonged exposure to sulphur dioxide. *J. Ind. Hyg.*, **14**, 159

KEINANEN, H., RINTAMAA, R., OBERG, T., SARKIMO, M., TALJA, HELI and SAARENHIEMO, A. (1990). Evaluation of catastrophic failure risk in pressure vessels. Results of pressure vessel test with a large vessel. *Tutkimuksia- Valt. Tek. Tutkimuskeskus*, **708**

KEISTER, R.G., PESETSKY, B.I. and CLARK, S.W. (1971). Butadiene explosion at Texas City − 3. *Loss Prevention*, **5**, 67

KEITER, A.G. (1989). Emergency pressure relief discharge control by passive quenching. *Runaway Reactions*, p. 425

KEITER, A.G. (1992). Emergency pressure relief discharge control by passive quenching − update. *Plant/Operations Prog.*, **11**, 157

KEITH, L.H. (1980). Analysis of toxic chemicals in the environment. In Haque, R. (1980a), *op. cit.*, p. 41

KEITH, L.H. (ed.) (1982). *Energy and the Environment*, vol. 2, *Acid Rain* (Ann Arbor, MI: Ann Arbor Science)

KEITH, L.H. and WALTERS, D.B. (1985−). *Compendium of Safety Data Sheets for Research and Industrial Chemicals* (Cambridge: VCH)

KEITH, P.C. (1965). The gas turbine as a new feature in a large air separation plant. *Ammonia Plant Safety*, **7**, 63

KELETI, C. (ed.) (1985). *Nitric Acid and Fertilizer Nitrates* (Basel: Dekker)

KELLARD, B. (1993). *Croner's Substances Hazardous to Health* (New Malden, Surrey: Croner)

KELLEHER, T.W. (1989). *Piper Alpha Public Inquiry. Transcript, Days 146, 147*

KELLEHER, T. and FAIR, J.R. (1996). Distillation studies in a high-gravity contactor. *Ind. Eng. Chem. Res.* **35**, 4646−55

KELLER, A., HEINXLE, E. and HUNGERBUHLER, K. (1996). Development and assessment of inherently safe processes in the fine chemical industry. *Int. Conf. and Workshop on Process Saf. Manag. and Inherently Safer Process.*, Oct. 8−11, 1996, Orlando, FL, 213−223. (New York: American Institute of Chemical Engineers)

KELLER, A.Z. (1977). The need for reliability training. *First Nat. Conf. on Reliability. NRC6/3.* (Culcheth, Lancashire: UK Atom. Energy Auth.)

KELLER, A.Z. and LAMB, R. (1987). Quantify the risks from chemical spills into rivers. *Process Engng*, Jul., 49

KELLER, A.Z. and STIPHO, N.A. (1981). Availability studies of two similar chlorine production plants. *Reliab. Engng*, **2**(4), 271

KELLER, A.Z. and WILSON, H.C. (1991). An evaluation of chemically related disasters using the Bradford Disaster Scale. *Hazards*, **XI**, p. 1

KELLER, A.Z. *et al.* (1990). The Bradford Disaster Scale. *Disaster Management*, **2**(4)

KELLER, C.L. (1987). Future requirements for unrestricted entry into confined spaces. *Plant/Operations Prog.*, **6**, 142

KELLER, C.L., KERLIN, D.J. and LOESER, R. (1986). Motor vehicles. *Fire Protection Handbook*, 16th ed. (ed. A.E. Cote and J.L. Linville), pp. 18−27

KELLER, H.N. and HOELSCHER, H.E. (1957). Development of static charges in a non-conducting system. *Ind. Engng Chem.*, **49**, 1433

KELLERMAN, O. (1966). *Present Views on Recurring Inspection of Reactor Pressure Vessels in the Federal Republic of Germany. Recurring Inspection of Nuclear Reactor Steel Pressure Vessels. STI/Doc/10/81 1968.* Int. Atom. Energy Agency, Vienna

KELLERMAN, O. and SEIPEL, H.G. (1967). Analysis of the improvement in safety obtained by a containment and by other safety devices for water cooled reactors. *Int. Atom. Energy Agency, Symp. on Containment and Siting of Nuclear Power Reactors* (Vienna)

KELLEY, C.R. (1968). *Manual and Automatic Control* (New York: Wiley)

KELLEY, C.S. (1975). Pilot ignition time of wood exposed to time-dependent radiation from pools of burning fuel. *Fire Technol.*, **11**, 119

KELLEY, H.H. (1973). The processes of causal attribution. *Am. Psychol.*, **28**(2), 107

M.W. KELLOGG CO. (1956). *Design of Piping Systems* (New York: Wiley)

M.W. KELLOGG CO. (1967). *Design Specifications – Process Plant Layout* (New York)

KELLUM, G.B. (1973). You can predict ball and roller bearing failures. *Hydrocarbon Process.*, **52**(1), 85

KELLY, A. (1974). The control of industrial maintenance, Pt 2. *Plant Engr*, **18**(2), 17

KELLY, A. (1980). A review of the maintenance management problem. *Terotechnica*, **1**, 243

KELLY, A. (1981). A maintenance plan for a continuously operating plant. *Terotechnica*, **2**, 193

KELLY, A. (1986). *Maintenance Planning and Control* (Oxford: Butterworth-Heinemann)

KELLY, A. and HARRIS, M.J. (1971). Simulation: an aid to maintenance decisions. *Plant Engr*, **15**(11), 43

KELLY, A. and HARRIS, M.J. (1978). *Management of Industrial Maintenance* (London: Newnes-Butterworths)

KELLY, A.B. and HARRIS, M.J. (1983). *Management of Industrial Maintenance* (Oxford: Butterworth-Heinemann)

KELLY, A.B. (1971). Transportation accidents and the law. *Loss Prevention*, **5**, 15

KELLY, B.A. and OAKES, T.W. (1980). *The Spill Prevention and Control Countermeasure Plan for Oak Ridge National Laboratory. Rep. ORNL/TM-7677.* Oak Ridge Nat. Lab., Oak Ridge, TN

KELLY, B.E. (1982). Fault propagation in process plant: some advances in computer aided fault tree synthesis. *Reliability of Large Scale Systems, IEE Coll. Dig. 1982/41 (London)*, p. 4/1

KELLY, B.E. and LEES, F.P. (1986a). The propagation of faults in process plants: 1, Modelling of fault propagation. *Reliab. Engng*, **16**, 1

KELLY, B.E. and LEES, F.P. (1986b). The propagation of faults in process plants: 2, Fault tree synthesis. *Reliab. Engng*, **16**, 39

KELLY, B.E. and LEES, F.P. (1986c). The propagation of faults in process plants: 3, An interactive, computer-based facility. *Reliab. Engng*, **16**, 63

KELLY, B.E. and LEES, F.P. (1986d). The propagation of faults in process plants: 4, Fault tree synthesis of a pump system changeover sequence. *Reliab. Engng*, **16**, 87

KELLY, C.S. (1973). The transfer of radiation from a flame to its fuel. *J. Fire Flammability*, **4**, 56

KELLY, G.N. and HEMMING, C.R. (1983). *The Radiological Consequences of Degraded Core Accidents for the Sizewell PWR. The Impact of Adopting Revised Frequencies of Occurrence. Rep. NRPB R160.* Nat. Radiol. Prot. Board, Harwell

KELLY, G.N., JONES, J.A. and HUNT, B.W. (1977). *An Estimate of the Radiological Consequences of Notional Accidental Releases of Radioactivity from a Fast Breeder Reactor. Rep. NRPB-R53* (London: HM Stationery Office)

KELLY, M.P. (1985). Applications of acoustic emission monitoring. *API 50th Midyear Mtg*, **64**, 156

KELLY, T.J. (1986). For energy savings – look at your boiler blowdown. *Plant/Operations Prog.*, **5**, 110

KELLY, W.J. (1991). Oversights and myths in a HAZOP program. *Hydrocarbon Process.*, **70**(10), 114

KELTNER, N.R., NICOLETTE, V.F., BROWN, N.N. and BAINBRIDGE, B.L. (1990). Test unit effects on heat transfer in large fires. *J. Haz. Materials*, **25**(1/2), 31

KEMBLE, J.V.H. (1983). The early management of burns. *Surgery*, **1**(3), 52

KEMENY, J. (chrmn). (1979). *The Need for Change – The Legacy of TMI. Report of the President's Commission on the Accident at Three Mile Island* (Washington, DC: US Govt Printing Office)

KEMENY, J.G. and SNELL, J.L. (1960). *Finite Markov Chains* (Princeton, NJ: van Nostrand)

KEMLER, H. (1975). French system for the control of tank-containers used for the transportation of hazardous materials. *Transport of Hazardous Cargoes*, **4**

KEMMER, F.N. and ODLAND, K. (1968). Water pollution control. Chemical treatment. *Chem. Engng*, **75**(Oct. 14), 83

KEMP, H.S. (1983). Formation and management of AIChE design institutes. *Chem. Engng Prog.*, **79**(6), 9

KEMP, H.S. (1987). Process safety – a commitment? *Chem. Engng Prog.*, **83**(5), 46

KEMP, I.C. (1991). Some aspects of the practical application of pinch technology methods. *Trans. Inst. Chem. Engrs*, **69A**, 471

KEMP, I.C. and DEAKIN, A.W. (1989). The cascade analysis for energy and process integration of batch processes. *Chem. Engng Res. Des.*, **67**, 495, 510 and 517

KEMP, I.C. and HART, D.R. (1987). Energy management – obligation or opportunity? *Chem. Engr, Lond.*, **437**, 26

KEMP, J.B. (1970). The thermal stress problem in pressure vessels. *Hydrocarbon Process.*, **49**(11), 195

KEMP, R. and GERARD, S. (1991). Hazard assessment techniques for UK landfill: HALO. *Integrated Environ. Mngmt*, **5**, 20

KEMP, R., O'RIORDAN, T. and PURDUE, M. (1984). Investigation as legitimacy: the maturing of the big public inquiry. *Geoforum*, **15**, 477

KEMPER, F.D. (1966). A near-fatal case of hydrogen sulphide poisoning. *Can. Med. Ass. J.*, **94**, 1130

KEMPER, M.J., MORLEY, J.I., MCWILLIAM, J.A. and SLATER, D. (1966–67). Pressure vessels of high proof strength warm worked stainless steel. *Proc. Instn Mech. Engrs*, **181**, pt 1, 137

KENDALL, H. (1975). *Nuclear Power Risks* (Cambridge, MA: Union of Concerned Scientists)

KENDALL, H.W. and MOGLEWER, S. (eds) (1974). *Preliminary Review of the AEC Reactor Safety Study.* Sierra Club/Union of Concerned Scientists, San Francisco, CA

KENDALL, J. (1938). Breathe Freely. *The Truth about Poison Gas* (London: Bell)

KENDALL, M.G. (1948). *Rank Correlation Methods* (London: Methuen)

KENDALL, M.G. (1955). Further contributions to the theory of paired comparisons. *Biometrics*, **11**, 43

KENDALL, M.G. (1970). *Rank Correlation Methods*, 4th ed. (London: Griffin)

KENDALL, M.G. and STUART, A. (1958–). *The Advanced Theory of Statistics*, vols 1–3 (London: Griffin)

KENDIK, E. (1986). Designing escape routes in buildings. *Fire Technol.*, **22**, 272

KENEFICK, J.F. and CHIRILLO, R.D. (1985). Photogrammetry and computer-aided piping design. *Chem. Engng*, **92** (Feb. 18), 173

KENNARD, H.J. (1984). Joint experts tackle waste. *Hazardous Cargo Bull.*, **5**(6), 3

KENNEDY, I.M., KOLLMANN, W. and CHEN, J.Y. (1990). A model for soot formation in a laminar diffusion flame. *Combust. Flame*, **81**, 73

KENNEDY, J. (1962). *Fire and Arson Investigation* (Chicago, IL: Investigations Inst.)

KENNEDY, R.H. (1983). Selecting temperature sensors. *Chem. Engng*, **90**(Aug. 8), 54

KENNEDY, R.P. (1976). A review of procedures for the analysis and design of concrete structures to resist missile impact effects. *Nucl. Engng Des.*, **37**, 183

KENNEDY, R.P. (1979). Above ground vertical tanks. In *Nuclear Regulatory Commission* (1979 *NUREG/CR-1161*), Sec. 2.2

KENNEDY, R.P. (1989). In Kennedy, R.P. *et al.* (1989), *op. cit.*, p. A-1

KENNEDY, R.P. and RAVINDRA, M.K. (1984). Seismic fragilities for nuclear power plant risk studies. *Nucl. Engng. Des.*, **79**, 47

KENNEDY, R.P., MURRAY, R.C., RAVINDRA, M.K., REED, J.W. and STEVENSON, J.D. (1989). *Assessment of Seismic Margin Calculation Methods. Rep. NUREG/CR-5270.* Nuclear Regulatory Comm., Washington, DC

KENNEDY, T.L. and JOHNSON, G.W. (1975). Civilian bomb injuries. *Br. Med. J.*, Feb. 15, 382

KENNEL, C., GÖTTGENS, J. and PETERS, N. (1991). The basic structure of lean propane flames. *Combustion*, **23**, 479

KENNEY, G.D. (1991). The double-edged sword of objective regulations. *Safety Developments in the Offshore Oil and Gas Industry* (London: Instn Mech. Engrs), p. 125

KENNEY, W.F. (1993). *Process Risk Management Systems* (New York: VCH)

KENNON, B.R. (1927). Report of a case of injury to the skin and eyes by liquid sulphur dioxide. *J. Ind. Hyg.*, **9**, 486

KENNY, T.M. (1991). Guard against freezing in water lines. *Chem. Engng Prog.*, **87**(9), 44

KENNY, T.M. (1992). Steam tracing: doing it right. *Chem. Engng Prog.*, **88**(8), 40

KENSON, R.E. and HOFFLAND, R.O. (1980). Control of toxic air emissions in chemical manufacture. *Chem. Engng. Prog.*, **76**(2), 80

KENSON, R.E. (1981). Controlling odors. *Chem. Engng*, **88**(Jan. 26), 94

KENT, G.R. (1964). Practical design of flare stacks. *Hydrocarbon Process.*, **43**(8), 121

KENT, G.R. (1968). Find radiation effect of flares. *Hydrocarbon Process.*, **47**(6), 119

KENT, G.R. (1972). Make profit from flares. *Hydrocarbon Process.*, **51**(10), 121

KENT, G.R. (1978a). Preliminary pipeline sizing. *Chem. Engng*, **85**(Sep. 25), 119

KENT, G.R. (1978b). Selecting gaskets for flanged joints. *Chem. Engng*, **85**(Mar. 27)

KENT, G.R. (1989). Troubleshooting vibration problems. *Chem. Engng*, **96**(2), 145

KENT, J.H. and BILGER, R.W. (1973). Turbulent diffusion flames. *Combustion*, **14**, 615

KENT, J.H. and HONNERY, D. (1987). Soot and mixture fraction in turbulent diffusion flames. *Combust. Sci. Technol.*, **54**, 383

KENT, J.H. and HONNERY, D.R. (1991). Soot formation in diffusion flames – a unifying trend. *Combust. Sci. Technol.*, **75**, 167

KENT, J.H. and WAGNER, H.G. (1985). Temperature and fuel effects in sooting diffusion flames. *Combustion*, **20**, 1007

KENT, T.D. (1978). Civil and criminal law liability of engineering executives. *Chem. Engng Prog.*, **74**(3), 11

KENTISH, D.N.W. (1982a). *Industrial Pipework* (New York: McGraw-Hill)

KENTISH, D.N.W. (1982b). *Pipework Design Data* (New York: McGraw-Hill)

KEPNER, C.H. and TREGOE, B.B. (1965). *The Rational Manager* (New York: McGraw-Hill)

KER WILSON, W. (1970–71) Some early experiences of mechanical vibration in marine engine systems. *Proc. Instn Mech. Engrs*, **185**(pt 1), 1023

KERAVNOU, E.T. and JOHNSON, L. (1986). *Competent Expert Systems* (London: Kogan Page)

KERKLO, P. (1982). Oil seals for centrifugal compressors. *Hydrocarbon Process.*, **61**(10), 113

KERKOFS, A.W.N. (1967). *Basic Electricity for Electronic Engineers* (London: Philips)

KERMAN, B.R. (1981). A note on the maximum ground level concentration for inversion-limited chimney plumes. *Atmos. Environ.*, **15**, 119

KERMAN, B.R. (1982). A similarity model of shoreline fumigation. *Atmos. Environ*, **16**, 467

KERMAN, B.R. (1983). Effective stratification for plume rise and boundary layer growth comparisons. *Atmos. Environ.*, **17**, 235

KERN, D.Q. (1950). *Process Heat Transfer* (New York: McGraw-Hill)

KERN, D.Q. (1966). Heat exchanger design for fouling service. *Chem. Engng Prog.*, **62**(7), 51

KERN, D.Q. and KRAUS, A.D. (1972). *Extended Surface Heat Transfer* (New York: McGraw-Hill)

KERN, J.L. (1972). Control systems for prevention of hazardous material spills in process plants. *Control of Hazardous Material Spills 1*, p. 19

KERN, R. (1960). How to design heat exchanger piping. *Petrol. Refiner*, **39**(2), 137

KERN, R. (1966). Plant layout and piping design for minimum cost systems. Hydrocarbon Process. *Petrol. Refiner*, **45**(10), 119

KERN, R. (1969). How to size process piping for two phase flow. *Hydrocarbon Process.*, **48**(10), 105

KERN, R. (1971). Pump piping design. *Chem. Engng*, **78**(Oct. 11), 85

KERN, R. (1972a). How discharge piping affects pump performance. *Hydrocarbon Process.*, **51**(3), 89

KERN, R. (1972b). How to size pump suction piping. *Hydrocarbon Process.*, **51**(4), 119

KERN, R. (1974–). Practical piping design. *Chem. Engng*, 1974, **81**(Dec. 23), 58; 1975, **82**(Jan. 6), 115; (Feb. 3), 72; (Mar. 3), 161; (Apr. 14), 85; (Apr. 28), 119; (May 26), 113; (Jun. 23), 145; (Aug. 4), 107; (Sep. 15), 129; (Oct. 13), 125; (Nov. 10), 209

KERN, R. (1975a). Control valves in process plants. *Chem. Engng*, **82**(Apr. 14), 85

KERN, R. (1975b). How to design piping for pump-suction conditions. *Chem. Engng*, **82**(Apr. 28), 119

KERN, R. (1975c). How to size piping for pump-discharge conditions. *Chem. Engng*, **82**(May 26), 113

KERN, R. (1977a). Arrangements of process and storage vessels. *Chem. Engng*, **84**(Nov. 7), 93

KERN, R. (1977b). How to find optimum layout for heat exchangers. *Chem. Engng*, **84**(Sep. 12), 169

KERN, R. (1977c). How to get the best process plant layouts for pumps and compressors. *Chem. Engng*, **84**(Dec. 5), 131

KERN, R. (1977d). How the manage plant design to obtain minimum cost. *Chem. Engng*, **84**(May 23), 130

KERN, R. (1977e). Layout arrangements for distillation columns. *Chem. Engng*, **84**(Aug. 15), 153

KERN, R. (1977f). Pressure relief valves for process plants. *Chem. Engng*, **84**(Feb. 28), 187

KERN, R. (1977g). Specifications are the key to successful plant design. *Chem. Engng*, **84**(Jul. 4), 123

KERN, R. (1978a). Arranging the housed chemical process plant. *Chem. Engng*, **85**(Jul. 17), 123

KERN, R. (1978b). Controlling the cost factors in plant design. *Chem. Engng*, **85**(Aug. 14), 141

KERN, R. (1978c). How to arrange the plot plan for process plants. *Chem. Engng*, **85**(May 8), 191

KERN, R. (1978d). Instrument arrangements for ease of maintenance and convenient operation. *Chem. Engng*, **85**(Apr. 10), 127

KERN, R. (1978e). Piperack design for process plants. *Chem. Engng*, **85**(Jan. 30), 105

KERN, R. (1978f). Space requirements and layout for process furnaces. *Chem. Engng*, **85**(Feb. 27), 117

KERNS, G.D. (1972). How to check process designs. *Hydrocarbon Process.*, **51**(1), 100

KERR, A.G. and BYRNE, J.E.T. (1975a). Blast injuries of the ear. *Br. Med. J.*, Mar. 8, 559

KERR, A.G. and BYRNE, J.E.T. (1975b). Concussive effects of bomb blast on the ear. *J. Laryngology Otology*, **89**, 131

KERR, H.W. (1976). A review of factors affecting toughness in welded steels. *Int. J. Pres. Ves. Piping*, **4**, 119

KERR, W. (1957). Complementary theories of safety psychology. *J. Soc. Psychol.*, **45**, 3

KERRIDGE, A.E. (1978). Improve CPM milestone networks. *Hydrocarbon Process.*, **57**(9), 277

KERRIDGE, A.E. (1979). Check project progress with Bell and 'S' curves. *Hydrocarbon Process.*, **58**(3), 189

KERRIDGE, A.E. (1981). When you initiate a project . . . *Hydrocarbon Process.*, **60**(12), 81

KERRIDGE, A.E. (1987). Operators can use expert systems. *Hydrocarbon Process.*, **66**(9), 97

KERRIDGE, A.E. (1992). Hard-dollar estimating. *Hydrocarbon Process.*, **71**(2), 135; **71**(3), 141

KERRIDGE, A.E. (1993). Plan for constructability. *Hydrocarbon Process.*, **72**(1), 135; **72**(2), 151

KERRY, F.G. (1985). Cryogenic control-valve selection is exacting. *Oil Gas J.*, **83**(Dec. 16), 96

KERRY, F.G. and HUGILL, J.T. (1961). Impurity control in air separation plants. *Ammonia Plant Safety*, **3**, 1. See also *Chem. Engng Prog.*, 1961, **57**(4), 37

KERSCHBAUM (1920). Die Gaskampfmittel. In Schwarte, M. (1920), *op. cit.*

KERSE, C.S. (1975). *The Law Relating to Noise* (London: Oyez)

KERSHAW, G.R. (1969). Commercial occupations. In Handley, W. (1969), *op. cit.*, p. 245

KERSHAW, G.R. and SYMONS, W.G. (1977). Commercial occupations. In Handley, W. (1977), *op. cit.*, p. 280

KERSHAW, R. and CULLEN, J.G. (1989). Ammonia plant prolonged operation between major overhauls. *Ammonia Plant Safety*, **29**, p. 49. See also *Plant/Operations Prog.*, 1989, **8**, 73

KERSKEN, M. and EHRENBERGER, W. (1981). A statistical assessment of reliability features of diverse programs. *Reliab. Engng*, **2**(3), 233

KESLER, M.G. (1988). Retrofitting refinery and petrochemical plants. *Chem. Engng Prog.*, **84**(6), 59

KESSEL, C.J. and WICKENS, C.D. (1982). The transfer of failure detection skills between monitoring and controlling dynamic systems. *Hum. Factors*, **24**, 49

KESSLER, D. (1986). Development of hazardous materials truck routes in the Dallas-Fort Worth metropolitan area. *Am. Inst. Chem. Engrs, Symp on Loss Prevention*

KETCHUM, B.H., KESTER, D.R. and PARK, P.K. (eds) (1981). *Ocean Dumping of Industrial Wastes* (New York: Plenum Press)

KETRON, P.H. (1980). Reliability of backup agitation for vessels containing RDX slurry. *Loss Prevention*, **13**, 154

KETTERINGHAM, P.J.A. and O'BRIEN, D.D. (1974). Simulation study of computer-aided soaking pit scheduling. In Edwards, E. and Lees, F.P. (1974), *op. cit.*, p. 260

KETTERINGHAM, P.J.A., O'BRIEN, D.D. and COLE, P.G. (1970). A computer-based interactive display system to aid steel plant scheduling. *Man–Computer Interaction* (London: Instn Elec. Engrs), p. 35

KEULEGAN, G.H. (1957). *An Experimental Study of the Motion of Saline Water from Locks into Fresh Water Channels. Rep. 5168.* Nat. Bur. Standards

KEVIL, C.G. (1967). Flare line rupture in an ethylene plant. *Ammonia Plant Safety*, **9**, p. 37

KEVORKOV, L.R., LUTOVINOV, S.Z. and TIKHONENKO, L.K. (1977). Influence of scale factors on the critical discharge of saturated water from straight tubes with a sharp inlet edge. *Teploenergetika*, **24**(7), 72

KEY, A. (chrmn) (1970). *Disposal of Solid Toxic Wastes* (London: Min. of Housing and Local Govt)

KEY, M. (1992). VOCs: vogue pollutants? *Chem. Engr, Lond.*, **518**, 3

KEYMED INDUSTRIAL (1983). *Guide to Visual Inspection of Plant* (Southend)

KEYWORTH, C.J., SMITH, D.G. and ARCHER, H.E. (1992). Emergency notification under SARA Title III: impacts on facility emergency planning. *J. Haz. Materials*, **31**, 241

KHADKE, H.Y., CHIDAMBARAM, M. and GOPALAN, M.N. (1989). Optimal redundancy allocation of a boiler-feedwater treatment plant. *J. Loss Prev. Process Ind.*, **2**(3), 171

KHADRA, F.F. (1980). Urea reactor bottom failure. *Ammonia Plant Safety*, **22**, p. 69

KHAJEHNAJAFI, S. and SHINDE, A. (1994). Prediction of discharge rate from pressurized vessel blowdown through sheared pipe. *Process Safety Prog.*, **13**, 75

KHALIL, M.F. (1983). Laminar outward flow of wet steam between two discs. In Veziroglu, T.N. (1983), *op. cit.*, p. 66

KHAN, A.A. (1990). Risk analysis of an LPG storage facility in India. *J. Loss Prev. Process Ind.*, **3**(4), 406

KHAN, A.R. and HUNT, A. (1989). The propagation of faults in process plants: integration of fault propagation technology into computer-aided design. *Computer Aided Process Engineering* (Rugby: Instn Chem. Engrs), p. 35

KHAN, F.I. and AMYOTTE, P.R. (2002). Inherent safety in offshore oil and gas activities: A Review of the present status and future directions. *J. of Loss Prev. in the Process Ind.* **15** (4), (July), 279–90

KHAN, F.I. and AMYOTTE, P.R. (2003a). Integrated inherent safety index (I2SI): A tool for inherent safety evaluation. *Proc. of the 37th Annu Loss Prev. Symp.*, March 31-April 3, 2003, New Orleans, LA (ed. J. F. Murphy and R.W. Johnson), Paper LPS-4d (New York: American Institute of Chemical Engineers)

KHAN, F.I. and AMYOTTE, P.R. (2003b). How to make inherent safety practice a reality. *Canadian J. of Chem. Eng.* **81**, 2–16

KHAN, F.I., SADIQ, R. and AMYOTTE, P.R. (2003). Evaluation of available indices for inherently safer design options. *Process Saf Prog.* **22** (2) (Junc), 83–97

KHAN, R.I. (1957). *The Dynamics of Interviewing* (New York: Wiley)

KHAN, S.U. (1980). Role of humic substances in predicting fate and transport of pollutants in the environment. In Haque, R. (1980), *op. cit.*, p. 215

KHANDELWAL, P.K. (1994). Use restriction orifices to maintain required flow rates. *Chem. Engng Prog.*, **90**(3), 34

KHARBANDA, O. (1985). Another Bhopal? Never again. *Chem. Engr, Lond.*, **416**, 40

KHARBANDA, O. (1988). Bhopal (letter). *Chem. Engr, Lond.*, **451**, 4

KHARBANDA, O.P. and STALLWORTHY, E.A. (1982). *How to Learn from Project Disasters* (Aldershot: Gower)

KHARBANDA, O.P. and STALLWORTHY, E.A. (1985). *Effective Project Cost Control* (Rugby: Instn Chem. Engrs)

KHARBANDA, O.P. and STALLWORTHY, E.A. (1986a). *A Guide to Project Implementation* (Rugby: Instn Chem. Engrs)

KHARBANDA, O.P. and STALLWORTHY, E. (1986b). *Management Disasters and How to Prevent Them* (Aldershot: Gower)

KHARBANDA, O.P. and STALLWORTHY, E.A. (1988). *Safety in the Chemical Industry: Lessons from Major Disasters* (London: Heinemann)

KHARBANDA, O.P. and STALLWORTHY, E.A. (1988). *Safety in the Chemical Industry.* (London: Heinemann Professional Publishing, Ltd.), Chapters 25 and 26

KHARBANDA, O.P. and STALLWORTHY, E.A. (1989). Planning for emergencies – lessons from the chemical industry. *Long Range Planning*, **22**, 83

KHATTAK, M.N. and WHITMORE, F.C. (1978). The Kepone incineration test, protocol, sampling and analysis and results. *Control of Hazardous Material Spills 4*, p. 268

KHENAT, B. and HASNI, T. (1977). The first years of operation of the Skikda LNG plant with a discussion of mercury corrosion of aluminium cryogenic exchangers. *LNG Conf.*, **5**

KHERADMAND-FARD, M. (1971). Intrinsically safe power supplies protected by zener diodes. *Electrical Safety*, **1**, 172

KHINNAVAR, R.S., AMINABHAVI, T.M., BALUNDGI, R.H., KUTAC, A., and SHUKLA, S.S. (1991). Resistance of barrier elastomers to hazardous organic liquids. *J. Haz. Materials*, **28**, 295

KHITRIN, L.N. and GOLDENBERG, S.A. (1957). Thermal theory of ignition of gas mixtures: limiting conditions. *Combustion*, **6**, 545

KHITRIN, L.N., MOIN, P.B., SMIRNOV, B.B. and SHEVCHUK, V.U. (1965). Peculiarities of laminar- and turbulent-flame flashbacks. *Combustion*, **10**, 1285

KHODA, T. and HENLEY, E.J. (1988). On digraphs, fault trees and cut sets. *Reliab. Engng*, **20**, 35

KHOGALI, M. (1977). Occupational lung diseases. *Chem. Engr, Lond.*, **323**, 571

KHOO, B.C. and CHEW, T.C. (1993). Transport of volatile pollutants across air–sea interface. *Health, Safety and Loss Prevention*, p. 347

KIANG, Y.H. (1976). Controlling vinyl chloride emissions. *Chem. Engng Prog.*, **72**(12), 37

KIANG, Y.H. and METRY, A.A. (1982). *Hazardous Waste Processing Technology* (Ann Arbor, MI: Ann Arbor Science)

KIDD, A.L. (ed.) (1987). *Knowledge Acquisition for Expert Systems* (New York: Plenum Press)

KIEFER, P.J., KINNEY, G.F. and STUART, M.C. (1954). *Principles of Engineering Thermodynamics* (New York: Wiley)

KIEFNER, J.F. and FORTE, T.C. (1985). Hydrostatic retesting. *Oil Gas J.*, **83**(Jan. 7), 93; (Jan. 14), 83

KIEFNER, J.F., MAXEY, W A., EIBER, R.J. and DUFFY, A.R. (1973). Failure stress levels of flaws in pressurized cylinders. *Progress in Flaw Growth and Fracture Toughness Testing. Pub. STP 536* (Washington, DC: Am. Soc. Testing and Materials), p. 461

KIELMAN, J.K. and FAWCETT, H.H. (1982). Radiation hazards: their prevention and control. In Fawcett, H.H. and Wood, W.S. (1982), *op. cit.*, p. 837

KIER, B. and MÜLLER, G. (1983). *Handbuch Störfalle* (Berlin: Erich Schmidt)

KIES, F.K., FRANSON, I.A. and COAD, B. (1970). Finding new uses for ferritic stainless steel. *Chem. Engr*, **77**(Mar. 23), 150

KIGUCHI, I., KUMAZAWA, T. and NAKAI, T. (1976). For EO: air and oxygen equal. *Hydrocarbon Process.*, **55**(3), 69

KIGUCHI, T. and SHERIDAN, T.B. (1979). Criteria for selecting measures of plant information with application to nuclear reactors. *IEEE Trans. Systems, Man, Cybernetics*, **SMC-9**, 165

KIHARA, H. and IKEDA, K. (1973). Technical application of fracture mechanics to thin-walled structures. *Int. J. Pres. Ves. Piping*, **1**, 317

KILBY, J.L. (1968). Flare system explosions. *Loss Prevention*, **2**, 110. See also *Chem. Engng Prog.*, 1968, **64**(6), 49

KILIAN, D.J. (1982). The relationship between safety management and occupational health programs. In Fawcett, H.H. and Wood, W.S. (1982), *op. cit.*, p. 173

KILLINGTON, A.J. (1983). Impact of legislation on the use and control of lead within the industry. *Chem. Ind.*, Apr. 4, 258

KIMBOROUGH, R.D. *et al.* (1977). Epidemiology and pathology of a tetrachlorodibenzodioxin poisoning episode. *Arch. Environ. Health*, **32**(2), 77

KIME, J.W., BOYLSTON, J.W. and VAN DYKE, J. (1980). The first United States LNG base load trade from Algeria – the Cove Point operation. *Gastech 79*, p. 151

KIMERLE, R.A. (1980). Aquatic hazard evaluation state of the art. In Haque, R. (1980a), *op. cit.*, p. 451

KIMMERLE, G. and EBEN, A. (1964). On the toxicity of methyl isocyanate and its quantitative determination in air. *Arch. Toxikol.*, **29**, 235

KIMMONS, R. (1979). Manage projects for results. *Hydrocarbon Process.*, **58**(4), 245

KINBARA, T. and AKITA, K. (1960). An approximate solution of the equation for self-ignition. *Combust. Flame*, **4**, 173

KINBARA, T., ENDO, H. and SEGA, S. (1966). Combustion propagation through solid materials. I – Downward propagation of smouldering along a thin vertical sheet of paper. *Combust. Flame*, **10**, 29

KINCHIN, G.H. (1982). The concept of risk. In Green, A.E. (1982b), *op. cit.*, p. 3

KING, B.G. (1949). High concentration-short time exposure and toxicity. *J. Ind. Hyg. Toxicol.*, **31**, 365

KING, C.F. and RUDD, D.F. (1972). Design and maintenance of economically failure-tolerant processes. *AIChE J.*, **18**, 257

KING, C.J. (1980). *Separation Processes*, 2nd ed. (New York: McGraw-Hill)

KING, C.J. (1993). Practical fire risk management. In Melchers, R.E. and Stewart, M.G. (1993), *op. cit.*, p. 227

KING, C.J. and ALEXANDER, J.M. (1992). Fire risk management planning. *CHEMECA 92*, p. 11–2

KING, D. (1973). Logic diagrams are easy. *Chem. Engng*, **80**(Aug. 20), 154

KING, D.G. (1975). Total accident control. In Hendry, W.D. (1975), *op. cit.*, paper B1

KING, D.L., USHIBA, K.K. and WHYTE, T.E. (1982). Make oxygenated chemicals from methanol. *Hydrocarbon Process.*, **61**(11), 131

KING, E. (1983). Assessment of occupational exposure to lead. *Chem. Ind.*, Apr. 4, 261

KING, G.G. (1984). Equation predicts buried pipeline temperatures. *Oil Gas J.*, **82**(Mar. 16), 65

KING, I.J. (1980). Noise and vibration. In Waldron, H.A. and Harrington, J.M. (1980), *op. cit.*, p. 144

KING, J. (1992). Management responsibility for offshore safety. *Major Hazards Onshore and Offshore*, p. 27

KING, J.C. (1988). Some measurements of turbulence length scale in the stably stratified surface layer. In Puttock, J.S. (1988b), *op. cit.*, p. 39

KING, J.R. (1971). *Probability Charts for Decision-Making* (London: Industrial Press)

KING, M. (1990). Technologies for controlling the emission of volatile organic compounds to the atmosphere. *Process Safety Environ.*, **68B**, 207

KING, N.J. and HINCHCLIFFE, P.R. (1981). The assessment of the environmental hazards posed by chemicals. *Chem. Ind.*, Sep. 5, 585

KING, N.K. (1973). The influence of water vapour on the emission spectra of flames. *Combust. Sci. Technol.*, **6**, 247

KING, R. (1975a). Flixborough – the role of water reexamined. *Process Engng*, Sep., 69

KING, R. (1975b). Major fire and explosion hazards in hydrocarbon processing plants. *Protection*, **12**(9), Nov. 11

KING, R. (1975c). A mechanism for transient pressure increase. *Instn Chem. Engrs, Symp. on Technical Lessons of Flixborough* (London)

KING, R. (1976a). Accidents caused by water in process plants. *Ind. Safety*, Jan., 6

KING, R. (1976b). Danger – beware of water. *Fire*, Feb., 464

KING, R. (1977). Latent superheat – a hazard of two phase liquid systems. *Chemical Process Hazards*, **VI**, p. 139

KING, R. (1983). Tackling fluids handling problems for the process industries. *Chem. Engr, Lond.*, **395**, 40

KING, R. (1985). Controlling major hazards – will CIMAH do the trick? *Process Engng*, May, 61

KING, R. (1986). New materials for old problems. *Chem. Engr, Lond.*, **425**, 17

KING, R. (1990). *Safety in the Process Industries* (London: Butterworth-Heinemann)

KING, R. (1991). Safety relief (letter). *Chem. Engr, Lond.*, **496**, 6

KING, R. and GILLES, E.D. (1986). Early detection of hazardous states in chemical reactors with model-based measuring techniques. *Loss Prevention and Safety Promotion 5*, p. 40.1

KING, R. and MAGID, J. (1979). *Industrial Hazard and Safety Handbook* (London: Newnes-Butterworths)

KING, R. and TAYLOR, M. (1977). Post accident investigation – in the aftermath of a catastrophe. *Chemical Engineering in a Hostile World* (London: Clapp & Poliak)

KING, R.A. (1986). Expert systems for materials selection and corrosion. *Chem. Engr, Lond.*, **431**, 42

KING, R.C. and CRACKER, S. (1967). *Piping Handbook*, 5th ed. (New York: McGraw-Hill)

KING, R.W. (1976). The dangers of sudden boiling of superheated liquids. *Fire Prev. Sci. Technol.*, **15**, 17

KING, R.W. (1984). Public inquiries (letter). *Chem. Engr, Lond.*, **406**, 3

KING, S.D.M. (1960). *Vocational Training in View of Technological Change. EPA Proj. 418*. Eur. Productivity Agency, Paris

KING, S.D.M. (1964). *Training within the Organisation* (London: Tavistock)

KING, W.R. (1965). The operator's nightmare. *Chem. Engng Prog.*, **61**(9), 28

KINGDON, R.S. (1969). Industrial training. In Handley, W. (1969), *op. cit.*, p. 3

KINGERY, C.N. (1966). *Air Blast Parameters versus Distance for Hemispherical TNT Surface Bursts. Rep. BRL 1344*. US Army, Ballistic Res. Lab., Aberdeen Proving Ground, MD

KINGERY, C.N. and BULMASH, G. (1984). *Air Blast Parameters from TNT Spherical Air Burst and Hemispherical Surface Burst. Tech. Rep. ARBRL-TR 02555*. US Army, Ballistic Res. Lab., Aberdeen Proving Ground, MD

KINGERY, C.N. and PANNILL, B.F. (1964). *Peak Overpressure versus Scaled Distance for TNT Surface Bursts (Hemispherical Charges). BRL Memo Rep. 1518*. US Army, Ballistic Res. Lab., Aberdeen Proving Ground, MD

KINGMAN, F.E.T., COLEMAN, E.H. and ROGOWSKI, Z.W. (1952). The ignition of inflammable gases from sparks from aluminium, paint and rusty steel. *J. Appl. Chem.*, **2**, 449

KINGSLEY, R.J. (1990). (Foreword). In Horsley, D.C.M. and Parkinson, J.S. (1990), *op. cit.*

KINGSLEY, R.J., KNEALE, M. and SCHWARTZ, E. (1968–69). Commissioning of medium-scale process plants. *Proc. Instn. Mech. Engrs*, **183**(pt 1), 205

KINGSHOTT, P.W. (1988). Boiler blowdown vessels. *Plant Engr*, **32**(3), 19

KINKEAD, A.K. (1982). Some aspects of redundancy in relationship to structural reliability. *Reliab. Engng*, **3**(5), 363

KINKEAD, A.N. (1983). Cargo protection afforded by hull structure of LPG carriers during grounding. In *Loss Prevention and Safety Promotion 4*, vol. 2, p. D1

KINLOCH, D.S. (1989). *Piper Alpha Public Inquiry. Transcript, Days 147*, 153

KINNEBROCK, W. (1980). Notes on experiments about the behaviour of liquid gases during release. In Hartwig, S. (1980), *op. cit.*, p. 173

KINNERSLEY, P. (1973). *The Hazards of Work* (London: Pluto Press)

KINNERSLEY, P. (1975). What really happened at Flixborough? *New Scientist*, **65**(938), 520

KINNERSLEY, P. (1976). Asbestos: unsafe at any concentration. *New Civil Engr*, May 20, 32; Jun. 17, 32

KINNEY, G.F. (1962). *Explosive Shocks in Air* (New York: Macmillan)

KINNEY, G.F. (1968). *Engineering Elements of Explosions. Rep. NWC TP-4654*. Naval Weapons Res. Center

KINNEY, G.F. and GRAHAM, K.J. (1985). *Explosive Shocks in Air*, 2nd ed. (Berlin: Springer)

KINSELLA, K. (1993). A rapid assessment methodology for the prediction of vapour cloud explosion overpressure. *Health, Safety and Loss Prevention*, p. 200

KINSLER, L.E. and FREY, A.R. (1972). *Fundamentals of Acoustics*, 2nd ed. (New York: Wiley)

KINSLEY, G.R. (1978). Controlling cross-connections keeps plant drinking water pure. *Chem. Engng*, **85**(Apr. 10), 121

KINSLEY, G.R. (1979). Specifying sound levels for new equipment. *Chem. Engng*, **86**(Jun. 18), 106

KINSLEY, G.R. (1980). Machine guarding: a key safety requirement. *Chem. Engng*, **87**(Feb. 25), 125

KINSLOW, R. (ed.) (1970). *High Velocity Impact Phenomena* (New York: Academic Press)

KINSMAN, P., WILDAY, J., NUSSEY, C. and MERCER, A. (1994). Users' views of model performance and expectations. *J. Loss Prev. Process Ind.*, **7**, 139

KINTOFF, W. (1939). *School Experiments in Chemical Warfare* (London: Massie)

KIPIN, P. (1985). Cleaning pipelines: a pigging primer. *Chem. Engng*, **92**(Feb. 4), 53

KIRBY, B.R. and PRESTON, R.R. (1988). High temperature properties of hot-rolled structural steels for use in fire engineering design studies. *Fire Safety J.*, **13**, 27

KIRBY, D.C.A. and DEROO, J.L. (1984). Water spray protection for a chemical processing unit: one company's view. *Plant/Operations Prog.*, **3**, 254

KIRBY, G.N. (1982). What you need to know about maintenance paints. *Chem. Engng*, **89**(Jul. 24), 75; (Aug. 23), 113

KIRBY, G.N. (1985a). Explosion pressure shock resistance. *Chem. Engng Prog.*, **81**(11), 48

KIRBY, G.N. (1985b). Selecting alloys for chloride service. *Chem. Engng*, **92**(Feb. 4), 81; (Mar. 4), 99

KIRBY, G.N. and SIWEK, R. (1986). Preventing failures of equipment subject to explosions. *Chem. Engng*, **93**(Jun. 23), 125

KIRBY, W.A. and WHEELER, R.V. (1931a). Explosions in closed cylinders. Pt IV, Correlation of flame movement and pressure development in methane-air explosions. *J. Chem. Soc.*, 847

KIRBY, W.A. and WHEELER, R.V. (1931b). Explosions in closed cylinders. Part V, The effect of restrictions. *J. Chem. Soc*, 2203

KIRCH, L.S., KARGOL, J.A., MAGEE, J.W. and STUPER, W.S. (1988). Stability of acrylic monomers. *Plant/Operations Prog.*, **7**, 270

KIRCH, L.S., MAGEE, J.W. and STUPER, W.W. (1987). Applying calorimetry to polymerization. *Chem. Engng. Prog.*, **83**(6), 22

KIRCH, L.S., MAGEE, J.W. and STUPER, W.W. (1990). Application of the DIERS methodology to the study of runaway polymerization. *Plant/Operations Prog.*, **9**, 11

KIRCHNER, R.W. (1967). Equipment testing: are you getting what you pay for? *Chem. Engng*, **74**(Aug. 28), 107

KIRCHOFF, E.O., NOVITSKY, G.H., O'CONNOR, P.J. and WILSON, J.D. (1983). Design and safe operation of a multifunctional monomer production facility. *Plant/Operations Prog.*, **2**, 238

KIRK, J. and TAYLOR, R.S. (1971). Design for safety of gas-cooled reactors. *Proc. Fourth Int. Conf. on Peaceful Uses of Atomic Energy* (New York: United Nations and Vienna: Int. Atom. Energy Agency), p. 151

KIRK, J.C. (1967). How water laws affect the HPI. *Hydrocarbon Process.*, **46**(7), 107

KIRK, P.L. (1969). *Fire Investigation* (New York: Wiley)

KIRK, R.E. and OTHMER, D.F. (eds) (1963–). *Encyclopaedia of Chemical Technology*, 2nd ed. (New York: Wiley-Interscience)

KIRK, R.E. and OTHMER, D.F. (eds) (1978–). *Encyclopaedia of Chemical Technology*, 3rd ed. (New York: Wiley-Interscience)

KIRK, R.E. and OTHMER, D.F. (eds) (1991–). *Kirk–Othmer Encyclopaedia of Chemical Technology*, 4th ed. (New York: Wiley)

KIRKHAM, D. (1987). Fire protection for cable tunnels. *Fire Surveyor*, **16**(5), 12

KIRKLAND, W. (1986). Cleaning in place. *Chem. Engr, Lond.*, **429**, 35

KIRKPATRICK, H.L. (1987). Program designs and prices vessels. *Hydrocarbon Process.*, **66**(3), 41

KIRKWOOD, J.G. and BRINCKLEY, S.R. (1945). *Theory of the Propagation of Shockwaves from Explosive Sources in Air and Water. OSRD Rep. 4814*

KIRKWOOD, R.L., LOCKE, M.H. and DOUGLAS, J.M. (1988). A prototype expert system for synthesizing chemical process flowsheets. *Comput. Chem. Engng*, **12**, 329

KIRLAN, P. (1977). Understanding flexible rotor critical speeds. *Hydrocarbon Process.*, **56**(5), 209

KIRLIK, A. (1993). Modeling strategic behavior in human–automation interaction: why an 'aid' can (and should) go unused. *Hum. Factors*, **35**, 221

KIRSCH, E. (1979). New developments in work safety law in respect of state supervision. In BASF (1979), *op. cit.*, p. 49

KIRSCH, M., VROLYK, J.J., MELVOLD, R.W. and LAFORNARA, J.P. (1976). A hazardous material spills warning system. *Control of Hazardous Material Spills 3*, p. 84

KIRSCH, M., MELVOLD, R.W., VROLYK, J.J. and LAFORNARA, J.P. (1978). A hazardous materials spill warning system. *Control of Hazardous Material Spills 4*, p. 84

KIRSCH-VOLDERS, M. (ed.) (1984). *Mutagenicity, Carcinogenicity and Teratogenicity of Industrial Pollutants* (New York: Plenum Press)

KIRSCHNER, M.J. (1991). Oxygen. *Ullmanns Encyclopaedia of Chemical Technology*, rev. ed., vol. A18 (Weinheim: VCH), p. 329

KIRVEN, J.B. and HANDKE, D.P. (1973). Plan safety for new refining facilities. *Hydrocarbon Process.*, **52**(6), 121

KIRWAN, B. (1988). A comparative evaluation of five human reliability assessment techniques. In Sayers, B.A. (1988), *op. cit.*, p. 87

KIRWAN, B. (1990). Human reliability assessment. In Wilson, J.R. and Corlett, E.N. (1990), *op. cit.*, p. 706

KIRWAN, B. and AINSWORTH, L.K. (eds) (1992). *A Guide to Task Analysis* (London: Taylor & Francis)

KIRWAN, B., MARTIN, B., RYCRAFT, H. and SMITH, A. (1990). Human error data collection and data generation. *Int. J. Qual. Reliab. Mngmt*, **7**(4), 34

KISTER, H.Z. (1979). When tower start-up has problems. *Hydrocarbon Process.*, **58**(2), 89

KISTER, H.Z. (1981a). How to prepare and test columns before start-up. *Chem. Engng*, **88**(Apr. 6), 97

KISTER, H.Z. (1981b). Inspection assures trouble-free operation. *Chem. Engng*, **88**(Feb. 9), 107

KISTER, H.Z. (1990). *Distillation Operation* (New York: McGraw-Hill)

KISTER, H.Z. (1992). *Distillation Design* (New York: McGraw-Hill)

KISTER, H.Z. and HOWER, T.C. (1987). Unusual operating histories of gas processing and olefins plant columns. *Plant/Operations Prog.*, **6**, 151

KISTER, H.Z. and TOWNSEND, R.W. (1984). Ethylene from NGL feedstocks, Pt 4. *Hydrocarbon Process.*, **63**(1), 105

KISTIAKOWSKY, G.B. (1949). Initiation of detonation of explosives. *Combustion*, **3**, p. 560

KITAGAWA, T. (1975). Presumption of causes of explosions in very large tankers. *Transport of Hazardous Cargoes*, **4**

KITAMURA, K., OKUMOTO, Y. and SHIBUE, T. (1978). Model testing of the strength of double bottom structures in grounding. *J. Soc. Naval Architects., Japan*, **143**, 356

KITAMURA, Y., MORIMITSU, H. and TAKAHASHI, T. (1986). Critical superheat for flashing of superheated liquid jets. *Ind. Engng Chem. Fund.*, **25**, 206

KITCHIN, J.B. and GRAHAM, A. (1961). Mental loading of process operators: an attempt to devise a method of analysis and assessment. *Ergonomics*, **4**, 1. See also Edwards, E. and Lees, F.P. (1974), *op. cit.*, p. 105

KITSON, J.C. and GUY, R. (1980). A guide to sprinkler pumps performance and testing. *Fire Surveyor*, **9**(3), 20

KITTLEMAN, T.A. and AKELL, R.B. (1978). The cost of controlling organic emissions. *Chem. Engng*, **74**(4), 87

KIVEN, J. (1966). Controlling thermal expansion in piping systems. *Chem. Engng*, **73**(Feb. 14), 153

KIVENSON, G. (1971). *Durability and Reliability in Engineering Design* (New York: Hayden)

KIWAN, A.R. (1970a). *Gas Flow During and After the Deflagration of a Spherical Cloud of Fuel-Air Mixture. Rep. 1511.* Ballistics Res. Lab., Aberdeen Proving Ground, MD

KIWAN, A.R. (1970b). *Self-similar Flow Outside an Expanding Sphere. Rep. 1495.* Ballistics Res. Lab., Aberdeen Proving Ground, MD

KIWAN, A.R. (1971). *FAE Flow Computations Using AFAMF Code. Rep. 1547.* Ballistics Res. Lab., Aberdeen Proving Ground, MD

KJELLEN, U. (1983). *The Deviation Concept in Occupational Accidents Control: Theory and Method. Report.* Occup. Accid. Group, R. Technol. Univ, Stockholm

KJELLEN, U. (ed.) (1984). *Occupational Accident Research* (Amsterdam: Elsevier)

KLAASSEN, C.D., AMDUR, M. and DOULL, J. (1986). *Casarett and Doull's Toxicology*, 3rd ed. (New York: Macmillan)

KLAASSEN, K.B. (1984). *Reliability of Analogue Electronic Systems* (Amsterdam: Elsevier)

KLAASSEN, P. (1969). Safety aspects in process design. *Symp. on Inherent Hazards of Manufacturing and Storage in the Process Industry*, The Hague, p. 135

KLAASSEN, P. (1971). Loss prevention aspects in process plant design. *Major Loss Prevention*, p. 111

KLAASSEN, P. (1979). Importance of software/hardware for safe processing. *Loss Prevention*, **12**, p. 118

KLAASSEN, P. (1980a). Vessel rupture in a hot oil heating system of a chemical plant. *Loss Prevention and Safety Promotion 3*, vol. 4, p. 1511

KLAASSEN, P. (1980b). Water drainage from vessels containing flashing hydrocarbons. *Loss Prevention and Safety Promotion 3*, vol. 4, p. 1513

KLAIS, O. (1987). Methane formation during hydrogenation reactions in the presence of Raney nickel catalyst. *Hazards from Pressure*, p. 25

KLAIS, O. and GREWER, T. (1983). Pressure increase in exothermic decomposition reactions – Part II. *Loss Prevention and Safety Promotion 4*, vol. 3, p. C24

KLAASSEN, M., GORE, J.P., SIVATHANU, Y.R., HAMINS, A. and KASHIWAGI, T. (1992). Radiative heat feedback in a toluene pool fire. *Combustion*, **24**, 1713

KLAUSER, W. (1980). CO_2 in the atmosphere – its risk and behaviour during release from the point of view of industry. In Hartwig, S. (1980), *op. cit.*, 255

KLAVER, R.F. (1973–74). *API Statics Res. Program, Pt 1* (Am. Petrol. Inst.)

KLEE, A.J. (1972). Utilization of expert opinion in decision-making. *AIChE J.*, **18**, 1107

KLEENE, S.C. (1967). *Mathematical Logic* (New York: Wiley)

DE KLEER, J. (1975). *Qualitative and Quantitative Knowledge in Classical Mechanics. Rep. TR-352.* AI Laboratory, MIT, Cambridge, MA

DE KLEER, J. (1984). How circuits work. *Artificial Intelligence*, **24**, 205

DE KLEER, J. (1985). How circuits work. In Bobrow, D.G. (1985) *op. cit.*

DE KLEER, J. (1992). Physics, qualitative. In Shapiro, S.C. (1992) *op. cit.*, p. 1149

DE KLEER, J. and BROWN, J.S. (1983). Assumptions and ambiguities in mechanistic mental models. In Gentner, D. and Stevens, A.L. (1983), *op. cit.*, p. 155

DE KLEER, J. and BROWN, J.S. (1984). A qualitative physics based on confluences. *Artificial Intelligence*, **24**, 7

DE KLEER, J. and BROWN, J.S. (1986). Theories of causal ordering. *Artificial Intelligence*, **29**, 33

KLEIN, H.H. (1986). Analysis of DIERS venting tests: validation of a tool for sizing emergency relief systems for runaway chemical reactions. *Plant/Operations Prog.*, **5**, 1

KLEIN, H.H. (1987). A computer model for sizing two-phase emergency relief systems, validated against DIERS large scale test data. *Hazards from Pressure*, p. 143

KLEIN, R.A. (1991a). Achieve total quality management. *Chem. Engng Prog.*, **87**(11), 83

KLEIN, R.A. (1991b). For 'quality' performance reviews... *Hydrocarbon Process.*, **70**(9), 193

KLEIN, R.C. (1993). Employ immersion heaters properly. *Chem. Engng Prog.*, **89**(8), 38

KLEIN, W. (1990). Fire resisting glass can help check flame spread. *Fire Int.*, **122**, 33

KLEINDORFER, P.R. and KUNREUTHER, H.C. (eds) (1987). *Insuring and Managing Hazardous Risk* (Berlin: Springer)

KLEINFELD, M., GIEL, C. and ROSSO, A. (1964). Acute hydrogen sulphide intoxication: an unusual source of exposure. *Ind. Med. Surg.*, **33**, 656

KLEINSTEIN, G. (1964). Mixing in turbulent axially symmetric free jets. *J. Spacecraft*, **1**(4), 403

KLEM, T.J. and SAUNDERS, J.C. (1986). Explosion in cold storage kills fire fighter. *Ammonia Plant Safety*, **26**, 145

KLEMENT, A.W. (ed.) (1982). *CRC Handbook of Environmental Radiation* (Boca Raton, FL: CRC Press)

KLEN, E.F. and GRIER, J.C. (1978). Better water treatment in cooling tower systems. *Ammonia Plant Safety*, **20**, 57

KLESNIL, M. and LUKAS, P. (1980). *Fatigue of Metallic Materials* (Amsterdam: Elsevier)

KLETZ, T.A. (1971). Hazard analysis – a quantitative approach to safety. *Major Loss Prevention*, p. 111

KLETZ, T.A. (1972a). Specifying and designing protective systems. *Loss Prevention*, **6**, 15

KLETZ, T.A. (1972b). In discussion of paper by Levine, R.Y. (1972b), *op. cit.*

KLETZ, T.A. (1972c). In discussion of paper by Ostroot, G. (1972), *op. cit.*

KLETZ, T.A. (1973a). Hazard analysis – a quantitative approach to safety. *Protection*, **10**(8), 19

KLETZ, T.A. (1973b). The uses, availability and pitfalls of data on reliability. *Process Technol.*, **18**(Mar.), 111

KLETZ, T.A. (1974a). Are safety valves old hat? *Chem. Process.*, **20**(9), 77

KLETZ, T.A. (1974b). Case histories in loss prevention. *Loss Prevention*, **8**, 126. See also *Chem. Engng Prog.*, 1974, **70**(4), 80

KLETZ, T.A. (1974c). Hazard analysis – a quantitative approach to safety. *Course on Process Safety – Theory and Practice.* Dept of *Chem. Engng Teesside Polytechnic, Middlesbrough*

KLETZ, T.A. (1974d). Over-pressure and other risks – some myths of the chemical industry. *Loss Prevention and Safety Promotion 1*, p. 309

KLETZ, T.A. (1974e). Protection of pressure vessels against fire. *Fire Prev.*, **103**, 17

KLETZ, T.A. (1975a). Accident data – the need for a new look as the sort of data that are collected and analysed. In Hendry, W.D. (1975), *op. cit.*, paper E2

KLETZ, T.A. (1975b). Emergency isolation valves for chemical plants. *Loss Prevention*, **9**, p. 134. See also *Chem. Engng Prog.*, 1975, **71**(9), 63

KLETZ, T.A. (1975c). The Flixborough cyclohexane disaster. *Loss Prevention*, **9**, 106

KLETZ, T.A. (1975d). Protection of pressure vessels against excessive temperature. Imperial Chemical Industries, London, *Process Safety Guide 2*

KLETZ, T.A. (1975e). Some of the wider questions raised by Flixborough. *Instn Chem. Engrs, Symp. on Technical Lessons of Flixbrough* (London)

KLETZ, T.A. (1976a). Accident data – the need for a new look as the sort of data that are collected and analysed. *J. Occup. Accid.*, **1**(1), 95

KLETZ, T.A. (1976b). Accidents caused by reverse flow. *Hydrocarbon Process.*, **55**(3), 187

KLETZ, T.A. (1976c). Accidents that will occur in the coming year. *Loss Prevention*, **10**, 151

KLETZ, T.A. (1976d). The application of hazard analysis to risks to the public at large. *World Cong. of Chemical Engineering*, Amsterdam

KLETZ, T.A. (1976e). Flixborough – the aftermath. *Prevention of Occupational Risks*, **3**, 289

KLETZ, T.A. (1976f). Preventing catastrophic accidents. *Chem. Engng*, **83**(8), 124

KLETZ, T.A. (1976g). The protection of pressure vessels against fire. *Fire Int.*, **53**, 18

KLETZ, T.A. (1976h). Some fires and explosions in liquids of high flash point. *J. Haz. Materials*, **1**(2), 165

KLETZ, T.A. (1976i). A three-pronged approach to plant modification. *Loss Prevention*, **10**, 91. See also *Chem. Engng Prog.*, 1976, **72**(11), 48

KLETZ, T.A. (1976j). In discussion of paper by Dailey, W.V. (1976), *op. cit.*

KLETZ, T.A. (1977a). The application of hazard analysis to risks to the public at large. *Occup. Safety Health*, **7**(10), 12; **7**(11), 12

KLETZ, T.A. (1977b). Evaluate risk in plant design. *Hydrocarbon Process.*, **56**(5), 297

KLETZ, T.A. (1977c). Explosive hazards in the process industries. *Chem. Australia*, **44**(9), 236

KLETZ, T.A. (1977d). Protect pressure vessels from fire. *Hydrocarbon Process.*, **56**(8), 98

KLETZ, T.A. (1977e). The risk equations. What risks should we run? *New Scientist*, **74**(1051), 320

KLETZ, T.A. (1977f). Risk evaluation in plant design. In *Health and Safety in the Oil Industry* (Heyden)

KLETZ, T.A. (1977g). The safety officer – past and future. *Occup. Safety Health*, **7**(7), 10

KLETZ, T.A. *et al.* (1977h). Safety surveys. *Instn Mech. Engrs, Conf. on Safety Audits – Essential for Every Factory?* (London)

KLETZ, T.A. (1977i). Some myths on hazardous materials. *J. Haz. Materials*, **2**(1), 1

KLETZ, T.A. (1977j). Unconfined vapour cloud explosions. *Loss Prevention*, **11**, 50

KLETZ, T.A. (1977k). What are the causes of change and innovation in safety? *Loss Prevention and Safety Promotion 2*, p. 1

KLETZ, T.A. (1977l). Private communication

KLETZ, T.A. (1977–78). Some myths on hazardous materials. *J. Haz. Materials*, **2**, 1

KLETZ, T.A. (1978). What you don't have, can't leak. *Chem. Ind.* 9124, (May 6), 287–92

KLETZ, T.A. (1978a). Practical applications of hazard analysis. *Chem. Engng Prog.*, **74**(10), 47

KLETZ, T.A. (1978b). Safety – the springs of action and innovation. *Occup. Safety Health*, **8**(3), 10

KLETZ, T.A. (1978c). What you don't have, can't leak. *Chem. Ind.*, May 6, 287

KLETZ, T.A. (1979). Is there a simpler solution? *The Chem. Eng* (March), 161–4

KLETZ, T.A. (1979a). Causes of hydrocarbon oxidation fires. In *Loss Prevention*, **12**, 96

KLETZ, T.A. (1979b). Consider vapor cloud dangers when you build a plant. *Hydrocarbon Processing*, **58**(10), 205

KLETZ, T.A. (1979c). A decade of safety lessons. *Hydrocarbon Processing*, **58**(6), 195

KLETZ, T.A. (1979d). The Health and Safety at Work Act – co-operation versus confrontation. *Instn Works Managers Mtg*, Merseyside

KLETZ, T.A. (1979e). Industrial Safety – The Shaking of the Foundations. Inaugural Lecture, Loughborough Univ. of Technology

KLETZ, T.A. (1979f). Is there a simpler solution? *Chem. Engr, Lond.*, **342**, 161

KLETZ, T.A. (1979g). 'Layered' accident investigation. *Hydrocarbon Processing*, **58**(11), 373

KLETZ, T.A. (1979h). Learn from these HPI plant fires. *Hydrocarbon Processing*, **58**(1), 243

KLETZ, T.A. (1979i). A matter of opinion – the case for safety by consent. *Works Mngmnt*, **32**(7), 89

KLETZ, T.A. (1979j). Practical applications of hazard analysis. *Loss Prevention*, **12**, 34. See also *Chem. Engng Prog.*, 1978, **74**(10), 47

KLETZ, T.A. (1979k). Processing chemicals safely. *Spectrum*, **163**, 11

KLETZ, T.A. (1979l). Reliability analysis and loss prevention. *Chem. Ind.*, Aug. 4, 519

KLETZ, T.A. (1979m). Safety in numbers. *Focus on Engineering* (Cambridge: Hobsons), p. 60

KLETZ, T.A. (1979n). Towards intrinsically safer plants. *Design 79* (London: Instn Chem. Engrs)

KLETZ, T.A. (1980a). Benefits and risks: their assessment in relation to human needs. *Endeavour*, 4(2), 46

KLETZ, T.A. (1980b). Benefits and risks: their assessment in relation to human needs. *Int. Atom. Energy Bull.*, **22**(5/6), 2

KLETZ, T.A. (1980c). *Human Factors in Process Operation and Design. Safety Note 80/11.* ICI, Petrochemicals Div

KLETZ, T.A. (1980d). Industry's responsibilities to the community – do they result in expenditure which is not cost effective? *Conf. on Industry and the Environment*, Centre for Extension Studies, Loughborough Univ. of Technology

KLETZ, T.A. (1980e). The man in the middle – some accidents caused by simple mistakes. *Loss Prevention and Safety Promotion 3*, vol. 2, p. 205

KLETZ, T.A. (1980f). Nitrogen – our most dangerous gas. *Loss Prevention and Safety Promotion 3*, vol. 3, p. 1518

KLETZ, T.A. (1980g). Organisations have no memory. *Loss Prevention*, **13**, 1

KLETZ, T.A. (1980h). Plant layout and location: methods for taking hazardous occurrences into account. *Loss Prevention*, **13**, 147

KLETZ, T.A. (1980i). Safety aspects of pressurised systems. *Fourth Int. Conf. on Pressure Vessel Technology* (London: Instn Mech. Engrs)

KLETZ, T.A. (1980j). Seek intrinsically safer plants. *Hydrocarbon Process.*, **59**(8), 137

KLETZ, T.A. (1980k). The shaking of the foundations. *Occup. Safety Health*, **10**(1), 8; **10**(2), 34

KLETZ, T.A. (1980–). How much safety can we afford? *The Safety Representative*, Dec., 6; Jan., 3

KLETZ, T.A. (1981a). Benefits and risks: their assessment in relation to human needs. In Griffiths, R.F. (1981), *op. cit.*, p. 36

KLETZ, T.A. (1981b). Case histories in transportation. *Loss Prevention*, **14**, 183

KLETZ, T.A. (1981c). Catalogue of disaster. *The Safety Representative*, Oct., 2

KLETZ, T.A. (1981d). Hazard analysis – the manager and the expert. *Reliab. Engng*, **2**, 35

KLETZ, T.A. (1981e). How far should we go in bringing old plants up to modern standards. *Loss Prevention*, **14**, 165

KLETZ, T.A. (1981f). Is plant safety too expensive? *Hydrocarbon Process.*, **60**(12), 171

KLETZ, T.A. (1981g). Minimize plant spills. *Hydrocarbon Process.*, **61**(3), 207

KLETZ, T.A. (1981h). Old plants and modern standards. *Health and Safety at Work*, **3**(6), 56

KLETZ, T.A. (1981i). Plant instruments: which ones don't work and why. *Loss Prevention*, **14**, 108. See also *Chem. Engng Prog.*, 1980, **76**(7), 68

KLETZ, T.A. (1981j). Pollution: does industry care? *Occup. Safety Health*, **11**(5), 20

KLETZ, T.A. (1981k). *Risk. Br. Ass. Advancement of Science*, Ann. Mtg, York

KLETZ, T.A. (1981l). Safety audits and surveys. *Health and Safety at Work*, **4**(3), 20

KLETZ, T.A. (1981m). The risk illiterates. *Occup. Safety Health*, **11**(11), 14

KLETZ, T.A. (1981n). *The Size of Fireballs and BLEVES. Health and Safety Executive Seminar on Thermal Radiation and Overpressure Calculations* (London: Health and Safety Executive)

KLETZ, T.A. (1982a). Accident investigations should inspire prevention. *Health and Safety at Work*, **4**(12), 20

KLETZ, T.A. (1982b). Beware the hazards of new technologies. *Hydrocarbon Process.*, **61**(5), 297

KLETZ, T.A. (1982c). Flame trap assembly for use with high melting point materials. *Plant/Operations Prog.*, **1**, 252

KLETZ, T.A. (1982d). Grey hairs cost nothing. *Universities Safety Ass. Conf.*, Edinburgh

KLETZ, T.A. (1982e). Hazard analysis – a review of criteria. *Reliab. Engng*, **3**(4), 325

KLETZ, T.A. (1982f). Hazards in chemical system maintenance: permits. In Fawcett, H.H. and Wood, W.S. (1982), *op. cit.*, p. 807

KLETZ, T.A. (1982g). Human problems with computer control. *Plant/Operations Prog.*, **1**, 209

KLETZ, T.A. (1982h). *Measurements on Large Pool Fires. Safety Note 82/3*. ICI, Petrochemicals Div.

KLETZ, T.A. (1982i). Minimize your product spillage. *Hydrocarbon Process.*, **61**(3), 207

KLETZ, T.A. (1982j). One organisation's memory – the use of a computerised system to store and retrieve information on loss prevention. *Fifteenth Loss Prevention Symp.* (New York: Am. Inst. Chem. Engrs)

KLETZ, T.A. (1982k). The safety adviser in the high technology industries: training and role. *Developments 82* (Rugby: Instn Chem. Engrs)

KLETZ, T.A. (1982l). Three Mile Island: lessons for the HPI. *Hydrocarbon Process.*, **61**(6), 187

KLETZ, T.A. (1983). Inherently safer plant – The concept, Its scope and benefits. *Loss Prev. Bull.* **051**(June), 1–8

KLETZ, T.A. (1983a). After the investigation of the fire. *Fire Prev.*, **162**, Sep., 16

KLETZ, T.A. (1983b). Chemstar lessons. *Health and Safety at Work*, Mar., 7

KLETZ, T.A. (1983c). English/American vocabulary (letter). *Chem. Engr, Lond.*, **391**, 4

KLETZ, T.A. (1983d). *Hazop and Hazan* (Rugby: Instn Chem Engrs)

KLETZ, T.A. (1983e). *How Safe Is It? Maths at Work* (ed. G. Howson and R. McLone) (London: Heinemann), p. 1

KLETZ, T.A. (1983f). Loss prevention – some future problems. *Chem. Engr, Lond.*, **389**, 33

KLETZ, T.A. (1983g). The mathematics of risk. *Interdisciplinary Sci. Revs.*, **8**, 115

KLETZ, T.A. (1983h). Now or later? A numerical comparison of short and long term hazards. *Loss Prevention and Safety Promotion 4*, vol. 1, p. A1. See also *Chem. Engr, Lond.*, **397**, 9

KLETZ, T.A. (1983i). Royal Society assesses risk assessment (book review). *Health and Safety at Work*, **5**(May), 42

KLETZ, T.A. (1983j). Three Mile Island: lessons on human factors. *Phys. Bull.*, **34**, 137

KLETZ, T.A. (1984). *Cheaper, safer plants, or wealth and safety at work*. (Rugby, Warwickshire, England: The Institution of Chemical Engineers), ISBN 0-85295-167-1

KLETZ, T.A. (1984a). Accident investigation: how far should we go? *Plant/Operations Prog.*, **3**, 1

KLETZ, T.A. (1984b). The analysis of hazards and the economy of risk prevention. *CHEMRAWN III*

KLETZ, T.A. (1984c). Anglo-American glossary. *Chem. Engr, Lond.*, **410**, 33

KLETZ, T.A. (1984d). *Cheaper, Safer Plants* (Rugby: Instn Chem Engrs)

KLETZ, T.A. (1984e). Flixborough. The view from ten years on. *Chem. Engr, Lond.*, **404**, 28

KLETZ, T.A. (1984f). The Flixborough explosion – ten years later. *Plant/Operations Prog.*, **3**, 133

KLETZ, T.A. (1984g). Grey hairs cost nothing. *Plant/Operations Prog.*, **3**, 210

KLETZ, T.A. (1984h). Hazardous consequences – prediction and measurement. In Petts, J.I. (1984), *op. cit.*

KLETZ, T.A. (1984i). The industrial approach to hazard identification and analysis and setting criteria for risk. In Petts, J.I. (1984), *op. cit.*

KLETZ, T.A. (1984j). *Myths of the Chemical Industry* (Rugby: Instn Chem Engrs)

KLETZ, T.A. (1984k). The prevention of major leaks – better inspection after construction? *Plant/Operations Prog.*, **3**, 19

KLETZ, T.A. (1984l). Talking about safety. *Chem. Engr, Lond.*, **399**, 33

KLETZ, T.A. (1984m). Talking about safety. *Chem. Engr, Lond.*, **406**, 29

KLETZ, T.A. (1985). Inherently safer plants. *Plant/Operations Prog.* **4** (3) (July), 164–67

KLETZ, T.A. (1985a). An atlas of safety thinking. *Chem. Engr, Lond.*, **414**, 47

KLETZ, T.A. (1985b). Avoid working in flammable atmospheres. *Health and Safety at Work*, Oct., 18

KLETZ, T.A. (1985c). The cost of saving a life. *Chem. Engr, Lond.*, **419**, 57

KLETZ, T.A. (1985d). Does data on reliability tell us about equipment or people? *Reliab. Engng*, **11**, 185

KLETZ, T.A. (1985e). Eliminating potential process hazards. *Chem. Engng*, **92**(Apr. 1), 48

KLETZ, T.A. (1985f). *An Engineer's View of Human Error.* (Rugby: Instn Chem Engrs)

KLETZ, T.A. (1985g). How did it happen? *Occup. Safety Health*, Feb., 25

KLETZ, T.A. (1985h). Inherently safer plants. *Plant/Operations Prog.*, **4**, 164

KLETZ, T.A. (1985i). Knowledge in the wrong place. *Chem. Engr, Lond.*, **417**, 51

KLETZ, T.A. (1985j). Lessons of Bhopal (letter). *The Listener*, Jun. 20, 22

KLETZ, T.A. (1985k). Make plants inherently safe. *Hydrocarbon Process.*, **64**(9), 172

KLETZ, T.A. (1985l). Making a chemical plant safe. *Chem. Britain*, Dec., 1091

KLETZ, T.A. (1985m). What is good practice? *Chem. Engr, Lond.*, **419**, 57

KLETZ, T.A. (1985n). What is a protective system? *Chem. Engr, Lond.*, **416**, 29

KLETZ, T.A. (1985o). What went wrong? case histories. *Hydrocarbon Process.*, **64**(12), 81

KLETZ, T.A. (1985p). *What Went Wrong?* (Houston, TX: Gulf)

KLETZ, T.A. (1985q). Why are fewer accident reports being published? *Chem. Engr, Lond.*, **410**, 39

KLETZ, T.A. (1986a). Accident reports and missing recommendations. *Loss Prevention and Safety Promotion 5*, 20–1

KLETZ, T.A. (1986b). Big losers. *Chem. Engr, Lond.*, **424**, 55

KLETZ, T.A. (1986c). Do engineers believe in Father Christmas? *Chem. Engr, Lond.*, **431**, 46

KLETZ, T.A. (1986d). *Hazop and Hazan*, 2nd ed. (Rugby: Instn Chem Engrs)

KLETZ, T.A. (1986e). Inherent safety and the nuclear industry. *Chem. Engr, Lond.*, **427**, 35

KLETZ, T.A. (1986f). Modification chains. *Plant/Operations Prog.*, **5**, 136

KLETZ, T.A. (1986g). Myths of the chemical industry. *New Zealand Engng*, Apr. 1, 29

KLETZ, T.A. (1986h). Package deals. *Chem. Engr, Lond.*, **421**, 37

KLETZ, T.A. (1986i). Tests should be like real life. *Chem. Engr, Lond.*, **429**, 39

KLETZ, T.A. (1986j). Transportation of hazardous substances: the UK scene. *Plant/Operations Prog.*, **5**, 160

KLETZ, T.A. (1986k). *What Went Wrong?* 2nd ed. (Houston, TX: Gulf)

KLETZ, T.A. (1986l). Will cold petrol explode in the open air? *Chem. Engr, Lond.*, **426**, 63

KLETZ, T.A. (1987a). Another pipe failure. *Chem. Engr, Lond.*, **432**, 32

KLETZ, T.A. (1987b). Artificial stupidity. *Chem. Engr, Lond.*, **432**, 32

KLETZ, T.A. (1987c). The best building has no walls. *Chem. Engr, Lond.*, **435**, 40

KLETZ, T.A. (1987d). Clap hands if your believe that . . . *Chem. Engr, Lond.*, **443**, 44

KLETZ, T.A. (1987e). Correlating causes and effects. *Chem. Engr, Lond.*, **437**, 41

KLETZ, T.A. (1987f). An engineer's approach to human error. *Preventing Major Chemical Accidents*, p. 6.23

KLETZ, T.A. (1987g). Nuclear safety (letter). *Chem. Engr, Lond.*, **432**, 4

KLETZ, T.A. (1987h). Old problems return in cycles. *Chem. Engr, Lond.*, **437**, 41

KLETZ, T.A. (1987i). Optimisation and safety. *Process Optimisation* (Rugby: Instn Chem. Engrs), p. 153

KLETZ, T.A. (1987j). Useless equipment and operations. *Chem. Engr, Lond.*, **435**, 40

KLETZ, T.A. (1988a). Ammonia can explode. *Chem. Engr, Lond.*, **445**, 41

KLETZ, T.A. (1988b). Computers and hazop. *Chem. Engr, Lond.*, **453**, 52

KLETZ, T.A. (1988c). Do you know what you don't know? *Chem. Engr, Lond.*, **448**, 47

KLETZ, T.A. (1988d). Fires and explosions of hydrocarbon oxidation plants. *Plant/Operations Prog.*, **7**, 226

KLETZ, T.A. (1988e). The hazards of insularity. *Chem. Engr, Lond.*, **448**, 47

KLETZ, T.A. (1988f). Indices of woe. *Chem. Engr, Lond.*, **445**, 41

KLETZ, T.A. (1988g). Investment in safety measures (letter). *Chem. Engr, Lond.*, **446**, 4

KLETZ, T.A. (1988h). *Learning from Accidents in Industry* (London: Butterworths)

KLETZ, T.A. (1988i). On the need to publish more case histories. *Plant/Operations Prog.*, **7**, 145

KLETZ, T.A. (1988j). The past and future of loss prevention. *Chem. Engr, Lond.*, **455**, 25

KLETZ, T.A. (1988k). Plants should be friendly. *Chem. Engr, Lond.*, **446**, 35

KLETZ, T.A. (1988l). Should undergraduates be instructed in loss prevention? *Plant/Operations Prog.*, **7**, 95

KLETZ, T.A. (1988m). Some hazards aren't obvious. *Chem. Engr, Lond.*, **451**, 34

KLETZ, T.A. (1988n). *What Went Wrong?* 2nd ed. (Houston, TX: Gulf)

KLETZ, T.A. (1989). Friendly plants. *Chemical Eng. Prog.*, **85** (7) (July), 18–26

KLETZ, T.A. (1989a). Abuse of relief systems. *Selection and Use of Pressure Relief Systems for Process Plant* (Manchester: Instn. Chem. Engrs, North-West Branch), p. 8.1

KLETZ, T.A. (1989b). The BP fires and explosion. *Chem. Engr, Lond.*, **463**, 43

KLETZ, T.A. (1989c). Friendly plants. *Chem. Engng Prog.*, **85**(7), 18

KLETZ, T.A. (1989d). Good safety procedures can prevent accidents: some examples. *Plant/Operations Prog.*, **8**, 1

KLETZ, T.A. (1989e). Static pipes and balls (letter). *Chem. Engr, Lond.*, **456**, 4

KLETZ, T.A. (1990). Plants should be friendly. *Saf. and Loss Prev. in the Chem. and Oil Process. Ind.*, Oct.23–27, 1989, Singapore, 423–35. IChemE Symposium Series, no. 120 (Rugby, Warwickshire, U. K.: The Institution of Chemical Engineers)

KLETZ, T.A. (1990a). Consequences and visions. *J. Occup. Accid.*, **13**(1–2), 93

KLETZ, T.A. (1990b). *Critical Aspects of Safety and Loss Prevention* (London: Butterworth)

KLETZ, T.A. (1990c). Ether, protoplasm and human error. *Chem. Engr, Lond.*, **479**, 19

KLETZ, T.A. (1990d). *Improving Chemical Engineering Practices* (New York: Hemisphere) (revision of Kletz (1984j))

KLETZ, T.A. (1990e). Lessons of another ethylene oxide explosion. *Chem. Engr, Lond.*, **469**, 15

KLETZ, T.A. (1990f). The need for friendly plants. *J. Occup. Accid.*, **13**(1–2), 3

KLETZ, T.A. (1990g). *Plant Design for Safety. A User-Friendly Approach.* (New York: Hemisphere) (Revision of Kletz (1984d))

KLETZ, T.A. (1990h). Plants should be friendly. *Safety and Loss Prevention in the Chemical and Oil Processing Industries*, p. 423

KLETZ, T.A. (1991a). Ammonia incidents. *J. Loss Prev. Process Ind.*, **4**, 207

KLETZ, T.A. (1991b). Are managers above the law? *Chem. Engr, Lond.*, **507**, 38

KLETZ, T.A. (1991c). Billiard balls and polo mints. *Chem. Engr, Lond.*, **495**, 21

KLETZ, T.A. (1991d). A change in feed composition can be dangerous. *Chem. Engr, Lond.*, **488**, 28

KLETZ, T.A. (1991e). *An Engineer's View of Human Error*, 2nd ed. (Rugby: Instn Chem. Engrs)

KLETZ, T.A. (1991f). The freeloaders. *Chem. Engr, Lond.*, **504**, 46

KLETZ, T.A. (1991g). Human problems with computer control: an update. *Plant/Operations Prog.*, **10**, 17

KLETZ, T.A. (1991h). Inherently safer plants – recent progress. *Hazards*, **XI**, p. 225

KLETZ, T.A. (1991i). Inherently safer plants: an update. *Plant/Operations Prog.*, **10**, 81

KLETZ, T.A. (1991j). The Phillips explosion – this is where I came in. Chem Engr, Lond., **493**(Mar. 28), 47

KLETZ, T.A. (1991k). Process safety – an engineering achievement. *Proc. Instn Mech. Engrs*, **205**, 11

KLETZ, T.A. (1991l). Reflection on Seveso, 15 years on. *Chem. Engr, Lond.*, **500**, 3

KLETZ, T.A. (1991m). Second class citizens. *Chem. Engr, Lond.*, **498**, 38

KLETZ, T.A. (1991n). Slaves of a lesser god. *Chem. Engr, Lond.*, **500**, 30

KLETZ, T.A. (1991o). Inherently safer plants: An Update. *Plant/Operations Prog.* **10** (2) (April), 81–4

KLETZ, T.A. (1991p). Inherently safer plants – recent progress. *IChemE Symp. Series* No. 124, 225–33

KLETZ, T.A. (1992a). Chemicals with 'use by' dates. *J. Loss Prev. Process Ind.*, **5**, 132

KLETZ, T.A. (1992b). *Hazop and Hazan*, 3rd ed. (Rugby: Instn Chem Engrs)

KLETZ, T.A. (1992c). Inherently safer design – a review. *Loss Prevention and Safety Promotion 7*, vol. 1, lecture

KLETZ, T.A. (1993). The unforeseen side effects of improving the environment, *Process Safety Progress* 12(3), (July) (American Institute of Chemical Engineers)

KLETZ, T.A. (1993a). Computer control – living with human error. *Reliab. Engng System Safety, 39*, 257

KLETZ, T.A. (1993b). *Lessons from Disaster* (Rugby: Instn Chem Engrs)

KLETZ, T.A. (1993c). Organizations have no memory when it comes to safety. *Hydrocarbon Process.*, **72**(6), 88

KLETZ, T.A. (1993d). Process safety: new solutions to old problems. *Health, Safety and Loss Prevention*, p. 1

KLETZ, T.A. (1993e). The unforeseen side-effects of improving the environment. *Process Safety Prog.*, **12**, 147

KLETZ, T.A. (1993f). What can we do to encourage the development of inherntly safer designs. *Safety and Health News*, (Oct.), 1

KLETZ, T.A. (1994). Warning from Accidents, 2nd ed. (Oxford: Butterworth-Heinumann)

KLETZ, T.A. (1995). Inherently safer design – The growth of an idea. *Proc. of the 29th Annu. Loss Prev. Symp.*, July 31–August 2, 1995, Boston, MA, (ed. E. D. Wixom and R. P. Benedetti,) Paper 1a. (New York: American Institute of Chemical Engineers)

KLETZ, T.A. (1996). Inherently safer design: achievements and prospects. *Int. Conf. and Workshop on Process Saf. Manage. and Inherently Safer Processes*, Oct. 8–11, 1996, Orlando, FL, 197–206. (New York: American Institute of Chemical Engineers)

KLETZ, T.A. (1997). Inherently safer design – The growth of an idea. *Conf. on Inherent SHE: The Cost effective route to improved safety, health and environmental performance*, June 16–17, 1997, London (London: IBC Conferences Limited)

KLETZ, T.A. (1998). *Process plants: A handbook for inherently safer design* (Philadelphia, PA: Taylor and Francis)

KLETZ, T.A. (1999a). Safer by design. *chemistry in britain* **35** (1), (January), 37–9

KLETZ, T.A. (1999b). The origins and history of loss prevention. *Trans IChemE* **77**, Part B, (May), 109–16

KLETZ, T.A. (2001). Some Problems and Opportunities that have been overlooked. *Proc. of the Mary Kay O'Connor Process Saf. Cent. Annu. Symp.* College Station, Texas,

KLETZ, T.A. (2002). Personal communications, January and April

KLETZ, T.A. and LAWLEY, H.G. (1982). Safety technology in industry. Chemical. In Green, A.E. (1982b) *op. cit.*, p. 317

KLETZ, T.A. and TURNER, E. (1979). *Is the Number of Serious Accidents in the Oil and Chemical Industries Increasing? Imperial Chemical Industries Paper* (London: Chem. Ind. Ass.)

KLETZ, T.A. and WHITAKER, G.D. (1973). *Human Error and Plant Operation. Rep. EDN 4099.* Safety and Loss Prev. Group, Petrochemicals Div., Imperial Chemical Industries

KLETZ, T.A. et al. (1977). Safety surveys. *Instn Mech. Engrs, Conf. on Safety Audits – Essential for Every Factory?* (London)

KLIEWER, V.D. (1983). Benefits of modular plant design. *Chem. Engng Prog.*, **79**(10), 58

KLIMISCH, H.J., BRETZ, R., DOE, J.E. and PURSER, D.A. (1987). Classification of dangerous substances and pesticides in the European Economic Community Directives: a proposed revision of criteria for inhalation toxicity. *Regul. Toxicol. Pharmacol.*, **7**, 21

KLINE, M.B. (1984). Suitability of the lognormal distribution for corrective maintenance repair times. *Reliab. Engng*, **9**(2), 65

KLINE, P.E., VOGEL, A.J., YOUNG, A.E., TOWNSEND, D.I., MAYER, M.P. and AERSTIN, F.G. (1974). Guidelines for process scaling. *Chem. Engng Prog.*, **70**(10), 67

KLING, A.L. (1965). Action stations and disaster planning. In Fawcett, H.H. and Wood, W.S. (1965) *op. cit.*, p. 527

KLINGEBIEL, W.J. and MOULTON, R.W. (1971). Analysis of flow choking of two-phase one-component mixtures. *AIChE J.*, **17**, 383

KLINGSCH, W. (1981/82). Fire resistance of solid steel columns. *Fire Safety J.*, **4**(4), 237

KLINKENBERG, A. (1957). Laboratory and plant-scale experiments on the generation and prevention of static electricity. *Am. Petrol. Inst. Proc., Ann. Mtg*

KLINKENBERG, A. (1959). *Genie chim.*, **82**, 149

KLINKENBERG, A. (1964a). Electrical charging of low conductivity liquids in turbulent flow. *Chem. Ing. Tech.*, **36**, 283

KLINKENBERG, A. (1964b). Theoretical aspects and practical implications of static electricity in the petroleum industry. *Advances in Petroleum Chemistry and Refining*, **8**, 87

KLINKENBERG, A. (1967a). Electrical conductivity of low dielectric constant liquids by DC measurement. *J. Inst. Petrol.*, **53**, 57

KLINKENBERG, A. (1967b). On the electric streaming current. *Electrochimica Acta*, **12**, 104

KLINKENBERG, A. (1967c). Static electricity in liquids. *Static Electrification*, **2**, p. 63

KLINKENBERG, A. (1970). The effect of ionic dissociation and recombination on the relaxation of charge in low conductivity dielectric constant liquids. *Chem. Engng Sci.*, **25**, 223

KLINKENBERG, A. (1971). Dimensional analysis in the field of static electricity. *Static Electrification*, **3**, 111

KLINKENBERG, A. and VAN DER MINNE, J.L. (1958). *Electrostatics in the Petroleum Industry* (Amsterdam: Elsevier)

KLINKENBERG, A. and MOOY, H. (1948). Dimensionless groups in fluid friction, heat, and mass transfer. *Chem. Engng Prog.*, **44**, 17

KLOOSTER, H.J., VOGT, G.A. and BERNHART, D.G. (1976). Refinery odor control. *Hydrocarbon Process.*, **55**(4), 121

KLOOSTER, H.J. *et al.* (1975). Optimizing the design of relief and flare systems. *Chem. Engng Prog.*, **71**(1), 39

KLOOTWIJK, J. (1976). Fire engineering. *Course on Loss Prevention in the Process Industries*. Dept of Chem. Engng, Loughborough Univ. of Technol.

KLOSEK, J. and MCKINLEY, C. (1968). Densities of liquefied natural gas and of low molecular weight hydrocarbons. *LNG Conf. 1*

KLOTZ, O. (1917). Acute death from chlorine poisoning. *J. Lab. Clin. Med.*, **2**, 889

KLOUMAN, G.H. (1964). Special purpose and high strength steels. *Chem. Engng Prog.*, **60**(7), 101

KLUG, W. (1969). A method for the determination of dispersion conditions from synoptic observations. *Staub*, **29**(4), 143

KLUG, W.R. (1989). Keeping the ghost in the bottle. *Loss Prevention and Safety Promotion 6*, p. 97.1

KLUMPAR, I.V. and SLAVSKY, S.T. (1985a). Updated cost factors: installation labor. *Chem. Engng*, **92**(Sep. 16), 85

KLUMPAR, I.V. and SLAVSKY, S.T. (1985b). Updated cost factors: Pt 1, Process equipment. Pt 1. *Chem. Engng*, **92**(Jul. 22), 73

KLUMPAR, I.V. and SLAVSKY, S.T. (1985c). Updated cost factors: Pt 2, Commodity materials. *Chem. Engng, Albany*, Aug. 19, 76

KLUNICK, C.H. (1977). Has AFFF agent come of age? *Hydrocarbon Process.*, **56**(9), 293

KLUWER PUBLISHERS (1992). *Chemical Safety Sheets* (Dordrecht)

KMETZ, C.J. (1990). An in-plant container labeling program. *Chem. Engng Prog.*, **86**(3), 33

KNEALE, M. (1984). The engineering of relief disposal. *The Protection of Exothermic Reactors and Pressurised Storage Vessels* (Rugby: Instn Chem. Engrs), p. 183

KNEALE, M. (1989). The safe disposal of relief discharges. *Selection and Use of Pressure Relief Systems for Process Plant* (Manchester: Instn. Chem. Engrs, North-West Branch), p. 7.1

KNEALE, M. and BINNS, J.S. (1977). Relief of runaway polymerisations. *Chemical Process Hazards*, **VI**, p. 47

KNEALE, M. and FORSTER, G.M. (1967). An emergency condensing system for a large propylene oxide polymerisation reactor. *Chemical Process Hazards*, **III**, p. 98

KNECHT, R.W. (1978). Coastal Zone Management; friend or foe? *API Proc., Refining Dept, 43rd Midyear Mtg*, **57**, 15

KNEE, H.E. (1991). The Centralized Reliability Data Organization (CREDO): an advanced nuclear reactor reliability, availability and maintainability data bank and data analysis center. In Cannon, A.G. and Bendell, A. (1991), *op. cit.*, p. 101

KNEEBONE, A. and BOYLE, G.J. (1973). *Laboratory Investigations into the Characteristics of LNG Spills on Water. Rep. TN G2.0012.73*. Shell Res. Ltd, Thornton Res. Centre, Chester, Cheshire

KNEEBONE, A. and BOYLE, G.J. (1973). *LNG Spillages on Water. Evaporation, Spreading and Vapour Dispersion. Rep. TNGR 00122.73*. Shell. Res. Ltd, Thornton Res. Centre, Chester

KNEEBONE, A. and PREW, L.R. (1974). Shipboard jettison tests of LNG onto the sea. *LNG Conf. 4*

KNEZEVIC, J. (1987). Condition parameter based approach to calculation of reliability characteristics. *Reliab. Engng*, **19**(1), 29

KNIEF, R.A. (1981). *Nuclear Energy Technology* (New York: McGraw-Hill)

KNIEL, L. (1974). Moving natural gas from the Arctic to markets. *LNG Conf.*, **4**

KNIEL, L., WINTER, O. and CHUNG-HU TSAI (1980). Ethylene. *Kirk–Othmer Encyclopaedia of Chemical Technology*, 3rd ed., vol. 9 (New York: Wiley), p. 393

KNIEL, L., WINTER, O. and STORK, K. (1980). *Ethylene: Keystone to the Petrochemical Industry* (New York: Dekker)

KNIES, H. (1979). Hazardous substances and reactions. In BASF (1979), *op. cit.*, p. 153

KNIFE, C.W. (1982–83). Tanker blaze on Interstate 70. *Fire Int.*, **78**, 52

KNIGHT, B. *et al.* (1984–85). The use of expert systems in industrial control. *J. Inst. Measmnt Control*, **17**, 489

KNIGHT, F. (1987). Basics for a CADD project manager. *Chem. Engr, Lond.*, **434**, 12

KNIGHT, H.A. (1973). Large plant reliability. *Chem. Engr, Lond.*, **278**, 464

KNIGHT, J. (ed.) (1976). *Operation of Instruments in Adverse Environments* (Bristol: Adam Hilger)

KNIGHT, P. and ROBSON, J.A.W. (1984). Empirical formulae for ground-wave field-strength calculation. *Electron. Letters*, **20**(18), 740

KNIGHT, W.P. (1978). Evaluate waste heat steam generators. *Hydrocarbon Process.*, **57**(7), 126

KNOBLOCH, K., SZENDZIKOWSKI, S., CZAJKOWSKA, T. and KRYSIAK, B. (1971). Experimental studies of the acute and subacute toxicity of acrylonitrile. *Med. Prac.*, **22**, 257

KNOLL, H. and TINNEY, S. (1971). Why use vertical in-line pumps? *Hydrocarbon Process.*, **50**(5), 131

KNOTH, R.J., LASKO, G.E. and MATEJKA, W.A. (1970). New Ni-free stainless bids to oust austenitic. *Chem. Engng*, **77**(May 18), 170

KNOTT, J.F. (1973). *Fundamentals of Fracture Mechanics* (London: Butterworths)

KNOTT, T. (1990). Offshore safety already along the Cullen road. *Chem. Engr, Lond.*, **487**, 15

KNOTT, T. (1994a). Born again Brent. *Offshore Engr*, Mar., 22

KNOTT, T. (1994b). Taking Ninian to market. *Offshore Engr*, Apr., 24

KNOWLES, A.E., TWEEDLE, F. and VAN DER POST, J.L. (1977). *The Background and Implications of IGE/TD/1 Edition 2. Comm. 1044*. Instn Gas Engrs, London

KNOWLTON, R.E. (1976). Hazard and operability studies and their initial applications in R&D. *R&D Mngmnt*, **7**, 1

KNOWLTON, R.E. (1979). Hazop and its contribution to plant operability. *Design 79* (London: Instn Chem. Engrs)

KNOWLTON, R.E. (1981). *An Introduction to Hazard and Operability Studies. The Guide Word Approach* (Vancouver, BC: Chemetics Int. Ltd)

KNOWLTON, R.E. (1988). Avoiding the process safety 'management gap'. *Preventing Major Chemical and Related Process Accidents*, p. 63

KNOWLTON, R.E. (1989). The widespread acceptability of hazard and operability studies. *Thirteenth Symp. on the Prevention of Occupational Risks in the Chemical Industry*, Budapest

KNOWLTON, R.E. (1990). Dealing with the process safety 'management gap'. *Plant/Operations Prog.*, **9**, 108

KNOWLTON, R.E. (1992). *A Manual of Hazard and Operability Studies* (Vancouver, BC: Chemetics Int. Ltd)

KNOWLTON, R.E. (1993). Process safety: life-cycle quality control during design and operations (poster item). *Process Safety Management*, p. 409

KNOX, A.C. (1984). Venting requirements for deaerating heaters. *Chem. Engng*, **91**(Jan. 23), 95

KNOX, B.J. (1990). Safety standards – a time for change. *Piper Alpha: Lessons for Life Cycle Management* (Rugby: Instn Chem. Engrs), p. 77

KNOX, E.G. (1975). Negligible risks to health. *Community Health*, **6**(5), 244

KNOX, J.B. (1984). The modeling of dispersion of heavy gases. In de Wispelaere, C. (1984), *op. cit.*, p. 285

KNOX, J.H. (1959). Rate constants of elementary reactions in hydrocarbon oxidation – competitive oxidation of lower hydrocarbons. (1959). *Combustion*, **7**, 122

KNOX, J.H. (1967a). The interpretation of cool flame and low-temperature combustion phenomena. In Ashmore, P.G., Dainton, F.S. and Sugden, T.M. (1967), *op. cit.*, p. 250

KNOX, J.H. (1967b). *Photochemistry and Reaction Kinetics* (Cambridge: Cambridge Univ. Press)

KNOX, J.H. and KINNEAR, C.G. (1971). The mechanism of combustion of pentane in the gas phase between 250 and 400°C. *Combustion*, **13**, 217

KNUDSEN, J.G. (1984). Fouling of heat exchangers: are we solving the problem? *Chem. Engng Prog.*, **80**(2), 63

KNUDSEN, S. and KROGSTAD, P.A. (1987). A wind tunnel simulation of the Thorney Island Phase II Trial 20. *J. Haz. Materials*, **16**, 173

KNÜLLE, H.R. (1972). Problems with heat exchangers in ethylene plants. *Chem. Engng Prog.*, **68**(7), 53

KNUTH, D.E. (1968). *The Art of Computer Programming*, vol. 1, *Fundamental Algorithms*; 1973, 2nd ed. (Reading, MA: Addison-Wesley)

KNUTH, D.E. (1969). *The Art of Computer Programming*, vol. 2, *Semi-Numerical Algorithms*; 1981, 2nd ed. (Reading, MA: Addison-Wesley)

KNUTH, D.E. (1973). *The Art of Computer Programming*, vol. 3, *Sorting and Searching* (Reading, MA: Addison-Wesley)

KNYSTAUTAS, R., GUIRAO, C., LEE, J.H. and SULMISTRAS, A. (1989). Measurements of cell size in hydrocarbon-air mixtures and predictions of critical tube diameter, critical initiation energy and detonability limits. In Bowen, J.R. *et al.* (1984b), *op. cit.*, p. 23

KNYSTAUTAS, R. and LEE, J.H. (1976). On the effective energy for direct initiation of gaseous detonations. *Combust. Flame*, **27**, 221

KNYSTAUTAS, R., LEE, J.H. and GUIRAO, C.M. (1982). The critical tube diameter for detonation failure in hydrocarbon-air mixtures. *Combust. Flame*, **48**, 63

KNYSTAUTAS, R., LEE, J.H., MOEN, I. and WAGNER, H.G. (1979). Direct initiation of spherical detonation by a hot turbulent gas jet. *Combustion*, **17**, p. 1247

KO, Y., ANDERSON, R.W. and ARPACI, V.S. (1991). Spark ignition of propane–air mixtures near the minimum ignition energy: Part I: An experimental study. *Combust. Flame*, **83**, 75

KO, Y., ARPACI, V.S. and ANDERSON, R.W. (1991). Spark ignition of propane–air mixtures near the minimum ignition energy: Part II: A model development. *Combust. Flame*, **83**, 88

KOBATAKE, K., TSUKADA, T., TAKEHASHI, S., AKATSU, M., IKEMOTO, Y. and BABA, O. (1975). Elevated temperature fatigue tests and inelastic stress analysis of bellows. *J. Pres. Ves. Technol.*, **97**, 223

KOBAYASHI, Y., KOBAYASHI, T. and COOMBS, K.M. (1986). Severe carryover phenomena in a 1,200 MTPD methanol plant. *Plant/Operations Prog.*, **5**, 103

KOBAYASI, K. (1955). An experimental study on the combustion of a fuel droplet. *Combustion*, **5**, 141

KOBER, D. and MARTIN, E. (1972). Safety aspects of LNG transportation with special consideration of inland waterways and coastal ports. *LNG Conf. 3*

KOBERSTEIN, E. (1973). Model reactor studies of the hydrogen cyanide synthesis from methane and ammonia. *Ind. Engng Chem. Proc. Des. Dev.*, **12**, 444

KOBERT, R. (1912). *Kompendium der Praktischen Toxikologie* (Stuttgart)

KOBORI, M., HANDA, T. and YUMOTO, T. (1981). Effect of tank height on fire spread between two model oil tanks. *J. Fire Flammability*, **12**, 157

KOBRIN, G. (1969). Reformer manifold cracking. *Ammonia Plant Safety*, **11**, 14. See also *Chem. Engng Prog., 1969*, **65**(3), 59

KOBRIN, G. (1970). Evaluate equipment condition by field inspection and tests. *Hydrocarbon Process.*, **49**(1), 115

KOBRIN, G. (1978). Metallurgical evaluation of a primary reformer. *Ammonia Plant Safety*, **20**, 103

KOBRIN, G. and KOPECKI, E.S. (1978). Choosing alloys for ammonia services. *Chem. Engng*, **85**(Dec. 18), 115

KOBUS, H. (ed.) (1981). *Hydraulic Modelling* (London: Pitman)

KOBYAKOV, A.I. (1993). Include heuristics in protection systems. *Hydrocarbon Process.*, **72**(2), 79

KOCH, E.R. and VAHRENHOLT, F. (1978). *Seveso ist überall* (Cologne: Kiepenheuer & Witsch)

KOCH, R.H. (1966). A guide to packing and gaskets. *Chem. Engng*, **73**(Mar. 28), 138

KOCH, T.A., KRAUSE, K.R. and MEHDIZADEH, M. (1996). Improved safety through distributed manufacturing of hazardous chemicals. *The 5th World Cong. of Chem. Eng.*, Jul. 14–18, 1996, San Diego, CA, Vol. II, 1095–98. (New York: American Institute of Chemical Engineers)

KOCHHAR, A.K. (1979). *Development of Computer Based Production Systems* (London: Arnold)

KOCHHAR, D.S. and ALI, H.M. (1979). Age as a factor in combined manual and decision tasks. *Hum. Factors*, **21**, 595

KOEHORST, L.J.B. (1989). *An Analysis of Accidents with Casualties in the Chemical Industry Based on Historical Facts* (Berlin: Springer)

KOEHORST, L.J.B. and BOCKHOLTS, P. (1991). FACTs: most comprehensive system for industrial safety. In Cannon, A.G. and Bendell, A. (1991), *op. cit.*, p. 53

KOELSCH (1920). *Zentralblatt Gewerbehyg.*, **8**(5)

KOEN, B.V. and CARNINO, A. (1974). Reliability calculations with a list processing technique. *IEEE Trans. Reliab.*, **R-23**, 43

KOENEN, H. *et al.* (1961). *Explosivestoffe*, **9**, 21

KOENIG, J.P. (1979). Strip-wound pressure vessels: manufacture and operating experience. *Ammonia Plant Safety*, **21**, 89

KOENIGSBERGER, M.D. (1986). Preventing pollution at source. *Chem. Engng Prog.*, **82**(5), 7

KOERNER, R.M. and LORD, A.E. (1987). Microwave system for locating faults in hazardous material dikes. *J. Haz. Materials*, **14**(3), 387

KOERNER, R.M., LORD, A.E. and DEISHER, J.N. (1976). Acoustic emission stress and leak monitoring to prevent spills from buried pipelines. *Control of Hazardous Material Spills*, **3**, 8

KOESTEL, A., GIDO, R.G. and LAMKIN, D.E. (1980). *Drop Size Estimates for a Loss of Coolant Accident. Rep NUREG/CR-1607* (Washington, DC: Nucl. Regul. Comm.)

KOETSIER, W.T. (1976). *Chemical Engineering in a Changing World* (Amsterdam: Elsevier)

KOFSTAD, P. (1988). *High Temperature Corrosion* (London: Elsevier Applied Science)

KOGARKO, S.M., ADUSHKIN, V.V. and LYAMIN, A.G. (1966). An investigation of spherical detonations of gas mixtures. *Int. Chem. Engng*, **6**, 393

KOGLER, R.D. (1971). Vinyl chloride tank car incident. *Loss Prevention*, **5**, 26

KOH, K.H. (1990). Control of hazardous chemicals. *Safety and Loss Prevention in the Chemical and Oil Processing Industries*, p. 399

KOHAN, A.L. (1987). *Pressure Vessel Systems* (New York: McGraw-Hill)

KOHAN, D. (1985). Criteria given for sizing and selection of control valves. *Oil Gas J.*, **83**(Dec. 16), 76

KOHAN, D. (1984). The design of interlocks and alarms. *Chem. Engng*, **91**(Feb. 20), 73

KOHDA, T. and HENLEY, E.J. (1988). On digraphs, fault trees and cut sets. *Reliab. Engng*, **20**, 35

KOHDA, T., HENLEY, E.J. and INOUE, K. (1989). Finding modules in fault trees. *IEEE Trans. Reliab.*, **38**, 165

KOHLBRAND, H.T. (1985). Reactive chemical screening for pilot-plant safety. *Chem. Engng Prog.*, **81**(4), 52

KOHLBRAND, H.T. (1987a). The relationship between theory and testing in the evaluation of reactive chemical hazards. *Preventing Major Chemical Accidents*, p. 4.15

KOHLBRAND, H.T. (1987b). Synopsis of research needs in reactive chemicals. *Preventing Major Chemical Accidents*, p. 4.139

KOHLBRAND, H.T. (1989). The use of Simusolv in the modeling of ARC (accelerating rate calorimetry) data. *Runaway Reactions*, p. 86

KOHLBRAND, H.T. (1990). An industrial reactive chemicals program. *Am. Inst Chem. Engrs, Spring Nat. Mtg*

KOHLBRAND, H.T. (1991). Case history of a deflagration involving an organic solvent/oxygen system below its flash point. *Plant/Operations Prog.*, **10**, 52

KOHLER, J. (1993). *Explosives*, 4th ed. (New York: VCH) – see Meyer, R. (1977)

KÖHLER, W. (1985). Firemen battle for 30 hours after petrol train crash. *Fire Int.*, **94**, 24

KOHLI, I.P. (1979). Steam tracing of pipelines. *Chem. Engng*, **86**(Mar. 26), 156

KOHLSCHMIDT, J. (1972). The determination of parameters characterising the flammability of dusts – both deposited and airborne – for the evaluation of properties constituting the risk. *Int. Symp. on Dust and Explosion Risks in Mines and Industry* (Heidelberg: Berufsgenossenschaft der *Chem. Ind.*), p. 156

KOHN, P.M. (1977a). No watering down sea for US effluent laws. *Chem. Engng*, **84**(Aug. 15), 95

KOHN, P.M. (1977b). The ring tightens on benzene emissions. *Chem. Engng*, **84**(Jul. 18), 64

KOHN, P.M. (1978a). CE cost indexes maintain 13-year ascent. *Chem. Engng*, **85**(May 8), 189

KOHN, P.M. (1978b). Ethylene diamine route eases pollution worries. *Chem. Engng*, **85**(Mar. 27), 90

KOHN, P.M. (1978c). Impasse over benzene while opponents bicker. *Chem. Engng*, **85**(Jul. 31), 41

KOHN, P.M. (1978d). Tough cleanup rules set for chemical dumps. *Chem. Engng*, **85**(Dec. 18), 56

KOHN, P.M. (1979a). Computers keep tabs on occupational health. *Chem. Engng*, **86**(Aug. 13), 98

KOHN, P.M. (1979b). Energy efficient motors gain wider interest. *Chem. Engng*, **86**(Jul. 16), 49

KOHN, P.M. (1982). New hope for Love Canal. *Chem. Engng*, **89**(Aug. 23), 55

KOHN, P.M. and HUGHSON, R.V. (1980). Engineering ethics. *Chem. Engng*, **87**(May 5), 96

KOHOUT, J., SLABY, O., BEJCKOVA, H. and SAMANOVA, M. (1988). Multiple hydrogen sulphide poisoning in a paper mill. *Prac. Lakar*, **40**, 249 (HSE Translation 13121)

KOIKE, M. (1975a). Guide to Japanese High Pressure Gas Control Law. *Hydrocarbon Process.*, **54**(12), 67

KOIKE, M. (1975b). Guide to Japanese Pressure Vessel Code. *Hydrocarbon Process.*, **54**(12), 65

KOIKE, M. (1975c). Guide to Japanese Std PV construction. *Hydrocarbon Process.*, **54**(12), 66

KOIKE, M. (1978a). Guide to Japanese High Pressure Gas Control Law. *Hydrocarbon Process.*, **57**(12), 98

KOIKE, M. (1978b). Guide to Japanese Pressure Vessel Code. *Hydrocarbon Process.*, **57**(12), 96

KOIKE, M. (1978c). Guide to Japanese Std PV construction. *Hydrocarbon Process.*, **57**(12), 97

KOIVISTO, R., KAKKO, R., DOHNAL, M. and JARVELAINEN, M. (1991). Accidental release. Fuzzy selection of adequate formal model. *J. Loss Prev. Process Ind.*, **4**, 317

KOIVISTO, R. and NIELSEN, D. (1994). FIRE – a database on chemical warehouse fires. *J. Loss Prev. Process Ind.*, **7**, 209

KOIVISTO, R., VAIJA, P. and DOHNAL, M. (1989). Intelligent interface of accident databases. *Loss Prevention and Safety Promotion 6*, p. 98.1

KOIVISTO, R. *et al.* (1992). Risk reasoning and revitalization of information poor knowledge. *Loss Prevention and Safety Promotion 7*, vol. 3, 140–1

KOKAWA, M., MIYAKI, S. and SHINGAI, S. (1983). Fault location using digraph and inverse direction search with application. *Automatica*, **190**, 729

KOKEMOR, F.G. (1980). Synthesis start-up heater failure. *Ammonia Plant Safety*, **22**, 159

KOKKINOWRACHOS, K. (1986). Engineering aspects of novel offshore production systems. *Gastech 85*, p. 168

KOKOZSKA, L. and FLOOD, J. (1985). A guide to EPA-approved PCB disposal methods. *Chem. Engng*, **92**(Jul. 8), 41

KOLATA, G.B. (1980). Love Canal: false alarm caused by botched study. *Science*, **208**, 1239

KOLB, J. and ROSS, S.S. (1980). *Product Safety and Liability* (New York: McGraw-Hill)

KOLB, W. and BOLTZE, H. (1987). The key to higher loading limits for cargo tanks on gas carriers – the LGA additional pressure relieving system. *Gastech 86*, p. 240

KOLB, W.B., PAPADIMITRIOU, A.A., CERRO, R.L., LEAVITT, D.L. and SUMMERS, J.C. (1993). The ins and outs of coating monolithic structures. *Chem. Engng Prog.*, **89**(2), 61

KOLBER, A.R., WONG, T.K., GRANT, L.D., DEWOSKIN, R.S. and HUGHES, T.J. (eds) (1983). *In Vitro Toxicity Testing of Environmental Agents, pts A and B* (New York: Plenum Press)

KOLFF, S.W. (1986). Corrosion of a CO_2 absorption tower. *Ammonia Plant Safety*, **26**, 1. See also *Plant/Operations Prog.*, 1986, **5**, 65

KOLLER, G., FISCHER, U. and HUNGERBUHLER, K. (2000). Assessing safety, health, and environmental impact early during process development. *Ind. Eng. Chem. Res.* **39**, 960–72

KOLFF, S.W. and MERTENS, P.R. (1984). The impact of effective quality control. *Ammonia Plant Safety*, **24**, 115. See also *Plant/Operations Prog.*, 1984, **3**, 117

KOLLIG, H.P., KITCHENS, B.E. and HAMRICK, K.J. (1993). FATE, the environmental fate constants information database. *J. Chem. Inf., Comput. Sci.*, **33**(1), 131

KOLLMAN, W. (ed.) (1980). *Prediction Methods for Turbulent Flows* (New York: McGraw-Hill)

KOLLURU, R.V. (1991). Understand the basics of risk assessment. *Chem. Engng Prog.*, **87**(3), 61

KOLODJI, B.P. (1993). Hazard resolutions in sulfur plants from design to start up. *Process Safety Prog.*, **12**, 127

KOLODNER, H. (1977a). A practical approach to vapour cloud protection. *Loss Prevention and Safety Promotion 2*, p. 333

KOLODNER, H. (1977b). Use a fault tree approach. *Hydrocarbon Process.*, **56**(9), 303

KOLODNER, H. (1980). Fresh hydrogen compressor (C-31) explosion in the hydrogenolysis unit. *Loss Prevention and Safety Promotion 3*, vol. 4, p. 1494

KOLODNER, H. and PRATT, T. (1980). A study of the cool flame phenomena and its relationship to explosion potential. *Loss Prevention and Safety Promotion 3*, vol. 2, p. 345

KOLSKY, H. (1963). *Stress Waves in Solids* (New York: Dover Press)

KOMAREK, R. (1993). Briquets used to reduce environmental hazards. *Pollution Eng.*, (December), 47–8

KOMAR, P.D. (ed.) (1983). *CRC Handbook of Coastal Processes and Erosion* (Boca Raton, FL: CRC Press)

KONDILI, E., PANTILIDES, C.C. and SARGENT, R.W.H. (1988). A general algorithm for scheduling batch operations. *Proc. Third Int. Symp on Process Systems Engineering*, p. 62

KONDO, J. (1975). Air-sea bulk transfer coefficients in diabatic conditions. *Boundary Layer Met.*, **9**, 91

KONDRASUOV, V. (1978). On the relationship of the hazard of poisoning by vapours and gases of toxic substances by skin and inhalation routes. *Gig. Tr. Prof. Zabol.*, Feb., 34

KONG, F.K., EVANS, R.H., COHEN, E. and ROLL, F. (eds) (1983). *Handbook of Structural Concrete* (London: Pitman)

KONGSO, H.E. (1970). *RELY4: a Monte Carlo Computer Program for System Reliability Analysis. Rep. Riso-M-1500.* Riso Nat. Lab., Riso, Denmark

KONGSO, H.E. (1992). Application of a guide to analysis of occupational hazards in the Danish iron and chemical industry. *Hazard Identification and Risk Analysis, Human Factors and Human Reliability in Process Safety*, p. 195

KONO, M., KUMAGAI, S. and SAKAI, T. (1976). Ignition of gases by two successive sparks with reference to frequency effect of capacitance sparks. *Combust. Flame*, **27**, 85

KONOKI, K. (1971). Redundancy system of chemical process plant. *J. Chem. Engng Japan*, **4**(1), 77

KONOKI, K., SHIONOHARA, T. and SHIBATA, K. (1982). Creep rupture of steam reforming tube due to thermal stress. *Plant/Operations Prog.*, **1**, 122

KONSTANTINOV, L.V. and GONZALEZ, A.J. (1989). *Nucl. Safety*, **30**, 53

KONTOLEON, J.M. (1980). Reliability analysis of a special redundant system. *Reliab. Engng*, **1**(2), 143

KONTOLEON, J.M. (1981). A general approach for determining reliability measures of repairable systems. *Reliab. Engng*, **2**(1), 57

KONTOLEON, J.M. (1982). An algorithm for maximising the overall network reliability. *Reliab. Engng*, **3**(2), 101

KONTOLEON, J.M. and KONTOLEON, N. (1974). Reliability analysis of a system subject to partial and catastrophic failures. *IEEE Trans Reliab.*, **R-23**(4), 277

KONWAY, M., MEAD, H.B. and PAGE, F.M. (1975). Fall of liquid metals into water. *Nature*, **254**, 321

KOOHYAR, A.N. (1967). Ignition of wood by flame radiation. Ph.D. Thesis, Univ. of Oklahoma, OK

KOOISTRA, L.F. and LEMCOE, M.M. (1962). Low cycle fatigue research on full-size pressure vessels. *Weld. J.*, **41**(res. suppl.), 297S

KOOLEN, J.L.A. ('1998). Simple and robust design of chemical plants. *Comp. Chem. Eng.*. **22**, S255–62

KOOLHAAS, J., RAMDAS, V. and PUTNAM, T.M. (1979). *Thermophysical Properties Research Literature Retrieval Guide* (New York: IFI/Plenum Data Co.)

KOOPMAN, R.P. and THOMPSON, D. (1986). Description of test-site for large-scale heavy gas dispersion and combustion experiments. In Hartwig, S. (1986), *op. cit.*, p. 299

KOOPMAN, R.P., BOWMAN, B.R. and ERMAK, D.R. (1980). Data and calculations on 5 m^3 LNG spill tests. *Liquefied Gaseous Fuels Safety and Environmental Control Assessment Program, Second Status Rep., Rep. P* (Dept of Energy)

KOOPMAN, R.P., CEDERWALL, R.T., ERMAK, D.L., GOLDWIRE, H.C., HOGAN, W.J., MCCLURE, J.W., MCRAE, T.G., MORGAN, D.L., RODEAN, H.C. and SHINN, J.H. (1982). Analysis of Burro series 40 m^3 LNG spills. *J. Haz. Materials*, **6**(1/2), 43. See also in Britter, R.E. and Griffiths, R.F. (1982), *op. cit.*, p. 43

KOOPMAN, R.P., MCRAE, T.G., GOLDWIRE, H.C., ERMAK, D.L. and KANSA, E.J. (1986). Results of recent large-scale NH_3 and N_2O_4 dispersion experiments. In Hartwig, S. (1986), *op. cit.*, p. 137

KOOPMAN, R.P., ERMAK, D.L. and CHAN, S.T. (1989). A review of recent field tests and mathematical modelling of atmospheric dispersion of large spills of denser-than air gases. *Atmos. Environ.*, **23**, 731

KOPALINSKY, E.M. and BRYANT, R.A.A. (1976). Friction coefficients for bubbly two-phase flow in horizontal pipes. *AIChE J.*, **22**, 82

KOPECEK, J.T. (1977). Risk assessment of a liquefied natural gas terminal. *Tenth Int. TNO Conf. on Risk Analysis: Industry, Government and Society*

KOPELMAN, B. (1959). *Materials for Nuclear Reactors* (New York: McGraw-Hill)

KOPETZ, H. (1979). *Software Reliability* (London: Macmillan)

KOPPEL, L.B. (1993). Improve information system master planning. *Hydrocarbon Process.*, **72**(5), 45

KORDYLEWSKI, W. (1979). Critical parameters of thermal explosion. *Combust. Flame*, **34**, 109

KORDYLEWSKI, W. (1980). Critical conditions for thermal ignition of porous bodies. *Combust. Flame*, **38**, 103

KORDYLEWSKI, W. and AMROGOWICZ, J. (1992). Comparison of $NaHCO_3$ and $NH_4H_2PO_4$ effectiveness as dust explosion suppressants. *Combust. Flame*, **90**, 344

KORDYLEWSKI, W. and KRAJEWSKI, Z. (1981). Thermal ignition of self-heating porous slab. *Combust. Flame*, **41**, 113

KORDYLEWSKI, W. and KRAJEWSKI, Z. (1985). Thermal ignition of self-heating porous slab: comparison with experiment. *Combust. Flame*, **62**, 135

KORDYLEWSKI, W. and WACH, J. (1986). Influence of ducting on the explosion pressure. *Combust. Flame*, **66**, 77

KORDYLEWSKI, W. and WACH, J. (1988). Influence of ducting on explosion pressure: small scale experiments. *Combust. Flame*, **71**, 51

KORMAN, H.F. (1970). Theoretical modeling of cool flames. *Combust. Sci. Technol.*, **2**, 149

KÖRNER, J.P. (1985). Design and construction of full-scale supercritical gas extraction plants. *Chem. Engng Prog.*, **81**(4), 63

KORNHAUSER, W. (1962). *Scientists in Industry* (Berkeley, CA: Univ. of California Press)

KOROLL, G.W., KUMAR, R.K. and BOWLES, E.M. (1993). Burning velocities of hydrogen-air mixtures. *Combust. Flame*, **94**, 330

KORTLANDT, D. and KRAGT, H. (1980a). Ergonomics in the struggle against 'alarm inflation' in process control systems. *Automation for Safety in Shipping and Offshore Petroleum Operations* (ed. A.B. Aune and J. Vlietstia) (New York: North Holland), p. 231

KORTLANDT, D. and KRAGT, H. (1980b). Process alarm systems as a monitoring tool for the operator. *Loss Prevention and Safety Promotion 3*, vol. 2, p. 804

KORTNER, H. (1992). Uncertainties in quantitative risk analyses. *Loss Prevention and Safety Promotion 7*, vol. 1, 31–1

KORZUCH, W.T. (1961). Are canned pumps for you? *Chem. Engng*, **68**(May 1), 95

KOSAR, R.P. and BLAHUT, N.W. (1986). Adapting expert systems to simulation training of process operators. *InTech*, **33**(9), 79

KOSDON, F.J., WILLIAMS, F.A. and BUMAN, C. (1969). Combustion of vertical cellulosic cylinders in air. *Combustion*, **12**, 253

KOSEKI, H. (1989). Combustion, properties of large liquid pool fires. *Fire Technol.*, **25**(3), 241

KOSEKI, H. and HAYASAKA, H. (1989). Estimation of the thermal balance in a heptane pool fire. *J. Fire Sciences*, **37**(4), 237

KOSEKI, H. and MULHOLLAND, G.W. (1991). The effect of diameter on the burning of crude oil pool fires. *Fire Technol.*, **27**, 54

KOSEKI, H. and YUMOTO, T. (1988). Air entrainment and thermal radiation from heptane pool fires. *Fire Technol*, **24**, 33

KOSHAL, D. (1993). *Manufacturing Engineers Reference Book* (Oxford: Butterworth-Heinemann)

KOSHLAND, D.E. (ed.) (1974). *How Safe is Safe?* (Washington, DC: Nat. Acad. Sci.)

KOSMIDER, S., ROGALA, E. and PACHOLEK, A. (1967). Electrocardiographic and histochemical studies of the heart muscle in acute experimental hydrogen sulphide poisoning. *Arch. Immunol. Ther. Exp.*, **15**, 731

KOSSEBAU, F. (1982). Explosion-proof bucket elevators. *The Control and Prevention of Dust Explosions* (London: Oyez/IBC), paper 17

KOSSOY, A., BELOCHVOSTOV, V. and GUSTIN, J.L. (1994). Methodological aspects of the application of adiabatic calorimetry for thermal safety investigation. *J. Loss Prev. Process Ind.*, **7**, 397

KOSTERING, S. and BECKER, H. (1987). European/USA petrochemical gases – a producer/consumer view. *Gastech*, **86**, p. 327

KOSZMAN, I. and GAVIS, J. (1962a). Development of charge on low conductivity liquids flowing past surfaces. Engineering predictions from the theory developed for tube flow. *Chem. Engng Sci.*, **17**, 1013

KOSZMAN, I. and GAVIS, J. (1962b). Development of charge on low conductivity liquids flowing past surfaces. Experimental verification and application of the theory developed for tube flow. *Chem. Engng Sci.*, **17**, 1023

KOTAS, T.J. (1985). *The Exergy Method of Thermal Plant Analysis* (London: Butterworths)

KOTAS, T.J. (1986). Exergy method of thermal and chemical plant analysis. *Chem. Engng Res. Des.*, **64**, 212

KOTCHARIAN, M. and SIMON, J.M. (1982). Safety of liquefied gases containment on land and at sea. In *Gastech 81*

KOTHARI, K.M. and MERONEY, R.N. (1979). *Building Effects on National Transonic Facility Exhaust Plume. Rep. CER-79-80-KMK-RNM-35*. Dept of Civil Engng, Colorado State Univ., Fort Collins, CO

KOTHARI, K.M., MERONEY, R.N. and NEFF, D.E. (1981). *LNG Plume Interaction with Surface Obstacles. Rep. CER-81-82-KMK-RNM-DEN22*. Dept of Civil Engng, Colorado State Univ., Fort Collins, CO

KOTHARI, K.M. and MERONEY, R.N. (1982). *Accelerated dilution of liquefied natural gas plumes with fences and vortex generators. Rep. CER-81–82–KMK-RNM-9*. Dept of Civil Engng, Colorado State Univ., Fort Collins, CO

KOTOYORI, T. (1989a). Critical ignition temperatures of chemical substances. *J. Loss Prev. Process Ind.*, **2**(1), 16

KOTOYORI, T. (1989b). Reaction processes up to explosion of chlorinated aromatic amines. *Loss Prevention and Safety Promotion*, **6**, p. 38.1

KOTOYORI, T. (1991). Investigation of a thermal runaway reaction involving a nitration process. *J. Loss Prev. Process Ind.*, **4**(2), 120

KOTOYORI, T. (1993). Critical temperatures for the thermal explosion of liquids. *Combust. Flame*, **95**, 307

KOTOYORI, T., TSURUMI, H., UENO, T. and NISHIHARA, A. (1976). Case history: explosion during distillation of 4-chloro-2-methylaniline. *J. Haz. Materials*, **1**(3), 253

KOTTOWSKI, H.M. and KASHIKI, K. (1991). Potential and limits of dimensional analysis for the design of chemical facilities. *J. Loss Prev. Process Ind.*, **4**, 194

KOTZERKE (1970). Catalytic decomposition of urea nitrate caused by heavy metal ions. *Prevention of Occupational Risks*, **1**

KOURGANOFF, V. (1963). *Basic Methods in Transfer Problems* (New York: Dover)

KOVACS, F. and HONTI, G. (1974). Secondary heat effect on LPG storage spheres in case of fire. *Loss Prevention and Safety Promotion 1*, p. 385

KOVAR, F.R., TRIGGER, K.R., GUYMON, L.G. and HARVEY, J.F. (1982). High explosive detonations in varying oxygen atmospheres. In Nellis, W.J., Seaman, L. and Graham, A.R. (1982), *op. cit.*, p. 563

KOWALCZYK, S.W. (1992). Root out the silent effects of electrical noise. *Chem. Engng*, **99**(6), 145

KOWALSKI, R. (1974). Predicate logic as a programming language. *Information Processing 74* (Amsterdam: North Holland), p. 569

KOWALSKI, R. (1979). *Logic for Problem-Solving* (New York: North Holland)

KOWITZ, T.A., REBA, R.C., PARKER, R.T. and SPICER, W.S. (1967). Effects of chlorine gas upon respiratory function. *Arch. Environ. Health*, **14**, 545

KOZACHENKO, L. (1962). The combustion mechanism and burning velocity in turbulent flow. *Combustion*, **8**, 567

KOZAK, J.J., HANCOCK, P.A., ARTHUR, E.J. and CHRYSLER, S.T. (1993). Transfer of training from virtual reality. *Ergonomics*, **36**, 777

KOZLOV, B.A. and USHAKOV, I.A. (1970). *Reliability Handbook* (New York: Holt, Rinehart & Winston)

KRAEMER, K. (1974). Are couplings the weak link in rotating machinery systems? *Hydrocarbon Process.*, **53**(2), 97

KRAFT, A. (1984). XCON: an expert configuration system at Digital Equipment Corporation. In Winston, P.H. and Prendergast, K.A. (1984), *op. cit.*

KRAFT, R.L. (1992). Incorporate environmental reviews into facility design. *Chem. Engng Prog.*, **88**(8), 46

KRAGELSKI, I.V. and DEMKIN, N.B. (1960). Contact areas of rough surfaces. *Wear*, **3**, 170

KRAGT, H. (1982). Process alarm systems. *Twenty-Sixth Ann. Mtg of Human Factors Soc.*, Seattle, WA

KRAGT, H. (1983). Operator tasks and annunciator systems. Doctoral thesis, Technische Hogeschool Eindhoven

KRAGT, H. (1984a). A comparative simulation study of annunciator systems. *Ergonomics*, **27**, 927

KRAGT, H. (1984b). A comparative simulation study of annunciator systems. *Ergonomic Problems in Process Operations* (Rugby: Instn Chem. Engrs), p. 115

KRAGT, H. and BONTEN, J. (1983). Evaluation of a conventional process-alarm system in a fertiliser plant. *IEEE Trans. Systems Man Cybernetics*, **SMC-13**, 586

KRAGT, H. and DANIELS, M.J.M. (1984). Some remarks on the process operator and his job: the human operator in the control room of tomorrow. *Ergonomic Problems in Process Operations* (Rugby: Instn Chem. Engrs), p. 191

KRAGT, H. and LANDEWEERD, J.A. (1974). Mental skills in process control. In Edwards, E. and Lees, F.P. (1974), *op. cit.*, p. 135

KRAMER, C.G. (1967). Chlorine. *J. Occup. Med.*, **9**(4), 193

KRAMER, H. (1981). Electrostatic charging of poorly conducting liquid systems – suspensions, emulsions and solutions – by agitating. *J. Electrostatics*, **10**, 89

KRAMER, H. (1987). Safety criteria for the assessment of the incendivity of gas discharges in explosible flammable gas or vapour–air mixtures. *Fortschritte der Sicherheitstechnik 1* (Frankfurt-am-Main: DECHEMA), p. 349

KRAMER, H. and ASANO, K. (1979). Incendivity of sparks from electrostatically charged liquids. *J. Electrostatics*, **6**, 361

KRAMER, H., ASANO, K. and SCHÖN, G. (1977). Criteria for the safe filling of tank vehicles – theoretical and experimental study to assess electrostatic hazards. *Proc. Third Int. Conf. on Static Electricity* (Paris: Soc. de Chimie Industrielle), p. 31a

KRAMER, H. and SCHÖN, G. (1973). Charge relaxation of highly charged fuels. *Static Electricity*, **2**

KRÄMER, H. and SCHÖN, G. (1975). Estimation of space charge and field strength in tanks during top filling with electrostatically chargeable fuels. *Proc. Ninth World Petrol. Cong.*, vol. 6 (London: Applied Science), p. 147

KRAMER, M.A. (1987a). Expert systems for process fault diagnosis: a general framework. In Reklaitis, G.V. and Spriggs, H.D. (1988), *op. cit.*

KRAMER, M.A. (1987b). Malfunction diagnosis using quantitative models with non-Boolean reasoning in expert systems. *AIChE J.*, **33**, 130

KRAMER, M.A. (1988). Automated diagnosis of malfunctions based on object-orientated programming. *J. Loss Prev. Process Ind.*, **1**(4), 226

KRAMER, M.A. and FINCH, F.E. (1989). Fault diagnosis of chemical processes. In *Knowledge-Based Systems, Supervision and Control* (ed. S.G. Tzafestas) (New York: Plenum Press)

KRAMER, M.A. and LEONARD, J.A. (1990). Diagnosis using backpropagation neural networks – analysis and criticism. *Comput. Chem. Engng*, **14**, 1323

KRAMER, M.A. and PALOWITCH, B.L. (1987). A rule-based approach to fault diagnosis using the signed directed graph. *AIChE J.*, **33**, 1067

KRAMER, O. (1917). Chlorine gas poisoning. *Vjscht Ger. Med. Off. Sanitatswesen*, **53**, 181

KRAMERS, H. and MEIJNEN, J.K.G. (1974). Safety organization. *Loss Prevention and Safety Promotion 1*, p. 45

KRAMERS, H. and WESTERTERP, K.R. (1963). *Elements of Chemical Reactor Design and Operation* (London: Chapman & Hall)

KRANENBERG, C. (1984). Wind-induced entrainment in a stably stratified fluid. *J. Fluid Mech.*, **145**, 253

KRASNER, L.M. (1974). The feasibility of quantitatively analyzing investments in loss prevention activities. *Loss Prevention*, **8**, 77

KRATSIOS, G. and LONG, W.P. (1968). Experiences with primary reformer furnaces and transfer lines. *Ammonia Plant Safety*, **10**, 8

KRATZ, A.P. and ROSECRANS, C.Z. (1922). *A Study of Explosions of Gaseous Mixtures*. Bulletin. Univ. of Illinois

KRAUS, M.N. (1965). Pneumatic conveying. *Chem. Engng*, **72**(Apr. 12), 167; (May 10), 149

KRAUS, M.N. (1966a). Pneumatic conveying. *Chem. Engng*, **73**(Jan. 3), 104

KRAUS, M.N. (1966b). Starting up pneumatic conveyors. *Chem. Engng*, **73**(Jan. 31), 94

KRAUS, M.N. (1979). Baghouses: selecting, specifying and testing industrial dust collectors. *Chem. Engng*, **86**(Apr. 23), 133

KRAUS, M.N. (1979). Baghouses: separating and collecting industrial dusts. *Chem. Engng*, **86**(Apr. 9), 94

KRAUS, M.N. (1980). How to protect mechanical equipment and motors. *Chem. Engng*, **87**(Dec. 15), 59

KRAUS, M.N. (1986). Pneumatic conveying systems. *Chem. Engng*, **93**(19), 50

KRAUS, N.J. (1980). FRP underground horizontal tanks for corrosive chemicals. *Chem. Engng*, **87**(Feb. 11), 125

KRAUSE, F.D., CHESTER, E.H. and GILLESPIE, D.G. (1968). The prevalence of chronic obstructive airway in chlorine gas workers. *Med. Abstr. Ann. Mtg Am. Thoracic Soc.*, p. 150

KRAUSE, R. (1985). A new method of applying protective insulation to the inside of double walled installations for cryogenic storage. *Gastech 84*, p. 406

KRAUSE, R. (1986). A comparison of various plastic materials concerning their behaviour at cryogenic temperatures. *Gastech 85*, p. 524

KRAUSE, T.R., HIDLEY, J.H. and HODSON, S.J. (1990). *The Behaviour-Based Safety Process* (New York: van Nostrand Reinhold)

KRAUTKRAMER, J. and KRAUTKRAMER, H. (1969). *Ultrasonic Testing of Materials*, 2nd ed. (London: Allen & Unwin)

KRAYBILL, H.F. and MEHLMAN, M.A. (ed.) (1977). *Occupational Carcinogenesis in Environmental Cancer* (New York: Wiley)

KREBS, T.M. (1962). New creep–rupture data on furnace tubes. *Hydrocarbon Process.*, 41(8), 135

KREBS, T.M. (1963). Casebook of a corrosion detective. *Chem. Engng*, 70(Feb. 4), 122; (Feb. 18), 186

KREITH, F. (1962). *Radiation Heat Transfer* (Scranton, PA: Int. Textbook)

KREITH, F. (1980). *Economics of Solar Energy and Conservation Systems*, vols 1–3 (Boca Raton, FL: CRC Press)

KREMER, G. (1981). The Accident Regulation. *Technische Überwachung*, 22(Apr.), 153

KREMER, G. (1987). Safety technology in the chemical industry – responsibility and obligation in the Federal Republic of Germany. *Fortschritte der Sicherheitstechnik 1* (Frankfurt-am-Main: DECHEMA), p. 1

KREMER, H. (1976). Leak rates of static and dynamic seal elements on chemical and petrochemical plants. *OGEW/DGMK Mtg*, Salzburg

KREMERS, J. (1983). Avoid water hammer. *Hydrocarbon Process.*, 62(3), 67

KRETZSCHMAR, J.G. and MERTENS, I. (1984). Influence of the turbulence typing scheme upon the cumulative frequency distributions of the calculated relative concentrations for different averaging times. *Atmos. Environ.*, 18, 2377

VAN KREVELEN, D.W. and CHERMIN, H.A.G. (1959). Generalized flame stability diagram for the prediction of the interchangeability of gases. *Combustion*, 7, 358

KREWSKI, D.R. and VAN RYZIN, J. (1981). Dose response models for quantal response toxicity data. In *Statistics and Related Topics* (ed. M. Csorgo, D. Dawson, J.N.K. Rao and E. Saleh) (Amsterdam: North Holland)

KREY, P.W. (1974). Washout ratios. In Beadle, R.W. and Semonin, R.G. (1974), *op. cit.*

KREYSZIG, E. (1967). *Advanced Engineering Mathematics* (New York: Wiley)

KRICK, E.V. (1969). *An Introduction to Engineering and Engineering Design* (New York: Wiley)

KRIENBERG, M. (1980). Pump selection for fuel oil tank farms. *Chem. Engng*, 87(Sep. 22), 179

KRIGMAN, A. (1984). Toxic gas monitoring: simple instruments – complex measure. *Instrum. Technol.*, 31(5), 487

KRIGMAN, A. (1985). Batch process control: sweet dreams or a nightmare? *InTech*, Oct., 9

KRIGMAN, A. and REDDING, R. (1974). Test for intrinsic safety. *Hydrocarbon Process.*, 53(10), 198

KRIKKE, R.H., HOVING, J. and SMIT, K. (1976). Estimate ethylene furnace tube life. *Hydrocarbon Process.*, 55(8), 151

KRIMSKY, S. and GOLDING, D. (eds) (1992). *Social Theories of Risk* (Westport, CT: Praeger)

KRISHER, A.S. (1965). Case histories of stainless steel use. *Chem. Engng Prog.*, 61(4), 99

KRISHER, A.S. (1978). Raw water treatment in the CPI. *Chem. Engng*, 85(Aug. 28), 79

KRISHER, A.S. (1987). Plant integrity programs. *Chem. Engng. Prog.*, 83(5), 20

KRISHNA, C.R. and BERLAND, A.L. (1980). A model for dust cloud autoignition. *Combust. Flame*, 37, 207

KRISHNAIAH, D., PRABHAKAR, A. and RAJU, D.C. (1990). Hazardous chemicals – handling, storage and transportation.

Safety and Loss Prevention in the Chemical and Oil Processing Industries, p. 409

KRISHNAMURTY, R. and HALL, J.G. (1987). Numerical and approximate solutions for plume rise. *Atmos. Environ.*, 21, 2083

KRISHNAN, M.R.V. and GANESH, J. (1980). Loss prevention and safety promotion – a typical case study (paints and dye industry). *Loss Prevention and Safety Promotion 3*, vol. 3, p. 1481

KRISHNASWAMI, S. (1979). Manage multiple projects better. *Hydrocarbon Process.*, 58(1), 229

KRISHR, A.S. (1986). Material requirements for anhydrous ammonia. *Process Industries Corrosion – the Theory and Practice* (Houston, TX: Nat. Ass. Corrosion Engrs), p. 311

KROEMER, K.H.E. (1992). Personnel training for safer material handling. *Ergonomics*, 35, 1119

KROGER (1979). Decomposition of n-butyricaldoxime during distillation with explosion of the distilling container. *Prevention of Occupational Risks*, 1

KROGH, M. (1914–15). The diffusion of gases through the lungs of man. *J. Physiol.*, 49, 271

KROGSTAD, P.A. and PETTERSEN, R.M. (1986). Wind tunnel modelling of a release of heavy gas near a building. *Atmos. Environ.*, 20, 867

KROHN, D.A. (1992). *Fiber Optic Sensors* (Pittsburgh, PA: Instrum. Soc. Am.)

KROHN, G.S. (1983). Flowcharts used for procedural instructions. *Hum. Factors*, 25, 573

KROHN, P.L., WHITTERIDGE, D. and ZUCKERMAN, S. (1942). Physiological effects of blast. *The Lancet*, Feb. 28, 252

KROLIKOWSKI, S.R. (1965). Safety in operating a nitrogen-wash plant. *Chem. Engr, Lond.*, 186, CE40

KRON, G. (1963). *Diakoptics – The Piecewise Solution of Large-Scale Systems* (London: MacDonald)

KRONVALL, J. (1978). Testing of houses for air leakage using a pressure method. *ASHRAE Trans.*, 84(pt 1), 72

KROPP, A. (1987). A comparison of published ground motion parameters. In Cermak, A.S. (1987), *op. cit.*, p. 493

KROUPA, R. (1993). Selecting valves for tough conditions. *Chem. Engng*, 100(2), 74

KRUPP, H. (1971). Physical models of the static electrification of solids. *Static Electrification*, 3, 1

KRUPPER, J. (1991). Fire behaviour of ducts. *Fire Safety J.*, 17, 135

KRUTZSCH, W.C. (1983). Careful piping design reduces pump wear. *Oil Gas J.*, 81(Sep. 19), 125

KRYGIELSKI, D.J. and BENNETT, G.F. (1978). Intergovernmental considerations in the disposal of Kepone – a municipal perspective. *Control of Hazardous Material Spills*, 4, p. 274

KRYSTOW, P.E. (1971). Materials and corrosion problems in urea plants. *Ammonia Plant Safety*, 13, p. 93. See also *Chem. Engng Prog.*, 1971, 67(4), 59

KRZYSZTOFORSKI, A., CIBOROWSKI, S., POHORECKI, R. and WOJCIK, Z. (1986). Chemical reaction engineering as a tool in process safety considerations. *Loss Prevention and Safety Promotion 5*, p. 45–1

KRYSZTOFORSKI, A., WOJCIK, Z., POHORECKI, R. and BALDYGA, J. (1986). Industrial contribution to the reaction engineering of cyclohexane oxidation. *Ind. Engng Chem. Process Des. Dev.*, 25, 894

KRZYSZTOFORSKI, A., GASIOROWSKI, P., MACIEJCZYK, S. and WOJCIK, Z. (1992). Cyclohexane oxidation using oxygen enriched air. *Loss Prevention and Safety Promotion 7*, vol. 2, 64–1

KUBIAS, F.O. (1982). Tools and techniques for chemical safety training. In Fawcett, H.H. and Wood, W.S. (1982), *op. cit.*, p. 2739

KUBIC, W.L. and STEIN, F.P. (1988). A theory of design reliability using probability and fuzzy sets. *AIChE J.*, **34**, 583

KUCHTA, J.M., BARTKOWIAK, A. and ZABETAKIS, M.G. (1965). Hot surface ignition temperatures of hydrocarbon fuel vapor−air mixtures. *J. Chem. Engng Data*, **10**, 282

KUCHTA, J.M., CATO, R.J. and ZABETAKIS, M.G. (1964). Comparison of hot surface and hot gas ignition temperatures (letter). *Combust. Flame*, **8**, 348

KUCHTA, J.M., FURNO, H.L., BARTKOWIAK, A. and MARTINDILL, G.H. (1968). Effect of pressure and temperature on flammability of chlorinated hydrocarbons in oxygen−nitrogen tetroxide−nitrogen atmospheres. *J. Chem. Engng Data*, **13**, 421

KUEHL, D.K. (1962). Laminar-burning velocities of propane−air mixtures. *Combustion*, **8**, 510

KUEHN, D.R. and DAVIDSON, H. (1961). Computer control. II, Mathematics of control. *Chem. Engng Prog.*, **57**(6), 44

KUGLER, E.L. and STEFFGEN, F.W. (eds) (1980). *Hydrocarbon Synthesis from Carbon Monoxide and Hydrogen* (Washington, DC: Am. Chem. Soc.)

KUHL, A.L., KAMEL, M.M. and OPPENHEIM, A.K. (1973). Pressure waves generated by steady flames. *Combustion*, **14**, 1201

KUHLMAN, R. (n.d.) *Professional Accident Investigation* (Loganville, GA: Institute Press, International Loss Control Institute)

KUHLMAN, RAYMOND, *Professional Accident Investigation − Methods and Techniques*, Atlanta: International Loss Control Institute (Now DNV), Library of Congress Catalog 77−77275

KUHN, A.T. (1971). *Industrial Electrochemical Processes* (Amsterdam: Elsevier)

KUHN, J.W. (1964). Safety in industrial use of radioactive sources. *Chem. Engng Prog.*, **60**(4), 41

KUHN, P.M. (1979). What's US PO and EO outlook? *Hydrocarbon Process.*, **58**(10), 123

KUHN, P.M. (1980a). Report says EPA control of toxics is lagging. *Chem. Engng*, **87**(Dec. 15), 39

KUHN, P.M. (1980b). What's ahead for propylene and ethylene oxide. *Chem. Engng Prog.*, **76**(1), 53

KUHNE, G. (1983). Fire protection design for off-shore process plants. *Loss Prevention and Safety Promotion 4*, vol. 2, p. D60

KÜHNEN, G. (1971). Experience gained from dust explosions and its practical significance. *Staub Reinhaltung der Luft*, **31**(3), 40 (English ed.)

KUIPERS, B. (1975). A frame for frames: representing knowledge for recognition. In Bobrow, D.G. and Collins, A. (1975), *op. cit.*, p. 151

KUIPERS, B. (1984). Commonsense reasoning about causality: deriving behaviour from structure. *Artificial Intelligence*, **24**, 169

KUIPERS, B. (1986). Qualitative simulation. *Artificial Intelligence*, **29**, 289

KUIPERS, B. (1989). Qualitative reasoning: modelling and simulation with incomplete knowledge. *Automatica*, **25**, 571

KUIST, B.B. and FIFE, H.E. (1971). Process continuity in polymer plants. *Chem. Engng Prog.*, **67**(1), 33

KUJAWSKI, R.A. (1985). Follow these rules to write good instructions. *Chem. Engng*, **92**(Apr. 15), 107

KUKKONEN, J. and NIKMO, J. (1992). Modeling heavy gas cloud transport in sloping terrain. *J. Haz. Materials*, **31**, 155

KUKKONEN, J. and VESALA, T. (1991). Aerosol dynamics in releases of hazardous materials. *Modeling and Mitigating the Consequences of Accidental Releases*, p. 53

KUKKONEN, J. and VESALA, T. (1992). Modelling aerosol releases of hazardous materials. *Loss Prevention and Safety Promotion 7*, vol. 3, 104.1

KUKKONEN, J., SAVOLAINEN, A.L., VALKAMA, I. and JUNTTO, S. (1992). Consequences of a major chemical accident in Dnava, Lithuania. *Loss Prevention and Safety Promotion 7*, vol. 1, 21.1

KUKKONEN, J., SAVOLAINEN, A.L., VALKAMA, I., JUNTTO, S. and VESALA, T. (1993). Long range transport of ammonia released in a major chemical accident at Ionova, Lithuania. *J. Haz. Materials*, **35**, 1

KULI, J.C. (1964). Aluminum and its alloys in the CPI. *Chem. Engng Prog.*, **60**(5), 45

KULIKOWSKI, C.A. and WEISS, S.M. (eds) (1982). Representation of expert knowledge for consultation: the CASNET and EXPERT projects. *Artificial Intelligence in Medicine* (ed. P. Slovitz) (Boulder, CO: Westview Press)

KULLENBERG, G.E.B. (1982). *Pollutant Transfer and Transport in the Sea* (Boca Raton, FL: CRC Press)

KULLERD, B.E. (1968). Interpreting and applying article 500 of NEC. *Loss Prevention*, **2**, 1

KUMAGAI, S. (1957). Combustion of fuel sprays. *Combustion*, **6**, 668

KUMAGAI, S. and KIMURA, I. (1957). Ignition of flowing gases by heated wires. *Combustion*, **6**, 554

KUMAGAI, S., SAKAI, T. and YASUGAHIRA, N. (1972). Calorimetry of spark energy. *Combust. Sci. Technol.*, **6**, 233

KUMAMOTO, H. and HENLEY, E.J. (1978). Protective system hazard analysis. *Ind. Engng Chem. Fund.*, **17**, 274

KUMAMOTO, H. and HENLEY, E.J. (1979). Safety and reliability synthesis of systems with control loops. *AIChE J.*, **25**, 108

KUMAMOTO, H. and HENLEY, E.J. (1986). Automated fault tree synthesis by disturbance analysis. *Ind. Engng Chem. Fund.*, **25**, 233

KUMAMOTO, H. and HENLEY, E.J. (1987). Application of expert system techniques to fault diagnosis. *Am. Inst. Chem. Engrs, Ann Mtg*, San Francisco, CA

KUMAMOTO, H., INOUE, K. and HENLEY, E.J. (1981). Computer-aided protective system hazard analysis. *Comput. Chem. Engng*, **5**, 93

KUMANA, J.D. and KOTHARI, S.P. (1982). Predict storage-tank heat transfer precisely. *Chem. Engng*, **89**(Mar. 22), 127

KUMAR, A. (1981). *Chemical Process Synthesis and Engineering Design* (New York: McGraw-Hill)

KUMAR, A., LUO, J. and BENNETT, G.F. (1993). Statistical evaluation of lower flammability distance (LFD) using four hazardous release models. *Process Safety Prog.*, **12**, 1

KUMAR, C.V., MOHAN, V., KANAPPAN, R. and CHANDRASEKHARAN, K. (1980). The role of conceptual thinking and process visualization in loss prevention in industries in developing countries. *Loss Prevention and Safety Promotion 3*, vol., 3, p. 1486

KUMAR, D.S.S.S., CHIDAMBARAM, M. and GOPALAN, M.N. (1989). Reliability analysis of protective systems for the ammonia-air mixing plant. *J. Loss Prev. Process Ind.*, **2**(2), 95

KUMAR, P. and GREWAL, P.S. (1988). Synthesis gas compressor failure: a case study. *Ammonia Plant Safety*, **28**, 107

KUMAR, R.K. (1989). Ignition of hydrogen−oxygen−diluent mixtures adjacent to a hot, nonreactive surface. *Combust. Flame*, **75**, 197

KUMAR, R.K., TAMM, H. and HARRISON, W.C. (1983). Combustion of hydrogen at high concentrations including the effect of obstacles. *Combust. Sci. Technol.*, **35**, 175

KUMAR, R.K., DEWIT, W.A. and GREIG, D.R. (1989). Vented explosion of hydrogen-air mixtures in a large volume. *Combust. Sci. Technol.*, **66**, 251

KUMAR, R.K., BOWLES, E.M. and MINTZ, K.J. (1992). Large-scale dust explosion experiments to determine the effects of scaling on explosion parameters. *Combust. Flame*, **89**, 320

KUMAR, S. (1986). Reactive scavenging of pollutants by rain: a modeling approach. *Atmos. Environ.*, **20**, 1015

KUMIG, J. and SCHULZ, K.H. (1957). *Dermatologica*, **115**, 540

KÜMPER, E. (1972). Safety of LNG storage tanks. *LNG Conf. 3*

KUNAR, R.R. (1985). Methods of seismic qualification for hazardous facilities in the UK. *Earthquake Engineering in Britain* (London: Instn Civil Engrs), p. 293

KUNESH, J.G. and HOLLENACK, W.R. (1983). Solving pilot-plant problems. *Chem. Engng Prog.*, **79**(8), 51

KUNG, H.C. and STAVRIANIDIS, P. (1982). Buoyant plumes of large-scale pool fires. *Combustion*, **19**, 905

KUNG, H.L. (1980). Man monitoring in occupation medicine. *Loss Prevention and Safety Promotion 3*, vol. 2, p. 241

KUNII, D. and LEVENSPIEL, O. (1991). *Fluidization Engineering*, 2nd ed. (Oxford: Butterworth-Heinemann)

KUNKEL, B.A. (1983). *A Comparison of Evaporative Source Strength Models for Toxic Chemical Spills. Rep. AFGLTR-83-0307.* Air Force Geophysics Lab., Hanscom AFB, MA

KUNKEL, B.A. (1985). *Development of an Atmospheric Diffusion Model for Toxic Chemical Releases. Rep. AFGL-TR-85-0338.* Air Force Geophysics Lab., Hanscom AFB, MA

KUNREUTHER, H. (1984). Compensation as a tool for siting hazardous facilities. *CHEMRAWN III*

KUNREUTHER, H. and LATHROP, J.W. (1981). Siting hazardous facilities: lessons from LNG. *Risk Analysis*, **1**, 289

KUNREUTHER, H.C. and LEY, E.V. (eds) (1982). *Decision Processes and Institutional Aspects of Risk* (Berlin: Springer)

KUNREUTHER, H., LINNEROOTH, J. and STARNES, R. (eds) (1981). Risk Analysis and Decision Processes. *The Siting of Liquefied Energy Gas Facilities in Four Countries* (Berlin: Springer)

KUNSTMANN, G. (1980). Leak rates at valve stems. *Verfahrenstechnik*, **14**, 260

KUNUGI, M. and JINNO, H. (1967). Determination of size and concentration of soot particles in diffusion flames by a light-scattering technique. *Combustion*, **11**, 257

KUNZ, F. and THALMANN, H. (1978). The response behaviour of smoke detectors in theory and practice. *Fire Int.*, **60**, 73

KUNZ, W. and RITTLE, H. (1970). *Issues as Elements of Information Systems. Working Paper 131.* Inst. of Urban and Regional Dev., Univ. of California, CA

KUNZI, H. (1980). Routine investigations of thermic hazards: some uneasiness remains with the examiner. *Loss Prevention and Safety Promotion 3*, vol. 2, p. 386

KUO, K. and SUMMERFIELD, M. (eds) (1983). Fundamentals of solid-propellant combustion. *Prog. Astronaut. Aeronaut.*, **90**

KUO, K.K. (1986). *Principles of Combustion* (New York: Wiley)

KUORINKA, I., LORTIE, M. and GAUTREAU, M. (1994). Manual handling in warehouses: the illusion of correct working postures. *Ergonomics*, **37**, 655

KURLAND, J.J. and BRYANT, D.R. (1987). Shipboard polymerization of acrylic acid. *Plant/Operations Prog.*, **6**, 203

KURTZ (1984). *Handbook of Engineering Economics* (New York: McGraw-Hill)

KURYLA, M. (1993). TSCA traps for the unwary plant manager. *Chem Engng*, **100**(9), 161

KURYLA, M.L. and YOHAY, S.C. (1992). New safety rules add to plant manager's worries. *Chem. Engng*, **99**(6), 153

KURZ, G.R. (1973). Inert-gas protection. *Chem. Engng*, **80**(Mar. 19), 112

KURZAWA, C.J. (1980). Get ready for mega projects. *Hydrocarbon Process.*, **59**(4), 283

KUSHA, A. (1977). Failure of ammonia plant air compressor. *Ammonia Plant Safety*, **19**, 151

KUSHA, A. (1978). Experience of Shahpur ammonia plant. *Ammonia Plant Safety*, **20**, 144

KUSHA, A. and LLOYD, D. (1976). Failure and repair of ammonia converter basket. *Ammonia Plant Safety*, **18**, 35

KUSNETZ, H.L. (1974). Industrial hygiene factors in design and operating practice. *Loss Prevention*, **8**, 20

KUSNETZ, H.L. and LYNCH, J. (1984). The role and impact of industrial hygiene expertise in reducing cancer risks. In Deisler, P.F. (1984b), *op. cit.*, p. 99

KUSNETZ, H.L. and PHILLIPS, C.F. (1983). Industrial hygiene evaluations for control design. *Plant/Operations Prog.*, **2**, 146

KUSSMAUL, K. and KRAEGELOH, E. (1973). The formation, significance and evaluation of welding defects in pressure vessels. *Int. J. Pres. Ves. Piping*, **1**, 263

KUSSMAUL, K., BLIND, D. and EWALD, J. (1977). Investigation methods for the detection and study of stress-relief cracking. *Int. J. Pres. Ves. Piping*, **5**, 159

KUSSMAUL, K., EWALD, J., MAIER, G. and SCHELLHAMMER, W. (1980). Enhancement of the quality of the reactor pressure vessel used in light water power plants by advanced material, fabrication and testing technologies. *Int. J. Pres. Ves. Piping*, **8**

KUTATELADZE, S.S. (1972). *Fluid Mechanics — Soviet Res.*, **1**, 29

KUTATELADZE, S.S. and BORISHANSKII, V.M. (1966). *A Concise Encyclopaedia of Heat Transfer* (New York: Pergamon Press)

KUTZ, M.P. (1986). *Mechanical Engineers Handbook* (New York: Wiley)

KUZNETSOV, N.M. and KOPOTEV, V.A. (1985). Detonation in a relaxing gas. *Combust. Flame*, **61**, 109

KVALSETH, T.O. (1979). A decision theoretic model of the sampling behavior of the human process monitor: experimental evaluation. *Hum. Factors*, **21**, 671

KVAMSDAL, R. (1973). Design aspects of the proposed IMCO code for the bulk carriage of liquefied gas. *Transport of Hazardous Cargoes*, **3**

KWANG SUB JEONG, SOON HEUNG CHANG and TAE WOON KIM (1987). Development of the dynamic fault tree using Markovian process and supercomponent. *Reliab. Engng*, **19**(2), 137

KYBURG, H. and SMOKLER, H. (eds) (1964). *Studies in Analytical Probability* (New York: Wiley)

KYDD, P.H. (1959). Hydrocarbon products from rich hydrocarbon flames. *Combust. Flame*, **3**, 133

KYDD, P.H. and FOSS, W.I. (1964). A comparison of the influence of heat losses and three-dimensional effects on flammability limits. *Combust. Flame*, **8**, 267

KYLE, S.R. (1989). *Piper Alpha Public Inquiry. Transcript, Days 152, 153*

KYTE, W. (1985). Dewpoint corrosion '85. *Chem. Engr, Lond.*, **418**, 29

VAN LAAR, G. (1993). Dust explosion hazards in extraction systems and practical solutions. *Health, Safety and Loss Prevention*, p. 335

LABADIBI, H.M.S. (1990). Hierarchical knowledge models in process engineering. Ph.D. Thesis, Univ. of Leeds

LABADIBI, H.M.S. and MCGREAVY, C. (1992). Model-based diagnostic expert system for chemical plants. *Inst. Elec. Engrs Colloquium on Intelligent Fault Diagnosis, Digest 1992/045* (London), p. 4/1

LABATH, N.A. and AMENDOLA, A. (1989). Analysis of 'domino' effect incident scenarios by the Dylam approach. *Loss Prevention and Safety Promotion 6*, p. 80.1

LABATH, N.A., REAL, M.C., HUESPE, A.J. and MASERA, M.J. (1986). A safety and risk assessment of a hydrogen sulphide storage plant sited at the boundary of a nuclear power plant. *Reliab. Engng, 14*, 223

LABINE, R.A. (1960). How the Air Force liquefies hydrogen. *Chem. Engng, 67*(Jan. 25), 86

LABINE, R.A. (1961). Chemical plants ride out Gulf Coast hurricane as good planning averts major disaster. *Chem. Engng, 68*(Oct. 16), 94

LABOUR INSPECTORATE (1982). *Operational Safety Report. Guideline for the Compilation. Draft Manual CP 3E* (Voorburg, Netherlands)

LABROW, S. (1968). Reciprocating compressor control. *Chem. Process Engng, 49*(12), 70

LACEY, B.G. (1969). Noise and its control in process plants. *Chem. Engng, 76*(Jun. 16), 74

LACEY, P.C.M. (ed.) (1965). *Symp. on Two-Phase Flow, Dept of Chem. Engng, Univ. of Exeter*

LACEY, S. (1922/23). Flow of gas in pipes. *Trans. Inst. Gas Engrs*, 246

LACOURSIERE, J.P., SARLIS, J.N. and RAVARY, P.M. (1998). The SO$_2$SAFE^{TM} Technology for storage and transport of sulfur dioxide. *Proc. of the 32nd Annu. Loss Prev. Symp.*, Mar. 8–12, 1998, New Orleans, LA, Paper 3e (New York: American Institute of Chemical Engineers)

LADD, R. (1966). Brittle fracture of steels. *Chem. Engng, 73*(Jul. 4), 140

LADERMAN, A.J. (1966). Detonability of hydrogen oxygen mixtures in large vessels at low initial pressures. *AIAA J., 4*(10), 1784

LADERMAN, A.J., URTIEW, P.A. and OPPENHEIM, A.K. (1963). On the generation of a shock wave by flame in an explosive gas. *Combustion, 9*, 265

LADICK, R. (1978). The involvement of total encapsulating suits used in the control of hazardous material spills. *Control of Hazardous Material Spills, 4*, 309

LADOU, J. (1981). *Occupational Health Law* (Basel: Dekker)

LAFAVE, V. and WILSON, C.D. (1983). The operation and maintenance of refrigerated LNG and LPG storage systems. *Gastech 82*, p. 339

LAFFITTE, P. and BOUCHET, R. (1959). Suppression of explosion waves in gaseous mixtures by means of fine powders. *Combustion, 7*, 504

LAFFITTE, P. and DELBOURGO, R. (1953). Ignition by condenser sparks. Regions of flammability of ethane, propane, n-butane and n-pentane. *Combustion, 4*, 114

LAFFITTE, P., DELBOURGO, R., COMBOURIEU, J. and DUMONT, J.C. (1965). The influence of particle diameter on the specificity of fine powders in the extinction of flames. *Combust. Flame, 9*, 357

LAFORNARA, J.P. (1978). Feasibility of a mobile hazardous materials spills water treatment system based on ozonolysis. *Control of Hazardous Material Spills, 4*, p. 344

LAFORNARA, J.P. (1982). Sampling, analysis and detection. In Bennett, G.F., Feates and Wilder (1982), *op. cit.*, p. 8.2

LAFORNARA, J.P. and HARSCH, J. (1976). EPA pumps Virginia pond to remove spilled toxaphene. *Control of Hazardous Material Spills, 3*, 404

LAFORNARA, J.P., POLITO, M. and SCHOLZ, R. (1976). Removal of spilled herbicide from a New Jersey Lake. *Control of Hazardous Material Spills, 3*, 378

LAFORNARA, J.P., FREESTOPNE, F.J. and POLITO, M. (1978). Spill clean-up at a defunct industrial waste disposal site. *Control of Hazardous Material Spills, 4*, 152

LAFORNARA, J.P., MARSHALL, M.D., MCGOFF, M.J. and GREER, J.S. (1978). Soil surface sealing to prevent penetration of hazardous material spills. *Control of Hazardous Material Spills, 4*, 296

LAFORNARA, J.P., FRANK, U. and WILDER, I. (1978). Environmental emergency response unit (EERU) operations at the Hopewell Virginia Kepone incident. *Control of Hazardous Material Spills, 4*, 236

LAFORNARA, J.P., LAMPL, H.J., MASSEY, T. and LORENZETTI, R. (1978). Hazardous materials removal during 1977 Johnstown, Pennsylvania, flood emergency. *Control of Hazardous Material Spills, 4*, 156

LAGADEC, P. (1982). *Major Technological Risk* (Oxford: Pergamon Press)

LAGIERE, M., MINISCLOUX, C. and ROUX, A. (1984). Computer two-phase flow model predicts pipeline pressure and temperature profiles. *Oil Gas J., 82*(Apr. 9), 82

LAGRON, J.P., BOULANGER, A. and LUYTEN, W. (1986). LPG and LNG terminals associated with underground storage. *Gastech 85*, p. 421

LAHDENPERA, A., KORHONEN, E. and NYSTROM, L. (1989). An expert system for selection of solid–liquid separation equipment. *Comput. Chem. Engng, 13*, 467

LAHR, P.T. (1989). Better standards, better pumps. *Chem. Engng, 96*(7), 96

LAI, Y.S. (1992). Performance of a safety relief valve under back pressure conditions. *J. Loss Prev. Process Ind., 5*, 55

LAING, I.G. (1986). Legislative developments in the field of health, safety and environmental control: the impact on the chemical industry. *Chem. Ind.*, Apr. 7, 231

LAING, J.S. and HENDERSON, L.B. (1965). Engineering ammonia storage tank insulation for safety. *Ammonia Plant Safety, 7*, 33

LAITINEN, H. (1982). Reporting noninjury accidents: a tool in accident prevention. *J. Occup. Accid., 4*(2/4), 275

LAITINEN, H. (1984). Estimation of potential seriousness of accidents and near accidents. *J. Occup. Accid., 6*(1/3), 167

LAKE, C. (1988). Steam turbines. *Plant Engr, 32*(6), 26

LAKE, I.A. (1986). *Hazard Warning and Fault Tree Analysis*. Warren Centre, Univ. of Sydney

LAKE, R.L., DEMOREY, F.W. and EIBER, R.J. (1968). Performance of 5083 and 7039 aluminium alloys in vessel burst tests and notched test panel tests at − 220°F. *LNG Conf. 1*

LAKEY, J.R.A. (1988). Major emergencies: fixed installations and transport (Conf. rep.). *J. Loss Prev. Process Ind., 1*(1), 52

LAKEY, R.J. (1971). Recent developments with respect to codes and regulations relating to bulk carriage of chemical liquids by sea (including requirements for cargoes with special hazards). *Transport of Hazardous Cargoes, 2*, paper 1

LAKEY, R.J. (1973). Present state of the proposed IMCO code for the bulk carriage of liquefied gases. *Transport of Hazardous Cargoes, 3*

LAKEY, R.J. (1976). The 1973 Marine Pollution Convention's impact on ships transporting hazardous materials. *J. Haz. Materials*, **1**(2), 113

LAKEY, R.J. and THOMAS, W.D. (1983). The LNG/LPG fleet record. *Gastech 82*, p. 119

LAKNER, A.A. and ANDERSON, R.T. (1984). *Reliability Engineering for Nuclear and Other High Technology Systems* (Amsterdam: Elsevier Applied Science)

LAKSHIMISHA, K.N., PAUL, P.J. and MUKUNDA, H.S. (1991). On the flammability limit and heat loss in flames with detailed chemistry. *Combustion*, **23**, 433

LAKSHIMISHA, K.N., ZHANG, Y., ROGG, B. and BRAY, K.N.C. (1992). Modelling auto-ignition in a turbulent medium. *Combustion*, **24**, 421

LAKSHMANAN, R. and STEPHANOPOULOS, G. (1988a). Synthesis of operating procedures for complete chemical plants: I, Hierarchical structured modeling for nonlinear planning. *Comput. Chem. Engng*, **12**, 985

LAKSHMANAN, R. and STEPHANOPOULOS, G. (1988b). Synthesis of operating procedures for complete chemical plants: II, A nonlinear planning methodology. *Comput. Chem. Engng*, **12**, 1003

LAKSHMANAN, R. and STEPHANOPOULOS, G. (1990). Synthesis of operating procedures for complete chemical plants: III, Planning in the presence of qualitative mixing constraints. *Comput. Chem. Engng*, **14**, 301

LALLANDE, A. (1981). Clean air debate begins. *Chem. Engng*, **88**(May 4), 37

LALLANDE, A. (1982). Clean Air Act reform looks better in 1982. *Chem. Engng*, **89**(Feb. 22), 43

LALANNE, M., BERTHIER, P. and DER HAGOPIAN, J. (1983). *Mechanical Vibrations for Engineers* (New York: Wiley)

LAM, V. and SANDBERG, C. (1992). Choose the right heat tracing system. *Chem. Engng Prog.*, **88**(3), 63

LAMANNA, C. (1959). The most poisonous poison. *Science*, **130**(Sep. 25), 763

LAMARSH, J.R. (1965). *Introduction to Nuclear Reactor Theory* (Reading, MA: Addison-Wesley)

LAMB, H. (1932). *Hydrodynamics* (New York: Dover Publs)

LAMB, M. (1980). Control of toxic atmospheres in the pharmaceutical industry. In Hughes, D.J. (1980), *op. cit.*, paper 2

LAMB, R.G. (1979). The effects of release height on material dispersion in the convective planetary boundary layer. *Fourth Symp. on Turbulence, Diffusion and Air Pollution* (Boston, MA: Am. Met. Soc.), p. 27

LAMB, R.G. (1982). Diffusion in the convective boundary layer. In Nieuwstadt, F.T.M. and van Dop, H. (1982), *op. cit.*, p. 159

LAMB, R.G. (1984). Air pollution models as descriptors of cause–effect relationships. *Atmos. Environ.*, **18**, 591

LAMBERT, H.E. (1975). *Fault Trees for Decision-making in Systems Analysis. Rep. UCRL-51829.* Lawrence Livermore Lab., Livermore, CA

LAMBERT, H.E. (1976). *Fault Trees for Location of Sensors in Chemical Processing Systems. Rep. UCRL-78442.* Lawrence Livermore Lab., Livermore, CA

LAMBERT, H.E. (1977). Fault trees for locating sensors in process systems. *Chem. Engn Prog.*, **73**(8), 81

LAMBERT, H.E. (1978a). *Failure Modes and Effects Analysis.* NATO Advanced Study Inst. on Synthesis and Analysis Methods for Safety and Reliability Studies, Urbino, Italy

LAMBERT, H.E. (1978b). *Probabilistic Importance and Applications to Fault Tree Analysis.* NATO Advanced Study Inst.

on Synthesis and Analysis Methods for Safety and Reliability Studies, Urbino, Italy

LAMBERT, H.E. (1979). Comments on the Lapp–Powers 'Computer aided synthesis of fault trees', *IEEE Trans. Reliab.*, **R-26**, 6

LAMBERT, H.E. and GILMAN, F.M. (1977a). *The IMPORTANCE computer code. Rep. UCRL-79269.* Lawrence Livermore Lab., Livermore, CA

LAMBERT, H.E. and GILMAN, F.M. (1977b). The IMPORTANCE computer code. *Am. Inst. Chem. Engrs, Eleventh Symp. on Loss Prevention* (New York)

LAMBERT, H.E. and YADIGAROGLU, G. (1976). *Fault Tree Diagnosis of System Fault Conditions. Rep. UCRL-78170.* Lawrence Livermore Lab., Livermore, CA

LAMBERT, P.G. and AMERY, G. (1989). Assessment of chemical reaction hazards in batch processing. *Runaway Reactions*, p. 525

LAMBERT, P.G., AMERY, G. and WATTS, D.J. (1992). Combine reaction calorimetry with gas evolution measurement. *Chem. Engng Prog.*, **88**(10), 53

LAMBERTON, W.J. and VAUGHAN, F.H. (1974). Will existing equipment fail by catastrophic brittle fracture? *Hydrocarbon Process.*, **53**(9), 217

LAMBERTSEN, C.J. (1974). Respiration. In Mountcastle, V.B. (1974), *op. cit.*, vol. 2, p. 1361

LAMBLE, J.H. (ed.) (1962). *Principles and Practice of Non-Destructive Testing* (London: Heywood)

LAMBOURN, R.F. (1985). The analysis of tachograph charts for road accident investigation. *Forensic Sci. Int.*, **28**, 181

LAMERSE, H. and BOSMAN, M. (1985). Data of gas compressors and instrumentation – hard to collect, easy to analyse. *Reliab. Engng*, **13**(2), 65

LAMIT, L.G. and ENGINEERING MODEL ASSOCIATES INC. (1981). *Engineering Model Building* (Englewood Cliffs, NJ: Prentice-Hall)

LAMMERTINK, E. (1992). Safe storage of propylene oxide, A new approach. *Int. Process Saf. Group Technical Meeting*, April 29 – May 1, 1992, Taormina, Italy, Paper 1.2

LAMOND, D.C. (1962). Hydrogen fire. *Ammonia Plant Safety*, **4**, p. 8

LAMPL, H.J., MASSEY, T. and FREESTONE, F.J. (1978). Assessment and control of a contamination incident, Haverford, Pennsylvania, November–December 1976. *Control of Hazardous Material Spills 4*, p. 145

LANCASTER, J.F. (1964). Pressure vessel steels – US vs Europe. *Hydrocarbon Process.*, **43**(10), 161

LANCASTER, J.F. (1969). Material selection for petroleum and petrochemical plant. *Chem. Engr, Lond.*, **226**, CE67

LANCASTER, J.F. (1970). US vs European reactor steels. *Hydrocarbon Process.*, **49**(6), 84

LANCASTER, J.F. (1971). UK and German material standards. *Hydrocarbon Process.*, **59**(1), 74

LANCASTER, J.F. (1973). Failures of boilers and pressure vessels: their causes and prevention. *Int. J. Pres. Ves. Piping*, **1**, 155

LANCASTER, J.F. (1975). What causes equipment to fail? *Hydrocarbon Process.*, **54**(1), 74

LANCASTER, J.F. (1976). Design against corrosion failure. *Soc. Chem. Ind., Symp. on Loss Prevention* (London)

LANCASTER, J.F. and HOYT, W.B. (1966). US vs British and European piping specifications. *Hydrocarbon Process.*, **45**(10), 127

LANCE, R.C., BARNARD, A.J. and HOOYMAN, J.E. (1979). Measurement of flash points: apparatus, methodology, applications. *J. Haz. Materials*, **3**(1), 107

VAN'T LAND, C.M. (1984). Selection of industrial dryers. *Chem. Engng*, **91**(Mar. 5), 53

LAND NORDRHEIN-WESTFALEN (1981). Administrative orders for the implementation of the Accident Ordnance. *Ministerialblatt fur das Land Nordrhein-Westfalen*, **34**(64), 1400

LANDAU, H.G. (1937a). The ignition of gases by local sources. *Chem. Rev.*, **21**, 245

LANDAU, H.G. (1937b). The ignition of gases by local sources. *Combustion*, **2**, 127

LANDAU, L.D. and LIFSHITZ, E.M. (1959). *Fluid Mechanics* (London: Pergamon Press)

LANDAU, R. (ed.) (1967). *The Chemical Plant from Process Selection to Commercial Operation* (New York: Reinhold)

LANDAU, R. (1981). Process innovation. *Chem. Ind.*, May 2, 321

LANDAU, R.N. and CUTRO, R.S. (1993). Assess risk due to desired chemistry. *Chem. Engng Prog.*, **89**(4), 66

LANDAU, R.N. and WILLIAMS, L.R. (1991). Reactor calorimetry: a powerful tool. *Chem. Engng Prog.*, **87**(12), 65

LANDER, C. (1942). Review of recent progress in heat transfer. *Proc. Instn Mech. Engrs*, **148**, 81

LANDEWEERD, J.A. (1979). Internal representation of a process, fault diagnosis and fault correction. *Ergonomics*, **22**, 1343

LANDEWEERD, J.A. and ROOKMAAKER, D.P. (1980). Ergonomic systems design in two maintenance departments. Theory and practice. *Ergonomics*, **23**, 727

LANDEWEERD, J.A., SEEGERS, H.J.J.L. and PRAAGMAA, J. (1981). Effects of instruction, visual imagery and educational background on process control performance. *Ergonomics*, **24**, 133

LANDIS, J.G., LINNEY, R.E. and HANLEY, B.F. (1994). *Dispersion of instantaneous jets. Process Safety Prog.*, **13**, 35

LANDIS, R.L. (1982). Solve flashing-fluid critical-flow problems. *Chem. Engng*, **89**(Mar. 8), 79

LANDIS, R.L. and HAMILTON, J.E. (1984). Technique evaluates engineering alternatives. *Chem. Engng*, **91**(Oct. 1), 91

LANDPHAIR, H.C. and MOTLOCH, J.L. (1985). *Site Reconnaissance and Engineering* (Amsterdam: Elsevier)

LANDROCK, A.H. (1983). *Handbook of Plastics Flammability and Combustion Toxicology* (Park Ridge, NJ: Noyes Data Corp.)

LANDRUM, H. and WATSON, M.B. (1980). Problems with synthesis loop heat exchangers. *Ammonia Plant Safety*, **22**, 218

LANDRUM, R.J. (1969). Designing for corrosion resistance. *Chem. Engng*, **76**(Feb. 24), 118; (Mar. 24), 172

LANDRUM, R.J. (1970). Designing for corrosion resistance. Equipment. *Chem. Engng*, **77**(Oct. 12), 75

LANDRY, W.A. and HILLER, A.E. (1975). Hazards involved in transportation of liquid sulfur: sulfur gases and their control. *Transport of Hazardous Cargoes*, **4**

LANDY, J.A. (1964a). Basic check points in fire protection layout. In Vervalin, C.H. (1964a), *op. cit.*, p. 215

LANDY, J.A. (1964b). Basic fire protection systems for chemical plant. *Chem. Engng*, **71**(Apr. 27), 139

LANDY, J.A. (1964c). Check these points on fire protection layout. *Hydrocarbon Process.*, **43**(3), 168

LANE, A.M. and HOLZBOCK, W.G. (1962). How to save with hydrostatic drives. *Chem. Engng*, **69**(Apr. 2), 111

LANE, D.A. and THOMSON, B.A. (1981). Monitoring a chlorine spill from a train derailment. *J. Air Poll. Control Ass.*, **31**, 122

LANE, J.R., RENWICK, P.D. and AL-HASSAN, T. (1994). The HSE offshore fire and explosion research programme. *J. Loss Prev. Process Ind.*, **7**, 376

LANE-SERFF, G.F., LINDEN, P.F. and HILLEL, M. (1993). Forced, angled plumes. *J. Haz. Materials*, **33**, 75

LANEY, T.J. (1963). Lessons from the Aramco tank fire. *Hydrocarbon Process.*, **42**(11), 247

LANEY, T.J. (1964). Lessons from the Aramco tank fire. In Vervalin, C.H. (1964a), *op. cit.*, p. 347

LANG, A. (1962). Air plant explosion. *Ammonia Plant Safety*, **4**, 71. See also *Chem. Engng Prog., 1962*, **58**(2), 71

LANG, A. (1965). Explosion of condensers in an oxygen plant. *Ammonia Plant Safety*, **7**, 60

LANG, R.Z.J., MOORHOUSE, J. and PAUL, G.J. (1980). Waterproof insulation materials. *Chemical Process Hazards*, **VII**, 275

LANGDON, R.M. (1983). Vibrating transducers for chemical plant. *Chem. Engrs, Lond.*, **397**, 33

LANGDON-THOMAS, G.J. (1967a). Fire regulations and the architect. *RIBA J.*, Jul., 286

LANGDON-THOMAS, G.J. (1967b). The Fire Regulations and their technical background. *The Building Regulations: Health, Administration, Fire and Safety* (London: R. Soc. Health)

LANGDON-THOMAS, G.J. (ca. 1967c). *The Fire Regulations and their Technical Background* (London: HM Stationery Office)

LANGDON-THOMAS, G.J. (1972). *Fire Safety in Buildings, Principles and Practice* (London: Black)

LANGEFORS, U. and KIHLSTROM, B. (1963). *The Modern Technique of* 2nd ed. (Stockholm: Almqvist & Wiksell); 1978, 3rd ed. (Stockholm: AWE/GEBERS)

LANGELUEDDECKKE, J. (1991). Safety measures for ammonia. The experience of the BASF works fire department. *Ki, Klima, Kaelte, Heiz.*, **19**(9), 343

VAN LANGENHOVE, H., LOOTENS, A. and SCHAMP, N. (1988). Objective evaluation of an odour nuisance problem based on inquiry results. *Atmos. Environ.*, **22**, 2509

LANGER, B.S. (1983). Wastewater reuse and recycle in petroleum refineries. *Chem. Engng Prog.*, **79**(5), 67

LANGEVELD, J.M. (1976). Structural design of control rooms. *Process Industry Hazards*, p. 205

LANGFORD, C.G. (1983). Installing, maintaining and troubleshooting control valves. *Chem. Eng.*, **90**(Sep. 5), 101

LANGHAAR, J.W. (1953). Cooling pond may answer your cooling water problem. *Chem. Engng*, **60**(8), 194

LANGLANDS, J.H.M., WALLACE, W.F.M. and SIMPSON, M.J.C. (1971). Insulation workers in Belfast, 2: Mortality in men still at work. *Br. J. Ind. Med.*, **28**, 217

LANGLEY, B.C. (ca. 1986). *Plant Maintenance* (Englewood Cliffs, NJ: Prentice-Hall)

LANGLEY, E. (1978). Health and safety aspects of the proposed notification scheme for new chemicals. *Chem. Ind.*, Jul. 15, 504

LANGVARDT, P.W. (1985). Acrylonitrile. *Ullmanns Encyclopaedia of Chemical Technology*, rev. ed., vol. A1 (Weinheim: VCH), p. 177

LANNOY, A. and LEROY, E. (1989). A practical way of assessing semi-confined explosion effect. *Loss Prevention and Safety Promotion 6*, p. 99.1

LANORE, J.M. and KALLI, H. (1977). Usefulness of the Monte Carlo method in reliability calculation. *Trans ANS*, **27**, 374

LANOUETTE, W.J. (1980). The Kemeny Commission report. *Bull. Atom. Scientists*, Jan., 20

LANS, H.J.D. and BJORDAL, E.N. (1983). Application of risk analysis techniques. *Loss Prevention and Safety Promotion 4*, vol. 1, p. G46

LANSDOWN, A.R. (1982). *A Practical Guide to Lubricant Selection* (Oxford: Pergamon Press)

LANTZY, R.J. (1992). Vapor cloud source modeling workshop chairman's report. *Plant/Operations Prog.*, **11**, 41

LANTZY, R.J., MYERS, R.D., PFENNUING, D.B. and MILLSAP, S.B. (1990). Atmospheric release tests of monomethylamine. *J. Loss Prev. Process Ind.*, **3**(1), 77

LAPEDES, D.N. (ed.) (1978). *McGraw-Hill Encyclopaedia of Science and Technology*, 4th ed. (New York: McGraw-Hill)

LAPIDUS, L. and AMUNDSON, N.R. (1977). *Chemical Reactor Theory* (Englewood Cliffs, NJ: Prentice-Hall)

LAPIN, A. and FOSTER, R.H. (1967). Oxygen diffusion in the atmosphere from liquid oxygen pools. *Cryogenic Engng*, **13**, 555

LAPINA, R.P. (1975). Can you rerate your centrifugal compressor. *Chem. Engng*, **82**(Jan. 20), 95

LAPINA, R.P. (1982). How to use the performance curves to evaluate behavior of centrifugal compressors. *Chem. Engng*, **89**(Jan. 25), 86

LAPORTE, T. (1982). On the design and management of nearly error-free organizational control systems. In Sills, D. (1982), *op. cit.*

LAPP, S. (1981). How to avoid problems with cold service pumps. *Hydrocarbon Processing*, **60**(1), 127

LAPP, S.A. (1990). The major risk index system. *Plant/Operations Prog.*, **9**, 176

LAPP, S.A. and POWERS, G.J. (1977a). Computer-aided synthesis of fault trees. *IEEE Trans Reliab.*, **R-26**, 2

LAPP, S.A. and POWERS, G.J. (1977b). Computer assisted generation and analysis of fault trees. *Loss Prevention and Safety Promotion 2*, p. 377

LAPP, S.A. and POWERS, G.J. (1977c). A method for the generation of fault trees. In Gangadharan, A.C. and Brown, S.J. (1977), *op. cit.*, p. 95

LAPP, S.A. and POWERS, G.J. (1979). Update of the Lapp–Powers fault-tree synthesis algorithm. *IEEE Trans Reliab.*, **R-29**, 12

LAPPLE, C.E. (1943). Isothermal and adiabatic flow of compressible fluids. *Trans Am. Soc. Chem. Engrs*, **39**, 385

LAPPLE, C.E. (1962). One page chart sums up characteristics of fine particles. *Chem. Engng*, **69**(Jun. 11), 205

LAPPLE, C.E. (1970). Electrostatic phenomena with particulates. *Adv. Chem. Engng*, **8**, 1

DE LARDEREI, J.A. (1988). Promoting awareness and preparation for emergencies at local level: a worldwide programme. *Preventing Major Chemical and Related Process Accidents*, p. 733

LARK, P.D., CRAVEN, B.R. and BOSWORTH, R.C.L. (1968). *The Handling of Chemical Data* (Oxford: Pergamon Press)

LARKEN, J. (1992). Command training for emergencies. *Health and Safety in the Offshore Oil and Gas Industries* (Rugby: Instn Chem. Engrs)

LAROCHE, G. (1972). Biological effects of short-term exposures to hazardous materials. *Control of Hazardous Material Spills 1*, p. 199

LARSEN, J.R. and OLSON, W. (1957). *Measurements of Air Blast Effects from Simulated Nuclear Core Excursions. Rep. BRLM-1102*. Atom. Energy Comm., Washington, DC

LARSEN, R.I. (1969). A new mathematical model of air pollution concentration averaging time and frequency. *J. Air Poll. Control Ass.*, **19**(1), 24

LARSEN, S.E. and GRYNING, S.E. (1986). Dispersion climatology in a coastal zone. *Atmos. Environ.*, **20**, 1325

LARSEN, T.E. (1984). Flame detectors: reducing false alarms. *Fire Surveyor*, **13**(1), 21

LARSON, D.W., REESE, R.T. and WILMOT, E.L. (1983). The Caldecott Tunnel fire – environments, regulatory considerations and probabilities. *Proc. Seventh Int. Symp. on Packaging and Transportation of Radioactive Materials, New Orleans, LA*, p. 1620

LARSON, J. and MICHALSKI, R.S. (1977). *Inductive Inference in the Variable Value Predicate Logic System VL2: Methodology and Computer Implementation. Rep. 869*. Dept of Comput. Sci., Univ. of Illinois, Urban, IL

LARSON, R.A. and WEBER, E.J. (1994). *Reaction Mechanisms in Environmental Organic Chemistry* (Boca Raton, FL: Lewis)

LARSSON, L.H. (1984). *Subcritical Crack Growth due to Fatigue, Stress Corrosion and Creep* (Amsterdam: Elsevier Applied Science)

DE LASA, H.I. (ed.) (1986). *Chemical Reactor Design and Technology* (Nijhoff)

LASATER, R.C. and HOPKINS, J.H. (1977). Removing particulates from stack gases. *Chem. Engng*, **84**(Oct. 17), 111

LASH, L.D. and KOMINEK, E.G. (1975). Primary-waste-treatment methods. *Chem. Engng*, **82**(Oct. 6), 49

LASHER, R.J. (1989). Integrity testing of process control systems. *Control Engng*, **36**(10)

LASHOVER, J.H. (1985). Computerized emergency system provides rapid warning of hazardous chemical leaks. *Chem. Processing*, Sep., 39

LASK, E. (1986). Naples fuel depot fire contained. *Chem. Engng*, **93**(Jan. 20), 20C

LASKA (1970). Safety distances in great storage zones of dangerous gases. *Prevention of Occupational Risks*, **1**

LASKIN, S. and GOLDSTEIN, B.D. (eds) (1977). *Benzene Toxicity* (New York: McGraw-Hill)

LASKOWSKI, J. (1984). Ferrography powerful maintenance technique. *Oil Gas J.*, **82**(Nov. 5), 85

LASSEIGNE, A.H. (1984). The NTSB Hazardous Material Spill Map program – an update. *Plant/Operations Prog.*, **3**, 94

LATHA, P., GAUTAM, G. and RAGHAVAN, K.V. (1992). Strategies for the quantification of thermally initiated cascade effects. *J. Loss Prev. Process Ind.*, **5**, 18

LATHAM, D.J., KIRBY, B.R. and THOMSON, G. (1987). The temperatures attained by unprotected structural steelwork in experimental natural fires. *Fire Safety J.*, **12**(2), 139

LATHROP, J. and LINEROOTH, J. (1983). The role of risk assessment in a political decision process. *Analysing and Aiding Decision Processes* (ed. P. Humphreys, O. Svenson and A. Vari) (Amsterdam: North Holland)

LATHROP, J.K. (1977). Fire fighter casualties in LP gas tank rupture incidents. *J. NFPA*, **82**

LATHROP, J.K. (1978). 54 killed in two grain elevator explosions. *Fire J.*, Sep

LATHROP, J.K. (1979). LP-gas truck fire near Mexico City, Mexico, kills twelve. *Fire J.*, Jan., 57

LATHROP, S.P. and LARSON, C.D. (1978). Spill of PVC latex on a public water supply watershed: a case history. *Control of Hazardous Material Spills 4*, p. 161

LATINO, C.J. (1987). Solving human-caused failure problems. *Chem. Engng. Prog.*, **83**(5), 42

LATOMBE, J. (ed.) (1978). *Artificial Intelligence and Pattern Recognition in Computer Aided Design* (New York: North Holland)

LATREILLE, P.J. (1985). On the structural details of gas tankers. *Gastech 84*, p. 359

LATTY, C. (1987). Nuclear valves – signed, sealed and delivered. *Process Engng*, Aug., 41

LATZKO, D.G.H., TURNER, C.E., LANDES, J.D., MCCABE, D.E. and HELLEN, T.K. (1984). *Post-Yield Fracture Mechanics*, 2nd ed. (Amsterdam: Elsevier Applied Science)

LAUBSCH, J. (1984). Advanced LISP programming. In T. O'Shea and M. Eisenstadt (1984), *op. cit.*, p. 110

LAUDERBACK, W. (1972). Unique flare system retards smoke. *Hydrocarbon Processing.*, **51**(1), 127

LAUDERBACK, W. (1975). Private communication (to N.A. Eisenberg *et al.*)

LAUER, J.D. and ANTAL, P.A. (1972). Electrostatic charges and the flow of organic fluids. *Loss Prevention*, **6**, 34

LAUGHLIN, R.G.W., GALLO, T. and ROBEY, H. (1983). Wet air oxidation for hazardous waste control. *J. Haz. Materials*, **8**(1), 1

LAUNDER, B.E. and SPALDING, D.B. (1972). *Mathematical Models of Turbulence*. (New York: Academic Press)

LAUR, C.E. (1970). Arzew liquefaction plant – assessment of five years of operation. *LNG Conf. 2*

LAURENDEAU, N.M. (1982). Thermal ignition of methane–air mixtures by hot surfaces: a critical examination. *Combust. Flame*, 1982, **46**, 29

LAURENDEAU, N.M. and CARON, R.N. (1982). Influence of hot surface size on methane–air ignition temperature. *Combust. Flame*, **46**, 213

LAUTKASKI, R. (1992). Validation of flame drag correlations with data from large pool fires. *J. Loss Prev. Process Ind.*, **5**, 175

LAUTKASKI, R. and FIEANDT, J. (1980). Risk assessments of the transportation of hazardous liquefied gases in bulk. *Loss Prevention and Safety Promotion 3*, vol. 3, p. 1052

LAUTKASKI, R. and MANKAMO, T. (1977). A risk assessment of the rail transportation of liquid chlorine. *Loss Prevention and Safety Promotion 2*, p. 407

LAUTZENHEISER, C.E. (1974). In-service inspection programmes in the United States. *Int. J. Pres. Ves. Piping*, **2**, 1

LAUWERYS, R.R. (1983). *Industrial Chemical Exposure: Guidelines for Biological Monitoring* (Chichester: Wiley)

VON LAVANTE, E. and STREHLOW, R.A. (1983). The mechanism of lean flame extinction. *Combust. Flame*, **49**, 123

LAVE, L.B. and SESKIN, E.P. (1977a). *Air Pollution and Human Health* (Baltimore, MD: John Hopkins Press)

LAVE, L.B. and SESKIN, E.P. (1977b). Does air pollution shorten lives? In Fairley, W.B. and Mosteller, F. (1977), *op. cit.*, p. 143

LAVERMAN, R.J. (1984). Evaporation loss control effectiveness for storage tanks. *API Proc., Refining Dept, 49th Midyear Mtg*, **63**, 287

LAVERY, T.F., HOLZWORTH, G.C., SCHIERMEIER, F.A., SNYDER, W.H., TRUPPI, L.E., GREENE, B.R. and EGAN, B.A. (1986). The full scale plume study: a summary of data collected and phenomena observed. *Fifth Joint Conf. on Applications of Air Pollution Meteorology* (Boston, MA: Am. Met. Soc), p. 211

LAVIRON, A. (1985). Simulation and reliability analysis of sequential complex systems with ESCAF. *Reliab. Engng*, **12**(3), 139

LAVIRON, A. (1986). Analysing critical components in complex system design: a comparison of methods with the ESCAF approach. *Reliab. Engng*, **15**(3), 209

LAVIRON, A. and HEISING, C.D. (1985). Error transmission in large complex fault trees using the ESCAF method. *Reliab. Engng*, **12**(3), 181

LAVIRON, A., CARNINO, A. and MANARANCHE, J.C. (1982). ESCAP – a new and cheap system for complex reliability analysis and computation. *IEEE Trans. Reliab.*, **R-31**, 339

LAW, C.K. (1975). Asymptotic theory for ignition and extinction in droplet burning. *Combust. Flame*, 24, 89

LAW, C.K. (1978). On the stagnation point ignition of a premixed combustible. *Int. J. Heat Mass Transfer*, **21**, 1363

LAW, C.K. (1989). Dynamics of stretched flames. *Combustion*, **22**, 1381

LAW, C.K. and EGOLFOPOULOS, F.N. (1991). A kinetic criterion of flammability limits: the C–H–O–inert system. *Combustion*, **23**, 413

LAW, C.K. and EGOLFOPOULOS, F.N. (1992). A unified chain-thermal theory of fundamental flammability limits. *Combustion*, **24**, 137

LAW, C.K. and LAW, H.K. (1979). Thermal ignition analysis in boundary layer flows. *J. Fluid Mechanics*, **92**, 97

LAW, C.K. and WILLIAMS, F.A. (1972). Kinetics and convection in the combustion of alkane droplets. *Combust. Flame*, **19**, 393

LAW, M. (1969). Structural fire protection in the process industry. *Building*, **217**(Jul. 18), 29/86

LAW, M. (1971). *A Relationship Between Fire Grading and Building Design and Contents. Fire Res. Note 877*. Fire Res. Station, Borehamwood

LAW, M. (1972). Nomograms for the fire protection of standard steelwork. *Fire Prev. Sci. Technol.*, **3**, 19

LAW, M. (1991). Fire grading and fire behaviour. *Fire Safety J.*, **17**, 147

LAW COMMISSION (1970a). *Civil Liability for Dangerous Things and Activities. Law Comm. 32* (London: HM Stationery Office)

LAW COMMISSION (1970b). *Strict Liability and the Enforcement of the Factories Act 1961. Law Comm. 30* (London: HM Stationery Office)

LAWLESS, E. (1974). *Technology and Social Shock – 100 Cases of Public Concern over Technology, Midwest Res. Inst., Kansas City, KS*

LAWLESS, J.F. (1982). *Statistical Models and Methods for Lifetime Data* (New York: Wiley)

LAWLEY, H.G. (1974a). Operability studies. *Course of Process Safety – Theory and Practice*. Dept of Chem. Engng, Teesside Polytechnic, Middlesbrough

LAWLEY, H.G. (1974b). Operability studies and hazard analysis. *Loss Prevention*, **8**, 105

LAWLEY, H.G. (1976). Size up plant hazards this way. *Hydrocarbon Process.*, **55**(4), 247

LAWLEY, H.G. (1980). Safety technology in the chemical industry: a problem in hazard analysis with solution. *Reliab. Engng.*, **1**(2), 89

LAWLEY, H.G. (1981). Private communication

LAWLEY, H.G. and KLETZ, T.A. (1975). High-pressure-trip systems for vessel protection. *Chem. Engng*, **82**(May 12), 81

LAWN, C.J., STREET, P.J. and BAUM, M.M. (1981). Spontaneous combustion in beds of small fuel particles. *Combustion*, **18**, 731

LAWRENCE, A. (1983). Environmental-rule changes will affect facility planning. *Chem. Engng*, **90**(Jan. 10), 85

LAWRENCE, D. (1996). *Quantifying inherent safety for the assessment of chemical process routes* (Loughborough, UK: Loughborough University)

LAWRENCE, D. and EDWARDS, D.W. (1994). Inherent safety assessment of chemical process routes by expert judgement. *The 1994 IChemE Research Event*, 886–88

LAWRENCE, D., EDWARDS, D.W. and RUSHTON, A.G. (1993). Quantifying inherent safety. *Proc. Inst. Mech. Eng. 4*, 1–8

LAWRENCE, D.J. (1980). International safety standards in a technological age – role of PDMS. *Interflow 80* (Rugby: Instn Chem. Engrs), p. 53

LAWRENCE, E.K. (1952). Analytical study of flame initiation. *SM Thesis, Massachusetts Inst. Technol., Cambridge, MA*

LAWRENCE, F.E. (1973). Safety in the chemical industry. *Chem. Engr, Lond.*, **272**, 211

LAWRENCE, G.M. (1984). Shell rupture of a secondary reformer. *Ammonia Plant Safety*, **24**, 83. See also *Plant/Operations Prog.*, 1984, **3**, 60

LAWRENCE, G.M. (1986). Multiple cracking and leakage of a hot synthesis gas pipe. *Ammonia Plant Safety*, **26**, 239. See also *Plant/Operations Prog.*, 1986, **5**, 175

LAWRENCE, G.M. (1989). Explosion in a nitric acid tail gas duct. *Ammonia Plant Safety*, **29**, 267. See also *Plant/Operations Prog.*, 1989, **8**, 33

LAWRENCE, G.N. and CROSS, S.R. (1994). Control emissions from marine vessel loading. *Chem. Engng Prog.*, **90**(3), 61

LAWRENCE, J. (1986). Towards more effective protection of the aquatic environment – an industry view. *Chem. Ind.*, Jun. 2, 380

LAWRENCE, J.A. (1966). Ammonia tank leak. *Ammonia Plant Safety*, **8**, p. 30

LAWRENCE, J.A. (1977). High temperature shift converter problems. *Ammonia Plant Safety*, **19**, 46

LAWRENCE, J.A. and BUSTER, A.A. (1972). Guide to trouble-free plant operation ... computer-process interface. *Chem. Engng*, **79**(Jun. 26), 102

LAWRENCE, J.C. (1991). The mortality of burns. *Fire Safety J.*, **17**, 205

LAWRENCE, J.C. and BULL, J.P. (1988). Fire casualty experience. *Fire Materials*, **13**, 109

LAWRENCE, L.A.J. (1965–66). High reliability servos. Development of redundancy techniques for the achievement of highly reliable autocontrol of nuclear reactors. *Proc. Instn. Mech. Engrs*, **180**(pt 1), 237

LAWRENCE, W.E. and JOHNSON, E.E. (1974). Design for limiting explosion damage. *Chem. Engng*, **81**(Jan. 7), 96

LAWRENCE, W.W. and COOK, S.E. (1967). The thermal decomposition of ethylene. *Loss Prevention*, **1**, p. 10

LAWRENCE, W.W. and COOK, S.E. (1970). Initiation and propagation of explosive reactions in chlorine–ethane mixtures. *Ind. Engng Chem. Process Des. Dev.*, **9**, 47

LAWSON, D.I. and SIMMS, D.L. (1952a). The ignition of wood by radiation. *Br. J. Appl. Phys.*, **3**(9), 288

LAWSON, D.I. and SIMMS, D.L. (1952b). The ignition of wood by radiation. *Br. J. Appl. Phys.*, **3**(12), 394

LAWSON, J.R. and QUINTIERE, J.G. (1985). *Slide Rule Estimates of Fire Growth. Rep. NSBRI 85-3196.* Nat. Bureau Standards, Washington, DC

LAWSON, E. and DENKOWSKI, W.J. (1971). Power actuators for valves. *Chem. Engng*, **78**(Oct. 11), 135

LAWSON, H.H. (1964). Aluminized steel protects insulation. *Chem. Engng*, **71**(Aug. 31), 94

LAWSON, R.E., SNYDER, W.H. and THOMPSON, R.S. (1989). Estimation of maximum surface concentrations from sources near to complex terrain in neutral flow. *Atmos. Environ.*, **23**, 321

LAWSON, S.G. and RASBASH, D.J. (1976). Heat stress in firefighting. *Fire Engng J.*, **36**(102), 16

LAWSON, V. (ed.) (1982). *Practical Experience of Machine Translation* (Amsterdam: North Holland)

LAWTHER, P.J. (1979). Epidemics of non-infectious disease. In Doll, Sir R. and McLean, A.E.M. (1979), *op. cit.*, p. 63. See also *Proc. R. Soc.*, 1979, *Ser. B*, **205**, 63

LAXTON, J. (1985). High-temperature insulation: making the right choice. *Process Engng*, Jul., 61

LAYFIELD, SIR F. (1987). *Sizewell B Public Inquiry* (London: HM Stationery Office)

LAYLAND, D.E., MCNAUGHTON, D.J. and BODNER, P.M. (1986). Assessing the reliability of dispersion models for hazardous materials releases. *Fifth Joint Conf. on Applications of Air Pollution Modeling* (Boston, MA: Am. Met. Soc.), p. 295

LAYTON, D. (1978). Some economic aspects of corrosion control. *Chem. Ind.*, Sep. 2, 647

LAZAR, E.C. (1976a). Damage incidents from improper land disposal. *J. Haz. Materials*, **1**(2), 157

LAZAR, E.C. (1976b). Summary of damage incidents from improper land disposal. *Control of Hazardous Material Spills 3*, p. 437

LAZARI, L., BURLEY, S.J. and AL-HASSANI, S.T.S. (1991). Containment of explosive blast in a spherical vessel with nozzles. *Hazards*, **XI**, p. 135

LAZARI, L., BURLEY, S.J. and AL-HASSANI, S.T.S. (1991). Containment of explosive blast in a spherical vessel with nozzles. *IChemE Symp. Series* No. 124, 135–53

LAZORKO, K. (1973). Now: air quality standards. *Chem. Engng*, **80**(Apr. 2), 28

LAZZERI, L. (1979). Application of plastic analysis to the pipe whip problem. *Int. J. Pres. Ves. Piping*, **7**, 183

LE, N.B., SANTAY, A.J. and ZABRENSKI, J.S. (1988). Laboratory safety design criteria. *Plant/Operations Prog.*, **7**, 87

LEA, F.C. (1930). *Hydraulics*, 5th ed. (London: Arnold)

LEA, R.F. (1977). Ambulance services. In Easton, K.D. (1977), *op. cit.* p. 410

LEACH, G. (1972). *The Biocrats* (Rickmansworth: Penguin)

LEACH, J.S.L. (1978). The predictability of corrosion problems. *Chem. Ind.*, Sep. 2, 643

LEACH, L.L. (1962). Insulation for cryogenic vessels. *Hydrocarbon Process.*, **41**(11), 241

LEACH, S.J. (1967). Stratification and mixing of fluids of different densities. *Chemical Process Hazards*, **III**, p. 17

LEACH, S.J. and BLOOMFIELD, D.P. (1973). Ventilation in relation to toxic and flammable gases in buildings. *Build. Sci.*, **8**, 289

LEAGUE OF RED CROSS SOCIETIES (1976). *Red Cross Disaster Relief Handbook* (Geneva)

LEAHEY, D.M. (1987). The vertical extent of the loglinear wind profile under stable stratification (letter). *Atmos. Environ.*, **21**, 261

LEAHEY, D.M. and DAVIES, M.J.E. (1984). Observations of plume rise from sour gas flares. *Atmos. Environ.*, **18**, 917

LEAHEY, D.M. and SCHROEDER, M.B. (1987). Observations and predictions of jet diffusion flame behaviour. *Atmos. Environ.*, **21**, 777

LEAR, J.B. (1993). Computer hazard and operability studies. *Sydney Univ. Chem. Engng Symp. on Safety and Reliability of Process Control Systems*

LEBLANC, J.B., SHAN, M.N. and BUIVIDAS, L.J. (1980). Reliability: key to NH₃ profits. *Hydrocarbon Process.*, **59**(3), 68–G

LEBLANC, L. (1994a). Industry marches into deepwater (1954–1994). *Offshore*, Apr., 36

LEBLANC, L. (1994b). Industry marches to the sea: the formative years (genesis – 1954). *Offshore*, Apr., 23

LEBLANC, R.W. (1986). Loss prevention through machinery vibration surveillance and analysis. *Plant/Operations Prog.*, **5**, 179

LEBUSER, U. and SCHECKER, H.G. (1986). Vaporization rates of liquids and liquefied gas. *Loss Prevention and Safety Promotion 5*, p. 33–1

LEBUSER, U. and SCHECKER, H.G. (1987). Vaporisation of liquids from open pools. *Fortschritte der Sicherheitstechnik 1* (Frankfurt-am-Main: DECHEMA), p. 331

LECHAT, J.F. and CAUDRON, S. (1987). Reception of different qualities of LNG in large storage tanks. *Gastech 86*, p. 388

LECK, M.J. and LOWE, D.J. (1985). Development and performance of the gas sensor system. *J. Haz. Materials*, 11(1/3), 65. See also in McQuaid, J. (1985), *op. cit.*, p. 65

LECKNER, B. (1971). The spectral and total emissivity of carbon dioxide. *Combust. Flame*, 17, 37

LECKNER, B. (1972). Spectral and total emissivity of water vapor and carbon dioxide. *Combust. Flame*, 19, 33

LECLERC-BATTIN, F., MARQUAIRE, P.M., COME, G.M. and BARONNET, F. (1991). Autoignitions of gas-phase mixtures of 1,4−dioxane and chlorine. *J. Loss Prev. Process Ind.*, 4, 170

LE CORNU, M.J.P. (1980). Countering the fire hazards of plastic components. *Loss Prevention*, 13, p. 70. See also *Chem. Engng Prog.*, 1980, 76(1), 81

LEDERBERG, J. (1972). The freedom from control of science: notes from the ivory tower. *Southern Calif. Law Rev.*, 45, 609

LEDERER, W. and FENSTERHEIM, R. (1983). *Arsenic* (New York: van Nostrand)

LEDERMAN, P.B. (1977). Physical and chemical methods. *Chem. Engng*, 84(Aug. 15), 135

LEDNICKY, R. and LINDLEY, N.L. (1991). Reduce leaks and corrosion with plastic-lined pipe. *Chem. Engng Prog.*, 87(3), 45

LEE, B.M. (1989). Fire protection systems for LPG/LNG process and storage facilities: the essential features. *Gastech 88*, paper 4.7

LEE, C.K. (1975). Estimates of luminous flame radiation from fires. *Combust. Flame*, 24, 239

LEE, C.K., CHAIKEN, R.F. and SINGER, J.M. (1979). Interaction between duct fires and ventilation flow: an experimental study. *Combust. Sci. Technol.*, 20, 59

LEE, D.A. and BROWNE, J.M. (1993). Implications of varying probability in fault tree analysis. In Melchers, R.E. and Stewart, M.G. (1993), *op. cit.*, p. 165

LEE, E.L. and HORNIG, H.C. (1969). Equation of state of detonation product gases. *Combustion*, 12, 493

LEE, E.S. and REKLAITIS, G.V. (1989). Intermediate storage and the operation of periodic processes under equipment failure. *Comput. Chem. Engng*, 13, 1235

LEE, G.K. (1989). Evaluation of jet dispersion models with data. *Loss Prevention and Safety Promotion 6*, p. 73.1

LEE, G.L. (1980). Sampling: principles, methods, apparatus, surveys. In Waldron, H.A. and Harrington, J.M. (1980), *op. cit.*, p. 39

LEE, H.H. (1985). *Heterogeneous Reactor Design* (Boston, MA: Butterworths)

LEE, H.N. (1979). Atmospheric turbulence and diffusion boundary layer transport model. *Fourth Symp. on Turbulence, Diffusion and Air Pollution* (Boston, MA: Am. Met. Soc.), p. 471

LEE, H.N. (1986). Three dimensional analytical models suitable for gaseous and particulate pollutant transport, diffusion, transformation and removal. *Atmos. Environ.*, 19, 1951

LEE, J. and MORAY, N. (1992). Trust, control strategies and allocation of function in human−machine systems. *Ergonomics*, 35, 1243

LEE, J.H.S. (1977). Initiation of gaseous detonation. *Ann. Rev. Phys. Chem.*, 28, 75

LEE, J.H.S. (1980). The mechanism of transition from deflagration to detonation in vapour cloud explosions. *Prog. Energy Combust Sci.*, 6, 359

LEE, J.H.S. (1983a). Explosion in vessels: recent results. *Plant/Operations Prog.*, 2, 84

LEE, J.H.S. (1983b). Gas cloud explosion − current status. *Fire Safety J.*, 5, 251

LEE, J.H.S. and GUIRAO, C.M. (eds) (1982a). *Fuel−Air Explosions* (Waterloo, Ont.: Univ. of Waterloo Press)

LEE, J.H.S. and GUIRAO, C.M. (1982b). Pressure development in closed and vented vessels. *Plant/Operations Prog.*, 1, 75

LEE, J.H.S. and MATSUI, H. (1977). A comparison of the critical energies for direct initiation of spherical detonations in acetylene-oxygen mixtures. *Combust. Flame*, 28, 61

LEE, J.H.S. and MOEN, I.O. (1980). The mechanism of transition from deflagration to detonation in vapour cloud explosions. *Prog. Energy Combust. Sci.*, 6, 359

LEE, J.H.S. and RAMAMURTHI, K. (1976). On the concept of the critical size of a detonation kernel. *Combust. Flame*, 27, 331

LEE, J.H.S., LEE, B.H.K. and SHANFIELD, I. (1965). Two-dimensional unconfined gaseous detonation waves. *Combustion*, 10, 805

LEE, J.H.S., KNYSTAUTAS, R. and GUIRAO, C.M. (1975). Critical power density for direct initiation of unconfined gaseous detonations. *Combustion*, 15, 53

LEE, J.H.S., KNYSTAUTAS, R., GUIRAO, C.M. and LAM, E. (1975). Direct initiation of explosions in flammable mixtures. *Transport of Hazardous Cargoes*, 4

LEE, J.H.S., GUIRAO, C.M., CHIU, K.W. and BACH, G.G. (1977). Blast effects from vapor cloud explosions. *Loss Prevention*, 11, 59

LEE, J.H.S., KNYSTAUTAS, R. and GUIRAO, C.M. (1982). The link between cell size, critical tube diameter, initiation energy and detonability limits. In Lee, J.H.S. and Guirao, C.M. (1982), *op. cit.*, p. 157

LEE, J.H.S., KNYSTAUTAS, R. and FREIMAN, A. (1984). High speed turbulent deflagrations and transition to detonation in H_2−air mixtures. *Combust. Flame*, 56, 227

LEE, J.H.S., KNYSTAUTAS, R. and CHAN, C.K. (1985). Turbulent flame propagation in obstacle-filled tubes. *Combustion*, 20, 1663

LEE, J.P. and CHOKSHI, N.C. (1978). Foundation design for rotating fans. *Hydrocarbon Process.*, 57(10), 131

LEE, J.W. (1980a). Heat transfer: pre-treatment of cooling water systems. *Ammonia Plant Safety*, 22, 106

LEE, J.W. (1980b). Pre-treatment of cooling water systems. *Chem. Engng Prog.*, 76(7), 56

LEE, K.S. and SOMMERFELD, J.T. (1994). Maximum leakage times through puncture holes for process vessels of various shapes. *J. Haz. Materials*, 38, 27

LEE, K.Y., ZHONG, Z.Y. and TIEN, C.L. (1985). Blockage of thermal radiation by the soot layer in combustion of condensed fuels. *Combustion*, 20, 1629

LEE, R. (1961). New developments in the bulk transport of chemical reagents. *Chem. Engr, Lond.*, 155, A51

LEE, R. (1972). Fast cure for failures. *Chem. Engng*, 79(Jan. 24), 118

LEE, R.C. and MOORE, J.M. (1967). CORRELAP − computerized relationship layout planning. *J. Ind. Engng*, 18(3), 195

LEE, R.H. (1969). Electrical grounding: safe or hazardous? *Chem. Engng*, 76(Jul. 28), 158

LEE, R.L., GRESHO, P.M., CHAN, S.T. and UPSON, C.D. (1983). A three dimensional, finite element model for simulating heavier-than-air gaseous releases over variable terrain. In de Wispelaere, C. (1983), *op. cit.*, p. 555

LEE, R.P. (1976a). Systematical failure analysis. *Chem. Engng*, **83**(Aug. 16), 105

LEE, R.P. (1976b). Tracing causes of metal failures in process equipment. *Chem. Engng*, **83**(Sep. 13), 213

LEE, R.P. (1977a). How poor design causes equipment failures. *Chem. Engng*, **84**(Jan. 31), 129

LEE, R.P. (1977b). Some unusual failure modes. *Chem. Engng*, **84**(Jan. 3), 107

LEE, R.W. (1984). Selection of equipment for removing particulates from gases. *Chem. Ind.*, Jul. 2, 476

LEE, S.C. and BANKOFF, S.G. (1986). Instantaneous pressure drop of gas-liquid two-phase flow in a sharp-edged orifice. *Proc. Fourth Int. Symp. on Multiphase Transport and Particulate Phenomena* (ed. T.N. Veziroglu)

LEE, S.C. and TIEN, C.L. (1981). Optical constants of soot in hydrocarbon flames. *Combustion*, **18**, 1159

LEE, S.D. and MUDD, J.B. (eds) (1979). *Assessing Toxic Effects of Environmental Pollutants* (Ann Arbor, MI: Ann Arbor Science)

LEE, S.L. and OTTO, F.W. (1975). Gross vortex activities in a simple simulated urban fire. *Combustion*, **15**, 157

LEE, S.T. and TIEN, J.S. (1982). A numerical analysis of flame flashback in a premixed laminar system. *Combust. Flame*, **48**, 273

LEE, T.G. and ROBERTSON, A.F. (1960). Extinguishing effectiveness of some powdered materials on hydrocarbon fires. *Fire Res. Abs Revs*, **2**(1)

LEE, T.R. (1981). The public's perception of risk and the question of irrationality. *Proc. R. Soc., Ser. A*, **376**, 5. See also in Warner, Sir F. and Slater, D.H. (1981), *op. cit.*, p. 5

LEE, T.R., BROWN, J. and HENDERSON, J. (1984). The public's attitudes towards nuclear power in the south-west. *Atom*, **336**, 8

LEE, T.Y., OKRENT, D. and APOSTOLAKIS, G. (1979). A comparison of background risks and the incremental seismic risk due to nuclear power plants. *Nucl. Engng Des.*, **53**, 141

LEE, V.W. and TRIFUNAC, M.D. (1987). *Selection of Earthquake Resistant Design Criteria for Nuclear Power Plants – Methodology and Technical Cases. Rep. NUREG/CR-4903*, vols 1–3. Nucl. Regul. Comm., Washington, DC

LEE, W. (1974). Solving the plant engineer's material problems. *Chem. Engng*, **81**(Apr. 29), 158

LEE, W.R. (1977). Lightning injuries and death. In Golde, R.H. (1977), *op. cit.*, p. 521

LEE, W.S., GROSH, D.L., TILLMAN, F.A. and LEE, C.H. (1985). Fault tree analysis, methods, and applications – a review. *IEEE Trans Reliab.*, **R-34**, 194

LEE, Y.T. and APOSTOLAKIS, G.E. (1976). *Probability Intervals for the Top Event Unavailability of Fault Trees. Rep. UCLA-ENG-7663*. Univ. of California, Los Angeles, CA

LEEAH, C.J. (1968–). Better safety planning. *Hydrocarbon Process.*, 1968, **47**(10), 157; **47**(11), 248; **47**(12), 136; 1969, **48**(1), 149

LEECH, D.J. (1972). *Management of Engineering Design* (New York: Wiley)

LEEMING, J. (1989). *Piper Alpha Public Inquiry. Transcript, Days 65, 66*

LEEPER, J.G. (1980). Mercury – LNG's problem. *Hydrocarbon Process.*, **59**(11), 237

LEEPER, J.H. (1973). Human error in merchant marine safety. *Conf. on Bulk Transportation of Hazardous Materials by Water in the Future* (Nat. Acad. Sci., Nat. Acad. Engng, Nat. Res. Coun.), p. 221

LEES, C.S. (1985). Planning for emergencies – the role of the police. *Chem. Ind.*, Jul. 1, 433

LEES, F.P. (1968). An SI conversion table for chemical engineering. *Chem. Engr, Lond.*, **222**, CE341

LEES, F.P. (1970). Functions performed by the human operator and by an on-line computer. *Chem. Engr, Lond.*, **241**, 263

LEES, F.P. (1972). The improvement of plant integrity using a process control computer. *Measmnt Control*, **5**, T118.

LEES, F.P. (1973a). Quantification of man–machine system reliability in process control. *IEEE Trans Reliab.*, **R-22**, 124

LEES, F.P. (1973b). Some data on the failure modes of instruments in the chemical plant environment. *Chem. Engr, London.*, **277**, 418

LEES, F.P. (1974a). The development of the loss prevention approach to safety in the process industries. *Occup. Safety Health*, **4**(7), 10

LEES, F.P. (1974b). Developments in loss prevention in the process industries. The Delft symposium. *Chem. Engr, London.*, **290**, 617

LEES, F.P. (1974c). Research on the process operator. In Edwards, E. and Lees, F.P. (1974), *op. cit.*, p. 386

LEES, F.P. (1974d). Some problems in the drying of organic liquids. *Chem. Ind.*, Jan. 5, 25

LEES, F.P. (1975). Future trends in plant engineering. *The Engineer, Suppl.*, Aug. 28/Sep. 4, 9

LEES, F.P. (1976a). Design for man–machine system reliability in process control. In Henley, E.J. and Lynn, J.W. (1976), *op. cit.*, p. 233

LEES, F.P. (1976b). The reliability of instrumentation. *Chem. Ind.*, Mar. 6, p. 195

LEES, F.P. (1976c). A review of instrument failure data. *Process Industry Hazards*, p. 73

LEES, F.P. (1976d). Visual displays and the man-machine interface. *Acta IMEKO* 1976

LEES, F.P. (1977a). The contribution of the control system to loss prevention. *Chemical Process Hazards*, **VI**, p. 13

LEES, F.P. (1977b). Developments in the control of major hazard installations in the UK. *Loss Prevention and Safety Promotion 2*, p. 25

LEES, F.P. (1977c). Some aspects of failure and other data for use in reliability studies. *Loss Prevention and Safety Promotion 2*, p. 617

LEES, F.P. (1980a). Accident fatality number – a supplementary risk criterion. In *Loss Prevention and Safety Promotion 3*, vol. 2, p. 426

LEES, F.P. (1980b). *Loss Prevention in the Process Industries*, vols 1–2 (London: Butterworths)

LEES, F.P. (1980c). Some aspects of hazard survey and assessment. *Chem. Engrs, Lond.*, **363**, 736

LEES, F.P. (1980d). The work of the UK Advisory Committee on Major Hazards. *CHEMECA 80* (Canberra: Instn Engrs, Aust.), p. 283

LEES, F.P. (1980e). The alarm problem in process control. In *Digital Process Control, Adelaide* (ed. by C.P. Jeffreson) (Adelaide, S. Australia: Dept of Chem. Engng, Univ. of Adelaide), p. 82

LEES, F.P. (1980f). The display problem in process control. In *Digital Process Control, Adelaide* (ed. by C.P. Jeffreson) (Adelaide, S. Australia: Dept of Chem. Engng, Univ. of Adelaide), p. 73

LEES, F.P. (1981a). Computer support for diagnostic tasks in the process industries. In Rasmussen, J. and Rouse, W.B. (1981), *op. cit.*, p. 369

LEES, F.P. (1981b). Hazard survey and assessment (letter). *Chem. Engr, Lond.*, **366**, 121

LEES, F.P. (1982a). A general relation for a single-channel trip system. *Reliab. Engng*, **3**(1), 1

LEES, F.P. (1982b). The hazard warning structure of major hazards. *Trans. Instn Chem. Engrs*, **60**, 211

LEES, F.P. (1982c). Quantitative assessment and reliability engineering of major hazard plants in the context of hazard control. *The Assessment of Major Hazards* (Rugby: Instn Chem. Engrs), p. 225

LEES, F.P. (1983a). The assessment of human reliability in process control. *Instn Chem. Engrs, Conf. on Human Reliability in the Process Control Centre* (London)

LEES, F.P. (1983b). Hazard warning structure: illustrative examples based on actual cases. *Reliability 83*, vol. 2, p. 4B/1

LEES, F.P. (1983c). Hazard warning structure: some implications and applications. *Loss Prevention and Safety Promotion 4*, vol. 1, p. J1

LEES, F.P. (1983d). Life cycle costing. Its applicability and characteristics in the process industries. *Plant Engr*, **27**(1), 31; **27**(2), 25

LEES, F.P. (1983e). Process computer alarm and disturbance analysis: review of the state of the art. *Comput. Chem. Engng*, **7**, 669

LEES, F.P. (1983f). The relative risk from materials in storage and in process. *J. Haz. Materials*, **8**(2), 185

LEES, F.P. (1984). Process computer alarm and disturbance analysis: outline of methods for the systematic synthesis of the fault propagation structure. *Comput. Chem. Engng*, **8**, 91

LEES, F.P. (1985). Hazard warning structure; some illustrative examples based on actual cases. *Reliab. Engng*, **10**, 65

LEES, F.P. (1987). The assessment of major hazards: use of the impact model for injury and damage around a hazard source to determine sensitivity to errors in the constituent models. *J. Haz. Materials*, **14**(2), 259

LEES, F.P. (1991). Piper Alpha and the plant engineer. *Plant Engr*, **35**(Jan./Feb.), 33

LEES, F.P. (1992a). Regulations for the control of major hazards with special reference to Piper Alpha. *Loss Prevention and Safety Promotion 7*, vol. 1, lecture

LEES, F.P. (1992b). Risk analysis and its contribution to health and safety. *Health and Safety in the Offshore Oil and Gas Industries* (Rugby: Instn Chem. Engrs)

LEES, F.P. (1994a). The assessment of major hazards: a model for fatal injury from burns. *Process Safety Environ.*, **72B**, 127

LEES, F.P. (1994b). Major accident inquiries – lawyers or engineers? *Process Safety Environ.*, **72B**, 10

LEES, F.P. and ANG, M.L. (eds) (1989). *Safety Cases* (London: Butterworths)

LEES, F.P. and SAYERS, B. (1976). The behaviour of process operators in emergencies. In Sheridan, T.B. and Johanssen, G. (1976), *op. cit.*

LEES, F.P. and WITHERS, R.M.J. (1992). The assessment of major hazards: the lethal toxicity of chlorine (letter). *J. Haz. Materials*, **32**, 113

LEES, F.P., ANDOW, P.K. and MURPHY, C.P. (1980). The propagation of faults in process plants: a review of the basic event/fault information. *Reliab. Engng*, **1**, 149

LEES, F.P., POBLETE, B.R. and SIMPSON, G.B. (1986). The assessment of major hazards: generalisation of the impact

model for the estimation of injury and damage around a hazard source. *J. Haz. Materials*, **13**(2), 187

LEES, N. (1990). *Hazardous Materials: Sources of Information on Their Transportation* (London: Br. Lib. Sci. Ref. and Inform. Service)

LEES, P., MCALLISTER, D., SMITH, J.R., BRITTON, L.G. and HUGHES, J.F. (1981). Experimental and computational investigation of electrostatic fields in plastic tanks during filling with low conductivity oils. *J. Electrostatics*, **10**, 267

LEES, T.A. (1966). Urea plant corrosion. *Ammonia Plant Safety*, **8**, p. 40

LEESE, W.L.B. and JONES, P. (1983). Epidemiological surveys in the oil industry. *Eleventh World Petrol. Cong.*, London

LEESLEY, M.E. (ed.) (1982). *Computer-Aided Process Plant Design* (Houston, TX: Gulf)

LEESLEY, M.E. and NEWELL, R.G. (1972). The determination of plant layout by interactive computer methods. *Decision, Design and the Computer* (London: Instn. Chem. Engrs), p. 2:20

LEFEBURE, V. (1921). *The Riddle of the Rhine* (London)

LEFEBURE, V. (1928). Chemical warfare. *J. R. United Services Inst.*, **73**, 492

LEFEBVRE, A.H. and REID, R. (1966). The influence of turbulence on the structure and propagation of enclosed flames. *Combust. Flame*, **10**, 355

LEFEBVRE, M.H., ORAN, E.S., KAILASANAITH, K. and VAN TIGGELEN, P.J. (1993). The influence of the heat capacity and diluent on detonation structure. *Combust. Flame*, **95**, 206

LEGENDRE, G.S. and SOLOMON, J.P. (1986). Leak in an ammonia plant reformer convection section natural gas feed preheat coil. *Ammonia Plant Safety*, **26**, 120

LEGGE, T. (1934). *Industrial Maladies* (Oxford: Oxford Univ. Press)

LEHMANN, E.L. (1975). Nonparametrics. *Statistical Methods Based on Ranks* (San Francisco, CA/New York: Holden Day/McGraw-Hill)

LEHMANN, K.B. (1886). Experimental studies on the effect of gases and vapours of importance in industry and hygiene on the organism: ammonia and hydrogen chloride. *Arch. Hyg.*, **5**, 1

LEHMANN, K.B. (1887). Experimental studies on the effect of gases and vapours of importance in industry and hygiene on the organism: chlorine and bromine. *Arch. Hyg.*, **7**, 231

LEHMANN, K.B. (1892). Experimental studies on the effect of gases and vapours of importance in industry and hygiene on the organism: hydrogen sulphide. *Arch. Hyg.*, **14**, 135

LEHMANN, K.B. (1893). Studies on the absorption of toxic gases and vapours by men. *Arch. Hyg.*, **17**, 324

LEHMANN, K.B. (1899a). Experimental studies of acquisition of tolerance to industrial gases (ammonia, chlorine, hydrogen sulphide). *Arch. Hyg.*, **34**, 272

LEHMANN, K.B. (1899b). How much chlorine does a dog take up in a chlorine atmosphere and by which route? *Arch. Hyg.*, **34**, 302

LEHMANN, K.B. (1919). *Kurzes Lehrbuch der Arbeits- und Gewerbehygiene*

LEHMANN, K.B. and FLURY, F. (1938). *Toxikologie und Hygiene des technischen Lösungsmittel* (Berlin: Springer)

LEHMANN, K.B. and SCHMIDT-KELH, L. (1936). The 13 most important chlorinated hydrocarbons of the saturated series in respect of industrial hygiene. *Arch. Hyg.*, **116**, 131

LEHNER, P.E. and ZIRK, D.A. (1987). Cognitive factors in user/expert system interaction. *Hum. Factors*, **29**, 97

LEHRER, I.H. (1981). A new model for free turbulent jets of miscible fluids of different density and a jet mixing time criterion. *Trans Instn Chem. Engrs*, **59**, 247

LEHTO, D.L. and LARSON, R.A. (1969). *Long Range Propagation of Spherical Shockwaves from Explosions in Air. Rep. NOLTR-69-88.* Naval Ordnance Lab., White Oak, MD

LEIBER, C.O. (1977). Betrachtungen zu Explosions- und Detonatationserscheinungen. *J. Occup. Accid.*, **1**(2), 159

LEIBER, C.O. (1980). Explosions of fluid tanks. Empirical results – typical accidents. *J. Occup. Accid.*, **3**(1), 21

LEIBER, C.O. (1985). Approximate quantitative aspects of hot spot. *J. Haz. Materials*, **12**(1), 43; **13**(3), 311

LEIBOVICI, C.F., VERNEUIL, V.S. and YANG, P. (1993). Improve prediction with data reconciliation. *Hydrocarbon Process.*, **72**(10), 79

LEICHT, R., WINGENDER, H.J. and RUPPERT, K.A. (1991). A computer aided approach for chemical plant safety assessment. *First ESReDA Seminar on The Use of Expert Systems in Safety Assessment and Management*

LEIGH, H.R. (1987). *Temperature Measurement and Control* (London: Inst. Elec. Engrs)

LEIGH, R. (1985). Rivals to the thermocouple. *Process Engng*, Dec., 48

LEIGHTON, P.A. (1961). *Photochemistry of Air Pollution* (New York: Academic Press)

LEINEMAN, H. (1970). *VFDB-Z*, **19**(1), 23

LEINGANG, W.J. and VICK, K.A. (1992). Troubleshooting flow-induced vibration on a syngas compressor. *Ammonia Plant Safety*, **32**, 289

LEISCH, S.O., KAUFFMAN, C.W. and SICHEL, M. (1985). Smouldering combustion in horizontal dust layers. *Combustion*, **20**, 1601

LEISER, H. (1992). Towards a goal-setting regime: plans and issues – general overview of developments and current plans. *Health and Safety in the Offshore Oil and Gas Industries* (Rugby: Instn Chem. Engrs)

LEITE, O.C. (1988). Predict flare noise. *Hydrocarbon Process.*, **67**(12), 51

LEITE, O.C. (1990). Structural design of stacks. *Hydrocarbon Process.*, **69**(11), 64

LEITES, R. (1929). Experimental studies on the simultaneous effect of heat and toxic gases on the organism. *Arch. f. Hyg.*, **102**, 91

LEMAY, R.E. and METCALFE, J.K. (1964). Safe handling of phosphorus. *Chem. Engng Prog.*, **60**(12), 69

LEMKE, D. (1976). Heat accumulation method and adiabatic storage for safe evaluation of thermally unstable products. *Prevention of Occupational Risks*, **3**, 103

LEMKOWITZ, S.M. (1992a). A unique program for integrating health, safety, environmental and social aspects into undergraduate chemical engineering education. *Loss Prevention and Safety Promotion 7*, vol. 2, 53.1

LEMKOWITZ, S.M. (1992b). A unique program for integrating health, safety, environment and social aspects into undergraduate chemical engineering education. *Plant/Operations Prog.*, **11**, 140

LEMMON, E.J. (1972). *Beginning Logic* (London: Nelson)

LEMOFF, T.C. (1989). BLEVEs without ignition (letter). *Chem. Engng Prog.*, **85**(11), 8

LEMOINE, V., PRESSOUYRE, G.M., CHARLES, J. and CADIOU, L. (1986). Stress corrosion cracking of 3.5 Ni steel in liquid ammonia. *Gastech 85*, p. 511

LEMONNIER, H., SELMER-OLSEN, S., AARVIK, A. and FJELD, M. (1991). The effect of relaxed mechanical nonequilibrium on gas–liquid critical flow modeling. *Am. Nucl. Soc., Proc. Twenty-Seventh Nat. Heat Transfer Conf.*

LEMOTT, S.R., PESKIN, R.L. and LEVINE, D.G. (1971). Effect of fuel molecular weight on particle ignition. *Combust. Flame*, **16**, 17

LENARD, P. (1882). *Ann. Phys.*, **46**, 584

LENAT, D.B. (1982). AM: an artificial intelligence approach to discovery in mathematics as heuristic search. In Davis, R. and Lenat, D.B. (1980), *op. cit.*

LENAT, D.B. (1983). EURISKO: a program that learns new heuristics and domain concepts: the nature of heuristics III: program design and results. *Artificial Intelligence*, **19**, 189

LENAT, D.B. and BROWN, J.S. (1984). Why AM and EURISKO appear to work. *Artificial Intelligence*, **23**

LENGEMAN, R.A. and HARDISON, L.C. (1964). Automate your pilot plants. *Chem. Engng*, **71**(Jun. 22), 129

LENIOR, T.M.J., RIJNSDORP, J.E. and VERHAGEN, L.H.J.M. (1980). From field operators to central control room operators. An integrated educational, research and consultancy approach. *Ergonomics*, **23**, 741

LENNART, R.P. (1964). Chemicals via pipeline. *Chem. Engng Prog.*, **60**(3), 23

LENOIR, E.M. and DAVENPORT, J.A. (1992). A survey of vapor cloud explosions- second update. *Am. Inst. Chem. Engrs, Twenty-Sixth Symp. on Loss Prevention* (New York), paper 74d

LENOIR, E.M. and DAVENPORT, J.L. (1993). A survey of vapor cloud explosions. *Process Safety Prog.*, **12**, 12

LENOIR, J.M. (1969). Furnace tubes: how hot? *Hydrocarbon Process.*, **48**(10), 97

LENOIR, J.M. (1975). Predict flash points accurately. *Hydrocarbon Process.*, **54**(1), 95

LENZ, L.E. (1970). Reliability design in process plants. *Chem., Engng Prog.*, **66**(12), 42

LENZ, R.R. (1965). *Explosives and Bomb Disposal Guide* (Springfield, IL: Thomas Books)

LEONARD, J.P. and FRANK, M.E. (1979). The coal based chemical complex. *Chem. Engng Prog.*, **75**(6), 68

LEONARD, J.T. (1976). *Pro-static Agents in Jet Fuels. NRL Rep. 8021*

LEONARD, J.T. (1981). Static electricity in hydrocarbon liquids and fuels. *J. Electrostatics*, **10**, 17

LEONARD, J.T. and CARHART, H.W. (1967). Electrical discharges from a fuel surface. *Static Electrification*, **2**, p. 100

LEONARD, J.T. and CARHART, H.W. (1970). Effect of conductivity on charge generation in hydrocarbon fuels flowing through fiber glass filters. *J. Colloid Interf. Sci.*, **32**, 383

LEONARD, J.T. and CLARK, R.C. (1975). Generation of static electricity by carbon dioxide in inerting and fire extinguishing systems. *Static Electrification*, **4**, 301

LEONARD, J.T. and CLARK, R.C. (1977). Ignition of flammable vapors by CO_2 fire extinguishers. *J. Fire Flammability*, **8**, 131

LEONARD, P.E. (1976). Must failure be the motivator? *Loss Prevention*, **10**, 111

LEONARD, R.B. (1969). New corrosion-resistant alloys. *Chem. Engng Prog.*, **65**(7), 84

LEONARD, S.C. (1993). Improve maintenance reliability. *Hydrocarbon Process.*, **72**(8), 162–C

LEONARDOS, G., KENDALL, D. and BARNARD, N. (1969). Odor threshold determinations of 53 odorant chemicals. *J. Air Poll. Control Ass.*, **19**(2), 91

LEONE, F.A. (1990). Hazard materials testing at the US Department of Energy's liquefied gaseous fuel spill facility. *Plant/Operations Prog.*, **9**, 226

LEONE, J.M., RODEAN, H.C. and CHAN, S.T. (1985). *FEM3 Phase Change Model. Rep. UCID-20353*. Lawrence Livermore Nat. Lab., Livermore, CA

LEONELLI, P., STRAMIGIOLI, C. and SPADONI, G. (1994). The modelling of pool vaporization. *J. Loss Prev. Process Ind.*, **7**, 443

LEONG, K.J., MACFARLAND, H.N. and SELLERS, E.A. (1961). Acute sulphur dioxide toxicity. *Arch. Environ. Health*, **3**, 66

LEPKOWSKI, W. (1973). Congress ready to move on control of toxic materials. *Chem. Engng*, **80**(Sep. 17), 90

LEPKOWSKI, W. (1985). Bhopal visited. *Chem. Ind.*, Mar. 4, 143

LEPKOWSKI, W. (1986). Bhopal . . . in the lingering aftermath. *Chem. Ind.*, Mar. 3, 156

LEPLAT, J. (1978a). Accident analyses and work analyses. *J. Occup. Accid.*, **1**, 331

LEPLAT, J. (1978b). Factors determining work-load. *Ergonomics*, **21**, 143

LEPLAT, J. (1981). Task analysis and activity analysis in situations of field diagnosis. In Rasmussen, J. and Rouse, W.B. (1981), *op. cit.*, p. 287

LEPLAT, J. (1985). *Erreur Humain, Fiabilité Humaine dans le Travail* (Paris: Colin)

LEPLAT, J. (1987). Accidents and incidents production: methods of analysis. In Rasmussen, J., Duncan, K. and Leplat, J. (1987), *op. cit.*

LERAY, M., PETIT, P. and PARADOWSKI, H. (1986a). Boil-off gas recovery using reliquefaction in LNG. *Gastech 85*, p. 391

LERAY, M., PETIT, P. and PARADOWSKI, H. (1986b). Boiloff gas vapors are recovered by liquefaction in LNG. *Oil Gas J.*, **84**(Feb. 24), 59

LERMAN, M.J. and CARRABOTTA, S. (1991). Maximize the life of glass-lined reactors. *Chem. Engng Prog.*, **87**(12), 37

LERMANT, J.C. and YIP, S. (1984–). A generalized Semenov model for thermal ignition in nonuniform temperature systems. *Combust. Flame*, **57**, 41; 58, 81, 87; 1986, **63**, 303, 305

LEROU, J.J. and NG, K.M. (1996). Chemical reaction engineering: A multiscale approach to a multiobjective task. *Chem. Eng. Sci.* **51**(10), 1595–614

LEROUX, L. (1933). *La Guerre Chimique* (Paris: E'ds Spes)

LEROY, N.R. and JOHNSON, D.M. (1969). The story of a reservoir fire. *Hydrocarbon Process.*, **48**(5), 129

LESINS, V. and MORITZ, J.J. (1991). Develop realistic safety procedures for pilot plants. *Chem. Engng Prog.*, **87**(1), 39

LESLIE, D.C. (1973). *Developments in the Theory of Turbulence* (Oxford: Oxford Univ. Press)

LESLIE, I.R.M. and BIRK, A.M. (1991). State of the art review of pressure liquefied gas container failure modes and associated projectile hazards. *J. Haz. Materials*, **28**, 329

LESLIE, V.J. and FERGUSON, D. (1985). Radioisotope techniques for solving ammonia plant problems. *Ammonia Plant Safety*, **25**, 105

LESPERANCE, T.W. (1968). Water pollution control. Biological treatment. *Chem. Engng*, **75**(Oct. 14), 89

LESSE, P.F. (1977). Mathematical simulation of a full scale fire test. *Fire Prev. Sci. Technol.*, **16**, 23

LESSENGER, J.E. (1985). Medical management of anhydrous ammonia emergencies. *Ammonia Plant Safety*, **25**, 177. See also *Plant/Operations Prog.*, 1985, **4**, 20

LESTER, H. (1992). Machinery Directive 89/392/EEC – article 100A. In Instn Mech. Engrs (1993), *op. cit.*, 1

LESTER, L.F. and BOMBACI, D.H. (1984). The relationship between personality and irrational judgment in civil pilots. *Hum. Factors*, **26**, 565

LESTER, R.K. (1986). Rethinking nuclear power. *Sci. American*, **254**(3), 23

LETCHFORD, P.W. (1975). Code for hot tapping. *Loss Prevention*, **9**, p. 34

LETT, JAMES, (1997). *A Field Guide to Critical Thinking* (New York: Rowman and Littlefield Publishers,)

LETTAU, H.H. (1959). Wind profile, surface stress and geostrophic drag coefficients in the atmospheric surface layer. *Adv. Geophys.*, **6**, 241

LETTAU, H.H. and DAVIDSON, B. (1957). *Exploring the Atmosphere's First Mile*, vols 1–2 (Oxford: Pergamon Press)

LEUBA, H.R. (1964). Quantification in man–machine systems. *Hum. Factors*, **6**, 555

LEUBE, G. and KREITER, H. (1971). Acute chlorine gas poisoning observations on 90 patients with acute intoxication. *Med. Klin.*, **66**, 354

LEUCHTER, H. (1977). Principles involved in the designing of machines for the chemical industry with regard to technical safety. *Prevention of Occupational Risks*, **5**, 25

LEUNG, J.C. (1986a). A generalised correlation for one-component homogeneous equilibrium flashing choked flow. *AIChE J.*, **32**, 1743

LEUNG, J.C. (1986b). Simplified vent sizing equations for emergency relief requirements in reactors and storage vessels. *AIChE J.*, **32**, 1622

LEUNG, J.C. (1987). Overpressure during emergency relief venting in bubbly and churn-turbulent flow. *AIChE J.*, **33**, 952

LEUNG, J.C. (1990a). Similarity between flashing and nonflashing two-phase flows. *AIChE J.*, **36**, 797

LEUNG, J.C. (1990b). Two-phase flow discharge in nozzles and pipes – a unified approach. *J. Loss Prev. Process Ind.*, **3**(1), 27

LEUNG, J.C. (1992a). Size safety relief valves for flashing liquids. *Chem. Engng Prog.*, **88**(2), 70

LEUNG, J.C. (1992b). Venting of runaway reactions with gas generation. *AIChE J.*, **38**, 723

LEUNG, J.C. and EPSTEIN, M. (1988). A generalized critical flow model for nonideal gases. *AIChE J.*, **34**, 1568

LEUNG, J.C. and EPSTEIN, M. (1990a). The discharge of two-phase flashing flow from an inclined duct. *J. Heat Transfer*, **112**, 524

LEUNG, J.C. and EPSTEIN, M. (1990b). A generalized correlation for two-phase non-flashing homogeneous choked. *J. Heat Transfer*, **112**, 528

LEUNG, J.C. and EPSTEIN, M. (1991). Flashing two-phase flow including the effects of non-condensible gases. *J. Heat Transfer*, **113**, 269

LEUNG, J.C. and FAUSKE, H.K. (1987). Runaway system characterization and vent sizing based on DIERS methodology. *Plant/Operations Prog.*, **6**, 77

LEUNG, J.C. and FISHER, H.G. (1989). Two-phase flow venting from reactor vessels. *J. Loss Prev. Process Ind.*, **2**(2), 78

LEUNG, J.C. and GROLMES, M.A. (1987–). The discharge of two-phase flashing flow in a horizontal duct. *AIChE J.*, 1987, **33**, 524; 1988, **34**, 1030

LEUNG, J.C. and GROLMES, M.A. (1988). A generalised correlation for flashing choked flow of initially sub-cooled liquid. *AIChE J.*, **34**, 688

LEUNG, J.C. and NAZARIO, F.N. (1990). Two-phase flashing flow methods and comparisons. *J. Loss Prev. Process Ind.*, **3**(2), 253

LEUNG, J.C., FAUSKE, H.K. and FISHER, H.G. (1986). Thermal runaway reactions in a low thermal inertia apparatus. *Thermochimica Acta*, **104**, 13

LEUNG, J.C., BUCKLAND, A.C., JONES, A.R. and PESCE, L.D. (1988). Emergency relief requirements for reactive chemical storage tanks. *Preventing Major Chemical and Related Process Accidents*, p. 169

LEUNG, J.C., CREED, M.J. and FISHER, H.G. (1989). Round-robin 'vent sizing package' results. *Runaway Reactions*, p. 264

LEUNG, J.C., STEPANIUK, N.J. and CANTRELL, G.L. (1989). Emergency relief considerations under segregation scenarios. *Plant/Operations Prog.*, **8**, 3

LEUNG, K.M., LINDSTEDT, R.P. and JONES, W.P. (1991). A simplified reaction mechanism for soot formation in nonpremixed flames. *Combust. Flame*, **87**, 289

LEUSCHKE, G. (1966). On the investigation of flammable dusts in relation to fire and explosion hazards. *Staub-Reinhaltung der Luft*, **26**, 40 (German ed.)

LEUSCHKE, G. (1980). Induction times of dust deposits stored at and above self-ignition temperatures. *Loss Prevention and Safety Promotion 3*, vol. 2, p. 647

LEUSCHKE, G. (1981). Experimental investigations on self-ignition of dust deposits in hot environments. *Runaway Reactions, Unstable Products and Combustible Products* (London: Instn Chem. Engrs), paper 2/G

LEV, Y. (1981a). A novel method for controlling LNG pool fires. *Fire Technol.*, **17**, 275

LEV, Y. (1981b). The use of foam for LNG fire fighting. *Fire Technol.*, **17**, 17

LEV, Y. (1991). Water protection of surfaces exposed to impinging LPG jet fires. *J. Loss Prev. Process Ind.*, **4**, 252

LEV, Y. and STRACHAN, D.C. (1989a). Dry testing of water spray systems. *J. Fire Sciences*, **7**(1), 3

LEV, Y. and STRACHAN, D.C. (1989b). A study of cooling water requirements for the protection of metal surfaces against thermal radiation. *Fire Technol.*, **25**(3), 213

LEVENE, G.M. and CALNAN, C.D. (1974). *A Colour Atlas of Dermatology* (Wolfe)

LEVENSPIEL, O. (1962). *Chemical Reaction Engineering* (New York: Wiley)

LEVENSPIEL, O. (1977). The discharge of gases from a reservoir through a pipe. *AIChE J.*, **23**, 402

LEVERENZ, F.L. and ERDMANN, R.C. (1975). *Critique of the AEC Reactor Safety Study (WASH 1400). Rep. EPRI 217-2-3.* Electric Power Res. Inst., Palo Alto, CA

LEVERENZ, F.L. and ERDMANN, R.C. (1979). *Comparison of the EPRI and Lewis Committee Review of the Reactor Safety Study. Rep. NP-1130.* Electric Power Res. Inst., Palo Alto, CA

LEVERENZ, F.L. and KIRCH, H. (1976). *User's Guide for the WAM-BAM Computer Code. Rep. ERPR-217-2-5.* Elec. Power Res. Inst., Palo Alto, CA

LEVESON, N.G. (1986). Software safety: what, why and how. *ACM Comput Surveys*, **18**(2), Jun.

LEVESON, N.G. (1987). Building safe software. In Littlewood, B. (1987b), *op. cit.*, p. 1

LEVESON, N.G. (1995). *Safeware: System Safety and Computers* (Reading, MA: Addison-Weley)

LEVESON, N.G. and HARVEY, P.R. (1983). Analyzing software safety. *IEEE Trans Software Engng*, **SE-9**, 589

LEVESON, N.G. and STOLZY, J.L. (1983). Safety analysis of ADA programs using fault trees. *IEEE Trans Reliab.*, **R-32**, 479

LEVIN, Z. and HOBBS, P.V. (1971). Splashing of water drops on solid and wetted surfaces: hydrodynamics and charge separation. *Phil. Trans. R. Soc., Ser. A*, **269**, 555

LEVINE, D. and BOYARS, C. (1965). The sensitivity of nitroglycerine to impact. *Combust. Flame*, **9**, 131

LEVINE, J.M. and SAMET, G. (1974). Information seeking with input pacing and multiple decision opportunities. *Hum. Factors*, **16**, 384

LEVINE, M.S. (1979). Respirator use and protection from exposure to carbon monoxide. *Am. Ind. Hyg. Ass. J.*, **40**(9), 832

LEVINE, R.D. (1989). The dynamics of elementary processes and combustion kinetics. *Combust. Flame*, **78**, 5

LEVINE, R.S. (1985). Workshop on flame radiation and soot. *Fire Technol.*, **21**, 41

LEVINE, R.S. and PAGNI, P.J. (1983). *Fire Science for Fire Safety* (New York: Gordon & Breach)

LEVINE, R.Y. (1965a). Electrical equipment in outside chemical and petroleum plants. *Fire J.*, **59**(3), 30

LEVINE, R.Y. (1965b). A new approach to electrical safety. *Hydrocarbon Process.*, **44**(2), 161

LEVINE, R.Y. (1968a). Criteria for electrical equipment for use in explosive environment. *Ann. N.Y. Acad. Sci.*, **152**(art. 1), 511

LEVINE, R.Y. (1968b). Electrical hazard equipment classification. *Loss Prevention*, **2**, 4

LEVINE, R.Y. (1972a). Classes and limits of hazardous areas. *Chem. Engng*, **79**(May 1), 50

LEVINE, R.Y. (1972b). Impact of environmental regulations on loss prevention. *Loss Prevention*, **6**, 130

LEVINE, R.Y. (1983). New concepts in the classification of Class II hazardous locations. *Plant/Operations Prog.*, **2**, 140

LEVINE, S. (1978). The role of risk assessment in the nuclear regulatory process. *Nucl. Safety*, **19**, 556

LEVINE, S. (1979). The role of risk assessment in the nuclear regulatory process. *Ann. Nucl. Energy*, **6**, 281

LEVINE, S., JOKSIMOVICH, V. and STETSON, F. (1983). Probabilistic risk assessment in the US. *Reliab. Engng*, **6**, 197

LEVINSON, D.W. (1977). Copper metallurgy as a diagnostic tool for analysis of the origin of building fires. *Fire Technol.*, **13**, 211

LEVINSON, R.Y. (ed.) (1975). *Work Hazard: Threshold Limit Values for Chemical Agents in the Workplace in the United States, Germany and Sweden* (Geneva: Int. Fed. Chem. and Gen. Workers)

LEVINSON, W.A. (1989). How to purge small vessels with nitrogen. *Chem. Engng*, **96**(11), 249

LEVITT, R. (1980). *Implementing Public Policy* (London: Croom Helm)

LEVITZKY, J.J. (1988). Safety, health and loss prevention in the undergraduate chemical engineering curriculum. *Plant/Operations Prog.*, **7**, 190

LEVY, A. (1955). The fast and slow reactions of hydrogen–oxygen–propane mixtures. *Combustion*, **5**, 495

LEVY, D.M., DIVERTIE, M.B., LITZOW, T.J. and HENDERSON, J.W. (1964). Ammonia burns of the face and respiratory tract. *J. Am. Med. Ass.*, **190**, 95

LEVY, L.B. (1987). Inhibitor-oxygen interactions in acrylic acid stabilization. *Plant/Operations Prog.*, **6**, 188

LEVY, L.B. (1993). The effect of oxygen in vinyl acetate and acrylic monomer stabilization. *Process Safety Prog.*, **12**, 47

LEVY, L.B. and LARKIN, M.B. (1993). Emergency response shortstop inhibition during the approach to an acrylic acid runaway polymerisation. *Process Safety Prog.*, **12**, 111

LEVY, L.B. and PENROD, J.D. (1989). The anatomy of an acrylic acid runaway polymerization. *Runaway Reactions*, p. 491. See also *Plant/Operations Prog.*, **8**, 105

LEVY, S. (1965). Prediction of two-phase critical flow rate. *J. Heat Transfer*, **87**, 53

LEVY, S. (1980a). Homogeneous non-equilibrium critical flow model. *Int. J. Heat Mass Transfer*, **25**, 759

LEVY, S. (1980b). Three Mile Island: a call for fundamentals and real-time analysis. *Heat Transfer Engng*, **1**(4), 47

S. LEVY INC. (1982). *Critical Flow Data Review and Analysis. Rep. EPRI NP-2192*. Elec. Power Res. Inst., Palo Alto, CA

LEWANDOWSKI, G.A. (1978). Specifying mechanical design of electrostatic precipitators. *Chem. Engng*, **85**(Jun. 19), 108

LEWELLEN, W.S. and SYKES, R.A. (1986). Analysis of concentration fluctuations from lidar analysis of atmospheric plumes. *J. Clim. Appl. Met.*, **25**, 1145

LEWIN (1920). *Die Kohlenoxydvergiftung* (Berlin)

LEWIN, J. (1977). Constraints for discharge to water. *Chemical Engineering in a Hostile World* (London: Clapp & Poliak)

LEWINS, J. (1988). United Kingdom and USSR reactor types. In Worley, N.G. and Lewins, J. (1988), *op. cit.*, 47

LEWIS, B. and VON ELBE, G. (1937). Theory of flame propagation. *Combustion*, **2**, 183

LEWIS, B. and VON ELBE, G. (1938). *Combustion, Flames and Explosions of Gases* (Cambridge: Cambridge Univ. Press)

LEWIS, B. and VON ELBE, G. (1943). Stability and structure of burner flames. *J. Chem. Phys.*, **11**, 75

LEWIS, B. and VON ELBE, G. (1947). Ignition of explosive mixtures by electric sparks. II, Theory of the propagation of flame from an instantaneous point source of ignition. *J. Chem. Phys.*, **15**, 803

LEWIS, B. and VON ELBE, G. (1951). *Combustion, Flames and Explosions of Gases* (New York: Academic Press)

LEWIS, B. and VON ELBE, G. (1961). *Combustion, Flames and Explosions of Gases*, 2nd ed. (New York: Academic Press)

LEWIS, B. and VON ELBE, G. (1987). *Combustion, Flames and Explosions of Gases*, 3rd ed. (Academic Press)

LEWIS, B. and FRIAUF, J.B. (1930). Explosions in detonating mixtures. I, Calculation of rates of explosions in mixtures of hydrogen and oxygen and the influence of rare gases. *J. Am. Chem. Soc.*, **52**, 3905

LEWIS, B.J. and LOW, L.M. (eds) (1973). *Readings in Maintenance Management* (Boston, MA: Cahners Books)

LEWIS, C.F. (1964). Why metals fail. *Chem. Engng*, **71**(Apr. 13), 214; (May 11), 188

LEWIS, C.R., EDWARDS, R.E. and SANTORO, M.A. (1976). Incineration of industrial wastes. *Chem. Engng*, **83**(Oct. 18), 115

LEWIS, D.A. and DAVIDSON, J.F. (1985). Pressure drop for bubbly gas–liquid flow through orifice plates and nozzles. *Chem. Engng Res. Des.*, **63**, 149

LEWIS, D.J. (n.d.). Event characterisation and ignition (paper)

LEWIS, D.J. (1971). Reducing explosion hazards. *Chem. Process.*, Oct.

LEWIS, D.J. (1979). The Mond Fire, Explosion and Toxicity Index – a development of the Dow Index. *Am. Inst. Chem. Engrs, Symp. on Loss Prevention* (New York)

LEWIS, D.J. (1980a). Autoignition temperature determinations and their relationship to other types of potential ignition sources and their application to practical situations. *Chemical Process Hazards, VII*, p. 257

LEWIS, D.J. (1980b). The Mond Fire, Explosion and Toxicity Index applied to plant layout and spacing. *Loss Prevention*, **13**, 20

LEWIS, D.J. (1980c). Unconfined vapour-cloud explosions: definitions of source of fuel. *Prog. Energy Combust. Sci.*, **6**, 121

LEWIS, D.J. (1980d). Unconfined vapour cloud explosions: historical perspective and predictive method based on incident record. *Prog. Energy Combust. Sci.*, **6**, 151

LEWIS, D.J. (1983). Beware heat of July. *Haz. Cargo Bull.*, **4**(8), 19

LEWIS, D.J. (1984a). Case histories of past accidents, the causes and consequences. In Petts, J.I. (1984), *op. cit.*

LEWIS, D.J. (1984b). 'Cold' causes vent block. *Haz. Cargo Bull.*, **5**(8), 21

LEWIS, D.J. (1984c). A fateful combination. *Haz. Cargo Bull.*, **5**(7), 44

LEWIS, D.J. (1984d). Hazards of scaling up. *Haz. Cargo Bull.*, **5**(4), 24

LEWIS, D.J. (1984e). Professional training and qualifications required for persons involved in health and safety in different industries (paper)

LEWIS, D.J. (1984f). Wisdom before the event. *Haz. Cargo Bull.*, **5**(7), 42

LEWIS, D.J. (1985). New definition for BLEVEs. *Haz. Cargo Bull.*, Apr., 28

LEWIS, D.J. (1986a). A review of gas and vapour explosion venting for process plant. *Am. Inst. Chem. Engrs, Symp on Loss Prevention*

LEWIS, D.J. (1986b). A review of some transportation accidents, identification of causes and minimization of consequences. *Am. Inst. Chem. Engrs, Symp on Loss Prevention*

LEWIS, D.J. (1989a). Loss estimation. *Chem. Engr, Lond.*, **462**, 4

LEWIS, D.J. (1989b). Plant layout and storage area spacing. *Loss Prevention and Safety Promotion 6*, p. 7.1

LEWIS, D.J. (1989c). Raising the roof. *Haz. Cargo Bull.*, **10**(10), 84

LEWIS, D.J. (1993). Case histories: accidents. *Dept of Chem. Engng, Loughborough Univ. of Technol., Course on Hazard Control and Major Hazards*

LEWIS, D.M. (1961). *Magnetic and Electrical Methods of Non-Destructive Testing* (London: Allen & Unwin)

LEWIS, E.E. (1979). *Nuclear Power Reactor Safety* (New York: Wiley)

LEWIS, E.E. and TU ZHUGUO (1986). Monte Carlo reliability modelling in inhomogeneous Markov processes. *Reliab. Engng*, **16**(4), 277

LEWIS, G. (1988). Accident management in the USSR and the United Kingdom. In Worley, N.G. and Lewins, J. (1988), *op. cit.*, p. 35

LEWIS, G.J. (1961). Bulk supply of hydrogen. *Chem. Engr, Lond.*, **155**, A56

LEWIS, G.J. (1975). Role of returnable packages in the distribution operations of the chemical industry. *Distribution Conf.* (London: Chem. Ind. Ass.)

LEWIS, H. (1971). Hazardous materials and the trucking industry. *Loss Prevention*, **5**, 13

LEWIS, H.W. (chrmn) (1978). *Risk Assessment Review Group Report to the US Nuclear Regulatory Commission. Rep. NUREG/CR-0400*. Nucl. Regul. Comm, Washington, DC

LEWIS, H.W. (1980). The safety of fusion reactors. *Sci. American*, **242**(Mar.), 33

LEWIS, K. (1979). Terotechnology in process plant design. *Design 79* (London: Instn Chem. Engrs)

LEWIS, J.H., COON, P.A., CLARE, V.R. and STURDIVAN, L.M. (1987). *An Empirical Mathematical Model to Estimate the Probability of Skin Penetration by Various Projectiles*. Chemical Systems Laboratory, US Army Res. Dev. Command, Aberdeen Proving Ground, MD

LEWIS, J.I. and TARSEY, A.R. (1978). EPA's hazardous spill control regulations. *Control of Hazardous Material Spills*, **4**, p. 1

LEWIS, M.J. (1993). Failure diagnosis. *Plant Engr*, **37**(5), 30; **37**(6), 29

LEWIS, M.W.J. (1984). Failure diagnosis. *Chem. Engng*, **91**(Jul. 9), 117; (Aug. 6), 87

LEWIS, P. (1974). Increase reliability by better compressor component design. *Hydrocarbon Process.*, **53**(5), 117

LEWIS, P. (1980). Controlling health hazards in the chemical industry. *Chem. Ind.*, Nov. 1, 860

LEWIS, P. (1981). Ports: safe, speedy and successful. *Chemy. Ind.*, Nov. 7, 754

LEWIS, P. (1985a). Occupational exposure limits: the new HSE guidance note EH 40 – an industrial view. *Chemy. Ind.*, Aug. 5, 519

LEWIS, P. (1985b). Proposed new legislation for the control of substances hazardous to health. *Chemy. Ind.*, Mar. 18, 181

LEWIS, P. (1986a). CIA approach to occupational exposure limits for toxic gas mixtures. *Chem. Ind.*, Apr. 21, 268

LEWIS, P. (1986b). Occupational exposure limits: EHG 40/86 principal changes in the new HSE guidance note. *Chem. Ind.*, Nov. 17, 762

LEWIS, R. (1970). *An Introduction to Reliability Engineering* (London: McGraw-Hill)

LEWIS, R. (1972). *The Nuclear Power Rebellion* (New York: Viking)

LEWIS, R.E. (1981). Estuary mixing. *Chem. Engr, Lond.*, **371/372**, 381

LEWIS, S. and DONEGANI, A.N. (1993). The benefits of risk analysis to goal-oriented safety management in the UK offshore industry. *Health, Safety and Loss Prevention*, p. 148

LEWITT, E.H. (1952). *Hydraulics*, 9th ed. (London: Pitman)

LEWITT, E.H. (1953). *Thermodynamics Applied to Heat Engines* (London: Pitman)

LEWORTHY, L.R. (1975). Automatic water sprinkler systems. *Protection*, **12**(9)

LEYEL, C.P. (1975). New tubes solve reformer heat problem. *Ammonia Plant Safety*, **17**, p. 41

LEYER, J.C. (1982). An experimental study of pressure fields by exploding cylindrical clouds. *Combust. Flame*, **48**, 251

LEYER, J.C. and MANSON, N. (1971). Development of vibratory flame propagation in short closed tubes and vessels. *Combustion*, **13**, p. 551

LEYER, J.C., GUERRAUD, C. and MANSON, N. (1975). Flame propagation in small spheres of unconfined and slightly confined flammable mixtures. *Combustion*, **15**, p. 645

LEYER, J.C., DESBORDES, D., SAINT-CLOUD, J.P. and LANNOY, A. (1993). Unconfined deflagrative explosion without turbulence: experiment and model. *J. Haz. Materials*, **34**, 123

LI, C.C. and MOORE, J.B. (1977). Estimating flash points of organic compounds. *J. Fire Flammability*, **8**, 38

LI, D. (1984). *Prolog Database System* (New York: Wiley)

LI, R. and OLSON, J.H. (1991). Fault detection and diagnosis in a closed loop nonlinear process: application of extended Kalman filter. *Ind. Engng Chem. Res.*, **30**, 898

LI, W.W. and MERONEY, R.N. (1983). Gas dispersion near a cubical model building. *J. Wind Engng Ind. Aerodynam.*, **12**, 15

LI XIAO-YUN, LEIJDENS, H. and OOMS, G. (1986). An experimental verification of a theoretical model for the dispersion of a stack plume heavier than air. *Atmos. Environ.*, **6**, 1087

LI ZONG-KAI and BRIGGS, G.A. (1988). Simple PDF models for convectively driven diffusion. *Atmos. Environ.*, **22**, 55

LIAM FINN, W.D. and ATKINSON, G.M. (1985). Probability of seismically induced liquefaction in British sector of North Sea. *Earthquake Engineering in Britain* (London: Instn Civil Engrs), p. 365

LIAN, H.Y., KAWAJI, M. and CHAN, A.M.C. (1993). Vibration effects of two-phase cross flow on heat exchangers. *Hydrocarbon Process.*, **72**(3), 53

LIANG, H. and TANAKA, T. (1990). Arrhenius parameters for the auto-oxidation of solid organic compounds. *Ind. Engng Chem. Res.*, **29**, 329

LIANG-SHIH FAN (1989). *Gas−Liquid−Solid Fluidization Engineering* (Oxford: Butterworth-Heinemann)

LICHT, W. (1980). *Air Pollution Control Engineering* (New York: Dekker)

LICHTENBERG, W.H. (1972). Overflow of an ammonia storage tank. *Ammonia Plant Safety*, **14**, 24

LICHTENBERG, W.H. (1977). Ammonia tank floor and foundation failure. *Ammonia Plant Safety*, **19**, 59

LICHTENBERG, W.H. (1987). Ammonia truck and rail loading. (discussion contribution). *Ammonia Plant Safety*, **27**, 179

LICHTENBERG, W.H. and MCKERLIE, T.H. (1979). Ammonia vapor detectors: their application and performance. *Ammonia Plant Safety*, **21**, 73

LICHTENSTEIN, I. (1964). Design cracking furnaces by computer. *Chem. Engng Prog.*, **60**(12), 64

LICHTENSTEIN, S., FISCHHOFF, B. and PHILLIPS, L.D. (1977). Calibration of probabilities: the state of the art. *Decision Making and Change in Human Affairs* (ed. J. Jungermann and G. de Zeeuw) (Dordrecht: Reidel)

LICHTENSTEIN, S., SLOVIC, P., FISCHHOFF, B., LAYMAN, M. and COMBS, B. (1978). Judged frequency of lethal events. *J. Exp. Psychol. (Human Learning and Memory)*, **4**, 551

LIDBETTER, H. and MONK-JONES, N. (1919). SSA14. Motor Ambulance Convoy (printed for private circulation only by J. Ellis Benson, Withington, Manchester)

LIDDELL, F.D.K., MCDONALD, J.C. and THOMAS, D.C. (1977). Methods of cohort analysis: appraisal by application to asbestos mining. *J. R. Stat. Soc., Ser. A*, **140**, 469

LIDE, D.R. (ed.) (1994). *CRC Handbook of Chemistry and Physics*, 75th ed. (Boca Raton, FL: CRC Press)

LIDIARD, A.B. (1984). *Rep AERA-M 3402*. UK Atom. Energy Authority

LIEBER, R.E. (1982). Process control graphics for petrochemical plants. *Chem. Engng Prog.*, **78**(12), 45

LIEBER, R.E. and HERNDON, T.R. (1965). Analog simulation spells safe start-up. *Hydrocarbon Process.*, **44**(2), 177

LIEBER, R.E. and HERNDON, T.R. (1973). Analog simulation of safe start-ups. In Vervalin, C.H. (1973a), *op. cit.*, p. 405

LIEBERMAN, N.P. (1979). Design processes for reduced maintenance. *Hydrocarbon Process.*, **58**(1), 89

LIEBERMAN, N.P. (1983). *Process Design for Reliable Operations* (Houston, TX: Gulf)

LIEBMAN, I., CONTI, R.S. and CASHDOLLAR, K.L. (1977). Dust cloud concentration probe. *Rev. Sci. Instrum.*, **48**, 1314

LIEBOWITZ, H. (ed.) (1968). *Fracture − An Advanced Treatise*, vols I−VII (New York: Academic Press)

LIEN, E. (1990). *Piper Alpha Public Inquiry. Transcript, Day 172*

LIEN, K.M. (1989). A framework for opportunistic problem-solving. *Comput. Chem. Engng*, **13**, 331

LIENHARD, J.H. (1981). *A Heat Transfer Textbook* (Englewood Cliffs, NJ: Prentice-Hall)

LIEPMAN, H.W. and ROSHKO, A. (1967). *Elements of Gas Dynamics* (New York: Wiley)

LIETZE, D. (1983). Method of operation and limitations of dry flame traps. *J. Occup. Accid.*, **5**(1), 17

LIEVENS, M.C. (1976). *Fiabilite des Systemes* (Toulouse: Cepadues Editions)

LIGHTHILL, J. (1978). *Waves in Fluids* (Cambridge: Cambridge Univ. Press)

LIGHTHILL, M.J. (1973). *Artificial Intelligence: a General Survey* (London: Science Research Council)

LIGHTLE, J. and HOHMAN, J. (1979). Keep pumps operating efficiently. *Hydrocarbon Process.*, **58**(9), 227

LIGI, J.L. (1964). Design pilot plants for safety. *Hydrocarbon Process.*, **43**(1), 181

LIGI, J.L. (1969). Processing with cryogenics. *Hydrocarbon Process.*, **48**(4), 93; **48**(5), 137; **48**(6), 118

LIGI, J.L. (1973). Design pilot plants for safety. In Vervalin, C.H. (1973a), *op. cit.*, p. 311

LIGTHART, V.H.M. (1980). Determination of probability of marine accidents with respect to gas carriers proceeding in Dutch coastal and inland waters. *J. Haz. Materials*, **3**(3), 233

LIHOU, D.A. (1977). Review of furnace design methods. *Trans. Instn Chem Engrs*, **55**, 225

LIHOU, D.A. (1979). Instrumentation scheme design to aid fault finding. *Design 79* (London: Instn Chem. Engrs)

LIHOU, D.A. (1980a). Efficient use of operability studies. *Safety Promotion and Loss Prevention in the Process Industries* (London: Oyez)

LIHOU, D.A. (1980b). Fault trees from operability studies. *Safety Promotion and Loss Prevention in the Process Industries* (London: Oyez)

LIHOU, D.A. (1981). Aiding process plant operators in fault finding and corrective action. In Rasmussen, J. and Rouse, W.B. (1981), *op. cit.*, p. 501

LIHOU, D.A. (1985a). Bhopal and beyond . . . *Chem. Engr, Lond.*, **414**, 15

LIHOU, D.A. (1985b). Computers in control? *Chem. Engr, Lond.*, **411**, 15

LIHOU, D.A. (1986).'Hazards IX' notebook. *Chem. Engr, Lond.*, **427**, 41

LIHOU, D.A. (1987). PESs and the HSE guidelines. *Chem. Engr, Lond.*, **439**, 16

LIHOU, D.A. (1988). A safe and operable plant at minimum cost. *Preventing Major Chemical and Related Process Accidents*, p. 27

LIHOU, D.A. (1990a). Management styles – the effects on loss prevention. *Safety and Loss Prevention in the Chemical and Oil Processing Industries*, p. 147

LIHOU, D.A. (1990b). The role of hazops in getting it right first time. *Piper Alpha: Lessons for Life Cycle Management* (Rugby: Instn Chem. Engrs), p. 47

LIHOU, D.A. (1993). Prediction of flammable limits for inerting and vacuum. *J. Loss Prev. Process Ind.*, **6**, 265

LIHOU, D.A. and JACKSON, P.P. (1985). Using task analysis to specify PLC software for batch processes. *Multistream 85* (Rugby: Instn Chem. Engrs), p. 279

LIHOU, D. and KABIR, Z. (1985). Sequential testing of safety systems. *Chem. Engr, Lond.*, **420**, 35; 1986, **423**, 2

LIHOU, D.A. and MAUND, J.K. (1982). Thermal radiation hazards from fireballs. *The Assessment of Major Hazards* (Rugby: Instn Chem. Engrs), p. 191

LIHOU, D.A. and SPENCE, G.D. (1988). Proper use of data with the Weibull distribution. *J. Loss Prev. Process Ind.*, **1**(2), 110

LIHOU, D.A., RAHIMI, R. and FLETCHER, J.P. (1980). Computer-aided operability studies for loss control. *Loss Prevention and Safety Promotion 3*, vol. 2, p. 579

LIJINSKY, W. (1980). Transport and fate of chemicals in evaluation of their carcinogenicity. In Haque, R. (1980a), *op. cit.*, p. 437

LIKERT, R. (1967). *The Human Organization* (New York: McGraw-Hill)

LILES, J.A. (1983). Computer aids for process engineering. *Chem. Engng Prog.*, **79**(6), 43

LILIS, R., ANDERSON, H., NICHOLSON, W.J., DAUM, S., FISCHBEIN, A.S. and SELIKOFF, I.J. (1975). Prevention of disease among vinyl chloride and polyvinyl chloride workers. *Ann. N. Y. Acad. Sci.*, **246**(Jan. 31), 22

LILLIS, E.J. and YOUNG, D. (1975). EPA looks at 'fugitive emissions'. *J. Air Poll. Control Ass.*, **25**(10), 1015

LILWALL, R.C. (1976). *Seismicity and Seismic Hazard in Britain* (London: HM Stationery Office)

LIM, P.K. (1994). Environmentally benign oil-water interfacial synthesis. *Preprints of Papers Presented at the 208th ACS National Meeting*, Aug. 21–25, 1994, Washington, DC, 238. Center for Great Lakes Studies, University of Wisconsin-Milwaukee, (Milwaukee, WI: Division of Environmental Chemistry, American Chemical Society)

LIMNIOS, N. (1987). Failure rate of non-repairable systems. *Reliab. Engng*, **17**(2), 83

LIMNIOS, N. (1987). A formal definition of fault tree graph models and an exhaustive test of their structural data. *Reliab. Engng*, **18**(4), 267

LIMNIOS, N. and JEANNETTE, J.P. (1987). Event trees and their treatment on PC computers. *Reliab. Engng*, **18**, 197

LIN, A., YU, J.P., YEI, W.Y. and YEI, M.Y. (1993). Workplace monitoring programme in Taiwan. *Health, Safety and Loss Prevention*, p. 544

LIN, C.T., BRULE, M.R., YOUNG, F.K., LEE, L.L., STARLING, K.E. and CHAO, J. (1980). Data bank for synthetic fuels. *Hydrocarbon Process.*, **59**(5), 229; **59**(8), 117; **59**(11), 225

LIN, D., MITTELMAN, A., HALPIN, V. and CANNON, D. (1994). *Inherently safer chemistry: A guide to current industrial processes to address high risk chemicals* (Washington, D.C.: Office of Pollution Prevention and Toxics, U. S. Environmental Protection Agency)

LIN, M. and AYUSMAN, S. (1994). Direct catalytic conversion of methane to acetic acid in aqueous medium. *Nature* **238**, (April 14), 613–15

LIN, Y.J. and SHROFF, G.H. (1991). Chlorine related accidental releases and their dispersion. *Modeling and Mitigating the Consequences of Accidental Releases*, p. 169

LINCH, A.L. (1974). *Biological Monitoring for Industrial Chemical Exposure Control* (Boca Raton, FL: CRC Press)

LINCH, A.L. (1981). *Evaluation of Ambient Air Quality by Personnel Monitoring*, 2nd ed. (Boca Raton, FL: CRC Press)

LINCOLN, R.S. (1960). Human factors in attainment of reliability. *IRE Trans Reliab. Qual. Control*, **RQC-9**, 97

LIND, C.D. (1974). *Explosion Hazards Associated with Spills of Large Quantities of Hazardous Materials. Phase 1. Rep. AD A001242*. Nat. Tech. Inf. Service, Springfield, VA

LIND, C.D. (1975). What causes unconfined vapour cloud explosions? *Loss Prevention*, **9**, 101

LIND, C.D. and STREHLOW, R.A. (1975). Unconfined vapour cloud explosion study. *Transport of Hazardous Cargoes*, **4**

LIND, C.D. and WHITSON, J. (1977). *Explosion Hazards Associated with Large Quantities of Hazardous Materials (Phase 3). Rep. CG-D-85-77.* US Coast Guard

LIND, M. (1981). The use of flow models for automated plant diagnosis. In Rasmussen, J. and Rouse, W.B. (1981), *op. cit.*, p. 411

LIND, N.C. (ed.) (1982). *Technological Risk* (Waterloo, Ont.: Univ. of Waterloo Press)

LINDAUER, G.C. and HAGERTY, D.J. (1983). Ethics simulation in the classroom. *Chem. Engng Prog.*, **79**(7), 17

LINDBERG, S., DOLATA, J. and MERCKE, U. (1987). Nasal exposure to airway irritants triggers a mucociliary defense reflex in the rabbit maxillary sinus. *Acta Otolaryngol.*, **104**, 552

LINDBLOM, U. (1989). Recent developments in LPG storage in shallow rock caverns. *Gastech 88*, paper 3.2

LINDBLOM, U. and QUAST, P. (1983). Methods of underground LPG storage. *Gastech 82*, p. 416

VAN DER LINDE, P. and HINTZE, N. (1978). *Time Bomb* (New York: Doubleday)

LINDEIJER, E.W. (1968). Explosions and explosion hazards. *Carriage of Dangerous Cargoes*, **1**, 14

LINDEIJER, E.W. and PASMAN, H.J. (1969). New test methods on explosion properties. *Symp. on Inherent Hazards of Manufacturing and Storage in the Process Industry*, The Hague, p. 23

LINDELL, M.K. and PERRY, R.W. (1980). Evaluation criteria for emergency response in radiological transportation. *J. Haz. Materials*, **3**(4), 335

LINDEMANN, P.F. and WHITEHORN, V.J. (1965). Safety and fire protection today. *Hydrocarbon Process. Petrol Refiner*, **44**(4), 125

LINDEN, P.F. and SIMPSON, J.E. (1988). Development of discontinuities in a turbulent fluid. In Puttock, J.S. (1988b), *op. cit.*, p. 97

LINDEN, P.F. and SIMPSON, J.E. (1990). Continuous two-dimensional releases from an elevated source. *J. Loss Prev. Process Ind.*, **3**(1), 82

LINDERMAN, R.B. (1974). A standard of missile protection. *Trans. Am. Nucl. Soc.*, **18**, 217

LINDLEY, A.L.G. and BROWN, F.H.S. (1958). Failure of a 60-MW steam turbo-generator at Uskmouth power station. *Proc. Inst. Mech. Engrs*, **172**, 627

LINDLEY, D.V. (1965). *Introduction to Probability and Statistics from a Bayesian Viewpoint*, vols 1–2 (Cambridge: Cambridge Univ. Press)

LINDLEY, D.V. and MILLER, J.C.P. (1964). *Cambridge Elementary Statistical Tables* (Cambridge: Cambridge Univ. Press)

LINDLEY, J. (1985). Centrifuges Pt 2: safe operation. *Chem. Engr, Lond.*, **411**, 41

LINDLEY, J. (1987). *User Guide for the Safe Operation of Centrifuges*, 2nd ed. (Rugby: Instn Chem. Engrs)

LINDLEY, N.L. and FLOYD, J.C. (1993). Piping systems: how installation costs stack up. *Chem Engng*, **100**(1), 94

LINDNER, H. and SEIBRING, H. (1967). Self-ignition of organic substances in lagging material. *Chem. Ing. Tech.*, **39**(11), 667

LINDQVIST, B.H. (1987). Monotone Markov models. *Reliab. Engng*, **17**(1), 47

LINDSAY, D. and ROGERS, R.L. (1994). Methodology for selecting disposal systems for chemical reactor relief. *Hazards*, **XII**, p. 481

LINDSAY, P.H. and NORMAN, D.A. (1972). *Human Information Processing: An Introduction to Psychology* (New York: Academic Press)

LINDSAY, R.K., BUCHANAN, B.G., FEIGENBAUM, E.A. and LEDERBERG, J. (1980). *Applications of Artificial Intelligence for Organic Chemistry: The DENDRAL Project* (New York: McGraw-Hill)

LINDSEY, A.W. (1975). Ultimate disposal of spilled hazardous materials. *Chem. Engng*, **82**(Oct. 27), 107

LINDSEY, M.N. (1967). Mechanical seals: carbon's key role. *Chem. Engng*, **74**(Feb. 27), 160

LINDSEY, P.G. and HENDERSHOT, D.C. (2003). The importance of inherently safer processes to site security. *2003 ASME International Mechanical Engineering Congress and RD&D Expo*, Nov. 15–21, 2003, Washington DC, IMECE2003–41962

LINDSTEDT, R.P. and MICHELS, H.J. (1989). Deflagration to detonation transitions and strong deflagrations in alkane and alkene air mixtures. *Combust. Flame*, **76**, 169

LINDZEN, R.S. (1990). *Dynamics in Atmospheric Physics* (Cambridge: Cambridge University Press)

LINE, L.E., RHODES, H.A. and GILMER, T.E. (1959). The spark ignition of dust clouds. *J. Phys. Chem.*, **63**, 290

LING, K.C. (1976). Loss control – a practical application of the Health and Safety at Work etc. Act 1974. *Process Industry Hazards*, p. 109

LING, R.C. (1979). Application of loss control to the fine chemicals industry. *Chem. Engr, Lond.*, **344**, 342

LINGELEM, M.N. and MAJEED, A.L. (1992). Hydrate formation and control in long distance submarine pipelines. *Chem. Engng Res. Des.*, **70A**, 38

LINHART, H.D. (1970). Recent advances in turbomachinery for the liquefaction, storage and transport of LNG. *LNG Conf. 2*

LINHART, V. and JELLINEK, E. (1977). Prediction of the fatigue life and damage accumulation in components at variable cyclic stresses. *Int. J. Pres. Ves. Piping*, **5**, 53

LINNEROOTH, J. (1975). *The Evaluation of Life Saving: a Survey. Res. Rep. PP-75-21*. Int. Inst. Appl. Systems Analysis

LINNETT, J.W. (1953). Methods of measuring burning velocity. *Combustion*, 4, 20

LINNETT, J.W. and HOARE, M.F. (1949). Burning velocities in ethylene–air mixtures. *Combustion*, **3**, 195

LINNETT, J.W. and HOARE, M.F. (1951). Burning velocity determinations, Pt III. Burning velocities of ethylene-air-carbon dioxide mixtures. *Trans. Faraday Soc.*, **47**, 179

LINNETT, J.W. and SIMPSON, J.S.M. (1957). Limits of inflammability. *Combustion*, **6**, 20

LINNHOFF, B. (1993). Pinch analysis – a state-of-the-art overview. *Chem. Engng Res. Des.*, **71A**, 503

LINNHOFF, B. and DHOLE, V.R. (1992). Shaftwork targeting for low temperature plants. *Chem. Engng Res. Des.*, **70A**, 180

LINNHOFF, B. and EASTWOOD, A.R. (1987). Overall site optimisation by pinch technology. *Chem. Engng Res. Des.*, **65**, 408

LINNHOFF, B. and FLOWER, J.R. (1978). Synthesis of heat exchanger networks. *AIChE J.*, **24**, 633, 642

LINNHOFF, B. and HINDMARSH, E. (1983). The pinch design method for heat exchanger networks. *Chem. Engng Sci.*, **38**, 745

LINNHOFF, B. and KOTJABASAKIS, E. (1986). Downstream paths for operable process design. *Chem. Engng Prog.*, **82**(5), 23

LINNHOFF, B. and POLLEY, G. (1988). Stepping beyond the pinch. *Chem. Engr, Lond.*, **445**, 25

LINNHOFF, B. and TURNER, J.A. (1980). Simple concepts in process synthesis given energy savings and elegant designs. *Chem. Engr, Lond.*, **363**, 742

LINNHOFF, B. and TURNER, J.A. (1981). Heat-recovery networks: new insights yield big savings. *Chem. Engng*, **88**(Nov. 2), 56

LINNHOFF, B. and VREDEVELD, D.R. (1984). Pinch technology has come of age. *Chem. Engng Prog.*, **80**(7), 33

LINNHOFF, B., MASON, D.R. and WARDLE, I. (1979). Understanding heat exchanger networks. *Comput. Chem. Engng*, 3, 295

LINNHOFF, B., TURNER, J.A. and BOLAND, D. (1980). The importance of setting targets in the design of heat exchanger networks. *Interflow 80* (Rugby: Instn Chem. Engrs), p. 217

LINNHOFF, B. *et al.* (1982). *User Guide on Process Integration for the Efficient Use of Energy* (Rugby: Instn Chem. Engrs)

LINNHOFF, B., POLLEY, G.T. and SAHDER, V. (1988). General process improvements through pinch technology. *Chem. Engng Prog.*, **84**(6), 51

LINNHOFF, B., SMITH, R. and WILLIAMS, J.D. (1990). The optimization of process changes and utility selection in heat integrated processes. *Chem. Engng Res. Des.*, **68**, 221

LINSTEDT, R.P. and MICHELS, H.J. (1988). Deflagration to detonation transition in mixtures of alkane LNG/LPG constituents with O_2/N_2. *Combust. Flame*, **72**, 63

LINSTONE, H. and TUROFF, M. (eds) (1975). *The Delphi Method* (Reading, MA: Addison Wesley)

LINSTROTH, R.L. (1991). Check for atmospheric corrosion when using stainless steel. *Chem. Engng Prog.*, **87**(7), 49

LINTON, F.L. and BRINK, J.A. (1967). Safer preheat air for ammonia plants. *Ammonia Plant Safety*, **9**, 42. See also *Chem. Engng Prog.*, 1967, **63**(2), 83

LINTON, W.H. and SHERWOOD, T.K. (1950). Mass transfer from solid shapes to water in streamline and turbulent flow. *Chem. Engng Prog.*, **46**, 258

LIPOWICZ, M.A. (1985). Fluid-flow and flow-network software. *Chem. Engng*, **92**(Nov. 25), 59

LIPPMANN, M. and SCHLESINGER, R.B. (1979). *Chemical Contamination in the Human Environment* (New York: Oxford Univ. Press)

LIPPOLD, O. (1968). *Human Respiration* (San Francisco, CA: W.H. Freeman)

LIPSETT, M. (1987). Causation in toxic tort litigation; the role of epidemiological evidence. *J. Haz. Materials*, **15**(1/2), 185

LIPSETT, S.G. (1966). Explosion from molten materials and water. *Fire Technol.*, **2**, 118

LIPSKA, A.E. (1966). *The Pyrolysis of Cellulosic and Synthetic Materials in a Fire Environment. Rep. USNRDL-TR- 1113.* Naval Radiolog. Def. Lab.

LIPTAK, B.G. (1964). Control valves for slurry and viscous services. *Chem. Engng*, **71**(Apr. 13), 185

LIPTAK, B.G. (1965). Rupture-disk systems provide protection for pressure vessels. *Chem. Engng*, **72**(Jul. 5), 127

LIPTAK, B.G. (1967). Process instrumentation for slurries and viscous materials. *Chem. Engng*, **74**(Jan. 30), 133; (Feb. 13), 151

LIPTAK, B.G. (1970). Costs of process instruments. *Chem. Engng*, **77**(Sep. 7), 60; (Sep. 21), 175; (Oct. 5), 83; (Nov. 2), 94

LIPTAK, B.G. (1983). Control valves in optimised systems. *Chem. Eng*, **90**(Sep. 5), 104

LIPTAK, B.G. (1987). Pressure regulators. *Chem. Engng*, **94**(5), 69

LIPTAK, B.G. (1993). Level measurement with problem liquids. *Chem. Engng*, **100**(3), 130

LIPTAK, B.G. and VENCZEL, K. (eds) (1982). *Instrument Engineers' Handbook – Process Measurement* (Radnor, PA: Chilton Book Co.)

LIPTON, M.A. and ROTARIU, G.J. (1941). In Geiling, E.M.K. and McLean, F.C. (1941), *op. cit.*

LIPTON, S. (1989). Fugitive emissions. *Chem. Engng Prog.*, **85**(6), 42

LIPTON, S. (1990). Fugitive emission control in packed valves. *Chem. Engng Prog.*, **86**(8), 70

LIPTON, S. (1992). Reduce fugitive emissions through improved process equipment. *Chem. Engng Prog.*, **88**(10), 61

LIPTON, S. and LYNCH, J. (1987). *Health Hazard Control in the Chemical Process Industry* (New York: Wiley)

LIPTON, S. and LYNCH, J. (1994). *Handbook of Health Hazard Control in the Chemical Process Industry* (New York: Wiley)

LIQUEFIED PETROLEUM GAS INDUSTRY TECHNICAL ASSOCIATION (n.d.). *An Introduction to Liquefied Petroleum Gas* (Shepperton, Middlesex)

LIQUEFIED PETROLEUM GAS INDUSTRY TECHNICAL ASSOCIATION (1988). *A Guide to the Writing of Safety Reports* (London)

LIQUEFIED PETROLEUM GAS INDUSTRY TECHNICAL ASSOCIATION – see also Appendices 27 and 28

LISK, D.J. (1974). Recent developments in the analysis of toxic elements. *Science*, **184**, 1137

LISS, V.M. (1987). Preventing corrosion under insulation. *Chem. Engng*, **94**(4), 97

LIST, E.J. (1982). Turbulent jets and plumes. *Ann. Rev. Fluid Mech.*, **14**, 189

LIST, G. and ABKOWITZ, M. (1988). Towards improved hazardous materials flow data. *J. Haz. Materials*, **17**(3), 287

LIST, H.L. (due 1986). *Petrochemical Technology* (Englewood Cliffs, NJ: Prentice-Hall)

LISTER, E. (1980). Improving plant performance. *Plant Engr*, **24**(9), 23

LISTON, D.M. (1965). Safety aspects of site selection, plant layout and unit plot planning. In Fawcett, H.H. and Wood, W.S. (1965), *op. cit.*, p. 35

LISTON, D.M. (1982). Safety aspects of site selection, plant layout, and unit plot planning. In Fawcett, H.H. and Wood, W.S. (1982), *op. cit.* p. 35

LITCHFIELD, E.L. (1965). *Flame Arresters* (Pittsburgh, PA: Instrum. Soc. Am.)

LITCHFIELD, E.L., HAY, M.H. and FORSHEY, D.R. (1963). Direct electrical initiation of freely expanding gaseous detonation waves. *Combustion*, **9**, 282

LITCHFIELD, J.T. and WILCOXON, F. (1949). A simplified method of evaluating dose-effect experiments. *J. Pharmacol. Exp. Therapeut.*, **96**, 99

LITCHFIELD, W., LEGRAND, W.H., MARTIN, M., SCHELINE, D.C. and HURD, V. (1959). Distribution and storage of ethylene. *Chem. Engng Prog.*, **55**(4), 68

A.D. LITTLE INC. (1960). *Handbook for Hydrogen Handling Equipment. WADC Tech Rep. 59–751* (New York)

A.D. LITTLE INC. (1968). *Cost Effectiveness in Traffic Safety* (New York: Praeger)

A.D. LITTLE INC. (1974a). *Hazard Assessment Computer System (HACS). User Manual* (Cambridge, MA)

A.D. LITTLE INC. (1974b). *A Modal Economic and Safety Analysis of the Transportation of Hazardous Substances in Bulk*. Rep. to Maritime Administration (Cambridge, MA)

A.D. LITTLE INC. (1974c). *Technology and Practices for Processing, Transferring and Storing Liquefied Natural Gas* (Cambridge, MA)

A.D. LITTLE INC. (1979). *Experiments Involving Pool and Vapor Fires from Spills of LNG on Water. Rep. C6–D-55–79* (Cambridge, MA)

A.D. LITTLE INC. (1984). *The Management of Technological Risks* (London)

LITTLEJOHN, I.J. (1990). *Piper Alpha Public Inquiry. Transcript, Day 168*

LITTLETON, C.T. (1962). *Industrial Piping*, 2nd ed. (New York: McGraw-Hill)

LITTLEWOOD, B. (1987a). How good are software reliability predictions. In Littlewood, B. (1987), *op. cit.*, p. 154

LITTLEWOOD, B. (ed.) (1987b). *Software Reliability: Achievement and Assessment* (Oxford: Blackwell)

LITTLEWOOD, B. (1989). Predicting software reliability. *Phil. Trans R. Soc., Ser. A*, **327**, 477

LITTLEWOOD, B. and MILLER, D. (eds) (1991). *Software Reliability and Safety* (London: Elsevier Applied Science)

LIU, D.D.S. and MACFARLANE, R. (1983). Laminar burning velocities of hydrogen−air and hydrogen−air−steam mixtures. *Combust. Flame*, **49**, 59

LIU, M., DURRAN, D., MUNDKUR, P., YOCKE, M. and AMES, J. (1976). *The Chemistry, Dispersion, and Transport of Air Pollutants Emitted from Fossil Fuel Plants in California: Data Analysis and Emission Impact Model. Air Resources Board Contract ARB 4-258* (Sacramento, CA)

LIU, Q. and KATSABANIS, P.D. (1993). Hazard evaluation of sulphide dust explosions. *J. Haz. Materials*, **33**, 35

LIU, Y.A. and LENZE, B. (1989). The influence of turbulence on the burning velocity of premixed CH_4−H_2 flames with different velocities. *Combustion*, **22**, 747

LIU, Y.A., MCGEE, H.A. and EPPERLY, W.R. (eds) (1987). *Recent Developments in Chemical Process and Plant Design* (New York: Wiley)

LIU, Y.A., MCGEE, H.A. and EPPERLY, W.R. (eds) (1987). *Recent Developments in Chemical Process and Plant Design* (New York: Wiley-Interscience)

LIVINGSTON, A.D. (1992). Effective management of the design and introduction of computer controlled systems. *Hazard Identification and Risk Analysis, Human Factors and Human Reliability in Process Safety*, p. 415

LIVINGSTON, A.D. and GREEN, M. (1992). Evaluation of incident investigation techniques and associated organisational issues. *Loss Prevention and Safety Promotion 7*, vol. 1, 30−1

LIVINGSTON-BOOTH, A., OAKES, F., SYLVESTER-EVANS, R. and WILLIAMS, J. (1985). Engineering decisions understress. *Reliab. Engng*, **13**(4), 211

LIVINGSTONE, E.H. (1984). Designing and maintaining safe, high-pressure facilities. *Chem. Engng Prog.*, **80**(3), 70

LIVINGSTONE, E.H. (1993). Build your working knowledge of process compressors. *Chem. Engng Prog.*, **89**(2), 27

LIVINGSTONE, J.G. (1977). Some problems in handling naphtha. *Ammonia Plant Safety*, **19**, 30

LIVINGSTONE, J.G. (1979). Production vs safety: how shall we allocate our resources. *Ammonia Plant Safety*, **21**, 29

LIVINGSTONE, J.G. (1989). How safe is your ammonia plant? *J. Loss Prev. Process Ind.*, **2**(3), 126

LIVINGSTONE, J.G. and PINTO, A. (1983). New ammonia process reduces costs. *Chem. Engng Prog.*, **79**(5), 62

LIVINGSTONE, W.L. (1980). The impact of Three Mile Island on process control computer technology. *Sixth Advanced Control Conf.*, Purdue Univ.

LIVSEY, W.O. and JUNEJO, A.A. (1976). The collapse of heat exchanger tubes with ovality and simulated defects. *Int. J. Pres. Ves. Piping*, **4**, 47

LLEWELLYN, F.J. (1947−48). Incendiary action of electric sparks in relation to their physical properties. *Trans. Inst. Rubber. Technol.*, **23**, 29

LLEWELYN, A.E. (1984). Tests reveal problems over explosion vent sizing (letter). *Process Engng*, Dec., 19

LLOYD, A.W. and ROBERTS, K.W.J. (1971). Process safety. *Major Loss Prevention*, p. 243

LLOYD, D.P. (1989). *Piper Alpha Public Inquiry. Transcript, Day 84*

LLOYD, D.R. and LIPOW, M. (1962). *Reliability: Management, Methods and Mathematics* (Englewood Cliffs, NJ: Prentice-Hall)

LLOYD, E.H., O'DONNELL, T. and WILKINSON, J.C. (eds) (1979). *The Mathematics of Hydrology and Water Resources* (London: Academic Press)

LLOYD, F.C. (1979). Electrostatic hazards in powder handling. *Fire Prev. Sci. Technol.*, **21**, 8

LLOYD, P.J. (ed.) (1987). *Particle Size Analysis* (New York: Wiley)

LLOYD, S.A. and CATCHPOLE, J.O. (1988). Realisation of reliability from design to operation for offshore facilities. *Reliab. Engng System Safety*, **21**, 27

LLOYD, W.D. (1983). Methanol synthesis gas explosion. *Plant/ Operations Prog.*, **2**, 120

LLOYD'S LIST (1974). *Entries Nov. 9−29. Yoko Maru No. 10* (London)

LLOYD'S REGISTER (n.d.). *Guidelines on Protection of Diesel Engines for Hazardous Zone 2* (London)

LLOYD'S REGISTER OF SHIPPING (n.d.). *Requirements for Design, Manufacture and Testing of Pressure Components. Land-based Nuclear Installations* (London)

LLOYD'S REGISTER OF SHIPPING (1992). Caveat emptor. *Plant Engr*, **36**(4), 37

LO, F. and OAKES, D. (1993). Leap from pilot- to full-scale production. *Chem Engng*, **100**(11), 110

LOADER, K. (1981). Flash points. *Fire Prev. Sci. Technol.*, **24**, 8

LOBANOFF, V.S. and ROSS, R.R. (1985). *Centrifugal Pumps: Design and Application*; 1992, 2nd ed. (Houston, TX: Gulf)

LOBEL, J. (1985). *Computer Security and Access Control* (Maidenhead: McGraw-Hill)

LOCHMAN, W.J. (1971). Corrosion specs increase costs. *Hydrocarbon Process.*, **50**(7), 101

LOCHMAN, W.J. (1972). Creep and creep−rupture life. *Hydrocarbon Process.*, **51**(8), 71

LOCK, B. (1976). Hazard and operability studies. *Course in Loss Prevention in the Process Industries*. Dept of Chem. Engng, Loughborough Univ. of Technol.

LOCK, J. (1979). Countering the rise in flammable risk. *Processing*, Apr., 25

LOCKE, J.M. (1977). Reflections on protecting society. *Sci. Publ. Policy*, **4**, 160

LOCKEMANN, W.F. (1974). Explosion in a water treating unit. *Ammonia Plant Safety*, **16**, p. 54

LOCKHART, R.W. and MARTINELLI, R.C. (1949). Proposed correlation of data for isothermal two-phase, two-component flow in pipes. *Chem. Engng Prog.*, **45**, 39

LOCKS, M.O. (1973). *Reliability, Maintainability and Availability Assessment* (New York: Hayden)

LOCKS, M.O. (1979). Synthesis of fault trees: an example of non-coherence. *IEEE Trans. Reliab.*, **R-28**, 2

LOCKS, M.O. (1980). Fault trees, prime implicants and non-coherence. *IEEE Trans. Reliab.*, **R-29**, 130

LOCKWOOD, C. (1989). *Piper Alpha Public Inquiry. Transcript, Days 17−20*

LOCKWOOD, F.C. (1977). The modelling of turbulent premixed and diffusion combustion in the computation of engineering flows. *Combust. Flame*, **29**, 111

LOCKWOOD, F.C. and MALALASEKERA, W.M.G. (1989). Fire computation: the 'flashover' phenomenon. *Combustion*, **22**, 1319

LOCKWOOD, F.C. and NAGUIB, A.S. (1975). The prediction of the fluctuations in the properties of free, round jet turbulent diffusion flames. *Combust. Flame*, **24**, 109

LOCKWOOD, F.C. and SYED, S.A. (1979). Consideration of the problem of combustion modelling for engineering applications. *Combust. Sci. Technol.*, **19**, 129

LOCKWOOD, J. (1993). Value of lessons learned for auditing purposes. *Health, Safety and Loss Prevention*, p. 71

LOCKWOOD, N.R. (1965). This new truck is a fire killer. *Hydrocarbon Process.*, **44**(6), 169

LOCKWOOD, N.R. (1973a). Amsterdam refinery designed for safety. *Hydrocarbon Process.*, **47**(7), 157

LOCKWOOD, N.R. (1973b). Mobil's two-agent fire-truck. In Vervalin, C.H. (1973a), *op. cit.*, p. 303

LOCKWOOD, N.R. (1986). Foam extinguishing agents and systems. In Cote, A.E. and Linville, J.L. (1986), *op. cit.*, p. 19–32

LODAL, P.N. (1996). Inherently safer processes using the low cost plant approach. *Int. Conf. and Workshop on Process Saf. Manage. and Inherently Safer Processes*, Oct. 8–11, 1996, Orlando, FL, Workshop E: Company Implementation Programs for Inherently Safer Processes: Incentives; Barriers

LODE, T. (1994). Can offshore experience help in reducing onshore losses? *J. Loss Prev. Process Ind.*, **7**, 382

LODGE, J.P., WAGGONER, A.P., KLODT, D.T. and CRAIN, C.N. (1981). Non-health effects of airborne particulate matter. *Atmos. Environ.*, **15**, 431. Also 1983, **17**, 431

LOEB, L.B. (1945). The basic mechanisms of static electrification. *Science*, **102**(Dec. 7), 573

LOEB, L.B. (1958). *Static Electrification* (Berlin: Springer)

LOEB, L.B. (1965). *Electrical Coronas* (Berkeley: Univ. of California Press)

LOEB, M.B. (1974). From liquid nitrogen to purge gas. *Chem. Engng*, **81**(Dec. 9), 134

LOEB, M.B. (1975). Determining chlorine gas flow rates from cylinders. *Chem. Engng*, **82**(Jan. 6), 126

LOEB, M. (1978). On the analysis and interpretation of vigilance: some remarks on two recent articles by Craig. *Hum. Factors*, **20**, 447

LOEHR, R.C. (1987). Land disposal of hazardous wastes. In Martin, E., J. and Johnson, J.H. (1987), *op. cit.*, p. 365

LOEN, R.O. (1966). Use 'engineered training' for better plant start-up. *Hydrocarbon Process.*, **45**(10), 199

LOEN, R.O. (1973). Engineered training for better plant start-ups. In Vervalin, C.H. (1973a), *op. cit.*, p. 414

LOESCHER, K.J. (1971). Metallurgical failure and the microscope. *Loss Prevention*, 5, p. 83

LOFBERG, E.K. (1965). Installation and maintenance of glass equipment. *Chem. Engng*, **72**(Apr. 26), 156

LOFQUIST, K. (1960). Flow and stress near an interface between stratified fluids. *Phys. Fluids*, **3**, 158

LOFTHOUSE, J.A. (1969). Large chemical plants and their problems. *Chem. Engr, Lond.*, **231**, 153

LOFTUS, J. (1970). Reliability in ethylene plants. *Chem. Engng Prog.*, **66**(12), 53

LOFTUS, J., SCHUTT, H.C. and SAROFIM, A.F. (1967). Design of furnaces for tubular reactors. *Chem. Engng Prog.*, **63**(7), 47

LOGAN, H.L. (1966). *The Stress Corrosion of Metals* (New York: Wiley)

LOGAN, J.D. and BDZIL, J.B. (1982). Self-similar solution of the spherical detonation problem. *Combust. Flame*, **46**, 253

LOGAN, P.J. (1976). A simplified approach . . . pressure vessel head design. *Hydrocarbon Process.*, **55**(11), 265

LOGAN, W.P.D. (1953). Mortality in the London fog incident, 1952. *The Lancet*, **264**, 366

LOGINOW, A.W. (1986). A review of stress corrosion cracking of steel in liquefied ammonia service. *Materials Performance*, **25**(Dec.), 18

LOGINOW, A.W. and PHELPS, E.H. (1962). Stress-corrosion cracking of steels in agricultural ammonia. *Corrosion*, **18**, 299t

LOHMEYER, A., MERONEY, R.N. and PLATE, E.J. (1981). Model investigations of spreading of heavy gases released from an instantaneous volume source at the ground. In de Wispelaere, C. (1981), *op. cit.*, p. 433

LOHRY, E.J. (1994). Small business: make the most of Process Safety Management. *Chem. Engng Prog.*, **90**(2), 60

LOIACONO, N.J. (1987). Time to drain a tank with piping. *Chem. Engng*, **94**, 164

LOING, M.B. and YIP, B. (1989). Measurement of three-dimensional concentrations in turbulent jets and flames. *Combustion*, **22**, 701

LOIS, E. and MODARRES, M. (1985). A comparative assessment of the safety for one decade of performance of operating US nuclear power plants. *Reliab. Engng*, **13**(4), 191

LOIS, E. and SWITHENBANK, J. (1979). Fire hazards in oil tank arrays in a wind. *Combustion*, **17**, 1087

LOISON, R. (1952). The mechanisms of explosions in compressed air pipe ranges. *Seventh Int. Conf of Directors of Safety in Mines Research*

LOISON, R., CHAINEAUX, L. and DELCLAUX, J. (1954). Study of some safety problems in fire damp drainage. *Eighth Int. Conf. of Directors of Safety in Mines Research* (Safety in Mines Res. Estab.), paper 37

LOM, W.L. (1974). *Liquefied Natural Gas* (Barking: Applied Science)

LOMBARD, J.F. and CULBERSON, R.A. (1973). Defining reformer performance. *Ammonia Plant Safety*, **15**, 29

LOMBARDO, J.M. (1967). The case for digital backup in direct digital control systems. *Chem. Engng*, **74**(Jul. 3), 79

LOMNICKI, Z.A. (1973). Some aspects of the statistical approach to reliability. *J. R. Statist. Soc.*, **136**(3), 395

LOMNITZ, C. (1974). *Global Tectonics and Earthquake Risk* (Amsterdam; Elsevier)

LOMNITZ, C. and ROSENBLUETH, E. (eds) (1976). *Seismic Risk and Engineering Decisions* (Amsterdam: Elsevier)

LONDERGAN, R.J. and BORENSTEIN, H. (1983). EPRI plume model validation project results for a plains site power plant. In de Wispelaere, C. (1983), *op. cit.*, p. 675

LONDON, E.S. (1982). Operational safety. In Green, A.E. (1982b), *op. cit.*, p. 111

LONDON, R.L. (1968). Computer programs can be proved correct. *Proc. Fourth Systems Symp. – Formal Systems and Non-Numerical Problem Solving by Computers*, Case Western Reserve Univ.

LONDON FIRE BRIGADE (1973). *Hazchem Code Classification for Hazardous Chemical Containers* (London)

LONG, A.B. (1980). Technical assessment of disturbance analysis systems. *Nucl. Safety*, **21**, 38

LONG, A.B. et al. (1980). *Summary and Evaluation of Scoping and Feasibility Studies for Disturbance Analysis and Surveillance System (DASS). Rep. EPRI NP-1684.* Elec. Power Res. Inst., Palo Alto, CA

LONG, F.A. and SCHWEITZER, G.E. (eds) (1982). *Risk Assessment of Hazardous Waste Sites* (Washington, DC: Am. Chem. Soc.)

LONG, G.E. (1978). Spraying theory and practice. *Chem. Engng*, **85**(Mar. 13), 73

LONG, J.J. (1975). Train an effective fire brigade. *Hydrocarbon Process.*, **54**(8), 69

LONG, N.A. (1968). Have you considered stainless steel condenser tubes? *Ammonia Plant safety 10*, 35

LONG, P.E. and PEPPER, D.W. (1976). A comparison of six numerical schemes for calculating the advection of atmospheric pollution. *Third Symp. on Atmospheric Turbulence, Diffusion and Air Quality* (Boston, MA: Am. Met. Soc.), p. 181

LONG, R. (ed.) (1967). *The Production of Polymer and Plastic Intermediates from Petroleum* (London: Butterworths)

LONG, R. (1984). Variable speed drives. *Plant Engr*, **28**(1), 24

LONG, V.D. (1963). Estimation of the extent of hazard areas around a vent. *Chemical Process Hazards*, **II**, p. 6

LONGBOTTOM, R. (1980). *Computer System Reliability* (Chichester: Wiley)

LONGINOW, A., OJDROVICH, G., BERTRAM, L. and WIEDERMAN, A. (1973). *People Survivability in a Direct Effects Environment and Related Topics*. IIT Res. Inst., Chicago, IL

LONGMATE, N. (1981). *The Doodlebugs: the Story of the Flying Bombs* (London: Hutchinson)

LONGMATE, N. (1985). *Hitler's Rockets* (London: Hutchinson)

LONSDALE, H. (1975). Ammonia tank failure – South Africa. *Ammonia Plant Safety*, **17**, p. 126

LONSDALE, J.T. and MUNDY, J.E. (1982). Estimating pipe heat-tracing costs. *Chem. Engng*, **89**(Nov. 29), 89

VAN DE LOO, P.J. (1988). Engineering method for measuring and predicting pipe noise in industrial plants. *Proc. Internoise 88*, p. 1547

VAN LOO, R. and OPSCHOOR, G. (1989). Methods for calculation of damage resulting from the physical effects of the accidental release of dangerous materials. *Loss Prevention and Safety Promotion 6*, p. 61.1

LOOMIS, A.G. and PERROTT, G.ST.J. (1928). Measurement of the temperatures of stationary flames. *Ind. Engng Chem.*, **20**, 1004

LOOMIS, T.A. (1974). *Essentials of Toxicology*, 2nd ed. (Philadelphia, PA: Lea & Febiger)

VAN LOOSEN, J. (1968). What you need to know about boilers. *Hydrocarbon Process.*, **47**(6), 95

LOOTVOET, R. (1986). Technical and economic aspects of Gaz Transport type LNG carriers. *Gastech 85*, p. 227

LOPEZ, A., PRIOR, M., YONG, S., ALBASSAM, M. and LILLIE, L.E. (1987). Biochemical and cytological alterations in the respiratory tract of rats exposed for 4 hours to hydrogen sulphide. *Fund. Appl. Toxicol.*, **9**, 753

LOPEZ, A., PRIOR, M.G., REIFFENSTEIN, R.J. and GOODWIN, L.R. (1989). Peracute toxic effects of inhaled hydrogen sulphide and injected sodium hydrosulphide on the lungs of rats. *Fund. Appl. Toxicol.*, **12**, 367

LOPEZ, J.P., BADIN, J., LIETO, J. and GROLLIER-BARON, R. (1990). Water spray and steam curtain design: a common research programme. *J. Loss Prev. Process Ind.*, **3**(2), 207

LOPEZ, J.-P., LIETO, J. and CROLLIER-BARRON, R. (1991). Dilution and absorption of unconfined hazardous gases by liquid spray curtains. *Modeling and Mitigating the Consequences of Accidental Releases*, p. 453

LOPINTO, L. (1982). Fog formation in low-temperature condensers. *Chem. Engng*, **89**(May 17), 111

LOPINTO, L. (1983). Designing and writing operating manuals. *Chem. Engng*, **90**(Jul. 11), 77

LOPINTO, L. (1993). Complying with OSHA's new safety law. *Chem Engng*, **100**(1), 108

LORD, A.E. and KOERNER, R.M. (1988). Nondestructive testing (NDT) techniques to detect contained subsurface hazardous waste (1988). *J. Haz. Materials*, **19**(1), 119

LORD, A.E., KOERNER, R.M. and FREESTONE, F.J. (1982). The identification and location of buried containers via non-destructive testing methods. *J. Haz. Materials*, **5**(3), 221

LORD, E.A. and HIRST, I.G. (1979). Safety considerations given to design and operation of a large scale VCM producing complex. *Design 79* (London: Instn Chem. Engrs)

LORD, K.W. (1983). *The Data Center Disaster Consultant*, 2nd ed. (Englewood Cliffs, NJ: Prentice-Hall)

LORD, R.C., MINTON, P.E. and SLUSSER, R.P. (1970a). Design of heat exchangers. *Chem. Engng*, **77**(Jan. 26, 96)

LORD, R.C., MINTON, P.E. and SLUSSER, R.P. (1970b). Guide to trouble-free heat exchangers. *Chem. Engng*, **77**(Jun. 1), 153

LORD, V.F. (1971). An area classification method. *Electrical Safety*, **1**, p. 11

LOREA-HERDNANDEZ, A. (1993). Evolution of process safety management in a Mexican company. *Process Safety Management*, p. 163

LORENTZ, F. (1967). Case study of incident involving acetylenic alcohol. *Loss Prevention*, 1, 1

LORENTZ, F. (1973). Acetylenic alcohol plant explosion. In Vervalin, C.H. (1973a), *op. cit.*, p. 21

LORENZO, D.K. (1990). *A Manager's Guide to Reducing Human Errors: Improving Human Performance in the Chemical Industry* (Washington, DC: Manufacturers Ass.)

LORIO, D. (1974). A new approach to turbomachinery analysis. *Hydrocarbon Process.*, **53**(1), 105

LORTZ, R.W. (1966). Insure long runs through better relief valve maintenance. *Hydrocarbon Process.*, **45**(5), 175

LOS ALAMOS SCIENTIFIC LABORATORIES (1950). *The Effects of Atomic Weapons* (Los Alamos)

LOSASSO, A.L. (1970). Pressure vessels fabricated in foreign countries. *Loss Prevention*, 4, 109

LOSCHKE, E. (1933). *Die wichtigsten Vergiftungen* (Munich: Verlag Lehmann)

LOSS PREVENTION BULLETIN — see end of reference list

LOTHROP, W.C. and HANDRICK, G.R. (1949). The relationship between performance and constitution of pure organic explosive compounds. *Chem. Revs.*, **44**, 419

LOTKA, A.J. (1920). Undamped oscillations derived from the law of mass action. *J. Am. Chem. Soc.*, **42**, 1595

LOTT, R.A. (1984). Case study of plume dispersion over elevated terrain. *Atmos. Environ.*, **18**, 125

LOTT, R.A. (1986). Model performance – plume dispersion over elevated terrain. *Atmos. Environ.*, **20**, 1547

LOTTES, P.A. and FLINN, W.S. (1956). A method of analysis of natural circulation boiling systems. *Nucl. Sci. Engng*, **1**, 461

LOU, S.C. (1973). Noise-control design for process plants. *Chem. Engng*, **80**(Nov. 26), 77

LOUBET, D. (1986). A methodology of operating procedures to increase, through self participation, the safety and reliability level of crude oil refining units. *Loss Prevention and Safety Promotion 5*, 5–1

LOUCKS, C.M. (1973). Boosting capacities with chemicals. *Chem. Engng*, **80**(Feb. 26), 79

LOUDON, D.E. (1963). Requirements for safe discharge of hydrocarbons to atmosphere. *API Proc., Div. of Refining*, **43**, 418

LOUGHAN, J.R. and YOKOMOTO, C.F. (1989). Computation of one-dimensional spread of a leaking flammable gas. *Fire Technol.*, **25**(4), 308

LOUHEVAARA, V., SMOLANDER, J., KORHONEN, O. and TUOMI, T. (1986). Maximal working times with self-contained breathing apparatus. *Ergonomics*, **29**, 77

LOUVAR, J.F. and KUBIAS, F.O. (1994). Safety and chemical engineering education. *Process Safety Prog.*, **13**, 72

LOVACHEV, L.A. (1971). The theory of limits on flame propagation in gases. *Combust. Flame*, **17**, 275

LOVACHEV, L.A. (1976). On the flammability limits of dust–air mixtures. *Fizika Gorenia i Vzryva*, **12**, 307

LOVACHEV, L.A. (1978a). On convective flame acceleration in long horizontal galleries. *Combust. Sci. Technol.*, **18**, 153

LOVACHEV, L.A. (1978b). On explosibility characteristics of dust clouds. *Fizika Gorenia i Vzryva*, **14**, 153

LOVACHEV, L.A. (1979). Flammability limits – a review. *Combust. Sci. Technol.*, **20**, 209

LOVACHEV, L.A. (1981). Chemical kinetics in flames – a review. *Combust. Sci. Technol.*, **25**, 49

LOVACHEV, L.A., BABKIN, V.S., BUNEV, V.A., V'YUN, A.V., KRIVULIN, V.N. and BARATOV, A.N. (1973). Flammability limits: an invited review. *Combust. Flame*, **20**, 259

LOVE, D.L. (1992). No hassle reboiler selection. *Hydrocarbon Process.*, **71**(10), 41

LOVE, F.S. (1972). Pumps – a troubleshooting guide. *Hydrocarbon Process.*, **51**(1), 91

LOVE, J. (1984). The management of instrumentation and control projects. *Chem. Engr, Lond.*, **403**, 17

LOVE, J. (1987a). Batch process control. *Chem. Engr, Lond.*, **437**, 34

LOVE, J. (1987b). Confidence in control. *Chem. Engr, Lond.*, **443**, 36

LOVE, J. (1987c). Strategies for batch process control. *Chem. Engr, Lond.*, **440**, 29

LOVE, J. (1988). Trends and issues in batch control. *Chem. Engr, Lond.*, **447**, 24

LOVE, R.A. (1993) Attain ISO-9000 certification. *Hydrocarbon Process.*, **72**(10), 82–B

LOVEJOY, G.R. and CLARK, I.M. (1983). Furnace safety systems. *Plant/Operations Prog.*, **2**, 13

LOVELACE, B.G. (1979). Safe sampling of liquid process streams. *Loss Prevention*, 13, 111. See also *Chem. Engng Prog.*, 1979, **75**(11), 51

LOVELACE FOUNDATION FOR MEDICAL EDUCATION AND RESEARCH – see Appendix 28

LOVELAND, R.J. (1981). Electrostatic ignition hazards in industry. *J. Electrostatics*, **11**, 3

LOVELESS, N.E. (1962). Direction-of-motion stereotypes: a review. *Ergonomics*, **5**, 357

LOVETT, K.M., SWIGGETT, G.E. and COBB, C.B. (1982). When you select a plant site . . . *Hydrocarbon Processing*, **61**(5), 285

LOVETT, O.P. (1971). Control valves. *Chem. Engng*, **78** (Oct. 11), 129

LOVING, F.A. (1957). Barricading hazardous reactions. *Ind. Engng Chem.*, **49**, 1744

LOVINS, A.B. (1975). *World Energy Strategies: Facts, Issues, and Options* (Cambridge, MA: Ballinger/Friends of the Earth)

LOVINS, A.B. (1977). *Soft Energy Paths* (Cambridge, MA: Ballinger/Friends of the Earth)

LØ VOLD, K. (1974). Safety improvements in the production of black powder. *Loss Prevention and Safety Promotion 1*, 331

LOVSTRAND, K.G. (1975). On the discharge of static electricity with Attain induction eliminators. *Static Electrification*, **4**, 246

LOVSTRAND, K.G. (1981). The ignition power of brush discharges. Experimental work on the energy density. *J. Electrostatics*, **10**, 161

LAW, T.A.W. and NOLTINGK, B.E. (1972). Quantitative aspects of reliability in process control systems. *IEEE Revs*, **119**(8R), 1033

LOWE, C.R. and KOSTRZEWSKI, J. (1973). *Epidemiology: A Guide to Teaching Methods* (London: Int. Epidemiology Ass.)

LOWE, D.R.T. (1980). Major incident criteria – what risk should society accept? *Eurochem Conf.*

LOWE, D.R.T. (1984). The hazards of risk analysis. *Reliab. Engng*, **9**, 243

LOWE, D.R.T. and SOLOMON, C.H. (1983). Hazard identification procedures. *Loss Prevention and Safety Promotion 4*, vol. 1, p. G8

LOWE, E.I. and HIDDEN, A.E. (1971). *Computer Control in Industrial Processes* (London: Peter Peregrinus)

LOWE, G. (1985). Use a safety advisory group. *Hydrocarbon Process.*, **64**(12)75

LOWE, H. and EHRFELD, W. (1999). State-of-the-art in microreaction technology: concepts, manufacturing and applications. *Electrochimica Acta* 44, 3679–89

LOWELL, V.W. (1967). *Airline Safety is a Myth* (New York: Bartholomew House)

LOWENHEIM, F.A. and MORAN, M.K. (1975). *Industrial Chemicals*, 4th ed. (New York: Wiley-Interscience) (4th ed. of book of same name by Faith, W.L., Keyes, D.B. and Clark, R.L. (1965))

LOWENSTEIN, J.G. (1958). Calculate adequate rupture disc size. *Chem. Engng*, **65**(1), 157

LOWENSTEIN, J.G. (1979). A pilot plant safety program. *Chem. Engng Prog.*, **75**(9), 51

LOWENSTEIN, J.G. (1985). The pilot plant. *Chem. Engng*, **92**(Dec. 9/23), 62

LOWES, T.M. and NEWALL, A.J. (1971). The emissivities of flame soot dispersions. *Combust. Flame*, **16**, 191

LOWRANCE, W.W. (1976). *Of Acceptable Risk* (Los Altos, CA: William Kaufmann)

LOWRANCE, W.W. (1984). Reducing industrial cancer: the strategic agenda. In Deisler, P.F. (1984b), *op. cit.*, p. 21

LOWRIE, R. (1987). Materials for oxygen service. *Chem. Engng*, **94**(6), 75

LOWRY, R.P. and AGUILO, A. (1974). Acetic acid today. *Hydrocarbon Process.*, **53**(11), 103

LU, G. (1989). *Unpublished PhD Thesis, Univ. of Cambridge*

LU, P.L., DABORA, E.K. and NICHOLLS, J.A. (1971). The role of explosion limit chemical kinetics in H_2–CO–O_2 detonations. *Combust. Flame*, **16**, 195

LUBERS, M.D. (1991). Representing design dependencies in an issue base style. *IEEE Software*, **81**

LUBINEAU, A. (1996). Making a splash in synthesis: water as a solvent. *Chem. and Ind.*, (Feb. 19), 123–26

LUBOW, M. (1987). *Working Out with Autocad* (Thousand Oaks, CA: New Riders)

LUCAS, D. and EMBREY, D.E. (1988). New directions in qualitative modelling. Sayers, B.A. (1988), *op. cit.*, p. 170

LUCAS, S., ALONSO, E., SANZ, J.A. and COCERO, M.J. (2003). Safety study in a supercritical extraction plant. *Chem. Eng. Technol.* **26** (4) 449–61

LUCAS, D.A. (1991). Organisational aspects of near miss reporting. In Van der Schaaf, T.W., Lucas, D.A. and Hale A.R. (1991), *op. cit.*, p. 127

LUCAS, D.M., ROE, D.R. and WATERLOW, D.K. (1983). British Gas Canvey LNG terminal: the analysis of the hazards of the ship-to-shore transfer. *Loss Prevention and Safety Promotion 4*, vol. 2, p. B65

LUCAS, W. (ed.) (1991). *Process Pipe and Tube Welding* (Abington, Cambridge: Abington)

LUCE, W.A. and PEACOCK, J.H. (1962). Stainless steel technology for the CPI. *Ind. Engng Chem.*, **54**(6), 63

LUCE, W.A. and PEACOCK, J.H. (1963). Stainless steels. *Ind. Engng Chem.*, **55**(7), 60

LUCK, C. (1984). Hazardous waste. In A.D. Little (1984), *op. cit.*

LUCK, C.J. (1977). The key role of supervisors in safe working. *Loss Prevention and Safety Promotion 2*, p. 73

LUCK, C.J. and HOWE, C. (1988). Assessment of safety, health and environment controls in devolved management systems. *Preventing Major Chemical and Related Process Accidents*, p. 43

LUCKETT, G.A. and POLLARD, R.T. (1973). The gaseous oxidation of isobutane (Part 1). *Combust. Flame*, **21**, 265

LUCKRITZ, R.T. (1977). *An Investigation of Blast Waves Generated by Constant Velocity Flames. Rep. UILU-Eng 770502.* Univ. of Illinois, Urbana, IL

LUCKRITZ, R.T. and SCHNEIDER, A.L. (1980). Decision making in hazardous materials transportation. *J. Haz. Materials*, **4**(2), 129

LUCKRITZ, R.T. and SCHWING (1973). Tank cleaning, an unrespected hazard. *Prevention of Occupational Risks*, **2**, p. 462

LUCQUIN, M. and ANTONIK, S. (1972). Multistage ignition in hydrocarbon combustion (letter). *Combust. Flame*, **19**, 311

LUCQUIN, M. and LAFFITTE, P. (1957). Cool flames of pentane–oxygen mixtures. *Combustion*, **6**, 130

LUDDEKE, D.E. (1975). Ammonia pipeline maintenance and repair. *Ammonia Plant Safety*, **17**, 99

LUDERER, F. (1987). Safety arrangements in electrostatic coating plants. *Fortschritte der Sicherheitstechnik 1* (Frankfurt-am-Main: DECHEMA), p. 375

LUDFORD, L. (1974). Propane blast kills four. *Fire Engng*, **127**(Jul.), 35

LUDWIG, C.B. *et al.* (1968). *Study on Exhaust Plume Radiation Protection. Rep. CR-61 222.* Nat. Aeronaut. Space Admin.

VON LUDWIG, D. (1975). *Supplementary Testimony to the US Federal Power Commission re: Distrigzas Corporation et al. (Docket CP 73–132 et al.) and Eascogas LNG Inc. et al. (Docket CP 73–47 et al.) October 27, 1975*

LUDWIG, E.E. (1964–). *Applied Process Design for Chemical and Petrochemical Plants*, vols 1–3; 1977–, 2nd ed.; 1994–, 3rd ed. (Houston, TX: Gulf)

LUDWIG, E.E. (1973a). Designing process plants to meet OSHA standards. *Chem. Engng*, **80**(Sep. 3), 88

LUDWIG, E.E. (1973b). Project managers should know OSHA. *Hydrocarbon Process.*, **52**(6), 135

LUDWIG, E.E. (1978). Inspection of equipment for high-pressure service. *Safety in Polyethylene Plants* **3**, 5

LUDWIG, F.L. (1981). A model for simulating the behavior of pollutants emitted at ground level under time-varying meteorological conditions. *Atmos. Environ.*, **15**, 989

LUDWIG, F.L. (1984). Transport and diffusion calculations during time-and space-varying meteorological conditions using a microcomputer. *J. Air Poll. Control Ass.*, **34**, 227

LUDWIG, F.L. (1986). A nonsteady-state adaptive volume plume model suitable for regions of strong vertical gradients of wind and stability. *Fifth Joint Conf. on Applications of Air Pollution Modeling* (Boston, MA: Am. Met. Soc.), p. 359

LUDWIG, F.L. and LIVINGSTON, J.M. (1983). Plume behavior in stable air. In de Wispelaere, C. (1983), *op. cit.*, p. 841

LUDWIG, F.L., LISTON, E.M. and SALAS, L.J. (1983). Tracer techniques for estimating emissions from inaccessible ground level sources. *Atmos. Environ.*, **17**, 2167

LUDWIG, P.W.P.H. and RUIJTERMAN, C. (1974). Calculation of pressure and flow transients in a system filled with liquid. *Loss Prevention and Safety Promotion 1*, p. 75

LUDWIG, R.C. and JOHNSON, J.E. (1972). The management of hazardous materials in the environment – a view by industry. *Control of Hazardous Material Spills*, **1**, p. 11

LUDWIGSON, J. (ed.) (1982). *1982 Hazardous Material Spills Conference* (Rockville, MD: Govt Insts)

LUECKE, R.H. and MCGUIRE, M.L. (1965). Chemical reactor stability by Liapunov's direct method. *AIChE J.*, **11**, 749

LUETZOW, D. (1989). Safe and effective operation of high-pressure plunger pumps for ammonia and carbamate in a urea plant. *Ammonia Plant Safety*, **29**, p. 269

LUFTI, M., EL-MIGHARBIL and HASABALLAH, E.M. (1980). Waste heat boiler failure and modifications. *Ammonia Plant Safety*, **22**, 114

LUGENHEIM, W. (1977). *Gesundheits-, Arbeits- und Brandschutz bei Chemischen Arbeiten in Laboratorien* (Berlin: Tribuene)

LUKASIEWICZ, J. (1951). *Aristotle's Syllogistics from the Standpoint of Modern Formal Logic* (Oxford: Clarendon Press)

LUKE, M.T. and GOLZ, L.A. (1970). Software self-monitoring of a process control computer system. *Advances in Instrumentation 25* (Pittsburgh, PA: Instrum. Soc. Am.)

LUKER, J.A. and LEIBSON, M.J. (1959). Dynamic loading of rupture discs with detonation waves. *J. Chem. Engng Data*, **4**, 133

LUKS, K.D. and KOHN, J.P. (1981a). Avoid freeze-up in LNG/LPG work. *Hydrocarbon Process.*, **60**(4), 135

LUKS, K.D. and KOHN, J.P. (1981b). LNG liquid–liquid immiscibility. *Hydrocarbon Process.*, **60**(9), 257

LUMLEY, J.L. and PANOFSKY, H.A. (1964). *The Structure of Atmospheric Turbulence* (New York: Interscience)

LUMLEY, K.P.S. (1976). A proportional study of cancer registrations in dockyard workers. *Br. J. Ind. Med.*, **33**, 108

LUNA, R.E. and CHURCH, H.W. (1972). A comparison of turbulence intensity and stability ratio measurements to Pasquill stability classes. *J. Appl. Met.*, **11**, 663

LUND, O.E. and WIELAND, H. (1966). Pathological–anatomical findings in experimental hydrogen sulphide poisoning. An investigation in Rhesus monkeys. *Int. Arch. Gewerbepath. Gewerbehyg.*, **22**, 46

LUND, S., AKTEN, H.T., PAGE, R.T. and BINNS, K.A. (1985). Cathodic protection, coatings protect Statpipe system. *Oil Gas J.*, **83**(Dec. 23), 47

LUNDBERG, W.C. (1979). Extending catalyst life. *Ammonia Plant Safety*, **21**, 105

LUNDE, K.E. (1984). Capacity exponents for field construction costs. *Chem. Engng*, **91**(Mar. 5), 71

LUNDE, L. (1984). Stress corrosion cracking of steels in ammonia: specially vapor-phase cracking. *Ammonia Plant Safety*, **24**, 154

LUNDE, L. and NYBORG, R. (1987). The effect of oxygen and water on stress corrosion cracking of mild steel in liquid and vaporous ammonia. *Ammonia Plant Safety*, **27**, 49. See also *Plant/Operations Prog.*, 1987, **6**, 11

LUNDE, L. and NYBORG, R. (1989). SCC of carbon steels in ammonia. Crack growth studies and means to prevent cracking. *Corrosion 89* (Houston, TX: Nat. Ass. Corrosion Engrs)

LUNDE, L. and NYBORG, R. (1990). Prevention of stress corrosion cracking of carbon steels in ammonia. *Ammonia Plant Safety*, **30**, 60

LUNDQUIST, S. and LAUFKE, H. (1977). A method for quantification of fire risk in industry. *Loss Prevention and Safety Promotion 2*, p. 399

LUNDSTROM, E.A. and OPPENHEIM, A.K. (1969). On the influence of non-steadiness on the thickness of detonation waves. *Proc. R. Soc., Ser. A*, **310**, 759

LUNN, G.A. (1982a). An apparatus for the measurement of maximum experimental safe gaps at standard and elevated temperatures. *J. Haz. Materials*, **6**(4), 329

LUNN, G.A. (1982b). The influence of chemical structure and combustion reactions of the maximum experimental safe gap of industrial gases and liquids. *J. Haz. Materials*, **6**(4), 341

LUNN, G.A. (1984a). The maximum experimental safe gap: the effects of oxygen enrichment and the influence of reaction kinetics. *J. Haz. Materials*, **8**(3), 261

LUNN, G.A. (1984b). *Venting Gas and Dust Explosions — A Review* (Rugby: Instn Chem. Engrs)

LUNN, G.A. (1988a). *Guide to Dust Explosion Prevention and Protection. Pt. 3: Venting of Weak Explosions and the Effect of Vent Ducts* (Rugby: Instn. Chem. Engrs)

LUNN, G.A. (1988b). A note on the lower explosibility limit of organic dusts. *J. Haz. Materials*, **17**(2), 207

LUNN, G.A. (1989a). Methods for sizing dust explosion event areas: a comparison when reduced pressures are low. *J. Loss Prev. Process Ind.*, **2**(4), 200

LUNN, G.A. (1989b). Venting requirements for weak dust-handling equipment. *Chem. Engr, Lond.*, **456**, 18

LUNN, G.A. (1990). Safety of powder handling systems: review and detailed discussion of venting. *Process Safety Environ.*, **68B**, **155**

LUNN, G.A. (1992). *Guide to Dust Explosion Prevention and Protection — Pt 1, Venting*, 2nd ed. (Rugby: Instn Chem. Engrs)

LUNN, G.A. and CAIRNS, F. (1985). The venting of dust explosions in a dust collector. *J. Haz. Materials*, **12**(1), 87

LUNN, G.A. and PHILLIPS, H. (1973). *A Summary of Experimental Data on the Maximum Safe Gap* (London: Safety in Mine Res. Estab.)

LUNN, G.A. and PRZYBYLSKI, J. (1985). A Schlieren investigation of ignition downstream of a flat-plate flame trap. *J. Haz. Materials*, **10**(1), 125

LUNN, G.A. and SANSONE, E.B. (1990). *Destruction of Hazardous Chemicals in the Laboratory* (New York: Wiley)

LUNN, G.A., BROOKES, D.E. and NICOL, A. (1988). Using the K_{st}-nomographs to estimate the venting requirements in weak dust-handling equipment. *J. Loss Prev. Process Ind.*, **1**(3), 123

LUNN, G.A., CROWHURST, D. and HEY, M. (1988). The effect of vent ducts on the reduced explosion pressures of vented dust explosions. *J. Loss Prev. Process Ind.*, **1**(4), 182

LUNN, H.P. (1971). The effects of containerisation on the transportation of dangerous chemicals. *Transport of Hazardous Cargoes*, **2**, (paper 14)

LUNN, H.P. (1975). The movement of chemicals in freight containers. *Distribution Conf.* (London: Chem. Ind. Ass.)

LUPINI, R. and TIRABASSI, T. (1985). Temporal behaviour of continuous releases in flows with wind direction reversal. *Atmos. Environ.*, **19**, 1237

LUPTON, H.R. (1953). Graphical analysis of pressure surge in pumping systems. *Instn Water Engrs*, **7**(2), 87

LURIE, M. and MICHAILOFF, M. (1936). Evaporation from free water surfaces. *Ind. Engng Chem.*, **28**, 345

LUSK, D.T. and DORNEY, D.C. (1972). LNG storage tank systems. *LNG Conf. 3*

LUSS, D. and AMUNDSON, N.R. (1968). Stability of batch catalytic fluidised beds. *AIChE J.*, **14**, 211

LÜTOLF, J. (1972). Equipment for the determination of the explosion characteristics of combustible dusts. *In. Symp. on Dust and Explosion Risks in Mines and Industry* (Heidelberg: Berufsgenossenschaft der Chem. Ind.), p. 106

LUTTGENS, G. (1985). Collection of accidents caused by static electricity. *J. Electrostatics*, **16**, 247

LÜTTGENS, G. and GLOR, M. (1989). *Understanding and Controlling Static Electricity* (Expert Verlag)

LUTZ, W.K. (1994). Consider chemistry and physics in all phases of chemical plant design. *American Institute of Chemical Engineers 1994 Summer National Meeting*, Aug. 14–17, 1994, Denver, CO, Paper No. 47c

LUTZ, W.K. (1995a). Take chemistry and physics into consideration in all phases of chemical plant design. *Process Safety Prog.* **14** (3), (July), 153–62

LUTZ, W.K. (1995b). Putting safety into chemical plant design. *Chem. Health and Saf.* **2** (6) (Nov./Dec.), 12–15

LUTZ, W.K. (1997). Advancing inherent safety into methodology. *Process Safety Prog* **16** (2) (Summer), 86–88

LÜTZKE, K. (1971). *Erdöl und Kohle*, **24**, 165, 231

LÜTZKE, K. (1974). The dispersion of liquefied gas in the open during discharge from the liquid phase — a comparison with meteorological parameters. *Staub Reinhaltung der Luft*, **34**(2), 33 (English ed.)

LÜTZOW, D. and HEMMER, G. (1983). Safe burning of explosive offgas. *Plant/Operations Prog.*, **2**, 150

LUTZSKY, M. and LEHTO, D. (1968). Shock propagation in spherically symmetric exponential atmospheres. *Phys. Fluids*, **11**, 1466

LUTZSKY, M. and LEHTO, D. (1968). Shock propagation in spherically symmetric exponential atmospheres. *Phys. Fluids*, **11**, 1466

LUUS, R. (1971). Determination of starting conditions for chemical reactor stability. *Ind. Engng Chem. Fund.*, **10**, 322

LUX, J.A. (1973). Water treatment corrosion, and cleaning of steam systems. *Ammonia Plant Safety*, **15**, p. 6

LUXENBERG, H.R. and KUEHN, R.L. (eds) (1968). *Display Systems Engineering* (New York: McGraw-Hill)

LUXON, S.G. (1991). Legal requirements (COSHH Regulations 1988). In Simpson, D. and Simpson, W.G. (1991), *op. cit.*, p. 34

LUYBEN, W.L. (1966). Stability of autorefrigerated chemical reactors. *AIChE J.*, **12**, 662

LUYBEN, W.L. (1968). Effect of imperfect mixing on autorefrigerated reactor stability. *AIChE J.*, **14**, 880

LUYBEN, W.L. (1973). *Process Modelling* (New York: McGraw-Hill)

LUYBEN, W.L. (1990). A rational approach to control valve sizing. *Ind. Engng Chem. Res.*, **29**, 700

LUYBEN, W.L. and HENDERSHOT, D.C. (2003). Dynamic disadvantages of intensification in inherently safer process design. *Mary Kay O'Connor Process Safety Center Symp. – 2003 Proc,*, Oct. 28–29, 2003, 357–377. (College Station, TX: Mary Kay O'Connor Process Safety Center)

LUYBEN, W.L. and MELCIC, M. (1978). Consider reactor control lags. *Hydrocarbon Process.*, **57**(3), 115. See also Kane, L.A. (1980), *op. cit.*, p. 216

LYCETT, J.E., MAUDSLEY, D. and MILNER, D.L. (1989). TREMEX: transport emergency expert system. In Guy, G.B. (1989), *op. cit.*, p. 160

LYDERSEN, A.L. (1979). *Fluid Flow and Heat Transfer* (New York: Wiley)

LYDERSEN, S. and RAUSAND, M. (1987). A systematic approach to accelerated life testing. *Reliab. Engng*, **18**(4), 285

LYLE, A.R. and STRAWSON, H. (1971). Estimation of electrostatic hazards in tank filling operations. *Static Electrification*, **3**, p. 234

LYLE, A.R. and STRAWSON, H. (1972). Electrostatic hazards in tank-filling operations. *Physics Bull.*, **23**, 453

LYLE, A.R. and STRAWSON, H. (1973). Electrostatic hazards in tank filling operations. *Fire Prev. Sci. Technol.*, **4**, 8

LYLE, O. (1947). *The Efficient Use of Steam* (London: HM Stationery Office)

LYMAN, W.J., REEHL, W.F. and ROSENBLATT, D.H. (1990). *Handbook of Chemical Property Estimation Methods – Environmental Behavior of Organic Compounds* (Washington, DC: Am. Chem. Soc.)

LYNCH, C.T. (ed.) (1974–). *Handbook of Materials Science*, vols 1–3 (Boca Raton, FL: CRC Press)

LYNCH, E.P. (1973). Process logic diagrams. *Chem. Engng*, **80**(May 28), 105

LYNCH, E.P. (1974). Using Boolean algebra and logic diagrams. *Chem. Engng*, **81**(Aug. 19), 107; (Sep. 16), 111; Oct. 14), 101

LYNCH, E.P. (1980). *Applied Symbolic Logic* (New York: Wiley)

LYNCH, J.W. (1980). The new hazardous waste regulations. *Chem. Engng*, **87**(Jul. 28), 55

LYNCH, M.E. (1967). How to investigate a plant disaster. *Hydrocarbon Process.*, **46**(8), 161

LYNCH, M.E. (1973). How to investigate a plant disaster. In Vervalin, C.H. (1973a), *op. cit.*, p. 444

LYNCH, P.J. (1977). Recent BR developments in the movement of hazardous goods. *Symp. Hazard 77 – Advances in the Safe Handing and Movement of Hazardous Substances*, Teesside Polytechnic, Middlesbrough

LYNSKEY, P.J. (1985). The development of an effective emergency procedure for a toxic hazard site. *The Assessment and Control of Major Hazards* (Rugby: Instn Chem. Engrs), p. 183

LYON, P.R., PYMAN, M.A.F. and SLATER, D.H. (1982). Port planning and management aspects of the safe shipment of LNG and LPG. *Gastech 81*, p. 271

LYONS, C.G. and GARDINER, P.C. (n.d.a). *The Damage Caused by the German Flying Bomb (Fly). Rep. REN 454.* Res. and Expts Dept, Ministry of Home Security

LYONS, C.G. and GARDINER, P.C. (n.d.b). *The Damage Caused by the German Long Range Rocket Bomb. Rep. REN 530.* Res. and Expts Dept, Ministry of Home Security

LYONS, C.G. and GARDINER, P.C. (n.d.c). *The Screening of Blast by Buildings. Rep. REN 523.* Res. and Expts Dept, Ministry of Home Security

LYONS, D.M. and HENDRICKS, A.J. (1992). Planning, reactive. In Shapiro, S.C. (1992), *op. cit.*, p. 1171

LYONS, J. (1977). *Semantics*, vols 1–2 (New York: Cambridge Univ. Press)

LYONS, J.E. (1980). Up petrochemical value by liquid phase catalytic oxidation. *Hydrocarbon Process.*, **59**(11), 107

LYONS, J.L. (1979). Simplified sizing of a poppet relief valve for optimum flow and pressure drop. In Haupt, R.W. and Meyer, R.A. (1979), *op. cit.*, p. 65

LYONS, R.A. (1989). *Piper Alpha Public Inquiry. Transcript, Days 160, 164*

LYONS, W.A. (1976). Turbulent diffusion and pollutant transport in shoreline environments. In Haugen, D.A. (1976), *op. cit.*, p. 136

MAAS, W. (1969). Summary and a few general considerations. *Symp. on Inherent Hazards of Manufacturing and Storage in the Process Industry*, The Hague, p. 217

MAAS, W. (1974). Emergency planning (EP). *Loss Prevention and Safety Promotion 1*, p. 127

MABE, W.J. and MULHOLLAND, R.A. (1990). Variable high-speed motors optimize sealless pumps. *Hydrocarbon Process*, **69**(2), 71

MABEY, R. (1974). *The Pollution Handbook* (Harmondsworth, Middlesex: Penguin Books)

MABILEAU, J.M. (1982). The demand for tonnage: will anticipated excess of supply necessarily lead to a shortage of tonnage? *Gastech 81*, p. 433

MCADAMS, W.H. (1954). *Heat Transmission*, 3rd ed. (New York: McGraw-Hill)

MCALISTER, J. (1984). Control disturbances? Check grounding. *Chem. Engng*, **91**(Dec. 24), 77

MCALISTER, J. (1986). Surveillance – the forgotten maintenance tool. *Chem. Engng*, **93**(Oct. 13), 87

MACALLAN, J.L. (1989). *Piper Alpha Public Inquiry. Transcript, Days 122–124.*

MCALLISTER, D.G. (1970). Process plant utilities. Air. *Chem. Engng*, **77**(Dec. 14), 138

MCALLISTER, D.G. (1973). How to select a system for instrument air. *Chem. Engng*, **80**(Feb. 26), 39

MACARTHUR, D.A. and PACKHAM, D.R. (1971). Radiation from an ethylene diffusion flame. *Combust. Sci. Technol.*, **2**, 299

MACARTHUR, J.G. (1972). Ammonia storage tank repair. *Ammonia Plant Safety*, **14**, p. 1

MCAULIFFE, C.A. (ed.) (1977). *The Chemistry of Mercury* (London: Macmillan)

MACAUSLAND, D.D. and WEBB, J.A. (1981). Development of materials and fabrication requirements for oilfield production valves in sour (H_2S/chloride) service. *Chem. Engr, Lond.*, **368**, 229

MCBEAN, E.A. (1993). Description of the mathematical structure of air pathway models. In Asante-Duah, D.K. (1993), *op. cit.*, p. 339

MCBEAN, R.A. (1982). *The Influence of Road Geometry at a Sample of Accident Sites. Rep. LR 1053.* Transport and Road Res. Lab., Crowthorne

MCBRIDE, A.C. (1980). The plant engineer and GRP. *Plant Engr*, **23**(7), Aug./Sep., 21

MCBRIDE, W.C. (1981). *A Denser than Air Vapour Dispersion Validation Study using the ZEPHYR and DISCO Models* (report). Energy Resources Co.

MCCABE, J.S. and HICKEY, K.J. (1969). Heavy reactors now site assembled. *Hydrocarbon Process.*, **48**(5), 133

MCCABE, L.C. and CLAYTON, G.D. (1952). Air pollution by hydrogen sulphide in Pozo Rica, Mexico. An evaluation of the incident of Nov. 24, 1950. *Arch. Ind. Hyg. Occup. Med.*, **6**, 199

MCCABE, W.C. and SMITH, J.C. (1967). *Unit Operations of Chemical Engineering*, 2nd ed. (New York: McGraw-Hill)

MCCABE, W.C., SMITH, J.C. and HARRIOTT, P. (1985). *Unit Operations of Chemical Engineering*, 4th ed. (New York: McGraw-Hill)

MCCAFFREY, B.J. (1989). Momentum diffusion flame characteristics and the effects of water spray. *Combust. Sci. Technol.*, **63**, 315

MCCAFFREY, B.J. and EVANS, D.D. (1986). Very large methane jet diffusion flames. *Combustion*, **21**, 25

MCCAFFREY, B.J. and HARKLEROAD, M. (1989). Combustion efficiency, radiation, CO and soot yield from a variety of gaseous, liquid, and solid fuelled buoyant diffusion flames. *Combustion*, **22**, 1251

MCCAFFREY, B.J., QUINTIERE, J.G. and HARKLEROAD, M. (1981). Estimating room temperatures and the likelihood of flashover using fire test data correlations. *Fire Technol.*, **17**, 98; **18**, 122

MACCALLAN, S.E.A. and SETTERSTROM, C. (1940). Toxicity of ammonia, chlorine, hydrogen cyanide, hydrogen sulphide and sulphur dioxide gases. I, General methods and correlations. *Contrib. Boyce Thompson Inst.*, **11**, 325

MCCALLISTER, R.A. (1971). Sources and cost of plant startup problems. *Oil Gas J.*, **69**(Jun. 28), 72

MCCAMY, C.S., SHOUB, H. and LEE, T.G. (1957). Fire extinguishment by means of dry powder. *Combustion*, **6**, p. 795

MCCANDLESS, J.B. and IMGRAM, A.G. (1976). Evaluate materials using new non-destructive test method. *Hydrocarbon Process.*, **55**(10), 159

MCCANN, D.P.J., THOMAS, G.O. and EDWARDS, D.H. (1985). Gasdynamics of vented explosions. *Combust. Flame* **59**, 233; **60**, 63

MCCANN, J., CHOI, E., YAMASAKI, E. and AMES, B.N. (1975–). Detection of carcinogens as mutagens in the Salmonella/microsome test: assay of 300 chemicals. *Proc. Nat. Acad. Sci, 1975*, **72**, 5135; 1976, **73**, 950

MCCAREY, J.J. (1966). Petrochemical fire threatens giant refinery. *Fire Engng*, **119**, Aug., 103

MCCARTHY, D. (1971). Standards and codes of practice in overseas countries – a comparison with UK practice. *Electrical Safety*, **1**, p. 77

MCCARTHY, J. (1960). Recursive functions of symbolic expressions and their computation by machine: Part 1. *CACM 3*, p. 184

MCCARTHY, J. (1978). History of LISP. *SIGPLAN Notices*, **13**, 217

MCCARTHY, J., ABRAHAMS, P.W., EDWARDS, D.J., HART, T.P. and LEVIN, M.I. (1962). *LISP 1.5 Programmers Manual* (Cambridge, MA: MIT Press)

MCCARTHY, A.J., DITZ, J.M. and GEREN, P.M. (1996). Inherently safer design review of a new alkylation process. *5th World Cong. of Chem. Eng.*, July 14–18, 1996, San Diego, CA, Vol. II, 1099–1106 (New York: American Institute of Chemical Engineers)

MCCARTHY, A.J. and MILLER, U.R. (1997). Inherently safer design principles are proven in expansions. *Hydrocarb. Process. 76* (4) (April), 122–25

MCCARTHY, J.E. (1978). Choosing a flue-gas desulfurization system. *Chem. Engng*, **85**(Mar. 13), 79

MCCARTHY, J.G. (ed.) (1990). *Plastic Surgery* (Philadelphia, PA: W.B. Saunders)

MCCARTHY, J.J. (1943). *The Science of Fighting Fires* (New York: Norton)

MACCARY, R.R. (1960a). How to select pressure-vessel size. *Chem. Engng*, **67**(Oct. 17), 187

MACCARY, R.R. (1960b). Pressure vessel design for high or cryogenic temperatures. *Chem. Engng*, **67**(Nov. 28), 131

MACCARY, R.R. (1975). A technical basis for characterising flaws detected by preservice and inservice examinations of nuclear power plant components. *J. Pres. Ves. Technol.*, **97**, 322

MCCAUGHEY, M. and FLETCHER, D.F. (1993). Calculations of wind-induced pressure distribution on a model building. *Fire Safety J.*, **21**, 189

MCCAUL, C. (1989). Keep the wear out of pump wear rings. *Chem. Engng*, **96**(10), 149

MCCHESNEY, M. and MCCHESNEY, P. (1981). Preventing burns from insulated pipes. *Chem. Engng*, **88**(Jul. 27), 58

MCCHESNEY, M. and MCCHESNEY, P. (1982). Insulation without economics. *Chem. Engng*, **89**(May 3), 70

MCCHRYSTAL, W.G. (1982). Computer assisted maintenance planning. *Plant Engr*, **26**(5), 27

MCCLELLAND, G.D. (1968). Maintenance and safe operation of high-pressure equipment. *Chem. Engng*, **75**(Sep. 23), 202

MCCLESKEY, F. (1988a). *Drag Coefficients for Irregular Fragments. Rep. NSWC TR 87–89*. Naval Surface Weapons Center

MCCLESKEY, F. (1988b). *Fragment Hazard Computer Program (FRAGHAZ). Rep. NSWC TR 87–59*. Naval Surface Weapons Center

MCCLESKEY, F. (1988c). *Quantity Distance Fragment Hazard Computer Program (FRAGHAZ). Rep. NSWC TR 87–459*. Naval Surface Weapons Center

MCCLESKEY, F. (1992). Fragmentation characteristics of horizontally stacked bombs. *Proc. Twenty-Fifth Explosives Safety Seminar* (Explosives Safety Board, Dept of Defense), p. 217

MCCLESKEY, F., NEADES, D.N. and RUDOLPH, R.R. (1990). A comparison of two personnel injury criteria based on fragmentation. *Proc. Twenty-Fourth Explosives Safety Seminar* (Explosives Safety Board, Dept of Defense), p. 2423

MCCLOSKEY, C.M. (1989). Safe handling of organic peroxides: an overview. *Plant/Operations Prog.*, **8**, 185

MCCLUER, P.E. and WHITTLE, D.K. (1992). Lessons learned from hazop reviews of FCCUs. *Hydrocarbon Process.*, **71**(8), 140–C

MCCLURE, J.M. (1994). Ease instrument rebuilds after plant explosions. *Chem. Engng Prog.*, **90**(3), 84

MCCLURE, K.A. (1927). Unpublished War Dept Rep. See Wheatley, C.J. *et al.* (1988 SRD R438)

MCCONNAUGHEY, W.E. (1971). Hazard evaluation. *Transport of Hazardous Cargoes*, **2**, paper 3

MCCONNAUGHEY, W.E., WELSH, M.E., LAKEY, R.J. and GOLDMAN, R.M. (1970). Hazardous materials transportation. *Chem. Engng Prog.*, **66**(2), 57

MCCONNELL, J.H. and BRADY, R.R. (1960). Austenitic stainless steels. Thousands of tons in 300 to 425° F service. *Chem. Engng*, **67**(Jul. 11), 125

MCCONNELL, R.A. (1992). Safety assurance program for existing operations. *Ammonia Plant Safety*, **32**, p. 104

MCCORD, C.P. (1926). Industrial poisoning from low concentrations of chlorine gas. *J. Am. Med. Ass.*, 1687

MCCORKLE, J.J. (1980). Safety designed into a cyclohexane processing facility. *Loss Prevention and Safety Promotion 3*, vol. 2, p. 770

MCCORMICK, D. (1974). The essential elements of a personal protection programme. *Chemy Ind.*, Apr. 20, 328

MCCORMICK, E.J. (1957a). *Human Engineering*, 2nd ed. (New York: McGraw-Hill)

MCCORMICK, E.J. (1957b). *Human Factors Engineering*, 2nd ed.; 1970, 3rd ed. (New York: McGraw-Hill)

MCCORMICK, E.J. (1976). *Human Factors in Engineering and Design*, 4th ed. (New York: McGraw-Hill)

MCCORMICK, E.J. and SANDERS, M.S. (1982). *Human Factors in Engineering and Design*, 5th ed. (New York: McGraw-Hill). See also Sanders, M.S. and McCormick, E.J. (1987)

MCCORMICK, J.W. and SCHMITT, C.R. (1974). Extinguishment of selected metal fires using carbon microspheres. *Fire Technol.*, **10**, 197

MCCORMICK, N.J. (1987). Asymptotic unavailability for an M-out-of-N system without testing. *Reliab. Engng*, **17**(3), 189

MCCORNACK, R.L. (1961). *Inspector Accuracy: a Study of the Literature. Rep. SC TM-53—61(14).* Sandia Labs, Albuquerque, NM

MCCOY, C.S., DILLENBACK, M.D. and TRUAX, D.J. (1986). Major fire in a steam-methane reformer furnace. *Ammonia Plant Safety*, **26**, p. 234. See also *Plant/Operations Prog.*, 1986, **5**, 165

MCCOY, R.C. (1969). Comparing flow-ratio control systems. *Chem. Engng*, **76**(Jun. 16), 92

MACCREADY, P.B., SMITH, T.B. and WOLF, M.A. (1961). *Vertical Diffusion from a Low Altitude Line Sourse* (report). Met. Res. Inc., Altadena, CA

MCCREANNEY, W.C. (1969). Skin care. In Handley, W. (1969), *op. cit.*, p. 399

MCCREANNEY, W.C. (1977). Skin care. In Handley, W. (1977), *op. cit.*, p. 361

MCCREEDY, T. (1999). Reducing the risks of synthesis. *Chem. & Ind.*, 15 (August 2), 588—90

MCCRINDLE, W.I. (1977). *A First Guide to Loss Prevention* (London: Instn Chem. Engrs)

MCCRINDLE, W.I. (1981). *A First Guide to Loss Prevention*, 2nd ed. (Rugby: Instn Chem. Engrs)

MCCULLOCH, D.R. (1981). Preventing failure of bellows expansion-joints. *Chem. Engng*, **88**(Nov. 16), 259

MCCULLOUGH, C.R. (1957). *Safety Aspects of Nuclear Reactors* (New York: van Nostrand)

MCCULLOUGH, D.G. (1968). *The Johnstown Flood* (New York: Simon & Schuster)

MCCULLOUGH, D.G. (1973). Yardsticks for maintenance performance. *Chem. Engng*, **80**(Oct. 29), 120

MCDANIEL, J.T. (1986). Explosion of a Benfield solution storage tank. *Ammonia Plant Safety*, **26**, p. 77. See also *Plant/Operations Prog.*, 1986, **5**, 45

MCDERMID, J.A. (1989a). Assurance in high-integrity software. In Sennet, C.T. (1989), *op. cit.*

MCDERMID, J.A. (ed.) (1989b). *The Theory and Practice of Refinement: Approaches to the Formal Development of Large Scale Systems* (London: Butterworths)

MCDERMID, J.A. (1991). *Software Engineers Reference Book* (London: Butterworth-Heinemann)

MCDERMID, J.A. (1993). Formal methods: use and relevance for the development of safety-critical systems. In Bennett, P. (1993), *op. cit.*, p. 96

MACDERMOD, E. (1965). Design and inspection of pressure vessels. In Fawcett, H.H. and Wood, W.S. (1965), *op. cit.*, p. 155

MACDERMOD, E. (1982). Design and inspection of pressure vessels. In Fawcett, H.H. and Wood, W.S. (1982), *op. cit.*, p. 119

MCDERMOTT, H.J. (1976). *Handbook of Ventilation for Contaminant Control* (Ann Arbor, MI: Ann Arbor Science)

MCDERMOTT, J. (1980). *R1: a Rule-based Configurer of Computer Systems. Rep. CMU-CS-80—119.* Dept of Comput. Sci., Carnegie-Mellon Univ., Pittsburgh, PA

MCDERMOTT, J. (1981). R1's formative years. *AI Magazine*, **2**

MCDERMOTT, J. (1982a). R1: a rule-based configurer of computer systems. *Artificial Intelligence*, **19**, 39

MCDERMOTT, J. (1982b). XCEL, a computer sales person's assistant. *Machine Intelligence 10* (ed. by J.E. Hayes, D. Michie and Y.H. Pao (Chichester: Ellis Horwood)

MACDERMOTT, P.E. (1976). Fires in oil soaked lagging. *Chem. Britain*, Feb., 69

MCDEVITT, C.A., STEWARD, F.R. and VENART, J.E.S. (1987). What is a BLEVE? *Proc. Fourth Tech. Seminar on Chemical Spills*, p. 137

MCDEVITT, C.A., CHAN, C.K., TENNANKORE, K.N. and STEWARD, F.R. (1990). Initiation step of boiling liquid expanding vapour explosions. *J. Haz. Materials*, **25**(1/2), 169

MACDIARMID, J.A. and NORTH, G.J.T. (1989). Lessons learned from an explosion in a process unit. *Plant/Operations Prog.*, **8**, 96

MCDONALD, A.G. (1989). *Piper Alpha Public Inquiry. Transcript, Day 93*

MACDONALD, G.A. (1972). *Volcanoes* (Englewood Cliffs, NJ: Prentice-Hall)

MACDONALD, G.L. (1972). *The Involvement of Tractor Design in Accidents. Rep. 3/72.* Dept of Mech. Engng, Univ. of Queensland, St Lucia

MACDONALD, H.F., DARLEY, P.J. and CLARKE, R.H. (1973). Recent developments in prediction of environmental consequences of radioactive releases from nuclear power plants. In *Physical Behaviour of Radioactive Contaminants in the Atmosphere* (Vienna: Int. Atom. Energy Agency), p. 337

MACDONALD, J.A. and WHITE, R.G. (1965). *Spontaneous Ignition of Kerosene (Avtur) Fuel Vapour: the Effect of Vessel Size. Tech. Rep. 65138.* R. Aircraft Estab.

MACDONALD, J.O.S. (1960). The manufacture of a highly toxic organophosphorus compound. *Chemical Process Hazards*, **I**, p. 107

MACDONALD, J.O.S. (1973). Efficient valve packing – the design of stuffing box glands. *Process Technol. Int.*, **18**, 367

MACDONALD, R.J. and HOWAT, C.S. (1988). Data reconciliation and parameter estimation in plant performance analysis. *AIChE J.*, **34**, 1

MACDONALD, R.W. (1977). The interaction of chlorine and seawater, with special reference to the four liquid chlorine tank cars lost in British Columbia coastal waters. *J. Haz. Materials*, **2**(1), 51

MCDONNELL, R.G., GLASS, J.A. and DAUES, G.W. (1967). Analyzing liquid O_2 for contaminants. *Ammonia Plant Safety*, **9**, p. 6

MCDOWELL, D.M. (1965). How sigma phase affects stainless. *Hydrocarbon Process.*, **44**(8), 143

MCDOWELL, D.M. (1967). Which steel for heaters, exchangers in heavy hydrogenation service? *Hydrocarbon Process.*, **46**(1), 245

MCDOWELL, D.M. (1973). Corrosion: back to basics. *Chem. Engng*, **80**(Oct. 1), 92

MCDOWELL, D.M. (1974). Materials of construction report. *Chem. Engng*, **81**(Nov. 11), 118

MCDOWELL, D.M. and O'CONNOR, B.A. (1977). *Hydraulic Behaviour of Estuaries* (London: Macmillan)

MCDOWELL, D.W. and MILLIGAN, J.D. (1965). Multilayer reactors resist hydrogen attack. *Hydrocarbon Process.*, **44**(12), 119

MCDOWELL, J.K. and DAVIS, J.F. (1991). Managing qualitative simulation in knowledge-based chemical diagnosis. *AIChE J.*, **37**, 569

MCDUFF, T. (1974). The probability of vessel collisions. *Ocean Ind.*, Sep.

MACEK, A. (1979). Flammability limits: a re-examination. *Combust. Sci. Technol.*, **21**, 43

MCELROY, J. (1969). A comparative study of urban and rural dispersion. *J. Appl. Met.*, **8**, 19

MCELROY, J.L. and POOLER, F. (1968). *The St Louis Dispersion Study. Rep. AP-53.* US Public Health Service, Nat. Air Poll. Control Admin.

MCELROY, R.E. (1965). *Accident Prevention Manual* (Chicago, IL: Nat. Safety Coun.)

MCELROY, R.E. and OLISHIFSKI, J.B. (1971). *Fundamentals of Industrial Hygiene* (Chicago, IL: Nat. Safety Coun.)

MACEWEN, J.D. and VERNOT, E.H. (1970). *Toxic Hazards Research Unit Annual Report.* Aerospace Med. Res. Lab., Wright Patterson Air Force Base, Dayton, OH

MACEWEN, J.D. and VERNOT, E.H. (1972). *Toxic Hazards Research Unit Annual Technical Report 1972. Rep. AMRL-TR-72–62 (AD 755–358).* SysteMed Corp., Dayton, OH

MACEWEN, J.D. and VERNOT, E.H. (1974). *Toxic Hazards Research Unit Annual Technical Report 1974. Rep. TR-74–78.* Aerospace Med. Res. Lab., Wright Patterson Air Force Base, Dayton, OH

MACEWEN, J.D. and VERNOT, E.H. (1976). *Toxic Hazards Research Unit Annual Technical Report 1976. Rep. AMRL series (AD A031–860).* SysteMed Corp., Dayton, OH

MACEWEN, J.D., THEODORE, J. and VERNOT, E.H. (1970). Human exposure to EEL concentrations of monomethylhydrazine. *Proc. First Ann. Conf. on Environmental Toxicology, Rep. AMRL-TR-70-102.* (Wright Patterson AFB, OH)

MCEWAN, M.J. and PHILLIPS, L.F. (1975). *Chemistry of the Atmosphere* (London: Arnold)

MACEY, M. (1989). *Piper Alpha Public Inquiry. Transcript, Day 149*

MACEYKA, T.D. (1983). Detect and evaluate surge in multistage compressors. *Hydrocarbon Process.,* **62**(11), 203

MCFARLAND, I. (1969). Preventing flange fires. *Ammonia Plant Safety,* **11**, p. 20. See also *Chem. Engng Prog.,* 1969, **65**(8), 59

MCFARLAND, I. (1970). Safety in pressure vessel design. *Ammonia Plant Safety,* **12**, p. 43. See also *Chem. Engng Prog.,* 1970, **66**(6), 56

MCFARLAND, I. (1974). New ways to fabricate large high pressure vessels. *Ammonia Plant Safety,* **16**, p. 113

MACFARLANE, D.R. and EWING, T.F. (1990). Acute health effects from accidental releases of high toxic hazard chemicals. *J. Loss Prev. Process Ind.,* **3**(1), 167

MACFARLANE, K. (1991). Development of plume and jet release models (poster item). *Modeling and Mitigating the Consequences of Accidental Releases,* p. 657

MACFARLANE, K., PROTHERO, A., PUTTOCK, J.S., ROBERTS, P.T. and WITLOX, H.W.M. (1990). *Development and Validation of Atmospheric Dispersion Models for Ideal Gases and Hydrogen Fluoride,* vol. I, *HGSYSTEM Technical Reference Manual.* (report). Shell Res. Ltd

MACFARLANE, P. (1986). Getting revamps into shape. *Chem. Engr, Lond.,* **421**, 38

MACFATTER, W.E. (1972). Reliability experiences in a large refinery. *Chem Engng Prog.,* **68**(3), 52

MACFIE, R.A.B. (1983). Safety responsibilities of plant operators and specialist contractor. *Plant/Operations Prog.,* **2**, 226

MCGARRY, J. (1983). Correlation and prediction of the vapor pressures of pure liquids over large pressure ranges. *Ind. Engng Chem. Process Des. Dev.,* **22**, 313

MCGARRY, J.F. (1958). A checklist for plant layout. *Petrol. Refiner,* **37**(10), 109

MCLAREN, W.M. (1989). *Piper Alpha Public Inquiry. Transcript, Day 79*

MCGETTRICK, A. (1982). *Program Verification Using ADA* (Cambridge: Cambridge Univ. Press)

MCGILL, J.C. and MCGILL, E.C. (1978). Plant vent and flare hydrocarbon recovery and reprocessing systems. *API Proc., Refining Dept, 43rd Midyear Mtg,* **57**, 46

MCGILL, J.C. and MCGILL, E.C. (1978). Save lost hydrocarbons. *Hydrocarbon Process.,* **57**(5), 158

MACGILL, S.M. (1982). *Mossmorran–Braefoot Bay: Aspects of a Risk Debate. Working Paper 327.* School of Geography, Univ. of Leeds

MCGILL, S.M. (ca. 1987). *The Politics of Anxiety: Sellafields Cancer-Link Controversy* (London: Pion)

MCGILLIVRAY, R.J. (1963a). Safety in sweetening operations. *Am. Petrol. Inst., Div. of Refining, Twenty Eighth Midyear Mtg,* paper 27–63

MCGILLIVRAY, R.J. (1963b). Take these steps to sweetening safety. *Hydrocarbon Process. Petrol. Refiner.,* **42**(5), 145

MCGINLEY, H. (1986). Motor vehicles. *Fire Protection Handbook,* 16th ed. (ed. A.E. Cote and J.L. Linville), pp. 13–20

MCGINTY, L. and ATHERLEY, G. (1977). Acceptability versus democracy. *New Scientist,* **323**

MCGLASHAN, R.S. and HVIDING, G. (1992). Troll onshore development. *Trans. Inst. Chem. Engrs,* **70A**, 3

MCGRATH, P.T. (1993). Thermal comfort. *Plant Engr,* **37**(3)

MCGRATH, R.V. (1963). Stability of API Standard 650 tank shells. *API Proc., Div. of Refining,* **43**, 458

MCGRATH, R.V. (1973). Improve your API tank specs. *Hydrocarbon Process.,* **52**(8), 79

MCGRATH, R.V. (1975). Highlights of revised API tank standards. *Hydrocarbon Process.,* **54**(5), 89

MCGRATH, R.V. (1976). New concepts for storage tank design. *Hydrocarbon Process.,* **55**(5), 97

MCGREAVY, C. and NEWMAN, M.W. (1974). How to reduce refractory spalling in steam reformer furnaces. *Ammonia Plant Safety,* **16**, p. 59

MACGREGOR, C.W. (1964). *Handbook of Analytical Design for Wear* (New York: Plenum Press)

MCGUFFEY, J.R., PALUZELLE, R. and MULDREW, W.E. (1962). Handling gaseous fluorine in industry. *Ind. Engng Chem.,* **54**(5), 46

MCGUINNESS, R.M. (1969). Ammonia in the eye. *Br. Med. J.,* 575

MCGUINESS, W.N. (1990). Block-type ball valve considerations. *Hydrocarbon Process.,* **69**(3), 101

MCGUIRE, G. (1980). Improving safety through training. *Gastech 79,* p. 251

MCGUIRE, G. and WHITE, B. (1986). *Liquefied Gas Handling Principles on Ships and in Terminals* (London: Witherby & Co.)

MCGUIRE, J.H. (1953). *Heat Transfer by Radiation. Fire Res. Special Rep. 2* (London: HM Stationery Office)

MCGUIRE, J.H. (1958). Fire danger from static electricity. *Br. Chem. Engng,* **3**, 136

MCGUIRE, J.H. (1964). *Large Scale Use of Inert Gas to Extinguish Building Fires. Res. Paper 246.* Nat Res. Council of Canada, Div. of Building Res., Ottawa

MCGUIRE, M.L. and LAPIDUS, L. (1965). On the stability of a detailed packed-bed reactor. *AIChE J.,* **11**, 85

MCGUIRE, R.K. (1976). Methodology for incorporating parameter uncertainties into seismic hazard analysis for low risk design intensities. *Int. Symp. on Earthquake Structural Engineering,* vol. 2, p. 913

MCGUIRK, J.J. and PAPADIMITROU, C. (1988). Stably stratified free surface shear layers with internal hydraulic jumps. In Puttock, J.S. (1988b), *op. cit.,* p. 385

MCHALE, E.T. (1975). Flame inhibition by potassium compounds. *Combust. Flame,* **24**, 277

MACHE, H. (1918). *Die Physik der Verbrennungserscheinungen* (Leipzig: Veit)

MACHIDA, S. (1990). ILO's programme on major hazard control. *Safety and Loss Prevention in the Chemical and Oil Processing Industries*, p. 69

MACHLE, W. and KITZMILLER, K. (1935). The effects of the inhalation of hydrogen fluoride. II, The response following exposure to low concentration. *J. Ind. Hyg.*, **17**, 223

MACHLE, W. and SCOTT, E.W. (1935). The effects of the inhalation of hydrogen fluoride. III, Fluorine storage following exposure to sub-lethal concentrations. *J. Ind. Hyg.*, **17**, 230

MACHLE, W., THAMANN, F., KITZMILLER, K. and CHOLAK, J. (1934). The effects of the inhalation of hydrogen fluoride. I, The response following exposure to high concentrations. *J. Ind. Hyg.*, **16**, 129

MACHLE, W., KITZMILLER, K.V., SCOTT, E.W. and TREON, J.F. (1942). The effect of the inhalation of hydrogen chloride. *J. Ind. Hyg. Toxicol.*, **24**, 222

MACHOL, R.E. and GRAY, P. (eds) (1962). *Recent Developments in Information and Decision Processes* (London: Macmillan)

MCHUGH, M. and KRUKONIS, V. (1986). *Supercritical Fluid Extraction – Principles and Practise* (London: Butterworths)

MCINTIRE, J.R. (1977a). A measure of refinery reliability. *API Proc., Refining Dept, 42nd Midyear Mtg*, **56**, 287

MCINTIRE, J.R. (1977b). Measure refinery reliability. *Hydrocarbon Process.*, **56**(5), 121

MCINTOSH, A.C. and TOLPUTT, T.A. (1990). Critical heat losses to avoid self-heating in coal piles. *Combust. Sci. Technol.*, **69**, 133

MCINTOSH, A.R. (1989). *Piper Alpha Public Inquiry. Transcript, Days 142*, 143

MCINTOSH, D.H. (1972). *Meteorological Glossary*, 5th ed. (London: HM Stationery Office)

MCINTOSH, D.H. and THOM, A.S. (1969). *Essentials of Meteorology* (London: Wykeham Publs)

MACINTYRE, D.R. (1980). How to prevent stress corrosion cracking in stainless steels. *Chem. Engng*, **87**(Apr. 7), 107; (May 5), 131

MCINTYRE, D.R. (1981). Estimating temperature extremes of accidental fires. *Chem. Engng*, **88**(Jun. 15), 123

MCINTYRE, D.R. (1982). Evaluating the cost of corrosion-control methods. *Chem. Engng*, **89**(Apr. 5), 127

MCINTYRE, P.J. (1992). Development of the operating procedures for Sizewell B PWR. *Int. Atom. Energy Agency, Specialists Mtg on Operating Procedures for Nuclear Power Plants and Their Presentation* (Vienna)

MACK, R.J. and SHULTZ, J.T. (1987). Failure of ammonia converter start-up heater tube. *Ammonia Plant Safety*, **27**, 127

MACK, W.A. (1975). VCM reduction and control. *Chem. Engng Prog.*, **71**(9), 41

MACK, W.C. (1971). How to predict remaining service life of tubing. *Chem. Engng*, **78**(Aug. 9), 124

MACK, W.C. (1976). Selecting steel tubing for high-temperature service. *Chem. Engng*, **83**(Jun. 7), 145

MACKAY, B. (1992). Avoid steam trap problems. *Chem. Engng Prog.*, **88**(1), 41

MACKAY, D. and MATSUGU, R.S. (1973). Evaporation rates of liquid hydrocarbon spills on land and water. *Can. J. Chem. Engng*, **51**, 434

MACKAY, D., SHIU, W.Y. and SUTHERLAND, R.J. (1980). Estimating volatilization and water column diffusion rates of hydrophobic contaminants. In Haque, R. (1980), *op. cit.*, p. 127

MACKAY, D.J., MURRAY, S.B., MOEN, I.O. and THIBAULT, P.A. (1988). Flame-jet ignition of large fuel–air clouds. *Combustion*, **22**, 1339

MCKAY, F.F., WORRELL, G.R., THORNTON, B.C. and LEWIS, H.L. (1977). If an ethylene pipe ruptures. *Hydrocarbon Process.*, **56**(11), 487

MCKAY, G., NORRIE, K.M., POOTS, V.J.P. and TURNER, J.M.C. (1975). Diffusion processes in the slow oxidation of isobutane. *Combust. Flame*, **25**, 219

MCKAY, R.S. (1993). Fine water spray – a halon replacement technology. *Fire Surveyor*, **22**(3), 16

MCKECHNIE, S. (1985). Health and safety at work. *Chem. Ind.*, Mar. 18, 176

MCKEE, H.C. (1976). Air-pollution instrumentation. *Chem. Engng*, **83**(Jun. 21), 155

MCKEE, R.E. (1990). *Piper Alpha Public Inquiry. Transcript, Days 171, 172*

MCKEEVER, D.J. and LAWRENSON, R. (1992). Safety management offshore requirements. *Major Hazards Onshore and Offshore*, p. 149

MCKELVEY, F.E. (1964). Estimate physical properties accurately. *Hydrocarbon Process.*, **43**(5), 145; **43**(6), 147; **43**(7), 146; **43**(8), 127

MCKENNA, F.P. (1975). Do safety measures really work? An examination of risk homeostasis theory. *Ergonomics*, **28**, 489

MCKENNA, F.P. (1985). Evidence and assumptions relevant to risk homeostasis. *Ergonomics*, **28**, 1539

MCKENNA, P. (1989). The fire safety factor in passenger and cargo ships. *Fire Int.*, **120**, 26

MCKENNA, S.P. and HALE, A.R. (1981). The effect of emergency first aid training on the incidence of accidents in factories. *J. Occup. Accid.*, **3**(2), 101

MCKENNA, S.P. and HALE, A.R. (1982). Changing behaviour towards danger: the effect of first aid training. *J. Occup. Accid.*, **4**(1), 47

MACKENTHUN, K.M. and GUARRAIA, L.J. (1978). The development and use of water quality criteria to protect aquatic integrity. *Control of Hazardous Material Spills* **4**, p. 80

MACKENZIE, J. (1990). Hydrogen peroxide without accidents. *Chem. Engng*, **97**(6), 84

MACKENZIE, J. (1991). Considerations for the safe design of processes using hydrogen peroxide and organics. *Plant/Operations Prog.*, **10**, 164

MACKENZIE, R.C. (1962). *The Differential Thermal Analysis Data Index* (London: Macmillan)

MACKENZIE, R.C. (1974). Highways and byways in thermal analysis. *Analysis*, **99**, 900

MACKENZIE, R.M. (1958). On the accuracy of inspectors. *Ergonomics*, **1**, 258

MACKEOWN, S.S. and WOUK, V. (1942). Electrical charges produced by flowing gasoline. *Ind. Engng Chem.*, **34**, 659

MACKEY, C.F. (1974). Pre-start-up safety checkout – prelude to a successful start-up. *Loss Prevention*, **8**, 40

MACKEY, W.M. (1895). Spontaneous combustion of oils spread on cotton – II. *J. Soc. Chem. Ind.*, **14**, 940

MACKEY, W.M. (1896). Apparatus for the determination of the relative liability to spontaneous combustion of oils spread on cotton wool. *J. Soc. Chem. Ind.*, **15**, 90

MACKIE, J.A. and NIESEN, K. (1984). Hazardous-waste management: the alternatives. *Chem. Engng*, **91**(Aug. 6), 51

MACKIE, R.R. (ed.) (1977). *Vigilance. Theory, Operational Performance and Physiological Correlates* (New York: Plenum Press)

MCKINLAY, A.F. (1982). Non-ionizing radiation. In Green, A.E. (1982), *op. cit.*, p. 527

MCKINLEY, C. (1968). Problems involved in the design of pressure relieving systems for permanently installed marine bulk liquid cargo containers. *Carriage of Dangerous Goods*, **1**, p. 89

MCKINLEY, C. (1973). Consideration of hazards associated with bulk sea transport of liquefied gases. *Transport of Hazardous Cargoes*, **3**

MCKINLEY, C. (1982). Pressure relief for chemical processes. In Fawcett, H.H. and Wood, W.S. (1982), *op. cit.*, p. 105

MCKINNEY, M.S. and OERLEMANS, J. (1985). The planning and building of Antwerp gas terminal. *Gastech 84*, p. 448

MCKINNON, R. (1974). VCM vapour – the 'panic' in perspective. *Occup. Safety Health, Sep.*, 12

MACKINVEN, R., HENSEL, J.G. and GLASSMAN, I. (1970). Influence of laboratory parameters on flame spread over liquid surfaces. *Combust. Sci. Technol.*, **1**, 293

MCKNIGHT, A.D., MARSTRAND, P.K. and SINCLAIR, I.D. (eds) (1974). *Environmental Pollution Control: Technical, Economic and Legal Aspects* (London: Allen & Unwin)

MCKOWN, R. (1974). *A Comprehensive Guide to Factory Law*, 5th ed. (London: Goodwin)

MACLACHLAN, S. (1985). Hydrogen sulphide in the oil industry. *Chem. Engr, Lond.*, **418**, 27

MCLAIN, L. (1975a). Basic lessons for the professionals. *The Engineer, Lond.*, May 15, 7

MCLAIN, L. (1975b). Instruments fail to detect fire at Shell refinery. *The Engineer, Lond., Jan. 2/9*, 15

MCLAREN, D.B. and UPCHURCH, J.C. (1970). Guide to trouble-free distillation. *Chem. Engng*, **77**(Jun. 1), 139

MCLAREN, W.M. (1989). *Piper Alpha Public Inquiry. Transcript, Days 79*

MCLARTY, T.E. (1976). Selecting silencers to suppress plant noise. *Chem. Engng*, **83**(Apr. 12), 104

MCLEAN, A.D. (1981). Chemical incidents handled by the United Kingdom public fire service in 1980. *J. Haz. Materials*, **5**(1/2), 3

MCLEAN, A.E.M. (1977a). Assessment of hazard and conflict of interest. *R. Soc. Mtg on Long-Term Hazards to Man from Man-Made Chemicals in the Environment* (London)

MCLEAN, A.E.M. (1977b). Risks and benefit in food and additives. *Proc. Nutr. Soc.*, **36**, 85

MCLEAN, A.E.M. (1979). Hazards from chemicals: scientific questions and conflicts of interest. *Proc. R. Soc., Ser. B*, **205**, 179. See also in Doll, Sir R. and McLean, A.E.M. (1979), *op. cit.*, p. 179

MCLEAN, A.E.M. (1981). Assessment and evaluation of risk to health from chemicals. *Proc. R. Soc., Ser. A*, **376**, 51. See also in Warner, Sir F. and Slater, D.H. (1981), *op. cit.*, p. 51

MACLEAN, D. (1982). Risk and consent: philosophical issues for centralized decisions. *Risk Analysis*, **2**, 59

MCLEAN, D. and CRIPPS, R.M. (1986). Design, construction and periodical surveys of ships for liquefied gases – a classification society's viewpoint. *Gastech 85*, p. 183

MCLEAN, J.A. (1992). Security of electricity supply systems including standby supplies. *Major Hazards Onshore and Offshore*, p. 275

MACLEAN, J.N. *et al.* (1962). The self-association of hydrogen fluoride vapour. *J. Inorg. Nucl. Chem.*, **24**, 1549

MCLEAN, M.G. (1982). How to select and apply flexible-impeller pumps. *Chem. Engng*, **89**(Sept. 20), 101

MCLEOD, L.M. (1969). Comments on Div. 2 vessel design. *Hydrocarbon Process.*, **48**(12), 125

MCLEOD, P.D. and MCCALLUM, I.R. (1975). Control strategies of experienced and novice controllers with a zero energy nuclear reactor. *Ergonomics*, **18**, 547

MCLINTOCK, I.S. (1968). The effect of various diluents on soot production in laminar ethylene diffusion flames. *Combust. Flame*, **12**, 217

MCLOUGHLIN, J. (1972). *The Law Relating to Pollution* (Manchester: Manchester Univ. Press)

MCLOUGHLIN, J. (1973). *The Water Act 1973* (London: Sweet & Maxwell)

MCLOUGHLIN, J. (1976). *The Law and Practice Relating to Pollution Control in the United Kingdom* (London: Graham & Trotman)

MCLOUGHLIN, J. (1980). Pollution control and law in practice. Methods of control used in different countries. *Chem. Ind.*, Apr. 19, 328

MCLOUGHLIN, J. (1981). Risk and legal liability. In Griffiths, R.F. (1981), *op. cit.*, p. 106

MCMAHON, T.A. and DENISON, P.J. (1979). Empirical atmospheric deposition parameters – a survey. *Atmos. Environ.*, **13**, 571

MCMARLIN, R.M. and GERRISH, R.W. (1992). Can your 'non-combustible' insulation burn? *Hydrocarbon Process.*, **71**(11), 101

MCMASTER, R.C. (1959). *Non-Destructive Testing Handbook*, vols 1–2 (New York: Ronald Press)

MCMILLAN, A. (1986). Flammable hazards: a difficult area to classify. *Process Engng*, Jul., 35

MACMILLAN, A.J. (1975). Electrical apparatus for Zones 2. *Electrical Safety*, **2**, p. 85

MACMILLAN, A.J. (1982). The use of instrumentation in potentially explosive atmospheres. *Flammable Atmospheres and Area Classification, IEE Coll. Dig. 1982/26* (London: Insn Elec. Engrs)

MACMILLAN, C.L. (1976). Service behavior of reformer outlet manifolds. *Ammonia Plant Safety*, **18**, 54

MACMILLAN, R.H. (ed.) (1962). *Non-Linear Control Systems* (Oxford: Pergamon Press)

MACMILLAN, W.J. (1974). Testing pipe insulation systems. *Loss Prevention*, **8**, p. 48

MCMORRIS, R.L. and GRAVLEY, R.J. (1993). Managing data from large-scale continuous monitoring projects. *Chem. Engng Prog.*, **89**(3), 111

MCMULLEN, G. (1976). *A Review of the 11th May Ammonia Truck Accident* (report). Health Dept, City of Houston, TX

JOHN J. MCMULLEN ASSOCIATES INC. (1977). *Draft Vessel Traffic Analysis Environmental Impact Report for Point Conception LNG Import Terminal*. Rep. to California Public Utilities Comm.

MCMULLEN, R.W. (1975). The change of concentration standard deviations with distance. *J. Air Poll. Control Ass.*, **25**, 1057

MCMULLEN, T.B. (1975). Interpreting the eight-hour national ambient air standard for carbon monoxide. *J. Air Poll. Control Ass.*, **25**(10), 1009

MCMURRAY, R. (1982). Flare radiation estimated. *Hydrocarbon Process.*, **61**(11), 175

MCMURRAY, T.R. (1962). Safety in disposal of large quantities of liquid oxygen. *Ammonia Plant Safety*, **4**, p. 12

MACNAB, J.I. (1981). The role of thermochemistry in chemical process hazards – catalytic nitro reduction processes. *Runaway Reactions, Unstable Products and Combustible Powders* (London: Instn. Chem. Engrs), paper 3/S

MCNALLY, L. (1979). Increase pump seal life. *Hydrocarbon Process.*, **58**(1), 107

MACNALTY, SIR A.S. and MELLOR, W.F. (eds) (1968). *The Medical Services in War* (London: HM Stationery Office)

MCNAMARA, B.P. (1976). *Estimates of the Toxicity of Hydrocyanic Acid Vapors in Man. Rep. EB-TR-76023.* Edgewood Arsenal

MCNAUGHTON, D.J. and BERKOWITZ, C.M. (1980). Overview of US research activities in the dispersion of gases. In Hartwig, S. (1980), *op. cit.*, p. 15

MCNAUGHTON, D.J., WORLEY, G.G. and BODNER, P.M. (1987). Evaluating emergency response models for the chemical industry. *Chem. Engng. Prog.*, **83**(1), 46

MCNAUGHTON, K.J. (1979). Relining deteriorated pipelines with polyolefin pipe. *Chem. Engng*, **86**(Nov. 19), 173

MCNAUGHTON, K.J. (1982). The ABCs of occupational skin disease. *Chem. Engng*, **89**(Mar. 22), 147; (Apr.), 149

MCNAUGHTON, K.J. (1983). Ethylene dichloride process. *Chem. Eng*, **90**(Dec. 12), 15

MCNEIL, C.S.L. (1976). Emergency response information systems at Environment Canada, *Control of Hazardous Material Spills*, **3**, p. 105

MCNEIL, J.P. (1989). *Piper Alpha Public Inquiry. Transcript, Days 61, 62*

MACNEISH, T.S. (1968). *Civil Engineering in the Process Industries* (London: Leonard Hill)

MCOMIE, W.A. (1949). Comparative toxicity of methacrylonitrile and acrylonitrile. *J. Ind. Hyg. Toxicol.*, **31**, 113

MCPHAIL, M.R. and WINDHORST, J.C.A. (1993). Integration of risk management into process safety standards and application of those standards. *Process Safety Management*, p. 101

MACPHERSON, G. (1981). *An Introduction to Electrical Machines and Transformers* (New York; Wiley)

MCPHERSON, J.H. (1967). To get safe performance, know people. *Hydrocarbon Process.*, **46**(3), 215

MCPHERSON, R.W., STARKS, C.M. and FRYAR, G.J. (1979). Vinyl chloride monomer...what you should know. *Hydrocarbon Process.*, **58**(3), 75

MCPHERSON, S.W. (1989). Columbia Nitrogen's spill alert computer system. *Ammonia Plant Safety*, **29**, p. 109. See also *Plant/Operations Prog.*, 1989, **8**, 206

MACPHERSON, MAJOR-GENERAL SIR W.G., HERRINGHAM, MAJOR-GENERAL SIR W.P., ELLIOTT, COLONEL T.R. and BALFOUR, COLONEL A. (1923). *History of the Great War. Medical Services, Diseases of the War*, vol. II (London: HM Stationery Office)

MCQUAID, J. (1975). Air entrainment into bounded axisymmetric sprays. *Proc. Inst. Mech. Engrs*, **189**, 197

MCQUAID, J. (1976a). The design of water-spray ventilators. *J. Occup. Accid.*, **1**(1), 1

MCQUAID, J. (1976b). *Some Experiments on the Structure of Stably-stratified Shear Flows. Tech. Paper P21.* Safety in Mines Res. Estab., Sheffield

MCQUAID, J. (1977). The design of water-spray barriers for chemical plants. *Loss Prevention and Safety Promotion 2*, p. 511

MCQUAID, J. (1979a). *Dispersion of Heavier-than-air Gases in the Atmosphere: Review of Research and Progress Report on HSE Activities. HSL Tech. Paper 8.* Health and Safety Executive, Sheffield

MCQUAID, J. (1979b). Field trials on the release of heavy gases. *Proc. Tenth Int. Tech. Mtg on Air Pollution Modelling and Its Application* (NATO/CCMS), p. 663

MCQUAID, J. (1980). Dispersion of heavy gases by the wind and by steam or water-spray barriers. *Course on Process Safety – Theory and Practice.* Teesside Polytechnic, Middlesbrough

MCQUAID, J. (1982a). *Application of Box Models (paper). Von Karman Inst., Lecture Series on Heavy Gas Dispersion*

MCQUAID, J. (1982b). Future directions of dense-gas dispersion research. *J. Haz. Materials*, **6**(1/2), 231. See also in Britter, R.E. and Griffiths, R.F. (1982), *op. cit.*, p. 231

MCQUAID, J. (1984a). The box model of heavy gas dispersion: a useful and practical tool. *J. Occup. Accid.*, **6**(4), 253

MCQUAID, J. (1984b). Large scale experiments on the dispersion of heavy gas clouds. In Ooms, G. and Tennekes, H. (1984), *op. cit.*, p. 129

MCQUAID, J. (1984c). Observations on the current status of field experiments on heavy gas dispersion. Ooms, G. and Tennekes, H. (1984), *op. cit.*, p. 241

MCQUAID, J. (ed.) (1985a). *Heavy Gas Dispersion Trials at Thorney Island* (Amsterdam: Elsevier)

MCQUAID, J. (1985b). Objectives and design of the Phase I Heavy Gas Dispersion Trials. *J. Haz. Materials*, **11**(1/3), 1. See also McQuaid, J. (1985a), *op. cit.*, p. 1

MCQUAID, J. (1985c). Trials on dispersion of heavy gas clouds. *Plant/Operations Prog.*, **4**, 58

MCQUAID, J. (1986). Some effects of obstructions on the dispersion of heavy gas clouds. *Refinement of Estimates of the Consequences of Heavy Toxic Vapour Release* (Rugby: Instn Chem. Engrs), p. 3.1

MCQUAID, J. (1987). Design of the Thorney continuous release trials. *J. Haz. Materials*, **16**, 1

MCQUAID, J. (1991). Know your enemy: the science of accident prevention. *Process Safety Environ.*, **69B**, 9

MCQUAID, J. (1991). Know your enemy: The science of accident prevention. *Trans. IChemE* **69**, Part B (Feb.), 9–19

MCQUAID, J. and FITZPATRICK, R.D. (1981). The uses and limitations of water spray barriers. In Greenwood, D.V. (1981), *op. cit.*, paper 1

MCQUAID, J. and FITZPATRICK, R.D. (1983). Air entrainment by water sprays: strategies for application to the dispersion of gas plumes. *J. Occup. Accid.*, **5**(2), 121

MCQUAID, J. and MOODIE, K. (1982). The scope for the reduction of the hazard of flammable or toxic gas clouds. *Seminar on Toxic and Flammable Clouds – Hazards and Protection* (London: Instn Mech. Engrs)

MCQUAID, J. and MOODIE, K. (1983). The scope for reduction of the hazard of flammable or toxic gas plumes. *J. Occup. Accid.*, **5**(2), 135

MCQUAID, J. and ROBERTS, A.F. (1982). Loss of containment: its effects and control. *Developments 82* (Rugby: Instn Chem. Engrs)

MCQUEEN, W. *et al.* (1972). *Spills of LNG on Water. Rep. EE61E-72.* Esso Res. and Engng Co.

MCRAE, T.G. (1983). Large-scale rapid phase transition explosions. In Veziroglu, T.N. (1983), *op. cit.*, p. 164

MCRAE, T.G. (1986). Evaluation and source strength and dispersion model predictions with data from large nitrogen tetroxide field experiments. In de Wispelaere, C., Schiermeier, F.A. and Gillani, N.V. (1986), *op. cit.*, p. 685

MCRAE, T.G., CEDERWALL, R.T., GOLDWIRE, H.C., HIPPLE, D.L., JOHNSON, G.W., KOOPMAN, R.P., MCCLURE, J.W. and MORRIS, L.K. (1984). *Eagle Series Data Report: 1983 Nitrogen Tetroxide Spills. Rep. UCID-20063.* Lawrence Livermore Lab., Livermore, CA

MCREYNOLDS, A.D. (1989). *Piper Alpha Public Inquiry. Transcript, Days 124–126*

MACRO, A. and BUXTON, J.N. (1987). *The Craft of Software Engineering* (Reading, MA: Addison-Wesley)

MACRORY (ed.) (1983). *Britain, Europe and the Environment* (London: Oyez)

MCSWEEN, T.E. (1993). Improve your safety program with a behavioural approach. *Hydrocarbon Process.*, **72**(8), 119

MCVEHIL, G.A. (1964). Wind and temperature profiles near the ground in stable stratification. *Quart. J. R. Met. Soc.*, **90**, 136

MCWHIRTER, D. (ed.) (1985). *The Guinness Book of Records 1986* (London: Guinness Superlatives Ltd)

MCWHIRTER, N. (ed.) (1976). *The Guinness Book of Records*, 23rd ed. (London: Guinness Superlatives Ltd)

MCWHORTER, E.W. (1978). Analyze steam tracing . . . do it right. *Hydrocarbon Process.*, **57**(12), 125. See also in Kane, L.A. (1980), *op. cit.*, p. 114

MCWILLIAMS, J. and STEEL, R.J. (1985). *Gas, the Battle for Ypres, 1915* (St Catherines, Ont.: Vanwell Publ.)

MADANSKY, A. and PEISAKOFF, M.P. (1960). *Thor and Atlas In-Flight Reliability Growth. Rep. RM-2598.*, Rand Corp., Santa Monica, CA

MADAYAG, E.F. (ed.) (1969). *Metal Fatigue: Theory and Design* (New York: Wiley)

MADDEN, J. (1987). CAD in process plant engineering. *Computer-Aided Engng J.*, 4(6), 237

MADDEN, J. (1993). Safety aspects of plant layout. *Course on Hazard Control and Major Hazards.* Dept of Chem. Engng, Loughborough Univ. of Technol.

MADDEN, J. and TAYLOR, V.T. (1979). Design data requirements in the process industries. *Computer Aided Design*, **11**(3), 142

MADDEN, J., PULFORD, C. and SHADBOLT, N. (1990). Plant layout — untouched by human hand? *Chem. Engr, Lond.*, **474**, 32

MADDISON, N. (1994). Explosion hazards in large scale solution purification of metal dust. *Hazards*, **XII**, p. 13

MADDOCK, M.J. (1967). Check furnace performance by computer. *Hydrocarbon Process.*, **46**(6), 161

MADDOCK, M.J. (1984). Todays technology helps keep ethylene plants competitive. *Oil Gas J.*, **82**(Sept. 3), 60

MADDOCKS, A.J. and JACKSON, G.A. (1981). Measurement of radio frequency voltage and power induced in structures on St Fergus gas terminals. *Radio Electron. Engr*, **51**, 187

MADDOX, R.N. and SAFTI, M.V. (1985). Slip velocity added to two phase pipeline calculations. *Oil Gas J.*, **83**(Jun. 17), 94

MADDOX, V. (1973). How to measure piston rod motion in reciprocating compressors. *Hydrocarbon Process.*, **52**(1), 81

MADDOX, V. (1977). Process machinery vibration data better. *Hydrocarbon Process.*, **56**(10), 179

MADER, C.L. (1979). *Numerical Modeling of Detonations* (London: Univ. of California Press)

MADHAVAN, S. (1984). Ammonia process simulator. *Ammonia Plant Safety* **24**, 167

MADHAVAN, S. and KIRSTEN, R.A. (1988). Anatomy of a plant audit. *Hydrocarbon Process.*, **67**(11), 123; **67**(12), 92

MADHAVAN, S. and KIRSTEN, R.A. (1989). Introduction to expert systems and potential ammonia plant applications. *Ammonia Plant Safety*, **29**, 57. See also *Plant/Operations Prog.,* 1989, **9**, 1

MADHAVAN, S. and LANDRY, B. (1988). Technical audits of existing ammonia plants. *Ammonia Plant Safety*, **28**, 1

MADHAVAN, S. and SATHE, S.Y. (1987). Inspection of ammonia plants. *Ammonia Plant Safety*, **27**, 106. See also *Plant/Operations Prog.,* 1987, **6**, 35

MADNI, A.M. (1988). The role of human factors in expert systems design and acceptance. *Hum. Factors*, **30**, 395

MADORSKY, S.L. (1964). *Thermal Degradation of Organic Polymers* (New York: Wiley)

MADRON, F. (1985). A new approach to the identification of gross errors in chemical engineering measurements. *Chem. Engng Sci.*, **40**, 1855

MADSEN, W.W. and WAGNER, R.C. (1994). An accurate methodology for modeling the characteristics of explosion effects. *Process Safety Prog.*, **13**, 171

MAEKAWA, M. (1975). Flame quenching by rectangular channels as a function of channel lengths for methane-air mixture. *Combust. Sci. Technol.*, **11**, 141

MAEKAWA, M. and TAKEICHI, M. (1979). Non-ignition of an ejected flame through a gap greater than the quenching distance. *Fire Res.*, **1**(6), 317

MAEKAWA, M., TAKEICHI, M. and KATO, S. (1978). Experimental determination of non-ignition diameter for various ignition positions. *Combust. Sci. Technol.*, **18**, 223

MAGEE, R.S. and MCALEVY, R.F. (1971). The mechanism of flame spread. *J. Fire Flammability*, **2**, 271

MAGILL, P.L., HOLDEN, F.R. and ACKLEY, C. (eds) (1956). *Air Pollution Handbook* (New York: McGraw-Hill)

MAGISON, E.C. (1966). *Electrical Instruments in Hazardous Locations*, 2nd ed. (New York: Plenum Press)

MAGISON, E.C. (1975). Electrical safety standards in the United States — a status report. *Electrical Safety*, **2**, 11

MAGISON, E.C. (1978). *Electrical Instruments in Hazardous Locations*, 3rd ed. (Pittsburg, PA: Instrum. Soc. Am.)

MAGISON, E.C. (1984). *Intrinsic Safety,* 3rd ed. (Pittsburgh, PA: Instrum. Soc. Am.)

DE MAGLIE, B. (1976). Oxygen best for EO. *Hydrocarbon Process.*, **55**(3), 78

MAGLIOZZI, T.L. (1967). Control system prevents surging in centrifugal-flow compressors. *Chem. Engng*, **74**(May 8), 139

MAGNUS, G. (1961). Tests on combustion velocity of liquid fuels and temperature distribution in flames and beneath surface of burning liquid. *Int. Symp. on The Use of Models in Fire Research 1959* (Washington, DC: Nat. Acad. Sci.)

MAGNUS, K. (1982). *Trends in Cancer Incidence: Causes and Practical Implications* (Washington, DC: Hemisphere Press)

MAGUIRE, B.A., SLACK, C. and WILLIAMS, A.J. (1962). The concentration limits for coal dust—air mixtures for upward propagation of flame in a vertical tube. *Combust. Flame*, **6**, 287

MAH, R.S.H. (1990). *Chemical Process Structures and Information Flows* (Oxford: Butterworth-Heinemann)

MAH, R.S.H., STANLEY, G.M. and DOWNING, D.M. (1976). Reconciliation and rectification of process flow and inventory data. *Ind. Engn Chem. Proc. Des. Dev.*, **15**, 175

MAHAJAN, K.K. (1975). Tall stack design simplified. *Hydrocarbon Process.*, **54**(9), 217

MAHAJAN, K.K. (1977a). Analyze tower vibration quicker. *Hydrocarbon Process.*, **56**(5), 217

MAHAJAN, K.K. (1977b). Size vessel stiffness quickly. *Hydrocarbon Process.*, **56**(4), 207

MAHALEC, V. and MOTARD, R.L. (1977). Procedures for the initial design of chemical processing systems. *Comput. Chem. Engng*, **1**, 57

MAHER, J.B. and VAN GELDER, L.R. (1972a). LNG tank dynamics. *LNG Conf. 3*

MAHER, J.B. and VAN GELDER, L.R. (1972b). Rollover and thermal overfill in flat bottom LNG tanks. *Pipeline Gas J.,* Sep., 46

MAHER, S.T., RODIBAUGH, R.K., SHARP, D.R. and DESAEDELEER, G. (1988). Relief valve testing interval optimisation program for the cost-effective control of major hazards. *Preventing Major Chemical and Related Process Accidents,* p. 117

MAHGEREFTEH, H., GIRI, N. and WONG, A. (1993). Standing blowdown on its head., *Chem. Engr, Lond.,* **548**, 12

MAHLEY, H.S. (1971). Static electricity in the handling of petroleum products. *Am. Inst. Chem. Engrs, Proc. Sixty-Fourth Ann. Mtg*

MAHLEY, H.S. (1972). Static electricity in handling of petroleum products. *Loss Prevention,* **6**, 51

MAHLEY, H.S. (1975). Fight tank fires subsurface. *Hydrocarbon Process.,* **54**(8), 72

MAHLEY, H.S. and WARREN, J.H. (1968). New fuel loading guides pinpoint danger areas. *Hydrocarbon Process. Petrol. Refiner,* **47**(5), 116

MAHMOOD, M.H., TURNER, A.K., MAN, M.C.M. and FOGARTY, P.O. (1984). Hydrogen as a by-product of chlorine/caustic production. *Chem. Ind.,* Jan. 16, 50

MAHONEY, D.G. (1990). *Large Property Damage Losses in the Hydrocarbon-Chemical Industries: A Thirty-Year Review,* 13th ed. (New York: M&M Protection Consultants)

MAHOOD, R.F. and MARTIN, J.E. (1979). Improve automatic control system reliability. *Hydrocarbon Process.,* **58**(5), 215. See also in Kane, L.A. (1980), *op. cit.,* p. 165

MAIER, D. (1983). *The Theory of Relational Databases* (New York: Computer Science Press)

MAINI, T. and STEPHENSON, S. (1989). Attenuation of radiant heat with free-standing water curtains: results of full-scale experiments. *Gastech 88,* paper 4.6

MAINSTONE, R.J. (1971). *The Breakage of Glass Windows by Gas Explosions. Current Paper CP26/71.* Building Res. Estab, Watford

MAINSTONE, R.J. (1972). The effect of explosions on buildings. *Symp. on Buildings and the Hazard of Explosion* (Watford: Building Res. Estab.)

DI MAIO, V.J.M. (1981). Penetration and perforation of skin by bullets and missiles. *Am. J. Forensic Med. Pathol.,* **2**(2), 107

DI MAIO, V.J.M., COPELAND, A.R., BESANT-MATTHEWS, P.E., FLETCHER, L.A. and JONES, A. (1982). Minimal velocities necessary for perforation of skin by air gun pellets and bullets. *J. Forensic Sci.,* **27**(4), 894

MAISEL, D.S. (1980). Hydrocarbons: methane, ethane, propane. *Kirk–Othmer Encyclopaedia of Chemical Technology,* 3rd ed., vol. 12 (New York: Wiley), p. 901

MAISEY, H.R. (1957). Explosion, suppression and protection. *Inst. Fire Engrs Quart.,* **17**(25), 35

MAISEY, H.R. (1965). Gaseous and dust explosion venting. *Chem. Process. Engng,* **46**(10), 527; **46**(12), 662

MAISEY, H.R. (1980). Explosion suppression and automatic explosion control. *Chemical Process Hazards,* **VII**, p. 171

MAJER, V. and SVOBODA, V. (eds) (1985). *Enthalpies of Vaporization of Organic Compounds* (Oxford: Blackwells)

MAJONE, G. (1989). *Evidence, Argument and Persuasion in the Policy Process* (New Haven, CT: Yale Univ. Press)

MAJOR HAZARDS ASSESSMENT PANEL (1987). *Chlorine Toxicity Monograph* (Rugby: Instn Chem. Engrs)

MAJOR HAZARDS ASSESSMENT PANEL (1988). *Ammonia Toxicity Monograph* (Rugby: Instn Chem. Engrs)

MAJOR HAZARDS ASSESSMENT PANEL (1989). *Overpressure Monograph* (Rugby: Instn Chem. Engrs)

MAJOR HAZARDS ASSESSMENT PANEL (1993). *Phosgene Toxicity Monograph* (Rugby: Instn Chem. Engrs)

MAJUMDAR, S.R. (1990). Regulatory requirements and hazardous materials. *Chem. Engng Prog.,* **86**(5), 17

MAK, C.-P., MUHLE, H. and ACHINI, R. (1997). Integrated solutions to environmental protection in process R&D. *Chimia* **51**, 184–88

MAKAY, E. (1976). How to avoid field problems with … boiler feed pumps. *Hydrocarbon Process.,* **55**(12), 79

MAKHAN, R.S. and HONSE, R.W. (1978). Carbon dioxide stripper explosion. *Ammonia Plant Safety,* **20**, 131

MAKHVILADZE, G.M. and PHILIPPOV, V.D. (1991). Unsteadystate zone model of free convection within a venting enclosure. *J. Loss Prev. Process Ind.,* **4**, 246

MAKHVILADZE, G.M. and ROGATYKH, D.I. (1991). Nonuniformities in initial temperature and concentration as a cause of explosive chemical reactions in combustible gases. *Combust. Flame,* **87**, 347

MAKRIS, J. (1988). Comments on the Emergency Planning and Community Right-to-Know Act of 1986. *Preventing Major Chemical and Related Process Accidents,* p. 729

MAKRIS, J.L. (1994). Opportunities for sound chemical accident prevention. *Process Safety Prog.,* **13**, 6

MAKSTENEK, M.I., LELAKIN, A.L., CHERNOPLEKOV, A.N. and CHERNYSHEV, K.O. (1992). The experience of realization of accident data base with flexible structure. *Loss Prevention and Safety Promotion 7,* vol. 3, 144–1

MALAGOLI, M., ZAPPELLINI, G., FONTANA, M. and PISCIOTTA, G. (1992). A new hazard identification procedure in the process plant analysis. *Loss Prevention and Safety Promotion 7,* vol. 1, 34–1

MALASKY, S. (1974). *System Safety – Planning, Engineering, Management* (Rochelle Park, NJ: Hayden)

MALATERRE, G., FERRANDEZ, F., FLEURY, D. and LECHNER, D. (1988). Decision making in emergency situations. *Ergonomics,* **31**, 643

MALCOLM, B. (1993). Software in safety-related systems: basic concepts and concerns. In Bennett, P. (1993), *op. cit.,* p. 1

MALCOLM, J.G. (1975). Implementation of CIA recommended safety procedures. *Distribution Conf.* (London: Chem. Ind. Ass.), p. 38

MALEY, G. (1970). *Manual of Logic Circuits* (Englewood Cliffs, NJ: Prentice-Hall)

DE MALHERBE, M.C., WING, R.D., LADERMAN, A.J. and OPPENHEIM, A.K. (1966). Response of a cylindrical shell to internal blast loading. *J. Mech. Engng Sci.,* **8**, 91

MALHOTRA, H.L. (1972). Building materials and fire. *Fire Prev. Sci. Technol.,* **1**, 4

MALHOTRA, H.L. (1977). The philosophy and principles of fire tests. *Fire Prev. Sci. Technol.,* **17**, 24

MALHOTRA, H.L. (1984a). Compartmentation. *Fire Surveyor,* **13**(5), 4

MALHOTRA, H.L. (1984b). Fire resistance and behaviour. *Fire Surveyor,* **13**(6), 13

MALINA, M.A. (1988). Plant capacity as a key to inventory. *Chem. Engng,* **95**(8) May 23, 139

MALINA, M.A. (1980). Storage capacity: how big should it be? *Chem. Engng,* **87**(Jan. 28), 121

MALL, R.D. (1988). Safety in multistream ammonia plants utilizing common utility services. *Ammonia Plant Safety,* **28**, 112. See also *Plant/Operations Prog.,* 1988, **7**, 136

MALLAMAD, S.M., LEVINE, J.M. and FLEISHMAN, E.A. (1980). Identifying ability requirements by decision flow diagram. *Hum. Factors,* **22**, 57

MALLARD, E. and LE CHATELIER, H. (1883). *Ann Mines, 8,* 274

MALLE, K.G. (1979). EEC 'Dangerous Substances' Directive. The German chemical industry view. *Chem. Ind.,* May 5, 308

MALLEK, J. (1969). Blast shatters chemical plant. *Mod. Manufacturing,* Aug., 104

MALLETT, R. (1992). Rate your risk management plans. *Hydrocarbon Process.,* **71**(8), 111

MALLETT, R. (1993). Evaluate the consequences of incidents. *Chem. Engng Prog.,* **89**(1), 69; **89**(2), 74

MALLICK, R.W. and GAUDREAU, A.T. (1951). *Plant Layout* (London: Chapman & Hall)

MALLINSON, J.H. (1966a). Plastic pipe: its performance and limitations. *Chem. Engng,* **73**(Feb. 14), 180

MALLINSON, J.H. (1966b). Reinforced plastic pipe: joining and supporting. *Chem. Engng,* **73**(Jan. 17), 168

MALLINSON, J.H. (1968). Rigging from a sky hook. *Chem. Engng,* **75**(Sep. 9), 150

MALLINSON, J.H. (1969). Grounding reinforced plastic process systems. *Chem. Engng,* **76** (Aug. 25), 136

MALLINSON, J.H. (1982). Abrasion of fiber-reinforced plastics in corrosive environments. *Chem. Engng,* **89** (May 31), 143

MALLORY, H.D. (1980). The fireball spectrum of a condensed explosive. *Combust. Sci. Technol.,* **24**, 161

MALLOW, J.E. (1964a). Air line explosion dangers. In Vervalin, C.H. (1964a), *op. cit.,* p. 122

MALLOW, J.E. (1964b). Danger! Air line explosions. *Hydrocarbon Process.,* **43**(3), 157

MALLOY, J.B. (1971). Risk analysis of chemical plants. *Chem. Engng Prog.,* **67**(10), 68

MALLOY, J.F. (1969). *Thermal Insulation* (New York: van Nostrand Reinhold)

MALONE, L.J.S.D. and WARIN, J.F. (1945). A domestic case of chlorine gas poisoning. *Br. Med. J.,* **281**, 1178

MALONE, R.J. (1980). Sizing external heat exchangers for batch reactors. *Chem. Engng,* **87**(Dec. 1), 95

MALONE, T. (ed.) (1951). *Compendium of Meteorology* (Boston, MA: Am. Met. Soc.)

MALONE, T., KIRKPATRICK, M., MALLORY, K., EIKE, D., JOHNSON, J. and WALKER, R. (1980). *Human Factors Evaluation of Control Room Design and Operator Performance at Three Mile Island 2. Rep. NUREG/CR-1270.* Nucl. Regul. Comm., Washington, DC

MALOW, M. (1980). Benzene or butane for MAN. *Hydrocarbon Process.,* **59**(11), 149

MALPAS, R. (1979). Meditations on maturity. *Chem. Ind.,* Dec. 1, 826

MALPAS, R. (1983). The plant after next. *Harvard Business Rev.,* Jul./Aug., 122

MALTONI, C. and LEFEMINE, G. (1975). Carcinogenicity bioassays of vinyl chloride – current results. *Ann N.Y. Acad. Sci.,* **246**(Jan. 31), 195

MALY, R. (1981). Ignition model for spark discharges and the early phase of flame front growth. *Combustion,* **18**, p. 1747

MALY, R. and VOGEL, M. (1979). Initiation and propagation of flame fronts in lean CH_4–air mixtures by the three modes of the ignition spark. *Combustion,* **17**, 821

MALYCHUK, M. and GOLLAHALLI, S.R. (1987). Ignition characteristics of petroleum and synthetic fuel pools exposed to external radiation. *Combust. Sci. Technol.,* **53**, 225

MAMDANI, A. and GAINES, B.R. (1981). *Fuzzy Reasoning and its Applications* (New York: Academic Press)

MANAHAN, S.E. (1991). *Environmental Chemistry,* 5th ed. (Chelsea, MI: Lewis)

MANAHAN, S.E. (1993). *Fundamentals of Environmental Chemistry* (Boca Raton, FL: Lewis)

MANAS, J.L. (1984). BLEVEs – their nature and prevention. *Fire Int.,* **87**, 27

MANCE, G. (1984). Derivation of environmental quality objectives and standards for black and grey list substances. *Chem. Ind.,* Jul. 16, 509

MANCINE, B.J. (1994). Succeed at ISO 9000 registration. *Chem. Engng Prog.,* **90**(2), 55

MANCINI, G. and AMENDOLA, A. (1983). Human models and the data problem. *Euredata* 4, paper 15.3

MANCINI, G. and VOLTA, G. (1983). Probabilistic risk assessment: use and misuse. *Euredata 4,* paper 1.1

MANCINI, R.A. (1988). The use (and misuse) of bonding for control of static ignition hazards. *Plant/Operations Prog.,* **7**, 23

MANCINI, R.A. (1992). Workshop on unconfined vapour cloud explosions. *Plant/Operations Prog.,* **11**, 27

MANCUSO, J.R. (1988). Retrofitting gear couplings with diaphragm couplings. *Hydrocarbon Process.,* **67**(10), 47

MANDELIK, B.G. and NEWSOME, D.S. (1981). Hydrogen. *Kirk–Othmer Encyclopaedia of Chemical Technology,* 3rd ed., vol. 9 (New York: Wiley), p. 950

MANDELIK, B.G. and TURNER, W. (1977). How to operate nitric acid plants. *Hydrocarbon Process.,* **56**(7), 175

MANDICH, N.V. and KRULIK, G.A. (1992). Substitution of non-hazardous for hazardous process chemicals in the printed circuit industry *Me. Finish.* **90** (11), 49–51

MANGIALAVORI, G. and RUBINO, F. (1992). Experimental tests on large hydrocarbon pool fires. *Loss Prevention and Safety Promotion 7,* vol. 2, 83–1

MANGNALL, K. (1971). *A Technical Guide to Vacuum and Pressure Producing Equipment* (Bolton, Lancashire: Hick Hargreaves)

MANI, S., SHOOR, S.K. and PETERSEN, H.K. (1990). Experience with a simulator for training ammonia plant operators. *Ammonia Plant Safety,* **29**, 244. See also *Plant/Operations Prog.,* 1989, **9**, 6

MANINS, P.C. (1984). Chimney plume penetration of the seabreeze inversion. *Atmos. Environ.,* **18**, 2339

MANKABADY, S. (1982). The control of movements of gas carriers in ports. *Gastech 81,* p. 279

MANN, L. (1977). Manage maintenance productivity. *Hydrocarbon Process.,* **56**(1), 82

MANN, N.R., SCHAFER, R.E. and SINGPURWALLA, N.D. (1974). *Methods for Statistical Analysis of Reliability and Life Data* (New York: Wiley)

MANN, R. (1985). So what future do you see for process fibre optics. *Process Engng,* Jun., 79

MANNA, Z. (1973). *Introduction to the Mathematical Theory of Computation* (New York: McGraw-Hill)

MANNA, Z. and PNUELI, A. (1969). Formalization of properties of recursively defined functions. *ACM Symp. on Theory of Computing,* p. 201

MANNAN, M.S., GENTILE, M. and O'CONNOR, T.M. (2001). Chemical Incident Data Mining and Application to Chemical Safety Analysis. *Proc. of the CCPS 2001 Int. Conf Workshop,* Toronto, Ontario, Canada, Oct. 2–5, pp. 137–56.

MANNING, D.H. (1976). *Disaster Technology* (Oxford: Pergamon)

MANNING, D.P. (1983). Deaths and injuries caused by slipping, tripping and falling. *Ergonomics,* **26**, 3

MANNING, W.R.D. (1957). Strength of cylinders. *Ind. Engng Chem.,* **49**(12), 1969

MANNING, W.R.D. (1960). High pressure joints. *Chem. Engr, Lond.,* **152**, A53

MANNING, W.R.D. (1978). Bursting pressure as the basis of cylinder design. *J. Pres. Ves. Technol.,* **100**, 374

MANNING, W.R.D. and LABROW, S. (1971). *High Pressure Engineering* (London: Leonard Hill)

MANNON, J.H. (1981). Infrared laser is key to British gas detector. *Chem. Engng*, **88**(Dec. 14), 47

MANNON, J.H. and JOHNSON, E.P. (1982). Caution: airwaves may be hazardous to your plant. *Chem. Engng*, **89**(Mar. 8), 37

MANNAN, M.S., ROGERS, W.J., GENTILE, M. and O'CONNOR, T.M. (2003). Inherently safer design. *Hydrocarbon Proc.* **82** (3), (March), 59–61

MANSFIELD, A.P. (1977). The emergency load transfer scheme. *Symp. Hazard 77 – Advances in the Safe Handling and Movement of Hazardous Substances,* Teesside Polytechnic, Middlesbrough

MANSFIELD, D. and CASSIDY, K. (1994). Inherently safer approaches to plant design: the benefits of an inherently safer approach and how this can be built into the design process. *Hazards,* **XII**, p. 285

MANSFIELD, D.P. (1992). Proposed offshore safety cases – a comparison with onshore CIMAH safety cases. *Major Hazards Onshore and Offshore,* p. 39

MANSFIELD, D.P. (1994). *Inherently Safer Approaches to Plant Design.* Warrington, (Cheshire, U.K.: United Kingdom Atomic Energy Authority) AEA/CS/HSE R1016

MANSFIELD, D.P. (1996). Viewpoints on implementing inherent safety. *Chem. Eng.* **103** (3), (Mar.), 78–80

MANSFIELD, D.P. (1997). The INSIDE Project. *Inherent SHE: The cost effective route to improved safety, health and environmental performance,* Jun. 16–17 1997, Kensington Hilton Hotel, London, (London: IBC UK Conf.s Limited)

MANSFIELD, D.P. and CASSIDY, K. (1994). Inherently safer approaches to plant design. *IChemE Symp. Series* **134**, 285–99

MANSFIELD, D.P. and HAWKSLEY, J.L. (1998). Improved business performance through inherent safety, health and environmental design and operation: A report of a London Conference. *J. Clean. Prod.* **6** (2), 147–50

MANSFIELD, D.P., MALMEN, Y. and SUOKAS, E. (1996). The development of an integrated toolkit for inherent SHE. *Int. Conf. and Workshop on Process Saf. Manage. and Inherently Safer Processes,* Oct. 8–11, 1996, Orlando, FL, 103–117. (New York: American Institute of Chemical Engineers)

MANSFIELD, D.P., TURNEY, R.D., ROGERS, R.L., VERWOERD, M. and BOTS, P. (1995). How to integrate inherent SHE in process development and plant design. *IChemE Conf., Major Haz. Onshore and Offshore.* UMIST'95, Manchester, UK

MANSFIELD, J.A. (1969). *Heat Transfer Hazards of Liquid Rocket Propellant Explosions.* URS Res. Co., Burlingame, CA

MANSFIELD, W.P. (1956). Crank case explosions: development of new protective devices. *Proc. Instn Mech. Engrs,* **170**, 825

MANSILLON, R. (1972). Study of possible accidents in storage of refrigerated hydrocarbons at atmospheric pressure. *LNG Conf. 3*

MANSON, N. (1949). *Revue de l'Institut Français du Petrole et Annales des Combustibles Liquides,* 4, 338

MANSON, N. and FERRI, F. (1953). Contribution to the study of spherical detonation waves. *Combustion,* 4, p. 486

MANSON, S.S. (1966). *Thermal Stress and Low Cycle Fatigue* (New York: McGraw-Hill)

MANSOT, J. (1989). Incendie du Depot Shell du Port Edouard Herriot a' Lyon les 2 et 3 Juin 1987. *Loss Prevention and Safety Promotion 6,* p. 41.1

MANSOURI, S.H. and HEYWOOD, J.B. (1980). Correlations for the viscosity and Prandtl number of hydrocarbon-air combustion products. *Combust. Sci. Technol.,* **23**, 251

MANT, W.D. (1993). Motorways and chemical plants – risk interaction. In Melchers, R.E. and Stewart, M.G. (1993), *op. cit.,* p. 43

MANTEL, N. and BRYAN, W.R. (1961). 'Safety' testing of carcinogenic agents. *J. Nat. Cancer Inst.,* **27**, 455

MANTELL, C.L. (ed.) (1975). *Solid Wastes* (New York: Wiley)

MANTON, J., VON ELBE, G. and LEWIS, B. (1953). Burning velocity measurements in a spherical vessel with central ignition. *Combustion,* 4, 358

MANUFACTURING CHEMISTS ASSOCIATION – see Appendix 28

MANYIK, R.M. (1978) Acetylene. *Kirk–Othmer Encyclopaedia of Chemical Technology,* 3rd ed., vol. 1 (New York: Wiley), p. 192

MANZER, L.E. (1993a). Toward catalysis in the 21st century chemical industry. *Catal. Today* **18**, 199–207

MANZER, L.E. (1993b). Environmentally safer processes: opportunities for catalysis and process R&D. *Chem. for Sustainable Dev.* **1**, 147–51

MANZER, L.E. (1994). Chemistry and catalysis. In *Benign by Design: Alternative Synthetic Design for Pollution Prevention,* (ed. P.T. Anastas, and C.A. Farris), pp 144–54. (Washington, D.C.: American Chemical Society)

MAPLES, R.E. and ADLER, P. (1978). Options for olefin feedstocks. *Hydrocarbon Process.,* **57**(7), 171

MAPSTONE, G. (1968). Find exponents for cost estimates. *Hydrocarbon Process.,* **48**(5), 165

MARA, P.B. (1967). Aluminum for nitrogen fertiliser equipment. *Ammonia Plant Safety,* **9**, 64

MARBLE, F.E. (1970). Dynamics of dusty gases. *Ann. Revs Fluid Mech.,* **2**, 397

MARCEL, O. (1991). Measurement of the velocity of natural convection movements in an LNG storage tank. *Gastech 90*, p. 7.5

MARCH (1989). *Managing Maintenance into the 1990's* (London: March Consulting Group)

MARCHAJ, T.J. (1985). Internal pressure equalising system for liquefied hydrocarbon storage tanks. *Gastech 84,* p. 481

MARCHAL, M. and DUC, J. (1959). The electronic computation of water hammer in hydraulic plants. *Sulzer Tech. Rev.,* **41**(2)

MARCHANT, E.W. (1984). Effect of wind on smoke movement and smoke control systems. *Fire Safety J.,* **7**, 55

MARCHELLO, J.M. and KELLY, J.J. (1975). *Gas Cleaning for Air Quality* (New York: Dekker)

MARCILLE, R. (1986). Improvement of operating instructions for nuclear power plants. *Loss Prevention and Safety Promotion 5,* 4–1

MARCIN, B.G. and SCHULTE, M.M. (1982). Modular-engineered process plants. *Chem. Engng Prog.,* **78**(11), 37

MARCOMBE, J.T., KRAUSE, T.R. and FINLEY, R.M. (1993). Behaviour-based safety at Monsanto's Pensacola plant. *Chem. Engr, Lond.,* **541**, 15

MARCOVITCH, M.J. (1979). Whither pneumatic and electronic instrumentation. *Chem. Engng,* **86**(Oct. 15), 45

DE LA MARE, R.F. (1975). The economic implications of plant reliability. *Chem. Ind.,* May 3, 366

DE LA MARE, R.F. (1976). The strategical and tactical implications of process plant reliability. *Advances in Reliability* (Culcheth: Lancashire: UK Atom. Energy Auth.), ART2

DE LA MARE, R.F. (1980). A study of mechanical valve reliability. In Moss, T.R. (1980), *op. cit.,* p. 73

DE LA MARE, R.F. (1982). *Manufacturing Systems Economics* (New York: Holt, Rinehart & Winston)

DE LA MARE, R.F. and ANDERSEN, O. (1981). *Pipeline Reliability* (Hovik, Norway: Det Norske Veritas and Bradford: Univ. of Bradford)

DE LA MARE, R.F. and BAKOUROS, Y.L. (1994). Predicting pipeline reliability using discriminant analysis: a comparison between North Sea and Gulf of Mexico performance. *Conf. on Valve and Pipeline Reliability.* Engng Dept, Univ of Manchester

DE LA MARE, R.F. and BALL, K.J. (1981). Early life availability of chemical process plant. *Reliab. Engng*, **2**(4), 289

MARESCA, J.W., STARR, J.W., WISE, R.F., HILLGER, R.W. and TAFURI, A.N. (1993). Evaluation of volumetric leak detection systems for large underground tanks. *J. Haz. Materials*, **34**, 335

MARGERISON, T. (1976). Dioxin clean-up: can it be done? *New Scientist*, **71**, 627

MARGERISON, T., WALLACE, M. and HALLENSTEIN, D. (1980). *The Superpoison* (London: Macmillan)

MARGETTS, A. (1986a). Design and test your PLC control logic this way. *Process Engng,* Mar., 41

MARGETTS, A. (1986b). PLCs for alarm and shutdown systems. *Chem. Engr, Lond.*, **427**, 35

MARGETTS, A.J. (1987). Procedures for the safety assessment of programmable logic controllers used for reactor trip systems. *Hazards from Pressure* (Rugby: Instn Chem. Engrs), p. 301

MARGOLIN, S.V. (1984). Why use outside R&D. *Chemtech*, **14**(2), 90

MARGOLIS, I. (1978). Plant electrical systems. *Chem. Engng*, **85**(Dec. 18), 81

MARGUS, E.A. (1982). Engineered plastics for pumps. *Chem. Engng Prog.*, **78**(12), 69

MARGUS, E.A. (1991). Choosing thermoplastic pumps. *Chem. Engng*, **98**(7), 106

MARIN, A. (1992). Cost and benefits of risk reduction. The Royal Society (1992), *op. cit.*, p. 192

MARIN, A. and PSACHAROPLOULOS, G. (1982). The reward for risk in the UK labor market: evidence from the UK and reconciliation with other studies. *J. Political Economy*, **90**, 827

MARINAK, M.J. (1967). Pilot plant procedures. Pilot plant prestart safety checklist. *Chem. Engng Prog.*, **63**(11), 58

MARION, L. (1978a). CPI fear data disclosure. *Chem. Engng*, **85**(Sept. 25), 80

MARION, L. (1978b). Debate over toxicity simmers in Washington. *Chem. Engng*, **85**(an. 2), 30

MARION, L. (1978c). Pollution takes its toll – in real money, that is. *Chem. Engng*, **85**(Jul. 31), 50

MARION, L. (1979). Monitoring carcinogens – with a badge. *Chem. Engng*, **86**(Oct. 8), 66

MARISTANY, B.A. (1968). Equipment decision: repair or replace. *Chem. Engng*, **75**(Nov. 4), 210

MARK, H.F. *et al.* (1989). *Encyclopaedia of Polymer Science and Engineering* (New York: Wiley)

MARK, R.K. and STUART, D.E. (1977). Disasters as a necessary part of cost–benefit analysis. *Science, N.Y.*, **197**, 1160

MARKE, P. (1991). Cable tunnels: an integrated fire detection/suppression system for rapid extinguishment. *Fire Technol.*, **27**, 219

MARKHAM, R. (1987a). Ammonia-handling procedure (discussion contribution). *Ammonia Plant Safety*, **27**, 180

MARKHAM, R.S. (1987b). Review of damages from ammonia spills. *Ammonia Plant Safety*, **27**, 136

MARKHAM, R.S. and HONSE, R.W. (1978). Carbon dioxide stripper explosion. *Ammonia Plant Safety*, **20**, 131

MARKOVITZ, R.E. (1971). Picking the best vessel jacket. *Chem. Engng*, **78**(Nov. 15), 156

MARKOVITZ, R.E. (1977). Process and project data pertinent to vessel design. *Chem. Engng*, **84**(Oct. 10), 123

MARKS, A. (1979). *Handbook of Pipeline Engineering Computations* (Tulsa, OK: Petrol. Publ. Co.)

MARKS, J.B. and HOLTON, K.D. (1974). Protection of thermal insulation. *Chem. Engng Prog.*, **74**(8), 46

MARKS, M.D. and CROSBY, E.J. (1975). Use of a centrifugal pressure nozzle as a chemical reactor. *Ind. Engng Chem. Process Des. Dev.*, **14**, 20

MARKS, T.E. (1991). *Handbook on the Electricity at Work Regulations* (Nottingham: William Ernest Publ.)

MARKSTEIN, G.H. (1957). A shock tube study of flame-front pressure wave interaction. *Combustion*, **6**, 387

MARKSTEIN, G.H. (ed.) (1964). *Nonsteady Flame Propagation* (Oxford: Pergamon Press)

MARKSTEIN, G.H. (1974). Radiative energy transfer from gaseous diffusion flames. *Combustion*, **15**, 1285

MARKSTEIN, G.H. (1976). Radiative energy transfer from turbulent diffusion flames. *Combust. Flame*, **27**, 51

MARKSTEIN, G.H. (1977). Scaling of radiative characteristics of turbulent diffusion flames. *Combustion*, **16**, 1407

MARKSTEIN, G.H. (1985). Relationship between smoke point and radiant emission from buoyant turbulent and laminar diffusion flames. *Combustion*, **20**, 1055

MARKSTEIN, G.H. (1989). Correlations for smoke points and radiant emission of laminar hydrocarbon diffusion flames. *Combustion*, **22**, 363

MARLAR, C.H. (1971). Lessons from hurricane Celia. *Hydrocarbon Process.*, **50**(3), 109

MARLIER, G. (1989). Managing the major accidents – a training for emergency managers of refineries and chemical plants. *Loss Prevention and Safety Promotion 6*, p. 19.1

MARLIER, G. and ROURE, R. (1986). How to handle disasters. An experienced preplanning, training and test procedure for fire fighting in the French petroleum industry. *Loss Prevention and Safety Promotion 5*, p. 48–1

MARLOW, R.E. (1975). Ultrasonics as a detection tool. *Ammonia Plant Safety*, **17**, 65

MARMARAS, N., LIOUKAS, S. and LAIOS, L. (1992). Identifying competences for the design of systems supporting complex decision-making tasks: a managerial planning application. *Ergonomics*, **35**, 1221

MARNEY, J.F. and FOORD, D.J. (1984). Diagnostics. Practical application of expert systems. *Plant Engr*, **28**(5), 23

MARNICIO, R.J., HAKKINEN, P.J., LUTKENHOFF, S.D., HERTZBERG, R.C. and MOSKOWITZ, P.D. (1991). Risk analysis software and data bases: review of Riskware 90 conference and exhibition. *Risk Analysis*, **11**(3), 545

MARPLE, L. and THROOP, L. (1993). Use of physical and chemical properties to assess environmental transport and fate. *J. Haz. Materials*, **35**, 331

MARPLES, D.R. (1988). *The Social Impact of the Chernobyl Disaster* (New York: St Martins Press)

MARPLES, D.R. (1991). *Chernobyl, Five Years Later, April 26, 1991* (Washington, DC: Comm. on Security and Cooperation in Europe)

MARQUARDT, W. (1992). An object-oriented representation of structured process models. Comput. *Chem. Engng*, **16**, S329

MARR, D. (1982). *Vision* (San Francisco, CA: W.H. Freeman)

MARRE, T. and REICHERT, M. (1979). Plant inspection and maintenance. In BASF (1979), *op. cit.*, p. 245

MARRIOTT, C.A. (1987). Hazards of aircraft crash. In Cullingford, M.C., Shah, S.M. and Gittus, J.H. (1987), *op. cit.,* p. 345

MARRIOTT, J.B., MERZ, M., NIHOUL, J. and WARD, J. (eds) (1988). *High Temperature Alloys* (London: Elsevier Applied Science)

MARRS, G.P. and LEES, F.P. (1989). Overpressure protection of batch chemical reactors: some quantitative data. *Loss Prevention and Safety Promotion 6,* p. 78.1

MARRS, G.P., LEES, F.P., BARTON, J. and SCILLY, N. (1989). Overpressure protection of batch chemical reactors. *Chem. Engng Res. Des.,* **67**, 381

MARSDEN, C. (1963). *Solvents Guide,* 2nd ed. (London: Cleaver-Hume)

MARSDEN, P., FERRARIO, M. and GREEN, M. (1992). A methodological approach to the development of an industrial emergency response system. *Major Hazards Onshore and Offshore,* p. 307

MARSH AND MACLENNAN (1992). *Large Property Damage Losses in the Hydrocarbon-Chemical Industries. A Thirty-year Review,* 14th ed. (New York: M&M Protection Consultants)

MARSH, K.J. *et al.* (ca. 1985). *Acoustic Fatigue of Pipes. Rep. 85/82.* CONCAWE, the Hague

MARSH, R.W. and OLIVO, C.T. (1979). *Basics of Refrigeration,* 2nd ed. (New York: van Nostrand Reinhold)

MARSHAL, J.R. (1968). The challenge of pollution control. Current status, future outlook. *Chem. Engng,* **75** (Oct. 14), 14

MARSHALL, A.L. (1989). Concrete marine storage of LNG — with particular reference to floating platforms. *Gastech 88,* paper 5.3

MARSHALL, B.K. and POSTILL, D.C. (1984). Wind tunnel modeling aids platform design. *Oil Gas J.,* **82**(Jun. 18), 114

MARSHALL, E.C. and SHEPHERD, A. (1981). A fault-finding programme for continuous plant operators. In Rasmussen, J. and Rouse, W.B. (1981), *op. cit.,* p. 575

MARSHALL, E.C., DUNCAN, K.D. and BAKER, S.M. (1981). The role of withheld information in the training of process plant fault diagnosis. *Ergonomics,* **24**, 711

MARSHALL, E.C., SCANLON, K.E., SHEPHERD, A. and DUNCAN, K.D. (1981). Panel diagnosis training for major-hazard continuous-process installations. *Chem. Engr, Lond.,* **365**, 66

MARSHALL, J., MUNDT, A., HULT, M., MCKEALVY, T.C., MYERS, P. and SAWYER, J. (1995). The relative risk of pressurized and refrigerated storage for six chemicals. *Process Safety Prog.* **14** (3), (July), 200–211

MARSHALL, J.G. (1977). Hazardous clouds and flames arising from continuous releases into the atmosphere. *Chemical Process Hazards,* **VI**, p. 99

MARSHALL, J.G. (1979). The dispersion in the atmosphere of upwards discharges from vents. In Minors, C. (1979), *op. cit.,* paper 8

MARSHALL, J.G. (1980). The size of flammable clouds arising from continuous releases into the atmosphere — Part 2. *Chemical Process Hazards,* **VII**, p. 11

MARSHALL, J.G. (1989). *Piper Alpha Public Inquiry. Transcript, Day 112*

MARSHALL, J.G. and RUTLEDGE, P.V. (1982). Fire and explosion. In Green, A.E. (1982b), *op. cit.,* p. 497

MARSHALL, J.R. (1964). Explosion testing becomes a booming business. *Chem. Engng,* **71**(Nov. 9), 98

MARSHALL, J.W. (1969). Catalyst safety in large ammonia plants. *Ammonia Plant Safety,* **11**, 54

MARSHALL, M.R. (1977). Calculation of gas explosion relief requirements: the use of empirical equations. *Chemical Process Hazards,* **VI**, p. 21

MARSHALL, M.R. (1980). Gaseous and dust explosion venting: determination of explosion relief requirements. *Loss Prevention and Safety Promotion 3,* vol. 2, p. 1210

MARSHALL, M.R. (1983). The effect of ventilation on the accumulation and dispersal of hazardous gases. *Loss Prevention and Safety Promotion 4,* vol. 3, p. E11

MARSHALL, M.R. and STEWART-DARLING, F.L. (1986). Buoyancy-driven natural ventilation of enclosed spaces. *Hazards,* **IX**, p. 7

MARSHALL, M.R., HARRIS, R.J. and MOPPETT, D.J. (1977). The response of glass windows to explosion pressures. *Chemical Process Hazards,* **VI**, p. 83

MARSHALL, N.R. and MORGAN, H.P. (1992). User's guide to BRE spill plume calculations. *Fire Surveyor,* **21**(6), 14

MARSHALL, P. (1984). *Austenitic Stainless Steels* (Amsterdam: Elsevier Applied Science)

MARSHALL, P.D. (1970). Aspects of spectra differences between earthquakes and underground explosions. *Geophys. J. R. Astro. Soc.,* **20**, 397

MARSHALL, S.P. and BRANDT, J.L. (1971). Installed cost of corrosion-resistant piping. *Chem. Engng,* **78**(Aug. 23), 68; (Oct 4), 111

MARSHALL, S.P. and BRANDT, J.L. (1974). 1974 installed cost of corrosion-resistant piping. *Chem. Engng,* **81**(Oct. 28), 94

MARSHALL, V.C. (1974). Flixborough (letter). *Chem. Engr, Lond.,* **287/288**, 483

MARSHALL, V.C. (1975a). The act and the engineer. *Chem. Engr, Lond.,* **303**, 661

MARSHALL, V.C. (1975b). The Health and Safety at Work Act or the major hazard and the law. *Instn Chem. Engrs, Symp. on Technical Lessons of Flixborough* (London)

MARSHALL, V.C. (1975c). Process-plant safety — a strategic approach. *Chem. Engng,* **82**(Dec. 22), 58

MARSHALL, V.C. (1975d). *Report on the Disaster. Rep. to Flixborough Court of Inquiry* (Univ. of Bradford)

MARSHALL, V.C. (1976a). Safety of control rooms — a strategic view. *Mtg of Belgian Petrol. Inst.,* Antwerp

MARSHALL, V.C. (1976b). Seveso and Flixborough. *Chem. Engr, Lond.,* **314**, 697

MARSHALL, V.C. (1976c). The siting and construction of control buildings — a strategic approach. *Process Industry Hazards,* p. 187

MARSHALL, V.C. (1976d). The strategic approach to safety. *Chem. Engr, Lond.,* **308**, 260

MARSHALL, V.C. (1977a). Chemical conurbations — the domino danger. *Chemical Engineering in a Hostile World* (London: Clapp & Poliak)

MARSHALL, V.C. (1977b). How lethal are explosions and toxic escapes? *Chem. Engr, Lond.,* **323**, 573

MARSHALL, V.C. (1977c). Safety in the laboratory. *Chem. Ind.,* Jul. 2, 521

MARSHALL, V.C. (1977d). What are major hazards? *Safety Surveyor,* **5**(3), 5

MARSHALL, V.C. (1978). Physical implications of major hazards. *Proc. Harwell Environmental Seminar on Major Chemical Hazards*

MARSHALL, V.C. (1979a). The hazards, the probable outcome and frequency. *Symp. on Bulk Storage and Handling of Flammable Gases and Liquids,* Oyez Communications, London

MARSHALL, V.C. (1979b). Legal obligations in relation to chemical engineering laboratories. *Instn Chem. Engrs Symp. on Safety in Chemical Engineering Departments* (London)

MARSHALL, V.C. (1979c). What happened at Harrisburg? *Chem. Engr, Lond.*, **346**, 479

MARSHALL, V.C. (1980a). An analysis of the Canvey Report. *Chem. Engr, Lond.*, **356**, 357

MARSHALL, V.C. (1980b). Historical and theoretical approaches to the prediction of hazard and risk. In *Loss Prevention and Safety Promotion,* vol. 3, p. 395

MARSHALL, V.C. (1980c). Seveso, an analysis of the official report. *Chem. Engr, Lond.*, **358**, 499

MARSHALL, V.C. (1980d). TNT equivalence – the Decatur anomaly. *Chem. Engr, Lond.*, **353**, 108

MARSHALL, V.C. (1981a). Hazard and risk (letter). *Chem. Engr, Lond.*, **368**, 219

MARSHALL, V.C. (1981b). Hazards . . . risk. . . which? *Health and Safety at Work,* March, 32

MARSHALL, V.C. (1982a). Assessment of hazard and risk. In Bennett, G.F., Feates, F.S. and Wilder, I. (1982), *op. cit.*, p. 5–1

MARSHALL, V.C. (1982b). Quantification as a means of control of toxic hazards. *The Assessment of Major Hazards* (Rugby: Instn Chem. Engrs), p. 69

MARSHALL, V.C. (1982c). Risk and hazard (letter). *Chem. Engr, Lond.*, **383/384**, 344

MARSHALL, V.C. (1982d). Unconfined-vapor-cloud explosions. *Chem. Engng*, **89**(Jun. 14), 149

MARSHALL, V.C. (1983a). Accrediting the safety professional. *Chem. Engr, Lond.*, **392**, 95

MARSHALL, V.C. (1983b). The Chemical Scythe (book review). *Chem. Engr, Lond.*, **392**, 129

MARSHALL, V.C. (1984). Shaping a strategy of control. *Chem. Engr, Lond.*, **404**, 26

MARSHALL, V.C. (1985). Major chemical hazards. Ph.D. Thesis, Univ. of Bradford

MARSHALL, V.C. (1986a). Sombre news from the Chief Inspector of Factories. *Chem. Engr, Lond.*, **430**, 7

MARSHALL, V.C. (1986b). Work permit systems. *Chem. Engr, Lond.*, **427**, 44

MARSHALL, V.C. (1987). *Major Chemical Hazards* (Chichester: Ellis Horwood)

MARSHALL, V.C. (1988a). Bhopal (letter). *Chem. Engr, Lond.*, **451**, 6

MARSHALL, V.C. (1988b). Nuclear power (letter). *Chem. Engr, Lond.*, **450**, 4

MARSHALL, V.C. (1988c). A perspective view of industrial disasters. *Atom*, **376**(Feb.), 2

MARSHALL, V.C. (1989a). Modes and consequences of the failure of road and rail tankers carrying liquefied gases and other hazardous liquids. In Guy, G.B. (1989), *op. cit.*, p. 136

MARSHALL, V.C. (1989b). The predictions of human mortality from chemical accidents with special reference to the acute toxicity of chlorine. *J. Haz. Materials*, **22**(1), 13

MARSHALL, V.C. (1989c). A review of major hazards and accidents. In Lees, F.P. and Ang, M.L. (1989), *op. cit.*, p. 1

MARSHALL, V.C. (1989d). *Piper Alpha Public Inquiry. Transcript, Days 155, 156*

MARSHALL, V.C. (1990). The social acceptability of the chemical and process industries. *Trans. IChemE* **68**, Part B (May), 83–93

MARSHALL, V.C. (1990a). Chlorine toxicity (letter). *Chem. Engr, Lond.*, **480**, 4

MARSHALL, V.C. (1990b). Fundamentals of fire and explosion. *Safety and Loss Prevention in the Chemical and Oil Processing Industries*, p. 185

MARSHALL, V.C. (1990c). Loss in a name. *Chem. Engr, Lond.*, **473**, 4

MARSHALL, V.C. (1990d). The social acceptability of the chemical and process industries: a proposal for an integrated approach. *Process Safety Environ.*, **68B**, 83

MARSHALL, V.C. (1991a). Judging the acceptability of risks. *Safety and Reliability*, **11**(2), 29

MARSHALL, V.C. (1991b). Modes and consequences of the failure of road and rail tankers carrying liquefied gases and other hazardous liquids. In Advisory Committee on Dangerous Substances (1991), *op. cit.*, p. 90

MARSHALL, V.C. (1991c). Teaching process safety. *Chem. Engr, Lond.*, **502**, 30

MARSHALL, V.C. and COCKRAM, M.D. (1979). Safety in batch reactors – a strategic view. In Minors, C. (1979), *op. cit.*, paper 2

MARSHALL, V.C. and TOWNSEND, A. (eds) (1991). *Safety in Chemical Engineering Research and Development* (Rugby: Instn Chem Engrs)

MARSHALL, W. (1982). *An Assessment of the Integrity of PWR Vessels. Second Report* (London: HM Stationery Office)

MARSHALL, SIR W. (ed.) (1984). *Nuclear Power Technology*, vols 1–3 (Oxford: Oxford Univ. Press)

MARSHALL, W. *et al.* (1976). *An Assessment of the Integrity of PWR Pressure Vessels* (report). UK Atom. Energy Auth.

MARSHALL, W. *et al.* (1982). Pressure vessel integrity. *Atom*, **308**(Jun.), 122

MARSHALL, W.F., PALMER, H.B. and SEERY, D.I. (1964). Particle size effects and flame propagation rate control in laminar coal dust flames. *J. Inst. Fuel*, **37**, 342

MARSHALL, W.W. (1981). Construction materials for the chemical process industries. *Chem. Engr, Lond.*, **368**, 221

MARSILI, G., VOLLONI, C. and ZAPPONI, G.A. (1992). Risk communication in preventing and mitigating consequences of major chemical industrial hazards. *Loss Prevention and Safety Promotion 7*, vol. 3, 168–1

MARTEL, J.T. (1984). Safety and reliability of microprocessor-based combustion safeguard systems. *Plant/Operations Prog.*, **3**, 80

MARTEL, Y., BOTTE, J.M. and REGAZZACCI, P. (1987). Reduce costs with metal bellows shaft seals. *Hydrocarbon Process.*, **66**(10), 39

MARTENSON, C. and SUPKO, G.M. (1983). Experiences with high-alloy stainless steels in the HPI. *Hydrocarbon Process.*, **62**(1), 69

MARTI, B. (1976). Current information on and prevention of accidents and occupational diseases in the chemical industry: introduction. *Prevention of Occupational Risks*, **3**, 283

MARTIN, A.M., NOLEN, S.L., GESS, P.S. and BAESEN, T.A. (1992). Control odors from CPI facilities. *Chem. Engng Prog.*, **88**(12), 53

MARTIN, B. (1966). How A&M fire school serves the HPI. *Hydrocarbon Process.*, **45**(12), 141

MARTIN, B.R. and WRIGHT, R.I. (1987). A practical method of common cause failure modelling. *Reliab. Engng*, **19**(3), 185

MARTIN, C. (1981). Fire detection devices. *Chem. Ind.*, Jul. 18, 490

MARTIN, C. (1985). The CIMAH Regulations. *Chem. Ind.*, Dec. 16, 810

MARTIN, C. (1986). To inspect or not that is the question. *Chem. Ind.*, Sep. 15, 598

MARTIN, C. and RANBY, P. (1982). The Sixth Amendment and all that. *Chem. Ind.*, Oct. 16, 796

MARTIN, D. (1977). Safety valves blow their top to save the plant. *Process Engng,* Oct.

MARTIN, D.G. (1956). Flame speeds of mixtures containing several combustible components or a known quantity of diluent. *Fuel*, **35**, 352

MARTIN, D.O. and TIKVART, J.A. (1968). A general atmospheric diffusion model for estimating the effects of air quality of one or more source. *Air Poll. Control Ass., Sixty-First Ann. Mtg*

MARTIN, E.J. and JOHNSON, J.H. (1987). *Hazardous Waste Management Engineering* (New York: van Nostrand Reinhold)

MARTIN, G. (1984). Emergency planning. *Fire Surveyor*, **13**(1), 5

MARTIN, G.F. (1979). A case study: designing new chemical engineering laboratories. *Instn Chem. Engrs, Symp. on Safety in Chemical Engineering Departments* (London)

MARTIN, H. and RUPPERT, K.A. (1993). Critical storage volume of self-reacting products. *J. Loss Prev. Process Ind.*, **6**, 303

MARTIN, H.W. (1983). An opportunity: less than design rate operations. *Plant/Operations Prog.*, **2**, 144

MARTIN, J. and OXMAN, S. (1988). *Building Expert Systems: A Tutorial* (Englewood Cliffs, NJ: Prentice-Hall)

MARTIN, J.B. and SYMONDS, P.S. (1966). Mode approximations for impulsively-loaded rigid-plastic structures. *Proc. Am. Soc. Civil Engrs, J. Engng Mechanics Div.*, **92**, 43

MARTIN, J.D. (1984). Hydrogen – the UK market and production methods. *Chem. Ind.,* Jan. 16, 46

MARTIN, L.D., YOUNG, J.O. and BANKS, W.P. (1977). How to simplify lube oil system cleanup. *Hydrocarbon Process.*, **56**(1), 88

MARTIN, M.K. and HEYWOOD, J.B. (1977). Approximate relationships for the thermodynamic properties of hydrocarbon–air combustion products. *Combust. Sci. Technol.*, **15**, 1

MARTIN, M.W. and SCHINZINGER, R. (1989). *Ethics in Engineering* (New York: McGraw-Hill)

MARTIN, P. (1976a). Mechanical reliability – a systematic approach to interactive failures. *Advances in Reliability Technology. ART7* (Culcheth, Lancashire: UK Atom. Energy Auth.)

MARTIN, P. (1976b). Reliability in mechanical design and production. In Henley, E.J. and Lynn, J.W. (1976), *op. cit.,* p. 267

MARTIN, P. (1980). A systematic approach to interactive failures. In Moss, T.R. (1980), *op. cit.,* p. 36

MARTIN, P. (1982). Consequential failures in mechanical systems. *Reliab. Engng*, **3**(1), 23

MARTIN, P., STRUTT, J.E. and KINEAD, N. (1983). A review of mechanical reliability modelling in relation to failure mechanisms. *Reliab. Engng*, **6**, 13

MARTIN, R.E. (1979). Hydratect – a new water presence detector for steam power plant. *Pulse,* Sep.

MARTIN, R.E. (1980). Hydratect – detection of the presence or absence of water in steam plant. *Chem. Engr, Lond.*, **362**, 687

MARTIN, R.J., SHEPHERD, D. and HAMLIN, C.C. (1986). Halon 1301 total flood systems. *Fire Surveyor*, **15**(3), 35; **15**(4), 16; **15**(5), 17

MARTIN, R.L. (1970). Detection of ball bearing malfunctions. *Instrum. Control Systems*, **43**(Dec.), 79

MARTIN, S. (1964). Ignition of organic materials by radiation. *Fire Res. Abst. Revs*, **6**(2), 85

MARTIN, S. (1965). Diffusion-controlled ignition of cellulosic materials by intense radiant energy. *Combustion*, **10**, 877

MARTIN, S.B. (1956). *The Mechanisms of Ignition of Cellulosic Materials by Intense Thermal Radiation. Rep. USNRDL-TR-102*. Naval Radiological Defense Lab

MARTIN, T. (1988a). Pump control. *Chem. Engr, Lond.*, **452**, 37

MARTIN, T. (1988b). Pump shutdown: watch your step. *Chem. Engr, Lond.*, **455**, 96

MARTIN, T. (1989a). Pumps in parallel, heater shutdowns. *Chem. Engr, Lond.*, **458**, 53

MARTIN, T. (1989b). Too many controls. *Chem. Engr, Lond.*, **460**, 59

MARTIN, T. (1989c). Unintentional vacuums. *Chem. Engr, Lond.*, **464**, 53

MARTIN-SOLIS, G.A., ANDOW, P.K. and LEES, F.P. (1977). An approach to fault tree synthesis for process plants. *Loss Prevention and Safety Promotion 2*, p. 367

MARTIN-SOLIS, G.A., ANDOW, P.K. and LEES, F.P. (1980). Synthesis of fault trees containing multi-state variables. In *Loss Prevention and Safety Promotion 3*, vol. 2, p. 554

MARTIN-SOLIS, G.A., ANDOW, P.K. and LEES, F.P. (1982). Fault tree synthesis for real time and design applications. *Trans. Instn Chem. Engrs*, **60**, 14

MARTINDALE, R. (1979). Charging for direct discharges. *Chem. Ind.,* Mar. 17, 195

MARTINEZ, E., BELTRAMINI, L., LEONE, H., RUIZ, C.A. and HUETE, E. (1992). Knowledge elicitation and structuring for a real-time expert system for monitoring a butadiene extraction system. *Comput. Chem. Engng.*, **16**, S345

MARTINEZ, O.A., MADHAVAN, S., KELLETT, D.J. and HAMROUNI, H. (1987). Damage to and replacement of an ammonia storage tank foundation. *Ammonia Plant Safety* **27**, 162

MARTINEZ, O.A., MADHAVAN, S. and KELLETT, D.J. (1987). Damage to and replacement of an ammonia storage tank foundation. *Plant/Operations Prog.*, **6**, 129

MARTINO, E. (1983). Materials for reciprocating process gas compressors. *Chem. Engng Prog.*, **79**(4), 77

MARTINO, R.L. (1963). Plain talk on critical path methods. *Chem. Engng*, **70**(Jun. 10), 221

MARTINO, R.L. (1964). How useful are critical path methods. *Chem. Engng*, **71**(Sep. 14), 220

MARTINOT, B. (1984). Reduction of explosion risks subsequent to ethylene decomposition on high pressure polyethylene plants. *The Protection of Exothermic Reactors and Pressurised Storage Vessels* (Rugby: Instn Chem. Engrs), p. 315

MARTINOVIC, A. (1983). Architectures of distributed digital control systems. *Chem. Engng Prog.*, **79**(2), 67

MARTINS, J.P. (1992). Truth maintenance systems. In Shapiro, S.C. (1992), *op. cit.,* p. 1613

MARTINS, K. (1988). Responding properly to hazardous waste spills. *Chem. Engng*, **95**(1), Jan. 18, 87

MARTINSEN, W.E. (1982). Propane spill and fire control methods. *Am. Petrol. Inst., Committee on Safety and Fire Protection, Spring Mtg*

MARTINSEN, W.E. (1984). BLEVEs once more (letter). *Haz. Cargo Bull.*, **5**(7), 50

MARTINSEN, W.E. (1992). Notes on a few vapor suppression systems. *Plant/Operations Prog.*, **11**, 22

MARTINSEN, W.E. and MUHLENKAMP, S.P. (1977). Disperse LNG vapors with water. *Hydrocarbon Process.*, **56**(7), 261

MARTINSEN, W.E., JOHNSON, D.W. and TERRELL, W.F. (1986). BLEVEs: their causes, effects and prevention. *Hydrocarbon Process.*, **65**(11), 141

MARTINSEN, W.E., JOHNSON, D.W. and MILLSAP, S.B. (1989). Determining spacing by radiant heat limits. *Plant/Operations Prog.*, **8**, 25

MARTINSON, A.M. (1964). Weight via waterways. *Chem. Engng Prog.*, **60**(3), 26

MARTON, F.D. (1965). Compressor surge control. *Instrum. Control Systems*, **38**(5), 149

MARTZ, H.F. (1984). On broadening failure rate distributions in PRA uncertainty analysis. *Risk Analysis*, **4**, 15

MARTZ, J.W. and PFAHLER, R.R. (1975). How to troubleshoot large industrial fans. *Hydrocarbon Process*, **54**(6), 57

MARX, W.B. (1973). Private communication (to R.R. Maccary, AEC Directorate of Licensing)

MARY KAY O'CONNOR PROCESS SAFETY CENTER (2002). *Challenges in Implementing Inherent Safety Principles in New and Existing Chemical Processes* (College Station, TX: Texas A&M University)

MARY KAY O'CONNOR PROCESS SAFETY CENTER (2003). *Center-line*, Newsletter Spring 2003, College Station, Texas: Mary Kay O'Connor Process Saf. Cent.

MARY KAY O'CONNOR PROCESS SAFETY CENTER, (2003). *Center-line*, Newsletter Summer 2003, College Station, Texas: Mary Kay O'Connor Process Saf. Cent.

MARYON, R.H., WHITLOCK, J.B.G. and JENKINS, G.J. (1986). An analysis of short-range dispersion experiments on the windward slope of an isolated hill. *Atmos. Environ.*, **20**, 2157

MARZAIS, J.J.L. (1973). A water injecting device for high pressure gases. *Safety in Polyethylene Plants*, **1**, p. 42

MARZI, M. (1982). Solving problems of varying heat transfer areas in batch processes. *Chem. Engng*, **89**(May 17), 101

MARZO, L. and FERNANDEZ, L. (1980). Destroy NO_x catalytically. *Hydrocarbon Process.*, **59**(2), 87

MASAITIS, G.J. and TORNAY, E.G. (1982). LNG transfer ship-to-ship following 'LNG Libra' tailshaft failure. *Gastech 81*, p. 263

MASCHIO, G. and ZANELLI, S. (1992). Analysis of runaway phenomena in polymerization reactors. *Loss Prevention and Safety Promotion 7*, vol. 1, 12-1

MASCONE, C.F. (1986). In case of emergency, call your nearest chemical plant. *Chem. Engng*, **93**(21), 22

MASCONE, C.F., SANTAQUILANI, A.G. and BUTCHER, C. (1991a). Engineering ethics: how ChEs respond. *Chem. Engng Prog.*, **87**(10), 73

MASCONE, C.F., SANTAQUILANI, A.G. and BUTCHER, C. (1991b). Engineering ethics – what are the right choices? *Chem. Engng Prog.*, **87**(4), 61

MASEK, J.A. (1964a). Thermowell design for process piping. *Hydrocarbon Process.*, **43**(2), 139; **43**(3), 119; **43**(4), 165

MASEK, J.A. (1964b). Tips on piping layout for flexibility. *Hydrocarbon Process.*, **43**(7), 143

MASEK, J.A. (1968). Metallic piping. *Chem. Engng*, **75** (Jun. 17), 215

MASEK, J.A. (1972). Where to locate thermowells. *Hydrocarbon Process.*, **51**(4), 147

MASEK, J.A. (1978). Weld thermocouples: its faster and cheaper. *Hydrocarbon Process.*, **57**(1), 106

MASEK, J.A. (1981). Designing pressure-gage connections. *Chem. Engng*, **88**(May 4), 71

MASEK, J.A. (1982). Installing pressure switches. *Chem. Engng*, **89**(Nov. 15), 113

MASEK, J.A. (1983). Improved instrument valves for pressure switches and pressure gages. *Chem. Engng*, **90** (Apr. 18), 89

MASLIYAH, J.H. and STEWARD, F.R. (1969). Radiative heat transfer from a turbulent diffusion buoyant flame with mixing controlled combustion. *Combust. Flame*, **13**, 613

MASON, B.J. (1971). *The Physics of Clouds,* 2nd ed. (Oxford: Clarendon Press)

MASON, D.G., GUPTA, M.K. and SCHOLZ, R.C. (1972). A mobile multipurpose' treatment system for processing hazardous material contaminated waters. *Control of Hazardous Material Spills 1*, p. 153

MASON, D.P. (1989). Highlights of FM inspection guidelines on emergency relief systems. *Runaway Reactions,* p. 722

MASON, G.S. (1982). Decanting without interface control. *Chem. Engng*, **89**(Sep. 20), 129

MASON, G.S. and ARNOLD, C. (1984). Contain liquid spills and improve safety with a flooded stormwater sewer. *Chem. Engng*, **91**(Sep. 17), 105

MASON, J.F. (1963). Ductile iron safe for refinery service. *Hydrocarbon Process.*, **42**(2), 165

MASON, J.F. (1979). The technical blow-by-blow. *IEEE Spectrum*, **16**(11), 33

MASON, P.I. (1975). Electrostatic hazards in liquid transport. *Univ. of Southampton, Fifth Electrostatics Industrial Course*

MASON, P.I. and REES, W.D. (1981). Hazards from plastic pipes carrying insulating liquids. *J. Electrostatics*, **10**, 137

MASON, W. and WHEELER, R.V. (1917). The 'uniform movement' during the propagation of flame. *J. Chem. Soc.*, 1044

MASON, W. and WHEELER, R.V. (1920a). The propagation of flame in mixtures of methane and air. Part I, Horizontal propagation. *J. Chem. Soc.*, **36**

MASON, W. and WHEELER, R.V. (1920b). The propagation of flame in mixtures of methane and air. Pt II, Vertical propagation. *J. Chem. Soc.*, 1227

MASON, W. and WHEELER, R.V. (1920c). The propagation of flame in mixtures of methane and air. Pt III, Propagation in currents of the mixtures. *J. Chem. Soc.*, 1237

MASON, W. and WHEELER, R.V. (1922). The ignition of gases. II, Ignition by a heated surface. Mixtures of methane and air. *J. Chem. Soc.*, **121**, 2079

MASON, W. and WHEELER, R.V. (1924). The ignition of gases. IV, Ignition by a heated surface. Mixtures of paraffins with air. *J. Chem. Soc.*, **125**, 1869

MASON, W.A. (1964). Oxidizing salt explosion. *Ammonia Plant Safety*, **6**, p. 45

MASON, W.E. and WILSON, M.J.G. (1967). Laminar flames of lycopodium dust in air. *Combust. Flame*, **11**, 195

MASSA, R.J. (1987). The SECUREPLAN bomb utility – a PC-based analytic tool for bomb defence. *Nucl. Materials Mngmt*, **16**, 299

MASSEY, B.S. (1968–). *Mechanics of Fluids;* 1970, 2nd ed. (London: van Nostrand); 1975, 3rd ed.; 1979, 4th ed. (New York: van Nostrand); 1983, 5th ed. (Wokingham: van Nostrand); 1989, 6th ed. (London: Chapman & Hall)

MASSEY, P.S. (1964). Frozen-earth storage: how Phillips built it. *Hydrocarbon Process.*, **43**(3), 147

MASSMEYER, K., GEISS, H., HEINEMAN, K., MÖLLMANN, M. and POLSTER, G. (1986). Atmospheric dispersion around an isolated hill – programme and preliminary results. In de Wispelaere, C., Schiermeier, F.A. and Gillani, N.V. (1986), *op. cit.*, p. 581

MASSO, A.H. and RUDD, D.F. (1968). Disaster propagation. Propagation of explosive-like violence. *Ind. Engng Chem. Fund.*, **7**, 131

MASSO, A.H. and RUDD, D.F. (1969). The synthesis of system designs. II, Heuristic structuring. *AIChE J.*, **15**, 10

MASSON, M. (1991). Understanding, reporting and preventing human fixation. Van der Schaaf, T.W., Lucas, D.A. and Hale, A.R. (1991), *op. cit.*, p. 35

MASTERS, K. (1972). *Spray Drying* (London: Leonard Hill); 1976, 2nd ed. (London: Godwin)

MASTERS, K. (1979). *Spray Drying Handbook,* 3rd ed.; 1985, 4th ed. (London: Godwin); 1991, 5th ed. (Burnt Hill, Harlow: Longmans)

MASUDA, H., KOMATSU, T. and IINOYA, K. (1976). The static electrification of particles in gas-solids pipe flow. *AIChE J.,* **22**, 558

MASUDA, S. and SCHÖN, G. (1967). Space–charge distribution in liquids after passage through a filter. *Static Electrification,* **2**, p. 112

MATEN, S. (1967). New vibration velocity standards. *Hydrocarbon Process.,* **46**(1), 137

MATEN, S. (1970). Program maintenance by measuring vibration velocity. *Hydrocarbon Process.,* **49**(9), 291

VAN MATER, P.R., EDINBERG, D.L., ORSERO, P. and FINIFTER, D. (1982). A comparison of collision resistance of membrane tank-type and spherical tank-type LNG tankers. *Gastech 81*

MATERIALS TECHNOLOGY INSTITUTE OF THE CHEMICAL PROCESS INDUSTRIES, Cortest Labs, Cypress, TX 77429

MATHAY, W.L. and MCKNIGHT, A. (1966). Stainless steel jacketing for insulation. *Chem. Engng Prog.,* **62**(1), 95

MATHER, J.S.B. (1993). Noise control in rotating machinery. In Instn Mech. Engrs (1993), *op. cit.,* 21

MATHESON, A.B. (1989). *Piper Alpha Public Inquiry. Transcript, Day 149*

MATHESON COMPANY INC. (1961). *Matheson Gas Data Book* (East Rutherford, NJ)

MATHESON, D. (1960). Some significant chemical accidents. *Chemical Process Hazards,* **I**, p. 1

MATHEWS, T. (1983). Bursting discs for overpressure protection. *Chem. Engr, Lond.,* **395**, 21

MATHIESEN, T.C., BAKKE, H.J. and BØE, C. (1973). Safety aspects of ships for bulk carriage of chlorine. *Transport of Hazardous Cargoes,* **3**

MATHIESEN, T.C., FLATSETH, H.H., SOLBERG, D.M. and TVEIT, O.J. (1977). Risk management in marine transportation of LNG. *LNG Conf. 5*

MATHIESEN, T.C., KOREN, J., VALSGARD, S., SKRAMSTAD, E. and THYGESEN, B. (1991). Useful service life of LNG carriers – 20, 30, 40 years? *Gastech 90,* p. 4.7

MATHIS, J.J. and GROSE, W.L. (1973). *Review of Methods for Predicting Air Pollution Dispersion. Rep. NASA SP-322.* Nat. Aeronaut. Space Admin., Washington, DC

MATHUR, J. (1973). Steam traps. *Chem. Engng,* **80**(Feb. 26), 47

MATLEY, J. (1965). How to join dissimilar steels. *Chem. Engng,* **72**(Nov. 8), 230

MATLEY, J. (1969). Keys to successful plant start-ups. *Chem. Engng,* **76**(Sep. 8), 110

MATLEY, J. (1970). Trends in maintenance. *Chem. Engng,* **77**(Mar. 23), 112

MATLEY, J. (1982). CE Plant Cost Index – revised. *Chem. Engng,* **89**(Apr. 19), 153

MATLEY, J. and CHEMICAL ENGINEERING STAFF (1979). *Pumps, Compressors, Fans and Blowers* (New York: McGraw-Hill)

MATLEY, J. and GREENE, R. (1987). Ethics of health, safety, and environment. What's right. *Chem. Engng,* **94**(3), 40

MATLEY, J., GREENE, R. and MCCAULEY, C. (1987). Health, safety and environment – CE readers say what's 'right'. *Chem. Engng,* **94**(13), 108

MATSON, A.F. and DUFOUR, R.E. (1950). *The Lower Limit of Flammability and the Autogenous Ignition Temperature of Certain Common Solvent Vapours Encountered in Ovens. Bull. Res. 43.* Underwriters Lab.

MATSUBARA, Y. (1991). Numerical modelling for potential relaxation of charged oil inside containers. *Electrostatics,* **8**, 293

MATSUDA, T. and NAITO, M. (1975). *Pressure Relief Equipment for Dust Explosions. Doc. RIIS-SD-75–1.* Research Institute of Industrial Safety (Japan)

MATSUI, H. and LEE, J.H. (1976). Influence of electrode geometry and spacing on the critical energy for direct initiation of spherical gaseous detonations. *Combust. Flame,* **27**, 217

MATSUI, H. and LEE, J.H. (1979). On the measure of the relative detonation hazards of gaseous fuel-oxygen and air mixtures. *Combustion,* **17**, 1269

MATT, L. (1889). Experimental contributions to the theory of the effects of poisonous gases on human beings. Dissertation, Univ. of Wurzburg

DE MATTEIS, U. (1982). Planning and budgeting for maintenance. *Chem. Engng,* **89**(Nov. 15), 105

MATTHEWS, L.G. (1961). Oxygen plant explosion. *Ammonia Plant Safety,* **3**, 12. See also *Chem. Engng Prog.,* 1961, **57**(4), 48

MATTHEWS, R.R. (1982). The development of safety targets for CEGB nuclear stations. *Nucl. Energy,* **21**, 323

MATTHEWS, S.D. (1977). *MOCARS: a Monte Carlo Code for Determining the Distribution and Simulation Limits. Rep. TREE-1138.* EG&G Idaho, Idaho Falls, ID

MATTHEWS, T. (1981). Explosion and fire at Salford (letter). *Chem. Engr, Lond.,* **388**, 5

MATTHEWS, W.D. and OWEN, G.G. (1963). Safety aspects of reconstructed ICI tonnage oxygen plant. *Ammonia Plant Safety,* **5**

MATTHIAS, C.S. (1990). Dispersion of a dense cylindrical cloud in calm air. *J. Haz. Materials,* **24**(1), 39

MATTHIAS, C.S. (1992). Dispersion of a dense cylindrical cloud in a turbulent atmosphere. *J. Haz. Materials,* **30**, 117

MATTHIESSEN, C. (1994). An overview of EPA's chemical Accident Release Prevention programs. *Process Safety Prog.,* **13**, 61

MATTHIESSEN, R.C. (1986). Estimating chemical exposure levels in the workplace. *Chem. Engng Prog.,* **82**(4), 30

MATTHIESSEN, R.C. (1992). Chemical accident prevention under the Clean Air Act Amendments. *Plant/Operations Prog.,* **11**, 99

MATTOO, B.N., WANI, A.K. and ASKEGAR, M.D. (1974). Casualty criteria for wounds from firearms with special reference to shot penetration, Pt II. *J. Forensic Sci.,* **19**, 585

MATTSON, R.E. (1974). Improve electrical systems safety and reliability. *Hydrocarbon Process.,* **53**(6), 146

MATUZAWA, T. (1964). *Study of Earthquakes* (Tokyo: Uno Shoten)

MAUCHLY, J.W. (1962). Critical path scheduling. *Chem. Engng,* **69**(Apr. 16), 139

MAUCK, J.W. and TOMITA, N. (1974). Vibration and strain in high pressure piping. *Safety in Polyethylene Plants,* **2**, 22

MAUGHAN, R.G. (1971). Modern railway track inspection. *Loss Prevention,* **5**, 41

MAUKONEN, D. and VEST, G. (1973). Heat treating welds. *Chem. Engng,* **80**(Jul. 9), 100

MAUL, P.R. (1980). Atmospheric transport of sulphur compound pollutants. Ph.D. Thesis, Imperial Coll., London

MAUNDER, A. (1985). Pneumatic conveying of hazardous materials. *Chem. Engr, Lond.,* **412**, 11

MAURER, B. (1979). Discharges due to charges of particles in large storage silos. *Chem. Ing. Tech.,* **51**, 98. See also *Ger. Chem. Engng,* **4**, 189

MAURER, B. (1984). *VDI-Berichte,* **494**, 119

MAURER, B., GLOR, M., LUTTGENS, G. and POST, L. (1987). Hazards associated with propagating brush discharges on flexible intermediate bulk containers, compounds and coated materials. *Electrostatics*, **7**, 217

MAURER, B. *et al.* (1977). Modeling of vapour cloud dispersion and deflagration after bursting of tanks filled with liquefied gas. *Loss Prevention and Safety Promotion in the Process Industries 2*, p. 305

MAVROVIC, I. (1974). Improved urea process is developed. *Ammonia Plant Safety* **16**, 127

MAVROVOUNIOTIS, M.L. and STEPHANOPOULOS, G. (1988). Formal order-of-magnitude reasoning in process engineering. *Comput. Chem. Engng*, **12**, 867

MAX, D.A. and JONES, S.T. (1983). Flareless ethylene plant. *Hydrocarbon Process.*, **62**(12), 89

MAXEY, W.A., KOCH, F.O., PEECK, A., JUNKE, R.G., VOGT, G.H., PETERS, P.A., KUGLER, J. and SEIFERT, K. (1985). Transition temperature tests show fracture behaviour. *Oil Gas J.*, **83**(Apr. 22), 75

MAX-LINO, R.A. (1989). Human factors in the design of programmable electronic systems. *Loss Prevention and Safety Promotion 6*, p. 53.1

MAXWELL, J.B. (1950). *Data Book on Hydrocarbons* (Princeton, NJ: van Nostrand)

MAXWELL, J.H. (1980). Vibration analysis pinpoints coupling problems. *Hydrocarbon Process.*, **59**(1), 95

MAXWELL, J.H. (1981). Diagnosing induction motor vibration. *Hydrocarbon Process.*, **60**(1), 117

MAXWELL, M.A. (1981). Consider baghouses for product recovery. *Chem. Engng*, **88**(Dec. 28), 54

MAY, D. and RUDD, D.F. (1976). Development of solvay clusters of chemical reactions. *Chem Engng. Sci.*, **31**, 59

MAY, D.L. (1985). Does your plant instrumentation work properly? *Chem. Engng*, **92**(Mar. 4), 67

MAY, D.L. (1988). How much backup for process control computers? *Chem. Engng*, **95**(6), Apr. 25, 64

MAY, D.L. and PAYNE, J.T. (1992). Validate process data automatically. *Chem. Engng*, **99**(6), 112

MAY, F.G. (1958). The washout by rain of lycopodium spores. *Quart J. R. Met. Soc.*, **84**, 451

MAY, F.G.J. (1978). Australian test procedures for determination of compatibility and stability of military explosives. *J. Haz. Materials*, **2**(2), 127

MAY, G. (1973). Chloracne from the accidental production of tetrachlorodibenzodioxin. *Br. J. Ind. Med.*, **30**, 276

MAY, J. (1989). *The Greenpeace Book of the Nuclear Age* (London: Gollancz)

MAY, P.J. and WILLIAMS, W. (1986). *Disaster Policy Implementation* (New York: Plenum Press)

MAY, W.G. and MCQUEEN, W. (1973). Radiation from large liquefied natural gas fires. *Combust. Sci. Technol.*, **7**, 51

MAY, W.G., MCQUEEN, W. and WHIPP, R.H. (1973). Dispersion of LNG spills. *Hydrocarbon Process.*, **52**(5), 105

MAY, W.G. and PERUMAL, P. (1974). The spreading and evaporation of LNG on water. *Am. Soc. Mech. Engrs, Ann. Winter Mtg* (New York)

MAY, W.W. (1982). $s for lives: ethical considerations in the use of cost/benefit analysis by for-profit firms. *Risk Analysis*, **2**, 35

MAYAN, M.H. and MERILAN, C.P. (1972). Effects of ammonia inhalation on respiration rate of rabbits. *J. Anim. Sci.*, **34**, 448

MAYANARI, T. (1986). The IHI SPB LNG system – application of advanced Al-alloy welding technology. *Gastech 85*, p. 285

MAYER, E. (1957). A theory of flame propagation limits due to heat loss. *Combust. Flame*, **1**, 438

MAYERS, M.R. (1969). *Occupational Health: Hazards of the Work Environment* (Baltimore, MD: Williams & Wilkins)

MAYES, H.A. and YALLOP, H.J. (1965). The explosive properties of liquid acetylene. *Chem. Engr, Lond.*, **185**, CE25

MAYINGER, F. (1981). Stand der thermohydraulischen Kenntnisse bei Druckentlastungsvorgängen. *Chem. Ing. Tech.*, **53**, 424

MAYLOR, J.B. (1969). Corrosion problems in the chemical industry. *Chem. Engr, Lond.*, **266**, CE74

MAYNARD, H.B. (1963). *Industrial Engineering Handbook* (New York: McGraw-Hill)

MAYO, H.C. (1968). Large plant operation. *Ammonia Plant Safety*, **10**, p. 109

MAYO, L.H. (1970). *Scientific Method, Adversarial System and Technology Assessment. Monogr. 5*. Program of Policy Studies in Science and Technology, George Washington Univ.

MAZUR, A. (1973). Disputes between experts. *Minerva*, **11**, 243

MAZZOCCHI, A., MORTARINO, C. and PICCININI, N. (1983). Incident report collection and analysis – method of classification and lessons to be learned. *Euredata 4*, paper 7.4

MAZZONCINI, R. (1975). Guide to Italian Pressure Vessel Code. *Hydrocarbon Process.*, **54**(12), 63

MAZZONCINI, R. (1978). Guide to Italian Pressure Vessel Code. *Hydrocarbon Process.*, **57**(12), 95

MEADE, W.G. and PRESS, L. (1977). Vinyl chloride – as mirror of the future? *Loss Prevention*, **11**, 120

MEADOR, L. and SHAH, A. (1969). Steam lines designed for two phase. *Hydrocarbon Process.*, **48**(1), 143

MEADOWCROFT, A.E. (1975). Hazard classification – the UN system. *Distribution Conf.* (London: Chem. Ind. Ass.), p. 4

MEADOWS, D.G. (1976). Hazard analysis. *Processing*, Apr. 13

MEAKINS, J.C. and PRIESTLEY, J.G. (1919). The after-effects of chlorine gas poisoning. *Can. Med. Ass. J.*, **9**, 968

MEALEY, J.D. (1970). Costing of occupational injuries. *Ammonia Plant Safety*, **12**, 19

MEANS, R.E. and BECK, T.R. (1984). A techno/economic review of chlor-alkali technology. *Chem. Engng*, **91**(Oct. 29), 46

MEARNS, A.B. (1965). Fault tree analysis: the study of unlikely events in complex systems. In Boeing Company (1965), *op. cit.*

MEARS, R.B. (1960). Plant site, layout minimize corrosion. *Chem. Engng*, **67**(Jan. 11), 144

MECHANICAL HANDLING ENGINEERS ASSOCIATION (1962). *Troughed Belt Conveyors* (London)

MECKLENBURGH, J.C. (1973). *Plant Layout* (London: Leonard Hill)

MECKLENBURGH, J.C. (1976). Plant layout and loss prevention. *Course on Loss Prevention in the Process Industries*. Dept. of Chem. Engng, Loughborough Univ. of Technol.

MECKLENBURGH, J.C. (1977a). The investigation of major process disasters. *Chem. Engr, Lond.*, **323**, 578

MECKLENBURGH, J.C. (1977b). *Lessons from the Post-enquiry Discussion of Flixborough. Design Developments Related to Safety, Cost and Solid Systems* (London: Instn Chem. Engrs), paper A-2–1

MECKLENBURGH, J.C. (1982). Hazard assessment of plant layout. *The Assessment of Major Hazards* (Rugby: Instn Chem Engrs), p. 321

MECKLENBURGH, J.C. (ed.) (1985). *Process Plant Layout* (London: Godwin)

MECKLENBURGH, J.C. (1986). Hazard zone sizes within buildings. *Loss Prevention and Safety Promotion 5*, p. 52–1

MEDARD, L. (1970). Rupture of an ammonia road tanker. *Ammonia Plant Safety,* 12, 17

MEDARD, L. (1979). *Les Explosifs Occasionnels,* vols 1–2 (Paris: Technique et Documentation)

MEDHEKAR, S.R., BLEY, D.C. and GEKLER, W.C. (1993). Prediction of vessel and piping failures rates in chemical process plant using the Thomas model. *Process Safety Prog.,* 12, 123

MEDHEKAR, S.R., GEKLER, W.C. and BLEY, D.C. (1993). Frequency estimates for transport-related hydrofluoric and sulfuric acid release scenarios. *Process Safety Prog.,* 12, 166

MEDICAL RESEARCH COMMITTEE (1918). *Atlas of Gas Poisoning* (France: BEF)

MEDICAL RESEARCH COUNCIL (1956). *The Hazards to Man of Nuclear and Allied Radiations* (London: HM Stationery Office)

MEDICAL RESEARCH COUNCIL (1975a). *Criteria for Controlling Radiation Doses to the Public after Accidental Escape of Radioactive Material* (London: HM Stationery Office)

MEDICAL RESEARCH COUNCIL (1975b). *The Toxicity of Plutonium* (London: HM Stationery Office)

MEDICI, M. (1974). *The Natural Gas Industry* (London: Newnes-Butterworths)

MEDLAM, J.W.N. (1990). Developments in reliability and availability of fire and gas and ESD systems. *Safety and Loss Prevention in the Chemical and Oil Processing Industries,* p. 219

MEDLAND, W.G.R. (1986). Comments on the mathematical modelling of collision risk. *Inst. of Mathematics and its Applications, Mtg on Mathematics in Major Accident Risk Assessment*

MEDLOCK, R.S. (1980). State of the art and trends in industrial instrumentation. *Chem. Engr, Lond.,* 362, 679

MEEHL, P.E. (1954). *Clinical versus Statistical Prediction* (Minneapolis, MN: Univ. of Minnesota Press)

MEEK, J.A.E. and EYSTER, W.J. (1920). Experiments on the pathology and physiology of acute phosgene poisoning. *Am. J. Physiol.,* 51, 303

MEEK, J.M. and GRAGGS, J.D. (1953). *Electric Breakdown of Gases* (London: Oxford Univ. Press)

MEEKER, J. (1985). Applying condition monitoring: a case history. *Process Engng,* Jun., 71

MEEKS, J. and CAMPION, S.F. (1977). Designing safety into new plants. *Chemical Engineering in a Hostile World* (London: Clapp & Poliak)

VAN DER MEER, D. (1971). Electrostatic charge generation during washing of tanks with water sprays – I: General introduction. *Static Electrification* 3, p. 153

VAN MEERBEKE, R.C. (1982). Accident at Cove Point LNG facility. *Chem. Engng Prog.,* 78(1), 39

MEETHAM, A.R. (1968). *Atmospheric Pollution – Its Origins and Prevention* (Oxford: Pergamon Press)

MEGANCK (1973). Explosion due to thermal decomposition of a high boiling organic chemical. *Prevention of Occupational Risks,* 2, 453

MEGARIDIS, C.M. and DOBBINS, R.A. (1989). Comparison of soot growth in smoking and non-smoking ethylene diffusion flames. *Combust. Sci. Technol.,* 66, 1

MEGAW, W.J. (1962). The penetration of iodine into buildings. *Int. J. Air Water Poll.,* 6, 121

MEGOW, L. and DAWSON, J.D. (1977). You can field repair a multi-layer vessel. *Hydrocarbon Process.,* 56(3), 153

MEGUERIAN, G.H. and RADOWSKY, F.W. (1967). Hydrocarbon ignition in the presence of metal oxides. *Combust. Flame,* 11, 135

MEGYESY, E.F. (1978). Size vessel nozzles faster. *Hydrocarbon Process.,* 57(12), 100

MEGYESY, E.F. (1983). *Pressure Vessel Handbook,* 6th ed. (Tulsa, OK: Pressure Vessel Handbook Publ. Co.)

MEHLMAN, M.A. (1987). Health effects and toxicity of phosgene: a scientific review. *Def. Sci. J.,* 37, 269

MEHLMAN, M.A., SHAPIRO, R. and BLUMENTHAL, H. (eds) (1976). *Advances in Modern Toxicology: New Concepts in Safety Evaluation* (Washington, DC: Hemisphere)

MEHNE, P.H. (1984). Reducing risks in exploratory research at high pressures and temperatures. *Plant/Operations Prog.,* 3, 37

MEHRA, D.K. (1983). Shell-and-tube heat exchangers. *Chem. Engng,* 90(Jul. 25), 47

MEHRA, Y.R. (1978). How to estimate power and condenser duty for ethylene refrigeration systems. *Chem. Engng,* 85(Dec. 18), 97

MEHRA, Y.R. (1979a). Charts for systems using ethane refrigerant. *Chem. Engng,* 86(Feb. 12), 95

MEHRA, Y.R. (1979b). Refrigerant charts for propylene systems. *Chem. Engng,* 86(Jan. 15), 1331

MEHRA, Y.R. (1979c). Refrigerant properties of propane. *Chem. Engng,* 86(Mar. 26), 165

MEHRA, Y.R. (1982). Refrigeration systems for low temperature processes. *Chem. Engng,* 89(Jul. 12), 94

MEHTA, A.K. and WONG, F. (1967). *Measurement of Flammability and Burn Potential of Fabrics. Rep. NTIS COM-73–10960.* Fuel Research Laboratory, MIT

MEHTA, A.K., WONG, F. and WILLIAMS, G.C. (1973). *Measurement of Flammability and Burn Potential of Fabrics. Summary Rep. on NSF-Grant GI-31881.* Fuels Res. Lab., Massachusetts Inst. Technol.

MEHTA, C. (1992). The business link. *Chem. Engng,* 99(5), 96

MEHTA, D.D. (1972). For process designers . . . compressors and converters. *Hydrocarbon Process.,* 51(6), 129

MEHTA, D.D. and ROSS, D.E. (1970). Optimize ICI methanol process. *Hydrocarbon Process.,* 49(11), 183

MEHTA, G.A. (1983). The benefits of batch process control. *Chem. Engng Prog.,* 79(10), 47

MEIDL, J.H. (1972). *Hazardous Materials Handbook* (Beverly Hills, CA: Glencoe Press)

MEIER, F.A. (1982). Is your control system ready to start up? *Chem. Engng,* 89(Feb. 22), 76

MEIERS, S. and JARMAN, M. (1993). VRJTankeat: a thermal model of deluge cooling water rates for atmospheric storage tanks. In Melchers, R.E. and Stewart, M.G. (1993), *op. cit.,* p. 85

MEIJER, C.H., FROGNER, B. and LONG, A.B. (1980). A disturbance analysis system for on-line power plant surveillance and diagnosis. In Gessellschaft für Reactorsicherheit (1980), *op. cit.,* p. 213

MEIJER, H.P. (1968). Development and future of the port of Rotterdam. *Carriage of Dangerous Goods,* 1, 75

MEINHARDT, D.P. (1981). Chlorine. *Chem. Engng,* 88 (Oct. 5), 125

MEINHOLD, G. and LAADING, G. (1983). Offshore processing – methanol plant on a semi-submersible platform. *Gastech* 82, p. 241

MEIROVITCH, L. (1975). *Elements of Vibration Analysis* (New York: McGraw-Hill)

MEISSNER, R.E. and SHELTON, D.C. (1992). Plant layout: minimizing problems. of plant layout. *Chem. Engng,* 99(4), 81

MEISTER, D. (1964). Methods of predicting human reliability in man–machine systems. *Hum. Factors,* 6, 621

MEISTER, D. (1967). *Tables for Predicting the Operational Performance of Personnel. Effectiveness of Display Subsystem Measurement and Prediction Techniques* (by J.J. Hornyak), *Rep. RADC-TR-67–292*, appendix A. Rome Air Dev. Center, Griffiths Air Force Base, NY

MEISTER, D. (1971). *Human Factors: Theory and Practice* (New York: Wiley)

MEISTER, D. (1987). Systems design, development, and testing. In Salvendy, G. (1987), *op. cit.*, p. 17

MEISTER, D. and RABIDEAU, G.F. (1965). *Human Factor Evaluation in System Development* (New York: Wiley)

MEJDELL, G.T. (1980). Modelling of pressure build-up in a runaway reactor. *Loss Prevention and Safety Promotion 3*, vol. 3, p. 1325

MEKKI, K., ATREYA, A., AGRAWAL, S. and WICHMAN, I. (1991).Wind-aided flame spread over charring and noncharring solids: an experimental investigation. *Combustion*, **23**, 1701

MELAMED, D. (1989). Fixing Superfund. *Chem. Engng*, **96**(11), 30

MELANCON, C.L. (1980). Improving emergency control and response systems. *Loss Prevention*, **13**, 43

MELCHERS, R.E. (1984). Human error in structural reliability assessment. *Reliab. Engng*, **7**, 61

MELCHERS, R.E. (1993a). On the treatment of uncertainty information in PRA. In Melchers, R.E. and Stewart, M.G. (1993), *op. cit.*, p. 13

MELCHERS, R.E. (1993b). Society, tolerable risk and the ALARP principle. In Melchers, R.E. and Stewart, M.G. (1993), *op. cit.*, p. 243

MELCHERS, R.E. and STEWART, M.G. (eds) (1993). *Probabilistic Risk and Hazard Assessment* (Rotterdam: Balkema)

MELDRUM, D.N. (1972). Aqueous film-forming foam – facts and fallacies. *Fire J.*, **66**(1), 57

MELDRUM, D.N. (1982). Fighting chemical fires with foam. In Fawcett, H.H. and Wood, W.S. (1982), *op. cit.*, p. 399

MELDRUM, D.N. and WILLIAMS, J.R. (1965). Dry chemical-compatible foam. *Fire J.*, **59**(6), 16

MELHAM, G.A., CROCE, P.A. and ABRAHAM, H. (1993). Data summary of the National Fire Protection Association's BLEVE tests. *Process Safety Prog.*, **12**, 76

MELHEM, G.A. (1993). Hazard reduction benefits from reduced storage temperature of pressurized liquids. *In Prevention and Control of Accidental Releases of Hazardous Gases*, (ed. V.M. Fthenakis, 411–37 (New York: Van Nostrand Reinhold)

MELINEK, S.J. (1974). A method of evaluating human life for economic purposes. *Accid. Analysis Prev.*, **6**, 114

MELINEK, S.J. (1989). Prediction of fire resistance of insulated steel. *Fire Safety J.*, **14**(3), 127

MELINEK, S.J. (1993a). The distribution of fire load. *Fire Safety J.*, **20**, 83

MELINEK, S.J. (1993b). Effectiveness of sprinklers in reducing fire severity. *Fire Safety J.*, **21**, 299

MELINEK, S.J. (1993c). Estimation of optimal safety levels. *Fire Safety J.*, **20**, 71

MELINEK, S.J. and THOMAS, P.H. (1987). Heat flow to insulated steel. *Fire Safety J.*, **12**(1), 1

MELIS, M., ZANI, F., LUISI, T. and PRELATI, M. (1983). Component event data collection in Enel IV-Caorso BWR nuclear power station for the European reliability data system. *Euredata 4*, paper 9.4

MELLAN, I. (1950). *Industrial Solvents,* 2nd ed. (New York: Reinhold)

MELLAN, I. (1957–). *Handbook of Solvents,* vols 1–3 (New York: Reinhold)

MELLAN, I. (1977). *Industrial Solvents Handbook,* 2nd ed. (Park Ridge, NJ: Noyes Data Corp.)

VAN MELLE, W. (1979). A domain-independent production-rule system for consultation programs. *Proc. Int. Joint Conf. on Artificial Intelligence 6*, p. 923

VAN MELLE, W. (1980). A domain-independent system that aids in constructing knowledge-based consultation programs. Ph.D. Thesis, Stanford Univ., Stanford, CA

VAN MELLE, W. (1981). *System Aids in Constructing Consultation Programs* (Ann Arbor, MI: Univ. of Michigan Press)

VAN MELLE, W., SHORTLIFFE, E.H. and BUCHANAN, B.G. (1981). EMYCIN: a domain-independent system that aids constructing knowledge-based consultation programs. *Machine Intelligence, Infotech State of the Art Rep. 9*, p. 281

MELLEY, B.W. (1989). Comparison of several computer hydraulic programs for IBM PCs and compatibles. *Fire Technol.*, **25**(4), 336

MELLISH, C.E. and LINNETT, J.W. (1953). The influence of inert gases on some flame phenomena. *Combustion*, 4, p. 407

MELLIT, B. (1995). Safety and loss control strategy and the London Underground safety case. *Safety Cases* (London: IBC Technical Services)

MELLOR, A.M., MANN, D.M., BOGGS, T.L., DICKINSON, C.W. and ROSE, W.E. (1987). Hazard initiation in energetic materials: status of technology assessment. *Combust. Sci. Technol.*, **54**, 203

MELLOR, A.M. *et al.* (1988). Research needs and plan for energetic material hazard mitigation. *Combust. Sci. Technol.*, **59**, 391

MELLOR, W. (ed.) (1972). *Casualties and Medical Statistics* (London: HM Stationery Office)

MELO, P.F.F.E., LIMA, J.E. and OLIVIERA, L.F. (1987). An initial appraisal of the event-tree methodology used in the Angra 1 nuclear power station probabilistic safety assessment. *Reliab. Engng*, **17**(3), 165

MELVILLE, G.S. (1994). When QRA is not quite the right approach. *J. Loss Prev. Process Ind.*, **7**, 387

MELVILLE, J. and FOSTER, G.G.A. (1975). A pictorial review of failures in conventional boiler plant. *Int. J. Pres. Ves. Piping*, **3**, 1

MELVILLE, W.K. and BRAY, K.N.C. (1979a). A model of the two-phase turbulent jet. *Int. J. Heat Mass Transfer*, **22**, 647

MELVILLE, W.K. and BRAY, K.N.C. (1979b). The two-phase turbulent jet. *Int. J. Heat Mass Transfer*, **22**, 279

MELVIN, A. (1966). Spontaneous ignition of methane-air mixtures at high pressure. *Combust. Flame*, **10**, 120, 129

MELVIN, A. (1969a. The development of cool flames in propane–oxygen mixtures (letter). *Combust. Flame*, **13**, 438

MELVIN, A. (1969b). Flame propagation on thermal ignition. *Combust. Flame*, **13**, 199

MELVIN, A. (1988). *Natural Gas: Basic Science and Technology* (Bristol: Adam Hilger)

MELVOID, R.W. and GIBSON, S.C. (1988). A guidance manual for selection and use of sorbents for liquid hazardous substance releases. *J. Haz. Materials*, **17**(3), 329

MEMEDLYAEV, Z.N., GLIKIN, M.A. and TYULPINOV, A.D. (1992). Efficiency and explosion safety of oxidation processes in fluidized catalyst bed. *Loss Prevention and Safety Promotion 7*, vol. 1, 20–1

MENASHE, J. and BERENBLUT, B.J. (1981). Fire hazards in chemical plants: an approach to their quantification. *The Recognition and Reduction of Ignition Hazards in the Chemical Industry* (London: Oyez)

MENCHER, S.K. (1967). Minimising waste in the petrochemical industry. *Chem. Engng Prog.*, **63**(10), 80

MENDEL, O. (1968). Cost comparisons for process piping. *Chem. Engng*, **75**(Jun. 17), 255

MENDENHALL, M.D. (1980). Shaft overlays proven effective. *Hydrocarbon Process.*, **59**(5), 191

MENDOZA, V.A., SMOLENSKY, V.G. and STRAITZ, J.F. (1993). Understand flame and explosion quenching speeds. *Chem. Engng Prog.*, **89**(5), 38

MENKES, J. (1962). Analytical solution of a flame with cylindrical symmetry. *Combustion*, **8**, p. 426

MENNEAR, J. (ed.) (1979). *Cadmium Toxicity* (New York: Dekker)

MENZEL, D.B. and SMOLKO, E.D. (1984). Interspecies extrapolation. In Rodricks, J.V. and Tardiff, R.G. (1984), *op. cit.*, p. 23

MENZIES, I.A. (1976). Pressure systems: materials of construction. *Course on Loss Prevention in the Process Industries.* Dept of Chem. Engng, Loughborough Univ. of Technol.

MENZIES, I.A. *et al.* (1965). *Corrosion and Protection of Metals* (London: Iliffe)

MENZIES, R.M. and STRONG, R. (1979). Some methods of loss prevention. *Chem. Engr, Lond.*, **342**, 151

MEPPELDER, F.M. (1977). The introduction of safety analysis reports for the chemical process industries in the Netherlands. *Loss Prevention and Safety Promotion 2*, p. 43

MERATLA, Z. (1990). Approach to a leak on an LNG tank bottom. *J. Haz. Materials*, **24**(1), 67

MERCALLI, G. (1902). *Boll. Soc. Sismologica Italiana*, **8**, 184

MERCER, A. (1988). Methods of validating models of dense gas dispersion. In Puttock, J.S. (1988b), *op. cit.*, p. 169

MERCER, A. and DAVIES, J.K.W. (1987). An analysis of the turbulence records from the Thorney Island continuous release trials. *J. Haz. Materials*, **16**, 21

MERCER, A. and NUSSEY, C. (1987). The Thorney Island continuous release trials: mass and flux balances. *J. Haz. Materials*, **16**, 9

MERCER, A. and PORTER, S.R. (1991). Land use planning the role of heavy gas dispersion modelling. *Fourth Symp. on Heavy Gases and Safety Analysis*, Wuppertal

MERCER, B.W., DAWSON, G.W., AMES, L.L., MCNEESE, J.A. and BAKER, E.G. (1978). Current methodology for disposal of spilled hazardous materials. *Control of Hazardous Material Spills 4*, p. 190

MERCER, J.R. (1969). Reliability of solenoid valves. *Safety and Failure of Components* (London: Instn Mech. Engrs), p. 89

MERCK COMPANY (1989). *The Merck Index of Chemicals and Drugs: An Encyclopaedia of Chemicals, Drugs and Biologicals*, 11th ed. (Rahway, NJ: Merck)

MERCK, E. (1889). *Merck's Index of Fine Chemicals and Drugs for the Materia Medica and the Arts* (Darmstadt: Merck)

MERCX, W.P.M. (1990). The effects of explosions on humans. *Europex Newsletter*, **13**, 2

MERCX, W.P.M. (1997). *Extended Modeling and Experimental Research into Gas Explosions*. Final Summary Report for EMERGE, EC Contract EV5V-CT-93-0274

MERCX, W.P.M., JOHNSON, D.M. and PUTTOCK, J.S. (1995). Validation of scaling techniques for experimental vapor cloud explosion investigations, *Process Safety Progr.*, **14**, p. 120,

MERCX, W.P.M., WEERHEIJM, J. and VERHAGEN, T.L.A. (1991). Some considerations on the damage criteria and safety distances for industrial explosions. *Hazards*, **XI**, p. 255

MERCX, W.P.M., VAN WINGERDEN, C.J.M. and PASMAN, H.J. (1992). Venting of gaseous explosions. *Major Hazards Onshore and Offshore*, p. 411

MERCX, W.P.M., VAN DEN BERG, A.C. and THE, H.G. (1993). Current research at TNO on vapour cloud explosion modelling. *Health, Safety and Loss Prevention*, p. 212

MERCX, W.P.M., VAN WEES, R.M.M. and OPSCHOOR, G. (1993). Current research at TNO on vapor cloud explosion modeling. *Process Safety Prog.*, **12**, 222

MERCX, W.P.M., VAN WINGERDEN, C.J.M. and PASMAN, H.J. (1993). Venting of gaseous explosions. *Process Safety Prog.*, **12**, 40

MEREWOOD, D.W. (1958). Industrial explosion protection. *Br. Chem. Engng*, **3**, 188

MERONEY, R.N. (1979a). Lift-off of buoyant gas initially on the ground. *J. Ind. Aerodyn.*, **5**, 1

MERONEY, R.N. (1979b). Physical modeling of atmospheric dispersion of heavy gases released at the ground from short stacks. *Proc. Tenth Int. Tech. Mtg on Air Pollution Modelling and Its Application* (NATO/CCMS), p. 689

MERONEY, R.N. (1981). Physical simulation of dispersion in complex terrain and valley drainage flow situations. In de Wispelaere, C. (1981), *op. cit.*, p. 489

MERONEY, R.N. (1982). Wind-tunnel experiments on dense gas dispersion. *J. Haz. Materials*, **6**(1/2), 85. See also in Britter, R.E. and Griffiths, R.F. (1982), *op. cit.* p. 85

MERONEY, R.N. (1984a). Transient characteristics of dense gas dispersion. Pt 1: A depth-averaged model. *J. Haz. Materials*, **9**(2), 139

MERONEY, R.N. (1984b). Transient characteristics of dense gas dispersion: Pt 2: Numerical experiments on dense cloud physics. *J. Haz. Materials*, **9**(2), 159

MERONEY, R.N. (1985). *Guidelines for Fluid Modeling of Dense Gas Cloud Dispersion. Rep. CER-85–86–RNM-18.* Dept of Civil Engng, Colorado State Univ., Fort Collins, CO

MERONEY, R.N. (1986). *Guideline for Fluid Modeling of LNG Dispersion. Support Document. Rep. CER-85–86–RNM-50.* Dept of Civil Engng, Colorado State Univ., Fort Collins, CO

MERONEY, R.N. (1987a). Guidelines for fluid modeling of dense gas cloud dispersion. *J. Haz. Materials*, **17**(1), 23

MERONEY, R.N. (1987b). Validation of fluid modeling techniques for assessing hazards of dense gas cloud dispersion. *J. Haz. Materials*, **15**(3), 377

MERONEY, R.N. (1991). Numerical simulation of the mitigation of HF cloud concentrations by means of vapor barriers and water spray curtains. *Modeling and Mitigating the Consequences of Accidental Releases*, p. 431

MERONEY, R.N. (1992). Dispersion in non-flat obstructed terrain and advanced modeling techniques. *Plant/Operations Prog.*, **11**, 6

MERONEY, R.N. and LOHMEYER, A. (1982). *Gravity Spreading and Dispersion of Dense Gas Clouds Released Suddenly into a Turbulent Boundary Layer. Rep. CER-82–83–RNM-AL-7.* Dept of Civil Engng, Colorado State Univ., Fort Collins, CO

MERONEY, R.N. and LOHMEYER, A. (1983). Statistical characteristics of instantaneous dense gas clouds released in an atmospheric boundary layer wind tunnel. *Boundary Layer Met.*, **28**, 1

MERONEY, R.N. and LOHMEYER, A. (1984). Prediction of propane cloud dispersion by a wind-tunnel-data calibrated box model. *J. Haz. Materials*, **8**(3), 205

MERONEY, R.N. and NEFF, D.E. (1980). Laboratory simulation of liquid natural gas vapor dispersion over land or water. In Cermak, J.E. (1980), *op. cit.*, vol. 2, p. 1139

MERONEY, R.N. and NEFF, D.E. (1981). Physical modeling of forty cubic meter LNG spills at China Lake, California. In de Wispelaere, C. (1981), *op. cit.*, p. 449

MERONEY, R.N. and NEFF, D.E. (1984). Wind tunnel simulation of cold gas spills. *Am. Soc. Mech. Engrs, Proc. 1984 Mtg*

MERONEY, R.N. and NEFF, D.E. (1985). Dispersion of vapor from liquefied natural gas spills – evaluation of simulation in a meteorological wind tunnel: five cubic metre China Lake spills. *J. Wind Engng Ind. Aerodyn.*, **10**, 1

MERONEY, R.N. and NEFF, D.E. (1986). Heat transfer and humidity effects during cold dense gas dispersion. In Hartwig, S. (1986), *op. cit.*, p. 169

MERONEY, R.N. and SEONG-HEE SHIN (1992). Depth-integrated dispersion model prediction of dense gas dispersion in the presence of barriers and water spray curtains. *Am. Soc. Mech. Engrs, Symp. on Measurement and Modeling of Environmental Flows*

MERONEY, R.N. and YANG, B.T. (1971). *Wind Tunnel Study on Gaseous Mixing due to Various Stack Heights and Injection Rates above an Isolated Structure. Rep. CER71–72RNM-BTY16.* Fluid Dynamics and Diffusion Lab., Colorado State Univ., Fort Collins, CO

MERONEY, R.N., CERMAK, J.E. and NEFF, D.E. (1976). Dispersion of vapour from LNG spills – simulation in a meteorological wind tunnel. *Third Symp. on Atmospheric Turbulence, Diffusion and Air Quality* (Boston, MA: Am. Met. Soc.), p. 243

MERONEY, R.N., NEFF, D.E., CERMAK, J.E. and MEGAHED, M. (1977). *Dispersion of Vapor from LNG Spills. Rep. CER-76–77–RNM-JEC-DEN-MM57.* Dept of Civil Engng, Colorado State Univ., Fort Collins, CO

MERONEY, R.N., LINDLEY, D. and BOWEN, A.J. (1980). Physical modeling of flow over complex terrain. In Cermak, J.E. (1980), *op. cit.*, vol. 2, p. 975

MERONEY, R.N., KOTHARI, K.M., NEFF, D.E. and ANDREIEV, G. (1983). *Model Study of Liquefied Natural Gas Vapor Cloud Dispersion with Water Spray Curtains. Rep. CER-82–83–RNM-KMK-DEN-GA41.* Dept of Civil Engng, Colorado State Univ., Fort Collins, CO

MERRIAM, R.L. (1972). *View Factor Between an Inclined Cylinder and a Plane – Numerical Technique* (report) (A.D. Little)

MERRICK, D. (1983). *Coal Combustion and Conversion Technology* (London: Macmillan)

MERRICK, R.C. (1986). A guide to selecting manual valves. *Chem. Engng*, **93**(17), Sep. 1, 52

MERRICK, R.D. and CIUFFREDA, A.R. (1983). Brittle fracture of a pressure vessel: study results and recommendations. *API Proc., Refining Dept, 48th Midyear Mtg*, **62**, 11

MERRICK, R.D. and MANTELL, C.L. (1965). Low-nickel stainless steel. *Chem. Engng*, **72**(Aug. 30), 144

MERRICKS, S.L. (1977). The anatomy of an incident in North Yorkshire. *Symp. Hazard 77 – Advances in the Safe Handling and Movement of Hazardous Substances,* Teesside Polytechnic, Middlesbrough

MERRIFIELD, R. (1985). The development of ball valves. *Chem. Engr, Lond.*, **418**, 36

MERRIFIELD, R. and ROBERTS, T.A. (1991). A comparison of explosion hazards associated with the transport of explosives and industrial chemicals with explosive properties. *Hazards*, **XI**, p. 209

MERRILL, J.W. and VALENTIN, J.E.C. (1942). Dust control in the plastics industry. *Trans. Am. Inst. Chem. Engrs*, **38**, 813

MERRIMAN, P.C. (1985). The Control of Industrial Major Accident Hazards Regulations 1984 – information to the public. *Chem. Ind.*, Jul. 1, 431

MERRISON, A.W. (1971). *Inquiry into the Basis of Design and Method of Erection of Steel Box Girder Bridges* (London: HM Stationery Office)

MERTE, H. and CLARK, J.A. (1964). Boiling heat transfer with cryogenic fluids at standard, fractional and near-zero gravity. *Trans. ASME, Ser. C*, **86**, 351

MERZ, H.A. and HALBRITTER, J.P. (1989). Computer-supported risk analysis for manufacturing, processing and storage in industries with explosive materials (EESP). *Loss Prevention and Safety Promotion 6*, p. 102.1

MERZHANOV, A.G. (1966). On critical conditions for thermal explosion of a hot spot. *Combust. Flame*, **10**, 341

MERZHANOV, A.G. (1967). Thermal explosion and ignition as a method for kinetic studies of exothermic reactions in the condensed phase. *Combust. Flame*, **11**, 201

MERZHANOV, A.G. and AVERSON, A.E. (1971). The present state of thermal ignition theory: an invited review. *Combust. Flame*, **16**, 89

MERZHANOV, A.G. and DUBOVITSKY, F.I. (1966). Present state of the theory of thermal explosions. *Russ. Chem. Revs*, **35**(4), 278

MERZHANOV, A.G., BARZYKIN, V.V. and GONTKOVSKAYA, V.T. (1963). The problem of a local thermal explosion. *Doklad. Acad. Nauk. USSR*, **48**(2), 380

MESLER, R. (1985). Flash vaporization from pools and jets. Private communication

MESLER, R.J. (1976). The Dow Chemical Company's distribution emergency/response system. *Control of Hazardous Material Spills 3*, p. 108

MESLIN, T.B. (1981). Assessment and management of risk in the transport of dangerous materials: the case of chlorine transport in France. *Risk Analysis*, **1**, 137

MESLOH, K. (1964). Air heater explosion. *Ammonia Plant Safety*, **6**, p. 39

MESNIAEFF, P.G. (1980). Timers are the key elements in process control systems. *Chem. Engng*, **87**(Dec. 1), 87

MESSER, J. (1979). Selecting and maintaining reciprocating compressor piston rods. *Chem. Engng*, **86**(May 21), 209

MESSING, A. (1980). Run a good safety/health program. *Hydrocarbon Process.*, **59**(10), 193

MESSINGER, M. and SHOOMAN, M.L. (1970). Techniques for optimum spares allocation: a tutorial review. *IEEE Trans Reliab.*, **R-19**, 156

DE MESTRE, N. (1990). *The Mathematics of Projectiles in Sport* (Cambridge: Cambridge Univ. Press)

MESZAROSE, E. (1981). *Atmospheric Chemistry* (Amsterdam: Elsevier)

METCALF AND EDDY INC. (1991). *Wastewater Engineering*, 3rd ed. (New York: McGraw-Hill)

METEOROLOGICAL OFFICE (1968). *Tables of Surface Wind Speed and Direction over the United Kingdom* (London: HM Stationery Office)

METEOROLOGICAL OFFICE (1972). *Meteorological Glossary*, 5th ed. (London: HM Stationery Office)

METEROLOGICAL OFFICE (1975). Map of mean and extreme temperature over the United Kingdom. *Climatology Memo 73* (London: HM Stationary Office)

METEOROLOGICAL OFFICE (1977). Private communication

METEOROLOGICAL OFFICE (1991). *Meteorological Glossary*, 6th ed. (London: HM Stationery Office)

METEOROLOGICAL OFFICE (c. 1994). *Climate of the British Isles. Leaflet ES4* (Bracknell)

METEOROLOGICAL RESEARCH INC. (1976). *Total Hydrocarbon Emission Measurements of Valves and Compressors at ARCO's Ellwood Facility. Rep. 911–115–1661*

METHNER, J.C. (1968). Hard sell for fire fighters. *Chem. Engng*, **75**(Feb. 12), 112

METROPOLIS, N. and ULAM, S. (1949). The Monte Carlo method. *J. Am. Statist. Soc.*, **44**(247), 335

METZGER, P.M. (1972). *The Atomic Establishment* (New York: Simon & Schuster)

METZGER, T.R., HANDERMANN, A.C., STOOKEY, D.J. and RYGG, A. (1985). On-site, on-demand nitrogen generation. *Plant/Operations Prog.*, **4**, 168

MEWIS, J. (1984). How much safety do we need in ChE education? *Chem. Engng Educ.*, Spring, 82

MEYER, A. (1938). *Les Gaz de Combat* (Paris: Charles-Lavauzelle)

MEYER, E. (1972). *Mechanical Seals*, 2nd ed. (London: Butterworths)

MEYER, E. (1989). *Chemistry of Hazardous Materials*, 2nd ed. (Englewood Cliffs, NJ: Prentice Hall)

MEYER, H.H. (1961). The selection of prime movers in the petroleum chemistry industry. *Chem. Engr, Lond.*, **155**, A37

MEYER, J. (1926). *Der Gaskampf und die chemischen Kampfstoffe*, 2nd ed. (Leipzig: S. Hirzel)

MEYER, J.W. and OPPENHEIM, A.K. (1971a). Coherence theory of the strong ignition limit. *Combust. Flame*, **17**, 65

MEYER, J.W. and OPPENHEIM, A.K. (1971b). On the shock-induced ignition of explosive gases. *Combustion*, **13**, 1153

MEYER, J.W., URTIEW, P.A. and OPPENHEIM, A.K. (1970). On the inadequacy of gasdynamic processes for triggering the transition to detonation. *Combust. Flame*, **14**, 13

MEYER, M.A. and BOOKER, J.M. (1990). *Eliciting and Analyzing Expert Judgment: a Practical Guide. Rep. NUREG/CR-5424.* Nucl. Regul. Comm., Washington, DC

MEYER, P.L. (1965). *Introductory Probability and Statistical Applications* (Palo Alto, CA: Addison-Wesley)

MEYER, M. (1943). *The Science of Explosives* (New York: Thomas Y. Crowell)

MEYER, R. (1973). *Explosivstoffe* (Weinhem: Verlag Chemie)

MEYER, R. (1977). *Explosives; 1981*, 2nd ed. (Weinheim: Verlag Chemie); 1987, 3rd ed. (Weinheim: VCH). See also Kohler, J. (1993)

MEYER, R.F. (1957). The impact of a shock wave on a moveable wall. *J. Fluid Mech.*, **3**, 309

MEYER, R.J. (1977). Solve vertical pump vibration problems. *Hydrocarbon Process.*, **56**(8), 145

MEYER, W.H. and CHURCH, F.W. (1961). Industrial hygiene aspects of mechanical operations in a petroleum industry. *Med. Bull.*, **21**(2), 256

MEYERS, G.J. (1976). *Software Reliability, Principles and Practice* (Chichester: Wiley)

MEYERS, R.A. (1984). *Handbook of Synfuels Technology* (New York: McGraw-Hill)

MEYERS, R.A. (ed.) (1986a). *Handbook of Chemicals Production Processes* (New York: McGraw-Hill)

MEYERS, R.A. (1986b). *Handbook of Petroleum Refining Processes* (New York: McGraw-Hill)

MEYERS, S. and SHANLEY, E.S. (1990). Industrial explosives – a brief history of their development and use. *J. Haz. Materials*, **23**(2), 183

VON MEYTHALLER, F. and GROSS, H. (1957). Acute ammonia poisoning. *Med. Klim.*, **52**, 165

MIANNAY, C.R. (1974). Improve bearing life with oil mist lubrication. *Hydrocarbon Process.* **53**(5), 113

MICHAELIS, M. (1986). Challenger and Chernobyl: why? *Interdisciplinary Sci. Rev.*, **11**, 324

MICHAELIS, P., MAVROTHALASSITIS, G. and PINEAU, J.P. (1992). Contribution to boilover and frothover quantification. *Loss Prevention and Safety Promotion 7*, vol. 2, 100–1

MICHAELSEN, R.H., MICHIE, D. and BOULANGER, A. (1985). The technology of expert systems. *Byte*, Apr., 303

MICHAELSON, M., TAITELMAN, U., and BURSTEIN, S. (1984). Management of crush syndrome. *Resuscitation*, **12**, 141

MICHAELSON, M., TAITELMAN, U., BSHOUTY, Z., BAR-JOSEPH, G. and BURSTEIN, S. (1984). Crush syndrome: experience from the Lebanon war 1982. *Israel J. Med. Sci.*, **20**, 305

MICHALOVIC, J.G., AKERS, C.K., BAIER, R.E. and PILIE, R.J. (1976). Optimization of universal gelling agent and development of means of applying to spilled hazardous materials. *Control of Hazardous Material Spills 3*, p. 259

MICHALOVIC, J.G., AKERS, C.K., KING, R.W. and PILIE, R.J. (1978). Development of means for applying multi-purpose gelling agent to spilled hazardous materials. *Control of Hazardous Material Spills 4*, p. 378

MICHALSKI, R.S., CARBONELL, J. and MITCHELL, T.M. (eds) (1983). *Machine Learning: An Artificial Intelligence Approach* (Palo Alto, CA: Tioga Press)

MICHEL, L., BEATTIE, R.D. and GOODGAME, T.H. (1954). Census of equipment scale up practice. *Chem. Engng Prog.*, **50**, 332

MICHELINI, M. (1978). High-pressure piping pulsations and vibration. In *Safety in Polyethylene Plants, 3*, 41

MICHELS, H.J. and RASHIDI, F. (1992). Shock temperature as a criterion for the detonability of LNG/LPG constituents. *Combust. Flame*, **91**, 364

MICHELS, H.J., MUNDAY, G. and UBBELOHDE, A.R. (1970). Detonation limits in mixtures of oxygen and homologous hydrocarbons. *Proc. R. Soc.*, Ser. A, **319**(1539), 461

MICHELS, H.J., RICHARDSON, S.M. and SHARIFI, T. (1988). Catastrophic failure of large liquid storage tanks – theory and experiment. *Preventing Major Chemical and Related Process Accidents*, p. 687

MICHELS, J.M. (1965). Computer evaluation of the safety fault tree modes. In *Boeing Company* (1965), *op. cit.*

MICHIE, D. (1974). *On Machine Intelligence* (Edinburgh: Edinburgh Univ. Press)

MICHIE, D. (ed.) (1979). *Expert Systems in the Micro Electronic Age* (Edinburgh: Edinburgh Univ. Press)

MICHIE, D. (1980). Expert systems. *Computer J.*, **23**, 369

MICHIE, D. (1981). Expert knowledge re-engineered. *Computer Bull.*, Jun., 6

MICHIE, D. (1982a). Bayes, Turing and the logic of corroboration. In Michie, D. (1982c), *op. cit.*, p. 38

MICHIE, D. (1982b). Heuristic search. In Michie, D. (1982c), *op. cit.*, p. 44

MICHIE, D. (ed.) (1982c). *Introductory Readings in Expert Systems* (New York: Gordon & Breach)

MICHIE, D. (1982d). *Machine Intelligence and Related Topics* (New York: Gordon & Breach)

MICHIE, D. (1983). The state of the art in machine learning. In Hayes, J.E. and Michie, D. (1983), *op. cit.*, p. 208

MICHOT, C., BIGOURD, J. and PINEAU, J.P. (1992). INERIS's action in the field of dangerous substances. *Loss Prevention and Safety Promotion 7*, vol. 2, 84–1

MIDDENDORF, W.H. (1969). *Engineering Design* (Boston, MA: Allyn & Bacon)

MIDDLEDITCH, B.S. (ed.) (1981). *Environmental Effects of Offshore Oil Production. The Buccaneer Gas and Oil Field Study* (New York: Plenum Press)

MIDDLETON, A.A. and LLOYD, P.A. (1984). Computer aided relief network design. *The Protection of Exothermic Reactors and Pressurised Storage Vessels* (Rugby: Instn Chem. Engrs), p. 75

MIDDLETON, A.H. (1976). Noise from chemical plant. *Chem. Engr, Lond.*, **306**, 115

MIDDLETON, C.J. (1989). *Piper Alpha Public Inquiry. Transcript, Days 148, 149*

MIDDLETON, G.V. (1966). Experiments on density and turbidity currents. 1, Motion of the head. *Can. J. Earth Sci.*, **3**, 523

MIDDLETON, J.C. and REVILL, B.K. (1983). The intensification of chemical reactors for fluids. *Instn Chem. Engrs, Research Mtg, Manchester* (Rugby: Instn Chem. Engrs).

MIDDLETON, J.F. (1983). Developments in flame detectors. *Fire Safety J.*, **6**, 175

MIDGLEY, P.M. (1989). The production and release to the atmosphere of 1,1,1-trichloroethane (methyl chloroform). *Atmos. Environ.*, **23**, 2663

MIE, G. (1908). *Ann. Phys.*, **25**, 377

MIEDEMA, H.M.E. and HAM, J.M. (1988). Odour annoyance in residential areas. *Atmos. Environ.*, **22**, 2501

MIEZE, G. (1970). Reliability analysis of complex systems with the aid of the fault tree method. *Kerntechnik*, **12**(9), 377

MIGIHON, R.D., RAGONESE, F.P. and THERIAULT, M.J. (1993). Development and implementation of a process safety management system. *Health, Safety and Loss Prevention*, p. 468

MIKASINOVIC, M. and DAUTOVICH, D.R. (1977). Designing steam transmission lines without steam traps. *Chem. Engng*, **84**(Mar. 14), 137

MIKASINOVIC, M. and MARCUCCI, P.A. (1984). Sizing pipe for external pressure. *Chem. Engng*, **91**(Apr. 30), 61

MIKASINOVIC, M. and TUNG, P.C. (1981). Sizing centrifugal pumps for safety service. *Chem. Engng*, **88**(Feb. 23), 83

MIKESELL, J.L., BUCKLAND, A.C., DIAZ, V. and KIVES, J.J. (1991). Evaporation of contained spills of multicomponent nonideal solutions. *Modeling and Mitigating the Consequences of Accidental Releases*, p. 103

MIKKELSEN, T., LARSEN, S.E. and THYKIERG-NIELSEN, S. (1984). Description of the Riso puff diffusion model. *Nucl. Technol.*, **67**, 56

MIKLOUCICH, F.J. and NORONHA, J.A. (1982). Heat transfer analysis of fire tests on water-filled drums. *Plant/ Operations Prog.*, **1**, 65

MILBURN, M.W. and CAMERON, I.T. (1992). *Planning for Hazardous Industrial Activities in Queensland* (report). Australian Ins. for Urban Studies, St Lucia, Queensland

MILBY, T.H. (1962). Hydrogen sulphide intoxication. Review of the literature and report of two unusual accidents resulting in two cases of non-fatal poisoning. *J. Occup. Med.*, **4**, 431

MILES, M.E. (1991). Valve and piping safety factors. *Hydrocarbon Process.*, **70**(8), 103

MILES, P.C. and GOULDIN, F.C. (1992). Mean reaction rates and flamelet statistics for reaction rate modelling in premixed turbulent flames. *Combustion*, **24**, 477

MILGRAM, J. and ERB, P.R. (1984). How floaters respond to subsea blowouts. *Petrol. Engr*, Jun., 64

MILHE, M. (1986). Water model simulation of hazardous gas diffusion in the atmosphere. *Loss Prevention and Safety Promotion 5*, p. 29–1

MILJURE, R. (1992). Infrared thermography finds process hot spots. *Chem. Engng Prog.*, **88**(3), 81

MILKE, J.A. (1986). Practical applications: procedure for quantitative analysis. *Fire Safety J.*, **11**(1), 135

MILL, J. (1985). Expert systems made simple. *Computing*, Feb. 21, 13

MILL, R.C. (1986). Some views on the management of safe operations. *Loss Prevention and Safety Promotion 5*, 16–1

MILL, R.C. (1992). *Human Factors in Process Operations* (Rugby: Instn Chem. Engrs)

MILL, T. (1980). Data needed to predict the environmental fate of organic chemicals. In Haque, R. (1980), *op. cit.*, p. 297

MILLAR, F.C. (1978). Use of turbo expanders with FCC. *Hydrocarbon Process.*, **57**(7), 111

MILLER, B.M. *et al.* (1986). *Laboratory Safety* (Washington, DC: Am. Soc. Microbiology)

MILLER, C. and FRICKER, C. (1993). Planning and hazards. *Progress in Planning*, **40**, 167

MILLER, C.A. (1979). Converting construction costs from one country to another. *Chem. Engng*, **86**(Jul. 2), 89

MILLER, D.E. (1976). *Occupational Safety, Health and Fire Index* (New York: Dekker)

MILLER, D.J. and WEBB, C.W. (1967). Lower flammability limits of hydrogen sulfide and carbon sulfide mixtures. *J. Chem. Engng Data*, **12**, 568

MILLER, D.K. (1979). Sizing dual-suction risers in liquid overfeed refrigeration systems. *Chem. Engng*, **86** (Sep. 24), 117

MILLER, D.R., BEGEMAN, S.R. and LINTNER, E.L. (1983). Current and future applications of on-line surveillance and monitoring systems in the petroleum industry. *Chem. Ind.*, Oct. 17, 782

MILLER, D.S. (1971). *Internal Flow Systems: A Guide to Losses in Pipe and Duct Systems;* 1990, 2nd ed. (Cranfield: Br. Hydromech. Res. Ass.)

MILLER, E.J., WILSON, K.D. and LEWIS, C.R. (1988). Expert system shells – do they deliver what they promise? *Chem. Engng Prog.*, **84**(10), 37

MILLER, F.J., MENZEL, D.B. and COFFIN, D.L. (1978). Similarity between man and laboratory animals in regional pulmonary deposition of ozone. *Environ. Res.*, **17**, 84

MILLER, G.M. (1972). Building the plant. *Chem. Engng Prog.*, **68**(2), 46

MILLER, H. (1973). *Medicine and Society* (Oxford: Oxford Univ. Press)

MILLER, I. and FREUND, J.E. (1965). *Probability and Statistics for Engineers* (Englewood Cliffs, NJ: Prentice-Hall)

MILLER, I.R. (1979). A code to build trust in FRP equipment. *Chem. Engng*, **86**(Jun. 18), 78

MILLER, J.A. (1970). Carcinogenesis by chemicals – an overview. *Cancer Res.*, **30**, 559

MILLER, J.A., MITCHELL, R.E., SMOOKE, M.D. and KEE, R.J. (1982). Toward a comprehensive chemical kinetic mechanism for the oxidation of acetylene: comparison of model predictions with results from flame and shock tube experiments. *Combustion*, **19**, 181

MILLER, J.E. and BLOOD, J.M. (1963). *Modern Maintenance Management* (New York: Am. Mgmt Ass.)

MILLER, J.T. (1964). *The Revised Course in Industrial Instrument Technology* (London: United Trade Press)

MILLER, J.W., SCHNEIDER, J. and VARCHOK, F. (1976). *Ship Operations in Restricted Waterways*. Science Appl. Inc., Rep. to Department of Commerce Maritime Administration

MILLER, K. (1994). An offshore gas field safety strategy and quantitative risk assessment model. *J. Loss Prev. Process Ind.*, **7**, 331

MILLER, M.C. (1990). Drag coefficients measurements for typical bomb and projectile fragments. Explosives Safety Board, Dept of Defense, *Proc. Twenty-Fourth Explosives Safety Seminar*, p. 2393

MILLER, M.J. (1974). Reliability of fire protection systems. *Loss Prevention*, **8**, 85

MILLER, M.J. (1977). Risk management and reliability. *Third Int. System Safety Conf.*, Washington, DC

MILLER, M.L. (ed.) (1977). *Toxic Substances Control Conf. 2* (Washington, DC: Govt Insts Inc.)

MILLER, M.T. (1971). Transportation emergency plans in Canada. *Loss Prevention*, **5**, 37

MILLER, N.H. (1968). Modern lubrication practice. *Chem. Engng*, **75**(Feb. 26), 155; (Mar. 11), 193; (Mar. 25), 105

MILLER, D.P. and SWAIN, A.D. (1987). Human error and human reliability. In Salvendy, G. (1987), *op. cit.*, p. 219

MILLER, P.D., HIBSHMAN, H.J. and CONNELL, J.R. (1958). The design of smokeless, nonluminous flares. *API Proc.*, **38**, 276

MILLER, R. (1968). Matching materials to temperatures. *Chem. Engng*, **75**(May 6), 210

MILLER, R.A., POPLE, H.E. and MYERS, J.D. (1982). INTERNISTI: an experimental computer-based diagnostic consultant for general internal medicine. *New England J. Med.*, **307**(8), 468

MILLER, R.E. (1959). Some factors governing the ignition delay of a gaseous fuel. *Combustion*, **7**, 417

MILLER, R.G. and KING, R.A. (1987). Mothballing your plant. *Chem. Engr, Lond.*, **434**, 41

MILLER, R.K. and MCINTYRE, P. (1987). *Non-Destructive Testing Handbook*, 2nd ed., vol. 5 (Philadelphia, PA: Am. Soc. Non-Destructive Testing)

MILLER, R.L. (1967). Solving practical problems of big ammonia plants. *Chem. Engng*, **74**(Jun. 5), 125

MILLER, R.L. *et al.* (1970). Process technology for sale or licence. *Chem. Engng*, **77**(Apr. 20), 114

MILLER, R.L. and HOWARD, W.B. (1971). Management tools in loss prevention. *Major Loss Prevention*, p. 203

MILLER, R.T. (1968). Construction and operating plant experience. *LNG Conf. 1*

MILLER, R.W. and CASERTA, L.V. (1971). Failure of a low temperature ammonia line. *Ammonia Plant Safety*, **13**, 5

MILLER, S.A. (1964). *Acetylene – Its Properties, Manufacture and Use* (London: Benn)

MILLER, S.A. (ed.) (1969). *Ethylene and its Industrial Derivatives* (London: Benn)

MILLER, S.A. and PENNY, E. (1960). Hazards in handling acetylene in chemical processes particularly under pressure. *Chemical Process Hazards*, **I**, p. 87

MILLER, W.A. (1993). How to choose a fluoropolymer. *Chem. Engng*, **100**(4), 163

MILLER, W.E. (ed.) (1965). *Digital Computer Applications to Process Control* (New York: Plenum Press)

MILLS, C.A. (1957). Respiratory and cardiac deaths in the Los Angeles smog. *Am. J. Med. Sci.*, Apr., 379

MILLS, D. (1990). *Pneumatic Conveying Design Guide* (Oxford: Butterworth-Heinemann)

MILLS, D.A. (1983). National hazardous waste management practices in western Europe. *Chem. Ind., Jun. 6*, 421

MILLS, J.S. and HAIGHTON, E.J. (1983). Prevention of electrostatic hazards associated with shipboard inert gas operation. *Electrostatics*, **6**, p. 13

MILLS, J.S. and OLDHAM, R.C. (1983). Electrostatic hazards on oil tankers. *Marine Eng. Rev.*, Jun., 10

MILLS, P.G. (1984). *Blowout Prevention* (Dordrecht: Reidel)

MILLS, P.L. and CHAUDHARI, R.V. (1997). Multiphase catalytic reactor engineering and design for pharmaceuticals and fine chemicals. *Catalysis Today* **37**, 367–404

MILLS, R.A. (1984). Implementation of Part II of the Control of Pollution Act. *Chem. Ind.*, Jul. 16, 496

MILLS, R.G. and HATFIELD, S.A. (1974). Sequential task performance: task module relationships, reliabilities and times. *Hum. Factors*, **16**, 117

MILLS, M.T. (1987). Modeling the release and dispersion of toxic combustion products from chemical fires. *Vapor Cloud Modeling*, p. 801

MILNE, I., AINSWORTH, R.A., DOWLING, A.R. and STEWART, A.T. (1986). *Assessment of the Integrity of Structures Containing Defects. Rep. R/H/R6, rev. 3.* Central Electricity Generating Board, Barnwell

MILNE, I., AINSWORTH, R.A., DOWLING, A.R. and STEWART, A.T. (1988a). Assessment of the integrity of structures containing defects. *Int. J. Pres. Ves. Piping*, **32**, 3

MILNE, I., AINSWORTH, R.A., DOWLING, A.R. and STEWART, A.T. (1988b). Background to and validation of CEGB Report R/H/R6 – Revision 3. *Int. J. Pres. Ves. Piping*, **32**, 105

MILNE, J. (1939). *Earthquakes and Other Earth Movements*, 7th ed. (rev. by A.L. Lee) (London: Kegan Paul)

MILNE-THOMSON, L.M. (1938–). *Theoretical Hydrodynamics* (London: Macmillan); 1941, 2nd ed. (New York: Macmillan); 1955, 3rd ed. (London: Macmillan); 1960, 4th ed.; 1968, 5th ed. (New York: Macmillan)

MILNE-THOMSON, L.M. (1947–). *Theoretical Aerodynamics* (New York: van Nostrand); 1958, 3rd ed. (London: Macmillan); 1966, 4th ed. (New York: Dover)

MILNES, M.H. (1971). Formation of 2,3,7,8-tetrachlorodibenzo-*p*-dioxin by thermal decomposition of sodium 2,4,5-trichlorophenate. *Nature*, **232**, 395

MIN, T.C., FAUSKE, H.K. and PETRICK, M. (1966). Effect of flow passages on two-phase critical flow. *Ind. Engng Chem. Fund.*, **5**, 50

MINAMI *et al.* (1960). Experimental study on cavitation in centrifugal pumps. *Bull. JSME*, **3**(9), 19

MINARICK, J.W. and KUKIELKA, C.A. (1982). *Precursors to Potential Severe Core Damage Accidents: 1969–1979: A Status Report. Rep. NUREG/CR-2497.* Nucl. Regul. Comm., Washington, DC

MINARICK, J.W., HARRIS, J.D., AUSTIN, P.N., CLETHER, J.W. and HAGEN, E.W. (1985). *Precursors to Potential Severe Core Damage Accidents: 1985: A Status Report. Rep. NUREG/CR-4674.* Nucl. Regul. Comm., Washington, DC

MINGLE, J.O. (1989). Inchemacy: an ethical challenge for chemical engineers. *Chem. Engng Prog.*, **85**(8), 19

MINGLE, J.O. and REAGAN, C.E. (1980). Legal and moral responsibilities of the engineer. *Chem. Engng Prog.*, **76**(12), 15

MINGLE, J.O., CHAWLA, T.O. and PERSON, L.W. (1977). Probabilistic safety analysis methods with application to nuclear and chemical systems. *Eleventh Loss Prevention Symp.* (New York: Am. Inst. Chem. Engrs)

MINICH, R.N. (1979). How to prepare operations manuals. *Hydrocarbon Process.*, **58**(11), 343

MINISTRY OF DEFENCE, EXPLOSIVES STORAGE AND TRANSPORT COMMITTEE (n.d.). *Leaflet no. 5*

MINISTRY OF DEFENCE (1972). *Explosive Hazard Assessment. Manual of Tests. Waltham Abbey Estab., SCC 3*

MINISTRY OF DEFENCE (1983). *MoD Practices and Procedures for Reliability and Maintainability, Part 2, Reliability Apportionment, Modelling and Calculation, Defence Std 0041 (Pt 2)/1* (London)

MINISTRY OF DEFENCE (1987). *Requirements for the Development of Safety-Critical Software for Airborne Systems. Interim Defence Std 0031/1* (London)

MINISTRY OF DEFENCE (1989a). *The Light Rescue Handbook* (London: HM Stationery Office)

MINISTRY OF DEFENCE (1989b). *Requirements for the Analysis of Safety Critical Hazards. Interim Defence Std 00–56* (London)

MINISTRY OF DEFENCE (1989c). *Requirements for the Procurement of Safety Critical Software in Defence Equipment. Interim Defence Std 00–55* (London)

MINISTRY OF THE ENVIRONMENT, CANADA (1983). *Portable Computing System for Use in Toxic Gas Emergencies. Rep. ARB-162–83–ARSP.* Min. of the Environ., Ont., Canada

MINISTRY OF THE ENVIRONMENT, CANADA (1986). *Review and Modification of Heavy Gas Model* (report). Min. of the Environ., Ont., Canada

MINISTRY OF HEALTH (1962). *Report of the Subcommittee on Accident and Emergency Services* (London: HM Stationery Office)

MINISTRY OF HOME SECURITY (n.d.a). *The Damage Caused to Houses by the Silvertown Explosion, 19th January, 1917. Rep. REN 190*

MINISTRY OF HOME SECURITY (n.d.b). *House Damage by HE Weapons Acting by Blast, Res. and Expts Dept, Rep. REN 214 rev.* (undated but not earlier than 1944)

MINISTRY OF HOME SECURITY (1945). *'Big Ben'. Public Record Office, File HO 191/198,* (Kew, London), *Rep. S104*

MINISTRY OF HOME SECURITY — see Appendix 28

MINISTRY OF HOUSING (1985). *Environmental Program for the Netherlands 1986–90* (The Hague)

MINISTRY OF HOUSING AND LOCAL GOVERNMENT (1959). *Cmnd 1027* (London: HM Stationery Office)

MINISTRY OF HOUSING AND LOCAL GOVERNMENT (1970). *Taken for Granted. Report of the Working Party on Sewage Disposal* (London: HM Stationery Office)

MINISTRY OF INTERNATIONAL TRADE AND INDUSTRY (1976). *A Report on the Experimental Results of Explosion and Fires of Liquid Ethylene Facilities* (Tokyo) (English translation: Insurance Tech. Bureau)

MINISTRY OF LABOUR (1962). *Organisation of Industrial Health Services. Health Services Safety, Health and Welfare Bklts, New Series 21* (London: HM Stationery Office)

MINISTRY OF LABOUR (1964). *Duties of Local Authorities. The Factories Act 1961*, 2nd ed. (London: HM Stationery Office)

MINISTRY OF LABOUR (1968). *Works Safety Committees in Practice – Some Case Studies* (London: HM Stationery Office)

MINISTRY OF POWER (n.d.a). *Test and Certification of the Flame-proof Enclosures of Electrical Apparatus. Testing Memo 4* (London)

MINISTRY OF POWER (n.d.b). *Test and Certification of Intrinsically Safe Apparatus and Circuits. Testing Memo 10* (London)

MINISTRY OF POWER (1944). *The Efficient Use of Fuel* (London: HM Stationery Office)

MINISTRY OF SOCIAL AFFAIRS (1972). *Experiments with acrylonitrile* (Voorburg, the Netherlands)

MINISTRY OF SOCIAL AFFAIRS (1975). *Experiments with chlorine* (Voorburg, the Netherlands)

MINISTRY OF SOCIAL AFFAIRS (1976). *Report on the Explosion at DSM at Beek 7 November 1975, Gas Explosion: Naphtha Cracker II* (Voorburg, the Netherlands)

MINISTRY OF SOCIAL AFFAIRS, DIRECTORATE GENERAL OF LABOUR (1979a). *Hazard and Operability Study. Why? When? How?* (Voorburg, the Netherlands)

MINISTRY OF SOCIAL AFFAIRS, DIRECTORATE GENERAL OF LABOUR (1979b). *Provisional Guidelines for the Safeguarding of Fuel Burning Installations with a Maximum Load of 600 kW in the Process Industry and Fired with Gaseous or Liquid Fuels* (Voorburg, the Netherlands)

MINISTRY OF SOCIAL AFFAIRS AND PUBLIC HEALTH (1968). *Report concerning an Inquiry into the Cause of the Explosion on 20 January 1968 at the Premises of Shell Nederland Raffinaderijin Pernis* (The Hague: State Publ. House)

MINISTRY OF SUPPLY (1952). *Textbook of Air Armaments*, ch. 12, Vulnerability of human targets to fragmenting and blast weapons (London) (now declassified)

MINISTRY OF TECHNOLOGY (1969a). *Report of the Committee of Enquiry on Pressure Vessels* (London: HM Stationery Office)

MINISTRY OF TECHNOLOGY (1969b). *A Study of Engineering Maintenance in Manufacturing Industry* (London)

MINISTRY OF TRANSPORT (n.d). *Carriage of Dangerous Goods and Explosives in Ships* (London: HM Stationery Office)

MINISTRY OF WORKS (1946). *Fire Grading of Buildings, pt 1, General Principles and Structural Precautions. Post-War Building Studies 20* (London: HM Stationery Office)

MINKOFF, G.J. and TIPPER, C.F.H. (1962). *Chemistry of Combustion Reactions* (London: Butterworths)

MINNING, C.P. (1977). Calculation of shape factors between rings and inverted cones sharing a common axis. *J. Heat Transfer,* **99**, 492

MINODA, K., FUJITANI, T., IRISAWA, T., OKUI, N., IINO, N., NAGAOKA, H., HANADA, N. and MAYANARI, T. (1986). The IHI SPB LNG system – application of advanced Al-alloy welding technology. *Gastech 85*, p. 285

MINORS, C. (1979). *The Safe Venting of Chemical Reactors* (Rugby: Instn Chem. Engrs)

MINORSKY, V.U. (1959). An analysis of ship collision with reference to protection of nuclear power plants. *J. Ship Res.,* **3**(2), 1

MINSKY, M. (1967). *Computation: Finite and Infinite Machines* (Englewood Cliffs, NJ: Prentice-Hall)

MINSKY, M. (ed.) (1968). *Semantic Information Processing* (Cambridge, MA: MIT Press)

MINSKY, M. (1975). A framework for representing knowledge. *The Psychology of Computer Vision* (ed. P. Winston) (New York: McGraw-Hill)

MINSKY, M.L. and PAPERT, S. (1969). *Perceptions* (Cambridge, MA: MIT Press)

MINTON, L.A. and JOHNSON, R.W. (1992). Fault tree faults. *Hazard Identification and Risk Analysis, Human Factors and Human Reliability in Process Safety,* p. 59

MINTSCHEW, M., MARKOW, C., NANOVSKA, S. and BELEW, B. (1972). Research into the explosion and ignition properties of dust deposits in forage plants, mills, and premises for the storage of bruise. *Int. Symp. on Dust and Explosion Risks in Mines and Industry* (Heidelberg: Berufsgenossenschaft der Chem. Ind.), p. 193

MINTZ, K.J. (1993). Upper explosive limit of dusts: experimental evidence for its existence under certain circumstances. *Combust. Flame,* **94**, 125

MIR, A. and SIDDIQUI, M.I.R. (1990). Vibration-free rod-baffle design of tube bundle for boiler feed water preheater. *Ammonia Plant Safety,* **30**, 122

MIRON, Y. (1980). Thermal decomposition of monomethylamine nitrate. *J. Haz. Materials,* **3**(4), 301

MIRZAI, A.R. (ed.) (1990). *Artificial Intelligence: Concepts and Applications in Engineering* (London: Chapman & Hall)

MISCHIATTI, M. and RIPAMONTI, B. (1985). Burst test yields vessel data. *Hydrocarbon Process.,* **64**(11), 83

MISONO, M. and OKUHARA, T. (1993). Solid superacid catalysts. *Chemtech* (November), 23–29

MISRA, K.B. (1992). *Reliability Analysis and Prediction* (Amsterdam: Elsevier)

MISRA, K.B. (1994). *New Trends in System Reliability Evaluation* (Amsterdam: Elsevier)

MISRA, K.B. and GADANI, J.P. (1982). Confidence limits assessment using sensitivity calculations. *Reliab. Engng,* **3**(4), 279

MISRA, P.K. (1985). Importance of concentration fluctuations in LNG and heavy gas modeling. *Heavy Gas Workshop* (Downsview, Ont.: Concord Sci. Corp.)

MISSEN, R.W. (1962). Pressure drop in vapour relief systems. *Chem. Engng,* **69**(Oct. 29), 101

MISSISSAUGA NEWS (1979). *Special Edition,* Nov. 12

MITAL, A. (1984a). Comprehensive maximum accepted weight of lift database for regular 8-hour work shifts. *Ergonomics,* **27**, 1127

MITAL, A. (1984b). Prediction of maximum weights of lift acceptable to male and female industrial workers. *J. Occup. Accid.,* **5**(4), 223

MITAL, A., KUO, T. and FAARD, H.F. (1994). A quantitative evaluation of gloves used with non-powered hand tools in routine maintenance tasks. *Ergonomics,* **37**, 333

MITANI, T. (1980). Asymptotic theory for extinction of curved flames with heat loss. *Combust. Sci. Technol.,* **23**, 93

MITANI, T. (1981). A flame inhibition theory by inert dust and spray. *Combust. Flame,* **43**, 243

MITANI, T. (1982). A study of thermal and chemical effects of heterogeneous flame suppressants. *Combust. Flame,* **44**, 247

MITANI, T. (1983). Flame retardant effects of CF_3Br and $NaHCO_3$. *Combust. Flame,* **50**, 177

MITANI, T. and NIIOKA, T. (1982). Comparison of experiments and theory on heterogeneous flame suppressants. *Combustion,* **19**, 869

MITCALF, J. (1975). Failure in start-up heater tube. *Ammonia Plant Safety* **17**, 123

MITCHELL, A.E. (1982). A comparison of short-term dispersion estimates resulting from various atmospheric stability classification methods. *Atmos. Environ,* **16**, 765

MITCHELL, A.R. and GRIFFITHS, D.F. (1980). *The Finite Difference Method in Partial Differential Equations* (Chichester: Wiley)

MITCHELL, C.M. and MILLER, R.A. (1983). Design strategies for computer based information displays in real-time control systems. *Hum. Factors,* **25**, 353

MITCHELL, C.W. and YANT, W.P. (1925). *Correlations of the Data Obtained from Refinery Accidents with a Laboratory Study of H_2S and its Treatment* (report). Bureau of Mines, Washington, DC

MITCHELL, E. (1975). *The Employer's Guide to the Law on Health, Safety and Welfare at Work* (London: Business Books)

MITCHELL, G.H. (1972). *Operational Research* (London: English Univ. Press)

MITCHELL, H.A. (1963). Ammonia tolerance of the Californian leaf-nosed bat. *J. Mammology,* **44**, 543

MITCHELL, N.D. (1951). New light on self-ignition. NFPA Quart., **45**(2), 3

MITCHELL, P.R. (1982). Peculiarities of gas tanker operations: crews and qualifications, safety and training. *Gastech 81,* p. 431

MITCHELL, R. (1982). Risk and hazard (letter). *Chem. Engr, Lond.,* **383/384**, 344

MITCHELL, R.A. (1967). *Introduction to Weibull Analysis. Rep. 3001.* Pratt and Whitney Aircraft, East Hartford, CT

MITCHELL, R.C., KIRSCH, M., HAMERMESH, C.L. and SINOR, J.E. (1972). Methods for plugging leaking chemical containers. *Control of Hazardous Material Spills,* **1**, 103

MITCHELL, R.C., VROLYK, J.J. and MELVOLD, R.W. (1976). Prototype system for plugging leaks in ruptured containers. *Control of Hazardous Material Spills,* **3**, 225

MITCHELL, R.C., VROLYK, J.J., COOK, R.L. and WILDER, I. (1978). System for plugging leaks of hazardous materials. *Control of Hazardous Material Spills,* **4**, 332

MITCHELL, R.L. (1971). Chlorine emergency procedures. *Loss Prevention,* **5**, 32

MITCHELL, R.L. (1972). A contingency plan for response to a chlorine spill. *Control of Hazardous Material Spills,* **1**, 71

MITCHELL, R.L. (1982). Chlorine. In Bennett, G.F., Feates, F.S. and Wilder, I. (1982), *op. cit.,* pp. 10–53

MITCHELL, T.J. and SMITH, G.M. (1931). *Casualties and Medical Statistics of the Great War* (London: HM Stationery Office)

MITCHELL, W.W. (1983). Earthquake design considerations for fixed offshore platforms. *Oil Gas J.,* **81**(Jun. 20), 87

MIT-GRI SAFETY AND RESEARCH WORKSHOP (1982). *LNG Fire – Combustion and Radiation, Workshop Proc.,* vol. 3 *GRI 82/0019.3* (Boston, MA: MIT)

MITLER, H.E. (1984). Zone modeling of forced ventilation. *Combust. Sci. Technol.,* **39**, 83

MITLER, H.E. (1991). Predicting the spread rates of fires on vertical surfaces. *Combustion,* **23**, 1715

MITLER, R.K. (ca. 1989). *Industrial Robot Handbook* (New York: van Nostrand Reinhold)

MITROVANOV, V.V. and SOLOUKHIN, R.I. (1964). The diffraction of multifront detonation waves. *Soviet Phys. Dokl.,* **9**, 1055

MITRUKA, B.M. and RAWNSLEY, H.M. (1977). *Clinal, Biochemical and Hematological Reference Values in Normal Experimental Animals* (New York: Masson)

MITSUI, R. and TANAKA, T. (1973). Simple models of dust explosion. Predicting ignition temperature and minimum explosive limit in terms of particle size. *Ind. Engng Chem. Proc. Des. Dev.,* **12**, 384

MITTLEMAN, A., and LIN, D. (1995) Inherently safer chemistry. *Chem. Ind.* (September 4), 694–96

MITTELMAN, A., LIN, D., CANNON, D. and HO, C., K. (1994). Inherently safe chemistry guide. *Preprints of Papers Presented at the 208th ACS Nat. Meet.,* Aug. 21–25, 1994, Washington DC, 271–272. Center for Great Lakes Studies, University of Wisconsin-Milwaukee, Milwaukee, WI: Division of Environmental Chemistry, American Chemical Society

MITTER, D.A. (1991). A methodology for quantifying expert system usability. *Hum. Factors,* **33**, 233

MIX, K.H. (1989). The evaluation of thermal instabilities using ARC and advanced DTA methods. *Runaway Reactions,* p. 176

MIXON, G.M. (1967). Chemical plant lighting. *Chem. Engng,* **74**(Jun. 5), 113

MIXTER, G. (1959). Thermal radiation burns beneath fabric systems. *Ann. N.Y. Acad. Sci,* **82**, 701

MIYAKE, A. and OGAWA, T. (1992). Non-ideal detonation properties of prilled ammonium nitrate. *Loss Prevention and Safety Promotion 7,* vol. 2, 85–1

MIYASAKI, K. and MIZUTANI, Y. (1975). Ignition of sprays by an incident shock. *Combust. Flame,* **25**, 177

MIZEREK, P. (1996). Disinfection techniques for water and wastewater. *The Nat. Environ. J.,* (January/February), 22–28

MIZNER, G.A. (1981). Large-scale liquefied gas pool fires. *Health and Safety Executive, Seminar on Thermal Radiation and Overpressure Calculations* (London: Health and Safety Executive)

MIZNER, G.A. and EYRE, J.A. (1982). Large-scale LNG and LPG pool fires. *The Assessment of Major Hazards* (Rugby: Instn Chem. Engrs), p. 147

MIZNER, G.A. and EYRE, J.A. (1983). Radiation from liquefied gas fires on water. *Combust. Sci. Technol.*, **35**, 33

MIZOGUCHI, F. (1983). PROLOG based expert system. *New Generation Computing*, **1**, 99

MIZSEY, P. and FONYO, Z. (1990). Toward a more realistic overall process synthesis – the combined approach. *Comput. Chem. Engng*, **14**, 1213

MIZUNO, T. and YOKOYAMA, O. (1986). On the vertical diffusion from a ground level source in a developing mixing layer. *Atmos. Environ.*, **20**, 21

MIZUTA, H., TAKAHASHI, K., KAMIMURA, K., YAMAGUCHI, T. and MASUDA, S. (1991). The fabrication reliability data analysis system DANTE-QC1. In Cannon, A.G. and Bendell, A. (1991), *op. cit.*, p. 145

MIZUTANI, Y. (1972a). Amplification of turbulence level by a flame and turbulent flame velocity. *Combust. Flame*, **19**, 203

MIZUTANI, Y. (1972b). Turbulent flame velocities in premixed sprays. Part II, Theoretical analysis. *Combust. Sci. Technol.*, **6**, 11

MIZUTANI, Y. and NAKAJIMA, A. (1973). Combustion of fuel vapor-drop-air systems. *Combust. Flame*, **20**, 343, 351

MIZUTANI, Y. and NISHIMOTO, T. (1972). Turbulent flame velocities in premixed sprays. Part I, Experimental study. *Combust. Sci. Technol.*, **6**, 1

MOBLEY, KEITH, (1999). *Root Cause Failure Analysis* (New York: Butterworth-Heinemann), ISBN 0-7506-7158-0

MODAK, A.T. (1975). Nonluminous radiation from hydrocarbon-air diffusion flames. *Combust. Sci. Technol.*, **10**, 245

MODAK, A.T. (1977). Thermal radiation from pool fires. *Combust. Flame*, **29**, 177

MODAK, A.T. (1978). *Radiation from Products of Combustion. Tech. Rep. 0A0E6, BU-1,* Factory Mutual Res. Corp., Norwood, MA

MODAK, A.T. (1978/79). Radiation from products of combustion. *Fire Res.*, **1**(b), 339

MODAK, A.T. (1981). The burning of large pool fires. *Fire Safety J.*, **3**(3), 177

MODAK, A.T. and CROCE, P.A. (1977). Plastic pool fires. *Combust. Flame*, **30**, 251

MODAK, A.T. and ORLOFF, L. (1980). Fires of insulations on tank exteriors. *Fire Technol.*, **16**, 118

MODARRES, M. and CADMAN, T. (1986). A method for alarm system analysis for process plants. *Comput. Chem. Engng*, **10**, 557

MODDREL, J.F.R. (1977). The CIA CHEMSAFE scheme – its principles and practice. *Symp. Hazards 77 – Advances in the Safe Handling and Movement of Hazardous Substances,* Teesside Polytechnic, Middlesbrough

MODER, J.J. and PHILLIPS, C.R. (1970). *Project Management with CPM and PERT,* 2nd ed. (New York: van Nostrand Reinhold)

MODEST, M.F. (1993). *Radiation Heat Transfer* (New York: McGraw Hill)

MODY, V. and JAKHETE, R. (1988). *Dust Control Handbook* (Park Ridge, NJ: Noyes Data Corp.)

MOEIN, G.J., SMITH, A.J. and STEWART, P.L. (1976). Follow-up study of the distribution and fate of polychlorinated biphenyls and benzenes in soil and ground water samples after an accidental spill of transformer fluid. *Control of Hazardous Material Spills,* **3**, 368

MOEIN, G.J. (1978). Magnitude of the chemical spill problem. A regional overview. *Control of Hazardous Material Spills,* **4**, 9

MOELWYN-HUGHES, E.A. (1961). *Physical Chemistry,* 2nd ed. (Oxford: Pergamon Press)

MOEN, I.O. (1982a). The influence of turbulence on flame propagation in obstacle environments. In Lee, J.H.S. and Guirao, C.M. (1982), *op. cit.*, p. 101

MOEN, I.O. (1982b). Large scale experiments. In Lee, J.H.S. and Guirao, C.M. (1982), *op. cit.*, p. 991

MOEN, I.O. (1993). Transition to detonation in fuel-air explosive clouds. *J. Haz. Materials*, **33**, 159

MOEN, I.O., DONATO, M., KNYSTAUTAS, R. and LEE, J.H. (1980). Flame acceleration due to turbulence produced by obstacles. *Combust Flame*, **39**, 21

MOEN, I.O., DONATO, M., KNYSTAUTAS, R., LEE, J.H. and WAGNER, H.G. (1980). Turbulent flame propagation and acceleration in the presence of obstacles. *Prog. Astronaut. Aeronaut.*, **75**, 33

MOEN, I.O., DONATO, M., KNYSTAUTAS, R. and LEE, J.H. (1981). The influence of confinement on the propagation of detonations near the detonability limits. *Combustion*, **18**, p. 1615

MOEN, I.O., LEE, J.H.S., HJERTAGER, B.H., FUHRE, K. and ECKHOFF, R.K. (1982). Pressure development due to turbulent flame propagation large-scale methane–air explosions. *Combust. Flame*, **47**, 31

MOEN, I.O., MURRAY, S.B., BJERKETVEDT, D., RINNAN, A., KNYSTAUTAS, R. and LEE, J.H. (1982). Diffraction of detonation from tubes into large fuel-air explosive cloud. *Combustion*, **19**, 635

MOEN, I.O., FUNK, J.W., WARD, S.A., RUDE, G.M. and THIBAULT, P.A. (1984). Detonation length scales for fuel-air explosives. In Bowen, J.R. *et al.,* (1984b), *op, cit.*, p. 55

MOEN, I.O., WARD, S.A., THIBAULT, P.A., LEE, J.H., KNYSTAUTAS, R., DEAN, T. and WESTBROOK, C.K. (1985). The influence of diluents and inhibitors on detonations. *Combustion*, **20**, 1717

MOEN, I.O., BJERKETVEDT, D., JENSSEN, A. and THIBAULT, A. (1985). Transition to detonation in a large fuel-air cloud. *Combust. Flame*, **61**, 285

MOEN, I.O., SULMISTRAS, A., HJERTAGER, B.H. and BAKKE, J.R. (1986). Turbulent flame propagation and transition to detonation in large fuel–air clouds. *Combustion*, **21**, 1617

MOEN, I.O., BJERKETVEDT, D., ENGEBRETSE, N.T., JENSSEN, A., HJERTAGER, B.H. and BAKKE, J.R. (1989). Transition to detonation in a flame jet. *Combust. Flame*, **75**, 297

MOGHISSI, A.A., MARLAND, R.E., CONGEL, F.J. and ECKERMAN, K.F. (1980). Methodology for environmental human exposure and health risk assessment. In Haque, R. (1980a), *op. cit.*, p. 471

MOHAN, V.K., BECKER, K.R. and HAY, J.E. (1982). Hazard evaluation of organic peroxides. *J. Haz. Materials*, **5**(3), 197

MOHLMAN, F.W. (1950). Waste disposal as a factor in plant location. *Chem. Engng Prog.*, **46**(7), 321

MOIENI, P., APOSTOLAKIS, G. and CUMMINGS, G.E. (1981). On random and systematic failures. *Reliab. Engng*, **2**(3), 199

MOIR, D.N. (1984). Size reduction. *Chem. Engng*, **91** (Apr. 16), 54

MOISIO, T. (1972). Brittle fracture in failed ammonia plant. *Metal Constr.*, Jan., 3

MOL, A. and WESTENBRINK, J.J. (1974). Steam cracker quench coolers. *Hydrocarbon. Process.*, **53**(2), 83

MOLE, N. and CHATWIN, P.C. (1987). Assessing and modeling variability in dispersing vapour clouds. *Vapor Cloud Modeling,* p. 779

VANDER MOLEN, R. and NICHOLLS, J.A. (1979). Blast wave initiation energy for the detonation of methane–ethane–air mixtures. *Combust. Sci. Technol.*, **21**, 75

MOLICH, K. (1980). Consider gas turbines for heavy loads. *Chem. Engng*, **87**(Aug. 25), 79

MOLINA, M. and ROWLAND, S. (1974). Stratospheric sink for chlorofluoromethanes: chlorine catalysed destruction of ozone. *Nature*, **249**(Jun. 28), 810

MOLLE, W. and WEVER, E. (1984). *Oil Refineries and Petrochemical Industries in Western Europe* (Aldershot: Gower)

MÖLLER, W., REDEKER, T. and SCHULTZ, P. (1989). CHEMSAFE – a database for safety characteristics. *Loss Prevention and Safety Promotion 6*, p. 105.1

MOLLERHOJ, J.B. and ROBERT, M. (1983). Accident investigation. In Parmeggiani, L. (1983), *op. cit.*, p. 18

MOLNAR, I.J. (1971). Foundations for elevated towers. *Hydrocarbon Process.*, **50**(4), 129

MOLSTAD, M.C. and GUNTHER, H. (1967). Hydrogen attack of carbon and low-alloy steels. *Ammonia Plant Safety*, **9**, p. 29

MOLTON, L.S. (1988). Apportionment of tort liability among multiple polluters in the USA. *J. Haz. Materials*, **18**(3), 229

MONGER, J.M., SELLO, H. and LEHWALDER, D.C. (1961). Explosion limits of liquid systems containing hydrogen peroxide, water, and oxygenated compounds. *J. Chem. Engng Data*, **6**(1), 23

MONGER, J.M., BAUMGARTNER, H.J., HOOD, G.C. and SANBORN, C.E. (1964). Detonations in hydrogen peroxide vapor. *J. Chem. Engng Data*, **9**(1), 124

MONIN, A.S. (1962). Empirical data on turbulence in the surface layer of the atmosphere. *J. Geophys. Res.*, **67**, 3103

MONIN, A.S. (1970). The atmospheric boundary layer. *Ann. Rev. Fluid Mech.*, **2**, 225

MONJI, N. and BUSINGER, J.A. (1972). Stability dependence of temperature, humidity and wind velocity variances in the atmospheric surface layer. *J. Met. Soc. Japan, ser. II*, **50**

MONIN, A.S. and OBUKHOV, A.M. (1954). Basic laws of turbulent mixing in the ground layer of the atmosphere. *Trans. Geophyz. Inst. Akad. Nauk. SSSR*, **24**(151), 163

MONIN, A.S. and YAGLOM, A.M. (1971). *Statistical Fluid Mechanics: Mechanics of Turbulence* (Cambridge, MA: MIT Press)

MONIZ, B.J. (1985). Performance of primary reformer materials of construction after 63,500 hours. *Ammonia Plant Safety*, **25**, 183

MONK, D. (1988). The fine art of safety audits. *Chem. Engng Prog.*, **84**(9), 34

MONK, R.G. and DAVIS, R.A. (1979). Noise aspects of venting from chemical plant. In Minors, C. (1979), *op. cit.*, paper 9

MONROE, E.S. (1968). Burning waste waters. *Chem. Engng*, **75**(Sep. 23), 215

MONROE, E.S. (1975). How to test steam traps. *Chem. Engng*, **82**(Sep. 1), 99

MONROE, E.S. (1976a). How to size and rate steam traps. *Chem. Engng*, **83**(Apr. 12), 119

MONROE, E.S. (1976b). Install steam traps correctly. *Chem. Engng*, **83**(May 10), 121

MONROE, E.S. (1976c). Selecting the right steam trap. *Chem. Engng*, **83**(Jan. 5), 129

MONROE, E.S. (1983a). Condensate recovery systems. *Chem. Engng*, **90**(Jun. 13), 117

MONROE, E.S. (1983b). Quicker, cheaper testing of incinerator performance. *Chem. Engng*, **90**(Feb. 21), 69

MONROE, E.S. (1985). Steam-trap capacities. *Chem. Engng*, **92**(Apr. 15), 73

MONROE, L.R. (1970). Process plant utilities. *Chem. Engng*, **77**(Dec. 14), 130

MUNROY, A.D. and MAJUL, V.M.G. (1977). Stop fires in EO plants. *Hydrocarbon Process.*, **56**(9), 175

MONSANTO CORPORATION (ca. 1990). *Procedure for Acoustic Emission Testing of Tanks/Vessels. Issue 6*

MONSANTO RESEARCH CORP. (1978). *Source assessment – Fugitive Hydrocarbon Emissions from Petrochemical Plants. Rep. EPA 600–4–78–004*

MONSON, R.R. (1980). *Occupational Epidemiology* (Boca Raton, FL: CRC Press)

MONTAGNA, J., LEONE, H., MELLIT, T., VECCHIETTI, A. and CERRO, R. (1987). SIMBAD: a process simulator linked to a DBMS – I. Comput. *Chem. Engng*, **11**, 63

MONTAGUE, T.J. and MACNEIL, A.R. (1980). Mass ammonia inhalation. *Chest*, **77**, 496

MONTET, A., PRZYDROZNY, M. and INQUIMBERT, R. (1977). LNG transportation by road tank trailers. *LNG Conf.*, **5**

MONTGOMERY, D.R. (1967). Integrated system for plant wastes combats stream pollution. *Chem. Engng*, **74**(Feb. 27), 108

MONTGOMERY, G. (1980). Perform pipe stress analysis on a minicomputer. *Hydrocarbon Processing*, **59**(12), 87

MONTGOMERY, G.J. (1991). Optimize process design via A* search. *Chem. Engng Prog.*, **87**(6), 70

MONTGOMERY, R. (1967). Centrifugal pumps troubleshooting. *Chem. Engng*, **74**(Apr. 24), 182

MONTGOMERY, T.L. *et al.* (1973). A simplified technique used to evaluate atmospheric dispersion from emissions from large power plants. *J. Air Poll. Control Ass.*, **23**, 395

MONTLE, J.F. and MAYHAN, K.G. (1974). Role of magnesium oxychloride cements in fire loss prevention. *Loss Prevention*, **8**, p. 52

MOODIE, G.C. and STUDDERT-KENNEDY, G. (1970). *Opinion, Public and Pressure Groups* (London: Allen & Unwin)

MOODIE, K. (1981). Experimental assessment of a full scale water spray barrier for dispersing dense gases. In Greenwood, D.V. (1981), *op. cit.*, paper 5

MOODIE, K. (1982). *Assessment of Failure Sequence for Volatile Liquids Stored Without Pressure Relief, with Particular Reference to Liquid Propane* (report). R&LS Div., Health and Safety Executive, Sheffield

MOODIE, K. (1985). The use of water spray barriers to disperse spills of heavy gases. *Plant/Operations Prog.*, **4**, 234

MOODIE, K. (1988). Experiments and modelling: an overview with particular reference to fire engulfment. *J. Haz. Materials*, **20**, 149

MOODIE, K., BILLINGE, K. and CUTLER, D.P. (1985). The fire engulfment of LPG storage tanks. *The Assessment and Control of Major Hazards* (Rugby: Instn Chem. Engrs), p. 87

MOODIE, K., COWLEY, L.T., DENNY, R.B., SMALL, L.M. and WILLIAMS, I. (1988). Fire engulfment tests on a 5-tonne LPG tank. *J. Haz. Materials*, **20**, 55

MOODIE, K. and EWAN, B.C.R. (1990). Jets discharging to atmosphere. *J. Loss Prev. Process Ind.*, **3**(1), 68

MOODIE, K. and JAGGER, S.F. (1987). Flow through pressure relief devices and the dispersion of discharge. *Hazards from Pressure* (Rugby: Instn Chem. Engrs), p. 215

MOODIE, T.W. (1971). Measurement comes to hazardous dust area classification. *ISA Trans.*, **10**(30), 224

MOODY, D.W. (1977). The new edition of ANSI B31.3 – some significant changes. *API Proc.*, Refining Dept, 42nd Midyear Mtg, **56**, 135

MOODY, F.J. (1965). Maximum flow rate of a single component two-phase mixture. *J. Heat Transfer*, **87**, 134

MOODY, F.J. (1966). Maximum two-phase vessel blowdown from pipes. *J. Heat Transfer*, **87**, 285

MOODY, F.J. (1969). *Prediction of Blowdown Thrust and Jet Forces. ASME Paper 69–HT-31*

MOODY, F.J. (1973). Time dependent pipe forces caused by blowdown and flow stoppage. *J. Fluids Engng*, 422

MOODY, F.J., WHEELER, A.J. and WARD, M.G. (1979). The role of various parameters in safety relief valve pipe forces. In Haupt, R.W. and Meyer, R.A. (1979), *op. cit.*, p. 97

MOODY, G.B. (1969). Mechanical design of a tall stack. *Hydrocarbon Process.*, **48**(9), 173

MOODY, G.B. (1970). Tubing: an ideal pipe support. *Hydrocarbon Process.*, **49**(4), 117

MOODY, G.B. (1972). How to design saddle supports for horizontal vessels. *Hydrocarbon Process.*, **51**(11), 157

MOODY, H.S. (1974). Optimisation and control of shutdowns on large continuous stream chemical plants using critical path analysis. *Plant Engr*, **18**(3), 39

MOODY, L.F. (1944). Friction factors for pipe flow. *Trans Am. Soc. Mech. Engrs*, **66**, 671

MOON, G.B. (1978). A law primer for the chemical engineer. *Chem. Engng*, **85**(Apr. 10), 114

MOON, P. and SPENCER, D.E. (1961). *Field Theory Handbook* (Berlin: Springer)

MOON, P. and TASKER, G. (1979). Continuous monitoring of energy in a large chemical plant. *Chem. Ind.*, Sep. 1, 585

MOON, R.A., MUNDY, S.C. and RICH, L.L. (1981). TVA ammonia from coal project: update. *Ammonia Plant Safety* **23**, 62

MOONEY, D.G. (1991). An overview of the Shell fluoro aromatics plant explosion. *Hazards XI*, p. 381

MOONEY, G.H. (1977). *The Valuation of Human Life* (London: Macmillan)

MOORE, A. (1984). Pressure-relieving systems. *Chem. Engr, Lond.*, **407**, 13; **410**, 5

MOORE, A. (1986). Selecting a flowmeter. *Chem. Engr, Lond.*, **424**, 39

MOORE, A. (1987). Design/construct decision support using expert systems. *Process Engng*, Jan., **25**

MOORE, A.D. (1954). In *Introduction to Electric Fields* (by W.E. Rogers) (1954), *op. cit.*, appendix

MOORE, A.D. (1973). *Electrostatics and its Applications* (New York: Wiley-Interscience)

MOORE, A.H. and ELONKA, S.M. (1971). *Electrical Systems and Equipment for Industry* (New York: van Nostrand Reinhold)

MOORE, C.A. (1992). *Quality Management System Audit* (London: HM Stationery Office)

MOORE, C.V. (1967). The design of barricades for hazardous pressure systems. *Nucl. Engng Des.*, **5**, 81

MOORE, D.A. (2002). Regulating inherent safety. *17th Annu. Int. Conf. and Workshop: Risk, Reliability, and Security*, Oct. 8–11, 2002, Jacksonville, FL (New York: American Institute of Chemical Engineers)

MOORE, D.A. and ROGERS, J.E.L. (2003). The challenge of inherent safety. *Cent for Chem Process Saf. 18th Annu. Int. Conf.: Managing Chemical Reactivity Hazards and High Energy Release Events*, Sep. 23–25, 2003, Scottsdale, Arizona, 121–128 (New York: American Institute of Chemical Engineers)

MOORE, D.J. (1969). The distribution of SO_2 emitted from tall chimneys. *Phil. Trans. R. Soc., Ser. A*, **265**, 245

MOORE, D.J. (1975). A simple boundary-layer model for predicting time mean ground-level concentrations of material emitted from tall chimneys. *Proc. Instn Mech. Engrs*, **189**, 33

MOORE, D.W. (1986). Halogenated agents and systems. In Cote, A.E. and Linville, J.L. (1986), *op. cit.*, p. 19–11

MOORE, H.F. and BALLINGER, L.H. (1970). The ddc-supervisory control computer in a centralised control room. *Am. Petrol. Inst., Div. of Refin., Thirty-Fifth Midyear Mtg*, Houston, TX

MOORE, J.A. (1973). Development of a low maintenance heat recovery exchanger. *Chem. Engng Prog.*, **69**(1), 43

MOORE, J.A. and HERB, S.M. (1983). *Understanding Distributed Process Control*, rev. ed. (Pittsburgh, PA: Instrum. Soc. Am.)

MOORE, J.C. (1960). Research reactor fault analysis. *Nucl. Engng*, **5**(52), Sep., 385

MOORE, J.C. (1966). Research reactor fault analysis. *Nucl. Engng*, **11**(118), Mar., 195; **11**(119), Apr., 306

MOORE, J.C. (1969). Electric motors in the CPI. *Loss Prevention*, **3**, p. 59. See also *Chem. Engng Prog.*, 1969, **65**(4), 36

MOORE, J.C. (1974). Fault analysis in complex plant. *Plant Engr*, **18**(3), 26

MOORE, J.C. (1975). Electric motor drives for centrifugal compressors. *Hydrocarbon Process.*, **54**(5), 133

MOORE, J.M. (1962). *Plant Layout and Design* (New York: Macmillan)

MOORE, K.L. and BIRD, D.B. (1965). How to reduce hydrogen plant corrosion. *Hydrocarbon Process.*, **44**(5), 179

MOORE, M.J. and VISCUSI, W.K. (1988). The quantity-adjusted value of life. *Economic Inquiry*, **XXVI**, 369

MOORE, M.J. and SIEVERDING, C.H. (1976). *Two-Phase Steam Flow in Turbines and Separators* (New York: McGraw-Hill)

MOORE, P.A.C. and REES, W.D. (1981). Forced dispersion of gases by water and steam. In Greenwood, D.V. (1981), *op. cit.*, paper 4

MOORE, P. (1979a). Characterisation of dust explosibility: comparative study of test methods. *Chem. Ind.*, Jul. 7, 430

MOORE, P.E. (1979b). Explosion suppression in industry. *Phys. Technol.*, **10**, 202

MOORE, P.E. (1980). Dust explosion hazard assessment. *Chemical Process Hazards VII*, p. 77

MOORE, P.E. (1981). Propagation and suppression of gas and dust explosions. Ph.D. Thesis, Univ. of Surrey

MOORE, P.E. (1982a). Industrial explosion protection. *Fire Int.*, **77**, 26

MOORE, P.E. (1982b). Industrial explosion suppression – system design criteria. *The Control and Prevention of Dust Explosions* (London: Oyez/IBC), paper 9

MOORE, P. (1984). Explosion suppression trials. *Chem. Engr, Lond.*, **410**, 23

MOORE, P. (1986). Towards large volume explosion suppression systems. *Chem. Engr, Lond.*, **430**, 43

MOORE, P.E. (1990). Industrial explosion protection – venting or suppression? *Process Safety Environ.*, **68B, 167**

MOORE, P.E. and BARTKNECHT, W. (1986). Extending the limits of explosion suppression systems. *Loss Prevention and Safety Promotion 5*, p. 63–1

MOORE, P.E. and COOPER, S.P. (1993). Developments in industrial explosion suppression. *Health, Safety and Loss Prevention*, p. 255

MOORE, P.E. and SIWEK, R. (2002). An update on the European explosion suppression and explosion venting standards. *Proc. Safety Prog.* **21**(1) 74

MOORE, R.B. (1967). Fouling is a safety problem. *Ammonia Plant Safety,* **9**, 62

MOORE, R.E. (1979). Selecting materials to meet environmental conditions. *Chem. Engng,* **86**(Jul. 2), 101; (Jul. 30), 91

MOORE, R.L. (1985). Adding real-time expert system capabilities to large distributed control systems. *Control Engng,* **32**(4), 118

MOORE, R.V. (ed.) (1972). *Nuclear Power* (London: Peter Peregrinus)

MOORE, S., SAMDANI, S., ONDREY, G. and PARKINSON, G. (1994). New roles for supercritical fluids. *Chem. Eng.,* (March), 32–5

MOORE, S.R. and WEINBERG, F.J. (1981). High propagation rates of explosions in large volumes of gaseous mixtures. *Nature,* **290**(Mar. 5), 39

MOORE, W.J. (1962). *Physical Chemistry,* 4th ed. (London: Longmans)

MOORE, W.W. (1963). Engineering foundations of tanks on soft soils for minimum cost. *API Proc., Div. of Refining,* **43**, 482

MOORES, C.W. (1972). Wastewater biotreatment. What it can and cannot do. *Chem. Engng,* **79**(Dec. 25), 63

MOORHOUSE, J. (1982). Scaling criteria for pool fires derived from large scale experiments. *The Assessment of Major Hazards* (Rugby: Instn Chem. Engrs), p. 165

MOORHOUSE, J. and CARPENTER, R.J. (1986). Factors affecting vapour evolution rates from liquefied gas spills. *Refinement of Estimates of the Consequences of Heavy Toxic Vapour Release* (Rugby: Instn Chem. Engrs), p. 4.1

MOORHOUSE, J. and PRITCHARD, M.J. (1982). Thermal radiation hazards from large pool fires and fireballs – a literature review. *The Assessment of Major Hazards* (Rugby: Instn Chem. Engrs), p. 123

MOORHOUSE, J., WILLIAMS, A. and MADDISON, T.E. (1974). An investigation of the minimum ignition energies of some C1 to C7 hydrocarbons. *Combust. Flame,* **23**, 203

MOOSEMILLER, M.D. and WHIPPLE, T. (2002). Inherent safety of dikes against catastrophic failure of storage tanks. *5th Annu. Symp.: Beyond Regulatory Compliance, Making Safety Second Nature,* Oct. 29–30, 2002, (College Station, TX, Track I: Inherent Safety Session. College Station, TX: Mary Kay O'Connor Process Safety Center)

MORAAL, J. (1992). Human factors in loss prevention. *Hazard Identification and Risk Analysis, Human Factors and Human Reliability in Process Safety,* p. 239

MORAIN, W.A. (1964). Reciprocating expansion engines with unlubricated cylinders. *Ammonia Plant Safety,* **6**, 27

MORAN, E.P. (1992). Explosive accident summary: World War II. *Proc. Twenty-Fifth Explosives Safety Seminar.* (Explosives Safety Board, Dept of Defense), p. 113

MORAN, P.A. (1984). *An Introduction to Probability Theory* (Oxford: Oxford Univ. Press)

MORAN, R.D. (1979). Prepare for an OSHA inspection. Hydrocarbon Process., **58**(10), 185

MORAND, P., CLAUDE, J. and HERBRETEAU, A. (1985). A new concept for large and safe LPG storage terminals. *Gastech 84,* p. 527

MORAY, N. (ed.) (1979). *Mental Workload. Its Theory and Measurement* (New York: Plenum Press)

MORAY, N. (1981). The role of attention in the detection of errors and the diagnosis of failures in man–machine

systems. In Rasmussen, J. and Rouse, W.B. (1980), *op. cit.,* p. 185

MORAY, N. and ROTENBERG, I. (1989). Fault management in process control: eye movements and action. *Ergonomics,* **32**, 1319

MORAY, N., DESSOUKY, M.I., KIJOWSKI, B.A. and ADAPATHYA, R. (1991). Strategic behaviour, workload and performance in task scheduling. *Hum. Factors,* **33**, 607

MORDACQ, GENERAL (1933). *Le Drame de l'Yser* (Paris: Editions Portique)

MORELAND, P.J. (1976). State of the art of electrical resistance and potential monitoring methods. *Diagnostic Techniques* (London: Instn Chem. Engrs)

MORELAND, P.J. and HINES, J.G. (1978). Corrosion monitoring ... selecting the right system. *Hydrocarbon Process.,* **57**(11), 251. See also Kane, L.A. (1980), *op. cit.,* p. 179

MORES, S., NOLAN, P.F. and O'BRIEN, G. (1994). Determination of the self-accelerating decomposition temperature (SADT) from thermal stability data generated using accelerating rate calorimetry. *Hazards,* **XII**, p. 609

MORETON, M.D. (1989). *Piper Alpha Public Inquiry. Transcript, Days 64, 65*

MORETTI, E.C. and MUKHOPADHYAY, N. (1993). VOC control: current practices and future trends. *Chem. Engng Prog.,* **89**(7), 20

MORGAN, C. (1990). *Deriving Programs from Specifications* (Englewood Cliffs, NJ: Prentice-Hall)

MORGAN, C.K.W.M. and SEATON, A. (1975). *Occupational Lung Diseases* (Philadelphia, PA: W.B. Saunders)

MORGAN, C.T., COOK, J.S., CHAPANIS, A. and LUND, M.W. (1963). *Human Engineering Guide to Equipment Design* (New York: McGraw-Hill)

MORGAN, D.L. (1984). *Dispersion Phenomenology of LNG Vapor in the Burro and Coyote LNG Spill Experiments. Rep. UCRL-91741.* Lawrence Livermore Nat. Lab., Livermore, CA

MORGAN, D.L., MORRIS, L.K. and ERMAK, D.L. (1983). *SLAB: a Time-dependent Computer Model for the Dispersion of Heavy Gases Released in the Atmosphere. Rep. UCRL-53383.* Lawrence Livermore Nat. Lab., Livermore, CA

MORGAN, D.L., KANSA, E.J. and MORRIS, L.K. (1984). Simulations and parameter variation studies of heavy gas dispersion using the slab model. In Ooms, G. and Tennekes, H. (1984), *op. cit.,* p. 83

MORGAN, D.L. *et al.* (1984). *Phenomenology and Modelling in Liquefied Natural Gas Vapour Dispersion, REP. UCRL-53581.* Lawrence Livermore Nat. Lab., Livermore, CA

MORGAN, G.O. and REED, J.D. (1965). Storage of liquefied ammonia in large tanks at atmospheric pressure. *Ammonia Plant Safety,* **7**, p. 33

MORGAN, J. (1992). Towards a stronger safety culture. *Health and Safety in the Offshore Oil and Gas Industries* (Rugby: Instn Chem. Engrs)

MORGAN, J.F. (1974). Establishing in-plant standards of exposure to toxic materials in chemical plant design. *Loss Prevention,* **8**, 35

MORGAN, J.M. (1985). Cost effective fault tree analysis for safer plants. *Process Engng,* **66**(1), 28

MORGAN, J.M. and ANDREWS, J.D. (1984). Assessment of safety systems using fault tree analysis. Paper presented at *Instn Gas Engrs, Fiftieth Autumn Mtg* (London)

MORGAN, N. (1992a). Contaminated land: a problem in everybody's backyard. *Chem. Engr, Lond.,* **529**, s19

MORGAN, N. (1992b). Knowledge is power. *Chem. Engr, Lond.,* Mar. 12, 17

MORGAN, P. (1961). *Glass Reinforced Plastics* (London: Iliffe)

MORGAN, R.D. (1979). Prepare for an OSHA inspection. *Hydrocarbon Process.*, **58**(10), 185

MORGAN, S.W. (1992). Use process integration to improve process designs and the design process. *Chem. Engng Prog.*, **88**(9), 62

MORGAN, V.T., COLLOCOTT, S.J. and MORROW, R. (1981). The electrification of operating portable CO_2 fire extinguishers. *J. Electrostatics*, **9**, 201

MORGENEGG, E.E. (1978). New design rules for frangible tanks. *Hydrocarbon Process.*, **57**(5), 167

MORGENEGG, E.E. (1982). Tank overpressure: an uplifting experience. *API Proc., Refining Dept, 47th Midyear Mtg*, **61**, 188

MORGESTER, J.J., FRISK, D.L., ZIMMERMANN, G.L., VINCENT, R.C. and JORDAN, G.H. (1979). Control of emissions from refinery valves and flanges. *Chem. Engng Prog.*, **75**(8), 40

MORIARTY, F. (1986). Measurement and prediction of bioaccumulation in aquatic organisms. *Chem. Ind.*, Nov. 3, 737

MORIKAWA, T. (1976). An improved sodium carbonate- based extinguishant for sodium fires. *Fire Technol.*, **12**, 124

MORITZ, A.R. and HENRIQUES, F.C. (1947). *Am. J. Path.*, **23**, 695

MORLEY, P. and HEASMAN, W. (1987). Selecting a cryogenic valve. *Chem. Engr, Lond.*, **433**, 13

MORLEY, P.G. (1989a). Sizing pressure safety valves for flashing liquid duty. *Chem. Engr, Lond.*, **465**, 47. See also Selmer-Olsen, S. (1992), *op. cit.*, p. 234

MORLEY, P.G. (1989b). Sizing pressure safety valves for gas duty. *Chem. Engr, Lond.*, **463**, 21

MORLOCK, W.J. (1978). Inline vs horizontal . . . which is best? *Hydrocarbon Process.*, **57**(6), 99

MORONEY, M.J. (1956). *Facts from Figures*, 3rd ed. (Harmondsworth: Penguin)

MOROS, A., HOWELLS, P. and RYALL, D. (1993). Gas-freeing techniques for crude oil storage tanks during cleaning. *Health, Safety and Loss Prevention*, p. 242

MOROZZO, COUNT (1795). *The Repertory of Arts and Manufactures*, 2, 416

MORRAN, D. (1993). How to design stressless steam lines. *Chem. Engng*, **100**(10), 88

MORRIS, A. (1986). Time for a fresh approach to plant instrumentation. *Process Engng*, Jan., **32**

MORRIS, A.J. and CALLADINE, C.R. (1971). Simple upper-bound calculations for the indentation of cylindrical shells. *Int. J. Mech. Sci.*, **13**, 331

MORRIS, A.J., MONTAGUE, G.A. and WILLIS, M.J. (1994). Artificial neural networks. Studies in process modelling and control. *Chem. Engng Res. Des.*, **72A**, 3

MORRIS, B.C. and STARILE, D.J. (1993). Process safety management program considerations to reduce future risk management plan efforts. *Process Safety Management*, p. 1

MORRIS, D.H.A. (1974). Fawley compressor house: gas dispersal and explosion venting. *Loss Prevention and Safety Promotion 1*, p. 369

MORRIS, G. (1955–). Testing the reliability of machines. Engineering, 1955, Nov. 25; 1957, Jul. 12

MORRIS, H. (1949). Ignition of gas mixtures by electric sparks. *Combustion*, **3**, 361

MORRIS, H. (1977). Inspect tank cars before shipping them. *Chem. Engng*, **84**(Nov. 7), 109

MORRIS, M., MILES, A. and COOPER, J. (1994). Quantification of escalation effects in offshore quantitative risk assessment. *J. Loss Prev. Process Ind.*, **7**, 337

MORRIS, N.M. and ROUSE, W.B. (1985). Review and evaluation of empirical research in troubleshooting. *Hum. Factors*, **27**, 503

MORRIS, P.J.T. (1983). The industrial history of acetylene: the rise and fall of a chemical feedstock. *Chem. Ind.*, Sep. 19, 710

MORRIS, P.W.G. and HOUGH, G.H. (1987). *The Anatomy of Major Projects* (Chichester: Wiley)

MORRIS, R. (1976). The role of simulation in tracing and loss prevention. *Diagnostic Techniques* (London: Instn Chem. Engrs)

MORRIS, S.D. (1988a). Choking conditions for flashing one-component flows in nozzles and valves – a simple estimation method. *Preventing Major Chemical and Related Process Accidents*, p. 281

MORRIS, S.D. (1988b). *Offshore Relief Systems: Some Guidelines in Designing for Two-phase Flow. Rep. ETU-027.* Energy Technol. Unit, Heriot-Watt Univ., Edinburgh

MORRIS, S.D. (1990a). Discharge coefficients for choked gas-liquid flow through nozzles and orifices and applications to safety devices. *J. Loss Prev. Process Ind.*, **3**(3), 303

MORRIS, S.D. (1990b). Flashing flow through relief lines, pipe breaks and cracks. *J. Loss Prev. Process Ind.*, **3**(1), 17

MORRIS, S.D. and WHITE, G. (1984). The vapour generation term for flashing flow with particular reference to chlorine and ethylene. *The Protection of Exothermic Reactors and Pressurised Storage Vessels* (Rugby: Instn Chem. Engrs), p. 173

MORRISH, M.G. (1986). Software and hardware aspects of industrial controller implementation. In Warwick, K. and Rees, D. (1986), *op. cit.*, p. 267

MORRISON, D.L. and DUNCAN, K.D. (1988). Strategies and tactics in fault diagnosis. *Ergonomics*, **31**, 761

MORRISON, D.M. (1987). Prestressed concrete/steel storage tanks for liquefied natural gas storage in the Arabian Gulf. *Gastech 86*, p. 409

MORRISON, D.M. and MARSHALL, H.T. (1970). Frost heaves and storage vessel foundations. *Ammonia Plant Safety*, **12**, p. 11

MORRISSETTE, M.D. (1979). Hazard evaluation of chemicals for bulk marine shipment. *J. Haz. Materials*, **3**(1), 33

MORRISON, R. and ROSS, D. (1978). Monitoring for groundwater contamination at hazardous waste disposal sites. *Control of Hazardous Material Spills*, 4, 281

MORRISON, R.V. (1984). Analyzing batch process cycles. *Chem. Engng*, **91**(Jun. 25), 175

MORRISON, W.G. (1978). Improving the reliability of electrical distribution systems. *Chem. Engng*, **85**(Sep. 25), 102

MORROW, C.T. (1963). *Shock and Vibration Engineering* (New York: Wiley)

MORROW, D.R. (1968). Performance curves in plant operation. *Chem. Engng Prog.*, **64**(12), 56

MORROW, L.C. (ed.) (1957). *Maintenance Engineering Handbook* (New York: McGraw-Hill)

MORROW, N.L., BRIEF, R.S. and BERTRAND, R.R. (1972a). Air sampling and analysis. *Chem. Engng*, **79**(May 8), 125

MORROW, N.L., BRIEF, R.S. and BERTRAND, R.R. (1972b). Sampling and analyzing air pollution sources. *Chem. Engng*, **79**(Jan. 24), 84

MORROW, R.S. (1975). Why use minicomputer systems for vibration monitoring? *Hydrocarbon Process*, **54**(10), 125

MORROW, T.B. (1980). A wind tunnel investigation of the dispersion of chemical vapors emitted from marine chemical carriers. In Cermak, J.E. (1980), *op. cit.*, vol. 2, p. 1127

MORROW, T.B. (1982a). *Development of an Improved Pipeline Breakflow Model. Rep. SWRI-6917/3.* Southwestern Res. Inst., San Antonio, TX

MORROW, T.B. (1982b). *Prediction of Flammable Vapour Contours for Ground Level Releases of Negatively Buoyant Gases and Vapours* (paper). Southwestern Res. Inst., San Antonio, TX

MORROW, T.B. (1982c). When an LPG line breaks, how does product release/disperse? *Pipeline Gas J.,* **290**(7), 31

MORROW, T.B. (1985). *Investigation of the Hazards Posed by Chemical Vapor Releases in Marine Operations – Phase II. SWRI Proj. 06-5686.* Southwestern Res. Inst., San Antonio, TX

MORROW, T.B., ASTLEFORD, W.J., BASS, R.L. and MAGOTT, R.J. (1980). The dispersion of chemical vapours emitted from marine chemical carriers. *Loss Prevention and Safety Promotion 3,* vol. 3, p. 1040

MORROW, T.B., BASS, R.L. and LOCK, J.A. (1982). An LPG pipeline break flow model. *Am. Soc. Mech. Engrs, Fifth Ann. Energy Sources Technology Conf., ASME paper I00149 (New York)*

MORROW, T.B., ASTLEFORD, W.J., MAGOTT, R.J. and BASS, R.L. (1983). Analysis and measurement of chemical vapour concentration during cargo transfer operations. *Loss Prevention and Safety Promotion 4,* p. C22

MORROW, T.B., BUCKINGHAM, J.C. and DODGE, F.T. (1984). Water channel tests of dense plume dispersion in a turbulent boundary layer. In Ooms, G. and Tennekes, H. (1984), *op. cit.,* p. 323

MORSE, A.R. (n.d.). See Cross, J. and FARRER, D. (1982), *op. cit.,* p. 39

MORSE, A.R. (1958). *Investigation of the Ignition of Grain Dust Clouds by Mechanical Sparks. Rep. NRC 4968 ERB 494.* Nat. Res. Coun. of Canada

MORSE, P.M. (1936). *Vibration and Sound* (New York: McGraw-Hill); 1976, 2nd ed. (New York: Am. Inst. Phys)

MORT, E. (1965). A nurse's role in chemical plant safety. In Fawcett, H.H. and Wood, W.S. (1965), *op. cit.,* p. 240

MORTON, A.J. (1973). Engineering and administration in advanced plant projects. *Chem. Engr, Lond.,* **278**, 457

MORTON, B.R. (1959). Forced plumes. *J. Fluid Mech.,* **5**, 151

MORTON, B.R. (1961). On the momentum–mass flux diagram for turbulent jets, plumes and wakes. *J. Fluid Mech.,* **101**, 10

MORTON, B.R. (1965). Modeling fire plumes. *Combustion,* **10**, 973

MORTON, B.R., TAYLOR, G.I. and TURNER, J.S. (1956). Turbulent gravitational convection from maintained and instantaneous sources. *Proc. R. Soc., Ser. A,* **234**, 1

MORTON, D.S. (1962). Heat exchangers dominate process heat transfer. *Chem, Engng,* **69**(Jun. 11), 170

MORTON, F. (1970). *Report of the Inquiry into the Safety of Natural Gas as a Fuel* (London: HM Stationery Office)

MORTON, F. (1975). *Assessment of evidence. Rep. to Flixborough Court of Inquiry* (Manchester: Univ. of Manchester Inst. Sci. and Technol.)

MORTON, J. and HENDRY, A.W. (1973). A theoretical investigation of the lateral strength of brick walls with precompression. *Proc. Br. Ceram. Soc.,* **21**, 181

MORTON, J., DAVIES, S.R. and HENDRY, A.W. (1971). The stability of load-bearing brickwork structures following accidental damage to a major bearing wall or pier. *Proc. Second Int. Brick Masonry Conf.* (Br. Ceramic Ass.)

MORTON, V.M. and NETTLETON, M.A. (1977). Pressures and their venting in spherical expanding flames. *Combust. Flame,* **30**, 111

MORTON, W.I. (1976). Safety techniques for workers handling hazardous materials. *Chem. Engng,* **83**(Oct. 18), 127

MOSCHINI, G.P. and SCHROEDER, E.H. (1968). Tips on starting up and installing centrifugal compressors. *Loss Prevention,* **2**, 74

MOSER, J., HARGENS, D. and WILDER, I. (1976). Sources of information and assistance for hazardous material spills. *Control of Hazardous Material Spills,* **3**, 88

MOSER, J.H. (1991). CADM: a computer program for modeling the dispersion of heavy gas in air (poster item). *Modeling and Mitigating the Consequences of Accidental Releases,* p. 689

MOSER, J.H., BLEWITT, D.N., STEINBERG, K.W. and PETERSEN, R.L. (1991). HF-DEGADIS simulation of dense gas dispersion from a wind tunnel-modeled oil refinery. *Modeling and Mitigating the Consequences of Accidental Releases,* p. 289

MOSER, W.A., MOEL, M.P. and HECKARD, J.M. (1976). Risk analysis and oil spill expectation for deepwater ports. *Eighth Ann. Conf. on Offshore Technology,* p. 285

MOSEY, D. (1990). *Reactor Accidents. Nuclear Safety and the Role of Institutional Failure* (Oxford: Butterworth-Heinemann)

MOSIG, U. (1977). Safety by means of instrumentation and control equipment with examples from petrochemical plants. *Loss Prevention and Safety Promotion 2,* p. 415

MOSKOWITZ, D.B. (1986). Aerial surveillance upheld in court. *Chem. Engng,* **93**(Jul. 21), 28

MOSKOWITZ, L. (1986). Rule-based programming. *Byte,* **11**(2)

MOSLEH, A. and APOSTOLAKIS, G. (1982). Models for the use of expert opinions. *Workshop on Low-Probability/High Consequence Risk Analysis.* Soc. for Risk Analysis, Arlington, VA

MOSLEH, A. and APOSTOLAKIS, G. (1985). The development of a generic data base for failure rates. *Int. ANS/ENS Topical Mtg on Probabilistic Safety Methods and Applications,* San Francisco, CA

MOSLEH, A. and APOSTOLAKIS, G. (1986). The assessment of probability distributions from expert opinions with an application to seismic fragility curves. *Risk Analysis,* **6**, 447

MOSLEH, A., BIER, V.M. and APOSTOLAKIS, G. (1987). *Methods for the Elicitation and use of Expert Opinion in Risk Assessment. Phase 1, a Critical Evaluation and Directions for Future Research. Rep. NUREG/CR-4962.* Nucl. Regul. Comm., Washington, DC

MOSLEH, A., BIER, V.M. and APOSTOLAKIS, G. (1988). A critique of current practice for the use of expert opinions in probabilistic risk assessment. *Reliab. Engng System Safety,* **20**, 63

MOSLEH, A. *et al.* (1988). *Procedures for Treating Common Cause Failures in Safety and Reliability Studies. Rep. EPRI NP-5613.* Elec. Power Res. Inst., Palo Alto, CA

MOSLEY, W.H. and BUNGEY, J.H. (1982). *Reinforced Concrete Design,* 2nd ed. (London: Macmillan)

MOSS, A. (1977). The disposal of effluent to tidal waters by pipeline. *Chemical Engineering in a Hostile World* (London: Clapp & Poliak)

MOSS, A.A.H. (1980). The chemist and the environment. *Chem. Ind.,* Oct. 18, 807

MOSS, D.R. (1987). *Pressure Vessel Design Manual* (Houston, TX: Gulf)

MOSS, J.B. (1989). Theoretical models of combustion phenomena relevant to risk assessment. In Cox, R.A. (1989a), *op. cit.,* p. 171

MOSS, T.H. and SILLS, D.L. (eds) (1981). The Three Mile Island accident. *Ann. N.Y. Acad. Sci.*, 365

MOSS, T.R. (ed.). (1980). *Mechanical Reliability* (Guildford, Surrey: IPC Sci. and Technol. Press)

MOSTELLER, F. (1977). Assessing unknown numbers: order of magnitude estimation. In Fairley, W.B. and Mosteller, F. (1977), *op. cit.*, p. 163

MOSTERT, N. (1974). *Supership* (Harmondsworth: Penguin Books)

MOTARD, R.L. (1989). Integrated computer-aided process engineering. *Comput. Chem. Engng*, 13, 1199

MOTIEE, M. (1991). Estimate possibility of hydrates. *Hydrocarbon Process.*, 70(7), 98

MOTT, N.F. (1947). Fragmentation of shell cases. *Proc. R. Soc., Ser. A*, 189, 300

MOTTER, M.J. (1991). Is your transformer a time bomb? *Chem. Engng Prog.*, 87(12), 60

MOTTERSHEAD, A. (1977). Is it cost effective to provide fire protection equipment on computers? *First Nat. Conf. on Reliability. NRC4/7* (Culcheth, Lancashire: UK Atom. Energy Auth.)

MOUBRAY, J. (1991). *Reliability-Centred Maintenance* (Oxford: Butterworth-Heinemann)

MOULD, R.F. (1988). *Chernobyl. The Real Story* (Oxford: Pergamon Press)

MOULDING, K. (1990). Offshore progress. *Chem. Engr, Lond.*, 470, 4

MOULTON, R.S. (1944). Cleveland gas explosion and fire. *NFPA Quart.*, Nov. 15

MOUNTCASTLE, V.B. (ed.) (1974). *Physiology*, 13th ed. (Saint Louis, MO: C.V. Mosby Co.)

MOUSSA, N.A., CARON, R.N., JEFFREYS, D.J. and ALLAN, D.S. (1982). *Ignition sources of LNG vapour clouds. Draft Rep.* A.D. Little, Cambridge, MA

MOVILLIAT, P. and GILTAIRE, M. (1979). Measurement of the ignition energy of gaseous mixtures by capacitive discharge − ignition by discharge from a person charged with static electricity. *J. Electrostatics*, 6, 307 (in French)

MOVILLIAT, P. and MONOMAKOFF, A. (1977). Ignition of gas mixtures by discharge of a person charged with static electricity. *Int. Conf. on Safety in Mines, Res. Paper* C3

MOWRER, F. (1990). Lag times associated with fire detection and suppression. *Fire Technol.*, 26, 244

MOWRER, F. (1993). Software review: 'THE' Sprinkler Program. *Fire Technol.*, 29, 60

MOY, S.S.J. (1981). *Plastic Methods for Steel and Concrete Structures* (London: Macmillan)

MOYSEN, F.D.H. (1985). Control of industrial processes to prevent loss of containment of flammable or toxic substances. *Chem. Ind.*, Jul. 15, 472

MRAZ, G.J. and NISBETT, E.G. (1979). Safety aspects of design, manufacture and operation of forged vessels for high pressure service. In Vodar, B. and Marteau, P. (1979), *op. cit.*, p. 1063

MUDAN, K.S. (1984a). Gravity spreading and turbulent dispersion of pressurized releases containing aerosols. In Ooms, G. and Tennekes, H. (1984), *op. cit.*, p. 73

MUDAN, K.S. (1984b). Hazard models and the personal computer. In A.D. Little (1984), *op. cit.*

MUDAN, K.S. (1984c). Thermal radiation hazards from hydrocarbon pool fires. *Prog. Energy Combust. Sci.*, 10, 59

MUDAN, K.S. (1986). Blast waves in partially confined spaces − a numerical simulations model. In Hartwig, S. (1986), *op. cit.* p. 287

MUDAN, K.S. (1987). Geometric view factors for thermal radiation hazard assessment. *Fire Safety J.*, 12(2), 89

MUDAN, K.S. (1989a). Acute inhalation toxicity of hydrogen fluoride. *Am. Inst. Chem. Engrs, Summer Nat. Mtg* (New York)

MUDAN, K.S. (1989b). Evaluation of fire and flammability hazards. *Encyclopaedia of Environmental Control Technology* (Houston, TX: Gulf), ch. 14

MUDAN, K.S. (1989c). *FIRST − A Facility Initial Risk Screening Technique for Hazard Ranking in the Chemical Process Industries* (report). A.D. Little, Cambridge, MA

MUDAN, K.S. and CROCE, P.A. (1984). Thermal radiation model for LNG trench fires. *Am. Soc. Mech. Engrs, Winter Ann. Mtg* (New York)

MUEHLBERGER, C.W. (1928). The constant-boiling mixture of hydrogen fluoride and water. *J. Phys. Chem.*, 32, 1888

MUELLER, J. (1974). How to prepare an EIR. *Chem. Engng*, 81(Oct., 21), 39

MUELLER, J.F. (1980). Is your power distribution system reliable? *Hydrocarbon Process.*, 59(5), 169

MUIR, D.C.F. (1976). Health hazards of thermal insulation products. *Ann. Occup. Hyg.*, 19, 139

MUIR, D.M. (1985). *Dust and Fume Control*, 2nd ed. (Rugby: Instn Chem. Engrs)

MUIR, G.D. (1971). *Hazards in the Chemical Laboratory;* 1977, 2nd ed. (London: Chem. Soc.). See also Bretherick, L. (1981)

MUIR, I.F.K. and BARCLAY, T.L. (1974). *Burns and Their Treatment* (London: Lloyd Luke)

MUIR, W.R. (1980). Chemical selection and evaluation: implementing the Toxic Substances Control Act. In Haque, R. (1980). *op. cit.*, p. 15

MUIR WOOD, R. (1985). Problems of seismic hazard at British and Mediterranean petrochemical installations. *The Assessment and Control of Major Hazards* (Rugby: Instn Chem. Engrs), p. 221

MUKADDAM, W. (1982). Pneumatic memory smooths valve opening. *Chem. Engng*, 89(Jun. 28), 115

MUKERJI, A. (1977). Unloading and storage technique for vinyl chloride monomer. *Chem. Engng*, 84(Sep. 12), 155

MUKERJI, A. (1980). How to size relief valves. *Chem. Engng*, 87(Jun. 2), 79

MUKHERJEE, S.K., JAITELY, R.C., CHATTERJEE, S. and AGGARWAL, K.K. (1988). Failure of secondary catalyst and air burner catalyst. *Ammonia Plant Safety* 28, 144

MUKHERJEE, S.K., GHOSH, S.R. and CHATTERJEE, S. (1992). Failure of waste heat boiler downstream of secondary reformer. *Ammonia Plant Safety*, 32, 143

MULDER, D.E. *et al.* (1986). *Review of Literature Data on Phosgene.* Notox Toxicol. Research and Consultancy, Amersfoort, Netherlands

MULDER, J.S. and VAN DER ZALM, H.O. (1967). A fatal case of ammonia poisoning. *Tijdscher Soc. Geneeskd*, 45, 458

MULESKI, G.E., ARIMAN, T. and AUMEN, C.P. (1979). A shell model of a buried pipe in a seismic environment. *J. Pres. Ves. Technol.*, 101, 44

MULET, A., CORRIPIO, A.B. and EVANS, L.B. (1981). Estimate costs of pressure vessel via correlations. *Chem. Engng*, 88(Oct. 5), 145

MULHEARN, P.J. (1977). Turbulent flow over a very rough surface. *Sixth Australian Hydraulics and Mechanics Conf.*

MULHOLLAND, K.L., BENDIXEN, L.M. and KEEPORTS, G.L. (2001). Making EHS an integral part of process design. *Int. Conf. and Workshop: Making Process Saf. Pay − The Business*

Case, Oct. 2–5, 2001, Toronto, Canada, 19–27 (New York: American Institute of Chemical Engineers)

MULHOLLAND, K.L., SYLVESTER, R.W. and DYER, J.A. (2000). Sustainability: waste minimization, green chemistry and inherently safer processing. *Environ Prog.* **19** (4), (Winster), 260–8

MULHOLLAND, M. and JURY, M.R. (1987). System design for a real-time atmospheric dispersion model. *Atmos. Environ.*, **21**, 1059

MULLEN, J.W., FENN, J.B. and IRBY, M.R. (1949). The ignition of high velocity streams of combustible gas by heated cylindrical rods. *Combustion*, **3**, 317

MULLER, B.W., BRODD, A.R. and LEO, J.P. (1983). Hazardous waste remedial action – Picillo farm, Coventry, Rhode Island: an overview. *J. Haz. Materials*, **7**(2), 113

MULLER, K.E. and BARTON, C.N. (1987). A nonlinear version of the Coburn, Forster and Kane model of blood carboxyhemoglobin. *Atmos. Environ.*, **21**, 1963

MULLER, R. (1910). On the toxicological effects of phosgene. *Z. Angew. Chemie*, 1489

MULLER, U. (1932). *Die chemische Waffe* (Berlin: Verlag Chemie)

MÜLLER-DETHLEFS, K. and SCHLADER, A.F. (1976). The effect of steam on flame temperature, burning velocity and carbon formation in hydrocarbon flames. *Combust. Flame*, **27**, 205

MULLER-HILDEBRAND, D. (1963). *Electrostatically Initiated Explosion of Sugar Dust in Air. Investigation of Kopingebro in 1961. Socker Handlingar II 18*, no. 3

MULLHAUPT, R. (1986). Update: the national quick response sprinkler research project. *Am. Inst. Chem. Engrs, Symp on Loss Prevention* (New York)

MULLHI, J.S., ANG, M.L., LEES, F.P. and ANDREWS, J.D. (1988). The propagation of faults in process plants: Pt 5, Fault tree synthesis for a butane vaporiser system. *Reliab. Engng*, **23**, 31

MULLICK, S. (1993). On-line models bring business benefits. *Chem. Engr, Lond.*, **551**, s13

MULLIER, A., RUSTIN, A. and VAN HECKE, F. (1984). Fire in a compressor house. *Ammonia Plant Safety*, **24**, 79. *Plant/Operations Prog.*, 1984, **3**, 46

MULLIGAN, T.J. and FOX, R.D. (1976). Treatment of industrial waste waters. *Chem. Engng*, **83**(Oct. 18), 49

MULLIN, J. (1986). (Bar)baric standard broached. *Chem. Engr, Lond.*, **422**, 41

MULLIN, J.W. (1967). SI units in chemical engineering. *Chem. Engr, Lond.*, **211**, 176

MULLINS, B.P. (1949). Combustion in vitiated atmospheres. *Fuel*, **28**, 205

MULLINS, B.P. (1953). Studies on the spontaneous ignition of fuels injected into a hot air stream. *Fuel*, 211, 218, 234, 327, 343, 451, 467, 481

MULLINS, B.P. (1955). *Spontaneous Ignition of Liquid Fuels* (London: Butterworths)

MULLINS, B.P. and PENNER, S. (1959). Spontaneous ignition. In *Explosions, Detonations, Flammability and Ignition* (New York: Pergamon Press)

MULLINS, M. (1977). Safe transportation of hazardous chemicals. *Chemical Engineering in a Hostile World* (London: Clapp & Poliak)

MULTER, R.K. (1982). Solar ponds collect sun's heat. *Chem. Engng*, **89**(Mar. 8), 87

MUMFORD, C.J. and LIHOU, D.A. (1979a). New projects – hazard assessment and control. Part 1 – Hazard identification. *Design 79* (London: Instn Chem. Engrs), p. P2.1

MUMFORD, C.J. and LIHOU, D.A. (1979b). New projects – hazard assessment and control. Part 2 – Assessment and engineering control. *Design 79* (London: Instn Chem. Engrs), p. P3.1

MUMFORD, C.J. and LIHOU, D.A. (1979c). New projects – hazard assessment and control. Part 3 – Planning controls and minimising risk. *Design 79* (London: Instn Chem. Engrs), p. P4.1

MÜNCHENER RÜCK (1978). *World Map of Natural Hazards* (Munich: Münchener Rückversicherungs Gesellschaft)

MUNDAY, G. (1963). The calculation of venting areas for pressure relief of explosions in vessels. *Chemical Process Hazards*, **II**, p. 46

MUNDAY, G. (1971a). Basis of design for safe high pressure equipment. *Engineering Solids under Pressure* (ed. H.L.D. Pugh) (London: Instn Mech. Engrs)

MUNDAY, G. (1971b). Detonations in vessels and pipelines. *Chem. Engr, Lond.*, **248**, 135

MUNDAY, G. (1974). Design of explosion reliefs. *Fire Prev. Sci. Technol.*, **9**, 23

MUNDAY, G. (1975a). *The Sizes of Possible Flammable Vapour Cloud Discharged From Section 25A. Rep. to Flixborough Court of Inquiry, Cremer and Warner Rep. 9* (London: Imperial Coll.)

MUNDAY, G. (1975b). Unconfined vapour cloud explosions. *Instn. Chem. Engrs, Symp. on Technical Lessons of Flixborough* (London)

MUNDAY, G. (1976a). Unconfined vapour cloud explosions. *Chem. Engr, Lond.*, **308**, 278

MUNDAY, G. (1976b). Unconfined vapour cloud explosions: a reappraisal of TNT equivalence. *Process Industry Hazards*, p. 19

MUNDAY, G. (1977). On-line monitoring and analysis of the hazard of chemical plant. *Loss Prevention and Safety Promotion 2*, p. 273

MUNDAY, G. (1979). *Chances of Ignition Leading to Unconfined Explosions in Flammable Gas and Vapour Clouds.* Insurance Tech. Bur., London

MUNDAY, G. (1980). Initial velocities attained by plant-generated missiles. *Nucl. Engng*, **19**, 165

MUNDAY, G. and CAVE, L. (1974). Evaluation of blast wave damage from very large vapour cloud explosions. *Int. Atom. Energy Agency, Symp. on Siting of Nuclear Facilities* (Vienna)

MUNDAY, G., UBBELOHDE, A.R. and WOOD, I.F. (1968). Fluctuating detonation in gases. *Proc. R. Soc., Ser. A*, **306**(1485), 171

MUNDAY, G., PHILLIPS, H.C.D., SINGH, J. and WINDEBANK, C.S. (1980). Instantaneous fractional annual loss – a measure of the hazard of an industrial operation. *Loss Prevention and Safety Promotion 3*, vol. 3, p. 1460

MUNDO, K.J. (1979). Reliability of chemical plants. In Froment, G.F. (1979), *op. cit.*, p. 55

MUNDY, N. (1992). Preventative maintenance. *Plant Engr*, **36**(2), 26

MUNICH RE (1991). *Losses in the Oil, Petrochemical and Chemical Industries – A Report* (Munich)

MUNKMAN, J. (1971). *Employer's Liability at Common Law*, 7th ed. (London: Butterworths)

MUNN, A. (1963). Occupational tumours of the bladder. *Chemical Process Hazards*, **II**, p. 59

MUNN, A. (1978). Health supervision of the chemical worker. *Chem. Ind.*, Jul. 15, 507

MUNN, P. (1988). Protection against corrosion. *Plant Engr*, **32**(3), 27

MUNN, R.E. (1966). *Descriptive Micometeorology* (New York: Academic Press)

MUNN, R.E. (ed.) (1979). *Environmental Impact Assessment – Principles and Procedures* (New York: Wiley)

MUNNS, D.D.B.H. (1977). Practical aspects of sampling particulate emissions to the atmosphere. *Filtration Separation*, 14, 656

MUNRO, C.S.H. (1963). Obsolescence of chemical process equipment. *Chem. Engr, Lond.*, 167, CE81

MUNRO, H.N. (ed.) (1969). *Mammalian Protein Metabolism*, vol. 3 (New York: Academic Press)

MUNRO, H.P. (1967). Uses of critical path planning. *J. Inst. Petrol.*, 53, 385

MUNRO, H.P., MARTIN, F.W. and ROBERTS, M.C. (1968). How to use simulation techniques to determine optimum manning levels for continuous process plants. *Chem. Engr, Lond.*, 222, 355. See also Edwards, E. and Lees, F.P. (1974), *op. cit.*, p. 320

MUNSON, J.S. (1968). Air pollution control. Dry mechanical collectors. *Chem. Engng*, 75(Oct. 14), 147

MUNSON, R.E. (1980). Safety considerations for layout and design of processes housed indoors. *Loss Prevention*, 13, p. 15

MUNSON, R.E. (1985). Process hazards management in Du Pont. *Plant/Operations Prog.*, 4, 13

MUNTSCH, O. (1935). *Leitfaden der Pathologie und Therapie der Kampfgaserkrankungen* (Leipzig: George Thieme)

MURAD, R.J., LAMENDOLA, J., ISODA, H. and SUMMERFIELD, M. (1970). A study of some factors influencing the ignition of a liquid fuel pool. *Combust. Flame*, 15, 289

MURALI, J. (1983). API steels and SSC. *Oil gas J.*, 81(Aug. 1), 94; (Aug. 8), 120

MURASZEW, A., FEDELE, J.B. and KUBY, W.C. (1979). The fire whirl phenomenon. *Combust. Flame*, 34, 29

MURATA, M., NARITOMI, M., YOSHIDA, Y. and KOKUBU, M. (1974). Behaviour of sodium aerosol in atmosphere. *J. Nucl. Sci. Technol. (Japan)*, 11(2), 65

MURATA, S., KANO, M., OKAMOTO, T., NISHIMO TO, T., TATEISHI, M. and HORI, T. (1983). The Hitachi Zosen/CBI spherical LNG tank ship system. *Gastech 82*, p. 129

MURDOCH, P. (1971). The interdependence of port and ship and the need for proper liaison and appropriate bye-laws. *Transport of Hazardous Cargoes*, 2, paper 9

MURGATROYD, R.A. and TATE, J.F. (1987). *A Case of the Application of the SHERPA Technique to the Dinorwig Casing Ultrasonic System* (report). Human Reliability Associates, Wigan, Lancashire

MURGATROYD, R.A. *et al.* (1986). *A Systematic Approach to Optimizing Human Performance in Ultrasonic Inspection* (report). Safety and Reliability Directorate, Culcheth, Warrington

MURPHEY, J.L. (1971). Start-up made easier. Cold service unit. *Hydrocarbon Process.*, 50(12), 92

MURPHY, A.J. (1982). The role of an occupational health nurse in a chemical surveillance program. In Fawcett, H.H. and Wood, W.S. (1982), *op. cit.*, p. 195

MURPHY, B.D. and NELSON, C.B. (1983). The treatment of ground deposition, species decay and growth and source height effects in a Lagrangian trajectory model. *Atmos. Environ.*, 17, 2545

MURPHY, J.F. (1988). A burner management emphasis program. *Plant/Operations Prog.*, 7, 173

MURPHY, M.J. (1985). Flame spread rates over methanol fuel spills. *Combust. Sci. Technol.*, 44, 223

MURPHY, R.F. (1974). Extrusion and pelleting operations. *Safety in Polythylene Plants* 2, 6

MURPHY, R.F. (1981). Fire fighting foam systems. *Hydrocarbon Process.*, 60(10), 213

MURPHY, S.D. (1980). Toxicological dynamics. In Haque, R. (1980a), *op. cit.*, p. 425

MURPHY, T.S. (1944a). Determining needed relieving capacity for rupture diaphragms. *Chem. Metall. Engng*, 51(12), 99

MURPHY, T.S. (1944b). Rupture diaphragms: characteristics, uses and size calculations. *Chem. Metall. Engng*, 51(11), 108

MURRAY, F.W., JAQUETTE, D.L. and KING, W.S. (1976). *Hazards Associated with the Importation of Liquefied Natural Gas. Rep. R-1845-RC*. Rand Corp., Santa Monica, CA

MURRAY, J. (1989). *Piper Alpha Public Inquiry. Transcript, Day 108*

MURRAY, M.G. (1973). How to specify better pump baseplates. *Hydrocarbon Process.*, 52(9), 157

MURRAY, M.G. (1974). All the facts about machinery alignment. *Hydrocarbon Process.*, 53(10), 139

MURRAY, M.G. (1975). Improve machinery maintenance by lube oil analysis. *Hydrocarbon Process.*, 54(1), 65

MURRAY, M.G. (1976). Select O-rings carefully. *Hydrocarbon Process.*, 55(4), 191

MURRAY, M.G. (1977). Electric motors – a mechanical viewpoint. *Hydrocarbon Process.*, 56(2), 127

MURRAY, M.G. (1979). Out of room? Use minimum movement machinery alignment. *Hydrocarbon Process.*, 58(1), 112

MURRAY, M.G. (1980a). Easy alignment measurements for gear-type couplings. *Hydrocarbon Process.*, 59(9), 245

MURRAY, M.G. (1980b). Flange fitup problems. Try a dutchman. *Hydrocarbon Process.*, 59(1), 105

MURRAY, M.G. (1983). Machinery shaft/coupling alignment strategy cuts manhours and downtime. *Oil Gas J.*, 81(July 11), 55

MURRAY, N.W. (1984). *Introduction to the Theory of Thin-Walled Structures* (Oxford: Oxford Univ. Press)

MURRAY, R.M. and WRIGHT, J.E. (1967). Trouble-free start-up of distillation columns. *Chem. Engng Prog.*, 63(12), 40

MURRAY, R.W. and O'NEILL, K.J. (1983). Automatic fire detection – application and installation. *Plant/Operations Prog.*, 2, 67

MURRAY, S.B. and LEE, J.H. (1984). The influence of yielding confinement on large-scale ethylene–air detonations. In Bowen, J.R. *et al.*, (1984b), *op, cit.*, p. 80

MURRELL, J.V. (1988). Recent advances in sprinkler technology and the implications for high-rack storage in major warehousing. *Preventing Major Chemical and Related Process Accidents*, p. 269

MURRELL, J.V. and FIELD, P. (1990). Sprinkler protection of post pallet storage in high racks. *Fire Surveyor*, 19(1), 16

MURRELL, K.F.H. (1952a). The design of instrument scales. *Instrum. Pract.*, 6, 225

MURRELL, K.F.H. (1952b). The use and arrangement of dials. *Instrum. Pract.*, 6, 250

MURRELL, K.F.H. (1965a). *Ergonomics: Man in his Working Environment* (London: Chapman & Hall)

MURRELL, K.F.H. (1965b). *Human Performance in Industry* (New York: Reinhold)

MURTY, A.S.R. and VERMA, A.K. (1986). Inverse normal and lognormal distributions for reliability design. *Reliab. Engng*, 15(1), 55

MURTY, K.A. and BLACKSHEAR, P.L. (1967). Pyrolysis effects in the transfer of heat and mass in thermally decomposing organic materials. *Combustion*, 11, 517

MURZA, J.C., GENTNER, D.H. and MCMAHON, C.J. (1981). Compositional effects in the temper embrittlement of bainitic 2 1/4 Cr−1Mo steel. *API Proc., Refining Dept, 46th Midyear Mtg*, **60**, 248

MUSA, J. *et al.* (1987). *Software Reliability* (New York: McGraw-Hill)

MUSCHELKNAUTZ, S. and MAYINGER, F. (1986). Fluid separation in two-phase flow depressurization. *1986 European Two-Phase Mtg*, Munich, paper

MUSCHELKNAUTZ, S. and MAYINGER, F. (1990). Separation of liquid from vapour upon pressure relief. *J. Loss Prev. Process Ind.*, **3**(1), 125

MUSSELWHITE, N. (1985). A chief fire officers view of intelligent fire alarm systems. *Fire Surveyor*, **14**(4), 5

MUTH, E. (1967). Stochastic theory of repairable systems. PhD Thesis, Polytechnic Inst. of Brooklyn, New York

MUTHER, R. (1955). *Practical Plant Layout* (New York: McGraw-Hill)

MUTHER, R. (1961). *Systematic Layout Planning* (Boston, MA: Ind. Educ. Inst.)

MUTHER, R. (1973). *Systematic Layout Planning*, 2nd ed. (Boston, MA: Cahners)

MUZZALL, C.E. (ed.) (1964). *Compendium of Gas Autoclave Engineering Studies. Rep. Y-1478*. Engng Div., Nucl. Div., Union Carbide Corp., Oak Ridge, TN

MYATT, D.O. (1975). Information sources/systems for hazardous materials control. *Transport of Hazardous Cargoes*, 4

MYERS, D.R., DAVIS, J.F. and HERMAN, D.J. (1988). A task oriented approach to knowledge-based systems for process engineering design. *Comput. Chem. Engng*, **12**, 959

MYERS, J.F. and WOOD, L.E. (1965). Enhancing accuracy of rupture disks. *Chem. Engng*, **72**(Nov. 8), 209

MYERS, K. (1995). *The view from the European Commission. Safety Cases* (London: IBC Technical Services)

MYERS, L.C. (1980). The chemical reactivity test − a compatibility screening test for explosives. *J. Haz. Materials*, **4**(1), 77

MYERS, P., MUDAN, K. and HACHMUTH, H. (1992). The risks of HF and sulfuric acid alkylation. *Hazard Identification and Risk Analysis, Human Factors and Human Reliability in Process Safety*, p. 127

MYERS, R.H., WONG, K.L. and GORDY, H.M. (1964). *Reliability Engineering for Electronic Systems* (New York: Wiley)

MYERS, W. (1986). Introduction to expert systems. *IEEE Expert*, **1**(1), 100

MYERSON, E.B. (1991). Calculate saturated-gas loads for vacuum systems. *Chem. Engng Prog.*, **87**(3), 92

VAN MYNEN, R. (1990). Risk management concepts in the chemical industry: one large manufacturer's approach. *Plant/Operations Prog.*, **9**, 191

NABERT, K. and SCHÖN, G. (1970). *Sicherheitstechnische Kennzahlen brennbarer Gase und Dampfe*, 2nd ed. (Marburg: Deutcher Eichverlag)

NACHMIAS, C. (1992). *Research Methods in the Social Sciences* (New York: St Martin Press)

NACHMIAS, D. (1976). *Research Methods in the Social Sciences* (New York: St Martin Press)

NADEAU, J. (1982). Bioassay. In Bennett, G.F., Feates, F.S. and Wilder, I. (1982), *op. cit.*, pp. 8−14

NADEAU, R.J. and DEWLING, R.T. (1972). Hazardous material spills vs oil spills common biological denominator. *Control of Hazardous Materials Spills 1*, p. 211

NADER, R. (1965). *Unsafe at any Speed* (New York: Grossman)

NADER, R. (1971). *Whistle Blowing: The Report of the Conference on Professional Responsibility* (New York: Bantam Books)

NAESHEIM, T. (1981). *The Alexander L. Kielland Accident* (report). Ministry of Justice and Police, Oslo

NAEVE, K.L. (1968). Operating experiences of a small LNG plant. *LNG Conf. 1*

NAGAISHI, T., YONEDA, K. and HIKITA, T. (1971). On the structure of detonation waves in gases. *Combust. Flame*, **16**, 35

NAGAMOTO, R., USHIJIMA, M., SAKAI, D., HAGIWARA, K., TAKAHASHI, T. and KURAMOTO, Y. (1982). On the study of the tank system of the 125,000 cu m MRV type LNG carrier (loads and stress analysis). *Gastech 81*, p. 187

NAGAMOTO, R., MATOBA, M., KURAMOTO, Y., USHIJAMA, M. and SAKAI, D. (1983). On the study of the tank system of a 125,000 m^3 MRV type LNG carrier (thermal leak before failure and buckling analysis). *Gastech 82*, p. 249

NAGEL, O., SCHOEMER, E. and SPÄHN, H. (1979). The combined contribution of technical rules and technical developments for safety in chemical plants. BASF (1979), *op. cit.*, p. 209

NAGL, G.J. (1982). Effects of insulation on refractory structures. *Chem. Engng*, **89**(Oct. 18), 127

NAGOA, S., STROUD, J.D., HAMADA, T., PINKUS, H. and BIRMINGHAM, D.J. (1972). The effect of sodium hydrooxide and hydrochloric acid on human epidermis. *Acta. Dermatol.* (Stockholm), **52**, 11

NAGY, G.Z. (1991). The odor impact model. *J. Air Waste Mngmt Ass.*, **41**, 1360

NAGY, J. and VERAKIS, H.C. (1983). *Development and Control of Dust Explosions* (New York: Dekker)

NAIDUS, E.S. (1976). Tests of smoke and explosion device. *Loss Prevention*, **10**, 122

NAIDUS, E.S. (1981a). Fire venting and sprinklers in single storey buildings. *Fire Surveyor*, **10**(2), 32

NAIDUS, E.S. (1981b). Full-scale explosion study of relief vents for protecting large structures. *Loss Prevention*, **14**, 35

NAIK, M.D. and NAIR, K.P.K. (1965). Multistage replacement strategies. *Operations Res.*, **279**, 13

NAILEN, R.L. (1970a). Guide to trouble-free electric motor drives. *Chem. Engng*, **77**(Jun. 1), 181

NAILEN, R.L. (1970b). Specifying motors for a quiet plant. *Chem. Engng*, **77**(May 8), 157

NAILEN, R.L. (1973a). Are your motors too hot? *Hydrocarbon Process.*, **52**(8), 91

NAILEN, R.L. (1973b). Is that motor really overloaded? *Hydrocarbon Process.*, **52**(9), 117

NAILEN, R.L. (1973c). Pick motors by their 'START'. *Hydrocarbon Process.*, **52**(1), 117

NAILEN, R.L. (1974). How to improve your motor specs. *Hydrocarbon Process.*, **53**(11), 211

NAILEN, R.L. (1975). How to apply electric motors in explosive atmospheres. *Hydrocarbon Process.*, **54**(2), 101

NAILEN, R.L. (1978). Protect motors by heat sensing. *Hydrocarbon Processing*, **57**(4), 175

NAIR, M.R.S. and GUPTA, M.C. (1973). Measurement of flame quenching distances in constant volume combustion vessels. *Combust. Flame*, **21**, 321

NAIR, M.R.S. and GUPTA, M.C. (1974). Burning velocity measurement by bomb method. *Combust. Flame*, **22**, 219

NAIR, P.K. and CANFIELD, F.B. (1992). Consider computer integrated manufacturing for continuous process plants. *Chem. Engng Prog.*, **88**(11), 71

NAKAKUKI, A. (1973). Flame velocity on the surface of volatile liquids (letter). *Combust. Flame*, **20**, 135

NAKAKUKI, A. (1974a). Blow-off and flame spread in liquid fires. *Combustion Sci. Technol.*, **9**, 71

NAKAKUKI, A. (1974b). Liquid fuel fires in the laminar flame region. *Combust. Flame*, **23**, 337

NAKAKUKI, A. (1976). Flame spread over solid and liquid fuels. *J. Fire Flammability*, **7**, 19

NAKANISHI, E. and REID, R.C. (1971). Liquid natural gas—water reactions. *Chem. Engng Prog.*, **67**(12), 36

NAKANO, R., SUGAWARA, Y., YAMAGATA, S. and NAKAMURA, J. (1983). An experimental study on the mixing of stratified layers using liquefied Freon. *Gastech 82*, p. 383

NAKANO, T. (1978). Safety design of ammonia plants. *Ammonia Plant Safety*, **20**, p. 37

NAKANO, Y., HIRAI, E. and HAYASHI, Y. (1990). Estimation of flash point of emulsified fuels. *Safety and Loss Prevention in the Chemical and Oil Processing Industries*, p. 437

NAM, S. and BILL, R.G. (1993). Numerical simulation of thermal plumes. *Fire Safety J.*, **21**, 231

NAMUR COMMITTEE (1985). Development of sequence controls for charge processes with variable formulation from pretailored functional modules

NANCEKIEVILL, D.C. (1980). Three major accidents. In Frey, R. and Safar, P. (1980), *op. cit.*, p. 128

NAPADENSKY, H.S. (1968). Sensitivity of explosive systems to detonation and subdetonation reactions. *Ann N.Y. Acad. Sci*, **152**(art. 1), 220

NAPIER, D.H. (1971). Static electrification in the process industries. *Major Loss Prevention*, p. 170

NAPIER, D.H. (1972). Natural gas and the environment: before combustion. In Hepple, P. (1973), *op. cit.*, p. 199

NAPIER, D.H. (1974a). *Aspects of Pollution Control Associated with the Process Industries* (London: Keith Shipton Developments Ltd)

NAPIER, D.H. (1974b). Generation of static electricity in steam screens and water curtains. *Chemical Process Hazards*, V, p. 244

NAPIER, D.H. (1977a). *Major Hazards* (London: Keith Shipton Developments Ltd)

NAPIER, D.H. (1977b). Minimum ignition energy of dust suspensions. *Chemical Process Hazards*, VI, p. 73

NAPIER, D.H. (1978). Educational and professional aspects of health and safety. *Chem. Ind.*, May 6, 303

NAPIER, D.H. (ed.) (1979a). *Carriage of Dangerous Goods Overland* (London: Oyez)

NAPIER, D.H. (1979b). Types of hazards in chemical engineering laboratories. Instn Chem. Engrs, Symp. on Safety in Chemical Engineering Departments (London)

NAPIER, D.H. (1980). The growing technology of industrial safety. *Chem. Ind.*, Jul. 19, 560

NAPIER, D. (1981). Risk assessment in the chemical and process industries. *The Hazards of Industrial Explosion from Dusts* (London: Oyez/IBC), paper 6

NAPIER, D.H. (1983). Static charging in powder conveying systems. *Loss Prevention and Safety Promotion 4*, vol. 3, p. E53

NAPIER, D.H. and PALMER, P.J. (ca. 1988). *Computer Aided Fault Tree Synthesis* (paper). Dept of Chem. Engng, Univ. of Toronto

NAPIER, D.H. and ROOPCHAND, D.R. (1986). An approach to hazard analysis of LNG spills. *J. Occup. Accidents*, **7**(4), 251

NAPIER, D.H. and ROSSELL, D.A. (1973). Spray electrification of xylene. *J. Phys. D, Appl. Phys.*, **6**, L118

NAPIER, D.H. and ROSSELL, D.A. (1974). Hazard assessment and critical parameters relating to static electrification in the process industries. *Loss Prevention and Safety Promotion 1*, p. 55

NAPIER, D.H. and ROSSELL, D.A. (1977). Hazard aspects of static electrification in dispersions of organic liquids. *Loss Prevention and Safety Promotion 2*, p. 157

NAPIER, D.H. and VLATIS, J. (1981). Safety considerations in the self-heating of two-component systems. *Runaway Reactions, Unstable Products and Combustible Powders* (Rugby: Instn Chem. Engrs), paper 2/F

NAPPO, C.J. (1974). A method for evaluating the accuracy of air pollution prediction models. *First Symp. on Atmospheric Diffusion and Air Pollution* (Boston, MA: Am. Met. Soc.), p. 325

NAPPO, C.J. (1976). The simulation of atmospheric transport using observed and estimated winds. *Third Symp. on Atmospheric Turbulence, Diffusion and Air Quality* (Boston, MA: Am. Met. Soc.), p. 100

NAPPO, C.J. (1981). Atmospheric turbulence and diffusion estimates derived from observations of a smoke plume. *Atmos. Environ.*, **15**, 541

NARASIMHAN, K.S. and FOSTER, P.J. (1965). The rate of growth of soot in turbulent flow with combustion products and methane. *Combustion*, **10**, p. 253

NARASIMHAN, N.D. (1986). Predict flare noise. *Hydrocarbon Process.*, **65**(4), 133

NARAYANAN, A. and SHARKEY, N.E. (1985). *An Introduction to Lisp* (Chichester: Ellis Horwood)

NARESKY, J.J. (1967). Data systems – where they stand today. *Ann. Reliab. Maintainab.*, **6**, 113

NARSIMHAN, G.N. (1993). Private communication.

NARISIMHAN, S. and MAH, R.S.H. (1987). Generalized likelihood ratio methods for gross error identification. *AIChE J.*, **33**, 1514

NASH, F.R. (1993). *Estimating Device Reliability* (Boston: Kluwer)

NASH, G. (1966). New concept in the supply of foam compound to oil fires. *Fire Int.*, **1**(12), 40

NASH, J.R. (1976). *Darkest Hours* (Chicago, IL: Nelson Hall)

NASH, P. (1969). Fire control: manual. In Handley, W. (1969), *op. cit.*, p. 72

NASH, P. (1972a). A high-speed sprinkler system for the protection of high-racked storage. *Fire Prev. Sci. Technol.*, **3**, 4

NASH, P. (1972b). Towards an international standard for fire statistics. *Fire Prev. Sci. Technol.*, **1**, 19

NASH, P. (1973a). And now – the in-line sprinkler. *Fire Prev. Sci. Technol.*, **5**, 20

NASH, P. (1973b). The performance of fire-fighting foams on flammable liquid fires. *Fire Prev. Sci. Technol.*, **5**, 4

NASH, P. (1974a). The fire protection of flammable liquid storages with water sprays. *Chemical Process Hazards*, V, p. 51

NASH, P. (1974b). The performance of water sprays on flammable liquid fires. *Fire Prev. Sci. Technol.*, **10**, 4

NASH, P. (1975a). Some design aspects of fire extinguishers. *Fire Prev. Sci. Technol.*, **12**, 24

NASH, P. (1975b). The testing of sprinkler installations. *Fire Surveyor*, **4**(4), 23

NASH, P. (1975c). The testing of sprinklers. *Fire Surveyor*, **4**(2), 23

NASH, P. (1977a). A non-electric resetting zoned sprinkler system for high-racked storage. *Fire Prev. Sci. Technol.*, **18**, 14

NASH, P. (1977b). The selection of fixed protection in chemical process plants. *Chemical Process Hazards*, **VI**, p. 127

NASH, P. (1984). Foams and foam systems. *Fire Surveyor*, **13**(4), 13

NASH, P. (1990). Sprinkler systems: a review of the new British Standard 5306 Part 2: 1990. *Fire Surveyor*, **19**(6), 14

NASH, P. (1991). Portable fire extinguishers: standards, design, and use. *Fire Surveyor*, **20**(3), 22

NASH, P. and THEOBALD, C.R. (1976a). The use of automatic sprinklers as fire sensors in a chemical plant. *Fire Prev. Sci. Technol.*, **15**, 11

NASH, P. and THEOBALD, C.R. (1976b). The use of automatic sprinklers as fire sensors in chemical plant. *Instn. Chem. Engrs, Symp on Technical Lessons of Flixborough* (London)

NASH, P. and WHITTLE, J. (1978). Fighting fires in oil storage tanks using base injection of foam. *Fire Technol.*, **14**, 15, 147

NASH, P. and YOUNG, R. (1976). The reliability of sprinkler systems. *Fire Surveyor*, **5**(2), 21

NASH, R. (1984). Conservation, environment and energy. *Encyclopaedia of American Political History* (ed. J.P. Greene), vol. 1 (New York: Charles Scribners Sons), vol. 1, p. 342

NASH, T. (1982). Asbestos. Its use must be controlled. *Health and Safety at Work*, Aug., 22

NASR, A.M. (1986). When to select a sealless pump. *Chem. Engng*, **93**(May 26), 85

NASR, A. (1992). Prevent failures of magdrive pumps. *Chem. Engng Prog.*, **88**(11), 31

NASR, A. (1992). Prevent failures of mag drive pumps. *Chem. Engng. Prog.* (Nov.), 31–34

NASR, A. (1994). Sealless pumps: limitations and developments. *Hydrocarbon Processing* (July), 63–66

NASS, L.I. (ed.) (1976–). *Encyclopaedia of PVC*, 1976, vol. 1; 1977, vols 2–3 (Basel: Dekker)

NASS, L.I. and HEIBERGER, C.A. (1986). *Encyclopaedia of PVC*, 2nd ed. (New York: Dekker)

NASSIKAS, J.N. (1970). A regulator's view of the growing LNG industry. *LNG Conf. 2*

NASSOPOULES, G.P. (1982). Safety of liquefied gas containment systems on land and at sea. *Gastech 81*, p. 347

NATHAN, I. (1966). Systems analysis and the reliability and maintainability function. *Ann. Reliab. Maintainab.*, **5**, 284

NATHAN, M.F. (1978). Choosing a process for chloride removal. *Chem. Engng*, **85**(Jan. 30), 93

NATIONAL ACADEMY OF ENGINEERING (1970). *Public Safety* (Washington, DC)

NATIONAL ACADEMY OF ENGINEERING (1972). *Perspectives in Benefit-Risk Decision Making* (Washington, DC)

NATIONAL ACADEMY OF ENGINEERING (1986). *Hazards: Technology and Fairness* (Washington, DC)

NATIONAL ACADEMY OF SCIENCES (1975). *Principles for Evaluating Chemicals in the Environment* (New York)

NATIONAL ACADEMY OF SCIENCES (1984). Guidance Levels for Emergency and Continuous Exposure to Selected Airborne Contaminants (Washington, DC)

NATIONAL ACADEMY OF SCIENCES/NATIONAL RESEARCH COUNCIL, COMMITTEE ON TOXICOLOGY (1972a). *Guides for Short-term Exposure of the Public to Air Pollutants: IV: Guide for Ammonia* (Washington, DC)

NATIONAL ACADEMY OF SCIENCES/NATIONAL RESEARCH COUNCIL (1972b). *Lead* (New York)

NATIONAL ACADEMY OF SCIENCES/NATIONAL RESEARCH COUNCIL, ADVISORY COMMITTEE ON THE BIOLOGICAL EFFECTS OF RADIATION (1972c). *The Effects on Populations of Exposure to Low Levels of Ionizing Radiation* (New York)

NATIONAL ACADEMY OF SCIENCES/NATIONAL RESEARCH COUNCIL, COMMITTEE ON TOXICOLOGY (1973a). *Guides for Short-term Exposure of the Public to Air Pollutants: VIII: Guide for Chlorine* (Washington, DC)

NATIONAL ACADEMY OF SCIENCES/NATIONAL RESEARCH COUNCIL (1973b). *Risk Assessment in the Federal Government: Managing the Process* (Washington, DC: Nat. Acad. Press)

NATIONAL ACADEMY OF SCIENCES/NATIONAL RESEARCH COUNCIL (1975). *Principles for Evaluating Chemicals in the Environment* (New York)

NATIONAL ACADEMY OF SCIENCES/NATIONAL RESEARCH COUNCIL (1976). *Medical and Biological Effects of Environmental Pollutants, Chlorine and Hydrogen Chloride* (Washington, DC)

NATIONAL ACADEMY OF SCIENCES/NATIONAL RESEARCH COUNCIL (1983). Risk Assessment in the Federal Government: Managing the Process (Washington, DC)

NATIONAL AERONAUTICAL AND SPACE ADMINISTRATION (1972–). *ASRDI Oxygen Technology Survey* (Washington, DC)

NATIONAL AERONAUTICAL AND SPACE ADMINISTRATION (1979). On the experiments with 7.5 m and 15 m JP-4 fuel pool fire measurements. Quoted in Mudan (1984c), *op. cit.*, as then unpublished work by Mansfield (Ref. 10)

NATIONAL ARP ANIMAL COMMITTEE (1939). *Animals in Chemical Warfare* (London: HM Stationery Office)

NATIONAL ASSOCIATION OF CORROSION ENGINEERS (1984). *Corrosion Basics* (Houston, TX)

NATIONAL BOARD OF FIRE UNDERWRITERS (1948). *The Texas City Disaster, Facts and Lessons* (New York)

NATIONAL BOARD OF FIRE UNDERWRITERS (1956). Fire and Explosion Hazards of Organic Peroxides (report)

NATIONAL BOARD OF FIRE UNDERWRITERS and FIRE INSURANCE RATING ORGANIZATION OF NEW JERSEY (1952). *Warren Petroleum Corporation Propane Fire and Explosion 7 July 1951*

NATIONAL BUREAU OF STANDARDS (1947). *Selected Values of Properties of Hydrocarbons. Circ. 461* (Washington, DC: US Govt Printing Office)

NATIONAL BUREAU OF STANDARDS (1951). *Monte Carlo Method. Appl. Mech. Ser. 12* (Washington, DC)

NATIONAL BUREAU OF STANDARDS (1952). *Selected Values of Chemical Thermodynamic Properties. Circ. 500* (Washington, DC: US Govt Printing Office)

NATIONAL BUREAU OF STANDARDS (1969). *Selected Values of Chemical Thermodynamic Properties. Circ. 270–3* (Washington, DC: US Govt Printing Office)

NATIONAL BUREAU OF STANDARDS (ca. 1986). *Examination of a Pressure Vessel that Ruptured in the Chicago Refinery of the Union Oil Company on July 23, 1984. Rep. NBSIR 86–3049* (Washington, DC)

NATIONAL BUREAU OF STANDARDS (1988). *Investigation into the Ashland Oil Storage Tank Collapse on January 2, 1988* (Washington, DC)

NATIONAL BUREAU OF STANDARDS, CRYOGENICS DIVISION (1978). *LNG Materials and Fluids – A User's Manual of Property Data in Graphic Format* (Boulder, CO)

NATIONAL BUREAU OF STANDARDS – see also Appendix 28

NATIONAL CENTRE FOR SYSTEMS RELIABILITY – see Appendix 28

NATIONAL COAL BOARD (n.d.a). *The Maintenance of Flameproof Apparatus* (London)

NATIONAL COAL BOARD (n.d.b). *The Maintenance of Intrinsically Safe Apparatus* (London)

NATIONAL COMPUTER CENTRE (1987). *The STARTS Guide*, vols 1–2, 2nd ed. (NCC Publs)

NATIONAL COMPUTER CENTRE (1989). *The STARTS Purchasers Handbook*, 2nd ed. (NCC Publs)

NATIONAL DEFENSE RESEARCH COMMITTEE (1946). *Effects of Impact and Explosion. Summary Tech. Rep. of Div. 2*, vol. 1

NATIONAL EARTHQUAKE INFORMATION CENTER (1970). *World Seismicity 1961–1969* (Washington, DC)

NATIONAL FEDERATION OF BUILDING TRADES EMPLOYERS (1975). *Construction Safety*, rev. ed. (London)

NATIONAL FIRE PROTECTION ASSOCIATION (1950). *Static Electricity. Bull. 77* (Boston, MA)

NATIONAL FIRE PROTECTION ASSOCIATION (1957). *Report of Important Dust Explosions* (Boston, MA)

NATIONAL FIRE PROTECTION ASSOCIATION (1973). In Vervalin, C.H. (1973a), *op. cit.*, p. 85

NATIONAL FIRE PROTECTION ASSOCIATION – see also Appendices 27 and 28

NATIONAL FIRE PROTECTION ASSOCIATION – *Standard 921 Investigating Fires and Explosions* Boston: NFPA

NATIONAL FUEL EFFICIENCY SERVICE LTD (1989). *The Boiler Operators Handbook*, 2nd ed. (London: Graham & Trotman)

NATIONAL INFORMATION SERVICE FOR EARTHQUAKE ENGINEERING (1982). *Computer Software for Earthquake Engineering. Earthquake Engng Center, Univ. of Calif., Berkeley, Calif.*

NATIONAL INSTITUTE OF INDUSTRIAL PSYCHOLOGY (n.d.). *2000 Accidents* (London)

NATIONAL INSTITUTE OF OCCUPATIONAL SAFETY AND HEALTH – see Appendix 28

NATIONAL LIBRARY OF MEDICINE (1974). *Toxline Chemical Dictionary* (Washington, DC). See also *Chem. Engng News*, 1974, **52**(4), 22

NATIONAL MATERIALS ADVISORY BOARD, NATIONAL ACADEMY OF SCIENCES (ca. 1974). *Safety Aspects of Liquefied Natural Gas in the Marine Environment. Publ. NMAB 354* (Washington, DC)

NATIONAL OCEANIC AND ATMOSPHERIC ADMINISTRATION (1972). *Earthquakes* (Washington, DC: US Govt Printing Office)

NATIONAL PHYSICAL LABORATORY (1955). *Noise Measurement Techniques* (London: HM Stationery Office)

NATIONAL PORTS COUNCIL (1976). *Analysis of Marine Accidents in Ports and Harbours* (London)

NATIONAL RADIOLOGICAL PROTECTION BOARD (1989). *Living with Radiation* (London: HM Stationery Office)

NATIONAL RADIOLOGICAL PROTECTION BOARD (1993). *Radiological Protection Bulletin 141* (Didcot)

NATIONAL RADIOLOGICAL PROTECTION BOARD – see also Appendix 28

NATIONAL RESEARCH COUNCIL (1979). *Hydrogen Sulfide* (Baltimore, MD: Univ. Park Press)

NATIONAL RESEARCH COUNCIL (1981). *Prudent Practices for Handling Hazardous Chemicals in Laboratories* (New York: Nat. Acad. Press)

NATIONAL RESEARCH COUNCIL (1983). *Risk Assessment in the Federal Government* (Washington, DC: Nat. Acad. Press)

NATIONAL RESEARCH COUNCIL (1984). *Ocean Disposal Systems* (Washington, DC: Nat. Acad. Press)

NATIONAL RESEARCH COUNCIL (1989). *Improving Risk Communication* (Washington, DC: Nat. Acad. Press)

NATIONAL RESEARCH COUNCIL, COMMITTEE ON ASSESSMENT OF SAFETY OF OUTER CONTINENTAL SHELF ACTIVITIES (1981). *Safety and Offshore Oil* (Washington, DC: Nat. Acad. Press)

NATIONAL RESEARCH COUNCIL, TRANSPORTATION RESEARCH BOARD (1988). *Pipelines and Public Safety* (Washington, DC)

NATIONAL SAFETY COUNCIL (1974). *Accident Prevention Manual for Industrial Operations* (Chicago, IL)

NATIONAL SAFETY COUNCIL – see also Appendix 28

NATIONAL SAFETY COUNCIL (US) (1995). *Accident Investigation*, 2nd Edition. Chicago, IL: ISBN 0-87912-187-4

NATIONAL SOCIETY FOR CLEAN AIR (1978). *NSCA Reference Book*; 1988, 3rd ed. (Brighton, Sussex)

NATIONAL SUPPLY COMPANY (UK) LTD (1979). *Chiksan Truck and Tank Car Loading Systems* (Stockport, Cheshire)

NATIONAL TECHNICAL INFORMATION SERVICE (1963). *Nuclear Reactors and Earthquakes. Publ. TID-7024* (Springfield, VA)

NATIONAL TRANSPORTATION SAFETY BOARD (Annual Reports)

NATIONAL TRANSPORTATION SAFETY BOARD (1971). *Special Study of Risk Concepts in Dangerous Goods Transportation Regulation* (report) (Washington, DC)

NATIONAL TRANSPORTATION SAFETY BOARD (1978). *Special Study – Safe Service Life for Liquid Petroleum Pipelines. Rep. NTSB-PSS-78-1* (Washington, DC)

NATIONAL TRANSPORTATION SAFETY BOARD (1979). *Annual Report* (Washington, DC)

NATIONAL TRANSPORTATION SAFETY BOARD (1981). *Review of Rotorcraft Accidents 1977–1979* (Washington, DC)

NATIONAL TRANSPORTATION SAFETY BOARD (1983). *Annual Report* (Washington, DC)

NATIONAL TRANSPORTATION SAFETY BOARD – see also Appendix 28

NATIONAL VULCAN (1977). It's hard to believe! *Plant Engr*, **21**(3), 13

NAUGHTON, D.A. (1968). Turbine accidents which should not have happened. *Loss Prevention*, **2**, 54

NAUJOKAS, A.A. (1979). Preventing carbon bed combustion problems. *Loss Prevention*, **12**, 128

NAUJOKAS, A.A. (1985). Spontaneous combustion of carbon beds. *Plant/Operations Prog.*, **4**, 120

NAUMAN, E.B. (1987). *Chemical Reactor Design* (Chichester: Wiley)

NAUR, P. (1966). Proof of algorithms by general snapshots. *BIT*, **6**, 310

NAVAZ, M.C. (1987). An approach towards establishing a safety survey of a liquefied gas terminal. *Gastech 86*, p. 129

NAWAR, G. and SAMSUDIN, R. (1993). Evaluation and control of human error in the management of engineering systems. In Melchers, R.E. and Stewart, M.G. (1993), *op. cit.*, p. 179

NAWAR, G. and SALTER, R. (1993). Probability of death and quantification of the value of life. In Melchers, R.E. and Stewart, M.G. (1993), *op. cit.*, p. 157

NAWROCKI, L.H. (1981). Computer-based maintenance training in the military. In Rasmussen, J. and Rouse, W.B. (1981), *op. cit.*, p. 605

NAYLOR, C. (1983). *Build Your Own Expert System* (Wilmslow, Cheshire: Sigma Tech. Press)

NAYLOR, C. (1984). How to build an inferencing engine. In Forsyth, R. (1984), *op. cit.*, p. 63

NAYLOR, C.A. and WHEELER, R.V. (1931). The ignition of gases. VI, Ignition by a heated surface. Mixtures of methane with oxygen and nitrogen, argon or helium. *J. Chem. Soc.*,

2456 (Note: authors state on later paper that this paper, actually numbered VII, should bear number VI)

NAYLOR, C.A. and WHEELER, R.V. (1933). The ignition of gases. VIII, Ignition by a heated surface. a) Mixtures of ethane, propane or butane with air; b) Mixtures of ethylene, propylene or butylene with air. *J. Chem. Soc.*, **1240**

NAZ, P. (1989). Simplified prediction and numerical simulation of the functioning of fragmentation explosives. Inst. Franco-Allemand de Recherches, Saint-Louis, *Rep. PB91–110858*

NAZARIO, F.N. (1988). Preventing or surviving explosions. *Chem. Engng*, **95**(Aug. 15), 102

NDUBIZU, C.C. and DURBETAKI, P. (1978). Modeling radiative ignition of fabrics in air. *Fire Res.*, **1**(4/5), 281

NDUBIZU, C.C., RAMAKER, D.E., TATEM, P.A. and WILLIAMS, F.W. (1983). A model of freely burning pool fires. *Combust. Sci. Technol.*, **31**, 233

NEADES, D.N. and RUDOLPH, R.R. (ca. 1983). *An Examination of Injury Criteria for Potential Application to Explosive Safety Studies*. Explosives Safety Board, Department of Defense, Washington, DC

NEAL, H.G. (1976). Safety methods utilized in plant operations. *Ammonia Plant Safety*, **18**, 7

NEAL, R.A. and GIBSON, J.E. (1984). The uses of toxicology and epidemiology in identifying and assessing carcinogenic risks. In Deisler, P. F. (1984), *op. cit.*, p. 39

NEAL, W.G. (1971). Safe operation of bulk liquid chemical cargoes. *Transport of Hazardous Cargoes*, **2**, paper 7

NEALE, D. (1974). The monitoring of critical equipment. *Chem. Engng Prog.*, **70**(10), 53

NEALE, D.F. (1971). How to specify centrifugal compressors. Inspection, shipping and erection. *Hydrocarbon Process.*, **50**(1), 81

NEALE, D.F. (1973). Better mechanical testing can improve compressor reliability. *Hydrocarbon Process.*, **52**(9), 165

NEALE, D.F. (1976). Specify shop testing to . . . improve compressor reliability. *Hydrocarbon Process.*, **55**(6), 135

NEALE, M.J. (ed.) (1973). *Tribology Handbook* (London: Butterworths)

NEALE, M.J. (1980). The reduction of process hazards by plant condition monitoring. *Chemical Process Hazards*, **VII**, p. 43

MICHAEL NEALE AND ASSOCIATES (1979). *A Guide to Condition Monitoring of Machinery* (London: HM Stationery Office)

NEEDHAM, M.A. (1992). Human factors onshore and offshore. *Major Hazards Onshore and Offshore*, p. 319

NEEDLE, G.J. (1974). An introduction to power stations. *Plant Engr*, **18**(10), 17

NEEDLE, G.J. (1979). Flow induced vibration in heat exchangers and boilers. *Plant Engr*, **23**(10), Dec., 13

NEELY, W.B. (1976). Predicting the flux of organics across of the air/water interface. *Control of Hazardous Material Spills 3*, p. 197

NEELY, W.B. (1980). A method for selecting the most appropriate environmental experiments on a new chemical. In Haque, R. (1980), *op. cit.*, p. 287

NEELY, W.B. (1981). *Chemicals in the Environment* (Basel: Dekker)

NEELY, W.B. and BLAU, G.E. (eds) (1985). *Environmental Exposure to Chemicals* (Boca Raton, FL: CRC Press)

NEELY, W.B. and LUTZ, R.W. (1985). Estimating exposure from a chemical spilled into a river. *J. Haz. Materials*, **10**(1), 33

NEELY, W.B. and MESLER, R.J. (1978). An emergency response system, for handling accidental discharges of chemicals into rivers. *Control of Hazardous Material Spills*, 4, p. 19

NEERKEN, R.F. (1974). Pump selection for the chemical process industries. *Chem. Engng*, **81**(Feb. 18), 104

NEERKEN, R.F. (1975). Compressor selection for the chemical process industries. *Chem. Engng*, **82**(Jan. 20), 78

NEERKEN, R.F. (1980a). How to select and apply positive displacement rotary pumps. *Chem. Engng*, **87**(Apr. 7), 76

NEERKEN, R.F. (1980b). Use steam turbines as process drivers. *Chem. Engng*, **87**(Aug. 25), 63

NEERKEN, R.F. (1987). Progress in pumps. *Chem. Engng*, **94**(12), 76

NEFF, D.E. and MERONEY, R.N. (1981). *The Behaviour of LNG Vapor Clouds: Wind Tunnel Simulation of 40 m^3 Spill Tests at China Lake Naval Weapons Center, California. Rep. CER-81-82-DEN-RNM-1*. Dept of Civil Engng, Colorado State Univ., Fort Collins, CO

NEFF, D.E. and MERONEY, R.N. (1982). *The Behaviour of LNG Vapor Clouds: Wind Tunnel Tests on the Modeling of Heavy Plume Dispersion. Rep. CER-81-82-DEN-RNM-25*. Dept of Civil Engng, Colorado State Univ., Fort Collins, CO. Also Rep. GRI/0145, Gas Res. Inst. Chicago, IL

NEFF, D.E., MERONEY, R.N. and CERMAK, J.E. (1976). *Wind Tunnel Study of the Negatively Buoyant Plume due to an LNG Spill. Rep. CER-81-82-DEN-RNM-25*. Dept of Civil Engng, Colorado State Univ., Fort Collins, CO. Also *Rep. CER76-77-DEN-RNM-JEC22*, Gas Res. Inst. Chicago, IL

NEFF, W.A. (1990). An industrial perspective on environmental and health legislation in Canada. *Plant/Operations Prog.*, **9**, 259

NEGRON, R.M. (1994). Using ultraviolet disinfection in place of chlorination. *The Nat. Environ. J.*, (March/April), 48–50

NEIL, G.D. (1971). The practicalities of insuring a petrochemical complex in the UK. *Major Loss Prevention*, p. 13

NEIL, G.D. (1978). Private communication

NEILL, E.T. and BETHEL, M.B. (1962). Protect your plant against flooding. *Hydrocarbon Process.*, **41**(3), 144

NEILSON, A.J. (1980). Missile impact on metal structures. *Nucl. Engng*, **19**, 191

NEILSON, I.D., KEMP, R.N. and WILKINS, H.A. (1979). *Accidents Involving Heavy Goods Vehicles in Great Britain. Rep. SR 470*. Transport and Road Res. Lab., Crowthorne

NEIMAN, M.B. and GAL, D. (1968). On the sequence of elementary steps in gas phase hydrocarbon oxidation. *Combust. Flame*, **12**, 371

NEJEDLY, J., TOPINKA, F. and SKARKA, J. (1992). Risk evaluation of a molten white phosphorus storage tank farm. *Loss Prevention and Safety Promotion 7*, vol. 3, 166–1

NELKIN, D. and POLLAK, M. (1979). Consensus and conflict resolution: the politics of assessing risk. *Sci. Publ. Policy*, **6**, 307

NELLIS, W.J., SEAMAN, L. and GRAHAM, A.R. (eds) (1982). *Shock Waves in Condensed Matter* (New York: Am. Inst. Physics)

NELMS, C.R. (1988). Computerized hazards analysis. *J. Loss Prev. Process Ind.*, **1**(3), 168

NELSON, A.J. and RASMUSON, D.M. (1982). A correction to S. Kaplan's method of discrete probability distribution. *Risk Analysis*, **2**, 205; 207

NELSON, A.L., BATTS, J.R. and BEADLES, R.L. (1970). A computer program for approximating system reliability. *IEEE Trans Reliab*, **R-19**, 61

NELSON, D.B. (1982). Safety and health awareness – what does industry expect from new chemical engineering graduates? *Plant/Operations Prog.*, **1**, 114

NELSON, E. (1961). Kinetics of drug absorption, distribution, metabolism, and excretion. *J. Pharmaceut. Sci.*, **50**, 181

NELSON, G.A. (1965). Use curves to predict steel life. *Hydrocarbon Process.*, 44(5), 195

NELSON, G.A. (1966). When to use low alloy steels for hydrogen service. *Hydrocarbon Process.*, 45(5), 201

NELSON, G.A. (1974). *Corrosion Data Survey*, 5th ed. (Nat. Ass. Corrosion Engrs)

NELSON, H.E. (1987). Concepts for life safety analysis. *Fire Safety J.*, 12(3), 167

NELSON, H.N. (1979). The need for full-function test features in smoke detectors. *Fire Technol.*, 15, 10

NELSON, H.P. (1964). Handling chlorine – 2 Barge and pipeline safety. *Chem. Engng Prog.*, 60(9), 88

NELSON, L.S. (1967). Weibull probability paper. *Evaluation Engng*, May/Jun., 24

NELSON, N. (1984). Use of toxicity test data in the estimation of risks to human health. In Rodricks, R.V. and Tardiff, R.G. (1984a), *op. cit.*, p. 13

NELSON, P.A. and ELLIOTT, S.J. (1992). *Active Control of Sound* (New York: Academic Press)

NELSON, R. (1979). Material properties in SI units. *Chem. Engng Prog.*, 75(10), 95; 75(11), 70; 1980, 76(4), 86; 76(5), 83

NELSON, R.F. (1978). Natural resources and national needs – a commentary on CZM. *API Proc., Refining Dept, 43rd Midyear Mtg*, 57, 12

NELSON, R.W. (1972). Protection of enclosures handling combustible dusts. *Loss Prevention*, 6, 100

NELSON, R.W. (1977). Know your insurer's expectations. *Hydrocarbon Process.*, 56(8), 103

NELSON, W. (1973a). A new theory to explain physical explosions. *Combustion*, May, 31

NELSON, W. (1973b). A new theory to explain physical explosions. *Tappi*, 56(3), 121

NELSON, W.E. (1974). Improve machinery maintenance. *Hydrocarbon Process.*, 53(1), 85

NELSON, W.E. (1977). Compressor seal fundamentals. *Hydrocarbon Process.*, 56(12), 91

NELSON, W.E. (1980). Maintenance techniques for turbomachinery. *Hydrocarbon Process.*, 59(1), 71

NELSON, W.E. (1993). Improve relief valve reliability. *Hydrocarbon Process.*, 72(1), 59

NELSON, W.L. (1948). Safety in the use and handling of ammonia. *Oil Gas J.*, (Sep. 16), 1053

NELSON, W.L. (1958). *Petroleum Refinery Engineering*, 4th ed. (New York: McGraw-Hill)

NELSON, W.L. (1974). Refinery cycle vs stream-time efficiency. *Oil Gas J.*, 72(Feb. 18), 93

NELSON, W.L. (1976). *Guide to Refinery Operating Costs*, 3rd ed. (Tulsa, OK: Pennwell)

NELSON, W.R. (1982). REACTOR – an expert system for diagnosis and treatment of nuclear reactor accidents. *Nat. Conf. on Artificial Intelligence*, Pittsburgh

NELSON, W.R. (1984). *Response Trees and Expert Systems for Nuclear Reactor Operations. Rep. NUREG/CR 3631.* Nucl. Regul. Comm., Washington, DC

NELSON, W.R. and BLACKMAN, H.S. (1985). *Response Tree Evaluation: Implications for Use of Artificial Intelligence in Control Rooms. Rep. NUREG/CR 3631.* Nucl. Regul. Comm., Washington, DC

NELSON, W.R. and JENKINS, J.P. (1985). Expert system for operator problem solving in process control. *Chem. Engng Prog.*, 81(12), 25

NEMA, A. and TARE, V. (1989). Atmospheric dispersion under non-homogeneous and unsteady conditions. *Atmos. Environ.*, 23, 851

NEMEROW, N.L. (1971). *Liquid Waste of Industry* (New York: Addison-Wesley)

NEMEROW, N.L. (1974). *Scientific Stream Pollution Analysis* (New York: McGraw-Hill)

NEMEROW, N.L. and DASGUPTA, A. (1991). *Industrial and Hazardous Waste Treatment* (New York: van Nostrand Reinhold)

NEMES, I., DANOCZY, E., VIDOCZY, T. and GAL, D. (1976). A new method for determination of reactivity ratios in hydrocarbon oxidation processes. *Combust. Flame*, 27, 285

NEMETH, T., VARGA, E.I., HANGOS, K.M. and JORGENSEN, S.B. (1992). A tool for development of qualitative and approximate quantitative process models. *Comput. Chem. Engng*, 16, S491

DI NENNO, P.J. (1982). Simplified radiation heat transfer calculations from large open hydrocarbon fires. *Soc. Fire Prot. Engrs Ann. Mtg*, paper

NEOWHOUSE, M.M. (1993). The use of software based qualitative risk assessment methodologies in industry. In Melchers, R.E. and Stewart, M.G. (1993), *op. cit.*, p. 147

NERTNEY, R.J. (1976). *A Contractor Guide to Advance Preparation for Accident Investigation. Rep. SSDC 3.* Dept of Energy, Washington, DC

NERTNEY, R.J. (1978). *Management Factors in Accident and Incident Prevention. Rep. SSDC 13.* Dept of Energy, Washington, DC

NERTNEY, R.J. (1978). *Safety Considerations in Evaluation of Maintenance Programs. Rep. SSDC 12.* Dept of Energy, Washington, DC

NESMEYANOV, A.N. (1963). *Vapour Pressure of the Chemical Elements* (Amsterdam: Elsevier)

NESMITH, J. (1970). Getting the facts – the reporter's job. *Loss Prevention*, 4, 93

NETTERVILLE, D.D.J. (1990). Plume rise, entrainment and dispersion in turbulent winds. *Atmos. Environ.*, 24A, 1061

NETTLETON, M.A. (1975). Pressure as a function of time and distance in a vented vessel. *Combust. Flame*, 24, 65

NETTLETON, M.A. (1976). Alleviation of blast waves from large vapour clouds. *J. Occup. Accidents*, 1(1), 3

NETTLETON, M.A. (1977a). Design of explosion relief for dust cloud explosions in large vessels. *Chemical Process Hazards*, VI, unbound paper

NETTLETON, M.A. (1977b). Shock wave chemistry in dusty gases and fogs: a review. *Combust. Flame*, 28, 3

NETTLETON, M.A. (1977c). Some aspects of vapour cloud explosions. *J. Occup. Accid.*, 1(2), 149

NETTLETON, M.A. (1978a). Dust clouds and fogs – flammability, explosibility and detonation limits, burning velocities and minimum ignition analysis. *Fire Prev. Sci. Technol.*, 20, 12

NETTLETON, M.A. (1978b). Some features influencing the venting of vessels in which pressure gradients initially exist. *Fire Prev. Sci. Technol.*, 19, 4

NETTLETON, M.A. (1979). Detonation and flammability limits. *Fire Prev. Sci. Technol.*, 21, 23

NETTLETON, M.A. (1980a). Comments on 'Concentration limits to the initiation of detonation in fuel/air mixtures' by D.C. *Bull. Trans. Instn Chem. Engrs*, 60, 281

NETTLETON, M.A. (1980b). Detonation and flammability limits of gases in confined and unconfined situations. *Fire Prev. Sci. Technol.*, 23, 29

NETTLETON, M.A. (1987). *Gaseous Detonations* (London: Chapman & Hall)

NETTLETON, M.A. and STIRLING, R. (1973). Detonations in suspensions of coal dust in oxygen. *Combust. Flame*, **21**, 307

NEUBER, H. (1937). *Kerbspannungslehre* (Berlin: Springer)

VON NEUMANN, J. (1942). *Progress Report on the Theory of Detonation Waves. OSRD Rep. 549*

NEUMAN, T.W. and BENASHKI, R.C. (1981). Toxicology terminology (letter). *Chem Engng*, **88**(Jan. 26), 5

NEUMANN, F. (1954). *Earthquake Intensity and Related Ground Motion* (Seattle, WA: Univ. of Washington Press)

VON NEUMANN, J. (1942). *OSRD Rep. 549*

VON NEUMANN, J. and RICHTMYER, R.D. (1950). A method for the numerical calculation of hydrodynamic shocks. *J. Appl. Phys.*, **21**, 232

DE NEVERS, N. (1984a). Pipe rupture by sudden fluid heating. *Ind. Engng Chem. Process Des. Dev.*, **23**, 669

DE NEVERS, N. (1984b). Spread and down-slope flow of negatively buoyant clouds. *Atmos. Environ.*, **18**, 2023

DE NEVERS, N. (1988). Chemical engineers as expert witnesses in accident cases. *Chem. Engng Prog.*, **84**(6), 22

DE NEVERS, N. (1992). Cloud formation on vapour or liquid mixing with air. *J. Loss Prev. Process Ind.*, **5**, 205

NEVILL, K. (1985). Glandless centrifugal pumps for critical process duties. *Process Engng*, Jul., 56

NEVILLE, D. and WHITE, G.C. (1991). Research into the structural integrity of LNG tanks. *Hazards*, **XI**, p. 175

NEVILLE, J. (1990). *The Blitz* (London: Hodder & Stoughton)

NEW, J.H. (1983). Computer designed fire protection systems. *Fire Surveyor*, **12**(3), 17

NEW SOUTH WALES GOVERNMENT (1988). *Industry Emergency Planning Guidelines. Haz. Ind. Planning Guideline 1*. Dept of Planning, Sydney

NEW SOUTH WALES GOVERNMENT (1989a). *Fire Safety Study Guidelines. Haz. Ind. Planning Guideline 2*. Dept of Planning, Sydney

NEW SOUTH WALES GOVERNMENT (1989b). *Environment Risk Impact Assessment Guidelines. Haz. Ind. Planning Guideline 3*. Dept of Planning, Sydney

NEW SOUTH WALES GOVERNMENT (1990). *Risk Criteria for Land Use Planning Safety. Haz. Ind. Planning Guideline 4*. Dept of Planning, Sydney

NEW SOUTH WALES GOVERNMENT (1991). *Hazard Audit Guidelines. Haz. Ind. Planning Guideline 5*. Dept of Planning, Sydney

NEWALL, J. (1969). Fire control: automatic. In Handley, W. (1969), *op. cit.*, p. 92

NEWBOLD, E.M. (1926). *A Contribution to the Study of the Human Factor in the Causation of Accidents. Ind. Fatigue Res. Board, Rep. 19* (London: HM Stationery Office)

NEWBORN, M. (1975). *Computer Chess* (New York: Academic Press)

NEWBROUGH, E.T. (1967). *Effective Maintenance Management* (New York: McGraw-Hill).

NEWBY, M. (1986). Applications of concepts of ageing in reliability data analysis. *Reliab. Engng*, **14**(4), 291

NEWBY, T. (1988). Sealing options for pumps. *Chem. Engr, Lond.*, **447**, 33

NEWBY, T. and FORTH, D. (1991). Glandless pumps and valves – a technical update. *Hazards*, **XI**, p. 119

NEWBY, T. and FORTH, D. (1991). Glandless pumps and valves – a technical update. *IChemE Symp. Series* No. 124, 119–34

NEWCOMBE, H.B. (1974). *A Method of Monitoring Nationally for Possible Delayed Effects of Various Occupational Environments. Rep. NRCC 13686*. Nat. Res. Coun. Canada

NEWELL, A. (1973). Production systems: models of control structures. *Visual Information Processing* (ed. W.G. Chase) (New York: Academic Press), p. 463

NEWELL, A. and SIMON, H.A. (1963). GPS, a program that simulates human thought. In Feigenbaum, E.A. and Feldman, J. (1963), *op. cit.*, p. 279

NEWELL, A. and SIMON, H.A. (1972). *Human Problem-Solving* (Englewood Cliffs, NJ: Prentice-Hall)

NEWELL, A., SHAW, J.C. and SIMON, H.A. (1958). Chess-playing programs and the problem of complexity. *IBM J. Res. Dev.*, **2**, 320

NEWELL, A., SHAW, J.C. and SIMON, H.A. (1960). Report on a general problem-solving program for a computer. *Proc. Int. Conf. on Information Processing* (Paris: UNESCO), p. 256

NEWELL, R.G. (1974). Algorithms for the design of chemical plant layout and pipe routing. Ph.D. Thesis, Univ. of London

NEWITT, D.M. (1937). The oxidation of hydrocarbons at high pressure. *Combustion*, **2**, p. 157

NEWITT, D.M. (1940). *The Design of High Pressure Plant and Properties of Fluids at High Pressure* (New York: Oxford Univ. Press)

NEWITT, D.M. and THORNES, L.S. (1937). The oxidation of propane. Part I, The products of the slow oxidation at atmospheric and at reduced pressures. *J. Chem. Soc.*, **1656**

NEWITT, E.J. (1955). The slow oxidation of some aliphatic alcohols. *Combustion*, **5**, 558

NEWLAND, D.E. (1975). *Report of an Investigation of Possible Causes of Failure of the 20 inch Bypass Pipe Assembly at Flixborough. Rep. to Flixborough Court of Inquiry* (Univ. of Sheffield)

NEWMAN, D.R., RYDEN, J. and LAMB, L.T. (1970). Ultrasonic thickness measurements. Instrum. *Control Systems*, Dec., 71

NEWMAN, J.S. and TEWARSON, A. (1983). Flame propagation in ducts. *Combust. Flame*, **51**, 347

NEWMARK, N.M. (1956a). An engineering approach to blast resistant design. *Trans Am. Soc. Civil Engrs*, **121**, 45

NEWMARK, N.M. (1956b). *An Engineering Approach to Blast Resistant Design* (Urbana, IL: Univ of Illinois)

NEWMARK, N.M. and HALL, W.J. (1973). A rational approach to seismic design standards for structures. *Proc Fifth World Conf. on Earthquake Engineering*, vol. 2, p. 2266

NEWMARK, N.M. and HALL, W.J. (1975). Pipeline design to resist large fault displacements. *Proc. US Nat. Conf. on Earthquake Engng* (California: Earthquake Engng Res. Inst.)

NEWMARK, N.M. and HALL, W.J. (1978). Development of criteria for seismic review of selected nuclear power plants. Nucl. Regul. Comm., Washington, DC, *Rep. NUREG/CR-0098*

NEWMARK, N.M. and ROSENBLUETH, E. (1971). *Fundamentals of Earthquake Engineering* (Englewood Cliffs, NJ: Prentice-Hall)

NEWTON, C.L. (1979). Cryogenics. *Kirk−Othmer Encyclopaedia of Chemical Technology*, 3rd ed., vol. 7 (New York: Wiley), p. 227

NEWSOM, G. and SHERRATT, J.C. (1972). *Water Pollution* (London: John Sherratt)

NG, H.J. and ROBINSON, D.B. (1976). The role of n-butane in hydrate formation. *AIChE J.*, **22**, 656

NG, H.J. and ROBINSON, D.R. (1976). The measurement and prediction of hydrates formation in liquid hydrocarbon systems. *Ind. Engng. Chem. Fund.*, **15**, 293

NG, L.K., ENG, C.N. and ZACK, R. (1983). Ethylene from NGL feedstocks, Pt 3. *Hydrocarbon Process.*, **62**(12), 99

NG TONG LENG (1990). Plant maintenance in the small and medium chemical factories. *Safety and Loss Prevention in the Chemical and Oil Processing Industries*, p. 627

NGUYEN, K. and BRANCH, M.C. (1987). Ignition temperature of bulk 6061 aluminum, 302 stainless steel and 1018 carbon steel in oxygen. *Combust. Sci. Technol.*, **53**, 277

NGUYEN, T.V. and POPE, S.B. (1984). Monte Carlo calculations of turbulent diffusion flames. *Combust. Sci. Technol.*, **42**, 13

VAN NICE, L.J. and CARPENTER, H.J. (1965). *Thermal Radiation from Saturn Fireballs*. *Rep. NAS 9-4810*. TRW Systems, Redondo Beach, CA

NICHOLAS, J. (1983). Pressure dam bearing design for optimum turbomachinery stability. *Hydrocarbon Process.*, **62**(4), 91

NICHOLLS, C. (1974). Vibration monitoring and analysis of critical machinery. *Plant Engr*, **18**(6), 13

NICHOLLS, G. (1981). Basis and development of sprinkler systems. *Fire Surveyor*, **10**(3), 12

NICHOLLS, H.J. (1984). *Induction Heating* (London: Electricity Council)

NICHOLLS, J.A., SICHEL, M., GABRIJEL, Z., OZA, R.D. and VANDER-MOLEN, R. (1979). Detonability of unconfined natural gas–air clouds. *Combustion*, **17**, 1223

NICHOLS, F.P. (1994). Pressure relief – the next major challenge? *Hazards*, **XII**, p. 457

NICHOLS, J. (1982). Eye safety in chemical operations. In Fawcett, H.H. and Wood, W.S. (1982), *op. cit.*, p. 573

NICHOLS, R.W. (chrmn) (1969). *Report of the Committee of Inquiry on Pressure Vessels* (London: Min. of Technol.)

NICHOLS, R.W. (1973). Some applications of fracture mechanics in power engineering. *Int. J. Pres. Ves. Piping*, **1**, 281

NICHOLS, R.W. (ed.) (1976a). *Acoustic Emission* (Barking: Applied Science)

NICHOLS, R.W. (ed.) (1976b). *Pressure Vessel Engineering Technology* (Barking: Applied Science)

NICHOLS, R.W. (ed.) (1979a). *Developments in Pressure Vessel Technology. 1, Flaw Analysis* (London: Applied Science)

NICHOLS, R.W. (ed.) (1979b). *Developments in Pressure Vessel Technology. 2, Inspection and Testing* (London: Applied Science)

NICHOLS, R.W. (ed.) (1980). *Developments in Pressure Vessel Technology. 3, Materials and Fabrication* (London: Applied Science)

NICHOLS, R.W. (ed.) (1983). *Developments in Pressure Vessel Technology. 4, Design for Specific Applications* (London: Applied Science)

NICHOLS, R.W. (ed.) (1987). *Developments in Pressure Vessel Technology. 5, Pressure Vessel Standards and Codes* (London: Elsevier)

NICHOLS, R.W. (ed.) (1984). *Non-Destructive Examination for Pressurised Components* (Amsterdam: Elsevier Applied Science)

NICHOLS, S. and READINGS, C.J. (1979). Aircraft observations of the structure of the lower boundary layer over the sea. *Quart. J. Roy. Met. Soc.*, **105**, 785

NICHOLSON, A.S. (1989). A comparative study of methods for establishing load handling capabilities. *Ergonomics*, **32**, 1125

NICHOLSON, C.S. (1969). Personal equipment and protection. In Handley, W. (1969), *op. cit.*, p. 381

NICHOLSON, G.D. (1974). Why the swing to wafer type check valves. *Hydrocarbon Process.*, **53**(6), 93

NICHOLSON, K.R. (1976). Protective clothing for Coast Guard response personnel. *Control of Hazardous Material Spills*, **3**, 173

NICHOLSON, K.W. (1988). The dry deposition of small particles: a review of experimental measurements. *Atmos. Environ.*, **22**, 2653

NICHOLSON, K.W. (1993). Wind-tunnel experiments on the resuspension of particulate material. *Atmos. Environ.*, **27A**, 181

NICHOLSON, M. (1967). *The System* (London: Hodder & Stoughton)

NICHOLSON, M. (1970). *The Environmental Revolution* (London: Hodder & Stoughton)

NICHOLSON, M.K., AZIZ, K. and GREGORY, G.A. (1978). Intermittent two-phase flow in horizontal pipes: predictive models. *Can. J. Chem. Engng*, **56**, 653

NICHOLSON, W.J., HAMMOND, E.C., SEIDMAN, H. and SELIKOFF, I.J. (1975). Mortality experience of a cohort of vinyl chloride-polyvinyl chloride workers. *Ann. N.Y. Acad. Sci.*, **246**, 225

NICKEL, H. (1980). Talking back to the anti-nukes. *Fortune*, Jan. 28, 108

NICKERSON, M.H. (1954). Water flow characteristics of sprinkler systems. *Nat. Fire Prot. Ass., Proc. Fifty-Eighth Ann. Mtg*, p. 140

NICKOLA, P.W. (1977). *The Hanford 67–Series; a Volume of Atmospheric Field Diffusion Measurements. Rep. PNL-2433.* Battelle Pacific Northwest Labs, Richland, WA

NICOLAESCU, D. (1977). Problems relating to industrial safety and safety engineering in the rubber and plastics industry. *Prevention of Occupational Risks*, **5**, p. 139

NICOLESCU, T. and WEBER, R. (1981). Reliability of systems with various functions. *Reliab. Engng*, **2**(2), 147

NICOLSON, A. (1991). The effect of O_2 concentration on methacrylic acid stability. *Plant/Operations Prog.*, **10**, 171

NICOLSON, N.J. (1979). A review of the nitrate problem. *Chem. Ind.*, Mar. 17, 189

NIDEN, A. (1968). Effect of ammonia inhalation of the terminal airways. *Eleventh Aspen Emphysema Conf.*, **11**, 41

NIEH, C.D. and ZENGYAN, H. (1986). Estimate exchanger vibration. *Hydrocarbon Process.*, **65**(4), 61

NIELSEN, A. (1964). Safety aspects of ammonia catalyst reduction and catalyst disposal. *Ammonia Plant Safety*, **6**, p. 7

NIELSEN, A. (1971). Crack formation in storage tanks. *Ammonia Plant Safety*, **13**, 103

NIELSEN, A., HANSEN, J.B., HOUKEN, J. and GAM, E.A. (1982). Revamp of ammonia plants. *Plant/Operations Prog.*, **1**, 186

NIELSEN, D.S. (1971). *The Cause–Consequence Diagram Method as a Basis for Quantitative Accident Analysis. Rep. Riso-M-1374.* Atom. Energy Comm., Res. Estab., Riso, Denmark

NIELSEN, D.S. (1974). *Use of Cause–Consequence Charts in Practical Systems Analysis. Rep. Riso-M-1743.* Atom. Energy Comm., Res. Estab., Riso, Denmark

NIELSEN, D.S. (1975). Use of cause-consequence charts in practical systems analysis. In Barlow, R.E., Fussell, J.B. and Singpurwalla, N.D. (1975), *op. cit.*

NIELSEN, D.S. (1977). Benefits of a reliability analysis of a proposed instrument air system. *First Nat. Conf. on Reliability NRC4/2.* (Culcheth, Lancashire: UK Atom. Energy Auth.)

NIELSEN, D.S. (1991). Validation of two-phase outflow model. *J. Loss Prev. Process Ind.*, **4**, 236

NIELSEN, D.S. and PLATZ, O. (1983). An assessment of the rate of occurrence of high hydrogen content in a chlorine plant. *Reliab. Engng*, **4**, 4

NIELSEN, D.S., PLATZ, O. and KONGSO, H.E. (1977). *Reliability Analysis of Proposed Instrument Air System. Rep. Riso-M-1903.* Atom. Energy Comm., Res. Estab., Riso, Denmark

NIELSEN, D.S., PLATZ, O. and RUNGE, B. (1977). A cause—consequence chart of a redundant protection system. *IEEE Trans Reliab.*, **R-24**, 8

NIELSEN, J.N. (1960). *Missile Aerodynamics* (New York: McGraw-Hill)

NIELSEN, L.B., PRAHM, L.P., BERKOWICZ, R. and CONRADSEN, K. (1981). Net radiation estimated from standard meteorological data. In de Wispelaere, C. (1981), *op. cit.*, p. 385

NIELSEN, L.R. (1978). *Predicting the Properties of Mixtures: Mixture Rules in Science and Engineering* (New York: Dekker)

NIELSEN, M. (1991). Dense gas field experiments with obstacles. *J. Loss Prev. Process Ind.*, **4**(1), 29

NIELSEN, M. and JENSEN, N.O. (1989). Research on continuous and instantaneous gas clouds. *J. Haz. Materials*, **21**, 101

NIEMEYER, C.E. and LIVINGSTON, G.N. (1993). Choose the right flare system design. *Chem. Engng Prog.*, **89**(12), 39

NIEMKIEWICZ, I.J. (1962). Canned pumps solve process problems. *Chem. Engng*, **69**(Feb. 19), 117

NIESS, J. (1980). Particulate removal from NO_x abatement off-gas in a nitric acid plant. *Ammonia Plant Safety*, **21**, p. 171

NIEUWHOF, G.W.E. (1975). An introduction to fault tree analysis with emphasis on failure rate evaluation. *Microelectron. Reliab.*, **14**, 105

NIEUWHOF, G.W.E. (1981). Reliability notes. *Reliab. Engng*, **2**(4), 315

NIEUWHOF, G.W.E. (1983a). Human error. *Reliab. Engng*, **6**, 191

NIEUWHOF, G.W.E. (1983b). Some comments on the terminology used in the availability analysis of a continuously operating repairable system. *Reliab. Engng*, **5**, 117

NIEUWHOF, G.W.E. (1984). The concept of failure in reliability engineering. Reliab. Engng, **7**, 53

NIEUWHOF, G.W.E. (1985a). A graphical interpretation of the general equation of reliability. *Reliab. Engng*, **11**, 121

NIEUWHOF, G.W.E. (1985b). Matching a failure probability distribution to test data. *Reliab. Engng*, **11**(3), 163

NIEUWHOF, G.W.E. (1985c). Risk: a probabilistic concept. *Reliab. Engng*, **10**, 183

NIEUWHOF, G.W.E. (1986). An analytical interpretation of the general equation of reliability. *Reliab. Engng*, **14**(2), 149

NIEUWSTADT, F.T.M. (1980a). An analytic solution of the time-dependent, one-dimensional diffusion equation in the atmospheric boundary layer. *Atmos. Environ.*, **14**, 1361

NIEUWSTADT, F.T.M. (1980b). Application of mixed layer similarity to the observed dispersion from a ground-level source. *J. Appl. Met.*, **19**, 157

NIEUWSTADT, F.T.M. (1984a). Some aspects of the turbulent stable boundary layer. *Boundary Layer Met.*, **30**, 31

NIEUWSTADT, F.T.M. (1984b). The turbulent structure of the stable nocturnal boundary layer. *J. Atmos. Sci.*, **41**, 2202

NIEUWSTADT, F.T.M. (1992). A large-eddy simulation of a line source in a convective atmospheric boundary layer. *Atmos. Environ.*, **26A**, 485; 497

NIEUWSTADT, F.T.M. and VAN DOP, H. (eds) (1982). *Atmospheric Turbulence and Air Pollution Modelling* (Dordrecht: Reidel)

NIEUWSTADT, F.T.M. and VAN DUUREN, H. (1979). Dispersion experiments with SF6 from the 213 m high meteorological mast at Cabauw in the Netherlands. *Fourth Symp. on Turbulence, Diffusion and Air Pollution* (Boston, MA: Am. Met. Soc.), p. 34

NIEUWSTADT, F.T.M. and VAN ULDEN, A.P. (1978). A numerical study on the vertical dispersion of passive contaminants

from a continuous source in the atmospheric surface layer. *Atmos. Environ.*, **12**, 2119

NIGHTINGALE, P.J. (1989a). Investigation into an explosion that occurred during a welding operation. *Ammonia Plant Safety*, **29**, p. 14. See also *Plant/Operations Prog.*, 1989, **8**, 29

NIGHTINGALE, P.J. (1989b). Severe cracking of CO_2 absorber exit line. *Ammonia Plant Safety*, **29**, p. 95

NIGHTINGALE, P.J. (1990). Cause of cracking in a high pressure synthesis gas pipe. *Ammonia Plant Safety*, **30**, p. 100

NII, H.P. and AIELLO, N. (1979). AGE (Attempt to Generalize): a knowledge-based program for building knowledge-based programs. *Int. Joint Conf. on Artificial Intelligence* **6**, p. 645

NIIDA, K., ITOH, J., UMEDA, T., KOBAYASHI, S. and ICHIKAWA, A. (1986). Some expert system experiments in process engineering. *Chem. Engng Res. Des.*, **64**, 372

NIKMO, J. and KUKKONEN, J. (1992). Heavy gas cloud transport in complex terrain. *Loss Prevention and Safety Promotion* **7**, vol. 1, 17–1

NIKODEM, H.J. (1978). *Risk Assessment Study for an Assumed LNG Terminal in the Lysekil Area.* Battelle Institute.V., Frankfurt-am-Main, Rep. to Swedish Energy Comm.

NIKODEM, H.J. (1980). Risk analysis for storage and transport of LNG. In Hartwig, S. (1980), *op. cit.*, p. 177

NIKODEM, H.J. (1983). Examples of analyses of gas cloud explosion hazards. In Hartwig, S. (1983), *op. cit.*, p. 305

NIKURADSE, J. (1933). Laws of flow in rough pipes. *Forschungsheft VDI*, 361

NILLSON, E. (1983a). Authorities aspects. *Loss Prevention and Safety Promotion 4*, vol. 1, p. H19

NILLSON, E. (1983b). The propane explosion in Gothenburg 8th May 1981. *Loss Prevention and Safety Promotion*, vol. 1, p. L23

NILSEN, J.M. (1972a). OSHA: acronym for trouble. *Chem. Engng*, **79**(Mar. 20), 58

NILSEN, J.M. (1972b). Tackling pollution control. *Chem. Engng*, **79**(Feb. 7), 34

NILSEN, J.M. (1974). VCM — exposure standards: too tough too soon? *Chem. Engng*, **80**(Jun. 24), 110

NILSSON, B. (1989). Structured modelling of chemical processes with control systems. *Am. Inst. Chem. Engrs, Ann Mtg, San Francisco* (New York), paper 27g

NILSSON, N. (1971). *Problem-Solving Methods in Artificial Intelligence* (New York: McGraw-Hill)

NILSSON, N. (1980). *Principles of Artificial Intelligence* (Palo Alto, CA: Tioga)

NILSSON, N.J. (1965). *Learning Machines* (New York: McGraw-Hill)

NILSSON, N.J. (1988). Action networks. In *From Formal Systems to Practical Systems* (ed. J. Weber, J. Tenenberg and J. Allen). Dept of Comput. Sci., Univ. of Rochester, Rochester, NY

VON NIMITZ, W., WACHEL, J.C. and SZENASI, F.R. (1974). Case histories of specialised turbomachinery problems. *Hydrocarbon Process.*, **53**(4), 141

NIMMO, I. (1992). Make the most of instrumentation modernization projects. *Chem. Engng Prog.*, **88**(6), 26

NIMMO, I. (1993). Start up plants safely. *Chem. Engng Prog.*, **89**(12), 66

NIMMO, I., NUNN, S.R. and EDDERSHAW, B.W. (1987). Lessons learned from the failure of a computer system controlling a nylon polymer plant. In Daniels, B.K. (1987), *op. cit.*

NIMPTSCH, P. (1979). Emergency planning and transport safety. In BASF (1979), *op. cit.*, p. 295

NING, D.T. and YAP, D. (1986). Climatology of convective boundary-layer parameters over Ontario, Canada. *Atmos. Environ.*, **20**, 2315

NINO, L.E. (1974). Predict failures and overhaul interval to minimise downtime. *Hydrocarbon Process.*, **53**(1), 108

NIPPES, P.I. (1981). Electromagnetic shaft currents in turbomachinery: an update, Pt 2. *Ammonia Plant Safety*, **23**, 196

NISBET, D.F. (1971). Case history: failures in a steam-methane reformer furnace. *Hydrocarbon Process.*, **50**(5), 103

NISBET, T.S. (1974). *Rolling Bearings* (Oxford: Oxford Univ. Press)

NISENFELD, A.E. (1972). Shutdown features of in-line process control. *Loss Prevention*, **6**, 1. See also *Chem. Engng Prog.*, 1972, **68**(4), 54

NISENFELD, A.E. *et al.* (1975). For easier compressor control. *Hydrocarbon Process.*, **54**(4), 153

NISENFELD, A.E. and CHO, C.H. (1978). Parallel compressor control . . . what should be considered. *Hydrocarbon Process.*, **57**(2), 147. See also in Kane, L.A. (1980), *op. cit.*, p. 126

NISENFELD, A.E. and SEEMAN, R.C. (1981). *Distillation Columns* (Pittsburgh, PA: Instrum. Soc. Amer.)

NISHIDA, N., STEPHANOPOULOS, G. and WESTERBERG, A.W. (1981). A review of process synthesis. *AIChE J.*, **27**, 321

NISHIDA, S. (1991). *Failure Analysis in Engineering Applications* (London: Butterworth-Heinemann)

NISHIKAWA, K. (1993). Case study – who is guilty? *Health, Safety and Loss Prevention*, p. 575

NISHIO, G. and MACHIDA, S. (1987). Pool fires under atmosphere and ventilation in steady-state burning. *Fire Technol.*, **23**, 146; 186

NISSLER, K.H. (1991). Keeping turbocompressors in top shape. *Chem. Engng*, **98**(3), 104

NITTINGER, R.H. (1964). Vibration analysis can keep your plant humming. *Chem. Engng*, **71**(Aug. 17), 152

NITTINGER, R.H. (1966). Non-destructive testing. *Chem. Engng*, **73**(Feb. 28), 87

NITZ, K.C., ENDLICH, R.M. and LUDWIG, F.L. (1986). On-line wind field calculations for routine and emergency response applications. In de Wispelaere, C., Schiermeier, F.A. and Gillani, N.V. (1986), *op. cit.*, p. 123

NIVOLIANITOU, Z., AMENDOLA, A. and REINA, G. (1986). Reliability analysis of chemical processes by the DYLAM approach. *Reliab. Engng*, **14**, 163

NIXON, F. (1971). *Managing to Achieve Quality and Reliability* (New York: McGraw-Hill)

NIXON, J. (1971). 'Risk point' determination as an alternative to hazardous area classification: Part 2. *Electrical Safety*, **1**, 150

NIXON, W., HILL, M.D. and NAIR, S. (1989). An overview of CONDOR – a new UK probabilistic accident consequence analysis code. *Proc. PSA 89*

NOBLES, E.J. (1967). Problems in shutting down and starting up. *Ammonia Plant Safety*, **9**, 84

NOCK, E. (1985). Welding fumes and gases. *Plant Engr*, **29**(5), 21

NOGAJ, R.J. (1972). Selecting wastewater aeration equipment. *Chem. Engng*, **79**(Apr. 17), 95

NOGITA, S. (1972). Statistical test and adjustment of process data. *Ind. Engng Chem. Process Des. Dev.*, **11**, 197

NOGITA, S. and UCHIYAMA, Y. (1973). Multibalance measurement of flow systems. *Proc. 1973 Ann. Reliability and Maintainability Symp.*, p. 590

NOIA (1994). Status report on legislation, regulations impacting US offshore. *Offshore*, May, 72

NOISE ADVISORY COUNCIL (1975). *Noise Units* (London: HM Stationery Office)

NOISE ADVISORY COUNCIL (1978). *A Guide to Measurement and Prediction of the Equivalent Continuous Sound Level* (London: HM Stationery Office)

DI NOLA, A., SESSA, S., PEDRYCZ, W. and SANCHEZ, E. (1989). *Fuzzy Relation Equations and Their Application to Knowledge Engineering* (Dordrecht: Kluwer)

NOLAN, B. (1979). Westwego, Louisiana: 36 die in grain elevator blast. *Firehouse*, **3**, 42

NOLAN, M.E. (1973). A simple model for the detonation limits of gas mixtures. *Combustion Sci. Technol.*, **7**, 57

NOLAN, P.F. (1989). Safety and loss prevention teaching. *J. Loss Prev. Process Ind.*, **2**(1), 3

NOLAN, P.F. and BARTON, J.F. (1987). Some lessons from thermal-runaway incidents. *J. Haz. Materials*, **14**(2), 233

NOLAN, P.F. and BRADLEY, W.J. (1987). A simple technique for the optimization of layout and location for chemical plant safety. *Plant/Operations Prog.*, **6**, 57

NOLAN, P.F. and BROWN, D.M. (1986). The interaction of shock waves and cylindrical structures. *Am. Inst. Chem. Engrs, Symp. on Loss Prevention* (New York)

NOLAN, P.F. and PROCTOR, P. (1988). *A Review of Known Methods for the Vent Sizing of Chemical Reactor Plant. HSE Contract Res. Rep. 9/1988*. Health and Safety Executive, London

NOLAN, P.F., PETTITT, G.N., HARDY, N.R. and BETTIS, R.J. (1990). Release conditions following loss of containment. *J. Loss Prev. Process Ind.*, **3**(1), 97

NOLAN, P.F., HARDY, N.R. and PETTITT, G.N. (1991). The physical modelling of two-phase releases following sudden failure of vessels. *Modeling and Mitigating the Consequences of Accidental Releases*, p. 125

NOLAN, P.F., PETTITT, G.N. and HARDY, N.R. (1991). Characterization of two-phase releases. *Probabilistic Safety Assessment Management* (ed. G. Apostolakis) (New York: Elsevier)

NOMA, E., BERGLUND, B., BERGLUND, U., JOHANSSON, I. and BAIRD, J.C. (1988). Joint representation of physical location and volatile organic compounds in indoor air from a healthy and a sick building. *Atmos. Environ.*, **22**, 451

NOMURA, S. and TANAKA, T. (1980). Prediction of maximum rate of pressure rise due to dust explosion in closed spherical and non-spherical vessels. *Ind. Engng Chem. Process Des. Dev.*, **19**, 451

NOMURA, S., TORIMOTO, M. and TANAKA, T. (1984). Theoretical upper limit of dust explosion in relation to oxygen concentration. *Ind. Engng Chem. Process Des. Dev.*, **23**, 420

NONHEBEL, G. (1958). The technical provisions of the British Clean Air Act. *Int. J. Air Pollution*, **1**, 120

NONHEBEL, G. (1960). Recommendations on heights for new industrial chimneys. *J. Inst. Fuel*, **33**, 479

NONHEBEL, G. (1972). *Gas Purification Processes for Air Pollution Control* (London: Newnes-Butterworths)

NONHEBEL, G. (1975). Gases vented to relief valves (letter). *Chem. Engr, Lond.*, **293**, 48

NONHEBEL, G. (1981). Industrial air pollution: British progress – a review. *Atmos. Environ.*, **15**, 213

NOON, RANDALL K. (2000). *Forensic Engineering Investigation* Chapter 23 (New York: CRC Press), (Managements Role), ISBN 0-8493-0911-5

DE NORA, O. and GALLONE, P. (1968). Chlorine. In *The Encyclopaedia of the Chemical Elements* (ed. C.A. Hampel) (New York: Reinhold)

NORCROSS, G. (1981). A company's experience of environmental protection requirements at Continental European coastal sites. In Cairns, W.J. and Rogers, P.M. (1981), *op. cit.*, p. 249

NORDA, T. (1976). On-line infra-red inspection and petrochemical plants. *Diagnostic Techniques* (London: Instn Chem. Engrs)

NORDA, T. (1977). Use infra-red scanning to find equipment hot spots. *Hydrocarbon Process.*, **56**(1), 109

NORDEN, R.B. (1960a). Evaluating wear-resistant materials. *Chem. Engng*, **67**(Jul. 11), 162

NORDEN, R.B. (1960b). New on-stream corrosion check. *Chem. Engng*, **67**(May 30), 123

NORDEN, R.B. (1960c). Reinforced plastic better than steel for storage tanks. *Chem. Engng*, **67**(Apr. 18), 214

NORDEN, R.B. (1960d). Tankers: stainless to plastic gamut. *Chem. Engng*, **67**(Jan. 25), 130

NORDEN, R.B. (1965). Modern engineering design practices. *Chem. Engng*, **72**(Aug. 2), 91

NORDGARD, T. (1990). *Piper Alpha Public Inquiry. Transcript, Day 165*

NORDIC COUNCIL OF MINISTERS (1987). *Personal Safety in Microprocessor Control Systems. Rep. NORD 1987:189*

NORDIC LIAISON COMMITTEE FOR ATOMIC ENERGY — see Appendix 28

NORDMO, F. and EMBLEM, K. (1989). Hydrogen explosion in a hydrogen gasometer. *Loss Prevention and Safety Promotion 6*, p. 40.1

NORDQUIST, J.M. (1945). Theory of largest values applied to earthquake motions. *Trans. Am. Geophys. Union*, **26**, 29

NORMAN, C. (1975). Candle power at Brown's Ferry. *Nature*, **257**(Oct. 16), 525

NORMAN, C. (1987). A guide to the use of foam on hazardous material spills. *Haz. Materials Waste Mngmt*, Sep./Oct.

NORMAN, D.A. and RUMELHART, D.E. and the LNR Research Group (1975). *Explorations in Cognition* (San Francisco, CA: Freeman)

NORMAN, E.C. and DIMAIO, L.R. (1989). Aqueous foams for vapor mitigation – practical considerations. *Loss Prevention and Safety Promotion 6*, p. 106.1

NORMAN, E.C. and DOWELL, H.A. (1978). The use of foams to control vapor emissions from hazardous material spills. *Control of Hazardous Material Spills 4*, p. 399

NORMAN, E.C. and DOWELL, H.A. (1980). Using aqueous foams to lessen vaporization from hazardous chemical spills. *Loss Prevention*, **13**, 27

NORMAN, E.C. and SWIHART, T.M. (1990). Foam system design for ammonia storage tank areas. *Ammonia Plant Safety*, **30**, 84. See also *Plant/Operations Prog.*, **10**, 22

NORMAN, P.W. and NAVEED, S. (1988). Knowledge acquisition analysis for the construction of real-time supervisory expert systems. *Chem. Engng Res. Des.*, **66**, 470

NORMAN, R.S. (1973). Predict plant noise problems. *Hydrocarbon Process.*, **52**(10), 89

NORMAN, W.S. (1961). *Absorption, Distillation and Cooling Towers* (London: Longmans)

NORONHA, J.A. (1986). Use of the DIERS bench-scale apparatus for restabilization and venting runaway reactions. *Loss Prevention and Safety Promotion 5*, pp. 41–1

NORONHA, J.A. and JUBA, M.R. (1980). An update: kinetic model and tests for runaway reactions. *Loss Prevention and Safety Promotion 3*, vol. 3, p. 892

NORONHA, J.A., JUBA, M.R., LOW, H.M., PASCOE, W.E., SCHIFFHAUER, E.J. and SIMPSON, B.L. (1979). Kinetic model and tests for runaway thermally initiated styrene polymerization. *J. Haz. Materials*, **3**(1), 91

NORONHA, J.A., JUBA, M.R., BROWN, C.S. and SCHMITT, J.G. (1980). The kinetics of runaway reactions. *Loss Prevention*, **13**, p. 50

NORONHA, J.A., MERRY, J.T. and REID, W.C. (1982). Deflagration pressure containment (DPC) for vessel safety design. *Plant/Operations Prog.*, **1**, 1

NORONHA, J.A., SEYLER, R.J. and TORRES, A.J. (1989). Simplified chemical and equipment screening for emergency venting safety reviews based on the DIERS technology. *Runaway Reactions*, p. 660

NORRIS, C.H., HANSEN, R.J., HOLLEY, M.J., BIGGS, J.M., NAYMET, S. and MINAMI, J.K. (1959). *Structural Design for Dynamic Loads* (New York: McGraw-Hill)

NORRISH (1935). A theory of the combustion of hydrocarbons. *Proc. R. Soc., Ser. A*, **150**, 36

NORRISH, R.G.W. and FOORD, V.G. (1936). The kinetics of the combustion of methane. *Proc. R. Soc., Ser. A*, **157**, 503

NORSTROM, G.P. (1982a). Fire/explosion losses in the CPI. *Chem. Engng Prog.*, **78**(8), 80

NORSTROM, G.P. (1982b). An insurer's perspective of the chemical industry. In Fawcett, H.H. and Wood, W.S. (1982), *op. cit.*, p. 509

NORSTROM, G.P. (1986). Property insurance considerations in loss prevention expenditures. *Plant/Operations Prog.*, **5**, 209

NORTH, J.D. (1949). Some aspects of the relationship between airworthiness and safety. *J. R. Aeronaut. Soc.*, **53**, 915

NORTH, M.A. (1974). *Some Statistics of Fire in Road Vehicles. Fire Res. Note 1011*. Fire Res. Station, Boreham Wood

NORTH, P.W.T. and PARR, J.G. (1968). Analysis of a high speed coupling failure. *Ammonia Plant Safety*, **10**, 75

NORTHCOTT, J.A. (1969). International co-operation and research. In Handley, W. (1969), *op. cit.*, p. 456

NORTHCROFT, L.G. and BARBER, W.M. (1968). *Steam Trapping and Air Venting* (London: Hutchinson)

NORTHEY, J.W. (1988). Fire detection. The new code and standards. *Fire Surveyor*, **17**(4), 75

NORTHRUP, R.P. (1964). Explosion-proof electrical systems. In Vervalin, C.H. (1964a), *op. cit.*, p. 75

NORTON, A. (1984). Where IT may not build jobs. *Computing*, Dec. 6, 13

NORTON, G., PILKINGTON, D.F. and CARR, J.B. (1994). The strength and mode of failure of a square duct entering a square plate under internal pressure, by experiment, classical and non-linear finite element analysis. *Hazards*, **XII**, p. 75

NORTON, J.F. (ed.) (1984). *High Temperature Materials Corrosion in Coal Gasification Atmospheres* (Amsterdam: Elsevier Applied Science)

NORTON, M.P. and PRUITI, A. (1991). Universal prediction scheme for estimating flow-induced industrial pipeline noise and vibration. *Appl. Acoustics*, **33**, 313

NOSKOFF, A.A. (1935). The German attack against Russian positions at Baranowitschi in the night of 24 to 25 September 1916. *Gasschutz und Luftschutz*, **5**, 128, 153

NOSS, H. (1971). The explosion disaster in the Nordhafen silo in Kiel. *Brandschutz*, **25**, 285

NOTMAN, J., GERRARD, A.M. and DE LA MARE, R. (1981). Reliability of solids handling equipment. *Reliable Production in the Process Industries* (Rugby: Instn Chem. Engrs), p. 171

NOTT, S.L. (1988). Implementing the new superfund. *J. Haz. Materials*, **18**(3), 229

NOTTAGE, H.B., SLABY, J.G. and GOJSZA, W.P. (1952). Isothermal ventilation-jet fundamentals. *Heating, Piping and Air Conditioning*, **24**(1), 165

NOVACEK, D.A. (1983). Pump and motor failure in a hot potassium-carbonate system. *Plant/Operations Prog.*, **2**, 209

NOVAK, C.F., SCHECHTER, R.S. and LAKE, L.W. (1988). Rule-based mineral sequences in geochemical flow processes. *AIChE J.*, **34**, 1607

NOVAK, P. and CABELKA, J. (1981). *Models in Hydraulic Engineering: Physical Principles and Design Applications* (London: Pitman)

NOVAK, R.G. (1970). Eliminating or disposing of industrial solid wastes. *Chem. Engng*, **77**(Oct. 5), 78

NOVAK, R.G. (1972). Hazardous chemicals disposal in a large chemical complex. *Loss Prevention*, **6**, 135

NOVAK, R.G. and PFROMMER, C. (1982). Incineration on land. In G.F. Bennett, G.F., Feates and Wilder (1982), *op. cit.*, p. 14–20

NOVICK, S. (1970). *The Careless Atom* (New York: Delta)

NOVOTNY, M., PANTOFLICEK, J. and LEBR, F. (1972). Some problems of determination of ignition temperature in a powder-air mixture. *Int. Symp on Dust and Explosion Risks in Mines and Industry* (Heidelberg: Berufsgenossenschaft der Chem. Ind.), p. 181

NOWAK, J.D. and JOYE, D.D. (1981). Analytical equations for head vs flow rate characteristics of branch/recycle piping systems. *Ind. Engng Chem. Process Des. Dev.*, **20**, 460

NOWICKI, P.L. (1991). Improve reactor control. *Chem. Engng Prog.*, **87**(12), 32

NOWLAN, F.S. (1972). Planning and operational aspects of 'on condition' philosophies. *Aircraft Engng*, **44**(Mar.), 26

NOWLAN, F.S. and HEAP, H. (1978). *Reliability-centred Maintenance* (report). Nat. Tech. Inf. Service, Springfield, VA

NOWLEN, S.P. (1992). Nuclear power plants: a unique challenge. *Fire Safety J.*, **19**, 3

NOZAKI, N., BESSYO, K., KAWAGUCHI, Y., OZAWA, H. and NAKANISHI, M. (1983). Fracture-safe steel plate for LPG tanks. *Gastech 82*, p. 437

NUCKOLLS, A.H. (1929). *Underwriters Laboratory Method for the Classification of the Hazards of Liquids* (Chicago, IL: Nat. Board Fire Underwriters)

NUCLEAR ENERGY AGENCY (1977). *Proceedings of Meeting of a Task Force on Problems of Rare Events in the Reliability Analysis of Nuclear Power Plants. CSNI Rep. 23* (Paris: OECD)

NUCLEAR ENERGY AGENCY (1984). *International Comparison Study on Reactor Accident Consequence Modelling* (Paris: OECD)

NUCLEAR ENERGY AGENCY (1987). *The Radiological Impact of the Chernobyl Accident in OECD Countries* (Paris: OECD)

NUCLEAR INSTALLATIONS INSPECTORATE – see Appendix 28 under Health and Safety Executive

NUCLEAR REGULATORY COMMISSION (1972). *Onsite Meteorological Programs. Regulatory Guide 1.23* (Washington, DC)

NUCLEAR REGULATORY COMMISSION (1973). *Design Response Spectra for Seismic Design of Nuclear Power Plants. Regulatory Guide 1.60* (Washington, DC)

NUCLEAR REGULATORY COMMISSION (1974). *Methods for Estimating Atmospheric Transport and Dispersion of Gaseous Effluents in Routine Releases from Light-water-cooled Reactors. Regulatory Guide 1.111* (Washington, DC)

NUCLEAR REGULATORY COMMISSION (1976). *Recommendations Related to Browns Ferry Fire. Rep. NUREG-0050* (Washington, DC)

NUCLEAR REGULATORY COMMISSION (1977a). *Analysis of the Risk to the Public from Possible Damage to the Diablo Canyon Nuclear Power Station from Seismic Events. Docket Nos. 50–275 and 50–323* (Washington, DC)

NUCLEAR REGULATORY COMMISSION (1977b). *Estimating Aquatic Dispersion of Effluents from Accidents and Routine Reactor Releases. Regulatory Guide 1.113* (Washington, DC)

NUCLEAR REGULATORY COMMISSION (1977c). *Methods for Estimating Atmospheric Transport and Dispersion of Gaseous Effluents in Routine Releases from Light Water Cooled Reactors. Regulatory Guide 1.111* (Washington, DC)

NUCLEAR REGULATORY COMMISSION (1979a). *Atmospheric Dispersion Models for Potential Accident Consequence Assessments at Nuclear Power Plants. Regulatory Guide 1.145* (Washington, DC)

NUCLEAR REGULATORY COMMISSION (1979b). *Investigation into the March 28 1979 Three Mile Island Accident by Office of Inspection and Enforcement* (Washington, DC)

NUCLEAR REGULATORY COMMISSION (1979c). *Recommended Revisions to Nuclear Regulatory Commission Seismic Design Criteria. Rep. NUREG/CR-1161* (Washington, DC)

NUCLEAR REGULATORY COMMISSION (1979d). *TMI-2. Lessons Learnt. Task Force – Final Report* (Washington, DC)

NUCLEAR REGULATORY COMMISSION (1982a). *PRA Procedures Guide. A Guide to the Performance of Probabilistic Risk Assessments for Nuclear Power Plants. Draft Rep. NUREG/CR-2300* (Washington, DC)

NUCLEAR REGULATORY COMMISSION (1982b). *Safety Goals for Nuclear Power Plants: a Discussion Paper. NUREG-OBBO* (Washington, DC)

NUCLEAR REGULATORY COMMISSION (1987a). *Reactor Risk Reference Document (draft). Rep. NUREG/CR-1150* (Washington, DC)

NUCLEAR REGULATORY COMMISSION (1987b). *Report on the Accident at Chernobyl Nuclear Power Station. Rep. NUREG-1250* (Washington, DC)

NUCLEAR REGULATORY COMMISSION (1987c). *Severe Accident Risks: an Assessment for Five US Nuclear Power Plants, Second Draft. Rep. NUREG/CR-1150* (Washington, DC). (Formerly entitled *Reactor Risk Reference Document, First Draft*)

NUCLEAR REGULATORY COMMISSION (1989). *Reactor Risk Reference Document. Rep. NUREG-1150* (Washington, DC)

NUCLEAR REGULATORY COMMISSION – see also Appendix 28

NUCLEAR REGULATORY COMMISSION, STEAM EXPLOSION REVIEW GROUP (1985). *A Review of the Current Understanding of the Potential for Containment Failure from In-vessel Steam Explosions. Rep. NUREG/CR-1116* (Washington, DC)

NUSSER J.A. (1982). Dispersion modeling. In Bennett, G.F., Feates and Wilder (1982), *op. cit.*, p. 8–22

NUSSER, J.A., GALLAGHER, T.W. and ST JOHN, J.P. (1978). Mathematical modeling for impact and control of hazardous material spills. *Control of Hazardous Material Spills*, **4**, p. 422

NUSSEY, C. (1983). Uncertainty in risk assessment. *Euredata 4*, paper 2.3

NUSSEY, C. (1994). Research to improve the quality of hazard and risk assessment for major chemical hazards. *J. Loss Prev. Process Ind.*, **7**, 175

NUSSEY, C. and PAPE, R.P. (1987). The significance of vapour cloud modeling in the assessment of major toxic hazards. *Vapor Cloud Modeling*, p. 889

NUSSEY, C., DAVIES, J.K.W. and MERCER, A. (1985). The effect of averaging time on the statistical properties of sensor records. *J. Haz. Materials*, **11**(1/3), 125. See also in McQuaid, J. (1985), *op. cit.*, p. 125

NUSSEY, C., MERCER, A. and FITZPATRICK, R.D. (1986). The effect of uncertainty in chlorine toxicity data on risk estimation. In Hartwig, S. (1986), *op. cit.*, p. 197

NUSSEY, C., MERCER, A. and CLAY, G.A. (1990). Consequences: toxic aspects of two-phase releases. *J. Loss Prev. Process Ind.*, **3**(1), 156

NUSSEY, C., PANTONY, M.F. and SMALLWOOD, R.J. (1992). Health and Safety Executive's risk assessment tool RISKAT. *Major Hazards Onshore and Offshore*, p. 607

NUSSEY, C., PANTONY, M.F. and SMALLWOOD, R.J. (1993). Health and Safety Executive's Risk Assessment Tool RISKAT. *Process Safety Environ.*, **71B**, 29

NUTTLI, O.W. and HERRMANN, R.B. (1987). Ground motion relations for Eastern North American earthquakes. In Cermak, A.S. (1987), *op. cit.*, p. 231

NYE, W.B. (1972). The hazardous material spill experience in Shawnee Lake, Ohio – a case history. *Control of Hazardous Material Spills*, **1**, 217

NYLUND, J. (1984). Fire survival of process vessels containing gas. *The Protection of Exothermic Reactors and Pressurised Storage Vessels* (Rugby: Instn Chem. Engrs), p. 137

NYPRO (UK) LTD (1975). *Site Investigation Report. Rep. to Flixborough Court of Inquiry*

NYREN, A. (1989). Tests by releasing liquid and gaseous SO_2 from a pressure tank. *Loss Prevention and Safety Promotion 6*, p. 71.1

NYREN, K. and WINTER, S. (1983). Two phase discharge of liquefied gases through pipes. Field experiments with ammonia and theoretical model. *Loss Prevention and Safety Promotion 4*, vol. 1., p. E1

NYREN, K. and WINTER, S. (1987). Discharge of condensed sulfur dioxide: a field test study of the source behaviour with different release geometries. *J. Haz. Materials*, **14**(3), 365

OAKLAND, J.S. (1981). Management tools in the manufacture of chemicals: statistical quality control. *Chem. Ind.*, Aug. 15, 562

OAKLAND, J.S. (1982). Management tools in the manufacture of chemicals: decision networks. *Chem. Ind.*, Nov. 20, 898

OAKLAND, J.S. (1989). *Total Quality Management* (Oxford: Heinemann)

OAKLAND, J.S. (1993). *Total Quality Management, 2nd ed.* (Oxford: Butterworth-Heinemann)

OBERENDER, W. and BUNG, W. (1984). The reliability of controlled safety valves in conventional power plants. *Reliab. Engng*, **9**, 33

OBERG, E., JONES, F.D., HORTON, H.L. and RYFFEL, H.H. (1992). *Machinery's Handbook, 24th ed.* (ed. R.E. Green) (New York: Ind. Press)

OBERHAGEMANN, D. and SCHECKER, H.G. (1989). Theoretical and experimental investigations of the autoignition temperatures of multicomponent fuel/air mixtures. *Loss Prevention and Safety Promotion 6*, p. 107.1

OBERHANSBERG, J. (1977). Dismantling of chemical plants handling hazardous substances. *Loss Prevention and Safety Promotion 2*, p. 77

OBERMESSER, C.F. (1960). Instrumentation and control of nuclear power plants. *Chem. Engng Prog.*, **56**(9), 60

O'BRIEN, G.J., GORDON, M.D., HENSLER, C.J. and MARCALI, K. (1982). Thermal stability hazards analysis. *Chem. Engng Prog.*, **78**(1), 46

O'BRIEN, J.J. (1969). *Scheduling Handbook* (New York: McGraw-Hill)

O'BRIEN, N.G. and PORTER, H.F. (1984). Design of gas–liquid contactors. *Chem. Engng Prog.*, **80**(5), 44

O'BRIEN, T.H. (1955). *Civil Defence* (London: HM Stationery Office)

OBUKHOV, A.M. (1971). Turbulence in an atmosphere with a non-uniform temperature. *Boundary-Layer Met.*, **2**, 7

OCCUPATIONAL HEALTH AND SAFETY ADMINISTRATION (1977). *Identification, Classification and Regulation of Occupational Carcinogens* (Washington, DC)

OCCUPATIONAL HEALTH AND SAFETY ADMINISTRATION (1983). Occupational exposure to inorganic arsenic. *Fed. Register*, **48**(Jan. 14), 1867

OCCUPATIONAL HEALTH AND SAFETY ADMINISTRATION (1990a). *Phillips 66 Company Houston Chemical Complex Explosion and Fire* (Washington, DC)

OCCUPATIONAL SAFETY AND HEALTH ADMINISTRATION (1990b). Process safety management of highly hazardous chemicals: notice of proposed rulemaking. *Federal Register*, **29**(Jul. 17), CFR Pt 1910

OCCUPATIONAL SAFETY AND HEALTH ADMINISTRATION – see also Appendix 28

OCCUPATIONAL SAFETY and HEALTH ADMINISTRATION (US) *Process Safety Management Standard 29CFR 1910.119*(0)(6) Washington, D.C.: US Department of Labor

OCKERBLOOM, N.E. (1972). World benzene outlook. *Hydrocarbon Process.*, **51**(11), 117

O'CONNELL, W.L. (1976). How to attack air pollution control problems. *Chem. Engng*, **83**(Oct. 18), 97

O'CONNOR, C.J. and LIRTZMANN, S.I. (1984). *Handbook of Chemical Industry Labeling* (Park Ridge, NJ: Noyes)

O'CONNOR, P.D.T. (1977). Practical reliability assessment using Weibull analysis. *First Nat. Conf. on Reliability. NRC3/3* (Culcheth, Lancashire; UK Atom. Energy Auth.)

O'CONNOR, P.D.T. (1981). *Practical Reliability Engineering* (London: Heyden)

O'CONNOR, P.D.T. (1984). *Practical Reliability Engineering*, 2nd ed. (Chichester: Wiley)

O'DEA, D.M. (1975). User experience with computerized machinery vibration analysis. *Hydrocarbon Process.*, **54**(12), 81

ODEEN, K. (1985). Fire resistance of wood structures. *Fire Technol.*, **21**, 34

O'DELL, L. (1993). Keep moisture out of compressed air. *Chem. Engng Prog.*, **89**(2), 37

O'DELL, L.R. (1992). Beware of group trapping. *Chem. Engng Prog.*, **88**(1), 37

ODI, T.O. and KARIMI, I.A. (1988). Sizing of intermediate storage for variabilities in noncontinuous processes with parallel units. *Comput. Chem. Engng*, **12**, 561

O'DOGHERTY, M.J. (1965). *The Shock Hazard Associated with the Extinction of Fires involving Electrical Equipment. Fire Res. Tech. Paper 13* (London: HM Stationery Office)

O'DOGHERTY, M.J., NASH, P. and YOUNG, R.A. (1966). *A Study of the Performance of Automatic Sprinkler Systems. Fire Res. Tech. Paper 17* (London: HM Stationery Office)

O'DONOVAN, K.H. and RALLIS, C.J. (1959). A modified analysis for the determination of the burning velocity of a gas mixture in a spherical constant volume combustion vessel. *Combust. Flame*, **3**, 201 and 419

O'DRISCOLL, J.J. (1965). Permit systems. In Fawcett, H.H. and Wood, W.S. (1965), *op. cit.*, p. 445

O'DRISCOLL, J.J. (1972). Methods for prevention of spills in the rail transportation of hazardous materials. *Control of Hazardous Materials Spills*, **1**, 47

O'DRISCOLL, J.J. (1975a). Handling derailment accidents involving ammonia. *Ammonia Plant Safety,* **17**, 119

O'DRISCOLL, J.J. (1975b). Southern's Hazardous Materials Program. *Loss Prevention,* **9**, 71

O'DRISCOLL, J.J. (1976). Protection of personnel in hazardous materials spill incidents. *Control of Hazardous Materials Spills,* **3**, 168

O'DRISCOLL, J.J. (1978). Development of a transportation emergency action guide for hazardous material incidents. *Control of Hazardous Material Spills,* **4**, 419

OEHLERT, G.W. (1983). Some statistics for wind roses. *Atmos. Environ.,* **17**, 2473

OELERT, H.H. and FLORIAN, T. (1972). Detection and evaluation of odor nuisance from diesel exhaust gas. *Staub Reinhaltung der Luft,* **32**(10), 20 (English ed.)

OELLRICH, L.R. (1987). Nitrogen-producing inert gas plants onboard liquefied gas tankers – aspects and experience. *Gastech 86,* p. 251

OENBRING, P.R. and SIFFERMAN, T.R. (1980a). Flare design based on full-scale plant data. *API Proc., Refining Dept, 45th Midyear Mtg,* **59**, 220

OENBRING, P.R. and SIFFERMAN, T.R. (1980b). Flare design – are current methods too conservative? *Hydrocarbon Process.,* **59**(5), 124

OERLEMANS, J. (1986). Linear theory of the urban heat island circulation. *Atmos. Environ.,* **20**, 447

VON OETTINGEN, W.F. (1937). The halogenated hydrocarbons: their toxicity and potential dangers. *J. Ind. Hyg. Toxicol.,* **19**, 349

OFFICE OF CIVIL DEFENSE (1965). *Critical Industry Repair Analysis: Petroleum Industry. Advance Res. Inc. Rep.* (Wellesley Hill, MA)

OFFICE OF EMERGENCY PREPAREDNESS (1972). *Disaster Preparedness* (Washington, DC)

OFFICE OF POPULATION CENSUS AND SURVEYS (1980). *General Household Survey 1978* (London: HM Stationery Office)

OFFICE OF POPULATION CENSUS AND SURVEYS (1981a). *1981 Census* (London: HM Stationery Office)

OFFICE OF POPULATION CENSUS AND SURVEYS (1981b). *1981 Census Users Guide* (London: HM Stationery Office)

OFFICE OF POPULATION CENSUS AND SURVEYS (1981c). *1981 Labour Force Survey* (London: HM Stationery Office)

OFFICE OF SCIENCE AND TECHNOLOGY POLICY (1979). Identification, Characterization and Control of Potential Carcinogens. *A Framework for Federal Decision Making* (Washington, DC). See also *J. Nat. Cancer Inst.,* 1980, **61**, 169

OFFICE OF TECHNOLOGY ASSESSMENT – see US Congress, Office of Technology Assessment

OFFICE OF TOXIC SUBSTANCES – see Environmental Protection Agency, Office of Toxic Substances

OFFORD, D.C. (1983). Can HSE prevent another Flixborough? In Health and Safety Executive (1983d), *op. cit.,* p. 57

OGASAWARA, H. (1969). A theoretical approach to two-phase critical flow. 3rd report, The critical condition including interphasic slip. *Bull. JSME,* **12**, 827

OGATA, M. (1884). On the toxicity of sulphur dioxide. *Arch. f. Hyg.,* **2**, 223

OGAWA, A. (1984). *Separation of Particles from Air and Gases* (Boca Raton, FL: CRC Press)

OGAWA, N., UMEKAWA, N., AKIBA, T. and KOBAYASHI, T. (1986). Future LNG carriers. *Gastech 85,* p. 256

OGAWA, T., MIYAKE, A. and MATSUO, H. (1992). Explosion characteristics of ethylene oxide mixed with nitrogen or/and propylene oxide. *Loss Prevention and Safety Promotion 7,* vol. 1, 14–1

OGAWA, Y. and OIKAWA, S. (1982). A field investigation of the flow and diffusion around a model cube. *Atmos. Environ.,* **16**, 207

OGAWA, Y., DIOSEY, P.G., UEHARA, K. and UEDA, H. (1982). Plume behaviour in stratified flows. *Atmos. Environ.,* **16**, 1419

OGAWA, Y., OIKAWA, S. and UEHARA, K. (1983). Field and wind tunnel study of the flow and diffusion around a model cube. *Atmos. Environ.,* **17**, 1145 and 1161

OGILVIE, C.M. *et al.* (1957). A standardised breath holding technique for the clinical measurement of the diffusing capacity of carbon monoxide. *J. Clin. Invest.,* **36**, 1

OGISO, C., FUJITA, A. and UEHARA, Y. (1986). Experimental study to assess the characteristics of vapor explosion. *Loss Prevention and Safety Promotion 5,* pp. 36–1

OGISO, C., TAKAGI, N. and KITAGAWA, T. (ca 1986). On the mechanism of vapor explosion. *Proc. First Pacific Chemical Engineering Conf.,* pt II, paper 9–4

OGIWARA, R., YOKAWA, T., NAKANO, H., SHIRAHA, M. and KITAYAMA, Y. (1986). A new generation of LNG carriers for economy and operational flexibility. *Gastech 85,* p. 246

OGIWARA, R., HASHIGUCHI, H., KAMOI, N., MURAKAMI, A. and KAKU, I. (1991). Development and experience of a low boiloff panel insulation system for LNG carriers. *Gastech 90,* p. 4.1

OGLE, R.A., BEDDOW, J.K., CHEN, L.D. and BUTTER, P.B. (1988). An investigation of aluminium dust explosions. *Combust. Sci. Technol.,* **61**, 75

OGLESBY, S. and NICHOLS, G.B. (1978). *Electrostatic Precipitation* (New York: Dekker)

OGNEDAL, M. (1990). *Piper Alpha Public Inquiry. Transcript, Days 169, 170*

OGNEDAL, M. (1995). The Norwegian experience and views. *Safety Cases* (London: IBC Technical Services)

O'GRADY, P.J. (1986). *Controlling Automated Manufacturing Systems* (London: Kogan Page)

OGURA, Y. and PHILLLIPS, N. (1962). Scale analysis of deep and shallow convection in the atmosphere. *J. Atm. Sci.,* **19**, 173

OH, J.I.H. and ALBERS, H.J. (1989). Procedure aspects of safety in the process industry. *Loss Prevention and Safety Promotion 6,* p. 2.1

OH, J.I.H. and HUSMANN, C.A.W.A. (1986). Process safety analysis: the approach and organization. *Loss Prevention and Safety Promotion 5,* p. 67–1

OH, J.I.H. and HUSMANN, C.A.W.A. (1988). Major hazard regulation in the Netherlands: the organizational safety aspects. *Preventing Major Chemical and Related Process Accidents,* p. 415

O'HANLON, J.F. (1980). *A User's Guide to Vacuum Technology* (New York: Wiley)

O'HARA, J.B., JENTZ, N.E., NEWTLON, R.J. and TEEPLE, R.V. (1978). Feedstocks from coal: how? when? *Hydrocarbon Process.,* **57**(11), 117

O'HARA, K.L. and POOLE, M.B. (1973). Initial and operational inspection in polyethylene plants. *Safety in Polyethylene Plants,* **1**, 53

OHBA, R., OKABAYASHI, K. and OKAMOTO, H. (1989). Prediction of gas diffusion in complicated terrain by a potential flow model. *Atmos. Environ.,* **23**, 71

OHBA, R., UKEGUCHI, N., KAKISHIMA, S. and LAMB, B. (1990). Wind tunnel experiment of gas diffusion in stably stratified flow over complex terrain. *Atmos. Environ.,* **24A**, 1987

OHE, S. (1978). *Computer-Aided Data Book of Vapour Pressures* (Japan: Data Book)

OHIWA, N., TANAKA, K. and YAMAGUCHI, S. (1993). Noise characteristics of turbulent diffusion flames with coherent structure. *Combust. Sci. Technol.*, **90**, 61

OHNISHI, W. (1980). Computer-aided application of safety laws and regulations. *Loss Prevention and Safety Promotion 2*, vol. 3, p. 791

OHRUI, T., OHKUBO, K. and IMAI, O. (1980). Technical improvements in the strong nitric acid process. *Ammonia Plant Safety*, **21**, 164

OHSAKI, K., MUKAI, K., KANEZASHI, H. and KANAMARU, T. (1986). Improvement of weldability in Fe−36% Ni alloy for liquefied natural gas transportation tanks. *Gastech 85*, p. 517

OHYAGI, S., YOSHIHASHI, T. and HARIGAYA, Y. (1984). Direct initiation of planar detonation waves in methane/oxygen/nitrogen mixtures. In Bowen, J.R. *et al.*, (1984b), *op. cit.*, p. 3

OIL AND CHEMICAL PLANT CONSTRUCTORS ASSOCIATION (1974). *Safety Manual for Mechanical Plant Construction*, 2nd ed. (London: Kluwer-Harrap)

OIL COMPANIES INTERNATIONAL MARINE FORUM − see Appendix 27

OIL COMPANIES MATERIALS ASSOCIATION (1972). *Procedural Specification for Limitation of Noise in Plant and Equipment for the Petroleum Industry* (London)

OIL COMPANIES ASSOCIATION (1977). *Recommendation for the Protection of Diesel Engines Operating in Hazardous Areas* (London)

OIL INDUSTRY ADVISORY COMMITTEE (1986). *A Guide to the Principles and Operation of Permit-to-work Procedures as Applied to the UK Petroleum Industry* (London: HM Stationery Office)

OIL INDUSTRY ADVISORY COMMITTEE (1991). *Guidance on Permit-to-work Systems in the Petroleum Industry*, rev. ed. (London: HM Stationery Office)

OIL INSURANCE ASSOCIATION (1973). In Vervalin, C.H. (1973a), *op. cit.*, p. 92

OIL INSURANCE ASSOCIATION − see also Appendix 28

OKAMOTO, S. (1973). *Introduction to Earthquake Engineering* (Tokyo: Univ. of Tokyo Press)

O'KANE, G.J. (1983). Inhalation of ammonia vapour. A report on the management of eight patients during the acute stages. *Anaesthesis*, **38**, 1208

OKASAKI, S., LEYER, J.C. and KAGEYAMA, T. (1981). Effets depression induits par l'explosion de charges combustibles cylindriques non confinees. *Combustion Inst, First Speciliasts Mtg*, Bordeaux

OKAYAMA, Y. (1991). A primitive study of a fire detection method controlled by artificial neural net. *Fire Safety J.*, **17**, 535

OKE, T.R. (1978). *Boundary Layer Climates* (London: Methuen)

OKEY, R.W., DIGREGORIO, D. and KOMINEK, E.G. (1979). Waste-sludge treatment in the CPI. *Chem. Engng*, **86** (Jan. 29), 86

OKOUCHI, S. and SASAKI, S. (1984). Volatility of mercury in water. *J. Haz. Materials*, **8**(4), 341

OKRENT, D. (1975). *A Survey of Expert Opinion on Low Probability Earthquakes. Rep. UCLA-ENG-7515.* Univ. of California, Los Angeles, CA

OKRENT, D. (1977). *A General Evaluation Approach to Risk−Benefit for Large Technological Systems and its Application to Nuclear Power. Rep. UCLA-ENG-7777.* Univ. of California, Los Angeles, CA

OKRENT, D. (1981). Industrial risks. *Proc. R. Soc., Ser. A*, **376**, 133. See also in Warner, Sir F. and Slater, D.H. (1981), *op. cit.*, p. 133

OKRENT, D. and BALDEWICZ, W.L. (1982). On the development of threshold criteria for action for light water reactors. *Risk Analysis*, **2**, 149

OKRENT, D. and WILSON, R. (1982). Safety regulations in the USA. In Green, A.E. (1982), *op. cit.*, p. 601

OKRENT, D., APOSTOLAKIS, G. and OKRENT, N.D. (1985). On the usefulness of quantitative safety goals for state regulation of energy systems. *J. Haz. Materials*, **10**(2/3), 279

DE OLIVEIRA, D.B. (1973). This paraffin chlorination unit exploded. *Hydrocarbon Process.*, **52**(3), 112

OLIVER, E. and WILSON, J. (1972). *Practical Security in Commerce and Industry* (London: Gower)

OLIVER, G.W.B. (1993). 40 rules of thumb for trouble-free projects. *Chem Engng*, **100**(4), 118

OLIVER, R.M. and SMITH, J.A. (eds) (ca. 1990). *Influence Diagrams, Belief Nets and Decision-Making* (New York: Wiley)

OLIVER, T. (1902). *Dangerous Trades* (London: Murray)

OLIVERIO, M., ALP, E., BOURQUE, C.P. and MATTHIAS, S. (1986). *COBRA II Enhancements. Concord Sci. Corp., Toronto*

OLIVI, L. (1983). Uncertainty analysis in probabilistic risk assessment. *Euredata 4*, paper 2.4

OLIVIER, J.R. and SCHEUBER, K. (1978). Gas pulsations and vibrations in LPDE plants. *Safety in Polyethylene Plants*, **3**, 37

OLIVIERA, E. (1984). Developing expert systems builders in logic programming. *New Generation Comput.*, **2**, 187

DE OLIVIERA, L.F.S. and DO AMARAL NETTO, J.D. (1987). Influence of the demand rate and repair rate on the reliability of a single-channel protective system. *Reliab. Engng*, **17**(4), 267

OLIVO, J. (1994). Loss prevention in a modern ethylene plant. *J. Loss Prev. Process Ind.*, **7**, 403

OLIVO, J. (1994). Loss prevention in a modern ethylene plant. *Process Safety Prog.*, **13**, 159

OLKKOLA, E., SALO, R., VALANTO, T., RIIHIMAKI, S. and FIEANDT, J. (1986). Experiences in supporting plant design and commissioning by hazard and risk analysis. *Loss Prevention and Safety Promotion 5*, p. 25−1

OLMOS, J. and WOLF, L. (1977). A modular approach to fault tree and reliability analysis, Dept of Nucl. Engng, Massachusetts Inst. Technol., MA, *Rep. MITNE-209*

OLORUNMAIYE, J.A. and IMIDE, N.E. (1993). Computation of natural gas pipeline rupture problems using the method of characteristics. *J. Haz. Materials*, **34**, 81

ÖLSCHLAGER, F.R. (1986). Improvement of gas tanker operating efficiency. *Gastech 85*, p. 504

OLSEN, E.A. (1969). The new rules for ammonia highway tank transports. *Ammonia Plant Safety*, **11**, 46

OLSEN, H.L., GAYHART, E.L. and EDMONSON, R.B. (1953). Propagation of incipient spark-ignited flames in hydrogen-air and propane-air mixtures. *Combustion*, **4**, p. 144

OLSEN, K.R. (1987). Hazard communication. *Chem. Engng*, **94**(5), 107

OLSEN, R.E. (1994). Process safety and quality management: how they fit together. *Process Safety Prog.*, **13**, 59

OLSON, C.R. (1963). Speed adjustment in drives for the chemical industry. *Ind. Engng Chem.*, **55**(3), 24

OLSON, C.R. (1979). Select motors to save energy. *Hydrocarbon Process.*, **58**(8), 73

OLSON, C.R. and MCKELVY, E.S. (1967). How to specify and cost estimate electric motors. *Hydrocarbon Process.*, **46**(10), 118

OLSON, M. (1971). *The Logic of Collective Action* (Cambridge, MA: Harvard Univ. Press)

OLTHOF, M. and OLESKIEWICZ, J. (1982). Anaerobic treatment of industrial wastewaters. *Chem. Engng*, **89**(Nov. 15), 121

OLUJIC, Z. (1985). Predicting two-phase-flow friction loss in horizontal pipes. *Chem. Engng*, **92**(Jun. 24), 45

O'MARA, R.L. and BURNS, C.C. (1984). Balancing risk and availability in toxic chemical production. *Chem. Engng Prog.*, **80**(9), 49

O'MARA, M.M., CRIDER, L.B. and DANIEL, R.L. (1971). Combustion products from vinyl chloride monomer. *Am. Ind. Hyg. Ass J.*, **32**(3), 153

O'MEAGHER, A.J. and BENNETTS, I.D. (1991). Modelling of concrete walls in fire. *Fire Safety J.*, **17**, 315

ONDERDANK, J.K. (1986). Understanding shutdown systems. *Chem. Engng*, **93**(Jul. 7), 45

ONDREY, G. (1992). Ethylene oxide in the safety spotlight. *Chem. Engng*, **99**(12), 37

ONDREY, G. (1993a). Havoc at Hoechst. *Chem. Engng*, **100**(4), 55

ONDREY, G. (1993b). Hoechst's accident streak continues. Chem Engng, **100**(5), 50

ONDREY, G. and FOUHY, K. (1991). Plasma arcs sputter new waste treatment. *Chem. Engng*, **98**(12), 32

ONDREY, G., HOFFMANN, P. and MOORE, S. (1992). 'Green' laws spark hydrogen technology. *Chem. Engng*, **99**(5), 30

O'NEAL, L.R., ELWONGER, C.M. and HUGHES, W.T. (1973). Fatigue-vibration problems in piping systems. *Safety in Polyethylene Plants*, **1**, 35. See also *Chem. Engng Prog.*, 1972, **68**(11), 68

O'NEIL, R. (1980). The PISC programme. *Atom*, Jul. 181

O'NEIL, R. AND JORDAN, G. (1972). *Periodic Inspection of Pressure Vessels* (London: Inst. Mech. Engrs)

O'NEILL, M.J. (1975). Measurement of exothermic reactions by DSC. *Anal. Chem.*, 47, 630

O'NEILL, P.A. (1993). *Industrial Compressors* (Oxford: Butterworth-Heinemann)

O'NEILL, P.S. and TERBOT, J.W. (1972). Impact of new heat exchanger technology on large LNG plants. *LNG Conf. 3*

ONG, J.O.Y. and HENLEY, E.J. (1979). Reliability parameters for process applications (abstract). *Comput. Chem. Engng*, **3**, 496

ONISHI, Y., OLSEN, A.R., PARKHURST, M.A. and WHELAN, G. (1985). Computer-based environmental exposure and risk assessment methodology for hazardous materials. *J. Haz. Materials*, **10**(2/3), 389

ONO, S., KAWANO, H., NIHO, H. and FUKUYAMA, G. (1976). Ignition in a free convection from vertical hot plate. *J. JSME*, **19**(132), 676

D'ONOFRIO, E.J. (1980). Cool flame autoignition in glycols. *Loss Prevention*, **13**, 89

ONUMA, Y. and OGASAWARA, M. (1977). An approach to the theoretical description of turbulent jet diffusion flames. 1. *Combust. Flame*, **30**, 163

OOMS, G. (1972). A new method for the calculation of the plume path of gases emitted by a stack. *Atmos. Environ.*, **6**, 899

OOMS, G. and DUIJM, N.J. (1984). Dispersion of a stack plume heavier than air. In Ooms, G. and Tennekes, H. (1984), *op. cit.*, p. 1

OOMS, G. and TENNEKES, H. (eds) (1984). *Atmospheric Dispersion of Heavy Gases and Small Particles* (Berlin: Springer)

OOMS, G., MAHIEU, A.P. and ZELIS, F. (1974). The plume path of vent gases heavier than air. *Loss Prevention and Safety Promotion 1*, p. 211

OOSTERKAMP, W.J. (1993). Inherent and passive safety aspects in the design of advanced natural circulation

boiling water reactors. In *Proc. ASME/JSME Nucl. Eng. Jt. Conf., Vol. 1*, (New York: ASME) pp. 713–728.

OOSTERLING, J.O. and ORBONS, H.G. (1981). The small leak that caused a big leak. *Ammonia Plant Safety*, **23**, 116

OPEN UNIVERSITY (1975a). *Air Pollution* (Milton Keynes: Open Univ. Press)

OPEN UNIVERSITY (1975b). *Clean and Dirty Water* (Milton Keynes: Open Univ. Press)

OPEN UNIVERSITY (1975c). *Control of the Acoustic Environment* (Milton Keynes: Open Univ. Press)

OPEN UNIVERSITY (1975d). *Toxic Wastes* (Milton Keynes: Open Univ. Press)

OPENSHAW, S. (1982). The geography of reactor siting policies in the UK Trans. *Inst. Br. Geogr.*, NS **7**, 150

OPIE, R. (1985). Will concrete barges help exploit North Sea 'marginal' fields? *Process Engng*, Dec., 33

OPILA, R.L. (1993). Carefully plan dust collection systems. *Chem. Engng Prog.*, **89**(5), 22

OPPENHEIM, A.K. (1953). Gasdynamic analysis of the development of gaseous detonation and its hydraulic analogy. *Combustion*, **4**, 471

OPPENHEIM, A.K. (1969). The past, present and future of discussions on gaseous detonation phenomena. *Combustion*, **12**, 795

OPPENHEIM, A.K. and ROSCISZEWSKI, J. (1963). Determination of the detonation wave structure. *Combustion*, **9**, 424

OPPENHEIM, A.K., KUHL, A.L. and KAMEL, M.M. (1972a). On flame-generated self-similar blast waves. *Conf on Fuel–Air Explosions*, p. 147

OPPENHEIM, A.K., KUHL, A.L. and KAMEL, M.M. (1972b). On self-similar blast waves headed by Chapman-Jouguet detonation. *J. Fluid Mech.*, **55**, 257

OPPENHEIM, A.K., KURYLO, J., COHEN, L.M. and KAMEL, M.M. (1977). Blast waves generated by exploding clouds. *Proc. Eleventh Int. Symp. on Shock Tubes and Waves*, p. 465

OPPENHEIM, N.A. (1966). *Questionnaire Design and Attitude Measurement* (New York: Basic Books)

OPSCHOOR, G. (1974). *Investigation of the Explosive Vaporization of LNG Spreading on Water*. Ref. 74-03386. TNO Central Technish Instituut

OPSCHOOR, G. (1975a). *Investigations into the Spreading and Evaporation of Burning LNG Spills on Water and the Heat Radiation from LNG-fire on Water*. Rep. 75-03777. TNO, Apeldoorn

OPSCHOOR, G. (1975b). *Investigations into the Spreading and Evaporation of Burning and Non-Burning LNG-Spills on Land and the Heat Radiation from LNG-Fires*. Rep. 75-02050. TNO, Apeldoorn

OPSCHOOR, G. (1977). Investigation into the spreading and evaporation of LNG spilled on water. *Cryogenics*, **17**, 629

OPSCHOOR, G. (1978). *Rep. 78-0834*. TNO, Apeldoorn

OPSCHOOR, G. (1979). *Methods for the Calculation of the Physical Effects of the Escape of Dangerous Materials* (report). TNO, Voorburg, The Netherlands

OPSCHOOR, G. (1980). The spreading and evaporation of LNG-and burning LNG-spills on water. *J. Haz. Materials*, **3**, 249

OPSCHOOR, G. and HOFTIJZER, G.W. (1980). Methods for the calculation of the physical effects of the accidental release of dangerous materials (liquids and gases). *Loss Prevention and Safety Promotion 3*, vol. 3, p. 1148

OPSCHOOR, G. and SCHECKER, H.G. (1983). Consequence analysis. *Loss Prevention and Safety Promotion 4*, vol. 1, p. G25

OPSCHOOR, G., VAN LOO, R.O.M. and PASMAN, H.J. (1992). Methods for calculation of damage resulting from physical effects of the accidental release of dangerous materials. *Hazard Identification and Risk Analysis, Human Factors and Human Reliability in Process Safety*, p. 21

ORAN, E.S., BORIS, J.P., YOUNG, T., FLANI GAN, M., BURKS, T. and PICONE, M. (1981). Numerical simulations of detonations in hydrogen-air and methane-air mixtures. *Combustion*, **18**, p. 1641

ORAN, E.S., BORIS, J.P., YOUNG, T.R., FRITTS, M.J., PICONE, M.J. and FYFE, D. (1982). Numerical simulation of fuel-air explosions: current methods and capabilities. In Lee, J.H.S. and Guirao, C.M. (1982), *op. cit.*, p. 447

ORAN, E.S., YOUNG, T.R., BORIS, J.P., PICONE, J.M. and EDWARDS, D.H. (1982). A study of detonation structure: the formation of unreacted gas pockets. *Combustion*, **19**, p. 573

ORANJE, L. (1983). Handling two-phase gas condensate flow in offshore pipeline systems. *Oil Gas J.*, **81**(Apr. 18), 128

ORBONS, H.G. (1988). Weld cracking in reformer outlet parts after 12 years in service. *Ammonia Plant Safety*, **28**, 45

ORBONS, H.G. and HUURDEMANN, T.L. (1985). Stress corrosion cracking in syngas heat exchangers. *Ammonia Plant Safety*, **25**, p. 89. See also *Plant/Operations Prog.*, 1985, **4**, 49

OREDA (1984). *Offshore Reliability Data Handbook* (Hovik, Norway)

OREDA (1992). OREDA-92. *Offshore Reliability Data*, 2nd ed. (Hovik, Norway)

OREFFICE, P.F. (1984). Law and the threat it poses to the US chemical industry. *Chem. Ind.*, Jan. 2, 15

O'REILLY, A. (1986). Experiences in process integration. *Chem. Engr, Lond.*, **425**, 56

O'REILLY, B.M. (1972). Dust explosions in factories. *Chemy. Ind.*, Jan. 1

O'REILLY, B.M. (1976). *Flixborough. The lessons to be learned* (London: Health and Safety Executive)

O'REILLY, J.N. and GLOYNE, S.R. (1941). Blast injury of the lungs. *The Lancet*, Oct. 11, 423

O'REILLY, J.T. (1987). *Emergency Response to Chemical Accidents: Planning and Co-ordinating Solutions* (New York: McGraw-Hill)

ORELL, A. and REMBRAND, R. (1986). A model for gas liquid slug flow in a vertical tube. *Ind. Engng Chem. Fund.*, **25**, 196

OREY, W.J. (1972a). Incidents provide lessons in plant fire protection. *Fire Engng*, **125**(Jun.), 52

OREY, W.J. (1972b). Protection of hot dip tanks holding flammable liquids. *Fire Engng*, **125**(Jul.), 45

OREY, W.J. (1972c). Selecting standpipe hose, nozzles for plant needs. *Fire Engng*, **125**(Mar.), 22

OREY, W.J. (1973). Precautions for handling flammable liquids safely. *Fire Engng*, **126**(Oct.), 56

ORGANISATION FOR ECONOMIC COOPERATION AND DEVELOPMENT (1974). *Mercury and the Environment* (Paris)

ORGANISATION FOR ECONOMIC COOPERATION AND DEVELOPMENT (1992). *Guiding Principles for Chemical Accident Prevention, Preparedness and Response. Rep. OCDE/GD(92) 43* (Paris)

ORIOLO, D.J. (1958). How clear are those instructions? *Oil Gas J.*, **56**(Nov. 24), 93

O'RIORDAN, T. (1982). Risk-perception studies and policy priorities. *Risk Analysis*, **2**, 95

O'RIORDAN, T. (1989). Air pollution legislation and regulation in the European Community: a review essay. *Atmos. Environ.*, **23**, 293

O'RIORDAN, T. (1990). Hazard and risk in the modern word: political models for programme design. In Handmer, J. and Penning-Rowsell, E.C. (1990), *op. cit.*

O'RIORDAN, T., KEMP, R. and PURDUE, M. (1988). *Sizewell B: An Anatomy of an Inquiry* (London: Macmillan)

ORLEMAN, J.A. *et al.* (1983) *Fugitive Dust Control Technology* (Park Ridge, NJ: Noyes Data Corp.)

ORLOFF (1973). Report on safety precautions in the chemical industry of the USSR. *Prevention of Occupational Risks*, **2**, 520

ORLOFF, L. (1981). Simplified radiation modeling of pool fires. *Combustion*, **18**, 549

ORLOFF, L. and DE RIS, J. (1982). Froude modeling of pool fires. *Combustion*, **19**, 885

ORLOFF, L., MODAK, A.T. and MARKSTEIN, G.H. (1979). Radiation from smoke layers. *Combustion*, **17**, 1029

ORLOFF, L., DE RIS, J. and DELICHATSIOS, M.A. (1992). Radiation from buoyant turbulent diffusion flames. *Combust. Sci. Technol.*, **84**, 177

VAN ORMER, H. (1979). Include power cost in air compressor selection. *Hydrocarbon Process.*, **58**(9), 219

VAN ORMER, H. (1980). Better service from rotary screw air compressor packages. *Hydrocarbon Process.*, **59**(5), 181

ORMOND, A.L. and ISHERWOOD, J. (1992). Potential for environmental incidents. approach to assessment. *Safety and Reliab Soc. Symp.*, Oct.

ORMSBY, R.W. (1982). Process hazards control at Air Products. *Plant/Operations Prog.*, **1**, 141

ORMSBY, R.W. (1990). Implementation of effective safety programs. *Plant/Operations Prog.*, **9**, 166

ORMSBY, R.W. and LE, N.B. (1988). Societal risk curves for historical hazardous chemical transportation accidents. *Preventing Major Chemical and Related Process Accidents*, p. 133

ORNELLAS, D.L. (1967). The heat and products of detonation of HMX, TNT, NM and FEFO. *Am. Chem. Soc., Symp. on Detonation and Reactions in Shock Waves* (Washington, DC)

ORNELLAS, D.L. (1968). The heat and products of detonation of cyclotetramethylene tetramine 2,4,6—trinitrotoluene, nitromethane and bis[2,2-dinitro-2-fluoroethyl]formal. *J. Phys. Chem.*, **72**(7), 2390

ORNELLAS, D.L., CARPENTER, J.H. and GUNN, S.R. (1966). Detonation calorimeter and results obtained with pentaerythritol tetranitrate (PETN). *Rev. Sci. Instrum.*, **37**, 907

O'ROURKE, J.F. (1974). The use of intumescent coatings for fire protection of structural steel. *Loss Prevention*, **8**, p. 69

ORR, D. (1977). Social risks and the energy option. *Sci. Publ. Policy*, **4**, 179

ORRELL, B. and CRYAN, J. (1987). CIMAH safety cases: getting rid of the hazard. *Chem. Engr, Lond.*, **439**, 14

ORRELL, W. and CRYAN, J. (1987). Getting rid of the hazard. *Chem. Eng.* (August), 14—15

ORSINI, B. (chrmn) (1977). *Parliamentary Commission of Inquiry on the Escape of Toxic Substances on 10 July 1976 at the ICMESA Establishment and the Consequent Potential Dangers to Health and the Environment due to Industrial Activity. Final Report* (Rome)

ORSINI, B. (chrmn) (1980). *Seveso* (translation of official Italian report by Health and Safety Executive) (London: Health and Safety Executive)

ORUM, P. (1998). Time to reduce hazards. *Chem. Ind.* (June 15), 500

OSBORN, G.R. (1943). Findings in 262 fatal accidents. *The Lancet*, Sep. 4, 277

OSBORNE, P., MORGAN, N. and VAREY, P. (1992). Switzerland in focus. *Chem. Engr, Lond.*, **523**, 21

OSBURN, H.G. (1987). Personnel selection. In Salvendy, G. (1987), *op. cit.*, p. 911

OSER, B.L. (ed.) (1965). *Hawk's Physiological Chemistry*, 14th ed. (New York: Blakiston)

OSER, B.L. (1981). The rat as a model for human toxicological evaluation. *J. Toxicol. Environ. Health*, **8**, 521

OSER, M.G. (1990). Implementing programmable safety shutdown. *Hydrocarbon Process.*, **69**(10), 83

OSHAS (18001 − 1999). *Occupational Health and Safety Management Systems − Specification*, British Standards Institute

O'SHEA, P. (1982). Flammable gases and liquids − a discussion of the parameters affecting their release and dispersion. *Flammable Atmospheres and Area Classification, IEE Coll. Dig 1982/26* (London), p. 8/1

O'SHEA, S. (1985). Fire alarm systems for hazardous areas. *Fire Surveyor*, **14**(1), 24

O'SHEA, S.M. (1983). Nitrogen blanketing of centrifuges reduces fire hazards. *Chem. Engng*, **90**(April 4), 73

O'SHEA, T. and EISENSTADT, M. (1984). *Artificial Intelligence: Tools, Techniques and Applications* (London: Harper & Row)

O'SHELL, H.E. and BIRD, F.E. (1969). Incident recall. *Nat. Safety News*, Oct., 58

O'SHIMA, E. (1983). Computer aided plant operation. *Comput. Chem. Engng*, **7**, 311

O'SHIMA, E. (1993). Regulation of chemical process safety in Japan (workshop). *Process Safety Management*, p. 375

O'SHIMA, E. (1994a). Japanese regulations. *Process Safety Prog.*, **13**, 14

O'SHIMA, E. (1994b). Safety management in Japan. *Process Safety Prog.*, **13**, 7

OSMAN, R.M. (1975). Ammonia plant heat exchanger problems. *Ammonia Plant Safety*, **17**, p. 44

OSMAN, R.M. and RUZISKA, P.A. (1976). Correlating tube failures with operating severity. *Ammonia Plant Safety*, **18**, p. 42

OSMOND, A.H. and TALLENTS, C.J. (1968). Ammonia attacks. *Br. Med. J.*, 740

OSTEBO, R., MUSAEUS, S.I., BELLAMY, L.J. and GEYER, T.A.W. (1989). Successfulness of shallow gas seismic predictions − a case history. *Norwegian Petrol. Soc., Conf. on Shallow Gas and Leaky Reservoirs* (Oslo)

OSTER, R., BELL, K. and KOTTOWSKI, H.M. (1990). Hydrodynamic considerations of venting with high viscosity non-reacting fluids. *J. Loss Prev. Process Ind.*, **3**(1), 50

OSTERWALDER, U. (1996). Continuous process to fit batch operation: safe phosgene production on demand. *Symp. Pap. − Inst. Chem. Eng., North West. Branch*, 6.1–6.6. (Rugby, Warwickshire, England: IchemE)

OSTRANDER, V.P. (1971). Spacecraft reliability techniques for industrial plants. *Chem. Engng Prog.*, **67**(1), 49

OSTROFSKY, B. (1968). Non-destructive tests for on- and off-stream inspections. *Chem. Engng*, **75**(May 20), 174

OSTROFSKY, B. (1980−). Materials identification in the plant. *Chem. Engng*, **1980, 87**(Dec. 15), 91; 1981, **88**(Jan. 12), 141; (Feb. 9), 119

OSTROFSKY, B. and HECKLER, N.B. (1970). Ultrasonics checks reformer headers. *Hydrocarbon Process.*, **49**(7), 99

OSTROVSKY, G.M. and BEREZINSKY, T.A. (1991). About one approach to solving the problem of the synthesis of chemical plants. *Comput. Chem. Engng*, **15**, 369

OSTROM, E. (1986). A method of institutional analysis. In Kaufman, F.X., Majone, G. and Ostrom, V. (1986), *op. cit.*

OSTROOT, G. (1967). Lessons learned from a coupling failure and lube oil fire. *Loss Prevention*, **1**, 13

OSTROOT, G. (1972). Explosions in gas- or oil-fired furnaces. *Loss Prevention*, **6**, 112

OSTROOT, G. (1973). Coupling failure and lube oil fire. In Vervalin, C.H. (1973a), *op. cit.*, p. 16

OSTROOT, G. (1976a). Ammonia plant design safety procedure. *Ammonia Plant Safety*, **18**, 4

OSTROOT, G. (1976b). Case history: a two burner boiler explosion. *Hydrocarbon Process.*, **55**(12), 85

O'SULLIVAN, T.F., MCCHESNEY, H.B. and POLLOCK, W.H. (1978). Coal-fired process heaters? *Hydrocarbon Process.*, **57**(7), 95

OSWATITSCH (1965). *Gas Dynamics* (New York: Academic Press)

OTSUKA, E., INOUE, S. and JOJIMA, T. (1976). What's new in urea technology? *Hydrocarbon Process.*, **55**(11), 160

OTT, W.R. (1978). *Environmental Indices − Theory and Practice* (Ann Arbor, MI: Ann Arbor Science)

OTTAWAY, M.R. (1986). Thermal hazard evaluation by accelerating rate calorimetry. *Anal. Proc.*, **23**, 116

OTTER, R.J. (1979). EEC 'Environmental Substances' Directive. Environmental quality objectives. *Chem. Ind.*, May 5, 302

OTTERMAN, B. (1975). Analysis of large LNG spills on water. Pt 1: Liquid spread and evaporation. *Cryogenics*, Aug., 455

OTTEWELL, S. (1991). Chernobyl five years on. *Chem. Engr, Lond.*, **495**, 15

OTTEWELL, S. (1993). Be safe: use CONDAM. *Chem. Engr, Lond.*, **548**, 14

OTTMERS, D.M., JONES, D.C., KEITH, L.H. and HALL, R.C. (1979). Instruments for environmental monitoring. *Chem. Engng*, **86**(Oct. 15), 85

OTTOSON, J. (1974). Human contribution to fire origins. *Fire J.*, **68**(Jul.), 19

OTWAY, H.J. (1975). *Risk Assessment. IAEA-IIASA Res. Proj. Rep.* (Vienna)

OTWAY, H.J. and ERDMAN, R.L. (1970). Reactor siting and design from a risk viewpoint. *Nucl. Engng Des.*, **13**(2), 365

OTWAY, H. and PELTU, M. (eds) (1985). *Regulating Industrial Risks* (London: Butterworths)

OTWAY, H.J. and THOMAS, K. (1982). Reflections on risk perception and policy. *Risk Analysis*, **2**, 69

OTWAY, H.J. and WINTERFELDT, D. (1982). Beyond acceptable risk: on the social acceptability of technologies. *Policy Sci.*, **14**, 247

OTWAY, H.J. and WYNNE, B. (1989). Risk communication: paradigm and paradox. *Risk Analysis*, **9**, 141

OUDAR, J. and WISE, H. (1985). *Deactivation and Poisoning of Catalysts* (New York: Dekker)

OUDRY, COLONEL (1924). 45th Algerian Division − the Second Battle of Ypres (April–May 1915). *Ypres Times*, **2**(2), Apr., 47

OUELLETTE, P., HOA, S.V. and LI, L. (1990). Nondestructive evaluation of fiberglass reinforced plastic road tankers subjected to internal pressure using acoustic emission monitoring. *J. Haz. Materials*, **25**(1/2), 49

OUELLETTE, R.P., DILKS, C.F., THOMPSON, W.C. and CHEREMISINOFF, P.N. (1986). *Asbestos Hazard Management: Guidebook to Abatement* (Lancaster, PA: Technomic)

OVE ARUP AND PARTNERS (1977). *Design Guide for Fire Safety of Bare Exterior Structural Steel* (London)

OVENU, H. (1969). Inspection service − vital factor in securing maximum plant availability. *Proc. Am. Power Conf.*

OVERCAMP, T.J. (1976). A general Gaussian diffusion deposition model for elevated point sources. *J. Appl. Met.*, **7**, 1167

OVERCAMP, T.J. and KU, T. (1986). Effect of a virtual origin correction on entrainment coefficients as determined from observations of plume rise. *Atmos. Environ.*, **20**, 293

OVERCASH, M.R. and PAL, D. (1980). *Design of Land Treatment Systems for Industrial Wastes* (New York: Wiley)

OVERFIELD, J.L. and RICHARD, J.W. (1976). Hazardous materials Separation system. *Control of Hazardous Material Spills 3*, p. 382

OVERHOFF, K.H., SCHECKER, H.G., FELLENSIEK, J. and ONKEN, U. (1989). Investigation for the design of a new water trap flame arrester. *Loss Prevention and Safety Promotion 6*, p. 54.1

OWEN, A.J. and VON ZAHN, E. (1974). The activities of the international group of experts on the explosion risks of unstable substances (IGUS). *Loss Prevention and Safety Promotion 1*, p. 235

OWEN, D.B. (1962). *Handbook of Statistical Tables* (Reading, MA: Addison-Wesley)

OWEN, F. and MAIDMENT, D. (eds) (1993). *Quality Assurance of Process Plant* (Rugby: Instn Chem. Engrs)

OWEN, L.T. (1974). Selecting in-plant dust control systems. *Chem. Engng*, **81**(Oct. 14), 120

OWEN, N.L. (1965). How to rig and lift heavy vessels. *Hydrocarbon Process.*, **44**(12), 129

OWEN, P.R. (1965). Buffeting excitation of boiler tube vibration. *J. Mech. Engng Sci.*, **7**(4), 431

OWEN, W.L., SARTOR, J.D. and VAN HORN, W.H. (1960). *Performance Characteristics of Wet Decontamination Procedures. Rep. USNRDL-TR-335*. Nav. Radiol. Def. Lab

OWENS, J.E. (1965). *Static Electricity. ISA Monogr. 110* (Pittsburgh, PA: Instrum. Soc. Am.), p. 113

OWENS, J.E. (1988). Spark ignition hazards caused by charge induction. *Plant/Operations Prog.*, **7**, 37

OWENS, M.J. and SPOUGE, J.R. (1991). The effect of improved platform design on emergency evacuation requirements. *Safety Developments in the Offshore Oil and Gas Industry* (London: Instn Mech. Engrs), p. 99

OXMAN, R. and GERO, J.S. (1987). Using an expert system for design diagnosis and design synthesis. *Expert Systems*, **4**(1), 4

OYELEYE, O.O. and KRAMER, M.A. (1988). Qualitative simulation of chemical process systems: steady state analysis. *AIChE J.*, **34**, 1441

OYEZ SCIENTIFIC AND TECHNICAL SERVICES – see Appendix 28

OZERO, B.J. and PROCELLI, J.V. (1984). Can developments keep ethylene oxide viable? *Hydrocarbon Process.*, **63**(3), 55

OZISIK, M.N. (1973). *Radiative Transfer and Interactions with Convection and Conduction* (New York: Wiley-Interscience)

OZMORE, R.L. (1978). Safety control for a waste heat boiler. *Loss Prevention*, **12**, p. 49. See also *Chem. Engng Prog.*, 1978, **74**(10), 45

OZOG, H. (1985). Hazard identification, analysis and control. *Chem. Engng*, **92** Feb. 18, 161

OZOG, H. (1993). Auditing performance-based process safety management systems. *Process Safety Management*, p. 213

OZOG, H. and BENDIXEN, L.M. (1987). Hazard identification and quantification. *Chem. Engng. Prog.*, **83**(4), 55

OZOG, H., and STICKLES, R.P. (1992). What to do about process safety audits. *Chem. Engng*, **99**(9), 173

PACKER, S. (1985). In the midst of life . . . *Chem. Engr, Lond.*, **415**, 56

PACKMAN, R. (1968). *A Guide to Industrial Safety and Health* (London: Longmans)

PACYNA, J.M. and OTTAR, B. (1989). *Control and Fate of Atmospheric Trace Metals.* (Dordrecht: Kluwer)

PADDLE, G.M. (1980). Epidemiological and animal data as a basis for risk assessment. *Arch. Toxicol., Suppl.*, **3**, p. 263

PAESCHKE, W. (1937). Experimental investigations on the roughness and stability problem in the layer of air near the earth. Beitrag. *Phys. freien Atmos.*, **24**, 163

PAFF, R. (1993). Safety consideration on LPG storage tanks. *Health, Safety and Loss Prevention*, p. 296

PAGE, G.A. (1987). The hazard assessment program at a large chemical company. *Preventing Major Chemical Accidents*, p. 1.61

PAGE, G.A. (1990). Risk management program. *Safety and Loss Prevention in the Chemical and Oil Processing Industries*, p. 157

PAGE, J. and LEBENS, R. (eds) (1986). *Climate in the United Kingdom* (London: HM Stationery Office)

PAGE, L.B. and PERRY, J.E. (1986). A simple approach to fault tree probabilities. *Comput. Chem. Engng*, **10**, 249

PAGE, R.C. (1971). Training personnel for chemical tankers and the importance of skilled operators as opposed to the provision of elaborate equipment. *Transport of Hazardous Cargoes*, **2**, paper 8

PAGE, R.C. and GARDNER, A.W. (1971). *Petroleum Tankship Safety* (London: Maritime Press)

PAGE, T. (1979). Keeping score: an actuarial approach to zero-infinity dilemma. In *Energy Risk Management* (ed. Goodman and Roux) (New York: Academic Press), p. 177

PAGELLA, C. and DE FAVERI, D.M. (1993). Fire in industrial buildings: simulation of fire growth, smoke production and venting. *Process Safety Environ.*, **71B**, 180

PAGET, E. (1978). Some current problems in toxicology. *Chem. Ind.*, Apr. 15, 260

PAGET, G.E. (ed.) (1970). *Methods in Toxicology* (Oxford: Blackwell)

PAGNI, P.J. and OKOH, C.I. (1985). Soot generation within radiating diffusion flames. *Combustion*, **20**, 1045

PAGNOTTO, L.D., ELKINS, H.B. and BRUGSCH, H.G. (1979). Benzene exposure in the rubber coating industry – a follow up. *Am. J. Ind. Hyg.*, **40**, 137

PAHLOW, E.M. and DENDY, F.M. (1993). Beyond Hazop and Hazan – the never-ending challenge of hazard management. *Health, Safety and Loss Prevention*, 492

PAI, C.C.D. and HUGHES, R.R. (1987). Strategies for formulating and solving two-stage problems for process design under uncertainty. *Comput. Chem. Engng*, **11**, 695

PAIGE, P.M. (1967). How to estimate the pressure drop of flashing mixtures. *Chem. Engng*, **74**(Aug. 14), 159

PAINE, R.J., INSLEY, E.M. and EBERHARD, W.L. (1986). Evaluation of plume rise estimation methods at the EPA's full scale plume study site. *Fifth Joint Conf. on Applications of Air Pollution Meteorology* (Boston, MA: Am. Met. Soc), p. 323

PAINE, R.J., PLEIM, J.E., HEINOLD, D.W. and EGAN, B.A. (1986). Physical processes in the release and dispersion of toxic air contaminants. *Air Poll. Control Ass., Seventy- Ninth Mtg*, paper 86–76.3

PAINE, R.J., SMITH, D.G. and EGAN, B.A. (1987). Use of meteorological data in assessing potential impacts of accidental releases of vapor clouds. *Vapor Cloud Modeling*, p. 293

PAKLEPPA, H. (1991). Super performance of future LEG/LPG carriers. *Gastech 90*, p. 4.3

PAKLEPPA, H. and VAN BERKEL, P.B. (1987). Offshore terminals for low temperature liquefied gases. *Gastech 86*, p. 358

PAL, T., GRIFFIN, G.D., MILLER, G.H., WATSON, A.P., DAUGHERTY, M.L. and VO-DINH, T. (1993). Permeation measurements of chemical agent simulants through protective clothing materials. *J. Haz. Materials*, **33**, 123

PALANIAPPAN, C. (2001). *Expert System for Design of Inherently Safer Chemical Processes* (Singapore: National University of Singapore)

PALANIAPPAN, C., SRINIVASAN, R. and HALIM, I. (2002). A material-centric methodology for developing inherently safer environmentally benign processes. *Comput. Chem. Eng.* **26**, 757–74

PALANIAPPAN, C., SRINIVASAN, R. and TAN, R.B.H. (2000a). An intelligent support system for design of inherently safer process flowsheets. *AIChE Inst. of Chem. Eng. Annual Meeting*, Nov.12–17, 2000, Los Angeles, CA, Paper 249b

PALANIAPPAN, C., SRINIVASAN, R. and TAN R.B.H. (2000b). ISafe – An intelligent system for design of inherently safer process: basic chemistry and route selection. *Reg. Symp. Chem. Eng.*, Dec. 11–13, 2000, Singapore

PALANIAPPAN, C., SRINIVASAN, R. and TAN, R.B.H. (2000c). Development of inherently safer processes: A decision support system for synthesis route selection. *Third Int. Conf. on Loss Prev. (Saf., Health, Environ.) in Oil, Chem. and Process Ind.*, December, 4–8, 2000, Singapore

PALANIAPPAN, C., SRINIVASAN, R. and TAN R.B.H. (2002a). Expert system for the design of inherently safer processes. 1. route selection stage. *Ind. Eng. Chem. Res.* **41**, 6698–710

PALANIAPPAN, C., SRINIVASAN, R. and TAN, R.B.H. (2002b). Expert system for the design of inherently safer processes. 2. Flowsheet development stage. *Ind. Eng. Chem. Res.* **41**, 6711–22

PALAS, R.F. (1989). Dynamic simulation in system analysis and design. *Chem. Engng Prog.*, **85**(2), 21

PALAZZI, E. (1992). Safety problems in operations concerning the manipulation and storage of chemically unstable mixtures of wastes. *Loss Prevention and Safety Promotion 7*, vol. 3, 154–1

PALAZZI, E. and FERRAIOLO, G. (1989). A runaway criterion for nonhomogeneous exothermically reacting fluids subject to free convection. *Loss Prevention and Safety Promotion 6*, p. 108.1

PALAZZI, E., DE FAVERI, M., FUMAROLA, G. and FERRAIOLO, G. (1982). Diffusion from a steady source of short duration. *Atmos. Environ.*, **16**, 2785

PALAZZI, E., FUMAROLA, G., DE FAVERI, D.M. and FERRAIOLO, G. (1984). Flammability limits with short duration gas releases. *Plant/Operations Prog.*, **3**, 159

PALAZZI, E., ZILLI, M., CONVERTI, A. and FERRAIOLO, G. (1989). Safety problems in biotechnology – the case of citric acid. *Loss Prevention and Safety Promotion 6*, p. 6.1

PALEN, J.W. (1980). Designing heat exchangers by computer. *Chem. Engng Prog.*, **82**(7), 23

PALLEN, J.J. (1990). Lessons learned from an anhydrous ammonia release incident. *Ammonia Plant Safety*, **30**, p. 11

PALLES-CLARK, P.C. (1982). History of area classification and the basic content of BS 5345: Part 2. *Flammable Atmospheres and Area Classification, IEE Coll. Dig. 1982/26* (London), p. 1/1

PALLES-CLARK, P.C. and JAMES, R.L. (1971). An approach to electrical installations in refrigerating plants using ammonia. *Electrical Safety*, **1**, p. 59

PALLUZI, R.P. (1987). Testing for leaks in pilot plants. *Chem. Engng*, **94**(Nov. 9), 81

PALLUZI, R.P. (1990). Piloting processes into commercial plants. *Chem. Engng*, **97**(3), 76

PALLUZI, R.P. (1991). Understand your pilot plant options. *Chem. Engng. Prog.*, **87**(1), 21

PALLUZI, R.P. (1992). *Pilot Plant Design, Construction and Operation* (New York: McGraw-Hill)

PALM, J.W. (1972). Watch these trends in sulfur plant design. *Hydrocarbon Process.*, **51**(3), 105

PALM-LEIS, A. and STREHLOW, R.A. (1969). On the propagation of turbulent flames. *Combust. Flame*, 13, 111

PALMER, A. (1962). Survey of battle casualties, Eighth Air Force, June, July and August 1944. In Beyer, J.C. (ed.) (1962), *op. cit.*

PALMER, A.C. (1989). *Piper Alpha Inquiry. Transcript, Day 82*

PALMER, H.B. and BEER, J.M. (eds) (1973). *Combustion Technology* (New York: Academic Press)

PALMER, J. (1980). Explosion protection for fabric dust collectors. *Plant Engr*, **23**(3) Apr., 23

PALMER, J.R. (1971). US pressure vessel codes. *Hydrocarbon Process.*, **50**(6), 99

PALMER, K. (1979). Plant design; safety; insurance. *Design 79* (London: Instn Chem. Engrs)

PALMER, K.N. (1956). A review of information on selected aspects of gas and vapour explosions. *J. Inst. Fuel*, **29**, 293

PALMER, K.N. (1957). Smouldering combustion in dusts and fibrous materials. *Combust. Flame*, 1, 129

PALMER, K.N. (1959). The quenching of flame by wire gauzes. *Combustion*, **7**, 497

PALMER, K.N. (1960). The quenching of flame by perforated sheeting and block flame arresters. *Chemical Process Hazards*, **I**, p. 51

PALMER, K.N. (1967a). The explosibility of dusts in small-scale tests and large-scale industrial plant. *Chemical Process Hazards*, **III**, p. 66

PALMER, K.N. (1967b). Explosion suppression and relief venting. *Chem. Ind.*, Jun. 10, 936

PALMER, K.N. (1971). The relief venting of dust explosions in process plant. *Major Loss Prevention*, p. 142

PALMER, K.N. (1973a). *Dust Explosions and Fires* (London: Chapman & Hall)

PALMER, K.N. (1973b). Effect of diluent dusts on the explosibility of some plastic dusts. *Prevention of Occupational Risks*, **2**, 148

PALMER, K.N. (1974a). Explosion protection of a dust extraction system. *Chemical Process Hazards*, **V**, p. 211

PALMER, K.N. (1974b). Relief venting of dust explosions. *Loss Prevention*, **8**, 10. See also *Chem. Engng Prog.*, 1974, **70**(4), 57

PALMER, K.N. (1974c). Relief venting of dust explosions. *Loss Prevention and Safety Promotion 1*, p. 175

PALMER, K.N. (1975a). Dust explosion hazards in pneumatic transportation. *Fire Prev. Sci Technol.*, **11**, 4

PALMER, K.N. (1975b). Explosions in dust collection plant. *Chem. Engr, Lond.*, **295**, 136

PALMER, K.N. (1975/76). Recent advances in the relief venting of dust explosions. *J. Haz. Materials*, **1**(2), 97

PALMER, K.N. (1976a). Flammable gas detection. *Course on Fire and Explosion Safety in the Chemical Industry.* Dept of Fire Safety Engng, Univ. of Edinburgh

PALMER, K.N. (1976b). Ignition sources and their protection. *Course on Fire and Explosion Safety in the Chemical Industry.* Dept of Fire Safety Engng, Univ. of Edinburgh

PALMER, K.N. (1976c). Portable flammable gas detectors and some problems in their use. *Prevention of Occupational Risks*, **3**, p. 435

PALMER, K.N. (1977). Relief venting of dust explosions in large volumes. *Chemical Process Hazards*, **VI**, unbound paper

PALMER, K.N. (1981a). Dust explosions: basic characteristics, ignition sources and methods of protection against explosions. *The Hazards of Industrial Explosion from Dusts* (London: Oyez/IBC), paper 1

PALMER, K.N. (1981b). Fire problems with powders. *The Hazards of Industrial Explosion from Dusts* (London: Oyez/IBC), paper 12

PALMER, K.N. (1982). Dust explosion hazards in the bulk handling of polymers. *Developments 82* (Rugby: Instn Chem. Engrs)

PALMER, K.N. (1983). Explosions in flammable smokes from smouldering fires. *Loss Prevention and Safety Promotion 4*, vol. 3, p. D11

PALMER, K.N. (1990). Dust explosions: initiation, characteristics, and protection. *Chem. Engng Prog.*, **86**(3), 24

PALMER, K.N. and ROGOWSKI, Z.W. (1967). The use of flame arresters for protection of enclosed equipment in propane–air atmospheres. *Chemical Process Hazards*, **III**, p. 76

PALMER, K.N. and TONKIN, P.S. (1957). The ignition of dust layers on a hot surface. *Combust. Flame*, **1**, 14

PALMER, K.N. and TONKIN, P.S. (1961). The flammability limits of moving mixtures of propane and air. *J. Appl. Chem.*, **11**, 5

PALMER, K.N. and TONKIN, P.S. (1963). The quenching of propane–air explosions by crimped-ribbon flame arresters. *Chemical Process Hazards*, **II**, p. 15

PALMER, K.N. and TONKIN, P.S. (1980). External pressures caused by venting gas explosions in a large chamber. *Loss Prevention and Safety Promotion 3*, vol. 3, p. 1274

PALMER, O.J. (1965). Which: air or electric instruments? *Hydrocarbon Process.*, **44**(5), 241

PALMER, R.K. (1957). *Hydrocarbon Losses from Flanges and Valves* (report). Joint District, Federal and State Project for the Evaluation of Refinery Emissions, Air Pollution Control District, County of Los Angeles, CA

PALMER, T. (1992). Recent advances in process control. *Chem. Engr, Lond.*, **520**, 17

PALMER, T.Y. (1981). Large fire winds, gases and smoke. *Atmos. Environ.*, **15**, 2079

PAN, H.Y., MINET, R.G., BENSON, S.W. and TSOTSIS, T.T. (1994). Process for converting hydrogen chloride to chlorine. *Ind. Eng. Chem. Res.* **33**, 2996–3003

PAN, J.J. (1970). Reliability engineering for the process plant industry. M.Sc. Thesis, Loughborough Univ. of Technol.

PANA, P. (1976). Berechnung der stationären Massenstromdichte von Wasserdampfgemischen und der auftretenden Rückstosskräfte. *Wiss. Bericht., IRSW-24*

PANCHARATNAM, S. (1972a). Transient behavior of a solar pond through Fourier analysis. *Ind. Engng Chem. Process Des. Dev.*, **11**, 626

PANCHARATNAM, S. (1972b). Transient behavior of a solar pond and prediction of evaporation rates. *Ind. Engng Chem. Process Des. Dev.*, **11**, 287

PANDE, P.K., SPECTOR, M.E. and CHATTERJEE, P. (1975). *Computerized Fault Tree Analysis, TREEL and MICSUP. Rep. ORC 75-3.* Operations Res. Centre, Univ. of Calif., Berkely, CA

PANESAR, K.S. (1979a). Include maintenance cost in pump selection. *Hydrocarbon Process.*, **58**(1), 104

PANESAR, K.S. (1979b). Match motor to driven machine. *Hydrocarbon Process.*, **58**(8), 79

PANESAR, K.S. (1980). Choose the right seal. *Hydrocarbon Process.*, **59**(1), 107

PANESAR, K.W. (1978). Consider suction specific speed. *Hydrocarbon Process.*, **57**(6), 107

PANGESTU, W. (1991). Studies on the behaviour of floating LPG plant at sea. *Gastech 90*, p. 4.2

PANKOWSKI, R.A. (1984). Electrical circuits in hazardous locations. *Chem. Engng*, **91**(Feb. 20), 93

PANOFSKY, H.A. (1968). In Hansen, F.V. and Shreve, J.D. (1968), *op. cit.*, p. 47

PANOFSKY, H.A. (1973). Tower micrometeorology. In Haugen, D.A. (1973), *op. cit.*, p. 151

PANOFSKY, H.A. (1974). The atmospheric boundary layer below 150 m. *Ann. Rev. Fluid Mech.*, **6**, 147

PANOFSKY, H.A. (1978). Matching in the convective planetary boundary layer. *J. Atmos. Sci.*, **35**, 272

PANOFSKY, H.A. and DUTTON, J.A. (1984). *Atmospheric Turbulence. Models and Methods for Engineering Applications* (New York: Wiley)

PANOFSKY, H.A. and PRASAD, B. (1965). Similarity theories and diffusion. *Int. J. Air Water Poll.*, **9**, 419

PANOFSKY, H.A., BLACKADAR and MCVEHIL, G.E. (1960). The diabatic wind profile. *Quart. J. R. Met. Soc.*, **86**, 390

PANOFSKY, H.A., TENNEKES, H., LENSCHOW, D.H. and WYNGAARD, J.C. (1977). The characteristics of turbulent velocity components in the surface layer under convective conditions. *Boundary Layer Met.*, **11**, 355

PANTELIDES, C.C. (1988). SPEEDUP – recent advances in process simulation. *Comput. Chem. Engng*, **12**, 745

PANTON, R. (1971). Effects of structure on average properties of two-dimensional detonations. *Combust. Flame*, **16**, 75

PANTON, R.L. and RITTMANN, J.G. (1971). Pyrolysis of a slab of porous material. *Combustion*, **13**, 881

PANTONY, D.A. (1961). *A Chemist's Introduction to Statistics, Theory of Error and Design of Experiments* (London: R. Soc. Chem.)

PANTONY, M.F. (1981). Limits of exposure to thermal radiation. *Health and Safety Executive Seminar on Thermal Radiation and Overpressure Calculations* (London: Health and Safety Executive)

PANTONY, M.F. and FULLAM, B.W. (1989). A strategy for LPG safety – a view from the Health and Safety Executive. *Gastech 88*, paper 4.8

PANTONY, M.F. and SMITH, L.M. (1982). An example of HSE's assessment of major hazards as an aid to planning control by local authorities. *The Assessment of Major Hazards* (Rugby: Instn Chem. Engrs), p. 377

PANTONY, M.F., SCILLY, N.F. and BARTON, J.A. (1989). Safety of exothermic reactions: a UK strategy. *Runaway Reactions*, p. 504. See also *Plant/Operations Prog.*, 1989, **8**, 113

PANTRY, S. (1986). HSELINE: a data base for occupational safety and health. *J. Occup. Accid.*, **7**(4), 285

PAPA, D.M. (1983a). Back pressure considerations for safety relief valves. *API Proc., Refining Dept, 48th Midyear Mtg*, **62**, 97.

PAPA, D.M. (1983b). How back pressure affects safety relief valves. *Hydrocarbon Process.*, **62**(5), 79

PAPA, D.M. (1991). Clear up pressure relief valve sizing methods. *Chem. Engng Prog.*, **87**(8), 81

PAPADOURAKIS, A., CARAM, H.S. and BARNER, C.L. (1991). Upper and lower bounds of droplet evaporation in two-phase jets. *J. Loss Prev. Process Ind.*, **4**(2), 93

PAPAGIANNES, G.J. (1992). Select the right dryer. *Chem. Engng Prog.*, **88**(12), 20

PAPAS, K.K. and SOMMERFELD, J.T. (1991). Comparisons of tank drainage times: orifice drains versus piping. *Am. Inst. Chem. Engrs, Summer Nat. Mtg* (New York)

PAPAZOGLOU, I.A. and KOOPMAN, R.P. (1990). Reliability of online versus standby safety systems in process plants: comparative assessment of two systems for controlling accidental gas releases. *J. Loss Prev. Process Ind.*, **3**(2), 212

PAPZOGLOU, I.A., CHRISTOU, M., NIVOLIANITOU, Z. and ANEZIRIS, O. (1992). On the management of severe chemical accidents DECARA: a computer code for consequence analysis in chemical installations. Case study: ammonia plant. *J. Haz. Materials*, **31**, 135

PAPAZOGLOU, I.A., NIVOLIANITOU, Z., ANEZERIS, O. and CHRISTOU, M. (1992). Probabilistic safety analysis in chemical installations. *J. Loss Prev. Process Ind.*, **5**, 181

PAPE, R.P. (1984). General considerations in producing a development control policy around a Notifiable Installation. In Petts, J.I. (1984), *op. cit.*

PAPE, R.P. (1992). The role of QRA in offshore safety cases. *Major Hazards Onshore and Offshore*, p. 593

PAPE, R.P. (1989a). HSE assessments for consultation distances for major hazard installations. In Lees, F.P. and Ang, M.L. (1989), *op. cit.*, p. 90

PAPE, R.P. (1989b). *Piper Alpha Public Inquiry. Transcript, Day 133*

PAPE, R.P. and NUSSEY, C. (1985). A basic approach for the analysis of risks from major toxic hazards. *The Assessment and Control of Major Hazards* (Rugby: Instn Chem. Engrs), p. 367

PAPE, R.P. and NUSSEY, C. (1989). Quantitative assessment of major hazard installations: 3, The HSE Risk Assessment Tool (RAT). In Lees, F.P. and Ang, M.L. (1989), *op. cit.*, p. 197

PAPENFUSS, R.H. (1965). Safe handling of oxygen in hydrocarbon systems. *Ammonia Plant Safety 7*, p. 68

PAPOULIS, A. (1965). *Probability, Random Variables and Stochastic Processes* (New York: McGraw-Hill)

PAPPAS, J.A. (1994). Safety and risk management on offshore process installations during design and construction. *J. Loss Prev. Process Ind.*, **7**, 345

PAQUET JR., D.A. and RAY W.H. (1994a). Tubular reactors for emulsion polymerization: I. experimental investigation. *AIChE J.* **40**, 1 (January), 73–87

PAQUET JR., D.A. and RAY, W.H. (1994b). Tubular reactors for emulsion polymerization: II. Model comparisons with experiments. *AIChE J.* **40**, 1 (January), 88–96

PARADIES, M., LIGHT, T. and UNGER, L. (1993). Using feedback to improve your PSM (poster item). *Process Safety Management*, p. 429

PARADIES, M., UNGER, L. and RAMEY-SMITH, A. (1991). NRC Human Performance Investigation Process (HPIP). *Proc. Thirty-Fifth Ann. Mtg of the Human Factors Soc.*

PARADIES, M., UNGER, L. and RAMEY-SMITH, A. (1992). Development and testing of NRC Human Performance Investigation Process (HPIP). *Hazard Identification and Risk Analysis, Human Factors and Human Reliability in Process Safety*, p. 253

PARADINE, C.G. and RIVETT, B.H.P. (1980). *Statistical Methods for Technologists* (London: English Univ. Press)

PARASURAMAN, R. (1987). Human computer monitoring. *Hum. Factors*, **29**, 695

PARAKRAMA, R. (1985). Improving batch chemical processes. *Chem. Engr, Lond.*, **417**, 24

PARDOE, J.G.M. (1968). The safety of air transport in the future. *Symp. at Fire Res. Station, Borehamwood*, p. 84

PAREKH, R. (1975). Equipment for controlling gaseous pollutants. *Chem. Engng*, **82**(Oct. 6), 129

PARET, G. (1992). Process development for plant with high hazard potential. *Loss Prevention and Safety Promotion 7*, vol. 1, 26–1

PARFITT, A. and ANDREASSEN, L. (1992). Get those applications right. *Chem. Engr, Lond.*, **529**, s6

PARIAG, A.F., WELCH, I.E. and KERNS, G.E. (1986). Failure and repair of the shell of a primary waste heat boiler on a 1,100 ton/day ammonia plant. *Ammonia Plant Safety*, **26**, 150

PARIDA, F.C., RAO, S.R.M. and MITRAGOTRI, D.S. (1985). Effect of humidity on the ignition behaviour of a liquid sodium pool. *J. Haz. Materials*, **12**(3), 303

PARIS MEMORANDUM OF UNDERSTANDING ON PORT STATE CONTROL (1993). *Annual Report* (Rijswijk, The Netherlands)

PARIS, P.C. and ERDOGAN, F. (1963). A critical analysis of crack propagation laws. *J. Basic Engng*, **85**, 528

PARIS, P.C. and SIH, G.C. (1965). Stress analysis of cracks. In *Fracture Toughness Testing and Its Applications, ASTM STP 381*, p. 30

PARIS, P.C., TADA, H., ZAHOOR, A. and ERNST, H. (1978). *The Theory of Instability of the Tearing Mode of Elastic–Plastic Crack Growth. ASTM STP 668* (Philadelphia, PA: Am. Soc. Testing and Mats), p. 5

PARK, D.E. (1996). Inherently safer chemical processes – A life cycle approach. *Int. Conf. and Workshop on Process Saf. Manage. and Inherently Safer Processes*, Oct. 8–11, 1996, Orlando, FL, 316–319. (New York: American Institute of Chemical Engineers)

PARK, K.S. (1987). *Human Reliability* (Amsterdam: Elsevier)

PARK, S.W. and HIMMELBLAU, D.M. (1980). Error in the propagation of error formula. *AIChE J.*, **26**, 168

PARK, S.W. and HIMMELBLAU, D.M. (1983). Fault detection and diagnosis via parameter estimation in lumped dynamic systems. *Ind. Engng Chem. Process Des. Dev.*, **22**, 482

PARK, S.W. and HIMMELBLAU, D.M. (1987). Structural design for systems fault diagnosis. *Comput. Chem. Engng*, **11**, 713

PARKE, D.V. and SMITH, R.L. (eds) (1977) *Drug Metabolism: From Microbe to Man* (London: Taylor & Francis)

PARKER, A. (ed.) (1978). *Industrial Air Pollution Handbook* (London: McGraw-Hill)

PARKER, A. (1982). Landfill. In Bennett, G.F., Feates and Wilder (1982), *op. cit.*, pp. 14–11

PARKER, A.P. (1979). *Mechanics of Fracture and Fatigue in some Common Structural Configurations. RMCS Tech. Note MAT/18.* Roy. Coll. Mil. Sci., Shrivenham

PARKER, A.P. (1981). *The Mechanics of Fatigue and Fracture* (London: Spon)

PARKER, C.M. (1966). Steels for the CPI. *Chem. Engng*, **73** (Nov. 7), 187

PARKER, C.M. and SULLIVAN, J.W.W. (1963). Steels for low temperatures. *Ind. Engng Chem.*, **55**(5), 18

PARKER, D.J. and HANDMER, J. (eds) (1992). *Hazard Management and Emergency Planning Perspectives in Britain* (London: James & James)

PARKER, E.R. (1957). *Brittle Behaviour of Engineering Structures* (New York: Wiley)

PARKER, H.W. (1975). *Waste Water Systems Engineering* (Englewood Cliffs, NJ: Prentice-Hall)

PARKER, J.U. (1967). Anatomy of a plant safety inspection. *Hydrocarbon Process.*, **46**(1), 197

PARKER, J.U. (1973). Anatomy of a plant safety inspection. In Vervalin, C.H. (1973a), *op. cit.*, p. 425

PARKER, N.H. (1960). Guides to reducing agitator failures. *Chem. Engng*, **67**(Dec. 12), 190

PARKER, N.H. (1964). Equipment and performance guarantees. *Chem. Engng*, **71**(Sep. 28), 113

PARKER, R. (1915a). Poisonous gases (letter). *Br. Med. J.*, Jun. 12, 1027

PARKER, R. (1915b). Poisonous gases (letter). *Br. Med. J.*, Jun. 19, 1066

PARKER, R.C. (1988). *Looking Good in Print* (Ventana Press)

PARKER, R.J. (chrmn) (1975). *The Flixborough Disaster. Report of the Court of Inquiry* (London: HM Stationery Office)

PARKER, R.J. (chrmn) (1978). *The Windscale Inquiry* (London: HM Stationery Office)

PARKER, R.O. (1974). Calculating thermal radiation hazards in large fires. *Fire Technol.*, **10**, 147

PARKER, R.O. and SPATA, J.K. (1968). Downwind travel of vapors from large pools of cryogenic liquids. *LNG Conf. 1*

PARKER, R.V. (1984). Safety systems for a continuous petrochemical process under vacuum. *Plant/Operations Prog.*, **3**, 119

PARKER, V. (1991). *Licensing Technology and Patents* (Rugby: Instn Chem. Engrs)

PARKER, W.G. (1958). A review of some unusual stationary flame reactions. *Combust. Flame*, 2, 69

PARKER, W.J. (1972). Flame spread model for cellulosic materials. *J. Fire Flammability*, **3**, 254

PARKES, D. (ed.) (1978). *Terotechnology Handbook* (London: HM Stationery Office)

PARKES, E.W. (1955). The permanent deformation of a cantilever struck transversely at its tip. *Proc. Roy. Soc., Ser. A*, **228**, 462

PARKES, W.R. (1974). *Occupational Lung Diseases* (London: Butterworths)

PARKHOUSE, J.G. (1987). Damage accumulation in structures. *Reliab. Engng*, **17**(2), 97

PARKIN, P.H. and HUMPHREYS, H.R. (1969). *Acoustics, Noise and Buildings* (London: Faber & Faber)

PARKINS, R.N. (1982). *Corrosion Processes* (Amsterdam: Elsevier Applied Science)

PARKINS, R.N. (1989). The application of stress corrosion crack growth kinetics to predicting lifetimes of structures. *Corrosion Sci.*, **29**, 1019

PARKINSON, D.B. (1987). Quadric reliability indices. *Reliab. Engng*, **17**(1), 23

PARKINSON, G. (1978). Save pollution credits for a rainy day. *Chem. Engng*, **85**(Oct. 23), 106

PARKINSON, G. (1980). New models to control old pollution source. *Chem. Engng*, **87**(Dec. 1), 38

PARKINSON, G. (1986). Makers of gaskets, packing, try asbestos replacements. *Chem. Engng*, **93**(20), Oct. 27, 23

PARKINSON, G. (1987). Safety concern for underground pipes is growing. *Chem. Engng*, **94**(18), 22

PARKINSON, G. (1988). CPI girds for Proposition 65. *Chem. Engng*, **95**(3), Mar. 14, 39

PARKINSON, G. (1990). HF's future is up in the air. *Chem. Engng*, **97**(5), 39

PARKINSON, G. (1991). Holistic maintenance. *Chem. Engng*, **98**(4), 30

PARKINSON, G. and ONDREY, G. (1992). Vacuum pumps are running dry. *Chem. Engng*, **99**(6), 121

PARKINSON, G. and BASTA, N. (1991). Water supply and disposal update. *Chem. Engng*, **98**(4), 37

PARKINSON, G. and JOHNSON, E. (1989). Surging interest in leakproof pumps. *Chem. Engng*, **96**(6), 30

PARKINSON, G., SHORT, H. and USHIO, S. (1982). For more plant projects the word is: go modular. *Chem. Engng*, **89** (Nov. 15), 66

PARKINSON, J.S. (1979). Assessment of plant pressure relief systems. *Design 79* (London: Instn Chem. Engrs), p. K3.1

PARKINSON, J.S. (1980). The engineer's liability to prosecution (letter). *Chem. Engr, Lond.*, Dec., 728

PARKINSON, S.T. (1982). The role of the user in successful new product development. *R&D Management*, **12**(3), 123

PARKS, J.C. and HINKLE, E.A. (1963). Analysis of hydrocarbon impurities in oxygen plant streams by gas chromatography. *Ammonia Plant Safety*, **5**

PARKS, J.R. (1961). Criticality criteria for various configurations of a self-heating chemical as functions of activation energy and temperature of assembly. *J. Chem. Phys.*, **34**, 46

PARKYN, B. (1970). *Glass Reinforced Plastics* (London: Iliffe)

PARLIAMENTARY COMMISSIONER (1975–76). *Third Report of the Parliamentary Commissioner for Administration* (London: HM Stationery Office)

PARLOW, CHARLES, (1999). *Normal Accidents Living with High-Risk Technology* (Princeton, NY: Princeton University Press) ISBN 0-691-00412-9

PARMAKIAN, J. (1955). *Waterhammer Analysis* (London: Prentice-Hall)

PARMAR, J.C. and LEES, F.P. (1987a). The propagation of faults in process plants: hazard identification. *Reliab. Engng*, **17**(4), 277

PARMAR, J.C. and LEES, F.P. (1987b). The propagation of faults in process plants: hazard identification for a water separator system. *Reliab. Engng*, **17**(4), 303

PARMEGGIANI, L. (ed.). (1983). *ILO Encyclopaedia of Occupational Health and Safety, 3rd ed.* (Geneva: Int. Labour Office)

PARMELE, C.S., O'CONNELL, W.L. and BASDEKIS, H.S. (1979). Absorption, pollution, recovers solvents. *Chem. Engng*, **86**(Dec. 31), 59

PARMENTIER, N. and NENOT, J.C. (1989). Radiation damage aspects of the Chernobyl accident. *Atmos. Environ.*, **23**, 771

PARNAROUSKIS, M.C., PERRY, W.W. and ARTICOLA, W.P. (1980). Applications of the US Coast Guard vulnerability model for hazard assessment. *Loss Prevention and Safety Promotion 3*, vol. 3, p. 1158

PARNELL, A.C. (1979). Automated high bay warehouses. Confusion amongst authorities? *Fire Surveyor*, **8**(3), 42

PARNELL, A.C. (1977). Why protect steelwork? *Fire Surveyor*, **6**(1), 44

PARNELL, D. (1981). Questions on sulfur. *Hydrocarbon Process.*, **60**(4), 127

PARNELL, D.C. (1973). Claus sulfur recovery unit start-up. *Chem. Engng Prog.*, **69**(8), 89

PARNIANPOUR, M., BEJJANI, F.J. and PAVLIDIS, L. (1987). Worker training: the fallacy of a single, correct lifting technique. *Ergonomics*, **30**, 331

PAROS, S.V. (1976). Choosing cold insulation for your piping and storage? *Hydrocarbon Process.*, **55**(11), 257

PARRATT, N.J. (1972). *Fibre Reinforced Materials Technology* (New York: van Nostrand Reinhold)

PARRISH, C.B. (1966). Onstream inspection increases run length. *Hydrocarbon Process.*, **45**(5), 170

PARRISH, R.W. and SIEDEL, R.O. (1978). Environmental considerations in acid gas removal. *Ammonia Plant Safety 20*, p. 116

PARRISH, W.R. and PRAUSNITZ, J.M. (1972). Dissociation pressures of gas hydrates formed by gas mixtures. *Ind. Engng Chem. Process Des. Dev.*, **11**, 26

PARRY, C.F. (1989). Selecting the relief system and its components. *Selection and Use of Pressure Relief Systems for Process Plant* (Manchester: Instn. Chem. Engrs, North-West Branch), p. 3.1

PARRY, C.F. (1992). *Relief Systems Handbook* (Rugby: Instn Chem. Engrs)

PARRY, D.L. (1976). Qualify vessel integrity with acoustic emission analysis. *Hydrocarbon Process.*, **55**(12), 132

PARRY, D.L. (1980). Detection and analysis of vessel discontinuities by acoustic emission inspection. *Ammonia Plant Safety*, **22**, 140

PARRY, I. (1990). Managing your lubricant. *Plant Engr*, Jan./Feb., 31

PARSAYE, K. and CHIGNELL, M. (1988). *Expert Systems for Experts* (Chichester: Wiley)

PARSON, K.C. (1988). Protective clothing; heat exchange and physiological objectives. *Ergonomics*, **31**, 991

PARSONS, C.L. (1918). The American chemist in warfare. *Science*, **48**(1242), 377. See also *Ind. Engng Chem.*, **10**, 776

PARSONS, J.D.W. (1977). Some aspects of the data collection from field reliability studies. *First Nat. Conf. on Reliability. NRC5/7* (Culcheth, Lancashire: UK Atom. Energy Auth.)

PARSONS, J.R., OGLESBY, M.W. and SMITH, D.L. (1970). *In Advances in Instrumentation 25* (Pittsburgh, PA: Instrum. Soc. Am.), p. 567

PARSONS, R. (1971a). The electrical double layer, electrode reactions and static. *Static Electrification*, **3**, p. 124

PARSONS, R. (1971b). Guidelines for plant start-up. *Chem. Engng Prog.*, **67**(12), 29

PARTINGTON, E.V. (1980). Applying terotechnology to chemical plant maintenance. *Chem. Engr, Lond.*, **356**, 312

PARTRIDGE, E.P. and PAULSON, E.G. (1963). Design and operate for water economy. *Chem. Engng*, **70**, Jun. 10, 175

PARTRIDGE, H. (1977). How to minimize fork truck accidents and reduce operating costs. In Handley, W. (1977), *op. cit.*, p. 260

PARTRIDGE, L.J. (1976). Coal-based ammonia plant operation. *Ammonia Plant Safety*, **18**, 73

PARUIT, B. and KIMMEL, W. (1979). Control blowdown to the flare. *Hydrocarbon Process.*, **58**(10), 117. See also in Kane, L.A. (1980), *op. cit.*, p. 147

PARUNGO, F.P., PUESCHEL, R.F. and WELLMAN, D.L. (1980). Chemical characteristics of oil refinery plumes in Los Angeles. *Atmos. Environ.*, **14**, 509

PARZEN, E. (1960). *Probability Theory and its Applications* (New York: Wiley)

PASMAN, H.J. and LEMKOWITZ, S.M. (2000). Hierarchy of concepts for risk reduction in chemical plant. *Prog. Saf. Sci. Technol. [Int. Symp. Saf. Sci. Technol.]* **2**, 122–32

PASMAN, H.J. and WAGNER, H.G. (1986). Explosion. A review of the state of understanding and predictive capabilities to safeguard against industrial explosion incidents. *Loss Prevention and Safety Promotion 5*, vol. I, pp. 8–1; vol. III, p. 176

PASMAN, H.J., GROOTHUIZEN, T.M. and DE GOOIJER, H. (1974). Design of pressure relief vents. *Loss Prevention and Safety Promotion 1*, p. 185

PASMAN, H.J., DUXBURY, H.A. and BJORDAL, E.N. (1992). Major hazards in the process industries: achievements and challenges in loss prevention. *J. Haz. Materials*, **30**, 1

PASQUA, P.F. (1953). Metastable flow of Freon 12. *Refrig. Engng*, **61**, 1084A

PASQUILL, F. (1943). Evaporation from a plane free-liquid surface into a turbulent air stream. *Proc. R. Soc., Ser. A*, **182**, 75

PASQUILL, F. (1956). Meteorological research at Porton. *Nature*, **177**, 1148

PASQUILL, F. (1961). The estimation of the dispersion of windborne materials. *Met. Mag.*, **90**(1063), 33

PASQUILL, F. (1962a). *Atmospheric Diffusion* (London: van Nostrand)

PASQUILL, F. (1962b). Some observed properties of medium-scale diffusion in the atmosphere. *Quart. J. R. Met. Soc.*, **88**(375), 70

PASQUILL, F. (1965). Meteorological aspects of the spread of windborne contaminants. *World Health Organisation, Conf. on Protection of the Public in the Event of Radiation Accidents*, p. 43

PASQUILL, F. (1970). Prediction of diffusion over an urban area – current practice and future prospects. *Proc. Symp. on Multiple Source Diffusion Models* (Environ. Prot. Agency), p. 3.1

PASQUILL, F. (1972). Some aspects of boundary layer description. *Quart. J. R. Met. Soc.*, 469

PASQUILL, F. (1974). *Atmospheric Diffusion*, 2nd ed. (New York: Wiley)

PASQUILL, F. (1975). Atmospheric dispersion models for environmental pollution applications. *Am. Met. Soc., Lectures on Air Pollution and Environmental Impact Analysis* (Boston, MA), p. 350

PASQUILL, F. (1976a). *Atmospheric Dispersion Parameters in Gaussian Plume Modeling, Pt II: Possible Requirements for Change in Turner Workbook Values. Rep. EPA-600/4-76-030b.* Environ. Prot. Agency

PASQUILL, F. (1976b). The dispersion of material in the atmospheric boundary layer – the basis for generalization. In Haugen, D.A. (1976), *op. cit.*, p. 1

PASQUILL, F. and SMITH, F.B. (1971). The meteorological basis for the estimation of the dispersion of windborne material. *Proc Second Int. Clean Air Congress* (New York: Academic Press), p. 1067

PASQUILL, F. and SMITH, F.B. (1983). *Atmospheric Diffusion*, 3rd ed. (New York: Wiley)

PAÄSSLER, P. *et al.* (1985). Acetylene. *Ullmanns Encyclopaedia of Chemical Technology*, rev. ed., vol. A1 (Weinheim: VCH), p. 97

PASSOW, N.R. (1978). US environmental regulations affecting the chemical process industries. *Chem. Engng*, **85**(Nov. 20), 173

PASSOW, N.R. (1980). Preparing the Environmental Impact Statement. *Chem. Engng*, **87**(Dec. 15), 69

PASTORINI, R., DE FAVERI, D., PELOSO, A. and FERRAIOLO, G. (1980). High level of labour accidents in small companies. *Loss Prevention and Safety Promotion 3*, vol. 3, p. 1382

PASTORINO, R., DE FAVERI, D.M., ZONATO, C., MAGA, L. and FERRAIOLO, G. (1986). Industrial safety and factories dimension. A sample investigation on 18 small companies: results and work perspective. *Loss Prevention and Safety Promotion 5*, p. 26–1

PASTORINI, R., DE FAVERI, D.M., DEALBERTI, P., VIDILI, A. and FERRAIOLO, G. (1989). The fire prevention system in an oil refinery – optimization of the plant fire time. *Loss Prevention and Safety Promotion 6*, p. 62.1

PASTORINO, R., PAGELLA, C. and DE FAVERI, D.M. (1992). Occupational accidents trend in Italy (1885–1980) and some correlations with industrial protection trend. *Loss Prevention and Safety Promotion 7*, vol. 1, 27–1

PATÉ-CORNELL, E. (1987). Probabilistic risk assessment and safety regulations in the chemical industry. *J. Haz. Materials*, **15**(1), 97

PATEL, J., KAPADIA, C.M. and OWEN, D.B. (1976). *Handbook of Statistical Distributions* (New York: Dekker)

PATEL, J.K. and READ, C.B. (1982). *Handbook of the Normal Distribution* (New York: Dekker)

PATERNOTTE, P.H. (1978). The control performance of operators controlling a continuous distillation column. *Ergonomics*, **21**, 671

PATERSON and ROWE (1970). In Astbury N.F. *et al.* (1970), *op. cit.*, Appendix

PATERSON, A.G. (1969). Compressed air. In Handley, W. (1969), *op. cit.*, p. 57

PATERSON, A.G. (1977). Compressed air. In Handley, W. (1977), *op. cit.*, p. 198

PATERSON, E.A. (1989). *Piper Alpha Public Inquiry. Transcript, Day 88*

PATERSON, S. (1939–). The ignition of inflammable gases by hot moving particles. *Phil. Mag.*, 1939, **28**, 1; 1940, **30**, 437

PATERSON, T.T. (1950). The theory of the social threshold. *The Sociological Rev.*, **42**, sec. 3

PATHAK, V.K. and RATTAN, I.S. (1985). Forestalling construction problems. *Chem. Engng*, **92**(Nov. 25), 69

PATHE, D.C. (1985). Process training simulators save startup and operations. *Oil Gas J.*, **83**(Feb. 11), 75

PATIENCE, K. (1989). The Bombay explosion. *After the Battle*, **64**, 1

PATIL, L.R.S., SMITH, R.G., VORWALD, A.J. and MOONEY, T.F. (1970). The health of diaphragm cell workers exposed to chlorine gas. *Am. Ind. Hyg. Ass. J.*, **31**, 678

PATRICK, C.P. (1983). Harmonisation of toxicity testing in a chemical safety evaluation. *Chem. Ind.*, Jan. 17, 55

PATRICK, E.A. (1979). *Decision Analysis in Medicine. Methods and Applications* (Boca Raton, FL: CRC Press)

PATRICK, J. (1975). *The Psychology of Training* (London: Methuen)

PATRICK, J. (1992). *Training* (London: Academic Press)

PATRICK, J. (1993). Cognitive aspects of fault-finding training and transfer. *Le Travail humain*, **56**, 187

PATRICK, J. and HAINES, B. (1988). Training and transfer of fault finding skill. *Ergonomics*, **31**, 193

PATRICK, J. and STAMMERS, R.B. (1981). The role of computers in training for problem diagnosis. In Rasmussen, J. and Rouse, W.B. (1981), *op. cit.*, p. 589

PATRICK, J., BARWELL, F., TOMS, M. and DUNCAN, K.D. (1986). *Fault-finding Skills – An Appraisal of Training Methods.* (Sheffield: Manpower Services Comm.)

PATRICK, J., HAINES, B., MUNLEY, G. and WALLACE, A. (1989). Transfer of fault-finding between simulated chemical plants. *Hum. Factors*, **31**, 503

PATRICK, P.K. (1975). Treatment and disposal of hazardous wastes in Western Europe. *J. Haz. Materials*, **1**(1), 45

PATTERNOTTE, P.H. and VERHAGEN, L.H.J.M. (1979). Human operator research with a simulated distillation process. *Ergonomics*, **22**, 19

PATTERSON, A.M., KINGERY, C.N., ROWE, R.D., PETES, J. and DEWEY, J.M. (1972). Fireball and shock wave anomalies observed in chemical explosion. *Combust. Flame*, **19**, 25

PATTERSON, D. (1968). Application of a computerised alarm analysis system to a nuclear power station. *Proc. IEE*, **115**, 1858

PATTERSON, D.W. (1990). *Introduction to Artificial Intelligence and Expert Systems* (Englewood Cliffs, NJ: Prentice Hall)

PATTERSON, L.B. (1979). Ammonia separator accident. *Ammonia Plant Safety*, **21**, 95

PATTERSON, L.W. and CLARK, J.M. (1971). Process plant reliability in Puerto Rico. *Chem. Engng Prog.*, **67**(1), 54

PATTERSON, W.C. (1976). *Nuclear Power* (Harmondsworth: Penguin)

PATTERSON, W.C. (1985). *Going Critical* (London: Paladin)

PATTERSON, W.W. (1973). Failure of a boiler pressure shell. *Ammonia Plant Safety*, **15**, 89

PATTERSON, W.W. (1977). Replacing a 1,000 ton/day ammonia converter. *Ammonia Plant Safety*, **19**, p. 97

PATTISON, D.A. (1975). Practical lubrication for process plants. *Chem. Engn, 82*(Apr. 28), 98

PATTISON, D.A. (1969). Oil spill cleanup: a matter of $s and methods. *Chem. Engng*, **76**(Feb. 10), 50

PATTON, P.W. (1983). Vacuum systems in the CPI. *Chem. Engng Prog.*, **79**(12), 56

PATTY, F.A. (ed.) (1948–). *Industrial Hygiene and Toxicology*; 1958, 2nd ed. (New York: Interscience); 1963, 2nd rev. ed.; 1978, 3rd ed. (New York: Wiley)

PATTY, F.A. (1962). *Industrial Hygiene and Toxicology*, 2nd ed. (New York: Wiley)

PATZLER, K. (1992). Electronics enter the electric motor. *Chem. Engng*, **99**(7), 71

PAU, L.F. (1975). *Diagnostique des Pannes dans les Syst³mes* (Toulouse: Cepadues-Édition)

PAU, L.F. (1981a). Application of pattern recognition to failure analysis and diagnosis. In Rasmussen, J. and Rouse, W.B. (1981), *op. cit.*, p. 389

PAU, L.F. (ca. 1981b). *Failure Diagnosis and Performance Modelling* (New York: Dekker)

PAUL, B.O. (1994). Mag-Drive pumps eliminate VOC emissions. *Chemi. Process.* (Feb.), 66–8

PAUL, E.L. (1988). design of reaction systems for specialty organic chemicals. *Chem. Eng. Sci.* **43**(8), 1773–82

PAULA, H.M. and ROBERTS, M.W. (1991). Reliability performance of fault-tolerant digital control systems. *Plant/Operations Prog.*, **10**, 115

PAULSEN, R.L. (1984). Human behaviour in fires: an introduction. *Fire Technol.*, **20**(2), 15

PAULSON, C.A. (1970). The mathematical representation of wind speed and temperature profiles in the unstable atmospheric surface layer. *J. Appl. Met.*, **9**, 857

PAULSON, E.G. (1977). How to get rid of toxic organics. *Chem. Engng*, **84**(Oct. 17), 21

PAUSTENBACH, D.J. (ed.) (1989). *The Risk Assessment of Environmental Hazards: A Textbook of Case Studies* (New York: Wiley)

PAUTHENIER (1961). *Proc. Conf. on La Physique des Forces Electrostatiques* (Paris: CNRS)

PAUTHENIER, M.M. and MOREAU-HANOT, M. (1932). The charge of spherical particles in an ionised field. *J. Phys. Radium, Ser.*, **7**, 3, 590

PAWEL, D., VAN TIGGELEN, P.J., VASATKO, H. and WAGNER, H.G. (1970). Initiation of detonation in various gas mixtures. *Combust. Flame*, **15**, 173

PAYMAN, W. (1928). The 'normal' propagation of flame in gaseous mixtures. *Combustion*, **1**, 51

PAYMAN, W. and SHEPHERD, W.C.F. (1937). Explosion waves and shock waves. IV, Quasi detonation in mixtures of methane and air. *Proc. R. Soc., Ser. A*, **158**, 348

PAYMAN, W. and SHEPHERD W.C.F. (1946). Explosion waves and shock waves. VI, The disturbance produced by bursting diaphragms with compressed air. *Proc. R. Soc., Ser. A*, **186**, 293

PAYMAN, W. and TITMAN, H. (1935). Explosion waves and shock waves. III, The initiation of detonation in mixtures of ethylene and oxygen and of carbon monoxide and oxygen. *Proc. R. Soc., Ser. A*, **152**, 418

PAYMAN, W. and WHEELER, R.V. (1922). *Fuel*, **1**, 185

PAYNE, A.M.M. (1941). Some air raid casualty experience. *Br. Med. J.*, Apr. 12

PAYNE, B.J. (1981). Dealing with hazard and risk in planning (1). In Griffiths, R.F. (1981), *op. cit.*, **77**

PAYNE, D. and ALTMAN, J.W. (1962). *An Index of Electronic Equipment Operability: Report of Development. Rep. AIR-C43-1/62/FR*. Am. Inst. Res.

PAYNE, G. (1964). Safe handling and storage of liquefied gases, in particular chlorine and chlorofluoro-hydrocarbon refrigerants. *Chem. Engr, Lond.*, **178**, CE92

PAYNE, J.R. (1980). The PVRC's gasket test program II: a status report. *API Proc., Refining Dept, 41st Midyear Mtg*, **59**, 85

PAYNE, J.R. and BAZERGUI, A. (1981). More progress in gasket testing; the PVRC program. *API Proc., Refining Dept, 46th Midyear Mtg*, **60**, 271

PAYNE, M.P., DELIC, J. and TURNER, R.M. (1990). *Toxicology of Substances in Relation to Major Hazards: Ammonia* (London: HM Stationery Office)

PAYNE, S. (1971). Living with bomb threats. *Chem. Engng*, **78**(Feb. 22), 80

PAYNE, W.R. (1978). Toxicology and process design. *Chem. Engng*, **85** Apr. 24, 83

PAYNE, W.R.S. (1986). In-house management of wastes. *Chem. Ind.*, Jun. 16, 411

PEACOCK, R.D. and BABRAUSKAS, V. (1991). Analysis of large scale fire test data. *Fire Safety J.*, **17**, 387

PEACOCK, R.D. and BUKOWSKI, R.W. (1990). A prototype methodology for fire hazard analysis. *Fire Technol.*, **26**(1), 15

PEACOCK, S.T. and WATSON, I.A. (1982). Effect of site locations on the reliability of automatic fire detection systems. *Reliab. Engng*, **3**(5), 379

PEACOCK, S.T., KAMATH, A.R.R. and KELLER, A.Z. (1982). Reliability appraisal of fire detection systems. *Sixteenth Loss Prevention Symp.* (New York: Am. Inst. Chem. Engrs)

PEACOCK, W.S. and WHITE, I.L. (1992). Site investigation and risk analyses. *Proc Instn Civ. Engrs, Civ. Engng*, **92**, 74

PEARCE, D. (1981). Nuclear power and the public interest. *Sci. Public Policy*, **8**, 23

PEARCE, D.W. (1981). Risk assessment: use and misuse. *Proc. R. Soc., Ser. A*, **376**, 181. See also in Warner, Sir F. and Slater, D.H. (1981), *op. cit.*, p. 181

PEARCE, I.D. (1985). The effect of process duty on relief valve reliability. *Reliability*, **85**, p. 5B/R/1

PEARCE, K. (1977). Deep sea disposal of wastes. *Chemical Engineering in a Hostile World* (London: Clapp & Poliak)

PEARCE, S.J. (1957). *Bureau of Mines Approval System for Respiratory Protective Devices* (Pittsburgh, PA: Bureau of Mines)

PEARCE, S.W. (1977). Disposal to land. *Chemical Engineering in a Hostile World* (London: Clapp & Poliak)

PEARL, J. (1984). *Heuristics* (Reading, MA: Addison-Wesley)

PEARSALL, I.S. (1965–66). The velocity of water hammer waves. *Proc. Instn Mech. Engrs*, **180**, pt 3E, 12

PEARSON, C.R. (1982). Notification of new chemicals: some practical implications. *Chem. Ind.*, Oct. 16, 801

PEARSON, C.R. (1988). Assessment of risk of significant environmental contamination from storage and distribution sites. *Preventing Major Chemical and Related Process Accidents*, p. 471

PEARSON, E.S. and HARTLEY, H.O. (1956). *Biometrika Tables for Statisticians* (Cambridge: Cambridge Univ. Press)

PEARSON, G.D.M. (1975). Availability evaluation of chemical plant systems. Ph.D. Thesis, Univ. of Nottingham

PEARSON, G.D.M. (1977). Computer program for approximating the reliability characteristics of acyclic directed graphs. *IEEE Trans. Reliab.*, **R-26**, 32

PEARSON, J. (1991). *Computers and Safety in the Process Industries. The HSE Guidelines* (paper)

PEARSON, J. (1992). Emergency shutdown systems in onshore and offshore process operations. *Major Hazards Onshore and Offshore*, p. 671

PEARSON, J. and BRAZENDALE, J. (1988). Computer control chemical plant – design and assessment framework. *Preventing Major Chemical and Related Process Accidents*, p. 195

PEARSON, J.E., NONHEBEL, G. and ULANDER, P.H.N. (1935). Removal of smoke and acid constituents from flue gases by non-effluent water processes. *J. Inst. Fuel*, **8**(Feb.), 119; (Apr.), 183

PEARSON, K. and HARTLEY, H.O. (1954). *Biometrika Tables for Statisticians* (Cambridge: Cambridge Univ. Press)

PEARSON, L. (1977a). Practical aspects of hazards in plant commissioning. *Course Chemical Plant Commissioning*. Dept of Chem Engng, Univ. of Leeds

PEARSON, L. (1977b). When its time for start-up. *Hydrocarbon Process.*, **58**(8), 116

PEARSON, R. (1984). Reconsidering the Law of the Sea Convention. *Chem. Engng*, **91**(Jul. 9), 20E

PEARSON, T.W. (1978). Response capacity – the first man in. *Control of Hazardous Material Spills*, **4**, 447

PEASE, J. (1935). The mechanism of slow oxidation of propane. *Am. Chem. Soc.*, **57**, 2296

PEASE, R.N. (1937). The slow combustion of gaseous paraffins. *Combustion*, **2**, p. 146

PEATE, J. (1984). Using electricity for industrial process heating. *Process Engng*, Dec., 34

PEBWORTH, L.A. (1977). Catastrophic failure in process lines. *Ammonia Plant Safety*, **19**, 148

PEBWORTH, L.A. and WICKES, J.R. (1975). Optimizing process control and data systems. *Ammonia Plant Safety*, **17**, 75

PECK, G.C. and PALLES-CLARK, P.C. (1982). *The Development of Hazardous Area Classification* (London: Insurance Tech. Bur.)

PECK, J.C. and JEAN, P. (1982). Prediction of sloshing loads in LNG ships. *Gastech 81*, p. 253

PECK, T.P. (1974). *Occupational Safety and Health – Guide to Information Sources* (Detroit, MI: Gale Res. Co.)

PECKNER, D. and BERNSTEIN, I.M. (1977). *Handbook of Stainless Steels* (New York: McGraw-Hill)

PEDERSEN, E.S. (1978). *Nuclear Power*, vols 1–2 (Ann Arbor, MI: Ann Arbor Science)

PEDERSEN, F. and SELIG, R.S. (1989). Predicting the consequences of short-term exposure to high concentrations of gaseous ammonia. *J. Haz. Materials*, **21**, 143

PEDERSEN, G.H. and ECKHOFF, R.K. (1987). Initiation of grain dust explosions by heat generated by single impact between solid bodies. *Fire Safety J.*, **12**(2), 153

PEDERSEN, O.M. (1974). *An Analysis of Operator's Information and Display Requirements during Power Plant Boiler Start-up. Rep. Riso-M-1738*. Atom. Energy Comm. Res. Estab., Riso, Denmark

PEDERSEN, O.M. (1984). *Human Errors in Test and Calibration: Analysis of Actual Events for the Evaluation of Coverage* and Applicability of Search Strategies and Risk Management. *Rep. Riso-M-2470*. Riso Nat. Lab., Riso, Denmark

PEDERSEN, O.M. (1985). *Human Risk Contributions in the Process Industry: Guides for their Pre-identification in Well-structured Activities and for Post-incident Analysis. Rep. Riso-M-2513*. Riso. Nat. Lab, Riso, Denmark

PEDLEY, J.B., NAYLOR, R.D. and KIRBY, S.P. (1986). *Thermochemical Data of Organic Compounds*, 2nd ed. (London: Chapman & Hall)

PEEBLES, M.W.H. (1977). World LNG trade: present status and long term prospects. *LNG Conf. 5*. See also *Hydrocarbon Process.*, 1978, **57**(1), 68–D

PEEK, F.W. (1929). *Dielectric Phenomena in High Voltage Engineering* (New York: McGraw-Hill)

PEELER, J. (1975). European surface transport. *Distribution Conf.* (London: Chem. Ind. Ass.), p. 42

PEGRAM, N. (1990). Value engineering in the process industries. *Chem. Engr, Lond.*, **482**, 37

PEINE, H.G. (1979). Safety in BASF. In BASF (1979), *op. cit.*, p. 63

PEISSARD, W.G. (1982). Corrosion from halon: the true extent of the problem. *Fire Int.*, **74**, 88

PEKELNEY, D. (1990). Hazardous waste generation, transportation, reclamation, and disposal: California's manifest system and the case of halogenated solvents. *J. Haz. Materials*, **23**(3), 293

PEKNY, J.F. (1993). Communications: profiting from an information explosion. *Chem. Engng Prog.*, **89**(11), 51

PEKROL, P.J. (1975). Vibration monitoring increases equipment availability. *Chem. Engng*, **82**(Aug. 18), 109

PELLINI, W.S. (1971). Principles of fracture-safe design. *Welding J.*, Mar. (suppl.), 91–S; Apr., 147–S

PELLMONT, G. (1979). Explosion and ignition behaviour of hybrid mixtures of flammable dusts and gases. Dissertation, ETH Zurich, No. 6498

PELLMONT, G. (1982). Explosion characteristics and hybrid mixtures. *The Control and Prevention of Dust Explosions* (London: Oyez/IBC), paper 1

PELOSI, P.F. and CAPPABIANCA, C.J. (1985). Corrosion control in steam and condensate lines. *Chem. Engng*, **92**(Jun. 24), 61

PELTI, T.F., KLEIN, J. and DHARJATI, P.S. (1990). Diagnostic model processor: using deep knowledge for process fault diagnosis. *AIChE J.*, **36**, 565

PELTO, P.J. and PURCELL, W.L. (1977). *MFAULT – a Computer Program for Analyzing Fault Trees. Rep. BNWL-2145*. Battelle Northwest Lab., Richland, WA

PEMBERTON, A.W. (1974). *Plant Layout and Materials Handling* (London: Macmillan)

DE LA PENA, M. (1990). *Piper Alpha Public Inquiry. Transcript, Days 167, 172*

PENCE, T.G. (1969). Transportation. *Chem. Engng*, **76** (Apr. 14), 23

PENDERGAST, M.M. and CRAWFORD, T.V. (1974). Actual standard deviations of vertical and horizontal wind direction compared to estimates from other measurements. *First Symp. on Atmospheric Diffusion and Air Pollution* (Boston, MA: Am. Met. Soc.), p. 1

PENDERGRASS, W. and ARYA, S.P.S. (1984). Dispersion in neutral boundary layer over a step change in surface roughness. *Atmos. Environ.*, **18**, 1267, 1281; 1986, **20**, 411

PENG, L.C. (1979). Toward more consistent pipe stress analysis. *Hydrocarbon Process.*, **58**(5), 207

PENGELLY, D. (1978). Select expansion joints properly. *Hydrocarbon Process.*, **57**(3), 141

PENINGTON, A.H. (1954). War gases and chronic lung disease. *Med. J. Austr.*, **510**

PENKETT, S. (1989). The changing atmosphere – ozone, holes and greenhouses. *Chem. Engr, Lond.*, **463**, 14

PENLAND, J.R. (1967). Large scale inerting. *Loss Prevention*, **1**, 56

PENMAN, J. and FRASER, J.R. (1983). The efficient calculation of electrostatic fields in large systems. *Electrostatics*, **6**, 243

PENNELL, A. (1963). Hazards in the electrolysis of brine in mercury cells. *Chemical Process Hazards*, **II**, p. 35

PENNER, S.S. and MULLINS, B.P. (1959). *Explosions, Detonations, Flammability and Ignition* (London: Pergamon Press)

PENNEY, LORD, SAMUELS, D.E.J. and SCORGIE, G.C. (1970). The nuclear explosive yields at Hiroshima and Nagasaki. *Phil. Trans. R. Soc., Lond., Ser. A*, **266**(1177), 357

PENNEY, W.G. and THORNILL, C.K. (1952). The dispersion, under gravity, of a column of fluid supported on a rigid horizontal plane. *Phil. Trans. R. Soc., Lond., Ser. A*, **244**, 285

PENNEY, W.R. (1970). Guide to trouble-free mixers. *Chem. Engng*, **77**(Jun. 1), 171

PENNEY, W.R. (1978). Inert gas in liquid mars pump performance. *Chem. Engng*, **85**(Jul. 3), 63

PENNIALL, T.H. (1980). Trends in graphics. *Ergonomics*, **23**, 921

PENNINGER, J.L.M. and OKAZAKI, J.K. (1980). Designing a high pressure laboratory. *Chem. Engng Prog.*, **76**(6), 65

PENNINGER, J.M.L. *et al.* (eds) (1985). *Supercritical Fluid Technology* (New York: Elsevier)

PENNINGTON, J. (1992). Information key to on-line sealing. *Hydrocarbon Process.*, **71**(1), 63

PENNINGTON, J.C.A. (1968). Simple plants cause trouble, too. *Ammonia Plant Safety*, **10**, 108

PENNY, E. (1956). The velocity of detonation of compressed acetylene. *Faraday Soc. Disc.*, **22**, 157

PENNY, J.J. (1986). The role of flame detection in AFD. *Fire Surveyor*, **15**(2), 22

PENROD, J.W. (1962). Operating safety and dependability with dry-ring compression. *Ammonia Plant Safety*, **4**, 21

PENTREATH, R.J. (1980). *Nuclear Power, Man and the Environment* (London: Taylor & Francis)

PERALDI, O., KNYSTAUTAS, R. and LEE, J.H. (1986). Criteria for transition to detonation in tubes. *Combustion*, **21**, 1629

PERBAL, G. (1983). Safety aspects of urea-ammonium nitrate solutions. *Loss Prevention and Safety Promotion 4*, vol. 1, p. M10

PERBAL, R. (1992). Transient flow phenomena and reaction forces during blowdown of gas at high pressures through relief valves behind rupture discs. *Loss Prevention and Safety Promotion 7*, vol. 2, 61–1

PERBAL, R. (1993). Transient flow phenomena and reaction forces during blowdown of gas at high pressures through relief lines behind rupture discs. *Process Safety Prog.*, **12**, 232

PERCHE, A., PEREZ, A. and LUCQUIN, M. (1970). On the Lotka–Frank–Kamenetskii isothermal theory of cool flames. *Combust. Flame*, **15**, 89

PERCHE, A., PEREZ, A. and LUCQUIN, M. (1971). Simulation by an analog computer of normal explosions and cool flames. *Combust. Flame*, **17**, 179

PERCIFUL, J.C. and EDWARDS, R.S. (1967). Pilot plant procedures. Transfer of information from the chemist. *Chem. Engng Prog.*, **63**(11), 69

PERDIKARIS, G.A. and MAYINGER, F. (1994). Numerical simulation of heavy gas cloud dispersion within topographically complex terrain. *J. Loss Prev. Process Ind.*, **7**, 391

PERGAMENT, H.S. and FISHBURNE, E.S. (1978). Influence of buoyancy on turbulent hydrogen/air diffusion flames. *Combust. Sci. Technol.*, **18**, 127

PERKET, C. (1978). An assessment of hazardous waste disposal landfills: the state of the art. *Control of Hazardous Material Spills*, vol. 4, p. 208

PERKIN, W.H. (1882). Some observations on the luminous incomplete combustion of ether and other organic bodies. *J. Chem. Soc.*, 363

PERKINS, C. (1980). The vulnerability of electronics equipment to damage and its impact on process plant loss. *Loss Prevention and Safety Promotion 3*, vol. 2, p. 754

PERKINS, E.S. and STUHLBARG, D.S. (1971). US rotating equipment standards. *Hydrocarbon Process.*, **50**(6), 104

PERKINS, R.L. (1975). Maintenance costs and equipment reliability. *Chem. Engng*, **82**(Mar. 31), 126

PERLMUTTER, D.D. (1965). *Introduction to Chemical Process Control* (New York: Wiley)

PERNOT, C., HURIET, C., MIDON, A. and GRUN, G. (1972). Acute industrial ammonia intoxication. A discussion of four cases. *Arch. Mal. Prof.*, **33**, 5

PERREGAARD, J., PEDERSEN, B.S. and GANI, R. (1992). Steady state and dynamic simulation of complex chemical processes. *Chem. Engng Res. Des.*, **70A**, 99

PERRI, J.M. (1953). Fire fighting foams. In Bikerman, J.J. (1953), *op. cit.*, p. 189

PERRIMAN, R. (1986). Emissions from tall chimneys. *Chem. Engr, Lond.*, **431**, 19

PERRIN, D.A. (1996). Inherent safety in processes: Softer aspects of process safety. *Int. Conf. and Workshop on Process Safety Manage. and Inherently Safer Processes*, Oct. 8–11, 1996, Orlando, FL, 207–12. (New York: American Institute of Chemical Engineers)

PERROTT, I.R. (1989). *Piper Alpha Public Inquiry. Transcript, Day 163*

PERROW, C. (1984). *Normal Accidents* (New York: Basic Books)

PERRY, A.H. (1981). *Environmental Hazards in the British Isles* (London: Allen & Unwin)

PERRY, B.G. (1990). Operational safety program. *Plant/Operations Prog.*, **9**, 246

PERRY, B.G. (1994). You need more than technology. *Process Safety Prog.*, **13**, 3

PERRY, E.S. and WEISSBERGER, A. (eds) (1979). *Techniques of Chemistry*, vol. XIII, *Laboratory Engineering and Manipulations*, 3rd ed. (New York: Wiley)

PERRY, J.A. (1949). Critical flow through sharp-edged orifices. *Trans. ASME*, **71**, 757

PERRY, J.H. (1934–). *Chemical Engineers Handbook*; 1941, 2nd ed.; 1950, 3rd ed.; 1963, 4th ed. (New York: McGraw-Hill). See also first Perry, R.H. and Chilton, C.H. and then Perry, R.H. and Green, D.W.

PERRY, J.H. (1950). *Chemical Business Handbook* 3rd ed. (New York: McGraw-Hill)

PERRY, J.H. (ed.) (1954). *Chemical Business Handbook* (New York: McGraw-Hill)

PERRY, R. (1980). How to prevent valve problems. *Hydrocarbon Process.*, **59**(8), 67

PERRY, R. and YOUNG, R.J. (eds) (1977). *Handbook of Air Pollution Analysis* (London: Chapman & Hall)

PERRY, R.G. (1995). The paradox of inherent safety. *Int. Symp. on Runaway Reactions & Pressure Relief Design*, Aug. 2–4, 1995, Boston, MA

PERRY, R.H. and CHILTON, C.H. (1973). *Chemical Engineers Handbook*, 5th ed. (New York: McGraw-Hill)

PERRY, R.H. and GREEN, D.W. (1984). *Perry's Chemical Engineers Handbook*, 6th ed. (New York: McGraw-Hill)

PERRY, W. (1985). Dinosaurs are alive and living in corporate standards. *Chem. Engr, Lond.*, **415**, 84

PERRY, W.W. and ARTICOLA, W.P. (1980). *Study to Modify the Vulnerability Model of the Risk Management System. Rep. CG-D-22-80.* Enviro Control Inc., Rockville, MD

PERRYMAN, R. (1984). Guidelines for the implementation of odour treatment. *Chem. Ind.*, May 7, 327

PESATA, V. *et al.* (1977). Some new methods for the determination of safety characteristics. *Loss Prevention and Safety Promotion 2*, p. 183

PESCHEL, H. and FETTING, F. (1972). The laminar flame speed of methane-oxygen mixtures and some remarks on flame pressure. *Combust. Flame*, **19**, 136

PESETSKY, B. and BEST, R.D. (1980). Methane diluent requirements for ethylene oxide storage and handling. *Loss Prevention*, **13**, 132

PESETSKY, B. and FISHER, J.J.A. (1975). Static problems in handling charged liquids. *Loss Prevention*, **9**, p. 114

PESETSKY, B., CAWSE, J.N. and VYN, W.T. (1980). Liquid phase decomposition of ethylene oxide. *Loss Prevention*, **13**, p. 123

PESUIT, D.R. (1979). Consider air absorption in plant noise control. *Chem. Engng*, **86**(Sep. 24), 125

PETER, H.K. (1980). Testing of tank flame traps against static and dynamic ignition break. *Loss Prevention and Safety Promotion 3*, vol. 3, p. 1399

PETERS, G.A. and HANSEL, J.G. (1992). Risk assessment of tornado wind and pressure phenomena on storage tanks. *Hazard Identification and Risk Analysis, Human Factors and Human Reliability in Process Safety*, p. 33

PETERS, J.M. (1973). Relating flares to air quality. *Pollution Engng*, Oct., 25

PETERS, J.S. (1994). Type-testing isolation valves under abrasive flow conditions. *Engng Dept, Univ of Manchester, Conf. on Valve and Pipeline Reliability*

PETERS, M.S. (1948). *Plant Design and Economics for Chemical Engineers* (New York: McGraw-Hill)

PETERS, M.S. and TIMMERHAUS, K.D. (1968). *Plant Design and Economics for Chemical Engineers*, 2nd ed.; 1980, 3rd ed.; 1991, 4th ed. (New York: McGraw-Hill)

PETERS, N. and GÖTTGENS, J. (1991). Scaling of buoyant turbulent jet diffusion flames. *Combust. Flame*, **85**, 206

PETERS, N. and SMOOKE, M.D. (1985). Fluid dynamic–chemical interactions at the lean flammability limit. *Combust. Flame*, **60**, 171

PETERS, N. and WILLIAMS, F.A. (1983). Liftoff characteristics of turbulent jet diffusion flames. *AIAA J.*, **21**, 423

PETERS, R.A. (1984). Computer-aided engineering. *Chem. Engng*, **91**(Jun. 11), 91; (Jul. 9), 107

PETERS, R.A. Integrated computer-aided engineering. *Chem. Engng*, **92**(May 13), 95

PETERSEN, D. (1971). *Techniques of Safety Management*; 1978, 2nd ed. (New York: McGraw-Hill)

PETERSEN, D. (1975). *OSHA Compliance Manual* (New York: McGraw-Hill)

PETERSEN, D. (1980). *Human Error Reduction and Safety Management* (Goshen, NY: Aloray)

PETERSEN, D. (1984). *Safety Supervision* (Goshen, NY: Aloray)

PETERSEN, D. (1988a). *Safe Behavior Reinforcement* (Goshen, NY: Aloray)

PETERSEN, D. (1988b). *Safety Management – A Human Approach*, 2nd ed. (Goshen, NY: Aloray)

PETERSEN, D. (1989). *Techniques of Safety Management*, 3rd ed. (Goshen, NY: Aloray)

PETERSEN, H.J.S. (1989). Acceptability of risk – a Danish case. *J. Loss Prev. Process Ind.*, **2**(1), 53

PETERSEN, K. (1992). Testing sensors for accuracy and speed. *Chem. Engng*, **99**(2), 131

PETERSEN, K.E. (1983). Analysis of pipe failures in Swedish nuclear power plants. *Euredata 4*, paper 8.1

PETERSEN, K.E. (1994). European model evaluation activity. *J. Loss Prev. Process Ind.*, **7**, 130

PETERSEN, M.E. (1986a). The role of extinguishers in fire protection. In *Fire Protection Handbook*, 16th ed. (ed. A.E. Cote and J.L. Linville), p. 20–2

PETERSEN, M.E. (1986b). Selection, operation and distribution of fire extinguishers. In *Fire Protection Handbook*, 16th ed. (ed. A.E. Cote and J.L. Linville), p. 20–9

PETERSEN, R.L. (1986). Statistical evaluation of integral and regulatory plume rise algorithms. *Fifth Joint Conf. on Applications of Air Pollution Meteorology* (Boston, MA: Am. Met. Soc), p. 319

PETERSEN, R.L. and DIENER, R. (1990). Vapour barrier assessment programme for delaying and diluting heavier-than-air HF vapour clouds. A wind tunnel modeling evaluation. *J. Loss Prev. Process Ind.*, **3**(2), 187

PETERSEN, R.L. and RATCLIFF, M.A. (1989). *Effect of Homogeneous and Heterogeneous Surface Roughness on HTAG Dispersion. Proj. 87-0417.* CPP Wind Engineering Consultants, Cermak Peterka Petersen Inc., Fort Collins

PETERSEN, W.B. (1986). INPUFF, a multiple source puff model. *Fifth Joint Conf. on Applications of Air Pollution Meteorology* (Boston, MA: Am. Met. Soc), p. 242

PETERSEN, W.B. and LAVDAS, L.G. (1986). *INPUFF 2.0 – a Multiple Source Gaussian Puff Dispersion Algorithm. Users Guide.* Environ. Prot. Agency, Research Triangle Park, NC

PETERSON DAN (1984). *Human-Error Reduction and Safety Management* (New York: Deer Park Aloray Publishing) ISBN 0-913690-09-0

PETERSON, A. and GROSS, E. (1967). *Handbook of Noise Measurement* (Bourne End, Buckinghamshire: General Radio Co. (UK) Ltd)

PETERSON, C. and MILLER, A. (1964). Mode, median and mean as optimal strategies. *J. Exptl Psychol.*, **68**, 363

PETERSON, C.R. and BEACH, L.R. (1967). Man as an intuitive statistician. *Psychol. Bull.*, **68**, 29

PETERSON, D.J. (1991). Better piping and expansion joint design. *Hydrocarbon Process.*, **70**(1), 89

PETERSON, E.R. and MADDOX, R.N. (1986). Vaporizer system smoothens flow from gas cylinders. *Chem. Engng*, **93**(18), Sep. 29, 115

PETERSON, G.D. (1979). Halon: advantages and disadvantages. *Fire Int.*, **64**, 76

PETERSON, J. (1981). *Petri Net Theory and the Modeling of Systems* (Englewood Cliffs, NJ: Prentice-Hall)

PETERSON, J.B., MORIZUMI, S.J. and CARPENTER, H.J. (1968). Thermal radiation from stored LNG release. *LNG Conf. 1*

PETERSON, J.F. and MANN, W.L. (1985). Steam-system design: how it evolves. *Chem. Engng*, **92**(Oct. 14), 62

PETERSON, J.L. (1977). Petri Nets. *ACM Comput. Surveys*, **9**(3)

PETERSON, J.N., CHEN, C.-C. and EVANS, L.B. (1978). Computer programs for chemical engineers. *Chem. Engng*, **85** (Jun. 5), 145; (Jul. 3), 69; (Jul. 31), 79; (Aug. 28), 107

PETERSON, J.R. and DAVIDSE, G.J. (1993). Consider pump cast refractory for critical linings. *Hydrocarbon Process.*, **72**(1), 55

PETERSEN, K.E. (1982). *Analysis of Pipe Failures in Swedish Nuclear Power Plants. Rep. SAK-1-D(82)8.* Risö Nat. Lab., Risö, Denmark

PETERSON, K.R. (1985). A nomographic solution of the Gaussian diffusion equation. *Atmos. Environ.*, **19**, 87

PETERSON, P. (1967). Explosions in flare stacks. *Loss Prevention*, **1**, p. 85. See also *Chem. Engng Prog.*, 1967, **63**(8), 67

PETERSON, P. and CUTLER, H.R. (1973). Explosion protection for centrifuges. *Loss Prevention*, **7**, 88. See also *Chem. Engng Prog.*, 1973, **69**(4), 42

PETES, J. (1968). Blast and fragmentation characteristics. *Ann. N.Y. Acad. Sci.*, **152**, art. 1, 283

PETHE, S., SINGH, R. and KNOPF, F.C. (1989). A simple technique for locating loops in heat exchanger networks. *Comput. Chem. Engng*, **13**, 859

PETHERBRIDGE, R.J.S. (1982). Designing operations into computer controlled plant. *Design 82* (Rugby: Instn Chem. Engrs)

PETHERBRIDGE, R.J.S. and KINDER, M.P. (1983). Chemical packages: safety in storage and transportation of hazardous materials. *Chem. Engr, Lond.*, **398**, 39

PETHICK, A.J. and WOOD, J. (1991). Designing a safer and more efficient drill floor: the human factors contribution. *Safety Developments in the Offshore Oil and Gas Industry* (London: Instn Mech. Engrs), p. 93

PETINO, G. (1981). An experimental method for measuring the electrostatic charges on falling and sliding particles. *The Hazards of Industrial Explosion from Dusts* (London: Oyez/IBC), paper 5

PETKAR, D.V. (1980). Setting up reliability goals for systems. *Reliab. Engng*, **1**, 43

PETKAR, D.V. (1981). Safety standard for chemical plant. *Reliab. Engng*, **2**, 53

PETKUS, J.J. (1963). Oily insulation can cause plant fires. *Hydrocarbon Process.*, **42**(11), 251

PETKUS, J.J. (1964). Oily insulation can cause plant fires. In Vervalin, C.H. (1964a), *op. cit.*, p. 290

PETO, J. (1978). The hygiene standard for chrysotile asbestos. *The Lancet*, **1**(8062), Mar. 4, 484

PETO, R. (1979). Detection of risk of cancer to man. *Proc. R. Soc., Ser. B*, **205**, 111. See also Doll, Sir R. and McLean, A.E.M. (1979), *op. cit.*, p. 111

PETO, R. (1980). Distorting the epidemiology of cancer: the need for a more balanced overview. *Nature*, **284** (Mar.), 297

PETRIE, J.R. (1983). Fire safety regulations for offshore oil rigs. *Fire Surveyor*, **12**(1), 17

PETRIE, J.R. (1989). *Piper Alpha Public Inquiry. Transcript, Days 104, 105, 158, 159*

PETRIE, J.R. (1990). *Piper Alpha Public Inquiry. Transcript, Days 172–175*

PETROLEUM INDUSTRY TRAINING BOARD – see Appendix 28

PETROPOULOS, H. and BREBNER, J. (1981). Stereotypes of direction-of-movement of rotary controls associated with linear displays: the effects of scale presence and position, of pointer direction, and distances between the control and display. *Ergonomics*, **24**, 143

PETROULAS, T. and REKLAITIS, G.V. (1984). Computer-aided synthesis and design of plant utility systems. *AIChE J.*, **30**, 69

PETROW, K. (1968). *The Black Tide in the Wake of the Torrey Canyon* (London: Hodder and Stoughton)

PETRY, CAPITAINE (1910). *Monographes de Systémes d'Artillerie* (Brussels: Polleunis)

PETSINGER, R.E. and MARSH, H.W. (1966). High strength steels for lower cost tanks. *Chem. Engng*, **73**(May 9), 182

PETTERSSON, O. (1988). Practical need for scientific material models for structural design – general review. *Fire Safety J.*, **13**, 1

PETTERSSON, O., MAGNUSSON, S.E. and THOR, J. (1976). *Fire Engineering Design of Structures. Publ. 50.* Swedish Inst Steel Construction

PETTI, T.F., KLEIN, J. and DHURJATI, P.S. (1990). Diagnostic model processor: using deep knowledge for process fault diagnosis. *AIChE J.*, **36**, 565

PETTITT, A.B. (1945). *Protection of Vessels Exposed to Fire. Safety Memo. 123.* Rubber Reserve Co., Washington, DC

PETTITT, G.N., HARDY, N.R., NOLAN, P.F. and JONES, C.D. (1992). Characterisation of aerosols formed by the flashing process following catastrophic vessel failure. *Proc. Nat. Heat Transfer Conf.*, p. 325

PETTITT, G.N., SCHUMACHER, R.R. and SEELEY, L.A. (1993). Evaluating the probability of major hazardous incidents as a result of escalation events. *J. Loss Prev. Process Ind.*, **6**, 37

PETTMAN, M.J. and HUMPHREYS, G.C. (1975). Improve designs to save energy. *Hydrocarbon Process.*, **54**(1), 77

PETTS, J.I. (1984a). Controlled explosion. *Planning*, **593** (Nov. 2), 10

PETTS, J.I. (ed.) (1984b). *Major Hazard Installations: Planning and Assessment* (seminar proc.). Dept of Chem. Engng, Loughborough Univ. of Technol.

PETTS, J.I. (ed.) (1984c). *Major Hazard Installations: Planning Implications and Problems* (seminar proc.). Dept of Chem. Engng, Loughborough Univ. of Technol.

PETTS, J.I. (1985a). *The Canvey Public Inquiries 1975–1982. Hazard and Risk Issues. Working Paper MHC/85/1.* Dept of Chem. Engng, Loughborough Univ. of Technol.

PETTS, J.I. (1985b). *Planning and Major Hazard Control. Working Paper MHC/85/2.* Dept of Chem. Engng, Loughborough Univ. of Technol.

PETTS, J.I. (1987). Planning control for hazardous installations – the effectiveness of HSE advice. *The Planner*, Mar., 22

PETTS, J.I. (1988a). Major hazard control: the local community and risk decisions. *Preventing Major Chemical and Related Process Accidents*, p. 497

PETTS, J.I. (1988b). Planning and hazardous installation control. *Progress in Planning*, **29**(1), 1

PETTS, J.I. (1989). Planning and major hazard installations. In Lees, F.P. and Ang, M.L. (1989), *op. cit.*, p. 57

PETTS, J.I. (1992). Land-use planning and major hazards: an overview of issues and problems. *Conf. on Major Hazards Aspects of Land-Use Planning*

PETTS, J.I. and EDULJEE, G. (1994). *Environmental Impact Assessment for Waste Treatment and Disposal Facilities* (Chichester: Wiley)

PETTS, J.I., WITHERS, J. and LEES, F. (1986). Expert evidence at inquiries into major hazards. *Project Appraisal*, **1**, 1

PETTS, J.I., WITHERS, R.M.J. and LEES, F.P. (1987). The assessment of major hazards: the density and other characteristics of the exposed population around a hazard source. *J. Haz. Materials*, **14**(3), 337

PEW, R.W. (1974). Human perceptual-motor performance. *Human Information Processing* (ed. B.H. Kantowitz) (New York: Erlbaum)

PEW, R.W., MILLER, D.C. and FEEHLER, C.E. (1981). *Evaluation of Proposed Control Room Improvements through Analysis of Critical Operator Decisions. Rep. EPRI NP-1982.* Elec. Power Res. Inst., Palo Alto, CA

PEYTON, V.F. (1984). Better tank fire protection in use. *Oil Gas J.*, **82**(Jul. 23), 70

PFEFFER, H.A., BHALLA, S.K. and DORE, J.C. (1984). Pilot design – how to avoid an expensive mistake. *Plant/Operations Prog.*, **3**, 98

PFEIFER, J.B. (1974). *Sulfur Removal and Recovery for Industrial Processes* (New York: Am. Chem. Soc.)

PFENNING, D.B. and CORNWELL, J.B. (1985). Computerized processing of Thorney Island trial data for comparison with model predictions. *J. Haz. Materials*, **11**(1/3), 347. See also in McQuaid, J. (1985), *op. cit.*, p. 347

PFENNING, D.B., MILLSAP, S.B. and JOHNSON, D.W. (1987). Comparison of turbulent jet model predictions with small-scale pressurized releases of ammonia and propane. *Vapor Cloud Modeling*, p. 84

PHADKE, P.S. (1983). Calculator program for rupture disk design. *Hydrocarbon Process.*, **62**(12), 34–B

PHALEN, R.F. (1984). *Inhalation Studies: Foundations and Techniques* (Boca Raton, FL: CRC Press)

PHAM, D.T. (ed.) (1988). *Expert Systems in Engineering* (Kempston, Bedfordshire: IFS Publs)

PHAM, Q.T. (1979). Explicit equations for the solution of turbulent pipe-flow problems. *Trans. Instn Chem. Engrs*, **57**, 281

PHELPS, D.R. (1970). Flange loads in a pumping system. *Chem. Engng Prog.*, **66**(5), 43

PHELPS, E.H. (1972). Stress corrosion cracking in ammonias. *Ammonia Plant Safety*, **14**, 109

PHELPS, E.H. (1974). Causes of stress-corrosion cracking of steel in ammonia. *Ammonia Plant Safety*, **16**, 32

PHELPS, P.J. and JUREIDINI, R. (1992). The computer modelling of tank spills and containment overflows. *Loss Prevention and Safety Promotion 7*, vol. 3, 141–1

PHIBBS, E. and KUMAMOTO, S.H. (1974). Fault tree analysis (letter). *IEEE Trans. Reliab.*, **R-23**, 266

PHILIP, E.B. (ca. 1945). *Empirical Formulae for the Average Circle Radii of A, B and C Damage. Rep. REN 558.* Chief Scientific Adviser's Department, Ministry of Works

PHILIPSON, L.L. (1974a). *Investigation of the Feasibility of the Delphi Technique for Estimating Risk Analysis Parameters.* (rep. to Dept of Transportation) Inst. for Safety and Systems Management, Univ. of Southern Calif., CA

PHILIPSON, L.L. (1974b). *Risk Analysis in Hazardous Materials Transportation: a Mechanism for Interfacing the Risk Analysis Model with the Hazardous Materials Incident Reporting System.* (rep. to Dept of Transportation). Inst. for Safety and Systems Management, Univ. of Southern Calif., CA

PHILIPSON, L.L. (1976). A revised system safety glossary. *Hazard Prev.*, **8**(2)

PHILIPSON, L.L. (1978a). The LNG decision problem. *Am. Inst. Chem. Engrs, Twenty-First Ann. Mtg* (New York)

PHILIPSON, L.L. (1978b). The operational reliability of LNG systems. *Proc. Ann. Reliability and Maintainability Symp.*

PHILIPSON, L.L. (1980). LNG operations risk analyses: evaluation and comparison of techniques and results. In Apostolakis, G., Garribba, S. and Volta, G. (1980), *op. cit.*, p. 401

PHILIPSON, L.L. and NAPADENSKY, H.S. (1982). The methodologies of hazardous materials transportation risk assessment. *J. Haz. Materials*, **6**(4), 361

PHILIPSON, L.L. and SCHAEFFER, M.S. (1975). Hazardous materials shipment data in risk assessment. *Proc. Ann. Reliability and Maintainability Symp.*, p. 519

PHILLEY, J.O. (1992a). Acceptable risk: an overview. *Plant/ Operations Prog.*, **11**, 218

PHILLEY, J. (1992b). Investigate incidents with MRC. *Hydrocarbon Process.*, **71**(9), 77

PHILLEY, JACK (2000). Application of critical thinking concepts to determining the most likely scenario, *Mary Kay O'Connor Process Saf. Cent. Conf.,* Texas A&M University, College Station, TX

PHILLEY, JACK, Investigate Incidents with MRC. *Hydrocarbon Processing Magazine,* September 1992, Gulf Publishing Co.

PHILLIPS, A.L. (1986). Computer risk – is it necessary? *Inst. of Mathematics and its Applications, Mtg on Mathematics in Major Accident Risk Assessment*

PHILLIPS, B.R. (1979). Failure at the inlet nozzle weldment of an ammonia synthesis converter. *Ammonia Plant Safety,* **21**, 21

PHILLIPS, D.W. (1982). Seismic structural analysis of a typical plant item. Appendix II of Alderson, M.A.H.G. (1982), *op. cit.*, p. 44

PHILLIPS, E.A. (1971). A railroad tank car safety program. *Ammonia Plant Safety,* **13**, 10

PHILLIPS, H. (1965). Flame in a buoyant methane layer. *Combustion,* **10**, 1277

PHILLIPS, H. (1971a). The basic theory of the flameproof enclosure. *Electrical Safety,* **1**, 99

PHILLIPS, H. (1971b). *The Mechanism of Flameproof Protection. SMRE Res. Rep. 275.* Safety in Mines Res. Estab.

PHILLIPS, H. (1972a). A nondimensional parameter characterizing mixing processes in a model of thermal gas ignition. *Combust. Flame,* **19**, 181

PHILLIPS, H. (1972b). Theory of suppression of explosions by narrow gaps. *Chemical Process Hazards,* **IV**, p. 17

PHILLIPS, H. (1973). The use of a thermal model of ignition to explain aspects of flameproof enclosure. *Combust. Flame,* **20**, 121

PHILLIPS, H. (1980). *Decay of Spherical Detonations and Shocks* (London: HM Stationery Office)

PHILLIPS, H. (1981a). Differences between determinations of maximum experimental safe gap. *J. Haz. Materials,* **4**(3), 245

PHILLIPS, H. (1981b). Unconfined vapour cloud explosions: a new look at Gugan's book. *Chem. Engr, Lond.,* **369**, 286

PHILLIPS, H. (1982). UK research into unconfined vapour cloud explosions. In Lee, J.H.S. and Guirao, C.M. (1982), *op. cit.*, p. 737

PHILLIPS, H. (1984). Maximum explosion pressure in flameproof enclosures: the effects of the vessel and the ambient temperature. *J. Haz. Materials,* **8**(3), 251

PHILLIPS, H. (1986). Flame acceleration in vapour cloud explosions. *Modelling and Simulation in Engineering* (ed. B. Wahlstroem and K. Leiviskae) (Amsterdam: Elsevier)

PHILLIPS, H. (1987). Two-fluid modeling of vapour cloud explosions. *Vapor Cloud Modeling,* p. 667

PHILLIPS, H. (1989). Prediction of blast from unconfined vapour cloud explosions. *Loss Prevention and Safety Promotion 6,* p. 27.1

PHILLIPS, H. and PRITCHARD, D.K. (1986). Performance requirements of flame arresters in practical applications. *Hazards,* **IX**, p. 47

PHILLIPS, H.C.D. (1990). Loss reduction proposals – determining priorities for implementation. *Safety and Loss Prevention in the Chemical and Oil Processing Industries,* p. 103

PHILLIPS, J. (1979). The solution to difficult shaft sealing problems. *Chem. Engr, Lond.,* **341**, 87

PHILLIPS, K.G. and HENSHELWOOD, F.W. (1991). A discussion of the detonation potential of ethylene vapor clouds. *Modeling and Mitigating the Consequences of Accidental Releases,* p. 529

PHILLIPS, L.D., EMBREY, D.E., HUMPHREYS, P. and SELBY, D.L. (1990). A socio-technical approach to assessing human reliability. In Oliver, R.M. and Smith, J.A. (1990), *op. cit.*

PHILLIPS, L.D., HUMPHREYS, P. and EMBREY, D.E. (1983). *A Sociotechnical Approach to Assessing Human Reliability. Tech. Rep. 83-4.* Decision Analysis Unit, London School of Economics

PHILLIPS, L.K. and GRAY, H. (1996). *Accidents do happen: Toxic chemical accident patterns in the United States* (Boston, MA: National Environmental Law Center)

PHILLIPS, M.J. (1981). A preventive maintenance plan for a system subject to revealed and unrevealed faults. *Reliab. Engng,* **2**(3), 221

PHILLIPS, R. (1981). Fire detection and alarm systems: introduction to evaluation and testing. *Fire Surveyor,* **10**(1), 12

PHILLIPS, R.F. (1967). Petrochemicals. *Chem. Engng,* **74**(May 22), 153

PHILLIPS, S.G. (1990). Skid-mounted pilot-plant facilities. *Chem. Engng Prog.,* **86**(6), 11

PHILLIPS, W.E. (1981). Safety considerations and management from a grain operator's point of view. *The Hazards of Industrial Explosion from Dusts* (London: Oyez/IBC), paper 19

PHILLIPS, Y.Y. (1986). Primary blast injuries. *Ann. Emergency Med.,* **15**, 1446

PHILPOTT, J.E. (1963). Bursting disc survey. *Engng Mater. Des.,* **6**, 24

PHONG, S. (1990). Disaster preparedness planning. *Safety and Loss Prevention in the Chemical and Oil Processing Industries,* p. 41

PHOON, W.H. (1993). Occupational health in the oil, chemical and process industries. *Health, Safety and Loss Prevention,* p. 521

PHOON, W.H. and TAN, K.T. (1990). Occupational health hazards from chemicals and their control. *Safety and Loss Prevention in the Chemical and Oil Processing Industries,* p. 337

PHYLAKTOU, H. and ANDREWS, G.E. (1991a). The acceleration of flame propagation in a tube by an obstacle. *Combust. Flame,* **85**, 363

PHYLAKTOU, H. and ANDREWS, G.E. (1991b). Gas explosions in long closed vessels. *Combust. Sci. Technol.,* **77**, 27

PHYLAKOTOU, H. and ANDREWS, G.E. (1993). Gas explosions in linked vessels. *J. Loss Prev. Process Ind.,* **6**, 15

PHYLAKTOU, H.N., ANDREWS, G.E. and HERATH, P. (1990). Fast flame speeds and rates of pressure rise in the initial period of gas explosions in large L/D cylindrical enclosures. *J. Loss Prev. Process Ind.,* **3**(4), 355

PHYLAKTOU, H., ANDREWS, G.E., MOUNTER, N. and KHAMIS, K.M. (1992). Spherical explosions aggravated by obstacles. *Major Hazards Onshore and Offshore,* p. 525

PHYLAKTOU, H., FOLEY, M. and ANDREWS, G.E. (1993). Explosion enhancement through a 90° curved bend. *J. Loss Prev. Process Ind.,* **6**, 21

PHYLAKTOU, H., ANDREWS, G.E. and LIU, Y. (1994). Turbulent explosions: a study of the influence of the obstacle scale. *Hazards,* **XII**, p. 269

PICARD, D.J. and BISHNOI, P.R. (1988). The importance of real-fluid behaviour and non-isentropic effects in modeling decompression characteristics of pipeline fluids for application in ductile fracture propagation analysis. *Can. J. Chem. Engng*, **66**, 3

PICARD, D.J. and BISHNOI, P.R. (1989). The importance of real-fluid behaviour in predicting release rates from high pressure sour gas pipeline ruptures. *Can. J. Chem. Engng*, **67**, 3

PICCIANO, D. (1980). Love Canal chromosome study (letter). *Science*, **209**, 754

PICCININI, N. and ANATRA, U. (1983). Intermediate tank as a way of increasing plant availability. *Ind Engng Chem. Fund.*, **22**, 206

PICCININI, N. and LEVY, G. (1984). Ethylene oxide reactor: safety according to operability analysis. *Can. J. Chem. Engng*, **62**, 547

PICCININI, N., ANATRA, U., MALADNRINO, G. and BARONE, D. (1980). A comparison of different techniques used to quantify fault trees. *Loss Prevention and Safety Promotion 3*, vol. 2, p. 614

PICCININI, N., ANATRA, U., MALANDRINO, G. and BARONE, D. (1982). Safety analysis of an allyl chloride plant. *Plant/Operations Prog.*, **1**, 69

PICCININI, N., SCARRONE, M. and CIARAMBINO, I. (1994). Probabilistic analysis of transient events by an event tree directly extracted from operability analysis. Evaluation of accidental releases of gas from a pressure-regulating installation. *J. Loss Prev. Process Ind.*, **7**, 23

PICCIOLO, G. and METZGER, B. Environmental impact assessment for a major petrochemical plant in Italy – policy considerations. *Loss Prevention and Safety Promotion 7*, vol. 1, 5–1

PICCIOTTI, M. (1978). Design for ethylene plant safety. *Hydrocarbon Process.*, **57**(3), 191; **57**(4), 261

PICCIOTTI, M. (1980). Why an LPG olefins plant? *Hydrocarbon Process.*, **59**(4), 223

PICKARD, J.M. (1983). Utilization of accelerating rate calorimetry for the characterization of energetic materials. *J. Haz. Materials*, **7**(3), 291

PICKARD, LOWE AND GARRICK INC. (1983). *Seabrook Station Probabilistic Safety Assessment*. Rep. to Public Service Co. of New Hampshire and Yankee Atomic Electric Co.

PICKBOURNE, D.E. (1975). A review of the BCISC lost time/injury accident statistics. In Hendry, W. (1975), *op. cit.*, paper E1

PICKERING, E.E. and BOCKHOLT, J.L. (1971). *Probabilistic Air Blast Criteria for Urban Structures* (report). Standford Res. Inst., Menlo Park, CA

PICKETT, D.J. (1979). *Electrochemical Reactor Design*, 2nd ed. (Amsterdam: Elsevier)

PICKFORD, J. (1969). *Analysis of Surge* (London: Macmillan)

PICKLES, J.H. (1986). A stochastic domino model for a sequential failure process. *Reliab. Engng*, **16**(3), 219

PICKLES, J.H. and BITTLESTON, S.H. (1983). Unconfined vapour cloud explosions – the asymmetrical blast from an elongated cloud. *Combust. Flame*, **51**, 45

PICKLES, R.G. (1971). Hazard reduction in the formaldehyde process. *Major Loss Prevention*, p. 57

PICKNETT, R.G. (1978a). *Field Experiments on the Behaviour of Dense Clouds. Pt 1, Main Report. Rep. Ptn IL 1154/78/1*. Chemical Defence Estab., Porton Down

PICKNETT, R.G. (1978b). *Field Experiments on the Behaviour of Dense Clouds. Pt 2, Detailed Results of Experiments 1–21 (and preliminary). Rep. Ptn IL 1154/78/2*. Chemical Defence Estab., Porton Down

PICKNETT, R.G. (1978c). *Field Experiments on the Behaviour of Dense Clouds. Pt 3, Detailed Results of Experiments*

22–39. Rep. Ptn IL 1154/78/3. Chemical Defence Estab., Porton Down

PICKNETT, R.G. (1981). Dispersion of dense gas puffs released in the atmosphere at ground level. *Atmos. Environ.*, **15**, 509

PICKRELL, J.A. (ed.) (1981). *Lung Connective Tissue: Location, Metabolism and Response to Injury* (Boca Raton, FL: CRC Press)

PICKUP, A. (1983). A method for evaluating the reliability of complex multiple redundant systems. *Reliab. Engng*, **5**, 53

PICONE, J.M., ORAN, E.S., BORIS, J.P. and YOUNG, T.R. (1984). Theory of vorticity by shock wave and flame interactions. In Bowen, J.R. *et al.*, (1984b), *op. cit.*, p. 429

PIDGEON, N.F., BLOCKLEY, D.I. and TURNER, B.A. (1986). Design practice and snow loading: lessons from a roof collapse. *Struct. Engr*, **64A**(3), 67

PIEHL, R.L. (1975). Survey: air cooler corrosion. *Hydrocarbon Process.*, **54**(7), 169

PIELA, P.C., EPPERLY, T.G., WESTERBERG, K.M. and WESTERBERG, A.W. (1991). ASCEND: an object-oriented computer environment for modeling and analysis: the modeling language. *Comput. Chem. Engng*, **15**, 53

PIEPER, C.J., ZICK, L.P. and LA FAVE, I.V. (1982). State-of-the-art assessment of refrigerated liquefied gas storage systems using flat-bottom tanks. *API Proc., Refining Dept, 47th Midyear Mtg*, **61**, 176

PIER, S.M., ALLISON, R.C., CUNNINGHAM, E.M. and WARD, C.H. (1978). Methods for the categorization of hazardous materials. *Control of Hazardous Material Spills 4*, p. 27

PIERCE, R.R. (1968). Protecting concrete floors from chemicals. *Chem. Engng*, **75**(Dec. 16), 118

PIERCE, T.D. and TURNER, D.B. (1980). *User's Guide for MPTER. Rep. EPA-600/8-80-016*. Environ. Prot. Agency, Research Triangle Park, NC

PIERCE, T.E. (1986). Estimating surface concentrations from an elevated, buoyant plume in a limited-mixed convective boundary layer. *Fifth Joint Conf. on Applications of Air Pollution Meteorology* (Boston, MA: Am. Met. Soc), p. 331

PIEROT, M. (1968). Operating experiences of the Arzew plant. *LNG Conf. 1*

PIERSON, K.L. (1984). How safe is safe enough? *Plant/Operations Prog.*, **3**, 71

PIERUSCHKA, E. (1963). *Principles of Reliability* (Englewood Cliffs, NJ: Prentice-Hall)

PIETERS, H.A.J. and CREIGHTON, J.W. (1957). *Safety in the Chemical Laboratory*, 2nd ed. (London: Academic Press)

PIETERSEN, C.M. (1985). *Analysis of the LPG Incident in San Juan Ixhuatepec, Mexico City, 19 November 1984. Rep. 85-0222*. TNO, Apeldoorn, The Netherlands

PIETERSEN, C.M. (1986a). Analysis of the LPG disaster in Mexico City, 19 November 1984. *Gastech 85*, p. 112

PIETERSEN, C.M. (1986b). Analysis of the LPG disaster in Mexico City. *Loss Prevention and Safety Promotion 5*, p. 21–1

PIETERSEN, C.M. (1986c). Calculation of intoxication effects in a risk analysis associated with toxic gases. In Hartwig, S. (1986), *op. cit.* p. 183

PIETERSEN, C.M. (1988a). Analysis of the LPG-disaster in Mexico City. *J. Haz. Mats*, **20**, 85

PIETERSEN, C.M. (1988b). Emergency management and support systems. *Preventing Major Chemical and Related Process Accidents* (Rugby: Instn Chem. Engrs), p. 657

PIETERSEN, C.M. (1990). Consequences of accidental releases of hazardous materials. *J. Loss Prev. Process Ind.*, **3**(1), 136

PIETERSEN, C.M. (1993). Off-site emergency management for industrial areas. *Health, Safety and Loss Prevention*, p. 394

PIETERSEN, C.M. and VAN HET VELD, B.F.P. (1992). Risk assessment and risk contour mapping. *J. Loss Prev. Process Ind.*, **5**, 60

PIGGOTT, M.R. (1980). *Load Bearing Fibre Composites* (Oxford: Pergamon Press)

PIGNATARO, L.J. (1973). *Traffic Engineering: Theory and Practice* (Englewood Cliffs, NJ: Prentice-Hall)

PIGNATO, J.A. (1983). Enhancement of fire protection with directional cooling spray nozzles. *Fire Technol.*, **19**, 5

PIKAAR, M.J. (1983). Unconfined vapour cloud dispersion and combustion – an overview of theory and experiments. *Loss Prevention and Safety Promotion 4*, vol. 1, p. C20.

PIKAAR, M.J. (1985). Unconfined vapour cloud dispersion and combustion – an overview of theory and experiments. *Chem. Engng Res. Dev.*, **63**, 75

PIKAAR, M.J., BRAITHWAITE, J.M. and COX, A.P. (1986). Process safety assessment of new and existing plants. *Loss Prevention and Safety Promotion 5*, 10–1

PIKAAR, R.N., THOMASSEN, P.A.J., DEGELING, P. and VAN ANDEL, H. (1990). Ergonomics on control room design. *Ergonomics*, **33**, 589

PIKE, D. (1981). Ethylene. *Chem. Engng*, **88**(Oct. 5), 134

PIKE, H. (1975a). Air transport of hazardous materials – safety is the key. *Instn Chem. Engrs, Symp. on Transport of Hazardous Materials* (London)

PIKE, H. (1975b). The movement of chemicals by air. *Distribution Conf.* (London: Instn Chem. Engrs)

PIKULIK, A. (1976). Selecting and specifying valves for new plants. *Chem. Engng*, **83**(Sept. 13), 168

PIKULIK, A. and DIAZ, H.E. (1977). Cost estimating for major process equipment. *Chem. Engng*, **84**(Oct. 10), 107

PILARSKI, L. (1985). The Swiss initiate a chemical-safety plan. *Chem. Engng*, **92**(Apr. 15), 20C

PILBOROUGH, L. (1971). *Inspection of Chemical Plant* (London: Leonard Hill)

PILBOROUGH, L. (1989). *Inspection of Industrial Plant*, 2nd ed. (Aldershot, Hampshire: Gower Technical)

PILIE, R.J., BAIER, R.E., ZIEGLER, R.C., LEONARD, R.P., MICHALOVIC, J.G., PELE, S.L. and BOCK, D. (1975). *Methods to Treat, Control and Monitor Spilled Hazardous Materials. EPA-670/2-75-042.* Environ. Prot. Agency

PILKEY, W.D. and CHANG, P.Y. (1978). *Modern Formulas for Statics and Dynamics* (New York: McGraw-Hill)

PILKINGTON, D.F., PLATT, G. and NORTON, G. (1994). Design and development of a rig for the pressure testing of weak vessels and subsequent work relating to the strength of flat plates. *Hazards*, **XII**, p. 57

PILL, J. (1971). The Delphi method: substance, content, a critique and an annotated bibliography. *Socioecon. Planning Sci*, **5**, 57

PILLANS, G.P. (1989). *Piper Alpha Public Inquiry. Transcript, Day 79*

PILLOY, M. and RICHARD, L.L. (1968). Three years experience with the methane ship Jules Verne and the terminal at Le Havre. *LNG Conf. 1*

PILZ, V. (1977). Auflegung, Einsatz und Wirkungsweise von Blowdown-Systemen als Sicherheitsrichtung bei Chemieanlagen. *Chem. Ing. Tech.*, **49**, 873

PILZ, V. (1980a). Fault tree analysis: how useful? *Hydrocarbon Process.*, **59**(5), 275

PILZ, V. (1980b). What is wrong with risk analysis? The benefits and drawbacks on risk analysis. *Loss Prevention and Safety Promotion 3*, vol. 2, p. 448

PILZ, V. (1982). On the meaning of the breakdown ordinance. *Chem. Ing. Tech.*, **54**, 712

PILZ, V. (1986). Systematic investigating procedures for the detection of thermic hazards and the prevention of runaway chemical reactions. *Control and Prevention of Chemical Reaction Hazards* (London: IBC Tech. Services Ltd)

PILZ, V. (1987). Comparison of the systems of technological safety regulation for the chemical industry in Europe. *Fortschritte der Sicherheitstechnik 1* (Frankfurtam-Main: DECHEMA), p. 25

PILZ, V. (1989). Safety-related regulations for chemical plant in various European countries. *Loss Prevention and Safety Promotion 6*, p. 67.1

PILZ, V. and VAN HERCK, W. (1976). Chemical engineering investigations with respect to the safety of large chemical plants. *Eur. Fed. Chem. Engng, Third Symp. on Large Chemical Plants*, Antwerp

PIMBLE, J. and O'TOOLE, S. (1982). Analysis of accident reports. *Ergonomics*, **25**, 967

PINDER, J.N. (1985). *The Study of Noise from Steel Pipes. Rep. 84/55.* CONCAWE, the Hague

PINDER, J.N. (1993). Noise radiation from gas-filled pipes. In Instn Mech. Engrs (1993), *op. cit.*, 13

PINDERA, M.Z. and BRZUSTOWSKI, T.A. (1984). A model for the ignition and extinction of liquid fuel droplets. *Combust. Flame*, **55**, 79

PINEAU (1973). Dust explosions in elongated containers – protection by discharge vents. *Prevention of Occupational Risks*, **2**, 364

PINEAU, J. (1982). Propagation of dust explosions in ducts. *The Control and Prevention of Dust Explosions* (London: Oyez/IBC), paper 3

PINEAU, J., GILTAIRE, M. and DANGREAUX, J. (1974). *Effectiveness of Explosion Venting. Cahiers de Notes Documentaires, no. 74, 1er trimestre* (Inst. Nat. de Recherche et Securite), p. 75

PINEAU, J., GILTAIRE, M. and DANGREAUX, J. (1976). *Effectiveness of de Notes Documentaires, no. 83, 2ie'me trimestre* (Inst. Nat. de Recherche et Securite), p. 33 *(Note 1005–83–76) Venting. Cahiers*

PINEAU, J., GILTAIRE, M. and DANGREAUX, J. (1978). *Effectiveness of Venting in the Case of Dust Explosions. Cahiers de Notes Documentaires, no. 90, 1er trimestre* (Inst. Nat. de Recherche et Securite), p. 37 *(Note 1095–90–79)*

PINEAU, J.P., CHAINEAUX, J. and RONCHAIL, G. (1986). Influence on gas and dust explosion development of lengthening and presence of obstacles in closed or vented vessels. *Loss Prevention and Safety Promotion 5*, p. 61–1

PINEAU, J.P., CHAINEUX, J., LEFIN, Y. and MAVROTHALASSITIS, G. (1991). Learning from critical analysis of hazard studies and from accidents in France. *Modeling and Mitigating the Consequences of Accidental Releases*, p. 563

PINEY, M. and CARROL, C. (1983). How to use fume cupboards effectively. *Health and Safety at Work*, Oct.

PINKER, S. (1984). *Language Learnability and Language Development* (Cambridge, MA: Harvard Univ. Press)

PINKNEY, D.R. (1986). Programmable electronic systems. *SRS Quart. Digest*, Jul., 21

PINKNEY, D.R. and HIGNETT, K.C. (1987). PLCs and reliable shutdown (letter). *Chem. Engr, Lond.*, **433**, 5

PINNEY, G.G. (1962). Oxygen trailer fire. *Ammonia Plant Safety*, **4**, p. 49

PINNINGTON, J. (1989). Bellows valves. *Plant Engr*, Nov./Dec., 33

PINSKY, M.L., VICKERY, T.P. and FREEMAN, K.P. (1990). Investigation of a drum explosion of 30% aqueous KOCN. *J. Loss Prev. Process Ind.*, **3**(4), 345

PINTO, Y. and DAVIS, E.J. (1971). The motion of vapor bubbles growing in uniformly heated liquids. *AIChE J.*, **17**, 1452

PIOTROWSKI, T.C. (1991). Specification of flame arresting devices for manifolded low pressure storage tanks. *Plant/Operations Prog.*, **10**, 102

PIPELINE INDUSTRIES GUILD (1984). *Pipelines: Design, Construction and Operation* (London: Construction Press)

PIPER, D.F. (1976). A practical guide to the introduction of preventive maintenance. *Plant Engr*, **20**(1/2), 9

PIPKIN, O.A. (1959). Detonation – old processes are not immune. *Proc. Am. Petrol. Inst.*, **39**(3), 21

PIPKIN, O.A. (1964). Coking unit explosion. In Vervalin, C.H. (1964a), *op. cit.*, p. 339

PIPKIN, O.A. and SLIEPCEVICH, C.M. (1964). Effect of wind on buoyant diffusion flames. *Ind. Engng Chem. Fund.*, **3**, 147

PIRANI, M.A. and YARWOOD, J. (1961). *Principles of Vacuum Technology* (London: Chapman & Hall)

PIRIE, J.M. (ed.) (1960–) – see notes on *Chemical Process Hazards* series at head of the reference list

PIRUMOV, U.G. and ROSLYAKOV, G.S. (1986). *Gas Flow in Nozzles* (Berlin: Springer)

PISO, E. (1981). Task analysis for process-control tasks: the method of Annett *et al.* applied. *J. Occup. Psychol.*, **54**, 247

PITBLADO, G. (1989). COSHH. *Plant Engr*, Jul./Aug., 25

PITBLADO, R.M. (1986a). *Consequence Models for BLEVE Incidents. Major Ind. Hazards Proj. NSW 2006.* Warren Center, Sydney Univ.

PITBLADO, R.M. (1986b). Hazard identification and risk control: review and recommendations. *CHEMECA 86*, p. 271

PITBLADO, R. (1994a). Quality and offshore quantitative risk assessment. *J. Loss Prev. Process Ind.*, **7**, 360

PITBLADO, R.M. (1994b). *Risk Analysis in the Process Industries*, 2nd ed. (Rugby: Instn Chem. Engrs)

PITBLADO, R.M. and LAKE, I.A. (1986). A spreadsheet approach for fault tree analysis and hazard warning. *CHEMECA 86*, 290

PITBLADO, R.M. and LAKE, I.A. (1987). Guidelines for the application of hazard warning. *Chem. Engng Res. Dev.*, **65**, 334

PITBLADO, R.M. and NALPANIS, P. (1989). Quantitative assessment of major hazard installations: 2, Computer programs. In Lees, F.P. and Ang, M.L. (1989), *op. cit.*, p. 179

PITBLADO, R.M., BELLAMY, L. and GEYER, T. (1989). Safety assessment of computer controlled process plants. *Loss Prevention and Safety Promotion 6*, p. 46.1

PITBLADO, R.M., SHAW, S.J. and STEVENS, G.C. (1990). The SAFETI risk assessment package and case study application. *Safety and Loss Prevention in the Chemical and Oil Processing Industries*, p. 51

PITBLADO, R.M., WILLIAMS, J. and SLATER, D.H. (1990). Quantitative assessment of process safety programs. *Plant/Operations Prog.*, **9**, 169

PITCHER, J.H. *et al.* (1976). Stainless steels: CPI work horses. *Chem. Engng*, **83**(Nov. 22), 119

PITKIN, D.W. (1972). Total cost evaluation of mobile equipment. *Chem. Engng*, **79**(Oct. 16), 128

PITT, G.J. and MILLWARD, G.R. (1979). *Coal and Modern Coal Processing: An Introduction* (New York: Academic Press)

PITT, M.J. (1978). The disposal of waste in the chemical laboratory. *Chem. Ind.*, Dec. 2, 903

PITT, M.J. (1982). A vapour hazard index for volatile chemicals. *Chem. Ind.*, Oct. 16, 804

PITT, M.J. (1987). Security with chemicals. *Chem. Engr, Lond.*, **442**, 38

PITT, M.J. and PITT, E. (1985). *Handbook of Laboratory Waste Disposal* (Chichester: Ellis Horwood)

PITT, M.J. *et al.* (1991). Teaching safety (letter). *Chem. Engr, Lond.*, **504**, 4

PITT, P.L., CLEMENTS, R.M. and TOPHAM, D.R. (1991). The early phase of spark ignition. *Combust. Sci. Technol.*, **78**, 289

PITTMAN, E.D. (1985). Specifying rotary valves. *Chem. Engng*, **92**(Jul. 8), 89

PITTMAN, J.F. (1972a). *Blast and Fragment Hazards from Bursting High Pressure Tanks. Rep. NOLTR 72-102.* Naval Ordnance Lab., White Oak, Silver Spring, MD

PITTMAN, J.F. (1972b). Pressures, fragments and damage from bursting pressure tanks. *Dept of Defense, Explosives Safety Board, Minutes of Fourteenth Explosives Safety Seminar*, p. 1117

PITTMAN, J.F. (1976). *Blast and Fragments from Superpressure Vessel Rupture. Rep. NSWC/WOL/TR 75-87.* Naval Surface Weapons Center, White Oak, MD

PITTOM, A. (1978). The control of toxic hazards: the way ahead. *Chem. Ind.*, Feb. 4, 77

PITTS, G., MOSS, T.R., STRUTT, J.E. and SAYLES, R.S. (1980). A new approach to reliability data classification. In Moss, T.R. (1980), *op. cit.*, p. 127

PITTS, S.F. and GOWAN, P.C. (1977). Learning safety from near misses. *Loss Prevention*, **11**, 9

PITTS, W.M. (1989). Assessment of theories for the behavior and blowout of lifted turbulent jet diffusion flames. *Combustion*, **22**, 809

PIVER, W.T. (1984). Simplified estimation technique for organic contaminant transport in groundwater. *J. Haz. Materials*, **8**(4), 331

PIZATELLA, T.J., PUTZ-ANDERSON, V., BOBRICK, T.G., MCGLOTHLIN, J.D. and WATERS, T.R. (1992). Understanding and evaluating manual handling injuries: NIOSH research studies. *Ergonomics*, **35**, 945

PIZZI, F.P. (1982). Assess environmental risks. *Hydrocarbon Process.*, **61**(4), 253

PLAMPING, K. (1986). The design of process plant alarm systems. Ph.D. Thesis, Loughborough. Univ. of Technol.

PLAMPING, K. and ANDOW, P.K. (1983). The design of process alarm systems. *Trans. Inst. Meas. Control*, **5**, 161

PLANER, R.G. (1979). *Fire Loss Control* (Basel: Dekker)

PLANNING RESEARCH CORP. (1979). *Deepwater Port Approach/Exit Hazard and Risk Assessment.* Rep. to US Coast Guard

PLANT SAFETY LTD (1990). Safety first at work. *Plant Engr*, May/Jun., 28

PLAPPERT, J.F. (1975). Selecting large electric drivers for the HPI. *Hydrocarbon Process.*, **54**(12), 103

PLATE, E.J. (1971). *Aerodynamic Characteristics of Atmospheric Boundary Layers. Rep. TID-25465.* Atom. Energy Comm., Washington, DC

PLATE, E.J. (ed.) (1982). *Engineering Meteorology* (Amsterdam: Elsevier)

PLATT, E.A. and FORD, J.A. (1966). Handling and testing unstable materials. *Chem. Engng Prog.*, **62**(3), 98

PLATT, G. (1966). Describe interlock sequences. *ISA J.*, **13**(1), 51

PLATT, H. (1972). Drivers for process pumps. *Hydrocarbon Process.*, **51**(6), 95

PLATTEN, C.F. (1971). *Economies of Scale in Manufacturing Industry* (London: Cambridge Univ. Press)

PLATZ, O. (1976). Availability of a renewable, checked system. *IEEE Trans. Reliab.*, **25**, 56

PLATZ, O. (1977). Methods for calculating down time distributions. *First Nat. Conf. on Reliability. NRC3/6.* (Culcheth, Lancashire: UK Atom. Energy Auth.)

PLATZ, O. (1980). Asymptotic distribution of downtime for a cold standby system. *IEEE Trans. Reliab.*, **29**, 79

PLATZ, O. (1984). A Markov model for common-cause failures. *Reliab. Engng*, **9**(1), 25

PLATZ, O. and OLSEN, J.V. (1976). *FAUNET − a Program Package for Evaluation of Fault trees and Networks. Rep. Riso-348.* Riso Nat. Lab., Riso, Denmark

PLAVSIC, B. (1993). Estimate costs of plants world-wide. *Chem Engng*, **100**(8), 100

PLEHIERS, P.M., REYNIERS, G.G. and FROMENT, G.F. (1990). Simulation of the run length of an ethane cracking furnace. *Ind. Engng Chem. Res.*, **29**, 636,

PLETCHER, D. (1984). *Industrial Electrochemistry* (London: Chapman & Hall)

PLETT, E.G. (1984). Flammability of gaseous mixtures of ethylene oxide, nitrogen and air. *Plant/Operations Prog.*, **3**, 190

PLEWES, L. and HINDLE, J. (1977). Hospital based schemes: Luton and Dunstable. In Easton, K.D. (1977), *op. cit.*, p. 234

PLIICKEBAUM, J.W., STRAUSS, W.A. and EDSE, R. (1964). Propagation of spherical combustion waves. *Combust. Flame*, **8**, 89

PLOUGH, A. and KRIMSKY, S. (1987). The emergence of risk communication studies: social and political context. *Sci. Technol. Human Values*, **12**(3/4), 4

PLOUM, A.J.W. (1977). The Brunei liquefied natural gas plant. *LNG Conf. 5*

PLOWDEN, W. (1971). *The Motor Car and Politics 1896−1970* (London: Bodley Head)

PLUDEK, V.R. (1977). *Design and Corrosion Control* (London: Macmillan)

PLUMBE, C.C. (1953) *Factory Health, Safety and Welfare Encyclopaedia* (London: Nat. Trade Press)

PLUMMER, R. (1986). A lot of hot air. *Process Engng*, Apr., 73

PLUNKETT, E.R. (1976). *Handbook of Industrial Toxicology* (New York: Chem. Publ. Co.)

POBLETE, B.R., LEES, F.P. and SIMPSON, G.B. (1984). The assessment of major hazards: estimation of injury and damage around a hazard source using an impact model based on inverse square law and probit relations. *J. Haz. Materials*, **9**(3), 355

POCCHIARI, F. (1978). Accidental TCDD contamination at Seveso (Italy): epidemiological aspects. *World Health Organisation, Conf. on Co-ordination of Epidemiological Studies on Long Term Hazards of Chlorinated Dibenzodioxins and Chlorinated Dibenzofurans* (Lyon)

POCHIN, E.E. (1973). Occupational and other fatality rates. *Soc. Occup. Med., Symp. on The Assessment of Exposure and Risk*, p. 35

POCHIN, E.E. (1975). The acceptance of risk. *Br. Med. Bull.*, **31**(3), 184

POCHIN, E.E. (1981a). Quantification of risk in medical procedures. *Proc. R. Soc., Ser. A*, **376**, 87. See also in Warner, Sir F. and Slater, D.H. (1981), *op. cit.*, p. 87

POCHIN, E.E. (1981b). Risk−benefit in medicine. *Risk−Benefit Analysis in Drug Research* (ed. J.F. Cavella) (Lancaster: MTP)

POCHIN, E. (1983). *Nuclear Radiation: Risks and Benefits* (Oxford: Oxford Univ. Press)

POCOCK, R. and DOCHERTY, A. (1978). Industrial environmental control. Legislative, industrial and social aspects. *Chem. Ind.*, Jul. 15, 512

POCOCK, R.F. (1977). *Nuclear Power, Its Development in the United Kingdom* (London: Unwin Brothers/Instn Nucl. Engrs)

POCOCK, S. and ALLEN, J. (1986). Machine condition monitoring − are you convinced yet? *Process Engng*, Aug., 41

PODA, G.A. (1966). Hydrogen sulphide can be handled safely. *Arch. Environ. Health*, **12**, 795

PODIO, A.L., FOSDICK, M.R. and MILLS, J. (1983). Gulf Coast blowouts. *Oil Gas J.*, **81**(Oct. 31), 119; (Nov. 7), 125

VAN POELGEEST, F.M. (1978). *Substandard Tankers*. Navigation Res. Centre, Netherlands Maritime Inst.

POGANY, G.A. (1979). On the safe use of alkylene oxides in high pressure bench-scale experiments. *Chem. Ind.*, Jan. 6, 16

POHL, H.A. *Dielectrophoresis* (Cambridge: Cambridge Univ. Press)

POHL, J.H., LEE, J., PAYNE, R. and TICHENOR, B.A. (1986). Combustion efficiency of flares. *Combust. Sci. Technol.*, **50**, 217

POHTO, H.A. (1979). Energy release from rupturing high-pressure vessels: a possible code consideration. *J. Pres. Ves. Technol.*, **101**, 165

POJASEK, R.B. (1979a). Solid-waste disposal: solidification. *Chem. Engng*, **86**(Aug. 13), 141

POJASEK, R.B. (ed.) (1979b). *Toxic and Hazardous Waste Disposal, vol. 1* (New York: Wiley)

POJE, G.V. (1990). Rising American expectations for a larger public role to prevent chemical accidents. *Plant/Operations Prog.*, **9**, 186

POLAK, E. (1983). Is odour quantifiable? *Chem. Ind.*, Jan. 3.30

POLAK, P. (1982). *Designing for Strength* (London: Macmillan)

POLAND, J., HINDASH, I.O. and KASSOY, D.R. (1982). Ignition processes in confined thermal explosions. *Combust. Sci. Technol.*, **27**, 215

POLENTZ, L.M. (1980). Calculate the energy in a pressurized vessel. *Chem. Engng*, **87**(Dec. 15), 95

POLI, A.A. and CIRILLO, M.C. (1993). On the use of the normalized mean square error in evaluating dispersion model performance. *Atmos. Environ.*, **27**A, 2427

POLITAKIS, P. (1985). *Empirical Analysis for Expert Systems* (London: Pitman)

POLITZ, F.C. (1985). Poor relief valve piping design results in crude unit fire. *API 50th Midyear Mtg*, **64**, 192

POLITZ, F.C. (1988). Managing overpressure − more than safety relief protection. *J. Haz. Materials*, **20**, 265

POLK, M. (1987). Better couplings reduce pump maintenance. *Hydrocarbon Process.*, **66**(1), 62

POLLACK, W.I. and STEELY, C.N. (eds) (1990). *Corrosion 89: Corrosion under Wet Thermal Insulation: New Techniques for Solving Old Problems* (Houston, TX: Nat. Ass. Corrosion Engrs)

POLLAK, H.M. (1963). How to select centrifugal pumps. *Chem. Engng*, **70**(Feb. 4), 81

POLLAK, P. (1991). Make or buy? A key question in new products development. *Speciality Chemicals*, Jun./Jul., 303

POLLAK, R. (1973). Selecting fan sand blowers. *Chem. Engng*, **80**(Jan. 22), 86

POLLARD, A. (1971). *Process Control* (London: Heinemann)

POLLARD, E.I. (1980). Synchronous motors . . . avoid torsional vibration problems. *Hydrocarbon Process.*, **59**(2), 97

POLLARD, F.B. and ARNOLD, J.H. (eds) (1966). *Aerospace Ordnance Handbook* (Englewood Cliffs, NJ: Prentice-Hall)

POLLEY, G.T. and SHAHI, M.H.P. (1991). Interfacing heat exchanger network synthesis and detailed heat exchanger design. *Trans. Inst. Chem. Engrs*, **69A**, 445

POLLEY, G.T., SHAHI, M.H.P. and NUNEZ, M.P. (1991). Rapid design algorithms for shell-and-tube and compact heat exchangers. *Trans. Inst. Chem. Engrs*, **69A**, 435

POLLINA, V. (1962). Safety sampling. *J. ASSE*, Aug.

POLLOCK, A.A. (1986). Acoustic emission capabilities and application in monitoring corrosion. In *Corrosion Monitoring in Industrial Plants Using Non-Destructive Testing and Electrochemical Methods* (ed. G.C. Moran and P. Labine) (New York: Am. Soc. Testing Mats), p. 30

POLLOCK, A.J.P., CAMPBELL, D. and REID, W.H. (1986). Smoke inhalation clinical and laboratory studies. *Fire Safety J.*, **10**(2), 155

POLLOCK, D.D. (1981–). *Physical Properties of Materials for Engineers*, vols 1–3 (Boca Raton, FL: CRC Press)

POLLOCK, G.E. (1975). Lessons from a dust collector fire. *Ammonia Plant Safety*, **17**, 145

POLLOCK, W.I. (1992). What's hot in metal alloys. *Chem. Engng*, **99**(10), 78

POLLOCKS, I. and RICHARDS, E.T. (1964). Failure rate data (FARADA) program conducted by the US Ordnance Laboratory, Corona, CA. *Proc. Third Reliability and Maintainability Conf.* (New York: Am. Inst. Mech. Engrs)

POLOVKO, A.M. (1968). *Fundamentals of Reliability Theory* (New York: Academic Press)

POLYA, G. (1954). *Mathematics and Plausible Reasoning*, vol. 2 (New York: Wiley)

POLYA, G. (1957). *How to Solve It*, 2nd ed. (New York: Doubleday Anchor)

POLYMEROPOULOS, C.E. (1974). Flame propagation in a one-dimensional liquid fuel spray. *Combust. Sci. Technol.*, **9**, 197

POLYMEROPOULOS, C.E. (1984). Flame propagation in aerosols of fuel droplets, fuel vapor and air. *Combust. Sci. Technol.*, **40**, 217

POLYMEROPOULOS, C.E. and DAS, S. (1975). The effect of droplet size on the burning velocity of kerosene-air sprays. *Combust. Flame*, **25**, 247

POLYMEROPOULOS, C.E. and PESKIN, R.L. (1969). Ignition and extinction of liquid fuel drops in numerical computations. *Combust. Flame*, **13**, 166

POLYTECHNIC INT. (1980). *Inert Gas Systems Manual*, 2nd ed. (Luton)

PONG CHUNG and SMITH, E.E. (1982). Emissivity of smoke. *J. Fire Flammability*, **13**, 104

PONTECORVO, A.B. (1965). A method of predicting human reliability. *Ann Reliab. Maintainab.*, **4**, 340

PONTIGGIA, R. (1992). Hazard analysis techniques adopted by 'TECNIMONT' in the development of process plant projects. *Loss Prevention and Safety Promotion 7*, vol. 3, 142–1

PONTIUS, P.W., VAN TASSEL, P.A. and FIELD, J.H. (1959). Operator training on new plant start-up. *Chem. Engng Prog.*, **55**(8), 38

PONTON, J.W. (1980). Crack-opening displacement as a method of measuring crack length in compact tension specimens. *J. Strain Anal. Engng Design*, **15**, 31

PONTON, J.W. (1996). The disposable batch plant. *5th World Con. of Chem. Eng.*, Jul. 14–18, 1996, San Diego, CA, Vol. II, 1119–23. (New York: American Institute of Chemical Engineers)

PONTON, J.W. and LAING, D.M. (1993). A hierarchical approach to the design of process control systems. *Chem. Engng Res. Des.*, **71A**, 181

PONTON, J.W. and VASEK, V. (1986). A two-level approach to chemical plant and process simulation. *Comput. Chem. Engng*, **10**, 277

POOK, L.P. (1975a). Analysis and application of fatigue crack growth. *J. Strain Anal.*, **10**(4), 242

POOK, L.P. (1975b). *Basic Statistics of Fatigue Crack Growth. NEL Rep. 595.* Nat. Engng Lab., East Kilbride

POOK, L.P. (1977). Practical implications of fatigue crack threshold. *Metal Sci.*, **11**(8), 382

POOK, L.P. (1979). Approximate stress intensity factors obtained from simple plate bending theory. *Engng Fracture Mechanics*, **12**, 505

POOLE, G. (1985). Improving the design of emergency relief systems. *Process Engng*, May, 67

POOLE, P.C. (1975). Debugging and testing. In Bauer, F.L. (1975a), *op. cit.*, p. 278

POPAT, N., CATLIN, C.A., ANTZEN, B., LINDSTEDT, R.P., HJERTAGER, B.H., SOLBERG, T., SAETER, R. and VAN DEN BERG, A.C. (1996). *J. of Hazard. Mater.*, **45**, 1

POPE, W.C. and CRESSWELL, T.J. (1965). Safety programs management. *J. ASSE*, Aug.

POPLE, H.E. (1982). Heuristic methods for imposing structure on ill-structured problems: the structuring of medical diagnostics. In Szolovits, P. (1982), *op. cit.*

POPOV, E.P. (1976). *Mechanics of Materials*, 2nd ed. (Englewood Cliffs, NJ: Prentice-Hall)

POPPER, H. (1963). Last year's major explosions prod this year's safety push. *Chem. Engng*, Jan. 7, 91

POPPER, H. (1971). The CPI's cost of meeting environmental standards. *Chem. Engng.*, **78**(Aug. 23), 106

POPPLESTONE, R. (1992a). POPLOG. In Shapiro, S.C. (1992), *op. cit.*, p. 1181

POPPLESTONE, R. (1992b). POP11. In Shapiro, S.C. (1992), *op. cit.*, p. 1182

POPPLESTONE, R.J. (1967). The design philosophy of POP-2. *Machine Intelligence 3* (ed. D. Michie) (Edinburgh: Edinburgh Univ. Press), p. 393

PORCELLI, R.V. and JURAN, B. (1986). Selecting the process for your next MMA plant. *Hydrocarbon Process.*, **65**(3), 37

PORT OF LONDON AUTHORITY (1977). *Notice to Mariners: Precautions to be Observed by Vessels (River Thames)* (London)

PORTEOUS, A. (1985). *Hazardous Waste Management Handbook* (London: Butterworths)

PORTEOUS, W. and BLANDER, M. (1975). Limits of superheat and explosive boiling of light hydrocarbons, halocarbons and hydrocarbon mixtures. *AIChE J.*, **21**, 560

PORTEOUS, W.M. and REID, R.C. (1976). Light hydrocarbon vapor explosions. *Chem. Engng Prog.*, **72**(5), 83

PORTER, A.L., ROSSINI, F.A., CARPLENTER, S.R. and ROPER, A.T. (1980). *A Guidebook for Technology Assessment and Impact Analysis* (New York: North Holland)

PORTER, J.W. (1988). The major concerns of the Environmental Protection Agency. *Preventing Major Chemical and Related Process Accidents*, p. 765

PORTER, J.W. (1989). Hazardous waste cleanup programs: a critical review. *Chem. Engng Prog.*, **85**(4), 16

PORTER, L. and LAWLER, E. (1968). *Managerial Attitudes and Performance* (Homewood, IL: Dorsey Press/R.D. Irwin)

PORTER, R.J. (1974). *Analytical Calorimetry*, vol. 3 (New York: Plenum Press)

PORTER, R.L. (1962). Unusual pressure relief devices. *Ind. Engng Chem.*, **54**(1), 24

PORTER, R.L. *et al.* (1956). Design and construction of barricades. *Ind. Engng Chem.*, **48**(5), 841

PORTER, S.R. (1991). Risk mitigation in land use planning: indoor releases of toxic gases. *Modeling and Mitigating the Consequences of Accidental Releases*, p. 463

PORTER, W.H.L. (1980). Generation of missiles and destructive shockfronts and their consequences. *Nucl. Engng*, **19**, 171

PORTER, W.H.L. (1982). The thrust on the supports of a two chamber vessel when the bursting disc in the duct connecting the chambers is ruptured. *The Assessment of Major Hazards* (Rugby: Instn Chem. Engrs), p. 109

PORTMANN, J.E. and NORTON, M.G. (1981). Disposal to sea of toxic materials. *Chem. Ind.*, Apr. 18, 285

PORZEL, F.B. (1972). *Introduction to a Unified Theory of Explosions (UTE). Rep. NOLTR 72-209.* Naval Ordnance Lab., White Oak, Silver Spring, MD

PORZEL, F.B. (1974). Damage potential from real explosions: total head and prompt energy. *Explosives Safety Board, Dept of Defense, Minutes Sixteenth Explosives Safety Seminar*, vol. 1, p. 633

PORZEL, F.B. (1980). Technology base of the Navy Explosives Safety Management Program. *Explosives Safety Board, Dept of Defense, Minutes Nineteenth Explosives Safety Seminar*, vol. 2, p. 1423

POST OFFICE (n.d.). *Engineering Instructions, Petroleum Premises, Lines General, Z3902*

POST, E. (1943). Formal reductions of the general combinatorial decision problem. *Am. J. Math.*, **65**, 197

POSTLETHWAITE, B.C. (1989). Noise. *Plant Engr*, Sep./Oct., 26

POSTLETHWAITE, B. (1993). Noise control. *Plant Engr*, **37**(2), 32

POSTNIKOV, S.N. (1978). *Electrophysical and Electrochemical Phenomena in Friction, Cutting and Lubrication* (New York: van Nostrand)

POSTON, T. and STEWART, I. (1978). *Catastrophe Theory and its Applications* (London: Pitman)

POTASH, L.M. (1977). Design of maps and map-related research. *Hum. Factors*, **19**, 139

POTEN AND PARTNERS (1982). *Liquefied Gas Ship Safety: Analysis of the Record 1964 to 1981* (New York)

POTHANIKAT, J.J. (1976). New approach to . . . piping system weight analysis. *Hydrocarbon Process.*, **55**(4), 195

POTTER, A.E. (1960). Flame quenching. *Prog. Combust. Sci. Technol.*, **1**, 145

POTTER, A.E. and BERLAD, A.L. (1957). The effect of fuel type and pressure on flame quenching. *Combustion*, **6**, p. 27

POTTER, A.E., HEIMEL, S. and BUTLER, J.N. (1962). Apparent flame strength. *Combustion*, **8**, p. 1027

POTTER, J. and BOUCH, W.E. (1970). The use of satellite LNG storage plants with ballasting for the control of the Wobbe Index. *LNG Conf. 2*

POTTER, N. (1994). Post-Piper spending toll equals six new field developments. *Offshore*, May, 142

POTTS, A.E. (1993). An overview of the applicability of PRA techniques to the design of offshore platforms and associated marine operations. In Melchers, R.E. and Stewart, M.G. (1993), *op. cit.*, p. 27

POTTS, C. and BRUNS, G. (1988). Recording the reasons for design decisions. *Proc. Tenth Conf. on Software Engineering* (Los Alamos, CA: CS Press), p. 418

POUCET, A. (1983). *Computer Aided Fault Tree Synthesis. Rep. EUR 8707 EN.* Comm. Eur. Comm., Directorate General for Sci. Res and Dev., Joint Res. Centre, Ispra, Italy

POUCET, A. (ed.) (1989). *HF-RBE: Human Factors Reliability Benchmark Exercise: Summary Contributions of Participants* (report). Comm. Eur. Comm., Directorate General for Sci. Res and Dev., Joint Res. Centre, Ispra, Italy

POUCET, A. (1990). STARS: knowledge-based tools for safety and reliability analysis. *Reliability Engineering and System Safety*, **30**, 379

POUCET, A. and DE MEESTER, P. (1981). Modular fault tree synthesis – a new method for computer aided fault tree construction. *Reliab. Engng*, **2**(1), 65

POUDRET, J., HUTHER, M., JEAN, P. and VAUGHAN, H. (1981). Grounding of a membrane tanker: correlation between damage predictions and observations. *Soc. Naval Architects Marine Engrs* (New York)

POULTON, E.C. (1971). *Environment and Human Efficiency* (Springfield, IL: Thomas)

POWELL, B. (1967). Safety and the large ammonia plant. *Ammonia Plant Safety*, **9**, p. 50

POWELL, F. (1969). Ignition of gases and vapours. *Ind. Engng Chem.*, **61**(12), 29

POWELL, F. (1978). Ignition of flammable gases and vapours by friction between footwear and flooring materials. *J. Haz. Materials*, **2**(4), 309

POWELL, F. (1984). Ignition of gases and vapours by hot surfaces and particles – a review. *Ninth Int. Symp on Prevention of Occupational Accidents and Diseases in the Chemical Industry*, Lucerne, p. 267

POWELL, H.N. (1955). The height of diffusion flames and the relative importance of mixing and reaction rates. *Combustion*, **5**, 290

POWELL, J. and CANTER, D. (1984). Studies in development of method to assess the management factor in industrial loss. *Ergonomic Problems in Process Operations* (Rugby: Instn Chem. Engrs), p. 211

POWELL, J.H. (1981). Estimation of the deficiencies of parallel-redundant groups of components under regular replacement policies. *Reliab. Engng*, **2**(4), 303

POWELL, R. (1984). Estimating worker exposure to gases and vapors leaking from pumps and valves. *Am. Ind. Hyg. Ass. J.*, **45**, A7

POWELL, R.H. (ed.) (1979). *Handbooks and Tables in Science and Engineering* (New York: Mansell)

POWELL, R.W. (1940). Further experiments on the evaporation of water from saturated surfaces. *Trans. Instn Chem. Engrs*, **18**, 36

POWELL, R.W. and GRIFFITHS, E. (1935). The evaporation of water from plane and cylindrical surfaces. *Trans. Instn Chem. Engrs*, **13**, 175

POWELL, T.E. (1985). A review of recent developments in project evaluation. *Chem. Engng*, **92**(Nov. 11), 187

POWELL-SMITH, V. (1974). *Questions and Answers on the Health and Safety at Work Act (A Protection Handbook)* (London: Osborne)

POWER, J.M. and KRIER, H. (1986). Attenuation of blast waves when detonating explosives inside barriers. *J. Haz. Materials*, **13**(2), 121

POWER, V. (1983). Attention to documentation yields unified drawings. *Chem. Eng*, **90**(Nov. 28), 73

POWERS, G.J. (1975). Private communication

POWERS, G.J. (1977). Comment on 'Computer aided synthesis of fault trees'. *IEE Trans. Reliab.*, **R-26**, 316

POWERS, G.J. and JONES, R.L. (1973). Reaction path synthesis strategies. *AIChE J.*, **19**, 1204

POWERS, G.J. and LAPP, S.A. (1976). Computer-aided fault tree synthesis. *Chem. Engng Prog.*, **72**(4), 89

POWERS, G.J. and LAPP, S.A. (1977). Computer-aided synthesis of fault trees. *IEEE Trans. Reliab.*, **R-26**, 2

POWERS, G.J. and TOMPKINS, F.C. (1974a). Fault tree synthesis for chemical processes. *AIChE J.*, **20**, 376

POWERS, G.J. and TOMPKINS, F.C. (1974b). A synthesis strategy for fault trees in chemical processing systems. *Loss Prevention*, **8**, 91

POWERS, G.J. and TOMPKINS, F.C. (1976). Computer aided synthesis of fault trees for complex processing systems. In Henley, E.J. and Lynn, J.W. (1976), *op. cit.*, p. 307

POYNTON, J.P. (1979). Basics of reciprocating metering pumps. *Chem. Engng*, **86**(May 21), 156

PRADHAN, S. (1993). Apply reliability centred maintenance to sealless pumps. *Hydrocarbon Process.*, **72**(1), 43

PRADO, G.P., LEE, M.L., HITES, R.A., HOULT, D.P. and HOWARD, J.B. (1977). Soot and hydrocarbon formation in a turbulent diffusion flame. *Combustion*, **16**, p. 649

PRAHL, J. and EMMONS, H.W. (1975). Fire induced flow through an opening. *Combust. Flame*, **25**, 369

PRANDTL, L. (1933). New results in turbulence research. *Z. Ver. Deutsch. Ing.*, **77**, 105

PRASEK, F. (1988). Acoustic emission test of a 12,000 t refrigerated ammonia storage tank. *Ammonia Plant Safety*, **28**, 66

PRATER, B.E. (1978). Industrial environmental control. The measurement of environmental impact. *Chem. Ind.*, Jul. 15, 517

PRATT, D.B. (1986). The fire and explosion hazards of hydraulic accumulators. *Hazards*, **IX**, p. 227

PRATT, F.L. (ed.) (1966). *Fracture* (London: Chapman & Hall)

PRATT, M. (ed.) (1993). *Remedial Processes for Contaminated Land* (Rugby: Instn Chem. Engrs)

PRATT, M.W.T. (1974). *The Cost of Safety. Course of Process Safety – Theory and Practice*. Dept of Chem Engng, Teesside Polytechnic, Middlesbrough

PRATT, N.R. (1965). From laboratory to production. *Chem. Engng Prog.*, **61**(2), 41

PRATT, T.H. (1992). Electrostatic hazards for automatic tank gauges. *J. Loss Prev. Process Ind.*, **5**, 104

PRATT, T.H. (1993). Electrostatic ignition in enriched oxygen atmospheres. *Process Safety Prog.*, **12**, 203

PRATT, T.H. (1994). Static electricity in pneumatic transport systems: three case histories. *Process Safety Prog.*, **13**, 109

PRATT, T.H., ANDERSON, S.E., WILLIAMS, G.M. and HULBURT, D.A. (1989). Electrostatic hazards in liquid pumping operations. *Loss Prevention and Safety Promotion 6*, p. 57.1

PRATT, V. (1979). *LISP. Encyclopaedia of Computer Science and Technology*, vol. 10 (ed. by J. Belzer, A.G. Holzman and A. Kent) (New York: Dekker)

PREDDY, D.L. (1969). Guidelines for safety and loss prevention. *Chem. Engng*, **76**(Apr. 21), 94

PREECE, J.A. (1982). Public relations in the United Kingdom. In Green, A.E. (1982), *op. cit.*, p. 235

PREECE, K. (1988). Health and safety at work. *Prospect*, Summer issue, 15

PREECE, P. (1986). The making of PFG and PIG. *Chem. Engr, Lond.*, **426**, 87

PREECE, P. (1990). Graphical user interfaces for process design. *Chem. Engr, Lond.*, **482**, 30

PREECE, P.E. and STEPHENS, M.B. (1989). PROCEDE – Opening windows on the design process. *Computer Aided Process Engineering* (Rugby: Instn Chem. Engrs), p. 211

PREECE, P.E., AL BADER, S.A., DAVILA PERNIA, J.L., EVANS, J.A.C. and GILLER, M.K. (1987). The development of a 2–D graphical process flowsheet generator (PFG) and a piping and instrument generator (PIG). *Comput. Chem. Engng*, **11**, 279

PRENGLE, H.W. and BARONA, N. (1970). Make petrochemicals by liquid phase oxidation. *Hydrocarbon Process.*, **49**(3), 106; 49(11), 159

PRENTICE, J.S., SMITH, S.E. and VIRTUE, L.S. (1974). Safety in polyethylene plant compressor areas. *Safety in Polyethylene Plants*, **2**, 1

PRENTICE, R.L., WILLIAMS, B.J. and PETERSON, A.V. (1981). On the regression analysis of multivariate failure time data. *Biometrika*, **68**, 373

PRENTISS, A.M. (1937). *Chemicals in War* (New York: McGraw-Hill)

PRENTISS, A.M. (1951). *Civil Defense in Modern War* (New York: McGraw-Hill)

PRESCOTT, G.R. (1982). Cracking and nitriding in ammonia converters. *Plant/Operations Prog.*, **1**, 94

PRESCOTT, G.R. and BADGER, F.W. (1980). Cracking problems in ammonia converters. *Ammonia Plant Safety*, **22**, 146

PRESCOTT, G.R., BLOMMAERT, P. and GRISOLIA, L. (1986). Failure of high pressure synthesis pipe. *Ammonia Plant Safety*, **26**, p. 228. See also *Plant/Operations Prog.*, 1986, **5**, 155

PRESCOTT, G.R., PODHORSKY, M. and BLOMMAERT, P. (1988). Improvements in boiler and tubejoint design. *Ammonia Plant Safety*, **28**, 135

PRESCOTT, J.H. (1966). Explosion leads to a safer ethylene plant design. *Chem. Engng*, **73**(Feb. 14), 96

PRESCOTT, J.H. (1967). Control houses are girded to survive boom. *Chem. Engng*, **74**(Mar. 13), 96

PRESCOTT, J.H. (1968). Identify pipes on pipe racks easily. *Chem. Engng*, **75**(Jul. 15), 148

PRESCOTT, J.H. (1970). Gas-pipeline code revised. *Chem. Engng*, **77**(Nov. 30), 28

PRESCOTT, J.H. (1971). More options to methanol. *Chem. Engng*, **78**(Apr. 5), 60

PRESCOTT, J.H. (1972). Thermography: new CPI tool to forewarn trouble. *Chem. Engng*, **79**(Sep. 18), 178

PRESIDENTS SCIENCE ADVISORY COMMITTEE (1963). *Use of Pesticides* (Washington, DC)

PRESIDENTS SCIENCE ADVISORY COMMITTEE, PANEL ON CHEMICALS AND HEALTH (1973). *Chemicals and Health* (Washington, DC)

PRESLES, H.N., BAUER, P., HEUZE, O. and BROCHET, C. (1985). Investigations on detonations of gaseous explosive mixtures at high initial pressure. *Combust. Sci. Technol.*, **43**, 315

PRESSURE GAUGE MANUFACTURERS ASSOCIATION (1980). Pressure gauge safety. *Measmnt Control*, **13**, 283

PRESTON, C.A. (1983). Overall view of safety in operation of tankers carrying hazardous cargoes and planning for contingencies and emergencies. *Loss Prevention and Safety Promotion 4*, vol. 2, p. A1

PRESTON, F.W. (1942). The mechanical properties of glass. *J. Appl. Phys.*, **13**, 623

PRESTON, M. (1993). The application and use of computers in planning and training for emergency response. *Health, Safety and Loss Prevention*, p. 403

PRESTON, M.L. (1994). The use of computers in emergency response planning. *Hazards*, **XII**, p. 385

PRESTON, M.L. (1997). Inherent SHE – 20 Years of Evolution. *IChemE Symp. Series*, No. 141, 11–23

PRESTON, M.L. and FRANK, G.W.H. (1985). A new tool for batch process engineers – ICI's Batchmaster. *Plant/Operations Prog.*, **4**, 217

PRESTON, M.L. and SINCLAIR, P.A. (1987). Gas dispersion modeling in ICI and the calibration of DISP2 burst model against Thorney Island data. *Vapor Cloud Modeling*, p. 357

PRESTON, M.L. and TURNEY, R.D. (1991). The process systems contribution to reliability engineering and risk assessment. In *Computer-Oriented Process Engineering*, (Puigjaner, L.//Espuna, A. Amsterdam: Elsevier Science Publishers) pp. 249–57.

PRETT, D.M. and GARCIA, C.E. (1988). *Fundamental Process Control* (Oxford: Butterworth-Heinemann)

PRETT, D.M. and MORARI, M. (1987). *The Shell Process Control Workshop* (Boston, MA: Butterworths)

PRETTRE, M. (1949). Researches on the combustion of mixtures of normal pentane and oxygen between 250° and 360°C. *Combustion*, **3**, p. 397

PRETTRE, M., DUMANOIS, P. and LAFITTE, P.C. (1930). *R. Heb. Séanc. Acad. Sci., Paris*, **191**, 329, 414

PREUSSNER, R.D. and BROZ, L.D. (1977). Hydrocarbon emission reduction systems. *Chem. Engng Prog.*, **73**(8), 69

PREW, L.R. (1976). LNG ship cargo systems – some design and operating considerations. *Trans. Inst. Marine Engrs*, **88**, 8

PRICE, A. (1980). Containment of uranium hexafluoride at Springfield Works. In Hughes, D.J. (1980), *op. cit.*, paper 5

PRICE, B.C. (1991). Know the range and limitations of screw compressors. *Chem. Engng Prog.*, **87**(2), 50

PRICE, C. and HUNT, J. (1991). Automating FMEA through multiple models. British Comput. Soc. *ESG Conf.*

PRICE, C., HUNT, J., LEE, M. and ORMSBY, A. (1991). Automating failure mode effects analysis. *Artificial Intelligence Appl. Inst., AI in Safety Colloquium* (Edinburgh)

PRICE, D. and WEHNER, J.F. (1965). The transition from burning to detonation in cast explosives. *Combust. Flame*, **9**, 73

PRICE, D.J. and BROWN, H.H. (1922). *Dust Explosions* (Boston, MA: Nat. Fire Prot. Ass.)

PRICE, D.J. DE S. (1963). *Big Science, Little Science* (New York: Columbia Univ. Press)

PRICE, D.K. (1965). *The Scientific Estate* (Cambridge, MA: Harvard Univ. Press)

PRICE, E.B. (1970). Safe handling of solvents in centrifuge operations. *Ind. Engng Chem.*, **62**(1), 48

PRICE, F. (1985). *Right First Time* (London: Gower)

PRICE, F.C. (1960). Explosions rip chemical plants. *Chem. Engng*, **67**(Nov. 14), 114

PRICE, F.C. (1961). Ball valve story: lusty newcomer looms large. *Chem. Engng*, **68**(Oct. 30), 56

PRICE, F.C. (1965). Long distance pipelines: shouldn't they be safer? *Chem. Engng*, **72**(Nov. 22), 131

PRICE, F.C. (1974a). Are long-distance pipelines getting safer? *Chem. Engng*, **81**(Oct. 14), 94

PRICE, F.C. (1974b). Citizens group questions safety of a pipeline. *Chem. Engng.* **71**(Jan. 6), 28

PRICE, H.A. and MCALLISTER, D.G. (1970). Process plant utilities. Inert gas. *Chem. Engng*, **77**(Dec. 14), 142

PRICE, H.E. (1985). The allocation of function in systems. *Hum. Factors*, **27**, 33

PRICE, J.A. (1962). Safety in HF alkylation requires two steps. *Hydrocarbon Process.*, **41**(5), 156

PRICE, J.W.H. (1983). Schematic method for presenting structural reliability arguments. *Reliab. Engng*, 4(4), 235

PRICE, L.H. (1965). The right lubrication for compressors. *Hydrocarbon Process.*, **44**(7), 145

PRICE, M.H. (1977). Computer assisted vibration monitoring successful. *Hydrocarbon Process.*, **56**(12), 85

PRICE, S.K., HUGHES, J.E., MORRISON, S.C. and POTGEITER, P.D. (1983). Fatal ammonia inhalation. A case report with autopsy. *S. Africa Med. J.*, **64**, 952

PRICE, W. (1978). CPI burglaries on the rise. *Chem. Engng*, **85**(Jul. 17), 78

PRICE, W.D.J. (1979/80). Methods of risk identification. *Fire Safety J.*, **2**(2), 105

PRICKETT, P.W. (1984). The performance of bursting discs at various rates of pressurisation. *The Protection of Exothermic Reactors and Pressurised Storage Vessels* (Rugby: Instn Chem. Engrs), p. 83

PRIDDLE, R.J. (1990). *Piper Alpha Public Inquiry. Transcript, Days 175, 176*

PRIDGEN, T.D. and GARCIA, N. (1967). What chemical engineers should know about structural design. *Chem. Engng*, **74**(May 8), 143; (May 22), 187

PRIEL, V. (1974). *Systematic Maintenance Organisation* (London: Macdonald & Evans)

PRIEST, D. (1992). Measuring corrosion rates fast. *Chem. Engng*, **99**(11), 169

PRIESTLEY, C.H.B. (1959). *Turbulent Transfer in the Lower Atmosphere* (Chicago, IL: Univ. of Chicago Press)

PRIESTLEY, C.H.B. and BALL, F.K. (1955). Continuous convection from an isolated source of heat. *Quart. J. R. Met. Soc.*, **81**, 144

PRIESTLY, M.J.N. *et al.* (1986). Seismic design of storage tanks. *Bull. N.Z. Nat. Soc. Earthquake Engng*, **19**(4)

PRIETO, R.F. (1985). Shortcut methods for estimating solids inventory. *Chem. Engng*, **92**(Nov. 11), 229

PRIJATEL, J. (1984). Failure analysis of ammonia plant shutdown instrumentation and control. *Ammonia Plant Safety*, **24**, 104. See also *Plant/Operations Prog.*, 1984, **3**, 25

PRIJATEL, J. (1983). Accidental venting of liquid ammonia. *Plant/Operations Prog.*, **2**, 131

PRIMACK, J. and VON HIPPEL, F. (1974). *Advice and Dissent: Scientists in the Political Arena* (New York: Basic Books)

PRIMACK, J. (1975). Nuclear reactor safety. An introduction to issues. *Bull. Atom. Scientists*, Sep.

PRINCE, A. (1966). *Alloy Phase Equilibria* (New York: Elsevier)

PRINCE, C.E. and ODINGA, K. (1986). Problems incurred during catalyst removal in ammonia plants. *Ammonia Plant Safety*, **26**, 82

PRINE, B. (1992). Analysis of titanium/carbon steel heat exchanger fire. *Plant/Operations Prog.*, **11**, 113

PRINGLE, D.R.S. and BROWN, A.E. (1990). International Safety Rating System. New Zealand's experience with a successful strategy (abstract). *J. Occup. Accid.*, **12**(1–3), 41

PRIOR, M.G., YONG, S., SHARMA, A. and LOPEZ, A. (1988). Concentration–time interactions in hydrogen sulphide toxicity in rats. *Can. J. Vet. Res.*, **4**, 375

PRITCHARD, B.N. (1983). Deterioration of structures supporting refinery equipment. *Loss Prevention and Safety Promotion 4*, vol. 3, p. G10

PRITCHARD, D.K. (1981). Breakage of glass windows by explosions. *J. Occup. Accid.*, **3**, 69

PRITCHARD, D.K. (1983). An experimental and theoretical study of blast effects on simple structures (cantilevers). *Loss Prevention and Safety Promotion 4*, vol. 3, p. D23

PRITCHARD, D.K. (1989). A review of methods for predicting blast damage from vapour cloud explosions. *J. Loss Prev. Process Ind.*, **2**(4), 187

PRITCHARD, K. (1989). Applying simulation to the control industry. *Control Engng*, **36**(5), 70

PRITCHARD, M.J. and BINDING, T.M. (1992). FIRE 2: a new approach for predicting thermal radiation levels from hydrocarbon pool fires. *Major Hazards Onshore and Offshore*, p. 491

PRITCHARD, R., GUY, J.J. and CONNOR, N.E. (1978). *Handbook of Industrial Gas Utilization — Engineering Principles and Practices* (NewYork: van Nostrand Reinhold)

PRITCHETT, D.H. (1972). Electric motors: a guide to standards. *Chem. Engng*, **79**(Dec. 11), 88

PROBER, R. (ed.) (1972). *CRC Handbook of Environmental Control* (Boca Raton, FL: CRC Press)

PROCACCIA, H. (1991). Reliability data banks at Electricite de France (EDF). In Cannon, A.G. and Bendell, A. (1991), *op. cit.*, p. 169

PROCESS CONTROL TRAINING (1981). Process Plant Fault Diagnosis (Cardiff)

PROCTOR, J.A. (1963). Noise control in air separation plants. *Ammonia Plant Safety*, **5**

PROCTOR, J.P. (1969). Glass as an aid to safety. In Handley, W. (1969), *op. cit.*, p. 315

PROCTOR, L.D. and WARR, A.J. (2002). Development of a continuous process for the industrial generation of diazomethane. *Org. Process Res. & Dev.* **6**, 884–92

PROCTOR, S.I. (1983). The FLOWTRAN simulation system. *Chem. Engng Prog.*, **79**(6), 49

PROES, N. (1977). Is LNG still a viable option? *Hydrocarbon Process.*, **56**(4), 93

PROGRAMMES ANALYSIS UNIT (1972). *An Economic and Technical Appraisal of Air Pollution in the United Kingdom* (London: HM Stationery Office)

PROJECT SOFTWARE AND DEVELOPMENT INC. (1991). Maintenance system. *Plant Engr*, **35**(Jan./Feb.), 30

PROKOP, J. (1988). The Ashland tank disaster. *Hydrocarbon Process.*, **67**(5), 105

PROKOPETZ, A.T. and WALTERS, D.B. (1987). Handling and management of hazardous research chemicals. In Young, J.A. (1987A), *op. cit.*, p. 139

PROKOPEVA, A.S. and YUSHKOV, G. (1975). *Gig. Tr. Prof. Zabol.*, Sep., 54

PROKOPEVA, A.S., YUSHKOV, G.G. and UBASHEEV, O. (1973). Toxicological characteristics of the one-time action of ammonia on animals during brief exposure. *Gig. Tr. Prof. Zabol.*, Jun., **56**

PROKOPOVITSH, A.S. (1986). Combustible metal agents and application techniques. In Cote, A.E. and Linville, J.L. (1986), *op. cit.*, p. 19–49

PRONIKOV, A.S. (1973). *Dependability and Durability of Engineering Products* (London: Butterworths)

PROTHERO, A. (1969). Computing with thermochemical data. *Combust. Flame*, **13**, 399

PROUST, C. and VEYSSIERE, B. (1988). Fundamental properties of flames propagating in starch dust–air mixtures. *Combust. Sci. Technol.*, **62**, 149

PROUZA, Z. (1970). Group poisoning with hydrogen sulphide in an unusual situation on a viscose plant. *Prak. Lekar*, **50**, 27

PROVINCIALE WATERSTAAT GRONINGEN (1979). *Pollution Control and Use of Norms in Groningen — Criteria for Risk Related to Dangerous Goods* (Groningen, Netherlands)

PRUDNIKOV, A.G. (1959). Flame turbulence. *Combustion*, **7**, 575

PRUGH, R.W. (1967). Pilot plant procedures. Preliminary evaluation of safety hazards. *Chem. Engng Prog.*, **63**(11), 49

PRUGH, R.W. (1981). Application of fault tree analysis. *Loss Prevention*, **14**, p. 1. See also *Chem. Engng Prog.*, 1980, **76**(7), 59

PRUGH, R.W. (1982). Practical application of fault tree analysis. In Fawcett, H.H. and Wood, W.S. (1982), *op. cit.*, p. 789

PRUGH, R.W. (1985–). Mitigation of vapor cloud hazards. *Plant/Operations Prog.*, 1985, **4**, 95; 1986, **5**, 169

PRUGH, R.W. (1987a). Evaluation of unconfined vapor cloud explosion hazards. *Vapor Cloud Modeling*, p. 712

PRUGH, R.W. (1987b). Guidelines for vapor release mitigation. *Plant/Operations Prog.*, **6**, 171

PRUGH, R.W. (1987c). Post-release mitigation design for mitigation of releases. *Preventing Major Chemical Accidents*, p. 5.1

PRUGH, R.W. (1989). Semi-quantitative risk analysis. *Loss Prevention and Safety Promotion 6*, p. 48.1

PRUGH, R.W. (1991). Quantify BLEVE hazards. *Chem. Engng Prog.*, **87**(2), 66

PRUGH, R.W. (1992). Hazardous fluid releases: prevention and protection by design and Operation. *J. Loss Prev. in the Process Ind.* **5**(2), 67–72

PRUGH, R.W. (1992a). Computer-aided HAZOP and fault tree analysis. *J. Loss Prev. Process Ind.*, **5**, 3

PRUGH, R.W. (1992b). Hazardous fluid releases: prevention and protection by design and operation. *J. Loss Prev. Process Ind.*, **5**, 67

PRUGH, R.W. (1992c). Improved F/N graph presentation and criteria. *J. Loss Prev. Process Ind.*, **5**, 239

PRUGH, R.W. (1992d). Pre-release mitigation. *Plant/Operations Prog.*, **11**, 19

PRUGH, R.W. (1994). Quantitative evaluation of fireball hazards. *Process Safety Prog.*, **13**, 83

PRUGH, R.W., HOWARD, A. and WINDSOR, D.G. (1993). High pressure operations: managing safety. *Chem Engng*, **100**(9), 99

PRUPPACHER, H.R., SEMONIN, R.G. and SLINN, W.G.N. (eds). (1983). *Precipitation Scavenging, Dry Deposition and Resuspension*, vols 1–2 (Amsterdam: Elsevier)

PU, Y.K., JAROSINSKI, J., TAI, C.S., KAUFFMAN, C.W. and SICHEL, M. (1989). The investigation of the feature of dispersion induced turbulence and its effects on dust explosions in closed vessels. *Combustion*, **22**, 1777

PU, Y.K., MAZURKIEWICZ, JAROSINSKI, J. and KAUFFMAN, C.W. (1989). Comparative study of the influence of obstacles on the propagation of dust and gas flames. *Combustion*, **22**, 1789

PU, Y.K., JAROSINKSI, J., JOHNSON, V.G. and KAUFFMAN, C.W. (1991). Turbulence effects on dust explosions in the 20 litre spherical vessel. *Combustion*, **23**, 843

PUBLIC HEALTH SERVICE (1941–). *Survey Compounds which have been Tested for Carcinogenic Activity. Publ. 149*

PUBLIC RECORD OFFICE (1945a). 'Big Ben'. File HO 191/198, Rep. S104 (Kew, London)

PUBLIC RECORD OFFICE (1945b). *A Comparison of the Standardised Casualty Rates for People in Unprotected Parts of Dwellings Exposed to Rocket Bombs (V2), Flying Bombs (V1) and Parachute Mines. File HO 191/198, Rep. S118* (Kew, London)

PUBLIC RECORD OFFICE (1945c). *The Damage Caused by the German Long Range Rocket Bomb. File HO 191/198, Rep S111* (Kew, London)

PUBLIC RECORD OFFICE (1945d). *File AIR 20/3439* (Kew, London)

PUBLIC RECORD OFFICE (1945e). *File PREM 3/111* (Kew, London)

PUCCINI, M. (1983). Videographic aids for Bayesian inference. *Euredata 4*, paper 12.2

PUCILL, P.M. (1973). Fire protection systems for computers: some special requirements and features. *Fire Prev. Sci. Technol.*, **6**, 13

PUCKETT, D.B. (1976). What you should know about fiberglass-reinforced plastic tanks. *Chem. Engng*, **83**(Sep. 27), 129

PUCKORIUS, P.R. (1978). Proper start-up protects cooling tower systems. *Chem. Engng*, **85**(Jan. 2), 101

PUGET SOUND COUNCIL OF GOVERNMENTS (1975). *Regional Disaster Mitigation Plan for the Central Puget Sound Region*

PUGH, H.D.L. (1970). *The Mechanical Behaviour of Materials under Pressure* (Amsterdam: Elsevier)

PUGH, T., YORK, B., LAXTON, G. and SIMPSON, G.D. (1993). Improve heat exchanger leak detection. *Hydrocarbon Process.*, **72**(11), 85

PUGSLEY, SIR A. (1966). *The Safety of Structures* (London: Arnold)

PUIGJANER, L. and ESPUNA, A. (eds) (1991). *Computer-Oriented Process Engineering* (Amsterdam: Elsevier)

PUISEAUX, L. (1977). *La Babel Nucléaire: Energie et Développement* (Paris: Editions Galilé)

PUJADO, P.R., VORA, B.V. and KRUEDING, A.P. (1977). Newest acrylonitrile process. *Hydrocarbon Process.*, **56**(5), 169

PUKLAVEC, V. and LINDENAU, D.P. (1985). Optimisation of LPG carrier design and its influence on long term operating costs. *Gastech 84*, p. 265

PULLY, A.S. (1993). Utilization and results of hazard and operability studies in a petroleum refinery. *Process Safety Prog.*, **12**, 106

PURANIK, S.A., HATHI, K.K. and SENGUPTA, R. (1990). Prevention of hazards through technological alternatives. *Saf. and Loss Prev. in the Chem. and Oil Processing Ind.*, Oct. 23–27, 1989, Singapore, 581–7. IChemE Symp. Series, no. 120. Rugby, Warwickshire (U.K.: The Institution of Chemical Engineers)

PURARELLI, C. (1982). Optimum design of horizontal liquid–vapor separators. *Chem. Engng*, **89**(Nov. 15), 127

PURCELL, V.P. (1968). Operating main turbine drives. *Loss Prevention*, **2**, 77

PURCHASE, I.F.H. (1980). Inter-species comparison of carcinogenicity. *Br. J. Cancer*, **41**, 454

PURDUE EUROPE (1977). *The Operator Interface in Process Control in European Industries.* (Trondheim, Norway: SINTEF)

PURDY, G. (1988). A practical application of quantified risk analysis. In Sayers, B.A. (1988), *op. cit.*, p. 139

PURDY, G. (1993). Risk analysis of the transportation of dangerous goods by road and rail. *J. Haz. Materials*, **33**, 229

PURDY, G. and DAVIES, P.C. (1985). Toxic gas incidents – some important considerations for emergency planning. *Multistream 85* (Rugby: Instn Chem. Engrs), p. 257

PURDY, G. and WASILIEWSKI, M. (1994). Risk management strategies for chlorine installations. *J. Loss Prev. Process Ind.*, **7**, 147

PURDY, G., CAMPBELL, H.S., GRINT, G.C. and SMITH, L.M. (1987). An analysis of the risks arising from the transport of liquefied gases in Great Britain. *Int. Specialist Mtg on Major Hazards in the Transport and Storage of Pressure Liquefied Gases*, Univ. of New Brunswick

PURDY, G., CAMPBELL, H.S., GRINT, G.C. and SMITH, L.M. (1988). An analysis of risks arising from the transport of liquefied gases in Great Britain. *J. Haz. Materials*, **20**, 335

PURDY, G., PITBLADO, R.M. and BAGSTER, D.F. (1992). Tank fire escalation – modelling and mitigation. *Loss Prevention and Safety Promotion 7*, vol. 3, 163–1

PURINGTON, R.G. (1986). Fire streams. *Fire Protection Handbook*, 16th ed. (ed. A.E. Cote and J.L. Linville), p. 17–90

PUROHIT, G.P. (1983). Estimating costs of shell-and-tube heat exchangers. *Chem. Engng*, **90**(Aug. 22), 56

PURVIN, R.L., WITHINGTON, H.W. and SMITH, C. (1974). The Arctic air/sea LNG project. *LNG Conf. 4*

PUSCH, W. and WAGNER, H.G. (1962). Investigation of the dependence of the limits of detonability on tube diameter. *Combust. Flame*, **6**, 157

PUTNAM, A.A. (1974). Effect of recirculated products on burning velocity and critical velocity gradient. *Combust. Flame*, **22**, 277

PUTNAM, A.A. and GRINBERG, I.M. (1965). Axial temperature variation in a turbulent buoyancy-controlled, diffusion flame (letter). *Combust. Flame*, **9**, 420

PUTNAM, A.A., BALL, D.A. and LEVY, A. (1980). Effect of fuel composition on relation of burning velocity to product of quenching distance and flashback velocity gradient. *Combust. Flame*, **37**, 193

PUTNAM, B.R. (1980). Cost benefit analysis for regulatory policy making. *Chem. Engng Prog.*, **76**(2), 20

PUTNEY, B. (1981). *WAMCOM Common Cause Methodologies using Large Fault Trees. Rep. EPRI NP-1851*. Elec. Power Res. Inst., Palo Alto, CA

VAN DE PUTTE, T. (1981). Purpose and framework of a safety study in the process industry. *J. Haz. Materials*, **4**, 225

VAN DE PUTTE, T. (1983). The safety report legislation and its application in the Netherlands. *J. Haz. Materials*, **7**, 131

VAN DE PUTTE, T. and MEPPELDER, F.H. (1980). Identification of major hazard installations in the process industries. *Loss Prevention and Safety Promotion 3*, vol. 2, p. 291

PUTTOCK, J.S. (1985). Thorney Island data and dispersion modelling. *J. Haz. Mats*, **11**(1/3), 381. See also McQuaid, J. (1985a), *op. cit.*, p. 381

PUTTOCK, J.S. (1987a). Analysis of meteorological data for the Thorney Island Phase I trials. *J. Haz. Materials*, **16**, 43

PUTTOCK, J.S. (1987b). Comparison of Thorney Island data with predictions of HEGABOX/HEGADAS. *J. Haz. Materials*, **16**, 439

PUTTOCK, J.S. (1987c). The development and use of HEGABOX/HEGADAS dispersion models for hazard analysis. *Vapor Cloud Modeling*, p. 317

PUTTOCK, J.S. (1988a). A model for gravity-dominated dispersion of dense-gas clouds. In Puttock, J.S. (1988b), *op. cit.*, p. 233

PUTTOCK, J.S. (ed.) (1988b). *Stably Stratified Flow and Dense Gas Dispersion* (Oxford: Clarendon Press)

PUTTOCK, J.S. (1989). Dispersion of dense vapour clouds in contact with the ground – theory and experiment. In Cox, R.A. (1989a), *op. cit.*, p. 145

PUTTOCK, J.S. (1999). A phenomenological model (SCOPE) in detail, *AIChE, CCPS conference*, San Francisco, Oct.

PUTTOCK, J.S. and COLENBRANDER, G.W. (1985a). Dense gas dispersion – experimental research. *Heavy Gas Workshop* (Downsview, Ont.: Concord Sci. Corp.)

PUTTOCK, J.S. and COLENBRANDER, G.W. (1985b). Thorney Island data and dispersion modelling. *J. Haz. Materials*, **11**(1/3), 381. See also in McQuaid, J. (1985), *op. cit.*, p. 381

PUTTOCK, J.S., BLACKMORE, D.R. and COLENBRANDER, G.W. (1982). Field experiments on dense gas dispersion. *J. Haz. Materials*, **6**(1/2), 13. See also in Britter, R.E. and Griffiths, R.F., (1982), *op. cit.*, p. 13

PUTTOCK, J.S., COLENBRANDER, G.W. and BLACKMORE, D.R. (1983). Maplin Sands experiments 1980: dispersion results from continuous releases of refrigerated liquid propane. In Hartwig, S. (1983), *op. cit.*, p. 147

PUTTOCK, J.S., COLENBRANDER, G.W. and BLACKMORE, D.R. (1984). Maplin Sands experiments 1980: dispersion results from continuous releases of refrigerated liquid propane and LNG. In Wispelaere, C. (1984), *op. cit.*, p. 353

PUTTOCK, J.S., MCFARLANE, K., PROTHERO, A., REES, F.J., ROBERTS, P.T., WITLOX, H.W.M. and BLEWITT, D.N. (1991). Dispersion models and hydrogen fluoride predictions. *J. Loss Prev. Process Ind.*, 4(1), 16

PUTTOCK, J.S., MCFARLANE, K., PROTHERO, A., ROBERTS, P.T., REES, F.J., WITLOX, H.W.M. and BLEWITT, D.N. (1991). The HGSYSTEM dispersion modelling package development and predictions. *Modeling and Mitigating the Consequences of Accidental Releases*, p. 223

PUYEAR, R.B. (1992). Pick the right material for process hardware. *Chem. Engng*, 99(10), 90

PUYEAR, R.B. and CONLISK, P.J. (1980). Improving fiber-reinforced plastic equipment reliability. In *Loss Prevention*, 13, 62

PUZAK, P.P. and LOSS, F.J. (1979). Metallurgical and mechanical considerations in selection of fracture-safe explosives containment vessel. *J. Pres. Ves. Technol.*, 101, 242

PYBUS, D.S. (1981). Boiler feed water treatment. *Plant Engr*, 25(3), 27

PYE, A.M. (1979). A review of asbestos substitute materials in industrial applications. *J. Haz. Materials*, 3(2), 125

PYE, D. (1964). *The Nature of Design* (New York: Reinhold)

PYLE, I.C. (1987). Designing for safety using ADA packages. In Daniels, B.K. (1987), *op. cit.*

PYLE, I.C. (1991). *Developing Safety Systems: A Guide Using ADA* (Englewood Cliffs, NJ: Prentice-Hall)

PYLE, I.C. (1993). Use of ADA in safety critical systems. In Bennett, P. (1993), *op. cit.*, p. 154

PYMAN, M.A.F. and MITCHELL, F.R. (1982). Identifying major process hazards at the conceptual design stage. *Design 82* (Rugby: Instn Chem. Engrs)

PYMAN, M.A.F. and GJERSTAD, T. (1983). Experience in applying hazard assessment techniques offshore in the Norwegian sector. *Loss Prevention and Safety Promotion 4*, vol. 2, p. B35

PYPER, P.C. and GRAHAM, W.J.H. (1982–83). Analysis of terrorist injuries treated at Craigavon Area hospital, Northern Ireland. *Injury*, 14, 332

PYY, P. (1992). Human factors in scheduled production outages. *Loss Prevention and Safety Promotion 7*, vol. 2, 51–1

QIAN, D.Q. (1990). An improved method for fault location of chemical plants. *Comput. Chem. Engng*, 14, 41

QUAM, G.N. (1963). *Safety Practice for Chemical Laboratories* (Villanova, PA: Villanova Press)

QUANT, J.T., VAN DAM, J., ENGEL, W.F. and WATTIMENA, F. (1963). The Shell chlorine process. *Chem. Engr, Lond.*, 170, CE224

QUANTRILL, T.E. and LIU, Y.A. (1991). *Artificial Intelligence in Chemical Engineering* (San Diego, CA: Academic Press)

QUARANTELLI, E.L. (ed.) (1978). *Disasters Theory and Research* (London: Sage)

QUARANTELLI, E.L. (1984). Chemical disaster preparedness at the local community level. *J. Haz. Materials*, 8, 239

QUARANTELLI, E.L. (1991). Disaster planning for transportation accidents involving hazardous materials. *J. Haz. Materials*, 27, 49

QUARANTELLI, E.L., LAWRENCE, C., TIERNEY, K. and JOHNSON, T. (1979). Initial findings from a study of socio-behavioral

preparations and planning for acute chemical hazard disasters. *J. Haz. Materials*, 3(1), 77

QUARANTELLI, E.L., DYNES, R.R. and WENGER, D.E. (1986). *The Disaster Research Center: its History and Activities*. Disaster Res. Center, Univ. of Delaware, Newark, NJ

QUARTULLI, O.J., FLEMING, J.B. and FINNERAN, J.A. (1968). Best pressure for NH_3 plants. *Hydrocarbon Process.*, 47(11), 153

QUEENER, J.S. (1970). The close relationship between employee safety and public safety in modern chemical plant design. In National Academy of Engineering (1970), *op. cit.*, p. 77

QUEINNEC, C. (1984). *LISP* (London: Macmillan)

QUELCH, J. and CAMERON, I.T. (1993). Dealing with model uncertainty in quantified risk assessment. In Melchers, R.E. and Stewart, M.G. (1993), *op. cit.*, p. 171

QUELCH, J. and CAMERON, I.T. (1994). Uncertainty representation and propagation in quantified risk assessment using fuzzy sets. *J. Loss Prev. Process Ind.*, 7, 463

QUENNE, R. and SIGNORET, J.P. (1977). Calculation of the reliability of nuclear reactor control and protection systems by computer simulation. *First Nat. Conf. on Reliability. NRC3/12.* (Culcheth, Lancashire: UK Atom. Energy Auth.)

QUIGLEY, H.A. (1966). Economics of multiple units. *Chem. Engng*, 73(Aug. 29), 97

QUINLAN, J.R. (1979). Discovering rules by induction from large collections of examples. In Michie, D. (1979), *op. cit.*, p. 168

QUINLAN, J.R. (1983a). Fundamentals of the knowledge engineering problem. In Hayes, J.E. and Michie, D. (1983), *op. cit.*, p. 33

QUINLAN, J.R. (1983b). Learning efficient classification procedures and their application to chess end games. In Michalski, R.S., Carbonell, J.G. and Mitchell, T.M. (1983), *op. cit.*, p. 463

QUINLAN, J.R. (ed.) (1987). *Applications of Expert Systems* (Addison-Wesley)

QUINLAN, J.R., COMPTON, P.J., HORN, K.A. and LAZARUS, L. (1986). Inductive knowledge acquisition: a case study. *Proc. Second Austral. Conf. on Applications of Expert Systems*, Sydney

QUINN, M.E., WEIR, E.D. and HOPPE, T.F. (1984). The design of emergency fail safe systems for the continuous sulfonation of an aromatic compound. *The Protection of Exothermic Reactors and Pressurised Storage Vessels* (Rugby: Instn Chem. Engrs), p. 31

QUINTIERE, J.G. (1975). Some observations on building corridor fires. *Combustion*, 15, 173

QUINTIERE, J.G. (1979). The spread of fire from a compartment: a review. In *Design of Buildings for Fire Safety. ASTM STP 685.* (ed. E.E. Smith and T.Z. Harmathy) (Philadelphia, PA: Am. Soc. Testing Materials), p. 139

QUINTIERE, J.G. (1981). A simplified theory for generalising results for a radiant panel rate of flame spread apparatus. *Fire and Materials*, 5, 52

QUINTIERE, J.G. (1983). *A Simple Correlation for Predicting Temperature in a Room Fire. Rep. NBSIR 83-2712.* Nat. Bureau Standards, Washington, DC

QUINTIERE, J. (1988). Analytical methods for fire safety design. *Fire Technol.*, 24, 333

QUINTIERE, J., MCCAFFREY, B.J. and KASHIWAGI, T. (1978). A scaling study of a corridor subject to a room fire. *Combust. Sci. Technol.*, 18, 1

QUINTON, P.J. (1962). The flow resistance of flame arresters. *Br. Chem. Engng*, 7, 914

QUIRK, W.J. (1985). *Verification and Validation of Real-Time Software* (Berlin: Springer)

QURAIDIS, S.I. (1982). Urea autoclave failure and repair. *Plant/Operations Prog.*, **1**, 159

QURESHI, A.R. (1988). The role of hazard and operability study in risk analysis of major hazard plant. *J. Loss Prev. Process Ind.*, **1**(2), 104

RABALD, E. (1968). *Corrosion Guide* (Amsterdam: Elsevier)

RABINKOV, V.A. (1988). The distribution of flammable gas concentrations in rooms. *Fire Safety J.*, **13**, 211

RACINE, W.J. (1972). Plant designed to protect environment. *Hydrocarbon Process.*, **51**(3), 115

RACKWITZ, R. (1984). Failure rates of general systems including structural components. *Reliab. Engng*, **9**(4), 229

RADANT, S. (1982). Dust explosions in silos. *The Control and Prevention of Dust Explosions* (London: Oyez/IBC), paper 15

RADEMEYER, D.J.E. (1990). Near failure of a 50 bar nitrogen supply tank. *Ammonia Plant Safety*, **30**, p 33

RADIAN CORPORATION (1975). *Control of Hydrocarbon Emissions from Petroleum Liquids. Rep. EPA 600/2-75-042* (Austin, TX)

RADIAN CORPORATION (1979). *Emission Factors and Frequency of Leak Occurrence for Fittings in Refinery Process Units. Rep EPA 600/2-79-044* (Austin, TX)

RADIAN CORPORATION (1986). *Description of the Radian Complex Hazardous Air Release Model (CHARM)* (report). (Austin, TX)

RADIER, H.H. (1939). Flame arresters. *J. Inst. Petrol.*, **25**, 377

RADIUS, N. and DOUWES, C. (1971). *Advances in Static Electricity*, **1**, 240

RADLE, J. (1992). Select the right steam trap. *Chem. Engng Prog.*, **88**(1), 30

RAE, D. (1961). A measurement of the temperature of some frictional sparks. *Combust. Flame*, **5**, 341

RAE, D. (1965). *The Ignition of Gas by the Impact of Brass against Magnesium Anodes and other Parts of a Ship's Tank. Res. Rep. 229.* Safety in Mines Estab., Buxton

RAE, D. (1971). Coal dust explosions in large tubes. In *Shock Tube Research* (ed. J.L. Stollery, A.D. Gaydon and P.R. Owen) (London: Chapman & Hall), p. 1

RAE, D. and THOMPSON, W. (1979). Experiments on prevention and suppression on coal-dust explosions by bromochlorofluoromethane and on prevention by carbon tetrachloride. *Combust. Flame*, **35**, 131

RAE, D., SINGH, B. and DANSON, R. (1964). *The Size and Temperature of a Hot Square in a Cold Surface necessary for the Ignition of Methane. Res. Rep. 224.* Safety in Mines Res. Estab., Buxton

RAES, H. (1977). The influence of a building's construction and fire load on the intensity and duration of a fire. *Fire Prev. Sci. Technol.*, **16**, 4

RAEZER, S.D. (1961). The relationship between burning velocity and space velocity of a spherical combustion wave in a closed spherical chamber. *Combust. Flame*, **5**, 77

RAEZER, S.D. and OLSEN, H.L. (1962). Measurement of laminar flame speeds of ethylene-air and propane-air by the double kernel method. *Combust. Flame*, **6**, 227

RAFAAT, H.M.N. and ABDOUNI, A.H. (1987). Development of an expert system for human reliability analysis. *J. Occup. Accidents*, **9**, 137

RAFFERTY, P. (1980). Voluntary chlorine inhalation: a new form of self abuse. *Br. Med. J.*, **211**, 1178

RAGHAVEN, K.V. (1992). Temperature runaway in fixed bed reactors: online and Offline Checks for Intrinsic Safety. *J. of Loss Prev. Process Ind.* **5**(3), 153–9

RAFTERY, M. (1972). Ignition properties of dust/air mixtures – the development of explosibility tests. *Int. Symp. on Dust and Explosion Risks in Mines and Industry* (Heidelberg: Berufsgenossenschaft der Chem. Ind.), p. 134

RAGHAVEN, K.V. (1992). Temperature runaway in fixed bed reactors: on-line and off-line checks for intrinsic safety. *J. Loss Prev. Process Ind.*, **5**, 153

RAGHAVAN, R., COLES, E. and DIETZ, D. (1991). Cleaning excavated soil using extraction agents: a state-of-the-art review. *J. Haz. Materials*, **26**(1), 81

RAGLAND, K. (1972). Multiple box model for dispersion of air pollutants from aerial sources. *Atmos. Environ.*, **7**, 1017

RAGSDALE, K.H. (1969). Lighting for safety. In Handley, W. (1969), *op. cit.*, p. 321

RAHA, K. and CHHABRA, J.S. (1993). Static charge development and impact sensitivity of high explosives. *J. Haz. Materials*, **34**, 385

RAHEJA, M.M., CHANDRA, S., MISRA, A. and BHARGAVA, R.N. (1992). Explosion in purge gas recovery unit cold box. *Ammonia Plant Safety*, **32**, 209

RAHN, F.J., ADAMANTIADES, A.G., KENTON, J.E. and BRAUN, C. (1984). *A Guide to Nuclear Power Technology* (Chichester: Wiley)

RAIFFA, H. (1968). *Decision Analysis* (Reading, MA: Addison Wesley)

RAIFFA, H. and SCHLAIFER, R. (1961). *Applied Statistical Decision Theory* (Cambridge, MA: Harvard Univ. Press)

RAINE, A.J. (1986). Fuel storage protection. *Fire Surveyor*, **15**(6), 9

RAINS, W.A. (1975). Test on fire protection systems for steel. *Loss Prevention*, **9**, 2

RAINS, W.A. (1977). Fire resistance – how to test for it. *Chem. Engng*, **84**(Dec. 19), 97

RAINS, W.A. (1985). Accelerated ageing tests for evaluating fire proofing materials. *Plant/Operations Prog.*, **4**, 246

RAJ, P.P.K. (1977). Calculation of thermal radiation hazards from LNG fires – a review of the state-of-the-art. *Am. Gas Ass., Transmission Conf.*, p. T.135

RAJ, P.P.K. (1979). A criterion for classifying accidental liquid spills into instantaneous and continuous types. *Combust. Sci. Technol.*, **19**, 251

RAJ, P.K. (1981). Models for cryogenic liquid spill behaviour on land and water. *J. Haz. Materials*, **5**(1/2), 111

RAJ, P.K. (1982). Ammonia. In Bennett, G.F., Feates, F.S. and Wilson, I. (1982), *op. cit.*, pp. 10–34

RAJ, P.K. (1985). Summary of heavy gas spills modeling research. *Heavy Gas Workshop* (Downsview, Ont.: Concord Sci. Corp.), p. 51

RAJ, P.K. (1986). Dispersion of hazardous reactive chemical vapours in the atmosphere. *Proc. 1986 Hazardous Material Spills Conf.*, p. 305

RAJ, P.K. (1990). Chemical behaviour models and computerized hazard assessment tools for use in contingency planning and emergency response. *Int. Workshop on Hazard Assessment and Disaster Mitigation in Petroleum and Chemical Process Industries*, Madras

RAJ, P.K. (1991). Chemical release/spill source models – a review. *Modeling and Mitigating the Consequences of Accidental Releases*, p. 87

RAJ, P. and ATALLAH, S. (1975). Thermal radiation from LNG spill fires. *Advances in Cryogenic Engineering*, **20**, 143

RAJ, P.P.K. and EMMONS, H.W. (1975). On the burning of a large flammable vapor cloud. *Combustion Inst., Joint Tech. Mtg of the Western and Central State Sections* (Pittsburgh, PA)

RAJ, P.K.P. and KALELKAR, A.S. (1973). Fire hazard presented by a spreading burning pool of LNG on water. *Combustion Inst., Western States Section, Fall Mtg*, paper

RAJ, P.K. and KALELKAR, A.S. (1974). *Assessment Models in Support of the Hazard Assessment Handbook. MA: Rep. CG-D-65-74.* (Cambridge, MA: A. D. Little)

RAJ, P.K. and MORRIS, J.A. (1990). Computerized spill hazard evaluation models. *J. Haz. Materials.* **25**(1/2). 77

RAJ, P.K. and REID, R.C. (1978a). Fate of liquid ammonia spilled onto water. *Environ. Sci. Technol.,* **12**, 1422

RAJ, P.P.K. and REID, R.C. (1978b). Underwater release of LNG. *Control of Hazardous Material Spills 4*, p. 432

RAJ, P.K., HAGOPIAN, J. and KALELKAR, A.S. (1974). *Prediction of Hazards of Spills of Anhydrous Ammonia on Water. Rep. NTIS AD 779400* (Cambridge, MA: A. D. Little)

RAJ, P.K., HAGOPIAN, J.H., KALELKAR, A.S. and CECE, J. (1975). Predicting hazards from ammonia spills. *Ammonia Plant Safety,* **17**, 102

RAJ, P.K. *et al.* (1979). LPG spill tests on water. An overview of results. *Am. Gas Ass., Transmission Conf.*, New Orleans, paper

RAJ, P.P.K., MOUSSA, N.A. and ARAVAMUDAN, K. (1979a). *Experiments Involving Pool and Vapour Fires from Spills of Liquefied Natural Gas on Water. Rep. CG-D-55-79* (Cambridge, MA: A. D. Little)

RAJ, P.K., MOUSSA, N.A. and ARAVAMUDAN, K. (1979b). LNG spill fire tests on water − an overview of results. *Am. Gas. Ass., Transmission Conf.*, New Orleans

RAJ, P.K., VENKATARAMANA, M. and MORRIS, J. (1987). A hybrid box−Gaussian dispersion model including considerations of chemical reactions and liquid aerosol effects. *Vapor Cloud Modeling*, p. 453

RAJARATNAM, N. (1976). *Turbulent Jets* (Amsterdam: Elsevier)

RAJU, K.S.N. and RATTAN, V.K. (1979). Heat-pipe construction. *Chem. Engng,* **86**(Dec. 17), 99

RAJU, M.S. and STREHLOW, R.A. (1984). Numerical investigations of non-ideal explosions. *J. Haz. Materials,* **9**(3), **265**

RAK, G.R. (1978). Corrosion monitoring equipment. *Chem. Engng Prog.,* **74**(3), 46

RAKACZKY, J.A. (1975). *The Suppression of Thermal Hazards from Explosions of Munitions: a Literature Survey. BRL Interim Memo. Rep. 377.* Ballistics Res. Lab., Aberdeen Proving Ground, MD,

RAKER, D. and RICE, H. (1988). *Inside Autocad*, 4th ed. (Thousand Oaks, CA: New Riders)

RAKESTRAW, J.A. and HILDREW, B. (1975). Flixborough. *Mag. of Lloyds Register of Shipping*, Sep., 11

RAKITA, P.E., AULTMAN, J.F. and STAPLETON, L. (1990). Handle commercial Grignard reagents safely. *Chem. Engng,* **97**(3), 110

RAKOCZYNSKI, R.W. and LONG, B.A. (1993). Evaluate the economics of secondary containment options for tanks. *Chem. Engng Prog.,* **89**(9), 74

RALEIGH, LORD (1882). *Phil. Mag.,* **14**, 184

RALL, D.F. (1970). Animal models for pharmacological studies. *Proc. Symp. on Animal Models for Biomedical Research III* (New York: Nat Acad. Sci.)

RALL, W. and FROMM, D. (1986). Damage at syngas turbine by excessive steam superheating. *Ammonia Plant Safety,* **26**, 124. See also *Plant/Operations Prog.*, 1986, **5**, 90

RALL, W. and SPAEHN, H. (1985). Rupture of the start-up heater coil in the synthesis loop of the new ammonia plant. *Ammonia Plant Safety,* **25**, 77

RALLIS, C.J., GARFORTH, A.M. and STEINZ, J.A. (1965). *The Determination of Laminar Burning Velocity with Particular Reference to the Constant Volume Method. Part 3, Experimental Procedure and Results. Rep. 26.* Univ. of Witwatersrand Johannesburg

RAMACHANDRAN, G. (1982). Properties of extreme order statistics and their application to fire protection and insurance problems. *Fire Safety J.,* **5**, 59

RAMACHANDRAN, G. (1988). Probabilistic approach to fire risk evaltin. *Fire Technol.,* **24**, 204

RAMACHANDRAN, G. (1990a). Human behaviour in fires − a review of research in the United Kingdom. *Fire Technol.,* **26**, 149

RAMACHANDRAN, G. (1990b). Risk of explosion from trains carrying LPG and other dangerous goods. *Fire Surveyor,* **19**(3), 8

RAMACHANDRAN, G. (1990c). Tradeoffs between fire safety measures: probabilistic evaluation. *Fire Surveyor,* **19**(2), 4

RAMACHANDRAN, G. (1991). Informative fire warning systems. *Fire Technol.,* **27**, 66

RAMADORAI, R. (1980). Explosion in the flare stack separation drum of 600 t ammonia plant during commissioning period. *Loss Prevention and Safety Promotion 3*, vol. 2, p. 161

RAMACKERS, L. (1970). Charging of silica powder during pneumatic transport and neutralisation by inductive countercharge generation. *Advances in Static Electricity,* **1**, 370

RAMAGE, J., SWITHENBY, S. and ROSS, S. (1982) *Electrostatics* (Milton Keynes: Open Univ.)

RAMALHO, R.S. (1978). Principles of activated sludge treatment. *Hydrocarbon Process.,* **57**(10), 112; **57**(11), 275; **57**(12), 147

RAMALHO, R.S. (1979). Design of aerobic treatment units. *Hydrocarbon Process.,* **58**(10), 99; **58**(11), 285

RAMAN, J.R., STEPHENS, G.K. and HADDAD, S.G. (1986). Major industrial hazards: risk assessment, public perception and government policies. *CHEMECA 86*, p. 281

RAMAN, R. (1985). *Chemical Process Computations* (Amsterdam: Elsevier)

RAMAN, R. (1993). Source term modelling of chlorine release inside buildings. In Melchers, R.E. and Stewart, M.G. (1993), *op. cit.*, p. 69

RAMAN, R. and GROSSMAN, I.E. (1991). Relation between MILP modelling and logical inference for chemical process synthesis. *Comput. Chem. Engng,* **15**, 73

RAMANATHAN, P., KANNAN, S. and DAVIS, J.F. (1993). Use knowledge-based system programming toolkits to improve plant troubleshooting. *Chem. Engng Prog.,* **89**(6), 75

RAMAZZINI, B. (1713). *De Morbis Artificum* (English translation by W.C. Wright) (New York: Hafner)

RAMBOUSEK (1913). *Industrial Poisoning* (London)

RAMESH, T.S., SHUM, S.K. and DAVIS, J.F. (1988). A structured framework for efficient problem solving in diagnostic expert systems. *Comput. Chem. Engng,* **12**, 891

RAMESH, T.S., DAVIS, J.F. and SCHWENZER, G.M. (1992). Knowledge-based diagnostic systems for continuous process operations based upon the task framework. *Comput. Chem. Engng,* **16**, 109

RAMEY, J.T. (1970). Providing for public safety in the nuclear industry − the engineering approach. In National Academy of Engineering (1970), *op. cit.*, p. 96

RAMIREZ, W.F. (1976). *Process Simulation* (Lexington, MA: D.C. Heath)

RAMIREZ, W.F. (1989). *Computational Methods for Process Simulation* (Oxford: Butterworth-Heinemann)

RAMIREZ, W.F. and TURNER, B.A. (1969). The dynamic modelling, stability and control of a continuous stirred tank reactor. *AIChE J.*, **15**, 853

RAMIREZ-DOMINGUEZ, E. and BANARES-ALCANTARA, R. (1993). A representation of engineering design intent. *Proc. Sixth Int. Conf. on Industrial and Engineering Applications of Artificial Intelligence and Expert Systems* (Reading: MA: Gordon & Breach), p. 229

RAMMAH, N.A. (1985). Emergency shutdown pressure surge predictions limit LPG loading. *Oil Gas J.*, **83**(Nov. 25), 87

RAMOHALLI, K., JONES, S. and BASHAR, R. (1984). Some experimental data on lift-off characteristics of turbulent jet flames. *Combust. Sci. Technol.*, **36**, 199

RAMSAT, A. and BARRETT, R. (1987). *AI in Practice: Examples in POP-11* (Chichester: Ellis Horwood)

RAMSAY, C.G., SYLVESTER-EVANS, R. and ENGLISH, M.A. (1982). Siting and layout of major hazardous installations. *The Assessment of Major Hazards* (Rugby: Instn Chem. Engrs), p. 335

RAMSAY, C.G., BOLSOVER, A.J., JONES, R.J. and MEDLAND, W.G. (1994). Quantitative risk assessment applied to offshore process installations. Challenges after the Piper Alpha disaster. *J. Loss Prev. Process Ind.*, **7**, 317

RAMSAY, C.W. (1990). Plant modifications: maintain your mechanical integrity. *Plant/Operations Prog.*, **9**, 117

RAMSDELL, J.V. (1990). Diffusion in building wakes for ground level releases. *Atmos. Environ*, **24B**, 377; **26B**, 515

RAMSDELL, J.V. and HINDS, W.T. (1971). Concentration fluctuations and peak-to-mean concentration ratios in plumes from a ground-level continuous point source. *Atmos. Environ.*, **5**, 483

RAMSDELL, J.V., GLANTZ, C.S. and KERNS, R.E. (1985). Hanford atmospheric dispersion data: 1959–1974. *Atmos. Environ.*, **19**, 83

RAMSDEN, J.H. (1972). Pumps: inspection, shipping and installation. *Hydrocarbon Process.*, **51**(6), 105

RAMSDEN, J.H. (1978). How to choose and install mechanical seals. *Chem. Engng*, **85**(Oct. 9), 97

RAMSDEN, J.M. (1985). World airline safety audit. *Flight Int.*, Jan. 16, 29

RAMSEY, C.E. (1972). Methods for prevention of spills in the truck transportation of hazardous materials. *Control of Hazardous Material Spills 1*, p. 49

RAMSEY, J.D. (1973). Identification of contributory factors in occupational injury and illness. In Widner, J.T. (1973), *op. cit.*, p. 328

RAMSEY, M.E., BRENNAN, R.B. and DANAHER, W.J. (1993). Computer assisted management of hazardous substances. *Health, Safety and Loss Prevention*, p. 504

RAMSEY, W.G. (ed.) (1974). The V-weapons. *After the Battle*, **6**, 2

RAMSEY, W.G. (ed.) (1983). The atomic bomb. *After the Battle*, **41**, 1

RAMSEY, W.G. (ed.) (1987). *The Blitz, Then and Now*, 1987, vol. 1; 1988, vol. 2; 1990, vol. 3 (London: Battle of Britain Prints Int.)

RAMSEY, W.D. and ZOLLER, G.C. (1976). How the design of shafts, seals affects agitator performance. *Chem. Engng*, **83**(Aug. 30), 101

RAMSHAW, C. (1983). 'HiGee' distillation. An example of process intensification. *Chem. Engr, Lond.*, **389**, 13

RAMSHAW, C. (1985). Process intensification: a game for *n* players. *Chem. Engr, Lond.*, **416**, 30

RAMSHAW, C. (1987). Opportunities for exploiting centrifugal fields. *Chem. Engr, Lond.*, **437**, 17

RAMSHAW, C. (1999). Process intensification and green chemistry. *Green Chem.* (Feb.), G15-G17

RAMSKILL, P.K. (1988). A description of the 'ENGULF' computer codes – codes to model the thermal response of an LPG tank either fully or partially engulfed by fire. *J. Haz. Materials*, **20**, 177

RANA, S.A.Z. (1985). Understand multistage compressor antisurge control. *Hydrocarbon Process.*, **64**(3), 69

RANADE, S.M., ROBERT, W.E. and ZAPATE-SUAREZ, A. (1988). Evaluate electric drive applications quickly. *Hydrocarbon Process.*, **67**(10), 41

RANBY, E.B. and HEWITT, F. (1982). Emergency planning. In Green, A.E. (1982), *op. cit.*, p. 217

RAND, C. (1987). Operations support using expert systems. *Process Engng*, Jan., 28

RANDALL, P.N. and GINSBURGH, I. (1961). Bursting of tubular specimens by gaseous detonation. *J. Basic Engng*, **83**, 519

RANDALL, P.N., BLAND, J., DUDLEY, W.M. and JACOBS, R.B. (1957). Effects of gaseous detonations upon vessels and piping. *Chem. Engng Prog.*, **53**(12), 574

RANDERSON, D. (1979). Review panel on sigma computations. *Bull. Am. Met. Soc.*, **60**, 682

RANDHAVA, R. and CALDERONE, S. (1985). Hazard analysis of supercritical extraction. *Chem. Engng Prog.*, **81**(6), 59

RANDHAVA, R., LO, R.N. and KRONSEDER, J.G. (1982). Modular pilot-plant technology. *Chem. Engng Prog.*, **78**(11), 76

RANKIN, A.D. (1989). *Piper Alpha Public Inquiry. Transcript, Days 50–52*

RANKIN, J.P. (1984). Sneak circuits – a class of random, unrepeatable glitch. *Plant/Operations Prog.*, **3**, 175

RANNEY, M.W. (1979). *Offshore Oil Technology* (Park Ridge, NJ: Noyes Data Corp.)

RANNILL, J. (1969). Shushing remery noises. *Chem. Engng*, **76**(Mar. 24), 84

RANNILL, J. (1971). Co-operative fire fighting. *Chem. Engng*, **78**(May 31), 84

RANZ, W.E. and MARSHALL, W.R. (1952). Evaporation from drops. *Chem. Engng Prog.*, **48**, 141, 173

RAO, K.V.K. (1982). Head losses in fittings. *Chem. Engng, Albany*, **89**(Feb. 8), 132

RAO, K.V.K. (1983). Physical properties of common heat transfer fluids. *Chem. Eng*, **90**(Sep. 19), 89

RAO, K.V.L. and LEFEBVRE, A.H. (1976). Minimum ignition energies in flowing kerosine-air mixtures. *Combust. Flame*, **27**, 1

RAO, P.V. and RAJ, P.K. (1990). Expert system interface to hazard prediction models. *J. Haz. Materials*, **25**(1/2), 93

RAO, R.B.K.N. (1991). Comadem: condition monitoring and diagnostic engineering management *Plant Engr*, **35**(Sep./Oct.), 28

RAO, S., WILTZEN, R.C.A. and JACOBS, W. (1986). Investigation of damage and repair of a 1000 MTD horizontal ammonia converter. *Ammonia Plant Safety*, **26**, 128. See also *Plant/Operations Prog.*, 1986, **5**, 116

RAO, S.S. (1992). *Reliability Based Design* (New York: McGraw-Hill)

RAPAROTTE, J.L. and BRZUSTOWSKI, T.A. (1986). A new approach to modelling radiation from open flames. *Combust. Sci. Technol.*, **49**, 251

RAPEAN, J.C., PEARSON, D.L. and SELLO, H. (1959). A test for hazardous chemical decomposition. *Ind. Engng Chem.*, **51**(2), 77A

RAPHAEL, B. (1968). SIR, a computer program for semantic information retrieval. In Minsky, M. (1968), *op. cit.*, p. 33

RAPHAEL, B. (1976). *The Thinking Computer: Mind inside Matter* (San Francisco, CA: W.H. Freeman)

RAPOPORT, A. (1961). Some comments on accident research. *Behavioural Approaches to Accident Research* (New York: Ass for the Aid of Crippled Children)

RAPP, R.W. (1984). Good tank seal design pays off. *Oil Gas J.*, **82**(May, 7), 170

RAPPAPORT, M.W. (1975). Information services and assistance for transhippers of hazardols materials. *Transport of Hazardous Cargoes*, 4

RAPPINI, G.E. *et al.* (1977). The role of terotechnology in the industrial loss prevention. *Loss Prevention and Safety Promotion 2*, p. 63

RASBASH, D.J. (1956). Properties of fires of liquids. *Fuel*, **35**, 94

RASBASH, D.J. (1960). Mechanisms of extinction of liquid fires with water spray. *Combust. Flame*, **4**(3), 223

RASBASH, D.J. (1962). The extinction of fires by water sprays. *Fire Res. Abstr. Rev.*, **4**(122), 28

RASBASH, D.J. (1968). Liquefied gases as fire-fighting agents. *Chem. Engr, Lond.*, **217**, CE89

RASBASH, D.J. (1969a). Review of latest developments in fire protection. *Instn Fire Engrs Quart.*, **29**(7), 261

RASBASH D.J. (1969b). Smoke and toxic gases. *Rev. Tech. Gen.*, **87**(10), 25

RASBASH, D.J. (1969c). The relief of gas and vapour explosions in domestic structures. *Struct. Engr*, **47**(10), 404

RASBASH, D.J. (1970a). *Evidence of Explosive Effects Accompanying the Propagation of Flame over a Spillage of Flammable Liquid in Air. Fire Res. Memo 24*. Fire Res. Station, Borehamwood

RASBASH, D.J. (1970b). Fire protection engineering with particular reference to chemical engineering. *Chem. Engr, Lond.*, **243**, CE385

RASBASH, D.J. (1972). The role of detection in fire protection. *Rev. Tech. Feu*, **13**(20), 23

RASBASH, D.J. (1974a). The fire problem. *Building Res. Estab., Symp. on Cost Effectiveness and Fire Protection, Ref. B448/74* (Watford)

RASBASH, D.J. (1974b). Present and future design philosophy for fire and explosion hazards in the chemical industry. *Fire Prev. Sci. Technol.*, **8**, 16

RASBASH, D.J. (1975a). Interpretation of statistics on the performance of sprinkler systems. *Control*, May, 63

RASHBASH, D.J. (1975b). Relevance of firepoint theory to the assessment of fire behaviour of combustible materials. *Int. Symp. on Fire Safety of Combustible Materials* (Edinburgh: Dept of Fire Safety Engng, Univ. of Edinburgh)

RASBASH, D.J. (1976a). Automatic installations for fire and explosion protection. *Course on Fire and Explosion Safety in the Chemical Industry*. Dept of Fire Safety Engng Univ. of Edinburgh

RASBASH, D.J. (1976b). Explosions in enclosures. *Course on Fire and Explosion Safety in the Chemical Industry*. Dept of Fire Safety Engng, Univ. of Edinburgh

RASBASH, D.J. (1976c). A flame extinction criterion for fire spread. *Combust. Flame*, **26**, 411

RASBASH, D.J. (1976d). Gas and vapour explosions in enclosed spaces. *Course on Fire and Explosion Safety in the Chemical Industry*. Dept of Fire Safety Engng, Univ. of Edinburgh

RASBASH, D.J. (1977a). The definition and evaluation of fire safety. *Fire Prevention Sci. Technol.*, **16**, 17

RASBASH, D.J. (1977b). *Fire and Explosion Hazard to Dalgety Bay and Aberdour Associated witll the Proposed Fife NGL Plant* (report). Dept of Fire Safety Engng, Univ. of Edinburgh

RASBASH, D.J. (1979/80). Review of Explosion and fire hazard for liquefied petroleum gas. *Fire Safety J.*, **2**(4), 223

RASBASH, D.J. (1980a). Analytical approach to fire safety. *Fire Surveyor*, **9**(4), 20

RASBASH, D.J. (1980b). Criteria for decisions on acceptability of major fire and explosion hazards with particular reference to the chemical and fuel industries. *Chemical Process Hazards*, **VII**, p. 25

RASBASH, D.J. (1980c). Review of explosion and fire hazard for liquefied petroleum gas. *Fire Safety J.*, **2**(4), 223

RASBASH, D.J. (1982a). Canvey Inquiry (letter). *Chem. Engr, Lond.*, **376**, 27

RASBASH, D.J. (1982b). Risk and hazard (letter). *Chem. Engr, Lond.*, **381**, 253

RASBASH, D.J. (1985). Criteria of acceptability for use with quantitative approaches to fire safety. *Fire Safety J.*, **8**, 141

RASBASH, D.J. (1986a). Credibility and truth: a view of risk appraisal. *Fire Safety J.*, **11**, 243

RASBASH, D.J. (1986b). Quantification of explosion parameters for combustible fuel-air mixtures. *Fire Safety J.*, **11**(1), 113

RASBASH, D.J. (1991). Major fire disasters involving flashover. *Fire Safety J.*, **17**, 85

RASBASH, D.J. and DRYSDALE, D.D. (1982). Fundamentals of smoke production. *Fire Safety J.*, **5**, 77

RASBASH, D.J. and LANGFORD, B. (1968). Burning of wood in atmospheres of reduced oxygen concentration. *Combust. Flame*, **12**, 33

RASBASH, D.J. and ROGOWSKI, Z.W. (1957). Extinction of fires in liquids by cooling with water sprays. *Combust. Flame*, **1**, 453

RASBASH, D.J. and ROGOWSKI, Z.W. (1960a). Gaseous explosion in vented ducts. *Combust Flame*, **4**, 301

RASBASH, D.J. and ROGOWSKI, Z.W. (1960b). Relief of explosions in duct systems. *Chemical Process Hazards*, **I**, p. 58

RASBASH, D.J. and STARK, G.W.V. (1960). Extinction of fires in burning liquids with water sprays from hand lines. *Instn Fire Engrs Quart.*, **20**(3), 55

RASBASH, D.J. and STARK, G.W.V. (1962). Some aerodynamic properties of sprays. *Chem. Engr, Lond.*, Dec., A83

RASBASH, D.J., ROGOWSKI, Z.E. and STARK, G.W.V. (1956). Properties of fires in liquids. *Fuel*, **35**, 94

RASBASH, D.J., ROGOWSKI, Z.W. and STARK, G.W.V. (1960). Mechanisms of extinction of liquid fires with water sprays. *Combust. Flame*, **4**, 223

RASBASH, D.J., DRYSDALE, D.D. and KEMP, N. (1976). Design of an explosion relief system for a building handling liquefied fuel gases. *Process Industry Hazards*, p. 145

RASBASH, D.J. DRYSDALE, D.D. and KEMP, N. (1976e). Design of an explosion relief system for a building handling liquefied fuel gases, *I. Chem. E. Symp.* Ser No. 47

RASBASH, D.J., DRYSDALE, D.D. and DEEPAK, D. (1986). Critical heat and mass transfer at pilot ignition and extinction of a material. *Fire Safety J.*, **10**(1), 1

RASE, H.F. (1963). *Piping Design for Process Plants* (New York: Wiley)

RASE, H.F. (1977). *Chemical Reactor Design for Process Plants*, vols 1 and 2 (New York: Wiley)

RASE, H.F. (1990). *Fixed-Bed Reactor Design and Diagnostics* (Oxford: Butterworth-Heinemann)

RASE, H.F. and BARROW, M.H. (1957). *Project Engineering of Process Plants* (New York: Wiley)

RASHIDI, F. and MICHELS, H.J. (1992). Marginal detonation of aliphatic natural gas constituents with O_2N_2. *Combust. Flame*, **91**, 373

RASMUSON, D.M. and MARSHALL, N.H. (1978). FATRAM — a core efficient cut set algorithm. *IEEE Trans. Reliab.*, **R-27**, 250

RASMUSON, D.M., WILSON, J.R. and BURDICK, G.J. (1979). *Design Optimisation of a Prototype Large Breeder Reactor Safety System. Rep. CONF 790816-28.* Idaho Nat. Energy Lab., Idaho Falls, ID

RASMUSON, D.M. *et al.* (1978). *COMCAN II — a Computer Program for Common Cause Failure Analysis. Rep. TREE-1289.* EG&G Idaho Inc., Idaho Falls, ID

RASMUSSEN, B. (1988). Occurrence and impact of unwanted chemical reactions. *J. Loss Prev. Process Ind.*, **1**(2), 92

RASMUSSEN, B. and SMITH-HANSEN, L. (1989). Cost-effective and uniform procedure for performing risk analysis of similar chemical plants. *Loss Prevention and Safety Promotion 6*, p. 66.1

RASMUSSEN, E.J. (1975). Alarm and shutdown devices protect process equipment. *Chem. Engng*, **82**(May 12), 74

RASMUSSEN, J. (1968a). *Characteristics of the Operator, Automation Equipment and Designer in Plant Automation. Rep. Risö-M-808.* Atom. Energy Comm., Res. Estab., Risö, Denmark,

RASMUSSEN, J. (1968b). *On the Communication between Operators and Instrumentation in Automatic Process Plant. Rep. Risö* Atom. Energy Comm., Res. Estab., Risö, Denmark. See also Edwards, E. and Lees, F.P. (1974), *op. cit.*, p. 222

RASMUSSEN, J. (1968c). *On the Reliability of Process Plants and Instrumentation Systems.* Atom. Energy Comm., Res. Estab., Risö, Denmark

RASMUSSEN, J. (1969). Man—machine communication in the light of accident records. *Man—Machine Systems, IEEE Conf. Rec., 69C58-MMS*

RASMUSSEN, J. (1971). Man as information receiver in complex tasks. *Displays* (London: Instn. Elec. Engrs)

RASMUSSEN, J. (1973). *The Role of the Man—Machine Interface in Systems Reliability. Rep. Risö-M-1673.* Atom. Energy Comm., Res. Estab., Risö, Denmark

RASMUSSEN, J. (1974). *The Human Data Processor as a System Component, Bits and Pieces of a Model. Rep. Risö-M-1722.* Atom. Energy Comm., Res. Estab., Risö, Denmark

RASMUSSEN, J. (1976a). Outlines of a hybrid model of the process plant operator. In Sheridan, T.R. and Johanssen, G. (1976), *op. cit.*

RASMUSSEN, J. (1976b). The role of the man—machine interface in systems reliability. In Henley, E.J. and Lynn, J.W. (1976), *op. cit.*, p. 315

RASMUSSEN, J. (1977). Man as a system component. *Man—Computer Research* (ed. H. Smith and T. Green) (New York: Academic Press)

RASMUSSEN, J. (1978). Human reliability. *NATO Advanced Study Int. on Synthesis and Analysis Methods for Safety and Reliability*, Urbino, Italy

RASMUSSEN, J. (1980a). Notes on human error analysis and prediction. In Apostolakis, G., Garribba, S. and Volta, G. (1980), *op. cit.*, p. 357

RASMUSSEN, J. (1980b). What can be learned from human error reports? In Duncan, K.D., Gruneberg, M.M. and Willis, D. (1980), *op. cit.*, p. 97

RASMUSSEN, J. (1981a). Human systems design criteria. *Comput. Ind.*, **2**, 297

RASMUSSEN, J. (1981b). Models of mental strategies in process plant diagnosis. In Rasmussen, J. and Rouse, W.B. (1981), *op. cit.*, p. 241

RASMUSSEN, J. (1982a). Human errors. A taxonomy for describing human malfunction in industrial installations. *J. Occup. Accid.*, **4**(2/4), 311

RASMUSSEN, J. (1982b). Human reliability in risk analysis. In Green, A.E. (1982), *op.cit.*, p. 143

RASMUSSEN, J. (1983). Skills, rules, knowledge: signals, signs and symbols and other distinctions in human performance models. *IEEE Trans. Syst. Man Cybern.*, **SMC-13**, 257

RASMUSSEN, J. (1985). Trends in human reliability analysis. *Ergonomics*, **28**, 1185

RASMUSSEN, J. (1986). *Information Processing and Human—Machine Interaction* (Amsterdam: Elsevier)

RASMUSSEN, J. (1987). Approaches to the control of the effects of human error on chemical plant safety. *Preventing Major Chemical Accidents*, p. 6.1

RASMUSSEN, J. (1988). Safety control and risk management: topics for cross-disciplinary research and development. *Preventing Major Chemical and Related Process Accidents*, p. 523

RASMUSSEN, J. (1990). Human error and the problem of causality in accidents. *Phil. Trans. R. Soc. Lond., Ser. B*, **327**, 449

RASMUSSEN, J. (1991). Safety control: some basic distinctions and research issues in high hazard, low risk operation. *Workshop on Risk Management*, Bad Homburg, paper

RASMUSSEN, J., DUNCAN, K. and LEPLAT, J. (eds) (1987). *New Technology and Human Error* (Chichester: Wiley)

RASMUSSEN, J. and GOODSTEIN, L.P. (1972). *Experiments on Data Presentation to Process Operators in Diagnostic Task. Rep. Risö-M-256.* Atom. Energy Comm., Res. Estab., Risö, Denmark

RASMUSSEN, J. and JENSEN, A. (1973). *A Study of Mental Procedures in Electronic Trouble-Shooting. Rep. Risö-M-1582.* Atom. Energy Comm., Res. Estab., Risö, Denmark

RASMUSSEN, J. and JENSEN, A. (1974). Mental procedures in real-life tasks: a case study of electronic trouble shooting. *Ergonomics*, **17**, 293

RASMUSSEN, J. and LIND, M. (1982). A model of human decision making in complex systems and its use for design of system control strategies. *Proc. Am. Control Conf.*, p. 270

RASMUSSEN, J. and PEDERSEN, O.M. (1982). *Formalized Search Strategies for Human Risk Contributions: a Framework for Further Development. Rep. Risö-M-2351.* Risö Nat. Lab., Risö, Denmark

RASMUSSEN, J. and PEDERSEN, O.M. (1983). Human factors in probabilistic risk analysis and in risk management. *Instn Chem. Engrs, Conf. on Human Reliability in the Process Control Centre* (Rugby)

RASMUSSEN, J. and ROUSE, W.B. (eds) (1981). *Human Detection and Diagnosis of System Failures* (New York: Plenum Press)

RASMUSSEN, J. and TAYLOR, J.R. (1976). *Notes on Human Factors Problems in Process Plant Reliability and Safety Prediction. Rep. Risö-M-1894.* Atom. Energy Comm., Res. Estab., Risö, Denmark

RASMUSSEN, K. and GOW, H.B.F. (1992). The importance of information on industrial risk: a new documentation centre. *J. Haz. Materials*, **30**, 355

RASMUSSEN, N.C. (1974). The AEC study on the estimation of risks to the public from potential accidents in nuclear power plants. *Nucl. Safety*, **15**, 375

RASMUSSEN, N.C. (1981). Methods of hazard analysis and nuclear safety engineering. *Ann. N.Y. Acad. Sci*, **365**, 20

RASSMAN, F.H. (1971). How to specify centrifugal compressors. Design, specifications and accessories. *Hydrocarbon Process.*, **50**(10), 72

RASSOKHIN, N.G., KUZEVANOV, V.S. and TSIKLAURI, G.V. (1980). Calculation of unsteady two-phase single-component discharge from a rupture (case of loss of coolant accident in a power plant). *Heat Transfer – Soviet Research*, **12**(6), 97

RATA, J.M. (1993). Standardisation efforts world-wide. In Bennett, P. (1993), *op. cit.*, p. 56

RATCLIFF, M.A., PETERSEN, R.L., SCHATZ, K.W. and HESKESTAD, G. (1993). Physical modeling of water sprays – comparison against field operations. *Process Safety Prog.*, **12**, 143

RATCLIFF, T.A. (1975). How NIOSH handles quality control. *Hydrocarbon Process.*, **54**(12), 157

RATCLIFFE, K.B. (1993). STATAS: development of an HSE audit scheme for loss of containment incidents (poster item). *Process Safety Management*, p. 533

RATHBONE, T. (1939). Vibration tolerance. *Power Plant Engng*, Nov.

RATHBONE, T.C. (1963). A proposal for standard vibration limits. *Production Engng*, **34**(Mar.), 68

RATTAN, I.S. and PATHAK, V.K. (1985). Start-up of centrifugal pumps in flashing or cryogenic liquid service. *Chem. Engng*, **92**(Apr. 1), 95

RAU, J.G. (1970). *Optimization and Probability in Systems Engineering* (New York: van Nostrand Reinhold)

RAUPACH, M.R., THOM, A.S. and EDWARDS, I. (1980). A wind tunnel study of turbulent flow close to regularly arrayed surfaces. *Boundary Layer Met.*, **18**, 373

RAUSCH, A.H., CHI, K.TSAO and ROWLEY, R.M. (1977). *Third Stage Development of the Vulnerability Model. A Simulation System for Assessing Damage Resulting from Marine Spills. Rep. CG-D-5-78.* Enviro Control Inc., Rockville, MD

RAUSCH, A.H., EISENBERG, N.A. and LYNCH, C.J. (1977). *Continuing Develoment of the Vulnerability Model. A Simulation System for Assessing Damage Resulting from Marine Spills. Rep. CG-D-53-77.* Enviro Control Inc., Rockville, MD

RAUSCH, D.A. (1986). Operating discipline: a key to safety. *Chem. Engng Prog.*, **82**(6), 13

RAUSCH, D.A. (1990). Product stewardship: a supplied/customer partnership in safety. *Plant/Operations Prog.*, **9**, 154

RAVETZ, J.R. (1971). *Scientific Knowledge and Its Social Problems* (Oxford: Clarendon Press)

RAVETZ, J.R. (1977a). Turning points in the relations of science with society. *Sci. Publ. Policy*, **4**, 445

RAVETZ, J.R. (1977b). The use of mathematics in the assessment of risk. In Council for Science and Society (1977), *op. cit.*, p. 99

RAVETZ, J.R. (1979). Public perceptions of acceptable risks. *Sci. Public Policy*, **6**, 298

RAVINDRA, M.K. (1992). Seismic assessment of chemical facilities under California: Risk Management and Prevention Program. *Hazard Identification and Risk Analysis, Human Factors and Human Reliability in Process Safety*, p. 11

RAWILL, J. (1971). Co-operative fire fighting. *Chem. Engng*, **78**(May 31), 84

RAWLINS, J.S.P. (1977–78). Physical and pathophysiological effects of blast. *Injury*, **9**, 313

RAWLS, J. (1971). *Theory of Justice* (Cambridge, MA: Harvard Univ. Press)

RAWSON, H. (1980). *Properties and Applications of Glass* (Amsterdam: Elsevier)

RAWSON, K. (ed.) (1994). *Carriage of Bulk Oil and Chemicals at Sea* (Rugby: Instn Chem. Engrs)

RAWSON, K.J. and TUPPER, E.C. (1968). *Basic Ship Theory;* 1976, 2nd ed.; 1983, 3rd ed. (London: Longmans); 1994, 4th ed. (Harlow: Longmans)

RAWSON, R.W. (1978). The HPI can prevent cancer. *Hydrocarbon Process.*, **57**(6), 193

RAWSTHORNE, K.H. and WILLIAMS, J.A. (1961). Bulk handling and storage of hydrogen peroxide. *Chem. Engr, Lond.*, **156**, A71

RAY, B. (1977). Flame sensing with ultraviolet radiation detectors. *Fire Prev. Sci. Technol.*, **17**, 17

RAY, B. (1978). Detection of explosions. *Phys. Technol.*, **9**, 235

RAY, G.C., SPINEK, R.A. and STOBAUGH, R.B. (1970). Ethylene oxide: how, where, who – future. *Hydrocarbon Process.*, **49**(10), 105

RAY, J., CARY, R. and BELGER, P. (1992). Fast-track DCS retrofitting. *Hydrocarbon Process.*, **71**(1), 45

RAY, S.R. and GLASSMAN, I. (1983). The detailed processes involved in flame spread over solid fuels. *Combust. Sci. Technol.*, **32**, 33

RAY, M.S. and JOHNSTON, D.W. (1990). *Chemical Engineering Design Project: A Case Study Approach* (New York: Gordon & Breach)

RAYLEIGH, LORD (1932). On the instability of jets. Cited in Lamb, H. (1932), *op. cit.*, pp. 472–473

RAYMENT, P. (1986). Operator interface in batch process control. *Chem. Ind.*, May 5, 315

RAYMUS, G.J. (1973). Evaluating the options for packaging chemical products. *Chem. Engng*, **80**(Oct. 8), 67

RAYNESFORD, J.E. (1975). Use dynamic absorbers to reduce vibration. *Hydrocarbon Process.*, **54**(4), 167

READ, J. (1942). *Explosives* (Harmondsworth: Penguin)

READ, P.P. (1993). *Ablaze* (London: Secker)

READINGS, C.J., HAUGEN, D.A. and KAIMAL, J.C. (1974). The 1973 Minnesota atmospheric boundary layer experiment. *Weather*, **29**, 309

READY, D.F. and SCHWAB, R.F. (1981). Incinerator problems and how to prevent them. *Loss Prevention* 14, p. 66. See also *Chem. Engng Prog.*, 1980, **76**(10), 52

REALE, M.J. and YOUNG, J.A. (1987). Precautionary labels and material safety data sheets. In Young, J.A, (1987a), *op. cit.*, p. 7

REARICK, J.S. (1969). How to design pressure relief systems. *Hydrocarbon Process.*, **48**(8), 104; **48**(9), 161

REASON, J.T. (1977). Skill and error in everyday life. *Adult Learning* (ed. M.J.A. Howe) (London: Wiley), p. 21

REASON, J.T. (1979). Actions not as planned: the price of automation. *Aspects of Consciousness*, vol. 1 (ed. G. Underwood and R. Stevens) (London: Academic Press), p. 67

REASON, J. (1986). Recurrent errors in process environments: some implications for the design of intelligent support systems. In Hollnagel, E. *et al.* (1986), *op. cit.*, p. 255

REASON, J. (1987a). The Chernobyl errors. *Bull. Br. Psychol. Soc.*, **40**, 201

REASON, J. (1987b). A framework for classifying errors. In Rasmussen, J., Duncan, K. and Leplat, J. (1987), *op. cit.*, p. 5

REASON, J. (1987c). The psychology of mistakes: a brief review of planning failures. In Rasmussen, J., Duncan, K. and Leplat, J. (1987), *op. cit.*, p. 45

REASON, J. (1990). *Human Error* (Cambridge: Cambridge Univ. Press)

REASON, J. (1991). Too little and too late: a commentary on accident and incident reporting schemes. In Van der Schaaf, T.W., Lucas, D.A. and Hale A.R. (1991), *op. cit.*, p. 9

REASON – DECISION SYSTEMS INC., LONGVIEW, TX, www.rootcause.com

REASON, J.T., MANSTEAD, A., STRADLING, S., BAXTER, J. and CAMPBELL, K. (1990). Errors and violations on the roads: a real distinction? *Hum. Factors*, **33**, 1315

REAY, D. (1977). *User Guide to Fire and Explosion Hazards in Drying of Particulate Materials* (Rugby: Instn Chem. Engrs)

REBENSTORF, M.A. (1961). Designing a high pressure laboratory. *Ind. Engng. Chem.*, **32**(1), 40A

REBOH, R. (1980). Using a matcher to make an expert consultation system behave intelligently. *Proc. Am. Ass. Artificial Intelligence – 80*, p. 231

REBOH, R. (1981). *Knowledge Engineering Techniques and Tools in the Prospector Environment. Tech. Note 243.* Stanford Res. Inst., Stanford, CA

REBSCH (1976). Decomposition of the aqueous solution of an organic nitrate. *Prevention of Occupational Risks*, **3**, p. 485

REBSDAT, S. and MAYER, D. (1987). Ethylene oxide. *Ullmanns Encyclopaedia of Chemical Technology*, rev. ed., vol. A10 (Weinheim: VCH), p. 117

RECHT, J.L. (1965). Systems safety analysis: an introduction. *Nat. Safety News*, Dec., 37

RECHT, J.L. (1966a). Systems safety analysis: error rates and costs. *Nat. Safety News*, Jun. 26

RECHT, J.L. (1966b). Systems safety analysis: failure mode and effect. *Nat. Safety News*, Feb., 24

RECHT, J.L. (1966c). Systems safety analysis: the fault tree. *Nat. Safety News, Apr. 37*

RECHT, R.F. (1971). Containing ballistic fragments. *Engineering Solids under Pressure* (ed. H.L.D. Pugh) (London: Instn. Mech. Engrs)

RECIO, M. et al. (1979). Chemical makers feel full impact of TSCA. *Chem. Engng*, **86**(Jul. 2), 36

RECIO, M. and RHEIN, R. (1985). EPA's air toxics policy arouses opposition. *Chem. Engng*, **92**(Jul. 8), 29

RECORD, F.A., BUBENICK, D.V. and KINDYA, R.J. (1982). *Acid Rain Information Book* (Park Ridge, NJ: Noyes Data Corp.)

REDDEN, P.A. (1987). Using audiovisual materials in safety training. In Young, J.A, (1987a), *op. cit.*, p. 323

REDDI, S.V. (1992). Thermal insulation safety. *Hydrocarbon Process.*, **71**(10), 122-C; **71**(11), 100-C

REDDING, C. (1984). Seveso (letter). *Chem. Engr, Lond.*, **407**, 3

REDDING, J.H. (1980). Can a computer reduce your maintenance? *Hydrocarbon Process.*, **59**(1), 79

REDDING, R.J. (1969). Intrinsic safety: A European view. *Chem. Engng*, **76**(Jun. 30), 137

REDDING, R.J. (1971a). *Intrinsic Safety* (London: McGraw-Hill)

REDDING, R.J. (1971b). The intrinsic safety of instrument signals and power supplies by the zener barrier method. *Electrical Safety*, **1**, p. 85

REDDING, R.J. (1972). Intrinsic safety and its significance to instrumentation. *Instrum. Technol.*, **19**(10), 33

REDDING, R.J. (1975). The planning and monitoring of the electrical safety of large instrumentation projects. *Electrical Safety*, **2**, p. 80

REDDING, R.J. (1977). Reliability criteria. *Measmnt Control*, **10**, 201

REDDY, K.V. and HUSAN, A. (1982). Modeling and simulation of an ammonia synthesis loop. *Ind. Engng Chem. Process Des. Dev.*, **21**, 359

REDEKER, T. and SCHEBSDAT, F. (1977). Characteristic safety data of diethylether – an example for the significance of characteristic safety data in explosion protection application. *Prevention of Occupational Risks*, **5**, p. 193

REDINGTON, J.M. (1965). The safety dollar. *Chem. Engng Prog.*, **61**(2), 56

REDLICH, O. (1976). *Thermodynamics: Fundamentals, Applications* (Amsterdam: Elsevier)

REDMAN, J. (1986a). BHRA launches pipeline test facility. *Chem. Engr, Lond.*, **429**, 17

REDMAN, J. (1986b). Mossmorran on line ahead of schedule and below budget. *Chem. Engr, Lond.*, **421**, 10

REDMAN, J. (1989a). Control of NO_X emission from large combustion plant. *Chem. Engr, Lond.*, **458**, 33

REDMAN, J. (1989b). The early returns from a toxic poll. *Chem. Engr, Lond.*, **462**, 17

REDMAN, J. (1989c). Pollution is waste. *Chem. Engr, Lond.*, **461**, 16

REDMAN, J. (1990). Orimulsion – a new world fuel. *Chem. Engr, Lond.*, **472**, 12

REDMAN, J. (1991). Polyethylene. *Chem. Engr, Lond.*, **504**, 26

REDMAN, J. and KLETZ, T.A. (1991). Unknown reaction in Shell explosion. *Chem. Engr, Lond.*, **491**, 15

REDMAN, J., and SMITH, H. (1991). Dutch chemical industry. *Chem. Engr, Lond.*, **508**, 22

REDMAN, J. et al. (1990). Italy in focus. *Chem. Engr, Lond.*, **486**, 21

REDMAN, J. et al. (1991). France in focus. *Chem. Engr, Lond.*, **501**, 22

REDMILL, F.J. (ed.) (1988). *Dependability of Critical Computer Systems 1* (London: Elsevier Applied Science)

REDMILL, F.J. (ed.) (1989). *Dependability of Critical Computer Systems 2* (London: Elsevier Applied Science)

REDMILL, F. and ANDERSON, T. (eds) (1993). *Directions in Safety Critical Systems* (London: Springer)

REDMOND, J.D. (1986). Selecting second-generation duplex stainless steels. *Chem. Engng*, **93**(20), **93**(22), 103

REDMOND, J.D. and MISKA, K.H. (1982). The basics of stainless steels. *Chem. Engng*, **89**(Oct. 18), 79

REDMOND, J.D. and MISKA, K.H. (1983). High-performance stainless steels for high-chloride service. *Chem. Engng*, **90**(Jul. 25), **93**(Aug. 22), 91

REDMOND, T. (1984). Hazard assessment for insurance purposes. *Chem. Engr, Lond.*, **406**, 17

REDMOND, T.C. (1990). Piper Alpha – the cost of the lessons. *Piper Alpha: Lessons for Life Cycle Management* (Ruby: Instn Chem. Engrs), p. 113

REDONDO, J.M. (1987). Effects of ground proximity on dense gas entrainment. *J. Haz. Materials*, **16**, 381

REDPATH, P.G. (1976). Fire: occurrence and effects. *Course on Loss Prevention the Process Industries* Dept of Chem. Engng, Loughborough Univ. of Technol.

REDSICKER, D.R. and O'CONNOR J.J. (1997). *Practical Fire and Arson Investigation*, 2nd Edition (New York: CRC Press), ISBN 0-8493-8155-X

REED (1920). *J. Pharm Exp. Therapeut.*, **15**, 301

REED, C.L. and KUHRE, C.J. (1979). Make syn gas by partial oxidation. *Hydrocarbon Process.*, **58**(9), 191

REED, J. (1977). Largest wartime explosions: 21 Maintenance Unit, RAF, Fauld, Staffs, November 27, 1944. *After the Battle*, **18**, 35

REED, J.D. (1967). Nitric acid plant explosion. *Ammonia Plant Safety*, **9**, p. 82

REED, J.D. (1974). Containment of leaks from vessels containing liquefied gases with particular reference to ammonia. *Loss Prevention and Safety Promotion 1*, p. 191

REED, J.W. (1973). Distant blast predictions for explosions. *Dept of Defense, Explosives Safety Board, Minutes of Fifteenth Explosives Safety Seminar*, vol. II, p. 1403

REED, J.W. (1980). *Predictions of Nuisance Damage and Hazard from Accidental Explosions During Trident Missile Test Flights Rep. SAND 79-0626*. Sandia Nat. Labs, Albuquerque, NM

REED, J.W. (1992). Airblast damage to windows. *Explosives Safety Board, Dept of Defense, Washington, DC, Minutes of Twenty-Fifth Explosives Safety Seminar*

REED, J.W., PAPE, B.J., MINOR, J.C. and DEHART, R.C. (1968). Evaluation of window pane damage intensity in San Antonio resulting from Medina Facility explosion on November 13, 1963. *Ann. N.Y. Acad. Sci.*, **152**, art. 1, 565

REED, P. (1939). Calse of pipeline explosion being sought. *Oil Gas J.*, Dec., 14

REED, R. (1961). *Plant Layout* (Homewood, IL: Irwin)

REED, R. (1967). *Plant Location* (Homewood, IL: Irwin)

REED, R. (1969). *Effective Material Handling, Planning and Management* (London: Ind. & Commercial Techniques Ltd)

REED, R.D. (1968). Design and operation of flare systems. *Loss Prevention*, **2**, p. 114. See also *Chem. Engng Prog.*, **64**(6), 53

REED, R.D. (1972). What is flare's proper purge rate? *Oil Gas J.*, Feb. 14, 91

REED, R.D. (1973). *Furnace Operations* (Houston, TX: Gulf)

REED, R.D. (1977). Nitrogen oxide problems in industry. *Chem. Engng*, **84**(Oct. 17), 153

REED, R.D. and NARAYAN, B.C. (1979). Mixing fluids under turbulent flow conditions. *Chem. Engng*, **86**(Jun. 4), 131

REED, R.P. and CLARK, A.F. (eds) (1983). *Materials at Low Temperatures* (Metals Park, OH: Am. Soc. for Metals)

REED, S.B. (1967). Flame stretch – a connecting principle for blow off data. *Combust. Flame*, **11**, 177

REED, S.B. (1971). Comment on the paper 'Blowoff and the flame stretch concept' by Edmonson and Heap. *Combust. Flame*, **17**, 105

REES, F.J. (1990). *Development and Validation of Atmospheric Dispersion Models for Ideal Gases and Hydrogen Fluoride, vol. III, HGSYSTEM Systems Manuall* (report). Shell Res. Ltd

REES, J.M. (1976). The identification of hazard areas. *Course on Fire and Explosion Safety in the Chemical Industry*. Dept of Fire Safety Engng, Univ. of Edinburgh

REES, R. (1982). *Wyre Borough Council. Application by Broseley Estates Ltd. Public Inquiry Rep. D/80/SP/P, File PNW/5063/2/9/16 of January 25, 1982*

REES, W. (1975). Labelling for transport. *Distribution Conf.* (London: Chem. Ind. Ass.), p. 22

REES, W.D. (1967). Electrostatics in the oil industry – hazards, remedies and applications. *Chem. Engr*, **205**, CE7

REES, W.D. (1971). Flow rate limitations imposed by static electricity on tankers. *Symp. on the International Oil Tanker and Terminal Safety Guide* (London: Inst. Petrol.)

REES, W.D. (1975). In discussion of Strawson, H. and Lyle, A.R. (1975b), *op. cit.*

REES, W.D. (1981). Static hazards during top loading of road tankers with highly insulating liquids: flow rate limitation proposals to minimize risk. *J. Electrostatics*. **11**, 13

REEVES, A.B. and MENNELL, S.J. (1989). The application of expert systems technology to risk assessment. *Loss Prevention and Safety Promotion 6*, p. 103.1

REEVES, A.B., WELLS, G.L. and LINKENS, D.A. (1989). A description of GOFA. *Loss Prevention and Safety Promotion 6*, p. 28.1

REEVES, A.L. (ed.) (1981). *Toxicology: Principles and Practice* (New York: Wiley)

REEVES, G.G. (1987). Avoiding self-priming centrifugal pump problems. *Hydrocarbon Process.*, **66**(1), 60

REEVES, P. (1987). Direct in-process analysis using optical fibre technology. *Process Engng*, Nov., 67

REGENASS, W. (1976). Hazard studies applying the preparative heat flow calorimeter. *Prevention of Occupational Risks*, **3**, p. 157

REGENASS, W. (1984). The control of exothermic reactors. *The Protection of Exothermic Reactors and Pressurised Storage Vessels* (Rugby: Instn Chem. Engrs), p. 1

REGENASS, W., OSTERWALDER, U. and BROGLI, F. (1984). Reactor engineering for inherent safety. *Eighth Int. Symp. on Chemical Reaction Engineering* (Rugby: Instn Chem. Engrs)

REGENCZUK, T.J. (1967). Selecting temperature monitoring and control systems. *Chem. Engng*, **74**(Dec. 4), 202

REGISTRAR GENERAL (1978). *Occupational Mortality 1970–72* (London: HM Stationery Office)

DE LA REGUERA, R.F., RIVAS, J.L., LEGATOS, N.A. and MARCHAJ, J. (1983). An 80,000 m³ double prestressed concrete wall LNG tank: its design, construction and commissioning. *Gastech 82*, p. 362

DE REGULES, S.E. (1967). A production engineer's provision for start-up. *Chem. Engng Prog.*, **63**(12), 61

REGULINSKI, T.L. (1973). On modeling human performance reliability. *IEEE Trans. Reliab.*, **R-22**, 114

REHRIG, P. (1981). Selecting centrifugal compressor materials for harsh environments. *Hydrocarbon Process.*, **60**(10), 137

REIBLE, D.D., SHAIR, F.H. and KAUPER, E. (1981). Plume dispersion and bifurcation in directional shear flows associated with complex terrain. *Atmos. Environ.*, **15**, 1165

REICH, P. (1976). Air or oxygen for VCM. *Hydrocarbon Process.*, **55**(3), 85

REICHENBACH, R., SQUIRES, D. and PENNER, S.S. (1962). Flame propagation in liquid-fuel droplet arrays. *Combustion*, **8**, p. 1068

REICHMANN, W.J. (1961). *Uses and Abuses of Statistics* (Harmondsworth: Penguin)

REID, D. (1992). Grinding slowly, grinding small. *Chem. Engr, Lond.*, **529**, s4

REID, E.C. and RENSHAW, J.C. (1977). *Steam* (Melbourne: Cole)

REID, I.A.B., ROBINSON, C. and SMITH, D.B. (1985). Spontaneous ignition of methane: measurement and chemical model. *Combustion*, **20**, p. 1833

REID, J.D. (1979). Markov chain simulations of vertical dispersion in the neutral surface layer for surface and elevated releases. *Boundary Layer Met.*, **16**, 3

REID, R.C. (1976). Superheated liquids. *Am. Scientist*, **64**, 146

REID, R.C. (1978). Superheated liquids. A laboratory curiosity and, possibly, an industrial clrse. *Chem. Engng Educ.*, **12**(2), 60; **12**(3), 108

REID, R.C. (1979). Possible mechanism for pressurized-liquid tank explosions or BLEVEs. *Science*, **203**, 1263

REID, R.C. (1980). Some theories on boiling liquid expanding vapour explosions. *Fire*, Mar., 525

REID, R.C. and SHERWOOD, T.K. (1958). *The Properties of Gases and Liquids*, 1st ed.; 1966, 2nd ed. (New York: McGraw-Hill). See also Reid, R.C., Prausnitz, J.M. and Sherwood, T.K. and then Reid, R.C., Prausnitz, J.M. and Poling B.E.

REID, R.C. and SMITH, K.A. (1978). Behaviour of LPG on water. *Hydrocarbon Process.*, **57**(4), 117

REID, R.C. and WANG, R. (1978). The boiling rates of LNG on typical dike floor materials. *Cryogenics*, July

REID, R.C., PRAUSNITZ, J.M. and SHERWOOD, T.K. (1977). *The Properties of Gases and Liquids*, 3rd ed. (New York: McGraw-Hill)

REID, R.C., PRAUSNITZ, J.M. and POLING, B.E. (1987). *The Properties of Gases and Liquids*, 4th ed. (New York: McGraw-Hill)

REIDENBACH, B. (1988). Spray treatment procedure improves cooling system efficiency and reliability. *Plant/ Operations Prog.*, **7**, 209

REIDER, H.L. (ca. 1968). Private communication (to A.G. Hirsch)

REIDER, R., OTWAY, H.J. and KNIGHT, H.T. (1965). An unconfined large volume hydrogen/air explosion. *Pyrodynamics*, **2**, 249

REIFF, D.D. (1965). Liquefaction and storage of hydrogen. *Chem. Engng*, **72**(Sep. 13), 191

VAN REIJENDAM, J.W., VERBRUGGE, H., TEN BRINK, J.V., COOLEGEM, J., HEIJMEN, H.J. and WINTER, M. (1992). Explosion in the Shell fluoroaromatics plant at Stanlow 1990. An investigation of the underlying chemistry. *J. Loss Prev. Process Ind.*, **5**, 211

REIJNHART, R. and ROSE, R. (1980). Vapour cloud dispersion and the evaporation of volatile liquids in atmospheric wind fields – II. Wind tunnel experiments. *Atmos. Environ.*, **14**, 759

REIJNHART, R., PIEPERS, J. and TONEMAN, L.H. (1980). Vapour cloud dispersion and the evaporation of volatile liquids in atmospheric wind fields – I. Theoretical model. *Atmos. Environ.*, **14**, 751

REILLY, M.J. and SCHMITZ, R.A. (1966–). Dynamics of a tubular reactor with recycle. *AIChE J.*, 1966, **12**, 153; 1967, **13**, 519

REIN, R.G., SLIEPCEVICH, C.M. and WELKER, J.R. (1970). Radiation view factors for tilted cylinders. *J. Fire Flammability*, **1**, 140

REINARTZ, S.J. (1993). An empirical study of team behaviour in a complex and dynamic problem-solving context: a discussion of methodological and analytical aspects. *Ergonomics*, **36**, 1281

REINAUER, T.V. (1981). Application of fabric filters to explosive conditions. *The Hazards of Industrial Explosion from Dusts* (London: Oyez/IBC), paper 13

REINSCHKE, K. (1973). *Zuverlässigkeit von Systeme* (Berlin: Technik)

REIS, T. (1966). Compare vinyl acetate processes. *Hydrocarbon Process.*, **45**(11), 171

REISER, B. (1985a). A remark on Ichikawa's upper bound of probability of failure. *Reliab. Engng*, **13**(3), 181

REISER, B. (1985b). There is no such thing as a free lunch: a comment on a new method for reliability demonstration. *Reliab. Engng*, **13**(3), 175

REISLER, R., PETTIT, B. and KENNEDY, L. (1976). *Air Blast Data from Height-of-burst Studies in Canada. Vol. I, HOB 5.4 to 71.9 feet. BRL Memo Rep. 1950.* US Army, Ballistic Res. Lab., Aberdeen Proving Ground, MD

REISLER, R., PETTIT, B. and KENNEDY, L. (1977). *Air Blast Data from Height-of-Burst Studies in Canada. Vol. II, HOB 45.4 to 144.5 feet. BRL Memo Rep. 1990.* US Army, Ballistic Res. Lab., Aberdeen Proving Ground, MD

REITER, J. (1980). *AL/X: an Expert System using Plausible Inference* (report). Machine Intelligence Res. Unit, Univ. of Edinburgh

REITER, W.M. and SOBEL, R. (1973). Waste control management. *Chem. Engng*, **80**(Jun. 18), 59

REITER, W.M., SOBEL, R. and SULLIVAN, W.L. (1980). The pollution control review: a tool in risk identification. *Loss Prevention*, **13**, 175

REITMAN, W. (ed.) (1984). *Artificial Intelligence Applications for Business* (Norwoocl, NJ: Ablex)

REKLAITIS, G.V. (1983). *Introduction to Material and Energy Balances* (New York: Wiley)

REKLAITIS, G.V. and SPRIGGS, D. (eds) (1988). *Foundations of Computer Aided Process Operations* (Amsterdam: Elsevier)

RELIABILITY ANALYSIS CENTER (1985). *Nonelectronic Parts Reliability Data. Rep. NPRD-3.* Rome Air Development Center, Griffiths Air Force Base, NY

RELIO, M. (1978). Data on toxics – a thorn in the side of the CPI. *Chem. Engng*, **85**(Oct. 23), 114

RELY, A. (1987). Antwerp 'City of Sudden Death'. *After the Battle*, **57**, 43

REMBOLD, U., ARMBRUSTER, K. and ÜLZMANN, W. (1983). *Interface Technology for Computer-Controlled Manufacturing Processes* (New York: Dekker)

REMER, D.S. and CHAI, L.H. (1990). Estimate costs of scaled up process plants. *Chem. Engng*, **97**(4), 138

REMIREZ, R. (1967). Starch dust blamed in New Jersey disaster. *Chem. Engng*, **74**(Apr. 10), 112

REMIREZ, R. (1968a). Ethylene or acetylene route to vinyl acetate? *Chem. Engng*, **75**(Aug. 12), 94

REMIREZ, R. (1968b). Naphtha-to-ethylene route is source of benzene. *Chem. Engng*, **75**(Sep. 9), 92

REMIREZ, R. (1968c). Thermal pollution: hot issue for industry. *Chem. Engng*, **75**(Mar. 25), 48

REMIREZ, R. (1982). Incineration at sea – are the odds now in favor? *Chem. Engng*, **89**(Mar. 22), 49

RENDELL, J. (1988a). On or off-site disposal. *Process Engng*, Mar., 71

RENDELL, J. (1988b). Pump those problem fluids. *Process Engng*, Apr., 49

RENDOS, J.J. (1963). Oil separator explosion in an air separation plant. *Ammonia Plant Safety*, **5**

RENDOS, J.J. (1964). Air separation plant fire. *Ammonia Plant Safety*, **6**, p. 41

RENDOS, J.J. (1967). Liquid oxygen pump failures. *Ammonia Plant Safety*, **9**, p. 5

RENFRO, E.M. (1975). Foundation repair techniques. *Hydrocarbon Process.*, **54**(1), 68

RENFRO, E.M. (1978). Better ways to repair machinery foundations. *Hydrocarbon Process.*, **57**(1), 95

RENFRO, E.M. (1979). Good foundations reduce machinery maintenance. *Hydrocarbon Process.*, **58**(1), 123

RENN, O. (1992). Risk communication: towards a rational discourse with the public. *J. Haz. Materials*, **29**, 465

RENOLD LTD (1980). Developments in drives and power transmission. *Plant Engr*, **23**(6), 15

RENOWDEN, D.G. and WOOLHEAD, F.E. (1990). Fire loss risk management. *Fire Surveyor*, **19**(3), 17; **19**(4), 9; **19**(5), 14

RENSHAW, F.M. (1990). A major accident prevention program. *Plant/Operations Prog.*, **9**, 194

RENSHAW, W.G. (1964). Corrosion resistant steels for petrochemicals. *Chem. Engng Prog.*, **60**(9), 98

RENTON, T. (1984). Specification and realisation of industrial control alarm system. *Chem. Engr, Lond.*, **402**, 33

RENTSCH, H. (1971). *Handbuch für Explosionsschutz* (Essen: W. Girardet)

RENWICK, P. and TOLLOCZKO, J. (1992). Offshore safety. *Chem. Engr, Lond.*, **525**, 17

DE RENZO, D.J. (1986). *Solvents Safety Handbook* (Park Ridge, NJ: Noyes)

REPPE, W. (1952). *Chemie und Technik der Acetylendruckreactionen* (Weinheim: Verlag Chemie)

REROLLE, E. (1968). Arzew masters LNG operating problems. *Hydrocarbon Process.*, **47**(7), 115

RESEARCH TASK FORCE ON RISK-BASED INSPECTION GUIDELINES (1992). Risk-based inspection – development of guidelines, vol. 2, Pt 1 – Light water reactor (LWR) nuclear power plant components. *Pub CRTD, vol. 20-2*. Am. Soc. mech. Engrs, New York

RESEN, L. (1984). Superfund a headache for the US chemical industry. *Chem. Engr, Lond.*, **409**, 11

RESEN, L. (1985a). Bhopal through the eyes of the US press. *Chem. Engr, Lond.*, **413**, 15

RESEN, L. (1985b). Is the message lost? *Chem. Engr, Lond.*, **416**, 35

RESEN, L. (1986). The community is concerned. The chemical industry is too. *Chem. Engr, Lond.*, **429**, 12

RESNICK, W. (1981). *Process Analysis and Design for Chemical Engineers* (New York: McGraw-Hill)

RESOURCE PLANNING ASSOCIATES (1977). *Alternative Site Study. Northeast Coast Liqueled Natural Gas Conversion Facility* (Cambridge, MA)

RESPLANDY, A. (1967). Etude expérimentale des propriétés de l'ammoniac conditionnant les mesures à prendre pour la sécurité du voisinage des stockages industriels. *Genie. Chem.*, **102**(6), 691

RESS, J.F. (1982). Safe production of bis(chloromethyl)ether. *Plant/Operations Prog.*, **1**, 248

VAN REST, D. (1982). Environmental impact assessment: an aid to design for environmental acceptability. *Design 82* (Rugby: Instn Chem. Engrs)

REUILLON, M. *et al.* (1977). New developments on the extinction of metal fires and of sodium fires in particular. *Loss Prevention and Safety Promotion 2*, p. 481

REUTER (1994)

REYNOLDS, C.E. (1971). *Reinforced Concrete Designers Handbook*, 7th ed. (London: Cement and Concrete Ass.)

REYNOLDS, D.J. (1956). The cost of road accidents. *J. R. Statist. Soc., Ser. A*, **119**, 393

REYNOLDS, J. and TRIMARKE, C.R. (1962). Four routes to urea by Toyo Koatsu. *Hydrocarbon Process.*, **41**(2), 109

REYNOLDS, J.A. (1971). Pump installation and maintenance. *Chem. Engng*, **78**(Oct. 11), 67

REYNOLDS, J.A. (1989). New ANSI pump standards. *Chem. Engng*, **96**(7), 98

REYNOLDS, J.M. (1978). Controlling the use of hazardous materials in research and development laboratories. *J. Haz. Materials*, **2**(4), 299

REYNOLDS, R.P. (1974). Better information and maintenance costs. *Inst. of Cost and Management Accountants Summer School*, Oxford

REYNOLDS, R.P. (1976). Why not a maintenance accountant? *Seventh Nat. Maintenance Engineering Conf.*, London

REYNOLDS, T.W. and GERSTEIN, M. (1949). Influence of molecular structure of hydrocarbons on rate of flame propagation. *Combustion*, **3**, 190

ROX, J. (1977). The use of equipment life history data in reliability and availability assessment studies. *First Nat. Conf. on Reliability. NRC5/2*. Culcheth, Lancashire: UK Atom. Energy Auth.)

REZA, F.M. (1961). *An Introduction to Information Theory* (New York: McGraw-Hill)

RHEE, K.T. (1982). Optimum clearance for rapid flame propagation. *Combustion*, **19**, p. 607

RHEE, K.T. and CHANG, S.L. (1985). Empirical equations for adiabatic flame temperature for some fuel–air combustion systems. *Combust. Sci. Technol.*, **44**, 75

RHEIN, R. (1989). Report lambasts Superfund work. *Chem. Engng*, **96**(3), 58

RHEIN, R.A. (1964). Preliminary experiments on ignition of carbon dioxide (letter). *Combust. Flame*, **8**, 346

RHIND, D. (ed.) (1983). *A Census Users Handbook* (London: Methuen)

RHOADS, D.E. (1985). Lockouts save lives. *Chem. Engng*, **92**(Sept. 2), 95

RHOADS, T.W., COLE, D.G. and NORTON, R.L. (1984). Optimizing fugitive VOC emission control. *Chem. Engng Prog.*, **80**(9), 39

RHODES, T. (1982). Shutdown systems for better managed process operations. *Hydrocarbon Process.*, **61**(5), 151

RHYNE, W.R. (1990). Evaluating routing alternatives for transporting hazardous materials using simplified risk indicators and complete probabilistic analysis. *Transport Res. Record 1264*. (Transport. Res. Board, Nat. Res. Coun.) p. 1

RIANCE, X.P. (1988). More learning the hard way. *Chem. Engng*, **95**(2) Feb. 15, 103

RIAZI, M.R. and DAUBERT, T.E. (1987). Predict flash and pour points. *Hydrocarbon Process.*, **66**(9), 81

RIBOVICH, J., MURPHY, J. and WATSON, R. (1977). Detonation studies with nitric oxide, nitrous oxide, nitrogen tetroxide, carbon monoxide and ethylene. *J. Haz. Materials*, **1**(4), 275

RICCA, P.M. (1983). IR techniques in fire-suppression systems. *Plant/Operations Prog.*, **2**, 99

RICCA, P.M. and BRADSHAW, P.M. (1984). TAPS machinery monitoring program. *Oil Gas J.*, **82**(Aug. 13), 96; (Aug. 20), 100

RICCI, L.J. (1974). Cancer questions plague. *Chem. Engng*, **81**(Dec. 23), 28

RICCI, L.J. (1976a). Cancer-causing chemicals defy easy detection. *Chem. Engng*, **83**(Apr. 12), 72

RICCI, L.J. (1976b). Kepone tragedy may spur tough toxic-chemicals-control law. *Chem. Engng*, **83**(Mar. 15), 53

RICCI, L.J. (1976c). US toxic-substances-control law, real impact fuzzy until 1980. *Chem. Engng*, **83**(Nov. 22), 81

RICCI, L.J. (1977a). Chemicals spawn reproductive woes. *Chem. Engng*, **84**(Aug. 1), 30

RICCI, L.J. (1977b). Mixing NO_X. *Chem. Engng*, **84**(Feb. 14), 33; (Apr. 11), 84; (Apr. 25), 70

RICCI, L.J. (1978a). CPI niche for heat pipes. *Chem. Engng*, **85**(Jan. 16), 84

RICCI, L.J. (1978b). Hazardous waste disposers feel down in the dumps. *Chem. Engng*, **85**(May 22), 54

RICCI, L.J. (1978c). Innovative R&D. Gone with the wind *Chem. Engng*, **85**(Sep. 25), 73

RICCI, L.J. (1979a). New health monitors: small but accurate. *Chem. Engng*, **86**(Dec. 31), 28

RICCI, L.J. (1979b). Tightening the loop in ammonia manufacture. *Chem. Engng*, **86**(Jan. 29), 54

RICCI, L.J. (1981). Oceanborne plants: finding a CPI berth. *Chem. Engng*, **88**(Oct. 19), 83

RICCI, P.F. and COX, L.A. (1987). De minimis considerations in health risk assessment. *J. Haz. Materials*, **15**(1/2), 77

RICCI, P.F., SAGAN, L.A. and WHIPPLE, C.G. (eds) (1984). *Technological Risk Assessment* (Dordrecht: Martinus Nijhoff)

RICE, A.P. (1982). Seveso accident: dioxin. In Bennett, G.F., Feates, F.S. and Wilder, I. (1982), *op. cit.*, pp. 11–18

RICE, G.B., NICHOLS, F.D., SNYDER, H.J. and SHUJINS, J. (1976). Case history – cleanup of Little Mountain salvage yard acid sludge and oil lagoon. *Control of Hazardous Material Spills*, **3**, 386

RICE, J.R. (1968a). Mathematical analysis in the mechanics of fracture. In Liebowitz, H. (ed.) (1968), *op. cit.*, vol. II

RICE, J.R. (1968b). A path independent integral and the approximate analysis of strain concentrations by notches and cracks. *J. Appl. Mech.*, **35**, 379

RICE, J.R. (1976). *Mechanics of Crack Tip Deformation and Extension by Fatigue. ASTM STP 415.* Am. Soc. Testing and Mats, Philadelphia, PA

RICH, E. (1983). *Artificial Intelligence* (New York: McGraw-Hill)

RICH, L.A. (1976). Where do environmental laws and regulation stand today? *Chem. Engng*, **83**(Oct. 18), 9

RICH, S.H. and VENKATASUBRAMANIAN, V. (1988). Model-based reasoning in diagnostic expert systems for chemical process plants. *Comput. Chem. Engng*, **11**, 111

RICH, S.H. and VENKATASUBRAMANIAN, V. (1989). Causality-based failure-driven learning in diagnostic expert systems. *AIChE J.*, **35**, 943

RICH, S.H., VENKATASUBRAMANIAN, V., NASRALLAH, M. and MAIEO, C. (1989). Development of a diagnostic expert system for a whipped toppings process. *J. Loss Prev. Process Ind.*, **2**(3), 145

RICHARD, D., BOULEY, G. and BOUDENE, C. (1978). *Bull. Eur. Physiopathol. Respir.*, **14**, 573

RICHARD, D., JOUNAY, J.M. and BOUDENE, C. (1978). Acute 61 toxicity by pulmonary penetration of ammonia gas in the rabbit. *C.R. Acad. Sci., Ser. D*, **287**, 375

RICHARDS, E.J. (1969). Noise generation in process plant. *Chem. Engr, Lond.*, **229**, CE223

RICHARDS, E.T. (1980). The successful exchange of reliability data between government and industry. *Reliab. Engng*, **1**(1), 49

RICHARDS, S.G. (1989). *Piper Alpha Public Inquiry. Tanscript, Days 110, 111*

RICHARDSON, E.G. (1960). Aerodynamic capture of particles. Introduction. *Int. J. Air Poll.*, **3**, 3

RICHARDSON, J.A. (1976). Terotechnology – the key to loss prevention? *Soc. Chem. Ind., Symp. on Loss Prevention* (London)

RICHARDSON, J.A. and TEMPLETON, L.R. (1973). The impact of plant reliability upon cost. *Chem., Engr, Lond.*, **280**, 625

RICHARDSON, J.J. (ed.) (1985). *Artificial Intelligence in Maintenance* (Park Ridge, NJ: Noyes)

RICHARDSON, J.M. (1956). Mathematical theory of turbulent flames. *Proc. Symp. on Aerothermodynamics Gasdynamics* (Evanston, IL: Northwestern Univ. Press), p. 169

RICHARDSON, J.W. (ed.) (1975). *Disaster Planning* (Bristol: Wright)

RICHARDSON, L.F. (1920). The supply of energy from and to atmospheric eddies. *Proc. R. Soc., Ser. A*, **97**, 354

RICHARDSON, L.F. (1925). Turbulence and vertical temperature difference near trees. *Phil. Mag.*, **49**, 81

RICHARDSON, L.F. (1926). Atmospheric diffusion shown on a distance neighbour graph. *Proc. R. Soc., Ser. A*, **110**, 709

RICHARDSON, L.F. and PROCTOR, D. (1925). Diffusion over distances ranging from 3 km to 86 km. *Memoirs R. Met. Soc.*, **1**(1)

RICHARDSON, M. (1994). *Chemical Safety* (New York: VCH Publs)

RICHARDSON, S.M. (1989a). *Piper Alpha Public Inquiry. Transcript, Day 89*

RICHARDSON, S.M. (1989b). *Piper Alpha Public Inquiry. Transcript, Days 90, 91*

RICHARDSON, S.M. (1989c). *Piper Alpha Public Inquiry. Transcript, Days 107, 108*

RICHARDSON, S.M. (1989d). *Piper Alpha Public Inquiry. Transcript, Day 127*

RICHARDSON, S.M. and SAVILLE, G. (1992). Blowdown of vessels and pipelines. *Major Hazards Onshore and Offshore*, p. 195

RICHARDSON, S.M., SAVILLE, G. and GRIFFITHS, J.F. (1990). Autoignition – occurrence and effects. *Piper Alpha: Lessons for Life Cycle Management* (Rugby: Instn Chem. Engrs), p. 13. See also *Process Safety Environ.*, **68B**, 239

RICHARDSON, T. (1991). Learn from the Phillips explosion. *Hydrocarbon Process.*, **70**(3), 83

RICHARDSON, W.H. (1962). Inspecting the design for missing horse shoe nails. *Chem. Engng*, **69**(Apr. 30), 126

RICHARDSON, W.H. (1963). How to foresee operating difficulties. *Chem. Engng*, **70**(Oct. 14), 216

RICHARDSON, W.H. (1966). Who should have the last word on operating variables? *Chem. Engng*, **73**(Dec. 5), 156

RICHINGS, W.V. (1967). Noise control in chemical processing. *Chem. Process. Engng*, **48**(10), 77; **48**(11), 66

RICHMANN, H.B. (1964). Now proved: phenol from air-oxidation of toluene. *Chem. Engng*, **71**(Sep. 24), 106

RICHMOND, D. (1965). Instrumentation for safe operation. In Fawcett, H.H. and Wood, W.S. (1965), *op. cit.*, p. 132

RICHMOND, D. (1982). Instrumentation for safe operation. In Fawcett, H.H. and Wood, W.S. (1982), *op. cit.*, p. 87

RICHMOND, D.R. (1971). Blast effects inside structures. *Probleme des Baulichen Schutzes* (Ernst Mach Inst., Germany), p. 9

RICHMOND, D.R. (1972). Recent advances in developing blast criteria for personnel inside structures. *Stanford Res. Inst., Fifth Mtg of Ad Hoc Subcommittee on Initial Effects and Structures to Protect against Them*, Appendix B (Menlo Park, CA)

RICHMOND, D.R. (1974). Whole-body impact studies with sheep. *Defense Civil Preparedness Agency, Proc. DCPA All Effects Research Contractors Conf.* (Washington, DC), p. 41

RICHMOND, D.R. and FLETCHER, E.R. (1971). Blast criteria for personnel in relation to quantity-distance. *Explosives Safety Board, Dept of Defense, Proc. Thirteenth Explosives Safety Seminar*, p. 401

RICHMOND, D.R. and KILGORE, D.E. (1971). *Blast Effects Inside Structures. Rep. DNA-2775P.* Defense Nucl. Agency, Dept of Defence, Washington, DC

RICHMOND, D.R. and WHITE, C.S. (1962a). *A Tentative Estimation of Man's Tolerance to Overpressures from Air Blast. Rep. DASA-1335.* Defense Atom. Support Agency, Dept of Defence, Washington, DC

RICHMOND, D.R. and WHITE, C.S. (1962b). *Biological Effects of Blast and Shock. Rep. DASA-1777.* Defense Atom. Support Agency, Dept of Defence, Washington, DC

RICHMOND, D.R., WETHERBE, M.B., TABORELLI, R.V., CHIFFELLE, T.L. and WHITE, C.S. (1957). The biological response to overpressure. I, Effects on dogs of five- to ten-second duration overpressures having various times to pressure rise. *J. Aviation Med.*, **28**, 447

RICHMOND, D.R., TABORELLI, R.V., SHERPING, F., WETHERBE, M.B., SANCHEZ, R.T., GOLDIZEN, V.C. and WHITE, C.S. (1957). *Shock Tube Studies of the Effects of Sharp-rising, Long-duration Overpressures on Biological Systems. Rep. TID 6065.* US

Atom. Energy Comm., Office of Tech. Services, Dept of Commerce, Washington, DC

RICHMOND, D.R., CLARE, V.R., GOLDIZEN, V.C., PRATT, D.E., SANCHEZ, T. and WHITE, C.S. (1961). Biological effects of overpressure. II, A shock tube utilized to produce sharp rising overpressure of 400 milliseconds duration and its employment in biomedical experiments. *Aerospace Med.*, **32**, 997

RICHMOND, D.R., DAMON, E.G., FLETCHER, E.R., BOWEN, I.G. and WHITE, C.S. (1968). *The Relationship Between Selected Blast Wave Parameters and the Response of Mammals Exposed to Blast. Rep. DASA-1860.* Defense Atom. Support Agency, Dept of Defence, Washington, DC. See also *Ann. N.Y. Acad. Sci, 1968,* **152**(art. 1), 103

RICHMOND, J.K., SINGER, J.M., COOK, E.B., OXENDINE, J.R., GRUMER, J. and BURGESS, D.S. (1957). Turbulent burning velocities of natural gas-air flames with pipe-flow turbulence. *Combustion,* **6**, 303

RICHTER, C.F. (1958). Elementary Seismology (San Francisco, CA: W.H. Freeman)

RICHTER, S.H. (1978a). Relief and blowdown for two-phase flow. *API Proc., Refining Dept, 43rd Midyear Mtg,* **57**, 23

RICHTER, S.H. (1978b). Size relief systems for two-phase flow. *Hydrocarbon Process.,* **57**(7), 145. See also in Kane, L.A. (1980), *op. cit.,* p. 152

RICHTERS, C.E. VETERINARY GENERAL (1936a). Animals in chemical warfare. *J. R. Army Vet. Corps,* Feb., 78

RICHTERS (1936b). *Die Tiere im Chemischen Krieg* (Berlin: Schoetz)

RICKENBACH, H. (1982). Ventex explosion barrier valves. *The Control and Prevention of Dust Explosions* (London: Oyez/IBC), paper 13

RICKETT, M.A. (1977). Marking of vehicles carrying cylinders and drums. *Symp. Hazard 77 — Advances in the Safe Handling and Movement of Hazardous Substances,* Teesside Polytechnic, Middlesbrough

RICKLES, R.N. (1965). Waste recovery and pollution abatement. *Chem. Engng,* **72**(Sep. 27), 133

RICKMANN, W.S., HOLDER, N.D. and YOUNG, D.T. (1985). Circulating bed incineration of hazardous wastes. *Chem. Engng Prog.,* **81**(3), 34

RICOU, F.P. and SPALDING, D.B. (1961). Measurements of entrainment by axisymmetrical turbulent jets. *J. Fluid Mech.,* **11**(1), 21

RIDDELL, F. (1977). Personal protection and equipment. In Handley, W. (1977), *op. cit.,* p. 366

RIDDICK, J.A. and BUNGER, W.B. (1970). *Organic Solvents,* 3rd ed. (New York: Wiley-Interscience)

RIDDLESTONE, H.G. (1957). *The Effect of Circuit Resistance on the Dischare Energy Required for Ignition of Methane Air Mixtures. Rep. D/T105.* Elec. Res. Ass., Leatherhead

RIDDLESTONE, H.G. (1958). *The Spontaneous Ignition Temperature of Inflammable Gases and Vapours. A Critical Resume. Rep. D/T110.* Elec. Res. Ass., Leatherhead

RIDDLESTONE, H.G. (1967). Comparison of national requirements for electrical apparatus for use in hazardous atmospheres. *Chemical Process Hazards,* **III**, p. 55

RIDDLESTONE, H.G. (1975). 1971–5 – a review of developments and problems outstanding. *Electrical Safety,* **2**, 1

RIDDLESTONE, H.G. (1979). Electrical equipment for use in flammable atmospheres. *Fire Prev. Sci. Technol.,* **21**, 18

RIDDLESTONE, H.G. and BARTELS, A. (1965). The relative hazards of ferrous and non-sparking tools in the petroleum industry. *J. Inst. Petrol.,* **51**(495), 106

RIDE, D.J. (1984a). An assessment of the effects of fluctuations on the severity of poisoning by toxic vapours. *J. Haz. Materials,* **9**(2), 235

RIDE, D.J. (1984b). A probabilistic model for dosage. In Ooms, G. and Tennekes, H. (1984), *op. cit.,* p. 267

RIDE, D.J. (1988). A model for the observed intermittency of a meandering plume. *J. Haz. Materials,* **19**, 131

RIDEAL, E.K. and ROBERTSON, A.J.B. (1948). The sensitiveness of solid high explosives to impact. *Proc. R. Soc., Ser. A,* **195**, 135

RIDENHOUR, L.W. (1982). Process and equipment monitoring by computer. *Plant/Operations Prog.,* **1**, 236

RIDEOUT, T.B., RANKIN, W.L., SHIKIAR, R. and DESTEESE, J.G. (1985). Human engineering in LNG facility design. *Chem. Engng Prog.,* **81**(12), 40

RIDEY, M. (1991). Relieving overpressure. *Chem. Engng,* **98**(12), 98

RIDLEY, J.R. (ed.) (1983). *Safety at Work;* 1986, 2nd ed.; (London: Butterworths); 1990, 3rd ed. (London: Butterworth-Heinemann); 1994, 4th ed. (Oxford: Butterworth-Heinemann)

RIEGEL, E.R. (1953). *Chemical Process Machinery* (New York: Reinhold)

TE RIELE, P.H.M. (1977). The atmospheric dispersion of heavy gases emitted at or near ground level. *Loss Prevention and Safety Promotion 2,* p. 347

RIESCHL, U. (1979). Water fog stream heat radiation attenuation. *Fire Technol.,* **15**, 262

RIETHMANN, J. *et al.* (1976). Mechanism of decomposition of nitrobenzene solutions of aluminum chloride. *Chem. Ing. Tech.,* **48**(8), 729

RIETHMULLER, M.L. (1985). Computer processing of visual records from the Thorney Island large scale gas trials. *J. Haz. Materials,* **11**(1/3), 179. See also in McQuaid, J. (1985), *op. cit.,* p. 179

RIETHMULLER, M.L. (1986). Wind tunnel simulation of chlorine spills. *Loss Prevention and Safety Promotion 5,* p. 28-1

RIEZEL, Y. (1996). Inherently Safer Process: An integral part of PSM implementation. *Int. Conf. and Workshop on Process Saf. Manage. and Inherently Safer Processes,* Oct. 8–11, 1996, Orlando, FL, 320–328 (New York: American Institute of Chemical Engineers)

RIGARD, J. and VADOT, L. (1979). Evaluate LNG's storage hazard. *Hydrocarbon Process.,* **58**(7), 267

RIGBY, H.F. (1977). Glass as an aid to safety. In Handley, W. (1977), *op. cit.,* p. 327

RIGBY, L.V. (1967). *The Sandia Human Error Rate Bank (SHERB). Rep.* Albuquerque, NM

RIGBY, L.V. (1971a). Motivation: its origins and nature. *Chem. Tech.,* Jun. 348

RIGBY, L.V. (1971b). The nature of human error, *Chem. Tech.,* Dec., 712

RIGBY, L.V. and EDELMAN, D.A. (1968a). *An Analysis of Human Variability in Mechanical Inspection. Rep. SCRR-68-282.* Sandia Nat. Labs, Albuquerque, NM

RIGBY, L.V. and EDELMAN, D.A. (1968b). A predictive scale of aircraft emergencies. *Hum. Factors,* **10**, 475

RIGBY, L.V. and SWAIN, A.D. (1968). Effects of assembly error on product acceptability and reliability. *Am. Soc. Mech. Engrs, Proc. Seventh Ann. Reliability and Maintainability Conf.* (New York), p. 3-12

RIGBY, L.V. and SWAIN, A.D. (1975). Some human-factor applications to quality control in a high technology industry. In Drury and Fox, J.G. (1975), *op. cit.,* p. 201

RIGGS, J.L. (1976). *Production Systems: Planning, Analysis and Control*, 2nd ed. (New York: Wiley)

RIJKELS, D.F. (1984). Cancer risk reduction in Western European industry. In Deisler, P.F. (1984b), *op. cit.*, p. 197

RIJNMOND PUBLIC AUTHORITY (1982). *Risk Analysis of Six Potentially Hazardous Industrial Objects in the Rijnmond Area, A Pilot Study* (Reidel: Dordrecht)

RIJNSDORP, J.E. (1986). Impact of operator tasks on process automation systems. *Chem. Ind.*, May 5, 304

RILEY, B.S. and BATES, H.J. (1980). *Fatal Accidents in Great Britain in 1976 Involving Heavy Goods Vehicles. Rep. SR 586.* Transport and Road Res. Lab, Crowthorne

RILEY, J.F. (1983). Selection and application of special extinguishing agents in industrial hazards. *Plant/ Operations Prog.*, **2**, 101

RILEY, R.V. (1979). Accidents with pipelines in the USA: selected case histories and a review of the activities of the National Transportation Safety Board. *Third Int. Conf. on Internal and External Protection of Pipes* (Cranfield, Bedfordshire: Br. Hydromechanics Res. Ass.)

RILEY, W.J. (1975). Combustible gas detectors. *Instrum. Technol.*, **22**(Aug.), 33

RIMINGTON, J.D. (1990). *Piper Alpha Public Inquiry. Transcript, Days 157, 158*

RIMINGTON, J.D. (1993). Overview of risk assessment. *Process Safety Environ.*, **71B**, 112

RINARD, I.H. (1982). A handbook to control-system design. *Chem. Engng*, **89**(Nov. 29), 46

RINARD, J.W. and STONE, D.B. (1977). Maintainability of inline pumps. *Hydrocarbon Process.*, **56**(1), 101

RINDFLEISCH, H.N. and SCHECKER, H.G. (1983). SYSP – a computer program for simulating safety and controlled systems in chemical plant. *Loss Prevention and Safety Promotion 4*, vol. l, paper

RINDNER, R.M. (1968). Response of explosives to fragment impact. *Ann. N.Y. Acad. Sci*, **152**(art. 1), 250

RINDNER, R.M. and WACHTELL, S. (1973). Establishment of design criteria for safe processing of hazardous materials. *Loss Prevention*, **7**, 92

RINEHART, J.S. (1959). Fracturing under explosive loading. *Chem. Engng Prog.*, **55**(2), 59

RING, T.A. and FOX, J.M. (1974). Staek gas cleanup progress. *Hydrocarbon Process.*, **53**(10), 119

RINGER, M.W. (1985). Detonation tests and response analysis of vessels and piping containing gas and aerated liquid. *Plant/Operations Prog.*, **4**, 26

RINGLE, M. (1979). Philosophy and artificial intelligence. *Philosophical Perspectives in Artificial Intelligence* (ed. M. Ringle) (Brighton: Harvester)

RINKER, F.G. and SCHULTZ, T.J. (1978). The safe operation of a commercial seale Kepone incineration system. *Control of Hazardous Material Spills 4*, p. 260

RINNAN, A. (1982). Transmission of detonation through tubes and orifices. In Lee, J.H.S. and Guirao, C.M. (1982), *op. cit.*, p. 553

RINNE, J.E. (1963). Economics and design problems associated with tanks on soft soils. *API Proc., Div. of Refining*, **43**, 490

RIORDAN, C. (1980). Toxic chemical transport and fate research in US Environmental Protection Agency. In Haque, R. (1980a), *op. cit.*, p. 21

RIOU, C. and ZERMATI, C. (1982). Commissioning of the 120,000 m3 storage tanks of the Gaz de France LNG terminal. *Gastech 81*

RIOU, Y. (1986). Methods for describing heavy gas dispersion in the environment of industrial sites. *Loss Prevention and Safety Promotion 5*, vol. I, p. 30–1; vol. III, p.216

RIOU, Y. (1987). Comparison between MERCURE-GL code calculations, wind tunnel measurements and Thorney Island field trials. *J. Haz. Materials*, **16**, 247

RIOU, Y. (1988). The use of a three dimensional model for simulating Thorney Island field trials. In Puttock, J.S. (1988b), *op. cit.*, p. 323

RIOU, Y.J. and SAAB, A.E. (1986). A three-dimensional numerical model for the dispersion of heavy gases over complex terrain. In de Wispelaere, C., Schiermeier, F.A. and Gillani, N.V. (1986), *op. cit.*, p. 743

RIP, A. (1988). Should social amplification of risk be counteracted? *Risk Analysis*, **8**, 193

RIPLEY, K.D. (1972). Monitoring of industrial effluents. *Chem. Engng*, **79**(May 8), 119

RIPPIN, D.W.T. (1983). Simulation of single- and multi- product batch chemical plants for optimal design and operation. *Comput. Chem. Engng*, **7**, 137

RIPPLE, D.F. (1987). Hazardous release mitigation developments in Europe. *Preventing Major Chemical Accidents*, p. 5.111

RIPPS, D.L. (1965). Adjustment of experimental data. *Chem. Engng Prog., Symp Ser.*, **61**, 8

DE RIS, J. (1969). The spread of a laminar diffusion flame. Combustion, **12**, 241

DE RIS, J. (1970). Duct fires. *Combust. Sci. Technol.*, **2**, 239

DE RIS, J. (1979). Fire radiation – a review. *Combustion*, **17**, 1003

DE RIS, J. (1986). Chemistry and physics of fire. *Fire Protection Handbook*, 16th ed.(ed. A.E. Cote and J.M. Linville), p. 4.1

DE RIS, J. and ORLOFF, L. (1972). A dimensionless correlation of pool burning data. *Combust. Flame*, **18**, 381

DE RIS, J., KANURY, A.M. and YUEN, M.C. (1973). Pressure modeling of fires. *Combustion*, **14**, 1083

RISINGER, J.L. (1962). Fire protection handbook. *Hydrocarbon Process.*, **41**(1), 182; **41**(2), 188; **41**(3), 204; **41**(4), 178; **41**(5), 209; **41**(7), 141

RISINGER, J.L. (1964a). Burning characteristics of hydrocarbon products. In Vervalin, C.H. (1964a), *op. cit.*, p. 25

RISINGER, J.L. (1964b). Controlling floating roof tank fires. In Vervalin, C.H. (1964a), *op. cit.*, p. 309

RISINGER, J.L. (1964c). Controlling high vapor-pressure fires. In Vervalin, C.H. (1964a), *op. cit.*, p. 320

RISINGER, J.L. (1964d). Flame propagation of hydrocarbon products. In Vervalin, C.H. (1964a), *op. cit.*, p. 30

RISINGER, J.L. (1964e). How oil acts when it burns. In Vervalin, C.H. (1964a), *op. cit.*, p. 1

RISINGER, J.L. (1964f). How to control and prevent crude oil tank fires. In Vervalin, C.H. (1964a), *op. cit.*, p. 313

RISINGER, J.L. (1964g). How to fight refinery fires. In Vervalin, C.H. (1964a), *op. cit.*, p. 300

RISINGER, J.L. (1964h). How to handle and store jet fuels. In Vervalin, C.H. (1964a), *op. cit.*, p. 272

RISINGER, J.L. (1964i). Selection and layout of plant site. In Vervalin, C.H. (1964a), *op. cit.*, p. 13

RISINGER, J.L. (1964j). What you should know about fires. In Vervalin, C.H. (1964a), *op. cit.*, p. 8

RISINGER, J.L. and VERVALIN, C.H. (1964). Sources of ignition. In Vervalin, C.H. (1964a), *op. cit.*, p. 35

RISK RESEARCH COMMITTEE (1980). *Accident Risks in Norway: How Do We Perceive and Handle Risks* (Oslo)

RISO NATIONAL LABORATORY – see Appendix 28

RISTON, J. (1976). Health and Safety at Work Act and the plant engineer. *Plant Engr*, **20**(8/9), Sep./Oct., 26; (10/11) Nov./Dec., 11

RITCHIE, C.B. (1989). *Piper Alpha Public Inquiry. Transcript, Days 86, 87*

RITTER, K. (1972). Technical and safety parameters for the assessment of the explosion risk of dusts. *Int. Symp. on Dust and Explosion Risks in Mines and Industry* (Heidelberg; Berufsgenossenschaft der Chem. Ind., p. 27

RITTER, K. (1982). Preventive measures for the avoidance of dust explosions. *The Control and Prevention of Dust Explosions* (London: Oyez/IBC), paper 4

RITTMAN, B.E. (1982). Application of two-thirds law to plume rise from industrial-sized sources. *Atmos. Environ.*, **16**, 2575; 1983, **17**, 1034

RIVAS, R.J. and RUDD, D.F. (1974). Synthesis of failure-safe operations. *AIChE J.*, **20**, 320

RIVAS, R.J., RUDD, D.F. and KELLY, L.R. (1974). Computer-aided safety interlock systems. *AIChE J.*, **20**, 311

RIVERA, F.P. (1975). Problems of static electricity in the oil industry. *Fire Int.*, **50**, 73

ROACH, J.R. and LEES, F.P. (1981). Some features of and activities in hazard and operability (hazop) studies. *Chem. Engr, Lond.*, Oct, 456

ROACH, S.A. (1970). Hygiene standards for asbestos. *Ann. Occup. Hyg.*, **13**, 7

ROARK, R.J. and YOUNG, W.C. (1975). *Formulas for Stress and Strain* (New York: McGraw-Hill)

ROBBINS, L.A. (1979). The miniplant concept. *Chem. Engng Prog.*, **75**(9), 48

ROBBINS, T. (1974). Use of the computer in planning and progressing commissioning operations. *Inst. Chem. Engrs, Symp. on Commissioning of Chemical and Allied Plant* (London)

ROBENS, LORD (chrmn) (1972). *Safety and Health at Work. Cmnd 5034* (London: HM Stationery Office)

ROBERSON, C.G., O'HEARNE, M.D. and HARKINS, B.L. (1993). Benefits of standardised advanced controls. *Hydrocarbon Process.*, **72**(7), 63

ROBERTS, A.A. (1915). *The Poison War* (London: Heinemann)

ROBERTS, A.F. (1964). Calorific values of partially decomposed wood samples (letter). *Combust. Flame*, **8**, 245

ROBERTS, A.F. (1966). Spread of flame on a liquid surface. Ph.D. Thesis, Imperial Coll., London

ROBERTS, A.F. (1967a). An analogue method of estimating wood pyrolysis rates. *Combustion*, **11**, p. 561

ROBERTS, A.F. (1967b). Calorific values of the volatile pyrolysis products of wood (letter). *Combust. Flame*, **11**, 439

ROBERTS, A.F. (1969/70) Fires in ducts under forced ventilation conditions. *Fire Technol.*, **5/6**, 13

ROBERTS, A.F. (1970). A review of kinetics data for the pyrolysis of wood and related substances. *Combust. Flame*, **14**, 261

ROBERTS, A.F. (1971a). Comments on 'Duct fires' by J. de Ris. *Combust. Sci. Technol.*, **3**, 263

ROBERTS, A.F. (1971b). Comments on 'The ignition of a liquid fuel pool' by Murad *et al.* (letter). *Combust. Flame*, **17**, 440

ROBERTS, A.F. (1971c). The heat of reaction during the pyrolysis of wood. *Combust. Flame*, **17**, 79

ROBERTS, A.F. (1971d). Problems associated with the theoretical analysis of the burning of wood. *Combustion*, **13**, p. 893

ROBERTS, A.F. (1975a). Extinction phenomena in liquids. *Combustion*, **15**, p. 305

ROBERTS, A.F. (1975b). *A Survey of Certain Aspects of the Blast Damage. Rep. to the Flixborough Court of Inquiry* (Safety in Mines Res. Estab.)

ROBERTS, A.F. (1980). The collection of data from chemical plant incidents as an aid to the improvement of hazard analysis techniques. *Loss Prevention and Safety Promotion 3*, p. 1443

ROBERTS, A.F. (1981a). Methods for estimating the harmful consequences of fires and explosions following a release of LPG. *Health and Safety Executive Seminar on Thermal Radiation and Overpressure Calculations* (London: Health and Safety Executive)

ROBERTS, A.F. (1981b). Thermal radiation hazards from releases of LPG from pressurised storage. *Health and Safety Executive Seminar on Thermal Radiation and Overpressure Calculatiolls* (London: Health and Safety Executive)

ROBERTS, A.F. (1981/82). Thermal radiation hazards from releases of LPG from pressurised storage. *Fire Safety J.*, **4**(3), 197

ROBERTS, A.F. (1982). The effect of conditions prior to loss of containment on fireball behaviour. *The Assessment of Major Hazards* (Rugby: Instn Chem. Engrs), p. 181

ROBERTS, A.F. (1987). HSE's research programme on loss prevention. *Chem. Engng Res. Dev.*, **65**, 291

ROBERTS, A.F. and CLOUGH, G. (1963). Thermal decomposition of wood in an inert atmosphere. *Combustion*, **9**, 158

ROBERTS, A.F. and CLOUGH, G. (1967a). Model studies of heat transfer in mine fires. *Safety in Mines Res. Estab., Res. Rep. 247* (London: HM Stationery Office)

ROBERTS, A.F. and CLOUGH, G. (1967b). The propagation of fires in passages lined with flammable material. *Combust. Flame*, **11**, 365

ROBERTS, A.F. and KENNEDY, M. (1965). Modelling of mine roadway fires. *Safety in Mines Res. Estab., Res. Rep. 239* (London: HM Stationery Office)

ROBERTS, A.F. and PRITCHARD, D.K. (1982). Blast effect from unconfined vapour cloud explosions. *J. Occup. Accidents*, **3**(2), 231

ROBERTS, A.F. and QUINCE, B.W. (1973). A limiting condition for the burning of flammable liquids. *Combust. Flame*, **20**, 245

ROBERTS, A.F., CLOUGH, G. and BLACKWELL, J.R. (1966). A model duct for mine fire research. *Safety in Mines Res. Estab., Res. Rep. 243* (London: HM Stationery Office)

ROBERTS, A.F., CUTLER, D.P. and BILLINGE, K. (1983). Fire engulfment trials with LPG tanks with a range of fire protection methods. *Loss Prevention and Safety Promotion 4*, vol. 3, p. D1

ROBERTS, E.L.M. (1977). Chemical handling. In Handley, W. (1977), *op. cit.*, p. 34

ROBERTS, E.M. (1979). Review of statistics of extreme values with applications to air quality data. *J. Air Poll. Control Ass.*, **29**, 632, 733

ROBERTS, G.K. (1970). *Political Parties and Pressure Groups in Britain* (London: Weidenfeld & Nicolson)

ROBERTS, J.H.T. (1913). The disintegration of metals at high temperatures. Condensation nuclei from hot wires. *Phil. Mag.*, **25**, 270

ROBERTS, S.K. (1989). *Piper Alpha Public Inquiry. Transcript, Day 65*

ROBERTS, K.C. (1976). A fast response spill recovery system. *Control of Hazardous Material Spills 3*, p. 152

ROBERTS, K.H. (1989). New challenges in organizational research: high reliability organizations. *Ind. Crisis Quart.*, **3**(2), 111

ROBERTS, K.H. and GARGANO, G. (1989). Managing a high reliability organization: a case for interdependence. *Managing Complexity in High Technology Organizations* (New York: Oxford Univ. Press)

ROBERTS, L.E.J. (1989). The challenge facing risk assessment. In Cox, R.A. (1989a) *op. cit.*, p. 1

ROBERTS, N.H. (1964). *Mathematical Methods in Reliability Engineering* (New York: McGraw-Hill)

ROBERTS, N.W. and STEWART, M.J. (1973). Economics of long-distance high-pressure gas transmission. In Hepple, P. (1973), *op. cit.*, p. 61

ROBERTS, O.F.T. (1923). The theoretical seattering of smoke in a turbulent atmosphere. *Proc. R. Soc., Ser. A*, **104**, 640

ROBERTS, P. (1990). Application of hazard and risk analysis to gas making and gas storage facilities. *Safety and Loss Prevention in the Chemical and Oil Processing Industries*, p. 527

ROBERTS, P., SMITH, D.B. and WARD, D.R. (1980). Flammability of paraffin hydrocarbons in confined and unconfined conditions. *Chemical Process Hazards*, **VII**, p. 157

ROBERTS, P. and SMITH, D.B. (1983). Purge limits for paraffin hydrocarbons in air and nitrogen. *Loss Prevention and Safety Promotion 4*, vol. 3, p. E1

ROBERTS, P.H. (1961). Analytical theory of turbulent diffusion. *J. Fluid Mech.*, **11**, 259

ROBERTS, P.T., PUTTOCK, J.S. and BLEWITT, D.N. (1990). Gravity spreading and surface roughness effects in dispersion of dense gas plumes. *Am. Inst. Chem. Engrs Symp. on Health and Safety*, Orlando, FA

ROBERTS, P.T. and HALL, D.J. (1994). Wind-tunnel simulation. Boundary layer effects in dense diffusion experiments. *J. Loss Prev. Proeess Ind.*, **7**, 106

ROBERTS, R.H. and HANDMAN, S.E. (1986). Minimize ammonia releases. *Hydrocarbon Process.*, **65**(3), 58

ROBERTS, R.K. (1973). The IMCO code for the construction and equipment of ships carrying dangerous liquid chemicals in bulk. Experience during the first year of implementation. *Transport of Hazardous Cargoes*, **3**

ROBERTS, R.K. (1975). International transport of dangerous goods in portable tanks. *Distribution Conf.* (London: Chem. Ind. Ass.), p. 34

ROBERTS, R.K. (1983). The International Bulk Chemical and Gas Codes: an assessment of aspects of design and construction requiring clarification or investigation. *Loss Prevention and Safety Promotion 4*, vol. 2, p. A11

ROBERTS, T.A., MERRIFIELD, R. and THARMALINGHAM, S. (1990). Thermal radiation hazards from organic peroxides. *J. Loss Prev. Process Ind.*, **3**(2), 244

ROBERTS, T.A. and ROYLE, M. (1991). Classification of energetic industrial chemicals for transport. *Hazards*, **XI**, p. 191

ROBERTSON, A. (1981). Seveso facts (letter). *Chem. Ind.*, Apr. 4, 196

ROBERTSON, A.J.B. (1949). The thermal initiation of explosion in liquid explosives. *Combustion*, **3**, p. 545

ROBERTSON, C.A. (1970). Materials engineering of hydro-treating equipment. *Loss Prevention*, **4**, p. 46. See also *Chem. Engng Prog.*, 1970, **66**(9), 33

ROBERTSON, C.A. (1972). Materials selection for pumps. *Hydrocarbon Process.*, **51**(6), 93

ROBERTSON, G.G. (1989). *Piper Alpha Public Inquiry. Transcript, Days 96, 97*

ROBERTSON, J.H., COWEN, W.F. and LONGFIELD, J.Y. (1980). Water pollution control. *Chem. Engng*, **87**(Jun. 30), 102

ROBERTSON, R.B. (1974a). Fire engineering in relation to process plant design. *Fire Int.*, **45, 69**

ROBERTSON, R.B. (1974b). Fire engineering in relation to plant design and layout. *Course on Process Safety – Theory and Practice*. Dept of Chem. Engng, Teeside Polytechnic, Middlesbrough

ROBERTSON, R.B. (1976a). Fire safety related to process plant layout. *Course on Fire and Explosion Safety in the Chemical Industry*. Dept of Fire Safety Engng, Univ. of Edinburgh

ROBERTSON, R.B. (1976b). Spacing in chemical plant design against loss by fire. *Process Industry Hazards*, p. 157

ROBERTSON, S.S.J. and LOVELAND, R.J. (1981). Radio frequency ignition hazards: a review. *IEE Proc.*, **128**, 607

ROBERTSON, W.S. (ed.) (1981). *Boiler Efficiency and Safety* (London: Macmillan)

ROBERTSON, W.S. (ed.) (1983). *Lubrication in Practice*, 2nd ed. (London: Macmillan)

ROBERTSON, W.W., MAGEE, E.M., FAIN, J. and MATSEN, F.A. (1955). Self-combustion of acetylene. Pt I, Pre-ignition kinetics. *Combustion*, **5**, 628

ROBERTSON, W.W. and MATSEN, F.A. (1957). A cool flame phenomenon in the combustion of acetylene. *Combust. Flame*, **1**, 99

ROBINS, A.G. and FACKRELL, J.E. (1983). *Mean Concentration Levels around Buildings due to Nearby Low Level Emissions. Rep. TPRD/M/1261/N82*. Central Elec. Generating Board, Leatherhead

ROBINSON, A.G. (1971). The control of ships carrying dangerous cargoes in harbours and their approaches. *Transport of Hazardous Cargoes*, **2**, paper 11

ROBINSON, B.W. (1977). Factors influencing the limitations to fractional dead time. *First Nat. Conf. on Reliability* (Culcheth, Lancashire: UK Atom. Energy Auth.), NRC3/1

ROBINSON, C. and SMITH, D.B. (1984). The auto-ignition temperature of methane. *J. Haz. Materials*, **8**(3), 199

ROBINSON, C.S. (1944). *Explosions. Their Anatomy and Destructiveness* (New York: McGraw-Hill)

ROBINSON, C.S. and GILLILAND, E.R. (1950). *Elements of Fractional Distillation*, 4th ed. (New York: McGraw-Hill)

ROBINSON, E.R. (1975). *Time Dependent Chemical Processes* (London: Applied Science)

ROBINSON, F.A. and AMIES, F.A. (1967). *Chemists and the Law* (London: Spon)

ROBINSON, G. (1977). Assess and control HPI plant noise. *Hydrocarbon Process.*, **56**(6), 223

ROBINSON, H. and WHEELER, R.V. (1933). Explosions of methane and air: propagation through a restricted tube. *J. Chem. Soc.*, 758

ROBINSON, H.S. (1967a). Loss risks in large integrated plants. *Loss Prevention*, **1**, p. 18

ROBINSON, H.S. (1967b). Process control and design case histories. *Loss Prevention*, **1**, p. 32

ROBINSON, H.S. (1968). Fire and explosion risks in the modern ammonia plant. *Ammonia Plant Satety*, **10**, p. 68

ROBINSON, H.S. (1978). An underwriter looks at LDPE plant safety. *Safety in Polyethylene Plants*, **3**, p. 59

ROBINSON, J.A. (1979). The logical basis of programming by assertion and query. In Michie, D. (1979), *op. cit.*, p. 105

ROBINSON, J.A. (1983). Logic programming – past present and future. *New Generation Comput.*, **1**, 107

ROBINSON, J.E. (1993). Underground storage tanks: removal vs abandonment. *Hydrocarbon Process.*, **72**(8), 149

ROBINSON, J.N. and TETELMAN, A.S. (1974). *Measurement of K_{IC} on Small Specimens using Critical Crack Tip Opening Displacement. Fracture Toughness and Slow Stable Cracking. STP 559* (Philadelphia, PA: Am. Soc. Testing Mats)

ROBINSON, J.O. (1993). Clean boiler systems chemically. *Chem. Engng Prog.*, **89**(4), 53

ROBINSON, J.P. (1971). *The Problem of Chemical and Biological Warfare*, vol. 1, *The Rise of CB Weapons* (Stockholm)

ROBINSON, J.P. and CONVERSE, P.E. (1966). *Summary of US Time Use Survey* (report). Inst. for Social Res., Univ. of Michigan, Ann Arbor, MI

ROBINSON, J.R. (1973). Start-up of a Bender process unit. *Chem. Engng Prog.*, **69**(8), 101

ROBINSON, M.J. and STRUTT, J.E. (1983). The assessment and prediction of plant corrosion using an on-litle monitoring system. In Butcher, D.W. (1983), *op. cit.*, p. 79

ROBINSON, N. (ed.) (1966). *Solar Radiation* (Amsterdam: Elsevier)

ROBINSON, T. (1993). Compare non-metal and metal valves. *Chem. Engng Prog.*, **89**(12), 25

ROBINSON, T. and YODAIKEN, R. (1989). The evolution of chronic hazard evaluation. *J. Haz. Materials*, **21**, 201

ROBINSON, W.F. (ed.) (1986). *The Solid Waste Handbook* (New York: Wiley)

ROBNETT, J.D. (1979). Engineering approaches to energy conservation. *Chem. Engng Prog.*, **75**(3), 59

ROBSON, R.E. (1983). On the theory of plume trapping by an elevated inversion. *Atmos. Environ.*, **17**, 1923

ROBURN, J. (1968). Hazard evaluation of new chemicals. *Carriage of Dangerous Goods*, **1**, p. 146

ROCHE, B. *et al.* (1977). Safety rules concerning fire and explosion hazards in a research centre equipped with pilot plants. *Loss Prevention and Safety Promotion 2*, p. 87

ROCHESTER, A. (1969). Fork lift trucks. In Handley, W. (1969), *op. cit.*, p. 122

ROCHINA, V. (1986). A practical risk analysis tool use in the case of hydrogen production unit. *Loss Prevention and Safety Promotion 5*, 18-1

ROCK, N.I.C. and BUTCHER, N.A. (1991). The role of predictive analysis in offshore satety. *Satety Developments in the Offshore Oil and Gas Industry* (London: Instn Mech. Engrs), p. 109

ROCKETT, J.A. (1976). Fire induced gas flow in an enclosure. *Combust. Sci. Technol.*, **12**, 165

ROCKLEY, J.C. (1964). *Introduction to Industrial Radiography* (London: Butterworths)

ROCKWELL, T. and BHISE, V. (1970). Two approaches to a non-accident measure for continuous assessment of safety performance. *J. Safety Res.*, Sep.

RODEAN, H.C. (1982). *Analysis of Turbulent Wind Velocity and Gas Concentration Fluctuations during the Burro Series 40-m³ LNG Spill Experiments. Rep. UCRL 53353.* Lawrence Livermore Nat. Lab., Livermore, CA

RODEAN, H.C. (1984). Effects of a spill of LNG on mean flow and turbulence under low wind speed, slightly stable atmospheric conditions. In Ooms, G. and Tennekes, H. (1984), *op. cit.*, p. 157

RODEAN, H.C., HOGAN, W.J., URTIEW, P.A., GOLDWIRE, H.C., MCRAE, T.G. and MORGAN, D.L. (1984). *Vapor Burn Analysis for the Coyote Series LNG Spill Experiments. Rep. 53530.* Lawrence Livermore Nat. Lab., Livermore, CA

RODEAN, H.C. and CHAN, S.T. (1986). *Generalized Phase Change Models in the FEM3 Model for Gas Dispersion. Rep. UCID-20817.* Lawrence Livermore Nat. Lab., Livermore, CA

RODGERS, S.J. (1976). Commercially available monitors for airborne hazardous chemicals – an overview. *Control of Hazardous Material Spills*, **3**, p. 208

RODGERS, T.H. (1974). *Marine Corrosion* (London: Butterworths)

RODGERS, W.H. (1987). Guerilla decision-making: judicial review of risk assessments. *J. Haz. Materials*, **15**(1/2), 205

RODGERS, W.P. (1971). *Introduction to System Safety Engineering (New York: Wiley)*

RODI, W. (1980). Turbulence models for environmental problems. In Kollman, W. (1980), *op. cit.*, p. 259

RODI, W. (ed.) (1982). *Turbulent Buoyant Jets and Plumes* (Oxford: Pergamon Press)

RODIN, J. (1978). Predicting man-made hazards – state of the art. *First Conf. on Predicting and Designing for Natural and Man-Made Hazards* (New York: Am. Soc. Civil Engrs), p. 21

RODRICKS, J.V. and TARDIFF, R.G. (1984a). *Assessment and Management of Chemical Risks* (Washington, DC: Am. Chem. Soc.)

RODRICKS, R.V. and TARDIFF, R.G. (1984b). Conceptual basis for risk assessment. In Rodricks, R.V. and Tardiff, R.G. (1984a), *op. cit.*, p. 3

RODRICKS, J.V. and TAYLOR, M.R. (1989). Comparison of risk management in US regulatory agencies. *J. Haz. Materials*, **20**, 239

RODRIGUEZ, A.L.M. and CORONEL, F.J.M. (1992). A better way to price plants. *Chem. Engng*, **99**(3), 124

RAE, L.B. (1978). *Practical Electrical Project Engineering* (New York: McGraw-Hill)

RAE, R.A. (1969). Can large motor failures be cut? *Hydrocarbon Process.*, **48**(5), 126

ROEBUCK, A.H. (1978). Safe chemical cleaning – the organic way. *Chem. Engng*, **85**(Jul. 31), 107

ROEBUCK, B. (1985). The presentation and availability of the data and plans for future analysis. *J. Haz. Materials*, **11** (1/3), 373. See also in McQuaid, J. (1985), *op. cit.*, p. 373

ROEHLICH, F., HILTZ, R.H. and BRUGGER, J.E. (1976). An emergency collection system for spilled hazardous materials. *Control of Hazardous Material Spills*, **3**, p. 314

ROEHRS, R.J. (1967). Equipment testing: leak testing assures sound vessels. *Chem. Engng*, **74**(Aug. 28), 120

VAN ROECKEL, L. (1985). Thermal runaway reactions: hazard evaluation. In Hoffman, J.M. and Maser, D.C. (1985), *op. cit.*, p. 69

ROGACZEWSKA, T. (1975). Absorption of acrylonitrile vapours through the skin in animals. *Med. Prac.*, **26**, 459 (HSE Translation 12869)

ROGALSKI, J. and SAMURCAY, R. (1993). Analysing communication in complex distributed decision-making. *Ergonomics*, **36**, 1329

ROGERS, A.J. (1991). Radio communication fire detector system. *Fire Surveyor*, **20**(5), 19

ROGERS, G.F.C. and MAYHEW, Y.R. (1957). Engineering Thermodynamics: Work and Heat Transfer; 1967, 2nd ed.; 1980, 3rd ed. (London: Longmans); 1992, 4th ed. (Harlow: Longmans)

ROGERS, G.O. and SORENSEN, J.H. (1989). Warning and response in two hazardous materials transportation accidents in the USA. *J. Haz. Materials*, **22**(1), 57

ROGERS, G.O. and SORENSEN, J.H. (1991). Adoption of emergency planning practices for chemical hazards in the United States. *J. Haz. Materials*, **27**, 3

ROGERS, H.J., SPECTO, R.G. and TROUNCE, J.R. (1981). *A Textbook of Clinical Pharlllacology* (London: Hodder & Stoughton)

ROGERS, J.A. (1965). Explosion in nitrogen gas compressors. *Ammonia Plant Safety*, **7**, p. 11

ROGERS, J.K. (1988). Safety services. *Plant Engr*, **32**(3), 31

ROGERS, J.M. (1988). Detecting toxic vapors early. *Chem. Engng*, **95**(5), Apr. 11, 81

ROGERS, L.M. (1976). *Temperature and Strain Field Measurements in NDT. Diagnostic Techniques* (London: Instn. Chem. Engrs)

ROGERS, L.M. (1986). The detection and monitoring of cracks in structures, process vessels and pipework by acoustic emission. *Hazards*, **IX**, p. 201

ROGERS, L.M. and ALEXANDER, F. (1983). Detection and monitoring of propagating cracks in steel structures by acoustic emission analysis. In Butcher, D.W. (1983), *op. cit.*, p. 11

ROGERS, P. (1980). Seal safety. *Plant Engr*, **24**(3), 21

ROGERS, R.L. (1989). The advantages and limitations of adiabatic Dewar calorimetry in chemical hazard testing. *Runaway Reactions*, p. 281. See also *Plant/Operations Prog.*, 1989, **8**, 109

ROGERS, R.L. (1989). The systematic assessment of chemical reaction hazards. *Runaway Reactions*, p. 578

ROGERS, R.L. and HALLAM, S. (1991). A chemical approach to inherent safety. *Hazards*, **XI**, p. 235. See also *Process Safety Environ.*, **69B**, 149

ROGERS, R.L. and HALLAM, S. (1991). A chemical approach to inherent safety. *IChemE Symp. Series* No. 124, 235–41

ROGERS, R.L., MANSFIELD, D.P., MALMEN, Y., TURNEY, R.D. and VERWOERD, M. (1995). The INSIDE project: Integrating inherent safety in chemical process development and plant design. *Inter. Symp. on Runaway React. and Pressure Relief Design*, Aug. 2–4, 1995, Boston, MA (ed. G.A. Melhem and H.G. Fisher) 668–89 (New York: American Institute of Chemical Engineers)

ROGERS, T.M. (1988). Improving valve quality. *Hydrocarbon Process.*, **67**(8), 37

ROGERS, W.E. (1954). *Introduction to Electric Fields* (New York: McGraw-Hill)

ROGERSON, J.E. (1982). Fire tests on NFPA IIIA combustible liquids stored in drums. *Plant/Operatiolls Prog.*, **1**, 33

ROGERSON, J.E. (1982). Entrance of dust into pressurized enclosures. *Plant/Operations Prog.*, **1**, 48

ROGERSON, J.H. (1986). *Quality Assurance in Process Plant Manufacture* (London: Elsevier Applied Science)

ROGOJINA, A. (1977). Some aspects of the safety engineering for the processing and conditioning machines in the varnishes and paints industry. *Prevention of Occupational Risks*, **5**, p. 133

ROGOSHENSKI, P., BOYSON, H. and WAGNER, K. (1983). *Remedial Action Technology for Waste Disposal Sites* (Park Ridge, NJ: Noyes Data Corp.). See also Wagner, K. *et al.* (1986)

ROGOVIN, M. (1980). *Three Mile Island – a report to the Commissioners and to the Public* (Washington, DC: Nucl. Regul. Comm.)

ROGOWSKI, Z.W. (1975). Protection of electrical equipment using flame arresters as pressure reliefs. *Electrical Safety*, **2**, p. 107

ROGOWSKI, Z.W. (1977). Modern methods of explosion protection. *Chemical Process Hazards*, **VI**, p. 63

ROGOWSKI, Z.W. (1980). Flame arresters in industry. *Chemical Process Hazards*, **VII**, p. 53

ROGOWSKI, Z.W. and RASBASH, D.J. (1963). Relief of explosions in propan–air mixtures moving in a straight unobstructed duct. *Chemical Process Hazards*, **II**, p. 21

ROHLEDER, G.V. (1969). Considerations in an ammonia pipeline system. *Ammonia Plant Safety*, **11**, 35

ROHLEDER, G.V. (1971). Driver loaded terminals for anhydrous ammonia. *Ammonia Plant Safety*, **13**, p. 18

ROHLFING, R.L. (1984). Ammonia plant design for minimum manpower requirements. *Plant/Operations Prog.*, **3**, 173

ROHNE, H. (1895). Studies on shrapnel from field artillery. *Archiv der Artillerie- und Ingenieur-Offiziere*, **101**, 385

ROHSENHOW, W.M., HARTNETT, J.P. and GANIC, E.N. (eds) (1985). *Handbook of Heat Transfer Fundamentals* (New York: McGraw-Hill)

ROLFE, S.T. and BARSOM, J.M. (1977). *Fracture and Fatigue Control in Structures* (Englewood Cliffs, NJ: Prentice-Hall)

ROLFE, S.T. and BARSOM, J.M. (1987). *Fracture and Fatigue Control in Structures*, 2nd ed. (Englewood Cliffs, NJ: Prentice-Hall)

ROLL, H.U. (1965). *Physics of the Marine Atmosphere* (New York: Academic Press)

ROLLET, A.P. and BOUAZIZ, R. (1972). *L'Analyse Thermique*, vol. 2, *L'Examen des Processus Chimiques* (Paris: Gauthier-Villars)

ROLLINS, D.K. and DAVIS, J.F. (1992). Unbiased estimation of gross errors in process measurements. *AIChE J.*, **38**, 563

ROLLINS, D.K. and DAVIS, J.F. (1993). Gross error detection when variance – covariance matrices are unknown. *AIChE J.*, **39**, 1335

ROLLINS, J.W. (1972). Exchangers nitriding in a loop. *Ammonia Plant Safety*, **14**, p. 28

ROLSTON, J.A. (1980). Fiberglass composite materials and fabrication processes. *Chem. Engng*, **87**(Jan. 28), 96

ROLT, L.T.C. (1982). *Red for Danger*, 4th ed. (Newton Abbott: David & Charles)

ROMAGNOLI, J.A. and STEPHANOPOULOS, G. (1981). Rectification of process measurement data in the presence of gross errors. *Chem. Engng Sci.*, **36**, 1849

ROMANO, A., DOSI, W. and BELLO, G.C. (1983). Probabilistic risk assessment of a large marine terminal of liquefied petroleum gases. *Loss Prevention and Safety Promotion 4*, vol. 2, p. B21

ROMANO R. (1983). Control emissions with flare efficiency. *Hydrocarbon Process.*, **62**(10), 78

RÖMCKE, O. and EVENSEN, O.K. (1940). The chlorine gas poisoning in Mjondalen. *Nord. Med.*, **7**, 1224

ROMER, C.J. (1987). The importance of accident and injury prevention. *Ergonomics*, **30**, 173

RØMER, H., BROCKHOFF, L., HAASTRUP, P. and STYHR PETERSEN, H.J. (1993). Marine transport of dangerous goods. Risk assessment based on historical accident data. *J. Loss Prev. Process Ind.*, **6**, 219

ROMIG, R.P., ROTHFUS, R. and KERMODE, R.I. (1966). Partial vaporization in orifices and valves. *Ind. Engng Chem. Proc. Des. Dev.*, **5**, 342

ROMMEL, J.C. and TRAIFOROS, S.A. (1983). Comparison of analytical and experimental results tor a PWR pressurizer safety valve discharge. In Veziroglu, T.N. (1983), *op. cit.*, p. 198

ROMPP, H. (1966). *Chemie Lexikon*, 6th ed. (Stuttgart: Frankhsche Verlags Handlung W. Keller)

RON, D., TAITELMAN, U., MICHAELSON, M., BAR-JOSEPH, G., BURSTEIN, S. and BETTER, O.S. (1984). Prevention of acute renal failure in traumatic rhabdomyolisis. *Arch. Intern. Med.*, **144**, 277

RONAN, W.W. (1953). *Training for Emergency Procedures in Multiengine Aircraft. Rep. AIR-153-53-FR-44,.* Am. Inst. Res., Pittsburgh, PA

RONEN, Y. (ed.) (1986). *CRC Handbook of Nuclear Reactors Calculations*, vols 1–3 (Boca Raton, FL: CRC Press)

RONEY, J.C. (1975). Coupling failure between compressors. *Ammonia Plant Safety*, **17**, p. 143

RONEY, J.C. and ACREE, R.E. (1973). Process isolation using stopples. *Ammonia Plant Safety*, **15**, p. 56

RONEY, J.C. and PERSSON, J.P. (1984). Philosophy of primary reformer repairs. *Ammonia Plant Safety*, **24**, 58

RONNEY, P.D. (1985). Effect of gravity on laminar premixed gas combustion. II, Ignition and extinction Phenomena. *Combust. Flame*, 62, 121

RONNEY, P.D. and WACHMAN, H.Y. (1985). Effect of gravity on laminar premixed gas comblstion. I, Flammability limits and burning velocities. *Combust. Flame*, **62**, 107

RONOLD, A. (1992). Computer simulations of wind and ventilation aid platform design and safety. *Major Hazards Onshore and Offshore*, p. 375

RONZANI, E. (1908). On the influence of inhalation of industrial irritant gases on the organism's powers of resistance to infectious diseases, Pt 1. *Arch. Hyg.* **70**

RONZANI, E. (1909). On the influence of inhalations of industrial irritant gases on the organisms powers of resistance to infectious diseases. H, Hydrogen fluoride, ammonia, hydrogen chloride. *Arch. Hyg.*, **70**, 217

ROOD, A.P. (1992a). Design and performance of the 'HSE' monitor. In Stanley-Wood, N.G. and Lines, R.W. (1992), *op. cit.*, p. 276

ROOD, A.P. (1992b). Particle characterisation for health and safety auditing. In Stanley-Wood, N.G. and Lines, R.W. (1992), *op. cit.*, p. 262

ROODMAN, R.G. (1982). Operations: a critical factor often neglected in plant design. *Chem. Engng*, **89**(May 17), 131

DE ROOIJ, C.G., VAN EICK, A.J. and VAN DE MEENT, D. (1981). *Glucocorticosteroids in the Therapy of Acute Phosgene poisoning in Rats and Mice. Rep. MBL-1981-6*

ROOK, L.W. (1962). *Reduction of Human Error in Industrial Production. Rep. SCTM 93-62(14)*. Sandia Nat. Labs, Albuquerque, NM

ROOK, L.W. (1964). Evaluation of system performance from rank order data. *Hum. Factors*, **6**, 533

ROOK, P. (ed.) (1990). *Software Reliability Handbook* (New York: Elsevier Applied Science)

ROOK, P. (1991). Project planning and control. In McDermid, J.A. (1991), *op. cit.*

ROOKE, D.E. and FILSTEAD, C.G. (1970). Six years operational experience with the 'Methane Princess' and 'Methane Progress'. *LNG Conf. 2*

ROOKE, D.P. and CARTWRIGHT, D.J. (1976). *Compendium of Stress Intensity Factors* (London: HM Stationery Office)

ROONEY, I.W. (1986). Advances in detection and reliability of atmospheric chlorine detection systems. In Wall, K. (1986), *op. cit.*, p. 82

ROONEY, J.J. and FUSSELL, J.B. (1978). *BACKFIRE II – a Computer Program for Common Cause Failure Analysis of Complex Systems*, Univ. of Tennessee

ROONEY, J.J., TURNER, J.H. and ARENDT, J.S. (1988). A preliminary hazards analysis of a fluid catalytic cracking unit complex. *J. Loss Prev. Process Ind.*, **1**(2), 96

ROONEY, J.J., SMITH, M.F. and ARENDT, J.S. (1993). A practical approach for integrating process safety and total quality management programs. *Process Safety Management*, p. 147

ROONEY, J.P. (1983). Predict reliability of instrumentation. *Hydrocarbon Process.*, **62**(1), 89

ROOPCHAND, D.R. (1983). Prediction of the effects of spills from LNG. *Thirty-Third CSChE Conf.*, Toronto

ROOPCHAND, D.R. and MODEREGGER, M. (1993). Risk based guidelines for the use of remote impounding basins. In Melchers, R.E. and Stewart, M.G. (1993), *op. cit.*, p. 77

ROOS, C.W. (1984). *Computer Systems for Occupational Safety and Health Management* (New York: Dekker)

ROOT, R.B. and SCHMITZ, R.A. (1969). An experimental study of steady state multiplicity in a loop reactor. *AIChE J.*, **15**, 670

ROOTSAERT, T. and HARRINGTON, J. (1992). A knowledge based computer program 'COMHAZOP' as aid for hazard and operability studies in process plants. *Loss Prevention and Safety Promotion 7*, vol. 3, 156–1

ROPER, D.L. and RYANS, J.L. (1989). Selecting the right vacuum gage. *Chem. Engng*, **96**(3), 125

ROPER, F., ARNO, J. and JAGGERS, H.C. (1991). The effect of release velocity and geometry on burning times for non-premixed fuel gas clouds. *Combust. Sci. Technol.*, **78**, 307

ROPER, F.G., JAGGERS, H.C., FRANKLIN, D.P., SLAVEN, N. and CAMPBELL, A. (1986). Factors controlling scaling laws for buoyancy controlled combustion of spherical clouds. *Combustion*, **21**, 1609

RØREN, E.M.Q. (1987). Development of frontier gas fields: the technological challenge. *Gastech 86*, p. 176

RORTY, M. and MCLEARN, M.E. (1993). Test, monitor, and maintain aboveground tanks properly. *Chem. Engng Prog.*, **89**(12), 45

ROSA, R.R. and BONNET, M.H. (1993). Performance and alertness on 8 h and 12 h rotating shifts at a natural gas utility. *Ergonomics*, **36**, 1177

ROSAK, J. and SKARKA, J. (1980). Methods for estimation of the effects of accidental release of liquefied gases. *Loss Prevention and Safety Promotion 3*, vol. 3, p. 1173

ROSALER, R.C. and RICE, J.O. (1983). *Standard Handbook of Plant Engineering* (New York: McGraw-Hill)

ROSE, A. (1985). Some do's and don't of parametric cost estimating. *Process Engng, Dec., 60*

ROSE, H.E. (1970). Dust explosion hazards. *Br. Chem. Engng*, **15**, 371

ROSE, H.E. (1975). Useful approach to dust explosion hazards. *Loss Prevention*, **9**, 82

ROSE, H.E. and PRIEDE, T. (1959a). Ignition phenomena in hydrogen-air mixtures. *Combustion 7*, p. 436

ROSE, H.E. and PRIEDE, T. (1959b). An investigation of the characteristics of spark discharges as employed in ignition experiments. *Combustion*, 7, p. 454

ROSE, H.E. and WOOD, A.J. (1966). *An Introduction to Electrostatic Precipitation in Theory and Practice*, 2nd ed. (London: Constable)

ROSE, J.C., WELLS, G.L. and YEATS, B.H. (1978). *A Guide to Project Procedure* (London: Instn. Chem. Engrs)

ROSE, J.W. and COOPER, J.R. (1977). *Technical Data on Fuel* (London: Br. Nat. Cttee, World Energy Conf.)

ROSE, L.M. (1976). *Engineering Investment Decisions* (Amsterdam: Elsevier)

ROSE, L.M. (1981). *Chemical Reactor Design in Practice* (Amsterdam: Elsevier)

ROSE, L.M. (1985). *Distillation Design in Practice* (Amsterdam: Elsevier)

ROSE, P. (1990). A model-based system for fault diagnosis of chemical process plants. M.S. Thesis, Massachusetts Inst. of Technol.

ROSE, P. and KRAMER, M.A. (1991). *Qualitative Analysis of Causal Feedback. AAAI-91*

ROSE, R.G. and HOWELL, T.E. (1962). Performance of protective barriers. *Chem. Engng Prog.*, **58**(6), 75

ROSE, S.P. (1968). Reliability in large plants. *Chem. Process.*, *suppl.*, Feb., 514

ROSE, T.J. and NOURISH, R. (1982). Occupational hygiene considerations: an HSE viewpoint. *Chem. Ind.*, Nov. 20, 892

ROSEBROOK, D.D. (1977). Fugitive hydrocarbon emissions. *Chem. Engng*, **84**(Oct. 17), 145

ROSEBROOK, D.D. and WORM, G. (1993). Personal exposures, indoor and outdoor relationships, and breath levels of toxic air pollutants measured for 355 persons in New Jersey (discussion). *Atmos. Environ.*, **27A**, 2243, 2247. See also Wallace, L.A. *et al.* (1985), *op. cit.*

ROSEMAN, W.W. (1975). How to define maintenance work load in the HPI. *Hydrocarbon Process.*, **54**(4), 187

ROSEN, E. and HENLEY, E.J. (1974). Reliability optimization using intermediate storage tanks. *Proc. AIChE/GVC Mtg* (New York: Am. Inst. Chem. Engrs)

ROSENBERG, D. (1987). Joint and several liability for toxic torts. *J. Haz. Materials*, **15**(1/2), 219

ROSENBERG, J., MAH, R.H.S. and IORDACHE, C. (1987). Evaluation of schemes for detecting and identifying gross errors in process data. *Ind. Engng Chem. Res.*, **26**, 555

ROSENBERG, R.H. (1968). Low-pressure alarm for gas cylinders in hazardous areas. *Chem. Engng*, **75**(Jul. 29), 172

ROSENBLATT, M. and HASSIG, P.J. (1986). Numerical simulation of the combustion of an uncollfilled LNG vapor cloud at high constant burning velocity. *Combust. Sci. Technol.*, **45**, 245

ROSENFELD, J.L.J., STRACHAN, D.C., TROMANS, P.S. and SEARSON, P.A. (1981). Experiments on the incendivity of radiofrequency, breakflash discharges (1.8–21 Mhz c.w.). *Radio Electron. Engr*, **51**, 175

ROSENHAH, A.K. and COLE, R. (1973–). Derailment proves CHEMTREC works. *Fire Engng*, **126**(Jul.), 55

ROSENHAHN, A.K. (1986). Fire department apparatus, equipment, and facilities. *Fire Protection Handbook*, 16th ed. (ed. A.E. Cote and J.L. Linville), pp. 15–59

ROSENHOLTZ, M.J., CARSON, T.R., WEEKS, M.H., WILINSKI, F., FORD, D.F. and OBERST, F.W. (1963). A toxicopathological study in animals after single brief exposures to hydrogen fluoride. *Ind. Hyg. J.*, 253

ROSENHOUSE, G., LIN, I.J. and ZIMMELS, Y. (1982). The use of sound-level maps in acoustic design. *Chem. Engng*, **89**(Mar. 8) 72

ROSENOF, H.P. (1982a). Building batch control systems around recipes. *Chem. Engng Prog.*, **78**(9), 59

ROSENOF, H.P. (1982b). Successful batch control planning: a path to plant-wide automation. *Control Engllg*, **29**(4), 107

ROSENOF, H.P. and GHOSH, A. (1987). *Batch Process Automation* (New York: van Nostrand Reinhold)

ROSENTHAL, I. (1993). The corporation and the community: credibility, legitimacy, and imposed risk. *Process Safety Prog.*, **12**, 209

ROSENTHAL, J. and LEWIS, P.G. (1994). The real costs of risk analysis. *Process Safety Prog.*, **13**, 92

ROSENTHAL, L.A. (1988). Static electricity and plastic drums. *Plant/Operations Prog.*, **7**, 51

ROSENZWEIG, M. and BUTCHER, C. (1992a). Can you use that knowledge? *Chem. Engng Prog.*, **88**(4), 76

ROSENZWEIG, M. and BUTCHER, C. (1992b). Should you use that knowledge *Chem. Engllg Prog.*, **88**(10), 85

DES ROSIERS, P.E. (1987). Evaluation of technology for wastes and soils contaminated with dioxins, furans and related substances. *J. Haz. Materials*, **14**(1), 119

ROSKILL, O.W. (1976). The location of chemical plants. *Chem. Engr, Lond.*, **305**, 29

ROSS, C. (1986). Good vibes from anti-sound. *Chem. Engr, Lond.*, **422**, 22

ROSS, C.W. (1984). *Computer Systems of Occupational Safety and Health Management* (New York: Dekker)

ROSS, E.W. and MCHENRY, H.T. (1963). High temperature metals. *Chem. Engng*, **70**(Nov. 25), 97

ROSS, J.E. (1973). More evidence in support of Murphy's law. *Loss Prevention*, **7**, p. 7

ROSS, L.L. (1991). CWRT pushes waste reduction. *Hydrocarbon Process.*, **70**(5), 121

ROSS, L.W. (1966). Data sources for calculating free energy and heat of reaction. *Chem. Engng*, **73**(Oct. 10), 205

ROSS, L.W. (1967). Sizing up anti-pollution legislation. *Chem. Engng*, **74**(Jul. 17), 19

ROSS, L.W. (1968). Predict pollutants dispersion. *Hydrocarbon Process.*, **47**(8), 144

ROSS, L.W. (1991). Audit for plant safety. *Hydrocarbon Process.*, **70**(8), 137

ROSS, P.A. (1993). Survey equipment for leakage. *Chem Engng*, **100**(2), EE suppl., 14

ROSS, R.C. (1973). How much tankage is enough? *Hydrocarbon Process.*, **52**(8), 75

ROSS, R.D. (1982). Disposal of hazardous materials. In Fawcett, H.H. and Wood, W.S. (1982), *op. cit.*, p. 649

ROSS, R.R. (1983). Centrifugal pipeline pumps. *Oil Gas J.*, **81**(Apr. 4), 123; (Apr. 11), 78

ROSS, S. (1971). Current legislation. *Chem. Engng*, **78**(Jun. 21), 9

ROSS, S.M. (1972–). Introduction to Probability Models; 1980, 2nd ed. (New York: Academic Press); 1985, 3rd ed. (Orlando, FL: Academic Press); 1989, 4th ed.; 1993, 5th ed. (Boston: Academic Press)

ROSS, S.S. (1972a). Designing your plant for easier emission testing. *Chem. Engng*, **79**(June 26), 112

ROSS, S.S. (1972b). Federal laws and regulations. *Chem. Engng*, **79**(May 8), 9

ROSS, S.S. and WHITE, L.J. (1972). International pollution control. *Chem. Engng*, **79**(May 8), 137

ROSS, T.K. and FRESHWATER, D.C. (1957). *Chemical Engineering Data Book* (London: Leonard Hill)

ROSSANO, A.T. and COOPER, H.B.H. (1968). Air pollution control. Sampling and analysis. *Cheml. Engng*, **75**(Oct. 14), 142

ROSSBERG, M. *et al.* (1986). Chlorinated hydrocarbons. *Ullmans Encyclopaedia of Chemical Technology*, rev. ed. vol. A6 (Weinheim: VCH), p. 233

VAN ROSSEN, P. (1975). Guide to Dutch Pressure Vessel Code. *Hydrocarbon Process.*, **54**(12), 61

VAN ROSSEN, P. (1978). Guide to Dutch Stoomwezen Code. *Hydrocarbon Process.*, **57**(12), 93

ROSSER, W.A. (1967). The shape and size of a drop diffusion flame (letter). *Combust. Flame*, **11**, 442

ROSSER, W.A. and RAJAPAKSE, Y. (1969). Thermal stability of a reactive spherical shell. *Combust. Flame*, **13**, 311

ROSSER, W.A., INAMI, S.H. and WISE, H. (1966). The quenching of premixed flames by volatile inhibitors. *Combust. Flame*, **10**, 287

ROSSI, M.J. (1985). Non-electronic Parts Reliability Data. Rep. NPRD-3. Reliab. Analysis Center, Rome Air Dev. Centerl

ROSSI, P.H., WRIGHT, J.D., WEBER-BURDIN, E. and PEREIRA, J. (1983). *Victims of the Environment. Loss from Natural Hazards in the United States, 1970–1980* (New York: Plenum Press)

ROSSINI, F.D. (1952). *Selected Values of Physical and Thermodynamic Properties of Hydrocarbons and Related Compounds API Res. Proj. 44*. Carnegie Inst. Technol., Pittsburgh, P A

ROSSINI, F.D. *et al. (1947). Selected Values of Properties of Hydrocarbons. Nat. Bur. Standards Circ. 461* (Washington, DC: US Govt Printing Office)

ROSSITER, A.P., SPRIGGS, H.D. and KLEE, H. (1993). Apply process integration to waste minimization. *Chem. Engng Prog.*, **89**(1), 30

ROST, M. and VISICH, E.T. (1969). Pumps. *Chem. Engng*, **76**(Apr. 14), 45

ROSTEN, H.I. and SPALDING, D.B. (n.d.) *The PHOENICS Reference Manual* (London: CHAM Ltd.)

ROTA, R., RUBINO, F., MESSINA, S., CANU, P., CARRA, S. and MORBIDELLI, M. (1987). Mathematical modeling of deflagrations in closed and vented vessels. *Vapor Cloud Modeling*, p. 869

ROTA, R., CANU, P., CARRA, S. and MORBIDELLI, M. (1991). Vented gas deflagration modeling: a simplified approach. *Combust. Flame*, **85**, 319

ROTA, R., CANU, P., CARRA, S. and MORBIDELLI, M. (1992). Diagrams for vent panels design. *Loss Prevention and Safety Promotion 7*, vol. 3, 118–1

ROTE, R. and PROCTOR, J.H. (1960). A-300 alloy steels serve the −50° to −50°F range. *Chem Engng*, **67**(Jun. 27), 119

ROTH, A. (1967). *Vacuum Sealing Techniques* (Oxford: Pergamon Press)

ROTH, A. (1990). *Vacuum Technology*, 3rd ed. (Amsterdam: Elsevier)

ROTH, E.M. (1964). *Space − Cabin Atmospheres, Pt II Fire and Blast Hazards. Spec. Pub. 48.* Nat. Aeronalt. Space Admin.

ROTH, J. and FIEDLER, H. (1971). Problems of reliability and availability. *Chem. Tech.*, **32**(11), 644

ROTH, J.A., BRANDES, W.F., HARBISON, R.D. and MIGUEZ, W.S. (1978). Training the on-scene co-ordinator for hazardous substances spill response. *Control of Hazardous Material Spills*, 4, p. 451

ROTH, J.E. (1979). Controlling construction costs. *Chem. Engng*, **86**(Oct. 8), 88

ROTH, S.E. (1965). Instrument maintenance cost trends. *Hydrocarbon Process.*, **44**(5), 199

ROTHE, J.P. (1969). *La Seismicité du Globe* (Paris: UNESCO)

ROTHMAN, R.A. (1973). Gaskets and packings. *Chem. Engng*, **80**(Feb. 26), 29

ROTHSCHILD, LORD (1978). Risk. *The Listener*, Nov. 30, 715

ROTHSCHILD, M. (1993). Applications of risk contours in quantative risk analysis (poster item). *Process Safety Management*, p. 397

ROTHWELL, G.P. (1979). Measuring corrosion as it happens. *Chem. Engng*, **86**(May 7), 107

ROTSTEIN, G.E., LAVIE, R. and LEWIN, D.R. (1992). A qualitative process-oriented approach for cheimical plant operations − the generation of feasible operation procedures. *Comput. Chem. Engng*, **16**, S337

ROTTMAN, J.W. and SIMPSON, J.E. (1984a). Gravity currents produced by instantaneous releases of a heavy fluid in a rectangular channel. *J. Fluid Mech.*, **135**, 95

ROTTMAN, J.W. and SIMPSON, J.E. (1984b). The initial development of gravity currents from fixed volume releases of heavy fluids. In Ooms, G. and Tennekes, H. (1984), *op. cit.*, p. 347

ROTTMAN, J.W., HUNT, J.C.R. and MERCER, A. (1985). The initial and gravity-spreading phases of heavy gas dispersion: comparison of models with Phase I data. *J. Haz. Materials*, 11(1/3), 261. See also in McQuaid, J. (1985), *op. cit.*, p. 261

ROTTMAN, J.W., SIMPSON, J.E., HUNT, J.C.R. and BRITTER, R.E. (1985). Unsteady gravity current flows over obstacles: some observations and analysis related to the Phase II trials. *J. Haz. Materials*, **11**(1/3), 325. See also in McQuaid, J. (1985), *op. cit.*, p. 325

ROTZLER, R.W., GLASS, J.A., GORDON, W.E. and HESLOR, W.R. (1960). Oxygen plant reboiler explosion. *Ammonia Plant Safety*, **2**, 31. See also *Chem. Engng Prog.*, 1960, **56**(6), 68

ROUGHLEY, J. (1963). Maintenance planning in relation to production. *Chem. Engr*, Lond., **172**, A61

ROUHIATNEN, V. (1982). Inadvertent starts causing accidents. *J. Occup. Accid.*, **4**, 165

ROUHIAINEN, V. (1990). The quality assessment of safety analysis. Doctoral thesis, Technical Research Centre of Finland (VTT), Espoo, Finland

ROUHIAINEN, V. (1992). Identification and assessment of environmental risks. *Loss Prevention and Safety Promotion 7*, vol. 1, 3-1

ROUHIAINEN, V. and SUOKAS, J. (1989). Planning and evaluation of the quality of safety and risk analysis. *Loss Prevention and Safety Promotion 6*, p. 47.1

ROULIER, M.H., LANDRETH, R.E. and CARNES, R.A. (1975). Current research on the disposal of hazardous wastes. *J. Haz. Materials*, **1**(1), 59

ROUSE, H., YIH, C.S. and HUMPHREYS, H.W. (1952). Gravitational convection from a boundary source. *Tellus*, **4**, 201

ROUSE, W.B. (1977). Human-compllter interaction in multitask situations. *IEEE Trans. Syst. Man Cybern.*, **SMC- 7**, 384

ROUSE, W.B. (1979). Problem solving performance for maintenance trainees in a fault diagnosis task. Hum. Factors, **21**, 195

ROUSE, W.B. (1979). Problem solving performance of first semester maintenance trainees in two fault diagnosis tasks. *Hum. Factors*, **21**, 611

ROUSE, W.B. (1981). Experimental studies and mathematical models of human problem solving performance in fault diagnosis tasks. In Rasmussen, J. and Rouse, W.B. (1981), *op. cit.*, p 199

ROUSE, W.B. and GOPHER, D. (1977). Estimation and control theory: application to modeling human behaviour. *Hum. Factors*, **19**, 315

ROUSE, W.B. and HUNT, R.M. (1982). *Human Problem-solving in Fault Diagnosis Tasks. Rep. 82-2.* Center for Man− Maclline Svstems Res., Georgia Inst. Technol., Atlanta, GA

ROUSSAKIS, N. and LAPP, K. (1991). A comprehensive test method for inline flame arresters. *Plant/Operations Prog.*, **10**, 85

ROUSSEAU, R.W. (ed.) (1987). *Handbook of Separation Process Technology* (New York: Wiley)

ROUSSEAUX, C.G. (1987). The use of animal models for predicting the potential chronic effects of highly toxic chemicals. *J. Haz. Materials*, **14**(3), 283

ROUTH, E.J. (1882). *The Elementary Part of a Treatise on the Dynamics of a System of Rigid Bodies*, 4th ed.; 1905, 7th ed. (London: Macmillan)

ROUTH, E.J. (1905). *The Advanced Part of a Treatise on the Dynamics of a System of Rigid Bodies*, 6th ed. (London: Macmillan)

ROVISON, J.M., GARCIA, A.A. and COLLINS, D.A. (2001). Minimizing off-Site consequences of anhydrous ammonia systems. *ammonia Plant Saf. Relat. Facil.* **41**, 273−83

ROWAN, D.A. (1986). Chemical plant fault diagnosis using expert systems technology. *Int. Fed. Autolllat. Control Workshop*, Kyoto

ROWE, G. (1990). Setting safety priorities: a technical and social process. *J. Occup. Accid.*, **12**, 31

ROWE, G.D. (1973). Essential of good industrial lighting. *Chem. Engng*, **80**(Dec. 10), 113; (Dec. 24), 50

ROWE, P.N. (1963). The correlation of engineering data. *Chem. Engr, Lond.*, **166**, CE69

ROWE, P.N. (1978). Some simple considerations in the design of a fluidised bed chemical reactor. *Chem. Ind.*, Jun. 17, 474

ROWE, R.D. and TAS, I. (1986). A simple technique for stack design in complex terrain. *Fifth Joint Conf. on Applications of Air Pollution Meteorology* (Boston, MA: Am. Met. Soc.), p. 207

ROWE, S.M., STARKIE, A.J. and NOLAN, P.F. (1994). The control of runaway polymerisation reactions by inhibition techniques. *Hazards*, XII, p. 575

ROWE, W.D. (1977). *The Anatomy of Risk* (New York: Wiley)

ROWE, W.D. (ed.) (1983). *Evaluation Methods for Environmental Standards* (Boca Raton, FL: CRC Press)

ROWEK, A.L. and COOK, K.S. (1986). US Coast Guard programme for new and existing liquefied gas ships. *Gastech 85*, p. 197

ROWLAND, D. (1993). Regulatory issues. In Bennett, P. (1993), *op. cit.*, p 19

ROWLANDS, J. and MORELAND, P.J. (1976). The instrumentation and use of linear polarisation measurements for corrosion monitoring. *Diagnostic Techniques* (London: Instn Chem. Engrs)

ROWLEY, R.M. and RAUSCH, A.H. (1977). *Vulnerability Model User's Guide* (rep. to US Coast Guard). Enviro Control, Rockville, MD

RAY, J.O. (1984). Explain those safety rules. *Hydrocarbon Process.*, **63**(9), 196

RAY, P., ROSE, J.C. and PARVIN, R. (1984). A case study: protection of a batch reactor making detergent ethoxylates. *The Protection of Exothermic Reactors and Pressurised Storage Vessels* (Rugby: Instn Chem. Engrs), p. 23

RAY, R. (1990). *A Primer on the Taguchi Method* (New York: van Nostrand Reinhold)

ROYAL, M.J. and NIMMO, N.M. (1969). Why LP methanol costs less. *Hydrocarbon Process.*, **48**(3), 147

ROYAL COMMISSION ON ENVIRONMENTAL POLLUTION — see Appendix 28

THE ROYAL INSTITUTE OF CHEMISTRY (1976). *Code of Practice for Chemical Laboratories* (London)

THE ROYAL SOCIETY (1978). *Long-Term Toxic Effects* (London)

THE ROYAL SOCIETY (1983). *Risk Assessment* (London)

THE ROYAL SOCIETY (1992). *Risk: Analysis, Perception and Management* (London)

THE ROYAL SOCIETY FOR THE PREVENTION OF ACCIDENTS — see Appendix 28

ROYALTY, C.A. and WOOSLEY, B.D. (1973). Design criteria for high pressure piping. *Safety in Polyethylene Plants*, **1**, p. 10. See also *Chem. Engng Prog.*, 1972, **68**(11), 72

ROYCE, S. (1988a). Getting the most out of corrosion control. *Process Engng*, Jan., 37

ROYCE, S. (1988b). Meeting the chlorine specification. *Process Engng, Jan.*, 47

ROYLANCE, H.M. (1968). Quantity–distance protection. *Ann N.Y. Acad. Sci*, **152** art. 1, 10

ROYLE, J.K. and BOUCHER, R.F. (1980). Survey of the use of power fluidics as valves in fluid flow systems. *Interflow 80* (Rugby: Instn Chem. Engrs), p. 273

ROYSE, S. (1988). Expert advice on control valve selection. *Chem. Engr, Lond.*, **444**, 27

ROZENFELD, I.L. (1981). *Corrosion Inhibitors* (New York: McGraw-Hill)

ROZLOVSKY, A.I. (1970). Doklad. Akad. *Nauk. SSSR*, **192**, 135. (English translation: *Proc. Acad. Sci. USSR*, 1970, **192**, 369)

ROZLOVSKY, A.I. and ZAKAZNOV, V.F. (1971). Effect of gas expansion during combustion on the possibility of using flame arresters. *Combust. Flame*, **17**, 215

RUAN KEQIANG (1986). A method of transformation of the disjoint sum in fault tree analysis. *Reliab. Engng*, **16**(3), 207

RUBACH, T., SCHECKER, H.G. and ONKEN, U. (1992). Investigations for the increase of the maxi gas flow rate through a water trap flame arrester. *Loss Prevention and Safety Promotion 7*, vol. 1, 19-1

RUBERT, A. and SCHAUMANN, P. (1988). Critical temperatures of steel columns exposed to fire. *Fire Safety J.*, **13**, 39

RUBIN, A. (1975). The role of hypothesis in medical diagnosis. *Proc. Int. Joint Conf. on Artificial Intelligence*, **4**, 856

RUBIN, F.L. (1960). Design of air cooled heat exchangers. *Chem. Engng*, **67**(Oct. 31), 91

RUBIN, F.L. (1961). Testing for leaks in heat exchangers. *Chem. Engng*, **68**(Jul. 24), 160

RUBIN, F.L. (1980). What's the difference between TEMA exchanger classes *Hydrocarbon Process.*, **59**(5), 96

RUBIN, F.L. and GAINSBORO, N.R. (1979). Latest TEMA standards for shell-and-tube heat exchallgers. *Chem. Engng*, **86**(Sep. 24), 111

RUBIN, H.H. and ARIEFF, A.J. (1945). Carbon disulfide and hydrogen sulfide clinical study of chronic low-grade exposures. *J. Ind. Hyg. Toxicol.*, **27**, 123

RUBIN, I.I. (1990). *Handbook of Plastic Materials and Technology* (New York: Wiley)

RUBIN, J.N. (1987). Synopsis of research needs in improving plant safety. *Preventing Major Chemical Accidents*, p. 3.209

RUBINSTEIN, E. (1979a). The accident that shouldn't have happened. *IEEE Spectrum*, **16**(11), 33

RUBINSTEIN, E. (1979b). Three Mile Island and the future of nuclear power. *IEEE Spectrum*, **16**(11), 30

RUBINSTEIN, R.Y. (1981). *Simulation and the Monte Carlo Method* (New York: Wiley)

RUCKSTUHL, R.E. (1972). Specifying compressor lube and seal systems. *Hydrocarbon Process.*, **51**(5), 127

RUDD, D.F. (1962). Reliability theory in chemical system design. *Ind. Engng Chem. Fund.*, **1**, 138

RUDD, D.F. (1968). The synthesis of system designs. *AIChE J.*, **14**, 343

RUDD, D.F., POWERS, G.J. and SIIROLA, J.J. (1973). *Process Systems* (Englewood Cliffs, NJ: Prentice-Hall)

RUDD, D.F. and WATSON, C.C. (1968). *Strategy of Process Design* (New York: Wiley)

RUDD, D.T. (1989). *Piper Alpha Public Inquiry. Transcript, Day 150*

RUDDY, E.N. and CARROLL, L.A. (1993). Select the best VOC control strategy. *Chem. Engng Prog.*, **89**(7), 28

RUDINGER, G. (1959). *Wave Diagrams for Nonsteady Flow in Ducts* (New York: van Nostrand)

RUDNICKI, M.I. and VANDER WALL, E.M. (1980). Gelled LNG – a safer approach *Gastech 79*, p. 247

RUDOLPH, A.W. (1974). Federal information sources. *Chem. Engng*, **81**(Oct. 21), 33

RUDOLPH, K. (1977). Reliability assessment for not-safety-related systems in nuclear power plants. *First Nat. Conf on Reliability. NRC4/4.* (Culcheth, Lancashire: UK Atom. Energy Auth.)

RUDRAM, D.A. and LAMBOURN, R.F. (1981). The scientific investigation of road accidents. *J. Haz. Materials*, **3**, 177

RUDSTON, S.G. (1989). Plant fitness through lubrieant monitoring. *Plant Engr*, Jan./Feb., 20

RUE, J.R. (1977). Non-chromate treatment of cooling water. *Ammonia Plant Safety*, **19**, 133

RUFF, M., ZUMSTEG, F. and FANNELOP, T.K. (1988). Water eontent and energy balance for gas cloud emanating from a cryogenic spill. *J. Haz. Materials*, **19**(1), 51

RUGGIERO, F., MACCHI, G., and MORICI, A. (1992). Offsite emergency planning and public structure for intervention in Italy. *Loss Prevention and Safety Promotion 7*, vol. 2, 70-l

RUHEMANN, M. (1949). *The Separation of Gases*, 2nd ed. (Oxford: Clarendon Press)

DE RUITER, P.A. and VAN LOOKEREN CAMPAGNE, N. (1980). The development of a safety policy in the Rijnmond region. *Loss Prevention and Safety Promotion 3*, vol. 2, p. 284

RULEO, P.A. (1964). Relief valve or rupture disc In Vervalin, C.H. (1964a), *op. cit.*, p. 124

RULKENS, P.F.M., BONGERS, A.J.P., TEN BRINK, G.P. and GULDEMOND, C.P. (1983). The application of gas curtains for diluting flammable gas clouds to prevent their ignition. *Loss Prevention and Safety Promotion 4*, vol. 1, p. F15

RUMAR, K. (1990). The basic driver error: late detection. *Hum. Factors*, **33**, 1281

RUMFORD, F. (1960). *Chemical Engineering Materials* (London: Constable)

RUMPF, F. (1990). *Particle Technology* (London: Chapman & Hall)

RUNCA, E. (1982). A practical numerical algorithm to compute steady-state ground level concenration by a K-model. *Atmos. Environ*, **16**, 753

RUNCA, E. (1992). Basic Lagrangian and Eueleerian modelling of atmospheric diffusion. *Atmos. Environ.*, **26A**, 513

RUNCA, E. and SARDEI, F. (1975). Numerical treatment of time dependent advection and diffusion of air pollutants. *Atmos Environ.*, **9**, 69

RUNES, E. (1971). Explosion venting. *Am. Inst. Chem. Engrs, Sixty-Fourth Ann. Mtg* (New York)

RUNES, E. (1972). Explosion venting. *Loss Prevention*, **6**, p. 63

RUNES, E. and KAMINSKY, E.M. (1965). Steam snuffing system assures safety. *Hydrocarbon Process*, **44**(4), 175

RUNES, E. and KAMINSKY, E.M. (1973). Steam snuffing system assures safety. In Vervalin, C.H. (1973a), *op. cit.*, p. 310

RUPP, H. and HENSCHLER, D. (1967). Effects of low chlorine and bromine concentrations on man. *Int. Arch. Gewerbepathol. Gewerbehyg.*, **23**, 79

RUPPERT, K.A., MUSCHELKNAUTZ, S. and KLUG, F. (1992). Design of liquid separation during two-phase venting. *Loss Prevention and Safety Promotion 7*, vol. 1, 42-1

RUSCH, G.M. (1993). The history and development of emergency response planning guidelines. *J. Haz. Materials*, **33**, 193

RUSCHENA, L.J. (1980). The need for toxicity data. *CHEME-CA 80* (Canberra: Instn Engrs Aust.)

RUSCK, S. (1977). Protection of distribution systems. In Golde, R.H. (1977), *op. cit.*, p. 747

RUSHBROOK, F. (1979). *Fire Aboard*, 2nd ed. (Glasgow: Brown Son & Ferguson)

RUSHDI, A.M. (1987). Efficient computation of k-to-l-out-of-m system reliability. *Reliab. Engng*, **17**(3), 157

RUSHFORD, R. (1977). Haard and operability studies in the chemical industries, *Trans North East Inst. Engrs Shipbuilders*, **93**(5), 117

RUSHMERE, J.D. (1954). *The Self-association of Hydrogen Fluoride. A Short Review of the Published Literature. Rep. SCS-TN-31*

RUSHTON, A.G. (1989). The integration of safety considerations. *Computer Aided Process Engineering* (Rugby: Instn Chem. Engrs), p. 27

RUSHTON, A.G. (1991a). The selection of trip system configuration. *Hazards*, **XI**, p. 329

RUSHTON, A.G. (1991b). The selection of trip system configuration including common cause failure among system channels. *Process Safety Environ.*, **69B**, 200

RUSHTON, A.G. (1992). Protective device faults – vulnerability to management. *Major Hazards Onshore and Offshore*, p. 183

RUSHTON, A.G., EDWARDS, D.W. and LAWRENCE, D. (1994). Inherent safety and computer aided design. *Trans. IChemE* **72**, Part B, (May), 83–87

RUSTON, A.G. (1995). Quality assurance of hazard and operability study performance in the context of offshore safety. Rep. to HSE, in press (referenced in Rushton, A.G. *et al.* (1994), *op. cit.*)

RUSHTON, A.G., EDWARDS, D.E. and LAWRENCE, D. (1994). Inherent safety and computer aided process design. *Fourth Eur. Symp. on Computer Aided Process Engineering* (Rugby: Instn Chem. Engrs), p. 413

RUSHTON, A.G., GOWERS, R.E., EDMONDSON, J.N. and AL-HASSAN, T. (1994). Hazard and operability study of offshore installations – a survey of variations in practice. *Hazards*, **XII**, p. 341

RUSHTON, L. and ALDERSON, M.R. (1981). A case-control study to investigate the association between exposure to benzene and deaths from leukaemia in oil refinery workers. *Br. J. Cancer*, **43**, 77

RUSSEL, A. (1909). Coefficients of capacity and mutual attraction and repulsion of two electrified spherical conductors when close together. *Proc. R. Soc., Ser. A*, **82**, 524

RUSSELL, D.L. (1980). Monitoring and sampling liquid effluents. *Chem. Engng*, **87**(Oct. 20) 109

RUSSELL, D.L. (1985). Managing your environmental audit. *Chem. Engng*, **92**(Jun. 24), 37

RUSSELL, D.L. and HART, S.W. (1987). Underground storage tanks. *Chem. Engng*, **94**(4), 61; **94**(13), 5

RUSSELL, G. (1977). The value of dry powders. *Fire*, **70**(Oct.), 248

RUSSELL, H. (1982). Tempered water for safety showers and eye baths. *Chem. Engng*, **89**(Nov. 29), 85

RUSSELL, H.W. (1985). Equipment selection; how to really assess costs. *Chem. Engng*, **92**(Aug. 5), 103

RUSSELL, H.W. (1989). Photography in engineering, Pt 1. *Chem. Engng*, **96**(1), 137

RUSSEL, J.B. (1989). *Piper Alpha Public Inquiry. Transcript, Day 35*

RUSSELL, L.H. and CANFIELD, J.A. (1973). Experimental measurement of heat transfer to a cylinder immersed in a large aviation-fuel fire. *J. Heat Transfer*, **95**, 397

RUSSELL, L.R. and SCHUELLER, G.I. (1974). Probabilistic models for Texas Gulf Coast hurricane occurrences. *J. Petrol. Technol.*, Mar. 279

RUSSELL, P.C. and HERBERT, P.A. (1974). Preparation for commissioning. *Instn Chem Engrs, Symp. on Commissioning of Chemical and Allied Plant* (London)

RUSSELL, R. (1977). FEA discount or giveaway? *Fire Surveyor*, **6**(5), 21

RUSSELL, S. and FERGUSON, R.A.D. (1980). Assessing the health costs of fuel systems. *Sci. Publ. Policy*, **7**, 365

RUSSELL, T.W.F., ETCHELLS, A.W., JENSEN, R.H. and ARRUDA, P.J. (1974). Pressure drop and hold-up in stratified gas-liquid flow. *AIChE J.*, **20**, 664

RUSSELL, W.W. (1973). *Monsanto Report on Explosion Venting* (St Louis, MO: Monsanto Ltd)

RUSSELL, W.W. (1974). Safety in flange joints. *Loss Prevention*, **8**, p. 100

RUSSELL, W.W. (1976). Hazard control of plant process changes. *Loss Preventing*, **10**, 80

RUSSO, E.P. and HAYDEL, B.G. (1983). Lap-joint flange slip connection eases differential settling in storage-tank piping. *Oil Gas J.*, **81**(Sept. 19), 113

RUSSO, E.P., HAYDEL, B.G. and EPTON, M.A. (1984). Stress analysis improved for steam traced piping. *Oil Gas J.*, **82**(Oct. 1), 172

RUSSO, M.F. and PESKIN, R.L. (1987). Knowledge-based systems for the engineer. *Chem. Engng. Prog.*, **83**(9), 38

RUSSO, T.J. and TORTORELLA, A.J. (1992). Plant layout: the contribution of CAD. *Chem. Engng*, **99**(4), 97

RUST, E.A. (1979). Explosion venting of low pressure equipment. *Chem. Engng*, **86**(Nov. 5), 102

RUST, E. and EBERT, A. (1947). *Unfälle beim chemischen Arbeiten*, 2nd ed. (Zurich: Rascher)

RUST, E. and GRATTON, J. (1980). Can offshore siting solve the LNG terminal impasse? *Gastech 79*, p. 285

RUST, J.H. and WEAVER. E. (eds) (1976). *Nuclear Power Safety* (New York: Pergamon Press)

RUTAN, C.R. (1993). Turn your attention to rotating machinery. *Chem. Engng Prog.*, **89**(7), 67

RUTCH, C., ANSELMO, K.J., LINNEY, R.E. and SHEESLEY, J.H. (1992). Tornado risk analysis. *Plant/Operations Prog.*, **11**, 134

RUTGERS, A.J. and DE SMET, M. (1952). Elecrokinetic potentials and electrical conductance in solution of low dielectric constant. *Trans. Faraday Soc.*, **48**, 635

RUTGERS, A.J., DE SMET, M. and DE MYER, G. (1957). Influence of turbulence upon electrokinetic phenomena. Experimental determination of the thickness of the diffuse part of the double layer. *Trans. Faraday Soc.*, **53**, 393

RUTH, J.H. (1986). Odor thresholds and irritation levels for several chemical substances: a review. *J. Am. Ind. Hyg. Ass.*, **47**, A-142

RUTHERFORD, D. (1980). *Tanker Cargo Handling* (London: Griffin)

RUTHERFORD, J.B. (1975). Road transport of hazardous chemicals. *Instn Chem. Engrs, Symp. on Transport of Hazardous Materials* (London)

RUTHERFORD, W.H. (1972–73). Experience in the Accident and Emergency Department of the Royal Victoria Hospital with patients from civil disturbances in Belfast., 1969–72, with a review of disasters in the United Kingdom, 1951–1971. *Injury*, **4**, 189

RUTHERFORD, W.H. (1975). The surgery of violence. II, Disaster procedures. *Br. Med. J.*, **1**(5955), Feb. 22, 443

RUTHVEN, D.M. (1984). *Principles of Adsorption and Adsorption Processes* (Chichester: Wiley)

RUTSTEIN, R. (1979a). The effectiveness of automatic fire detection systems. *Fire Surveyor*, **8**(4), 37

RUTSTEIN, R. (1979b). The estimation of the fire hazard in different occupancies. *Fire Surveyor*, **8**(2), 21

RUTSTEIN, R. and CLARKE, M.B.J. (1979). The probability of fire in different sectors of industry. *Fire Surveyor*, **8**(1), 20

RUYTERS, P.J.M. (1968). The reaction of structures and structural components to blast loading (discussion). *Ann. N.Y. Acad. Sci.*, **152**(art., 1), 491

RUZISKA, P.A. (1972). Automatic trip systems. *Ammonia Plant Safety*, **14**, p. 129

RUZISKA, F.A. and WORLEY, A.C. (1974). An improved header/transfer line. *Ammonia Plant Safety*, **16**, p. 85

RUZISKA, P.A. and BAGNOLI, D.L. (1974). Steam reformer tube testing – a status report. *Ammonia Plant Safety*, **16**, p. 10

RUZISKA, P.A., SONG, C.C., WILKINSON, R.A. and UNRUH, W. (1985). Exxon Chemical low energy amonia process start-up experience. *Ammonia Plant Safety*, **25**, p. 22. See also *Plant/Operations Prog.*, 1985, **4**, 79

RUZZO, W.P. (1991). Prepare for stormwater regulations. *Hydrocarbon Process.*, **70**(12), 105

RYAN, C.S. (1981). Review of API epidemiology research. *API Proc., Refining Dept, 46th Midyear Mtg*, **60**, 362

RYAN, D.L. (1979–). *Computer-Aided Graphics and Design*; 1985, 2nd ed.; 1994, 3rd ed. (New York: Dekker)

RYAN, G.T. (1972). Managing the project start-up. *Chem. Engng Prog.*, **68**(12), 65

RYAN, W., LAMB, B. and ROBINSON, E. (1984). An atmospheric tracer investigation of transport and diffusion around a large, isolated hill. *Atmos. Environ.*, **18**, 2003

RYANS, J.L. (1984). Advantages of integrated vacuum pumping systems. *Chem. Engng Prog.*, **80**(6), 59

RYANS, J.L. and CROLL, S. (1981). Selecting vacuum systems, pt 1. *Chem. Engng*, **88**(Dec. 14), 72

RYANS, J.L. and ROPER, D.L. (1986). *Process Vacuum System Design and Operation* (New York: McGraw-Hill)

RYASON, P.R. and HIRSCH, E. (1974). Cool flame quench distances. *Combust. Flame*, **22**, 131

RYCHENER, M.D. (1979). A semantic network of production rules in a system for describing computer structures. *Proc. Int. Joint Conf. on Artificial Intelligence*, **6**, 743

RYCKMAN, M.D. and PETERS, J.L. (1982). Toxic corridor projection models for emergency response. *Transportation Res. Board, Sixty First Ann. Mtg*, Washington, DC

RYCKMAN, M.D., WIESE, T.J. and RYCKMAN, D.W. (1978). REACT's response to hazardous nlaterial spills. *Control of Hazardous Material Spill*, **4**, p. 24

RYCROFT, R.J.G. (1979). Occupational dermatoses. *Current Approaches to Occupational Medicine* (ed. A.W. Gardner) (Bristol: John Wright), p. 124

RYCROFT, R.J.G. (1980). Industrial dermatitis. *Chemical Process Hazards*, **VII**, p. 121

RYDER, G.H. (1975). *Mechanics of Machines* (London: Macmillan)

RYDER, M. (1991). Containment of airborne hazards in the workplace. *Process Ind. J.* Sep., 37

RYER-POWDER, J.E. (1991). Health effects of ammonia. *Plant/Operations Prog.*, **10**, 228

RYERSON, C.M. (1973). Relating factory test failure results to field reliability, required field maintenance, and to total life cycle costs. *Microelectron. Reliab.*, **12**, 357

RYIN, L.A. *et al.* (1990). Radiocontamination patterns and possible health consequences of the accident at Chernobyl nuclear power station. *J. Soc. Radiol. Prot.*, **10**, 3

RYLANDER, P.N. (ed.) (1985). *Hydrogenation Methods* (Orlando, FL: Academic Press)

RYMAN, D.J. and STEENBERGEN, J.E. (1976). Field balancing of rotary equipment. *Chem. Engng Prog.*, **72**(2), 46

RYMARZ, T.M. and KLIPSTEIN, D.H. (1975). Removing particulates from gases. *Chem. Engllg*, **82**(Oct. 6), 113

SAARI, J. (1977). Characteristics of tasks associated with the occurrence of accidents. *J. Occup Accid.*, **1**(3), 273

SAATY, T.L. (1977). A scaling method for priorities in hierarchical structures. *J. Math. Psychol*, **15**, 234

SAATY, T.L. (1980). *The Analytic Hierarchy Process* (New York; McGraw-Hill)

SAATY, T.L. (1982). *Decision Making for Leaders* (Belmont, CA: Lifetime Learning)

SABATIER, P.A. and MAZMANIAN, D.A. (1983). *Can Regulation Work?* (New York: Plenum Press)

SABLON, F. (1976). Humidification, a process for the reduction of hazards in the production of explosives. *Prevention of Occupational Risks*, **3**, p. 467

SABNIS, J., SIMMONS, W. and TEUSCHER, L. (1982). Two-phase blowdown from high pressure liquid pipelines. *The Assessment of Major Hazards* (Rugby: Instn *Chem. Engrs)*, p. 39

SABO, D.S. and DRANOFF, J.S. (1966). On the stability of a continuous flow stirred tank reactor. *AIChE J.*, **12**, 1223

SABO, D.S. and DRANOFF, J.S. (1970). Stability analysis of a continuous flow stirred tank reactor with consecutive reactions. *AIChE J.*, **16**, 211

SABORN, J.W. (1970). *Piper Alpha Public Inquiry. Transcript, Day 56*

SACCOMANNO, F.F. and ALLEN, B. (1988). Locating emergency response capability for dangerous goods incidents on a road network. *Transport. Res. Record 1193* (Transport. Res. Board, Nat. Res. Coun.), p. 1

SACCOMANNO, F.F. and EL-HAGE, S. (1989). Minimizing derailments of railcars carrying dangerous commodities through effective marshalling strategies. *Transport. Res. Record 1245* (Transport. Res. Board, Nat. Res. Coun.) p. 34

SACCOMANNO, F.F., SHORTREED, J.H., VAN AERDE, M. and HIGGS, J. (1989). Comparison of risk measures for the transport of dangerous commodities by truck and rail. *Transport. Res. Record 1245* (Transport. Res. Board, Nat. Res. Coun.) p. 1

SACCOMANNO, F.F., SHORTREED, J.H. and MEHTA, R. (1990). Fatality risk curves for transporting chlorine and liquefied petroleum gas by truck and rail. *Transport. Res. Record 1264* (Transport. Res. Board, Nat. Res. Coun.) p. 29

SACERDOTI, E.D. (1975). The non-linear nature of plans. *Fourth Int. Joint Conf. on Artificial Intelligence*, p. 206

SACERDOTI, E.D. (1977). *A Structure for Plans and Behaviour* (New York: Elsevier)

SACERDOTI, E. (1982). The organization of expert systems: a tutorial. *Artificial Intelligence*, **18**, 135

SACHERE, R.M. (1971). Responsibilities of the plant owner and the contractor. *Loss Prevention*, **5**, p. 76

SACHS, G. (1970). Economics and technical factors in chemical plant layout. *Chem. Engr, Lond.*, **242**, CE304

SACHS, G. (1978). (Letter) *Chem. Engr, Lond.*, **329**, 87

SACHS, P. (1978). *Wind Forces in Engineering*, 2nd ed. (Oxford: Pergamon Press)

SACHS, P.A., PATERSON, A.M. and TURNER, M.H.M. (1985). *ESCORT — an Expert System for Complex Operations* (paper). PA Computers and Telecommunications, London

SACHS, R.G. (1944). *The Dependence of Blast on Ambient Pressure and Temperature. Rep. 466*. Ballistics Res. Lab., Aberdeen, MD

SACKMAN, H. (1967). *Computers, System Science and Evolving Society* (New York: Wiley)

SACKMAN, H. (1970). *Man—Computer Problem-Solving* (New York: Auerbach)

SADEE, C. (1973a). Investigation and simulation of secondary space explosions. *Prevention of Occupational Risks*, **2**, p. 266

SADEE, C. (1973b). Spontaneous ignition and decomposition of hydrocarbons in chemical installations. *Prevention of Occupational Risks*, **2**, p. 438

SADEE, C.P.M. (1975). *Estimation of the TNT Equivalent of the Amount of Reacted Cyclohexane, and of the Dimensions and Shape of the Cloud in Relation to the Explosion which Occurred on the Flixborough Site of Nypro (UK) Ltd on 1st June 1974. Rep. to Flixborough Court of Inquiry* (Dutch State Mines)

SADEE, C., SAMUELS, D.E. and O'BRIEN, T.P. (1976–77). The characteristics of the explosion of cyclohexane at the Nypro (UK) Flixborough plant on June 1st 1974. *J. Occup. Accid.*, **1**, 203

SADLER, D.L. and MATSUZ, B.T. (1994). A comprehensive program for preventing cyclohexane oxidation process piping failure. *Process Safety Prog.*, **13**, 45

SAEKS, R., and LIBERTY, S.R. (eds) (1977). *Rational Fault Analysis* (New York: Dekker)

SAELID, S., MJAAVATTEN, A. and FJALESTAD, K. (1992). An object oriented operator support system based on process models and an expert system shell. *Comput. Chem. Engng*, **16**, S97

SAFER EMERGENCY SYSTEMS INC. (n.d.). *SAFER System for Managing Chemical Hazards* (Westlake Village, CA)

SAFETY IN MINES RESEARCH BOARD — see Appendix 28

SAFETY IN MINES RESEARCH ESTABLISHMENT — see Appendix 28

SAFETY AND RELIABILITY DIRECTORATE (1990). *MHIDAS* (Culcheth, Warrington)

SAFETY AND RELIABILITY DIRECTORATE — see also Appendix 28

SAFFELL, J. (1992). Water quality. *Plant Engr*, **36**(3), 31

SAGALOVA, S.L. and RESNICK, V.A. (1962). *Teploenergetika*, **12**, 19

SAGAN, L.A. (1981). Have the nuclear power risk assessors failed? In Berg, G.G. and Maillie, H.D. (1981), *op. cit.*, p. 321

SAGAN, L.A. and WHIPPLE, C.G. (1984). Cancer control in the workplace: some examples from the utility industry. In Deisler, P.F. (1984b), *op. cit.*, p. 159

SAGE, A.P. (1977). *Methodology for Large Systems* (New York: McGraw-Hill)

SAGE, J.A. and WEINBERG, F.J. (1959). An attempt at measuring homogeneous ignition temperatures. *Combustion*, **7**, p. 464

SAGENDORF, J. (1975). *Diffusion under Low Wind Speed and Inversion Conditions. ERL-ARL Tech. Memo. 52*. NOAA, Environ. Res. Labs

SAGUES, A.A. and DAVIS, B.H. (1982). Monitoring corrosion through metal levels of streams. *Hydrocarbon Process.*, **61**(1), 98

SAIA, S.A. (1976). Vapor clouds and fires in a light hydrocarbon unit. *Loss Prevention*, **10**, p. 23. See also *Chem. Engng Prog.*, 1976, **72**(11), 56

SAIBEL, E. (1961a). Failure of a thick walled pressure vessel. *Ind. Engng Chem.*, **53**(7), 56A

SAIBEL, E. (1961b). Failure of thick walled pressure vessels. *Ind. Engng Chem.*, **53**(12), 975

SAIFUTDINOUV, M.M. (1966). Maximum permissible concentration of ammonia in the atmosphere. *Hyg. Sanit.*, **176**, 171

SAIGA, M. *et al.* (1991). An advanced LNG leakage monitoring system. *Gastech 90*, p. 3.5

SAINSBURY, F. (1977). The largest wartime explosions: Silvertown, London, January 19, 1917. *After the Battle*, **18**, 30

ST CLAIR, W.H. (1973). Combination unit pumphouse fire. In Vervalin, C.H. (1973a), *op. cit.*, p. 6

SAINT-DIZIER, P. (1987). *Initiation a la Programmation en PROLOG* (Paris: Editions Eyrolles)

ST-GEORGES, M., BUCHLIN, J.M., RIETHMULLER, M.L., LOPEZ, J.P., LIETO, J. and GRIOLET, F. (1992a). Fundamental multi-disciplinary study of liquid sprays for absorption of pollutant or toxic clouds. *Loss Prevention and Safety Promotion 7*, vol. 2, 65–1

ST-GEORGES, M., BUCHLIN, J.M., REITHMULLER, M.L., LOPEZ, J.P., LIETO, J. and GRIOLET, F. (1992b). Fundamental multi-disciplinary study of liquid sprays for absorption of pollutant or toxic clouds. *Process Safety Environ.*, **70B**, 205

ST JOHN, K. (1978). Use combustible gas analyzers. *Hydrocarbon Process.*, **57**(5), 282

THE ST JOHN AMBULANCE ASSOCIATION (1961). *Occupational First Aid*, 2nd ed. (London)

THE ST JOHN AMBULANCE ASSOCIATION (1992). *First Aid Manual: the Authorized Manual of the St John Ambulance, St Andrew's Ambulance Association and the British Red Cross*, 6th ed. (London: Dorling Kindersley)

ST PIERRE, A.F. (1975). Plant instrument zeroing is now a function of process computers. *Chem. Engng*, **82**(Nov. 10), 223

SAITO, K. and NAKANO, R. (1988). Medical diagnostic expert system based on PDP model. *Proc. IEEE Int. Conf. on Neural Networks* (New York: Instn Electrical and Electronic Engrs)

SAKAI, Y. (1964). This ammonia unit exploded. In Vervalin, C.H. (1964a), *op. cit.*, p. 326

SAKAMOTO, K. (1985). An epidemiological study of the incidence of hearing loss by exposure to an industrial noisy environment. *J. Occup. Accid.*, **7**(2), 101

SAKURAI, T. (1989). Toxic gas tests with several pure and mixed gases using mice. *J. Fire Sci.*, **7**(1), 22

SALAMITOU, J. and ROBIN, A. (1980). Pollution control and law in practice. The situation in France. *Chem. Ind.*, Apr. 19, 335

SALCEDO, R.N., CROSS, F.L. and CHRISMON, R.L. (1989). *Environmental Impacts of Hazardous Waste Treatment, Storage and Disposal Facilities* (London: Technomic Publ.)

SALDANHA, PERES, J.E.C. and ROGERSON, H. (1984). The use of probablistic fracture mechanics in devising quality control policies in the fabrication industries. *Reliab. Engng*, **8**(3), 149

SALE, J.W. (1979a). Safety valve mass flow rate calculations and correction factors. In Haupt, R.W. and Meyer, R.A. (1979), *op. cit.*, p. 1

SALE, J.W. (1979b). Safety valve vent pipe sizing for open discharge systems. In Haupt, R.W. and Meyer, R.A. (1979), *op. cit.*, p. 79

SALEM, A. (1981). Estimate capital costs fast. *Hydrocarbon Process.*, **60**(9), 199

SALEM, S.L. and APOSTOLAKIS, G. (1980). The CAT methodology for fault tree construction. In Apostolakis, G., Garribba, S. and Volta, G. (1980). *op. cit.*, p. 109

SALEM, S.L., APOSTOLAKIS, G.E. and OKRENT, D. (1975a). On the automatic construction of fault trees. *Trans. Am. Nucl. Soc.*, **22**, 475

SALEM, S.L., APOSTOLAKIS, G.E. and OKRENT, D. (1975b). *A Computer-oriented Approach to Fault-tree Construction. Rep. UCLA-ENG-7635.* Univ. of California, Los Angeles, CA

SALEM, S.L., APOSTOLAKIS, G.E. and OKRENT, D. (1977). A new methodology for the computer-aided construction of fault trees. *Ann. Nucl. Energy*, **4**, 417

SALEM, S.L., WU, J.S. and APOSTOLAKIS, G. (1979). Decision table development and application to the construction of fault trees. *Nucl. Technol.*, **42**, 51

SALETAN, D.I. (1959a). How to calculate and combat static electricity hazards in process plants. *Chem. Engng*, **66**(Jun. 29), 101

SALETAN, D.I. (1959b). The theory behind static electricity hazards in process plants. *Chem. Engng*, **66**(Jun. 1), 99

SALISBURY, A. (1985). Requirements for integrated pipeline systems. *Process Engng*, Feb., 27

SALISBURY, R.E. (1990). Acute care of the burned hand. In McCarthy, J.G. (1990), *op. cit.*, p. 5399

SALKOVITCH, S.R. (1982). Effective document control. *Chem. Engng*, **89**(Dec. 27), 67

SALLENBACH, H.G. (1980). Stepwise solution to vibrating equipment foundation design. *Hydrocarbon Process.*, **59**(3), 93

SALLET, D.W. (1978). *On the Sizing of Pressure Relief Valves for Pressure Vessels which are Used in the Transport of Liquified Gases. ASME paper 78–WA/HT-39* (New York: Am. Soc. Mech. Engrs)

SALLET, D.W. (1979a). Pressure relief valve sizing for vessels containing compressed liquefied gasses. *Inst. Mech. Engrs, Conf. on Pressure Relief Devices* (London), paper C274/19

SALLET, D.W. (1979b). Two-phase flow aspects in sizing pressure relief valves. In Haupt, R.W. and Meyer, R.A. (1979), *op. cit.*, p. 31

SALLET, D.W. (1984). Thermal hydraulics of valves for nuclear application. *Nucl. Sci. Engng*, **88**, 220

SALLET, D.W. (1990a). Critical two-phase mass flow rates of liquefied gases. *J. Loss Prev. Process Ind.*, **3**(1), 38

SALLET, D.W. (1990b). On the prevention of two-phase fluid venting during safety valve action. *J. Loss Prev. Process Ind.*, **3**(1), 53

SALLET, D.W. (1990c). Subcooled and saturated liquid flow through valves and nozzles. *J. Haz. Materials*, **25**(1/ 2), 181

SALLET, D.W. and SOMERS, G.W. (1985). Flow capacity and response of safety relief valves to saturated water flow. *Plant/Operations Prog.*, **4**, 207

SALLET, D.W., WESKE, J.R. and GUHLER, M. (1979). The flow of liquids and gases through pressure relief valves. In Haupt, R.W. and Meyer, R.A. (1979), *op. cit.*, p. 13

SALOOJA, K.C. (1960). Studies of combustion processes leading to ignition in hydrocarbons. *Combust. Flame*, **4**, 117

SALOOJA, K.C. (1961). Effect of temperature on the ignition characteristics of hydrocarbons. *Combust. Flame*, **5**, 243

SALOOJA, K.C. (1964a). Factors responsible for the differences in the oxidation behaviour of hydrocarbons in the liquid and gaseous phases: studies in cyclohexene and cyclohexane. *Combust. Flame*, **8**, 311

SALOOJA, K.C. (1964b). Influence of surface-to-volume ratio of quartz reaction vessels on preflame and ignition characteristics of hydrocarbons. *Combust. Flame*, **8**, 203

SALOOJA, K.C. (1966). Studies of combustion processes leading to ignition of some oxygen derivatives of hydrocarbons. *Combust. Flame*, **10**, 11

SALOOJA, K.C. (1967). Studies of oxidation processes leading to ignition of butanes, butenes and cyclobutane. *Combust. Flame*, **11**, 320

SALOOJA, K.C. (1968a). Combustion studies of olefins and of their influence on hydrocarbon combustion processes. *Combust. Flame*, **12**, 401

SALOOJA, K.C. (1968b). Ignition behaviour of mixtures of hydrocarbons. *Combust. Flame*, **12**, 597

SALOT, W.J. (1967). Safe and economical aluminum storage tanks. *Chem. Engng*, **74**(May 8), 172

SALOT, W.J. (1972). Anticipating reformer tube life in normal service. *Ammonia Plant Safety*, **14**, 119

SALOT, W.J. (1973). Catalyst tube performance. *Ammonia Plant Safety*, **15**, p. 1

SALOT, W.J. (1975). High pressure reformer tube operating problems. *Ammonia Plant Safety*, **17**, p. 34

SALOT, W.J. (1978). Mystery leaks in a waste-heat boiler. *Chem. Engng*, **85**(Sept. 11), 177

SALOT, W.J. (1982). Bent bayonets are bad for boilers. *Hydrocarbon Process.*, **61**(1), 77

SALTER, J. (1994). The legal position of the new regime. *Offshore Safety Case Management* (Rugby: Instn *Chem. Engrs*), p. js1

SALTER, J. and THOMAS, P. (eds) (1981). *Planning Law for Industry* (London: Int. Bar Ass.)

SALTER, M. (1992). Managing the unexpected. *Health and Safety in the Offshore Oil and Gas Industries* (Rugby: Instn *Chem. Engrs*)

SALTER, R.F., FIKE, L.L. and HANSEN, F.A. (1963). How to size rupture discs. *Hydrocarbon Process. Petrol. Reiner*, **42**(5), 159

SALTZ, E.X. (1980). Refinery on skids speeds start-up. *Hydrocarbon Process.*, **59**(5), 161

SALVENDY, G. (ed.) (1987). *Handbook of Human Factors* (New York: Wiley)

SAMANS, C.H. (1963). Which steel for refinery service? *Hydrocarbon Process.*, **42**(10), 169; **42**(11), 241

SAMANS, C.H. (1966). Which material for process plant piping? *Hydrocarbon Process.*, **45**(10), 135

SAMANS, C.H. (1968). Making the most of contemporary seals. *Chem. Engng*, **75**(Feb. 12), 150

SAMBETH, J. (1983). What really happened at Seveso. *Chem. Engng*, **90**(May 16), 44

SAMDAL, U.N., KORTNER, H. and GRAMMELTVEDT, J.A. (1992). A user's view of human reliability. *Hazard Identification and Risk Analysis, Human Factors and Human Reliability in Process Safety*, p. 281

SAMDANI, G. (1990a). Neural nets. *Chem. Engng*, **97**(8), 37

SAMDANI, G. (1990b). Nonpoint emissions get pinpoint scrutiny. *Chem. Engng*, **97**(5), 42

SAMDANI, G. and FOUHY, K. (1992). Smart software. *Chem. Engng*, **99**(4), 31

SAMID, G. (1990). *Computer Organized Cost Engineering* (Basel: Dekker)

SAMOILOFF, A.A. (1968). Mechanical seals: longer runs, less maintenance. *Chem. Engng*, **75**(Jan. 29), 130

SAMPSON, J.D. (1991). The employee's standpoint. The Fellowship of Engineering (1991b), *op. cit.*, p. 101

SAMUEL, A.L. (1959). Some studies in machine learning using the game of checkers. *IBM J. Res. Dev.*, **3**, 211

SAMUEL, A.L. (1960). Programming computers to play games. In *Advances in Computers* (edited by F.L. Alt) (London: Academic Press), p. 165

SAMUEL, A.L. (1963a). Some studies in machine learning using the game of checkers, II, recent developments. *IBM J. Res. Dev.*, **11**, 601

SAMUEL, A.L. (1963b). Some studies in machine learning using the game of checkers. In Feigenbaum, E.A. and Feldman, J. (1963) *op. cit.*

SAMUEL, A.L. (1968). Some studies in machine learning using the game of checkers: recent progress. *IBM J. Res. Devel.*, **11**, 601

SAMUELS, B. (1992). Explosion hazard assessment of off-shore modules using 1/12 scale models. *Major Hazards Onshore and Offshore*, p. 543

SAMUELS, B. (1993). Explosion hazard assessment of off-shore modules using 1/12th scale models. *Process Safety Environ.*, **71B**, 21

SAMUELS, E.J. and O'BRIEN, T.P. (1975). *An Estimate of the Energy of Explosion at Nypro. Rep. to Flixborough Court of Inquiry* (Foulness: Atom. Weapons Res. Estab.)

SAMUELS, H. (1967). *Industrial Law* (London: Pitman)

SAMUELS, H. (1969). *Factory Law*, 8th ed. (London: Knight)

SAMUELS, J.P. (1991). Application in a research environment (COSHH Regulations 1988). In Simpson, D. and Simpson, W.G. (1991) *op. cit.*, p. 130

SAMUELY, F.J. and HAMANN, C.W. (1939). *Civil Protection* (London: Architect. Press)

SAN JOSE, R., CASANOVA, J.L., VILORIA, R.E. and CASANOVA, J. (1986). Evaluation of the turbulent parameters of the unstable surface boundary layer outside Businger's range. *Atmos. Environ.*, **19**, 1555

SANCAKTAR, S. (1982). An illustration of matrix formulation for a probabilistic risk-assessment study. *Risk Analysis*, **2**, 137

SANCAKTAR, S. (1983). Financial risk assessment at the plant design stage. *Plant/Operations Prog.*, **2**, 176

VAN DE SANDE, E. and SMITH, J.M. (1976). Jet break-up and air entrainment by low velocity water jets. *Chem. Engng Sci.*, **31**, 219

SANDEFUR, M. (1980). How to implement digital control. *Hydrocarbon Process.*, **59**(7), 115

SANDERS, A.F. (1991). Simulation as a tool in the measurement of human performance. *Ergonomics*, **34**, 995

SANDERS, A.J. and ALDWINCKLE, D.S. (1987). The analysis of marine accidents affecting industrial installation safety. *Preventing Major Chemical Accidents*, p. 1.67

SANDERS, K.W. (1971). Explosion in feed-gas section. *Ammonia Plant Safety*, **13**, p. 8

SANDERS, M.S. (1982). *Workbook for Human Factors in Design and Engineering*, 2nd ed.; 1987, 3rd ed. (Dubuque, IA: Kendal Hunt)

SANDERS, M.S. (1993). *Human Factors in Design and Engineering*, 7th ed. (New York: McGraw-Hill)

SANDERS, M.S. and MCCORMICK, E.J. (1987). *Human Factors in Design and Engineering*, 6th ed. (New York: McGraw-Hill). See also Sanders, M.S. (1993)

SANDERS, R.E. (1983). Plant modifications; troubles and treatment. *Chem. Engng Prog.*, **79**(2), 73

SANDERS, R.E. (1992). Small, quick changes can create bad memories. *Chem. Engng Prog.*, **88**(5), 78

SANDERS, R.E. (1993a). Don't become another victim of vacuum. *Chem. Engng Prog.*, **89**(9), 54

SANDERS, R.E. (1993b). *Management of Change in Chemical Plants: Learning from Case Histories* (Oxford: Butterworth-Heinemann)

SANDERS, R.E., HAINES, D.L. and WOOD, J.H. (1990). Stop tank abuse. *Plant/Operations Prog.*, **9**, 61

SANDERS, R.G., PANTAZELOS, T.G. and RICH, S.R. (1972a). A mobile physical-chemical treatment system for hazardous material contaminated waters. *Control of Hazardous Material Spills*, **3**, p. 412

SANDERS, R.G., PANTAZELOS, T.G. and RICH, S.R. (1972b). A short contact time physical-chemical treatment system for hazardous material contaminated waters. *Control of Hazardous Material Spills*, **1**, p. 145

SANDERS, R.W. and NODORP, R.W. (1963). Nitrogen compression system explosion. *Chem. Engng Prog.*, **59**(3), 72

SANDERSON, H. (1981). BR and chemical industry. *Chem. Ind.*, Nov. 7, 762

SANDERSON, P.G. (1969). Welding operations. In Handley, W. (1969), *op. cit.*, p. 234

SANDERSON, P.G. (1977). Welding operations. In Handley, W. (1977), *op. cit.*, p. 178

SANDERSON, P.M., VERHAGE, A.G. and FULD, R.B. (1989). State–space and verbal protocol methods for studying the human operator in process control. *Ergonomics*, **32**, 1343

SANDLER, G. (1963). *System Reliability Engineering* (Englewood Cliffs, NJ: Prentice-Hall)

SANDLER, H.J. and LUCKIEWICZ, E.T. (1987). *Practical Process Engineering* (New York: McGraw-Hill)

SANDLIN, S.G. (1963). *Piper Alpha Public Inquiry. Transcript, Days 63, 64*

SANDNER, B.W. (1978). Alignment monitoring of hyper compressor plungers. *Safety in Polyethylene Plants*, **3**, p. 47. See also *Chem. Engng Prog.*, 1977, **73**(12), 50

SANDS, D.F. and BULKLEY, W.L. (1967). Implementing process safety policies. *Loss Prevention*, **1**, p. 53

SANDTORY, H. and EDWARDS, L. (1980). *Failure Characteristics of Components. Rep. OR 230 55232.01.01*. Ship Res. Inst. of Norway, Trondheim

SANDTORY, H. and RAUSAND, M. (1991). RCM – Closing the loop between design reliability and operational reliability. *Maintenance*, **6**(1), 13

SANEMATSU, H.S. (1969a). Turbulent flame propagation in a homogeneous gas mixture. *Combust. Flame*, **13**, 1

SANEMATSU, H.S. (1969b). Turbulent flame propagation parameters (letter). *Combust. Flame*, **13**, 91

SANGERHAUSEN, C.R. (1981). Reduce seal failures in light hydrocarbon service. *Hydrocarbon Process.*, **60**(1), 111; **60**(3), 119

SANG HOON HAN, TAE WOON KIM and YOUNGE CHOI (1989). Development of a computer code AFTC for fault tree construction using decision tables and supercomponent concept. *Reliab. Engng System Safety*, **25**, 15

SANGIOVANNI, J. and ROMANS, H.C. (1987). Expert systems in industry: a survey. *Chem. Engng. Prog.*, **83**(9), 52

SANGIOVANNI, J.J. and KESTEN, A.S. (1977). A theoretical and experimental investigation of the ignition of fuel droplets. *Combust. Sci. Technol.*, **16**, 59

SANKARAN, N. (1993). Management of change – the systematic use of hazard evaluation procedures and audits. *Process Safety Prog.*, **12**, 181

SANQUIST, T.F. (1992). A model for training cognitive skills for severe accident management. *Hazard Identification and Risk Analysis, Human Factors and Human Reliability in Process Safety*, p. 327

SANSOM, D.F. and GUST, G. (1975). Flameproof components in hazardous areas. *Electrical Safety*, **2**, p. 113

SANTI, E. (1979). Finding volume in partially filled tanks. *Chem., Engng*, **86**(June 18), 144

SANTINI, D.G. (1961). Factors influencing the design and operation of a peracetic acid unit. *Chem. Engng Prog.*, **57**(12), 61

DE SANTIS, G.J. (1976). How to select a centrifugal pump. *Chem. Engng*, **83**(Nov. 22), 163

DE SANTOLI, L. and LO GIUDICE, G.M. (2000). Natural convection in inherently safe passive emergency systems. *Heat Technol. (Pisa)* **18**(1), 85–92

SANTO, G. and DELICHATSIOS, M.A. (1984). Effects of vitiated air on radiation and completeness of combustion in propane pool fires. *Fire Safety J.*, **7**, 159

SANTOLERI, J.J. (1973). Chlorinated hydrocarbon waste disposal and recovery systems. *Chem. Engng Prog.*, **69**(1), 68

SANTOLERI, J.J. (1983). Energy savings for process heaters. *Plant/Operations Prog.*, **2**, 233

SANTOMAURO, L., MAESTRO, V. and BARBERIS, C. (1979). Experimental determination of diffusion coefficients above an urban area. *Proc. Tenth Int. Tech. Mtg on Air Pollution Modelling and Its Application* (NATO/CCMS), p. 295

SANTON, R.C. (1976). Autoignition temperatures (letter). *Chem. Engr, Lond.*, **309**, 391; **313**, 611

SANTON, R.C. (1977). *Private communication* (referring to work done for the Insurance Technical Bureau)

SANTON, R.C. (1992). An expert system for the design of dust explosion relief vents. *Loss Prevention and Safety Promotion 7*, vol. 3, 152–1

SANTON, R.C. and WRIGHTSON, I. (1989). The explosion hazards of halogenated hydrocarbons. *Loss Prevention and Safety Promotion 6*, p. 111.1

SANTON, R.C., POSTILL, A., FURMAN, T.T., VADERA, S., BYERS-BROWN, W. and PALMER, K.N. (1991). A feasibility study into the use of expert systems for explosion relief vent design. *Hazards*, **XI**, p. 317

SANTORO, R.J. and SEMERJIAN, H.G. (1985). Soot formation in diffusion flames: flow rate, fuel species and temperature effects. *Combustion*, **20**, p. 997

SANTORO, R.J., SEMERJIAN, H.G. and DOBBINS, R.A. (1983). Soot particle measurements in diffusion flames. *Combust. Flame*, **51**, 203

SARA, M.N. (1994). *Standard Handbook for Solid and Hazardous Waste Facility Assessment* (Boca Raton, FL: Lewis Publs)

DE SARAM, H. (1985). *Programming in Micro-Prolog* (Chichester: Ellis Horwood)

SARAOJA, E.K. (1977). Lightning earths. In Golde, R.H. (1977), *op. cit.*, p. 577

SARAPPO, J.W. (1969). Contract maintenance: its place in chemical plants. *Chem. Engng*, **76**(Nov. 17), 264

SARGENT, G.D. (1969). Dust collection equipment. *Chem. Engng*, **76**(Jan. 27), 131

SARGENT, H.B. (1957). How to design a hazard free system to handle acetylene. *Chem. Engng*, **64**(2), 250

SARGENT, R. (1980). Process control and the impact of microelectronics. *Chem. Ind.*, Dec. 20, 926

SARGENT, R.W.H. (1983). Advances in modelling and analysis of chemical process systems. *Comput. Chem. Engng*, **7**, 219

SARGENT, R.W.H. and WESTERBERG, A.W. (1964). 'SPEED-UP' in chemical engineering design. *Trans. Instn Chem. Engrs*, **42**, T190

SARGENT, R.W.H., PERKINS, J.D. and THOMAS, S. (1982). SPEED-UP: Simulation program for the economic evaluation and design of unified processes. In Leesley, M.D. (1982), *op. cit.*

SARKES, L.A. and MAN, D.B. (1974). A survey of LNG technological needs in the USA – 1974 to beyond 2000. *LNG Conf. 4*

SARSTEN, J.A. (1972). LNG stratification and rollover. *Conf Proc. on LNG Importation and Terminal Safety* (Boston, MA: Committee on Hazardous Materials), p. 27

SARSTEN, J.A. (1972). LNG stratification and rollover. *Pipeline Gas J.*, **199**(Sep.), 37

SARTORELLI, E. (1969). Clinical and pathological aspects of acute bronchial pneumonia from the inhalation of gas irritants. *Lab. Med.*, **60**, 561

SARTORI, M. (1939). *The War Gases* (New York: van Nostrand)

SASANOW, S. (1989). Wellheads on the seabed. *New Scientist*, Mar. 18, 43

SASONOW, S. (1994). Reflections on a dizzy decade. *Offshore Engr*, Apr., 49

SASS, H.M. (1986). Technical and cultural aspects of risk perception. In Hartwig, S. (1986), *op. cit.*, p. 25

SASS, R. and ECKERMANN, R. (1992). CHEMSAFE – a user friendly information system of safety values. *Loss Prevention and Safety Promotion 7*, vol. 3, 145–1

SATHE, S.Y. and O'CONNOR, T.M. (1988). Review of nitriding in ammonia plants. *Ammonia Plant Safety*, **28**, 115

SATHER, H.N. (1974). *Biostatistical Aspects of Risk–Benefit: the Use of Competing Risk Analysis. Rep. UCLA-ENG-7477.* Univ. of California, Los Angeles, CA

SATHER, H. (1976). *Statistical Models for Competing Risks Analysis. Rep. UCLA-ENG-7676.* Univ. of California, Los Angeles, CA

SATO, K. (1989). Gas screening effect caused by steam curtains. *J. Loss Prev. Process Ind.*, **2**(4), 209

SATO, K., AOKI, M. and NOYORI, R. (1998). A 'green' route to adipic acid: direct oxidation of cyclohexenes with 30 percent hydrogen peroxide. *Science* **281**, (September 11 1998), 1646–47

SATO, T., KUNIMOTO, T., YOSHI, S. and HASHIMOTO, T. (1969). On the monochromatic distribution of the radiation from luminous flame. *JSME Bull-12*, **1135**

SATTERFIELD, C.N. (1970). *Mass Transfer in Heterogeneous Catalysis* (Cambridge, MA: MIT Press)

SATTLER, F.J. (1989a). Finding below-surface defects with nondestructive testing. *Chem. Engng*, **96**(9), 161

SATTLER, F.J. (1989b). Finding surface defects with nondestructive testing. *Chem. Engng*, **96**(7), 125

SATTLER, F.J. (1989c). Improving plant maintenance via nondestructive testing. *Chem. Engng*, **96**(12), 115

SATTLER, F.J. (1989d). Nondestructive testing via optical and acoustical methods. *Chem. Engng*, **96**(11), 191

SATTLER, F.J. (1990). Nondestructive testing methods can aid plant operation. *Chem. Engng*, **97**(10), 177

SATYANARAYANA, K., BORAH, M. and RAO, P.G. (1991). Prediction of thermal hazards from fireballs. *J. Loss Prev. Process Ind.*, **4**, 344

SAUNBY, J.B. and KIFF, B.W. (1976). Liquid phase oxidation . . . hydrocarbons to petrochemicals. *Hydrocarbon Process.*, **55**(11), 247

SAUNDERS, E.A.D. (1980). The current approach to the design of shell and tube heat exchangers. *Interflow 80* (Rugby: Instn *Chem. Engrs)*, p. 259

SAUNDERS, G.T. (1987). Laboratory hoods. In Young, J.A, (1987a), *op. cit.*, p. 287

SAUNDERS, P.T. (1980). *An Introduction to Catastrophe Theory* (Cambridge: Cambridge Univ. Press)

SAVAGE, K.M. (1975). Marine gas hazards control (cleaning and gas-freeing shipboard tanks). *Transport of Hazardous Cargoes*, **4**

SAVAGE, L.D. (1962). The enclosed laminar diffusion flame. *Combust. Flame*, **6**, 77

SAVAGE, L.J. (1962a). Bayesian statistics. In Machol, R.E. and Gray, P. (1962), *op. cit.*, p. 161

SAVAGE, L.J. (1962b). *The Foundations of Statistical Inference* (London: Methuen)

SAVAGE, L.J. (1964). The foundations of statistics reconsidered. In Kyburg, H. and Smokler, H. (1964), *op. cit.*

SAVAGE, L.J. (1971). Elicitation of personal probabilities and expectations. *J. Am. Statist. Ass.*, **66**(336), 783

SAVAGE, P. and MARSHALL, T. (1978). A cease-fire looming for UK contractors, unions? *Chem. Engng*, **85**(Feb. 13), 54C

SAVAGE, P., PORTNOY, K. and PARKINSON, G. (1988). Demothballing is back in fashion. *Chem. Engng*, **95**(8), 26

SAVAGE, P.E., GOPALAN, S., MIZAN, T.I., MARTINO, C.J. and BROCK, E.E. (1995). Reactions at supercritical conditions: applications and fundamentals. *AIChE J.* **41**, 7 (July), 1723–78

SAVAGE, P.E.A. (1971). Disaster planning: a review. *Injury*, **3**(1), 49

SAVAGE, P.E.A. (1977). Hospital disaster planning. In Easton, K. (1977), *op. cit.*, 247

SAVAGE, P.E.A. (1979). Future disaster planning in the UK. *Disaster*, **3**(1), 75

SAVAGE, P.R. (1978). Asbestos: will rules tighten still further? *Chem. Engng*, **85**(Dec. 4), 60I

SAVAGE, P.R. (1979). Legislative puzzle for chemical shippers. *Chem. Engng*, **86**(Apr. 9), 54J

SAVAGE, P.R. *et al.* (1979). Harrisburg clouds world nuclear plans. *Chem. Engng*, **86**(May 7), 28J

SAVAS, E.S. (1965). *Computer Control of Industrial Processes* (New York: McGraw-Hill)

SAVI, L. (1971). Packing standards in relation to performance testing. *Transport of Hazardous Cargoes*, **2**, paper 12

SAVIANO, F., LAGANA, V. and BISI, P. (1981). Integrate recovery systems for low energy ammonia production. *Hydrocarbon Process.*, **60**(7), 99

SAVILLE, G. (1989a). *Piper Alpha Inquiry. Transcript, Day 87*

SAVILLE, G. (1989b). *Piper Alpha Inquiry. Transcript, Day 107*

SAVILLE, G. (1989c). *Piper Alpha Inquiry. Transcript, Day 119*

SAVILLE, G. (1989d). *Piper Alpha Inquiry. Transcript, Day 127*

SAVILONIS, B. and RICHARDS, R. (1988). Survey and evaluation of existing smoke movements models. *Fire Safety J.*, **13**, 87

SAVKOVIC-STENANOVIC, J. (1994). A qualitative model for estimation of plant behaviour. *Comput. Chem. Engng*, **18**(Suppl.), S713

SAVOIE, D.L. (1988). Effect of normal standard errors on lognormal distributions. *Atmos. Environ.*, **22**, 1957

SAVORY, S. (ed.) (1988). *Artificial Intelligence and Expert Systems* (Chichester: Ellis Horwood)

SAVVIDES, C.N. and TAM, V.H.Y. (1991). The modeling of gas explosions: a comparison of 3D numerical model with simpler models. *Modeling and Mitigating the Consequences of Accidental Releases*, p. 599

SAVY, J.B., BERNREUTER, D.L. and CHEN, J.C. (1987). A methodology to correct for effect of the local site characteristics in seismic hazard assessment. In Cermak, A.S. (1987), *op. cit.*, p. 243

SAWYER, J.G. and WILLIAMS, G.P. (1974). Turndown efficiency of a single-train plant. *Ammonia Plant Safety*, **16**, p. 68

SAWYER, J.G., WILLIAMS, G.P. and CLEGG, J.W. (1972). Causes of shutdowns in ammonia plants. *Ammonia Plant Safety*, **14**, p. 62

SAWYER, P. (1989). Process simulation, integration and control – challenges and responses. *Chem. Engr, Lond.*, **464**, 47

SAWYER, P. (1991a). Language, loops and logic. *Chem. Engr, Lond.*, **489**, 24

SAWYER, P. (1991b). Simulation on the move. *Chem. Engr, Lond.*, **501**, 31

SAWYER, P. (1991c). Towards the paperless project. *Chem. Engr, Lond.*, **496**, 36

SAWYER, P. (1992a). Batch standards on the move. *Chem. Engr, Lond.*, **524**, 28

SAWYER, P. (1992b). Software for safety. *Chem. Engr, Lond.*, **526**, 32

SAWYER, P. (1993a). *Computer-Controlled Batch Processing* (Rugby: Instn *Chem. Engrs*)

SAWYER, P. (1993b). STEP steps in. *Chem. Engr, Lond.*, **540**, 20

SAWYER, R.L. (2002). Implementation of inherently safer systems regulations. *17th Annu. Int. Conf. and Workshop: Risk, Reliability, and Security*, Oct. 8–11, 2002, Jacksonville, FL, Regulation of Inherent Safety Workshop

SAX, N.I. (1957–). *Dangerous Properties of Materials*; 1963, 2nd ed.; 1968, 3rd ed. (New York: Reinhold) 1975, 4th ed.; 1984, 6th ed.; 1989, 7th ed. (New York: van Nostrand Reinhold). See also Lewis, R.J.

SAX, N.I. (ed.) (1974). *Industrial Pollution* (New York: van Nostrand)

SAX, N.I. (1981). *Cancer Causing Chemicals* (New York: van Nostrand Reinhold)

SAX, N.I. (1986). *Hazardous Chemicals Information Annual*, No. 1 (Wokingham: van Nostrand Reinhold)

SAX, N.I. and LEWIS, R.J. (1986). *Rapid Guide to Hazardous Chemicals in the Workplace* (New York: van Nostrand Reinhold)

SAXTON, J.C. and NARKUS-KRAMER, M. (1975). EPA findings on solid wastes from industrial chemicals. *Chem. Engng*, **82**(Apr. 28), 98

SAYERS, B. (1976). Safety assessment of a hazardous chemical plant. *Soc. Chem. Ind., Symp. on Loss Prevention* (London)

SAYERS, B.A. (ed.) (1988). *Human Factors and Decision Making* (London: Elsevier Applied Science)

SAYLES, R.S. (1983). The use of discriminant function techniques in reliability assessment and data classification. *Reliab. Engng*, **6**, 103

SAYLES, R.S. and MACPHERSON, P.B. (1980). Tribological aspects of reliability. In Moss, T.R. (1980), *op. cit.*, p. 55

SAYYED, S. (1976). How compressor components affect performance. *Hydrocarbon Process.*, **55**(9), 273

SAYYED, S. (1978). Compressor shop tests: do they ensure reliability? *Hydrocarbon Process.*, **57**(4), 201

SAYYED, S. (1985). Selecting high-performance centrifugal compressors. *Hydrocarbon Process.*, **64**(10), 57

SCACCIA, C. and THEOCLITUS, G. (1980). Heat exchangers. Types, performance and applications. *Chem. Engng*, **87**(Oct. 6), 121

SCALI, C., SPINAZZOLA, D.A. and ZANELLI, S. (1980). Stability and control of batch reactor for the polymerisation of vinyl chloride. *Loss Prevention and Safety Promotion 3*, vol. 3, p. 1348

SCANLON, T.J. (1989). *Piper Alpha Public Inquiry. Transcript, Days 72, 162*

SCANNELL, G.F. (1971). How industry handles a transportation emergency. *Loss Prevention*, **5**, p. 1

SCARRONE, M., PICCININI, N. and MASSOBRIO, C. (1989). A reliability data bank for the natural gas distribution industry. *J. Loss Prev. Process Ind.*, **2**(4), 235

VAN DER SCHAAF, J. (1986). Development of a device to diminish the risks of torches at LPG-refuelling stations. *Loss Prevention and Safety Promotion 5*, p. 69–1

VAN DER SCHAAF, J.B.R. and OPSCHOOR, G. (1980). Methodology for the execution of risk analyses; applied to the storage and handling of LPG in the Netherlands. *Loss Prevention and Safety Promotion 3*, vol. 2, p. 315

VAN DER SCHAAF, J.B.R. and STEUNENBERG, C.F. (1983). Investigations into the circumstances surrounding an accident in Nijmegen involving a tanker. *Loss Prevention and Safety Promotion 4*, vol., 1, p. N14

VAN DER SCHAAF, T.W. (1988). Critical incidents and human recovery: some examples of research techniques. *Human Recovery* (ed. L.H.J. Goossens) (Delft: Delft Univ. of Technology)

VAN DER SCHAAF, T.W. (1991a). Development of a near miss management system at a chemical process plant. In Van der Schaaf, T.W., Lucas, D.A. and Hale A.R. (1991), *op. cit.*, p. 57

VAN DER SCHAAF, T.W. (1991b). A framework for designing near miss management systems. In Van der Schaaf, T.W., Lucas, D.A. and Hale A.R. (1991), *op. cit.*, p. 27

VAN DER SCHAAF, T.W. (1991c). Introduction. In Van der Schaaf, T.W., Lucas, D.A. and Hale A.R. (1991), *op. cit.*, p. 1

VAN DER SCHAAF, T.W. (1993). Developing and using cognitive task typologies. *Ergonomics*, **36**, 1439

VAN DER SCHAAF, T.W., LUCAS, D.A. and HALE, A.R. (eds) (1991). *Near Miss Reporting as a Safety Tool* (Oxford: Butterworth-Heinemann)

SCHAAFSTAL, A. (1993). Knowledge and strategies in a diagnostic skill. *Ergonomics*, **36**, 1305

SCHACHER, G.E., FAIRALL, C.W. and ZANNETTI, P. (1982). Comparison of stability methods for parameterizing coastal overwater dispersion. *First Int. Conf. on Meteorology and Air/Sea Interaction of the Coastal Zone* (Boston, MA: Am. Met. Soc.)

SCHACK, C.J., PRATSINIS, S.E. and FRIEDLANDER, S.K. (1985). A general correlation for deposition of suspended particles from turbulent gases to completely rough surfaces. *Atmos. Environ.*, **19**, 953

SCHÄCKE, G. and SAWAI, G. (1990). Occupational health prevention in landfills for hazardous waste. *Safety and Loss Prevention in the Chemical and Oil Processing Industries*, p. 363

SCHÄCKE, H. and FALKE, K.O. (1986). Safe operation of a pneumatic conveying (flash) dryer. *Loss Prevention and Safety Promotion 5*, p. 49–1

SCHAEFER, A.T. and BRITTAIN, E.L. (1988). *Autocad Productivity Book* (Chapel Hill, NC: Ventana Press)

SCHAEFFER, H.K. (1979). *Sicherheit in der Chemie. Sicherheitstechnik, Arbeitsschutz, Gefaehrliche Stoffe* (Munich: Carl Hansen)

SCHAEFFER, J. (1980). Use flammable vapor sensors? *Hydrocarbon Process.*, **59**(1), 211

SCHÄFER, SIR E. (1915). On the immediate effects of the inhalation of chlorine gas. *Br. Med. J.*, **11**(Aug. 19), 245

SCHÄFER, H.K. (1976). Problems of the secure direction of chemical reactions: introduction. *Prevention of Occupational Risks*, **3**, p. 63

SCHÄFER (1973a). The accident as a psychophysical phenomenon. *Prevention of Occupational Risks*, **2**, p. 46

SCHÄFER (1973b). The drying of an explosible dust in a nitrogen cycle. *Prevention of Occupational Risks*, **2**, p. 162

SCHAGEN, I.P. (1987). A new model for software failure. *Reliab. Engng*, **18**, 205

SCHAGEN, I.P. and SALLIH, M.M. (1987). Fitting software failure data with stochastic models. *Reliab. Engng*, **17**(2), 111

SCHAICH, J.R. (1991). Estimate fugitive emissions from process equipment. *Chem. Engng Prog.*, **87**(8), 31

SCHALL, W.C. (1962). Selecting sensible sampling systems. *Chem. Engng*, **69**(May 14), 157

SCHALLA, R.L. and MCDONALD, G.E. (1955). Mechanism of smoke formation in diffusion flames. *Combustion,* **5,** p. 316

SCHALLER, L.C. (1990). Off site risk assessment and risk minimization. *Plant/Operations Prog.,* **9,** 50

SCHAMPEL, K. and STEEN, H. (1975). Flame arresting high velocity valves on cargo tanks of tankers for inflammable liquids. *Transport of Hazardous Cargoes,* **4**

SCHAMPEL, K. and STEEN, H. (1976). Flame arresting high velocity valves on cargo tanks of tankers for inflammable liquids. *J. Haz. Materials,* **1**(3), 223

SCHAMPEL, K. and STEEN, H. (1977). Explosion protection in systems for combustion of industrial exhaust air with combustible components (after-burn systems). *Prevention of Occupational Risks,* **5,** p. 233

SCHANK, R.C. and ABELSON, R.P. (1977). *Scripts, Plans, Goals and Understanding* (Hillsdale, NJ: Lawrence Erlbaum)

SCHANK, R.C. and COLBY, K.M. (eds) (1973). *Computer Models of Thought and Language* (San Francisco, CA: Freeman)

SCHANK, R.C. and RIESBECK, C.K. (1981). *Inside Computer Understanding* (Hillsdale, NJ: Lawrence Erlbaum)

SCHANZENBACH, G.P. (1975). Sensors for machinery monitoring. *Hydrocarbon. Process.,* **54**(2), 85

SCHAPER (1973). Technical measures for the prevention of acro-osteolysis with autoclave cleaners. *Prevention of Occupational Risks,* **2**

SCHARFSTEIN, L.R. (1964). Special stainless steels. *Chem. Engng Prog.,* **60**(11), 90

SCHARLE, W.J. (1965) The safe handling of liquid hydrogen. *Chem. Engr, Lond.,* **185,** CE16

SCHARLE, W., SALOT, W.J. and HARDY, O.F. (1971). A review of primary reformer catalysts. *Ammonia Plant Safety,* **13,** p. 73

SCHATZ, K.W. and FTHENAKIS, V.M. (1994). Mitigation of hydrogen fluoride aerosols by dry powders. *J. Loss Prev. Process Ind.,* **7,** 451

SCHATZ, K.W. and KOOPMAN, R.P. (1990). Water spray mitigation of hydrofluoric acid releases. *J. Loss Prev. Process Ind.,* **3**(2), 222

SCHATZMANN, M., KOENIG, G. and LOHMEYER, A. (1986). Dispersion modeling of hazardous materials in the wind tunnel. In de Wispelaere, C., Schiermeier, F.A. and Gillani, N.V. (1986), *op. cit.,* p. 641

SCHATZMANN, M., LOHMEYER, A. and ORTNER, G. (1987). Flue gas discharge from cooling towers. Wind tunnel investigation of building downwash effects on ground level concentrations. *Atmos. Environ.,* **21,** 1713

SCHATZMANN, M., SNYDER, W.H. and LAWSON, R.I. (1993). Experiments with heavy gas jets in laminar and turbulent crossflows. *Atmos. Environ.,* **27A,** 1105

SCHAYES, G. (1982). Direct determination of diffusivity profiles from synoptic reports. *Atmos. Environ.,* **16,** 1407

SCHEEL, L.F. (1967a). Compressor pulse dampers — when are they necessary? *Hydrocarbon Process.,* **46**(2), 149

SCHEEL, L.F. (1967b). New piston compressor rating method. *Hydrocarbon Process.,* **46**(12), 133

SCHEEL, L.F. (1968). New ideas on centrifugal compressors. *Hydrocarbon Process.,* **47**(9), 253; **47**(10), 161; **47**(11), 238

SCHEEL, L.F. (1969a). New ideas on fan selection. *Hydrocarbon Process.,* **48**(6), 125

SCHEEL, L.F. (1969b). New ideas on rotary compressors. *Hydrocarbon Process.,* **48**(11), 271

SCHEEL, L.F. (1970). What you need to know about gas expanders. *Hydrocarbon Process.,* **49**(2), 105

SCHEEL, L.F. (1971). How to specify centrifugal compressors. Air compressor rotor and components. *Hydrocarbon Process.,* **50**(10), 74

SCHEFER, R.W., NAMAZIAN, M. and KELLY, J. (1989). Comparison of turbulent-jet and bluff-body stabilized flames. *Combust. Sci. Technol.,* **67,** 123

SCHEFFLER, N.E. (1994). Improved Fire and Explosion Index hazard classification. *Process Safety Prog.,* **13,** 214

SCHEFFLER, N.E. (1995). Inherently safer latex plants. *Proc. of the 29th Annu. Loss Prev. Symp.,* July 31—August 2, 1995, Boston, MA, (ed. E. D. Wixom and R. P. Benedetti,) Paper 1b (New York: American Institute of Chemical Engineers)

SCHEFFLER, N.E., GREENE, L.S. and FRURIP, D.J. (1993). Vapor suppression of chemicals using foams. *Process Safety Prog.,* **12,** 151

SCHEID, D.C. (1987). Health, safety and environmental protection. *Plant/Operations Prog.,* **6,** 211

SCHEIDEMANTEL, V.R. (1933). Delayed consequences of hydrogen sulphide poisoning (disorders of cardiac activity and lesion of the trunkal ganglia). *Munch. Med. Wochen,* **80,** 1494

SCHEIDWEILER, A. (1983). The distribution of intelligence in future fire detection systems. *Fire Safety J.,* **6,** 209

SCHEIMANN, W.E. (1990). Port: a process safety audit system. *Plant/Operations Prog.,* **9,** 156

SCHEINSON, R.S., PENNER-HAHN, J.E. and INDRITZ, D. (1989). The physical and chemical action of fire suppressants. *Fire Safety J.,* **15,** 437

SCHEIRMAN, W.L. and ROGERS, T.F. (1985). How to calculate storage-tank heat loss. *Oil Gas J.,* **83**(Jan. 14), 96

SCHEITHE, W. (1992). Better bearing vibration analysis. *Hydrocarbon Process.,* **71**(7), 57

SCHELL, F.R. and MATLOCK, B. (1979). *Industrial Welding Procedures* (New York: van Nostrand Reinhold)

SCHELLEKENS, P.L. (1984). Alarm management in distributed control systems. *Control Engng,* **31**(12)

SCHELLING, T. (1968). The life you save may be your own. In *Problems in Public Expenditure Analysis* (ed. S.B. Chase) (Brookings Inst.), p. 127

SCHENKEL, J.R.H. (1977). Design and the identification of potential hazards. *Design Developments Related to Safety, Costs and Solid Systems* (London: Instn Chem. Engrs), paper A-3—1

SCHERBERGER, R.F. (1977). Industrial hygiene control methods. *Loss Prevention,* **11,** p. 114

SCHERFF, R.E. and NOVACEK, D.A. (1985). Fire in a primary reformer penthouse. *Ammonia Plant Safety,* **25,** p. 84

SCHEUERMANN, K.P. (1985). Studies about the influence of turbulence on the course of explosions. *Process Safety Prog.,* **13,** 219

SCHEWE, G.J. and CARVITTI, J. (1986). Human exposure estimates using guideline models: an integrated approach. *Fifth Joint Conf. on Applications of Air Pollution Meteorology* (Boston, MA: Am. Met. Soc), p. 281

SCHIAPPA, C.A. and WINEGARDNER, D.K. (1994). Pressure relief system alternatives for vessel jackets. *J. Loss Prev. Process Ind.,* **7,** 7

SCHIAVELLO, B. (1992). Troubleshoot centrifugal pumps. *Chem. Engng Prog.,* **88**(11), 35

SCHICHT, H.H. (1969). Flow patterns for an adiabatic two-phase flow of water and air within a horizontal tube. *Verfahrenstechnik,* **3**(4), 153

SCHIEFER, H.M. and PAPE, P.G. (1982). Uses of silicones in process plants. *Chem. Engng,* **89**(Feb. 8), 123

SCHIELER, L. and PAUZE, D. (1976). *Hazardous Materials* (New York: Delmar)

SCHIERMEIER, F.A., LAVERY, T.F. and DICRISTOFARO, D.C. (1986). Meteorological events that produced the highest ground-level concentrations during complex terrain/ field experiments. In de Wispelaere, C., Schiermeier, F.A. and Gillani, N.V. (1986), *op. cit.*, p. 99

SCHIERWATER, F.W. (1971). The safe operation of exothermic reactions, especially in the liquid phase. *Major Loss Prevention*, p. 47

SCHIERWATER, F.W. (1979). Safety at work today – a task for both the company and the professional association. In BASF (1979), *op. cit.*, p. 35

SCHIFFHAUER, D.J. (1984). Hydrogen compression with reciprocating compressors. *Hydrocarbon Process.*, 63(1), 75

SCHIFFMAN, D.A. (1986). Analyzing failure data with unknown scale: applying the Pearson–Fisher theorem. *Reliab. Engng*, 16(2), 121

SCHILDKNECHT, J. (1977). Development and application of a 'mini-pilot reaction calorimeter'. *Loss Prevention and Safety Promotion 2*, p. 139

SCHILDKNECHT, M. (1984). *Versuche zur Freistrahlung von Wasserstoff-Luft-Gemischen im Hinblick auf den Übergang Deflagration-Detonation. Rep. BIeV-R-65.769– 1.* Battelle Inst., Frankfurt

SCHILDKNECHT, M. and GEIGER, W. (1982). *Detonationsahnliche Explosionsformen-Mogliche Intiierung Detonations ahnlicher Explosionsformen durch partiellen Einschluss. Teilaufgabe 1 des Teilforschungsprogramm Gasexplosionen. Rep. BIeV-R- 64.176–2.* Battelle Inst., Frankfurt

SCHILDKNECHT, M. and GEIGER, W. (1983). Explosion development in unconfined and partially confined ethylene–air mixtures due to jet initiation. *Loss Prevention and Safety Promotion 4*, vol. 1, p. D30

SCHILDKNECHT, M., GEIGER, W. and STOCK, M. (1984). Flame propagation and pressure build-up in a free gas–air mixture due to jet ignition. In Bowen, J.R. *et al.* (1984b), *op. cit.*, p. 474

SCHILDT, B.E. (1972). Mortality rate in quantified combined injuries. *Strahlentherapie*, 144, 40

SCHILDWACHTER, J. (1961). Water injection solves dirty-gas problems in centrifugal compressors. *Chem. Engng*, 68(Nov. 13), 254

SCHILLER, L. (1932). *Handbuch der Experimentalphysik*, vol. 4, pt 4 (Leipzig)

SCHILLER, W. (1933). In *Überkritische Entspannung kompressibler Flüssigkeiten. Forschung auf dem Geibiet des Ingenieurwesens*, 4, 128

SCHILLING, R.S.F. (1973). *Occupational Health Practice* (London: Butterworths)

SCHILLINGS, R.C. (1970). Protection for chlorine load cells. *Chem. Engng Prog.*, 66(2), 68

SCHILLMOELLER, N.H. (1974). Technology, its control and the engineer. *Sci. Publ. Policy*, 1, 150

SCHILLMOLLER, C.M. (1979). Alloy selection for VCM plants. *Hydrocarbon Process.*, 58(3), 89

SCHILLMOLLER, C.M. (1986). Amine stress cracking: causes and cures. *Hydrocarbon Process.*, 65(6), 37

SCHILLMOLLER, C.M. and ALTHOFF, H.J. (1984). How to avoid failure of stainless steel. *Chem. Engng*, 91(May 28), 119

SCHILLMOLLER, C.M. and VAN DEN BRUCK, U.W. (1984). Furnace alloys update. *Hydrocarbon Process.*, 63(12), 55

SCHILLY, R.C. (1962). Hydrocarbon accumulation in an air box. *Ammonia Plant Safety*, 4, p. 1

SCHILPEROORD, A.A. (1986). The safe packaging of dangerous goods. *Loss Prevention and Safety Promotion 5*, p. 50–1

SCHIRM, A.C. (1971). How to specify centrifugal compressors. Bearings, seals and testing. *Hydrocarbon Process.*, 50(10), 76

SCHIRRIPA, J.T. (1977). Carcinogens – how to deal with them. *Loss Prevention*, 11, p. 125

SCHLAGBAUER, M. and HENSCHLER, D. (1967). Toxicity of chlorine and bromine with single and repeated inhalation. *Int. Arch. Gewerbepath. Gewerbehyg.*, 23, 91

SCHLAPOWSKY, A.F. (1967). *Bull. N.Y. Acad. Med.*, 43(8)

SCHLATTER, R.G. and NOEL, C.J. (1972). Axial compressor control in a liquefaction unit. *LNG Conf. 3*

SCHLEGEL, W.F. (1972). Design and scale-up of polymerization reactor. *Chem. Engng*, 79(Mar. 20), 88

SCHLEICH, K. (1976). Reliability of chemical reactions. *Prevention of Occupational Risks*, 3, p. 117

SCHLEIFENBAUM, R. (1981). Industrial investment in Germany in the context of German planning and pollution law. In Salter, J. and Thomas, P. (1981), *op. cit.*, paper 19

SCHLEY, J.R. (1960). New surfacing methods view with old for high temperature process applications! *Chem. Engng*, 67(Mar. 21), 184

SCHLICHTHARLE, G. (1984). Reformer inlet pigtail repair. *Ammonia Plant Safety*, 24, 86

SCHLICHTHARLE, G. and HUBERICH, T. (1983). Tank-car loading station for liquid and aqueous ammonia. *Plant/Operations Prog.*, 2, 165

SCHLICHTLING, H. (1936). *Ingen-Arch.*, 7, 1

SCHLICHTLING, H. (1960). *Boundary Layer Theory* (New York: McGraw-Hill)

SCHLIEPHAKE, V., GIESBRECHT, H. and LOFFLER, U. (1992). Formation and self-ignition of polyglycoles in insulation materials: a source of danger for ethylene oxide plants. *Loss Prevention and Safety Promotion*, 7, vol. 2, 86–1

SCHLIMMER, J.A. and LANGLEY, P. (1992). Learning, machine. In Shapiro, S.C. (1992), *op. cit.*, p. 785

SCHLOBOHM, D.A. (1984). Introduction to Prolog. Pt II: a simple expert system. *Robotics Age*, Dec., 24

SCHLUNDER, E.U. (ed.) (1982). *Heat Exchanger Design Handbook* (New York: Hemisphere)

SCHMALZ, F. (1982). Grinding and mixing plant, spray driers. *The Control and Prevention of Dust Explosions* (London: Oyez/IBC), paper 14

SCHMEAL, W.R., MACNAB, A.J. and RHODES, P.R. (1978). Corrosion in amine/sour gas treating contactors. *Chem. Engng. Prog.*, 74(3), 37

SCHMER, L.D. and GOODMAN, M. (1990). Using ultrasonics to inspect bellows and towers. *Hydrocarbon Process.*, 69(1), 61

SCHMID, C.F. (1966). *Scaling Techniques in Sociological Research* (Englewood Cliffs, NJ: Prentice-Hall)

SCHMID, R. (1980). Maintenance and safety of the process units. *Loss Prevention and Safety Promotion 3*, vol. 2, p. 735

SCHMIDLI, J., BANNERJEE, S. and YADIGAROGLU, G. (1990). Effects of vapour/aerosol and pool formation on rupture of vessels containing superheated liquid. *J. Loss Prev. Process Ind.*, 3(1), 104

SCHMIDT, E.H.W., STEINICKE, H. and NEUBERT, U. (1953). Flame and schlieren photographs of combustion waves in tubes. *Combustion*, 4, p. 658

SCHMIDT, H., HABERLE, K. and RECKLINGHAUSEN, M.K. (1955). *Tech. Uberwach.*, 7, 432

SCHMIDT, F.H. (1965). On the rise of hot plumes in the atmosphere. *Int. J. Air Water Poll.*, **9**, 175

SCHMIDT, H. (1971). Protective measures and experience in acetylene decomposition in piping and equipment. *Major Loss Prevention*, p. 165

SCHMIDT, T.R. (1977a). Ground level detector tames flare stack flames. *Chem. Engng*, **84**(Apr. 11), 121

SCHMIDT, T.R. (1977b). A smokeless flare system. *Ecolibrium* (Shell Oil Co.), **6**(1)

SCHMIDT, W. (1925). *Der Massenaustausch in freier Luft und verwandte Erscheinungen* (Hamburg: Verlag von Henri Grand)

SCHMIT, K.H., LEE, T.E., MITCHEN, J.H. and SPITERI, E.E. (1978). How to cope with aluminum alklyls. *Hydrocarbon Process.*, **57**(11), 341

SCHMITT, D.W. (1964). Compressor test facility explosion. In Vervalin, C.H. (1964a), *op. cit.*, p. 330

SCHMITT, S.A. (1969). *Measuring Uncertainty. An Elementary Introduction to Bayesian Statistics* (Reading, MA: Addison-Wesley)

SCHMITT, W. and WELLEIN, R. (1980). Methods of determining the influence of quality assurance on the reliability of primary components of a PWR. *Int. J. Pres. Ves. Piping*, **8**, 187

SCHMITTINGER, P. (1986). Chlorine. *Ullmanns Encyclopaedia of Chemical Technology, rev.* ed., vol. A6 (Weinheim: VCH), p. 399

SCHMITZ, R.A. (1967). A further study of diffusion flame stability. *Combust. Flame*, **11**, 49

SCHNATZ, G. (1986). Major results of the BMFT-HSWE heavy gas project. In Hartwig, S. (1986), *op. cit.*, p. 35

SCHNATZ, G. and FLOTHMANN, D. (1980). A 'K'-model and its modification for the dispersion of heavy-gases. In Hartwig, S. (1980), *op. cit.*, p. 125

SCHNATZ, G., KIRSCH, J. and HEUDORFER, W. (1983). Investigation of energy fluxes in heavy gas dispersion. In Hartwig, S. (1983), *op. cit.*, p. 53

SCHNEEWEISS, W. (1973). *Zuverlassigkeitstheorie* (Berlin: Springer)

SCHNEIDER, A. (1981). Three Mile Island: the chemical engineering aspects. *Loss Prevention*, **14**, p. 96. See also *Chem. Engng Prog.*, 1980, **76**(9), 13

SCHNEIDER, A.L. (1980). Liquefied natural gas spills on water: fire modelling. *J. Fire Flammability*, **11**, 302

SCHNEIDER, A.L., LIND, C.D. and PARNAROUSKIS, M. (1980). US Coast Guard liquefied natural gas research at China Lake. *Gastech 79*, p. 215

SCHNEIDER, D.F. (1993). How to calculate purge gas volumes. *Hydrocarbon Process.*, **72**(11), 89

SCHNEIDER, G.G. *et al.* (1975a). Electrostatic precipitators: how they are used in the CPI. *Chem. Engng*, **82**(Aug. 18), 97

SCHNEIDER, G.G. *et al.* (1975b). Selecting and specifying electrostatic precipitators. *Chem. Engng*, **82**(May 26), 94

SCHNEIDER, H. and PFORTNER, H. (1981). *Spread of the Flame and Shock Waves in the Deflagration of Hydrogen–Air Mixtures.* Frauenhofer Inst., Pfinztal-Berghaven, Germany

SCHNEIDER, M.E. and KENT, L.A. (1989). Measurements of gas velocities and temperatures in a large open pool fire. *Fire Technol.*, **25**(1), 51

SCHNEIDER, U. (1988). Concrete at high temperatures – a general review. *Fire Safety J.*, **13**, 55

SCHNEIDER, U. (ed.) (1990). Repairability of fire damaged structures. *Fire Safety J.*, **16**, 251 (special issue)

SCHNEIDER, W. and DILLER, W. (1991). Phosgene. *Ullmanns Encyclopaedia of Chemical Technology*, rev. ed., vol. A19 (Weinheim: VCH), p. 411

SCHNEIDERMAN, P.M. (1978). Amortization of pollution control facilities: how the election works. *Control of Hazardous Material Spills*, 4, p. 77

SCHNEIDMAN, D. and STROBEL, L. (1974). Indiana gas leak capped: 1700 return. *Chicago Daily Tribune*, Sep. 15

SCHOCH, W. (1979). Self-regulation of the industry in the field of technical monitoring. In BASF (1979), *op. cit.*, p. 167

SCHOEN, W. and DROSTE, B. (1988). Investigations of water spraying systems for LPG storage tanks by full scale fire tests. *J. Haz. Materials*, **20**, 73

SCHOEN, W., PROBST, U. and DROSTE, B. (1989). Experimental investigations of fire protection measures for LPG storage tanks. *Loss Prevention and Safety Promotion 6*, p. 51.1

SCHOFIELD, C. (1982). *Guide to Safety in Mixing* (Rugby: Instn Chem. Engrs)

SCHOFIELD, C. (1983). The IChemE guide to safety in mixing. *Chem. Engr, Lond.*, **391**, 23

SCHOFIELD, C. (1984). *Guide to Dust Explosion Prevention and Protection. Pt 1, Venting* (Rugby: Instn *Chem. Engrs*)

SCHOFIELD, D.C. and ABBOTT, J.A. (1988). *Guide to Dust Explosion Prevention and Protection. Pt. 2: Ignition Prevention, Containment, Inerting, Suppression and Isolation* (Rugby: Instn. *Chem. Engrs*)

SCHOFIELD, F. (1976). Chemical hazards: the screening process. *Prevention of Occupational Risks*, 3, p. 175

SCHOFIELD, K.G. (1975). *A Literature Survey of Strain and Restraint Measurements in Welded Joints* (Cambridge: Welding Inst.)

SCHOFIELD, M. (1988). Corrosion horror stories. *Chem. Engr, Lond.*, **446**, 39

SCHOFIELD, M. and KING, R. (1988). Stocking fillers for plant engineers. *Chem. Engr, Lond.*, **455**, 83

SCHOLEFIELD, D.A. and GARSIDE, J.E. (1949). The structure and stability of diffusion flames. *Combustion*, 3, p. 102

SCHOLL, E.W. (1982). Explosion pressure relief. *The Control and Prevention of Dust Explosions* (London: Oyez/IBC), paper 7

SCHOLL, E.W., REEH, D., WIEMANN, W., FABER, M., KUEHNEN, G., BECK, H. and GLIENKE, N. (1979). *Combustion and Explosion Parameters of Dusts. STF-Rep* (2)

SCHOLZ, R.C. (1982). Field implemented measures. In Bennett, G.F., Feates, F.S. and Wilder, I. (1982), *op. cit.*, p. 9–2

SCHON, D.A. (1983). *The Reflective Practitioner; How Professionals Think in Action* (New York: Basic Books)

SCHON, G. (1962a). Electrostatic charging processes and their ignition hazards. *Chem. Ing. Tech.*, **34**, 432

SCHON, G. (1962b). Safety in industry: ignition hazards due to electrostatics. *J. Electrostatics*, **11**, 309

SCHON, G. (1965). In Freytag, H.H. (1965), *op. cit.*

SCHON, G. (1968). Some problems of hazards of static electricity in the carriage of dangerous goods. *Carriage of Dangerous Goods*, 1, p. 103

SCHONN, G. (1967). In discussion of Gibson, N. and Lloyd, F.C. (1967), *op. cit.*, p. 99

SCHON, G. (1973). Basic principles of explosion protection. *Prevention of Occupational Risks*, 2, p. 88

SCHON, G. (1977). Basic aspects for setting up safety recommendations to prevent explosion hazards. *Loss Prevention and Safety Promotion*, 2, p. 35

SCHON, G. and KRAMER, H. (1971). On the size of stationary space charge clouds in streaming media. *Static Electrification*, 3, p. 138

SCHON, G. and MASUDA, S. (1967). On the expansion of a space–charge cloud. *Static Electrification*, **2**, p. 123

SCHON, G. and MASUDA, S. (1969). On the expansion and shrinkage of space charge clouds. *Br. J. Appl. Phys (J. Phys D), Series 2: Appl. Phys*, **2**, 115

SCHONBUCHER, A., GOCK, D. and FIALA, R. (1992). Prediction of the heat radiation and safety distances for large fires with the model OSRAMO. *Loss Prevention and Safety Promotion 7*, vol. 2, 68–1

SCHONFELD, L. (ca. 1993). *Kommunikation in der Krise* (Frankfurt-am-Main: Hoechst AG)

SCHÖNEBURG, E. (1992). Lernfähiges Expertensystem. *Sonderausdruck aus CAV 10/92*

SCHOOFS, G.R. (1992). Fire and explosion hazards induced by repressurisation of air dryers. *AIChE J.*, **38**, 1385

SCHOPPERS, M.J. (1987). Universal plans for reactive robots in unpredictable environments. *Proc. Tenth Int. Joint Conf. on Artificial Intelligence* (San Mateo, CA: Morgan Kaufmann), p. 1039

SCHORLING, M. and KARDELS, D. (1993). Accidental release of pollutants in an industrial complex. *Health, Safety and Loss Prevention*, p. 382

SCHOLTE, T.G. and VAAGS, P.B. (1959). The burning velocity of hydrogen-air mixtures and mixtures of some hydrocarbons with air. *Combust. Flame*, **3**, 495

SCHOLTE, T.G. and VAAGS, P.B. (1959). Burning velocities of mixtures of hydrogen, carbon monoxide and methane with air. *Combust. Flame*, **3**, 511

SCHOTTE, W. (1987). Fog formation of hydrogen fluoride in air. *Ind. Engng Chem. Res.*, **26**, 300

SCHOTTE, W. (1988). Thermodynamic model for HF fog formation. Private communication to C.A. Soczek

SCHOTTLE, E. and HADER, J. (1977). The development of process control for a multicharge pharmaceutical production unit. *Loss Prevention and Safety Promotion 2*, p. 199

SCHRADER, R.F. and MOWINCKEL, P.M. (1986). Offshore loading of LNG: a review of methods, procedures and constructions with emphasis on safety and operability. *Gastech 85*, p. 401

SCHRADER-FRECHETTE, K.S. (1981). *Nuclear Power and Public Policy* (Dordrecht: Reidel)

SCHRADER-FRECHETTE, K.S. (1984a). *Risk Analysis and Scientific Method* (Dordrecht: Reidel)

SCHRADER-FRECHETTE, K.S. (1984b). *Science Policy, Ethics, and Economic Methodology* (Dordrecht: Reidel)

SCHRAMEK, H. (1984). Flexible processing by multipurpose plants in the transformation to chemical speciality end products. *CHEMRAWN III*

SCHREIBER, A.M. (1982). Using event trees and fault trees. *Chem. Engng*, **89**(Oct. 4), 115

SCHREIBER, S. (1990). The Center for Chemical Process Safety. *Plant/Operations Prog.*, **9**, 143

SCHREIBER, S. (1991). Guidelines for technical management of chemical process safety. *Plant/Operations Prog.*, **10**, 65

SCHREIBER, S. (1994). Measuring performance and effectiveness of process safety management. *Process Safety Prog.*, **13**, 64

SCHREIBER, S. and SWEENEY, J.C. (1989). Engineering visions of process safety management. *Loss Prevention and Safety Promotion 6*, p. 34.1

SCHREURS, P. and MEWIS, J. (1987). Development of a transport phenomena model for accidental releases of heavy gases in an industrial environment. *Atmos. Environ.*, **21**, 765

SCHREURS, P., MEWIS, J. and HAVENS, J. (1987). Numerical aspects of a Lagrangian particle model for atmospheric dispersion of heavy gases. *J. Haz. Materials*, **17**(1), 61

SCHREURS, P., VERBRUGGEN, A., DE WEL, H. and DENTENEER, A. (1992). Quality assurance of safety and environmental impact studies. *Loss Prevention and Safety Promotion 7*, vol. 1, 4–1

SCHRODER, F.S. (1982). Calculating heat loss or gain by an insulated pipe. *Chem. Engng*, **89**(Jan. 25), 111

SCHROTER, R.C. and LEVER, M.J. (1980). The deposition of materials in the respiratory tract. In Waldron, H.A. and Harrington, J.M. (1980), *op. cit.*, p. 76

SCHROY, J.M. (1979). Prediction of workplace contaminant levels. *NIOSH Symp. on Control Techniques in the Plastics and Resins Industry* (Washington, DC)

SCHUBER, G. (1989). Ignition breakthrough behaviour of dust/air and hybrid mixtures through narrow gaps. *Loss Prevention and Safety Promotion 6*, p. 14.1

SCHUCHART, P. and SCHLICHTHARLE, G. (1985). Cracks in the outlet section of steam reformer tubes. *Ammonia Plant Safety*, **25**, p. 54

SCHUDER, C.B. (1970). Coping with control valve noise. *Chem. Engng*, **77**(Oct. 19), 149

SCHUELER, R.C. (1972). Metal dusting. *Hydrocarbon Process.*, **51**(8), 73

SCHUELLER, G.I. (1982). External hazards. In Green, A.E. (1982), *op. cit.*, p. 483

SCHULEIN, P., KLOET, D. and STOLK, D.J. (1993). Designing, developing and operating crisis management training systems (poster item). *Process Safety Management*, p. 445

SCHULLER, M.R., MURPHY, J.C. and GLASSER, K.F. (1974). LNG storage tanks for metropolitan areas. *LNG Conf. 4*

SCHULMAN, L.L. and SCIRE, J.S. (1980). *Buoyant Line and Point Source (BLP) Dispersion Model User's Guide. Doc. P-7304B.* Environ. Res. and Technol. Inc., Concord, MA

SCHULMEYER, C.G. (1990). *Zero Defect Software* (New York: McGraw-Hill)

SCHULMEYER, G.G. and MACMANUS, J.L. (eds) (1987). *Handbook of Software Quality Assurance* (New York: van Nostrand Reinhold)

SCHULTZ, M.A. (1961). *Control of Nuclear Reactors and Power Plants* (London: McGraw-Hill)

SCHULTZ, N. (1985). *Fire and Flammability Handbook* (New York: van Nostrand Reinhold)

SCHULTZ, N. (1986). Fire protection for cable trays in petrochemical facilities. *Plant/Operations Prog.*, **5**, 35

SCHULZ, N. and SCHOFT, H. (1986). Pressure relief of foaming or highly viscous liquids. *Loss Prevention and Safety Promotion 5*, p. 44–1

SCHULZ, N., PILZ, V. and SCHACKE, H. (1983). Controlling thermal instabilities in chemical processes. *Loss Prevention and Safety Promotion 4*, vol. 3, p. B1

SCHULZ, T. and BRAUN, M. (1992). Fatigue-crack leakage prevention by operational vibration simulation tests of piping. *Loss Prevention and Safety Promotion 7*, vol. 2, 74–1

SCHULTZ, T.J. and SHAH, J.K. (1978). Personnel safety and toxic containment during Kepone incineration tests. *Control of Hazardous Material Spills*, **4**, p. 265

SCHULTZ, W.H. (1918a). The reaction of the heart towards chlorine. *J. Pharmacol. Exp. Therapeut.*, **11**, 179

SCHULTZ, W.H. (1918b). The reaction of the respiratory mechanism to chlorine gas. *J. Pharmacol. Exp. Therapeut.*, **11**, 180

SCHULTZ, W.H. and HUNT, H.R. (1918). Some pathological phenomena following inhalation of chlorine gas. *J. Pharmacol. Exp. Therapeut.*, **11**, 181

SCHULZ, K.H. (1957). Clinical and experimental investigations concerning the etiology of chloracne. *Arch. Klin. Exp. Derm.*, **206**, 589

SCHULZ-FORBERG, B., DROSTE, B. and CHARLETT, H. (1984). Failure mechanisms of propane-tanks under thermal stresses including fire engulfment. *Transport and Storage of LPG and LNG* (Antwerp: Koninklijke Vlaamse Ingenieursvereniging), p. 295

SCHULZE, R.H. (1990). Dispersion modeling using personal computers. *Atmos. Environ.*, **24A**, 2051

SCHUMACHER, J.C. (1960). *Perchlorates, their Properties and Uses* (New York: Reinhold)

SCHUMACHER, M. (1979). *Seawater Corrosion Handbook* (Park Ridge, NJ: Noyes Data Corp.)

SCHUMACHER, W.J. (1977). Wear and galling can knock out equipment. *Chem. Engng*, **84**(May 9), 155

SCHUMB, W.C., SATTERFIELD, C.N. and WENTWORTH, R.L. (1955). *Hydrogen Peroxide* (New York: Reinhold)

SCHUPP, B.A., LEMKOWITZ, S.M., GOOSSENS, L., PASMAN, H.J. and HALE, A.R. (2000). Improving on chemical process safety through distributed computer assisted knowledge analysis of preliminary design. *Eur. Symp. on Comp. Aided Process Eng.-10* S. Pierucci, 751–56. Elsevier Science B.V.

SCHURINGA, A. and LUTTIK, C. (1960). Measurement of space density of charge in flowing liquids. *J. Sci. Instrum.*, **37**, 332

SCHUTZE, W. (1927). On the dangers to men and animals from high concentrations of some toxic gases taken through the skin. *Arch. Hyg.*, **98**, 70

SCHWAB, R. (1929). The effect of glucose on the course of poisoning. *Z. Exp. Med.*, **67**, 513

SCHWAB, R.F. (1968). Interpretation of dust explosion data. *Loss Prevention*, **2**, p. 37

SCHWAB, R.F. (1971). Chlorofluorohydrocarbon reaction with aluminum rotor. *Loss Prevention*, **5**, p. 111

SCHWAB, R.F. (1986). Dusts. *Fire Protection Handbook*, 16th ed. (ed. A.E. Cote and J.M. Linville), p. 5–97

SCHWAB, R.F. (1987). Recent developments in deflagration venting design. *Preventing Major Chemical Accidents*, p. 3.101

SCHWAB, R.F. (1988). Consequences of solvent ignition during a drum filling operation. *Plant/Operations Prog.*, **7**, 242

SCHWAB, R.F. (1988). Phenol plant explosion – March 9, 1982. *Preventing Major Chemical and Related Process Accidents*, p. 683

SCHWAB, R.F. and DOYLE, W.H. (1967). Hazards of electrolytic chlorine plants. *Electrochem. Technol.*, **5**(5–6), 228

SCHWAB, R.F. and DOYLE, W.H. (1970). Hazards in phthalic anhydride plants. *Loss Prevention*, **4**, p. 25. See also *Chem. Engng Prog.*, 1970, **66**(9), 49

SCHWAB, R.F. and LAWLER, J.B. (1974). Laboratory evaluation tests of fireproofing materials. *Loss Prevention*, **8**, p. 42

SCHWAB, R.F. and OTHMER, D.F. (1964). Dust explosions. *Chem. Process Engng*, **45**, 165

SCHWANEKE, R. (1965). Air contamination by small leaks on chemical process equipment and methods of prevention. *Wasser Luft Betrieb*, **8**(11), 725

SCHWANEKE, R. (1968). Air quality: problems of relief and safety valves. *Wasser Luft Betrieb*, **12**(7), 426

SCHWANEKE, R. (1970). Air contamination by leaks on chemical process equipment. *Luftverunreinigung*, Dec., **9**

SCHWARTE, M. (1920). *Die Technik im Weltkrieg* (Berlin: Mittler & Sohn)

VON SCHWARTZ, E. (1904). *Fire and Explosion Risks*; 1918, 2nd ed.; 1926, 3rd ed.; 1940, 4th ed. (London: Griffin)

SCHWARTZ, H.E. (1976). Accidents caused by handling of chlorine. *Prevention of Occupational Risks*, **3**, p. 429

SCHWARTZ, K.J. and LIE, T.T. (1985). Investigating unexposed surface temperature criteria of ASTM E119. *Fire Technol.*, **21**, 169

SCHWARTZ, L., TULIPAN, L. and BIRMINGHAM, D.J. (1957). *Occupational Diseases of the Skin*, 3rd ed. (Philadelphia, PA: Lea & Ferbiger)

SCHWARTZ, M. (1982a). Building codes for process plants. *Chem. Engng*, **89**(Dec. 27), 57

SCHWARTZ, M. (1982b). Forces acting in and on structural materials. *Chem. Engng*, **89**(Nov. 1), 89

SCHWARTZ, M. (1982c). Structural design of chemical process plants. *Chem. Engng*, **89**(May 17), 96

SCHWARTZ, M. (1983a). Masonry structures. *Chem. Engng*, **90**(May 2), 67

SCHWARTZ, M. (1983b). Reinforced concrete for structures and equipment. *Chem. Engng*, **90**(Mar. 7), 165

SCHWARTZ, M. (1983c). Steel and concrete structures for supporting process equipment. *Chem. Eng*, **90**(Oct. 31), 51

SCHWARTZ, M. (1983d). Steel-frame construction for process structures. *Chem. Engng*, **90**(Jul. 11), 79

SCHWARTZ, M. (1983e). Supports for process vessels and storage equipment. *Chem. Eng*, **90**(Sep. 5), 119

SCHWARTZ, M. (1984). Wall and roof construction for process and plant buildings. *Chem. Engng*, **91**(Jan. 23), 77

SCHWARTZ, M. (1991). *Composite Materials Handbook*, 2nd ed. (New York: McGraw-Hill)

SCHWARTZ, M. and KOSLOV, J. (1984). Piping and instrumentation diagrams. *Chem. Engng*, **91**(Jul. 9), 85

SCHWARTZ, M. and SPAHN, H. (1988). Piping failure in a high-pressure steam system. *Ammonia Plant Safety*, **28**, 18

SCHWARTZ, M.M. (1979). *Metals Joining Manual* (New York: McGraw-Hill)

SCHWARTZ, R. and KELLER, M. (1977). Environmental factors vs flare application. *Loss Prevention*, **11**, p. 31. See also *Chem. Engng Prog.*, 1977, **73**(9), 41

SCHWARTZ, S. (1993). Process hazard screening – a method for identifying batch pharmaceutical processes with the greatest risk of a catastrophic accident (poster item). *Process Safety Management*, p. 475

SCHWARTZHOFF, J.A. and SOMMERFELD, J.T. (1988). How fast do spheres drain? *Chem. Engng*, **95**, 158

SCHWARTZMAN, H. (1981). Proctor and Gamble environmental incident program loss prevention aspects. *Loss Prevention*, **14**, p. 73

SCHWARTZMAN, H. and WIESE, R.J. (1976). Spill protection engineering for industry – an integrated approach. *Control of Hazardous Material Spills*, **3**, p. 33

SCHWARZ, G.W. (1976). Preventing vibration in shell-and-tube heat exchangers. *Chem. Engng*, **83**(Jul. 19), 134

SCHWARZ, H.E. (1976). Damages caused by handling of chlorine. *Prevention of Occupational Risks*, **3**, p. 429

SCHWARZ, R.J. (1980). Establishing and managing a reliable plant-wide inerting system. *Loss Prevention*, **13**, p. 142

SCHWEIKERT, L.E. (1986). Tests prove two-phase efficiency for offshore pipeline. *Oil Gas J.*, **84**(Feb. 3), 39

SCHWEINFURTH (1974). Industrial fire protection with high pressure installations. *Fire Int.*, **45**, 79

SCHWEITZER, J.W. and HANSON, D.N. (1971). Stability limit of charged drops. *J. Colloid Interace Sci.*, **35**, 417

SCHWEITZER, O.R. (1982). Systematic trial-and-error problem solving. *Chem. Engng*, **89**(May 17), 109

SCHWEITZER, P.A. (1969). *Handbook of Corrosion Resistant Piping* (New York: Ind. Press)

SCHWEITZER, P.A. (1976). *Corrosion Resistance Tables* (Basel: Dekker)

SCHWEITZER, P.A. (ed.) (1979). *Handbook of Separation Techniques for Chemical Engineers* (New York: McGraw-Hill)

SCHWEITZER, P.A. (ed.) (1983). *Corrosion and Corrosion Protection Handbook* (New York: Dekker)

SCHWEITZER, P.A. (ed.) (1989). *Corrosion and Corrosion Protection Handbook*, 2nd ed. (New York: Dekker)

SCHWEIZER, J.W. and HANSON, D.N. (1971). Stability limit of charged droplets. *J. Coll. Int. Sci.*, **35**, 417

SCHWENDEMAN, T.G. and WILCOX, H.K. (1987). *Underground Storage Systems. Leak Detection and Monitoring* (Chichester: Wiley)

SCHWENDTNER, A.H. (1977). LNG ocean transport costs. *Pipeline Gas J.*, Aug., 52

SCHWEPPE, J.L. and FOUST, A.S. (1953). Effect of forced circulation rate on boiling heat transfer and pressure drop in a short vertical tube. *Heat Transfer – Atlantic City* (New York: Am. Inst. Chem. Engrs), p. 77

SCHWERDTEL, W. (1970). Make acetic acid from n-butenes. *Hydrocarbon Process.*, **49**(11), 117

SCHWETZ, B.A. *et al.* (1973). *Environ. Health Perspectives*, exptl issue (**5**), 87

SCHWEYER, H.M. (1955). *Process Engineering Economics* (New York: McGraw-Hill)

SCHWING, R.C. and ALBERS, W.A. (eds) (1980). *Societal Risk Assessment. How Safe is Safe Enough?* (New York: Plenum Press)

SCHWINN, V. and STREISSELBERGER, A. (1993). Avoid H_2S corrosion in refinery vessels. *Hydrocarbon Process.*, **72**(1), 103

SCIENCE APPLICATIONS INT. (1974). *Risk Assessment of Storage and Transport of Liquefied Natural Gas and LP Gas. Rep. PB-247-415*

SCIENTIFIC INSTRUMENT RESEARCH ASSOCIATION (1970). Instruments in Working Environments (London: Adam Hilger)

SCILLY, N.F. (1981). Blast waves from unconfined vapour cloud explosions. *Health and Safety Executive Seminar on Thermal Radiation and Overpressure Calculations* (London: Health and Safety Executive)

SCILLY, N.F. (1982). *Report of incident at 'Los Alfaques' camping site at San Carlos de la Rapita.* (internal report). Health and Safety Executive

SCILLY, N.F. (1989). *Piper Alpha Public Inquiry. Transcript, Day 75*

SCILLY, N.F. and CROWTHER, J.H. (1992). Methodology for predicting domino effects from pressure vessel fragmentation. *Hazard Indentification and Risk Analysis, Human Factors and Human Reliability in Process Safety*, p. 1

SCILLY, N.F. and HIGH, W.G. (1986). The blast effects of explosions. *Loss Prevention and Safety Promotion 5*, p. 39–1

SCIRE, J.S. and LURMANN, F. (1983). Development of the MESOPUFF II dispersion model. *Am. Met. Soc., Proc. Sixth Symp. on Turbulence and Diffusion*, p. 32

SCISSON, S.E. (1969). Mined underground storage for anhydrous ammonia. *Ammonia Plant Safety*, **11**, p. 30

SCONCE, J.S. (ed.) (1972). *Chlorine, Its Manufacture, Properties and Uses* (Huntington, NY: Krieger)

SCORER, R.S. (1958). *Natural Aerodynamics* (London: Pergamon Press)

SCORER, R.S. (1968). *Air Pollution* (Oxford: Pergamon Press)

SCORER, R.S. (1973). *Pollution in the Air. Problems, Policies and Priorities* (London: Routledge & Kegan Paul)

SCORER, R.S. (1978). *Environmental Aerodynamics* (Chichester: Ellis Horwood)

SCORER, R.S. and BARRETT, C.F. (1962). Gaseous pollution from chimneys. *Int. J. Air Water Poll.*, **6**, 49

SCOTHERN, E. (1978). *Piper Alpha Public Inquiry. Transcript, Days 23–25*

SCOTT, A.C., CLANCEY, W.J., DAVIS, R. and SHORTLIFFE, E.H. (1977). Explanation capabilities of knowledge-based production systems. *Am. J. Comput. Linguistics, Microfiche*, **62**

SCOTT, B. (1992). Identify alkylation hazards. *Hydrocarbon Process.*, **71**(10), 77

SCOTT, D. (1975). Debris examination – a prognostic approach to failure prevention. *Wear*, **34**, 15

SCOTT, D. (1980). Some seldom considered aspects of pressure relief systems. *Instn Chem. Engrs, Conf. on Explosion, Fire Hazards and Relief Venting in Chemical Plants* (Rugby)

SCOTT, D. (1995). Safety cases – a view from the insurance industry. Safety Cases (London: IBC Technical Services)

SCOTT, D. and CRAWLEY, F. (1992). *Process Plant Design and Operation* (Rugby: Instn Chem. Engrs)

SCOTT, D. and SMITH, A.I. (1975). *Improvement of Design and Materials by Failure Analysis and the Prognostic Approach to Reliability. Rep. 594.* Nat. Engng Lab., East Kilbride

SCOTT, D.A. (1940). Hazards with butadiene and its peroxides. *Inst. Petrol. J.*, **26**(199), 272

SCOTT, D.S. (1963). Properties of concurrent gas-liquid flow. *Adv. Chem. Engng*, **4**, 199

SCOTT, J.N. (1992). Succeeding at emergency response. *Chem. Engng Prog.*, **88**(12), 62

SCOTT, P.M. and KIEFNER, J.F. (1984). In-service pipeline repairs. *Oil Gas J.*, **82**(Dec. 3), 108 Pt 1

SCOTT, P.P. (1979). Attitudes towards regulating control of carcinogens in food. *Chem. Ind.*, Feb. 3, 83

SCOTT, P.P. (1983). *Variations in Two-vehicle Accident Frequencies, 1970–1978. Rep. LR 1086.* Transport and Road Res. Lab., Crowthorne

SCOTT, R. and MACLEOD, N. (1991). *Process Design Case Studies* (Rugby: Instn Chem. Engrs)

SCOTT, R.B., DENTON, W.H. and NICHOLLS, C.M. (1964). *Technology and Uses of Liquid Hydrogen* (Oxford: Pergamon Press)

SCOTT, R.L. (1971). *A Review of Safety-related Occurrences in Nuclear Power Reactors from 1967–1970. Rep. ORNL-TM-3435.* Oak Ridge Nat. Lab., Oak Ridge

SCOTT, R.M. (1989). *Chemical Hazards in the Workplace* (Chelsea, MI: Lewis Publs)

SCOTT, R.W. (1961). Electrical safety guide to hazardous process areas. *Chem. Engng*, **68**(Jul. 10), 123

SCOTT, R.W. (1964). Electrical safety in atmospheres containing hydrogen. *Chem. Engng*, **71**(Mar. 2), 79

SCOTT, S. (1992). Management of safety – permit-to-work systems. *Major Hazards Onshore and Offshore*, p. 171

SCOTT, W.J. (1982). Handling cooling-water systems during a low-pH excursion. *Chem. Engng*, **89**(Feb. 22), 121

SCULLEY, J.C. (1966). *The Fundamentals of Corrosion* (London: Commonwealth & Int. Library)

SCULLY, W.A. (1981). Safety-relief-valve malfunctions: symptoms, causes and cures. *Chem. Engng*, **88**(Aug. 10), 111

SEADER, J.D. (1985). *Computer Modeling of Chemical Processes* (New York: Am. Inst. *Chem. Engrs*)

SEADER, J.D. and EINHORN, I.N. (1977). Some physical, chemical, toxicological and physiological aspects of fire smokes. *Combustion*, 16, p. 1423

SEAGLE, S.R. and BANNON, B.P. (1982). Titanium: its properties and uses. *Chem. Engng*, 89(Mar. 8), 111

SEAKINS, M. and HINSHELWOOD, SIR C.N. (1963). Some correlations in the kinetics of gas-phase hydrocarbon oxidations. *Proc. R. Soc., Ser. A*, 276, 324

SEARBY, G. (1992). Acoustic instability in premixed flames. *Combust. Sci. Technol.*, 81, 221

SEARGEANT, R.J. and ROBINETT, F.E. (1968). *An Experimental Investigation of the Atmospheric Diffusion and Ignition of Boil-off Vapors Associated with a Spillage of Liquefied Natural Gas. Rep. 08072-7.* TRW Systems

SEARLE, C.E. (ed.) (1976). *Chemical Carcinogens*; 1984, 2nd ed. (Washington, DC: Am. Chem. Soc.)

SEARSON, A.H. (1972). Fire in a catalytic reforming unit. *Loss Prevention*, 6, p. 58. See also *Chem. Engng Prog.*, 1972, 68(5), 65

SEARSON, A.H. (1977). Control of major fire emergencies in refineries and petrochemical plants: action checklist for emergency procedures. *Loss Prevention and Safety Promotion*, 2, p. 495

SEARSON, A.H. (1982). Learning from incident experience. *Developments 82* (Rugby: Instn Chem. Engrs)

SEARSON, A.H. (1983). Explosion in cone-roof gas oil tank. *Loss Prevention and Safety Promotion 4*, vol. 1, p. L15

SEATON, W.H. (1991). Group contribution method for predicting the lower and upper flammability limits of vapors in air. *J. Haz. Materials*, 27, 169

SEATON, W.H. and HARRISON, B.K. (1990). A new general method for estimation of heats of combustion for hazard evaluation. *J. Loss Prev. Process Ind.*, 3(3), 311

SEATON, W.H., FREEDMAN, E. and TREWEEK, D.N. (1974). *CHETAH: the ASTM chemical thermodynamic and energy release evaluation program. ASTM Data Ser. Pub. DS51* (Philadelphia, PA: Am. Soc. Testing Mats)

SEAVER, D.A. and SITWELL, W.G. (1983). *Procedures for using Expert Judgement to Estimate Human Error Probabilistics in Nuclear Power Plant Operations. Rep. NUREG/CR-2743.* Nucl. Regul. Comm., Washington, DC

SEBASTIAN, F.P. and CARDINAL, P.J. (1968). Water pollution control. Solid waste disposal. *Chem. Engng*, 75(Oct. 14), 112

SEDDON, G.N.D. and KELLY, A. (1981). An investigation into the causes and characteristics of maintenance work in a batch-processing chemical plant, with a view to developing maintenance control systems. *Terotechnica*, 2, 35

SEDDON, R.H. (1989). *Piper Alpha Public Inquiry. Transcript, Day 86*

SEDEFIAN, L. and BENNETT, E. (1980). A comparison of turbulence classification schemes. *Atmos. Environ.*, 14, 741

SEDILLE, M. (1965–66). Centrifugal compressors for the chemical industry: recent progress. *Proc. Instn Mech. Engrs*, 180(pt 1), 1195

SEDRIKS, A.J. (1979). *Corrosion of Stainless Steels* (New York: Wiley)

SEEBERGER, J.J. and WIERWILLE, W.W. (1976). Estimating the amount of eye movement data required for panel design and instrument placement. *Hum. Factors*, 18, 281

SEEBOLD, J.G. (1972a). Flare noise: causes and cures. *Hydrocarbon Process.*, 51(10), 143

SEEBOLD, J.G. (1972b). How to reduce control valve and furnace combustion noise. *Hydrocarbon Process.*, 51(3), 97

SEEBOLD, J.G. (1973). Smooth piping reduces noise – fact or fiction. *Hydrocarbon Process.*, 52(9), 189

SEEBOLD, J.G. (1975a). Control plant noise this way. *Hydrocarbon Process.*, 54(8), 80

SEEBOLD, J.G. (1975b). Reduce noise from pulsating combustion in elevated flares. *Hydrocarbon Process.*, 54(9), 225

SEEBOLD, J.G. (1982). Reduce noise in process piping. *Hydrocarbon Processing*, 61(10), 75

SEEBOLD, J.G. (1984). Practical flare design. *Chem. Engng*, 91(Dec. 10), 69

SEEBOLD, J.G. and HERSH, A.S. (1971). Control flare steam noise. *Hydrocarbon Process.*, 50(2), 140

SEEGER, P.G. (1974). Heat transfer by radiation from fires of liquid fuels in tanks. In Afgan, N. and Beer, J.M. (1974), *op. cit.*, p. 431

SEEHOF, J.M. and EVANS, W.O. (1967). Automated layout design program. *J. Ind. Engng*, 18(12), 195

SEELS, F.H. (1973). Industrial water pretreatment. *Chem. Engng*, 80(Feb. 26), 27

SEEMAN, G.R., HARRIS, G.L., SOUTER, N.K., SCHERB, M.V. and GUSTAFSON, H.A. (1978). Development of a remotely operated vehicle to control hazardous material spills. *Control of Hazardous Material Spills*, 4, p. 356

SEFTON, A.D. (1989). *Piper Alpha Public Inquiry. Transcript, Days 154, 155*

SEGALOFF, L. (1961). *Task Sirocco. Community Reaction to an Accidental Chlorine Exposure. Proj. Summit. Contract DA-18-064-Cml-2757.* Inst. for Coop. Res., Univ. of Pennsylvania, Philadelphia, PA

SEGERSTAHL, B. (ed.) (1991). *Chernobyl. A Policy Response Study* (Berlin: Springer)

SEGRAVES, R.O. and WICKERSHAM, D. (1991). Learn lessons from PEPCON explosions. *Chem. Engng Prog.*, 87(6), 65

SEHMEL, G.A. (1980). Particle and dry gas deposition: a review. *Atmos. Environ.*, 14, 983

SEHMEL, G.A. and HODGSON, W.H. (1978). *A Model for Predicting Dry Deposition of Particles and Gases to Environmental Surfaces. Rep. BNWL-2000–3.* Battelle Pacific Northwest Lab., Richland, WA

SEHMEL, G.A. and HODGSON, W.H. (1980). A model for predicting dry deposition of particles and gases to environmental surfaces. *Implications of the Clean Air Amendments 1977 and of Energy Considerations for Air Pollution Control* (ed. W. Licht) (New York: Am. Inst. Chem. Engrs), p. 218

SEIDENBERGER, J.W. and BARNARD, A.J. (1976). A new look at cleanup of laboratory chemical spills. *Control of Hazardous Material Spills*, 3, p. 288

SEIFERT, H. and GIESBRECHT, H. (1986). Safer design of inflammable gas vents. *Loss Prevention and Safety Promotion*, 5, p. 70.1

SEIFERT, H., MAURER, B. and GIESBRECHT, H. (1983). Steam curtains – effectiveness and electrostatic hazards. *Loss Prevention and Safety Promotion 4*, vol. 1, p. F1

SEIFERT, W.F., JACKSON, L.L. and SECH, C.E. (1972). Organic fluids for high temperature heat transfer systems. *Chem. Engng*, 79(Oct. 30), 96

SEIGNEUR, C. (1992). Understand the basics of air quality modeling. *Chem. Engng Prog.*, 88(3), 68

SEINFELD, J.H. (1975). *Air Pollution: Physical and Chemical Fundamentals* (New York: McGraw-Hill)

SEITH, R.W. and HEENAN, W.A. (1986). Gross error detection and data reconciliation in steam-metering systems. *AIChE J.*, 32, 733

SEITZ, D.R. (1983). Profile of a successful computer control project. *Chem. Engng Prog.*, **79**(10), 44

SELA, U. (1978). Design a mobile machinery analysis laboratory. *Hydrocarbon Process.*, **57**(8), 115

SELA, U. (1980). Analysis of cat plant blower couping failure. *Chem. Engng Prog.*, **76**(11), 43

SELBY, K. (1982). Controlled disposal of special wastes. A water authority's viewpoint. *Chem. Ind.*, Apr. 17, 251

SELIKOFF, I.J. (1979). Polybrominated biphenyls in Michigan (abstract). *Proc. Roy. Soc., SER. B*, **205**, 153. See also in Doll, Sir R. and McLean, A.E.M. (1979), *op. cit.*, p. 153

SELIKOFF, I.J. and HAMMOND, E.C. (eds) (1975). Toxicity of vinyl chloride-polyvinyl chloride. *Ann. N.Y. Acad. Sci.*, **246** (special issue)

SELIKOFF, I.J., CHURG, J. and HAMMOND, E.C. (1976). The occurrence of asbestosis among insulation workers in the US. *Ann. N.Y. Acad. Sci.*, **132**, 139

SELIM, F.E. (1981). Liquefied petroleum gas. *Kirk–Othmer Encyclopaedia of Chemical Technology*, 3rd ed., vol. 14 (New York: Wiley), p. 383

SELL, P.S. (1985). *Expert Systems – A Practical Introduction* (London: Macmillan)

SELL, R.G. (1980). Success and failure in implementing changes in job design. *Ergonomics*, **23**, 809

SELL, R.G., CROSSMAN, E.R.F.W. and BOX, A. (1962). An ergonomic method of analysis applied to hot strip mills. *Ergonomics*, **5**, 203

SELLERS, J.G. (1976). Quantification of toxic gas emission hazards. *Process Industry Hazards*, p. 127

SELLERS, J.G. (1984). Are your risks being properly managed? *The Management of Technological Risks* (London: A.D. Little)

SELLERS, J.G. (1988). Mossmorran – the role of the independent safety assessment. *Preventing Major Chemical and Related Process Accidents*, p. 633

SELLERS, J.G. and BENDIG, L.L. (1989). Risk comparison of LPG distribution by truck or pipeline. *Gastech 88*, paper 4.1

SELLERS, J.G. and BENDIG, L.L. (1990). Risk comparison of LPG distribution by truck or pipeline. *Safety and Loss Prevention in the Chemical and Oil Processing Industries*, p. 385

SELLERS, G. and LUCK, C.J. (1986). Are the risks in your gas plant being properly managed? *Gastech 85*, p. 133

SELLERS, J.G. and PICCIOLO, G. (1988). Safety analysis of the Priolo ethylene plant. *Preventing Major Chemical and Related Process Accidents*, p. 459

SELLERS, J.G., LUCK, C.J. and PANTONY, M. (1985). An independent hazard and operability audit during the design and construction of a major natural gas liquids facility. *Gastech 84*, p. 177

SELM, R.F. and HULSE, B.T. (1960). Deep well disposal of industrial wastes. *Chem. Engng Prog.*, **56**(5), 138

SELMAN, A.C. (1982). Systems reliability assessment in health care engineering. In Green, A.E. (1982), *op. cit.*, p. 367

SELMER-OLSEN, S. (1991). Etude théorique et expérimentale des ecoulements diphasiques en tuyere convergen tedivergente. Doctoral Thesis, Institut Nat. Polytechnique de Grenoble

SELMER-OLSEN, S. (1992). Pressure relief and two-phase flow. *Major Harards Onshore and Offshore*, p. 229

SELOVER, T.B. (1990). Use of DIPPR data in reactive chemical hazardassessment. *Plant/Operations Prog.*, **9**, 207

SELOVER, T.B. and BUCK, E. (1987). DIPPR in its seventh year. *Chem. Engng. Prog.*, **83**(7), 18

SELVA, R.A. and HEUSER, A.H. (1990). Structural integrity of a 12,000 ton refrigerated ammonia stroage tank in the presence of stress corrosion cracking. *Ammonia Plant Safety*, **30**, p. 39

SELVIDGE. J. (1972). Assigning probabilities to rare events. Ph.D. Thesis, Graduate School of Buiness Admin., Harvard Univ.

SEMANDERES, S.M. (1971). ELRAFT – a computer program for the efficient logic reduction analysis of fault trees. *IEEE Trans. Nucl. sci.*, **NS-18**, 481

SEMENOV, N. (1928). On the theory of combustion processes. *Z. Phys. Chem.*, **48**, 571

SEMENOV, N.N. (1935). *Chemical Kinetics and Chain Reactions* (Oxford: Oxford Univ. Press)

SEMENOV, N.N. (1940). *Usp. Fiz. Nauk.* (USSR), **23**, 251

SEMENOV, N.N. (1942). *Thermal Theory of Combustion and Explosion. Tech. Memo. 1024*. Nat. Advisory Cttee Aeronaut

SEMENOV, N.N. (1959). *Some Problems of Chemical Kinetics and Reactivity* (London: Pergamon Press)

SEMINARA, J.L., GONZALES, W.R. and PARSONS, S.O. (1976). *Human factors review of nuclear power plant control room design. Rep. EPRI NP-309*. Electric Power Res. Inst.,. Palo Alto, CA

SEMONELLI, C.T. (1990). Secondary containment of underground storage tanks. *Chem. Engng Prog.*, **86**(6), 78

SEMPRUCCI, L.B. and HOLBROOK, B.P. (1979). The application of RELAP 4/REPIPE to determine force time histories on relief valve discharge piping. In Haupt, R.W. and Meyer, R.A. (1979), *op. cit.*, p. 127

SEN, A.K. and LUDFORD, G.S.S. (1982). Maximum flame temperature and burning rate of combustible mixtures. *Combustion*, **19**, p. 267

SENCE, L.H. (1970). Reducing pump noise. *Chem. Engng Prog.*, **66**(5), 55

SENDERS, J.W. (1964). The human operator as a monitor and controller of multi-degree of freedom systems. *IEEE Trans. Hum. Fact. Electron.*, **HFE-5**, 2

SENDERS, J.W. (1980). Is there a cure for human error? *Psychology Today*, Apr., 52

SENDERS, J.W. and POSNER, M.J.M. (1976). A queuing model of monitoring and supervisory behaviour. In Sheridan, T.B. and Johannsen, G. (1976), *op. cit.*

SENECAL, J.A. (1989a). Deflagration suppression of high K_{st} dusts. *Plant/Operations Prog.*, **8**, 147

SENECAL, J.A. (1989b). Deflagration suppression of high-K_{st} dusts and dust-vapor systems. *Loss Prevention and Safety Promotion 6*, p. 55.1

SENECAL, J.A. (1991). Manganese mill dust explosion. *J. Loss Prev. Process Ind.*, **4**, 332

SENECAL, J.A. (1992a). Halon replacement chemicals: perspectives on the alternatives. *Fire Technol.*, **28**, 332

SENECAL, J.A. (1992b). Halon replacement: the law and the options. *Plant/Operations Prog.*, **11**, 182

SENGUPTA, M. and STAATS, F.Y. (1978a). A new approach to relief valve load calculations. *Hydrocarbon Process.*, **57**, 160

SENGUPTA, M. and STAATS, F.Y. (1978b). Relief valve load calculations. *API Proc., Refining Dept*, **57**, 37

SENIOR, R. (1978). Reliability theory for the plant engineer. *Plant Engr*, **22**(5), 13

SENKAN, S.M., ROBINSON, J.M. and GUPTA, A.K. (1983). Sooting limits of chlorinated hydrocarbon-methane-air premixed flames. *Combust. Flame*, **49**, 305

SENKBEIL, E.G. (1994). Laboratory safety course in the chemistry curriculum. *J. Haz. Materials*, **36**, 159

SENNET, C.T. (ed.) (1989). *High Integrity Software* (London: Pitman)

SEONG-HEE SHIN, MERONEY, R.N. and WILLIAMS, T. (1991). Wind tunnel model validation for vapor dispersion from vapor detention system. *Modeling and Mitigating the Consequences of Accidental Releases*, p. 369

SEPPA, W.O. (1964). Fundamentals of sewer design. *Hydrocarbon Process. Petrol. Refiner*, **43**(10), 171

SERGHIDES, T.K. (1984). Estimate friction factor accurately. *Chem. Engng*, **91**(Mar. 5), 63

SERRA, A. and BARLOW, R.E. (eds) (1986). *Theory of Reliability* (Amsterdam: Elsevier)

SERRANT (1924). *La Guerre Chimique*

DE SERRES, F.J. and SHELBY, M.D. (eds) (1982). *Comparative Chemical Mutagenesis* (New York: Plenum Press)

SERRIDGE, M. (1990). Better machinery monitoring. *Hydrocarbon Process.*, **69**(7), 53

SERVICE IN INFORMATICS AND ANALYSIS LTD. (1982). CHA-FINCH. *The Chemical Accidents, Failure Incidents and Chemical Hazards Databank* (London)

SESHADRI, K., BERLAD, A.L. and TANGIRALA, V. (1992). The structure of premixed particle-cloud flames. *Combust. Flame*, **89**, 333

SESSA, T., PECORA, L., VECCHIONE, G. and MOLE, R. (1970). Cardiorespiratory function in bronchopneumopathies caused by irritant gases. *Lung Heart*, **26**(9), 1097

SESSELBERG, F. (1926). *Der Stellungskampf* (Berlin: Mittler & Sohn)

SESTAK, E.J. (1965). *Venting of Chemical Plant Equipment. Engng Bull. N-53*. Factory Insurance Ass., Hartford, CT

SETARO, D.M. (1980). High pressure boiler water treatment. *Ammonia Plant Safety*, **22**, p. 109

SETCHKIN, N.P. (1954). Self-ignition temperature of combustible liquids. *J. Res. Nat. Bur. Standards*, **53**(1), 49

SETHURAMAN, S. and BROWN, R.M. (1976). Validity of the log−linear profile relationship over a rough terrain during stable conditions. *Boundary Layer Met.*, **10**, 489

SETHURAMAN, S. and RAYNOR, G.S. (1979). Effects of changes in upwind surface characteristics on mean wind speed and turbulence near a coastline. *Fourth Symp. on Turbulence, Diffusion and Air Pollution* (Boston, MA: Am. Met. Soc.), p. 48

SETHURAMAN, S., BROWN, R.M. and TICHLER, J. (1974). Spectra of atmospheric turbulence over the sea during stably stratified conditions. *First Symp. on Atmospheric Diffusion and Air Pollution* (Boston, MA: Am. Met. Soc.), p. 71

SETHURAMAN, S., MEYERS, R.E. and BROWN, R.M. (1976). Ratio of Lagrangian to Eulerian energy dissipation scales over water during stable conditions. *Third Symp. on Atmospheric Turbulence, Diffusion and Air Quality* (Boston, MA: Am. Met. Soc.), p. 264

SETON, W., FITTERER, R.S. and HARRIS, T. (1992). Step up to the challenge: hazmat storage. *Chem. Engng*, **99**(6)

SETOYAMA, I., WADA, E. and FUNAKOSHI, Y. (1980). Performance and maintenance of an ammonia plant using the Braun purifier process. *Ammonia Plant Safety*, **21**, p. 82

SETTERLUND, R.B. (1991). Selecting process piping materials. *Hydrocarbon Process.*, **70**(8), 93

SETTLES, J.E. (1968). Deficiencies in the testing and classification of dangerous materials. *Ann. N.Y. Acad. Sci.*, **152**(art 1), 199

SEUNG WOOK BAEK and CHAN LEE (1989). Heat transfer in a radiating medium between flame and fuel surface. *Combust. Flame*, **75**, 153

SEUNG WOOK BAEK, KOOK YOUNG AHN and JONG UCK KIM (1994). Ignition and explosion of carbon particle clouds in a confined geometry. *Combust. Flame*, **96**, 121

SEURAT, S., HOSTACHE, G. and GROS, A. (1978). Technology transfer at Arzew. *Hydrocarbon Process.*, **57**(1), 199

SEVERA, J.E. (1973). Startup of a vacuum distillation unit. *Chem. Engng Prog.*, **69**(8), 85

SEVERNS, G. and HEDRICK, J. (1983). Planning control methods for batch processes. *Chem. Engng*, **90**(Apr. 18), 69

SEWELL, M.G. (1984a). Dangerous goods by road (letter). *Chem. Engr, Lond.*, **402**, 3

SEWELL, M.G. (1984b). Recent developments in legislation on road conveyance of dangerous substances. *Chem. Engr, Lond.*, **405**, 20

SEWELL, M.G. (1987). (Letter). *Chem. Engr, Lond.*, **440**, 7

SEWELL, R.S.G. and KINNEY, G.F. (1968). Response of structures to blast: a new criterion. *Ann. N.Y. Acad. Sci.*, **152**(art. 1), 532

SEWELL, W.R.D. (1971). Environmental perceptions and attitudes of engineers and public health officials. *Environ. Behaviour*, **3**, 23

SEXTON, K., SPENGLER, J.D. and TREITMAN, R.D. (1984). Personal exposure to respirable particles: a case study in Waterbury, Vermont. *Atmos. Environ.*, **18**, 1385

SEYLER, R.J. (1980). Application of pressure DTA (DSC) to thermal hazard evaluation. *Thermochimica Acta*, **39**, 171

SEYMOUR, T. (1992). OSHA's interest in plant safety. *Plant/Operations Prog.*, **11**, 164

SEYNHAEFVE, J.M., LOMBRE, R., DUCROCQ, J. and BOLLE, L. (1994). Physical modeling of rapid transients in long pipes, in case of vaporization: an efficient tool for safety management. *Process Safety Prog.*, **13**, 95

SHA, P.Y. and BOHINSKY, J.A. (1985). Structural design procedure for plant retrofit. *Chem. Engng Prog.*, **81**(6), 20

SHABICA, A.C. (1963). Evaluating the hazards in chemical processing. *Chem. Engng Prog.*, **59**(9), 57

SHACKEL, B. (1980). Factors influencing the application of ergonomics in practice. *Ergonomics*, **23**, 817

SHAEIWITZ, J.A., LAPP, S.A. and POWERS, G.J. (1977). Fault tree analysis of sequential systems. *Ind. Engng Chem. Process Des. Dev.*, **16**, 529

SHAFAGHI, A. and GIBSON, S.B. (1985). Hazard and operability study: a flexible technique for process system safety and reliability analysis. In Hoffman, J.M. and Maser, D.C. (1985), *op. cit.*, p. 33

SHAFAGHI, A., ANDOW, P.K. and LEES, F.P. (1984a). Fault tree synthesis based on control loop structure. *Trans. Instn Chem. Engrs*, **62**, 101

SHAFAGHI, A., LEES, F.P. and ANDOW, P.K. (1984b). An illustrative example of fault tree synthesis based on control loop structure. *Reliab. Engng*, **8**(4), 193

SHAFER, G. (1976). *A Mathematical Theory of Evidence*. (Princeton, NJ: Princeton Univ. Press)

SHAH, A.P., SHAH, H.M. and MODY, A.J. (1993). Design guidelines for storage and handling facilities for hazardous chemicals. *Health, Safety and Loss Prevention*, p. 509

SHAH, G.C. (1978). Trouble-shooting distillation columns. *Chem. Engng*, **85**(Jul. 31), 70

SHAH, J. (1987). Inspection of ammonia storage tanks (discussion contribution). *Ammonia Plant Safety*, **27**, p. 177

SHAH, S., FISCHER, U. and HUNGERBUHLER, K. (2003). Evaluating the degree of inherent safety in early phases of chemical process design: A hierarchical approach. *Cent. for Chem. Process Safety 18th Annu. Int. Conf: Managing Chem. React. Hazards and High Ener. Release Events*, Sept. 23−25, 2003, Scottsdale, Arizona, 103−19. (New York: American Institute of Chemical Engineers)

SHAH, J.M. (1982). Refrigerated ammonia storage tanks. *Plant/Operations Prog.*, **2**, 90

SHAH, M.J. (1967). Computer control of ethylene production. *Chem. Tech.*, May, 70

SHAH, Y.T. (1978). *Gas—Liquid—Solid Reactor Design* (London: McGraw-Hill)

SHANER, R.L. (1978). Energy scarcity: a process design incentive. *Chem. Engng Prog.*, **74**(5), 47

SHANLEY, A. (1999). Solvents: less is more. *Chem. Eng.* **106**(13) (Dec.), 57–60

SHANLEY, A., SAVAGE, P. and FOUHY, K. (1992). Bridging the trust gap. *Chem. Engng*, **99**(6), 26

SHANMUGAM, C. (1980). Estimating delivery times. *Chem. Engng.*, **87**(Nov. 3), 166

SHANMUGAM, C. (1981). Instrument-installation manhours. *Chem. Engng*, **88**(Jan. 26), 124

SHANNON, C.E. (1950a). Automatic chess player. *Sci. American*, **182**

SHANNON, C.E. (1950b). Programming a digital computer for playing chess. *Phil. Mag.*, **41**, 356

SHANNON, H.S. (1986). Road accident data: interpreting British experience with particular reference to risk homeostasis theory. *Ergonomics*, **29**, 1005

SHANNON, R.W.E. (1973). Crack growth monitoring by strain sensing. *Int. J. Pres. Ves. Piping*, **1**, 61

SHANNON, R.W.E. (1974a). The failure behaviour of line pipe defects. *Int. J. Pres. Ves. Piping*, **2**, 243

SHANNON, R.W.E. (1974b). Stress intensity factors for thick-walled cylinders. *Int. J. Pres. Ves. Piping*, **2**, 19

SHANNON, W.B., CAMPBELL, W.W., CARSTAIRS, R., DOLLIN, F., FRANKEL, A., HIRST, A.C.W. and SHAO-LIN LEE and HELLMAN, J.M. (1969). Study of firebrand trajectories in a turbulent swirling natural convection plume. *Combust. Flame*, **13**, 645

SHAO-LIN LEE and HELLMAN, J.M. (1970). Firebrand trajectory study using an empirical velocity-dependent burning law. *Combust. Flame*, **15**, 265

SHAPIRO, A. (1953–). *The Dynamics and Thermodynamics of Compressible Fluid Flow*, 1953, vol. 1; 1954, vol. 2 (New York: Ronald Press)

SHAPIRO, A. (1987). *Structured Induction in Expert Systems* (Reading, MA: Addison-Wesley)

SHAPIRO, A. and NIBLETT, T. (1982). Automatic induction of classification rules for a chess endgame. *Advances in Computer Chess 3* (ed. M.R.B. Clarke) (Oxford: Pergamon Press)

SHAPIRO, A.H. (1961). *Shape and Flow* (London: Heinemann)

SHAPIRO, J. (1972). *Radiation Protection* (Cambridge, MA: Harvard Univ. Press)

SHAPIRO, S.C. (1992). *Encyclopaedia of Artificial Intelligence*, vols 1–2 (New York: Wiley)

SHAPO, M.S. (1979). *A Nation of Guinea Pigs. The Unknown Risks of Chemical Technology* (New York: Free Press)

SHARANGPANI, R. and TAY, S.P. (2001). Growth and characterization of rapid themal chlorinated oxides grown using in situ generated HCl. *J. of the Electrochem. Society* **148** (1)F5–F8

SHARIT, J. and SALVENDY, G. (1982). Occupational stress: review and appraisal. *Hum. Factors*, **24**, 129

SHARKEY, J.J., CUTRO, R.S., FRASER, W.J. and WILDMAN, G.T. (1992). Process safety testing program for reducing risks associated with large scale chemical manufacturing operations. *Plant/Operations Prog.*, **11**, 238

SHARLAND, I.J. (1990). *Woods Practical Guide to Noise Control*, 5th ed. (Colchester: Woods)

SHARLIN, H.I. (1989). Risk perception: changing the terms of the debate. *J. Haz. Materials*, **21**, 261

SHARMA, G.N. (1981). Hot redundant versus cold redundant systems. *Reliab. Engng*, **2**(3), 193

SHARMA, O.P. and SIRIGNANO, W.A. (1969). Ignition of stagnation point flow by a hot body. *Combust. Sci. Technol.*, **1**, 95

SHARMA, O.P. and SIRIGNANO, W.A. (1970). On the ignition of a pre-mixed fuel by a hot projectile. *Combust. Sci. Technol.*, **1**, 481

SHARMA, T.P., LAL, B.B. and SINGH, J. (1987). Metal fire extinguishment. *Fire Technol.*, **23**, 205

SHARMA, T.P., CHIMOTE, R.S., LAL, B.B. and SINGH, J. (1992). Experiments on extinction of liquid hydrocarbon fires by a particulate mineral. *Combust. Flame*, **89**, 229

SHARMA, T.P., VARSHNEY, B.S. and KUMAR, S. (1993). Studies on the burning behaviour of metal powder fires and their extinguishment, Part II. *Fire Safety J.*, **21**, 153. See also Varshney, B.S., Sharma, T.P. and Kumar, S. (1990)

SHARP, D.H. and WEST, T.F. (ed.) (1982). *The Chemical Industry* (Chichester: Ellis Horwood)

SHARP, E.C. (1977). Operating and maintenance records for heating equipment. *Chem. Engng*, **84**(Apr. 25), 141; (May 23), 171

SHARPE, R.A. (1976). Gasify coal for syngas. *Hydrocarbon Process.*, **55**(11), 171

SHARRATT, P.N. (1990). Computational fluid dynamics and its application in the process industries. *Chem. Engng Res. Des.*, **68**, 13

SHARRY, J.A. (1975). Chemical explosion, Los Angeles, California. *Fire J.*, **69**(Jan.), 62

SHARRY, J.A. and WALLS, W.L. (1974). LP gas distribution plant fire. *Fire J.*, **68**(Jan.), 52

SHATTES, W. (1986). Design and use of containers for cryogenic liquids. *Plant/Operations Prog.*, **5**, 17

SHAVER, E.M. and FORNERY, L.J. (1988). Properties of industrial dense plumes. *Atmos. Environ.*, **22**, 833

SHAW, B. (1961). A simple installation for handling molten phthalic anhydride. *Chem. Engr, Lond.*, **156**, A73

SHAW, B.J. (1981). Characterization study of temper embrittlement of chromium-molybdenum steels. *API Proc., Refinery Dept, 46 Midyear Mtg*, **60**, 225

SHAW, C.P. (1971). Training: key to mechanical seal life. *Hydrocarbon Process.*, **50**(2), 121

SHAW, C.P. (1977). Better performance from anti-friction bearings. *Hydrocarbon Process.*, **56**(4), 185

SHAW, C.R. (ed.) (1981). *Prevention of Occupational Cancer* (Boca Raton, FL: CRC Press)

SHAW, D., SWARTOUT, W. and GREEN, C. (1975). Inferring LISP programs from examples. *Proc. Int. Joint Conf. on Artificial Intelligence 4*, 260

SHAW, D.A. (1990). SAFIRE: program for design of emergency pressure relief systems. *Chem. Engng Prog.*, **86**(7), 14

SHAW, J. (1967). *Reactor Operation* (Oxford: Pergamon Press)

SHAW, J. (1978). EPA and OSHA: gearing for change. *Chem. Engng*, **85**(Oct. 9), 70

SHAW, J. (1979). OSHA performance faulted as accident rate rises. *Chem. Engng*, **86**(Feb. 12), 56

SHAW, J.A. (1985). Smart alarm systems: where do we go from here? *InTech*, Dec.

SHAW, J.A. (1991). Design your DCs to reduce operator error. *Chem. Engng Prog.*, **87**(2), 24

SHAW, J.C. (1985). Human factors aspects of advanced process control. *Plant/Operations Prog.*, **4**, 111

SHAW, J.G.F. (1992). The integration of process simulation and engineering design. Comput. *Chem. Engng*, **16**, S465

SHAW, J.S. (1978). How safe is safety gear? *Chem. Engng*, **85**(Apr. 10), 75

SHAW, J.S. (1980a). EPA spells out the law on hazardous wastes. *Chem. Engng*, **87**(Jun. 2), 44

SHAW, J.S. (1980b). Superfund proves, if not painless, at least bearable. *Chem. Engng*, **87**(Dec. 29), 24

SHAW, L. and SICHEL, H.S. (1971). *Accident Proneness: Research in the Occurrence, Causation and Prevention of Road Accidents* (Oxford: Pergamon Press)

SHAW, M. (1987). Weibull analysis of component failure data from accelerated testing. *Reliab. Engng*, **19**(3), 237

SHAW, M.C. (1971). Fundamentals of wear. *Chem. Tech.*, Jul., 432

SHAW, M.W. (1980). Love Canal chromosome study (letter). *Science*, **209**, 751

SHAW, R.J., SYKES, J.A. and ORMSBY, R.W. (1980). Plant-test manual. *Chem. Engng*, **87**(Aug. 11), 126

SHAW, S. and GOWER, C. (1976). Extending the use of ADAPTS to hazardous materials. *Control of Hazardous Material Spills*, **3**, p. 230

SHAW, S.J. (1992). Use of risk assessment as an offshore design tool. *J. Loss Prev. Process Ind.*, **5**, 10

SHAW, S.J. and KRISTOFFERSEN, T.A.J. (1993). Cost effective QRA used in preparation of offshore safety cases. *Health, Safety and Loss Prevention*, p. 135

SHCHELKIN, K.I. (1943). *J. Tech. Phys.* (USSR), **13**, 520

SHCHELKIN, K.I. (1959). Two cases of unstable combustion. *Zhur. Eksp. Teor. Fiz.*, **36**, 600

SHCHELKIN, K.I. and TROSHIN, Y.A.T. (1965). *Gas Dynamics of Combustion* (Moscow: Acad. Sci. USSR)

SHEA, D.M. and JELINEK, S.F. (1986). An analysis of atmospheric dispersion of toxic contaminants resulting from a spill of a highly reactive acid. *Fifth Joint Conf. on Applications of Air Pollution Meteorology* (Boston, MA: Am. Met. Soc), p. 291

SHEAR, R.E. and MCCANE, P. (1960). *Normally Reflected Shock Front Pressures. Memo Rep. 1273.* US Army, Ballistic Res. Lab., Aberdeen Proving Ground, MD

SHEAR, R.P. and DAY, B.D. (1959). *Tables of Thermodynamic and Shock Front Parameters for Air. BRL Memo Rep. 1206.* US Army, Ballistic Res. Lab., Aberdeen Proving Ground, MD

SHEARER, K. (1969). Some experience in the prediction and control of chemical plant noise. *Chem. Engr, Lond.*, **227**, CE115

SHEARING, J. (1991). Building models, analyzing processes. *Chem. Engng*, **98**(9), 219

SHEARMAN, J. (1992). Computer aided design builds a better plant. *Chem. Engng*, **99**(4), 177

SHEARON, W.H. (1962). Chemical safety. An evaluation. *AIChE J.*, **54**(3), 43

SHEEHAN, E. (1986). Heat transfer oil boilers. *Plant/Operations Prog.*, **5**, 193

SHEEHY, N.P. (1981). The interview in accident investigation. Methodological pitfalls. *Ergonomics*, **24**, 437

SHEEN, MR JUSTICE (1987). *MV Herald of Free Enterprise. Report of Court* (London: HM Stationery Office)

SHEETS, H.D., O'HARA, M.J. and SNYDER, M.J. (1967). Refractory ceramics for the CPI. *Chem. Engng Prog.*, **64**(2), 35

SHEFFIELD, R.E. (1992). Integrate process and control system design. *Chem. Engng Prog.*, **88**(10), 30

SHEFFY, T.B. (1981). Investigation of polychlorinated biphenyls (PCBs) in natural gas pipelines and distribution systems. *J. Haz Materials*, **5**(1/2), 93

SHEI, W.F. and CONRADI, L.L. (1990). Hazard evaluation of two industrial parks in Taiwan, Republic of China. *Safety and Loss Prevention in the Chemical and Oil Processing Industries*, p. 27

SHEID, D.C. (1987). Health, safety and environmental protection: a new audit program. *Plant/Operations Prog.*, **6**, 211

SHEILAN, M. and SMITH, R.F. (1984). Hydraulic flow effect on amine plant corrosion. *Oil Gas J.*, **82**(Nov. 19), 138

SHELBY, S.E., DAVIS, G.M. and GAY, J.P. (1991). Meet new NPDES permit limits. *Hydrocarbon Process.*, **70**(8), 67

SHELDON, C.W. and SCHUDER, C.B. (1965). Sizing control valves for liquid–gas mixtures. *Instrum. Control Systems*, **38**(1), 134

SHELDON, J.H. (1960). On the natural history of falls in old age. *Br. Med. J.*, Dec. 10, 1685

SHELDON, R.A. (1997). Catalysis and pollution prevention. *Chem. Ind.*, (Jan. 6), 12–15

SHELDON, R.A. and KOCHI, J.K. (1973). Mechanisms of metal catalysed oxidation of organic compounds in the liquid phase. *Oxid. Combust. Revs*, **5**, 135

SHELDRIK, M.G. (1968). Waterway pollution policies begin to evolve. *Chem. Engng*, **75**(Mar. 22), 54

SHELDRIK, M.G. (1969). Deep-well disposal: are safeguards being ignored? *Chem. Engng*, **76**(Apr. 7), 74

SHELL (1983). *The Petroleum Handbook*, 6th ed. (Amsterdam: Elsevier)

SHELL CHEMICAL CO. LTD (1963). *Safety in fuel handling: ASA-3* (London)

SHELL CHEMICAL CORP. (1959). *Concentrated Hydrogen Peroxide. Summary of Research Data on Safety Limitations, Bull. SC59–44*

SHELL INTERNATIONAL (1988). *Shell Safety Management Program* (The Hague)

SHELL INTERNATIONAL PETROLEUM CO. (1933–). *The Petroleum Handbook*; 1959, 4th ed.; 1966, 5th ed. (London)

SHELL RESEARCH LTD (1980). *Refrigerated Gas Safety. Spill Tests in LNG and Refrigerated Liquid Propane on the Sea, Maplin Sands, UK, 1980.* (report). Shell Res. Centre, Thornton, Cheshire

SHELL UK EXPLORATION AND PRODUCTION (1976). *Hazard Evaluation. Proposed Natural Gas Liquids Plant, Peterhead* (London)

SHELL UK EXPLORATION AND PRODUCTION (1982). *North Sea fields. Facts and Figures*, 2nd ed. (London)

SHELLEY, J. (1985). *Essentials of FORTRAN 77* (Chichester: Wiley)

SHELLEY, P.G. and SILLS, E.J. (1969). Monomer storage and protection. *Loss Prevention*, **3**, p. 83. See also *Chem. Engng Prog.*, 1969, **65**(4), 29

SHELLEY, S. (1990). Making inroads with modular construction. *Chem. Engng*, **97**(8), 30

SHELLEY, S. (1991). Out of thin air. *Chem. Engng*, **98**(6), 30

SHELLEY, S. (1993). Keeping fire at bay. *Chem Engng*, **100**(11), 71

SHELLEY, S. (1994). Re-engineering ethylene's cold train. *Chem. Engng*, **101**(1), 37

SHEN, T.T. (1981). Estimating hazardous air emissions from disposal sites. *Poll. Engng*, Aug.

SHEPHERD, A. (1976). *Fault Location Tasks in Chemical Process Operation. Diagnostic Techniques* (London: Instn *Chem. Engrs*)

SHEPHERD, A. (1979). Information requirements of chemical plant operators. *Design 79* (London: Instn *Chem. Engrs*)

SHEPHERD, A. (1982a). Phenol. In Bennett, G.F., Feates and Wilder (1982), *op. cit.*, p. 12–12

SHEPHERD, A. (1982b). Task analysis, training development and some implications for process plant design. *Design 82* (Rugby: Instn Chem., Engrs)

SHEPHERD, A. (1986). Issues in the training of process operators. *Int. J. Ind. Ergonomics*, **1**, 49

SHEPHERD, A. (1992). Training process control skills. *Major Hazards Onshore and Offshore*, p. 293

SHEPHERD, A. (1993). An approach to information requirements specification for process control tasks. *Ergonomics*, **36**, 1425

SHEPHERD, A. and DUNCAN, K.D. (1980). Analysing a complex planning task. In Duncan, K.D., Gruneberg, M.M. and Willis, D. (1980), *op. cit.*, p. 137

SHEPHERD, A., MARSHALL, E.C., TURNER, A. and DUNCAN, K.D. (1977). Diagnosis of plant failure from a control panel: a comparison of three training methods. *Ergonomics*, **20**, 347

SHEPHERD, J.E., MELHEM, G.A. and ATHENS, P. (1991). Unconfined vapor cloud explosions: a new perspective. *Modeling and Mitigating the Consequences of Accidental Releases*, p. 613

SHEPHERD, R. (1970). LNG is coming of age. *Chem. Engng*, **77**(Dec. 14), 94

SHEPHERD, W.C.F. (1949). The ignition of gas mixtures by impulsive pressures. *Combustion*, **3**, p. 301

SHEPPARD, B. (1992). Determine bearing inner race defect length. *Hydrocarbon Process.*, **71**(12), 101

SHEPPARD, C.M. (1992). Disengagement predictions via drift flux correlation vertical, horizontal and spherical vessels. *Plant/Operations Prog.*, **11**, 229

SHEPPARD, C.M. (1993). DIERS churn-turbulent disengagement correlation extended to horizontal cylinders and spheres. *J. Loss Prev. Process Ind.*, **6**, 177

SHEPPARD, C.M. (1994). DIERS bubbly disengagement correlation extended to horizontal cylinders and spheres. *J. Loss Prev. Process Ind.*, **7**, 3

SHEPPARD, J.M. (1983). Problems in using reliability growth models for planning and monitoring development programmes. *Euredata 4*, paper 10.2

SHEPPARD, J.M. (1985). Are reliability growth models practical planning and monitoring tools for development? *Reliab. Engng*, **11**, 1

SHEPPARD, P.A. (1958). Turbulent transfer through the earth's surface and through the air above. *Quart. J. R. Met. Soc.*, **84**, 205

SHEPPARD, R.A. (1990). *Piper Alpha Public Inquiry. Transcript, Day 171*

SHEPPARD, R.S. and GEGNER, P.J. (1965). Experiences with titanium equipment. *Chem. Engng Prog.*, **61**(3), 114

SHEPPARD, W.L. (1984). Failure analysis of chemically resistant monolithic surfacings. *Chem. Engng*, **91**(Jul. 23), 123

SHER, E. (1982). A theoretical study of the combustion of liquids at a free surface. *Combust. Flame*, **47**, 109

SHER, E. and OZDOR, N. (1992). Laminar burning velocities of n-butane/air mixtures enriched with hydrogen. *Combust. Flame*, **89**, 214

SHER, E. and REFAEL, S. (1982). Numerical analysis of the early phase development of spark-ignited flames in CH4–air mixture. *Combustion*, **19**, p. 251

SHERIDAN, T.B. (1981). Understanding human error and aiding human diagnostic behaviour in nuclear power plants. In Rasmussen, J. and Rouse, W.B. (1981), *op. cit.*, p. 19

SHERIDAN, T.B. and FERRELL, W.R. (1974). *Man/Machine Systems* (Cambridge, MA: MIT Press)

SHERIDAN, T.B. and JOHANNSEN, G. (eds) (1976). *Monitoring Behaviour and Supervisory Control* (New York: Plenum Press)

SHERIF, Y.S. (1981). Reliability – risks, resources, rewards. *Reliab. Engng*, **2**(3), 167

SHERIF, Y.S. (1982a). Inverse truncated normal distribution as a failure model. *Reliab. Engng*, **3**(3), 209

SHERIF, Y.S. (1982b). An optimal maintenance model for life cycle costing analysis. *Reliab. Engng*, **3**(3), 173

SHERIF, Y.S. (1982c). Reliability analysis of utility and stand-by electric power systems under uncertainty. *Reliab. Engng*, **3**(3), 203

SHERMAN, A.T. (1960). Rate instruments for maintenance. *Chem. Engng*, **67**(Mar. 21), 178

SHERMAN, J.L. (1970). Storage of anhydrous ammonia in a mined cavern. *Ammonia Plant Safety*, **12**, p. 8

SHERMAN, M.P., TIEZSEN, S.R., BENEDICK, W.B., FISK, W. and CARCASSI, M. (1985). The effect of transverse venting on flame acceleration and transition to detonation in a large channel. *Tenth Int. Coll. on Dynamics of Explosions and Reactive Systems*, Berkeley, CA

SHERMAN, W.F. and STADTMULLER, A.A. (1987). *Experimental Techniques in High Pressure Research* (Chichester: Wiley)

SHERRARD, C.P. (1992). An approach to the quantitative evaluation of safety on existing mobile drilling units. *Major Hazards Onshore and Offshore*, p. 639

SHERRELL, S.J. (1987). Fitting box models to the Thorney Island Phase I data set. *J. Haz. Materials*, **16**, 395

SHERRINGRON, D.C. (1991). Polymer supported systems: towards clean chemistry? *Chem. Ind.*, (Jan. 7), 15–19

SHERRY, A. (1969–70). Availability of power stations and their equipment. *Proc. Instn Mech. Engrs*, **184**(pt 1), 311

SHERWIN, D.J. (1978). Hyperexponentially distributed failure to process plant. *Fifth Symp. on Reliability Technology, Univ. of Bradford*

SHERWIN, D.J. (1983). Failure and maintenance data analysis at a petrochemical plant. *Reliab. Engng*, **5**, 197

SHERWIN, D.J. (1984). Availability and productiveness of systems of redundant and limited stages of like items in terms of item availability. *Int. J. Prod. Res.*, **22**, 109

SHERWIN, D.J. and BOSSCHE, A. (1993). *The Reliability, Availability and Productiveness of Systems* (London: Chapman & Hall)

SHERWIN, D.J. and LEES, F.P. (1980). An investigation of the application of failure data analysis to decision-making in maintenance of process plants. *Proc. Instn Mech. Engrs*, **194**, 301; 308

SHERWIN, M.B. (1981). Chemicals from methanol. *Hydrocarbon Process.*, **60**(3), 79

SHERWOOD, E.M. (1963). Less common metals for process equipment. *Ind. Engng Chem.*, **55**(10), 59

SHERWOOD, M. (1978). Toxicology, environmental monitoring and contract research. *Chem. Ind.*, Apr. 15, 254

SHERWOOD, M. (1982). Setting up SHOP at Stanlow. *Chem. Ind.*, Dec. 18, 994

SHERWOOD, R.J. (1989). Material verification and inspection program. *Ammonia Plant Safety*, **29**, p. 152. See also *Plant/Operations Prog.*, 1989, **8**, 88

SHERWOOD, T.K. (1940). Mass transfer and friction in turbulent flow. *Trans. Am. Inst. Chem. Engrs*, **36**, 817

SHERWOOD, T.K. (1963). *A Course in Process Design* (Boston, MA: MIT Press)

SHERWOOD, T.K. and PIGFORD, R.L. (1952). *Absorption and Extraction*, 2nd ed. (New York: McGraw-Hill)

SHERWOOD, T.K., PIGFORD, R.L. and WILKE, C.R. (1975). *Mass Transfer* (New York: McGraw-Hill)

SHETINKOV, E.S. (1959). Calculation of flame velocity in turbulent stream. *Combustion*, **7**, p. 583

SHEVCHENKO, N.F. et al. (1982). *Fundamentals of Explosion Protection of Electrical Systems* (Moscow: Energoizdat)

SHEVENELL, L. and HOFFMAN, F.O. (1993). Necessity of uncertainty analyses in risk assessment. *J. Haz. Materials*, **35**, 369

SHI, B. and SEINFELD, J.H. (1991). On mass transport limitation to the rate of reaction of gases in liquid droplets. *Atmos. Environ.*, **25A**, 2371

SHIBASAKI, T., TAKEMURA, K., KAWAI, T. and MOHRI, T. (1987). Experiences of niobium-containing alloys for steam reformers. *Ammonia Plant Safety*, **27**, 56

SHICK, T. and VAUGHN, L. *How to Think about Weird things, Critical Thinking for a New Age*, 2nd Edition, (Mountain View, CA: Mayfield Publishing CO.), ISBN 0-7674-0013-5

SHIEL, H.H., SIEGELL, J. and JONES, A.L. (1991). Modelling plant hydrocarbon emissions. *Chem. Engr, Lond.*, **502**, 21

SHIELD, B.H. (1967). Battery failure causes major failure of gas turbine. *Loss Prevention*, **1**, p. 75. See also *Chem. Engng Prog.*, 1967, **63**(8), 58

SHIELD, B.H. (1973). Major failure of gas turbine. In Vervalin, C.H. (1973a), *op. cit.*, p. 12

SHIELDS, S.E. and VANDER KLAY, A.P. (1977). Amoco's new generation of pilot plants. *Chem. Engng Prog.*, **73**(9), 73

SHIELDS, T.J. and SILCOCK, G.W. (1987). *Building Fires* (London: Longman)

SHIELDS, T.J., SILCOCK, G.W.H., DONEGAN, H.A. and BELL, Y.A. (1987). Methodological problems associated with the use of the Delphi technique. *Fire Technol.*, **23**, 175

SHIELDS, VUDENGAARD, N.B. and BERZINS, V. (1984). Revamping hydrogen plants of improved yield and energy efficiency. *Plant/Operations Prog.*, **3**, 108

SHIELER, L. and PAUZE, D. (1976). *Hazardous Materials* (New York: van Nostrand)

SHIGLEY, J.E. (1977). *Mechanical Engineering Design*, 3rd ed. (New York: McGraw-Hill)

SHIH, C.F., GERMAN, M.D. and KUMAR, V. (1981). An engineering approach for examining crack growth stability in flawed structures. *Int. J. Pres. Ves. Piping*, **9**, 159

SHILLITO, D. (1991). Setting new standards. *Chem. Engr, Lond.*, **499**, 14

SHIMAGAKI, M., MIYASHITA, K., ILYAS, A. and KALYUBI, A. (1987). Ammonia converter operation. *Plant/Operations Prog.*, **6**, 118

SHIMATANI, T. and OKUDA, O. (1992). 'Pipeless' batch chemical plants offer a new approach. *Chem. Engng*, **99**(10), 181

SHIMPO, Y. (1979). A revolutionary pulp plant construction method. *Chem. Engng Prog.*, **75**(10), 70

SHINDO, A. and UMEDA, T. (1977). A systems approach to the safety design of chemical processing systems. *Loss Prevention and Safety Promotion 2*, p. 433

SHINNAR, R. (1973). Sizing of storage tanks for off-grade material. *Ind. Engng Chem. Process Des. Dev.*, **6**, 263

SHINNAR, R., DOYLE, F.J., BUDMAN, H.M. and MORARI, M. (1992). Design considerations for tubular reactors with highly exothermic reactions. *AIChE J.* **38** (11), (November), 1729–43

SHINOTAKE, A., KODA, S. and AKITA, K. (1985). An experimental study of radiative properties of pool fires of an intermediate scale. *Combust. Sci. Technol.*, **43**, 85

SHINOZUKA, M., TAKADA, S. and ISHAKAWA, H. (1979). Some aspects of seismic risk analysis of underground lifeline systems. *J. Pres. Ves. Technol.*, **101**, 31

SHINSKEY, F.G. (1967). *Process Control Systems* (New York: McGraw-Hill)

SHINSKEY, F.G. (1977). *Distillation Control* (New York: McGraw-Hill)

SHINSKEY, F.G. (1978). Control systems can save energy. *Chem. Engng Prog.*, **74**(5), 43

SHINSKEY, F.G. (1979). *Process Control Systems–Application/Design/Adjustment* (New York: McGraw-Hill)

SHINSKEY, F.G. (1983). Uncontrollable processes and what to do about them. *Hydrocarbon Process.*, **62**(11), 179

SHIOZAKI, J., MATSUYAMA, H., OSHIMA, E. and IRI, M. (1985). An improved algorithm for diagnosis of system failures in the chemical process. *Comput. Chem. Engng*, **9**, 285

SHIPES, K. (1983). Correcting air-cooled heat exchanger problems. *Chem. Engng Prog.*, **79**(7), 56

SHIPLEY, E.E. (1973). Testing can reduce gear failures. *Hydrocarbon Process.*, **52**(12), 61

SHIPP, M.P. (1984). Hydrocarbon fire standard for offshore installations. *Fire Surveyor*, **13**(6), 4

SHIRAI, Y. and TSUJII, J.I. (1985). *Artificial Intelligence* (Chichester: Wiley)

SHIRLEY, R.S. (1987). Some lessons learned using expert systems for process control. *Proc. 1987 Am. Control Conf.*

SHIRLEY, R.S. and FORTIN, D.A. (1986). Developing an expert system for process fault detection and analysis. *InTech*, **33**(4), 51

SHIRVILL, L.C. (1992). Performance of passive fire protection in jet fires. *Major Hazards Onshore and Offshore*, p. 111

SHOCAIR, M. (1978). Plant layout. Ph.D. Thesis, Univ. of Sheffield

SHOEMAKER, W.C. (ed.) (1995). *Textbook of Clinical Care* (New York: W.B. Saunders)

SHONNARD, D.R., HERLEVICH, J., PARIKH, P., STARK, D., BARNA, B.A., ROGERS, T.N., CROWL, D.A. and KLINE, A.A. (1996). Pollution assessment software as chemical industry process simulator enhancements. *The 5th World Cong. of Chem. Eng.*, Jul. 14–18, 1996, San Diego, CA, Vol. III, 329–37 (New York: American Institute of Chemical Engineers)

SHOOMAN, M.L. (1968a). *Probabilistic Reliability: An Engineering Approach* (New York: McGraw-Hill)

SHOOMAN, M.L. (1968b). Reliability physics models. *IEEE Trans. Reliab.*, **R-17**, 14

SHOOMAN, M.L. (1976). Software reliability analysis and prediction. In Henley, E.J. and Lynn, J.W. (1976), *op. cit.*, p. 343

SHOOMAN, M.L. (1983). *Software Engineering: Reliability, Development and Management* (Maidenhead: McGraw-Hill)

SHOOTER, D., LYMAN, W.J. and SINCLAIR, J.R. (1981). Treatment techniques for spills of hazardous water-soluble chemicals – a feasibility study. *J. Haz. Materials*, **5**(1/2), 131

SHOOTER, R.A. (chrmn) (1980). *Investigation into the Cause of the 1978 Birmingham Smallpox Occurrence* (London: HM Stationery Office)

SHORE, D. (1973). Towards quieter flaring. *Chem. Engng Prog.*, **69**(10), 60

SHORT, H. (1983a). British meeting explores topic of intensification. *Chem. Engng*, **90**(Jun. 13), 27; (Jul. 25), 5 (letter)

SHORT, H. (1983b). New mass-transfer find is a matter of gravity. *Chem. Engng*, **90**(Feb. 21), 23

SHORT, H. (1986). The talk of Europe: Mossmorran's new cracker. *Chem. Engng*, **93**(Jan. 20), 31

SHORT, H. (1989). NH3 break thru: small plants. *Chem. Engng*, **96**(7), 41

SHORT, H. et al. (1984). Dioxin cleanup methods get worldwide attention. *Chem. Engng*, **91**(Nov. 26), 22

SHORT, H.C. and BOLTON, L. (1985). New styrene process pares production costs. *Chem. Engng*, **92**(Aug. 19), 30

SHORT, J.M., HELM, F.H., FINGER, M. and KAMLET, M.J. (1981). Chemistry of detonations. VII, A simplified method for predicting explosive performance in the cylinder test. *Combust. Flame*, **43**, 99

SHORT, W. (ed.) (1979). *Fuel Economy Handbook*, 2nd ed. (London: Graham & Trottman)

SHORT, W.A. (1972). Electrical equipment for hazardous locations. *Chem. Engng*, **79**(May 1), 59

SHORT, W.A. (1982). Electrical equipment used in hazardous locations. *Plant/Operations Prog.*, **1**, 14

SHORT, W.A. (1985). International standards: do they affect our industry? *Chem. Engng Prog.*, **81**(11), 14z

SHORTER, D.N. (1985). The Alvey RESCU project. *Inst. Elec. Engrs, Computing and Control Div. Colloquium on Real-Time Expert Systems in Process Control* (London)

SHORTHOUSE, B.O. (1983). Boiler circulation during chemical cleaning. *Chem. Engng*, **90**(Aug. 22), 75

SHORTLIFFE, E.H. (1976). *Computer Based Medical Consultation: MYCIN* (New York: Elsevier)

SHORTLIFFE, E.H. and BUCHANAN, B.G. (1975). A model of inexact reasoning in medicine. *Math. Biosci.*, **23**, 351

SHORTLIFFE, E.H., BUCHANAN, B.G. and FEIGENBAUM, E.A. (1979). Knowledge engineering for medical decision-making: a review of computer-based clinical decision aids. *Proc. IEEE*, **67**, 1207

SHORTREED, J. (1990). Recent advances in research and future requirements. *Plant/Operations Prog.*, **9**, 198

SHORTREED, J.H. and STEWART, A. (1988). Risk assessment and legislation. *J. Haz. Materials*, **20**, 315

SHOUMAN, A.R. and DONALDSON, A.B. (1975). The stationery problem of thermal ignition in a reactive slab with unsymmetric boundary temperatures. *Combust. Flame*, **24**, 203

SHOUMAN, A.R. and DONALDSON, A.B. (1977). Prediction of critical conditions for thermal explosion problems by a series method. *Combust. Flame*, **29**, 213

SHOUMAN, A.R., DONALDSON, A.B. and TSAO, H.Y. (1974). Exact solution to the one-dimensional stationary energy equation for a self-heating slab. *Combust. Flame*, **23**, 17

SHREIDER, Y.A. (ed.) (1964). *Method of Statistical Testing. Monte Carlo Method* (Amsterdam: Elsevier)

SHREIR, L.L. (ed.) (1976). *Corrosion*, vols 1–2 (London: Newnes)

SHREVE, R.N. (1967). *Chemical Process Industries* (New York: McGraw-Hill)

SHRIVASTAVA, P. (1987). *Bhopal. Anatomy of a Crisis* (Cambridge, MA: Ballinger)

SHRIVASTAVA, S.K. (1991). Fault tolerant system structuring concepts. In McDermid, J.A. (1991), *op. cit.*

SHRIVER, D.F. (1969). *Manipulation of Air Sensitive Compounds.* (London: McGraw-Hill)

SHRIVER, E.L., FINK, C.D. and TREXLER, R.C. (1964). *FORECAST Systems Analysis and Training Methods for Electronics Maintenance. Res. Rep. 13.* Hum. Resources Res. Office, Alexandria, VA

SHTAYIEH, S. (1983a). Planning and care mark repair of 14–year-old leak in Kuwait Oil Co. LPG tank. *Oil Gas J.*, **81**(Jan. 10), 95

SHTAYIEH, S. (1983b). Refrigerated propane tank opening for inspection and leak repairs. *Gastech 82*, p. 397

SHTAYIEH, S., DURR, C.A., MCMILLAN, J.C. and COLLINS (1982). Operation of the world's largest LPG plant. *Gastech 81*, p. 417

SHTERN, V.Y. (1964). *The Gas Phase Oxidation of Hydrocarbons* (Oxford: Pergamon Press)

SHUBIN, J.A. and MADEHEIM, H. (1951). *Plant Layout* (New York: Prentice-Hall)

SHUCKROW, A.J., MERCER, B.W. and DAWSON, G.W. (1972). The application of sorption process for in situ treatment of hazardous material spills. *Control of Hazardous Material Spills 1*, p. 173

SHUCKROW, A.J., TOUHILL, C.J. and PAJAK, A.P. (1987). Hazardous-waste leachate management. In Martin, E.J. and Johnson, J.H. (1987), *op. cit.*, p. 441

SHUKER, F.S. (1983). When to use refractory metals and alloys in the plant. *Chem. Engng*, **90**(May 2), 81

SHULL, W.W. and CHURCH, M.L. (1991). Field performance test: key to pump savings. *Hydrocarbon Process.*, **70**(1), 77

SHUM, S.K., DAVIS, J.F., PUNCH, W.F. and CHANDRASEKARAN, B. (1988). An expert system approach to malfunction diagnosis in chemical plants. *Comput. Chem. Engng*, **12**, 27

SHUMAKER, F.E. (1980). Ship-to-ship transfer of LNG, the 'El Paso Kayser'/'El Paso Sonatrach' gas transfer operation. *Gastech 79*, p. 173

SHYVERS, H.C. (1981). The offshore scene. *Reliable Production in the Process Industries* (Rugby: Instn Chem. Engrs), p. 41

SIBLEY, H.W. (1983). Selecting refrigerants for process systems. *Chem. Engng*, **90**(May 16), 71

SIBULKIN, M. (1973). Estimates of the effect of flame size on radiation from flames. *Combust. Sci. Technol.*, **7**, 141

SIBULKIN, M. (1980). Pressure rise generated by combustion of a gas pocket. *Combust. Flame*, **38**, 329

SICCAMA, E.H. (1971). The environmental risk arising from the bulk storage of dangerous goods in port areas. *Transport of Hazardous Cargoes*, **2**, p. 10

SICCAMA, E.H. (1973). The environmental risk arising from the bulk storage of dangerous chemicals. *De Ingenieur*, **85**(24), 502

SICHEL, M. (1982a). The detonation of sprays: recent results. In Lee, J.H.S. and Guirao, C.M. (1982), *op. cit.*, p. 265

SICHEL, M. (1982b). Liquid spray detonation. In Lee, J.H.S. and Guirao, C.M. (1982), *op. cit.*, p. 983

SICHEL, M. and PALANISWAMY, S. (1985). Sheath combustion of sprays. *Combustion*, **20**, p. 1789

SICHEL, M., RAO, C.S. and NICHOLLS, J.A. (1971). A simple theory for the propagation of film detonations. *Combustion*, **13**, p. 1141

SICKLES, R.W. (1968). Air pollution control. Electrostatic precipitators. *Chem. Engng*, **75**(Oct. 14), 156

SIDDALL, E. (1959). Statistical analysis of nuclear reactor safety standards. *Nucleonics*, **17**(2), 64

SIDDALL, R.G. and MCGRATH, I.A. (1963). The emissivity of luminous flames. *Combustion*, **9**, p. 102

SIDDLE, J. (1986a). Engineering insurance. *Plant Engr*, **30**(1), 18

SIDDLE, J. (1986b). Robots and loss control. *Plant Engr*, **30**(4), 15

SIDDLE, J.F. (1986). Fire hazards of flammable liquids in small containers. *Hazards*, **IX**, p. 111

SIDDONS, D.J. (1977). Setting availability targets for power station systems during their early design stages. *First Nat. Conf. on Reliability. NCR2/3* (Culcheth, Lancashire: UK Atom. Energy Auth.)

SIDE, J. (1989). *Piper Alpha Public Inquiry. Transcript, Days 144, 145*

SIDLEY and AUSTIN (1987). *Superfund Handbook*, 2nd ed. (Concorde, MA: ERT Inc.); 1989, 3rd ed. (Concorde, MA: ENSR Corp.). See also Casler, J. (1985)

SIDWICK, J.M. and BARNARD, R. (1981). Treatment before discharge. *Chem. Ind.*, Apr. 18, 277

SIEBERT, O.W. (1970). Failure analysis: a materials engineering tool. *Chem. Engng Prog.*, **66**(9), 63

SIEGEL, A.I. and WOLF, J.A. (1969). *Man-Machine Simulation Models* (New York: Wiley)

SIEGEL, A.I. *et al.* (1984). *Maintenance Personnel Performance Simulations (MAPPS) Model. Rep. NUREG/CR-3626*, vols 1–2. Nucl. Regul. Comm., Washington, DC

SIEGEL, M. (1989). Explaining risk to the public. *Chem. Engng Prog.*, **85**(5), 20

SIEGEL, R. and HOWELL, J.R. (1972–). *Thermal Radiation Heat Transfer* (New York: McGraw-Hill); 1981, 2nd ed. (New York: Hemisphere)

SIEGEL, R. and HOWELL, J.R. (1991). *Thermal Radiation Heat Transfer*, 3rd ed. (New York: Hemisphere)

SIEGEL, R.D. (1981). Controls developing for fugitive emissions. *Hydrocarbon Process.*, **60**(10), 105

SIEGEL, R.D., FOYE, C.M., PETRILLO, J.L. and ROSENFELD, M.F. (1979). The impact of the Clean Air Act amendments on new and expanded plants. *Chem. Engng Prog.*, **75**(8), 13

SIEGEL, S. (1956). *Non-Parametric Statistics for the Behavioural Sciences* (New York: McGraw-Hill)

SIEGEL, W. (1978). Safer low-density polyethylene units. *Safety in Polyethylene Plants*, **3**, p. 80

SIEGHART, P. (ed.) (1979). *The Big Public Inquiry* (London: Outer Circle Policy Unit)

SIEKEVITZ, P. (1972). The social responsibilities of scientists. *Ann. N.Y. Acad. Sci.*, **196**(art. 4)

SIEUR (1915). *French Mil. Arch., Vincennes, Rep. Box 16, no. 826*

SIEVERING, H. (1982). Profile measurements of particle dry deposition velocity at an air–land interface. *Atmos. Environ*, **16**, 301

SIEVERING, H. (1989). The dry deposition of small particles a review of experimental measurement. *Atmos. Environ.*, **23**, 2863

SIGALES, B. (1975). How to design settling drums. *Chem. Engng*, **82**(Jun. 23), 141

SIGMOND, R.S. (1978). Corona discharges. *Electrical Breakdown in Gases* (ed. J.M. Meek and J.D. Craggs) (New York: Wiley)

SIGNORET, J.P. and LEROY, A. (1985). The 1800 m water depth drilling project: risk analysis. *Reliab. Engng*, **11**(2), 83

SIGNORET, J.P., MURON, O. and COHEN, G. (1983). Optimal testing of stand-by systems. *Reliab. Engng*, **4**(3), 127

SIGNORINI, G. (1961). Find optimum frequency of inspection. *Chem. Engng*, **68**(Mar. 6), 138

SIH, G.C. (1966). On the Westergaard method of crack analysis. *Int. J. Frac. Mech.*, **2**, 628

SIH, G.C. (1973a). *Handbook of Stress Intensity Factors for Researchers and Engineers*. Inst. for Fracture and Solid Mechanics, Lehigh Univ., Bethlehem, PA

SIH, G.C. (ed.) (1973b). *Methods of Analysis and Solutions of Crack Problems* (Leiden: Noordhoff)

SIIROLA, J.J., GROSSMAN, I.E. and STEPHANOPOULOS, G. (eds) (1990). *Foundations of Computer Aided Process Design* (Amsterdam: Elsevier)

SIIROLA, J.J., POWERS, G.J. and RUDD, D.F. (1971a). Synthesis of system designs. III, Towards a process concept generator. *AIChE J.*, **17**, 667

SIIROLA, J., POWERS, G. and RUDD, D.F. (1971b). Synthesis of chemical process designs. *Ind. Engng Chem. Fund.*, **10**, 353

SIKLOSSY, L. (1976). *Let's Talk LISP* (Englewood Cliffs, NJ: Prentice-Hall)

SILBAUGH, S.A., WOLFF, R.K., JOHNSON, W.K., MAUDERLY, J.L. and MACKEN, C.A. (1981). Effects of sulphuric acid aerosols on pulmonary function of guinea pigs. *J. Toxicol. Environ. Health*, **7**, 339

SILBERBERG, M.A. and MECLOT, M. (1983). EdF nuclear power plant event file. *Euredata 4*, paper 11.1

SILLETT, C. (1988). Flowsheeting for food product safety. *Chem. Engr, Lond.*, **450**, 21

SILLS, D.L., WOLF, C.P. and SHELANSKI, V.B. (1982). *Accident at Three Mile Island: the Human Dimensions* (Boulder, CO: Westview Press)

SILSBEE, F.B. (1942). *Static Electricity. Nat. Bur. Standards, Circ. C-438* (Washington, DC)

SILSBY, R.I. and OCKERBLOOM, N.C. (1971). Can olefins meet 70's needs? *Hydrocarbon Process.*, **50**(3), 83

DE SILVA, R. (1988). In these days of energy conservation flares remain in fashion. *Process Engng*, Feb., 75

SILVER, L. (1967). Screening for chemical process hazards. *Loss Prevention*, **1**, p. 58. See also *Chem. Engng Prog.*, 1967, **63**(8), 43

SILVER, R.S. (1937). The ignition of gas mixtures by hot particles. *Phil. Mag.*, **23**, 633

SILVER, S.D. and MCGRATH, F.P. (1942). *Chlorine. Median Lethal Concentration for Mice. Rep. EATR 351*. War Dept, Chemical Warfare Center, Edgwood Arsenal, MD

SILVER, S.D. and MCGRATH, F.P. (1948). A comparison of acute toxicities of ethylene imine and ammonia to mice. *J. Ind. Hyg. Toxicol.*, **30**, 7

SILVER, S.D., MCGRATH, F.P. and FERGUSON, R.L. (1942). *Chlorine. Median Lethal Concentration for Mice. Rep. EATR 373*. War Dept, Chemical Warfare Center, Edgwood Arsenal, MD

SILVER PLATTER INFORMATION SERVICES (1990a). *Hazardous Chemical Databanks: CHEM-BANK* (London)

SILVER PLATTER INFORMATION SERVICES (1990b). *Occupational Safety and Health; OSH-ROM* (London)

SILVERMAN, D. (1964a). Electrical design: distribution. *Chem. Engng*, **71**(May 25), 131

SILVERMAN, D. (1964b). Electrical design: heating. *Chem. Engng*, **71**(Jul. 20), 161

SILVERMAN, D. (1964c). Electrical design: lighting. *Chem. Engng*, **71**(Jul. 6), 121

SILVERMAN, D. (1964d). Electrical design: motors and controls. *Chem. Engng*, **71**(Jun. 22), 133

SILVERMAN, L., WHITTENBERGER, J. and MULLER, J. (1949). Physiological response of man to ammonia at low concentrations. *J. Ind. Hyg. Toxicol.*, **31**, 74

SILVERSTEIN, J.L., WOOD, B.H. and LESHAW, S.A. (1981). Case study in reactor design for hazards prevention. *Loss Prevention*, **14**, p. 78

SILVERSTEIN, R.M. and CURTIS, S.D. (1971). Cooling water. *Chem. Engng*, **78**(Aug. 9), 84

SILVESTRI, A., GOODMAN, A., MCCORMACK, L.M., RAZULIS, M., JONES, A.R. and DAVIS, M.E.P. (1976). Detection of hazardous substances. *Control of Hazardous Material Spills*, **3**, p. 78

SILVESTRI, A., RAZULIS, M., GOODMAN, A., VASQUEZ, A., STRAUCH, L. and JONES, A.R. (1978). Identification of hazardous substances. *Control of Hazardous Material Spills*, **4**, p. 91

SIMINSKI, V.J. (1987). Application of national consensus standards to pilot plant safety. *Chem. Engng. Prog.*, **83**(1), 60

SIMIU, E. and SCANLAN, R.H. (1978). *Wind Effects on Structures: An Introduction to Wind Engineering* (New York: Wiley)

SIMKIN, T. and FISKE, R.S. (1983). *Krakatoa 1883* (Washington, DC: Smithsonian Inst.)

SIMMON, C.A. (1972). Allowable pump piping loads. *Hydrocarbon Process.*, **51**(6), 98

SIMMONDS, D. (1993). Clean fire extinguishing agents, Part 1. *Fire Surveyor*, **22**(5), 15

SIMMONDS, W.A. and CUBBAGE, P.A. (1960). The design of explosion reliefs for industrial drying ovens. *Chemical Process Hazards*, I, p. 69

SIMMONS, J.A. (1975). Risk assessment method for volatile toxic and flammable materials. *Transport of Hazardous Cargoes*, 4

SIMMONS, J.A. and ERDMANN, R.C. (1975). Risk assessment strategy for marine transportation. *Transport of Hazardous Cargoes*, 4

SIMMONS, J.A., ERDMANN, R.C. and NAFT, B.N. (1973). *The Risk of Catastrophic Spills of Toxic Chemicals. Rep. UCLA ENG-7425.* Univ. of California, Los Angeles, CA

SIMMONS, J.A., ERDMANN, R.C. and NAFT, B.N. (1974). Risk assessment of large spills of toxic materials. *Nat. Conf. on Control of Hazardous Material Spills, San Francisco, CA*

SIMMONS, P.E. (1978). For turbomachinery . . . new acceptance criteria proposed. *Hydrocarbon Process.*, **57**(1), 169

SIMMONS, R.F. (1973). Semantic networks; their computation and use for understanding English sentences. In Schank, R.C. and Colby, K.M. (1973), *op. cit.*, p. 63

SIMMONS, R.F. and WRIGHT, N. (1972). The burning velocities of near limit mixtures of propane, air, and hydrogen bromide. *Combust. Flame*, **18**, 203

SIMMROCK, K.H. (1978). Compare propylene oxide routes. *Hydrocarbon Process.*, **57**(11), 105

SIMMS, D.L. (1960). Ignition of cellulosic materials by radiation. *Combust. Flame*, **4**, 293

SIMMS, D.L. (1961). Experiments on the ignition of cellulosic materials by thermal radiation. *Combust. Flame*, **5**, 369

SIMMS, D.L. (1962). Damage to cellulosic materials by thermal radiation. *Combust. Flame*, **6**, 303

SIMMS, D.L. (1963). On the pilot ignition of wood by radiation. *Combust. Flame*, **7**, 253

SIMMS, D.L. and LAW, M. (1967). Ignition of wet and dry wood by radiation. *Combust. Flame*, **11**, 377

SIMMS, D.L., HIRD, D. and WRAIGH, H. (1960). The temperatures and duration of fires. Some experiments with models with restricted ventilation. *Fire Res. Note 412* (London: Joint Fire Res. Org.)

SIMMS, R.K. (1965). Suffocation of workers in a CO converter. *Ammonia Plant Safety*, **7**, p. 10

SIMO, F.E. (1973). Which flow control: valve or pump? *Hydrocarbon Process.*, **52**(7), 103

SIMON, D.M., BELLES, F.E. and SPAKOWSKI, A.E. (1953). Investigation and interpretation of the flammability region of some lean hydrocarbon-air mixtures. *Combustion*, **4**, p. 126

SIMON, E. (1976). Inertizing techniques and design modifications for the prevention of explosions and fire in filter centrifuges. *Prevention of Occupational Risks*, **3**, p. 367

SIMON, H. and THOMSON, S.J. (1972). Relief system optimization. *Loss Prevention*, **6**, p. 74. See also *Chem. Engng Prog.*, 1972, **68**(5), 52

SIMON, H. and WHELAN, T.W. (1970). Use computer to select optimum control valves. *Hydrocarbon Process.*, **49**(7), 103

SIMON, H.A. (1969). *The Sciences of the Artificial* (Cambridge, Mass.: MIT Press)

SIMON, H.A. (1975). *A Students Introduction to Engineering Design* (New York: Pergamon Press)

SIMON, H.A. (1983). Why should machines learn? In R.S. Michalski, J. Carbonell and T.M. Mitchell (1983), *op. cit.*

SIMONDS, R.H. and GRIMALDI, J.V. (1963). *Safety Management*, 2nd ed. (Homewood, IL: Richard D. Irwin)

SIMONS, E.A. (1976). A minor catastrophe averted. *Control of Hazardous Material Spills 3*, p. 342

SIMONS, G.L. (1984). *Introducing Artificial Intelligence* (Manchester: National Computing Centre)

SIMONS, G. (1988). *Evolution of the Intelligent Machine* (Manchester: National Computing Centre)

SIMONS, J.H. (1931). Hydrogen fluoride and its solutions. *Chem. Rev.*, **8**, 213

SIMPSON, D. and SIMPSON, W.G. (1991). *The COSHH Regulations: A Practical Guide* (London: R. Soc. Chem.)

SIMPSON, E.A. (1980). The European Commission's philosophy in respect of monitoring the aquatic environment. *Chem. Ind.*, Aug. 2, 584

SIMPSON, H.C., ROONEY, D.H. and GRATTAN, E. (1979). Two-phase pressure drops in valve and orifice restrictions. *Eur. Two-Phase Flow Group Mtg*, Ispra, Italy

SIMPSON, H.G. (1971). Design for loss prevention – plant layout. *Major Loss Prevention*, p. 105

SIMPSON, H.G. (1974). The ICI vapour barrier. *Power Works Engng*, May 8

SIMPSON, J.E. (1982). Gravity currents in the laboratory, atmosphere and ocean. *Ann. Rev. Fluid Mech.*, **14**, 213

SIMPSON, J.E. (1987). *Gravity Currents: In the Environment and the Laboratory* (Chichester: Ellis Horwood)

SIMPSON, J.E. and BRITTER, R.E. (1979). The dynamics of the head of a gravity current advancing over a horizontal surface. *J. Fluid Mech.*, **94**, 477

SIMPSON, L.L. (1968). Sizing piping for process plant. *Chem. Engng*, **75**(Jun. 17), 192

SIMPSON, L.L. (1969). Process piping: functional design. *Chem. Engng*, **76**(Apr. 14), 167

SIMPSON, L.L. (1972). Tubing rupture in liquid filled exchangers. *Loss Prevention*, **6**, p. 92

SIMPSON, L.L. (1986). Equations for the VDI and Bartknecht nomograms. *Plant/Operations Prog.*, **5**, 49

SIMPSON, L.L. (1991). Estimate two-phase flow in safety devices. *Chem. Engng*, **98**(8), 98

SIMPSON, M. and JONES, F.R. (1992). *Evaluation of the Safety of Composite Gas Containers. HSE Contract Res. Rep. 37/1992* (London: HM Stationery Office)

SIMS, E.R. (1987). Safely store hazardous and flammable materials. *Chem. Engng*, **94**(13), 129

SIMS, I.G. (1980). Condition monitoring of chemical plants. *Chem. Engr, Lond.*, **356**, 308

SIMSON, R.E. and SIMPSON, G.R. (1971). Fatal hydrogen sulphide poisoning associated with industrial waste exposure. *Med. J. Aust.*, **1**, 331

SINAIKO, H.W. (ed.) (1961). *Selected Papers on Human Factors in the Design and Use of Control Systems* (New York: Dover)

SINCLAIR, C.G. and SWEIS, F.L. (1980). The influence of oxygen concentration on dust explosibility. *Chemical Process Hazards*, VII, p. 103

SINCLAIR, I.A.C., SELL, R.G., BEISHON, R.J. and BAINBRIDGE, L. (1966). Ergonomic study of an LD waste heat boiler control room. *J. Iron Steel Inst.*, **204**, 434

SINCLAIR, J.D., PSOTA-KELTY, L.A. and WESCHLER, C.J. (1985). Indoor/outdoor concentrations and indoor surface accumulations of ionic substances. *Atmos. Environ.*, **19**, 315

SINCLAIR, J.R. and BAUER, W.H. (1976). Containment and recovery of floating hazardous chemicals with commercially available devices. *Control of Hazardous Material Spills*, **3**, p. 272

SINCLAIR, T.C. (1963). *Control of Hazards in Nuclear Reactors* (London: Temple Press)

SINCLAIR, T. CRAIG (1972). *A Cost-Effectiveness Approach to Industrial Safety* (London: HM Stationery Office)

SINCLAIR, T. CRAIG, MARSTRAND, P. and NEWICK, P. (1972a). *Human Life and Safety in Relation to Technical Change.* (report). Sci. Policy Res. Unit, Univ. of Sussex

SINCLAIR, T. CRAIG, MARSTRAND, P. and NEWICK, P. (1972b). *Innovation and Human Risk* (London: Centre for the Study of Ind. Innovation)

SINGER, D. (1990). Fault tree analysis based on fuzzy logic. *Comput. Chem. Engng*, **14**, 259

SINGER, I.A. and SMITH, M.E. (1953). Relation of gustiness to other meteorological parameters. *J. Met.*, **10**, 121

SINGER, I.A. and SMITH, M.E. (1966). Atmospheric dispersion at Brookhaven National Laboratory. *Int. J. Air Water Poll.*, **10**, 125

SINGER, J.M. and VON ELBE, G. (1957). Flame propagation in cylindrical tubes near the quenching limit. *Combustion*, **6**, p. 127

SINGER, S.F. (ed.) (1975). *The Changing Global Environment* (Dordrecht: Reidel)

SINGH, B. (1991). Selecting nonasbestos sheet gasketing. *Hydrocarbon Process.*, **70**(4), 79

SINGH, B. (1993). Specifying clad CRA piping. *Hydrocarbon Process.*, **72**(1), 51

SINGH, B.R. and THOMAS, F.H. (1980). A method for predicting turbine blade lives using unfailed specimens and analytical techniques. In Moss, T.R. (1980), *op. cit.*, p. 105

SINGH, C. (1974). Reliability analysis of large repairable systems. *Microelectron. Reliab.*, **13**, 487

SINGH, C. and BILLINTON, R. (1977). *System Reliability Modelling and Evaluation* (London: Hutchinson)

SINGH, C.P.P. and SARAF, D.N. (1981). Process simulation of ammonia plant. *Ind. Engng Chem. Process Des. Dev.*, **20**, 425

SINGH, J. (1977). Gas explosions in single and compartmented vessels. Ph.D. Thesis, Univ. of London

SINGH, J. (1979a). Sizing of relief for exothermic reactions. In Minors, C. (1979), *op. cit.*, paper 5

SINGH, J. (1979b). Sizing vents for gas explosions. *Chem. Engng*, **86**(Sep. 24), 103

SINGH, J. (1979c). Suppression of internal explosions in process plant. *J. Occup. Accid.*, **2**, 113

SINGH, J. (1981). Selecting heat-transfer fluids for high-temperature service. *Chem. Engng*, **88**(Jun. 1), 53

SINGH, J. (1984). Gas explosions in compartmented vessels: pressure piling. *Trans. Instn Chem. Engrs*, **62**, 351

SINGH, J. (1988a). Design of process plant handling highly unstable and flammable materials by the use of test data. *Preventing Major Chemical and Related Process Accidents*, p. 229

SINGH, J. (1988b). Explosion propagation in non-spherical vessels: simplified equations and applications. *J. Loss Prev. Process Ind.*, **1**(1), 39

SINGH, J. (1989). PHI-TEC: enhanced vent sizing calorimeter – application and comparison with existing devices. *Runaway Reactions*, p. 313

SINGH, J. (1990a). Safety system design for runaway reaction explosions: use of data from bench-scale equipment. *Safety and Loss Prevention in the Chemical and Oil Processing Industries*, p. 233

SINGH, J. (1990b). Sizing relief vents for runaway reactions. *Chem. Engng*, **97**(8), 104

SINGH, J. (1992a). Runaway reaction hazard and venting of a cyanide reaction. *J. Loss Prev. Process Ind.*, **5**, 140

SINGH, J. (1992b). Safe disposal of reactive chemicals following emergency venting. *Major Hazards Onshore and Offshore*, p. 249

SINGH, J. (1993). Assessing semi-batch reaction hazards. *Chem. Engr, Lond.*, **537**, 21

SINGH, J. (1993). Assessing semi-batch reaction hazards. *Chemical Eng.*, **537**, 21 (Feb. 5), 23–5

SINGH, J. (1994a). Gas explosions in interconnected vessels: pressure piling. *Hazards*, **XII**, p. 195

SINGH, J. (1994b). Vent sizing for gas-generating runaway reactions. *J. Loss Prev. Process Ind.*, **7**, 481

SINGH, J. and BOEY, D. (1991). Design of quench system for runaway reactions using bench-scale equipment. *Plant/Operations Prog.*, **10**, 58

SINGH, J. and BOEY, D. (1992). Modelling of mass transfer controlled exothermic reactions using adiabatic calorimetry. *Loss Prevention and Safety Promotion 7*, vol. 3, 127–1

SINGH, J. and MCBRIDE, M. (1990). Successfully model complex chemical hazard scenarios. *Chem. Engng Prog.*, **86**(10), 71

SINGH, J. and MUNDAY, G. (1979). IFAL: a model for the evaluation of chemical process losses. *Design 79* (London: Instn Chem. Engrs)

SINGH, J., CAVE, L. and MCBRIDE, M. (1990). Loss of containment – some practical aspects. *J. Loss Prev. Process Ind.*, **3**(1), 146

SINGH, K.P. (1976). Design of skirt-mounted supports. *Hydrocarbon Process.*, **55**(4), 199

SINGH, K.P. (1990). Pressure vessels: design, operation and maintenance. *Chem. Engng*, **97**(7), 62

SINGH, M.G., HINDI, K.S., SCHMIDT, G. and TZAFESTAS, S. (eds) (1987). *Fault Detection and Reliability* (Oxford: Pergamon Press)

SINGH, M.P. (1990). Vulnerability analysis for airborne release of extremely hazardous substances. *Atmos. Environ.*, **24A**, 769

SINGH, M.P. and GHOSH, S. (1987). Bhopal gas tragedy: model simulation of the dispersion scenario. *J. Haz. Materials*, **17**(1), 1

SINGH, M.P., KUMARI, M. and GHOSH, S. (1990). A mathematical model for the recent oleum leakage in Delhi. *Atmos. Environ.*, **24A**, 735

SINGHAL, S.N., DELICHATSIOS, M.A. and DE RIS, J. (1989). Offshore stack-enclosed gas flares. *Fire Safety J.*, **15**, 211, 227

SINGPURWALLA, N.D. (1972). Extreme values from a log normal law with applications to air pollution problems. *Technometrics*, **14**, 703

SINGLETON, B. (1989). CIMAH safety cases: 3, toxics and ultratoxics safety cases. In Lees, F.P. and Ang, M.L. (1989), *op. cit.*, p. 243

SINGLETON, W.T. (1969). Display design: principles and procedures. *Ergonomics*, **12**, 543

SINGLETON, W.T. (1976a). Ergonomics: where have we been and where are we going: IV. *Ergonomics*, **19**, 307

SINGLETON, W.T. (1976b). *Human Aspects of Safety* (London: Keith Shipton Developments)

SINGLETON, W.T. (1982). Accidents and progress of technology – a plenary report. *J. Occup. Accid.*, **4**(2/4), 91

SINGLETON, W.T. (1984). Application of human error analysis to occupational accident research. *J. Occup. Accid.*, **6**(1/3), 107

SINGLETON, W.T. (1986). Ergonomics and its application to Sizewell B. *The Pressurised Water Reactor and the United*

Kingdom (ed. D. Weaver and J. Walker) (Birmingham: Univ. of Birmingham)

SINGLETON, W.T., EASTERBY, R.S. and WHITFIELD, D.C. (eds) (1967). *The Human Operator in Complex Systems* (London: Taylor & Francis)

SINGLETON, W.T., FOX, J.G. and WHITFIELD, D.C. (eds) (1971). *Measurement of Man at Work* (London: Taylor & Francis)

SINGO SHIGEO (1986). *Zero Quality Control: Source Inspection and the Poka-Yoke System* (Stamford, CT: Productivity Press)

SINHA, B.P. and HENDRY, A.W. (1970). Further tests on model brick wall and piers. *Proc. Br. Ceram. Soc.*, Feb., 83

SINHA, S.K. and KALE, B.K. (1980). *Life Testing and Reliability Estimation* (New York: Wiley)

SINNOTT, J.E. (1988). Minimizing lightning damage to data and control circuits. *Chem. Engng*, **95**(1), 127

SINNOTT, R.K. (1983). Safety and loss prevention. In *Chemical Engineering* (ed. J.M. Coulson and J.F. Richardson) (Oxford: Pergamon Press), Ch. 9

SIOUI, R.H. and ROBLEE, L.H.S. (1969). The prediction of the burning constants of suspended hydrocarbon fuel droplets. *Combust. Flame*, **13**, 447

SIRIGNANO, W.A. (1972). A critical discussion of theories of flame spread across solid and liquid fuels. *Combust. Sci. Technol.*, **6**, 95

SIRIGNANO, W.A. and GLASSMAN, I. (1970). Flame spreading across liquid fuels: surface tension-driven flows. *Combust. Sci. Technol.*, **1**, 307

SISMAN, T. and GEORGHIU (1977). Improvements to the machines and systems used in the pharmaceutical industry with a view to avoiding the danger of accidents and occupational diseases. *Prevention of Occupational Risks*, **5**, p. 89

SITARAMAN, V., SANTOSO, E. and SATHE, S. (1988). Failure and repair of a primary waste heat boiler. *Ammonia Plant Safety*, **28**, 27

SITTIG, M. (1961–). Combine oxygen and hydrocarbons for profit. *Hydrocarbon Process.*, 1961, **40**(11), 309; **40**(12), 147; 1962, **41**(1), 163; **41**(2), 169; **41**(3), 177; **41**(4), 157; **41**(5), 169; **41**(6), 175; **41**(7), 119; **41**(8), 129; **41**(9), 245; **41**(10), 159

SITTIG, M. (1962). *Combining Oxygen and Hydrocarbons for Profit* (Houston, TX: Gulf)

SITTIG, M. (1966). *Nitrogen in Industry* (London: van Nostrand)

SITTIG, M. (1974). *Pollution Detection and Monitoring Handbook* (Park Ridge, NJ: Noyes Data Corp.)

SITTIG, M. (1978). *Vinyl Chloride and PVC Manufacture* (Park Ridge, NJ: Noyes Data Corp.)

SITTIG, M. (1979). *Hazardous and Toxic Effects of Industrial Chemicals* (Park Ridge, NJ: Noyes Data Corp.)

SITTIG, M. (1981). *Handbook of Toxic and Hazardous Chemicals*; 1985, 2nd ed.; 1991, 3rd ed. (Park Ridge, NJ: Noyes)

SITTIG, M. (1994). *World-wide Limits for Toxic and Hazardous Chemicals in Air, Water and Soil* (Park Ridge, NJ: Noyes)

SIU, N.O. (1982). Physical models for compartment fires. *Reliab. Engng*, **3**(3), 229

SIU, N.O. and APOSTOLAKIS, G.E. (1982). Probabilistic models for cable tray fires. *Reliab. Engng*, **3**(3), 213

SIU, N. and APOSTOLAKIS, G.E. (1985). On the quantification of model uncertainties. *Eighth Int. Conf. on Structural Mechanisms in Reactor Technology*, Brussels

SIU, N. and APOSTOLAKIS, G. (1986). Modeling the detection rates of fires in nuclear plants: development and appli-

cation of a methodology for treating imprecise evidence. *Risk Analysis*, **6**

SIVALLS, C.R. (1985). Hydrogen sulphide service dictates special equipment design above requirements. *Oil Gas J.*, **83**(Jul. 8), 56

SIVASHINSKY, G.A. (1974). On a converging spherical flame front. *Int. J. Heat Mass Transfer*, **17**, 1499

SIVASHINSKY, G.I. (1988). Cascade-renormalisation theory of turbulent flame speed. *Combust. Sci. Technol.*, **62**, 77

SIVATHANU, Y.R. and GORE, J.P. (1993). Total radiative heat loss in jet flames from single point radiative flux measurements. *Combust. Flame*, **94**, 265

SIVERTSEN, B. (1984). Estimation of diffuse hydrocarbon leakages from petrochemical factories. In de Wispelaere, C. (1984), *op. cit.*, p. 693

SIWEK, R. (1982). Explosion characteristics and influencing factors. *The Control and Prevention of Dust Explosions* (London: Oyez/IBC), paper 2

SIWEK, R. (1989a). Dust explosion venting for dusts pneumatically conveyed into vessels. *Plant/Operations Prog.*, **8**, 129

SIWEK, R. (1989b). New knowledge about rotary air locks in preventing dust ignition breakthrough. *Plant/Operations Prog.*, **8**, 165

SIWEK, R. (1994). New revised VDI Guideline 3673 Pressure Release of Dust Explosions. *Process Safety Prog.*, **13**, 190

SIWEK, R. and CESANA, C. (1993). Assessment of fire and explosion hazards of combustible products for unit operations. *Health, Safety and Loss Prevention*, p. 322

SIWEK, R. and ROSENBERG, E. (1989). Ethylene oxide decomposition. *Loss Prevention and Safety Promotion 6*, p. 52.1

SIWEK, R. and WEGMULLER, R. (1993). Explosion protection in industry. *Health, Safety and Loss Prevention*, p. 264

SIWEK, R., GLOR, M. and TORREGIANI, T. (1992). Dust explosion venting at elevated initial pressure. *Loss Prevention and Safety Promotion 7*, vol. 2, 57–1

SIXSMITH, T. (1993). Compare plastic fusion methods for ultrapure piping systems. *Chem. Engng Prog.*, **89**(4), 62

SJOSTROM, H. (1972). *Thalidomide and the Power of the Drug Companies* (Harmondsworth: Penguin)

SKABO, R.R. (1970). Internal maintenance of vessels. *Chem. Engng Prog.*, **66**(1), 33

SKALA, V. (1974). Improving instrument service factors. *Instrum. Technol.*, **21**(11), 27

SKANDIA INTERNATIONAL (1985) BLEVE! The Tragedy of San Juanico. *First*, Nov. 2

SKANS, S. (1978). *Safety Implications of Human Error. Rep. TA 838–R5*. LUTAB, Bromma, Sweden

SKANS, S. (1980). *Ergonomic Investigations of Operation and Maintenance of Personnel's Work Situation in Nuclear Power Plants: a Main Study. Rep. TA 895–R6*. LUTAB, Bromma, Sweden

SKARKA, J. (1987). Considerations on the maximum hazard limits originating from LPG processing and handling. *Vapor Cloud Modeling*, p. 943

SKAUDAHL, R.E. and ZEBROSKI, E.L. (1969). Finding materials for fast reactors of the future. *Chem. Engng*, **76**(Aug. 11), 114

SKELTON, R.P. (1983). *Fatigue at High Temperatures* (London: Elsevier Applied Science)

SKELTON, R.P. (ed.) (1987). *High Temperature Fatigue* (London: Elsevier Applied Science)

SKIBIN, D. (1983). Analysis of a finite release of a pollutant followed by an abrupt change of wind direction. In de Wispelaere, C. (1983), *op. cit.*, p. 591

SKIBIN, D. and BUSINGER, J.S. (1985). The vertical extent of the loglinear wind profile under stable stratification. *Atmos. Environ.*, **19**, 27

SKILLERN, M.R. (1976). The future of LPG as a petrochemical feedstock. *Chem. Engng Prog.*, **72**(10), 34

SKINNER, A.C. (1985). Retrofitting increases energy recovery. *Chem. Engng Prog.*, **81**(2), 28

SKINNER, D.R. (1982). *Introduction to Petroleum Production*, vols 1–3 (Houston, TX: Gulf Publ. Co.)

SKINNER, G.B., MUELLER, G., GRIMM, U. and SCHELLER, K. (1968). Initiation of detonation by incident shock waves in hydrogen–oxygen–argon mixtures. *Combust. Flame*, **12**, 436

SKINNER, J.B. (1962). Tabulating the toxics. *Chem. Engng*, **69**(Jun. 11), 183

SKINNER, S.J. (1981). Explosive evolution of gas in the manufacture of ethyl polysilicate. *Loss Prevention*, **14**, p. 178

SKINNER, S.J. (1985). TLV and toxicity. *Chem. Engr, Lond.*, **417**, 4

SKINNER, S.J. (1989). Case histories of dust explosions. *Plant/Operations Prog.*, **8**, 211

SKIPPON, S.M. and SHORT, R.T. (1993). Suitability of flame detectors for offshore applications. *Fire Safety J.*, **21**, 1

SKITT, J. (1982). Controlled disposal of special wastes. A waste disposal authority's viewpoint. *Chem. Ind.*, Apr. 17, 262

SKLJANSKAJA, R.M. and RAPPOPORT, T.L. (1935). Experimental studies on chronic poisoning and the development of the offspring of chlorine-poisoned rabbits. *Arch. Exp. Pathol. Pharmacol.*, **177**, 276

SKLJANSKAJA, R.M., KLAUS, K.M. and SSIDOROWA, L.M. (1935). On the effect of chlorine on the female organism. *Arch. Hyg. Bakt.*, **114**, 103

SKOLE, B. and PILARSKI, L. (1978). Listing of chemicals is underway in Sweden. *Chem. Engng*, **85**(Apr. 10), 70L

SKOLE, R. *et al.* (1978). Do dynamite workers face a chronic health threat? *Chem. Engng*, **85**(May 22), 44C

SKOULOUDIS, A.N. (1992). Benchmark exercises on the emergency venting of vessels. *J. Loss Prev. Process Ind.*, **5**, 89

SKOULOUDIS, A.N. and KOTTOWSKI, H.M. (1991). Hydrodynamic aspects of venting vessels containing viscous fluids. *Plant/Operations Prog.*, **10**, 107

SKOULOUDIS, A.N., KOTTOWSKI, H.M., BELL, K.I. and OSTER, R. (1989). Towards the prediction of venting characteristics with the codes SAFIRE and DIERS. *Runaway Reactions*, p. 247

SKOULOUDIS, A.N., BELL, K. and KOTTOWSKI, H.M. (1990). Venting of vessels containing reacting fluids: a parametric study with SAFIRE and DIERS. *J. Loss Prev. Process Ind.*, **3**(1), 13

SKOWRONSKI, M.J. (1980). Susceptibility of floating roofs in wind and rain storms. *API Proc., Refining Dept, 41st Midyear Mtg*, **59**, 72

SKRAMSTAD, E. (1974). NV613. *Analysis of Temperature Fields in the Steady and Unsteady State. Rep. 74–46–M*. Det norske Veritas

SKUM (1986). Application of foam in the petroleum industry. *Fire Int.*, **98**, 55

SKUPNIEWICZ, C.E. and SCHACHER, G.E. (1984). *Measured Plume Dispersion Parameters over Water. Rep. NPS-61–84–012*. Naval Postgraduate School, Monterey, CA

SKUPNIEWICZ, C.E. and SCHACHIER, G.E. (1986). Parameterization of plume dispersion over water. *Atmos. Environ.*, **20**, 1333

SLACK, A.V. and JAMES, G.R. (eds) (1973). *Ammonia*, vols 1–4 (New York: Dekker)

SLACK, C. and WOODHEAD, D.W. (1966). Correlation of ignitability of gases and vapours by a spark at a flange gap. *Proc. IEE*, **113**(2), 297

SLACK, J.H. (1971). Explosions in buildings – the behaviour of reinforced concrete frames. *Concrete*, **5**, 109

SLADE, D.H. (ed.) (1968). *Meteorology and Atomic Energy 1968* (report). Office of Inf. Services, Atom. Energy Comm., Washington, DC

SLATER, D.H. (1974). Industrial safety factors of ozones. *Chemical Process Hazards*, V(suppl.), paper B

SLATER, D.H. (1978a). Vapour clouds. *Chem. Ind.*, May 6, 295

SLATER, D.H. (1978b). *Private communication*

SLATER, D.H. (1979). Siting of hazardous plant. *Health and Safety at Work*, Jan.

SLATER, D.H. (1984). An introduction to hazard assessment – fault trees, event trees, hazard models. Presentation of results in terms of individual and societal risks, evaluation of the results. In Petts, J.I. (1984b), *op. cit.*

SLATER, D.H. and COX, R.A. (1984). Learning from near-miss experience. *1984 European Major Hazards Conf.* (London: Oyez)

SLATER, D.H., RAMSAY, C.G. and COX, R.A. (1981). Hazard analysis: what can it achieve? In Cairns, W.J. and Rogers, P.M. (1981), *op. cit.*, p. 235

SLATER, D.H., CORRAN, E.R. and PITBLADO, R.M. (eds) (1986). *Major Industrial Hazards Project Report*. Warren Centre for Advanced Engng, Univ. of Sydney

SLATER, E.C. (1950). The components of the dihydroenzymase oxidase system. *Biochem. J.*, **46**, 484

SLATTER, P.A. (1987). *Building Expert Systems* (Chichester: Ellis Horwood)

SLATTERY, J.C. (1972). *Momentum Energy and Mass Transfer in Continua* (New York: McGraw-Hill)

SLAWSON, P.R. and CSANADY, G.T. (1971). Effect of atmospheric conditions on plume rise. *J. Fluid Mech.*, **47**, 33

SLEAR, D.G., LONG, R.L. and JONES, J.D. (1983). Repair of TMI OTSG tube failures. *Plant/Operations Prog.*, **2**, 173

SLEDGE, D.J. (1982). Specification and achievement of reliability in Royal Air Force equipment. *Reliab. Engng*, **3**(5), 345

SLEEMAN, D. and BROWN, J.S. (1982). *Intelligent Tutoring Systems* (London: Academic Press)

SLEIGHT, R.B. (1948). The effect of instrument dial shape on legibility. *J. Appl. Psychol.*, **32**, 170

SLINEY, D. and WOLBARSHT, M.L. (1980). *Safety with Lasers and Other Optical Sources* (New York: Plenum Press)

SLOAN, C.E. (1967). Drying systems and equipment. *Chem. Engng*, **74**(Jun. 19), 169

SLOAN, S.A. and NETTLETON, M.A. (1975). A model for the axial decay of a shock wave in a large and abrupt area change. *J. Fluid Mech.*, **71**, 769

SLOAN, S.A. and NETTLETON, M.A. (1978). A model for the decay of a wall shock in a large and abrupt area change. *J. Fluid Mech.*, **88**, 259

SLOANE, B.A. (1992). Maximize your energy savings with proper insulation size. *Hydrocarbon Process.*, **71**(10), 67

SLOANE, T.M. (1990a). Energy requirements for spherical ignitions in methane-air mixtures at different equivalence ratios. *Combust. Sci. Technol.*, **73**, 351

SLOANE, T.M. (1990b). Numerical simulation of electric spark ignition in atmospheric pressure methane–air mixtures. *Combust. Sci. Technol.*, **73**, 367

SLOANE, T.M. (1992). Ignition and flame propagation modeling with an improved propane oxidation mechanism. *Combust. Sci. Technol.*, **83**, 77

SLOANE, T.M. and RONNEY, P.D. (1993). A comparison of ignition phenomena modelled with detailed and simplified kinetics. *Combust. Sci. Technol.*, **88**, 1

SLOANE, T.M. and SCHOENE, A.Y. (1989). Energy requirements for spherical ignitions in atmospheric-pressure methane-air mixtures: a computational study. *Combustion*, **22**, 1669

SLOMAN, A., HARDY, S. and GIBSON, J. (1983). POPLOG: a multilanguage program development environment. *Inf. Technol. Res. Dev.*, **2**, 109

SLOPIANKA, G. and MIEZE, G. (1968). *Failure Rates of Pressure Vessels, Part I: Evaluation of VdTUV Statistics. Rep. IRS-I-34 (FRG)*

SLOT, G.M.J. (1938). Ammonia gas burns: account of six cases. *The Lancet*, **235**, 1356

SLOVIC, P. (1986). Information and educating the public about risk. *Risk Analysis*, **6**(4), 403

SLOVIC, P., FISCHHOFF, B. and LICHTENSTEIN, S. (1976). Cognitive processes and societal risk taking. *Cognition and Social Behaviour* (ed. J.S. Carrol and J.W. Payne) (New York: Wiley), p. 165

SLOVIC, P., FISCHHOFF, B. and LICHTENSTEIN, S. (1980). Facts and fears: understanding perceived risk. In Schwing, R.C. and Albers, W.W. (1980), *op. cit.*

SLOVIC, P., FISCHHOFF, B. and LICHTENSTEIN, S. (1982). Why study risk perception? *Risk Analysis*, **2**, 83

SLOVIC, P., LICHTENSTEIN, S. and FISCHHOFF, B. (1984). Modelling the societal impact of fatal accidents. *Mngmt Sci.*, **30**, 464

SLOVIC, P. *et al.* (1984). Behavioral decision theory perspectives on risk and safety. *Research Perspectives on Decision Making under Uncertainty* (ed. K. Borcherding) (Amsterdam: North-Holland)

SLOVIC, P., FLYNN, J.H. and LAYMAN, M. (1991). Perceived risk, trust and the politics of nuclear waste. *Science*, **254**, 1603

VAN SLYKE, W.J. and GRIFFING, D.W. (1975). *ALLCUTS – a Fast Comprehensive Fault Tree Analysis Code. Rep. ARH-ST-112.* Atlantic Richfield Hanford Co.

SMAIL, G.G., BROOKSBANK, R.J. and THORNTON, W.M. (1931). Electrical resistance of moisture films on glazed surfaces. *J. Instn Elec. Engrs*, **69**, 427

SMALL, D.J. (1972). *The Noise of Gas Regulator Valves. Comm. 883.* (Instn Gas Engrs, London). See also *Instn Gas Engrs J.*, **12**, 272

SMALL, F.H. and SNYDER, G.E. (1974). Controlling in-plant toxic spills. *Loss Prevention*, **8**, p. 24

SMALL, W.M. (1968). Not enough fouling? You're fooling! *Chem. Engng Prog.*, **64**(3), 82

SMART, R.C. (1947). *The Technology of Industrial Fire and Explosion Hazards* (London: Chapman & Hall)

SMEDLEY, J.M., WILLIAMS, A. and BARTLE, K.D. (1992). A mechanism for the formation of soot particles and soot deposits. *Combust. Flame*, **91**, 71

SMEGO, R.A. (1966). Increased profits through materials management. *Chem. Engng*, **73**(Dec. 5), 118

SMELT, A.B. and JAMES, D.P. (1973). *Testing and Evaluation of Pipeline and Control Equipment. Comm. 908.* (Instn Gas Engrs, London). See also *Instn Gas Engrs J.*, **13**, 262

SMET, E.J. (1986). Sequential completion of plant construction to allow early production. *Plant/Operations Prog.*, **5**, 108

SMILLIE, R.J. and AYOUB, M.A. (1976). Accident causation theories: a simulation approach. *J. Occup. Accid.*, **1**(1), 47

SMIT, W. (1971). Electrostatic charge generation during washing of tanks with water sprays – III: Mathematical methods. *Static Electrification*, **3**, p. 178

SMITH (1973). Monitoring and control of combustible gas concentrations in the atmosphere. *Prevention of Occupational Risks*, **2**, p. 201

SMITH, A. (1991). Quality – an integrated approach. In Ince, D. (1991a), *op. cit.*, p. 16

SMITH, A.G., SHAH, M.N., NORTHING, R.A. and SENOUSSAOUI, F. (1985). Synthesis gas compressor/turbine problems. *Ammonia Plant Safety*, **25**, p. 201

SMITH, A.J. (1982). Polychlorinated biphenyls. In Bennett, G.F., Feates and Wilder (1982), *op. cit.*, p. 12–2

SMITH, A.J. (1992). *The Development of a Model to Incorporate Management and Organizational Influences in Quantified Risk Assessment. HSE Contract Res, Rep. 38/1992.* WS Atkins Ltd, Warrington

SMITH, A.J., PENINGTON, J., LYDIATE, C.W. and JACKSON, A.R.G. (1992). Human factors in management and organization. *Hazard Identification and Risk Analysis, Human Factors and Human Reliability in Process Safety*, p. 371

SMITH, A.M. and WATSON, I.A. (1980). Common cause failures – a dilemma in perspective. *Reliab. Engng*, **1**, 127

SMITH, A.P. (1979a). Rid pumps of pipe stress. *Chem. Engng*, **86**(Jul. 16), 121

SMITH, A.P. (1979b). Unexplained wear in large centrifugal pumps. *Chem. Engng*, **86**(Aug. 13), 153

SMITH, B. (1992). Infrared thermography. *Plant Engr*, **36**(1), 31

SMITH, B.D. and SRIVASTAVA, R. (1986). *Thermodynamic Data for Pure Compounds* (Amsterdam: Elsevier)

SMITH, B.P. (1987). Exposure and risk assessment. In Martin, E.J. and Johnson, J.H. (1987), *op. cit.*, p. 37

SMITH, C.A. (1984). Covered mesh protection for industrial plant and steelwork. *Fire Surveyor*, **13**(6), 8

SMITH, C.L. and BRODMANN, M.T. (1976). Process control trends for the future. *Chem. Engng*, **83**(Jun. 21), 129

SMITH, C.L. (1979). Fundamentals of control theory. *Chem. Engng*, **86**(Oct. 15), 11

SMITH, C.O. (1976). *Introduction to Reliability Design* (New York: McGraw-Hill)

SMITH, C.R. (1981). *A User Guide to Dust and Fume Control* (Rugby: Instn. Chem. Engrs)

SMITH, C.R. (1985). Seismic design approach for the Sizewell B nuclear power plant. *Earthquake Engineering in Britain* (London: Instn Civil Engrs), p. 305

SMITH, C.S. (1969). *Quality and Reliability: An Integrated Approach* (London: Pitman)

SMITH, C.W. (1977). Toxic substances control is here. *Hydrocarbon Process.*, **56**(1), 213

SMITH, D. (1970). Pneumatic transport and its hazards. *Loss Prevention*, **4**, p. 58. See also *Chem. Engng Prog.*, 1970, **66**(9), 41

SMITH, D. (1986). *Welding Skills and Technology* (New York: McGraw-Hill)

SMITH, D. and AGNEW, J.T. (1957). The effect of pressure on the laminar burning velocity of methane−oxygen−nitrogen mixtures. *Combustion*, **6**, p. 83

SMITH, D.A. (1992). Measurements of flame length and flame angle in an inclined trench. *Fire Safety J.*, **18**, 231

SMITH, D.A. and COX, G. (1992). Major chemical species in buoyant turbulent diffusion flames. *Combust. Flame*, **91**, 226

SMITH, D.J. (1972). *Reliability Engineering* (London: Macmillan)

SMITH, D.J. (1981). *Reliability and Maintainability in Perspective* (London: Macmillan)

SMITH, D.J. (1985a). *Reliability and Maintainability in Perspective*, 2nd and 3rd ed. (London: Macmillan)

SMITH, D.J. (1985b). Reliability of software-based fire and gas detection control systems. *Reliab. Engng*, **12**(3), 161

SMITH, D.J. (1991). *Reliability, Maintainability, and Risk* (Oxford: Butterworth-Heinemann)

SMITH, D.J. and BABB, A.H. (1973). *Maintainability Engineering* (London: Pitman)

SMITH, D.J. and EDGE, J.S. (1991). *Quality Procedures for Hardware and Software* (London: Elsevier Applied Science)

SMITH, D.J. and WOOD, K.B. (1989). *Engineering Quality Software* (New York: Elsevier)

SMITH, D.R. (1976). Clean water cuts boiler downtime. *Processing*, Aug., **15**

SMITH, D.T. (1965a). Handling hazardous chemicals. *Chem. Engng Prog.*, **61**(2), 51

SMITH, D.T. (1965b). Proper maintenance prevents accidents. In Fawcett, H.H. and Wood, W.S. (1965), *op. cit.*, p. 472

SMITH, D.T. (1965c). Services and facilities. In Fawcett, H.H. and Wood, W.S. (1965), *op. cit.*, p. 63

SMITH, D.T. (1967). The Chambers works emergency control plan. *Loss Prevention*, **1**, p. 31

SMITH, D.T. (1982a). Maintenance of chemical equipment. Fawcett, H.H. and Wood, W.S. (1982), *op. cit.*, p. 231

SMITH, D.T. (1982b). Services and facilities. In Fawcett, H.H. and Wood, W.S. (1982), *op. cit.*, p. 211

SMITH, D.W. (1982). Runaway reactions and thermal explosion. *Chem. Engng*, Dec. 13, 79

SMITH, D.W. (1984). Assessing the hazards of runaway reactions. *Chem. Engng*, **91**(May 14), 54

SMITH, D.W., TAYLOR, M.C., YOUNG, R. and STEPHENS, T. (1980). Accelerating rate calorimetry. *Am. Lab.*, Jun., 51

SMITH, E. and HARRIS, M.J. (1990). Managing the maintenance of high reliability, major hazard plant. *Seventh Int. Symp. on Reliability and Maintainability, Brest*

SMITH, E.H. (1993). *Mechanical Engineers Reference Book*, 12th ed. (Oxford: Butterworth-Heinemann)

SMITH, E.L. (1966). Texas Eastern plans LNG concrete tank. *Hydrocarbon Process.*, **45**(8), 145

SMITH, F.A. (1973). Water transportation of petroleum and petroleum products in the future. *Conf. on Bulk Transportation of Hazardous Materials by Water in the Future* (Washington, DC: Nat. Acad. Sci., Nat. Acad. Engng, Nat. Res. Coun.), p. 41

SMITH, F.B. (1956). The diffusion from a continuous elevated point source into a turbulent atmosphere. *J. Fluid Mech.*, **2**, 49

SMITH, F.B. (1962). The problem of deposition in atmospheric diffusion of particulate matter. *J. Atmos. Sci.*, **19**, 429

SMITH, F.B. (1973). *A Scheme for Estimating the Vertical and Horizontal Dispersion of a Neutral Plume from a Source Near Ground Level. Turbulence and Diffusion Note 40*. Boundary Layer Research Branch, Meteorological Office, London

SMITH, F.B. (1979). The relation between the Pasquill stability P and Kazanski-Monin stability μ (in neutral and unstable conditions). *Atmos. Environ.*, **13**, 879

SMITH, F.B. and HAY, S. (1961). The expansion of clusters of particles in the atmosphere. *Quart. J. R. Met. Soc.*, **87**(371), 82

SMITH, F.G.W. (1974). *CRC Handbook of Marine Science*, sec. A, *Oceanography*, vol. 1, *Physical* (Boca Raton, FL: CRC Press)

SMITH, G. and PATEL, A. (1987). Step by step through the pinch. *Chem. Engr, Lond.*, **442**, 26

SMITH, G.N. (1959). Thermal stress analysis of refinery piping system. *Chem. Engng Prog.*, **55**(2), 122

SMITH, H. and GRACE, D.W. (1961). Application of truncated distributions in process start-ups and inventory control. *Technometrics*, **3**, 429

SMITH, H.M. (1975). *Principles of Holography*, 2nd ed. (New York: Wiley)

SMITH, J. (1978a). EEC to pull plug on ground-water polluters. *Chem. Engng*, **85**(Apr. 24), 36I

SMITH, J. (1978b). Europe's toxicity rule does not copy TSCA. *Chem. Engng*, **85**(Dec. 4), 76

SMITH, J. (1978c). Brussels brews tough new rules on chemicals. *Chem. Engng*, **85**(Dec. 4), 60C

SMITH, J. (1994). Ship classification rules and international conventions – their relevance to design. In Rawson, K. (1994), *op. cit.*, p. 83

SMITH, J. *et al.* (1979). Toxics solidification: under fire yet picking up. *Chem. Engng*, **86**(Jul. 2), 28E

SMITH, J.B. (1949). *Explosion Pressures in Industrial Piping System*. Factory Mutual Insurance Ass., Hartford, CT

SMITH, J.B. (1987). System approach for plant reliability. *Chem. Engng Prog.*, **83**(4), 47

SMITH, J.B. and PAULSON, G.G. (1989). Synthesis gas compressor coupling corrosion fatigue failures. *Ammonia Plant Safety*, **29**, p. 65

SMITH, J.D. (1983). *Gears and Their Vibration* (London: Macmillan)

SMITH, J.F. (1968). *Literature of Chemical Technology* (New York: Am. Chem. Soc.)

SMITH, J.H.F. (1956). Explosions and poisons: a comparison of American and British precautions. *Trans Instn Chem. Engrs*, **34**, 127

SMITH, J.J. (1976). Integrating safe marine transportation systems for handling hazardous materials in bulk. *Control of Hazardous Material Spills*, **3**, p. 24

SMITH, J.M. (1956). *Chemical Engineering Kinetics* (New York: McGraw-Hill)

SMITH, J.M. and VAN DOORN, M. (1981). Water sprays in confined applications – mixing and release from enclosed spaces. In Greenwood, D.V. (1981), *op. cit.*, paper 2

SMITH, J.T. (1978). Fire service evaluation and field tests of EPA developed hazardous spills control devices. *Control of Hazardous Material Spills*, **4**, p. 444

SMITH, K. (1992). *Environmental Hazards* (London: Routledge)

SMITH, K.A. and GERMELES, A.E. (1974). LNG tank stratification consequent to filling procedures. *LNG Conf. 4*

SMITH, K.A., LEWIS, J.P., RANDALL, G.A. and MELDON, J.H. (1975). Mixing and rollover in LNG storage tanks. *Advances in Cryogenic Engineering*, **20**, 124

SMITH, L.R. (1972). Submerged LNG pumps – notes on recent significant projects. *LNG Conf. 3*

SMITH, M.A. (1985). *Contaminated Land: Reclamation and Treatment* (New York: Plenum Press)

SMITH, M.E. (1951). The Forecasting of Micrometeorological Variables. *Met. Monogr.*, **4**, 50

SMITH, M.E. (1956). The variation of effluent concentrations from an elevated point source. *Arch. Ind. Health*, **14**, 56

SMITH, M.E. (1979). Solid-waste disposal: deepwell injection. *Chem. Engng*, **86**(Apr. 9), 107

SMITH, M.E. (1986). Expectations and experience in diffusion model validation. In de Wispelaere, C., Schiermeier, F.A. and Gillani, N.V. (1986), *op. cit.*, p. 403

SMITH, M.E., CARLSON, J.H., MARTIN, J.R. and LAWRENCE, F.J. (1983). Atmospheric modeling for emergencies. *Plant/Operations Prog.*, **2**, 61

SMITH, M.T.E., BIRCH, A.D., BROWN, D.R. and FAIRWEATHER, M. (1986). Studies of ignition and flame propagation in turbulent jets of natural gas, propane and a gas with high hydrogen content. *Combustion*, **21**, p. 1403

SMITH, P.E. (1983). Welded metal bellows seals: applications and problems. *Hydrocarbon Process.*, **62**(1), 87

SMITH, P.J., GRIFFIN, W.C., ROCKWELL, T.H. and THOMAS, M. (1986). Modeling fault diagnosis as the activation and use of a frame system. *Hum. Factors*, **28**, 703

SMITH, P.R. and VAN LAAN, T.J. (1987). *Piping and Pipe Support Systems* (New York: McGraw-Hill)

SMITH, R. (1986). Condition monitoring. *Plant Engr*, **30**(6), 24

SMITH, R. (1992). *Catastrophes and Disasters* (Edinburgh: Chambers)

SMITH, R. and HARTLEY, G. (1982). A view from the nuclear fuel reprocessing industry. *Chem. Ind.*, Nov. 20, 894

SMITH, R. and LINNHOFF, B. (1988). The design of separators in the context of overall processes. *Chem. Engng Res. Des.*, **66**, 195

SMITH, R. and PETALA, E. (1991—). Waste minimization in the process industries. *Chem. Engr, Lond.*, **506**, 24; **509/510**, 17; 1992, **513**, 24; **517**, 21; **523**, 32

SMITH, R.A. (1987). *Vaporisers: Selection, Design and Operation* (London: Longman)

SMITH, R.A. and MILLER, G.R. (1988). *Chemical Exposure Index. Guidelines for Safe Storage and Handling of High Toxic Hazard Materials* (New York: Am. Inst. Chem. Engrs), Appendix B

SMITH, R.C. and SMITH, P. (1968). *Mechanics* (London: Macmillan)

SMITH, R.E. (1985). Pollution legislation and control. *Plant Engr*, **29**(4), 15

SMITH, R.J. (1982). Square one — instrument and control checklist. *Measmnt Control*, **15**, 249

SMITH, R.K. (1967). Radiation effects on large fire plumes. *Combustion*, **11**, p. 507

SMITH, R.L. (1986). Extreme value statistics and reliability applications. *Reliab. Engng*, **15**(3), 161

SMITH, R.P. (1981). Boredom: a review. *Hum. Factors*, **23**, 329

SMITH, R.P. and GOSSELIN, R.E. (1979). Hydrogen sulphide poisoning. *J. Occup. Med.*, **21**, 93

SMITH, R.S. (1972). Control stack gas pollution. *Hydrocarbon Process.*, **51**(9), 223

SMITH, S.F. (1969). Running nips. In Handley, W. (1969), *op. cit.*, p. 204

SMITH, S.F. (1977). Running nips. In Handley, W. (1977), *op. cit.*, p. 186

SMITH, S.F. (1984). Adaptive learning systems. In Forsyth, R. (1984), *op. cit.*, p. 168

SMITH, T. (1987). Selecting timber for corrosive environments. *Process Engng*, Jul., 69

SMITH, T.A. (1985). An analysis of a 100 te storage vessel. *The Assessment and Control of Major Hazards* (Rugby: Instn Chem. Engrs), p. 107

SMITH, T.A. and WARWICK, R.G. (1974). A survey of defects in pressure vessels built to high standards of construction and its relevance to nuclear primary circuits. *Int. J. Pres. Ves. Piping*, **2**, 283

SMITH, T.A. and WARWICK, R.G. (1978). Survey of defects in pressure vessels built to high standards of construction. *Am. Soc., Mech. Engrs, New York, Winter Ann. Mtg*, paper PVP-PB-032

SMITH, T.B., KAUPER, E.K., BERMAN, S. and VUKOVICH, F. (1964). *Micrometeorological investigation of Naval Missile Facility, Point Arguello, California. Rep. MR164—FR167*, vols I–II, Met. Res. Inc., Altadena, CA

SMITH, W. (1975). The Court of Inquiry. *Instn Chem. Engrs, Symp. on Technical Lessons of Flixborough* (London)

SMITH, W.E. and SMITH, A.M. (1975). *Minimata* (New York: Holt, Rinehart & Winston)

SMITH, W.K. and KING, J.B. (1970). Surface temperature of materials during radiant heating to ignition. *J. Fire Flammability*, **1**, 272

SMITH, W.N. (1968). Cost of being out? Inventory is key. *Hydrocarbon Process.*, **47**(7), 155

SMITH, W.N. and SANTANGELO, J.G. (eds) (1980). *Hydrogen: Production and Marketing* (Washington, DC: Am. Chem. Soc.)

SMITH, W.Q. and LIEN, D.A. (1968). *Automated Fault Tree Program. Doc. AS-2740*. The Boeing Company, Seattle, WA

SMITH-HANSEN, L. (1988). Risk analysis of a warehouse for the mixing, repackaging and distribution of organic chemicals. *J. Loss Prev. Process Ind.*, **1**(4), 233

SMITH-HANSEN, L. (1994). Pesticide warehouse fire consequences: experiences from a Danish risk analysis study. *J. Loss Prev. Process Ind.*, **7**, 157

SMITH-HANSEN, L. and JORGENSEN, K.H. (1993). Characterisation of fire products from organophosphorus pesticides using the DIN 53436 method. *J. Loss Prev. Process Ind.*, **6**, 227

SMITHELLS, C.J. (1973). *Metals Reference Book* (London: Butterworths)

SMITHIES, J.N., BURRY, P.E. and SPEARPOINT, M.J. (1991). Background signals from fire detectors: measurement, analysis, application. *Fire Safety J.*, **17**, 445

SMITHSON, M. (1989). *Ignorance and Uncertainty: Emerging Paradigms* (Berlin: Springer)

SMOLEN, A.M. and MASE, J.R. (1982). ASME pressure-vessel code: which division to choose? *Chem. Engng*, **89**(Jan. 11), 133

VON SMOLUCHOWKSI, M. (1921). *Theorie der Kataphoretischen Strome*, pt 5, **22**, 385

SMOOT, L.D., HECKER, W.C. and WILLIAMS, G.A. (1976). Prediction of propagating methane–air flames. *Combust. Flame*, **26**, 323

SMULDERS, A.J.M. (1980). The role of the ambulance in disaster. In De Boer, J. and Baillie, T.W. (1980), *op. cit.*, p. 59

SMYLLIE, R.J. (1989). *Piper Alpha Public Inquiry. Transcript, Day 94*

SMYTH, D. (1978). *Alternatives to Animal Experiments* (London: Scholar Press)

SMYTH, M.I. (1944). Injuries from flying glass. *Br. Med. J.*, Dec. 16, 799; also *Br. Med. J.*, Nov. 23, 1940, 719

SMYTH, H.F. and CARPENTER, C.P. (1949). Further experience with the range finding test in the industrial toxicology laboratory. *J. Ind. Hyg. Toxicol.*, **30**, 63

SMYTHE, W.R. (1939). *Static and Dynamic Electricity*; 1950, 2nd ed.; 1967, 3rd ed. (New York: McGraw-Hill); 1989, 3rd rev. ed. (New York: Hemisphere)

SNAITH, E.R. (1982). Reliability evaluation of large power systems. *Reliability of Large Scale Systems, IEE Colloquium Dig. 1982/41* (London: IEE), p. 3/1

SNEDDON, I.N. (1946). The distribution of stress in the neighbourhood of a crack in an elastic solid. *Proc. R. Soc., Ser. A*, **187**, 229

SNEDDON, I.N. (1973). Integral transform methods. In Sih, G.C. (1973), *op. cit.*

SNEDDON, I.N. and LOWENGRUB, M. (1969). *Crack Problems in the Classical Theory of Elasticity* (New York: Wiley)

SNEE, T.J. (1980). A method of measuring the concentration of flammable gases. *Chemical Process Hazards*, **VII**, p. 89

SNEE, T.J. (1987). Incident investigation and hazard evaluation using differential scanning calorimetry and accelerating rate calorimetry. *J. Occup. Accid.*, **8**(4), 261

SNEE, T.J. (1989). The influence of a constant power heat source on the critical conditions for thermal explosion. *Runaway Reactions*, p. 112

SNEE, T.J. and GRIFFITHS, J.F. (1989). Criteria for spontaneous ignition in exothermic, autocatalytic reactions: chain branching and self-heating in the oxidation of cyclohexane in closed vessels. *Combust. Flame*, **75**, 381

SNEE, T.J. and HARE, J.A. (1992). Development and application of a pilot scale facility for studying runaway exothermic reactions. *J. Loss Prev. Process Ind.*, **5**, 46

SNEE, T.J. and HARE, J.A. (1994). Laboratory scale investigation of runaway reaction venting. *Hazards*, **XII**, p. 541

SNEE, T.J., BASSANI, C. and LIGTHART, J.A.M. (1993). Determination of the thermokinetic parameters of an exothermic reaction using isothermal, adiabatic and temperature-programmed calorimetry in conjunction with spectrophotometry. *J. Loss Prev. Process Ind.*, **6**, 87

SNELLINK, G. (1977). Hazard assessment of LNG supply and storage. *LNG Conf. 5*

SNOOK, R. (1977). The accident flying squad – resuscitation and release. In Easton, K.D. (1977), *op. cit.*, p. 218

SNOOK, S.H. and CIRIELLO, V.M. (1991). The design of manual handling tasks: revised tables of maximum acceptable weights and forces. *Ergonomics*, **34**, 1197

SNOW, D.A. (ed.) (1991). *Plant Engineer's Reference Book* (Oxford: Butterworth-Heinemann)

SNOW, D.W. (1981). FRP dust fire incident. *Loss Prevention*, **14**, p. 104

SNOW, K. (1978). Shipping hazardous material. *Chem. Engng*, **85**(Nov. 6), 102

SNOW, M. (1979). Maintaining pressure vessels. *Chem. Engng*, **86**(Jan. 1), 109

SNYDER, I.G. (1974). Implementing a good safety program. *Chem. Engng*, **81**(May 27), 112

SNYDER, J.S. (1965). Testing reactions and materials for safety. In Fawcett, H.H. and Wood, W.S. (1965), *op. cit.*, p. 296

SNYDER, J.S. (1982). Testing reactions and materials for safety. In Fawcett, H.H. and Wood, W.S. (1982), *op. cit.*, p. 285

SNYDER, P.G. (1988). Brittle fracture of a high pressure heat exchanger. *Plant/Operations Prog.*, **7**, 148

SNYDER, P.G. (1996). Is your plant inherently safer. *Hydrocarb. Process.* (July), 77–82

SNYDER, R. (1984). Basic concepts of the dose-response relationship. In Rodricks, R.V. and Tardiff, R.G. (1984a), *op. cit.*, p. 37

SNYDER, W.H. (1981). *Guidelines for Fluid Modeling of Atmospheric Diffusion. Rep. EPA-600/8–81–009*. Environ. Prot. Agency, Res. Triangle Park, NC

SNYDER, W.H. (1990). Fluid modeling applied to atmospheric diffusion in complex terrain. *Atmos. Environ.*, **24A**, 2071

SNYDER, W.H. and BRITTER, R.E. (1987). A wind tunnel study of the flow structure and dispersion from sources upwind of three-dimensional hills. *Atmos. Environ.*, **21**, 735

SNYDER, W.H. and HUNT, J.C.R. (1984). Turbulent diffusion from a point source in stratified and neutral flows around a three-dimensional hill – II. *Atmos. Environ.*, **18**, 1969

SOBEL, R. (1979). How industry can prepare for RCRA. *Chem. Engng*, **86**(Jan. 29), 82

SOBER, R.F., and PAUL, D. (1992). Less subjective odor assessment. *Chem. Engng*, **99**(9), 130

SOBOL, I.M. (1974). *The Monte Carlo Method* (Chicago, IL: Univ. of Chicago Press)

SOBONYA, R. (1977). Fatal anhydrous ammonia inhalation. *Hum. Pathol.*, **8**, 293

SOCHA, G.E. (1974). Industrial hygiene considerations in process design. *Loss Prevention*, **8**, p. 29

SOCIAL AUDIT (1974). *The Alkali Inspectorate. Rep. 4*

SOCIETY OF CHEMICAL INDUSTRY (1957). *Disposal of Industrial Waste Materials* (London)

SOCIETY OF CHEMICAL INDUSTRY (1962). *Physics and Chemistry of High Pressures* (London)

SOCIETY OF CHEMICAL INDUSTRY (1967). *Biological Treatment of Industrial Waste* (London)

SOCIETY OF CHEMICAL INDUSTRY (1968). *Occupational Health Services* (London) (reprints)

SOCIETY OF CHEMICAL INDUSTRY (1974). *Report of the Occupational Safety Conference* (London)

SOCIETY OF CHEMICAL INDUSTRY (1978). Evidence to the Royal Commission on Environmental Pollution on the impact of agricultural chemicals on the environment. *Chem. Ind.*, May 30, 322

SOCIETY OF CHEMICAL INDUSTRY (1979). *Corrosion of Steel Reinforcements in Concrete* (London)

SOCIETY OF CHEMICAL INDUSTRY (1980). *Reclamation of Contaminated Land* (London)

SOCIETY OF CHEMICAL INDUSTRY (1981a). *Marine Corrosion of Offshore Structures* (London)

SOCIETY OF CHEMICAL INDUSTRY (1981b). *Sulphur Emissions and the Environment* (London)

SOCIETY OF INDUSTRIAL EMERGENCY SERVICES OFFICERS (1986) *Guide to Emergency Planning* (Borehamwood: Paramount)

SOCIETY OF INTERNATIONAL GAS TANKER AND TERMINAL OPERATORS – see Appendices 27 and 28

SOCIETY OF LABOUR LAWYERS (1970). *Occupational Accidents and the Law* (London: Fabian Soc.)

SOCIETY OF THE PLASTICS INDUSTRY (1973). *Plastics Industry Safety Handbook* (Boston, MA: Cahners)

SOCIETY OF THE PLASTICS INDUSTRY (1987). *Recommended Practice for Acoustic Emission Testing of Fiberglass Tanks/ Vessels*

SOCIETY FOR UNDERWATER TECHNOLOGY (1985). *Design and Installation of Subsea Systems* (London: Graham & Trotman)

SOCIO-ECONOMIC SYSTEMS (1977). *Environmental Impact Report for the Proposal Oxnard LNG Facilities. Draft EIR*, Appendix B, Safety

SODEN, J.D. (1985). Basics of fire protection design. *Hydrocarbon Process.*, **64**(5), 157; **64**(6), 101

SODEN, J.E. and JOHNSON, J.C. (1978). Burial and other high-potential response techniques for spills of hazardous chemicals that sink. *Control of Hazardous Material Spills*, **4**, p. 202

SODERMAN, J.V. (ed.) (1982). *CRC Handbook of Identified Carcinogens and Noncarcinogens: Carcinogenicity– Mutagenicity Database*, vols 1–2 (Boca Raton, FL: CRC Press)

SOESAN, J.M. and FFOOKS, R.C. (1973). LNG carriers – the new liners. In Hepple, P. (1973), *op. cit.*, p. 99

DE SOETE, G.G. (1971). The influence of isotropic turbulence on the critical ignition energy. *Combustion*, **13**, p. 735

SOFYANOS, T. (1981). *Causes and Consequences of Fires and Explosions on Offshore Platforms: Statistical Survey of Gulf of Mexico Data. Rep. 81–0057.* Det norske Veritas, Oslo

SOHNKE, F. (1971). Safety measures and plans for emergency action in case of accidents in the Federal German waterways and coastal waters. *Transport of Hazardous Cargoes*, **2** (paper 6)

SOHRE, J.S. (1970). You can predict reliability of turbomachinery. *Hydrocarbon Process.*, **49**(1), 100

SOHRE, J.S. (1977). Turbo-machinery problems: causes and cures. *Hydrocarbon Process.*, **56**(12), 77

SOHRE, J.S. (1979). Are magnetic currents destroying your machinery? *Hydrocarbon Process.*, **58**(4), 207

SOHRE, J.S. (1981). Electromagnetic shaft currents in turbomachinery: an update, Pt 1. *Ammonia Plant Safety*, **23**, 185

SOJKA, S.A. (2002). The spinning tube-in-tube reactor system. *6th Ann. Green Chem. and Eng. Conf.*, Jun. 24–27, 2002, Washington, DC

SOKOLIK, A.S. (1960). *Self-Ignition: Flame and Detonation in Gases* (Moscow). (English translation: Israel Program for Scientific Translations, 1963)

SOKOLOV, A. (1989). Diesel engines in hazardous areas on offshore installations (paper). *Mtg of Joint Offshore Group of Inst. Marine Engrs and Roy. Instn Nav. Architects*

DE SOLA, J.H. (1985). European gas terminals and inland distribution systems. *Gastech 84*, p. 99

SOLBERG, D.M. (1982a). Gas explosion research related to safety of ships and offshore platforms. In Lee, J.H.S. and Guirao, C.M. (1982), *op. cit.*, p. 787

SOLBERG, D.M. (1982b). Industrial gas explosion problems. *Plant/Operations Prog.*, **1**, 243

SOLBERG, D.M. and BORGNES, O. (1983). Thermal response of process equipment to hydrocarbon fires. *Plant/Operations Prog.*, **2**, 50

SOLBERG, D.M. and SKRAMSTAD, E. (1982). Assessment of consequences from accidental release of liquefied gases. *Gastech 81*, p. 299

SOLBERG, D.M., PAPPAS, J. and SKRAMSTAD, E. (1980). Experimental investigations on flame acceleration and pressure rise phenomena in large scale vented gas explosions. *Loss Prevention and Safety Promotion 3*, vol. 3, p. 1295

SOLBERG, D.M., PAPPAS, J.A. and SKRAMSTAD, E. (1981). Observations of flame instabilities in large scale vented gas explosions. *Combustion*, **18**, p. 1607

SOLIMAN, R. (1984). Two-phase pressure drop computed. *Hydrocarbon Process.*, **63**(4), 155

SOLOMON, C.H. (1980). The fire event chain. *Loss Prevention and Safety Promotion 3*, vol. 4, p. 1500

SOLOMON, K.A. (1976). *State by State Chemical Risk Assessment Survey. Rep. UCLA-ENG-7732.* Univ. of California, Los Angeles, CA

SOLOMON, K.A. (1987). Comparing risk management practices at the local levels of government with those at the state and federal levels. *J. Haz. Materials*, **15**(1/2), 265

SOLOMON, K.A. and ALESCH, K.A. (1989). The index of harm: a measure for comparing occupational risk across industries. *J. Occup. Accid.*, **11**(1), 19

SOLOMON, K.A. and KASTENBERG, W.E. (1988). Estimating emergency planning zones for the Shoreham nuclear reactor. A review of four safety analyses. *J. Haz. Materials*, **18**(3), 269

SOLOMON, K.A., OKRENT, D. and KASTENBERG, W.E. (1975a). Pressure vessel integrity and weld inspection procedure. *Nucl. Engng Des.*, **35**(1), 87

SOLOMON, K.A., OKRENT, D. and KASTENBERG, W.E. (1975b). *Pressure Vessel Integrity and Weld Inspection Procedure. Rep. UCLA-ENG-7496.* Univ. of California, Los Angeles, CA

SOLOMON, K.A., RUBIN, M. and OKRENT, D. (1976). *On Risks from the Storage of Hazardous Chemicals. Rep. UCLA-ENG-76125.* Univ. of California, Los Angeles, CA

SOLOUKHIN, R.I. (1966). *Shock Waves and Detonations in Gases* (Baltimore, MD: Mono)

SOLOUKHIN, R.I. (1969). Nonstationary phenomena in gaseous detonation. *Combustion*, **12**, p. 799

SOLTER, R.L., FIKE, L.L. and HANSEN, F.A. (1963a). How to size rupture discs. *Hydrocarbon Process.*, **42**(5), 159

SOLTER, R.L., FIKE, L.L. and HANSEN, F.A. (1963b). Selection and sizing of rupture units for the petroleum industry. *API Proc. Div. of Refining*, **43**, 448

SOMEAH, K. (1992). On-line tube cleaning: the basics. *Chem. Engng Prog.*, **88**(7), 39

SOMERVILLE (1981). Emergency guides slammed. *Haz. Cargo Bull.*, Jun., 14

SOMERVILLE, I. (1989). *Software Engineering*, 3rd ed. (Reading, MA: Addison-Wesley)

SOMERVILLE, R.L. (1990). Reduce risks of handling liquified toxic gas. *Chem. Engng Prog.*, **86**(12), 64

SOMERVILLE, R.L. (1993). Control of liquefied toxic gas releases. *Prev. and Control of Accidental Releases of Hazard. Gases* (ed. V. M. Fthenakis) 438–47 (New York: Van Nostrand Reinhold)

SOMMER, E.C. (1965). How Esso stores refrigerated LPG. *Hydrocarbon Process.*, **44**(5), 195

SOMMER, M., COOKE, E.F., GOEHRING, J.E. and HAYDEL, J.J. (1971). Process control in a petrochemical complex. *Chem. Engng Prog.*, **67**(10), 54

SOMMERFELD, J.T. (1990). Drainage of conical tanks with piping. *Chem. Engng. Educ.*, **24**, 145

SOMMERFELD, J.T. and STALLYBRASS, M.P. (1993). Elliptic integral solutions for fluid discharge rates from punctured horizontal cylindrical vessels. *J. Loss Prev. Process Ind.*, **6**, 11

SOMMERS, H.A. (1965). The chlor-alkali industry. *Chem. Engng Prog.*, **61**(3), 94

SOMMERS, W.P. (1961). Gaseous detonation wave interactions with non-rigid boundaries. *ARS J.*, **31**, 1780

SONG, C.C. (1986). Low-cost start-up of a large ammonia plant. *Ammonia Plant Safety*, **26**, p. 134

SONG, C.C. and UNRUH, W. (1974). Solving a steam generating tube failure. *Ammonia Plant Safety*, **16**, p. 28

SONG, L., BLACK, J.A. and DUNNE, M.C. (1993). Probabilistic analysis in traffic accident risk assessment. In Melchers, R.E. and Stewart, M.G. (1993), *op. cit.*, p. 47

SONJU, O.K. and HUSTAD, J. (1984). An experimental study of turbulent jet diffusion flames. In Bowen, J.R. *et al.*, *op. cit.*, p. 320

SONLEY, J.M. (1982). Recent studies relating to the development of poison resistant catalytic flammable gas sensors for use on LNG/LPG offshore installations. *Gastech 81*, p. 343

SONNENFELDT, R.W. (1964). Reliability and direct digital control. *Instrumentation in the Chemical and Petroleum Industries 5* (Pittsburgh, PA: Instrum. Soc. Am.), p. 15

SONTAG, J.M. (ed.) (1981). *Carcinogens in Industry and the Environment* (New York: Dekker)

SONTI, R.S. (1984). Practical design and operation of vapor-depressuring systems. *Chem. Engng*, **91**(Jan. 23), 66

SOON HEUNG CHANG, MYUNG KI KIM and JOO YOUNG PARK (1985). Analytical and numerical unavailability analysis of redundant systems with dependent human error. *Reliab. Engng*, 10(1), 49

SOON H. CHANG, JOO Y. PARK and MYUNG K. KIM (1985). The Monte-Carlo method without sorting for uncertainty propagation analysis in PRA. *Reliab. Engng*, 10, 233

SORELL, G. (1965). Catastrophic oxidation. *Ammonia Plant Safety*, 7, p. 7

SORELL, G. (1968). Controlling corrosion by process design. *Chem. Engng*, 75(Jul. 29), 164

SORELL, G. (1970). Designing for corrosion resistance. Process control. *Chem. Engng*, 77(Oct. 12), 83

SORELL, G. and ZEIS, L.A. (1967). The brittle fracture hazard. *Ammonia Plant Safety*, 9, p. 18

SORENSEN, F. and PETERSEN, H.J.S. (1992). Substitution of organic solvents. *Staub − Reinhaltund Der Luft* 52, 113−18

SORENSEN, J.H. (1987). Evacuations due to off-site releases from chemical accidents: experience from 1980 to 1984. *J. Haz. Materials*, 14(2), 247

SORENSEN, J.H., CARNES, S.A. and ROGERS, G.O. (1992). An approach for deriving emergency planning zones for chemical munitions emergencies. *J. Haz. Materials*, 30, 223

SORENSON, K.G. and BESUNER, P.M. (1977). A workable approach to extending the life of expensive life-limited components. In Gangadharan, A.C. and Brown, S.J. (1977), *op. cit.*, p. 77

SORENSON, S.C., SAVAGE, L.D. and STREHLOW, R.A. (1975). Flammability limits − a new technique. *Combust. Flame*, 24, 347; 25, 284

SORKIN, R.D., KANTOWITZ, B.H. and KANTOWITZ, S.C. (1988). Likelihood alarm displays. *Hum. Factors*, 30, 445

SOROTZKIN, J. and LOCK, J.A. (1983). Simulation prepares operators for steam upsets in refining/petrochemical complex. *Oil Gas. J.*, 81(Jan. 24), 55

SÖTEBIER, D.L. and RALL, W. (1986). Problems with leakages at large flanges in the BASF new ammonia plant. *Plant/Operations Prog.*, 5, 86

SOULE, L.M. (1969−). Basic concepts of industrial process control. *Chem. Engng*, 1969, 76(Sept. 22), 133; (Oct. 20), 115; (Dec. 1), 101; 1970, 77(Jan. 12), 103; (Jan. 26), 120; (Feb. 23), 113; (Mar. 9), 142

SOUNDARAJAN, R. (1990). An overview of present day immobilization technologies. *J. Haz. Materials*, 24(2/3), 199

SOUND RESEARCH LABORATORIES (1991). *Noise Control in Industry*, 3rd ed. (Sudbury)

SOUSA, A.C.M. and VENART, J.E.S. (1983). Thermal modelling of LPG rail tank cars exposed to fire environment. *Fourth Int. Conf. on Mathematical Modelling*, Zurich

SOUTHWELL, K.B. (1977). Grinding. In Handley, W. (1977), *op. cit.*, p. 147

SOUTTER, J. (1993). Private communication

SOUTTER, J.K. and CHUNG, P.W.H. (1993). Desirable plans and safe plans. *Twelfth UK Planning/Scheduling SIG*, Cambridge

SOUTTER, J. and CHUNG, P.W.H. (1995a). Partial order planning with goals of prevention. *Int. Joint Conf. on Artificial Intelligence*, Canada

SOUTTER, J. and CHUNG, P.W.H. (1995b). Planning architectures for operating procedure synthesis. *Int. Conf. on Industrial and Engineering Applications of Artificial Intelligence and Expert System*, Australia

SOWA, J.F. (1984). *Conceptual Structures: Information Processing in Mind and Machine* (New York: Addison-Wesley)

SOWBY, F.D. (1964). *In Symp. on Transport of Radioactive Materials* (ed. F.J. Neary) (London: HM Stationery Office)

SOWBY, F.D. (1965). Radiation and other risks. *Health Phys.*, 11(9), 879

SOYLEMEZ, S. and SEIDER, W.D. (1973). A new technique for precedence-ordering chemical process equation sets. *AIChE J.*, 19, 934

SOZZI, G.L. and SUTHERLAND, W.A. (1975). Critical flow of saturated and subcooled water at high pressure. *Am. Soc. Mech. Engrs, Symp. on Non-Equilibrium Two-Phase Flow* (New York), p. 19

SPAAR, R. and SUTER, G. (1992). A simplified hazard analysis scheme for use in process development. *Loss Prevention and Safety Promotion 7*, vol. 2, 55−1

SPADONI, G., PASQUALI, G., MACCHI, G. and RICCHIUTI, A. (1992). 'Source term' models to evaluate the dispersion of heavy gases. *Loss Prevention and Safety Promotion 7*, vol. 2, 94−1

SPAIN, I.L. and PAAUWE, J. (eds) (1977). *High Pressure Technology*, vols 1−2 (New York: Dekker)

SPAKOWSKI, A.E. (1952). *Pressure Limit of Flame Propagation of Pure Hydrocarbon−Air Mixtures at Reduced Pressures. Res. Memo., E52H15*. Nat. Advis. Comm. Aeronaut.

SPALDING, D.B. (1955). *Some Fundamentals of Combustion* (London: Butterworths)

SPALDING, D.B. (1956). A mixing rule for laminar flame speed. *Fuel, Lond.*, 35, 347

SPALDING, D.B. (1957a). Ends and means in flame theory. *Combustion*, 6, p. 12

SPALDING, D.B. (1957b). One-dimensional laminar flame theory for temperature−explicit reaction rates. *Combust. Flame*, 1, 296

SPALDING, D.B. (1957c). Predicting the laminar flame speed in gases with temperature−explicit reaction rates. *Combust. Flame*, 1, 287

SPALDING, D.B. (1957d). A theory of inflammability limits and flame-quenching. *Proc. R. Soc., SER. A*, 240, 83

SPALDING, D.B. (1960). The theory of steady laminar spherical flame propagation: equations and numerical solution. *Combust. Flame*, 4, 51

SPALDING, D.B. (1962). The burning rate of liquid fuels from open trays by natural convection. *Fire Res. Abs. Revs*, 4, 234

SPALDING, D.B. (1963a). The art of partial modelling. *Combustion*, 9, p. 833

SPALDING, D.B. (1963b). Contribution to the theory of gaseous detonation waves. *Combustion*, 9, p. 417

SPALDING, D.B. (1971). Concentration fluctuations in a round turbulent free jet. *Chem. Engng Sci.*, 26, 95

SPALDING, D.B. (1976). Mathematical models of turbulent flames: a review. *Combust. Sci. Technol.*, 13, 3

SPALDING, D.B. (1983). Chemical reaction in turbulent fluids. *Physico Chemical Hydrodynamics*, 4, 323

SPALDING, D.B. and JAIN, V.K. (1961). The theory of steady laminar spherical flame propagation: analytical solutions. *Combust. Flame*, 5, 11

SPALDING, D.B. and YUMLU, V.S. (1959). Experimental demonstration of the existence of two flame speeds (letter). *Combust. Flame*, 3, 553

SPALDING, D.B., JAIN, V.K. and SAMAIN, M.D. (1961). The theory of steady laminar spherical flame propagation: analogue solutions. *Combust. Flame*, 5, 19

SPALDING, D.B., STEPHENSON, P.L. and TAYLOR, R.G. (1971). A calculation procedure for the prediction of laminar flame speeds. *Combust. Flame*, 17, 55

SPANGLER, H.D. (1972). Stainless steels in a waste heat boiler. *Ammonia Plant Safety*, **14**, p. 26

SPANGLER, M.B. (1982). The role of interdisciplinary analysis in bridging the gap between the technical and human sides of risk assessment. *Risk Analysis*, **2**, 101

SPANGLER, T.C. and JOHNSON, V.C. (1989). The dispersion of offshore pollutant plumes into moderately complex coastal terrain. *Atmos. Environ.*, **23**, 2133

SPARKS, C.R. and WACHTEL, J.C. (1977). Pulsation in centrifugal pumps and piping systems. *Hydrocarbon Process.*, **56**(7), 183

SPARROW, E.M. (1962). Heat radiation between simply arranged surfaces having different temperatures and emissivities. *AIChE J.*, **8**, 12

SPARROW, E.M. and CESS, R.D. (1966). *Radiation Heat Transfer* (Brooks Cole)

SPARROW, E.M. and CESS, R.D. (1978). *Radiation Heat Transfer*, rev. ed. (New York: McGraw-Hill)

SPARROW, E.M., MILLER, G.B. and JONSSON, V.K. (1962). Radiating effectiveness of annular-finned space radiators, including mutual irradiation between radiator elements. *J. Aerospace Sci.*, **29**, 1291

SPARROW, R.E. (1986). Firebox explosion in a primary reformer furnace. *Ammonia Plant Safety*, **26**, p. 49. See also *Plant/Operations Prog.*, 1986, **5**, 122

SPATZ, R., DREES, S. and SCHOFT, H. (1992). Direct condensation of released vapour in blow-down systems during venting. *Loss Prevention and Safety Promotion 7*, vol. 2, 62–1

SPEAKER, D.M., THOMPSON, S.R. and LUCKAS, W.J. (1982). *Identification and Analysis of Human Errors underlying Pump and Valve related Events reported by Nuclear Power Plant Licensees. Rep. NUREG/CR-2417.* Nucl. Regul. Comm., Washington, DC

SPEARING, M.H. (1974). The instrument specialist's role in commissioning with special reference to process analysers. *Instn. Chem. Engrs, Symp. on Commissioning of Chemical and Allied Plant* (London)

SPECTOR, W.S. (ed.) (1956a). *Handbook of Toxicology* (Philadelphia, PA: Saunders)

SPECTOR, W.S. (1956b). *Handbook of Biological Data* (Washington, DC: Nat. Acad. Sci.)

SPEECHLY, D., THORNTON, R.E. and WOODS, W.A. (1979). Principles of total containment system design. *North Western Branch Papers No. 2, Institution of Chemical Engineers*, 7.1–7.21

SPEED, D.W. (1966). Incoloy alloys. *Ammonia Plant Safety*, **8**, p. 46

SPEEDIE, I. (1985). Engineering for successful revamps. *Chem. Engr, Lond.*, **413**, 49

SPEIDEL, S.R. (1985). An experimental study on the behaviour of the outer concrete wall of a double wall LNG storage facility under extreme thermal loads. *Gastech 84*, p. 434

SPEIGHT, J.G. (1990). *Fuel Science and Technology Handbook* (New York: Dekker)

SPEIGHT, J.G. (1993). *Gas Processing: Environmental Aspects and Methods* (Oxford: Butterworth-Heinemann)

SPEIR, W.B. (1970). Choosing and planning industrial sites. *Chem. Engng*, **77**(Nov. 30), 69

SPENCE, D. and MCHALE, E.T. (1975). The role of negative halogen ions in hydrocarbon flame inhibition. *Combust. Flame*, **24**, 211

SPENCE, E. (1994). Safety case implementation a case study. *Offshore Safety Case Management* (Rugby: Instn Chem. Engrs), p. es1

SPENCE, J. and CARLSON, W.B. (1967–68). A study of nozzles in pressure vessels under pressure fatigue loading. *Proc. Instn Mech. Engrs*, **182**(pt 1), 657

SPENCE, J. and TOOTH, A.S. (1994). *Pressure Vessel Design* (London: Spon)

SPENCE, J.P. and NORONHA, J.A. (1988). Reliable detection on runaway reaction precursors in liquid phase reactions. *Plant/Operations Prog.*, **7**, 231

SPENCE, J.P., HUFF, W.J., NORONHA, J.A. and TORRES, A.J. *Reliable detection of runaway reaction precursors.* Am. Inst. Chem. Engrs, Symp. on Loss Prevention (New York)

SPENCE, K. and TOWNEND, D.T.A. (1949). The two-stage process in the combustion of higher hydrocarbons and their derivatives. *Combustion*, **3**, p. 404

SPENCE, L.A. (1972). Methods of obtaining stable control. *Chem. Engng Prog.*, **68**(3), 50

SPENCER, D.B. and HANSEN, J.S. (1985). A better way to monitor bearings. *Hydrocarbon Process.*, **64**(1), 75

SPENCER, E. (1967). Steam vacuum refrigeration. *Hydrocarbon Process.*, **46**(6), 136

SPENCER, G.S. (1968). *An Introduction to Plasticity* (London: Chapman & Hall)

SPENCER, J. (1962). An investigation of process control skill. *Occup. Psychol.*, **36**, 30

SPENCER, H. (1979). Health and safety in the laboratory. *Chem. Ind.*, Nov. 3, 728

SPENCER, H.M. (1945). Empirical heat capacity equations for various gases. *J. Am. Chem. Soc.*, **67**, 1859

SPENCER, H.M. and FLANNAGAN, G.N. (1942). Empirical heat capacity equations of gases. *J. Am. Chem. Soc.*, **64**, 2511

SPENCER, H.M. and JUSTICE, J.L. (1934). Empirical heat capacity equations for single gases. *J. Am. Chem. Soc.*, **56**, 2311

SPENCER, R.A. (1978). Design concept and use of plastic-lined steel pipe. *Chem. Ind.*, Jun. 3, 369

SPENCER, W.F. and FARMER, W.J. (1980). Assessment of the vapor behaviour of toxic organic chemicals. In Haque, R. (1980a), *op. cit.*, p. 143

SPERRAZZA, J.A. and KOKINAKIS, W. (1967). *Ballistic Limits of Tissue and Clothing. Tech. Note 1645.* US Army, Ballistic Res. Lab.

SPERRAZZA, J. and KOKINAKIS, W. (1968). Ballistic limits of tissue and clothing. *Ann. N.Y. Acad. Sci.*, **152**(art 1), 163

SPETZLER, C.S. and STAEL VON HOLSTEIN, C.A.S. (1975). Probability encoding in decision analysis. *Mngmt Sci.*, **22**, 340

SPICER, T.O. (1985). Mathematical modeling and experimental investigation of heavier-than-air gas dispersion in the atmosphere. Ph.D. Thesis, Univ. of Arkansas

SPICER, T.O. and HAVENS, J.A. (1985). Modeling the Phase I Thorney Island experiments. *J. Haz. Materials.*, **11**, 237. See also in McQuaid, J. (1985), *op. cit.*, p. 237

SPICER, T.O. and HAVENS, J.A. (1986). Development of a heavier-than-air dispersion model for the US Coast Guard Hazard Assessment Computer System. In Hartwig, S. (1986), *op. cit.*, p. 73

SPICER, T.O. and HAVENS, J.A. (1987). Field test validation of the DEGADIS model. *J. Haz. Materials*, **16**, 231

SPICER, T.O., HAVENS, J.A., TEBEAN, P.A. and KEY, L.E. (1986). DEGADIS – a heavier-than-air gas atmospheric dispersion model developed for the US Coast Guard. *Am. Poll. Control Ass., Proc. Ann. Conf.*, paper 86–42.2

SPICER, T.O., HAVENS, J.A. and KAY, L.E. (1987). Extension of DEGADIS for modeling aerosol releases. *Vapor Cloud Modeling*, p. 416

SPIEGELMAN, A. (1969). Risk evaluation of chemical plants. *Loss Prevention*, **3**, p. 1

SPIEGELMAN, A. (1980). Hazard control in the chemical and allied industries. In *Loss Prevention and Safety Promotion 3*, vol. 2, p. 129

SPIEGELMAN, A. (1981). Risk management from an insurer's standpoint. *The Hazards of Industrial Explosion from Dusts* (London: Oyez/IBC), paper 7

SPIEGELMAN, A. (1987). Gas plant problems and the insurance industry. *Plant/Operations Prog.*, **6**, 190

SPIERS, H.M. (ed.) (1950). *Technical Data on Fuel*, 5th ed. (London: Br. Nat. Cttee, World Power Conf.). See also Rose, J.W. and Cooper, J.R.

SPIERS, J.A.L. (1976). Fire safety – management. *Dept of Fire Safety Engng, Univ. of Edinburgh, Symp. on Fire and Explosion Safety in the Chemical Industry*

SPIKER, V.A., HARPER, W.R. and HAYES, J.F. (1985). The effect of job experience on the maintenance proficiency of army automotive mechanics. *Hum. Factors*, **27**, 301

SPILLER, R. (1994). The employee response. *Offshore Safety Case Management* (Rugby: Instn Chem. Engrs), p. rsl

SPINKS, W.S. (1963). *Vacuum Technology* (London: Chapman & Hall)

SPIRAX-SARCO LTD (ca. 1993). *Steam and Steam Trapping* (Cheltenham)

SPITZ, P.H. (1965). How to evaluate licensed processes. *Chem. Engng*, **72**(Dec. 20), 91

SPITZ, P.H. (1988). *Petrochemicals. The Rise of an Industry* (Chichester: Wiley)

SPITZGO, C.R. (1976). Guidelines for overall chemical plant layout. *Chem. Engng*, **83**(Sep. 27), 103

SPIVEY, M. (1992). Liquid storage woes? Intermediate bulk containers may be the answer. *Chem. Engng*, **99**(9), 145

SPOCK, T. (1993). Control fugitive emissions from valves. *Chem Engng*, **100**(2), 82

SPOUGE, J.R. (1989). *Piper Alpha Public Inquiry. Transcript, Days 140, 141*

SPRANZA, F.G. (1981). Improve your plant security system. *Hydrocarbon Processing*, **60**(9), 331

SPRANZA, F.G. (1982). Set high security standards. *Hydrocarbon Process.*, **61**(4), 269

SPRANZA, F.G. (1991). Will terrorists hit your plant? *Hydrocarbon Process.*, **70**(7), 102

SPRANZA, F.G. (1992). Plant security checklist. *Hydrocarbon Process.*, **71**(9), 71

SPRINGER, S.T. (1987). How chemicals harm us. In Young, J.A. (1987a), *op. cit.*, p. 115

SPRINGMANN, H. (1977). Plan large O_2 and N_2 plants. *Hydrocarbon Process.*, **56**(2), 97

SPRINGMANN, H. (1985). Cryogenics: principles and applications. *Chem. Engng*, **92**(May 13), 58

SPROESSER, W.D. (1981). Miniature models help to clarify design ideas. *Chem. Engng*, **88**(Apr. 20), 145

SPROSTON, J.L. (ed.) (1987). *Electrostatics 87* (Bristol: Adam Hilger)

SQUELLATI, G. (1980). *Critical Review of the CAT Algorithms for Automated Fault Tree Construction. Tech. Note 1.06.01.80.84 PER 392/80*. Comm. Eur. Commun., Joint Res. Centre, Ispra, Italy

SQUIRE, R.A. (1981). Ranking animal carcinogens: a proposed regulatory approach. *Science*, **214**, 877

SQUIRE, R.H. (1991). An ammonia storage tank study. *Plant/Operations Prog.*, **10**, 212

SQUIRES, A.M. (1967). Air pollution: the control of SO_2 from power stacks. *Chem. Engng*, **74**(Nov. 6), 260; (Nov. 20), 133; (Dec. 4), 189; (Dec. 18), 101

SREEVES, M. (1989). Enclosure testing for halon: benefits of NFPA method. *Fire Surveyor*, **18**(6), 4

SRI INT. (1988a). *Directory of Chemical Producers: Canada* (San Antonio, TX)

SRI INT. (1988b). *Directory of Chemical Producers: United States* (San Antonio, TX)

SRI INT. (1988c). *Directory of Chemical Producers: Western Europe* (San Antonio, TX)

SRINIVASAN, D., MOHAN, V. and DEVOTTA, S. (1980). Management of environmental risks and prevention of accidents in chemical industries. *Loss Prevention and Safety Promotion 3*, vol. 3, p. 1451

SRINIVASAN, R. and MEIBAO, X. (2003). *Technical Report: ISafe Index for Methyl Methacrylate Process* (Singapore: Laboratory for Intelligence Applications in Chemical Engineering, Department of Chemical and Environmental Engineering, National University of Singapore)

SRIRAM, M. and SAHASTRABUDHE, V.D. (1988). Bhopal (letter). *Chem. Engr, Lond.*, **451**, 4

SRIRAM, M. and SAHASTRABUDHE, V.D. (1989). Bhopal postscript. *Chem. Engr, Lond.*, **461**, 4

SRIRAMULU, V., PADIYAR, K.S. and SHET, U.S.P. (1978). Correlation of burning velocity data for enclosed flames. *Combust. Sci. Technol.*, **19**, 71

STACEY, F.D. (1969). *Physics of the Earth* (New York: Wiley)

STACEY, R. (1984). An overview of powder valve design. *Chem. Engr, Lond.*, **404**, 45

STADTHERR, M.A., COFER, H.N. and CAMARDA, K.V. (1993). Hardware: developing the metacomputer. *Chem. Engng Prog.*, **89**(11), 27

STAFFORD, A.E. and WHITTLE, J.C. (1983). Corrosion control. *Plant Engr*, **27**(4), 19

STAFFORD, J. (1977). Safety balance: VCM case study. *Sci. Publ. Policy*, **4**, 134

STAFFORD, J.D. (1971). The high pressure centrifugal compressor loop. *Ammonia Plant Safety*, **13**, p. 60. See also *Chem. Engng Prog.*, 1971, **67**(4), 55

STAGG, K.G. and ZIENKIEWICZ, O.C. (1968). *Rock Mechanics in Engineering Practice* (New York: Wiley)

STAHL, E. (1988). Basic elements of the individual systems of safety related standards and nature and responsibility of controlling bodies. The situation in the Federal Republic of Germany. *DECHEMA Workshop*, Bad Soden (Frankfurt-am-Main)

STAHL, G. (1968). Transport of LNG from the Atlantic Coast to Switzerland by road. *LNG Conf. 1*

STAHL, Q.R. (1969a). *Air Pollution Aspects of Chlorine Gas* (report). Litton Systems Inc., Bethesda, MD

STAHL, Q.R. (1969b). *Air Pollution Aspects of Hydrochloric Acid* (report). Nat. Tech. Inf. Service, Springfield VA

STAHL, Q.R. (1969c). *Preliminary Air Pollution Survey of Hydrochloric Acid: a Literature Review* (report). Nat. Tech. Inf. Service, Springfield VA

STAHL, W.R., STRASSMAN, F. and RICHARD, G. (1949). *Catastrophe at the BASF on 28 July 1948*. Mediation Court of Land Rhineland-Palatinate, Neustad Office (in German).

STAHLKOPF, K.E. and STEELE, L.E. (eds) (1984). *Light Water Reactor Structural Integrity* (Amsterdam: Elsevier Applied Science)

STAINTHORP, F.P. and WEST, B. (1974). Computer controlled plant start-up. *Chem. Engr, Lond.*, **289**, 526

STAIRMAND, C.J. (1951). Sampling of dust-laden gases. *Trans. Instn Chem. Engrs*, **29**, 15

STALDER, H. (1982). Explosion protection slide valves – Wey system. *The Control and Prevention of Dust Explosions* (London: Oyez/IBC), paper 11

STALDER, H. (1987). Flame barrier valves. *Plant/Operations Prog.*, **6**, 175

STALLWORTHY, E.A. (1987). Safety first and always: management needs foresight. *Chem. Engr, Lond.*, **438**, 44

STALLWORTHY, E.A. (1989). Performance auditing. *Chem. Engr, Lond.*, **462**, 51

STAMBULEANU, A. (1976). *Flame Combustion Processes in Industry* (Tonbridge Wells: Abacus Press)

STAMMERS, R.B. (1978). Human factors in airfield air traffic control. *Ergonomics*, **21**, 483

STANCLIFFE, I. (1977). Colour and environment. In Handley, W. (1977), *op. cit.*, p. 305

STANDEN, R. (1989). *Piper Alpha Public Inquiry. Transcript, Day 114*

STANDIFORD, F.C. (1979). Effect of non-condensibles on condenser design and heat transfer. *Chem. Engng Prog.*, **75**(7), 59

STANEK, W.M. (1973). Railway transport of hazardous materials. *Conf. on Bulk Transportation of Hazardous Materials by Water in the Future* (Washington, DC: Nat. Acad. Sci., Nat. Res. Coun., Nat. Acad. Engng), p. 33

STANFILL, I.C. (1968). Startup experiences and special features at Memphis LNG plant. *LNG Conf. 1*

STANIAR, W. (ed.) (1959). *Plant Engineering Handbook* (New York: McGraw-Hill)

STANKIEWICZ, A.I. and MOULIJN, J.A. (2000). Process intensification: transforming chemical engineering. *Chem. Eng. Prog.* **96**(1) (Jan.), 22–34

STANKIEWICZ, A.I. and MOULIJN, J.A. (2003). *Re-engineering the chemical processing plant: process intensification.* (New York: Marcel Dekker, Inc.)

STANLEY, G.M. and MAH, R.S.H. (1977). Estimation of flows and temperatures in process networks. *AIChE J.*, **23**, 642

STANLEY, P. (ed.) (1977). *Fracture Mechanics in Engineering Practice* (Barking: Applied Science)

STANLEY-WOOD, N.G. and LINES, R.W. (eds) (1992). *Particle Size Analysis* (London: R. Soc. Chem.)

STANNARD, J.H. (1977). Thermal radiation hazards associated with marine LNG spills. *Fire Technol.*, **13**, 35

STANTON, T.E. and PANNELL, J.R. (1914). The similarity of motion in relation to the surface friction of fluids. *NPL Collected Researches*, **11**, 293; 320

STAPCZYNSKI, J.S. (1982). Blast injuries. *Ann. Emergency Med.*, **11**, 687

STAPLES, B.G., PROCOPIO, J.M. and SU, G.J. (1970). Vapour-pressure data for common acids at high temperatures. *Chem. Engng*, **77**(Nov. 16), 113

STAPLETON, R.N. (1962). Developing the better operator. *Chem. Engng*, **69**(Oct. 15), 198

STAPLETON, R.N. (1963). Unsafe acts inspection. *Chem. Engng*, **70**(Aug. 19), 185

STARK, G.W.V. (1972). Liquid spillage fires. *Chemical Process Hazards*, **IV**, p. 71

STARK, G.W.V. (1976). The control of gas and vapour fires. *Fire Surveyor*, **5**(4), 48

STARK, V. and WASHIE, A.E. (1968). LNG for small utilities and satellite plants. *LNG Conf. 1*

STARKE, R. and ROTH, P. (1986). An experimental investigation of flame behavior during cylindrical vessel explosions. *Combust. Flame*, **66**, 249

STARKE, R. and ROTH, P. (1989). An experimental investigation of flame behavior during explosions in cylindrical enclosures with obstacles. *Combust. Flame*, **75**, 111

STARKMAN, E.S., SHROCK, V.E., NEUSEN, K.F. and MANEELY, D.J. (1964). Expansion of very low-quality two-phase fluid through a convergent–divergent nozzle. *J. Basic Engng*, **68**(2), 247

STARLING, K.E. *et al.* (1971–). Thermo data refined for LPG. *Hydrocarbon Process.*, **50**(3), 101; **50**(4), 139; **50**(4), 140; **50**(6), 116; **50**(7), 115; **50**(9), 170; **50**(10), 90; 1972, **51**(2), 86; **51**(5), 129; **51**(6), 107

STAROSELSKY, N. and CARTER, D. (1990). Protecting multicase compressors. *Hydrocarbon Process.*, **69**(3), 85

STAROSELSKY, N. and LADIN, L. (1979). Improved surge control for centrifugal compressors. *Chem. Engng*, **86**(May 21), 175

STARR, C. (1969). Social benefit vs technological risk. *Science*, **165**, 1232

STARR, C. (1970). An overview of the problems of public safety. In National Academy of Engineering (1970), *op. cit.*, p. 3

STARR, C. (1972). Benefit cost studies in sociotechnical systems. In National Academy of Engineering (1972), *op. cit.*, p. 17

STARR, C. (1981). The Three Mile Island accident: the other lesson. *Ann. N.Y. Acad. Sci.*, **365**, 292

STARR, C. (1985). Risk management, assessment and acceptability. *Risk Analysis*, **5**(2), 57

STARR, C. and RITTERBUSH, P.C. (eds) (1980). *Science, Technology and the Human Prospect* (New York: Pergamon Press)

STARR, C., RUDMAN, R. and WHIPPLE, C. (1976). Philosophical basis for risk analysis. *Ann. Rev. Energy 1.* Annual Reviews Inc., Palo Alto, CA

STARR, T.B. (1983). Mechanisms of formaldehyde toxicity and risk evaluation. In Clary, J.J., Gibson, J.E. and Waritz, R.S. (1974), *op. cit.*

STATESIR, W.A. (1973). Explosive reactivity of organics and chlorine. *Loss Prevention*, **7**, p. 114. See also *Chem. Engng Prog.*, 1973, **69**(4), 52

STATHARAS, J.C., BARTZIS, J.G., VENETSANOS, A. and WURTZ, J. (1993). Prediction of ammonia releases using the ADREA-HF code. *Process Safety Prog.*, **12**, 118

STAUFFER, E.E. (1975). Extinguishing toluene diisocyanate fires with dry chemicals. *Fire Technol.*, **11**, 255

STAUNTON, M.A. (1982). The legal implications of chemical engineering. *Chem. Engng Prog.*, **78**(9), 7

STEAD, J.P. (1992). A desktop data system to support hazard identification and risk analysis studies. *Hazard Identification and Risk Analysis, Human Factors and Human Reliability in Process Safety*, p. 459

STEBBING, L.S. (1942). *A Modern Introduction to Logic* (London: Methuen)

STECKLER, K.D., BAUM, H.R. and QUINTIERE, J.G. (1985). Fire induced flows through room openings – flow coefficients. *Combustion*, **20**, p. 1591

STEEL, B.G. (1971). Fire detectors for use on chemical plants. *Major Loss Prevention*, p. 133

STEEL, D. (1987). *MV Herald of Free Enterprise. Formal Investigation* (London: HM Stationery Office)

STEEL, G. (1984). *Common LISP* (Bedford, MA: Digital Press)

STEEL, J.G. (1975a). Intermediate bulk containers. *Distribution Conf.* (London: Chem. Ind. Ass.), p. 35

STEEL, J.G. (1975b). Package performance tests – the concept and the reality. *Distribution Conf.* (London: Chem. Ind. Ass.), p. 13

STEEL, M.A.B., FARIDANY, E.K. and FFOOKS, R.C. (1983). Multilobe tanks for storage of gas liquids. *Gastech 82*, p. 403

STEELE, A.B. and DUGGAN, J.J. (1959). Safe handling of 'reactive' chemicals. *Chem. Engng*, **66**(Apr. 20), 157

STEELE, C.H. and NOLAN, P.F. (1989). The design and operation of a reflux heat flow calorimeter for studying reactions at boiling. *Runaway Reactions*, p. 198

STEELE, C.H., NOLAN, P.F., HIRST, A., STARKIE, A.J. and LOWE, J.L. (1989). Scale-up and heat transfer data for safe reactor operation. *Runaway Reactions*, p. 597

STEELE, C.H., NOLAN, P.F., LOWE, J.L., HIRST, A. and STARKIE, A.J. (1993). Monitoring reaction exotherms in pilot scale batch reactors. *J. Loss Prev. Process Ind.*, **6**, 115

STEELE, E.P. (1975). Air steam coil failure. *Ammonia Plant Safety*, **17**, p. 147

STEEN (1973). Use of gas warning devices in rooms and outside. *Prevention of Occupational Risks*, **2**, p. 248

STEEN, H. (1971). On the measurement of the propagation of explosive mixtures in free atmosphere. *Electrical Safety*, **1**, p. 64

STEEN, H. (1974). Inflammability of liquid mixtures with inflammable and non-inflammable components. *Loss Prevention and Safety Promotion*, **1**, p. 293

STEEN, H. and SCHAMPEL, K. (1983). Experimental investigations on the run-up distance of gaseous detonations in large pipes. *Loss Prevention and Safety Promotion*, **4**, vol. 3, p. E23

VAN STEEN, J.F.J. (1992). A perspective on structured expert judgment. *J. Haz. Materials*, **29**, 365

VAN STEEN, J.F.J. (1996). *The Inside Project, Phase 1, Task 1f Report: An assessment of hurdles and solutions concerning the adoption of inherently safer approaches in industry* (Apeldoorn, Netherlands: TNO)

VAN STEEN, J.F.J. and KOEHORST, L.J.B. (1993). The Smart project: a framework and tools for addressing and improving management of safety. *Process Safety Management*, p. 251

VAN STEEN, J.F.J., GOOSSENS, L.H.J. and COOKE, R.M. (1989). Protocols for expert opinion use in risk analysis. *Loss Prevention and Safety Promotion*, **6**, p. 42.1

STEENKIST, R. and NIEUWSTADT, F.T.M. (1981). Dispersion near to a tall stack. In de Wispelaere, C. (1981), *op. cit.*, p. 357

STEENSLAND, O., ASKHEIM, N.E. and VOSSGARD, R. (1975). Reactivity of diluted cargo remains. *Transport of Hazardous Cargoes*, 4

STEENSMA, M. and WESTERERP, K.R. (1988). Thermally safe operation of a cooled semibatch reactor. slow liquid-liquid reactions. *Chem. Eng. Sci.* **43**(8), 2125–32

STEENSMA, M. and WESTERERP, K.R. (1990). Thermally safe operation of a semibatch reactor for liquid-liquid reactions. Slow reactions. *Ind. Eng. Chem. Res.* **29**, 1259–70

STEERE, N.V. (ed.) (1971). *Handbook of Laboratory Safety*, 2nd ed. (Boca Raton, FL: CRC Press)

STEFFEN, H. (1975). Guide to German Pressure Vessel Code. *Hydrocarbon Process.*, **54**(12), 62

STEFFEN, H. (1978). Guide to A.D. Merkblatt Code. *Hydrocarbon Process.*, **57**(12), 92

STEFFENS, L. (1931). *Autobiography* (New York: Harcourt Brace)

STEFFY, J. (1994). Accessing industrial fire and explosion information in the CA files on STN. *J. Haz. Materials*, **36**, 165

STEFIK, M. (1981a). Planning with constraints (MOLGEN: Part 1). *Artificial Intelligence*, **16**, 111

STEFIK, M. (1981b). Planning with constraints (MOLGEN: Part 1). *Artificial Intelligence*, **16**, 141

STEFIK, M. and BOBROW, D.G. (1986). Object-oriented programming themes and variations. *AI Mag.*, **6**, 40

STEFIK, M., AIKINS, J., BALZER, R., BENOIT, J., BIRNBAUM, L., HAYESROTH, F. and SACERDOTI, E. (1982). The organization of expert systems: a tutorial. *Artificial Intelligence*, **18**, 135

STEFIK, M.J. and DE KLEER, J. (1983). Prospects for expert systems in CAD. *Comput. Des.*, Apr. 21, 65

STEGELMAN, A.F. and RENFFTLEN, R. (1983). On line mechanical cleaning of heat exchangers. *Hydrocarbon Process.*, **82**(1), 95

STEGGERDA, J.J. (1965). Thermal stability: an extension of the Frank–Kamenetsky theory. *J. Chem. Phys.*, **43**, 4446

STEIGER, F.H. (1971). Practical applications of the Weibull distribution function. *Chem. Tech.*, Apr., 225

STEIGERWALD, B.J. (1958). *Emissions of Hydrocarbons to the Atmosphere from Seals on Pumps and Compressors. Joint District, Federal and State Project for the Evaluation of Refinery Emissions, Air Pollution Control District, County of Los Angeles, CA, Rep. 6*

STEIMER, J.G. (1983). Liquefied energy gas storage tanks – construction considerations. *Gastech 82*, p. 343

STEIN, T.N. (1980). Analyzing and controlling noise in process plants. *Chem. Engng*, **87**(Mar. 10), 129

STEINBACH, J. (1987). Practically oriented representation of modelling criteria for semi-batch reactors. *Fortschritte der Sicherheitstechnik* 1 (Frankfurt-am-Main: DECHEMA), p. 133

STEINBERG, S.L. (1964). Corrosion resistant floors. *Chem. Engng*, **71**(Mar. 16), 178

STEINBERG, T.A., WILSON, D.B. and BENZ, F. (1992). The combustion phase of burning metals. *Combust. Flame*, **91**, 200; **93**, 343

STEINBRECHER, L. (1986). Analysis of a tank fire 'classic'. *Hydrocarbon Process.*, **65**(5), 85

STEINHERZ, H.A. (1962). High vacuum technology and equipment. *Chem. Engng*, **69**(Aug. 20), 117

STEINHOFF, R.C. (1959). Development of safe procedures. *Am. Petrol. Inst, API Proc.* **39**(3)

STEINHOFF, R.C. (1973). Development of safe procedures. In Vervalin, C.H. (1973a), *op. cit.*, p. 422

STEINKIRCHNER, K.K. (1978). 1985 LPG supply/demand promising. *Hydrocarbon Process.*, **57**(8), 60A

STELLMAN, J.M. and DAUM, S.M. (1973). *Work is Dangerous to Your Health* (New York: Random House)

STEN, T. and ULLEBERG, T. (1992). Top down approach to human factors. *Hazard Identification and Risk Analysis, Human Factors and Human Reliability in Process Safety*, p. 389

STENDER, J.H. (1974). What the HPI can expect from OSHA. *Hydrocarbon Process*, **53**(7), 190

STEPANOFF, A.J. (1957). *Centrifugal and Axial Flow Pumps: Theory, Design and Application*, 2nd ed. (New York: Wiley)

STEPHAN, D.G. and ATCHESON, J. (1989). The EPA's approach to pollution prevention. *Chem. Engng Prog.*, **85**(6), 53

STEPHAN, R.J. (1994). Fire department response to hazardous materials incidents. *J. Haz. Materials*, **36**, 183

STEPHANOPOULOS, G. (1983). Synthesis of control systems for chemical plants – a challenge for creativity. *Comput. Chem. Engng*, **7**, 331

STEPHANOPOULOS, G. (1984). *Chemical Process Control* (Englewood Cliffs, NJ: Prentice-Hall)

STEPHANOPOULOS, G. (1987). The future of expert systems in chemical engineering. *Chem. Engng Prog.*, **83**(9), 44

STEPHANOPOULOS, G. (1990). Artificial intelligence in process engineering – current state and future trends. *Comput. Chem. Engng*, **14**, 1259

STEPHANOPOULOS, G. and TOWNSEND, D.W. (1986). Synthesis in process development. *Chem. Engng Res. Des.,* **64**, 160

STEPHANOPOULOS, G., JOHNSTON, J., KRITICOS, T., LAKSHMANAN, R., MAVROVOUNIOTIS, M. and SILETTI, C. (1987). DESIGN-KIT: an object-oriented environment for process engineering. *Comput. Chem. Engng*, **11**, 655

STEPHANOPOULOS, G. *et al.* (1987). *An Intelligent System for Planning Plant-wide Process Control Strategies*. Int. Fed. Automat. Control Cong., Munich

STEPHANOPOULOS, G., HANNING, G. and LEONE, H. (1990a). MODEL.LA. A modeling language for process engineering. I, The formal framework. *Comput. Chem. Engng*, **14**, 813

STEPHANOPOULOS, G., HANNING, G. and LEONE, H. (1990b). MODEL.LA. A modeling language for process engineering. II, Multifaceted modeling of processing systems. *Comput. Chem. Engng*, **14**, 847

STEPHENS, D.K. (1977). Training for safe operations of ethylene plants. *Am. Inst. Chem. Engrs, Symp. on Loss Prevention* (New York)

STEPHENS, D.R. and MCCONNELL, D.P. (1985). Riser design codes. *Oil Gas J.*, **83**(Jul. 15), 128; (Jul. 29), 139

STEPHENS, J.D. and VIDALIN, F. (1988). Stress corrosion cracking in ammonia field storage tanks. *Ammonia Plant Safety*, **28**, 9

STEPHENS, M. (1980). *Three Mile Island* (London: Junction Books)

STEPHENS, M.M. (1970). *Minimising Damage to Refineries from Nuclear Attack, Natural and Other Disasters* (report). Office of Oil and Gas, Dept of Interior

STEPHENS, M.M. (1973). *Vulnerability of Total Petroleum Systems. Rep. DAHC20-7-CV-0216*. Defense Civil Preparedness Agency

STEPHENS, M.M. and STEPHENS, O.F. (1957). *Petroleum and Natural Gas Production* (University Park, PA: Pennsylvania State Univ.)

STEPHENS, T.J.R. and LIVINGSTON, C.B. (1973). Explosion of a chlorine distillate receiver. *Loss Prevention*, **7**, p. 104. See also *Chem. Engng Prog.*, 1973, **69**(4), 45

STEPHENSON, J.J. (1981). Condition monitoring: how to establish an effective programme. *Plant Engr*, **25**(4), 23

STEPHENSON, D. (1981a). *Pipeline Design for Water Engineers*, 2nd ed. (Amsterdam: Elsevier)

STEPHENSON, D. (1981b). *Stormwater Hydrology and Drainage* (Amsterdam: Elsevier)

STEPHENSON, D. (1984). *Pipeflow Analysis* (Amsterdam: Elsevier)

STEPHENSON, M.E. (1980). Setting toxicological and environmental testing priorities for commercial chemicals. In Haque, R. (1980), *op. cit.*, p. 35

STEPHENSON, P.L. and TAYLOR, R.G. (1973). Laminar flame propagation in hydrogen, oxygen, nitrogen mixtures. *Combust. Flame*, **20**, 231

STEPHENSON, S. and COWARD, M.J. (1987). Attenuation of radiant heat on LNG/LPG carriers with freestanding water curtains. *Gastech 86*, p. 157

STERLING, L. and SHAPIRO, E. (1986). *The Art of Prolog* (Cambridge, MA: MIT Press)

STERLING, M.B. (1977). Ammonia pipeline rupture. *Ammonia Plant Safety*, **19**, p. 77

STERLING, M.B. and MOON, A.J. (1975). Failure in secondary reformer vessel. *Ammonia Plant Safety*, **17**, p. 135

STERN, A.C. (ed.) (1962–). *Air Pollution*, vols 1–2 (New York: Academic Press)

STERN, A.C. *et al.* (1973). *Fundamentals of Air Pollution* (New York: Academic Press)

STERN, A.L. (1962). A guide to crushing and grinding. *Chem. Engng*, **69**(Dec. 10), 129

STERN, A.M. (1980). Environmental testing of toxic substances. In Haque, R. (1980), *op. cit.*, p. 459

STERN, E. (1986). The portable computing system for use in gas emergencies. *Risk Analysis*, **6**, 381

STERN, O.T. (1924). Theory of the electrolytic double layer. *Z. Electrochem.*, **30**, 508

STERNBERG, R.J. (ed.) (1982). *Handbook of Human Intelligence* (New York: Cambridge Univ. Press)

STERNGLASS, E. (1973). *Low Level Radiation* (New York: Ballantine)

STERNLICHT, B., LEWIS, P. and RIEGER, N. (1968). Failure prevention is everybody's problem. *Loss Prevention*, **2**, p. 62

STEVE, E.H. (1971). Logic diagram boosts process control efficiency. *Chem. Engng*, **78**(Mar. 8), 96

STEVENS, A. (1984). How shall we judge an expert system? In Forsyth, R. (1984), *op. cit.*, p. 37

STEVENS, B. (1983). Event, reliability, accidental and abnormal occurrences data banking using computer bureau services. *Euredata 4*, paper 11.2

STEVENS, D. (1986). Self regulation and the role of the safety officer. *Chem. Ind.*, Sep. 15, 600

STEVENS, F.W. (1928). The gaseous explosive reaction at constant pressure. *Combustion*, **1**, p. 36

STEVENS, G. and STICKLES, R.P. (1992). Prioritization of safety related plant modifications using cost-risk benefit analysis. *Hazard Identification and Risk Analysis, Human Factors and Human Reliability in Process Safety*, p. 71

STEVENS, G.C. and HUMPHREYS, A.M. (1992). Calibrating hazop studies. *Major Hazards Onshore and Offshore*, p. 483

STEVENS, G.C., LEGRAND, P. and MOKHTAR, A.A. (1992). Cost-effective implementation of safety and environmental projects. *Loss Prevention and Safety Promotion 7*, vol. 2, 56–1

STEVENS, J.C., CAIN, W.S. and WEINSTEIN, D.E. (1987). Ageing impairs the ability to detect gas odor. *Fire Technol.*, **23**, 198

STEVENS, J.J. (1980). What happened at Seveso. *Chem. Ind.*, Jul. 19, 564

STEVENS, J.J. (1988). Controlling environmental liability in property transactions through the use of environmental assessments in the state of California. *J. Haz. Materials*, **18**(3), 245

STEVENS, R. (1983). Emergency generators: a reliability study based on an analysis of failures. *Plant/Operations Prog.*, **2**, 203

STEVENS, R. (1985). A couple of unusual occurrences. *Plant/Operations Prog.*, **4**, 68

STEVENS, T.C. and LITTLEWOOD, M.J. (1979). Plastics pipes for the transport and distribution of potable water in the UK. *Chem. Ind.*, Mar. 17, 205

STEVENS, V.L. (1979). Chlorocarbons and chlorohydrocarbons: other chloroethanes. *Kirk-Othmer Encyclopaedia of Chemical Technology*, 3rd ed., vol. 5 (New York: Wiley), p. 722

STEVENS-GUILLE, P.D. and CRAGO, W.A. (1975). Application of spiral wound gaskets for leak-tight joints. *J. Pres. Ves. Technol.*, **97**, 29, 142

STEVENSTON, L.G. (1959). Experience in high pressure technology. *Chem. Engng Prog.*, **55**(4), 94

STEVERDING, B. and NIEBERLEIN, V.A. (1965). Ablation for heat shielding. *Chem. Engng*, **72**(Jul. 19), 163

STEWARD, A.M. and VAN AERDE, M. (1990a). An empirical analysis of Canadian gasoline and LPG truck releases. *J. Haz. Materials*, **25**(1/2), 205

STEWARD, A.M. and VAN AERDE, M. (1990b). Enhancements and updates to the RISKMOD risk analysis model. *J. Haz. Materials*, **25**(1/2), 107

STEWARD, F.R. (1964). Linear flame heights for various fuels. *Combust. Flame*, **8**, 171

STEWARD, F.R. (1970). Prediction of the height of turbulent diffusion buoyant flames. *Combust. Sci. Technol.*, **2**, 203

STEWARD, F.R. (1971). A mechanistic fire spread model. *Combust. Sci. Technol.*, **4**, 177

STEWARD, H.F. (1974). *The Technology of Caprolactam Manufacture* (report). Res. Dev. Unit, Manchester Business School, Manchester

STEWARD, P. (1988). Subsea separation and transport. *Chem. Engr, Lond.*, **444**, 22

STEWARD, P. (1989). Subsea highlights from Oslo. *Chem. Engr, Lond.*, **460**, 25

STEWARD, P. (1990). Offshore progress report. *Chem. Engr, Lond.*, **468**, 12

STEWART, C.D., SYED, K.J. and MOSS, J.B. (1991). Modelling soot formation in non-premixed kerosene−air flames. *Combust. Sci. Technol.*, **75**, 211

STEWART, D. and KULLOCH, D. (1968). *Principles of Corrosion and Protection* (London: Macmillan)

STEWART HUGHES LTD (1981). *Gearbox Monitoring Manual. Rep. SHL/18/81*

STEWART, H.V.M. (1963). *Guide to Efficient Maintenance Management* (London: Business Publs)

STEWART, J.D. (1958). *British Pressure Groups* (Oxford: Oxford Univ. Press)

STEWART, N.E. and MCVEY, J.W. (1994). Design concepts for a hydrogen fluoride emergency de-inventory system. *Process Safety Prog.*, **13**, 104

STEWART, N.G. et al. (1954). *AERA Rep. HP/R 1452*. UK Atom. Energy Auth., Harwell

STEWART, N.G., GALE, J.J. and CROOKS, R.N. (1958). The atmospheric diffusion of gases discharged from the chimney of the Harwell reactor BEPO. *Int. J. Air Poll.*, **1**, 87

STEWART, O.W., RUSSELL, C.K. and CONE, W.V. (1941). Injury to the central nervous system by blast. *The Lancet*, Feb. 8, 172

STEWART, P.H., ROTHEM, T. and GOLDEN, D.M. (1989). Tabulation of rate constants for combustion modeling. *Combustion*, **22**, 943

STEWART, R.M. (1971). High integrity protective systems. *Major Loss Prevention*, p. 99

STEWART, R.M. (1974a). The design and operation of high integrity protective systems. *Chem. Engr, Lond.*, **290**, 622

STEWART, R.M. (1974b). The design and operation of high integrity protective systems. *Course on Process Safety − Theory and Practice*. Dept of Chem. Engng, Teesside Polytechnic, Middlesbrough

STEWART, R.M. (1983). A systematic approach towards automating the management of machinery complexes. In Butcher, D.W. (1983), *op. cit.*, p. 293

STEWART, R.M. and HENSLEY, G. (1971). High integrity protective systems in hazardous chemical plants. *CRST Specialist Mtg on Applicability of Quantitative Reliability Analysis of Complex Systems and Nuclear Plants in its Relation to Safety*

STEYMANN, E.H. (1966). Handling materials outside the plant. *Chem. Engng*, **73**(Dec. 5), 127

STEYMANN, E.H. (1976). Industry's views regarding pollution-control laws. *Chem. Engng*, **83**(Oct. 18), 33

STICH, H.F. (ed.) (1982−). *Carcinogens and Mutagens in the Environment* (Boca Raton, FL: CRC Press)

STIKVOORT, W.J. (1986). Piping reactions on pressure-vessel nozzles. *Chem. Engng*, **93**(Jul. 7), 51

STILES, A.B. (1983). *Catalyst Manufacture: Laboratory and Commercial Preparation* (New York: Dekker)

STILLMAN, T.G. (1971). Foam attack on tank fire. *Fire Command*, **38**(10), 26

STINCHCOMB, W.W. (1980). *Mechanics of Nondestructive Testing* (New York: Plenum Press)

STINDT, W.H. (1971). Pump selection. *Chem. Engng*, **78**(Oct. 11), 43

STINTON, H.G. (1978). Taking stock after the tanker explosion in Spain. *Fire Int.*, **62**, 43

STINTON, H.G. (1979). Spanish campsite disaster. *J. Ass. Petrol. Explosives Act Admin*, **18**(1), 17

STINTON, H.G. (1983). Spanish camp site disaster. *J. Haz. Materials*, **7**(4), 393

STIPPICK, J. (1979). Design weld-neck flanges fast. *Hydrocarbon Process.*, **58**(5), 201

STIPPICK, J.A. (1992). How to check welding procedures without being an expert. *Hydrocarbon Process.*, **71**(11), 93

STIRLING, A.G. (1972). Methods for prevention of spills in the maritime industry − government considerations. *Control of Hazardous Material Spills 1*, p. 37

STIRLING, L.A. and SHAPIRO, E. (1986). *The Art of Prolog* (Cambridge, MA: MIT Press).

STITT, E.H. (2002). Alternative multiphase reactors for fine chemicals: A world beyond stirred tanks? *Chem. Eng. J.* **90**, 47−60

STOBAUGH, R.B. (1965a). Benzene: how, where, who and the future. *Hydrocarbon Process.*, **44**(9), 209

STOBAUGH, R.B. (1965b). Cyclohexane: how, where, who and the future. *Hydrocarbon Process.*, **44**(1), 157

STOBAUGH, R.B. (1965c). Styrene: how, where, who . . . future. *Hydrocarbon Process.*, **44**(12), 137

STOBAUGH, R.B. (1966a). Acetylene: how, where, who − future. *Hydrocarbon Process.*, **45**(8), 125

STOBAUGH, R.B. (1966b). Aromatics and derivatives: how, where, who − future. *Hydrocarbon Process.*, **45**(5), 205

STOBAUGH, R.B. (1966c). Ethylene: how, where, who − future. *Hydrocarbon Process.*, **45**(10), 143

STOBAUGH, R.B. (1966d). Naphthalene: how, where, who − future. *Hydrocarbon Process.*, **45**(3), 149

STOBAUGH, R.B. (1966e). Phenol: how, where, who − future. *Hydrocarbon Process.*, **45**(1), 143

STOBAUGH, R.B. (1966f). Toluene: how, where, who − future. *Hydrocarbon Process.*, **45**(2), 137

STOBAUGH, R.B. (1966g). Xylene: how, where, who − future. *Hydrocarbon Process.*, **45**(4), 149

STOBAUGH, R.B. (1967a). Butadiene: how, where, who − future. *Hydrocarbon Process.*, **46**(6), 141

STOBAUGH, R.B. (1967b). Propylene: how, where, who − future. *Hydrocarbon Process.*, **46**(1), 143

STOBAUGH, R.B. (1988). *Innovation and Competition: The Global Management of Petrochemical Products* (Boston, MA: Harvard Business School Press)

STOBAUGH, R.B., ALLEN, W.C. and VAN STERNBERGH, R.H. (1972). Vinyl acetate: how, where, who − future. *Hydrocarbon Process.*, **51**(5), 153

STOBAUGH, R.B. et al. (1973). Propylene oxide: how, where, who − future. *Hydrocarbon Process.*, **52**(1), 99

STOCK, A.J. (1966). Solid lubricants for process plants. *Chem., Engng*, **73**(Jul. 18), 176

STOCK, M. (1987). In *Fortschritte der Sicherheitstechnik II* (Frankfurt: DECHEMA)

STOCK, M. and GEIGER, W. (1984). Assessment of vapor cloud explosion hazards based on recent research results. *Ninth Int. Symp. on Prevention of Occupational Accidents and Diseases in the Chemical Industry, Luzern*

STOCK, M., SCHILDKNECHT, M. and GEIGER, W. (1984). Flame acceleration by a postflame local explosion. In Bowen, J.R. *et al.* (1984b), *op. cit.*, p. 491

STOCK, M., GEIGER, W. and GIESBRECHT, H. (1989). Scaling of vapor cloud explosion hazards based on recent research results. *Twelfth Int. Symp. on Dynamics of Explosions and Reactive Systems*, Ann Arbor, MI

STOCKBURGER, D.A. and KUHNER, H. (1979). Safety considerations in the planning of chemical plants. In BASF (1979), *op. cit.*, p. 183

STOCKHAM, J.J. and FOCHTMAN, E.G. (eds) (1979). *Particle Size Analysis* (Ann Arbor, MI: Ann Arbor Science)

STOCKHOLM INTERNATIONAL PEACE RESEARCH INSTITUTE (1978). *Anti-Personnel Weapons* (London: Taylor & Francis)

STOECKER, G. (1976). Behaviour-modifying techniques for the prevention of occupational risks. *Prevention of Occupational Risks*, **3**, p. 35

STOEHR, L.A. *et al.* (1977). *Spill Risk Analysis Program: Methodology Development and Demonstration*, vols 1–2 (rep. to US Coast Guard). Operations Res. Inc.

STOESS, H.A. (1970). *Pneumatic Conveying*; 1983, 2nd ed. (New York: Wiley-Interscience)

STOESSEL, F. (1989). Experimental study of thermal hazards during the hydrogenation of aromatic nitro compounds. *Loss Prevention and Safety Promotion 6*, p. 77.1

STOESSEL, F. (1993a). Experimental study of thermal hazards during the hydrogenation of aromatic nitro compounds. *J. Loss Prev. Process Ind.*, **6**, 79

STOESSEL, F. (1993b). What is your thermal risk? *Chem. Engng Prog.*, **89**(10), 68

STOKER, J.J. (1957). *Water Waves* (New York: Interscience)

STOKES (1849). On some points on the received theory of sound. *Phil. Mag., ser. 3*, **XXXIV**, p. 52

STOKES, K.J. (1979). Compression systems for ammonia plants. *Ammonia Plant Safety*, **21**, p. 119. See also *Chem. Engng Prog.*, 1979, **75**(7), 88

STOKES, J.W., HOLLY, L. and MAYER, K.H. (1974). The ASME Boiler and Pressure Vessel Code. *Non-Destructive Testing*, Jun., **145**

STOKINGER, H.E. (1949). In *Pharmacology and Toxicology of Uranium Hexafluoride* (ed. Vaegtlin, C. and Hodge, H.C.) (New York: Academic Press), ch. 17

STOKOE, J.N., POMPHREY, D.M. and BUTCHARD, A.C. (1981). Failure analysis as an aid to reliable production. *Reliable Production in the Process Industries* (Rugby: Instn Chem. Engrs), p. 159

STOKOE, J.N., POTTS, J.W. and MARRON, V.I.J. (1969). Service failures of coils in pressure vessels. *Chem. Engr, Lond.*, **226**, CE60

STOLL, A.M. and CHIANTA, M.A. (1969). Heat transfer through fabrics as related to thermal injury. *Aerospace Med.*, **40**, 1232

STOLL, A.M. and CHIANTA, M.A. (1971). Heat transfer through fabrics as related to thermal injury. *Trans N.Y. Acad. Sci., ser. 2*, **33**, 649

STOLL, A.M. and GREENE, L.C. (1958). *The Production of Burns by Thermal Radiation of Medium Intensity. ASME paper 58-A-219*. Am Soc. Mech. Engrs, New York

STOLL, A.M. and GREENE, L.C. (1959). Relationship between pain and tissue damage due to thermal radiation. *J. Appl. Physiol.*, **14**, 373

STOLL, A.P. (1965). The storage and handling of hydrogen with safety. *Chem. Engr, Lond.*, **185**, CE11

STOLZ, D.R. *et al.* (1974). *Toxicol. Appl. Pharmacol.*, **29**, 157

STONE, D. and TWEED, C. (1991). Representation issues in the design of an intelligent system for building standards. In Kahkonen, K. and Bjork (1991), *op. cit.*

STONE, J.P., WILLIAMS, F.W. and HAZLETT, R.N. (1976). Some fire considerations for transportation of self-heating chemicals. *J. Fire Flammability*, **7**, 303

STONE, L.K., HILL, R.F. and NEEDELS, T.S. (1974). Considerations for the safety of LNG storage terminals. *LNG Conf. 4*

STONE, R., GRAY, H. and TICKNER, J. (1995). *Opportunities for promoting inherent safety* (Boston, MA: National Environmental Law Center)

STONE, R.B. (1987). Underground storage of hazardous waste. *J. Haz. Materials*, **14**(1), 23

STONEBREAKER, J., FREESTONE, F.J. and PELTIER, W.H. (1978). Cleanup of an oil and mixed chemical waste spill in Kernersville, NC, June, 1977. *Control of Hazardous Material Spills*, **4**, p. 182

STORCH, O. (1979). *Industrial Separators for Gas Cleaning* (Amsterdam: Elsevier)

STOREY, P.D. (1980). Calorimetric studies of the thermal explosion properties of aromatic diazonium salts. *Loss Prevention and Safety Promotion 3*, vol. 3, p. 1435

STOREY, P.D. (1981). Calorimetric studies of the thermal explosion properties of aromatic diazonium salts. *Runaway Reactions, Unstable Products and Combustible Powders* (London: Instn Chem. Engrs), paper 3/P

STOREY, P.D. (1992). State of European research on major industrial hazards. *Loss Prevention and Safety Promotion 7*, vol. 1, lecture

STORIE, V.J. (1984). *Involvement of Goods Vehicles and Public Service Vehicles in Motorway Accidents. Rep. LR 1113*. Transport and Road Res. Lab., Crowthorne

STORK, K., ABRAHAMS, M.A. and RHOE, A. (1974). More petrochemicals from crude. *Hydrocarbon Process.*, **53**(11), 157

STORK, K., HANISIAN, J. and BAC, J. (1976). Recover acetylene in olefins plants. *Hydrocarbon Process.*, **55**(11), 151

STOTT, A.L. (1969). The computer control system on the ICI Mond Division Paraquat plant. *Direct Digital Control of Processes, IEE Colloquium Dig. 69/2* (London: IEE), p. 6/1

STOUFFS, P. and GIOT, M. (1993). Pipeline leak detection based on mass balance: importance of the packing term. *J. Loss Prev. Process Ind.*, **6**, 307

STOUT, H.P. (1952). Ignition of wood by radiation (letter). *Br. J. Appl. Phys.*, **3**, 394. See also Lawson, D.I. and Simms, D.L. (1952a)

STOUT, H.P. and JONES, E. (1949). The ignition of gaseous explosive media by hot wires. *Combustion*, **3**, p. 329

STOUT, J.B. (1992). Develop a strategic automation plan. *Hydrocarbon Process.*, **71**(5), 57

STOVER, G. (1972). Getting the best from stationary batteries. *Chem. Engng*, **79**(Jun. 26), 126

STRACHAN, D.C., DENNIS, P.N.J., BRADLEY, D. DANN, R. and BULL, D.C. (1989). Imaging of hydrocarbon vapour and gas leakages by absorption thermography. *Loss Prevention and Safety Promotion 6*, p. 113.1

STRADER, N.R. (1973). CRT consoles: new look in control rooms. *Chem. Engng*, **80**(Apr. 30), 83

STRAHLER, A.H. and STRAHLER, A.N. (1974). *Introduction to Environmental Science* (Santa Barbara, CA: Hamilton)

STRAITZ, J.F. (1977). Make the flare protect the environment. *Hydrocarbon Process.*, **56**(10), 131

STRAITZ, J.F. (1987). Flare technology safety. *Chem. Engng. Prog.*, **83**(7), 53

STRAITZ, J.F. and O'LEARY, J.A. (1977). Flare radiation. *Loss Prevention and Safety Promotion 2*, p. 281

STRAITZ, J.F., O'LEARY, J.A., BRENNAN, J.E. and KARDAN, C.J. (1977). Flare testing and safety. *Loss Prevention*, **11**, p. 23

STRAJA, S. (1994). The importance of the pollutant dispersion along the nominal wind direction. *Atmos. Environ.*, **28A**, 371

STRAMIGIOLI, C. and SPADONI, G. (1992). Vaporization from multicomponent liquid spills. *Loss Prevention and Safety Promotion 7*, vol. 3, 102–1

STRASHOK, P. and UNRUH, W. (1972). Failure of a reformer outlet header. *Ammonia Plant Safety*, **14**, p. 43

STRATMEYER, R.J. (1964). Do you know what you're getting for your repair dollar? *Chem. Engng*, **71**(Mar. 2), 120

STRATTA, J.L. (1987). *Manual of Seismic Design* (Englewood Cliffs, NJ: Prentice-Hall)

STRAUS, D.B. (1981). Lessons for the resolution of technical disputes. *Ann. N.Y. Acad. Sci.*, **365**, 334

STRAUSS, R. (1985). Filtering samples to on-line analyzers. *Hydrocarbon Process.*, **64**(1), 69

STRAUSS, W. (ed.) (1971–). *Air Pollution Control* (New York: Wiley-Interscience)

STRAUSS, W. (1975). *Industrial Gas Cleaning*, 2nd ed. (Oxford: Pergamon Press)

STRAUSS, W. and WOODHOUSE, G. (1958). Air pollution from tall chimneys. *Br. Chem. Engng*, **3**, 620

STRAUSS, W.A. and EDSE, R. (1959). Burning velocity measurements by the constant-pressure bomb method. *Combustion*, **7**, p. 377

STRAWSON, H. (1973). Electrostatic explosions and fires. *Chart. Mech. Engr*, Nov., 91

STRAWSON, H. and LYLE, A.R. (1975a). The avoidance of electrostatic hazards during the filling of road and rail tanker wagons. *Electrical Safety*, **2**, p. 26

STRAWSON, H. and LYLE, A.R. (1975b). Safe charge densities for road and rail tank car filling. *Static Electrification*, **4**, p. 276

STRAWSON, J.W. (1978). Guide to British Code BS 5500. *Hydrocarbon Process.*, **57**(12), 91

STREET, B.F. (1991). Plant design, operation and modification. The Fellowship of Engineering (1991b), *op. cit.*, p. 65

STREET, B.F., EVANS, T.J. and HANCOCK, B.M. (1984). The role of the Institution of Chemical Engineers in international safety. *Plant/Oper. Prog.*, **3**, 63

STREET, H. and FRAME, F.R. (1966). *Law Relating to Nuclear Energy* (London: Butterworths)

STREET, T.T., WILLIAMS, F.W. and ALEXANDER, J.I. (1980). Logic aided fire detection system. *J. Fire Flammability*, **11**, 212

STREETER, H.W. and PHELPS, E.B. (1925). *A Study of the Pollution and Natural Purification of the Ohio River. Bull. 146.* Public Health Service, Washington, DC

STREETER, V.L. (1951–). *Fluid Mechanics*; 1958, 2nd ed.; 1962, 3rd ed.; 1966, 4th ed. (New York: McGraw-Hill). See also Streeter, V.L. and Wylie, E.B.

STREETER, V.L. (1964). Waterhammer analysis of pipelines. *Proc. Am. Soc. Civil Engrs, Hydr. Div.*, **90**(Hy 4), paper 152

STREETER, V.L. and WYLIE, E.B. (1967). *Hydraulic Transients* (New York: McGraw-Hill)

STREETER, V.L. and WYLIE, E.B. (1975–). *Fluid Mechanics*, 6th ed.; 1979, 7th ed.; 1985, 8th ed. (New York: McGraw-Hill)

STREHLOW, R.A. (1968a). *Fundamentals of Combustion* (Scranton, PA: Int. Textbook)

STREHLOW, R.A. (1968b). Gas phase detonations: recent developments. *Combust. Flame*, **12**, 81

STREHLOW, R.A. (1971). Detonation structure and gross properties. *Combust. Sci. Technol.*, **4**, 65

STREHLOW, R.A. (1973a). *Equivalent Yield of the Explosion in the Alton and Southern Gateway Yard, East St Louis, Illinois, January 22, 1972. Rep. AAE TR-73–3.* Univ. of Illinois

STREHLOW, R.A. (1973b). Unconfined vapour cloud explosions – an overview. *Combustion*, **14**, p. 1189

STREHLOW, R.A. (1975). Blast waves generated by constant velocity flames: a simplified approach. *Combust. Flame*, **24**, 257

STREHLOW, R.A. (1981). Blast wave from deflagrative explosions: an acoustic approach. *Loss Prevention*, **14**, p. 145

STREHLOW, R.A. and BAKER, W.E. (1975). *The Characterisation and Evaluation of Accidental Explosions. Rep. CR 134779, 1975.* Aerospace Safety Res. and Data Inst., Lewis Research Center, NASA, Cleveland, OH

STREHLOW, R.A. and BAKER, W.E. (1976). The characterisation and evaluation of accidental explosions. *Prog. Energy Combust. Sci.*, **2**, 27

STREHLOW, R.A. and BILLER, J.R. (1969). On the strength of transverse waves in gaseous detonations. *Combust. Flame*, **13**, 577

STREHLOW, R.A. and ENGEL, C.D. (1969). Transverse waves in detonations: II, Structure and spacing in H_2-O_2, $C_2H_2-O_2$, C_2H_4-O2, and CH_4-O_2 systems. *AIAA J.*, **7**, 3

STREHLOW, R.A. and FERNANDES, F.D. (1965). Transverse waves in detonations. *Combust. Flame*, **9**, 109

STREHLOW, R.A. and RICKER, R.E. (1976). The blast wave from a bursting sphere. *Loss Prevention*, **10**, p. 115

STREHLOW, R.A. and SAVAGE, L.D. (1978). The concept of flame stretch. *Combust. Flame*, **31**, 209

STREHLOW, R.A., CROOKER, A.J. and CUSEY, R.E. (1967). Detonation initiation behind an accelerating shock wave. *Combust. Flame*, **11**, 339

STREHLOW, R.A., LIAUGMINAS, R., WATSON, R.H. and EYMAN, J.R. (1967). Transverse wave structure in detonations. *Combustion*, **11**, p. 683

STREHLOW, R.A., SAVAGE, L.D. and VANCE, G.M. (1973). On the measurement of energy release rates in vapour cloud explosions. *Combust. Sci. Technol.*, **6**, 307

STREHLOW, R.A., LUCKRITZ, R.T., ADAMCZYK, A.A. and SHIMPI, S.A. (1979). The blast wave generated by spherical flames. *Combust. Flame*, **35**, 297

STREHLOW, R.A., NICHOLLS, J.A., MAGISON, E.C. and SCHRAM, P.J. (1979). An investigation of the maximum experimental safe gap anomaly. *J. Haz. Materials*, **3**(1), 1

STREICH, H.J. and FEELEY, F.G. (1972). Design and operation of process waste heat boilers. Loss Prevention 6, p. 118. See also *Chem. Engng Prog.*, 1972, **68**(7), 57

STRELZOFF, S. (1964). Explosion during breaking in of a compressor. *Ammonia Plant Safety*, **6**, p. 17

STRELZOFF, S. (1981). *Technology and Manufacture of Ammonia* (Malibar, FL: Krieger)

STRELZOFF, S. and PAN, L.C. (1968). Designing pressure vessels. *Chem. Engng*, **75**(Nov. 4), 191

STRELZOFF, S., PAN, L.C. and MILLER, E.J. (1968). Trends in pressure vessels and closures. *Chem. Engng,* **75**(Oct. 21), 143

STRELZOFF, S.Z. (1970). Ethylene and propylene: booming building blocks. *Chem. Engng,* **77**(Aug. 24), 75

STRETCH, D. (1986). The dispersion of slightly dense contaminants in a turbulent boundary layer. Ph.D. Thesis, Univ. of Cambridge

STRETCH, D.D., BRITTER, R.E. and HUNT, J.C.R. (1984). The dispersion of slightly dense contaminants. In Ooms, G. and Tennekes, H. (1984), *op. cit.,* p. 333

STRETCH, K.L. (1969). The relationship between containment characteristics and gaseous reactions. *Struct. Engr,* **47**, 408

STRICKLAND-CONSTABLE, R.F. (1949). The burning velocity of gas in relation to the ignition delay period. *Combustion,* **3**, p. 229

STRICOFF, R.S. (1990). Implementing process safety management systems. *Plant/Operations Prog.,* **9**, 236

STRICOFF, R.S. (1994). Implementing process safety management systems. *Plant/Operations Prog.,* **13**, 21

STRICOFF, R.S. and WALTERS, D.B. (1990). *Laboratory Health and Safety Handbook: A Guide for the Preparation of a Chemical Hygiene Plan* (New York: Wiley)

STRIMAITIS, D.G. and SNYDER, W.H. (1986). An evaluation of the complex terrain dispersion model against laboratory observations: neutral flow over 2–D and 3–D hills. *Fifth Joint Conf. on Applications of Air Pollution Modeling* (Boston, MA: Am. Met. Soc.), p. 215

STRIPLING, T.E., KHAN, A.U. and DILLON, K. (1985). Hydraulic analysis aids high-pressure system design. *Oil Gas J.,* **83**(Jan. 28), 125

STRNAD, J. (1980). Preventive measures against static electricity hazard in the explosives production. *Loss Prevention and Safety Promotion 3,* vol. 2, p. 631

STROCK, O.J. (1988). *Telemetry Computer Systems – the New Generation* (Pittsburgh, PA: Instrum. Soc. Am.)

STROFER and NICKEL (1989). Safe operation and design of distillation columns. *Loss Prevention and Safety Promotion 6,* p. 114.1

STROM, G.H. and HALITSKY, J. (1955). Important considerations in the use of the wind tunnel for pollution studies of power plants. *Trans. Am. Soc. Mech. Engrs,* **77**, 789

STROM, G.H., HACKMAN, M. and KAPLIN, E.J. (1957). Atmospheric dispersal of industrial stack gases determined by concentration measurements in scale model wind tunnel measurements. *J. Air Poll. Control Ass.,* **7**, 198

STRONG, B.R. and BASCHIERE, R.J. (1979). Steam hammer design loads for safety/relief valve discharge piping. In Haupt, R.W. and Meyer, R.A. (1979), *op. cit.,* p. 145

STRONG, R. and MENZIES, R.M. (1980). Safety appraisal system applied by British Nuclear Fuels Limited. *Loss Prevention and Safety Promotion 3,* vol. 2, p. 280

STROTHER-SMITH, N.C. (1971). The general philosophy of the Fire Protection Association. *Major Loss Prevention,* p. 24

STROUD, F.B., WILKERSON, R.T. and SMITH, A. (1978). Treatment and stabilization of PCB contaminated water and waste oil: a case study. *Control of Hazardous Material Spills,* 4, p. 135

STROUD, J. (1981). What to do about leaks. *Hydrocarbon Process.,* **60**(1), 108

STRUB, R.A. and MATILE, C. (1975). Centrifugal hypercompressors for ethylene service. *Hydrocarbon Process.,* **54**(6), 67

STRUCK, W.G. and REICHENBACH, H.W. (1967). Investigation of freely expanding spherical combustion waves using methods of high-speed photography. *Combustion,* **11**, p. 677

STRUCTURAL ENGINEERS ASSOCIATION OF CALIFORNIA, SEISMOLOGY COMMITTEE (1975). *Recommended Lateral Force Requirements and Commentary* (San Francisco, CA)

STRUTT, J.E., ROBINSON, R.J. and TURNER, W.H. (1981). Recent developments in electrochemical corrosion monitoring techniques. *Chem. Engr, Lond.,* **375**, 567

STRUTT, J.E., ALLSOP, K. and OUCHET, L. (1994). Reliability prediction of pipes and valves. *Engng Dept, Univ of Manchester, Conf. on Valve and Pipeline Reliability*

STUART, L.R. (1974). You can have safer construction. *Hydrocarbon Process.,* **53**(98), 106

STUBBERUD, G. (1971). Classification, and testing and standards for for dangerous chemicals. *Transport of Hazardous Cargoes,* **2**(paper 4)

STUBBERUD, G. (1973). Pollution of the sea by dangerous chemicals. *Transport of Hazardous Cargoes,* **3**

STUBBLEFIELD, F. (1993). Get topnotch performance from jacketed pipes. *Chem Engng,* **100**(6), 110

STUBENRAUCH, E.H. (1962). Getting the most out of gaskets. *Chem. Engng,* **69**(Sep. 3), 123

STUCKER, T. (1991). Tracking ever-shrinking emissions. *Chem. Engng,* **98**(10), 90

STUDER, D.W., COOPER, B.A. and DOELP, L.C. (1988). Vaporization and dispersion modeling of contained refrigerated liquid spills. *Plant/Operations Prog.,* **7**, 127

STUEBEN, W.J. and BALL, B.L. (1979). The Pensacola ammonia accident. *Ammonia Plant Safety,* **21**, p. 76

STUHLBARG, D.S. (1971). US heat exchanger standards. *Hydrocarbon Process.,* **50**(6), 106

STULL, D.R. (1970). Identifying chemical reaction hazards. *Loss Prevention,* 4, p. 16

STULL, D.R. (1973). Linking thermodynamics and kinetics to predict real chemical hazards. *Loss Prevention,* **7**, p. 67

STULL, D.R. (1976). Linking thermodynamics and kinetics to predict real chemical hazards. *Prevention of Occupational Risks,* **3**, p. 187

STULL, D.R. (1977). *Fundamentals of Fire and Explosion* (New York: Am. Inst. Chem. Engrs)

STULL, D.R. and PROPHET, H. (1971). JANAF *Thermochemical Tables. Publ. NSRDS-NBS37.* Nat. Bur. Standards, Washington, DC

STULL, J.O. (1987a). Considerations for design and selection of chemical-protective clothing. *J. Haz. Materials,* **14**(2), 165

STULL, J.O. (1987b). Protective suits for chemical spill response. *Chem. Engng. Prog.,* **83**(11), 34

STUMPE, B. (1984). Control centres for nuclear research. *Ergonomic Problems in Process Operations* (Rugby: Instn Chem. Engrs), p. 121

STUNDER, M. and SETHURAMAN, S. (1986). A statistical evaluation and comparison of coastal point source dispersion models. *Atmos. Environ.,* **20**, 301

STUPFEL, M., ROMANY, F., MAGNIER, M. and POLIANSKI, J. (1971). Comparative acute toxicity in male and female mice of some air pollutants. Automobile gas, nitrogen oxides, sulphur dioxide, ozone, ammonia, and carbon dioxide. *C.R. Soc. Bid.,* **165**, 1869

STURRIDGE, H. (1987). How to shift information from an expert's brain over to a PC. *Computer News,* Mar. 12, 21

STURROCK, T. (1981). Reliable production from a continuous process plant. *Reliable Production in the Process Industries* (Rugby: Instn Chem. Engrs), p. 29

STYER, R.F. and WIER, J.T. (1973). Revised piping code highlights. *Hydrocarbon Process.*, **53**(3), 101

SU, F.Y. and PATNIAK, P.C. (1981). *SIGMET-N: Near Field Solution of Contaminant Dispersion Phenomena. Model Description and User's Guide* (report). Science Applications Inc.

SU, Y.P., HOMAN, H.S. and SIRIGNANO, W.A. (1979). Numerical predictions of conditions for ignition of a combustible gas by a hot, inert particle. *Combust. Sci. Technol.*, **21**, 65

SUAYSOMPOL, K. and WOOD, R.M. (1991). The flexible pinch design method for heat exchanger networks. *Trans. Inst. Chem. Engrs*, **69A**, 458; 465

SUBRAMANIAM, T.K. and CANGELOSI, J.V. (1989). Predict safe oxygen in combustible gases. *Chem. Engng*, **96**(12), 108

SUBRAMANIAN, V., GALAKRISHNAN, G., CHANDRASHEEKAR, V. and PANDIT, A. (1988). Configuration of a two-phase jet. *Combust. Sci. Technol.*, **57**, 71

SUBRAMANYAM, K. (1981). *Scientific and Technical Information Resources* (Basel: Dekker)

SUBRAMANYAM, N., PANDIAN, G. and GANAPATHY, V. (1979). How to size waste heat boilers. *Hydrocarbon Process.*, **58**(9), 261

SUCHAN (1973). Summary of the International Symposium on Dust Explosion Risks in Mines and Industry, Karlovy Vary 11–13, October 1972. *Prevention of Occupational Risks*, **2**, p. 401

SUCHMAN, E.A. (1961). A conceptual analysis of the accident problem. *Social Problems*, **8**(3), 241

SUCHMAN, E.A. and SCHERZER, A.L. (1960). Accident proneness. *Current Research in Childhood Accidents* (New York: Ass. for the Aid of Crippled Children)

SUCKLING, C.J., SUCKLING, K.E. and SUCKLING, C.W. (1978). *Chemistry through Models* (New York: Cambridge Univ. Press)

SUESS, H. and LERNER, R. (1956). *Z. Praventivmed.*, **2**, 64

SUEYAMA, T. and TAKAMI, K. (1978). Tube damage in ammonia plant exchangers. *Ammonia Plant Safety*, **20**, p. 134

SUEYASU, S. and HIKITA, T. (1965). Study of low temperature flames on n-hexane. *Combust. Flame*, **9**, 1

SUGAI, K. (1970). Low cost, high yield for forming ethylene. *Chem. Engng*, **77**(Jun. 15), 126

SUGAWARA, Y. and MINEGISHI, K. (1985). Design of an in-ground storage tank for refrigerated propane. *Gastech 84*, p. 466

SUH, N.P. and SAKA, N. (eds) (1980). *Fundamentals of Tribology* (Cambridge, MA: MIT Press)

SULLIVAN, F.M. and BARLOW, S.M. (1979). Congenital malformations and other reproductive hazards from environmental chemicals. *Proc. R. Soc., Ser. B*, **205**, 91. See also In Doll, Sir R. and McLean, A.E.M. (1979), *op. cit.*, p. 91

SULLIVAN, F.M. and BARLOW, M. (1982). *Reproductive Hazards of Industrial Chemicals* (London: Academic Press)

SULLIVAN, H.C. (1989). Safety and loss prevention concerns for engineers in polymer facilities. *Plant/Operations Prog.*, **8**, 195

SULLIVAN, P.J.E. and SHARSHAR, R. (1992). The performance of concrete at elevated temperatures. *Fire Technol.*, **28**, 240

SULZER-AZAROFF, B. (1987). The modification of occupational safety behaviour. *J. Occup. Accid.*, **9**(3), 177

SUMARIA, V.H., ROVNAK, J.A., HEITNER, I. and HERBERT, R.J. (1976). Model to predict transient consequences of a heat exchanger tube rupture. *API, Div. of Refining*, **55**, 631

SUMATHAIPALA, K., VENART, J.E.S. and STEWARD, F.R. (1990a). Two-phase critical discharge through a pressure relief valve: an experimental study. *Safety and Loss Prevention in the Chemical and Oil Processing Industries*, p. 569

SUMATHAIPALA, K., VENART, J.E.S. and STEWARD, F.R. (1990b). Two-phase swelling and entrainment during pressure relief valve discharges. *J. Haz. Materials*, **25**(1/2), 219

SUMBRY, L.C. and BOGNER, B.R. (1993). Lining tank bottoms to extend service life. *Chem. Engng*, **100**(1), 131

SUMI, K. and TSUCHIYA, Y. (1976). *Assessment of Relative Toxicity of Materials – Toxicity Index. NRCC15367, DBR paper 685*. Nat. Res. Coun. of Canada

SUMMER, E.C., FEELY, F.J., MARSHALL, B.T. and PALMER, A.J. (1980). Safety audit of refrigerated liquefied gas facilities. *Am. Petrol. Inst., Forty-Fifth Midyear Refinery Mtg*, Houston, TX

SUMMER, W. (1963). *Methods of Air Deodorisation* (Amsterdam: Elsevier)

SUMMER, W. (1971). *Odour Pollution of Air* (London: Leonard Hill)

SUMMERELL, H.M. (1981). Consider axial-flow fans when choosing a gas mover. *Chem. Engng*, **88**(Jun. 1), 59

SUMMERFELD, J.T. and WHITE, R.H. (1979). Estimate energy consumption from heat of reaction. *Chem. Engng*, **86**(Nov. 19), 140

SUMMERS, P.D. (1977). A methodology for LISP program construction from examples. *J. ACM*, **24**, 161

SUMMERS, ANGELA E. and ZACHARY, BRYAN A. (2000). Partial stroke testing of safety block valves, *Control Engineering*, November 2000

SUMMERS, ANGELA E. (2002). High integrity pressure protective systems, *Instrum. Engineers Handb.*, Volume 3 Chapter 2.6 (Boca Raton, FL: CRC Press)

SUMMERS, ANGELA E. and ZACHARY, BRYAN A. (2002). Improve facility SIS performance and reliability, Hydrocarbon Process., October 2002

SUMMERS, ANGELA E. (2003a). Flare reduction with high integrity protection systems (HIPS), *ISA Tech. Symp. 2003*, Houston, TX, Mar. 19–20 (Research Triangle Park, NC: Instrumentation, Systems, and Automation Society)

SUMMERS, ANGELA E. (2003b). High integrity protective systems for reactive processes, *Center for Chem. Process Saf. Symp.*, Scottsdale, Arizona, Sep. 23–25 (New York: American Institute of Chemical Engineers)

SUMMERS-SMITH, D. (1969). *An Introduction to Tribology in Industry* (London: Machinery Publ.)

SUMMERS-SMITH, D. (1982). Leakage of flammable vapour through dynamic seals. *Flammable Atmospheres and Area Classification, IEE Colloquium Dig. 1982/26* (London: IEE), p. 7/1

SUMMERS-SMITH, D. and LIVINGSTONE, J.G. (1976). Compressor oil system problems. *Ammonia Plant Safety*, **18**, p. 70

SUMMERS-SMITH, J.D. (ed.) (1988). *Mechanical Seal for Improved Performance*; 1992, 2nd ed. (London: Mech. Engng Publs)

SUMMITT, R. and SLIKER, A. (eds) (1980). *Handbook of Materials Science*, vol. 4 (Boca Raton, FL: CRC Press)

SUMNER, C.A. and PFANN, J.R. (1976). Sulfur trioxide spill control. *Control of Hazardous Material Spills*, **3**, p. 192

SUNAVALA, P.D., HULSE, C. and THRING, M.W. (1957). Mixing and combustion in free and enclosed turbulent jet diffusion flames. *Combust Flame*, **1**, 179

SUNDELL, M.J. and NASMAN, J.H. (1993). Anchoring catalytic functionality on a polymer. *Chemtech* (Dec.), 16–23

SUNDERSTROM, G.A. (1993). Towards models of tasks and task complexity in supervisory control applications. *Ergonomics*, **36**, 1413

SUNDIN, D. (1991). New options in transformer oils. *Chem. Engng*, **98**(12), 125

SUNDSTROM, D.W. and KLEI, H.E. (1979). *Wastewater Treatment* (Englewood Cliffs, NJ: Prentice-Hall)

SUNDUKOV, I.N. and PREDVODITELEV, A.S. (1959). Flame propagation in two-phase mixtures. *Combustion*, **7**, p. 352

SUNSHINE, I. (ed.) (1969). *Handbook of Analytical Toxicology* (Boca Raton, FL: CRC Press)

SUNSHINE, I. (1978). *Toxicology* (West Palm Beach, FL: CRC Press) SOUKAS, J. (1981). *Experience of Work Safety Analysis. Seventh SCRATCH Seminar on Methods for Risk and Safety Analysis* (Stockholm: NORDFORSK)

SUOKAS, J. (1982). Safety analysis of a liquefied gas storage and loading system. *J. Occup. Accid.*, **4**, 347

SUOKAS, J. (1986). Evaluation of the coverage and validity of hazard and operability study. *Loss Prevention and Safety Promotion 5*, 12–1

SUOKAS, J. (1988). The limitations of safety and risk analysis. *Preventing Major Chemical and Related Process Accidents*, p. 483

SUOKAS, J. (1989). The identification of human contribution to chemical accidents. *Loss Prevention and Safety Promotion 6*, p. 32.1

SUOKAS, J. and KAKKO, R. (1989). On the problems and future of safety and risk analysis. *J. Haz. Materials*, **21**, 105

SUOKAS, J. and ROUHIAINEN, V. (1984). *Work Safety Analysis. Method Description and User's Guide. Res. Rep. 314.* VTT, Espoo, Finland

SUOKAS, J. and ROUHIAINEN, V. (1989). Quality control in safety and risk analyses. *J. Loss Prev. Process Ind.*, **2**(2), 67

SUOKAS, J., HEINO, P. and KARVONEN, I. (1990). Expert systems in safety management. *J. Occup. Accid.*, **12**(1/3), 63

SUPP, E. (1981). Improved methanol process. *Hydrocarbon Process.*, **60**(3), 71

SUPPES, P. (1957). *An Introduction to Logic* (Princeton, NJ: van Nostrand Reinhold)

SURDI, V.L. and ROMAINE, D. (1964). Fundamentals of low temperature piping design. *Hydrocarbon Process.*, **43**(6), 116

SURGEON GENERAL (1950). *German Aviation Medicine in World War II* (Washington, DC: US Govt Printing Office)

SURGEON GENERAL'S ADVISORY COMMITTEE ON SMOKING AND HEALTH (1964). *Smoking and Health* (Washington, DC)

SURIS, A.L., FLANKIN, E.V. and SHORING, S.N. (1977). Length of free diffusion flames. *Combust. Explosion Shock Waves*, **13**

SURMAN, P.G., BODERO, J. and SIMPSON, R.W. (1987). The prediction of the numbers of violations of standards and the frequency of air pollution episodes using extreme value theory. *Atmos. Environ.*, **21**, 1843

SURPRENANT, N. (1990). Shutting off fugitive emissions. *Chem. Engng*, **97**(9), 199

SURRIDGE, A.D. and GOLDREICH, Y. (1988). On the spatial characteristics of the nocturnal stable boundary layer over a complex urban terrain. *Atmos. Environ.*, **22**, 1

SURRY, J. (1969a). *An Annotated Bibliography for Industrial Accident Research and Related Fields* (report). Labour Safety Coun. of Ontario, Toronto, Ont.

SURRY, J. (1969b). *Industrial Accident Research: a Human Engineering Appraisal* (report). Dept of Ind. Engng, Univ. of Toronto

SURTEES, L.S. and ROONEY, P. (1977). Tips on specifying FRP piping. *Chem. Engng*, **84**(Nov. 21), 215

SURUDA, A.J., PETTIT, T.A., NOONAN, G.P. and RONK, R.M. (1994). Deadly rescue: the confined space hazard. *J. Haz. Materials*, **36**, 45

SUSS, K.H. (1977). Non-contacting safety devices on machines for the chemical industry. *Prevention of Occupational Risks*, **5**, p. 75

SUSSAMS, J.E. (1977). *Industrial Logistics* (London: Gower Press)

SUSSMAN, G.J. and STALLMAN, R. (1975). Heuristic techniques in computer-aided circuit analysis. *IEEE Trans. Circuits and Systems*, **CAS-22**, 857

SUTCLIFFE, J.R. (1980). Shipping risks at Braefoot Bay. *Sci. Public Policy*, **7**, 356

SUTER, G., KUPR, T. and ANGLY, H. (1992). Reaction calorimetry at minus 80°C. *Loss Prevention and Safety Promotion 7*, vol. 2, 87–1

SUTERLIN, R. (1966). Les projectiles. *Memorial de l'Artillerie*, **40**(3)

SUTHERLAND, A.B. (1985). Burn injury – later management. *Surgery*, **1**(18), 423

SUTHERLAND, M.E. and WEGERT, H.W. (1973). An acetylene decomposition incident. *Loss Prevention*, **7**, p. 99. See also *Chem. Engng Prog.*, 1973, **69**(4), 48

SUTHERLAND, V. (1992). Behaviour under stress. *Health and Safety in the Offshore Oil and Gas Industries* (Rugby: Instn Chem. Engrs)

SUTHERLAND, W.H. (1990). Episodic ammonia release reduction in ammonia derivative units. *Ammonia Plant Safety*, **30**, p. 76

SUTTON, I.S. (1992). Writing operating procedures. *Hazard Identification and Risk Analysis, Human Factors and Human Reliability in Process Safety*, p. 293

SUTTON, I.S. (1992). *Process Reliability and Risk Management* (New York: van Nostrand Reinhold)

SUTTON, O.G. (1932). A theory of eddy diffusion in the atmosphere. *Proc. R. Soc., Ser. A*, **135**, 143

SUTTON, O.G. (1934). Wind structure and evaporation in a turbulent atmosphere. *Proc. R. Soc., Ser. A*, **146**, 701

SUTTON, O.G. (1947). The theoretical distribution of airborne pollution from factory chimneys. *Quart. J. R. Met. Soc.*, **73**, 426

SUTTON, O.G. (1949). *Atmospheric Turbulence* (London: Methuen)

SUTTON, O.G. (1950). The dispersion of hot gases in the atmosphere. *J. Met.*, **7**, 307

SUTTON, O.G. (1953). *Micrometeorology* (New York: McGraw-Hill)

SUTTON, O.G. (1956). Dispersal of airborne effluents from stacks. *Br. Chem. Engng*, **1**, 202

SUTTON, O.G. (1962). *The Challenge of the Atmosphere* (London: Hutchinson)

SUTTON, P. (1968). The design of a noise specification for process plant. *J. Sound Vib.*, **8**, 33

SUTTON, P. (1975). Process plant noise: evaluation and control. *Proc. Ninth World Petrol. Cong.*, vol. 6 (London: Applied Science), p. 243

SUTTON, P. (1977a). *The Protection Handbook of Industrial Noise Control* (London: Alan Osborne)

SUTTON, P. (1977b). *The Protection Handbook of Pollution Control* (London: Alan Osborne)

SUZUKI, M., YAMOMOTO, T., FUTAGAMI, N. and MAEDA, S. (1978). Atomisation of superheated liquid jet. *Fuel Soc., of Japan, First Int. Conf. on Liquid Atomisation and Spray Systems* (Tokyo), p. 37

SUZUKI, M., TAKAHASHI, H. and SAITO, M. (1979). Experimental determination of the stress intensity factor for a part-through thickness crack. *Int. J. Pres. Ves. Piping*, **7**, 424

SUZUKI, T. and HIRANO, T. (1982). Flame propagation across a liquid fuel in an air stream. *Combustion*, **19**, p. 877

SVANES, T. and DELANEY, J.R. (1981). SCAT: System Control Analysis and Training Simulator. In Rasmussen, J. and Rouse, W.B. (1981), *op. cit.*, p. 659

SVAROVSKY, L. (ed.) (1990). *Solid–Liquid Separation* (Oxford: Butterworth-Heinemann)

SVEHLA, R.A. and MCBRIDE, B.J. (1973). *Fortran IV Computer Program for Calculation of Thermodynamic and Transport Properties of Complex Chemical Systems. Rep. TN D-7056.* Nat. Aeronaut. Space Admin.

SVENSON, O. (1989). On expert judgements in safety analyses in the process industries. *Reliab. Engng System Safety*, **25**, 291

SVENSON, O. (1991a). The Accident Evolution and Barrier Function (AEB) model. *Risk Analysis*, **11**(3), 499

SVENSON, O. (1991b). Analysis of incidents and accidents in the processing industries. In Apostolakis, G.E. (1991), *op. cit.*, p. 271

SWAID, H. (1991). Nocturnal variation of air-surface temperature gradients for typical urban and rural surfaces. *Atmos. Environ.*, **25B, 333**

SWAIN, A.D. (1964a). Some problems in the measurement of human performance in man–machine systems. *Hum. Factors*, **6**, 687

SWAIN, A.D. (1964b). *THERP. Rep. SC-R-64–1338.* Sandia Labs, Albuquerque, NM

SWAIN, A.D. (1969). *Human Reliability Assessment in Nuclear Reactor Plants. Rep. SC-R-69–1236.* Sandia Labs, Albuquerque, NM

SWAIN, A.D. (1968). *Some Limitations in using the Simple Multiplicative Model in Behaviour Quantification. Rep. SC-R-68–1697.* Sandia Labs, Albuquerque, NM

SWAIN, A.D. (1970). *Development of a Human Error Rate Data Bank. Rep. SC-R-70–4286.* Sandia Labs, Albuquerque, NM

SWAIN, A.D. (1972). *Design Techniques for Improving Human Performance in Production* (London: Ind. & Commercial Techniques Ltd)

SWAIN, A.D. (1973a). Design of industrial jobs a worker can and will do. *Hum. Factors*, **15**, 129

SWAIN, A.D. (1973b). An Error–Cause Removal program for industry. *Hum. Factors*, **15**, 207

SWAIN, A.D. (1982). A note on the accuracy of predictions using THERP. *Hum. Factors Soc. Bull.*, **25**(4), 1

SWAIN, A.D. (1987a). *Accident Sequence Evaluation Program Human Reliability Procedure. Rep. SAND86–1996.* Sandia Labs, Albuquerque, NM. See also *NUREG/CR-4772*

SWAIN, A.D. (1987b). Relative advantage of people and machines in process industries. *Preventing Major Chemical Accidents*, p. 6.97

SWAIN, A.D. (1987c). Synopsis of human factors research needs. *Preventing Major Chemical Accidents*, p. 6.179

SWAIN, A.D. and GUTTMAN, H.E. (1980). *Handbook of Human Reliability Analysis with Emphasis on Nuclear Power Plant Applications. Draft Report for Interim Use and Comment. Rep. NUREG/CR-1278 (draft).* Nucl. Regul. Comm, Washington, DC

SWAIN, A.D. and GUTTMAN, H.E. (1983). *Handbook of Human Reliability Analysis with Emphasis on Nuclear Power Plant Applications. Final Report. Rep. NUREG/CR-1278.* Nucl. Regul. Comm, Washington, DC

SWAIN, R.T. and HOPPER, B.J. (1971). Contractor problems during start-up. *Chem. Engng Prog.*, **67**(12), 32

SWALM, R.O. (1973). Future risk analysis techniques. *Conf. on Bulk Transportation of Hazardous Materials by Water in the Future* (Washington, DC: Nat. Acad. Sci., Nat. Acad. Engng, Nat. Res. Coun.)

SWANDBY, R.K. (1962). Corrosion charts: guides to materials selection. *Chem. Engng*, **69**(Nov. 12), 186

SWANDER, A.J. and VAIL, R.E. (1992). Building procedures for repeatability and consistency in job performance. *Hazard Identification and Risk Analysis, Human Factors and Human Reliability in Process Safety*, p. 339

SWANDER, W. and POTTS, J.D. (1986). Flare system design. *Am. Petrol. Inst., Seminar on Fundamental Applications of Loss Prevention* (New York)

SWANDER, W. and POTTS, J.D. (1989). *Safer Flare System Design. Tech. Paper S490.* John Zink Co., Tulsa, OK

SWANN, H.W. (1953). *Analysis* harmful static electrification. *Br. J. Appl. Phys.*, **4**(suppl. 2), S68

SWANN, H.W. (1957). The use of electricity under inflammable conditions. *Elec. J.*, Aug., 965

SWANN, H. (1959). *Electrical Safety* (London: Macmillan)

SWANSON, C.G. (1980). Carbon dioxide removal in ammonia synthesis gas by Selexol. *Ammonia Plant Safety*, **21**, p. 152

SWANSON, K. (1983). An advanced combustion control system. *Plant/Oper. Prog.*, **2**, 30

SWARTOUT, W.R. (1983). XPLAIN: a system for creating and explaining expert system consulting programs. *Artificial Intelligence*, **21**, 285

SWEARINGEN, J.J., MCFADDEN, E.B., GARNER, J.D. and BLETHROW, J.G. (1960). Human voluntary tolerance to vertical impact. *Aerospace Med.*, **31**, 989

SWEARINGEN, J.S. (1970). Engineer's guide to turbo expanders. *Hydrocarbon Process.*, **49**(4), 97

SWEARINGEN, J.S. (1972). Turboexpanders and processes that use them. *Chem. Engng Prog.*, **68**(7), 95

SWEAT, M.E. (1983). An ammonia tank failure. *Plant/Operations Prog.*, **2**, 114

SWEENEY, J.C. (1992). Measuring process safety management. *Plant/Operations Prog.*, **11**, 89

SWEENEY, J.C. (1993). ARCO Chemicals HAZOP experience. *Process Safety Prog.*, **12**, 83

SWEET, C. (ed.) (1980). *The Fast Breeder Reactor. Need? Cost? Risk?* (London: Macmillan)

SWEET, S.W. (1976). A look at two surge control systems. *ISA Trans*, **14**(2), 172

SWEETLAND, K. (1993). Dealing with lead in the workplace. *Chem. Engng*, **100**(1), 127

SWEIS, F.K. (1987). The effect of mixtures of particle sizes on the minimum ignition temperature of a dust cloud. *J. Haz. Materials*, **14**(2), 241

SWEIS, F.K. and SINCLAIR, C.G. (1985). The effect of particle size on the maximum permissible oxygen concentration to prevent dust explosions. *J. Haz. Materials*, **10**(1), 59

SWERN, D. (ed.) (1970–). *Organic Peroxides, 1970,* vol. 1; 1971, vol. 2; 1972, vol. 3 (London: Wiley-Interscience)

SWETT, C.C. (1955). *NACA Res. Memo. E55 I 16*

SWETT, C.C. (1956). *Spark ignition of flowing gases. NACA Rep. 1287*

SWETT, C.C. (1957). Spark ignition of flowing gases using long-duration discharges. *Combustion*, **6**, p. 523

SWIERZAWSKI, T.J. and GRIFFITH, P. (1990). Preventing water hammer in large horizontal pipes passing steam and water. *J. Heat Transfer*, **112**, 523

SWIFT, I. (1981). Developments in dust explosibility testing: the effects of test variables. *The Hazards of Industrial Explosion from Dusts* (London: Oyez/IBC), paper 3

SWIFT, I. (1982). Developments in dust explosibility testing; the effect of test variables. In Lee, J.H.S. and Guirao, C.M. (1982), *op. cit.*, p. 375

SWIFT, I. (1983). Gaseous combustion venting — a simplified approach. *Loss Prevention and Safety Promotion 4*, vol. 3, p. F21

SWIFT, I. (1984). Developments in emergency relief system design. *Chem. Engr, Lond.*, **406**, 30

SWIFT, I. (1985). ERS in context. *Chem. Engr, Lond.*, **416**, 38

SWIFT, I. (1988a). Design of deflagration protection systems. J. Loss Prev. Process Ind., 1(1), 5

SWIFT, I. (1988b). Designing explosion vents easily and accurately. *Chem. Engng*, **95**(5), 65

SWIFT, I. (1988c). Developments in explosion protection. *Plant/Operations Prog.*, **7**, 159

SWIFT, I. (1989). NFPA 68 guide for venting of deflagrations: what's new and how it affects you. *J. Loss Prev. Process Ind.*, **2**(1), 5

SWIFT, I. and EPSTEIN, M. (1987). Performance of low explosion vents. *Plant/Operations Prog.*, **6**, 98

SWIFT, I., BATZ, G. and DEGOODE, B. (1986). The performance of deflagration venting systems and their effect on vent size requirements. *Loss Prevention and Safety Promotion 5*, p. 60–1

SWIFT, I., FAUSKE, H.K. and GROLMES, M.A. (1983). Emergency relief systems for runaway reactions. *Plant/Operations Prog.*, **2**, 116

SWIFT, P. (1984). A user's guide to dust control. *Chem. Engr, Lond.*, **403**, 22

SWIFT, R.L. (1963). Detection of hazardous atmospheres. NFPA Quart., **57**(2), 168

SWINBANK, C. (1973). Packaging of Chemicals (London: Newnes-Butterworths)

SWINBANK, C. (1975). Packaged goods lost at sea. Distribution Conf. (London: Chem. Ind. Ass.), p. 50

SWINBANK, W.C. (1964). The exponential wind-profile. Quart. J. R. Met. Soc., **90**, 119

SWINBANK, W.G. (1968). A comparison between predictions of dimensional analysis for the constant flux layer and observations in unstable conditions. *Quart. J. R. Met. Soc.*, **94**, 460

SWINDELLS, I., NOLAN, P.F. and PRATT, D.B. (1986). Safety aspects of the storage of heated bitumen, Hazards IX, p. 91

SWINSON, P.S.G. (1983). An introduction to Prolog. *Computer-Aided Design*, **15**(4), 241

SWISDAK, M.M. (1975). *Explosion effects and properties. Pt 1 — Explosion effects in air. Rep. NSWC/WOL/TR 75–116* Naval Surface Weapons Center, White Oak, Silver Spring, MD

SWITHENBANK, J. (1972). Ecological aspects of combustion devices (with reference to hydrocarbon flaring). *AIChE J.*, **18**, 553

SWITHENBANK, J., SAYAS, D. and BOYSEN, F. (1991). Fluid dynamics on a PC. *Chem. Engr, Lond.*, **503**, 16

SYED, K.J., STEWART, C.D. and MOSS, J.B. (1991). Modelling soot formation and thermal radiation in buoyant turbulent diffusion flames. *Combustion*, **23**, p. 1533

SYKES, B.A. and BROWN, D. (1975). *A review of the technology of high pressure system. Commun. 973.* Instn Gas Engrs, London. See also *Instn Gas Engrs J.*, **15**, 262

SYKES, R.I. and HENN, D.S. (1992). Large-eddy simulation of concentration fluctuations in a dispersing plume. *Atmos. Environ.*, **26A**, 3127

SYKES, W.G. (1966). Hazards of potentially explosive materials. *Chem. Engng Prog.*, **62**(12), 49

SYKES, W.G., JOHNSON, R.H. and HAINER, R.M. (1963). Ammonium nitrate explosion hazards. *Chem. Engng Prog.*, **59**(1), 66

SYKES, W.G., MEYERS, S., PARKS, J.R. and CHANDLER, S.S. (1963). The effect of low concentrations of organic materials on the properties of ammonium nitrate. *Ammonia Plant Safety*, **5**

SYLVANDER, N.E. and KATZ, D.L. (1948). Design and construction of pressure relieving systems. *Engng Res. Bull. 31* (Univ. of Michigan Press)

SYLVESTER-EVANS, R. (1990a). 'ackground» — to the Piper Alpha tragedy. Piper Alpha: Lessons for Life Cycle Management (Rugby: Instn Chem. Engrs), p. 121

SYLVESTER-EVANS, R. (1990b). Piper Alpha — the gas leak and its tragic consequences. *Chem. Engr, Lond.*, **485**, 3

SYLVESTER-EVANS, R. (1991). 'Background' — to the Piper Alpha tragedy. *Process Safety Environ.*, **69B**, 3

SYLVIA, D. (1972a). HI system labels and placards proposal. *Fire Engng*, **125**, Dec., 22

SYLVIA, D. (1972b). Positioning pumpers and men for flammable liquid fires. *Fire Engng*, **125**, Mar., 12

SYLVIA, D. (1972c). Standard pressures for pump operations. *Fire Engng*, **125**, Dec., 16

SYLVIA, D. (1972d). Symposium evaluates halons. *Fire Engng*, **125**, Sept., 32

SYLVIA, D. (1972e). Using foam to extinguish flammable liquid fires. *Fire Engng*, **125**, Feb., 14

SYLVIA, D. (1973). Fire Commission report ask US administration, academy. *Fire Engng*, **126**, Jun., 36

SYLVIA, D. (1974). HI system near adoption by DOT for identifying hazardous materials. *Fire Engng*, **127**, Aug., **43**

SYMALLA, M. (1984). How to effectively select fire-safe valves and actuators. *Hydrocarbon Process.*, **63**(7), 83

SYMES, W.L. (1915a). Note on the treatment of the symptoms arising from inhalation of irritant gases and vapours. *Br. Med. J.*, Jul. 3, 12

SYMES, W.L. (1915b). Poisonous gases (letter). *Br. Med. J.*, Jun. 12, 1028

SYMONDS, P.S. (1968). Plastic shear deformations in dynamic load problems. *Proc. Engineering Plasticity Conf.* (Cambridge; Cambridge Univ. Press), p. 647

SYMONS, W.G. (1969). The role of HM Factory Inspectorate. In Handley, W. (1969), *op. cit.*, p. 407

SYNGE, J.L. and GRIFFITHS, B.A. (1960). *Principles of Mechanics*, 3rd ed. (New York: McGraw-Hill)

SYRBE, M. (1981). Automatic error detection and error recording of a distributed, fault-tolerant process computer system. In Rasmussen, J. and Rouse, W.B. (1981), *op. cit.*, p. 475

SYSTEM ENGINEERING COMMITTEE (1961). Application of probability methods to generating capacity problems. AIEE Trans., **80**, 1165

SZABO, Z.G. (1959). Some problems of elementary reactions in gas-phase slow combustion. *Combustion*, **7**, p. 118

SZARGUT, J., MORRIS, D.R. and STEWARD, F.R. (1988). *Energy Analysis for Thermal, Chemical and Metallurgical Processes* (New York: Hemisphere Publ.)

SZEKELY, J. and APELIAN, D. (eds) (1984). *Plasma Processing and Synthesis of Materials* (Amsterdam: Elsevier)

SZELINSKI, B.A. (1980). Pollution control and law in practice. The German approach. *Chem. Ind.*, April 19, 342

SZERI, A.Z. (1980). *Tribology, Friction, Lubrication and Wear* (Washington, DC: Hemisphere Publ. Co.)

SZILAS, A.P. (1975). *Production and Transport of Oil and Gas* (Amsterdam: Elsevier)

SZOLOVITS, P. (ed.) (1982). *Artificial Intelligence in Medicine* (Boulder, CO: Westview Press)

TABACK, H.J. *et al.* (1977). *Control of Hydrocarbon Emissions from Stationary Sources in the California South Coast Air Basin.* (prelim. rep.). KVB Engng, Tustin. CA

TABAR, D.C. (1989). Foam-water sprinkler protection for flammable liquids. *Plant/Operations Prog.*, **8**, 218

TABATA, Y., KODAMA, T. and KOTOYORI, T. (1987). Explosion hazards of chlorine drying towers. *J. Haz. Materials*, **17**(1), 47

TABERSHAW, I.R. (1982). The impact of TSCA on the practice of occupational medicine. In Fawcett, H.H. and Wood, W.S. (1982), *op. cit.*, p. 189

TABERSHAW, I.R. and LAMM, S.H. (1977). Benzene and leukaemia (letter). *The Lancet*, **2**(8043), 867

TABOREK, J., AOKI, T., RITTER, R.B., PALEN, J.W. and KNUDSEN, J.G. (1972a). Fouling: the major unresolved problem in heat transfer. *Chem. Engng Prog.*, **68**(2), 59

TABOREK, J. AOKI, T., RITTER, R.B., PALEN, J.W. and KNUDSEN, J.G. (1972b). Predictive methods for fouling behaviour. *Chem. Engng Prog.*, **68**(7), 69

TABORELLI, R.V. and BOWEN, I.G. (1957). *Tertiary Effects of Blast – Displacement. Rep. ITR-1469.* Civil Effects Test Group, Atom. Energy Comm., Office of Tech. Services, Dept of Commerce, Washington, DC

TABORELLI, R.V., BOWEN, I.G. and FLETCHER, E.R. (1959). *Tertiary Effects of Blast – Displacement. Rep. WT- 1469.* Civil Effects Test Group, Atom. Energy Comm., Office of Tech. Services, Dept of Commerce, Washington, DC

TACHAKRA, S.S. (1988). The human cost of Bhopal – 'an estimate'. *J. Loss Prevention Process Ind.*, **1**(1), 3

TADA, H., PARIS, P.C. and IRWIN, G.R. (eds) (1973). *Stress Analysis of Cracks Handbook;* 1985, 2nd ed. (St Louis, MO: Del Res. Corp.)

TADAO, N. (1993). Pipeless plants boost batch processing. *Chem. Engng*, **100**(6), 102

TAFT, J.R. (1981). *Simulations of Experimental Spills Using the MARIAH Model* (report). Deygon-Ra Inc.

TAFT, J.R., RYNE, M.S. and WESTON, D.A. (1983). MARIAH: a dispersion model for evaluating realistic heavy gas spill scenarios. *Proc. AGA Transmission Conf.* (Arlington, VA: Am. Gas Ass.)

TAGLIAZUCCA, M. and NANNI, T. (1983). An atmospheric diffusion classification scheme based on the Kazanski–Monin stability parameter. *Atmos. Environ.*, **17**, 2205

TAGOE, C. and RAMHARRY, W. (1993). Are there contaminants in your feedstream? *Hydrocarbon Process.*, **72**(1), 117

TAGUCHI, G. (1979). *Introduction to Off-Line Quality Control* (Magaya, Japan: Japan Quality Control Assn)

TAGUCHI, G. (1981). *On-Line Quality Control during Production* (Tokyo: Japanese Standards Association)

TAGUCHI, G., ELSAYED, E.A. and HSIANG, T.C. (1989). *Quality Engineering in Production Systems* (New York: McGraw-Hill)

TAIG, A.R. (1980). Radioactive materials packages and the British railway accident environment. *Sixth Int. Symp. on Packaging and Transportation of Radioactive Materials*, p. 190

TAILLET, J. (1991). Investigation procedures applied to major electrostatic accidents in complex systems. *Electrostatics*, **8**, p. 63

TAILLET, J., POIDRAS, P., GLOR, M. and MAURER, B. (1992). Testing a new device for suppressing electrostatic charges on polymer granules. *J. Loss Prev. Process Ind.*, **5**, 235

TAIT, N. (1976). *Asbestos Kills* (London: The Silbury Fund)

TAIT, S.R. (1981). Dust dispersion systems – which is best? *The Hazards of Industrial Explosion from Dusts* (London: Oyez/IBC), paper 15

TAIT, S.R., REPUCCI, R.G. and TOU, J.C. (1980). The effects of fumigants on grain dust explosions. *J. Haz. Materials*, **4**(2), 177

TAITEL, Y. and DUKLER, A.E.M. (1976). *Int. J. Multiphase Flow*, **2**, 591

TAKAHASHI, F. and GLASSMAN, I. (1984). Sooting correlations for premixed flames. *Combust. Sci. Technol.*, **37**, 1. See also 1987, **54**, 407, 410

TAKAHASHI, F., MIZOMOTO, M. and IKAI, S. (1982). Transition from laminar to turbulent free jet diffusion flames. *Combust. Flame*, **48**, 85

TAKAHASHI, K., MIYAKE, T., INOMATA, T., MORIWAKI, T. and OKAZAKI, S. (1991). Influence of obstacles on flame propagation and gas movement ahead of flame in an open space – effects of nets laid on floor. *Combust. Flame*, **84**, 110

TAKAI, R., YONEDA, K. and HIKITA, T. (1975). Study of detonation wave structure. *Combustion*, **15**, p. 69

TAKAO, S. and NARUSAWA, U. (1980). An experimental study of heat and mass transfer across a diffusive interface. *Int. J. Heat Mass Transfer*, **23**, 1283

TAKAO, S. and SUZUKI, O. (1983). Prediction of LNG tank roll-over. *Gastech, 82*, p. 392

TAKEDA, H. (1988). Model experiments of ship fire. *Combustion*, **22**, 1311

TAKEMURA, K., SHIBASAKI, T. and KAWAI, T. (1981). Stress corrosion cracking of heat-resisting alloys. *Ammonia Plant Safety*, **23**, 99

TAKENO, K. and HIRANO, T. (1989). Behaviour of combustible liquid soaked in porous beds during flame spread. *Combustion*, **22**, 1223

TAKENO, K., ICHINOSE, T., HYODO, Y. and NAKAMURA, H. (1994). Evaporation rates of liquid hydrogen and liquid oxygen spilled onto the ground. *J. Loss Prev. Process Ind.*, **7**, 425

TAKENO, T. (1977). Ignition criterion by thermal explosion theory. *Combust. Flame*, **29**, 209

TAKENO, T. and SATO, K. (1980). Effect of oxygen diffusion on ignition and extinction of self-heating in porous bodies. *Combust. Flame*, **38**, 75

TAKI, S. and FUJIWARA, T. (1984). Numerical simulations on the establishment of gaseous detonation. In Bowen, J.R. *et al.* (1984b), *op, cit.*, p. 186

TALBOT, J.P.P. (1977). The bath tub myth. *Quality Assurance*, **3**(Dec.)

TALBOT, J.S. (1968). Water pollution control. Deep wells. *Chem. Engng*, **75**(Oct. 14), 108

TALBOT, J.S. and BEARDON, P. (1964). The deep well method of industrial waste disposal. *Chem. Engng Prog.*, **60**(1), 49

TALLEY, D.G. (1992). Universal definitions of mixture fractions and progress variables for arbitrary states of pre-mixing. *Combust. Flame*, **88**, 83

TALMAGE, W.P. (1971). Safe vapor phase oxidation. *Chem. Tech.*, Feb., 117

TALTY, J.T. (1986). Integrating safety and health issues into engineering school curricula. *Chem. Engng Prog.*, **82**(10), 13

TALWAR, M. (1983). Analyzing centrifugal-pump circuits. *Chem. Engng*, **90**(Aug. 22), 69

TAM, R.Y. and LUDFORD, G.S.S. (1988). The lean flammability limit: a four step model. *Combust. Flame*, **72**, 35

TAM, V.H.Y. (1989). A simple transient release rate model for releases of LPG from pipelines. *Loss Prevention and Safety Promotion 6*, p. 115.1

TAM, V.H.Y. and COWLEY, L.T. (1989). Consequence of pressurised LPG release Å a full-scale experiment. *Gastech 88*, paper 4.3

TAM, V.H.Y. and HIGGINS, R.B. (1990). Simple transient release rate models for releases of pressurised liquid petroleum gas from pipelines. *J. Haz. Materials*, **25**(1/2), 193

TAMANINI, F. (1976). A study of the extinguishment of vertical wood slabs in self-sustaining burning by water spray application. *Combust. Sci. Technol.*, **14**, 1

TAMANINI, F. (1977). Reaction rates, air entrainment and radiation in turbulent fire plumes. *Combust. Flame*, **30**, 85

TAMANINI, F. (1982). Dust explosion research at FMRC. In Lee, J.H.S. and Guirao, C.M. (1982), *op. cit.*, p. 645

TAMANINI, F. (1990). Turbulence effects on dust explosion venting. *Plant/Operations Prog.*, **9**, 52

TAMANINI, F. (1995). An improved correlation of experimental data on the effects of ducts in vented dust explosions, *Loss Prev. Process Ind.*, **1**, 243

TAMANINI, F. (1996). Vent sizing in partial-volume deflagrations and its application to the case of spray driers. *J. Loss Prev. Process Ind.*, **9**(5), 339

TAMANINI, F. (2002). Development of an engineering tool to quantify the explosion hazard of flammable liquid spills. *AIChE 36th Loss Prev. Symp.* New Orleans, Paper LPS-4c Mar. 10–13,

TAMANINI, F. and CHAFFEE, J.L. (1992a). Combined turbulence and flame instability effects in vented gas explosions. *Loss Prevention and Safety Promotion 7*, vol. 1, 38–1

TAMANINI, F. and CHAFFEE, J.L. (1992b). Turbulent vented gas explosions with and without acoustically-induced instabilities. *Combustion*, **24**, p. 1845

TAMANINI, F. and VALIULIS, J.V. (1996). Improved guidelines for the sizing of vents in dust explosions, *J. Loss Prev. Process Ind.* **9**(1), 105

TAMANINI, F., URAL, E.A., CHAFFEE, J.L. and HOSLER, J.F. (1989). Hydrogen combustion experiments in a 1/4 scale model of a nuclear power plant. *Combustion*, **22**, 1715

TAMAS, J. (1986). Determining physical properties for multicomponent hydrocarbon mixtures. *Chem. Engng*, **93**(18), Sep. 29, 103

TAMIR, A. (1994). *Impinging-Stream Reactors* (Amsterdam: Elsevier)

TAMM, H., HARRISON, W.C. and KUMAR, R.K. (1982). Hydrogen combustion research at WNRE. In Lee, J.H.S. and Guirao, C.M. (1982), *op. cit.*, p. 893

TAMMANI, B. (1987). Avoid partial condenser flooding. *Hydrocarbon Process.*, **66**(8), 43

TAMPLIN, A.B. (1973). The BEIR report: a focus on issues. *Bull. Atom. Scientists/Sci. Publ. Affairs*, Mar., 19, 47

TAMURA, M., ISHIDA, H., ITOH, M., YOSHIDA, T., WATANABE, M., MURANAGA, K., ABE,T. and MORISAKI, S. (1987). Evaluation of the deflagration hazards of organic peroxides by the revised time-pressure test. *J. Haz. Materials*, **17**(1), 89

TAN, C.C. and DAWSON, B. (1983). Application of adaptive noise cancellations to the condition monitoring of rolling bearing elements. In Butcher, D.W. (1983), *op. cit.*, p. 167

TAN, M.A., KUMAR, R.P. and KUILANOFF, G. (1984). Modular design and construction. *Chem. Engng*, **91**(May 28), 89

TAN, S.H. (1967a). Flare system design simplified. *Hydrocarbon Process.*, **46**(1), 172

TAN, S.H. (1967b). Simplified flare system sizing. *Hydrocarbon Process.*, **46**(10), 149

TANAKA, M. (1974). Fatigue life estimation of bellows based on elastic-plastic calculations. *Int. J. Pres. Ves. Piping*, **2**, 51

TANAKA, N., WADA, Y., TAMURA, M. and YOSHIDA, T. (1990). Performance of pressure vessel test concerned with heating of pressure vessel and bursting pressure of rupture disk. *J. Haz. Materials*, **23**(1), 89

TANAKA, T. and UMEKAWA, N. (1985). The prediction of sloshing pressure in prismatic tanks of LNG carriers. *Gastech 84*, p. 275

TANAKA, Y. (1989). Numerical simulations for combustion of quiescent and turbulent mixtures in confined vessels. *Combust. Flame*, **75**, 123

TANCREDE, M., WILSON, R., ZEISE, L. and CROUCH, E.A.C. (1987). The carcinogenic risk of some organic vapors indoors: a theoretical survey. *Atmos. Environ.*, **21**, 2187

TANDY, T.F. and THOMAS, L.R. (1979). Start-up damage to a carbon dioxide absorber. *Ammonia Plant Safety*, **21**, p. 9

TANFORD, C. (1947). Theory of burning velocity. I, Temperature and free radical concentrations near the flame front, relative importance of heat conduction and diffusion. *J. Chem. Phys.*, **15**, 433

TANFORD, C. and PEASE, R.N. (1947). Theory of burning velocity. II, The square root law for burning velocity. *J. Chem. Phys.*, **15**, 861

TANG, S.S. (1968). Short cut method for calculating tower deflection. *Hydrocarbon Process.*, **47**(11), 230

TANGREN, R.F., DODGE, C.H. and SEIFERT, H.S. (1949). Compressibility effects in two-phase flow. *J. Appl. Phys.*, **20**, 637

TANI, M., YAMASHI, K., TANIKOSHI, K., FUTAKAWA, M. and TANIFUJI, S. (1992). Object oriented video: interaction with real-world objects through live video. *Assoc. Comput. Machinery, Proc. Conf. on Human Factors in Computing Systems (CHI)*, p. 593

TANIMOTO, S. (1974). Ethylene explosion experiment provides useful data. *Safety in Polyethylene Plants*, **2**, p. 14

TANIMOTO, S.L. (1990). *The Elements of Artificial Intelligence Using Common LISP* (New York: Computer Sci. Press)

TANK, R.W. (1973). *Focus on Environmental Geology* (Oxford: Oxford Univ. Press)

TANKER SAFETY GROUP (1977). Tanker safety. Interim Report of the Tanker Safety Group. *Trade Ind.*, **26**(8), Feb. 25, 483

TANNE (1970). Accidents caused by the transfusion of liquid oxygen. *Prevention of Occupational Risks*, **1**

TANNER, A.C. (1976). Fighting fires in very large crude carriers. *Fire Int.*, **52**, 28

TANNER, R.K. (1994). Computer-based aids for pressure relief contingency assessment. *Hazards XII*, p. 467

TANSY, M.F., KENDALL, F.M., FANTASIA, J., LANDIN, W.E. and OBERLY, R. (1981). Acute and subchronic toxicity studies of rats exposed to vapours of methyl mercaptan and other sulphur-reduced compounds. *J. Toxicol. Environ. Health*, **8**, 71

TANZER, E.K. (1963). Comparing refrigeration systems. *Chem. Engng*, **70**(Jun. 10), 215; (Jun. 24), 105

TAO, T.M. and WATSON, A.T. (1987). A data fitting algorithm with shape preserving features. *AIChE J.*, **33**, 1567

TAO, T.M. and WATSON, A.T. (1988). An adaptive algorithm for fitting with splines. *AIChE J.*, **34**, 1722

TAPLIN, G.V., CHOPRA, S., YANDA, R.L. and ELAM, D. (1976). Radionuclide lung-imaging procedures in the assessment of injury due to ammonia inhalation. *Chest*, **69**, 582

TAPROOT — SYSTEMS IMPROVEMENTS INC., Knoxville, Tn, www.taproot.com

TARAKAD, R.R., DURR, C.A. and HUNT, R. (1987). Modular engineering — applications in liquefaction plant design. *Gastech 86*, p. 403

TARHAN, M.O. (1983). *Catalytic Reactor Design* (New York: McGraw-Hill)

TARLAS, H.D. (1970). The selection and use of protective coatings. *Chem. Engng Prog.*, **66**(8), 38

TARPA, A. (1977). Clinical biological changes due to the effect of vinyl chloride. *Prevention of Occupational Risks*, **5**, p. 249

TARR, J. and DAMME, C. (1978). The special relevance of cancer. *Chem. Engng*, **85**(Apr. 24), 86

TARRANTS, W. (1965). Applying measurement concepts to the appraisal of safety performance. *J. ASSE*, May

TARRANTS, W.E. (1967). *A Selected Bibliography of Reference Materials in Safety Engineering and Related Uses* (Chicago, IL: Am. Soc. Safety Engrs)

TARRANTS, W. (1970). A definition of the safety measurement problem. *J. Safety Res.*, Sep.

TARRANTS, W.E. (1973). The evaluation of safety program effectiveness. In Widner, J.T. (1973), *op. cit.*, p. 387

TARVER, C.M. (1982). Chemical energy release in one-dimensional detonation waves in gaseous explosives. *Combust. Flame*, **46**, 111

TASK FORCE OF PAST PRESIDENTS OF THE SOCIETY OF TOXICOLOGY (1982). Animal data in hazard evaluation: paths and pitfalls. *Fundam. Appl. Toxicol.*, **2**, 101

VAN TASSEL, G.W. (1983). Training of LNG carrier personnel. *Gastech 82*, p. 271

TASUCHER, W.A. and STREIFF, F.A. (1979). Static mixing of gases. *Chem. Engng*, **75**(4), 61

TATARELLI, G. (1946). Cumulative chlorine poisoning on board a submarine. *Ann. Naval Colonial Med.*, **51**, 337

TATE, A. (1976). Project planning using a hierarchi nonlinear planner. Rep. 25. Artificial Intelligence Dept, Edinburgh Univ.

TATE, A. (1977). Generating project networks. *Proc. Fifth Int. Joint Conf. on Artificial Intelligence* (San Mateo, CA: Morgan-Kaufmann), p. 888

TATE, R.W. (1970). Estimating liquid discharge from pressurised containers. *Chem. Engng*, **77**(Nov. 2), 126

TATEM, P.A., GANN, R.G. and CARHART, H.W. (1973a). Pressurization with nitrogen as an extinguishant for fires in confined spaces. *Combust. Sci. Technol.*, **7**, 213

TATEM, P.A., GANN, R.G. and CARHART, H.W. (1973b). Pressurization with nitrogen as an extinguishant for fires in confined spaces. II, Cellulosic and fabric fuels. *Combust. Sci. Technol.*, **9**, 255

TATEM, P.A., WILLIAMS, F., NDUBIZU, C.C. and RAMAKER, D.E. (1986). Influence of complete enclosure on liquid pool fires. *Combust. Sci. Technol.*, **45**, 185

TATHAM, F., JENNINGS, D. and KLAHN, J. (1986). Mossmorran NGL fractionation plant computer-based operation and control system. *Oil Gas J.*, **84**(Jan. 13), 60

TATHAM, M. and MILES, J.E. (1920). The Friends Ambulance Unit 1914–1919 (London: Swarthmore Press)

TATKEN, R.L. and LEWIS, R.J. (1984). *Registry of Toxic Effects of Chemical Substances*, 1981–1982 ed. (Cincinnati, OH: National Inst. Occup. Safety Health)

TATOR, K.B. (1970). Trends in materials applications. *Chem. Engng*, **77**(Oct.), 14, 137

TAUBE, A.T. (1974). Decoke furnace tubes faster. *Hydrocarbon Process.*, **53**(4), 151

TAVEL, N.G., MARAVEN, S.A. and TAYLOR, J.R. (1989). Major hazards assessment for contingency planning. *Loss Prevention and Safety Promotion 6*, p. 64.1

TAVENER, S.J. and CLARK, J.H. (1997). Recent highlights in phase transfer catalysis. *Chem. Ind.* (Jan. 6), 22–4

TAVERNA, M. and CHITI, M. (1970). Compare routes to caprolactam. *Hydrocarbon Process.*, **49**(11), 137

TAVLARIDES, L.L. (1985). *Process Modifications for Industrial Pollution Source Reduction* (Chelsea, MD: Lewis Publs)

TAYLER, C. (1984a). Fife ethylene: countdown to completion. *Process Engng*, July, 21

TAYLER, C. (1984b). How can we improve the plant's economic performance? *Process Engng*, Nov., 27

TAYLER, C. (1984c). Valve manufacture and the lure of BS 5351. *Process Engng*, Aug., 31

TAYLER, C. (1985a). 'As built' piping database means simpler certification. *Process Engng*, Nov., 33

TAYLER, C. (1985b). Do you have a mind for 'distributed intelligence'? *Process Engng*, Aug., 36

TAYLER, C. (1985c). Inspection and QA procedures still causing problems. *Process Engng*, May, 55

TAYLER, C. (1985d). Integrating process control and production management. *Process Engng*, July, 40

TAYLER, C. (1985e). Progress and obstacles along the CAD/CAE path. *Process Engng*, June, 39

TAYLER, C. (1985f). Public attitudes to pollution and waste management. *Process Engng*, May, 32

TAYLER, C. (1986a). All change in the explosives business. *Process Engng*, Oct., 83

TAYLER, C. (1986b). From panel boards to computer screens. *Process Engng*, Oct., 59

TAYLER, C. (1986c). Using triple redundancy to increase plant availability. *Process Engng*, Aug., 30

TAYLER, C. (1986d). Where do you put the computer in today's control system? *Process Engng*, Dec., 29

TAYLER, C. (1986e). Will radio transmitters now replace instrument cabling? *Process Engng*, Jan., 37

TAYLER, C. (1987a). Reduce instrument costs: go talk to your piping engineer. *Process Engng*, July, 33

TAYLER, C. (1987b). What benefit expert systems for supervisory control? *Process Engng*, Apr., 45

TAYLER, C. (1988a). Let's get physical. *Process Engng*, Feb., 57

TAYLER, C. (1988b). Moving towards more expert process management. *Process Engng*, Apr., 63

TAYLOR, LORD (1967). *First Aid in the Factory*, 3rd ed. (London: Longmans Green)

TAYLOR, A. (1979). Comparison of predicted and actual hazard rates. *Second Nat. Reliab. Conf.*, paper 4B/1

TAYLOR, A. (1981). Comparison of predicted and actual reliabilities on a chemical plant. *Reliable Production in the Process Industries* (Rugby: Instn Chem. Engrs), p. 105

TAYLOR, A.H. (1981). Oxygen. *Kirk-Othmer Encyclopaedia of Chemical Technology*, 3rd ed., vol. 16 (New York: Wiley), p. 653

TAYLOR, B.G. (1989). *Piper Alpha Public Inquiry. Transcript, Day 131*

TAYLOR, C. (1985). The ins and outs of bellows design. *Chem. Engr, Lond.*, **417**, 40

TAYLOR, C.C., SPIEGELHALTER, D.J. and MICHIE, D. (1994). *Evaluating the Quality of Machine Learning Methods* (Chichester: Ellis Horwood)

TAYLOR, D. (1978). *Mercury as an Environmental Pollutant. A Bibliography*, 5th ed. (Brixham, Derbyshire: Imperial Chemical Industries Ltd)

TAYLOR, D.E. and PRICE, C.F. (1971). Velocity of fragments from bursting gas reservoirs. *J. Engng Ind.*, **93B**, 981

TAYLOR, F. (1975). A fire service point of view. *Instn Chem. Engrs, Symp. on Transport of Hazardous Materials* (London)

TAYLOR, G. (chrmn) (1991). *Report of the Halons Technical Options Committee.* Technology and Economics Review Panel, UN Environment Programme

TAYLOR, G.F. (1954). *Elementary Meteorology* (Englewood Cliffs, NJ: Prentice-Hall)

TAYLOR, G.I. (1915). Eddy motion in the atmosphere. *Phil Trans R. Soc., Ser. A*, **215**, 1

TAYLOR, G.I. (1921). Diffusion by continuous movements. *Proc. London Math. Soc.*, **20**, 196

TAYLOR, G.I. (1927). Turbulence. *Quart. J. R. Met. Soc.*, **53**, 201

TAYLOR, G.I. (1946). The air wave surrounding an expanding sphere. *Proc. R. Soc., Ser. A*, **186**, 273

TAYLOR, G.I. (1950). The dynamics of the combustion products behind plane and spherical detonation fronts in explosives. *Proc. R. Soc., Ser. A*, **200**, 235

TAYLOR, G.I. SIR (1963a). Analysis of the explosion of a long cylindrical bomb detonated at one end. *Scientific Papers of Sir Geoffrey Ingram Taylor* (ed. G.K. Batchelor), vol. 3 (Cambridge: Cambridge Univ. Press)

TAYLOR, G.I. SIR (1963b). The fragmentation of tubular bombs. *Scientific Papers of Sir Geoffrey Ingram Taylor* (ed. G.K. Batchelor), vol. 3 (Cambridge: Cambridge Univ. Press)

TAYLOR, H.D. (1976). What happens when loss prevention fails. *Chem. Engr, Lond.*, **308**, 263

TAYLOR, H.D. (1977). *Flixborough: The Implications for Management* (London: Keith Shipman Developments)

TAYLOR, H.D. and REDPATH, P.G. (1971). Incidents resulting from process design and operating deficiencies. *Major Loss Prevention*, p. 41

TAYLOR, H.D. and REDPATH, P.G. (1972). Incidents resulting from process design and operating deficiencies. *Fire Prevention Sci. Technol.*, **1**, 12

TAYLOR, I. (1967). What NPSH for dissolved gases. *Hydrocarbon Process.*, **46**(8), 133

TAYLOR, I. (1987). Pump bypasses now more important. *Chem. Engng*, **94**(7), 53

TAYLOR, J. (1952). *Detonation in Condensed Explosives* (Oxford: Clarendon Press)

TAYLOR, J.F., GRIMMETT, H.L. and COMINGS, E.W. (1951). Isothermal free jets of air mixing with air. *Chem. Engng Prog.*, **47**(4), 175

TAYLOR, J.H. (ed.) (1965). *Project Sand Storm. An Experimental Program in Atmospheric Diffusion. Rep. AFCRL-65−649*, vol. II. Air Force Cambridge Res. Labs

TAYLOR, J.R. (1973). *A Formalisation of Failure Mode Analysis of Control Systems. Rep. Risö-M-1654.* Atom. Energy Comm., Res. Estab, Risö, Denmark

TAYLOR, J.R. (1974a). *Design Errors in Nuclear Power Plant. Rep. Risö-M-1742.* Atom. Energy Comm., Res. Estab, Risö, Denmark

TAYLOR, J.R. (1974b). *Proving Correctness for a Real Time Operating System. Rep. Risö-M-1708.* Atom. Energy Comm., Res. Estab, Risö, Denmark

TAYLOR, J.R. (1974c). *A Semiautomatic Method for Qualitative Failure Mode Analysis. Rep. Risö-M-1707.* Atom. Energy Comm., Res. Estab, Risö, Denmark

TAYLOR, J.R. (1974d). *Sequential Effects in Failure Mode Analysis. Rep. Risö-M-1740.* Atom. Energy Comm., Res. Estab, Risö, Denmark

TAYLOR, J.R. (1975a). Sequential effects in failure mode analysis. In Barlow, R.E., Fussell, J.B. and Singpurwalla, N.D. (1975), *op. cit.,* p. 881

TAYLOR, J.R. (1975b). *Some Assumptions Underlying Reliability Prediction. Rep. Risö-M-1794.* Atom. Energy Comm., Res. Estab, Risö, Denmark

TAYLOR, J.R. (1975c). *A Study of Abnormal Occurrence Reports. Rep. Risö-M-1837.* Atom. Energy Comm., Res. Estab, Risö, Denmark

TAYLOR, J.R. (1976a). *Common Mode and Coupled Failure. Rep. Risö-M-1826.* Atom. Energy Comm., Res. Estab, Risö, Denmark

TAYLOR, J.R. (1976b). *Experience with a Safety Analysis Procedure. Rep. Risö-M-1878.* Atom. Energy Comm., Res. Estab, Risö, Denmark

TAYLOR, J.R. (1976c). *Interlock Design using Fault Tree and Cause Consequence Analysis. Rep. Risö-M-1890.* Atom. Energy Comm., Res. Estab, Risö, Denmark

TAYLOR, J.R. (1978a). *Cause−Consequence Diagrams.* NATO Advanced Study Inst. on Synthesis and Analysis Methods for Safety and Reliability Studies, Urbino, Italy

TAYLOR, J.R. (1978b). In discussion of paper by Fussell, J.B. (1978a), *op. cit.*

TAYLOR, J.R. (1979). *A Background to Risk Analysis,* vols 1−4. (report). Risö Nat. Lab., Risö, Denmark

TAYLOR, J.R. (1982). An algorithm for fault tree construction. *IEEE Trans. Reliab.*, **R-31**, 137

TAYLOR, J.R. (1984). Practical utilization of safety analysis results (abstract). *J. Occup. Accid.*, **6**(1/3), 213

TAYLOR, J.R. (1992). A comparative evaluation of safety features based on risk analysis for 25 plants. *Loss Prevention and Safety Promotion 7*, vol. 1, 9−1

TAYLOR, J.R. (1993). Use of simulation for emergency planning and training. *Health, Safety and Loss Prevention*, p. 387

TAYLOR, J.R. (1994). *Risk Analysis for Process Plant, Pipelines and Transport* (London: Spon)

TAYLOR, J.R. and HOLLO, E. (1977a). *Algorithms and Programs for Consequence Diagram and Fault Tree Construction. Rep. Risö-M-1907.* Atom. Energy Comm., Res. Estab, Risö, Denmark

TAYLOR, J.R. and HOLLO, E. (1977b). Experience with algorithms for automatic failure analysis. *Soc. Ind. Appl. Math., Conf. on Nuclear Systems Reliability Engineering and Risk Assessment* (Philadelphia, PA)

TAYLOR, J.R. and OLSEN, J.V. (1983). A comparison of automatic fault tree construction with manual methods of hazard analysis. *Loss Prevention and Safety Promotion 4*, vol. 1, p. J28

TAYLOR, J.R., SELIG, R.S., NIELSEN, H., PETERSEN, C.G. and ANDERSEN, A. (1986). A design review approach to safety analysis. *Loss Prevention and Safety Promotion 5*, 11−1

TAYLOR, K. (1988). The engineer and liability. *Plant Engr*, **32**(4), 28

TAYLOR, L.S. (1971). *Radiation Protection Standards* (London: Butterworths)

TAYLOR, M. (1991). Maintenance management. *Plant Engr*, **35**, Mar./Apr., 37

TAYLOR, M.W., LIND, C.D. and CECE, J.M. (1978). Vapor cloud explosion study. *Control of Hazardous Material Spills*, 4, p. 406

TAYLOR, P.B. and FOSTER, P.J. (1974). The total emissivities of luminous and non-luminous flames. *Chem. Engng Sci.*, **17**, 1591

TAYLOR, P.B. and FOSTER, P.J. (1975). Some gray gas weighting coefficients for $CO_2−H_2O−$soot mixtures. *Chem. Engng. Sci.*, **18**, 1331

TAYLOR, P.H. (1985a). On the role of partial confinement in the generation of fast flames. *Combust. Flame,* **44,** 161

TAYLOR, P.H. (1985b). Vapor cloud explosions in the directional blast wave from an elongated cloud with edge ignition. *Combust. Sci. Technol.,* **44,** 207

TAYLOR, P.H. (1986). Fast flames in a vented duct. *Combustion,* **21,** 1601

TAYLOR, P.H. and BIMSON, S.J. (1989). Flame propagation along a vented duct containing grids. *Combustion,* **22,** 1355

TAYLOR, P.W. (1964). How to operate direct fired heaters safely. In Vervalin, C.H. (1964a). *op. cit.,* p. 252

TAYLOR, R.J., WARNER, J. and BACON, N.E. (1970). Scale length in atmospheric turbulence as measured from an aircraft. *Quart. J. R. Met. Soc.,* **96,** 750

TAYLOR, R.M. and LEWIS, W.I. (1987). Using robots to reduce personnel radiation exposure at the Savannah River plant. *Plant/Operations Prog.,* **6,** 165

TAYLOR, T.E. (1968–69). High temperature testing of pressure vessels. *Proc. Instn. Mech Engrs,* **183,** pt 1, 51

TAYLOR, T.H. (1971). Liquefied petroleum gases. Paper presented at Cranfield Inst. Technol.

TAYLOR, W.A. (1988). *What Every Engineer Should Know about Artificial Intelligence* (Cambridge, MA: MIT Press)

TAYLOR, W.I. (1968). Blast wave behavior in confined regions. *Ann. N.Y. Acad. Sci.,* **152** (art. 1), 339

TAYLOR, W.K. and PINTO, A. (1987). Commissioning of CIL's Ammonia II plant. *Ammonia Plant Safety,* **27,** 43. See also *Plant/Operations Prog.,* 1987, **6,** 106

TAYYABKHAN, M.T. and RICHARDSON, T.C. (1965). Monte Carlo techniques. *Chem. Engng Prog.,* **61** (1), 78

TEA, D.C. (1989). *Piper Alpha Public Inquiry. Transcript, Day 78*

TECHNICA LTD (1985). *Prediction of the Proportions of Errant Vessels in Traffic Lanes on the United Kingdom Continental Shelf* (London: HM Stationery Office)

TECHNICA LTD. (1985). *Manual of Industrial Hazard Assessment* (London) (WHAZAN)

TECHNICA LTD (1987). *TECJET* (London)

TECHNICA LTD (1988). *Safety of Computer Controlled Process Systems.* Pub. Rep. C970. (rep. for Dienst Centraal Milieubeheer, Rijnmond) (London)

TECHNICA LTD (1991). What is the ORA toolkit? *Offshore Risk Analysis,* **1** (Nov.)

TECHNICAL RESEARCH CENTRE OF FINDLAND (VTT) — see Appendix 28

TECHNISCHE öBERWACHUNGSVEREIN (1979). *Deutsche Risikostudie Kernkraftwerke* (Cologne: Verlag TUV)

TEDDER, D.W. and POHLAND, F.G. (eds) (1990). *Emerging Technologies in Hazardous Waste Management* (Washington, DC: Am. Chem. Soc.)

TEDESCHI, R.J. (1982). *Acetylene-Based Chemicals from Coal and Other Natural Resources* (Basel: Dekker)

TEDESCHI, R.J., CLARK, G.S., MOORE, G.L., HALFON, A. and IMPROTA, J. (1967). Liquid acetylene. *Ind. Engng Chem. Proc. Des. Dev.,* **7,** 303

TEGROTENHUIS, W.E., CAMERON, R.J. and BUTCHER, M.G. (1998). Microchannel devices for efficient contacting of liquids in solvent extraction. *Sep. Sci. and Technol.*

TEICHMANN, T. (1986). Joint probabilities of partially coupled events. *Reliab. Engng,* **14** (2), 133

TEITELMAN and MASINTER, L. (1981). The INTERLISP programming environment. *IEEE Trans. Comput.,* **C-14** (4), 25

TEISSINGER, J. and SOUCEK, B. (1949). Absorption and elimination of carbon disulphide in man. *J. Ind. Hyg. Toxicol.,* **31,** 67

TEJA, A.S. (ed.) (1981). *Chemical Engineering and the Environment* (Oxford: Blackwells)

TEJA, A.S. (1989). University-industry research in thermophysical properties. *Chem. Engng Prog.,* **85** (9), 20

TELESMANIC, M.J. (1968). Axial flow compressor accidents. *Loss Prevention,* **2,** p. 59. See also *Chem. Engng Prog.,* 1968, **64** (6), 68

TELLER, A.J. (1972). Air pollution control. *Chem. Engng,* **79** (May 8), 93

TELLER, E., TALLEY, W.K., HIGGINS, G.H. and JOHNSON, G.W. (1968). *The Constructive Uses of Nuclear Explosives* (New York: McGraw-Hill)

TEMPERLEY, H.N.V. (1981). *Graph Theory and Its Applications* (Chichester: Ellis Horwood)

TEMPLE, C.J. (1977a). Insurance, safety and the chemical engineer. *Chemical Engineering in a Hostile World* (London: Clapp & Poliak)

TEMPLE, C.J. (1977b). *Seveso – the Issues and the Lessons* (London: Keith Shipton Developments)

TEMPLE, J.A.G. (1983). Time-of-flight inspection: theory. *Nucl. Energy,* **22,** 335

TEMPLE, J.A.G. (1985). An empirical approach to fatigue crack growth data for ferritic steels in a pressurized water reactor environment. *Nucl. Energy,* **24,** 53

TEMPLE, M.A.P. (1980). Repair or replacement. *Plant Engr,* **24** (6), 21

TEMPLE, R.E., GOODING, W.T., WOERNER, P.F. and BENNETT, G.F. (1978). A new universal sorbent for hazardous spills. *Control of Hazardous Material Spills,* 4, p. 382

TEMPLE, R.E., ESTERHAY, R.J., GOODING, W.T. and BENNETT, G.F. (1980). A new universal sorbent for hazardous spills. *J. Haz. Materials,* **4** (2), 185

TEMPLE, R.W. (1973). Underground vs surface storage. *Hydrocarbon Process.,* **52** (8), 85

TEMPLETON, H.C. (1971). Valve installation, operation and maintenance. *Chem. Engng,* **78** (Oct. 11), 141

TENDOLKAR, N.B., SITARAMAN, V. and PONNUSWAMY, A. (1976). Corrosion in a naphtha reformer. *Ammonia Plant Safety,* **18,** p. 65

TENNECO OIL COMPANY (1964a). Electricity in the plant. In Vervalin, C.H. (1964a), *op. cit.,* p. 161

TENNECO OIL COMPANY (1964b). Emergency systems in the plant. In Vervalin, C.H. (1964a), *op. cit.,* p. 225

TENNEKES, H. (1973a). The logarithmic wind profile. *J. Atmos. Sci.,* **30,** 234

TENNEKES, H. (1973b). Similarity laws and scale relations in planetary boundary layers. In Haugen, D.A. (1973), *op. cit.,* p. 177

TENNEKES, H. (1976). Observations on the dynamics and statistics of simple box models with a variable inversion lid. *Third Symp. on Atmospheric Turbulence, Diffusion and Air Quality* (Boston, MA: Am. Met. Soc.), p. 397

TENNEKES, H. (1982). Similarity relations, scaling laws and spectral dynamics. In Nieuwstadt, F.T.M. and van Dop, H. (1982), *op. cit.,* p. 37

TENNEKES, H. and LUMLEY, J.L. (1972). *A First Course in Turbulence* (Cambridge, MA: MIT Press)

TENNEKES, H. and VAN ULDEN, A.P. (1974). Short-term forecasts of temperature and mixing height on sunny days. *First Symp. on Atmospheric Diffusion and Air Pollution* (Boston, MA: Am. Met. Soc.), p. 35

TENNISWOOD, C.F., SHARP, T. and CLARK, D.G.N. (1993). Case studies in probabilistic risk assessment. In Melchers, R.E. and Stewart, M.G. (1993), *op. cit.*, p. 111

TENZER, R., FORD, B., MATTOX, W. and BRUGGER, J.E. (1979). Characteristics of the mobile field use system for the detoxification/incineration of residuals from oil and hazardous material spill clean-up operations. *J. Haz. Materials*, **3**(1), 61

TEODORCZYK, A., LEE, J.H.S. and KNYSAUTAS, R. (1989). Propagation mechanism of quasi-detonations. *Combustion*, **22**, 1723

TEPPER, H.J. (1987). Inert gas systems onboard liquefied gas tankers. *Gastech 86*, p. 255

TERASHIMA, Y., MAEHARA, J., IMANISHI, K., NAKAGAWA, H., YAMADA, T., MATSUMURA, Y. and IZUMI, Y. (1987). Development of a large bore two-cycle duel fuel diesel engine for LNG carriers. *Gastech 86*, p. 284

TERDRE, N. (1994). Compact manifold cuts deepwater production costs by one-third. *Offshore*, May, 109

TERKONDA, P.K. (1983). Energy conservation and modeling of indoor pollution. *Fourth Int. Conf. on Mathematical Modelling in Science and Technology* (ed. Avula, J.R. *et al.*) (New York: Pergamon Press), p. 981

TERWIESCH, B. (1976). Suspension PVC in large reactors. *Hydrocarbon Process.*, **55**(11), 117

TESMEN, A.B. (1973). Materials of construction for process plants. *Chem. Engng*, **80**(Feb. 19), 140; (Mar. 19), 126; (May 14), 158

TESNER, P.A., ROBINOVITCH, H.J. and RAFALKES, I.S. (1962). The formation of dispersed carbon in hydrocarbon diffusion flames. *Combustion*, **8**, p. 801

TETELMAN, A.S. and MCEVILY, A.J. (1967). *Fracture of Structural Materials* (New York: Wiley)

TETLOW, J.A. (1979). Effect of new legislation and EEC directives on the water industry. *Chem. Ind.*, Mar. 17, 183

TEUSCHER, L., SABNIS, J. and DRIVAS, P. (1983). *Oil Gas J.*, Sep., 162

TEWARSON, A. (1985). Fully developed enclosure fires of wood cribs. *Combustion*, **20**, p. 1555

TEWARSON, A. and PION, R.F. (1976). Flammability of plastics — I. Burning intensity. *Combust. Flame*, **26**, 85

TEWARSON, A. and STECIAK, J. (1983). Fire ventilation. *Combust. Flame*, **53**, 123

TEZEKJIAN, E.A. (1963). How to control centrifugal compressors. *Hydrocarbon Process.*, **4297**, 169

THACKARA, A.D., ROBB, R.M. and HALL-CRAGGS, R.B.T. (1960). An approach to the problems of process hazards in plant design. *Chemical Process Hazards*, **I**, p. 5

THACKER, C.M. (1970). What's ahead for carbon disulfide. *Hydrocarbon Process.*, **49**(4), 124; **49**(5), 137

THAIN, W. (1976). Measurement of exposure to toxic gases by personal monitors. *Prevention of Occupational Risks*, **3**, p. 451

THAMBYNAYAGAM, R.K.M., WOOD, R.K. and WINTER, P. (1981). DPS — an engineer's tool for dynamic process analysis. *Chem. Engr, Lond.*, **365**, 58

THARMALINGAM, S. (1989a). Assessing runway reactions and sizing vents. *Chem. Engr, Lond.*, **463**, 33

THARMALINGHAM, S. (1989b). The evaluation of self-heating in bulk handling of unstable solids. *Runaway Reactions*, p. 293

THEDIE, J. and ABRAHAM, C. (1969). The economic cost of road accidents. *Traffic Engng Control*, **2**, 589

THEIMER, O.F. (1973). Cause and prevention of dust explosions in grain elevators and flour mills. *Powder Technol.*, **8**(3–4), 137

THEOBALD, C.R. (1977). Studies of fires in industrial buildings — Part 1: The growth and development of fire. *Fire Prev. Sci. Technol.*, **17**, 4

THEOBALD, C. (1981). The effect of nozzle design on the stability and performance of turbulent water jets. *Fire Safety J.*, **4**(1), 1

THEOBALD, C.R. (1984). The design of a general purpose firefighting jet and spray branch. *Fire Safety J.*, **7**, 177

THEOBALD, C.R. (1987a). Heated wind tunnel testing for fast acting sprinklers. *Fire Surveyor*, **16**(2), 9

THEOBALD, C.R. (1987b). Thermal response of sprinklers. Pt I, FRS heated wind tunnel. *Fire Safety J.*, **12**(1), 51

THEOBALD, C.R., WESTLEY, S.A. and WHITBREAD, S. (1988). Thermal response of sprinklers. Pt 2, Characteristics and test methods. *Fire Safety J.*, **13**, 99

THEODORE, L. and BUONICORE, A.J. (1976). *Industrial Air Pollution Control Equipment for Particulates* (Boca Raton, FL: CRC Press)

THEODORE, L. and BUONICORE, A.J. (eds) (1982). *Air Pollution Control Equipment: Selection, Design, Operation and Maintenance* (Englewood Cliffs, NJ: Prentice-Hall)

THEODORE, L., SOSA, F.J. and FAJARDO, P.L. (1980). *Particulate Air Pollution: Problems and Solutions* (Boca Raton, FL: CRC Press)

THEODORE, L., REYNOLDS, J. and TAYLOR, F. (1989). *Accident and Emergency Management* (New York: Wiley)

THEOFANOUS, T.G. (1981). A physicochemical mechanism for the ignition of the Seveso accident. *Nature*, **291**(Jun. 25), 640

THEOFANOUS, T.G. (1983). The physicochemical origins of the Seveso accident. *Chem. Engng Sci.*, **38**, 1615, 1631

THIBAULT, L. (1993). Taking care of enclosed gear drives. *Chem. Engng*, **100**(5), 145

THIBAULT, P., LIU, Y.K., CHAN, C., LEE, J.H., KNYSTAUTAS, R., GUIRAO, C., HJERTAGER, B. and FUHRE, K. (1982). Transmission of an explosion through an orifice. *Combustion*, **19**, p. 599

THIBODEAUX, L.J. (1979). Chemodynamics (New York: Wiley)

THIBODEAUX, L.J. and HWANG, S.T. (1982). Land farming of petroleum wastes — modeling the air emissions problem. *Environ. Prog.*, Feb.

THIBODEAUX, L.J., SPRINGER, C. and RILEY, L.M. (1982). Models of mechanisms for the vapor phase emission of hazardous chemicals from landfills. *J. Haz. Materials*, **7**(1), 63

THIELSCH, H. (1965). *Defects and Failures in Pressure Vessels and Piping* (New York: Reinhold)

THIELSCH, H. (1968). Fabrication problems and failures in vessels and piping in ammonia and related plants. *Ammonia Plant Safety*, **10**, p. 55

THIESS, A.M. and ZAPP, H. (1979). Health and safety measures in BASF from the viewpoint of the company medical staff. In BASF (1979), *op. cit.*, p. 95

THIMBLEBY, H. (1991). Can humans think? *Ergonomics*, **34**, 1269

THIRUVENGADAM, A. (1966). Now there's a way to work out erosion strength of materials. *Product Engng*, **37**(Aug. 15), 55

THIYAGARAJAN, R. and HERMANCE, C.E. (1975). Prediction of ignition conditions for flammable mixtures drifting over heated planar surfaces. *Combust. Inst., Joint Spring Mtg* (Pittsburgh, PA)

THODOS, G. (1964). Hazards with ammonia and mercury. *AIChE J.*, **10**, 274

THOENES, H.W. (1964). Fire and explosion processes in compressor with oil lubricated cylinders. *Tech.-berwach.*, **5**(1), 12

THOFT-CHRISTENSEN, P. and SO/RESEN, J.D. (1987). Reliability analysis of tubular joints in offshore structures. *Reliab. Engng*, **19**(3), 171

THOM, N.G. and DOUGLAS, R.T. (1976). A study of hydrogen sulphide levels in the geothermal areas of Rotorua, New Zealand. *Fourth Int. Clean Air Cong.*, Tokyo, p. 565

THOMA, F.A. (1973). AGMA views on shop testing and vendor required data. *Hydrocarbon Process.*, **52**(12), 56

THOMA, G.J., HILDEBRAND, G., VALSARAJ, K.T., THIBODEAUX, L.J. and SPRINGER, C. (1992). Transport of chemical vapors through soil: a landfill cover experiment. *J. Haz. Materials*, **30**, 333

THOMAS, A. (1951). An isothermal calorimeter for measuring low rates of heat evolution over long periods. *Trans. Faraday Soc.*, **47**, 569

THOMAS, A. (1962). Carbon formation in flames. *Combust. Flame*, **6**, 46

THOMAS, D. (1985). Condition monitoring. *Plant Engr*, **29**(4), 20

THOMAS, D.E., JAMES, R.A. and SPARKS, C.R. (1972). Some plant noise problems and solutions. *Hydrocarbon Process.*, **51**(10), 149

THOMAS, F.H. (1977). Data analysis on the RB211. *First Nat. Conf on Reliability. NRC5/1.* (Culcheth, Lancashire: UK Atom Energy Auth.)

THOMAS, G.O. and OAKLEY, G.L. (1993). On practical difficulties encountered when testing flame and deto- nation arresters to BS 7244. *Process Safety Environ.*, **71B**, 187

THOMAS, G.O., EDWARDS, M.J. and EDWARDS, D.H. (1990). Studies of detonation quenching by water sprays. *Combust. Sci. Technol.*, **71**, 233

THOMAS, G.O., SUTTON, P. and EDWARDS, D.H. (1991). The behavior of detonation waves at concentration gradients. *Combust. Flame*, **84**, 312

THOMAS, G.O., JONES, A. and EDWARDS, M.J. (1991). Influence of water sprays on explosion development in fuel–air mixtures. *Combust. Sci. Technol.*, **80**, 47

THOMAS, G.R. (1987). K-301 syngas compressor train: near catastrophic failure of speed increaser pinion shaft. *Ammonia Plant Safety* **27**, 4

THOMAS, H.A. (1948). Pollution load capacity of streams. *Water Sewage Works*, **95**, 409

THOMAS, H.M. (1977). A probability model for the failure of pressure containing parts. *First Nat. Conf on Reliability. NRC3/2. (Culcheth, Lancashire: UK Atom Energy Auth.)*

THOMAS, H.M. (1981). Pipe and vessel failure probability. *Reliab. Engng*, **2**(2), 83

THOMAS, I. (1991). Coode Island – vapour recovery to blame? *Chem. Engr, Lond.*, **506**, 17

THOMAS, I.F. (1993). Official incident reports, can they be relied on – the local scene. *CHEMECA 93*, p. 133–2

THOMAS, I.F. (1995). The politics of safety. *Chem Engng Aust.*, **20**(1), 14

THOMAS, I.F. and GUGAN, K., (1993). Official incident reports, can they be relied on – Flixborough. *CHEMECA 93*, p. 141–2

THOMAS, K. (1981). Comparative risk perception: how the public perceives the risks and benefits of energy systems. *Proc. R. Soc., Ser. A*, **376**, 35. See also in Warner, Sir F. and Slater, D.H. (1981), *op. cit.*, p. 35

THOMAS, L. (1980). *Fundamentals of Heat Transfer* (Englewood Cliffs, NJ: Prentice-Hall)

THOMAS, L. (1992). Fired heaters in the oil and gas industries. *Chem. Engr, Lond.*, **514**, 17

THOMAS, L.C. (1986). A survey of maintenance and replacement models for maintainability and reliability of multiitem systems. *Reliab. Engng*, **16**(4), 297

THOMAS, M.W. (1948). *The Early Factory Legislation* (London: Thames Bank)

THOMAS, P.D. (1972). Embrittlement. *Hydrocarbon Process.*, **51**(8), 65

THOMAS, P.H. (1958). On the thermal conduction equation for self-heating materials with surface cooling. *Trans Faraday Soc.*, **54**, 60

THOMAS, P.H. (1960). Some approximations in the theory of self-heating and thermal explosion. *Trans Faraday Soc.*, **56**, 833

THOMAS, P.H. (1961). Effect of reactant consumption on the induction period and critical condition for a thermal explosion. *Proc. R. Soc., Ser. A*, **262**, 192

THOMAS, P.H. (1963). The size of flames from natural fires. *Combustion*, **9**, p. 844

THOMAS, P.H. (1965a). A comparison of some hot spot theories. *Combust. Flame*, **9**, 369

THOMAS, P.H. (1965b). *Fire Spread in Wooden Cribs, Part III: the Effect of Wind. Fire Res. Note 600.* Fire Res. Station, Borehamwood

THOMAS, P.H. (1972). Self-heating and thermal ignition – a guide to its theory and application. *Ignition, Heat Release and Non-Combustibility of Materials. ASTM STP 502.*(New York: Am. Soc. for Testing and Materials)

THOMAS, P.H. (1973a). An approximate theory of 'hot spot' criticality. *Combust. Flame*, **21**, 99

THOMAS, P.H. (1973b). Behavior of fires in enclosures – some recent progress. *Combustion*, **14**, p. 1007

THOMAS, P.H. (1974). *Fire in Model Rooms.* (report). Building Res. Estab., Borehamwood

THOMAS, P.H. (1980). Fire: triangles jumps and falls. *Fire Prev. Sci. Technol.*, **23**, 3

THOMAS, P.H. (1981a). Fire modeling and fire behavior in rooms. *Combustion*, **18**, p. 503

THOMAS, P.H. (1981b). Testing products and materials for their contribution to flashover in rooms. *Fire and Materials*, **5**, 103

THOMAS, P.H. (1982). Modelling of compartment fires. *Sixth Int. Fire Protection Seminar* (Verein. zur Forderung des Deutschen Brandschutzes), p. 29

THOMAS, P.H. (1984). On the surface ignition of self-heating materials. *Combust. Sci. Technol.*, **36**, 263

THOMAS, P.H. (1992). Fire modelling: a mature technology? *Fire Safety J.*, **19**, 125

THOMAS, P.H. (1993). Comparisons between plume theories. *Fire Safety J.*, **20**, 289

THOMAS, P.H. and BOWES, P.C. (1961a). Some aspects of the ignition and self-heating of solid cellulosic materials. *Br J. Appl. Phys.*, **12**, 222

THOMAS, P.H. and BOWES, P.C. (1961b). Thermal ignition in a slab with one face at a constant high temperature. *Trans. Faraday Soc.*, **57**, 2007

THOMAS, P.H. and BOWES, P.C. (1967). Note on thermal ignition in the hollow circular cylinder heated at the inner face. *Pyrodynamics*, **5**, 29

THOMAS, P.H. and BULLEN, M.L. (1979/80). Burning of fuels in fully developed room fires. *Fire Safety J.*, **2**(4), 275

THOMAS, P.H. and HESELDEN, A.J.M. (1962). Behaviour of fully developed fire in an enclosure. *Combust. Flame*, **6**, 133

THOMAS, P.H. and HINCKLEY, P.L. (1964). *Design of Roof Venting Systems for Single Storey Buildings. Fire Res. Tech. Paper 10* (London: HM Stationery Office)

THOMAS, P.H. and LAW, M. (1972). *The Projection of flames from Burning Buildings. Fre Res. Note 921.* Fire Res. Station, Borehamwood

THOMAS, P.H. and LAW, M. (1974). The projection of flames from buildings on fire. *Fire Prev. Sci. Technol.*, **10**, 19

THOMAS, P.H., HINCKLEY, P.L., THEOBALD, C.R. and SIMMS, D.L. (1963). *Investigations into the Flow of Hot Gases in Roof Venting. Fire Res. Tech. Paper 70* (London: HM Stationery Office)

THOMAS, P.H., PICKARD, R.W. and WRAIGHT, H.G.H. (1963).*On the Size and Orientation of Buoyant Diffusion Flames and the Effect of Wind. Fire Res. Note 516.* Fire Res. Station, Borehamwood

THOMAS, P.H., SIMMS, D.L. and WRAIGHT, H. (1964). *Fire Spread in Wooden Cribs. Part 1. Fire Res. Note 537.* Fire Res. Station, Borehamwood

THOMAS, P.H., BALDWIN, R. and HESELDEN, A.J.M. (1965). Buoyant diffusion flames: some measurements of air entrainment, heat transfer, and flame merging. *Combustion*, **10**, p. 983

THOMAS, P.H., SIMMS, D.L. and WRAIGHT, H. (1965). *Fire Spread in Wooden Cribs. Part 2. Fire Res. Note 799.* Fire Res. Station, Borehamwood

THOMAS, P.H. and THEOBALD, C.R. (1977). Studies of fires in industrial buildings – The burning rates and duration of fires. *Fire Prev. Sci. Technol.*, **17**, 15

THOMAS, P.H., WEBSTER, C.T. and RAFTERY, M.M. (1961). Some experiments in buoyant diffusion flames. *Combust. Flame*, **5**, 359

THOMAS, P.W. (1984). Multi-stage overpressure protection and product containment on high pressure polymerisation reactors. *The Protection of Exothermic Reactors and Pressurised Storage Vessels* (Rugby: Instn Chem. Engrs), p. 229

THOMAS, R.G. (1986). Tube bundle modification to the 124–C synthesis gas aftercooler. *Ammonia Plant Safety*, **26**, p. 177

THOMAS, T.J., SRINAVASAN, S. and GOWER, C.C. (1976). The influence of environmental factors on hazardous spill amelioration. *Control of Hazardous Material Spills,* vol. 3, p. 303

THOMAS, V.M. (1982). Safety technology in industry. Coal- mining. In Green, A.E. (1982), *op. cit.*, p. 383

THOMAS, W.A. (1978). Floating production and subsea completions. In Institute of Petroleum (1978), *op. cit.*, p. 79

THOMAS, W.B. (1980). Improve water treatment systems. *Hydrocarbon Process.*, **59**(1), 99

THOMAS, W.B. (1981). Are you prepared for utility emergencies? *Hydrocarbon Process.*, **60**(8), 73

THOMAS, W.G. (1962). *Coll. Engng*, **39**, 377

THOMASON, R. (1969). *Introduction to Reliability and Quality* (London: Machinery Publ. Co.)

THOMASSON, J.M. and LEBOSSE, M. (1982). Experience with CAD. *Hydrocarbon Process.*, **61**(6), 125: **61**(7), 161

THOME, R.J., CLINE, M.W. and GRILLO, J.A. (1979). Batch process automation. *Chem. Engng Prog.*, **75**(5), 54

THOMPSON, A. (1965). Operating experience with direct digital control. In Miller, W.E. (1965), *op. cit.*, p. 55. See also *Chem. Engr, Lond., 1965*, **188**, CE96

THOMPSON, A.A. (1969). *Fatigue Failures Induced in Heat Exchanger Tubes by Vortex Shedding. ASME paper 69-Pet.-6.* Am. Soc. Mech. Engrs, New York

THOMPSON, A.R. (1968). Water pollution control. Cooling towers. *Chem. Engng*, **75**(Oct. 14), 100

THOMPSON, A.W. (1978). Hydrogen embrittlement of stainless steels and carbon steels. *API Proc., Refining Dept, 43rd Midyear Mtg*, **57**, 317

THOMPSON, D.A. (1981). Commercial air crew detection of system failures: state of the art and future trends. In Rasmussen, J. and Rouse,W.B. (1981), *op. cit., p. 37*

THOMPSON, D.W. (1994). Relief valve reliability data. *Engng Dept, Univ of Manchester, Conf. on Valve and Pipeline Reliability*

THOMPSON, G. (1994). The comparative design evaluation of pipe joints with respect to reliability. *Engng Dept, Univ of Manchester, Conf. on Valve and Pipeline Reliability*

THOMPSON, J.C. (1985). When a floating roof sinks. *Hydrocarbon Processing*, **64**(7), 119

THOMPSON, J.E. and TRICKLER, C.J. (1983). Fans and fan systems. *Chem. Engng*, **90**(Mar. 21), 48

THOMPSON, J.H. (1990). Support insulation of a fully-integrated refrigerated ammonia tank. *Ammonia Plant Safety*, **30**, p. 241

THOMPSON, J.M.T. (1981). *Instabilities and Catastrophes in Science and Engineering* (Chichester: Wiley)

THOMPSON, J.N. (1978). The UK approach to the control of noise from industrial sources. *Chem. Ind.*, Oct. 21, 790

THOMPSON, J.R. and CARNEGIE, R.N. (1989). Building a high-integrity ammonia storage tank – 1988. *Ammonia Plant Safety*, **29**, p. 116. See also *Plant/Operations Prog.*, 1989, **8**, 92

THOMPSON, J.R., KAISER, G., BROOKS, J.C., HEOK, M., MCKELVEY, T., RAJ, A. and GREINER, W. (1990). Ammonia barging risk assessment. *Ammonia Plant Safety*, **30**, p. 1

THOMPSON, J.R., SEKULA, L.J. and WHITSON, P.E. (1992). Successful application of ammonia detectors. *Ammonia Plant Safety*, **32**, p. 77

THOMPSON, L. (1982). EPRI research on hydrogen combustion and control for nuclear reactor safety. In Lee, J.H.S. and Guirao, C.M. (1982), *op. cit.*, p. 919

THOMPSON, L. and BUXTON, O.E. (1979a). Maximum isen-tropic flow of dry saturated steam through pressure relief valves. *J. Pres. Ves. Technol.*, **101**, 113

THOMPSON, L. and BUXTON, O.E. (1979b). Maximum isen-tropic flow of dry saturated steam through pressure relief valves. In Haupt, R.W. and Meyer, R.A. (1979), *op. cit.*, p.43

THOMPSON, M. (1980). *An Outline of the Cultural Theory of Risk.* Int. Inst. for Appl. Systems Analysis, Laxenburg, Austria

THOMPSON, M., ELLIS, R. and WILDAVSKY, A. (1990). *Cultural Theory* (Boulder, CO: Westview Press)

THOMPSON, N.J. (1929). Auto-ignition temperatures of flammable liquids. *Ind. Engng Chem.*, **21**(2), 134

THOMPSON, N.J. and COUSINS, E.W. (1949). Explosion tests on glass windows: effect on glass breakage of varying the rate of pressure application. *J. Am. Ceramic Soc.*, **32**, 313

THOMPSON, R.G. (1976). Water-pollution instrumentation. *Chem. Engng*, **83**(Jun. 21), 151

THOMPSON, R.L. (1973). Catalyst handling in a nitrogen atmosphere. *Ammonia Plant Safety*, **15**, p. 42

THOMPSON, R.L. and BROOKS, J.B. (1977). Ammonia plant converter basket failure. *Ammonia Plant Safety*, **19**, p. 100

THOMPSON, R.S. (1993). Building amplification factors for sources near buildings: a wind-tunnel study. *Atmos. Environ.*, **27A**, 2313

THOMPSON, R.S. and SHIPMAN, M.S. (1986). Streamlines in stratified flow over a three-dimensional hill. *Fifth Joint Conf. on Applications of Air Pollution Meteorology* (Boston, MA: Am. Met. Soc.), p. 273

THOMPSON, R.W. and KING, C.J. (1972). Systematic synthesis of separation schemes. *AIChE J.*, **18**, 941

THOMPSON, S.M. (1990). Liquefied petroleum gas. *Ullmanns Encyclopaedia of Chemical Technology,* rev. ed., vol. A15 (Weinheim: VCH), p. 347

THOMPSON, T. and BECKERLEY, J.G. (1973). *The Technology of Nuclear Reactor Safety* (Cambridge, MA: MIT Press)

THOMPSON, T.J. (1971). Nuclear power – today and tomorrow. *Chem. Tech.*, Aug., 495

THOMPSON, W. (1991). Portrait of an integrated plant. *Hydrocarbon Process.*, **70**(6), 51

THOMPSON, W.A. (1981). Some probabilistic aspects of safety assessment. In Berg, G.G. and Maillie, H.D., (1981), op. cit., p. 285

THOMPSON, W.S. (1972). Formula to find fog nozzle reaction. *Fire Engng*, **125**(Oct.), 51

THOMPSON, W.S. (1973). Nozzles that keep their head. *Fire Engng*, **126**(Aug.), 40

THOMSON, B.J. and VON ZAHN-ULLMANN, S (1988). OECDIGUS: an example of good international laboratory cooperation. (1988). *J. Loss Prev. Process Ind.*, **1**(1), 48

THOMSON, J.H. (1941). *Guide to Explosives Acts 1875 and 1923*, 4th ed. (London: HM Stationery Office)

THOMSON, J.R. (1987). *Engineering Safety Assessment* (Harlow, Essex: Longman)

THOMSON, J.R. (1994). Quantified risk assessment – a nuclear industry viewpoint. *J. Loss Prev. Process Ind.*, **7**, 419

THOMSON, J.R. and NIGHTINGALE, A.P.M. (1988). A simple method for determining the maximum consequences of notional toxic and radiotoxic gas discharges. *J. Haz. Materials*, **17**(3), 239

THOMSON, M.S. (1989). *Piper Alpha Public Inquiry. Transcript, Days 106, 107*

THOMSON, W.F. (1951). Transmission of pressure waves in liquid-filled tubes. *Proc. First US Nat. Cong. on Applied Mechanics*, p. 933

THOMSON, W.T. and DEANS, N.D. (1983). Condition monitoring for induction motors for availability assessment in offshore installations. *Euredata 4*, paper 17.1

THOMSON, W.T., LEONARD, R.A., MILNE, A.J. and PENMAN, J. (1984). Failure identification of offshore induction motors using on-condition monitoring. *Reliab. Engng*, **9**(1), 49

THONGREN, J.T. (1970). Predict exchanger tube damage. *Hydrocarbon Process.*, **49**(4), 129

THORLEY, A.R.D. (1980). Surge suppression and control in slurry pipelines. *Chem. Engr, Lond.*, **355**, 222

THORLEY, D. (1984). The dynamic response of check valves. *Chem. Engr, Lond.*, **402**, 12

THORNE, P.F. (1978a). The dilution of flammable polar solvents by water for safe disposal. *J. Haz. Materials*, **2**(4), 321

THORNE, P.F. (1978b). The flame inhibiting properties of trichlorofluoromethane and dichlorofluoromethane compared with Halons, carbon dioxide and nitrogen. *Fire Res.*, **1**(4/5), 237

THORNE, P.F. (1978c). The toxicity of halon extinguishing agents – a new appraisal. *Fire Int.*, **60**, 41

THORNE, P.F. (1982). Explosion research carried out at the Fire Research Station, UK. In Lee, J.H.S. and Guirao, C.M. (1982), op. cit., p. 999

THORNE, P.F. (1986). Estimating the hazard around a vent: a case study. *Hazards IX*, p. 19

THORNE, P.F., ROGOWSKI, Z.W. and FIELD, P. (1983). Performance of low inertia explosion reliefs fitted to a 22 m³ cubical

chamber. *Loss Prevention and Safety Promotion 4*, vol. 3., p. F1

THORNE, P.F., THEOBALD, C.R. and MELINEK, S.J. (1988). The thermal performance of sprinkler heads. *Fire Safety J.*, **14**(1/2), 89

THORNHILL, C.K. (1960). *Explosions in Air. ARDE Memo. (B) 57/60*. Armament Res. Dev. Estab

THORNLEY, F. (1985). Ambulance service arrangements for dealing with incidents involving toxic and chemical hazards. *Chem. Ind.*, Jul. 1, 438

THORNTON, D.P. (1970). Corrosion free HF alkylation. *Chem. Engng*, **77**(Jul. 13), 108

THORNTON, D.P., WARD, D.J., and ERICKSON, R.A. (1972). MRG process for SNG. *Hydrocarbon Process.*, **51**(8), 81

THORNTON, N.C. and SETTERSTROM, C. (1940). Toxicity of ammonia, chlorine, hydrogen cyanide, hydrogen sulphide, and sulphur dioxide. III, Green plants. *Contrib. Boyce Thompson Inst.*, **11**, 343

THORNTON, W.M. (1914a). The electrical ignition of gaseous mixtures. *Proc. R. Soc., Ser. A*, **90**, 272

THORNTON, W.M. (1914b). The ignition of gases by condenser discharge sparks. *Proc. R. Soc., Ser. A*, **91**, 17

THORNTON, W.M. (1915a). The ignition of gases by impulsive electrical discharge. *Proc. R. Soc., Ser. A*, **92**, 381

THORNTON, W.M. (1915b). The reaction between gas and pole in the electrical ignition of gaseous mixtures. *Proc. R. Soc., Ser. A*, **92**, 9

THORNTON, W.M. (1919). The ignition of gases by hot wires. *Phil. Mag.*, **38**, 613

THOROGOOD, J.L. and CRAWLEY, F.K. (1990). Design and operation of a single vent diverter system. *Process Safety Environ.*, **68B**, 94

THOROGOOD, R.M., DAVEY, B. and HENDRY, W. (1973). Elements in the cost of liquefaction of natural gas. In Hepple, P. (1973), op. cit., p. 77

THORPE, J.J. (1974). Epidemiological survey of leukaemia in persons potentially exposed to benzene. *J. Occup. Med*, **16**, 375

THORPE, J.J. (1978a). Benzene – the current controversy. *API Proc., Refining Dept*, **57**, 408

THORPE, J.J. (1978b). How much benzene causes cancer. *Hydrocarbon Process.*, **57**(5), 172

THORSEN, M.A. and MØLHAVE, L. (1987). Elements of a standard protocol for measurements in the indoor atmospheric environment. *Atmos. Environ.*, **21**, 1411

THRING, M.W. (1952). *The Science of Flames and Furnaces* (London: Chapman & Hall)

THRING, M.W. (1980). *The Engineer's Conscience* (Bury St Edmunds, Suffolk: Mech. Engng Publs)

THRING, M.W. and LOWES, T.M. (1972). Luminous radiation from industrial flames. *Combust. Sci. Technol.*, **5**, 251

THRING, M.W. and NEWBY, M.P. (1953). Combustion length of enclosed turbulent jet flames. *Combustion*, **4**, p. 789

THRODAHL, M.C. (1978). The technical basis for regulatory decision-making. *Chem. Ind.*, Apr. 1, 221

THRUSH, R.E. and HONEYCHURCH, R.W. (1969). How to specify centrifuges. *Hydrocarbon Process.*, **48**(10), 81

THUILLIER, J. and PONS, F. (1978). Advances in reformer tube metallurgy and production. *Ammonia Plant Safety*, **20**, p. 89

THUMA, N.K., O'NEILL, P.E., BROWNLEE, S.G. and VALENTINE, R.S. (1978). Biodegradation of spilled hazardous materials. *Control of Hazardous Materials Spills 4*, p. 217

THUMANN, A. (1974). Interdisciplinary plant-noise control. *Chem. Engng*, **81**(Aug. 19), 120

THURLBECK, S.D. and DEOGAN, M.S. (1994). Passive fire protection for offshore pipeline risers. *Hazards*, **XII**, p. 441

THURLOW, C. (1965). Pumps and the chemical plant. *Chem. Engng*, **72**(May 24), 117: (Jun. 7), 213: (Jun. 21), 133

THURLOW, C. (1971). Centrifugal pumps. *Chem. Engng*, **78**(Oct. 11), 29

THURSTON, D.B. (1980). *Design for Safety* (New York: McGraw-Hill)

THURSTONE, L.L. (1927). A law of comparative judgement. *Psychol. Rev.*, **34**, 273

THURSTONE, L.L. (1931). Rank order as a psychophysical method. *J. Exp. Psychol.*, **14**, 187

THWAITES, D. (1993). Awareness, preparedness for emergency at local level. *Health, Safety and Loss Prevention*, p. 434

THYGERSON, A.L. (1972). *Safety:* 1976, 2nd ed.: 1986, 2nd rev. ed. (Englewood Cliffs, NJ: Prentice-Hall)

THYGERSON, A.L. (1977). *Accidents and Disasters* (Englewood Cliffs, NJ: Prentice Hall)

TIAN, B. and WANG, S. (1990). Improved safe operation of an old polybutadiene rubber plant after modification. *Safety and Loss Prevention in the Chemical and Oil Processing Industries*, p. 639

TICKNER, J. (1994). The case for inherent safety. *Chem. and Ind.* (Oct. 3), 796.

TIEN, C.L. (1968). Thermal radiation properties of gases. *Advances in Heat Transfer 5* (ed. J.P. Harnett and T.F. Irvine) (New York: Academic Press), p. 300

TIEN, C.L. and LEE, S.C. (1982). Flame radiation. *Prog. Energy Combust. Sci.*, **8**, 41

TIEN, J.S. (1973). The effects of perturbations on the flammability limits. *Combust. Sci. Technol.*, **7**, 185

TIERNEY, K.J. (1981). Community and organizational awareness of and preparedness for acute chemical emergencies. *J. Haz. Materials*, 4(4), 331

TIERNEY, K.J. (1982). Developing a community-preparedness capability for sudden emergencies involving hazardous materials. In Fawcett, H.H. and Wood, W.S. (1982), *op. cit.*, p. 759

TIESZEN, S.R., STAMPS, D.W., WESTBROOK, C.K. and PITZ, W.J. (1991). Gaseous hydrocarbon-air detonations. *Combust. Flame*, **84**, 376

TIETZE, L.F. (1995). Domino reactions in organic synthesis. *Chem. Ind.* (Jun. 19), 453–7

VAN TIGGELEN, A. and BURGER, J. (1964). Considerations on the theoretical existence of fundamental flammability limits in gaseous mixtures (letter). *Combust. Flame*, **8**, 343

TIHANSKY, D.P. (1975). A conceptual framework for risk-benefit assessments of hazardous materials transport. *Transport of Hazardous Cargoes*, 4

TIJSSEN, P. (1977). Optimizing ammonia plant operation by computer. *Ammonia Plant Safety*, **19**, p. 155

TILLEY, A.J., WILKINSON, R.T., WARREN, P.S.G., WATSON, B. and DRUD, M. (1982). The sleep and performance of shift workers. *Hum. Factors*, **24**, 629

TILLEY, K.W. (1967). Fault diagnosis for maintenance personnel. *Ergonomics*, **10**, 206

TILLMAN, W.A. and HOBBS, G.E. (1949). The accident-prone automobile driver. *Am. J. Psychiat.*, **106**(5), 321

TILTON, J.N. and FARLEY, C.W. (1990). Predicting liquid jet break-up and aerosol formation during the accidental release of pressurised hydrogen fluoride. *Plant/Operations Prog.*, **9**, 120

TILTON, J.N., SQUIRE, R.H., SAFFLE, C.S. and ATKINS, C.R. (1992). Ammonia storage tank study. *Ammonia Plant Safety*, **32**, p. 63

TILY, P. (1986). Practical options for on-line flow measurement. *Process Engng*, May, 85

TIMAR, J. (1978). Platforms – steel structures. In Institute of Petroleum (1978), *op. cit.*, p. 57

TIMBERLAKE, D.L. and GOVIND, R. (1994). Expert system for solvent substitution. *Preprints of Papers Presented at the 208th ACS National Meeting*, Aug. 21–25, 1994, Washington, D.C., 215–17. Center for Great Lakes Studies, University of Wisconsin-Milwaukee (Milwaukee, WI: Division of Environmental Chemistry, American Chemical Society)

TIMBERWOLF CONSULTING SERVICES (1991). *WBD2 Wound Ballistics Simulation Model* (Hosingen, Luxembourg)

TIMMERMANN, D.N. (1989). How to specify and select gear drives. *Chem. Engng Prog.*, **85**(3), 15

TIMMERMANN, P. (1968). *Fault Data for the Prediction of Reliability of Electronic and Mechanical Systems and Equipment. Rep. Risö-M-709*. Atom. Energy Comm., Res. Estab, Risö, Denmark

TIMMERMANS, J. (1950). *Physicochemical Constants of Pure Organic Compounds*, vol. 1, 1950; vol. 2, 1965 (Amsterdam: Elsevier)

TIMMERMANS, W.J. (1984). Deepwater pipelining. *Oil Gas J.*, **82**(Dec. 17), 78: (Dec. 24), 57

TIMMINS, P.F. (1983). Guidelines for detecting hydrogen blisters and similar damage in sour-gas service. *Oil Gas J.*, **81**(Feb. 14), 116

TIMMINS, P.F. (1984). Assessing hydrogen damage in sour-service lines and vessels is key to inspection. *Oil Gas J.*, **82**(Nov. 5), 103

TIMOSHENKO, S. (1936). *Theory of Elastic Stability* (New York: McGraw-Hill)

TIMOSHENKO, S. and YOUNG, D.H. (1956). *Engineering Mechanics* (New York: McGraw-Hill)

TIMOSHENKO, S. and GERE, J.M. (1973). *Mechanics of Materials* (New York: van Nostrand)

TIMOSHENKO, S.P. and GOODIER, J.N. (1970). *Theory of Elasticity*, 3rd ed. (New York: McGraw-Hill)

TIMOSHENKO, S., YOUNG, D.H. and WEAVER, W. (1974). *Vibration Problems in Engineering*, 4th ed. (New York: Wiley)

TING, Y.C. (1950). The toxicity of ammonia. *Science*, **112**(Jul. 21), 91

TINGEY, F.H. (1962). Plant material balance. Construction and interpretation. *Ind. Engng Chem.*, **54**(4), 36

TINKER, J. (1974). Could Britain's nuclear know-how have saved Nypro? *New Scientist*, June 13, 680

TINNEY, W.S. (1978). How to obtain trouble-free performance from centrifugal pumps. *Chem. Engng*, **85**(Jun. 5), 139

TINNEY, W.S., KNOLL, H.J. and DIEHL, C.H. (1973). Loop extends mechanical seal life. *Hydrocarbon Process.*, **52**(1), 123

TINSLEY, I.J. (1979). *Chemical Concepts in Pollutant Behaviour* (New York: Wiley-Interscience)

TINSON, R. (1967). The electrostatic hazard during loading of petroleum products into glass-reinforced plastic tanks. *J. Inst. Petrol.*, **53**, 303

TIPLER, B.A. (1979). Spark erosion damage to compressor bearings in ammonia plants. *Ammonia Plant Safety*, **21**, p. 124

TIPPER, C.F. (1957). Gas phase slow oxidation. *Quart. Rev.*, **11**, 313

TIPPER, C.F. (1962). *The Brittle Fracture Story* (New York: Cambridge Univ. Press)

TIPTON, C.R. *et al.* (1960). *Reactor Handbook*, vols 1–4, 2nd ed. (New York: Interscience)

TIRATSOO, E.N. (1967). *Natural Gas* (Beaconsfield: Scientific Press)

TISSERAND, M. (1985). Progress in the prevention of falls by slipping. *Ergonomics*, **28**, 1027

TITE, J.P., GREENING, K. and SUTTON, P. (1989). Explosion hazard of air ingress into flare stacks. *Chem. Engng Res. Des.*, **67**, 373

TITMAN, H. (1955/56). *Trans. Instn Min. Engrs*, **115**, 535

TITMAN, H. and WYNN, A.H.A. (1954). *The Ignition of Explosive Gas Mixtures by Friction. Rep. 95*. Safety in Mines Res. Estab., Sheffield

TITZE, H. (1967). *Elemente des Apparatebaus,* 2nd ed. (Berlin: Springer)

TIZARD, H.T. and PYE, D.R. (1922). Experiments on the ignition of gases by sudden compression. *Phil. Mag.*, **44**, 79

TJOA, I.B. and BIEGLER, L.T. (1991). Simultaneous strategies for data reconciliation and gross error detection of non-linear systems. *Comput. Chem. Engng*, **15**, 679

TJOE, T.N. and LINNHOFF, B. (1986). Using pinch technology for process retrofit. *Chem. Engng*, **93**(Apr. 28), 47

TNO (1979). *Methods for the Calculation of the Physical Effects of the Escape of Dangerous Materials (Liquids and Gases),* vols 1–2. Prins Maurits Lab., Rijswijk, The Netherlands

TNO – see also Appendix 28

TOA, Y.X. and KAVIANY, M. (1991). Burning rate of liquid supplied through a wick. *Combust. Flame*, **86**, 47

TOBIN, P.S. (1994). Risk management through inherently safe chemistry. *Preprints of Papers Presented at the 208th ACS National Meeting,* Aug. 21–25, 1994, Washington, D.C., 268–270. Center for Great Lakes Studies, University of Wisconsin-Milwaukee (Milwaukee, WI: Division of Environmental Chemistry, American Chemical Society)

TOCA, F.M. (1977). Anticipating toxic materials and exposures. *Loss Prevention*, **11**, p. 106

TODD, A.C.B. (1989). *Piper Alpha Public Inquiry. Transcript Days 70,* 71

TODD, C. (1985). Intelligent fire alarm systems. *Fire Surveyor*, **14**(2), 5

TOGNOTTI, L. and PETARCA, L. (1992). Safety problems in the production of organic hydroperoxides. *Loss Prevention and Safety Promotion 7*, vol. 2, 88–1

TOKHI, M.O. and LEITCH, R.R. (1992). *Active Noise Control* (Oxford: Clarendon Press)

TOLAND, J. (1957). *Ships in the Sky* (New York: Holt)

TOLSON, P. (1972). The ignition of flammable atmospheres by small amounts of metal vapor and particles. *Combust. Flame*, **18**, 19

TOLSON, P. (1980). The energy needed to ignite methane by discharges from a charged person. *J. Electrostatics*, **8**, 289

TOMASKO, D., WEIGAND, G.G. and THOMPSON, S.L. (1981a). A model for the free jet expansion and target stagnation pressure along an axis of symmetry. *Workshop on Jet Impingement and Pipe Whip,* Genoa, p. 187

TOMASKO, D., WEIGAND, G.G. and THOMPSON, S.L. (1981b). A model for two-phase jet loading. *Workshop on Jet Impingement and Pipe Whip,* Genoa, p. 165

TOMBS, S. (1988). The management of safety in the process industries: a redefinition. *J. Loss Prev. Process Ind.*, **1**(4), 179

TOMBS, S. (1990). Piper Alpha – a case study in distorted communication. *Piper Alpha: Lessons for Life Cycle Management* (Rugby: Instn Chem. Engrs), p. 99

TOMECEK, D.V. and MILKE, J.A. (1993). A study of the effect of partial loss of protection on the fire resistance of steel columns. *Fire Technol.*, **29**, 3

TOMFOHRDE, J.H. (1985). Design for process safety. *Hydrocarbon Process.*, **64**(12), 71

TOMIC, B. and LEDERMAN, L. (1991). IAEA's experience in compiling a generic component reliability data base. In Cannon, A.G. and Bendell, A. (1991), *op. cit.*, p. 197

TOMITA, S., HWANG, K. and O'SHIMA, E. (1989). On the development of an AI-based system for synthesising plant operating procedures. *Computer Aided Process Engineering* (Rugby: Instn Chem. Engrs), p. 303

VON TOMKEWITSCH, R. (1983). Fire detector systems with 'distributed intelligence'. The pulse polling system. *Fire Safety J.*, **6**, 225

TOMKINS, B. (1973). Fatigue failure criteria for thick-walled cylindrical pressure vessels. *Int. J. Pres. Ves. Piping*, **1**, 37

TOMKINS, B. (1983). Life prediction at elevated temperatures. *J. Press. Ves. Technol.*, **105**, 269

TOMKINS, B., LIDBURY, D.P.G. and HARROP, L.P. (1989). Phenomena of failure, failure modes of plant materials and probabilistic failure mechanics. In Cox, R.A. (1989), *op. cit.*, p. 13

TOMLINSON, H.C. (1985). Improve your operator training. *Hydrocarbon Process.*, **64**(2), 115

TOMLINSON, M.J. (1980). *Foundation Design and Construction,* 4th ed. (London: Pitman)

TOMLINSON, T., FINN, A. and LIMB, D. (1990a). Exergy analysis – a case study. *Chem. Engr, Lond.*, **484**, 39

TOMLINSON, T., FINN, A. and LIMB, D. (1990b). Exergy analysis in process development – introduction. *Chem. Engr, Lond.*, **483**, 25

TOMLINSON, W.R. (1958). *Properties of Explosives of Military Interest. Tech. Rep. 1740, rev. 1* (revised by O.E. Sheffield)

TOMPKINS, B. and RIFFLE, D. (1983). Careful safety evaluation identifies fire hazards on offshore facilities. *Oil Gas J.*, **81**(Oct. 3), 98

TOMPKINS, B. (1984). Offshore facility design must consider egress of personnel during emergency situations. *Oil Gas J.*, **82**(May 7), 154

TOMPKINS, B.G. and HOWELL, A.L. (1985). Analysis picks best offshore cable fireproofing. *Oil Gas J.*, **83**(Nov. 11), 65

TOMS, M. and PATRICK, J. (1987). Some components of fault-finding. *Hum. Factors*, **29**, 587

TOMS, M. and PATRICK, J. (1989). Components of fault-finding: symptom interpretation. *Hum. Factors*, **31**, 465

TONG, D.A. (1983). Human reaction to fire alarms. *Fire Surveyor*, **12**(4), 4

TONG, G.E. (1978). Fermentation routes to C3 and C4 chemicals. *Chem. Engng Prog.*, **74**(4), 70

TONG, W.R., SEAGRAVE, R.L. and WIEDERHORN, R. (1977). 3, 4–Dichloroaniline autoclave incident. *Loss Prevention*, **11**, p. 71

TONGUE, H. (1934). *The Design and Construction of High Pressure Chemical Plant* (London: Chapman & Hall)

TONKOVICH, A.Y., ZILKA, J.L., LAMONT, M.J., WANG, R.S. and WEGENG, R.S. (1999). Microchannel reactors for fuel processing applications. I. Water gas shift reactor. *Chem. Eng. Sci.* **54**, 2947–51

TOOGOOD, J.T. (1972). *Legal Requirements for Periodic Examination of Pressure Vessels under the Factories Act 1961. Paper C42/72.* Instn Mech. Engrs, London

TOOLA, A. (1992). Plant level safety analysis. *J. Loss Prev. Process Ind.*, **5**, 119

TOONG, T. (1957). Ignition and combustion in a laminar boundary layer over a hot surface. *Combustion*, **6**, p. 532

TOPMILLER, D.A., ECKEL, J.S. and KOZINSKY, E.J. (1982). *Human Reliability Data Bank for Nuclear Power Plant*

Operations, vol. 1, *A Review of Existing Human Reliability Data Banks. Rep. NUREG/CR-2744.* Nucl. Regul. Comm., Washington, DC

TORGERSON, W.S. (1967). *Theory and Method of Scaling* (New York: Wiley)

TORNAY, E.G., GILMORE, R.A. and FESKOS, J. (1991). Service longevity of the LNG 'Aquarius' class vessels. *Gastech 90,* p. 4.8

TORRANCE, K.E. (1971). Subsurface flows preceding flame spread over a liquid fuel. *Combust. Sci. Technol., 3,* 133

TORRANCE, K.E. and MAHAJAN, R.L. (1975). Fire spread over liquid fuels: liquid phase parameters. *Combustion, 15,* p. 281

TORRENT, J.G., FUCHS, J.C. and FERNANDEZ, S.M. (1990). K_{st} variation due to presence of inflammable gases. *Europex Newsletter, 13,* 7

TORRENT, J.G., FUCHS, J.C. and BORRAJO, J.L. (1991). On the combustion mechanism of coal dust in the presence of fire damp. *Combust. Flame, 87,* 371

TORREY, L. (1980). The hazards of cleaning up Three Mile Island. *New Scientist,* Sep. 11, 3

TORTOISHELL, G. (1986). Are fibre optics safe in flammable atmospheres? *Process Engng,* May, 111

TOSIC, N.M. (1993). Integration of a German parent company's process safety policies at a US subsidiary: an example of Bayer AG & Miles Inc. (poster item). *Process Safety Management,* p. 387

TOTH, W.C. (1990). Process safety management implementation at small locations. *Plant/Operations Prog., 9,* 244

TOTTER, J.R. (1982). Some problems in determining risk from cancer. *Risk Analysis, 2,* 19

TOUCHARD, G. (1978). Streaming currents generated by laminar and turbulent flow. *J. Electrostatics, 5,* 463

TOUCHARD, G. and DUMARQUE, P. (1978). Transport of electric charges by convection of a dielectric liquid in a cylindrical metal tube. *J. Electroanal. Chem. Interfacial Chem., 88,* 387

TOUCHARD, G., BENMADDA, M., HUMEAU, P., SAUNIERE, J. and BORZEIX, J. (1987). Static electrification by dusty gas flowing through polyethylene pipes. *Electrostatics, 7,* p. 97

TOULOUKIAN, Y.S., GERRITSEN, J.K. and MOORE, N.Y. (1967). *Thermophysical Properties Research Literature Retrieval Guide,* 2nd ed. (London: Macdonald)

TOUMA, J.S. (1977). Dependence of the wind profile power law on stability for various locations. *J. Air Poll. Control Ass., 27,* 863

TOUMA, J.S. and STROUPE, K.T. (1991). TSCREEN: a personal computer system for screening toxic air pollution impacts (poster item). *Modeling and Mitigating the Consequences of Accidental Releases,* p. 723

TOUMA, J.S., ZAPERT, J.G., THISTLE, H., LONDERGAN, R.J. and TOPAZICO, R. (1991). Performance evaluation of toxics dispersion models for simulating heavier-than-air releases (poster item). *Modeling and Mitigating the Consequences of Accidental Releases,* p. 735

TOURETSZKY, D.S. and HINTON, G.E. (1985). Symbols among the neurons: details of a connectionist inference architecture. *Proc. Inst. Joint Conf. on Artificial Intelligence*

TOVEY, H. (1974). The development of the National Fire Data System. *Fire J.,* Nov., 91

TOWILL, D.R. (1976). Transfer functions and learning curves. *Ergonomics, 19,* 623

TOWLE, L.C. (1975a). Earthing in hazardous atmospheres. *Electrical Safety, 2,* p. 96

TOWLE, L.C. (1975b). Instrumentation requirements in Zone 2 locations. Electrical *Safety, 2,* p. 91

TOWNE, D.M. (1981). A general-purpose system for simulating and training complex diagnosis and troubleshooting tasks. In Rasmussen, J. and Rouse, W.B. (1981), *op. cit.,* p. 621

TOWNEND, D.T.A. (1937). Ignition regions of hydrocarbons. *Combustion, 2,* p. 134

TOWNEND, D.T.A. and HSIEH, M.S. (1939). *J. Chem. Soc., 332*

TOWNEND, D.T.A. and MANDELKAR (1933). The influence of pressure on the spontaneous ignition of inflammable gas–air mixtures. I, Butane–air mixtures. *Proc. R. Soc., Ser. A, 141,* 484

TOWNSEND, A. (1992). *Maintenance of Process Plant,* 2nd ed. (Rugby: Instn Chem. Engrs)

TOWNSEND, A.A. (1956). *The Structure of Turbulent Shear Flow* (London: Cambridge Univ. Press)

TOWNSEND, C.A. and PANTONY, M.F. (1979). The control of runaway reactions – the view of H.M. Factory Inspectorate. In Minors, C. (1979), *op. cit.,* paper 1

TOWNSEND, D.I. (1977). Hazard evaluation of self-accelerating reactions. *Chem. Engng Prog., 73*(9), 80

TOWNSEND, D.I. (1981). Accelerating rate calorimetry. *Runaway Reactions, Unstable Products and Combustible Powders* (London: Instn Chem. Engrs), paper 3/Q

TOWNSEND, D.I. and TOU, J.C. (1980). Thermal hazard evaluation by an accelerating rate calorimeter. *Thermochmica Acta, 37,* 1

TOWNSEND, D.W. (1980). Second law analysis in practice. *Chem. Engr, Lond., 301,* 628

TOWNSEND, D.W. and LINNHOFF, B. (1982). Designing total energy systems by systematic methods. *Chem. Engr, Lond., 378,* 91

TOWNSEND, I. and VALDER, C.E. (1993). Modification of an accelerating rate calorimeter for operation from sub-ambient temperatures. *J. Loss Prev. Process Ind., 6,* 75

TRACHT, J.W. (1964). Communications between maintenance, design and production. *Chem. Engng, 71*(Apr. 27), 174

TRACY, A.W. (1962a). Corrosion of copper alloys. *Chem. Engng, 69*(Jan. 8), 130

TRACY, A.W. (1962b). Use of copper for corrosion resistant equipment. *Chem. Engng, 69*(Jan. 22), 152

TRACY, A.W. (1963). Where and when to use copper alloys. *Chem. Engng Prog., 59*(9), 78

TRACY, K.D. and ZITRIDES, T.G. (1979). Mutant bacteria aid Exxon waste system. *Hydrocarbon Process., 58*(10), 95

TRADE AND TECHNICAL PRESS (1986). *Seals and Sealing Handbook,* 2nd ed. (Morden)

TRADES UNION CONGRESS (1975). *Health and Safety at Work* (London)

TRADES UNION CONGRESS (1978). *Handbook on Health and Safety at Work,* 2nd ed. (London)

TRADES UNION CONGRESS (1986). *Health and Safety at Work: Know Your Rights* (London)

TRADES UNION CONGRESS (1987). *Nuclear Energy. General Council Report* (London)

TRAIN, D. (1983). Treatment and disposal of toxic wastes. *Chem. Engr, Lond., 390,* 35

TRAMBOUZE, P. (1979). Reactor scale-up philosophy. *Chem. Engng, 86*(Sep. 10), 122

TRAMBOUZE, P., VAN LANDEGHEM, H. and WAUQUIER, J.P. (1988). *Chemical Reactors: Design/Engineering/Operation* (Paris: Editions Techniques)

TRANSPORTATION SYSTEMS CENTER (1978). *Offshore Vessel Traffic Management (OVTM) System* (rep. to US Coast Guard)

TRASK, M.N. (1990). Operating discipline. *Plant/Operations Prog.*, **9**, 158

TRAVERSARI, A. and BENI, P. (1974). Design of a safe secondary compressor. *Safety in Polyethylene Plants*, **2**, p. 8

TRAVIS, C.C., RICHTER, S.A., CROUCH, E.A.C., WILSON, R. and KLEMA, E.D. (1987). Risk and regulation. *CHEMTECH*, Aug.

TRBOJEVIC, V.M. and GJERSTAD, T. (1989). How relevant is your treatment of accidental loads. *Loss Prevention and Safety Promotion 6*, p. 63.1

TRBOJEVIC, V.M. and MAINI, Y.N.T. (1985). An approach to the assessment of double containment RLG storage tanks under abnormal liquid loads. *The Assessment and Control of Major Hazards* (Rugby: Instn Chem. Engrs), p. 124

TRBOJEVIC, V.M. and SLATER, D.H. (1989). Tank failures modes and their consequences. *Plant/Operations Prog.*, **8**, 84

TRBOJEVIC, V.M., BELLAMY, L.J., BRABAZON, P.G., GUDMESTAD, T. and RETTDAL, W.K. (1994). Methodology for the analysis of risks during the construction and installation phases of an offshore platform. *J. Loss Prev. Process Ind.*, **7**, 350

TREGELLES, P.G. and WORTHINGTON, B. (1982). Reliability assessment as an aid to the development of mining equipment. *Mining Engr,* Aug., 95

TRELEAVEN, W.D. (1983). Acoustic emission testing for chemical plants. *Chem. Engng*, **90**(Feb. 7), 87

TRELIVING, L. (1985). Thermal image cameras: the new life savers. *Fire Int.*, **92**, 81

TRENCHARD, A.J. (1990). Process plant alarm diagnosis using fault tree knowledge. Ph.D. Thesis, Loughborough. Univ. of Technol.

TREON, J.F., DUTRA, F.R., CAPPEL, J., SIGMON, H. and YOUNKER, W. (1950). Toxicity of sulphuric acid mist. *Arch. Ind. Hyg. Occup. Med.*, **2**, 716

TRESEDER, R.S. (1981). Guarding against hydrogen embrittlement. *Chem. Engng*, **88**(June 29), 105

TREVAN, J.W. (1927). The error of determination of toxicity. *Proc. R. Soc., Ser. B*, **101**, 483

TREVANA, D.H. (1961). *Static Fields in Electricity and Magnetism* (London: Butterworths)

TREVETHICK, R.A. (1976). *Environmental Industrial Health Hazards,* rev. ed. (London: Heinemann)

TREVINO, C. and MENDEZ, F. (1992). Reduced kinetic mechanism for methane ignition. *Combustion*, **24**, p. 121

TREVINO, C. and SEN, M. (1981). Effect of plate thermal resistance on boundary layer ignition. *Combust. Flame*, **43**, 121

TREWARTHA, G.T. (1968). *An Introduction to Climate,* 4th ed. (New York: McGraw-Hill)

TREWEEK, D.N., CLAYDON, C.R. and SEATON, W.H. (1973). Appraising energy hazard potentials. *Loss Prevention*, **7**, p. 21

TREWEEK, D.N., HOYLAND, J.R., ALEXANDER, C.A. and PARDUE, W.M. (1976). Use of simple thermodynamic and structural parameters to predict self-reactivity hazard ratings of chemicals. *J. Haz. Materials*, **1**(3), 173

TREWHITT, J. (1981). Final rules lag, but PMN remains a hot topic. *Chem. Engng*, **88**(Jul. 13), 41

TREWHITT, J. (1982). 'Good science' is the new byword at EPA. *Chem. Engng*, **89**(Mar. 22), 57

TREWHITT, J. (1985). EPA takes solid action on risk-assessment guidelines. *Chem. Engng*, Jan. 21, 47

TREWHITT, J. and CAHAN, V. (1983). Hazy US cancer policy is causing dissension. *Chem. Engng*, **90**(Apr. 4), 35

TREYBAL, R.E. (1951 –). *Liquid Extraction:* 1963, 2nd ed. (New York: McGraw-Hill)

TREYBAL, R.E. (1955). *Mass Transfer Operations:* 1967, 2nd ed.: 1980, 3rd ed. (New York: McGraw-Hill)

TREZEK, G.J. and BALCERZAK, M.J. (1969). Some analytical and experimental results for large balloon detonations (letter). *Combust. Flame*, **13**, 431

TRIBOLET, R.O. (1964). Ultrasonic testing of critical equipment. *Ammonia Plant Safety*, **6**, p. 20

TRIBUS, M. (1969). *Rational Descriptions, Decisions and Designs* (New York: Pergamon Press)

TRIGGS, L.E. (1979). Improving boiler reliability. *Loss Prevention*, **12**, p. 60. See also *Chem. Engng Prog.*, 1978, **74**(10), 53

TRIGGS, T.J. (1988). The ergonomics of decision-making in large scale systems: information displays and expert knowledge elicitation. *Ergonomics*, **31**, 771

TRIMM, D.L. (1980). *Design of Industrial Catalysts* (Amsterdam: Elsevier)

TRIVEDI, B.C. and CULBERTSON, B.M. (1982). Maleic Anhydride (New York: Plenum Press)

TRIVEDI, K.K., ROACH, J.R. and O'NEILL, B.K. (1987). Shell targeting in heat exchanger networks. *AIChE J.*, **33**, 2087

TROE, J. (1989). Towards a quantitative understanding of elementary combustion reactions. *Combustion*, **22**, 843

TROOP, P.A. (1979). Co-generation in a changing regulatory environment. *Chem. Engng*, **86**(Feb. 26), 111

TROPP, R.I. (1986). More efficient turnarounds. *Hydrocarbon Process.*, **65**(1), 55.

TROSCINSKI, E.S. and WATSON, R.G. (1970). Controlling deposits in cooling water systems. *Chem. Engng*, **77**(Mar. 9), 125

TROTTER, C. (1983). Circumstances surrounding emissions of ammonia – a case history. *Loss Prevention and Safety Promotion 4*, vol. 1, p. L1

TROTTER, J.A. (1970) Techniques of predictive maintenance. *Chem. Engng*, **77**(Aug., 24), 66

TROTTER, J.A. (1973). New tools for modern maintenance. *Chem. Engng*, **80**(Feb. 26) (Deskbook issue), 87

TROTTER, J.A. (1979). Reduce maintenance cost with computers. *Hydrocarbon Process.*, **58**(1), 94

TROWBRIDGE, M.E. (1979). Government-industry interface. A European industry view. *Chem. Ind.*, Oct. 20, 699

TROWBRIDGE, T.D. (1987). Permit space standards proposed by OSHA. *Chem. Engng. Prog.*, **83**(6), 68

TROXLER, R. (1977). Elektronische Systeme in der Sicherheitstechnik. *J. Occup. Accid.*, **1**(3), 245

TROYAN, J. (1960). How to prepare for plant startups in the chemical industries. *Chem. Engng*, **67**(Sep. 5), 107

TROYAN, J.E. (1961a). Elements of operating manuals. *Chem. Engng*, **68**(Mar. 6), 134

TROYAN, J.E. (1961b). Hints for plant startup. Pt II, Troubleshooting new equipment. *Chem. Engng*, **68**(Mar. 20), 147

TROYAN, J.E. (1961c). Hints for plant startup. Pt III, Pumps, compressors and agitators. *Chem. Engng*, **68**(May 1), 91

TROYAN, J.E. (1963). Using common senses in plant operation. *Chem. Engng*, **70**(Mar. 4), 120

TROYAN, J.E. and LEVINE, R. (1968). Ethylene oxide explosion at Doe Run. *Loss Prevention*, **2**, p. 125

TRUAX, D.J. (1978). Are hydrogen permeation models accurate? *Hydrocarbon Process.*, **57**(8), 149

TRUE, W.R. (1985). US pipeline mileage up a bit in 1984. *Oil Gas J.*, **83**(Nov. 25), 55

TRUMAN, J.E. and HAIGH, P.M. (1980). Stainless steels and fluid handling. *Interflow 80* (Rugby: Instn Chem. Engrs), p. 187

TRUMAN, J.E. and KIRKBY, H.W. (1965). The possibility of service failure of stainless steel by stress corrosion cracking. *Metallurgia*, Aug.

TRUMBORE, D.C., WILKINSON, C.R. and WOLFSBERGER, S. (1991). Evaluation of techniques for in situ determination of explosion hazards in asphalt tanks. *J. Loss Prev. Process Ind.*, **4**, 230

TRUNKEY, D.D. (1983). Trauma. *Sci. Amer.*, **248**(2), 20

TRUSCOTT, J.M. and LIVINGSTONE, J.G. (1978). Inspection of a 12,000 ton ammonia storage tank. *Ammonia Plant Safety*, **20**, p. 153

TRUSCOTT, W.T. (1970). *Analytical Methods in Reliability Engineering* (London: Ind. Commercial Techniques)

TSAHALIS, D.T. (1978). Mitigation of chemical spills, RIDIS: a river dispersion model. *Control of Hazardous Material Spills*, **4**, p. 427

TSAI, M.J. (1982). Accounting for dissolved gases in pump design. *Chem. Engng*, **89**(Jul. 24), 65

TSAI, T.C. (1985). Sizing a vertical separator by microcomputer. *Oil Gas J.*, **83**(Feb. 25), 113

TSAI, T.S., LANE, J.W. and LIN, C.S. (1986). *Modern Control Techniques for the Processing Industries* (New York: Dekker)

TSANG, G. (1971). Laboratory study of line thermals. *Atmos. Environ.*, **5**, 445

CHI K. TSAO and PERRY, W.W. (1979). *Modifications to the Vulnerability Model: A Simulation System for Assessing Damage Resulting from Marine Spills. Rep. CG-D-38-79.* Enviro Control Inc., Rockville, MD

TSCHIRLEY, F.H. (1986). Dioxin. *Sci. American*, **254**(2)

VON TSCHISCHWITZ *et al.* (1934). The German attack at Ypres on 22.4.1915. *Gasschutz and Luftschutz*, **9**, 233

TSE, R.S. (1972). The oxidation of 2–butenes during the induction period. *Combust. Flame*, **18**, 357

TSICHRITZIS, D. (1975a). Project management. In Bauer, F.L. (1975a), *op. cit.*, p. 374

TSICHRITZIS, D. (1975b). Reliability. In Bauer, F.L. (1975a), *op. cit.*, p. 319

TSO, P.O. and DI PAULO, J.A. (1974). *Chemical Carcinogenesis* (New York: Dekker)

TSOURIS, C. and PORCELLI, J.V. (2003). Process intensification – has its time finally come? *Chem. Eng. Prog.* **99** (10), (Oct.), 50–5

TSUCHIYA, K., OKUBO, T. and ISCHIZU, S. (1975). An epidemiological study of occupational bladder tumours in the dye industry of Japan. *Br. J. Ind. Med.*, **32**, 203

TSUCHIYA, M., YAMAJI, K., YAMAMOTO, H., FUKASAWA, I. and HASHIM, Z. (1990). Mitigation plan for gas processing plant. *Safety and Loss Prevention in the Chemical and Oil Processing Industries,* p. 471

TSUCHIYA, Y. (1982). Methods of determining heat release rate: state-of-the-art. *Fire Safety J.*, **5**, 49

TSUCHIYA, Y. and NAKAYA, I. (1986). Numerical analysis of fire toxicity: mathematical predictions and experimental results. *J. Fire Sci.*, 4(2), 126

TSUDA, S. (1970). Tri- and per-chloroethylene made via simple new route. *Chem. Engng*, **77**(May 4), 74

TSUGE, Y., MATSUYAMA, H. and MCGREAVY, C. (1989). Qualitative simulator for estimation of plant behaviour during abnormal situations. *Computer Aided Process Engineering* (Rugby: Instn Chem. Engrs), p. 293

TSUGE, Y., SHIOZAKI, J., MATSUYAMA, H. and O'SHIMA, E. (1985). *Fault Diagnosis Algorithms based on the Signed Directed Graph and its Modifications. PSE 85* (Rugby: Instn Chem. Engrs), p. 133

TU, A.T. (ed.) (1981 –). *Survey of Contemporary Toxicology*, vol. 1, 1981: vol. 2, 1982 (New York: Wiley)

TUBULAR EXCHANGERS MANUFACTURERS ASSOCIATION – see Appendix 28

TUCKER, COOK, IYER and STACEY (1970). *Global Geophysics* (London: English Univ. Press)

TUCKER, A. (1972). *The Toxic Metals* (London: Earth Island)

TUCKER, A.J.P. (1972). Reformer tube and weld inspection. *Ammonia Plant Safety* **14**, p. 75

TUCKER, D.M. (1975). The explosion and fire at Nypro (UK) Ltd Flixborough on 1 June 1974. *Instn Chem. Engrs, Symp on Technical Lessons of Flixborough* (London)

TUCKER, D.M. (1989). *Piper Alpha Inquiry. Transcript, Day 95*

TUCKER, D.M., DRYSDALE, D.D. and RASBASH, D.J. (1981). The extinction of diffusion flames burning in various oxygen concentrations by inert gases and bromotrifluoromethane. *Combust. Flame*, **41**, 293

TUCKER, J.F. (1969). Ethylene liquefaction: a case history. *Hydrocarbon Process.*, **48**(2), 99

TUCKER, K. and LETTIN, A. (1975). The Tower of London bomb explosion. *Br. Med. J.*, 287

TUCKER, R.E., YOUNG, A.L. and GRAY, A.P. (1983). *Human and Environmental Risks of Chlorinated Dioxins and Related Compounds* (New York: Plenum Publ Press)

TUCKER, W.E. and CLINE, W.E. (1970). Plant reliability and its impact on large plant design and economics. *Am. Inst. Chem. Engrs and Puerto Rican Inst. Chem. Engrs, Third Joint Mtg* (New York: Am. Inst. Chem. Engrs)

TUCKER, W.E. and CLINE, W.E. (1971). Large plant reliability. *Chem. Engng Prog.*, **86**(3), 52

TUCKER, W.H. (1983). Dilemmas in teaching engineering ethics. *Chem. Engng Prog.*, **79**(4), 20

TUDOR-HART, J. (1972). Data on occupational mortality 1959–63. *The Lancet,* Jun. 22, 192

TUEY, D.B. (1980). Toxicokinetics. In Hodgson, E. and Guthrie, F.E. (1980), *op. cit.,* p. 40

TUFANO, V. (1992). On the design of inerting systems against chemical explosions. *Loss Prevention and Safety Promotion 7,* vol. 3, 119–1

TUFANO, V., CRESCITELLI, S. and RUSSO, G. (1981). On the design of venting systems against gaseous explosions. *J. Occup. Accid.*, **3**(3), 143

TUFANO, V., CRESCITELLI, S. and RUSSO, G. (1983). Overall kinetic parameters and laminar burning velocity from pressure measurements in closed vessels. *Combust. Sci. Technol.*, **31**, 119

TUHTAR, D. (1989). *Fire and Explosion Protection: A Systems Approach* (Chichester: Ellis Horwood)

TULACZ, J. and SMITH, R.E. (1980). Assessment of missiles generated by pressure component failure and its application to recent gas-cooled nuclear plant design. *Nucl. Energy*, **19**, 151

TULER, S., KASPERSON, R.E. and RATICK, S. (1989). Human reliability and risk management in the transportation of spent nuclear fuel. In Guy, G.B. (1989), *op. cit.*, p. 169

TULIS, A.J. and SELMAN, J.R. (1982). Detonation tube studies of aluminium particles dispersed in air. *Combustion*, **19**, p. 655

TULIS, A.J. and SELMAN, J.R. (1984). Unconfined aluminum particle two-phase detonation in air. In Bowen, J.R. *et al.*, (1984b), *op, cit.,* p. 277

TULLY, T.J. (1972). Instrumentation for safe handling of hazardous materials. *Instrum. Technol.*, **19**(10), 43

TUNC, M. and VENART, J.E.S. (1984/85a). Incident radiation from an engulfing pool fire to a horizontal cylinder – Part I. *Fire Safety J.*, **8**, 81

TUNC, M. and VENART, J.E.S. (1984/85b). Incident radiation from a flare to a horizontal cylinder – Part II. *Fire Safety J.*, **8**, 89

TUNKEL, S.J. (1983). Barricade design criteria. *Chem. Engng Prog.*, **79**(9), 50

TUNNICLIFFE, M.F. (1973). Emissions to the air – the interpretation of best practicable means. *Chem. Engr, Lond.*, **271**, 121

TUNNICLIFFE, M.F. (1975). The United Kingdom approach and its application by central government. Standards of emission for scheduled processes. *Nat. Soc. for Clean Air, Int. Clean Air and Pollution Control Conf.*, Brighton

TURING, A.M. (1950). Computing machinery and intelligence. *Mind*, **59**(236)

TURING, A.M. (1953). In Bowden, B.V. (1953), *op. cit.*, p. 288

TURING, A.M. (1963). Can a machine think? In Feigenbaum, E.A. and Feldman, J. (1963), *op. cit.*, p. 11

TURK, A. (1969). Industrial odor control and its problems. *Chem. Engng*, **76**(Nov. 3), 70

TURK, A., JOHNSTON, J.W. and MOULTON, D.G. (eds) (1974). *Human Responses to Environmental Odors* (New York: Academic Press)

TURKENBURG, W.C. (1974). *De Ingenieur*, **86**(10), 189

TURNBULL, J.N., BUYERS, T.B. and SMITH, W.J.W. (1974). The organisation of supporting services for a major petrochemical plant commissioning. *Instn Chem. Engrs, Symp. on Commissioning of Chemical and Allied Plant* (London)

TURNER, A.E. (1969). Intrinsically safe controls for Mobil. *Hydrocarbon Process.*, **48**(6), 149

TURNER, B. (1973). How reliable is your high pressure seal oil system? *Hydrocarbon Process.*, **52**(11), 205

TURNER, B. (1974). Why clean turbomachines on stream *Hydrocarbon Process?*, **53**(1), 102

TURNER, B. (1977). Learn from equipment failure. *Hydrocarbon Process.*, **56**(11), 317

TURNER, B. (1984). Design audit. *Engng Des. Educ.*, Autumn, 28

TURNER, B.A. (1978). *Man-Made Disasters* (London:Wykeham)

TURNER, B.A. (1991). The development of a safety culture. *Chem. Ind.*, Apr. 241

TURNER, D. (1974). Industrial hygiene and works safety. *Chem. Ind.*, Apr. 20, 332

TURNER, D. (1978). The control of environmental hazards in the chemical industry. *Chem. Ind.*, July 15, 510

TURNER, D.B. (1961). Relationships between 24-hour mean air quality measurements and meteorological factors in Nashville. *J. Air Poll. Control Ass.*, **11**, 483

TURNER, D.B. (1964). A diffusion model for an urban area. *J. Appl. Met.*, **3**, 83

TURNER, D.B. (1970). *Workbook of Atmospheric Dispersion Estimates*. Office of Air Programs, Environ. Prot. Agency, Research Triangle Park, NC

TURNER, D.B. (1985). Proposed pragmatic methods for estimating plume rise and plume penetration through atmospheric layers. *Atmos. Environ.*, **19**, 1215

TURNER, D.B. and NOVAK, J.H. (1978). *User's Guide for RAM*, vols A and B. *Rep. EPA–600/8–78–016*. Environ. Prot. Agency, Research Triangle Park, NC

TURNER, J. (1992). Put process engineering in the fast lane. *Chem. Engng Prog.*, **88**(8), 59

TURNER, J.S. (1965). The coupled turbulent transports of salt and heat across a sharp density interface. *Int. J. Heat Mass Transfer*, **8**, 759

TURNER, J.S. (1966). Jets or plumes with negative or reversing buoyancy. *J. Fluid Mech.*, **26**, 779

TURNER, J.S. (1973). *Buoyancy Effects in Fluids* (Cambridge: Cambridge Univ. Press)

TURNER, M. (1987). Limitations of materials. *Chem. Engr, Lond.*, **442**, 17

TURNER, M. (1988). What every chemical engineer should know about corrosion tests. *Chem. Engr, Lond.*, **449**, 52

TURNER, M. (1989a). What every chemical engineer should know about crevice corrosion. *Chem. Engr, Lond.*, **464**, 43

TURNER, M. (1989b). What every chemical engineer should know about stress corrosion cracking. *Chem. Engr, Lond.*, **460**, 52

TURNER, M. (1990a). Corrosion at welds. *Chem. Engr, Lond.*, **473**, 25

TURNER, M. (1990b). Hoare revisited. *Chem. Engr, Lond.*, **479**, 34

TURNER, M. (1990c). What every chemical engineer should know about galvanic corrosion. *Chem. Engr, Lond.*, **468**, 28

TURNER, M. and KING, R. (1986). Should you worry about corrosion monitoring? *Chem. Engr, Lond.*, **422**, 17

TURNER, M.J. (1980). The safety evaluation of cross country pipelines. Chemical Process Hazards, **7**, p. 245

TURNER, N. (1975). Will Canvey be the next Flixborough? *New Scientist*, Nov. 27, 520

TURNER, P.J. and CHESTERS, D.A. (1982). The development of coanda flares. *Developments 82* (Rugby: Instn Chem. Engrs)

TURNER, R. (1984). *Logics for Artificial Intelligence* (Chichester: Ellis Horwood)

TURNER, R.M. and FAIRHURST, S. (1989a). *Assessment of the Toxicity of Major Hazard Substances. Specialist Inspectors Rep. 21* (Bootle: Health and Safety Executive)

TURNER, R.M. and FAIRHURST, S. (1989b). *Toxicology of Substances in Relation to Major Hazards: Acrylonitrile* (London: HM Stationery Office)

TURNER, R.M. and FAIRHURST, S. (1990a). *Toxicology of Substances in Relation to Major Hazards: Chlorine* (London: HM Stationery Office)

TURNER, R.M. and FAIRHURST, S. (1990b). *Toxicology of Substances in Relation to Major Hazards: Hydrogen Fluoride* (London: HM Stationery Office)

TURNER, R.M. and FAIRHURST, S. (1990c). *Toxicology of Substances in Relation to Major Hazards: Hydrogen Sulphide* (London: HM Stationery Office)

TURNER, R.M. and FAIRHURST, S. (1992). *Toxicology of Substances in Relation to Major Hazards: Sulphuric Acid Mist* (London: HM Stationery Office)

TURNER, W.C. (1966). How to support insulated pipe. *Chem. Engng*, **73**(Jun. 20), 211

TURNER, W.C. (1974). Criterion for installing insulation systems in petrochemical plants. *Chem. Engng Prog.*, **74**(8), 41

TURNER, W.C. and MALLOY, J.F. (1981). *Thermal Insulation Handbook* (New York: McGraw-Hill)

TURNEY, R.D. (1990a). Designing plants for 1990 and beyond: procedures for the control of safety, health and environmental hazards in the design of chemical plant. *Safety and Loss Prevention in the Chemical and Oil Processing Industries*, p. 557

TURNEY, R.D. (1990b). Designing plants for 1990 and beyond: procedures for the control of safety, health and environmental hazards in the design of chemical plant. *Process Safety Environ.*, **68B**, 12

TURNEY, R.D. (1991). The application of Total Quality Management to hazard studies and their recording. *Hazards*, **XI**, p. 299

TURNEY, R.D. (1993). Prevention of environmental incidents. *Health, Safety and Loss Prevention,* p. 412

TURNEY, R.D., MANSFIELD, D.P., MALMEN, Y., ROGERS, R.L., VERWOERD, M., SOUKAS, E. and PLAISIER, A. (1997). The INSIDE project on inherent SHE in Process Development and Design – The Toolkit and Its Application. *IChemE Symp. Series,* No. 141, 203–15

TURNS, S.R. and LOVETT, J.A. (1989). Measurements of oxides of nitrogen emissions from turbulent propane jet diffusion flames. *Combust. Sci. Technol.,* **66,** 233

TURPIN, M.P. (1982). Application of computer methods to reliability prediction and assessment in a commercial company. *Reliab. Engng,* **3**(4), 295

TURPIN, M.P. and KAMATH, A.R.R. (1986). The development of an equipment availability reporting database and analysis package. *Reliab. Engng,* **15**(2), 95

TURTON, K.J. (1984). *Principles of Turbomachinery* (London: Spon)

TUSS, B.M. (1985). Include thermal imaging in your maintenance program. *Hydrocarbon Process.,* **64**(1), 77

TUSTIN, W. (1971). Measurement and analysis of machinery vibration. *Loss Prevention,* **5,** p., 90. See also *Chem. Engng Prog.,* 1971, **67**(6), 62

TUTTLE, F.C. and COULTER, K.E. (1977). Experience in operation of high-pressure furnaces. *Loss Prevention,* **11,** p. 1

TUTTON, R.C. (1965). Liquid natural gas: storage and sea transport. *Chem. Engr,* **186,** CE36

TUVE, G.L. (1953). Air velocities in ventilating jets. *Heating, Piping Air Conditioning,* **25**(1), 181

TUVE, R.L. (1964). Light water and potassium bicarbonate – a new two-agent extinguishing system. *NFPA Quart.,* **58**(1), 64

VAN TUYEN, N. and REGNAUD, P. (1983). Problems and solutions for pipelining LNG. *Oil Gas J.,* **81**(Jan. 24), 69

TVEIT, O.J. (1990). *Piper Alpha Public Inquiry. Transcript, Days 165, 166*

TVEIT, O.J. (1994). Safety issues on offshore process installations. *J. Loss Prev. process Ind.,* **7,** 267

TVEIT, O., MOSS, T.R. and OSTBY, E. (1983). The OREDA project offshore reliability data collection and analysis. *Euredata 4,* paper 6.1

TVER, D.F. (1964). Try this route to industrial safety. *Hydrocarbon Process.,* **43**(1), 131

TVERSKY, A. and KAHNEMAN, D. (1974). Judgment under uncertainty: heuristics and biases. *Science,* **185,** 1124

TVRDY, M., HYSPECKA, L., HAVEL, S. and MAZANEC, K. (1980). Fracture toughness and other mechanical metallurgy characteristics of pressure vessel steels. *Int. J. Pres. Ves. Piping,* **8,** 91

TVRDY, M., HAVEL, S., HYSPECKA, L. and MAZABEC, K. (1981). Hydrogen embrittlement of CrMo and CrMoV pressure vessel steels. *Int. J. Pres. Ves. Piping,* **9**

TWEEDDALE, J.G. (1964). *Mechanical Properties of Metals* (London: Allen & Unwin)

TWEEDDALE, H.M. (1985). Improving process safety and loss control software in existing process plants. *Multistream 85* (Rugby: Instn Chem. Engrs), p. 269

TWEEDDALE, H.M. (1990a). Plant safety audits: their place, content and style. *Australian Inst. Petrol., Safety Audit Seminar*

TWEEDDALE, H.M. (1990b). Toward safety assurance with major hazards. *FUTURESAFE 90*

TWEEDDALE, H.M. (1990c). Upgrading safety management in process plant. *Safety and Loss Prevention in the Chemical and Oil Processing Industries,* p. 125

TWEEDDALE, H.M. (1992). Balancing quantitative and non-quantitative risk assessment. *Process Safety Environ.,* **70B,** 70

TWEEDDALE, H.M. (1993a). Alternatives to risk assessment. *National Ecological Risk Assessment Conf.,* Melbourne

TWEEDDALE, M. (1993b). Increasing the relevance of QRA. *Health, Safety and Loss Prevention,* p. 124

TWEEDDALE, H.M. (1993c). Maximising the usefulness of risk assessment. In Melchers, R.E. and Stewart, M.G. (1993), *op. cit.,* p. 1

TWEEDDALE, M. (1993d). The safety audit steeplechase. *Chem. Engr, Lond.,* **553,** 15

TWEEDDALE, H.M. and WOOD, S. (1990). Safety assessment of industrial facilities on Rosebank Peninsula. *CHEMECA 90*

TWIGG, R.J. (1985). Preventing corrosion during mothballing. *Chem. Engng,* **92**(Sep. 16), 91

TWILT, L. (1988). Strength and deformation properties of steel at elevated temperatures: some practical implications. *Fire Safety J.,* **13,** 9

TWISDALE, L.A. and VICKERY, P.J. (1992). Comparison of debris trajectory models for explosives safety hazard analysis. *Explosives Safety Board, Dept of Defense, Proc. Twenty-Fifth Ann. Explosives Safety Seminar,* p.513

TWISDALE, A., SUES, R.H., LAVELLE, F.M. and DRAKE, J.L. (1992). Uncertainties and probabilistic risk assessment of explosives. *Explosives Safety Board, Dept of Defense, Proc. Twenty-Fifth Explosives Safety Seminar,* p. 129

TWISDALE, L.A. *et al.,* (1978). *Tornado Missile Risk Analysis. Reps. NP–768 and NP–769.* Elec. Power Res. Inst., Palo Alto, CA

TYAGI, S., LORD, A.E. and KOERNER, R.M. (1983a). Use of a metal detector to detect buried drums in sandy soil. *J. Haz. Materials,* **7**(4), 353

TYAGI, S., LORD, A.E. and KOERNER, R.M. (1983b). Use of a proton precession magnetometer to detect buried drums in sandy soil. *J. Haz. Materials,* **8**(1), 11

TYAGI, S., LORD, A.E. and KOERNER, R.M. (1983c). Use of a very-low-frequency electromagnetic method at 9.5 Hz to detect buried drums in sandy soil. *J. Haz. Materials,* **7**(4), 353

TYE, J. (1969). Communicating the safety message. In Handley, W. (1969), *op. cit.,* p. 15

TYE, J. (1970). *Management Introduction to Total Loss Control* (London: Br. Safety Coun.)

TYE, J. and ULLYETT, K. (1971). *Safety – Censored* (London: Corgi Books)

TYE, W. (1970). Airworthiness and the Air Registration Board. *Aeronaut. J.,* **74**(719), 885

TYLER, B.J. (1985). Using the Mond Index to measure inherent hazards. *Plant/Operations Prog.,* **4,** 172

TYLER, B.J. (1985). Using the Mond index to measure inherent hazards. *Plant/Operations Prog.* **4** (3) (July), 172–5

TYLER, B.J. and HENDERSON, D.K. (1987). Spontaneous ignitions in dust layers: comparisons of experimental and calculated values. *Hazards from Pressure,* p. 45

TYLER, B.J. and JONES, D.R. (1981). Thermal ignition in an asymmetrically heated slab: effects of reactant consumption. *Combust. Flame,* **42,** 147

TYLER, B.J. and WESLEY, T.A.B. (1967). Numerical calculations of the critical conditions in thermal explosion theory with reactant consumption. *Combustion,* **11,** p. 1115

TYLER, B.J., THOMAS, A.R., DORAN, P. and GREIG, T.R. (1994). A toxicity hazard index. *Hazards,* **XII,** p. 351

TYLER, D.A. (1969). How noisy is a refinery? *Hydrocarbon Process.*, **48**(7), 173

TYUGU, E. (1988). *Knowledge-Based Programming* (Wokingham: Addison-Wesley)

TZENG, L.S., ATREYA, A. and WICHMAN, I.S. (1990). A one-dimensional model of piloted ignition. *Combust. Flame*, **80**, 94

UBBELOHDE (1935). Investigations on the combustion of hydrocarbons. I, The influence of molecular structure on combustion. *Proc. R. Soc., Ser. A,* **152**, 354

UBBELOHDE (1936). *Z. Elektrochem.*, **42**, 468

UBBELOHDE, A.R. (1949). Transition from deflagration to detonation: the physico-chemical aspects of stable detonation. *Combustion*, **3**, p. 566

UBBELOHDE, A.R. (1953). The possibility of weak detonation waves. *Combustion*, **4**, p. 464

UBBELOHDE, A.R. and MUNDAY, G.M. (1969). Some current problems in the marginal detonation of gases. *Combustion*, **12**, p. 809

UBBELOHDE, A.R. *et al.* (1955). Status of the theory of the kinetics of combustion reactions (panel discussion). *Combustion*, **5**, p. 799

UBEROI, T. (1992). Effectively select in-line filters. *Chem. Engng Prog.*, **88**(3), 75

UCHIDA, H. and NARIAI, H. (1966). Discharges of saturated water through pipes and orifices. *Proc. Third Int. Heat Transfer Conf.*, vol. 5

UCHIYAMA, T. (1982). Effective pressure relief of offsite piping. *Hydrocarbon Process.*, **61**(5), 213

UEDA, S., ONISHI, H., OKUBO, M. and TAKEGAWA, T. (1978). Corrosion of ammonia plant heat exchanger. *Ammonia Plant Safety*, **20**, p. 98

UEDA, T. (1980). Hi-yield DMT by Mitsui. *Hydrocarbon Process.*, **59**(11), 143

UEHARA, Y. and HASEGAWA, H. (1986). Analysis of causes of accidents at factories dealing with hazardous materials. *Loss Prevention and Safety Promotion 5*, p. 23–1

UEHARA, Y. and KITAMURA, A. (1990). Evaluation of potential hazards of oxidizing substances by burning tests. *Safety and Loss Prevention in the Chemical and Oil Processing Industries*, p. 485

UEHARA, Y. and NAKAJIMA, T. (1985). Proposal of a new test method for the classification of oxidizing substances. *J. Haz. Materials*, **10**(1), 89

UEHARA, Y. and SEINO, N. (1992). Experimental study on change of critical ignition temperature of material in varying ambient temperature. *Loss Prevention and Safety Promotion 7*, vol. 2, 90–1

UEHARA, Y., UEMATSU, H. and SAITO, Y. (1978). Thermal ignition of calcium hypochlorite. *Combust. Flame*, **32**, 85

UFFORD, P.S. (1972). Equipment reliability analysis for large plants. *Chem. Engng Prog.*, **68**(3), 47

UGUCCIONI, G. and MESSINA, S. (1986). Experimental test on large hydrocarbon fires. Results and comparison with existing fire models. *Loss Prevention and Safety Promotion 5*, p. 68–1

UHLIG, H.H. (ed.) (1951). *Corrosion Handbook* (New York: Wiley)

UITENHAM, L. and MUNJAL, R. (1991). Choose the right control scheme for pilot plants. *Chem. Engng Prog.*, **87**(1), 35

UK ATOMIC ENERGY AUTHORITY — see also Appendix X, 28

UK ATOMIC ENERGY AUTHORITY (1979). *Precis of Parts of Legal Documents and Dispositions made in January 1979 regarding the Tanker Explosion in Spain 11/7/78* (Culcheth, Lancashire)

UK ATOMIC ENERGY AUTHORITY (1987). *Nuclear Power Reactors* (London)

UK ATOMIC ENERGY AUTHORITY and BRITISH NUCLEAR FUELS LTD (1985). *Environmental Impact Assessment Relating to the Proposed Siting of the European Demonstration Fast Reactor Fuel Reprocessing Plant (EDRP) at Dounreay, Caithness* (London and Risley)

UK AUTOMATION COUNCIL (1965). *Advances in Automatic Control* (London: Instn Mech. Engrs)

ULBRECHT, J.J. and PATTERSON, G.K. (eds) (1985). *Mixing of Liquids by Mechanical Means* (New York: Gordon & Breach)

VAN ULDEN, A.P. (1974). On the spreading of a heavy gas released near the ground. *Loss Prevention and Safety Promotion*, **1**, p. 221

VAN ULDEN, A.P. (1979). The unsteady gravity spread of a dense cloud in a calm atmosphere. *Proc. Tenth Int. Tech. Mtg on Air Pollution Modelling and Its Application* (NATO/CCMS), p. 645

VAN ULDEN, A.P. (1984). A new bulk model for dense gas dispersion: two-dimensional spread in still air. In Ooms, G. and Tennekes, H. (1984), *op. cit.*, p. 419

VAN ULDEN, A.P. (1987). The heavy gas mixing process in still air at Thorney Island and the laboratory. *J. Haz. Materials*, **16**, 411

VAN ULDEN, A.P. (1988). The spreading and mixing of a dense cloud in still air. In Puttock, J.S. (1988b), *op. cit.*, p. 141

VAN ULDEN, A.P. (1992). A surface-layer similarity model for the dispersion of a skewed passive puff near the ground. *Atmos. Environ.*, **26A**, 681

VAN ULDEN, A.P. and DE HAAN, B.J. (1983). Two-dimensional unsteady gravity currents. In de Wispelaere, C. (1983), *op. cit.*, p. 603

ULERICH, N.H. and POWERS, G.J. (1988). On-line hazard aversion and fault diagnosis in chemical processes: the digraph + fault-tree method. *IEEE Trans Reliab.*, **R37**, 171

ULLMAN, J.D. (1982). *Principles of Database Systems,* 2nd ed. (Rockville: Computer Sci. Press)

ULLMANN (1969–). *Ullmanns Encyklopädie der technischen Chemie* (Munich: Urban & Schwarzenberg)

ULLMAN (1985–). *Ullmanns Encyclopedia of Chemical Technology* (Weinheim, Germany: VCH Publs)

ULM, R.C. (1963). Steel storage tanks give generations of service. API Proc., *Div. of Refining*, **43**, 470

ULRICH, G.D. (1984). *A Guide to Chemical Engineering Process Design and Economics* (New York: Wiley)

ULRICH, G.D. (1992). How to calculate utility costs. *Chem. Engng*, **99**(2), 110

ULRICH, J.S. (1973). Furnace tubes resist overheating. *Hydrocarbon Process.*, **52**(4), 184

UMBERS, I.G. (1979). A study of the control skills of gas grid control engineers. *Ergonomics*, **22**, 557

UMBERS, I.G. (1981). A study of control skills in an industrial task, and in a simulation, using the verbal protocol technique. *Ergonomics*, **24**, 275

UMBERS, I.G. and RIERSON, C.S. (1995). Task analysis in support of the design and development of a nuclear power plant safety system. *Ergonomics*, **8**, 443

UMEDA, T. (1982). Computer aided process synthesis. *Proc. Process Systems Engng Symp.*, Kyoto, Japan, p. 79

UMEDA, T., NISHIO, M. and KOMATSU, S. (1971). A method for plant data analysis and parameters estimation. *Ind. Engng Chem. Process Des. Dev.*, **10**, 236

UMEDA, T., KURIYAMA, T., O'SHIMA, E. and MATSUYAMA, H. (1980). A graphical approach to cause and effect analysis of chemical processing systems. *Chem. Engng Sci.*, **35**, 2379

UMEDA, Y., KURIYAMA, T., O'SHIMA, E. and MATSUYAMA, H. (1979). A graphical approach to causes and effects of chemical processing systems. *Comput. Chem. Engng*, **3**, 495

UMLAND, A.W. (1954). Explosive limits of hydrogen-chlorine mixtures. *J. Electrochem. Soc.*, **101**, 626

UMMARINO, G. (1979). Computerize seal coolant flow calculations. *Hydrocarbon Process.*, **58**(8), 137

UNCLES, R.F. (1966). What the engineer should know about containers. *Chem. Engng*, **73**(Jul. 4), 113: (Jul. 18), 167; (Aug. 15), 151

UNDERHILL, F.P. (1920). *The Lethal War Gases* (New Haven, CN: Yale Univ. Press)

UNDERWOOD, A.A. (1992). Switch to stainless steel – beneficial or harmful? *Chem. Engng Prog.*, **88**(6), 69

UNDERWOOD, B.Y. (1987). Dry deposition to a uniform canopy: evaluation of a first-order-closure mathematical model. *Atmos. Environ.*, **21**, 1573

UNDERWOOD, G.W. (1971). *Practical Fire Precautions* (London: Gower Press): 1979, 2nd ed. (Farnborough: Gower Press). See also Hirst, R.

UNDERWOOD, H.C., SOURWINE, R.E. and JOHNSON, C.D. (1976). Organize for plant emergencies. *Chem. Engng*, **83**(Oct. 11), 118

UNDERWOOD, J.G. (1978). *Hazards from Toxic Materials Released During Fires* (report). Insurance Technical Bureau, London

UNDERWOOD, J.H. (1975), A note on cylinder stress- intensity factors. *Int. J. Pres. Ves. Piping*, **3**, 229

UNDERWRITERS LABORATORIES (n.d.a). *Miscellaneous Hazard 2375* (Chicago, IL), p. 26

UNDERWRITERS LABORATORIES (n.d.b). *UL Handbook for Fire Ratings* (Chicago, IL)

UNDERWRITERS LABORATORIES (1944). *A New Type of Bomb for Investigation of Pressures Developed by Dust Explosions. Bull. Res. 30* (Chicago, IL)

UNDERWRITERS LABORATORIES (1950). *The Lower Limit of Flammability and the Autogeneous Ignition Temperature of Certain Common Solvent Vapours Encountered in Ovens. Bull. Res. 43* (Chicago, IL)

UNDERWRITERS LABORATORIES (1951). *An Investigation of Large Electric Motors and Generators of the Explosion Proof Type for Hazardous Locations, Class 1, Group D. Bull. Res. 46* (Chicago, IL)

UNDERWRITERS LABORATORIES (1953). *The Spontaneous Ignition and Dust Explosion Hazards of Certain Soybean Products. Bull. Res. 47* (Chicago, IL)

UNDERWRITERS LABORATORIES (1963). *The Compatibility Relationship between Mechanical Foam and Dry Chemical Fire Extinguishing Agents.* Bull. Res., 54 (Chicago, IL)

UNDERWRITERS LABORATORIES – see also Appendix 27

UNGAR, L.F. (1985). A framework for fault analysis. *Am. Inst. Chem. Engrs, Ann. Mtg,* Chicago (New York)

UNGAR, L.H., POWELL, B.A. and KAMENS, S.N. (1990). Adaptive networks for fault diagnosis and process control. *Comput. Chem. Engng*, **14**, 561

UNGER, E. (1983). Problems of operating reliability in control and indicating equipment. *Fire Safety J.*, **6**, 241

ÜNGÜT, A. and SHUFF, P.J. (1989). Deflagration to detonation transition from a venting pipe. *Combust. Sci. Technol.*, **63**, 75

UNGUT, A., SHUFF, P.J. and EYRE, J.A. (1984). Initiation of unconfined gaseous detonation by diffraction of a detonation front emerging from a pipe. In Bowen, J.R. *et al.* (1984b), *op. cit.*, p. 523

UNION CARBIDE CORPORATION (1985). *Bhopal Methyl Isocyanate Incident. Investigation Team Report* (Danbury, CT)

UNION OF CONCERNED SCIENTISTS (1977). *The Risks of Nuclear Power Reactors* (Cambridge, MA)

UNITED NATIONS (1993). *Recommendations on the Transport of Dangerous Goods,* 8th rev. ed. (New York)

UNITED NATIONS (1978). *The Gas Industry and the Environment* (Oxford: Pergamon Press)

UNITED NATIONS (1986). *Recommendations on the Transport of Dangerous Goods: Tests and Criteria* (New York)

UNITED NATIONS (1988a). *Recommendations on the Transport of Dangerous Goods,* 7th ed. (New York)

UNITED NATIONS (1988b). *Recommendations on the Transport of Dangerous Goods: Tests and Criteria, Addendum 1* (New York)

UNITED NATIONS (1990). *Recommendations on the Transport of Dangerous Goods: Tests and Criteria,* 2nd ed. (New York)

UNITED NATIONS (1991). *Recommendations on the Transport of Dangerous Goods,* 7th rev. ed. (New York)

UNITED NATIONS SCIENTIFIC COMMITTEE ON THE EFFECTS OF ATOMIC RADIATIONS (1977). *Sources and Effects of Ionizing Radiations* (New York)

UNITED NATIONS, SCIENTIFIC COMMITTEE ON THE EFFECTS OF ATOMIC RADIATION (1982). *Ionizing Radiations, Sources and Biological Effects of Atomic Radiation* (New York)

UNIVERSITY ENGINEERS INC. (1974). *An Experimental Study on the Mitigation of Flammable Vapor Dispersion and Fire Hazards Immediately Following LNG Spills on Land. AGA Proj. IS-100–1*

UNRUG, H.F. (1979). Choosing among plant fabrication methods. *Chem. Engng Prog.*, **75**(11), 40

UNWIN, A.J.T., ROBINS, J.A. and PAGE, S.J.R. (1974). Commissioning of a major materials handling offsite facility for a petrochemical complex. *Instn Chem. Engrs, Symp. on Commissioning of Chemical and Allied Plant* (London)

UPFOLD, A.T. (1971). Man-hour rating standardised for instrument maintenance. *Instrum. Technol.*, **18**(2), 46

UPMANN, J. (1878). *Traite sur la Pourde* (Paris: Dunod)

UPPAL, K.B. (1980). Insulating tanks: what it costs and how it's done. *Hydrocarbon Process.*, **59**(4), 165

UPPULURI, V.R.R. (1983). *Expert Opinion and Ranking Methods. Rep. NUREG/CR-3115.* Nucl. Regul. Comm., Washington, DC

URAGODA, C.G. (1970). A case of chlorine poisoning of occupational origin. *Ceylon Med. J.*, **15**, 223

URAL, E.A. (1992). Dust entrainability and its effect on explosion propagation in elongated structures. *Plant/Operations Prog.*, **11**, 176

URAL, E.A. (1993). A simplified method for predicting the effect of ducts connected to explosion vents. *J. Loss Prev. Process Ind.*, **6**, 3

URAL, E.A. and ZALOSH, R.G. (1985). A mathematical model for lean-hydrogen-air-steam mixture combustion in closed vessels. *Combustion*, **20**, p. 1727

URBAN, M. and LOSCHE, R. (1978). Development and use of a mobile chemical laboratory for hazardous material spill response activities. *Control of Hazardous Material Spills*, **4**, p. 311

URBANIK, T., DESROSIERS, A., LINDELL, M.K. and SCHULER, C.R. (1980). *Analysis of Techniques for Estimating Evacuation*

Times for Emergency Planning Zones. Rep. NUREG/CR-1745. Nucl. Regul. Comm.,Washington DC

URBANSKI, T. (1964–). *Chemistry and Technology of Explosives,* vols 1–4 (London: Macmillan)

URBANSKI, T. (1983). *Chemistry and Technology of Explosives,* vols 1–4 (Oxford: Pergamon Press)

URQUHART, L.C. (1959). *Civil Engineering Handbook,* 4th ed. (NewYork: McGraw-Hill)

URSENBACH, W.O. (1980). Unusual explosion and fire investigations. *Loss Prevention and Safety Promotion 3,* vol. 2, p. 138

URSENBACH, W.O. (1983). Hazard assessment from industrial explosion investigation. *Loss Prevention and Safety Promotion 4,* vol. 1, p. N4

URTIEW, P.A. (1981). *Flame Propagation in Gaseous Fuel Mixtures in Semi-Confined Geometries. Rep. UCID- 19000.* Lawrence Livermore Nat. Lab., Livermore, CA

URTIEW, P.A. (1982a). *Overpressure Resulting from Combustion of Explosive Gas in an Unconfined Geometry* (report). Lawrence Livermore Lab., Livermore, CA

URTIEW, P.A. (1982b). Recent flame propagation experiments at LLNL within the liquefied gaseous fuels spill safety program. In Lee, J.H.S. and Guirao, C.M. (1982), *op. cit.,* p. 929

URTIEW, P.A. and OPPENHEIM, A.K. (1965). Onset of detonation. *Combust. Flame,* **9**, 405

URTIEW, P.A. and OPPENHEIM, A.K. (1966). Experimental observations of the transition to detonation in an explosive gas. *Proc. R. Soc., Ser. A,* **295**, 13

URTIEW, P.A. and OPPENHEIM, A.K. (1967). Detonative ignition induced by shock merging. *Combustion,* **11**, p. 665

URTIEW, P.A. and OPPENHEIM, A.K. (1968). Transverse flame-shock interactions in explosive gas. *Proc. R. Soc., Ser. A,* **304**, 379

URTIEW, P.A. and TARVER, C.M. (1981). Effects of the cellular structure on the behaviour of gaseous detonation waves in transient conditions. *Prog. Astronaut. Aeronaut.,* **75**, 370

URTIEW, P.A., LADERMAN, A.J. and OPPENHEIM, A.K. (1965). Dynamics of the generation of pressure waves by accelerating flames. *Combustion,* **10**, p. 797

URTIEW, P.A., BRANDEIS, J. and HOGAN, W.J. (1983). Experimental study of flame propagation in semi- confined geometries with obstacles. *Combust. Sci. Technol.,* **30**, 105

US AIR FORCE, AIR FORCE SYSTEM COMMAND (n.d.). *System Safety Design Handbook. Hndbk DHI-6* (Washington, DC)

US ARMED SERVICES – see Appendix 27

US ARMY, CORPS OF ENGINEERS (1957). *Design of Structures to Resist the Effects of Atomic Weapons. Engineering and Design Manuals* (Washington, DC: US Govt Printing Office)

US ARMY, DEPARTMENT OF DEFENSE (1965). *Reliability Stress and Failure Rate Data for Electronic Equipment* (Washington, DC)

US ARMY MATERIAL DEVELOPMENT AND READINESS COMMAND (1977). *Handbook of Chemical Hazard Predictions* (Arlington,VA)

US ARMY, NAVY AND AIR FORCE (1990). *Structures to Resist the Effects of Accidental Explosions. Rep TM 5–1300, NAV-FAC P-397, AFR 88–22, rev. 1* (the TriServices Manual)

US ARMY, ORDNANCE CORPS (1958). *Ordnance Safety Manual. Publ. ORDM-7–224 1951* Sec. 17, May 15, 68

US ARMY, ORDNANCE CORPS (1960). *Properties of Explosions of Military Interest. Pamphlet ORD P20–177*

US COAST GUARD (1973). *Vessel Traffic System Analysis of Port Needs* (Washington, DC)

US COAST GUARD (1976a). *Analyzing Marine Safety Systems. Request for Proposal CG-60-351-A,* (Washington, DC)

US COAST GUARD (1976b). *Final Environmental Impact Statement, Loop Deepwater Port Licence Application* (Washington, DC)

US COAST GUARD (1979). *Analytical and Experimental Study to Improve Computer Models for Mixing and Dilution of Water Soluble Hazardous Chemicals. Request for Proposal CG-920622-A* (Washington, DC)

US CONGRESS (1973). *Staten Island Explosion: Safety Issues Concerning LNG Storage Facilities. Special Sub-committee on Investigations of the Committee on Interstate and Foreign Commerce, Hearings July 10–12,* 1973, Ser. 93– 42 (Washington, DC: US Govt Printing Office)

US CONGRESS (1979). *Accident at the Three Mile Island Nuclear Power Plant. Subcommittee on Energy and the Environment, Hearings May 9, 10, 11 and 15, 1979, Ser. 96–8, pt I* (Washington, DC: US Govt Printing Office)

US CONGRESS, JOINT COMMITTEE ON ATOMIC ENERGY (1960). *Radiation Protection Criteria and Standards: their Basis and Use. Summary-analysis of Hearings May 24–25 and June 1–3*

US CONGRESS, OFFICE OF TECHNOLOGY ASSESSMENT (1977). *Transportation of Liquefied Natural Gas* (Washington, DC)

US CONGRESS, OFFICE OF TECHNOLOGY ASSESSMENT (1981). *Assessment of Technologies for Determining Cancer Risks from the Environment* (Washington, DC: US Govt rinting Office)

US CONGRESS, OFFICE OF TECHNOLOGY ASSESSMENT (1983). *The Information Content of Pre-Manufacture Notices* (Washington, DC: US Govt Printing Office)

US EPA/OSHA JOINT REPORT – "*Chemical Accident Investigation Report*", (1998), EPA document 550-R-98-005 (Washington, D.C.: US Environmental Protection Administration)

US GEOLOGICAL SURVEY (1975). *The Interior of the Earth* (Washington, DC: US Govt Printing Office)

US NAVY, BUREAU OF NAVAL WEAPONS (n.d.). *Failure Rate Data Handbook* (FARADA) (Corona, CA)

US SENATE (1970). *Effects of 2,4,5-T on Man and the Environment. Subcommittee on Energy, Natural Resources and the Environment of the Committee of Commerce, Hearings April 7 and 15* (Washington, DC: US Govt Printing Office)

US SENATE (1978). *Liquefied Natural Gas: Safety, Siting and Policy Concerns. Rep. GPO-26–349* (Washington, DC)

USSR STATE COMMITTEE ON THE UTILIZATION OF ATOMIC ENERGY (1986). *The Accident at the Chernobyl Nuclear Power Plant and its Consequences, pts 1–2* (Vienna: Int. Atom. Energy Agency)

VACCARO, D.J. (1969). Power and control system reliability. *Loss Prevention,* **3**, p. 70

VADALA, A.J. (1930). Effects of gun explosions on the ear and hearing mechanism. *Milit. Surg.,* **66**, 710

VADERA, S. (ed.) (1989). *Expert System Applications* (Wilmslow: Sigma Press)

VAHAVIOLOS, S.J., POLLOCK, A. and LEW, N. (1991). Pinpoint structural defects with acoustic emissions. *Chem. Engng Prog.,* **87**(1), 60

VAIJA, P., JÄVELÄINEN, M. and DOHNAL, M. (1985). Fuzzy based expert system for analysis of accidents. *The Assessment and Control of Major Hazards* (Rugby: Instn Chem. Engrs), p. 397

VAIJA, P., JÄRVELÄINEN, M. and DOHNAL, M. (1986a). Expert system and fuzzy clustering of accident data. *Loss Prevention and Safety Promotion 5*, vol. I, p. 65–1: vol. III, p. 228

VAIJA, P., JÄRVELÄINEN, M. and DOHNAL, M. (1986b). Failure diagnosis of complex systems by a network of expert bases. *Reliab. Engng*, **16**(3), 237

VAILLANT, M. (1977). Fault tree based on past accidents. Factorial analysis of events. *Loss Prevention and Safety Promotion 2*, p. 425

VALCKENAERS, M. (1983). Risk assessment of LNG installations. *Gastech 82*, p. 301

VALDES, E.C. and SVOBODA, K.J. (1985). Estimating relief loads for thermally blocked-in liquids. *Chem. Engng*, **92**(Sep. 2), 77

VALENTIN, F.H.H. (1990). Making chemical-process plants odor-free. *Chem. Engng*, **97**(1), 112

VALENTIN and MERRILL (1942). Dust control in the plastics factory. *Trans. Am. Inst. Chem. Engrs*, **38**

VALENTIN, F.H.H. and NORTH, A.A. (eds) (1980). *Odour Control* (Stevenage: Warren Spring Lab.)

VALENTINE, R. (1988). *Preliminary Report on the Concentration–Time Response Relationship for Hydrogen Fluoride*. Rep. 281–88. Haskell Res. Lab., EI Dupont de Nemours

VALENZUELA, D.P. and MYERS, A.L. (1988). *Adsorption Equilibrium Data Handbook* (Englewood Cliffs, NJ: Prentice-Hall)

VALK, A. and SYLVESTER-EVANS, R. (1985). Safety in the design of gas terminals. *Gastech 84*, p. 157

VALKO, E.I. and TESORO, G.S. (1964). Antistatic additive. *Encyclopaedia of Polymer Science and Technology*, vol. 2 (New York: Interscience), p. 204

VALPIANA (1970a). Explosion caused by decomposition of a quaternary ammonium salt. *Prevention of Occupational Risks*, 1

VALPIANA (1970b). Explosive decomposition of 6-chlor-2,4-dinitro-diazo-bene in concentrated sulphuric acid. *Prevention of Occupational Risks*, 1

VALSARAJ, K.T. and THIBODEAUX, L.J. (1988). Equilibrium adsorption of chemical vapors on surface soils, landfills and landfarms – a review. *J. Haz. Materials*, **19**(1), 79

VANCE, F.P. (1966). An economic basis for setting confidence limits. *Ind. Engng Chem.*, **58**(2), 30

VANCE, G.M. and KRIER, H. (1974). A theory for spherical and cylindrical laminar premixed flames. *Combust. Flame*, **22**, 365

VANCHUK, J.T. (1978). Making safe compressed respiratory air. *Chem. Engng*, **85**(Apr. 24), 109

VANCINI, C.A. (1971). *Synthesis of Ammonia* (London: Macmillan)

VANCINI, C.A. (1982). Program calculates flame temperature. *Chem. Engng*, **89**(Mar. 22), 133

VANDELL, C. and FOERG, W. (1993). The pluses of positive displacement. *Chem Engng*, **100**(1), 74

VANDEGRIFT, G.F., REED, D.T. and TASKER, I.R. (eds) (1992). *Environmental Remediation* (Washington, DC: Am. Chem. Soc.)

VANDER AREND, P.C. (1961). Large scale liquid hydrogen production. *Chem. Engng Prog.*, **57**(10), 62

VANDER SCHRAAF, E. and STRAUSS, W.I. (1964). Direct digital control – an emerging technology. *Oil Gas J.*, **62**(46), 167

VANDERBECK, R.W. (1960). Special carbon steels, tough and ductile at subzero temperatures. *Chem. Engng*, **67**(May 30), 103

VANDERBORGHT, B. and KRETSCHMAR, J. (1984). A literature survey on tracer experiments for atmospheric dispersion modelling studies. *Atmos. Environ.*, **18**, 2395

VANDERGRIFT, A.E., SHANNON, L.J. and GORMAN, P.G. (1973). Controlling fine particles. *Chem. Engng*, **80**(Jun. 18), 107

VANDERGRIFT, E.F. (1980). Meeting OSHA regulations for toxic exposure. *Chem. Engng*, **87**(June 2), 69

VANDERLINDE, L.G. (1974). State of the art – smokeless flares. *Hydrocarbon Process.*, **53**(10), 99

VANDERLINDER, L.S. (1970). Trends in materials applications. Plastics. *Chem. Engng*, **77**(Oct. 12), 119

VANDERMEIREN, M. and VAN TIGGELEN, P.J. (1984). Cellular structure in detonation of acetylene–oxygen mixture. In Bowen, J.R. *et al.* (1984b), *op. cit.*, p. 104

VANDERVEEN, J.W., LUSS, D. and AMUNDSON, N.R. (1968). Stability of adiabatic packed bed reactors. Effect of flow variations and coupling between the particles. *AIChE J.*, **14**, 636, 993

VANDERWATER, R.G. (1989). Case history of an ethylene tank car explosion. *Chem. Engng Prog.*, **85**(12), 16

VANLANINGHAM, F.L. (1977). Gear case distortion causes failures. *Hydrocarbon Process.*, **56**(1), 91

VANMATRE, J. (1977). Clean heat exchange equipment on-stream. *Hydrocarbon Process.*, **56**(7), 115

VANNAH, W.E. and CALDER, W. (1981). Effects of service conditions on electronic instrumentation. *Chem. Engng*, **88**(Sep. 7), 143

VANPEE, M. (1993). On the cool flames of methane. *Combust. Sci. Technol.*, **93**, 363

VANPEE, M. and GRARD, F. (1955). The kinetics of the slow combustion of methane at high temperature. *Combustion*, **5**, p. 484

VANPEE, M. and SHIRODKAR, P.P. (1979). A study of flame inhibition by metal compounds. *Combustion*, **17**, p. 787

VANPEE, M. and WOLFHARD, H.G. (1959). *J. Am. Rocket Soc.*, **29**, 517

VANTINE, H.C., CHAN, J. and ERICKSON, L.M. (1982). Thin pulse initiation of PBX-9404. In Nellis. W.J., Seaman, L. and Graham, A.R. (1982), *op. cit* p. 558

VARADARAJAN, Heat transfer experiments of monodispersed vertically impacting sprays

VAREY, P. (1987). Pressurised water goes against the trend (editorial). *Chem. Engr, Lond.*, **434**, 3

VAREY, P. (1988). Fire and the valve. *Chem. Engr, Lond.*, **454**, 33

VAREY, P. (1989). Setting up sub-sea separation. *Chem. Engr, Lond.*, **456**, 31

VAREY, P. (1992a). Of chemistry and canaries. *Chem. Engr, Lond.*, **524**, 19

VAREY, P. (1992b). Fighting Forties. *Chem. Engr, Lond.*, **527**, 12

VARGAS, K.J. (1979). Managing plant turnarounds. *Chem. Engng*, **86**(Jun. 4), 116

VARGAS, K.J. (1980a). Maintenance records. *Chem. Engng*, **87**(Apr. 21), 161

VARGAS, K.J. (1980b). Setting up a spare parts program. *Chem. Engng*, **87**(Jan. 28), 133

VARGAS, K.J. (1982). Trouble-shooting compression refrigeration systems. *Chem. Engng*, **89**(Mar. 22), 137

VARMA, R.S. (1999). Solvent-free organic syntheses using supported reagents and microwave irradiation. *Green Chem.*, (Feb.), 43–55

VARNERIN, R.E. (1987). Safe disposal of hazardous waste. In Young, J.A, (1987a), *op. cit.*, p. 253

VARSHNEY, B.S., KUMAR, S. and SHARMA, T.P. (1990). Studies on the burning behaviour of metal powder fires and their extinguishment. *Fire Safety J.*, **16**, 93. See also Sharma, T.P., Varshney, B.S. and Kumar, S. (1993)

VARTIALA, H. (1982). Electronic machine guards. *J. Occup. Accid.*, 4(2/4), 171

VASEY, M.W. (1989). *Piper Alpha Public Inquiry. Transcript, Days 134, 135*

VASILE, C. and SEYMOUR, R.B. (eds) (1993). *Handbook of Polyolefins* (New York: Dekker)

VASILIEV, A.A., GAVRILENKO, T.P. and TOPCHIAN, M.E. (1972). On the Chapman–Jouguet surface in multiheaded gaseous detonations. *Astronaut. Acta*, **17**, 499

VATAVUK, W.M. and NEVERIL, R.B. (1980–). Estimating costs of air-pollution control equipment. *Chem. Engng, 1980*, **87**(Oct. 6), 165; (Nov. 3), 157; (Dec. 1), 111; (Dec. 30), 71; 1981, **88**(Jan. 26), 127; (Mar. 23), 223; (May 18), 171; (Jun. 15), 129; (Sep. 7), 139; (Nov. 30), 193; 1982, **89**(Mar. 22), 153; (Jul. 12), 129; (Oct. 14), 135; 1983, **90**(Jan. 24), 131; (Feb. 21), 89; (May 16), 95; 1984, **91**(Apr. 2), 97; (Apr. 30), 95

VATER, R.A. (1985). Dynamic load attenuation for double wall tanks. *Gastech 84*, p. 391

VAUGHAN, H. (1978). Bending and tearing of plate with application to ship bottom damage. *Naval Architect*, May

VAUGHAN, J.R. (1984). The assessment of the expert of the hazardous area around ignited fires. *SRS Quart. Dig.*, Jul.

VAUGHN, C.B. and MANCINI, R.A. (1993). Ignition of maleic anhydride soaked insulation. *J. Loss Prev. Process Ind.*, **6**, 61

VAUGHN, J.R. (1986). A reliability approach to the selection of fire detection systems. *SRS Quart. Digest*, Jul., 3

VAURIO, J.K. (1985). On exact and asymptotic unavailability expressions for standby components. *Reliab. Engng*, **10**(3), 151

VAURIO, J.K. (1986). On robust methods for failure rate estimation. *Reliab. Engng*, **14**(2), 123

VAZQUES-SIERRA, J.M., MARTI, J. and MOLINA, R. (1988). A multistep approach for evaluation of pipe impact effects. *Int. J. Press. Ves. Piping*, **31**, 15

VEAZEY, J.A. (1979). Protecting cooling towers from overpressure. *Ammonia Plant Safety*, **21**, p. 11. See also *Chem. Engng Prog.*, 1979, **75**(7), 73

VEAZEY, J.A. and WINGET, J.M. (1989). Rupture of a process line exit the high-temperature shift converter in an ammonia plant. *Ammonia Plant Safety*, **29**, p. 1

VEDDER, E.B. (1925). *The Medical Aspects of Chemical Warfare* (Baltimore, MD: Williams & Wilkins)

VEEVERS, A. (1989/90). The bathtub curve revisited. *Safety Reliab.*, **9**(4), 6

VEGA, J.M. and LINAN, A. (1984). Large activation energy analysis of the ignition of self-heating porous bodies. *Combust. Flame*, **57**, 247

VEGTER, H.J. (1969). Dust explosions in the starch industry. *Symp. on Inherent Hazards of Manufacturing and Storage in the Process Industry*, The Hague, p. 187

VEJTASA, S.A. and SCHMITZ, R.A. (1970). An experimental study of steady state multiplicity and stability in an adiabatic stirred reactor. *AIChE J.*, **16**, 410

VELETSOS, A.S. (1984). Seismic response and design of liquid storage tanks. *Guidelines for the Seismic Design of Oil and Gas Pipeline Systems* (Am. Soc. Civil Engrs), Ch. 7

VELETSOS, A.S. and YANG, J.Y. (1976). Earthquake response of liquid storage tanks. *Advances in Civil Engineering through Earthquake Mechanics* (New York: Am. Soc. Civil Engrs), p. 1

VELETSOS, A.S. and YU TANG (1986a). Dynamics of vertically excited liquid storage tanks. *J. Struct. Engng*, **112**, 1228

VELETSOS, A.S. and YU TANG (1986b). Interaction effects in vertically excited steel tanks. *Dynamic Response of Structures* (New York: Am. Soc. Civil Engrs), p. 636

VELIOTIS, P.T. (1982). A submarine LNG tanker concept for the Arctic. *Gastech 81*, p. 153

VELLENGA, S.J. (1961). Estimating the electric field inside a rectangular tank with boundaries at zero potential. *Appl. Sci. Res.*, **B9**, 35

VELLENGA, S.J. and KLINKENBERG, A. (1965). On the rate of discharge of electrically charged hydrocarbon liquids. *Chem. Engng Sci.*, **20**, 923

VELO, E., BOSCH, C.M. and RECASENS, F. (1996). Thermal safety of batch reactors and storage tanks. development and validation of runaway boundaries. *Ind. Eng. Chem. Res.* **35**, 1288–99

VELTMAN, G., LANGE, C.E., JÜHE, S., STEIN, G. and BACHNER, U. (1975). Clinical manifestations and course of vinyl chloride disease. *Ann. N.Y. Acad. Sci.*, **246**, 6

VELZ, C.J. (1984). *Applied Stream Sanitation* (New York: Wiley)

VENART, J.E.S. (1990a). The anatomy of a boiling liquid expanding vapor explosion (BLEVE). *Am. Inst. Chem. Engrs, Twenty Fourth Ann. Symp. on Loss Prevention* (New York)

VENART, J.E.S. (1990b). In-vessel transient thermohydraulics. *Proc. Eur. Seminar 14 on Heat Transfer and Major Technological Hazards* (ed. L. Bolle), p. IL1/1

VENART, J., SOUSA, A., STEWARD, F. and PRASAD, R. (1984). Experiments on the physical modelling of LPG tank cars under accident conditions. *Transport and Storage of LPG and LNG* (Antwerp: Koninklijke Vlaamse Ingenieursvereniging), p. 385

VENART, J.E., SUMATHIPALA, U.K., STEWARD, F.R. and SOUSA, A.C.M. (1988). Experiments on the thermo-hydraulic response of pressure liquefied gases in externally heated tanks with pressure relief. *Plant/Operations Prog.*, **7**, 139

VENART, J.E.S., SUMATHIPALA, K., HADJISOPHOCLEOUS, G. and SOUSA, A.C.M. (1989). Pressure relief of externally heated pressure liquefied gas vessels. *Loss Prevention and Safety Promotion 6*, p. 116.1

VENART, J.E.S., RUTLEDGE, G.A., SUMATHIPALA, K. and SOLLOWS, K. (1993). To BLEVE or not to BLEVE: anatomy of a boiling liquid expanding vapour explosion. *Process Safety Prog.*, **12**, 67

VENEZIANO, D., CORNELL, C.A. and O'HARA, T. (1984). *Historical Method of Seismic Hazard Analysis. Rep. NP–3438.* Elec. Power Res. Inst., Palo Alto, CA

VENKATASUBRAMANIAN, V. (1988). Towards a specialized shell for diagnostic expert systems for chemical plant. *J. Loss Prev. Process Ind.*, **1**(2), 84

VENKATASUBRAMANIAN, V. and KING CHAN (1989). A neural network methodology for process fault diagnosis. *AIChE J.*, **35**, 1993

VENTAKASUBRAMANIAN, V. and MCAVOY, T.J. (1992). Neural network applications in chemical engineering (editorial). *Comput. Chem. Engng*, **16**, v

VENTAKASUBRAMANIAN, V. and RICH, S.H. (1988). An object-oriented two-tier architecture for integrating compiled and deep-level knowledge for process diagnosis. *Comput. Chem. Engng*, **12**, 903

VENKATASUBRAMANIAN, V. and VAIDHYANATHAN, R. (1994). A knowledge-based framework for automating HAZOP analysis. *AIChE J.*, **40**, 496

VENTAKASUBRAMANIAN, V., VAIDHYANATHAN, R. and YAMAMOTO, Y. (1990). Process fault detection and diagnosis using neural networks. I, Steady state processes. *Comput. Chem. Engng*, 14, 699

VENKATESWARLU, C., GANGIAH, K. and RAO, M.B. (1992). Two-level methods for incipient fault diagnosis in non-linear chemical processes. *Comput. Chem. Engng*, 16, 463

VENKATRAM, A. (1978). Estimating the convective velocity scale for diffusion applications. *Boundary Layer Met.*, 15, 447

VENKATRAM, A. (1980a). Dispersion from an elevated source in a convective boundary layer. *Atmos. Environ.*, 14, 1

VENKATRAM, A. (1980b). Estimating the Monin–Obukhov length in the stable boundary layer for dispersion calculations. *Boundary Layer Met.*, 19, 481

VENKATRAM, A. (1980c). The relationship between the convective boundary layer and dispersion from tall stacks. *Atmos. Environ.*, 14, 763

VENKATRAM, A. (1981a). Estimation of turbulence velocity scales in the stable and unstable boundary layer for dispersion applications. In de Wispelaere, C. (1981), *op. cit.*, p. 169

VENKATRAM, A. (1981b). Model predictability with reference to concentrations associated with point sources. *Atmos. Environ*, 15, 1571

VENKATRAM, A. (1988a). Inherent uncertainty in air quality modelling. *Atmos. Environ.*, 22, 1221

VENKATRAM, A. (1988b). An interpretation of Taylor's statistical analysis of particle dispersion. *Atmos. Environ.*, 22, 865

VENKATRAM, A. and KURTZ, J. (1979). Models to study the short-term impact of large point sources. *Fourth Symp. on Turbulence, Diffusion and Air Pollution* (Boston, MA: Am. Met. Soc.), p. 79

VENKATRAM, A. and PAINE, R. (1985). A model to estimate dispersion of elevated releases into a shear-dominated boundary layer. *Atmos. Environ.*, 19, 1797

VENKATRAM, A. and VET, R. (1981). Modeling of dispersion from tall stacks. *Atmos. Environ*, 15, 1531

VENKATRAM, A., STRIMAITIS, D. and DICRISTOFARO, D. (1984). A semi-empirical model to estimate vertical dispersion of elevated releases in the stable boundary layer. *Atmos. Environ.*, 18, 923

VENN, F.A.H. (1991). Increasing safety awareness through high profile legislation – a Norwegian continental shelf offshore project experience. *Safety Developments in the Offshore Oil and Gas Industry* (London: Instn Mech. Engrs), p. 15

VENNARD, J.K. and STREET, R.L. (1982). *Elementary Fluid Mechanics, 6th ed.* (New York: Wiley)

VENNEN, H. (ca. 1993). *Störfälle in Serie?* (Frankfurt-am-Main: Hoechst AG)

VENNER, L. (1979). Fire damages Swedish nuclear power station. *Fire Int.*, 65, 78

VENTON, A.O.F. (1977). Data requirements for mechanical reliability. *First Nat. Conf. on Reliability* (Culcheth, Lancashire: UK Atom. Energy Auth.), NRC5/5

VENTON, A.O.F. (1980). Pursuing reliability in mechanical equipment design. In Moss, T.R. (1980), *op. cit.*, p. 20

VENTRONE, T.A. (1969). Batch nitration experiences. *Loss Prevention*, 3, p. 38

VERBAND DER CHEMISCHEN INDUSTRIE (1978). *Seveso ist nicht überall* (Frankfurt)

VERBERK, M.M. (1977). Effects of ammonia on volunteers. *Int. Arch. Occup. Environ. Health*, 39, 73

VERCAMER, P., SAUVE, P. and LOOTVOET, E. (1987). Fatigue tests on an LNG carrier tank corner assembly of Gas Transport Technique. *Gastech 86*, p. 221

VERDE, L. and LEVY, G. (1979). Safety strategy for plant design. *Hydrocarbon Process.*, 58(3), 215

VERDE, L., MORENO, S. and RICCARDI, R. (1977). Integrate refinery/ethylene production. *Hydrocarbon Process.*, 56(10), 165

VERDIN, A. (1973). *Gas Analysis Instrumentation* (London: Macmillan)

VERDIN, A. (1980). Process analysers. *Chem. Engr, Lond.*, 362, 683

VERDUIJN, W.D. (1979). Operating experience with a cryogenic syngas purifier. *Ammonia Plant Safety*, 21, p. 130

VERE, S.A. (1983). Planning in time: windows and durations for activities and goals. *IEEE Trans Pattern Anal. Mach. Intell.*, PAMI-5, 246

VERE, S. (1992). Planning. In Shapiro, S.C. (1992), *op. cit.*, p. 1159

VEREIN DEUTSCHER INGENIEURE – see Appendix 27

VERGISON, E., VAN DIEST, J. and BASLER, J.C. (1989). Atmospheric dispersion of toxic gases in a complex environment. *J. Haz. Materials*, 22(3), 331

VERHAGEN, H. (1989). Safety management. *Loss Prevention and Safety Promotion 6*, p. 1.1

VERHAGEN, T.L.A. and BUYTENEN, C.J.P. (1986). Validation of the TNO heavy gas dispersion model. In Hartwig, S. (1986), *op. cit.* p. 157

VERHAEGEN, P., VANHALST, B., DERIJCKE, H. and VANHOECKE, M. (1976). The value of some psychological theories on industrial accidents. *J. Occup. Accid.*, 1(1), 39

VERHEIJ, F.J. and DUIJM, N.J. (1991). *Wind Tunnel Modelling of Torch Fire. Rep. 91-422.* TNO-IMET, Apeldoorn, The Netherlands

VERHEIJ, Z.C. (1988). Experimental validation and/or support of vulnerability analysis. In Groothuizen, T.M. (1988), *op. cit.*, p. 206

VERHOEFF, J. (1981). Explosion hazards of tertiary butyl hydroperoxide (TBHP). *Runaway Reactions, Unstable Products and Combustible Powders* (London: Instn Chem. Engrs), paper 3/T

VERHOEFF, J. (1983a). Experimental study of the thermal explosion of liquids. Ph.D. Thesis, Technische Hogschool, Delft

VERHOEFF, J. (1983b). Thermal explosion of liquids. *Loss Prevention and Safety Promotion 4*, vol. 3, p. C5

VERHOEFF, J. (1988). Gas production and heat generation data concerning the thermal runaway of liquid reaction systems. In Groothuizen, T.M. (1988), *op. cit.*, p. 291

VERHOEFF, J. and HEEMSKERK, A.H. (1984). Gas production and heat generation data concerning the thermal runaway of liquid reaction systems. *The Protection of Exothermic Reactors and Pressurised Storage Vessels* (Rugby: Instn Chem. Engrs), p. 41

VERHOEFF, J. and JANSWOUDE, J.J. (1980). Test methods in process safety analysis. *Loss Prevention and Safety Promotion 3*, vol. 2, p. 323

VERHOLEK, M.G. (1986). CARE – modelling hazardous airborne releases. *Proc. Nat. Conf. on Hazardous Wastes and Hazardous Materials*

VERKADE, M. and CHIOTTI, P. (1978a). *A Bibliography of Topics Related to the Study of Grain Dust Fire and Explosion* (report). Energy Mineral Resources Inst., Iowa State Univ.

VERKADE, M. and CHIOTTI, P. (1978b). *Literature Survey of Dust Explosions in Grain Handling Facilities: Causes and*

Prevention (report). Energy Mineral Resources Inst., Iowa State Univ.

VERMEER, D.J., MEYER, J.W. and OPPENHEIM, A.K. (1972). Auto-ignition of hydrocarbons behind reflected shock waves. *Combust. Flame*, **18**, 327

VERMEIREN, G. (1989). LPG contaminations by interactions of sulphur compounds. *Gastech 88*, paper 5.11

VERMEULEN, J. (1987). Effect of functionally or topographically presented process schemes on operator performance. *Hum. Factors*, **29**, 383

VERMEULEN, J. and HANDS, P. (1993). The Seveso Directive – its background and implementation in Europe. *Health, Safety and Loss Prevention*, p. 102

VERNEUIL, V.S., PIN YANG and MADRON, F. (1992). Banish bad plant data. *Chem. Engng Prog.*, **88**(10), 45

VERNEY, P. (1979). *The Earthquake Handbook* (London: Paddington Press)

VERNICK, A.S. and WALKER, E.C. (eds) (1982). *Handbook of Wastewater Treatment Processes* (Basel: Dekker)

VERNOT, E.H., MACEWEN, J.D., HAUN, C.C. and KINKEAD, E.R. (1977). Acute toxicity and skin corrosion data for some organic and inorganic compounds and aqueous solutions. *Toxicol. Appl. Pharmacol.*, **42**, 417

VERRETT, J. (1970). In US Senate (1970), *op. cit.*, p. 190

VERSCHUEREN, K. (1977). *Handbook of Environmental Data on Organic Chemicals* (New York: van Nostrand Reinhold)

VERSEPUT, J.P. (1965). How Shell maintains its world-wide refineries. *Hydrocarbon Process.*, **44**(1), 78

VERSINO, B. and OTT, H. (eds) (1982). *Physico-Chemical Behavior of Atmospheric Pollutants* (Dordrecht:Reidel)

VERSTEEG, M.F. (1988). External safety policy in the Netherlands: an approach to risk management. *J. Haz. Materials*, **17**(2), 215

VERSTOEP, N.D.L. and SCHLUNK, R.W. (1978). Modern instrumentation in chemical plants. *Chem. Ind.*, May 20, 335

VERVALIN, C.H. (1962–). Fire protection handbook. *Hydrocarbon Process.*, 1962, **41**(8), 141; **41**(9), 282; **41**(10), 149; **41**(12), 137; 1963, **42**(1), 153; **42**(2), 171; **42**(3), 187

VERVALIN, C.H. (ed.) (1964a). *Fire Protection Manual for Hydrocarbon Processing Plants* (Houston, TX: Gulf)

VERVALIN, C.H. (1964b). Cationic and ampholytic surfactants, CO_2 and inhibition of combustion. In Vervalin, C.H. (1964a), *op. cit.*, p. 89

VERVALIN, C.H. (1964c). General case histories of fires. In Vervalin, C.H. (1964a), *op. cit.*, p. 357

VERVALIN, C.H. (1964d). Role of foam in fighting fires. In Vervalin, C.H. (1964a), *op.cit.*, p. 83

VERVALIN, C.H. (1964e). Use of water and dry chemicals. In Vervalin, C.H. (1964a), *op. cit.*, p. 76

VERVALIN, C.H. (1966). How to get fire-safety information. *Hydrocarbon Process.*, **45**(5), 227

VERVALIN, C.H. (1970). Responsibility of the press and the company. *Loss Prevention* 4, p. 89

VERVALIN, C.H. (1971–). Keep up with OSHA. *Hydrocarbon Process.*, 1971, **50**(12), 118: 1972, **51**(2), 107

VERVALIN, C.H. (1972a). Fire and safety information. *Hydrocarbon Process.*, **51**(4), 163: **51**(5), 162: **51**(7), 117

VERVALIN, C.H. (1972b). Know where OSHA is taking you. *Hydrocarbon Process.*, **51**(12), 61

VERVALIN, C.H. (1972c). Learn from HPI plant fires. *Hydrocarbon Process.*, **51**(12), 49

VERVALIN, C.H. (1972d). What governments expect . . . *Hydrocarbon Process.*, **51**(12), 85

VERVALIN, C.H. (1973a). *Fire Protection Manual for Hydrocarbon Processing Plants*, 2nd ed. (Houston, TX: Gulf)

VERVALIN, C.H. (1973b). Contact these sources for environmental information. *Hydrocarbon Process.*, **52**(10), 71

VERVALIN, C.H. (1973c). Environmental management resources: how to keep up with what's happening. *Hydrocarbon Process.*, **52**(2), 113

VERVALIN, C.H. (1973d). Learn from HPI tank fires. *Hydrocarbon Process.*, **52**(8), 81

VERVALIN, C.H. (1973e). Miscellaneous incidents. In Vervalin, C.H. (1973a), *op. cit.*, p. 80

VERVALIN, C.H. (1973f). Steps to note in tank safety. In Vervalin, C.H. (1973a), *op. cit.*, p. 301

VERVALIN, C.H. (1973g). Who's publishing in fire and safety. *Hydrocarbon Process.*, **52**(1), 128

VERVALIN, C.H. (1974a). API/NFPA report on HPI fire losses. *Hydrocarbon Process.*, **53**(5), 182

VERVALIN, C.H. (1974b). Some case histories of hydro-carbon fires. *Fire Int.*, **45**, 45

VERVALIN, C.H. (1975a). Fire/safety information resources. *Hydrocarbon Process.*, **54**(8), 65

VERVALIN, C.H. (1975b). The HPI looks at toxicity. *Hydrocarbon Process.*, **54**(4), 225

VERVALIN, C.H. (1975c). Latest API/NFPA fire loss report. *Hydrocarbon Process.*, **54**(6), 149

VERVALIN, C.H. (1975d). Trends in HPI fire/safety training. *Hydrocarbon Process.*, **54**(2), 146: **54**(3), 169

VERVALIN, C.H. (1975e). What the best fire schools teach. *Hydrocarbon Process.*, **54**(7), 202

VERVALIN, C.H. (1975f). Will your plant burn tomorrow? *Hydrocarbon Process.*, **54**(12), 145

VERVALIN, C.H. (1976a). Curtail vinyl chloride exposure. *Hydrocarbon Process.*, **55**(2), 182

VERVALIN, C.H. (1976b). Fire losses reported by API/ NFPA. *Hydrocarbon Process.*, **55**(5), 291

VERVALIN, C.H. (1976c). Loss prevention in the HPI. *Hydrocarbon Process.*, **55**(7), 205: **55**(8), 182: **55**(9), 321: **55**(10), 215: **55**(11), 305

VERVALIN, C.H. (1977). Fire losses reported by NFPA. *Hydrocarbon Process.*, **56**(2), 166

VERVALIN, C.H. (1978a). Fire-loss report for the HPI. *Hydrocarbon Process.*, **57**(1), 251

VERVALIN, C.H. (1978b). HPI loss – incident case histories. *Hydrocarbon Process.*, **57**(2), 183

VERVALIN, C.H. (1979). Use these 'keep up' resources for environmental management. *Hydrocarbon Process.*, **58**(10), 89

VERVALIN, C.H. (1981a). Know loss-prevention information sources. *Hydrocarbon Process.*, **60**(3), 221

VERVALIN, C.H. (1981b). Put loss experiences to work. *Hydrocarbon Process.*, **60**(8), 175

VERVALIN, C.H. (1981c). Train for better fire protection. *Hydrocarbon Process.*, **60**(4), 289: **60**(7), 239

VERVALIN, C.H. (1983). Define loss prevention problems and remedies. *Hydrocarbon Process.*, **62**(12), 111

VERVALIN, C.H. (1984). Training by simulation. *Hydrocarbon Process.*, **63**(12), 41

VERVALIN, C.H. (ed.) (1985a). *Fire Protection for Hydrocarbon Processing Plants, 3rd ed.* (Houston, TX: Gulf)

VERVALIN, C.H. (1986a). Environmental control. *Hydrocarbon Process.*, **64**(12), 77

VERVALIN, C.H. (1985b). CAER about emergency response. *Hydrocarbon Process.*, **64**(12), 77

VERVALIN, C.H. (1986a). Environmental Control. *Hydrocarbon Process.*, **65**(10), 41

VERVALIN, C.H. (1986b). Hazard evaluation. *Hydrocarbon Process.*, **65**(12), 35

VERVALIN, C.H. (1987). Explosion at Union Oil. *Hydrocarbon Process.*, **66**(1), 83

VERVALIN, C.H. (1990a). To incinerate waste . . . *Hydrocarbon Process.*, **69**(8), 61

VERVALIN, C.H. (1990b). Recent HPI loss incidents not result of operator error. *Hydrocarbon Process.*, **69**(10), 27

VERVER, G.H.L. and DE LEEUW, F.A.A.M. (1992). An operational puff dispersion model. *Atmos. Environ.*, **26A**, 3179

VERVISCH, P., PUECHBERTY, D. and MOHAMED, T. (1981). The spectral transmission of 0.4–4.5 mm of fire smoke. *Combust. Flame*, **41**, 179

VERVISCH, P. and COPPALLE, A. (1983). Fire flame radiation. *Combust. Flame*, **52**, 127

VERWIJS, J.W., VAN DEN BERG, H. and WESTERTERP, K.R. (1996). Startup strategy design and safeguarding of industrial adiabatic tubular reactor systems. *AIChE J.* **42**, (2), (Feb.), 503–15

VESELIND, P.A., PEIRCE, J.J. and WEINER, R.F. (1990). *Environmental Pollution and Control* (Oxford: Butterworth-Heinemann)

VESELY, W.E. (1969). *Analysis of Fault Trees by Kinetic Tree Theory. Rep. IN-1330.* Idaho Nucl. Corp., Idaho Falls, ID

VESELY, W.E. (1970a). *Reliability and Fault Tree Aplications at NRTS* (report). Idaho Nucl. Corp., Idaho Falls, ID

VESELY, W.E. (1970b). A time-dependent methodology for fault tree evaluation. *Nucl. Engng Res.*, **13**(2), 337

VESELY, W.E. (1977a). *Estimating CCF Probabilities in Reliability and Risk Analyses: Marshall–Olkin Specialisation. NSRE and RA* (Philadelphia, PA: Soc. Ind. Appl. Maths)

VESELY, W.E. (1977b). Failure data and risk analysis. In Gangadharan, A.C. and Brown, S.J. (1977), *op. cit.*, p. 61

VESELY, W.E. (1983). Minimal cut set failure and repair. *Reliab. Engng*, **5**, 65

VESELY, W.E. and GOLDBERG, F.F. (1977a). *FRANTIC – Computer Code for Time-dependent Availability Analysis. Rep. NUREG-0193.* Nucl. Regul. Comm., Washington, DC

VESELY, W.E. and GOLDBERG, F.F. (1977b). Time-dependent unavailability analysis for nuclear safety systems. *IEEE Trans Reliab.*, **R-26**, 257

VESELY, W.E. and NARUM, R.E. (1970). *PREP and KITT: Computer Codes for the Automatic Evaluation of a Fault Tree. Rep. IN-1349.* Idaho Nucl. Corp., Idaho Falls, ID

VESELY, W.E. *et al.* (1981). *Fault Tree Handbook. Rep. NUREG-0492.* Nucl. Regul. Comm., Washington, DC

VESTERGAARD, N.K. and RASMUSSEN, B. (1988). The use of risk analysis for design improvement. *J. Loss Prev. Process Ind.*, **1**(2), 113

VESTRUCCI, P. (1992). Human reliability analysis in accident conditions for process industry: the discrete convolution approach. *Loss Prevention and Safety Promotion 7*, vol. 2, 52–1

VETTER, G. and HERING, L. (1980). Leakfree pumps for the chemical process industries. *Chem. Engng*, **87**(Sep. 22), 149

VETTER, G., SCHLÜCKER, E., JAROSCH, J. and HORN, W. (1993). Know your diaphragm pump internals. *Chem. Engng*, **100**(11), 130

VEYSSIERE, B. (1984). 'Double-front' detonations in gas–solid particles mixtures. In Bowen, J.R. *et al.* (1984b), *op. cit.*, p. 264

VEZIROGLU, T.N. (ed.) (1983). *Proc. Third Multiphase Flow and Heat Transfer Symposium-Workshop* (Coral Gables, FL: Clean Energy Res. Inst., Univ. of Miami)

VIALLON, J. (1977). Study of flammability or decompositions of gases or vapours. *Loss Prevention and Safety Promotion 2*, p. 115

VICK, F.A. (1953). Theory of contact electrification. *Br. J. Appl. Phys.*, **4**(Suppl. 2), S1

VICK, K.A. (1987). Inspection and repair of a secondary reformer. *Ammonia Plant Safety*, **27**, 113

VICK, K.A., WITTHAUS, J.B. and MAYO, H.C. (1980). Steam purging and foundation repair of an atmospheric ammonia storage tank. *Ammonia Plant Safety*, **22**, p. 54

VICKERS, SIR G. (1968). *The Art of Judgment: A Study in Policy-Making* (London:Methuen)

VICTORY, G. and ROBERTS, R.K. (1971). Fire prevention and fire fighting for vessels carrying liquid chemicals in bulk. *Transport of Hazardous Cargoes*, **2**, paper 5

VIDALIN, K.E. and BERTRAM, D.W. (1974). Repairing ammonia tank insulation. *Ammonia Plant Safety*, **16**, p. 50

VIERA, G.A., SIMPSON, L.L. and REAM, B.C. (1993). Lessons learned from the ethylene oxide explosion at Seadrift, Texas. *Chem. Engng Prog.*, **89**(8), 66

VIERGAS, D.X., CAMPOS, J.L. and OLIVIERA, L.A. (1984). Tank fire in an atmospheric boundary layer. *Combust. Sci. Technol.*, **41**, 31

VIGEANT, S.A. and MAZZOLA, C.A. (1987). Modeling the impact of an accidental hazardous chemical release. *Vapor Cloud Modeling*, p. 923

VIK, I. and KJERSEM, G.L. (1987). A pre-feasibility study on onshore production and loading of LNG at Tromso/flaket. *Gastech 86*, p. 349

VILBRANDT, F.C. and DRYDEN, C.E. (1959). *Chemical Engineering Plant Design* (New York: McGraw-Hill)

VILCHEZ, J.A. and CASAL, J. (1991). Hazard index for runaway reactions. *J. Loss Prev. Process Ind.*, **4**(2), 125

DE VILDER, P. (1982). Waterborne polyethylene plant. *Chem. Engng Prog.*, **78**(11), 29

VILLERMAUX, E. and DUROX, D. (1992). On the physics of jet diffusion flames. *Combust. Sci. Technol.*, **84**, 279

VILLERMAUX, J. (1993). Basic chemical engineering research. Where are we going? *Chem. Engng Res. Des.*, **71A**, 45

VILKUS, R. and SEVERIN, M. (1989). Problems with a process gas cooler. *Ammonia Plant Safety*, **29**, p. 26

VILYUNOV, V.N. and ZARKO, V.E. (1989). *Ignition of Solids* (Amsterdam: Elsevier)

VINCENT, G.C. (1971). Rupture of a nitroaniline reactor. *Loss Prevention*, **5**, p. 46. See also *Chem. Engng Prog.*, 1971, **67**(6), 51

VINCENT, G.C. and GENT, C.W. (1978). Safety and environmental features in the startup. *Ammonia Plant Safety*, **20**, p. 22

VINCENT, G.C. and HOWARD, W.B. (1976). Hydrocarbon mist explosions – part I. Prevention by explosion suppression. *Loss Prevention*, **10**, p. 43

VINCENT, G.C., NELSON, R.C., HOWARD, W.B. and RUSSELL, W.W. (1976a). Hydrocarbon mist explosions – part II. Prevention by water fog. *Loss Prevention*, **10**, p. 55

VINCENT, G.C., NELSON, R.C., HOWARD, W.B. and RUSSEL, W.W. (1976b). Hydrocarbon mist explosions – prevention by water fog. *Prevention of Occupational Risks*, **3**, p. 511

VINCENT, J.H. (1977). Model experiments on the nature of air pollution transport near buildings. *Atmos. Environ.*, **11**, 765

VINCENT, J.H. (1978). Scalar transport in the near aerodynamic wakes of surface-mounted cubes. *Atmos. Environ.*, **12**, 1319

VINCENT, J.H. (1980). The physics of gases and aerosols. In Waldron, H.A. and Harrington, J.M. (1980), *op. cit.*, p. 1

VINCENT, J.H. and MARK, D. (1981). The basis of dust sampling in occupational hygiene. *Ann. Occup. Hyg.*, **24**, 375

VINCK, W. (1982). Europe. In Green, A.E. (1982), *op. cit.*, p. 577

VINCKIER, J. and VAN TIGGELEN, A. (1968). Structure and burning velocity of turbulent premixed flames. *Combust. Flame*, **12**, 561

VINCOLI, J.W. (1993). *Basic Guide to System Safety* (New York: van Nostrand Reinhold)

VINET (1919). Gas warfare and the work of the French chemical services. *Chimie et Industrie*, **2**, 1377

VINEY, M. (1990). Regulation of existing chemicals. *J. Loss Prev. Process Ind.*, **3**(3), 266

VINNEM, J.E. (1983). Quantitative risk analysis in the design of offshore installations. *Reliab. Engng*, **6**, 1

VIOLA, P.L. (1970). Cancerogenic effect of vinyl chloride (abstract). *Proc. Xth Int. Cancer Congress*, vol. 29

VIRGILS, R.E. (1971). Preventive maintenance today. *Hydrocarbon Process.*, **50**(1), 91

VISANUVIMOL, V. and SLATER, D.H. (1974). The detection and monitoring of ozone in industrial environments. *Chemical Process Hazards*, V, p. 16

VISICK, D. (1986). Human operators and their role in automated plant. *Chem. Ind.*, May 5, 310

VISVANATHAN, T.R. (1976). Fatal accident in CO_2 removal system. *Ammonia Plant Safety*, **18**, p. 93

VISWANATH, D.S. and NATARAJAN, G. (1989). *Data Book on the Viscosity of Liquids* (London: Taylor & Francis)

VIVIAN, B. (1985). Identifying and evaluating valve problems. *Process Engng*, Oct., 33

VIVIAN, B. (1988). Selecting process valves for in-service reliability. *Process Engng*, Apr. 33

VAN VLACK, L.H. (1970). *Materials Science for Engineers* (New York: Addison-Wesley)

VLAD, A. (1977). The working place as 'patient'clinician and therapist. *Prevention of Occupational Risks*, **5**, p. 283

VLAMING, D.J. (1984). Analysis of cavitation provides advanced NPSH estimates for centrifugal pumps. *Oil Gas J.*, **82**(Nov. 19), 143

VLEK, C.J.H. and KEREN, G. (1991). Behavioural decision theory and environmental risk management: what we have learned and what has been neglected. *Thirteenth Res. Conf on Subjective Probability, Utility and Decision-Making* Fribourg, Switzerland

VAN VLIET, J.T., DE GOOIJER, H., HOFTIJZER, G.W. and LOGTENBERG, T.M. (1980). The development of a method to visualize the hazards for the benefit of a safety policy in a province of Holland (South Holland). *Loss Prevention and Safety Promotion 3*, vol. 3, p. 1087

VODAR, B. and MARTEAU, P. (eds) (1980). *High Pressure Science and Technology* (Oxford: Pergamon Press)

VODSEDALEK, J. and BIELAK, O. (1977). Service life of steam boiler drums. *Int. J. Pres. Ves. Piping*, **5**, 23

VODYANIK, V.I. (1990). Prevention of explosion of process units. *Brandschutz, Explosionsschutz, Havarieschutz*, **2**, 75

VOEGELEN, J.F. (1965). Storage and disposal of dangerous chemicals. *J. Chem. Educ.*, **43**, A151

VOELKER, C.H. (1964). Profiting from stress relief improvements. *Chem. Engng*, **71**(Dec. 21), 91

VOELKER, C.H. (1965). Modern practice in field welding. *Chem. Engng*, **72**(Jan. 4), 67

VOELKER, C.H. (1973). Practical repair welding. *Hydrocarbon Process.*, **52**(1), 68, 71: **52**(2), 95

VOELKER, C.H. and ZEIS, L.A. (1972). How to repair weld HK-40 furnace tubes. *Hydrocarbon Process.*, **51**(4), 121

VOEVODSKY, V.V. (1959). On some reactions occurring during the explosion induction period. *Combustion*, **7**, p. 34

VOGEL, F. (1979). 'Our load of mutation' reappraisal of an old problem. *Proc. R. Soc., Ser. B*, **205**, 77. See also in Doll, Sir R. and McLean, A.E.M. (1979), *op. cit.*, p. 77

VOGEL, G.A. and MARTIN, E.J. (1983). Equipment sizes and integrated facility costs: hazardous waste incineration. *Chem. Engng*, **90**(Sep. 5), 143; (Oct. 17), 75; (Nov. 28), 87; 1984, **91**(Jan. 9), 97; (Feb. 6), 121

VOGES, U. (ed.) (1987). *Software Diversity in Computerised Control Systems* (New York: Springer)

VOGLER, T.C. and WEISSMAN, W. (1988). Thermodynamic availability analysis for maximising a system's efficiency. *Chem. Engng Prog.*, **84**(3), 35

VOGLER, W. (1986). Safety audits as a means of ensuring safety performance in a trans-national company. *Loss Prevention and Safety Promotion 5*, 19–1

VOGT, K.J., STRAKA, J. and GEISS, H. (1979). An extension of the Gaussian plume model for the case of changing weather conditions. *Proc. Tenth Int. Tech. Mtg on Air Pollution Modelling and Its Application* (NATO/ CCMS), p. 361

VOGTH-ERIKSEN, J. (1983). Cargo claims/cargo losses. *Gastech 82*, p. 456

VOIGHT, B. (ed.) (1978–). *Rockslides and Avalanches* (Amsterdam: Elsevier)

VOIGTLÄNDER, H.J. (1965). Accident prevention in the chemical industry. *Chemie für Labor und Betribe*, **16**(1), 18

VOIGTSBERGER (1973). Technical safety criteria for the assessment of works and working procedures dealing with gases. *Prevention of Occupational Risks 2*, p. 112

VOISON, C., GUERRIN, F., ROBIN, H., FUREN, D. and WATTEL, F. (1970). Respiratory function sequelae of intoxication by ammonia (summary of cases). *Le Poumon et le Coeur*, **26**, 1079

VOLK, W. (1979a). Calculator programs to correlate data. *Chem. Engng*, **86**(Apr. 23), 129

VOLK, W. (1979b). Calculator programs to fit a family of lines. *Chem. Engng*, **86**(Jun. 4), 133

VOLKMAN, Y. (1977). A simple method for safety factor evaluation. *AIChE J.*, **23**, 203

VOLLAND, H. (ed.) (1982). *CRC Handbook of Atmospherics*, vols 1–2 (Boca Raton, FL: CRC Press)

VOLLER, V. and KNIGHT, B. (1985). Expert systems. *Chem. Engng*, **92**(Jun. 10), 93

VONDRAN, J. and KARDOS, J. (1975). Determination of availability of a configuration with parallel capacity and internal storage. *Hungarian J. Ind. Chem.*, **3**, 565

VOOGD, J. and TIELROOY, J. (1967). Improvement in making hydrogen. *Hydrocarbon Process.*, **46**(9), 115

VÖRÖS, M. and HONTI, G. (1974). Explosion of a liquid CO_2 storage vessel in a carbon dioxide plant. *Loss Prevention and Safety Promotion 1*, p. 337

VORTHMAN, J.E. (1982). Facilities for the study of shock induced decomposition of high explosives. In Nellis. W.J., Seaman, L. and Graham, A.R. (1982), *op. cit.*, p. 680

VOS, B. (1971). Electrostatic charge generation during washing of tanks with water sprays – IV: Mechanical studies. *Static Electrification*, **3**, p. 184

VOS, B., DOUWES, C., RAMACKERS, L. and VAN DE WEERD, J.M. (1974). Electrostatic charging of suspensions during agitation. *Loss Prevention and Safety Promotion 1*, p. 381

VOS, J.G. (1973). *Toxicol Appl. Pharmacol.*, **29**, 229

VOSIKOVSKY, O. and COOKE, R.J. (1978). An analysis of crack extension by corrosion fatigue in a crude oil pipeline. *Int. J. Pres. Ves. Piping*, **6**, 113

VOSS, M. and LUDBROOK, A.C. (1974). Commissioning of chemical plants in developing countries. *Instn Chem. Engrs, Symp. on Commissioning of Chemical and Allied Plant* (London)

VOSSBRINCK, W.J.H. (1973). Nuts – how do you tighten them? *Hydrocarbon Process.*, **52**(1), 89

VOSSENAAR, B. (1975). The fire protection of Rotterdam: dealing with an exceptional problem. *Fire Int.*, **47**, 53

VOUETS, W. (1992). Preventative maintenance. *Plant Engr*, **36**(1), 34

VOUK, V.B., BUTLER, G.C., HOEL, D.G. and PEAKALL, D.B. (eds) (1985). *Methods for Estimating Risk of Chemical Injury: Human and Non-Human Biota and Ecosystems* (Chichester: Wiley)

VOVELLE, C. and DELBOURGO, R. (1973). Concentration and temperature profiles for cool and second-stage diethyl ether–air flames. *Combustion*, **14**, p. 267

VOVELLE, C., MELLOTTE, E.H. and DELBOURGO, R. (1982). Kinetics of the thermal degradation of cellulose and wood in inert and oxidative atmospheres. *Combustion*, **19**, p. 797

VOVELLE, C., AKRICH, R. and DELFAU, J.L. (1984). Mass loss rate measurements on solid materials under radiative heating. *Combust. Sci. Technol.*, **36**, 1

VOZNYAK, V.Y., KOVALENKO, A.P. and TROITSKY, S.N. (1989). Chernobyl: Events and Lesson (Moscow: Politizdat)

VRANCKEN, P.L.L. and MCHUGH, J. (1977). Twelve years operating experience with 'Methane Princess' and 'Methane Progress'. *LNG Conf. 5*

VREEDENBURGH, H. (1979). Steel or prefab concrete for inshore plants? *Chem. Engng Prog.*, **75**(10), 82

DE VRIES, R.C. (1990). An automated methodology for generating a fault tree. *IEEE Trans.*, **39**, 76

DE VRIES-GRIEVER, A.H.G. and MEIJMAN, T.F. (1987). The impact of abnormal hours of work on various modes of information processing: a process model on human costs of performance. *Ergonomics*, **30**, 1287

VUKADINOVIC, M. (1980). What to do when heat exchangers plug. In *Ammonia Plant Safety*, **22**, p. 154. See also *Chem. Engng Prog., 1980*, **76**(7), 36

VUMBACO, B.J. (ed.) (1983). *Transportation of Hazardous Materials: Towards a National Strategy*, vol. 1. *Special Rep. 197* (Washington, DC: Nat. Acad. Sci./Nat. Res. Coun., Transportation Safety Board)

WACHEL, J.C. (1973). Turbine and compressor vibrations. *Ammonia Plant Safety*, **15**, p. 69

WACHEL, J.C. and BATES, C. (1976). Escape piping vibrations while designing. *Hydrocarbon Process.*, **55**(10), 152

WACHTEL, C. (1941). *Chemical Warfare* (London: Chapman & Hall)

WACHTER, A. (1960). Proper design voids equipment corrosion. *Chem. Engng*, **76**(Feb. 22), 162

WADA, Y., PERLMAN, M.M. and KOKADO, H. (1979). *Charge Storage, Charge Transport and Electrostatics with Their Applications* (Amsterdam: Elsevier)

WADDAMS, A.L. and CANN, G. (1974). Shipping liquid ethylene. *Hydrocarbon Process.*, **53**(11), 135

WADE, D.E. (1987). Reduction of risks by reduction of toxic material inventories. *Preventing Major Chemical Accidents*, p. 2.1

WADE, D.E. (1990). Manage for process safety. *Hydrocarbon Process.*, **69**(9), 44

WADE, J.J. (1963). Hazards in processing light hydrocarbons. *Chemical Process Hazards*, **II**, p. 55

WADE, L.E., GENGELBACH, R.B., TRUMBLEY, B.L. and HALLBAUER, W.L. (1981). Methanol. *Kirk–Othmer Encyclopaedia of Chemical Technology*, 3rd ed. (New York: Wiley), vol. 15, p. 400

WADE, S.H. (1942). Evaporation of liquids in currents of air. *Trans. Instn Chem. Engrs*, **20**, 1

WADI, I. (1993). Take an integrated approach to refinery automation. *Hydrocarbon Process.*, **72**(9), 155

WAFELMAN, H.R. and BURHMANN, H. (1977). Maintain safer tank storage. *Hydrocarbon Process.*, **56**(1), 229

WAGENAAR, G., BOESMANS, B. and SCHRIJNEN, L.M. (1989). AUTOPES, the development of a knowledge based system for process control. *Loss Prevention and Safety Promotion 6*, p. 84.1

WAGENAAR, W.A. and REASON, J.T. (1990). Types and tokens in road accident causation. Errors and violations on the roads: a real distinction? *Hum. Factors*, **33**, 1365

WAGENER, P.E. (1956). Charge separation at contacts. *J. Appl. Phys.*, **27**, 1300

WAGLE, M.P. (1985). Predict saturation temperature as a function of vapor pressure. *Chem. Engng*, **92**(Jun. 10), 77

WAGLE, U.D., HOUSTON, G.W., ANDERSON, A.J. and MAGELI, O.L. (1978). Organic peroxide decomposition characteristics. *Safety in Polyethylene Plants*, **3**, p. 62

WAGMAN, D.D. (ed.) (1953). Selected Values of Chemical Thermodynamic Properties. Nat. Bur. Standards (Washington, DC: US Govt Printing Office)

WAGMAN, D. *et al.* (1969). *Selected Values of Chemical Thermodynamic Properties. Nat. Bur. Standards, Tech. Notes 270-3 and 270-4* (Washington, DC: US Govt Printing Office)

WAGNER, D.P. and FUSSELL, J.B. (1977). *Common Cause Failure Analysis for Complex Systems*. Mtg at Gatlinburg, TN

WAGNER, D.P., CATE, C.L. and FUSSELL, J.B. (1977). Computer aided common cause failure analysis for complex systems. *Am. Inst. Chem. Engrs, Eleventh Symp. on Loss Prevention*

WAGNER, H.G. (1963). Reaction zone and stability of gaseous detonations. *Combustion*, **9**, p. 454

WAGNER, H.G. (1979). Soot formation in combustion. *Combustion*, **17**, p. 3

WAGNER, H.G. (1982). Some experiments about flame acceleration. In Lee, J.H.S. and Guirao, C.M. (1982), *op. cit.*, p. 77

WAGNER, H.M. (1975). *Principles of Operational Research*, 2nd ed. (Englewood Cliffs, NJ: Prentice-Hall)

WAGNER, J.A. (1979). Selecting the proper flame detector. *Loss Prevention*, **12**, p. 68

WAGNER, J.C. (1960). Some pathological aspects of asbestosis in the Union of South Africa. *Proc. Conf. on Pneumoconiosis 1959* (ed. A.J. Grensten) (London Churchill)

WAGNER, J.F. (1975). Modelling of electrostatic phenomena. *Loss Prevention*, **9**, p. 119

WAGNER, J.G. (1975). Do you need a pharmacokinetic model, and, if so, which one? *J. Pharmacokinetics Biopharmaceut.*, **3**, 457

WAGNER, K. *et al.* (1986). *Remedial Action Technology for Waste Disposal Sites*, 2nd ed. (Park Ridge, NJ: Noyes Data Corp.). See also Rogoshenski, P., Boyson, H. and Wagner, K. (1983)

WAHL, H.J. and NEEB, E.F. (1977). Cause and effect of explosion in ammonia converter. *Ammonia Plant Safety*, **19**, p. 105

WAI-BU CHENG and MAH, R.S.H. (1976). Optimal design of pressure relieving piping network by discrete merging. *AIChe J.*, **22**, 471

WAINWRIGHT, L. (1915). What is the gas? *The Lancet,* Jul. 24, 198

WAINWRIGHT, M. (1992). How much is too much? *Chem. Engr, Lond.,* **529**, s9

WAITE, P.J., WHITEHOUSE, R.J., WINN, E.B. and WAKEHAM, W.A. (1983). The spread and vaporisation of cryogenic liquids on water. *J. Haz. Materials,* **8**(2), 165

WAITE, P.J., SHILLITO, D.E. and SYLVESTER-EVANS, R. (1988). Safety assessment and technology transfer. *Preventing Major Chemical and Related Process Accidents,* p. 1

WAITT, A.H. (1942). *Gas Warfare* (New York: Duell, Sloan & Pearce)

WAKABAYASHI, T. (1977). Major accidents in petrochemical industries and safety regulations in Japan. *Loss Prevention and Safety Promotion 2,* p. 473

WAKABAYASHI, T. (1978). New pollutant treating process for ammonia plants. *Ammonia Plant Safety,* **20**, p. 17

WAKE, G.C. (1971). An improved bound for the critical explosion condition for an exothermic reaction in an arbitrary shape. *Combust. Flame,* **17**, 171

WAKE, G.C. (1973). Estimation of critical parameters in thermal ignition. *Combust. Flame,* **21**, 119

WAKE, G.C. and RAYNER, M.E. (1973). Variational methods for nonlinear eigenvalue problems associated with thermal ignition. *J. Differential Equations,* **13**, 247

WAKE, G.C. and WALKER, I.K. (1964). Calorimetry of oxidation reactions. *N.Z. J. Sci.,* **7**, 227

WAKELING, H.M. (1988). International risk management within a fixed price contract. *Preventing Major Chemical and Related Process Accidents,* p. 13

WAKEMAN, C.R. and LAUGHTON, M.A. (1976). Some design aspects of electrical power distribution system reliability. In Henley, E.J. and Lynn, J.W. (1976), *op. cit.,* p. 417

WAKESHIMA, H. and TAKATA, K. (1958). On the limit of superheat. *J. Phy. Soc. Japan,* **13**, 1398

WAKOH, H. and HIRONA, T. (1991). Diffusion of leaked flammable gas in soil. *J. Loss Prev. Process Ind.,* **4**, 260

WALAS, S.M. (1959). *Reaction Kinetics for Chemical Engineers* (New York: McGraw-Hill)

WALAS, S.M. (1985). Chemical reactor data. *Chem. Engng,* **92**(Oct. 14), 79

WALAS, S.M. (1990). *Chemical Process Equipment. Selection and Design* (Boston, MA: Butterworth-Heinemann)

WALD, A. (1947). *Sequential Analysis* (New York: Wiley)

WALDBOTT, G.L. (1973). *Health Effects of Environmental Pollutants* (St Louis: C.V. Mosby)

WALDE, R.A. (1965). The relationship of the chemical structure of cutting oils to their oxygen compatibility. *Ammonia Plant Safety,* **7**, p. 21

WALDHEIM, P.A., FINNERAN, J.A. and WHITTINGTON, E.L. (1983). Productivity improvement in process design – a look to the future. *Plant/Operations Prog.,* **2**, 222

WALDIE, B. (1986). Chemical engineering at sea. *Chem. Engr, Lond.,* **426**, 17

WALDINGER, R. (1977). Achieving several goals simultaneously. *Machine Intelligence,* **8** (Chichester, Ellis Horwood)

WALDMAN, S. (1967). Fireproofing in chemical plants. *Loss Prevention,* **1**, p. 90. See also *Chem. Engng Prog.,* 1967, **63**(8), 71

WALDO, A. (1989). *The Consolidated Chemical Regulation Guidebook* (New York: Executive Enterprises)

WALDRAM, S.P. (1994). Toll manufacturing: rapid assessment of reactor relief systems for exothermic batch reactions. *Hazards,* **XII**, p. 525

WALDRON, H.A. (1979). *Lecture Notes on Occupational Medicine,* 2nd ed. (Oxford: Blackwells)

WALDRON, H.A. (1980a). The effects of inhaled materials. In Waldron, H.A. and Harrington, J.M. (1980), *op. cit.,* p. 96

WALDRON, H.A. (1980b). Environmental control: current concepts. In Waldron, H.A. and Harrington, J.M. (1980), *op. cit.,* p. 335

WALDRON, H.A. (ed.) (1989). *Occupational Health Practice* (London: Butterworths)

WALDRON, H.A. and HARRINGTON, J.M. (eds) (1980). *Occupational Hygiene: An Introductory Text* (Oxford: Blackwell)

WALDRON, R.D., ERSTFIELD, T.E. and CRISWELL, D.R. (1979). The role of chemical engineering in space manufacturing. *Chem. Engng,* **86**(Feb. 12), 80

WALFORD, P. (1986). Crisis management. *CHEMECA 86,* p. 277

WALIGORA, C.L. and MOTARD, R.L. (1977). Data management on engineering and construction projects. *Chem. Engng Prog.,* **73**(12), 62

WALK, C.J. (1993). Early safety, health and environmental process guidelines. *Process Safety Management,* p. 117

WALKER, A. (1987). *The Hazardous Loads Handbook* (London: Kogan Page)

WALKER, A.W. (1979). Noise and vibration effects. *Plant Engr,* **23**(4), 25

WALKER, C. (1971). *Environmental Pollution by Chemicals* (London: Hutchinson Educational)

WALKER, D.A., WILLIAMS, T.R., NAWROCKI, D.F., BRIDGES, W.G. and ARENDT, J.S. (1993). Managing safety. *Chem Engng,* **100**(3), 90

WALKER, F.E. (1969). *Estimating Production and Repair Effort in Blast-damaged Petroleum Refineries. SRI Proj. MU 6300–620* (report). Stanford Res. Inst., Menlo Park, CA

WALKER, G., COULTER, D.M. and NORRIE, D.H. (1972). *Steel pipe for the British Gas Industry. Commun. GC 194.* Instn Gas Engrs, London. See also *Instn Gas Engrs J.,* **12**, 71

WALKER, I.K. (1961). The heat balance in spontaneous ignition. 1, The critical state. *N.Z. J. Sci.,* **4**, 309

WALKER, I.K. (1967). *Fire Res. Abstrs. Revs,* **9**, 5

WALKER, J. (1973). *Disasters* (London: Studio Vista)

WALKER, J.J. (1970). Sizing relief areas for distillation columns. *Loss Prevention,* **4**, p. 104. See also *Chem. Engng Prog.,* 1970, **66**(9), 38

WALKER, J.L. (1982). Dealing with the media in a chemical emergency. *Chem. Engng,* **89**(Dec. 27), 61

WALKER, M.E. (1983). The analysis of defect to achieve realistic assessment of achieved reliability. *Euredata 4,* paper 12.3

WALKER, W. (1985). The new British Standard BS 2915: 1984 for bursting discs. *Chem. Engr, Lond.,* **411**, 18

WALL, I.B. (1976). Private communication (to T.A. Kletz)

WALL, I.B. (1982). Government sponsored assessment. In Green, A.E. (1982), *op. cit.,* p. 247

WALL, J. (1973). NG/LNG/SNG handbook. *Hydrocarbon Process.,* **52**(4), 87

WALL, J.D. (1975). LNG goes world-wide. *Hydrocarbon Process.,* **54**(4), 141

WALL, K. (ed.) (1986). *Modern Chlor-Alkali Technology,* vol. 3 (Chichester: Ellis Horwood)

WALLACE, A.E. and WEBB, W.P. (1981). Cut vessel costs with realistic corrosion allowances. *Chem. Engng,* **88**(Aug. 24), 123

WALLACE, A.F.C. (1980). *An Exploratory Study of Individual and Community Behaviour in an Extreme Situation* (Washington, DC: Nat. Acad. Sci.)

WALLACE, D.P. (1979). Atmospheric emissions and control. *Ammonia Plant Safety,* **21,** p. 51

WALLACE, I.G. (1989). *Piper Alpha Public Inquiry. Transcript, Days 146, 162, 163*

WALLACE, I.G. (1990). Safety auditing in the offshore industry. *Piper Alpha: Lessons for Life Cycle Management* (Rugby: Instn Chem. Engrs), p. 85

WALLACE, I.G. (1992). The assessment of evacuation, escape and rescue provisions of offshore installations. *Major Hazards Onshore and Offshore,* p. 713

WALLACE, L.A., PELLIZARI, E.R., HARTWELL, T.D., SPARACINO, C.M., SHELDON, L.S. and ZELON, H. (1985). Personal exposures, indoor–outdoor relationships and breath levels of toxic pollutants measured for 355 persons in New Jersey. *Atmos. Environ.,* **19,** 1651. See also Rosebrook, D.D. and Worm, G. (1993), *op. cit.*

WALLACE, M.J. (1978). Basic concepts of toxicology. *Chem. Engng,* **85**(Apr. 24), 72

WALLACE, M.J. (1979). Controlling fugitive emissions. *Chem. Engng,* **86***(Aug. 27),* 79

WALLACE, N.M. and DAVID, J. (1985). The design and application of seals for light hydrocarbon service. Br. *Pump Manufacturers Ass., Ninth Tech. Conf. on Reliability – The User-Maker Partnership* (London)

WALLACE, R.H. (1970). *Understanding and Measuring Vibrations* (London: Wykeham)

WALLACE, W.F.M. and LANGLANDS, J.H.M. (1971). Insulation workers in Belfast, 1: Comparison of a random sample with a control population. *Br. J. Ind. Med.,* **28,** 211

WALLBRIDGE, N. and GATES, E. (1984). Software for control valve sizing with your desktop computer. *Chem. Engr, Lond.,* **403,** 42

WALLER, R.A. and COVELLO, V.T. (eds) (1984). *Low Probability, High Consequence Risk Analysis* (New York: Plenum Press)

WALLEY, K.H. and ROBINSON, S.J.Q. (1972). The effects of scale on the economics of olefin production – the difference between theory and reality. *Eur Comm., Symp. on Olefins Production,* Bratislava

WALLICK, F. (1972). *The American Worker – An Endangered Species* (New York: Ballantine Books)

WALLINGTON, T.J., SCHNEIDER, W.F., WORSNOP, D.R., NIELSEN, O.J., SEHESTED, J., DEBRUYN, W.J. and SHORTER, J.A. (1994). The environmental impact of CFC replacements – HFCs and HCFCs. *Environ. Sci. Technol.* **28** (7), 320–25

WALLIS, B. (1984). Current trends in seals materials. *Chem. Engr, Lond.,* **408,** 10

WALLIS, G.B. (1969). *One Dimensional Two-Phase Flow* (New York: McGraw-Hill)

WALLIS, G.B. (1970). Minimum critical velocity for one-phase flow of liquids. *Ind. Engng Chem. Process Des. Dev.,* **9,** 483

WALLIS, G.B. (1980). Critical two-phase flow. *Int. J. Multiphase Flow,* **6,** 97

WALLIS, G.B. and RICHTER, H.J. (1978). An isentropic streamline model for flashing two-phase vapor–liquid flow. *J. Heat Transfer,* **100,** 595

WALLS, E.L. and METZNER, W.P. (1962). Fume hoods – safety vs cost. *Ind. Engng Chem.,* **54**(4), 42

WALLS, L.A. and BENDELL, A. (1985). The structure and exploration of reliability field data: what to look for and how to analyse it. *Reliability 85,* p. 5B/5/1. See also *Reliab. Engng,* **15,** 115

WALLS, L.A. and BENDELL, A. (1987). Time series methods in reliability. *Reliab. Engng,* **18**(4), 239

WALLS, W.L. (1963–64). LP gas tank truck accident and fire – Berlin. *NFPA Quart.,* **57,** 9

WALLS, W.L. (1964a). LP gas tank truck accident and fire. *Fire Int.,* **3,** 12

WALLS, W.L. (1964b). Vinyl chloride explosion. *NFPA Quart.,* Apr., 352

WALLS, W.L. (1971). Fire protection for LNG plants. *Hydrocarbon Process.,* **50**(9), 205

WALLS, W.L. (1973). Coping with LNG plant fires. In Vervalin, C.H. (1973a), *op. cit.,* p. 377

WALLS, W.L. (1978). Just what is a BLEVE? *Fire J.,* Nov.

WALLS, W.L. (1979). The BLEVE. *Fire Command,* May, 22: Jun., 35

WALLS, W.L. (1986). *Gases. Fire Protection Handbook,* 16th ed. (ed. A.E. Cote and J.M. Linville), p. 5–39

WALLSGROVE, R. (1988). CRYSTAL. *Personal Comput. World,* Nov., 172

WALMSLEY, F.H. (1973). Safety aspects of the transport of hazardous chemicals. *Chem. Engr, Lond.,* **270,** 95

WALMSLEY, F.H. (1974). Safety aspects of the transport of hazardous chemicals. *Course on Process Safety – Theory and Practice.* Dept of Chem. Engng, Teesside Polytechnic, Middlesbrough

WALMSLEY, H.L. (1982). The generation of electric currents by turbulent flow of dielectric liquids. I: Long pipes. *J. Phys D: Appl. Phys,* **15,** 1907

WALMSLEY, H.L. (1983a). Charge generation in low dielectric constant liquids. *Electrostatics,* **6,** p. 45

WALMSLEY, H.L. (1983b). Charge relaxation in pipes. *Electrostatics,* **6,** p. 209

WALMSLEY, H.L. (1983c). The generation of electric currents by turbulent flow of dielectric liquids. II: Pipes of finite length. *J. Phys D: Appl. Phys,* **16,** 1907

WALMSLEY, H.L. and WOODFORD, G. (1981a). The generation of electric currents by laminar flow of dielectric liquids. *J. Phys D: Appl. Phys,* **14,** 1761

WALMSLEY, H.L. and WOODFORD, G. (1981b). The polarity of current generated by the laminar flow of a dielectric liquid. *J. Electrostatics,* **10,** 283

WALSH, D. (ed.) (1988). *Chemical Safety Data Sheets,* vol. 1 (London: R. Soc. Chem.). See also Allen, A. (1988)

WALSH, F. and MILLS, G. (1993). Electrochemical techniques for a cleaner environment. *Chem. and Ind.,* (Aug. 2), 576–79

WALSH, J.J. (1984). Drying of temperature sensitive solids. *Plant/Operations Prog.,* **3,** 74

WALSH, P.J., DUDNEY, C.S. and COPENHAVER, E.D. (eds) (1983). *Indoor Air Quality* (Boca Raton, FL: CRC Press)

WALSHAM, G. (1974). A computer simulation model applied to design problems of a petrochemical complex. *Opl Res. Quart.,* **25**(3), 399

WALTER, D.L. (1983). Ethylene from NGL feedstocks, Pt 2. *Hydrocarbon Process.,* **62**(11), 187

WALTERS, D.B. and JAMESON, C.W. (eds) (1984). *Health and Safety for Toxicity Testing* (Boston, MA: Butterworths)

WALTERS, J.K. and WINT, A. (eds) (1981). *Industrial Effluent Treatment* (London: Applied Science)

WALTERS, P. (1977). Aids to rescue. *Fire Surveyor,* **6**(3), 33

WALTERS, W.J. (1973). *Production and Supply and Safety* (London: Instn Chem. Engrs)

WALTERS, W.J., DEAN, F.E. and CARNE, M. (1974). Environmental and safety aspects of LNG storage. *LNG Conf. 4*

WALTON, D.C. (1925). Naval medical aspects of chemical warfare. In Vedder, E.B. (1925), *op. cit.,* p. 264

WALTON, D.C. and ELDRIDGE, W.A. (ca. 1928). The action of chlorine on men poisoned by toxic smokes. *J. Pharm. Exp. Therapeut.*, **35**, 241

WALTON, D.C. and WITHERSPOON, M.G. (1928). Skin absorption of certain gases. *J. Pharm. Exp. Therapeut.*, **26**, 315

WALTON, J.A. (1980). Detector tubes for atmospheric monitoring. In Hughes, D.J. (1980), *op. cit.*, paper 7

WALTON, M. (1973). Industrial ammonia gassing. *Br. J. Ind. Med.*, **30**, 76

WALTON, N.H. (1963). Hydrogen compressor cylinder explosion. *Ammonia Plant Safety*, **5**

WALTON, N.H. (1965). Urea plant safety problems. *Ammonia Plant Safety*, **7**, p. 56

WALTON, N.H. (1966). Urea plant safety. *Ammonia Plant Safety*, **8**, p. 36

WALTON, N.H. (1967). Safety problems of urea plants. *Ammonia Plant Safety*, **9**, p. 73

WALTON, P. (1993). Assessment of the safety integrity of programmable electronic systems. In Melchers, R.E. and Stewart, M.G. (1993), *op. cit.*, p. 213

WALTON, W. (1964). Tankcar not a hollow shell. *Chem. Engng Prog.*, **60**(3), 29

WALTON, W.H. (ed.) (1971). *Inhaled Particles and Vapours*, vol. 3 (Old Woking: Unwin Bros). See also Davies, C.N. (1961, 1967)

WALTON, W.H. (ed.) (1977). *Inhaled Particles and Vapours*, vol. 4 (Oxford: Pergamon Press)

WAMPLER, F.M. (1988). Formation of diacrylic acid during acrylic acid storage. *Plant/Operations Prog.*, **7**, 183

WAMSLEY (1983). On-stream inspection – a contributor to increased equipment reliability and improved refinery productivity. *API Proc., Refining Dept, 48th Midyear Mtg*, **62**, 226

WAN, X.Z., HARDMAN, J.S., NEVELL, T.G., HARPER, D.J. and ROGERS, R.L. (1994). Experimental and theoretical studies of the temperature ramped screening test for the thermal stability of powders. *Hazards*, **XII**, p. 43

WANDS, R.C. (1972). Acute and chronic toxicity aspects of chemical spills. *Control of Hazardous Material Spills*, **1**, p. 207

WANDS, R.C. (1981). *Patty's Industrial Hygiene and Toxicology*, 3rd ed. (New York: Wiley)

WANG, C.C.K. (1984). Minimizing worker exposure during solid material sampling. *Chem. Engng Prog.*, **80**(9), 53

WANG, C.C.K. (1985). A 3-E quantitative decision model of toxic substance control through control technology use in the industrial environment. *J. Haz. Materials*, **12**(1), 65

WANG, C.C.K. and PARKER, H.E. (1984). How to be an effective expert witness. *Chem. Engng*, **91**(Feb. 20), 87

WANG, F.S. and PERLMUTTER, D.D. (1968a). Bounding chemical reactor transients by estimation of growth on decay rates. *AIChE J.*, **14**, 934

WANG, F.S. and PERLMUTTER, D.D. (1968b). Comparison of two methods used in analysis of reactor dynamics. *AIChE J.*, **14**, 976

WANG, F.S. and PERLMUTTER, D.D. (1968c). A composite phase plane for tubular reactor stability studies. *AIChE J.*, **14**, 335

WANG, L.R.L. and CHENG, K.M. (1979). Seismic response behaviour of buried pipelines. *J. Pres. Ves. Technol.*, **101**, 21

WANG, P.S. (1994). *C++ with Object Oriented Programming* (Boston: PWS)

WANG, S.I. and PATEL, N.M. (1984). Creative ways of revamping NH_3 plants can improve profitability. *Plant/Operations Prog.*, **3**, 101

WANG, S.S.Y., SAN KIANG and MERKL, W. (1994). Investigation of a thermal runaway hazard – drum storage of thionyl chloride/ethyl acetate mixture. *Process Safety Prog.*, **13**, 153

WANG, T.S., ZHANG, Y.F., ZHAO, H.F. QI, Z.M. and REN, A.F. (1981). A study of ignition by flame torches. *Combustion*, **18**, p. 1729

WANKEL, R.A. (1967). Purpose and function of Tennessee Eastman's reactive chemical program. *Loss Prevention*, **1**, p. 50

WANNER, D.F. (1977). Improve productivity with maintenance audits. *Hydrocarbon Process.*, **56**(5), 127

WANNER, J.C. (1986). Human factors and systems safety. Ergonomic control room design. *Loss Prevention and Safety Promotion 5*, 2–1

WARBURTON, G.B. (1976). *The Dynamical Behaviour of Structures* (Oxford: Pergamon Press)

WARBURTON, R. (1974). Case study of maintenance engineering. *Plant Engr*, **18**(7/8), 11

WARD, A.F.H. (1969). Training in workshops and laboratories in schools, colleges and universities. In Handley, W. (1969), *op. cit.*, p. 8

WARD, B. (1992). Halon replacements, alternatives and recycling. *Fire Surveyor*, **21**(6), 25

WARD, C. (1973). *Vandalism* (London: Architect. Press)

WARD, J.A. and EGAN, P.C. (1969). Experience with frozen inground storage units for liquefied natural gas – Canvey Island (UK). *Proc. Int. Conf. on Liquefied Natural Gas* (London)

WARD, J.A. and HILDREW, R.H. (1968a). Importation of liquefied natural gas from Algeria to the United Kingdom. *LNG Conf. 1*

WARD, J.A. and HILDREW, R.H. (1968b). Latest developments at Canvey Island. *LNG Conf. 1*

WARD, J.R. (1968). Lined-pipe systems. *Chem. Engng.*, **75**(Jun. 17), 238

WARD, K., MURRAY, B. and COSTELLO, G.P. (1983). Acute and long-term pulmonary sequelae of acute ammonia inhalation. *Ir. Med. J.*, **76**, 279

WARD, M. (1989). Dusts: testing for explosibility. *Plant Engr*, Mar./Apr., 26

WARD, R. (1972). *Power*, **116**(Feb.), 154

WARD, GENERAL SIR R. (1981). *Reopened Exploratory Inquiry into the Desirability of Revoking the Planning Permission Granted to United Refineries Ltd on 28th March 1973 for the Construction of an Oil Refinery on Canvey Island* (London: Dept of the Environment)

WARD, R. (1993a). Management creep: is it a real risk in the process industries? *Process Safety Management*, p. 59

WARD, R.B. (1993b). Management, safety and serendipity in education. *Health, Safety and Loss Prevention*, p. 565

WARD, R. (1993c). Mixing management, mystery and safety: an innovative seminar. *Process Safety Management*, p. 69

WARD, T.J. (1984). Predesign estimating of plant capital costs. *Chem. Engng*, **91**(Sep. 17), 121

WARD, W. (1978). Tanker fire contained. *Fire Command*, **45**(7), 28

WARD, W.D. and FRICKE, J.E. (1969). *Noise as a Public Health Hazard* (Washington, DC: Am. Speech and Hearing Ass.)

WARD, W.J., LEE, J.W. and FREYMARK, S.G. (1978). Advantages of chlorine dioxide as biocide. *Ammonia Plant Safety*, **20**, p. 64

WARD, W.J., LABINE, P.U. and REDFIELD, D.A. (1980). Organometallurgically hydrazine for maximum protection of steam generating systems. *Ammonia Plant Safety,* **21**, p. 57

WARDALE, J.K.S. and FRANK, E.H. (1970). Failure of liquid ethylene storage tanks. *LNG Conf. 2*

WARDALE, J.K.S. and TODD, G. (1967). The bulk storage of liquefied gases. *Chem. Engr, Lond.,* **213**, CE247

WARDE, E. (1991). Which super-duplex? *Chem. Engr, Lond.,* **502**, 32

WARDLAW, G. (1983). Human behaviour in emergencies – is widespread panic a myth? *Fire Int.,* **80**, 48

WARDLE, A. (n.d.). LPG carriers collision study. *Lloyds Register/Hull Structures/A and P Rep. 030*

WAREING, T.H. (1969). Entry into confined spaces. In Handley, W. (1969), *op. cit.,* p. 61

WARING, R.H. (1974). *Pumps – Selection, Systems and Application* (London: Trade & Technical Press)

WARLIN, L. and SVENSSON, T. (2000). Compact heat exchangers: design and service *Pet. Technol. Q.,* (Autumn), 119–27

WARNATZ, J. (1985). Chemistry of high temperature combustion of alkanes up to octane. *Combustion,* **20**, p. 845

WARNCKE, H. (1977). Monitoring systems for toxic gases. *Loss Prevention and Safety Promotion 2*, p. 487

WARNER, D.L. (1965). Deep-well disposal of industrial wastes. *Chem. Engng,* **72**(Jan. 4), 73

WARNER, SIR. F. (1975a). The Flixborough disaster. *Chem. Engng Prog.,* **71**(9), 77

WARNER, SIR F. (1975b). The Flixborough disaster – how it happened. *BSI News, Jun.,* 6

WARNER, SIR F. (1976). *Engineering Safety and the Environment* (London: Exchange Publs)

WARNER, SIR F. (1979). Sources and extent of pollution. *Proc. Roy. Soc., Ser. B,* **205**, 5. See also in Doll, Sir R. and McLean, A.E.M. (1979), *op. cit.,* p. 5

WARNER, SIR.F. (1980). Risk and response. *Sci. Publ. Policy,* **7**, 402

WARNER, SIR F. (1981a). The engineer's conscience (book review). *Chem. Engr, Lond.,* **366**, 123

WARNER, SIR F. (1981b). The foundations of risk assessment (foreword). In Griffiths, R.F. (1981), *op. cit.* p. ix

WARNER, SIR F. (1981c). Hazard and risk (letter). *Chem. Engr, Lond.,* **370**, 335

WARNER, SIR F. and SLATER, D.H. (1981). *The Assessment and Perception of Risk* (London: The Royal Society)

WARNER, K. (1981). Carriage of chemicals and other dangerous goods by air – new regulations. *Chem. Ind.,* Nov. 7, 759

WARREN, A.H. (1966). The protection of oil storage tanks. *Fire Int.,* **1**(12), 66

WARREN, C.W. (1975a). Hot tapping under pressure. *Loss Prevention,* **9**, p. 29

WARREN, C.W. (1975b). How to read instrument flowsheets. *Hydrocarbon Process.,* **54**(7), 163: **54**(9), 191

WARREN, D., PERREIRA, L.M. and PERREIRA, F. (1977). PROLOG – the language and its implementation compared with LISP. *Proc. ACM SIGART-SIGPLAN Symp. on AI and Programming Languages,* Rochester, NY

WARREN, D.V. (1977). Accident rates and acceptable risk levels for aeroplanes and their systems. In Council for Science and Society (1977), *op. cit.,* p. 60

WARREN, F.H., JOYCE, T.J., DAVIS, R.J. and FIRSTENBERG, H. (1974). Environmental factors in siting LNG facilities. *LNG Conf. 4*

WARREN, J.H. and CORONA, A.A. (1975). This method tests fire protective coatings. *Hydrocarbon Process.,* **54**(1), 121

WARREN, J.H., LANGE, R.F. and RHYNARD, D.L. (1974). Conductivity additives are best. *Hydrocarbon Process.,* **53**(12), 111

WARREN, M.C., CONRAD, J.P., BOCIAN, J.J. and HAYES, M. (1963). Clothing-borne epidemic. *J. Am. Med. Ass.,* **184**, 266

WARREN, P.J. (ed.) (1985). *Dangerous Chemicals Spillage Guide* (New Malden: Walters Sampson)

WARREN, P.R. (1992). *Private communication*

WARREN, P.R. and WEBB, B.C. (1980a). The relationship between tracer gas and pressurisation techniques in buildings. *Proc. First Air Infiltration Centre Conf.*

WARREN, P.R. and WEBB, B.C. (1980b). Ventilation measurements in housing. *Chart. Inst. Building Services, Conf. on Natural Ventilation by Design* (London)

WARREN, R.E. (1978). Control of hazardous material spills within urban areas by proper drainage management. *Control of Hazardous Material Spills,* **4**, p. 36

WARREN, W.R. (1957). *An Analytical and Experimental Study of Compressible Free Jets. Rep. 381.* Aeronaut. Engng Lab, Univ. of Princeton

WARREN SPRING LABORATORY (1972). *Oil pollution of the Sea and Shore* (London: HM Stationery Office)

WARRING, R.H. (ed.) (1970). *Handbook of Noise and Vibration Control* (Morden: Trade & Technical Press)

WARRING, R.H. (ca. 1973). *Handbook of Noise and Vibration Control, 2nd ed.* (Morden, Surrey: Trade & Technical Press)

WARRING, R.H. (1981). *Seals and Sealing Handbook* (Morden: Trade & Technical Press)

WARRING, R.H. (1987). *Handbook of Valves, Piping and Pipelines* (Houston, TX: Gulf)

WARWICK, I. (1977). The workings of the Health and Safety at Work Act – a contractor's view. *Chemical Engineering in a Hostile World* (London: Clapp & Poliak)

WARWICK, K. and REES, D. (ed.) (1986). *Industrial Digital Control Systems* (London: Peter Peregrinus)

WARWICK, R.G. (1967). An insurance company's view on the safety aspects of chemical plant design. *Chem. Ind.,* Jun. 10, 943

WARWICK, R.G. (1969). Steam boilers and pressure vessels. In Handley, W. (1969), *op. cit.,* p. 216

WASHBURN, E.W. (ed.). (1926–). *International Critical Tables* (New York: McGraw-Hill)

WASHIMI, K. and ASAKURA, M. (1966). Computer control of a vinyl chloride plant provides process optimization. *Chem. Engng,* **73**(Oct. 24), 133

WASHIO, T., KITAMURA, M. and SUGIYAMA, K. (1987). Development of failure diagnosis method based on transient information of nuclear power plant. *J. Nucl. Sci. Technol.,* **24**, 30

WASP, E.J., THOMPSON, T.L. and SNOEK, P.E. (1971). The era of slurry pipelines. *Chem. Tech.,* Sep., 552

WASS, H.S. (1983). *Sprinkler Hydraulics* (White Plains, NY: IRM Insurance)

WASSERMAN, S., SMITH, M.A. and MOTTU, S. (eds) (1989). *Encyclopaedia of Physical Sciences and Engineering Information Sources* (Detroit, MI: Gale Res.)

WASSOM, R.W. (1982). High-pressure heat exchanger ignition. *Chem. Engng Prog.,* **78**(7), 55

WATANABE, J. and MURAKAMI, Y. (1981). Prevention of temper embrittlement of chromium-molybdenum steel vessels by use of low-silicon forged shells. *API Proc., Refining Dept, 46th Midyear Mtg,* **60**, 216

WATANABE, K. and HIMMELBLAU, D.M. (1983). Fault diagnosis in nonlinear chemical processes. *AIChE J.,* **29**, 243: 250

WATANABE, K. and HIMMELBLAU, D.M. (1984). Incipient fault diagnosis of nonlinear processes with multiple causes of faults. *Chem. Engng Sci.*, **39**, 491

WATANABE, K. and HIMMELBLAU, D.M. (1986). Detection and location of a leak in a gas transport pipeline by a new acoustic method. *AIChE J.*, **32**, 1690

WATANABE, K., MATSUURA, I., ABE, M., KUBOTA, M. and HIMMELBLAU, D.M. (1989). Incipient fault diagnosis of chemical processes via artificial neural networks. *AIChE J.*, **35**, 1803

WATANABE, S., BOUHADEF, A., KINOSHITA, K., SUZUKI, Y., HAYASHI, N. and TOUCHARD, G. (1991). A new method of assessing streaming charge in liquids. *Electrostatics*, **8**, p. 85

WATANABE, Y. (1987). Errors in availability estimation by 2−state models of 3−state systems. *Reliab. Engng*, **18**, 223

WATCHORN, N. (1966). Ammonium nitrate stability. *Ammonia Plant Safety*, **8**, p. 32

WATERMAN, D.A. (1978). A rule-based approach to knowledge acquisition for man-machine interface programs. *Int. J. Man−Machine Studies*, **10**, 693

WATERMAN, D.A. (1979). User-oriented systems for capturing expertise: a rule-based approach. In Michie, D. (1979), *op. cit.*, p. 26

WATERMAN, D.A. (1981). Rule-based expert systems. Machine Intelligence, Infotech State of the Art Rep. 9

WATERMAN, D. (1985). *A Guide to Expert Systems* (Reading, MA: Addison-Wesley)

WATERMAN, D.A. and HAYES-ROTH, F. (eds) (1978). *Pattern-Directed Inference Systems* (New York: Academic Press)

WATERMAN, L. (1987). What the COSHH Regs will mean in practice. *Health and Safety at Work*, Nov., 21

WATERMAN, T.E. (1968). Room flashover − criteria and synthesis. *Fire Technol.*, **4**, 25

WATERS, A. and PONTON, J.W. (1989). Qualitative simulation and fault propagation in process plants. *Chem. Engng Res. Des.*, **67**, 407

WATERS, A. and PONTON, J.W. (1992). Managing constraints in design: using an AI toolkit as a DMBS. *Comput. Chem. Engng*, **16**, 987

WATERS, A., CHUNG, P.W.H.C. and PONTON, J. (1989). Representing safety constraints upon chemical plant designs. In Vadera, S. (1989), *op. cit.*, p. 199

WATERS, A.W. (1978). Design and application of a low cost pneumatic composition transmitter. *Chem. Engng Prog.*, **74**(6), 80

WATERS, J.R. and NIELSEN, N.R. (1988). *Crafting Knowledge-Based Systems* (Chichester: Wiley)

WATERS, T. (1988a). Human factors reliability benchmark exercise: report of SRD participation. In Sayers, B.A. (1988), *op. cit.*, p. 110

WATERS, T.L. (1988b). Human reliability assessment − a case study involving the analysis of operator actions during a loss of coolant accident. *SRS Quart Dig.*, Apr., 19

WATERS, T. (1989). Human Factors Reliability Benchmark Exercise: summary report. In Poucet, A. (1989), *op. cit.*, p. 367. See also Safety and Reliability Directorate, Culcheth, Lancashire, *Rep. RTS/123, 1988*

WATERS, T.R., PUTZ-ANDERSEN, V., GARG, A. and FINE, L.J. (1993). Revised NIOSH equation for the design and evaluation of manual lifting tasks. *Ergonomics*, **36**, 749

WATERWORTH, T.A. and CARR, M.J.T. (1975). Report on injuries sustained by patients treated at the Birmingham General Hospital following the recent bomb explosions. *Br. Med. J.*, Apr. 5, 25

WATKINS, R.A. (1969). Preventing ammonia plant fires. *Ammonia Plant Safety*, **11**, p. 23. See also *Chem. Engng Prog.*, 1969, **65**(3), 69

WATKINS, T.F., CACKETT, J.C. and HALL, R.G. (1968). *Chemical Warfare, Pyrotechnics and the Chemical Industry* (Oxford: Pergamon Press)

WATROUS, L.D. (1970). Freight train derailment and fire Crescent City, Illinois. *Fire J.*, **10**

WATSON, B. (1984). Fast response sprinkler heads. *Fire Surveyor*, **13**(4), 25

WATSON, E.H. (1974). Autoclave reactors for polyethylene production. *Safety in Polyethylene Plants*, **2**, p. 35

WATSON, I.A. (1980). Common mode/cause failures − an overall review (1980). In Moss T.R. (1980), *op. cit.*, p. 136

WATSON, I.A. (1985). Review of human factors in reliability and risk assessment. *The Assessment and Control of Major Hazards* (Rugby: Instn Chem. Engrs), p. 323

WALTON, I.A. and OAKES, F. (1988). Management in high risk industries. In Sayers, B.A. (1988), *op. cit.*, p. 202

WATSON, S.R. (1981). On risks and acceptability. *J. Soc. Radiolog. Prot.*, **1**(4), 21

WATT, W. and PEROV, B.V. (1985). Handbook of composites (Amsterdam: North Holland)

WATT COMMITTEE ON ENERGY (1986). *The Chernobyl Accident and Nuclear Safety in the United Kingdom* (London)

WATT COMMITTEE ON ENERGY (1991). *Five Years after Chernobyl: 1986−1991* (London)

WATTERS, G.Z. (1979). *Modern Analysis and Control of Unsteady Flow in Pipelines* (Ann Arbor, MI: Ann Arbor Science)

WATTERS, P. (1977). Road routing of dangerous substances. Symp. Hazard 77—Advances in the Safe Handling and Movement of Hazardous Substances, Teesside Polytechnic, Middlesbrough

WATTNER, K.W. (1976). Failure analysis of high speed couplings. *Hydrocarbon Process.*, **55**(1), 77

WATTON, B.K. and BRODIE, G.W. (1984). Reverse domed buckling discs, a pressure relief device: their performance and problems. *The Protection of Exothermic Reactors and Pressurised Storage Vessels* (Rugby: Instn Chem. Engrs), p. 51

WATTS, C. (1984). Emergency lighting. *Fire Surveyor*, **13**(5), 12

WATTS, H. (1951). *Storage of Petroleum* (London: Griffin)

WATTS, H. (1954). *Law Relating to Explosives* (London: Griffin)

WATTS, H. (1956). *The Law Relating to Petroleum Mixtures, Acetylene, Calcium Carbide, etc.* (London: Griffin)

WATTS, H.E. (1951). *Report on Explosion and Fire at Regent Oil Co Ltd Premises, Royal Edward Dock, Avonmouth, Bristol on 7th Sept. 1951* (London: HM Stationery Office)

WATTS, J.M. (1986). Dealing with uncertainty: some applications in fire protection engineering. *Fire Safety J.*, **11**(1), 127

WATTS, J.M. (1987). Computer models for evacuation analysis. *Fire Safety J.*, **12**(3), 237

WATTS, J.W. (1976). Effects of water spray on unconfined flammable gas. *Loss Prevention*, **10**, p. 48

WATTS, P. (1985). Vent system design on or offshore. *Chem. Engr, Lond.*, **410**, 34

WATTS, P.C.A. (1989). *Piper Alpha Inquiry. Transcript, Day 94*

WAUGH, W.K. (1966). Fire hazards in the petrochemical industry. *Fire Int.*, **1**(12), 71

WAY, D. (1969). Fire protection engineering. *Loss Prevention* **3**, p. 23

WAY, D.H. and HILADO, C.J. (1968). Fire tests on thermal insulation systems for pipes. *Fire Tech. Note 4*, 271

WAYNE, F.D. (1991). An economical formula for calculating atmospheric infrared transmissivities. *J. Loss Prev. Process Ind.*, **4**(2), 86

WAYNE, W.S., BÖCKENHAUER, M. and GRAY, R.C. (1991). Filling limits of liquefied gas cargo tanks. *Gastech 90*, p. 3.2

WEALE, K.E. (1967). *Chemical Reactions at High Pressures* (London: Spon)

WEARNE, S.H. (1979). A review of reports of failures. *Proc. Instn Mech. Engrs*, **193**, 125

WEAST, R.C. (ed.) (1985). *CRC Handbook of Data on Organic Compounds*, vols 1–2 (Boca Raton, FL: CRC Press)

WEATHERBURN, A.C. and CLINCK, F.J. (1976). Automatic ultrasonic weld scanner and recording system for examination of butt and T-butt welds in pressure vessels. *Diagnostic Techniques* (London: Instn. *Chem. Engrs*)

WEATHERBY, J.H. (1952). Chronic toxicity of NH₃ fumes by inhalation. *Proc. Soc. Explt Biol. Med.*, Oct., 300

WEATHERBY, J.J. (1964). Here's how one plant had high safety at low cost. In Vervalin, C.H. (1964a), *op. cit.*, p. 292

WEATHERFORD, W.D. and SHEPPARD, D.M. (1965). Basic studies of the mechanism of ignition of cellulosic materials. *Combustion*, **10**, p. 897

WEATHERFORD, W.D. and VALTIERRA, M.C. (1966). Piloted ignition of convection heated cellulose slabs. *Combust. Flame*, **10**, 279

WEATHERHEAD, D. (1977). Intrinsic safety. *Measamt Control*, **10**, 341

WEATHERHEAD, R.G. (1980). *FRP Technology* (London: Applied Science)

WEATHERILL, T. and CAMERON, I. (1988). Preliminary hazop studies using expert systems. *Third Int. Symp. on Process Systems Engng*, Sydney

WEATHERILL, T. and CAMERON, I.T. (1989). A prototype expert system for hazard and operability studies. *Comput. Chem. Engng*, **13**, 1229

WEAVER, D. (1987). Avoid those vibration problems on heat exchanger tubes. *Process Engng*, Nov., 33

WEAVER, F.L. (1972). Rotor design and vibration response. *Ammonia Plant Safety* **14**, p. 10

WEAVER, R. (1973). *Process Piping Design*, vols 1–2 (Houston, TX: Gulf)

WEAVER, W.W. (1980). Deterministic criteria versus probabilistic analysis: examining the single failure and separation criteria. *Nucl. Technol.*, **47**(2), 234

WEBB, E. (1993). Contractor management – a New Zealand perspective. *Health, Safety and Loss Prevention*, p. 92

WEBB, E.K. (1970). Profile relationships: the log–linear range and extension to strong stability. *Q. J. Roy. Met. Soc.*, **96**, 67

WEBB, H.E. (1963). Coping with the fire menace. *Chem. Engng*, **70**(Dec. 9), 196

WEBB, H.E. (1969). Accident investigation. *Chem. Engng*, **76**(Feb. 24), 88

WEBB, H.E. (1977). What to do when disaster strikes. *Chem. Engng*, **84**(Aug. 1), 117

WEBB, J.D. (1978). *Noise Control in Industry* (Sudbury: Sound Res. Lab.). See also Sound Research Laboratory (1991)

WEBB, J.T. (1991). The role of verification and validation tools in the production of critical software. In Ince, D. (1991a), *op. cit.* p. 33

WEBB, L.C. (1974). Development of correct feedwater treatment. *Ammonia Plant Safety*, **16**, p. 92

WEBB, M.E. (1962). Be hard-headed about safety gear. *Chem. Engng*, **69**(Jul. 23), 136

WEBB, W.P. and GUPTA, S.C. (1984). Metals for hydrogen service. *Chem. Engng*, **91**(Oct. 1), 113

WEBBER, D. (1985). Carbide speeds rationalisation plans. *Chem. Ind.*, Oct. 21, 673

WEBBER, D. (1986). Bhopal victims wait for compensation. *Chem. Ind.*, Jul. 7, 442

WEBBER, D.M. (1984). Gravity spreading in dense gas dispersion models. In Ooms, G. and Tennekes, H. (1984), *op. cit.*, p. 397

WEBBER, D.M. (1989). Evaporation and boiling of liquid pools – a unified treatment. In Cox, R.A. (1989a), *op. cit.*, p. 131

WEBBER, D.M. (1990). Modelling two-phase jets for hazard analysis. *J. Haz. Materials*, **23**(2), 167

WEBBER, D.M. (1991). Source terms. *J. Loss Prev. Process Ind.*, **4**(1), 5

WEBBER, D.M. and BRIGHTON, P.W.M. (1986). A mathematical model of a spreading, vaporising liquid pool. In Hartwig, S. (1986), *op. cit.* p. 223

WEBBER, D.M. and JONES, S.J. (1987). A model of spreading vaporising pools. *Vapor Cloud Modeling*, p. 226

WEBBER, D.M. and KUKKONEN, J.S. (1990). Modelling two-phase jets for hazard analysis. *J. Haz. Materials*, **23**(2), 167

WEBBER, D.M. and WHEATLEY, C.J. (1987). The effect of initial potential energy on the dilution of a heavy gas cloud. *J. Haz. Materials*, **16**, 357

WEBBER, D.M., JONES, S.J. and MARTIN, D. (1992). Advances in gas cloud dispersion modelling: heavy clouds on sloping ground. *Major Hazards Onshore and Offshore*, p. 351

WEBBER, D.M., JONES, S.J. and MARTIN, D. (1993). A model of the motion of a heavy gas cloud released on a uniform slope. *J. Haz. Materials*, **33**, 101

WEBBER, D.M., MERCER, A. and JONES, S.J. (1994). Hydrogen fluoride source terms and dispersion. *J. Loss Prev. Process Ind.*, **7**, 94

WEBBER, G.M.B. and HALLAN, P.J. (1988). Photoluminescence for aiding escape. *Fire Surveyor*, **17**(6), 17

WEBBER, W.O. (1979). Retrofit insulation for profit. *Hydrocarbon Process.*, **58**(7), 141

WEBER, A.H. (1976). *Atmospheric Dispersion Parameters in Gaussian Plume Modelling. Rep. EPA-600/4-76-030a*. Environ. Prot. Agency

WEBER, C.G. (1955). How to protect your pressure vessels. *Chem. Engng*, **62**(10), 170

WEBER, E. (1950). *Electromagnetic Fields* (New York: Wiley)

WEBER, E. (1965). *Electromagnetic Theory: Static Fields and Their Mapping* (New York: Dover)

WEBER, E. (ed.) (1982). *Air Pollution* (New York: Plenum Press)

WEBER, E.C. (1976). Development of Maryland's spill control containment and clean-up program. *Control of Hazardous Material Spills 3*, p. 126

WEBER, H. and FALBE, J. (1970). Oxo synthesis technology. *Ind. Engng Chem.*, **62**(4), 33

WEBER, J.H. (1980a). Calculate equation-of-state variables. *Chem. Engng*, **87**(Feb. 25), 93

WEBER, J.H. (1980b). Predict latent heat of vaporisation. *Chem. Engng*, **87**(Jan. 14), 105

WEBER, J.H. (1980c). Predict properties of gas mixtures. *Chem. Engng*, **87**(May 19), 151

WEBER, J.H. (1980d). Predict thermodynamic properties of pure gases and binary mixtures. *Chem. Engng*, **87**(Sep. 22), 155

WEBER, J.H. (1982). Predict equation-of-state variables. *Chem. Engng*, **89**(Jan. 11), 111

WEBER, J.P., SAVITT, J., KRC, J. and BROWNE, H.C. (1961). Detonation tests evaluate high pressure cells. *Ind. Engng Chem.*, **53**(11), 128A

WEBER, M.J. (ed.) (1982). *CRC Handbook of Laser Science and Technology* (Boca Raton, FL: CRC Press)

WEBER, R. (1992). A comparison of different estimators for the standard deviation of wind direction based on persistence. *Atmos. Environ.*, **26A**, 983

WEBER, R.O. (1989). Analytical models of fire spread due to radiation. *Combust. Flame*, **78**, 398

WEBER, W.J., SHERRILL, J.D., PIRBAZARI, M., UCHRIN, C.G. and LO, T.Y. (1980). Transport and differential accumulation of toxic substances in river–harbor–lake systems. In Haque, R. (1980), *op. cit.*, p. 191

WEBSTER'S II NEW RIVERSIDE DICTIONARY, OFFICE EDITION (1984). (New York: Houghton Mifflin Co., Berkley Publishing,)

WEBSTER, G.R. (1979). The canned pump in the petrochemical environment. *Chem. Engr, Lond.*, **341**, 91

WEBSTER, J.M. (1984). Multipurpose fire extinguishers. *Fire Surveyor*, **13**(2), 6

WEBSTER, T.J. (1974). Safety is good business. *Loss Prevention and Safety Promotion 1*, p. 41

WEBSTER, T.J. (1976). Total loss control. *Course on Loss Prevention in the Process Industries*. Dept of Chem. Engng, Loughborough Univ. of Technol.

WEBSTER, T.J. (1977). When safety is dangerous. *Loss Prevention and Safety Promotion 2*, p. 15

WECHSLER, M. (1970). *The Radiation-embrittlement of Pressure-vessel Steels and the Safety of Nuclear Reactor Pressure Vessels* (report). Oak Ridge Nat. Lab.

WEDDERBURN, A.A.I. (1992). How fast should the night shift rotate? *Ergonomics*, **35**, 1447

WEDDERBURN, K. (1965). *The Worker and the Law* (Harmondsworth: Penguin)

WEEDON, F.R., HARTZELL, A. and SETTERSTROM, C. (1939). Effect on animals of prolonged exposure to sulfur dioxide. *Contrib. Boyce Thompson Inst.*, **10**, 281

WEEDON, F.R., HARTZELL, A. and SETTERSTROM, C. (1940). Toxicity of ammonia, chlorine, hydrogen cyanide, hydrogen sulfide, and sulfur dioxide gases. V. *Animals*. *Contrib. Boyce Thompson Inst.*, **11**, 365

WEEKMAN, V.W. (1974). Laboratory reactors and their limitations. *AIChE J.*, **20**, 833

WEEKS, R.J. and HODGES, N.W. (1969–70). Component failures in generating plant. *Proc. Instn Mech. Engrs*, **184**(pt 3B), 3

WEEMS, S., BALL, D.H. and GRIFFIN, D.E. (1979). Computer control stabilizes ammonia operations. *Ammonia Plant Safety*, **21**, 39

VAN DE WEERD, J.M. (1971). Electrostatic charge generation during washing of tanks with water sprays – II: Measurements and interpretation. *Static Electrification*, **3**, p. 158

VAN DE WEERD, J.M. (1972). *Can Probe to Cloud Discharges be an Ignition Source in Tanker Explosions? External Rep. AMSR 001672*. Shell Res. NV

VAN DE WEERD, J.M. (1974). Generation and prevention of electrostatic charges in pneumatic transport of plastic powder. *Loss Prevention and Safety Promotion 1*, p. 71

VAN DE WEERD, J.M. (1975). Electrostatic charge generation during tank washing. Spark mechanisms in tanks filled with charged mist. *J. Electrostatics*, **1**, 295

WEERHEIJM, J. (1988). The strength of concrete under dynamic loading. In Groothuizen, T.M. (1988), *op. cit.*, p. 302

VAN WEES, R.M.M. (1989). *Explosion Hazards of Storage Vessels: Estimation of Explosion Effects. Rep. PML 1989–C61*. TNO, Prinz Maurits Lab, Rijswijk

VAN WEES, R.M.M. and MERCX, W.P.M. (1992). Uncertainties of explosion damage criteria. *Loss Prevention and Safety Promotion 7*, vol. 3, 162–1

WEGENG, R.S. and DROST, M.K. (1998). Opportunities for distributed processing using micro chemical systems. *AIChE 1998 Spring Nat. Meet.*, Mar. 1998, New Orleans, LA

WEGENG, R.S., DROST, M.K. and BRENCHLEY, D.L. (1999). Process intensification through miniaturization of chemical and thermal systems in the 21st Century. *3rd Int. Con. on Microreact. Technol.*, February, 1999, Frankfurt, Germany

DE WEGER, D., PIETERSEN, C.M. and REUZEL, P.G.J. (1991). Consequences of exposure to toxic gases following industrial disasters. *J. Loss Prev. Process Ind.*, **4**, 272

WEHMEIER, G., WESTPHAL, F. and FRIEDEL, L. (1994). Pressure relief system design for vapour or two-phase flow? *Hazards*, **XII**, p. 491

WEHNER, J.F. and PHILLIPS, T.D. (1962). The initiation of detonation by the impact of explosively driven solids. *Combustion*, **8**, p. 767

WEHRUM, W.L. (1993). Case study of a hydrogen peroxide related deflagration in a wastewater treatment tank. *Process Safety Prog.*, **12**, 199

WEI, J., FRASER, T.W. and SWARTZLANDER, M.W. (1978). *Structure of the Chemical Processing Industries* (New York: McGraw-Hill)

WEIBEL, E.R. (1963). *Morphometry of the Human Lung* (New York: Academic Press)

WEIBEL, E. (1973). *Morphometry of the Human Lung* (Berlin: Springer Verlag)

WEIBULL, H.R.W. (1968). Pressures recorded in partially closed chambers at explosion of TNT charges. *Ann. N.Y. Acad. Sci.*, **152**(art. 1), 357

WEIBULL, W. (1950). *Explosion of Spherical Charges in Air: Travel Time, Velocity of Front and Duration of Shock Waves. Rep. X-127*. Ballistics Res. Lab., Aberdeen Proving Ground, MD

WEIBULL, W. (1951). A statistical distribution function of wide applicability. *J. Appl. Mech.*, **18**, 293

WEIBY, P. and DICKINSON, K.R. (1976). Monitoring work areas for explosive and toxic hazards. *Chem. Engng*, **83**(Oct. 18), 139

WEICK, K. (1989). Mental models of high reliability systems. *Ind. Crisis Quart.*, **3**(2), 127

WEIDENBAUM, S.S. (ed.) (1980). *Hazardous Chemicals – Spills and Waterborne Transportation* (New York: Am. Inst. Chem. Engrs)

WEIGHELL, S.C. (1980). The performance of relief valves – an equipment users view. *Interflow 80* (Rugby: Instn Chem. Engrs), p. 1

WEIGHTMAN, R.T. (1993). ISO 9000: back to basics. *Hydrocarbon Process.*, **72**(8), 197

WEIHS, D. and SMALL, R.D. (1993). An approximate model of atmospheric plumes produced by large area fires. *Atmos. Environ.*, **27A**, 73

WEIL, G.L. (n.d.). *Nuclear Energy: Promises, Promises* (published by the author, 1730 M Street, Washington, DC)

WEIL, J.C., CORIO, L.A. and BROWER, R.P. (1986). Dispersion of buoyant plumes in the convective boundary layer. *Fifth Joint Conf. on Applications of Air Pollution Meteorology* (Boston, MA: Am. Met. Soc.), p. 335

WEILL, H., GEORGE, R., SCHWARTZ, M. and ZISKIND, M. (1969). Late evaluation of pulmonary function after acute exposure to chlorine gas. *Am. Rev. Resp. Dis.*, **99**, 374

WEIMER, J.C., FAETH, G.M. and OLSON, D.R. (1973). Penetration of vapour jets submerged in subcooled liquids. *AIChE J.*, **19**, 552

WEINBERG, A. (1985). Science and its limits: the regulator's dilemma. *Issues Sci. Technol.*, **II**(1), 59

WEINBERG, A.M. (1963). Criteria of scientific choice. *Minerva*, **1**, 159

WEINBERG, A.M. (1967). *Reflections on Big Science* (Oxford: Pergamon Press)

WEINBERG, A.M. (1976). Is nuclear power acceptable? *Sci. Publ. Policy*, **3**, 455

WEINBERG, A.M. (1972a). Science and trans-science. *Minerva*, **10**, 209. See also *Science*, **177**, 211

WEINBERG, A.M. (1972b). Social institutions and nuclear energy. *Science*, **177**(Jul. 7), 27

WEINBERG, F.J. (1986). *Gaseous Detonations* (London: Academic Press)

WEINBERGER, A.J. (1963). Calculating manufacturing costs. *Chem. Engng*, **70**(Dec. 23), 81

WEINER, A.L. (1966). How to clean and dry compressed air. *Hydrocarbon Process.*, **45**(2), 125

WEINER, A.L. (1974). Dynamic fluid drying. *Chem. Engng*, **81**(Sep. 16), 92

WEINER, R.J. and ROGERS, H.C. (1979). Dynamic failure phenomena in high-strength steels. *J. Appl. Phys.*, **50**, 8025

WEINER, R.T., MERCER, W.L. and GIBSON, R.B. (1972). Steel pipe for the British gas industry. Instn Gas Engrs, London, Commun. See also *Instn Gas Engrs, J.*, **12**, 275

WEINER, S. (1985). Use of programmable devices for safety. *Plant/Operations Prog.*, **4**, 47

WEIR, A. and MORRISON, R.B. (1954). Equilibrium temperature and compositions behind a detonation wave. *Ind. Engng Chem*, **46**, 1056

WEIR, E.D., GRAVENSTINE, G.W. and HOPPE, T.F. (1986). Thermal runaways: problems with agitation. *Plant/Operations Prog.*, **5**, 142

WEIR, R. (1980). High availability instrumentation systems. *Eurochem 1980* (Rugby: Instn Chem. Engrs), paper 6A-4

WEIR, W.T. (1968). Bayesian reliability evaluation. *Ann. Assurance Sci.*, **1**(2), 344

WEIRICK, M.L., FARQUHAR, S.M. and CHISMAR, B.P. (1994). Spill Containment and destruction of a reactive, volatile chemical. *Process Saf. Prog.* **13** (2), (April), 69–71

WEISAETH, L., (1988). Industrial disaster as a human crisis. *Preveting Major Chemical and Related Process Accidents*, p. 585

WEISAETH, L. (1992). Technological disasters: psychological and psychiatric aspects. *Loss Prevention and Safety Promotion 7*, vol. 1, lecture

WEISBERG, E. (1973). Impact of 1972 water laws. *Chem. Engng*, **80**(Jun. 18), 7

WEISBURGER, E.K. (1994). Lowering threshold limit values (TLVs) has additional benefits. *J. Haz. Materials*, **36**, 143

WEISE, E. (1988). Company policy on preventing chemical accidents with special emphasis on managerial aspects. *Preventing Major Chemical and Related Process Accidents*, p. 581

WEISMAN, J. (1968). Engineering design optimization under conditions of risk. Ph.D. Thesis, Univ. of Pittsburgh

WEISMANTEL, G.E. (1969a). Floods catch industry short. *Chem. Engng*, **75**(Jul. 14), 50

WEISMANTEL, G.E. (1969b). A fresh look at intrinsic safety. *Chem. Engng*, **76**(Jun. 30), 132

WEISMANTEL, G.E. (1970). Reducing NO_x emissions. *Chem. Engng*, **77**(Oct. 19), 82

WEISMANTEL, G.E. (1974). Stemming hazardous spills. *Chem. Engng*, **81**(Oct. 14), 56

WEISMANTEL, G.E. (1977). Plant siting barriers grow. *Chem. Engng.* **84**(Jun. 20), 69

WEISMANTEL, G.E. (1978). Underground storage: moving closer to real pay dirt. *Chem. Engng*, **85**(Jan. 16), 81

WEISMANTEL, G.E. (1981). Paints and coatings for CPI plants and equipment. *Chem. Engng*, **88**(Apr. 20), 130

WEISMANTEL, G.E. (1990).'Quality' is a major focus next year and beyond. Hydrocarbon Process., **69**(12), 60

WEISMANTEL, G.E. and PARKINSON, G. (1980).What's under the EPA bubble? *Chem. Engng*, **87**(Mar. 10), 77

WEISMANTEL, G. and RAMIREZ, R. (1978). CPI warm up to infrared. *Chem. Engng*, **85**(Dec. 4), 68

WEISMANTEL, G.E. and RICCI, L. (1979). Partial oxidation in comeback. *Chem. Engng*, **86**(Oct. 8), 57

WEISS, G. (ed.) (1981). *Hazardous Chemicals Data Book* (Park Ridge, NJ: Noyes)

WEISS, S.M. and KULIKOWSKI, C.A. (1979). EXPERT: A system for developing consultation models. *Proc. Sixth Int. Joint Conf. on Artificial Intelligence,* Tokyo, Japan, p. 942

WEISS, S.M. and KULIKOWSKI, C.A. (1981). *The EXPERT and CASNET Projects. Machine Intelligence, Infotech State of the Art Rep. 9*

WEISS, S.M. and KULIKOWSKI, C.A. (1983). *A Practical Guide to Designing Expert Systems* (London: Chapman & Hall)

WEISS, S.M., KULIKOWSKI, C.A., AMAREL, S. and SAFIR, A. (1977). A model-based consultation system for the long-term management of glaucoma. *Proc. Fifth Int. Joint Conf. on Artificial Intelligence* (San Mateo, CA: Morgan- Kaufmann), p. 826

WEISS, S.M., KULIKOWSKI, C.A. and SAFIR, A. (1978). A model-based method for computer-aided medical decision-making. *Artificial Intelligence*, **11**, 145

WEISSBACH, D. (1976). Methods for the handling of toxic intermediates generated in the laboratory and the pilot plant. *Prevention of Occupational Risks*, **3**, p. 497

WEISSMAN, C. (1967). *LISP 1.5 Primer* (Belmont, CA: Dickenson)

WEISSELBERG, E. (1967). Troubleshooting chemical solids handling. *Chem. Engng Prog.*, **63**(11), 72

WEISWEILER, F.G. and SERGEEV, G.N. (1987). *Nondestructive Testing of Large-Diameter Pipe for Oil and Gas Transmission Lines* (Weinheim: VCH)

WEITZMAN, L. (1990). Factors for selecting appropriate solidification/stabilization methods. *J. Haz. Materials*, **24**(2/3), 157

WEIZENBAUM, J. (1976). *Computer Power and Human Reason: From Judgment to Calculation* (San Francisco, CA: W.H. Freeman)

WELBOURNE, D. (1965). Data processing and control at Wylfa nuclear power station. *Advances in Automatic Control* (London: Instn. Mech. Engrs), p. 92

WELBOURNE, D. (1968). Alarm analysis and display at Wylfa nuclear power station. *Proc. IEE*, **115**, 1726

WELBOURNE, D. (1974). Computers for reactor safety. *Nucl. Engng Int.*, **946**

WELCH, D.L. (1992). Human factors in nuclear power and process safety. *Hazard Identification and Risk Analysis, Human Factors and Human Reliability in Process Safety*, p. 353

WELCH, R.I., MARMELSTEIN, A.D. and MAUGHAN, P.M. (1972). An aerial surveillance spill prevention program. *Control of Hazardous Material Spills 1*, p. 137

WELD, D. (1987). Comparative analysis. *Proc. IJCAI*, **959**

WELD, D. (1988a). Exaggeration. *Proc. AAAI,* **291**

WELD, D. (1988b). *Theories of Comparative Analysis. MIT Rep. AI–TRI–35.* Massachusetts Inst. Technol.

WELD, D.S. and DE KLEER, J. (eds) (1990). *Readings in Qualitative Reasoning about Physical Systems* (San Mateo, CA: Morgan-Kaufman)

WELDING INSTITUTE (1969). *Significance of Defects in Welds* (Cambridge)

WELDING INSTITUTE – see also Appendix 28

WELDING, T.V. (1984). Operational experience with total containment systems. *The Protection of Exothermic Reactors and Pressurised Storage Vessels* (Rugby: Instn Chem. Engrs), p. 251

WELDON, G.E. (1971). An insurance introductory view. *Loss Prevention,* **5**, p. 109

WELDON, G.E. (1972). Damage limiting construction for chemical plants. *Loss Prevention,* **6**, p. 105

WELFORD, A.T. (1976). Ergonomics: where have we been and where are we going: I. *Ergonomics,* **19**, 275

WELFORD, A.T. (1978). Mental work-load as a function of demand, capacity, strategy and skill. *Ergonomics,* **21**, 151

WELKER, J.R. (1965). The effect of wind on uncontrolled buoyant diffusion flames from burning liquid. Ph.D. Thesis, Univ. of Oklahoma

WELKER, J.R. (ca. 1973). *Radiant Heating from LNG Fires. Proj. IS-3–1, paper F-8–13.* Am. Gas Ass.

WELKER, J.R. (1982). Radiation from LPG fires. *Combustion Inst., Western States Section Mtg,* paper WSCI 82–38

WELKER, J.R. and SCHORR, H.P. (1979). LNG plant experience data base. *Am. Gas Ass., Transmission Conf.,* paper 79– T-21

WELKER, J.R. and SLIEPCEVICH, C.M. (1965). *The Effect of Wind on Flames. Tech. Rep. 2, Contract OCD-OS-62-89.* Res. Inst., Univ. of Oklahoma, Norman, OK

WELKER, J.R. and SLIEPCEVICH, C.M. (1966). Bending of wind-blown flames from liquid pools. *Fire Technol.,* **2**, 127

WELKER, J.R. and SLIEPCEVICH, C.M. (1970). *Susceptibility of Potential Target Components to Defeat by Thermal Action. Rep. OURI-1578–FR.* Res. Inst., Univ. of Oklahoma, Norman, OK

WELKER, J.R., PIPKIN, O.A. and SLIEPCEVICH, C.M. (1965). The effect of wind on flames. *Fire Technol.,* **1**(2), 122

WELKER, J.R., WESSON, H.R. and SLIEPCEVICH, C.M. (1969). LNG spills: to burn or not to burn. *Am. Gas Ass., Operating Section, Distribution Conf.,* paper 69–D-23

WELKER, J.R., WESSON, H.R. and BROWN, L.E. (1974). Use foam to disperse LNG vapors. *Hydrocarbon Process.,* **53**(2), 119

WELKER, J.R. *et al.* (1976). *Fire Safety Aboard LNG Vessels. Rep. AD/A-303 619.* Nat. Tech. Inf. Service, Springfield, VA

WELKER, J.R., MARTINSEN, W.E. and JOHNSON, D.W. (1986). Effectiveness of fire control agents for hexane fires. *Fire Technol.,* **22**, 329

WELLAND, R. (1981). *Decision Tables and Computer Programming* (London: Heyden)

WELLMAN, F. (1975). Some fundamental considerations in the collection and interpretation of accident data. In Hendry, W. (1975), *op. cit.,* paper C4

WELLS, A.A. (1961). Unstable crack propagation in metals – cleavage and fast fracture. *Cranfield Crack Propagation Symp. 1,* p. 210

WELLS, A.A. (1969). Crack opening displacement from elastic–plastic analyses of externally notched tension bars. *Engng Fracture Mech.* **1**(3), 399

WELLS, C. (1991). Inquests, inquiries and indictments: the official reception of death by disaster. *Legal Studies,* **11**(1), 71

WELLS, C. (1992). *Corporations and Criminal Responsibility* (Oxford: Oxford Univ. Press)

WELLS, C. (1993). Are you flexible in selecting mechanical seals? *Chem. Engng Prog.,* **89**(4), 58

WELLS, D.D. and WHITE, J.D. (1980). Shop or field fabrication? Transportation is often the key. *Hydrocarbon Process.,* **59**(1), 153

WELLS, G.L. (1980). *Safety in Process Plant Design* (London: Godwin)

WELLS, G.L. (1981). Safety reviews and plant design. *Hydrocarbon Process.,* **60**(1), 241

WELLS, G.L. and PHANG, C. (1992). A modified form of hazop with emphasis on operability. *Loss Prevention and Safety Promotion 7,* vol. 3, 139–1

WELLS, G.L. and ROSE, L.M. (1986). *The Art of Chemical Process Design* (Amsterdam: Elsevier).

WELLS, G.L., SEAGRAVE, C.J. and WHITEWAY, R.N.C. (1976). *Flowsheeting for Safety* (London: Instn Chem. Engrs)

WELLS, G.L., SEAGRAVE, C.J. and WHITEWAY, R.N.C. (1977). A sample safety check for use during plant design. *Design Developments Related to Safety, Costs and Solid Systems* (London: Instn Chem. Engrs), paper A-5–1

WELLS, G.L., HURST, N.W. and REEVES, A.B. (1990). A checklist for identifying the root causes of faults and errors on chemical plant. *Safety and Loss Prevention in the Chemical and Oil Processing Industries,* p. 175

WELLS, G.L., PHANG, C. and REEVES, A.B. (1991). Hazcheck and the development of major incidents. *Hazards,* **XI**, p. 305

WELLS, G.L., PHANG, C., WARDMAN, M. and WHETTON, C. (1992). Incident scenarios: their identification and evaluation. *Process Safety Environ.,* **70B**, 179

WELLS, G., WARDMAN, M. and WHETTON, C. (1993). Preliminary safety analysis. *J. Loss Prev. Process Ind.,* **6**, 47

WELLS, G., PHANG, C. and WARDMAN, M. (1994). Improvements in process safety reviews prior to HAZOP. *Hazards,* **XII**, p. 301

WELSH, C.R. (1977). A risk assessment method for deriving a test frequency for the pressure relief system on a marine nuclear reactor. *First Nat. Conf. on Reliability. NRC4/12.* (Culcheth, Lancashire; UK Atom. Energy Auth.)

WELSH, C.R. and DEBENHAM, A.A. (1986). A rationale and procedure for pursuing completeness in probabilistic safety studies. *Reliab. Engng,* **14**, 51

WELSH, C.R. and LUNDBERG, P.K. (1980). Methods used to assess the safety and reliability of a nuclear reactor rod control system. *Reliab. Engng,* **1**, 59

WELSH, S. (1992). Assessment and management of risks to the environment. *Major Hazards Onshore and Offshore,* p. 85

WELSH, S. (1993). Assessment and management of risks to the environment. *Process Safety Environ.,* **71B**, 3

WELTGE, C. and CLEMENT, H. (1964). How Conoco trains operators with PILOT. *Hydrocarbon Process.,* **43**(9), 253

WELTER, T.R. (1991). The quest for safe substitutes. *Ind. W.,* (Feb. 4), 38–43

WELTY, J.R., WICKS, C.E. and WILSON, R.E. (1976). *Fundamentals of Momentum, Heat and Mass Transfer, 2nd ed.* (New York: Wiley)

WENDER, P.A., HANDY, S.T. and WRIGHT, D.L. (1997). Towards the ideal synthesis. *Chem. Ind.,* (Oct. 6), 765–9

WENDES, M.J. (1992). One company's experience of formal safety assessment and preparation of offshore safety cases. *Major Hazards Onshore and Offshore,* p. 49

WENDLANDT (1986). In *Chemical Analysis,* vol. 19 (ed. P.J. Irvine and J.D. Winefordner) (New York: Wiley)

WENDLANDT, W.W. (1974). *Thermal Methods of Analysis,* 2nd ed. (New York: Wiley)

WENTZEL, R., FOUTCH, R.H., HARWARD, W.E. and JONES, W.E. (1981). *Restoring Hazardous Spill-damaged Areas: Technique Identification/Assessment. Rep. EPA–600/2– 81–208.* Environ. Prot. Agency, Cincinnati, OH

WERCHAN, R. (1976). Noise control in ammonia plants. *Ammonia Plant Safety,* **18**, p. 87

WERCHAN, R.E. and BRUCE, R.D. (1973). Process plant noise can be controlled. *Chem. Engng Prog.,* **69**(10), 51

WERSBORG, B.L., HOWARD, J.B. and WILLIAMS, G.C. (1973). Physical mechanisms in carbon formation in flames. *Combustion,* **14**, p. 929

WERTHENBACH, H.G. (1971a). Flame lengths in fires of liquid storage tanks. *Verfahrenstechnik,* **5**(3), 115

WERTHENBACH, H.G. (1971b). Spread of flames on cylindrical tanks for hydrocarbon fluids. *Gas und Ergas,* **112**(8), 383

WERTHENBACH, H.G. (1973). Fires of oil products in tanks – experiments and model. *VDI-Zeitschrift,* **115**(5), 383

WESSELS, E.C. (1980). Rating techniques for risk assessment. *Fire Int.,* **67**, 80

WESSON AND ASSOCIATES INC. (1977). *Compilation of Wesson and Associates Key Personnel Major Experiences in LNG Technology–Safety–Fire Protection, Industrial LNG Fire Training School and Comparison of NFPA No. 59 with the Proposed OPSO LNG Facility Federal Safety Regulations* (report). (Norman, OK)

WESSON, H.R. (1970). The piloted ignition of wood by radiant heat. Ph.D. Thesis, Univ. of Oklahoma, Norman, OK

WESSON, H.R. (1972). Studies of the effect of particle size on the flow characteristics of dry chemical. *Fire Technol.,* **8**(3), 173

WESSON, H.R. (1976). Propane tank failure mechanisms. *Butane–Propane News, Jul.*

WESSON, H.R., WELKER, J.R. and SLIEPCEVICH, C.M. (1971). The piloted ignition of wood by thermal radiation. *Combust. Flame,* **16**, 303

WESSON, H.R., WELKER, J.R. and BROWN, L.E. (1972). Control LNG-spill fires. *Hydrocarbon Process.,* **51**(12), 61

WESSON, H.R., WELKER, J.R., BROWN, L.E. and SLIEPCEVICH, C.M. (1973a). Fight LNG fires with foam. *Hydrocarbon Process.,* **52**(10), 165

WESSON, H.R., WELKER, J.R., BROWN, L.E. and SLIEPCEVICH, C.M. (1973b). Fight LNG spill fires with dry chemicals. *Hydrocarbon Process.,* **52**(11), 234

WESSON, H.R., WELKER, J.R. and BROWN, L.E. (1975). Control of LNG spill fires on land. *Advances in Cryogenic Engineering,* **20**, 151

WEST, A.S. (1979). The impact of TSCA. *Chem. Engng Prog.,* **75**(4), 26

WEST, A.S. (1982). Risk–benefit trade-offs under the Toxic Substances Control Act. *Chem. Engng Prog.,* **78**(8), 88

WEST, A.S. (1986). Safety evaluation of chemicals: the regulatory framework. *Plant/Operations Prog.,* **5**, 11

WEST, A.S. (1991). Worker hazard communications: the regulatory framework. *Plant/Operations Prog.,* **10**, 13

WEST, A.S. (1993a). Chemical reactivity evaluation: the CCPS program. *Process Safety Prog.,* **12**, 55

WEST, A.S. (1993b). Safety testing of new chemicals: the regulatory framework world-wide. *Process Safety Prog.,* **12**, 34

WEST, B. and CLARK, J.A. (1974). Operator interaction with a computer controlled distillation column. In Edwards, E. and Lees, F.P. (1974), *op. cit.,* p. 206

WEST, E.J., DIEZ, L.J., KENNEDY, A.B. and WILLIAMS, G.W. (1978). Life support system for ammonia plant work. *Ammonia Plant Safety,* **20**, p. 140

WEST, H. and BROWN, L.E. (1977). Analyze fire protection systems. *Hydrocarbon Process.,* **56**(8), 89

WEST, H.H., BROWN, L.E. and WELKER, J.R. (1975). Vapour dispersion, fire control and fire extinguishment for LNG spills. *Transport of Hazardous Cargoes,* **4**

WEST, H.H., PFENNING, D.B. and BROWN, L.E. (1980). A systems approach to LNG fire safety. In *Gastech 79,* p. 193

WEST, H.W.H. (1973). A note on the resistance of glass windows to pressures generated by gaseous explosions. *Proc. Br. Ceram. Soc.,* **21**, 213

WEST, H.W.H., HODGKINSON, H.R. and WEBB, W.F. (1973). The resistance of brick walls to lateral loading. *Proc. Br. Ceram. Soc.,* **21**, 141

WEST, J.B. (1965). *Ventilation/Blood Flow and Gas Exchange* (Oxford: Blackwell)

WEST, J.B. (ed.) (1977). *Bioengineering Aspects of the Lung* (New York: Dekker)

WEST, J.M. (1980). *Basic Corrosion and Oxidation* (Chichester: Ellis Horwood)

WESTBROOK, C.K. (1982a). Chemical kinetics in gaseous detonations. In Lee, J.H.S. and Guirao, C.M. (1982), *op. cit.,* p. 189

WESTBROOK, C.K. (1982b). Chemical kinetics of hydrocarbon oxidation in gaseous detonations. *Combust. Flame,* **46**, 191

WESTBROOK, C.K. (1982c). Inhibition of hydrocarbon oxidation in laminar flames and detonations by halogenated compounds. *Combustion,* **19**, p. 127

WESTBROOK, C.K. and DRYER, F.L. (1981). Chemical kinetics and modeling of combustion processes. *Combustion,* **18**, p. 749

WESTBROOK, C.K. and PITZ, W.J. (1984). Prediction of laminar flame properties of propane–air mixtures. In Bowen, J.R. *et al., op, cit.,* p. 211

WESTBROOK, C.K., and URTIEW, P.A. (1982). Chemical kinetic prediction of critical parameters in gaseous detonations. *Combustion,* **19**, p. 615

WESTBROOK, C.K., PITZ, W.J. and URTIEW, P.A. (1984). Chemical kinetics of propane oxidation in gaseous detonations. In Bowen, J.R. *et al.* (1984b), *op. cit.,* p. 151

WESTBROOK, E.A., HELLWIG, K. and ANDERSON, R.C. (1955). Self-combustion of acetylene. Pt II, Reactions in flame propagation. *Combustion,* **5**, p. 631

WESTBROOK, G.W. (1974). The bulk distribution of toxic substances: a safety assessment of the carriage of liquid chlorine. *Loss Prevention and Safety Promotion 1,* p. 197

WESTENBERG, A.A. (1957). Present status of information on transport properties applicable to combustion research. *Combust. Flame,* **1**, 346

WESTENBERG, A.A. and FAVIN, S. (1960). The theory of a spherical premixed laminar flame. *Combust. Flame,* **4**, 161

WESTERBERG, A.W. (1980). A review of process synthesis. *Computer Applications to Chemical Engineering* (ed. R.G. Squires and G.V. Reklaitis) (Washington, DC: Am. Chem. Soc.)

WESTERBERG, A.W. (1985). The synthesis of distillation-based separation systems. *Comput. Chem. Engng,* **9**, 421

WESTERBERG, A.W. (1989). Synthesis in engineering design. *Comput. Chem. Engng,* **13**, 365

WESTERBERG, A.W. (1993). Human aspects: redefining the role of chemical engineers. *Chem. Engng Prog.,* **89**(11), 60

WESTERBERG, A.W. and CHIEN, H.H. (1984). *Foundations of Computer-Aided Process Design* (Amsterdam: Elsevier)

WESTERBERG, A.W., HUTCHINSON, H.P., MOTARD, R.L. and WINTER, P. (1979). *Process Flowsheeting* (Cambridge: Cambridge Univ. Press)

WESTERGAARD, H.M. (1939). Bearing pressures and cracks. *J. Appl. Mech.*, **6**, A49

WESTERLING, D. and KILBOM, A. (1981). Physical strain in the handling of gas cylinders. *Ergonomics*, **24**, 623

WESTERMAN, M. (1981). Reliability in plant commissioning and operation. *Reliable Production in the Process Industries* (Rugby: Instn Chem. Engrs), p. 117

WESTERN GEOPHYSICAL (1994). Gulf of Mexico 3–D surveys. *Offshore*, Mar.

WESTERN OIL AND GAS ASSOCIATION (1977). *Hydrocarbon Emissions from Floating Roof Storage Tanks* (Los Angeles, CA)

WESTERTERP, K.R., VAN SWAAIJ, W.P.M. and BEENACKERS, A.A.C.M. (1984). *Chemical Reactor Design and Operation* (Chichester: Wiley)

WESTINE, P.S. (1972). R–W plane analysis for vulnerability of targets to air blast. *Shock and Vibration Bull.*, **42**(pt 5), p. 173

WESTINE, P.S. (1977). *Constrained Secondary Fragment Modeling.* (Final report of US Army Ballistic Res. Lab.) Southwest Res. Inst., San Antonio, TX

WESTINE, P.S. and BAKER, W.E. (1974). Energy selections for predicting deformations in blast loaded structures. *Dept of Defense, Explosives Safety Board, Minutes of Sixteenth Explosives Safety Seminar*

WESTINGHOUSE ELECTRIC COMPANY (1982). *Sizewell B Probabilistic Safety Study* (Monroeville)

WESTON, A.F. (1984). Obtaining reliable priority-pollutant analyses. *Chem. Engng*, **91**(Apr. 30), 54

WESTON, C.W. (1992). Ammonium nitrate. *Kirk–Othmer Encyclopaedia of Chemical Technology*, 4th ed., vol. 2 (New York: Wiley), p. 698

WESTON, F.C. (1974a). Design and select pneumatic conveying systems. *Hydrocarbon Process.*, **53**(3), 75

WESTON, F.C. (1974b). Improve solids handling instrumentation systems. *Hydrocarbon Process.*, **53**(3), 87

WESTON, H.B. (1975). The efficient use of fuel. *Plant Engr*, **19**(4), 33

WESTON, H.B. (1979). Energy savings in plant utilities and services. *Chem. Ind.*, Sep. 1, 589

WESTON, H.B. (1980). The role of energy targets. *Plant Engr*, **24**(7), 17

WESTON, L.M. (1983). Use of expert opinion in probabilistic risk assessment. In Swain, A.D. and Guttmann, H.E. (1983), *op. cit.*, Ch. 8

WESTON, R.E. and KAREL, L. (1946). An application of the dosimetric method for biologically assaying inhaled substances: the determination of the retained lethal dose, percentage retention and respiratory response in dogs exposed to different concentrations of phosgene. *J. Pharmac. Exp. Ther.*, **88**, 195

WESTON, R.E. and KAREL, L. (1947). An adaptation of the dosimetric method for use in smaller animals: the retained lethal dose and respiratory response in normal unanaesthetised rhesus monkeys (Macaca mulatta) exposed to phosgene. *J. Ind. Hyg. Toxicol.*, **29**, 29

WETHERILL, F.E. and WALLSGROVE, C.S. (1986). Operator training for startup. *Plant/Operations Prog.*, **5**, 78

WETHEROLD, R.G., ROSEBROOK, D.D. and TICHENOR, B.A. (1981). *Assessment of Atmospheric Emissions from Petroleum Refining. Rep. EPA–600/S2–80–075.* Ind. Environ. Res. Lab., Environ. Prot. Agency, Research Triangle Park, NC

WETHEROLD, R.G., HARRIS, G.E., STEINMENTZ, J.L. and KAMAS, J.W. (1983). Economics of controlling fugitive emissions. *Chem. Engng Prog.*, **79**(11), 43

WETT, T. (1985). Capacity down, production up: Rx for ethylene's 84 outlook. *Oil Gas J.*, **82**(Sep. 3), 55

WETTERICH, W. (1982). Optical flame barriers. *The Control and Prevention of Dust Explosions* (London: Oyez/IBC), paper 10

WHALEN, J.J. (1962). Select and apply gaskets effectively. *Chem. Engng*, **69**(Oct. 1), 83

WHALEN, J.J. (1963). Selecting and maintaining packings. *Chem. Engng*, **70**(Nov. 11), 256

WHALEY, H. and LEE, G.K. (1982). The behaviour of buoyant plumes in neutral and stable conditions in Canada. *Atmos. Environ.*, **16**, 2555

WHALLEY, D. (1989). *Piper Alpha Public Inquiry. Transcript, Days 84, 85*

WHALLEY, S. (1987). What can cause human error? *Chem. Engr, Lond.*, **433**, 37

WHALLEY, S. (1989). Displays made friendly. *Chem. Engr, Lond.*, **459**, 49

WHALLEY, S.P. and KIRWAN, B. (1989). An evaluation of five human error identification techniques. *Loss Prevention and Safety Promotion 6*, p. 31.1

WHALLEY, S. and LIHOU, D. (1988). Management factors and system safety. In Sayers, B.A. (1988), *op. cit.*, p. 172

WHALEY, T.P. and LONG, G.M. (1982). Pipelines. *Kirk–Othmer Encyclopaedia of Chemical Technology*, 3rd ed. (New York: Wiley), vol. 17, p. 906

WHARRY, D. and HIRST, R. (1974). *Fire Technology: Chemistry and Combustion* (Leicester: Instn Fire Engrs)

WHARTON, R.K. (1990). A medium scale test method for assessing the fire hazard of flammable solid materials. *J. Loss Prev. Process Ind.*, **3**(4), 349

WHEATCROFT, S. (1964). Licensing British air transport. *J. R. Aeronaut. Soc.*, **68**(639), 171

WHEATLEY, A.D. (1982). Effluent treatment. *Chem. Ind.*, Aug. 7, 512

WHEATLEY, C.J. (1986). Factors affecting cloud formation from releases of liquefied gases. *Refinement of Estimates of the Consequences of Heavy Toxic Vapour Release* (Rugby: Instn Chem. Engrs), p. 5.1

WHEATLEY, C.J. and HUNNS, D.M. (1981). Exact hazard rate formulae for r-from-n protective configurations. *Third Nat. Reliability Conf.*, p. B2/2/1

WHEATLEY, C.J. and PRINCE, A.J. (1987). Translational cloud speeds in the Thorney Island trials: mathematical modelling and data analysis. *J. Haz. Materials*, **16**, 185

WHEATLEY, C.J., PRINCE, A.J. and BRIGHTON, P.W.M. (1985). Comparison between data from the Thorney Island Heavy Gas Trials and predictions of simple dispersion models. *NATO/CCMS Int. Tech. Mtg on Air Pollution*, St Louis

WHEATLEY, C.J., BRIGHTON, P.W.M. and PRINCE, A.J. (1986). Comparison between data from heavy gas dispersion experiments at Thorney island and predictions of simple models. In de Wispelaere, C., Schiermeier, F.A. and Gillani, N.V. (1986), *op. cit.*, p. 707

WHEATLEY, W.C. (1971). Latest practices in the field of signalling and communications in hazardous environments. *Electrical Safety*, **1**, p. 140

WHEATON, E.L. (1948). *Texas City Remembers* (San Antonio, TX: Naylor)

WHEELER, A.J. and MOODY, F.J. (1979). A method for computing transient pressures and forces in safety relief valve discharge lines. In Haupt, R.W. and Meyer, R.A. (1979), *op. cit.*, p. 169

WHEELER, D.A. (1988). Atmospheric dispersal and deposition of radioactive material from Chernobyl. *Atmos. Environ.*, **22**, 853

WHEELER, D.B. *et al.* (1977). Fault tree analysis using bit manipulation. *IEEE Trans. Reliab.*, **R-26**, 95

WHEELER, E.M. (1976). Ammonia – food – safety. *Ammonia Plant Safety*, **18**, p. 1

WHEELER, R.N. and SUTHERLAND, M.E. (1975). Control of intransit VCM. *Chem. Engng Prog.*, **71**(9), 48

WHEELER, R.V. (1920). The ignition of gases. I, Ignition by the impulsive electrical discharge. Mixtures of methane and air. J. *Chem. Soc.*, **117**, 903

WHEELER, R.V. (1924). The ignition of gases. III, Ignition by the impulsive electrical discharge. Mixtures of paraffins with air. *J. Chem. Soc.*, **125**, 1858

WHEELER, R.V. (1935a). Explosion of carbonaceous dust. *Ind. Chem.*, **11**(122), 110

WHEELER, R.V. (1935b). Ignition of firedamp by compression. *Trans. Instn Min. Engrs*, **88**, 447; 459

WHEELER, T.A., HORA, S.C., CRAMOND, W.R. and UNWIN, S.D. (1989). *Analysis of Core Damage Frequency from Internal Events: Expert Judgment Elicitation. Rep. NUREG/CR-4550.* Nucl. Regul. Comm., Washington, DC

WHEELER, W.H. (1980). Chemical and engineering aspects of low NO_x concentration. *Chem. Engr, Lond.*, **362**, 693

WHELAN, J. (1977). Qatar fire. The big one everyone was afraid of. *Middle East Current Dig.*, Apr. 15

WHELAN, T.W. and THOMSON, S.J. (1975). Reduce relief system costs. *Hydrocarbon Process.*, **54**(8), 83

WHENRAY, J.J. (1972). *International Requirements for the Periodic Inspection of Pressure Vessels. Paper C56/72.* Instn Mech. Engrs, London

WHERRY, R.J. (1938). Orders for the presentation of pairs in the method of paired comparisons. *J. Exp. Psychol.*, **23**, 651

WHETTON, C. (1993). Maintainability and its application to process plant. *Process Safety Prog.*, **12**, 158

WHETTON, C.P. (1993). Sneak analysis of process systems. *Process Safety Environ.*, **71B**, 169

WHICKER, F.W. and SCHULTZ, V. (1982). *Radioecology: Nuclear Energy and the Environment,* vols 1–2 (Boca Raton, FL: CRC Press)

WHILLIER, A. (1967). Pump efficiency determination in chemical plant from temperature measurements. *Ind. Engng Chem. Proc. Des. Dev.*, **7**, 194

WHINCUP, M.H. (1963). *The General Principles of Employers' Liability at Common Law under the Factories Act 1961* (IIS)

WHINCUP, M.H. (1968). *Industrial Law* (London: Heinemann)

WHINCUP, M.H. (1976). *Modern Employment Law: A Guide to Job Security and Safety at Work* (London: Heinemann)

WHIPP, B.J. and WIBERG, D.M. (eds) (1983). *Modelling and Control of Breathing* (New York: Elsevier Biomedical)

WHISTON, J. (1977). When you commission a plant. *Hydrocarbon Process.*, **56**(11), 410; **56**(12), 169

WHISTON, J. (1991). Responsible Care – another dimension of the quality process. *Chem. Engr, Lond.*, **503**, 35

WHISTON, J. (ed.) (1991). *Safety in Chemical Production* (Oxford: Blackwell)

WHISTON, J. (1993). Corporate SHE policy and implementation of international SHE standards. *Health, Safety and Loss Prevention,* p. 19

WHISTON, J. and EDDERSHAW, B. (1989). Quality and safety – distant cousins or close relatives? *Chem. Engr, Lond.*, **461**, 97

WHITACRE, C.G., GRINER, J.H., MYIRSKI, M.M. and SLOOP, D.W. (1986). *Personal Computer Program for Chemical Hazard Prediction (D2PC).* US Army, Chem. Res. Dev. Center, Aberdeen Proving Ground, MD

WHITAKER, E. (1980). Mississauga's train of disaster. *Fire,* Jun, 15

WHITAKER, F. (1981). Reducing plant downtime by avoiding valve failures. *Plant Engr*, **25**(1), 24

WHITAKER, G.D. (1973a). Statistical reliability models for chemical process plant. *Design for Reliability* (Manchester: Instn Chem. Engrs, North-West Branch)

WHITAKER, G.D. (1973b). Statistical reliability models for chemical process plant. *Chem. Engr, Lond.*, **278**, 471

WHITAKER, G.W. (1993). Contractor hazard identification and control. *Process Safety Prog.*, **12**, 133

WHITAKER, N.R. (1972). Guide to trouble-free plant operation . . . instrumentation. *Chem. Engng*, **79**(Jun. 26), 96

WHITAKER, R. (1984). Onshore modular construction. *Chem. Engng*, **91**(May 28), 81

WHITBREAD, E.G. (1975). The regulation of the conveyance of dangerous goods. *Distribution Conf.* (London: Chem. Ind. Ass.), p. 3

WHITCRAFT, P.K. (1992). High-tech steels to the rescue. *Chem. Engng*, **99**(1), 116

WHITE, A.G. (1922). Limits for the propagation of flame in vapour-air mixtures. Pt 1, Mixtures of air and one vapour at the ordinary temperature and pressure. *J. Chem. Soc.*, p. 1244

WHITE, A.G. (1924). Limits for the propagation of flame in gas–air mixtures. Pt 1, Mixtures of air and one gas at the ordinary temperature and pressure. *J. Chem. Soc.*, 2387

WHITE, A.T. (1984). *Graphs, Groups and Surfaces,* rev. ed. (Amsterdam: North Holland)

WHITE, B. (1980). Safety considerations in the bulk transport of hazardous cargoes with particular reference to the ship/jetty interface. *Loss Prevention and Safety Promotion 3,* vol. 3, p. 1369

WHITE, B. and COOKE, CAPTAIN P. (1983). Training for gas handling operations at terminals. *Gastech 82,* p. 281

WHITE, B.A. (1993). Rethink the role of control valves. *Chem. Engng Prog.*, **89**(12), 31

WHITE, C.S. (1957). *Interior Missile and Dust Hazard. Rep. ITR-1420,* Appendix E. Civil Effects Test Group, US Atom. Energy Comm., Office of Tech. Services, Dept of Commerce, Washington, DC

WHITE, C.S. (1959). *Biological Blast Effects. Rep. TID-5564.* Atom. Energy Comm., Office of Tech. Services, Dept of Commerce, Washington, DC

WHITE, C.S. (1961). *Biological Effects of Blast. Rep DASA-1271.* Defense Atom. Support Agency, Dept of Defence, Washington, DC

WHITE, C.S. (1963). *Tentative Biological Criteria for Assessing Potential Hazards from Nuclear Explosions. Rep DASA-1462.* Defense Atom. Support Agency, Dept of Defence, Washington, DC

WHITE, C.S. (1965). Tentative biological criteria for estimating blast hazards. *Nat. Acad. Sci.–Nat. Res. Coun., Proc. Symp. on Protective Structures for Civilian Populations* (Washington, DC), p. 41

WHITE, C.S. (1968a). *The Nature of the Problems Involved in Estimating the Immediate Casualties from Nuclear Explosions. Rep. LF-1242-1.* Lovelace Foundation, Albuquerque,

NM, Rep. See also revised edition, *Rep. CEX-71.1,* Civil Effects Test Group, Atom. Energy Comm., Office of Tech. Services, Dept of Commerce, Washington, DC

WHITE, C.S. (1968b). *The Scope of Blast and Shock Biology and Problem Areas in Relating Physical and Biological Parameters. Rep DASA-1856.* Defense Atom. Support Agency, Dept of Defence, Washington, DC. See also *Ann. NY Acad. Sci, 1968,* **152**(art. 1), 89

WHITE, C.S. (1971). *The Nature of the Problems involved in Estimating the Immediate Casualties from Nuclear Explosions,* rev. version. *Rep. CEX-71.1.* Civil Effects Test Group, Atom. Energy Comm., Office of Tech. Services, Dept of Commerce, Washington, DC. See also White, C.S. (1968a)

WHITE, C.S. (1974). Selected biomedical effects. *Defense Civil Preparedness Agency, Proc. DCPA All Effects Research Contractors Conf.* (Washington, DC,), p. 33

WHITE, C.S. and RICHMOND, D.R. (1960). Blast biology. In *Clinical Cardiopulmonary Physiology* (ed. B.L. Gordon and R.C. Kory) (New York: Grune&Stratton), p. 974

WHITE, C.S., BOWEN, I.G. and RICHMOND, D.R. (1964). *A Comparative Analysis of some of the Immediate Environmental Effects at Hiroshima and Nagasaki. Rep. CEX-63.7.* Civil Effects Test Operations, US Atom. Energy Comm., Office of Tech. Services, Dept of Commerce, Washington, DC. See also *Health Phys.,* 1964, **10**, 89

WHITE, C.S., BOWEN, I.G. and RICHMOND, D.R. (1965). *Biological Tolerance to Air Blast and Related Biomedical Criteria. Rep. CEX-65.4.* Civil Effects Test Operations, US Atom. Energy Comm., Office of Tech. Services, Dept of Commerce, Washington, DC

WHITE, C.S., BOWEN, I.G. and RICHMOND, D.R. (1970). The relationship between eardrum rupture and blast-induced pressure variations. *Space Life Sci.,* **2**, 158

WHITE, C.S., JONES, R.K., DAMON, E.G., FLETCHER, E.R. and RICHMOND, D.R. (1971). *The Biodynamics of Air Blast. Rep. DNA-2738-T.* Defense Nucl. Agency, Dept of Defense, Washington, DC. See also *NATO AGARD Conf. Proc. on Linear Acceleration of Impact Type,* p. 14–1

WHITE, C.S., JONES, R.K., DAMON, E.G., FLETCHER, E.R. and RICHMOND, D.R. (1972). The biodynamics of air blast. *Nucl. Sci. Abstracts,* **26**, 1179

WHITE, C.S. and RICHMOND, D.R. (1959). *Blast biology. Rep. TID-5764.* US Atom. Energy Comm., Office of Tech. Services, Dept of Commerce, Washington, DC

WHITE, D.J.D. and WELL, D.D. (1979). Move vessels from shop to site. *Hydrocarbon Process.,* **58**(5), 195

WHITE, E.N. (1973). *Maintenance Planning, Control and Documentation* (Epping: Gower)

WHITE, E.S. (1971). A case of near fatal ammonia gas poisoning. *J. Occup. Med.,* **13**, 550

WHITE, G.F. (ed.) (1974). *Natural Hazards* (New York: Oxford Univ. Press)

WHITE, H.J. (1963). *Industrial Electrostatic Precipitation* (Oxford: Pergamon Press)

WHITE, H.L. (1986). *Introduction to Industrial Chemistry* (New York: Wiley)

WHITE, I.F., DIBBLE, G.J., HARKER, J.E. and O'BRIEN (1986). Safety considerations in the design of chlor-alkali plants. In Wall, K. (1986), *op. cit.,* p. 97

WHITE, J. (1927). The effect of pressure on the limits of propagation of flame in ether–air. *J. Chem. Soc.,* **498**

WHITE, L.A. (1963). Diagnose equipment troubles with gamma radiation. *Hydrocarbon Process.,* **42**(4), 153

WHITE, L.J. (1972). State laws and enforcement. *Chem. Engng,* **79**(May 8), 13

WHITE, L.J. (1973a). Are air standards workable? *Chem. Engng,* **80**(Aug. 20), 80

WHITE, L.J. (1973b). Occupational Safety and Health Act. *Chem. Engng,* **80**(Feb. 26), 13

WHITE, L.J. (1974). The laws and their enforcement. Progress amid problems. *Chem. Engng,* **81**(Oct. 21), 7

WHITE, M.C., INFANTE, P.F. and CHU, K.C. (1982). A quantitative estimate of leukaemia mortality associated with occupational exposure to benzene. *Risk Analysis,* **2**, 195

WHITE, M.H. (1972). Surge control for centrifugal compressors. *Chem. Engng,* **79**(Dec. 25), 54

WHITE, M.J. (1969). Trade Unions and safety. In Handley, W. (1969), *op. cit.,* p. 428

WHITE, M.J. (1977). Safety administration. In Handley, W. (1969), *op. cit.,* p. 442

WHITE, P.A.F. and SMITH, S.E. (1962). *Inert Atmospheres* (London: Butterworths)

WHITE, R.F. (1982). Analysis of risk. In Green, A.E. (1982b), *op. cit.,* p. 31

WHITE, R.F. (1986). A suggested method for the treatment of human error in the assessment of major hazards. *Reliab. Engng,* **15**, 171

WHITE, R.L. (1987). On-line troubleshooting of chemical plants. *Chem. Engng. Prog.,* **83**(5), 33

WHITE, R.P. (1934). *Dermatergoses or Occupational Affections of the Skin* (London: H.K. Lewis)

WHITE, R.W. and BOTSFORD, N.B. (1963). *Containment of Fragments from a Runaway Reactor. Rep. SRIA-113.* Stanford Res. Inst., Stanford, CA

WHITE, W.C.J. (1985). The Chemical Emergency Agency Service. *Chem. Ind.,* Jan. 21, 51

WHITEBLOOM, S.W. (1976). Oil and hazardous materials cannot be effectively removed from large bodies of waters or busy river systems unless... *Control of Hazardous Material Spills,* **3**, p. 123

WHITEHEAD, L.W. (1987). Planning considerations for industrial plants emphasizing occupational and environmental health and safety issues. *Appl. Ind. Hyg.,* **2**

WHITEHORN, V.J. and BROWN, H.W. (1967). How to handle a safety inspection. *Hydrocarbon Process.,* **46**(4), 125; **46**(5), 227

WHITEHORN, V.J. and BROWN, H.W. (1973). How to handle a safety inspection. In Vervalin, C.H. (1973a), *op. cit.,* p. 432

WHITEHOUSE, H.B. (1985). IFAL – a new risk analysis tool. *The Assessment and Control of Major Hazards* (Rugby: Instn Chem. Engrs), p. 309

WHITEHOUSE, R.C. (1990). Ball valves and their development. *Plant Engr,* Nov./Dec., 38

WHITELEY, J.R. and DAVIS, J.F. (1992). Knowledge-based interpretation of sensor patterns. *Comput. Chem. Engng,* **16**, 329

WHITELEY, R.A. (1989). Integrity testing for halon. *Fire Surveyor,* **18**(5), 17

WHITESELL, L.G. and BOWLES, E.H. (1965). Train power station operators by analog. *Hydrocarbon Process.,* **44**(1), 127

WHITESIDE, C.C. (1987). Overheating and loss of production of the syngas compressor in ammonia plant after a turnaround. *Ammonia Plant Safety,* **27**, 17

WHITE-STEVENS, D.T. and ELLIOT, D.G. (1986). Cooper Basin liquids project design and start-up. *Gastech 85,* p. 456

WHITFIELD, D. (1969). *A Pilot Survey of Human Factors Aspects of Power Reactor Safety and Control.* Ergonomics Dev. Unit, Univ. of Aston in Birmingham

WHITFIELD, D., BALL, R.G. and ORD, G. (1980). Some human factors aspects of computer-aiding concepts for air traffic controllers. *Hum. Factors,* **22**, 569

WHITFIELD, D.J.C. (1995). Ergonomics in the design and operation of Sizewell 'B' nuclear power station. *Ergonomics*, **38**, 455

WHITHAM, G.B. (1950). The propagation of spherical blast. *Proc. R. Soc., Ser. A*, **203**, 571

WHITHAM, G.B. (1974). *Linear and Non-Linear Waves* (London: Wiley Interscience)

WHITING, M.J.L. (1992). The benefits of process intensification for caro's acid production. *Trans. IChemE* **70** (March), 195–96

WHITING, L.D. and SHAFFER, R.E. (1978). Feasibility study of hazardous vapor amelioration techniques. *Control of Hazardous Materials Spills*, **4**, p. 384

WHITING, M.J.L. (1992). The benefits of process intensification for Caro's acid production. *Chem. Engng Res. Des.*, **70A**, 195

WHITLOCK, D.J. (1977). Imported LNG interchangeability poses problems. *Oil Gas J.*, Apr. 11, 62

WHITMAN, K. (1967). Starting a pump under computer control. *Chem. Engng*, **74**(Jan. 2), 81

WHITMAN, K.A. (1972). Computer monitoring and surveillance of process equipment. *Instrum. Technol.*, **19**(7), 50

WHITMAN, R.V. (1973). *Damage Probability Matrices for Prototype Buildings. Seismic Design Decision Analysis Rep. 8.* Massachusetts Inst. Technol., Cambridge, MA

WHITMORE, J.B. and GRAY, R.C. (1987). The SIGTTO recommendations and guidelines for linked ship/ shore emergency shutdown. *Gastech 86*, p. 121

WHITMORE, M.W. (1994). Estimation of stability temperatures from differential thermal analysis and thermal activity monitor data in combination. *Hazards*, **XII**, p. 597

WHITMORE, M.W. and WILBERFORCE, J.K. (1993). Use of the accelerating rate calorimeter and the thermal activity monitor to estimate stability temperatures. *J. Loss Prev. Process Ind.*, **6**, 95

WHITMORE, M.W., GLADWELL, J.P. and RUTLEDGE, P.V. (1993). Report of an explosion during the manufacture of an azodicarbonamide formulation. *J. Loss Prev. Process Ind.*, **6**, 169

WHITMORE, M.W., CUTLER, D.P. and GLADWELL, J.P. (1994). Venting of bottom-initiated decompositions in azodicarbamide and its formulations. *J. Loss Prev. Process Ind.*, **7**, 49

WHITNEY, J.B. (1978). Detecting faults in glass-lined reactors. *Chem. Engng Prog.*, **74**(10), 81

WHITNEY, M.G., BARKER, D.D. and SPIVEY, K.H. (1992). Ultimate capacity of blast loaded structures common to chemical plants. *Plant/Operations Prog.*, **11**, 205

WHITTAKER, D.M. (1975). Improve valve life in reciprocating compressors. *Hydrocarbon Process.*, **54**(5), 143

WHITTAKER, H. (1981). State perspectives on hazardous materials management. *J. Haz. Materials*, 4(4), 367

WHITTAKER, J.D., ANGLE, R.P., WILSON, D.J. and CHOUKALOS, M.G. (1982). Risk-based zoning for toxic gas pipelines. *Risk Analysis*, **2**, 163

WHITTINGHAM, R.B. (1988). The application of the combined THERP/HCR model in human reliability assessment. In Sayers, B.A. (1988), *op. cit.*, p. 126

WHITTOW, J. (1980). *Disasters* (London: Allen Lane)

WHITWORTH, D.P.D. (1987). Application of operator error analysis in the design of Sizewell B. *Reliab. Engng*, **19**(4), 299

WHYTE, B.A.C. (1970). *Safety on the Site* (London: United Trade Press)

WIBERG, J. (1980). Investigation of 'ageing' of pressure vessel during the growth of a fatigue crack. *Int. J. Pres. Ves. Piping*, **8**, 79

WICHER, M.W. (1975). Safety considerations in canning of valves. *Ammonia Plant Safety*, **17**, p. 111

WICHMAN, I.S. and AGRAWAL, S. (1991). Wind-aided flame spread over thick solids. *Combust. Flame*, **83**, 127

WICHMAN, I.S. and ATREYA, A. (1987). A simplified model for the pyrolysis of charring materials. *Combust. Flame*, **68**, 231

WICHMANN, B.A. (1992). A note on the floating point in critical systems. *Comput. J.*, **35**(1), 41

WICHTERLE, I., LINEK, J. and HALA, E. (eds) (1973–). *Vapor–Liquid Equilibrium Data Bibliography* (Amsterdam: Elsevier)

WICKELGREN, W.A. (1974). *How to Solve Problems* (San Francisco, CA: W.H. Freeman)

WICKENS, C.D. and KESSEL, C. (1981). Failure detection in dynamic systems. In Rasmussen, J. and Rouse, W.B. (1980), *op. cit.*, p. 155

WICKS, K.M. (1983). Risk assessment techniques for LNG/ LPG shipping terminals. *Gastech 82*, p. 295

WICKSTRÖM, U. (1981/82). Temperature calculation of insulated steel columns exposed to natural fire. *Fire Safety J.*, **4**(4), 219

WICKSTRÖM, U. (1985). Temperature analysis of heavily insulated steel structures exposed to fire. *Fire Safety J.*, **9**, 281

WIDERA, G.E.O. and LOGAN, D.L. (1979). Layered cylindrical pressure vessels. *J. Pres. Ves. Technol.*, **101**, 80

WIDGINTON, D.W. (1982). Ignition thresholds for pulsed microwave sources. *Electrical Safety*, **3**, p. 139

WIDMER, H.M. (1980). Control of hazardous material in industrial air. *Loss Prevention and Safety Promotion, 3*, vol. 2, p. 268

WIDNER, J.T. (ed.) (1973). *Selected Readings in Safety* (Macon, GA: Academy Press)

WIDROW, B. *et al.* (1975). Adaptive noise cancelling: principles and application. Proc. *IEEE*, **63** (12), 1692

WIEBELT, J.A. (1966). *Engineering Radiation Heat Transfer* (New York: Holt, Rinehart & Winston)

WIEBERSON, W.E., ZWISLER, W.H., SEELY, L.B. and BRINKLEY, S.B. (1968). *TIGER Computer Program Documentation.* Vol. 1, *Theoretical and Mathematical Formulations of the TIGER Computer Program.* Contact DA-04–200-AMC-3226. US Army, Ballistic Res. Lab., Aberdeen Proving Ground, MD

WIEDEMANN, P. (1992). An effective tool for feedback from past experience with steam crackers. *Loss Prevention and Safety Promotion 7*, vol. 1, 32–1

WIEDERMAN, A.H. (1986a). Air-blast and fragment environments produced by the bursting of vessels filled with very high pressure gases. *Advances in Impact, Blast Ballistics and Dynamic Analysis of Structures* (New York: Am. Soc. Mech. Engrs)

WIEDERMAN, A.H. (1986b). Air-blast and fragment environments produced by the bursting of vessels filled with two phase fluids. *Advances in Impact, Blast Ballistics and Dynamic Analysis of Structures* (New York: Am. Soc. Mech. Engrs)

WIEGEL, R.L. (1970). *Earthquake Engineering* (Englewood Cliffs, NJ: Prentice-Hall)

WIEHLE, C.K. and BOCKHOLT, J.L. (1971). *Blast Response of Five NFSS Buildings* (report). Standford Res. Inst., Menlo Park, CA

WIEKEMA, B.J. (1980). Vapour cloud explosion model. *J. Haz. Materials*, **3**(3), 221

WIEKEMA, B.J. (1983a). An analysis of vapour cloud explosions based on accidents. In Hartwig, S. (1983), *op. cit.*, p. 237

WIEKEMA, B.J. (1983b). Analysis of vapour cloud incidents. *Euredata 4*, paper 1.2

WIEKEMA, B.J. (1984). Vapour cloud incidents – an analysis based on accidents. *J. Haz. Materials*, **8**, 295, 313

WIEKEMA, B.J., PASMAN, H.J. and GROOTHUIZEN, T.M. (1977). The effect of tubes connected with pressure relief vents. *Loss Prevention and Safety Promotion 2*, p. 223

WIEMANN, W. (1982). *Desensitization. The Control and Prevention of Dust Explosions* (London: Oyez/IBC), paper 5

WIENER, E.L. (1974). An adaptive vigilance task with knowledge of results. *Hum. Factors*, **16**, 333

WIENER, E.L. (1987). Application of vigilance research: rare, medium or well done? *Hum. Factors*, **29**, 725

WIENER, E.L. and CURRY, R.E. (1980). Flight deck automation: promises and problems. *Ergonomics*, **23**, 995

WIENER, S.A. and HEARD, D.B. (1976). Spill prevention and material sensing devices. *Control of Hazardous Material Spills*, **3**, p. 16

WIER, J.T. (1975). Selecting valves for the CPI. *Chem. Engng*, **82**(Nov. 24), 62

WIERINGA, J. (1980). Representativeness of wind observations at airports. *Bull. Am. Met. Soc.*, **51**, 962

WIERINGA, J. (1981). Estimation of meso-scale and local-scale roughness for atmospheric transport modelling. In de Wispelaere, C. (1981), *op. cit.*, p. 279

WIERSMA, S.J. (1978). Flow characteristics of Halon 1301 in pipes. *Fire Technol.*, **14**, 5

WIERZBA, I., KARIM, G.A. and CHENG, H. (1988). The flammability of rich gaseous fuel mixtures including those containing propane in air. *J. Haz. Materials*, **20**, 303

WIERZBA, I., HARRIS, K. and KARIM, G.A. (1990). Effect of low temperature on the rich flammability limits of some gaseous fuels and their mixtures. *J. Haz. Materials*, **25**(1/2), 257

WIERZBICKI, T. and MYUNG SUNG SUH (1986). *Denting Analysis of Tubes Under Combined Loading. Rep. MITSG-5*. Undersea Grant Coll. Program, Massachusetts Inst. Technol., Cambridge, MA

VON WIESENTHAL, P. and COOPER, H.W. (1970). Guide to economics of fired heater design. *Chem. Engng*, **77**(Apr. 6), 164

WIESNER, C.J. (1964). Hydrometeorology and flood estimation. *Proc. Instn Civil Engrs*, **27**, 153

WIEST, J.D. and LEVY, F.K. (1969). *Management Guide to PERT-CPM* (Englewood Cliffs, NJ: Prentice-Hall)

WIESTLING, C. (1982). Improve HPI loss investigation. *Hydrocarbon Process.*, **61**(9), 345

WIGGINS, J.H., HASSELMAN, T.K. and CHROSTOWSKI, J.D. (1976). Seismic risk analysis for offshore platforms in the Gulf of Alaska. *Eighth Ann. Conf. on Offshore Technology*, p. 577

WIGGLESWORTH, E.C. (1974). Occupational eye protection. *Occup. Safety Health*, **4**(May), 22

WIGHTMAN, D.W. and BENDELL, A. (1986). The practical application of proportional hazards modelling. *Reliab. Engng*, **15**(1), 29

WIGHUS, R. (1983). Simulation of propane dispersion from spills in a petrochemical plant using a water tunnel technique. *Loss Prevention and Safety Promotion 4*, vol. 1, pF. 26

WIGHUS, R. (1994). Fires on offshore process installations. *J. Loss Prev. Process Ind.*, **7**, 305

WIGHUS, R. and MEDLAND, O. (1989). Smoke venting of liquid pool fires in a large industrial hall. *Loss Prevention and Safety Promotion 6*, p. 56.1

WIGLEY, D.A. (1979). *Materials for Low-Temperature Use* (Oxford: Oxford Univ. Press)

WIGNAL, G. and LEE, J.K. (1980). 240,000 forced to flee chlorine released in Canadian rail wreck. *Fire Engng,* Jul., **47**

WIJKMAN, A. and TIMBERLAKE, L. (1984). *Natural Disasters* (London: Earthscan)

WILBERFORCE, J.K. (1982). Comparison of methods of determination of adiabatic times to maximum rate of exothermic reactions. *J. Thermal Analysis*, **25**, 593

WILBERFORCE, J.K. (1984). Applications of the accelerating rate calorimeter in thermal hazards evaluation. In *The Protection of Exothermic Reactors and Pressurised Storage Vessels* (Rugby: Instn Chem. Engrs), p. 329

WILBY, R. (1987). Programmable controllers in distributed control. *Chem. Engr, Lond.*, **435**, 13

WILCOX, D.E. and BROMLEY, L.A. (1963). Computer estimation of heat and free energy of formation of simple inorganic compounds. *Ind. Engng Chem.*, **55**(7), 32

WILCOX, J.C. (1978). Improving boiler efficiency. *Chem. Engng*, **85**(Oct. 9), 127

WILCOX, J.R. (1983). The teaching of engineering ethics. *Chem. Engng Prog.*, **79**(5), 15

WILCOX, P.H.A. (1970). *The Detective Physician – The Life and Work of Sir William Wilcox 1870–1941* (London: Heinemann Medical)

WILCOX, R.J. and BAKER, A. (1986). Boiler combustion. *Plant Engr*, **30**(1), 22

WILD, A.G. (1991). How to specify a progressing cavity pump. *Chem. Engr, Lond.*, **506**, 20

WILD, J. (1989). Fans for fire smoke venting. *Fire Surveyor*, **18**(2), 5

WILDAVSKY, A. (1985). *Trial Without Error: Anticipation vs Resilience as Strategies for Risk Reduction*. Centre for Independent Studies, Sydney

WILDAVSKY, A. (1988). *Searching for Safety* (New Brunswick: Transaction Books)

WILDAY, A.J. (1987). The use of DIERS methods for two phase relief of vaporisers. *Hazards from Pressure*, p. 187

WILDAY, A.J. (1988). Pressure relief system sizing for external fire. *The Pressurised Storage of Flammable Liquids (Fire Problems and Fire Protection)* (London: IBC Technical Services)

WILDAY, A.J. (1989). Sizing the relief system for runaway reactions. *Selection and Use of Pressure Relief Systems for Process Plant* (Manchester: Instn. Chem. Engrs, North-West Branch), p. 6.1

WILDAY, A.J. (1991). The safe design of chemical plants with no need for pressure relief systems. *Hazards*, **XI**, p. 243

WILDAY, A.J. (1991). The safe design of chemical plants with no need for pressure relief systems. *IChemE Symp. Series* No. 124, 243–53

WILDE, D.G. (1972). Combustion of polyurethane foam in an experimental mine roadway. *Safety in Mines Res. Estab., Res. Rep. 282* (London: HM Stationery Office)

WILDE, D.G. and CURZON, G.E. (1971). Ignition and burning of dispersions of flammable oil in the form of pools. *Major Loss Prevention*, p. 148

WILDE, G.J.S. (1978). Theory of risk compensation for accident causation and practical consequences for accident prevention. *Heft zur Unfallheilkunde*, **130**, 134

WILDE, G.J.S. (1982b). Critical issues in risk homoeostasis theory: a reply. *Risk Analysis*, **2**, 249

WILDE, G.J.S. (1982a). The theory of risk homoeostasis: implications for safety. *Risk Analysis*, **2**, 209, 227, 235, 239, 243, 249

WILDE, G.J.S. (1985). Assumptions necessary and unnecessary to risk homoeostasis. *Ergonomics*, **28**, 1531

WILDE, G.J.S. (1988). Risk homeostasis theory and traffic accidents: propositions, deductions and discussion of dissension in recent reactions. *Ergonomics*, **31**, 441

WILDER, I. and BRUGGER, J.E. (1972). Present and future technology requirements for the containment of hazardous material spills. *Control of Hazardous Material Spills*, **1**, p. 77

WILDER, D.R. (1963). Brittle engineering materials. *Chem. Engng*, **70**(Nov. 11), 209

WILDI, T. (1981). *Electrical Power Technology* (New York: Wiley)

WILENSKY, R. (1983). *Planning and Understanding* (Reading, MA: Addison-Wesley)

WILES, C.C. (1987). A review of solidification/stabilization technology. *J. Haz. Materials*, **14**(1), 5

WILEY, S.K. (1973). Controlling vapor losses. *Chem. Engng*, **80**(Sep. 17), 116

WILEY, S.K. (1987). Incinerate your hazardous wastes. *Hydrocarbon Process.*, **66**(6), 51

WILHELM, S.M. (1991). The effect of elemental mercury on engineering materials used in ammonia plants. *Plant/Operations Prog.*, **10**, 189

WILHELMI, A.R. and KNOPP, P.V. (1979). Wet air oxidation – an alternative to incineration. *Chem. Engng Prog.*, **75**(8), 46

WILK, R.D., CERNANSKY, N.P., PITZ, W.J. and WESTBROOK, C.K. (1989). Propene oxidation at low and intermediate temperatures: a detailed chemical kinetic study. *Combust. Flame*, **77**, 145

WILKE, T. (1974). Fire and explosions onboard ships. *Veritas*, **80**(Sep.), 12

WILKES, S.H. (1948). Dust explosions in factories. *Trans Instn Chem. Engrs*, **26**, 77

WILKIE, F. (1985a). Alarm system reduces risk of brittle fracture. *Process Engng, Oct., 41*

WILKIE, F. (1985b). Keeping a high priority on planned maintenance. *Process Engng,* Nov., 59

WILKIE, F. (1986). A shaky future for the PWR? *Process Engng,* May, 61

WILKIE, F. (1987a). The answer lies in the oil. *Process Engng,* May, 85

WILKIE, F. (1987b). Flame arresters – not simply a matter of choice. *Process Engng,* Mar, 53

WILKIE, F. and BERKOVITCH, I. (1985). Refuelling for a fast future. *Process Engng,* Dec., 38

WILKINS, M.J. (1992). Simplify batch automation projects. *Chem. Engng Prog.*, **88**(4), 61

WILKINSON, H.L. (1989). A study of severe air crash environments with particular reference to the carriage of radioactive materials by air. In Guy, G.B. (1989), *op. cit.,* p. 109

WILKINSON, J. (1981). . . . And now for the bad news. *Fell and Rock J.*, **XXII**(2/67), 135

WILKINSON, J.A. (1986). PLCs for alarm and shutdown systems. *Chem. Engr, Lond.*, **429**, 3

WILKINSON, J.A. and BALLS, B.W. (1985). Microprocessor- based safety systems designed for fire and gas and emergency shutdown *Advances in Instrumentation 40*, pt 2 (Pittsburgh, PA: Instrum. Soc. Am.), p. 1211

WILKINSON, L. (1979). Justify olefin expansions. *Hydrocarbon Process.*, **58**(9), 209

WILKINSON, M. and GEDDES, K. (1993). An award winning process. *Chem. in Britain* (December), 1050–52

WILKINSON, N.B. (1966). *Explosions in History* (Wilmington, DE: The Hagley Museum)

WILKINSON, R.T. (1992a). How fast should the night shift rotate? *Ergonomics*, **35**, 1425

WILKINSON, R.T. (1992b). Reply to Folkard (on shiftwork). *Ergonomics*, **35**, 1465

WILKINSON, W.L. (1981). A case for nuclear energy. *Chem. Engr, Lond.*, **370**, 345

WILKS, Y.A. (1975). A preferential pattern-seeking semantics for natural language inference. *Artificial Intelligence*, **6**, 53

WILKSON, E.J. and DENGLER, H.P. (1972). Higher alcohols grow in the 1970s *Hydrocarbon Process.*, **51**(11), 69

WILLA, J.L. and CAMPBELL, J.C. (1983). Common misconceptions about cooling towers. *Chem. Engng Prog.*, **79**(12), 23

WILLATT, R. (1989). *Piper Alpha Public Inquiry. Transcript, Day 141*

WILLATT, R.M. (1991). The functional and safety aspects of the location of submarine pipeline risers. *Safety Developments in the Offshore Oil and Gas Industry* (London: Instn Mech. Engrs), p. 31

WILLCOCK, G.E. (1986). Benefits of an effective emergency plan. *Refinement of Estimates of the Consequences of Heavy Toxic Vapour Release* (Rugby: Instn Chem. Engrs), p. 7.1

WILLEMS, H.J. (1980). Disaster organisation and chemical works. In De Boer, J. and Baillie, T.W. (1980), *op. cit.,* p. 41

WILLETTE, D.J. (1982). Profile: safety system in action. *Hydrocarbon Process.*, **61**(5), 120

WILLEY, J. (1984). Human reliability – the search for real data. *Chem. Engr, Lond.*, **405**, 33

WILLEY, M. (1992). Line type heat detectors. *Fire Surveyor*, **21**(2), 8

WILLIAMS *et al.* (n.d.). *Measurement of Flammability and Burn Potential of Fabrics. Summary Rep. DSR Proj. 73884.* Massachusetts Inst. Technol.

WILLIAMS, A. (1973). Combustion of droplets of liquid fuels. *Combust. Flame*, **21**, 1

WILLIAMS, A. (1976). *Combustion of Sprays of Liquid Fuel* (London: Elek Science)

WILLIAMS, A. (1977). Metal toxicity. *Chem. Engr, Lond.*, **323**, 570

WILLIAMS, A. (1990). *Combustion of Liquid Fuel Sprays* (London: Butterworths)

WILLIAMS, A.D. and CATALAN, D. (1976). Contingency planning. One railroad's approach to handling hazardous material spills. *Control of Hazardous Material Spills*, **3**, p. 146

WILLIAMS, A.F. and LOM, W.L. (1974). *Liquefied Petroleum Gases* (Chichester: Ellis Horwood)

WILLIAMS, A.J. (1992). Major hazards – the development of European and UK legislation over 20 years. *Major Hazards Onshore and Offshore*, p. 15

WILLIAMS, A.J. and MALONE, M.O. (1977). How to field weld piping. *Hydrocarbon Process.*, **56**(1), 111

WILLIAMS, B.C. (1984a). Qualitative analysis of MOS circuits. *Artificial Intelligence*, **24**, 281

WILLIAMS, B.C. (1984b). The use of continuity in a qualitative physics. *Proc. Fourth Int. Conf. on Artificial Intelligence* (Menlo Park, CA: Am. Ass. Artificial Intelligence), p. 350

WILLIAMS, C. (1986). Expert systems, knowledge engineering and AI tools – an overview. *IEEE Expert*, **1**(4), 66

WILLIAMS, C.C. (1953). *Damage Initiation in Organic Materials Exposed to High Intensity Thermal Radiation. Tech. Rep. 2.* Fuels Res. Lab, Massachusetts Inst. of Technol., Cambridge, MA

WILLIAMS, C.C. (ed.) (1975) *Selected Values of Properties of Hydrocarbons and Related Compounds. API Res. Prog. 44.* Texas A&M Univ., TX

WILLIAMS, C.P. (1970). Explosion clad titanium steel plate. *Chem. Engng Prog.*, **66**(10), 48

WILLIAMS, D. (1971). Safety audits. *Major Loss Prevention*, p. 220

WILLIAMS, D. (1984). Closure fuels dioxin scare. *Chem. Ind.*, Aug. 20, 564

WILLIAMS, D. (1985). The 'Greening' of Hesse. *Chemy Ind.*, Dec. 2, 776

WILLIAMS, D. (1986). Germany copes with Rhine disaster. *Chem. Ind.*, Dec. 1, 802

WILLIAMS, D.J. (1979). Major disasters: disaster planning in hospitals. *Br. J. Hospital Med.*, **22**(4), 308

WILLIAMS, D.T. (1963). Cracks under the microscope. *Chem. Engng*, **70**(May 27), 154

WILLIAMS, D.T. and BOLLINGER, L.M. (1949). The effect of turbulence on flame speeds of bunsen-type flames. *Combustion*, **3**, p. 176

WILLIAMS, E.G. (1965). Logic diagram in control systems design. *Elec. Times*, Nov. 11, 729; Nov. 18, 771

WILLIAMS, F.A. (1959). Spray combustion theory. *Combust. Flame*, **3**, 215

WILLIAMS, F.A. (1965). *Combustion Theory* (Reading, MA: Addison-Wesley)

WILLIAMS, F.A. (1976a). Qualitative theory of non-ideal explosions. *Combust. Sci. Technol.*, **12**, 199

WILLIAMS, F.A. (1976b). Quenching thickness for detonations. *Combust. Flame*, **26**, 403

WILLIAMS, F.A. (1977). Mechanisms of fire spread. *Combustion* **16**, p. 1281

WILLIAMS, F.A. (1982). Laminar flame instability and turbulent flame propagation. In Lee, J.H.S. and Guirao, C.M. (1982), *op. cit.*, p. 69

WILLIAMS, F.A. (1983). Estimation of pressure fields in combustion of vapour clouds. *J. Fluid Mech.*, **127**, 429

WILLIAMS, F.A. (1992). The role of theory in combustion science. *Combustion*, **24**, p. 1

WILLIAMS, F.W., INDRITZ, D. and SHEINSON, R.S. (1975). Concentration limits for n-butane low temperature flames. *Combust. Sci. Technol.*, **11**, 67

WILLIAMS, G.C., HOTTEL, H.C. and SCURLOCK, A.C. (1949). Flame stabilisation and propagation in high velocity streams. *Combustion*, **3**, p. 21

WILLIAMS, G.M. and PRATT, T.H. (1990). Characteristics and hazards of electrostatic discharges in air. *J. Loss Prev. Process Ind.*, **3**(4), 381

WILLIAMS, G.P. (1978). Causes of ammonia plant shutdowns. *Ammonia Plant Safety* **20**, p. 123. See also *Chem. Engng Prog.*, **74**(9), 88

WILLIAMS, G.P. and SAWYER, J.G. (1974). What causes ammonia plant shutdowns. *Ammonia Plant Safety* **16**, p. 4

WILLIAMS, G.P. and HOEHING, W.W. (1983). Causes of ammonia plant shutdowns: survey IV. *Chem. Engng Prog.*, **79**(3), 11; **79**(5), 28

WILLIAMS, G.P., HOEHING, W.W. and BYINGTON, R.G. (1987). Causes of ammonia plant shutdowns: survey V. *Ammonia Plant Safety*, **27**, 72. See also *Plant/ Operations Prog.*, 1988, **7**, 99

WILLIAMS, H.B. (1964). Human factors in warning-and-response systems. In Grosser, G.H. *et al.* (1964), *op. cit.*, p. 80

WILLIAMS, H.D. (1972). Methods for prevention of spills in the maritime industry – industrial considerations. *Control of Hazardous Material Spills*, **1**, p. 41

WILLIAMS, H.D. (1973). More detail on the hazards of liquefied natural gas in marine transportation. *Proc. Marine Safety Coun.*, Oct., **205**

WILLIAMS, H.D. (1975). The national strike force concept in hazardous substance discharge response activities. *Transport of Hazardous Cargoes*, 4

WILLIAMS, H.D. (1976). Hazardous materials spills: stop them at source. *Control of Hazardous Material Spills 3*, vol. 2, p. 114

WILLIAMS, H.L. and RUSSELL, B.H. (1970). NASA reliability techniques in the chemical industry. *Chem. Engng Prog.*, **66**(12), 45

WILLIAMS, I. (1984). Fire protection of pressurized LPG tanks. *Transport and Storage of LPG and LNG* (Antwerp: Koninklijke Vlaamse Ingenieursvereniging), p. 315

WILLIAMS, I.K.G. (1979). Nuclear wastes: a problem in perspective? *Sci. Publ. Policy*, **6**, 134

WILLIAMS, J.A. (1982). Reprocessing irradiated nuclear fuel. In Green, A.E. (1982), *op. cit.*, p. 467

WILLIAMS, J.C. (1982). Cost-effectiveness of human factor recommendations in relation to equipment design. *Design 82* (Rugby: Instn Chem. Engrs)

WILLIAMS, J.C. (1983). Validation of human reliability assessment techniques. *Reliability 83*, vol. 1, p. 2B/2/1

WILLIAMS, J.C. (1985a). Quantification of human error in maintenance for process plant probabilistic risk assessment. *The Assessment and Control of Major Hazards* (Rugby: Instn Chem. Engrs), p. 353

WILLIAMS, J.C. (1985b). Validation of human reliability assessment techniques. *Reliab. Engng*, **11**(3), 149

WILLIAMS, J.C. (1986). HEART – a proposed method for assessing human error. *Ninth Symp. on Advances in Reliability Technology*

WILLIAMS, J.C. (1988a). A data-based method for assessing and reducing human error to improve operational experience. *Proc. IEEE Fourth Conf. on Human Factors in Power Plants* (New York)

WILLIAMS, J.C. (1988b). A human factors data base to influence safety and reliability. In Sayers, B.A. (1988), *op. cit.*, p. 223

WILLIAMS, J.C. (1992). Improved evaluation tool for users of HEART. *Hazard Identification and Risk Analysis, Human Factors and Human Reliability in Process Safety*, p. 261

WILLIAMS, J.C. (1993). Incorporating human performance variability in process safety assessment. *Process Safety Management*, p. 187

WILLIAMS, J.C. and HURST, N.W. (1992). A comparative study of the management effectiveness of two technically similar major hazards sites. *Major Hazards Onshore and Offshore*, p. 73

WILLIAMS, J.R. (1970). Polar solutions and fire fighting foam. *Loss Prevention*, 4, p. 53

WILLIAMS, O.A. (1983). *Pneumatic and Hydraulic Conveying of Solids* (New York: Dekker)

WILLIAMS, P.R. (1976). Tribology in action. *Plant Engr*, **20**(3), 35

WILLIAMS, P.T. (1990). The formation and control of dioxins and furans from incineration of municipal waste. *Process Safety Environ.*, **68B**, **122**

WILLIAMS, R.L. (1977). Systems reliability analysis using the GO methodology. In Gangadharan, A.C. and Brown, S.J. (1977), *op. cit.*, p. 103

WILLIAMS, R.M. (1982a). A model for the dry deposition of particles to natural water surfaces. *Atmos. Environ.*, **16**, 1933

WILLIAMS, R.M. (1982b). Uncertainties in the use of box models for estimating dry deposition velocity. *Atmos. Environ.*, **16**, 2707; 1983, **17**, 193

WILLIAMS, R.T. (1959). *Detoxification Mechanisms,* 2nd ed. (New York: Wiley)

WILLIAMS, T.J. (1983). Robots in the CPI: an appraisal and a note of caution. *Chem. Eng,* **90**(Dec. 26), 29

WILLIAMS, T.J. (ed.) (1989). *A Reference Model for Computer Integrated Manufacturing, a Description Viewpoint of Industrial Automation.* Instrum Soc. Am., Res. Triangle Park, NC

WILLIAMS, V.C. (1970). Cryogenics. *Chem. Engng,* **77**(Nov. 16), 92

WILLIAMSON, B.R. and MANN, L.R.B. (1981). Thermal hazards from propane (LPG) fireballs. *Combust. Sci. Technol.,* **25**, 141

WILLIAMSON, G.E. (1978). Labelling of dangerous substances. *Chem. Ind.,* May 6, 307

WILLIAMSON, G.E. (1985). The Classification, Packaging and Labelling of Dangerous Substances Regulations 1984. *Chem. Ind.,* Jan. 21, 40

WILLIAMSON, H.V. (1976). Halon 1301 flow in pipelines. *Fire Technol.,* **12**, 18

WILLIAMSON, H.V. (1986). Carbon dioxide and application systems. In Cote, A.E. and Linville, J.L. (1986), *op. cit.,* p. 19–1

WILLIAMSON, J.J. (1966). *The Chartered Insurance Institute Handbook* (London: Pitman)

WILLIAMSON, J.R. and KINGSLEY, W.K. (1987). Storage of laboratory chemicals. In Young, J.A, (1987), *op. cit.,* p. 205

WILLIAMSON, K.D. and EDESKUTY, F.J. (1984). *Liquid Cryogens* (Boca Raton, FL: CRC Press)

WILLIAMSON, M.M., HACKEL, R.A. and WRIGHT, J.W.D. (1978). Better training by vendors. *Hydrocarbon Process.,* **57**(1), 99

WILLIAMSON, R.B. and DEMBSEY, N.A. (1993). Advances in assessment methods for fire safety. *Fire Safety J.,* **20**, 15

WILLIS, B.H. and HENEBRY, W.M. (1976). What you need to know about environmental impact statements. *Hydrocarbon Process.,* **55**(10), 87

WILLIS, B.H. and HENEBRY, W.M. (1982–). The EIS: a current perspective. *Hydrocarbon Process.,* 1982, **61**(11), 223; 1983, **62**(1), 135

WILLIS, C.D. and CECE, J.M. (1975). Implementation of a Chemical Hazards Response Information System (CHRIS). *Transport of Hazardous Cargoes,* **4**

WILLIS, C.H.H. (1981). Method of identifying fuels after a fire or explosion. *J. Occup. Accid.,* **3**(3), 217

WILLIS, G.E. (1986). Dispersion of buoyant stack plumes in a convective boundary layer. *Fifth Joint Conf. on Applications of Air Pollution Meteorology* (Boston, MA: Am. Met. Soc.), p. 327

WILLIS, G.E. and DEARDORFF, J.W. (1976). A laboratory model of diffusion into the convective planetary boundary layer. *Quart. J. R. Met. Soc.,* **102**, 427

WILLIS, G.E. and DEARDORFF, J.W. (1978). A laboratory study of dispersion from an elevated source in a convective mixed layer. *Atmos. Environ.,* **12**, 1305

WILLIS, G.E. and DEARDORFF, J.W. (1983). On plume rise within a convective boundary layer. *Atmos. Environ.,* **17**, 2435

WILLIS, G.E. and DEARDORFF, J.W. (1987). Buoyant plume dispersion and inversion entrapment in and above a laboratory mixed layer. *Atmos. Environ.,* **21**, 1725

WILLIS, R.P. (1981). Reliable production in the process industries – the view of the contractor. *Reliable Production in the Process Industries* (Rugby: Instn Chem. Engrs), p. 1

WILLMAN, J.C. (1977). Case history: PCB transformer spill Seattle, Washington. *J. Haz. Materials,* **1**(4), 361

WILLMANN, J.C., BLAZEVICH, J. and SNYDER, H.J. (1976). PCB spill in the Duwamish-Seattle, WA. *Control of Hazardous Material Spills,* **3**, p. 351

WILLMOTH, B.M. (1978). Procedures to reduce contamination of groundwater by hazardous materials. *Control of Hazardous Material Spills,* **4**, p. 293

WILLRICH, M. and TAYLOR, T. (1974). *Nuclear Theft, Risks and Safeguards* (Cambridge, MA: Ballinger)

WILLS, J.S. (1970). Size vapor piping by computer. *Hydrocarbon Process.,* **49**(5), 149

WILLSKY, A.S. (1976). A survey of design methods for failure detection in dynamic systems. *Automatica,* **12**, 601

WILMOT, D.A. and LEONG, A.P. (1976). Another way to detect agitation. *Loss Prevention,* **10**, p. 19

WILSE, T. (1974). Fire and explosion on board ships. *Veritas,* **80**(Sep.), 12

WILSON, SIR A. (1963). *Report of the Committee on the Problem of Noise* (London: HM Stationery Office)

WILSON, A. (1987). Make batch process improvement using models. *Process Engng, Jun,* 45

WILSON, B.J. and MYERS, T.J. (1979). Risk reduction in adipic acid manufacture. *Loss Prevention,* **12**, p. 27

WILSON, B.R.W. (1976). Control and instrumentation aspects of loss prevention. *Course on Loss Prevention in the Process Industries.* Dept of Chem. Engng, Loughborough Univ. of Technol.

WILSON, C.E. (1989). *Noise Control* (London: Harper & Row)

WILSON, C.W. and BONI, A.A. (1980). Spherical piston initiation model of gas phase detonation. *Combust. Sci. Technol.,* **21**, 183

WILSON, D.C. (1980). Flixborough versus Seveso – comparing the hazards. *Chemical Process Hazards,* VII, p. 193

WILSON, D.C. (1981). *Waste Management: Planning, Evaluation, Technologies* (Oxford: Clarendon Press)

WILSON, D.C. (1982). Controlled disposal of special wastes. The definition of special wastes. *Chem. Ind.,* Apr. 17, 253

WILSON, D.C., SMITH, E.T. and PEARCE, K.W. (1981). Uncontrolled hazardous waste sites: a perspective of the problem in the UK. *Chem. Ind.,* Jan. 3, 18

WILSON, D.J. (1976). *Contamination of Building Air Intakes from Nearby Vents. Rep. 1.* Dept of Mech. Engng, Univ. of Alberta, Edmonton, Alberta

WILSON, D.J. (1979a). Flow patterns over flat-roofed buildings and application to exhaust stack design. *ASHRAE Trans.,* **85**(pt 2), 284

WILSON, D.J. (1979b). *The Release and Dispersion of Gas from Pipeline Ruptures. Rep. Alberta Environment Contract 790686.* Dept of Mech. Engng, Univ. of Alberta

WILSON, D.J. (1981a). Along-wind diffusion of source transients. *Atmos. Environ.,* **15**, 489

WILSON, D.J. (1981b). *Expansion and Plume Rise of Gas Jets from High Pressure Pipeline Ruptures. Rep. Alberta Environment Contract 810786.* Dept of Mech. Engng, Univ. of Alberta

WILSON, D.J. (1982). *Predicting Risk of Exposure to Peak Concentrations in Gas Emissions. Fluctuating Plumes. Rep. Alberta Environment Contract 820847.* Dept of Mech. Engng, Univ. of Alberta

WILSON, D.J. (1986). *Plume Dynamics and Concentration Fluctuations in Gas Emissions Rep. Alberta Environment Contract 840702.* Dept of Mech. Engng, Univ. of Alberta

WILSON, D.J. (1987). Stay indoors or evacuate to avoid exposure to toxic gas? *Emergency Preparedness Canada, Emergency Preparedness Digest,* **14**(1), 19

WILSON, D.J. (1990). *Model Development for EXPOSURE-1 and SHELTER-1: Toxic Load and Adverse Biological Effects for Outdoor, Indoor and Evacuation Exposures in Dispersing Toxic Gas Plumes. Engng Rep. 73.* Dept of Mech. Engng, Univ. of Alberta

WILSON, D.J. (1991a). Accounting for peak concentration in atmospheric dispersion for worst case hazard assessments. *Modeling and Mitigating the Consequences of Accidental Releases,* p. 385

WILSON, D.J. (1991b). Effectiveness of indoor sheltering during long duration toxic gas releases. *ER 91 Emergency Response Conf. on Technological Response to Dangerous Substances Accidents,* Calgary, Alberta

WILSON, D.J. (1995). The effectiveness of safety cases – a major contractor's view. *Safety Cases* (London: IBC Technical Services)

WILSON, D.J. and BRITTER, R.E. (1982). Estimates of building surface concentrations from nearby point sources. *Atmos. Environ.,* **16**, 2631

WILSON, D.J. and NETTERVILLE, D.D.J. (1978). Interaction of a roof-level plume with a downwind building. *Atmos. Environ.,* **12**, 1051

WILSON, D.J. and SIMMS, B.W. (1985). *Exposure Time Effects on Concentration Fluctuations in Plumes. Environment Contract 830990.* Dept of Mech. Engng, Univ. of Alberta

WILSON, D.J. and ZELT, B.W. (1990). *Technical Basis for EXPOSURE-1 and SHELTER-1: Models for Predicting Outdoor and Indoor Exposure Hazards from Toxic Gas Releases. Engng Rep. 72.* Dept of Mech. Engng, Univ. of Alberta

WILSON, D.J., FACKRELL, J.E. and ROBINS, A.G. (1982). Concentration fluctuations in an elevated plume: a diffusion–dissipation approximation. *Atmos. Environ.,* **16**, 2581

WILSON, D.J., ROBINS, A.G. and FACKRELL, J.E. (1982). Predicting the spatial distribution of concentration fluctuations from a ground level source. *Atmos. Environ,* **16**, 497

WILSON, D.J., ROBINS, A.G. and FACKRELL, J.E. (1985). Intermittency and conditionally averaged concentration fluctuation statistics in plumes. *Atoms. Environ.,* **19**, 1053; 1986, **20**, 1831

WILSON, D.K. (1988). Failure mode management: a loss prevention philosophy for programmable logic controllers. *Plant/Operations Prog.,* **7**, 254

WILSON, D.S. (1969). *The Modern Gas Industry* (London: Arnold)

WILSON, D.S., LEWIS, P., BRODERICK, J.J. and STERNLICHT, B. (1970). Failure analysis, detection, prevention, and control of turbo machinery. *Loss Prevention,* **4**, p. 38

WILSON, E. (1981). A selected annotated bibliography and guide to sources of information on planning for and responses to chemical emergencies. *J. Haz. Materials,* **4**(4), 373

WILSON, E.E. (1915). *A Basis for the Rational Design of Heat Transfer Apparatus* (New York: Am. Soc. Mech. Engrs)

WILSON, F.W. (1962). The challenge of higher pressures. *Chem. Engng,* **69**(Oct. 15), 159

WILSON, H.S. (1978). Future trends in instrumentation – where are we going? *Measmt Control,* **11**, 202

WILSON, J.G. (1971). Use of rhesus monkeys in teratological studies. *Am. Soc. Exptl Biol, Federation Proc.,* **30**, 104

WILSON, J.M. and TYRER, P.G. (1980). Explosion protection for diesel engines offshore. *Eur. Offshore Petrol Conf.,* paper EUR 226

WILSON, J.Q. (ed.) (1980). *The Politics of Regulation* (New York: Basic Books)

WILSON, J.R. (1968). Disaster threatens refinery tank field. *Ann. N.Y. Acad. Sci.,* **152**(art. 1), 604

WILSON, J.R. (1990). *Evaluation of Human Work* (London: Taylor & Francis)

WILSON, J.R. and RUTHERFORD, A. (1989). Mental models: theory and application in human factors. *Hum. Factors,* **31**, 617

WILSON, K.B. (1946). Calculations and analysis of longitudinal temperature gradients in tubular reactors. *Trans Instn Chem. Engrs,* **24**, 77

WILSON, K.C. (1990a). Process control, plant safety and the human factor: lessons to be learned from a major refinery incident. *Process Safety Environ.,* **68B**, 31

WILSON, K.C. (1990b). Process control, plant safety and the human factor: lessons to be learned from a major refinery incident. *Safety and Loss Prevention in the Chemical and Oil Processing Industries,* p. 615

WILSON, L.G. and CAVANAGH, P. (1972). *Electrostatic Hazards due to Clothing. Rep. 665.* Defence Res. Estab, Ottawa

WILSON, M. (1993). Protecting against storage tank bund fires. *Fire Surveyor,* **22**(4), 8

WILSON, N. (1977). The risk of fire or explosion due to static charges on textiles and clothing. *J. Electrostatics,* **4**, 67

WILSON, N. (1979). The nature and incendiary behaviour of spark discharges from the body. *Electrostatics,* **5**, p. 73

WILSON, N. (1983). The ignition of natural gas by spark discharges from the body. *Electrostatics,* **6**, p. 21

WILSON, R. (1964a). Here's fire protection Monsanto style. *Hydrocarbon Process.,* **43**(4), 185

WILSON, R. (1964b). How a large chemical plant handles fire-water layout. In Vervalin, C.H. (1964a), *op. cit.,* p. 296

WILSON, R. (1975). The costs of safety. *New Scientist,* **68**(973), 274

WILSON, R. (1992). Offshore emergency management – communications and control. *Health and Safety in the Offshore Oil and Gas Industries* (Rugby: Instn Chem. Engrs)

WILSON, R. and CROUCH, E.A. (1987). Risk assessment and comparisons: an introduction. *Science,* **236**(Apr. 17, 267

WILSON, R.A.M. and DANCKWERTS, P.V. (1964). Studies in turbulent mixing – II, A hot air jet. *Chem. Engng. Sci.,* **19**, 885

WILSON, R.F. (1960). *Colour in Industry Today* (London: Allen & Unwin)

WILSON, R.H. (1944). Health hazards encountered in the manufacture of synthetic rubber. *JAMA,* **124**, 701

WILSON, R.H., HOUGH, G.V. and MCCORMICK, W.E. (1948). Medical problems encountered in the manufacture of American-made rubber. *Ind. Med.,* **17**, 199

WILSON, R.P. and ATALLAH, S. (1975). A minimum effective length criterion for flame arrestors. *Transport of Hazardous Cargoes,* **4**

WILSON, R.P. and FLESSNER, M.F. (1979). Design criteria for flame arresters. *Loss Prevention,* **12**, p. 86

WILSON, R.T. (1978). How to pack high temperature valves. *Hydrocarbon Process.,* **57**(1), 91. See also in Kane, L.A. (1980), *op. cit.,* p. 172

WILSON, R.W. (1974). Diagnosis of engineering failures. *Br. Corros. J.,* **9**(3), 134

WILSON, S. (1992). Develop an effective crisis communications plan. *Chem. Engng Prog.,* **88**(3), 96

WILSON, S. (1993). Planning for a crisis. *Chem. Engng,* **100**(1), 125

WILSON, S.A. (1972). *Estimating Pipe Reliability by the Time-to-damage Method. Rep. GEAP-10452.* General Electric Co.

WILSON, S.A. (1976). Distribution of time-to-damage method in reliability of pipes. In Henley, E.J. and Lyhnn, J.W. (1976), *op. cit.*, p. 463

WILSON, W.E., O'DONOVAN, J.T. and FRISTROM, R.M. (1969). Flame inhibition by halogen compounds. *Combustion*, **12**, p. 929

WILTEN, H.M. (1962). In refinery service, look at today's stainless steel failures by stress corrosion. *Hydrocarbon Process.*, **41**(1), 122

WILTEN, H.M. (1963). Why steels fail in refinery service. *Hydrocarbon Process.*, **42**(2), 15

WILTEN, H.M. (1965). Basic guide to refinery corrosion. *Hydrocarbon Process.*, **44**(11), 301; **44**(12), 131

WILTSHIRE, W.J. (ed.) (1957). *A Further Handbook of Industrial Radiology* (London: Arnold)

WIMER, W.E. (1976). Oxygen gives lowest cost VCM. *Hydrocarbon Process.*, **55**(3), 81

WIMPRESS, R.N. (1963). Rating fired heaters. *Hydrocarbon Process.*, **42**(10), 115

WINBURN, D.C. (1987). *What Every Engineer Should Know about Lasers* (New York: Dekker)

WINBURN, D.C. (1990). *Practical Laser Safety*, 2nd ed. (New York: Dekker)

WINDEBANK, C.S. (1976). The economics of loss prevention a preview. *Course on Loss Prevention in the Process Industries.* Dept of Chem. Engng, Loughborough Univ. of Technol.

WINDEBANK, C.S. (1983). The application of risk assessment/ hazard analysis to marine operations – introduction. *Loss Prevention and Safety Promotion 4*, vol. 2, p. B1

WINDEBANK, E. (1983a). A Monte Carlo simulation method versus a general analytical method for determining reliability measures of repairable systems. *Reliab. Engng*, **5**(2), 73

WINDEBANK, E. (1983b). Reliability network analysis by PLANT computer program. *Euredata 4*, paper 5.1

WINDER, C., TOTTSZER, A., NAVRATIL, J. and TANDON, R. (1992). Hazardous materials incidents reporting: results of a nation-wide trial. *J. Haz. Materials*, **31**, 119

WINDHORST, J.C.A. (1995). Application of inherently safe design concepts, fitness for use and risk driven design process safety standards to an LPG project. *Loss Prev. and Saf. Prom. in the Process Ind.*, (ed. J.J. Mewis, H.J. Pasman, and E.E. De Rademacker), 543–4. (Amsterdam: Elsevier Science B.V)

WINDSCALE LOCAL LIAISON COMMITTEE (1979). *Notice to Local Householders concerning Windscale Emergency Arrangements* (Windscale: British Nuclear Fuels Ltd)

WINEGAR, B.H. (1980). Partial collapse of an atmospheric storage tank. *Ammonia Plant Safety*, **22**, 226

WINGENDER, H.J. (1986). *Reliability Data Collection and Use in Risk Assessment* (Berlin: Springer)

WINGENDER, H.J. (1991). Reliability data collection in process plants. In Cannon, A.G. and Bendell, A. (1991), *op. cit.*, p. 81

WS ATKINS, Contract Research Report # 325/2001. *Root Cause Analysis Literature Review* UK (London: HSE), ISBN 0-7176-19664

VAN WINGERDEN, C.J.M. (1989a). Experimental investigation into the strength of blast waves generated by vapour cloud explosions in congested areas. *Loss Prevention and Safety Promotion 6*, p. 26.1

VAN WINGERDEN, C.J.M. (1989b). On the scaling of vapour explosion experiments. *Chem. Engng Res. Des.*, **67**, 339

VAN WINGERDEN, C.J.M. (1989c). On the venting of large-scale methane–air explosions. *Loss Prevention and Safety Promotion 6*, p. 25.1

VAN WINGERDEN, C.J.M. and ZEEUWEN, J.P. (1983). Flame propagation in the presence of repeated obstacles: influence of gas reactivity and degree of confinement. *J. Haz. Materials*, **8**, 139

VAN WINGERDEN, C.J.M. and ZEEUWEN, J.P. (1986). Vapour cloud explosion experiments for use in hazard assessment. *Loss Prevention and Safety Promotion 5*, p. 37–1

VAN WINGERDEN, C.J.M. and ZEEUWEN, J.P. (1988). On the role of acoustically driven flame instabilities in vented gas explosions and their elimination. In Groothuizen, T.M. (1988), *op. cit.*, p. 254

VAN WINGERDEN, C.J.M., VAN DEN BERG, A.C. and OPSCHOOR, G. (1989). Vapour cloud explosion blast prediction. *Plant/ Operations Prog.*, **8**, 234

VAN WINGERDEN, C.J.M., OPSCHOOR, G. and PASMAN, H.J. (1990). Parameter study into blast generated by vapour cloud explosions. *Safety and Loss Prevention in the Chemical and Oil Processing Industries*, p. 277

VAN WINGERDEN, K. (1993). Prediction of pressure and flame effects in the direct surroundings of installations protected by dust explosion venting. *J. Loss Prev. Process Ind.*, **6**, 241

VAN WINGERDEN, K. (1994). Course and strength of accidental explosions on offshore installations. *J. Loss Prev. Process Ind.*, **7**, 295

VAN WINGERDEN, K., VISSER, J. and PASMAN, H. (1991). Combustion in obstructed diverging and non-diverging flow fields (poster). *Thirteenth ICDERS Conf*, Nagoya, Japan

VAN WINGERDEN, K., PEDERSEN, G.H. and WILKINS, B.A. (1994). Turbulent flame propagation in gas mixtures. *Hazards*, **XII**, p. 249

WINKLER, R.L. (1968). The consensus of subjective probability distributions. *Mngmt Sci.*, **15**(2), B61

WINKLER, R.L. (1981). Combining probability distributions from dependent information sources. *Mngmt Sci.*, **27**, 987

WINKLER, R.L. (1986). Expert resolution. *Mngmt Sci.*, **32**(3), 298

WINKLER, R.L. and MURPHY, A.H. (1978). 'Good' probability assessors. *J. Appl. Met.*, **7**, 751

WINN, LORD JUSTICE (chrmn) (1968). *Report of the Committee on Personal Injuries Litigation, Cmnd 3691* (London: HM Stationery Office)

WINNERLING, H.A. (1973). Quieting of process machinery. *Chem. Engng Prog.*, **69**(6), 96

WINOGRAD, T. (1972). *Understanding Natural Language* (New York: Academic Press)

WINOGRAD, T. (1975). Frame representations and the declarative/procedural controversy. In Bobrow, D.G. and Collins, A. (1975), op. cit., p. 185

WINOGRAD, T. (1978). On primitives, prototypes, and other semantic anomalies. *Workshops on Theoretical Issues in Natural Language Processing – 2*, p. 25

WINOGRAD, T. (1983). *Language as a Cognitive Process* (Reading, MA: Addison-Wesley)

WINSETT, W.G. (1981). Dust control in grain elevators. *The Hazards of Industrial Explosion from Dusts* (London: Oyez/IBC), paper 17

WINSTON, P.H. (1977). *Artificial Intelligence* (Reading, MA: Addison-Wesley)

WINSTON, P.H. (1980). Learning and reasoning by analogy. *Comm. ACM*, **23**(12)

WINSTON, P.H. (1982). Learning new principles from precedents and exercises. *Artificial Intelligence*, **19**(3)

WINSTON, P.H. (1984). *Artificial Intelligence,* 2nd ed. (Reading, MA: Addison-Wesley)

WINSTON, P.H. (1992). *Artificial Intelligence,* 3rd ed. (Reading, MA: Addison-Wesley)

WINSTON, P.H. and HORN, B.K.P. (1981). *LISP* (Reading, MA: Addison-Wesley)

WINSTON, P.H. and PRENDERGAST, K.A. (eds) (1984). *The AI Business* (Cambridge, MA: MIT Press)

WINSTON, W.P. (1984). *Petroleum Law Guide* (London: Applied Science)

WINTER (1973). Mechanism of extension and mixture of gases. *Prevention of Occupational Risks,* **2**, p. 174

WINTER, E.F. (1962). The electrostatic problem in aircraft fuelling. *J. R. Aero. Soc.,* **66**, 429

WINTER, P. (1985). Process flowsheeting. *Process Engng,* Mar., 29

WINTER, P. (1986). Computer aided process design at a watershed. *Chem. Engng Res. Des.,* **64**, 329

WINTER, P. (1990). Perspectives on computer aided process engineering. *Chem. Engng Res. Des.,* **68A**, 403

WINTER, P. (1992). Computer aided process engineering: the evolution continues. *Chem. Engng Prog.,* **88**(2), 77

WINTER, P. and NEWELL, R.G. (1977). Data bases in design management. *Chem. Engng Prog.,* **73**(6), 97

WINTER, P., BRANCH, S.J. and SMITH, R.E. (1982). Database technology and process design. ACHEMA Conf., Frankfurt-am-Main

VON WINTERFELDT, D. JOHN, R.S. and BORCHERDING, K. (1981). Cognitive components of risk ratings. *Risk Analysis,* **1**, 277

WINTERNITZ, M.C., LAMBERT, R.A., JACKSON, L. and SMITH, C.H. (1920). The pathology of chlorine poisoning. *Collected Studies on the Pathology of War Gas Poisoning* (New Haven, CT: Yale Univ. Press)

WINTERNITZ, M.C., SMITH, G.H. and MCNAMARA, F.P. (1920). Effect of interbronchial insufflation of acid. *J. Exp. Med.,* **32**, 199

WINTERNITZ, P.F. (1949). A method for calculating simultaneous, homogeneous gas equilibria and flame temperatures. *Combustion,* **3**, p. 623

WINTERS, K.H. and CLIFFE, K.A. (1985). The prediction of critical points for thermal explosions in a finite volume. *Combust. Flame,* **62**, 13

WINTERS, R.G. (1978). Lubricating air compressors. *Chem. Engng,* **85**(Aug. 14), 157

WIRTH, G.F. (1975). Preventing and dealing with in-plant hazardous spills. *Chem. Engng,* **82**(Aug. 18), 82

WIRTH, W. (1935). Experimental studies in the treatment of poisoning by inhalation of hydrogen cyanide. *Arch. Exp. Pathol.,* **179**, 558

WIRTH, W. (1937). On the treatment of hydrogen cyanide poisoning. *Zbl. Gewerbehyg.,* **14**, 11

WIRTH, W. and LAMMERHIRT (1934). On the question of detoxification from hydrogen cyanide. *Biochem. Z.,* **270**, 455

WISE, H. and AGOSTON, G.A. (1958). *Literature of the Combustion of Petroleum* (Washington, DC: Am. Chem. Soc.), p. 125

WISNIAK, J. and KERSKOWITZ, M. (1984). *Solubility of Gases and Solids* (Amsterdam: Elsevier)

WISNIAK, J. and TAMIR, A. (1978). *Mixing and Excess Thermodynamic Properties: A Literature Source Book* (Amsterdam: Elsevier)

WISNIAK, J. and TAMIR, A. (1980–). Liquid–Liquid Equilibrium Extraction: A Literature Source Book (Amsterdam: Elsevier)

WISOTSKI, J. and SNYER, W.H. (1965). *Characteristic of Blast Waves obtained from Cylindrical High Explosive Charges,* Contract N960921–7013. Univ. of Denver, CO

DE WISPELAERE, C. (ed.) (1981). *Air Pollution Modelling and its Application,* vol. 1 (New York: Plenum Press)

DE WISPELAERE, C. (ed.). (1983). *Air Pollution Modeling and its Application,* vol. 2 (New York: Plenum Press)

DE WISPELAERE, C. (ed.) (1984). *Air Pollution Modeling and Its Application,* vol. 3 (New York: Plenum Press)

DE WISPELAERE, C. (ed.) (1985). *Air Pollution Modeling and Its Application,* vol. 4 (New York: Plenum Press)

DE WISPELAERE, C., SCHIERMEIER, F.A. and GILLANI, N.V. (eds) (1986). *Air Pollution Modelling and its Application,* vol. 5 (New York: Plenum Press)

WISS, J., FLEURY, C. and FUCHS, V. (1995). Modelling and optimization of semi-Batch and continuous nitration of chlorobenzene from safety and technical viewpoints. *J. of Loss Prev. in the Process Ind.* **8**(4), 205–13

WISSMILLER, I.L. (1970). New relief valve gas equations. *Hydrocarbon Process.,* **49**(5), 123 LE WITA (1925) *La Guerre Chimique* (Paris)

WITHERILL, C.E. (1991). Forestalling failure in the plant. *Chem. Engng,* **98**(10), 126

WITHERS, R.M.J. (1985). Chlorine toxicity (letter). *Chem. Engr, Lond.,* **419**, 4

WITHERS, R.M.J. (1986a). The lethal toxicity of ammonia. A report to the 'Major Hazards Advisory Panel'. *Refinement of Estimates of the Consequences of Heavy Toxic Vapour Release* (Rugby: Instn Chem. Engrs), p. 6.1

WITHERS, J. (1986b). Toxic effects of chlorine. First report of the toxicity working party. In Wall, K. (1986), *op. cit.,* p. 52

WITHERS, J. (1988). *Major Industrial Hazards* (Aldershot: Gower)

WITHERS, R.M.J. and LEES, F.P. (1985a). The assessment of major hazards: the lethal toxicity of chlorine. Part 1, Review of information on toxicity. *J. Haz. Materials,* **12**(3), 231

WITHERS, R.M.J. and LEES, F.P. (1985b). The assessment of major hazards: the lethal toxicity of chlorine. Part 2, Model of toxicity to man. *J. Haz. Materials,* **12**(3), 283

WITHERS, R.M.J. and LEES, F.P. (1986a). The assessment of major hazards: the factors affecting lethal toxicity estimates and the associated uncertainties. *Hazards,* **IX**, p. 185

WITHERS, R.M.J. and LEES, F.P. (1986b). The assessment of major hazards: the lethal toxicity of bromine. *J. Haz. Materials,* **13**(3), 279

WITHERS, R.M.J. and LEES, F.P. (1987a). The assessment of major hazards: the lethal toxicity of bromine (letter). *J. Haz. Materials,* **14**, 274

WITHERS, R.M.J. and LEES, F.P. (1987b). The assessment of major hazards: the lethal toxicity of chlorine: 3, Cross-checks from gas warfare. *J. Haz. Materials,* **15**(3), 301

WITHERS, R.M.J. and LEES, F.P. (1991). The assessment of major hazards: the lethal effects of a condensed phase explosion in a built-up area. *Hazards,* **XI**, p. 37

WITHERS, R.M.J. and LEES, F.P. (1992). The assessment of major hazards: the lethal toxicity of chlorine (letter). *J. Haz. Materials,* **32**, 113

WITKIN, D.E. (1968). A new code worth its weight in metal. *Chem. Engng,* **75**(Aug. 26), 124

WITKIN, D.E. and MRAZ, G.J. (1978). Pressure vessel codes and reactor design. *Safety in Polyethylene Plants,* **3**, p. 20

WITLOX, H.W.M. (1990). Mathematical modelling of three-dimensional dense-gas dispersion problems. *Proc. Third*

IMA Conf. on Stably Stratified Flows (Inst. of Mathematics and its Applications)

WITLOX, H.W.M. (1991). Recent development of heavy-gas dispersion modelling. *Modeling and Mitigating the Consequences of Accidental Releases,* p. 183

WITLOX, H.W.M., MACFARLANE, K., REES, F.J. and PUTTOCK, J.S. (1990). *Development and Validation of Atmospheric Dispersion Models for Ideal Gases and Hydrogen Fluoride, vol. II, HGSYSTEM Program User's Manual* (report). Shell Res. Ltd

WITSCHI, H. and NETTHESHEIM, P. (eds) (1981–). *Mechanisms in Respiratory Toxicology,* vols 1–2 (Boca Raton, FL: CRC Press)

WITSCHAKOWSKI, W. (1974). *Hochdrucktechnik* (Ludwigshafen: BASF)

WITT, H.H. (1985). The difficulty of testing redundancy systems (validation of a reactor containment systems). *Reliab. Engng,* **13**(3), 185

WITT, P.A. (1971a). Applications of acoustic emission technology. *Loss Prevention,* **5**, p. 106. See also *Chem. Engng Prog.,* 1972, **68**(1), 56

WITT, P.A. (1971b). Disposal of solids wastes. *Chem. Engng,* **78**(Oct. 4), 62

WITT, P.A. (1972). Solid waste disposal. *Chem. Engng,* **79**(May 8), 109

WITTE, L.C. and COX, J.E. (1978). The vapour explosion: a second look. *J. Met.,* **30**(10), 29

WITTE, L.C., COX, J.E. and BOUVIER, J.E. (1970). The vapour explosion. *J. Met.,* **22**(2), 39

WITTER, R.E. (1982). Safety program payoff. *Plant/ Operations Prog.,* **1**, 139

WITTER, R.E. (1992). Guidelines for Hazard Evaluation Procedures: second edition. *Plant/Operations Prog.,* **11**, 50

WITTEVEEN, J. and TWILT, L. (1981/82). A critical view on the results of standard fire resistance tests on steel columns. *Fire Safety J.,* **4**(4), 259

WITTIG, S.L.K. (1970). High volume submillisecond pressure relief emergency system. *Ind Engng Chem. Process Des. Dev.,* **9**, 605

WOGALTER, M.S., GODFREY, S.S., FONTENELLE, G.A., DESAULNIERS, D.R., ROTHSTEIN, P.R. and LAUGHERY, K.R. (1987). The effectiveness of warnings. *Hum. Factors,* **29**, 599

WOHL, K. (1953). Quenching, flash-back, blow-off – theory and experiment. *Combustion,* 4, p. 68

WOHL, K. (1955). Burning velocity of unconfined turbulent flames. Theory for turbulent burning velocity. *Ind. Engng Chem.,* **47**, 825

WOHL, K. and SHORE, L. (1955). Burning velocity of unconfined turbulent flames. Experiments with butane-air and methane-air flames. *Ind. Engng Chem.,* **47**, 828

WOHL, K., GAZLEY, C. and KAPP, N. (1949). Diffusion flames. *Combustion,* **3**, p. 288

WOHL, K., KAPP, N.M. and GAZLEY, C. (1949). The stability of open flames. *Combustion,* **3**, p. 3

WOHL, K., SHORE, L., VON ROSENBERG, H. and WEIL, C.W. (1953). The burning velocity of turbulent flames. *Combustion,* 4, p. 620

WOHLLEBEN, W. and VAHRENHOLT, F. (1981). Precautions against accidents in chemical facilities. *J. Haz. Materials,* **5**(1/2), 41

WOHLSLAGEL, J., DIPASQUALE, L.C. and VERNOT, E.H. (1976). Toxicity of solid rocket motor exhaust: effects of HCl, HF and alumina on rodents. *J. Combust. Toxicol.,* **3**, 61

WOINSKY, S.G. (1972). Predicting flammable material classification. *Chem. Engng,* **79**(Nov. 27), 81

WOISIN, G. (1982). Comments on Vaughan: 'The tearing strength of mild steel plate'. *J. Ship Res.,* **26**, 50

WOLANSKI, P. (1982a). Dust explosion. In Lee, J.H.S. and Guirao, C.M. (1982), *op. cit.,* p. 979

WOLANSKI, P. (1982b). Fundamental problems of dust explosions. In Lee, J.H.S. and Guirao, C.M. (1982), *op. cit.,* p. 349

WOLANSKI, P., KAUFFMAN, C.W., SICHEL, M. and NICHOLLS, J.A. (1981). Detonation in methane-air mixtures. *Combustion,* 18, p. 1651

WOLANSKI, P., LEE, D., SICHEL, M., KAUFFMAN, C.W. and NICHOLLS, J.A. (1984). The structure of dust detonations. In Bowen, J.R. *et al.* (1984b), *op. cit.,* p. 241

WOLF, C.P. (1981). Public interest groups. *Ann. N.Y. Acad. Sci.,* **365**, 244

WOLF, K. (1985). Hazardous substances and cancer incidence: introduction to the special issue on risk assessment and risk management. *J. Haz. Materials,* **10**(2/3), 167

WOLF, K., HOLLAND, R. and RAJARATNAM, A. (1987). Vinyl chloride contamination: the hidden threat. *J. Haz. Materials,* **15**(1/2), 163

WOLFF, D. (2001). Producing hydrogen on-site. *Chem. Process.* **64** (3) (March), 51–53

WOLFF, F., EYRE, D.V. and GRENIER, M.R. (1979). Oxygen plants: 10 years of development and operation. *Chem. Engng Prog.,* **75**(7), 83

WOLFF, H.J. (1990). Medical surveillance and biological monitoring of workers in the chemical industry in the Federal Republic of Germany. *Safety and Loss prevention in the Chemical and Oil Processing Industries, p.325*

WOLFF, H.S. (1970). The hospital ward – a technological desert. In Scientific Instrument Research Association (1970), *op. cit.,* p. 90

WOLFF, M.F., CARRIER, G.F. and FENDELL, F.E. (1991). Wind- aided firespread across arrays of discrete fuel elements, Pt 2. *Combust. Sci. Technol.,* **77**, 261. See also Carrier, G.F., Fendell, F.E. and Wolff, M.F. (1991)

WOLFF, R.K., SILBAUGH, S.A., BROWNSTEIN, D.G., CARPENTER, R.L. and MAUDERLY, J.L. (1979). Toxicity of 0.4 and 0.8 μm sulphuric acid aerosols in the guinea pig. *J. Toxicol. Environ. Health,* **5**, 1037

WOLFHARD, H.G. (1958). The ignition of combustible mixtures by hot gases. *Jet Propulsion,* **28**, 798

WOLFHARD, H.G. and BRUSZAK, A.E. (1960). The passage of explosions through narrow cylindrical channels. *Combust. Flame,* 4, 149

WOLFHARD, H.G. and VANPEE, M. (1959). Ignition of fuel–air mixtures by hot gases and its relationship to firedamp explosions. *Combustion,* 7, p. 446

WOLFRUM, J. (1985). Chemical kinetics in combustion systems: the specific effects of energy, collisions, and transport processes. *Combustion,* 20, p. 559

WOLFSON, B.T. and DUNN, R.C. (1956). Generalized charts of detonation parameters for gaseous mixtures. *J. Chem. Engng Data,* **1**, 77

WOLFSON ELECTROSTATIC UNIT (1991). Guidelines on the electrostatic hazards during pneumatic conveying and storage of flammable substances. *J. Loss Prev. Process Ind.,* **4**, 211

WOLLENBERG, B.F. (1986). Feasibility study for an energy management system intelligent alarm processor. *IEEE Trans Power Systems,* **PWRS-1**(2), 241

WOLSTENHOLME, A.W. (1962). A general appraisal of the role of work study in the design process. *Chem. Engr, Lond.,* **163**, A88

WONG, K.K. (1982). Computer security. In Green, A.E. (1982), *op. cit., p. 353*

WONG, R.F. (1992). Troubleshooting rotating equipment. *Hydrocarbon Process.*, **71**(10), 123

WONG, S. and STEWARD, F.R. (1988). Radiative interchange factors between flames and tank car surfaces. *J. Haz. Materials*, **20**, 137

WONG, W. and ANSLEY-WATSON, P. (1985). Seal specification to improve life and reliability. *Process Engng*, Jun., 29

WONG, W.W. (1989). Safer relief valve sizing. *Chem. Engng*, **96**(5), 137

WONG, W.Y. (1992a). Capacity credit calculation for exchanger tube rupture. *Hydrocarbon Process.*, **71**(12), 89

WONG, W.Y. (1992b). PRV sizing for exchanger tube rupture. *Hydrocarbon Process.*, **71**(2), 59

WONG, W.Y. (1992c). Size relief valves more accurately. *Chem. Engng*, **99**(6), 163

WOOD, M. and GREEN, A. (1998). A methodological approach to process intensification. *IChemE Symp. Ser.* **144**, 405–16

WOOD, B.D. and BLACKSHEAR, P.L. (1969). *Some Observations on the Mode of Burning of a One and One-half Meter Diameter Pan of Fuel. Rep.* (NTIS AD704713). Univ. of Minnesota

WOOD, B.D., BLACKSHEAR, P.L. and ECKERT, E.R.G. (1971). Mass fire model: an experimental study of the heat transfer to liquid fuel burning from a sand-filled pan burner. *Combust. Sci. Technol.*, **4**, 113

WOOD, C.M. (1961). 'Automatic' Sprinkler Hydraulic Data. 'Automatic' Sprinkler Corp. of America, Cleveland, OH

WOOD, D.R., MUEHL, E.J. and LYON, A.E. (1974). Determining process plant reliability. *Chem. Engng Prog.*, **70**(10), 62

WOOD, E.F. (1984). *Groundwater Contamination from Hazardous Wastes* (Englewood Cliffs, NJ: Prentice-Hall)

WOOD, J.L. (1969). Explosion problems in chlorine manufacture. *Loss Prevention*, **3**, p. 45

WOOD, J.R. (1994). Chemical warfare – a chemical and toxicological review. *Am. J. Public Health*, **34**, 455

WOOD, K. (1971). Application of safety in major plant expansion. *Major Loss Prevention*, p. 61

WOOD, L. (1981). Metal rupture discs. *Fire Prev. Sci. Technol.*, **24**, 20

WOOD, L.E. (1965). Rupture discs. *Chem. Engng Prog.*, **61**(2), 93

WOOD, S. (1986). Expert systems in theoretically ill-formulated domains. In *Research and Development in Expert Systems* (ed. M.A. Bramer), p. 132

WOOD, S.T. (1984). Documentation for operating procedures and maintenance systems. *Plant Engr*, **27**(5) WOOD, W.P. (1989) *Piper Alpha Publuc Inquiry. Transcript, Day 107*

WOOD, W.S. (1963). How to handle additives safely. *Hydrocarbon Process.*, **42**(7), 183

WOOD, W.S. (1965). Flammable materials. In Fawcett, H.H. and Wood, W.S. (1965), *op. cit.*, p. 330

WOOD, W.S. (1973). Transporting, loading and unloading of hazardous materials. *Chem. Engng*, **80**(Jun. 25), 72

WOOD, W.S. (1982a). Safe handling of flammable and combustible materials. In Fawcett, H.H. and Wood, W.S. (1982), *op. cit.*, p. 339

WOOD, W.S. (1982b). Transportation of hazardous materials. In Fawcett, H.H. and Wood, W.S. (1982), *op. cit.*, p. 681

WOOD, W.S. (1988). Full hydraulic calculations for sprinkler systems. *Fire Surveyor*, **17**(4), 64

WOODARD, A.M. (1973). How to design a plant fire-water system. *Hydrocarbon Process.*, **52**(10), 103

WOODELL, M.L. and Ex, B. (1993). Process safety management and product stewardship: re-engineering engineers (poster item). *Process Safety Management*, p. 435

WOODHEAD, F. (1992). Laboratory fire safety. *Fire Surveyor*, **21**(5), 16

WOODHOUSE, G., SAMOLS, D. and NEWMAN, J. (1974). The economics and technology of large ethylene projects. *Chem. Engng*, **81**(Mar. 18), 72

WOODLEY, D.R. (ed.) (1964). *Encyclopaedia of Materials Handling* (Oxford: Pergamon Press)

WOODMAN, A.L., RICHTER, H.P., ADICOFF, A. and GORDON, A.S. (1978). AFFF spreading properties at elevated temperatures. *Fire Technol.*, **14**, 265

WOODMAN, R.C. (1967). Replacement policies for components that deteriorate. *Opl Res. Q.*, **18**, 3

WOODRUFF, D.M. (1993). A common sense approach to quality management. *Hydrocarbon Process.*, **72**(4), 143

WOODS, D. (1984). Some results on operator performance in emergency events. *Ergonomic Problems in Process Operations* (Rugby: Instn Chem. Engrs), p. 21

WOODS, D.D., O'BRIEN, J.F. and HANES, L.F. (1987). Human factors challenges in process control: the case of nuclear power plants. In Salvendy, G. (1987), *op. cit.*, p. 1724

WOODS, J.T. (1981). Preventing flammable atmosphere in electrical equipment rooms. *Loss Prevention*, **14**, p. 90

WOODS LTD (1960). *Woods Practical Guide to Fan Engineering* (London: Constable)

WOODS, P. (1993). Lightning – don't get caught. *Chem. Engr, Lond.*, **558**, 17

WOODS, W.A. and SPEECHLY, D. (1979). Principles of total containment system design. In Minors, C. (1979), *op. cit.*, paper 7

WOODS, W.A. and THORNTON, R.E. (1967). Calculation of transient forces during emergency escape of gases from an autoclave, with special reference to design methods. *Chemical Process Hazards*, **III**, p. 86

WOODSON, W.E. (1987). *Human Factors Reference Guide for Process Plants* (New York: McGraw-Hill)

WOODSON, W.E. and CONOVER, D.W. (1964). *Human Engineering Guide for Equipment Designers,* 2nd ed. (Los Angeles, CA: Univ. of California Press)

WOODWARD, C.D. (1989). The industrial fire problem. *Fire Safety J.*, **15**, 348

WOODWARD, E.C. and DREW, W.B. (1971). Flame quenching device using non-parallel walls (letter). *Combust. Flame*, **16**, 203

WOODWARD, J.L. (1989a). Dispersion modelling of an elevated high momentum release forming aerosols. *J. Loss Prev. Process Ind.*, **2**(1), 22

WOODWARD, J.L. (1989b). Does separation of hazmat cars in a railroad train improve safety from derailments. *J. Loss Prev. Process Ind.*, **2**(3), 176

WOODWARD, J.L. (1990). An integrated model for discharge rate, pool spread, and dispersion from punctured vessels. *J. Loss Prev. Process Ind.*, **3**(1), 33

WOODWARD, J.L. (1993). Expansion zone modeling of two-phase and gas discharges. *J. Haz. Mats*, **33**, 307

WOODWARD, J.L. (2000). A model of indoor releases with recirculating ventilation. *Process Safety Prog.*, **19**(3), 160

WOODWARD, J.L. and KETCHUM, D.K. (2001). Investigation and modeling of an explosion in a propane absorption column, *J. Loss Prev. Process Ind.* **14**(4), 251

WOODWARD, J.L. and THOMAS, J.K. (2000). Modeling indoor dispersion of aerosols or vapors and subsequent vented fire or explosion, *Mary Kay O'Connor 2000 Annu. Symp.,* College Station, Texas Oct. 24–25

WOODWARD, J.L., THOMAS, J.K. and KELLY, B.D. (2002). Lessons learned from an explosion in a large fractionator, *AIChE, 36th Loss Prev. Symp.* New Orleans, pp. 583–597, Mar. 10–14. *Process Safety Prog.* Feb. 2003

WOODWARD, J.L., HAVENS, J.A., MCBRIDE, W.C. and TAFT, J.R. (1982). A comparison with experimental data of several models for dispersion of heavy vapor clouds. *J. Haz. Materials,* **6**(1/2), 161. See also in Britter, R.E. and Griffiths, R.F. (1982), *op. cit.,* p. 161

WOODWARD, J.L. and MUDAN, K.S. (1991). Liquid and gas discharge rates through holes in process vessels. *J. Loss Prev. Process Ind.,* 4, 161

WOODWARD, J.L. and PAPADOURAKIS, A. (1991). Modeling of droplet entrainment and evaporation in a dispersing jet. *Modeling and Mitigating the Consequences of Accidental Releases,* p. 147

WOODWARD, J.L. and SILVESTRO, F.B. (1988). Application of risk assessment to a titanium dioxide plant which uses chlorine as a process fluid. *Preventing Major Chemical and Related Process Accidents,* p. 349

WOODWARD, J.L., HAVENS, J.A., MCBRIDE, W.C. and TAFT, J.R. (1983). A comparison with experimental data of several models for dispersion of heavy vapor clouds. In de Wispelaere, C. (1983), *op. cit.,* p. 477

WOODWARD, M.R. (1991). Aspects of path testing and mutation testing. In Ince, D. (1991a), *op. cit.* p. 1

WOODWARD-LUNDGREN AND ASSOCIATES (1973). *Cargo Spill Probability Analysis for the Deepwater Port Project* (rep. to Corps of Engineers)

WOODWORTH, M.E. (1955). Whiting refinery fire. *NFPA Quart.,* **49**(2), 79

WOODWORTH, M.E. (1958). Oil froth fire at Signal Hill refinery. *NFPA Quart.,* **52**(2), 85

WOODWORTH, M.E. (1968). The size factor in hazard evaluation. *Carriage of Dangerous Goods,* **1**, p. 63

WOODWORTH, M. (1973). Refinery fire at Shell Netherlands. In Vervalin, C.H. (1973a), *op. cit.,* p. 36

WOOLEY, R.J. (1986). Calculator program for finding values of physical properties. *Chem. Engng,* **93**(Mar. 31), 109

WOOLF, A.D. (1977). Asbestos hazards and standards. A study in the arbitrary acceptance of hazards. In Council for Science and Society (1977), *op. cit.,* p. 91

WOOLFOLK (1971). *Correlation of the rate of explosion with blast effects for non-ideal explosions. Contract 0017–69–C-4432, Final Rep.* Stanford Res. Inst., Menlo Park, CA

WOOLFOLK, R.W. and ABLOW, C.M. (1973). Blast waves from non-ideal explosions. *Proc. Conf on Mechanisms of Explosions and Blast Waves* (ed. J. Alstor), paper IV

WOOLFOLK, W. and SANDERS, R. (1984). Process safety relief valve testing. *Chem. Engng Prog.,* **80**(3), 60

WOOLFOLK, W.H. and SANDERS, R.E. (1987). Dynamic testing and maintenance of safety relief valves. *Chem. Engng,* **94**(15), 119

WOOLHEAD, F. (1989). Preventing and controlling cable fires. *Fire Surveyor,* **18**(3), 24

WOOLHEAD, F. (1990). The threat to business due to fire and electrical cable terminations. *Fire Surveyor,* **19**(2), 17

WOOLHEAD, G. (1981). Computerised medical and occupational system. *Chem. Ind.,* Sep. 5, 587

WOOLHOUSE, R.A. and SAYERS, D.R. (1973). Monnex compared with other potassium-based dry chemicals. *Fire J.,* **671**(1), 85

WOOLLATT, D. (1993). Factors affecting reciprocating compressor performance. *Hydrocarbon Process.,* **72**(6), 57

WOOLLEY, W.D. and FARDELL, P.J. (1982). Basic aspects of combustion toxicology. *Fire Safety J.,* **5**, 29

WOOLLEY, W.D. and RAFTERY, M.M. (1976). Smoke and toxicity hazards of plastics in fires. *J. Haz. Materials,* **1**(3), 215

WOOTTON, G. (1970). *Interest Groups* (Englewood Cliffs, NJ: Prentice-Hall)

WORLD BANK, OFFICE OF ENVIRONMENTAL AND SCIENTIFIC AFFAIRS (1985). *Manual of Hazard Assessment Techniques* (Washington, DC)

WORLD HEALTH ORGANIZATION (1972). *Health Hazards of the Human Environment* (Geneva)

WORLD HEALTH ORGANIZATION (1977). *Monographs on the Carcinogenic Risk of Chemicals to Man, vol. 15,* Some Fumigants, the Herbicides 2,4-D and 2,4,5-T, Chlorinated Dibenzodioxins and Miscellaneous Industrial Chemicals (Lyon)

WORLD HEALTH ORGANIZATION (1978). *Principles and Methods for Evaluating the Toxicity of Chemicals,* pt I (Geneva)

WORLD HEALTH ORGANIZATION (1983a). *Acrylonitrile. Environmental Health Criteria 28* (Geneva)

WORLD HEALTH ORGANIZATION (1983b). *Hydrogen sulphide. Environmental Health Criteria 19* (Geneva)

WORLD HEALTH ORGANIZATION (1986a). *Ammonia. Environmental Health Criteria 54* (Geneva)

WORLD HEALTH ORGANIZATION (1986b). *Asbestos and Other Natural Mineral Fibres. Environmental Health Criteria 53* (Geneva)

WORLD HEALTH ORGANIZATION — see also Appendix 28

WORLD OFFSHORE ACCIDENTS DATABANK (1988). *WOAD Statistical Report* (Hovik, Norway: Veritas Technology and Services A/S)

WORLEDGE, D.H. (1993). Industry advances in reliability centred maintenance. *Process Safety Environ.,* **71B**, 155

WORLEDGE, D.H. *et al.* (1988). EDF/EPRI collaborative program on operator reliability experiments. *Int. ENS/ ANS Conf. on Thermal Reactor Safety* (New York: Am. Nucl. Soc.)

WORLEY, N.G. and LEWINS, J. (eds) (1988). *The Chernobyl Accident and its Implications for the United Kingdom* (London: Watt Cttee on Energy)

WORRALL, E.M. and MERT, B. (1980). Application of dynamic scheduling rules in maintenance planning and scheduling. *Int. J. Prod. Res.,* **18**, 57

WORRELL, G.R. (1979). If ethylene decomposes in a pipe... *Hydrocarbon Process.,* **58**(4), 255

WORRELL, R.B. (1974). *Set Equation Transformation System (SETS). Rep.* Albuquerque, NM

WORRELL, R.B. and STACK, D.W. (1977). *A SETS User's Manual for the Fault Tree Analyst. Rep. SAND 77–2051.* Sandia Labs, Albuquerque, NM

WORRELL, R.B. (1985). *SETS Reference Manual. Rep. SAND 83–2675.* Sandia Nat. Lab., Albuquerque, NM, See also *NUREG/CR-4213*

WORSTER, R.C. (1959). *The Closing of Reflux Valves. Rep. SP626.* Br. Hydromech. Res. Ass., Cranfield

WORTHINGTON, D.R.E. (1992). Recent computer modelling advances for process hazard analysis. *Plant/Operations Prog.,* **11**, 106

WOTTGE, K.R. (1989). *Piper Alpha Public Inquiry. Transcript, Days 1–9, 85, 90, 112, 113*

WRAITH, R.E. and LAMB, E.B. (1971). *Public Enquiries as an Instrument of Government* (London: Allen & Unwin)

WRATT, D.S. (1987). An experimental investigation of some methods of estimating turbulence parameters for use in dispersion models. *Atmos. Environ.*, **21**, 2599

WREATHALL, J. (1993). Human error and process safety. *Health, Safety and Loss Prevention*, p. 82

WREATHALL, J. and REASON, J. (1993). Managing safety through controlling the causes of human error. *Process Safety Management*, p. 241

WRENN, C. (1986). Inerting for safety. *Plant/Operations Prog.*, **5**, 225

WRIGHT, C. (1986). Routine deaths: fatal accidents in the oil industry. *Sociological Rev.*, **34**, 265

WRIGHT, C.E. (1968). Nonmetallic pipe: promise and problems. *Chem. Engng*, **75**(Jun. 17), 23

WRIGHT, D. (1971). Practical hints on vibration. *Hydrocarbon Process.*, **50**(1), 83

WRIGHT, D. (1993). *A Guide to the IChemE's Model Forms of Conditions of Contract* (Rugby: Instn Chem. Engrs) (the '*Purple Book*')

WRIGHT, G.T. (1961). Oxygen plant vaporizer explosion. *Ammonia Plant Safety*, **3**, p. 9. See also *Chem. Engng Prog.*, 1961, **57**(4), 45

WRIGHT, H.C. (1973). *Infrared Techniques* (Oxford: Oxford Univ. Press)

WRIGHT, J. (1975). Which shaft coupling is best – lubricated or non-lubricated? *Hydrocarbon Process.*, **54**(4), 191

WRIGHT, J.B. (1981). Fire testing of valves. *API Proc., Refining Dept, 46th Midyear Mtg*, **60**, 267

WRIGHT, J.B. (1982). Cast austenitic alloys for valves. *Chem. Engng Prog.*, **78**(12)

WRIGHT, J.B. (1993). Avoid valve leaks. *Chem. Engng Prog.*, **89**(6), 62

WRIGHT, J.K. (1961). *Shock Tubes* (London: Methuen)

WRIGHT, J.M. and FRYER, K.C. (1982). Alternative fire protection for LPG vessels. *Gastech 81*, p. 325

WRIGHT, K. (1973). Plant communications. *Ammonia Plant Safety*, **15**, p. 37

WRIGHT, K. (1975). Accident involving hot water. *Ammonia Plant Safety*, **17**, p. 142

WRIGHT, K. and HANEY, G. (1972). Catalyst fusion. *Ammonia Plant Safety*, **14**, p. 34

WRIGHT, K.G. (1972). *Cost Effective Security* (New York: McGraw-Hill)

WRIGHT, L. and GINSBURGH, I. (1963). Take a new look at static electricity. *Hydrocarbon Process.*, **42**(10), 175

WRIGHT, L. and GINSBURGH, I. (1964). What experimentation shows about static electricity. In Vervalin, C.H. (1964a), *op. cit.* p. 153

WRIGHT, L.T. (1968). Piping codes and standards. *Chem. Engng*, **75**(Jun. 17), 247

WRIGHT, L.T. (1962). Try this analog for solving your fire-water problems. *Hydrocarbon Process.*, **41**(9), 274

WRIGHT, L.T. (1964). Use this analog for solving your fire-water problems. In Vervalin, C.H. (1964a), *op. cit.*, p. 285

WRIGHT, M., BELLAMY, L. and COX, R.A. (1993). Recent developments in chemical plant QRA. *Health, Safety and Loss Prevention*, p. 112

WRIGHT, M.R. (1985). Legislation for the control of hazardous materials during storage, transportation and use. *Chem. Ind.*, Jul. 18, 470

WRIGHT, R.G. (1971). Startup made easier. Computer systems. *Hydrocarbon Process.*, **50**(12), 94

WRIGHT, R.I. (1980). Instrument reliability. *J. Phys. E: Sci. Instrum.*, **13**, 8

WRIGHT, R.E., STEVERSON, J.A. and ZUROFF, W.F. (1987). Pipe break frequency estimation for nuclear power plants. *Nucl. Regul. Comm., Rep.* NUREG/CR-4407, (Washington, DC)

WRIGHT, R.V. (1962). Fire protection handbook. *Hydrocarbon Process.*, **41**(10), 175

WRIGHT, R.V. (1964). Fundamentals of static electricity. In Vervalin, C.H. (1964a), *op. cit.*, p. 140

WRIGHT, T.K. and BUTTERWORTH, C.W. (1987). Isothermal heat flow calorimeter. *Hazards from Pressure*, p. 85

WRIGHT, T.K. and ROGERS, R.L. (1986). Adiabatic Dewar calorimeter. *Hazards*, **IX**, p. 121

WRINCH, D. and JEFFREYS, H. (1923). On the seismic waves from the Oppau explosion. *MNRAS Geophys.*, **1**, (suppl.), 1

WROTNOWSKI, A.C. (1965). Oil mist removal in a synthetic ammonia plant. *Ammonia Plant Safety*, **7**, p. 24

WU, F.H. (1984). Drum separator design. A new approach. *Chem. Engng*, **91**(Apr. 2), 74

WU, J.M. and SCHROY, J.M. (1979). Emissions from spills. *Am. Air Poll. Control Ass., Special Conf. on Control of Specific (Toxic) Pollutants*, Pittsburgh

WU, J.S. and APOSTOLAKIS, G.E. (1992). Experience with probabilistic risk assessment in the nuclear power industry. *J. Haz. Materials*, **29**, 313

WU, R.S.H. (1985). Dynamic thermal analyzer for monitoring batch processes. *Chem. Engng Prog.*, **81**(9), 57

WU, S.S. (1965). A study of heat transfer coefficients in the lowest 400 meters of the atmosphere. *J. Geophys. Res.*, **70**, 1801

WU, Y. and SWITHENBANK, J. (1992). Experimental studies on gas-dynamics of venting explosions. *Chem. Engng Res. Des.*, **70A**, 200

WU-CHIEN, J.S. and APOSTOLAKIS, G. (1981). On risk aversion in acceptance criteria. *Reliab. Engng*, **2**(1), 45

WULFF, W. (1973). Fabric ignition. *Texas Res. J.*, Oct.

WULFF, W. and DURBETAKI, P. (1974). Fabric ignition and the burn injury hazard. In Afgan, N. and Beer, J.M. (1974), *op. cit.*, p. 451

WULFF, W., ZUBER, N., ALKIDAS, A. and HESS, R.W. (1973). Ignition of fabrics under radiative heating. *Combust. Sci. Technol.*, **6**, 321

WULPI, D.J. (1985). *Understanding How Components Fail* (Metals Park, OH: Am. Soc. for Metals)

WUTZ, M., ADAM, H. and WALCHER, W. (1989). *Theory and Practice of Vacuum Technology* (Braunschweig/Wiesbaden: Vieweg)

WYATT, L.M. (1976). *Materials of Construction for Steam Power Plant* (Barking: Applied Science)

WYLDE, L.E. and SHAW, T. (1983). Predictive monitoring of process equipment by statistical analysis of ultrasonic survey data. In Butcher, D.W. (1983), *op. cit.*, p. 137

WYMA, B.H. (1962). Uses of aluminum alloys expand. *Ind. Engng Chem.*, **54**(12), 46

WYNDER, E.L. and GORI, G.B. (1977). Contribution of the environment to cancer incidence: an epidemiological exercise (editorial). *J. Nat. Cancer Inst.*, **58**, 825

WYNGAARD, J.C. (1973). On surface layer turbulence. In Haugen, D.A. (1973), *op. cit.*, p. 101

WYNGAARD, J.C. (1975). Modeling the planetary boundary layer. Extension to the stable case. *Boundary Layer Met.*, **9**, 441

WYNGAARD, J.C. (1982). Boundary layer modelling. In Nieuwstadt, F.T.M. and van Dop, H. (1982), *op. cit.*, p. 69

WYNGAARD, J.C. and COTE, O.R. (1974). A planetary boundary layer model. *First Symp. on Atmospheric Diffusion and Air Pollution* (Boston, Mass.: Am. Met. Soc.), p. 23

WYNGAARD, J.C., ARYA, S.P.S. and COTE, O.R. (1974). Some aspects of the structure of the planetary boundary layer. *J. Atmos. Sci.*, **31**, 747

WYNGAARD, J.C., COTE, O.R. and RAO, K.S. (1974). Modeling the atmospheric boundary layer. *Adv. Geophys.*, **18A**, 193

WYNN, A.H.A. (1950). *Applications of Probability Theory to the Study of Mining Accidents. Rep. 7.* Safety in Mines Res. Estab.

WYNNE, B. (1980). Technology, risk and participation: on the social treatment of uncertainty. In Conrad, J. (1980), *op. cit.*

WYNNE, B. (1982). *Rationality and Ritual: the Windscale Inquiry and Nuclear Decisions in Britain* (Chalfont St Giles: Br. Soc. for the History of Sci.)

WYNNE, B. (1989). Frameworks of rationality in risk management: towards the testing of a naive sociology. In Brown, J. (1989), *op. cit.*

WYNNE, B. (1992). Risk and social learning: reification to engagement. In Krimsky, S. and Golding, D. (1992), *op. cit.*

WYNNE, D. (1986). Brigades role in fighting ship fires. *Fire Int.*, **97**, 39

XIAO-JING WANG (1989). Chaotic oscillations of cool flames. *Combust. Flame,* **75**, 107

YAGI, S. and IINO, H. (1962). Radiation from soot particles in luminous flames. *Combustion*, **8**, p. 288

YAGLOM, A.M. (1977). Comments on wind and temperature flux-profile relationships. *Boundary Layer Met.*, **11**, 89

YAKI, S.J. and CARPENTER, R. (1970). Pump packings: a case history. *Hydrocarbon Process.*, **9**(7), 136

YAKOVLEV, Y.S. (1963). *The Hydrodynamics of an Explosion. Rep. FTD–TT–63–381/1 +2.* US Air Force, Wright Patterson Air Force Base

YALLOP, H.J. (1980). *Explosion Investigation* (Edinburgh: Forensic Sci. Soc. Scottish Acad. Press)

YAMAGUCHI, T. (1975). Systems approach to pollution-free petroleum refinery. *Proc. Ninth World Petrol. Cong.,* vol. 6 (London: Applied Science), p. 271

YAMAKAWA, T.M. (1982). Test tank programme for liquefied gas storage using the GT/MDC containment system. *Gastech 81*, p. 387

YAMANE, T. (1967). *Statistics,* 2nd ed. (New York: Harper & Row)

YAMANOUCHI, N. and NAGASAWA, H. (1979). Using LNG cold for air separation. *Chem. Engng Prog.*, **75**(7), 78

YAMAOKA, I. and TSUJI, H. (1979). An experimental study of flammability limits using counterflow flames. *Combustion*, **17**, p. 843

YAMAOKA, I. and TSUJI, H. (1985). Determination of burning velocity using counterflow flames. *Combustion*, **20**, p. 1883

YAMASHIRO, C.E., ESPIELL, L.G.S. and FARINA, I.H. (1986). Program determines two-phase flow. *Hydrocarbon Processing*, **65**(12), 46

YAMASHITA, Y., SHOJI, S.S. and SUZUKI, M. (1990). A fault-diagnosis system based on qualitative reasoning. *Int. Chem. Engng*, **30**, 151

YAMAZAKI, K. and TSUJI, H. (1962). An experimental investigation on the stability of turbulent burner flames. *Combustion*, **8**, p. 543

YAMPOLSKY, M.L., ADAM, J.A. and KARAHALIOS, P. (1985). Optimization of fault tree analysis using the factorial

design approach. *Am. Inst. Chem. Engrs, Summer Nat. Mtg* (New York)

YANG, B.T. and MERONEY, R.N. (1972). *On Diffusion from an Instantaneous Point Source in a Neutrally Stratified Boundary Layer with a Light Scattering Probe. ONR Rep. N0014–68–A–0493–0001.* Dept of Civil Engng, Colorado State Univ., Fort Collins, CO

YANG, C.H. (1961). Burning velocity and the structure of flames near the extinction limits. *Combust. Flame*, **5**, 163

YANG, C.H. (1962). Theory of ignition and autoignition. *Combust. Flame*, **6**, 215

YANG, C.H. and GRAY, B.F. (1967). The determination of explosion limits from a unified thermal and chain theory. *Combustion*, **11**, p. 1099

YANG, J., KOIRANEN, T., KRASLAWSKI, A. and NYSTROM, L. (1993). Object oriented knowledge based systems for process equipment selection. *Comput. Chem. Engng*, **17**, 1181

YANG, K. (1973). Explosive interaction of liquefied natural gas and organic liquids. *Nature*, **243**, 221

YANG, R.T. (1987). *Gas Separation by Adsorption Processes* (Oxford: Butterworth-Heinemann)

YANT, W.P. (1930). Hydrogen sulphide in industry occurrence, effects and treatment. *Am. J. Public Health,* 598

YAO, C. (1974). Explosion venting of low-strength equipment and structures. *Loss Prevention*, **8**, p. 1

YAO, C. (1982). Venting. In Lee, J.H.S. and Guirao, C.M. (1982), *op. cit.*, p. 987

YAO, C. (1988). Early suppression of fast response sprinkler systems. *Chem. Engng Prog.*, **84**(9), 38

YAO, C., DE RIS, J., BAJPAI, S.N. and BUCKLEY, J.L. (1969). *Evaluation of Protection from Explosion Overpressure in AEC Glove Boxes. FMR Serial 16215.* FM Res. Corp.

YAO, H.C. and RUOF, C.H. (1964). Evidence for two temperature regimes in the slow oxidation of methane. *Combust. Flame*, **8**, 179

YAO, J.T.P. (1985). *Safety and Reliability of Existing Structures* (London: Pitman)

YAO, J.T.P. and MUNSE, W.H. (1962). Low cycle fatigue of metals – literature review. *Weld. J.*, **41**(res. suppl.), 182s

YAO, S.C. and CHOI, K.J. (1987). *Int. J. Multiphase Flow*, **13**, 639

YAO-CHUNG TSAO, DRURY, C.G. and MORAWSKI, T.B. (1979). Human performance in sampling inspection. *Hum. Factors*, **21**, 99

YARO, E.D. (1973). Reliability and profitability analysis of parallel processing trains. M.S. Thesis, Univ. of Arizona

YAROCH, G.N. (1978). Government roles in controlling hazardous materials. *Control of Hazardous Material Spills 4*, p. 69

YASUHARA, K., KITA, S. and HIKI, F. (1978). Preventing ethylene plant explosions with an extruder. *Safety in Polyethylene Plants,* **3**, p. 76. See also *Chem. Engng Prog.,* 1977, **73**(12), 40

YATES, H.G. (1949). Vibration diagnosis in marine geared turbines. *Trans N.E. Coast Instn Engrs and Shipbuilders*

YAWAK, J.D. (1978). Transfer technology effectively. *Hydrocarbon Processing*, **57**(6), 167

YAWS, C.L. and XIANG PAN (1992). Liquid heat capacity for 300 organics. *Chem. Engng*, **99**(4), 130

YAWS, C.L., NI, H.M. and CHIANG, M. (1988). Heat capacities for 700 compounds. *Chem. Engng*, **95**(7) May 9, 91

YAWS, C.L., YANG, H.-C. and CAWLEY, W.A. (1990). Predict enthalpy of vaporisation. *Hydrocarbon Processing*, **69**(6), 87

YAWS, C.L., YANG, H.-C., HOPPER, J.R. and HANSEN, K.C. (1990a). Hydrocarbons: water solubility data. *Chem. Engng*, **97**(4), 177

YAWS, C.L., YANG, H.-C., HOPPER, J.R. and HANSEN, K.C. (1990b). Organic chemicals: water solubility data. *Chem. Engng*, **97**(7), 115

YAWS, C.L., HAUR-CHUNG YANG and XIANG PAN (1991). 633 organic chemicals surface tension data. *Chem. Engng*, **98**(3), 140

YAWS, C.L., XIANG PAN and XIAOYAN LIN (1993). Water solubility data for 151 hydrocarbons. *Chem Engng*, **100**(2), 108

YAWS, C.L. *et al.* (1974–). Physical and thermodynamic properties. *Chem. Engng*, 1974, **81**(Jun. 10), 70; (Jul.8), 85; (Aug. 19), 99; (Sep. 30), 115; (Oct. 28), 113; (Nov. 25), 91; (Dec. 23), 67; 1975, **82**(Jan. 20), 99; (Feb. 17), 87; (Mar. 31), 101; (May 12), 89; (Jul. 21), 113; (Sep. 1), 107; (Sep. 29), 73; (Oct. 27), 119; (Dec. 8), 119; 1976, **83**(Jan. 19), 107; (Mar. 1), 107; (Apr. 12), 129; (Jun. 7), 114; (Jul. 5), 81; (Aug. 16), 79; (Oct. 25), 127; (Nov. 22), 153

YAZDANI, M. (1984). Knowledge engineering in PROLOG. In Forsyth, R. (1984), *op. cit.*, p. 91

YAZDANI, M. (1985). *Prolog Programming* (Oxford: Blackwells Sci. Publs)

YAZDANI, M. (1986). *Artificial Intelligence* (London: Chapman & Hall)

YEDIDIAH, S. (1974). How to improve pump performance. *Hydrocarbon Process.*, **53**(4), 165

YEDIDIAH, S. (1977). Diagnosing troubles of centrifugal pumps. *Chem. Engng*, **84**(Oct. 24), 125; (Nov. 21), 193; (Dec. 5), 141

YELLAND, A.E.J. (1966). Design considerations when assessing safety in chemical plants. *Chem. Engr, Lond.*, **201**, CE214

YELLIN, J. (1976) The NRC's Reactor Safety Study. *Bell J. Economics*, **7**(1)

YELLMAN, T.W. (1979). Comment on 'Comment on computer aided synthesis of fault trees'. *IEEE Trans. Reliab.*, **R-28**, 10

YEN WANG (ed.) (1969). *CRC Handbook of Radioactive Nuclides* (Boca Raton, FL: CRC Press)

YEN-HSIUNG KIANG (1981). *Waste Energy Utilization Technology* (Basel: Dekker)

YERGES, L.F. (1978). *Sound, Noise and Vibration Control,* 2nd ed. (New York: van Nostrand Reinhold)

YERSEL, M., GOBLE, R. and MORRILL, J. (1983). Short range experiments in an urban area. *Atmos. Environ.*, **17**, 275

YETTER, R.A., DRYER, F.L. and RABITZ, H. (1991). A comprehensive reaction mechanism for carbon monoxide/hydrogen/oxygen kinetics. *Combust. Sci. Technol.*, **79**, 97

YIH, C.S. (1951). Free convection due to a point source of heat. *Proc. First U.S. Nat. Cong. on Applied Mechanics,* p. 941

YIH, C.S. (1965). *Dynamics of Non-Homogenous Fluids* (London: Macmillan)

YIH, C.S. (1980). *Stratified Flows,* 2nd ed. (New York: Academic Press)

YIH-YUAN HSU and CHENG-CHIN YU (1992). A self-learning fault diagnosis system based on reinforcement learning. *Ind. Engng Chem. Res.*, **31**, 1937

YINON, J. and ZITRIN, S. (1987). *The Analysis of Explosives* (Oxford: Pergamon Press)

YIP, H.T., WELLER, G.C. and ALLAN, R.N. (1984). Reliability evaluation of protection devices in electrical power systems. *Reliab. Engng*, **9**(4), 191

YOCOM, J.E. (1962). Air pollution regulations. *Chem. Engng*, **69**(Jul. 23), 103

YOCOM, J.E. and DUFFEE, R.A. (1970). Controlling industrial odors. *Chem. Engng*, **77**(Jun. 15), 160

YOCOM, J.E. and WHEELER, D.H. (1963). How to get the most from air-pollution control systems. *Chem. Engng*, **70**(Jun. 24), 126

YOCOM, J.E., COLLINS, G.F. and BOWNE, N.E. (1971). Plant site selection. *Chem. Engng*, **78**(Jun. 21), 164

YODER, D. (1968). Mechanical aspects of piping design. *Chem. Engng*, **75**(Jun. 17), 251

YOKELL, S. (1983). Troubleshooting shell-and-tube heat exchangers. *Chem. Engng*, **90**(Jul. 25), 57

YOKELL, S. (1986). Understanding the pressure vessel codes. *Chem. Engng*, **93**(May 12), 75

DE YONG, L.V. and CAMPENELLA, G. (1989). A study of blast characteristics of several primary explosive and pyrotechnic compositions. *J. Haz. Materials*, **20**, 125

YOON, B.S., OH, J.K. and YOON, E.S. (1991). Knowledge base representation for the fault diagnosis expert system using fault-consequence digraph. *Hwahak Konghak*, **29**(1), 19

YOSHIDA, T., MURANAGA, K., MATSUNAGA, T. and TAMURA, M. (1985). Evaluation of explosive properties of organic peroxides in a modified Mk III ballistic mortar. *J. Haz. Materials*, **12**(1), 27

YOSHII, S. (1986). Research and development of offshore marginal field development technology. *Gastech 85*, p. 169

YOSHIKAWA, T., KIMURA, F., KOIDE, T. and KURATA, S. (1990). An emergency computation model for the wind field and diffusion during accidental nuclear pollutants release. *Atmos. Environ.* **24A**, 2739

YOSHIMURA, J.S. (1993). Computer integrated manufacturing/processing in the CPI. *Hydrocarbon Process.*, **72**(5), 65

YOSHIZAWA, Y. and KUBOTA, H. (1982). Experimental study on gas-phase ignition of cellulose under thermal radiation. *Combustion*, **19**, p. 787

YOST, F.G. and HALL, I.J. (1976). Failure by the processes of nucleation and growth. *IEEE Trans Reliab.*, **R-25**, 45

YOTHERS, M.T. (ed.) (1977). *Standards and Practices for Instrumentation,* 5th ed. (Pittsburgh, PA: Instrum. Soc. Am.)

YOU, H.Z. and KUNG, H.C. (1985). Strong buoyant plumes of growing rack storage fires. *Combustion*, **20**, p. 1547

YOUNAS, M. and SHEIKH, A.K. (1987). Characteristics of Birnbaum and Saunders Model. *Reliab. Engng*, **19**(3), 201

YOUNG, A.D. (1989). *Boundary Layers* (Oxford: BSP Professional Books)

YOUNG, A.J. (1955). *Process Control System Design* (London: Longmans Green)

YOUNG, C.A.J. (1973). The chemical and petrochemical industries in the 1980s. *Phil. Trans R. Soc., Ser. A*, **275**, 329

YOUNG, C.R. (1990). Local participation in industrial accident prevention and emergency preparedness. *Safety and Loss Prevention in the Chemical and Oil Processing Industries,* p. 13

YOUNG, D.D. (1979). Water quality issues: the water industry view. *Chem. Ind.*, Mar. 17, 199

YOUNG, E.K. (1986). Controlling and eliminating perceived risks. *Chem. Engng Prog.*, **82**(7), 14

YOUNG, G.W. (1953). *The Grace of Forgetting* (London: Country Life)

YOUNG, J.A. (1987a). *Improving Safety in the Chemical Laboratory* (New York: Wiley)

YOUNG, J.A. (1987b). Safety inspections, safety audits. In Young, J.A. (1987a), *op. cit.*, p. 47

YOUNG, J.A. (1987c). The 95 percent solution. In Young, J.A, (1987a), *op. cit.*, p. 41

YOUNG, P.J. and PARKER, A. (1984). Origin and control of landfill odours. *Chem. Ind.*, May 7, 329

YOUNG, P.W. and CLARK, C.B. (1973). Why shift catalysts deactivate. *Ammonia Plant Safety*, **15**, p. 18

YOUNG, R. (1981). Automatic on-off sprinkler systems. *Fire Surveyor*, **10**(4), 12

YOUNG, R.A. and NASH, P. (1977). The fire protection of modern high bay storages. *Fire Prev. Sci. Technol.*, **18**, 4

YOUNG, R.E. (1977). *Supervisory Remote Control Systems* (London: Peter Peregrinus)

YOUNG, R.E. (1982). *Control in Hazardous Environments* (London: Peter Peregrinus)

YOUNGER, A.M. and RUITER, J.L. (1961). Selecting pumps and compressors. *Chem. Engng*, **68**(Jun. 26), 117

YOUNESS, A. (1970). New approach to tower deflection. *Hydrocarbon Process.*, **49**(6), 121

YOWELL, R.L. (1968). Bisphenol-A dust explosions. *Loss Prevention*, **2**, p. 29. See also *Chem. Engng Prog.*, 1968, **64**(6), 58

YU, G., LAW, C.K. and WU, C.K. (1986). Laminar flame speeds of hydrocarbon + air mixtures with hydrogen addition. *Combust. Flame*, **63**, 339

YU-PEN SU and SIRIGNANO, W.A. (1981). Ignition of a combustible mixture by an inert hot particle. *Combustion*, **18**, p. 1719

YUAN, J. (1987). A Bayes approach to reliability assessment for systems with dependent components. *Reliab. Engng*, **17**(1), 1

YUASA, S. (1985). Spontaneous ignition of sodium in dry and moist air streams. *Combustion*, **20**, p. 1869

YUE, M.H., SHARKEY, J.J. and LEUNG, J.C. (1994). Relief vent sizing for a Grignard reaction. *J. Loss Prev. Process Ind.*, **7**, 413

YUEN, M.C. and CHEN, L.W. (1976). On drag of evaporating liquid droplets. *Combust. Sci. Technol.*, **14**, 147

YUEN, M.C. and CHEN, L.W. (1978). Heat transfer measurements of evaporating liquid droplets. *J. Heat Mass Transfer*, **21**, 537

YUEN, M.H. (1970). Process plant utilities. Electricity. *Chem. Engng*, **77**(Dec. 14), 144

YUEN, W.W. and TIEN, C.L. (1977). A simple calculation scheme for the luminous-flame emissivity. *Combustion*, **16**, p. 1481

YUKIMACHI, T., NAGASAKA, A. and SASOU, K. (1992). A method for system reliability analysis with change-over operation. *Ergonomics*, **35**, 499

YULE, A.J., EREAUT, P.R. and UNGUT, A. (1983). Droplet sizes and velocities in vaporizing sprays. *Combust. Flame*, **54**, 15

YUMLU, V.S. (1967a). Prediction of burning velocities of carbon monoxide-hydrogen-air flames. *Combust. Flame,* **11**, 190

YUMLU, V.S. (1967b). Prediction of burning velocities of saturated carbon monoxide-air flames by application of mixing rules. *Combust. Flame*, **11**, 389

YUMLU, V.S. (1968). The effects of additives on the burning velocities of flames and their possible prediction by the mixing rule. *Combust.* **Flame, 12**, 14

YUMOTO, T. (1971a). *An Experimental Study on Heat Radiation from Oil Tank Fire. Rep. 33.* Fire Res. Inst. of Japan

YUMOTO, T. (1971b). Heat transfer from flame to fuel surface in large pool fires (letter). *Combust. Flame*, **17**, 108

YUMOTO, T. (1977). Fire spread between two oil tanks. *J. Fire Flammability*, **8**, 494

YUMOTO, T., TAKAHASHI, A. and HANDA, T. (1977). Combustion behaviour of liquid fuel in a small vessel: effect of convective motion in the liquid on burning rate of hexane in the early stages of combustion. *Combust. Flame*, **30**, 33

YUN, W.Y. and BAI, D.S. (1986). Optimal numbers of redundant units for parallel systems with common mode failures. *Reliab. Engng*, **16**(3), 201

YURKANIN, R.M. and CLAUSSEN, E.O. (1970). Safety aspects of electrical heating systems. *Chem. Engng*, **77**(Dec. 14), 164

ZABAGA, J.G. (1980). New floating-roof tank loss/emission factors. *API Proc., Refining Dept, 41st Midyear Mtg*, **59**, 81

ZABEL, N.R. (1958). *Containment of Fragments from a Runaway Reactor. Rep. SRI-1.* Stanford Res. Inst.

ZABETAKIS, M.G. (1965). Fire and explosion hazards at temperature and pressure extremes. *Am. Inst. Chem. Engrs-Instn Chem Engrs, Joint Mtg* (London: Instn Chem. Engrs)

ZABETAKIS, M.G. and JONES, G.W. (1953). Flammability of carbon disulphide in mixtures of air and water vapour. *Ind. Engng Chem.*, **45**, 2079

ZABETAKIS, M.G. and JONES, G.W. (1955). Prevention of industrial gas explosion disasters. *Chem. Engng Prog.*, **51**(9), 411

ZABETAKIS, M.G. and RICHMOND, J.K. (1953). The determination and graphic representation of the limits of flammability of complex hydrocarbon fuels at low temperatures and pressures. *Combustion*, **4**, p. 121

ZABETAKIS, M.G., SCOTT, G.S. and JONES, G.W. (1951). Limits of flammability of paraffin hydrocarbons in air. *Ind. Engng Chem.*, **43**, 2120

ZACK, R. and SKAMSER, R.O. (1983). Ethylene from NGL feedstocks, Pt 1. *Hydrocarbon Process.*, **62**(10), 99

ZADEH, L.A. (1965). Fuzzy sets. *Inf. Control*, **8**, 338

ZADEH, L. (1968). Probability measures of fuzzy sets. *J. Math. Anal. Appl.*, **23**, 421

ZADEH, L. (1975a). The concept of a linguistic variable and its application to approximate reasoning. *Inf. Sci.*, 1975, **8**, 199, 301; **9**, 43

ZADEH, L. (1975b) Fuzzy logic and approximate reasoning. *Synthese*, **30**, 407

ZADEH, L. (1978). Fuzzy sets as a basis for a theory of possibility. *Fuzzy Sets Syst.*, **1**, 3

ZADEH, L.A. (1979). A theory of approximate reasoning. *Machine Intelligence 9* (ed. J.E. Hayes, D. Michie and L.I. Mikulich) (Chichester: Ellis Horwood), p. 149

ZADEH, L. (1983a). A computational approach to fuzzy quantifiers in natural languages. *Comput. Math. Appl.*, **9**, 149

ZADEH (1983b). The role of fuzzy logic in the management of uncertainty in expert systems. *Fuzzy Sets Syst.*, **11**, 199

ZADEH (1985). Syllogistic reasoning in fuzzy logic and its application to usuality and reasoning with dispositions. *IEEE Trans Syst. Man Cybernet.*, **SMC-15**, 754

ZADEH, L.A. (1992). Fuzzy sets and fuzzy logic: an overview. In Shapiro, S.C. (1992), *op. cit.,* p. 507

ZADEH, L.A. and FUKANAKA, K. (eds) (1975). *Fuzzy Sets and their Applications to Cognitive Decision Processes* (New York: Academic Press)

ZADEH, L.A. and KACPRZYK, J. (eds) (1992). *Fuzzy Logic for the Management of Uncertainty* (New York: Wiley)

ZADIC, J.E. and HIMMELMAN, W.A. (1978). *Highly Hazardous Materials Spills and Emergency Planning* (New York: Dekker)

ZAFAR, R.S.A. (1986). Design for compressors with direct quench recycle. *Chem. Engr, Lond.*, **423**, 35

ZAFFT, R.W. (1967). Absorption refrigeration. *Hydrocarbon Process.*, **46**(6), 131

ZAHODIAKIN, P. (1990). Puzzling out the new Clean Air Act. *Chem. Engng*, **97**(12), 24

ZAHORSKY, J.R. (1985). Degradation of pressure relief valve performance caused by inlet piping configuration. *Am. Soc. Mech. Engrs, Proc. Mtg on Testing and Analysis of Safety/Relief Valve Performance* (New York), p. 23

ZAIRI, M. (1991). *Total Quality Management for Engineers* (Cambridge: Woodhead Publ.)

ZAJAC, L.J. and OPPENHEIM, A.K. (1969). Thermodynamic computations for the gasdynamic analysis of explosion phenomena. *Combust. Flame*, **13**, 537

ZAJIC, J.E. and HIMMELMAN, W.A. (1978). *Highly Hazardous Materials Spills and Emergency Planning* (Basel: Dekker)

ZAKAUSKAS, A., ULINSKAS, R. and KATINAS, V. (1988). *Fluid Dynamics and Flow-Induced Vibration of Tube Banks* (London: Taylor & Francis)

ZAKER, T.A. (1971). Trajectory calculations in fragment hazard analysis. *Explosives Safety Board, Dept of Defense, Washington, DC, Minutes of Thirteenth Explosives Safety Seminar*

ZAKER, T.A. (1975a). *Computer Program for Predicting Casualties and Damage from Accidental Explosions. Tech. Paper 11.* Explosives Safety Board, Dept of Defense, Washington, DC

ZAKER, T.A. (1975b). *Fragment and Debris Hazards. Rep. DDESSB TP 12.* Explosives Safety Board, Dept of Defense, Washington, DC

ZALDIVAR, J.M., HERNANDEZ, H., NIEMAN, H., MOLGA, E. and BASSANI, C. (1993). The FIRES project: experimental study of thermal runaway due to agitation problems. *J. Loss Prev. Process Ind.*, **6**, 319

ZALEWSKI, T. (1906). Experimental investigations on the resistance of the eardrum. *Z. Ohren*, **52**, 109

ZALOSH, R.G. (1980a). A dispersal and blast wave analysis for vapour cloud explosions. *Loss Prevention and Safety Promotion 3*, p. 542

ZALOSH, R.G. (1980b). Gas explosion tests in room-size vented enclosures. *Loss Prevention*, **13**, p. 98

ZALOSH, R. (1981). Dispersal of LNG vapour clouds with water spray curtains. In Greenwood, D.V. (1981), *op. cit.*, paper 3

ZALOSH, R.G. (1982). Gas explosion research at Factory Mutual. In Lee, J.H.S. and Guirao, C.M. (1982), *op. cit.*, p.771

ZALOUDEK, F.R. (1961). *The Low Pressure Critical Discharge of Steam/Water Mixtures from Pipes. Rep. HW-68934*

ZAMBON, D.M. and HULL, G.B. (1982). A look at five modular projects. *Chem. Engng Prog.*, **78**(11), 53

ZANETTI, R. (1984a). Advice on safety: play it safe with consultants. *Chem. Engng*, **91**(Nov. 26), 25

ZANETTI, R. (1984b). Good breaks for rupture discs. *Chem. Engng*, **91**(Oct. 15), 22

ZANETTI, R. (1986a). CPI alter their ways of dealing with toxics. *Chem. Engng*, **93**(21), 27

ZANETTI, R. (1986b). Remote sensors zero in on toxic-gas targets. *Chem. Engng*, **93**(Mar. 3), 14

ZANGGER, H. (1932). *Die Gasschutzfrage* (Born: Ruber)

ZANKER, A. (1976). Simple way to develop empirical equations. *Chem. Engng*, **83**(Jan. 19), 101

ZANKER, A. (1977). Estimate tank breathing loss. *Hydrocarbon Process.*, **56**(1), 117

ZANNETTI, P. (1981). A new Gaussian puff algorithm for non-homogeneous, non-stationary dispersion in complex terrain. In de Wispelaere, C. (1981), *op. cit.*, p. 537

ZANNETTI, P. (1986). A new mixed segment-puff approach to dispersion modeling. *Atmos. Environ.*, **20**, 1121

ZANNETTI, P. (1990). *Air Pollution Modeling. Theories, Computational Models and Available Software* (New York: van Nostrand Reinhold)

ZAPATKA, T.F., HANN, R.W. and ZAPATKA, M.C. (1976). Assessment and evaluation of a safety factor with respect to ocean disposal of waste materials. *Control of Hazardous Material Spills*, **3**, p. 429

ZAPERT, J.G., LONDERGAN, R.J. and THISTLE, H. (1990). *Evaluation of Dense Gas Simulation Models. Rep. EPA-450/4–90.* Environ. Prot. Agency, Research Triangle Park, NC

ZAPP, K.H. (1985). Ammonium compounds: ammonium nitrate. *Ullmanns Encyclopedia of Chemical Technology*, rev. ed., vol. A12 (Weinheim: VCH), p. 243

ZAPPE, R.W. (1991). *Valve Selection Handbook*, 3rd ed. (Houston, TX: Gulf)

ZARUBA, Q. and MENCL, V. (1982). *Landslides and Their Control* (Amsterdam: Elsevier)

ZARYTKIEWICZ, E.D. (1975). Federal environmental laws and regulations. *Chem. Engng*, **82**(Oct. 6), 9

ZATEK, J.E. (1975). Meeting OSHA's requirements for TLV monitoring of airborne hazards. *Plant Engng, Ill.*, **29**(Aug. 21), 77

ZATURSKA, M.B. (1974). The interaction of hot spots. *Combust. Flame*, **23**, 313

ZATURSKA, M.B. (1975). Thermal explosion of interacting hot spots. *Combust. Flame*, **25**, 25

ZATURSKA, M.B. (1978). An elementary theory for the prediction of critical conditions for thermal explosion. *Combust. Flame*, **32**, 277

ZATURSKA, M.B. (1980). Prediction of criticality for multi-dimensional explosives. *Combust. Sci. Technol.*, **23**, 231

ZATURSKA, M.B. (1981). Comments on 'The Frank–Kamenetskii problem revisited. Part I: Boundary conditions of the first kind' by W. Gill, A.B. Donaldson and A.R. Shooman. *Combust. Flame*, **43**, 217

ZATURKSA, M.B. (1983). A note on the nature of the disappearance of criticality in thermal explosion theory. *Combust. Flame*, **49**, 151

ZATURSKA, M. (1984). Thermal explosion theory for partially insulated reactants. *Combust. Flame*, **56**, 97

ZATURSKA, M.B. and BANKS, W.H.H. (1982). A note on thermal explosion: the spherical geometry. *Combust. Flame*, **47**, 315

ZATURSKA, M.B. and BANKS, W.H.H. (1990). Thermal explosion for multidimensional reactants with partial insulation. *Combust. Flame*, **79**, 220

ZDONIK, S.B., BASSLER, E.J. and HALLEE, L.P. (1974). How feedstocks affect ethylene. *Hydrocarbon Process.*, **53**(2), 73

ZDONIK, S.B. and MEILUN, E.C. (1983). Olefin feedstock and product flexibility. *Chem. Engng Prog.*, **79**(9), 56

ZECH, W.A. (1968). Mechanical failures of centrifugal compressors. *Loss Prevention*, **2**, p. 80

ZEDDE, G. (1977). Transportation of hazardous substances in Europe. *Symp. Hazard 77 – Advances in the Safe Handling and Movement of Hazardous Substances,* Teesside Polytechnic, Middlesbrough.

ZEEHUISEN, H. (1922). On the action of some toxic gases and vapours on guinea pigs and white rats. *Arch. Neerl. Physiol.*, **7**, 146

ZEELENBERG, A.P. and DE BRUIJN, H.W. (1965). Kinetics, mechanism and products of the gaseous oxidation of cyclohexane. *Combust. Flame*, **9**, 281

ZEEMAN, E.C. (1972). A catastrophe machine. *Towards a Theoretical Biology, 4 Essays* (ed. C.H. Waddington) (Edinburgh: Edinburgh Univ. Press)

ZEEMAN, E.C. (1977). *Catastrophe Theory: Selected Papers 1972–1977* (Reading, MA: Addison-Wesley)

ZEEUWEN, J.P. (1981). Pressure piling (letter). *Chem. Engr, Lond.*, **373**, 467

ZEEUWEN, J.P. (1982). Current review of research at TNO into gas and dust explosions. In Lee, J.H.S. and Guirao, C.M. (1982), *op. cit.*, p. 687

ZEEUWEN, J.P. (1988). Dust explosion investigation. In Groothuizen, T.M. (1988), *op. cit.*, p. 269

ZEEUWEN, J.P. and VAN WINGERDEN, C.J.M. (1988). Experimental investigation into the blast effect produced by unconfined gas cloud explosions. In Groothuizen, T.M. (1988), *op. cit.*, p. 259

ZEEUWEN, J.P., VAN WINGERDEN, C.J.M. and DAUWE, R.M. (1983). Experimental investigation into the blast effect produced by unconfined vapour cloud explosions. *Loss Prevention and Safety Promotion 4*, vol. 1, p. D20

ZEEUWEN, J.P. and SCHIPPERS, J. (1980). Unconfined vapour cloud explosion modelling and estimation of structural damage as the consequence of explosions in hazard analysis and risk evaluation. *Loss Prevention and Safety Promotion 3*, vol. 2, p. 515

ZEHR (1970). Accidents due to unsuspected chemical reactions. *Prevention of Occupational Risks*, **1**

ZEHRI, M., HUTHER, M., GILBERT, J.P. and GIRIBONE, R. (1986). Burn-out risk analysis in gas storage plants. *Gastech 85*, p. 122

ZEIDENBERG, M. (1990). *Neural Networks in Artificial Intelligence* (Chichester: Ellis Horwood)

ZEIS, L.A. (1975). Stress-corrosion prevention. *Ammonia Plant Safety*, **17**, p. 56

ZEIS, L.A. and ESCHENBRENNER, G.P. (1980). Expected life of pressure equipment. *Ammonia Plant Safety*, **22**, p. 86

ZEIS, L.A. and HEINZ, E. (1970). Catalyst tubes in primary reformer furnaces. *Ammonia Plant Safety*, **12**, p. 55

ZEIS, L.A. and LANCASTER, J.F. (1974). Corrosion at elevated temperatures. *Ammonia Plant Safety*, **16**, p. 117

ZEIS, L.A. and PAUL, G.T. (1975). How to minimize stress cracking. *Hydrocarbon Process.*, **54**(9), 229

ZEITLIN, L.R. (1994). Failure to follow safety instructions: faulty communication or risky decisions? *Hum. Factors*, **36**, 172

ZELDOVICH, J.B. (1939). *J. Exp. Tech. Phys. (USSR)*, **9**, 1530

ZELDOVICH, Y.B. (1940). On the theory of the propagation of detonation in gaseous systems. *J. Exp. Theor. Phys. (USSR)*, **10**, 542. (English translation: NACA Tech. Memo 1261, 1950)

ZELDOVICH, Y.B. (1941). *J. Exp. Tech. Phys. (USSR)*, **11**, 159. (English translation: SMBE 4310, 1960)

ZELDOVICH, Y.B. and BARENBLATT, G.I. (1959). Theory of flame propagation. *Combust. Flame*, **3**, 61

ZELDOVICH, Y.B. and FRANK-KAMENETSKY, D.A. (1938). *Acta Physicochim.*, **9**, 341

ZELDOVICH, Y.B. and KOMPANEETS, S.A. (1960). *Theory of Detonation* (New York: Academic Press)

ZELDOVICH, Y.B. and RAIZA, Y.P. (1968). *Elements of Gas Dynamics and the Classical Theory of Shock Waves* (New York: Academic Press)

ZELDOVICH, Y.B., KOGARKO, S.M. and SEMEONOV, N.N. (1956). An experimental investigation of spherical detonation in gases. *Soviet Phys. Tech. Phys*, **1**(8), 1689

VAN ZELE, R.L. and DIENER, R. (1990). On the road to HF mitigation. *Hydrocarbon Process.*, **69**(6), 92; **69**(7), 77

ZELEN, M. (1963). *Statistical Theory of Reliability.* (Maddison, WI: Univ. of Wisconsin Press)

ZELEZNIK, F.J. and GORDON, S. (1968). Calculation of complex equilibria. *Ind. Engng Chem.*, **60**(6), 27

ZELLER, H. (1979). Safety from the viewpoint of the toxicologist. In BASF (1979), *op. cit.*, p. 131

ZELLER, J.R. (1985). The nitration of 5-chloro-1,3-dimethyl-1*H*-pyrazole: risk assessment before pilot plant scale-up. In Hoffman, J.M. and Maser, D.C. (1985), *op. cit.*, p. 107

ZELNICK, A.A. (1981). Avoiding oversized flange problems. *Chem. Engng*, **88**(Nov. 16), 253

ZELT, B.W. and WILSON, D.J. (1990). *User's Manual for EXPOSURE-1 and SHELTER-1 Software for Toxic Gas Exposure Hazard Estimates. Engng Rep. 74.* Dept of Mech. Engng, Univ. of Alberta

ZEMAN, O. (1982a). The dynamics and modeling of heavier-than-air cold gas releases. *Atmos. Environ.*, **16**, 741

ZEMAN, O. (1982b). *The Dynamics and Modeling of Heavier-air Cold Gas Releases. Rep. UCRL-15224.* Lawrence Livermore Nat. Lab., Livermore, CA

ZEMAN, O. (1984). Entrainment in gravity currents. In Ooms, G. and Tennekes, H. (1984), *op. cit.*, p. 53

ZEMANSKY, M.W. (1937). *Heat and Thermodynamics;* 1943, 2nd ed.; 1951, 3rd ed.; 1957, 4th ed.; 1968, 5th ed.; 1981, 6th ed. (New York: McGraw-Hill)

ZENG, Y. and OKRENT, D. (1991). Assessment of off-normal emissions from hazardous waste incinerators. *J. Haz. Materials*, **26**(1), 47, 63

ZENNER, G.H. (1960). Low temperature metals: a roundup. *Chem. Engng*, **67**(Oct. 31), 117

ZENZ, C. (1975). *Occupational Medicine* (Chicago, IL: Yearbook Publs)

ZEPP, R.G. (1980). Assessing the photochemistry of organic pollutants in aquatic environments. In Haque, R. (1980a), *op. cit.*, p. 69

ZERATSKY, E.D. (1964). How to fight aluminum alkyl fires. In Vervalin, C.H. (1964a), *op. cit.*, p. 324

ZERCHER, J.C. (1975). CHEMTREC up to date. *Loss Prevention*, **9**, p. 66

ZERCHER, J.C. (1976). CHEMTREC – for chemical transportation accident assistance. *Control of Hazardous Material Spills*, **3**, p. 110

ZERILLI, F.J., DOHERTY, R.M., SHORT, J.M., MCGUIRE, R.R. and KAMLET, M.J. (1988). A comparison of calculations of detonation pressures for a homologous series of polynitroaliphatic explosives by different equations of state. *Combust. Flame*, **74**, 295

ZERKANI, H. and RUSHTON, A.G. (1992). Computer emulation of hazard identification. *Interactions between Process Design and Process Control* (ed. J.D. Perkins) (Oxford: Pergamon Press), p. 221

ZERKANI, H. and RUSHTON, A.G. (1993). Computer aid for hazard identification. *Sixth Int. Conf. on Industrial and Engineering Applications of Artificial Intelligence* (New York: Gordon & Breach)

ZERMATI, C. (1982). Commissioning of the 120,000 m^3 storage tanks of the Gaz de France LNG terminal. *Gastech 81*, p. 403

ZERNIK, F. (1933). On irritant gas and material. *Gasschutz Luftschutz*, **2**, 101

ZHANG, B.Q. (1988). Using AE monitoring on on-line crack detection. *Hydrocarbon Process.*, **67**(1), 35

ZHANG, F. and GRÖNIG, H. (1991). Transition to detonation in corn starch dust—oxygen and —air mixtures. *Combust. Flame*, **86**, 21

ZHANG, D.K. and WALL, T.F. (1993). An analysis of the ignition of coal dust clouds. *Combust. Flame*, **92**, 475

ZHANG, G.S. and TANG, M.J. (1993). The investigation of a devastating accident — an accidental explosion of 40 tons of TNT. *J. Haz. Materials*, **34**, 225

ZHENG, L. (1990). Safety distance for a person under action of air shock wave. *Explosives Safety Board, Dept of Defense, Proc. Twenty-Fourth Explosives Safety Seminar*, p. 949

ZICK, L.P. and CLAPP, M.B. (1964). How to specify low temperature storage vessels. *Hydrocarbon Process.*, **43**(6), 125

ZICK, L.P. and LA FAVE, I.V. (1981). State-of-the-art assessment of refrigerated liquefied gas storage systems using flat bottom tanks. *Gastech 81*

ZICK, L.P. and MCGRATH, R.V. (1968). New design approach for large storage tanks. *Hydrocarbon Process.*, **47**(5), 143

ZIEBOLZ, H. (1964). Predictor instrumentation for selective control. *Instrum. Control Systems*, **37**(12), 84

ZIEBS, J., AURICH, D., ERBE, H., VEITH, H. and HELMS, R. (1981). The influence of the stress state on fracture toughness. *Int. J. Pres. Ves. Piping*, **9**

ZIEFLE, R.G. (1973). Designing safe high pressure polyethylene plants. *Safety in Polyethylene Plants*, **1**, p. 5. See also *Chem. Engng Prog.*, 1972, **68**(11), 51

ZIEGER, A. (1969). The chemical and physical properties of acetylene and safety measures for industrial use. *Symp. on Inherent Hazards of Manufacturing and Storage in the Process Industry*, The Hague, p. 65

ZIEGLER, A. (1992). Chemical risk assessment — a tool for disaster prevention. *J. Haz. Materials*, **31**, 233

ZIEGLER, J.G. (1957). Sizing valves for flashing fluids. *ISA J.*, **4**(3), 92

ZIEGLER, R.C. and LAFORNARA, J.P. (1972). In situ treatment methods for hazardous material spills. *Control of Hazardous Material Spills*, 1, p. 157

ZIELHUIS, R.L. (1970). Tentative emergency exposure limits for sulfur dioxide, sulfuric acid, chlorine and phosgene. *Ann. Occup. Hyg.*, **13**, 171

ZIELINSKI, R.M. (1967). Case study of a reactor fire. *Loss Prevention*, **1**, p. 65. See also *Chem. Engng Prog.*, 1967, **63**(8), 50

ZIELINSKI, R.M. (1973). Reactor fire at Avisun. In Vervalin, C.H. (1973a), *op. cit.*, p. 27

ZIELINSKI, M. (1978). Microcomputer systems: where do you begin trouble-shooting? *Hydrocarbon Process.*, **57**(1), 109

ZIENTARA, D.E. (1972). Measuring process variables. *Chem. Engng*, **79**(Sep. 11), 19

ZIKA, I. and MATYAS, R. (1992). Modern technologies and efficient loss-prevention systems in Czechoslovak chemical industry are inevitable. *Loss Prevention and Safety Promotion 7*, vol. 3, 130—1

ZILITINKEVICH, S.S. (1972a). On the determination of the height of the Ekman boundary layer. *Boundary Layer Met.*, **3**, 141

ZILITINKEVICH, S.S. (1972b). *Dynamics of the Atmospheric Boundary Layer* (Leningrad: Gidroometeoizdat)

ZILKA, M.I. (1982). The chemical engineer as super sleuth. *Chem. Engng*, **89**(May 17), 121

ZIMMER, M.F., MCDEVITT, J.A. and DALE, C.B. (1970). Assessment of the explosive hazards of large solid rocket motors. *J. Spacecraft Rockets*, **7**, 636

ZIMMERMANN, H.J. (1984). *Fuzzy Sets and Decision Analysis* (Amsterdam: North Holland)

ZIMMERMANN, H.J. (ed.) (1986). *Fuzzy Sets Theory and its Applications* (Dordrecht: Reidel)

ZIMMERMANN, H.J. (1987). *Fuzzy Sets, Decision-Making, and Expert Systems* (Boston, MA: Kluwer)

ZIMMERMAN, J.J. (1968). Machinery failure prevention: a fundamental aspect of plant safety. *Ammonia Plant Safety*, **10**, p. 91

ZIMMERMAN, R. (1990). *Governmental Management of Chemical Risk* (Chelsea, MI: Lewis)

ZIMOLONG, B. (1992). The HEPs of HEP experts: empirical evaluation of THERP and SLIM. *Hazard Identification and Risk Analysis, Human Factors and Human Reliability in Process Safety*, p. 273

ZINK, J.S. and REED, R.D. (n.d.). Flare stack gas burner. *US Patent 2 779 399*

ZINN, J. (1962). Initiation of explosions by hot spots. *J. Chem. Phys.*, **36**(7), 1949

ZINN, J. and MADER, C.L. (1960). Thermal initiation of explosives. *J. Appl. Phys.*, **31**, 323

ZINSSTAG (1973). A serious accident with sodium. *Prevention of Occupational Risks*, **2**, p. 428

ZIOMAS, I.C., ZEREFOS, C.S. and BAIS, A.F. (1989). Design of a system for real-time modelling of the dispersion of hazardous gas releases in industrial plants: 2, Accidental releases from storage installations. *J. Loss Prev. Process Ind.*, **2**(4), 194

ZIPF, G. (1984). Computation of minimal cut sets of fault trees: experiences with three different methods. *Reliab. Engng*, **7**(2), 159

ZIPPLER, D.B. (1976). Respiratory and protective equipment at a large nuclear facility. *Control of Hazardous Material Spills 3*, p. 163

ZIRSCHKY, J. and GILBERT, R.O. (1984). Detecting hot spots at hazardous-waste sites. *Chem. Engng*, **91**(Jul. 9), 97

ZISSOS, D. (1976). *Problems and Solutions in Logic Design* (Oxford: Oxford Univ. Press)

ZLOKARNIK, M. (1991). *Dimensional Analysis and Scale-Up in Chemical Engineering* (Berlin: Springer)

ZOHAR, D., COHEN, A. and AZAR, N. (1980). Promoting increased use of ear protectors in noise through information feedback. *Hum. Factors*, **22**, 69

ZOHRUL ABIR, A.B.M. (1987). Inspection policy based on minimization of fractional dead time. *Reliab. Engng*, **19**(2), 77

ZOLIN, B.I. (1970). Protective lining performance. *Chem. Engng Prog.*, **66**(8), 31

ZOLLER, L. and ESPING, J.P. (1993). Use 'What if' method for process hazard analysis. *Hydrocarbon Process.*, **72**(1), 132—B

ZONATO, C., VIDILI, A., PASTORINO, R. and DE FAVERI, D.M. (1993). Plume release of smoke coming from free burning fires. *J. Haz. Materials*, **34**, 69

ZOOK, R.J. (1976). Rupture disks for low-burst pressures. *Chem. Engng*, **83**(Mar. 1), 131

ZUBER, K. (1976). LNG facilities — engineered fire protection systems. *Fire Technol.*, **12**, 41

ZUBER, N. and FINDLAY, J. (1965). Average volumetric concentration in two-phase flow systems. *J. Heat Transfer*, **87**, 453

ZUCKERMAN, S. (1940). Experimental study of blast injuries to the lungs. *The Lancet*, Aug. 24, 219

ZUCKERMAN, S. (1941a). The problem of blast injuries (discussion). *Proc. R. Soc. Med.*, **34**, 171

ZUCKERMAN, S. (1941b). *Rupture of Eardrums by Blast. Rep. BPC 148/WS 13.* Min. of Home Security, Oxford

ZUCKERMANN, LORD (1980). The risks of a no-risk society. *Yearbook of World Afairs* (London: Stevens & Sons), p. 7

ZUKOSKI, E.E., CETEGEN, B.M. and KUBOTA, T. (1985). Visible structure of buoyant diffusion flames. *Combustion*, **20**, p. 361

ZUDANS, Z., YEN, T.C. and STEIGELMAN, W.H. (1965). *Thermal Stress Techniques in the Nuclear Industry* (New York: Elsevier)

ZUDKEVITCH, D. (1975). Imprecise data impact plant design and operation. *Hydrocarbon Process.*, **54**(3), 97

ZUMSTEG, F. and FANNELOP, T.K. (1991). Experiments on the evaporation of liquefied gas over water. *Modeling and Mitigating the Consequences of Accidental Releases,* p. 35

ZURER, P.S. (1986). Ocean incineration of hazardous wastes: US policy adrift. *Chem. Ind.*, Feb. 17, 116

ZWAGA, H.J.G. and VELDKAMP, M. (1984). Evaluation of integrated control and supervision systems. *Ergonomic Problems in Process Operations* (Rugby: Instn Chem. Engrs), p. 133

ZWART, A. (1987). *Acute (One Hour) Inhalation Toxicity Study of Phosgene in Rats. Rep. V 87.029/260831.* TNO, Rijswijk, the Netherlands

ZWART, A. and WOUTERSEN, R.A. (1988). Acute inhalation toxicity of chlorine in rats and mice: time-concentration-mortality relationships and effects on respiration. *J. Haz. Materials*, **19**(2), 195

ZWART, A., AERTS, J.H.E. and KLOKMAN-HOUWELING (1988). Determination of concentration-time-mortality relation ships to replace LC50 values. *Inhalation Toxicol.*, **2**, 105

ZWETSLOOT, G.I.J.M. and ASKOUNES-ASHFORD, N. (1999). *Towards Inherently Safer Production: A Feasibility Study on Implementation of an Inherent Safety Opportunity Audit and Technology Options Analysis in European Firms.* (Hoofddorp, The Netherlands: TNO Work and Employment) TNO Report R990341

ZYGADLOWSKI, J. (1968). Acute burns with ammonia vapours. *Otolaryngol. Pol.*, **22**, 773

Loss Prevention Bulletin

(Institution of Chemical Engineers)

LOSS PREVENTION BULLETIN (Institution of Chemical Engineers)

Sample issue 1975 – Reactor thermal stability, p. 2; Storage tank collapse, p. 2; Explosion and fire in ethylene plant, p. 3; How strong is your storage tank?, p. 3; Hammerblow in pipelines, p. 6

1   1975 – Are your plant modifications safe?, p. 1; More cases of water hammer, p. 7; Exothermic reactor, p. 10; Air eliminator explosion, p. 15

2   1975 – Static electricity as a fire hazard, p. 1; Investigation of electrostatic ignition hazards, p. 12; How can we tell if the accident record is really better?, p. 15

3   1975 – Explosive reactions caused by impurities, p. 1; Practising what is preached, p. 8; Hydrogen sulphide, H_2S, p. 17

4   1975 – Total loss control, p. 1; Engineering maintenance, p. 9

5   1975 – Chemical and physical explosions in furnaces and boilers, p. 1; Bellows again (see also LPB 6), p. 9; Reaction kinetics, p. 16

6   1975 – Hazards of corrosion, p. 1; Relief valves and bursting discs, p. 12; More explosions in plant equipment (see also LPB 7), p. 17

7   1976 – Instrumentation and control, p. 11; Nitrogen, p. 11; Human error, p. 15

8   1976 – Second chance and fail safe design, p. 1; Explosions and fire involving dusts, p. 7; Information for plant operators, p. 19

9   1976 – Vessel overflows, p. 1; Mishap miscellany, p. 5; Potential hazards due to seal failures, p. 12; Unexpected material properties, p. 26

10   1976 – Fin fan coolers, p. 1; Ball valves, p. 3; Radios (see also LPB 11), p. 4; Pressure relief valve chattering, p. 5; The spreading and dispersion of a large release of heavy gases, p. 9

11   1976 – A major incident, p. 2; The risk of hydrocarbons entering steam and condensate systems, p. 5; Mishap miscellany, p. 10; Control room design and location – one company's viewpoint (see also LPB 13), p. 16; Reliability assessment Pt 1, p. 18

12   1976 – Hazards of pyrophoric iron sulphide, p. 1; Safe design and operation of hot oil tanks, p. 7; Neutralisation of phosphorous trichloride, p. 11; Inadequate physical and chemical data – source of major hazard, p. 12; Reliability assessment Pt 2, p. 17

13   1977 – Chemical reactor hazards, p. 2; Hose and coupling incidents, p. 5; Laboratory accidents, p. 14; Coaxial pipes, p. 17; Phosphorus chlorides, p. 21

14   1977 – Operating, maintenance and modification procedures (software), p. 2; Excess flow valves – one company's view, p. 15; Diagnostic techniques, p. 17

15   1977 – Hydrogen (see also LPB 18), p. 2; Building ventilation, p. 10; Liquid metal embrittlement (see also LPB 17), p. 23; Safety systems for tanker loading and unloading, p. 30; Road tanker hose and coupling incidents, p. 33; Hose and coupling incidents, p. 35

16   1977 – The hazards of mineral acid and alkali handling (see also LPB 16), p. 3; Protection for bulk storage of corrosive liquids, p. 21; Control room design and location – a further viewpoint, p. 24

17   1977 – Lagging fires (see also LPB 19), p. 2; Relief valve and venting systems – more hazards associated with maintenance and modification, p. 15

18   1977 – Plant commissioning hazards (see also LPB 19, 20), p. 2; Laboratory accidents, p. 17; Co-axial pipes, p. 2; Excess flow valve incident, p. 22

19   1978 – The transport of chemicals by road, p. 1; Safety down the drain, p. 10; Compression ignition – a case history, p. 24

20   1978 – More software failures, p. 41; Hazards of gas cylinders, p. 45; A new approach to the layout of storage tanks, p. 55

21   1978 – Instrumentation, p. 68; Sparks from steam, p. 81; Hydrogen, p. 85

22   1978 – Hazards of improvisation, p. 101; Fire extinguishing agents and their uses, p. 107; Reflections on the rubbish bin, p. 114; A new approach to the layout of storage tanks (letter), p. 114

23   1978 – Loss prevention in the design and selection of bursting discs, p. 125; Material selection – a factor for loss prevention, p. 136; Hazards of radio frequency ignition, p. 147

24   1978 – Engineering maintenance, p. 155; Liquid lightness of tank dikes, p. 164; Ancillary equipment incidents, p. 167; Corrosion/erosion – some more accidents, p. 172

25   1979 – Fire and explosion hazards in dryers by D. Reay; Loss prevention: a one company view by D.J. Kerr; Reactive chemicals: cyanuric chloride, p. 19; Dimethyl sulphate (letter), p. 24

26   1979 – Bulk storage of corrosive liquids – catastrophic failure of phosphoric acid tanks (see also LPB 33), p. 31; Pressure testing, p. 39; Hazards of comminution, p. 48; Explosions in seal oil reservoirs of centrifugal gas compressors, p. 51; Liquid tightness of tank dikes (letter), p. 52

27   1979 – Acoustic emission testing, p. 59; Hazards involving the use and storage of drums, p. 68; Report on the International Symposium on Grain Elevator Explosions, p. 76; Engineering maintenance – further incidents, p. 80; Academic use of Hazard Workshop Modules by G.S.G. Beveridge

28   1979 – Thermal properties of reactors and some instabilities, p. 91; Flare systems, p. 97; Stress corrosion cracking of stainless steel, p. 104; Design recommendations for outdoor drum storage, p. 115, Guide Notes on the Safe Use of Stainless Steel in Chemical Process Plant (letter), p. 118

29   1979 – Some hazards due to layering of liquids, p. 124; Static electricity – further case histories, p. 134; Dead head hazards of centrifugal pumps, p. 139; Embrittlement of stainless steel by zinc, p. 149; A 'nitrogen explosion' (letter), p. 151

30   1979 – Fire in ducting, p. 155; Thermal decomposition tests, p. 159; Aniline toxicology, p. 165; 'Small is beautiful', p. 171; Embrittlement of stainless steel by zinc, p. 175

31   1980 – The Second Report of the Advisory Committee on Major Hazards – a review by G.S.G. Beveridge; Fire resistance and flame retardants, p. 5; Oxygen deficiency in nitrogen atmospheres and related problems, p. 11; Valve limitations, p. 17; Flare systems – pump out of knockout drum (letter), p. 27

32   1980 – Atmospheric storage tanks for flammable liquids, p. 1; Shell/bottom junction failure in a cone roof tank, p. 13; HSE Guides: tanker marking – packaging and labelling – hot work, p. 15; Legal obligations in relation to chemical engineering laboratories by V.C. Marshall; Hazards in chemical engineering laboratories by D.H. Napier; People in laboratories by MA Buttolph; Designing of chemical engineering laboratories by

(near) Caracas, Venezuela by W.M. Garrison (see also LPB 59, 60, 61)

58 1984 – One hundred largest losses – a thirty year review of property damage losses in the hydrocarbon-chemical industries by F.A. Manuele; Storage of liquid ammonia (see also LPB 59), p. 13; Accidents of the past No.5 – major fire at a marketing terminal by T.A. Kletz; The HSE heavy gas dispersion trials project at Thorney Island by B. Roebuck; The Human Factors in Reliability group by I.A. Watson

59 1984 – Intelligent fire detection systems by M. Finucane; Rapid ranking of hazards by C.J. Mumford and D.A. Lihou; Safe disposal of relief discharges by M. Kneale; Entry into a confined space – a retrospective look by J. Bond; Pressure reducers, p. 28; Lessons from the past No. 6 – reluctance to recognise new hazards by T.A. Kletz; Nitrogen – a gassing incident and a near miss, p. 30

60 1984 – Incidents 1982, p. 1; odour as an aid to chemical safety: odour thresholds compared with threshold limit values by P.J. Lynskey (see also LPB 61); Personal protection: respiratory protection – the importance of facepiece fit by M.J. Evans and P. Leinster; Weak spots in a refinery, p. 40

61 1985 – Emergency procedures by P.J. Lynskey; Update on the Spanish campsite disaster by L. Hymes; The phosphorus railcar explosion near Brownson, Nebraska, on Sunday, 4 February 1978 by I. Hymes; Major disaster averted in massive escape of vinyl chloride by J. Kilmartin (see also LPB 65); Accidents of today, p. 33

62 1985 – Toxic gas incidents – some important considerations for emergency planning by G. Purdy and P.C. Davies (see also LPB 63); Interactive video, p. 13; A matter of some agitation by M. Kneale; Safety aspects of the use of microcomputers and programmable controllers for chemical process control by S.J. Skinner Guide to reducing human error in process operation by P.W. Ball; A compressor overspeed, p. 27

63 1985 – Bhopal – the company's report, p. 1; The toxic hazards of hydrogen sulphide by P.J. Lynskey; Some H_2S incidents by HSE; The hazards of nitrogen by J. Bond, Improving process safety and loss control software in existing process plant by H.M. Tweedale; Preparation for entry into a confined space – one company's view (see also LPB 64), p. 31

64 1985 – Analysis of the LPG disaster at Mexico City by C.M. Pietersen; Refinery fire contained, p. 13; The toxic effects of chlorine by MHAP; Accidents of today, p. 27

65 1985 – Loss prevention implications of arson: Pt 1, Case histories and prevention by P.A. Carson, C.J. Mumford and R.B. Ward; Acetylene cylinder fire, p. 15; Violent polymerisation by J. Bond; An ethylene reflux pump fire by J. Bond, Spectators and other vulnerable exposes to major accident hazards: an emergency services headache by I. Hymes

66 1985 – Dust explosibility testing by P. Field (see also LPB 66); Some recent British initiatives on dust explosions prevention and protection by G. Butters; Principles for the safe transfer of technology by EFCE; Hydrogen cyanide release from a failed bonnet gasket, p. 23; The need to be blamed (letter) by D.A. Lihou (see also LPB 67)

66 suppl 1985 – Incidents 1983

67 1986 – Industrial health hazards: Pt 1, Sources of exposure to substances hazardous to health by P.A. Carson

and C.J. Mumford; Lessons from the past No.7 – flying saucers have landed by T.A. Kletz; Ludwigshafen – two case histories by V.C. Marshall

68 1986 – Unconfined vapour cloud explosions involving hydrogen rich gases – estimating the blast effects by J.L. Hawksley; The Spanish campsite disaster – a new scenario by H. Ens; Reactor incident, p. 33; Experience with a temperature monitoring system, p. 35; The effects of explosions in the process industries by the MHAP

69 1986 – Some problems with valves by J. Bond; Leak sealing using compound injection by J. Bond; Refinery fire, p. 17; Fatal accident on a hydrocarbon recovery plant, p. 25; The rupture of a tank of benzene, p. 29; Power station fire, p. 33

70 1986 – Handling the vapour cloud after failure of a flare line, p. 1; Polymer lined GRP tank failure, p. 7; Loss prevention implications of arson Pt 2 – Investigation by P.A. Carson and C.J. Mumford; Flashing liquid flow calculations by D.A. Carter

71 1986 – One hundred largest losses – a thirty year review of property damage losses in the hydrocarbon-chemical industries by F.A. Manuele; Maintaining your sprinkler system by Factory Mutual Engineering Corporation; Explosion in an oxidised bitumen tank, p. 71; Ship/shore transfer of liquefied gases by J. Bond; Fire in a coker unit, p. 27; Three injured in extruder explosion, p. 31; The Spanish campsite disaster (letter) by J. Foxcroft

72 1986 – Sizing pressure vacuum valves for cone roof tanks, p. 1; The first tanker explosion by P. Watson; Inadvertent pressure-testing of a storage tank to destruction, p. 7; The Spanish campsite disaster – a third view by V.C. Marshall; Comments on 'The Spanish campsite disaster – a third view' by H. Ens; Causes of accidents (the results of a study), p. 21; Some problems with valves – another case history, p. 28

73 1987 – Some essentials of lightning protection (in electrical systems) by Factory Mutual Engineering Corporation; Explosion in an isopropyl alcohol tank, p. 9; Is pipework given the attention it deserves by R.F. Towndrow; A tale of woe by C.D. Plummer; Spanish campsite disaster (letter) by J. Bond (see also LPB 72); Extruder explosion (letter) by L. Carter (see LPB 71)

74 1987 – An oleum leakage, p. 1; Explosion in a superheater, p. 15; Fire protection systems – how to cope with impairments, p. 17; A pump explodes, p. 23

75 1987 – Autoignition temperatures (AITs) – how hot is dangerous? by F.S. Ashmore and D. Blumson; The Sandoz warehouse fire, p. 11; Overpressuring of a storage tank, p. 19; Explosion in a heat exchanger during demolition, p. 22; Hydrogen sulphide incident, p. 23; The two safeguard approach for minimising human failure injuries by J. Bond and J.W. Bryans

76 1987 – Industrial health hazards: Pt 2, Development of UK legislation by P.A. Carson and C.J. Mumford; The Chernobyl accident by A.N. Hall, S.F. Hall and W. Nixon; HSE warning on road tanker discharge valves, p. 30; Failures of glass reinforced plastic tanks by T.E. Maddison

77 1987 – The Feyzin disaster, p. 1; Major LPG fire at US refinery, p. 11; Sabotage causes propane release, p. 17; Fire at LPG recovery installation of a refinery, p. 27

78 1987 – Static electricity hazards in industry by F.J. Pay; Electrostatic ignition of volatile liquid vapour, p. 9; Hydrogen explosion/detonation in an ammonia plant, p. 11; Temporary manhole covers by J. Bond; Three year

survey of accidents and dangerous occurrences in the UK chemical industry by B.J. Robinson; Molten sulphur spillage, p. 23; Heat of dilution causes violent eruption in storage tank, p. 26; Computerised chemical engineering industry information by S. Pantry

79 1988 – Flame arresters and the chemical tanker by P.B. Watson; The velocity flame stopper by W.B. Howard Faulty ducts cause switchroom fires, p. 23; Safety audit of refrigerated gas facilities, p. 25; Static electricity hazards in industry (letter) by P.A. Kalisvaart (see also LPB 78)

80 1988 – Protective clothing design by L.E. Jones; Operator burned whilst using incorrect tanker connection, p. 3; Gloves as protective equipment by I.P. Hoyle; Protection from protective clothing by I.P. Hoyle; Hazards on a rubber injection moulding machine, p. 14; Expect the unexpected, p. 14; Acetylene cylinders can be dangerous by F.S. Ashmore; Shattering rotameter cuts safety glasses in half p. 19; Dizzy heights, p. 20; Some problems with breathing apparatus, p. 21; The International Safety Rating System by J. Bond

81 1988 – Chernobyl – was it the worst accident? by V.C. Marshall; Glasnost now applies to the Soviet chemical industry, p. 15; Incinerator explosion, p. 17; Cool flames by R.D. Coffee; Autoignition, slow combustion and the variation of minimum ignition temperature with vessel size by T.J. Snee; Lessons from the past – entry into vessels, p. 29; Qatar tank failure (letter) by T.A. Kletz

82 1988 – Calculation of the intensity of thermal radiation from large fires by MHAP; 'Pops' in tanks, p. 12; Chemical risk: living up to public expectations by H.J. Corbett Tank farm burns for a whole week, p. 19; The Boston, Mass., incident of January 15th, 1919 by V.C. Marshall

83 1988 – Failure of a flexible pipe in an H_2 installation area, p. 1; Seizure of boiler relief valves, p. 3; Failure of articulated bellows by A. Smith; Another non-return valve failure, p. 9; Damaged flange seal face leads to ammonia release, p. 11; Fire on a gasoline treater unit, p. 13; Process worker sustains an acid burn to his right eye, p. 15; Release of UF_6 in a vaporiser room, p. 17; Some incidents caused by failure of screwed joints, p. 19; An explosion involving 1,1,1-trichlorethane by I. Wrightson and R.C. Santon; Three incidents, three lessons, p. 25; Explosion and fire at a phenol plant by R.F. Schwab; 'Pops' in tanks (letter) by T.A. Kletz (see also LPB 82)

84 1988 – Protection of warehouses against fire by A. Mackintosh; The forgotten hazards: services in warehouses by R. Heels; Materials selection for composite fire damper by B.D. Baines; Lack of segregation causes rapid spread of fire, p. 19; Fire in Auckland, New Zealand by B.S. Amstrong; Molten sulphur spillage (letter) by P.G. Ayers (see also LPB 78)

85 1989 – HF burn to fitter's forearm, p. 1; Serious accident during valve replacement caused by incorrect procedure, p. 3; Release of UF_6, vapour, p. 5; If it didn't quite happen ... tell somebody, p. 7; A programme of safety assurance of liquid carbon dioxide storage vessels by A.R. Coleman; UK and European hazard information systems – a personal view by B. Cahill; Caustic potash spillage in tank farm, p. 27; Protection of warehouses against fire (letter) by B. Johnston (see LPB 84)

86 1989 – A case history involving the release of chlorine, p. 1; Pressure testing and its hazards by R. Dooner and V. Marshall; Exposure to acetone cyanhydrin, p. 17; A reappraisal of safety in chlorine handling, p. 19; Release of HF fume due to valve fracture, p. 22; Wrongly labelled gasket causes chlorine release, p. 23; Release of chlorine from an expansion pot, p. 27; Protection of warehouses against fire (letter) by K.N. Palmer (see also LPB 85); Nitrous oxide: no laughing matter (letter) by L. Bretherick

87 1989 – Two men gassed by H_2S p. l; Three injured in extruder explosion, p. 3; Corrosion under insulation causes refinery fire, p. 5; Explosion when refuelling a liquid gas-driven stacker, p. 7; Pressure unrestrained, p. 9; Engineering for health by J.R. Jackson; Inherently safer design, p. 21; Emergency planning – in practice by J.M. Mullins; Incident during distillation of nitrobenzaldehyde, p. 27; Release of chlorine from an expansion pot (letter) by T.A. Kletz; Reporting and analysis of industrial incidents (letter) by D.J. Lewis

88 1989 – Operator burned during sampling operations, p. 1; Programmable electronic systems by B.W. Eddershaw; Warehousing of chemicals, p. 9; Storage tanks – a vendetta, p. 11; Explosion in a dinitrotoluene pipeline, p. 13; Raw glycide collecting tank explodes, p. 17; Lessons of the past, p. 18; Bitumen tank fire, p. 19; Electrostatic hazards from pouring powdered chemicals into vessels from plastic bags by J. Bond; Analysis of a tank fire 'classic' by L. Steinbrecher, Decomposition of nitric oxide (letter) by J.L. Currie (see also LPB 86); Explosion of 2-nitro-benzaldehyde (letter) by P.G. Urben (see also LPB 87)

89 1989 – Loss prevention in pilot plants: Pt 1 – The hazards by P.A. Carson and C.J. Mumford; Accident to a scaffolder in a plenum chamber, p. 9; Hydrocracker accident, p. 13; Operator fall from scaffold, p. 18; LPG storage cracking experience by J.E. Cantwell; Stress corrosion cracking results in hydrofluoric acid spill, p. 27; Sizing the relief system for runaway reactions by A.J. Wilday

90 1989 – Pavadena explosion, p. 0; The hazards of purging a line, p. 1, Loss control in pilot plants by P.A. Carson and C.J. Mumford: Pt 2, Control measures; LPG pipeline and Trans-Siberian Railway explosion and fire by D.J. Lewis; Investigation into glacial acetic acid spillage, p. 13; List of major incidents 1987, p. 15; Severe burns resulting in fatality caused by dust explosion during pallet shrink-wrapping, p. 27; Two reactor incidents, p. 29; Failure of a reverse buckling disc, p. 30; Incident during vacuum distillation of nitrobenzaldehyde (letter) by J.L. Cronin

91 1990 – Which foam? by D. Waters; Industrial health hazards Pt 3, Some lessons from experience by P.A. Carson and C.J. Mumford; Incompatible materials causes explosion in waste container, p. 15; Compressor incident causes propylene release, p. 16; Disconnecting is sometimes not enough, p. 17; Multiple accident due to release of CO gas in refinery compressor hall, p. 19; Drum of sodium hydrosulphite fire, p. 22; Relief valves on liquid gas tankers (letter) by J.L. Hawksley; Loss prevention in the 90s (letter) by F.K. Crawley

92 1990 – Safety valves: taking the pressure off productivity and improperly closed valves, p. 1; Fatal accident following a toluene leak from a stuffing box, p. 7; Incident involving an engineering foreman scalded by hot water, p. 9; Contractor began repair to the plant pipework without a permit-to-work p. 10; Report relating to a leak of anhydrous hydrofluoric acid (AHF) vapour, p. 11; Overflow of caustic solution from vent scrubber system to the stock tanks, p. 13; Bursting of a cylinder cover on a compressor releasing a large quantity of ammonia, p. 15; Gas leakage that taught a valuable lesson by M. Huvinen; Caustic

Acronyms

AAE	Additional Accommodation East (Piper Alpha)
AAM	accident anatomy method
AAR	Association of American Railroads
AAW	Additional Accommodation West (Piper Alpha)
ABC	general purpose dry chemical
ABCM	Association of British Chemical Manufacturers
ABI	Association of British Insurers
ABL	atmospheric boundary layer
ABMA	American Boiler Manufacturers Association
ABPI	Association of the British Pharmaceutical Industry
ACA	Advisory Committee on Asbestos
ACDS	Advisory Committee on Dangerous Substances
ACE	Army Corps of Engineers
ACGIH	American Conference of Government Industrial Hygienists
ACH	Acetone cyanohydrinbased route
ACMH	Advisory Committee on Major Hazards
ACOP	Approved Code of Practice
ACR	Average circle radius
ACRS	Advisory Committee on Reactor Safety
ACSNI	Advisory Committee on the Safety of Nuclear Installations
ACTS	Advisory Committee on Toxic Substances
ADI	acceptable daily intake
ADN	European Agreement for the International Transport of Dangerous Goods
ADNR	European Agreement for the International Transport of Dangerous Goods (River Rhine)
ADR	European Agreement concerning the International Carriage of Dangerous Goods by Road
ADT	adiabatic storage test
AE	acoustic emission
AEA	action error analysis
AEC	Atomic Energy Commission
AEL	accessible emission level
AFFF	aqueous film forming foam
AFNOR	Association Francaise de Normalisation
AFPA	Australian Fire Protection Association
AGA	American Gas Association
AGR	advanced gas-cooled reactor
AHM	acutely hazardous material
AHP	analytic hierarchy process
AI	artificial intelligence
AIA	American Insurance Association
AIB	Accident Investigation Branch
AIChE	American Institute of Chemical Engineers
AID	Aeronautical Inspection Directorate
AIHA	American Industrial Hygiene Association
AIR	American Institute for Research
AISI	American Iron and Steel Institute
AIT	autoignition temperature
AL	action level
AIARP	as low as reasonably practicable
Amoco	American Oil Company
AN	acrylonitrile
AN	ammonium nitrate
ANA	ammonium nitrate–ammonia

ANAP	action-not-as-planned
ANC	adaptive noise cancellation
ANL	Argonne National Laboratory
ANN	artificial neural network
ANSI	American National Standards Institute
AOD	argon–oxygen decarburization
APC	air pollution control
APCA	Air Pollution Control Association
APAU	Accident Prevention Advisory Unit
API	American Petroleum Institute
APJ	absolute probability judgement
APS	automatic protective system
AQ	air quality
AQD	Aeronautical Quality Assurance Directorate
AQL	acceptable quality level
AQP	average lot quality protection
AR	alcohol resistant
ARC	accelerating rate calorimetry
ASCE	American Society of Civil Engineers
ASHRAE	American Society of Healing Refrigerating and Air-conditioning Engineers
ASM	Algebraic stress model
ASME	American Society of Mechanical Engineers
ASQC	American Society of Quality Control
ASTM	American Society for Testing and Materials
ATMS	assumption-based truth maintenance system
ATN	augmented transition network
ALIT	Association of University Teachers
AWS	American Welding Society
BA	breathing apparatus
BACT	best available control technology
BAM	Bundesanstalt für Materialprüfung
BASEEFA	British Approvals Service for Electrical Equipment in Flammable Atmospheres
BASF	Badische Anilin und Soda Fabrik
BATNEC	best available technology not entailing excessive cost
BCC	British Cryogenic Council
BCF	bromochlorodifluoromethane
BCGA	British Compressed Gases Association
BCISC	British Chemical Industry Safety Council
BCME	bis(chloromethyl)ether
BCS	Beilby, Cottrell and Swinden (model)
BD	bursting disc
BFR	binomial failure rate
BG	British Gas
BGC	British Gas Corporation
BG/C&W	British Gas/Cremer and Werner (model)
BHEP	basic human error probability
BI	business interruption
BIA	Berufsgenossenschaftlisches Institut für Arbeitssicherheit
BITC	Bureau International Technique du Chlore
BLEVE	boiling liquid expanding vapour explosion
BLV	biological limit value
BM	Bureau of Mines
BMCS	Bureau of Motor Carrier Safety
BNF	Backus–Naur form
BNFL	British Nuclear Fuels Ltd
BNL	Brookhaven National Laboratory
BNP	back-propagation network
BOCA	Building Official and Code Administration
BOD	biological oxygen demand
BOHS	British Occupational Hygiene Society

BOP	blowout preventer
BOR	boil-off rate
BP	bulk property
BPCS	basic process control system
BPEO	best practical environmental option
BPF	British Plastics Federation
BPM	best practicable means
BR	British Rail
BRE	Building Research Establishment
BRL	Ballistics Research Laboratory
BSC	bench scale calorimeter
BSC	British Safety Council
BSC	British Steel Corporation
BSI	British Standards Institute
BTC	British Transport Commission
BTM	bromotrifluoromethane
BUA	built-up area
BWR	boiling water reactor
CA	Certifying Authority
CAA	Civil Aviation Authority
CAA	Clean Air Act
CAAA	Clean Air Act Amendments
CAD	computer-aided design
CADA	critical action and decision approach
CADE	computer aided design and engineering
CADET	critical action and decision evaluation technique
CAER	Community Awareness and Emergency Response
CAF	compressed asbestos fibre
CAPE	computer aided process engineering
CAPITB	Chemical and Allied Products Industry Training Board
CAWR	Control of Asbestos at Work Regulations 1987
CBI	Central Bureau of Investigation (India)
CBI	Confederation of British Industry
CBL	convective boundary layer
CCA	copper—chrome—arsenate
CCD	cause—consequence diagram
CCF	common cause failure
CCPS	Center for Chemical Process Safety
CCS	centralized control system
CCT	cause—consequence tree
CCW	Chester—Chisnall—Whitham
CD	conceptual dependency
CDE	Chemical Defence Establishment
CDFM	conservative deterministic failure margin
CE	critical examination
CEC	Commission of the European Communities
CEFIC	European Federation of Chemical Engineering
CEGB	Central Electricity Generating Board
CEI	Chemical Exposure Index (Dow)
CELD	cause—effect logic diagram
CEN	Comité de Normalisation Européen
CENELEC	Comité de Normalisation Européen Électrotechnique
CEP	Chemical Engineering Progress
CEP	Chemical Exposure Index (DOW)
CEQ	Commission on Environmental Quality
CERCLA	Comprehensive Environmental Response, Compensation and liability Act (Superfund)
CF	certainty factor

CF	credit factor (Dow)
CFC	Chlorofluorocarbons
CFC	chlorofluorohydrocarbon
CFD	computational fluid dynamics
CGA	Compressed Gas Association
CHA	concept hazard analysis
chazop	computer HAZOP
CHEMSAFE	Chemical Industry Scheme for Assistance in Freight Emergencies
CHEMTREC	Chemical Transportation Emergency Centre
CHEP	conditional human error probability
CHETAH	chemical thermodynamics and energy hazard
CHIP	Chemicals (Hazard Information and Packaging) Regulations 1993
CHL	classification of hazardous locations
CHP	cumene hydroperoxide
CI	Chlorine Institute
CIA	Chemical Industries Association
CIB	Consell Intenrationale de Batiment
CIBS	Chartered Institute of Building Services
CIBSE	Chartered Institute of Building Services Engineers
CIIT	Chemical Industry Institute of Toxicology
CIM	computer integrated manufacturing
CIM	Uniform Rules concerning the Contract for International Carriage of Goods by Rail
CIMAH	Control of Industrial Major Accident Hazards Regulations 1984
CIP	cleaning in place
CISHC	Chemical Industry Safety and Health Council
CJ	Chapman—Jouguet
CLAW	Control of Lead at Work Regulations 1980
CLER	Classification and Labelling of Explosives Regulations 1983
CM	chemical manufacturing
CM	condition monitoring
CMA	Chemical Manufacturers Association
CME	chloromethyl ether
CMF	common mode failure
CMI	Christian Michelsen Institute
COD	chemical oxygen demand
COD	crack opening displacement
COER	Control of Explosives Regulations 1991
CONCAWE	oil companies organization
CONDAM	Construction (Design and Management) Regulations
COP	Code of Practice
COP	coefficient of performance
COPA	Control of Pollution Act
COPEP	Chlorine Out-Plant Emergency Plan
COSHH	Control of Substances Hazardous to Health Regulations 1988
COTIF	Convention concerning International Carriage by Rail
COV	collision with other vehicle
CP	cathodic protection
CP	conceptual dependency
CPA	Canadian Petroleum Association
CPD	Committee for the Prevention of Disasters
CPL	Classification, Packaging and Labelling (Regulations) 1984

CPLR	Classification Packaging and Labelling of Dangerous Substances Regulations 1984	DSHA	Dangerous Substances in Harbour Areas (Regulations) 1987
CPR	critical pressure ratio	DSIR	Department of Scientific and Industrial Research
CPS	control and protection system		
CPS	critical path scheduling	DSM	Dutch State Mines
CPSC	Consumer Product Safety Commission	DTA	differential thermal analysis
CRHC	Chemical Reactor Hazard Centre	DTG	differential thermogravimetry
CRMR	complete and rigorous model-based reconciliation	DTI	Department of Trade and Industry
		DTIC	Defense Technical Information Center
CSE	concept safety evaluation	DTL	dangerous toxic load
CSF	cancer slope factor	DTp	Department of Transport (UK)
CSO	Central Statistical Office	DTT	de-energize-to-trip
CSR	concept safety review	E/E/PES	Electrical/Electronic/Programmable Electronic Systems
CSTR	continuous stirred tank reactor		
CTM	causal tree method	EC	error cause (Whalley)
CTOD	crack tip opening displacement	EC	European Community
CTU	cargo transport unit	ECCS	emergency core cooling system
CV	control valve	ECCS	European Convention for Constructional Steelwork
CVCP	Committee of Vice Chancellors and Principals		
		ECETOC	European Chemical Industry Ecology and Toxicology Centre
CVN	Charpy V-notch (test)		
CW	continuous wave	ECOIN	European Core Inventory of Existing Substances
CWA	closed world assumption		
CWR	Chemical Works Regulations 1922	ED	enumeration district
CWRT	Center for Waste Reduction Technology	EDC	ethylene dichloride
DAP	detergent alkylate plant	EDF	Electricité de France
DAS	disturbance analysis system	EDF	equivalent discrete failure
DBMS	database management system	EDP	Emergency depressurization
DCF	discounted cash flow	EECS	Electrical Equipment Certification Service
DCL	distributed control logic	EEGL	emergency exposure guidance level
DCS	digital control systems	EEI	Edison Electric Institute
DCS	Distributed Control Systems	EEI	emergency exposure index
DDB	dependency-directed backtracking	EEL	emergency exposure limit
DDC	direct digital control	EEMUA	Engineering Equipment and Materials Users Association
DDP	defect detection probability		
DDT	deflagration to detonation transition	EEPG	Emergency Exposure Planning Guidelines
DEA	diethanolamine	EER	evacuation, escape and rescue
DF	Damage Factor (Dow)	EEUA	Engineering Equipment Users Association
DF	dependent failure		
DFA	2,4-difluoroaniline	EF	error factor
DFNB	2,4-difluoronitrobenzene	EF	Escalation Factor (Dow)
DG	diesel generator	EFS	emergency feedwater system
DGL	Directorate General of Labour	EFV	excess flow valve
DHSV	down-hole safety valve	EHS	effective stack height
DIERS	Design Institute for Emergency Relief Systems	EHS	extraordinarily hazardous substance
		EIA	environmental impact assessment
DIMP	Design Institute for Multiphase Processing	EIDAS	Explosion Incidents Data Service
DIN	Deutsches Institut für Normung	EINECS	European Inventory of Commercial Chemical Substances
DMAC	N,N-dimethylacetamide		
DMV	diaphragm motor valve (Bhopal)	EIS	environmental impact statement
DO	damage only (accident)	EIV	emergency isolation valve
DO	dissolved oxygen	EJMA	Expansion Joint Manufacturers Association
DoE	Department of the Environment	ELD	engineering line diagram
DoEm	Department of Employment	ELSA	European laboratory for Structural Assessment
DoEn	Department of Energy		
DoI	Department of Industry	EM	error mechanism (Whalley)
DoT	Department of Trade	EMAS	Employment Medical Advisory Service
DOT	Department of Transportation (USA)	EML	estimated maximum loss
DPLU	Disaster Prevention and Limitation Unit	EMS	Emergency Medical Services
DPT	decomposition pressure test	EMS	environmental management system
DPT	dynamically positioned tanker	EnvIDAs	Environmental Incidents Data Service
DR	discounted premium rate	EO	equation oriented
DRI	Distribution Ranking Index (Dow)	EO	ethylene oxide
DRM	deterministic reference model	EOQ	economical order quantity
DSC	differential scanning calorimetry	EP	environmental protection
		EPA	Environmental Protection Act

EPA	Environmental Protection Agency		FMD	fragment mass distribution
EPC	error producing condition		FMEA	failure mode and effect analysis
EPCRA	Emergence Planning & Community Right to Know Act 1986		FMEC	Factory Mutual Engineering Corporation
EPRI	Electrical Power Research Institute		FMECA	failure modes, effects and criticality analysis
EPSC	European Process Safety Centre		FMRC	Factory Mutual Research Corporation
EQO	environmental quality objective		FN	frequency–number (curve)
EQS	environmental quality standard		FOD	Field Operations Division
ER	equivalent radiator		FoF	figure of friction
ER	Exposure Radius (Dow)		FoI	figure of insensitiveness
ERA	Electrical Research Association		FOPL	first order predicate logic
ERC	Emergency Response Commission		FP	fluoroprotein
ERCO	Energy Resources Company		FPA	Fire Precautions Act 1971
ERDS	European Reliability Data System		FPA	Fire Protection Association
ERG	emergency response guidelines		FPC	Federal Power Commission
ERM	equilibrium rate model		FPS	floating production system
ERPG	Emergency Response Planning Guidelines		FPSO	floating production, storage and offloading (system)
ERQ	East Replacement Quarters (Piper Alpha)		FR	filling ratio
ERS	emergency relief system		FRC	fast rescue craft
ERT	Environmental Research and Technology Inc.		FRG	Federal Republic of Germany
			FRP	fibre reinforced plastic
ERV	expiratory reserve volume		FRS	Fire Research Station
ES	environmental statement		FRV	functional residual volume
ES	explosion severity		FSA	formal safety assessment
ESA	European Space Agency		FSM	fault–symptom matrix
ESD	emergency shut-down		FTA	fault tree analysis
ESD,ESS	Emergency Shutdown System		FTE	fracture transition elastic (point)
ESH	electrical surface heating		FTP	fracture transition plastic (point)
ESP	equivalent standard plant		FTS	fault tree synthesis
ESTC	Explosives Storage and Transport Committee		FWPCA	Federal Water Pollution Control Act 1972
			GAO	General Accounting Office
ESV	emergency shut-down valve		GCM	Gas Conservation Module (Piper Alpha)
ESV	emergency shut-off valve		GD	Germeles–Drake
ET	error type (Whalley)		GDA	goal-directed activity
ETA	expected time of arrival		GdF	Gaz de France
ETT	energize-to-trip		GDO	general development order
FA	fragility analysis		GDR	German Democratic Republic
FAD	failure analysis diagram		GFM	goodness-of-fit measure
FAD	failure assessment diagram		GFRP	glass fibre reinforced plastic
FAE	fuel–air explosion		GIS	general incident scenario
FAFR	failure accident frequency rate		GMAW	gas-metal arc welding
FAR	fatal accident rate		GNP	Gross National Product
FB	fractional bias		GOFA	goal-oriented failure analysis
FBL	fracture-before-leak		GoI	Government of India
FCAW	fluxed core arc welding		GOM	Gulf of Mexico
FCT	flux-corrected transport		GP	general purging (nitrogen)
FDA	Food and Dreg Administration		GPH	General Process Hazard (Dow)
FDG	flue gas desulphuriration		GRP	glass reinforced plastic
FDM	finite difference method		GRS	Gesselschaft für Reaktorsicherheit
FDT	fractional dead time		GRT	Gross registered tonnage
FECA	Federal Emission Control Act 1974		GS(I&U)R	Gas Safety (Installations and Use) Regulations 1984
FEMA	Federal Emergency Management Agency			
FERC	Federal Energy Regulatory Commission		GSR	Gas Safety Regulations 1972
FET	Flammability, explosiveness and toxicity		GT	Gaz Transport
FFFP	film forming fluoroprotein (foam)		GTAW	gas-tungsten arc welding
FFP	flame front projection		GTST	goal tree–success tree
F&EI	Fire and Explosion Index		GVW	gross vehicle weight
F&G	fire and gas		HAC	hazardous area classification
FHA	Federal Highways Authority		HALO	hazard assessment of landfill
FIA	Factory Insurance Association		HANES	Health and Nutrition Examination Survey
FIBC	flexible intermediate bulk container		HAZ	heat affected zone
FICA	Federal Emission Control Act (Germany)		HAZAN	hazard analysis
FL	frequency–loss		HAZMAT	hazardous material
FM	Factory Mutual			
FMA	Fertilizer Manufacturers Association			

HAZOP	hazard and operability (study)	IAPH	International Association of Ports and Harbours
HCFC	hydrochlorfluorocarbon	IARC	International Agency for Research on Cancer
HCLPF	high confidence of low probability of failure		
HCR	human cognitive reliability	IATA	International Air Transport Association
HD	hazard distance	IBC	intermediate bulk container
HD	Hazard Division	IC	inspiratory capacity
HDPE	high density polyethylene	ICAO	International Civil Aviation Organization
HDR	high discharge rate	ICBO	International Conference of Building Officials
HEART	human error assessment and reduction technique	ICD	inherently cleaner design
HED	Human Error Database	ICE	Institution of Civil Engineers
HEH	homogeneous equilibrium model	ICFTU	International Confederation of Free Trade Unions
HEP	human error probability		
HER	human error rate	IChemE	Institution of Chemical Engineers
HF	hydrogen fluoride	ICHMAP	International Co-operative Hydrogen fluoride Mitigation and Assessment Program
HFC	hydrofluorocarbon		
HF/E	human factors/engineering		
HFLR	Highly Flammable Liquids and Liquefied Petroleum Gases Regulations 1972	ICI	Imperial Chemical Industries
		ICRP	International Committee on Radiological Protection
HFO	heavy fuel oil		
HFRG	Human Factors in Reliability Group	ICS	International Chamber of Shipping
HGDT	heavy gas dispersion trial	ID	identification number
HGT	heavy goods train	IDA	influence diagram analysis
HGV	heavy goods vehicle	IDLH	Immediately Dangerous to life and Health (limit)
HI	hazard index		
HIO	Hazardous Incident Ordinance (Germany)		
HIPS	high integrity protective system	IE	index of explosibility
HIS	Health Interview Survey	IEC	International Electrotechnical Commission
HISS	high integrity shutdown system	IEE	Institution of Electrical Engineers
HIT	hazardous installations and transport	IET	insulated exotherm test
HITI	high integrity trip initiator	IFAL	instantaneous fractional annual loss
HITS	high integrity trip system	IFE	Institution of Fire Engineers
HIVE	high integrity voting equipment	IGE	Institution of Gas Engineers
HM	hazardous material	IHVE	Institution of Heating and Ventilation Engineers
HMA	Human reliability assessment		
HMAP	Hydrogen Fluoride Mitigation and Assessment Program	IISO	Institution of Industrial Safety Officers
		IL	integrity level
HMCIF	HM Chief Inspector of Factories	ILCI	International Loss Control Institute
HMFI	HM Factory Inspectorate	ILO	International Labour Office
HMIP	HM Inspectorate of Pollution	IMechE	Institution of Mechanical Engineers
HMIR	Hazardous Materials Incident Reporting	IMAS	influence modelling and assessment system
HMV	hydraulic master valve	IMCO	Intergovernmental Maritime Consultative Organization
HND	Higher National Diploma		
HORATIO	Human Response Analyzer and Timer for Infrequent Occurrences	IMDG	International Maritime Dangerous Goods (Code)
HP	high pressure	IMO	International Maritime Organization
HPI	high pressure injection	INRS	Institute National de Recherche sur la Securité
HPIM	high purity inert medium		
HPIP	human performance investigation process	IOI	International Oil Insurers
HQ	hazard quotient	IP	industrial platform
HRA	human reliability analysis	IP	Institute of Petroleum
HRC	hydrocarbon release (database)	IPC	integrated pollution control
HRS	hazard ranking scheme	IPIECA	International Petroleum Industry Environment Conservation Association
HSB	Hartford Steam Boiler		
HSC	Health and Safety Commission	IPL	independent layer of protection
HSE	Health and Safety Executive	IPSG	International Process Safety Group
HSF	hardware, software and fire fighting	IR	infrared
HSWA	Health and Safety at Work etc. Act 1974	IRI	Industrial Risk Insurers
HTA	hierarchical task analysis	IRIS	incident reporting and investigation system
HTF	heat transfer fluid	IRLG	Interagency Regulatory Liaison Group
HTF	high temperature fluid	IRV	inspiratory reserve volume
HTGR	high temperature gas-cooled reactor	IS	ignition severity
HTHM	high toxic hazard materials	ISA	Instrument Society of America
HV	high velocity	ISA	Instrumentation, Systems and Automation Society
IAEA	International Atomic Energy Agency		

ISGHO	International Study Group on Hydrocarbon Oxidation	LPB	*Loss Prevention Bulletin*
ISGRA	International Study Group on Risk Assessment	LPG	liquefied petroleum gas
ISI	Inherent Safety Index	LPGA	liquefied Petroleum Gas Association
ISLO	Installations Subject to Licensing Ordinance (Germany)	LPGITA	Liquefied Petroleum Gas Industry Technical Association
ISO	International Standards Organisation	LPI	low pressure injection
ISR	incoming solar radiation	LPR	linear polarization resistance
ISRS	International Safety Rating System	LQW	Living Quarters West (Piper Alpha)
IST	isothermal storage test	LR	Lloyds Register
ITB	Industry Training Board	LRA	least replaceable assembly
ITB	Insurance Technical Bureau	LS	limit switch
ITC	Interagency Testing Committee	LS	logical sufficiency
ITOPF	International Tanker Owners Pollution Federation	LTA	lost time accident
ITSA	Independent Tank Storage Association	LTI	lost time injury
IUR	International Union of Railways	LTMS	logic-based truth maintenance system
JFRO	Joint Fire Research Organization	LTPD	lot tolerance per cent defective
JHEP	joint human error probability	LTSR	long-term safety review
JIT	'just-in-time'	LVHV	low volume, high velocity
JRC	Joint Research Centre	LWR	light water reactor
JSA	Job safety analysis	MAC	maximum allowable concentration
JT	Joule–Thomson	MACT	maximum achievable control technology
JTMS	justification-based truth maintenance system	MAPP	major accident prevention plan
KAF	knowledge acquisition facility	MAPP	methyl acetylene/propadiene/propylene
KASP	Kjeller Ammonia Stress Corrosion Project	MAPCO	Mid-America Pipeline Company
KWU	Kraftwerk Union	MAPPS	maintenance personnel performance simulator
LA	Local authority	MARPOL	International Conference on Revision of the International Regulations for Preventing Collisions at Sea
LAAPC	local authority air pollution control	MAUD	multi-attribute utility decomposition
LBB	leak-before-break	MAWP	maximum allowable working pressure
LBF	leak-before-fracture	MB	measure of belief
LCn	lethal concentration ($n\%$ level)	MC	motor cycle
LCC	life cycle costing	MCA	Manufacturing Chemists Association
$LCTn$	lethal toxic concentration ($n\%$ level)	MCC	methylcarbamoyl chloride
LCV	level control valve	MCL	maximum credible loss
LDn	lethal dose ($n\%$ level)	MCSOII	multiple-cause, systems-oriented incident investigation
LDC	London Dumping Convention	MCSP	Model Code of Safe Practice (IP)
LDG	List of Dangerous Goods and Conditions of Acceptance	MD	measure of disbelief
LDPE	low density polyethylene	MDA	modellers data archive
LEFM	linear elastic fracture mechanics	MDF	multiple degree-of-freedom
LEL	lower explosive limit	MDI	4,4-Diphenylmethane diisocyanate
LEPC	Local Emergency Preparedness Committee	ME	multienergy (model)
LER	licensee event report	MEA	monoethanolamine
LEV	local exhaust ventilation	MEHQ	methyl ether of hydroquinone
LF	landfill	MEL	maximum exposure limit
LFG	liquefied flammable gas	MEPC	Marine Environment Protection Committee
LFL	lower flammability limit	MES	multilinear events sequencing
LGC	liquefied gas carrier	MESG	maximum experimental safe gap
LGF	Liquefied Gaseous Fuels (Spill Effects Program)	MF	management factor
		MF	material factor
LGFSTF	Liquefied Gaseous Fuels Spill Test Facility	MFCC	multiple forced circulation circuit
LGV	light goods vehicle	MGL	multiple Greek letter
LITTS	large inventory top tier site	MHAP	Major Hazards Assessment Panel
LLNL	Lawrence Livermore National Laboratory	MHAU	Major Hazards Assessment Unit
LN	logical necessity	MHIDAS	Major Hazard Incidents Data Service
LNG	liquefied natural gas	MIBE	methyl-t-butyl ether
LNH3	liquefied ammonia	MIBK	methylisobutyl ketone
LOCA	loss of coolant accident	MIC	methyl isocyanate
LOPA	Layer of protection analysis	MIC	minimum igniting current
LOV	locally operated valve	MIE	minimum ignition energy
LOX	liquid oxygen	MIG	metal inert gas
LPA	local planning authority	MILNP	mixed integer non-linear programming
		MILP	mixed integer linear programming

MITI	Ministry of International Trade and Industry	NG	nitroglycerine
MLS	marine LNG system	NGL	natural gas liquids
MM	methyl methacrylate	NIG	National Interest Group
MM	modified Mercalli	NIHHS	Notification of Installations Handling
M&M	Marsh and McLennan		Hazardous Substances Regulations 1982
M&M	monitoring and maintenance	NII	Nuclear Installations Inspectorate
MMA	Methyl Methacrylate	NIOSH	National Institute of Occupational Safety
MO	Marshall−Olkin		and Health
MOC	Management of change	NIST	National Institute of Standards and
MOC	method of characteristics		Technology
MoD	Ministry of Defence	NML	Normal Maximum Loss
MOL	main oil line	NMRS	near miss reporting system
MOR	modification of risk	NMS	Dangerous Substances (Notification and
MORT	Management Oversight and Risk Tree		Marking of Sites) Regulations 1990
MOV	motor operated valve	NMSE	normalized mean square error
MPA	mean presented area	NMMS	near miss management system
MPD	minimum premarketing data	NMTMS	Non-monotonic truth maintenance system
MPDO	maximum probable days outage	NNS	Norwegian North Sea
MPE	maximum permissible exposure	NOAEL	no observable adverse effect level
MPI	magnetic particle inspection	NPD	Norwegian Petroleum Directorate
MPPD	Maximum Probable Property Damage	NPDES	National Pollutant Discharge Elimination
MPS	multiple point source		System
MPST	maximum permissible surface temperature	NPL	National Physical Laboratory
MRC	Medical Research Council	NPP	nuclear power plant
MRS	MIC refining still (Bhopal)	NPS	Naval Postgraduate School
MRS	Midlands Research Station	NPSH	net positive suction head
MSA	Marine Safety Agency	NPV	Net Present Value
MSC	Maritime Safety Committee	NRA	National Rivers Authority
MSDS	material safety data sheet	NRC	National Research Council
MSS	MIC storage system (Bhopal)	NRC	Nuclear Regulatory Commission
MTTF	mean time to failure	NRPB	National Radiological Protection Board
MTBF	mean time between failures	NRTS	National Reactor Test Station
MTTFF	mean time to first failure	NRV	non-return valve
MV	medium velocity	NSAC	Nuclear Safety Analysis Center
MV	munitions vehicle	NSC	National Safety Council
MW	molecular weight	NSCA	National Society for Clean Air
MWA	Mineral Workings Act 1971	NSPS	New Source Performance Standards
N/A	not applicable	NSR	New Source Reviews
NAAQS	National Ambient Air Quality Standards	NSW	New South Wales
NACE	National Association of Corrosion Engineers	NTIS	National Technical Information Service
NADOR	Notification of Accidents and Dangerous	NTP	National Toxicology Program
	Occurrences Regulations 1980	NTS	Nevada Test Site
NAMAS	National Measurement Accreditation	NTSB	National Transportation Safety Board
	Scheme	NWC	Naval Weapons Center
NAS	National Academy of Sciences	OAET	operator action event tree
NASA	National Aeronautical and Space	OARU	Occupational Accident Research Unit
	Administration	OAT	operator action tree
NBFU	National Board of Fire Underwriters	OB	oxygen balance
NBS	National Bureau of Standards	OBE	operating basis earthquake
NCB	National Coal Board	OBO	ore/bulk/oil
NCEC	National Chemical Emergency Centre	OC	operating characteristic
NCSR	National Centre for Systems Reliability	OCIMF	Oil Companies International Marine Forum
NCTR	National Center for Toxicological Research	OCM	on-condition maintenance
NDE	Non-destructive evidence	OD	oxygen demand
NDE	non-destructive examination	ODCB	o-dichlorobenzene
NDRC	National Defense Research Committee	OECD	Organization for Economic Cooperation and
NDT	non-ductility transition		Development
NDT	null ductility transition	OEL	Occupational Exposure Limit
NEA	Nuclear Energy Agency	OES	Occupational Exposure Standard
NEIC	National Earthquake Information Center	OFG	over-the-top foam generation
NEL	National Engineering Laboratory	OIA	Oil Industry Association
NEPA	National Environmental Policy Act	OIAC	Oil Industry Advisory Committee
NESHAP	National Exposure Standards for Hazardous	OHSAS	Occupational Health and Safety
	Air Pollutants		Management systems
NFPA	National Fire Protection Association	OII	Oil Industry Insurance

OILPOL	International Convention on the Prevention of Pollution of the Sea by Oil		P&I(D)	piping and instrument (diagram)
OIM	offshore installation manager		PIF	performance influencing factor
ONAN	*o*-nitroaniline		PIIS	Prototype index of inherent safety
ONCB	*o*-nitrochlorobenzene		PIS	Protective instrumented Systems
OOA	observable operational attribute		PISC	Plate Inspection Steering Committee
OOP	object oriented programming		PITB	Petroleum Industry Training Board
OOS	object oriented system		PIUS	process-inherent ultimate safety
OPCS	Office of Population Census and Surveys		PL	propositional logic
OPS	operating procedure synthesis		PLA	Port of London Authority
OPSO	Office of Pipeline Safety Operations		PLC	programmable logic control
OREDA	Offshore Reliability Data (bank)		P/M	peak-to-mean
ORNL	Oak Ridge National Laboratory		PM	preventive maintenance
ORR	overall risk rating		PML	Probable Maximum Loss
OSD	Offhore Safety Division		PMMA	polymethylmethacrylate
OSHA	Occupational Safety and Health Act 1970		PMN	premanufacture notice
OSHA	Occupational Safety and Health Administration		PMN	premarketing notification
			PNEUROP	European Committee of Manufacturers of Compressors, Vacuum Pumps and Pneumatic Tools
OS&RP	Office, Shops and Railway Premises (Act) 1963			
OSTP	Office of Science and Technology Policy		POIC	planning, organization implementation and control
OSW	Office of Solid Waste			
OT	overturning		PORV	power-operated relief valve
OTA	Office of Technology Assessment		PP	polypropylene
OTS	Office of Toxic Substances		PPA	potential problem analysis
OV	Original Value (Dow)		PPE	peak pressure enhancement
P/A	peak-to-average		PPE	personal protective equipment
Paris MOU	Paris Memorandum		PRA	probabilistic risk assessment
PASS	process and storage ship		PRP	potentially responsible person
PAW	Petroleum Administration for War		PRV	pressure relief valve
PAW	plasma arc welding		PS	point source
PB	pushbutton		PS	polystyrene
PBB	polybrominated biphenyl		PSA	process safety management
PBL	planetary boundary layer		PSA	pressure swing adsorption
PC	paired comparison		PSA	probabilistic safety assessment
PCA	preliminary consequence analysis		PSAT	Pre-start-up acceptance test
P(C)A	Petroleum (Consolidation) Act		PSD	prevention of significant deterioration
PCB	polychlorinated biphenyl		PSDI	Project Software and Development Inc.
PCM	parametric correlation method		PSF	performance shaping factor
PCV	pressure control valve		PSII	process safety incident investigation
PD	Property Damage		PSM	process safety management
PDF	probability density function		PSV	pressure safety valve
PDS	potentiodynamic scan		PT	passenger train
PE	programmable electronic (hardware)		PTB	Physikalische Technische Bundesanstalt
PEG	polyethylene glycol		PTFE	polytetrafluoroethene
PEEL	Public Emergency Exposure Limit		PTSD	post-traumatic stress disorder
PEL	permissible exposure limit		PTW	permit-to-work
PERT	project evaluation and review technique		PUF	polyurethane foam
PES	Programmable electronic system		PUHF	Process Unit Hazards Factor (Dow)
PETN	pentaerythritol tetranitrate		PV	pressure/vacuum
PF	potential failure		PVC	polyvinyl chloride
PFC	perfluorocarbon		PVH	process vent header (Bhopal)
PFD	Probability of failure on demand		PVQAB	Pressure Vessel Quality Assurance Board
PFD	process flow diagram		PVTC	Pressure Vessel Technical Committee
PG	Pasquill−Gifford		PVM	population vulnerability model
PGA	peak ground acceleration		PWG	Provinciale Waterstaat Groeningen
PGR	Road Traffic (Carriage of Dangerous Substances in Packages) Regulations		PWHT	post-weld heat treatment
			PWR	pressurized water reactor
PH	precipitation hardening		QA	quality assurance
PHA	preliminary hazard analysis		QHRDCS	quantitative human reliability data collection system
PHA	process hazard analysis			
PHEA	predictive human error analysis		QUASA	quality assessment assurance of safety analysis
PHSA	Planning (Hazardous Substances) Act 1990			
PI	personal injury (accident)		QC	quality control
PI	pressure indicator		QC	quality circle
			QF	quality factor

QPT	qualitative process theory		SAT	Site acceptance test
QRA	quantitative risk assessment		SAW	submerged arc welding
RAG	risk appraisal group		SBC	Standard Building Code
RARDE	Royal Armament Research and Development Establishment		SCA	small-scale cook-off bomb
			SCA	sneak circuit analysis
RAT	risk assessment tool		SCAT	systematic cause analysis technique
RBE	relative biological efficiency		SCBA	self-contained breathing apparatus
RBV	remotely-operated block valve		SCC	stress corrosion cracking
RC	resistance–capacitance		SCCM	subjective cause–consequence model
RCAS	root cause analysis system		SCI	Society of Chemical Industry
RCEP	Royal Commission on Environmental Pollution		SCM	short-cut classical method
			SCRA(M)	Short-cut risk assessment (method)
RCM	reliability-centred maintenance		SCS	safety critical system
RCRA	Resource Conservation and Recovery Act 1976		SCT	stress concentration theory
			SDF	single degree of freedom
RCS	reactor cooling system		SEAOC	Structural Engineers Association of California
RF	radio frequency			
RF	Rossi–Forelli		SEDEX	type of calorimeter
RfD	reference dose		SEDF	standard equivalent discrete failure
RH	relative humidity		SEER	Surveillance, Epidemiology and End Results Program
R–H	Rankin–Hugoniot			
RHI	Reaction Hazard Index		SEM	scanning electron microscopy
RIA	Robot Institute of America		SEM	slip equilibrium model
RIC	Royal Institute of Chemistry		SF	solid flame
RID	International Regulations Concerning the Carriage of Dangerous Goods by Rail		SFG	signal flow graph
			SGHWR	steam generating heavy water reactor
RIDDOR	Reporting of Injuries, Diseases and Dangerous Occurrences Regulations 1985		SGT	subgoal template
			SHE	safety, health and environment
RII	relative incapacitation index		SHERB	Sandia Human Error Rate Bank
RLA	remanent life assessment		SHERPA	systematic human error reduction and prediction approach
RME	reasonable maximum exposure			
RMEE	relative mass extinguishing efficiency		SHH	substance hazardous to health
RMP	risk management planning		SI	safety interlock
RMPP	Risk Management and Prevention Program (California)		SI	Statutory Instrument
			SIC	Standard Industrial Classification
RoRo	roll-on/roll-off		SIGTTO	Society of International Gas Tanker and Terminal Operators
RoSPA	Royal Society for the Prevention of Accidents			
			SIKAREX	type of calorimeter
ROV	remotely-operated valve		SIL	Safety integrity level
RPE	respiratory protective equipment		SIPRI	Stockholm International Peace Research Institute
RPI	Retail Price Index			
RPT	rapid phase transition		SIRA	Scientific Instrument Research Association
RPV	reactor pressure vessel		SIS	Safety Instrumented System
RRL	Road Research Laboratory		SIS	safety interlock system
RSPA	Research and Special Programs Administration		SIT	spontaneous ignition temperature
			SITTS	small inventory top tier site
RSS	Reactor *Safety Study* (*Rasmussen Report*)		SKHCR	standardized skilled and hospitalized casualty rate
RSSG	Royal Society Study Group			
RSST	Reactor Systems Screening Tool		SKR	skills–rules–knowledge (model)
RT	response type (Whalley)		SL	Sandia National Laboratory
RTJ	ring-type joint		SL	support list
RTR	Road Traffic (Carriage of Dangerous Substances in Road Tankers and Tank Containers) Regulations 1992		SLI(M)	success likelihood index (method)
			SLOT	Specified Level of Toxicity
			SLP	safety and loss prevention
RV	Replacement Value (Dow)		SLT	superheat limit temperature
RV	residual volume		SM	sequential modulator
RVVH	relief valve vent header (Bhopal)		SMAW	shielded metal arc welding
SA	sneak analysis		SMR	Seismic margin earthquake
SADT	self-accelerating decomposition temperature		SMRB	Safety in Mines Research Board
SAI	Science Applications Inc.		SMRE	Safety in Mines Research Establishment
SARA	Superfund Amendment and Reauthorization Act 1986		SMS	safety management system
			SNG	synthetic natural gas
SARAH	systematic approach to the reliability assessment of humans		SOAP	spectrometric oil analysis programme
			SOLAS	Safety of life at Sea
SAS	small area statistics		SOP	Standard Operating Procedure

SOT	Society of Toxicology
SPEAR	system for predictive error analysis and reduction
SPEGL	Short-term Public Emergency Guidance Level
SPH	Special Process Hazard (Dow)
SPI	Society of the Plastics Industry
SPM	shock pulse meter
SPS	Safety protection systems
SPS	submerged production station
SRD	Safety and Reliability Directorate
SRI	Stanford Research Institute
SRS	Systems Reliability Service
SRV	safety relief valve
SSD	Safety Shutdown System
SSE	safe shut-down earthquake
SSEB	South of Scotland Electricity Board
SSI	soil–structure interaction
SSIV	subsea isolation valve
SSMRP	Seismic Safety Margins Research Program
STAR	stability array (method)
STATAS	structured audit technique for the assessment of safety management systems
STCW	International Convention on Training, Certification and Watchkeeping for Seafarers
STEL	short-term exposure limit
STEP	sequentially timed events plotting procedure
STEP	sequentially timed events plot
STPL	short-term public exposure limit
SVA	single vehicle accident
SVO	collision of vehicle with stationary object
SWACER	shock-wave amplified coherent energy release
SWAN	stress wave analyser
TAFF	theoretical adiabatic flash fraction
TBC	p-t-butyl catechol
TBN	total base number
TCB	1,2,4,5-trichlorobenzene
TCDD	2,3,7,8-tetrachlorodibenzo-p-dioxin
TCP	2,4,5-trichlorophenol
TCPA	Town and Country Planning Act
TCPA	Toxic Catastrophe Prevention Act
TCSP	tank centre space potential
TDI	toluene diisocyanate
TEMA	Tubular Exchanger Manufacturers Association
TESEO	Tecnica Empirica Stima Errori Operatori
TFI	The Fertilizer Institute
TGZ	Technigaz
THERP	Technique for Human Error Rate Prediction
TIG	tungsten inert gas
TLC	total lung capacity
TLV	threshold limit value
TMI	Three Mile Island
TMR	triple modular redundant
TMS	truth maintenance system
TNO	Toegepast-Natuurwetenschappelijk Onderzoek
TNT	trinitrotoluene
TOA	technical options analysis
TOPS	Total Operations Processing System
TPA	third party activity
TPQ	threshold planning quantity
TQM	total quality management
TRA	total risk analysis
TRC	Thornton Research Centre
TRC	time–reliability correlation
TRI	toxic release inventory
TRRL	Transport and Road Research Laboratory
TS	turbulence scheme
TSCA	Toxic Substances Control Act
TSDF	treatment storage and disposal facility
TSR	temporary safe refuge
TT	task type (Whalley)
TUV	Technische berwachungs-Verein
TV	Tidal volume
TVA	Tennessee Valley Authority
TWA	time-weighted average
UAS	user-approved safety
UBC	Uniform Building Code
UCB	uncertainty bound
UCC	Union Carbide Corporation
UCIL	Union Carbide India Ltd
UCO	Use Classes Order
UCS	Union of Concerned Scientists
UEL	upper explosive limit
UFL	upper flammability limit
UF	uniform flux
UKAEA	UK Atomic Energy Authority
UKCIC	UK Chemical Information Centre
UKCP	UK Chlorine Producers
UKOOA	UK Oil Operators Association
UL	Underwriters Laboratories
UMC	underwater manifold centre
UN	United Nations
UNEP	UN Environment Programme
UNS	Unified Numbering System
UPS	Uninterrupted power supplies
UPS	uninterruptible power supply
UR(L)	United Refineries (Ltd)
USCG	US Coast Guard
USPHS	US Public Health Service
UST	underground storage tank
UV	ultraviolet
UVCE	unconfined vapour cloud explosion
VAE	Value of Area Exposed (Dow)
VC	vinyl chloride
VC	vital capacity
VCE	vapour cloud explosion
VCF	vapour cloud fire
VCM	vinyl chloride monomer
VDI	Verein Deutscher Ingenieure
VDU	visual display unit
VEEB	Vapour escape into, and explosion in, a building
VGS	vent gas scrubber (Bhopal)
VIM	vacuum induction melting
VKI	Von Karman Institute
VLCC	very large crude carrier
VM	Vulnerability model
VOC	volatile organic compound
VOD	vacuum–oxygen decarburization
VOD	velocity of detonation
VOF	velocity of fragment
VPM	value of production per month
VS	volumetric source
VSP	Vent Sizing Package
V&V	Verification and validation

WA	water authority	WOAD	World Offshore Accident Database
WAO	wet air oxidation		
WATCH	Working Group on the Assessment of Toxic Chemicals	WSA	work safety analysis
		WSL	Warren Spring Laboratory
WE	Western Europe	WSSN	Worldwide Standard Seismographic Network
WHO	World Health Organization		
WI	Welding Institute	ZND	Zeldovitch−von Neumann−Döring

Index

Computer Codes Index